Land Rov

Discovery

Parts Catalogue

1989 to 1998 MY

This Parts Catalogue covers all Land Rover Discovery vehicles from model year 1989 up to and including 1998, fitted with 2.0 Mpi, 3.5 & 3.9 V8 petrol engines and 200 Tdi & 300 Tdi diesel engines.

Part. No. RTC9947CF

Whilst every effort is made to ensure the accuracy of the particulars contained in the Catalogue, modifications and vehicle specification changes may affect the information specified. No responsibility is accepted for the incorrect supply of parts or any other consequence that may arise as a result of information in the Catalogue not being in accord with modifications or vehicle specification changes which are subsequent to the date of this Catalogue. Also no responsibility is accepted for the incorrect supply of parts or any other consequence that may arise as a result of any misinterpretation of the information specified in this Catalogue.

Brooklands Books Ltd., PO Box 146, Cobham,
Surrey KT11 1LG, England.
E-mail: sales@brooklands-books.com www.brooklands-books.com

Part Number: RTC 9947CF

ISBN 9781855206144 Ref: LR89PH 10T6/2399

CONTENTS

	Group	Page
Introduction	A	5
Engine	B	45
Gearbox	C	141
Transfer Box	D	191
Axles and Suspension	E	205
Steering	F	231
Vehicle and Engine Controls	G	243
Brakes	H	253
Fuel and Emission Systems	J	267
Exhaust Systems	K	281
Cooling and Heating	L	291
Body Electrics	M	321
Body and Chassis	N	403
Interior Trim	P	467
Seats	R	511
Stowage	S	543
Miscellaneous and Accessories	T	545
Special Vehicles Options	V	571
Numerics Index	X	575

DISCOVERY 1989MY UPTO 1999MY

GROUP A

MASTER INDEX

Section	Group	Section	Group
Engine	B	Fuel & Emission Systems	J
Gearbox	C	Exhaust System	K
Transfer Box	D	Cooling & Heating	L
Axles & Suspension	E	Body Electrics	M
Steering	F	Body & Chassis	N
Vehicle & Engine Controls	G	Interior Trim	P
Brakes	H	Seats	R

VIN NUMBER IDENTIFICATION

FROM 21st APRIL 1995
LAND ROVER INTRODUCED
AN ADDITIONAL SERIES OF VIN NUMBERS
TO THOSE ALREADY EXISTING
WHICH WILL COMMENCE FROM
MA500000

STATION WAGON 3 DOOR

VIN CODE	ENGINE		GEARBOX	STEERING
SALLJGBV7 ••	3.5 PETROL		5 SPEED	RH STG
SALLJGBV8 ••	3.5 PETROL		5 SPEED	LH STG
SALLJGBF7 ••	2.5 DIESEL	TDI	5 SPEED	RH STG
SALLJGBF8 ••	2.5 DIESEL	TDI	5 SPEED	LH STG
SALLJGBL7 ••	3.5 PETROL	PI	5 SPEED	RH STG
SALLJGBL8 ••	3.5 PETROL	PI	5 SPEED	LH STG
SALLJGBM7 ••	3.9 PETROL	PI	5 SPEED	RH STG
SALLJGBM8 ••	3.9 PETROL	PI	5 SPEED	LH STG

HOME/ROW

•• GA - 90 MODEL YEAR
•• HA - 91 MODEL YEAR
•• JA - 92 MODEL YEAR
•• KA - 93 MODEL YEAR
•• LA - 94 MODEL YEAR
•• MA - 95 MODEL YEAR
•• TA - 96 MODEL YEAR
•• VA - 97 MODEL YEAR

USA ONLY

•• RA - 94 MODEL YEAR
•• SA - 95 MODEL YEAR
•• TA - 96 MODEL YEAR
•• VA - 97 MODEL YEAR

NOTE (1) Leisure Models are fitted with Lo-Line Instrument Packs, Grey Carpets, Plain Door Trim Pads and are less Forward Facing Rear Seats

VIN NUMBER IDENTIFICATION

FROM 21st APRIL 1995
LAND ROVER INTRODUCED
AN ADDITIONAL SERIES OF VIN NUMBERS
TO THOSE ALREADY EXISTING
WHICH WILL COMMENCE FROM
MA500000

STATION WAGON 5 DOOR (HA ONWARDS)

VIN CODE	ENGINE		GEARBOX	STEERING
SALLJGML7 ••	3.5 PETROL	PI	5 SPEED	RH STG
SALLJGML8 ••	3.5 PETROL	PI	5 SPEED	LH STG
SALLJGMF7 ••	2.5 DIESEL	TDI	5 SPEED	RH STG
SALLJGMF8 ••	2.5 DIESEL	TDI	5 SPEED	LH STG
SALLGMY8 ••	2.0 PETROL	MPI	5 SPEED	LH STG
SALLGMY7 ••	2.0 PETROL	MPI	5 SPEED	RH STG
SALLJEMM7 ••	3.9 PETROL	PI	5 SPEED	RH STG
SALLJEMM8 ••	3.9 PETROL	PI	5 SPEED	LH STG
SALLJRMM7 ••	3.9 PETROL	PI	5 SPEED Rover Japan Nov 94 onwards	LH STG

HOME/ROW

•• GA - 90 MODEL YEAR
•• HA - 91 MODEL YEAR
•• JA - 92 MODEL YEAR
•• KA - 93 MODEL YEAR
•• LA - 94 MODEL YEAR
•• MA - 95 MODEL YEAR
•• TA - 96 MODEL YEAR
•• VA - 97 MODEL YEAR

USA ONLY

•• RA - 94 MODEL YEAR
•• SA - 95 MODEL YEAR
•• TA - 96 MODEL YEAR
•• VA - 97 MODEL YEAR

NOTE (1) Leisure Models are fitted with Lo-Line Instrument Packs, Grey Carpets, Plain Door Trim Pads and are less Forward Facing Rear Seats

SERIAL NUMBER IDENTIFICATION

ENGINES PETROL - MPI

20T4HG63 2.0 P.1 MPI Aircon (Greece only)
20T4HG62 2.0 P.1 MPI Aircon
20T4HG91 2.0 P.1 MPI Non Aircon

ENGINES - DIESEL

12L00001 4 Cylinder 2.5 200 TDI
17L00001 4 Cylinder 2.5 300 TDI EDC Man
18L00001 4 Cylinder 2.5 300 TDI EDC DETOX Man
19L00001 4 Cylinder 2.5 300 TDI EDC Auto
20L00001 4 Cylinder 2.5 300 TDI EDC DETOX Auto
21L00001 4 Cylinder 2.5 300 TDI EGR Man
22L00001 4 Cylinder 2.5 300 TDI EGR Auto

ENGINES - V8 PETROL

27G00001 3.5 Carburettered, 8. 13: 1. C/R Manual Gearbox

ENGINES - V8 PETROL EFI

22L00001 3.5 EFI 8. 13: 1. C/R Manual
23L00001 3.5 EFI 8. 13: 1. C/R Auto
24L00001 3.5 EFI 9. 35: 1. C/R
35L00001 3.9 EFI 9. 35: 1. C/R Manual
36L00001 3.5 EFI 9. 35: 1. C/R Auto
37L00001 3.9 EFI 8. 13: 1. C/R Manual
38L00001 3.5 EFI 8. 13: 1. C/R Auto

SERIAL NUMBER IDENTIFICATION

AXLE SERIAL NUMBERS
FRONT

18L00001B RHS V8
19L00001B LHS V8
34L00001B RHS DIESEL
35L00001B LHS DIESEL
38L00001B RHS V8)
39L00001B LHS V8) ASBESTOS FREE
42L00001B RHS DIESEL) BRAKES
43L00001B LHS DIESEL)
71L00001B RHS 4 CYL))
72L00001B LHS 4 CYL) NO)
73L00001B RHS V8) ABS)
74L00001B LHS V8)) MA081992
75L00001B RHS 4 CYL)) 95MY
76L00001B LHS 4 CYL) ABS) ONWARDS
77L00001B RHS V8) BRAKES)
78L00001B LHS V8))

AXLE SERIAL NUMBERS
REAR

28S00001B
32S00001B ASBESTOS FREE BRAKE
46S00001B NON ABS) MA081992 95MY ONWARDS
46S00001B ABS) 3BOLT FLANGE

GEARBOX - LT77 MANUAL TYPE

53A00001 5 SPEED, V8 EFI
57A00001 5 SPEED, V8 CARB
55A00001 5 SPEED, 4 Cylinder TDI
59A00001 5 SPEED, Petrol PI
62A00001 5 SPEED, Petrol 2.0 Greece
63A00001 5 SPEED, Petrol 2.0

GEARBOX - R380 MANUAL

53A00001IJ 5 SPEED, V8 EFI
55A00001IJ 5 SPEED, TDI
63A00001IJ 5 SPEED, MPI

GEARBOX - AUTOMATIC

ZF - 4 SPEED

TRANSFER GEARBOX - LT230T

28D00001 V8 Petrol and 4 Cylinder TDI
34D00001 V8 2.0 MPI

TECHNICAL INFORMATION

LAND-ROVER

CIRCULATE TO:	
SERVICE MGR.	X
RECEPTION	X
WORKSHOP	X
PARTS	X

SUBJECT	BULLETIN No.
Handbrake Gaiter Adrift	074/92/EN
MODEL: **DISCOVERY**	AFFECTED VEHICLES: All Discovery derivatives before VIN LJ 024361

DETAIL

Instances of the handbrake gaiter coming adrift have been found to be caused by the gaiter being caught by the clevis pin during operation.

An improved clevis pin and clip were introduced on vehicles after the above VIN to resolve this issue.
The new pin and clip can be fitted to vehicles before the above VIN to stop the handbrake gaiter coming adrift.

ACTION REQUIRED

In the event of customer complaint of the handbrake gaiter coming adrift, replace the clevis pin and clip with the parts listed below.

PARTS INFORMATION

Clevis pin	SYT 10004
Clip	EDP 7864

Note: These parts must be replaced as a set.
DO NOT reuse old pin or clip.

ILLUSTRATION

PROCEDURE

Refer to Workshop Manual Section 70 for replacing the clevis pin and clip.

The clevis pin **MUST** be fitted as shown in the illustration. The pin must have the oval head and not the round head of the old pin.

The retention clip is always fitted on the side of the handbrake nearest the transmission tunnel, on both right and left hand drive vehicles.

Use long nosed pliers to fit the clip, holding only the end shown uppermost in the illustration.
Push the clip onto the pin and turn anticlockwise so that the leg of the clip slides over the pin and rests behind it as shown on the diagram.

WARRANTY
Normal warranty policy and procedure applies.

COMP. CODE
8E5B

SRO
70-35-37 Replace clevis pin, clip 00.24 Hr and readjust handbrake.

TECHNICAL INFORMATION

LAND-ROVER

CIRCULATE TO:	
SERVICE MGR.	X
RECEPTION	X
WORKSHOP	X
PARTS	X

SUBJECT	03/12/93
Axle Swivel and Hub Seals - New Part Information	170/93/EN
MODEL: **Range Rover Discovery Defender**	AFFECTED VEHICLES: **All derivatives from LH 632889 - LJ 045306 - LD925688**

DETAIL

To improve axle swivel and hub sealing a new range of seals have been introduced at the above VIN numbers. These seals are directly interchangeable with the previous parts.

The information provided on seal FTC 3401 supersedes that provided on Technical Bulletin 083/93.

ACTION REQUIRED
If the new type swivel/hub seals leak in service, please submit a Product Report accompanied by a copy of the illustration overleaf identifying the area of the leak. Should customers complain of leaks in this area fit the new seals.

PARTS INFORMATION

Swivel seal	FTC 3401 replaces FRC 2889
Hub seal	FTC 2783 replaces FRC 8221
Stub axle seal	FTC 3145 replaces FRC 0951

ILLUSTRATION

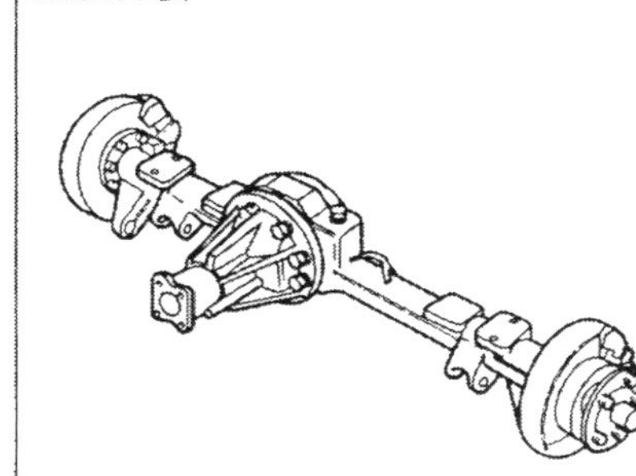

PROCEDURE

It must be noted that the new seals have changed orientation and the following guidelines should be followed:-

Hub oil seal FTC 2783 must be fitted with the lip facing inwards ie. with the seal manufacturers name and seal part number facing the bearings.
See illustration on following page.

Stub axle oil seal FTC 3145 on front axle application must be fitted with the spring to the outside, see illustration on following page.

Rear hub oil seal FTC 3145 must be fitted with the spring to the inside. See illustration on following page.

Current Special tools for these seals remain unchanged.

WARRANTY
Normal warranty policy and procedure applies.

	COMP. CODE
Front swivel hub oil seal	**5H0N**
Front hub seals	**5V3N**
Rear hub seals	**5W3N**

SRO

Refer to the Service Operation Repair Manual for individual SRO times.

TECHNICAL INFORMATION

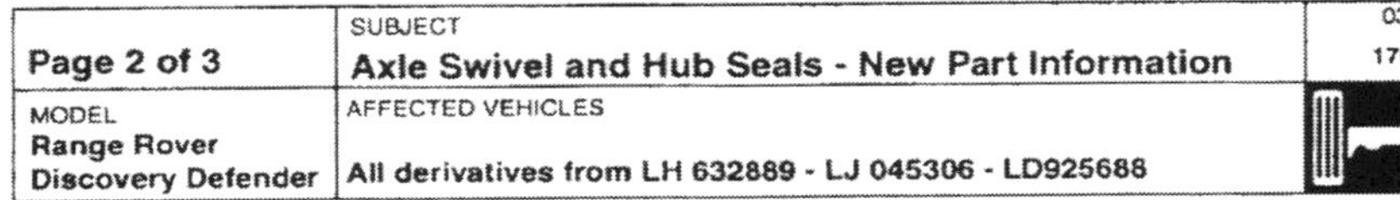

Page 2 of 3	SUBJECT Axle Swivel and Hub Seals - New Part Information	03/12/93 170/93/EN
MODEL Range Rover Discovery Defender	AFFECTED VEHICLES All derivatives from LH 632889 - LJ 045306 - LD925688	

REAR AXLE HUB ASSEMBLY

A - Rear axle hub assembly - oil seal FTC 3145 fitted with spring inwards.
B - Rear axle hub assembly - oil seal FTC 2783 (Springless type)

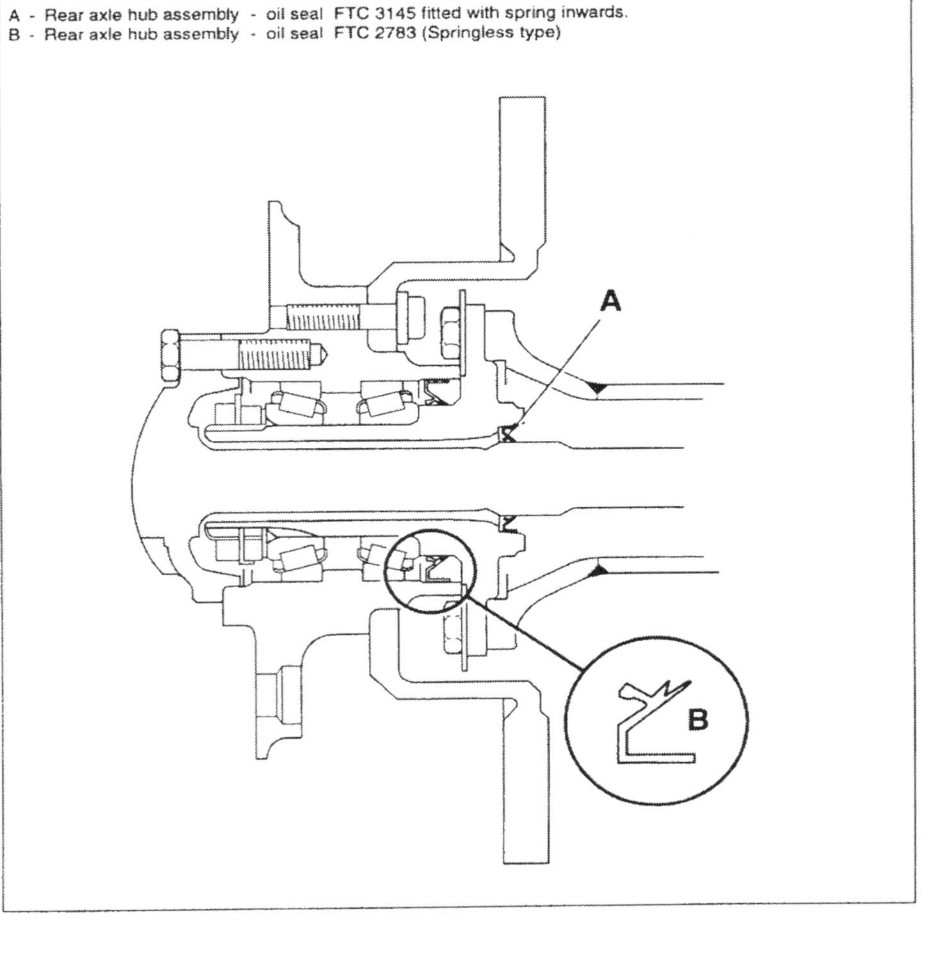

TECHNICAL INFORMATION

Page 3 of 3	SUBJECT Axle Swivel and Hub Seals - New Part Information	03/12/93 170/93/EN
MODEL Range Rover Discovery Defender	AFFECTED VEHICLES All derivatives from LH 632889 - LJ 045306 - LD925688	

FRONT AXLE HUB ASSEMBLY

C - Front axle oil seal - FTC3145 fitted with spring outwards.
D - Swivel housing oil seal - FTC 3401
E - Front axle hub oil seal - FTC 2783 (Springless type)

A post-rh ... N 22
Absorber-crankshaft vibration ... B 80
Absorber-crankshaft vibration ... B112
Abutment-cable accelerator ... B 98
Actuator assembly cruise control ... B 65
Actuator assembly cruise control ... B140
Actuator-central door locking ... T 33
Actuator-central door locking-driver ... N 78
Actuator-central door locking-passenger ... N 84
Actuator-central door locking-passenger ... N 82
Actuator-vacuum blower-heater ... L 20
Actuator-waste gate turbocharger ... B 30
Actuator-waste gate turbocharger ... B 67
Adapter-crankcase breather ... B 91
Adaptor assembly-oil filter ... B 15
Adaptor assembly-oil filter ... B 87
Adaptor assembly-oil filter ... B122
Adaptor-air conditioning compressor hose ... L 42
Adaptor-carburettor ... B 98
Adaptor-exhaust ... K 5
Adaptor-exterior mirror assembly motor ... N117
Adaptor-exterior mirror assembly motor ... N113
Adaptor-exterior mirror assembly motor ... N118
Adaptor-exterior mirror assembly motor ... N114
Adaptor-inlet manifold ... B 94
Adaptor-inlet manifold bypass ... B129
Adaptor-oil drain pipe turbocharger ... B 31
Adaptor-oil drain pipe turbocharger ... B 67
Adaptor-oil filter ... B 15
Adaptor-oil filter ... B 53
Adaptor-oil filter ... B122
Adaptor-pipe coolant ... B 17
Adaptor-pipe coolant ... B 55
Adaptor-pipe coolant ... L 2
Adaptor-pipe coolant ... L 5
Adaptor-pipe coolant ... L 7
Adaptor-pipe-dual male ... B 28
Adaptor-pipe-dual male ... C 42
Adaptor-pipe-dual male ... C 93
Adaptor-pipe-dual male ... C 95
Adaptor-pipe-dual male ... C 97
Adaptor-pipe-dual male ... L 42
Adaptor-transmission mounting plate ... B 40
Adaptor-transmission mounting plate ... C 71
Adaptor-water pump drive ... B 85
Adaptor-water pump drive ... B 86
Adaptor-water pump drive ... B120
Adjuster assembly-transfer box brake shoe ... D 22
Adjuster seatbelt height ... R 55
Adjuster seatbelt height ... R 54
Adjuster unit-headlamp ... M 3
Aerial front fender ... M132
Aerial front fender ... M133
Aerial front fender ... M136
Aerial front fender ... M138
Aerial-receiver central door locking & alarm ... M106
Aerial.-electric aerial ... T 46
Air cleaner assembly ... J 14
Air cleaner assembly ... J 15
Air cleaner assembly ... J 16
Air cleaner assembly ... J 17
Air cleaner assembly ... J 18
Air cleaner assembly ... J 19
Air conditioning sub assembly ... L 50
Air conditioning sub assembly-rhd ... L 39
Alternator assembly-127/65 amp-new-lh ... B 36
Alternator assembly-a127-100 amp-new ... B155
Alternator assembly-a127-65 amp-new-rh- ... B105
Alternator assembly-a127-65 amp-new-rh- ... B152
Alternator assembly-a127-72amp-new ... B152
Alternator assembly-a127i-100 amp-new ... B 73
Alternator assembly-a127i-100 amp-new ... B155
Alternator assembly-a127i-100 amp-new ... B186
Alternator assembly-a127i-85 amp-new ... B153
Alternator assembly-a133-80 amp-new ... B154
Amplifier assembly-audio power ... M137
Amplifier assembly-audio power ... M135
Amplifier assembly-audio power-footwell ... M140
Amplifier assembly-audio power-footwell ... M143
Amplifier audio power-rear end door ... M136
Amplifier audio power-rear end door ... M134
Amplifier audio power-rear end door ... M142
Amplifier audio power-rear end door ... M140
Amplifier-audio antenna-am/fm-pioneer ... M141
Amplifier-audio antenna-lh rear quarter ... M143
Amplifier-sounder unit catalyst overheat ... M 90
Antifreeze-205ltr ... T 28
Arm & bush assembly-exhaust valve rocker- ... B 18
Arm & bush assembly-inlet valve rocker-lh ... B 18
Arm assembly-backlight wiper ... M151
Arm assembly-backlight wiper ... M152
Arm assembly-radius front suspension ... E 43
Arm link - sunroof assembly-rh ... N 35
Arm link - sunroof assembly-rh ... N 36
Arm-clutch release clutch ... C 2
Arm-clutch release clutch ... C 24
Arm-clutch release clutch ... C 44
Arm-clutch release clutch ... C 45
Arm-clutch release clutch ... C 46
Arm-glovebox stay ... P 10
Arm-radius front suspension-rhd ... E 44
Arm-rotor distributor ... B101
Arm-rotor distributor ... B147
Arm-transfer box differential lock ... D 20
Arm-transfer box differential lock connecting ... D 20
Arm-transfer box gearchange ... D 16
Arm-windscreen wiper-rhd ... M144
Arm-windscreen wiper-rhd ... M145
Armrest assembly-rear door-dark granite ... P 38
Armrest assembly-rear door-sonar blue ... P 36
Armrest assembly-rear door-sonar blue ... P 41
Armrest-front door-rh-bahama beige ... P 31
Armrest-front door-rh-bahama beige ... P 34
Armrest-front door-rh-sonar blue ... P 29
Armrest-split rear seat-leather-dark granite ... R 43
Armrest-split rear seat-leather/cloth-dark ... R 45
Ashtray assembly-facia console ... P 14
Ashtray assembly-facia-ash grey ... P 12
Ashtray assembly-tunnel console ... P 19
Ashtray assembly-tunnel console ... P 20
Ashtray-door-ash grey ... P 28
Axle assembly-stub front suspension ... E 21
Axle-rear suspension hub stub ... E 40
Axle-stub-hub front suspension ... E 21
Axle-stub-hub front suspension ... E 19
Axle/drive unit assembly 4x4-front-rhd- ... E 3
Axle/drive unit assembly 4x4-front-rhd- ... E 4
Axle/drive unit assembly 4x4-front-rhd- ... E 5
Axle/drive unit-4x4 rear-asbestos free ... E 26
Axle/drive unit-4x4 rear-asbestos-less splash ... E 25
Backplate-transmission brake assembly ... D 22
Backplate-transmission brake assembly ... D 23
Badge ... N106
Badge steering wheel-sonar blue ... F 1
Badge-camel trophy-large ... V 5
Badge-gs model-mid silver ... N106
Badge-land rover ... N 42
Badge-land rover ... N 86
Badge-land rover-grey ... N122
Badge-land rover-grille radiator-silver ... N 43
Badge-premium ... N106
Badge-safari derivative ... N106
Baffle-air vent-rear ... N 31
Baffle-engine oil ... B 89
Baffle-oil separator ... B 21
Baffle-oil separator ... B 59
Bag-assembly-roof rack rails stowage ... S 1
Bag-console-bahama beige ... P 18
Bag-stowage-key & extractor ... E 42
Ball-detent manual transmission ... C 18
Ball-detent manual transmission ... C 20
Ball-detent manual transmission ... C 38
Ball-detent manual transmission ... C 40
Ball-detent manual transmission ... C 47
Ball-detent manual transmission ... C 66
Ball-detent manual transmission ... D 15
Ball-detent manual transmission ... D 14
Ball-selector yoke ... C 38
Ball-selector yoke ... C 66
Ball-selector yoke ... C 19
Ball-towing attachment-50mm ... T 20
Bar assembly-anti-roll rear suspension ... E 53
Bar assembly-front bumper centre ... T 10
Bar assembly-nudge-front ... T 10
Bar assembly-nudge-front-poly coated ... T 8
Bar assembly-nudge-front-poly coated ... T 9
Bar-anti-roll front suspension ... E 47
Bar-gearbox snubbing ... C 22
Bar-roof sports ... T 39
Bar-tie straight ... T 23
Bar-tie-bent forging-rhd ... F 12
Bar-underride protection ... N 11
Bar-vehicle recovery rear-lh ... T 21
Bar-vehicle recovery rear-rh ... T 20
Battery ... M102
Battery ... M103
Battery ... M104
Battery-072-bbms-wet ... M 20
Battery-072-bbms-wet ... M 21
Battery-091-bbms-standard-wet ... M 16
Battery-091-bbms-standard-wet ... M 17
Battery-091-bbms-standard-wet ... M 18
Battery-091-bbms-standard-wet ... M 19
Battery-torch key ... M107
Battery-torch key ... M109
Battery-transmitter burglar alarm ... T 35
Bc post-rh ... N 22
Bearing bush-connecting rod small end ... B 8
Bearing bush-connecting rod small end ... B 46
Bearing intermediate shaft assembly ... D 10
Bearing steering gear-power assisted steering ... F 14
Bearing-alternator front ... B 36
Bearing-alternator front ... B105
Bearing-alternator front ... B152
Bearing-alternator front ... B153
Bearing-ball ... B 14
Bearing-ball ... B119
Bearing-ball ... B144
Bearing-ball ... D 11
Bearing-ball ... D 12
Bearing-camshaft front ... B 4
Bearing-camshaft front ... B 43
Bearing-camshaft rear ... B 4
Bearing-camshaft rear ... B 43
Bearing-clutch release ... B 39
Bearing-clutch release ... B108
Bearing-clutch release ... B160
Bearing-clutch release ... C 2
Bearing-clutch release ... C 24
Bearing-clutch release ... C 44
Bearing-clutch release ... C 45
Bearing-clutch release ... C 46
Bearing-front/rear axle hub ... E 23
Bearing-front/rear axle hub ... E 24
Bearing-front/rear axle hub ... E 39
Bearing-front/rear axle hub ... E 40
Bearing-front/rear axle hub ... E 21
Bearing-front/rear axle hub ... E 19
Bearing-needle roller ... C 12
Bearing-needle roller ... C 13

Bearing-needle roller .... C 31
Bearing-needle roller .... C 32
Bearing-needle roller .... C 65
Bearing-needle roller .... E 21
Bearing-needle roller caged-rear B165
Bearing-needle roller manual transmission .... C 13
Bearing-needle roller manual transmission .... C 32
Bearing-needle roller manual transmission .... C 58
Bearing-needle roller manual transmission .... C 60
Bearing-needle roller manual transmission .... C 61
Bearing-needle roller manual transmission .... C 14
Bearing-needle roller manual transmission .... C 62
Bearing-needle roller manual transmission .... C 34
Bearing-taper roller .... C 15
Bearing-taper roller .... C 36
Bearing-taper roller .... C 47
Bearing-taper roller .... C 56
Bearing-taper roller .... C 14
Bearing-taper roller .... C 30
Bearing-taper roller .... C 11
Bearing-taper roller .... C 35
Bearing-taper roller .... D 15
Bearing-taper roller .... D 14
Bearing-taper roller .... E 11
Bearing-taper roller .... E 36
Bearing-taper roller .... E 10
Bearing-taper roller .... E 32
Bearing-taper roller differential .... E 11
Bearing-taper roller differential .... E 36
Bearing-taper roller-lower .... E 20
Bearing-taper roller-lower .... E 21
Bearing-taper roller-lower .... E 19
Bearing-taper roller-man trans .... C 64
Bearing-taper roller-man trans .... C 65
Bearing-taper roller-man trans .... C 62
Bellcrank-door .... N 80
Bellcrank-door .... N 78
Belt-coolant pump vee .... B 99
Belt-coolant pump vee B142
Belt-engine timing .... B 11
Belt-engine timing .... B 50
Belt-engine timing B169
Belt-polyvee air conditioning compressor .... B 35
Belt-polyvee air conditioning compressor .... B 71
Belt-polyvee alternator .... B 36
Belt-polyvee alternator .... B 86
Belt-polyvee alternator B100
Belt-polyvee alternator B105
Belt-polyvee alternator B121
Belt-polyvee alternator B144
Belt-polyvee alternator B145
Belt-polyvee alternator B154
Belt-polyvee alternator B186
Belt-polyvee alternator B153
Belt-polyvee coolant pump .... B 52

Belt-vee power assisted steering .... B 13
Belt-vee power assisted steering .... B 34
Bezel assembly windscreen demist-ash grey .... P 8
Bezel-facia headlamp levelling switch-rhd- .... M 66
Bezel-facia instrument-black .... P 14
Bezel-manual sunroof-mist grey .... P 72
Bezel-side window demist facia-ash grey-lh .... P 7
Bezel-side window demist facia-rh-ash .... P 4
Bezel-sunroof switch-mist grey .... M 76
Bezel-sunroof switch-mist grey .... P 72
Bimetal assembly-carburettor choke .... B 97
Bin assembly-front door-rh-bahama beige .... P 31
Bin assembly-front door-rh-dark granite .... P 34
Bin assembly-front door-rh-sonar blue .... P 28
Bin assembly-tunnel console stowage-ash .... P 14
Bin assembly-tunnel console stowage-beige .... T 19
Blade-backlight wiper M151
Blade-backlight wiper M152
Blade-windscreen wiper M144
Blade-windscreen wiper M145
Blank-centre panel switch-tunnel console-rh .... M 77
Blank-centre panel switch-tunnel console-rh .... P 21
Blank-electric window lift switch-tunnel .... M 77
Blank-facia headlamp levelling .... M 66
Blank-facia headlamp levelling .... M 10
Blank-facia radio aperture M133
Blank-facia radio aperture-ash M132
Blank-facia switch .... M 69
Blank-hole air conditioning-sonar blue .... P 1
Blank-instrument bezel switch .... M 65
Blank-instrument bezel switch .... M 72
Blank-non alarmed radio-ash grey .... P 12
Block-load space cover-retaining-ash grey .... P 22
Blower assembly heater-rhd .... L 27
Blower assembly heater-rhd .... L 38
Blower assembly-air conditioning .... L 49
Blower assembly-air conditioning-rhd .... L 46
Board-armrest.-leather-dark granite .... R 45
Board-armrest.-leather-dark granite .... R 43
Body assembly-multi point injection throttle B180
Body shell .... N 16
Body shell .... N 17
Body shell .... N 18
Body side-rear-lh .... V 2
Body side-rear-rh .... N 29
Body-carburettor-lh .... B 97
Body-transmission oil pump .... C 6
Body-transmission oil pump .... C 28
Bolt .... B 4
Bolt .... B 18
Bolt .... B 19
Bolt .... B 33
Bolt .... B 43
Bolt .... B 57
Bolt B184

Bolt .... C 5
Bolt .... C 27
Bolt .... C 53
Bolt .... D 22
Bolt .... D 23
Bolt .... E 11
Bolt .... E 23
Bolt .... E 36
Bolt .... E 39
Bolt .... E 49
Bolt .... E 44
Bolt .... F 15
Bolt .... G 9
Bolt .... G 6
Bolt .... G 8
Bolt .... H 15
Bolt .... H 18
Bolt .... J 13
Bolt & washer assembly-sump-5/16unc B115
Bolt & washer assembly-sump-sump to .... B 82
Bolt-1/2unf x 6 1/4 .... N 11
Bolt-1/4unc x 1 1/4 .... B 86
Bolt-1/4unc x 2 .... B119
Bolt-1/4unc x 7/8 .... B 87
Bolt-1/4unc x 7/8 .... B122
Bolt-3/8-unf .... E 1
Bolt-3/8-unf .... E 2
Bolt-3/8unc x 1 1/4 .... D 11
Bolt-3/8unc x 1 1/4 .... F 16
Bolt-3/8unc x 1 7/8 .... C 71
Bolt-3/8unc x 2.79 .... B 89
Bolt-3/8unf .... N 12
Bolt-3/8unf x 1 3/4 .... D 12
Bolt-3/8unf x 5 .... B100
Bolt-5/16unc x 3 .... B117
Bolt-5/16unc x 3.8 .... B118
Bolt-5/16unf x 1 5/8 .... B 80
Bolt-5/16unf x 1 5/8 .... B113
Bolt-7 seats-m8 x 25 .... L 43
Bolt-7/16unc x 1 1/8 .... B 92
Bolt-7/16unc x 2.71 .... B 88
Bolt-7/16unc x 2.71 .... B125
Bolt-7/16unc x 2.71 .... B124
Bolt-7/16unf x 2 .... F 12
Bolt-alternator to bracket-flanged head-m8 x .... B 73
Bolt-banjo .... B 31
Bolt-banjo .... B 62
Bolt-banjo .... B 66
Bolt-banjo .... B 67
Bolt-banjo .... B 94
Bolt-banjo .... B 65
Bolt-banjo .... D 3
Bolt-banjo .... E 6
Bolt-banjo .... E 27
Bolt-banjo fuel lines .... B 64

Bolt-banjo-6mm .... B 29
Bolt-banjo-boost capsule .... B 28
Bolt-banjo-pipe fuel to fuel filter .... J 11
Bolt-bracket mounting to mounting engine .... N 4
Bolt-bracket mounting-1/4unc x 1 1/2- B142
Bolt-bracket to pump-flanged head-m8 x 50 .... B 64
Bolt-caliper fixing-7/16"-unf .... H 16
Bolt-camshaft retaining plate-m12 x 45 .... B 60
Bolt-compressor to bracket-3/8unf x 5 .... B 35
Bolt-compressor-3/8unf x 5 .... B144
Bolt-connecting rod .... B 8
Bolt-connecting rod .... B 46
Bolt-connecting rod B166
Bolt-connecting rod-11/32uns-3a .... B 81
Bolt-connecting rod-11/32uns-3a B114
Bolt-coolant pump to front cover-1/4unc x 2 .... B 85
Bolt-coolant pump to front cover-m8 x 55 .... B 12
Bolt-crankshaft pulley .... B 7
Bolt-crankshaft pulley .... B 45
Bolt-crankshaft pulley .... B 80
Bolt-crankshaft pulley B165
Bolt-crankshaft pulley B113
Bolt-cylinder head fixing B172
Bolt-cylinder head fixing B173
Bolt-cylinder head fixing-7/16unc x 2 1/4 .... B 88
Bolt-cylinder head fixing-7/16unc x 2 1/4 .... B123
Bolt-cylinder head fixing-7/16unc x 2 1/4 .... B125
Bolt-cylinder head fixing-7/16unc x 3.91 .... B124
Bolt-cylinder head fixing-short-m12 x 100 .... B 16
Bolt-cylinder head fixing-short-m12 x 100 .... B 54
Bolt-dog point B113
Bolt-fan to viscous coupling-m6 x 14 .... B 13
Bolt-fixing ecu-m6 x 65 .... M111
Bolt-fixing flywheel crankshaft B189
Bolt-fixing flywheel crankshaft-long .... B 39
Bolt-fixing flywheel crankshaft-short .... B 40
Bolt-fixing flywheel crankshaft-short .... B 75
Bolt-fixing flywheel crankshaft-short .... B 76
Bolt-flanged head B148
Bolt-flanged head .... L 51
Bolt-flanged head .... M 81
Bolt-flanged head-3/8unc x 5 1/4 .... B143
Bolt-flanged head-7/16unf x 1 1/8 .... E 21
Bolt-flanged head-7/16unf x 1 1/8 .... E 19
Bolt-flanged head-alternator-m8 x 110 .... B155
Bolt-flanged head-breather chamber-m6 x 55 .... B 59
Bolt-flanged head-cover to block-m8 x 35 .... B 10
Bolt-flanged head-elbow to housing-m6 x 40 .... B 23
Bolt-flanged head-front cover to block-m8 x .... B 48
Bolt-flanged head-jacket cover-1/4unc x 1 .... B136
Bolt-flanged head-m10 x 100 .... B183
Bolt-flanged head-m10 x 40 .... D 4
Bolt-flanged head-m10 x 45 .... B171
Bolt-flanged head-m10 x 45 .... D 4
Bolt-flanged head-m10 x 75 .... B170

Bolt-flanged head-m6 x 25 .... B174
Bolt-flanged head-m6 x 40 .... H 6
Bolt-flanged head-m6 x 40 .... H 8
Bolt-flanged head-m6 x 50 .... B165
Bolt-flanged head-m6 x 60 .... B 28

Bolt-flanged head-m8 .... L 47
Bolt-flanged head-m8 x 120 .... B 51
Bolt-flanged head-m8 x 125 .... B 34
Bolt-flanged head-m8 x 125 .... B170
Bolt-flanged head-m8 x 25 .... B 65

Bolt-flanged head-m8 x 35 .... B 49
Bolt-flanged head-m8 x 40 .... C 44
Bolt-flanged head-m8 x 40 .... C 45
Bolt-flanged head-m8 x 40 .... C 46
Bolt-flanged head-m8 x 40 .... C 55

Bolt-flanged head-m8 x 50 .... B172
Bolt-flanged head-m8 x 50 .... B176
Bolt-flanged head-m8 x 50 .... B187
Bolt-flanged head-m8 x 50 .... B 49
Bolt-flanged head-m8 x 50 .... C 2

Bolt-flanged head-m8 x 50 .... K 11
Bolt-flanged head-m8 x 50 .... K 13
Bolt-flanged head-m8 x 50 .... N 7
Bolt-flanged head-m8 x 55 .... C 67
Bolt-flanged head-m8 x 55 .... C 55

Bolt-flanged head-m8 x 55 .... D 18
Bolt-flanged head-m8 x 55 .... D 17
Bolt-flanged head-m8 x 60 .... C 28
Bolt-flanged head-m8 x 60 .... C 7
Bolt-flanged head-m8 x 85-long .... B 62

Bolt-flanged head-m8 x 95 .... B145
Bolt-flanged head-rail to inlet manifold-m6 x B179
Bolt-flanged head-side cover to block-m8 x .... B 21
Bolt-flanged head-squab to frame-m8 .... R 8
Bolt-flanged head-valve to body .... H 9

Bolt-flanged head-valve to body .... H 10
Bolt-flywheel to crankshaft B189
Bolt-front cover to block-5/16unc x 3 .... B 84
Bolt-front cover to block-flanged head-m8 x .... B 48
Bolt-gear to camshaft-m12 x 30 .... B128

Bolt-hexagonal head-3/8unc x 2 .... B 95
Bolt-hexagonal head-3/8unc x 2 .... C 52
Bolt-hexagonal head-5/16unc x 4 3/4 .... B120
Bolt-hexagonal head-7/16" .... E 18
Bolt-hexagonal head-m10 x 100 .... N 1

Bolt-hexagonal head-m10 x 110 .... E 52
Bolt-hexagonal head-m10 x 35 .... E 20
Bolt-hexagonal head-m6 x 30 .... C 6
Bolt-hexagonal head-m6 x 30 .... H 11
Bolt-hexagonal head-m8 .... F 8

Bolt-hexagonal head-m8 x 100 .... M114
Bolt-hexagonal head-m8 x 100 .... N 2
Bolt-hexagonal head-m8 x 100 .... N 9
Bolt-hexagonal head-m8 x 55 .... N 53
Bolt-hexagonal head-m8 x 70 .... H 21

Bolt-hexagonal head-m8 x 70 .... K 10
Bolt-hexagonal head-m8 x 70 .... K 12
Bolt-hexagonal head-m8 x 70 .... N121
Bolt-hexagonal head-m8 x 90 .... D 3
Bolt-hexagonal socket .... E 48

Bolt-hexagonal socket-m12 x 100 .... F 11
Bolt-hexagonal socket-m12 x 50 .... E 7
Bolt-hexagonal socket-m12 x 50 .... E 28
Bolt-hexagonal socket-m12 x 50 .... E 8
Bolt-hexagonal socket-m12 x 50 .... E 29

Bolt-hexagonal socket-m12 x 50 .... E 30
Bolt-hexagonal socket-m12 x 50 .... E 31
Bolt-housing to block-flanged head-m10 x .... B 75
Bolt-housing to block-hexagonal head- .... C 72
Bolt-housing to block-m10 x 45 .... B 39

Bolt-inlet manifold to cylinder head-m8 x 90 .... B 24
Bolt-ladder frame to block-flanged head-m8 .... B 9
Bolt-m10 .... H 14
Bolt-m10 .... H 17
Bolt-m10 x 45 .... D 5

Bolt-m10 x 45 .... E 24
Bolt-m10 x 45 .... E 40
Bolt-m10 x 55 .... B 40
Bolt-m10 x 60-long .... D 7
Bolt-m10 x 60-long .... G 1

Bolt-m10-hexagonal .... E 13
Bolt-m10-hexagonal .... E 14
Bolt-m10-hexagonal .... E 16
Bolt-m12 x 45 .... C 4
Bolt-m12 x 45 .... C 26

Bolt-m5 .... D 20
Bolt-m6 x 20 .... N 27
Bolt-m6 x 20 .... N 20
Bolt-m6 x 20 .... N 30
Bolt-m6 x 30 .... B168

Bolt-m8 x 25 .... L 22
Bolt-m8 x 25 .... L 27
Bolt-m8 x 25 .... L 38
Bolt-m8 x 25 .... L 39
Bolt-m8 x 30 .... D 6

Bolt-m8 x 35 .... F 9
Bolt-m8 x 50 .... K 1
Bolt-manifold-hexagonal head-3/8unc x 1 1/2 .... B130
Bolt-output shaft to mainshaft-m10 x 130 .... C 85
Bolt-pawl to housing .... C 84

Bolt-plate cover/front cover to block-flanged .... B 10
Bolt-plate to body .... T 20
Bolt-pulley to camshaft-hexagonal head-m10 .... B 22
Bolt-ram pipes housing-hexagonal head- B138
Bolt-rocker shaft to head-flanged head-m8 x .... B 56

Bolt-rocker shaft-3/8unc x 2.79 .... B126
Bolt-sear belt lower anchorage .... R 54
Bolt-seat base to plinth .... R 20
Bolt-seat base to plinth .... R 18
Bolt-shear head .... B 64

Bolt-shear-lock steering column-m8 x 12 .... N 73
Bolt-shear-lock steering column-m8 x 12 .... N 74
Bolt-shear-lock steering column-m8 x 12 .... N 75
Bolt-shear-lock steering column-m8 x 20 .... F 6
Bolt-shear-lock steering column-m8 x 20 .... N 71

Bolt-shear-lock steering column-m8 x 20 .... N 72
Bolt-short-high tensile .... E 43
Bolt-shouldered-hexagonal head-m12 x 50 .... E 21
Bolt-shouldered-hexagonal head-m12 x 50 .... E 19
Bolt-shouldered-hexagonal head-m12 x 50 .... H 14

Bolt-split cable to pedal-m10 x 60-long .... G 4
Bolt-starter motor to bell housing-m10 x 45 .... B 37
Bolt-starter ring-5/16unf x 0.58 .... B161
Bolt-steering shaft-hexagonal head-m8 .... F 7
Bolt-strut alternator B157

Bolt-support bracket-flanged head-m8 x 80 .... B182
Bolt-tensioner to cover-flanged head-m10 x .... B 11
Bolt-tensioner to cover-flanged head-m10 x .... B 50
Bolt-terminal cable-battery .... M 16
Bolt-terminal cable-battery .... M 17

Bolt-terminal cable-battery .... M 18
Bolt-terminal cable-battery .... M 19
Bolt-terminal cable-battery .... M 20
Bolt-terminal cable-battery .... M 21
Bolt-timing belt tensioner pedestal B169

Bolt-torx-pan-seat belt lower anchorage .... R 55
Bolt-tow ball to plate .... T 21
Bolt-u .... B 31
Bolt-u .... K 10
Bolt-u .... K 11

Bolt-u .... K 12
Bonnet assembly .... N 41
Box assembly-brush starter motor .... B 37
Box assembly-brush starter motor .... B 38
Box assembly-brush starter motor .... B 74

Box assembly-brush starter motor B159
Box assembly-steering-rhd-new .... F 10
Box-gun .... T 18
Box-key pad radio M141
Box-key pad radio M139

Box-luggage .... T 41
Bracket .... B 6
Bracket .... B 14
Bracket .... B 37
Bracket .... B 91

Bracket .... B 55
Bracket .... F 16
Bracket .... G 11
Bracket .... N 50
Bracket N110

Bracket N123
Bracket air conditioning .... L 46
Bracket assembly accelerator B139
Bracket assembly power assisted steering .... F 17
Bracket assembly power assisted steering .... F 18

Bracket assembly power assisted steering .... L 2
Bracket assembly power assisted steering .... L 8
Bracket assembly-electric control unit M111
Bracket assembly-facia centre support-rear .... P 3
Bracket assembly-facia upper end support .... P 3

Bracket assembly-mounting air flow meter .... L 17
Bracket assembly-mounting alternator B187
Bracket assembly-rear damper-upper .... E 52
Bracket assembly-rear seat mounting-rh- .... R 51
Bracket assembly-sensor diesel .... L 17

Bracket assembly-throttle wire mounting B139
Bracket condenser-upper .... L 40
Bracket exhaust system .... K 11
Bracket exhaust system .... K 12
Bracket exhaust system-centre .... K 6

Bracket exhaust system-centre .... K 8
Bracket exhaust system-centre .... K 9
Bracket exhaust system-centre .... K 4
Bracket exhaust system-centre .... K 7
Bracket exhaust system-front .... K 2

Bracket exhaust system-front .... K 13
Bracket exhaust system-front .... K 15
Bracket fuel tank-front .... J 1
Bracket fuel tank-front .... J 2
Bracket fuel tank-front .... J 3

Bracket fuel tank-front .... J 4
Bracket fusebox .... M 82
Bracket harness .... B 39
Bracket harness .... B 69
Bracket harness .... B 75

Bracket harness .... B 76
Bracket harness B160
Bracket harness .... B 55
Bracket harness .... B 65
Bracket harness B120

Bracket harness M154
Bracket harness .... M 40
Bracket mounting-for edc engine harness .... M 51
Bracket mounting-for edc engine harness .... M 52
Bracket mounting-for edc engine harness .... M 53

Bracket mounting-rh .... N 55
Bracket power assisted steering pump & .... B 34
Bracket speedometer cable M114
Bracket- compressor/tensioner assembly air .... B 10
Bracket- compressor/tensioner assembly air .... F 7

Bracket- compressor/tensioner assembly air .... H 24
Bracket-5th gear selector fork support .... C 19
Bracket-5th gear selector fork support .... C 39
Bracket-actuator assembly cruise control .... B 65
Bracket-air bag mounting .... M 61

Bracket-air bag mounting .... M 62
Bracket-air charge/body air cooler .... L 44
Bracket-air charge/body air cooler .... L 51
Bracket-air compressor B100
Bracket-air compressor B144

Bracket-air conditioning duct-7 seats L 43
Bracket-alarm switch mounting M108
Bracket-alarm switch mounting M110
Bracket-alarm switch mounting M106
Bracket-ancillary mounting B145

Bracket-ancillary mounting B156
Bracket-ancillary mounting N 5
Bracket-body/chassis mounting-rh N 2
Bracket-bonnet stay pivot N 69
Bracket-brake pipe tee connector H 13

Bracket-brake pipe tee connector H 12
Bracket-cable accelerator B139
Bracket-charcoal canister J 20
Bracket-charcoal canister J 21
Bracket-charcoal canister J 22

Bracket-clamp roll bar mounting E 47
Bracket-clamp roll bar mounting E 53
Bracket-clipping-steel M154
Bracket-compact disc mounting M137
Bracket-compact disc mounting M135

Bracket-compact disc mounting-rh M140
Bracket-compact disc mounting-rh M143
Bracket-compressor air conditioning B 35
Bracket-compressor air conditioning B 71
Bracket-compressor air conditioning S 1

Bracket-compressor air conditioning-sanden B184
Bracket-coolant & power assisted steering B170
Bracket-coolant hose support B 51
Bracket-coolant pump support B 51
Bracket-coolant pump support B 85

Bracket-courtesy lamp M 12
Bracket-courtesy lamp M 13
Bracket-damper clutch G 13
Bracket-drive end c/w bearing alternator e- B152
Bracket-drive end c/w bearing alternator e- B153

Bracket-drive end c/w bearing alternator e-lh B 36
Bracket-drive end c/w bearing alternator e- B105
Bracket-electric control unit mounting M 97
Bracket-electronic control unit mounting M 91
Bracket-engine mounting-lh B 83

Bracket-engine mounting-lh N 4
Bracket-engine mounting-rh B 83
Bracket-engine mounting-rh N 5
Bracket-engine mounting-rh N 6
Bracket-engine mounting-rh N 7

Bracket-exhaust gas recirculation valve B 25
Bracket-exhaust mounting K 10
Bracket-exhaust mounting K 4
Bracket-exhaust mounting/lashing hanger K 1
Bracket-expansion tank L 12

Bracket-expansion tank L 13
Bracket-expansion tank L 14
Bracket-expansion tank L 15
Bracket-facia end support P 3
Bracket-facia grab handle reinforcement-rhd P 6

Bracket-facia instrument mounting P 11
Bracket-facia instrument mounting-lower P 5
Bracket-facia support P 3
Bracket-fender support-front N 28
Bracket-front bumper support N 13

Bracket-front door pull-front P 31
Bracket-front door pull-front P 34
Bracket-front door pull-front P 29
Bracket-front seat belt console mounting P 14
Bracket-front spring mounting-upper E 46

Bracket-front-rh N110
Bracket-fuel filter J 12
Bracket-fuel filter J 9
Bracket-fuel filter support J 13
Bracket-glovebox P 1

Bracket-handbrake cable mounting clip H 24
Bracket-handbrake cable relay lever H 22
Bracket-handbrake cable relay lever H 21
Bracket-handbrake cable relay lever H 24
Bracket-harness protector M 44

Bracket-harness protector M 45
Bracket-headlining support P 72
Bracket-heater blower mounting L 27
Bracket-heater blower mounting L 38
Bracket-heatshield exhaust system B 38

Bracket-heatshield exhaust system B 96
Bracket-heatshield exhaust system B131
Bracket-hose brake-lh-front H 8
Bracket-hose brake-rh-front E 22
Bracket-hose brake-rh-front H 5

Bracket-hose brake-rh-front H 7
Bracket-hose brake-rh-front H 9
Bracket-hose brake-rh-front H 10
Bracket-hose brake-rh-front H 6
Bracket-idler mounting ancillary drive B187

Bracket-injection pump diesel B 28
Bracket-injection pump diesel B 64
Bracket-injection pump diesel B 65
Bracket-inlet manifold support B 24
Bracket-inlet manifold support B 94

Bracket-inlet manifold support B177
Bracket-jack stowage mounting S 1
Bracket-lifting engine front B143
Bracket-lifting engine front B 17
Bracket-lifting engine front B 55

Bracket-lifting engine front B124
Bracket-lifting engine rear B 16
Bracket-lifting engine rear B 90
Bracket-lifting engine rear B125
Bracket-lifting engine rear B 17

Bracket-lifting engine rear B124
Bracket-load space cover retaining block P 22
Bracket-load/tail door N 83
Bracket-load/tail door-centre N 86
Bracket-long range driving lamp mounting T 10

Bracket-modulator antilock brakes H 3
Bracket-mounting M 91
Bracket-mounting air cleaner J 15
Bracket-mounting fuel lines-front B179
Bracket-mounting oil cooler C 42

Bracket-mounting oil cooler C 69
Bracket-mounting oil cooler C 92
Bracket-mounting oil cooler C 94
Bracket-mounting oil cooler C 96
Bracket-mounting oil pipe C 42

Bracket-mounting oil pipe C 69
Bracket-mounting oil pipe C 93
Bracket-mounting oil pipe C 95
Bracket-mounting oil pipe C 97
Bracket-mounting power assisted steering B 70

Bracket-mounting power assisted steering B 99
Bracket-mounting power assisted steering B142
Bracket-mounting power assisted steering B182
Bracket-mounting power assisted steering B143
Bracket-mounting-alternator-65 amp B106

Bracket-mounting-alternator-65 amp B156
Bracket-mounting-duct induction system J 17
Bracket-mounting-exhaust N 2
Bracket-mounting-facia P 3
Bracket-mounting-solenoid valve B 26

Bracket-mounting-steel M116
Bracket-mounting-steel M154
Bracket-pulley support belt tensioner B 10
Bracket-pulley support belt tensioner B 85
Bracket-pulley support belt tensioner B119

Bracket-radiator mounting-rh L 11
Bracket-radiator mounting-top L 1
Bracket-radiator mounting-top L 2
Bracket-radiator mounting-top L 4
Bracket-radiator mounting-top L 5

Bracket-radiator mounting-top L 8
Bracket-radiator upper mounting L 11
Bracket-radio housing M137
Bracket-radio housing M139
Bracket-radio housing-philips M142

Bracket-rear N 76
Bracket-rear N 80
Bracket-rear axle fulcrum E 49
Bracket-rear bumper mounting N 14
Bracket-rear lamp retaining M 7

Bracket-rear lamp retaining M 10
Bracket-rear seat belt lower anchorage R 59
Bracket-rear seat pivot-upper R 40
Bracket-rear seat pivot-upper R 38
Bracket-rear seat release catch-rh R 46

Bracket-rear-lh N111
Bracket-rear-rh N111
Bracket-relay M 84
Bracket-relay M 85
Bracket-relay M 86

Bracket-relay M 89
Bracket-relay mounting M 83
Bracket-relay mounting M 87
Bracket-relay mounting M 89
Bracket-relay mounting-electric seat M 88

Bracket-return spring-attachment accelerator B 98
Bracket-rh E 49
Bracket-rh R 51
Bracket-roof bow support N 34
Bracket-shield rear brake H 17

Bracket-shield rear brake H 18
Bracket-side-lh P 3
Bracket-speaker rear-lh P 62
Bracket-speaker/e post-lh L 43
Bracket-steering column-lhd-upper F 5

Bracket-steering column-without harness M 40
Bracket-sunroof mounting P 72
Bracket-support alternator B187
Bracket-support selector cable C 86
Bracket-support selector cable C 88

Bracket-support starter motor B 38
Bracket-support-lead ignition B151
Bracket-support-lead ignition-centre B103
Bracket-tail door door pull-rh P 39
Bracket-tail door door pull-rh P 42

Bracket-towing T 21
Bracket-towing-front & rear-less spoiler N 2
Bracket-towing-front & rear-less spoiler T 20
Bracket-transmission mounting C 22
Bracket-transmission mounting-rh C 68

Bracket-transmission mounting-rh C 90
Bracket-transmission mounting-rh N 8
Bracket-transmission mounting-rh N 10
Bracket-transmission mounting-rh-rear N 9
Bracket-transmission retaining D 6

Bracket-tunnel console mounting L 35
Bracket-vent assembly central facia P 4
Brake fluid T 28
Breather assembly-crankcase B 6
Breather assembly-crankcase B 21

Brush alternator B154
Buckle-front seat belt-ash grey-non audible R 53
Buckle-front seat belt-ash grey-non audible R 55
Buckle-lap belt-ash R 56
Buckle-lap belt-ash R 58

Buffer filler flap N 33
Buffer rubber-seat front R 18
Buffer rubber-seat front R 11
Buffer-bonnet N 70
Buffer-stop M 16

Buffer-stop M 18
Buffer-stop M 20
Bulb M 9
Bulb & holder assembly-black base-clear.- M128
Bulb & holder assembly-black base-clear.- M123

Bulb & holder assembly-clear. .... M 65
Bulb & holder assembly-clear.-black base M121
Bulb & holder assembly-clear.-black base M127
Bulb & holder assembly-grey M118
Bulb & holder assembly-orange .... L 28

Bulb & holder assembly-orange .... L 48
Bulb & holder assembly-orange .... M 70
Bulb & holder assembly-orange .... M 77
Bulb & holder assembly-orange .... M 71
Bulb & holder assembly-orange .... M 78

Bulb & holder assembly-orange .... M 72
Bulb & holder assembly-orange .... T 23
Bulb & holder assembly-with tachometer- .... P 13
Bulb instrument pack M121
Bulb instrument pack M127

Bulb-239-c11-5 watt-festoon .... M 13
Bulb-239-c11-5 watt-festoon .... M 15
Bulb-245-r19/10-10 watt .... M 12
Bulb-245-r19/10-10 watt .... M 13
Bulb-272-10 watt-festoon-36x10.5mm .... M 12

Bulb-272-10 watt-festoon-36x10.5mm .... M 13
Bulb-286-w5/1.2-1.2 watt-12v-1.2 watt .... L 28
Bulb-286-w5/1.2-1.2 watt-12v-1.2 watt .... L 48
Bulb-286-w5/1.2-1.2 watt-12v-1.2 watt .... M 63
Bulb-286-w5/1.2-1.2 watt-12v-1.2 watt M130

Bulb-286-w5/1.2-1.2 watt-12v-1.2 watt M131
Bulb-286-w5/1.2-1.2 watt-12v-1.2 watt .... M 68
Bulb-286-w5/1.2-1.2 watt-12v-1.2 watt .... P 13
Bulb-380-p25/2-21/5 watt .... M 8
Bulb-380-p25/2-21/5 watt .... M 9

Bulb-380-p25/2-21/5 watt-stop tail .... M 7
Bulb-382-p25/1-21 watt .... M 4
Bulb-382-p25/1-21 watt .... M 5
Bulb-382-p25/1-21 watt .... M 7
Bulb-382-p25/1-21 watt .... M 8

Bulb-382-p25/1-21 watt .... M 11
Bulb-382-p25/1-21 watt .... M 9
Bulb-501-w10/5-5 watt .... M 5
Bulb-501-w10/5-5 watt-side repeater-12v .... M 4
Bulb-501-w10/5-5 watt-sidelamp-12v .... M 3

Bulb-501-w10/5-5 watt-sidelamp-12v .... M 2
Bulb-headlamp-h3-55 watt .... M 6
Bulb-headlamp-h3-55 watt-type h3 .... T 23
Bulb-headlamp-h4-60/65 watt-main & .... M 3
Bulb-headlamp-h4-60/65 watt-quartz .... M 2

Bulb-headlamp-quartz halogen-cewe-type .... T 23
Bulb-wedge illumination .... P 16
Bulb-wedge illumination .... P 17
Bulkhead front assembly .... N 19
Bumper assembly-front-painted .... N 12

Bumper assembly-painted rear .... N 14
Bumper assembly-primed front .... T 7
Bumper cleaner-black-500ml .... T 28
Bush .... B 14
Bush .... B 94

Bush B106
Bush B156
Bush B165
Bush .... D 15
Bush .... D 16

Bush .... D 18
Bush .... D 20
Bush .... E 2
Bush .... F 3
Bush .... G 1

Bush .... G 4
Bush .... G 5
Bush .... G 7
Bush .... G 6
Bush .... G 8

Bush .... H 6
Bush .... H 8
Bush .... J 15
Bush .... M 83
Bush .... N 83

Bush .... N 82
Bush .... N 85
Bush .... N 79
Bush .... R 40
Bush .... R 38

Bush-1st gear selective-30.905 to 30.955mm .... C 59
Bush-1st gear selective-30.905 to 30.955mm .... C 62
Bush-1st gear selective-40.16 to 40.21mm .... C 13
Bush-1st gear selective-40.16 to 40.21mm .... C 33
Bush-3rd gear .... C 12

Bush-3rd gear .... C 31
Bush-anti-roll bar-front .... E 47
Bush-anti-roll bar-rear .... E 53
Bush-antilock brake system sensor .... E 20
Bush-antilock brake system sensor .... H 18

Bush-brake and clutch pedal .... G 9
Bush-change pivot manual transmission .... C 20
Bush-change pivot manual transmission .... C 40
Bush-change pivot manual transmission .... C 67
Bush-crankshaft .... B 7

Bush-crankshaft .... B 45
Bush-crankshaft .... B 80
Bush-crankshaft B112
Bush-cylinder head rocker arm .... B 18
Bush-cylinder head rocker arm .... B 56

Bush-damper mounting lower .... E 46
Bush-damper mounting lower .... E 51
Bush-damper mounting lower .... E 52
Bush-damper mounting upper-rear .... E 51
Bush-exhaust mounting rubber outer .... K 10

Bush-exhaust mounting rubber outer .... K 11
Bush-exhaust mounting rubber outer .... K 12
Bush-fixing ecu .... M 98
Bush-front suspension panhard rod .... E 43
Bush-front suspension radius arm to axle .... E 43

Bush-joint selector rod/yoke manual .... D 16
Bush-joint selector rod/yoke manual .... D 18
Bush-lower arm front/rear suspension .... E 47
Bush-lower arm front/rear suspension .... E 53
Bush-oil cooler to radiator mounting .... L 1

Bush-oil cooler to radiator mounting .... L 2
Bush-oil cooler to radiator mounting .... L 4
Bush-oil cooler to radiator mounting .... L 5
Bush-oil cooler to radiator mounting .... L 8
Bush-pedal pivot-brake .... G 9

Bush-rear suspension upper link .... E 49
Bush-rear-lower arm rear suspension .... E 48
Bush-retention M137
Bush-retention M139
Bush-retention M142

Bush-rubber-modulator antilock brakes- .... H 3
Bush-selector yoke spherical .... C 18
Bush-selector yoke spherical .... C 38
Bush-selector yoke spherical .... C 66
Bush-spherical B139

Bush-starter motor-centre .... B 37
Bush-starter motor-centre .... B 38
Bush-starter motor-centre .... B 74
Bush-starter motor-centre B159
Bush-throttle spindle tbi B135

Button sill-ash grey .... N 80
Button sill-ash grey .... N 83
Button sill-ash grey .... N 77
Button-lever release selector handle .... C 87
Button-lever release selector handle .... C 89

Button-rear seat squab release-winchester .... R 46
Buzzer passive restraint .... M 83
Cable assembly .... T 22
Cable assembly accelerator .... G 3
Cable assembly accelerator-lhd .... G 4

Cable assembly accelerator-lhd .... G 2
Cable assembly accelerator-rhd .... G 1
Cable assembly choke .... G 14
Cable assembly cruise control M112
Cable assembly handbrake .... H 22

Cable assembly handbrake .... H 23
Cable assembly handbrake .... H 21
Cable assembly starter & earth B107
Cable assembly starter & earth B158
Cable assembly starter & earth B164

Cable assembly starter & earth .... M 18
Cable assembly starter & earth .... M 19
Cable assembly starter & earth .... M 20
Cable assembly starter & earth .... M 21
Cable assembly-bonnet release assembly .... N 70

Cable assembly-kickdown automatic .... C 72
Cable speedometer-lhd M114
Cable speedometer-rhd M113
Cable-battery negative .... M 16
Cable-battery negative .... M 17

Cable-battery negative .... M 18
Cable-battery negative .... M 19
Cable-battery negative .... M 20
Cable-battery negative .... M 21
Cable-battery positive B158

Cable-battery positive .... M 16
Cable-battery positive .... M 17
Cable-battery positive .... M 18
Cable-battery positive .... M 19
Cable-battery positive .... M 20

Cable-battery positive .... M 21
Cable-control mode control-heater .... L 20
Cable-heater plug ignition .... B 16
Cable-heater plug ignition .... B 54
Cable-heater plug ignition-rhd .... M 50

Cable-selector automatic transmission .... C 86
Cable-selector automatic transmission .... C 88
Cable-slide assembly flexible drive-524mm .... R 21
Cable-wiper motor-rear end door .... M 42
Cable. .... M 16

Cable. .... M 17
Cable. .... M 19
Cable.-t post to starter .... M 18
Caliper assembly front brake-rh-asbestos- .... H 15
Caliper assembly rear brake-rh .... H 18

Camshaft assembly-engine .... B 22
Camshaft-engine .... B 60
Camshaft-engine .... B 92
Camshaft-engine B128
Camshaft-engine B175

Camshaft-reverse lock manual transmission .... C 55
Canister assembly charcoal .... J 21
Canister assembly charcoal .... J 22
Canister charcoal .... J 20
Canister charcoal .... J 21

Canister charcoal .... J 22
Cantrail assembly-rh .... N 24
Cap assembly-centre alloy wheel- .... E 41
Cap assembly-tank clutch .... G 11
Cap assembly-tank clutch .... G 12

Cap power assisted steering .... F 17
Cap-air box blanking .... J 19
Cap-backlight wiper spindle M151
Cap-backlight wiper spindle M152
Cap-clutch release lever pivot .... C 24

Cap-clutch release lever pivot .... C 46
Cap-cylinder head valve spring .... B 19
Cap-cylinder head valve spring .... B 57
Cap-cylinder head valve spring .... B 88
Cap-cylinder head valve spring B123

Cap-cylinder head valve spring B125
Cap-cylinder head valve spring B173
Cap-distributor ignition B101
Cap-distributor ignition B147
Cap-expansion tank pressure .... L 14

Cap-expansion tank pressure-15psi ... L 12
Cap-expansion tank pressure-15psi ... L 13
Cap-expansion tank pressure-15psi ... L 15
Cap-filler fuel filler-non-locking ... J 5
Cap-filler fuel filler-non-locking ... J 6

Cap-filter. ... B127
Cap-fluid level warning reservoir indicator ... H 1
Cap-front seat squab manual recline ... R 8
Cap-hub front suspension-without logo ... E 14
Cap-hub front suspension-without logo ... E 16

Cap-locking-nut road wheels-plastic-silver ... T 36
Cap-oil filler ... B 20
Cap-oil filler ... B 58
Cap-oil filler ... B 90
Cap-oil filler ... B127

Cap-oil filler ... B174
Cap-plenum moulding end-lh ... N 89
Cap-plenum pipe ... J 22
Cap-protective ... C 2
Cap-protective ... C 24

Cap-protective ... C 44
Cap-protective ... C 45
Cap-protective ... C 46
Cap-protective ... H 15
Cap-protective ... H 18

Cap-recliner-seat ... R 20
Cap-recliner-seat ... R 18
Cap-seat belt bolt head-ash ... R 56
Cap-seat belt bolt head-ash ... R 54
Cap-seat belt bolt head-ash grey ... P 79

Cap-stop lever ... B140
Cap-tamperproof ... B135
Cap-tamperproof ... B136
Cap-valve stem ... B 19
Cap-valve stem ... B 57

Cap-washer system ... M146
Cap-washer system ... M147
Cap-washer system ... M148
Cap-washer system ... M149
Cap-windscreen wiper spindle ... M144

Cap-windscreen wiper spindle ... M145
Capacitor alternator ... B 36
Capacitor alternator ... B105
Capacitor alternator ... B152
Capacitor alternator ... B154

Capacitor alternator-with lead ... B153
Capacitor ignition coil ... B104
Capacitor ignition coil ... B148
Capping-heater controls-wooden ... P 12
Capping-wood-facia ... V 3

Capping-wood-facia-rhd ... P 10
Capscrew-black ... P 62
Capscrew-dark granite ... P 63
Capscrew-dark granite ... R 25
Capsule-vacuum distributor ... B101

Capsule-vacuum distributor-2 pin ... B147
Car set-drop in mat-blue. ... T 32
Car set-seat cover-beige-waterproof ... T 12
Car set-seat cover-grey-waterproof ... T 11
Carburettor-engine-lh ... B 97

Carpet-centre floor wheelarch-rear-rh-sonar ... P 81
Carpet-centre floor wheelarch-rh-rear-flint ... P 82
Carpet-centre floor-charcoal ... P 79
Carpet-centre floor-sonar blue ... P 77
Carpet-front floor-rhd-charcoal ... P 78

Carpet-front floor-sonar blue ... P 76
Carpet-loadspace-blue ... T 32
Carpet-rear floor-sonar blue ... P 80
Carrier-gear automatic transmission ... C 76
Carrier-load/tail door spare wheel mounting ... N121

Carrier-sailboard-use with sports bars ... T 40
Carrier-ski roof rack-use with sports bars ... T 40
Carrier-trunk rack ski ... T 42
Cartridge-diesel fuel filter-heated ... J 12
Cartridge-engine oil filter ... B 15

Cartridge-engine oil filter ... B 53
Cartridge-engine oil filter ... B 87
Cartridge-engine oil filter ... B122
Cartridge-engine oil filter ... B171
Case assembly instrument pack ... M118

Case assembly instrument pack-less airbag ... M126
Case assembly instrument pack-less airbag ... M121
Case assembly-extension ... C 6
Case assembly-extension ... C 28
Case assembly-extension ... C 54

Case differential ... E 11
Case differential ... E 36
Case-extension-gearbox ... C 54
Case-extension-gearbox-rear ... C 6
Case-front axle ... E 6

Case-heater with air conditioning ... L 49
Case-heater with air conditioning-sonar blue ... P 15
Case-lower evaporator-air conditioning ... L 45
Case-lower evaporator-air conditioning ... L 49
Case-manual transmission ... C 3

Case-manual transmission ... C 25
Case-manual transmission ... C 47
Case-rear axle ... E 27
Case-remote plip ... M107
Case-remote plip ... M109

Case-upper evaporator-air conditioning ... L 45
Case-upper evaporator-air conditioning-rear ... L 49
Casing assembly-tail door-sonar blue ... P 39
Casing assembly-tail door-sonar blue-less ... P 40
Casing rear door-rh-sonar blue ... P 35

Casing rear door-rh-vinyl-dark granite ... P 37
Casing-front door-cloth-rh-dark granite ... P 30
Casing-front door-lh-sonar blue ... P 27
Casing-front door-lh-sonar blue-1 speaker ... P 25
Casing-front door-rh-cloth-dark granite- ... P 33

Casing-front door-rh-sonar blue ... P 26
Casing-front door-rh-sonar blue-1 speaker ... P 24
Casing-front door-rh-vinyl-dark granite ... P 32
Casing-transfer box ... D 5
Catalytic converter assembly engine exhaust ... K 1

Catalytic converter assembly engine exhaust ... K 3
Catch ... R 18
Catch ... R 11
Catch assembly-bonnet safety ... N 69
Catch assembly-side opening window-ash ... N 56

Catch overcentre ... J 18
Catcher-manual transmission oil ... C 3
Catcher-manual transmission oil ... C 25
Catcher-transfer box transmission brake oil ... D 22
Cd automatic changer-philips ... M136

Cd automatic changer-philips ... M135
Cd automatic changer-philips ... M140
Cd automatic changer-philips ... M143
Centre disc & logo-205 x 16 ... T 37
Chain-engine timing ... B 92

Chain-engine timing ... B128
Channel locating ... N 48
Channel locating-rh ... N 52
Channel-front door glass lift ... N 48
Channel-front door glass run-4.0mm ... N 48

Channel-rear door glass lift-lh ... N 51
Channel-rear door glass run-front-top-4.0mm ... N 51
Channel-rear door glass run-rubber ... N 52
Chassis-b/mtg-short ... N 3
Chassis-btm l/brk-lh ... N 3

Chassis-fnt body mtg ... N 3
Chassis-front end-lh ... N 3
Chassis-guss f/c/mem ... N 3
Chassis-no2 body mtg ... N 3
Chassis-no3 b/mtg-lh ... N 3

Chassis-no4 b/mtg-lh ... N 3
Chassis-no5 b/mtg-lh ... N 3
Cheater-exterior mirror-inner-rh ... N114
Cheater-exterior mirror-inner-rh ... N119
Cheater-exterior mirror-inner-rh-black ... N112

Cheater-exterior mirror-inner-rh-black ... N116
Check strap ... N 50
Check strap-rh ... N 46
Chock-spare wheel stowage ... S 1
Cigar lighter facia ... M130

Cigar lighter facia ... M131
Circlip ... B139
Circlip ... C 73
Circlip ... C 19
Circlip ... C 37

Circlip ... C 39
Circlip ... C 17
Circlip ... D 10
Circlip ... D 16
Circlip ... D 18

Circlip ... E 1
Circlip ... E 2
Circlip ... E 11
Circlip ... G 1
Circlip ... G 9

Circlip ... N 50
Circlip ... N 53
Circlip ... N 83
Circlip ... N 47
Circlip ... N 51

Circlip-10 spline shaft ... E 36
Circlip-front axle shaft to cv joint ... E 13
Circlip-front axle shaft to cv joint ... E 14
Circlip-front axle shaft to cv joint ... E 16
Circlip-large ... D 11

Circlip-large ... D 12
Circlip-man trans 1st & 2nd synchroniser hub ... C 13
Circlip-man trans 1st & 2nd synchroniser hub ... C 32
Circlip-man trans 2nd gear thrust washer ... C 14
Circlip-man trans 2nd gear thrust washer ... C 34

Circlip-man trans 5th gear synchroniser hub ... C 35
Circlip-man trans mainshaft/centre plate ... C 62
Circlip-piston ... B 8
Circlip-piston ... B 46
Cladding assembly-monoside-rh ... T 7

Cladding-door-front-lh ... T 7
Clamp ... C 42
Clamp ... C 69
Clamp ... C 84
Clamp ... C 93

Clamp ... C 95
Clamp ... C 97
Clamp ... F 15
Clamp ... K 1
Clamp ... K 10

Clamp ... L 6
Clamp ... L 18
Clamp ... L 3
Clamp ... L 17
Clamp air intake/duct-rear ... J 18

Clamp fuel injector ... B 29
Clamp fuel injector ... B 66
Clamp-2 pipes ... B 29
Clamp-cable ... C 86
Clamp-cable ... C 88

Clamp-cable ... N 35
Clamp-cable ... N 36
Clamp-distributor ... B101
Clamp-distributor ... B147
Clamp-exhaust downpipe to manifold ... K 13

Clamp-exhaust downpipe to manifold ... K 14
Clamp-exhaust intermediate pipe assembly ... K 11
Clamp-fixing battery ... M 16
Clamp-fixing battery ... M 17
Clamp-fixing battery ... M 18

Clamp-fixing battery ..... M 19
Clamp-fixing battery ..... M 20
Clamp-fixing battery ..... M 21
Clamp-fixing pump M148
Clamp-hose fuel lines B133
Clamp-hose fuel lines B134
Clamp-inlet manifold gasket seal ..... B 95
Clamp-inlet manifold gasket seal B130
Clamp-lock steering column ..... F 6
Clamp-lock steering column ..... N 71
Clamp-lock steering column ..... N 72
Clamp-lock steering column ..... N 73
Clamp-lock steering column ..... N 74
Clamp-lock steering column ..... N 75
Clamp-rocker shaft ..... B 18
Clamp-rocker shaft ..... B 56
Cleaner-alloy wheel-500ml ..... T 28
Cleaner-glazing-1ltr ..... T 28
Clip ..... B 67
Clip ..... B 91
Clip ..... B 97
Clip ..... B 98
Clip B116
Clip B133
Clip B134
Clip ..... C 2
Clip ..... C 24
Clip ..... C 44
Clip ..... C 45
Clip ..... C 46
Clip ..... D 20
Clip ..... D 14
Clip ..... G 14
Clip ..... H 10
Clip ..... H 23
Clip ..... H 6
Clip ..... H 8
Clip ..... H 12
Clip ..... H 21
Clip ..... J 5
Clip ..... J 6
Clip ..... J 14
Clip ..... J 20
Clip ..... K 4
Clip ..... L 5
Clip ..... L 8
Clip ..... L 21
Clip ..... N 33
Clip ..... N 41
Clip ..... N 71
Clip ..... N 76
Clip ..... N 83
Clip N112
Clip ..... N 15
Clip ..... N 70

Clip ..... N 72
Clip ..... N 84
Clip ..... N 91
Clip N111
Clip N116
Clip ..... N 31
Clip ..... N 73
Clip ..... N 85
Clip N114
Clip N119
Clip ..... N 75
Clip ..... P 4
Clip ..... P 11
Clip ..... P 51
Clip ..... P 64
Clip ..... P 8
Clip ..... P 31
Clip ..... P 36
Clip ..... P 44
Clip ..... P 10
Clip ..... P 38
Clip ..... P 28
Clip ..... P 34
Clip ..... P 62
Clip ..... R 46
Clip brake pipe ..... H 9
Clip brake pipe ..... H 10
Clip brake pipe ..... H 11
Clip brake pipe ..... H 13
Clip brake pipe ..... N 70
Clip-2 holes ..... H 9
Clip-5mm-6.5mm hole M158
Clip-7 seats ..... L 43
Clip-bonnet stay ..... N 69
Clip-brake pipe double ..... H 5
Clip-brake pipe double ..... H 7
Clip-brake pipe double ..... H 9
Clip-brake pipe double ..... H 10
Clip-brake pipe double ..... H 11
Clip-brake pipe double ..... H 13
Clip-brake pipe double ..... H 12
Clip-brake pipe treble ..... H 9
Clip-brake pipe treble ..... H 12
Clip-brake pipe treble-4.75mm ..... H 10
Clip-brake servo ..... H 1
Clip-cable ..... B 66
Clip-cable ..... B 90
Clip-cable ..... C 52
Clip-cable ..... E 19
Clip-cable ..... E 22
Clip-cable ..... G 14
Clip-cable-8mm hole ..... M 34
Clip-cable-8mm hole ..... M 57
Clip-cable-8mm hole M151
Clip-cable-8mm hole M158

Clip-cable-8mm hole ..... M 23
Clip-cable-8mm hole ..... M 42
Clip-cable-8mm hole ..... T 22
Clip-cable-rear ..... N 82
Clip-carpet to mudflap bolts ..... P 83
Clip-case-heater heater ..... L 21
Clip-clutch release-pivot post ..... C 24
Clip-clutch release-pivot post ..... C 46
Clip-cover to cushion ..... R 7
Clip-cradle-13mm-8mm hole M157
Clip-double fuel lines ..... J 11
Clip-double fuel lines-6 - 8mm ..... J 22
Clip-double fuel lines-6 - 8mm ..... J 9
Clip-drip finisher-front-black ..... N 90
Clip-e ..... C 86
Clip-e ..... C 88
Clip-e ..... C 87
Clip-e ..... C 89
Clip-edge harness ..... N 77
Clip-edge harness-12mm-clear. M158
Clip-edge pipe swivel ..... H 4
Clip-edge pipe swivel ..... H 2
Clip-fuel injector multi point injection B132
Clip-fuel injector multi point injection B133
Clip-fuel injector multi point injection B179
Clip-fuel injector multi point injection B134
Clip-harness M116
Clip-harness-14.5mm-6.5mm hole ..... M 59
Clip-harness-14mm-6mm hole M159
Clip-harness-seat buckle lead ..... M 60
Clip-hose ..... B 12
Clip-hose ..... B 23
Clip-hose ..... B 30
Clip-hose ..... B 62
Clip-hose ..... B 94
Clip-hose B137
Clip-hose B174
Clip-hose B180
Clip-hose ..... B 49
Clip-hose B136
Clip-hose ..... H 4
Clip-hose ..... J 6
Clip-hose ..... J 14
Clip-hose ..... J 15
Clip-hose ..... J 16
Clip-hose ..... J 17
Clip-hose ..... J 18
Clip-hose ..... J 20
Clip-hose ..... J 21
Clip-hose ..... J 22
Clip-hose ..... J 23
Clip-hose ..... J 9
Clip-hose ..... L 4
Clip-hose ..... L 5
Clip-hose ..... L 10

Clip-hose ..... L 13
Clip-hose ..... L 15
Clip-hose ..... L 16
Clip-hose ..... L 19
Clip-hose ..... L 30
Clip-hose ..... L 3
Clip-hose tee piece B136
Clip-hose-14.5mm-steel-25/64" hole ..... C 87
Clip-hose-14.5mm-steel-25/64" hole ..... G 3
Clip-hose-14.5mm-steel-25/64" hole M154
Clip-hose-18.3mm ..... B 91
Clip-hose-20mm ..... B 21
Clip-hose-20mm ..... B 59
Clip-hose-20mm ..... L 2
Clip-hose-20mm ..... L 12
Clip-hose-20mm ..... L 14
Clip-hose-20mm M150
Clip-hose-26mm ..... L 29
Clip-hose-26mm ..... L 31
Clip-hose-30mm ..... B 95
Clip-hose-30mm B129
Clip-hose-30mm B176
Clip-hose-30mm B130
Clip-hose-50mm-less spring assist ..... E 1
Clip-hose-50mm-less spring assist ..... L 9
Clip-hose-5mm/id M146
Clip-hose-5mm/id M148
Clip-hose-spring-band-60mm ..... J 5
Clip-hose-spring-band-60mm ..... L 9
Clip-hose-spring-band-60mm ..... L 10
Clip-hose-spring-band-60mm ..... L 16
Clip-hose-spring-band-60mm ..... L 19
Clip-hose-spring-band-60mm ..... L 3
Clip-hose-worm drive ..... J 19
Clip-hose-worm drive-25mm ..... B 21
Clip-hose-worm drive-25mm ..... B 24
Clip-hose-worm drive-25mm ..... B 59
Clip-hose-worm drive-25mm B176
Clip-hose-worm drive-25mm ..... F 19
Clip-hose-worm drive-25mm ..... F 21
Clip-hose-worm drive-25mm ..... F 20
Clip-hose-worm drive-25mm ..... L 29
Clip-insulation pad retention ..... N 41
Clip-linkage ..... N 76
Clip-linkage ..... N 80
Clip-linkage ..... N 77
Clip-omega-grey-10.5 x 8mm-6.5mm hole B153
Clip-omega-grey-10.5 x 8mm-6.5mm hole M146
Clip-omega-grey-10.5 x 8mm-6.5mm hole M147
Clip-omega-grey-10.5 x 8mm-6.5mm hole M148
Clip-omega-grey-10.5 x 8mm-6.5mm hole M149
Clip-omega-grey-10.5 x 8mm-6.5mm hole M152
Clip-omega-grey-10.5 x 8mm-6.5mm hole M153
Clip-p ..... B 59
Clip-p ..... B 79

Clip-p ... B101
Clip-p ... B174
Clip-p ... B 55
Clip-p ... B157
Clip-p ... C 87

Clip-p ... C 89
Clip-p ... F 21
Clip-p ... F 20
Clip-p ... G 3
Clip-p ... H 12

Clip-p ... J 21
Clip-p ... J 23
Clip-p ... J 24
Clip-p ... M 16
Clip-p ... M 17

Clip-p ... M 18
Clip-p ... M 19
Clip-p ... M 20
Clip-p ... M 21
Clip-p ... M 48

Clip-p ... M 51
Clip-p ... M 54
Clip-p ... M 55
Clip-p ... M 49
Clip-p ... M 52

Clip-p ... M 53
Clip-p-3/4" ... B151
Clip-p-3/4" ... M 95
Clip-p-5/16" ... C 3
Clip-p-5/16" ... C 25

Clip-p-5/16" ... D 3
Clip-p-5/16" ... J 22
Clip-p-black-13.6mm ... M153
Clip-p-m12 ... G 4
Clip-p-m12 ... G 10

Clip-p-pipe vacuum ... B147
Clip-p-plastic-adjustable-white.-4.0/12.0mm ... L 19
Clip-p-plastic-adjustable-white.-4.0/12.0mm ... M 28
Clip-panel fixing-5mm-4.8mm hole ... M158
Clip-pipe ... B 66

Clip-pipe ... B137
Clip-pipe ... C 4
Clip-pipe ... C 52
Clip-pipe ... E 6
Clip-pipe ... E 27

Clip-pipe ... F 21
Clip-pipe ... L 44
Clip-pipe ... L 51
Clip-pipe ... L 52
Clip-pipe-double ... H 24

Clip-pipe-rear ... M151
Clip-pipe-retaining speedo cable-single ... M113
Clip-pipe-single ... D 6
Clip-pipe-single ... K 4
Clip-pipe-tank to filter-double ... J 7

Clip-pivot-snap-13-17mm-7mm hole-grey ... B120
Clip-pivot-snap-13-17mm-7mm hole-grey ... M 46
Clip-pivot-snap-13-17mm-7mm hole-grey ... M 50
Clip-pivot-snap-13-17mm-7mm hole-grey ... M159
Clip-pivot-snap-13-17mm-7mm hole-grey ... M 23

Clip-plastic ... F 18
Clip-rear ... N 80
Clip-retainer pin ... N 76
Clip-retainer pin ... N 80
Clip-retainer pin ... N 83

Clip-retaining ... P 54
Clip-retaining-waist seal ... N 48
Clip-self adhesive-6mm ... M153
Clip-self adhesive-6mm ... M110
Clip-single fuel lines-8mm ... J 23

Clip-single fuel lines-8mm ... J 9
Clip-single fuel lines-8mm ... M 23
Clip-single fuel lines-separator to canister- ... J 24
Clip-snap-sack ... M152
Clip-snap-sack ... N 43

Clip-snap-sack ... N 91
Clip-snap-sack ... P 39
Clip-snap-sack ... P 31
Clip-snap-sack ... P 36
Clip-snap-sack ... P 41

Clip-snap-sack ... P 38
Clip-snap-sack ... P 60
Clip-snap-sack ... P 79
Clip-snap-sack ... P 28
Clip-snap-sack ... P 34

Clip-snap-sack ... P 55
Clip-snap-sack ... P 48
Clip-snap-sack ... P 62
Clip-snap-sack ... P 82
Clip-snap-sack ... R 1

Clip-snap-sack ... R 25
Clip-spacer ... B185
Clip-spacer ... F 21
Clip-spacer ... F 20
Clip-spacer-high tension lead ignition ... B 90

Clip-spacer-high tension lead ignition ... B103
Clip-spacer-high tension lead ignition ... B151
Clip-speedometer cable-retaining grommet & ... M113
Clip-spring steel ... B139
Clip-spring steel ... D 18

Clip-spring steel ... D 20
Clip-spring steel ... P 19
Clip-spring steel ... P 57
Clip-spring steel ... R 21
Clip-spring steel-switch panel to console ... P 21

Clip-spring-push on ... G 4
Clip-spring-push on ... G 3
Clip-spring-push on ... P 57
Clip-stowage bag to panel ... R 38
Clip-swivel ... B 26

Clip-swivel ... J 11
Clip-swivel ... L 19
Clip-swivel-10 x 10mm ... M157
Clip-swivel-15 x 10mm ... M 17
Clip-tail door lock link ... N 83

Clip-trim ... P 75
Clip-trim ... P 9
Clip-trim retention ... P 39
Clip-trim retention ... P 43
Clip-trim retention ... P 51

Clip-trim retention ... P 41
Clip-trim retention ... P 62
Clip-trim retention ... R 39
Clip-trim retention ... R 20
Clip-trim retention ... R 40

Clip-trim retention ... R 18
Clip-trim retention ... R 32
Clip-trim retention ... R 38
Clip-trim retention-black ... P 82
Clock digital-lcd ... M130

Clock digital-lcd-without hazard switch - ... M131
Closing panel ... N 20
Closing-rear crossmember ... N 32
Coil & bracket assembly ignition ... B185
Coil ignition ... B185

Coil ignition-lucas ... B104
Coil ignition-lucas ... B148
Coil-engine immobilizer ... M108
Coil-engine immobilizer ... M110
Cointray-facia console-ash grey ... P 12

Collar ... C 28
Collar ... C 7
Collar-distance manual transmission ... C 54
Collar-distance manual transmission ... C 14
Collar-distance manual transmission ... C 34

Collar-layshaft ... C 65
Collet-cylinder head valve ... B173
Column assembly steering system ... F 1
Column assembly-tilt steering system ... F 3
Compressor air conditioning ... B 71

Compressor air conditioning-sanden ... B183
Compressor assembly air conditioning ... B 35
Compressor assembly air conditioning ... B100
Compressor assembly air conditioning ... B144
Compressor assembly air conditioning ... B145

Condenser air conditioning ... L 40
Condenser assembly air conditioning ... L 47
Cone-synchronizer inner ... C 14
Cone-synchronizer inner ... C 34
Connecting rod assembly brake ... D 22

Connector brake ... H 13
Connector brake-rear ... H 11
Connector exhaust gas recirculation ... B 26
Connector fit ... C 81
Connector fit ... N 77

Connector-air duct centre ... L 29
Connector-inlet manifold adaptor ... B138
Connector-inlet manifold adaptor-1 outlet ... B136
Connector-outer air duct-rh ... L 29
Connector-wash system straight ... M148

Connector-wash system tee piece ... M147
Connector-wash system tee piece ... M149
Console assembly-facia-ash grey ... P 12
Console assembly-tunnel-dark granite ... P 19
Console assembly-tunnel-dark granite ... P 20

Console-tunnel base-sonar blue ... P 13
Container & bracket-windscreen wash ... M147
Container assembly-dual wash ... M147
Container assembly-dual wash ... M148
Container assembly-windscreen wash ... M146

Container-dual wash & headlamp wash ... M148
Container-dual wash & headlamp wash ... M149
Container-windscreen wash ... M146
Control & diagnostic unit air bag ... M 61
Control assembly heater ... L 20

Control assembly heater-air conditioning- ... L 48
Control assembly heater-less cables ... L 28
Control unit burglar alarm ... T 35
Control unit burglar alarm-433 mhz ... M107
Control unit burglar alarm-433 mhz ... M109

Control unit burglar alarm-type a&b ... T 34
Control unit dim dip-quartz halogen ... M 83
Control unit multi function ... M 81
Control unit passive restraint-sensor fitted ... M 62
Control unit sunroof ... M 97

Control unit-fan timer-green ... M 84
Control unit-fan timer-green ... M 89
Control unit-remote locking-315 mhz ... M103
Control unit-remote locking-315 mhz ... M104
Control unit-remote locking-418 mhz ... M102

Control-winch remote ... T 26
Converter assembly automatic transmission ... C 72
Coolbox-20 litre ... T 14
Cotter-valve ... B 19
Cotter-valve ... B 57

Cotter-valve ... B 88
Cotter-valve ... B123
Cotter-valve ... B125
Counter shaft-carb ... B 98
Coupler-rotary steering column ... M 61

Coupler-rotary steering column ... M 62
Coupler-rotary steering column ... M 64
Coupling ... B 15
Coupling-engine fan viscous ... B 13
Coupling-engine fan viscous ... B 52

Coupling-engine fan viscous ... B 86
Coupling-engine fan viscous-11 blade ... B121
Coupling-female ... L 49
Coupling-female ... L 54
Coupling-small ... L 53

Coupling-small ... L 55
Coupling-small ... L 46
Cover & driven plate assembly clutch ... B160
Cover & harness assembly-headlamp ... M 2
Cover alternator ... B154

Cover assembly-coverter automatic ... C 72
Cover assembly-cushion-large-light ... R 52
Cover assembly-engine timing gear ... B117
Cover assembly-front seat cushion-cloth- ... R 15
Cover assembly-front seat cushion-cloth- ... R 17

Cover assembly-front seat cushion-dark ... R 19
Cover assembly-front seat cushion-dark ... R 20
Cover assembly-front seat cushion-leather- ... R 23
Cover assembly-front seat cushion-light ... R 52
Cover assembly-front seat cushion-sonar ... R 7

Cover assembly-front seat squab-cloth-dark ... R 15
Cover assembly-front seat squab-cloth-dark ... R 16
Cover assembly-front seat squab-dark ... R 19
Cover assembly-front seat squab-dark ... R 20
Cover assembly-front seat squab-leather- ... R 23

Cover assembly-front seat squab-light ... R 52
Cover assembly-front seat squab-sonar blue ... R 4
Cover assembly-rear camshaft ... B173
Cover assembly-rear seat large split cushion- ... R 34
Cover assembly-rear seat large split cushion- ... R 42

Cover assembly-rear seat large split cushion- ... R 39
Cover assembly-rear seat large split cushion- ... R 44
Cover assembly-rear seat large split cushion- ... R 35
Cover assembly-rear seat large split squab- ... R 28
Cover assembly-rear seat large split squab- ... R 29

Cover assembly-rear seat large split squab- ... R 41
Cover assembly-rear seat large split squab- ... R 39
Cover assembly-rear seat large split squab- ... R 44
Cover assembly-rear seat small split cushion- ... R 34
Cover assembly-rear seat small split cushion- ... R 37

Cover assembly-rear seat small split cushion- ... R 42
Cover assembly-rear seat small split cushion- ... R 39
Cover assembly-rear seat small split cushion- ... R 45
Cover assembly-rear seat small split squab- ... R 28
Cover assembly-rear seat small split squab- ... R 41

Cover assembly-rear seat small split squab- ... R 39
Cover assembly-rear seat small split squab- ... R 44
Cover assembly-rear seat small split squab- ... R 31
Cover assembly-seat-front-headrestraint- ... R 24
Cover assembly-spare wheel-205 x 16- ... T 37

Cover assembly-spare wheel-quicksilver ... N121
Cover assembly-squab-large-light ... R 52
Cover assembly-valve rocker ... B 20
Cover assembly-valve rocker ... B 58
Cover assembly-valve rocker-lh ... B127

Cover assembly-valve rocker-rh ... B 90
Cover assembly-water jacket ... B136
Cover fusebox ... M 82
Cover fusebox ... P 8
Cover plate ... B 49

Cover plate ... N 70
Cover plate-ash grey ... N 60
Cover plate-ash grey ... N 63
Cover plate-side ... B 6
Cover steering wheel nut ... F 1

Cover-air conditioning belt tensioner ... B 72
Cover-armrest.-leather-dark granite ... R 43
Cover-armrest.-leather/cloth-dark granite- ... R 45
Cover-bell housing-bottom ... C 5
Cover-bell housing-bottom ... C 27

Cover-bell housing-bottom ... C 53
Cover-bell housing-bottom ... C 71
Cover-breather ... B 59
Cover-caliper brake-rh-top ... H 19
Cover-clutch assembly ... B 39

Cover-clutch assembly ... B 75
Cover-clutch assembly ... B108
Cover-clutch assembly ... B160
Cover-clutch assembly ... B189
Cover-control unit burglar alarm ... T 35

Cover-coolant pump ... B170
Cover-dust wheel cylinder ... D 22
Cover-end manual transmission-front ... C 56
Cover-end manual transmission-front-long ... C 8
Cover-end manual transmission-front-long ... C 29

Cover-end-camshaft cover-engine-rear ... B175
Cover-engine ... B 20
Cover-engine camshaft-exhaust ... B174
Cover-engine exhaust manifold ... K 9
Cover-engine exhaust manifold ... K 7

Cover-engine-front ... B 10
Cover-engine-front ... B 84
Cover-engine-front ... B117
Cover-engine-front ... B 49
Cover-engine-top ... B 58

Cover-facia lower-rh-rhd ... P 2
Cover-front door speaker-rh-ash grey ... P 31
Cover-front door speaker-rh-ash grey ... P 34
Cover-front seat belt adjuster mechanism- ... P 56
Cover-front seat belt upper anchorage-ash ... R 56

Cover-front seat belt upper anchorage-ash ... R 54
Cover-fuse ... M116
Cover-handle selector mechanism ... C 87
Cover-handle selector mechanism ... C 89
Cover-headlamp bulb insulation ... M 3

Cover-high mounted stop lamp intergal ... M 11
Cover-instrument pack rear-rhd-ash grey ... P 5
Cover-interlock solenoid ... D 6
Cover-lower front timing belt ... B168
Cover-lower rear timing belt ... B168

Cover-oil strainer ... B 82
Cover-oil strainer ... B115
Cover-protector-rear end door-ash grey- ... T 31
Cover-quarterlight pivot bracket catch ... N 56
Cover-rear lamp access-dark granite-5 seats ... P 63

Cover-rear lamp access-rh-sonar blue-5 ... P 61
Cover-rear lamp access-rh-sonar blue-7 ... P 61
Cover-seat-front-waterproof-beige ... T 12
Cover-seat-front-waterproof-grey ... T 11
Cover-seat-waterproof-front-grey ... T 11

Cover-spark plug ... B174
Cover-spring assembly detent arm ... C 20
Cover-spring assembly detent arm ... C 40
Cover-tail door handle-exterior ... N 86
Cover-terminal alternator ... B 73

Cover-terminal alternator ... B155
Cover-terminal alternator ... B157
Cover-terminal post ... M 16
Cover-terminal post ... M 17
Cover-terminal post ... M 18

Cover-terminal post ... M 19
Cover-terminal post ... M 20
Cover-terminal post ... M 21
Cover-terminal post ... M 28
Cover-timing belt-front ... B 10

Cover-timing belt-inner ... B 48
Cover-towing attachment ball ... T 20
Cover-transfer box ... D 5
Cover-transfer box front housing ... D 3
Cover-transfer box gear change housing ... C 67

Cover-transfer box power take off ... D 5
Cover-upper front timing belt ... B168
Cover-upper rear timing belt ... B168
Cover-waterproof-seat-rear inward facing ... T 12
Cover-waterproof-seat-rear inward facing ... T 11

Cover-wheel nut ... E 42
Cover-winch ... T 27
Cowl assembly instrument pack ... M117
Cowl assembly instrument pack ... M128
Cowl assembly instrument pack ... M123

Cowl assembly instrument pack-sonar blue ... P 5
Cowl assembly instrument pack-without ... P 11
Cowl-cooling system fan ... L 4
Cowl-cooling system fan ... L 3
Cowl-cooling system fan-upper ... L 5

Cowl-cooling system fan-upper ... L 8
Cowl-lower half left hand indicator system- ... F 4
Cowl-upper half left hand indicator system- ... F 4
Cradle-fuel tank support ... J 1
Cradle-fuel tank support ... J 2

Cradle-fuel tank support ... J 3
Cradle-fuel tank support ... J 4
Crank assembly-door bell ... N 83
Crankshaft handle-sunroof ... N 36
Crankshaft-engine ... B 7

Crankshaft-engine ... B 45
Crankshaft-engine ... B 80
Crankshaft-engine ... B112
Crankshaft-engine ... B165
Cross shaft differential ... E 11

Cross shaft differential-10 spline shaft ... E 36
Crossbrace ... N 20
Crossmember ... N 3
Crossmember assembly-chassis frame ... C 68
Crossmember assembly-chassis frame ... C 90

Crossmember assembly-chassis frame ... N 1
Crossmember assembly-chassis frame-front ... N 3
Crossmember assembly-rear floor ... N 32
Crossmember floor-rear ... N 24
Crossmember rear ... N 24

Crossmember-rear lower ... N 2
Crownwheel and pinion assembly ... E 11
Crownwheel and pinion assembly ... E 36
Cubby box lid-bahama beige ... T 19
Cubby box-stay ... T 19

Cushion-child restraint booster ... T 16
Cushion-large split rear seat-cloth-sonar ... R 33
Cushion-large split rear seat-kestral-cloth- ... R 34
Cushion-large split rear seat-leather-dark ... R 42
Cushion-large split rear seat-leather/cloth- ... R 39

Cushion-large split rear seat-leather/cloth- ... R 44
Cushion-small split rear seat-cloth-sonar ... R 36
Cushion-small split rear seat-kestral-cloth- ... R 34
Cushion-small split rear seat-leather-dark ... R 42
Cushion-small split rear seat-leather/cloth- ... R 39

Cushion-small split rear seat-leather/cloth- ... R 45
Cylinder ... C 78
Cylinder ... C 79
Cylinder ... C 81
Cylinder block casting assembly ... B164

Cylinder block-engine ... B 4
Cylinder block-engine ... B 43
Cylinder block-engine ... B 79
Cylinder block-engine ... B111
Cylinder block/ladder assembly ... B 9

Cylinder head assembly-engine ... B 16
Cylinder head assembly-engine ... B 88
Cylinder head assembly-engine ... B123
Cylinder head assembly-engine ... B125
Cylinder head assembly-engine ... B172

Cylinder head-engine-less core plugs ... B 54
D post-rh ... N 24
Damper assembly clutch ... G 13
Damper assembly front ... E 46
Damper assembly rear ... E 51

Damper assembly rear ... E 52
Damper assembly-carburettor ... B 97
Damper assembly-harmonic ... F 16
Damper assembly-steering ... F 16
Damper-swivel pin-rh ... E 21

Damper-swivel pin-rh ... E 19
De-icer-aerosol-500ml ... T 28
Decal catalyst-nimbus ... N 88
Decal front fender-rh-lower ... N104
Decal front fender-rh-upper ... N103

Decal fxi-grey N105
Decal v8-grey N105
Decal xs-grey N105
Decal-argyll N106
Decal-aviemore N106

Decal-b post blackout-rh ..... N 88
Decal-body side-rh-blue.-grey-silver N100
Decal-body side-rh-blue.-silver-grey ..... N 99
Decal-body side-rh-grey-silver/green N101
Decal-bonnet-mid silver N106

Decal-door blackout-rear-upper-rh ..... N 88
Decal-door-front-rh-blue.-silver-grey ..... N 99
Decal-door-front-rh-lower N104
Decal-door-front-rh-upper N103
Decal-door-lh-rear-blue. N102

Decal-door-rh-grey-silver/green N101
Decal-front fender rear-lh-blue. N102
Decal-front fender rear-rh-nimbus ..... N 87
Decal-front fender rear-silver ..... N 88
Decal-fuel filler door-blue.-grey-silver N100

Decal-nudge bar-black ..... T 9
Decal-special-large N106
Decal-tailgate-mid silver N106
Decal-tailgate-rear-nimbus ..... N 87
Decal-tonneau-rh-blue.-silver-grey ..... N 99

Decal-turbocharger-nimbus ..... N 87
Diaphragm-rear outer front seat ..... R 26
Diaphragm ..... R 26
Differential assembly differential-exchange- ..... E 7
Differential assembly differential-exchange- ..... E 8

Differential assembly-rear-new-10 spline ..... E 7
Differential assembly-rear-new-10 spline ..... E 28
Differential assembly-rear-new-24 spline ..... E 30
Differential assembly-rear-reconditioned-24 ..... E 31
Differential assembly-transfer box ..... D 7

Differential-new-24 spline shaft-4 bolt flange ..... E 28
Differential-new-24 spline shaft-4 bolt flange ..... E 8
Differential-reconditioned-24 spline shaft-4 ..... E 29
Diode-pektron ..... M 79
Diode-pektron ..... M 80

Diode-pektron ..... M 35
Dipstick automatic transmission-689mm ..... C 74
Dipstick-oil ..... B 9
Dipstick-oil ..... B 47
Dipstick-oil ..... B 79

Dipstick-oil B116
Dipstick-oil B167
Disc support-recline handle-seat ..... R 8
Disc synchro stop-mainshaft ..... C 30
Disc-brake spring ..... C 79

Disc-carburettor throttle B135
Disc-clutch automatic transmission-end ..... C 77
Disc-clutch automatic transmission-end ..... C 78
Disc-clutch automatic transmission-lined ..... C 79
Disc-clutch automatic transmission-lined ..... C 80

Disc-clutch automatic transmission-outer ..... C 81
Disc-solid brake-front ..... H 14
Disc-solid brake-rear ..... H 17
Disc-vented brake-front ..... H 14
Distributor assembly ignition-2 pin B146

Distributor ignition B101
Door assembly-front-rh ..... N 46
Door assembly-rear-rh ..... N 50
Door-outer front-rh ..... N 46
Dovetail female-rear door ..... N 85

Dovetail male-rear door ..... N 85
Dowel ..... B 10
Dowel ..... B 14
Dowel ..... B 40
Dowel ..... B 43

Dowel ..... B 48
Dowel ..... B 79
Dowel ..... B 84
Dowel ..... B 88
Dowel B108

Dowel B111
Dowel B117
Dowel B123
Dowel B125
Dowel B129

Dowel B145
Dowel B160
Dowel B164
Dowel B172
Dowel B189

Dowel ..... B 49
Dowel ..... C 2
Dowel ..... C 3
Dowel ..... C 4
Dowel ..... C 6

Dowel ..... C 24
Dowel ..... C 25
Dowel ..... C 26
Dowel ..... C 44
Dowel ..... C 45

Dowel ..... C 46
Dowel ..... C 47
Dowel ..... C 71
Dowel ..... C 19
Dowel ..... C 39

Dowel ..... C 85
Dowel ..... D 4
Dowel ..... D 5
Dowel-flywheel spigot alignment ..... B 40
Dowel-flywheel spigot alignment ..... B 76

Dowel-flywheel spigot alignment B161
Dowel-flywheel to crankshaft ..... B 7
Dowel-flywheel to crankshaft ..... B 39
Dowel-flywheel to crankshaft ..... B 45
Dowel-flywheel to crankshaft ..... B 75

Dowel-flywheel to crankshaft B165
Dowel-housing to gear case ..... C 52
Dowel-main bearing cap ..... B 4
Dowel-pin ..... B 60
Dowel-pin ..... C 75

Dowel-ring ..... B 4
Dowel-ring ..... B 16
Dowel-ring ..... B 29
Dowel-ring ..... B 40
Dowel-ring ..... B 43

Dowel-ring ..... D 4
Dowel-ring ..... D 6
Dowel-ring-main bearing cap B164
Dowel-stepped-backplate to block B164
Downpipe assembly exhaust system ..... B 31

Downpipe assembly exhaust system ..... K 2
Downpipe assembly exhaust system ..... K 3
Downpipe assembly exhaust system ..... K 6
Downpipe assembly exhaust system ..... K 8
Downpipe assembly exhaust system ..... K 9

Downpipe assembly exhaust system ..... K 13
Downpipe assembly exhaust system ..... K 14
Downpipe assembly exhaust system ..... K 7
Downpipe assembly exhaust system-rh ..... K 10
Drive assembly starter motor B107

Drive assembly starter motor B159
Drive assembly starter motor B188
Drive assembly starter motor-bosch ..... B 38
Drive assembly starter motor-bosch ..... B 74
Drive assembly starter motor-valeo ..... B 37

Drive assembly-freewheel-automatic ..... C 80
Drive assembly-oil pump ..... B 14
Driveshaft-front hub ..... E 23
Driveshaft-front-rh ..... E 16
Driveshaft-front-rh-10 spline shaft ..... E 14

Drum-transmission brake ..... D 22
Drum-transmission brake ..... D 23
Duct-air cleaner/air flow meter induction ..... J 19
Duct-air facia ..... P 4
Duct-air floor-rh ..... L 22

Duct-air floor-rh ..... L 35
Duct-air floor-rh ..... L 39
Duct-air inlet heater-rhd ..... L 33
Duct-air outlet foot-rh ..... L 35
Duct-air outlet foot-rh ..... L 21

Duct-air rear floor ..... L 34
Duct-cold air intake/air cleaner induction ..... J 19
Duct-elbow induction system ..... J 18
Duct-rear air conditioning-7 seats ..... L 43
Ecu-electric windows ..... M 93

Elbow assembly-duct air cleaner/throttle body ..... J 18
Elbow-carburettor air inlet ..... J 14
Elbow-engine coolant outlet ..... B 23
Elbow-engine coolant outlet ..... B 61
Elbow-engine coolant outlet ..... B 93

Elbow-engine coolant outlet B176
Elbow-engine coolant outlet B130
Electric control unit antilock brakes M111
Electric control unit cruise control M112
Electric control unit-air conditioning ..... L 26

Electric control unit-fuel control diesel M100
Electric control unit-fuel/ignition ..... M 96
Electric control unit-fuel/ignition ..... M 98
Electric control unit-fuel/ignition ..... M 99
Electric control unit-fuel/ignition-new ..... M 94

Electronic control unit engine ..... M 96
Electronic control unit-exhaust gas ..... M 91
Electronic control unit-exhaust gas ..... M 92
Element air cleaner ..... J 14
Element air cleaner ..... J 15

Element air cleaner ..... J 16
Element air cleaner ..... J 17
Element air cleaner ..... J 18
Element air cleaner ..... J 19
Element-camshaft cover oil separator B174

Element-inline fuel filter ..... J 7
End cap assembly-rear bumper-rh-black ..... N 14
End cap-front bumper-rh-black ..... N 12
End cap-front-rh-black ..... N 90
End cap-rear bumper-lh ..... T 7

End piece-drag link ..... F 15
Engine assembly - short ..... B 42
Engine base unit-part B163
Engine base unit-part-8.13:1 ..... B 78
Engine base unit-part-8.13:1 ..... B110

Engine base unit-part-short ..... B 3
Engine base unit-power unit ..... B 1
Engine unit-stripped-exchange B109
Engine unit-stripped-new ..... B 2
Engine unit-stripped-new ..... B 41

Engine unit-stripped-new B109
Engine unit-stripped-new B162
Engine unit-stripped-reconditioned ..... B 77
Escutcheon- bc post lower ..... N 23
Escutcheon- bc post lower-sonar blue ..... P 44

Escutcheon- bc post lower-sonar blue ..... P 53
Escutcheon- bc post upper adjuster ..... P 79
Escutcheon- bc post upper adjuster ..... P 56
Escutcheon- bc post upper adjuster ..... P 49
Escutcheon-door-lock-ash grey ..... N 84

Escutcheon-headrest-89mm ..... M 16
Escutcheon-headrest-89mm ..... M 17
Escutcheon-headrest-89mm ..... M 18
Escutcheon-headrest-89mm ..... M 19
Escutcheon-headrest-89mm ..... M 21

Escutcheon-headrest-89mm M159
Escutcheon-headrest-black ..... R 6
Escutcheon-headrest-dark granite ..... R 20
Escutcheon-headrest-dark granite ..... R 16
Escutcheon-headrest-dark granite ..... R 24

Escutcheon-rear door window regulator-ash .... P 36
Escutcheon-rear door window regulator-ash .... P 38
Escutcheon-rear quarter window latch-rh- .... P 58
Escutcheon-rear quarter window latch-rh- .... P 50
Escutcheon-rear seat belt-lh-sonar blue .... P 55

Escutcheon-rear seat belt-rh-sonar blue .... P 48
Escutcheon-window regulator-ash grey .... N 60
Escutcheon-window regulator-ash grey .... N 63
Evaporator air conditioning .... L 25
Evaporator air conditioning .... L 45

Evaporator air conditioning .... L 49
Evaporator assembly-air conditioning .... L 26
Excluder mud B117
Exhaust silencer-intermediate .... K 11
Expander assembly-transfer box brake shoe .... D 22

Extension-valance-rh .... N 20
Extinguisher-fire-2 kg .... T 17
Extractor blade fuse .... M 79
Extractor blade fuse .... M 80
Extractor blade fuse .... M 81

Extractor blade fuse .... M 82
Extrusion-side step .... T 44
Facia assembly-rhd-bahama beige .... P 7
Facia assembly-rhd-sonar blue .... P 1
Fan alternator-127/65 amp B153

Fan alternator-133/65 amp .... B 36
Fan alternator-133/65 amp B105
Fan alternator-80 amp B154
Fan assembly-cooling-7 blade .... B 86
Fan-cooling .... B 13

Fan-cooling .... B 52
Fan-cooling-11 blade-17"-alternative B121
Fan-cooling/air conditioning .... L 23
Fan-cooling/air conditioning .... L 26
Fastener-drive .... P 60

Fastener-drive .... P 55
Fastener-drive-ash grey .... R 1
Fastener-drive-black .... N 89
Fastener-drive-black .... P 83
Fastener-drive-dark granite .... P 63

Fastener-drive-dark granite .... P 79
Fastener-drive-dark granite .... P 56
Fastener-drive-dark granite .... R 25
Fastener-drive-sonar blue .... P 57
Fastener-drive-sonar blue .... P 64

Fastener-drive-sonar blue .... P 61
Fastener-drive-sonar blue .... P 48
Fastener-drive-sonar blue .... P 49
Fastener-drive-sonar blue .... R 48
Fastener-drive-winchester grey-plastic .... P 2

Fastener-expanding .... N 28
Fastener-expanding .... P 19
Fastener-expanding .... P 21
Fastener-fir tree .... P 9
Fastener-fir tree .... R 39

Fastener-fir tree .... R 32
Fastener-fir tree-ash grey .... P 44
Fastener-fir tree-ash grey .... R 18
Fastener-fir tree-ash grey .... R 11
Fastener-fir tree-bahama beige .... P 57

Fastener-fir tree-bahama beige .... P 63
Fastener-fir tree-bahama beige .... P 60
Fastener-fir tree-bahama beige .... P 48
Fastener-headlining fir tree-mist grey .... P 72
Fastener-quarter turn .... P 9

Fastener-quarter turn-ash grey .... S 1
Fastener-quarter turn-grey .... P 18
Filler-oil .... B 90
Filler-oil B127
Filter assembly-in line fuel lines .... B 91

Filter assembly-in line fuel lines-unheated .... J 12
Filter-container assembly filler neck M146
Filter-container assembly filler neck M147
Filter-container assembly filler neck M148
Filter-container assembly filler neck M149

Filter-crankcase breather B127
Filter-crankcase intake foam B127
Filter-in line fuel lines .... J 7
Filter-in line fuel lines .... J 9
Filter-in line fuel lines-heated .... J 12

Filter/sedimentor assembly-diesel .... J 13
Finisher assembly-a post upper-rh-mist .... P 43
Finisher assembly-a post upper-rh-mist .... P 51
Finisher assembly-pocket-rh-aspen grey .... P 67
Finisher top .... N 68

Finisher-a post lower-rh-bahama beige .... P 65
Finisher-a post upper-rh-sonar blue .... P 43
Finisher-a post upper-rh-sonar blue .... P 51
Finisher-bc post lower-rh-sonar blue .... P 44
Finisher-bc post lower-rh-sonar blue .... P 53

Finisher-bc post upper-rh-sonar blue .... P 52
Finisher-bc post-rh .... N 90
Finisher-bc post-sonar blue .... P 54
Finisher-body side sill-lh .... T 7
Finisher-body side sill-rh-black .... N 30

Finisher-d post upper-rh-sonar blue .... P 54
Finisher-d post-outer-lh .... N 91
Finisher-demist vent windscreen-rh-ash .... P 4
Finisher-e post upper-rh-dark granite .... P 61
Finisher-e post upper-rh-sonar blue .... P 60

Finisher-e post-dark granite .... P 63
Finisher-facia console radio-ash grey M136
Finisher-facia console radio-ash grey M138
Finisher-facia console radio-ash grey M141
Finisher-facia console radio-ash grey .... P 12

Finisher-floor cover carpet/crossbar-primer .... P 82
Finisher-front console automatic selector-ash .... P 16
Finisher-front console automatic selector-ash .... P 17
Finisher-glovebox-dark granite .... P 7
Finisher-headlamp .... N 43

Finisher-headlamp-rh-black .... N 42
Finisher-headlining-aspen grey .... P 71
Finisher-headlining-rear .... N 32
Finisher-lower screen-unheated .... N 68
Finisher-primed headlamp-rh .... N 43

Finisher-rear bumper .... T 7
Finisher-rear header-dark granite .... P 63
Finisher-rh-dark granite .... R 50
Finisher-roof drip rail-black-rear-lh-rh .... N 91
Finisher-roof drip rail-rh-side-black .... N 90

Finisher-sunroof-grey .... N 37
Finisher-sunroof-grey .... N 38
Finisher-sunroof-grey .... N 39
Finisher-sunroof-mist grey .... N 36
Finisher-sunroof-mist grey .... P 72

Finisher-tunnel console centre-ash grey .... P 19
Finisher-tunnel console centre-ash grey .... P 20
Finisher-tunnel console front-sonar blue .... P 14
Finisher-windscreen a post-rh .... N 68
Fixing clip jacket .... C 52

Flame trap-crankcase breather .... B 91
Flange assembly-front output .... D 11
Flange assembly-rear output .... D 12
Flange-driveshaft coupling differential-3 bolt .... E 34
Flange-driveshaft coupling differential-4 bolt .... E 10

Flap & seals assembly blower-heater .... L 21
Flasher unit electronic .... M 83
Flasher unit electronic-for towing .... M 81
Flinger-crankshaft pulley .... B 7
Flinger-crankshaft pulley .... B 45

Float-carburettor-lh .... B 97
Fluid-automatic transmission-dextron 2 .... T 28
Flywheel engine .... B 39
Flywheel engine .... B 75
Flywheel engine B108

Flywheel engine B160
Flywheel engine B189
Foam .... L 25
Foam .... P 10
Foam .... P 50

Foam cushion .... R 26
Foam cushion-front .... R 12
Foam cushion-front .... R 13
Foam-facia finisher assembly .... P 8
Foam-pad-self adhesive N118

Foam-pad-self adhesive .... P 12
Foam-pad-self adhesive .... P 72
Foam-squab .... R 12
Foam-squab .... R 26
Foam-squab-front .... R 13

Footrest assembly-front floor-rhd .... P 64
Fork assembly-3rd & 4th manual transmission .... C 38
Fork assembly-3rd & 4th manual transmission .... C 66
Fork assembly-3rd & 4th manual transmission .... C 19
Fork assembly-5th manual transmission .... C 66

Fork assembly-5th manual transmission .... C 19
Fork assembly-5th manual transmission .... C 39
Fork assembly-selector-1st/2nd manual .... C 18
Fork assembly-selector-1st/2nd manual .... C 38
Fork assembly-selector-1st/2nd manual .... C 66

Fork-high/low selector .... D 15
Fork-transfer box differential lock selector .... D 14
Frame assembly-chassis .... N 1
Frame assembly-manual front seat cushion- .... R 12
Frame assembly-manual front seat cushion- .... R 13

Frame assembly-manual front seat-rh-ash .... R 9
Frame assembly-manual lumbar support front .... R 12
Frame assy-sunroof-balanced .... N 35
Frame assy-sunroof-balanced .... N 36
Frame sub assembly-front seat squab-front- .... R 13

Frame-assembly-seat-front-elecric .... R 26
Frame-front door-rh .... N 49
Frame-front sunroof lower .... N 35
Frame-front sunroof lower .... N 36
Frame-front sunroof upper .... N 37

Frame-front sunroof upper .... T 47
Frame-radiator-top .... L 2
Frame-radiator-top .... L 7
Frame-rear air conditioning unit .... L 43
Frame-rear door upper-rh .... N 55

Frame-sunroof-lower .... N 37
Frame-sunroof-lower .... N 38
Frame-sunroof-lower .... N 39
Frame-sunroof-lower .... T 47
Fuel filler door assembly .... N 33

Fuse-20 amp-auto-standard .... M 81
Fuse-3 amp-auto-standard .... M 79
Fuse-3 amp-auto-standard .... M 80
Fuse-30 amp-auto-standard .... M 82
Fuse-electric seat-30 amp-auto-standard .... M 88

Fuse-heated windscreen-60 amp-maxi- M116
Fusebox assembly engine compartment .... M 82
Fusebox assembly passenger compartment .... M 81
Gaiter-gear lever assembly inner .... P 17
Gaiter-gear lever assembly inner .... P 75

Gaiter-gear lever assembly inner-sealing .... P 16
Gaiter-gear lever assembly-ash grey .... P 19
Gaiter-gear lever assembly-ash grey .... P 14
Gaiter-gear lever assembly-ash grey .... P 21
Gaiter-handbrake-ash grey .... H 22

Gaiter-handbrake-ash grey .... H 23
Gaiter-handbrake-grey .... H 20
Gaiter-hi/low lever-ash grey .... P 16
Gaiter-hi/low lever-ash grey .... P 19
Gaiter-hi/low lever-ash grey .... P 14

Gaiter-hi/low lever-ash grey .... P 21
Gaiter-propshaft sliding joint .... E 1
Gaiter-propshaft sliding joint .... E 2
Gasket .... B 5
Gasket .... D 3

Gasket ... D 21
Gasket ... N 46
Gasket ... N 50
Gasket exhaust system ... B 31
Gasket exhaust system ... K 5

Gasket exhaust system ... K 4
Gasket exhaust system-manifold to downpipe ... K 2
Gasket exhaust system-manifold to downpipe ... K 3
Gasket exhaust system-manifold to downpipe ... K 6
Gasket exhaust system-manifold to downpipe ... K 8

Gasket exhaust system-manifold to downpipe ... K 9
Gasket exhaust system-manifold to downpipe ... K 7
Gasket exhaust system-turbo outlet to ... K 13
Gasket exhaust system-turbo outlet to ... K 14
Gasket selector mechanism ... P 16

Gasket selector mechanism ... P 17
Gasket-access plate/torque convertor ... C 72
Gasket-adaptor ... B122
Gasket-axle shaft drive member-front ... E 23
Gasket-axle shaft drive member-front ... E 24

Gasket-axle shaft drive member-front ... E 39
Gasket-axle shaft drive member-front ... E 40
Gasket-barrel lock ... N 83
Gasket-bonnet hinge ... N 41
Gasket-bracket-pulley support belt tensioner ... B 10

Gasket-camshaft cover-exhaust ... B174
Gasket-carburettor to induction manifold- ... B 98
Gasket-carburettor-asbestos ... B 98
Gasket-centre plate to extension case ... C 3
Gasket-centre plate to extension case ... C 25

Gasket-clutch pedal box to bulkhead ... G 9
Gasket-converter housing lower access ... B 76
Gasket-coolant outlet elbow ... B 93
Gasket-coolant outlet elbow ... B 94
Gasket-coolant outlet elbow ... B130

Gasket-coolant pump body ... B 12
Gasket-coolant pump body ... B 51
Gasket-coolant pump body ... B119
Gasket-coolant pump body-to front cover- ... B 10
Gasket-coolant pump distance piece to block ... B 51

Gasket-cover front ... B 84
Gasket-cover front ... B117
Gasket-crankcase oil sump ... B 82
Gasket-crankcase oil sump ... B115
Gasket-crankcase oil sump ... B167

Gasket-crankcase rear oil seal housing ... B 43
Gasket-cylinder block coolant pump ... B 85
Gasket-cylinder block coolant pump ... B119
Gasket-cylinder block oil pump ... B 87
Gasket-cylinder block oil pump ... B122

Gasket-cylinder block oil pump ... B171
Gasket-cylinder block oil pump-asbestos free ... B 14
Gasket-cylinder block side cover ... B 6
Gasket-cylinder block side cover-asbestos ... B 21
Gasket-cylinder block side cover-asbestos ... B 59

Gasket-cylinder block to oil filter adaptor ... B 53
Gasket-cylinder block to oil filter adaptor- ... B 15
Gasket-cylinder block water gallery plate ... B164
Gasket-cylinder head ... B 88
Gasket-cylinder head ... B125

Gasket-cylinder head ... B124
Gasket-cylinder head ... B173
Gasket-cylinder head-1.30mm-1 hole ... B 16
Gasket-cylinder head-1.30mm-1 hole ... B 54
Gasket-differential housing to axle case ... E 12

Gasket-differential housing to axle case ... E 38
Gasket-door handle ... N 86
Gasket-end cover automatic transmission ... C 73
Gasket-engine inlet/exhaust manifold ... B 24
Gasket-engine rocker cover ... B 20

Gasket-engine rocker cover ... B 58
Gasket-engine rocker cover ... B 90
Gasket-engine rocker cover ... B127
Gasket-exhaust gas recirculating ... B 25
Gasket-exhaust gas recirculation ... L 16

Gasket-exhaust gas recirculation pipe to ... B 67
Gasket-exhaust gas recirculation valve to ... B 25
Gasket-exhaust gas recirculation valve to ... L 16
Gasket-exhaust manifold ... B 96
Gasket-exhaust manifold ... B131

Gasket-exhaust manifold ... B178
Gasket-exterior mirror ... N112
Gasket-exterior mirror ... N116
Gasket-exterior mirror ... N114
Gasket-exterior mirror ... N119

Gasket-external drain pipe ... B 44
Gasket-external drain pipe ... B 47
Gasket-flywheel/drive plate ... B 39
Gasket-front cover manual transmission ... C 8
Gasket-front cover manual transmission ... C 29

Gasket-front plate to timing chain cover ... B 49
Gasket-frt cover ... B 10
Gasket-fuel lift pump ... B 27
Gasket-fuel lift pump ... B 63
Gasket-gearchange ball pin housing ... C 21

Gasket-gearchange ball pin housing ... C 41
Gasket-headlamp lens-rh ... M 3
Gasket-injection pump-diesel engine ... B 28
Gasket-injection pump-diesel engine ... B 64
Gasket-inlet manifold ... B 62

Gasket-inlet manifold ... B 95
Gasket-inlet manifold ... B177
Gasket-inlet manifold ... B130
Gasket-lower swivel pin ... E 20
Gasket-lower swivel pin ... E 19

Gasket-oil drain turbocharger ... B 67
Gasket-oil separator ... B 21
Gasket-potentiometer-throttle multi point ... B135
Gasket-rear extension housing ... C 84
Gasket-rear lamp assembly ... M 15

Gasket-roof rack side rail front-rh ... N 40
Gasket-roof rack side rail rear-lh ... N 40
Gasket-sealing ... F 1
Gasket-stepping motor ... B136
Gasket-stub axle ... E 21

Gasket-stub axle ... E 19
Gasket-stub axle-asbestos free ... E 39
Gasket-stub axle-asbestos free ... E 40
Gasket-sump oil strainer ... B 82
Gasket-sump oil strainer ... B115

Gasket-sump/transmission case automatic ... C 74
Gasket-swivel pin bearing housing to axle ... E 13
Gasket-swivel pin bearing housing to axle ... E 14
Gasket-swivel pin bearing housing to axle ... E 16
Gasket-tail door hinge-upper ... N 53

Gasket-thermostat housing ... B176
Gasket-thermostat housing-cylinder head ... B 61
Gasket-thermostat housing-top-asbestos ... B 23
Gasket-throttle body ... B180
Gasket-timing gear cover plate ... B 10

Gasket-timing gear cover plate ... B 49
Gasket-timing gear housing drain plate ... B 10
Gasket-timing gear housing drain plate ... B 48
Gasket-transfer box bottom cover ... D 5
Gasket-transfer box bottom cover ... D 21

Gasket-transfer box cover to case ... D 3
Gasket-transfer box cover to case ... D 21
Gasket-transfer box gear change housing ... D 4
Gasket-transfer box gear change housing ... D 18
Gasket-transfer box gear change housing ... D 21

Gasket-transfer box gear change housing ... D 17
Gasket-transfer box gear change housing ... D 16
Gasket-transfer box gear change housing ... D 18
Gasket-transfer box gear change housing ... D 21
Gasket-transfer box power take off cover ... D 5

Gasket-transfer box power take off cover ... D 21
Gasket-transfer box side cover ... D 3
Gasket-transfer box side cover ... D 21
Gasket-transfer box speedometer housing ... D 5
Gasket-transfer box speedometer housing ... D 21

Gasket-turbocharger/exhaust manifold turbo ... B 31
Gasket-upper inlet manifold to lower inlet ... B177
Gasket-vacuum pump ... B 33
Gasket-vacuum pump ... B 69
Gasket-water jacket to plenum chamber ... B136

Gauge-fuel ... M120
Gauge-fuel ... M126
Gauge-fuel & temperature-without ... M117
Gauge-tachometer/fuel/temperature-with ... M117
Gauge-temperature ... M120

Gauge-temperature ... M126
Gauge-tyre pressure ... T 17
Gear & coupling assembly ... B101
Gear & coupling assembly ... B147
Gear assembly-planetary automatic ... C 82

Gear oil-ep90 ... T 28
Gear-1st driven manual transmission ... C 13
Gear-1st driven manual transmission ... C 32
Gear-1st driven manual transmission ... C 59
Gear-2nd driven manual transmission ... C 13

Gear-2nd driven manual transmission ... C 32
Gear-2nd driven manual transmission ... C 58
Gear-3rd driven manual transmission ... C 12
Gear-3rd driven manual transmission ... C 31
Gear-3rd driven manual transmission ... C 60

Gear-5th driven manual transmission ... C 61
Gear-5th speed countershaft manual ... C 37
Gear-5th speed countershaft manual ... C 65
Gear-5th speed countershaft manual ... C 17
Gear-5th speed main shaft manual ... C 14

Gear-5th speed main shaft manual ... C 34
Gear-automatic transmission sun ... C 82
Gear-crankshaft timing drive ... B165
Gear-drive-distributor ... B 92
Gear-drive-distributor ... B128

Gear-high output-1.2 : 1 ... D 15
Gear-intermediate shaft-1.2 : 1 ... D 10
Gear-low output ... D 15
Gear-mainshaft-26 teeth ... D 8
Gear-oil pump ... B 87

Gear-oil pump ... B122
Gear-oil pump ... C 6
Gear-oil pump ... C 28
Gear-oil pump ... C 7
Gear-oil pump idler-10 teeth ... B 14

Gear-oil pump-10 teeth ... B 14
Gear-parking automatic transmission ... C 84
Gear-reverse idler assembly manual ... C 65
Gear-reverse manual transmission ... C 37
Gear-reverse manual transmission ... C 62

Gear-reverse manual transmission ... C 65
Gear-reverse manual transmission ... C 17
Gear-reverse manual transmission ... C 63
Gear-ring automatic transmission-front ... C 82
Gear-ring-flywheel engine ... B 39

Gear-ring-flywheel engine ... B 40
Gear-ring-flywheel engine ... B 75
Gear-ring-flywheel engine ... B 76
Gear-ring-flywheel engine ... B108
Gear-ring-flywheel engine ... B160

Gear-ring-flywheel engine ... B189
Gear-speedometer drive ... D 12
Gearbox assembly-front seat fore & aft ... R 21
Gearbox assembly-recline-rh ... R 21
Gears & shaft assembly-transmission oil pump ... C 28

Glass door front-4.0mm-rh ... N 47
Glass panel-sunroof ... N 35
Glass panel-sunroof ... N 36
Glass-alpine light-clear.-rh ... N 67
Glass-exterior mirror assembly flat-rh ... N112

Glass-exterior mirror assembly flat-rh N115
Glass-exterior mirror assembly flat-rh N117
Glass-exterior mirror assembly flat-rh N113
Glass-exterior mirror assembly flat-rh N114
Glass-exterior mirror assembly-convex- N112

Glass-exterior mirror assembly-convex- N115
Glass-exterior mirror assembly-convex-lh N117
Glass-exterior mirror assembly-convex-rh N113
Glass-exterior mirror assembly-convex-rh N118
Glass-exterior mirror assembly-flat-rh N114

Glass-green backlight-with high-level stop .... N 67
Glass-green backlight-without high-level .... N 66
Glass-laminated-windscreen-green .... N 66
Glass-quarterlight tinted-lh-front .... N 56
Glass-rear door clear N120

Glass-rear quarter-lh-no aerial .... N 57
Glass-rear quarter-rh N120
Glass-rear quarter-rh .... N 52
Glass-rear quarter-rh-no aerial .... N 58
Glass-rear-door-4.0mm-rh .... N 51

Glazing rubber-5.0mm .... N 48
Glovebox assembly-facia-bahama beige .... P 7
Glovebox assembly-facia-dark granite .... P 7
Glovebox-inner-sonar blue .... P 1
Grease-375cc-one shot .... E 19

Grease-375cc-one shot .... E 22
Grille assembly-radiator .... N 43
Grille assembly-radiator-black .... N 42
Grille-front .... T 7
Grille-plenum .... N 89

Grip handbrake-ash grey .... H 22
Grip handbrake-ash grey .... H 23
Grip handbrake-grey .... H 20
Grommet B106
Grommet B156

Grommet .... D 18
Grommet .... D 17
Grommet .... G 14
Grommet .... G 6
Grommet .... G 8

Grommet .... G 3
Grommet .... H 24
Grommet .... J 1
Grommet .... J 2
Grommet .... J 3

Grommet .... J 4
Grommet .... J 5
Grommet .... J 6
Grommet .... J 15
Grommet .... J 16

Grommet .... J 17
Grommet .... J 19
Grommet .... J 23
Grommet .... L 7
Grommet .... L 22

Grommet .... L 39
Grommet .... L 47
Grommet .... L 46
Grommet .... M 3
Grommet .... M 34

Grommet M147
Grommet M149
Grommet M150
Grommet M152
Grommet .... N 41

Grommet .... N 42
Grommet .... N 43
Grommet .... N 69
Grommet .... N 20
Grommet .... N 70

Grommet .... N 82
Grommet .... N 85
Grommet .... N 79
Grommet .... P 22
Grommet .... R 59

Grommet-17/32" hole .... C 87
Grommet-17/32" hole .... C 89
Grommet-17/32" hole .... M 16
Grommet-17/32" hole .... M 17
Grommet-17/32" hole .... M 18

Grommet-17/32" hole .... M 19
Grommet-17/32" hole .... M 20
Grommet-17/32" hole .... M 21
Grommet-17/32" hole .... M 95
Grommet-ash grey .... P 31

Grommet-ash grey .... P 38
Grommet-ash grey .... P 34
Grommet-black-11 x 19mm .... G 10
Grommet-black-rubber-3 x 25.5mm .... L 36
Grommet-black-rubber-3 x 25.5mm .... L 37

Grommet-black-rubber-3 x 25.5mm M151
Grommet-blind .... J 16
Grommet-blind .... J 19
Grommet-blind .... N 21
Grommet-cable blanking M116

Grommet-cable-3 x 19mm-black M160
Grommet-container assembly pump M146
Grommet-container assembly pump M148
Grommet-convolute .... M 33
Grommet-convolute .... M 41

Grommet-convolute .... N 82
Grommet-convolute .... N 79
Grommet-for harness rear end door .... M 35
Grommet-grey .... P 36
Grommet-grey .... P 28

Grommet-hinge face-rh .... N 48
Grommet-mounting air cleaner .... J 19
Grommet-mounting air cleaner-front .... J 16
Grommet-mounting air cleaner-front .... J 17
Grommet-pipe evaporator-air conditioning .... L 47

Grommet-retaining speedo cable-rubber- M113
Grommet-rubber-17/64" hole B151
Grommet-sealing .... F 3
Grommet-sealing-airbag .... M 61
Grommet-sealing-airbag .... M 62

Grommet-wash hose connector M151
Grommet-wash hose connector M152
Guard-crankshaft pulley-front B168
Guard-dog-loadspace N123
Guard-dog-loadspace-mesh type .... T 13

Guard-fan alternator-65 amp B106
Guard-fan alternator-65 amp B156
Guide automatic transmission .... C 77
Guide automatic transmission .... P 16
Guide-carburettor jet needle .... B 97

Guide-clutch release bearing .... C 24
Guide-clutch release bearing .... C 46
Guide-clutch release bearing-aluminium .... C 44
Guide-clutch release bearing-aluminium .... C 45
Guide-clutch release bearing-cast iron .... C 2

Guide-cylinder head valve .... B 16
Guide-cylinder head valve .... B 54
Guide-cylinder head valve .... B 88
Guide-cylinder head valve B123
Guide-cylinder head valve B125

Guide-front door glass .... N 49
Guide-front door glass .... N 62
Guide-gear lever .... C 54
Guide-slide-nacelle selector mechanism-ash .... P 16
Guide-starter motor brush .... B 37

Guide-starter motor brush .... B 38
Guide-starter motor brush .... B 74
Gusset-headlamp side-rh .... N 21
Handle assembly-door .... N 83
Handle assembly-door-inner-rh-ash grey .... N 77

Handle assembly-front door-rh .... N 71
Handle assembly-front door-rh .... N 76
Handle assembly-front door-rh .... N 72
Handle assembly-front door-rh .... N 73
Handle assembly-front door-rh .... N 75

Handle assembly-front door-with keys-rh .... N 74
Handle assembly-grab-dark granite .... P 7
Handle assembly-grab-sonar blue .... P 6
Handle assembly-rear door-rh .... N 80
Handle assembly-roof trim grab-ash .... P 71

Handle assembly-sunroof .... N 37
Handle assembly-sunroof .... N 38
Handle assembly-sunroof .... N 39
Handle assembly-sunroof .... N 36
Handle assembly-sunroof .... T 47

Handle-door grab-inner-rh-ash grey/light .... N 81
Handle-door grab-inner-rh-ash grey/light .... N 84
Handle-e post grab-sonar blue .... P 62
Handle-front headrestraint grab-sonar blue .... R 6
Handle-front/rear door window regulator-ash .... N 60

Handle-front/rear door window regulator-ash .... N 63
Handle-jack .... S 1
Handwheel-front seat squab manual lumbar .... R 20
Handwheel-front seat squab manual lumbar .... R 16
Handwheel-front seat squab manual lumbar .... R 24

Handwheel-front seat squab manual recline .... R 20
Handwheel-front seat squab manual recline .... R 18
Handwheel-front seat squab manual recline .... R 8
Handwheel-front seat squab manual recline .... R 11
Hanger-exhaust front sub frame .... K 6

Hanger-exhaust front sub frame .... K 9
Hanger-exhaust front sub frame .... K 7
Hardtop assembly-c/w tailgate N120
Harness air bag-rhd .... M 61
Harness air bag-rhd .... T 46

Harness air conditioning .... M 58
Harness air conditioning-rear .... L 26
Harness air conditioning-rear .... P 15
Harness air conditioning-rhd .... T 46
Harness blower-air conditioning-front .... M 43

Harness blower-heater .... M 43
Harness body .... M 36
Harness body-rh .... M 34
Harness body-rh .... M 35
Harness body-with rear air-con .... M 37

Harness console-immobilisation M108
Harness console-immobilisation M110
Harness drivers door .... M 33
Harness earth bond .... B 37
Harness earth bond B104

Harness earth bond B148
Harness earth bond B157
Harness earth bond-front of engine to ignition .... M 47
Harness engine .... M 44
Harness engine .... M 45

Harness engine .... M 46
Harness engine .... M 48
Harness engine .... M 50
Harness engine .... M 51
Harness engine .... M 49

Harness engine .... M 52
Harness engine .... M 53
Harness facia .... M 38
Harness facia .... M 39
Harness front door .... M 33

Harness front door-driver .... M 41
Harness instrument pack M117
Harness interior light .... M 13
Harness main .... M 24
Harness main .... M 29

Harness main .... M 25
Harness main .... M 30
Harness main .... M 26
Harness main .... M 31
Harness main .... M 27

Harness main ..... M 32
Harness powerwash ..... M 60
Harness rear door ..... M 33
Harness rear door-rear door service ..... M 41
Harness steering column-with horn push only ..... M 40

Harness- tail door ..... M 34
Harness- tail door ..... M 35
Harness-central door locking-driver ..... M 33
Harness-central door locking-driver ..... N 79
Harness-crash sensor airbag-rhd ..... M 62

Harness-front wing-lh-rhd ..... M 23
Harness-front wing-rh-rhd ..... M 22
Harness-fuel injector engine ..... M 94
Harness-fuel injector engine ..... M 95
Harness-licence plate lamp-rear ..... M 15

Harness-link B107
Harness-link B158
Harness-link ..... C 54
Harness-link ..... L 21
Harness-link ..... M 82

Harness-link-3 pin B101
Harness-link-a post ..... M 33
Harness-link-amplifier remote M135
Harness-link-catalyst overheat ..... M 90
Harness-link-cd data lead M137

Harness-link-exhaust gas recirculation ..... M 91
Harness-link-front door speaker M132
Harness-link-fuel pump ..... M 57
Harness-link-gearbox-temperature ..... M 55
Harness-link-headlamp levelling ..... M 60

Harness-link-heated rear window ..... M 42
Harness-link-heated rear window/high ..... M 42
Harness-link-low tension ignition coil-2 pin B104
Harness-link-low tension ignition coil-2 pin B148
Harness-link-oil-temp warning-automatic ..... M 43

Harness-link-radio switches ..... M 68
Harness-link-selector illumiation selection ..... P 16
Harness-link-selector illumiation selection ..... P 17
Harness-link-with burglar alarm ..... T 35
Harness-link-with remote locking M106

Harness-link-with sunroof ..... M 42
Harness-pad wear-front ..... M 59
Harness-passenger door-door speakers M133
Harness-rear end door-less door speaker ..... M 42
Harness-supplementary audio system ..... M 41

Harness-transmission ..... M 54
Harness-transmission ..... M 55
Harness-transmission ..... M 56
Headlamp assembly-front lighting-rh-rhd-5 ..... M 1
Headlamp assembly-load levelling ..... T 25

Headlamp assembly-load levelling-rh-rhd- ..... M 3
Headlining-roof trim-aspen grey ..... P 66
Headrestraint assembly-blue ..... T 12
Headrestraint assembly-dark granite ..... R 20
Headrestraint assembly-dark granite ..... R 16

Headrestraint assembly-leather-dark granite ..... R 23
Headrestraint assembly-sonar blue ..... R 5
Heater assembly ..... L 20
Heater assembly non air conditioning-rhd ..... L 22
Heatshield alternator ..... B 36

Heatshield alternator ..... B 73
Heatshield alternator B186
Heatshield alternator-65 amp B153
Heatshield alternator-65 amp ..... K 4
Heatshield assembly air conditioning B145

Heatshield assembly exhaust-downpipe ..... B 38
Heatshield assembly exhaust-downpipe ..... C 68
Heatshield assembly exhaust-downpipe ..... C 90
Heatshield assembly exhaust-downpipe ..... K 2
Heatshield assembly exhaust-downpipe ..... K 3

Heatshield assembly exhaust-downpipe ..... K 10
Heatshield assembly exhaust-downpipe ..... N 10
Heatshield assembly exhaust-intermediate ..... K 4
Heatshield assembly steering gear ..... F 7
Heatshield bottom hose-bottom ..... L 19

Heatshield engine exhaust ..... K 4
Heatshield engine exhaust ..... L 17
Heatshield engine exhaust-rh ..... B 83
Heatshield engine exhaust-rh ..... N 6
Heatshield steering gear ..... F 9

Heatshield-bracket ..... D 6
Heatshield-bracket ..... K 2
Heatshield-bracket ..... K 3
Heatshield-bracket ..... K 6
Heatshield-bracket ..... K 8

Heatshield-bracket ..... K 9
Heatshield-bracket ..... K 7
Heatshield-catalyst bulkhead ..... K 3
Heatshield-clutch slave cylinder ..... K 3
Heatshield-exhaust elbow turbocharger ..... B 31

Heatshield-exhaust manifold ..... B 62
Heatshield-exhaust manifold ..... B 96
Heatshield-exhaust manifold B131
Heatshield-exhaust manifold ..... K 9
Heatshield-exhaust manifold ..... K 7

Heatshield-front engine mounting ..... K 6
Heatshield-front engine mounting ..... K 8
Heatshield-front engine mounting ..... N 7
Heatshield-front engine mounting-lh ..... B 83
Heatshield-front engine mounting-lh ..... K 2

Heatshield-front engine mounting-lh ..... N 6
Heatshield-starter motor B131
Heatshield-tunnel exhaust system ..... K 2
Heatshield-tunnel exhaust system ..... K 6
Heatshield-tunnel exhaust system ..... K 8

Heatshield-tunnel exhaust system ..... K 9
Heatshield-tunnel exhaust system ..... K 11
Heatshield-tunnel exhaust system ..... K 13
Heatshield-tunnel exhaust system ..... K 14
Heatshield-tunnel exhaust system ..... K 7

Hexsert ..... N 41
Hexsert ..... P 71
Hexsert-black ..... N 30
Hexsert-fixing speaker to rear end door M134
Hexsert-fixing speaker to rear end door M139

Hexsert-fixing speaker to rear end door M142
Hexsert-speaker to door M136
Hinge a post-rh-upper ..... N 47
Hinge assembly-bonnet-rh ..... N 41
Hinge assembly-front door-rh ..... N 46

Hinge assembly-front door-rh ..... N 51
Hinge assembly-fusebox lid ..... P 8
Hinge assembly-load/tail door-top ..... N 53
Hinge assembly-tailgate-upper N120
Hinge assembly-tunnel console lid-ash grey ..... P 19

Hinge assembly-tunnel console lid-ash grey ..... P 20
Hinge b post-rh-upper ..... N 50
Hinge-rear seat armrest-rh ..... R 45
Hinge-rear seat armrest-rh ..... R 43
Hinge-side opening window ..... N 56

Hinge-tilt & remove sunroof ..... N 38
Hinge-tilt & remove sunroof ..... N 39
Hinges-plastic-beige ..... T 19
Holder & bulb assembly clock M130
Holder & bulb assembly clock M131

Holder-double pole bulb ..... M 7
Holder-double pole bulb ..... M 8
Holder-double pole bulb ..... M 9
Holder-front lighting bulb ..... M 3
Holder-single pole bulb ..... M 7

Holder-single pole bulb ..... M 8
Holder-single pole bulb ..... M 9
Holder-single pole bulb-21w ..... M 5
Hook-front sub frame towing ..... N 12
Hook-loadspace cover-ash grey ..... P 23

Hook-roof trim coat/grab handle-ash grey ..... P 71
Horn assembly M108
Horn assembly M110
Horn assembly M106
Horn assembly-high note M101

Horn assembly-low note M101
Horn low note M101
Horn push assembly steering wheel ..... F 3
Horn push assembly steering wheel-ash grey ..... F 3
Hose assembly-breather fuel ..... J 2

Hose assembly-breather fuel ..... J 4
Hose assembly-breather fuel ..... J 5
Hose assembly-breather fuel ..... J 6
Hose assembly-breather-rocker cover to ..... B 91
Hose assembly-engine coolant valve ..... L 31

Hose assembly-intercooler turbocharger ..... L 10
Hose assembly-wash system-rear-rhd M152
Hose brake vacuum-pump to servo ..... H 2
Hose brake vacuum-to inlet manifold ..... H 4
Hose brake-front ..... H 5

Hose brake-front ..... H 7
Hose brake-front ..... H 9
Hose brake-front ..... H 10
Hose brake-rear ..... H 11
Hose brake-rear ..... H 13

Hose breather-rocker cover to flame trap-rh ..... B 91
Hose clutch ..... G 11
Hose clutch ..... G 13
Hose coolant valve/heater ..... L 30
Hose coolant valve/heater ..... L 31

Hose driver vent ..... L 29
Hose driver vent ..... P 4
Hose heater/coolant valve ..... L 30
Hose heater/coolant valve ..... L 31
Hose lift pump/filter ..... J 10

Hose vacuum..-cut to 18" from 34" hose ..... B 26
Hose vacuum..-cut to length required-6" ..... B 25
Hose vacuum..-cut to length required-6" B132
Hose vacuum..-cut to length required-6" B133
Hose vacuum..-cut to length required-6" B134

Hose wastegate control-90mm ..... B 30
Hose-air flow meter to turbocharger induction ..... J 19
Hose-air inlet turbocharger ..... J 18
Hose-air outlet turbocharger ..... J 19
Hose-bleed pipe assembly to expansion tank ..... L 13

Hose-bleed pipe assembly to expansion tank ..... L 15
Hose-bypass to inlet manifold coolant ..... B 95
Hose-bypass to pump coolant ..... B 12
Hose-bypass to pump coolant ..... B 51
Hose-camshaft cover to air intake breather ..... B 91

Hose-camshaft cover to inlet manifold B174
Hose-carburettor fuel pipe-carburettor to ..... B 91
Hose-compressor/condenser air conditioning ..... L 40
Hose-compressor/condenser air conditioning ..... L 44
Hose-compressor/condenser air conditioning ..... L 51

Hose-compressor/condenser air conditioning ..... L 53
Hose-compressor/condenser air conditioning ..... L 55
Hose-compressor/condenser air ..... L 41
Hose-condenser/evaporator air conditioning ..... L 53
Hose-condenser/evaporator air conditioning ..... L 55

Hose-condenser/evaporator air conditioning- ..... L 41
Hose-condenser/evaporator air conditioning- ..... L 45
Hose-condenser/receiver dryer air ..... L 41
Hose-cooling system bleed ..... L 2
Hose-cooling system bleed ..... L 12

Hose-cooling system bleed ..... L 19
Hose-crankcase breather ..... B 59
Hose-crankcase breather flexible ..... B 21
Hose-crankcase breather flexible-breather ..... B 91
Hose-drain pipe ..... L 22

Hose-duct to throttle body induction system B180
Hose-duct to throttle body induction system ..... J 16
Hose-duct to throttle body induction system ..... J 17
Hose-engine coolant valve ..... L 30
Hose-engine to heater valve coolant ..... L 30

Hose-engine to heater valve coolant ..... L 31
Hose-evapr/compressor air conditioning ..... L 44
Hose-evapr/compressor air conditioning ..... L 52
Hose-evapr/compressor air conditioning ..... L 55
Hose-evapr/compressor air conditioning ..... L 54

Hose-evapr/compressor air conditioning-rear ..... L 51
Hose-expansion tank overflow ..... L 14
Hose-filler fuel filler ..... J 5
Hose-filler fuel filler ..... J 6
Hose-flame trap t piece-plenum chamber B136

Hose-flexible ..... J 15
Hose-headlamp wash-10mm id x 5m M150
Hose-heater to inlet manifold coolant ..... B 95
Hose-heater to inlet manifold coolant B129
Hose-heater to pump coolant B129

Hose-heater-inlet ..... L 29
Hose-heater-inlet ..... L 32
Hose-heater-outlet ..... L 29
Hose-heater-outlet ..... L 32
Hose-intercooler ..... L 10

Hose-intercooler to inlet manifold air ..... L 10
Hose-intercooler to inlet manifold air ..... L 3
Hose-intercooler turbocharger-rear ..... L 3
Hose-intercooler-front ..... L 3
Hose-liquid ..... L 49

Hose-liquid ..... L 53
Hose-liquid ..... L 55
Hose-manifold B136
Hose-manifold to plenum de icer B137
Hose-modulator valve to air cleaner-white.- ..... L 16

Hose-oil separator drain ..... B 21
Hose-oil separator drain ..... B 59
Hose-pipe to heater ..... B 24
Hose-pipe to heater B176
Hose-pipe to oil cooler ..... C 94

Hose-pipe valve-air pipe/air valve B137
Hose-plenum B136
Hose-plenum chamber to starter motor B136
Hose-plenum-plenum chamber/air pipe B137
Hose-power assisted steering reservoir to ..... F 21

Hose-power assisted steering reservoir to ..... F 20
Hose-power assisted steering reservoir to ..... F 19
Hose-radiator bottom coolant ..... L 2
Hose-radiator bottom coolant ..... L 4
Hose-radiator bottom coolant ..... L 5

Hose-radiator bottom coolant ..... L 9
Hose-radiator bottom coolant ..... L 19
Hose-radiator to expansion tank ..... L 14
Hose-radiator top coolant ..... L 2
Hose-radiator top coolant ..... L 4

Hose-radiator top coolant ..... L 5
Hose-radiator top coolant ..... L 9
Hose-radiator top coolant ..... L 19
Hose-receiver dryer/evaporator air ..... L 45
Hose-receiver dryer/evaporator air ..... L 52

Hose-receiver dryer/evaporator air ..... L 53
Hose-receiver dryer/evaporator air ..... L 55
Hose-receiver dryer/evaporator air ..... L 41
Hose-regulator/filter multi point injection/tbi B179
Hose-return fuel B133

Hose-return fuel B134
Hose-solenoid valve B137
Hose-solenoid valve B136
Hose-suction. ..... L 49
Hose-suction. ..... L 53

Hose-suction. ..... L 55
Hose-throttle body to camshaft cover B174
Hose-turbocharger to pipe ..... L 10
Hose-vacuum cruise control-pump to t-piece ..... G 10
Hose-wash system M148

Hose-wash system M151
Hose-wash system jet to tee valve M146
Hose-wash system jet to tee valve M147
Hose-wash system jet to tee valve M148
Hose-wash system jet to tee valve M149

Hose-wash system pump to non return valve- M146
Hose-wash system pump to non return valve- M147
Hose-wash system pump to non return valve- M148
Hose-wash system pump to non return valve- M149
Hose-wash system-720mm/long M146

Hose-wash system-720mm/long M147
Hose-wash system-720mm/long M149
Hose. B136
Housing & bush assembly pinion housing ..... E 21
Housing assembly differential/final drive ..... E 30

Housing assembly differential/final drive ..... E 7
Housing assembly differential/final drive ..... E 28
Housing assembly differential/final drive ..... E 8
Housing assembly differential/final drive ..... E 29
Housing assembly differential/final drive ..... E 31

Housing assembly- gear change cross shaft ..... D 4
Housing assembly- gear change cross shaft ..... D 16
Housing assembly- gear change cross shaft ..... D 18
Housing assembly-clutch ..... B 39
Housing assembly-clutch ..... B 75

Housing assembly-clutch B189
Housing assembly-torque converter ..... C 72
Housing assembly-transfer box gear change ..... D 14
Housing-ball pin gear change-manual ..... C 20
Housing-ball pin gear change-manual ..... C 40

Housing-bearing shaft ..... D 5
Housing-centre column switches ..... M 63
Housing-centre column switches ..... M 64
Housing-change ball manual transmission ..... C 20
Housing-change ball manual transmission ..... C 40

Housing-clutch manual transmission ..... C 52
Housing-clutch manual transmission-manual ..... C 4
Housing-clutch manual transmission-manual ..... C 26
Housing-cross shaft ..... D 3
Housing-facia console switches-for 2 ..... M 65

Housing-facia vent louvre ..... P 4
Housing-facia vent louvre-black ..... P 15
Housing-gear selector ..... C 67
Housing-governor extension ..... C 85
Housing-governor extension ..... D 4

Housing-inner panel door release assembly- ..... N 81
Housing-inner panel door release assembly- ..... N 84
Housing-inner panel door release assembly- ..... N 78
Housing-manual transmission gearchange ..... D 4
Housing-output shaft front ..... D 3

Housing-ram pipes B138
Housing-speaker-sonar blue ..... P 39
Housing-speaker-sonar blue ..... P 41
Housing-speedometer driven gear ..... D 5
Housing-spindle ..... D 12

Housing-swivel pin bearing ..... E 13
Housing-swivel pin bearing ..... E 14
Housing-swivel pin bearing ..... E 16
Housing-swivel pin-rh ..... E 20
Housing-swivel pin-rh-rhd ..... E 17

Housing-swivel pin-rh-rhd ..... E 18
Housing-thermostat ..... B 23
Housing-thermostat ..... B 61
Housing-throttle linkage bearing-front B139
Housing-transmission thermostat ..... C 28

Housing-transmission thermostat ..... C 55
Housing-tunnel console switch ..... P 19
Housing-tunnel console switch ..... P 20
Housing-tunnel console switch-ash grey ..... P 21
Housing-tunnel console switch-rhd-ash ..... M 73

Housing-tunnel console switch-rhd-ash ..... M 74
Hub & sleeve assembly- transfer box ..... D 15
Hub assembly front ..... E 23
Hub assembly front ..... E 24
Hub assembly- rear ..... E 39

Hub assembly-front & rear ..... E 24
Hub assembly-front & rear ..... E 40
Hub-camshaft pulley ..... B 60
Hub-governor ..... C 84
Hub-selector automatic transmission ..... C 86

Idler-timing belt ..... B 11
Idler-timing belt ..... B 50
Illumination assembly selector mechanism ..... P 16
Illumination assembly selector mechanism-ash ..... P 17
Illumination assembly-instrument pack M121

Illumination assembly-instrument pack M127
Indicator-light emitting diode burglar alarm M128
Indicator-light emitting diode burglar alarm M106
Inhibitor connector ..... M 81
Injector assembly-diesel holder & nozzle ..... B 29

Injector assembly-diesel holder & nozzle-new ..... B 66
Injector-fuel multi point injection B132
Injector-fuel multi point injection B133
Injector-fuel multi point injection B179
Injector-fuel multi point injection B134

Insert-clutch lever ..... C 24
Insert-clutch lever ..... C 46
Insert-cylinder head exhaust valve seat ..... B 16
Insert-cylinder head exhaust valve seat ..... B 54
Insert-cylinder head exhaust valve seat B172

Insert-front headrestraint grab handle-dark ..... R 20
Insert-front headrestraint grab handle-dark ..... R 16
Insert-front headrestraint grab handle-sonar ..... R 6
Insert-inlet valve seat-cylinder head ..... B 16
Insert-inlet valve seat-cylinder head ..... B 54

Insert-inlet valve seat-cylinder head B172
Insert-inlet valve seat-cylinder head- ..... B 88
Insert-inlet valve seat-cylinder head- B123
Insert-inlet valve seat-cylinder head- B125
Insert-inlet/exhaust valve seat-cylinder head- B123

Insert-inlet/exhaust valve seat-cylinder head- B125
Insert-plain ..... R 39
Insert-plain-panel seat back-large ..... R 32
Insert-screw in-sensor to inlet manifold B177
Inspection hatch-seat plinth-ash grey ..... R 1

Instrument pack with warning lights-including M124
Instrument pack with warning lights-including M125
Instrument pack-uk police specification-less M129
Instrument pack-with drivers air bag M119
Insulator assembly-carburettor inlet manifold- ..... B 98

Intercooler assembly-coolant ..... L 2
Intercooler assembly-coolant ..... L 7
Interlock manual transmission ..... C 66
Interlock-5th/reverse ..... C 66
Intermediate assembly exhaust system ..... K 14

Intermediate/rear assembly exhaust system ..... K 1
Intermediate/rear assembly exhaust system ..... K 2
Intermediate/rear assembly exhaust system ..... K 6
Intermediate/rear assembly exhaust system ..... K 8
Intermediate/rear assembly exhaust system ..... K 9

Intermediate/rear assembly exhaust system ..... K 14
Intermediate/rear assembly exhaust system ..... K 7
Isolator upper ..... E 50
Isolator upper ..... E 52
J bolt-fixing battery-long ..... M 16

J bolt-fixing battery-long ..... M 18
J bolt-fixing battery-long ..... M 20
J bolt-fixing battery-short ..... M 17
J bolt-fixing battery-short ..... M 19
J bolt-fixing battery-short ..... M 21

Jack-hydraulic ..... S 1
Jacket insulation lower ..... C 52
Jacket insulation upper-manual ..... C 52
Jaw assembly-towing ..... T 20
Jet assembly-backlight wash single nozzle M151

Jet assembly-backlight wash single nozzle M152
Jet assembly-headlamp wash single M150
Jet assembly-nozzle-wash system M146
Jet assembly-nozzle-wash system M147
Jet assembly-nozzle-wash system M148

Jet assembly-nozzle-wash system M149
Jet assembly-piston cooling-no1-no3 ..... B 5
Jet assembly-piston cooling-no1-no3 ..... B 44
Jet-camshaft oil ..... B 43
Jet-carburettor-lh ..... B 97

Joint-air duct floor ..... L 34
Joint-ball B140
Joint-ball-lower arm front/rear suspension ..... E 47
Joint-ball-lower arm front/rear suspension ..... E 53
Joint-ball-outer steering gear ..... F 11

Joint-constant velocity ..... E 13
Joint-constant velocity ..... E 16
Joint-constant velocity-10 spline shaft rh ..... E 14
Joint-propshaft universal ..... E 1
Joint-propshaft universal ..... E 2

Joint-rear suspension upper link ball ..... E 49
Joint-universal steering linkage ..... F 7
Joint-universal steering linkage ..... F 9
Joint-universal steering linkage-upper ..... F 8
Junction-'y' piece exhaust system ..... K 10

Key ..... B 22
Key-blank owner ..... F 6
Key-blank owner ..... N 71
Key-blank owner ..... N 72
Key-blank owner N122

Key-blank owner ..... N 73
Key-blank owner ..... N 74
Key-blank owner ..... N 75
Key-camshaft location ..... B 92
Key-camshaft location B128

Key-crankshaft ..... B 7
Key-crankshaft ..... B 45
Key-crankshaft ..... B 80
Key-crankshaft B112
Key-crankshaft B165

Key-oil pump drive B171
Key-tool kit locking wheel nut-code a ..... E 42
Key-tool kit locking wheel nut-pin type ..... T 36
Kit air conditioning ..... L 40
Kit air conditioning ..... L 45

Kit burglar alarm ..... T 33
Kit burglar alarm ..... T 34
Kit burglar alarm ..... T 35
Kit decal ..... N 88
Kit interior wood trim-lhd ..... T 15

Kit service-side step ..... T 44
Kit-air conditioning-rhd ..... T 46
Kit-air intake duct pollen filter ..... N 89
Kit-air intake-air cleaner ..... J 19
Kit-alternator-a127-65 amp-new B105

Kit-alternator-a127-65 amp-new B152
Kit-anti roll bar ..... T 38
Kit-anti-roll bar sports suspension-front & ..... T 38
Kit-armrest-bahama beige ..... T 29
Kit-armrest-sonar blue ..... P 39

Kit-armrest-sonar blue ..... P 36
Kit-armrest-sonar blue ..... P 41
Kit-automatic transmission bearing overhaul ..... C 91
Kit-auxiliary wiring-safari 500 ..... T 24
Kit-ball joint-lh-l/h thread ..... F 11

Kit-ball joint-rh-r/h thread ..... F 15
Kit-body decals/badges-turquoise/light ..... T 5
Kit-body rubbing strip ..... T 6
Kit-body styling ..... T 7
Kit-bonnet locking ..... N 69

Kit-brake caliper piston seal ..... H 18
Kit-brake pad retaining ..... H 18
Kit-brake pad retaining-asbestos-front-solid ..... H 16
Kit-brake piston seal-front ..... H 15
Kit-brush starter motor B188

Kit-brush starter motor-with m78r starter B107
Kit-bulb ..... T 46
Kit-caliper set-rear brake-rh ..... H 15
Kit-caliper set-rear brake-rh ..... H 18
Kit-carburettor seal ..... B 97

Kit-clutch ..... B 39
Kit-clutch B108
Kit-clutch-asbestos free B160
Kit-clutch-automatic ..... C 76
Kit-console bag-bahama beige ..... P 18

Kit-cover-loadspace ..... P 22
Kit-differential gears ..... D 7
Kit-differential gears-10 spline shaft ..... E 11
Kit-differential gears-10 spline shaft ..... E 36
Kit-dog guard fitting ..... T 13

Kit-door key & lock cylinder-rh red cam ..... N 71
Kit-door key & lock cylinder-rh red cam ..... N 72
Kit-door key & lock cylinder-rh red cam ..... N 73
Kit-door key & lock cylinder-rh red cam ..... N 74
Kit-door key & lock cylinder-rh red cam ..... N 75

Kit-driving lamp wiring-cewe ..... T 23
Kit-electric winch-husky ..... T 17
Kit-engine balance weights B160
Kit-engine balance weights B161
Kit-eye bolts and bushes ..... R 59

Kit-first aid ..... T 17
Kit-fitting-gearbox r380 ..... C 1
Kit-fitting-gearbox r380 ..... C 23
Kit-fixing-screw ..... T 10
Kit-fixings alternator/vacuum pump ..... B 36

Kit-fixings alternator/vacuum pump-65amp B152
Kit-fixings alternator/vacuum pump-85amp B153
Kit-fixings alternator/vacuum pump-new B105
Kit-flange ..... E 10
Kit-flange ..... E 35

Kit-flange & mudshield ..... E 35
Kit-flange-front ..... D 11
Kit-flange-rear ..... D 12
Kit-foam gasket-light seals M123
Kit-front axle brackets ..... T 38

Kit-front lighting fog lamp-pair ..... T 24
Kit-fuel pump repair ..... B 27
Kit-fusible link ..... M 80
Kit-gaiter-gearbox ..... C 1
Kit-gaiter-gearbox ..... C 23

Kit-gasket ..... B 68
Kit-gasket ..... B 97
Kit-gasket and seal ..... D 21
Kit-gasket automatic transmission ..... C 91
Kit-gasket-overhaul ..... B 68

Kit-gears-1st/2nd ..... C 59
Kit-gears-3rd/4th ..... C 60
Kit-glass sunroof assembly-front-rear ..... T 47
Kit-gun box fixing ..... T 18
Kit-headlamp cover ..... M 2

Kit-headlamp fittings ..... M 3
Kit-heavy duty road spring-rear ..... E 51
Kit-heavy duty road spring-rear ..... T 38
Kit-high tension leads ignition-silicone B102
Kit-high tension leads ignition-silicone B149

Kit-igniter unit distributor B147
Kit-indicator trip/slip ring ..... F 3
Kit-insulation cover distributor B101
Kit-insulation cover distributor B147
Kit-lamp guard fixing ..... T 10

Kit-locking nut road wheels ..... E 42
Kit-locking nut road wheels-set of 5-alloy- ..... T 36
Kit-lumber mechanism-rh ..... R 12
Kit-lumber mechanism-rh ..... R 13
Kit-lumber mechanism-rh ..... R 26

Kit-lumber mechanism-rh ..... R 25
Kit-master cylinder/reservoir & seal brake ..... H 1
Kit-modulator-antilock brakes ..... H 3
Kit-mudflaps ..... V 1
Kit-oil pump repair ..... B 87

Kit-pipes etc heater-heater inlet ..... L 25
Kit-piston ring ..... B 46
Kit-piston ring B166
Kit-plate & seal-oil guide ..... D 6
Kit-power assisted steering box seal ..... F 13

Kit-power assisted steering box seal ..... F 14
Kit-power assisted steering pump seal ..... B 99
Kit-power assisted steering pump seal B142
Kit-propshaft coupling ..... E 2
Kit-rear axle brackets ..... T 38

Kit-rear spoiler ..... T 7
Kit-rear step fixing ..... T 43
Kit-repair brake master cylinder ..... H 1
Kit-repair brake servo ..... H 1
Kit-repair clutch master cylinder ..... G 11

Kit-repair clutch master cylinder ..... G 12
Kit-repair clutch slave cylinder ..... C 2
Kit-repair clutch slave cylinder ..... C 24
Kit-repair clutch slave cylinder ..... C 44
Kit-repair clutch slave cylinder ..... C 45

Kit-repair clutch slave cylinder ..... C 46
Kit-repair-lamp guards ..... T 8
Kit-roof rack conversion-c/w bag ..... N 40
Kit-seat belt fixing ..... R 54
Kit-seat belt fixing ..... R 57

Kit-seat-rear-inward-cloth-blue.-lh ..... T 29
Kit-seat-rear-inward-cloth-blue.-rh ..... R 49
Kit-seat-rear-inward-leather-bahama beige- ..... R 50
Kit-seat-rear-inward-left hand-leather-dark ..... R 50
Kit-seat-rear-inward-right hand-leather- ..... R 50

Kit-shaft-cross ..... D 7
Kit-side repeater harness ..... M 4
Kit-side repeater harness ..... M 5
Kit-side sills pair ..... T 7
Kit-side step ..... T 44

Kit-starter motor ..... B 37
Kit-starter motor sundry parts ..... B 37
Kit-starter motor sundry parts-with 2m100 ..... B107
Kit-steering box sector shaft seal ..... F 13
Kit-steering box sector shaft seal ..... F 14

Kit-steering box sector shaft seal ..... F 12
Kit-sunblind ..... T 47
Kit-sunroof roller blind ..... P 73
Kit-swivel pin-upper ..... E 21
Kit-synchroniser assembly-main shaft-5th & ..... C 63

Kit-throttle potentiometer multi point injection B180
Kit-tool & stowage assembly ..... T 17
Kit-towing ..... T 20
Kit-transducer speed-end fixings M114
Kit-transmission brake adjuster ..... D 22

Kit-transmission brake adjuster ..... D 23
Kit-transmission brake shoe ..... D 22
Kit-transmission brake shoe ..... D 23
Kit-transmission brake shoe retention ..... D 23
Kit-transmission brake springs ..... D 22

Kit-transmission brake springs ..... D 23
Kit-transmitter/receiver/interior lamp infra red- ..... T 34
Kit-trunk tidy net ..... P 71
Kit-vehicle label ..... P 13
Kit-wheelarch flare N109

Kit-wheelarch flare ..... T 6
Kit-winch accessories ..... T 26
Kit-winch accessories ..... T 27
Knob assembly-change manual transmission- ..... C 67
Knob assembly-change manual transmission- ..... C 21

Knob assembly-change manual transmission- ..... C 41
Knob assembly-transfer box change-ash ..... D 19
Knob assembly-transfer box change-grey ..... D 17
Knob heater/fan control ..... P 15
Knob trip reset M118

Knob trip reset M128
Knob trip reset M123
Knob-control heater ..... L 28
Knob-control heater ..... L 48
Knob-control heater ..... L 21

Knob-front seat belt adjuster mechanism-ash ..... R 55
Knob-front seat belt adjuster mechanism- ..... R 54
Knob-handbrake cable adjusting ..... H 23
Knob-handbrake cable adjusting ..... H 21
Knob-selector drive ..... C 88

Knob-selector drive ..... C 87
Knob-slider control-heater ..... L 28
Knob-slider control-heater ..... L 48
Knob-switch-hazard warning switch ..... M 63
Label anti freeze ..... N 44

Label-air conditioning ..... N 44
Label-body colour adhesive-davas white ..... N 45
Label-catalyst-fuel filler neck ..... K 1
Label-caution fuel filler-diesel ..... J 5
Label-caution fuel filler-diesel ..... J 6

Label-e mark ..... N 44
Label-fuel caution ..... J 5
Label-fuel caution ..... J 6
Label-fusebox ..... M 79
Label-fusebox ..... M 81

Label-fusebox-with air conditioning ..... M 80
Label-headlamp leveling information ..... M 3
Label-headlamp leveling information ..... M 2
Label-headlamp leveling information ..... M 75
Label-headlamp leveling information ..... N 44

Label-jack information ..... N 44
Label-jack information ..... S 1
Label-oil/diesel ..... N 44
Label-sunvisor warning ..... P 71
Label-transfer box ratio ..... N 67

Label-underbonnet asbestos ..... N 44
Label-underbonnet data ..... N 44
Label-underbonnet radiator caution B130
Label-underbonnet radiator caution ..... N 44
Label-underbonnet smoke absorbtion ..... N 44

Label-unleaded fuel only warning ..... J 5
Label-unleaded fuel only warning ..... J 6
Label-warning M143
Label-warning-cfc free ..... L 47
Label-warning-cfc free ..... N 44

Label-warning-manual ..... P 19
Label-warning-manual ..... P 20
Ladder-roof rack access ..... T 39
Lamp assembly-front lighting fog-safari 5000 ..... T 24
Lamp assembly-rear bumper-rh ..... M 9

Lamp assembly-rear-rh ..... M 7
Lamp assembly-rear-rh ..... M 8
Lamp assembly-rear-rh ..... M 9
Lamp guards-fixed ..... T 8
Lamp steady bars-pair ..... T 24

Lamp-auxiliary front lighting-pair ..... T 23
Lamp-centre headlining interior courtesy- ..... M 12
Lamp-driving-rally 1000 ..... T 23
Lamp-front direction indicator-rh ..... M 4
Lamp-front direction indicator-rh ..... M 5

Lamp-front lighting fog-for body styling kit ..... T 7
Lamp-front lighting fog-rally 1000 ..... T 23
Lamp-front lighting fog-rh ..... M 6
Lamp-front long range driving-cewe- ..... T 23
Lamp-front long range driving-safari 5000 ..... T 24

Lamp-glove box illumination ..... M 13
Lamp-hand-cewe ..... T 23
Lamp-interior courtesy-front ..... M 13
Lamp-interior courtesy-with on/off switch ..... M 12
Lamp-rear headlining interior courtesy ..... M 13

Lamp-rear high mounted stop ..... M 11
Lamp-rear licence plate ..... M 15
Lamp-stand alone warning-marker rear end ..... M 35
Lamp-trunk/loadspace interior courtesy-rear- ..... M 12
Lamp-trunk/loadspace interior courtesy-rear- ..... M 13

Latch assembly-front door-rh ..... N 76
Latch assembly-load/tail door ..... N 84
Latch assembly-rear door-rh ..... N 80
Latch assembly-rear seat squab-lh ..... R 46
Latch assembly-tunnel console lid-ash grey ..... P 19

Latch assembly-tunnel console lid-ash grey ..... P 20
Latch-glovebox ..... P 8
Lead antenna M136
Lead antenna M138
Lead antenna-am/fm M141

Lead antenna-fm M143
Lead-heated screen fuse M116
Lead-high tension ignition-no1 ..... B102
Lead-high tension ignition-no1 ..... B149
Lead-high tension ignition-no1 ..... B150

Lead-high tension ignition-no7 ..... B103
Lead-high tension ignition-no7 ..... B151
Lead-high tension ignition-to number one B185
Lead-king ignition B149
Lead-king ignition B103

Lead-king ignition B151
Lead-link radio M137
Lens-auxiliary lighting side repeater ..... M 4
Lens-auxiliary lighting side repeater ..... M 5
Lens-fog ..... T 24

Lens-front long range driving ..... T 24
Lens-replacement ..... T 24
Lens-replacement-rally 1000 ..... T 23
Lens-tunnel console face plate ..... P 16
Lens-warning instrument pack-diesel M117

Level-carburettor throttle ..... B 97
Level-carburettor throttle ..... B 98
Level-carburettor throttle B139
Lever assembly bell crank ..... D 16
Lever assembly bell crank ..... D 18

Lever assembly handbrake-ash grey ..... H 22
Lever assembly handbrake-ash grey ..... H 23
Lever assembly handbrake-grey ..... H 20
Lever assembly-operating automatic ..... C 86
Lever assembly-operating automatic ..... C 88

Lever assembly-operating/selector automatic ..... C 86
Lever assembly-operating/selector automatic ..... C 88
Lever assembly-selector selector mechanism ..... C 72
Lever assembly-throttle tbi B139
Lever driver vent ..... L 21

Lever-carburettor camshaft-lh ..... B 97
Lever-change manual transmission ..... C 67
Lever-detent automatic transmission ..... C 72
Lever-drop arm steering-rhd ..... F 11
Lever-handbrake cable relay ..... H 22

Lever-handbrake cable relay ..... H 21
Lever-kickdown automatic transmission ..... B 28
Lever-kickdown automatic transmission ..... B 65
Lever-lower manual transmission ..... C 21
Lever-lower manual transmission ..... C 41

Lever-selector selector mechanism ..... C 86
Lever-selector selector mechanism ..... C 37
Lever-selector selector mechanism ..... C 17
Lever-throttle multi point injection B140
Lever-tool kit nut cover removal ..... E 42

Lever-transfer box-bent ..... D 16
Lever-transfer box-bent ..... D 18
Lever-transmission brake ..... D 23
Lever-upper manual transmission ..... C 67
Lever-upper manual transmission ..... C 21

Lever-upper manual transmission ..... C 41
Lid assembly-facia fuse box-sonar blue ..... P 1
Lid assembly-tunnel console-dark granite ..... P 19
Lid assembly-tunnel console-dark granite ..... P 20
Lid assembly-tunnel console-granite ..... T 19

Lid-air cleaner ..... J 16
Lid-air cleaner ..... J 19
Lid-facia fuse box-bahama beige-rh ..... P 8
Liner-carpet protection-loadspace ..... T 30
Liner-fender-rh ..... N 28

Liner-front wheelarch-rh ..... N 28
Link - locks ..... N 76
Link - locks-lh ..... N 78
Link - locks-rh ..... N 81
Link adjusting ..... N 76

Link adjusting ..... N 80
Link adjusting ..... N 83
Link assembly-drag ..... F 15
Link assembly-lower-rear suspension ..... E 48
Link assembly-rear suspension upper-rh ..... E 49

Link assembly-windscreen wiper-rhd M144
Link cruise control B140
Link-adjuster alternator B106
Link-adjuster alternator B156
Link-carburettor/throttle connecting ..... B 98

Link-central door locking motor to latch ..... N 85
Link-compact disc automatic changer-philips M140
Link-compact disc automatic changer-philips M143
Link-front door lock/latch ..... N 77
Link-front door sill button/latch-rh ..... N 77

Link-fusible-100 amp-blue. ..... M 82
Link-handle ..... N 76
Link-lever accelerator ..... G 3
Link-load/tail door exterior release ..... N 83
Link-load/tail door exterior release ..... N 84

Link-load/tail door latch/crank ..... N 83
Link-load/tail door sill button/crank ..... N 83
Link-power assisted steering pump adjusting ..... B 34
Link-power assisted steering pump adjusting ..... B 99
Link-power assisted steering pump adjusting B142

Link-rear door crank/latch-rh ..... N 80
Link-rear door handle ..... N 80
Link-rear door handle-rh ..... N 81
Link-rear door interior release handle/latch ..... N 84
Link-rear door sill button/crank ..... N 80

Link-transfer box differential lock connecting ..... D 14
Load door assembly ..... N 53
Load stop adjustable-set of 4 ..... T 40
Lock & keys ..... T 19
Lock & keys ..... T 37

Lock assembly-load/tail door ..... N 84
Lock assembly-steering column ..... F 6
Lock assembly-steering column ..... N 71
Lock assembly-steering column ..... N 72
Lock assembly-steering column ..... N 73

Lock assembly-steering column ..... N 74
Lock assembly-steering column ..... N 75
Lock assembly-tailgate ..... N 71
Lock assembly-tailgate ..... N 83
Lock assembly-tailgate ..... N 72

Lock assembly-tailgate ..... N 73
Lock assembly-tailgate ..... N 74
Lock assembly-tailgate ..... N 75
Lock set ..... N 71
Lock set ..... N 73

Lock set ..... N 75
Lock set-rhd ..... N 72
Lock set-rhd ..... N 74
Lock-fuel filler flap ..... N 33
Lock-fuel filler flap ..... N 71

Lock-fuel filler flap ..... N 72
Lock-fuel filler flap ..... N 73
Lock-fuel filler flap ..... N 74
Lock-fuel filler flap ..... N 75
Locknut ..... C 25

Locknut ..... D 10
Locknut ..... E 23
Locknut ..... E 24
Locknut ..... E 39
Locknut ..... E 40

Locknut ..... E 49
Locknut ..... E 44
Locknut ..... E 46
Locknut ..... E 51
Locknut ..... E 52

Locknut ... F 1
Locknut ... H 21
Locknut ... N 80
Locknut-1/4-unf ... B 86
Locknut-1/4unf ... F 15

Locknut-10x32unf ... B 98
Locknut-10x32unf ... B103
Locknut-10x32unf ... B151
Locknut-7 seats ... L 43
Locknut-fixing inertia switch ... M 96

Locknut-fixing inertia switch ... M 99
Locknut-m10 ... C 20
Locknut-m10 ... C 40
Locknut-m10 ... G 9
Locknut-m16 ... E 47

Locknut-m16 ... E 53
Locknut-m8 ... B 34
Locknut-m8 ... F 7
Magazine-automatic changer compact disc ... M136
Magazine-automatic changer compact disc ... M135

Magazine-automatic changer compact disc ... M140
Magazine-automatic changer compact disc ... M143
Mainshaft bearing ... D 9
Manifold assembly-engine inlet ... B 94
Manifold-engine exhaust ... B 24

Manifold-engine exhaust ... B178
Manifold-engine exhaust-lh ... B 96
Manifold-engine exhaust-rh ... B131
Manifold-engine inlet ... B 24
Manifold-engine inlet ... B 62

Manifold-engine inlet ... B129
Manifold-engine lower inlet ... B177
Manifold-engine upper inlet ... B177
Mask assembly instrument pack ... M120
Master cylinder assembly brake ... H 1

Master cylinder clutch ... G 11
Master cylinder clutch ... G 12
Mat-centre console-automatic-black ... P 21
Mat-centre console-black ... P 15
Mat-centre console-black ... P 16

Mat-centre console-black ... P 14
Mat-centre console-black-manual ... P 19
Mat-facia-black ... P 1
Mat-facia-black ... P 7
Mat-loadspace-full-nylon ... T 30

Mat-loadspace-rubber ... T 31
Mat-rear bumper step-black ... N 15
Mat-rubber ... S 1
Matrix-heater heater ... L 20
Matrix-heater heater ... L 25

Mechanism assembly-transfer box gear ... D 16
Mechanism-manual front seat height ... R 18
Micro switch n.c contacts ... C 88
Micro switch n.o contacts ... C 88
Mirror assembly-electric control exterior- ... N115

Mirror assembly-electric control exterior-self ... N117
Mirror assembly-electric control exterior-self ... N118
Mirror assembly-hand set exterior-convex-rh ... N112
Mirror assembly-hand set self colour exterior- ... N113
Mirror assembly-hand set self colour exterior- ... N114

Mirror assembly-interior dipping-manual dim ... N 67
Modulator antilock brakes ... H 3
Module & cover assembly-driver air bag-ash ... M 61
Module & cover assembly-driver air bag-ash ... M 62
Module & cover assembly-passenger air bag- ... M 61

Module & cover assembly-passenger air bag- ... M 62
Module-distributor-2 pin ... B101
Module-distributor-2 pin ... B147
Monitor unit- overspeed-black ... M 83
Monobolt ... N 54

Motor & bracket assembly-windscreen wiper ... M144
Motor & bracket assembly-windscreen wiper- ... M145
Motor & fan balance assembly blower-heater ... L 20
Motor assembly blower-air conditioning ... L 24
Motor assembly blower-air conditioning ... L 26

Motor assembly blower-air conditioning ... L 42
Motor assembly-exterior mirror assembly ... N115
Motor assembly-sunroof ... N 36
Motor-backlight wiper ... M151
Motor-backlight wiper ... M152

Motor-cooling system fan ... L 1
Motor-front door window regulator-rh ... N 61
Motor-front seat ... R 21
Motor-headlamp-load levelling ... M 3
Motor-headlamp-load levelling ... M 2

Motor-multi point injection stepping ... B136
Motor-rear door window regulator-rh ... N 64
Motor-starter engine ... B188
Motor-starter engine-exchange-bosch ... B 38
Motor-starter engine-exchange-bosch ... B 74

Motor-starter engine-new ... B159
Motor-starter engine-new-lucas ... B107
Motor-starter engine-new-lucas ... B158
Motor-starter engine-new-valeo ... B 37
Motor-window regulator-front rh ... N 62

Motor-window regulator-rear rh ... N 65
Moulding assembly-rear end trim load space- ... P 63
Moulding-centre roof lamp-less sunroof ... P 72
Moulding-e post upper air vent ... N 31
Moulding-load space side casing-rh-sonar ... P 57

Moulding-plenum air intake extension-rh- ... N 89
Moulding-seat belt-upper-dark granite ... P 56
Mounting-engine rubber ... N 4
Mounting-engine rubber-3/8unf ... B 83
Mounting-engine rubber-body to engine ... N 6

Mounting-engine rubber-lh ... N 7
Mounting-gearbox ... C 22
Mounting-lower link flexible rubber ... E 48
Mounting-multi point injection throttle body ... B180
Mounting-rubber ... F 16

Mounting-rubber ... L 16
Mounting-rubber ... N 7
Mounting-rubber exhaust system ... K 1
Mounting-rubber exhaust system ... K 2
Mounting-rubber exhaust system ... K 6

Mounting-rubber exhaust system ... K 8
Mounting-rubber exhaust system ... K 9
Mounting-rubber exhaust system ... K 13
Mounting-rubber exhaust system ... K 4
Mounting-rubber exhaust system ... K 7

Mounting-rubber exhaust system ... K 15
Mounting-rubber flexible support ... J 18
Mounting-rubber-body to chassis frame ... N 8
Mounting-rubber-lh ... C 68
Mounting-rubber-lh ... C 90

Mounting-rubber-rh ... N 10
Mudflaps ... T 44
Mudflaps-kit front ... N110
Mudflaps-kit front ... T 45
Mudflaps-kit rear ... N111

Mudflaps-kit rear ... T 45
Mudshield-front & rear differential ... E 10
Mudshield-front & rear differential ... E 34
Mudshield-front & rear differential ... E 35
Mudshield-transfer box oil seal ... D 11

Mudshield-transfer box oil seal ... D 12
Needle & seat assembly-carburettor float ... B 97
Needle assembly-carburettor jet ... B 97
Net-carpet stowage ... T 30
Net-stowage-mist grey ... P 71

Nipple-cable automatic transmission ... C 86
Nipple-cable automatic transmission ... C 88
Nipple-grease-unf ... E 1
Nipple-grease-unf ... E 2
Nut ... B101

Nut ... C 87
Nut ... C 89
Nut ... D 15
Nut ... F 1
Nut ... F 13

Nut ... F 14
Nut ... F 3
Nut ... F 11
Nut ... G 1
Nut ... G 4

Nut ... G 9
Nut ... G 11
Nut ... G 6
Nut ... G 8
Nut ... H 5

Nut ... H 7
Nut ... H 11
Nut ... H 13
Nut ... J 23
Nut ... J 24

Nut ... M 7
Nut ... M107
Nut ... M144
Nut ... M145
Nut ... M152

Nut ... M137
Nut ... M 10
Nut ... N 35
Nut ... N 76
Nut ... N 36

Nut ... N 91
Nut ... P 4
Nut ... P 15
Nut ... P 19
Nut ... P 14

Nut ... P 10
Nut ... P 72
Nut ... S 1
Nut ... T 20
Nut ... T 21

Nut & washer set ... T 21
Nut & washer set-6mm ... B 73
Nut & washer set-alternator to bracket- ... B157
Nut & washer set-cover terminal-6mm ... B155
Nut-1/4unf ... B 96

Nut-1/4unf ... B131
Nut-20mm ... E 49
Nut-3/8-unf ... E 1
Nut-3/8-unf ... E 2
Nut-7/16unf ... B 97

Nut-alloy road wheels ... E 41
Nut-alternator-flanged head-m8 x 20 ... B155
Nut-bracket adjusting-hexagonal head-m8 ... B142
Nut-bracket mounting to mounting engine ... N 4
Nut-bracket mounting-hexagonal head- ... B144

Nut-castle ... E 10
Nut-castle ... E 34
Nut-castle-m12 ... E 47
Nut-castle-m12 ... E 53
Nut-castle-m12 ... F 15

Nut-castle-m12 ... F 11
Nut-clamp distributor ... B147
Nut-connecting rod ... B 8
Nut-connecting rod ... B 46
Nut-connecting rod ... B166

Nut-connecting rod-11/32uns-3b ... B 81
Nut-connecting rod-11/32uns-3b ... B114
Nut-console to floor ... P 21
Nut-fan guard-hexagonal head-m8 ... B156
Nut-fixing bracket egr-flanged head-m8 x 20 ... M 91

Nut-fixing bracket-flanged head-m8 ... M111
Nut-fixing bracket-hexagonal head-nyloc- ... M 45
Nut-fixing cd autochanger bracket ... M135
Nut-fixing control unit ... M109
Nut-fixing ecu-m4 ... M 93

Nut-flange L 1
Nut-flange-bracket mounting to mounting N 10
Nut-flange-bracket mounting to mounting N 5
Nut-flange-fixing ecu-m6 M111
Nut-flange-fuel rail to manifold-m6 B133

Nut-flange-fuel rail to manifold-m6 B134
Nut-flange-m10 B 24
Nut-flange-m10 B 31
Nut-flange-m10 B 62
Nut-flange-m10 C 68

Nut-flange-m10 C 90
Nut-flange-m10 K 13
Nut-flange-m10 K 14
Nut-flange-m6 B180
Nut-flange-m6 C 53

Nut-flange-m6 H 3
Nut-flange-m6 M 82
Nut-flange-m6 M145
Nut-flange-m6 M147
Nut-flange-m6 M149

Nut-flange-m6 N 83
Nut-flange-m6 N 86
Nut-flange-m8 B 25
Nut-flange-m8 L 27
Nut-flange-m8 L 38

Nut-flange-nyloc-m10 B178
Nut-flange-nyloc-m10 K 6
Nut-flange-nyloc-m10 K 8
Nut-flange-nyloc-m10 K 9
Nut-flange-nyloc-m10 K 7

Nut-flange-nyloc-m5 D 20
Nut-flange-nyloc-m5 D 17
Nut-flange-nyloc-m5 L 3
Nut-flange-nyloc-m5 L 17
Nut-flange-nyloc-m5 M 10

Nut-flange-nyloc-m6 C 69
Nut-flange-nyloc-m6 C 95
Nut-flange-nyloc-m6 C 97
Nut-flange-nyloc-m6 H 3
Nut-flange-nyloc-m6 H 9

Nut-flange-nyloc-m6 H 10
Nut-flange-nyloc-m6 H 18
Nut-flange-nyloc-m6 L 6
Nut-flange-nyloc-m6 L 18
Nut-flange-prevailing torque-m10 N 7

Nut-flange-valve to body-m6 H 9
Nut-flange-valve to body-m6 H 10
Nut-flanged head-m10 B 11
Nut-flanged head-m10 B 37
Nut-flanged head-m10 B 50

Nut-flanged head-m10 B 51
Nut-flanged head-m10 B 74
Nut-flanged head-m10 C 3
Nut-flanged head-m10 C 4
Nut-flanged head-m10 C 52

Nut-flanged head-m10 C 72
Nut-flanged head-m10 K 6
Nut-flanged head-m10 K 8
Nut-flanged head-m10 K 9
Nut-flanged head-m10 K 7

Nut-flanged head-m5 B186
Nut-flanged head-m5 M 13
Nut-flanged head-m5 N 43
Nut-flanged head-m8 B 12
Nut-flanged head-m8 B 29

Nut-flanged head-m8 B 56
Nut-flanged head-m8 B 73
Nut-flanged head-m8 x 20 B 24
Nut-flanged head-m8 x 20 B 28
Nut-flanged head-m8 x 20 B 34

Nut-flanged head-m8 x 20 B 62
Nut-flanged head-m8 x 20 B 64
Nut-flanged head-m8 x 20 B 66
Nut-flanged head-m8 x 20 B 71
Nut-flanged head-m8 x 20 B 99

Nut-flanged head-m8 x 20 B104
Nut-flanged head-m8 x 20 B148
Nut-flanged head-m8 x 20 B177
Nut-flanged head-m8 x 20 B157
Nut-flanged head-m8 x 20 H 24

Nut-flanged head-m8 x 20 J 15
Nut-flanged head-m8 x 20 K 12
Nut-flanged head-m8 x 20 L 1
Nut-flanged head-m8 x 20 L 2
Nut-flanged head-m8 x 20 L 4

Nut-flanged head-m8 x 20 L 19
Nut-flanged head-m8 x 20 L 17
Nut-hexagonal head-1/4unc x 1 B 85
Nut-hexagonal head-1/4unc x 1 B151
Nut-hexagonal head-3/8unf B 35

Nut-hexagonal head-3/8unf B100
Nut-hexagonal head-3/8unf H 14
Nut-hexagonal head-3/8unf N 9
Nut-hexagonal head-5/16unf B 80
Nut-hexagonal head-5/16unf B 84

Nut-hexagonal head-5/16unf B 98
Nut-hexagonal head-5/16unf B137
Nut-hexagonal head-5/16unf B113
Nut-hexagonal head-5/16unf B118
Nut-hexagonal head-5/16unf F 5

Nut-hexagonal head-5/16unf K 10
Nut-hexagonal head-coarse thread-m10 B 83
Nut-hexagonal head-coarse thread-m10 D 4
Nut-hexagonal head-coarse thread-m10 D 6
Nut-hexagonal head-coarse thread-m10 J 12

Nut-hexagonal head-coarse thread-m10 J 13
Nut-hexagonal head-coarse thread-m10 K 5
Nut-hexagonal head-coarse thread-m10 N 1
Nut-hexagonal head-coarse thread-m10 N 6
Nut-hexagonal head-m8 B106

Nut-hexagonal head-m8 B 65
Nut-hexagonal head-m8 D 20
Nut-hexagonal head-m8 G 3
Nut-hexagonal head-m8 L 5
Nut-hexagonal head-m8 L 8

Nut-hexagonal head-m8 M114
Nut-hexagonal head-m8 N 2
Nut-hexagonal head-nyloc-3/8-unf E 7
Nut-hexagonal head-nyloc-3/8-unf E 8
Nut-hexagonal head-nyloc-3/8-unf E 29

Nut-hexagonal head-nyloc-3/8-unf E 30
Nut-hexagonal head-nyloc-3/8-unf E 31
Nut-hexagonal head-nyloc-m10 C 22
Nut-hexagonal head-nyloc-m10 C 85
Nut-hexagonal head-nyloc-m10 E 47

Nut-hexagonal head-nyloc-m10 E 48
Nut-hexagonal head-nyloc-m10 E 53
Nut-hexagonal head-nyloc-m10 E 46
Nut-hexagonal head-nyloc-m10 J 1
Nut-hexagonal head-nyloc-m10 J 2

Nut-hexagonal head-nyloc-m10 J 3
Nut-hexagonal head-nyloc-m10 J 4
Nut-hexagonal head-nyloc-m16 E 10
Nut-hexagonal head-nyloc-m16 E 34
Nut-hexagonal head-nyloc-m16 E 35

Nut-hexagonal head-nyloc-m6 B 38
Nut-hexagonal head-nyloc-m6 C 5
Nut-hexagonal head-nyloc-m6 C 27
Nut-hexagonal head-nyloc-m6 C 42
Nut-hexagonal head-nyloc-m6 C 69

Nut-hexagonal head-nyloc-m6 C 92
Nut-hexagonal head-nyloc-m6 C 94
Nut-hexagonal head-nyloc-m6 C 96
Nut-hexagonal head-nyloc-m6 C 97
Nut-hexagonal head-nyloc-m6 D 5

Nut-hexagonal head-nyloc-m6 F 17
Nut-hexagonal head-nyloc-m6 F 18
Nut-hexagonal head-nyloc-m6 H 6
Nut-hexagonal head-nyloc-m6 H 8
Nut-hexagonal head-nyloc-m6 J 5

Nut-hexagonal head-nyloc-m6 J 6
Nut-hexagonal head-nyloc-m6 J 18
Nut-hexagonal head-nyloc-m6 J 20
Nut-hexagonal head-nyloc-m6 L 12
Nut-hexagonal head-nyloc-m6 L 13

Nut-hexagonal head-nyloc-m6 L 15
Nut-hexagonal head-nyloc-m6 L 16
Nut-hexagonal head-nyloc-m6 L 36
Nut-hexagonal head-nyloc-m6 L 3
Nut-hexagonal head-nyloc-m6 M 16

Nut-hexagonal head-nyloc-m6 M 17
Nut-hexagonal head-nyloc-m6 M 18
Nut-hexagonal head-nyloc-m6 M 19
Nut-hexagonal head-nyloc-m6 M 20
Nut-hexagonal head-nyloc-m6 M 21

Nut-hexagonal head-nyloc-m6 M 44
Nut-hexagonal head-nyloc-m6 M 47
Nut-hexagonal head-nyloc-m6 N110
Nut-hexagonal head-nyloc-m6 N 13
Nut-hexagonal head-nyloc-m6 N 15

Nut-hexagonal head-nyloc-m6 N 20
Nut-hexagonal head-nyloc-m6 N 70
Nut-hexagonal head-nyloc-m6 N111
Nut-hexagonal head-nyloc-m6 P 22
Nut-hexagonal head-nyloc-m6 P 64

Nut-hexagonal head-nyloc-m8 D 14
Nut-hexagonal head-nyloc-m8 E 43
Nut-hexagonal head-nyloc-m8 F 7
Nut-hexagonal head-nyloc-m8 F 8
Nut-hexagonal head-nyloc-m8 J 9

Nut-hexagonal head-nyloc-m8 K 1
Nut-hexagonal head-nyloc-m8 N 40
Nut-hexagonal-coarse thread-1/4unf B 83
Nut-hexagonal-coarse thread-1/4unf B132
Nut-hexagonal-coarse thread-1/4unf L 4

Nut-hexagonal-coarse thread-1/4unf M 16
Nut-hexagonal-coarse thread-1/4unf M 17
Nut-hexagonal-coarse thread-1/4unf M 18
Nut-hexagonal-coarse thread-1/4unf M 19
Nut-hexagonal-coarse thread-1/4unf M 20

Nut-hexagonal-coarse thread-1/4unf M 21
Nut-hexagonal-coarse thread-1/4unf N 6
Nut-hexagonal-coarse thread-1/4unf N 11
Nut-hexagonal-coarse thread-1/4unf N 69
Nut-hexagonal-coarse thread-1/4unf N 20

Nut-hexagonal-coarse thread-1/4unf N 51
Nut-hexagonal-coarse thread-1/4unf T 22
Nut-hexagonal-fixing control unit-m5 M105
Nut-hexagonal-m4 B 54
Nut-hexagonal-m4 R 21

Nut-hexagonal-m5 B 36
Nut-hexagonal-m5 B 90
Nut-hexagonal-m5 C 52
Nut-hexagonal-m5 G 4
Nut-hexagonal-m5 J 21

Nut-hexagonal-m5 M151
Nut-hexagonal-m5 M 28
Nut-hexagonal-m5 P 5
Nut-hexagonal-m5 P 6
Nut-hexagonal-m5 P 75

Nut-hexagonal-m5 S 1
Nut-hexagonal-nyloc E 48
Nut-hexagonal-nyloc E 49
Nut-hexagonal-nyloc-1/4" L 14
Nut-hexagonal-nyloc-m16 E 43

Nut-hexagonal-nyloc-m20 D 11
Nut-hexagonal-nyloc-m20 D 12
Nut-hexagonal-thin-1/2" E 19
Nut-hexagonal-thin-7/16unf G 11
Nut-hexagonal-thin-7/16unf G 13

Nut-hexagonal-thin-m12 .... E 21
Nut-hexagonal-thin-m12 .... H 14
Nut-linkage bearing-no10-unf B139
Nut-locking .... B 18
Nut-locking .... B 56

Nut-locking .... D 14
Nut-locking N122
Nut-locking-alloy road wheels-code a .... E 42
Nut-lokut .... N 42
Nut-lokut .... N 83

Nut-lokut .... N 84
Nut-lokut .... N 78
Nut-lokut .... N 82
Nut-lokut .... N 25
Nut-lokut .... N 79

Nut-lokut .... P 6
Nut-lokut .... P 62
Nut-lokut .... P 72
Nut-lokut .... R 48
Nut-m10 .... K 4

Nut-m16 .... N122
Nut-m4 .... M112
Nut-m4 .... N 78
Nut-m5 x 12 .... B140
Nut-m6 .... N 83

Nut-m6 .... P 6
Nut-m6 .... P 39
Nut-m6 .... P 42
Nut-m8 .... B 31
Nut-m8 .... C 86

Nut-m8 .... C 88
Nut-m8 .... K 2
Nut-m8 .... K 3
Nut-m8 .... K 11
Nut-m8 .... K 13

Nut-m8 .... K 14
Nut-m8 .... K 15
Nut-m8 .... L 14
Nut-m8 .... N121
Nut-master cylinder to servo-hexagonal .... H 1

Nut-narrow .... F 16
Nut-new-10 spline shaft-hexagonal head- .... E 28
Nut-no10-unf .... M 83
Nut-nyloc-7/16unf .... R 59
Nut-nyloc-upper .... F 12

Nut-nyloc-upper .... N 11
Nut-plastic .... P 1
Nut-plastic-fixing pump M148
Nut-plastic-m5 .... L 30
Nut-plastic-m5 .... L 31

Nut-plastic-push on-fixing ecu-m5 .... M 96
Nut-plastic-push on-fixing ecu-m5 .... M100
Nut-plastic-push on-m5 .... M 92
Nut-plastic-ratchet stud .... P 63
Nut-rear damper upper-nyloc-m12 .... E 52

Nut-spiralok-squab to panel .... R 39
Nut-spiralok-squab to panel .... R 32
Nut-spire .... B 91
Nut-spire .... N 27
Nut-spring-flat .... R 25

Nut-spring-j type .... L 21
Nut-spring-self thread .... N 30
Nut-spring-special-no10 .... N 13
Nut-spring-u type .... L 49
Nut-spring-u type .... L 21

Nut-spring-u type .... M 15
Nut-spring-u type .... P 1
Nut-spring-u type .... P 3
Nut-spring-u type .... P 4
Nut-spring-u type .... P 5

Nut-spring-u type .... P 13
Nut-spring-u type .... P 22
Nut-spring-u type .... P 14
Nut-spring-u type .... P 67
Nut-spring-u type .... P 47

Nut-spring-u type-fixing control unit- .... M 97
Nut-spring-u type-fixing solenoid egr .... M 91
Nut-stake .... C 37
Nut-stake .... C 17
Nut-stake-man trans countershaft .... C 65

Nut-steel road wheels-16mm .... E 41
Nut-tow loop to bumper-hexagonal head- .... N 12
Nut-tread plate to sill .... P 83
Nut-tube-female-m12 .... B 63
Nut-tube-female-m12 .... J 9

Nut-tube-female-m12 .... J 11
Nut-tube-female-spill return-m12 .... B 29
Nut-tube-female-tank to filter-m12 .... J 7
Nut-unf .... T 22
Nut-wing .... J 18

Nutsert-blind M144
Nutsert-blind .... P 3
Nutsert-blind-7 seats .... L 43
Nutsert-blind-fixing ecu .... M 98
Nutsert-blind-fixing ecu .... M 99

Nutsert-blind-fixing switch bonnet-m6 .... M106
Nutsert-blind-m6 .... C 69
Nutsert-blind-m6 .... C 95
Nutsert-blind-m6 .... C 97
Nutsert-blind-m6 .... F 21

Nutsert-blind-m6 .... F 20
O ring .... B 9
O ring .... B 14
O ring .... B 15
O ring .... B 22

O ring .... B 30
O ring .... B 33
O ring .... B 47
O ring .... B 53
O ring .... B 59

O ring .... B 61
O ring .... B 79
O ring .... B 90
O ring .... B 98
O ring B101

O ring B116
O ring B127
O ring B129
O ring B167
O ring B171

O ring B147
O ring .... C 28
O ring .... C 72
O ring .... C 75
O ring .... C 82

O ring .... C 84
O ring .... C 86
O ring .... C 88
O ring .... C 92
O ring .... C 7

O ring .... C 19
O ring .... C 39
O ring .... C 55
O ring .... C 87
O ring .... C 89

O ring .... C 93
O ring .... D 10
O ring .... D 12
O ring .... D 15
O ring .... D 16

O ring .... D 18
O ring .... D 6
O ring .... D 14
O ring .... D 19
O ring .... J 13

O ring .... J 15
O ring .... J 16
O ring .... L 7
O ring .... L 18
O ring .... L 3

O ring .... M 5
O ring air conditioning B100
O ring air conditioning B144
O ring-10mm-with cfc-free refrigerant .... L 46
O ring-11.2mm od-7.6mm id-1.8mm thick- .... L 45

O ring-14.5mm od-10.8mm id-1.8mm thick- .... L 41
O ring-17mm id-18.2mm od .... L 25
O ring-5/16"-less cfc free refrigerant .... L 49
O ring-5/8" diameter .... L 44
O ring-5/8" diameter .... L 51

O ring-6mm .... F 21
O ring-6mm .... F 20
O ring-anti rattle .... D 17
O ring-auto trans strainer to control unit .... C 73
O ring-crankshaft to toothed pulley .... B 45

O ring-fuel filter to pipe .... J 9
O ring-intermediate shaft-front .... D 21
O ring-large .... B 99
O ring-large B142
O ring-large .... L 54

O ring-less cfc free refrigerant .... L 40
O ring-less cfc free refrigerant .... L 53
O ring-less cfc free refrigerant .... L 55
O ring-oil cooler hose .... L 6
O ring-oil pump seal .... C 54

O ring-oil suction pipe seal .... B 47
O ring-oil suction pipe seal B167
O ring-petrol injector-small B179
O ring-radiator filler plug .... L 1
O ring-radiator filler plug .... L 4

O ring-radiator filler plug .... L 5
O ring-seal transducer M115
O ring-side repeater .... M 4
O ring-small B132
O ring-small B133

O ring-small B134
O ring-transmission oil cooler hose .... C 42
O ring-transmission oil cooler hose .... C 69
O ring-transmission oil cooler hose .... C 94
O ring-transmission oil cooler hose .... C 97

Oil cooler transmission .... C 42
Oil cooler transmission .... C 69
Oil cooler transmission .... C 92
Oil cooler transmission .... C 94
Oil cooler transmission .... C 96

Oil-1ltr-15w-40 .... T 28
Olive .... B 9
Olive .... B 63
Olive .... B 97
Olive .... J 9

Olive .... J 11
Olive-exhaust pipe joint flange .... K 1
Olive-exhaust pipe joint flange .... K 11
Olive-exhaust pipe joint flange .... K 13
Olive-spill return .... B 29

Olive-tank to filter .... J 7
Packing strip .... B 4
Packing strip .... B 43
Pad heated seat-squab .... R 26
Pad-5th gear selector fork .... C 19

Pad-5th gear selector fork .... C 39
Pad-anti rattle .... P 19
Pad-anti rattle .... P 21
Pad-anti rattle-loadspace-velour .... P 23
Pad-anti rattle-small-lh .... P 42

Pad-buffer .... N 2
Pad-cable sealing handbrake .... H 20
Pad-clutch release lever slipper .... C 2
Pad-clutch release lever slipper .... C 44
Pad-clutch release lever slipper .... C 45

Pad-detachable-key radio M138
Pad-detachable-key radio M141
Pad-front bulkhead insulation P 75
Pad-front bulkhead insulation-rhd P 74
Pad-front floor insulation P 74

Pad-heelboard insulation P 74
Pad-insulation bonnet N 41
Pad-insulation engine bay P 75
Pad-insulation-front tunnel P 16
Pad-interior mirror adhesive N 67

Pad-lever-pedal G 5
Pad-lever-pedal G 7
Pad-lever-pedal G 9
Pad-load floor insulation P 74
Pad-pedal lever accelerator G 4

Pad-pedal lever accelerator G 6
Pad-pedal lever accelerator G 8
Pad-pedal lever accelerator-rhd G 1
Pad-pedal lever accelerator-rhd P 64
Pad-reverse gear selector lever C 37

Pad-reverse gear selector lever C 17
Pad-tunnel insulation P 17
Paint-aerosol spray-aa yellow-(lrc584)-solid T 3
Paint-aerosol spray-zanzibar silver-(lrc391)- T 4
Paint-pencil-arken grey-(lrc445)-micatallic T 2

Paint-pencil-quicksilver-(lrc468) T 1
Pan assembly-oil transmission case- C 74
Panel assembly-facia console-sonar blue P 13
Panel assembly-tunnel console centre P 17
Panel battery mounting-rh N 21

Panel closing N 32
Panel closing-rh N 24
Panel floor-inner N 24
Panel floor-side-rh N 25
Panel header roof N 34

Panel header roof-outer N 32
Panel inner d post-rh N 25
Panel inner-rh N 24
Panel make up piece-lh N 25
Panel outer-outer-rh N 24

Panel wheelarch-lh N 25
Panel-access-plinth-ash grey R 1
Panel-access-rr floor N 25
Panel-alpine light trim-7 seats L 43
Panel-alpine light trim-dark granite P 63

Panel-alpine light trim-rh-sonar blue P 59
Panel-battery closing-rh N 21
Panel-bodyside trim-bahama beige P 50
Panel-bodyside trim-rh-bahama beige P 46
Panel-bodyside trim-rh-sonar blue P 45

Panel-closing-facia-rhd P 9
Panel-driver closing-rh-rhd P 9
Panel-floor extension N 25
Panel-floor to rear end N 32
Panel-headlamp mounting-lh N 19

Panel-longitudinal/dash side-rh N 22
Panel-mudflap mounting-rh-rear N111
Panel-passenger closure-rhd P 9
Panel-rear door facing-rh N 50
Panel-rear end-lower N 32

Panel-rear seat 1/3 R 40
Panel-rear seat 1/3-ash grey R 38
Panel-rear seat 2/3 R 40
Panel-rear seat 2/3-ash grey R 38
Panel-rear seat back-2/3-charcoal R 39

Panel-rear seat back-with out stowage bag- R 32
Panel-rear seat base-ash grey-1/3 R 38
Panel-rear seat base-sonar blue R 48
Pawl-parklock automatic transmission C 84
Pedal & bracket assembly brake/accelerator- G 7

Pedal & bracket assembly G 5
Pedal assembly accelerator G 1
Pedal assembly accelerator G 4
Pedal assembly-rhd G 9
Peg-air cleaner mtg-engine B 94

Pillar-cylinder head rocker support B 89
Pillar-cylinder head rocker support B126
Pin B 10
Pin C 67
Pin C 21

Pin C 37
Pin C 41
Pin C 17
Pin D 20
Pin E 47

Pin E 53
Pin E 19
Pin E 22
Pin G 6
Pin G 8

Pin N 50
Pin assembly-clevis H 22
Pin-clevis R 59
Pin-clevis-brake servo H 1
Pin-clevis/pivot B140

Pin-clevis/pivot C 19
Pin-clevis/pivot C 39
Pin-clevis/pivot D 16
Pin-clevis/pivot D 18
Pin-clevis/pivot D 20

Pin-clevis/pivot G 4
Pin-clevis/pivot G 11
Pin-clevis/pivot G 6
Pin-clevis/pivot G 8
Pin-clevis/pivot G 3

Pin-clevis/pivot H 23
Pin-clevis/pivot H 21
Pin-clevis/pivot H 24
Pin-clevis/pivot R 48
Pin-door check arm N 46

Pin-lower E 20
Pin-piston gudgeon B 8
Pin-piston gudgeon B 46
Pin-pivot check link N 54
Pin-rear axle drive flange centralising E 34

Pin-reverse gear selector lever pivot C 3
Pin-reverse gear selector lever pivot C 25
Pin-reverse gear selector lever pivot C 37
Pin-reverse gear selector lever pivot C 17
Pin-roll B101

Pin-roll B175
Pin-roll B147
Pin-roll D 14
Pin-roll G 6
Pin-roll G 8

Pin-roll-spring tension B 14
Pin-roll-spring tension B 66
Pin-roll-spring tension C 20
Pin-roll-spring tension C 40
Pin-roll-spring tension C 67

Pin-roll-spring tension C 72
Pin-roll-spring tension E 7
Pin-roll-spring tension E 8
Pin-roll-spring tension E 29
Pin-roll-spring tension E 30

Pin-roll-spring tension E 31
Pin-roll-spring tension-new-10 spline shaft E 28
Pin-selector lever/shaft manual transmission C 18
Pin-selector lever/shaft manual transmission C 38
Pin-selector lever/shaft manual transmission C 66

Pin-selector manual transmission C 18
Pin-selector manual transmission C 38
Pin-selector manual transmission C 66
Pin-split B 89
Pin-split B126

Pin-split B140
Pin-split C 86
Pin-split C 88
Pin-split D 16
Pin-split D 14

Pin-split E 47
Pin-split E 49
Pin-split E 53
Pin-split E 10
Pin-split E 34

Pin-split E 35
Pin-split F 15
Pin-split F 11
Pin-split G 4
Pin-split G 11

Pin-split G 3
Pin-split H 23
Pin-split H 21
Pin-split H 24
Pin-split R 59

Pin-spring B 80
Pin-spring B112
Pin-spring B129
Pin-spring B113
Pin-spring C 54

Pin-spring C 61
Pin-spring C 67
Pin-spring C 86
Pin-spring C 88
Pin-spring C 21

Pin-spring C 41
Pin-towing pintle T 20
Pin-upper E 18
Pipe K 13
Pipe assembly brake-front H 6

Pipe assembly brake-front H 8
Pipe assembly brake-master cylinder H 5
Pipe assembly brake-master cylinder H 7
Pipe assembly brake-master cylinder to valve H 9
Pipe assembly brake-rear H 11

Pipe assembly brake-rear-lh H 13
Pipe assembly brake-rear-rh H 12
Pipe assembly brake-valve to front hose-lh- H 10
Pipe assembly power assisted steering-rhd F 13
Pipe assembly-antilock brakes-modulator to H 9

Pipe assembly-canister to separator J 23
Pipe assembly-charcoal canister vacuum J 21
Pipe assembly-charcoal canister vacuum J 22
Pipe assembly-coolant L 4
Pipe assembly-engine coolant-diesel B 51

Pipe assembly-engine coolant-diesel B 61
Pipe assembly-evap/vs to canister J 21
Pipe assembly-evap/vs to canister-separator J 22
Pipe assembly-evap/vs to canister-separator J 24
Pipe assembly-filter to engine feed J 8

Pipe assembly-heater feed B129
Pipe assembly-intercooler turbocharger L 10
Pipe assembly-oil strainer B 47
Pipe assembly-pump to steering box F 20
Pipe assembly-pump to steering box-rhd F 21

Pipe assembly-purge J 21
Pipe assembly-seperator to atmosphere J 23
Pipe assembly-steering box to reservoir-rhd F 19
Pipe assembly-steering box to reservoir-rhd F 21
Pipe assembly-steering box to reservoir-rhd F 20

Pipe assembly-tank to filter J 8
Pipe assembly-tank to separator J 23
Pipe assembly-vacuum switch/water valve L 30
Pipe assembly-vacuum switch/water valve L 31
Pipe boost control B 62

Pipe clutch G 11
Pipe clutch-damper to slave cylinder G 13
Pipe clutch-master cylinder to damper G 12
Pipe fuel J 20
Pipe fuel J 23

Pipe fuel filler-leaded ... J 5
Pipe fuel filler-leaded ... J 6
Pipe fuel filler-separator to canister-unleaded ... J 24
Pipe fuel filler-unleaded ... K 1
Pipe fuel-tank to filter ... J 7
Pipe fuel-tank to sedimentor ... J 10
Pipe induction system ... B138
Pipe kit ... L 25
Pipe turbocharger oil feed ... B 31
Pipe turbocharger oil feed ... B 67
Pipe-adaptor turbocharger ... B 31
Pipe-adaptor turbocharger ... B 67
Pipe-breather fuel tank ... J 1
Pipe-breather fuel tank ... J 2
Pipe-breather fuel tank ... J 3
Pipe-breather fuel tank ... J 4
Pipe-breather fuel tank ... J 23
Pipe-coolant ... L 19
Pipe-cooler to pump oil ... L 3
Pipe-cylinder 1-high pressure fuel injection ... B 29
Pipe-cylinder 1-high pressure fuel injection ... B 66
Pipe-cylinder 2-high pressure fuel injection ... B 29
Pipe-cylinder 2-high pressure fuel injection ... B 66
Pipe-cylinder 3-high pressure fuel injection ... B 29
Pipe-cylinder 3-high pressure fuel injection ... B 66
Pipe-cylinder 4-high pressure fuel injection ... B 29
Pipe-cylinder 4-high pressure fuel injection ... B 66
Pipe-dipstick automatic transmission ... C 74
Pipe-engine oil suction ... B167
Pipe-engine to oil cooler ... L 6
Pipe-engine to oil cooler ... L 18
Pipe-evaporator/compressor air conditioning ... L 44
Pipe-exhaust gas recirculation ... L 16
Pipe-exhaust gas recirculation cooler ... L 17
Pipe-exhaust gas recirculation-to solenoid ... B 26
Pipe-feed-exhaust gas recirculation engine ... B 25
Pipe-feed-fuel lines ... B 97
Pipe-feed-fuel lines ... J 10
Pipe-filter to hose feed-tank to filter ... J 7
Pipe-fluid air conditioning ... L 44
Pipe-fluid air conditioning ... L 51
Pipe-fluid air conditioning-rear ... L 49
Pipe-front axle breather ... E 6
Pipe-fuel filter/fuel injection pump ... J 10
Pipe-heater to hose coolant ... B 24
Pipe-heater to hose coolant ... L 32
Pipe-heater to hose coolant-bottom ... B176
Pipe-heater to thermostat coolant ... B176
Pipe-inlet manifold vacuum ... B101
Pipe-inlet manifold vacuum ... M 98
Pipe-inlet manifold vacuum ... M 99
Pipe-oil cooler to engine ... L 6
Pipe-oil cooler to engine ... L 18
Pipe-oil cooler/transmission ... C 69
Pipe-oil cooler/transmission ... C 92
Pipe-oil cooler/transmission ... C 93
Pipe-oil cooler/transmission ... C 97
Pipe-oil cooler/transmission-front ... C 42
Pipe-oil cooler/transmission-intermediate ... C 94
Pipe-oil drain turbocharger ... B 31
Pipe-oil drain turbocharger ... B 44
Pipe-oil drain turbocharger ... B 67
Pipe-outlet stub ... B129
Pipe-outlet to heater coolant ... B129
Pipe-outlet to inlet manifold coolant ... B 94
Pipe-outlet to inlet manifold coolant ... B 95
Pipe-plenum chamber breather ... B137
Pipe-plenum chamber breather ... B136
Pipe-pump to cooler oil ... L 3
Pipe-ram ... B138
Pipe-rear axle breather ... E 27
Pipe-return fuel lines ... J 8
Pipe-return fuel lines ... J 10
Pipe-return fuel lines-tank to filter ... J 7
Pipe-sampling exhaust gas ... B 30
Pipe-sampling exhaust gas ... B 26
Pipe-solenoid valve/exhaust gas recirculation ... B 25
Pipe-spill return fuel injection ... B 29
Pipe-spill return fuel injection ... B 66
Pipe-spill return fuel injection ... B133
Pipe-spill return fuel injection ... B134
Pipe-suction air conditioning-rear ... L 49
Pipe-suction automatic transmission ... C 73
Pipe-sump assembly oil drain ... B 47
Pipe-transmission oil inlet ... C 6
Pipe-transmission oil inlet ... C 28
Pipe-transmission oil inlet ... C 54
Pipe-transmission oil inlet ... C 7
Pipe-transmission/oil cooler ... C 69
Pipe-transmission/oil cooler-front ... C 92
Pipe-transmission/oil cooler-rear ... C 94
Pipe-vacuum exhaust gas recirculation valve ... B 25
Pipe-vacuum heater ... L 36
Pipe-vacuum heater/air conditioning ... L 36
Pipe-vacuum heater/air conditioning ... L 37
Piston & connecting rod assembly-engine- ... B166
Piston front caliper ... H 15
Piston power assisted steering ... F 14
Piston-clutch-automatic ... C 77
Piston-clutch-automatic ... C 78
Piston-clutch-automatic ... C 79
Piston-clutch-automatic ... C 80
Piston-clutch-automatic ... C 81
Piston-engine-0.010" oversize ... B 46
Piston-engine-standard.-'x' ... B 8
Piston-engine-standard.-8.13:1 ... B 81
Piston-engine-standard.-8.13:1 ... B114
Piston-rear caliper brake ... H 18
Pivot assembly-exterior mirror assembly glass ... N112
Pivot assembly-exterior mirror assembly glass ... N113
Pivot assembly-exterior mirror assembly glass ... N114
Pivot-clutch release lever ... C 24
Pivot-clutch release lever ... C 46
Pivot-transfer box differential lock shaft/arm ... D 14
Plate ... B 28
Plate ... B 80
Plate ... B112
Plate ... B127
Plate ... C 2
Plate ... C 24
Plate ... C 44
Plate ... C 45
Plate ... C 46
Plate ... D 22
Plate ... L 46
Plate adjusting-upper ... N 70
Plate assembly-drive torque converter ... B 40
Plate assembly-drive torque converter ... B161
Plate assembly-weld bolt ... J 1
Plate assembly-weld bolt ... J 2
Plate assembly-weld bolt ... J 3
Plate assembly-weld bolt ... J 4
Plate mounting-rh ... N 8
Plate towing ... T 20
Plate- guide-guide-pawl ... C 84
Plate-5th gear synchroniser support ... C 14
Plate-5th gear synchroniser support ... C 34
Plate-anti vibration power assisted steering ... B 34
Plate-automatic transmission clutch end ... C 79
Plate-backing automatic transmission ... B 40
Plate-backing automatic transmission ... B 76
Plate-baffle ... C 47
Plate-blanking air conditioning switch ... L 28
Plate-blanking air conditioning switch ... L 48
Plate-camshaft pulley retaining ... B 22
Plate-camshaft pulley retaining ... B 60
Plate-camshaft pulley retaining ... B 64
Plate-camshaft retaining ... B 22
Plate-camshaft retaining ... B 60
Plate-camshaft retaining ... B128
Plate-centre ... C 80
Plate-charcoal canister bracket stiffening ... J 20
Plate-closure automatic transmission ... B 76
Plate-clutch-driven ... B 39
Plate-clutch-driven ... B 75
Plate-clutch-driven ... B108
Plate-clutch-driven ... B189
Plate-clutch-driven-asbestos free ... B160
Plate-corner ... C 74
Plate-crankshaft balancing ... B 80
Plate-crankshaft balancing ... B112
Plate-crankshaft balancing ... B113
Plate-cylinder head water outlet blanking ... B169
Plate-door actuator mounting ... N 79
Plate-drive assembly automatic transmission ... B 40
Plate-drive assembly automatic transmission ... B 76
Plate-drive assembly automatic transmission ... B161
Plate-exterior mirror retaining-rh ... N112
Plate-exterior mirror retaining-rh ... N116
Plate-exterior mirror retaining-rh ... N114
Plate-exterior mirror retaining-rh ... N119
Plate-forward clutch pressure ... C 78
Plate-front & rear stub axle locking ... E 39
Plate-front & rear stub axle locking ... E 40
Plate-front & rear stub axle locking ... E 19
Plate-front & rear stub axle locking ... E 22
Plate-gear lever retention ... D 16
Plate-gear lever retention ... D 18
Plate-gearbox mounting ... B 76
Plate-gearchange bias spring adjuster ... C 20
Plate-gearchange bias spring adjuster ... C 40
Plate-gearchange bias spring adjuster ... C 67
Plate-glovebox stay retaining ... P 10
Plate-graphics control-heater ... L 28
Plate-graphics control-heater ... L 48
Plate-guide-pawl ... C 84
Plate-inhibitor switch detent ... C 72
Plate-intermediate ... C 75
Plate-lever-gating selector mechanism ... C 19
Plate-lever-gating selector mechanism ... C 39
Plate-locking ... D 22
Plate-lower ... N 70
Plate-mist grey ... P 72
Plate-mounting distributor ... B101
Plate-mounting distributor ... B147
Plate-mounting fuel tank ... J 1
Plate-mounting fuel tank ... J 2
Plate-mounting fuel tank ... J 3
Plate-mounting fuel tank ... J 4
Plate-mounting pedal ... G 9
Plate-mudflap support-rh-rear ... N111
Plate-oil filler blanking ... B164
Plate-oil guide transmission case ... D 6
Plate-oil seal retainer ... D 12
Plate-oil seal retainer ... E 13
Plate-oil seal retainer ... E 16
Plate-oil seal retainer ... E 15
Plate-packing ... D 22
Plate-primed front bumper fog lamp blanking ... T 7
Plate-radiator bottom ... L 11
Plate-rear bumper tread-black ... N 15
Plate-rear door actuator-rh ... N 82
Plate-rear seat belt male anchor-twin ... R 56
Plate-reinforcement ... B 39
Plate-relay mounting ... M 83
Plate-retaining ... N 51
Plate-retaining ... N 48
Plate-striker ... N 69
Plate-support-drive plate automatic ... B161
Plate-switch pack blanking-mist grey ... P 72

Plate-synchroniser ... C 12
Plate-synchroniser ... C 13
Plate-synchroniser ... C 31
Plate-synchroniser ... C 32
Plate-synchroniser ... C 58
Plate-synchroniser ... C 60
Plate-synchroniser ... C 14
Plate-synchroniser ... C 59
Plate-synchroniser ... C 62
Plate-synchroniser ... C 34
Plate-synchroniser ... C 63
Plate-timing belt cover ... B 10
Plate-tow bar bracket slider ... T 20
Plate-trailer socket mounting ... T 22
Plate-transfer box anti rotation ... D 10
Plate-transfer box detent spring retaining ... C 72
Plate-transfer box detent spring retaining ... D 17
Plate-transfer box gearchange grommet ... D 18
Plate-transfer box gearchange grommet ... D 17
Plate-transfer box selector cross shaft end ... D 16
Plate-transfer box selector cross shaft end ... D 18
Plate-turbocharger blanking ... B 67
Plenum-inlet manifold ... B135
Plinth assembly-seat-rh-ash grey ... R 1
Plinth-licence plate-black-with bull bar ... N 13
Plinth-licence plate-black-with bull bar ... T 9
Plug ... B 4
Plug ... B 14
Plug ... B 15
Plug ... B 18
Plug ... B 43
Plug ... B 48
Plug ... B 56
Plug ... B 61
Plug ... B 76
Plug ... B 79
Plug ... B 87
Plug ... B111
Plug ... B122
Plug ... B138
Plug ... B172
Plug ... C 6
Plug ... C 18
Plug ... C 28
Plug ... C 38
Plug ... C 47
Plug ... C 66
Plug ... C 7
Plug ... D 3
Plug ... D 5
Plug ... D 6
Plug ... D 14
Plug ... H 1
Plug ... H 24
Plug and magnet assembly-transmission case ... C 25
Plug and magnet assembly-transmission case ... C 47
Plug and magnet assembly-transmission case ... C 74
Plug-7 pin ... T 22
Plug-blanking ... B 94
Plug-blanking ... B138
Plug-blanking ... B189
Plug-blanking ... C 54
Plug-blanking ... C 75
Plug-blanking ... J 13
Plug-blanking ... M150
Plug-blanking ... N 11
Plug-blanking ... N 89
Plug-blanking-13mm ... G 4
Plug-blanking-14mm ... G 10
Plug-blanking-25.5mm ... N 54
Plug-blanking-32mm ... M113
Plug-blanking-8mm ... M160
Plug-blanking-9.5mm ... N 47
Plug-blanking. ... B 79
Plug-blanking. ... B111
Plug-blanking. ... B178
Plug-blanking. ... F 14
Plug-camshaft blanking-rubber ... B 4
Plug-camshaft blanking-rubber ... B 43
Plug-coolant drain ... B 79
Plug-coolant drain ... B111
Plug-coolant drain ... C 26
Plug-coolant drain ... C 52
Plug-coolant drain ... S 1
Plug-core ... B 4
Plug-core ... B 16
Plug-core ... B 24
Plug-core ... B 43
Plug-core ... B 54
Plug-core ... B 62
Plug-core ... B 79
Plug-core ... B 84
Plug-core ... B 88
Plug-core ... B 89
Plug-core ... B 94
Plug-core ... B111
Plug-core ... B117
Plug-core ... B123
Plug-core ... B125
Plug-core ... B126
Plug-core ... B164
Plug-core ... B172
Plug-core ... C 72
Plug-core-6mm ... D 6
Plug-core-9mm ... C 47
Plug-core-transfer box-6mm ... C 3
Plug-core-transfer box-6mm ... C 25
Plug-cup ... B 79
Plug-cup ... B 88
Plug-cup ... B 94
Plug-cup ... B111
Plug-cup ... B123
Plug-cup ... B125
Plug-cup ... B135
Plug-cup ... B137
Plug-cylinder block oil way ... B 4
Plug-cylinder block oil way ... B 43
Plug-cylinder block oil way ... B 79
Plug-cylinder block oil way ... B111
Plug-cylinder block oil way-rubber ... B 17
Plug-drain ... B 5
Plug-drain ... B 24
Plug-drain ... B 44
Plug-drain ... B 62
Plug-drain ... B 55
Plug-drain ... C 3
Plug-drain ... C 25
Plug-drain ... C 47
Plug-drain ... D 4
Plug-drain ... D 5
Plug-drain ... E 20
Plug-drain ... E 18
Plug-drain ... J 13
Plug-drain/filler level ... B129
Plug-drain/filler level ... D 5
Plug-drain/filler level ... D 6
Plug-drain/filler level ... E 6
Plug-drain/filler level ... E 27
Plug-drain/filler level ... L 1
Plug-drain/filler level ... L 4
Plug-drain/filler level ... L 5
Plug-drain/filler level ... L 7
Plug-facia blanking ... P 8
Plug-heater ignition ... B 16
Plug-heater ignition ... B 54
Plug-level ... C 3
Plug-level ... C 25
Plug-level ... E 20
Plug-level ... E 18
Plug-oil pump ... B171
Plug-oil pump relief valve ... B 87
Plug-oil pump relief valve ... B122
Plug-oil pump relief valve ... B171
Plug-red ... J 15
Plug-roof trim blanking-aspen grey ... P 71
Plug-roof trim blanking-mist grey ... P 72
Plug-rubber ... B 10
Plug-rubber ... B 39
Plug-rubber ... B 40
Plug-rubber ... B 75
Plug-sparking ... B185
Plug-sparking-n9yc ... B151
Plug-sparking-rn11yc ... B103
Plug-sparking-rn11yc ... B151
Plug-square head ... E 20
Plug-square head ... E 18
Plug-sump assembly oil drain ... B 9
Plug-sump assembly oil drain ... B 47
Plug-sump assembly oil drain ... B 82
Plug-sump assembly oil drain ... B115
Plug-sump assembly oil drain ... B167
Plunger-locking manual transmission ... C 20
Plunger-locking manual transmission ... C 40
Plunger-oil pump relief valve ... B 14
Plunger-oil pump relief valve ... B 48
Plunger-oil pump relief valve ... B 87
Plunger-oil pump relief valve ... B122
Plunger-oil pump relief valve ... B171
Plunger-reverse switch ... C 21
Plunger-reverse switch ... C 41
Pocket-bodyside trim-rh-sonar blue ... P 47
Pocket-headlining-aspen grey ... P 67
Pocket-quarter trim-rear-sonar blue ... P 58
Pocket-tail door-rh-sonar blue ... P 39
Pocket-tail door-rh-sonar blue ... P 41
Pointer-engine timing ... B 84
Pointer-engine timing ... B118
Polish wax-500ml ... T 28
Pop out unit cigar lighter ... M130
Post-terminal ... M 16
Post-terminal ... M 17
Post-terminal ... M 18
Post-terminal ... M 19
Post-terminal ... M 20
Post-terminal ... M 21
Potentiometer-air con ... L 21
Potentiometer-throttle diesel ... B 64
Potentiometer-throttle multi point injection ... B135
Pressure control release valve-brake ... H 6
Pressure control release valve-brake ... H 8
Printed circuit board instrument pack ... M118
Printed circuit board instrument pack ... M127
Printed circuit board instrument pack ... P 13
Printed circuit board instrument pack-less ... M122
Propellor shaft-front ... E 1
Propellor shaft-rear ... E 2
Protector rear window ... N 54
Protector-driving lamp ... T 8
Protector-driving lamp-vinyl-pair-rally 1000 ... T 23
Protector-headlamp ... T 8
Protector-headlamp ... T 9
Protector-lamp-rh ... T 23
Protector-load space-stowable ... T 31
Protector-pipe fuel ... J 9
Protector-rear lamp-pair-lh ... T 8
Protector-tail door-aluminium ... P 42
Pulley alternator ... B 73
Pulley alternator-68mm-127/65 amp ... B105
Pulley alternator-68mm-127/65 amp ... B153
Pulley assembly-crankshaft ... B165

Pulley assembly-pump fuel injection pump ..... B 28
Pulley coolant pump ..... B 13
Pulley coolant pump ..... B 51
Pulley coolant pump ..... B 52
Pulley coolant pump ..... B 86
Pulley coolant pump B119
Pulley coolant pump B121
Pulley power assisted steering pump ..... B 34
Pulley power assisted steering pump ..... B 70
Pulley power assisted steering pump B182
Pulley power assisted steering pump-twin ..... B 99
Pulley power assisted steering pump-twin B142
Pulley-alternator ..... B 36
Pulley-alternator B154
Pulley-alternator B153
Pulley-assembly B143
Pulley-camshaft ..... B 22
Pulley-camshaft ..... B 60
Pulley-camshaft ..... B 64
Pulley-camshaft B175
Pulley-crankshaft ..... B 7
Pulley-crankshaft ..... B 45
Pulley-crankshaft-single groove B112
Pulley-crankshaft-water pump/alternator- ..... B 80
Pulley-driven air conditioning ..... B 86
Pulley-driven air conditioning B121
Pulley-idler ancillary drive ..... B 72
Pulley-idler ancillary drive B184
Pulley-idler ancillary drive B187
Pulley-polyvee alternator-61mm ..... B 73
Pulley-tensioner ancillary drive ..... B 35
Pulley-tensioner ancillary drive ..... B 72
Pulley-tensioner ancillary drive ..... B 85
Pulley-tensioner ancillary drive B100
Pulley-tensioner ancillary drive B119
Pulley-tensioner ancillary drive B144
Pulley-tensioner ancillary drive B145
Pump assembly power assisted steering ..... B 34
Pump assembly power assisted steering ..... B 70
Pump assembly power assisted steering ..... B 99
Pump assembly power assisted steering B142
Pump assembly power assisted steering B143
Pump assembly power assisted steering B182
Pump assembly-backlight wash M146
Pump assembly-backlight wash M148
Pump assembly-backlight wash-black M147
Pump assembly-backlight wash-black M149
Pump assembly-engine coolant ..... B 12
Pump assembly-engine coolant ..... B 51
Pump assembly-engine coolant ..... B 85
Pump assembly-engine coolant B119
Pump assembly-engine coolant B170
Pump assembly-engine oil ..... B 14
Pump assembly-engine oil B171
Pump assembly-fuel injection diesel-new ..... B 28
Pump assembly-headlamp wash M148
Pump assembly-headlamp wash M149
Pump assembly-windscreen wash-front M146
Pump assembly-windscreen wash-front M148
Pump assembly-windscreen wash-red M147
Pump assembly-windscreen wash-red M149
Pump brake vacuum ..... B 33
Pump brake vacuum ..... B 69
Pump-cruise control vacuum ..... G 10
Pump-expansion tank ejector ..... L 15
Pump-foot ..... T 17
Pump-fuel injection-new ..... B 64
Pump-fuel mechanical ..... B 27
Pump-fuel mechanical ..... B 63
Pump-fuel-unit assembly fuel tank ..... J 1
Pump-fuel-unit assembly fuel tank ..... J 2
Pump-transmission oil ..... C 54
Pump-transmission oil ..... C 75
Quarter assembly-rear-rh ..... N 29
Rack assembly-roof-camel l.e. only ..... V 4
Rack assembly-roof-standard ..... T 39
Rack-riding tack ..... T 13
Radiator & intercooler assembly-cooling ..... L 7
Radiator assembly ..... L 1
Radiator assembly ..... L 2
Radiator assembly ..... L 4
Radiator assembly ..... L 5
Radiator assembly ..... L 7
Radio cassette assembly- electronic-clarion- M136
Radio cassette assembly- electronic-philips- M138
Radio cassette assembly- electronic-philips- M141
Radio cassette electronic-auto/reversal-ash M132
Radio cassette electronic-auto/reversal-ash M133
Rail assembly-multi point injection fuel B132
Rail assembly-multi point injection fuel B133
Rail assembly-multi point injection fuel B134
Rail-multi point injection fuel distribution B179
Rail-roof rack side-rh ..... N 40
Rear assembly exhaust system ..... K 13
Rear assembly exhaust system-rear ..... K 4
Receiver dryer assembly air conditioning ..... L 23
Receiver dryer assembly r134a-air ..... L 41
Receiver-turnbuckle ..... P 9
Rectifier-alternator ..... B 36
Rectifier-alternator B105
Rectifier-alternator-65amp alternator-new B152
Rectifier-alternator-85 amp/100 amp B154
Rectifier-alternator-85amp alternator-new B153
Reel assembly-ash grey ..... R 55
Reflector ..... N 15
Reflector-rear-rectangular ..... M 35
Refridgerator ..... T 14
Regulator & brush box assembly alternator ..... B 36
Regulator & brush box assembly alternator B105
Regulator & brush box assembly alternator- B152
Regulator & brush box assembly alternator- B153
Regulator assembly alternator-85 amp/100 ..... B154
Regulator assembly-rear door electric glass- ..... N 64
Regulator-front door glass-electric-rh ..... N 61
Regulator-front door glass-manual-rh ..... N 60
Regulator-front door glass-rh ..... N 62
Regulator-fuel pressure B132
Regulator-fuel pressure B133
Regulator-fuel pressure B134
Regulator-multi point injection fuel pressure B179
Regulator-rear door glass-manual-rh ..... N 63
Regulator-rear door glass-rh ..... N 65
Reinforcement-roof ..... N 34
Reinforcement-valance member-outer rh ..... N 19
Relay ..... M 81
Relay assembly air conditioning ..... P 15
Relay assembly multi function ..... M 84
Relay assembly multi function ..... M 89
Relay blower-heater ..... L 24
Relay changeover ..... T 35
Relay normally open-dim dip-yellow ..... M 85
Relay normally open-dim dip-yellow ..... M 90
Relay normally open-electric seat-yellow ..... M 88
Relay normally open-horn-yellow ..... M 84
Relay normally open-power wash-yellow ..... M 86
Relay normally open-shift interlock-yellow ..... M 87
Relay normally open-yellow ..... M 83
Relay normally open-yellow M108
Relay normally open-yellow M110
Relay normally open-yellow M106
Relay-abs warning-green ..... M 85
Relay-abs warning-green ..... M 86
Relay-black ..... L 26
Relay-delay unit-rear wiper-blue ..... M 83
Relay-green ..... M 83
Relay-ignition heater plug-12v ..... M 83
Relay-ignition heater plug-12v ..... M 90
Relay-rear wiper-green ..... M 84
Relay-shift interlock-green ..... M 87
Relay-twin make-brown M106
Remote unit-locking system-315 mhz M103
Remote unit-locking system-315 mhz M104
Remote unit-locking system-418 mhz M102
Reservoir assembly power assisted steering ..... F 17
Reservoir assembly power assisted steering ..... F 18
Resistor B104
Resistor B148
Resistor ..... L 45
Resistor ..... L 42
Resistor ..... M 96
Resistor ..... M 95
Resistor pack-speed control-blower-heater ..... L 24
Resistor-dim dip ..... M 90
Resistor-pack heater ..... L 26
Resonator induction system ..... J 18
Restrictor-air flow ..... B 98
Restrictor-cylinder block oil feed B164
Retainer ..... E 51
Retainer ..... N 70
Retainer ..... S 1
Retainer-bearing automatic transmission ..... C 76
Retainer-front/rear carpet-dark granite ..... P 83
Retainer-gear lever assembly gaiter ..... P 75
Retainer-gear lever assembly gaiter-upper ..... P 19
Retainer-gear lever assembly gaiter-upper ..... P 14
Retainer-gear lever assembly gaiter-upper ..... P 21
Retainer-inner ..... E 52
Retainer-interlock ..... C 19
Retainer-interlock ..... C 39
Retainer-interlock spool ..... C 66
Retainer-interlock spool ..... C 39
Retainer-roof trim sunshade ..... N 35
Retainer-roof trim sunshade ..... N 36
Retainer-roof trim sunshade-aspen grey ..... P 71
Retainer-trim ..... P 60
Retainer-trim ..... P 62
Retainer-trim-7 seats ..... L 43
Retention-bearing clutch release ..... C 2
Retention-bearing clutch release ..... C 24
Retention-bearing clutch release ..... C 44
Retention-bearing clutch release ..... C 45
Retention-bearing clutch release ..... C 46
Ring power assisted steering ..... F 14
Ring set assembly-piston ..... C 79
Ring set assembly-piston-standard ..... B 8
Ring set assembly-piston-standard ..... B 81
Ring set assembly-piston-standard B114
Ring-backing automatic transmission ..... C 79
Ring-backing automatic transmission ..... C 80
Ring-backing automatic transmission ..... C 83
Ring-baulk ..... C 31
Ring-baulk ..... C 32
Ring-baulk ..... C 58
Ring-baulk ..... C 59
Ring-baulk ..... C 62
Ring-baulk ..... C 34
Ring-baulk ..... C 63
Ring-baulk-1st/2nd gear ..... C 13
Ring-baulk-3rd/4th gear ..... C 12
Ring-baulk-3rd/4th gear ..... C 60
Ring-baulk-5th gear ..... C 14
Ring-differential locking ..... D 13
Ring-differential locking ..... E 7
Ring-differential locking ..... E 8
Ring-differential locking ..... E 29
Ring-differential locking ..... E 30
Ring-differential locking ..... E 31
Ring-differential locking-new-10 spline shaft ..... E 28
Ring-drive automatic transmission B161
Ring-loadspace anti rattle ..... P 23

Ring-locking fuel tank-black J 3
Ring-locking fuel tank-black J 4
Ring-locking fuel tank-red J 2
Ring-locking fuel tank-yellow. J 1
Ring-mainshaft oil feed C 54

Ring-mainshaft oil feed C 61
Ring-mainshaft oil feed C 63
Ring-mainshaft oil feed-asbestos C 6
Ring-mainshaft oil feed-asbestos C 28
Ring-piston-forward shaft automatic C 79

Ring-pulsar antilock brakes H 18
Ring-retention C 61
Ring-retention C 63
Ring-retention E 46
Ring-self cancelling F 1

Ring-snap B 30
Ring-snap C 18
Ring-snap C 38
Ring-snap H 20
Ring-snap H 22

Ring-snap H 23
Ring-snap automatic transmission C 77
Ring-snap automatic transmission C 78
Ring-snap automatic transmission C 79
Ring-snap automatic transmission C 81

Ring-snap automatic transmission C 82
Ring-snap automatic transmission-large C 80
Ring-snap automatic transmission-small C 83
Ring-snap-manual C 66
Ring-snap-outer B 14

Ring-support C 77
Ring-synchroniser friction C 14
Ring-synchroniser friction C 34
Ring-synchroniser inner C 58
Ring-synchroniser inner C 60

Ring-synchroniser inner C 59
Ring-synchroniser intermediate C 58
Ring-synchroniser intermediate C 60
Ring-synchroniser intermediate C 59
Ring-towing-front T 7

Ring-transfer box differential retaining D 7
Rivet J 18
Rivet N 69
Rivet P 10
Rivet-3/16" x 0.45"lng S 1

Rivet-3/16" x 0.575"lng P 75
Rivet-3/16" x 0.575"lng R 59
Rivet-4.8mm diameter P 17
Rivet-7 seats L 43
Rivet-fixing bracket M106

Rivet-grey P 18
Rivet-liner to wing N 28
Rivet-plastic-drive N 30
Rivet-plastic-drive-black N 15
Rivet-pop-steel L 11

Rivet-pop-steel-fixing speaker rear e post M132
Road atlas T 46
Rocker assembly-cylinder head-rh B 56
Rocker assembly-cylinder head-rh B 89
Rocker assembly-cylinder head-rh B126

Rod assembly-carburettor choke B 98
Rod assembly-kickdown automatic C 72
Rod assembly-linkage connecting D 20
Rod assembly-push clutch master cylinder G 12
Rod assembly-push slave cylinder C 24

Rod assembly-push slave cylinder C 46
Rod-crankshaft connecting B 8
Rod-crankshaft connecting B 46
Rod-crankshaft connecting B 81
Rod-crankshaft connecting B114

Rod-engine push B 19
Rod-engine push B 57
Rod-engine push B 89
Rod-engine push B126
Rod-front suspension panhard E 43

Rod-push slave cylinder C 2
Rod-push slave cylinder C 44
Rod-push slave cylinder C 45
Rod-rear seat belt tie R 59
Rod-rear seat squab latch link-rh R 46

Rod-transmission support C 85
Roll-tool storage S 1
Roller P 10
Roller fairlead-winch T 26
Roller-tappet guide B 19

Roller-tappet guide B 57
Roof bow-centre N 34
Roof bow-rear N 34
Roof panel N 34
Rotor assembly-oil pump B 48

Rotor assembly-turbocharger B 30
Rotor blade-vacuum pump B 33
Rotor-oil pump inner B 48
Rotor-oil pump outer B 48
Rubbing strip assembly-front door-rh-black N 92

Rubbing strip assembly-front fender-rh- N 92
Rubbing strip-front door-black insert-rh N 96
Rubbing strip-front door-black insert-rh N 98
Rubbing strip-front door-rh-black N 93
Rubbing strip-front door-rh-with chrome N 94

Rubbing strip-front door-rh-with chrome N 97
Rubbing strip-front fender-black insert-rh N 96
Rubbing strip-front fender-black insert-rh N 98
Rubbing strip-front fender-rh-black N 93
Rubbing strip-front fender-rh-with chrome N 94

Rubbing strip-front fender-rh-with chrome N 97
Rubbing strip-rear door-black insert-rh N 96
Rubbing strip-rear door-black insert-rh N 98
Rubbing strip-rear door-rh-black N 92
Rubbing strip-rear door-rh-black N 93

Rubbing strip-rear door-rh-with chrome N 94
Rubbing strip-rear door-rh-with chrome N 97
Rubbing strip-rear fender-rh-black N 92
Rubbing strip-rear fender-rh-black N 93
Rubbing strip-rear quarter-black insert-rh N 96

Rubbing strip-rear quarter-black insert-rh N 98
Rubbing strip-rear quarter-rh-with chrome N 95
Rubbing strip-rear quarter-rh-with chrome N 97
Safety vest T 17
Screw B 58

Screw B 97
Screw B169
Screw B178
Screw B103
Screw B151

Screw C 79
Screw C 84
Screw C 74
Screw D 4
Screw D 15

Screw D 22
Screw F 1
Screw F 4
Screw F 13
Screw F 14

Screw G 6
Screw G 8
Screw H 14
Screw H 12
Screw H 21

Screw J 24
Screw L 22
Screw L 39
Screw L 21
Screw M 4

Screw M 5
Screw M 12
Screw M 13
Screw M144
Screw M145

Screw N 37
Screw N 38
Screw N 39
Screw N 41
Screw N 62

Screw N 65
Screw N 28
Screw N 36
Screw N 47
Screw N 51

Screw N 79
Screw P 3
Screw P 6
Screw P 63
Screw P 73

Screw P 23
Screw P 31
Screw P 10
Screw P 28
Screw P 34

Screw P 62
Screw P 71
Screw R 1
Screw R 20
Screw R 40

Screw R 48
Screw R 16
Screw T 47
Screw-1/2" P 11
Screw-1/4unf x 3/4 H 17

Screw-1/4unf x 5/8 N 20
Screw-10unf x 5/16 B 98
Screw-3.5 x 16 M 11
Screw-3.5 x 16 P 39
Screw-3.5 x 16 P 41

Screw-3/8unc x 1 1/2 B107
Screw-40mm long R 21
Screw-5/16unc x 7/8 B130
Screw-6 x 3/8 B 90
Screw-6 x 3/8 B127

Screw-7 seats L 43
Screw-adaptor to housing-hexagonal head- B 53
Screw-alternator bracket to block-flanged B 4
Screw-alternator to bracket-flanged head- B 73
Screw-anchor to body-hexagon socket-high R 54

Screw-bleed F 13
Screw-bleed brake H 15
Screw-bleed brake H 18
Screw-bleed clutch slave cylinder C 2
Screw-bleed clutch slave cylinder C 24

Screw-bleed clutch slave cylinder C 44
Screw-bleed clutch slave cylinder C 45
Screw-bleed clutch slave cylinder C 46
Screw-bleed-unheated J 12
Screw-block to gearbox-83mm-long C 73

Screw-body to engine block-hexagonal N 6
Screw-bracket adjusting-flanged head-m8 x B142
Screw-bracket mounting to chassis- N 5
Screw-bracket mounting-flanged head-m8 x B142
Screw-bracket mounting-hexagonal head- B140

Screw-bracket to block-flanged head-m10 x B 35
Screw-bracket to block-flanged head-m10 x B 71
Screw-bracket to body-flanged head-m6 x R 51
Screw-bracket to coolant pump-flanged B182
Screw-bracket to cylinder head-flanged B 16

Screw-bracket to cylinder head-flanged B 17
Screw-bracket to engine block-hexagonal B 83
Screw-bracket to gearbox-hexagonal head- C 90
Screw-bracket to injection pump-flanged B 28
Screw-bracket to manifold B179

Screw-breather chamber-hexagonal head- .... B 21
Screw-carburettor fast idle adjusting .... B 97
Screw-carburettor slow run adjusting .... B 97
Screw-carburettor throttle disc B135
Screw-clamp distributor-hexagonal head- B147

Screw-clutch damper to bracket-hexagonal .... G 13
Screw-clutch master cylinder to mounting .... G 12
Screw-clutch plate to flywheel-hexagonal B160
Screw-console to floor-m5 x 16 .... P 21
Screw-coolant pump to front cover-flanged .... B 85

Screw-coolant pump to front cover-m8 x 25 .... B 12
Screw-counter sunk .... E 21
Screw-counter sunk-m8 x 20 .... D 23
Screw-countershaft-10unf x 5/16 .... B139
Screw-cover bottom to plate-flanged head- .... C 71

Screw-cover to block-flanged head-m8 x 25 .... B 6
Screw-cover to block-flanged head-m8 x 25 .... B 59
Screw-cover to housing-hexagonal head-m6 .... C 72
Screw-cover-flanged head-1/4unc x 2 .... B119
Screw-cowl fixing M128

Screw-cylinder head-hexagonal head-3/8unc B125
Screw-cylinder head-hexagonal head- B124
Screw-drive assembly-3/8unc x 1 1/2 .... B158
Screw-engine tappet adjustment .... B 18
Screw-engine tappet adjustment .... B 56

Screw-fan to viscous coupling-flanged .... B 52
Screw-fan to viscous coupling-hexagonal .... B 52
Screw-fixing bracket ecu-hexagonal head .... M 91
Screw-fixing bracket egr-flanged head-m8 x .... M 91
Screw-fixing bracket-hexagonal head-m6 x M110

Screw-fixing cd autochanger-hexagonal M135
Screw-fixing cd autochanger-hexagonal M143
Screw-fixing cd bracket-hexagonal head-m5 .... M140
Screw-fixing cd bracket-hexagonal head-m5 .... M143
Screw-fixing control unit-hexagonal head- M105

Screw-fixing ecu-hexagonal head-m5 x 10 .... M 94
Screw-fixing ecu-hexagonal head-m5 x 12 .... M 99
Screw-fixing ecu-hexagonal head-m5 x 20 .... M 98
Screw-fixing horn-flanged head-m8 x 20 .... M106
Screw-fixing ignition heater plug-hexagonal .... M 90

Screw-fixing inertia switch-no10-self tapping .... M 95
Screw-fixing instrument cowl M123
Screw-fixing pump M148
Screw-fixing reflector .... M 35
Screw-fixing sounder unit-hexagonal head- M108

Screw-fixing sounder unit-hexagonal head- M110
Screw-fixing speaker front door-self M133
Screw-fixing speaker rear e post-self M136
Screw-fixing speaker rear e post-self M134
Screw-fixing speaker rear e post-self M139

Screw-fixing speaker rear e post-self M142
Screw-fixing speaker to rear end door- M140
Screw-fixing switch bonnet-hexagonal head- M106
Screw-flanged head-1/4unc x 1/2 .... B 95
Screw-flanged head-1/4unc x 5/8 .... B115

Screw-flanged head-m10 x 30 .... B 53
Screw-flanged head-m10 x 30 .... B 72
Screw-flanged head-m10 x 30 .... B 76
Screw-flanged head-m10 x 30 .... B184
Screw-flanged head-m10 x 30 .... B188

Screw-flanged head-m10 x 30 .... E 51
Screw-flanged head-m10 x 30 .... N 7
Screw-flanged head-m12 x 40 .... B184
Screw-flanged head-m12 x 40 .... C 52
Screw-flanged head-m6 x 10 .... B179

Screw-flanged head-m6 x 10 .... C 24
Screw-flanged head-m6 x 10 .... C 46
Screw-flanged head-m6 x 10 .... C 86
Screw-flanged head-m6 x 10 .... L 35
Screw-flanged head-m6 x 10 .... M 44

Screw-flanged head-m6 x 10 .... M 82
Screw-flanged head-m6 x 10 .... N 62
Screw-flanged head-m6 x 12 .... B180
Screw-flanged head-m6 x 12 .... B182
Screw-flanged head-m6 x 12 .... B173

Screw-flanged head-m6 x 12 .... E 16
Screw-flanged head-m6 x 12 .... E 15
Screw-flanged head-m6 x 12 .... P 38
Screw-flanged head-m6 x 16 .... K 4
Screw-flanged head-m6 x 16 .... L 12

Screw-flanged head-m6 x 16 .... L 15
Screw-flanged head-m6 x 16 .... N 65
Screw-flanged head-m6 x 20 .... B 28
Screw-flanged head-m6 x 20 .... B129
Screw-flanged head-m6 x 20 .... B168

Screw-flanged head-m6 x 20 .... B 65
Screw-flanged head-m6 x 20 .... C 28
Screw-flanged head-m6 x 20 .... C 7
Screw-flanged head-m6 x 20 .... F 17
Screw-flanged head-m6 x 20 .... N 53

Screw-flanged head-m6 x 20 .... N 89
Screw-flanged head-m6 x 20 .... N 13
Screw-flanged head-m6 x 25 .... B167
Screw-flanged head-m6 x 25 .... B170
Screw-flanged head-m6 x 25 .... B174

Screw-flanged head-m6 x 25 .... B177
Screw-flanged head-m6 x 25 .... C 5
Screw-flanged head-m6 x 25 .... C 27
Screw-flanged head-m6 x 25 .... C 53
Screw-flanged head-m6 x 25 .... F 18

Screw-flanged head-m6 x 25 .... K 2
Screw-flanged head-m6 x 25 .... K 3
Screw-flanged head-m8 x 12 .... B185
Screw-flanged head-m8 x 16 .... B 24
Screw-flanged head-m8 x 16 .... B 40

Screw-flanged head-m8 x 16 .... C 87
Screw-flanged head-m8 x 16 .... F 17
Screw-flanged head-m8 x 16 .... F 3
Screw-flanged head-m8 x 16 .... H 24
Screw-flanged head-m8 x 16 .... J 1

Screw-flanged head-m8 x 16 .... J 2
Screw-flanged head-m8 x 16 .... J 3
Screw-flanged head-m8 x 16 .... J 4
Screw-flanged head-m8 x 16 .... J 15
Screw-flanged head-m8 x 16 .... K 15

Screw-flanged head-m8 x 16 .... L 14
Screw-flanged head-m8 x 20 .... B 24
Screw-flanged head-m8 x 20 .... B 59
Screw-flanged head-m8 x 20 .... B 64
Screw-flanged head-m8 x 20 .... B 67

Screw-flanged head-m8 x 20 .... B 70
Screw-flanged head-m8 x 20 .... B 76
Screw-flanged head-m8 x 20 .... B143
Screw-flanged head-m8 x 20 .... B169
Screw-flanged head-m8 x 20 .... B176

Screw-flanged head-m8 x 20 .... B187
Screw-flanged head-m8 x 20 .... B189
Screw-flanged head-m8 x 20 .... B 55
Screw-flanged head-m8 x 20 .... E 43
Screw-flanged head-m8 x 20 .... E 49

Screw-flanged head-m8 x 20 .... G 11
Screw-flanged head-m8 x 20 .... J 9
Screw-flanged head-m8 x 20 .... L 1
Screw-flanged head-m8 x 20 .... L 2
Screw-flanged head-m8 x 20 .... L 4

Screw-flanged head-m8 x 20 .... L 5
Screw-flanged head-m8 x 20 .... L 8
Screw-flanged head-m8 x 20 .... N 7
Screw-flanged head-m8 x 20 .... N 69
Screw-flanged head-m8 x 25 .... B 10

Screw-flanged head-m8 x 25 .... B 61
Screw-flanged head-m8 x 25 .... B 49
Screw-flanged head-m8 x 25 .... C 2
Screw-flanged head-m8 x 25 .... C 8
Screw-flanged head-m8 x 25 .... C 44

Screw-flanged head-m8 x 25 .... C 45
Screw-flanged head-m8 x 25 .... C 46
Screw-flanged head-m8 x 25 .... C 56
Screw-flanged head-m8 x 25 .... C 67
Screw-flanged head-m8 x 25 .... D 3

Screw-flanged head-m8 x 25 .... D 10
Screw-flanged head-m8 x 25 .... J 21
Screw-flanged head-m8 x 25 .... J 22
Screw-flanged head-m8 x 25 .... N 14
Screw-flanged head-m8 x 30 .... B 34

Screw-flanged head-m8 x 30 .... B 51
Screw-flanged head-m8 x 30 .... B104
Screw-flanged head-m8 x 30 .... B177
Screw-flanged head-m8 x 30 .... B187
Screw-flanged head-m8 x 30 .... D 6

Screw-flanged head-m8 x 30 .... F 18
Screw-flanged head-m8 x 30 .... G 9
Screw-flanged head-no 10 x 1/2 .... L 35
Screw-flanged head-no10 .... N 13
Screw-front cover to block-flanged head-m8 .... B 48

Screw-front cover to block-hexagonal head- .... B 84
Screw-front door speaker fixing-self M132
Screw-fuel pressure regulator to fuel rail-m6 .... B133
Screw-fuel pressure regulator to fuel rail-m6 .... B134
Screw-fuel rail to manifold-flanged head- B132

Screw-gearbox extension housing front-m8 .... C 56
Screw-governer housing to gearbox- .... C 85
Screw-grab handle to squab-m8 .... R 6
Screw-grub .... C 6
Screw-grub .... C 18

Screw-grub .... C 38
Screw-grub .... C 66
Screw-grub .... D 15
Screw-grub .... F 13
Screw-heatshield-hexagonal head-m5 x 10 .... B 37

Screw-hexagon socket-5/16unc x 3/4 .... B129
Screw-hexagon socket-5/16unc x 3/4 .... B136
Screw-hexagon socket-fixing transducer M115
Screw-hexagon socket-m10 x 25 .... E 39
Screw-hexagon socket-m10 x 25 .... E 40

Screw-hexagon socket-m10 x 25 .... E 21
Screw-hexagon socket-m10 x 25 .... E 19
Screw-hexagon socket-m8 x 20 .... B 25
Screw-hexagon socket-m8 x 25 .... L 16
Screw-hexagon socket-plate to manifold-m8 .... B 67

Screw-hexagon socket-pump vacuum to .... B 33
Screw-hexagonal head .... N 55
Screw-hexagonal head .... N 52
Screw-hexagonal head-1/2unf x 2 .... E 19
Screw-hexagonal head-1/4unf x 1/2 .... H 19

Screw-hexagonal head-1/4unf x 3/4 .... L 4
Screw-hexagonal head-1/4unf x 3/4 .... M146
Screw-hexagonal head-1/4unf x 3/4 .... T 22
Screw-hexagonal head-1/4unf x 5/8 .... B 86
Screw-hexagonal head-1/4unf x 7/8 .... L 14

Screw-hexagonal head-3/8unc x 1 1/2 .... B101
Screw-hexagonal head-3/8unc x 1 1/8 .... B 96
Screw-hexagonal head-5/16unc x 1 1/8 .... B118
Screw-hexagonal head-5/16unc x 1 1/8 .... K 2
Screw-hexagonal head-5/16unf x 1 .... E 49

Screw-hexagonal head-5/16unf x 3/4 .... F 5
Screw-hexagonal head-5/8unc x 1 .... B 93
Screw-hexagonal head-5/8unc x 1 .... C 42
Screw-hexagonal head-5/8unc x 1 .... C 93
Screw-hexagonal head-5/8unc x 1 .... M 47

Screw-hexagonal head-7/16unc x 1 .... B108
Screw-hexagonal head-m10 x 20 .... J 1
Screw-hexagonal head-m10 x 20 .... J 2
Screw-hexagonal head-m10 x 20 .... J 3
Screw-hexagonal head-m10 x 20 .... J 4

Screw-hexagonal head-m10 x 25 .... C 22
Screw-hexagonal head-m10 x 25 .... E 46
Screw-hexagonal head-m10 x 25 .... J 12
Screw-hexagonal head-m10 x 25 .... N 1
Screw-hexagonal head-m10 x 30 .... D 5

Screw-hexagonal head-m10 x 30 .... E 48
Screw-hexagonal head-m10 x 35 .... E 47
Screw-hexagonal head-m10 x 35 .... E 53
Screw-hexagonal head-m10 x 35 .... H 24
Screw-hexagonal head-m12 x 25 .... C 68

Screw-hexagonal head-m12 x 25 .... N 4
Screw-hexagonal head-m12 x 25 .... N 10
Screw-hexagonal head-m12 x 25 .... N 9
Screw-hexagonal head-m12 x 30 .... C 4
Screw-hexagonal head-m12 x 30 .... C 26

Screw-hexagonal head-m4 x 10 .... M 82
Screw-hexagonal head-m5 x 10 .... B 38
Screw-hexagonal head-m5 x 10 .... B 62
Screw-hexagonal head-m5 x 10 .... L 17
Screw-hexagonal head-m5 x 10 .... M 83

Screw-hexagonal head-m5 x 10 .... P 3
Screw-hexagonal head-m5 x 12 .... G 4
Screw-hexagonal head-m5 x 16 .... B 26
Screw-hexagonal head-m5 x 16 .... C 52
Screw-hexagonal head-m5 x 16 .... M 57

Screw-hexagonal head-m5 x 20 .... M151
Screw-hexagonal head-m5 x 20 .... M137
Screw-hexagonal head-m5 x 25 .... M 28
Screw-hexagonal head-m6 x 10 .... C 88
Screw-hexagonal head-m6 x 12 .... N 46

Screw-hexagonal head-m6 x 12 .... N 86
Screw-hexagonal head-m6 x 12 .... N110
Screw-hexagonal head-m6 x 12 .... N111
Screw-hexagonal head-m6 x 14 .... B 15
Screw-hexagonal head-m6 x 14 .... K 10

Screw-hexagonal head-m6 x 16 .... B 99
Screw-hexagonal head-m6 x 16 .... B164
Screw-hexagonal head-m6 x 16 .... C 42
Screw-hexagonal head-m6 x 16 .... C 66
Screw-hexagonal head-m6 x 16 .... C 69

Screw-hexagonal head-m6 x 16 .... C 92
Screw-hexagonal head-m6 x 16 .... C 94
Screw-hexagonal head-m6 x 16 .... C 96
Screw-hexagonal head-m6 x 16 .... C 19
Screw-hexagonal head-m6 x 16 .... C 39

Screw-hexagonal head-m6 x 16 .... C 95
Screw-hexagonal head-m6 x 16 .... C 97
Screw-hexagonal head-m6 x 16 .... F 21
Screw-hexagonal head-m6 x 16 .... F 20
Screw-hexagonal head-m6 x 16 .... J 5

Screw-hexagonal head-m6 x 16 .... J 6
Screw-hexagonal head-m6 x 16 .... J 20
Screw-hexagonal head-m6 x 16 .... J 21
Screw-hexagonal head-m6 x 16 .... J 22
Screw-hexagonal head-m6 x 16 .... K 6

Screw-hexagonal head-m6 x 16 .... K 8
Screw-hexagonal head-m6 x 16 .... K 9
Screw-hexagonal head-m6 x 16 .... K 11
Screw-hexagonal head-m6 x 16 .... K 13
Screw-hexagonal head-m6 x 16 .... K 14

Screw-hexagonal head-m6 x 16 .... K 4
Screw-hexagonal head-m6 x 16 .... K 7
Screw-hexagonal head-m6 x 16 .... L 6
Screw-hexagonal head-m6 x 16 .... L 13
Screw-hexagonal head-m6 x 16 .... L 18

Screw-hexagonal head-m6 x 16 .... L 36
Screw-hexagonal head-m6 x 16 .... L 3
Screw-hexagonal head-m6 x 16 .... M 18
Screw-hexagonal head-m6 x 16 .... M 19
Screw-hexagonal head-m6 x 16 .... M144

Screw-hexagonal head-m6 x 16 .... M147
Screw-hexagonal head-m6 x 16 .... M149
Screw-hexagonal head-m6 x 16 .... M 47
Screw-hexagonal head-m6 x 16 .... M108
Screw-hexagonal head-m6 x 16 .... N 50

Screw-hexagonal head-m6 x 16 .... N 20
Screw-hexagonal head-m6 x 16 .... N 28
Screw-hexagonal head-m6 x 16 .... P 22
Screw-hexagonal head-m6 x 16 .... R 46
Screw-hexagonal head-m6 x 20 .... M152

Screw-hexagonal head-m6 x 25 .... F 21
Screw-hexagonal head-m6 x 30 .... N 70
Screw-hexagonal head-m6 x 35 .... C 69
Screw-hexagonal head-m6 x 35 .... C 95
Screw-hexagonal head-m6 x 35 .... C 97

Screw-hexagonal head-m6 x 45 .... C 5
Screw-hexagonal head-m6 x 45 .... C 27
Screw-hexagonal head-m8 x 15 .... L 17
Screw-hexagonal head-m8 x 16 .... C 89
Screw-hexagonal head-m8 x 16 .... F 3

Screw-hexagonal head-m8 x 16 .... J 15
Screw-hexagonal head-m8 x 16 .... K 13
Screw-hexagonal head-m8 x 16 .... K 15
Screw-hexagonal head-m8 x 25 .... B 99
Screw-hexagonal head-m8 x 25 .... C 29

Screw-hexagonal head-m8 x 25 .... C 54
Screw-hexagonal head-m8 x 25 .... C 21
Screw-hexagonal head-m8 x 25 .... C 41
Screw-hexagonal head-m8 x 25 .... D 14
Screw-hexagonal head-m8 x 25 .... M 16

Screw-hexagonal head-m8 x 25 .... M 17
Screw-hexagonal head-m8 x 25 .... M 18
Screw-hexagonal head-m8 x 25 .... M 19
Screw-hexagonal head-m8 x 25 .... M 20
Screw-hexagonal head-m8 x 25 .... M 21

Screw-hexagonal head-m8 x 35 .... C 39
Screw-hexagonal head-m8 x 40 .... C 24
Screw-hexagonal head-m8 x 40 .... K 6
Screw-hexagonal head-m8 x 40 .... K 8
Screw-hexagonal head-m8 x 40 .... K 9

Screw-hexagonal head-m8 x 40 .... K 7
Screw-hexagonal head-m8 x 45 .... B106
Screw-hinge to body-pan head-m5 x 12 .... N 56
Screw-housing thermostat to cyl head- .... B 23
Screw-housing to block-flanged head-m10 x .... B 39

Screw-housing to block-flanged head-m10 x .... B 75
Screw-housing to block-flanged head-m8 x .... B 43
Screw-housing to block-m10 x 14 .... C 72
Screw-idle adjustment tbi B140
Screw-jacket cover-1/4unc x 1 .... B136

Screw-jet to block-flanged head-m8 x 20 .... B 44
Screw-ladder frame to block-flanged head- .... B 9
Screw-lift pump fuel to side cover-flanged .... B 27
Screw-lift pump fuel to side cover-flanged .... B 63
Screw-lift pump fuel to side cover-m8 x 30 .... B 27

Screw-link adjusting-flanged head-m8 x 20 .... B156
Screw-link adjusting-flanged head-m8 x 20 .... B157
Screw-m10 x 30 .... D 5
Screw-m10 x 35 .... B175
Screw-m12 x 30 .... E 10

Screw-m12 x 30 .... E 34
Screw-m12 x 30 .... E 35
Screw-m12 x 45 .... B189
Screw-m4 x 12 .... P 5
Screw-m5 .... B 38

Screw-m5 .... M 6
Screw-m5 .... M114
Screw-m5 .... N 35
Screw-m5 .... P 22
Screw-m5 .... P 14

Screw-m5 x 10-black B174
Screw-m5 x 16 .... N 43
Screw-m5 x 16 .... P 19
Screw-m5 x 20 .... N114
Screw-m5 x 20 .... N119

Screw-m5 x 60 .... S 1
Screw-m6 .... L 33
Screw-m6 x 12 .... N 83
Screw-m6 x 12 .... P 29
Screw-m6 x 16 .... C 47

Screw-m6 x 16 .... D 17
Screw-m6 x 16 .... K 11
Screw-m6 x 16 .... N 15
Screw-m6 x 16 .... P 75
Screw-m6 x 18 .... B167

Screw-m6 x 20-counter sunk-recessed .... D 16
Screw-m6 x 20-counter sunk-recessed .... D 18
Screw-m6 x 20-counter sunk-recessed .... N 85
Screw-m6 x 25-patch lock .... D 16
Screw-m6 x 60 .... C 20

Screw-m6 x 60 .... C 40
Screw-m8 x 10 .... L 29
Screw-m8 x 10 .... P 13
Screw-m8 x 16 .... H 20
Screw-m8 x 16 .... N 33

Screw-m8 x 16 .... P 16
Screw-m8 x 20 .... B 25
Screw-m8 x 70 .... C 55
Screw-mounting bracket-hexagonal head- B156
Screw-no 6 x 12 .... N 76

Screw-no 6 x 12 .... N 81
Screw-no 6 x 12 .... N 84
Screw-paint clearing-m6 x 25 .... N111
Screw-pan .... M 82
Screw-pan .... N 43

Screw-pan .... N 51
Screw-pan head-m5 x 12 .... J 23
Screw-pan head-m5 x 12 .... N 69
Screw-pan head-m6 x 12 .... N 63
Screw-pan head-m6 x 12 .... N 64

Screw-pan head-m6 x 12 .... P 36
Screw-pan head-m6 x 12 .... P 42
Screw-pan head-m6 x 20 .... N 60
Screw-pan head-m6 x 20 .... N 61
Screw-pan-10mm .... N 42

Screw-panel to floor .... R 38
Screw-pipe-hexagonal head-m6 x 12 .... B137
Screw-plastite .... N 82
Screw-plate cover/front cover to block- .... B 10
Screw-plate mounting to pump-hexagonal .... B 34

Screw-plate mounting to pump-m6 x 28 .... C 75
Screw-plate thrust camshaft-hexagonal B128
Screw-plate thrust camshaft-hexagonal .... B 22
Screw-plate thrust camshaft-hexagonal .... B 60
Screw-plate-hexagonal head-3/8unc x 1 1/8 .... B131

Screw-plenum chamber-m8 x 80 .... B135
Screw-plinth to floor .... R 18
Screw-powerlok .... M 61
Screw-powerlok .... M 62
Screw-powerlok .... P 10

Screw-powerlok-pan head-m6 .... N 84
Screw-pulley to damper-flanged head-m8 x .... B 7
Screw-pulley to pump-flanged head-m8 x 16 .... B 65
Screw-pulley to pump-hexagonal head- .... B 13
Screw-pulley to pump-hexagonal head- B121

Screw-pulley-flanged head-m8 x 12 .... B143
Screw-pump to block-flanged head-m8 x 20 .... B 69
Screw-rack guide power assisted steering .... F 13
Screw-rocker cover to cylinder head-flanged .... B 20
Screw-seat base to plinth .... R 10

Screw-seat slide to floor .... R 25
Screw-sedimentor to bracket mounting .... J 13
Screw-self tapping .... L 28
Screw-self tapping .... L 48
Screw-self tapping .... P 5

Screw-self tapping .... P 19
Screw-self tapping .... R 1
Screw-self tapping ab .... M 15
Screw-self tapping ab .... P 1
Screw-self tapping ab-10 x 3/8 .... B 91

Screw-self tapping ab-10 x 3/8 .... K 4
Screw-self tapping ab-fixing inertia switch- .... M 96
Screw-self tapping ab-fixing inertia switch- .... M 99
Screw-self tapping ab-fixing p clip-10 x 3/8 .... M 95
Screw-self tapping ab-m8 x 12 .... B116

Screw-self tapping ab-m8 x 12 ..... B103
Screw-self tapping ab-m8 x 12 ..... B151
Screw-self tapping ab-m8 x 12 ..... N 25
Screw-self tapping ab-no 6 x 1/2 ..... C 88
Screw-self tapping ab-no 6 x 1/2 ..... L 4

Screw-self tapping ab-no 6 x 1/2 ..... M 79
Screw-self tapping ab-no 6 x 1/2 ..... M 80
Screw-self tapping ab-no 6 x 1/2 ..... P 5
Screw-self tapping ab-no 8 x 3/8 ..... L 28
Screw-self tapping ab-no 8 x 3/8 ..... L 21

Screw-self tapping ab-no 8 x 3/8 ..... M 34
Screw-self tapping ab-no 8 x 3/8 ..... M 89
Screw-self tapping ab-tube dipstick to ..... B 79
Screw-self tapping b-no 12 x 1" M144
Screw-self tapping b-no 4 x 1/2 ..... N 81

Screw-self tapping b-no 4 x 3/8 ..... P 47
Screw-self tapping b-no 4 x 5/8 ..... P 39
Screw-self tapping b-no 4 x 5/8 ..... P 41
Screw-self tapping b-no 8 x 1 1/4 ..... P 71
Screw-self tapping-fixing control unit- ..... M 97

Screw-self tapping-no 12 x 3/4 ..... N 27
Screw-self tapping-no 12 x 3/4 ..... N 30
Screw-self tapping-no 12 x 5/8 ..... P 75
Screw-self tapping-no 4 x 1/4 ..... M 77
Screw-self tapping-no 6 ..... B117

Screw-self tapping-no 6 x 1/2 ..... C 86
Screw-self tapping-no 6 x 3/8 ..... L 28
Screw-self tapping-no 6 x 3/8 ..... L 48
Screw-self tapping-no 8 x 1/2 ..... N 30
Screw-self tapping-no10 x 3/4 ..... B 91

Screw-self tapping-no10 x 3/4 ..... M 83
Screw-self tapping-no10 x 3/4 ..... N112
Screw-self tapping-no10 x 3/4 ..... N116
Screw-self tapping-no8 x 5/8 ..... N 62
Screw-self tapping-no8 x 5/8 ..... N 83

Screw-self tapping-no8 x 5/8 ..... N 84
Screw-self tapping-no8 x 5/8 ..... N 78
Screw-self tapping-no8 x 5/8 ..... N 82
Screw-self tapping-no8 x 5/8 ..... N 85
Screw-self tapping-no8 x 5/8 ..... N 79

Screw-self tapping-no8 x 5/8 ..... P 67
Screw-self tapping-no8 x 5/8 ..... P 71
Screw-self tapping-self tapping-m8 x 13 ..... N 81
Screw-self tapping-self tapping-m8 x 13 ..... P 4
Screw-self tapping-switch panel to console ..... P 21

Screw-side cover to block-flanged head-m8 ..... B 21
Screw-solenoid to starter motor-flanged ..... B 74
Screw-spacer crankshaft-7/16unf x 3/4 ..... B161
Screw-spindle housing to gear housing- ..... B 14
Screw-squab to frame-m8 ..... R 8

Screw-squab to panel-self tapping-no 14 x ..... R 39
Screw-squab to panel-self tapping-no 14 x ..... R 32
Screw-starter motor to bell housing-flanged ..... B 37
Screw-sunroof handle special ..... N 36
Screw-support bracket-m6 x 20-counter ..... R 46

Screw-thread forming-fixing amplifier-m6 x M141
Screw-torx drive ..... N 49
Screw-torx-countersunk ..... E 19
Screw-torx-countersunk-m8 ..... C 67
Screw-torx-flange-m5 x 30 ..... C 54

Screw-towing plate to tie bar-hexagonal ..... T 20
Screw-tread plate to sill-self tapping-no 8 x ..... P 83
Screw-tube dipstick to block-flanged head- ..... B 47
Screw-tube dipstick to block-hexagonal ..... B 9
Screw-washer-m5 x 20 ..... N 12

Screwdriver-tool kit ..... S 1
Scrim-load/tail door speaker fret-ash grey ..... P 39
Scrim-load/tail door speaker fret-ash grey ..... P 41
Seal ..... B 10
Seal ..... B 20

Seal ..... B 40
Seal ..... C 20
Seal ..... C 28
Seal ..... C 40
Seal ..... C 42

Seal ..... C 67
Seal ..... C 7
Seal ..... C 93
Seal ..... C 95
Seal ..... C 97

Seal ..... D 21
Seal ..... H 15
Seal ..... H 18
Seal ..... J 14
Seal ..... J 23

Seal ..... N 42
Seal ..... N 50
Seal ..... N 53
Seal ..... N 76
Seal ..... N 80

Seal ..... N 89
Seal ..... N 84
Seal ..... N 25
Seal ..... P 15
Seal ..... P 9

Seal ..... P 62
Seal ..... P 72
Seal ..... T 22
Seal air conditioning-7 seats ..... L 43
Seal air intake ..... L 20

Seal air intake ..... N 31
Seal assembly-front/rear hub ..... E 40
Seal assembly-front/rear hub ..... E 21
Seal assembly-front/rear hub-outer ..... E 23
Seal assembly-front/rear hub-outer ..... E 24

Seal assembly-front/rear hub-outer ..... E 39
Seal assembly-front/rear hub-outer ..... E 19
Seal induction system ..... J 14
Seal induction system ..... J 18
Seal selector mechanism ..... C 72

Seal-antilock brake system sensor ..... E 20
Seal-auto trans oil pump ..... C 75
Seal-auto trans output shaft ..... C 84
Seal-bearing cover ..... B 79
Seal-bearing cover B111

Seal-body to cooling pack air-top-bottom ..... L 11
Seal-bottom ..... C 5
Seal-bottom ..... C 27
Seal-bottom ..... C 53
Seal-bottom ..... C 71

Seal-camshaft oil ..... B 10
Seal-camshaft oil ..... B 48
Seal-camshaft oil B173
Seal-camshaft oil-rear B175
Seal-crankshaft front oil ..... B 10

Seal-crankshaft front oil ..... B 84
Seal-crankshaft front oil B117
Seal-crankshaft front oil B171
Seal-crankshaft front oil-inner ..... B 48
Seal-crankshaft front oil-outer ..... B 49

Seal-crankshaft rear oil ..... B 43
Seal-crankshaft rear oil ..... B 79
Seal-crankshaft rear oil B111
Seal-crankshaft rear oil B164
Seal-cylinder head valve stem ..... B 19

Seal-cylinder head valve stem ..... B 57
Seal-cylinder head valve stem B173
Seal-cylinder head valve stem oil B125
Seal-cylinder head valve stem oil-inlet ..... B 88
Seal-cylinder head valve stem oil-inlet B123

Seal-differential final drive pinion ..... E 10
Seal-differential final drive pinion ..... E 34
Seal-differential final drive pinion ..... E 35
Seal-door lock striker ..... N 80
Seal-door lock striker ..... N 77

Seal-door lock striker ..... N 85
Seal-duct central facia ..... P 4
Seal-duct central facia ..... P 12
Seal-duct driver/passenger ..... P 4
Seal-duct driver/passenger-outer ..... P 7

Seal-dust clutch housing B189
Seal-dust-gear change/plate manual ..... C 20
Seal-dust-gear change/plate manual ..... C 40
Seal-dust-gear change/plate manual ..... C 67
Seal-e post ..... P 10

Seal-engine oil filler cap B174
Seal-engine timing gear cover crankshaft ..... B 39
Seal-front ..... N 26
Seal-front door waist inner ..... N 47
Seal-front door-rh-sonar blue ..... N 49

Seal-front driveshaft ..... E 13
Seal-front driveshaft ..... E 14
Seal-front driveshaft ..... E 16
Seal-fuel pump unit tank ..... J 1
Seal-fuel pump unit tank ..... J 2

Seal-fuel pump unit tank ..... J 3
Seal-fuel pump unit tank ..... J 4
Seal-handbrake lever/body ..... H 23
Seal-inlet manifold gasket ..... B 95
Seal-inlet manifold gasket B130

Seal-load/tail door-sonar blue ..... N 54
Seal-man trans output shaft ..... C 54
Seal-man trans primary shaft ..... C 8
Seal-man trans primary shaft ..... C 29
Seal-man trans primary shaft ..... C 56

Seal-oil speedometer pinion ..... D 12
Seal-oil speedometer pinion ..... D 21
Seal-pedal box to bulkhead ..... G 6
Seal-pedal box to bulkhead ..... G 8
Seal-rear door-rh-sonar blue ..... N 55

Seal-rear quarter glass ..... N 52
Seal-rh-rear ..... M 7
Seal-rh-rear ..... M 10
Seal-rubber ..... J 13
Seal-rubber ..... N 37

Seal-rubber ..... N 38
Seal-rubber ..... N 39
Seal-rubber ..... T 47
Seal-scuttle panel-front ..... N 89
Seal-steering rack pinion ..... F 13

Seal-sunroof frame ..... N 37
Seal-sunroof frame ..... N 38
Seal-sunroof frame ..... N 39
Seal-sunroof frame ..... T 47
Seal-swivel pin housing-12.5mm ..... E 13

Seal-swivel pin housing-9mm ..... E 14
Seal-swivel pin housing-9mm ..... E 16
Seal-tailgate-outer ..... N 54
Seal-throttle spindle multi point injection B135
Seal-transfer box input ..... D 9

Seal-transfer box output ..... D 11
Seal-transfer box output ..... D 12
Seal-transfer box output ..... D 21
Seal-transfer box output flange spline ..... D 11
Seal-transfer box output flange spline ..... D 12

Seal-transfer box output flange spline ..... D 21
Seal-tunnel console automatic rubber ..... P 17
Seal-waist-front door-lh-outer ..... N 47
Seal-waist-rear door-outer-rh ..... N 52
Seal-wheelarch flare ..... N 31

Seal-windscreen glazing ..... N 67
Seal-windscreen wiper motor mounting M144
Seal-wiper linkage joint M144
Seal-wiper linkage joint M145
Sealant-hylomar 2000 ..... B 9

Sealant-hylomar 2000 ..... B 47
Sealant-pack ..... N 68
Sealant-silicone ..... D 3
Sealant-silicone ..... D 5
Sealant-silicone ..... D 18

Sealant-silicone ... D 17
Seat assembly-inward facing-cloth-sonar ... R 47
Seat assembly-inward facing-kestral-cloth- ... R 48
Seat belt assembly-double rear short end- ... R 57
Seat belt assembly-individual rear-rh-ash ... R 56

Seat belt assembly-short/long-rear-centre ... R 57
Seat belt-rh ... R 53
Seat-child restraint ... T 16
Seat-cylinder head exhaust valve-standard ... B 88
Seat-cylinder head inlet valve-0.010" ... B 88

Seat-cylinder head valve spring ... B 19
Seat-cylinder head valve spring ... B 57
Seat-electric front-rh-leather-dark granite ... R 22
Seat-electric front-rh-leather-light stone ... R 23
Seat-manual front-lh-leather/cloth-bahama ... R 20

Seat-manual front-lh-sonar blue ... R 3
Seat-manual front-rh-cloth-dark granite ... R 14
Seat-manual front-rh-kestral-cloth-dark ... R 15
Seat-manual front-rh-leather/cloth-dark ... R 19
Seat-manual front-rh-sonar blue ... R 2

Seat-spring lower-leveling ... E 46
Seat-spring lower-leveling ... E 51
Seatbelt-anchor ... R 54
Sensor assembly antilock brakes-front ... M111
Sensor assembly-airflow multi point injection ... L 17

Sensor assembly-crank engine ... B 76
Sensor assembly-crank engine ... B185
Sensor ultrasonic-aspen grey ... M106
Sensor ultrasonic-mink grey ... M108
Sensor ultrasonic-mink grey ... M110

Sensor unit fuel tank ... J 3
Sensor unit fuel tank ... J 4
Sensor-air inlet temperature engine-electric ... B177
Sensor-air temperature diesel ... B 62
Sensor-airflow-multi point inject ... J 15

Sensor-airflow-multi point inject ... J 16
Sensor-carburettor/multi point injection/tbi ... B 95
Sensor-carburettor/multi point injection/tbi ... B176
Sensor-carburettor/multi point injection/tbi ... B130
Sensor-coolant temperature diesel ... B 55

Sensor-coolant temperature diesel-orange ... B 23
Sensor-fuel temperature multi point ... B133
Sensor-fuel temperature multi point ... B179
Sensor-fuel temperature multi point ... B134
Sensor-fuel temperature multi point ... B132

Sensor-knock ignition ... B164
Sensor-oxygen multi point injection ... B178
Sensor-oxygen multi point injection ... K 4
Sensor-speed engine ... B 75
Sensor-temperature ... B130

Sensor-temperature-green ... B 61
Sensor-temperature-white ... B176
Sensor-thermostat air conditioning ... L 46
Sensor-turbocharger boost pressure ... L 17
Separator-vapour fuel filler ... J 23

Separtor assembly-crankcase breather oil ... B 21
Separtor assembly-crankcase breather oil ... B 59
Servo assembly brake ... H 1
Set-camshaft bearings ... B 79
Set-camshaft bearings ... B111

Set-connecting rod big end half bearing ... B165
Set-connecting rod big end half bearing- ... B 8
Set-connecting rod big end half bearing- ... B 46
Set-connecting rod big end half bearing- ... B 81
Set-connecting rod big end half bearing- ... B114

Set-coolant pump d ring ... B170
Set-crankshaft bearings-standard ... B 7
Set-crankshaft bearings-standard ... B 45
Set-crankshaft bearings-standard ... B 80
Set-crankshaft bearings-standard ... B112

Set-crankshaft bearings-standard ... B165
Set-cylinder head gasket ... B 32
Set-cylinder head gasket ... B181
Set-cylinder head gasket-asbestos ... B141
Set-engine gasket ... B141

Set-engine gasket ... B181
Set-engine gasket-asbestos ... B 32
Set-pad front brake system-asbestos ... H 16
Set-pad rear brake system-with sensor ... H 18
Set-ring automatic transmission ... C 82

Set-snow chains-205 x 16-600 x 16 ... T 36
Setscrew ... B135
Setscrew-1/4unc x 3/4 ... B 94
Setscrew-1/4unc x 3/4 ... B115
Setscrew-1/4unc x 3/4 ... B129

Setscrew-cylinder block main bearing cap ... B 43
Setscrew-cylinder block main bearing cap ... B 79
Setscrew-cylinder block main bearing cap ... B111
Setscrew-cylinder block main bearing cap ... B164
Setscrew-oil strainer to block-1/4unc x 3/4 ... B 82

Shaft & joint assembly- outer-rh ... E 13
Shaft & lever assembly-selector manual ... C 18
Shaft & lever assembly-selector manual ... C 38
Shaft assembly-front output ... D 13
Shaft assembly-high/low selector ... D 15

Shaft assembly-main automatic transmission ... C 85
Shaft assembly-primary manual transmission ... C 11
Shaft assembly-rear output ... D 12
Shaft assembly-rocker ... B 89
Shaft assembly-rocker ... B126

Shaft assembly-steering box sector-rhd ... F 13
Shaft automatic transmission ... C 72
Shaft automatic transmission ... C 84
Shaft primary manual transmission ... C 56
Shaft-camshaft reverse lock ... C 55

Shaft-counter automatic transmission ... C 76
Shaft-counter manual transmission ... C 36
Shaft-cross-transmission brake ... D 23
Shaft-front axle half-rh ... E 13
Shaft-front axle half-rh ... E 14

Shaft-front axle half-rh ... E 16
Shaft-gear change cross ... D 16
Shaft-gear change cross ... D 18
Shaft-intermediate transmission ... C 82
Shaft-lay ... C 15

Shaft-lay ... C 36
Shaft-lay ... C 64
Shaft-main manual transmission ... C 30
Shaft-main manual transmission ... C 57
Shaft-main manual transmission ... C 11

Shaft-oil pump ... B 14
Shaft-oil pump ... B 87
Shaft-oil pump ... B122
Shaft-pedal-accelerator ... G 1
Shaft-pinion differential ... C 30

Shaft-pinion differential ... C 11
Shaft-pivot pedal ... G 9
Shaft-rear axle half-lh-10 spline shaft ... E 40
Shaft-rear axle half-rh-10 spline shaft ... E 39
Shaft-reverse idler assembly-manual ... C 37

Shaft-reverse idler assembly-manual ... C 65
Shaft-reverse idler assembly-manual ... C 17
Shaft-rocker ... B 18
Shaft-rocker ... B 56
Shaft-rocker ... B 89

Shaft-rocker ... B126
Shaft-selector lever ... D 15
Shaft-selector manual transmission ... C 66
Shaft-spider ... C 82
Shaft-stub rear ... E 39

Shaft-sun gear ... C 82
Shaft-track rod adjusting ... F 15
Shaft-transfer box differential lock selector ... D 14
Shaft-transfer box-intermediate ... D 10
Shaft-universal joint steering ... F 7

Shaft-universal joint steering ... F 8
Shaft-universal joint steering ... F 9
Shampoo-500ml ... T 28
Shedder-rear door water-less subwoofer ... N 54
Shield-disc front brake-rh ... H 14

Shield-disc rear brake-rh ... H 17
Shield-disc rear brake-rh ... H 19
Shield-dust-idler ancillary drive-10mm id ... B187
Shield-inertia switch ... M 95
Shield-rear flange oil deflector ... B 80

Shield-rear flange oil deflector ... B113
Shim ... F 14
Shim ... N 14
Shim ... N 33
Shim ... N 77

Shim ... N 81
Shim ... N 85
Shim ... R 40
Shim ... R 38
Shim-0.003" ... E 20

Shim-0.003" ... E 18
Shim-0.038" ... E 37
Shim-1.1mm ... B 40
Shim-1.1mm ... B 76
Shim-1.25mm ... B161

Shim-1.51mm ... C 29
Shim-1.548mm ... E 12
Shim-1.548mm ... E 38
Shim-2.8mm ... C 75
Shim-2mm ... D 13

Shim-3.15mm ... D 9
Shim-bearing differential-0.060" ... E 32
Shim-bearing differential-2.155mm ... E 9
Shim-bearing differential-2.155mm ... E 33
Shim-door hinge ... N 47

Shim-door hinge ... N 48
Shim-door hinge-1.30mm ... N 46
Shim-door hinge-lower ... N 51
Shim-door hinge-lower ... N 52
Shim-front axle half shaft-0.45mm ... E 16

Shim-front axle half shaft-0.45mm ... E 15
Shim-inner rh-outer lh ... H 18
Shim-man trans 5th gear synchroniser hub- ... C 14
Shim-man trans 5th gear synchroniser hub- ... C 35
Shim-man trans countershaft bearing/centre ... C 50

Shim-man trans countershaft bearing/centre ... C 51
Shim-man trans countershaft bearing/front ... C 36
Shim-man trans countershaft bearing/front ... C 16
Shim-man trans mainshaft bearing/centre ... C 48
Shim-man trans mainshaft bearing/centre ... C 49

Shim-man trans primary shaft bearing/front ... C 9
Shim-man trans primary shaft bearing/front ... C 10
Shim-man trans reverse gear bush- ... C 63
Shroud air conditioning ... B 71
Silencer assembly exhaust system ... K 5

Silencer assembly exhaust system ... K 13
Silencer assembly exhaust system ... K 4
Sill-inner-rh ... N 24
Sill-outer-rh ... N 23
Slave cylinder clutch ... C 2

Slave cylinder clutch ... C 24
Slave cylinder clutch ... C 44
Slave cylinder clutch ... C 45
Slave cylinder clutch ... C 46
Sleeve ... E 13

Sleeve-air conditioning pipe protection-liquid ... L 49
Sleeve-guide clutch ... B 39
Sleeve-guide clutch ... C 2
Sleeve-rubber ... H 21
Sleeve-short ... C 73

Slide selector mechanism ... C 87
Slide-tappet guide ... B 19
Slide-tappet guide ... B 57
Socket-accessory power-12 s ... T 20
Socket-accessory power-12 s ... T 21

Socket-accessory power-fridge ..... T 14
Socket-electrical towing-7 pin ..... T 22
Socket-electrical towing-7 pin-12n ..... T 20
Soleniod-charcoal canister ..... J 20
Solenoid shut off-fuel injection pump ..... B 64

Solenoid starter motor B159
Solenoid starter motor B188
Solenoid starter motor-0 331 303 165-bosch ..... B 38
Solenoid starter motor-0 331 303 165-bosch ..... B 74
Solenoid starter motor-valeo ..... B 37

Solenoid starter motor-with m78r starter B107
Solenoid-control valve-with air conditioning B137
Solenoid-gear change lock ..... C 88
Solenoid-gear change lock ..... D 6
Solenoid-vacuum blower-heater ..... L 20

Solenoid-vacuum blower-heater ..... L 30
Solenoid-vacuum blower-heater ..... L 31
Sounder unit burglar alarm ..... T 35
Sounder unit burglar alarm-with battery back M108
Sounder unit burglar alarm-with battery back M110

Spacer ..... B 18
Spacer ..... B 21
Spacer ..... B 25
Spacer ..... B 36
Spacer ..... B 56

Spacer ..... B 73
Spacer ..... B 92
Spacer ..... B 94
Spacer B128
Spacer B142

Spacer ..... B 17
Spacer ..... C 13
Spacer ..... C 20
Spacer ..... C 32
Spacer ..... C 40

Spacer ..... C 67
Spacer ..... C 30
Spacer ..... C 37
Spacer ..... C 65
Spacer ..... C 17

Spacer ..... C 11
Spacer ..... D 10
Spacer ..... D 11
Spacer ..... D 12
Spacer ..... E 13

Spacer ..... E 14
Spacer ..... E 16
Spacer ..... E 23
Spacer ..... E 39
Spacer ..... G 9

Spacer ..... G 6
Spacer ..... G 8
Spacer ..... H 1
Spacer ..... J 1
Spacer ..... J 2

Spacer ..... J 3
Spacer ..... J 4
Spacer ..... K 9
Spacer ..... K 10
Spacer ..... K 11

Spacer ..... K 12
Spacer ..... K 7
Spacer ..... L 29
Spacer M144
Spacer M145

Spacer ..... N 41
Spacer ..... N 50
Spacer N121
Spacer ..... N 2
Spacer ..... N 36

Spacer ..... N 54
Spacer ..... P 73
Spacer ..... P 71
Spacer ..... R 40
Spacer ..... T 20

Spacer ..... T 21
Spacer-crank/drive plate automatic ..... B 40
Spacer-crank/drive plate automatic ..... B 76
Spacer-crank/drive plate automatic B161
Spacer-fan B154

Spacer-hinge to cushion ..... R 38
Spacer-lower ..... F 4
Spacer-mainshaft ..... C 57
Spacer-output shaft ..... C 85
Spacer-pinion bearing ..... E 10

Spacer-pinion bearing ..... E 32
Spacer-radio-ash grey M134
Spacer-reverse gear plunger ..... C 21
Spacer-reverse gear plunger ..... C 41
Spacer-set of 4 ..... N 35

Spanner-roof rack assembly ..... N 40
Speaker assembly rear-ash grey M136
Speaker assembly rear-e-post-ash grey M134
Speaker assembly rear-e-post-ash grey M139
Speaker assembly rear-e-post-ash grey M142

Speaker assembly rear-e-post-rh M132
Speaker assembly-front single cone-door- M139
Speaker assembly-front single cone-door- M142
Speaker-front door-ash grey-twin cone M136
Speaker-front single cone-door-ash M132

Speaker-front single cone-front door-ash M133
Speedometer-135mph M120
Speedometer-135mph M126
Speedometer-mph-with tachometer M117
Spigot-flywheel alignment ..... B 40

Spigot-flywheel alignment B161
Spindle-carburettor throttle B135
Spindle-oil pump ..... B 14
Spool-interlock ..... C 19
Spool-interlock ..... C 39

Sportbox-roof rack assembly ..... T 41
Spring B140
Spring ..... C 77
Spring ..... C 78
Spring ..... C 80

Spring ..... C 81
Spring ..... N 78
Spring 1st & 2nd synchroniser ..... C 12
Spring 1st & 2nd synchroniser ..... C 13
Spring 1st & 2nd synchroniser ..... C 31

Spring 1st & 2nd synchroniser ..... C 32
Spring 1st & 2nd synchroniser ..... C 14
Spring 1st & 2nd synchroniser ..... C 34
Spring clip ..... H 21
Spring ring ..... C 28

Spring ring ..... C 58
Spring ring ..... C 60
Spring ring ..... C 7
Spring ring ..... C 59
Spring ring ..... C 62

Spring ring ..... C 63
Spring-1st/2nd bias manual transmission ..... C 20
Spring-1st/2nd bias manual transmission ..... C 40
Spring-1st/2nd bias manual transmission ..... C 67
Spring-1st/2nd bias manual transmission ..... C 41

Spring-5th bias manual transmission ..... C 20
Spring-5th bias manual transmission ..... C 40
Spring-5th bias manual transmission ..... C 67
Spring-carburettor jet needle ..... B 97
Spring-carburettor throttle return ..... B 98

Spring-carburettor yellow piston ..... B 97
Spring-clutch assist ..... G 5
Spring-coil ..... B 37
Spring-coil ..... B 38
Spring-coil ..... B 74

Spring-coil ..... C 20
Spring-coil ..... C 40
Spring-coil-tension ..... R 48
Spring-cylinder head valve ..... B 19
Spring-cylinder head valve ..... B 57

Spring-cylinder head valve ..... B 88
Spring-cylinder head valve B123
Spring-cylinder head valve B125
Spring-cylinder head valve B173
Spring-detent manual transmission ..... C 18

Spring-detent manual transmission ..... C 38
Spring-detent manual transmission ..... C 47
Spring-detent manual transmission ..... C 66
Spring-detent manual transmission ..... D 17
Spring-front seat squab suspension-top ..... R 26

Spring-fuel fill door ..... N 33
Spring-glovebox return-rh ..... P 8
Spring-lever detent selector mechanism ..... C 72
Spring-oil pump relief valve ..... B 14
Spring-oil pump relief valve ..... B 48

Spring-oil pump relief valve ..... B 87
Spring-oil pump relief valve B122
Spring-oil pump relief valve B171
Spring-parking pawl automatic transmission ..... C 84
Spring-plunger manual transmission ..... C 20

Spring-plunger manual transmission ..... C 40
Spring-release button selector mechanism ..... C 87
Spring-return ..... G 1
Spring-return ..... G 4
Spring-return ..... N 69

Spring-return brake ..... G 9
Spring-return brake ..... G 6
Spring-return brake ..... G 8
Spring-return clutch ..... G 5
Spring-return-green ..... D 22

Spring-return-tension accelerator B139
Spring-return-tension accelerator B140
Spring-road rear coil ..... E 50
Spring-road-coil-blue stripe ..... E 45
Spring-rocker shaft ..... B 18

Spring-rocker shaft ..... B 56
Spring-rocker shaft ..... B 89
Spring-rocker shaft B126
Spring-short ..... C 73
Spring-timing belt tensioner B169

Spring-torsion ..... C 55
Spring-transfer box detent ..... D 15
Spring-transfer box detent ..... D 14
Spring-transfer box differential lock ..... D 14
Sprocket-camshaft ..... B 92

Sprocket-camshaft B128
Sprocket-crankshaft ..... B 80
Sprocket-crankshaft B112
Squab-large split rear seat-cloth-sonar blue ..... R 27
Squab-large split rear seat-kestral-cloth- ..... R 28

Squab-large split rear seat-leather-dark ..... R 41
Squab-large split rear seat-leather/cloth-dark ..... R 39
Squab-large split rear seat-leather/cloth-dark ..... R 44
Squab-small split rear seat-cloth-sonar blue ..... R 30
Squab-small split rear seat-kestral-cloth- ..... R 28

Squab-small split rear seat-leather-dark ..... R 41
Squab-small split rear seat-leather/cloth-dark ..... R 39
Squab-small split rear seat-leather/cloth-dark ..... R 44
Stay ..... N 28
Stay assembly-load/tail stop ..... N 53

Stay-bonnet ..... N 69
Stay-rear mudflap N111
Step rear retractable ..... T 43
Step-side runner-rh ..... T 44
Stop-bonnet cable-bonnet lock ..... N 70

Stop-front suspension rebound-lh ..... E 43
Stop-rebound rear suspension-2 holes ..... E 48
Stopper pads ..... G 6
Stopper pads ..... G 8
Strainer and pipe assembly-oil ..... B 14

Strainer and pipe assembly-oil ..... B 82
Strainer and pipe assembly-oil B115
Strainer-oil ..... B 87
Strainer-oil B122
Strainer-oil-automatic transmission fluid ..... C 73

Strainer-transmission oil ..... C 6
Strainer-transmission oil ..... C 28
Strainer-transmission oil ..... C 54
Strainer-transmission oil ..... C 7
Strap assembly fuel tank ..... J 1

Strap assembly fuel tank ..... J 2
Strap assembly fuel tank ..... J 3
Strap assembly fuel tank ..... J 4
Strap assy-towing ..... T 17
Strap-air conditioning duct-7 seats ..... L 43

Strap-load space cover securing-ash grey ..... P 23
Strap-roof rack lashing-pair-2 metres ..... T 40
Strap-tool kit retention ..... S 1
Striker-bonnet ..... N 69
Striker-centre console stowage box lid-ash ..... P 19

Striker-centre console stowage box lid-ash ..... P 20
Striker-door lock ..... N 80
Striker-door lock ..... N 77
Striker-door lock ..... N 85
Striker-facia glovebox ..... P 10

Striker-fuel filler door ..... N 33
Striker-rear seat squab-rh ..... R 40
Striker-rear seat squab-rh ..... R 38
Strip lamp fluoresent ..... M 14
Strip-anti rattle ..... P 21

Strip-anti rattle ..... P 10
Strip-edge protection-89mm ..... M 17
Strip-edge protection-89mm ..... M 19
Strip-edge protection-89mm ..... M 21
Strip-facia to windscreen sealing ..... N 66

Strut-alternator-65 amp B106
Strut-alternator-65 amp B156
Strut-tailgate gas N120
Stud ..... B 56
Stud ..... B 67

Stud ..... B 94
Stud ..... B 98
Stud ..... B 17
Stud ..... B 55
Stud ..... C 28

Stud ..... C 7
Stud ..... D 5
Stud ..... E 23
Stud ..... E 24
Stud ..... E 39

Stud ..... E 40
Stud ..... H 18
Stud ..... P 63
Stud ..... P 64
Stud ..... P 62

Stud-3/8unf x 1 1/8 ..... B 99
Stud-3/8unf x 1 3/4 ..... E 6
Stud-3/8unf x 1 3/4 ..... E 27
Stud-5/16unc x 1 3/8 ..... B111
Stud-5/16unc x 1 3/8 ..... B118

Stud-5/16unf x 1 1/2 ..... B 96
Stud-bolt ..... D 5
Stud-bracket mounting-3/8unf x 1 1/8 ..... B142
Stud-compressor to bracket-m8 x 90 ..... B 71
Stud-coolant pump to front cover-m8 x 110 ..... B 12

Stud-front cover to block-5/16unc x 1 3/8 ..... B 79
Stud-idler to cover-m10 x 20 ..... B 11
Stud-inlet manifold to cylinder head-m8 x 25 ..... B 24
Stud-inlet manifold-m8 x 25 ..... B172
Stud-lift pump fuel to side cover-m8 x 20 ..... B 6

Stud-lifting bracket ..... C 6
Stud-lifting bracket ..... C 85
Stud-lifting bracket ..... D 6
Stud-m10 x 25 ..... B178
Stud-m10 x 30 ..... B 39

Stud-m10 x 30 ..... B 75
Stud-m10 x 30 ..... B189
Stud-m10 x 30 ..... C 4
Stud-m10 x 30 ..... C 52
Stud-m10 x 35 ..... B 40

Stud-m10 x 35 ..... B 76
Stud-m10 x 75 ..... B 51
Stud-m6 x 22 ..... B129
Stud-m8 x 110 ..... B 10
Stud-m8 x 20 ..... B 34

Stud-m8 x 25 ..... B 62
Stud-m8 x 25 ..... B131
Stud-m8 x 30 ..... B 49
Stud-manifold to head-m10 x 25 ..... B 16
Stud-manifold to head-m8 x 25 ..... B 54

Stud-plastic ..... P 31
Stud-plastic ..... P 36
Stud-plastic ..... P 38
Stud-plastic ..... P 28
Stud-plastic ..... P 34

Stud-plastic ..... R 18
Stud-plastic ..... R 11
Stud-screw in-plastic-bahama beige ..... P 63
Stud-screw in-plastic-sonar blue ..... P 1
Stud-turbocharger to exhaust manifold-m8 x ..... B 31

Sump assembly-engine oil ..... B 9
Sump assembly-engine oil B167
Sump-engine oil ..... B 47
Sump-engine oil ..... B 82
Sump-engine oil B115

Sunroof assembly-electric operating glass- ..... T 47
Sunroof assembly-glass ..... N 35
Sunroof assembly-glass ..... N 37
Sunroof assembly-glass ..... N 38
Sunroof assembly-glass ..... N 39

Sunroof assembly-glass ..... N 36
Sunroof assembly-glass ..... T 47
Sunroof complete assy-rear-front ..... T 47
Sunshade assembly-sunroof sliding ..... N 35
Sunshade assembly-sunroof sliding ..... N 36

Sunvisor assembly-front header-less mirror- ..... P 70
Sunvisor assembly-front header-with mirror- ..... P 69
Sunvisor assembly-front header-with mirror- ..... P 68
Support bracket ..... N 32
Support-adjuster link alternator ..... B 85

Support-adjuster link alternator B106
Support-adjuster link alternator B120
Support-adjuster link alternator-link adjusting B156
Support-expansion tank mounting ..... L 13
Support-headlining front pocket moulding ..... P 71

Support-windscreen glazing ..... N 68
Suppressor assembly-noise distributor B101
Suppressor assembly-noise distributor B147
Switch assembly steering column ..... M 64
Switch assembly-transfer box low ratio ..... C 54

Switch assembly-vacuum control-heater ..... L 21
Switch handbrake unit ..... H 20
Switch heated seats-rh ..... M 78
Switch pack assembly seat-non memory-rh ..... M 77
Switch pack-air conditioning ..... L 50

Switch remote in car entertainment-volume up ..... M 71
Switch remote in car entertainment-waveband ..... M 68
Switch stop lamp ..... G 9
Switch stop lamp ..... G 6
Switch stop lamp ..... G 8

Switch-contact courtesy light ..... M 12
Switch-contact courtesy light ..... M 13
Switch-contact glovebox lamp ..... M 13
Switch-contact-bonnet burglar alarm M108
Switch-contact-bonnet burglar alarm M110

Switch-contact-bonnet burglar alarm M106
Switch-contact-bonnet burglar alarm ..... T 35
Switch-crash sensor air bag ..... M 61
Switch-cruise control enable/disable ..... M 65
Switch-cruise control pedal vent ..... G 10

Switch-cruise control set/resume ..... M 64
Switch-cruise control-resume/cancel ..... M 64
Switch-electronic diesel control ..... B 28
Switch-electronic diesel control ..... B 64
Switch-electronic diesel control M100

Switch-engine thermostat-green ..... B 23
Switch-engine thermostat-green ..... B 61
Switch-engine thermostat-yellow ..... B 93
Switch-engine thermostat-yellow B130
Switch-heater fuel filter-heated ..... J 12

Switch-ignition ..... F 6
Switch-ignition ..... N 71
Switch-ignition ..... N 72
Switch-ignition ..... N 73
Switch-ignition ..... N 74

Switch-ignition ..... N 75
Switch-inertia ..... M 95
Switch-inertia remote fuel pump ..... M 96
Switch-inertia remote fuel pump ..... M 99
Switch-inhibitor change selector ..... C 72

Switch-master lighting /indicator/headlamp ..... M 63
Switch-master lighting /indicator/headlamp ..... M 64
Switch-mirror ..... M 66
Switch-mirror ..... M 75
Switch-oil pressure engine ..... B 15

Switch-oil pressure engine ..... B 53
Switch-oil pressure engine ..... B 87
Switch-oil pressure engine B122
Switch-oil pressure engine B171
Switch-pressure air conditioning control ..... L 23

Switch-pressure air conditioning control-low ..... L 42
Switch-push push air conditioning ..... L 28
Switch-push push air conditioning ..... L 48
Switch-push push air conditioning-front ..... L 48
Switch-push push fog-front ..... M 65

Switch-push push fog-rear ..... M 67
Switch-push push fog-rear ..... M 70
Switch-push push fog-rear ..... T 23
Switch-push push hazard ..... M 65
Switch-push push heated rear window ..... M 67

Switch-push push heated rear window ..... M 70
Switch-push push heated windscreen ..... M 65
Switch-push push heated windscreen ..... M 67
Switch-push push isolator-electric window ..... M 77
Switch-push push isolator-electric window ..... M 75

Switch-push push-rear wash ..... M 67
Switch-push push-rear wash ..... M 70
Switch-rear wipe push push ..... M 67
Switch-rear wipe push push ..... M 70
Switch-reverse light manual transmission ..... C 20

Switch-reverse light manual transmission ..... C 40
Switch-reverse light manual transmission ..... C 54
Switch-reverse light manual transmission ..... D 14
Switch-rh-squab ..... M 77
Switch-rocker electric window lift ..... M 77

Switch-rocker electric window lift-driver-lhd ..... M 75
Switch-rocker electric window lift-rear door- ..... M 78
Switch-rocker sunroof-front ..... M 76
Switch-rotary headlamp leveling ..... M 66
Switch-rotary headlamp leveling ..... M 75

Switch-rotary heatr/fan control-2 position ..... L 21
Switch-rotary heatr/fan control-rear-with air ..... P 15
Switch-slide heatr/fan control-4 position ..... L 21
Switch-sunroof isolator push push-front ..... M 76
Switch-thumbwheel illumination control ..... M 72

Switch-transfer box audible warning ..... D 4
Switch-transfer box temperature sensor ..... D 6
Switch-transmission oil cooler temperature ..... C 42
Switch-transmission oil cooler temperature ..... C 92
Switch-transmission oil cooler temperature ..... C 95

Switch-transmission oil cooler temperature .... C 97
Switch-volt sensitive-black .... M 85
Switch-volt sensitive-cruise speed trip-black .... M 86
Switch-volt sensitive-switch volt sensitive- .... M 83
Switch-wash/wipe windscreen .... M 63

Switch-wash/wipe windscreen .... M 64
Synchroniser assembly-1st & 2nd mainshaft .... C 13
Synchroniser assembly-1st & 2nd mainshaft .... C 32
Synchroniser assembly-1st & 2nd mainshaft .... C 58
Synchroniser assembly-1st & 2nd mainshaft .... C 59

Synchroniser assembly-3rd & 4th mainshaft .... C 12
Synchroniser assembly-3rd & 4th mainshaft .... C 31
Synchroniser assembly-3rd & 4th mainshaft .... C 60
Synchroniser assembly-5th gear mainshaft .... C 14
Synchroniser assembly-5th gear mainshaft .... C 62

Synchroniser assembly-5th gear mainshaft .... C 34
Synchroniser assembly-5th gear mainshaft .... C 63
T connector-crankcase breather .... B 79
T connector-crankcase breather B136
T piece .... B 25

T piece B137
T piece .... G 10
T piece .... H 2
T piece .... L 46
T piece M146

T piece M148
T piece M150
T piece-rear .... H 11
T piece-rear .... H 6
T piece-rear .... H 8

Tab-differential locking ring .... E 7
Tab-differential locking ring .... E 8
Tab-differential locking ring .... E 29
Tab-differential locking ring .... E 30
Tab-differential locking ring .... E 31

Tab-differential locking ring-new-10 spline .... E 28
Tachometer-less airbag M120
Tachometer-less airbag M126
Tailgate assembly-upper-glass reinforced N120
Tailpipe assembly-exhaust system .... K 5

Tailpipe assembly-exhaust system .... K 12
Tailpipe exhaust system .... K 5
Tailpipe exhaust system .... K 4
Tank assembly fuel .... J 1
Tank assembly fuel .... J 2

Tank assembly fuel .... J 3
Tank assembly fuel .... J 4
Tank-radiator expansion .... L 12
Tank-radiator expansion .... L 13
Tank-radiator expansion .... L 14

Tank-radiator expansion .... L 15
Tap drain-cylinder block-alternative B111
Tape N107
Tape front door-rh-dark grey N108
Tape front door-rh-upper N107

Tape front wing-rh N107
Tape front wing-rh-dark grey N108
Tape rear quarter-lh-blue. N102
Tape rear quarter-rh-dark grey N108
Tape rear quarter-rh-lower N104

Tape rear quarter-rh-upper N103
Tape rear quarter-rh-upper N105
Tape rear quarter-rh-upper N107
Tape rear side door-rh-dark grey N108
Tape rear side door-rh-lower N107

Tape sidestripe-mid silver-10m N106
Tape-3.9i graphic N103
Tape-black-25mm-thin .... N 20
Tape-rh-front-upper N105
Tape-self adhesive-foam .... L 14

Tape-sonar blue .... N 54
Tape-xs graphic N104
Tappet-engine valve hydraulic .... B 19
Tappet-engine valve hydraulic .... B 57
Tappet-engine valve hydraulic .... B 89

Tappet-engine valve hydraulic B126
Tappet-engine valve hydraulic B173
Tappet-engine valve mechanical .... B 19
Tappet-engine valve mechanical .... B 57
Template-drilling .... P 10

Template-sunroof .... N 16
Template-sunroof .... N 17
Tensioner assembly B119
Tensioner-automatic ancillary drive .... B 51
Tensioner-automatic ancillary drive B184

Tensioner-automatic ancillary drive B187
Tensioner-timing belt .... B 11
Tensioner-timing belt .... B 50
Tensioner-timing belt B169
Thermocouple assembly-catalyst overheat .... M 90

Thermostat assembly-oil .... B 15
Thermostat assembly-oil .... B 53
Thermostat evaporator-air conditioning .... L 25
Thermostat evaporator-air conditioning .... L 26
Thermostat evaporator-air conditioning .... L 46

Thermostat-engine .... B 61
Thermostat-engine-88 degrees c .... B 23
Thermostat-engine-88 degrees c .... B 93
Thermostat-engine-88 degrees c B129
Thermostat-engine-88 degrees c B176

Thermostat-engine-88 degrees c .... L 19
Tie rod .... R 59
Tie-cable-4.8 x 270mm-inside serated B148
Tie-cable-4.8 x 270mm-inside serated .... E 6
Tie-cable-4.8 x 270mm-inside serated .... E 27

Tie-cable-4.8 x 270mm-inside serated .... J 23
Tie-cable-4.8 x 270mm-inside serated .... J 9
Tie-cable-4.8 x 270mm-inside serated .... M 20
Tie-cable-4.8 x 270mm-inside serated .... M 21
Tie-cable-4.8 x 270mm-inside serated .... M 45

Tie-cable-4.8 x 270mm-inside serated .... M 54
Tie-cable-4.8 x 270mm-inside serated .... M 95
Tie-cable-6 1/4" .... F 20
Tie-cable-6.5mm hole-8.0 x 155mm .... M 23
Tie-cable-6.5mm hole-8.0 x 155mm .... M 32

Tie-cable-8mm hole-4.6 x 215mm B157
Tie-cable-black-2.5 x 100mm-outside M155
Tie-cable-black-3.5 x 150mm-inside serated .... M 44
Tie-cable-black-3.5 x 150mm-inside serated .... M 48
Tie-cable-black-3.5 x 150mm-inside serated M132

Tie-cable-black-3.5 x 150mm-inside serated M133
Tie-cable-black-3.5 x 150mm-inside serated M150
Tie-cable-black-3.5 x 150mm-inside serated .... M 42
Tie-cable-black-3.5 x 150mm-inside serated .... N 79
Tie-cable-black-4.8 x 115mm-inside serated .... B 26

Tie-cable-fir tree-black-3.5 x 150mm-inside .... M 46
Tie-cable-fir tree-black-3.5 x 150mm-inside .... M 50
Tie-cable-fir tree-black-3.5 x 150mm-inside .... M 51
Tie-cable-fir tree-black-3.5 x 150mm-inside .... M 55
Tie-cable-fir tree-black-3.5 x 150mm-inside .... M 56

Tie-cable-fir tree-black-3.5 x 150mm-inside .... M 49
Tie-cable-fir tree-black-3.5 x 150mm-inside .... M 52
Tie-cable-fir tree-black-3.5 x 150mm-inside .... M 53
Tie-cable-plastic-adjustable-24mm M156
Tie-cable-separator to canister-4.8 x .... J 24

Tie-cable-white.-4.6 x 385mm-inside serated .... B 91
Tie-cable-white.-4.6 x 385mm-inside serated .... B 95
Tie-cable-white.-4.6 x 385mm-inside serated B176
Tie-cable-white.-4.6 x 385mm-inside serated B130
Tie-cable-white.-4.6 x 385mm-inside serated .... J 2

Tie-cable-white.-4.6 x 385mm-inside serated .... J 4
Tie-cable-white.-4.6 x 385mm-inside serated .... J 5
Tie-cable-white.-4.6 x 385mm-inside serated .... J 6
Tie-cable-white.-4.6 x 385mm-inside serated .... L 9
Tie-cable-white.-4.6 x 385mm-inside serated .... L 19

Tie-cable-white.-4.6 x 385mm-inside serated .... L 30
Tie-cable-white.-4.6 x 385mm-inside serated .... L 31
Tie-cable-white.-4.6 x 385mm-inside serated .... L 32
Tongue seat belt-stowage-ash grey .... R 58
Tongue-lap belt-ash .... R 56

Tongue-lap belt-ash grey .... R 58
Torsional vibration damper assembly-engine .... B 7
Torsional vibration damper-engine crankshaft .... B 45
Tow ball-50mm .... T 21
Tow bar kit .... T 20

Tow bar kit .... T 21
Tow bar-adjustable .... T 20
Towing attachment assembly .... N 2
Track rod power assisted steering .... F 15
Transducer speed .... G 6

Transducer speed .... G 8
Transducer speed M114
Transducer speed M115
Transfer gearbox assembly-1.2 : 1-new .... D 1
Transfer gearbox assembly-new-1.2 : 1 .... D 2

Transformer-mains-240v to 12v .... T 14
Transmission assembly automatic -new .... C 70
Transmission assembly manual-new .... C 1
Transmission assembly manual-new .... C 23
Transmission assembly manual-new .... C 43

Transmitter plip burglar alarm .... T 35
Transmitter plip burglar alarm-433 mhz M107
Transmitter plip burglar alarm-433 mhz M109
Tray-console picnic-ash grey-without .... P 12
Treadstrip-rh-front-ash .... P 82

Treadstrip-rh-rear-ash .... P 83
Triangle-warning .... T 17
Tube .... F 15
Tube .... N 70
Tube assembly-oil dipstick B167

Tube fluorescent .... M 14
Tube-breather transmission case .... C 3
Tube-breather transmission case .... C 25
Tube-breather transmission case .... C 55
Tube-breather transmission case .... C 85

Tube-breather transmission case .... D 3
Tube-carburettor vacuum ignition B147
Tube-drag link .... F 15
Tube-e post trim support .... P 62
Tube-inward facing rear seats support .... R 48

Tube-oil dipstick .... B 9
Tube-oil dipstick .... B 47
Tube-oil dipstick .... B 79
Tube-oil dipstick B116
Tube-sunroof drain-front .... N 36

Turbocharger assembly-new .... B 30
Turbocharger assembly-new .... B 67
Turnbuckle-fuse box lid-bahama beige .... P 8
Turnbuckle-fuse box lid-dark granite .... P 9
Tweeter assembly-a post audio M142

Tweeter audio system-front door-ash M132
Unit assembly-immobilization M108
Unit assembly-immobilization M110
Vacuum tank .... L 36
Valance front-lower .... N 12

Valance-front seat inner-rh-dark granite .... R 25
Valance-front seat outer-rh-ash grey .... R 18
Valance-front seat outer-rh-ash grey .... R 11
Valance-front seat outer-rh-dark granite .... R 25
Valve &pipe assembly-coolant heater .... L 30

Valve &pipe assembly-coolant heater .... L 31
Valve assembly air intake .... J 14
Valve assembly air intake .... J 15
Valve assembly air intake .... J 18
Valve assembly-solenoid .... L 26

Valve brake .... H 9
Valve brake .... H 10
Valve-air control .... B 91
Valve-bearing housing jet .... B 5
Valve-bearing housing jet .... B 44

Valve-block ... C 73
Valve-charging air conditioning ... L 40
Valve-charging air conditioning ... L 53
Valve-charging air conditioning ... L 41
Valve-charging air conditioning ... L 46

Valve-charging air conditioning ... L 54
Valve-check-vacuum control-heater ... L 37
Valve-coolant by passive heater ... L 30
Valve-coolant by passive heater ... L 31
Valve-cylinder head exhaust ... B 19

Valve-cylinder head exhaust ... B 57
Valve-cylinder head exhaust ... B 88
Valve-cylinder head exhaust ... B123
Valve-cylinder head exhaust ... B125
Valve-cylinder head exhaust ... B173

Valve-cylinder head inlet ... B 19
Valve-cylinder head inlet ... B 57
Valve-cylinder head inlet ... B 88
Valve-cylinder head inlet ... B123
Valve-cylinder head inlet ... B125

Valve-cylinder head inlet ... B173
Valve-drain blower-heater ... L 22
Valve-exhaust gas recirculation engine ... B 25
Valve-exhaust gas recirculation engine ... L 16
Valve-expansion evaporator-air conditioning ... L 25

Valve-expansion evaporator-air conditioning ... L 26
Valve-expansion evaporator-air conditioning ... L 45
Valve-expansion evaporator-air ... L 49
Valve-fuel cut fuel tank ... B 64
Valve-fuel cut fuel tank ... J 4

Valve-fuel cut fuel tank ... J 23
Valve-inlet manifold non return ... B 94
Valve-inlet manifold non return ... B138
Valve-inlet manifold non return ... H 4
Valve-non return brake vacuum ... H 1

Valve-solenoid-exhaust gas recirculation ... B 25
Valve-solenoid-exhaust gas recirculation ... B 26
Valve-solenoid-exhaust gas recirculation ... L 16
Valve-solenoid-exhaust gas recirculation ... M 91
Valve-vacuum delay-orange ... B101

Valve-vapour pressure relief ... J 6
Valve-vapour pressure relief ... J 24
Valve-wash system non return-5mm-front ... M146
Valve-wash system non return-5mm-front ... M147
Valve-wash system non return-5mm-front ... M148

Valve-wash system non return-5mm-front ... M149
Velcro-hook-mist grey ... P 72
Vent assembly-air conditioning duct-7 seats- ... L 43
Vent assembly-air extractor ... N 31
Vent assembly-face level facia-ash ... P 4

Vent assembly-face level facia-ash grey- ... P 7
Vent-rear air conditioning-7 seats-ash grey ... L 43
Washer ... B 11
Washer ... B 50
Washer ... B 80

Washer ... B 95
Washer ... B 98
Washer ... B139
Washer ... B113
Washer ... B130

Washer ... C 4
Washer ... C 25
Washer ... C 26
Washer ... C 68
Washer ... C 90

Washer ... D 11
Washer ... D 12
Washer ... D 16
Washer ... D 22
Washer ... D 23

Washer ... E 47
Washer ... E 48
Washer ... E 53
Washer ... E 51
Washer ... E 52

Washer ... F 16
Washer ... G 3
Washer ... J 13
Washer ... K 12
Washer ... L 4

Washer ... L 5
Washer ... L 22
Washer ... L 27
Washer ... L 38
Washer ... L 39

Washer ... M144
Washer ... N 10
Washer ... N 42
Washer ... N 62
Washer ... N 78

Washer ... P 1
Washer ... P 16
Washer ... P 8
Washer ... P 14
Washer ... P 9

Washer ... R 59
Washer ... T 20
Washer- shakeproof ... M146
Washer- shakeproof ... N 51
Washer- shakeproof ... N 52

Washer- shakeproof-fixing container ... M148
Washer- shakeproof-m4 ... B 54
Washer- shakeproof-m4 ... H 20
Washer-1/2" ... F 12
Washer-20mm ... E 49

Washer-3/8"-square ... B 35
Washer-3/8"-square ... B100
Washer-3/8"-square ... B106
Washer-3/8"-square ... B107
Washer-3/8"-square ... B108

Washer-3/8"-square ... B125
Washer-3/8"-square ... B124
Washer-3/8"-square ... E 23
Washer-3/8"-square ... E 24
Washer-3/8"-square ... E 39

Washer-3/8"-square ... E 40
Washer-3/8"-square ... E 46
Washer-3/8"-square ... G 9
Washer-3/8"-square ... H 14
Washer-3/8"-square ... H 17

Washer-3/8"-square ... N 9
Washer-7 seats-m10 ... L 43
Washer-8mm ... N 14
Washer-bracket mounting to alternator-3/8"- ... B156
Washer-clutch cover to flywheel-3/8"-square ... B160

Washer-copper ... B 9
Washer-copper ... B 15
Washer-copper ... B178
Washer-copper ... J 11
Washer-copper-m14 ... B 23

Washer-copper-sump drain plug ... B115
Washer-crankshaft thrust ... B165
Washer-crankshaft thrust-standard.-pair ... B 7
Washer-crankshaft thrust-standard.-pair ... B 45
Washer-cup ... E 43

Washer-cup ... H 3
Washer-cup ... R 40
Washer-cup ... R 38
Washer-cup-intermediate ... F 16
Washer-exhaust manifold to cylinder head tab ... B 96

Washer-exhaust manifold to cylinder head tab ... B131
Washer-fibre-self adhesive ... N 80
Washer-fibre-self adhesive ... N 85
Washer-fixing bracket ecu-m5 ... M 91
Washer-fixing bracket-m6 ... M 45

Washer-fixing cd autochanger-m4 ... M135
Washer-fixing cd autochanger-m4 ... M140
Washer-fixing cd autochanger-m4 ... M143
Washer-fixing control unit-m5 ... M105
Washer-fixing ecu-m5 ... M 94

Washer-fixing ecu-m5 ... M 98
Washer-fixing horn-m8 ... M106
Washer-fixing sounder unit-m8 ... M108
Washer-fixing sounder unit-m8 ... M110
Washer-fixing transducer-m6-standard. ... M115

Washer-foam ... P 8
Washer-front & rear hub nut tab ... E 23
Washer-front & rear hub nut tab ... E 24
Washer-front & rear hub nut tab ... E 39
Washer-front & rear hub nut tab ... E 40

Washer-front hub thrust ... E 21
Washer-front hub thrust ... E 19
Washer-lock ... B 14
Washer-lock ... E 18
Washer-lock ... E 19

Washer-lock ... E 22
Washer-lock ... M 83
Washer-lock ... N 60
Washer-lock ... N 61
Washer-lock ... N 63

Washer-lock ... N 64
Washer-lock ... N 76
Washer-lock ... N 81
Washer-lock ... N 84
Washer-lock ... P 29

Washer-lock-fixing ignition heater plug ... M 90
Washer-lock-lever stop ... B140
Washer-lock-outer ... E 24
Washer-lock-outer ... E 40
Washer-m10 ... B 37

Washer-m10 ... B 83
Washer-m10 ... C 3
Washer-m10 ... D 4
Washer-m10 ... D 6
Washer-m10 ... E 20

Washer-m10 ... E 19
Washer-m10 ... J 12
Washer-m10 ... N 1
Washer-m10 ... N 6
Washer-m4 ... M136

Washer-m5 ... C 52
Washer-m5 ... J 23
Washer-m5 ... M 6
Washer-m5 ... M 57
Washer-m5 ... M151

Washer-m5 ... M 28
Washer-m5 ... N112
Washer-m5 ... N 30
Washer-m5 ... N116
Washer-m5 ... N114

Washer-m5 ... N119
Washer-m5 ... P 63
Washer-m6 ... B 22
Washer-m6 ... B 33
Washer-m6 ... B 38

Washer-m6 ... B 90
Washer-m6 ... B127
Washer-m6 ... B137
Washer-m6 ... C 5
Washer-m6 ... C 27

Washer-m6 ... C 28
Washer-m6 ... C 7
Washer-m6 ... C 93
Washer-m6 ... C 95
Washer-m6 ... D 17

Washer-m6 ... F 15
Washer-m6 ... F 21
Washer-m6 ... F 20
Washer-m6 ... J 20
Washer-m6 ... M 44

Washer-m6 N 50
Washer-m6 N 55
Washer-m6 N 70
Washer-m6 R 40
Washer-m6 R 38
Washer-m6-oversize H 6
Washer-m6-oversize H 8
Washer-m6-oversize N 27
Washer-m6-oversize N 20
Washer-m6-standard. B 13
Washer-m6-standard. B 15
Washer-m6-standard. B 85
Washer-m6-standard. B 86
Washer-m6-standard. B119
Washer-m6-standard. B121
Washer-m6-standard. B132
Washer-m6-standard. C 42
Washer-m6-standard. C 69
Washer-m6-standard. C 92
Washer-m6-standard. C 94
Washer-m6-standard. C 96
Washer-m6-standard. C 85
Washer-m6-standard. C 97
Washer-m6-standard. D 5
Washer-m6-standard. D 20
Washer-m6-standard. D 14
Washer-m6-standard. E 13
Washer-m6-standard. E 16
Washer-m6-standard. E 15
Washer-m6-standard. F 17
Washer-m6-standard. F 18
Washer-m6-standard. G 13
Washer-m6-standard. H 9
Washer-m6-standard. H 10
Washer-m6-standard. H 11
Washer-m6-standard. H 19
Washer-m6-standard. J 5
Washer-m6-standard. J 6
Washer-m6-standard. J 18
Washer-m6-standard. J 21
Washer-m6-standard. J 22
Washer-m6-standard. K 2
Washer-m6-standard. K 6
Washer-m6-standard. K 8
Washer-m6-standard. K 9
Washer-m6-standard. K 13
Washer-m6-standard. K 14
Washer-m6-standard. K 4
Washer-m6-standard. K 7
Washer-m6-standard. L 12
Washer-m6-standard. L 13
Washer-m6-standard. L 15
Washer-m6-standard. L 18
Washer-m6-standard. L 36
Washer-m6-standard. L 3

Washer-m6-standard. M 47
Washer-m6-standard. N 60
Washer-m6-standard. N 61
Washer-m6-standard. N 63
Washer-m6-standard. N 64
Washer-m6-standard. N 86
Washer-m6-standard. N110
Washer-m6-standard. N 13
Washer-m6-standard. N 51
Washer-m6-standard. N111
Washer-m6-standard. N 52
Washer-m6-standard. P 6
Washer-m6-standard. P 22
Washer-m6-standard. P 64
Washer-m6-standard. R 46
Washer-m6-standard. R 51
Washer-m6-standard. S 1
Washer-m8 B 6
Washer-m8 B 14
Washer-m8 B 24
Washer-m8 B 39
Washer-m8 B 75
Washer-m8 B104
Washer-m8 B148
Washer-m8 B189
Washer-m8 C 2
Washer-m8 C 24
Washer-m8 C 19
Washer-m8 C 39
Washer-m8 E 21
Washer-m8 G 11
Washer-m8 G 12
Washer-m8 H 21
Washer-m8 H 24
Washer-m8 J 15
Washer-m8 K 1
Washer-m8 K 10
Washer-m8 K 11
Washer-m8 L 2
Washer-m8 M 16
Washer-m8 M 17
Washer-m8 M 18
Washer-m8 M 19
Washer-m8 M 20
Washer-m8 M 21
Washer-m8 N 41
Washer-oil pressure switch engine B 87
Washer-oil pressure switch engine B122
Washer-plain B 13
Washer-plain B 22
Washer-plain B 86
Washer-plain B 88
Washer-plain B 89
Washer-plain B 92
Washer-plain B 96

Washer-plain B 99
Washer-plain B106
Washer-plain B125
Washer-plain B126
Washer-plain B128
Washer-plain B131
Washer-plain B113
Washer-plain B124
Washer-plain C 42
Washer-plain C 93
Washer-plain E 11
Washer-plain E 13
Washer-plain E 14
Washer-plain E 16
Washer-plain E 36
Washer-plain E 43
Washer-plain E 21
Washer-plain E 10
Washer-plain E 34
Washer-plain F 1
Washer-plain F 5
Washer-plain F 16
Washer-plain F 18
Washer-plain F 3
Washer-plain H 12
Washer-plain J 13
Washer-plain J 23
Washer-plain J 24
Washer-plain K 2
Washer-plain K 10
Washer-plain K 11
Washer-plain K 12
Washer-plain K 4
Washer-plain K 15
Washer-plain L 1
Washer-plain N 12
Washer-plain N 33
Washer-plain N 40
Washer-plain N 76
Washer-plain N 51
Washer-plain N 81
Washer-plain N 25
Washer-plain P 3
Washer-plain P 39
Washer-plain P 41
Washer-plain P 47
Washer-plain R 59
Washer-plain-12mm E 2
Washer-plain-7 seats-m8 L 43
Washer-plain-7/16" F 12
Washer-plain-bracket adjusting B142
Washer-plain-bracket mounting to mounting N 4
Washer-plain-bracket mounting-m10- B144
Washer-plain-bracket mounting-m8 B140
Washer-plain-convertor drive plate-m10- B161

Washer-plain-fixing bracket ecu-m5- M 91
Washer-plain-fixing cd autochanger bracket- M135
Washer-plain-fixing control unit-m5- M105
Washer-plain-fixing ecu-m5-oversize M 94
Washer-plain-fixing ecu-m5-standard M 98
Washer-plain-fixing horn-m8 M106
Washer-plain-fixing sounder unit-m8 M108
Washer-plain-fixing sounder unit-m8 M110
Washer-plain-fixing speaker front door-m4 M133
Washer-plain-fixing speaker to rear end M134
Washer-plain-fixing speaker to rear end M142
Washer-plain-fixing speaker to rear end M140
Washer-plain-link adjusting B156
Washer-plain-m10 B175
Washer-plain-m10 B187
Washer-plain-m10-oversize B 83
Washer-plain-m10-oversize E 49
Washer-plain-m10-oversize E 52
Washer-plain-m10-oversize N 6
Washer-plain-m10-oversize N 14
Washer-plain-m10-oversize T 21
Washer-plain-m10-standard B 24
Washer-plain-m10-standard B 35
Washer-plain-m10-standard B 39
Washer-plain-m10-standard B 75
Washer-plain-m10-standard B100
Washer-plain-m10-standard B107
Washer-plain-m10-standard C 3
Washer-plain-m10-standard C 22
Washer-plain-m10-standard C 25
Washer-plain-m10-standard C 85
Washer-plain-m10-standard D 5
Washer-plain-m10-standard E 47
Washer-plain-m10-standard E 53
Washer-plain-m10-standard E 51
Washer-plain-m10-standard J 1
Washer-plain-m10-standard J 2
Washer-plain-m10-standard J 3
Washer-plain-m10-standard J 4
Washer-plain-m10-standard J 12
Washer-plain-m10-standard K 3
Washer-plain-m10-standard K 6
Washer-plain-m10-standard K 8
Washer-plain-m10-standard K 9
Washer-plain-m10-standard K 7
Washer-plain-m10-standard N 9
Washer-plain-m10-standard T 20
Washer-plain-m12 x 22 E 19
Washer-plain-m12 x 28 E 44
Washer-plain-m12 x 28 F 15
Washer-plain-m12 x 28 F 11
Washer-plain-m12 x 28 N 11
Washer-plain-m12-standard C 4
Washer-plain-m12-standard C 26
Washer-plain-m4 B 54

Washer-plain-m5 ... M 10
Washer-plain-m5-large ... P 14
Washer-plain-m5-oversize ... M 4
Washer-plain-m5-oversize ... M 5
Washer-plain-m5-oversize ... M151
Washer-plain-m5-oversize ... N 13
Washer-plain-m5-oversize ... P 75
Washer-plain-m5-oversize ... S 1
Washer-plain-m5-oversize-metric ... M137
Washer-plain-m5-standard ... B 26
Washer-plain-m5-standard ... C 52
Washer-plain-m5-standard ... D 17
Washer-plain-m5-standard ... G 4
Washer-plain-m5-standard ... J 21
Washer-plain-m5-standard ... L 3
Washer-plain-m5-standard ... M 57
Washer-plain-m5-standard ... M 28
Washer-plain-m5-standard ... N 69
Washer-plain-m5-standard ... N112
Washer-plain-m5-standard ... N 30
Washer-plain-m5-standard ... N116
Washer-plain-m5-standard ... N114
Washer-plain-m5-standard ... N119
Washer-plain-m5-standard ... P 5
Washer-plain-m5-standard ... P 6
Washer-plain-m5-standard ... P 62
Washer-plain-m6 ... C 6
Washer-plain-m6 ... L 6
Washer-plain-m6 ... L 16
Washer-plain-m6 ... L 33
Washer-plain-m6 ... M144
Washer-plain-m6 ... M147
Washer-plain-m6 ... M149
Washer-plain-m6 ... M152
Washer-plain-m6 ... N 49
Washer-plain-m6 ... N 20
Washer-plain-m6 ... R 39
Washer-plain-m6-extra large ... R 48
Washer-plain-m6-standard ... N 70
Washer-plain-m8 ... B 6
Washer-plain-m8 ... B 14
Washer-plain-m8 ... B 25
Washer-plain-m8 ... B 27
Washer-plain-m8 ... B 28
Washer-plain-m8 ... B 31
Washer-plain-m8 ... B 34
Washer-plain-m8 ... B 84
Washer-plain-m8 ... B 85
Washer-plain-m8 ... B104
Washer-plain-m8 ... B119
Washer-plain-m8 ... B135
Washer-plain-m8 ... B139
Washer-plain-m8 ... B148
Washer-plain-m8 ... B170
Washer-plain-m8 ... B118

Washer-plain-m8 ... B157
Washer-plain-m8 ... C 2
Washer-plain-m8 ... C 8
Washer-plain-m8 ... C 28
Washer-plain-m8 ... C 29
Washer-plain-m8 ... C 86
Washer-plain-m8 ... C 88
Washer-plain-m8 ... C 21
Washer-plain-m8 ... C 41
Washer-plain-m8 ... C 87
Washer-plain-m8 ... C 89
Washer-plain-m8 ... D 3
Washer-plain-m8 ... D 10
Washer-plain-m8 ... D 16
Washer-plain-m8 ... D 6
Washer-plain-m8 ... D 14
Washer-plain-m8 ... E 48
Washer-plain-m8 ... G 9
Washer-plain-m8 ... G 6
Washer-plain-m8 ... G 8
Washer-plain-m8 ... G 3
Washer-plain-m8 ... H 23
Washer-plain-m8 ... H 21
Washer-plain-m8 ... H 24
Washer-plain-m8 ... J 15
Washer-plain-m8 ... K 13
Washer-plain-m8 ... L 2
Washer-plain-m8 ... L 4
Washer-plain-m8 ... L 5
Washer-plain-m8 ... L 19
Washer-plain-m8 ... L 8
Washer-plain-m8 ... L 17
Washer-plain-m8 ... L 21
Washer-plain-m8 ... M 16
Washer-plain-m8 ... M 17
Washer-plain-m8 ... M 18
Washer-plain-m8 ... M 19
Washer-plain-m8 ... M 20
Washer-plain-m8 ... M 21
Washer-plain-m8 ... M114
Washer-plain-m8 ... M 47
Washer-plain-m8 ... N 41
Washer-plain-m8 ... N 53
Washer-plain-m8 ... N121
Washer-plain-m8 ... N 2
Washer-plain-m8 ... R 51
Washer-plain-master cylinder to servo-m8 ... H 1
Washer-plain-ram pipes housing-m8 ... B138
Washer-plain-squab to panel-m6 ... R 32
Washer-plain-standard ... H 19
Washer-plain-standard ... J 14
Washer-plain-standard ... N 55
Washer-plain-standard ... N 28
Washer-plain-starter motor-m10-standard ... B158
Washer-plastic ... M152

Washer-plastic ... N122
Washer-rubber ... F 16
Washer-rubber ... N 2
Washer-rubber ... P 8
Washer-rubber-7 seats ... L 43
Washer-sealing ... B 5
Washer-sealing ... B 14
Washer-sealing ... B 23
Washer-sealing ... B 28
Washer-sealing ... B 29
Washer-sealing ... B 31
Washer-sealing ... B 47
Washer-sealing ... B 53
Washer-sealing ... B 58
Washer-sealing ... B 62
Washer-sealing ... B 67
Washer-sealing ... B 75
Washer-sealing ... B 76
Washer-sealing ... B 87
Washer-sealing ... B 93
Washer-sealing ... B 94
Washer-sealing ... B 95
Washer-sealing ... B122
Washer-sealing ... B138
Washer-sealing ... B 17
Washer-sealing ... B 49
Washer-sealing ... B 55
Washer-sealing ... B 65
Washer-sealing ... B130
Washer-sealing ... C 6
Washer-sealing ... C 42
Washer-sealing ... C 75
Washer-sealing ... C 92
Washer-sealing ... C 7
Washer-sealing ... C 95
Washer-sealing ... C 97
Washer-sealing ... D 4
Washer-sealing ... D 21
Washer-sealing ... H 1
Washer-sealing ... H 4
Washer-sealing ... N 33
Washer-sealing-axle drain plug ... E 6
Washer-sealing-axle drain plug ... E 27
Washer-sealing-diesel injector spill return ... B 66
Washer-sealing-drain plug-automatic ... C 74
Washer-sealing-man trans drain plug ... C 3
Washer-sealing-man trans drain plug ... C 25
Washer-sealing-man trans drain plug ... C 47
Washer-sealing-man trans extension case ... C 28
Washer-sealing-man trans reverse light ... C 54
Washer-sealing-steering swivel pin housing ... E 20
Washer-sealing-steering swivel pin housing ... E 18
Washer-sealing-sump drain plug ... B 9
Washer-sealing-sump drain plug ... B 82
Washer-sealing-sump drain plug ... B167

Washer-sealing-transfer box breather pipe ... D 3
Washer-sealing-transfer box drain plug ... D 5
Washer-sealing-transfer box drain plug ... D 6
Washer-sealing-unheated ... J 12
Washer-special idler pulley ... B 60
Washer-spring ... F 1
Washer-spring ... L 36
Washer-spring ... M 18
Washer-spring ... M 19
Washer-spring ... N 13
Washer-spring-1/4 dia-square ... B 82
Washer-spring-1/4 dia-square ... B 84
Washer-spring-1/4 dia-square ... B 85
Washer-spring-1/4 dia-square ... B 90
Washer-spring-1/4 dia-square ... B 94
Washer-spring-1/4 dia-square ... B 95
Washer-spring-1/4 dia-square ... B115
Washer-spring-1/4 dia-square ... B127
Washer-spring-1/4 dia-square ... B129
Washer-spring-1/4 dia-square ... B118
Washer-spring-1/4 dia-square ... E 13
Washer-spring-1/4 dia-square ... E 16
Washer-spring-1/4 dia-square ... E 15
Washer-spring-1/4 dia-square ... H 14
Washer-spring-1/4 dia-square ... H 17
Washer-spring-1/4 dia-square ... H 19
Washer-spring-1/4 dia-square ... M 20
Washer-spring-1/4 dia-square ... N 69
Washer-spring-1/4 dia-square ... N 20
Washer-spring-1/4 dia-square ... T 22
Washer-spring-3/16 dia-square ... J 23
Washer-spring-3/16 dia-square ... J 24
Washer-spring-3/16 dia-square ... P 3
Washer-spring-3/8 ... C 72
Washer-spring-5/16" ... B 33
Washer-spring-5/16" ... B 35
Washer-spring-5/16" ... B 83
Washer-spring-5/16" ... B 93
Washer-spring-5/16" ... B 98
Washer-spring-5/16" ... B 99
Washer-spring-5/16" ... B119
Washer-spring-5/16" ... B137
Washer-spring-5/16" ... B124
Washer-spring-5/16" ... B130
Washer-spring-5/16" ... B136
Washer-spring-5/16" ... E 49
Washer-spring-5/16" ... E 46
Washer-spring-5/16" ... E 51
Washer-spring-5/16" ... E 52
Washer-spring-5/16" ... F 5
Washer-spring-5/16" ... M 47
Washer-spring-body to engine block-5/16" ... N 6
Washer-spring-bracket adjusting-5/16" ... B142
Washer-spring-double coil-1/4"-imperial ... L 4
Washer-spring-fuel rail to manifold-1/4 dia- ... B132

Washer-spring-m6-square ... C 71
Washer-spring-m6-square ... C 19
Washer-spring-m6-square ... C 39
Washer-spring-no 10-single coil-rectangular ... B103
Washer-spring-no 10-single coil-rectangular ... B140

Washer-spring-no 10-single coil-rectangular ... B151
Washer-spring-no 10-single coil-rectangular ... N 76
Washer-spring-no 10-single coil-rectangular ... N 80
Washer-spring-pulley-5/16" ... B144
Washer-sprung-7/16" ... H 18

Washer-sprung-7/16" ... H 16
Washer-squab to frame ... R 8
Washer-starlock ... E 21
Washer-starlock ... G 11
Washer-starlock ... G 13

Washer-starlock ... H 5
Washer-starlock ... H 7
Washer-starlock ... H 11
Washer-starlock ... H 13
Washer-starlock ... M 16

Washer-starlock ... M 17
Washer-starlock ... M 18
Washer-starlock ... M 19
Washer-starlock ... M 20
Washer-starlock ... M 21

Washer-starlock-m5 ... M114
Washer-starlock-m6 ... N 46
Washer-starlock-m8 ... N 49
Washer-starter motor-3/8"-square ... B158
Washer-starter ring ... B161

Washer-steering box drop arm nut tab- ... F 11
Washer-swivel pin thrust ... E 20
Washer-thrust ... D 7
Washer-thrust ... E 21
Washer-thrust manual transmission ... C 61

Washer-thrust manual transmission ... C 34
Washer-thrust manual transmission ... C 63
Washer-thrust manual transmission-1st/2nd ... C 14
Washer-waved ... B 89
Washer-waved ... B126

Washer-waved ... R 55
Washer-waved ... R 48
Washer_drain plug ... B 5
Washer_drain plug ... B 44
Water curtain-rh ... N 48

Water curtain-rh ... N 52
Waxstat adaptor ... B 15
Waxstat adaptor ... B 53
Weatherstrip-alpine light-rh ... N 67
Weatherstrip-quarterlight-rh ... N 56

Weatherstrip-rear door ... N 67
Weatherstrip-rear quarter-rh ... N 59
Weatherstrip-windscreen ... N 66
Weight-crankshaft balance-.17oz ... B 80
Weight-crankshaft balance-.17oz ... B113

Weight-crankshaft balance-1.00 oz ... B160
Weight-crankshaft balance-1.00 oz ... B161
Weight-flywheel balance-0.75 oz ... B160
Weight-flywheel balance-0.75 oz ... B161
Weight-governor ... C 84

Wheel assembly-steering soft feel-sonar blue ... F 1
Wheel assembly-steering-ash grey-vinyl ... F 2
Wheel-alloy road-7.0 x 16-styled- ... E 41
Wheel-alloy road-7.0 x 16-styled- ... T 36
Wheel-steel road-primer ... E 41

Wheelarch & valance-rh ... N 19
Wheelarch assembly-front-outer-lh ... T 7
Wheelarch assembly-rear-lh ... N 25
Wheelbrace ... S 1
Winch-electric-husky ... T 26

Winch-electric-xd9000i ... T 27
Wing front-rh ... N 27
Worm & valve assembly-power assisted ... F 14
Yoke assembly starter motor ... B 37
Yoke assembly-selector rod ... C 18

Yoke assembly-selector rod ... C 38
Yoke-shift rod manual transmission ... C 66

2000 - Petrol - T16 ..... B 162
- Air Conditioning Compressor ..... B 183
- Alternator ..... B 186
- Camshaft ..... B 175
- Crankshaft and Bearings ..... B 165
- Cylinder Head ..... B 172
- Engine Stripped ..... B 162
- Front Cover ..... B 168
- Manifold Exhaust ..... B 178
- Manifold Inlet ..... B 177
- Power Steering Pump ..... B 182
- Starter Motor ..... B 188

2500 - Diesel - Turbo - 200 Tdi ..... B 1
- Air Conditioning Compressor ..... B 35
- Alternator ..... B 36
- Camshaft ..... B 22
- Crankshaft and Bearings ..... B 7
- Cylinder Head ..... B 16
- Engine Stripped ..... B 2
- Front Cover ..... B 10
- Injection Pump ..... B 28
  - Washer-sealing ..... B 28
- Manifolds ..... B 24
- Oil Pump ..... B 14
- Power Steering Pump ..... B 34
- Starter Motor Lucas ..... B 37
- Turbocharger ..... B 30

2500 - Diesel - Turbo - 300 Tdi ..... B 41
- Air Conditioning Compressor ..... B 71
- Alternator ..... B 73
- Camshaft ..... B 60
- Crankshaft and Bearings ..... B 45
- Cylinder Head ..... B 54
- Engine Stripped ..... B 41
- Front Cover ..... B 48
- Injection Pump ..... B 64
- Manifolds ..... B 62
- Oil Filter ..... B 53
- Power Steering Pump ..... B 70
- Starter Motor ..... B 74
- Turbocharger ..... B 67

V8 Petrol - Carburetter ..... B 77
- Air Conditioning Compressor ..... B 100
- Alternator ..... B 105
- Crankshaft and Bearings ..... B 80
- Cylinder Head ..... B 88
- Distributor ..... B 101
- Engine Stripped ..... B 77
- Exhaust Manifold ..... B 96
- Front Cover ..... B 84
- Inlet Manifold ..... B 94
- Oil Pump ..... B 87
- Power Steering Pump & Mountings ..... B 99
- Starter Motor ..... B 107

V8 Petrol - EFi ..... B 109
- Air Conditioning Compressor V Belt Drive ..... B 144
- Alternator Lucas A133 ..... B 154
- Crankshaft and Bearings ..... B 112
- Cylinder Head 3.5 ..... B 123
- Cylinder Head 3.9 ..... B 125
- Distributor ..... B 146
- Engine Stripped ..... B 109
- Exhaust Manifold ..... B 131
- Front Cover ..... B 117
- Ignition Coil ..... B 148
- Inlet Manifold ..... B 129
- Oil Pump and Filter ..... B 122
- Power Steering Pump V Belt Drive ..... B 142
- Starter Motor Lucas ..... B 158

Illus	Part Number	Description	Quantity	Change Point	Remarks
1	RTC6637	Engine base unit-power unit	1		12L
1	RTC6638	Engine base unit-power unit	1		Switzerland

COMPLETE ENGINE INCLUDES FLYWHEEL,
CLUTCH AND ELECTRICS.
OVERHAUL GASKET KIT STC363
DECARBONISING GASKET KIT STC362

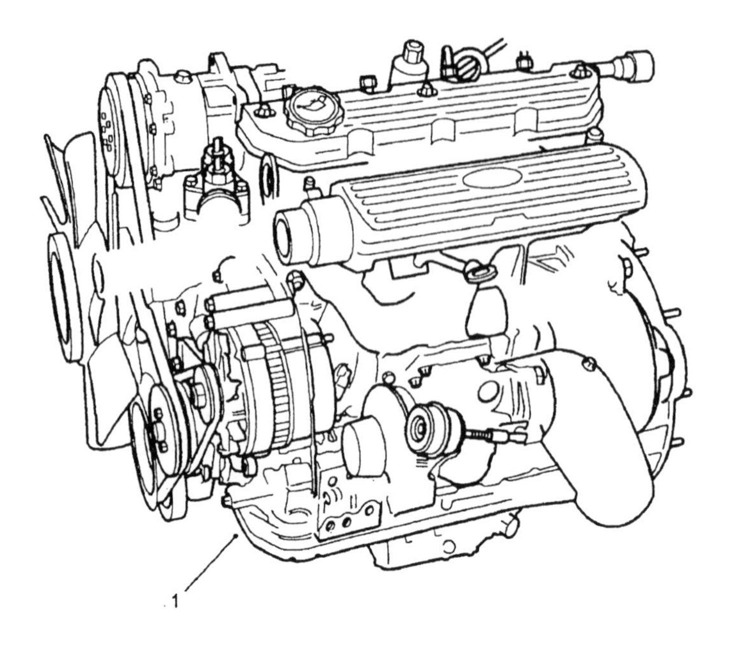

CABMAA1A

Illus	Part Number	Description	Quantity	Change Point	Remarks
1	STC1065N	Engine unit-stripped-new	1		12L
1	RTC6636R	Engine unit-stripped-reconditioned	1		

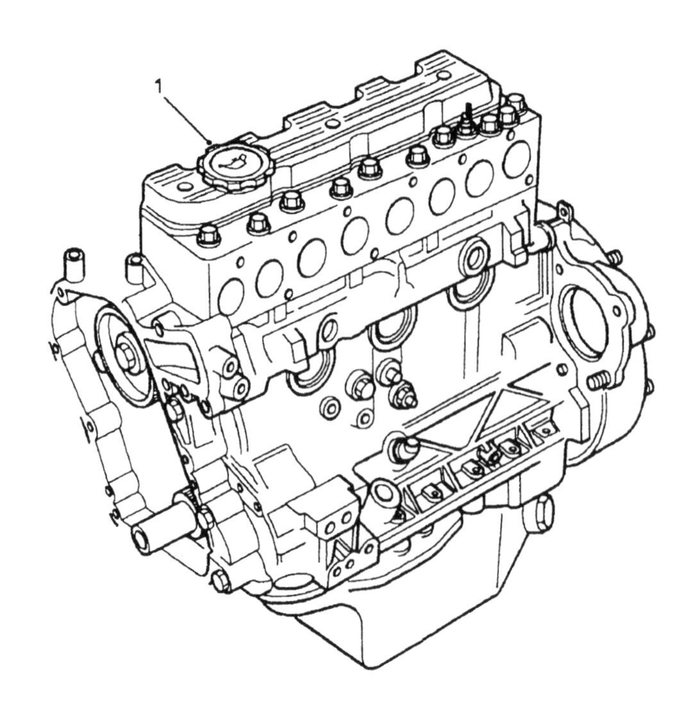

CABMAC1B

Engine Short

Illus	Part Number	Description	Quantity	Change Point	Remarks
1	RTC6635	Engine base unit-part-short	1		12L

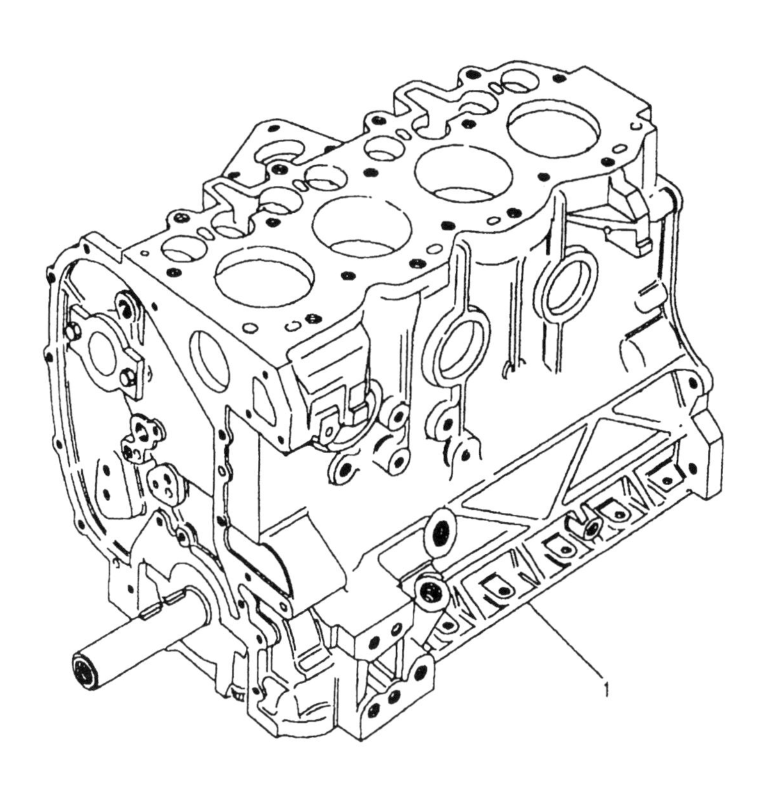

B 3

DISCOVERY 1989MY UPTO 1999MY | GROUP B | Engine 2500 - Diesel - Turbo - 200 Tdi

Cylinder Block

Illus	Part Number	Description	Quantity	Change Point	Remarks
1	ERR3	CYLINDER BLOCK-ENGINE	1		
2	501593	• Dowel-main bearing cap	10		
3	ERC4996	• Plug-core	7		
4	597586	• Plug-core	1		
5	ETC8442	• Bearing-camshaft front	1		
6	90519055	• Bearing-camshaft rear	3		
7	ETC4529	• Plug-core	1		
8	ETC8074	• Bolt	10		
9	FS108161L	• Screw-alternator bracket to block-flanged head-M8 x 16	1		
10	247127	Plug-cylinder block oil way	2		
11	ERC4644	Dowel-block to bell housing	2		
12	ETC4922	Plug	2		
13	ERR5034	Plug-camshaft blanking-rubber	3		
14	ETC8352	Dowel-ring	2		
15	ERR913	Packing strip	2		

B 4

DISCOVERY 1989MY UPTO 1999MY	GROUP B	Engine 2500 - Diesel - Turbo - 200 Tdi Jet Adaptors

Illus	Part Number	Description	Quantity	Change Point	Remarks
1	ETC6531	Jet assembly-piston cooling-No1-No3	2		
2	ETC6532	Jet assembly-piston cooling-No2-No4	2		
3	ETC5592	Valve-bearing housing jet	4		
4	AFU1879L	Washer-sealing	4		
5	AFU1887L	Washer-sealing	4		
6	ERC9410	Plug-drain	1		
7	AFU1882L	Washer_drain plug	1		
8	ERC8864	Gasket	1		

CABMBA2A

B 5

DISCOVERY 1989MY UPTO 1999MY	GROUP B	Engine 2500 - Diesel - Turbo - 200 Tdi Cover Side

Illus	Part Number	Description	Quantity	Change Point	Remarks
1	ETC7929	COVER PLATE-SIDE	1		
2	TE108041L	• Stud-lift pump fuel to side cover-M8 x 20	2		
3	ERR1475	Gasket-cylinder block side cover	1		
6	ERR506	Breather assembly-crankcase	1		
7	ERR2026	Gasket-cylinder block side cover-asbestos free	1		
		COVER TO BLOCK			
4	FS108257L	Screw-cover to block-flanged head-M8 x 25	9		2500 cc 4 Cylinder Diesel TDi 12L
5	WL108001L	Washer-M8	12		
9	WA108051L	Washer-plain-M8	3		
8	ERC9480	Bracket	1		

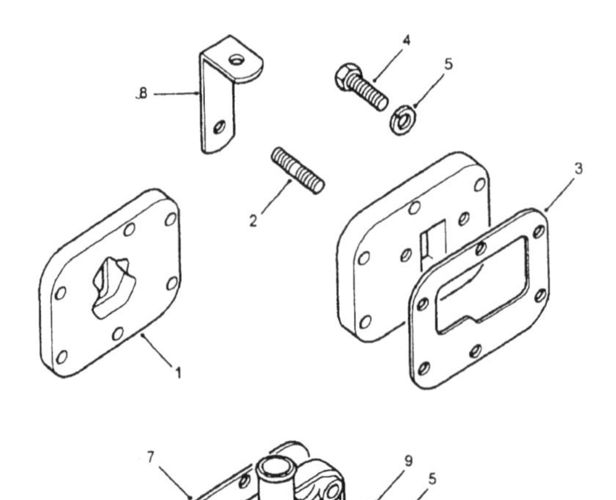

CABMBC1A

B 6

DISCOVERY 1989MY UPTO 1999MY

GROUP B

Engine
2500 - Diesel - Turbo - 200 Tdi
Crankshaft and Bearings

Illus	Part Number	Description	Quantity	Change Point	Remarks
1	ERR1181	Crankshaft-engine	1		
2	ERC4650	Dowel-flywheel to crankshaft	NLA		Use ERR1630.
2	ERR1630	Dowel-flywheel to crankshaft	1		
3	8566L	Bush-crankshaft	1		
4	235770	Key-crankshaft	2		
5	RTC4783	Set-crankshaft bearings-standard	NLA		Use STC3395.
5	STC3395	Set-crankshaft bearings-standard	1		
5	RTC478310	Set-crankshaft bearings-0.010" undersize	1		
6	RTC2825	Washer-crankshaft thrust-standard.-pair	A/R		
6	538131	Washer-crankshaft thrust-oversize.-0.0025"-pair	A/R		
6	538132	Washer-crankshaft thrust-oversize.-0.005"-pair	A/R		
6	538134	Washer-crankshaft thrust-oversize.-0.01"-pair	A/R		
7	ERR751	TORSIONAL VIBRATION DAMPER ASSEMBLY-ENGINE CRANKSHAFT	1		
13	ETC4390	• Flinger-crankshaft pulley	1		
8	ERR1642	Pulley-crankshaft	1		
9	ERR2351	Pulley-crankshaft	1		Except air con
9	ERR2352	Pulley-crankshaft	1		Air con
10	ERR605	Bolt-crankshaft pulley	1		
11	ERR705	Washer-crankshaft thrust	1		
12	FS108207L	Screw-pulley to damper-flanged head-M8 x 20	4		2500 cc 4 Cylinder Diesel TDi 12L

WHEN AIR CON IS FITTED USE ERR751 WITH CORRECT PULLEY

B 7

DISCOVERY 1989MY UPTO 1999MY

GROUP B

Engine
2500 - Diesel - Turbo - 200 Tdi
Piston, Connecting Rod and Bearings

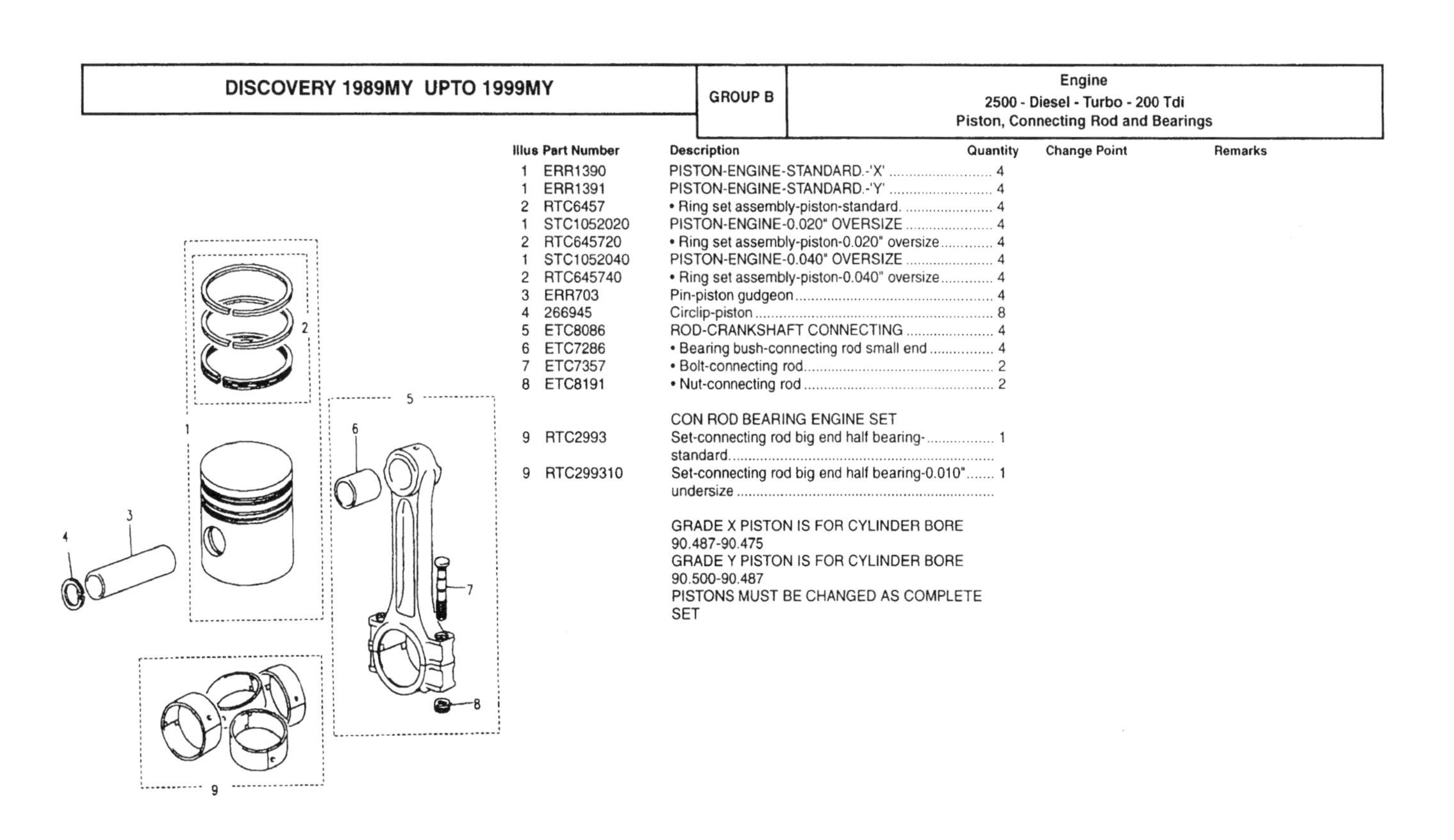

Illus	Part Number	Description	Quantity	Change Point	Remarks
1	ERR1390	PISTON-ENGINE-STANDARD.-'X'	4		
1	ERR1391	PISTON-ENGINE-STANDARD.-'Y'	4		
2	RTC6457	• Ring set assembly-piston-standard.	4		
1	STC1052020	PISTON-ENGINE-0.020" OVERSIZE	4		
2	RTC645720	• Ring set assembly-piston-0.020" oversize	4		
1	STC1052040	PISTON-ENGINE-0.040" OVERSIZE	4		
2	RTC645740	• Ring set assembly-piston-0.040" oversize	4		
3	ERR703	Pin-piston gudgeon	4		
4	266945	Circlip-piston	8		
5	ETC8086	ROD-CRANKSHAFT CONNECTING	4		
6	ETC7286	• Bearing bush-connecting rod small end	4		
7	ETC7357	• Bolt-connecting rod	2		
8	ETC8191	• Nut-connecting rod	2		
		CON ROD BEARING ENGINE SET			
9	RTC2993	Set-connecting rod big end half bearing-standard	1		
9	RTC299310	Set-connecting rod big end half bearing-0.010" undersize	1		

GRADE X PISTON IS FOR CYLINDER BORE 90.487-90.475
GRADE Y PISTON IS FOR CYLINDER BORE 90.500-90.487
PISTONS MUST BE CHANGED AS COMPLETE SET

B 8

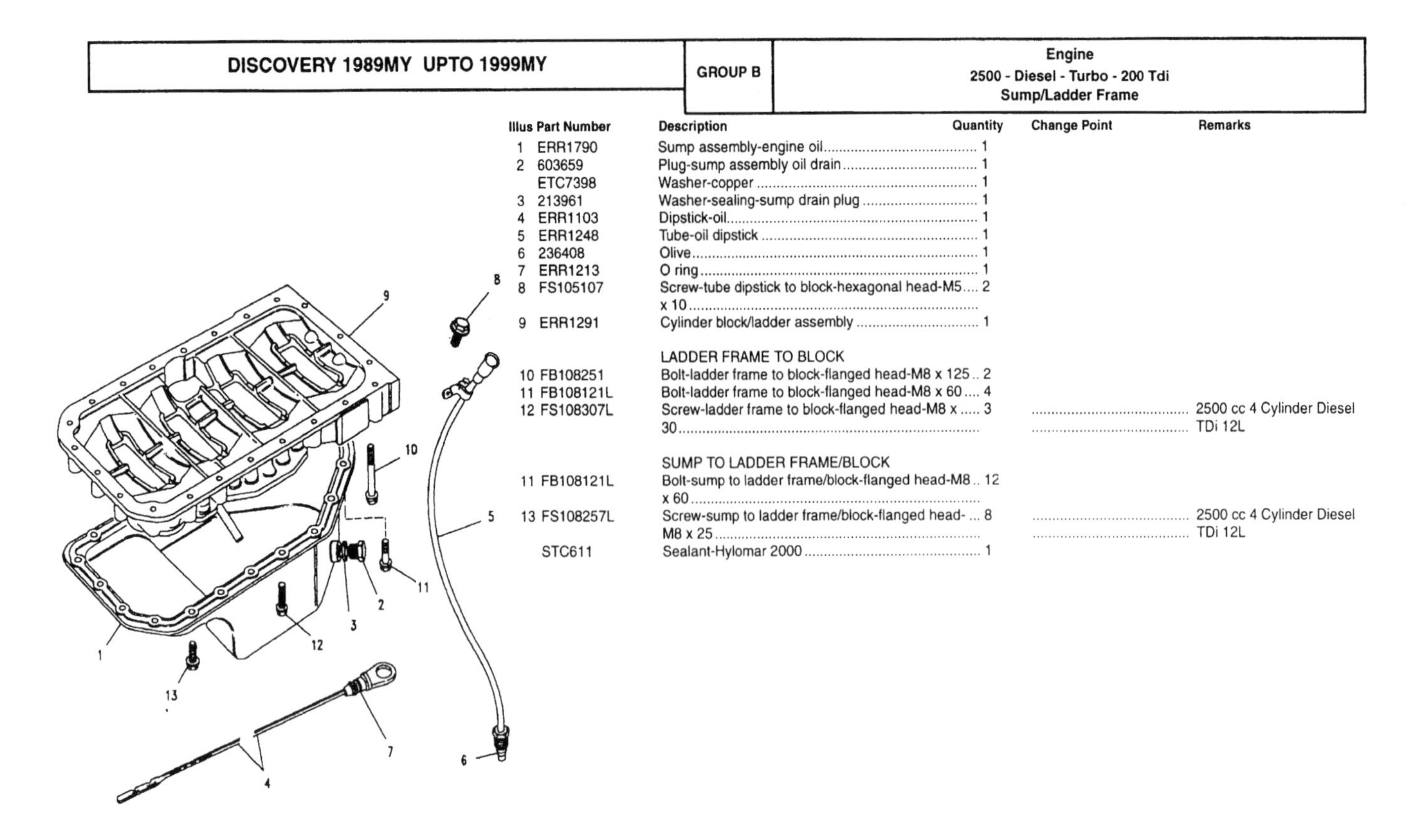

Illus	Part Number	Description	Quantity	Change Point	Remarks
1	ERR1790	Sump assembly-engine oil	1		
2	603659	Plug-sump assembly oil drain	1		
	ETC7398	Washer-copper	1		
3	213961	Washer-sealing-sump drain plug	1		
4	ERR1103	Dipstick-oil	1		
5	ERR1248	Tube-oil dipstick	1		
6	236408	Olive	1		
7	ERR1213	O ring	1		
8	FS105107	Screw-tube dipstick to block-hexagonal head-M5 x 10	2		
9	ERR1291	Cylinder block/ladder assembly	1		
		LADDER FRAME TO BLOCK			
10	FB108251	Bolt-ladder frame to block-flanged head-M8 x 125	2		
11	FB108121L	Bolt-ladder frame to block-flanged head-M8 x 60	4		
12	FS108307L	Screw-ladder frame to block-flanged head-M8 x 30	3		2500 cc 4 Cylinder Diesel TDi 12L
		SUMP TO LADDER FRAME/BLOCK			
11	FB108121L	Bolt-sump to ladder frame/block-flanged head-M8 x 60	12		
13	FS108257L	Screw-sump to ladder frame/block-flanged head-M8 x 25	8		2500 cc 4 Cylinder Diesel TDi 12L
	STC611	Sealant-Hylomar 2000	1		

Illus	Part Number	Description	Quantity	Change Point	Remarks
1	ERR1111	Cover-engine-front	1		
2	6395L	Dowel	1		
3	ETC5064	Seal-camshaft oil	1		
4	ETC5065	Seal-crankshaft front oil	NLA		Use ERR1632.
4	ERR1632	Seal-crankshaft front oil	NLA		Use ERR6490.
4	ERR6490	Seal-crankshaft front oil	1		
5	ERR1607	Gasket-coolant pump body-to front cover-asbestos free	1		
6	538039	Gasket-timing gear housing drain plate	1		
7	FS108257L	Screw-flanged head-M8 x 25	1		2500 cc 4 Cylinder Diesel TDi 12L
8	TD108201	Stud-M8 x 110	1		
9	FB108141L	Bracket- compressor/tensioner assembly air conditioning-cover to block-flanged head-M8 x 70	2		
10	FB108071L	Bolt-flanged head-cover to block-M8 x 35	1		
11	TE108061L	Stud-fuel injection pump-M8 x 30	3		
12	ERR390	Cover-timing belt-front	1		
13	ETC4154	Seal	1		
14	ERR635	Gasket-bracket-pulley support belt tensioner	1		
15	ETC8853	Bracket-pulley support belt tensioner	1		Air con
25	ERR25	Plate-timing belt cover	1		Except air con
16	ERC7295	Plug-rubber	1		
17	ERR1121M	Screw-plate to cover-M8 x 25	3		
18	ERR1122M	Bolt-plate cover/front cover to block-flanged head-M8 x 90	NLA		Use FB108181ML.
18	FB108181ML	Bolt-plate cover/front cover to block-flanged head-M8 x 90	2		
19	FB108161L	Bolt-plate cover/front cover to block-flanged head-M8 x 80	4		
20	SE106101L	Screw-plate cover/front cover to block-M6 x 10	NLA		Use FS106101L.
20	FS106101L	Screw-plate cover/front cover to block-flanged head-M6 x 10	3		
21	ERR14	Pin	1		
22	ERR1195	Gasket-timing gear cover plate	1		
23	ERR1108	Stud-alternator to cover-M8 x 1.25	2		
24	ERR394	Gasket-frt cover	1		

DISCOVERY 1989MY UPTO 1999MY

GROUP B

Engine
2500 - Diesel - Turbo - 200 Tdi
Timing Belt and Tensioner

Illus	Part Number	Description	Quantity	Change Point	Remarks
1	ETC8550	Belt-engine timing	1		
2	ERR2530	Tensioner-timing belt	1		
3	ERR900	Washer	1		
4	FB110141L	Bolt-tensioner to cover-flanged head-M10 x 70	1		
5	ETC8560	Idler-timing belt	1		
6	FN110041L	Nut-flanged head-M10	NLA		Use FN110047.
6	FN110047	Nut-flanged head-M10	1		
7	TD110041L	Stud-idler to cover-M10 x 20	1		

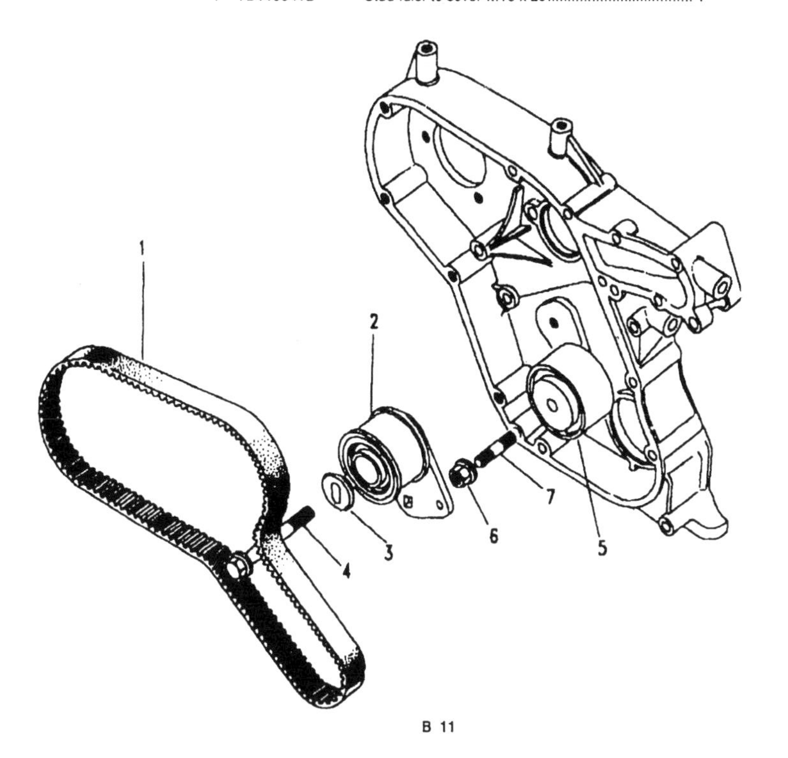

B 11

DISCOVERY 1989MY UPTO 1999MY

GROUP B

Engine
2500 - Diesel - Turbo - 200 Tdi
Coolant Pump

Illus	Part Number	Description	Quantity	Change Point	Remarks
1	RTC6395	PUMP ASSEMBLY-ENGINE COOLANT	1		
2	ERR388	• Gasket-coolant pump body	1		
3	FN108047L	Nut-flanged head-M8	1		2500 cc 4 Cylinder Diesel TDi 12L
4	ERR1120M	Bolt-coolant pump to front cover-M8 x 55	2		
5	ERR1121M	Screw-coolant pump to front cover-M8 x 25	3		
6	ERR1119M	Bolt-coolant pump to front cover-M8 x 110	2		
7	ERR39	Hose-bypass to pump coolant	1	Note (1)	
7	ERR1361	Hose-bypass to pump coolant	1	Note (2)	
8	TD108201	Stud-coolant pump to front cover-M8 x 110	1		
9	CN100408L	Clip-hose	2		

CHANGE POINTS:
(1) To (E)12L 07540A
(2) From (E)12L 07541A

B 12

Illus	Part Number	Description	Quantity	Change Point	Remarks
1	ERR3380	Fan-cooling	1		
2	ETC7238	Coupling-engine fan viscous	1		
3	4589L	Washer-plain	4		
4	ERC5709	Bolt-fan to viscous coupling-M6 x 14	4		
5	ETC5499	Pulley coolant pump	1		
6	ERR810	Belt-vee power assisted steering	1		
7	SH604051L	Screw-pulley to pump-hexagonal head-1/4UNF x 5/8	3		
8	WA106041L	Washer-M6-standard.	3		

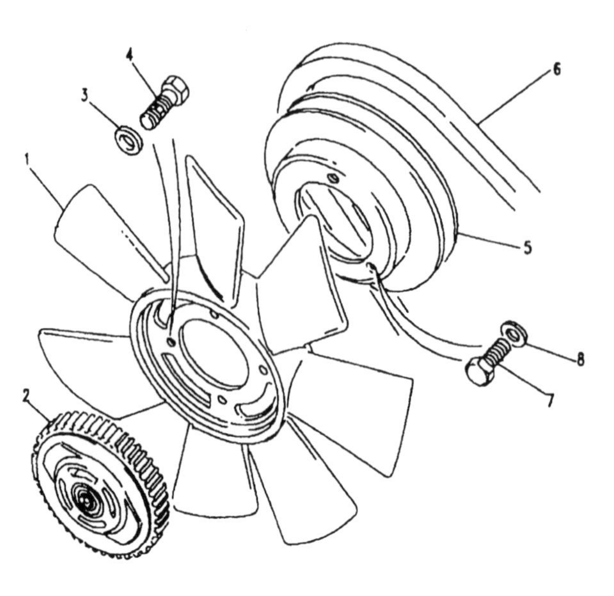

Illus	Part Number	Description	Quantity	Change Point	Remarks
1	ERR1178	PUMP ASSEMBLY-ENGINE OIL	1		
2	236257	• Dowel	2		
3	ERR1063	• Spindle-oil pump	1		
4	ERR1088	• Gear-oil pump idler-10 teeth	1		
5	ERC9706	• Gear-oil pump-10 teeth	1		
6	FS108201L	• Screw-spindle housing to gear housing-flanged head-M8 x 20	NLA		Use FS108207L.
7	WL108001L	• Washer-M8	4		
8	ETC4880	• Plunger-oil pump relief valve	1		
9	564456	• Spring-oil pump relief valve	1		
10	564455	• Plug	1		
11	232044	• Washer-sealing	1		
12	ERR1521	• Strainer and pipe assembly-oil	1		
13	ERR541	• Bracket	1		
14	WA108051L	• Washer-plain-M8	1		
15	WL108001L	• Washer-M8	1		
16	FS108201L	• Screw-bracket to strainer-flanged head-M8 x 20	NLA		Use FS108207L.
17	244488	• O ring	1		
18	244487	• Washer-lock	1		
19	FS108251L	• Screw-pump to block-flanged head-M8 x 25	NLA		Use FS108257L.
20	247665	• Washer-lock	2		
21	ERR850	Shaft-oil pump	1		
22	ERR3606	Gasket-cylinder block oil pump-asbestos free	1		
23	ERR928	DRIVE ASSEMBLY-OIL PUMP	1		
24	ERR309	• Pin-roll-spring tension	1		
25	ERR528	• Bearing-ball	1		
26	CR120305	• Ring-snap-outer	1		
27	ERR530	• Ring-snap-inner	1		
28	ERR531	• O ring	1		
29	ERR532	• O ring	1		
30	ERR500	• Bush	1		
31	ERR848	• Screw	1		

Filter and Adaptor

Illus	Part Number	Description	Quantity	Change Point	Remarks
1	ERR1299	ADAPTOR ASSEMBLY-OIL FILTER	1	Note (1)	
1	ERR2711	ADAPTOR ASSEMBLY-OIL FILTER	1	Note (2)	
2	ERC5923	• Thermostat assembly-oil	1		
3	ETC4022	• Adaptor-oil filter	1		Part of ERR1299.
4	ERC5913	• O ring	1		
5	ETC4021	• Plug	1		
6	SH106141L	• Screw-hexagonal head-M6 x 14	2		
7	WA106041L	• Washer-M6-standard	2		
3	ERR2623	• Adaptor-oil filter	1		Part of ERR2711.
	ERR2241	• Waxstat adaptor	1		Part of ERR2711.
8	ETC7398	Washer-copper	2		
9	ERR3340	Cartridge-engine oil filter	1		
10	FS110301L	Screw-flanged head-M10 x 30	2		
11	WL110001L	Washer-M10	2		
12	ERR3607	Gasket-cylinder block to oil filter adaptor-asbestos free	1		
13	PRC6387	Switch-oil pressure engine	1		
14	NRC8618	Coupling	2		

CHANGE POINTS:
(1) To (E)12L 51182A
(2) From (E)12L 51183A

Cylinder Head

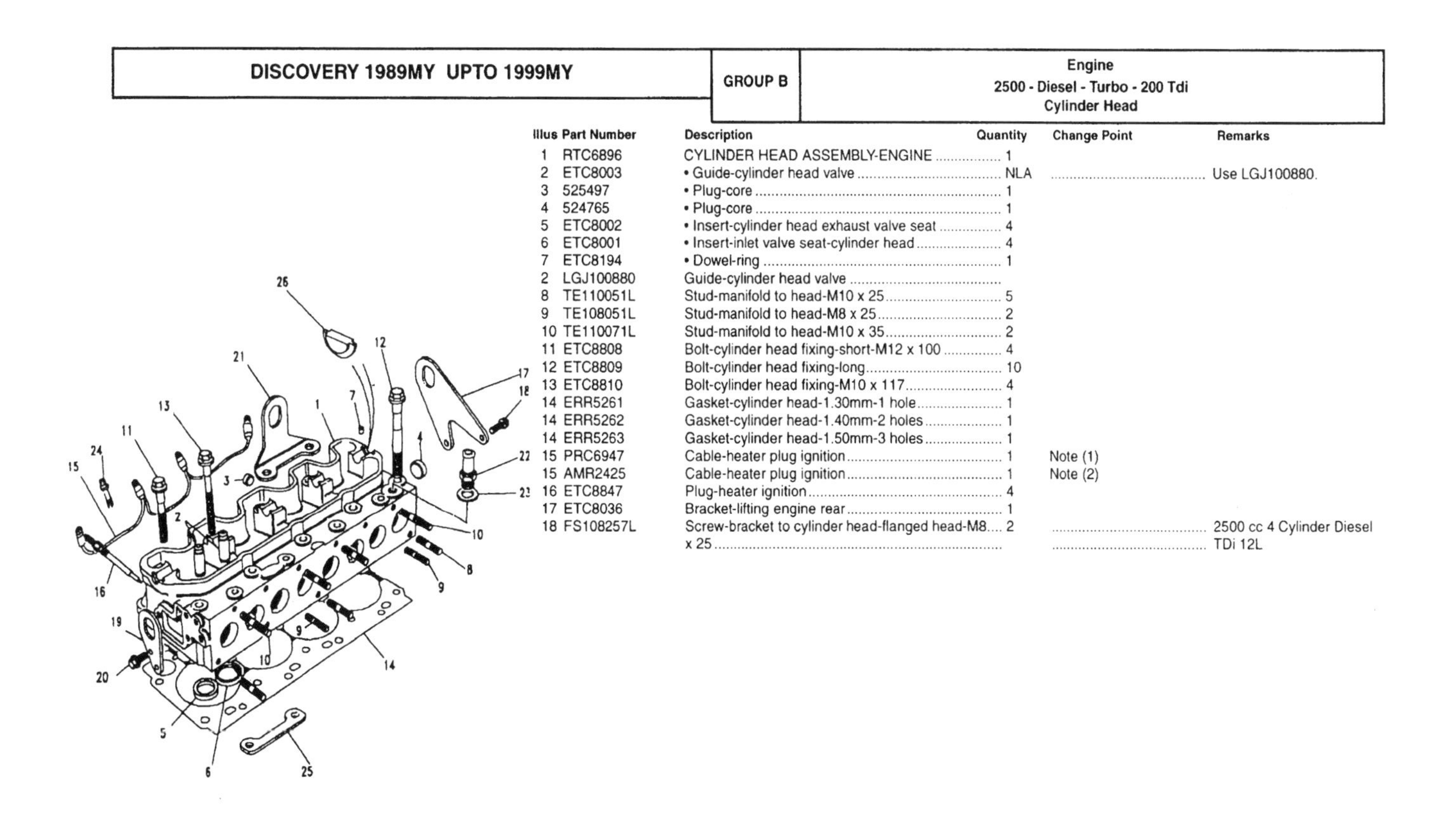

Illus	Part Number	Description	Quantity	Change Point	Remarks
1	RTC6896	CYLINDER HEAD ASSEMBLY-ENGINE	1		
2	ETC8003	• Guide-cylinder head valve	NLA		Use LGJ100880.
3	525497	• Plug-core	1		
4	524765	• Plug-core	1		
5	ETC8002	• Insert-cylinder head exhaust valve seat	4		
6	ETC8001	• Insert-inlet valve seat-cylinder head	4		
7	ETC8194	• Dowel-ring	1		
2	LGJ100880	Guide-cylinder head valve			
8	TE110051L	Stud-manifold to head-M10 x 25	5		
9	TE108051L	Stud-manifold to head-M8 x 25	2		
10	TE110071L	Stud-manifold to head-M10 x 35	2		
11	ETC8808	Bolt-cylinder head fixing-short-M12 x 100	4		
12	ETC8809	Bolt-cylinder head fixing-long	10		
13	ETC8810	Bolt-cylinder head fixing-M10 x 117	4		
14	ERR5261	Gasket-cylinder head-1.30mm-1 hole	1		
14	ERR5262	Gasket-cylinder head-1.40mm-2 holes	1		
14	ERR5263	Gasket-cylinder head-1.50mm-3 holes	1		
15	PRC6947	Cable-heater plug ignition	1	Note (1)	
15	AMR2425	Cable-heater plug ignition	1	Note (2)	
16	ETC8847	Plug-heater ignition	4		
17	ETC8036	Bracket-lifting engine rear	1		
18	FS108257L	Screw-bracket to cylinder head-flanged head-M8 x 25	2		2500 cc 4 Cylinder Diesel TDi 12L

CHANGE POINTS:
(1) To (E)12L 65883A
(2) From (E)12L 65884A

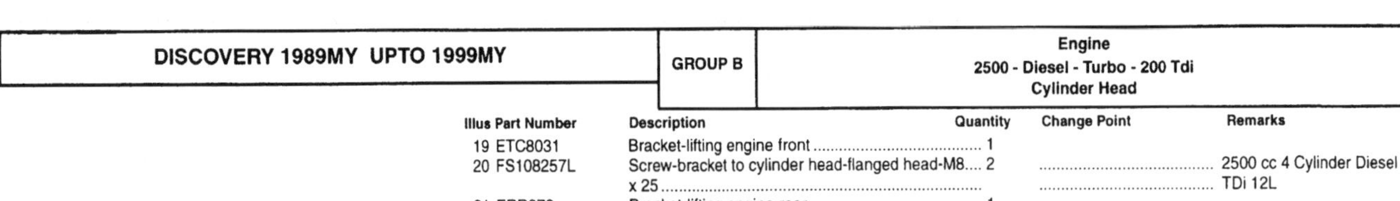

DISCOVERY 1989MY UPTO 1999MY	GROUP B	Engine 2500 - Diesel - Turbo - 200 Tdi Cylinder Head

Illus	Part Number	Description	Quantity	Change Point	Remarks
19	ETC8031	Bracket-lifting engine front	1		
20	FS108257L	Screw-bracket to cylinder head-flanged head-M8 x 25	2		2500 cc 4 Cylinder Diesel TDi 12L
21	ERR978	Bracket-lifting engine rear	1		
22	624091	Adaptor-pipe coolant	1		
23	243959	Washer-sealing	1		
24	ERR1019	Stud	4		
25	ERR1536	Spacer	2		
26	ERR765	Plug-cylinder block oil way-rubber	2		

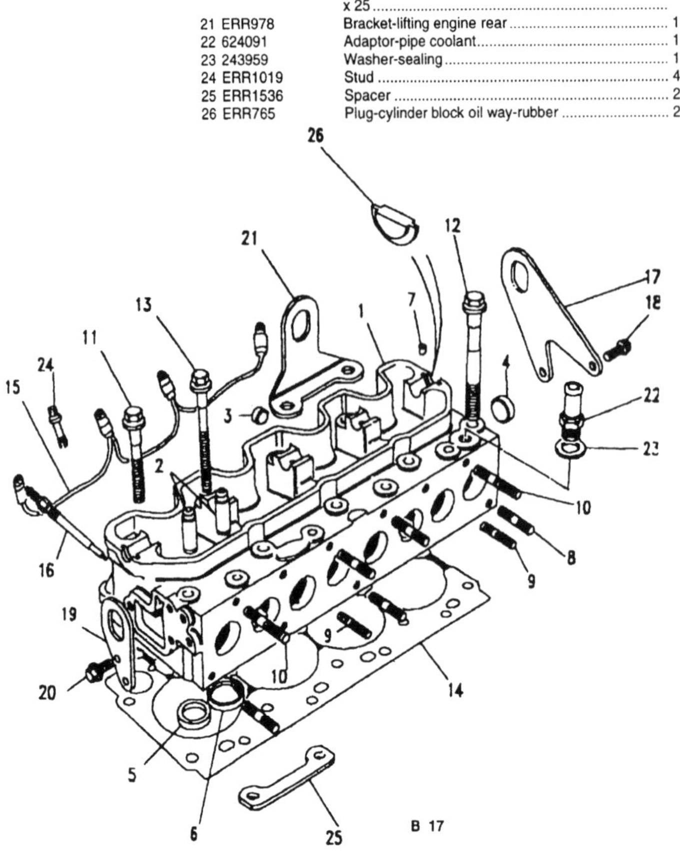

B 17

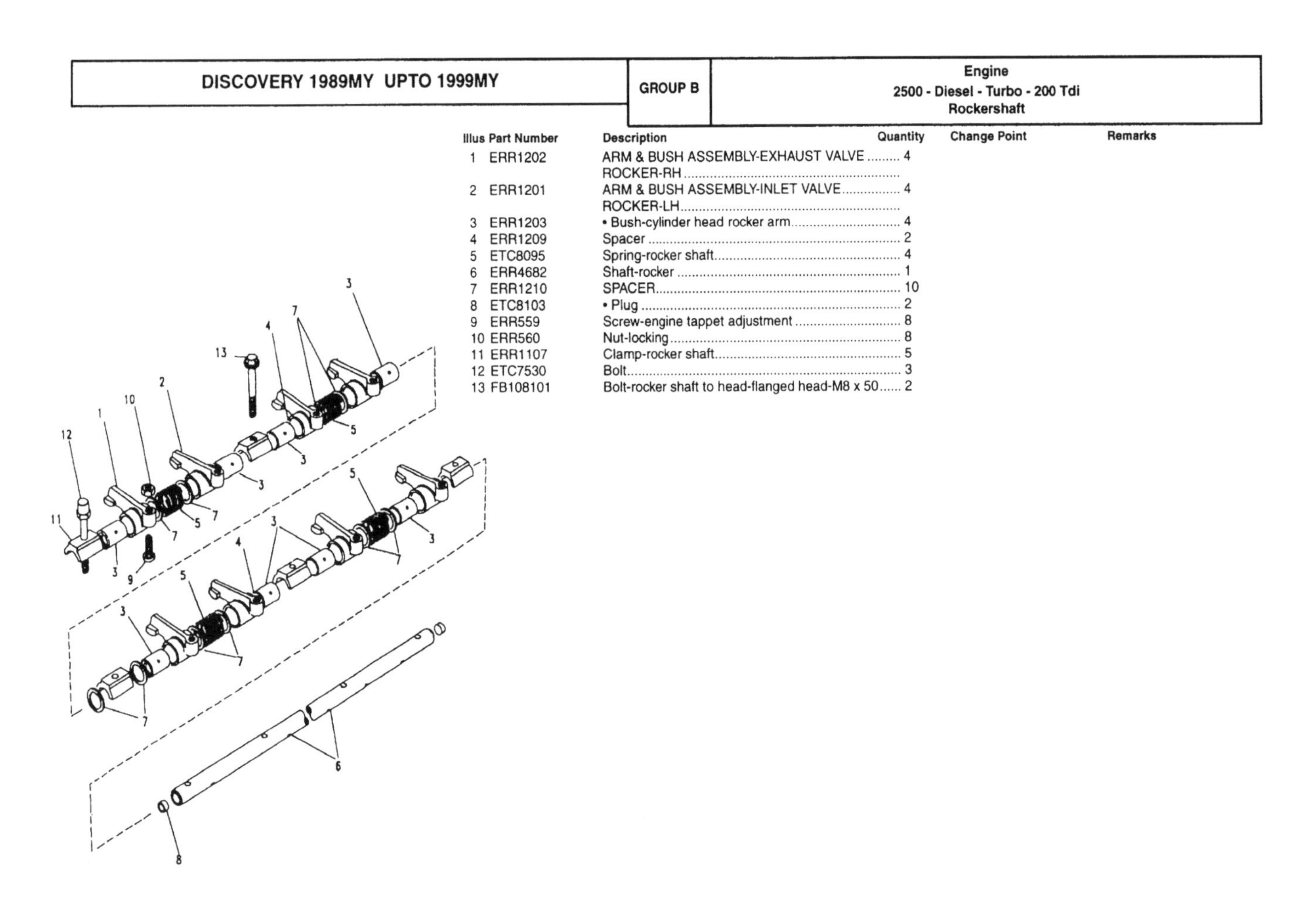

DISCOVERY 1989MY UPTO 1999MY	GROUP B	Engine 2500 - Diesel - Turbo - 200 Tdi Rockershaft

Illus	Part Number	Description	Quantity	Change Point	Remarks
1	ERR1202	ARM & BUSH ASSEMBLY-EXHAUST VALVE ROCKER-RH	4		
2	ERR1201	ARM & BUSH ASSEMBLY-INLET VALVE ROCKER-LH	4		
3	ERR1203	• Bush-cylinder head rocker arm	4		
4	ERR1209	Spacer	2		
5	ETC8095	Spring-rocker shaft	4		
6	ERR4682	Shaft-rocker	1		
7	ERR1210	SPACER	10		
8	ETC8103	• Plug	2		
9	ERR559	Screw-engine tappet adjustment	8		
10	ERR560	Nut-locking	8		
11	ERR1107	Clamp-rocker shaft	5		
12	ETC7530	Bolt	3		
13	FB108101	Bolt-rocker shaft to head-flanged head-M8 x 50	2		

B 18

Valves and Tappets

Illus	Part Number	Description	Quantity	Change Point	Remarks
1	RTC6564	TAPPET-ENGINE VALVE HYDRAULIC	8		
2	502473	• Slide-tappet guide	8		
3	ETC4246	• Bolt	8		
5	ERR561	• Roller-tappet guide	8		
6	ERR607	• Tappet-engine valve mechanical	8		
7	ERR1157	Valve-cylinder head inlet	4		
8	ERR1156	Valve-cylinder head exhaust	4		
9	ERR4640	Spring-cylinder head valve	8		
10	ETC4068	Cap-cylinder head valve spring	8		
11	ETC4069	Cotter-valve	16		
12	546799	Rod-engine push	8		
13	ETC8193	Seat-cylinder head valve spring	8		
14	ETC8663	Seal-cylinder head valve stem	8		
	LJC100270	Cap-valve stem	8		2500 cc 4 Cylinder Diesel TDi 12L

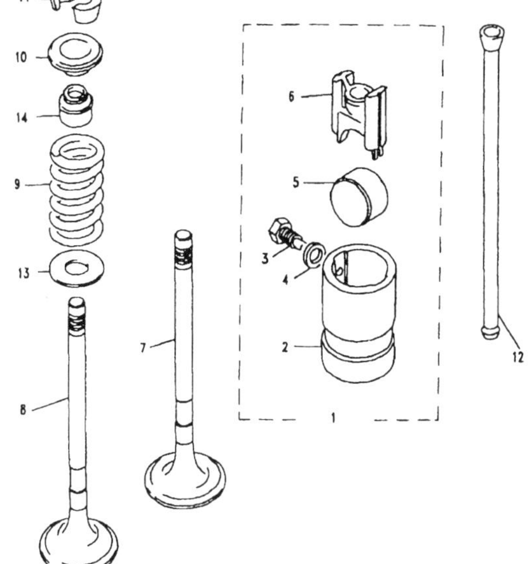

Rocker Cover

Illus	Part Number	Description	Quantity	Change Point	Remarks
1	ERR1530	Cover assembly-valve rocker	NLA	Note (1)	Brthr between the Cyl block and R/C is deleted.Use ERR3368, Replace ERR506 with 541010. Parts NLR ERR874, ERR877, CN100258, 79026.
1	ERR3368	Cover assembly-valve rocker	1	Note (2)	
2	ERR2393	Gasket-engine rocker cover	1		
3	ERR455	Cap-oil filler	1	Note (3)	
3	ERR2529	Cap-oil filler	NLA	Note (4)	Use ERR5041.
3	ERR5041	Cap-oil filler	1		
4	FS106255L	Screw-rocker cover to cylinder head-flanged head-M6 x 25	3		
5	ERR663	Seal	3		
	ERR4551	Cover-engine	1		Automatic

CHANGE POINTS:
(1) To (E)12L 45742A
(2) From (E)12L 45743A
(3) To (E)12L 46743A
(4) From (E)12L 46744A

Engine Breather

Illus	Part Number	Description	Quantity	Change Point	Remarks
1	ERR506	Breather assembly-crankcase	1	Note (1)	
1	541010	Breather assembly-crankcase	1	Note (2)	
2	FB108081L	Bolt-flanged head-side cover to block-M8 x 40	1		
3	FS108257L	Screw-side cover to block-flanged head-M8 x 25	5		2500 cc 4 Cylinder Diesel TDi 12L
4	ERR3605	Gasket-cylinder block side cover-asbestos	NLA		Use ERR2026. 2500 cc 4 Cylinder Diesel TDi 12L
5	ERR874	Baffle-oil separator	1		
6	79026	Screw-No 6 x 5/16	2		
7	ERR877	Hose-crankcase breather flexible	1		
8	CN100258L	Clip-hose-worm drive-25mm	2		
9	ERR509	Separtor assembly-crankcase breather oil	1	Note (3)	
9	ERR1471	Separtor assembly-crankcase breather oil	1	Note (4)	
10	FS106167L	Screw-breather chamber-hexagonal head-M6 x 16	2	Note (3)	
10	FB106121L	Bolt-flanged head-breather chamber-M6 x 60	2	Note (4)	
11	ERR1605	Hose-oil separator drain	1		
12	CN100208L	Clip-hose-20mm	2		
13	ERR875	Hose-crankcase breather flexible	1	Note (3)	
13	ERR1730	Hose-crankcase breather flexible	1	Note (4)	
14	CN100258L	Clip-hose-worm drive-25mm	2		
15	ERR878	Clip-hose	1		
16	ERR879	Spacer	1		
17	SH110201L	Screw-bracket to engine block-hexagonal head-M10 x 20	1		
18	ERR882	Gasket-oil separator	1		

CHANGE POINTS:
(1) To (E)12L 45742A
(2) From (E)12L 45743A
(3) To (E)12L 14074A
(4) From (E)12L 14075A

B 21

Camshaft

Illus	Part Number	Description	Quantity	Change Point	Remarks
1	ETC7128	Camshaft assembly-engine	1		
2	ERC1561	Plate-camshaft retaining	1		
3	FS106167L	Screw-plate thrust camshaft-hexagonal head-M6 x 16	2		
4	WL106001L	Washer-M6	2		
5	ERR666	Pulley-camshaft	1		
6	230313	Key	1		
7	BX110071M	Bolt-pulley to camshaft-hexagonal head-M10 x 35	1		
8	ETC4670	Washer-plain	1		
9	ERC8847	Plate-camshaft pulley retaining	1		
10	ERC8849	O ring	1		
11	ETC4076	O ring	1		

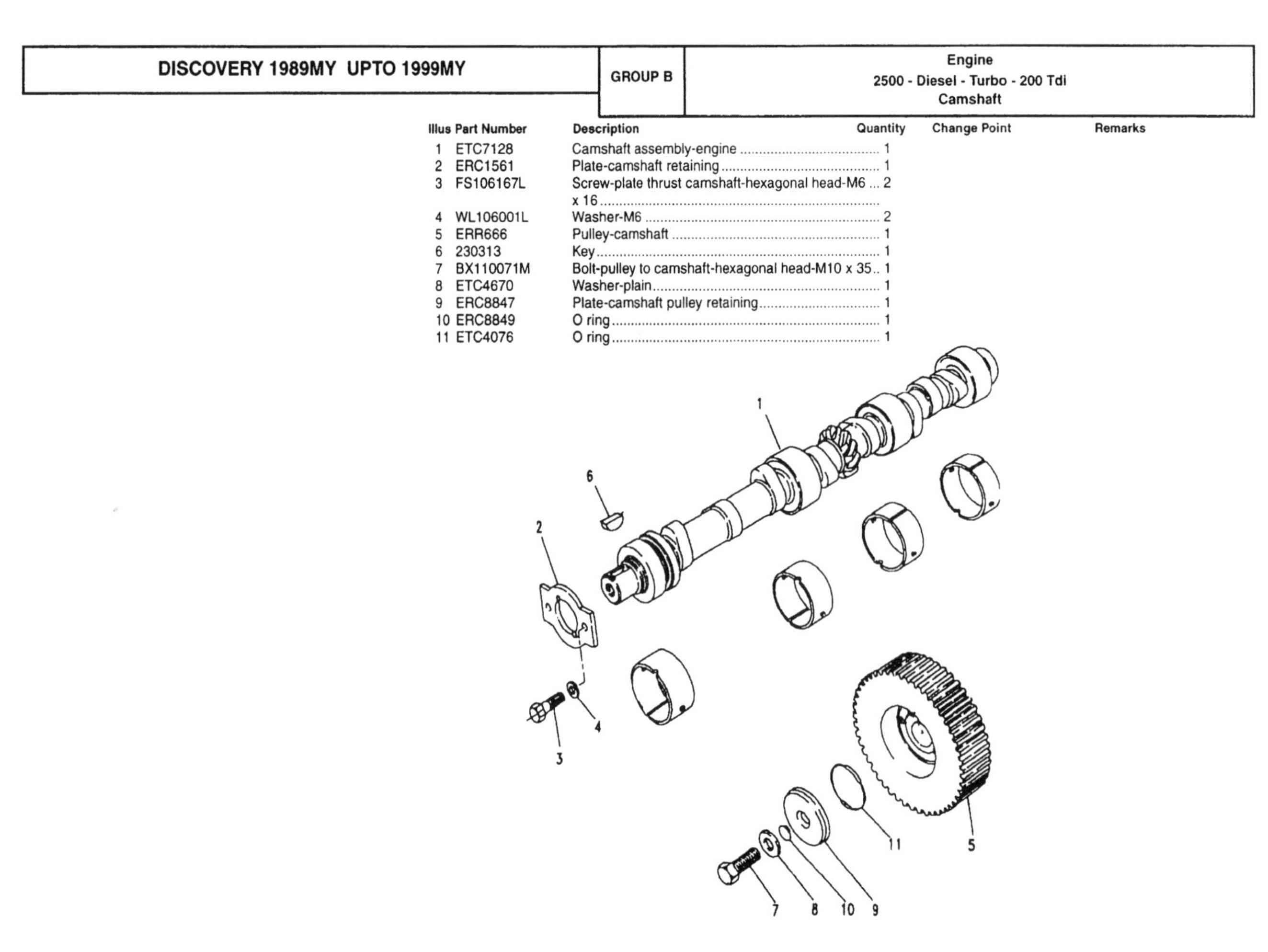

B 22

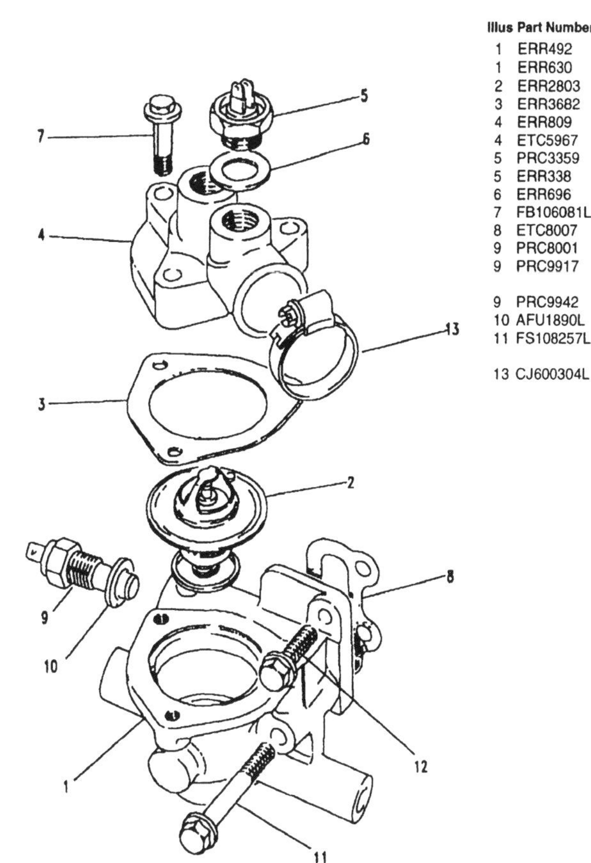

Illus	Part Number	Description	Quantity	Change Point	Remarks
1	ERR492	Housing-thermostat	1		
1	ERR630	Housing-thermostat	1		Switzerland
2	ERR2803	Thermostat-engine-88 degrees C	1		
3	ERR3682	Gasket-thermostat housing-top-asbestos free	1		
4	ERR809	Elbow-engine coolant outlet	1		Air con
4	ETC5967	Elbow-engine coolant outlet	1		Except air con
5	PRC3359	Switch-engine thermostat-Green	1		
5	ERR338	Switch-engine thermostat	1		Switzerland
6	ERR696	Washer-copper-M14	1		
7	FB106081L	Bolt-flanged head-elbow to housing-M6 x 40	3		
8	ETC8007	Gasket-thermostat housing-bottom	1		
9	PRC8001	Sensor-coolant temperature diesel-Orange	1		
9	PRC9917	Sensor-coolant temperature diesel-intercooler-Green	1		
9	PRC9942	Sensor-coolant temperature diesel	1		Gulf States
10	AFU1890L	Washer-sealing	1		
11	FS108257L	Screw-housing thermostat to cyl head-flanged head-M8 x 25	1		
13	CJ600304L	Clip-hose	1		

Illus	Part Number	Description	Quantity	Change Point	Remarks
1	ERR250	MANIFOLD-ENGINE INLET	1		
1	ERR537	MANIFOLD-ENGINE INLET	1		Switzerland
2	524765	• Plug-core	1		
3	ERR1208	Gasket-engine inlet/exhaust manifold	1		
4	FN108041L	Nut-flanged head-M8 x 20	NLA		Use FN108047L.
5	FB108181	Bolt-inlet manifold to cylinder head-M8 x 90	2		
7	TE108051L	Stud-inlet manifold to cylinder head-M8 x 25	2		
6	ERR789	MANIFOLD-ENGINE EXHAUST	1		
6	ERR788	MANIFOLD-ENGINE EXHAUST	1		Switzerland
7	ERR551L	• Stud-exhaust manifold-M8 x 25	4		
8	TE108041L	• Stud-M8 x 20	1		
9	FX110041L	Nut-flange-M10	2		
10	WA110061L	Washer-plain-M10-standard	11		
4	FN108041L	Nut-flanged head-M8 x 20	NLA		Use FN108047L.
12	ETC5577	Plug-drain	1		
13	ERR371	Pipe-heater to hose coolant	1		
14	ERR372	Hose-pipe to heater	1		
15	CN100258L	Clip-hose-worm drive-25mm	2		
9	FX110041L	Nut-flange-M10	5		Automatic
11	FN110041L	Nut-flanged head-M10	NLA		Use FN110047.
11	FN110047	Nut-flanged head-M10	2		
7	TE110051L	Stud-M10 x 25	5		Automatic
8	TE110071L	Stud-M10 x 35	2		Automatic
		MANIFOLD SUPPORT			
16	ERR1004	Bracket-inlet manifold support	1		
17	ERR545	Bracket-inlet manifold support-cylinder block	1		
		SCREW			
18	FS108161L	flanged head-M8 x 16	4		
18	FS108161L	flanged head-M8 x 16	2		Automatic
18	FS108201L	flanged head-M8 x 20	NLA		Use FS108207L. Automatic
18	FS108207L	flanged head-M8 x 20	2		
4	FN108041L	Nut-flanged head-M8 x 20	NLA		Use FN108047L. Automatic
19	WA108051L	Washer-plain-M8	1		
20	WL108001L	Washer-M8	1		

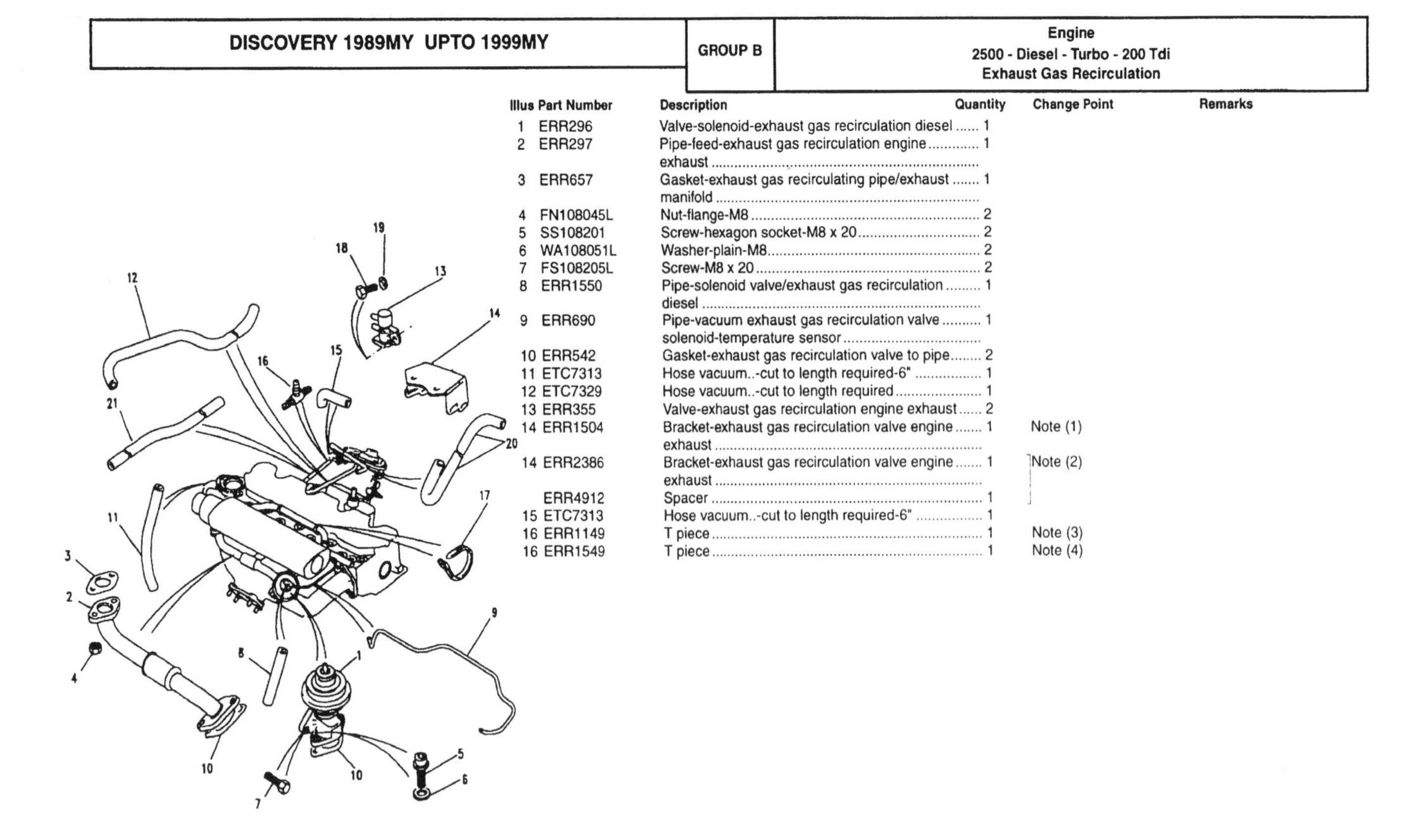

Illus	Part Number	Description	Quantity	Change Point	Remarks
1	ERR296	Valve-solenoid-exhaust gas recirculation diesel	1		
2	ERR297	Pipe-feed-exhaust gas recirculation engine exhaust	1		
3	ERR657	Gasket-exhaust gas recirculating pipe/exhaust manifold	1		
4	FN108045L	Nut-flange-M8	2		
5	SS108201	Screw-hexagon socket-M8 x 20	2		
6	WA108051L	Washer-plain-M8	2		
7	FS108205L	Screw-M8 x 20	2		
8	ERR1550	Pipe-solenoid valve/exhaust gas recirculation diesel	1		
9	ERR690	Pipe-vacuum exhaust gas recirculation valve solenoid-temperature sensor	1		
10	ERR542	Gasket-exhaust gas recirculation valve to pipe	2		
11	ETC7313	Hose vacuum..-cut to length required-6"	1		
12	ETC7329	Hose vacuum..-cut to length required	1		
13	ERR355	Valve-exhaust gas recirculation engine exhaust	2		
14	ERR1504	Bracket-exhaust gas recirculation valve engine exhaust	1	Note (1)	
14	ERR2386	Bracket-exhaust gas recirculation valve engine exhaust	1	Note (2)	
	ERR4912	Spacer	1		
15	ETC7313	Hose vacuum..-cut to length required-6"	1		
16	ERR1149	T piece	1	Note (3)	
16	ERR1549	T piece	1	Note (4)	

CHANGE POINTS:
(1) To (E)12L 29188A
(2) From (E)12L 29189A
(3) To (E)12L 20737A
(4) From (E)12L 20738A

B 25

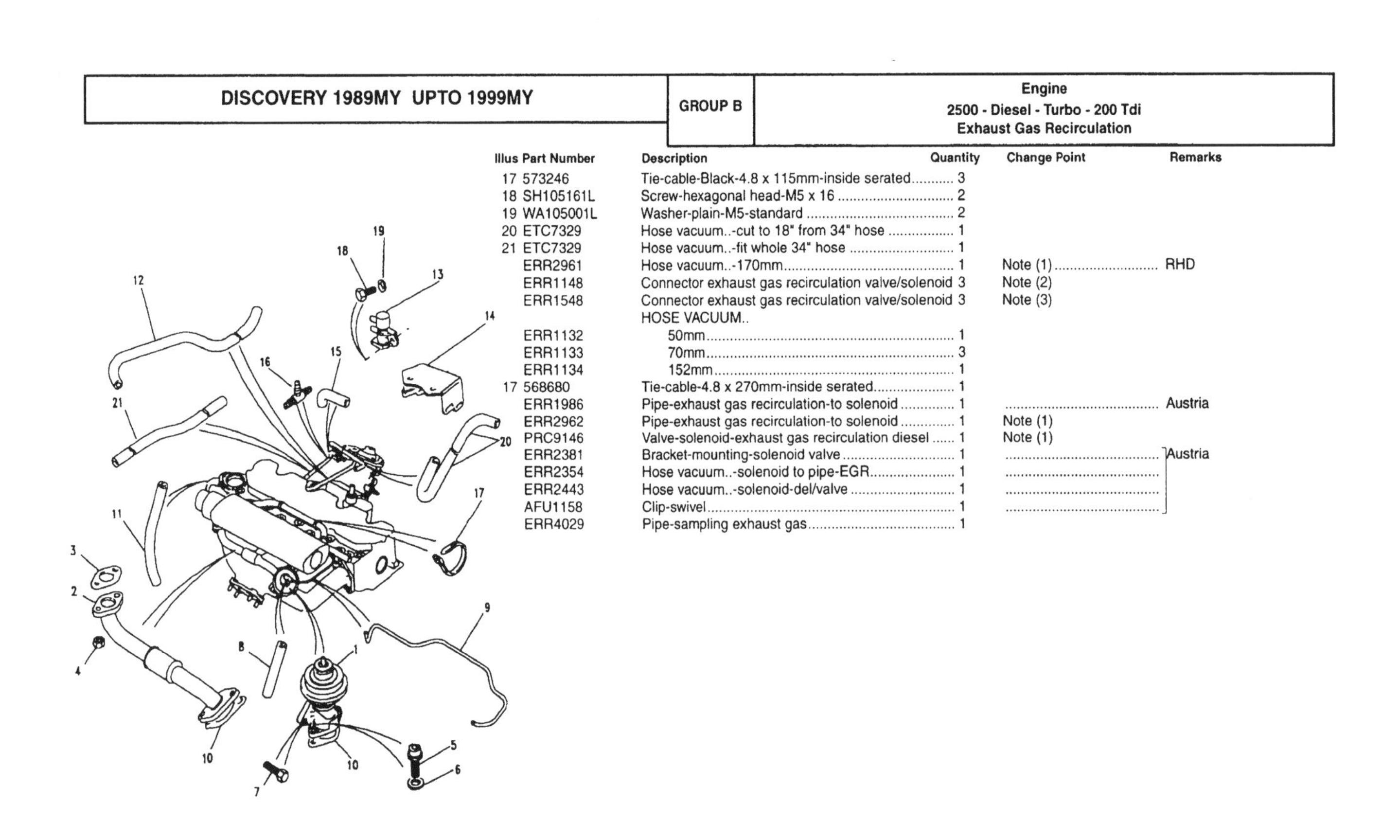

Illus	Part Number	Description	Quantity	Change Point	Remarks
17	573246	Tie-cable-Black-4.8 x 115mm-inside serated	3		
18	SH105161L	Screw-hexagonal head-M5 x 16	2		
19	WA105001L	Washer-plain-M5-standard	2		
20	ETC7329	Hose vacuum..-cut to 18" from 34" hose	1		
21	ETC7329	Hose vacuum..-fit whole 34" hose	1		
	ERR2961	Hose vacuum..-170mm	1	Note (1)	RHD
	ERR1148	Connector exhaust gas recirculation valve/solenoid	3	Note (2)	
	ERR1548	Connector exhaust gas recirculation valve/solenoid	3	Note (3)	
		HOSE VACUUM..			
	ERR1132	50mm	1		
	ERR1133	70mm	3		
	ERR1134	152mm	1		
17	568680	Tie-cable-4.8 x 270mm-inside serated	1		
	ERR1986	Pipe-exhaust gas recirculation-to solenoid	1		Austria
	ERR2962	Pipe-exhaust gas recirculation-to solenoid	1	Note (1)	
	PRC9146	Valve-solenoid-exhaust gas recirculation diesel	1	Note (1)	
	ERR2381	Bracket-mounting-solenoid valve	1		Austria
	ERR2354	Hose vacuum..-solenoid to pipe-EGR	1		
	ERR2443	Hose vacuum..-solenoid-del/valve	1		
	AFU1158	Clip-swivel	1		
	ERR4029	Pipe-sampling exhaust gas	1		

CHANGE POINTS:
(1) From (V) KA 034314
(2) To (E)12L 20737A
(3) From (E)12L 20738A

B 26

DISCOVERY 1989MY UPTO 1999MY	GROUP B	Engine 2500 - Diesel - Turbo - 200 Tdi Fuel Pump

Illus	Part Number	Description	Quantity	Change Point	Remarks
1	ETC7869	Pump-fuel mechanical	NLA		Use STC1190.
	STC1190	Kit-fuel pump repair	1		
2	ERR2028	Gasket-fuel lift pump	2		
3	SS108301L	Screw-lift pump fuel to side cover-M8 x 30	2		
3	FS108257L	Screw-lift pump fuel to side cover-flanged head-M8 x 25	2		Automatic 2500 cc 4 Cylinder Diesel TDi 12L
4	WA108051L	Washer-plain-M8	2		

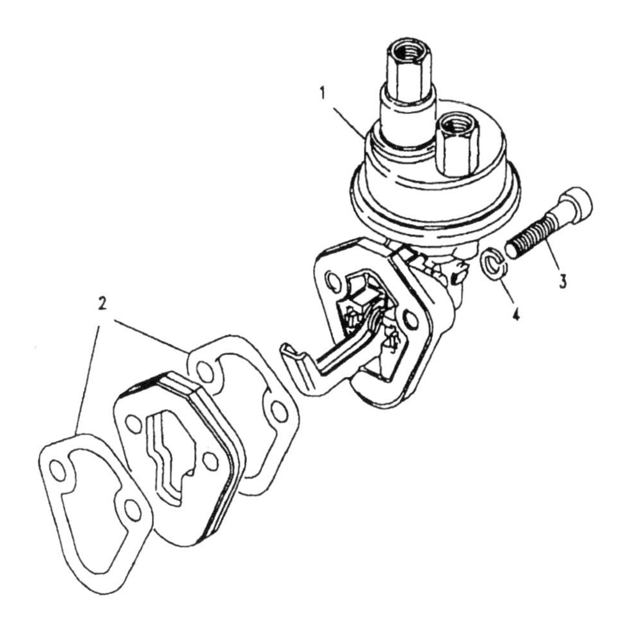

DISCOVERY 1989MY UPTO 1999MY	GROUP B	Engine 2500 - Diesel - Turbo - 200 Tdi Injection Pump

Illus	Part Number	Description	Quantity	Change Point	Remarks
1	ERR459	Pump assembly-fuel injection diesel-new	1		
1	ERR459E	Pump assembly-fuel injection diesel-exchange	1		
1	ERR1985	Pump assembly-fuel injection diesel-new	1		Austria
3	ERR631	Gasket-injection pump-diesel engine	1		
4	FN108041L	Nut-flanged head-M8 x 20	NLA		Use FN108047L.
5	ERR1661	Bracket-injection pump diesel	1		
5	ERR1604	Bracket-injection pump diesel	1		Automatic
6	FN108041L	Nut-flanged head-M8 x 20	NLA		Use FN108047L.
7	FS108207L	Screw-bracket to injection pump-flanged head-M8 x 20	2		2500 cc 4 Cylinder Diesel TDi 12L
8	ERR667	Pulley assembly-pump fuel injection pump drive	1		
9	ERR359	Plate	1		
10	FN108041L	Nut-flanged head-M8 x 20	NLA		Use FN108047L.
11	WA108051L	Washer-plain-M8	6		
12	ERR886	Bolt-banjo-boost capsule	1		
13	273069	Washer-sealing	2		
14	ERR1125	Bolt-banjo-fuel feed	1		
15	RTC6702	Switch-electronic diesel control	1		
16	STC262	Adaptor-pipe-dual male	4		
17	ERR4897	Lever-kickdown automatic transmission	1		Automatic
18	FS106121	Screw-M6 x 12	NLA		Use FB106121L part raised in error on TDB download. Automatic
18	FB106121L	Bolt-flanged head-M6 x 60	1		
19	FS106201L	Screw-flanged head-M6 x 20	1		Automatic

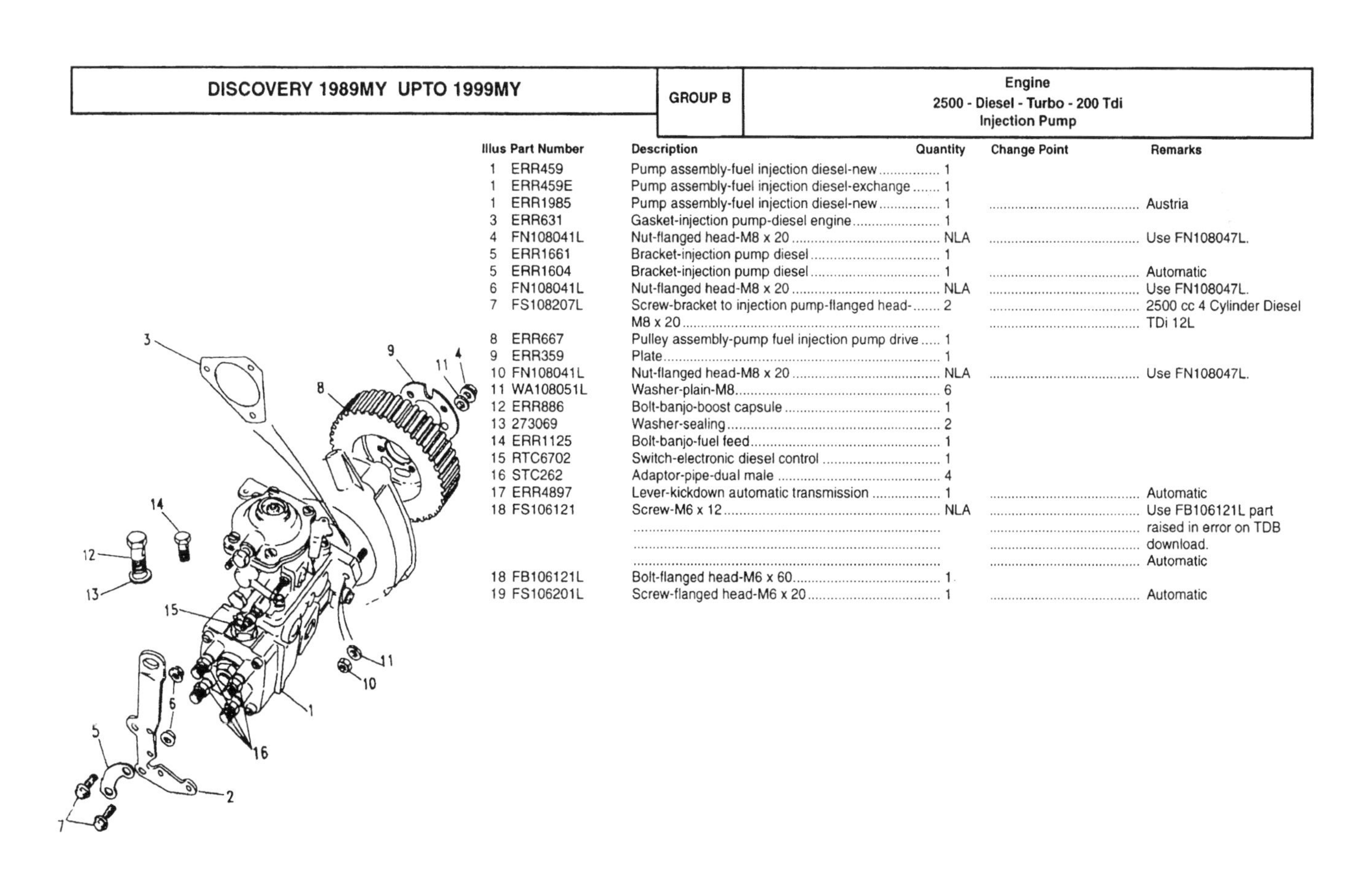

Illus	Part Number	Description	Quantity	Change Point	Remarks
1	ERR3651	Pipe-spill return fuel injection	1	Note (1)	
1	ERR3652	Pipe-spill return fuel injection	1	Note (2)	
2	RTC6732	Pipe-cylinder 1-high pressure fuel injection	1		
3	RTC6733	Pipe-cylinder 2-high pressure fuel injection	1		
4	RTC6734	Pipe-cylinder 3-high pressure fuel injection	1		
5	RTC6735	Pipe-cylinder 4-high pressure fuel injection	1		
6	ETC8412	Injector assembly-diesel holder & nozzle	4		
7	ETC8197	Clamp fuel injector	4		Manual
7	ERR1509	Clamp fuel injector	4		Automatic
8	ERR1200	Bolt-banjo-6mm	NLA		Use STC3297.
8	STC3297	BOLT-BANJO	1		
9	ERR1304	• Washer-sealing			
10	FN108047L	Nut-flanged head-M8	1		
11	AEU2129L	Clamp-2 pipes	4		
12	ETC8470	Dowel-ring	4		
13	ERR4621	Washer-sealing-diesel injector to cylinder head	4		
	NRC9770	Nut-tube-female-spill return-M12	1		
	NRC9771	Olive-spill return	1		

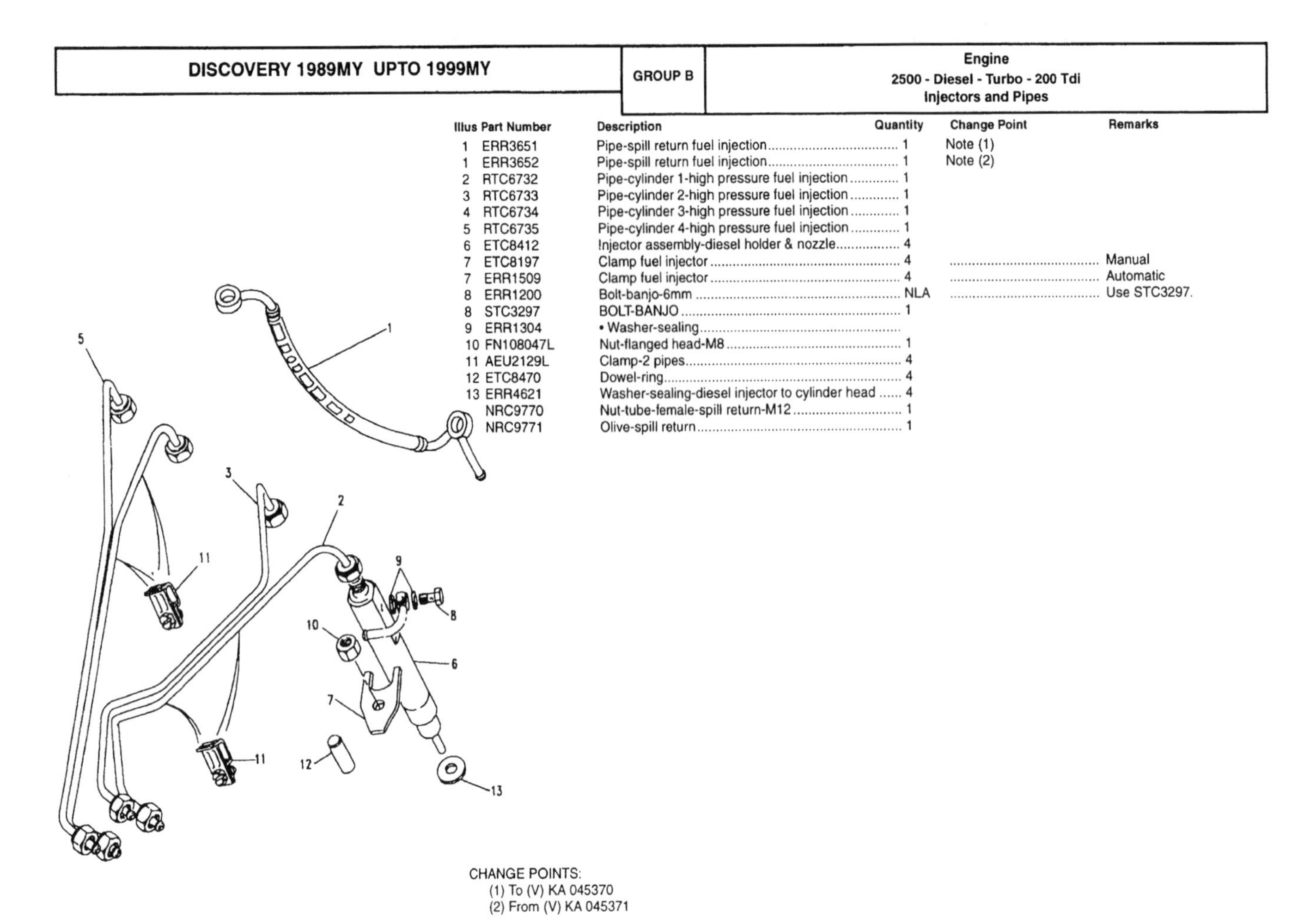

CHANGE POINTS:
(1) To (V) KA 045370
(2) From (V) KA 045371

Illus	Part Number	Description	Quantity	Change Point	Remarks
1	ETC7461	TURBOCHARGER ASSEMBLY-NEW	1		
1	ETC7461E	TURBOCHARGER ASSEMBLY-EXCHANGE	1		
2	RTC6535	• Rotor assembly-turbocharger	1		
3	RTC6536	• O ring	1		
4	RTC6537	• Ring-snap	1		
5	RTC6538	• Actuator-waste gate turbocharger	1		
6	UKC3803L	Clip-hose	2		
7	ERR4029	Pipe-sampling exhaust gas	1		
8	RTC6539	Hose wastegate control-90mm	1		
8	STC1897	Hose wastegate control-98mm-98mm	1		

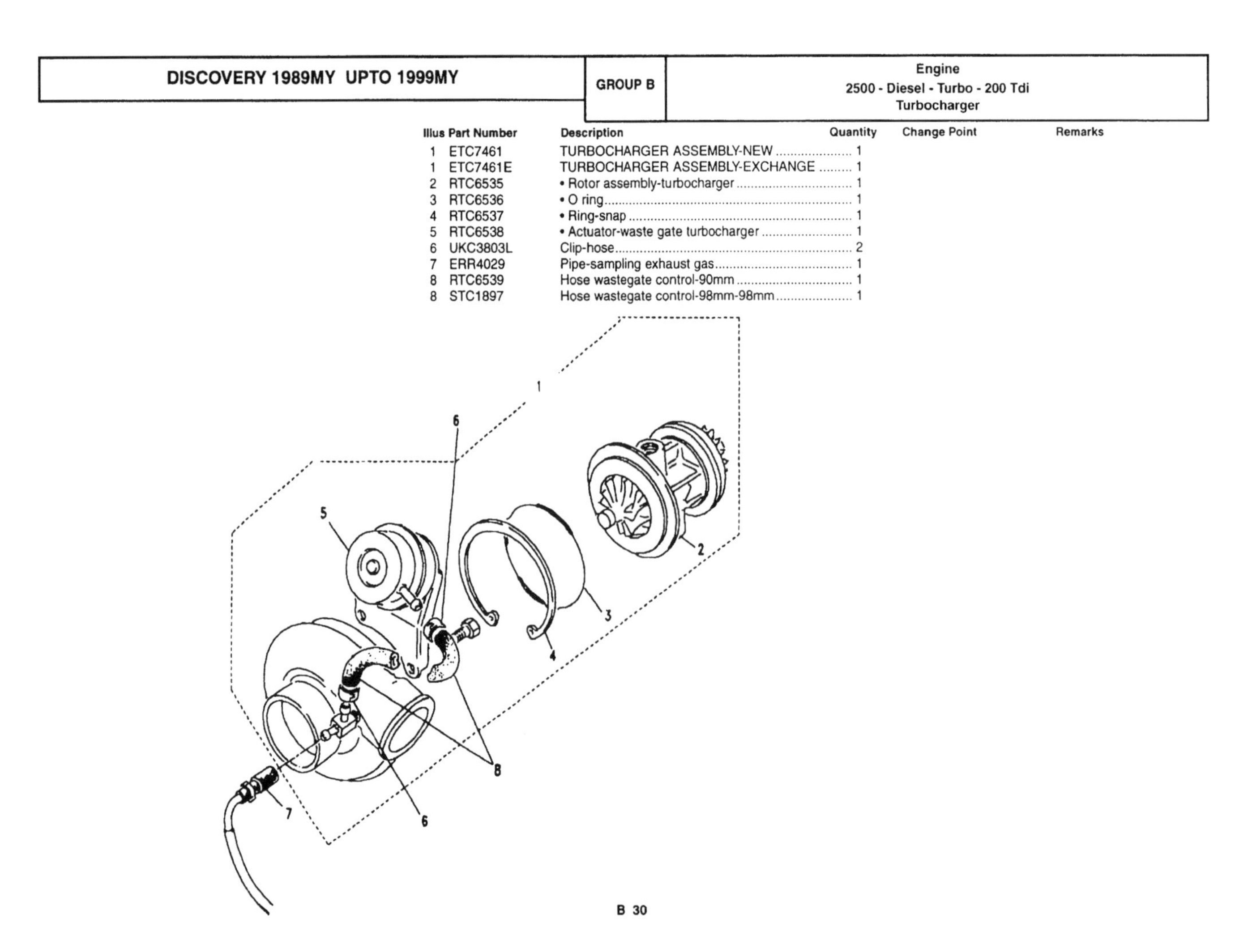

DISCOVERY 1989MY UPTO 1999MY

GROUP B

Engine
2500 - Diesel - Turbo - 200 Tdi
Turbo Charger Adaptors

Illus	Part Number	Description	Quantity	Change Point	Remarks
		OIL FEED			
1	ETC8820	Pipe-adaptor turbocharger	1		
2	ERR816	Pipe turbocharger oil feed	1	Note (1)	
2	ERR3493	Pipe turbocharger oil feed	1	Note (2)	
3	ERR1125	Bolt-banjo	1		
4	FRC4808	Washer-sealing	2		
5	ETC7463	Pipe-oil drain turbocharger	1	Note (1)	
5	ERR3492	Pipe-oil drain turbocharger	1	Note (3)	
6	ERR335	Adaptor-oil drain pipe turbocharger	1		
7	ERR896	Washer-sealing-turbocharger oil feed pipe	1		
		TURBO TO EXHAUST MANIFOLD			
8	ETC7514	Gasket-turbocharger/exhaust manifold turbo	1		
9	ERR551L	Stud-turbocharger to exhaust manifold-M8 x 25	2		
10	WA108051L	Washer-plain-M8	4		
11	ETC7184	Nut-M8	4		Manual
3	ERR1125	Bolt-banjo	1		Automatic
11	ERR4245	Nut-M8	4		Automatic
		TURBO TO DOWNPIPE			
12	ESR3260	Gasket exhaust system	1		
	ETC7513	Gasket exhaust system	2		Automatic
13	ERR544	Downpipe assembly exhaust system	1	Note (4)	
13	ERR1295	Downpipe assembly exhaust system	1	Note (5)	
13	ERR4530	Downpipe assembly exhaust system	1		Automatic
14	FX110041L	Nut-flange-M10	3		
14	FX110041L	Nut-flange-M10	2		Automatic
		HEATSHIELD			
15	ERR1293	Heatshield-exhaust elbow turbocharger	1		
16	STC313	Bolt-u	1		
17	ETC7184	Nut-M8	2		

CHANGE POINTS:
(1) To (E)12L 54592A
(2) From (E)12L 54592A
(3) From (E)12L 54593A
(4) To (E)12L 29384A
(5) From (E)12L 29385A

DISCOVERY 1989MY UPTO 1999MY

GROUP B

Engine
2500 - Diesel - Turbo - 200 Tdi
Gasket Kits

Illus	Part Number	Description	Quantity	Change Point	Remarks
1	STC363	Set-engine gasket-asbestos	NLA		Use STC1557.
1	STC1557	Set-engine gasket-asbestos free	1		
2	STC1172	Set-cylinder head gasket	1		

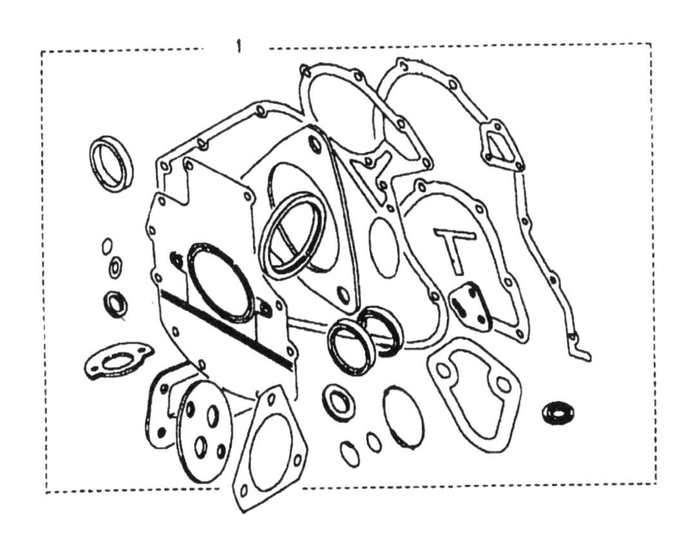

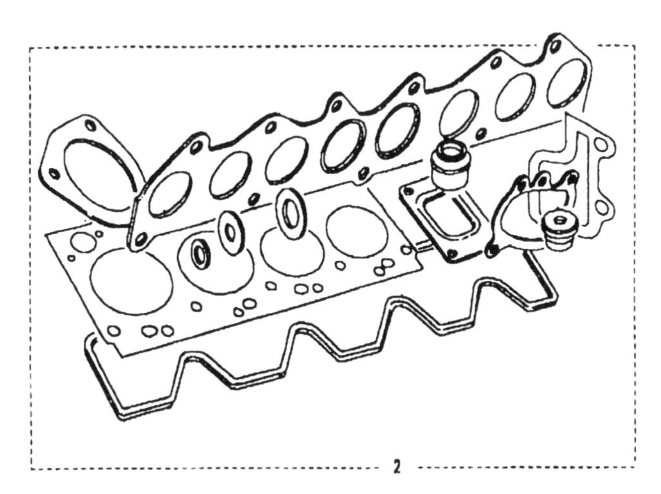

Engine
2500 - Diesel - Turbo - 200 Tdi
Vacuum Pump

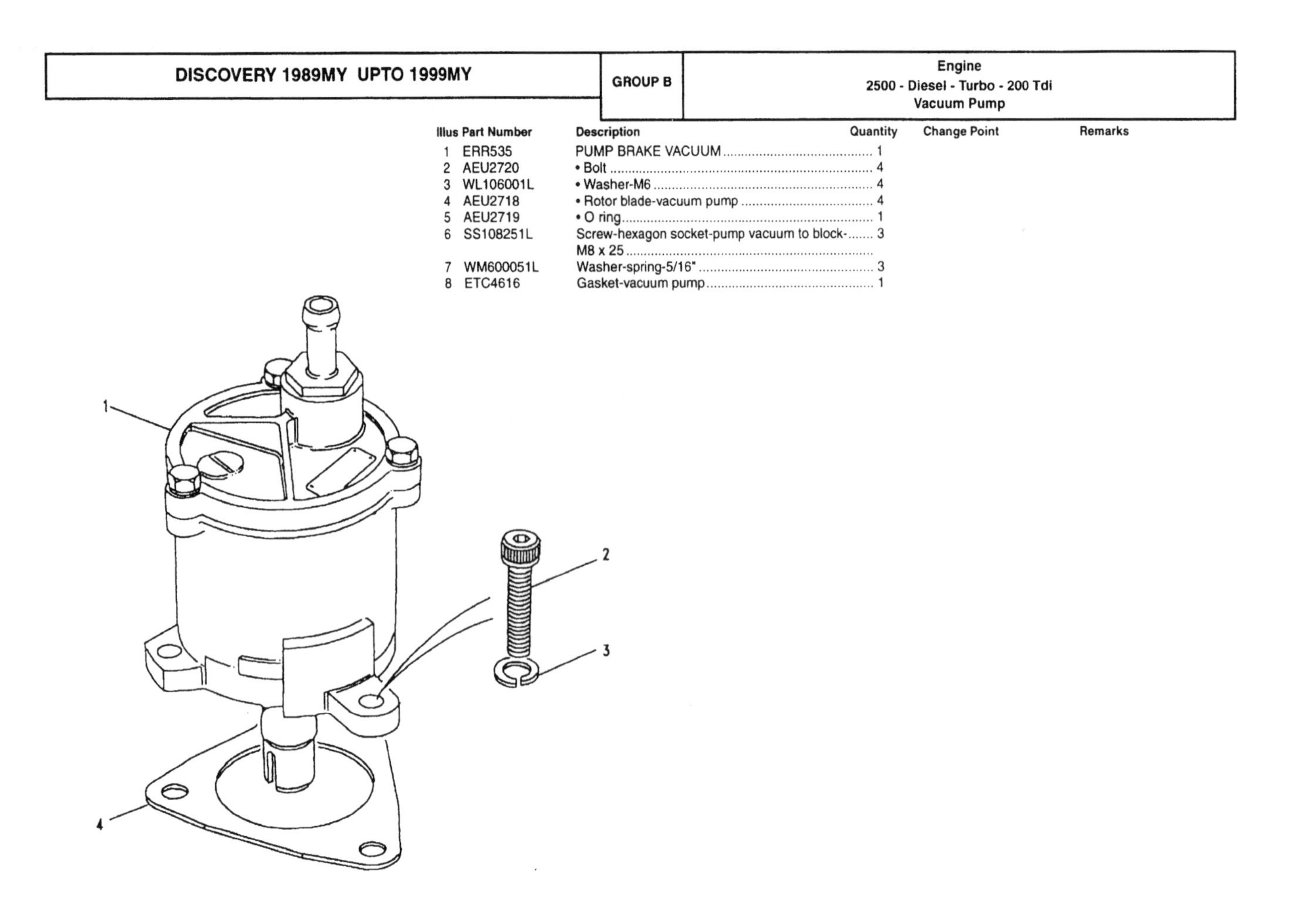

Illus	Part Number	Description	Quantity	Change Point	Remarks
1	ERR535	PUMP BRAKE VACUUM	1		
2	AEU2720	• Bolt	4		
3	WL106001L	• Washer-M6	4		
4	AEU2718	• Rotor blade-vacuum pump	4		
5	AEU2719	• O ring	1		
6	SS108251L	Screw-hexagon socket-pump vacuum to block-M8 x 25	3		
7	WM600051L	Washer-spring-5/16"	3		
8	ETC4616	Gasket-vacuum pump	1		

Engine
2500 - Diesel - Turbo - 200 Tdi
Power Steering Pump

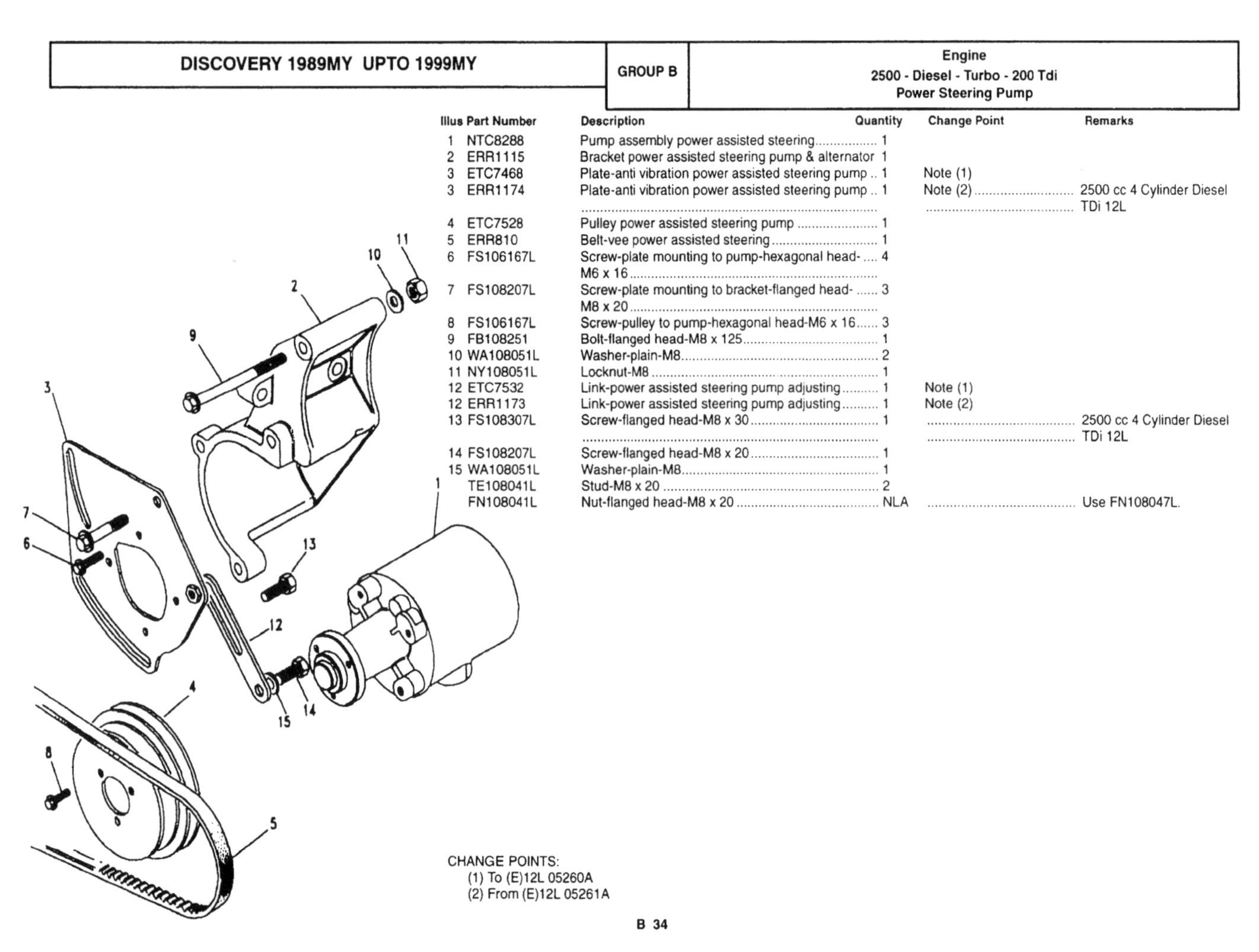

Illus	Part Number	Description	Quantity	Change Point	Remarks
1	NTC8288	Pump assembly power assisted steering	1		
2	ERR1115	Bracket power assisted steering pump & alternator	1		
3	ETC7468	Plate-anti vibration power assisted steering pump	1	Note (1)	
3	ERR1174	Plate-anti vibration power assisted steering pump	1	Note (2)	2500 cc 4 Cylinder Diesel TDi 12L
4	ETC7528	Pulley power assisted steering pump	1		
5	ERR810	Belt-vee power assisted steering	1		
6	FS106167L	Screw-plate mounting to pump-hexagonal head-M6 x 16	4		
7	FS108207L	Screw-plate mounting to bracket-flanged head-M8 x 20	3		
8	FS106167L	Screw-pulley to pump-hexagonal head-M6 x 16	3		
9	FB108251	Bolt-flanged head-M8 x 125	1		
10	WA108051L	Washer-plain-M8	2		
11	NY108051L	Locknut-M8	1		
12	ETC7532	Link-power assisted steering pump adjusting	1	Note (1)	
12	ERR1173	Link-power assisted steering pump adjusting	1	Note (2)	
13	FS108307L	Screw-flanged head-M8 x 30	1		2500 cc 4 Cylinder Diesel TDi 12L
14	FS108207L	Screw-flanged head-M8 x 20	1		
15	WA108051L	Washer-plain-M8	1		
	TE108041L	Stud-M8 x 20	2		
	FN108041L	Nut-flanged head-M8 x 20	NLA		Use FN108047L.

CHANGE POINTS:
(1) To (E)12L 05260A
(2) From (E)12L 05261A

DISCOVERY 1989MY UPTO 1999MY	GROUP B	Engine 2500 - Diesel - Turbo - 200 Tdi Air Conditioning Compressor

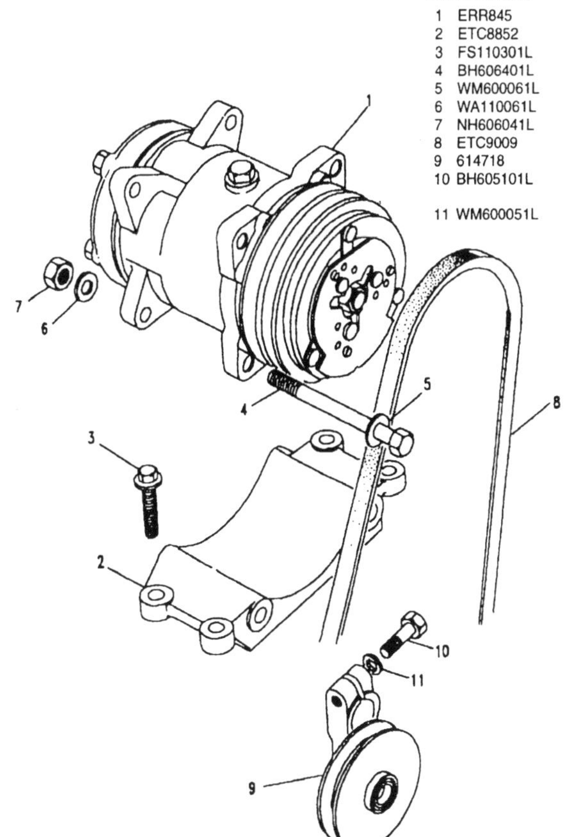

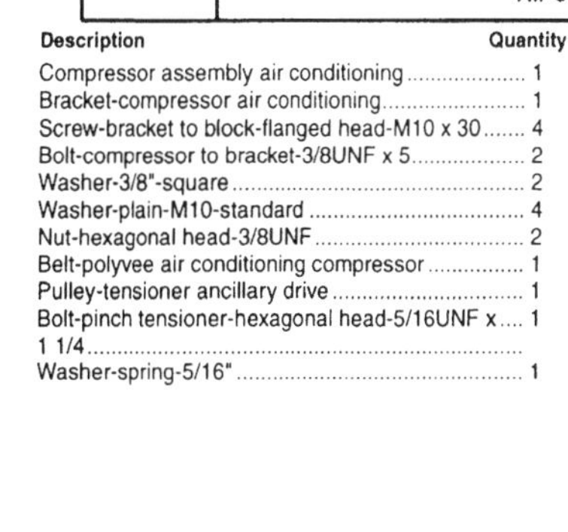

Illus	Part Number	Description	Quantity	Change Point	Remarks
1	ERR845	Compressor assembly air conditioning	1		
2	ETC8852	Bracket-compressor air conditioning	1		
3	FS110301L	Screw-bracket to block-flanged head-M10 x 30	4		
4	BH606401L	Bolt-compressor to bracket-3/8UNF x 5	2		
5	WM600061L	Washer-3/8"-square	2		
6	WA110061L	Washer-plain-M10-standard	4		
7	NH606041L	Nut-hexagonal head-3/8UNF	2		
8	ETC9009	Belt-polyvee air conditioning compressor	1		
9	614718	Pulley-tensioner ancillary drive	1		
10	BH605101L	Bolt-pinch tensioner-hexagonal head-5/16UNF x 1 1/4	1		
11	WM600051L	Washer-spring-5/16"	1		

B 35

DISCOVERY 1989MY UPTO 1999MY	GROUP B	Engine 2500 - Diesel - Turbo - 200 Tdi Alternator

Illus	Part Number	Description	Quantity	Change Point	Remarks
1	RTC5681N	ALTERNATOR ASSEMBLY-127/65 AMP-NEW-LH	1		
1	RTC5681E	ALTERNATOR ASSEMBLY-127/65 AMP-EXCHANGE-LH	1		
2	RTC5670	• Regulator & brush box assembly alternator	1		
3	RTC5671	• Rectifier-alternator	1		
4	RTC5926	• Bearing-alternator front	1		
5	RTC5688	• BRACKET-DRIVE END C/W BEARING ALTERNATOR E-LH	1		
6	RTC5925	• • Kit-fixings alternator/vacuum pump	1		
10	RTC6114	• • Capacitor alternator	1		
7	AAU2249L	Fan alternator-133/65 amp	1		
8	RTC5847	Pulley-alternator	1		
9	ETC7469	Belt-polyvee alternator	1		
	RTC5689	Spacer	1		
		ALTERNATOR HEATSHEILD			
11	ETC7519	Heatshield alternator	1	Note (1)	
11	UAM7498L	Heatshield alternator	1	Note (2)	
	NH105041L	Nut-hexagonal-M5	1	Note (2)	

ALTERNATOR IS MOUNTED
ON POWER STEERING PUMP BRACKET

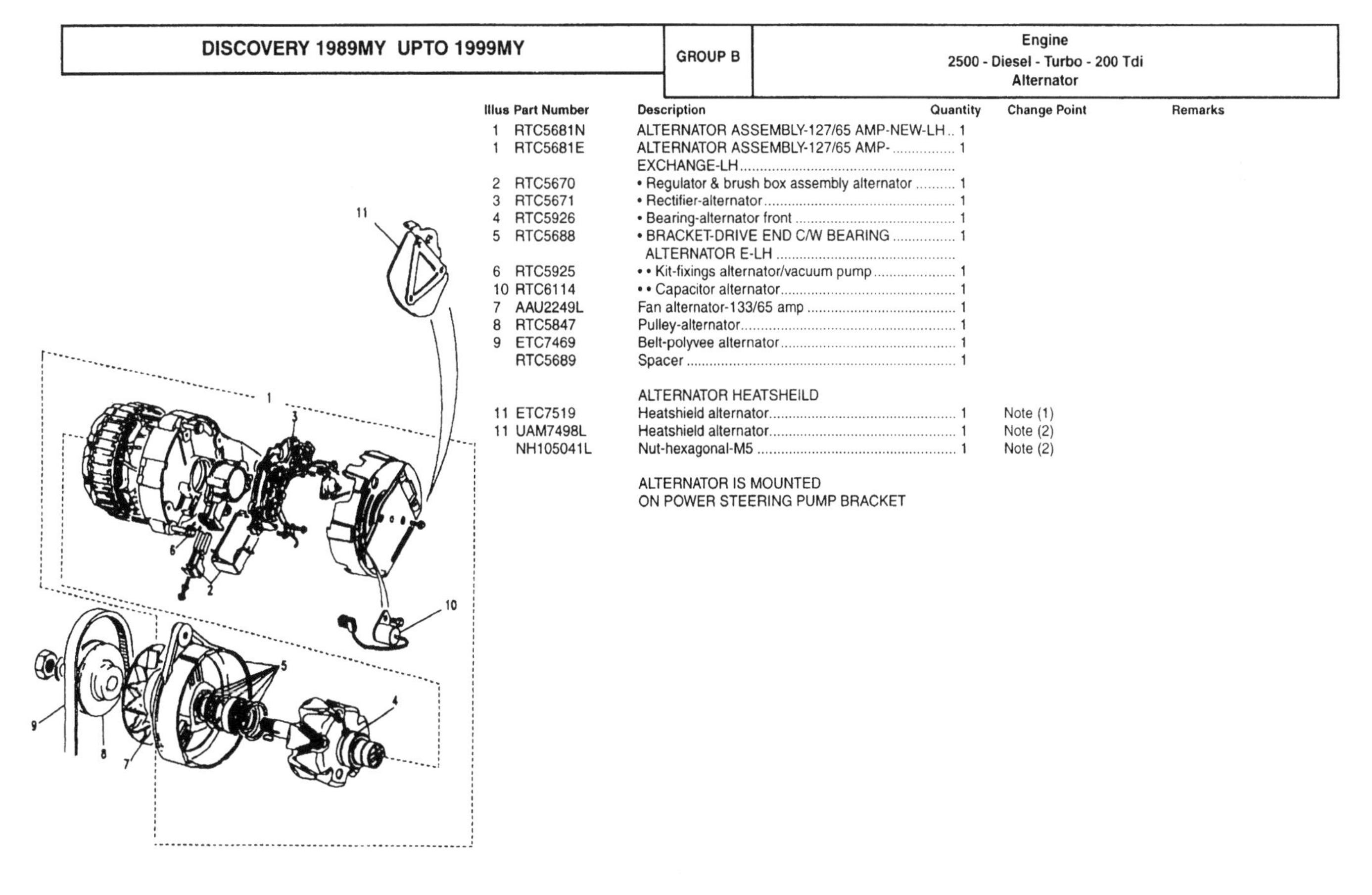

CHANGE POINTS:
(1) To (E)12L 36903A
(2) From (E)12L 36904A

B 36

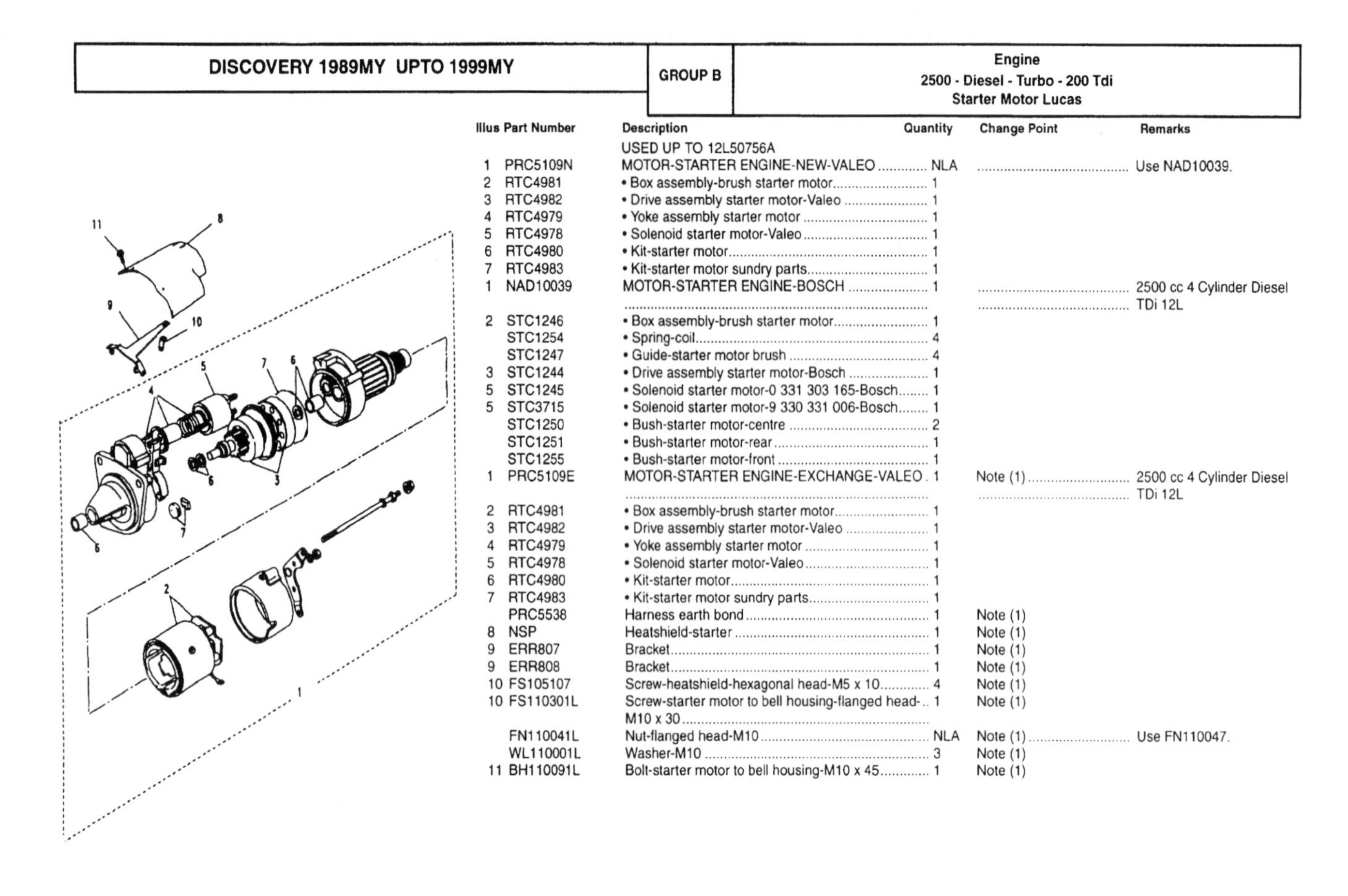

Illus	Part Number	Description	Quantity	Change Point	Remarks
		USED UP TO 12L50756A			
1	PRC5109N	MOTOR-STARTER ENGINE-NEW-VALEO	NLA		Use NAD10039.
2	RTC4981	• Box assembly-brush starter motor	1		
3	RTC4982	• Drive assembly starter motor-Valeo	1		
4	RTC4979	• Yoke assembly starter motor	1		
5	RTC4978	• Solenoid starter motor-Valeo	1		
6	RTC4980	• Kit-starter motor	1		
7	RTC4983	• Kit-starter motor sundry parts	1		
1	NAD10039	MOTOR-STARTER ENGINE-BOSCH	1		2500 cc 4 Cylinder Diesel TDi 12L
2	STC1246	• Box assembly-brush starter motor	1		
	STC1254	• Spring-coil	4		
	STC1247	• Guide-starter motor brush	4		
3	STC1244	• Drive assembly starter motor-Bosch	1		
5	STC1245	• Solenoid starter motor-0 331 303 165-Bosch	1		
5	STC3715	• Solenoid starter motor-9 330 331 006-Bosch	1		
	STC1250	• Bush-starter motor-centre	2		
	STC1251	• Bush-starter motor-rear	1		
	STC1255	• Bush-starter motor-front	1		
1	PRC5109E	MOTOR-STARTER ENGINE-EXCHANGE-VALEO	1	Note (1)	2500 cc 4 Cylinder Diesel TDi 12L
2	RTC4981	• Box assembly-brush starter motor	1		
3	RTC4982	• Drive assembly starter motor-Valeo	1		
4	RTC4979	• Yoke assembly starter motor	1		
5	RTC4978	• Solenoid starter motor-Valeo	1		
6	RTC4980	• Kit-starter motor	1		
7	RTC4983	• Kit-starter motor sundry parts	1		
	PRC5538	Harness earth bond	1	Note (1)	
8	NSP	Heatshield-starter	1	Note (1)	
9	ERR807	Bracket	1	Note (1)	
9	ERR808	Bracket	1	Note (1)	
10	FS105107	Screw-heatshield-hexagonal head-M5 x 10	4	Note (1)	
10	FS110301L	Screw-starter motor to bell housing-flanged head-M10 x 30	1	Note (1)	
	FN110041L	Nut-flanged head-M10	NLA	Note (1)	Use FN110047.
	WL110001L	Washer-M10	3	Note (1)	
11	BH110091L	Bolt-starter motor to bell housing-M10 x 45	1	Note (1)	

CHANGE POINTS:
(1) To (E)12L 50756A

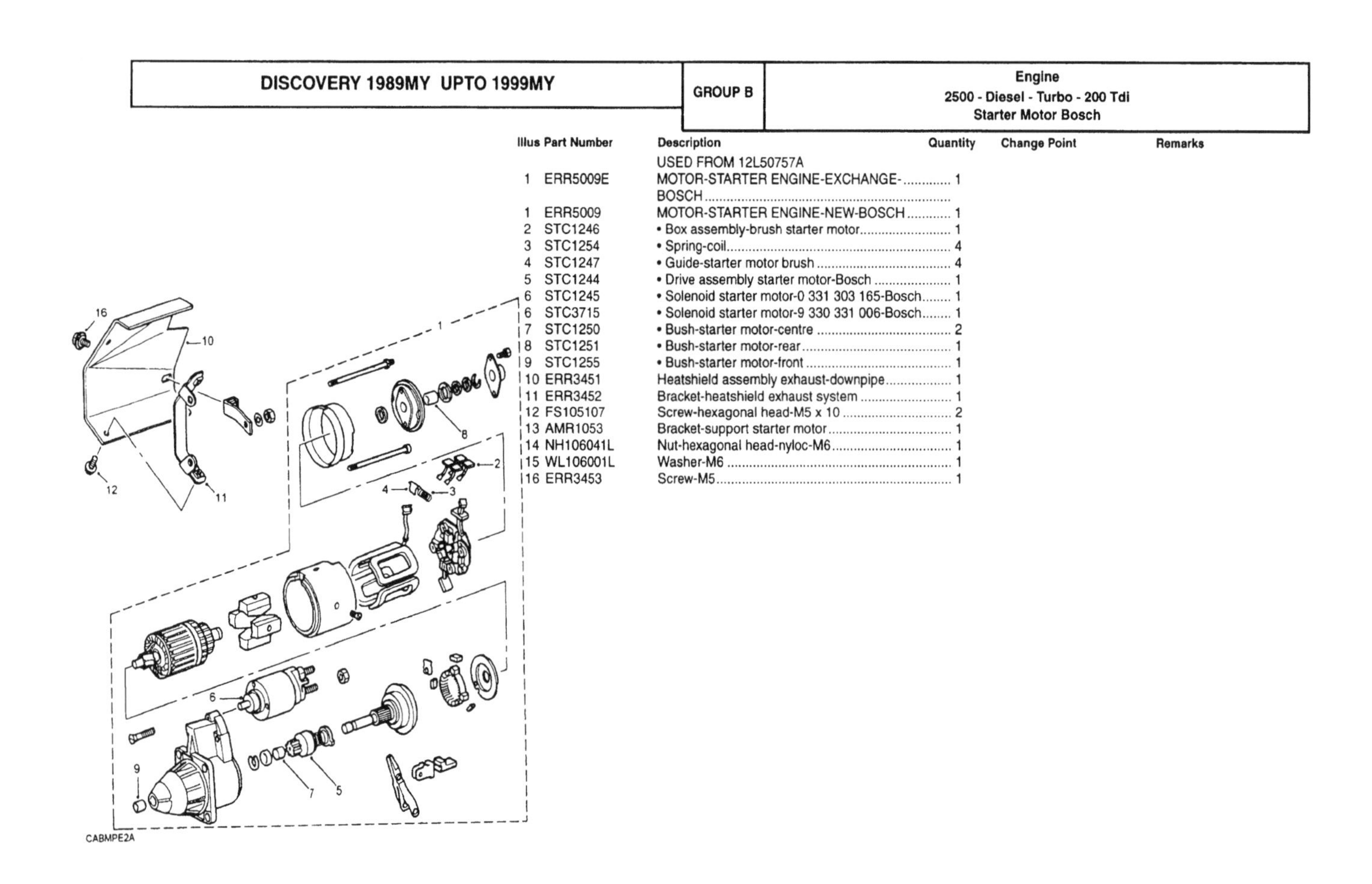

Illus	Part Number	Description	Quantity	Change Point	Remarks
		USED FROM 12L50757A			
1	ERR5009E	MOTOR-STARTER ENGINE-EXCHANGE-BOSCH	1		
1	ERR5009	MOTOR-STARTER ENGINE-NEW-BOSCH	1		
2	STC1246	• Box assembly-brush starter motor	1		
3	STC1254	• Spring-coil	4		
4	STC1247	• Guide-starter motor brush	4		
5	STC1244	• Drive assembly starter motor-Bosch	1		
6	STC1245	• Solenoid starter motor-0 331 303 165-Bosch	1		
6	STC3715	• Solenoid starter motor-9 330 331 006-Bosch	1		
7	STC1250	• Bush-starter motor-centre	2		
8	STC1251	• Bush-starter motor-rear	1		
9	STC1255	• Bush-starter motor-front	1		
10	ERR3451	Heatshield assembly exhaust-downpipe	1		
11	ERR3452	Bracket-heatshield exhaust system	1		
12	FS105107	Screw-hexagonal head-M5 x 10	2		
13	AMR1053	Bracket-support starter motor	1		
14	NH106041L	Nut-hexagonal head-nyloc-M6	1		
15	WL106001L	Washer-M6	1		
16	ERR3453	Screw-M5	1		

Illus	Part Number	Description	Quantity	Change Point	Remarks
1	ERR675	Housing assembly-clutch	1	Note (1)	
1	ERR3922	Housing assembly-clutch	1	Note (2)	
2	TE110061L	Stud-M10 x 30	9		
3	ERC9240	Stud-starter motor to housing-M10	1		
4	ERC7295	Plug-rubber	1		
		HOUSING TO BLOCK			
5	FS110301L	Screw-housing to block-flanged head-M10 x 30	8		
6	BH110091L	Bolt-housing to block-M10 x 45	2		
7	WA110061L	Washer-plain-M10-standard	8		
8	ERC9404	Bracket harness	2		
		FLYWHEEL AND CLUTCH			
9	ERR719	FLYWHEEL ENGINE	1		
10	568431	• Gear-ring-flywheel engine	1		
11	502116	• Dowel-flywheel to crankshaft	3		
12	ERC4658	Plate-reinforcement	NLA		Reinforcing plate is not needed. If one has been fitted and needs replacing omit plate and fit shorter bolts ERR4574.
13	ERC6551	Bolt-fixing flywheel crankshaft-long	8		
13	ERR4574	Bolt-fixing flywheel crankshaft-short	8		
	STC8358	KIT-CLUTCH	1		
14	FTC575	• Cover-clutch assembly	NLA		Use URB100760.
15	FTC2149	• Plate-clutch-driven	NLA		Use FTC4204.
	FRC9568	• BEARING-CLUTCH RELEASE	1		
	FRC4078	• • Sleeve-guide clutch	1		
14	URB100760	Cover-clutch assembly	1		
15	FTC4204	Plate-clutch-driven	1		
16	SX108201L	Bolt-cover clutch to flywheel-M8 x 20	6		
17	WL108001L	Washer-M8	6		
18	ERR2532	Seal-engine timing gear cover crankshaft	1		
19	ERR1440	Gasket-flywheel/drive plate housing/crankcase	1		
20	SS110801	Screw-M10 x 80	4		

CHANGE POINTS:
(1) To (E)12L 060160A
(2) From (E)12L 060161A

B 39

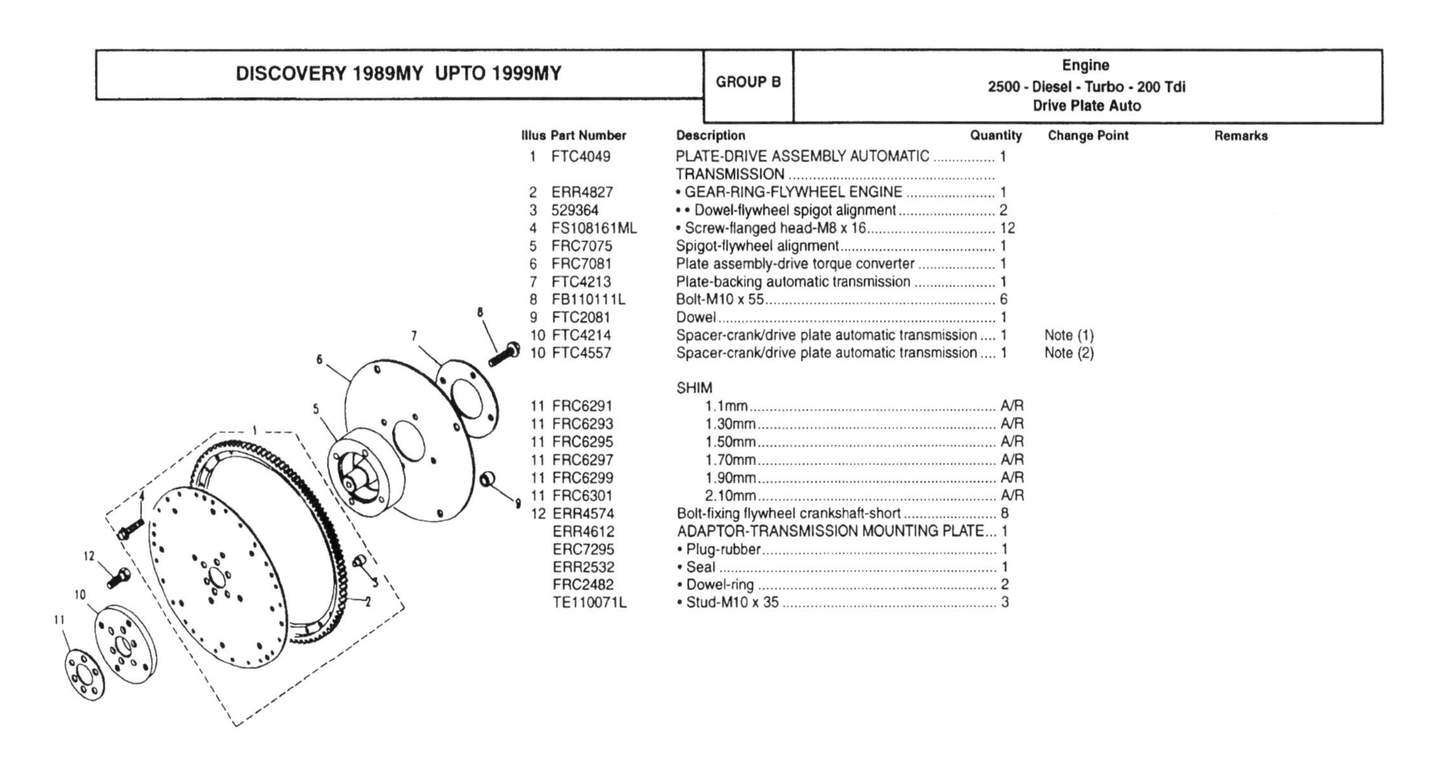

Illus	Part Number	Description	Quantity	Change Point	Remarks
1	FTC4049	PLATE-DRIVE ASSEMBLY AUTOMATIC TRANSMISSION	1		
2	ERR4827	• GEAR-RING-FLYWHEEL ENGINE	1		
3	529364	• • Dowel-flywheel spigot alignment	2		
4	FS108161ML	• Screw-flanged head-M8 x 16	12		
5	FRC7075	Spigot-flywheel alignment	1		
6	FRC7081	Plate assembly-drive torque converter	1		
7	FTC4213	Plate-backing automatic transmission	1		
8	FB110111L	Bolt-M10 x 55	6		
9	FTC2081	Dowel	1		
10	FTC4214	Spacer-crank/drive plate automatic transmission	1	Note (1)	
10	FTC4557	Spacer-crank/drive plate automatic transmission	1	Note (2)	
		SHIM			
11	FRC6291	1.1mm	A/R		
11	FRC6293	1.30mm	A/R		
11	FRC6295	1.50mm	A/R		
11	FRC6297	1.70mm	A/R		
11	FRC6299	1.90mm	A/R		
11	FRC6301	2.10mm	A/R		
12	ERR4574	Bolt-fixing flywheel crankshaft-short	8		
	ERR4612	ADAPTOR-TRANSMISSION MOUNTING PLATE	1		
	ERC7295	• Plug-rubber	1		
	ERR2532	• Seal	1		
	FRC2482	• Dowel-ring	2		
	TE110071L	• Stud-M10 x 35	3		

CHANGE POINTS:
(1) To (E)22L 03116A
(2) From (E)22L 03117A

B 40

DISCOVERY 1989MY UPTO 1999MY	GROUP B	Engine 2500 - Diesel - Turbo - 300 Tdi Engine Stripped

Illus	Part Number	Description	Quantity	Change Point	Remarks
1	STC1736N	Engine unit-stripped-new	1		
1	STC1736E	Engine unit-stripped-exchange	1		
		WHEN USED WITH AN AUTOMATIC GEARBOX REMOVE SPIGOT BEARING FROM REAR OF CRANKSHAFT			

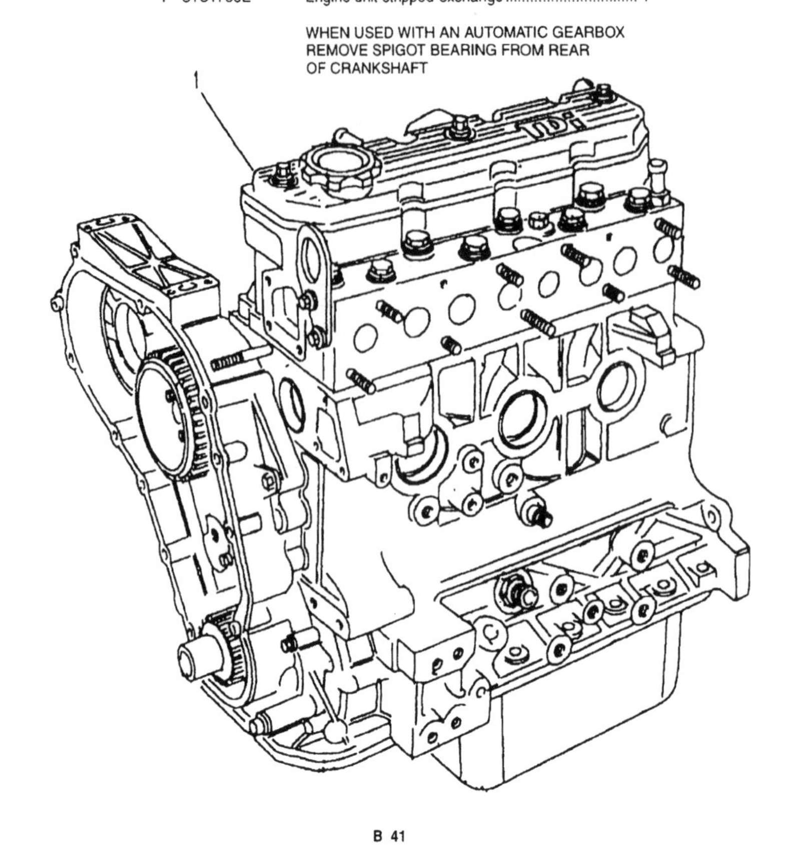

B 41

DISCOVERY 1989MY UPTO 1999MY	GROUP B	Engine 2500 - Diesel - Turbo - 300 Tdi Engine Short

Illus	Part Number	Description	Quantity	Change Point	Remarks
1	STC1675	Engine assembly - Short	1		

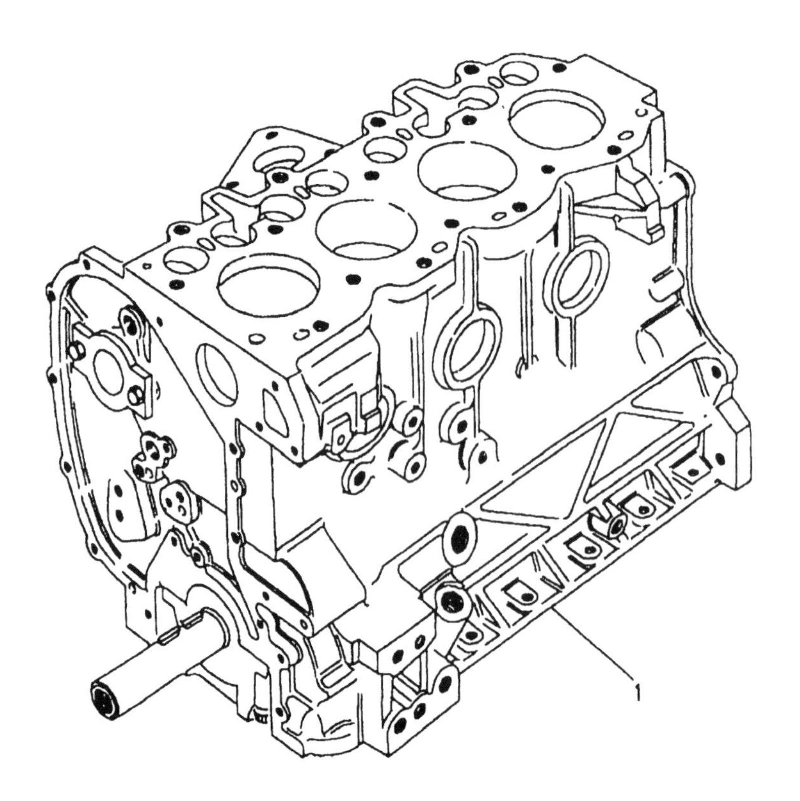

B 42

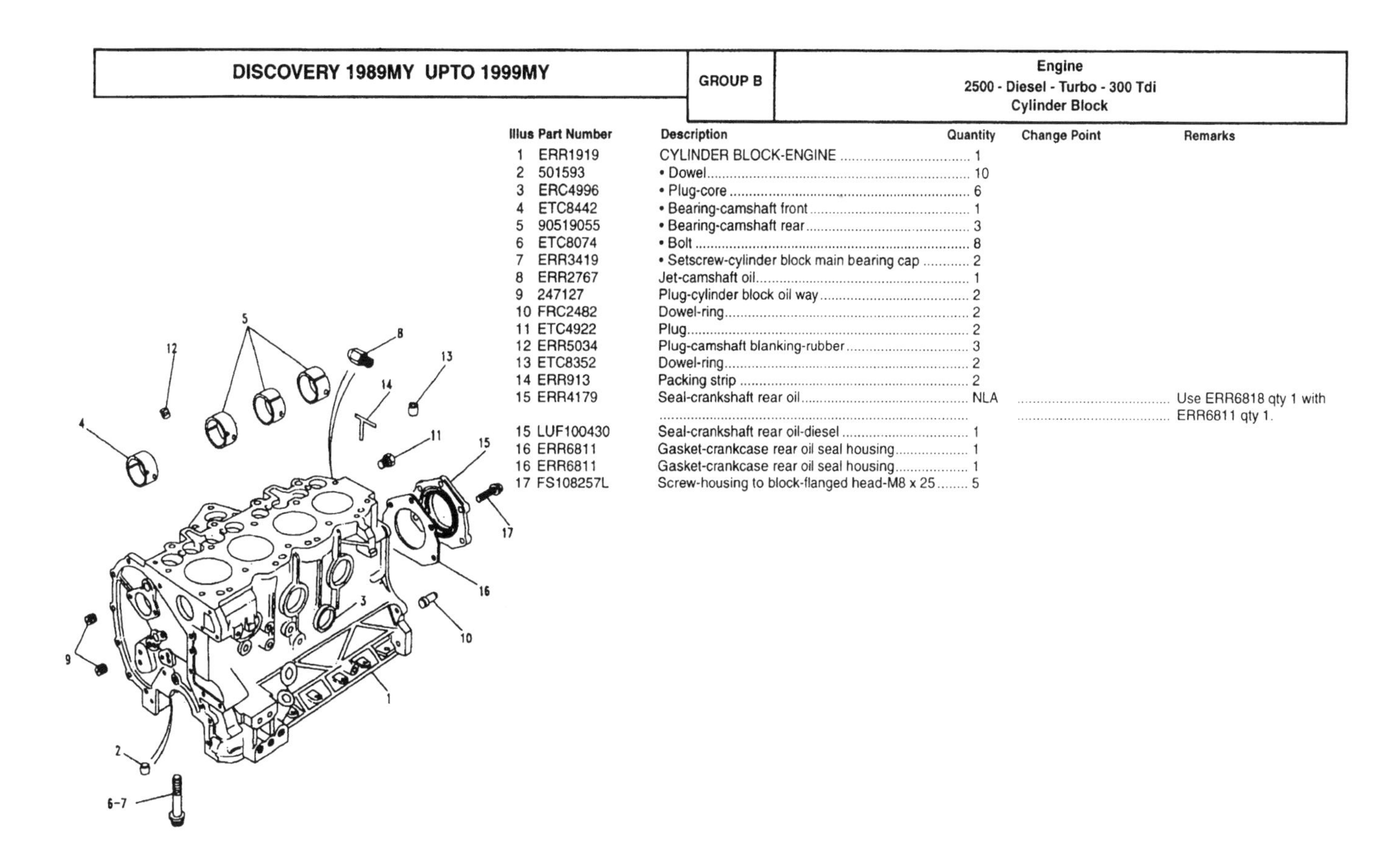

DISCOVERY 1989MY UPTO 1999MY	GROUP B	Engine 2500 - Diesel - Turbo - 300 Tdi Cylinder Block

Illus	Part Number	Description	Quantity	Change Point	Remarks
1	ERR1919	CYLINDER BLOCK-ENGINE	1		
2	501593	• Dowel	10		
3	ERC4996	• Plug-core	6		
4	ETC8442	• Bearing-camshaft front	1		
5	90519055	• Bearing-camshaft rear	3		
6	ETC8074	• Bolt	8		
7	ERR3419	• Setscrew-cylinder block main bearing cap	2		
8	ERR2767	Jet-camshaft oil	1		
9	247127	Plug-cylinder block oil way	2		
10	FRC2482	Dowel-ring	2		
11	ETC4922	Plug	2		
12	ERR5034	Plug-camshaft blanking-rubber	3		
13	ETC8352	Dowel-ring	2		
14	ERR913	Packing strip	2		
15	ERR4179	Seal-crankshaft rear oil	NLA		Use ERR6818 qty 1 with ERR6811 qty 1.
15	LUF100430	Seal-crankshaft rear oil-diesel	1		
16	ERR6811	Gasket-crankcase rear oil seal housing	1		
16	ERR6811	Gasket-crankcase rear oil seal housing	1		
17	FS108257L	Screw-housing to block-flanged head-M8 x 25	5		

B 43

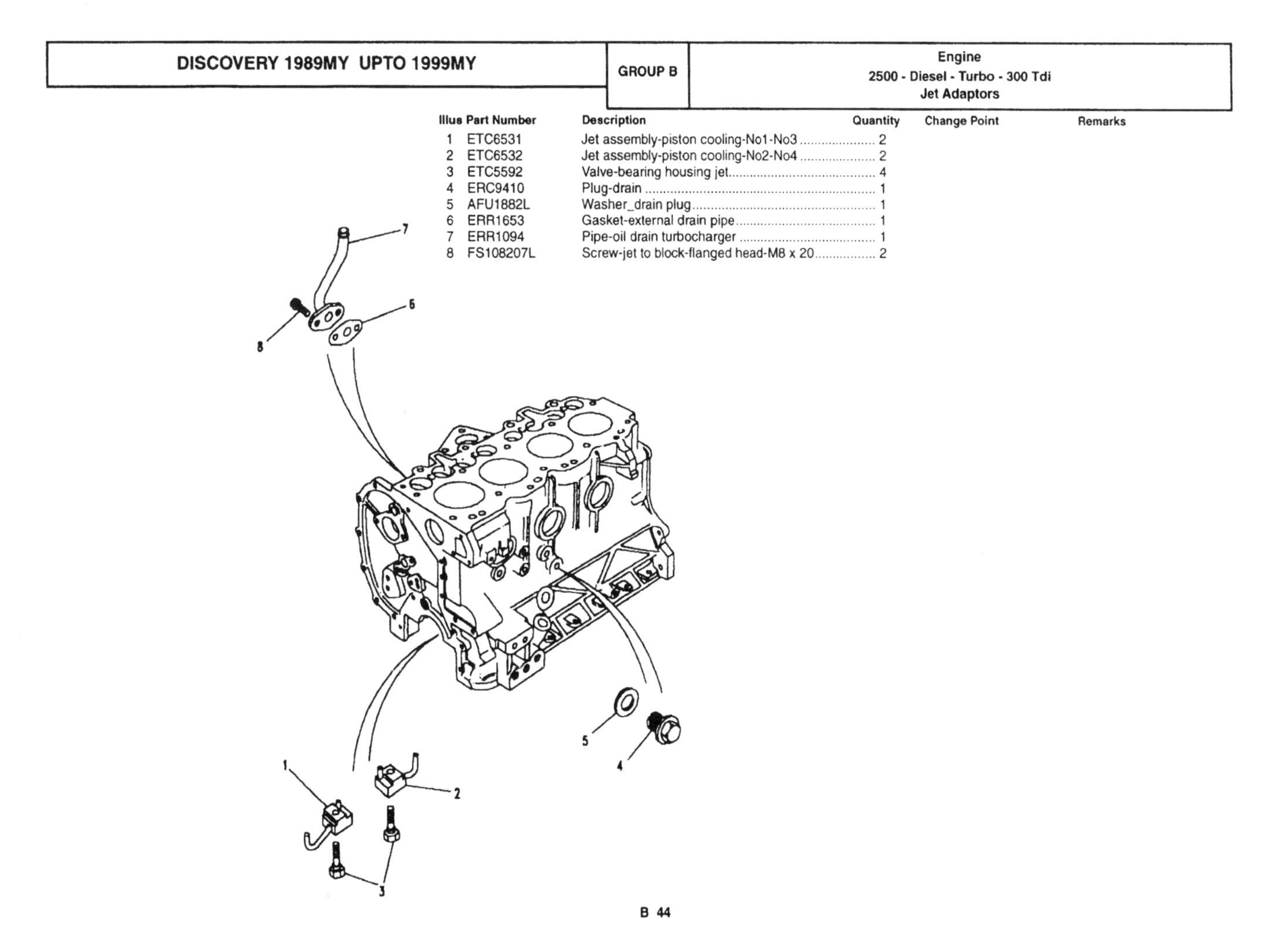

DISCOVERY 1989MY UPTO 1999MY	GROUP B	Engine 2500 - Diesel - Turbo - 300 Tdi Jet Adaptors

Illus	Part Number	Description	Quantity	Change Point	Remarks
1	ETC6531	Jet assembly-piston cooling-No1-No3	2		
2	ETC6532	Jet assembly-piston cooling-No2-No4	2		
3	ETC5592	Valve-bearing housing jet	4		
4	ERC9410	Plug-drain	1		
5	AFU1882L	Washer_drain plug	1		
6	ERR1653	Gasket-external drain pipe	1		
7	ERR1094	Pipe-oil drain turbocharger	1		
8	FS108207L	Screw-jet to block-flanged head-M8 x 20	2		

B 44

Crankshaft and Bearings

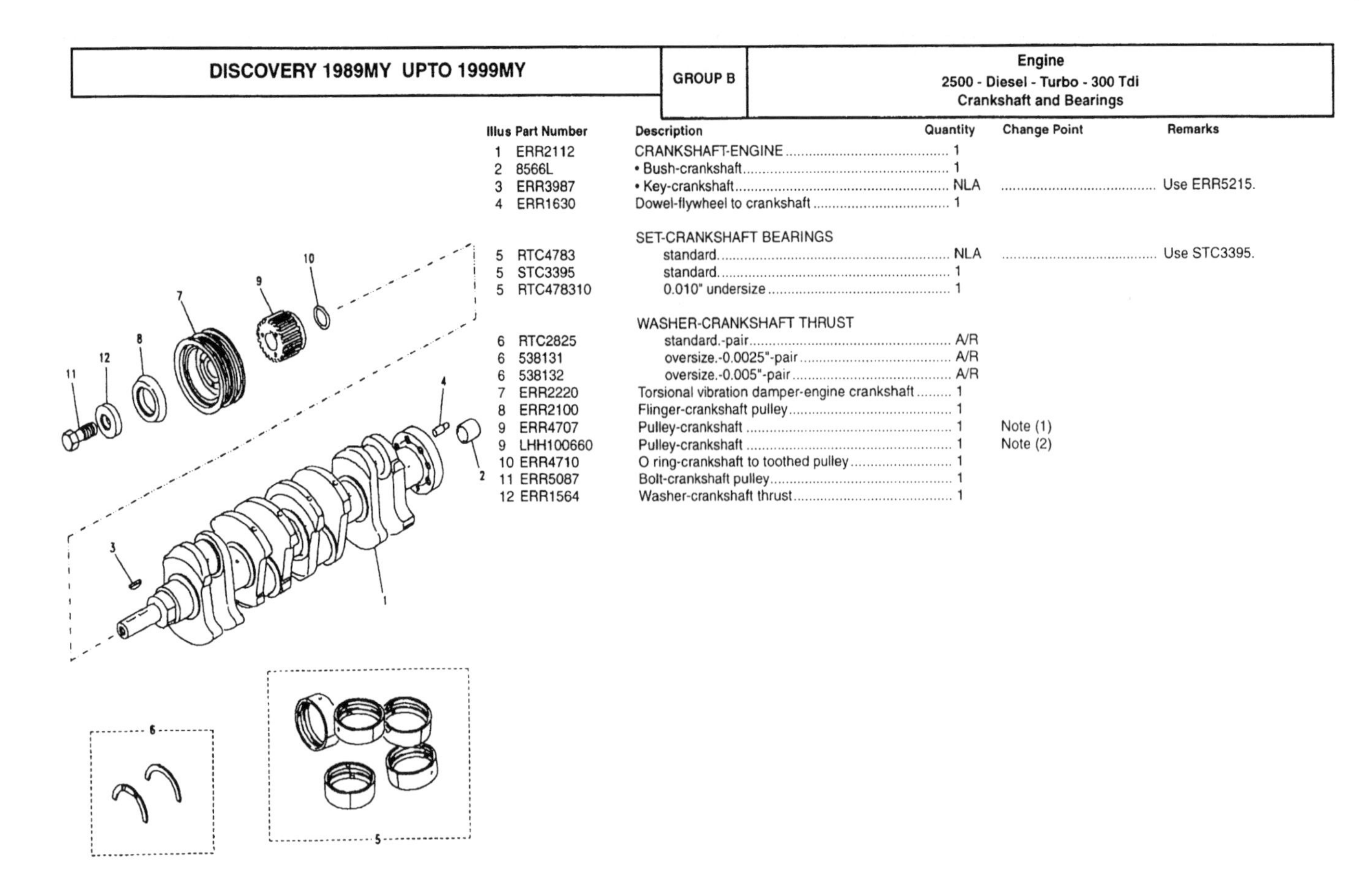

Illus	Part Number	Description	Quantity	Change Point	Remarks
1	ERR2112	CRANKSHAFT-ENGINE	1		
2	8566L	• Bush-crankshaft	1		
3	ERR3987	• Key-crankshaft	NLA		Use ERR5215.
4	ERR1630	Dowel-flywheel to crankshaft	1		
		SET-CRANKSHAFT BEARINGS			
5	RTC4783	standard	NLA		Use STC3395.
5	STC3395	standard	1		
5	RTC478310	0.010" undersize	1		
		WASHER-CRANKSHAFT THRUST			
6	RTC2825	standard.-pair	A/R		
6	538131	oversize.-0.0025"-pair	A/R		
6	538132	oversize.-0.005"-pair	A/R		
7	ERR2220	Torsional vibration damper-engine crankshaft	1		
8	ERR2100	Flinger-crankshaft pulley	1		
9	ERR4707	Pulley-crankshaft	1	Note (1)	
9	LHH100660	Pulley-crankshaft	1	Note (2)	
10	ERR4710	O ring-crankshaft to toothed pulley	1		
11	ERR5087	Bolt-crankshaft pulley	1		
12	ERR1564	Washer-crankshaft thrust	1		

CHANGE POINTS:
(1) To (V) VA 560897
(2) From (V) VA 560898

DISCOVERY 1989MY UPTO 1999MY | GROUP B | Engine
2500 - Diesel - Turbo - 300 Tdi

Piston, Connecting Rod and Bearings

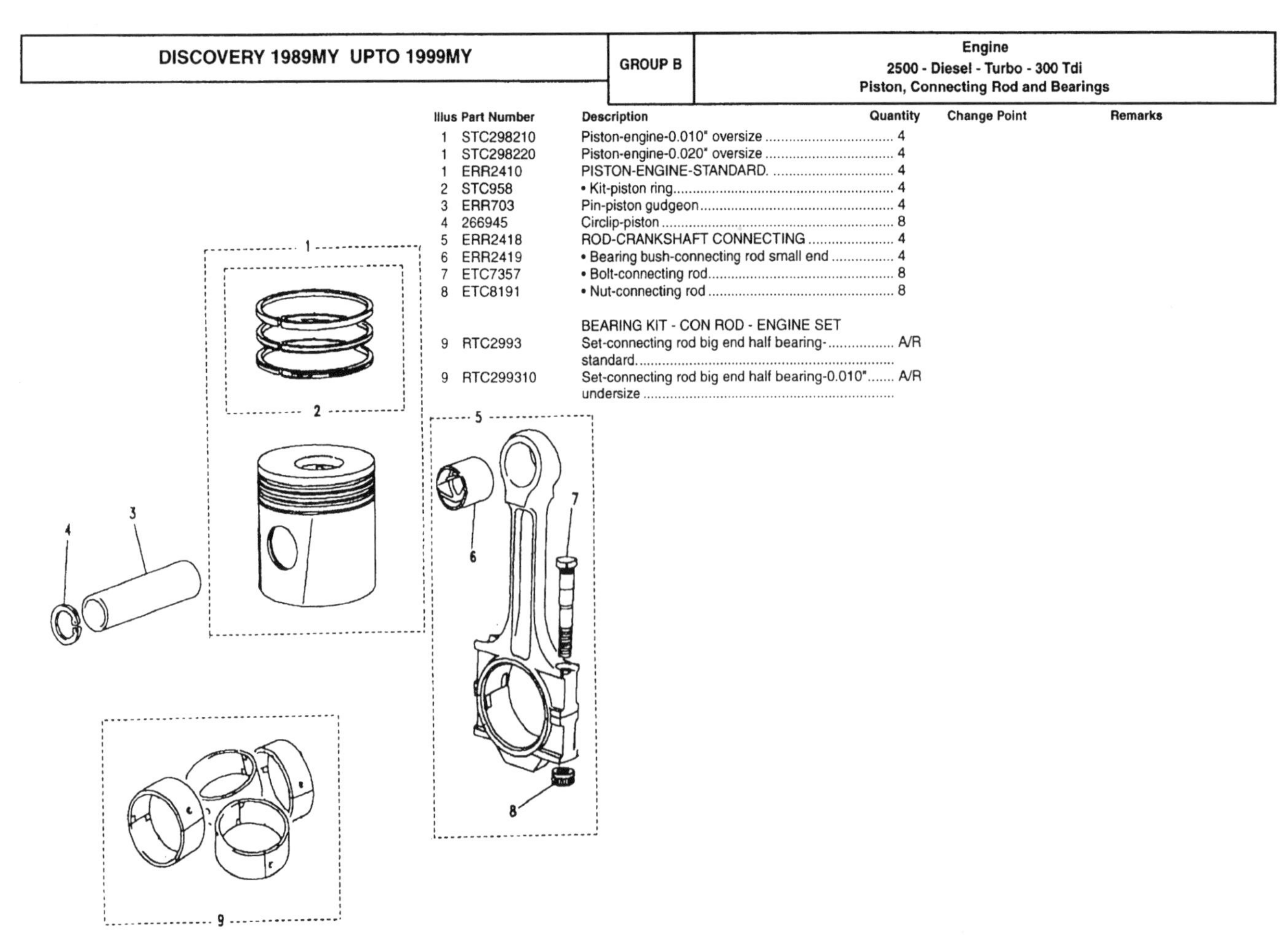

Illus	Part Number	Description	Quantity	Change Point	Remarks
1	STC298210	Piston-engine-0.010" oversize	4		
1	STC298220	Piston-engine-0.020" oversize	4		
1	ERR2410	PISTON-ENGINE-STANDARD.	4		
2	STC958	• Kit-piston ring	4		
3	ERR703	Pin-piston gudgeon	4		
4	266945	Circlip-piston	8		
5	ERR2418	ROD-CRANKSHAFT CONNECTING	4		
6	ERR2419	• Bearing bush-connecting rod small end	4		
7	ETC7357	• Bolt-connecting rod	8		
8	ETC8191	• Nut-connecting rod	8		
		BEARING KIT - CON ROD - ENGINE SET			
9	RTC2993	Set-connecting rod big end half bearing-standard	A/R		
9	RTC299310	Set-connecting rod big end half bearing-0.010" undersize	A/R		

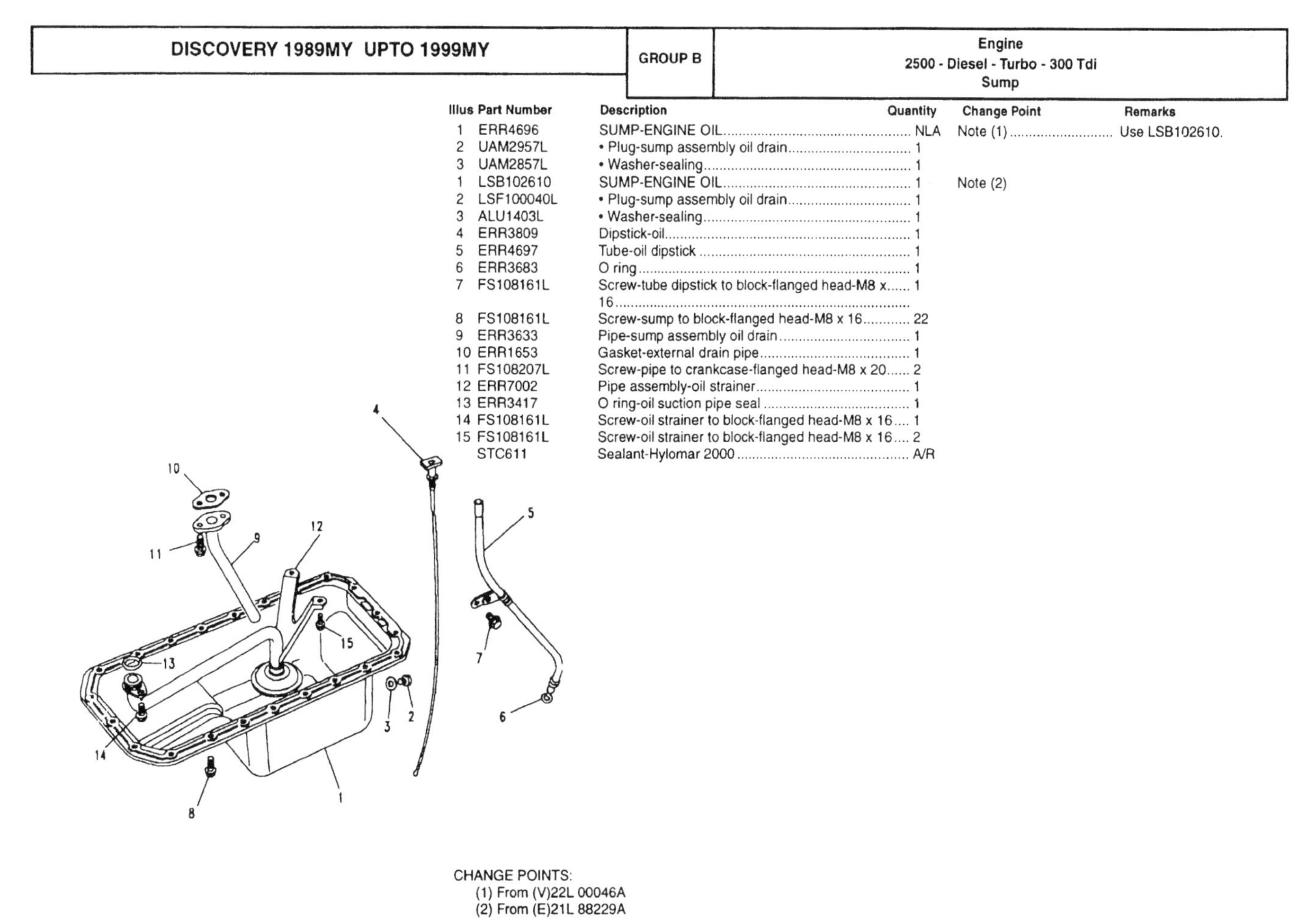

Illus	Part Number	Description	Quantity	Change Point	Remarks
1	ERR4696	SUMP-ENGINE OIL	NLA	Note (1)	Use LSB102610.
2	UAM2957L	• Plug-sump assembly oil drain	1		
3	UAM2857L	• Washer-sealing	1		
1	LSB102610	SUMP-ENGINE OIL	1	Note (2)	
2	LSF100040L	• Plug-sump assembly oil drain	1		
3	ALU1403L	• Washer-sealing	1		
4	ERR3809	Dipstick-oil	1		
5	ERR4697	Tube-oil dipstick	1		
6	ERR3683	O ring	1		
7	FS108161L	Screw-tube dipstick to block-flanged head-M8 x 16	1		
8	FS108161L	Screw-sump to block-flanged head-M8 x 16	22		
9	ERR3633	Pipe-sump assembly oil drain	1		
10	ERR1653	Gasket-external drain pipe	1		
11	FS108207L	Screw-pipe to crankcase-flanged head-M8 x 20	2		
12	ERR7002	Pipe assembly-oil strainer	1		
13	ERR3417	O ring-oil suction pipe seal	1		
14	FS108161L	Screw-oil strainer to block-flanged head-M8 x 16	1		
15	FS108161L	Screw-oil strainer to block-flanged head-M8 x 16	2		
	STC611	Sealant-Hylomar 2000	A/R		

CHANGE POINTS:
(1) From (V)22L 00046A
(2) From (E)21L 88229A

B 47

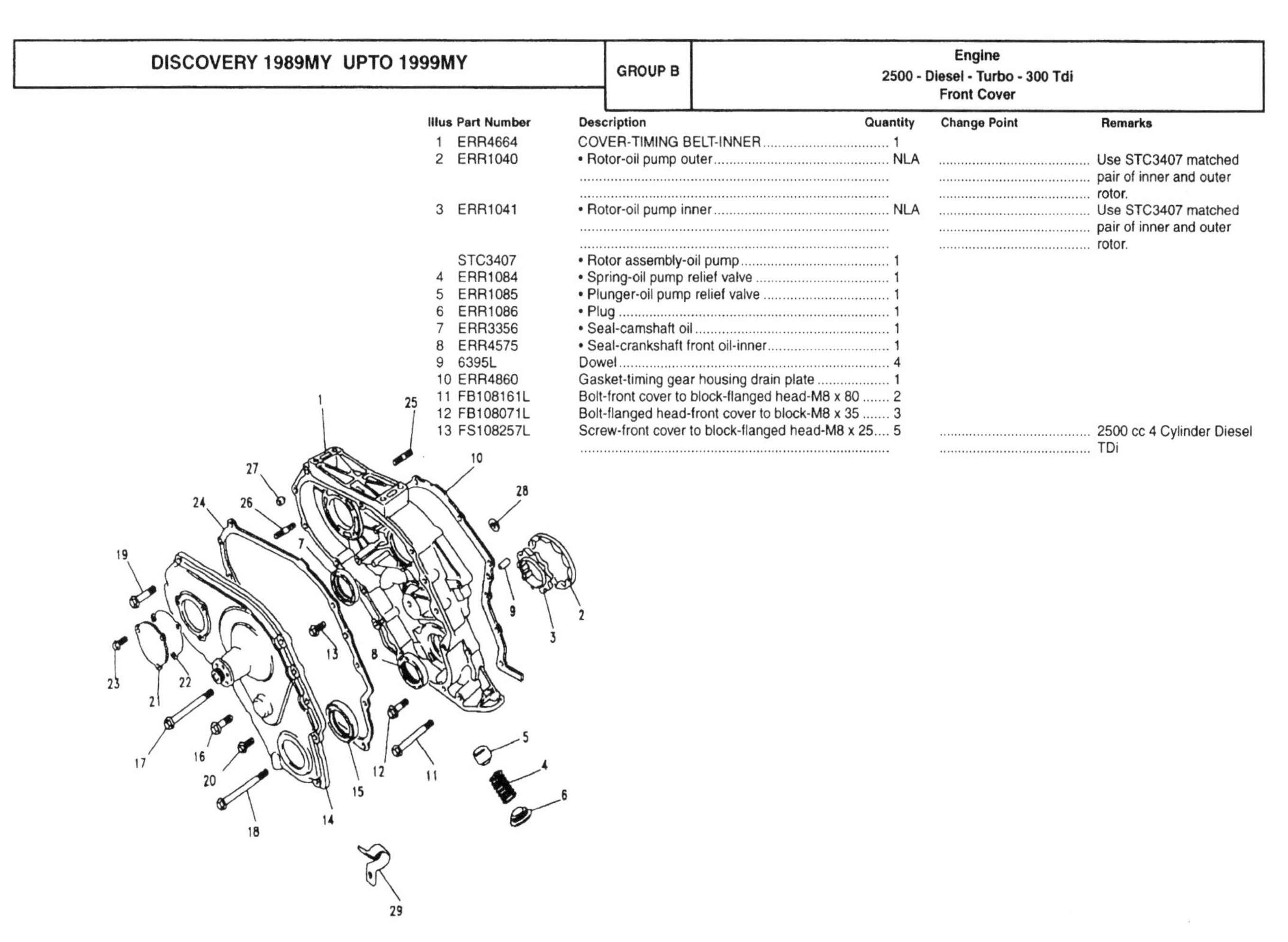

Illus	Part Number	Description	Quantity	Change Point	Remarks
1	ERR4664	COVER-TIMING BELT-INNER	1		
2	ERR1040	• Rotor-oil pump outer	NLA		Use STC3407 matched pair of inner and outer rotor.
3	ERR1041	• Rotor-oil pump inner	NLA		Use STC3407 matched pair of inner and outer rotor.
	STC3407	• Rotor assembly-oil pump	1		
4	ERR1084	• Spring-oil pump relief valve	1		
5	ERR1085	• Plunger-oil pump relief valve	1		
6	ERR1086	• Plug	1		
7	ERR3356	• Seal-camshaft oil	1		
8	ERR4575	• Seal-crankshaft front oil-inner	1		
9	6395L	Dowel	4		
10	ERR4860	Gasket-timing gear housing drain plate	1		
11	FB108161L	Bolt-front cover to block-flanged head-M8 x 80	2		
12	FB108071L	Bolt-flanged head-front cover to block-M8 x 35	3		
13	FS108257L	Screw-front cover to block-flanged head-M8 x 25	5		2500 cc 4 Cylinder Diesel TDi

B 48

DISCOVERY 1989MY UPTO 1999MY	GROUP B	Engine 2500 - Diesel - Turbo - 300 Tdi Front Cover

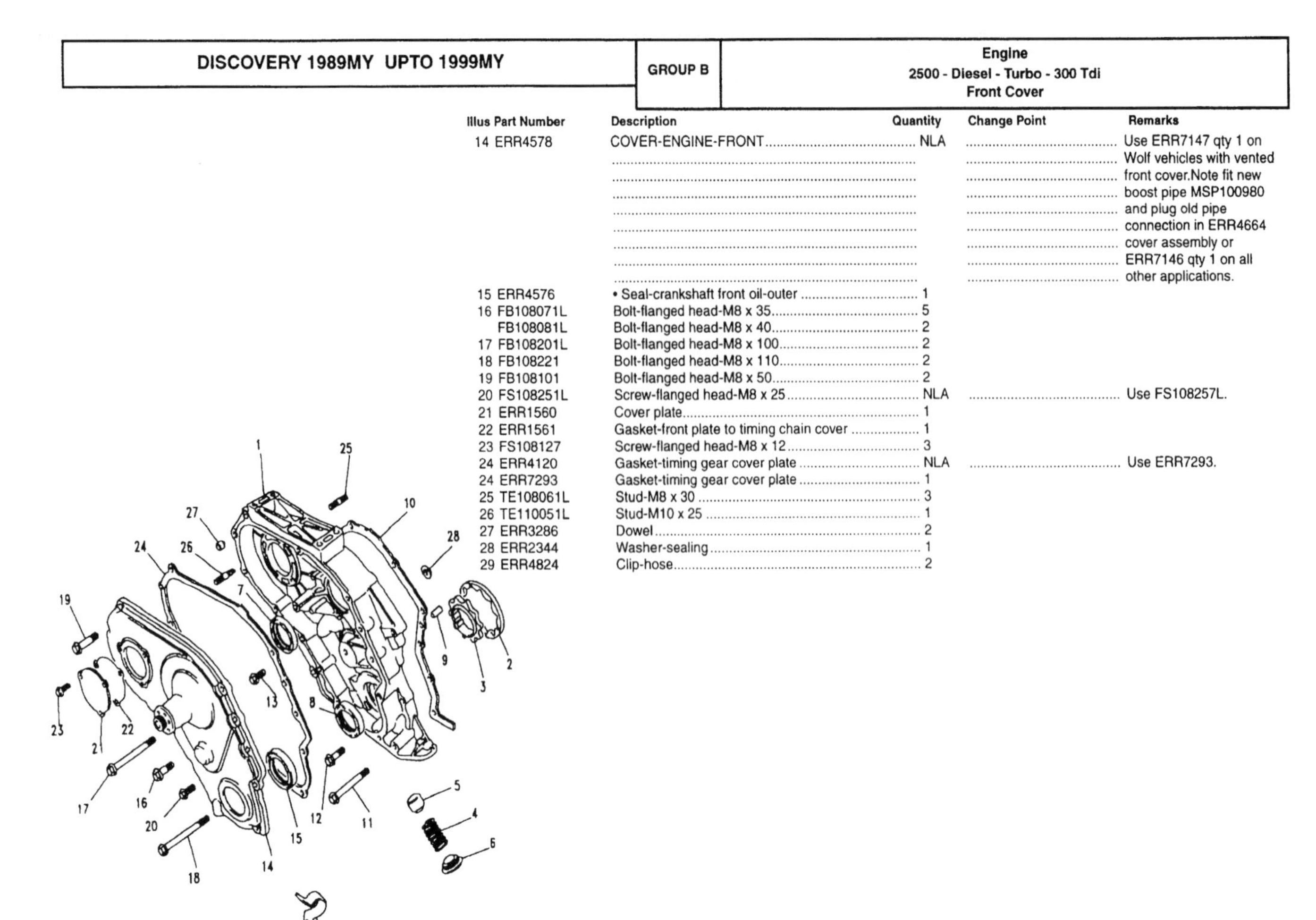

Illus	Part Number	Description	Quantity	Change Point	Remarks
14	ERR4578	COVER-ENGINE-FRONT	NLA		Use ERR7147 qty 1 on Wolf vehicles with vented front cover.Note fit new boost pipe MSP100980 and plug old pipe connection in ERR4664 cover assembly or ERR7146 qty 1 on all other applications.
15	ERR4576	• Seal-crankshaft front oil-outer	1		
16	FB108071L	Bolt-flanged head-M8 x 35	5		
	FB108081L	Bolt-flanged head-M8 x 40	2		
17	FB108201L	Bolt-flanged head-M8 x 100	2		
18	FB108221	Bolt-flanged head-M8 x 110	2		
19	FB108101	Bolt-flanged head-M8 x 50	2		
20	FS108251L	Screw-flanged head-M8 x 25	NLA		Use FS108257L.
21	ERR1560	Cover plate	1		
22	ERR1561	Gasket-front plate to timing chain cover	1		
23	FS108127	Screw-flanged head-M8 x 12	3		
24	ERR4120	Gasket-timing gear cover plate	NLA		Use ERR7293.
24	ERR7293	Gasket-timing gear cover plate	1		
25	TE108061L	Stud-M8 x 30	3		
26	TE110051L	Stud-M10 x 25	1		
27	ERR3286	Dowel	2		
28	ERR2344	Washer-sealing	1		
29	ERR4824	Clip-hose	2		

B 49

DISCOVERY 1989MY UPTO 1999MY	GROUP B	Engine 2500 - Diesel - Turbo - 300 Tdi Timing Belt and Tensioner

Illus	Part Number	Description	Quantity	Change Point	Remarks
1	ERR1092	Belt-engine timing	1		
2	ERR1972	Tensioner-timing belt	1	Note (1)	
2	LHP100860	Tensioner-timing belt	1	Note (2)	
3	ERR900	Washer	1		
4	FB110141L	Bolt-tensioner to cover-flanged head-M10 x 70	1		
5	FN110041L	Nut-flanged head-M10	NLA		Use FN110047.
5	FN110047	Nut-flanged head-M10	1		
6	ETC8560	Idler-timing belt	1		
7	ERR1973	Washer	1		

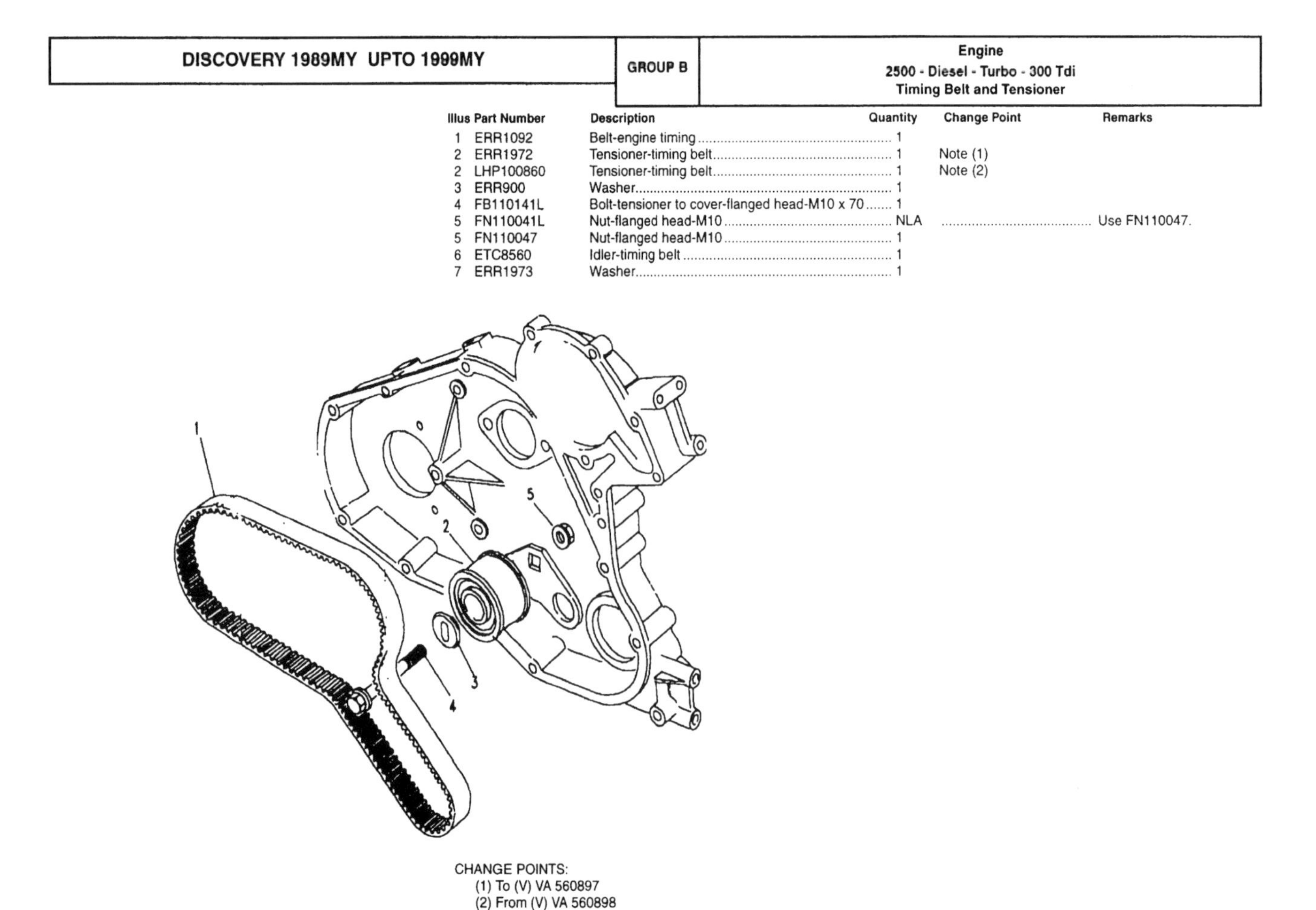

CHANGE POINTS:
(1) To (V) VA 560897
(2) From (V) VA 560898

B 50

DISCOVERY 1989MY UPTO 1999MY

GROUP B — Engine, 2500 - Diesel - Turbo - 300 Tdi, Coolant Pump

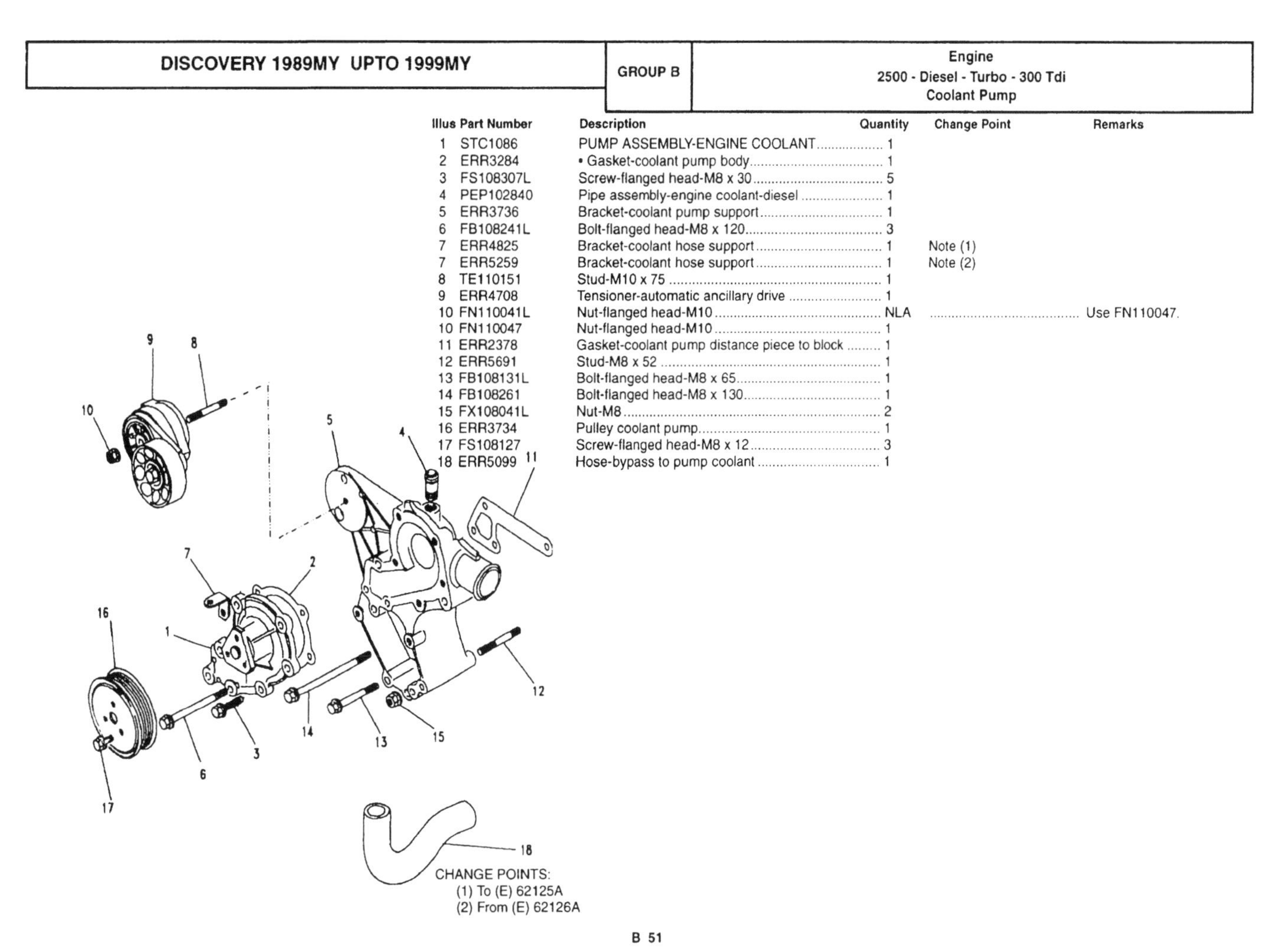

Illus	Part Number	Description	Quantity	Change Point	Remarks
1	STC1086	PUMP ASSEMBLY-ENGINE COOLANT	1		
2	ERR3284	• Gasket-coolant pump body	1		
3	FS108307L	Screw-flanged head-M8 x 30	5		
4	PEP102840	Pipe assembly-engine coolant-diesel	1		
5	ERR3736	Bracket-coolant pump support	1		
6	FB108241L	Bolt-flanged head-M8 x 120	3		
7	ERR4825	Bracket-coolant hose support	1	Note (1)	
7	ERR5259	Bracket-coolant hose support	1	Note (2)	
8	TE110151	Stud-M10 x 75	1		
9	ERR4708	Tensioner-automatic ancillary drive	1		
10	FN110041L	Nut-flanged head-M10	NLA		Use FN110047.
10	FN110047	Nut-flanged head-M10	1		
11	ERR2378	Gasket-coolant pump distance piece to block	1		
12	ERR5691	Stud-M8 x 52	1		
13	FB108131L	Bolt-flanged head-M8 x 65	1		
14	FB108261	Bolt-flanged head-M8 x 130	1		
15	FX108041L	Nut-M8	2		
16	ERR3734	Pulley coolant pump	1		
17	FS108127	Screw-flanged head-M8 x 12	3		
18	ERR5099	Hose-bypass to pump coolant	1		

CHANGE POINTS:
(1) To (E) 62125A
(2) From (E) 62126A

B 51

DISCOVERY 1989MY UPTO 1999MY

GROUP B — Engine, 2500 - Diesel - Turbo - 300 Tdi, Cooling Fan and Belt

Illus	Part Number	Description	Quantity	Change Point	Remarks
1	ERR2789	Fan-cooling	1		
2	ERR2266	Coupling-engine fan viscous	1		
4	SH108161L	Screw-fan to viscous coupling-hexagonal head-M8 x 16	NLA		Use FS108161L.
4	FS108161L	Screw-fan to viscous coupling-flanged head-M8 x 16	4		
3	ERR3735	Pulley coolant pump	1		
4	FS108127	Screw-pulley to pump-flanged head-M8 x 12	4		
5	ERR3287	Belt-polyvee coolant pump	1	Note (1)	
5	ERR5911	Belt-polyvee coolant pump	1	Note (2)	

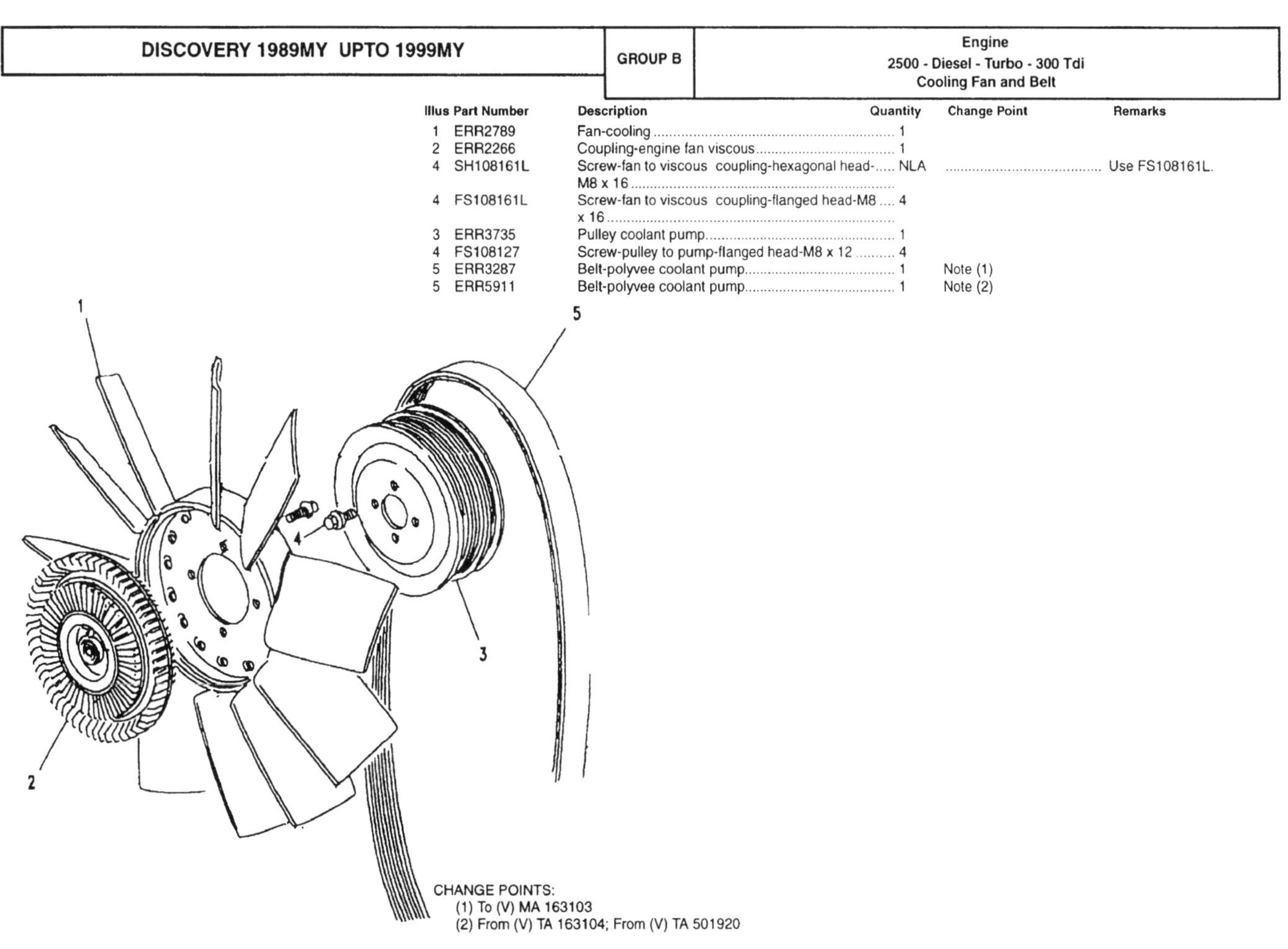

CHANGE POINTS:
(1) To (V) MA 163103
(2) From (V) TA 163104; From (V) TA 501920

B 52

DISCOVERY 1989MY UPTO 1999MY	GROUP B	Engine 2500 - Diesel - Turbo - 300 Tdi Oil Filter

Illus	Part Number	Description	Quantity	Change Point	Remarks
1	ERR2317	ADAPTOR-OIL FILTER	1		
2	ERC5923	• Thermostat assembly-oil	1		
3	ERR2241	• Waxstat adaptor	1		
4	ERC5913	• O ring	1		
5	PRC6387	• Switch-oil pressure engine	1		
6	FS106167L	• Screw-adaptor to housing-hexagonal head-M6 x 16	2		
7	AFU1887L	• Washer-sealing	1		
8	ERR3340	Cartridge-engine oil filter	1		
9	FS110301L	Screw-flanged head-M10 x 30	4		
10	ERR3283	Gasket-cylinder block to oil filter adaptor	1		

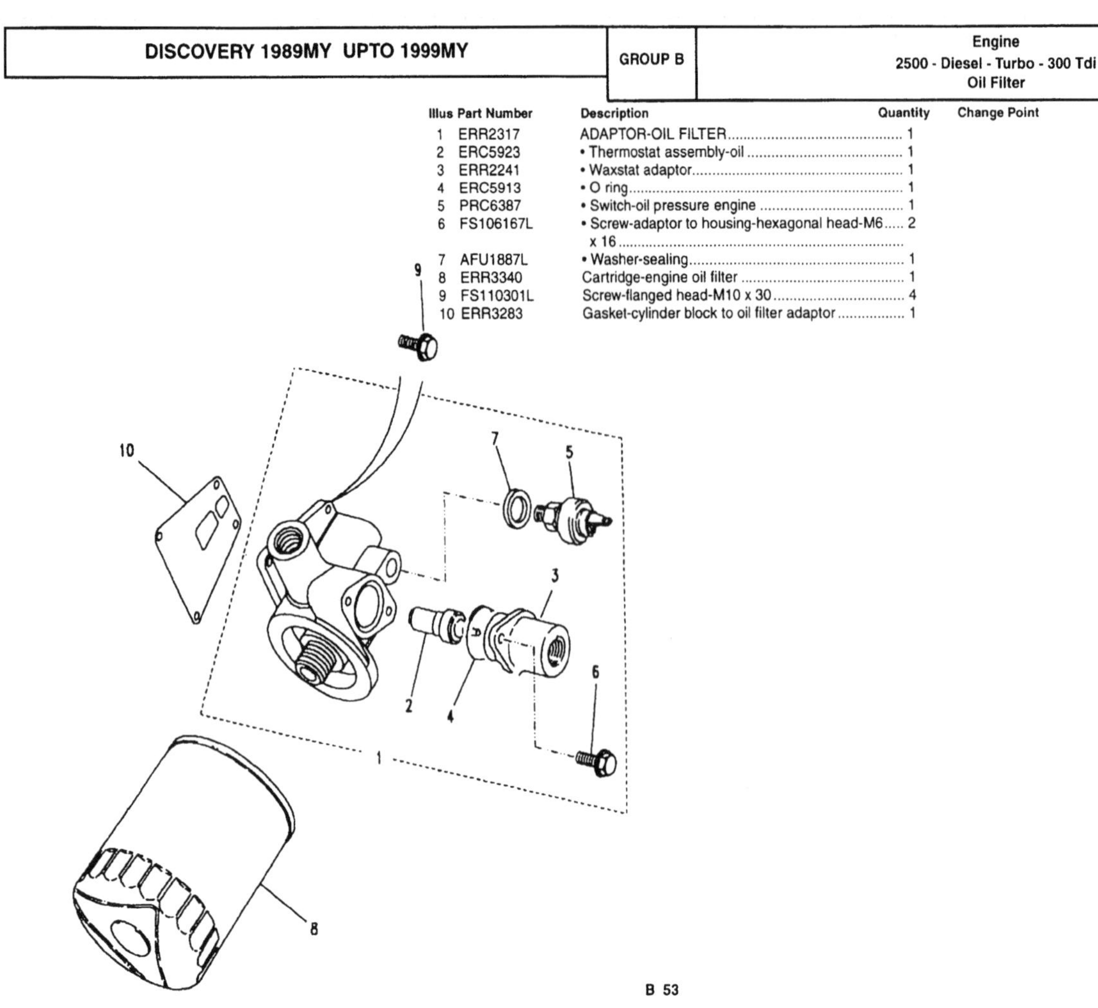

B 53

DISCOVERY 1989MY UPTO 1999MY	GROUP B	Engine 2500 - Diesel - Turbo - 300 Tdi Cylinder Head

Illus	Part Number	Description	Quantity	Change Point	Remarks
1	ERR5027	CYLINDER HEAD-ENGINE-LESS CORE PLUGS	1		
2	ETC8003	• Guide-cylinder head valve	NLA	Note (1); Note (2); Note (3)	Use LGJ100880.
2	LGJ100880	• Guide-cylinder head valve	8	Note (4); Note (5); Note (6)	
4	ERR3753	• Insert-cylinder head exhaust valve seat	4		
3	ERR2396	• Insert-inlet valve seat-cylinder head	4		
5	525497	Plug-core	1		
6	524765	Plug-core	1		
7	TE108051L	Stud-manifold to head-M8 x 25	2		
8	TE110071L	Stud-manifold to head-M10 x 35	7		
9	ETC8808	Bolt-cylinder head fixing-short-M12 x 100	4		
10	ERR1939	Bolt-cylinder head fixing-long-M12 x 140	10		
11	ETC8810	Bolt-cylinder head fixing-M10 x 117	4		
		GASKET-CYLINDER HEAD-SELECTIVE			
12	ERR5261	Gasket-cylinder head-1.30mm-1 hole	A/R		
12	ERR5262	Gasket-cylinder head-1.40mm-2 holes	A/R		
12	ERR5263	Gasket-cylinder head-1.50mm-3 holes	A/R		
12	ERR7154	Gasket-cylinder head-1.6mm	A/R		
13	AMR2425	Cable-heater plug ignition	1		
14	NH104041L	Nut-hexagonal-M4	4		
15	WA104001L	Washer-plain-M4	4		
16	WE104001L	Washer- shakeproof-M4	4		
17	ETC8847	Plug-heater ignition	4		

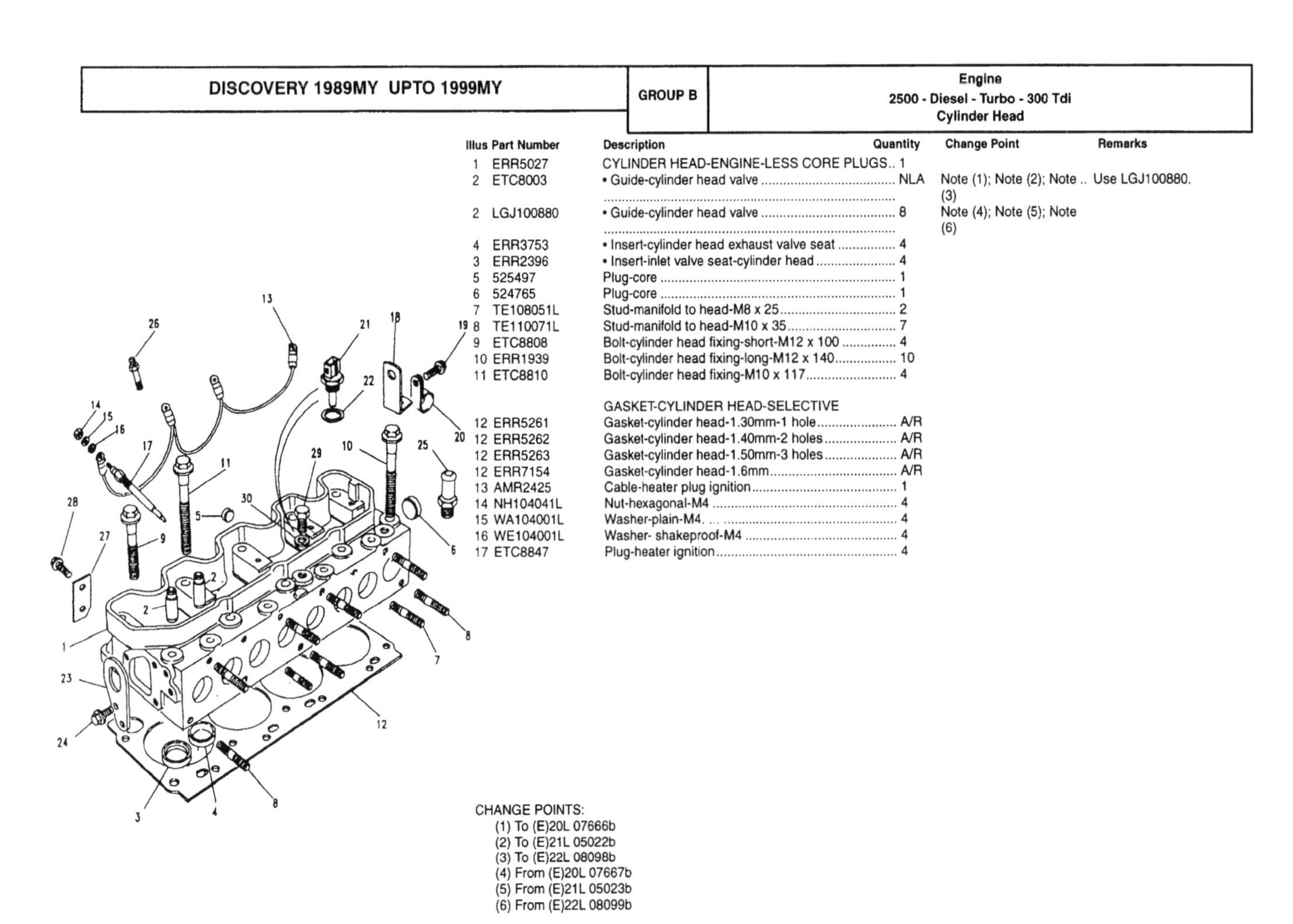

CHANGE POINTS:
(1) To (E)20L 07666b
(2) To (E)21L 05022b
(3) To (E)22L 08098b
(4) From (E)20L 07667b
(5) From (E)21L 05023b
(6) From (E)22L 08099b

B 54

DISCOVERY 1989MY UPTO 1999MY

GROUP B

Engine
2500 - Diesel - Turbo - 300 Tdi
Cylinder Head

Illus	Part Number	Description	Quantity	Change Point	Remarks
18	ERC9480	Bracket	1		EDC
19	FS108207L	Screw-flanged head-M8 x 20	2		
20	CP108251L	Clip-p	1		
21	ERR2081	Sensor-coolant temperature diesel	1		
22	ERR894	Washer-sealing	1		
23	ETC8031	Bracket-lifting engine front	1		
24	FS108207L	Screw-bracket to block-flanged head-M8 x 20	2		
25	624091	Adaptor-pipe coolant	1		
26	ERR1019	Stud	4		
27	ERC9404	Bracket harness	1		
28	FS108161L	Screw-flanged head-M8 x 16	1		
29	ETC5577	Plug-drain	1		Detoxed
30	FRC4808	Washer-sealing	1		Detoxed

B 55

DISCOVERY 1989MY UPTO 1999MY

GROUP B

Engine
2500 - Diesel - Turbo - 300 Tdi
Rockershaft

Illus	Part Number	Description	Quantity	Change Point	Remarks
1	ERR3343	ROCKER ASSEMBLY-CYLINDER HEAD-RH	4		
2	ERR3342	ROCKER ASSEMBLY-CYLINDER HEAD-LH	4		
3	ERR1203	• Bush-cylinder head rocker arm	8		
4	ERR2405	Spring-rocker shaft	4		
5	ERR4848	SHAFT-ROCKER	1		
6	ERR3457	• Plug	2		
7	ERR1209	Spacer	6		
8	ERR2732	Spacer	2		
9	ERR4883	Screw-engine tappet adjustment	8		
10	ERR560	Nut-locking	8		
11	ERR3779	Clamp-rocker shaft	5		
12	ERR4687	Stud	3		
13	FB108101	Bolt-rocker shaft to head-flanged head-M8 x 50	2		
14	FN108047L	Nut-flanged head-M8	3		

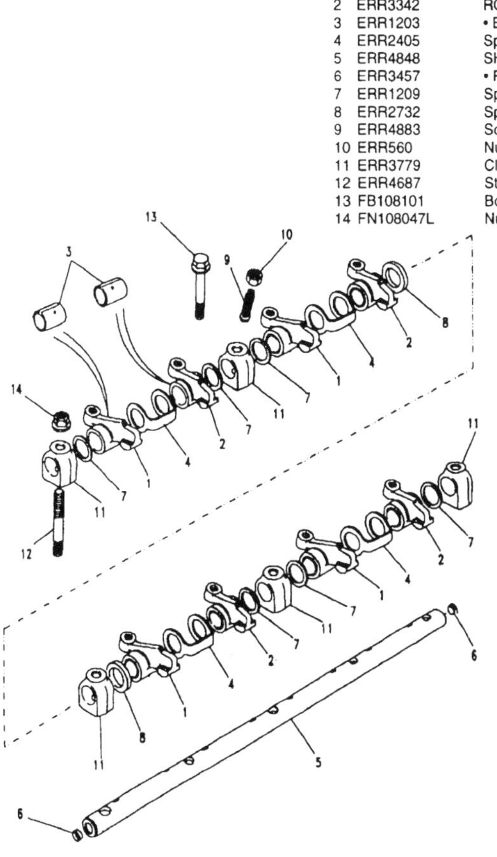

B 56

Illus	Part Number	Description	Quantity	Change Point	Remarks
1	RTC6564	TAPPET-ENGINE VALVE HYDRAULIC	8		
2	502473	• Slide-tappet guide	8		
3	ETC4246	• Bolt	8		
4	ERR561	• Roller-tappet guide	8		
5	ERR607	• Tappet-engine valve mechanical	8		
6	ERR3777	Valve-cylinder head inlet	4		
7	ERR1156	Valve-cylinder head exhaust	4		
8	ERR4640	Spring-cylinder head valve	8		
9	ETC4068	Cap-cylinder head valve spring	8		
10	ETC4069	Cotter-valve	16		
11	546799	Rod-engine push	8		
12	ETC8193	Seat-cylinder head valve spring	8		
13	ETC8663	Seal-cylinder head valve stem	8		
14	LJC100270	Cap-valve stem	8		

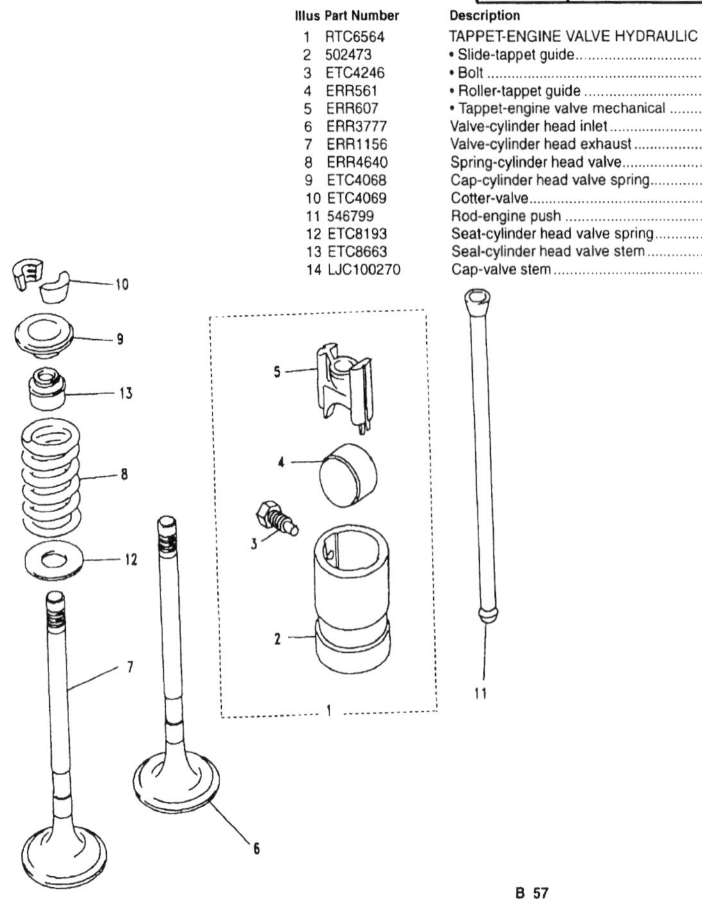

Illus	Part Number	Description	Quantity	Change Point	Remarks
1	ERR4691	Cover assembly-valve rocker	1		
2	ERR2409	Gasket-engine rocker cover	1		
3	ERR5041	Cap-oil filler	1		
4	ERR4834	Screw	3		
5	ERR3424	Washer-sealing	3		
6	ERR4632	Cover-engine-top	1		

DISCOVERY 1989MY UPTO 1999MY	GROUP B	Engine 2500 - Diesel - Turbo - 300 Tdi Engine Breather

Illus	Part Number	Description	Quantity	Change Point	Remarks
1	ERR4706	Cover-breather	1		
2	FS108257L	Screw-cover to block-flanged head-M8 x 25	3		
3	ERR2026	Gasket-cylinder block side cover-asbestos free	1		
4	ERR874	Baffle-oil separator	1		
5	79027	Screw-6 x 3/8	1		
6	ERR7340	Hose-crankcase breather	1		
7	CN100258L	Clip-hose-worm drive-25mm	2		
8	ERR1471	SEPARTOR ASSEMBLY-CRANKCASE BREATHER OIL	1		
	LLO100000	• O ring	1		
9	FB106111L	Bolt-flanged head-breather chamber-M6 x 55	1		
10	ERR1351	Hose-oil separator drain	1		
11	CN100208L	Clip-hose-20mm	2		
12	ERR4689	Hose-oil separator drain	1		
13	CN100258L	Clip-hose-worm drive-25mm	2		
14	CP108251L	Clip-p	1		
15	FS108207L	Screw-flanged head-M8 x 20	1		

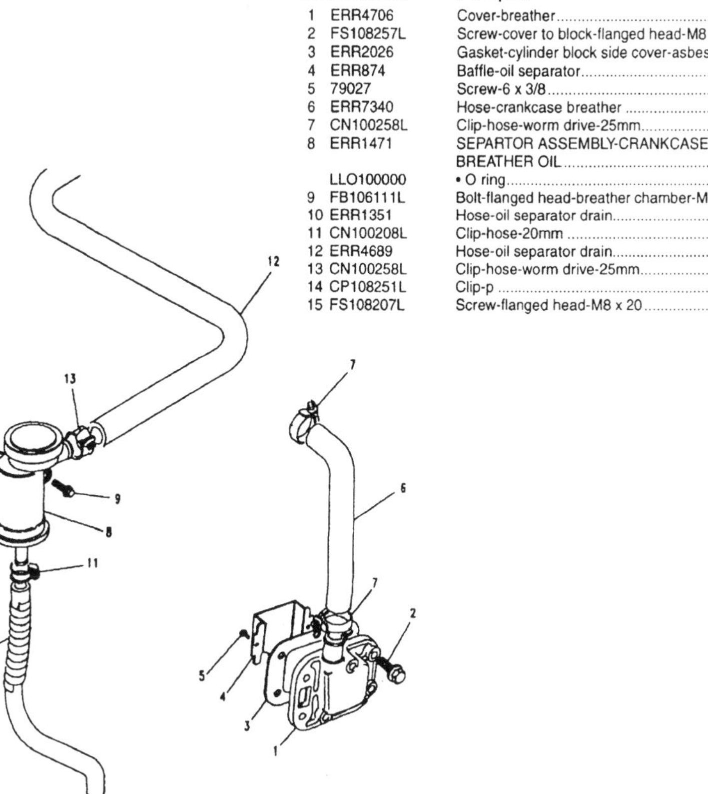

B 59

DISCOVERY 1989MY UPTO 1999MY	GROUP B	Engine 2500 - Diesel - Turbo - 300 Tdi Camshaft

Illus	Part Number	Description	Quantity	Change Point	Remarks
1	ERR3547	Camshaft-engine	1		
2	ERR3754	Plate-camshaft retaining	1		
3	FS106167L	Screw-plate thrust camshaft-hexagonal head-M6 x 16	2		
4	ERR3545	Pulley-camshaft	1		
5	BX112091	Bolt-camshaft retaining plate-M12 x 45	1	Note (1)	
5	FS112401	Screw-flanged head-M12 x 40	1	Note (2)	2500 cc L/R 4 Cylinder Diesel TDi
6	BDU1496L	Washer-special idler pulley	1	Note (1)	
7	ERR2216	Plate-camshaft pulley retaining	1		
8	ERR4709	Dowel-pin	1		
9	FS108161L	Screw-pulley to camshaft-flanged head-M8 x 16	3		
10	ERR3756	Hub-camshaft pulley	1		

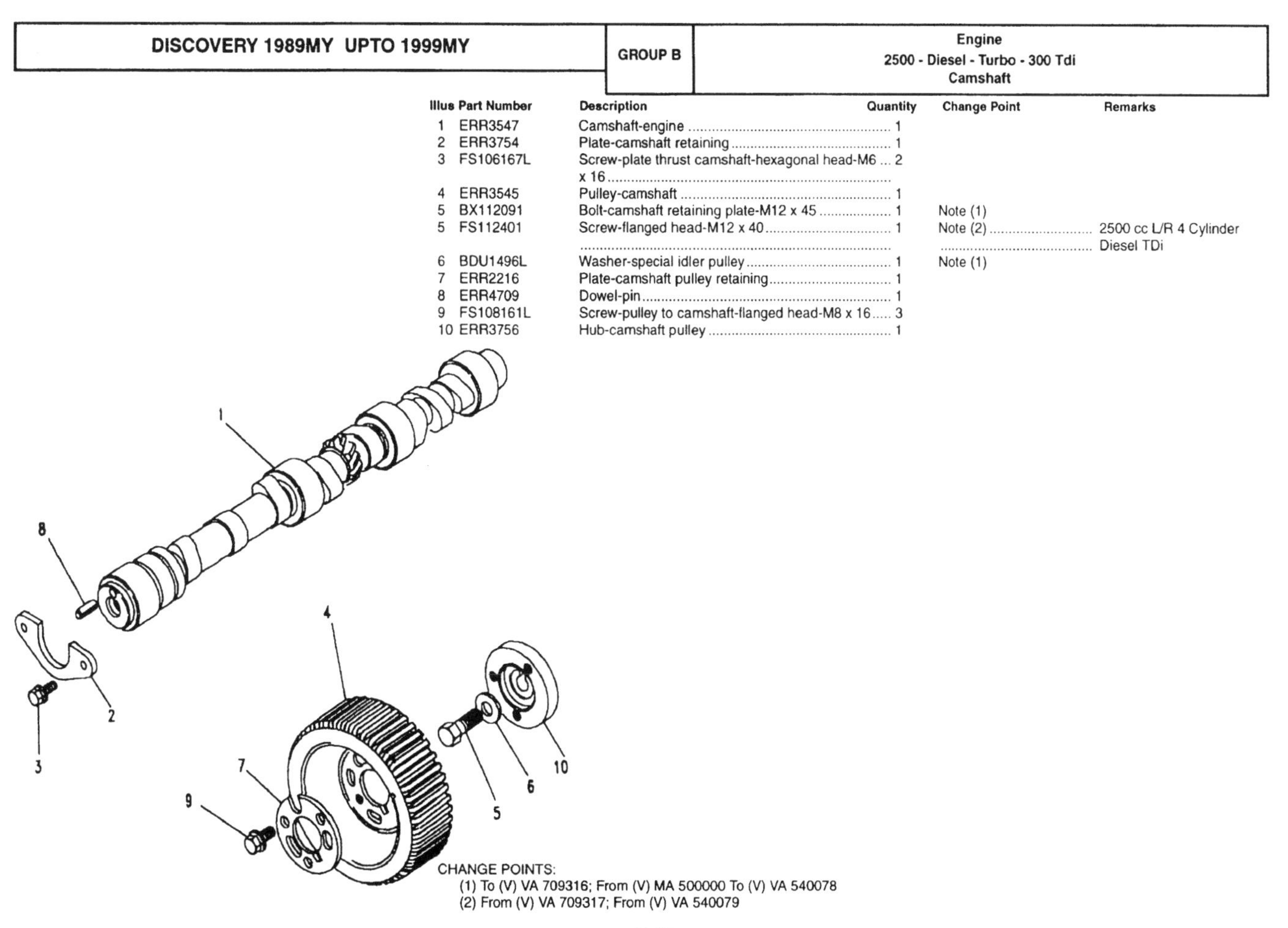

CHANGE POINTS:
(1) To (V) VA 709316; From (V) MA 500000 To (V) VA 540078
(2) From (V) VA 709317; From (V) VA 540079

B 60

Illus	Part Number	Description	Quantity	Change Point	Remarks
1	ERR3479	HOUSING-THERMOSTAT	1		Except air con
2	ERR3291	• Thermostat-engine	1		
3	ERR4685	• O ring	1		
4	ERR4686	• Plug	1		
5	ERR3622	• Pipe assembly-engine coolant-diesel	1	Note (1)	
5	PEP102840	• Pipe assembly-engine coolant-diesel	1		
6	ERR3737	• Elbow-engine coolant outlet	1		
7	FS108251L	• Screw-flanged head-M8 x 25	NLA		Use FS108257L.
1	ERR5098	HOUSING-THERMOSTAT	1		Air con
2	ERR3291	• Thermostat-engine	1		
3	ERR4685	• O ring	1		
4	ERR4686	• Plug	1		
5	ERR3622	• Pipe assembly-engine coolant-diesel	1	Note (1)	
5	PEP102840	• Pipe assembly-engine coolant-diesel	1		
6	ERR3738	• Elbow-engine coolant outlet	1		
7	FS108251L	• Screw-flanged head-M8 x 25	NLA		Use FS108257L.
8	ERR3490	Gasket-thermostat housing-cylinder head	1		
9	FS108257L	Screw-thermostat housing to cylinder head-flanged head-M8 x 25	4		2500 cc 4 Cylinder Diesel TDi
10	AMR1425	Sensor-temperature-Green	1		
11	PRC3359	Switch-engine thermostat-Green	1		
11	PRC3505	Switch-engine thermostat-Yellow.	1		

CHANGE POINTS:
(1) From (E)16L 00088A

Illus	Part Number	Description	Quantity	Change Point	Remarks
1	ERR3481	Manifold-engine inlet	1		Except EDC
1	ERR2284	Manifold-engine inlet	1		EDC
2	ETC5577	Plug-drain	1		Except EDC
3	524765	Plug-core	1		
4	ERR894	Washer-sealing	1		Except EDC
5	ERR3785	Gasket-inlet manifold	1		
6	TE108051L	Stud-M8 x 25	2		
7	FB108171L	Bolt-flanged head-M8 x 85-long	2		
8	FN108041L	Nut-flanged head-M8 x 20	NLA		Use FN108047L.
9	ERR4699	PIPE BOOST CONTROL	1		Except EDC
9	ERR4700	PIPE BOOST CONTROL	1		EDC
10	UKC3803L	• Clip-hose	1		
11	ERR886	Bolt-banjo	1		Except EDC
11	ERR1125	Bolt-banjo	1		EDC
12	273069	Washer-sealing	2		Except EDC
12	ERR894	Washer-sealing	2		EDC
15	TE110071L	Stud-M10 x 35	7		
16	FX110041L	Nut-flange-M10	7		
17	ESR2422	Heatshield-exhaust manifold	1		
18	FS105107	Screw-hexagonal head-M5 x 10	2		
19	ERR2082	Sensor-air temperature diesel	1		EDC
20	ERR894	Washer-sealing	1		EDC

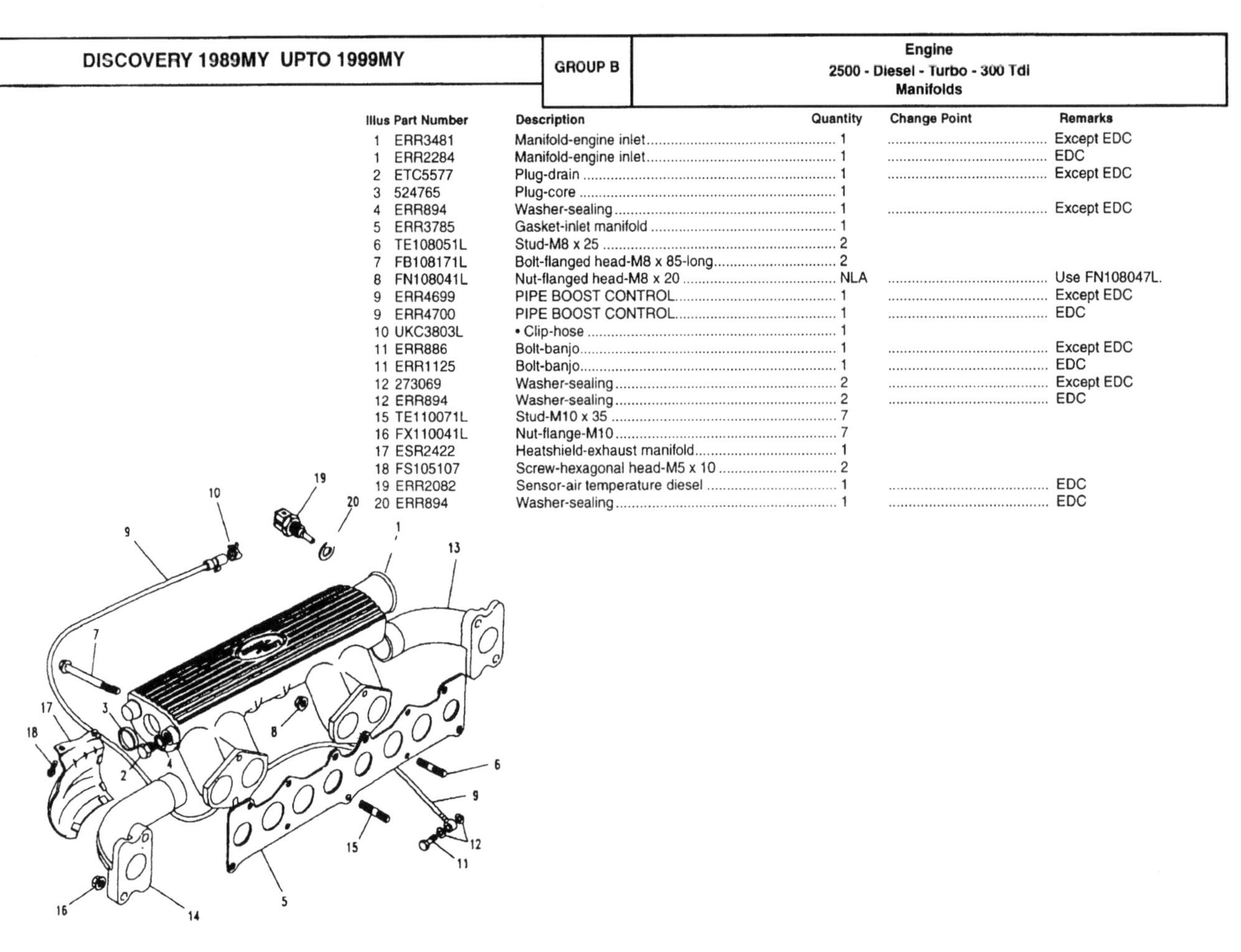

Illus	Part Number	Description	Quantity	Change Point	Remarks
1	ERR5057	Pump-fuel mechanical	1		
2	ERR2028	Gasket-fuel lift pump	1		
3	FS108307L	Screw-lift pump fuel to side cover-flanged head-M8 x 30	2		
4	NRC9770	Nut-tube-female-M12	2		
5	NRC9771	Olive	2		

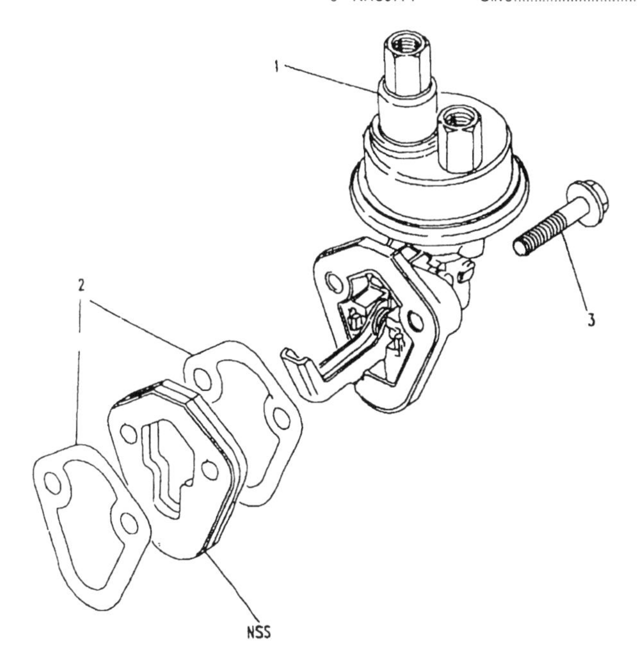

Illus	Part Number	Description	Quantity	Change Point	Remarks
1	ERR4419	PUMP-FUEL INJECTION-NEW	1		Detoxed
1	ERR4419E	PUMP-FUEL INJECTION-EXCHANGE	1		Detoxed
1	ERR3336	PUMP-FUEL INJECTION-NEW	1	Note (1)	EDC
1	ERR3336E	PUMP-FUEL INJECTION-EXCHANGE	1	Note (1)	EDC
1	ERR6727	PUMP-FUEL INJECTION-NEW	1	Note (2)	EDC
1	ERR4046	PUMP-FUEL INJECTION-NEW	1	Note (1)	EGR
1	ERR4046E	PUMP-FUEL INJECTION-EXCHANGE	1	Note (1)	EGR
	RTC6702	• Switch-electronic diesel control	1		
	STC3597	• Bolt-banjo fuel lines	1		
	STC3624	• Potentiometer-throttle diesel	1		
1	ERR6700	PUMP-FUEL INJECTION-NEW	1	Note (2)	EGR
	STC3254	• Solenoid shut off-fuel injection pump	1		
	STC3253	• Bolt-shear head	2		
	STC3252	• Valve-fuel cut fuel tank	1		
2	ERR2023	Gasket-injection pump-diesel engine	1		
3	FN108041L	Nut-flanged head-M8 x 20	NLA		Use FN108047L.
4	ERR6835	Bracket-injection pump diesel	1		
4	ERR6835	Bracket-injection pump diesel	1	Note (2)	
5	FN108041L	Nut-flanged head-M8 x 20	NLA		Use FN108047L.
6	FS108201L	Screw-flanged head-M8 x 20	NLA		Use FS108207L.
6	FS108207L	Screw-flanged head-M8 x 20	2		
7	FB108101	Bolt-bracket to pump-flanged head-M8 x 50	3		
8	ERR3545	Pulley-camshaft	1		
9	ERR2216	Plate-camshaft pulley retaining	1		

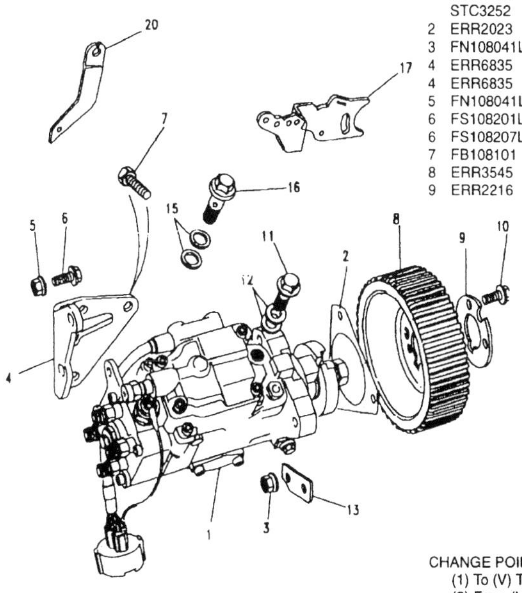

CHANGE POINTS:
(1) To (V) TA 703236; To (V) TA 534103
(2) From (V) VA 703237; From (V) VA 534104

Illus	Part Number	Description	Quantity	Change Point	Remarks
10	FS108161L	Screw-pulley to pump-flanged head-M8 x 16	3		
11	ERR886	Bolt-banjo	1		
12	ERR894	Washer-sealing	2		
13	ERC9404	Bracket harness	2		
14	FB108051	Bolt-flanged head-M8 x 25	3		
15	ERR894	Washer-sealing	2		
16	ERR1125	Bolt-banjo	1		
17	ERR4897	Lever-kickdown automatic transmission	1		Automatic Detoxed
18	FS106121	Screw-M6 x 12	NLA		Use FB106121L part raised in error on TDB download. Automatic Detoxed
19	FS106201L	Screw-flanged head-M6 x 20	1		Automatic Detoxed
19	FS106207L	Screw-flanged head-M6 x 20	1		
20	ERR1604	Bracket-injection pump diesel	1		Automatic Cruise control
	ANR4517	Actuator assembly cruise control	1		
5	NH108041L	Nut-hexagonal head-M8	1		
	ANR4518	Bracket-actuator assembly cruise control	1		
19	FS106161L	Screw-flanged head-M6 x 16	NLA		
19	FS106167L	Screw-hexagonal head-M6 x 16	1		

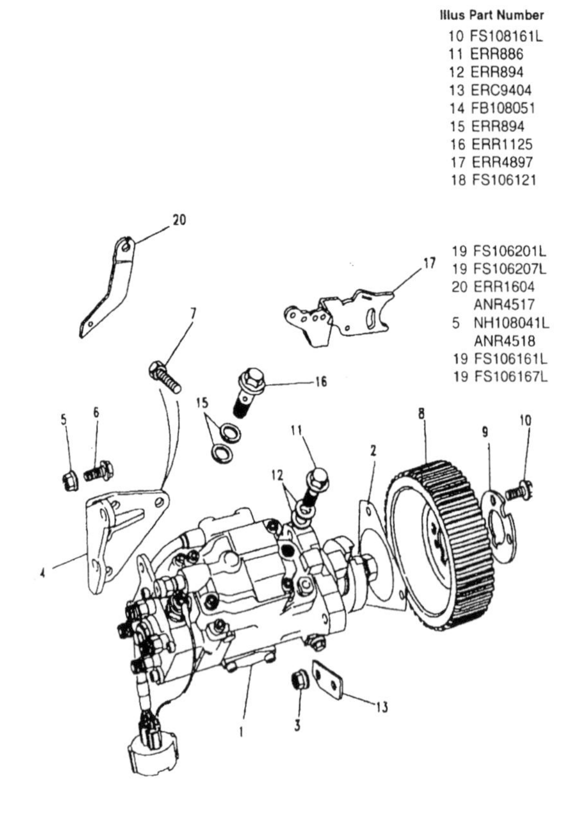

Illus	Part Number	Description	Quantity	Change Point	Remarks
1	ERR3306	Pipe-spill return fuel injection	1		
2	STC1694	Pipe-cylinder 1-high pressure fuel injection	1		Except EDC
3	STC1695	Pipe-cylinder 2-high pressure fuel injection	1		
4	STC1696	Pipe-cylinder 3-high pressure fuel injection	1		
5	STC1697	Pipe-cylinder 4-high pressure fuel injection	1		
2	STC1698	Pipe-cylinder 1-high pressure fuel injection	1		EDC
3	STC1699	Pipe-cylinder 2-high pressure fuel injection	1		
4	STC1700	Pipe-cylinder 3-high pressure fuel injection	1		
5	STC1701	Pipe-cylinder 4-high pressure fuel injection	1		
6	ERR3339	Injector assembly-diesel holder & nozzle-new	4		EGR
6	ERR3339E	Injector assembly-diesel holder & nozzle-exchange	4		
6	ERR3337	Injector assembly-diesel holder & nozzle	1		EDC
6	ERR3337E	Injector assembly-diesel holder & nozzle	1		
6	ERR3348	Injector assembly-diesel holder & nozzle	3		
6	ERR3348E	Injector assembly-diesel holder & nozzle	1		
7	ERR3780	Clamp fuel injector	4		
8	STC3297	BOLT-BANJO	1		
9	ERR1304	• Washer-sealing-diesel injector spill return pipe banjo	8		
10	FN108041L	Nut-flanged head-M8 x 20	NLA		Use FN108047L.
11	STC1738	Clip-pipe	2		
12	STC1739	Clip-pipe	2		
13	ERR3468	Pin-roll-spring tension	4		
14	ERR4621	Washer-sealing-diesel injector to cylinder head	4		
15	DCP7384L	Clip-cable	1		

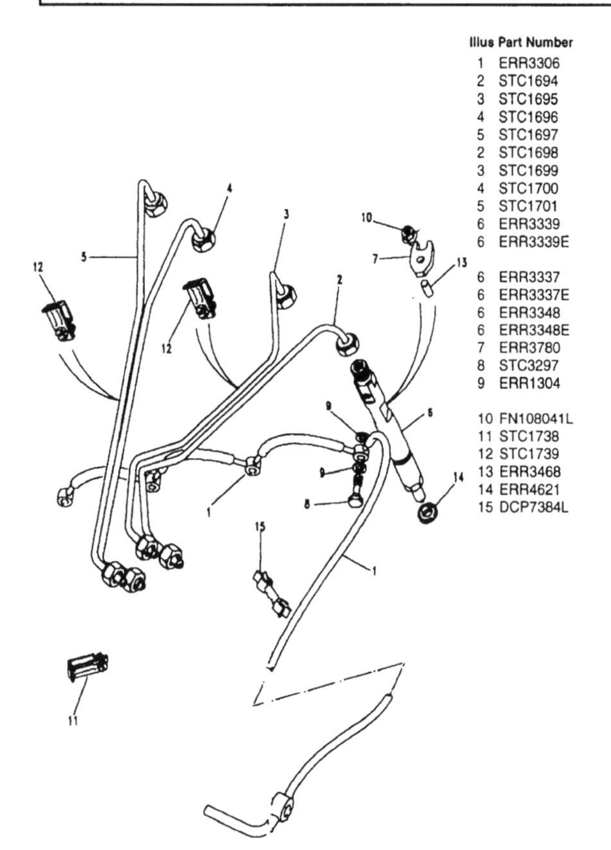

Illus	Part Number	Description	Quantity	Change Point	Remarks
1	ERR4802	TURBOCHARGER ASSEMBLY-NEW	1		
1	ERR4802E	TURBOCHARGER ASSEMBLY-EXCHANGE	1		
	STC3159	• Stud	3		
	STC3084	• Actuator-waste gate turbocharger	1		
	STC3592	• Clip	1		
2	ERR1125	Bolt-banjo	1		
3	FRC4808	Washer-sealing	2		
4	ERR4894	Pipe turbocharger oil feed	1		
5	ERR4895	Pipe-oil drain turbocharger	1		
6	ERR2109	Gasket-oil drain turbocharger	1		
7	ERR335	Adaptor-oil drain pipe turbocharger	1		
8	FS108207L	Screw-flanged head-M8 x 20	2		
9	ERR4698	Plate-turbocharger blanking	1		
10	ERR7173	Gasket-turbocharger blanking	1		
11	SS108201	Screw-hexagon socket-plate to manifold-M8 x 20	2		
11	SS108251L	Screw-hexagon socket-M8 x 25	1		
12	ETC8820	Pipe-adaptor turbocharger	1		

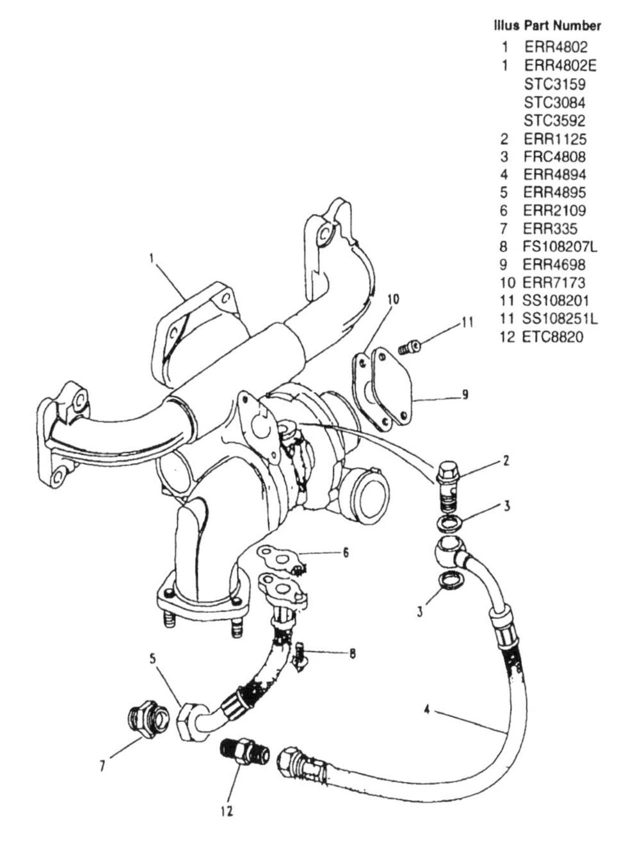

Illus	Part Number	Description	Quantity	Change Point	Remarks
1	STC2801	Kit-gasket-overhaul	1		
2	STC2802	Kit-gasket	1		

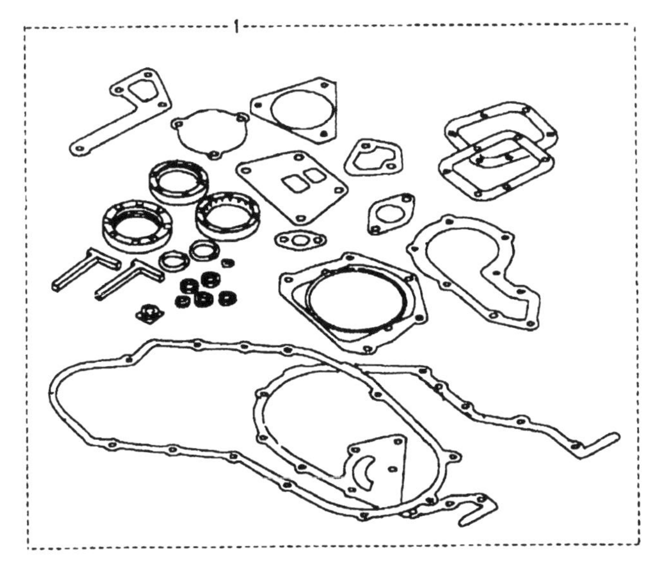

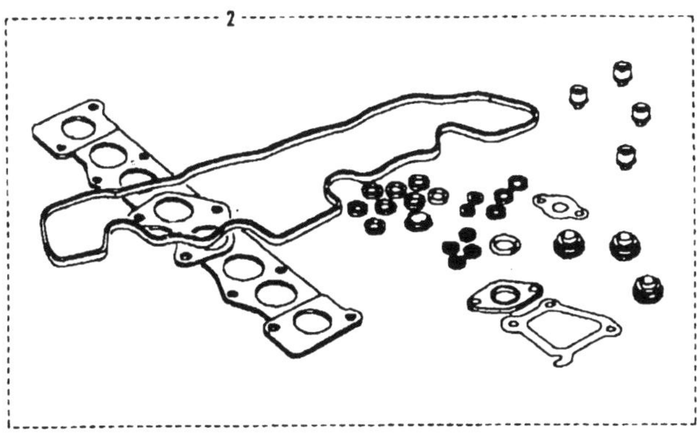

Illus	Part Number	Description	Quantity	Change Point	Remarks
1	ERR3539	Pump brake vacuum	1		
2	FS108201L	Screw-pump to block-flanged head-M8 x 20	NLA		Use FS108207L.
2	FS108207L	Screw-pump to block-flanged head-M8 x 20	5		2500 cc 4 Cylinder Diesel TDi
3	ERR2027	Gasket-vacuum pump	1		
4	ERC9404	Bracket harness	1		

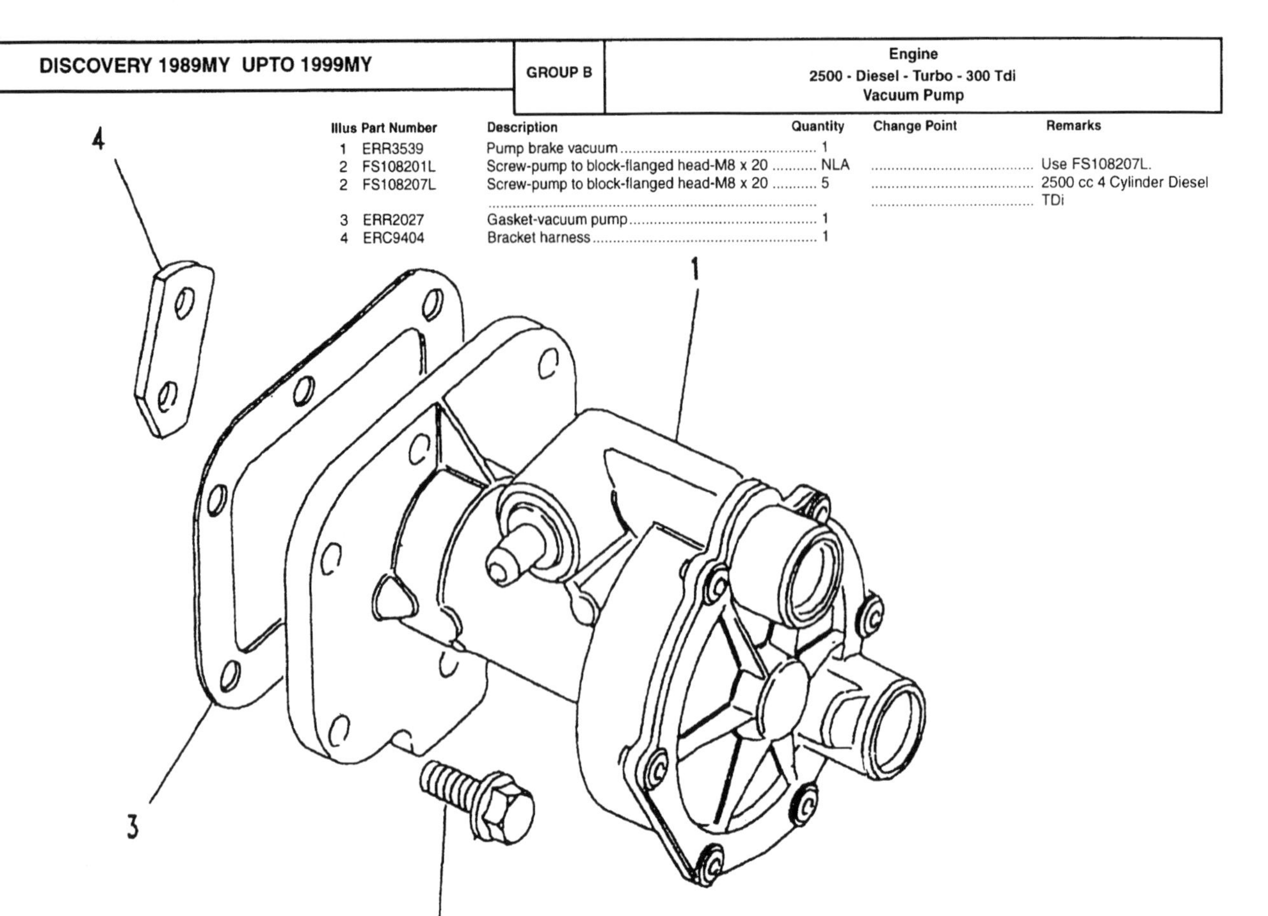

B 69

Illus	Part Number	Description	Quantity	Change Point	Remarks
1	ANR2157	Pump assembly power assisted steering	1		
2	ERR2228	Bracket-mounting power assisted steering pump	1		
4	FS108201L	Screw-flanged head-M8 x 20	NLA		Use FS108207L.
4	FS108207L	Screw-flanged head-M8 x 20	3		
5	ERR3733	Pulley power assisted steering pump	1		
6	FS108127	Screw-pulley to pump-flanged head-M8 x 12	3		

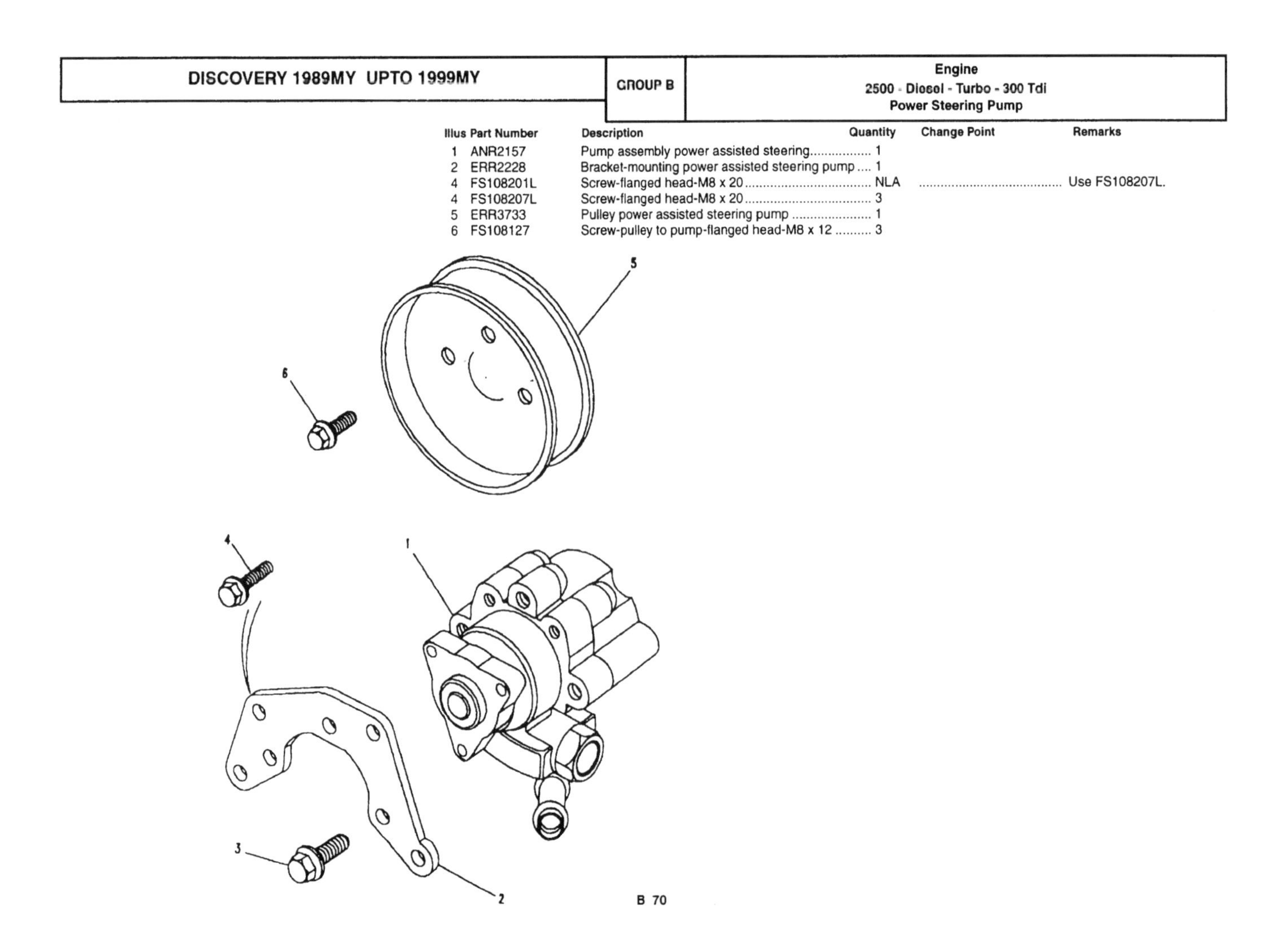

B 70

Air Conditioning Compressor

Illus	Part Number	Description	Quantity	Change Point	Remarks
1	BTR4717	Compressor air conditioning	1		
2	ERR2215	Belt-polyvee air conditioning compressor	1		
3	TE108181	Stud-compressor to bracket-M8 x 90	3		
4	FN108041L	Nut-flanged head-M8 x 20	NLA		Use FN108047L.
5	ERR4975	Bracket-compressor air conditioning	1		
6	FS110301L	Screw-bracket to block-flanged head-M10 x 30	4		
7	ERR4639	Shroud air conditioning	1		
8	FN108041L	Nut-flanged head-M8 x 20	NLA		Use FN108047L.

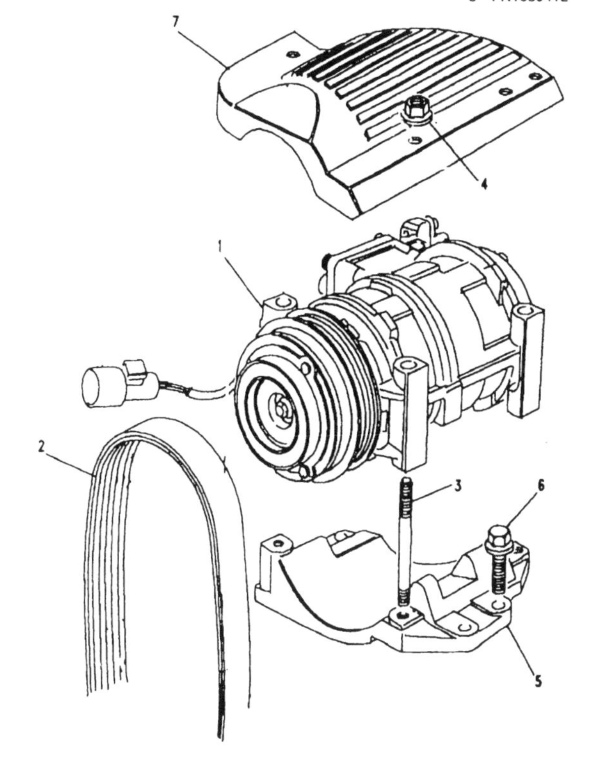

Air Conditioning Idlers

Illus	Part Number	Description	Quantity	Change Point	Remarks
1	ERR4543	PULLEY-TENSIONER ANCILLARY DRIVE	NLA		Use ERR7386 qty 1 with ERR7387 qty 1.
2	ERR3807	• Cover-air conditioning belt tensioner	1		
3	FS110301L	• Screw-flanged head-M10 x 30	1		
4	FS108251L	Screw-flanged head-M8 x 25	NLA		Use FS108257L.
4	FS108257L	Screw-flanged head-M8 x 25	3		
5	ERR2798	Pulley-idler ancillary drive	1	Note (1)	
5	ERR7295	Pulley-idler ancillary drive-with flinger and dust cover	1	Note (2)	
6	FS110301L	Screw-flanged head-M10 x 30	1		

CHANGE POINTS:
(1) To (E)20L 06398A; To (E)19L 11266A; To (E)25L 01117A
(2) From (E)20L 06399A; From (E)19L 11267A; From (E)25L 01118A

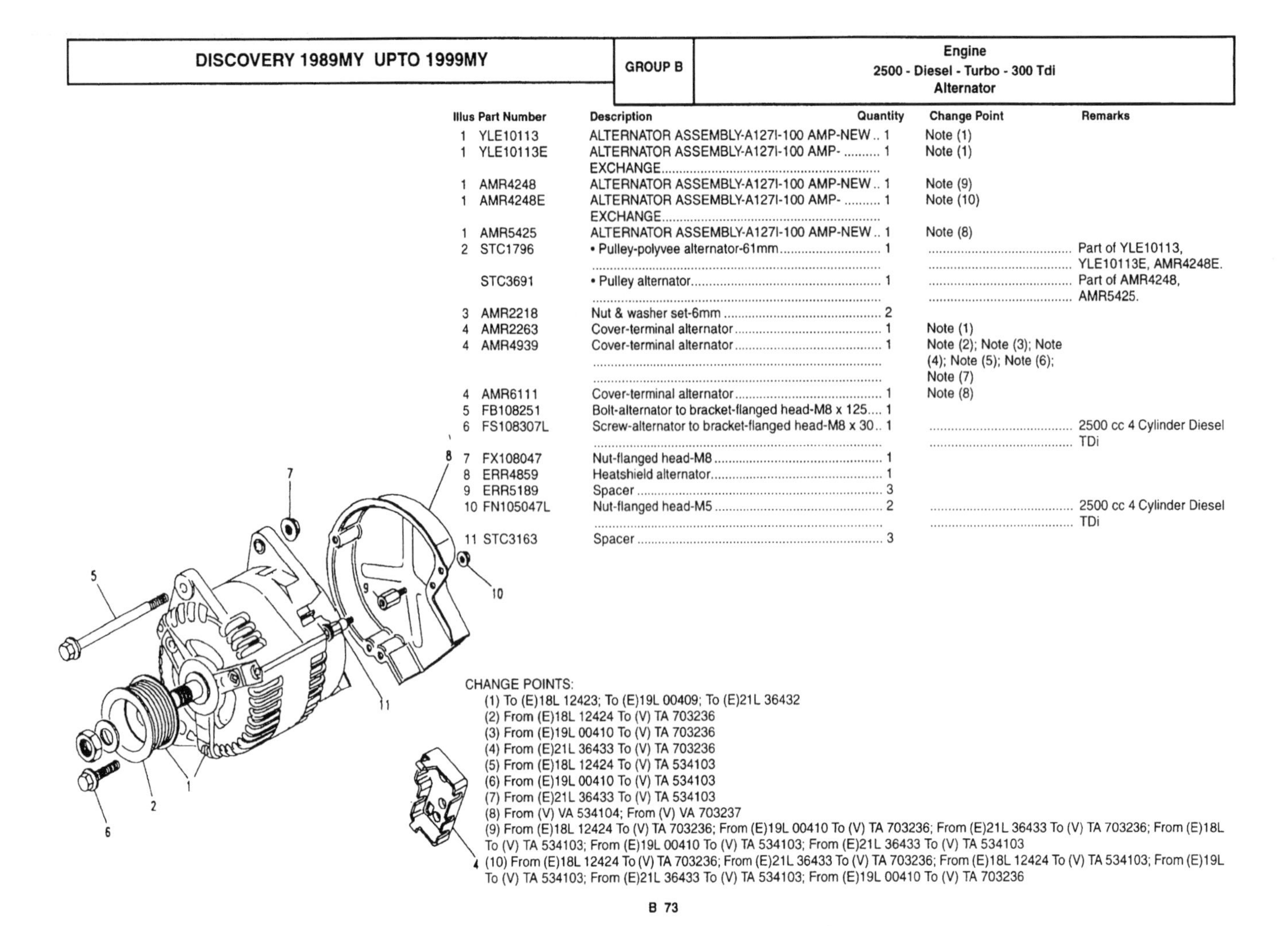

Illus	Part Number	Description	Quantity	Change Point	Remarks
1	YLE10113	ALTERNATOR ASSEMBLY-A127I-100 AMP-NEW	1	Note (1)	
1	YLE10113E	ALTERNATOR ASSEMBLY-A127I-100 AMP-EXCHANGE	1	Note (1)	
1	AMR4248	ALTERNATOR ASSEMBLY-A127I-100 AMP-NEW	1	Note (9)	
1	AMR4248E	ALTERNATOR ASSEMBLY-A127I-100 AMP-EXCHANGE	1	Note (10)	
1	AMR5425	ALTERNATOR ASSEMBLY-A127I-100 AMP-NEW	1	Note (8)	
2	STC1796	• Pulley-polyvee alternator-61mm	1		Part of YLE10113, YLE10113E, AMR4248E.
	STC3691	• Pulley alternator	1		Part of AMR4248, AMR5425.
3	AMR2218	Nut & washer set-6mm	2		
4	AMR2263	Cover-terminal alternator	1	Note (1)	
4	AMR4939	Cover-terminal alternator	1	Note (2); Note (3); Note (4); Note (5); Note (6); Note (7)	
4	AMR6111	Cover-terminal alternator	1	Note (8)	
5	FB108251	Bolt-alternator to bracket-flanged head-M8 x 125	1		
6	FS108307L	Screw-alternator to bracket-flanged head-M8 x 30	1		2500 cc 4 Cylinder Diesel TDi
7	FX108047	Nut-flanged head-M8	1		
8	ERR4859	Heatshield alternator	1		
9	ERR5189	Spacer	3		
10	FN105047L	Nut-flanged head-M5	2		2500 cc 4 Cylinder Diesel TDi
11	STC3163	Spacer	3		

CHANGE POINTS:

(1) To (E)18L 12423; To (E)19L 00409; To (E)21L 36432
(2) From (E)18L 12424 To (V) TA 703236
(3) From (E)19L 00410 To (V) TA 703236
(4) From (E)21L 36433 To (V) TA 703236
(5) From (E)18L 12424 To (V) TA 534103
(6) From (E)19L 00410 To (V) TA 534103
(7) From (E)21L 36433 To (V) TA 534103
(8) From (V) VA 534104; From (V) VA 703237
(9) From (E)18L 12424 To (V) TA 703236; From (E)19L 00410 To (V) TA 703236; From (E)21L 36433 To (V) TA 703236; From (E)18L To (V) TA 534103; From (E)19L 00410 To (V) TA 534103; From (E)21L 36433 To (V) TA 534103
(10) From (E)18L 12424 To (V) TA 703236; From (E)21L 36433 To (V) TA 703236; From (E)18L 12424 To (V) TA 534103; From (E)19L To (V) TA 534103; From (E)21L 36433 To (V) TA 534103; From (E)19L 00410 To (V) TA 703236

B 73

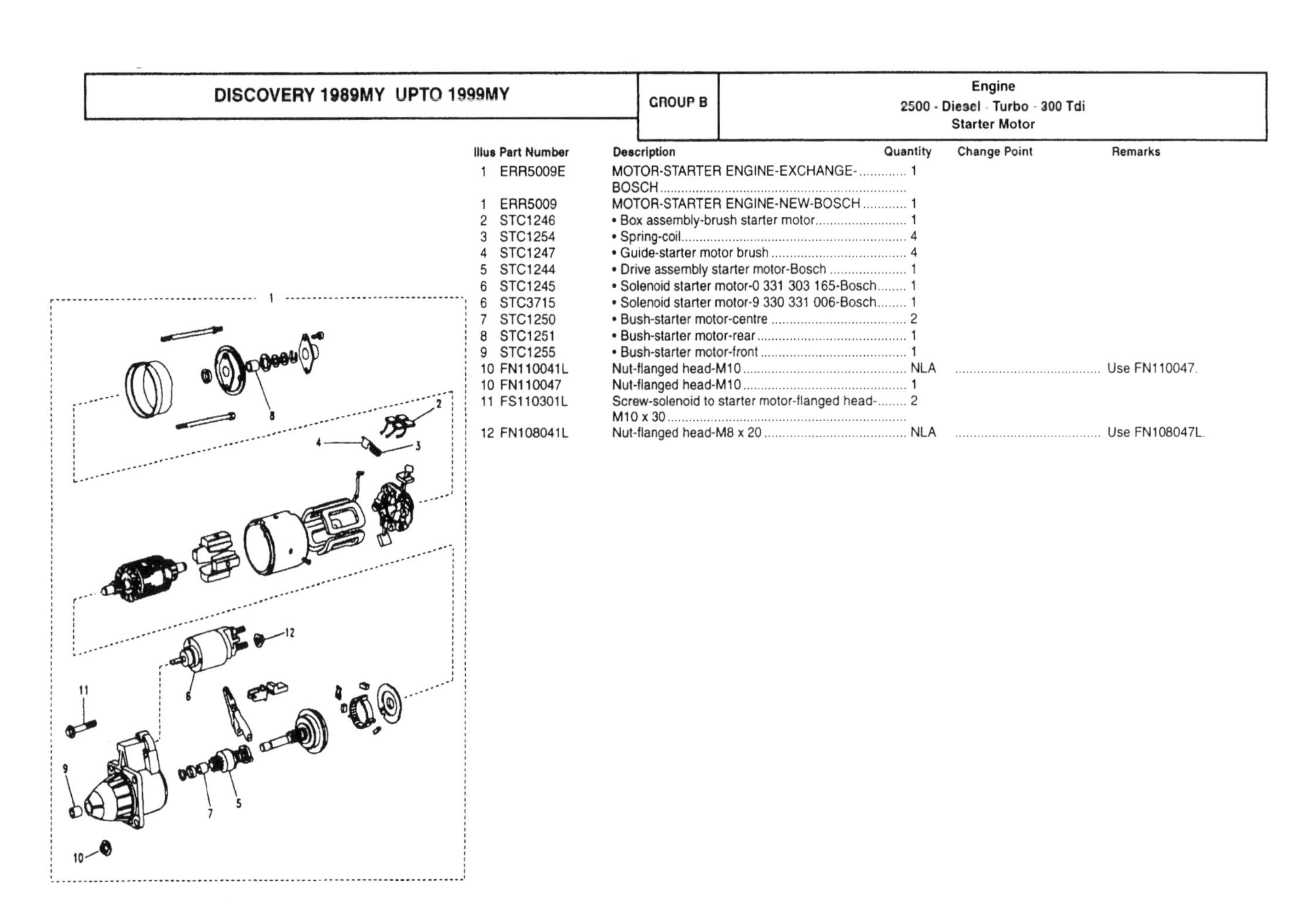

Illus	Part Number	Description	Quantity	Change Point	Remarks
1	ERR5009E	MOTOR-STARTER ENGINE-EXCHANGE-BOSCH	1		
1	ERR5009	MOTOR-STARTER ENGINE-NEW-BOSCH	1		
2	STC1246	• Box assembly-brush starter motor	1		
3	STC1254	• Spring-coil	4		
4	STC1247	• Guide-starter motor brush	4		
5	STC1244	• Drive assembly starter motor-Bosch	1		
6	STC1245	• Solenoid starter motor-0 331 303 165-Bosch	1		
6	STC3715	• Solenoid starter motor-9 330 331 006-Bosch	1		
7	STC1250	• Bush-starter motor-centre	2		
8	STC1251	• Bush-starter motor-rear	1		
9	STC1255	• Bush-starter motor-front	1		
10	FN110041L	Nut-flanged head-M10	NLA		Use FN110047.
10	FN110047	Nut-flanged head-M10	1		
11	FS110301L	Screw-solenoid to starter motor-flanged head-M10 x 30	2		
12	FN108041L	Nut-flanged head-M8 x 20	NLA		Use FN108047L.

B 74

DISCOVERY 1989MY UPTO 1999MY

GROUP B

Engine
2500 - Diesel - Turbo - 300 Tdi
Flywheel and Clutch

Illus	Part Number	Description	Quantity	Change Point	Remarks
1	ERR4723	HOUSING ASSEMBLY-CLUTCH	1		Detoxed
					EGR
2	TE110061L	• Stud-M10 x 30	13		
3	TE110051L	• Stud-M10 x 25	1		
1	ERR4722	HOUSING ASSEMBLY-CLUTCH	1		EDC
2	TE110061L	• Stud-M10 x 30	9		
3	TE110051L	• Stud-M10 x 25	1		
4	ERC7295	Plug-rubber	1		
5	FS110301L	Screw-housing to block-flanged head-M10 x 30	6		
6	FB110071L	Bolt-housing to block-flanged head-M10 x 35	2		
6	FB110141L	Bolt-housing to block-flanged head-M10 x 70	4		
7	WA110061L	Washer-plain-M10-standard	4		
8	AMR3755	Bracket harness	1		
9	ERR719	FLYWHEEL ENGINE	1		
10	568431	• Gear-ring-flywheel engine	1		
11	502116	• Dowel-flywheel to crankshaft	3		
12	ERR4574	Bolt-fixing flywheel crankshaft-short	8		
13	URB100760	Cover-clutch assembly	1		
14	FTC4204	Plate-clutch-driven	1		
15	SX108201L	Bolt-cover clutch to flywheel-M8 x 20	6		
16	WL108001L	Washer-M8	6		
17	ERR4645	Sensor-speed engine	1		
18	ERR894	Washer-sealing	1		

B 75

DISCOVERY 1989MY UPTO 1999MY

GROUP B

Engine
2500 - Diesel - Turbo - 300 Tdi
Drive Plate Auto

Illus	Part Number	Description	Quantity	Change Point	Remarks
1	ERR1616	Plate-gearbox mounting	1		
2	TE110071L	Stud-M10 x 35	3		
3	AMR3754	Bracket harness	1		
4	FS110301L	Screw-flanged head-M10 x 30	6		
5	ERR1618	Plate-closure automatic transmission	1		
6	ERR1617	Plug	1		
7	ERR1619	Gasket-converter housing lower access cover	1		
8	FS108207L	Screw-flanged head-M8 x 20	1		
9	ERC7295	Plug-rubber	1		
10	FTC4050	PLATE-DRIVE ASSEMBLY AUTOMATIC TRANSMISSION	1		EGR
					Detoxed
10	FTC4052	PLATE-DRIVE ASSEMBLY AUTOMATIC TRANSMISSION	1		EDC
11	ERR4829	• GEAR-RING-FLYWHEEL ENGINE	1		
12	529364	• • Dowel-flywheel spigot alignment	2		
13	FS108161ML	Screw-flanged head-M8 x 16	12		
13	SH108161M	Screw-hexagonal head-M8 x 16	12		
14	FTC4213	Plate-backing automatic transmission	1		
15	FTC4214	Spacer-crank/drive plate automatic transmission	1		
16	ERR4574	Bolt-fixing flywheel crankshaft-short	8		
17	ERR2079	Sensor assembly-crank engine	1		
18	ERR894	Washer-sealing	1		
19	FS110141M	Screw-M10 x 14	4		
		TORQUE CONVERTOR HEIGHT SHIM			
20	FRC6291	1.1mm	A/R		
20	FRC6293	1.30mm	A/R		
20	FRC6295	1.50mm	A/R		
20	FRC6297	1.70mm	A/R		
20	FRC6299	1.90mm	A/R		
20	FRC6301	2.10mm	A/R		

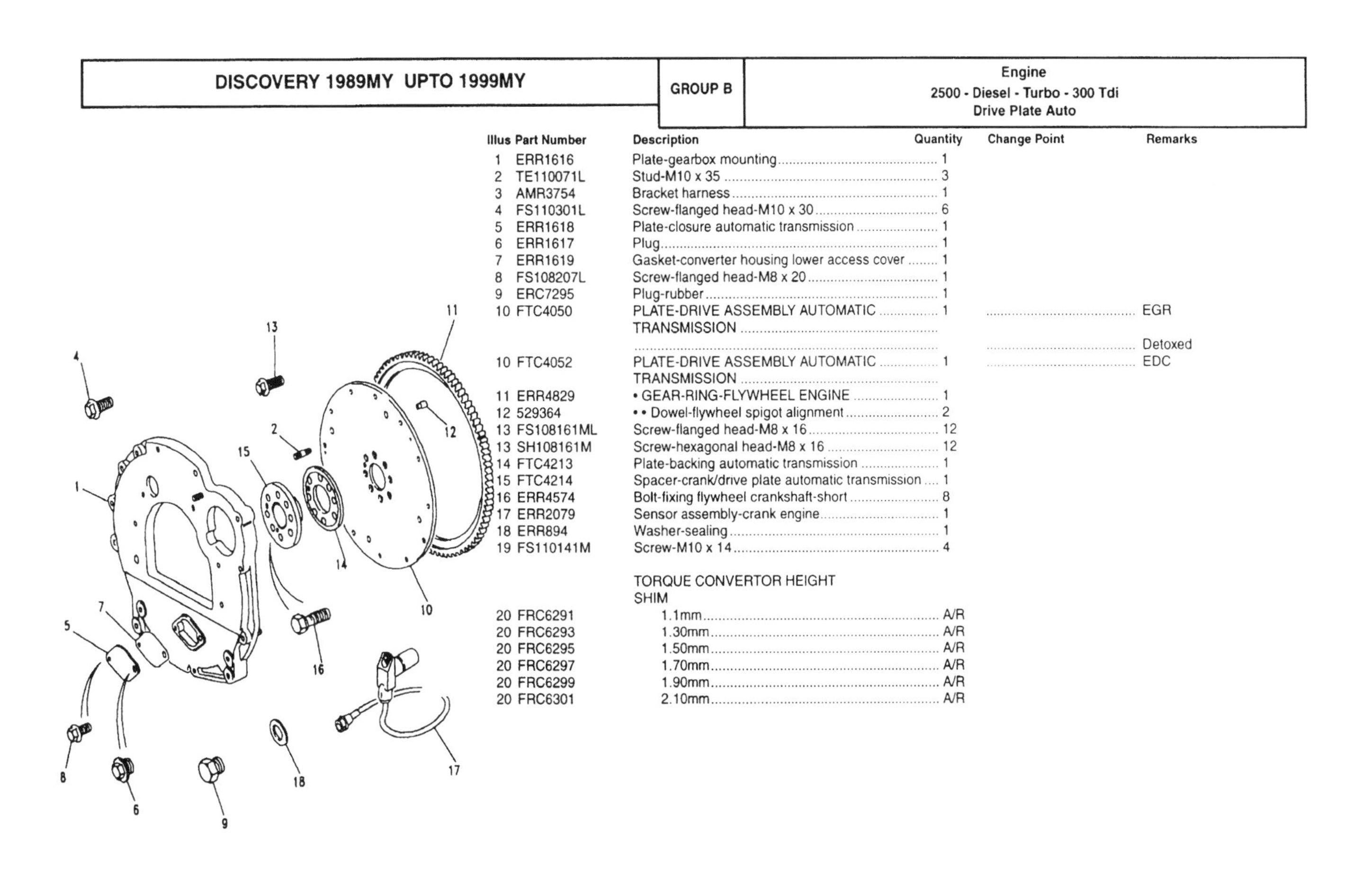

B 76

Illus	Part Number	Description	Quantity	Change Point	Remarks
1	RTC6682R	Engine unit-stripped-reconditioned	1		
1	RTC6682N	Engine unit-stripped-new	1	Note (1)	Note (1)

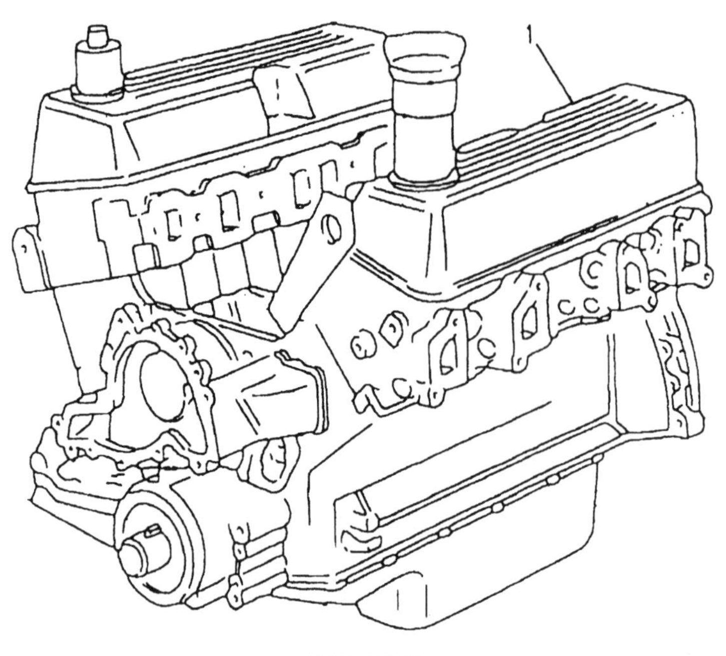

CHANGE POINTS:
(1) From (E)27G 00001

Remarks:
(1) Low Compression

B 77

Illus	Part Number	Description	Quantity	Change Point	Remarks
1	ETC7714	Engine base unit-part-8.13:1	1		Low Compression

COMPRISES BLOCK CRANKSHAFT
AND PISTONS

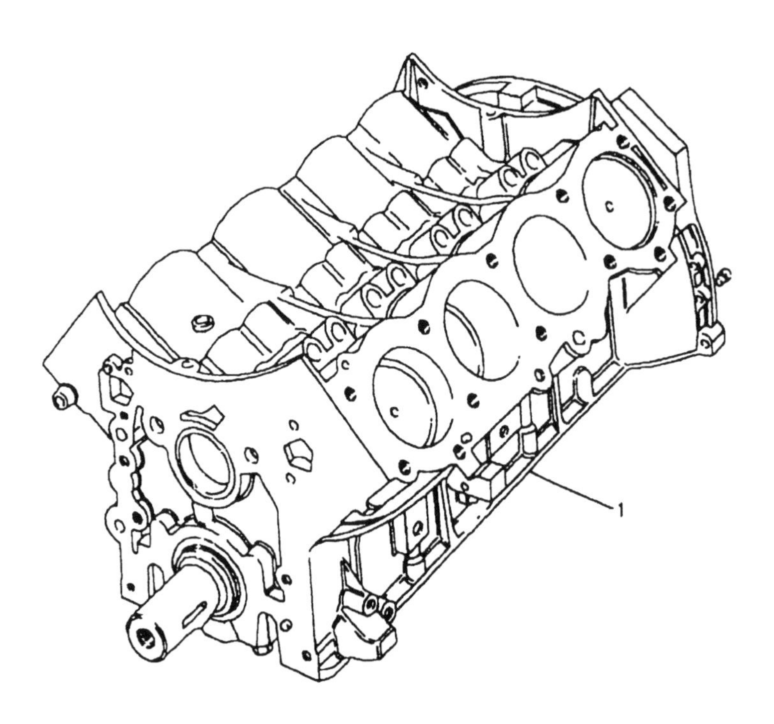

B 78

DISCOVERY 1989MY UPTO 1999MY

GROUP B — **Engine** / **V8 Petrol - Carburetter** / **Cylinder Block**

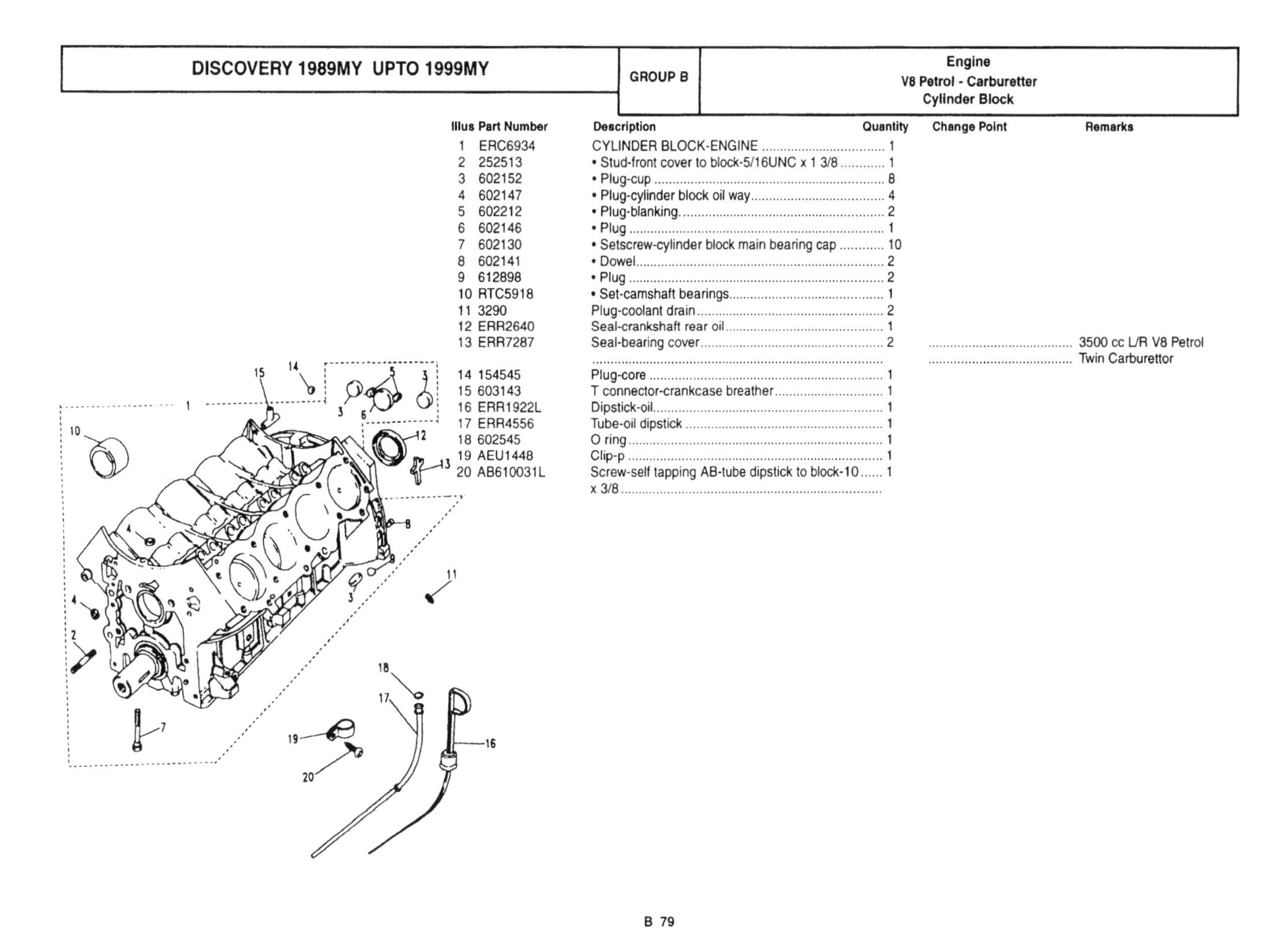

Illus	Part Number	Description	Quantity	Change Point	Remarks
1	ERC6934	CYLINDER BLOCK-ENGINE	1		
2	252513	• Stud-front cover to block-5/16UNC x 1 3/8	1		
3	602152	• Plug-cup	8		
4	602147	• Plug-cylinder block oil way	4		
5	602212	• Plug-blanking	2		
6	602146	• Plug	1		
7	602130	• Setscrew-cylinder block main bearing cap	10		
8	602141	• Dowel	2		
9	612898	• Plug	2		
10	RTC5918	• Set-camshaft bearings	1		
11	3290	Plug-coolant drain	2		
12	ERR2640	Seal-crankshaft rear oil	1		
13	ERR7287	Seal-bearing cover	2		3500 cc L/R V8 Petrol Twin Carburettor
14	154545	Plug-core	1		
15	603143	T connector-crankcase breather	1		
16	ERR1922L	Dipstick-oil	1		
17	ERR4556	Tube-oil dipstick	1		
18	602545	O ring	1		
19	AEU1448	Clip-p	1		
20	AB610031L	Screw-self tapping AB-tube dipstick to block-10 x 3/8	1		

DISCOVERY 1989MY UPTO 1999MY

GROUP B — **Engine** / **V8 Petrol - Carburetter** / **Crankshaft and Bearings**

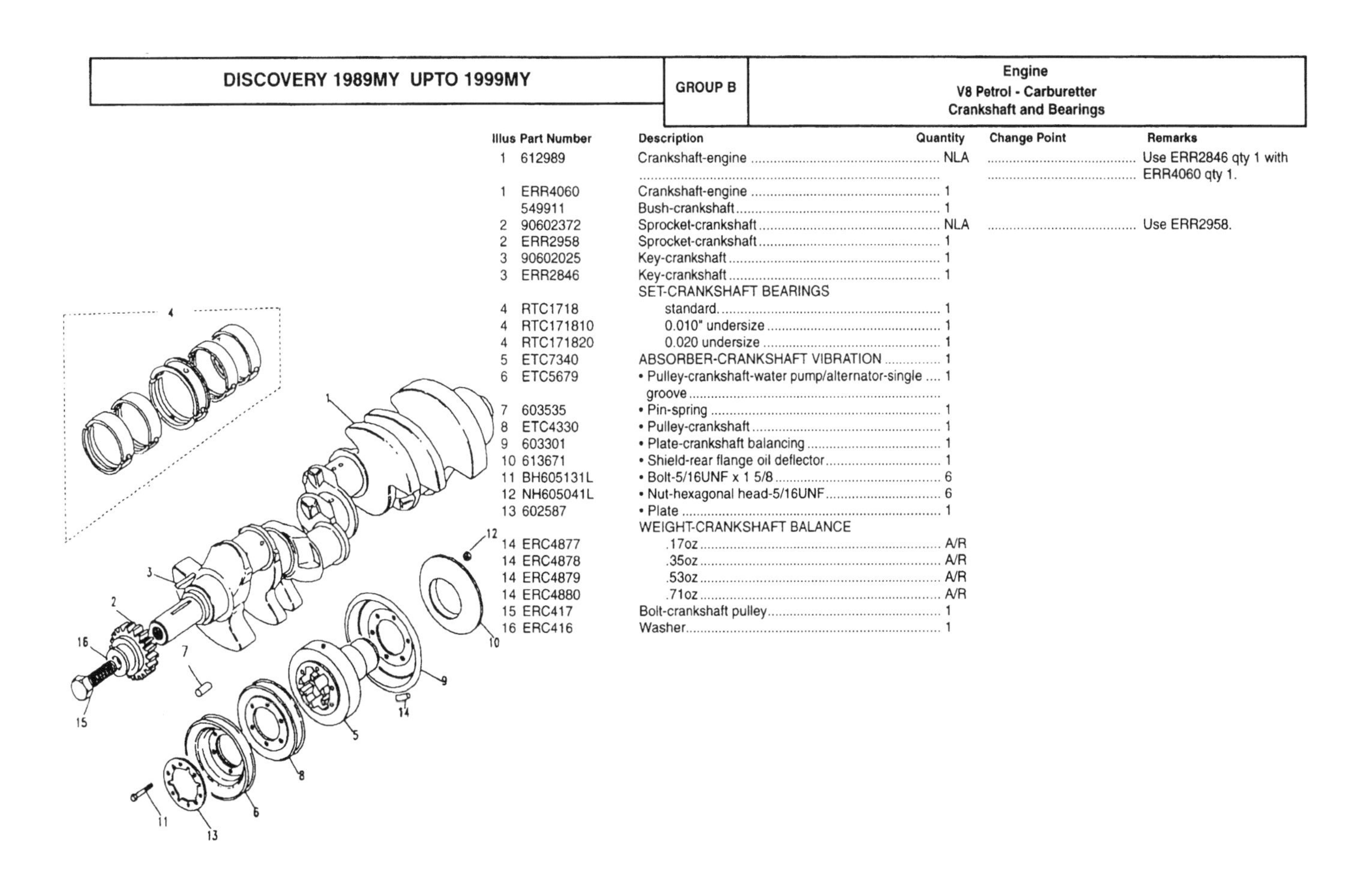

Illus	Part Number	Description	Quantity	Change Point	Remarks
1	612989	Crankshaft-engine	NLA		Use ERR2846 qty 1 with ERR4060 qty 1.
1	ERR4060	Crankshaft-engine	1		
	549911	Bush-crankshaft	1		
2	90602372	Sprocket-crankshaft	NLA		Use ERR2958.
2	ERR2958	Sprocket-crankshaft	1		
3	90602025	Key-crankshaft	1		
3	ERR2846	Key-crankshaft	1		
		SET-CRANKSHAFT BEARINGS			
4	RTC1718	standard	1		
4	RTC171810	0.010" undersize	1		
4	RTC171820	0.020 undersize	1		
5	ETC7340	ABSORBER-CRANKSHAFT VIBRATION	1		
6	ETC5679	• Pulley-crankshaft-water pump/alternator-single groove	1		
7	603535	• Pin-spring	1		
8	ETC4330	• Pulley-crankshaft	1		
9	603301	• Plate-crankshaft balancing	1		
10	613671	• Shield-rear flange oil deflector	1		
11	BH605131L	• Bolt-5/16UNF x 1 5/8	6		
12	NH605041L	• Nut-hexagonal head-5/16UNF	6		
13	602587	• Plate	1		
		WEIGHT-CRANKSHAFT BALANCE			
14	ERC4877	.17oz	A/R		
14	ERC4878	.35oz	A/R		
14	ERC4879	.53oz	A/R		
14	ERC4880	.71oz	A/R		
15	ERC417	Bolt-crankshaft pulley	1		
16	ERC416	Washer	1		

Illus	Part Number	Description	Quantity	Change Point	Remarks
1	RTC2186S	PISTON-ENGINE-STANDARD.-8.13:1	8		
2	RTC2408	• Ring set assembly-piston-standard.	8		
1	RTC218620	PISTON-ENGINE-0.020" OVERSIZE-8.13:1	8		
2	RTC240820	• Ring set assembly-piston-0.020" oversize	8		
3	602082	ROD-CRANKSHAFT CONNECTING	8		
4	602609	• Bolt-connecting rod-11/32UNS-3A	16		
5	602061	• Nut-connecting rod-11/32UNS-3B	16		
		SET-CONNECTING ROD BIG END HALF BEARING			
6	RTC2117	standard	1		
6	RTC211710	0.010" undersize	1		
6	RTC211720	0.020 undersize	1		

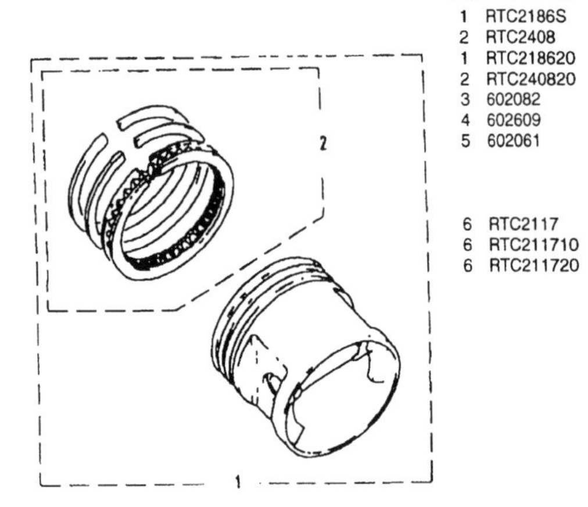

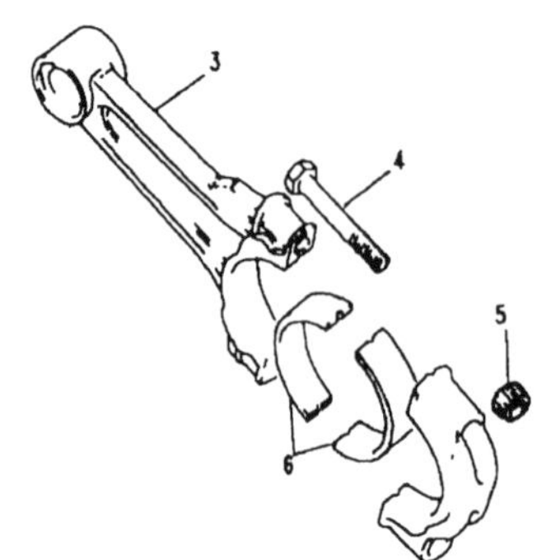

B 81

Illus	Part Number	Description	Quantity	Change Point	Remarks
1	ERR4633	Sump-engine oil	1		
2	213961	Washer-sealing-sump drain plug	1		
3	603659	Plug-sump assembly oil drain	1		
4	602087	Gasket-crankcase oil sump	1		
5	602199	Bolt & washer assembly-sump-sump to block-5/16UNC	NLA		Use LSO100000.
	LSO100000	Bolt & washer assembly-sump-sump to block	16		
6	ERR3788	Gasket-sump oil strainer	1		
7	SH504061L	Setscrew-oil strainer to block-1/4UNC x 3/4	2		
8	WM600041L	Washer-spring-1/4 dia-square	2		
9	ERR3677	Strainer and pipe assembly-oil	1		
10	602070	Cover-oil strainer	1		

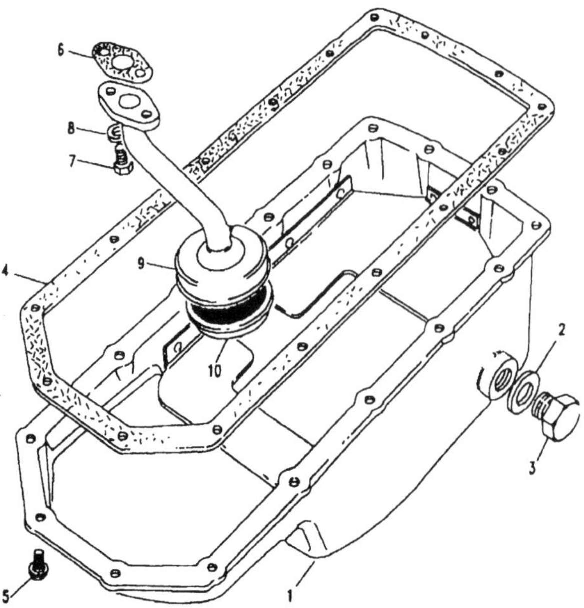

B 82

Engine
V8 Petrol - Carburetter
Engine Mountings

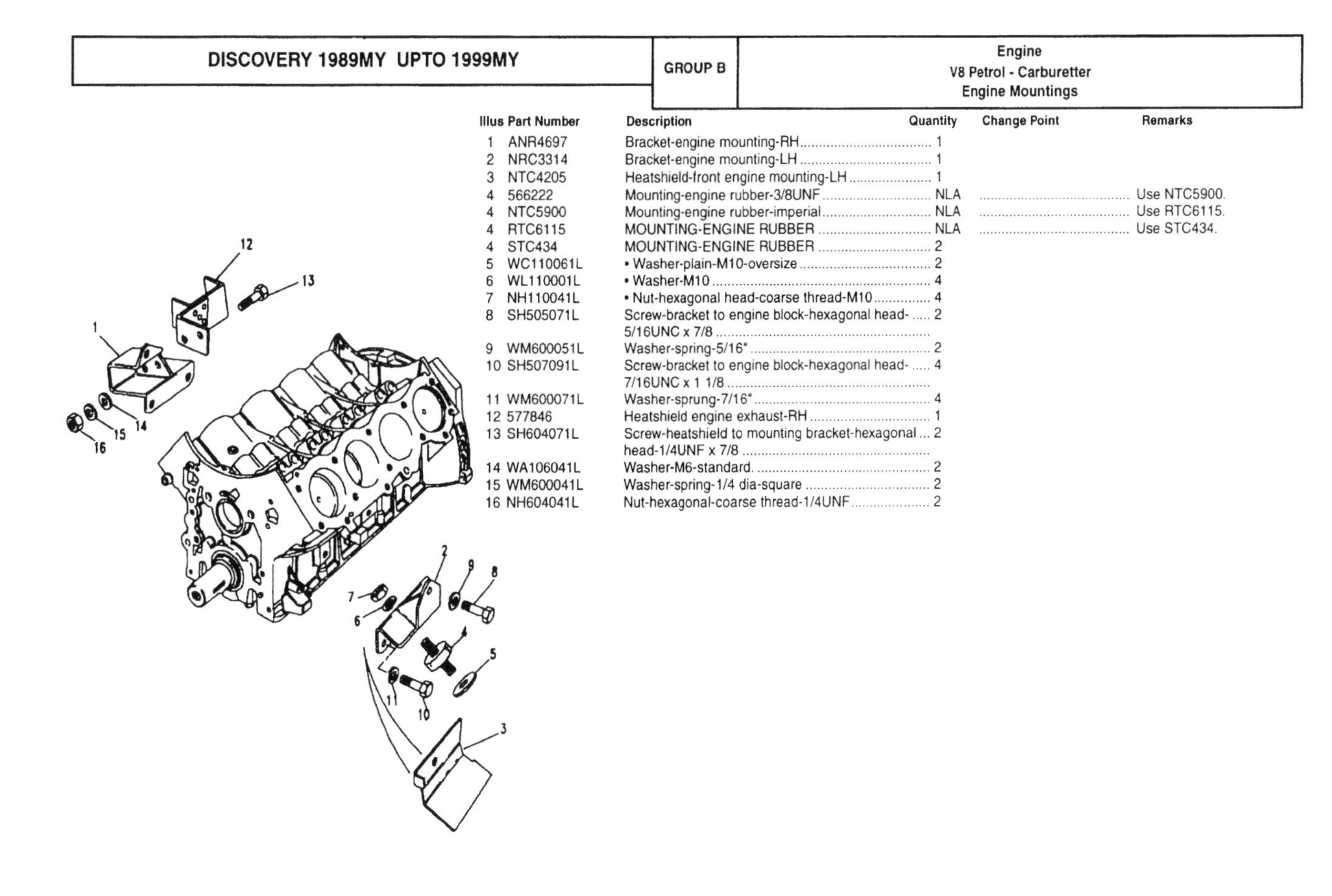

Illus	Part Number	Description	Quantity	Change Point	Remarks
1	ANR4697	Bracket-engine mounting-RH	1		
2	NRC3314	Bracket-engine mounting-LH	1		
3	NTC4205	Heatshield-front engine mounting-LH	1		
4	566222	Mounting-engine rubber-3/8UNF	NLA		Use NTC5900.
4	NTC5900	Mounting-engine rubber-imperial	NLA		Use RTC6115.
4	RTC6115	MOUNTING-ENGINE RUBBER	NLA		Use STC434.
4	STC434	MOUNTING-ENGINE RUBBER	2		
5	WC110061L	• Washer-plain-M10-oversize	2		
6	WL110001L	• Washer-M10	4		
7	NH110041L	• Nut-hexagonal head-coarse thread-M10	4		
8	SH505071L	Screw-bracket to engine block-hexagonal head-5/16UNC x 7/8	2		
9	WM600051L	Washer-spring-5/16"	2		
10	SH507091L	Screw-bracket to engine block-hexagonal head-7/16UNC x 1 1/8	4		
11	WM600071L	Washer-sprung-7/16"	4		
12	577846	Heatshield engine exhaust-RH	1		
13	SH604071L	Screw-heatshield to mounting bracket-hexagonal head-1/4UNF x 7/8	2		
14	WA106041L	Washer-M6-standard.	2		
15	WM600041L	Washer-spring-1/4 dia-square	2		
16	NH604041L	Nut-hexagonal-coarse thread-1/4UNF	2		

Engine
V8 Petrol - Carburetter
Front Cover

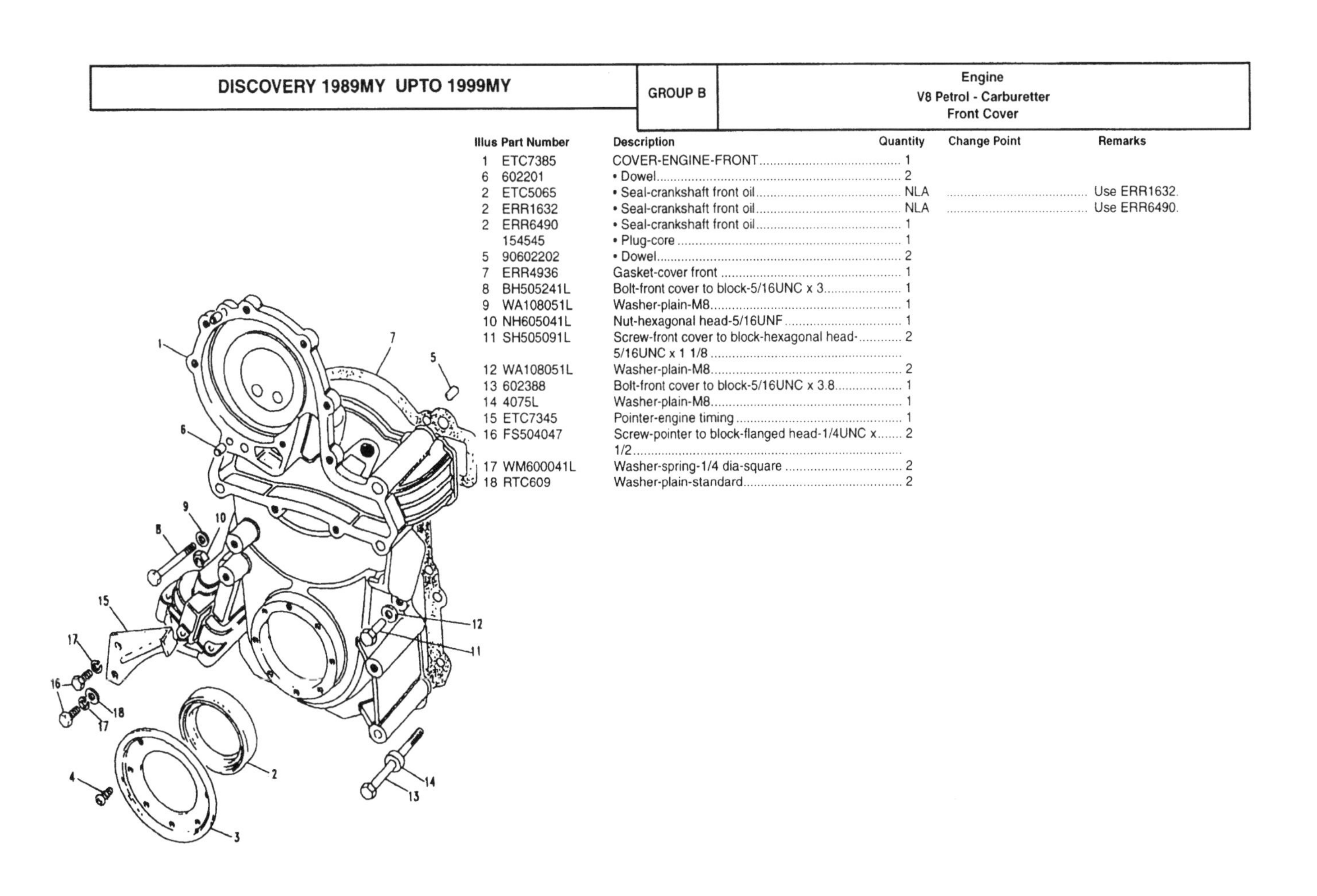

Illus	Part Number	Description	Quantity	Change Point	Remarks
1	ETC7385	COVER-ENGINE-FRONT	1		
6	602201	• Dowel	2		
2	ETC5065	• Seal-crankshaft front oil	NLA		Use ERR1632.
2	ERR1632	• Seal-crankshaft front oil	NLA		Use ERR6490.
2	ERR6490	• Seal-crankshaft front oil	1		
	154545	• Plug-core	1		
5	90602202	• Dowel	2		
7	ERR4936	Gasket-cover front	1		
8	BH505241L	Bolt-front cover to block-5/16UNC x 3	1		
9	WA108051L	Washer-plain-M8	1		
10	NH605041L	Nut-hexagonal head-5/16UNF	1		
11	SH505091L	Screw-front cover to block-hexagonal head-5/16UNC x 1 1/8	2		
12	WA108051L	Washer-plain-M8	2		
13	602388	Bolt-front cover to block-5/16UNC x 3.8	1		
14	4075L	Washer-plain-M8	1		
15	ETC7345	Pointer-engine timing	1		
16	FS504047	Screw-pointer to block-flanged head-1/4UNC x 1/2	2		
17	WM600041L	Washer-spring-1/4 dia-square	2		
18	RTC609	Washer-plain-standard	2		

DISCOVERY 1989MY UPTO 1999MY

GROUP B

Engine
V8 Petrol - Carburetter
Coolant Pump

Illus	Part Number	Description	Quantity	Change Point	Remarks
1	STC483	PUMP ASSEMBLY-ENGINE COOLANT	1		
2	ERR2428	• Gasket-cylinder block coolant pump	1		
3	WA108051L	Washer-plain-M8	4		
4	SH504161L	Screw-coolant pump to front cover-flanged head-1/4UNC x 2	1		
5	BH505121L	Screw-coolant pump to front cover-flanged head-5/16UNC x 1 1/2	2		
5	BH504161L	Bolt-coolant pump to front cover-1/4UNC x 2	1		
6	WA106041L	Washer-M6-standard.	6		
7	BH505441L	Bolt-coolant pump to front cover-hexagonal head-5/16UNC x 5 1/2	1		
8	BH505485L	Bolt-5/16UNC x 6	1		
8	BH504151L	Bolt-hexagonal head-1/4UNC x 1 7/8	2		
9	ETC6148	Bracket-pulley support belt tensioner	1		
10	614718	Pulley-tensioner ancillary drive	1		
11	ETC6104	Adaptor-water pump drive	1		
12	BH505401L	Bolt-hexagonal head-5/16UNC x 5	1		
13	BH505381L	Bolt-hexagonal head-5/16UNC x 4 3/4	NLA		Use BH505401L.
14	BH504111L	Bolt-1/4UNC x 1 3/8	1		
15	BH504121L	Bolt-1/4UNC x 1 1/2-hexagonal head	1		
16	BH504101L	Bolt-1/4UNC x 1 1/4	1		
17	ETC7208	Support-adjuster link alternator	1		
18	ERC1167	Bracket-coolant pump support	1		
19	NH504041L	Nut-hexagonal head-1/4UNC x 1	1		
20	WM600041L	Washer-spring-1/4 dia-square	1		

DISCOVERY 1989MY UPTO 1999MY

GROUP B

Engine
V8 Petrol - Carburetter
Viscous Drive & Fan

Illus	Part Number	Description	Quantity	Change Point	Remarks
1	ETC5422	Pulley coolant pump	1		Air con
1	ETC5499	Pulley coolant pump	1		Except air con
2	ETC5423	Pulley-driven air conditioning	1		Air con
3	SH604051L	Screw-hexagonal head-1/4UNF x 5/8	3		
4	WA106041L	Washer-M6-standard.	3		
5	ERC5708	Coupling-engine fan viscous	1		
6	ETC7553L	Fan assembly-cooling-7 blade	1		
7	BH604101L	Bolt-1/4UNC x 1 1/4	4		
8	NY604041	Locknut-1/4-UNF	4		
9	611612	Belt-polyvee alternator	1		
10	ETC6104	Adaptor-water pump drive	1		
11	4910L	Washer-plain	1		
12	ERC5709	Bolt-M6 x 14	1		

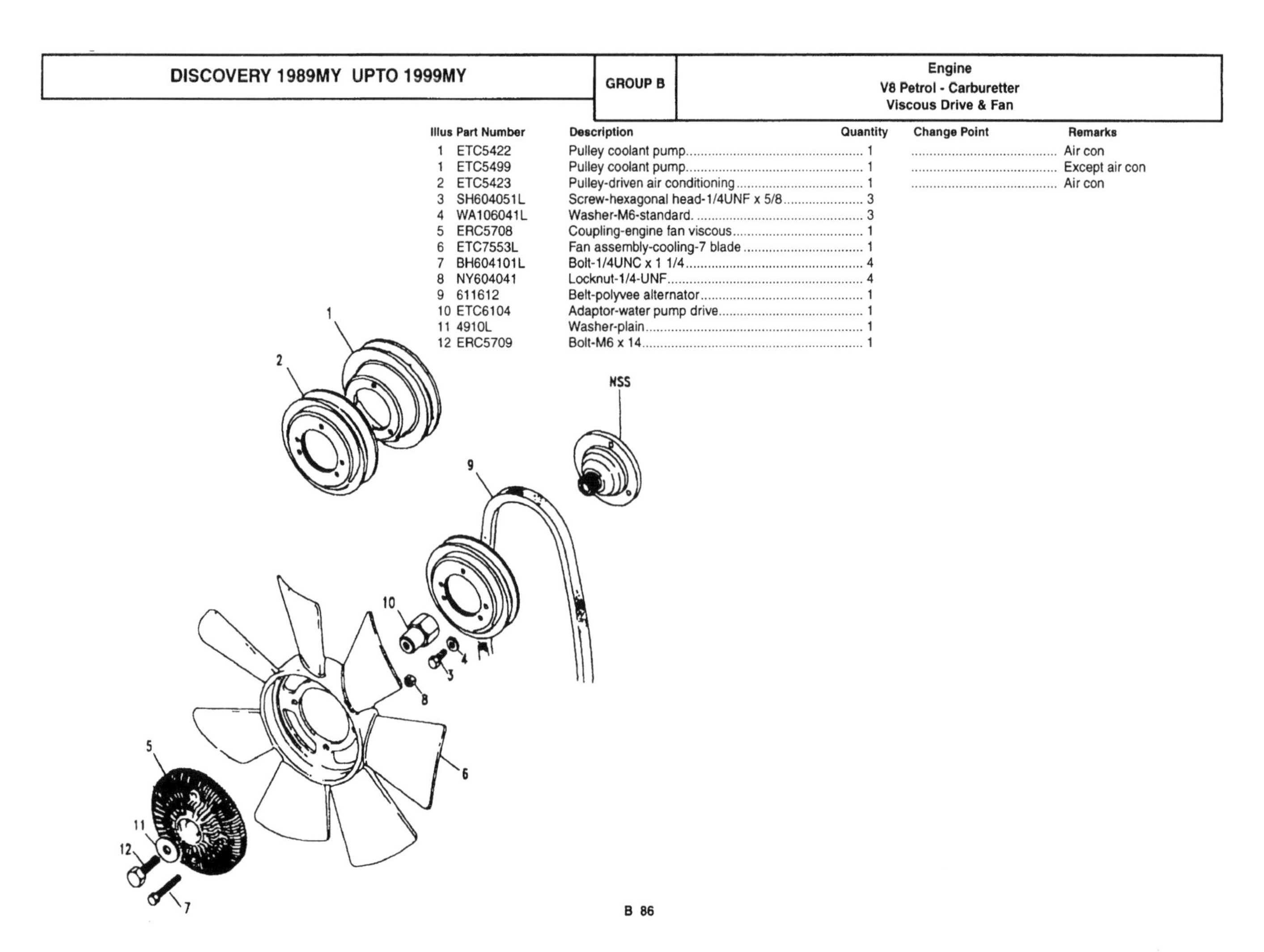

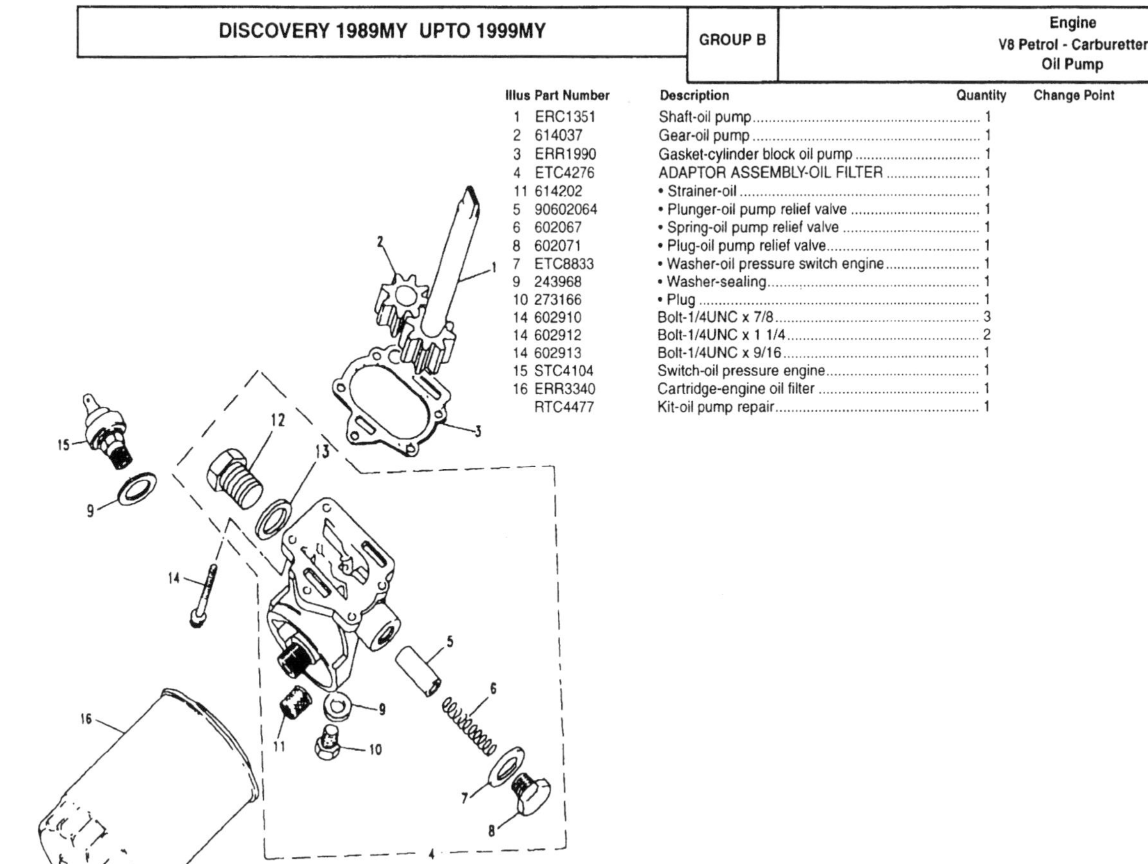

Illus	Part Number	Description	Quantity	Change Point	Remarks
1	ERC1351	Shaft-oil pump	1		
2	614037	Gear-oil pump	1		
3	ERR1990	Gasket-cylinder block oil pump	1		
4	ETC4276	ADAPTOR ASSEMBLY-OIL FILTER	1		
11	614202	• Strainer-oil	1		
5	90602064	• Plunger-oil pump relief valve	1		
6	602067	• Spring-oil pump relief valve	1		
8	602071	• Plug-oil pump relief valve	1		
7	ETC8833	• Washer-oil pressure switch engine	1		
9	243968	• Washer-sealing	1		
10	273166	• Plug	1		
14	602910	Bolt-1/4UNC x 7/8	3		
14	602912	Bolt-1/4UNC x 1 1/4	2		
14	602913	Bolt-1/4UNC x 9/16	1		
15	STC4104	Switch-oil pressure engine	1		
16	ERR3340	Cartridge-engine oil filter	1		
	RTC4477	Kit-oil pump repair	1		

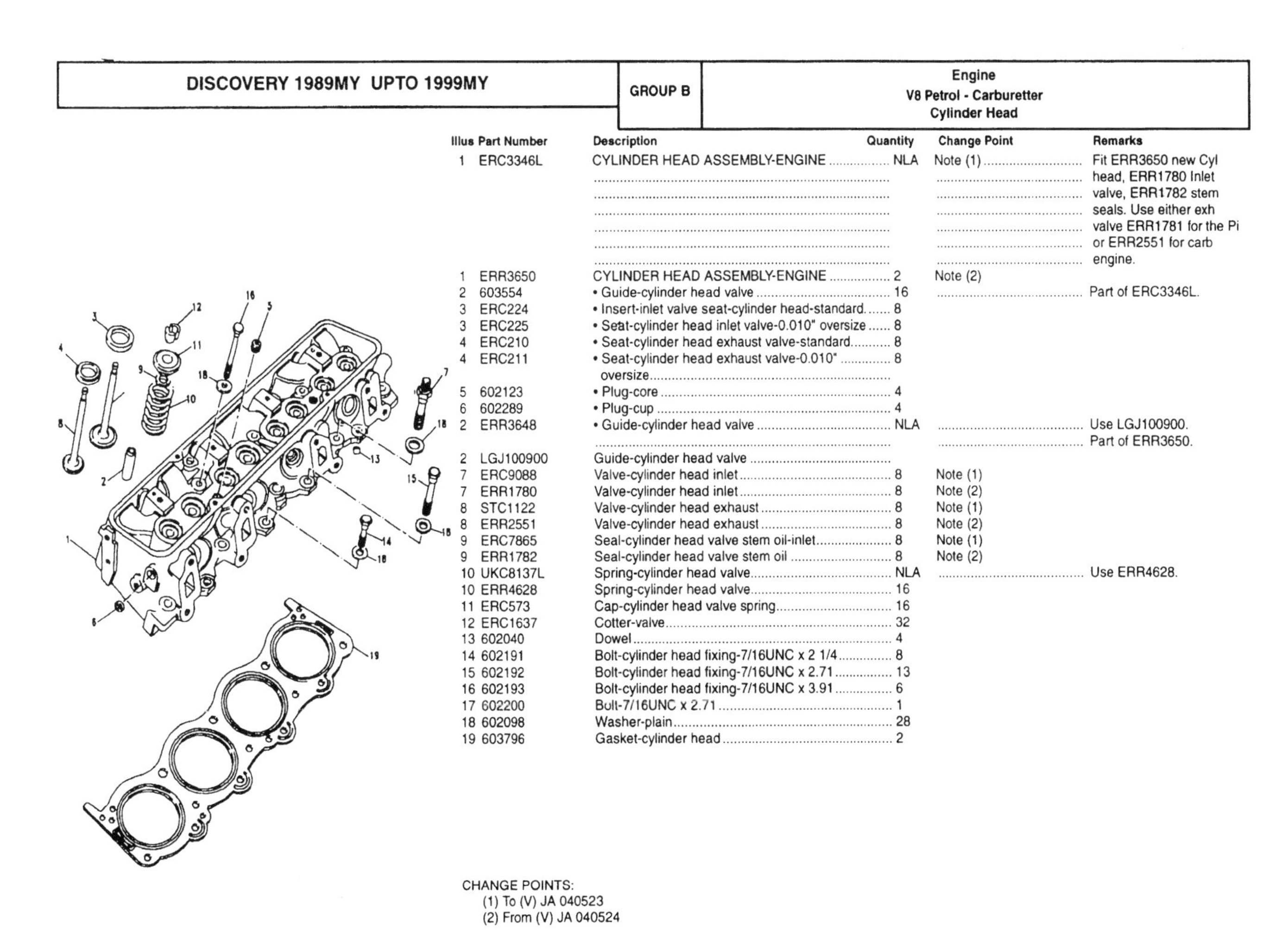

Illus	Part Number	Description	Quantity	Change Point	Remarks
1	ERC3346L	CYLINDER HEAD ASSEMBLY-ENGINE	NLA	Note (1)	Fit ERR3650 new Cyl head, ERR1780 Inlet valve, ERR1782 stem seals. Use either exh valve ERR1781 for the Pi or ERR2551 for carb engine.
1	ERR3650	CYLINDER HEAD ASSEMBLY-ENGINE	2	Note (2)	
2	603554	• Guide-cylinder head valve	16		Part of ERC3346L.
3	ERC224	• Insert-inlet valve seat-cylinder head-standard	8		
3	ERC225	• Seat-cylinder head inlet valve-0.010" oversize	8		
4	ERC210	• Seat-cylinder head exhaust valve-standard	8		
4	ERC211	• Seat-cylinder head exhaust valve-0.010" oversize	8		
5	602123	• Plug-core	4		
6	602289	• Plug-cup	4		
2	ERR3648	• Guide-cylinder head valve	NLA		Use LGJ100900. Part of ERR3650.
2	LGJ100900	Guide-cylinder head valve			
7	ERC9088	Valve-cylinder head inlet	8	Note (1)	
7	ERR1780	Valve-cylinder head inlet	8	Note (2)	
8	STC1122	Valve-cylinder head exhaust	8	Note (1)	
8	ERR2551	Valve-cylinder head exhaust	8	Note (2)	
9	ERC7865	Seal-cylinder head valve stem oil-inlet	8	Note (1)	
9	ERR1782	Seal-cylinder head valve stem oil	8	Note (2)	
10	UKC8137L	Spring-cylinder head valve	NLA		Use ERR4628.
10	ERR4628	Spring-cylinder head valve	16		
11	ERC573	Cap-cylinder head valve spring	16		
12	ERC1637	Cotter-valve	32		
13	602040	Dowel	4		
14	602191	Bolt-cylinder head fixing-7/16UNC x 2 1/4	8		
15	602192	Bolt-cylinder head fixing-7/16UNC x 2.71	13		
16	602193	Bolt-cylinder head fixing-7/16UNC x 3.91	6		
17	602200	Bolt-7/16UNC x 2.71	1		
18	602098	Washer-plain	28		
19	603796	Gasket-cylinder head	2		

CHANGE POINTS:
(1) To (V) JA 040523
(2) From (V) JA 040524

DISCOVERY 1989MY UPTO 1999MY

GROUP B — Engine, V8 Petrol - Carburetter, Rocker Shaft

Illus	Part Number	Description	Quantity	Change Point	Remarks
1	611660	SHAFT ASSEMBLY-ROCKER	2		
2	611659L	• SHAFT-ROCKER	2		
	154545	• • Plug-core	4		
3	602153	• Rocker assembly-cylinder head-RH	8		
4	602154	• Rocker assembly-cylinder head-LH	8		
5	603734	• Pillar-cylinder head rocker support	8		
6	602142	• Spring-rocker shaft	6		
7	602148	• Washer-waved	4		
8	602186	• Washer-plain	4		
9	PS606101L	• Pin-split	4		
11	602097	Bolt-3/8UNC x 2.79	8		
12	603378	Rod-engine push	16		
13	ERC4949	Tappet-engine valve hydraulic	16		
14	602172	Baffle-engine oil	2		

B 89

DISCOVERY 1989MY UPTO 1999MY

GROUP B — Engine, V8 Petrol - Carburetter, Rocker Cover

Illus	Part Number	Description	Quantity	Change Point	Remarks
1	ETC8681	COVER ASSEMBLY-VALVE ROCKER-RH	1		
2	79027	• Screw-6 x 3/8	6		
3	602512	• Gasket-engine rocker cover	NLA		Use ERR7288.
4	ETC8682	COVER ASSEMBLY-VALVE ROCKER-LH	1		
2	79027	• Screw-6 x 3/8	8		
3	602512	• Gasket-engine rocker cover	NLA		Use ERR7288.
5	ERC2989	Filler-oil	1		
6	625038	Cap-oil filler	1		
7	564258	O ring	1		
8	SY504072L	Screw-1/4UNC x 7/8	4		
9	603127	Screw-1/4UNC x 1 5/16	4		
10	WM600041L	Washer-spring-1/4 dia-square	8		
11	WB106041L	Washer-M6	8		
12	ERC614	Bracket-lifting engine rear	1		
13	SH506071L	Screw-hexagonal head-3/8UNC x 7/8	2		
14	WM600061L	Washer-3/8"-square	2		
15	SH105251L	Screw-hexagonal head-M5 x 25	1		
16	586440	Clip-cable	1		
17	78862	Screw-10UNC x 9/16	3		
18	610402	Clip-cable	1		
19	603672	Clip-spacer-high tension lead ignition	2		
20	NH105041L	Nut-hexagonal-M5	1		

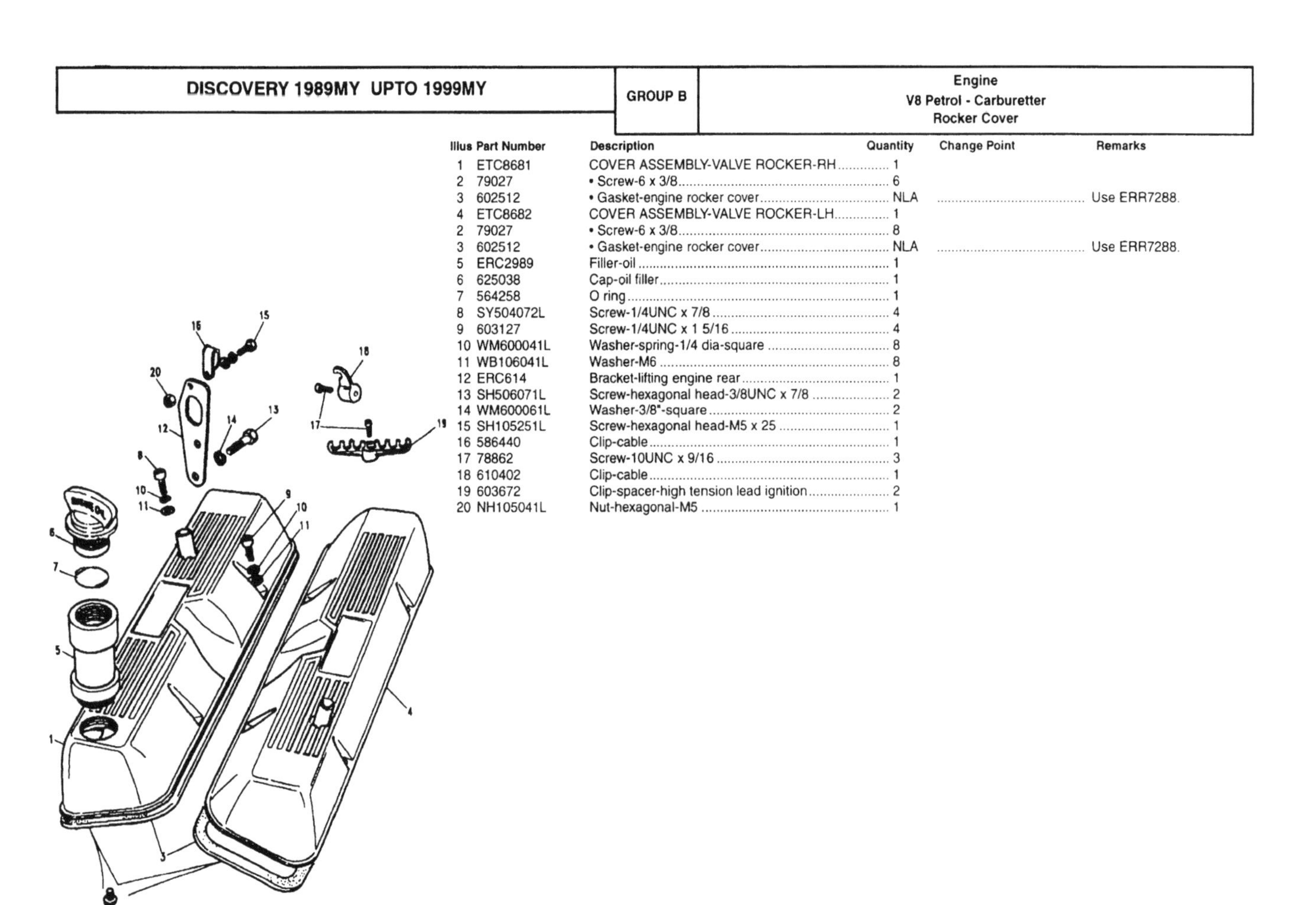

B 90

Breather System

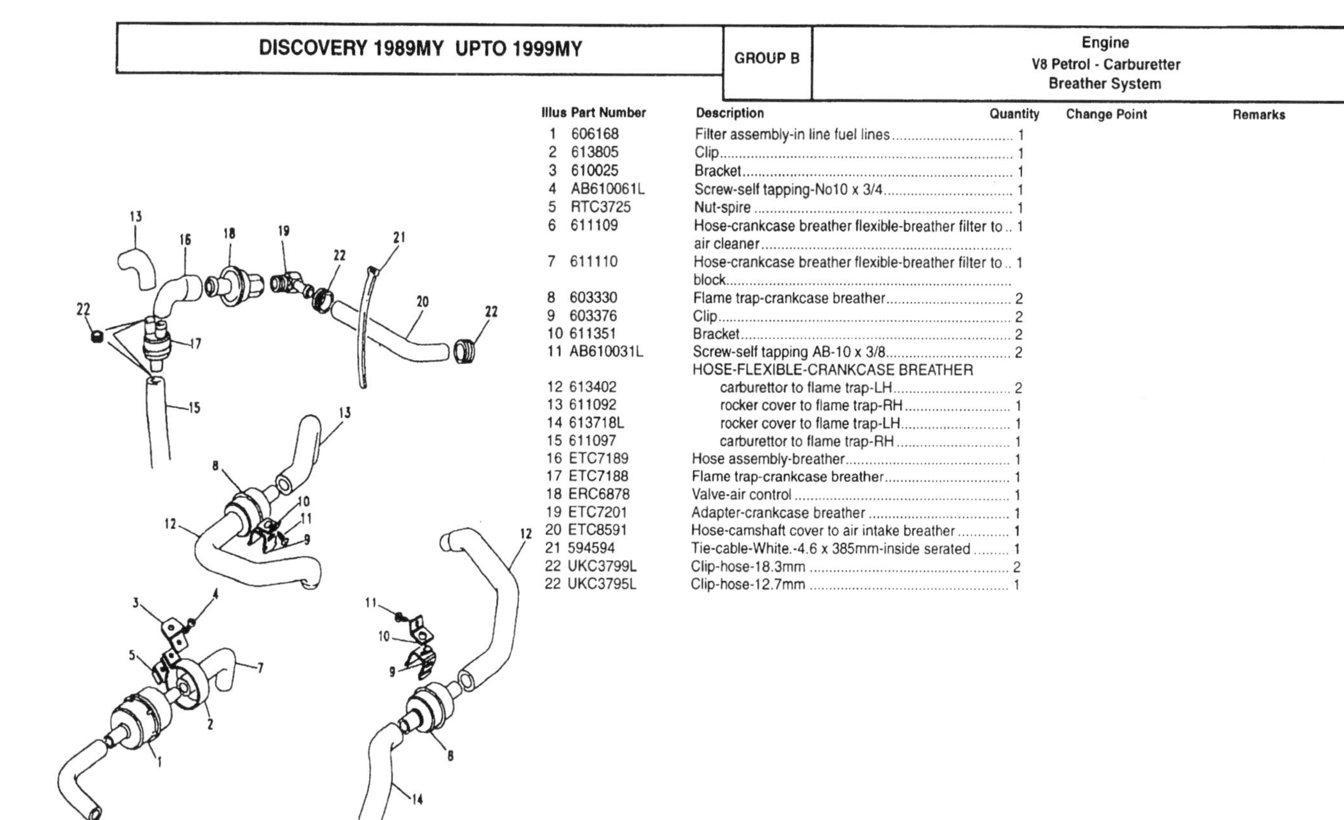

Illus	Part Number	Description	Quantity	Change Point	Remarks
1	606168	Filter assembly-in line fuel lines	1		
2	613805	Clip	1		
3	610025	Bracket	1		
4	AB610061L	Screw-self tapping-No10 x 3/4	1		
5	RTC3725	Nut-spire	1		
6	611109	Hose-crankcase breather flexible-breather filter to air cleaner	1		
7	611110	Hose-crankcase breather flexible-breather filter to block	1		
8	603330	Flame trap-crankcase breather	2		
9	603376	Clip	2		
10	611351	Bracket	2		
11	AB610031L	Screw-self tapping AB-10 x 3/8	2		
		HOSE-FLEXIBLE-CRANKCASE BREATHER			
12	613402	carburettor to flame trap-LH	2		
13	611092	rocker cover to flame trap-RH	1		
14	613718L	rocker cover to flame trap-LH	1		
15	611097	carburettor to flame trap-RH	1		
16	ETC7189	Hose assembly-breather	1		
17	ETC7188	Flame trap-crankcase breather	1		
18	ERC6878	Valve-air control	1		
19	ETC7201	Adapter-crankcase breather	1		
20	ETC8591	Hose-camshaft cover to air intake breather	1		
21	594594	Tie-cable-White.-4.6 x 385mm-inside serated	1		
22	UKC3799L	Clip-hose-18.3mm	2		
22	UKC3795L	Clip-hose-12.7mm	1		

B 91

Camshaft

Illus	Part Number	Description	Quantity	Change Point	Remarks
1	ETC6850L	Camshaft-engine	1		
2	610289	Sprocket-camshaft	1		
3	ERC7929	Chain-engine timing	1		
4	ERC2838	Key-camshaft location	1		
5	ERC2839	Spacer	1		
6	614188	Gear-drive-distributor	1		
7	ERC6552	Washer-plain	1		
8	602227	Bolt-7/16UNC x 1 1/8	1		

B 92

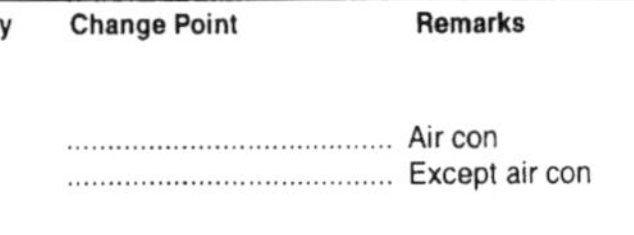

Engine
V8 Petrol - Carburetter
Thermostat Housing

Illus	Part Number	Description	Quantity	Change Point	Remarks
1	ETC4765	Thermostat-engine-88 degrees C	1		
2	611786	Gasket-coolant outlet elbow	1		
3	ETC6135	Elbow-engine coolant outlet	1		Air con
3	ETC4596	Elbow-engine coolant outlet	1		Except air con
4	SH505081L	Screw-hexagonal head-5/8UNC x 1	2		
5	WM600051L	Washer-spring-5/16"	2		
6	PRC3505	Switch-engine thermostat-Yellow.	1		
7	C457593	Washer-sealing	1		

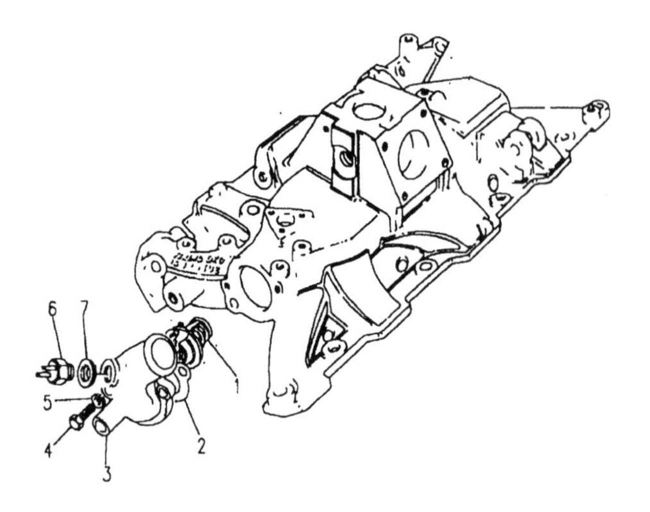

Engine
V8 Petrol - Carburetter
Inlet Manifold

Illus	Part Number	Description	Quantity	Change Point	Remarks
1	ETC7915	MANIFOLD ASSEMBLY-ENGINE INLET	1		
2	252514	• Stud	8		
3	ERC256	• Plug-core	2		
4	602152	• Plug-cup	1		
5	522932	• Bush	2		
6	603224	• Plug-blanking	1		
7	232043	• Washer-sealing	1		
8	603920	• Adaptor-inlet manifold	1		
9	534897	Bolt-banjo	1		
10	267604	Washer-sealing	2		
11	603277	Bolt-banjo	1		
12	603440	Pipe-outlet to inlet manifold coolant	1		
13	ERR4935	Gasket-coolant outlet elbow	1		
14	SH504061L	Setscrew-1/4UNC x 3/4	2		
15	WM600041L	Washer-spring-1/4 dia-square	2		
16	603561	Bracket-inlet manifold support	1		
17	ERC2297	Peg-air cleaner mtg-engine	1		
18	ERC2298	Spacer	1		
19	RTC5907	Valve-inlet manifold non return	1	Note (1)	
19	LKX100440	Valve-inlet manifold non return	1	Note (2)	
20	232043	Washer-sealing	1		
21	572548	Clip-hose	NLA		Use CN100168L.
21	CN100168L	Clip-hose	1		

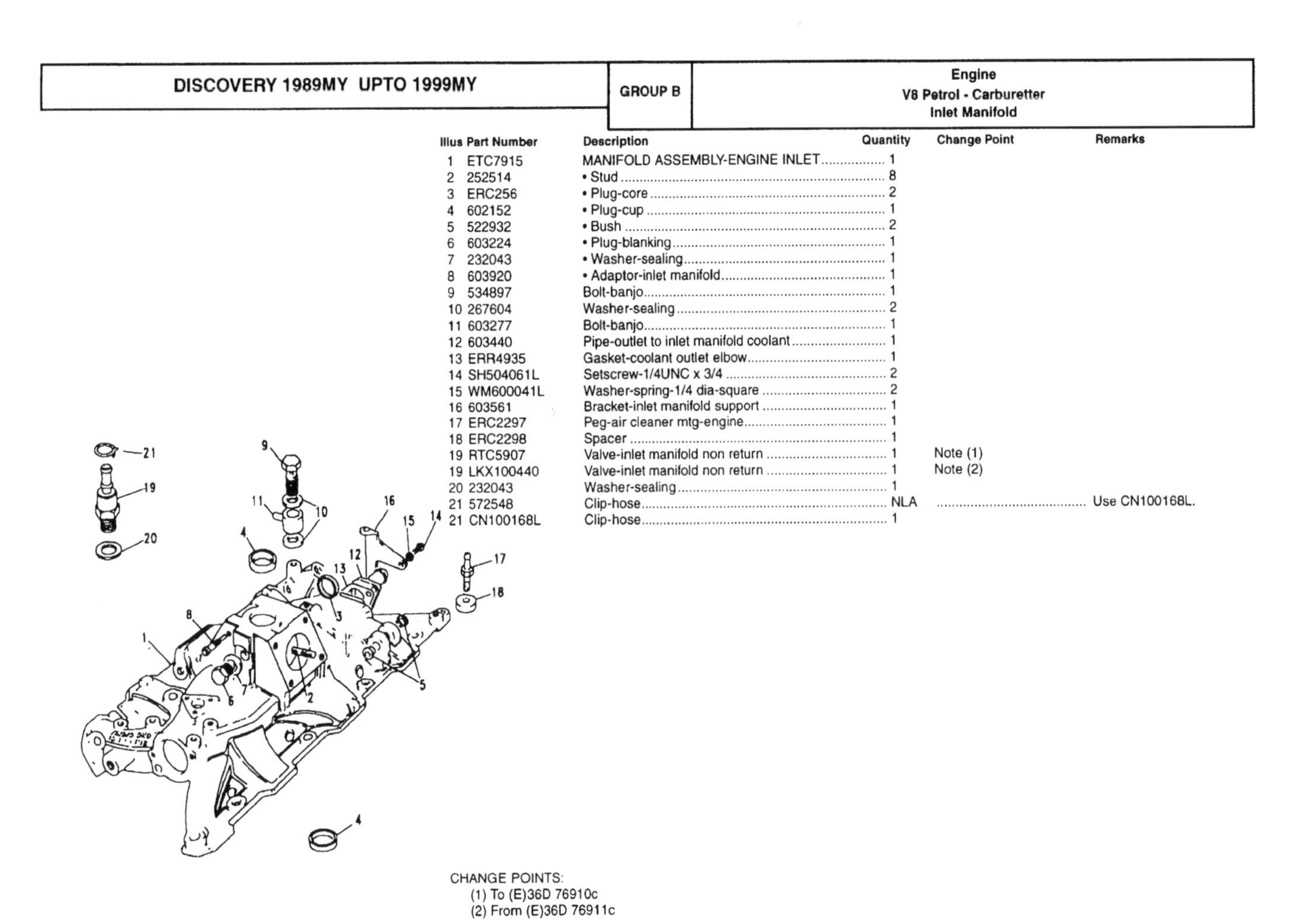

CHANGE POINTS:
(1) To (E)36D 76910c
(2) From (E)36D 76911c

Inlet Manifold-Coolant Hoses

Illus	Part Number	Description	Quantity	Change Point	Remarks
1	PRC8003	Sensor-carburettor/multi point injection/tbi coolant temperature	1		
2	90568054	Washer-sealing	1		
3	ERC2319	Hose-bypass to inlet manifold coolant	1		
4	CN100308L	Clip-hose-30mm	4		
5	BH506161L	Bolt-hexagonal head-3/8UNC x 2	2		
6	BH506121L	Bolt-hexagonal head-3/8UNC x 1 1/2	6		
7	90611504	Bolt-3/8UNC x 1 5/16	4		
8	2204L	Washer	12		
9	602099	Seal-inlet manifold gasket	NLA		Use ERR7283.
9	ERR7283	Seal-inlet manifold gasket	2		
10	602076	Clamp-inlet manifold gasket seal	2		
11	602236	Bolt-5/16UNC	2		
12	ERC3990	Gasket-inlet manifold	1		
13	ERC2143	Pipe-outlet to inlet manifold coolant	1		
14	FS504047	Screw-flanged head-1/4UNC x 1/2	2		
15	WM600041L	Washer-spring-1/4 dia-square	1		
16	ERC2320	Hose-heater to inlet manifold coolant	1		
	594594	Tie-cable-White.-4.6 x 385mm-inside serated	4		

Exhaust Manifold

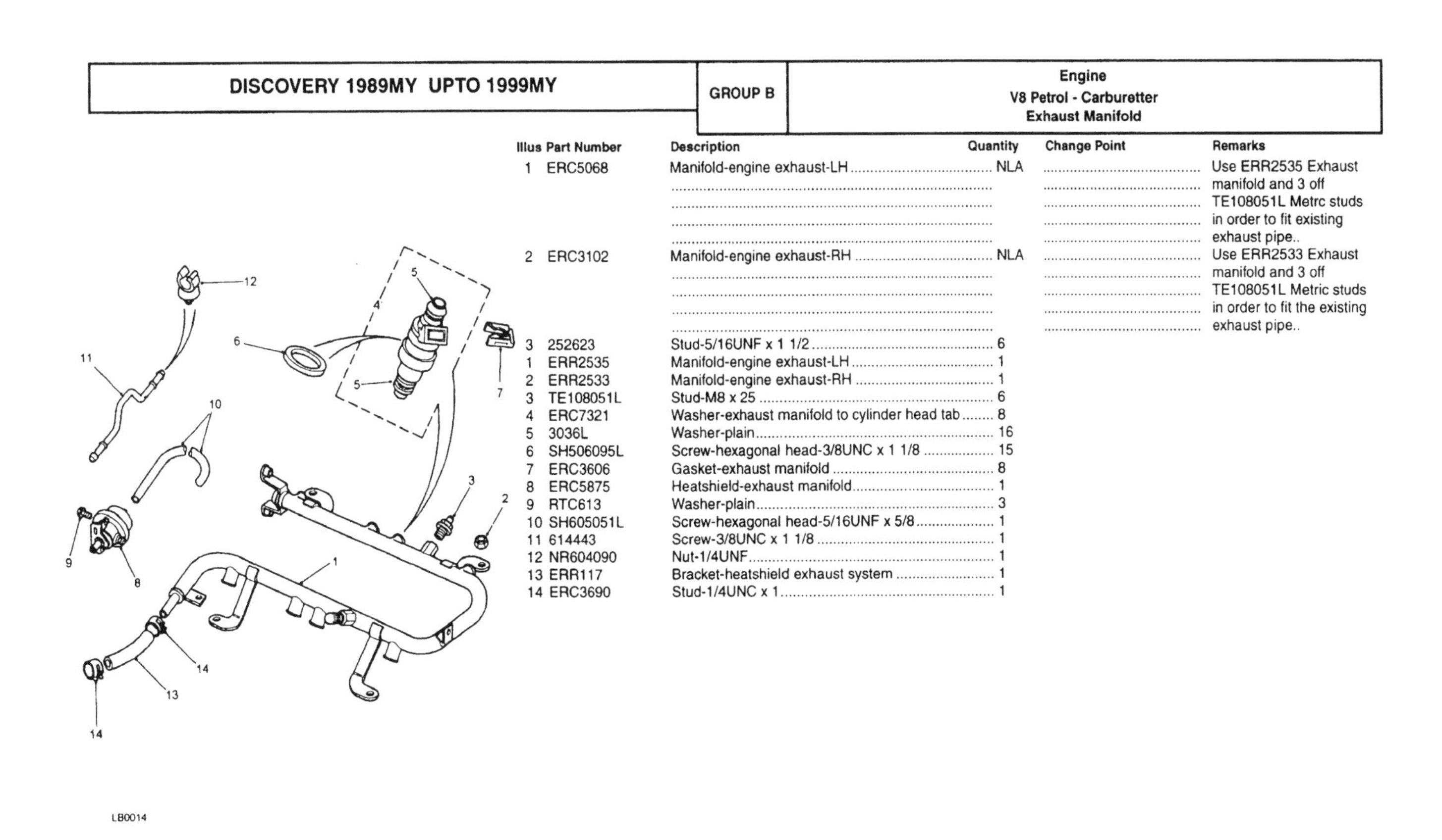

Illus	Part Number	Description	Quantity	Change Point	Remarks
1	ERC5068	Manifold-engine exhaust-LH	NLA		Use ERR2535 Exhaust manifold and 3 off TE108051L Metrc studs in order to fit existing exhaust pipe..
2	ERC3102	Manifold-engine exhaust-RH	NLA		Use ERR2533 Exhaust manifold and 3 off TE108051L Metric studs in order to fit the existing exhaust pipe..
3	252623	Stud-5/16UNF x 1 1/2	6		
1	ERR2535	Manifold-engine exhaust-LH	1		
2	ERR2533	Manifold-engine exhaust-RH	1		
3	TE108051L	Stud-M8 x 25	6		
4	ERC7321	Washer-exhaust manifold to cylinder head tab	8		
5	3036L	Washer-plain	16		
6	SH506095L	Screw-hexagonal head-3/8UNC x 1 1/8	15		
7	ERC3606	Gasket-exhaust manifold	8		
8	ERC5875	Heatshield-exhaust manifold	1		
9	RTC613	Washer-plain	3		
10	SH605051L	Screw-hexagonal head-5/16UNF x 5/8	1		
11	614443	Screw-3/8UNC x 1 1/8	1		
12	NR604090	Nut-1/4UNF	1		
13	ERR117	Bracket-heatshield exhaust system	1		
14	ERC3690	Stud-1/4UNC x 1	1		

DISCOVERY 1989MY UPTO 1999MY

GROUP B — Engine — V8 Petrol - Carburetter — Carburetter

Illus	Part Number	Description	Quantity	Change Point	Remarks
1	ERR489	CARBURETTOR-ENGINE-LH	1		
1	ERR490	CARBURETTOR-ENGINE-RH	1		
2	LZX2107L	• Level-carburettor throttle	1		
3	JZX1303L	• Screw-carburettor fast idle adjusting	2		
4	AUD4771L	• Clip	2		
5	JZX1181L	• Screw-carburettor slow run adjusting	2		
6	STC509	• Body-carburettor-LH	1		
6	STC510	• Body-carburettor-RH	1		
7	LZX1988L	• Lever-carburettor camshaft-LH	1		
7	LZX1989L	• Lever-carburettor camshaft-RH	1		
8	AUD4398L	• Spring-carburettor yellow piston	2		
9	RTC6436	• Needle assembly-carburettor jet	2		
10	AUD3306L	• Spring-carburettor jet needle	2		
11	JZX1039L	• Guide-carburettor jet needle	2		
12	JZX1394L	• Screw	6		
13	LZX1505	• Damper assembly-carburettor	2		
14	CUD2788L	• Jet-carburettor-LH	1		
14	CUD2785L	• Jet-carburettor-RH	1		
15	RTC3566	• Float-carburettor-LH	1		
15	LZX1600L	• Float-carburettor-RH	1		
16	CUD2399L	• Bimetal assembly-carburettor choke	2		
17	ETC7127	PIPE-FEED-FUEL LINES	1		
18	612064	• Olive	2		
19	534790	• Nut-7/16UNF	2		
20	STC205	Needle & seat assembly-carburettor float	2		
	RTC6072	Kit-carburettor seal	1		
	WZX1505	Kit-gasket	1		

DISCOVERY 1989MY UPTO 1999MY

GROUP B — Engine — V8 Petrol - Carburetter — Carburetter Adaptor and Linkages

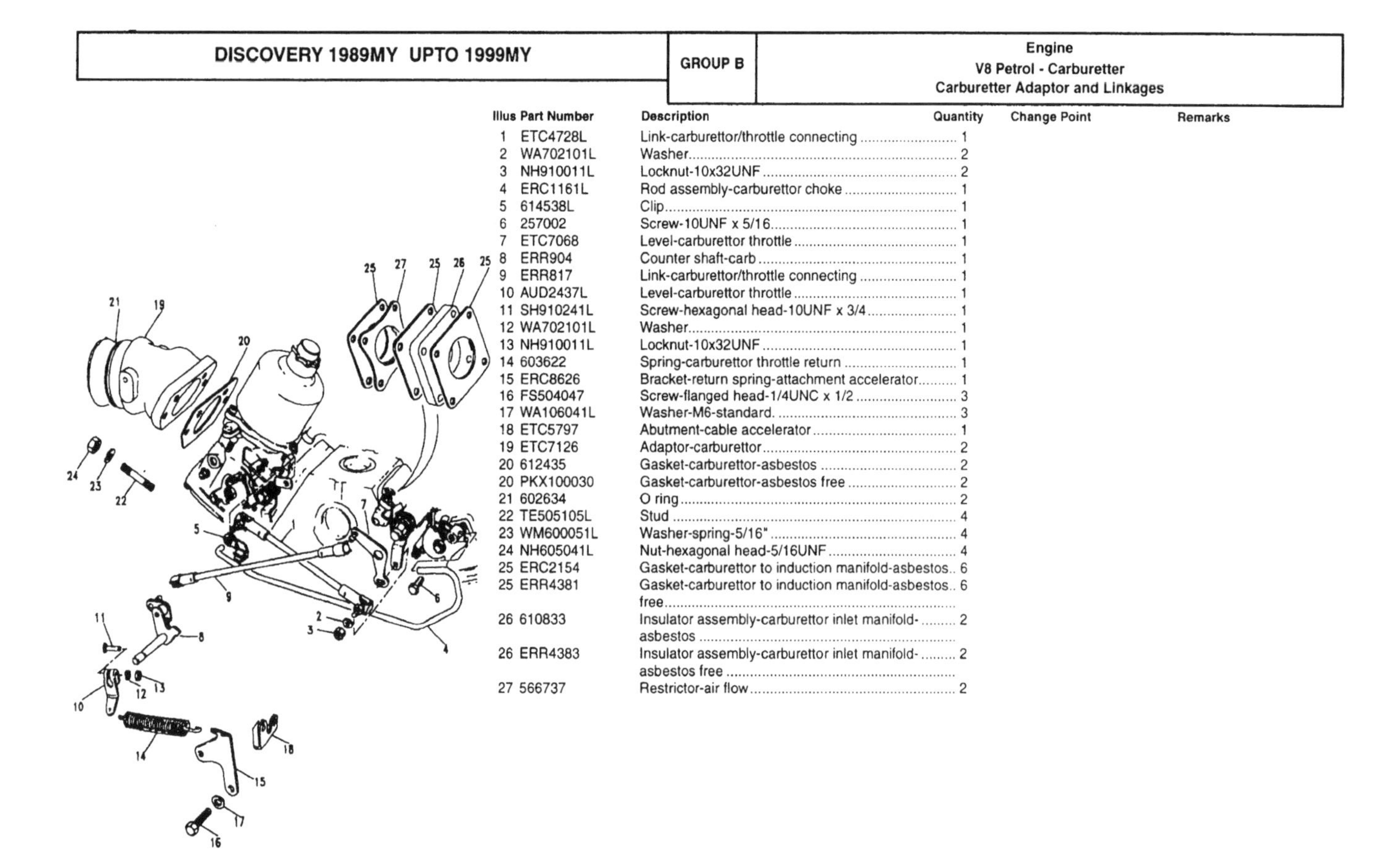

Illus	Part Number	Description	Quantity	Change Point	Remarks
1	ETC4728L	Link-carburettor/throttle connecting	1		
2	WA702101L	Washer	2		
3	NH910011L	Locknut-10x32UNF	2		
4	ERC1161L	Rod assembly-carburettor choke	1		
5	614538L	Clip	1		
6	257002	Screw-10UNF x 5/16	1		
7	ETC7068	Level-carburettor throttle	1		
8	ERR904	Counter shaft-carb	1		
9	ERR817	Link-carburettor/throttle connecting	1		
10	AUD2437L	Level-carburettor throttle	1		
11	SH910241L	Screw-hexagonal head-10UNF x 3/4	1		
12	WA702101L	Washer	1		
13	NH910011L	Locknut-10x32UNF	1		
14	603622	Spring-carburettor throttle return	1		
15	ERC8626	Bracket-return spring-attachment accelerator	1		
16	FS504047	Screw-flanged head-1/4UNC x 1/2	3		
17	WA106041L	Washer-M6-standard.	3		
18	ETC5797	Abutment-cable accelerator	1		
19	ETC7126	Adaptor-carburettor	2		
20	612435	Gasket-carburettor-asbestos	2		
20	PKX100030	Gasket-carburettor-asbestos free	2		
21	602634	O ring	2		
22	TE505105L	Stud	4		
23	WM600051L	Washer-spring-5/16"	4		
24	NH605041L	Nut-hexagonal head-5/16UNF	4		
25	ERC2154	Gasket-carburettor to induction manifold-asbestos	6		
25	ERR4381	Gasket-carburettor to induction manifold-asbestos free	6		
26	610833	Insulator assembly-carburettor inlet manifold-asbestos	2		
26	ERR4383	Insulator assembly-carburettor inlet manifold-asbestos free	2		
27	566737	Restrictor-air flow	2		

Power Steering Pump & Mountings

Illus	Part Number	Description	Quantity	Change Point	Remarks
1	ETC6496	PUMP ASSEMBLY POWER ASSISTED STEERING	NLA	Note (1)	Use NTC8286.
	RTC6074	• O ring-large	1		
	RTC6075	• O ring-small	1		
	RTC5935	• Kit-power assisted steering pump seal	NLA		Use STC1633.
	STC1633	Kit-power assisted steering pump seal			
1	NTC8286	PUMP ASSEMBLY POWER ASSISTED STEERING	1	Note (2)	
	RTC6074	• O ring-large	1		
	RTC6075	• O ring-small	1		
	STC1633	• Kit-power assisted steering pump seal	1		
2	ETC6408	Pulley power assisted steering pump-twin groove-A127 alternator	1		
3	ERC675	Belt-coolant pump vee	1		
4	ERR798	Bracket-mounting power assisted steering pump	1		
5	ETC6607	Link-power assisted steering pump adjusting	1		
6	4478	Washer-plain	1		
7	FN108041L	Nut-flanged head-M8 x 20	NLA		Use FN108047L.
8	FS106167L	Screw-hexagonal head-M6 x 16	1		
9	FS108207L	Screw-flanged head-M8 x 20	3		
10	SH108251L	Screw-hexagonal head-M8 x 25	NLA		Use FS108251L.
10	FS108251L	Screw-flanged head-M8 x 25	NLA		Use FS108257L.
11	TE506101L	Stud-3/8UNF x 1 1/8	1		
10	SS506060	Screw-3/8UNC x 3/4	1		
12	WA108051L	Washer-plain-M8	3		
13	WM600051L	Washer-spring-5/16"	1		

CHANGE POINTS:
(1) To (V)NOV 89 approx
(2) From (V)NOV 89 approx

B 99

DISCOVERY 1989MY UPTO 1999MY | GROUP B | Engine
V8 Petrol - Carburetter

Air Conditioning Compressor

Illus	Part Number	Description	Quantity	Change Point	Remarks
1	ETC7994	COMPRESSOR ASSEMBLY AIR CONDITIONING	1	Note (1)	
	STC3131	• O ring air conditioning	2		
1	ERR845	Compressor assembly air conditioning	1	Note (2)	
2	ERR4521	Bracket-air compressor	1		
3	611612	Belt-polyvee alternator	1		
4	614718	Pulley-tensioner ancillary drive	1		
5	WM600061L	Washer-3/8"-square	1		
6	WA110061L	Washer-plain-M10-standard	6		
7	WA108051L	Washer-plain-M8	1		
8	NH606041L	Nut-hexagonal head-3/8UNF	2		
9	BH606401L	Bolt-3/8UNF x 5	2		
10	BH605101L	Bolt-hexagonal head-5/16UNF x 1 1/4	1		
11	BH505341L	Bolt-5/16UNC x 4 1/4	1		
12	BH506131L	Bolt-3/8UNC x 1 7/8	1		

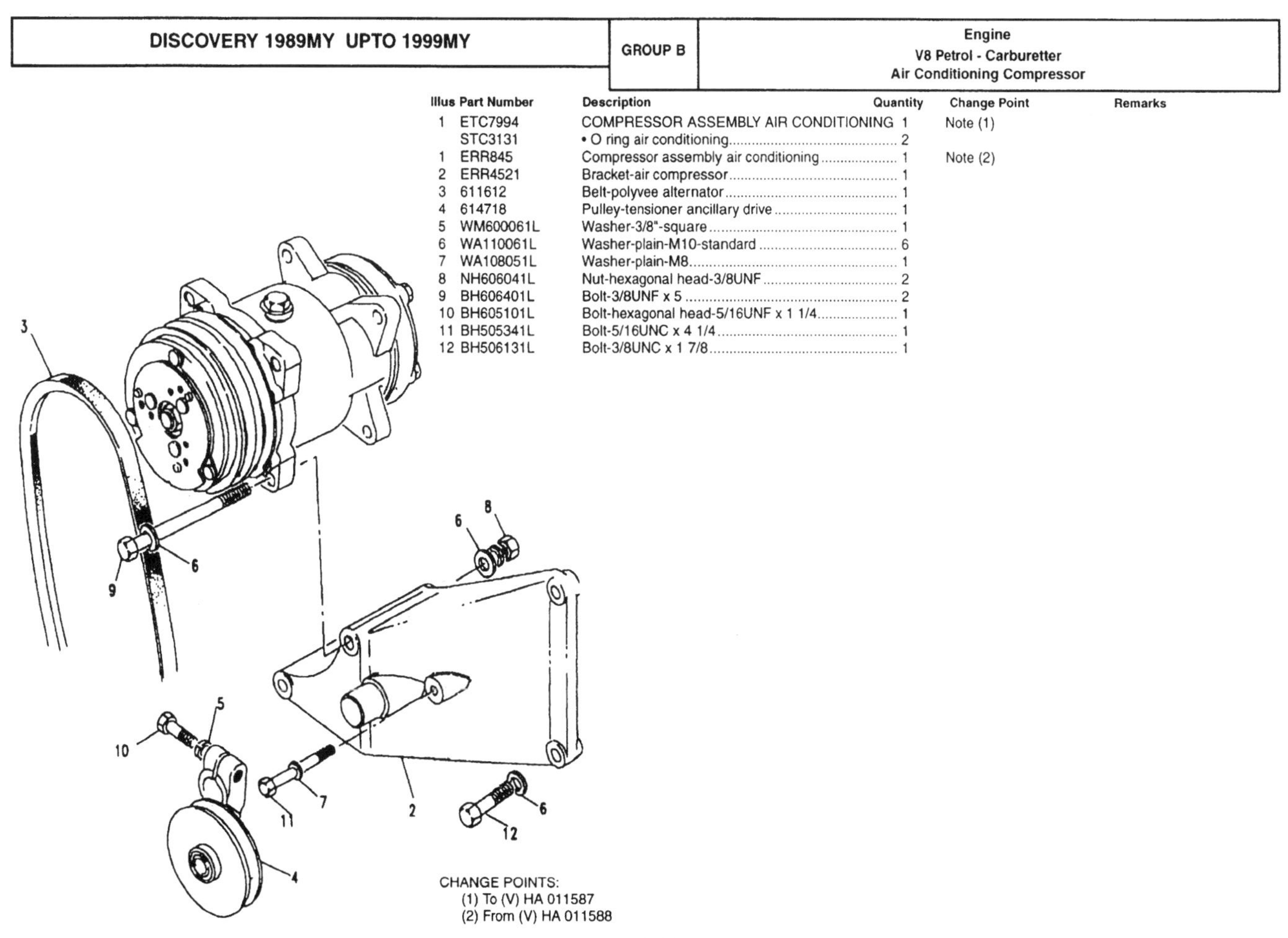

CHANGE POINTS:
(1) To (V) HA 011587
(2) From (V) HA 011588

B 100

DISCOVERY 1989MY UPTO 1999MY

GROUP B

Engine
V8 Petrol - Carburetter
Distributor

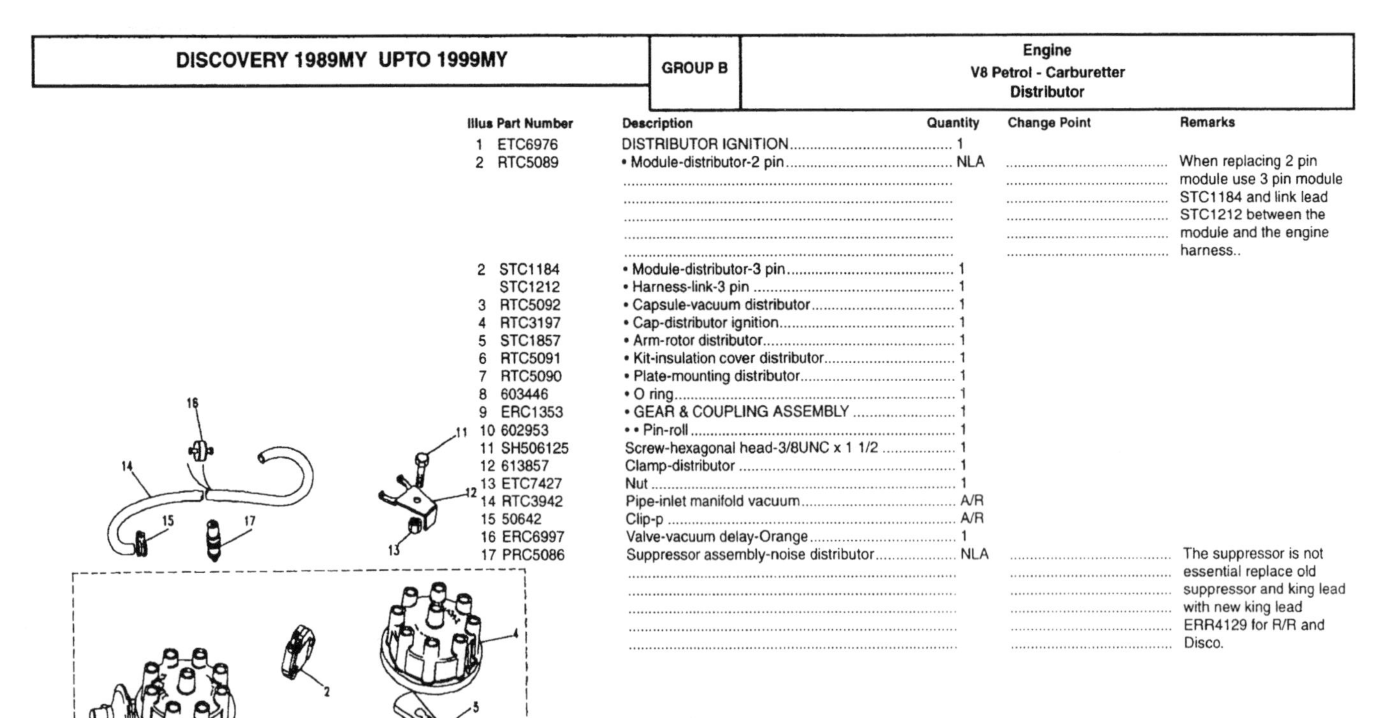

Illus	Part Number	Description	Quantity	Change Point	Remarks
1	ETC6976	DISTRIBUTOR IGNITION	1		
2	RTC5089	• Module-distributor-2 pin	NLA		When replacing 2 pin module use 3 pin module STC1184 and link lead STC1212 between the module and the engine harness..
2	STC1184	• Module-distributor-3 pin	1		
	STC1212	• Harness-link-3 pin	1		
3	RTC5092	• Capsule-vacuum distributor	1		
4	RTC3197	• Cap-distributor ignition	1		
5	STC1857	• Arm-rotor distributor	1		
6	RTC5091	• Kit-insulation cover distributor	1		
7	RTC5090	• Plate-mounting distributor	1		
8	603446	• O ring	1		
9	ERC1353	• GEAR & COUPLING ASSEMBLY	1		
10	602953	• • Pin-roll	1		
11	SH506125	Screw-hexagonal head-3/8UNC x 1 1/2	1		
12	613857	Clamp-distributor	1		
13	ETC7427	Nut	1		
14	RTC3942	Pipe-inlet manifold vacuum	A/R		
15	50642	Clip-p	A/R		
16	ERC6997	Valve-vacuum delay-Orange	1		
17	PRC5086	Suppressor assembly-noise distributor	NLA		The suppressor is not essential replace old suppressor and king lead with new king lead ERR4129 for R/R and Disco.

B 101

DISCOVERY 1989MY UPTO 1999MY

GROUP B

Engine
V8 Petrol - Carburetter
High Tension Leads

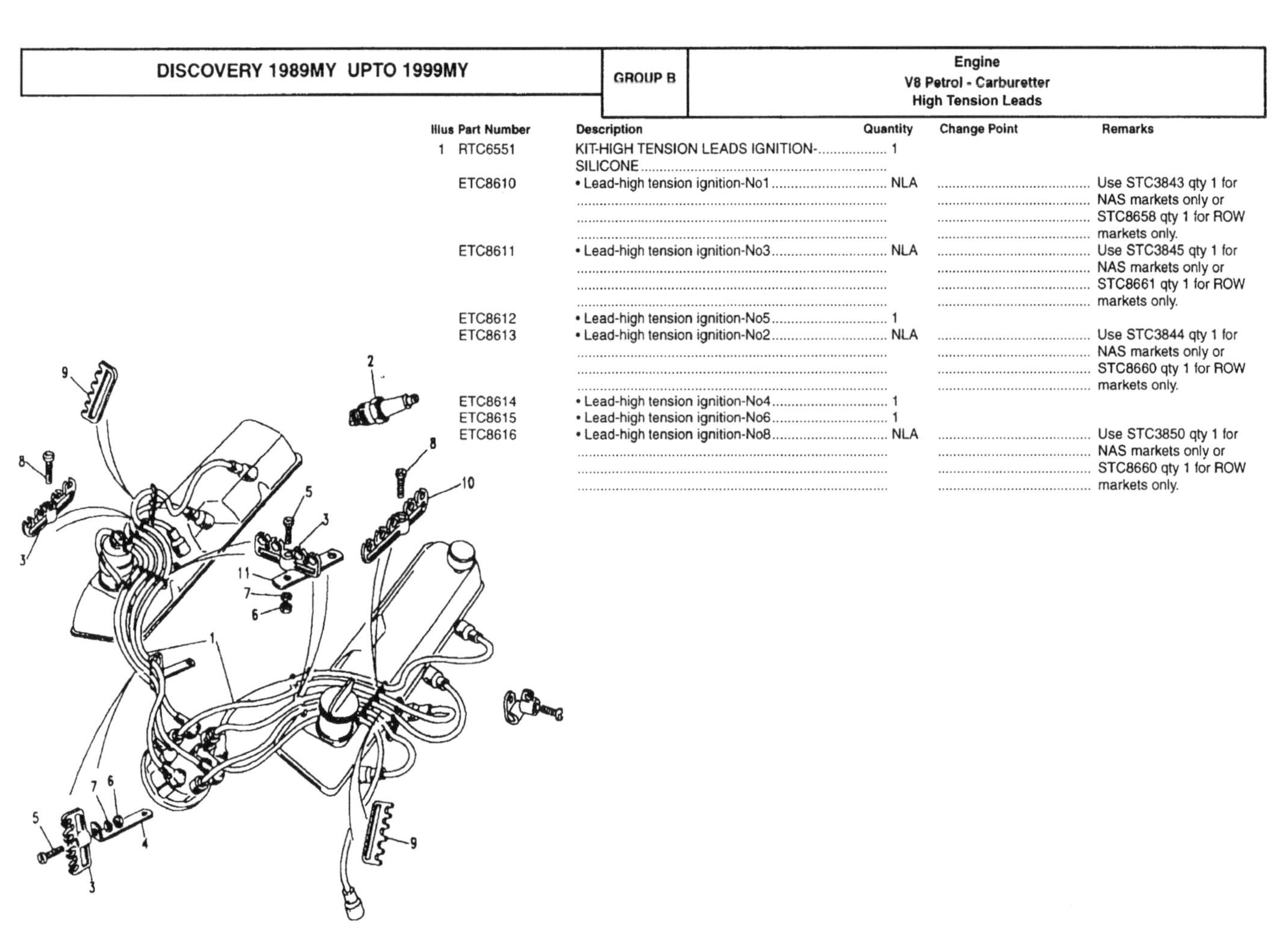

Illus	Part Number	Description	Quantity	Change Point	Remarks
1	RTC6551	KIT-HIGH TENSION LEADS IGNITION-SILICONE	1		
	ETC8610	• Lead-high tension ignition-No1	NLA		Use STC3843 qty 1 for NAS markets only or STC8658 qty 1 for ROW markets only.
	ETC8611	• Lead-high tension ignition-No3	NLA		Use STC3845 qty 1 for NAS markets only or STC8661 qty 1 for ROW markets only.
	ETC8612	• Lead-high tension ignition-No5	1		
	ETC8613	• Lead-high tension ignition-No2	NLA		Use STC3844 qty 1 for NAS markets only or STC8660 qty 1 for ROW markets only.
	ETC8614	• Lead-high tension ignition-No4	1		
	ETC8615	• Lead-high tension ignition-No6	1		
	ETC8616	• Lead-high tension ignition-No8	NLA		Use STC3850 qty 1 for NAS markets only or STC8660 qty 1 for ROW markets only.

B 102

Engine
V8 Petrol - Carburetter
High Tension Leads

Illus	Part Number	Description	Quantity	Change Point	Remarks
	ETC8617	• Lead-high tension ignition-No7	1		
	ETC8618	• Lead-king ignition	1		
2	ERR3799	Plug-sparking-RN11YC	8		
3	603672	Clip-spacer-high tension lead ignition	3		
4	ETC6237	Bracket-support-lead ignition-centre	1		
5	SE910161L	Screw	3		
6	NH910011L	Locknut-10x32UNF	3		
7	WL700101L	Washer-spring-No 10-single coil-rectangular	3		
8	AB610041L	Screw-self tapping AB-M8 x 12	3		
9	ETC8720	Clip-spacer-high tension lead ignition	2		
10	ETC7971	Clip-spacer-high tension lead ignition	1		
11	ETC6093	Bracket-support-lead ignition	2		

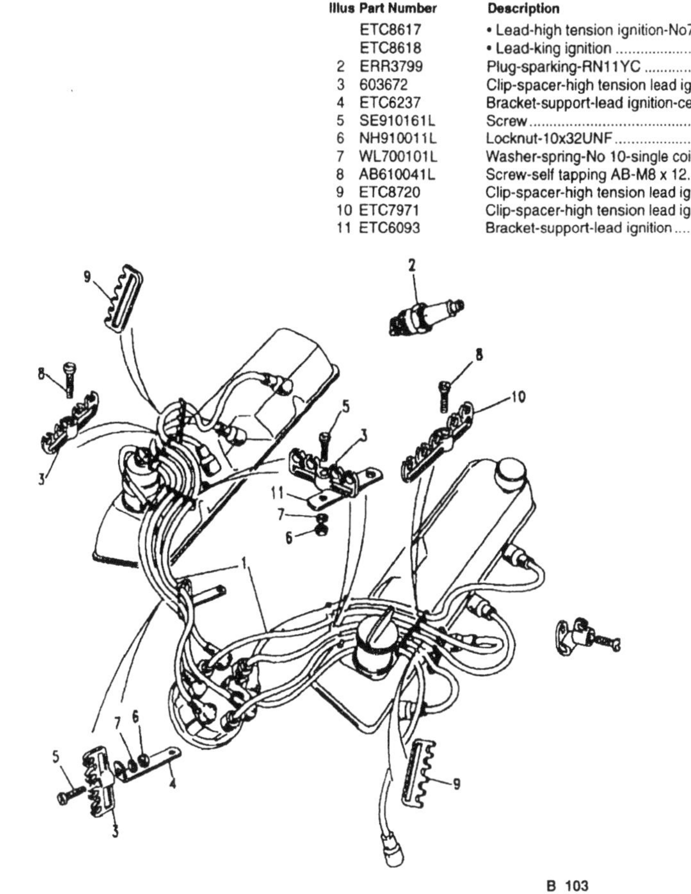

B 103

Engine
V8 Petrol - Carburetter
Coil

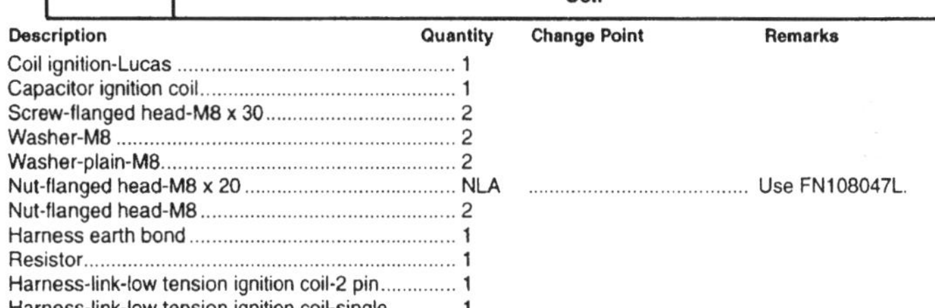

Illus	Part Number	Description	Quantity	Change Point	Remarks
1	RTC5628	Coil ignition-Lucas	1		
2	ADU7242L	Capacitor ignition coil	1		
3	FS108307L	Screw-flanged head-M8 x 30	2		
4	WL108001L	Washer-M8	2		
5	WA108051L	Washer-plain-M8	2		
6	FN108041L	Nut-flanged head-M8 x 20	NLA		Use FN108047L.
6	FN108047L	Nut-flanged head-M8	2		
	PRC4628	Harness earth bond	1		
7	DRC1752L	Resistor	1		
8	PRC6144	Harness-link-low tension ignition coil-2 pin	1		
8	PRC6141	Harness-link-low tension ignition coil-single connectors	1		

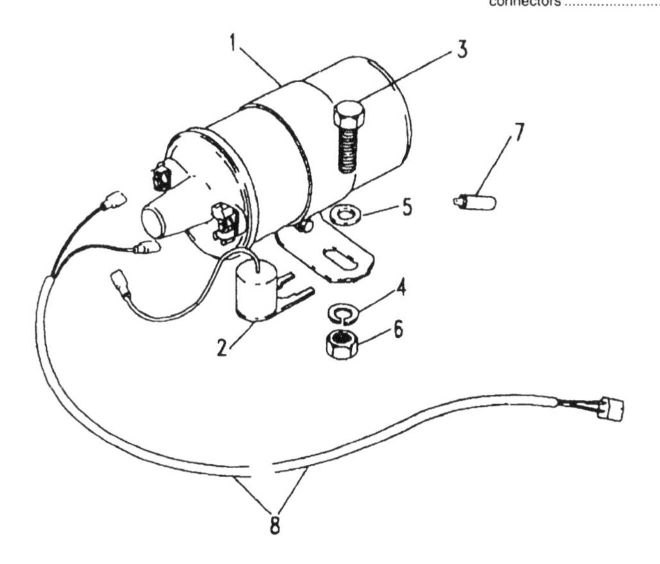

B 104

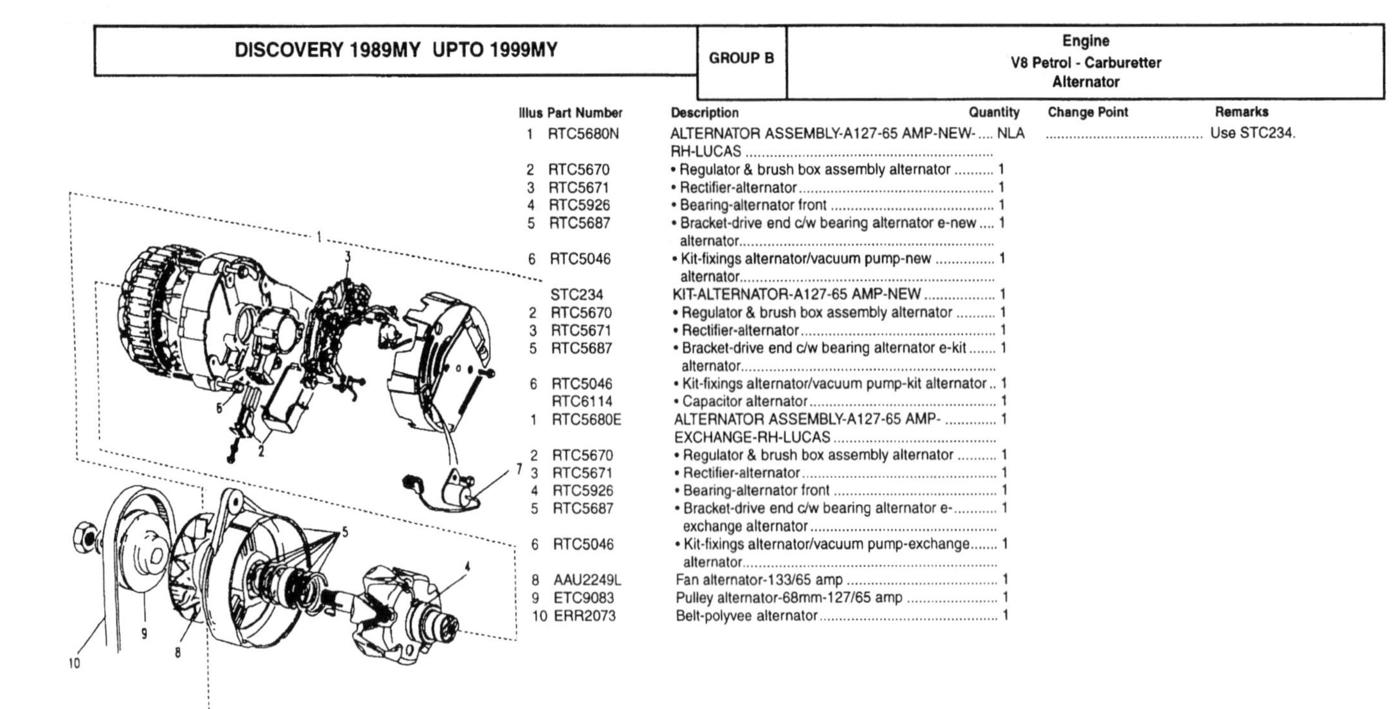

Illus	Part Number	Description	Quantity	Change Point	Remarks
1	RTC5680N	ALTERNATOR ASSEMBLY-A127-65 AMP-NEW-RH-LUCAS	NLA		Use STC234.
2	RTC5670	• Regulator & brush box assembly alternator	1		
3	RTC5671	• Rectifier-alternator	1		
4	RTC5926	• Bearing-alternator front	1		
5	RTC5687	• Bracket-drive end c/w bearing alternator e-new alternator	1		
6	RTC5046	• Kit-fixings alternator/vacuum pump-new alternator	1		
	STC234	KIT-ALTERNATOR-A127-65 AMP-NEW	1		
2	RTC5670	• Regulator & brush box assembly alternator	1		
3	RTC5671	• Rectifier-alternator	1		
5	RTC5687	• Bracket-drive end c/w bearing alternator e-kit alternator	1		
6	RTC5046	• Kit-fixings alternator/vacuum pump-kit alternator	1		
	RTC6114	• Capacitor alternator	1		
1	RTC5680E	ALTERNATOR ASSEMBLY-A127-65 AMP-EXCHANGE-RH-LUCAS	1		
2	RTC5670	• Regulator & brush box assembly alternator	1		
3	RTC5671	• Rectifier-alternator	1		
4	RTC5926	• Bearing-alternator front	1		
5	RTC5687	• Bracket-drive end c/w bearing alternator e-exchange alternator	1		
6	RTC5046	• Kit-fixings alternator/vacuum pump-exchange alternator	1		
8	AAU2249L	Fan alternator-133/65 amp	1		
9	ETC9083	Pulley alternator-68mm-127/65 amp	1		
10	ERR2073	Belt-polyvee alternator	1		

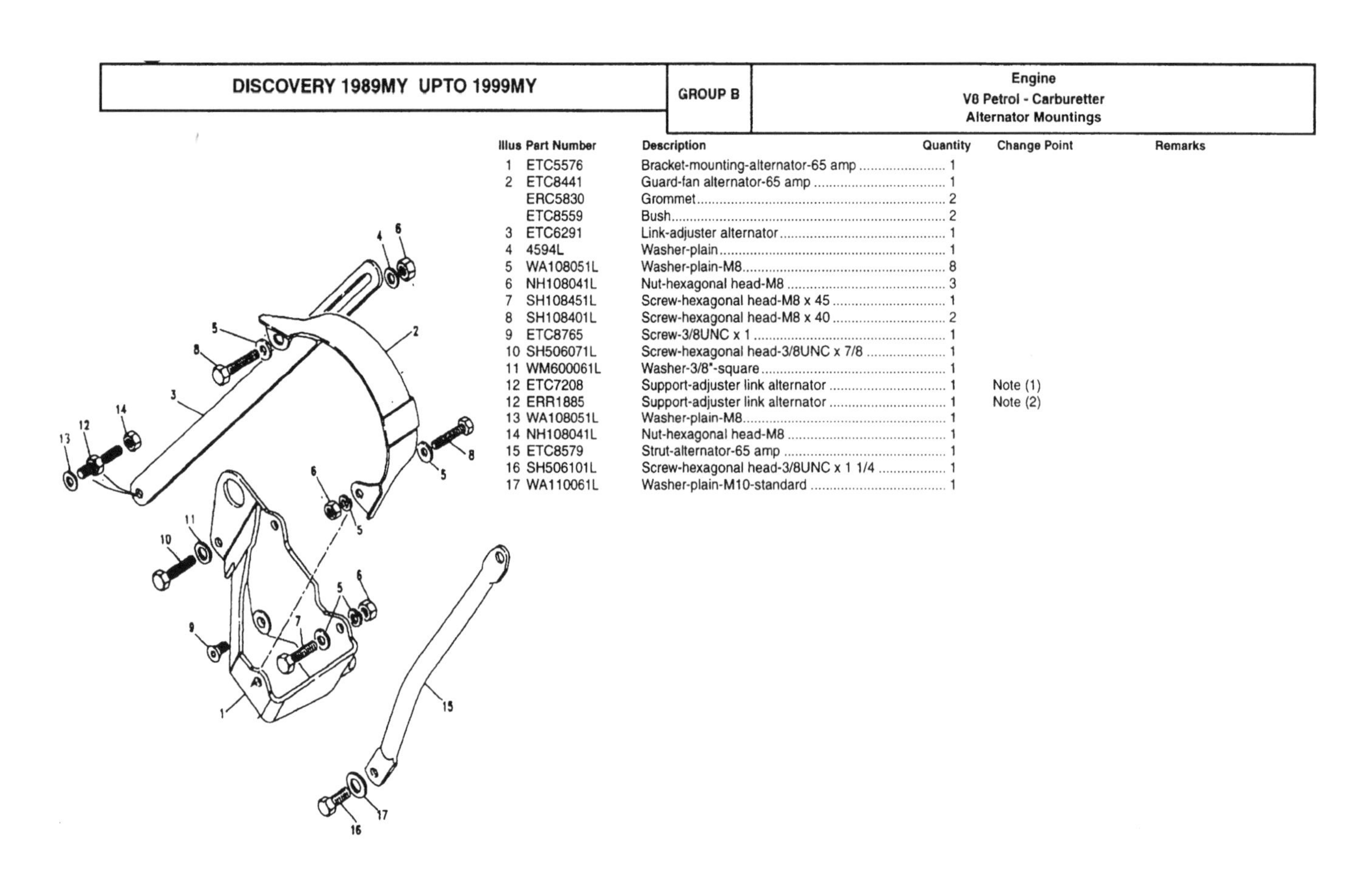

Illus	Part Number	Description	Quantity	Change Point	Remarks
1	ETC5576	Bracket-mounting-alternator-65 amp	1		
2	ETC8441	Guard-fan alternator-65 amp	1		
	ERC5830	Grommet	2		
	ETC8559	Bush	2		
3	ETC6291	Link-adjuster alternator	1		
4	4594L	Washer-plain	1		
5	WA108051L	Washer-plain-M8	8		
6	NH108041L	Nut-hexagonal head-M8	3		
7	SH108451L	Screw-hexagonal head-M8 x 45	1		
8	SH108401L	Screw-hexagonal head-M8 x 40	2		
9	ETC8765	Screw-3/8UNC x 1	1		
10	SH506071L	Screw-hexagonal head-3/8UNC x 7/8	1		
11	WM600061L	Washer-3/8"-square	1		
12	ETC7208	Support-adjuster link alternator	1	Note (1)	
12	ERR1885	Support-adjuster link alternator	1	Note (2)	
13	WA108051L	Washer-plain-M8	1		
14	NH108041L	Nut-hexagonal head-M8	1		
15	ETC8579	Strut-alternator-65 amp	1		
16	SH506101L	Screw-hexagonal head-3/8UNC x 1 1/4	1		
17	WA110061L	Washer-plain-M10-standard	1		

CHANGE POINTS:
(1) To (V) HA 002588
(2) From (V) HA 002589

DISCOVERY 1989MY UPTO 1999MY

GROUP B

Engine
V8 Petrol - Carburetter
Starter Motor

Illus	Part Number	Description	Quantity	Change Point	Remarks
1	RTC6061N	MOTOR-STARTER ENGINE-NEW-LUCAS	1		
2	RTC5048	• Kit-brush starter motor-with M78R starter motor-Lucas	1		
3	RTC5049	• Solenoid starter motor-with M78R starter motor-Lucas	1		
4	RTC5050	• Drive assembly starter motor	1		
5	SS506121L	Screw-3/8UNC x 1 1/2	2		
6	WM600061L	Washer-3/8"-square	2		
7	WA110061L	Washer-plain-M10-standard	2		
	PRC3623	Cable assembly starter & earth	1		
	PRC5208	Harness-link	1		
	90608178	Kit-starter motor sundry parts-with 2m100 starter motor	1		

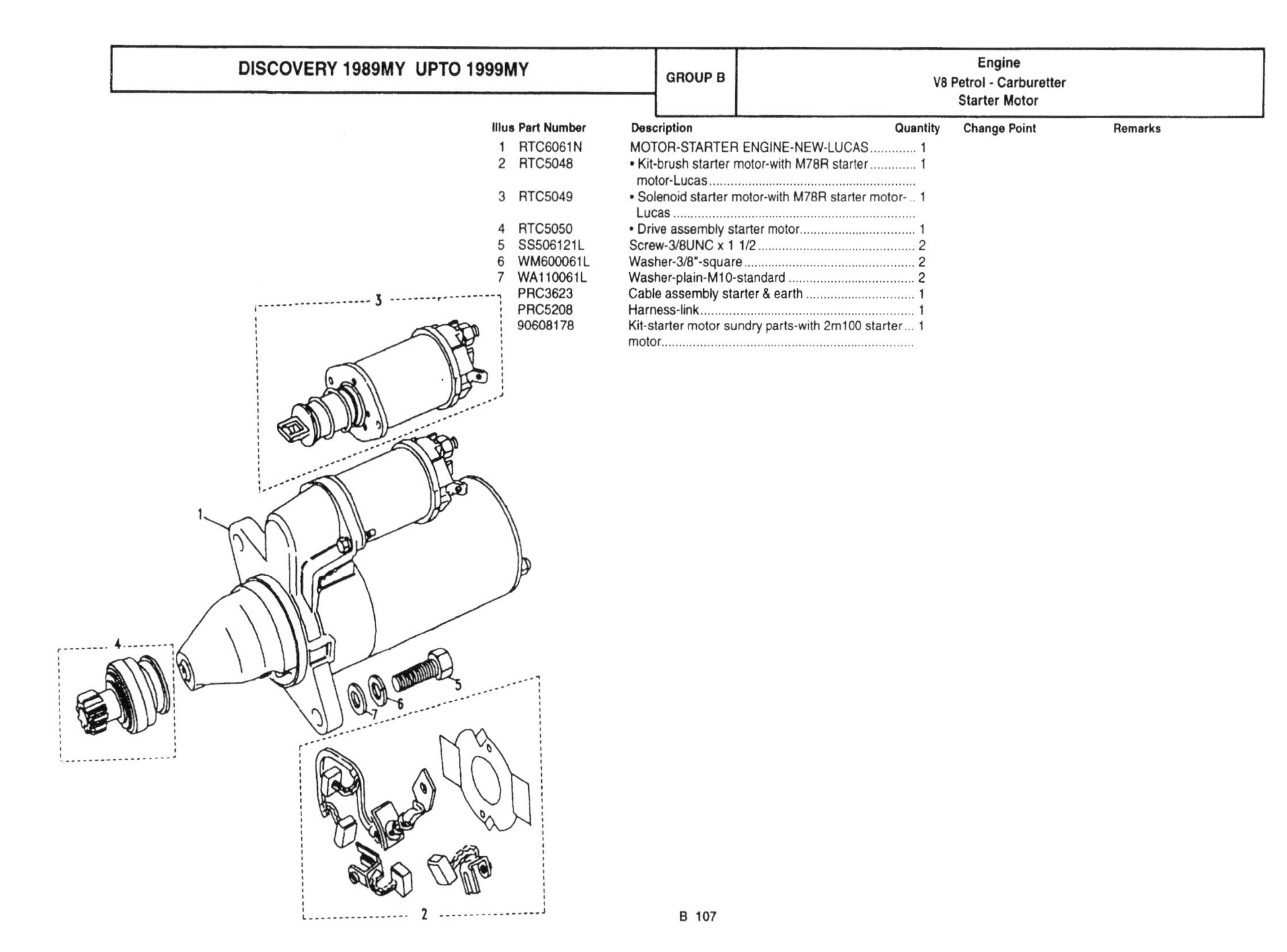

B 107

DISCOVERY 1989MY UPTO 1999MY

GROUP B

Engine
V8 Petrol - Carburetter
Flywheel and Clutch

Illus	Part Number	Description	Quantity	Change Point	Remarks
1	611324	FLYWHEEL ENGINE	NLA		Use ERR5575.
1	ERR5575	FLYWHEEL ENGINE	1		
2	611323	• Gear-ring-flywheel engine	1		
3	6395L	• Dowel	3		
4	SH607081L	Screw-hexagonal head-7/16UNC x 1	6		
5	FTC814	Plate-clutch-driven	1		
6	FTC813	Cover-clutch assembly	1		
	STC8360	KIT-CLUTCH	1		
5	FTC814	• Plate-clutch-driven	1		
6	FTC813	• Cover-clutch assembly	1		
	FRC9568	• Bearing-clutch release	1		
7	SH606061L	Screw-hexagonal head-3/8UNC x 3/4	6		
8	WM600061L	Washer-3/8"-square	6		

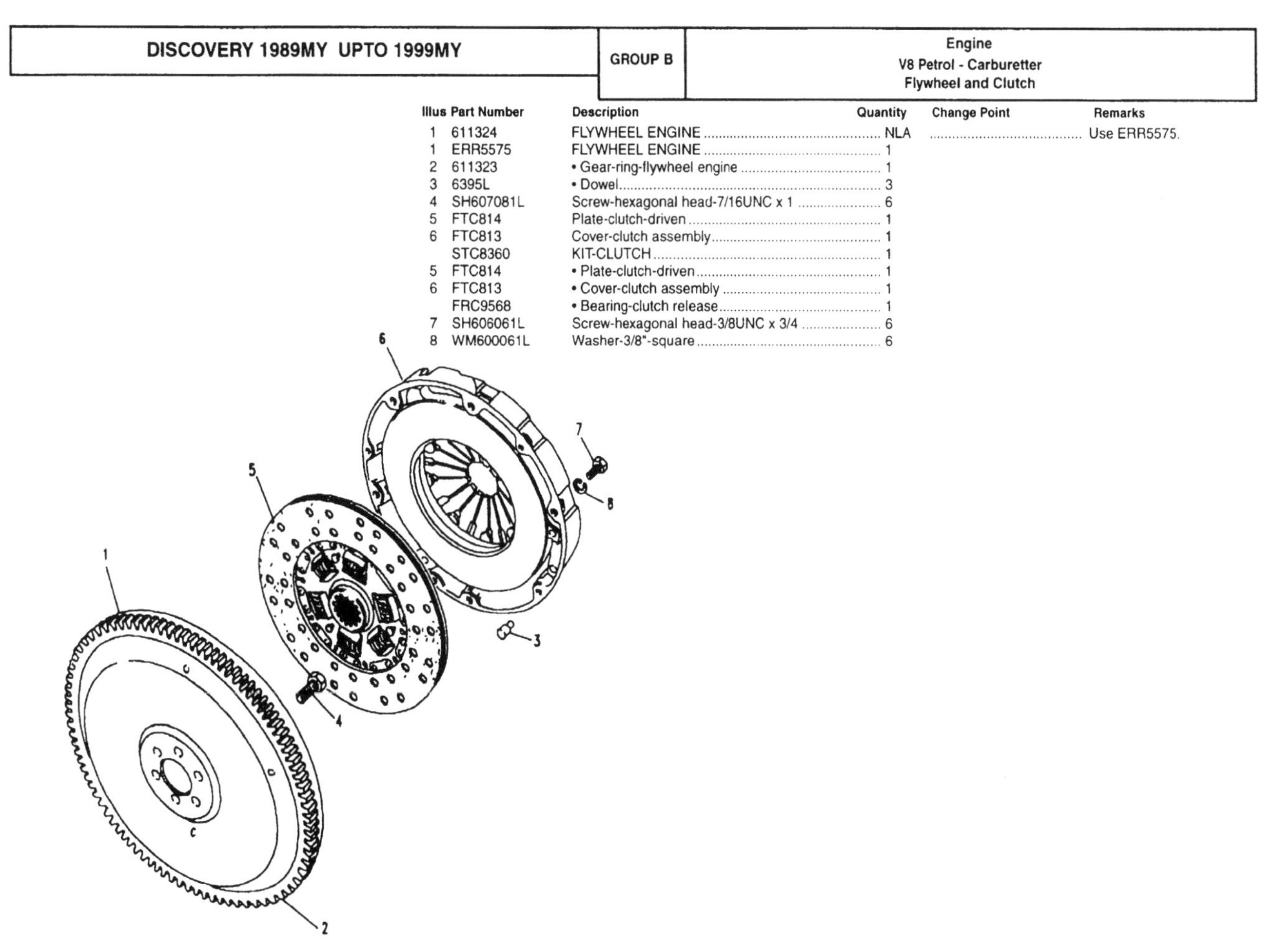

B 108

Illus	Part Number	Description	Quantity	Change Point	Remarks
		ENGINE UNIT-STRIPPED			
1	STC155N	new	1		24D
1	STC155R	reconditioned	1		24D
1	STC156N	new	1		22D
1	STC156R	reconditioned	1		22D
1	STC156N	new	1		23D
1	STC156R	reconditioned	1		23D
1	STC1600N	new	NLA	Note (1)	Use STC1102N.
1	STC1600N	new	NLA	Note (1)	Use STC1102N.
1	STC1102N	new	1	Note (1)	35D
1	STC1102N	new	1	Note (1)	36D
1	STC1102E	exchange	1	Note (1)	35D
1	STC1102E	exchange	1	Note (1)	36D
1	STC1101N	new	1	Note (1)	37D
1	STC1101N	new	1	Note (1)	38D
1	STC1861N	new	1	Note (2)	35D
1	STC1861N	new	1	Note (2)	36D
1	STC1861E	exchange	1	Note (2)	35D
1	STC1861E	exchange	1	Note (2)	36D
1	STC1862N	new	1	Note (2)	37D
1	STC1862N	new	1	Note (2)	38D
1	STC1862E	exchange	1	Note (2)	37D
1	STC1862E	exchange	1	Note (2)	38D

CHANGE POINTS:
(1) To (V) MA 081990
(2) From (V) MA 081991

B 109

Illus	Part Number	Description	Quantity	Change Point	Remarks
1	ETC7714	Engine base unit-part-8.13:1	1		22D
					23D
1	ETC7713	Engine base unit-part-9.35:1	1		24D
1	ERR514	Engine base unit-part-9.35:1	1	Note (1)	
1	ERR4165	Engine base unit-part	1	Note (2)	
1	ERR4170	Engine base unit-part	1	Note (3)	

COMPRISES BLOCK
CRANKSHAFT AND PISTONS

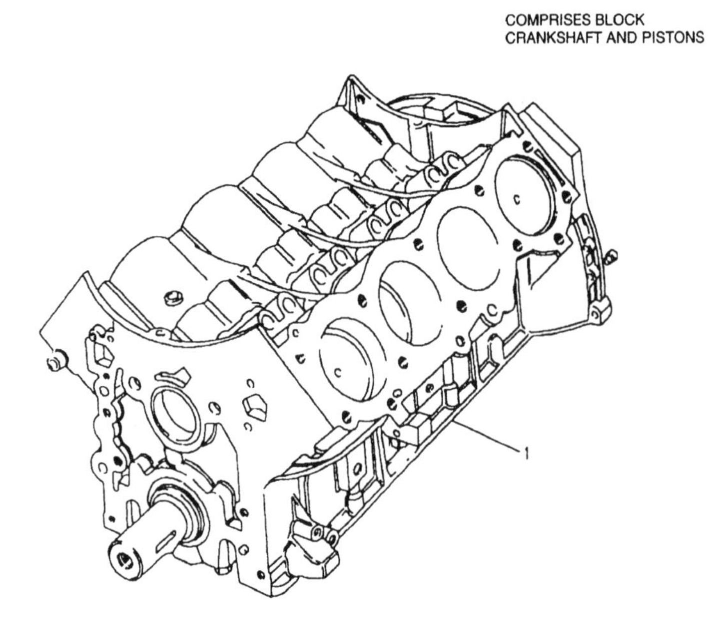

CHANGE POINTS:
(1) To (E)35D 08927; To (E)36D 25154
(2) From (E)35D 08928b; From (E)36D 25155b
(3) From (E)37D 02090b; From (E)38D 27238b

B 110

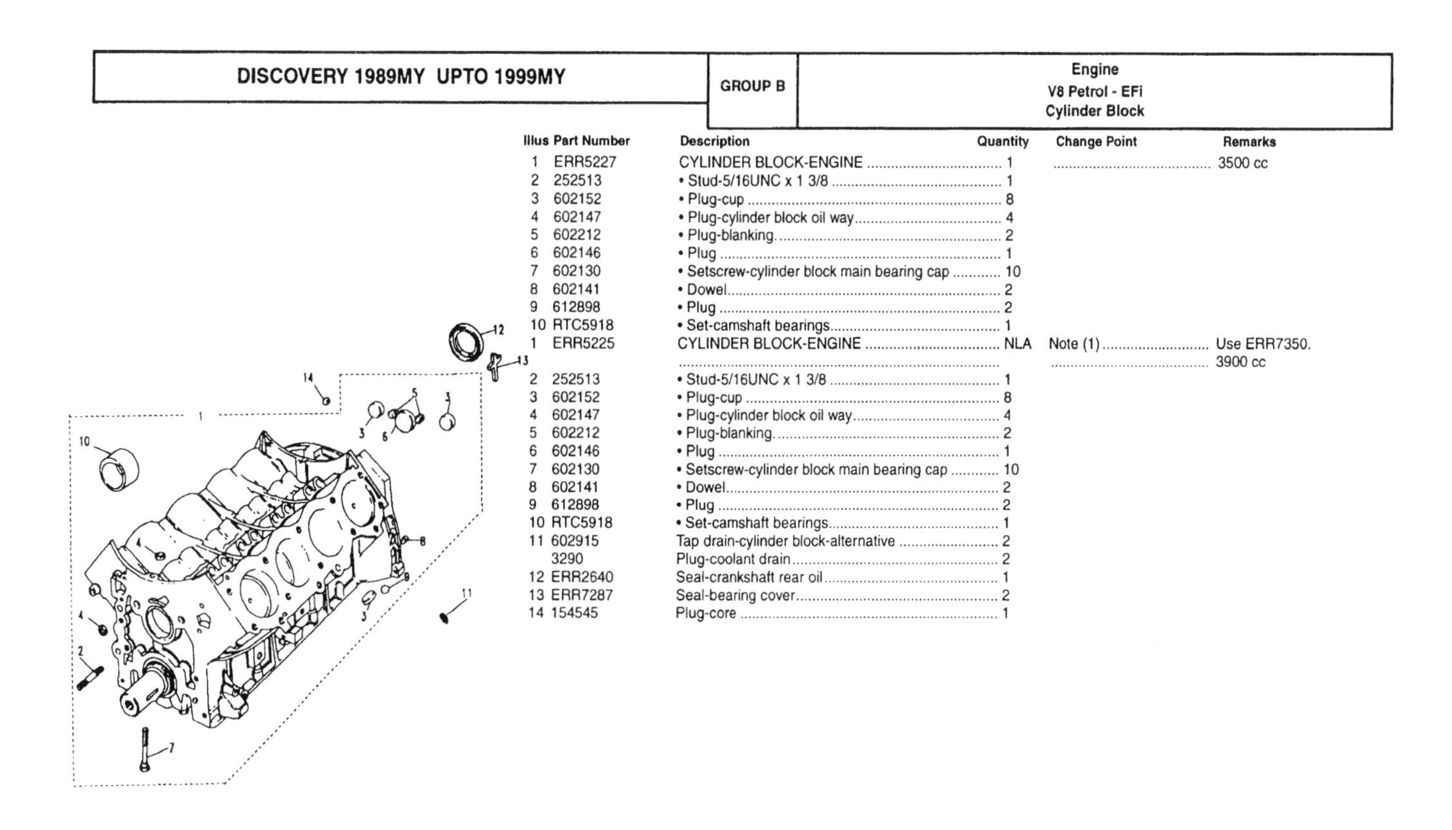

DISCOVERY 1989MY UPTO 1999MY

GROUP B

Engine
V8 Petrol - EFi
Cylinder Block

Illus	Part Number	Description	Quantity	Change Point	Remarks
1	ERR5227	CYLINDER BLOCK-ENGINE	1		3500 cc
2	252513	• Stud-5/16UNC x 1 3/8	1		
3	602152	• Plug-cup	8		
4	602147	• Plug-cylinder block oil way	4		
5	602212	• Plug-blanking	2		
6	602146	• Plug	1		
7	602130	• Setscrew-cylinder block main bearing cap	10		
8	602141	• Dowel	2		
9	612898	• Plug	2		
10	RTC5918	• Set-camshaft bearings	1		
1	ERR5225	CYLINDER BLOCK-ENGINE	NLA	Note (1)	Use ERR7350.
					3900 cc
2	252513	• Stud-5/16UNC x 1 3/8	1		
3	602152	• Plug-cup	8		
4	602147	• Plug-cylinder block oil way	4		
5	602212	• Plug-blanking	2		
6	602146	• Plug	1		
7	602130	• Setscrew-cylinder block main bearing cap	10		
8	602141	• Dowel	2		
9	612898	• Plug	2		
10	RTC5918	• Set-camshaft bearings	1		
11	602915	Tap drain-cylinder block-alternative	2		
	3290	Plug-coolant drain	2		
12	ERR2640	Seal-crankshaft rear oil	1		
13	ERR7287	Seal-bearing cover	2		
14	154545	Plug-core	1		

CHANGE POINTS:
(1) From (E)35D 08928b; From (E)36D 25524b; From (E)37D 01932b; From (E)38D 27238b

B 111

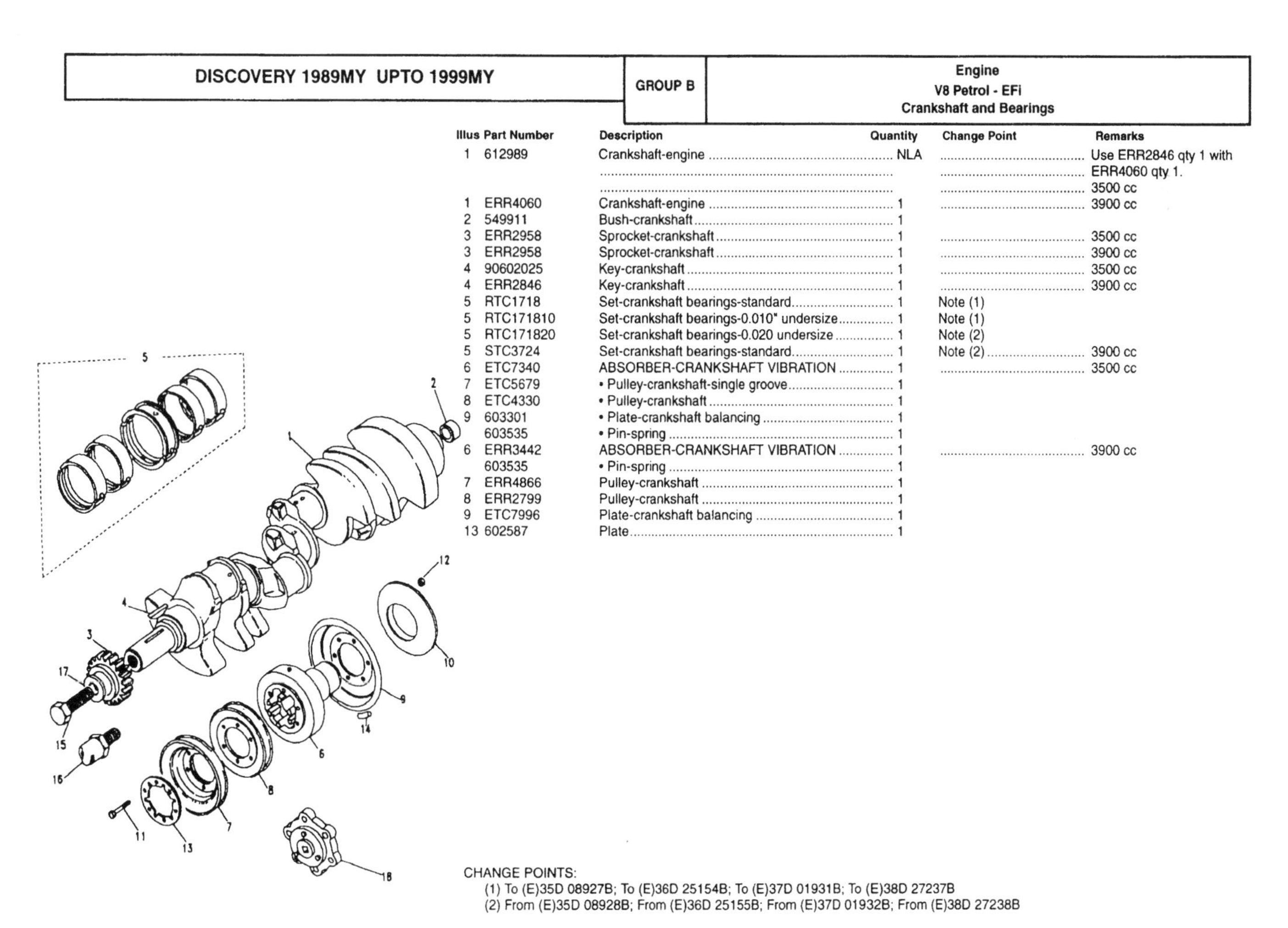

DISCOVERY 1989MY UPTO 1999MY

GROUP B

Engine
V8 Petrol - EFi
Crankshaft and Bearings

Illus	Part Number	Description	Quantity	Change Point	Remarks
1	612989	Crankshaft-engine	NLA		Use ERR2846 qty 1 with
					ERR4060 qty 1.
					3500 cc
1	ERR4060	Crankshaft-engine	1		3900 cc
2	549911	Bush-crankshaft	1		
3	ERR2958	Sprocket-crankshaft	1		3500 cc
3	ERR2958	Sprocket-crankshaft	1		3900 cc
4	90602025	Key-crankshaft	1		3500 cc
4	ERR2846	Key-crankshaft	1		3900 cc
5	RTC1718	Set-crankshaft bearings-standard	1	Note (1)	
5	RTC171810	Set-crankshaft bearings-0.010" undersize	1	Note (1)	
5	RTC171820	Set-crankshaft bearings-0.020 undersize	1	Note (2)	
5	STC3724	Set-crankshaft bearings-standard	1	Note (2)	3900 cc
6	ETC7340	ABSORBER-CRANKSHAFT VIBRATION	1		3500 cc
7	ETC5679	• Pulley-crankshaft-single groove	1		
8	ETC4330	• Pulley-crankshaft	1		
9	603301	• Plate-crankshaft balancing	1		
	603535	• Pin-spring	1		
6	ERR3442	ABSORBER-CRANKSHAFT VIBRATION	1		3900 cc
	603535	• Pin-spring	1		
7	ERR4866	Pulley-crankshaft	1		
8	ERR2799	Pulley-crankshaft	1		
9	ETC7996	Plate-crankshaft balancing	1		
13	602587	Plate	1		

CHANGE POINTS:
(1) To (E)35D 08927B; To (E)36D 25154B; To (E)37D 01931B; To (E)38D 27237B
(2) From (E)35D 08928B; From (E)36D 25155B; From (E)37D 01932B; From (E)38D 27238B

B 112

Illus	Part Number	Description	Quantity	Change Point	Remarks
10	613671	Shield-rear flange oil deflector	1		
11	BH605131L	Bolt-5/16UNF x 1 5/8	6		
12	NH605041L	Nut-hexagonal head-5/16UNF	6		
14	ERC4877	Weight-crankshaft balance-.17oz	A/R		
14	ERC4878	Weight-crankshaft balance-.35oz	A/R		
14	ERC4879	Weight-crankshaft balance-.53oz	A/R		
14	ERC4880	Weight-crankshaft balance-.71oz	A/R		
15	ERC417	Bolt-crankshaft pulley	1		
16	610178	Bolt-dog point	1		
	602411	Washer-plain	1		
17	ERC416	Washer	1		
	603535	Pin-spring	1		
18	ETC7419	Plate-crankshaft balancing	1		

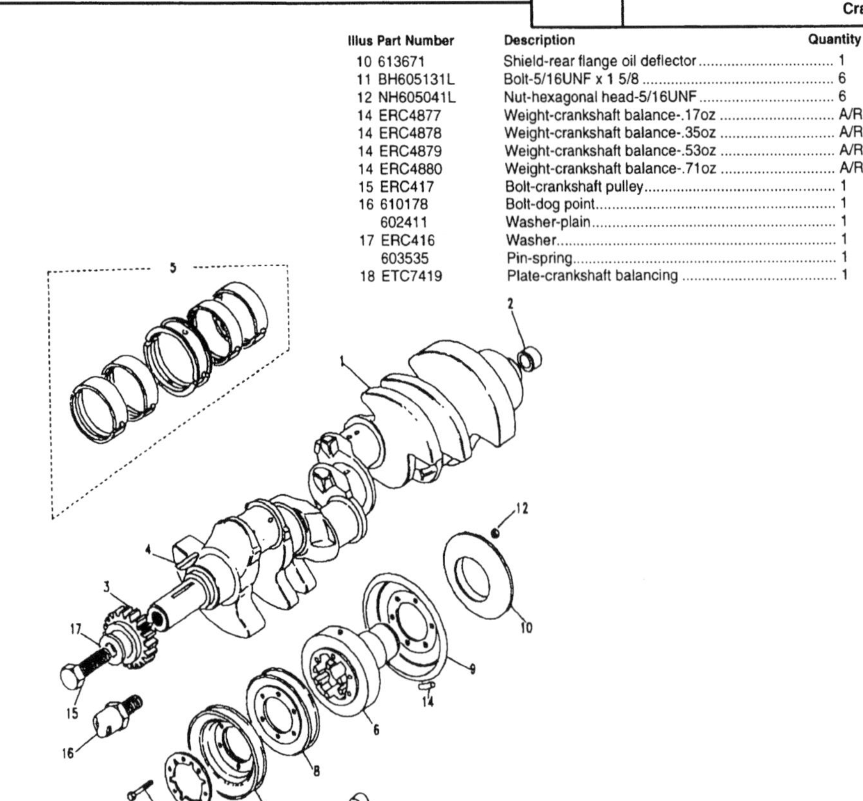

B 113

Illus	Part Number	Description	Quantity	Change Point	Remarks
		3.5 ENGINE LOW COMPRESSION			
1	RTC2186S	PISTON-ENGINE-STANDARD.-8.13:1	8		
1	RTC218620	PISTON-ENGINE-0.020" OVERSIZE-8.13:1	8		
2	RTC2408	• Ring set assembly-piston-standard.	8		Part of RTC2186S.
2	RTC240820	• Ring set assembly-piston-0.020" oversize	8		Part of RTC218620.
		3.5 ENGINE HIGH COMPRESSION			
1	RTC2295S	PISTON-ENGINE-STANDARD.-9.35:1	8		
1	RTC229520	PISTON-ENGINE-0.020" OVERSIZE-9.35:1	8		
2	RTC2408	• Ring set assembly-piston-standard.	8		Part of RTC2295S.
2	RTC240820	• Ring set assembly-piston-0.020" oversize	8		Part of RTC229520.
		3.9 ENGINE			
1	ERR2692	PISTON-ENGINE-STANDARD.	8		Low Compression
1	ERR2693	PISTON-ENGINE-STANDARD.	8		High Compression
2	RTC6066S	• Ring set assembly-piston-standard.	8		
		CON ROD AND BEARINGS			
3	602082	ROD-CRANKSHAFT CONNECTING	8		
4	602609	• Bolt-connecting rod-11/32UNS-3A	16		
5	602061	• Nut-connecting rod-11/32UNS-3B	16		
6	RTC2117	Set-connecting rod big end half bearing-standard.	1		
6	RTC211710	Set-connecting rod big end half bearing-0.010" undersize	1		
6	RTC211720	Set-connecting rod big end half bearing-0.020 undersize	1		

B 114

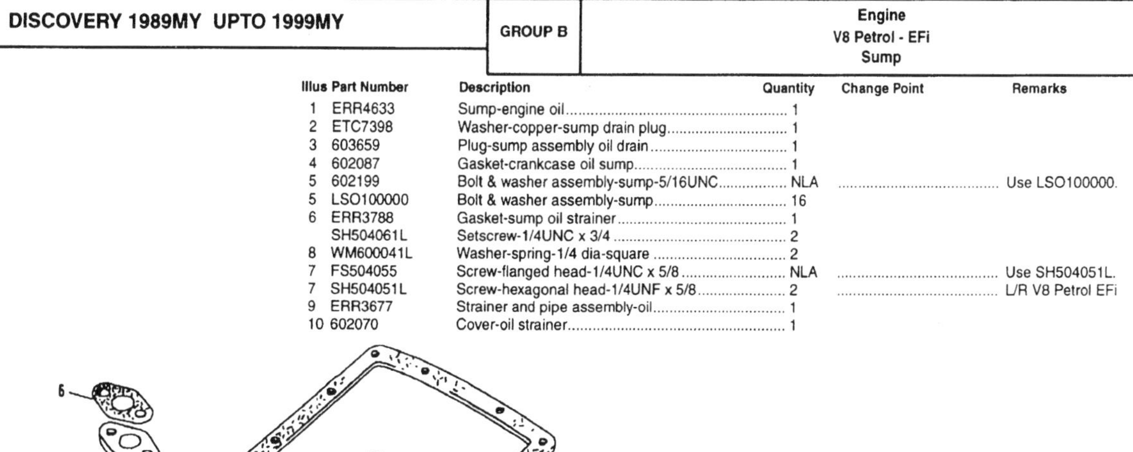

Illus	Part Number	Description	Quantity	Change Point	Remarks
1	ERR4633	Sump-engine oil	1		
2	ETC7398	Washer-copper-sump drain plug	1		
3	603659	Plug-sump assembly oil drain	1		
4	602087	Gasket-crankcase oil sump	1		
5	602199	Bolt & washer assembly-sump-5/16UNC	NLA		Use LSO100000.
5	LSO100000	Bolt & washer assembly-sump	16		
6	ERR3788	Gasket-sump oil strainer	1		
	SH504061L	Setscrew-1/4UNC x 3/4	2		
8	WM600041L	Washer-spring-1/4 dia-square	2		
7	FS504055	Screw-flanged head-1/4UNC x 5/8	NLA		Use SH504051L.
7	SH504051L	Screw-hexagonal head-1/4UNF x 5/8	2		L/R V8 Petrol EFi
9	ERR3677	Strainer and pipe assembly-oil	1		
10	602070	Cover-oil strainer	1		

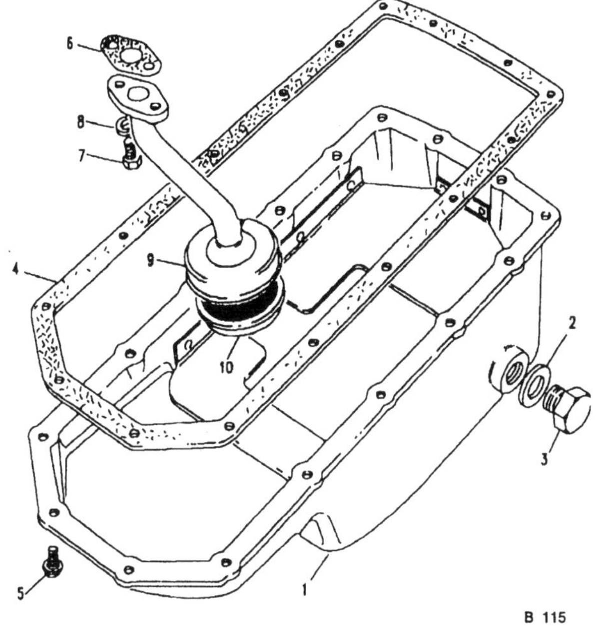

B 115

Illus	Part Number	Description	Quantity	Change Point	Remarks
1	ERR1922L	Dipstick-oil	1		
2	ERR4556	Tube-oil dipstick	1		
3	602545	O ring	1		
4	610489	Clip	1		
5	AB610041L	Screw-self tapping AB-M8 x 12	1		

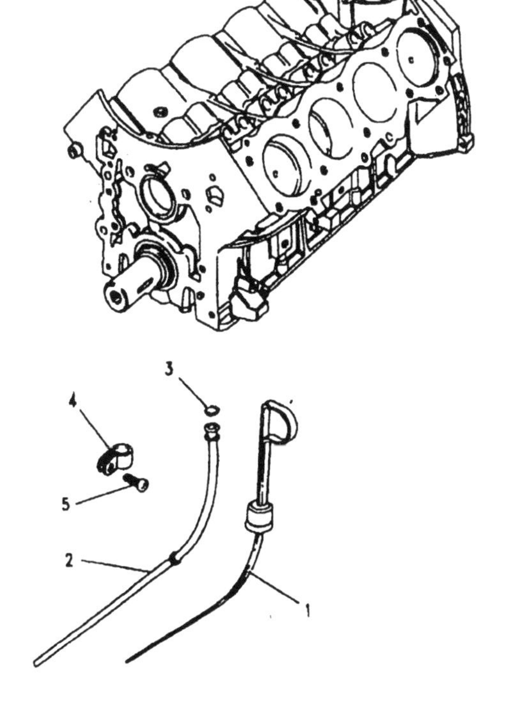

B 116

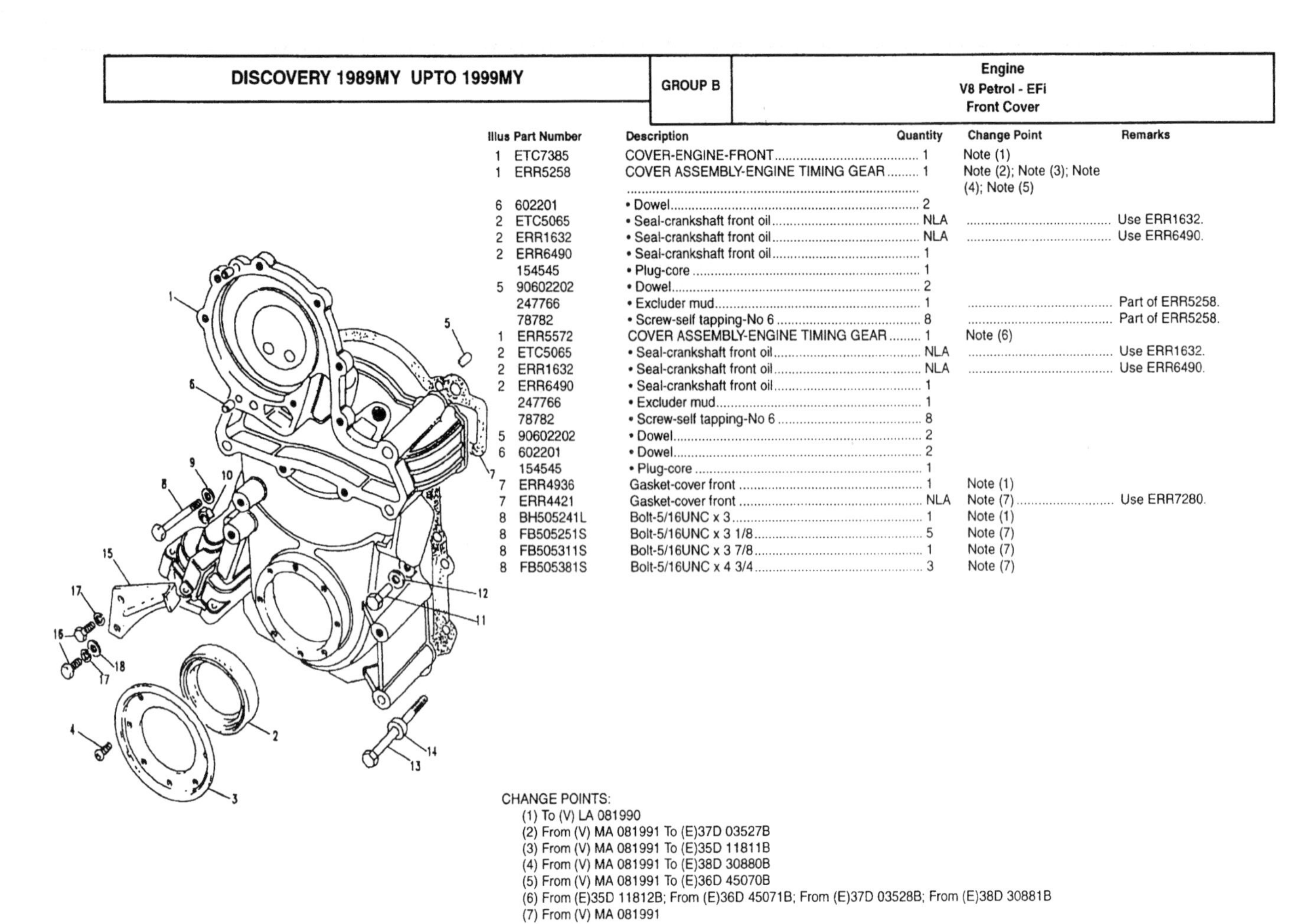

Illus	Part Number	Description	Quantity	Change Point	Remarks
1	ETC7385	COVER-ENGINE-FRONT	1	Note (1)	
1	ERR5258	COVER ASSEMBLY-ENGINE TIMING GEAR	1	Note (2); Note (3); Note (4); Note (5)	
6	602201	• Dowel	2		
2	ETC5065	• Seal-crankshaft front oil	NLA		Use ERR1632.
2	ERR1632	• Seal-crankshaft front oil	NLA		Use ERR6490.
2	ERR6490	• Seal-crankshaft front oil	1		
	154545	• Plug-core	1		
5	90602202	• Dowel	2		
	247766	• Excluder mud	1		Part of ERR5258.
	78782	• Screw-self tapping-No 6	8		Part of ERR5258.
1	ERR5572	COVER ASSEMBLY-ENGINE TIMING GEAR	1	Note (6)	
2	ETC5065	• Seal-crankshaft front oil	NLA		Use ERR1632.
2	ERR1632	• Seal-crankshaft front oil	NLA		Use ERR6490.
2	ERR6490	• Seal-crankshaft front oil	1		
	247766	• Excluder mud	1		
	78782	• Screw-self tapping-No 6	8		
5	90602202	• Dowel	2		
6	602201	• Dowel	2		
	154545	• Plug-core	1		
7	ERR4936	Gasket-cover front	1	Note (1)	
7	ERR4421	Gasket-cover front	NLA	Note (7)	Use ERR7280.
8	BH505241L	Bolt-5/16UNC x 3	1	Note (1)	
8	FB505251S	Bolt-5/16UNC x 3 1/8	5	Note (7)	
8	FB505311S	Bolt-5/16UNC x 3 7/8	1	Note (7)	
8	FB505381S	Bolt-5/16UNC x 4 3/4	3	Note (7)	

CHANGE POINTS:
(1) To (V) LA 081990
(2) From (V) MA 081991 To (E)37D 03527B
(3) From (V) MA 081991 To (E)35D 11811B
(4) From (V) MA 081991 To (E)38D 30880B
(5) From (V) MA 081991 To (E)36D 45070B
(6) From (E)35D 11812B; From (E)36D 45071B; From (E)37D 03528B; From (E)38D 30881B
(7) From (V) MA 081991

B 117

DISCOVERY 1989MY UPTO 1999MY | GROUP B | Engine V8 Petrol - EFi Front Cover

Illus	Part Number	Description	Quantity	Change Point	Remarks
9	WA108051L	Washer-plain-M8	1		
10	NH605041L	Nut-hexagonal head-5/16UNF	1		
11	SH505091L	Screw-hexagonal head-5/16UNC x 1 1/8	2		
12	WA108051L	Washer-plain-M8	2		
13	602388	Bolt-5/16UNC x 3.8	1		
14	4075L	Washer-plain-M8	1		
	252513	Stud-5/16UNC x 1 3/8	1		
15	ETC7345	Pointer-engine timing	1		
16	SH504051L	Screw-hexagonal head-1/4UNF x 5/8	2		
17	WM600041L	Washer-spring-1/4 dia-square	2		
18	RTC609	Washer-plain-standard	2		

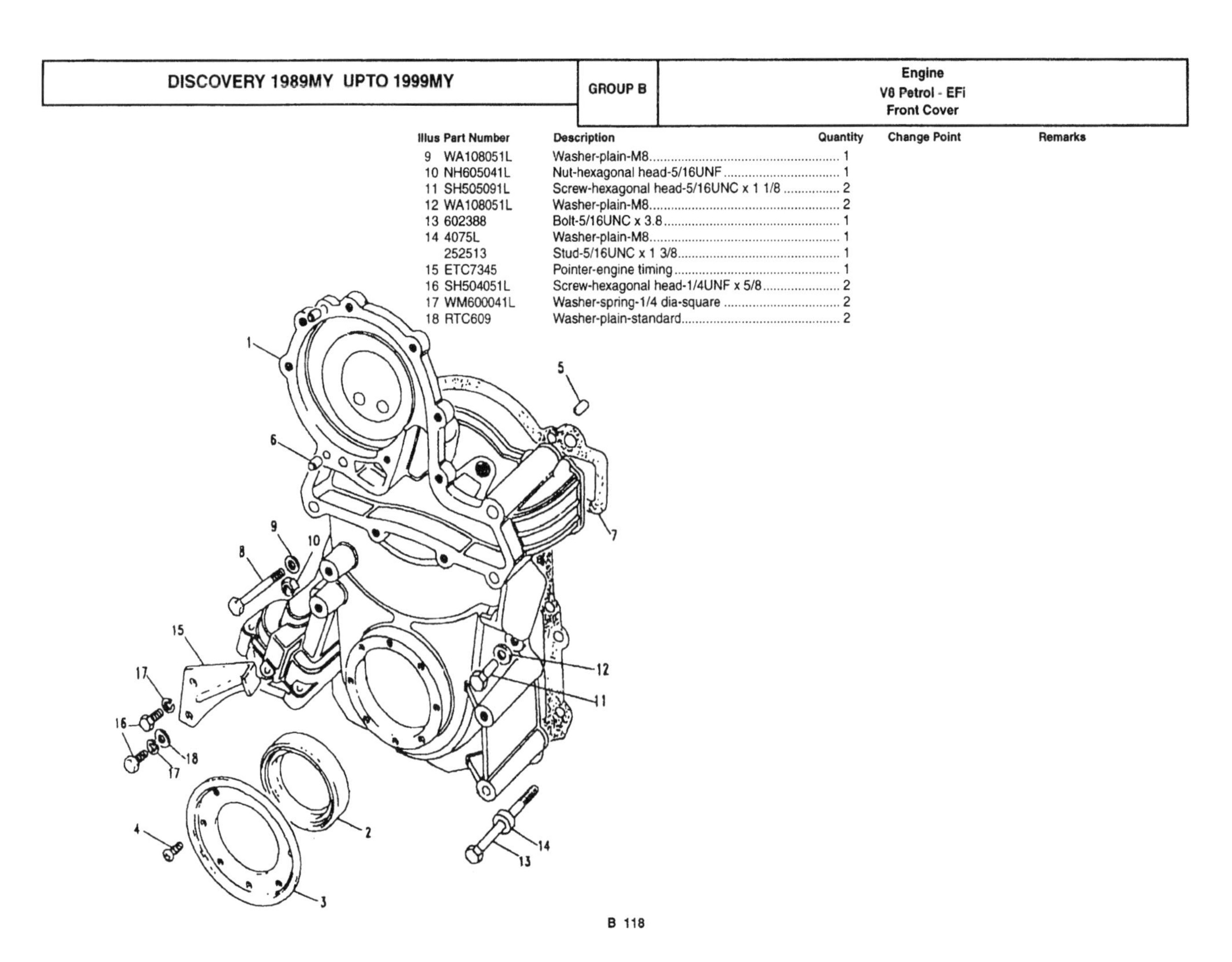

B 118

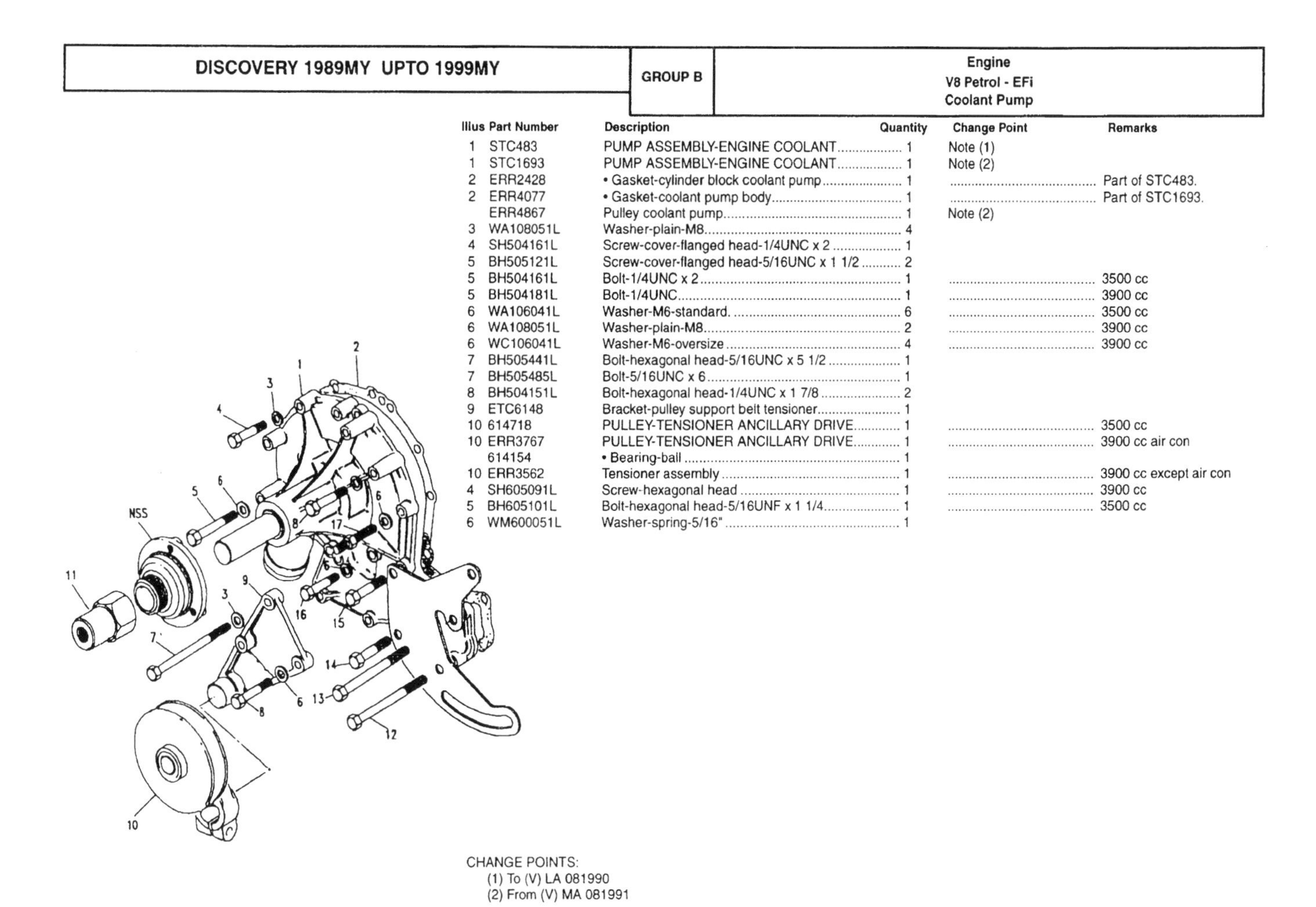

Illus	Part Number	Description	Quantity	Change Point	Remarks
1	STC483	PUMP ASSEMBLY-ENGINE COOLANT	1	Note (1)	
1	STC1693	PUMP ASSEMBLY-ENGINE COOLANT	1	Note (2)	
2	ERR2428	• Gasket-cylinder block coolant pump	1		Part of STC483.
2	ERR4077	• Gasket-coolant pump body	1		Part of STC1693.
	ERR4867	Pulley coolant pump	1	Note (2)	
3	WA108051L	Washer-plain-M8	4		
4	SH504161L	Screw-cover-flanged head-1/4UNC x 2	1		
5	BH505121L	Screw-cover-flanged head-5/16UNC x 1 1/2	2		
5	BH504161L	Bolt-1/4UNC x 2	1		3500 cc
5	BH504181L	Bolt-1/4UNC	1		3900 cc
6	WA106041L	Washer-M6-standard.	6		3500 cc
6	WA108051L	Washer-plain-M8	2		3900 cc
6	WC106041L	Washer-M6-oversize	4		3900 cc
7	BH505441L	Bolt-hexagonal head-5/16UNC x 5 1/2	1		
7	BH505485L	Bolt-5/16UNC x 6	1		
8	BH504151L	Bolt-hexagonal head-1/4UNC x 1 7/8	2		
9	ETC6148	Bracket-pulley support belt tensioner	1		
10	614718	PULLEY-TENSIONER ANCILLARY DRIVE	1		3500 cc
10	ERR3767	PULLEY-TENSIONER ANCILLARY DRIVE	1		3900 cc air con
	614154	• Bearing-ball	1		
10	ERR3562	Tensioner assembly	1		3900 cc except air con
4	SH605091L	Screw-hexagonal head	1		3900 cc
5	BH605101L	Bolt-hexagonal head-5/16UNF x 1 1/4	1		3500 cc
6	WM600051L	Washer-spring-5/16"	1		

CHANGE POINTS:
(1) To (V) LA 081990
(2) From (V) MA 081991

B 119

Illus	Part Number	Description	Quantity	Change Point	Remarks
11	ETC6104	Adaptor-water pump drive	1		
13	BH505381L	Bolt-hexagonal head-5/16UNC x 4 3/4	NLA		Use BH505401L.
12	BH505401L	Bolt-hexagonal head-5/16UNC x 5	1		
14	BH504111L	Bolt-1/4UNC x 1 3/8	1		
15	BH504121L	Bolt-1/4UNC x 1 1/2-hexagonal head	1		
16	BH504101L	Bolt-1/4UNC x 1 1/4	1		
17	ERR2673	Support-adjuster link alternator	1		3900 cc
	DBP8169L	Clip-pivot-snap-13-17mm-7mm hole-Grey	1		3900 cc
	ERC9404	Bracket harness	1		3900 cc
12	BH504161L	Bolt-1/4UNC x 2	3		3900 cc

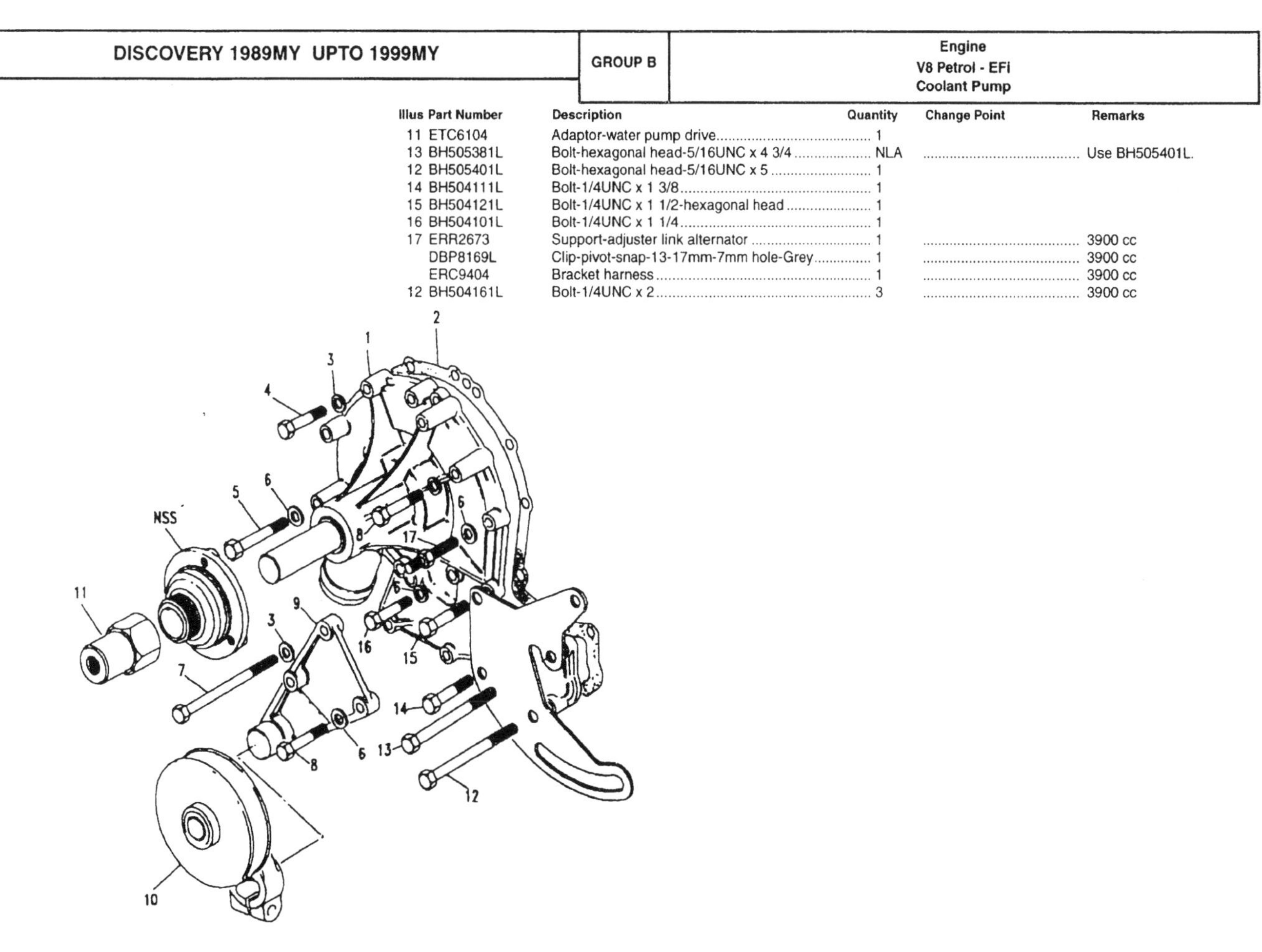

B 120

DISCOVERY 1989MY UPTO 1999MY	GROUP B	Engine V8 Petrol - EFI Viscous Drive and Fan

Illus	Part Number	Description	Quantity	Change Point	Remarks
1	ETC5499	Pulley coolant pump	1	Note (1)	Except air con
1	ERR4867	Pulley coolant pump	1	Note (2); Note (3)	Except air con
1	ETC5422	Pulley coolant pump	1		Air con
2	ETC5423	Pulley-driven air conditioning	1		Air con
3	SH604051L	Screw-pulley to pump-hexagonal head-1/4UNF x 5/8	3	Note (1)	
3	FS108127	Screw-pulley to pump-flanged head-M8 x 12	3	Note (3); Note (2)	
4	WA106041L	Washer-M6-standard.	3	Note (1)	
5	ETC1260	Coupling-engine fan viscous-11 blade	1		
5	ERR3443	Coupling-engine fan viscous-7 blade	1	Note (3); Note (2)	
6	ETC1275	Fan-cooling-11 blade-17"-alternative	1	Note (1)	
6	ERR3439	Fan-cooling-11 blade	1	Note (2); Note (3)	
7	SH505051L	Screw-fan to viscous coupling-hexagonal head-5/16UNC x 5/8	4	Note (1)	
7	ERR4194	Screw-fan to viscous coupling-hexagonal head-M8 x 14	4	Note (2); Note (3)	
8	611612	Belt-polyvee alternator	1	Note (1)	
8	ERR4461	Belt-polyvee alternator	1	Note (2); Note (3)	Except air con
8	ERR6191	Belt-polyvee alternator	1	Note (4); Note (5)	Except air con
8	ERR4623	Belt-polyvee alternator	1	Note (2); Note (3)	Air con
8	ERR5579	Belt-polyvee alternator	1	Note (4); Note (5)	Air con

CHANGE POINTS:
(1) To (V) LA 081990
(2) From (V) MA 081991 To (E)38D 32139
(3) From (V) MA 081991 To (E)36D 50529
(4) From (E)36D 50530C
(5) From (E)38D 32140C

B 121

DISCOVERY 1989MY UPTO 1999MY	GROUP B	Engine V8 Petrol - EFI Oil Pump and Filter

Illus	Part Number	Description	Quantity	Change Point	Remarks
		UP TO VIN LA081990			
1	ERC1351	Shaft-oil pump	1		
2	614037	Gear-oil pump	1		
3	ERR1990	Gasket-cylinder block oil pump	1		
4	ETC4276	ADAPTOR ASSEMBLY-OIL FILTER	1		
11	614202	• Strainer-oil	1		
5	90602064	• Plunger-oil pump relief valve	1		
6	602067	• Spring-oil pump relief valve	1		
8	602071	• Plug-oil pump relief valve	1		
7	ETC8833	• Washer-oil pressure switch engine	1		
9	243968	• Washer-sealing	1		
10	273166	• Plug	1		
14	602910	Bolt-1/4UNC x 7/8	3		
14	602912	Bolt-1/4UNC x 1 1/4	2		
14	602913	Bolt-1/4UNC x 9/16	1		
15	PRC7204	Switch-oil pressure engine	NLA		Use AMR2092.
15	AMR2092	Switch-oil pressure engine	NLA		Use STC4104.
16	ERR3340	Cartridge-engine oil filter	1		
17	ERC8501	Adaptor-oil filter	1		Manual
17	ERR2490	Adaptor-oil filter	1		Automatic
18	ERR852	Gasket-adaptor.	1		
19	ERC2226L	Adaptor-oil filter	1		
		FROM VIN MA081991			
16	ERR3340	Cartridge-engine oil filter	1		

B 122

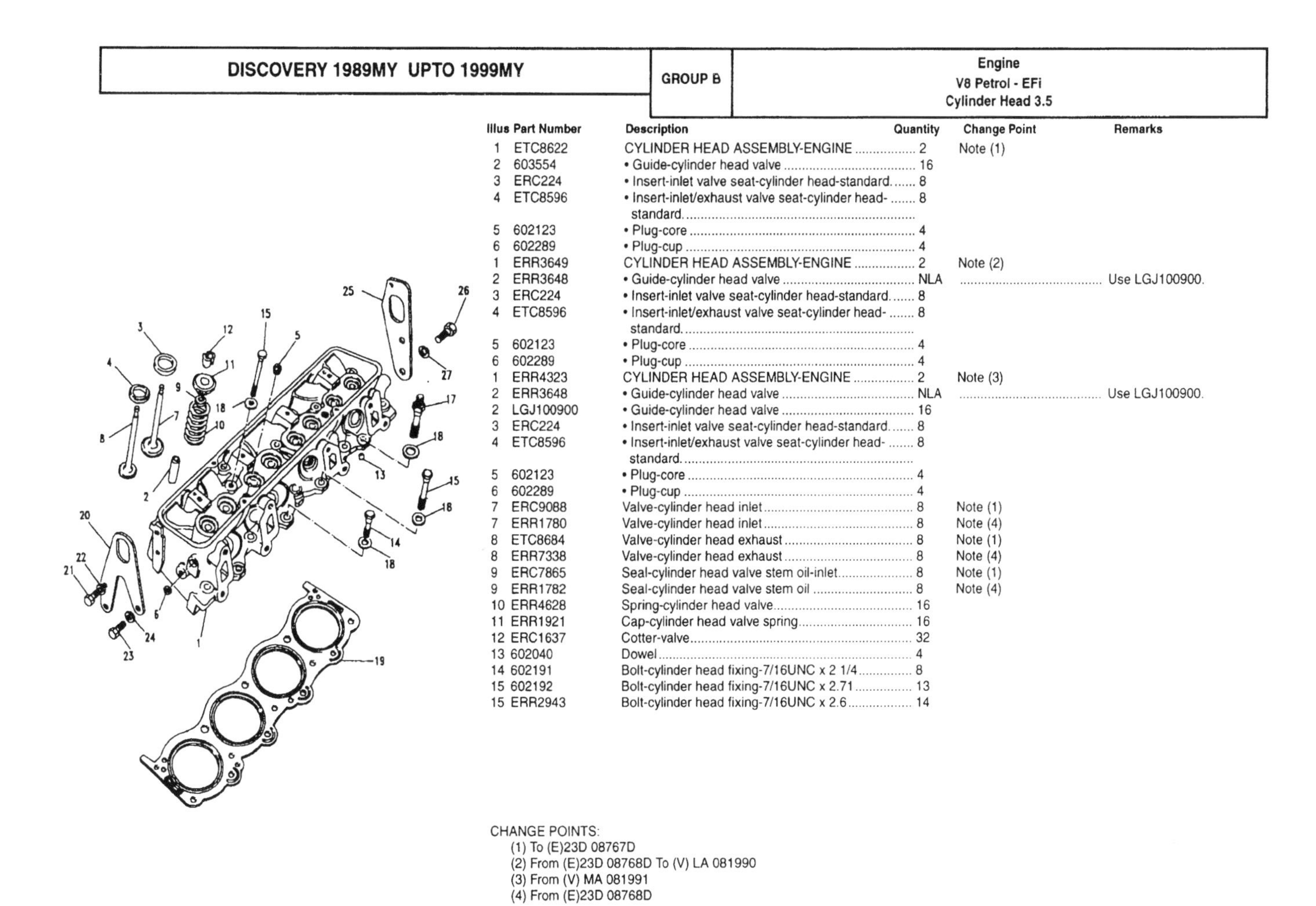

Illus	Part Number	Description	Quantity	Change Point	Remarks
1	ETC8622	CYLINDER HEAD ASSEMBLY-ENGINE	2	Note (1)	
2	603554	• Guide-cylinder head valve	16		
3	ERC224	• Insert-inlet valve seat-cylinder head-standard	8		
4	ETC8596	• Insert-inlet/exhaust valve seat-cylinder head-standard	8		
5	602123	• Plug-core	4		
6	602289	• Plug-cup	4		
1	ERR3649	CYLINDER HEAD ASSEMBLY-ENGINE	2	Note (2)	
2	ERR3648	• Guide-cylinder head valve	NLA		Use LGJ100900.
3	ERC224	• Insert-inlet valve seat-cylinder head-standard	8		
4	ETC8596	• Insert-inlet/exhaust valve seat-cylinder head-standard	8		
5	602123	• Plug-core	4		
6	602289	• Plug-cup	4		
1	ERR4323	CYLINDER HEAD ASSEMBLY-ENGINE	2	Note (3)	
2	ERR3648	• Guide-cylinder head valve	NLA		Use LGJ100900.
2	LGJ100900	• Guide-cylinder head valve	16		
3	ERC224	• Insert-inlet valve seat-cylinder head-standard	8		
4	ETC8596	• Insert-inlet/exhaust valve seat-cylinder head-standard	8		
5	602123	• Plug-core	4		
6	602289	• Plug-cup	4		
7	ERC9088	Valve-cylinder head inlet	8	Note (1)	
7	ERR1780	Valve-cylinder head inlet	8	Note (4)	
8	ETC8684	Valve-cylinder head exhaust	8	Note (1)	
8	ERR7338	Valve-cylinder head exhaust	8	Note (4)	
9	ERC7865	Seal-cylinder head valve stem oil-inlet	8	Note (1)	
9	ERR1782	Seal-cylinder head valve stem oil	8	Note (4)	
10	ERR4628	Spring-cylinder head valve	16		
11	ERR1921	Cap-cylinder head valve spring	16		
12	ERC1637	Cotter-valve	32		
13	602040	Dowel	4		
14	602191	Bolt-cylinder head fixing-7/16UNC x 2 1/4	8		
15	602192	Bolt-cylinder head fixing-7/16UNC x 2.71	13		
15	ERR2943	Bolt-cylinder head fixing-7/16UNC x 2.6	14		

CHANGE POINTS:
(1) To (E)23D 08767D
(2) From (E)23D 08768D To (V) LA 081990
(3) From (V) MA 081991
(4) From (E)23D 08768D

B 123

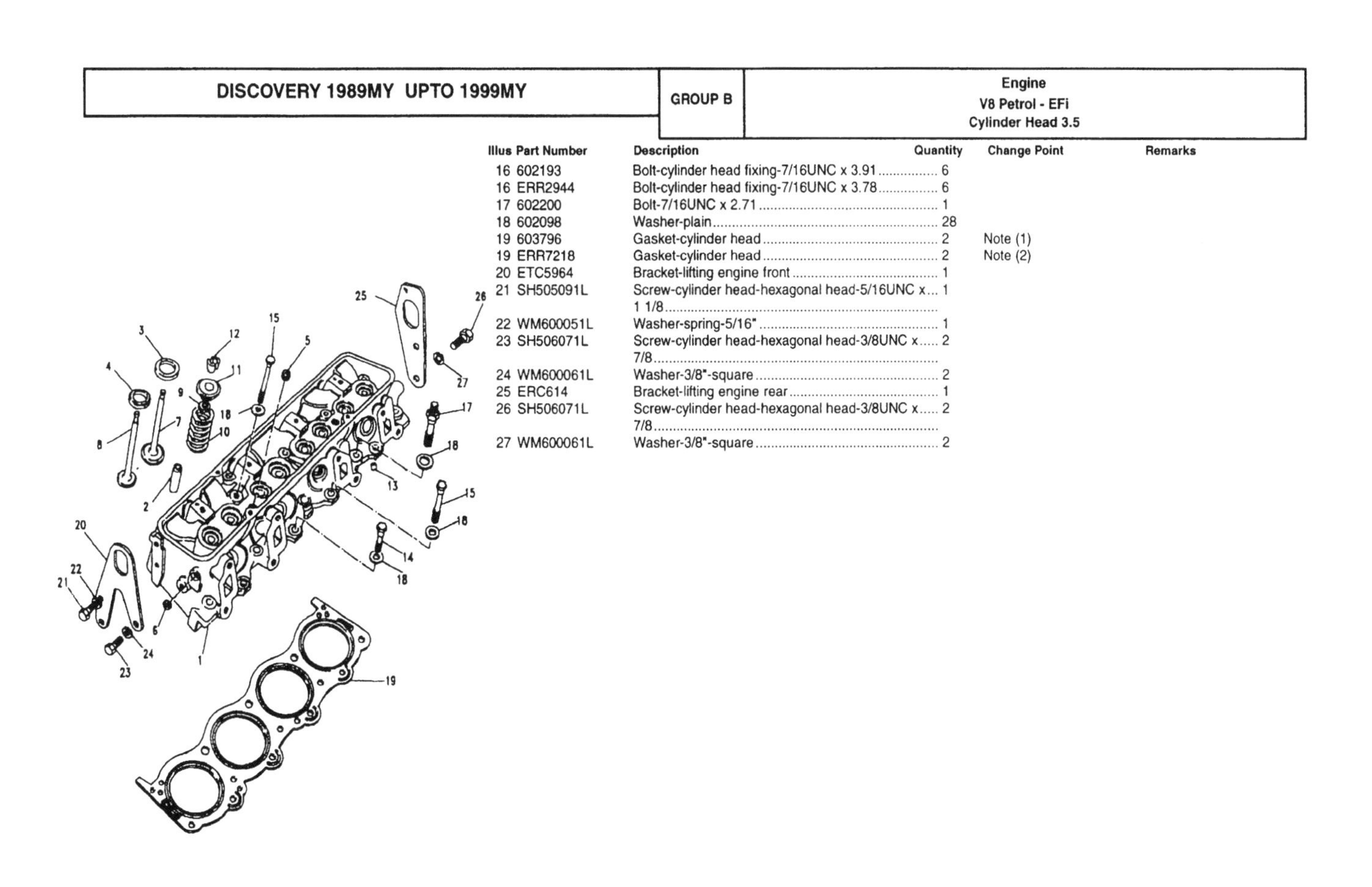

Illus	Part Number	Description	Quantity	Change Point	Remarks
16	602193	Bolt-cylinder head fixing-7/16UNC x 3.91	6		
16	ERR2944	Bolt-cylinder head fixing-7/16UNC x 3.78	6		
17	602200	Bolt-7/16UNC x 2.71	1		
18	602098	Washer-plain	28		
19	603796	Gasket-cylinder head	2	Note (1)	
19	ERR7218	Gasket-cylinder head	2	Note (2)	
20	ETC5964	Bracket-lifting engine front	1		
21	SH505091L	Screw-cylinder head-hexagonal head-5/16UNC x 1 1/8	1		
22	WM600051L	Washer-spring-5/16"	1		
23	SH506071L	Screw-cylinder head-hexagonal head-3/8UNC x 7/8	2		
24	WM600061L	Washer-3/8"-square	2		
25	ERC614	Bracket-lifting engine rear	1		
26	SH506071L	Screw-cylinder head-hexagonal head-3/8UNC x 7/8	2		
27	WM600061L	Washer-3/8"-square	2		

CHANGE POINTS:
(1) To (V) LA 081990
(2) From (V) MA 081991

B 124

DISCOVERY 1989MY UPTO 1999MY

GROUP B — **Engine** / **V8 Petrol - EFi** / **Cylinder Head 3.9**

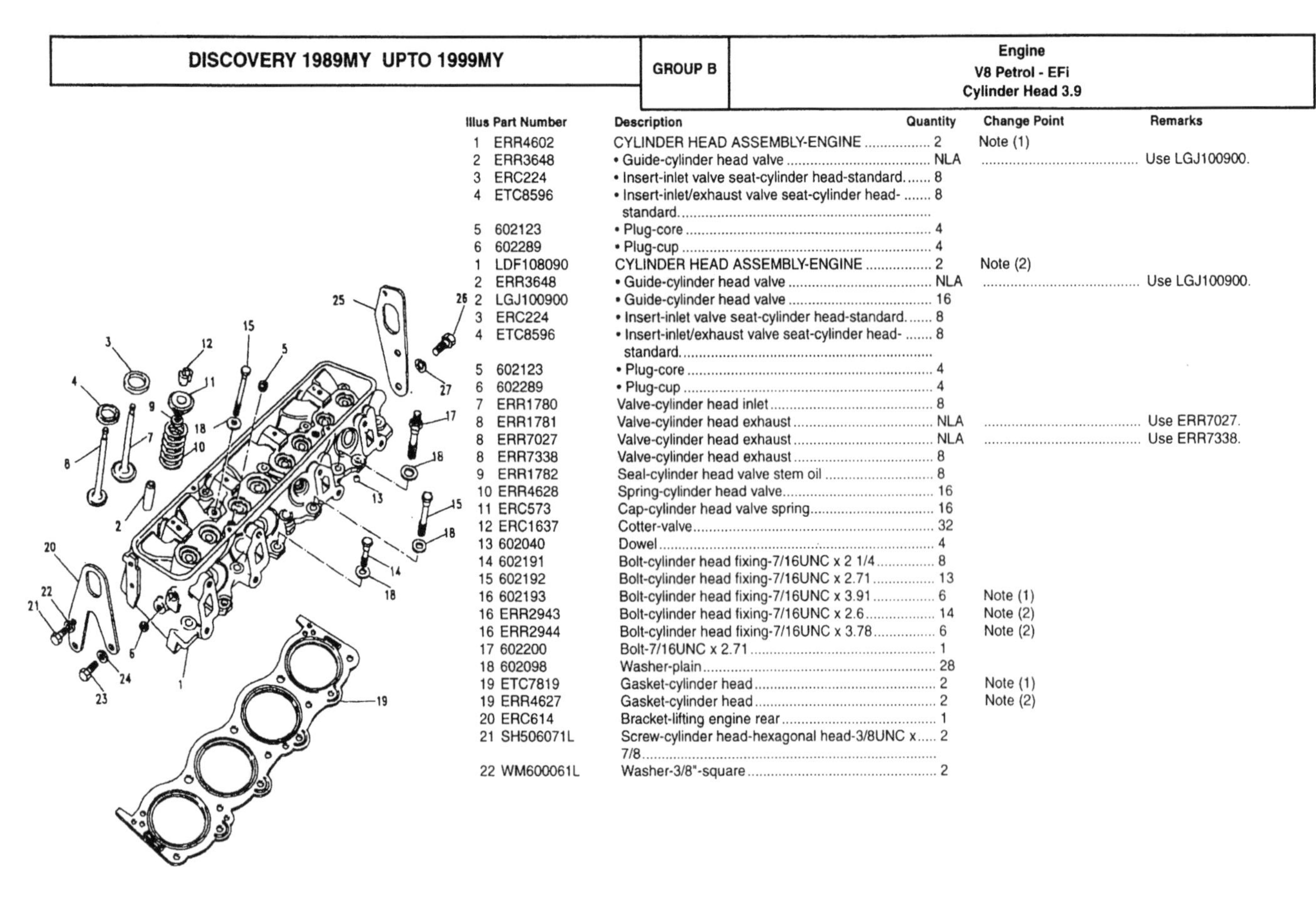

Illus	Part Number	Description	Quantity	Change Point	Remarks
1	ERR4602	CYLINDER HEAD ASSEMBLY-ENGINE	2	Note (1)	
2	ERR3648	• Guide-cylinder head valve	NLA		Use LGJ100900.
3	ERC224	• Insert-inlet valve seat-cylinder head-standard.	8		
4	ETC8596	• Insert-inlet/exhaust valve seat-cylinder head-standard.	8		
5	602123	• Plug-core	4		
6	602289	• Plug-cup	4		
1	LDF108090	CYLINDER HEAD ASSEMBLY-ENGINE	2	Note (2)	
2	ERR3648	• Guide-cylinder head valve	NLA		Use LGJ100900.
2	LGJ100900	• Guide-cylinder head valve	16		
3	ERC224	• Insert-inlet valve seat-cylinder head-standard.	8		
4	ETC8596	• Insert-inlet/exhaust valve seat-cylinder head-standard.	8		
5	602123	• Plug-core	4		
6	602289	• Plug-cup	4		
7	ERR1780	Valve-cylinder head inlet	8		
8	ERR1781	Valve-cylinder head exhaust	NLA		Use ERR7027.
8	ERR7027	Valve-cylinder head exhaust	NLA		Use ERR7338.
8	ERR7338	Valve-cylinder head exhaust	8		
9	ERR1782	Seal-cylinder head valve stem oil	8		
10	ERR4628	Spring-cylinder head valve	16		
11	ERC573	Cap-cylinder head valve spring	16		
12	ERC1637	Cotter-valve	32		
13	602040	Dowel	4		
14	602191	Bolt-cylinder head fixing-7/16UNC x 2 1/4	8		
15	602192	Bolt-cylinder head fixing-7/16UNC x 2.71	13		
16	602193	Bolt-cylinder head fixing-7/16UNC x 3.91	6	Note (1)	
16	ERR2943	Bolt-cylinder head fixing-7/16UNC x 2.6	14	Note (2)	
16	ERR2944	Bolt-cylinder head fixing-7/16UNC x 3.78	6	Note (2)	
17	602200	Bolt-7/16UNC x 2.71	1		
18	602098	Washer-plain	28		
19	ETC7819	Gasket-cylinder head	2	Note (1)	
19	ERR4627	Gasket-cylinder head	2	Note (2)	
20	ERC614	Bracket-lifting engine rear	1		
21	SH506071L	Screw-cylinder head-hexagonal head-3/8UNC x 7/8	2		
22	WM600061L	Washer-3/8"-square	2		

CHANGE POINTS:
(1) To (V) LA 081990
(2) From (V) MA 081991

DISCOVERY 1989MY UPTO 1999MY

GROUP B — **Engine** / **V8 Petrol - EFi** / **Rockershaft**

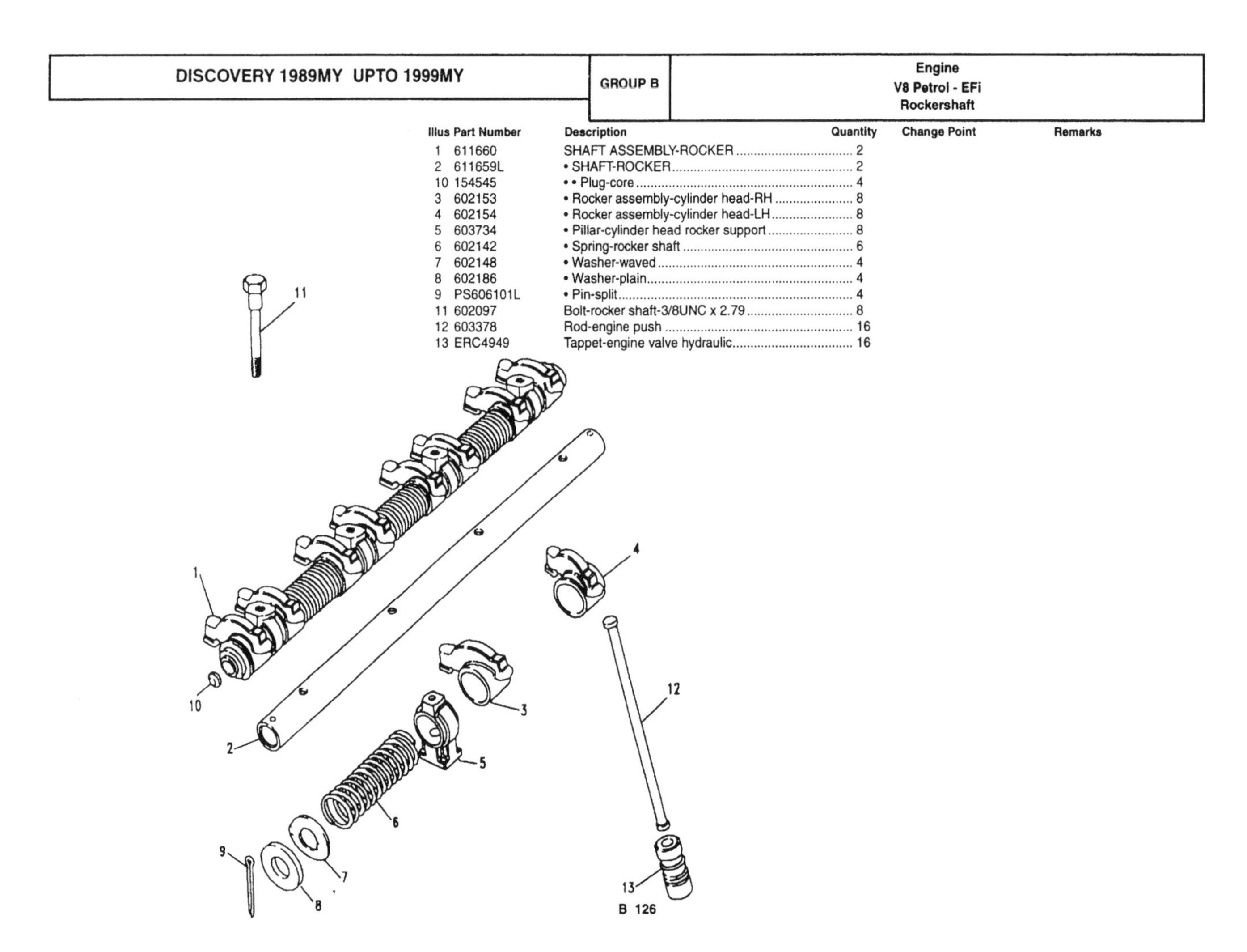

Illus	Part Number	Description	Quantity	Change Point	Remarks
1	611660	SHAFT ASSEMBLY-ROCKER	2		
2	611659L	• SHAFT-ROCKER	2		
10	154545	• • Plug-core	4		
3	602153	• Rocker assembly-cylinder head-RH	8		
4	602154	• Rocker assembly-cylinder head-LH	8		
5	603734	• Pillar-cylinder head rocker support	8		
6	602142	• Spring-rocker shaft	6		
7	602148	• Washer-waved	4		
8	602186	• Washer-plain	4		
9	PS606101L	• Pin-split	4		
11	602097	Bolt-rocker shaft-3/8UNC x 2.79	8		
12	603378	Rod-engine push	16		
13	ERC4949	Tappet-engine valve hydraulic	16		

DISCOVERY 1989MY UPTO 1999MY

GROUP B

Engine
V8 Petrol - EFi
Rocker Cover

Illus	Part Number	Description	Quantity	Change Point	Remarks
1	ETC8679	COVER ASSEMBLY-VALVE ROCKER-LH	1	Note (1)	
1	ERR4870	COVER ASSEMBLY-VALVE ROCKER-LH	1	Note (2)	
2	79027	• Screw-6 x 3/8	4		
3	ERC2989	• Filler-oil	1		
4	602512	• Gasket-engine rocker cover	NLA		Use ERR7288.
5	625038	Cap-oil filler	1		
6	564258	O ring	1		
7	ERC3933	Plate	1		
8	ERC3209	Filter-crankcase intake foam	1		
9	ERC3208	Cap-filter.	1		
10	ETC8680	COVER ASSEMBLY-VALVE ROCKER-RH	1	Note (1)	
10	ERR4672	COVER ASSEMBLY-VALVE ROCKER-RH	1	Note (2)	
11	79027	• Screw-6 x 3/8	8		
12	ERC248	• Filter-crankcase breather	1		
13	564258	• O ring-positive crankcase ventilation breather	1		
14	602512	• Gasket-engine rocker cover	NLA		Use ERR7288.
15	SY504072L	Screw-cover-1/4UNC x 7/8	4		
15	603127	Screw-cover-1/4UNC x 1 5/16	4		
16	WM600041L	Washer-spring-1/4 dia-square	8		
17	WB106041L	Washer-M6	8		

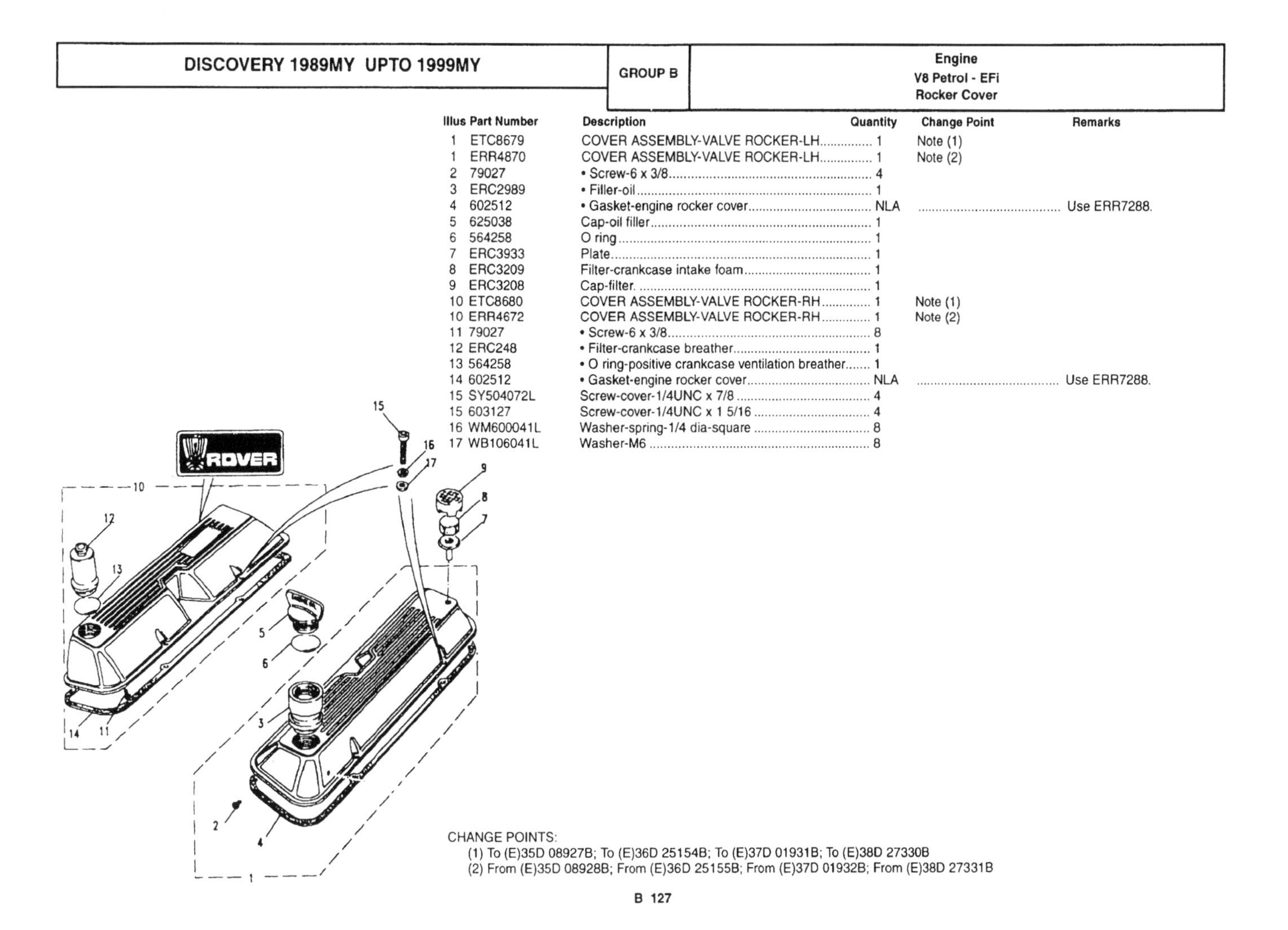

CHANGE POINTS:
(1) To (E)35D 08927B; To (E)36D 25154B; To (E)37D 01931B; To (E)38D 27330B
(2) From (E)35D 08928B; From (E)36D 25155B; From (E)37D 01932B; From (E)38D 27331B

B 127

DISCOVERY 1989MY UPTO 1999MY

GROUP B

Engine
V8 Petrol - EFi
Camshaft

Illus	Part Number	Description	Quantity	Change Point	Remarks
1	ETC6099	Camshaft-engine	1		3500 cc
1	ETC8686	Camshaft-engine	NLA	Note (1)	Use ERR5924 qty 1 with ERR5926 qty 1 with SH505061L qty 2. 3900 cc
1	ERR5924	Camshaft-engine	1	Note (2)	3900 cc
2	610289	Sprocket-camshaft	1		
3	ERC7929	Chain-engine timing	1		
4	ERC2838	Key-camshaft location	1		
5	ERC2839	Spacer	1		
6	614188	Gear-drive-distributor	1		
7	ERC6552	Washer-plain	1		
8	ERC5749	Bolt-gear to camshaft-M12 x 30	1		3500 cc
8	602227	Bolt-gear to camshaft-7/16UNC x 1 1/8	1		3900 cc
9	ERR5926	Plate-camshaft retaining	1	Note (2)	
10	SH505061L	Screw-plate thrust camshaft-hexagonal head-5/16UNC x 3/4	2	Note (2)	

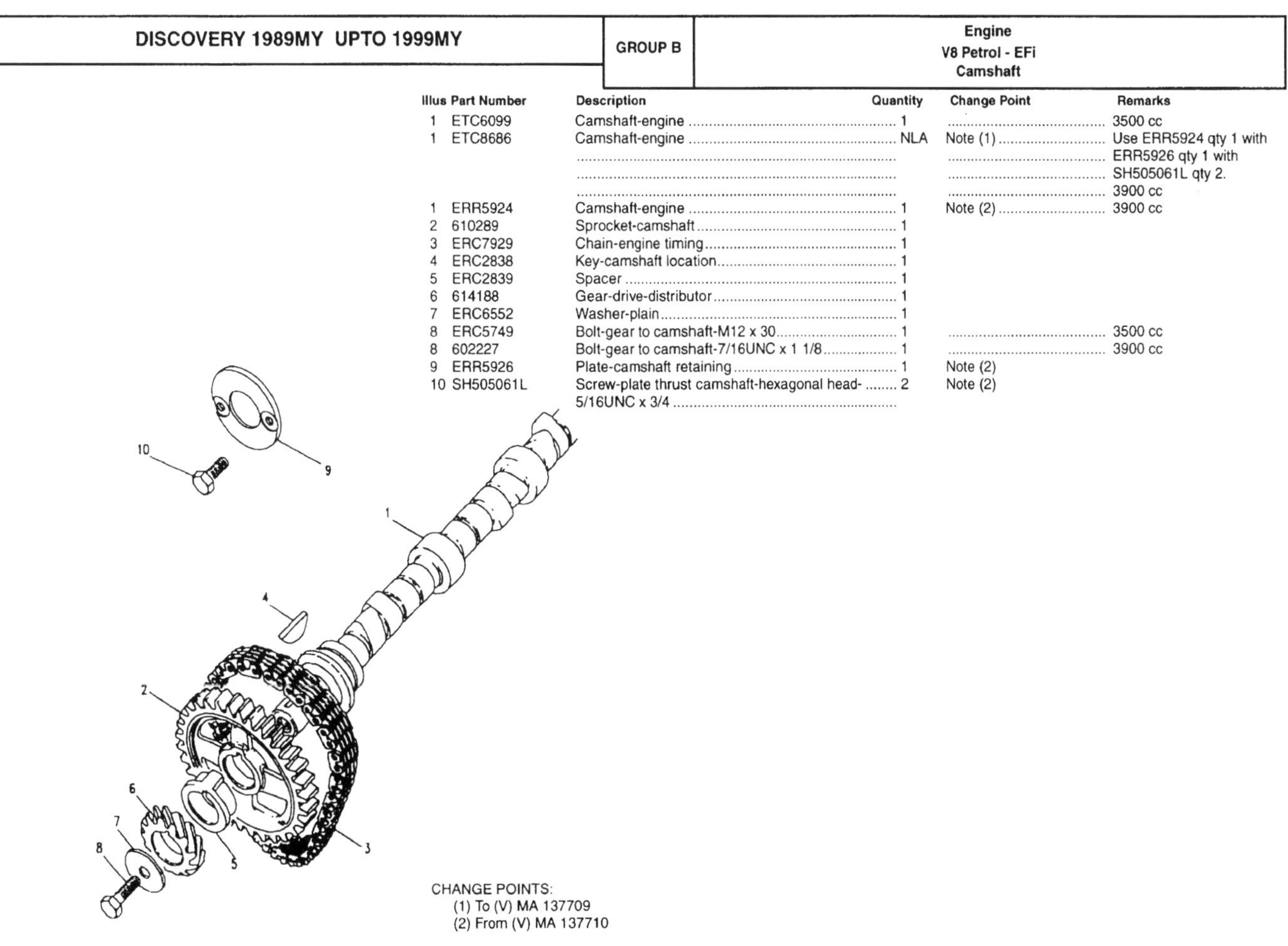

CHANGE POINTS:
(1) To (V) MA 137709
(2) From (V) MA 137710

B 128

DISCOVERY 1989MY UPTO 1999MY

GROUP B — **Engine** / **V8 Petrol - EFi** / **Inlet Manifold**

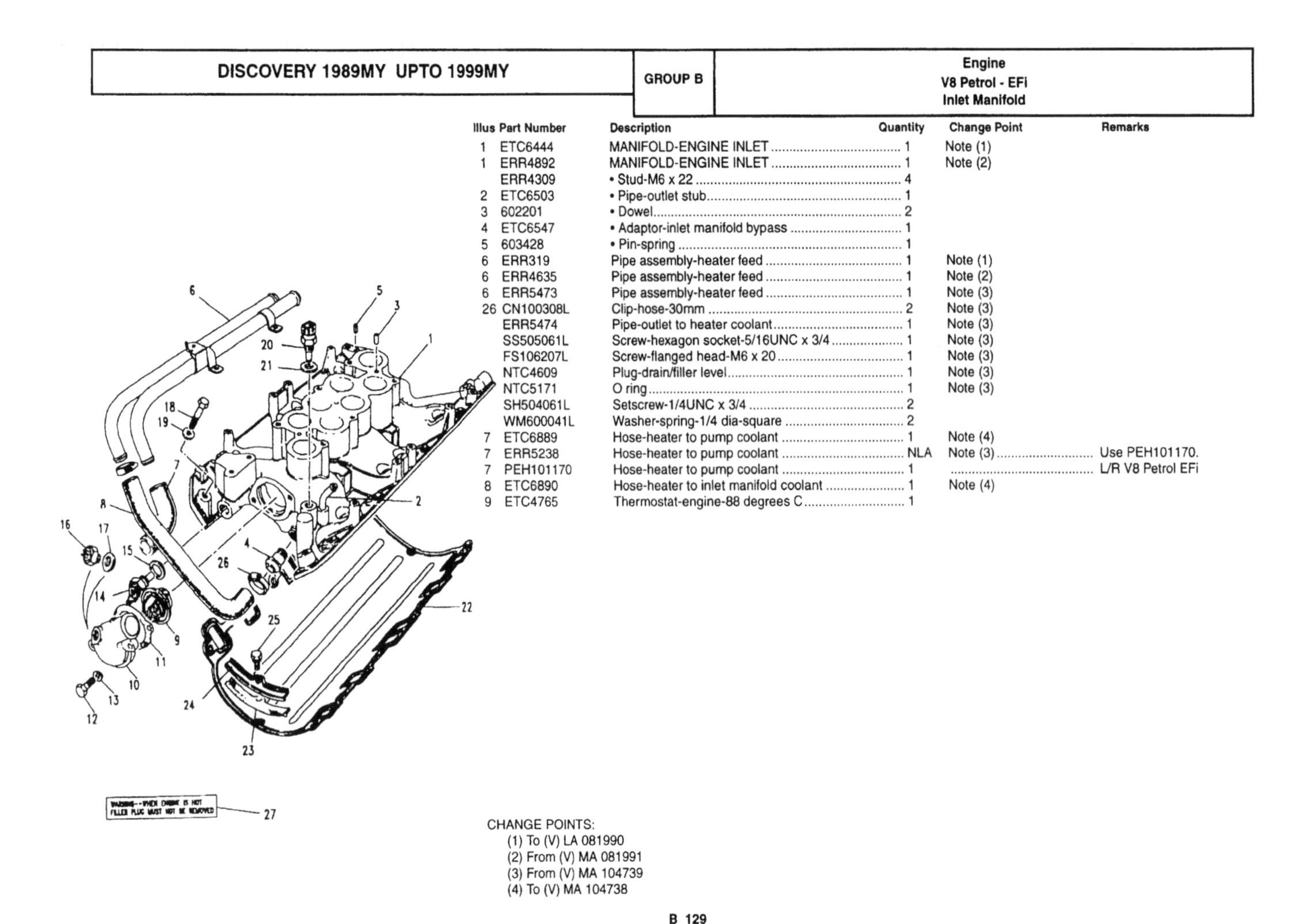

Illus	Part Number	Description	Quantity	Change Point	Remarks
1	ETC6444	MANIFOLD-ENGINE INLET	1	Note (1)	
1	ERR4892	MANIFOLD-ENGINE INLET	1	Note (2)	
	ERR4309	• Stud-M6 x 22	4		
2	ETC6503	• Pipe-outlet stub	1		
3	602201	• Dowel	2		
4	ETC6547	• Adaptor-inlet manifold bypass	1		
5	603428	• Pin-spring	1		
6	ERR319	Pipe assembly-heater feed	1	Note (1)	
6	ERR4635	Pipe assembly-heater feed	1	Note (2)	
6	ERR5473	Pipe assembly-heater feed	1	Note (3)	
26	CN100308L	Clip-hose-30mm	2	Note (3)	
	ERR5474	Pipe-outlet to heater coolant	1	Note (3)	
	SS505061L	Screw-hexagon socket-5/16UNC x 3/4	1	Note (3)	
	FS106207L	Screw-flanged head-M6 x 20	1	Note (3)	
	NTC4609	Plug-drain/filler level	1	Note (3)	
	NTC5171	O ring	1	Note (3)	
	SH504061L	Setscrew-1/4UNC x 3/4	2		
	WM600041L	Washer-spring-1/4 dia-square	2		
7	ETC6889	Hose-heater to pump coolant	1	Note (4)	
7	ERR5238	Hose-heater to pump coolant	NLA	Note (3)	Use PEH101170.
7	PEH101170	Hose-heater to pump coolant	1		L/R V8 Petrol EFi
8	ETC6890	Hose-heater to inlet manifold coolant	1	Note (4)	
9	ETC4765	Thermostat-engine-88 degrees C	1		

CHANGE POINTS:
(1) To (V) LA 081990
(2) From (V) MA 081991
(3) From (V) MA 104739
(4) To (V) MA 104738

B 129

DISCOVERY 1989MY UPTO 1999MY

GROUP B — **Engine** / **V8 Petrol - EFi** / **Inlet Manifold**

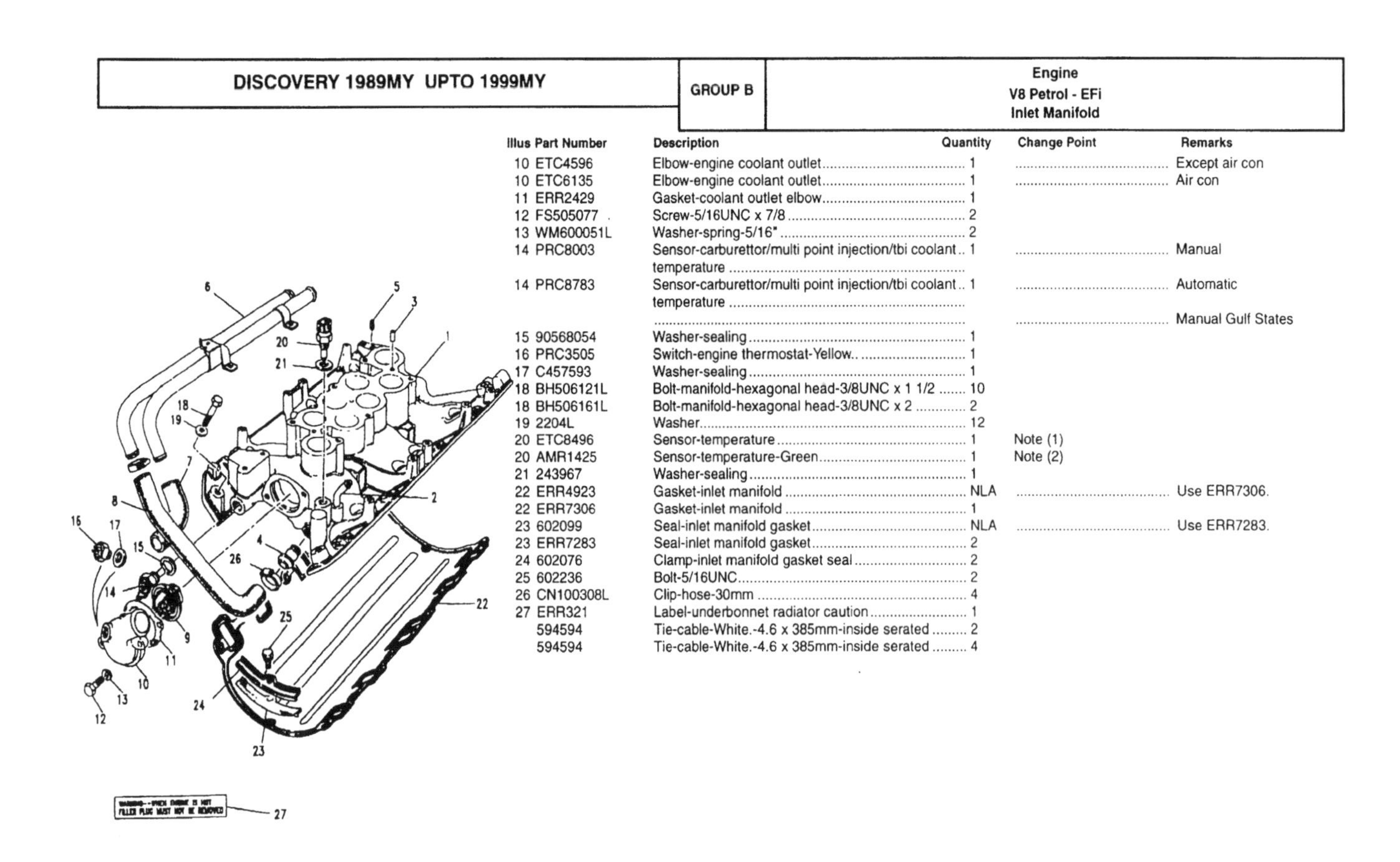

Illus	Part Number	Description	Quantity	Change Point	Remarks
10	ETC4596	Elbow-engine coolant outlet	1		Except air con
10	ETC6135	Elbow-engine coolant outlet	1		Air con
11	ERR2429	Gasket-coolant outlet elbow	1		
12	FS505077	Screw-5/16UNC x 7/8	2		
13	WM600051L	Washer-spring-5/16"	2		
14	PRC8003	Sensor-carburettor/multi point injection/tbi coolant temperature	1		Manual
14	PRC8783	Sensor-carburettor/multi point injection/tbi coolant temperature	1		Automatic
					Manual Gulf States
15	90568054	Washer-sealing	1		
16	PRC3505	Switch-engine thermostat-Yellow	1		
17	C457593	Washer-sealing	1		
18	BH506121L	Bolt-manifold-hexagonal head-3/8UNC x 1 1/2	10		
18	BH506161L	Bolt-manifold-hexagonal head-3/8UNC x 2	2		
19	2204L	Washer	12		
20	ETC8496	Sensor-temperature	1	Note (1)	
20	AMR1425	Sensor-temperature-Green	1	Note (2)	
21	243967	Washer-sealing	1		
22	ERR4923	Gasket-inlet manifold	NLA		Use ERR7306.
22	ERR7306	Gasket-inlet manifold	1		
23	602099	Seal-inlet manifold gasket	NLA		Use ERR7283.
23	ERR7283	Seal-inlet manifold gasket	2		
24	602076	Clamp-inlet manifold gasket seal	2		
25	602236	Bolt-5/16UNC	2		
26	CN100308L	Clip-hose-30mm	4		
27	ERR321	Label-underbonnet radiator caution	1		
	594594	Tie-cable-White.-4.6 x 385mm-inside serated	2		
	594594	Tie-cable-White.-4.6 x 385mm-inside serated	4		

CHANGE POINTS:
(1) To (V) LA 081990
(2) From (V) MA 081991

B 130

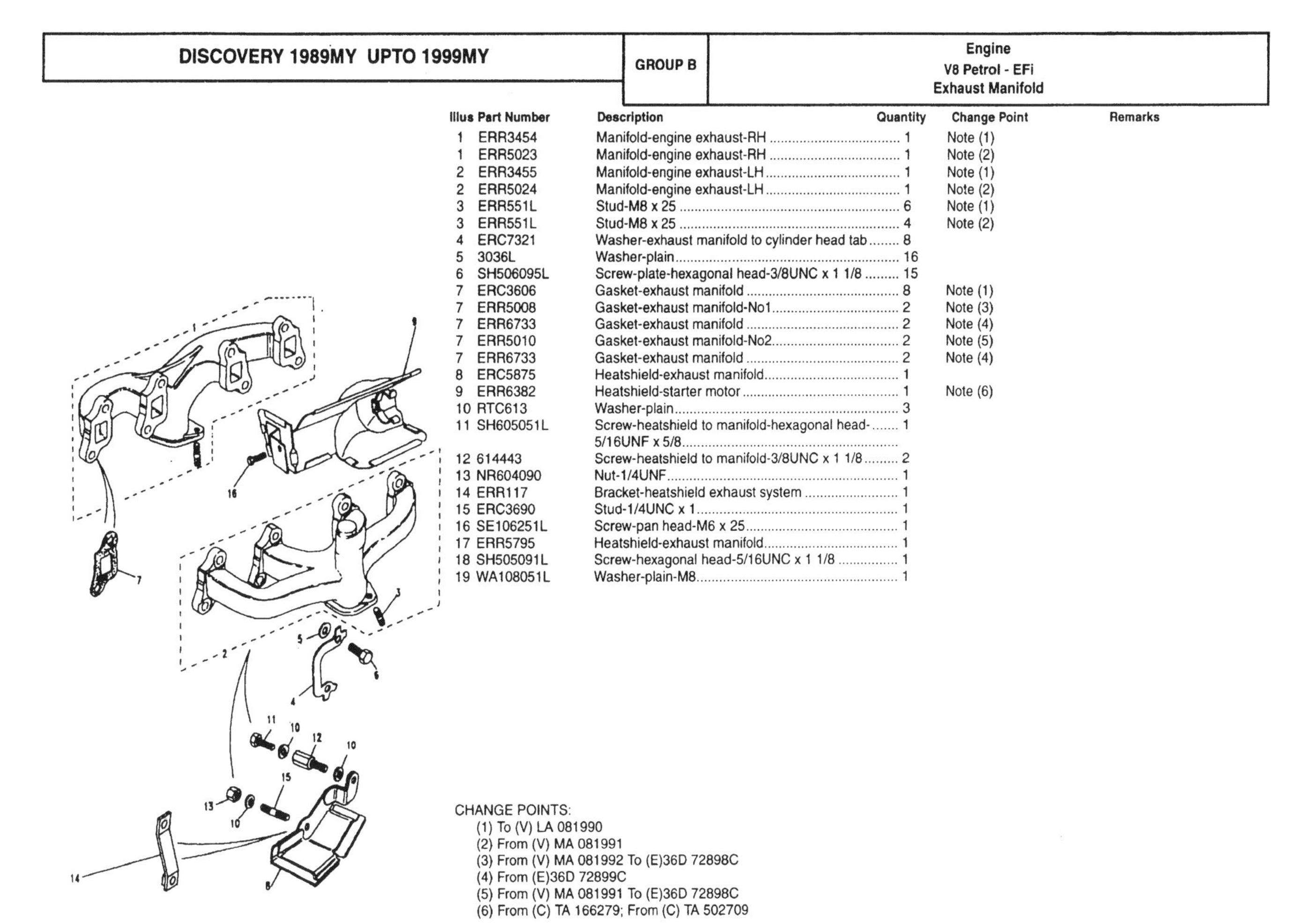

Illus	Part Number	Description	Quantity	Change Point	Remarks
1	ERR3454	Manifold-engine exhaust-RH	1	Note (1)	
1	ERR5023	Manifold-engine exhaust-RH	1	Note (2)	
2	ERR3455	Manifold-engine exhaust-LH	1	Note (1)	
2	ERR5024	Manifold-engine exhaust-LH	1	Note (2)	
3	ERR551L	Stud-M8 x 25	6	Note (1)	
3	ERR551L	Stud-M8 x 25	4	Note (2)	
4	ERC7321	Washer-exhaust manifold to cylinder head tab	8		
5	3036L	Washer-plain	16		
6	SH506095L	Screw-plate-hexagonal head-3/8UNC x 1 1/8	15		
7	ERC3606	Gasket-exhaust manifold	8	Note (1)	
7	ERR5008	Gasket-exhaust manifold-No1	2	Note (3)	
7	ERR6733	Gasket-exhaust manifold	2	Note (4)	
7	ERR5010	Gasket-exhaust manifold-No2	2	Note (5)	
7	ERR6733	Gasket-exhaust manifold	2	Note (4)	
8	ERC5875	Heatshield-exhaust manifold	1		
9	ERR6382	Heatshield-starter motor	1	Note (6)	
10	RTC613	Washer-plain	3		
11	SH605051L	Screw-heatshield to manifold-hexagonal head-5/16UNF x 5/8	1		
12	614443	Screw-heatshield to manifold-3/8UNC x 1 1/8	2		
13	NR604090	Nut-1/4UNF	1		
14	ERR117	Bracket-heatshield exhaust system	1		
15	ERC3690	Stud-1/4UNC x 1	1		
16	SE106251L	Screw-pan head-M6 x 25	1		
17	ERR5795	Heatshield-exhaust manifold	1		
18	SH505091L	Screw-hexagonal head-5/16UNC x 1 1/8	1		
19	WA108051L	Washer-plain-M8	1		

CHANGE POINTS:
(1) To (V) LA 081990
(2) From (V) MA 081991
(3) From (V) MA 081992 To (E)36D 72898C
(4) From (E)36D 72899C
(5) From (V) MA 081991 To (E)36D 72898C
(6) From (C) TA 166279; From (C) TA 502709

B 131

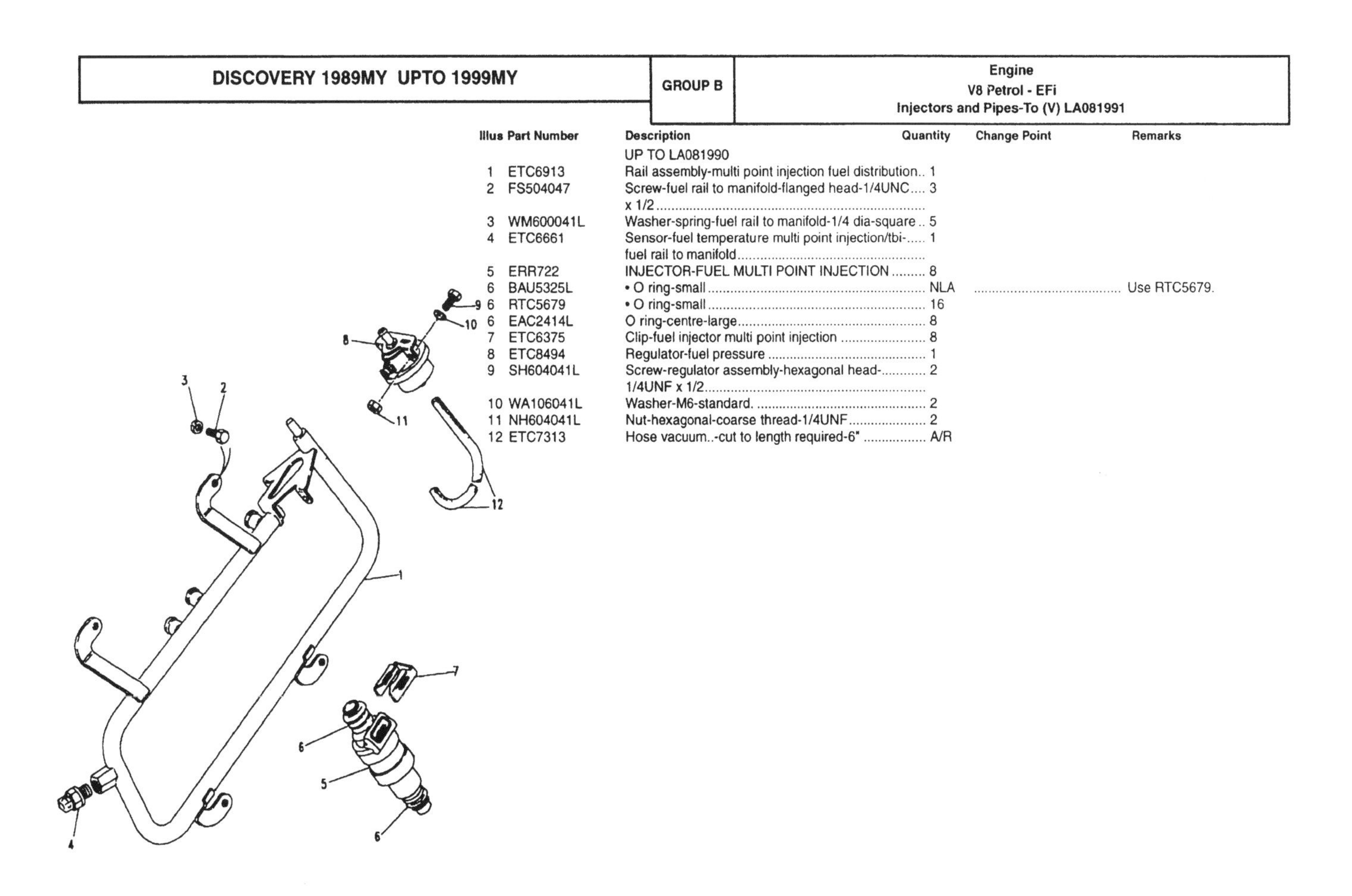

Illus	Part Number	Description	Quantity	Change Point	Remarks
		UP TO LA081990			
1	ETC6913	Rail assembly-multi point injection fuel distribution	1		
2	FS504047	Screw-fuel rail to manifold-flanged head-1/4UNC x 1/2	3		
3	WM600041L	Washer-spring-fuel rail to manifold-1/4 dia-square	5		
4	ETC6661	Sensor-fuel temperature multi point injection/tbi-fuel rail to manifold	1		
5	ERR722	INJECTOR-FUEL MULTI POINT INJECTION	8		
6	BAU5325L	• O ring-small	NLA		Use RTC5679.
6	RTC5679	• O ring-small	16		
6	EAC2414L	O ring-centre-large	8		
7	ETC6375	Clip-fuel injector multi point injection	8		
8	ETC8494	Regulator-fuel pressure	1		
9	SH604041L	Screw-regulator assembly-hexagonal head-1/4UNF x 1/2	2		
10	WA106041L	Washer-M6-standard.	2		
11	NH604041L	Nut-hexagonal-coarse thread-1/4UNF	2		
12	ETC7313	Hose vacuum..-cut to length required-6"	A/R		

B 132

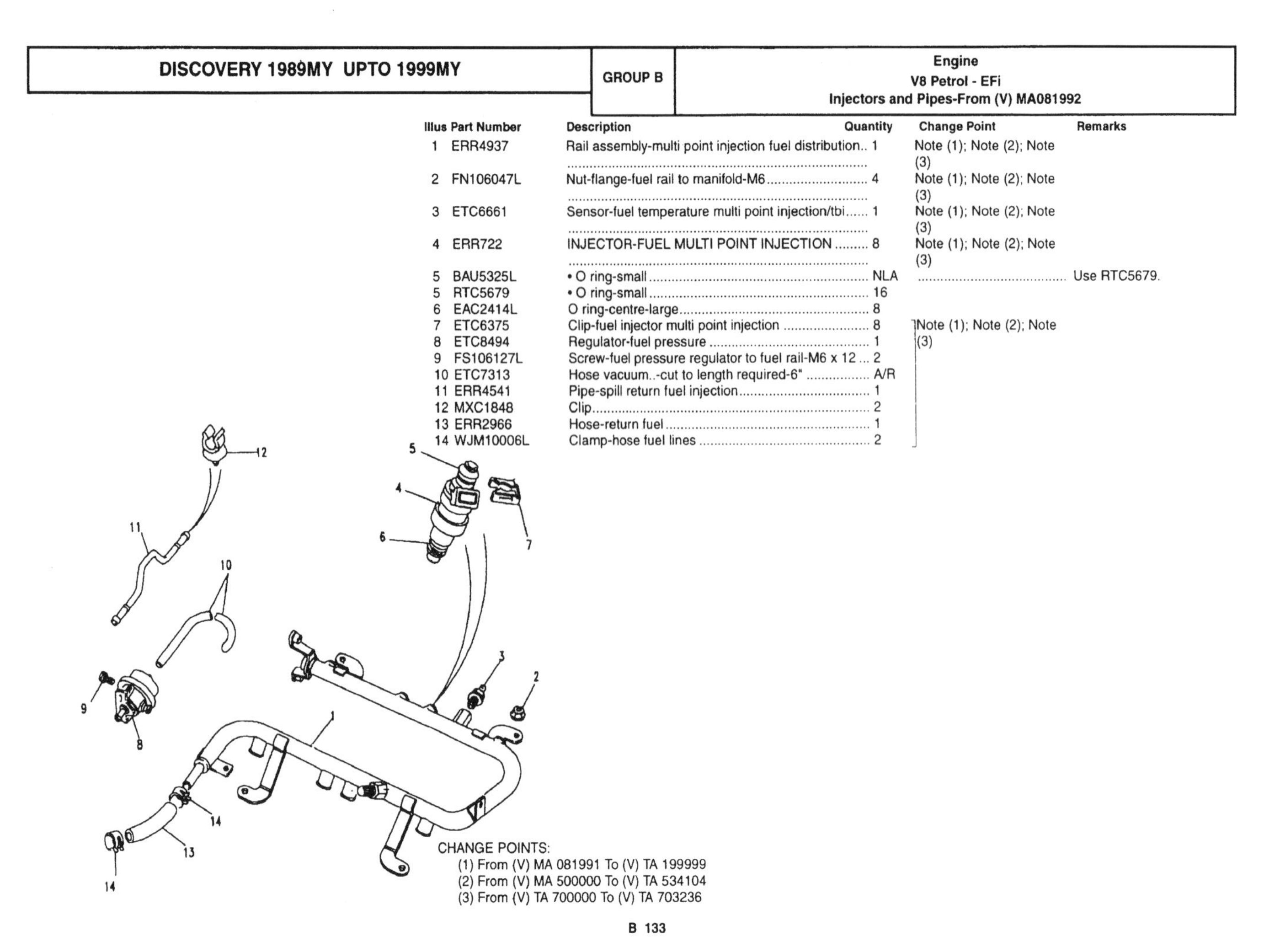

Illus	Part Number	Description	Quantity	Change Point	Remarks
1	ERR4937	Rail assembly-multi point injection fuel distribution..	1	Note (1); Note (2); Note (3)	
2	FN106047L	Nut-flange-fuel rail to manifold-M6	4	Note (1); Note (2); Note (3)	
3	ETC6661	Sensor-fuel temperature multi point injection/tbi	1	Note (1); Note (2); Note (3)	
4	ERR722	INJECTOR-FUEL MULTI POINT INJECTION	8	Note (1); Note (2); Note (3)	
5	BAU5325L	• O ring-small	NLA		Use RTC5679.
5	RTC5679	• O ring-small	16		
6	EAC2414L	O ring-centre-large	8		
7	ETC6375	Clip-fuel injector multi point injection	8	Note (1); Note (2); Note (3)	
8	ETC8494	Regulator-fuel pressure	1		
9	FS106127L	Screw-fuel pressure regulator to fuel rail-M6 x 12	2		
10	ETC7313	Hose vacuum..-cut to length required-6"	A/R		
11	ERR4541	Pipe-spill return fuel injection	1		
12	MXC1848	Clip	2		
13	ERR2966	Hose-return fuel	1		
14	WJM10006L	Clamp-hose fuel lines	2		

CHANGE POINTS:
(1) From (V) MA 081991 To (V) TA 199999
(2) From (V) MA 500000 To (V) TA 534104
(3) From (V) TA 700000 To (V) TA 703236

B 133

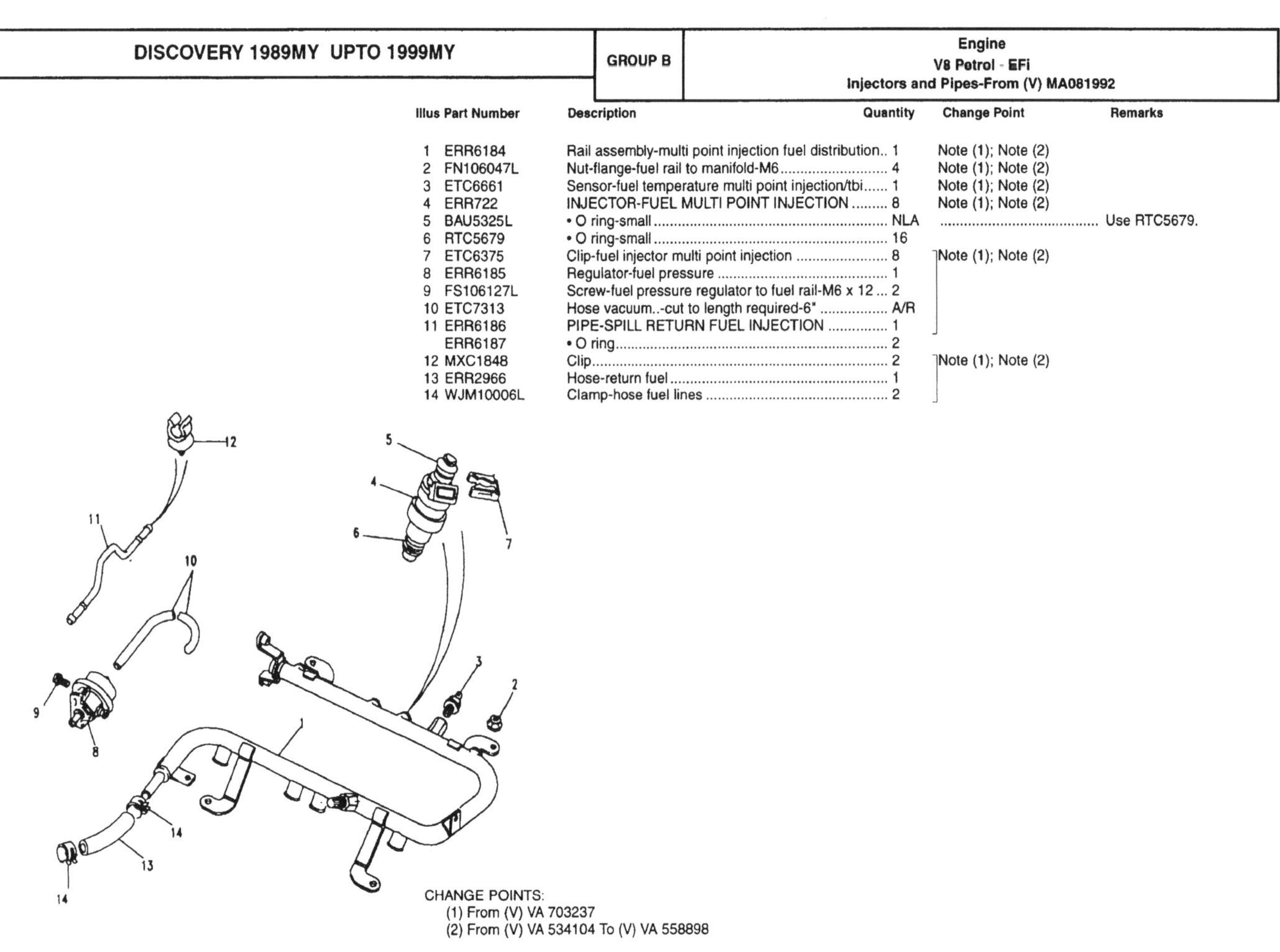

Illus	Part Number	Description	Quantity	Change Point	Remarks
1	ERR6184	Rail assembly-multi point injection fuel distribution..	1	Note (1); Note (2)	
2	FN106047L	Nut-flange-fuel rail to manifold-M6	4	Note (1); Note (2)	
3	ETC6661	Sensor-fuel temperature multi point injection/tbi	1	Note (1); Note (2)	
4	ERR722	INJECTOR-FUEL MULTI POINT INJECTION	8	Note (1); Note (2)	
5	BAU5325L	• O ring-small	NLA		Use RTC5679.
6	RTC5679	• O ring-small	16		
7	ETC6375	Clip-fuel injector multi point injection	8	Note (1); Note (2)	
8	ERR6185	Regulator-fuel pressure	1		
9	FS106127L	Screw-fuel pressure regulator to fuel rail-M6 x 12	2		
10	ETC7313	Hose vacuum..-cut to length required-6"	A/R		
11	ERR6186	PIPE-SPILL RETURN FUEL INJECTION	1		
	ERR6187	• O ring	2		
12	MXC1848	Clip	2	Note (1); Note (2)	
13	ERR2966	Hose-return fuel	1		
14	WJM10006L	Clamp-hose fuel lines	2		

CHANGE POINTS:
(1) From (V) VA 703237
(2) From (V) VA 534104 To (V) VA 558898

B 134

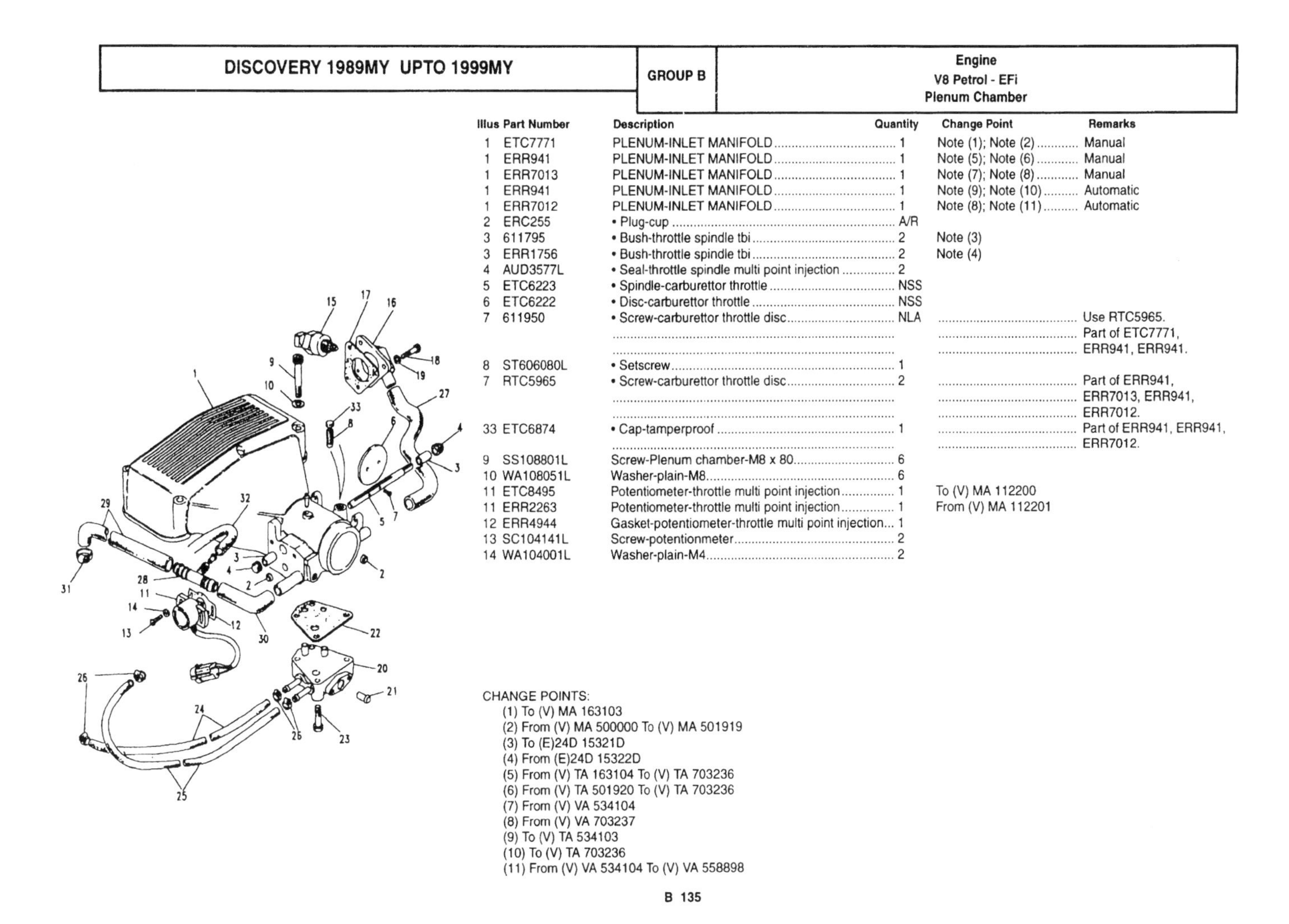

Illus	Part Number	Description	Quantity	Change Point	Remarks
1	ETC7771	PLENUM-INLET MANIFOLD	1	Note (1); Note (2)	Manual
1	ERR941	PLENUM-INLET MANIFOLD	1	Note (5); Note (6)	Manual
1	ERR7013	PLENUM-INLET MANIFOLD	1	Note (7); Note (8)	Manual
1	ERR941	PLENUM-INLET MANIFOLD	1	Note (9); Note (10)	Automatic
1	ERR7012	PLENUM-INLET MANIFOLD	1	Note (8); Note (11)	Automatic
2	ERC255	• Plug-cup	A/R		
3	611795	• Bush-throttle spindle tbi	2	Note (3)	
3	ERR1756	• Bush-throttle spindle tbi	2	Note (4)	
4	AUD3577L	• Seal-throttle spindle multi point injection	2		
5	ETC6223	• Spindle-carburettor throttle	NSS		
6	ETC6222	• Disc-carburettor throttle	NSS		
7	611950	• Screw-carburettor throttle disc	NLA		Use RTC5965. Part of ETC7771, ERR941, ERR941.
8	ST606080L	• Setscrew	1		
7	RTC5965	• Screw-carburettor throttle disc	2		Part of ERR941, ERR7013, ERR941, ERR7012.
33	ETC6874	• Cap-tamperproof	1		Part of ERR941, ERR941, ERR7012.
9	SS108801L	Screw-Plenum chamber-M8 x 80	6		
10	WA108051L	Washer-plain-M8	6		
11	ETC8495	Potentiometer-throttle multi point injection	1	To (V) MA 112200	
11	ERR2263	Potentiometer-throttle multi point injection	1	From (V) MA 112201	
12	ERR4944	Gasket-potentiometer-throttle multi point injection	1		
13	SC104141L	Screw-potentionmeter	2		
14	WA104001L	Washer-plain-M4	2		

CHANGE POINTS:
(1) To (V) MA 163103
(2) From (V) MA 500000 To (V) MA 501919
(3) To (E)24D 15321D
(4) From (E)24D 15322D
(5) From (V) TA 163104 To (V) TA 703236
(6) From (V) TA 501920 To (V) TA 703236
(7) From (V) VA 534104
(8) From (V) VA 703237
(9) To (V) TA 534103
(10) To (V) TA 703236
(11) From (V) VA 534104 To (V) VA 558898

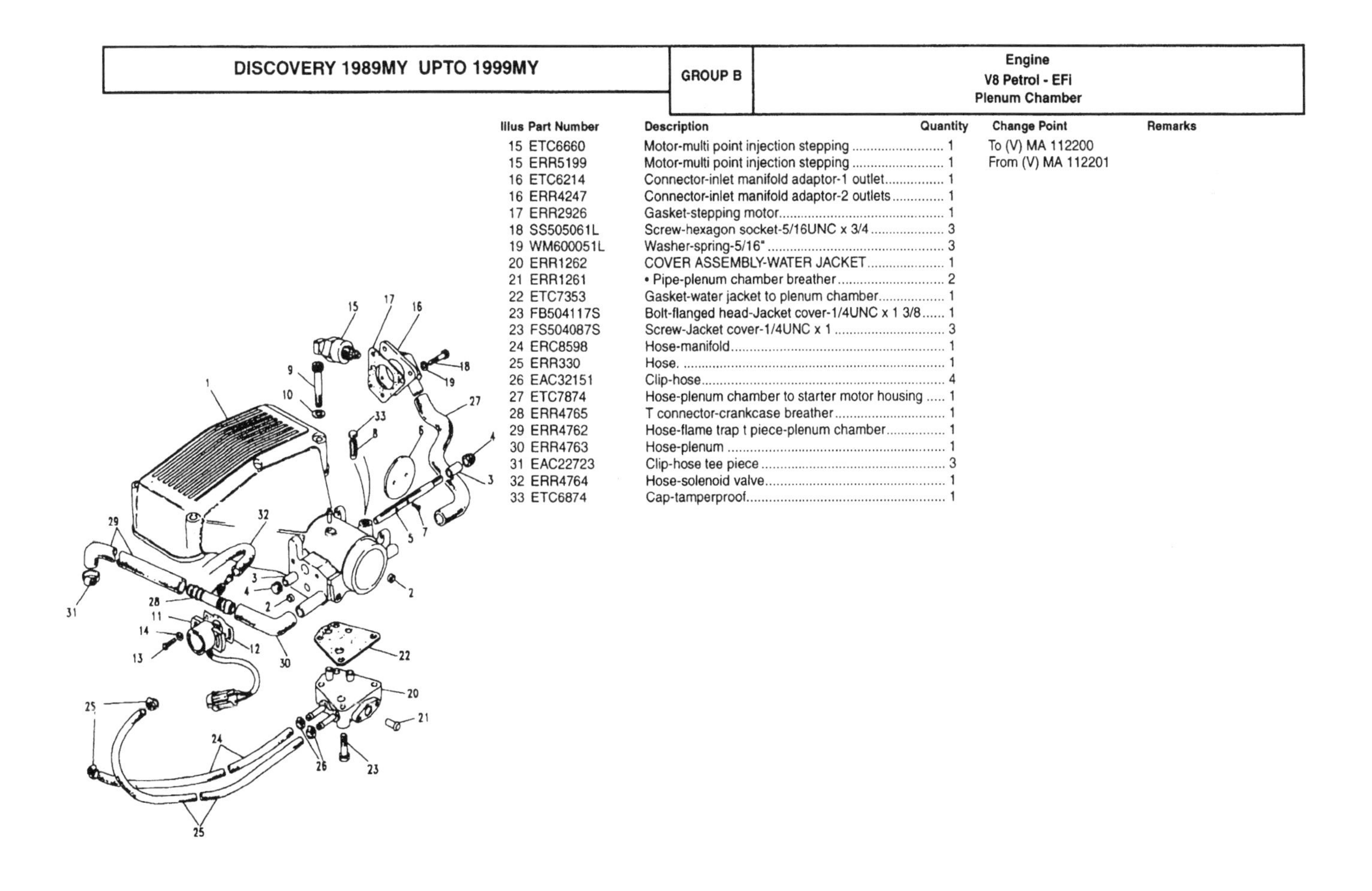

Illus	Part Number	Description	Quantity	Change Point	Remarks
15	ETC6660	Motor-multi point injection stepping	1	To (V) MA 112200	
15	ERR5199	Motor-multi point injection stepping	1	From (V) MA 112201	
16	ETC6214	Connector-inlet manifold adaptor-1 outlet	1		
16	ERR4247	Connector-inlet manifold adaptor-2 outlets	1		
17	ERR2926	Gasket-stepping motor	1		
18	SS505061L	Screw-hexagon socket-5/16UNC x 3/4	3		
19	WM600051L	Washer-spring-5/16"	3		
20	ERR1262	COVER ASSEMBLY-WATER JACKET	1		
21	ERR1261	• Pipe-plenum chamber breather	2		
22	ETC7353	Gasket-water jacket to plenum chamber	1		
23	FB504117S	Bolt-flanged head-Jacket cover-1/4UNC x 1 3/8	1		
23	FS504087S	Screw-Jacket cover-1/4UNC x 1	3		
24	ERC8598	Hose-manifold	1		
25	ERR330	Hose.	1		
26	EAC32151	Clip-hose	4		
27	ETC7874	Hose-plenum chamber to starter motor housing	1		
28	ERR4765	T connector-crankcase breather	1		
29	ERR4762	Hose-flame trap t piece-plenum chamber	1		
30	ERR4763	Hose-plenum	1		
31	EAC22723	Clip-hose tee piece	3		
32	ERR4764	Hose-solenoid valve	1		
33	ETC6874	Cap-tamperproof	1		

Plenum Hoses

Illus	Part Number	Description	Quantity	Change Point	Remarks
1	ERC9114L	Pipe-plenum chamber breather	1		Except air con
1	ERC9187	Pipe-plenum chamber breather	1		Air con
2	ERC3853L	Clip-pipe	2		
3	SH106121L	Screw-pipe-hexagonal head-M6 x 12	2		
4	WL106001L	Washer-M6	2		
5	602289	Plug-cup	1		
6	ERC9115	Hose-plenum-plenum chamber/air pipe	1		
7	ERC9118	Hose-pipe valve-air pipe/air valve	1		
8	ERC9117	Hose-plenum-air valve/plenum chamber	1		
9	EAC22723	Clip-hose	3		
10	ERC7508	Hose-pipe valve	1		Air con
11	ERC7536	Solenoid-control valve-with air conditioning	1		Air con
12	ERC7538	Screw-valve mounting-1/4UNC	1		
13	WM600051L	Washer-spring-5/16"	1		
14	NH605041L	Nut-hexagonal head-5/16UNF	1		
15	ERR4764	Hose-solenoid valve	1		
16	ERC9116	Hose-plenum	1		
	ETC7442	T piece	1		
	ERR4388	Hose-manifold to plenum de icer	1	Note (1)	

CHANGE POINTS:
(1) From (V) MA 106697

Ram Pipes Housing

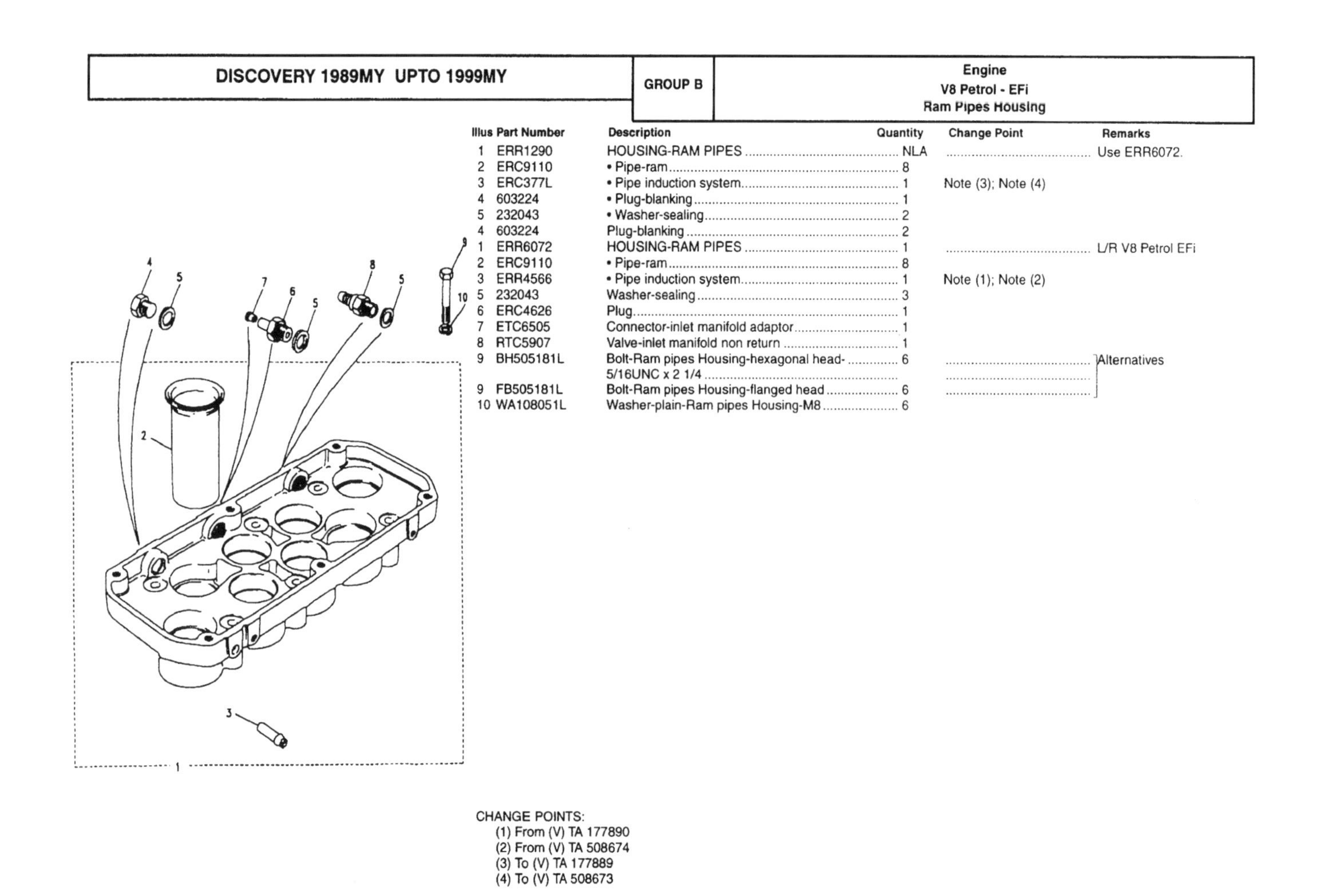

Illus	Part Number	Description	Quantity	Change Point	Remarks
1	ERR1290	HOUSING-RAM PIPES	NLA		Use ERR6072.
2	ERC9110	• Pipe-ram	8		
3	ERC377L	• Pipe induction system	1	Note (3); Note (4)	
4	603224	• Plug-blanking	1		
5	232043	• Washer-sealing	2		
4	603224	Plug-blanking	2		
1	ERR6072	HOUSING-RAM PIPES	1		L/R V8 Petrol EFi
2	ERC9110	• Pipe-ram	8		
3	ERR4566	• Pipe induction system	1	Note (1); Note (2)	
5	232043	Washer-sealing	3		
6	ERC4626	Plug	1		
7	ETC6505	Connector-inlet manifold adaptor	1		
8	RTC5907	Valve-inlet manifold non return	1		
9	BH505181L	Bolt-Ram pipes Housing-hexagonal head-5/16UNC x 2 1/4	6		Alternatives
9	FB505181L	Bolt-Ram pipes Housing-flanged head	6		
10	WA108051L	Washer-plain-Ram pipes Housing-M8	6		

CHANGE POINTS:
(1) From (V) TA 177890
(2) From (V) TA 508674
(3) To (V) TA 177889
(4) To (V) TA 508673

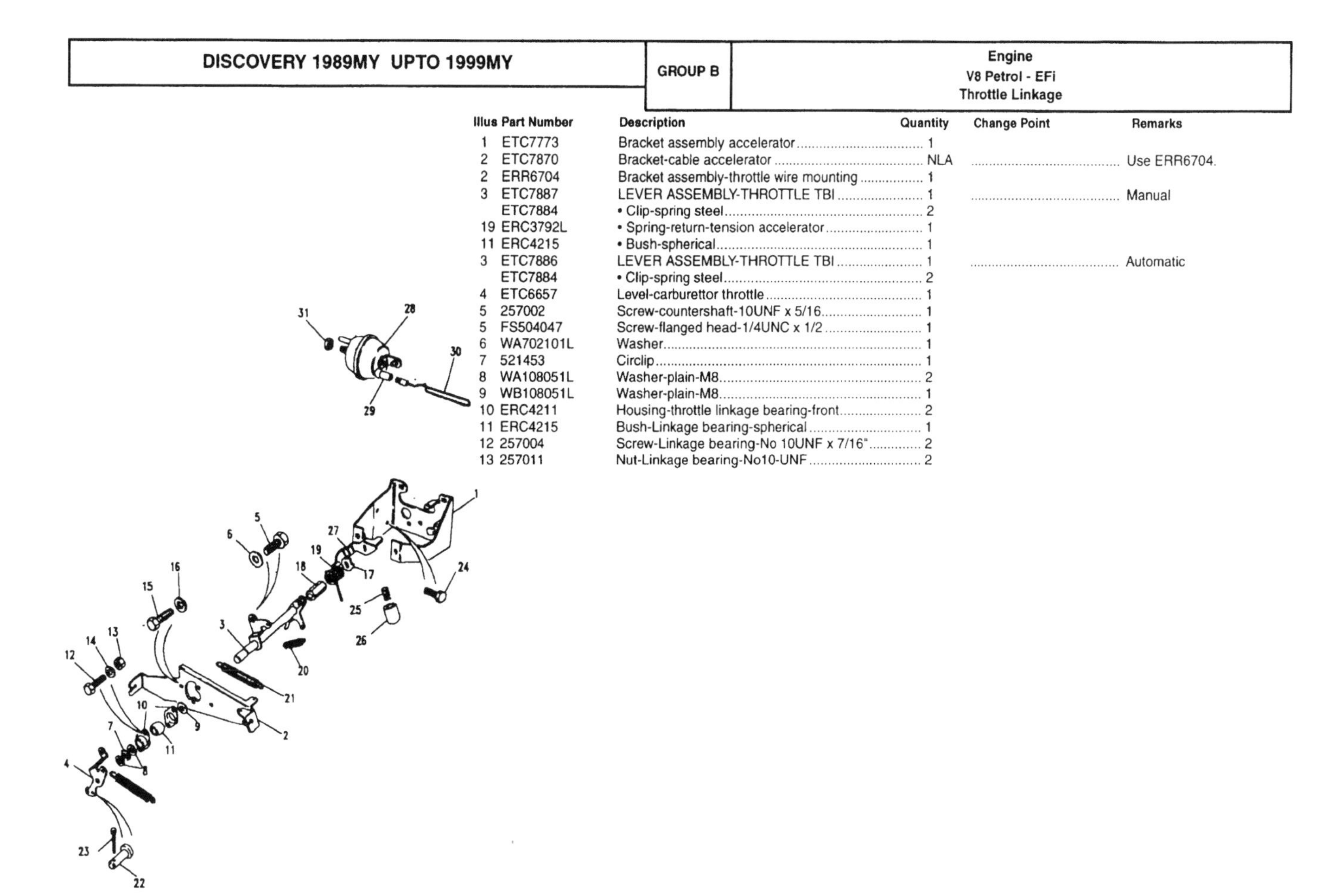

Illus	Part Number	Description	Quantity	Change Point	Remarks
1	ETC7773	Bracket assembly accelerator	1		
2	ETC7870	Bracket-cable accelerator	NLA		Use ERR6704.
2	ERR6704	Bracket assembly-throttle wire mounting	1		
3	ETC7887	LEVER ASSEMBLY-THROTTLE TBI	1		Manual
	ETC7884	• Clip-spring steel	2		
19	ERC3792L	• Spring-return-tension accelerator	1		
11	ERC4215	• Bush-spherical	1		
3	ETC7886	LEVER ASSEMBLY-THROTTLE TBI	1		Automatic
	ETC7884	• Clip-spring steel	2		
4	ETC6657	Level-carburettor throttle	1		
5	257002	Screw-countershaft-10UNF x 5/16	1		
5	FS504047	Screw-flanged head-1/4UNC x 1/2	1		
6	WA702101L	Washer	1		
7	521453	Circlip	1		
8	WA108051L	Washer-plain-M8	2		
9	WB108051L	Washer-plain-M8	1		
10	ERC4211	Housing-throttle linkage bearing-front	2		
11	ERC4215	Bush-Linkage bearing-spherical	1		
12	257004	Screw-Linkage bearing-No 10UNF x 7/16"	2		
13	257011	Nut-Linkage bearing-No10-UNF	2		

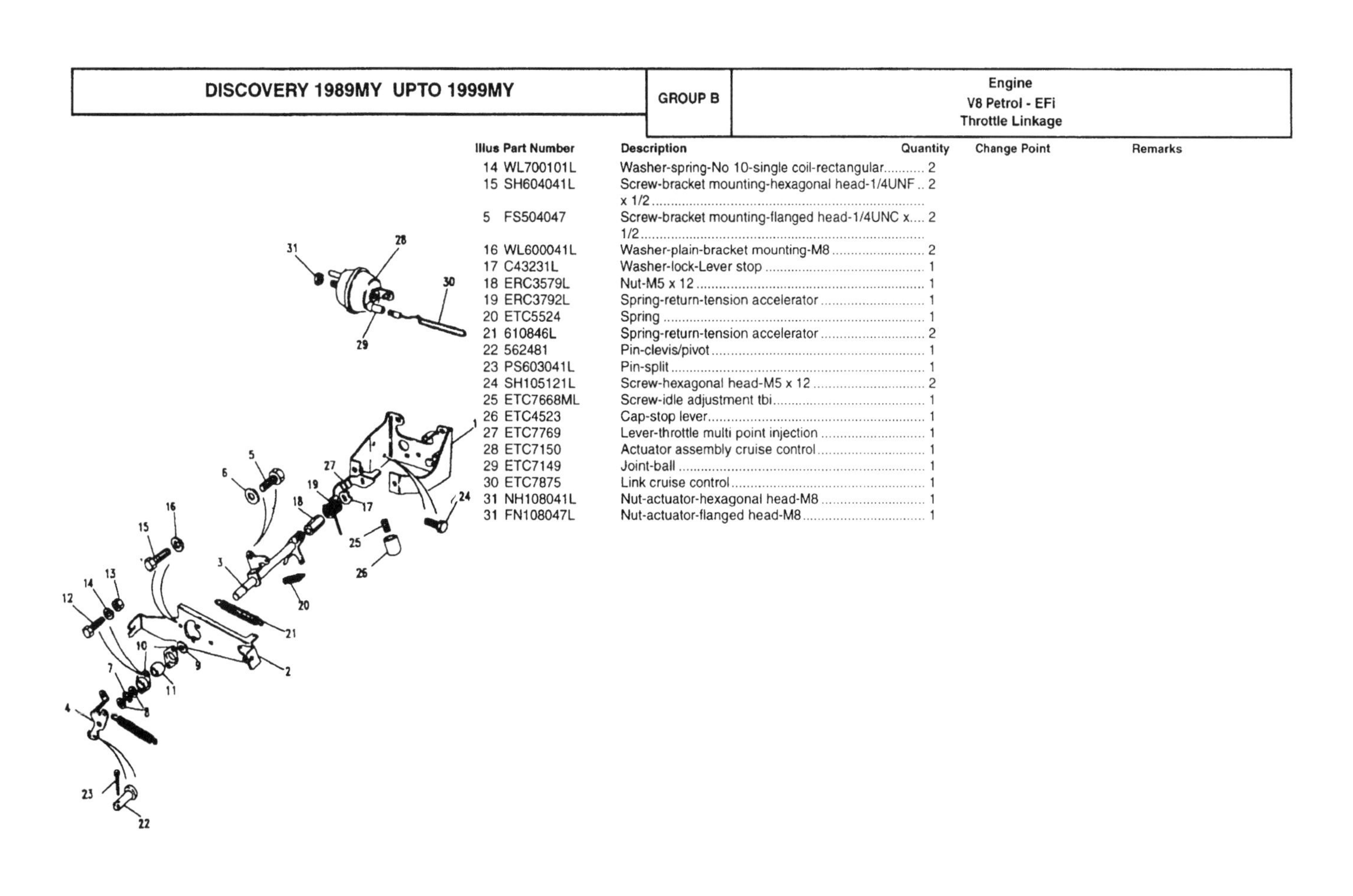

Illus	Part Number	Description	Quantity	Change Point	Remarks
14	WL700101L	Washer-spring-No 10-single coil-rectangular	2		
15	SH604041L	Screw-bracket mounting-hexagonal head-1/4UNF x 1/2	2		
5	FS504047	Screw-bracket mounting-flanged head-1/4UNC x 1/2	2		
16	WL600041L	Washer-plain-bracket mounting-M8	2		
17	C43231L	Washer-lock-Lever stop	1		
18	ERC3579L	Nut-M5 x 12	1		
19	ERC3792L	Spring-return-tension accelerator	1		
20	ETC5524	Spring	1		
21	610846L	Spring-return-tension accelerator	2		
22	562481	Pin-clevis/pivot	1		
23	PS603041L	Pin-split	1		
24	SH105121L	Screw-hexagonal head-M5 x 12	2		
25	ETC7668ML	Screw-idle adjustment tbi	1		
26	ETC4523	Cap-stop lever	1		
27	ETC7769	Lever-throttle multi point injection	1		
28	ETC7150	Actuator assembly cruise control	1		
29	ETC7149	Joint-ball	1		
30	ETC7875	Link cruise control	1		
31	NH108041L	Nut-actuator-hexagonal head-M8	1		
31	FN108047L	Nut-actuator-flanged head-M8	1		

DISCOVERY 1989MY UPTO 1999MY

GROUP B

Engine
V8 Petrol - EFi
Gasket Kit

Illus	Part Number	Description	Quantity	Change Point	Remarks
1	RTC3306	Set-engine gasket	1	Note (1)	
1	STC2823	Set-engine gasket	1	Note (2)	
2	RTC5254	Set-cylinder head gasket-asbestos	1	Note (1)	
2	STC1642	Set-cylinder head gasket-asbestos free	1	Note (1)	
2	STC2822	Set-cylinder head gasket	1	Note (2)	

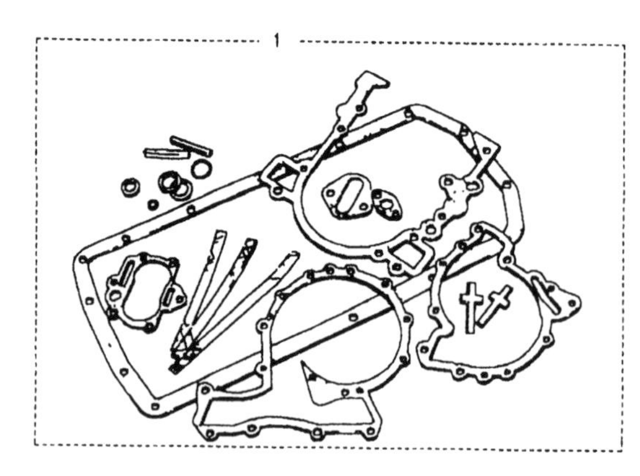

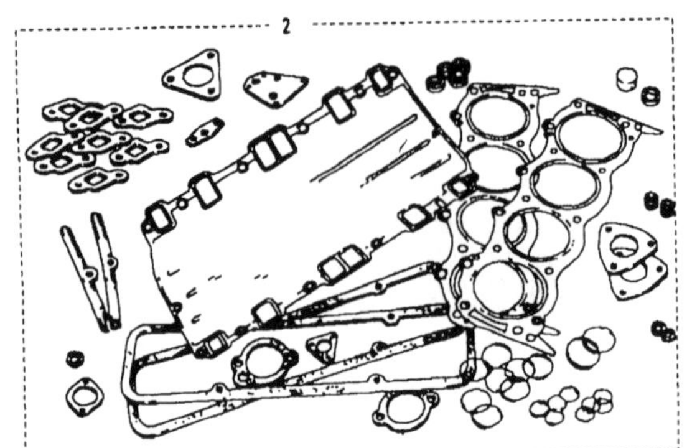

CHANGE POINTS:
(1) To (V) LA 081991
(2) From (V) MA 081992; From (V) MA 500000

B 141

DISCOVERY 1989MY UPTO 1999MY

GROUP B

Engine
V8 Petrol - EFi
Power Steering Pump V Belt Drive

Illus	Part Number	Description	Quantity	Change Point	Remarks
		UP TO VIN LA081990			
1	NTC8286	PUMP ASSEMBLY POWER ASSISTED STEERING	1	Note (1)	
1	NTC9198	PUMP ASSEMBLY POWER ASSISTED STEERING	1	Note (2)	
	RTC6074	• O ring-large	1		
	RTC6075	• O ring-small	1		
	STC1633	• Kit-power assisted steering pump seal	1		Part of NTC8286.
	RTC6760	• Spacer	1		Part of NTC9198.
2	ETC6408	Pulley power assisted steering pump-twin groove-A127 alternator	1		
3	ERC675	Belt-coolant pump vee	1		
4	ERR798	Bracket-mounting power assisted steering pump	1		
5	ETC6607	Link-power assisted steering pump adjusting	1		
6	2266L	Washer-plain-bracket adjusting	1		
7	NH108041L	Nut-bracket adjusting-hexagonal head-M8	1		
8	FS108207L	Screw-bracket mounting-flanged head-M8 x 20	3		3500 cc L/R V8 Petrol EFi
8	FS108207L	Screw-pulley-flanged head-M8 x 20	3		3500 cc L/R V8 Petrol EFi
8	FS108207L	Screw-bracket mounting-flanged head-M8 x 20	3		
9	FS108207L	Screw-pulley-flanged head-M8 x 20	3		
10	FS108251L	Screw-bracket adjusting-flanged head-M8 x 25	NLA		Use FS108257L.
10	FS108257L	Screw-bracket adjusting-flanged head-M8 x 25	1		
11	TE506101L	Stud-bracket mounting-3/8UNF x 1 1/8	1		
12	WA108051L	Washer-plain-pulley-M8	3		
13	WM600051L	Washer-spring-bracket adjusting-5/16"	1		
14	NH606041L	Nut-hexagonal head-3/8UNF	1		
15	BH504121L	Bolt-bracket mounting-1/4UNC x 1 1/2-hexagonal head	2	Note (3)	
16	BH505401L	Bolt-pulley-hexagonal head-5/16UNC x 5	1	Note (3)	
17	BH505381L	Bolt-hexagonal head-5/16UNC x 4 3/4	NLA	Note (3)	Use BH505401L.

CHANGE POINTS:
(1) To (V) JA 34313
(2) From (V) KA 34314
(3) To (V) LA 081990

B 142

Power Steering Pump Serpentine Belt Drive

Illus	Part Number	Description	Quantity	Change Point	Remarks
		FROM VIN MA081991			
1	ERR4066	Pump assembly power assisted steering	NLA	Note (1)	Use QVB101110.
2	ERR4054	Bracket-mounting power assisted steering pump-centre	1	Note (1)	
3	FS108207L	Screw-flanged head-M8 x 20	2		
4	ERR4427	Bracket-mounting power assisted steering pump-front	1	Note (1)	
5	ERR3062	Bolt-flanged head-3/8UNC x 5 1/4	1	Note (1)	
6	FS108207L	Screw-flanged head-M8 x 20	5		
7	ERR4679	Bracket-lifting engine front	1	Note (1)	
8	ERR4053	Bracket-mounting power assisted steering pump-rear	1	Note (1)	
9	ERR3063	Bolt-flanged head-Lifting Plate-3/8UNC x 2 5/8	1	Note (1)	
10	FB506267	Bolt-flanged head-Lifting Plate-3/8UNC x 3 1/4	1	Note (1)	
11	ERR4868	Pulley-assembly	1	Note (1)	
12	FS108127	Screw-pulley-flanged head-M8 x 12	3	Note (1)	

CHANGE POINTS:
(1) From (V) MA 081991

B 143

Air Conditioning Compressor V Belt Drive

Illus	Part Number	Description	Quantity	Change Point	Remarks
1	ETC7994	COMPRESSOR ASSEMBLY AIR CONDITIONING	1	Note (1); Note (2)	
	STC3131	• O ring air conditioning	2		
1	ERR845	Compressor assembly air conditioning	1	Note (3); Note (4)	
2	ERR4521	Bracket-air compressor	1		
3	611612	Belt-polyvee alternator	1		
4	ETC6159	PULLEY-TENSIONER ANCILLARY DRIVE	1		
	614154	• Bearing-ball	1		
5	WM600051L	Washer-spring-pulley-5/16"	1		
6	WA110061L	Washer-plain-bracket mounting-M10-standard	6		
7	WA108051L	Washer-plain-bracket mounting-M8	1		
8	NH606041L	Nut-bracket mounting-hexagonal head-3/8UNF	2		
9	BH606401L	Bolt-compressor-3/8UNF x 5	2		
10	BH605101L	Bolt-pulley-hexagonal head-5/16UNF x 1 1/4	1		
11	BH506345	Bolt-bracket mounting	1		
12	BH506131L	Bolt-bracket mounting-3/8UNC x 1 7/8	1		

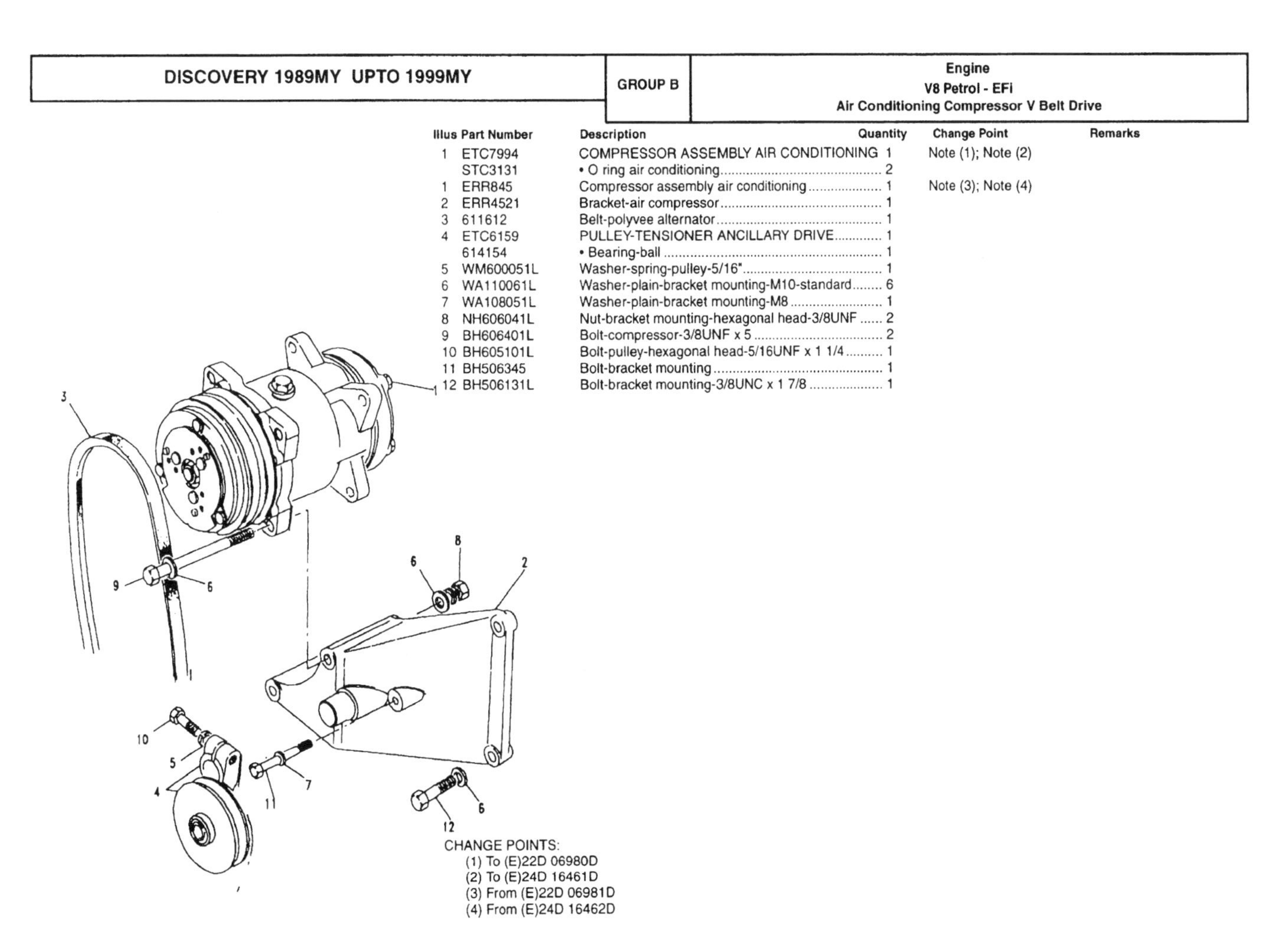

CHANGE POINTS:
(1) To (E)22D 06980D
(2) To (E)24D 16461D
(3) From (E)22D 06981D
(4) From (E)24D 16462D

B 144

DISCOVERY 1989MY UPTO 1999MY	GROUP B	Engine V8 Petrol - EFi Air-Con Compressor Serpentine Belt Drive

Illus	Part Number	Description	Quantity	Change Point	Remarks
		FROM VIN MA081991			
1	BTR5750	Compressor assembly air conditioning	1		
2	ERR3671	Heatshield assembly air conditioning compressor	1		
3	ERR4052	Dowel	2		
4	FB108197	Bolt-flanged head-M8 x 95	4		
5	ERR3440	Pulley-tensioner ancillary drive	1		
6	FB110151L	Bolt-flanged head-M10 x 75	1		
7	ERR4513	Bracket-ancillary mounting	1		
8	ERR3062	Bolt-flanged head-3/8UNC x 5 1/4	2		
9	FB506267	Bolt-flanged head-3/8UNC x 3 1/4	2		
10	ERR4623	Belt-polyvee alternator	1	Note (1); Note (2)	
10	ERR5579	Belt-polyvee alternator	1	Note (3); Note (4)	

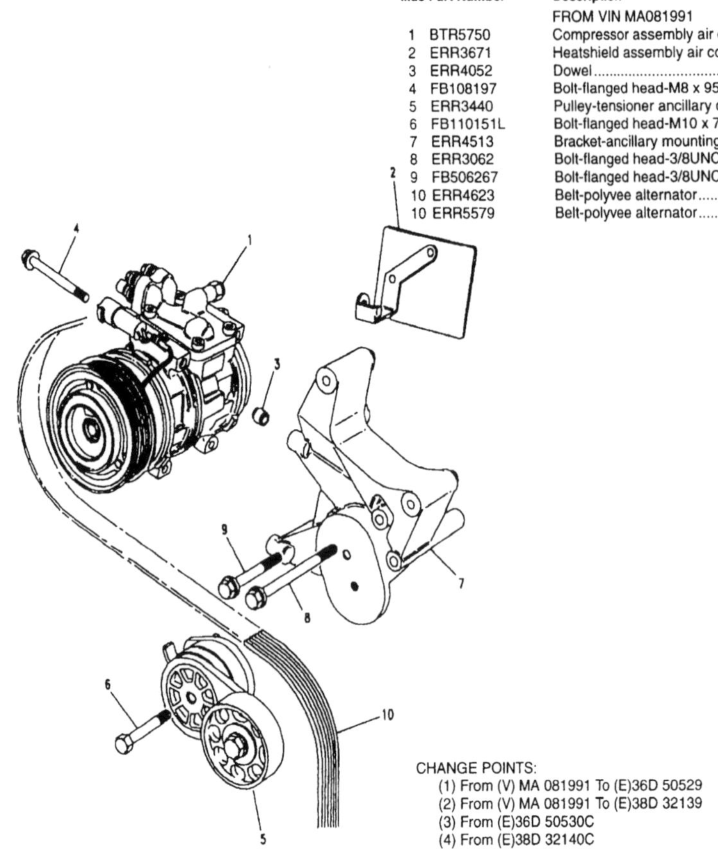

CHANGE POINTS:
(1) From (V) MA 081991 To (E)36D 50529
(2) From (V) MA 081991 To (E)38D 32139
(3) From (E)36D 50530C
(4) From (E)38D 32140C

B 145

DISCOVERY 1989MY UPTO 1999MY	GROUP B	Engine V8 Petrol - EFi Distributor

Illus	Part Number	Description	Quantity	Change Point	Remarks
		3.5 LOW COMPRESSION			
1	ERR1808	Distributor assembly ignition-2 pin	1		
1	ERR4254	Distributor assembly ignition-3 pin	1		
		3.5 LOW COMPRESSION (AUSTRALIA)			
1	ERR1810	Distributor assembly ignition-2 pin	1		
1	ERR4252	Distributor assembly ignition-3 pin	1		
		3.5 HIGH COMPRESSION			
1	ERR1809	Distributor assembly ignition-2 pin	1		
1	ERR4256	Distributor assembly ignition-3 pin	1		
		3.9 HIGH COMPRESSION			
1	ERR4738	Distributor assembly ignition	1	Note (1)	Catalyst
1	ERR4755	Distributor assembly ignition	1	Note (2)	
1	ERR5209	Distributor assembly ignition	1	Note (3)	
1	ERR4739	Distributor assembly ignition	1	Note (1)	Except Catalyst
1	ERR4754	Distributor assembly ignition	1	Note (2)	
1	ERR5208	Distributor assembly ignition	1	Note (3)	
		3.9 LOW COMPRESSION			
1	ERR4740	Distributor assembly ignition	1	Note (1)	
1	ERR4753	Distributor assembly ignition	1	Note (2)	
1	ERR5207	Distributor assembly ignition	1	Note (3)	

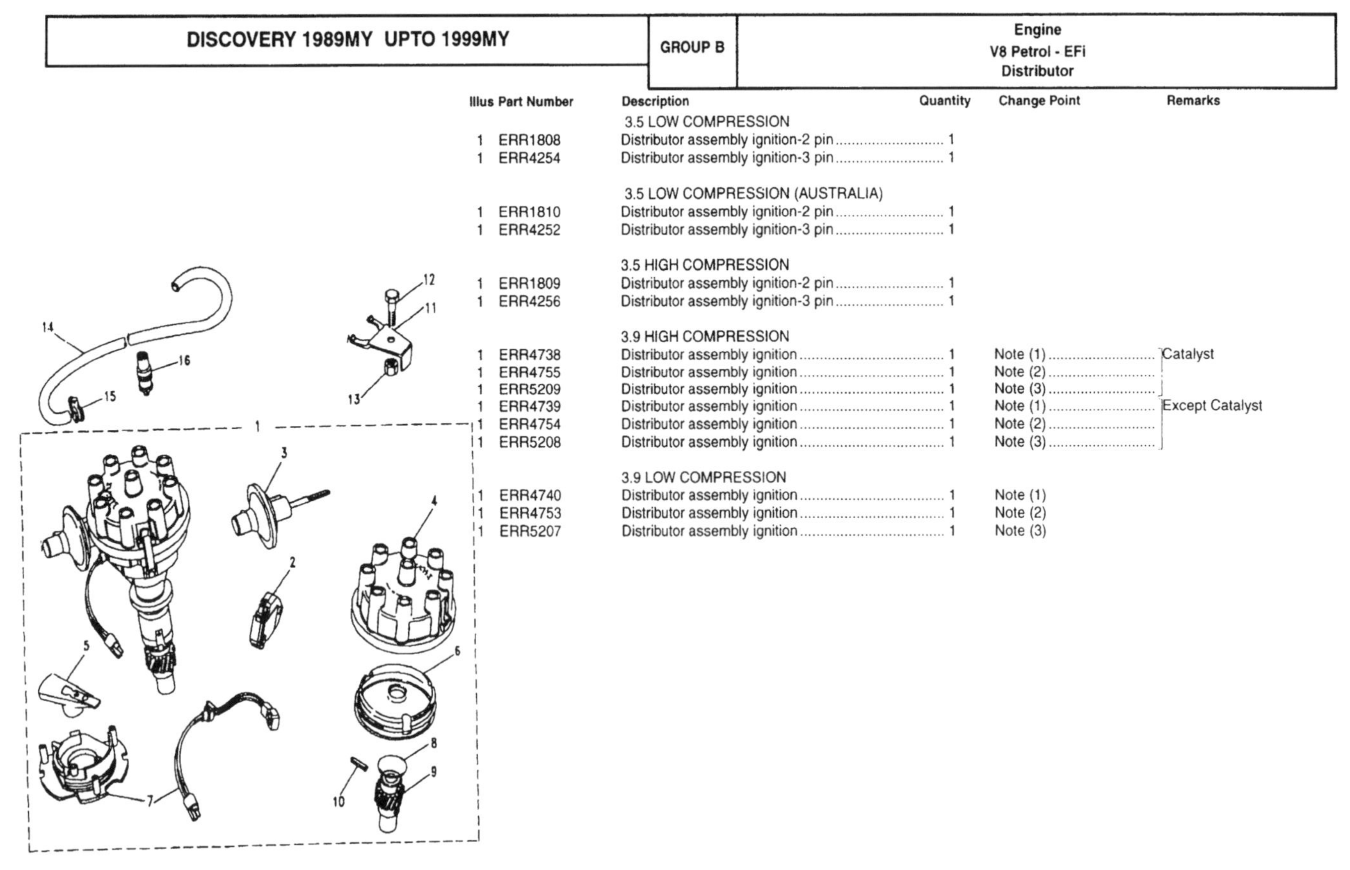

CHANGE POINTS:
(1) To (V) LA 081990
(2) From (V) MA 081991
(3) From (V) MA 094073

B 146

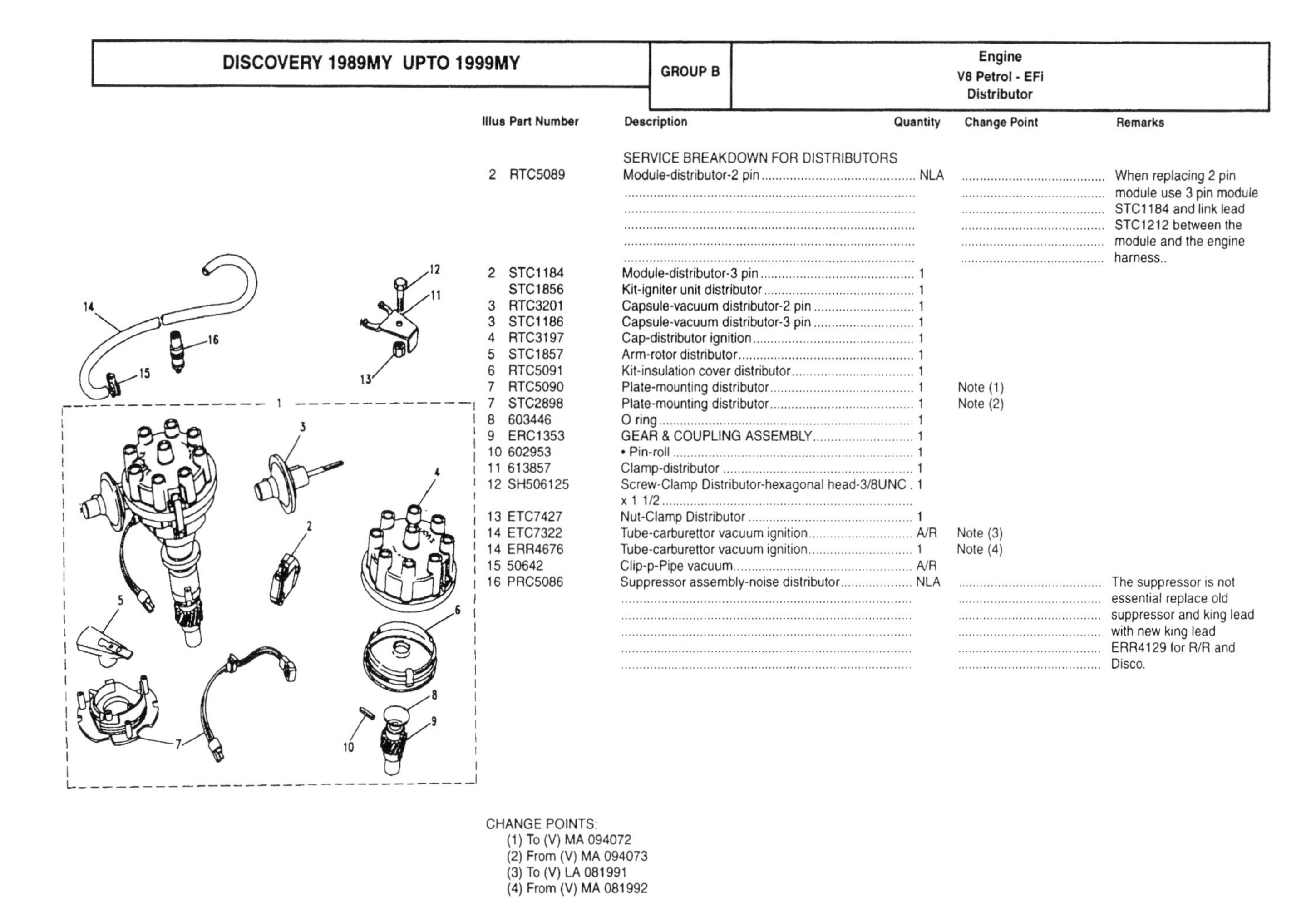

Illus	Part Number	Description	Quantity	Change Point	Remarks
		SERVICE BREAKDOWN FOR DISTRIBUTORS			
2	RTC5089	Module-distributor-2 pin	NLA		When replacing 2 pin module use 3 pin module STC1184 and link lead STC1212 between the module and the engine harness..
2	STC1184	Module-distributor-3 pin	1		
	STC1856	Kit-igniter unit distributor	1		
3	RTC3201	Capsule-vacuum distributor-2 pin	1		
3	STC1186	Capsule-vacuum distributor-3 pin	1		
4	RTC3197	Cap-distributor ignition	1		
5	STC1857	Arm-rotor distributor	1		
6	RTC5091	Kit-insulation cover distributor	1		
7	RTC5090	Plate-mounting distributor	1	Note (1)	
7	STC2898	Plate-mounting distributor	1	Note (2)	
8	603446	O ring	1		
9	ERC1353	GEAR & COUPLING ASSEMBLY	1		
10	602953	• Pin-roll	1		
11	613857	Clamp-distributor	1		
12	SH506125	Screw-Clamp Distributor-hexagonal head-3/8UNC x 1 1/2	1		
13	ETC7427	Nut-Clamp Distributor	1		
14	ETC7322	Tube-carburettor vacuum ignition	A/R	Note (3)	
14	ERR4676	Tube-carburettor vacuum ignition	1	Note (4)	
15	50642	Clip-p-Pipe vacuum	A/R		
16	PRC5086	Suppressor assembly-noise distributor	NLA		The suppressor is not essential replace old suppressor and king lead with new king lead ERR4129 for R/R and Disco.

CHANGE POINTS:
(1) To (V) MA 094072
(2) From (V) MA 094073
(3) To (V) LA 081991
(4) From (V) MA 081992

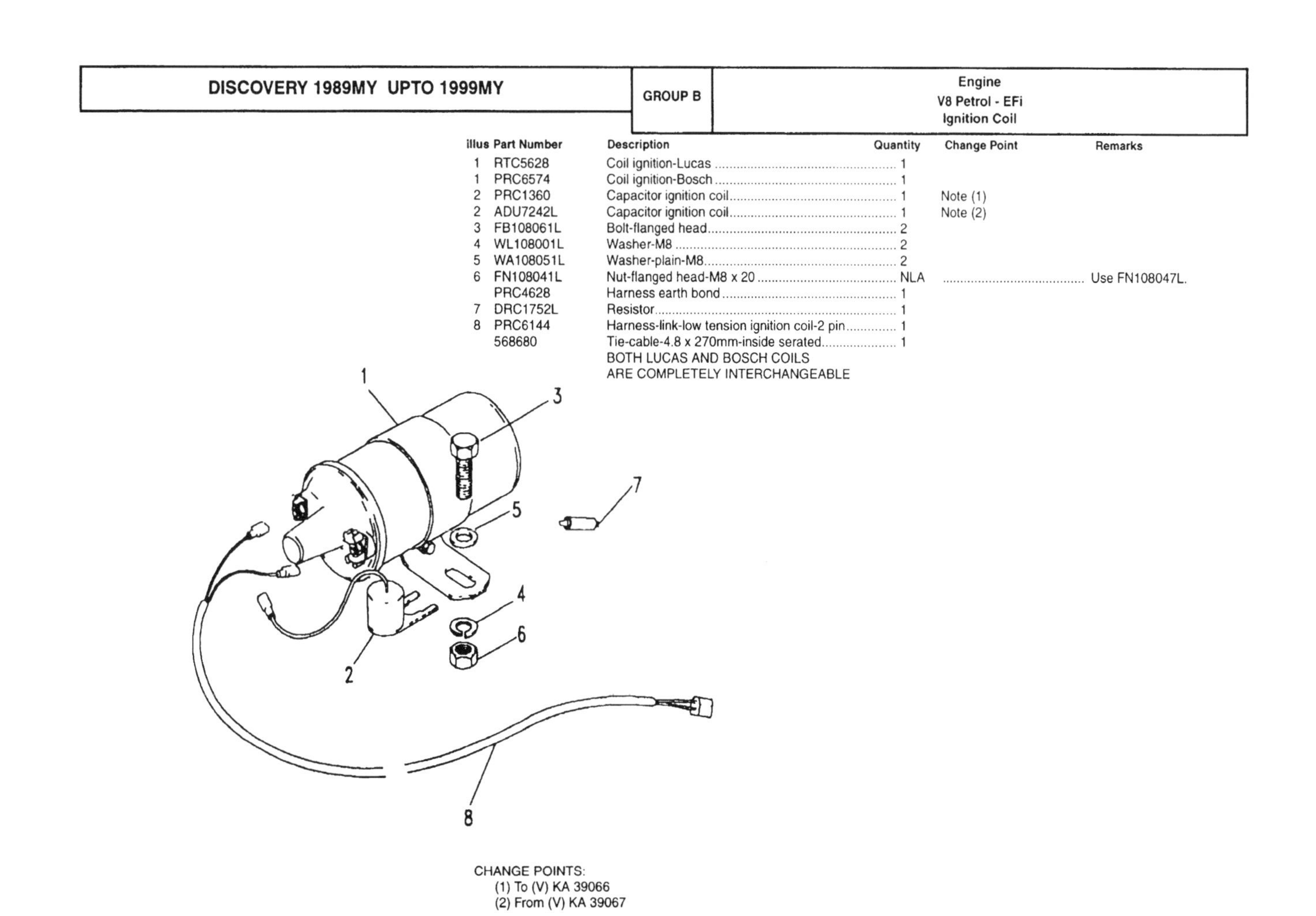

Illus	Part Number	Description	Quantity	Change Point	Remarks
1	RTC5628	Coil ignition-Lucas	1		
1	PRC6574	Coil ignition-Bosch	1		
2	PRC1360	Capacitor ignition coil	1	Note (1)	
2	ADU7242L	Capacitor ignition coil	1	Note (2)	
3	FB108061L	Bolt-flanged head	2		
4	WL108001L	Washer-M8	2		
5	WA108051L	Washer-plain-M8	2		
6	FN108041L	Nut-flanged head-M8 x 20	NLA		Use FN108047L.
	PRC4628	Harness earth bond	1		
7	DRC1752L	Resistor	1		
8	PRC6144	Harness-link-low tension ignition coil-2 pin	1		
	568680	Tie-cable-4.8 x 270mm-inside serated	1		
		BOTH LUCAS AND BOSCH COILS ARE COMPLETELY INTERCHANGEABLE			

CHANGE POINTS:
(1) To (V) KA 39066
(2) From (V) KA 39067

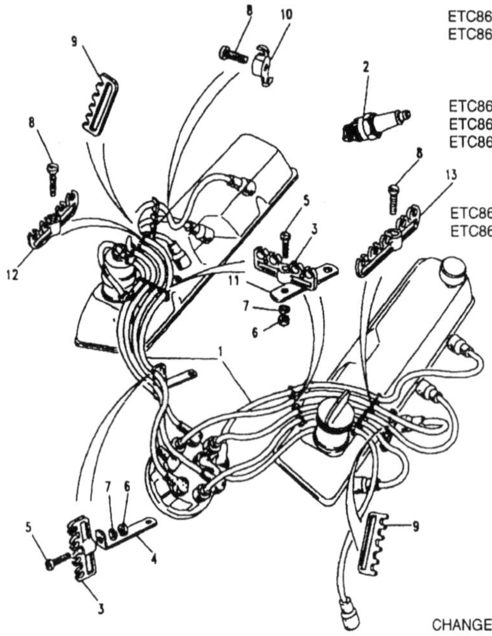

Illus	Part Number	Description	Quantity	Change Point	Remarks
		UP TO VIN LA081990			
1	RTC6551	KIT-HIGH TENSION LEADS IGNITION-SILICONE	1	Note (1)	
	ETC8610	• Lead-high tension ignition-No1	NLA		Use STC3843 qty 1 for NAS markets only or STC8658 qty 1 for ROW markets only.
	ETC8611	• Lead-high tension ignition-No3	NLA		Use STC3845 qty 1 for NAS markets only or STC8661 qty 1 for ROW markets only.
	ETC8612	• Lead-high tension ignition-No5	1		
	ETC8613	• Lead-high tension ignition-No2	NLA		Use STC3844 qty 1 for NAS markets only or STC8660 qty 1 for ROW markets only.
	ETC8614	• Lead-high tension ignition-No4	1		
	ETC8615	• Lead-high tension ignition-No6	1		
	ETC8616	• Lead-high tension ignition-No8	NLA		Use STC3850 qty 1 for NAS markets only or STC8660 qty 1 for ROW markets only.
	ETC8617	• Lead-high tension ignition-No7	1		
	ETC8618	• Lead-king ignition	1		

CHANGE POINTS:
(1) To (V) LA 081990

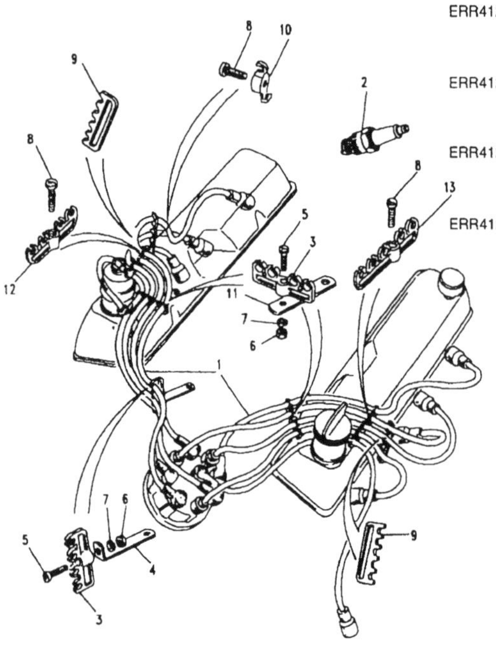

Illus	Part Number	Description	Quantity	Change Point	Remarks
	ERR4121	Lead-high tension ignition-No1	NLA	Note (1)	Use STC3835 qty 1 for ROW markets only. or STC8666 qty 1 for ROW markets only.
	ERR4122	Lead-high tension ignition-No2	NLA	Note (1)	Use STC3836 qty 1 for NAS markets only. or STC8665 qty 1 for ROW markets only..
	ERR4123	Lead-high tension ignition-No3	NLA	Note (1)	Use STC3837 qty 1 for NAS markets only. or STC8667 qty 1 for ROW markets only..
	ERR4124	Lead-high tension ignition-No4	NLA	Note (1)	Use STC3838 qty 1 for NAS markets only. or STC8664 qty 1 for ROW markets only..
	ERR4125	Lead-high tension ignition-No5	NLA	Note (1)	Use STC3839 qty 1 for NAS markets only. or STC8659 qty 1 for ROW markets only..
	ERR4126	Lead-high tension ignition-No6	NLA	Note (1)	Use STC3840 qty 1 for NAS markets only. or STC8664 qty 1 for ROW markets only..

CHANGE POINTS:
(1) From (V) MA 081991

DISCOVERY 1989MY UPTO 1999MY	GROUP B	Engine V8 Petrol - EFi High Tension Leads and Plugs

Illus	Part Number	Description	Quantity	Change Point	Remarks
	ERR4127	Lead-high tension ignition-No7	NLA	Note (1)	Use STC3841 qty 1 for NAS markets only. or STC8668 qty 1 for ROW markets only..
	ERR4128	Lead-high tension ignition-No8	NLA	Note (1)	Use STC3842 qty 1 for NAS markets only. or STC8665 qty 1 for ROW markets only..
	ERR4129	Lead-king ignition	NLA	Note (1)	Use STC8656.
	STC8656	Lead-king ignition	1	Note (1)	
	STC9124	Lead-high tension ignition	1		L/R V8 Petrol EFi
2	RTC3812	Plug-sparking-N9YC	8		
2	ETC7244	Plug-sparking-RN12YC	8		
2	ERR3799	Plug-sparking-RN11YC	8		
3	603672	Clip-spacer-high tension lead ignition	3		
3	603672	Clip-spacer-high tension lead ignition	5		Automatic
4	ERR395	Bracket-support-lead ignition	1		
5	SE910161L	Screw	3		
6	NH910011L	Locknut-10x32UNF	3		
7	WL700101L	Washer-spring-No 10-single coil-rectangular	3		
7	WM702001L	Washer-spring-3/16 dia-square	1		Automatic
7	WM600041L	Washer-spring-1/4 dia-square	3		Automatic
8	AB610041L	Screw-self tapping AB-M8 x 12	4		
9	603673	Clip-spacer-high tension lead ignition	2		
10	587477L	Clip-spacer-high tension lead ignition	1		
11	ETC6093	Bracket-support-lead ignition	2		
12	ETC7971	Clip-spacer-high tension lead ignition	1		
13	ETC7971	Clip-spacer-high tension lead ignition	1		
	6860L	Grommet-rubber-17/64" hole	1		
	AEU1446	Clip-p-3/4"	1		
	NH504041L	Nut-hexagonal head-1/4UNC x 1	1		Automatic

CHANGE POINTS:
(1) From (V) MA 081991

DISCOVERY 1989MY UPTO 1999MY	GROUP B	Engine V8 Petrol - EFi Alternator Lucas A127

Illus	Part Number	Description	Quantity	Change Point	Remarks
		ALTERNATOR ASSEMBLY			
1	RTC5680N	A127-65 AMP-NEW-RH-LUCAS	NLA		Use STC234.
1	STC234	A127-65 AMP-NEW	1		L/R V8 Petrol EFi
1	RTC5680E	A127-65 AMP-EXCHANGE-RH-LUCAS	1		
2	RTC5670	• Regulator & brush box assembly alternator-65Amp alternator-New	1		
3	RTC5671	• Rectifier-alternator-65Amp alternator-New	1		
4	RTC5926	• Bearing-alternator front	1		Part of RTC5680N, RTC5680E.
5	RTC5687	• Bracket-drive end c/w bearing alternator e-65Amp alternator-New	1		
6	RTC5046	• Kit-fixings alternator/vacuum pump-65Amp alternator-New	1		
7	RTC6114	• Capacitor alternator	1		Part of STC234.
1	YLE10088	ALTERNATOR ASSEMBLY-A127-72AMP-NEW	1	Note (1)	
1	YLE10088E	ALTERNATOR ASSEMBLY-A127-72AMP-EXCHANGE	1	Note (1)	
2	STC1604	• Regulator & brush box assembly alternator-72Amp alternator-New	1		
3	STC1605	• Rectifier-alternator-72Amp alternator-New	1		
4	RTC5047	• Bearing-alternator front	NLA		Use RTC5926.
5	RTC5687	• Bracket-drive end c/w bearing alternator e-72Amp alternator-New	1		
6	RTC5046	• Kit-fixings alternator/vacuum pump-72Amp alternator-New	1		
7	RTC6114	• Capacitor alternator	1		

CHANGE POINTS:
(1) From (V) KA 034314

DISCOVERY 1989MY UPTO 1999MY	GROUP B	Engine V8 Petrol - EFi Alternator Lucas A127

Illus	Part Number	Description	Quantity	Change Point	Remarks
1	YLE10099	ALTERNATOR ASSEMBLY-A127I-85 AMP-NEW	1		
1	YLE10099E	ALTERNATOR ASSEMBLY-A127I-85 AMP-EXCHANGE	1		
2	STC985	• Regulator & brush box assembly alternator-85Amp alternator-New	1		
3	STC986	• Rectifier-alternator-85Amp alternator-New	1		
4	RTC5047	• Bearing-alternator front	NLA		Use RTC5926.
5	RTC5687	• Bracket-drive end c/w bearing alternator e-85Amp alternator-New	1		
6	RTC5046	• Kit-fixings alternator/vacuum pump-85Amp-alternator-New	1		
7	RTC6114	• Capacitor alternator-with lead	1		
8	605410	Fan alternator-127/65 amp	1		
9	ETC9083	Pulley alternator-68mm-127/65 amp	1		3500 cc
9	STC862	Pulley-alternator	1		3900 cc
10	ETC7899	Belt-polyvee alternator	1	Note (1)	3500 cc
10	ERR2073	Belt-polyvee alternator	1	Note (2)	3500 cc
10	ERR2678	Belt-polyvee alternator	1		3900 cc
	ERR1608	Heatshield alternator-65 amp	1		
	ERR3764	Heatshield alternator-85 amp	1		
	PRC5273	Clip-omega-Grey-10.5 x 8mm-6.5mm hole	1		
7	AEU1616	Capacitor alternator-A series	1		

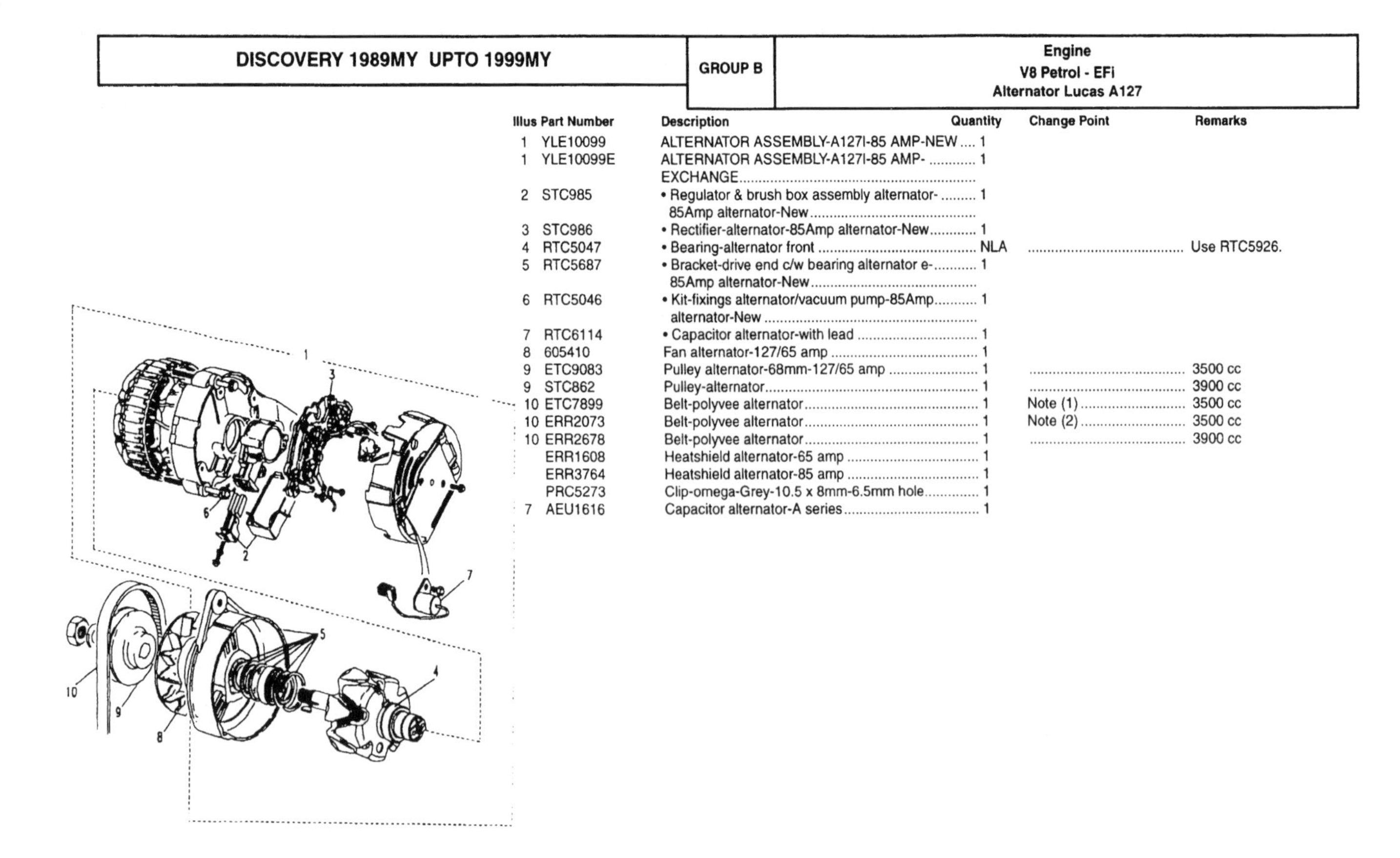

CHANGE POINTS:
(1) To (E)22D 06717D; To (E)24D 15830D
(2) From (E)22D 06718D; From (E)24D 15831D

B 153

DISCOVERY 1989MY UPTO 1999MY	GROUP B	Engine V8 Petrol - EFi Alternator Lucas A133

Illus	Part Number	Description	Quantity	Change Point	Remarks
1	RTC5053	ALTERNATOR ASSEMBLY-A133-80 AMP-NEW	NLA		Use STC1753.
1	STC1753	ALTERNATOR ASSEMBLY-A133-80 AMP-NEW	1		
1	RTC5053E	ALTERNATOR ASSEMBLY-A133-80 AMP-EXCHANGE	NLA		Exchange unit not available, use new unit STC1753.
2	RTC5610	• Cover alternator	1		
3	RTC5609	• Rectifier-alternator-85 amp/100 amp	1		
4	RTC5928	• Regulator assembly alternator-85 amp/100 amp	1		
5	RTC5611	• Brush alternator	1		
6	RTC5054	Fan alternator-80 amp	1		
7	ERR767	Pulley-alternator	1		
8	AEU1616	Capacitor alternator	2		
9	RTC6166	Spacer-fan	2		
10	ETC7899	Belt-polyvee alternator	1	Note (1)	RHD
10	ERR2073	Belt-polyvee alternator	1	Note (2)	
10	ERR914	Belt-polyvee alternator	1		LHD

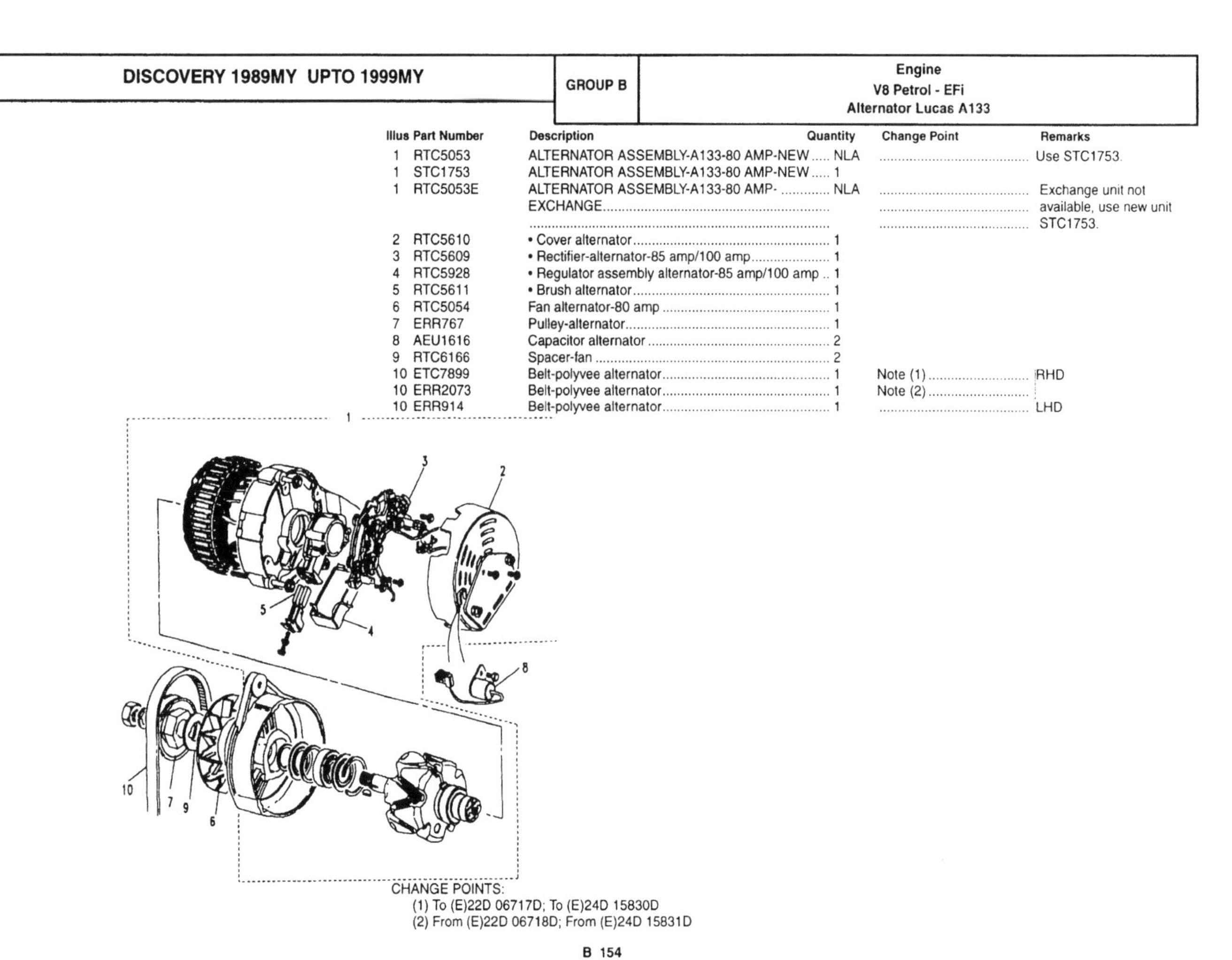

CHANGE POINTS:
(1) To (E)22D 06717D; To (E)24D 15830D
(2) From (E)22D 06718D; From (E)24D 15831D

B 154

Illus	Part Number	Description	Quantity	Change Point	Remarks
		FROM VIN MA081992			
1	AMR3107	Alternator assembly-A127i-100 amp-new	1	Note (1); Note (2)	
1	AMR3107E	Alternator assembly-A127i-100 amp-exchange	1	Note (1); Note (2)	
1	AMR4247	Alternator assembly-A127-100 amp-new	1	Note (3); Note (4)	
1	AMR4247E	Alternator assembly-A127-100 amp-exchange	1	Note (5); Note (4)	
2	AMR2218	Nut & washer set-Cover Terminal-6mm	1	Note (7)	
3	AMR2263	Cover-terminal alternator	1	Note (8); Note (2)	
3	AMR4939	Cover-terminal alternator	1	Note (9); Note (10)	
3	AMR6111	Cover-terminal alternator	1	Note (11)	
4	FB108227	Bolt-flanged head-alternator-M8 x 110	2	Note (7)	
5	FN108041L	Nut-alternator-flanged head-M8 x 20	NLA		Use FN108047L.

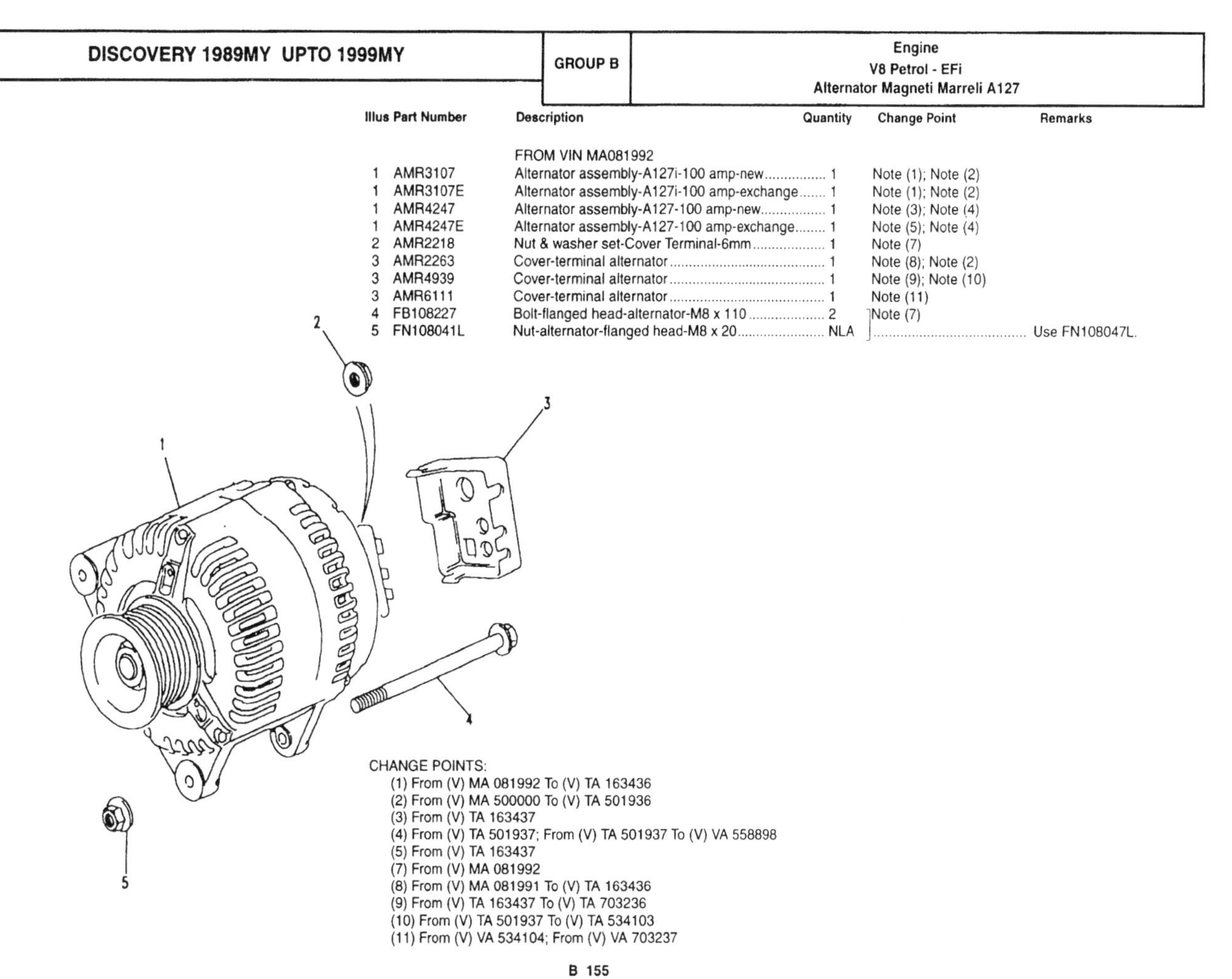

CHANGE POINTS:
(1) From (V) MA 081992 To (V) TA 163436
(2) From (V) MA 500000 To (V) TA 501936
(3) From (V) TA 163437
(4) From (V) TA 501937; From (V) TA 501937 To (V) VA 558898
(5) From (V) TA 163437
(7) From (V) MA 081992
(8) From (V) MA 081991 To (V) TA 163436
(9) From (V) TA 163437 To (V) TA 703236
(10) From (V) TA 501937 To (V) TA 534103
(11) From (V) VA 534104; From (V) VA 703237

B 155

Illus	Part Number	Description	Quantity	Change Point	Remarks
1	ETC5576	Bracket-mounting-alternator-65 amp	1		
1	ERR2676	Bracket-mounting-alternator-85 amp	1		
	ERR4513	Bracket-ancillary mounting	1		
2	ETC8441	Guard-fan alternator-65 amp	1		
2	ETC8440	Guard-fan alternator-80 amp	1		
	ERC5830	Grommet	2		
	ETC8559	Bush	2		
3	ETC6291	Link-adjuster alternator	1	Note (1)	
3	ERR1574	Link-adjuster alternator	1	Note (2)	
4	4594L	Washer-plain-link adjusting	1		
5	WA108051L	Washer-plain-fan guard-M8	8		
6	NH108041L	Nut-fan guard-hexagonal head-M8	3		
7	FS108207L	Screw-link adjusting-flanged head-M8 x 20	1		
7	SH108451L	Screw-mounting bracket-hexagonal head-M8 x 45	1		
8	SH108401L	Screw-hexagonal head-M8 x 40	2		
9	ETC8765	Screw-bracket mounting to alternator-3/8UNC x 1	1		
10	SH506071L	Screw-mounting bracket-hexagonal head-3/8UNC x 7/8	1		
10	SH506095L	Screw-bracket mounting to alternator-hexagonal head-3/8UNC x 1 1/8	2		
11	WM600061L	Washer-bracket mounting to alternator-3/8"-square	3		
12	ETC7208	Support-adjuster link alternator-link adjusting	1	Note (1)	
12	ERR1885	Support-adjuster link alternator	1	Note (2)	
13	WA108051L	Washer-plain-link adjusting-M8	1		
14	NH108041L	Nut-link adjusting-hexagonal head-M8	1		
15	ETC8579	Strut-alternator-65 amp	1		
15	ERR1589	Strut-alternator-85 amp	1		
16	SH506101L	Screw-strut to manifold-hexagonal head-3/8UNC x 1 1/4	1		
16	SH506095L	Screw-Strut alternator-hexagonal head-3/8UNC x 1 1/8	1		
17	WA110061L	Washer-plain-strut to manifold-M10-standard	1		

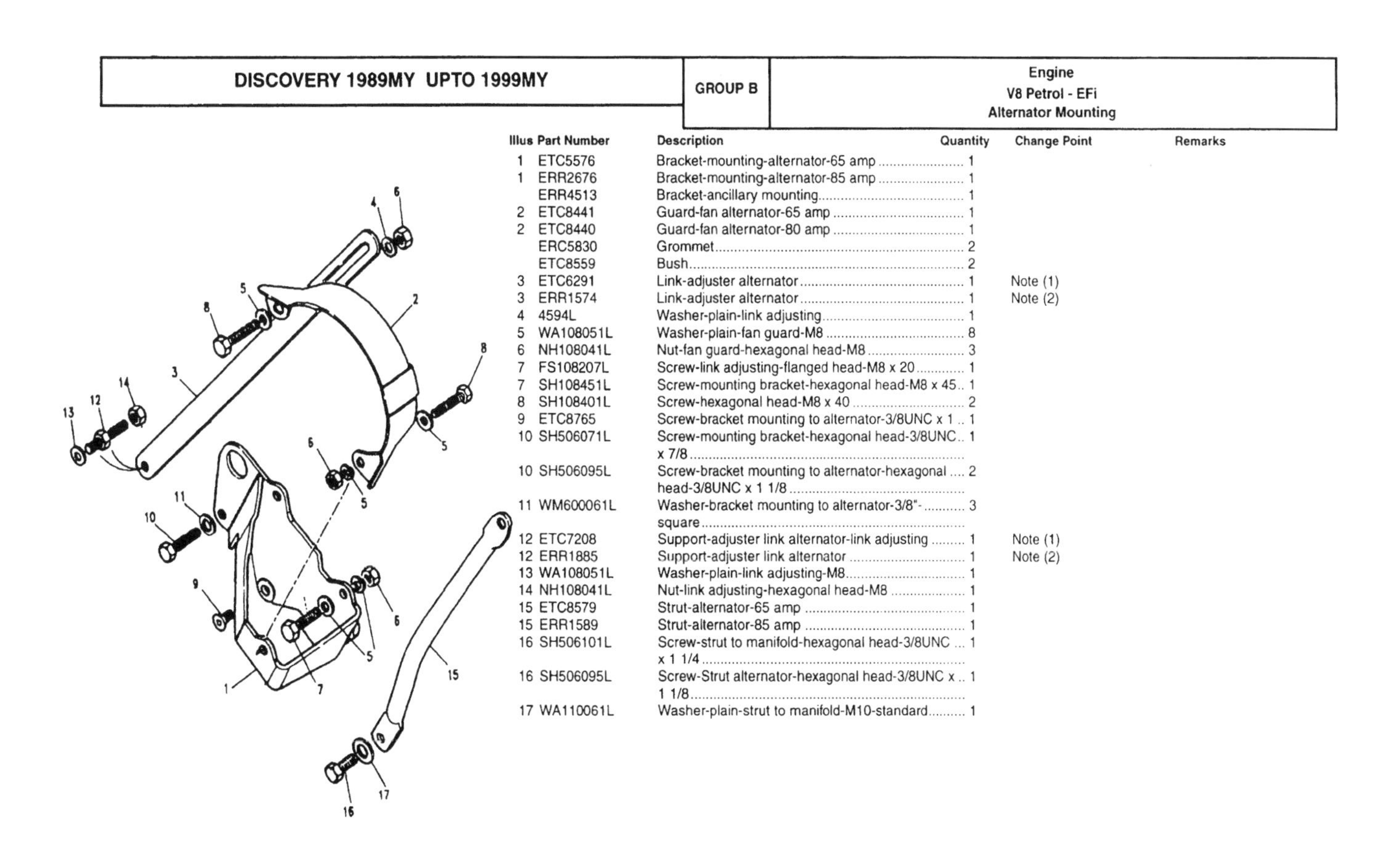

CHANGE POINTS:
(1) To (E)22D 06158D; To (E)24D 15546D
(2) From (E)22D 06159D; From (E)24D 15547D

B 156

DISCOVERY 1989MY UPTO 1999MY

GROUP B | **Engine** | **V8 Petrol - EFi** | **Alternator Mounting**

Illus	Part Number	Description	Quantity	Change Point	Remarks
8	FS108201L	Screw-link adjusting-flanged head-M8 x 20	NLA		Use FS108207L.
	ERR2677	Harness earth bond	1		
	AMR2263	Cover-terminal alternator	1		
	AMR2217	Nut & washer set-alternator to bracket-flanged head-5.0m	1		
	AMR2218	Nut & washer set-alternator to bracket-6mm	1		
	BH108281	Bolt-Strut alternator	1		
14	FN108041L	Nut-flanged head-M8 x 20	NLA		Use FN108047L.
16	FS108251L	Screw-flanged head-M8 x 25	NLA		Use FS108257L.
16	FS108257L	Screw-flanged head-M8 x 25	1		
13	WA108051L	Washer-plain-M8	4		
	YYD10003	Tie-cable-8mm hole-4.6 x 215mm	1		
	50641	Clip-p	1		

B 157

DISCOVERY 1989MY UPTO 1999MY

GROUP B | **Engine** | **V8 Petrol - EFi** | **Starter Motor Lucas**

UP TO VIN KA045926(3.5)

Illus	Part Number	Description	Quantity	Change Point	Remarks
1	RTC6061N	MOTOR-STARTER ENGINE-NEW-LUCAS	1	Note (1)	
1	RTC5228E	MOTOR-STARTER ENGINE-EXCHANGE	1		
5	SS506121L	Screw-Drive assembly-3/8UNC x 1 1/2	2	Note (1)	
6	WM600061L	Washer-starter motor-3/8"-square	2		
7	WA110061L	Washer-plain-starter motor-M10-standard	2		
	PRC3623	Cable assembly starter & earth	1		
	PRC5208	Harness-link	1		
	PRC7922	Cable-battery positive	1		

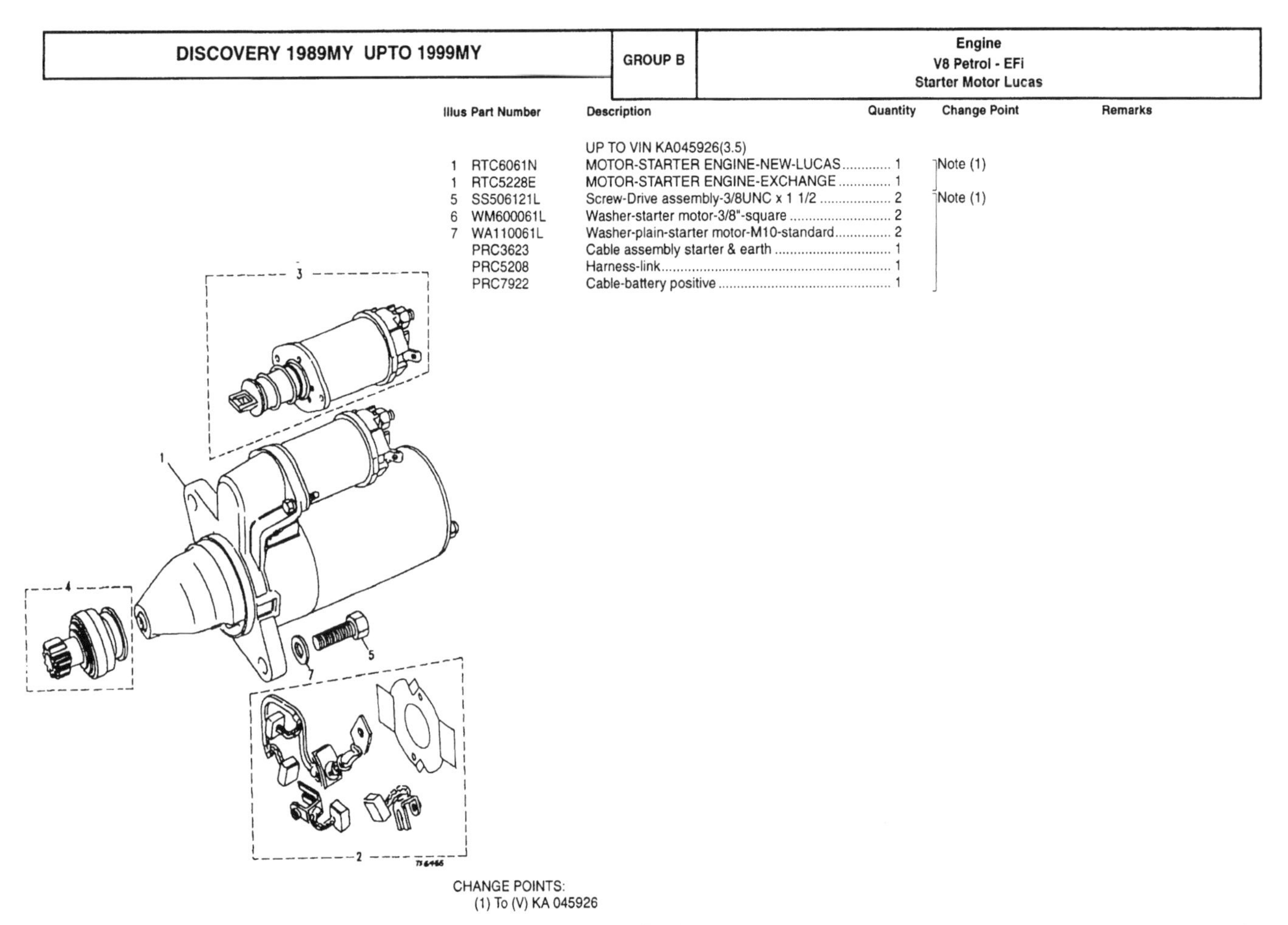

CHANGE POINTS:
(1) To (V) KA 045926

B 158

DISCOVERY 1989MY UPTO 1999MY — GROUP B

Engine
V8 Petrol - EFi
Starter Motor Bosch

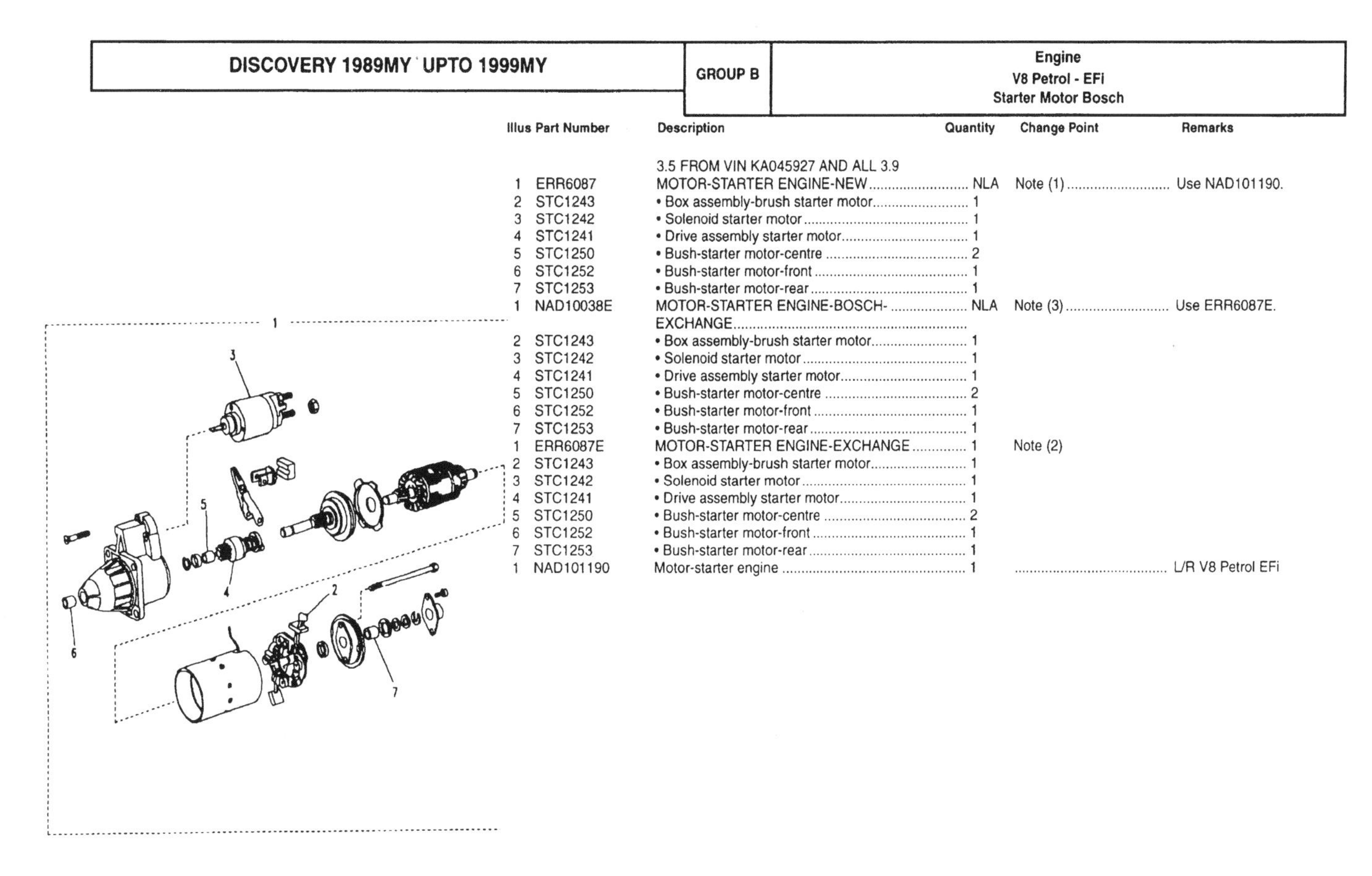

Illus	Part Number	Description	Quantity	Change Point	Remarks
		3.5 FROM VIN KA045927 AND ALL 3.9			
1	ERR6087	MOTOR-STARTER ENGINE-NEW	NLA	Note (1)	Use NAD101190.
2	STC1243	• Box assembly-brush starter motor	1		
3	STC1242	• Solenoid starter motor	1		
4	STC1241	• Drive assembly starter motor	1		
5	STC1250	• Bush-starter motor-centre	2		
6	STC1252	• Bush-starter motor-front	1		
7	STC1253	• Bush-starter motor-rear	1		
1	NAD10038E	MOTOR-STARTER ENGINE-BOSCH-EXCHANGE	NLA	Note (3)	Use ERR6087E.
2	STC1243	• Box assembly-brush starter motor	1		
3	STC1242	• Solenoid starter motor	1		
4	STC1241	• Drive assembly starter motor	1		
5	STC1250	• Bush-starter motor-centre	2		
6	STC1252	• Bush-starter motor-front	1		
7	STC1253	• Bush-starter motor-rear	1		
1	ERR6087E	MOTOR-STARTER ENGINE-EXCHANGE	1	Note (2)	
2	STC1243	• Box assembly-brush starter motor	1		
3	STC1242	• Solenoid starter motor	1		
4	STC1241	• Drive assembly starter motor	1		
5	STC1250	• Bush-starter motor-centre	2		
6	STC1252	• Bush-starter motor-front	1		
7	STC1253	• Bush-starter motor-rear	1		
1	NAD101190	Motor-starter engine	1		L/R V8 Petrol EFi

CHANGE POINTS:
(1) From (V) KA 045927
(2) From (V) KA 045927 To (V) MA 163587; To (V) MA 501978
(3) From (V) MA 163588; From (V) MA 501979

DISCOVERY 1989MY UPTO 1999MY — GROUP B

Engine
V8 Petrol - EFi
Flywheel and Clutch

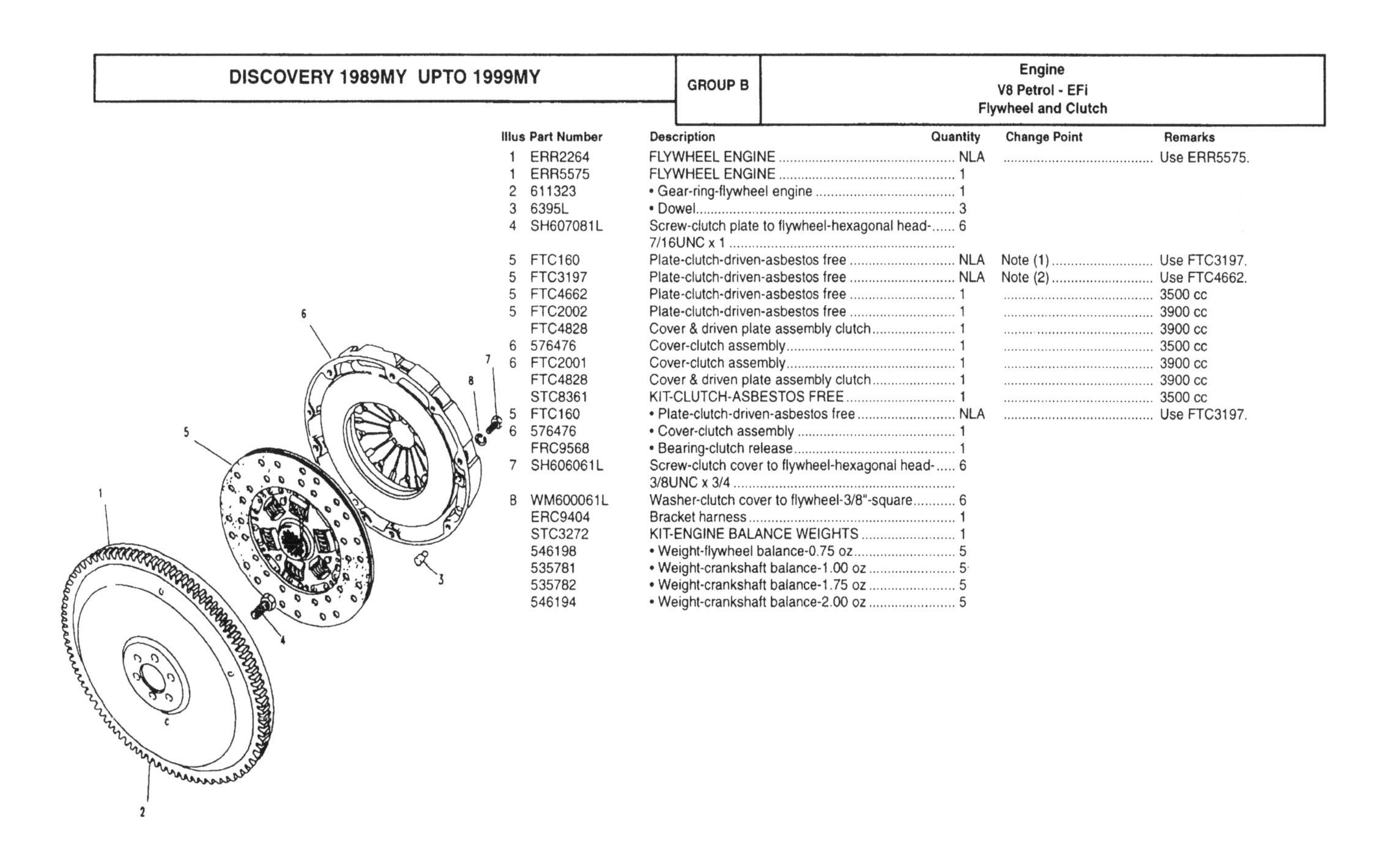

Illus	Part Number	Description	Quantity	Change Point	Remarks
1	ERR2264	FLYWHEEL ENGINE	NLA		Use ERR5575.
1	ERR5575	FLYWHEEL ENGINE	1		
2	611323	• Gear-ring-flywheel engine	1		
3	6395L	• Dowel	3		
4	SH607081L	Screw-clutch plate to flywheel-hexagonal head-7/16UNC x 1	6		
5	FTC160	Plate-clutch-driven-asbestos free	NLA	Note (1)	Use FTC3197.
5	FTC3197	Plate-clutch-driven-asbestos free	NLA	Note (2)	Use FTC4662.
5	FTC4662	Plate-clutch-driven-asbestos free	1		3500 cc
5	FTC2002	Plate-clutch-driven-asbestos free	1		3900 cc
	FTC4828	Cover & driven plate assembly clutch	1		3900 cc
6	576476	Cover-clutch assembly	1		3500 cc
6	FTC2001	Cover-clutch assembly	1		3900 cc
	FTC4828	Cover & driven plate assembly clutch	1		3900 cc
	STC8361	KIT-CLUTCH-ASBESTOS FREE	1		3500 cc
5	FTC160	• Plate-clutch-driven-asbestos free	NLA		Use FTC3197.
6	576476	• Cover-clutch assembly	1		
	FRC9568	• Bearing-clutch release	1		
7	SH606061L	Screw-clutch cover to flywheel-hexagonal head-3/8UNC x 3/4	6		
8	WM600061L	Washer-clutch cover to flywheel-3/8"-square	6		
	ERC9404	Bracket harness	1		
	STC3272	KIT-ENGINE BALANCE WEIGHTS	1		
	546198	• Weight-flywheel balance-0.75 oz	5		
	535781	• Weight-crankshaft balance-1.00 oz	5		
	535782	• Weight-crankshaft balance-1.75 oz	5		
	546194	• Weight-crankshaft balance-2.00 oz	5		

CHANGE POINTS:
(1) To (E)22D 08413D; To (E)24D 11300B
(2) From (E)22D 08414D; From (E)24D 11301B

Engine
V8 Petrol - EFi
Drive Plate Automatic

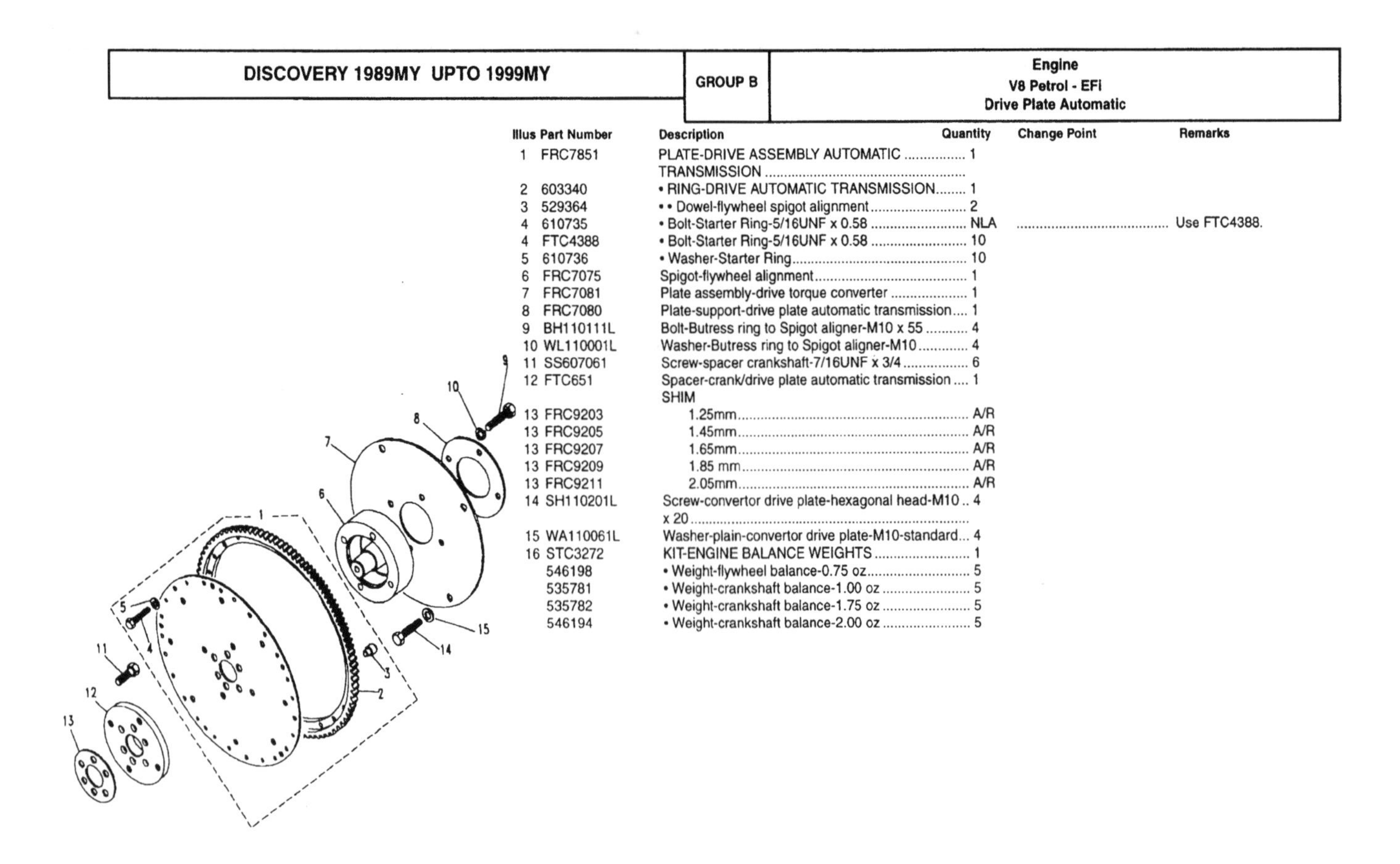

Illus	Part Number	Description	Quantity	Change Point	Remarks
1	FRC7851	PLATE-DRIVE ASSEMBLY AUTOMATIC TRANSMISSION	1		
2	603340	• RING-DRIVE AUTOMATIC TRANSMISSION	1		
3	529364	• • Dowel-flywheel spigot alignment	2		
4	610735	• Bolt-Starter Ring-5/16UNF x 0.58	NLA		Use FTC4388.
4	FTC4388	• Bolt-Starter Ring-5/16UNF x 0.58	10		
5	610736	• Washer-Starter Ring	10		
6	FRC7075	Spigot-flywheel alignment	1		
7	FRC7081	Plate assembly-drive torque converter	1		
8	FRC7080	Plate-support-drive plate automatic transmission	1		
9	BH110111L	Bolt-Butress ring to Spigot aligner-M10 x 55	4		
10	WL110001L	Washer-Butress ring to Spigot aligner-M10	4		
11	SS607061	Screw-spacer crankshaft-7/16UNF x 3/4	6		
12	FTC651	Spacer-crank/drive plate automatic transmission	1		
		SHIM			
13	FRC9203	1.25mm	A/R		
13	FRC9205	1.45mm	A/R		
13	FRC9207	1.65mm	A/R		
13	FRC9209	1.85 mm	A/R		
13	FRC9211	2.05mm	A/R		
14	SH110201L	Screw-convertor drive plate-hexagonal head-M10 x 20	4		
15	WA110061L	Washer-plain-convertor drive plate-M10-standard	4		
16	STC3272	KIT-ENGINE BALANCE WEIGHTS	1		
	546198	• Weight-flywheel balance-0.75 oz	5		
	535781	• Weight-crankshaft balance-1.00 oz	5		
	535782	• Weight-crankshaft balance-1.75 oz	5		
	546194	• Weight-crankshaft balance-2.00 oz	5		

B 161

Engine
2000 - Petrol - T16
Engine Stripped

Illus	Part Number	Description	Quantity	Change Point	Remarks
		GREECE UPTO KA338821			
1	YFM4509	Engine unit-stripped-new	1	To (V) KA 0338821	20T4HG63 Greece
		EUROPE FROM KA338822			
1	YFM4510	Engine unit-stripped-new	1	From (V) KA 338822	20T4HG91 except air con
1	YFM4571	Engine unit-stripped-new	1		20T4HJ58 except air con
1	YFM4591	Engine unit-stripped-new	1		20T4HK06 except air con
1	YFM4511	Engine unit-stripped-new	1		20T4HG92 air con
1	YFM4572	Engine unit-stripped-new	1		20T4HJ59 air con
1	YFM4590	Engine unit-stripped-new	1		20T4HK07 air con
1	STC8617R	Engine unit-stripped-reconditioned	1		

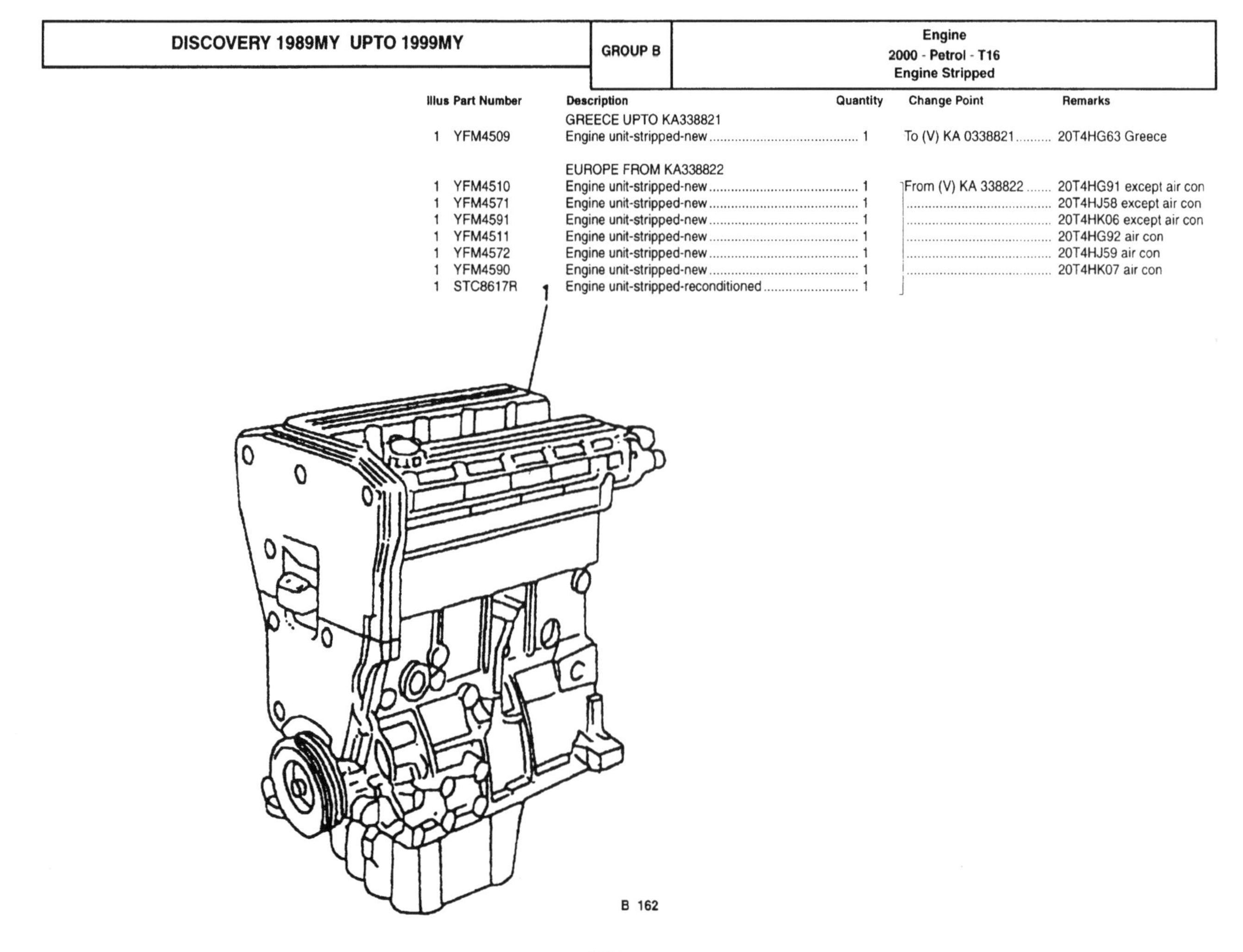

B 162

Illus	Part Number	Description	Quantity	Change Point	Remarks
1	LBB10468NL	Engine base unit-part	1		

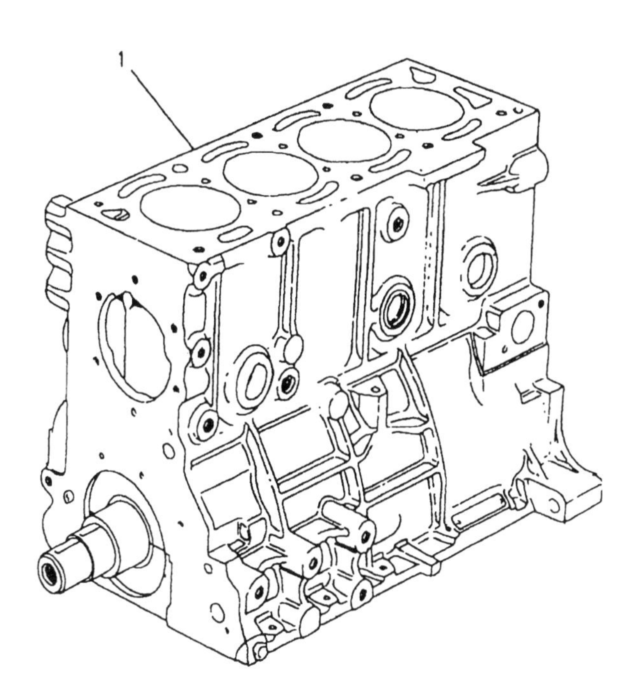

B 163

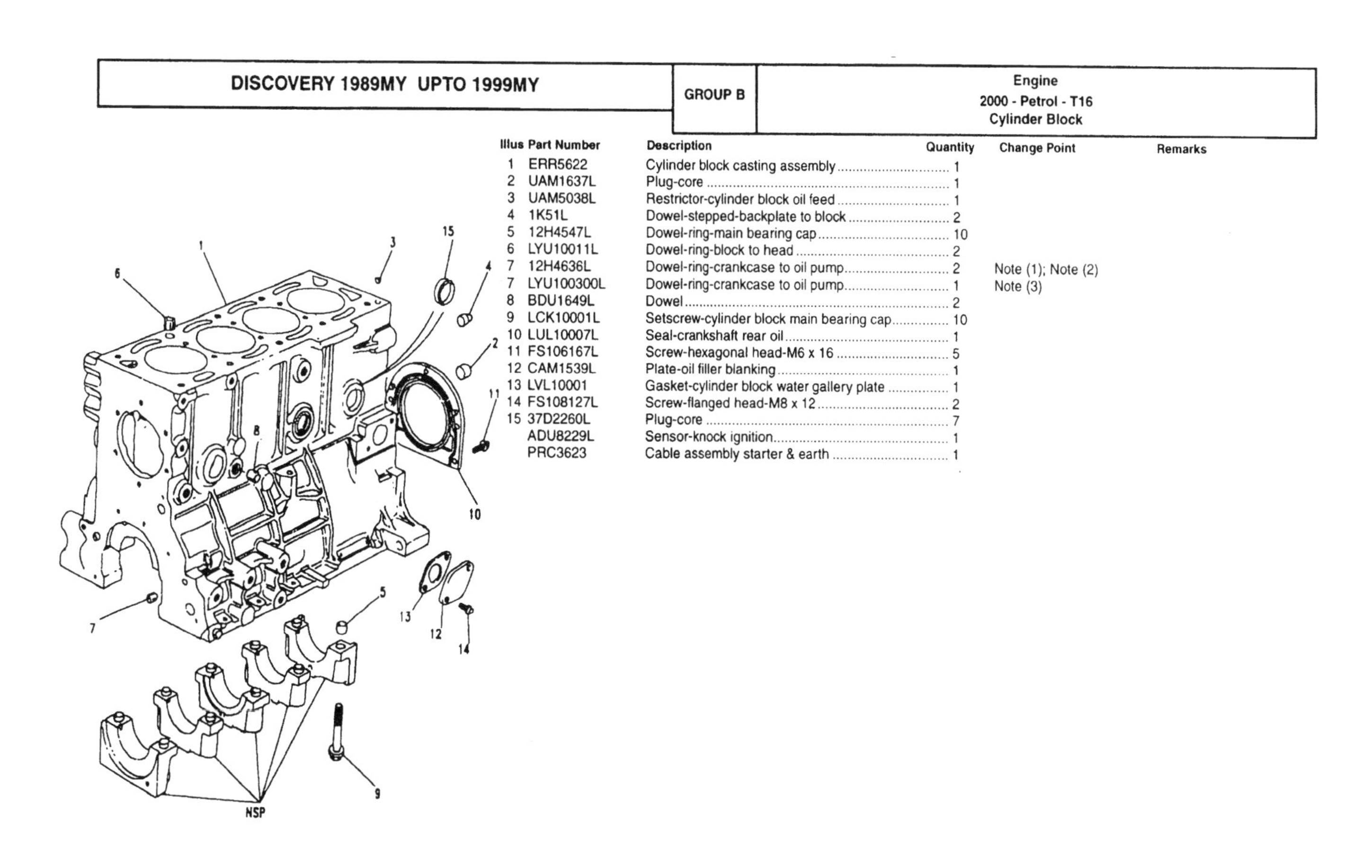

Illus	Part Number	Description	Quantity	Change Point	Remarks
1	ERR5622	Cylinder block casting assembly	1		
2	UAM1637L	Plug-core	1		
3	UAM5038L	Restrictor-cylinder block oil feed	1		
4	1K51L	Dowel-stepped-backplate to block	2		
5	12H4547L	Dowel-ring-main bearing cap	10		
6	LYU10011L	Dowel-ring-block to head	2		
7	12H4636L	Dowel-ring-crankcase to oil pump	2	Note (1); Note (2)	
7	LYU100300L	Dowel-ring-crankcase to oil pump	1	Note (3)	
8	BDU1649L	Dowel	2		
9	LCK10001L	Setscrew-cylinder block main bearing cap	10		
10	LUL10007L	Seal-crankshaft rear oil	1		
11	FS106167L	Screw-hexagonal head-M6 x 16	5		
12	CAM1539L	Plate-oil filler blanking	1		
13	LVL10001	Gasket-cylinder block water gallery plate	1		
14	FS108127L	Screw-flanged head-M8 x 12	2		
15	37D2260L	Plug-core	7		
	ADU8229L	Sensor-knock ignition	1		
	PRC3623	Cable assembly starter & earth	1		

CHANGE POINTS:
(1) To (V) MA 163103
(2) From (V) MA 500000 To (V) MA 501919
(3) From (V) TA 163104; From (V) TA 501920

B 164

Illus	Part Number	Description	Quantity	Change Point	Remarks
1	LFT10013L	CRANKSHAFT-ENGINE	1	Note (1)	
1	LEF10078	CRANKSHAFT-ENGINE	1	Note (2)	
	AHU1026L	• Bearing-needle roller caged-rear	1		Part of LFT10013L.
	22G2076L	• Bush	1		
2	BHM1153L	• Set-crankshaft bearings-standard.	1		
3	BHM1505L	• Set-connecting rod big end half bearing	1		
4	LFT10014L	• Washer-crankshaft thrust	1		Part of LFT10013L.
5	LEJ10002L	• Dowel-flywheel to crankshaft	1		
	DAM3653L	• Bearing-needle roller caged-rear	1		Part of LEF10078.
4	LEG10003	• Washer-crankshaft thrust-upper	1		Part of LEF10078.
4	LEG10004	• Washer-crankshaft thrust-lower	1		Part of LEF10078.
6	BDU1055L	Key-crankshaft	1		
7	LHG10036	Pulley assembly-crankshaft	1		
8	UAM7987L	Gear-crankshaft timing drive	1		
9	FB106105L	Bolt-flanged head-M6 x 50	4		
10	LYG10046	Bolt-crankshaft pulley	1		

CHANGE POINTS:
(1) To (V) KA 038821
(2) From (V) KA 038821

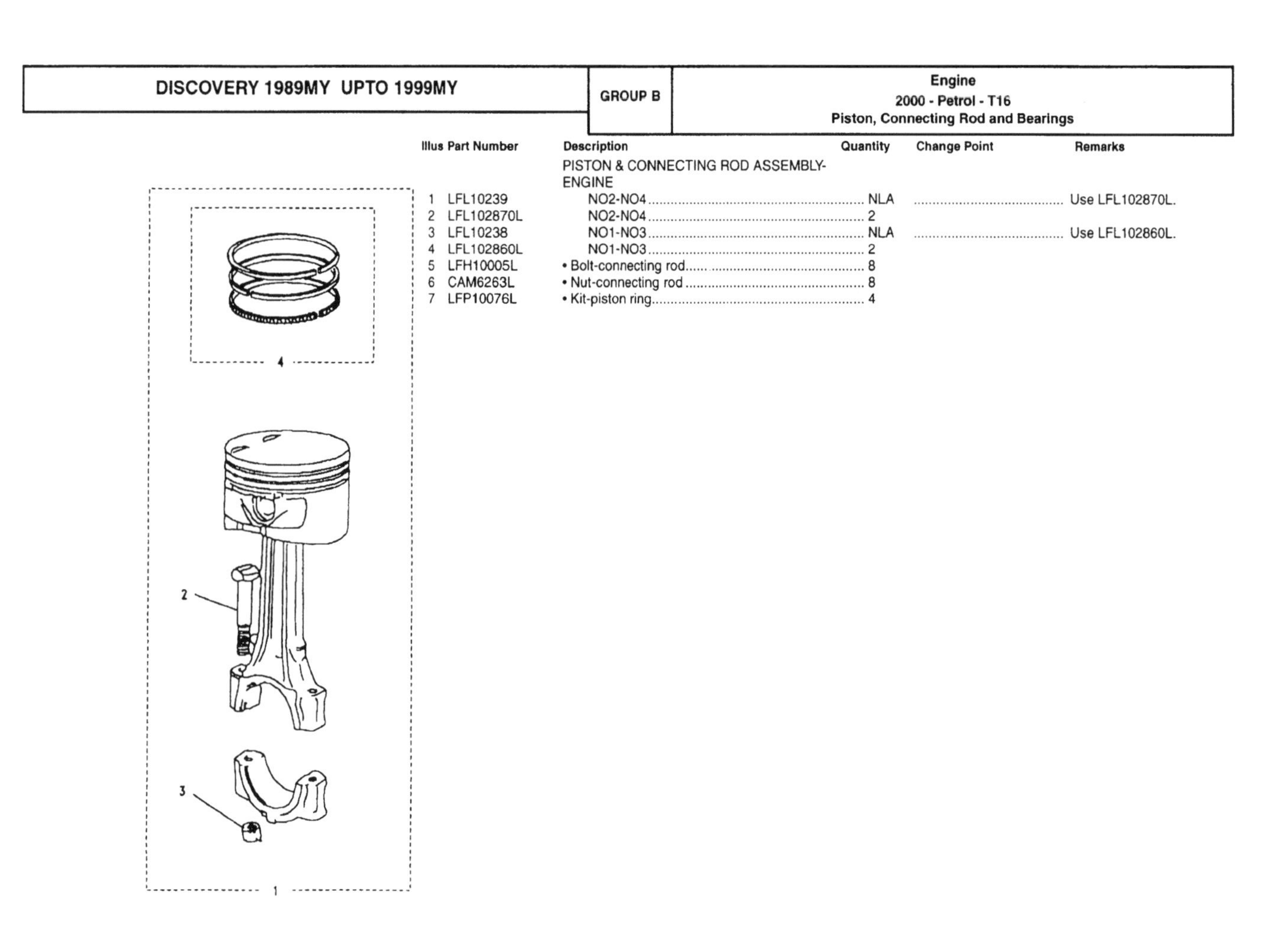

Illus	Part Number	Description	Quantity	Change Point	Remarks
		PISTON & CONNECTING ROD ASSEMBLY-ENGINE			
1	LFL10239	NO2-NO4	NLA		Use LFL102870L.
2	LFL102870L	NO2-NO4	2		
3	LFL10238	NO1-NO3	NLA		Use LFL102860L.
4	LFL102860L	NO1-NO3	2		
5	LFH10005L	• Bolt-connecting rod	8		
6	CAM6263L	• Nut-connecting rod	8		
7	LFP10076L	• Kit-piston ring	4		

Illus	Part Number	Description	Quantity	Change Point	Remarks
1	LSP10037	Pipe-engine oil suction	1		
2	CDU1344L	O ring-oil suction pipe seal	1		
3	FS106181L	Screw-M6 x 18	2		
4	FS106125L	Screw-flanged head-M6 x 12	1		
5	LSB10076	SUMP ASSEMBLY-ENGINE OIL	1		
6	UAM2957L	• Plug-sump assembly oil drain	1		
7	UAM2857L	• Washer-sealing-sump drain plug	1		
8	LVF10005	Gasket-crankcase oil sump	1		
9	FS106125L	Screw-flanged head-M6 x 12	17		
10	FS106255L	Screw-flanged head-M6 x 25	1		
11	LQM10026	DIPSTICK-OIL	1		
12	CDU4083	• O ring	NSS		
13	LQN10033L	Tube assembly-oil dipstick	1		
14	FS106125L	Screw-flanged head-M6 x 12	1		

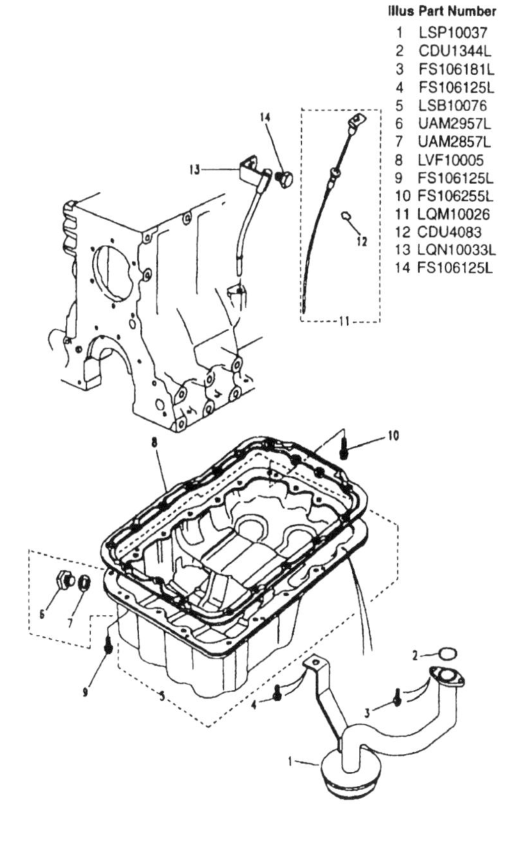

B 167

Illus	Part Number	Description	Quantity	Change Point	Remarks
1	LJR10163L	Cover-upper rear timing belt	1	Note (1)	
1	LJR102660	Cover-upper rear timing belt	1	Note (2)	
2	LJR10159L	Cover-lower rear timing belt	1		
3	FS106207L	Screw-flanged head-M6 x 20	9		
4	LJR10161L	Cover-upper front timing belt	1	Note (1)	
4	LJR102640	Cover-upper front timing belt	1	Note (2)	
5	LJR10169L	Cover-lower front timing belt	1	Note (1)	
5	LJR102670	Cover-lower front timing belt	1	Note (2)	
6	FS106207L	Screw-flanged head-M6 x 20	10		
7	LJR10165L	Guard-crankshaft pulley-front	1		
8	FB106065L	Bolt-M6 x 30	3		

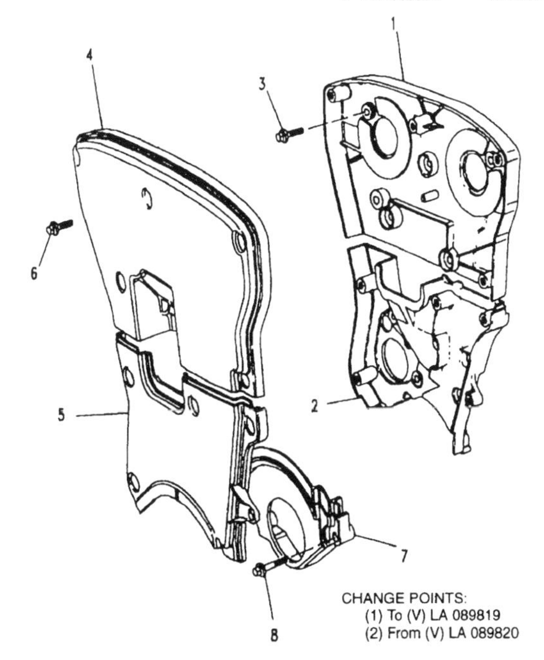

CHANGE POINTS:
(1) To (V) LA 089819
(2) From (V) LA 089820

B 168

DISCOVERY 1989MY UPTO 1999MY	GROUP B	Engine 2000 - Petrol - T16 Timing Belt and Tensioner

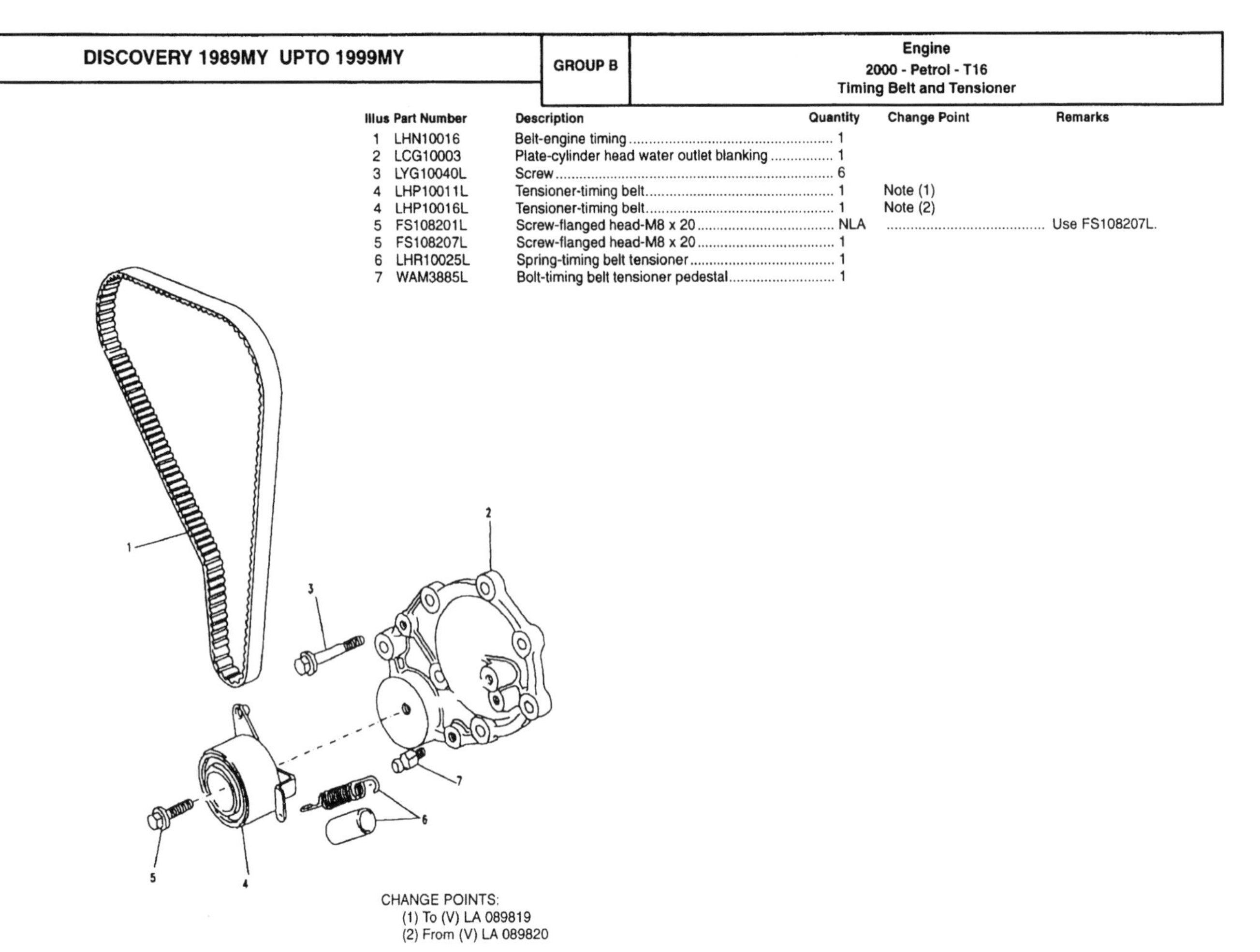

Illus	Part Number	Description	Quantity	Change Point	Remarks
1	LHN10016	Belt-engine timing	1		
2	LCG10003	Plate-cylinder head water outlet blanking	1		
3	LYG10040L	Screw	6		
4	LHP10011L	Tensioner-timing belt	1	Note (1)	
4	LHP10016L	Tensioner-timing belt	1	Note (2)	
5	FS108201L	Screw-flanged head-M8 x 20	NLA		Use FS108207L.
5	FS108207L	Screw-flanged head-M8 x 20	1		
6	LHR10025L	Spring-timing belt tensioner	1		
7	WAM3885L	Bolt-timing belt tensioner pedestal	1		

CHANGE POINTS:
(1) To (V) LA 089819
(2) From (V) LA 089820

B 169

DISCOVERY 1989MY UPTO 1999MY	GROUP B	Engine 2000 - Petrol - T16 Water Pump and Bracket

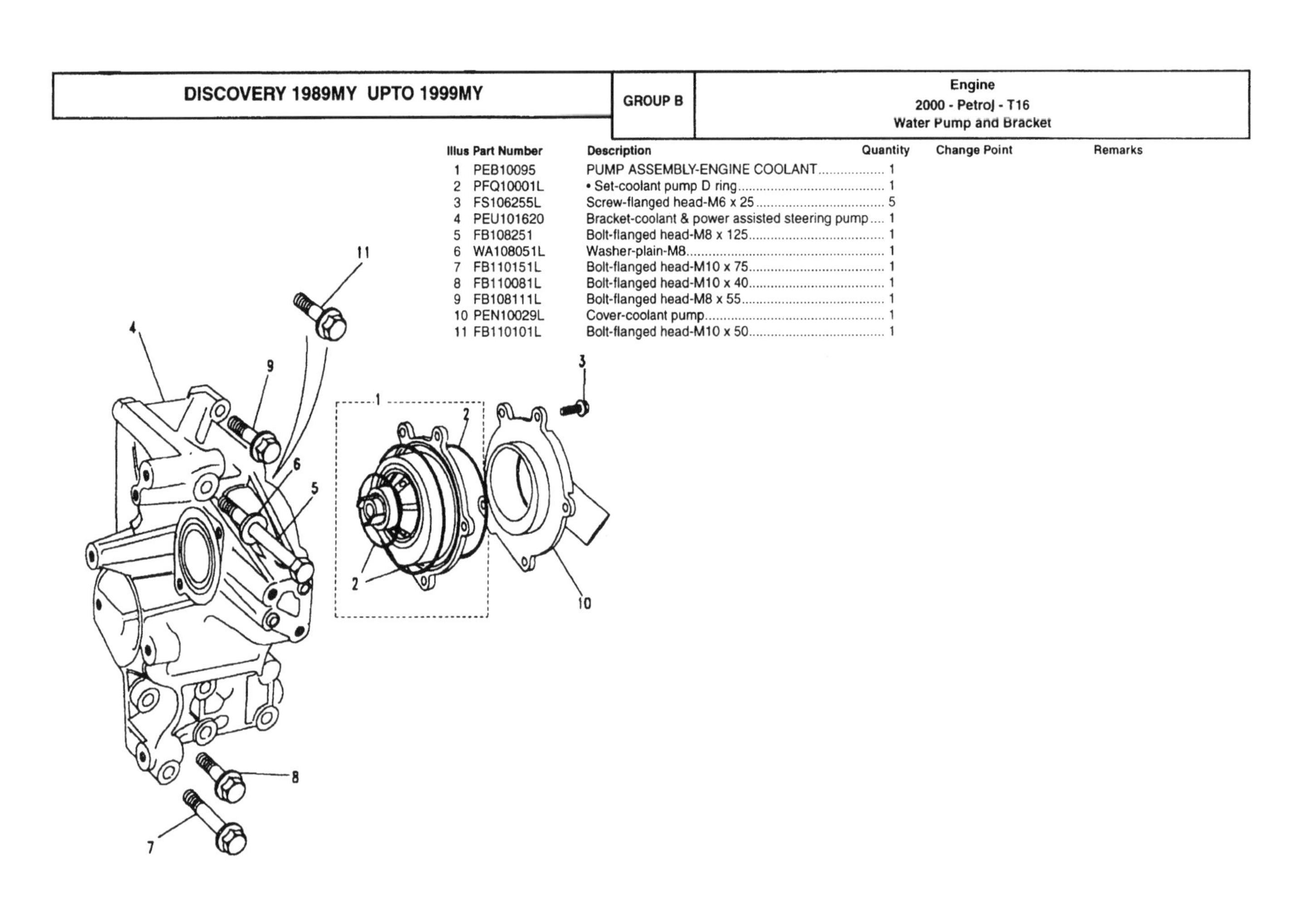

Illus	Part Number	Description	Quantity	Change Point	Remarks
1	PEB10095	PUMP ASSEMBLY-ENGINE COOLANT	1		
2	PFQ10001L	• Set-coolant pump D ring	1		
3	FS106255L	Screw-flanged head-M6 x 25	5		
4	PEU101620	Bracket-coolant & power assisted steering pump	1		
5	FB108251	Bolt-flanged head-M8 x 125	1		
6	WA108051L	Washer-plain-M8	1		
7	FB110151L	Bolt-flanged head-M10 x 75	1		
8	FB110081L	Bolt-flanged head-M10 x 40	1		
9	FB108111L	Bolt-flanged head-M8 x 55	1		
10	PEN10029L	Cover-coolant pump	1		
11	FB110101L	Bolt-flanged head-M10 x 50	1		

B 170

DISCOVERY 1989MY UPTO 1999MY

GROUP B

Engine
2000 - Petrol - T16
Oil Pump and Filter

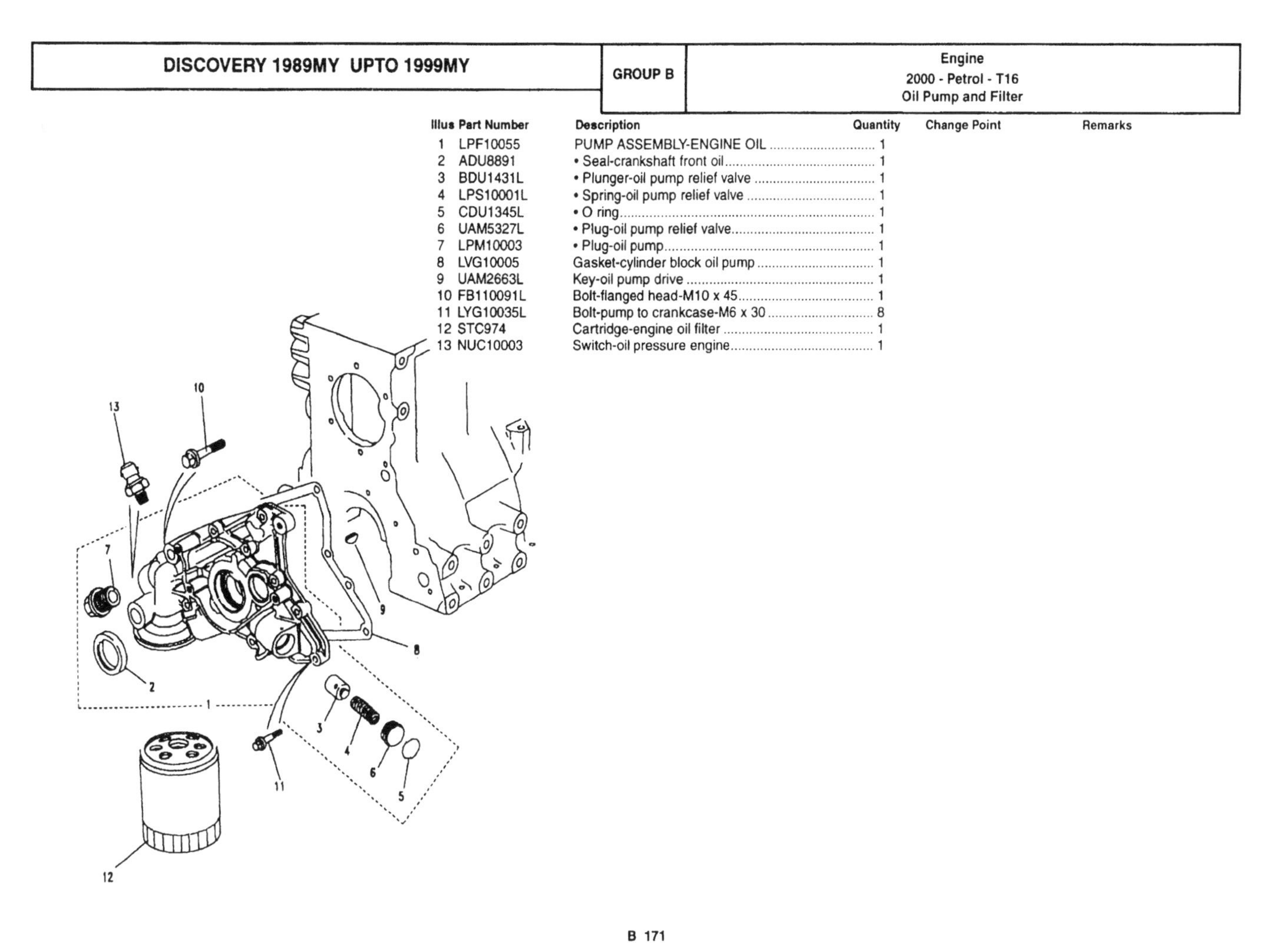

Illus	Part Number	Description	Quantity	Change Point	Remarks
1	LPF10055	PUMP ASSEMBLY-ENGINE OIL	1		
2	ADU8891	• Seal-crankshaft front oil	1		
3	BDU1431L	• Plunger-oil pump relief valve	1		
4	LPS10001L	• Spring-oil pump relief valve	1		
5	CDU1345L	• O ring	1		
6	UAM5327L	• Plug-oil pump relief valve	1		
7	LPM10003	• Plug-oil pump	1		
8	LVG10005	Gasket-cylinder block oil pump	1		
9	UAM2663L	Key-oil pump drive	1		
10	FB110091L	Bolt-flanged head-M10 x 45	1		
11	LYG10035L	Bolt-pump to crankcase-M6 x 30	8		
12	STC974	Cartridge-engine oil filter	1		
13	NUC10003	Switch-oil pressure engine	1		

DISCOVERY 1989MY UPTO 1999MY

GROUP B

Engine
2000 - Petrol - T16
Cylinder Head

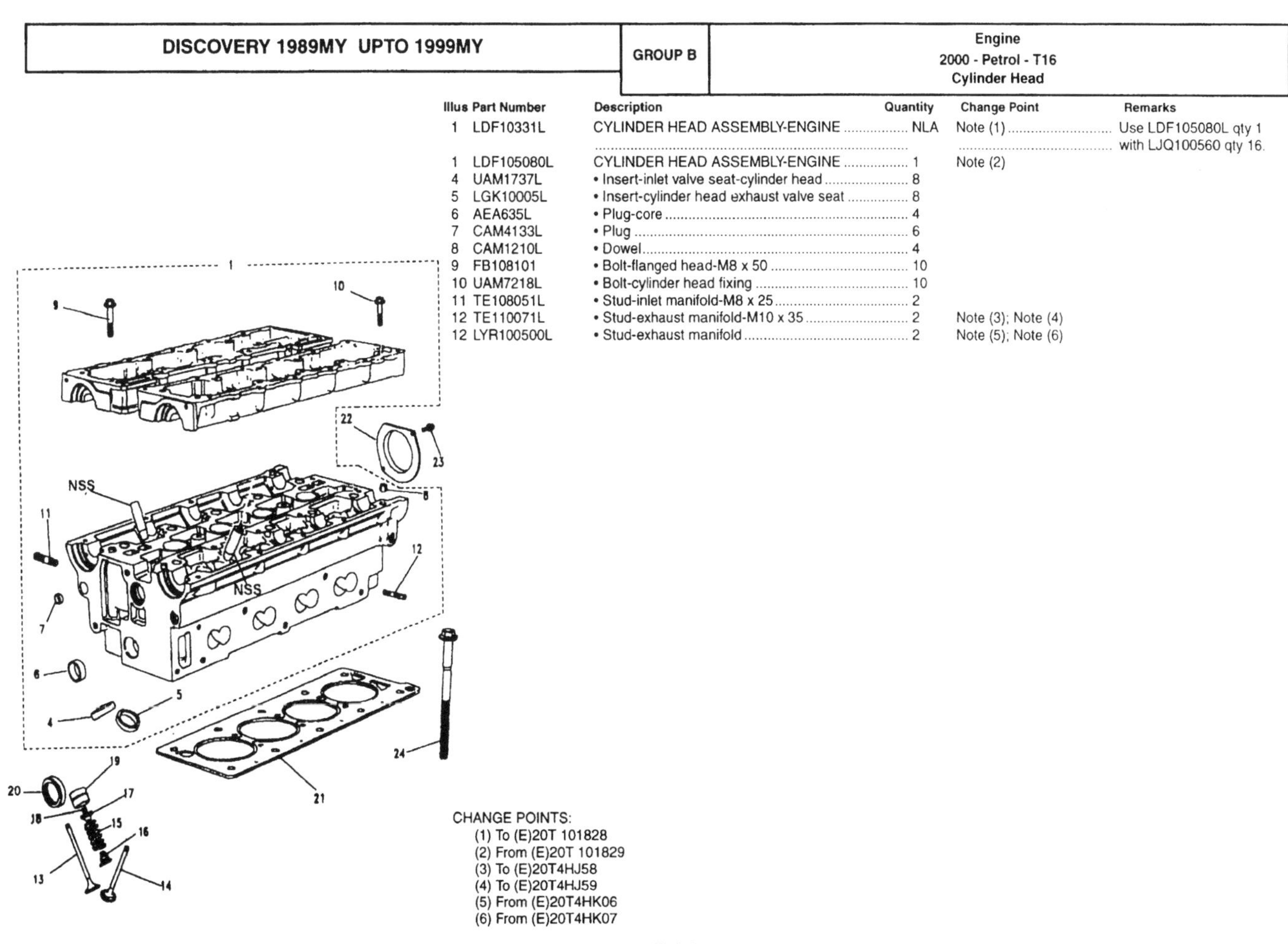

Illus	Part Number	Description	Quantity	Change Point	Remarks
1	LDF10331L	CYLINDER HEAD ASSEMBLY-ENGINE	NLA	Note (1)	Use LDF105080L qty 1 with LJQ100560 qty 16.
1	LDF105080L	CYLINDER HEAD ASSEMBLY-ENGINE	1	Note (2)	
4	UAM1737L	• Insert-inlet valve seat-cylinder head	8		
5	LGK10005L	• Insert-cylinder head exhaust valve seat	8		
6	AEA635L	• Plug-core	4		
7	CAM4133L	• Plug	6		
8	CAM1210L	• Dowel	4		
9	FB108101	• Bolt-flanged head-M8 x 50	10		
10	UAM7218L	• Bolt-cylinder head fixing	10		
11	TE108051L	• Stud-inlet manifold-M8 x 25	2		
12	TE110071L	• Stud-exhaust manifold-M10 x 35	2	Note (3); Note (4)	
12	LYR100500L	• Stud-exhaust manifold	2	Note (5); Note (6)	

CHANGE POINTS:
(1) To (E)20T 101828
(2) From (E)20T 101829
(3) To (E)20T4HJ58
(4) To (E)20T4HJ59
(5) From (E)20T4HK06
(6) From (E)20T4HK07

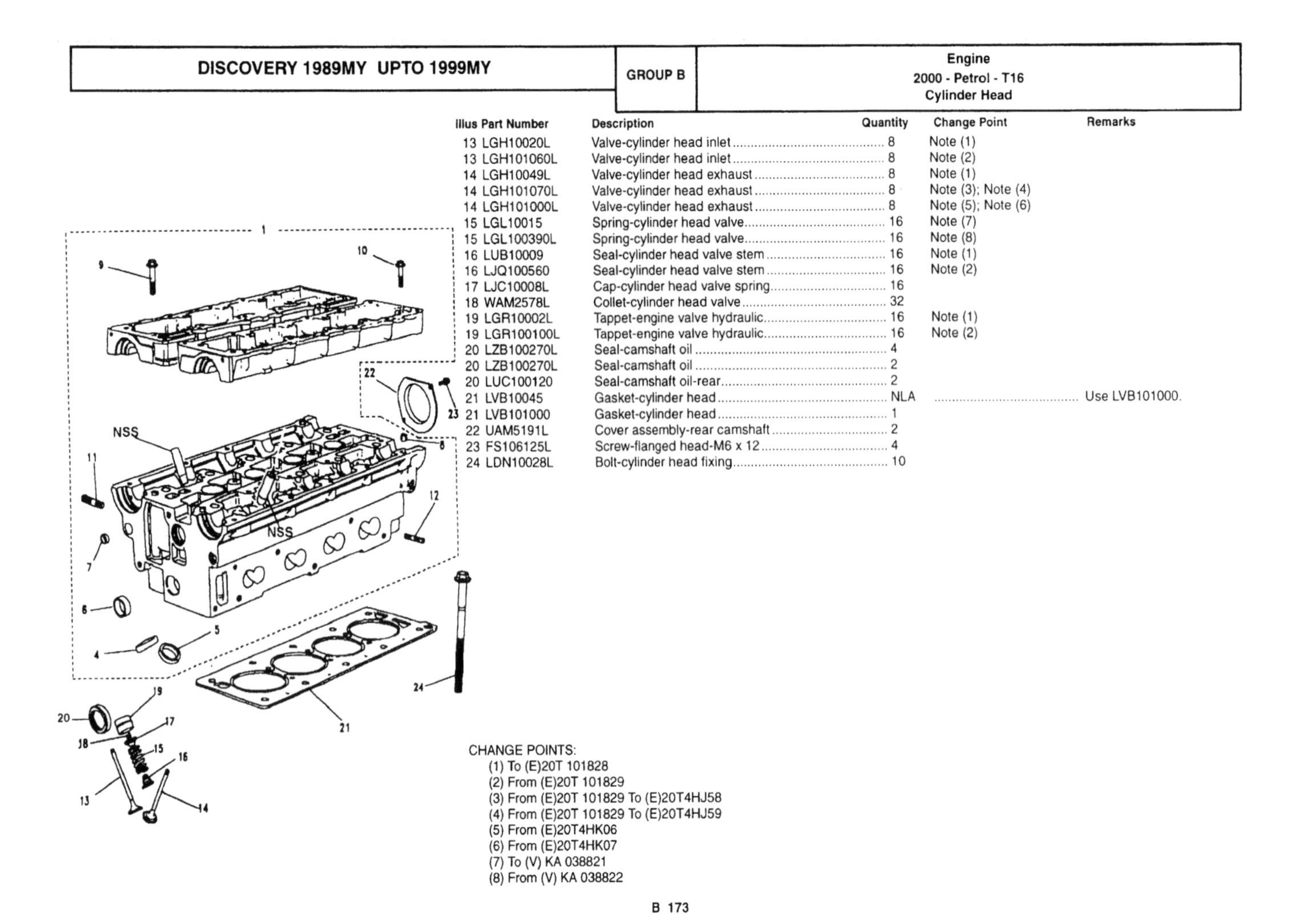

Illus	Part Number	Description	Quantity	Change Point	Remarks
13	LGH10020L	Valve-cylinder head inlet	8	Note (1)	
13	LGH101060L	Valve-cylinder head inlet	8	Note (2)	
14	LGH10049L	Valve-cylinder head exhaust	8	Note (1)	
14	LGH101070L	Valve-cylinder head exhaust	8	Note (3); Note (4)	
14	LGH101000L	Valve-cylinder head exhaust	8	Note (5); Note (6)	
15	LGL10015	Spring-cylinder head valve	16	Note (7)	
15	LGL100390L	Spring-cylinder head valve	16	Note (8)	
16	LUB10009	Seal-cylinder head valve stem	16	Note (1)	
16	LJQ100560	Seal-cylinder head valve stem	16	Note (2)	
17	LJC10008L	Cap-cylinder head valve spring	16		
18	WAM2578L	Collet-cylinder head valve	32		
19	LGR10002L	Tappet-engine valve hydraulic	16	Note (1)	
19	LGR100100L	Tappet-engine valve hydraulic	16	Note (2)	
20	LZB100270L	Seal-camshaft oil	4		
20	LZB100270L	Seal-camshaft oil	2		
20	LUC100120	Seal-camshaft oil-rear	2		
21	LVB10045	Gasket-cylinder head	NLA		Use LVB101000.
21	LVB101000	Gasket-cylinder head	1		
22	UAM5191L	Cover assembly-rear camshaft	2		
23	FS106125L	Screw-flanged head-M6 x 12	4		
24	LDN10028L	Bolt-cylinder head fixing	10		

CHANGE POINTS:
(1) To (E)20T 101828
(2) From (E)20T 101829
(3) From (E)20T 101829 To (E)20T4HJ58
(4) From (E)20T 101829 To (E)20T4HJ59
(5) From (E)20T4HK06
(6) From (E)20T4HK07
(7) To (V) KA 038821
(8) From (V) KA 038822

B 173

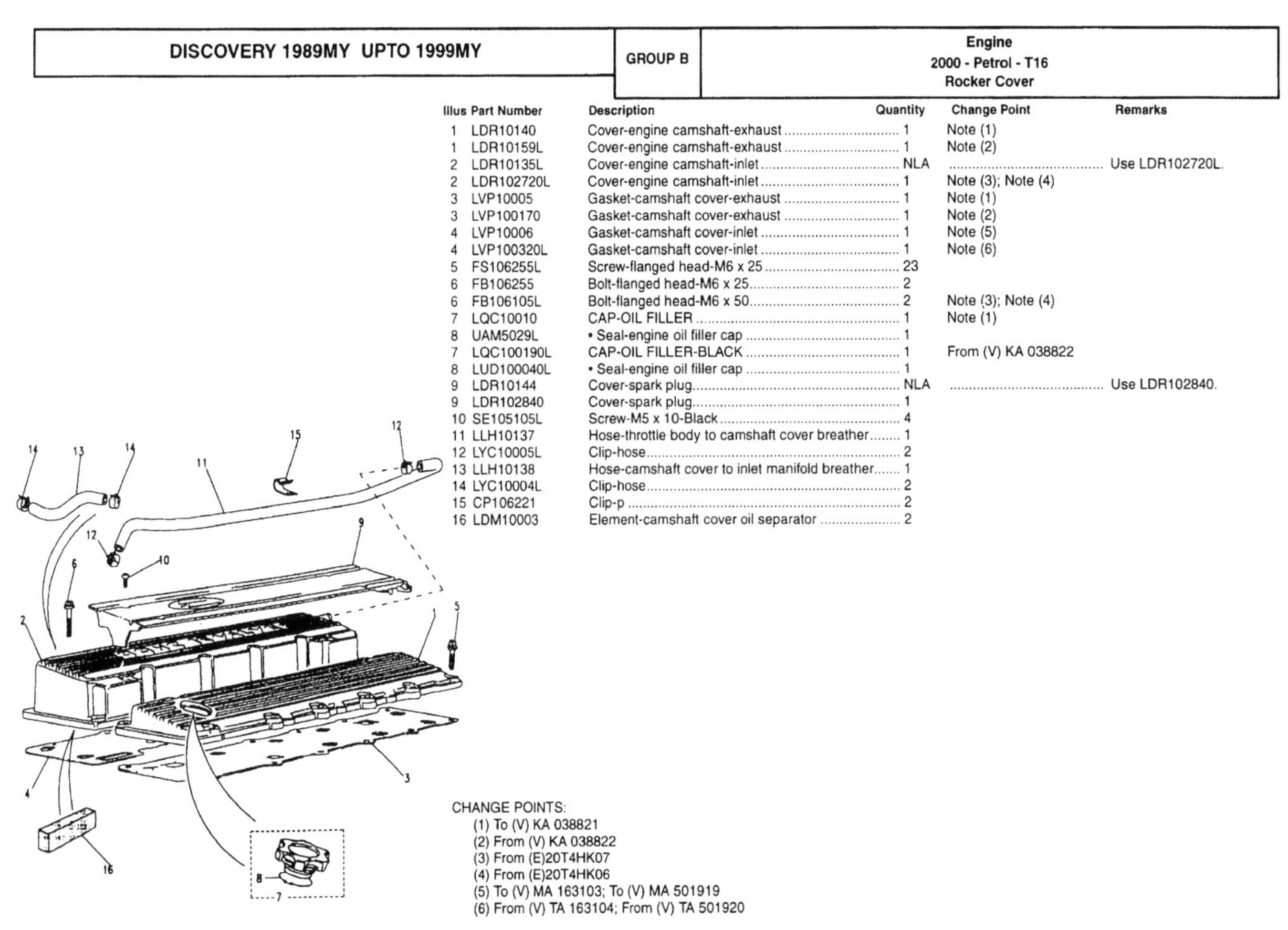

Illus	Part Number	Description	Quantity	Change Point	Remarks
1	LDR10140	Cover-engine camshaft-exhaust	1	Note (1)	
1	LDR10159L	Cover-engine camshaft-exhaust	1	Note (2)	
2	LDR10135L	Cover-engine camshaft-inlet	NLA		Use LDR102720L.
2	LDR102720L	Cover-engine camshaft-inlet	1	Note (3); Note (4)	
3	LVP10005	Gasket-camshaft cover-exhaust	1	Note (1)	
3	LVP100170	Gasket-camshaft cover-exhaust	1	Note (2)	
4	LVP10006	Gasket-camshaft cover-inlet	1	Note (5)	
4	LVP100320L	Gasket-camshaft cover-inlet	1	Note (6)	
5	FS106255L	Screw-flanged head-M6 x 25	23		
6	FB106255	Bolt-flanged head-M6 x 25	2		
6	FB106105L	Bolt-flanged head-M6 x 50	2	Note (3); Note (4)	
7	LQC10010	CAP-OIL FILLER	1	Note (1)	
8	UAM5029L	• Seal-engine oil filler cap	1		
7	LQC100190L	CAP-OIL FILLER-BLACK	1	From (V) KA 038822	
8	LUD100040L	• Seal-engine oil filler cap	1		
9	LDR10144	Cover-spark plug	NLA		Use LDR102840.
9	LDR102840	Cover-spark plug	1		
10	SE105105L	Screw-M5 x 10-Black	4		
11	LLH10137	Hose-throttle body to camshaft cover breather	1		
12	LYC10005L	Clip-hose	2		
13	LLH10138	Hose-camshaft cover to inlet manifold breather	1		
14	LYC10004L	Clip-hose	2		
15	CP106221	Clip-p	2		
16	LDM10003	Element-camshaft cover oil separator	2		

CHANGE POINTS:
(1) To (V) KA 038821
(2) From (V) KA 038822
(3) From (E)20T4HK07
(4) From (E)20T4HK06
(5) To (V) MA 163103; To (V) MA 501919
(6) From (V) TA 163104; From (V) TA 501920

B 174

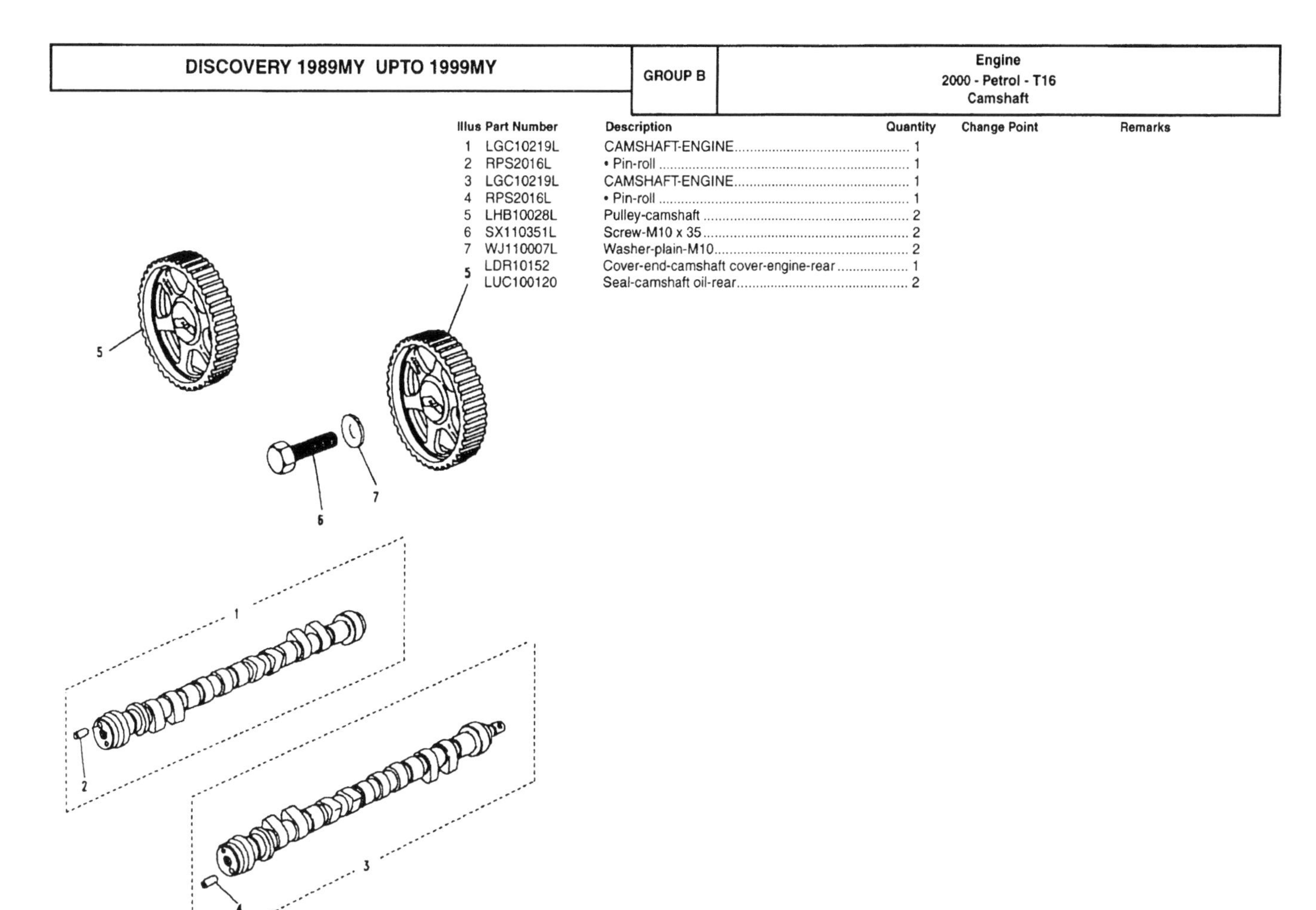

Illus	Part Number	Description	Quantity	Change Point	Remarks
1	LGC10219L	CAMSHAFT-ENGINE	1		
2	RPS2016L	• Pin-roll	1		
3	LGC10219L	CAMSHAFT-ENGINE	1		
4	RPS2016L	• Pin-roll	1		
5	LHB10028L	Pulley-camshaft	2		
6	SX110351L	Screw-M10 x 35	2		
7	WJ110007L	Washer-plain-M10	2		
	LDR10152	Cover-end-camshaft cover-engine-rear	1		
	LUC100120	Seal-camshaft oil-rear	2		

B 175

DISCOVERY 1989MY UPTO 1999MY | GROUP B | Engine 2000 - Petrol - T16 Thermostat Housing

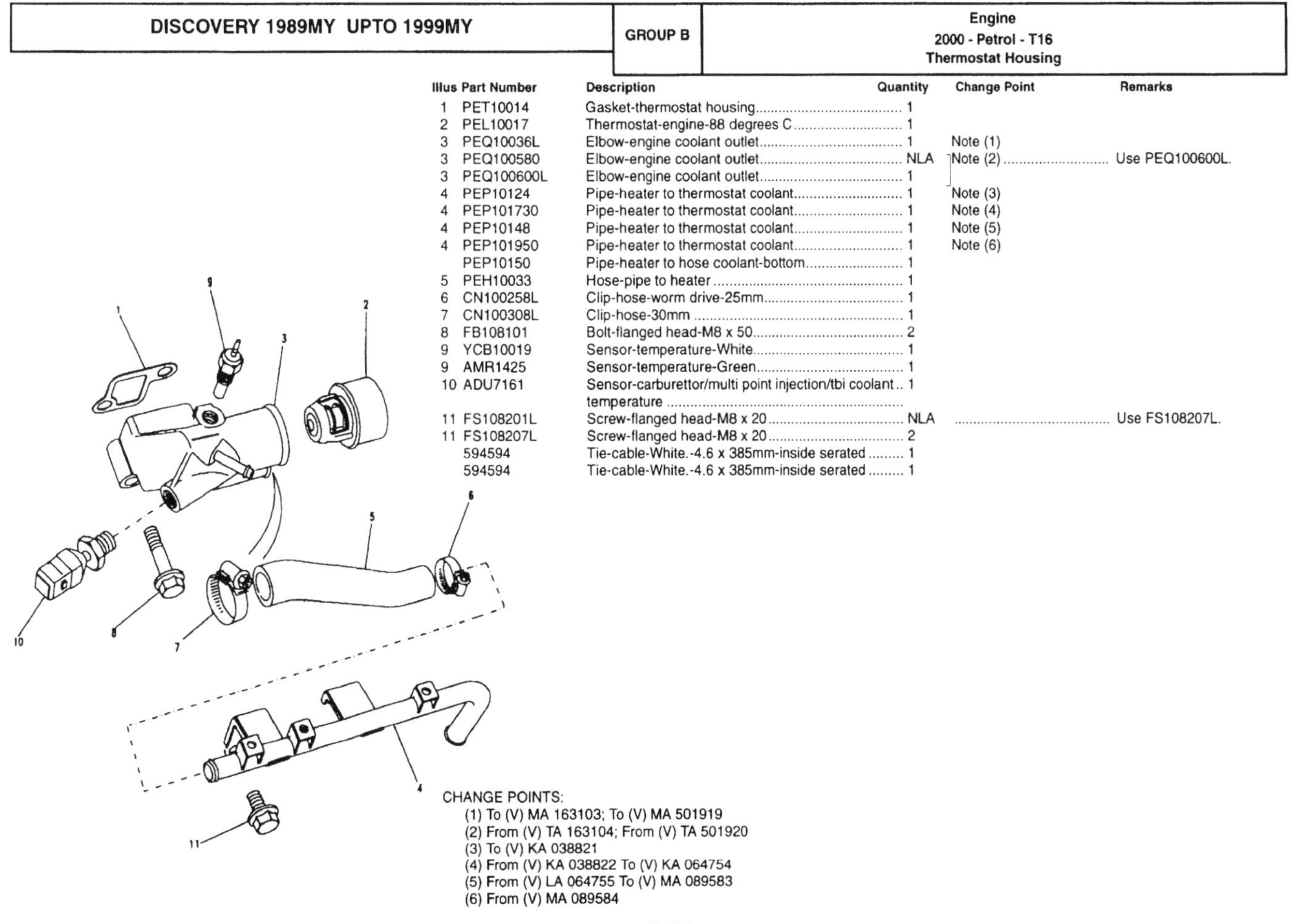

Illus	Part Number	Description	Quantity	Change Point	Remarks
1	PET10014	Gasket-thermostat housing	1		
2	PEL10017	Thermostat-engine-88 degrees C	1		
3	PEQ10036L	Elbow-engine coolant outlet	1	Note (1)	
3	PEQ100580	Elbow-engine coolant outlet	NLA	Note (2)	Use PEQ100600L.
3	PEQ100600L	Elbow-engine coolant outlet	1		
4	PEP10124	Pipe-heater to thermostat coolant	1	Note (3)	
4	PEP101730	Pipe-heater to thermostat coolant	1	Note (4)	
4	PEP10148	Pipe-heater to thermostat coolant	1	Note (5)	
4	PEP101950	Pipe-heater to thermostat coolant	1	Note (6)	
	PEP10150	Pipe-heater to hose coolant-bottom	1		
5	PEH10033	Hose-pipe to heater	1		
6	CN100258L	Clip-hose-worm drive-25mm	1		
7	CN100308L	Clip-hose-30mm	1		
8	FB108101	Bolt-flanged head-M8 x 50	2		
9	YCB10019	Sensor-temperature-White	1		
9	AMR1425	Sensor-temperature-Green	1		
10	ADU7161	Sensor-carburettor/multi point injection/tbi coolant temperature	1		
11	FS108201L	Screw-flanged head-M8 x 20	NLA		Use FS108207L.
11	FS108207L	Screw-flanged head-M8 x 20	2		
	594594	Tie-cable-White.-4.6 x 385mm-inside serated	1		
	594594	Tie-cable-White.-4.6 x 385mm-inside serated	1		

CHANGE POINTS:

(1) To (V) MA 163103; To (V) MA 501919
(2) From (V) TA 163104; From (V) TA 501920
(3) To (V) KA 038821
(4) From (V) KA 038822 To (V) KA 064754
(5) From (V) LA 064755 To (V) MA 089583
(6) From (V) MA 089584

B 176

Illus	Part Number	Description	Quantity	Change Point	Remarks
1	LKB105090	Manifold-engine upper inlet	1	Note (1)	
1	LKB105620	Manifold-engine upper inlet	1	Note (3)	
1	LKB107080	Manifold-engine upper inlet	1	Note (2)	
2	LKB10473L	Manifold-engine lower inlet	1		
3	LZN10001L	Insert-screw in-sensor to inlet manifold	1		
4	LKJ10030	Gasket-upper inlet manifold to lower inlet manifold	1		
5	FS106255L	Screw-flanged head-M6 x 25	2		
6	LKJ10027	Gasket-inlet manifold	1		
7	FS108307L	Screw-flanged head-M8 x 30	7		
8	FN108041L	Nut-flanged head-M8 x 20	NLA		Use FN108047L.
9	LKU10034L	Bracket-inlet manifold support	2		
10	FS106125L	Screw-flanged head-M6 x 12	2	Note (3)	
10	FS106167L	Screw-hexagonal head-M6 x 16	2	Note (2)	
	NNK10001L	Sensor-air inlet temperature engine-electric control unit	1		

CHANGE POINTS:
(1) To (V) KA 038821
(2) From (V) TA 163104; From (V) TA 501920
(3) To (V) MA 163103; To (V) MA 501919

B 177

Illus	Part Number	Description	Quantity	Change Point	Remarks
1	LKC10066	MANIFOLD-ENGINE EXHAUST	NLA		Use LKC10060 qty 1 with TE110051 qty 1.
1	LKC10060L	MANIFOLD-ENGINE EXHAUST	1		
5	TE110051L	• Stud-M10 x 25	4		
2	LKG10002	Gasket-exhaust manifold	1		
3	LYP10028L	Screw	3		
4	ESR2034	Nut-flange-nyloc-M10	2		
6	MHK10004L	Sensor-oxygen multi point injection	1		
	MYB100040L	Plug-blanking.	1		
	MYF100730L	Washer-copper	1		

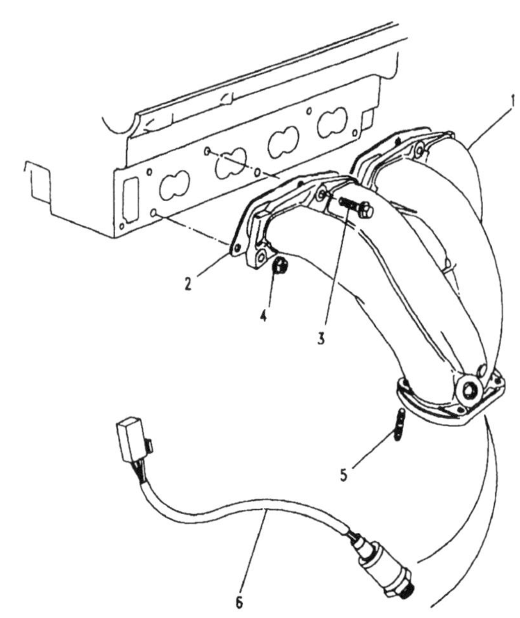

B 178

Illus	Part Number	Description	Quantity	Change Point	Remarks
1	MJN10026L	Rail-multi point injection fuel distribution	1		
2	FB106071	Bolt-flanged head-rail to inlet manifold-M6 x 35	2		
3	MHU10068	Bracket-mounting fuel lines-front	1		
4	FS106101L	Screw-flanged head-M6 x 10	2		
5	MYP10013L	Screw-bracket to manifold	2		
6	MKW10011L	Regulator-multi point injection fuel pressure	1		
7	MHU10069	Bracket-mounting fuel lines-rear	1		
8	FS106101L	Screw-flanged head-M6 x 10	2		
9	MYP10013L	Screw-regulator to rail	2		
10	MHH10020L	Hose-regulator/filter multi point injection/tbi	1		
11	MJY10029L	INJECTOR-FUEL MULTI POINT INJECTION	4		
12	BAU5325L	• O ring-petrol injector-small	NLA		Use RTC5679.
12	RTC5679	• O ring-petrol injector-small	8		
13	MHN10004L	Clip-fuel injector multi point injection	4		
14	MEK10002L	Sensor-fuel temperature multi point injection/tbi	1		
15	MYF10018L	O ring	1		
16	ESR1972	Rail-multi point injection fuel distribution	1		

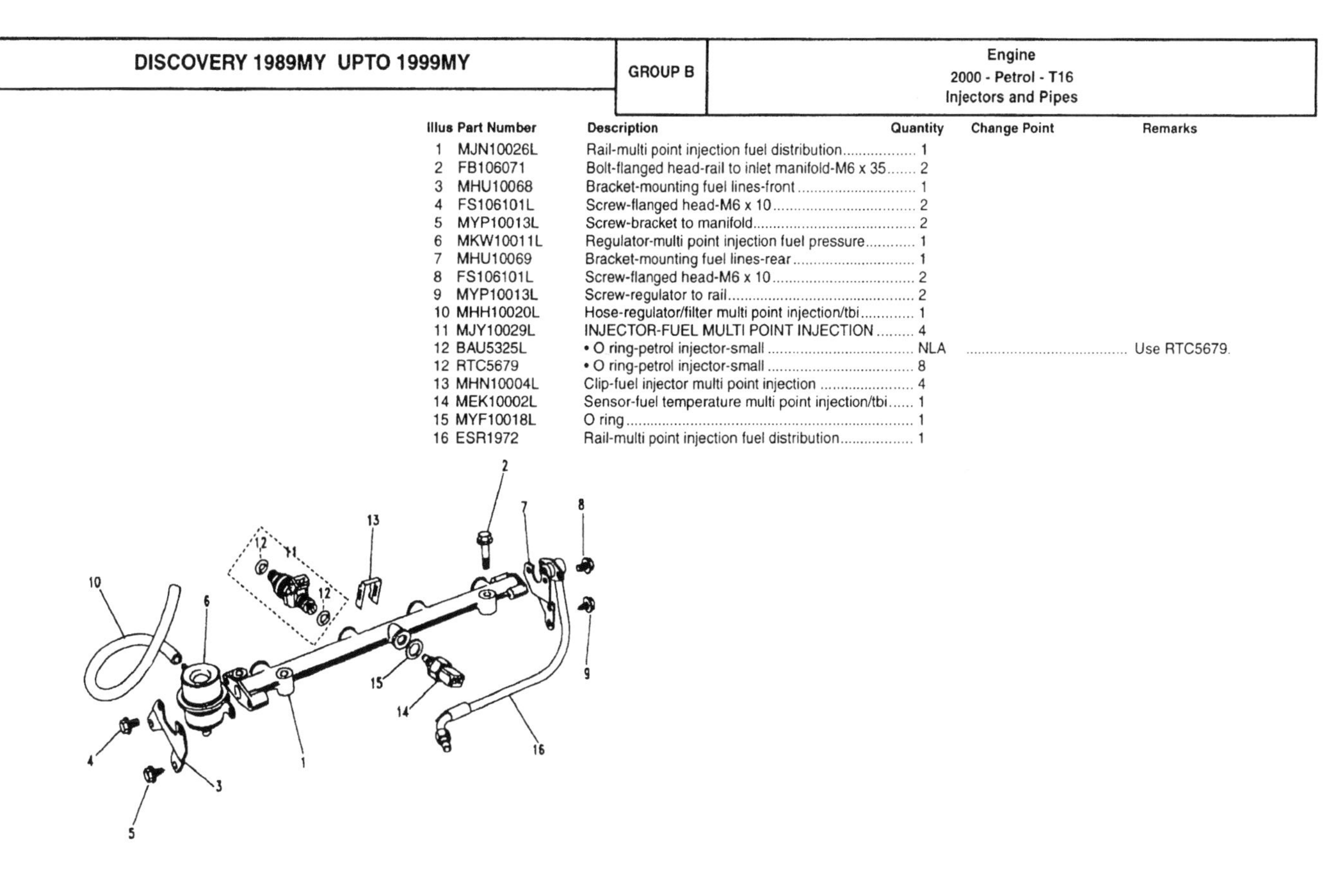

B 179

Illus	Part Number	Description	Quantity	Change Point	Remarks
1	MHB10085	BODY ASSEMBLY-MULTI POINT INJECTION THROTTLE	1	Note (1)	
1	MHB101140	BODY ASSEMBLY-MULTI POINT INJECTION THROTTLE	NLA		Use MHB101620.
1	MHB101620	BODY ASSEMBLY-MULTI POINT INJECTION THROTTLE	1	Note (2)	
	MHX10002L	• Gasket-throttle body	1		
	PEH100570	Hose-duct to throttle body induction system	1	Note (2)	
	CN100168L	Clip-hose	2	Note (2)	
2	MKG10004L	Kit-throttle potentiometer multi point injection	1		
3	FN106047L	Nut-flange-M6	4		
4	JZV1085L	Mounting-multi point injection throttle body flexible	1		
5	FS106125L	Screw-flanged head-M6 x 12	3		

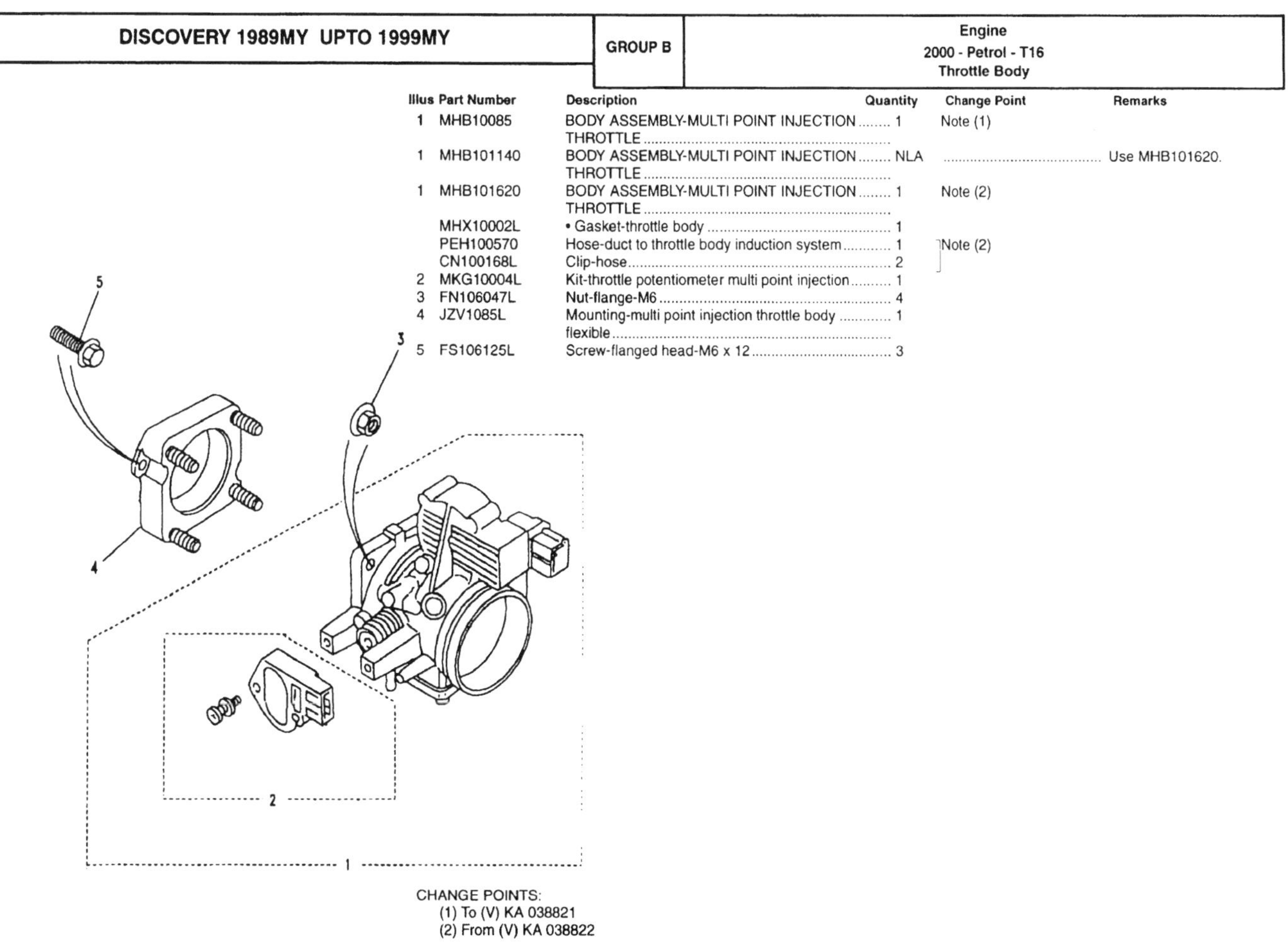

CHANGE POINTS:
(1) To (V) KA 038821
(2) From (V) KA 038822

B 180

Illus	Part Number	Description	Quantity	Change Point	Remarks
1	LVQ10040	Set-cylinder head gasket	1	Note (1)	
1	STC3160	Set-cylinder head gasket	1	Note (2)	
2	STC848	Set-engine gasket	1		

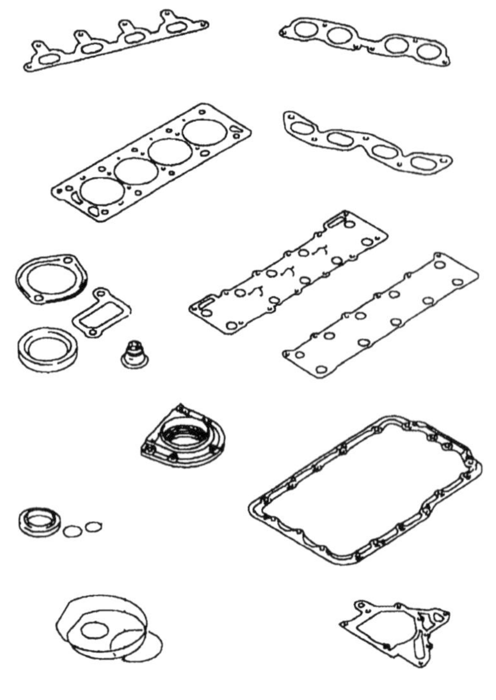

CHANGE POINTS:
(1) To (E) 101828
(2) From (E) 101829

B 181

Illus	Part Number	Description	Quantity	Change Point	Remarks
1	ANR1467	PUMP ASSEMBLY POWER ASSISTED STEERING	NLA		Use ANR4647.
1	ANR4647	PUMP ASSEMBLY POWER ASSISTED STEERING	1		
2	PQR10041L	• Pulley power assisted steering pump	1		
3	FS106125L	• Screw-flanged head-M6 x 12	3		
4	QVU10037	Bracket-mounting power assisted steering pump	1		
5	FS108201L	Screw-bracket to coolant pump-flanged head-M8 x 20	NLA		Use FS108207L.
5	FS108207L	Screw-bracket to coolant pump-flanged head-M8 x 20	2		
6	FS108161L	Screw-flanged head-M8 x 16	1		
7	FB108161L	Bolt-support bracket-flanged head-M8 x 80	2		

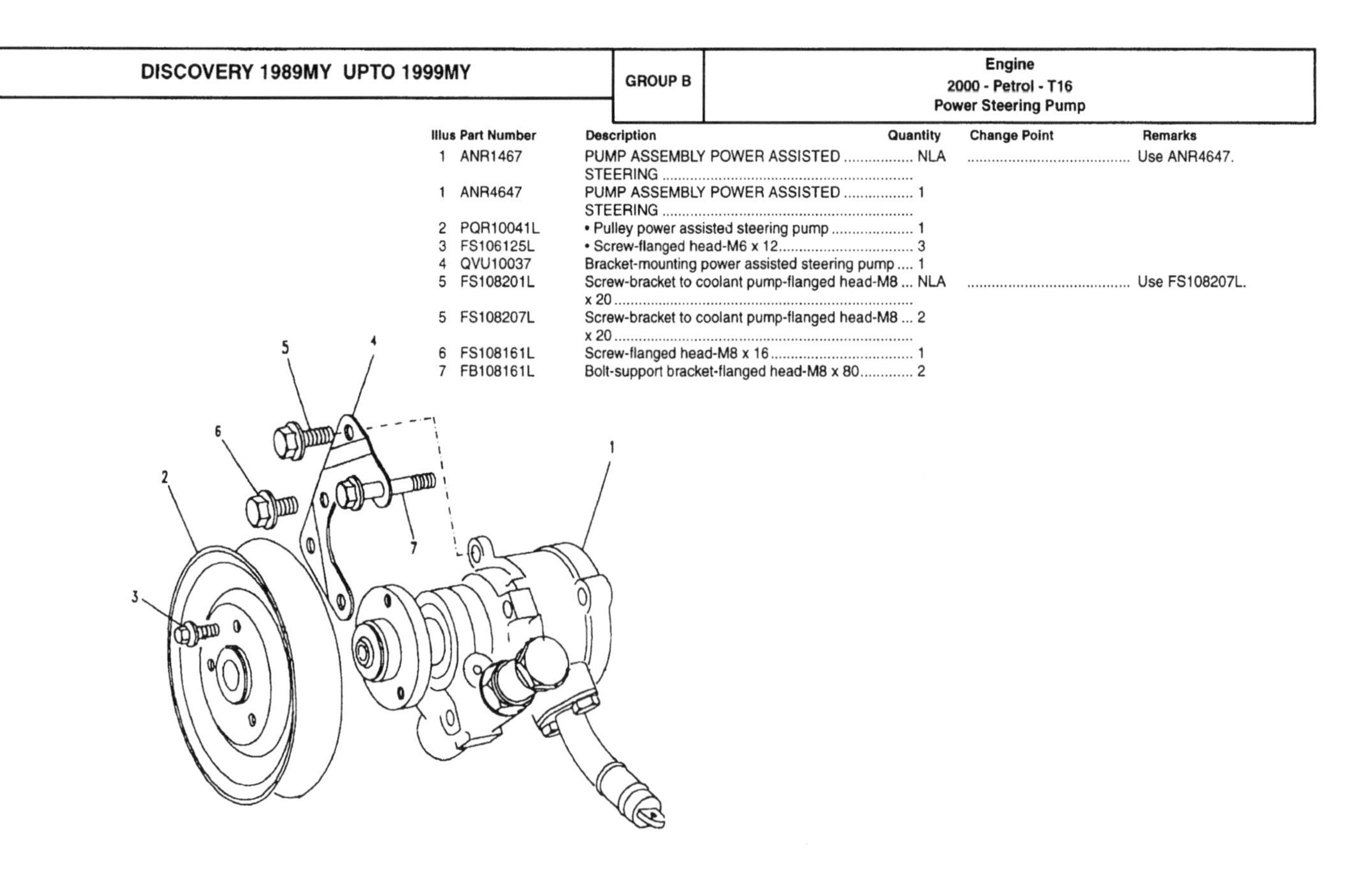

B 182

DISCOVERY 1989MY UPTO 1999MY	GROUP B	Engine 2000 - Petrol - T16 Air Conditioning Compressor

Illus	Part Number	Description	Quantity	Change Point	Remarks
1	BTR4187	Compressor air conditioning-Sanden	1	Note (1)	
1	BTR4719	Compressor air conditioning-Nippon Denso	1		
2	FB110201L	Bolt-flanged head-M10 x 100	4		

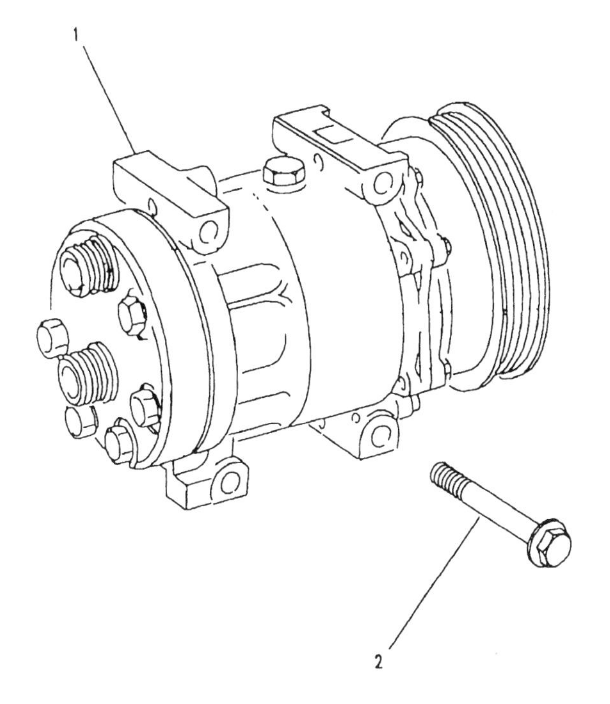

CHANGE POINTS:
(1) To (E)20T 4HG63

B 183

DISCOVERY 1989MY UPTO 1999MY	GROUP B	Engine 2000 - Petrol - T16 Compressor Bracket and Tensioner

Illus	Part Number	Description	Quantity	Change Point	Remarks
1	JPD10026	Bracket-compressor air conditioning-Sanden	1	Note (1)	
1	JPD10016	Bracket-compressor air conditioning-Nippon Denso	1		
2	PQG10008L	Tensioner-automatic ancillary drive	1		
3	PYG10006L	Bolt	1		
4	PQR10051	Pulley-idler ancillary drive	1		
5	FS112401L	Screw-flanged head-M12 x 40	1		
6	FS110301L	Screw-flanged head-M10 x 30	2		
7	SH110201L	Screw-bracket to block-hexagonal head-M10 x 20	2		
8	FB110121L	Bolt-bracket to block-M10 x 60	1		
9	FB110091L	Bolt-bracket to block-flanged head-M10 x 45	2		

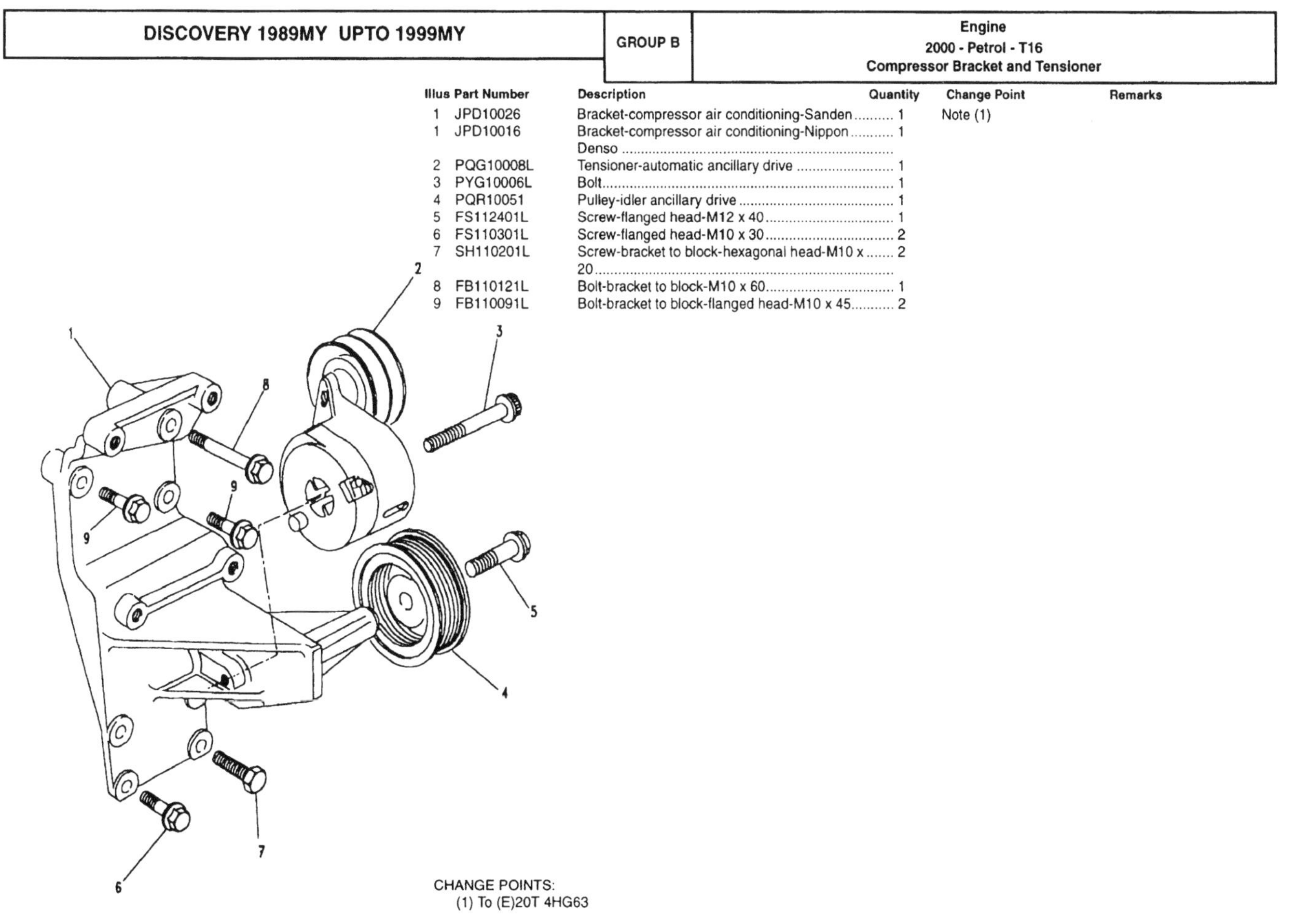

CHANGE POINTS:
(1) To (E)20T 4HG63

B 184

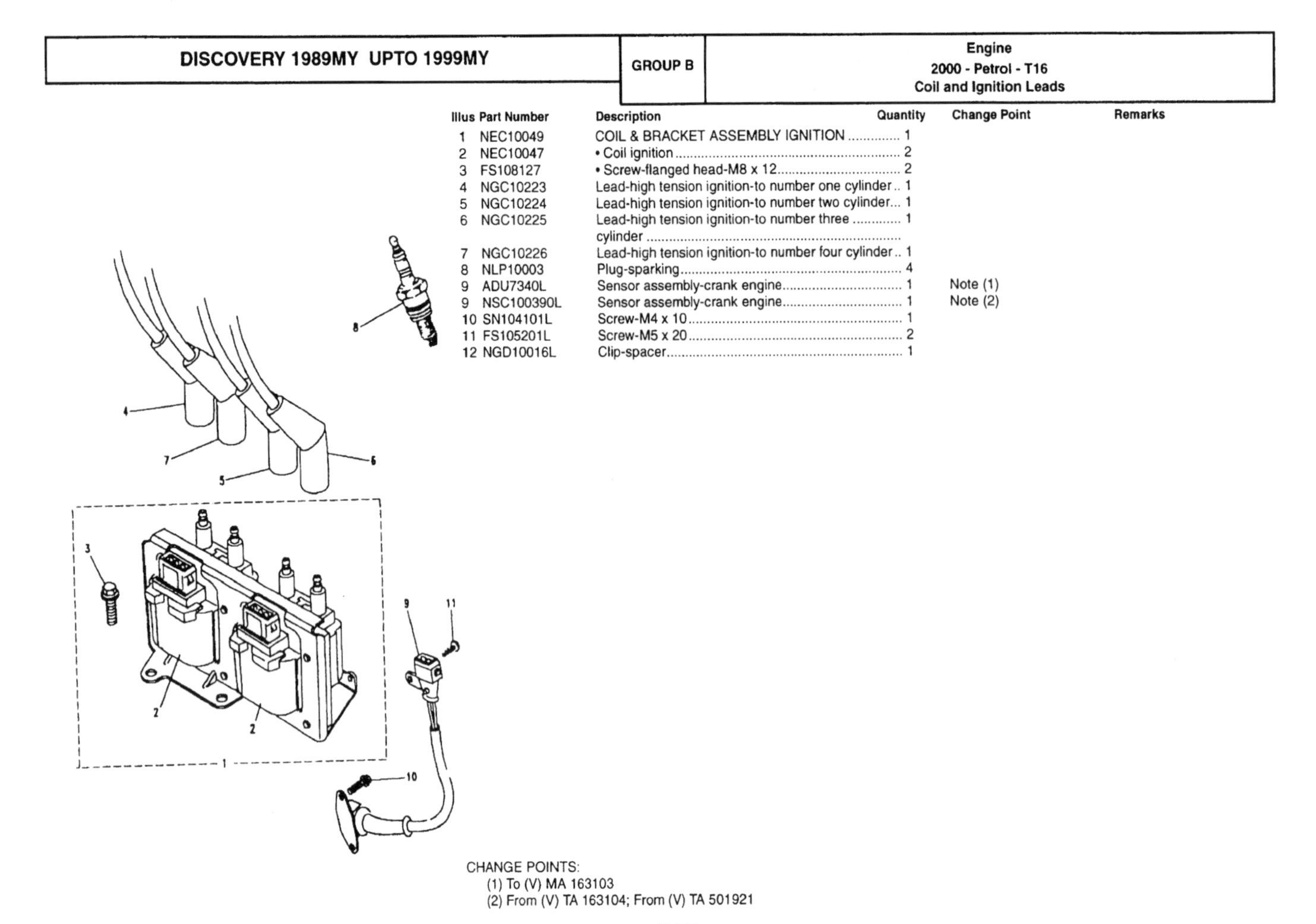

DISCOVERY 1989MY UPTO 1999MY

GROUP B — **Engine** — **2000 - Petrol - T16** — **Coil and Ignition Leads**

Illus	Part Number	Description	Quantity	Change Point	Remarks
1	NEC10049	COIL & BRACKET ASSEMBLY IGNITION	1		
2	NEC10047	• Coil ignition	2		
3	FS108127	• Screw-flanged head-M8 x 12	2		
4	NGC10223	Lead-high tension ignition-to number one cylinder	1		
5	NGC10224	Lead-high tension ignition-to number two cylinder	1		
6	NGC10225	Lead-high tension ignition-to number three cylinder	1		
7	NGC10226	Lead-high tension ignition-to number four cylinder	1		
8	NLP10003	Plug-sparking	4		
9	ADU7340L	Sensor assembly-crank engine	1	Note (1)	
9	NSC100390L	Sensor assembly-crank engine	1	Note (2)	
10	SN104101L	Screw-M4 x 10	1		
11	FS105201L	Screw-M5 x 20	2		
12	NGD10016L	Clip-spacer	1		

CHANGE POINTS:
(1) To (V) MA 163103
(2) From (V) TA 163104; From (V) TA 501921

B 185

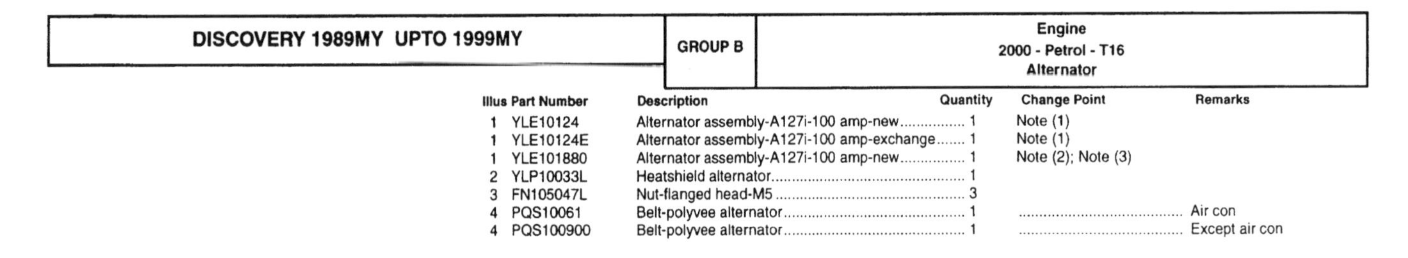

DISCOVERY 1989MY UPTO 1999MY

GROUP B — **Engine** — **2000 - Petrol - T16** — **Alternator**

Illus	Part Number	Description	Quantity	Change Point	Remarks
1	YLE10124	Alternator assembly-A127i-100 amp-new	1	Note (1)	
1	YLE10124E	Alternator assembly-A127i-100 amp-exchange	1	Note (1)	
1	YLE101880	Alternator assembly-A127i-100 amp-new	1	Note (2); Note (3)	
2	YLP10033L	Heatshield alternator	1		
3	FN105047L	Nut-flanged head-M5	3		
4	PQS10061	Belt-polyvee alternator	1		Air con
4	PQS100900	Belt-polyvee alternator	1		Except air con

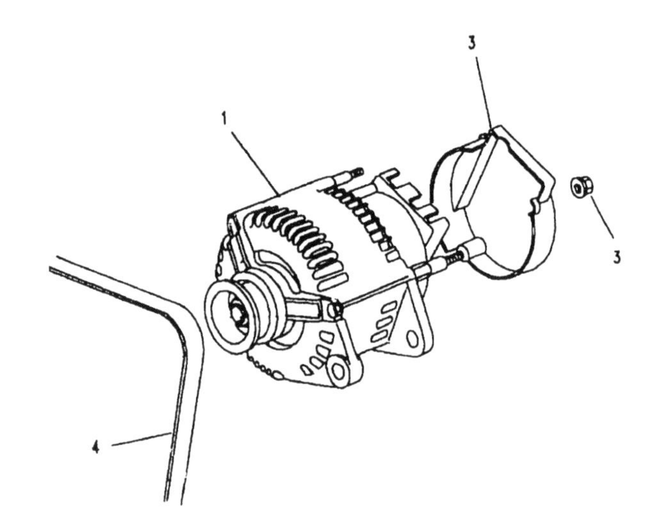

CHANGE POINTS:
(1) To (V) MA 163103
(2) From (V) TA 501920 To (V) VA 558898
(3) From (V) TA 163104

B 186

Alternator Mounting

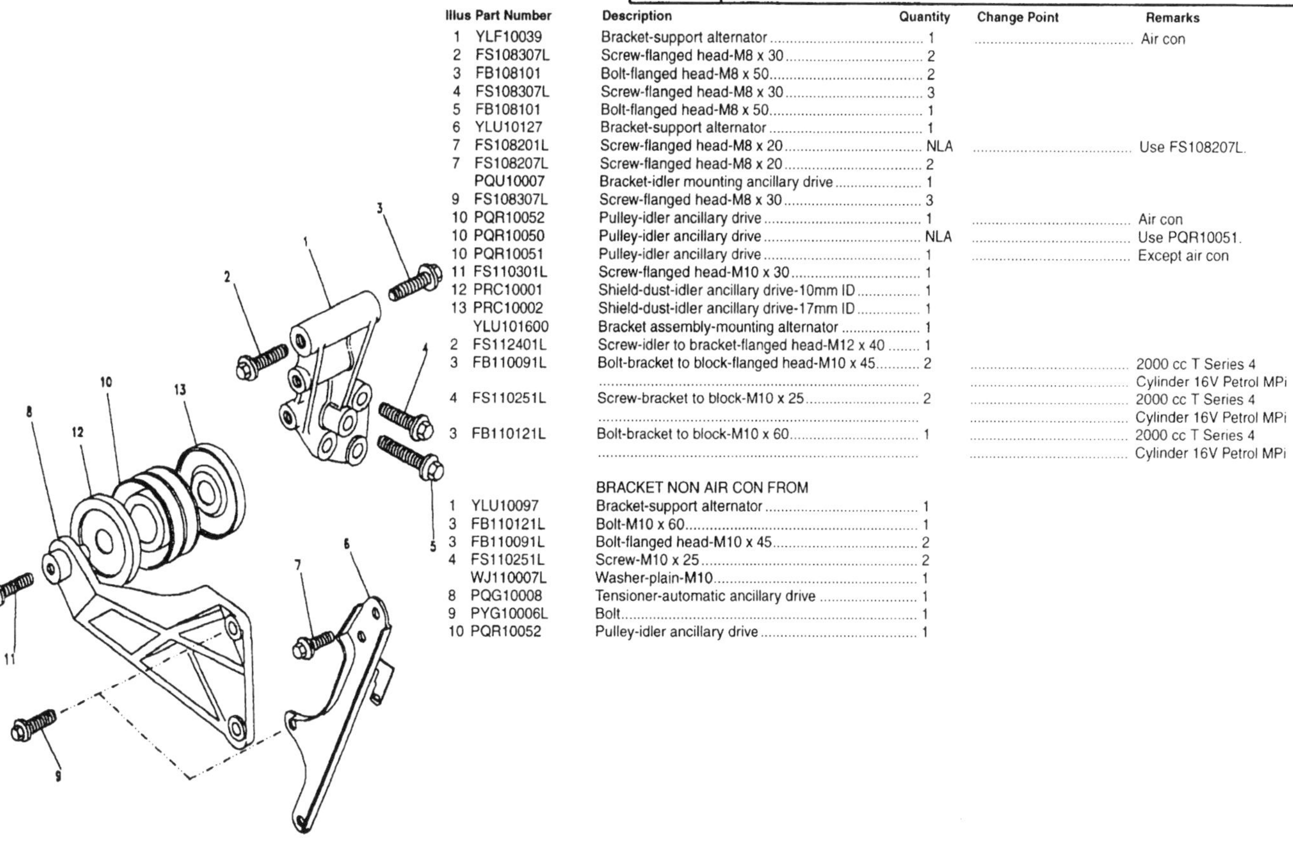

Illus	Part Number	Description	Quantity	Change Point	Remarks
1	YLF10039	Bracket-support alternator	1		Air con
2	FS108307L	Screw-flanged head-M8 x 30	2		
3	FB108101	Bolt-flanged head-M8 x 50	2		
4	FS108307L	Screw-flanged head-M8 x 30	3		
5	FB108101	Bolt-flanged head-M8 x 50	1		
6	YLU10127	Bracket-support alternator	1		
7	FS108201L	Screw-flanged head-M8 x 20	NLA		Use FS108207L.
7	FS108207L	Screw-flanged head-M8 x 20	2		
	PQU10007	Bracket-idler mounting ancillary drive	1		
9	FS108307L	Screw-flanged head-M8 x 30	3		
10	PQR10052	Pulley-idler ancillary drive	1		Air con
10	PQR10050	Pulley-idler ancillary drive	NLA		Use PQR10051.
10	PQR10051	Pulley-idler ancillary drive	1		Except air con
11	FS110301L	Screw-flanged head-M10 x 30	1		
12	PRC10001	Shield-dust-idler ancillary drive-10mm ID	1		
13	PRC10002	Shield-dust-idler ancillary drive-17mm ID	1		
	YLU101600	Bracket assembly-mounting alternator	1		
2	FS112401L	Screw-idler to bracket-flanged head-M12 x 40	1		
3	FB110091L	Bolt-bracket to block-flanged head-M10 x 45	2		2000 cc T Series 4 Cylinder 16V Petrol MPi
4	FS110251L	Screw-bracket to block-M10 x 25	2		2000 cc T Series 4 Cylinder 16V Petrol MPi
3	FB110121L	Bolt-bracket to block-M10 x 60	1		2000 cc T Series 4 Cylinder 16V Petrol MPi
		BRACKET NON AIR CON FROM			
1	YLU10097	Bracket-support alternator	1		
3	FB110121L	Bolt-M10 x 60	1		
3	FB110091L	Bolt-flanged head-M10 x 45	2		
4	FS110251L	Screw-M10 x 25	2		
	WJ110007L	Washer-plain-M10	1		
8	PQG10008	Tensioner-automatic ancillary drive	1		
9	PYG10006L	Bolt	1		
10	PQR10052	Pulley-idler ancillary drive	1		

B 187

Starter Motor

Illus	Part Number	Description	Quantity	Change Point	Remarks
1	NAD10048	MOTOR-STARTER ENGINE	1		
	STC852	• Solenoid starter motor	1		
	STC853	• Kit-brush starter motor	1		
	STC854	• Drive assembly starter motor	1		
2	FS110301L	Screw-flanged head-M10 x 30	3		

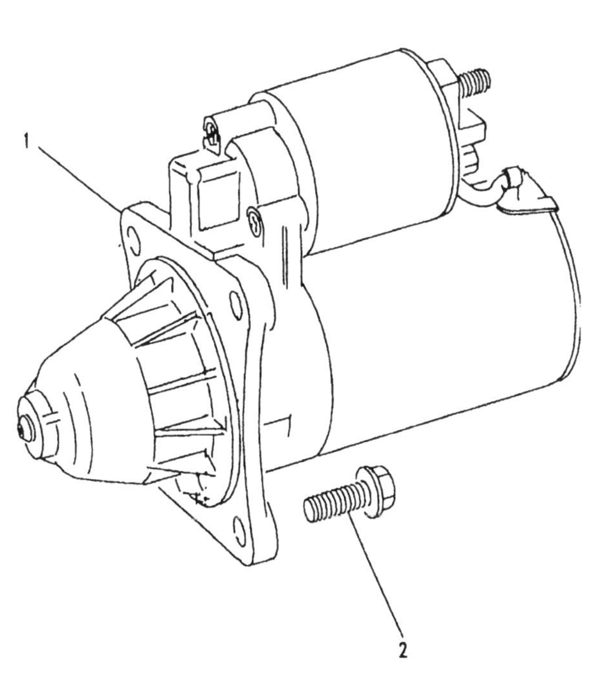

B 188

Illus	Part Number	Description	Quantity	Change Point	Remarks
1	PSD10102	FLYWHEEL ENGINE	1	Note (1)	
1	PSD102420	FLYWHEEL ENGINE	1	Note (2)	
2	DAM7638	• Gear-ring-flywheel engine	1		
3	1G2984	• Dowel	3		
4	LYG10012L	Bolt-flywheel to crankshaft	6		
5	FTC3317	Plate-clutch-driven	1	Note (3)	
5	FTC4034	Plate-clutch-driven	1	Note (4)	
6	576557	Cover-clutch assembly	1		
7	FS108207L	Screw-flanged head-M8 x 20	1		
8	WL108001L	Washer-M8	6		
9	KSP10093	Housing assembly-clutch	1	Note (3)	
9	KSP10094	Housing assembly-clutch	1	Note (4)	
	TE110061L	Stud-M10 x 30	3	Note (4)	
9	KSP101360	Housing assembly-clutch	1	Note (2)	
10	CDU1346L	Bolt-fixing flywheel crankshaft	6		
11	KYG10050	Screw-M12 x 45	2		
12	KSG10001	Seal-dust clutch housing	1		
13	KSG10002	Plug-blanking	1		

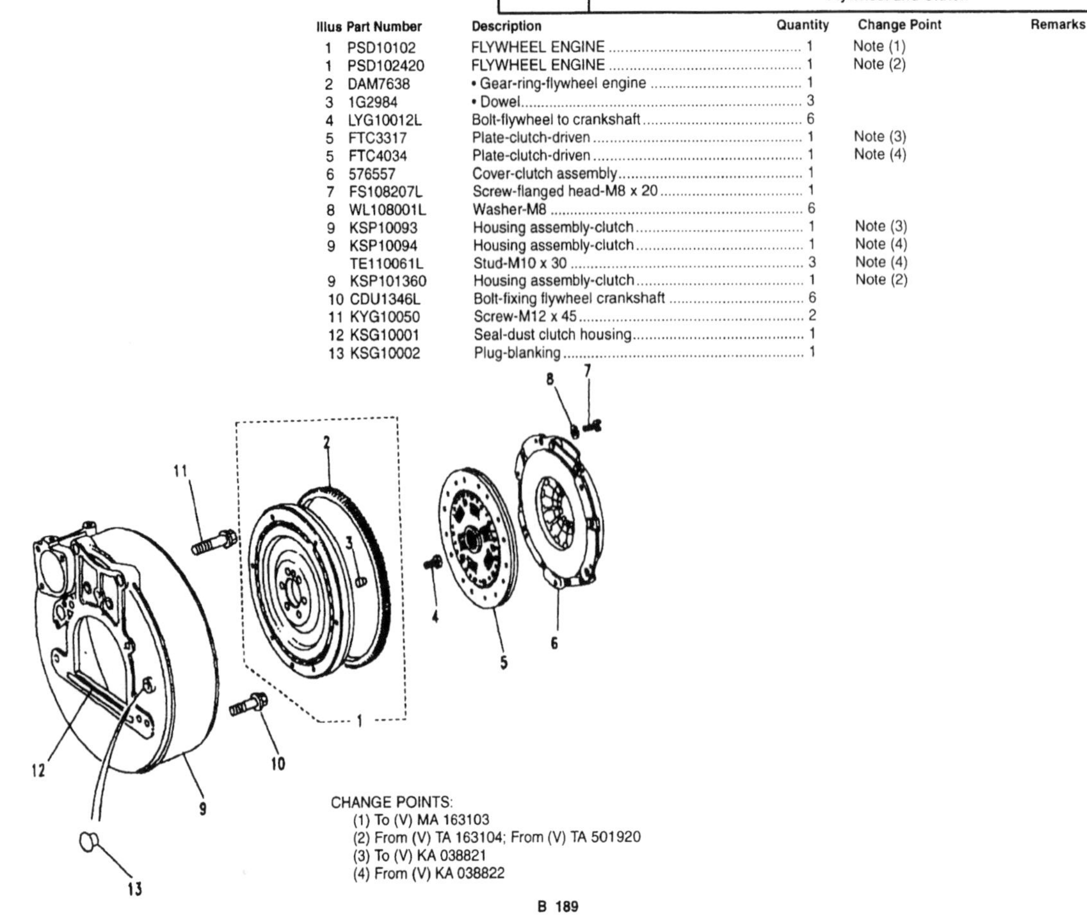

CHANGE POINTS:
(1) To (V) MA 163103
(2) From (V) TA 163104; From (V) TA 501920
(3) To (V) KA 038821
(4) From (V) KA 038822

B 189

77mm 5 Speed 4 Cylinder ... C 1
Bell Housing ... C 4
Bottom Cover ... C 5
Clutch Release Mechanism ... C 2
Extension Case ... C 6
Gearbox Assembly ... C 1
Gearcase ... C 3
Gearchange and Lever ... C 20
Layshaft ... C 15
Mainshaft ... C 8
Mainshaft Gears 1st/2nd/5th ... C 13
Mainshaft Gears 3rd/4th ... C 12
Selectors and Forks ... C 18
Transmission Mountings ... C 22
77mm 5 Speed V8 ... C 23
Bell Housing ... C 26
Bottom Cover ... C 27
Clutch Release-V8 ... C 24
Extension Case ... C 28
Gearbox Assembly ... C 23
Gearcase ... C 25
Gearchange and Lever ... C 40
Layshaft ... C 36
Mainshaft ... C 29
Mainshaft Gears 1st/2nd/5th ... C 32
Mainshaft Gears 3rd/4th ... C 31
Oil Cooler ... C 42
Selectors and Yoke Assembly ... C 38
R380 5 Speed ... C 43
Bell Housing ... C 52
Bottom Cover ... C 53
Clutch Release Mechanism-MPi ... C 45
Clutch Release Mechanism-V8 ... C 46
Extension Case ... C 54
Gearbox Assembly ... C 43
Gearcase ... C 47
Gearchange and Lever ... C 67
Layshaft ... C 64
Mainshaft ... C 56
Mainshaft Gears 1st/2nd ... C 58
Mainshaft Gears 3rd/4th ... C 60
Mainshaft Gears 5th/Reverse ... C 61
Oil Cooler-V8 ... C 69
Selectors and Shafts ... C 66
Transmission Mountings ... C 68
ZF Auto ... C 70
Adaptor Plate ... C 71
Brake 'C-C' ... C 80
Brake 'D' ... C 81
Brakes 'F' ... C 79
Casing and Sump ... C 73
Clutch 'A' ... C 76
Clutch 'B' ... C 77
Clutch 'E' ... C 78
Gearbox Assembly ... C 70
Gearchange ... C 86
Governor and Extension Housing ... C 84
Oil Cooler 300Tdi ... C 96
Oil Cooler-Automatic ... C 92
Oil Pump ... C 75
Service Kits ... C 91
Shift Interlock ... C 88
Torque Convertor and Housing ... C 72
Transmission Mounting ... C 90
Web Shaft and Fourth Gear ... C 82

Gearbox Assembly

Illus	Part Number	Description	Quantity	Change Point	Remarks
1	FTC4495	Transmission assembly manual-new	1	Note (1)	TDi
1	FTC4497	Transmission assembly manual-new	1		MPi
	STC2772	Kit-fitting-gearbox R380	1		
	STC2773	Kit-gaiter-gearbox	1		
		TRANSMISSION ASSEMBLY MANUAL			
1	RTC6563R	reconditioned	1		55A Suffix G
1	STC1093R	reconditioned	1		55A Suffix H
1	STC1514R	reconditioned	1		63A Suffix H

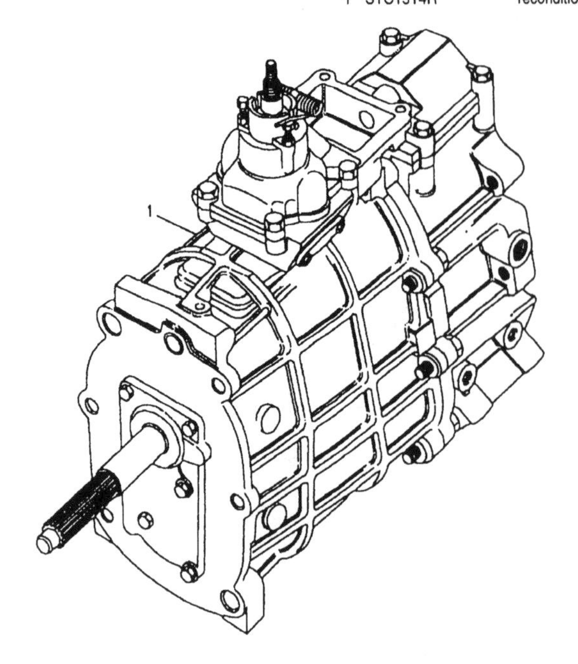

CHANGE POINTS:
(1) To (V) LA 081991

Clutch Release Mechanism

Illus	Part Number	Description	Quantity	Change Point	Remarks
1	FTC5218	Guide-clutch release bearing-cast iron	1	Note (1)	
2	FB108101	Bolt-flanged head-M8 x 50	2	Note (1)	
3	WA108051L	Washer-plain-M8	2	Note (1)	
4	FRC3326	Arm-clutch release clutch	1	Note (2)	
4	FTC2957	Arm-clutch release clutch	1	Note (3)	
5	FRC3416	Retention-bearing clutch release	1	Note (1)	
6	FRC9568	BEARING-CLUTCH RELEASE	1		
	FRC4078	• Sleeve-guide clutch	1		
6	FTC2772	Bearing-clutch release	1	Note (1)	MPi
7	FRC5255	Pad-clutch release lever slipper	2	Note (1)	
8	FRC3327	Clip	1		
9	FRC3417	Rod-push slave cylinder	1		
10	591988	Plate	1	Note (2)	
10	FRC2402	Plate	1	Note (3)	
11	TKC2786L	SLAVE CYLINDER CLUTCH	1	Note (2)	
	514244	• Kit-repair clutch slave cylinder	1		
11	FTC2498	Slave cylinder clutch	1	Note (3)	
12	FS108251L	Screw-flanged head-M8 x 25	NLA	Note (1)	Use FS108257L.
12	FS108257L	Screw-flanged head-M8 x 25	2	Note (1)	
13	WL108001L	Washer-M8	2	Note (1)	
14	594091	Cap-protective	1	Note (1)	
15	606733	Screw-bleed clutch slave cylinder	1	Note (1)	
16	FRC2481	Dowel	2	Note (1)	

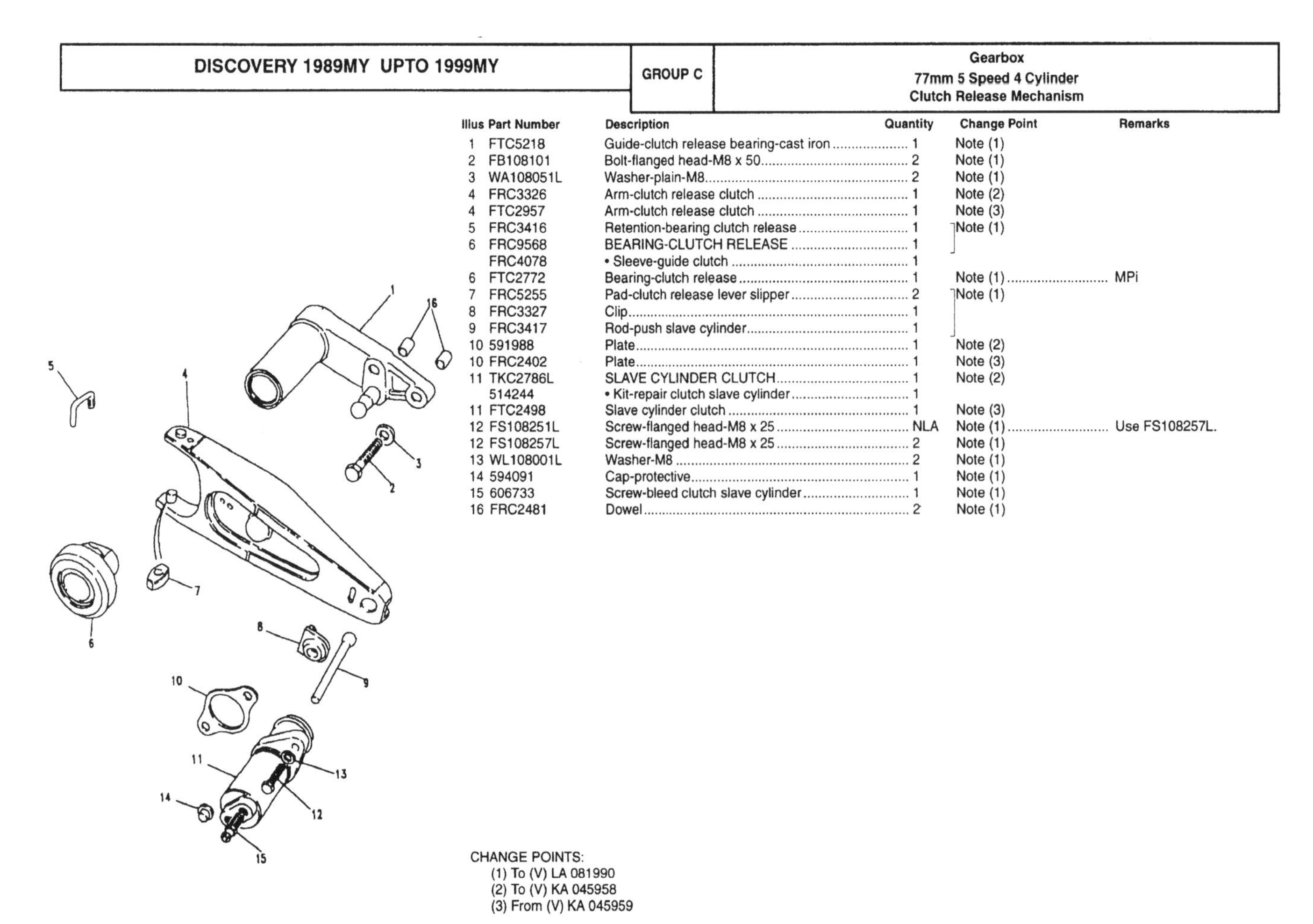

CHANGE POINTS:
(1) To (V) LA 081990
(2) To (V) KA 045958
(3) From (V) KA 045959

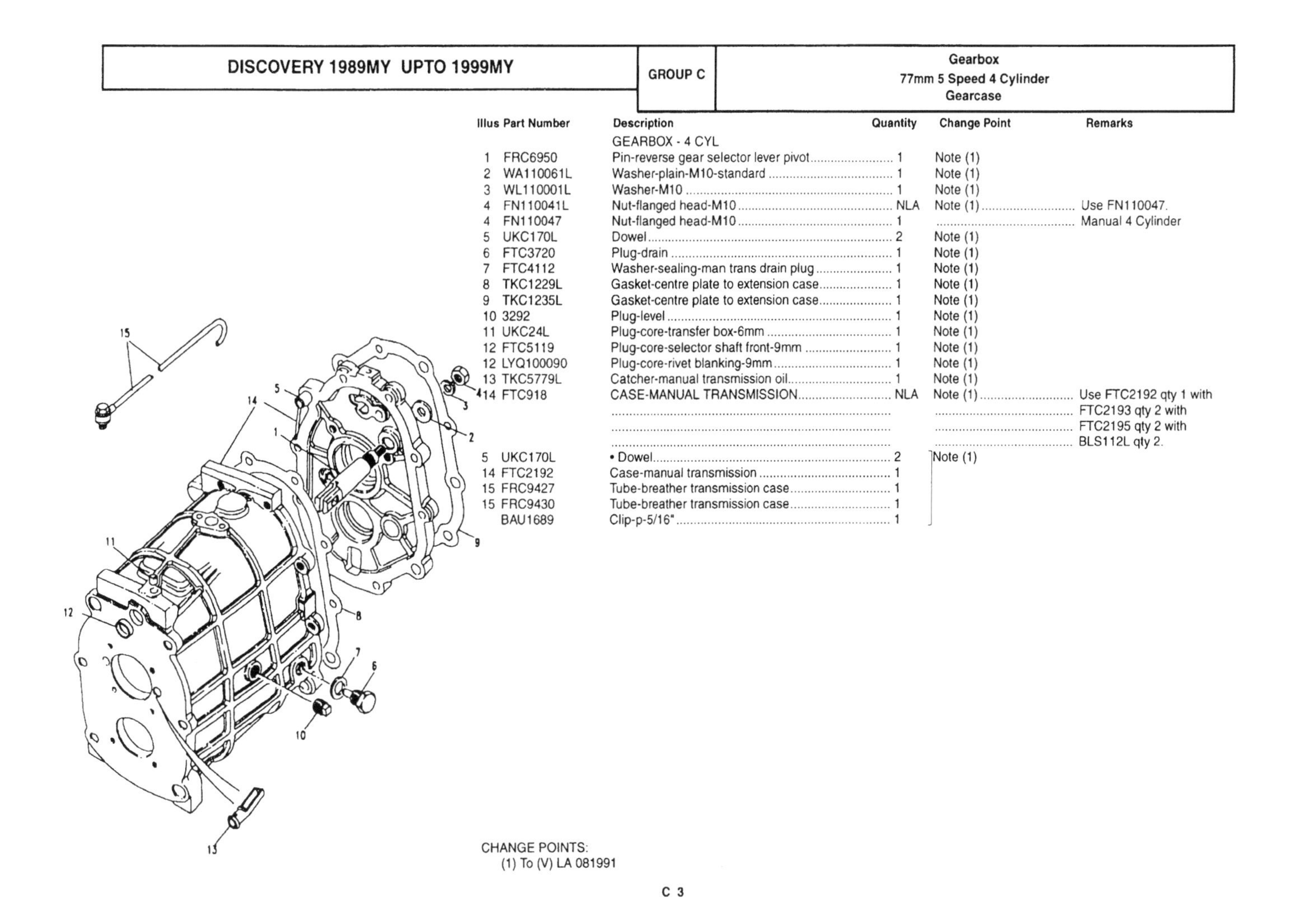

DISCOVERY 1989MY UPTO 1999MY	GROUP C	Gearbox 77mm 5 Speed 4 Cylinder Gearcase

Illus	Part Number	Description	Quantity	Change Point	Remarks
		GEARBOX - 4 CYL			
1	FRC6950	Pin-reverse gear selector lever pivot	1	Note (1)	
2	WA110061L	Washer-plain-M10-standard	1	Note (1)	
3	WL110001L	Washer-M10	1	Note (1)	
4	FN110041L	Nut-flanged head-M10	NLA	Note (1)	Use FN110047.
4	FN110047	Nut-flanged head-M10	1		Manual 4 Cylinder
5	UKC170L	Dowel	2	Note (1)	
6	FTC3720	Plug-drain	1	Note (1)	
7	FTC4112	Washer-sealing-man trans drain plug	1	Note (1)	
8	TKC1229L	Gasket-centre plate to extension case	1	Note (1)	
9	TKC1235L	Gasket-centre plate to extension case	1	Note (1)	
10	3292	Plug-level	1	Note (1)	
11	UKC24L	Plug-core-transfer box-6mm	1	Note (1)	
12	FTC5119	Plug-core-selector shaft front-9mm	1	Note (1)	
12	LYQ100090	Plug-core-rivet blanking-9mm	1	Note (1)	
13	TKC5779L	Catcher-manual transmission oil	1	Note (1)	
14	FTC918	CASE-MANUAL TRANSMISSION	NLA	Note (1)	Use FTC2192 qty 1 with FTC2193 qty 2 with FTC2195 qty 2 with BLS112L qty 2.
5	UKC170L	• Dowel	2	Note (1)	
14	FTC2192	Case-manual transmission	1		
15	FRC9427	Tube-breather transmission case	1		
15	FRC9430	Tube-breather transmission case	1		
	BAU1689	Clip-p-5/16"	1		

CHANGE POINTS:
(1) To (V) LA 081991

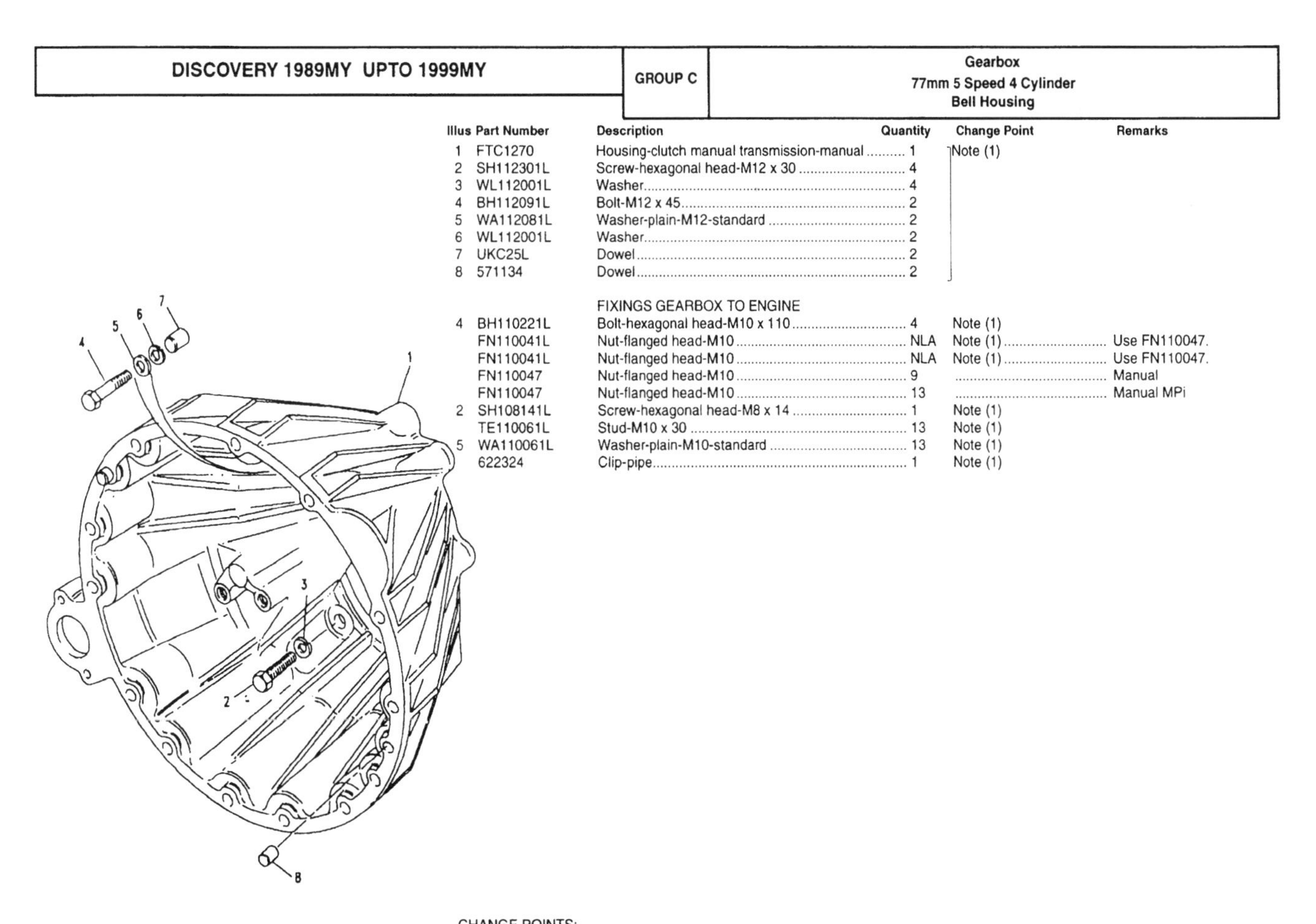

DISCOVERY 1989MY UPTO 1999MY	GROUP C	Gearbox 77mm 5 Speed 4 Cylinder Bell Housing

Illus	Part Number	Description	Quantity	Change Point	Remarks
1	FTC1270	Housing-clutch manual transmission-manual	1	Note (1)	
2	SH112301L	Screw-hexagonal head-M12 x 30	4		
3	WL112001L	Washer	4		
4	BH112091L	Bolt-M12 x 45	2		
5	WA112081L	Washer-plain-M12-standard	2		
6	WL112001L	Washer	2		
7	UKC25L	Dowel	2		
8	571134	Dowel	2		
		FIXINGS GEARBOX TO ENGINE			
4	BH110221L	Bolt-hexagonal head-M10 x 110	4	Note (1)	
	FN110041L	Nut-flanged head-M10	NLA	Note (1)	Use FN110047.
	FN110041L	Nut-flanged head-M10	NLA	Note (1)	Use FN110047.
	FN110047	Nut-flanged head-M10	9		Manual
	FN110047	Nut-flanged head-M10	13		Manual MPi
2	SH108141L	Screw-hexagonal head-M8 x 14	1	Note (1)	
	TE110061L	Stud-M10 x 30	13	Note (1)	
5	WA110061L	Washer-plain-M10-standard	13	Note (1)	
	622324	Clip-pipe	1	Note (1)	

CHANGE POINTS:
(1) To (V) LA 081991

Illus	Part Number	Description	Quantity	Change Point	Remarks
1	FRC2859	Cover-bell housing-bottom	1	Note (1)	
2	594087	Seal-bottom	1		
3	SH106451	Screw-hexagonal head-M6 x 45	5		
4	WL106001L	Washer-M6	5		
5	NH106041L	Nut-hexagonal head-nyloc-M6	5		
6	FS106255L	Screw-flanged head-M6 x 25	2		
7	WL106001L	Washer-M6	2		
8	594134	Bolt	2		
9	WL108001L	Washer-M8	2		
10	253023	Bolt	2		
11	232038	Washer	2		

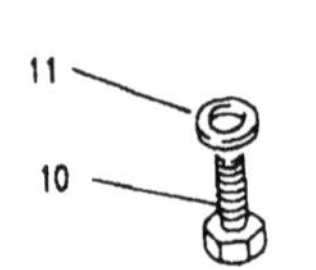

CHANGE POINTS:
(1) To (V) LA 081991

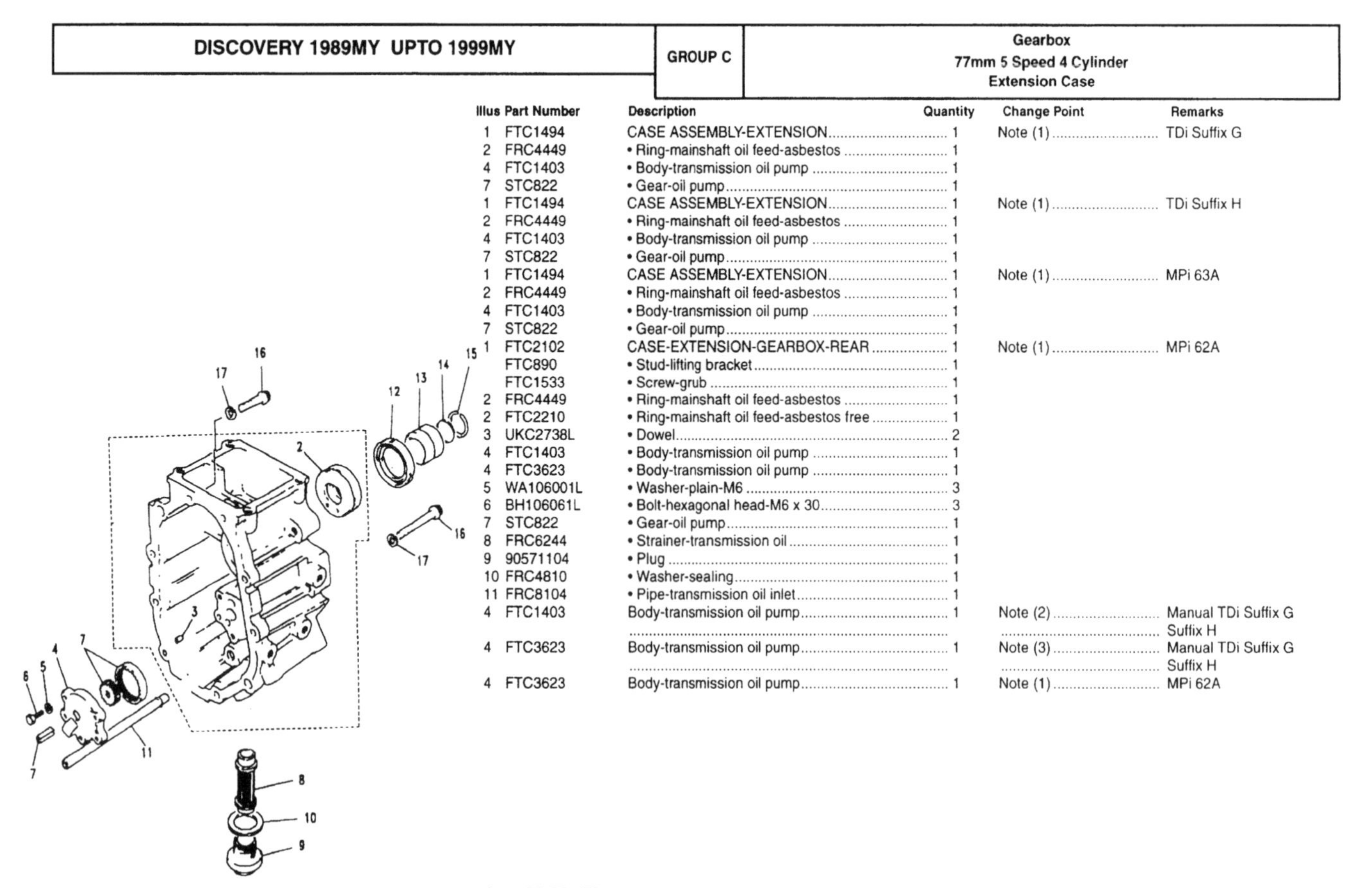

Illus	Part Number	Description	Quantity	Change Point	Remarks
1	FTC1494	CASE ASSEMBLY-EXTENSION	1	Note (1)	TDi Suffix G
2	FRC4449	• Ring-mainshaft oil feed-asbestos	1		
4	FTC1403	• Body-transmission oil pump	1		
7	STC822	• Gear-oil pump	1		
1	FTC1494	CASE ASSEMBLY-EXTENSION	1	Note (1)	TDi Suffix H
2	FRC4449	• Ring-mainshaft oil feed-asbestos	1		
4	FTC1403	• Body-transmission oil pump	1		
7	STC822	• Gear-oil pump	1		
1	FTC1494	CASE ASSEMBLY-EXTENSION	1	Note (1)	MPi 63A
2	FRC4449	• Ring-mainshaft oil feed-asbestos	1		
4	FTC1403	• Body-transmission oil pump	1		
7	STC822	• Gear-oil pump	1		
1	FTC2102	CASE-EXTENSION-GEARBOX-REAR	1	Note (1)	MPi 62A
	FTC890	• Stud-lifting bracket	1		
	FTC1533	• Screw-grub	1		
2	FRC4449	• Ring-mainshaft oil feed-asbestos	1		
2	FTC2210	• Ring-mainshaft oil feed-asbestos free	1		
3	UKC2738L	• Dowel	2		
4	FTC1403	• Body-transmission oil pump	1		
4	FTC3623	• Body-transmission oil pump	1		
5	WA106001L	• Washer-plain-M6	3		
6	BH106061L	• Bolt-hexagonal head-M6 x 30	3		
7	STC822	• Gear-oil pump	1		
8	FRC6244	• Strainer-transmission oil	1		
9	90571104	• Plug	1		
10	FRC4810	• Washer-sealing	1		
11	FRC8104	• Pipe-transmission oil inlet	1		
4	FTC1403	Body-transmission oil pump	1	Note (2)	Manual TDi Suffix G Suffix H
4	FTC3623	Body-transmission oil pump	1	Note (3)	Manual TDi Suffix G Suffix H
4	FTC3623	Body-transmission oil pump	1	Note (1)	MPi 62A

CHANGE POINTS:
(1) To (V) LA 081991
(2) To (G) 0207750H
(3) From (G) 0207751H To (V) LA 081991

Illus	Part Number	Description	Quantity	Change Point	Remarks
5	WL106001L	Washer-M6	3	Note (1)	TDi
6	FS106201L	Screw-flanged head-M6 x 20	3	Note (1)	TDi
7	STC822	Gear-oil pump	2	Note (1)	TDi
8	FRC6244	Strainer-transmission oil	1	Note (1)	TDi
9	90571104	Plug	1	Note (1)	TDi
10	FRC4810	Washer-sealing	1	Note (1)	TDi
11	FRC8104	Pipe-transmission oil inlet	1	Note (1)	TDi
12	FRC2365	Seal	1	Note (1)	TDi
13	FRC4493	Collar	1	Note (1)	TDi
14	FRC4501	O ring	1	Note (1)	TDi
15	FRC4494	Spring ring	1	Note (1)	TDi
16	FRC4282	Bolt-flanged head-M8 x 60	8	Note (1)	TDi
17	WL108001L	Washer-M8	8	Note (1)	TDi
	FTC890	Stud	1	Note (1)	TDi

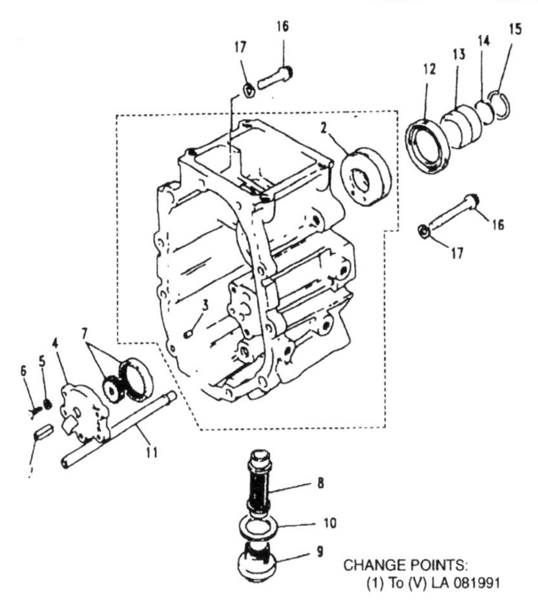

CHANGE POINTS:
(1) To (V) LA 081991

Illus	Part Number	Description	Quantity	Change Point	Remarks
1	FTC311	COVER-END MANUAL TRANSMISSION-FRONT-LONG	1	Note (1)	Twin Carburettor Suffix G
2	UKC1060L	• Seal-man trans primary shaft	NLA	Note (2); Note (3)	Use FTC5303.
2	FTC5303	• Seal-man trans primary shaft	1	Note (4); Note (5)	
1	FTC917	COVER-END MANUAL TRANSMISSION-FRONT-LONG	1	Note (1)	EFi Suffix G
2	UKC1060L	• Seal-man trans primary shaft	NLA	Note (2); Note (3)	Use FTC5303.
1	FTC2822	COVER-END MANUAL TRANSMISSION-FRONT-SHORT	1	Note (1)	Manual EFi Suffix H
2	UKC1060L	• Seal-man trans primary shaft	NLA		Use FTC5303.
2	FTC5303	• Seal-man trans primary shaft	1		
3	FS108251L	Screw-flanged head-M8 x 25	NLA	Note (1)	Use FS108257L. Suffix G
3	FS108257L	Screw-flanged head-M8 x 25	6	Note (1)	Suffix G
4	WA108051L	Washer-plain-M8	6	Note (1)	Suffix G
5	FTC316	Gasket-front cover manual transmission	1	Note (1)	Suffix G

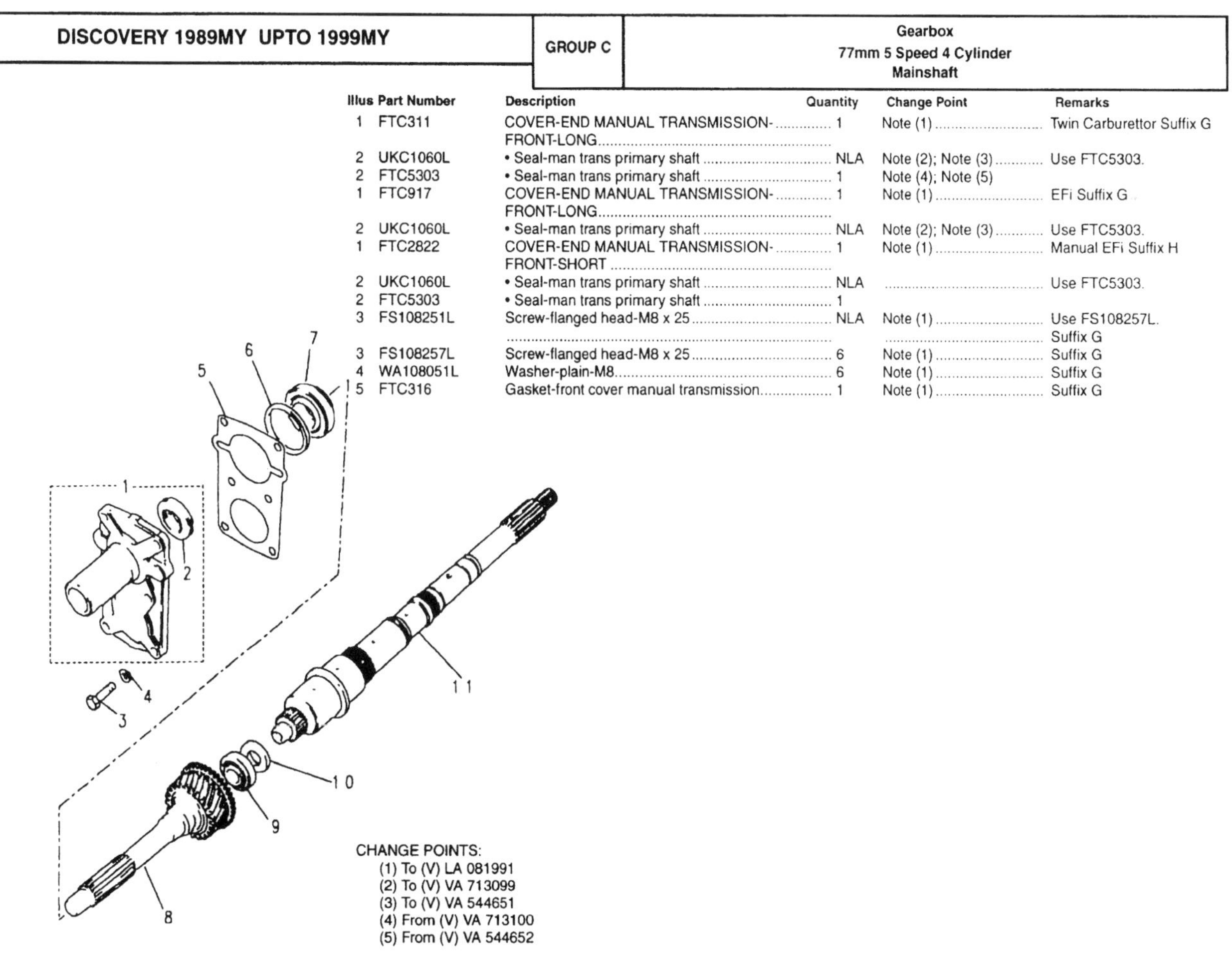

CHANGE POINTS:
(1) To (V) LA 081991
(2) To (V) VA 713099
(3) To (V) VA 544651
(4) From (V) VA 713100
(5) From (V) VA 544652

Illus	Part Number	Description	Quantity	Change Point	Remarks
		SHIM			
6	FRC4327	man trans primary shaft bearing/front cover-1.51mm	A/R	Note (1)	Suffix G
6	FRC4329	man trans primary shaft bearing/front cover-1.57mm	A/R		
6	FRC4331	man trans primary shaft bearing/front cover-1.63mm	A/R		
6	FRC4333	man trans primary shaft bearing/front cover-1.69mm	A/R		
6	FRC4335	man trans primary shaft bearing/front cover-1.75mm	A/R		
6	FRC4337	man trans primary shaft bearing/front cover-1.81mm	A/R		
6	FRC4339	man trans primary shaft bearing/front cover-1.87mm	A/R		
6	FRC4341	man trans primary shaft bearing/front cover-1.93mm	A/R		
6	FRC4343	man trans primary shaft bearing/front cover-1.99mm	A/R		
6	FRC4345	man trans primary shaft bearing/front cover-2.05mm	A/R		
6	FRC4347	man trans primary shaft bearing/front cover-2.11mm	A/R		

CHANGE POINTS:
(1) To (V) LA 081991

Illus	Part Number	Description	Quantity	Change Point	Remarks
		SHIM			
6	FRC4349	man trans primary shaft bearing/front cover-2.17mm	A/R	Note (1)	Suffix G
6	FRC4351	man trans primary shaft bearing/front cover-2.23mm	A/R		Suffix G
6	FRC4353	man trans primary shaft bearing/front cover-2.29mm	A/R		
6	FRC4355	man trans primary shaft bearing/front cover-2.35mm	A/R		
6	FRC4357	man trans primary shaft bearing/front cover-2.41mm	A/R		
6	FRC4359	man trans primary shaft bearing/front cover-2.47mm	A/R		
6	FRC4361	man trans primary shaft bearing/front cover-2.53mm	A/R		
6	FRC4363	man trans primary shaft bearing/front cover-2.59mm	A/R		
6	FRC4365	man trans primary shaft bearing/front cover-2.65mm	A/R		
6	FRC4367	man trans primary shaft bearing/front cover-2.71mm	A/R		
6	FRC4369	man trans primary shaft bearing/front cover-2.77mm	A/R		

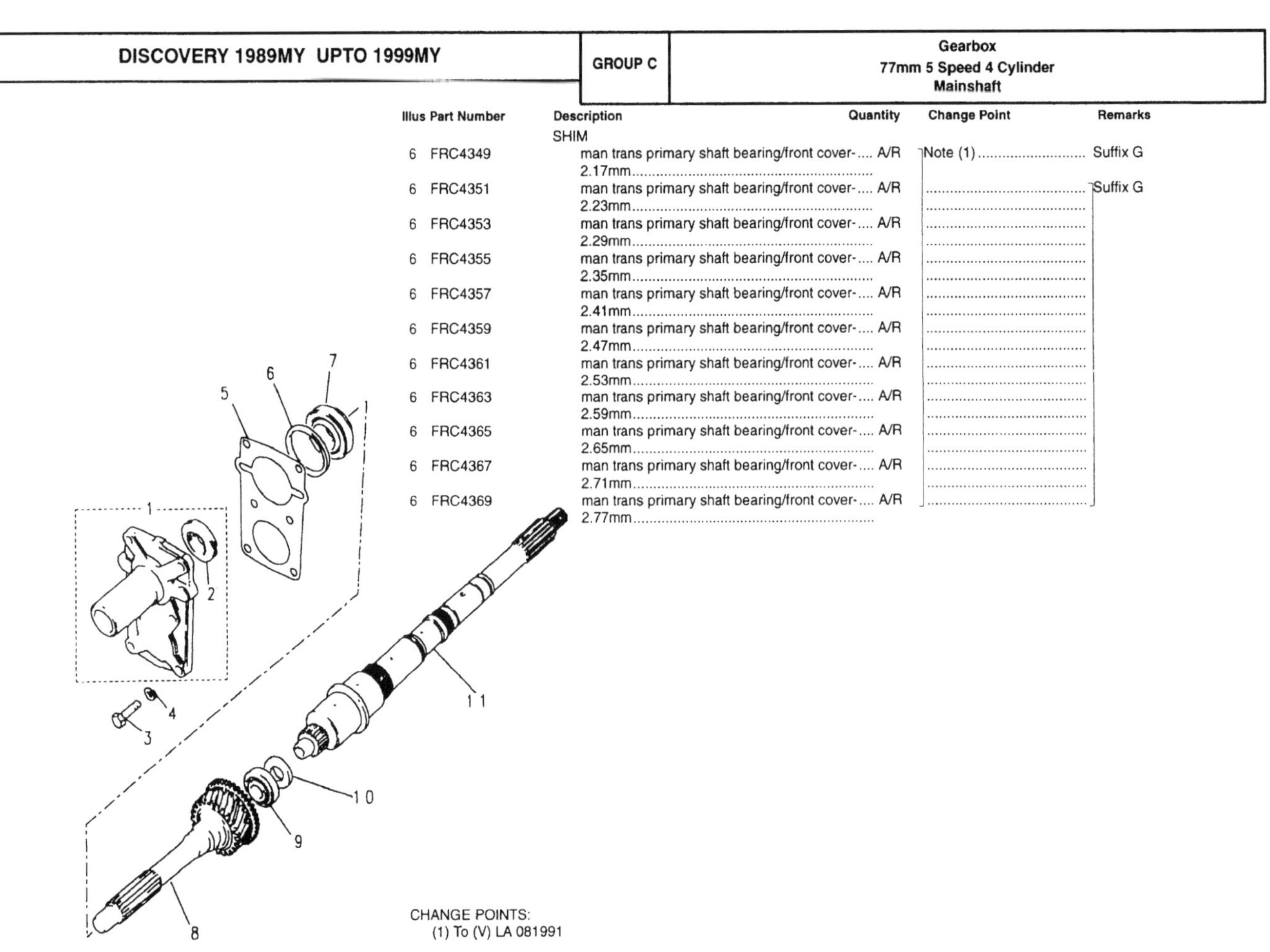

CHANGE POINTS:
(1) To (V) LA 081991

Illus	Part Number	Description	Quantity	Change Point	Remarks
7	RTC6751	Bearing-taper roller	1	Note (1)	
8	FTC1426	Shaft-pinion differential	1	Note (1)	Manual 4 Cylinder 55A
8	FTC2770	Shaft assembly-primary manual transmission	1	Note (1)	Manual 4 Cylinder 62A
8	FTC3664	Shaft assembly-primary manual transmission	1	Note (1)	Manual 4 Cylinder 63A
9	UKC8L	Bearing-taper roller	1	Note (1)	
10	UKC37L	Spacer	1	Note (1)	
11	FTC1446	Shaft-main manual transmission	1	Note (1)	Suffix G
11	FTC1282	Shaft-main manual transmission	NLA	Note (1)	Use STC1889. Suffix G
11	STC1889	Shaft-main manual transmission	1	Note (1)	Suffix H

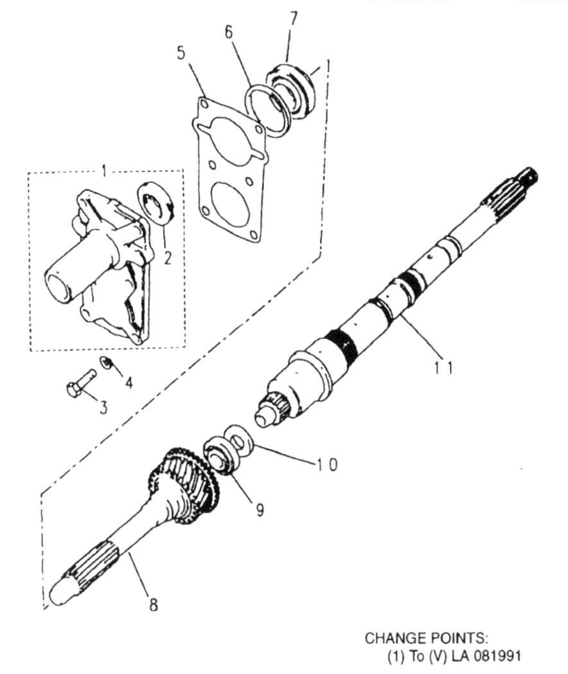

CHANGE POINTS:
(1) To (V) LA 081991

Illus	Part Number	Description	Quantity	Change Point	Remarks
		MANUAL - 4 CYL			
1	FRC9792L	SYNCHRONISER ASSEMBLY-3RD & 4TH MAINSHAFT ASSEMBLY	1	Note (1)	Suffix G Suffix H
2	UKC31L	• Spring 1st & 2nd synchroniser			
3	UKC3530L	• Plate-synchroniser			
4	FRC8232	Ring-baulk-3rd/4th gear	2	Note (1)	Suffix G Suffix H
5	FTC2505	Gear-3rd driven manual transmission	1		Suffix H except MPi
5	FTC2504	Gear-3rd driven manual transmission	1		MPi Suffix H
5	FTC358	Gear-3rd driven manual transmission	1		Suffix G
6	FRC5678	Bearing-needle roller	1	Note (1)	Suffix G
6	FTC1313	Bearing-needle roller	1		Suffix H
7	FTC1310	Bush-3rd gear	1		Suffix G Suffix H

CHANGE POINTS:
(1) To (V) LA 081991

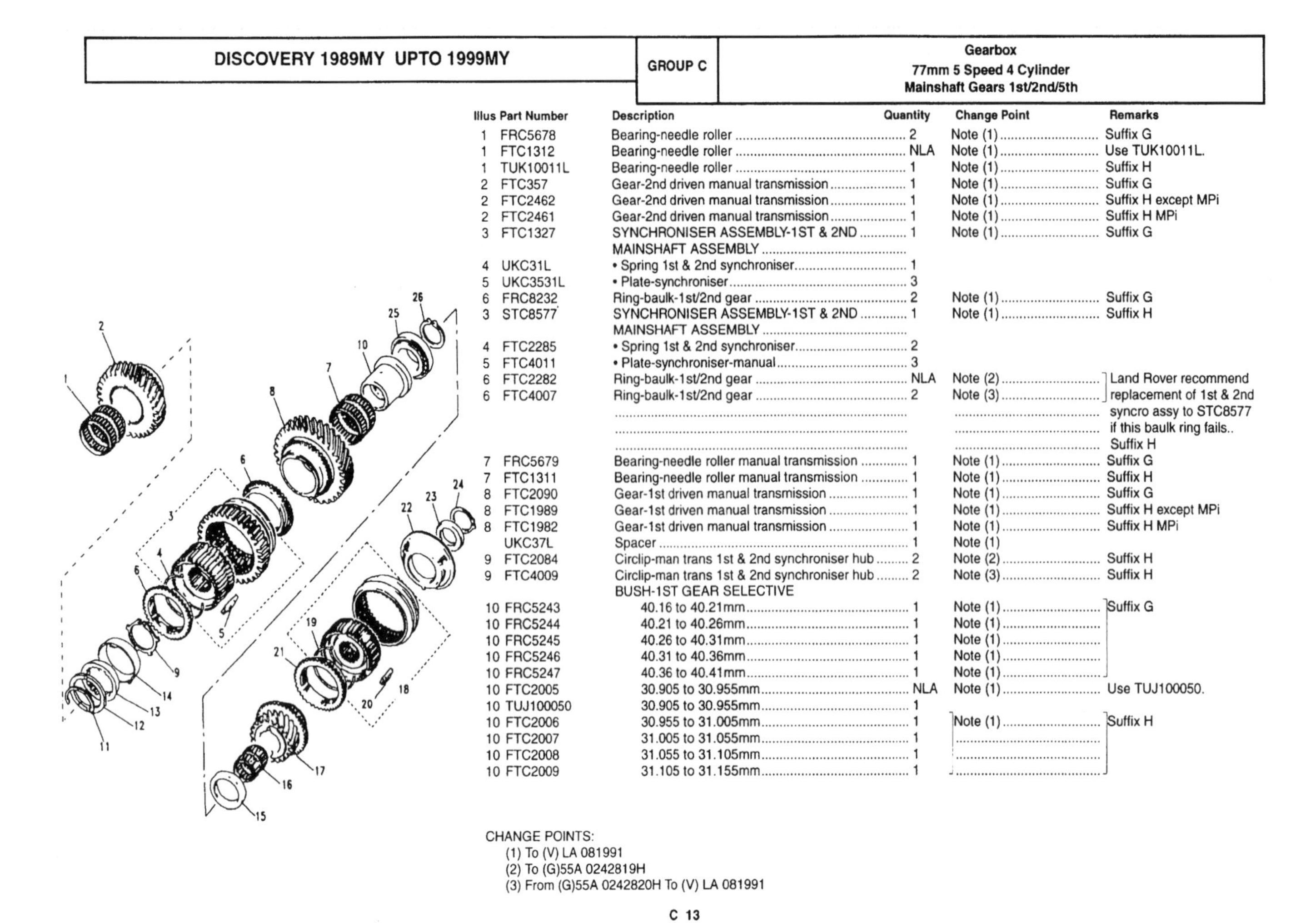

Illus	Part Number	Description	Quantity	Change Point	Remarks
1	FRC5678	Bearing-needle roller	2	Note (1)	Suffix G
1	FTC1312	Bearing-needle roller	NLA	Note (1)	Use TUK10011L.
1	TUK10011L	Bearing-needle roller	1	Note (1)	Suffix H
2	FTC357	Gear-2nd driven manual transmission	1	Note (1)	Suffix G
2	FTC2462	Gear-2nd driven manual transmission	1	Note (1)	Suffix H except MPi
2	FTC2461	Gear-2nd driven manual transmission	1	Note (1)	Suffix H MPi
3	FTC1327	SYNCHRONISER ASSEMBLY-1ST & 2ND MAINSHAFT ASSEMBLY	1	Note (1)	Suffix G
4	UKC31L	• Spring 1st & 2nd synchroniser	1		
5	UKC3531L	• Plate-synchroniser	3		
6	FRC8232	Ring-baulk-1st/2nd gear	2	Note (1)	Suffix G
3	STC8577	SYNCHRONISER ASSEMBLY-1ST & 2ND MAINSHAFT ASSEMBLY	1	Note (1)	Suffix H
4	FTC2285	• Spring 1st & 2nd synchroniser	2		
5	FTC4011	• Plate-synchroniser-manual	3		
6	FTC2282	Ring-baulk-1st/2nd gear	NLA	Note (2)	Land Rover recommend replacement of 1st & 2nd syncro assy to STC8577 if this baulk ring fails..
6	FTC4007	Ring-baulk-1st/2nd gear	2	Note (3)	Suffix H
7	FRC5679	Bearing-needle roller manual transmission	1	Note (1)	Suffix G
7	FTC1311	Bearing-needle roller manual transmission	1	Note (1)	Suffix H
8	FTC2090	Gear-1st driven manual transmission	1	Note (1)	Suffix G
8	FTC1989	Gear-1st driven manual transmission	1	Note (1)	Suffix H except MPi
8	FTC1982	Gear-1st driven manual transmission	1	Note (1)	Suffix H MPi
	UKC37L	Spacer	1	Note (1)	
9	FTC2084	Circlip-man trans 1st & 2nd synchroniser hub	2	Note (2)	Suffix H
9	FTC4009	Circlip-man trans 1st & 2nd synchroniser hub	2	Note (3)	Suffix H
		BUSH-1ST GEAR SELECTIVE			
10	FRC5243	40.16 to 40.21mm	1	Note (1)	Suffix G
10	FRC5244	40.21 to 40.26mm	1	Note (1)	
10	FRC5245	40.26 to 40.31mm	1	Note (1)	
10	FRC5246	40.31 to 40.36mm	1	Note (1)	
10	FRC5247	40.36 to 40.41mm	1	Note (1)	
10	FTC2005	30.905 to 30.955mm	NLA	Note (1)	Use TUJ100050.
10	TUJ100050	30.905 to 30.955mm	1		
10	FTC2006	30.955 to 31.005mm	1	Note (1)	Suffix H
10	FTC2007	31.005 to 31.055mm	1		
10	FTC2008	31.055 to 31.105mm	1		
10	FTC2009	31.105 to 31.155mm	1		

CHANGE POINTS:
(1) To (V) LA 081991
(2) To (G)55A 0242819H
(3) From (G)55A 0242820H To (V) LA 081991

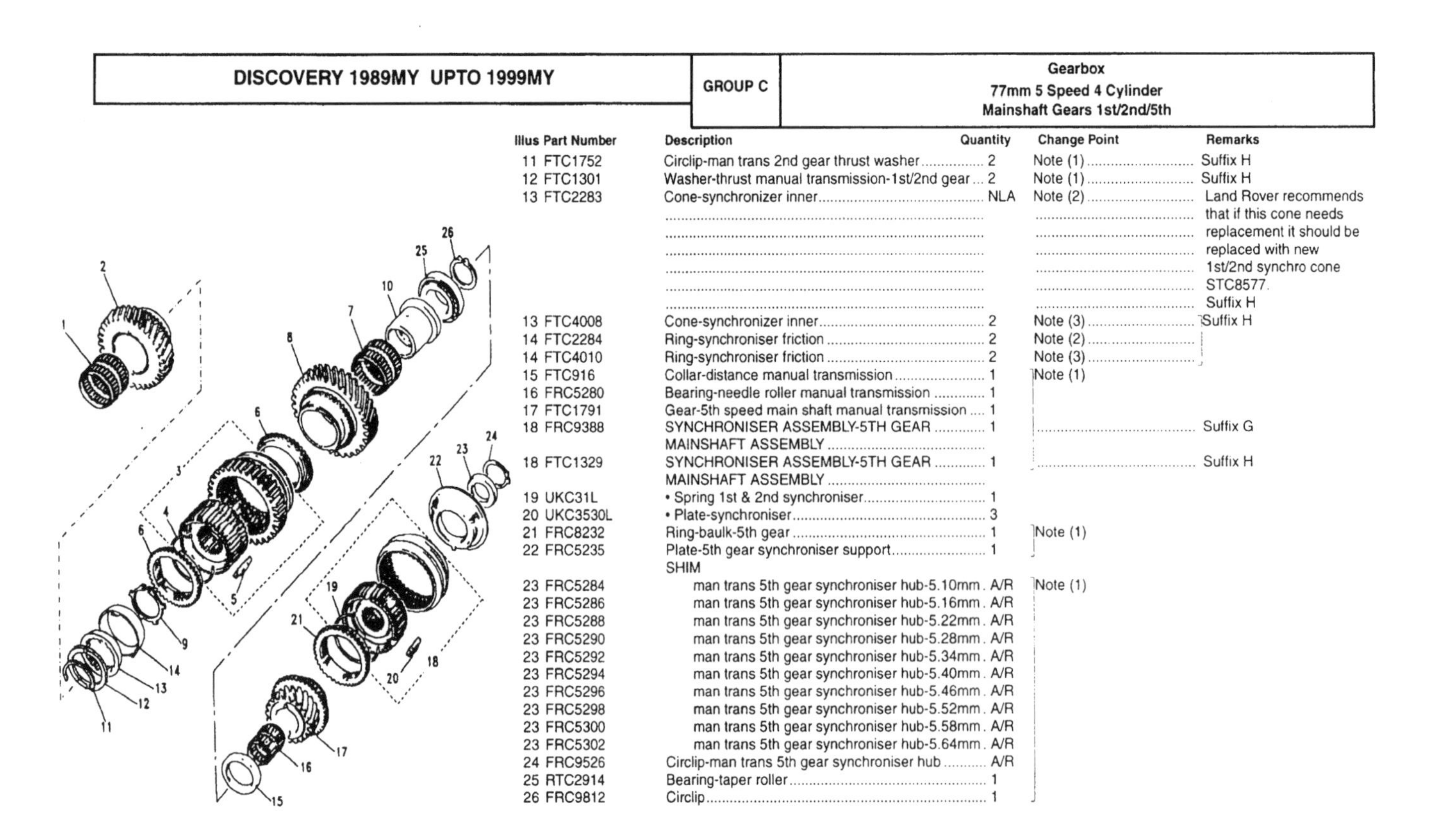

Illus	Part Number	Description	Quantity	Change Point	Remarks
11	FTC1752	Circlip-man trans 2nd gear thrust washer	2	Note (1)	Suffix H
12	FTC1301	Washer-thrust manual transmission-1st/2nd gear	2	Note (1)	Suffix H
13	FTC2283	Cone-synchronizer inner	NLA	Note (2)	Land Rover recommends that if this cone needs replacement it should be replaced with new 1st/2nd synchro cone STC8577. Suffix H
13	FTC4008	Cone-synchronizer inner	2	Note (3)	Suffix H
14	FTC2284	Ring-synchroniser friction	2	Note (2)	
14	FTC4010	Ring-synchroniser friction	2	Note (3)	
15	FTC916	Collar-distance manual transmission	1	Note (1)	
16	FRC5280	Bearing-needle roller manual transmission	1		
17	FTC1791	Gear-5th speed main shaft manual transmission	1		
18	FRC9388	SYNCHRONISER ASSEMBLY-5TH GEAR MAINSHAFT ASSEMBLY	1		Suffix G
18	FTC1329	SYNCHRONISER ASSEMBLY-5TH GEAR MAINSHAFT ASSEMBLY	1		Suffix H
19	UKC31L	• Spring 1st & 2nd synchroniser	1		
20	UKC3530L	• Plate-synchroniser	3		
21	FRC8232	Ring-baulk-5th gear	1	Note (1)	
22	FRC5235	Plate-5th gear synchroniser support	1		
		SHIM			
23	FRC5284	man trans 5th gear synchroniser hub-5.10mm	A/R	Note (1)	
23	FRC5286	man trans 5th gear synchroniser hub-5.16mm	A/R		
23	FRC5288	man trans 5th gear synchroniser hub-5.22mm	A/R		
23	FRC5290	man trans 5th gear synchroniser hub-5.28mm	A/R		
23	FRC5292	man trans 5th gear synchroniser hub-5.34mm	A/R		
23	FRC5294	man trans 5th gear synchroniser hub-5.40mm	A/R		
23	FRC5296	man trans 5th gear synchroniser hub-5.46mm	A/R		
23	FRC5298	man trans 5th gear synchroniser hub-5.52mm	A/R		
23	FRC5300	man trans 5th gear synchroniser hub-5.58mm	A/R		
23	FRC5302	man trans 5th gear synchroniser hub-5.64mm	A/R		
24	FRC9526	Circlip-man trans 5th gear synchroniser hub	A/R		
25	RTC2914	Bearing-taper roller	1		
26	FRC9812	Circlip	1		

CHANGE POINTS:
(1) To (V) LA 081991
(2) To (G)55A 0242819H
(3) From (G)55A 0242820H To (V) LA 081991

Illus	Part Number	Description	Quantity	Change Point	Remarks
		MANUAL - 4 CYLINDER			
1	FTC1074	Shaft-lay	1	Note (1)	55A Suffix G Suffix H
1	FTC1073	Shaft-lay	1	Note (1)	Manual 4 Cylinder MPi 63A
2	FTC317	Bearing-taper roller	1	Note (1)	Manual 4 Cylinder 55A Suffix G Suffix H except MPi
3	FTC248	Bearing-taper roller	1	Note (1)	Manual 4 Cylinder 55A Suffix H except MPi

CHANGE POINTS:
(1) To (V) LA 081991

DISCOVERY 1989MY UPTO 1999MY | GROUP C | Gearbox 77mm 5 Speed 4 Cylinder Layshaft

Illus	Part Number	Description	Quantity	Change Point	Remarks
		SHIM			
4	FTC263	man trans countershaft bearing/front cover-1.39mm	1	Note (1)	55A Suffix G Suffix H
4	FTC265	man trans countershaft bearing/front cover-1.45mm	1		
4	FTC267	man trans countershaft bearing/front cover-1.51mm	1		
4	FTC269	man trans countershaft bearing/front cover-1.57mm	1		
4	FTC271	man trans countershaft bearing/front cover-1.63mm	1		
4	FTC273	man trans countershaft bearing/front cover-1.69mm	1		
4	FTC275	man trans countershaft bearing/front cover-1.75mm	1		
4	FTC277	man trans countershaft bearing/front cover-1.81mm	1		
4	FTC279	man trans countershaft bearing/front cover-1.87mm	1		
4	FTC281	man trans countershaft bearing/front cover-1.93mm	1		
4	FTC283	man trans countershaft bearing/front cover-1.99mm	1		
4	FTC285	man trans countershaft bearing/front cover-2.05mm	1		
4	FTC287	man trans countershaft bearing/front cover-2.11mm	1		
4	FTC289	man trans countershaft bearing/front cover-2.17mm	1		
4	FTC291	man trans countershaft bearing/front cover-2.23mm	1		
4	FTC293	man trans countershaft bearing/front cover-2.29mm	1		
4	FTC295	man trans countershaft bearing/front cover-2.35mm	1		

CHANGE POINTS:
(1) To (V) LA 081991

Illus	Part Number	Description	Quantity	Change Point	Remarks
5	FTC419	Gear-5th speed countershaft manual transmission	1	Note (1)	Suffix G Suffix H
5	FTC1562	Gear-5th speed countershaft manual transmission	1		Suffix H Suffix G
6	FRC7214	Nut-stake	1		Suffix G Suffix H
7	FRC8285	Gear-reverse manual transmission	1		Suffix G Suffix H
8	FRC4947	Spacer	1		Suffix G Suffix H
9	FRC5186	Spacer-rear	1		Suffix G Suffix H
10	FRC5095	Shaft-reverse idler assembly-manual transmission	1		Suffix G Suffix H
11	UKC18L	Pin	1		Suffix G Suffix H
12	FRC8246	Lever-selector selector mechanism	1		Suffix G Suffix H
13	FRC8382	Pin-reverse gear selector lever pivot	1		Suffix G Suffix H
14	13H2023L	Circlip	1		Suffix G Suffix H
15	FTC1435	Pad-reverse gear selector lever	1		Suffix G Suffix H

CHANGE POINTS:
(1) To (V) LA 081991

C 17

Illus	Part Number	Description	Quantity	Change Point	Remarks
1	FTC1491	SHAFT & LEVER ASSEMBLY-SELECTOR MANUAL TRANSMISSION	NLA	Note (1)	Use FRC8127 qty 2 with FRC8475 qty 1 with FTC1490 qty 1. Suffix G
1	FTC1316	SHAFT & LEVER ASSEMBLY-SELECTOR MANUAL TRANSMISSION	1	Note (1)	Suffix H
	FRC8127	• Pin-selector lever/shaft manual transmission	1		
2	FTC1490	• Fork assembly-selector-1st/2nd manual transmission	1		Part of FTC1491.
2	FTC2450	• Fork assembly-selector-1st/2nd manual transmission	1		Part of FTC1316.
3	BLS112L	Ball-detent manual transmission	1	Note (1)	
4	FRC7195	Spring-detent manual transmission	1		Suffix G
5	UKC75	Plug	1		Suffix G
4	FTC2195	Spring-detent manual transmission	1		Suffix H
5	FTC2193	Plug	1		Suffix H
6	FTC4536	Screw-grub	1		
7	FTC1486	YOKE ASSEMBLY-SELECTOR ROD	1	Note (2)	
	FTC1488	• Pin-selector manual transmission	1		
7	STC1063	Yoke assembly-selector rod	1	Note (3)	
8	FRC9986	Bush-selector yoke spherical	1	Note (2)	Manual 4 Cylinder
8	FTC2203	Bush-selector yoke spherical	1	Note (3)	
9	FRC8882	Ring-snap	1	Note (1)	

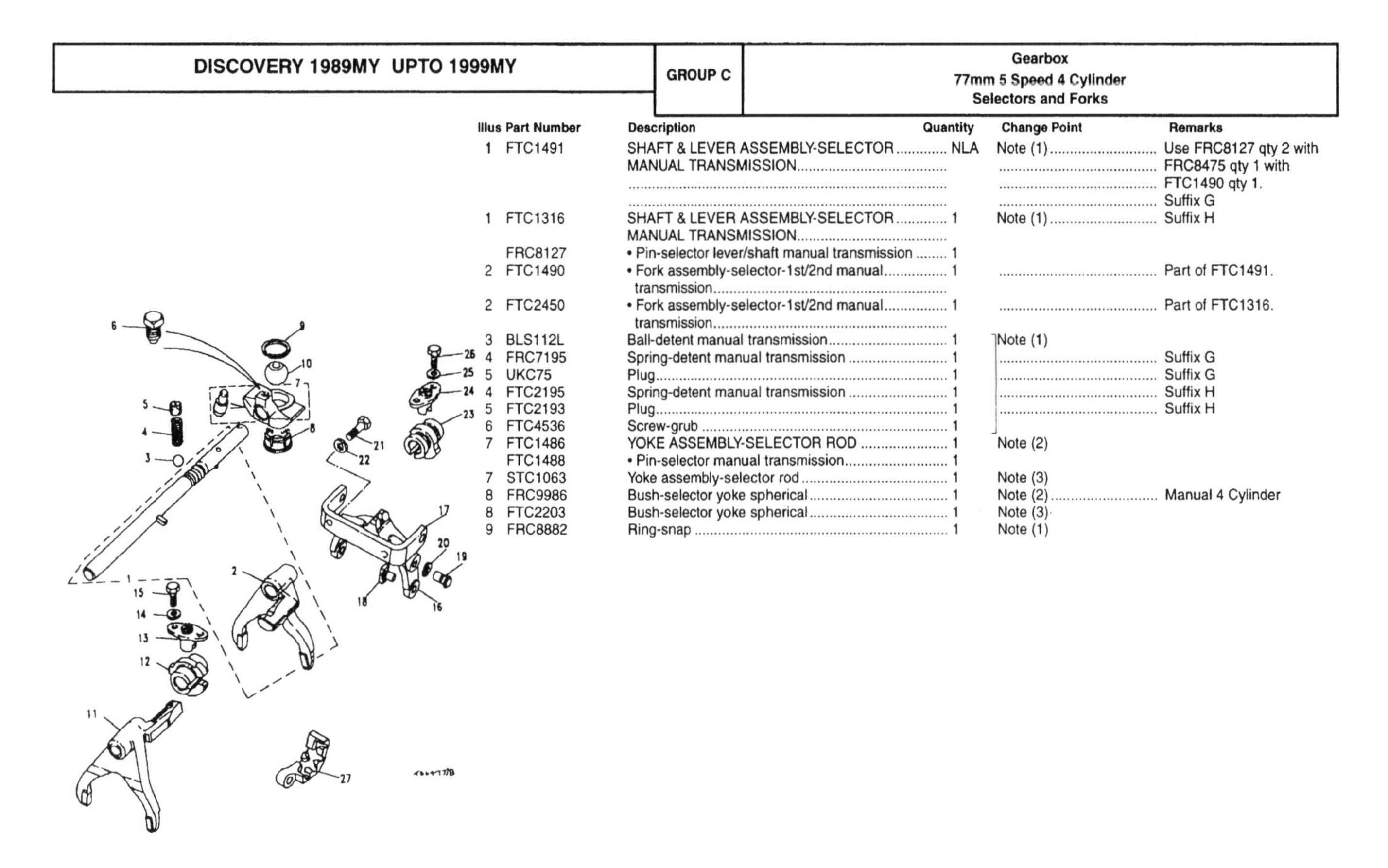

CHANGE POINTS:
(1) To (V) LA 081991
(2) To (G)55A 119840H
(3) From (G)55A 119841H To (V) LA 081991

C 18

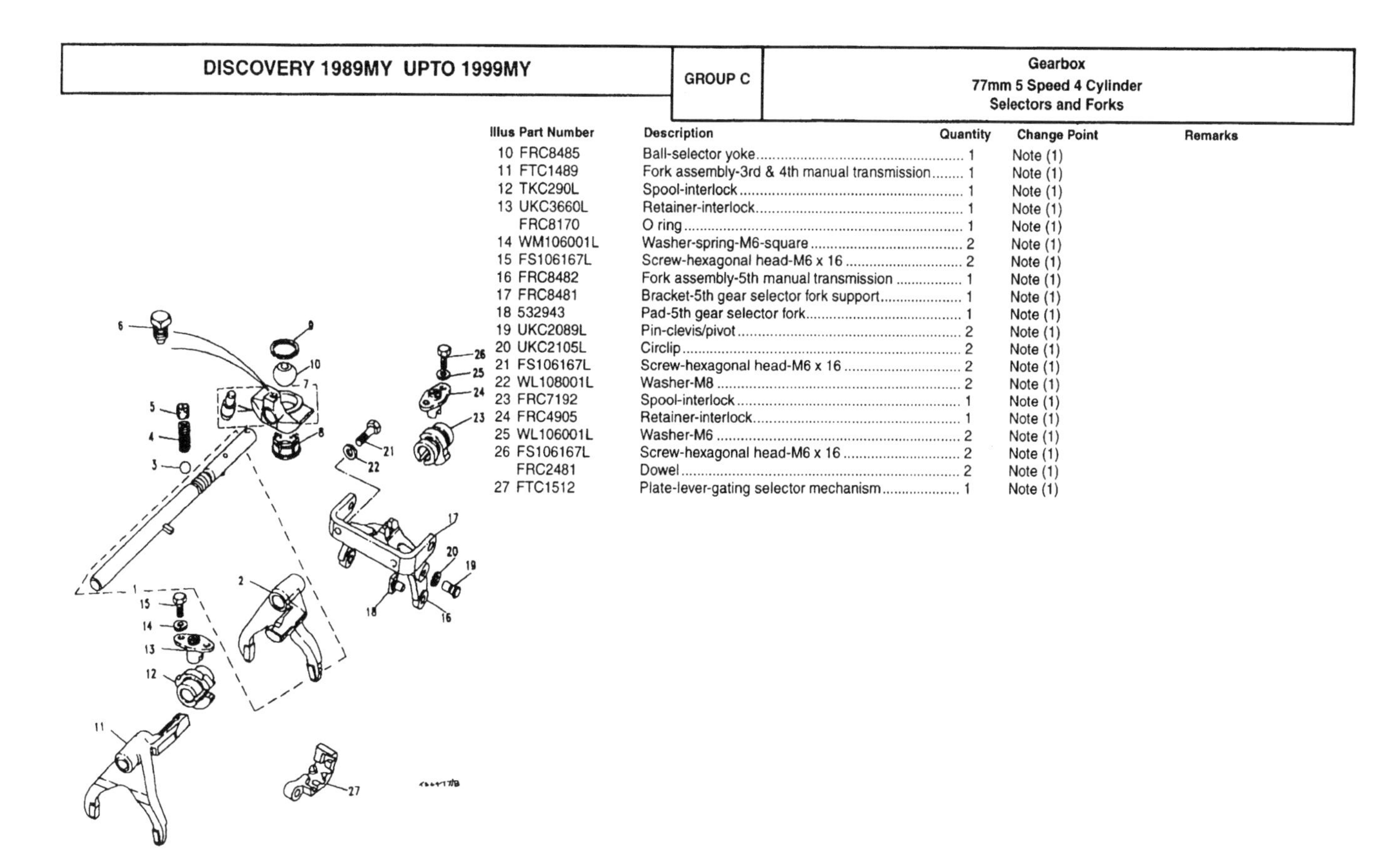

Illus	Part Number	Description	Quantity	Change Point	Remarks
10	FRC8485	Ball-selector yoke	1	Note (1)	
11	FTC1489	Fork assembly-3rd & 4th manual transmission	1	Note (1)	
12	TKC290L	Spool-interlock	1	Note (1)	
13	UKC3660L	Retainer-interlock	1	Note (1)	
	FRC8170	O ring	1	Note (1)	
14	WM106001L	Washer-spring-M6-square	2	Note (1)	
15	FS106167L	Screw-hexagonal head-M6 x 16	2	Note (1)	
16	FRC8482	Fork assembly-5th manual transmission	1	Note (1)	
17	FRC8481	Bracket-5th gear selector fork support	1	Note (1)	
18	532943	Pad-5th gear selector fork	1	Note (1)	
19	UKC2089L	Pin-clevis/pivot	2	Note (1)	
20	UKC2105L	Circlip	2	Note (1)	
21	FS106167L	Screw-hexagonal head-M6 x 16	2	Note (1)	
22	WL108001L	Washer-M8	2	Note (1)	
23	FRC7192	Spool-interlock	1	Note (1)	
24	FRC4905	Retainer-interlock	1	Note (1)	
25	WL106001L	Washer-M6	2	Note (1)	
26	FS106167L	Screw-hexagonal head-M6 x 16	2	Note (1)	
	FRC2481	Dowel	2	Note (1)	
27	FTC1512	Plate-lever-gating selector mechanism	1	Note (1)	

CHANGE POINTS:
(1) To (V) LA 081991

C 19

DISCOVERY 1989MY UPTO 1999MY | GROUP C | Gearbox 77mm 5 Speed 4 Cylinder Gearchange and Lever

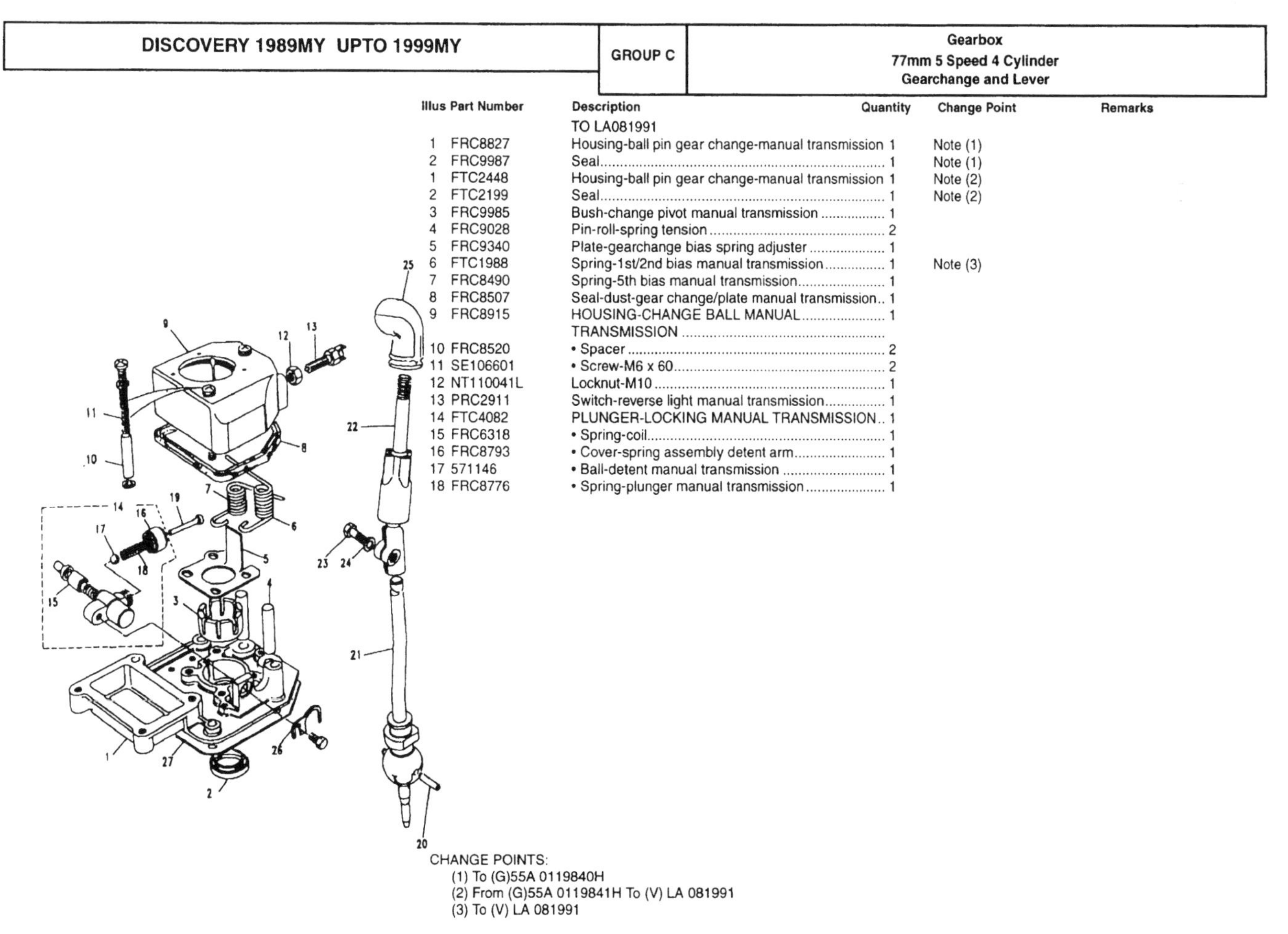

Illus	Part Number	Description	Quantity	Change Point	Remarks
		TO LA081991			
1	FRC8827	Housing-ball pin gear change-manual transmission	1	Note (1)	
2	FRC9987	Seal	1	Note (1)	
1	FTC2448	Housing-ball pin gear change-manual transmission	1	Note (2)	
2	FTC2199	Seal	1	Note (2)	
3	FRC9985	Bush-change pivot manual transmission	1		
4	FRC9028	Pin-roll-spring tension	2		
5	FRC9340	Plate-gearchange bias spring adjuster	1		
6	FTC1988	Spring-1st/2nd bias manual transmission	1	Note (3)	
7	FRC8490	Spring-5th bias manual transmission	1		
8	FRC8507	Seal-dust-gear change/plate manual transmission	1		
9	FRC8915	HOUSING-CHANGE BALL MANUAL TRANSMISSION	1		
10	FRC8520	• Spacer	2		
11	SE106601	• Screw-M6 x 60	2		
12	NT110041L	Locknut-M10	1		
13	PRC2911	Switch-reverse light manual transmission	1		
14	FTC4082	PLUNGER-LOCKING MANUAL TRANSMISSION	1		
15	FRC6318	• Spring-coil	1		
16	FRC8793	• Cover-spring assembly detent arm	1		
17	571146	• Ball-detent manual transmission	1		
18	FRC8776	• Spring-plunger manual transmission	1		

CHANGE POINTS:
(1) To (G)55A 0119840H
(2) From (G)55A 0119841H To (V) LA 081991
(3) To (V) LA 081991

C 20

Illus	Part Number	Description	Quantity	Change Point	Remarks
		TO (V) LA081991			
19	FRC8517	Plunger-reverse switch	1		
20	UKC3092	Pin	1		
21	FRC9669	Lever-lower manual transmission	1	Note (1)	
21	FTC4026	Lever-lower manual transmission	1	Note (2)	
22	FTC4129	Lever-upper manual transmission	1		
23	SH108251L	Screw-hexagonal head-M8 x 25	NLA		Use FS108251L.
23	FS108251L	Screw-flanged head-M8 x 25	NLA		Use FS108257L.
24	WA108051L	Washer-plain-M8	1		
25	MXC5322LUN	Knob assembly-change manual transmission-Ash Grey	1	Note (3)	
25	MXC5322LNF	Knob assembly-change manual transmission-Ash Grey	1	Note (4)	
26	FRC8496	Spacer-reverse gear plunger	1		
26	FRC8497	Spacer-reverse gear plunger	1		
27	FRC8505	Gasket-gearchange ball pin housing	1		
	FRC2626	Pin-spring	2		

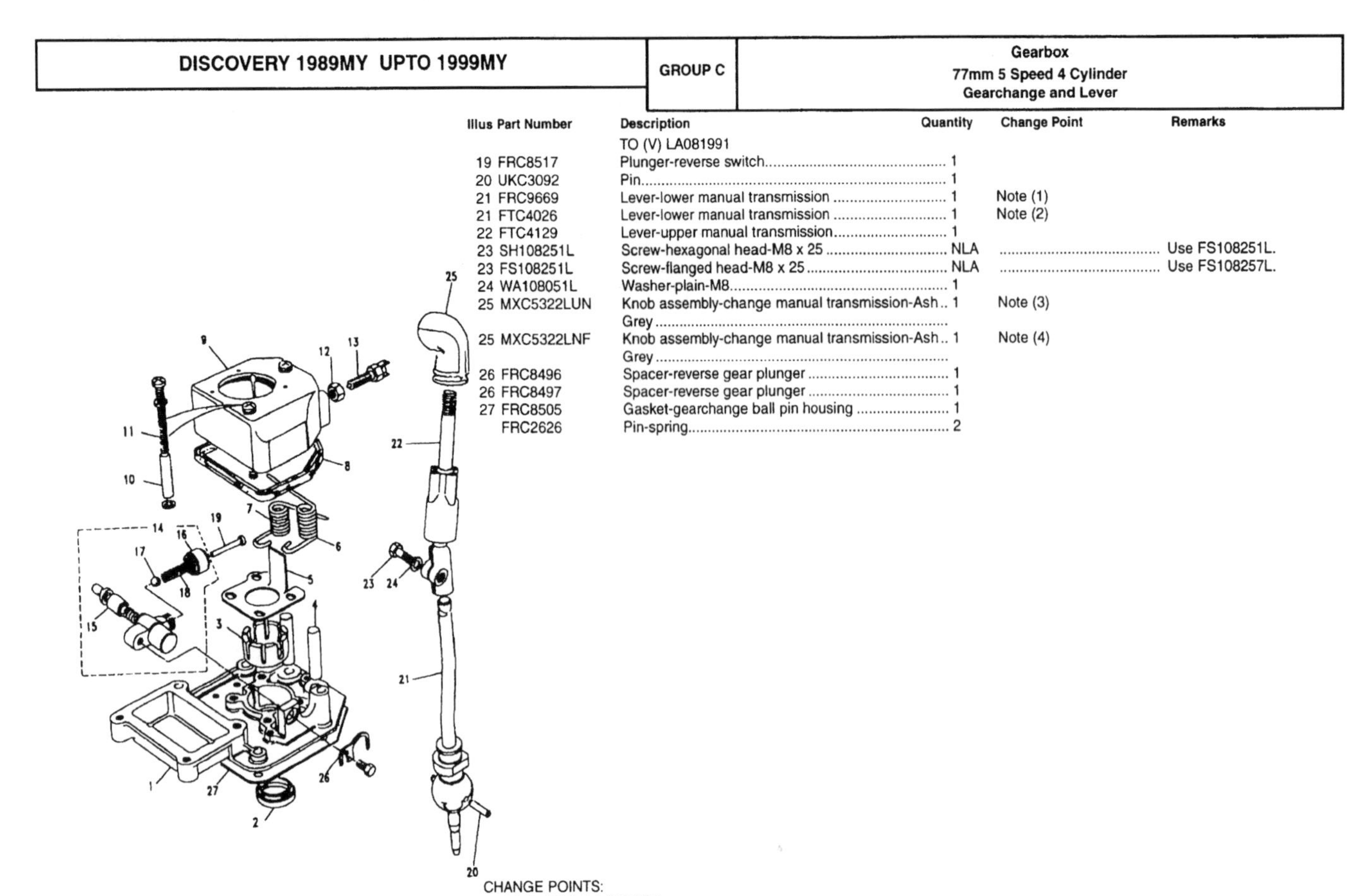

CHANGE POINTS:
(1) To (G)55A 0119840H
(2) From (G)55A 0119841H To (V) LA 081991
(3) To (V) JA 034313
(4) From (V) KA 034314; To (V) LA 081991

Illus	Part Number	Description	Quantity	Change Point	Remarks
		TDI			
1	NTC7558	Bracket-transmission mounting	1		
2	SH110251L	Screw-hexagonal head-M10 x 25	2		
3	WA110061L	Washer-plain-M10-standard	2		
4	SH108161L	Screw	2		
5	WA108051L	Washer-plain-M8	2		
6	SH108251L	Screw-hexagonal head-M8 x 25	NLA		Use FS108251L.
6	FS108251L	Screw-flanged head-M8 x 25	NLA		Use FS108257L.
6	FS108251L	Screw-flanged head-M8 x 25	NLA		Use FS108257L.
7	WA108051L	Washer-plain-M8	3		
8	NTC7853	Bar-gearbox snubbing	1		
9	SH110251L	Screw-hexagonal head-M10 x 25	2		
10	WA110061L	Washer-plain-M10-standard	4		
11	NY110047L	Nut-hexagonal head-nyloc-M10	2		
12	NTC6282	Mounting-gearbox	1		

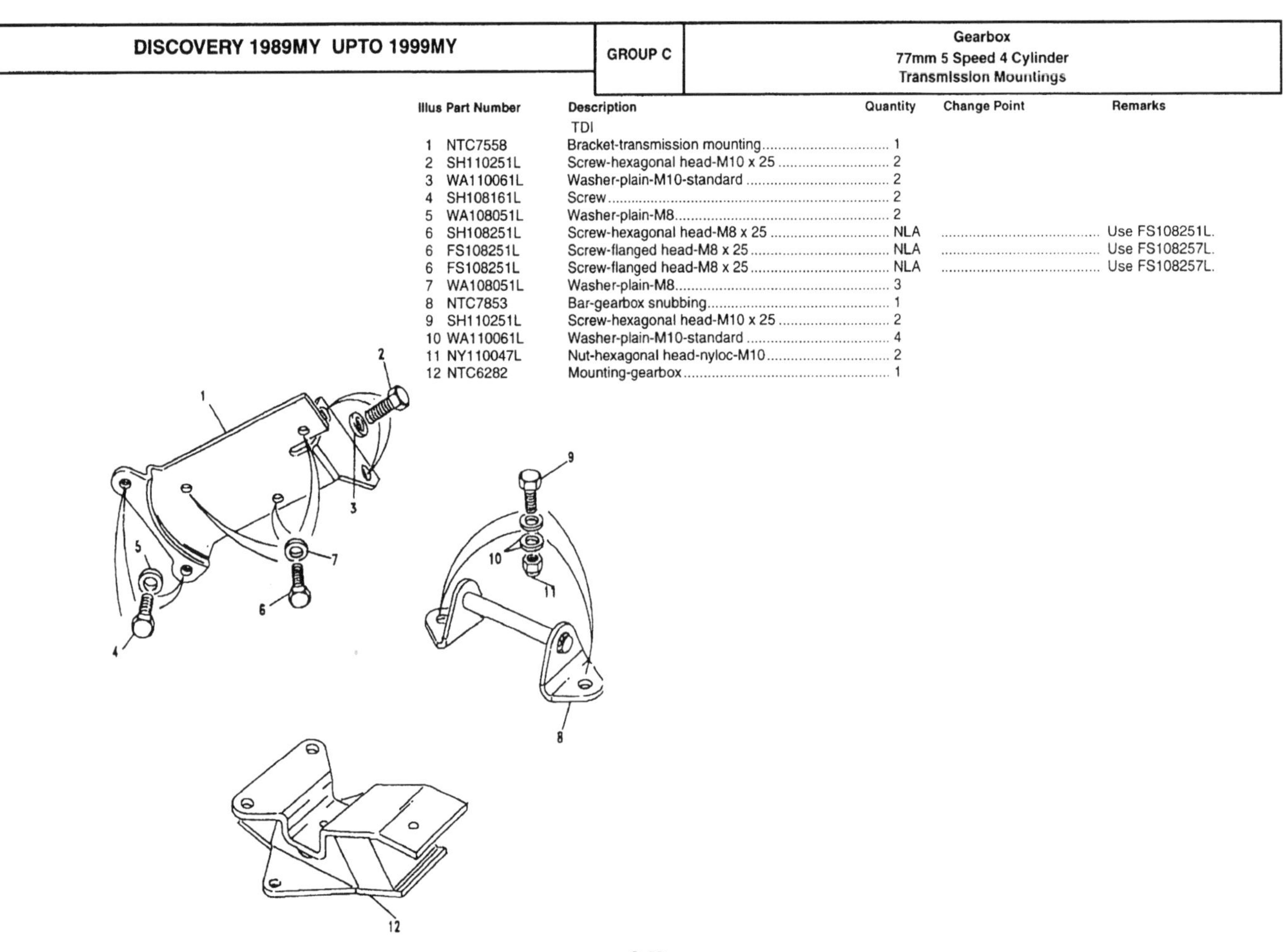

Illus	Part Number	Description	Quantity	Change Point	Remarks
		TO (V) LA081991			
1	RTC6722N	Transmission assembly manual-new	1		3500 cc Twin Carburettor
1	FTC4494	Transmission assembly manual-new	1		3500 cc EFi Suffix H
1	FTC4489	Transmission assembly manual-new	1		3900 cc EFi Suffix H
1	RTC6722R	Transmission assembly manual-reconditioned	1		Twin Carburettor
1	STC157R	Transmission assembly manual-reconditioned	1		EFi 59A Suffix G
1	STC1003R	Transmission assembly manual-reconditioned	1		EFi 59A Suffix H
1	STC1000R	Transmission assembly manual-reconditioned	1		EFi 53A Suffix H
	STC2772	Kit-fitting-gearbox R380	1		
	STC2773	Kit-gaiter-gearbox	1		

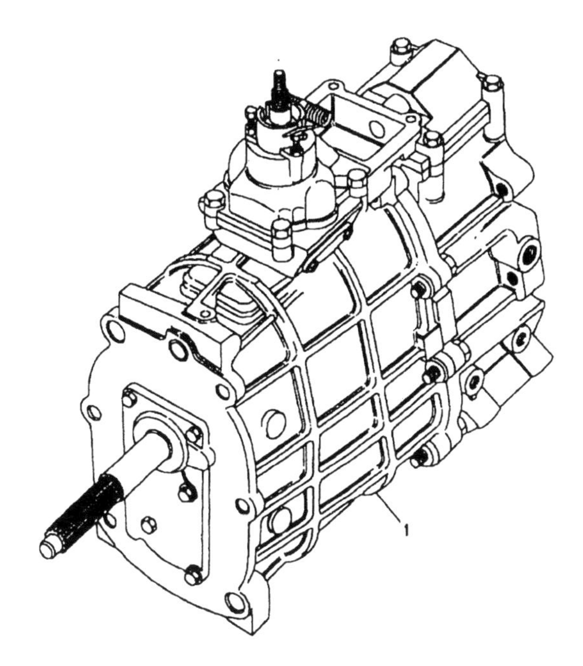

C 23

Illus	Part Number	Description	Quantity	Change Point	Remarks
		TO (V) LA081991			
1	FRC4803	Guide-clutch release bearing	1		
2	FRC2481	Dowel	2		
3	WL108001L	Washer-M8	1		
4	SH108401L	Screw-hexagonal head-M8 x 40	1		
5	FRC2528	Pivot-clutch release lever	1		
6	SH106101L	Screw-hexagonal head-M6 x 10	NLA		Use FS106101L.
6	FS106101L	Screw-flanged head-M6 x 10	1		Manual V8
7	WL106001L	Washer-M6	1		
8	571163	Clip-clutch release-pivot post	1		
9	FRC2975	Cap-clutch release lever pivot	1		
10	571161	Insert-clutch lever	1		
11	576203	Retention-bearing clutch release	1		
12	576137	Arm-clutch release clutch	1	To (V) KA 045958	
12	FTC2957	Arm-clutch release clutch	1	From (V) KA 045959	
13	576723	Clip	1		
14	571160	Rod assembly-push slave cylinder	1		
15	FRC9568	Bearing-clutch release	1		
16	591988	Plate	1	To (V) KA 045958	
16	FRC2402	Plate	1	From (V) KA 045959	
17	WL108001L	Washer-M8	2		
18	SH108251L	Screw-hexagonal head-M8 x 25	NLA		Use FS108251L.
18	FS108251L	Screw-flanged head-M8 x 25	NLA		Use FS108257L.
					Manual V8
19	TKC2786L	SLAVE CYLINDER CLUTCH	1	To (V) KA 045958	
	514244	• Kit-repair clutch slave cylinder	1		
19	FTC2498	Slave cylinder clutch	1	From (V) KA 045959	
20	606733	Screw-bleed clutch slave cylinder	1		
21	594091	Cap-protective	1		

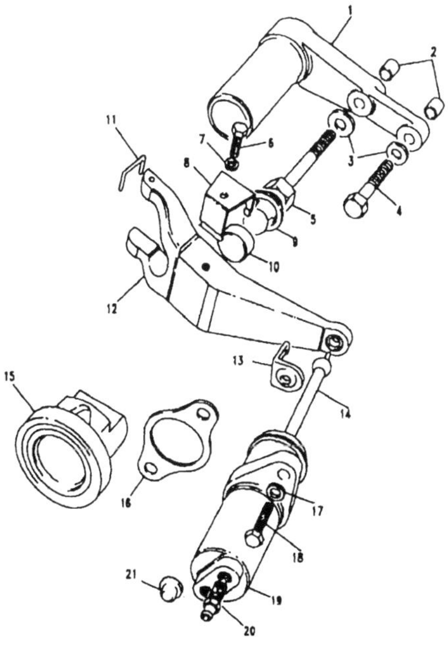

C 24

DISCOVERY 1989MY UPTO 1999MY	GROUP C	Gearbox 77mm 5 Speed V8 Gearcase

Illus	Part Number	Description	Quantity	Change Point	Remarks
1	FRC6950	Pin-reverse gear selector lever pivot	1	Note (1)	
2	WA110061L	Washer-plain-M10-standard	1	Note (1)	
3	WL100001L	Washer	1	Note (1)	Manual V8
4	NH100041L	Locknut	1	Note (1)	
5	UKC170L	Dowel	2	Note (1)	
6	FRC6145	Plug and magnet assembly-transmission case	1	Note (1)	
	FTC3720	Plug-drain	NLA	Note (1)	
7	FTC4112	Washer-sealing-man trans drain plug	1	Note (1)	
8	TKC1229L	Gasket-centre plate to extension case	1	Note (1)	
9	TKC1235L	Gasket-centre plate to extension case	1	Note (1)	
10	3292	Plug-level	1	Note (1)	
11	UKC24L	Plug-core-transfer box-6mm	1	Note (1)	
12	FTC5119	Plug-core-selector shaft front-9mm	1	Note (1)	
12	LYQ100090	Plug-core-rivet blanking-9mm	1	Note (1)	Manual V8
13	TKC5779L	Catcher-manual transmission oil	1	Note (1)	
14	STC1036	CASE-MANUAL TRANSMISSION	1	Note (1)	Twin Carburettor
5	UKC170L	• Dowel	2		
14	FTC918	Case-manual transmission	NLA	Note (1)	Use FTC2192 qty 1 with FTC2193 qty 2 with FTC2195 qty 2 with BLS112L qty 2.
14	FTC2192	Case-manual transmission	1	Note (1)	EFi
15	FRC9427	Tube-breather transmission case	1		
15	FRC9430	Tube-breather transmission case	1		
	BAU1689	Clip-p-5/16"	1		

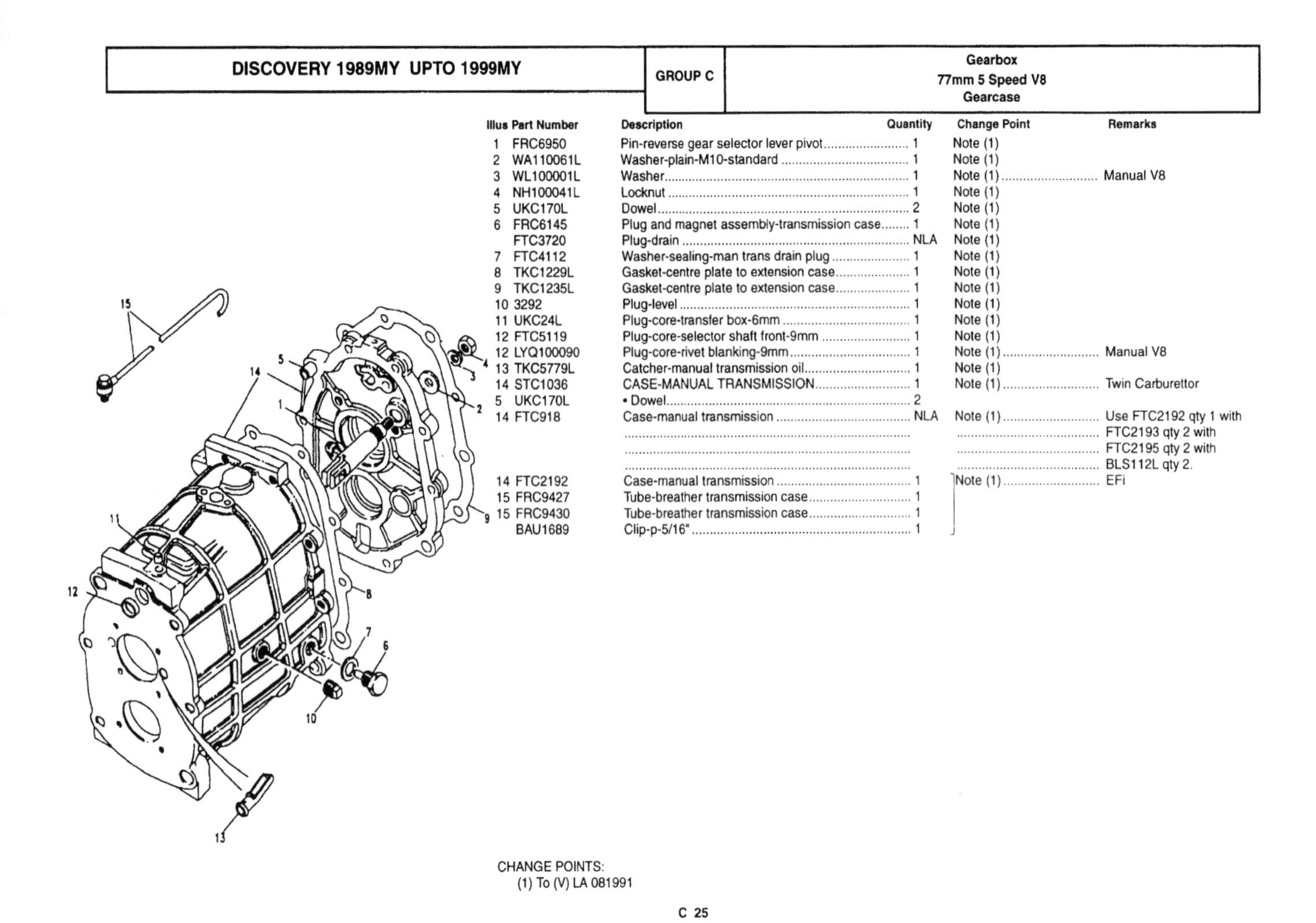

CHANGE POINTS:
(1) To (V) LA 081991

C 25

DISCOVERY 1989MY UPTO 1999MY	GROUP C	Gearbox 77mm 5 Speed V8 Bell Housing

Illus	Part Number	Description	Quantity	Change Point	Remarks
1	FRC6154	Housing-clutch manual transmission-manual	1	Note (1)	Suffix G Suffix H
2	SH112301L	Screw-hexagonal head-M12 x 30	4	Note (1)	
3	WL112001L	Washer	4	Note (1)	
4	BH112091L	Bolt-M12 x 45	2	Note (1)	
5	WA112081L	Washer-plain-M12-standard	2	Note (1)	
6	WL112001L	Washer	2	Note (1)	
7	UKC25L	Dowel	2	Note (1)	
8	3290	Plug-coolant drain	1	Note (1)	
9	BH506161L	Bolt-hexagonal head-3/8UNC x 2	8	Note (1)	
10	WM600061L	Washer-3/8"-square	8	Note (1)	
		FIXINGS GEARBOX TO ENGINE			
9	BH506161L	Bolt-hexagonal head-3/8UNC x 2	8	Note (1)	

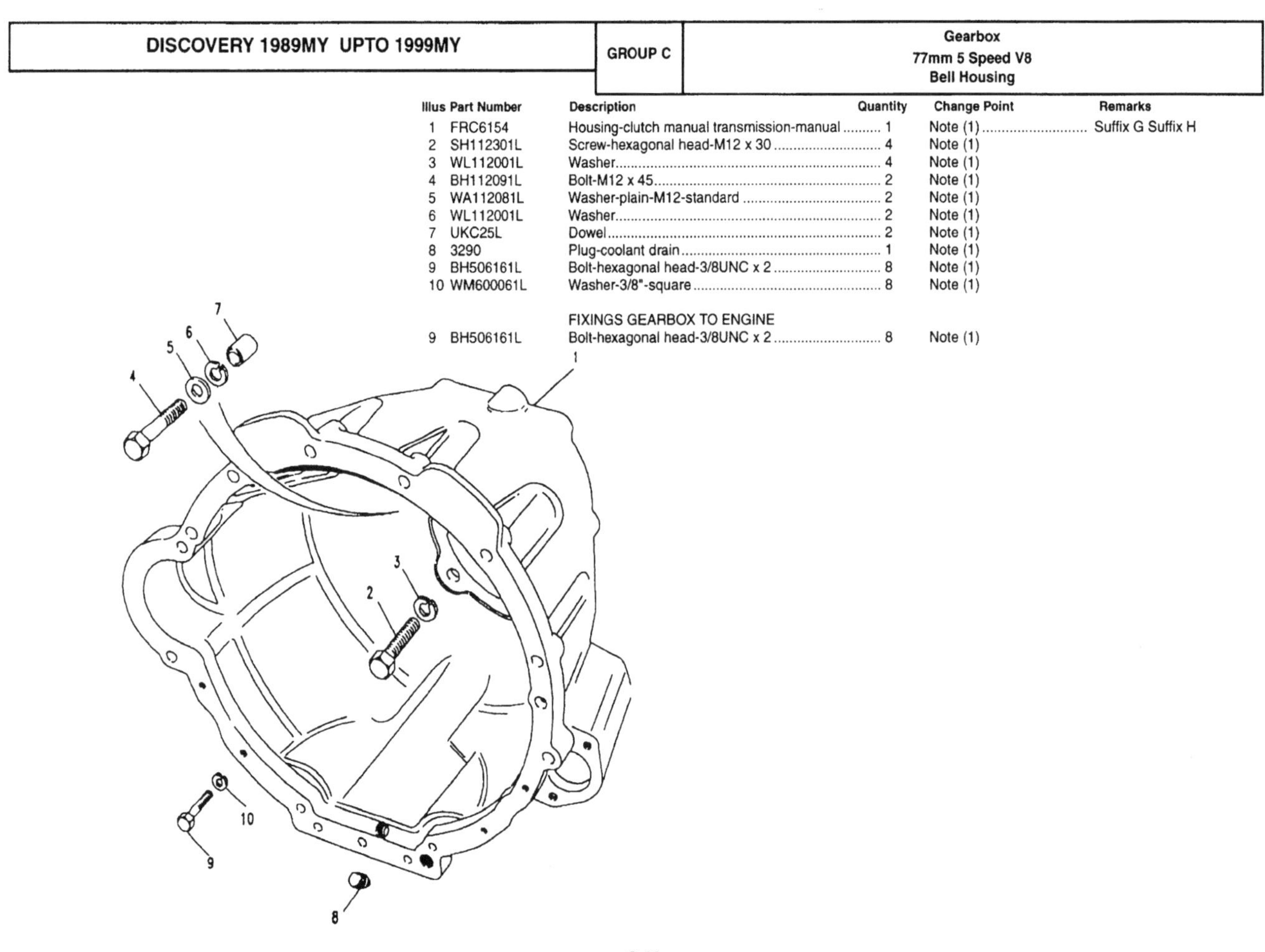

C 26

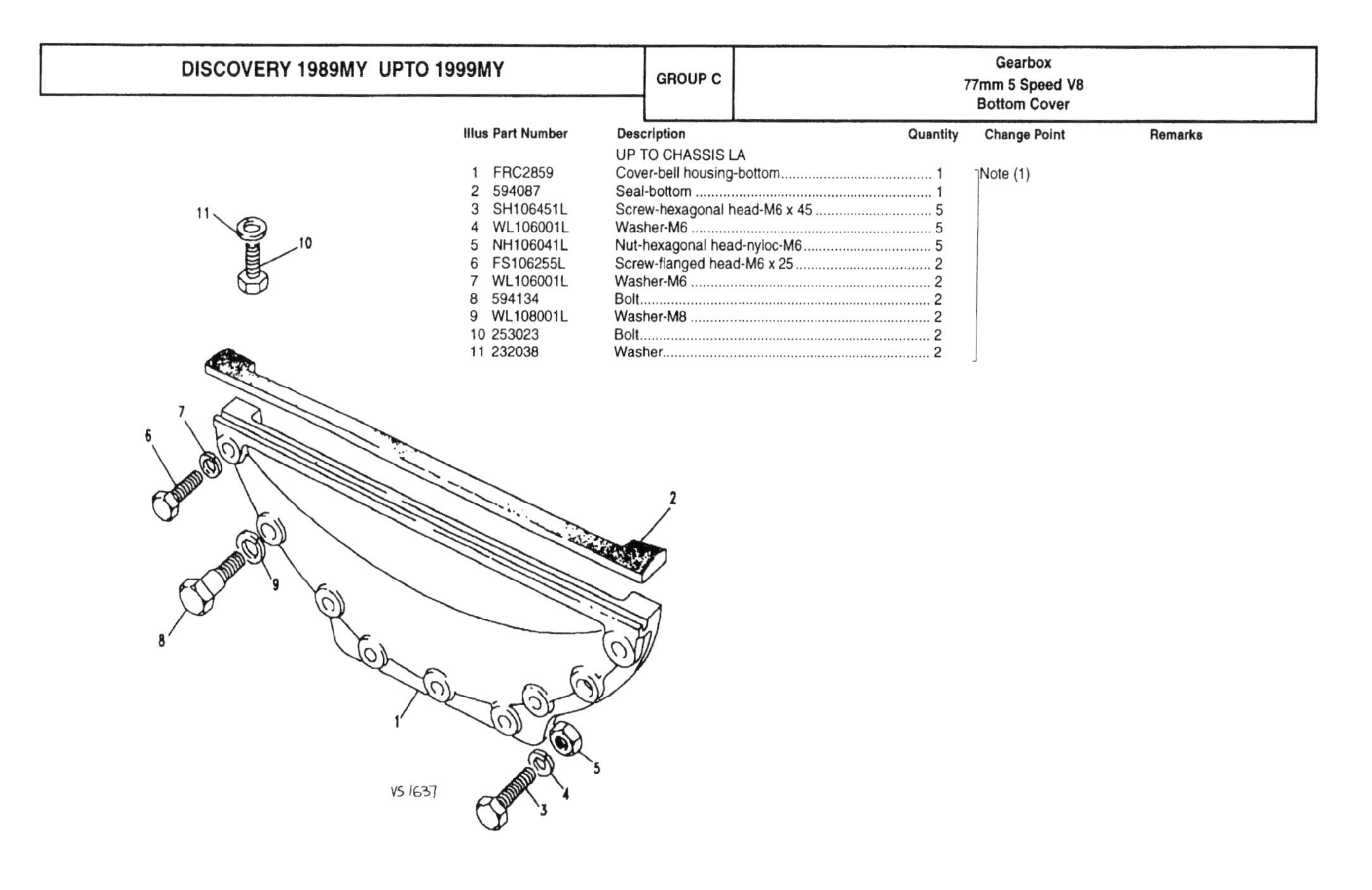

Illus	Part Number	Description	Quantity	Change Point	Remarks
		UP TO CHASSIS LA			
1	FRC2859	Cover-bell housing-bottom	1	Note (1)	
2	594087	Seal-bottom	1		
3	SH106451L	Screw-hexagonal head-M6 x 45	5		
4	WL106001L	Washer-M6	5		
5	NH106041L	Nut-hexagonal head-nyloc-M6	5		
6	FS106255L	Screw-flanged head-M6 x 25	2		
7	WL106001L	Washer-M6	2		
8	594134	Bolt	2		
9	WL108001L	Washer-M8	2		
10	253023	Bolt	2		
11	232038	Washer	2		

CHANGE POINTS:
(1) To (V) LA 081991

C 27

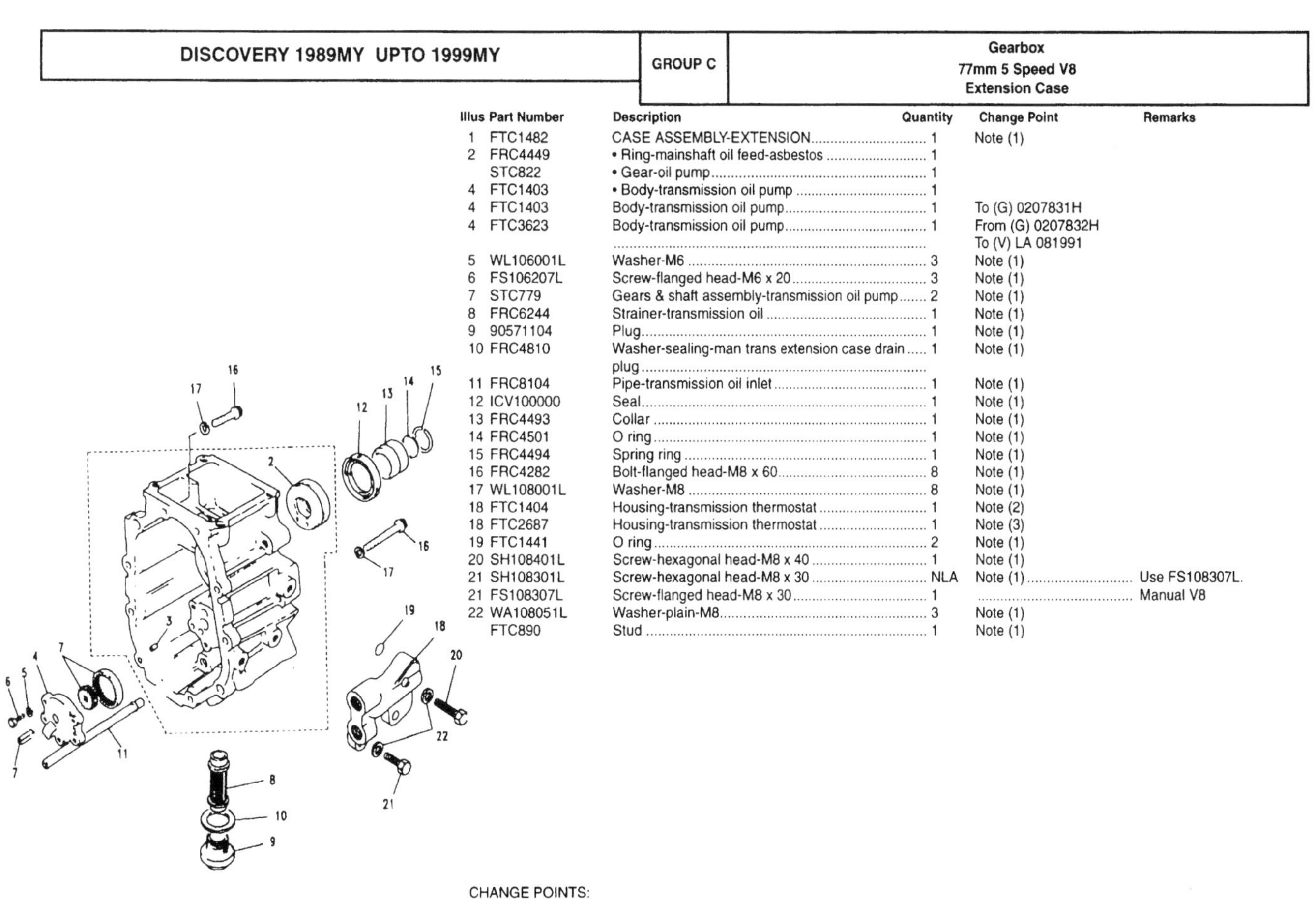

Illus	Part Number	Description	Quantity	Change Point	Remarks
1	FTC1482	CASE ASSEMBLY-EXTENSION	1	Note (1)	
2	FRC4449	• Ring-mainshaft oil feed-asbestos	1		
	STC822	• Gear-oil pump	1		
4	FTC1403	• Body-transmission oil pump	1		
4	FTC1403	Body-transmission oil pump	1	To (G) 0207831H	
4	FTC3623	Body-transmission oil pump	1	From (G) 0207832H To (V) LA 081991	
5	WL106001L	Washer-M6	3	Note (1)	
6	FS106207L	Screw-flanged head-M6 x 20	3	Note (1)	
7	STC779	Gears & shaft assembly-transmission oil pump	2	Note (1)	
8	FRC6244	Strainer-transmission oil	1	Note (1)	
9	90571104	Plug	1	Note (1)	
10	FRC4810	Washer-sealing-man trans extension case drain plug	1	Note (1)	
11	FRC8104	Pipe-transmission oil inlet	1	Note (1)	
12	ICV100000	Seal	1	Note (1)	
13	FRC4493	Collar	1	Note (1)	
14	FRC4501	O ring	1	Note (1)	
15	FRC4494	Spring ring	1	Note (1)	
16	FRC4282	Bolt-flanged head-M8 x 60	8	Note (1)	
17	WL108001L	Washer-M8	8	Note (1)	
18	FTC1404	Housing-transmission thermostat	1	Note (2)	
18	FTC2687	Housing-transmission thermostat	1	Note (3)	
19	FTC1441	O ring	2	Note (1)	
20	SH108401L	Screw-hexagonal head-M8 x 40	1	Note (1)	
21	SH108301L	Screw-hexagonal head-M8 x 30	NLA	Note (1)	Use FS108307L.
21	FS108307L	Screw-flanged head-M8 x 30	1		Manual V8
22	WA108051L	Washer-plain-M8	3	Note (1)	
	FTC890	Stud	1	Note (1)	

CHANGE POINTS:
(1) To (V) LA 081991
(2) To (V) JA 034313
(3) From (V) JA 034314 To (V) LA 081991

C 28

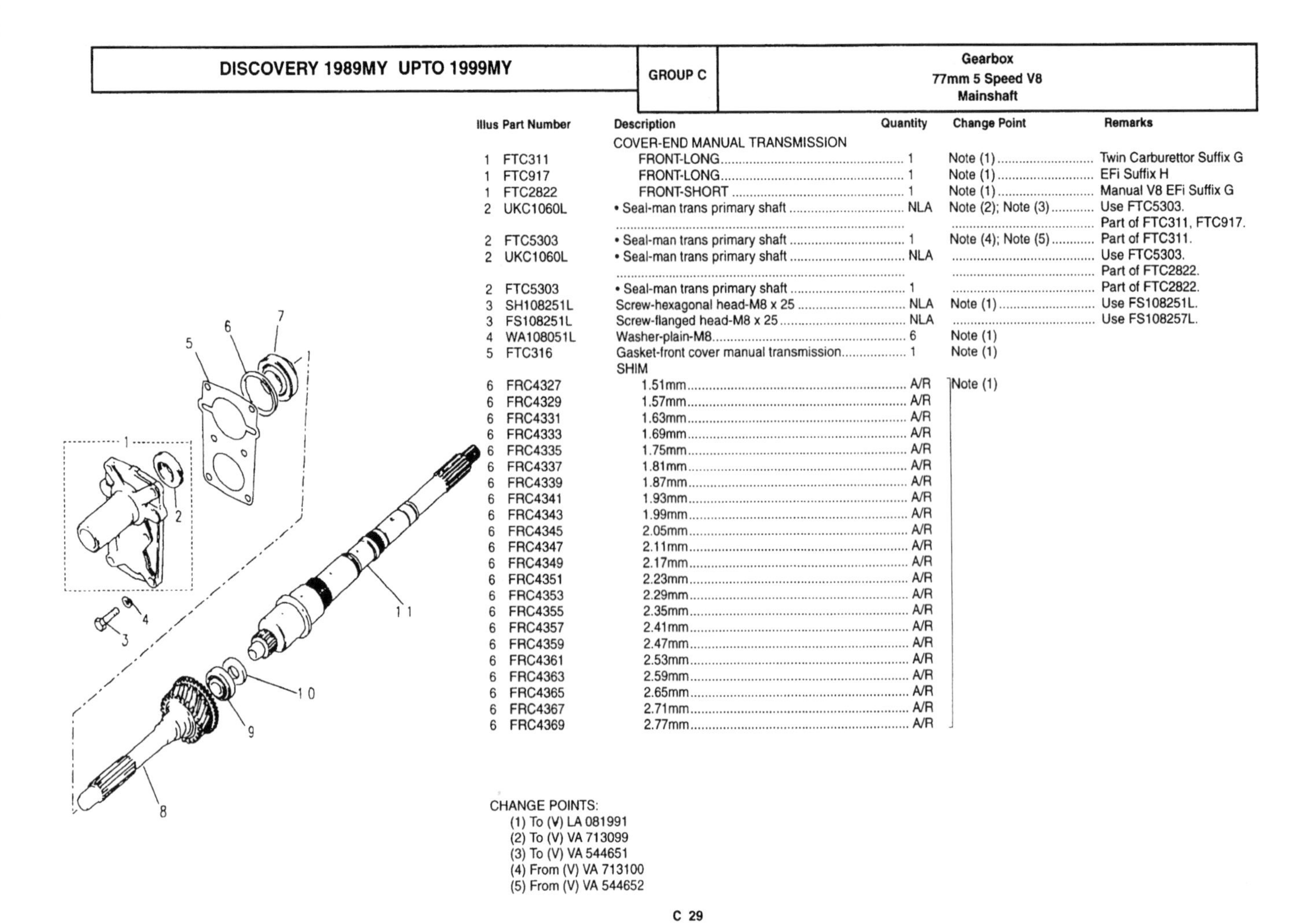

Illus	Part Number	Description	Quantity	Change Point	Remarks
		COVER-END MANUAL TRANSMISSION			
1	FTC311	FRONT-LONG	1	Note (1)	Twin Carburettor Suffix G
1	FTC917	FRONT-LONG	1	Note (1)	EFi Suffix H
1	FTC2822	FRONT-SHORT	1	Note (1)	Manual V8 EFi Suffix G
2	UKC1060L	• Seal-man trans primary shaft	NLA	Note (2); Note (3)	Use FTC5303. Part of FTC311, FTC917.
2	FTC5303	• Seal-man trans primary shaft	1	Note (4); Note (5)	Part of FTC311.
2	UKC1060L	• Seal-man trans primary shaft	NLA		Use FTC5303. Part of FTC2822.
2	FTC5303	• Seal-man trans primary shaft	1		Part of FTC2822.
3	SH108251L	Screw-hexagonal head-M8 x 25	NLA	Note (1)	Use FS108251L.
3	FS108251L	Screw-flanged head-M8 x 25	NLA		Use FS108257L.
4	WA108051L	Washer-plain-M8	6	Note (1)	
5	FTC316	Gasket-front cover manual transmission	1	Note (1)	
		SHIM			
6	FRC4327	1.51mm	A/R	Note (1)	
6	FRC4329	1.57mm	A/R		
6	FRC4331	1.63mm	A/R		
6	FRC4333	1.69mm	A/R		
6	FRC4335	1.75mm	A/R		
6	FRC4337	1.81mm	A/R		
6	FRC4339	1.87mm	A/R		
6	FRC4341	1.93mm	A/R		
6	FRC4343	1.99mm	A/R		
6	FRC4345	2.05mm	A/R		
6	FRC4347	2.11mm	A/R		
6	FRC4349	2.17mm	A/R		
6	FRC4351	2.23mm	A/R		
6	FRC4353	2.29mm	A/R		
6	FRC4355	2.35mm	A/R		
6	FRC4357	2.41mm	A/R		
6	FRC4359	2.47mm	A/R		
6	FRC4361	2.53mm	A/R		
6	FRC4363	2.59mm	A/R		
6	FRC4365	2.65mm	A/R		
6	FRC4367	2.71mm	A/R		
6	FRC4369	2.77mm	A/R		

CHANGE POINTS:
(1) To (V) LA 081991
(2) To (V) VA 713099
(3) To (V) VA 544651
(4) From (V) VA 713100
(5) From (V) VA 544652

C 29

Illus	Part Number	Description	Quantity	Change Point	Remarks
7	RTC6751	Bearing-taper roller	1	Note (1)	
8	FTC347	SHAFT-PINION DIFFERENTIAL	1	Note (1)	
	FTC1293	• Disc synchro stop-mainshaft	1		
9	UKC8L	Bearing-taper roller	1	Note (1)	
10	UKC37L	Spacer	1	Note (1)	
11	FRC5305	Shaft-main manual transmission	1	Note (1)	Suffix G
11	FTC1446	Shaft-main manual transmission	1	Note (1)	
11	FTC1282	Shaft-main manual transmission	NLA	Note (1)	Use STC1889. Suffix H
11	STC1889	Shaft-main manual transmission	1	Note (1)	Suffix H

CHANGE POINTS:
(1) To (V) LA 081991

C 30

Illus	Part Number	Description	Quantity	Change Point	Remarks
1	FRC9792L	SYNCHRONISER ASSEMBLY-3RD & 4TH MAINSHAFT ASSEMBLY	1	Note (1)	Suffix G Suffix H
2	UKC31L	• Spring 1st & 2nd synchroniser			
3	UKC3530L	• Plate-synchroniser			
4	FRC8232	Ring-baulk	2	Note (1)	Suffix G Suffix H
5	FTC358	Gear-3rd driven manual transmission	1		Suffix G Suffix H except EFi
5	FTC2504	Gear-3rd driven manual transmission	1		EFi Suffix G Suffix H
6	FRC5678	Bearing-needle roller	2		Manual V8 Suffix G
6	FTC1313	Bearing-needle roller	2		Suffix H
7	FTC1310	Bush-3rd gear	1		Suffix G Suffix H

CHANGE POINTS:
(1) To (V) LA 081991

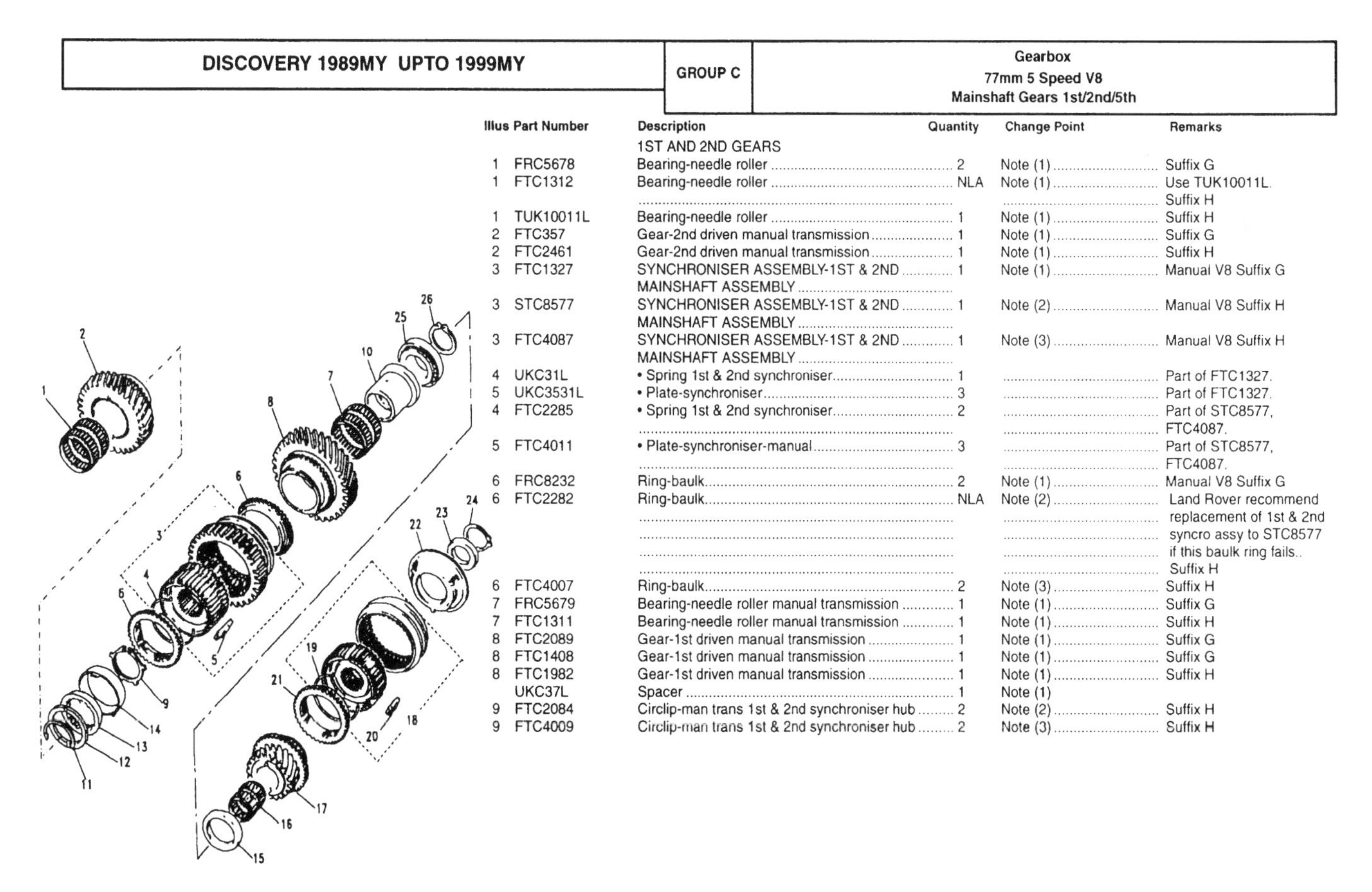

Illus	Part Number	Description	Quantity	Change Point	Remarks
		1ST AND 2ND GEARS			
1	FRC5678	Bearing-needle roller	2	Note (1)	Suffix G
1	FTC1312	Bearing-needle roller	NLA	Note (1)	Use TUK10011L. Suffix H
1	TUK10011L	Bearing-needle roller	1	Note (1)	Suffix H
2	FTC357	Gear-2nd driven manual transmission	1	Note (1)	Suffix G
2	FTC2461	Gear-2nd driven manual transmission	1	Note (1)	Suffix H
3	FTC1327	SYNCHRONISER ASSEMBLY-1ST & 2ND MAINSHAFT ASSEMBLY	1	Note (1)	Manual V8 Suffix G
3	STC8577	SYNCHRONISER ASSEMBLY-1ST & 2ND MAINSHAFT ASSEMBLY	1	Note (2)	Manual V8 Suffix H
3	FTC4087	SYNCHRONISER ASSEMBLY-1ST & 2ND MAINSHAFT ASSEMBLY	1	Note (3)	Manual V8 Suffix H
4	UKC31L	• Spring 1st & 2nd synchroniser	1		Part of FTC1327.
5	UKC3531L	• Plate-synchroniser	3		Part of FTC1327.
4	FTC2285	• Spring 1st & 2nd synchroniser	2		Part of STC8577, FTC4087.
5	FTC4011	• Plate-synchroniser-manual	3		Part of STC8577, FTC4087.
6	FRC8232	Ring-baulk	2	Note (1)	Manual V8 Suffix G
6	FTC2282	Ring-baulk	NLA	Note (2)	Land Rover recommend replacement of 1st & 2nd syncro assy to STC8577 if this baulk ring fails.. Suffix H
6	FTC4007	Ring-baulk	2	Note (3)	Suffix H
7	FRC5679	Bearing-needle roller manual transmission	1	Note (1)	Suffix G
7	FTC1311	Bearing-needle roller manual transmission	1	Note (1)	Suffix H
8	FTC2089	Gear-1st driven manual transmission	1	Note (1)	Suffix G
8	FTC1408	Gear-1st driven manual transmission	1	Note (1)	Suffix G
8	FTC1982	Gear-1st driven manual transmission	1	Note (1)	Suffix H
	UKC37L	Spacer	1	Note (1)	
9	FTC2084	Circlip-man trans 1st & 2nd synchroniser hub	2	Note (2)	Suffix H
9	FTC4009	Circlip-man trans 1st & 2nd synchroniser hub	2	Note (3)	Suffix H

CHANGE POINTS:
(1) To (V) LA 081991
(2) To (G)55A 0242819H
(3) From (G)55A 0242820H To (V) LA 081991

Illus	Part Number	Description	Quantity	Change Point	Remarks
		BUSH-1ST GEAR SELECTIVE			
10	FRC5243	40.16 to 40.21mm	1	Note (1)	Suffix G
10	FRC5244	40.21 to 40.26mm	1	Note (1)	
10	FRC5245	40.26 to 40.31mm	1	Note (1)	
10	FRC5246	40.31 to 40.36mm	1	Note (1)	
10	FRC5247	40.36 to 40.41mm	1	Note (1)	
10	FTC2005	30.905 to 30.955mm	NLA	Note (1)	Use TUJ100050.
10	TUJ100050	30.905 to 30.955mm	1		Suffix H
10	FTC2006	30.955 to 31.005mm	1	Note (1)	Suffix H
10	FTC2007	31.005 to 31.055mm	1	Note (1)	Suffix H
10	FTC2008	31.055 to 31.105mm	1	Note (1)	Suffix H
10	FTC2009	31.105 to 31.155mm	1	Note (1)	Suffix H

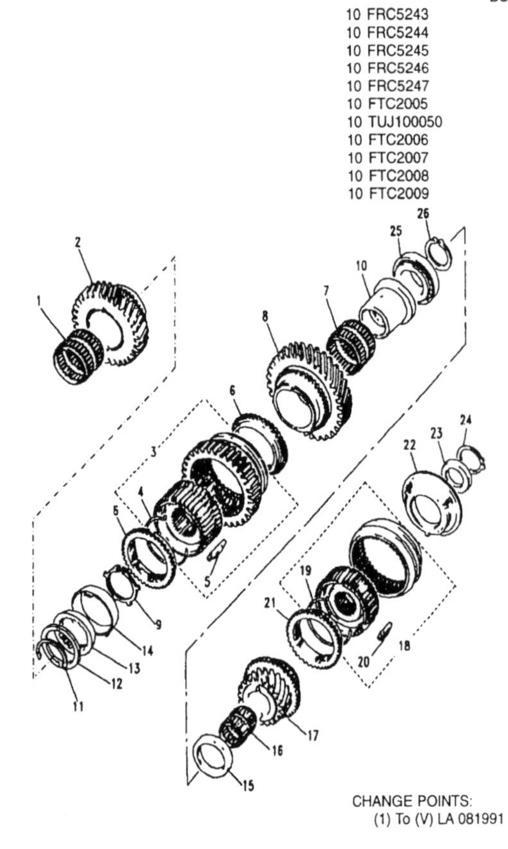

CHANGE POINTS:
(1) To (V) LA 081991

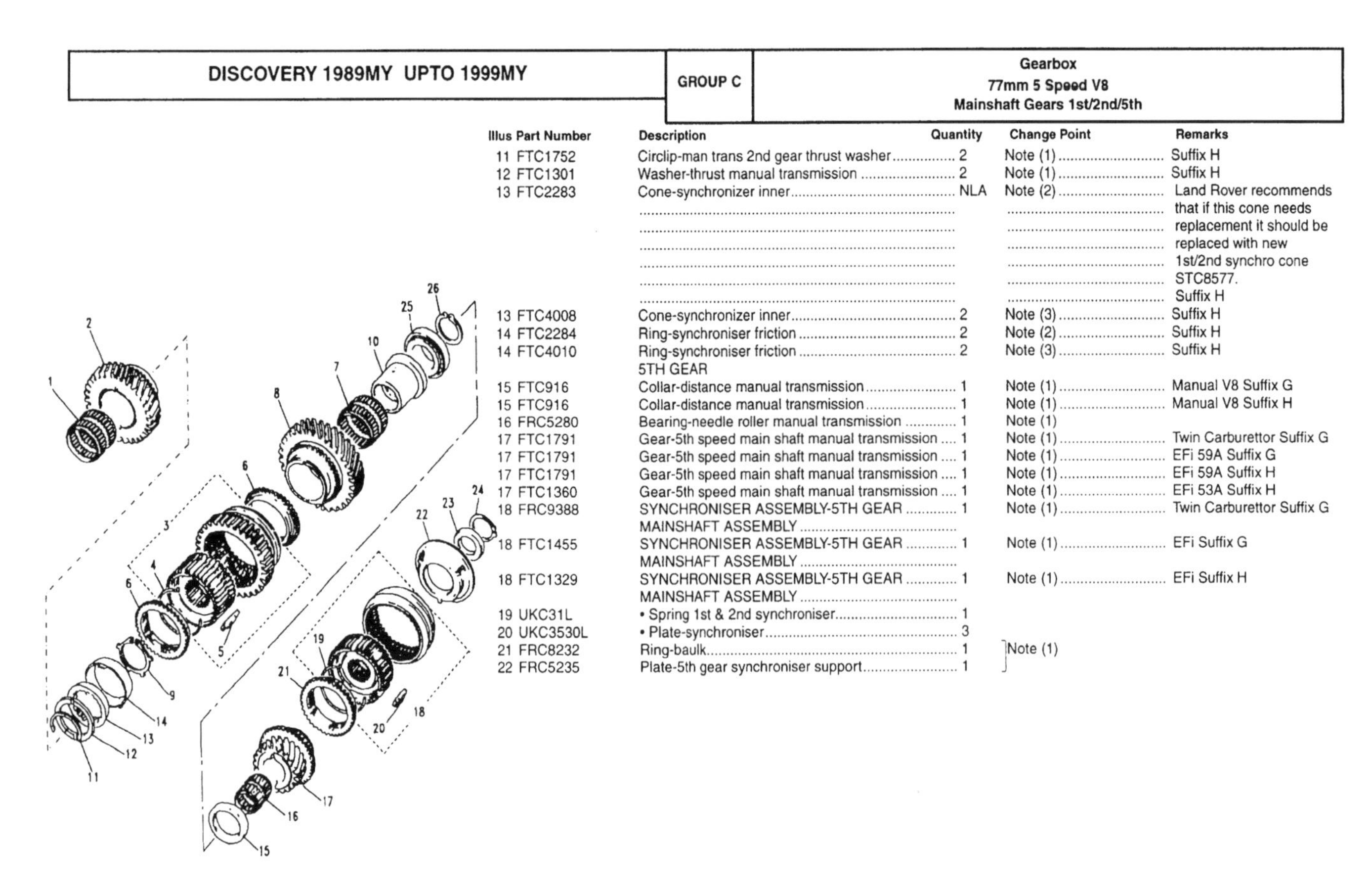

Illus	Part Number	Description	Quantity	Change Point	Remarks
11	FTC1752	Circlip-man trans 2nd gear thrust washer	2	Note (1)	Suffix H
12	FTC1301	Washer-thrust manual transmission	2	Note (1)	Suffix H
13	FTC2283	Cone-synchronizer inner	NLA	Note (2)	Land Rover recommends that if this cone needs replacement it should be replaced with new 1st/2nd synchro cone STC8577. Suffix H
13	FTC4008	Cone-synchronizer inner	2	Note (3)	Suffix H
14	FTC2284	Ring-synchroniser friction	2	Note (2)	Suffix H
14	FTC4010	Ring-synchroniser friction	2	Note (3)	Suffix H
		5TH GEAR			
15	FTC916	Collar-distance manual transmission	1	Note (1)	Manual V8 Suffix G
15	FTC916	Collar-distance manual transmission	1	Note (1)	Manual V8 Suffix H
16	FRC5280	Bearing-needle roller manual transmission	1	Note (1)	
17	FTC1791	Gear-5th speed main shaft manual transmission	1	Note (1)	Twin Carburettor Suffix G
17	FTC1791	Gear-5th speed main shaft manual transmission	1	Note (1)	EFi 59A Suffix G
17	FTC1791	Gear-5th speed main shaft manual transmission	1	Note (1)	EFi 59A Suffix H
17	FTC1360	Gear-5th speed main shaft manual transmission	1	Note (1)	EFi 53A Suffix H
18	FRC9388	SYNCHRONISER ASSEMBLY-5TH GEAR MAINSHAFT ASSEMBLY	1	Note (1)	Twin Carburettor Suffix G
18	FTC1455	SYNCHRONISER ASSEMBLY-5TH GEAR MAINSHAFT ASSEMBLY	1	Note (1)	EFi Suffix G
18	FTC1329	SYNCHRONISER ASSEMBLY-5TH GEAR MAINSHAFT ASSEMBLY	1	Note (1)	EFi Suffix H
19	UKC31L	• Spring 1st & 2nd synchroniser	1		
20	UKC3530L	• Plate-synchroniser	3		
21	FRC8232	Ring-baulk	1	Note (1)	
22	FRC5235	Plate-5th gear synchroniser support	1		

CHANGE POINTS:
(1) To (V) LA 081991
(2) To (G)55A 0242819H
(3) From (G)55A 0242820H To (V) LA 081991

DISCOVERY 1989MY UPTO 1999MY

GROUP C

Gearbox
77mm 5 Speed V8
Mainshaft Gears 1st/2nd/5th

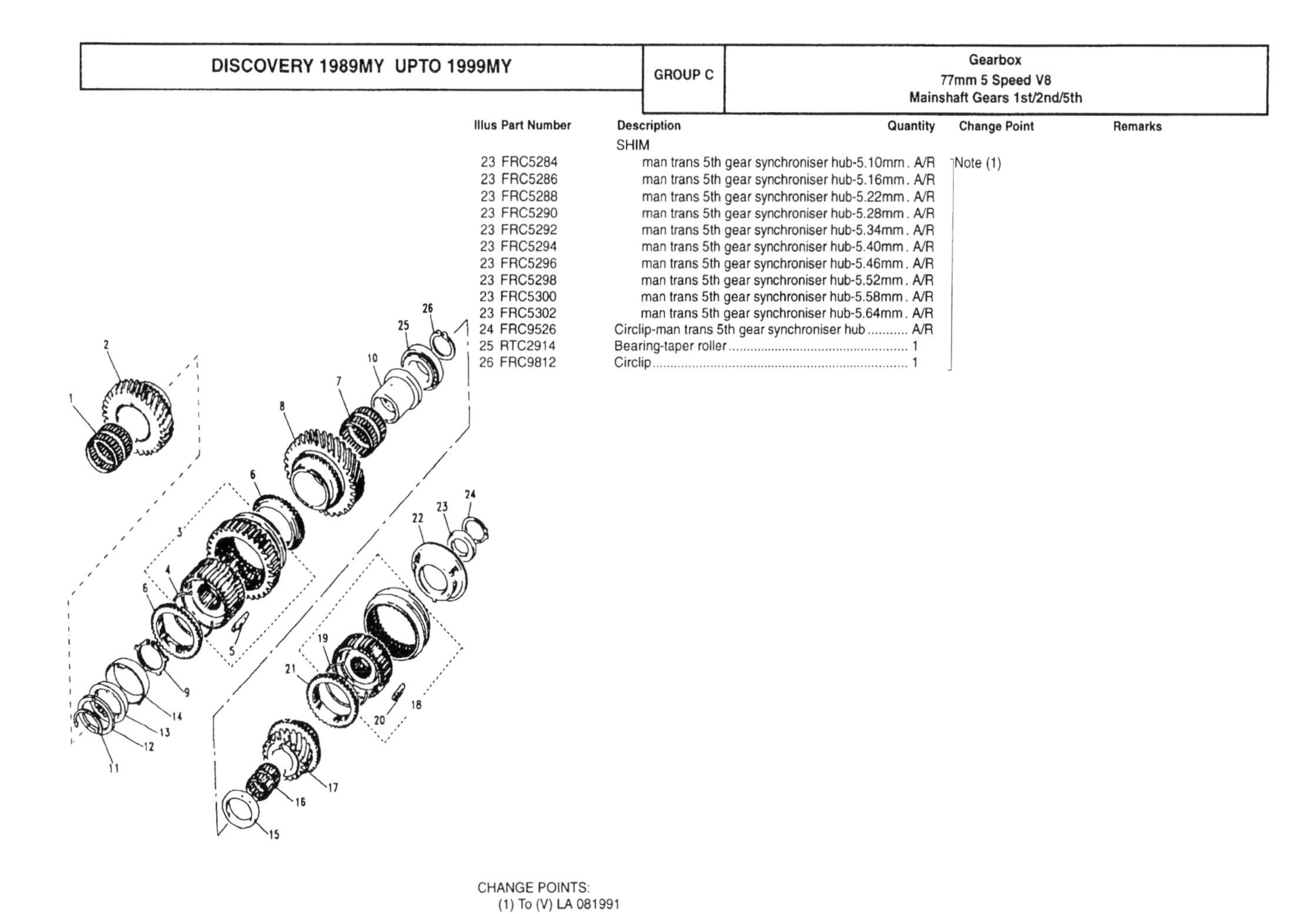

Illus	Part Number	Description	Quantity	Change Point	Remarks
		SHIM			
23	FRC5284	man trans 5th gear synchroniser hub-5.10mm	A/R	Note (1)	
23	FRC5286	man trans 5th gear synchroniser hub-5.16mm	A/R		
23	FRC5288	man trans 5th gear synchroniser hub-5.22mm	A/R		
23	FRC5290	man trans 5th gear synchroniser hub-5.28mm	A/R		
23	FRC5292	man trans 5th gear synchroniser hub-5.34mm	A/R		
23	FRC5294	man trans 5th gear synchroniser hub-5.40mm	A/R		
23	FRC5296	man trans 5th gear synchroniser hub-5.46mm	A/R		
23	FRC5298	man trans 5th gear synchroniser hub-5.52mm	A/R		
23	FRC5300	man trans 5th gear synchroniser hub-5.58mm	A/R		
23	FRC5302	man trans 5th gear synchroniser hub-5.64mm	A/R		
24	FRC9526	Circlip-man trans 5th gear synchroniser hub	A/R		
25	RTC2914	Bearing-taper roller	1		
26	FRC9812	Circlip	1		

CHANGE POINTS:
(1) To (V) LA 081991

C 35

DISCOVERY 1989MY UPTO 1999MY

GROUP C

Gearbox
77mm 5 Speed V8
Layshaft

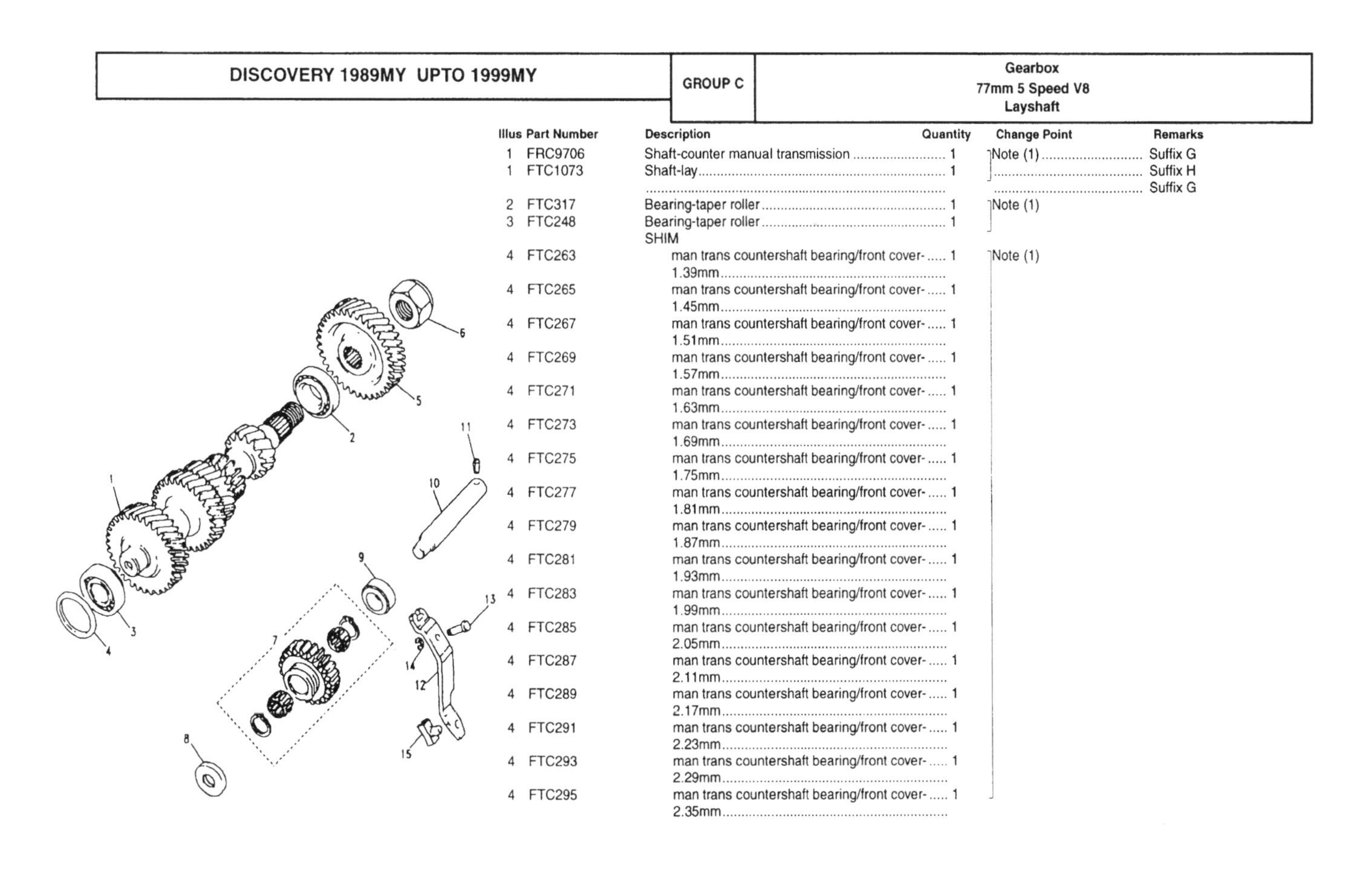

Illus	Part Number	Description	Quantity	Change Point	Remarks
1	FRC9706	Shaft-counter manual transmission	1	Note (1)	Suffix G
1	FTC1073	Shaft-lay	1		Suffix H
					Suffix G
2	FTC317	Bearing-taper roller	1	Note (1)	
3	FTC248	Bearing-taper roller	1		
		SHIM			
4	FTC263	man trans countershaft bearing/front cover-1.39mm	1	Note (1)	
4	FTC265	man trans countershaft bearing/front cover-1.45mm	1		
4	FTC267	man trans countershaft bearing/front cover-1.51mm	1		
4	FTC269	man trans countershaft bearing/front cover-1.57mm	1		
4	FTC271	man trans countershaft bearing/front cover-1.63mm	1		
4	FTC273	man trans countershaft bearing/front cover-1.69mm	1		
4	FTC275	man trans countershaft bearing/front cover-1.75mm	1		
4	FTC277	man trans countershaft bearing/front cover-1.81mm	1		
4	FTC279	man trans countershaft bearing/front cover-1.87mm	1		
4	FTC281	man trans countershaft bearing/front cover-1.93mm	1		
4	FTC283	man trans countershaft bearing/front cover-1.99mm	1		
4	FTC285	man trans countershaft bearing/front cover-2.05mm	1		
4	FTC287	man trans countershaft bearing/front cover-2.11mm	1		
4	FTC289	man trans countershaft bearing/front cover-2.17mm	1		
4	FTC291	man trans countershaft bearing/front cover-2.23mm	1		
4	FTC293	man trans countershaft bearing/front cover-2.29mm	1		
4	FTC295	man trans countershaft bearing/front cover-2.35mm	1		

CHANGE POINTS:
(1) To (V) LA 081991

C 36

DISCOVERY 1989MY UPTO 1999MY	GROUP C	Gearbox 77mm 5 Speed V8 Layshaft

Illus	Part Number	Description	Quantity	Change Point	Remarks
5	FTC346	Gear-5th speed countershaft manual transmission	1	Note (1)	Suffix G
5	FTC419	Gear-5th speed countershaft manual transmission	1	Note (1)	59A Suffix G
					Suffix H
5	FTC1359	Gear-5th speed countershaft manual transmission	1	Note (1)	53A Suffix H
6	FRC7214	Nut-stake	1		
7	FRC8285	Gear-reverse manual transmission	1		
8	FRC4947	Spacer	1		
9	FRC5186	Spacer-rear	1		
10	FRC5095	Shaft-reverse idler assembly-manual transmission	1		
11	UKC18L	Pin	1		
12	FRC8246	Lever-selector selector mechanism	1		
13	FRC8382	Pin-reverse gear selector lever pivot	1		
14	13H2023L	Circlip	1		
15	FTC1435	Pad-reverse gear selector lever	1		

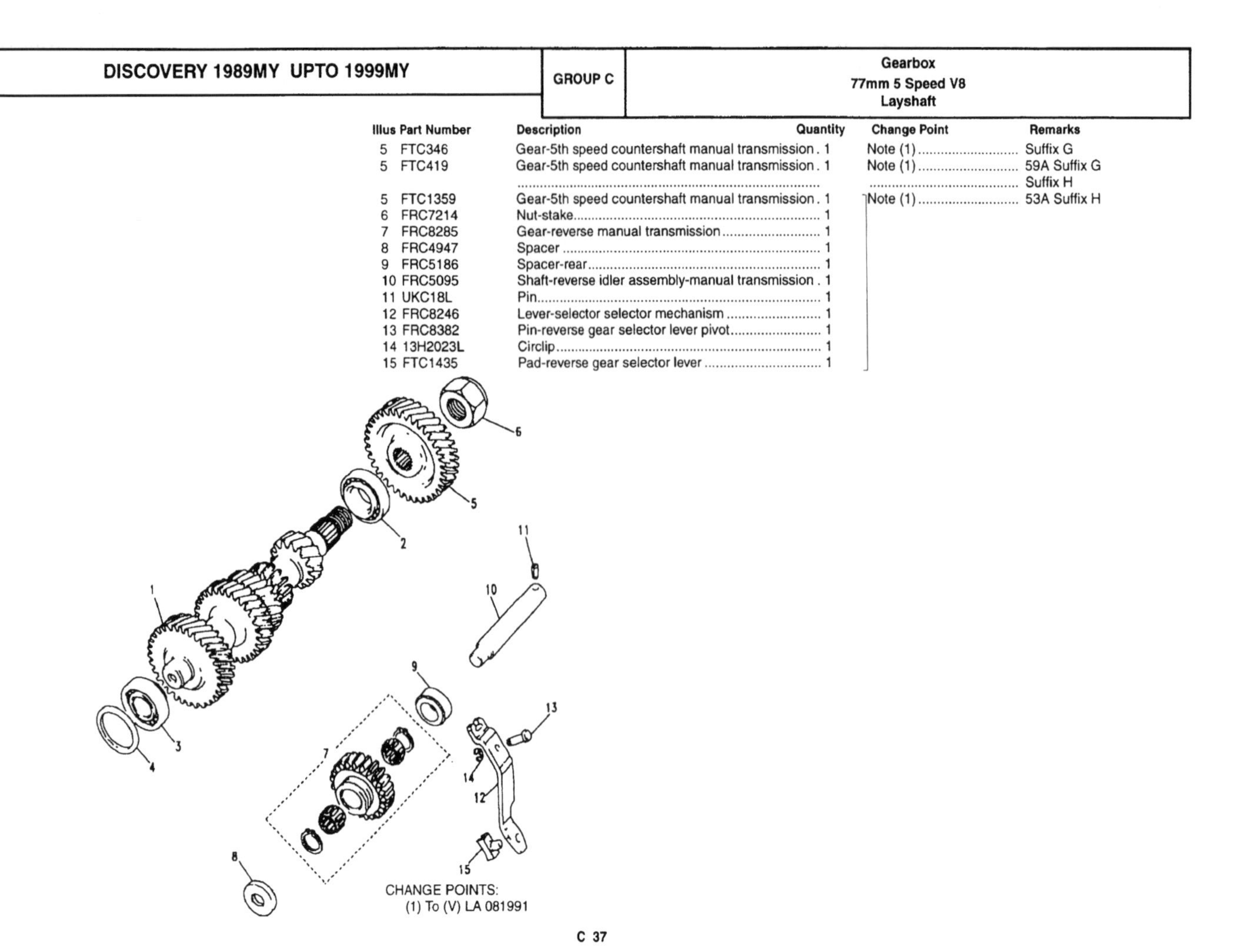

CHANGE POINTS:
(1) To (V) LA 081991

DISCOVERY 1989MY UPTO 1999MY	GROUP C	Gearbox 77mm 5 Speed V8 Selectors and Yoke Assembly

Illus	Part Number	Description	Quantity	Change Point	Remarks
1	FTC1491	SHAFT & LEVER ASSEMBLY-SELECTOR MANUAL TRANSMISSION	NLA	Note (1)	Use FRC8127 qty 2 with FRC8475 qty 1 with FTC1490 qty 1.
					Suffix G
	FRC8127	• Pin-selector lever/shaft manual transmission	1		
2	FTC1490	• Fork assembly-selector-1st/2nd manual transmission	1		
1	FTC1316	SHAFT & LEVER ASSEMBLY-SELECTOR MANUAL TRANSMISSION	1	Note (1)	Suffix H
	FRC8127	• Pin-selector lever/shaft manual transmission	1		
2	FTC2450	• Fork assembly-selector-1st/2nd manual transmission	1		
3	BLS112L	Ball-detent manual transmission	1	Note (1)	
4	FRC7195	Spring-detent manual transmission	1		Suffix G
5	UKC75	Plug	1		
4	FTC3382	Spring-detent manual transmission	1		Suffix H
5	FTC2193	Plug	1		
6	FTC4536	Screw-grub	1		
7	FTC1486	YOKE ASSEMBLY-SELECTOR ROD	1	Note (2)	
	FTC1488	• Pin-selector manual transmission	1		
8	FRC9986	Bush-selector yoke spherical	1	Note (2)	
7	STC1063	YOKE ASSEMBLY-SELECTOR ROD	1	Note (3)	
8	FTC2203	• Bush-selector yoke spherical	1		
9	FRC8882	Ring-snap	1	Note (1)	
10	FRC8485	Ball-selector yoke	1		
11	FTC1489	Fork assembly-3rd & 4th manual transmission	1		Suffix G
					Suffix H

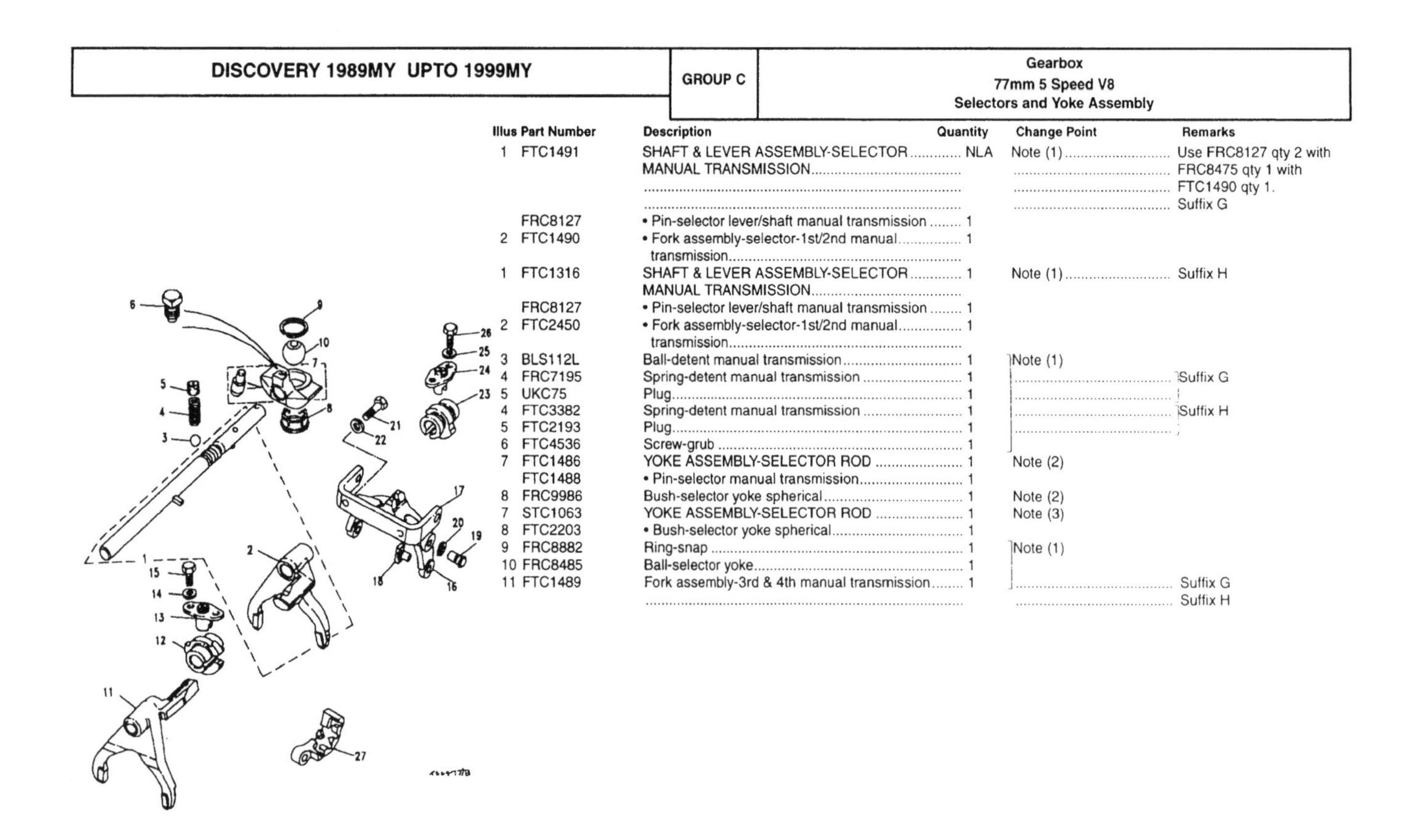

CHANGE POINTS:
(1) To (V) LA 081991
(2) To (V)59A 0120117H
(3) From (V)59A 0120118H To (V) LA 081991

DISCOVERY 1989MY UPTO 1999MY

GROUP C

Gearbox
77mm 5 Speed V8
Selectors and Yoke Assembly

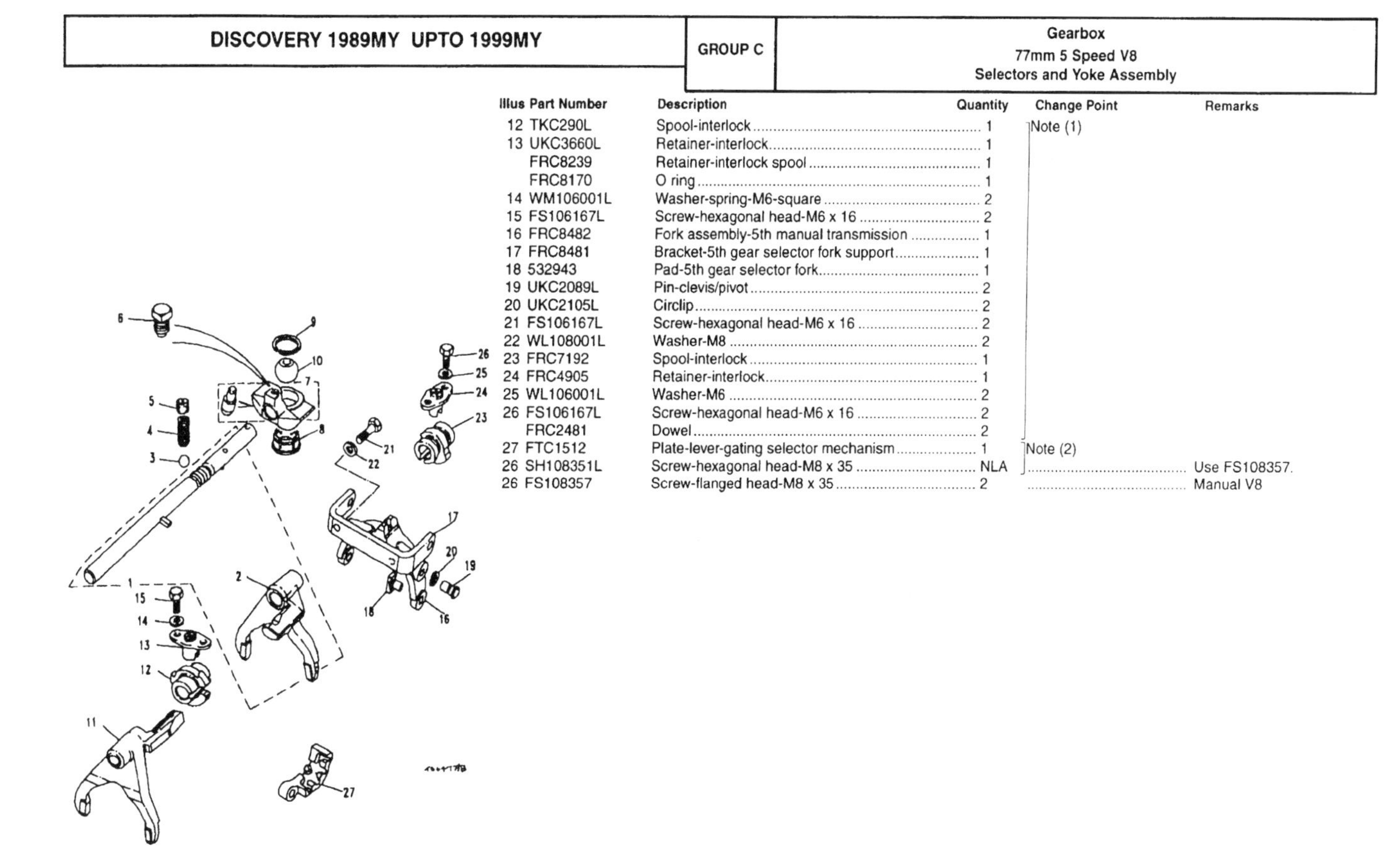

Illus	Part Number	Description	Quantity	Change Point	Remarks
12	TKC290L	Spool-interlock	1	Note (1)	
13	UKC3660L	Retainer-interlock	1		
	FRC8239	Retainer-interlock spool	1		
	FRC8170	O ring	1		
14	WM106001L	Washer-spring-M6-square	2		
15	FS106167L	Screw-hexagonal head-M6 x 16	2		
16	FRC8482	Fork assembly-5th manual transmission	1		
17	FRC8481	Bracket-5th gear selector fork support	1		
18	532943	Pad-5th gear selector fork	1		
19	UKC2089L	Pin-clevis/pivot	2		
20	UKC2105L	Circlip	2		
21	FS106167L	Screw-hexagonal head-M6 x 16	2		
22	WL108001L	Washer-M8	2		
23	FRC7192	Spool-interlock	1		
24	FRC4905	Retainer-interlock	1		
25	WL106001L	Washer-M6	2		
26	FS106167L	Screw-hexagonal head-M6 x 16	2		
	FRC2481	Dowel	2		
27	FTC1512	Plate-lever-gating selector mechanism	1	Note (2)	
26	SH108351L	Screw-hexagonal head-M8 x 35	NLA		Use FS108357.
26	FS108357	Screw-flanged head-M8 x 35	2		Manual V8

CHANGE POINTS:
(1) To (V) LA 081991
(2) To (G)55A 0183938H

C 39

DISCOVERY 1989MY UPTO 1999MY

GROUP C

Gearbox
77mm 5 Speed V8
Gearchange and Lever

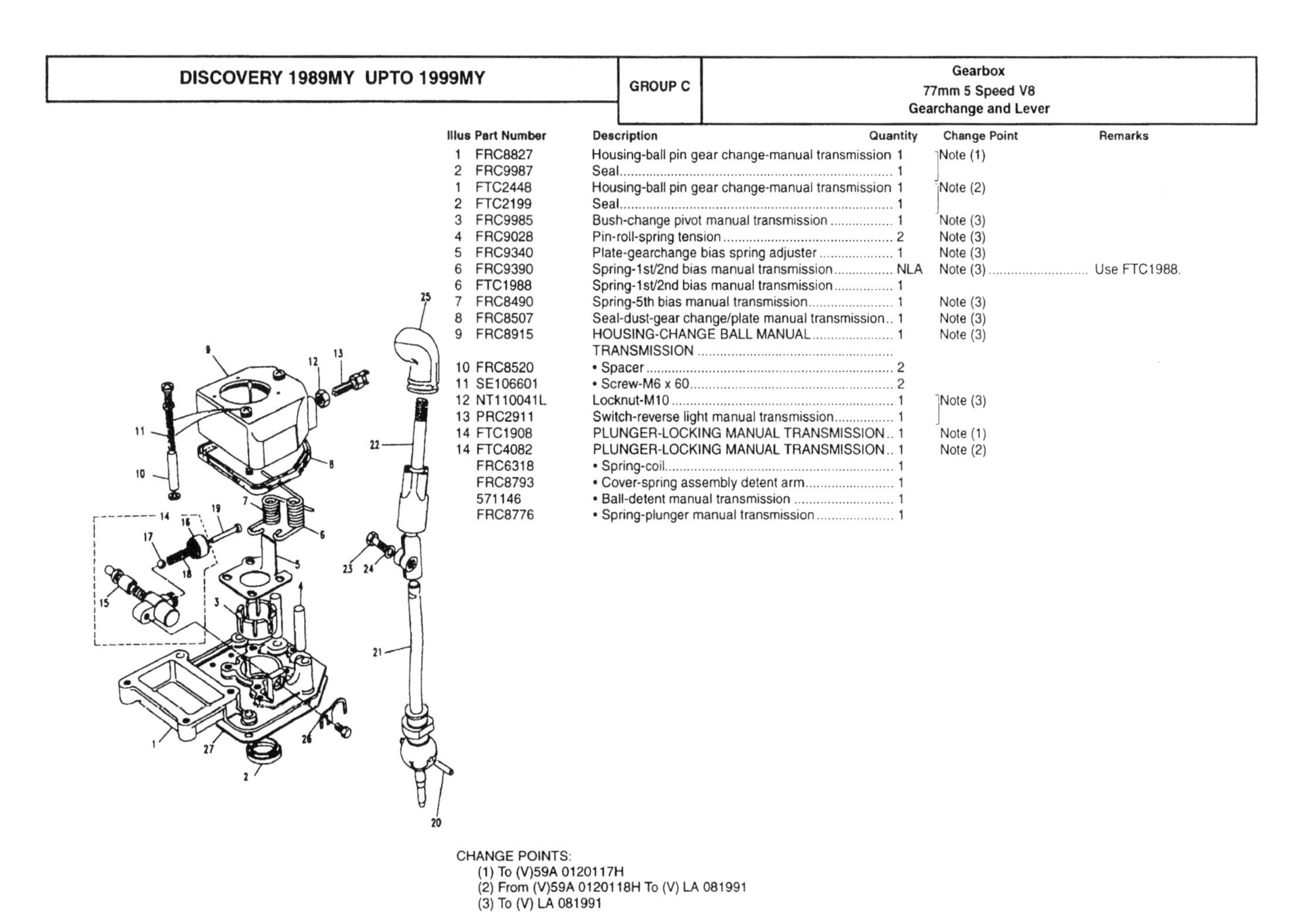

Illus	Part Number	Description	Quantity	Change Point	Remarks
1	FRC8827	Housing-ball pin gear change-manual transmission	1	Note (1)	
2	FRC9987	Seal	1		
1	FTC2448	Housing-ball pin gear change-manual transmission	1	Note (2)	
2	FTC2199	Seal	1		
3	FRC9985	Bush-change pivot manual transmission	1	Note (3)	
4	FRC9028	Pin-roll-spring tension	2	Note (3)	
5	FRC9340	Plate-gearchange bias spring adjuster	1	Note (3)	
6	FRC9390	Spring-1st/2nd bias manual transmission	NLA	Note (3)	Use FTC1988.
6	FTC1988	Spring-1st/2nd bias manual transmission	1		
7	FRC8490	Spring-5th bias manual transmission	1	Note (3)	
8	FRC8507	Seal-dust-gear change/plate manual transmission	1	Note (3)	
9	FRC8915	HOUSING-CHANGE BALL MANUAL TRANSMISSION	1	Note (3)	
10	FRC8520	• Spacer	2		
11	SE106601	• Screw-M6 x 60	2		
12	NT110041L	Locknut-M10	1	Note (3)	
13	PRC2911	Switch-reverse light manual transmission	1		
14	FTC1908	PLUNGER-LOCKING MANUAL TRANSMISSION	1	Note (1)	
14	FTC4082	PLUNGER-LOCKING MANUAL TRANSMISSION	1	Note (2)	
	FRC6318	• Spring-coil	1		
	FRC8793	• Cover-spring assembly detent arm	1		
	571146	• Ball-detent manual transmission	1		
	FRC8776	• Spring-plunger manual transmission	1		

CHANGE POINTS:
(1) To (V)59A 0120117H
(2) From (V)59A 0120118H To (V) LA 081991
(3) To (V) LA 081991

C 40

DISCOVERY 1989MY UPTO 1999MY

GROUP C

Gearbox
77mm 5 Speed V8
Gearchange and Lever

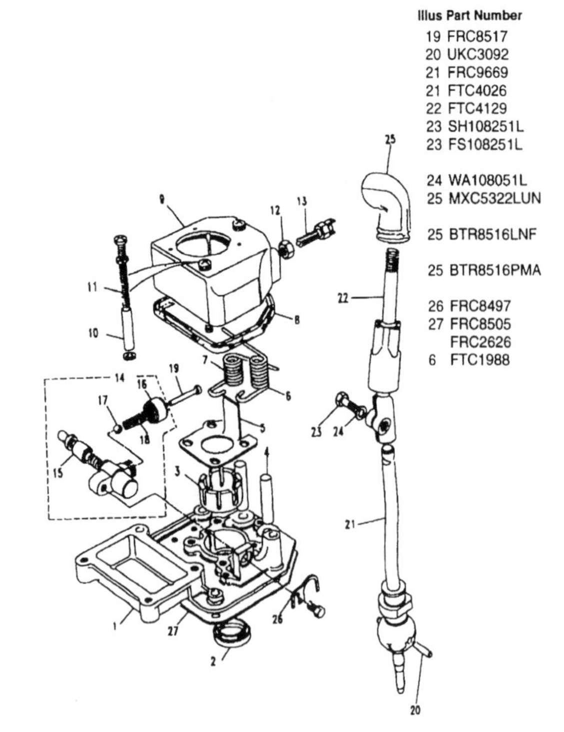

Illus	Part Number	Description	Quantity	Change Point	Remarks
19	FRC8517	Plunger-reverse switch	1	Note (1)	
20	UKC3092	Pin	2		
21	FRC9669	Lever-lower manual transmission	1	Note (2)	
21	FTC4026	Lever-lower manual transmission	1	Note (3)	
22	FTC4129	Lever-upper manual transmission	1	Note (1)	
23	SH108251L	Screw-hexagonal head-M8 x 25	NLA	Note (1)	Use FS108251L.
23	FS108251L	Screw-flanged head-M8 x 25	NLA		Use FS108257L.
					Manual V8
24	WA108051L	Washer-plain-M8	1	Note (1)	
25	MXC5322LUN	Knob assembly-change manual transmission-Ash Grey	1	Note (4)	
25	BTR8516LNF	Knob assembly-change manual transmission-Ash Grey	1	Note (5)	
25	BTR8516PMA	Knob assembly-change manual transmission-Ash Grey	1	Note (6)	
26	FRC8497	Spacer-reverse gear plunger	1	Note (1)	
27	FRC8505	Gasket-gearchange ball pin housing	1	Note (1)	
	FRC2626	Pin-spring	2	Note (1)	
6	FTC1988	Spring-1st/2nd bias manual transmission	1	Note (1)	

CHANGE POINTS:
(1) To (V) LA 081991
(2) To (V)59A 0120117H
(3) From (V)59A 0120118H To (V) LA 081991
(4) To (V) JA 034313
(5) From (V) KA 034314 To (V) LA 081990
(6) From (V) LA 081991

C 41

DISCOVERY 1989MY UPTO 1999MY

GROUP C

Gearbox
77mm 5 Speed V8
Oil Cooler

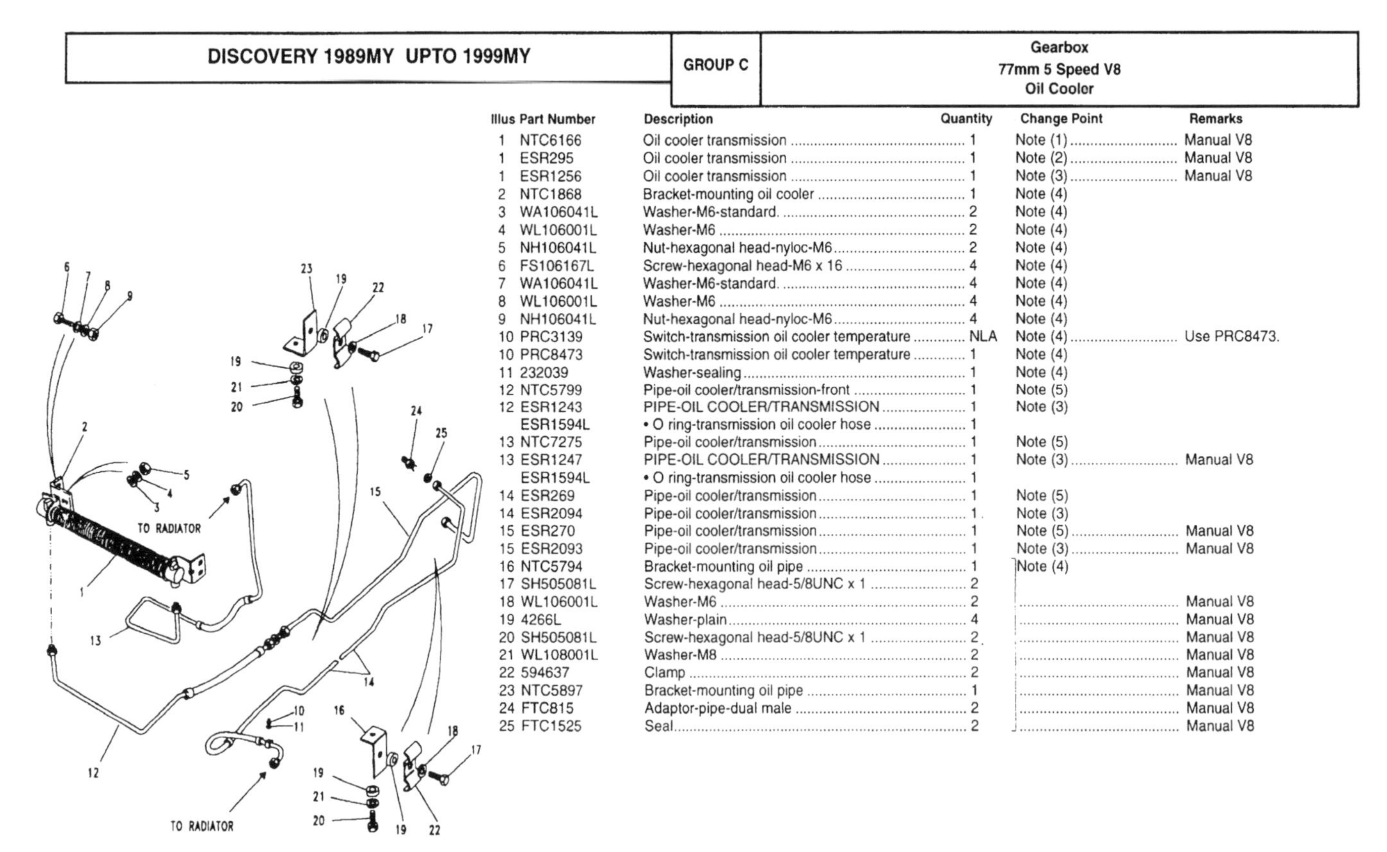

Illus	Part Number	Description	Quantity	Change Point	Remarks
1	NTC6166	Oil cooler transmission	1	Note (1)	Manual V8
1	ESR295	Oil cooler transmission	1	Note (2)	Manual V8
1	ESR1256	Oil cooler transmission	1	Note (3)	Manual V8
2	NTC1868	Bracket-mounting oil cooler	1	Note (4)	
3	WA106041L	Washer-M6-standard.	2	Note (4)	
4	WL106001L	Washer-M6	2	Note (4)	
5	NH106041L	Nut-hexagonal head-nyloc-M6	2	Note (4)	
6	FS106167L	Screw-hexagonal head-M6 x 16	4	Note (4)	
7	WA106041L	Washer-M6-standard.	4	Note (4)	
8	WL106001L	Washer-M6	4	Note (4)	
9	NH106041L	Nut-hexagonal head-nyloc-M6	4	Note (4)	
10	PRC3139	Switch-transmission oil cooler temperature	NLA	Note (4)	Use PRC8473.
10	PRC8473	Switch-transmission oil cooler temperature	1	Note (4)	
11	232039	Washer-sealing	1	Note (4)	
12	NTC5799	Pipe-oil cooler/transmission-front	1	Note (5)	
12	ESR1243	PIPE-OIL COOLER/TRANSMISSION	1	Note (3)	
	ESR1594L	• O ring-transmission oil cooler hose	1		
13	NTC7275	Pipe-oil cooler/transmission	1	Note (5)	
13	ESR1247	PIPE-OIL COOLER/TRANSMISSION	1	Note (3)	Manual V8
	ESR1594L	• O ring-transmission oil cooler hose	1		
14	ESR269	Pipe-oil cooler/transmission	1	Note (5)	
14	ESR2094	Pipe-oil cooler/transmission	1	Note (3)	
15	ESR270	Pipe-oil cooler/transmission	1	Note (5)	Manual V8
15	ESR2093	Pipe-oil cooler/transmission	1	Note (3)	Manual V8
16	NTC5794	Bracket-mounting oil pipe	1	Note (4)	
17	SH505081L	Screw-hexagonal head-5/8UNC x 1	2		
18	WL106001L	Washer-M6	2		Manual V8
19	4266L	Washer-plain	4		Manual V8
20	SH505081L	Screw-hexagonal head-5/8UNC x 1	2		Manual V8
21	WL108001L	Washer-M8	2		Manual V8
22	594637	Clamp	2		Manual V8
23	NTC5897	Bracket-mounting oil pipe	1		Manual V8
24	FTC815	Adaptor-pipe-dual male	2		Manual V8
25	FTC1525	Seal	2		Manual V8

CHANGE POINTS:
(1) From (V) FA 393361 To (V) GA 456980; From (V) HA 456981 To (V) HA 473001
(2) From (V) HA 473002 To (V) HA 486551; From (V) HA 000001 To (V) HA 012332; From (V) JA 012333
(3) From (V) KA 034314 To (V) LA 081990
(4) To (V) LA 081990
(5) To (V) JA 034313

C 42

Gearbox
R380 5 Speed
Gearbox Assembly

Illus	Part Number	Description	Quantity	Change Point	Remarks
1	TRC103040	Transmission assembly manual-new	1	Note (1)	Manual 4 Cylinder MPi 63A Suffix J
1	TRC103040	Transmission assembly manual-new	1		Manual 4 Cylinder MPi 63A Suffix K
1	STC1541E	Transmission assembly manual-exchange	1	Note (1)	Manual 4 Cylinder MPi 63A
1	TRC102940	Transmission assembly manual-new	1	Note (1)	Manual V8 53A Suffix J
1	TRC102940	Transmission assembly manual-new	1		Manual V8 53A Suffix K
1	STC1543E	Transmission assembly manual-exchange	1	Note (1)	Manual V8 53A
1	TRC102960	Transmission assembly manual-new	1	Note (1)	Manual 4 Cylinder TDi Suffix J
1	TRC102960	Transmission assembly manual-new	1		Manual 4 Cylinder TDi Suffix K
1	STC1544E	Transmission assembly manual-exchange	1	Note (1)	4 Cylinder TDi 55A

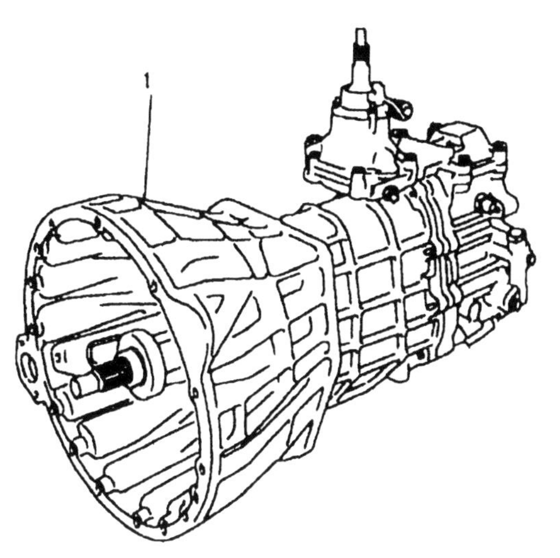

CHANGE POINTS:
(1) From (V) MA 081992

DISCOVERY 1989MY UPTO 1999MY | GROUP C

Gearbox
R380 5 Speed
Clutch Release Mechanism

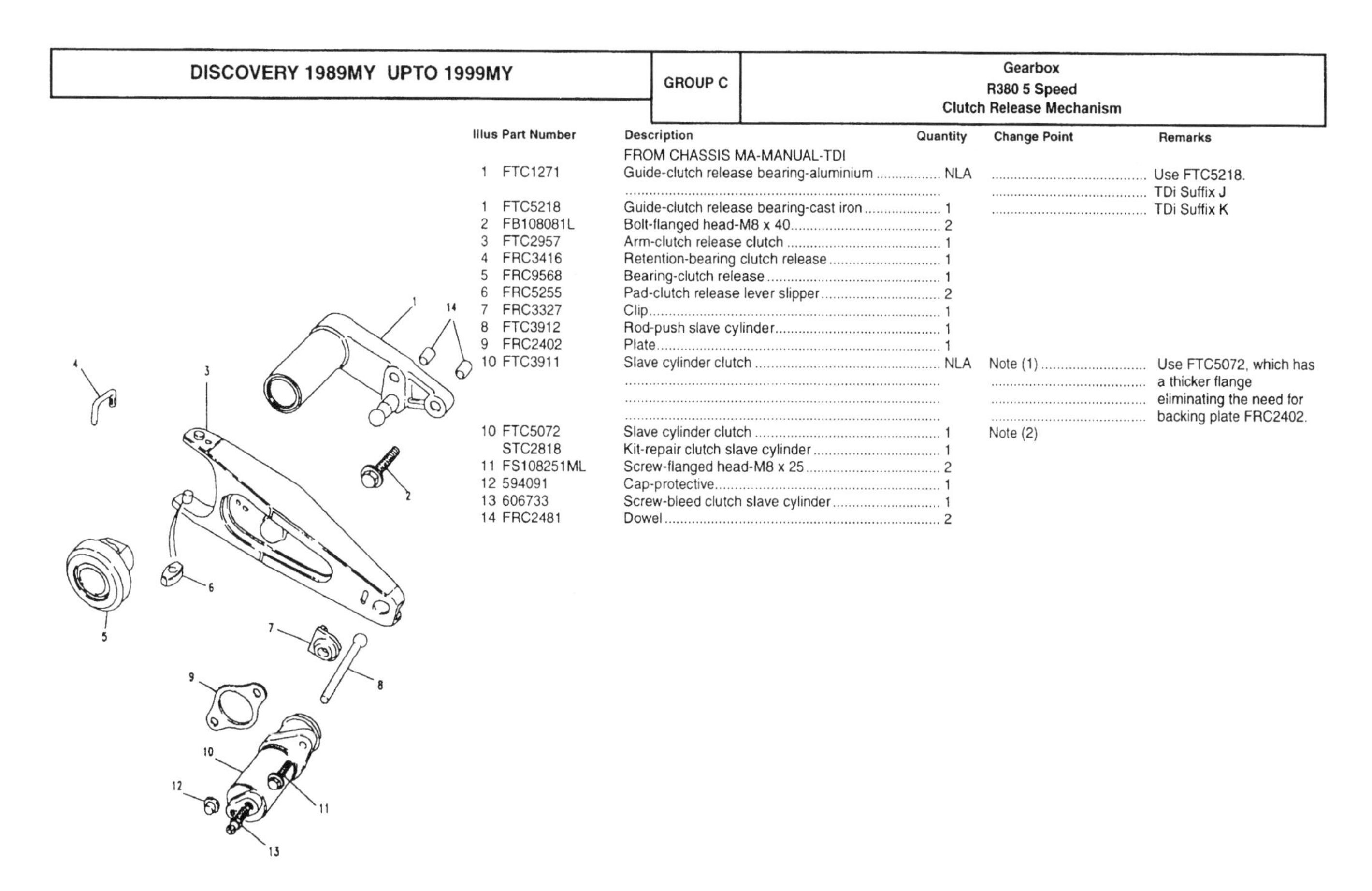

Illus	Part Number	Description	Quantity	Change Point	Remarks
		FROM CHASSIS MA-MANUAL-TDI			
1	FTC1271	Guide-clutch release bearing-aluminium	NLA		Use FTC5218. TDi Suffix J
1	FTC5218	Guide-clutch release bearing-cast iron	1		TDi Suffix K
2	FB108081L	Bolt-flanged head-M8 x 40	2		
3	FTC2957	Arm-clutch release clutch	1		
4	FRC3416	Retention-bearing clutch release	1		
5	FRC9568	Bearing-clutch release	1		
6	FRC5255	Pad-clutch release lever slipper	2		
7	FRC3327	Clip	1		
8	FTC3912	Rod-push slave cylinder	1		
9	FRC2402	Plate	1		
10	FTC3911	Slave cylinder clutch	NLA	Note (1)	Use FTC5072, which has a thicker flange eliminating the need for backing plate FRC2402.
10	FTC5072	Slave cylinder clutch	1	Note (2)	
	STC2818	Kit-repair clutch slave cylinder	1		
11	FS108251ML	Screw-flanged head-M8 x 25	2		
12	594091	Cap-protective	1		
13	606733	Screw-bleed clutch slave cylinder	1		
14	FRC2481	Dowel	2		

CHANGE POINTS:
(1) To (G) 437139J
(2) From (G) 437140J

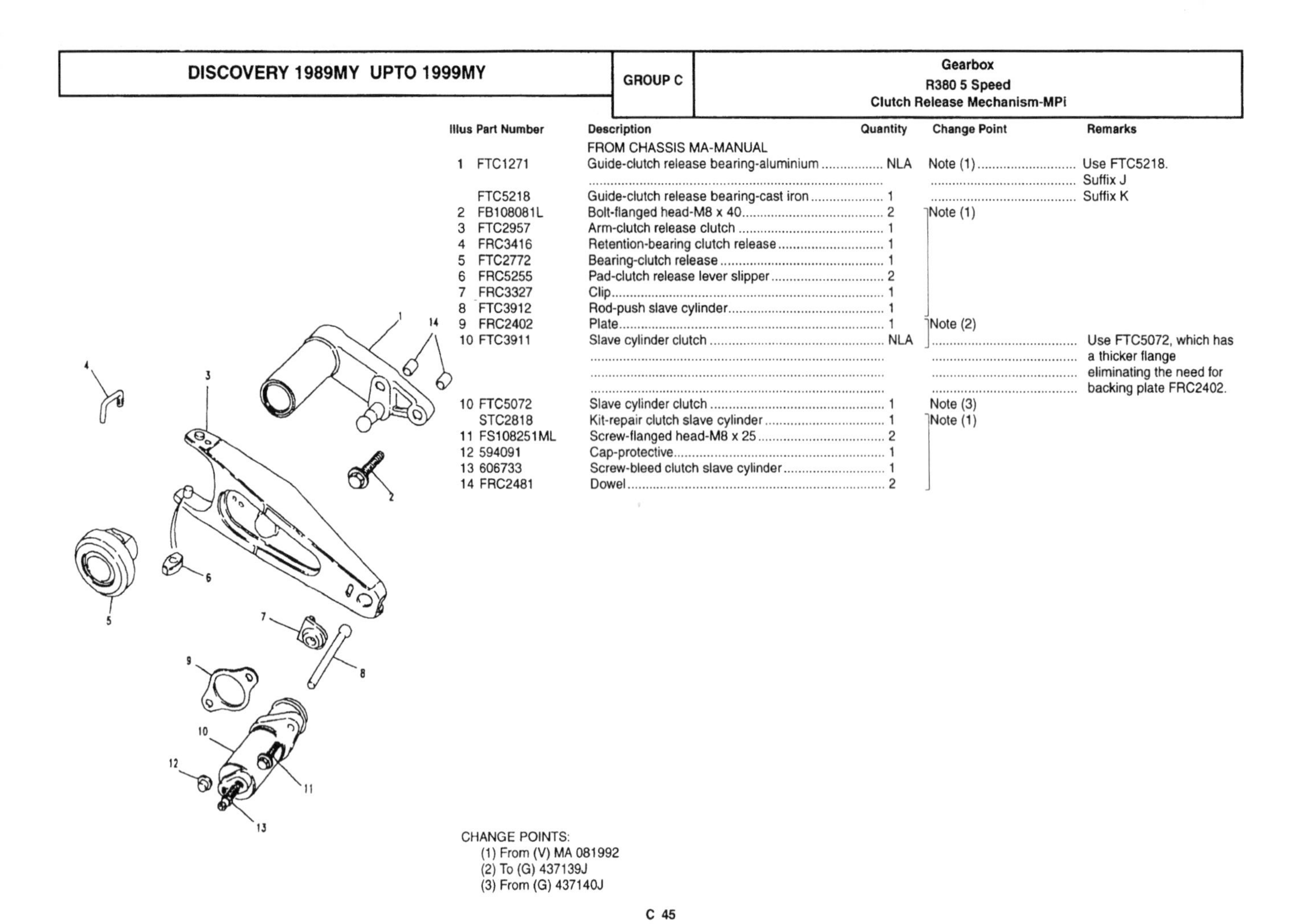

Illus	Part Number	Description	Quantity	Change Point	Remarks
		FROM CHASSIS MA-MANUAL			
1	FTC1271	Guide-clutch release bearing-aluminium	NLA	Note (1)	Use FTC5218. Suffix J
	FTC5218	Guide-clutch release bearing-cast iron	1		Suffix K
2	FB108081L	Bolt-flanged head-M8 x 40	2	Note (1)	
3	FTC2957	Arm-clutch release clutch	1		
4	FRC3416	Retention-bearing clutch release	1		
5	FTC2772	Bearing-clutch release	1		
6	FRC5255	Pad-clutch release lever slipper	2		
7	FRC3327	Clip	1		
8	FTC3912	Rod-push slave cylinder	1		
9	FRC2402	Plate	1	Note (2)	
10	FTC3911	Slave cylinder clutch	NLA		Use FTC5072, which has a thicker flange eliminating the need for backing plate FRC2402.
10	FTC5072	Slave cylinder clutch	1	Note (3)	
	STC2818	Kit-repair clutch slave cylinder	1	Note (1)	
11	FS108251ML	Screw-flanged head-M8 x 25	2		
12	594091	Cap-protective	1		
13	606733	Screw-bleed clutch slave cylinder	1		
14	FRC2481	Dowel	2		

CHANGE POINTS:
(1) From (V) MA 081992
(2) To (G) 437139J
(3) From (G) 437140J

C 45

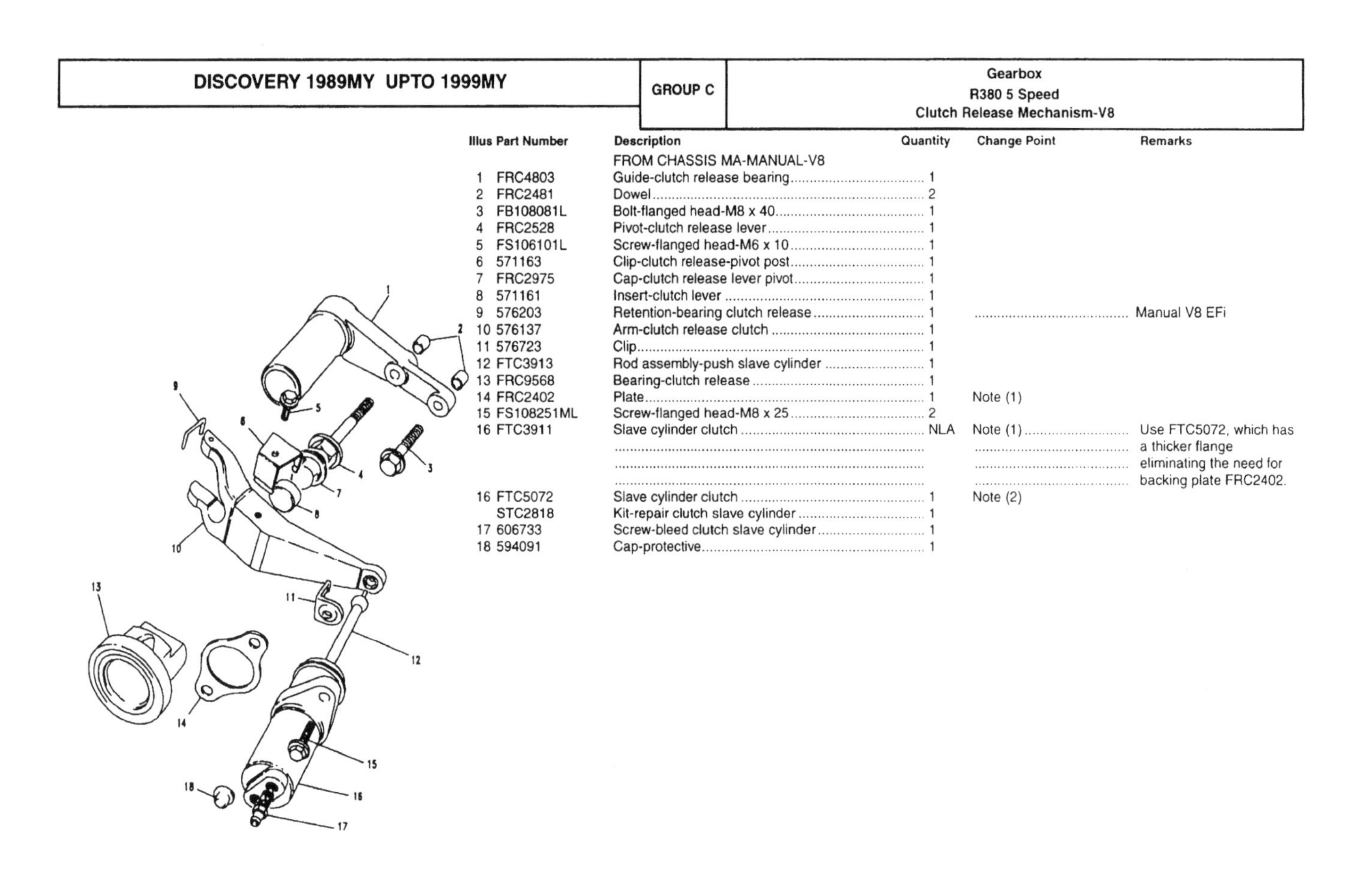

Illus	Part Number	Description	Quantity	Change Point	Remarks
		FROM CHASSIS MA-MANUAL-V8			
1	FRC4803	Guide-clutch release bearing	1		
2	FRC2481	Dowel	2		
3	FB108081L	Bolt-flanged head-M8 x 40	1		
4	FRC2528	Pivot-clutch release lever	1		
5	FS106101L	Screw-flanged head-M6 x 10	1		
6	571163	Clip-clutch release-pivot post	1		
7	FRC2975	Cap-clutch release lever pivot	1		
8	571161	Insert-clutch lever	1		
9	576203	Retention-bearing clutch release	1		Manual V8 EFi
10	576137	Arm-clutch release clutch	1		
11	576723	Clip	1		
12	FTC3913	Rod assembly-push slave cylinder	1		
13	FRC9568	Bearing-clutch release	1		
14	FRC2402	Plate	1	Note (1)	
15	FS108251ML	Screw-flanged head-M8 x 25	2		
16	FTC3911	Slave cylinder clutch	NLA	Note (1)	Use FTC5072, which has a thicker flange eliminating the need for backing plate FRC2402.
16	FTC5072	Slave cylinder clutch	1	Note (2)	
	STC2818	Kit-repair clutch slave cylinder	1		
17	606733	Screw-bleed clutch slave cylinder	1		
18	594091	Cap-protective	1		

CHANGE POINTS:
(1) To (G) 437139J
(2) From (G) 437140J

C 46

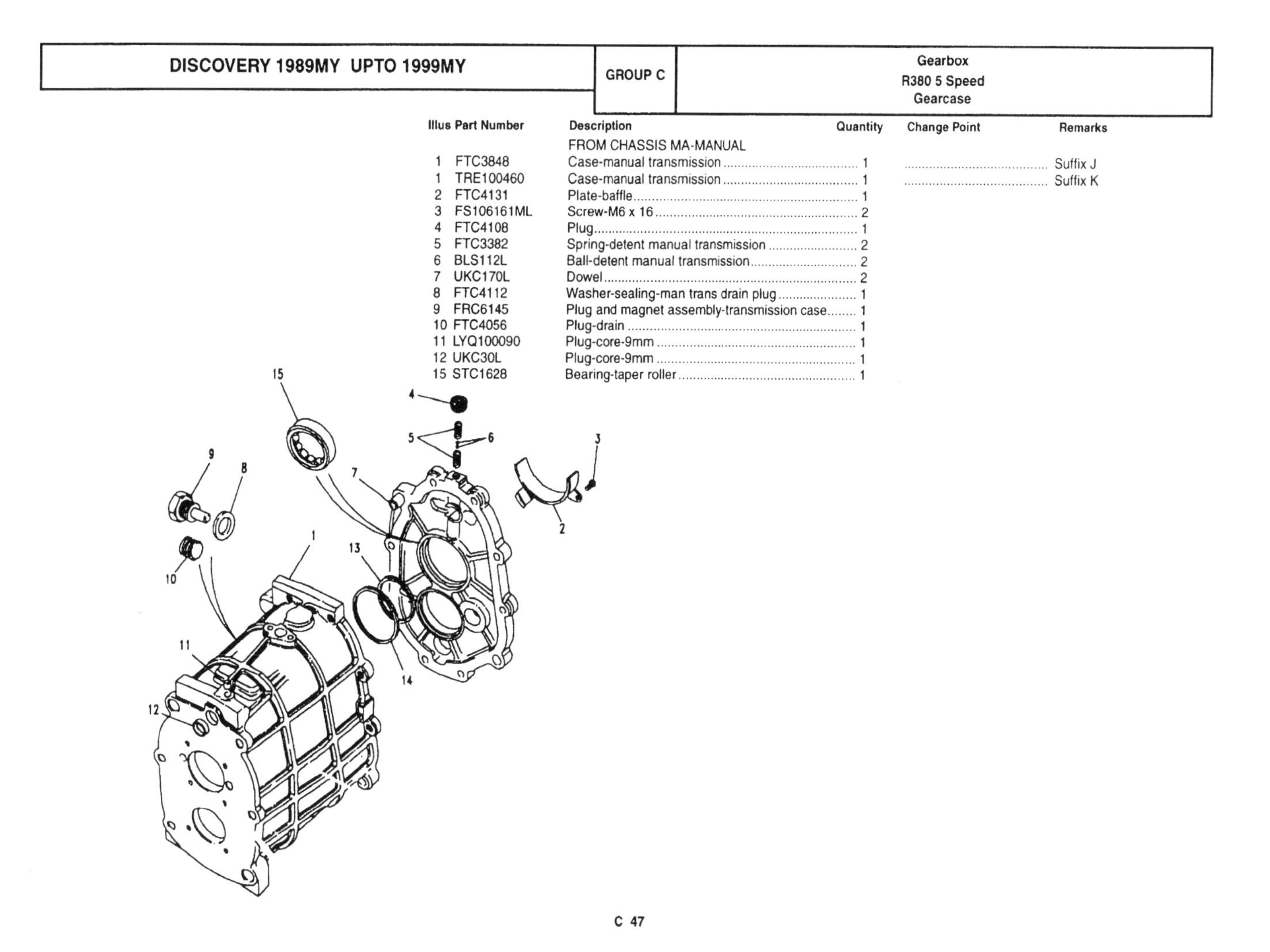

Illus	Part Number	Description	Quantity	Change Point	Remarks
		FROM CHASSIS MA-MANUAL			
1	FTC3848	Case-manual transmission	1		Suffix J
1	TRE100460	Case-manual transmission	1		Suffix K
2	FTC4131	Plate-baffle	1		
3	FS106161ML	Screw-M6 x 16	2		
4	FTC4108	Plug	1		
5	FTC3382	Spring-detent manual transmission	2		
6	BLS112L	Ball-detent manual transmission	2		
7	UKC170L	Dowel	2		
8	FTC4112	Washer-sealing-man trans drain plug	1		
9	FRC6145	Plug and magnet assembly-transmission case	1		
10	FTC4056	Plug-drain	1		
11	LYQ100090	Plug-core-9mm	1		
12	UKC30L	Plug-core-9mm	1		
15	STC1628	Bearing-taper roller	1		

DISCOVERY 1989MY UPTO 1999MY | GROUP C | Gearbox R380 5 Speed Gearcase

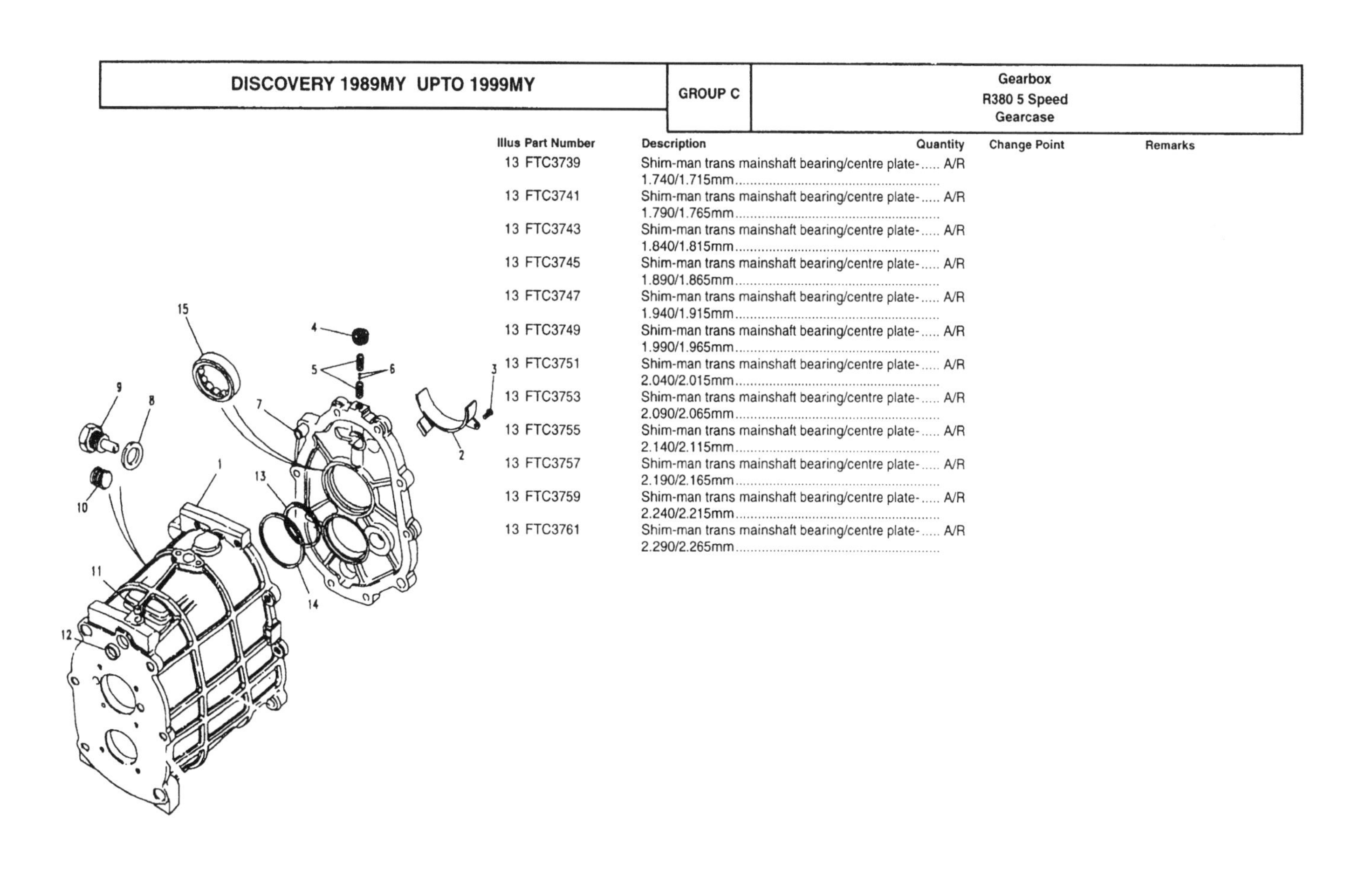

Illus	Part Number	Description	Quantity	Change Point	Remarks
13	FTC3739	Shim-man trans mainshaft bearing/centre plate-1.740/1.715mm	A/R		
13	FTC3741	Shim-man trans mainshaft bearing/centre plate-1.790/1.765mm	A/R		
13	FTC3743	Shim-man trans mainshaft bearing/centre plate-1.840/1.815mm	A/R		
13	FTC3745	Shim-man trans mainshaft bearing/centre plate-1.890/1.865mm	A/R		
13	FTC3747	Shim-man trans mainshaft bearing/centre plate-1.940/1.915mm	A/R		
13	FTC3749	Shim-man trans mainshaft bearing/centre plate-1.990/1.965mm	A/R		
13	FTC3751	Shim-man trans mainshaft bearing/centre plate-2.040/2.015mm	A/R		
13	FTC3753	Shim-man trans mainshaft bearing/centre plate-2.090/2.065mm	A/R		
13	FTC3755	Shim-man trans mainshaft bearing/centre plate-2.140/2.115mm	A/R		
13	FTC3757	Shim-man trans mainshaft bearing/centre plate-2.190/2.165mm	A/R		
13	FTC3759	Shim-man trans mainshaft bearing/centre plate-2.240/2.215mm	A/R		
13	FTC3761	Shim-man trans mainshaft bearing/centre plate-2.290/2.265mm	A/R		

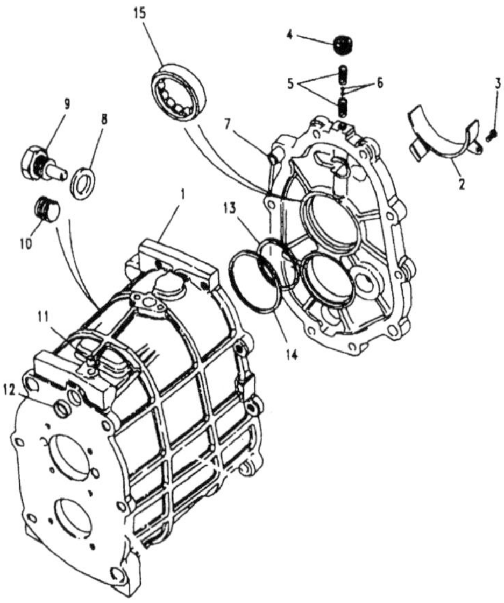

Illus	Part Number	Description	Quantity	Change Point	Remarks
13	FTC3763	Shim-man trans mainshaft bearing/centre plate-2.340/2.315mm	A/R		Manual Suffix J
13	FTC3765	Shim-man trans mainshaft bearing/centre plate-2.390/2.365mm	A/R		Manual Suffix J
13	FTC3767	Shim-man trans mainshaft bearing/centre plate-2.440/2.415mm	A/R		Manual Suffix J
13	FTC3769	Shim-man trans mainshaft bearing/centre plate-2.490/2.465mm	A/R		Manual Suffix J
13	FTC3771	Shim-man trans mainshaft bearing/centre plate-2.540/2.515mm	A/R		Manual Suffix J
13	FTC3773	Shim-man trans mainshaft bearing/centre plate-2.590/2.565mm	A/R		Manual Suffix J
13	FTC3775	Shim-man trans mainshaft bearing/centre plate-2.640/2.615mm	A/R		Manual Suffix J
13	FTC3777	Shim-man trans mainshaft bearing/centre plate-2.690/2.665mm	A/R		Manual Suffix J
13	FTC3779	Shim-man trans mainshaft bearing/centre plate-2.740/2.715mm	A/R		Manual Suffix J
13	FTC3781	Shim-man trans mainshaft bearing/centre plate-2.790/2.765mm	A/R		Manual Suffix J
13	FTC3783	Shim-man trans mainshaft bearing/centre plate-2.840/2.815mm	A/R		Manual Suffix J
13	FTC3785	Shim-man trans mainshaft bearing/centre plate-2.890/2.865mm	A/R		Manual Suffix J
13	FTC3787	Shim-man trans mainshaft bearing/centre plate-2.940/2.915mm	A/R		Manual Suffix J

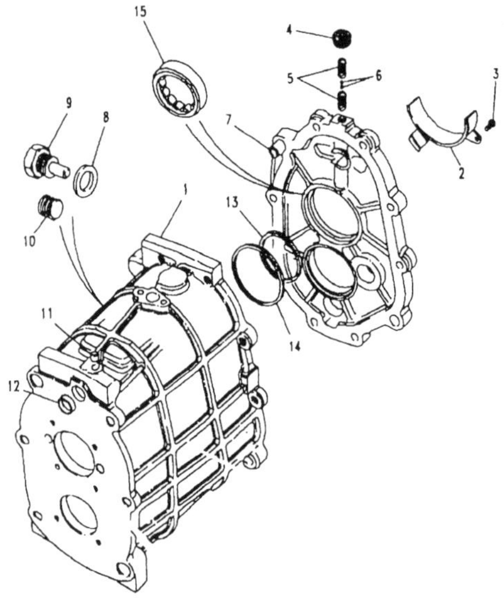

DISCOVERY 1989MY UPTO 1999MY | GROUP C | Gearbox R380 5 Speed Gearcase

Illus	Part Number	Description	Quantity	Change Point	Remarks
14	FTC4296	Shim-man trans countershaft bearing/centre plate-1.477/1.452mm	A/R		Manual Suffix J
14	FTC4298	Shim-man trans countershaft bearing/centre plate-1.527/1.502mm	A/R		
14	FTC4300	Shim-man trans countershaft bearing/centre plate-1.577/1.552mm	A/R		
14	FTC4302	Shim-man trans countershaft bearing/centre plate-1.627/1.602mm	A/R		
14	FTC4304	Shim-man trans countershaft bearing/centre plate-1.677/1.652mm	A/R		
14	FTC4306	Shim-man trans countershaft bearing/centre plate-1.727/1.702mm	A/R		
14	FTC4308	Shim-man trans countershaft bearing/centre plate-1.777/1.752mm	A/R		
14	FTC4310	Shim-man trans countershaft bearing/centre plate-1.827/1.802mm	A/R		
14	FTC4312	Shim-man trans countershaft bearing/centre plate-1.877/1.852mm	A/R		
14	FTC4314	Shim-man trans countershaft bearing/centre plate-1.927/1.902mm	A/R		
14	FTC4316	Shim-man trans countershaft bearing/centre plate-1.977/1.952mm	A/R		
14	FTC4318	Shim-man trans countershaft bearing/centre plate-2.027/2.002mm	A/R		
14	FTC4320	Shim-man trans countershaft bearing/centre plate-2.077/2.052mm	A/R		
14	FTC4322	Shim-man trans countershaft bearing/centre plate-2.127/2.102mm	A/R		
14	FTC4324	Shim-man trans countershaft bearing/centre plate-2.177/2.152mm	A/R		
14	FTC4326	Shim-man trans countershaft bearing/centre plate-2.227/2.202mm	A/R		

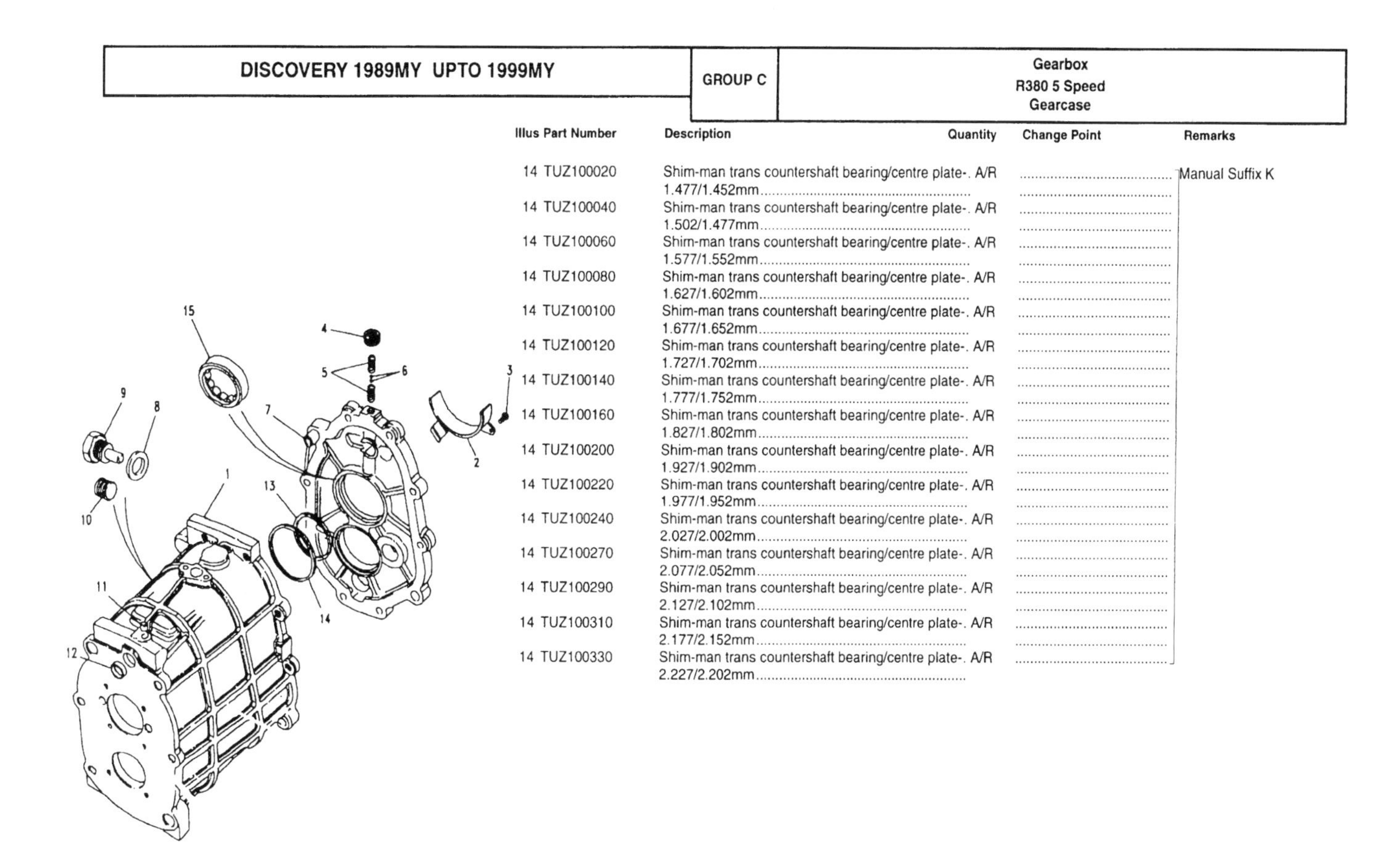

Illus	Part Number	Description	Quantity	Change Point	Remarks
14	TUZ100020	Shim-man trans countershaft bearing/centre plate-. 1.477/1.452mm	A/R		Manual Suffix K
14	TUZ100040	Shim-man trans countershaft bearing/centre plate-. 1.502/1.477mm	A/R		
14	TUZ100060	Shim-man trans countershaft bearing/centre plate-. 1.577/1.552mm	A/R		
14	TUZ100080	Shim-man trans countershaft bearing/centre plate-. 1.627/1.602mm	A/R		
14	TUZ100100	Shim-man trans countershaft bearing/centre plate-. 1.677/1.652mm	A/R		
14	TUZ100120	Shim-man trans countershaft bearing/centre plate-. 1.727/1.702mm	A/R		
14	TUZ100140	Shim-man trans countershaft bearing/centre plate-. 1.777/1.752mm	A/R		
14	TUZ100160	Shim-man trans countershaft bearing/centre plate-. 1.827/1.802mm	A/R		
14	TUZ100200	Shim-man trans countershaft bearing/centre plate-. 1.927/1.902mm	A/R		
14	TUZ100220	Shim-man trans countershaft bearing/centre plate-. 1.977/1.952mm	A/R		
14	TUZ100240	Shim-man trans countershaft bearing/centre plate-. 2.027/2.002mm	A/R		
14	TUZ100270	Shim-man trans countershaft bearing/centre plate-. 2.077/2.052mm	A/R		
14	TUZ100290	Shim-man trans countershaft bearing/centre plate-. 2.127/2.102mm	A/R		
14	TUZ100310	Shim-man trans countershaft bearing/centre plate-. 2.177/2.152mm	A/R		
14	TUZ100330	Shim-man trans countershaft bearing/centre plate-. 2.227/2.202mm	A/R		

Illus	Part Number	Description	Quantity	Change Point	Remarks
1	FTC3921	Housing-clutch manual transmission	1		Manual MPi
1	FTC3922	Housing-clutch manual transmission	1		V8 Twin Carburettor
1	FTC3921	Housing-clutch manual transmission	1		Manual Diesel TDi
1	FTC3922	Housing-clutch manual transmission	1		V8 Petrol EFi
2	FS112401L	Screw-flanged head-M12 x 40	6	From (C) MA	Manual TDi
3	UKC25L	Dowel-housing to gear case	2	From (C) MA	Manual TDi
	3290	Plug-coolant drain	1	From (C) MA	Manual TDi
	BH506161L	Bolt-hexagonal head-3/8UNC x 2	8		
	586440	Clip-cable	1		
2	SH105161L	Screw-hexagonal head-M5 x 16	1		
	NH105041L	Nut-hexagonal-M5	1		
	WA105001L	Washer-plain-M5-standard	1		
	WL105001L	Washer-M5	1		
	FB110071L	Bolt-flanged head-M10 x 35	4		Manual 4 Cylinder TDi
	FN110041L	Nut-flanged head-M10	NLA		Use FN110047. Manual 4 Cylinder TDi
	FN110047	Nut-flanged head-M10	9		Manual 4 Cylinder TDi
	FN110047	Nut-flanged head-M10	13		Manual 4 Cylinder MPi
2	SH108141L	Screw-hexagonal head-M8 x 14	1		Manual 4 Cylinder TDi
3	571134	Dowel	2		Manual 4 Cylinder TDi
	622324	Clip-pipe	1		Manual 4 Cylinder TDi
		FIXINGS GEARBOX TO ENGINE-MPI			
	TE110061L	Stud-M10 x 30	13		
	FN110041L	Nut-flanged head-M10	NLA		Use FN110047. Manual 4 Cylinder MPi
	WA110061L	Washer-plain-M10-standard	13		Manual 4 Cylinder MPi
3	571134	Dowel	2		Manual 4 Cylinder MPi
	622324	Clip-pipe	1 .		Manual 4 Cylinder MPi
	FTC4829	Jacket insulation upper-manual	1	From (V) TA	Manual V8
	FTC4830	Jacket insulation lower	1	From (V) TA	Manual V8
	FTC4900	Fixing clip jacket	4	From (V) TA	Manual V8

DISCOVERY 1989MY UPTO 1999MY

GROUP C

Gearbox
R380 5 Speed
Bottom Cover

Illus	Part Number	Description	Quantity	Change Point	Remarks
1	FRC2859	Cover-bell housing-bottom	1	Note (1)	
2	594087	Seal-bottom	1		
3	FS106255L	Screw-flanged head-M6 x 25	2		
4	FN106047L	Nut-flange-M6	3		
5	594134	Bolt	2		

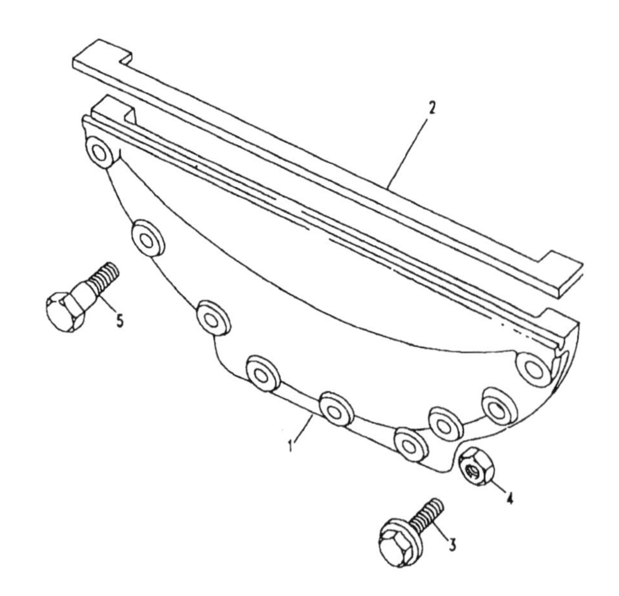

CHANGE POINTS:
(1) From (C) MA

C 53

DISCOVERY 1989MY UPTO 1999MY

GROUP C

Gearbox
R380 5 Speed
Extension Case

Illus	Part Number	Description	Quantity	Change Point	Remarks
1	FTC4241	CASE ASSEMBLY-EXTENSION	1	Note (1)	
1	FTC4522	CASE ASSEMBLY-EXTENSION	1	From (V) MA 109621	
	FTC3730	• Case-extension-gearbox	1		
2	LYQ100050	• Plug-blanking	1		
3	FRC2626	• Pin-spring	2		
4	FTC2392	• Guide-gear lever	1		
5	SH108251	• Screw-hexagonal head-M8 x 25	NLA		Use FS108257. Part of FTC4241.
6	CDU51	• Switch-reverse light manual transmission	1		Part of FTC4241.
7	ALU1403	• Washer-sealing-man trans reverse light switch	NLA		Use ALU1403L. Part of FTC4241.
7	ALU1403L	• Washer-sealing-man trans reverse light switch	1		
5	FS108257L	• Screw-flanged head-M8 x 25	2		Part of FTC4522.
	IGM100010	• Switch assembly-transfer box low ratio detect	1		Part of FTC4522.
	AMR3859	Harness-link	1	Note (1)	
8	FTC3701	Ring-mainshaft oil feed	1	Note (2); Note (3); Note (4)	
8	FTC4991	Ring-mainshaft oil feed	1	Note (5); Note (6); Note (7)	
9	FTC4206	PUMP-TRANSMISSION OIL	1		
	BZV1051	• O ring-oil pump seal	1		
10	FTC4449	Screw-torx-flange-M5 x 30	3		
11	FRC7855	Strainer-transmission oil	1		
12	FTC3387	Pipe-transmission oil inlet	1		
13	FTC2383	Seal-man trans output shaft	1		
14	FTC4021	Collar-distance manual transmission	1		

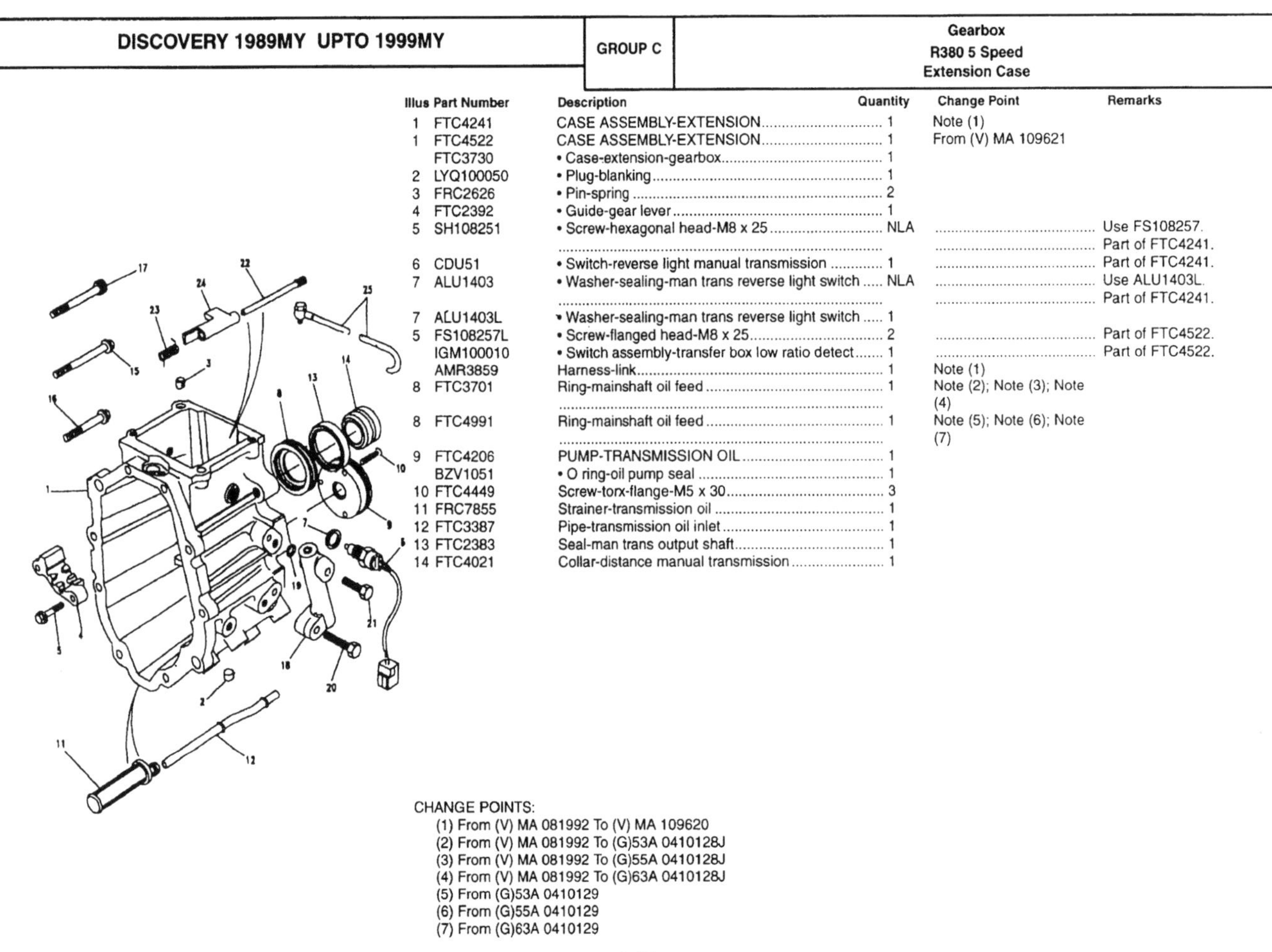

CHANGE POINTS:
(1) From (V) MA 081992 To (V) MA 109620
(2) From (V) MA 081992 To (G)53A 0410128J
(3) From (V) MA 081992 To (G)55A 0410128J
(4) From (V) MA 081992 To (G)63A 0410128J
(5) From (G)53A 0410129
(6) From (G)55A 0410129
(7) From (G)63A 0410129

C 54

DISCOVERY 1989MY UPTO 1999MY

GROUP C

Gearbox
R380 5 Speed
Extension Case

Illus	Part Number	Description	Quantity	Change Point	Remarks
15	FB108111L	Bolt-flanged head-M8 x 55	6		
16	FB108171L	Bolt-flanged head-M8 x 85-long	2		
17	FB108141S	Screw-M8 x 70	2		
18	FTC4053	Housing-transmission thermostat	1		MPi
18	FTC4053	Housing-transmission thermostat	1		TDi
18	FTC2687	Housing-transmission thermostat	1		Manual V8 except MPi
19	FTC1441	O ring	2		
20	FB108081L	Bolt-flanged head-M8 x 40	2		TDi
20	FB108081L	Bolt-flanged head-M8 x 40	2		MPi
20	FB108081L	Bolt-flanged head-M8 x 40	1		EFi
21	FS108301L	Screw-flanged head-M8 x 30	NLA		Use FS108307L. V8
21	FS108307L	Screw-flanged head-M8 x 30	2		V8
22	FTC3711	Shaft-camshaft reverse lock	1		
23	FTC4483	Spring-torsion	1		
24	FTC3713	Camshaft-reverse lock manual transmission	1		
25	FRC9430	Tube-breather transmission case	1		

DISCOVERY 1989MY UPTO 1999MY

GROUP C

Gearbox
R380 5 Speed
Mainshaft

Illus	Part Number	Description	Quantity	Change Point	Remarks
1	FTC3696	COVER-END MANUAL TRANSMISSION-FRONT	NLA		Use FTC5371. Suffix J
2	UKC1060L	• Seal-man trans primary shaft	NLA		Use FTC5303.
1	TVQ100130	COVER-END MANUAL TRANSMISSION	1		Suffix K
2	FTC5303	• Seal-man trans primary shaft	1		
3	FS108251ML	Screw-flanged head-M8 x 25	6		Suffix J
3	FS108301S	Screw-Gearbox extension housing front-M8 x 30	6		Suffix K
4	RTC6751	Bearing-taper roller	1		Suffix J
4	TZZ100190	Bearing-taper roller	1		Suffix K
5	FTC5048	Shaft primary manual transmission	1	Note (1)	MPi Suffix J
5	TUD102370	Shaft primary manual transmission	1		MPi Suffix K
5	FTC5046	Shaft primary manual transmission	1	Note (2)	Manual 4 Cylinder TDi Suffix J
5	TUD101970	Shaft primary manual transmission	1		TDi Suffix K
5	FTC5044	Shaft primary manual transmission	1	Note (1)	V8 Suffix J
5	TUD102340	Shaft primary manual transmission	1		V8 Suffix K
6	UKC8L	Bearing-taper roller-man trans primary shaft	1		

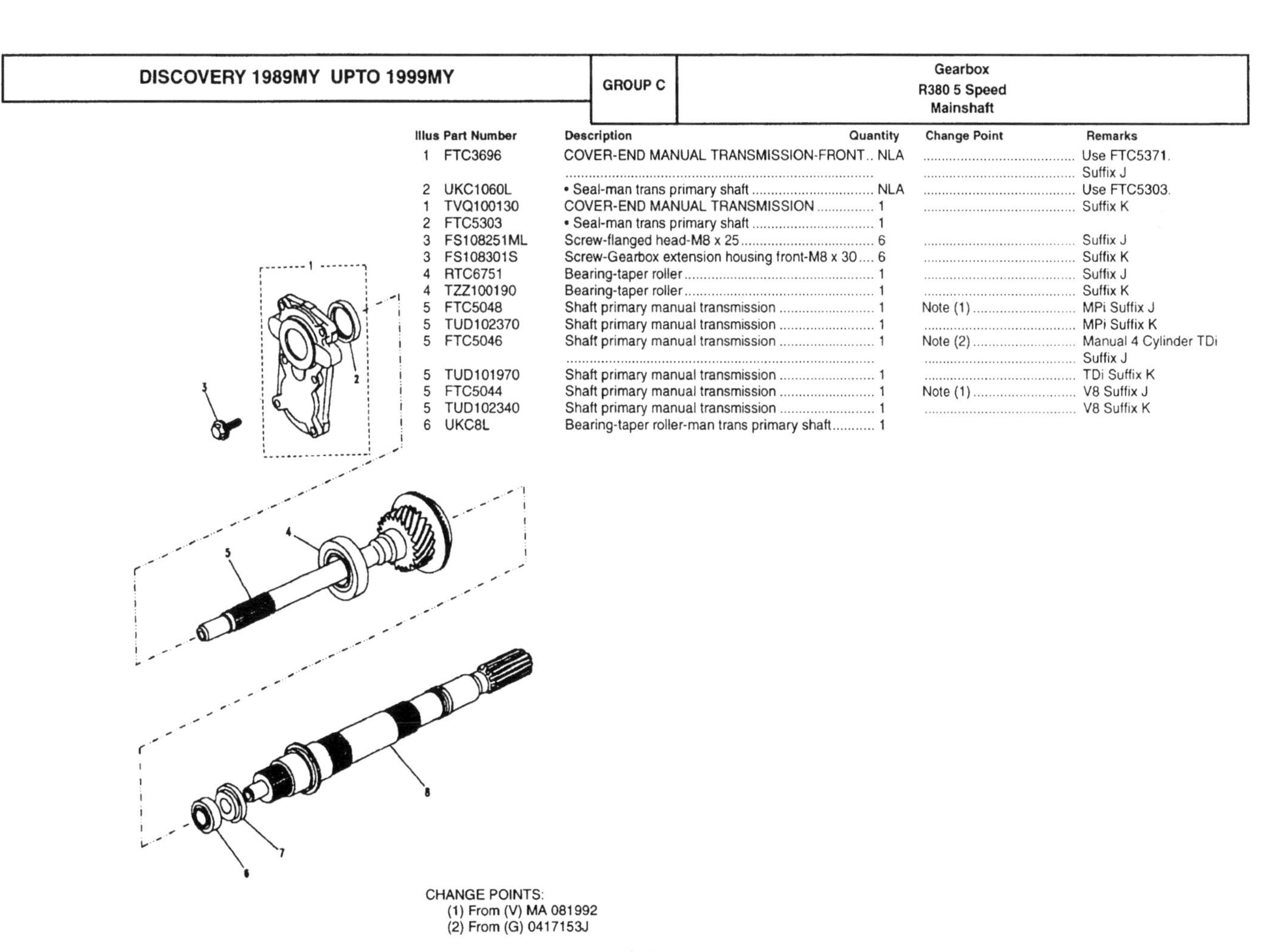

CHANGE POINTS:
(1) From (V) MA 081992
(2) From (G) 0417153J

Illus	Part Number	Description	Quantity	Change Point	Remarks
7	FTC2737	Spacer-mainshaft	1		
8	FTC3703	Shaft-main manual transmission	NLA		Use FTC5287 qty 1 Mainshaft strengthened, need to change mainshaft and fit new pair of thrust washer segments and with FTC5288 qty 1 with FTC4990 qty 2.
8	FTC5287	Shaft-main manual transmission	1		5 Speed
8	TUD101720	Shaft-main manual transmission	1		Suffix K

Illus	Part Number	Description	Quantity	Change Point	Remarks
1	FTC2714	Gear-2nd driven manual transmission	1	Note (1)	
1	FTC5038	Gear-2nd driven manual transmission	1	Note (2)	
2	FTC1311	Bearing-needle roller manual transmission	2		Suffix J
2	TUK100340	Bearing-needle roller manual transmission	2		Suffix K
3	FTC2397	Ring-synchroniser inner	NLA	Note (1)	Use FTC5019.
3	FTC5019	Ring-synchroniser inner	2	Note (2)	
4	FTC2396	Ring-synchroniser intermediate	2		
5	FTC3584	Ring-baulk	2		
6	FTC2733	SYNCHRONISER ASSEMBLY-1ST & 2ND MAINSHAFT ASSEMBLY	1	Note (1)	
6	FTC5100	SYNCHRONISER ASSEMBLY-1ST & 2ND MAINSHAFT ASSEMBLY	1	Note (2)	
7	FTC4172	• Spring ring	NLA		Use FTC4992. Part of FTC2733.
8	FTC4171	• Plate-synchroniser	NLA		
7	FTC4992	• Spring ring	2		Part of FTC5100.

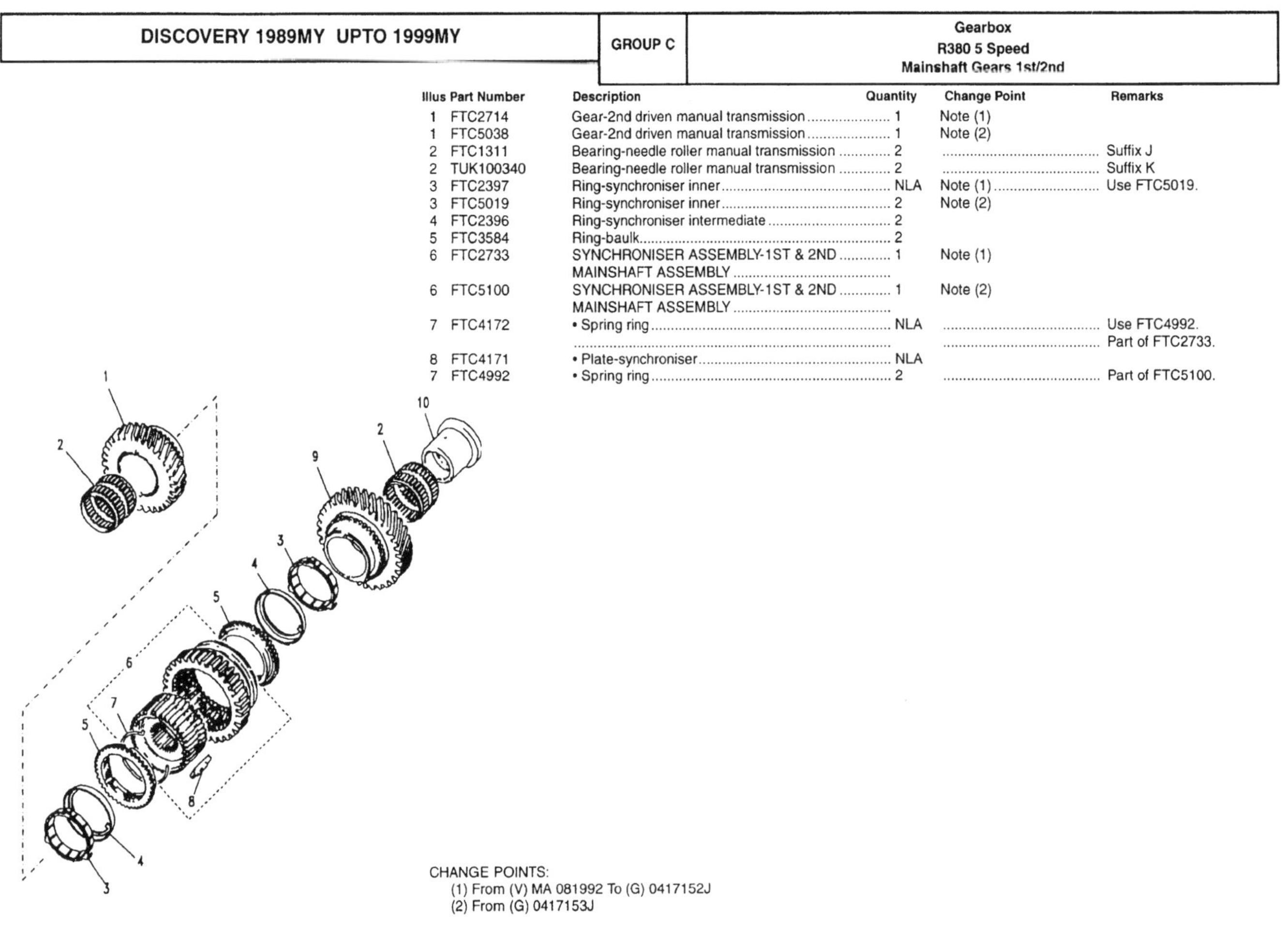

CHANGE POINTS:
(1) From (V) MA 081992 To (G) 0417152J
(2) From (G) 0417153J

Illus	Part Number	Description	Quantity	Change Point	Remarks
9	FTC2948	Gear-1st driven manual transmission	NLA	Note (1)	Use FTC5037 qty 1 FTC5038 and synchro kit STC3375 for diesel or FTC5040 qty 1 For Defender 4 cyl diesel MOD, BMC SANAYI Turkey and DAF 0 series & XUD9A fit FTC5037, FTC5040 and synchro kit STC3375. Diesel
9	FTC5037	Gear-1st driven manual transmission	1	Note (2)	Diesel Suffix J
9	TUB101800	Gear-1st driven manual transmission	1		Suffix K
9	FTC2716	Gear-1st driven manual transmission	NLA	Note (1)	Fit new 1st/2nd synchro kit STC3375 with 1st gear FTC5036 and 2nd gear FTC5038. Petrol
9	FTC5036	Gear-1st driven manual transmission	1	Note (2)	Petrol
10	FTC2005	Bush-1st gear selective-30.905 to 30.955mm	NLA		Use TUJ100050. Suffix J
10	TUJ100050	Bush-1st gear selective-30.905 to 30.955mm	1		Suffix K
	STC3375	KIT-GEARS-1ST/2ND	1		
3	FTC5019	• Ring-synchroniser inner	2		
4	FTC2396	• Ring-synchroniser intermediate	2		
5	FTC3584	• Ring-baulk	2		
6	FTC5100	• SYNCHRONISER ASSEMBLY-1ST & 2ND MAINSHAFT ASSEMBLY	1		
7	FTC4992	• • Spring ring	2		
8	FTC4171	• • Plate-synchroniser	NLA		

CHANGE POINTS:
(1) From (V) MA 081992 To (G) 0417152J
(2) From (G) 0417153J

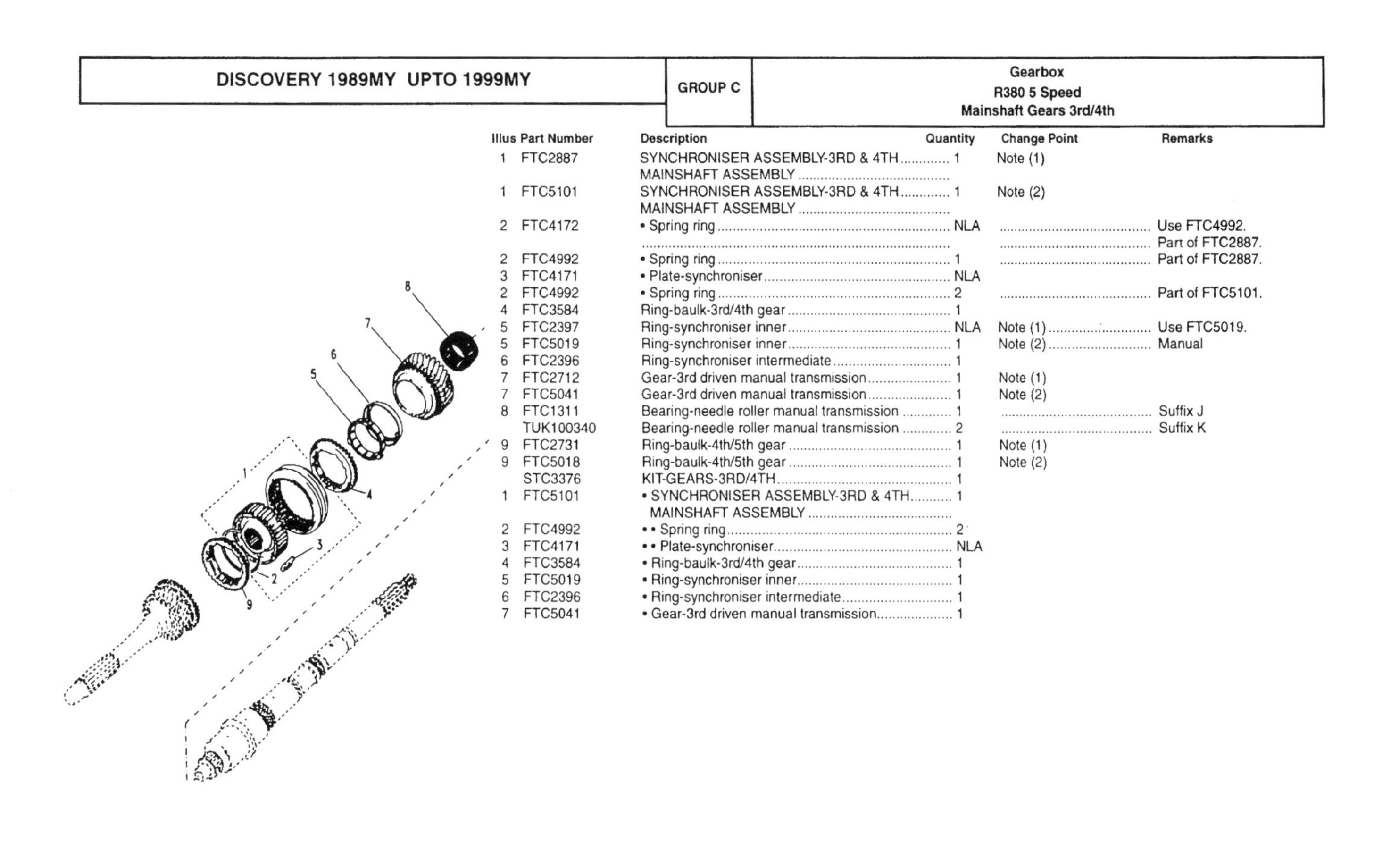

Illus	Part Number	Description	Quantity	Change Point	Remarks
1	FTC2887	SYNCHRONISER ASSEMBLY-3RD & 4TH MAINSHAFT ASSEMBLY	1	Note (1)	
1	FTC5101	SYNCHRONISER ASSEMBLY-3RD & 4TH MAINSHAFT ASSEMBLY	1	Note (2)	
2	FTC4172	• Spring ring	NLA		Use FTC4992. Part of FTC2887.
2	FTC4992	• Spring ring	1		Part of FTC2887.
3	FTC4171	• Plate-synchroniser	NLA		
2	FTC4992	• Spring ring	2		Part of FTC5101.
4	FTC3584	Ring-baulk-3rd/4th gear	1		
5	FTC2397	Ring-synchroniser inner	NLA	Note (1)	Use FTC5019.
5	FTC5019	Ring-synchroniser inner	1	Note (2)	Manual
6	FTC2396	Ring-synchroniser intermediate	1		
7	FTC2712	Gear-3rd driven manual transmission	1	Note (1)	
7	FTC5041	Gear-3rd driven manual transmission	1	Note (2)	
8	FTC1311	Bearing-needle roller manual transmission	1		Suffix J
	TUK100340	Bearing-needle roller manual transmission	2		Suffix K
9	FTC2731	Ring-baulk-4th/5th gear	1	Note (1)	
9	FTC5018	Ring-baulk-4th/5th gear	1	Note (2)	
	STC3376	KIT-GEARS-3RD/4TH	1		
1	FTC5101	• SYNCHRONISER ASSEMBLY-3RD & 4TH MAINSHAFT ASSEMBLY	1		
2	FTC4992	• • Spring ring	2		
3	FTC4171	• • Plate-synchroniser	NLA		
4	FTC3584	• Ring-baulk-3rd/4th gear	1		
5	FTC5019	• Ring-synchroniser inner	1		
6	FTC2396	• Ring-synchroniser intermediate	1		
7	FTC5041	• Gear-3rd driven manual transmission	1		

CHANGE POINTS:
(1) From (V) MA 081992 To (G) 0417152J
(2) From (G) 0417153J

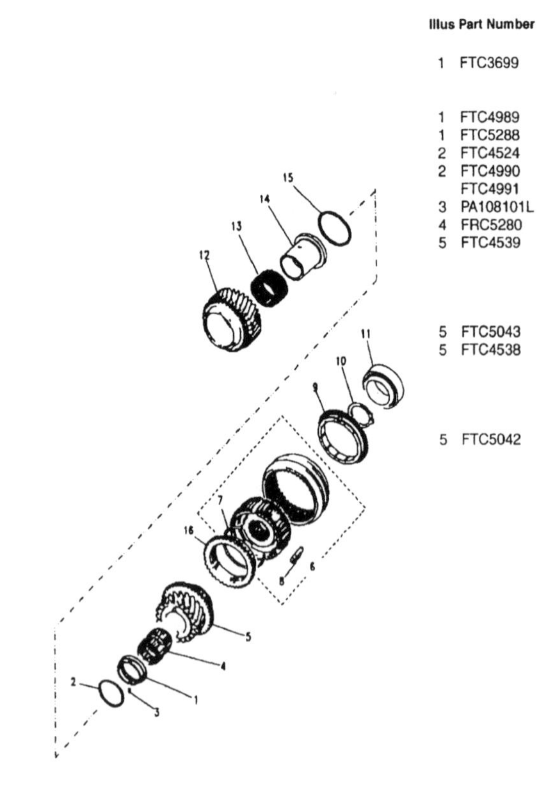

Illus	Part Number	Description	Quantity	Change Point	Remarks
		FROM VIN MA081992			
1	FTC3699	Washer-thrust manual transmission	NLA	Note (1)	Use FTC4989 qty 2 with FTC4990 qty 1 with FTC4991 qty 1.
1	FTC4989	Washer-thrust manual transmission	NLA	Note (2)	Use FTC5288.
1	FTC5288	Washer-thrust manual transmission	1	Note (2)	
2	FTC4524	Ring-retention	1	Note (1)	
2	FTC4990	Ring-retention	1	Note (2)	
	FTC4991	Ring-mainshaft oil feed	1	Note (2)	
3	PA108101L	Pin-spring	1	Note (1)	
4	FRC5280	Bearing-needle roller manual transmission	1		
5	FTC4539	Gear-5th driven manual transmission	NLA	Note (3)	Fit new 5th/reverse synchro & reverse gear kit STC3377 and 5th gear FTC5043.. Diesel
5	FTC5043	Gear-5th driven manual transmission	1	Note (4)	Diesel
5	FTC4538	Gear-5th driven manual transmission	NLA	Note (3)	Fit new 5th gear FTC5042 and 5th/reverse synchro & reverse gear kit STC3377.. Petrol
5	FTC5042	Gear-5th driven manual transmission	1	Note (4)	Petrol

CHANGE POINTS:
(1) From (V) MA 081992 To (G) 0410128J
(2) From (G) 0410129J
(3) From (V) MA 081992 To (G) 0417152J
(4) From (G) 0417153J

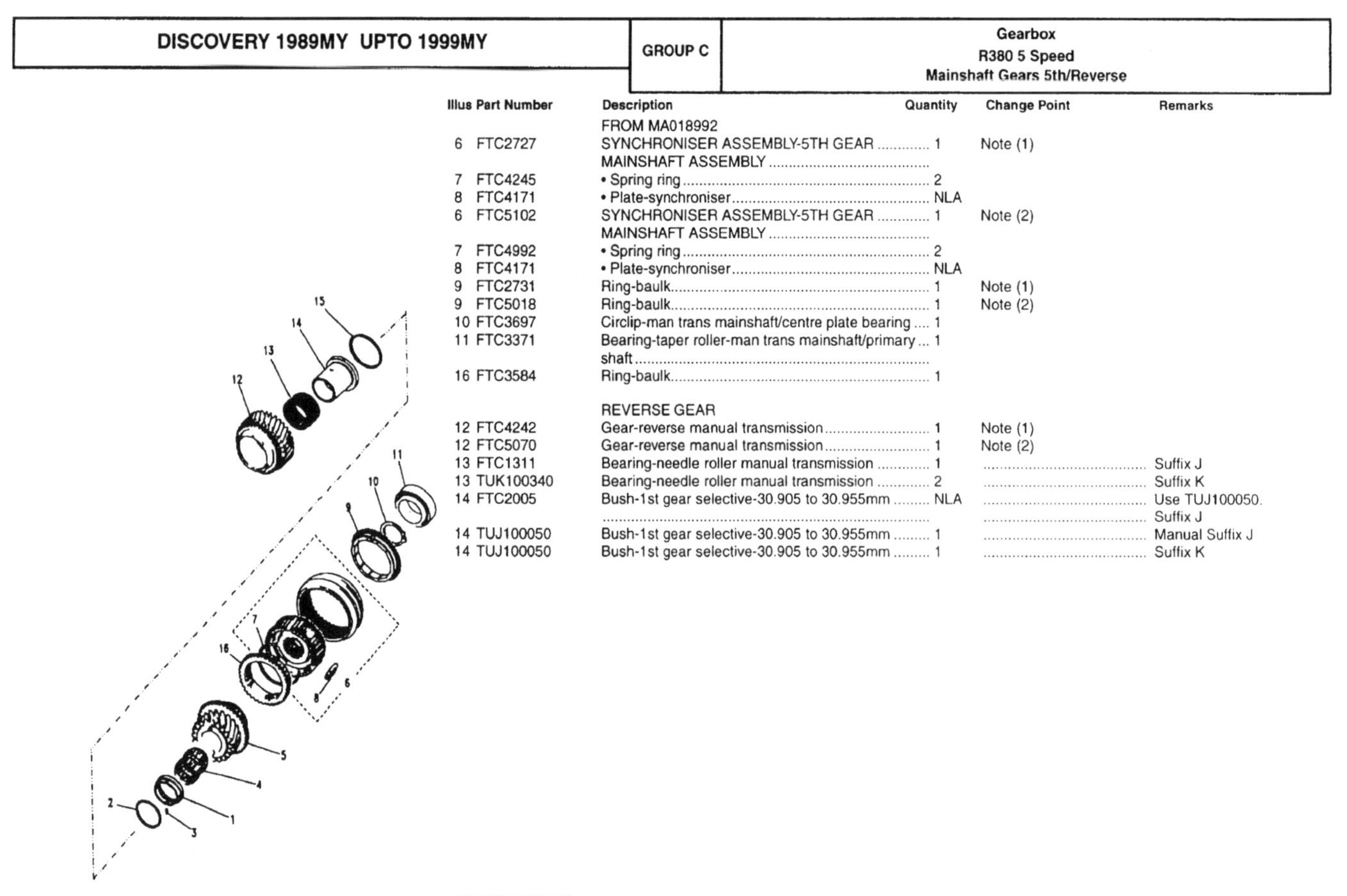

Illus	Part Number	Description	Quantity	Change Point	Remarks
		FROM MA018992			
6	FTC2727	SYNCHRONISER ASSEMBLY-5TH GEAR MAINSHAFT ASSEMBLY	1	Note (1)	
7	FTC4245	• Spring ring	2		
8	FTC4171	• Plate-synchroniser	NLA		
6	FTC5102	SYNCHRONISER ASSEMBLY-5TH GEAR MAINSHAFT ASSEMBLY	1	Note (2)	
7	FTC4992	• Spring ring	2		
8	FTC4171	• Plate-synchroniser	NLA		
9	FTC2731	Ring-baulk	1	Note (1)	
9	FTC5018	Ring-baulk	1	Note (2)	
10	FTC3697	Circlip-man trans mainshaft/centre plate bearing	1		
11	FTC3371	Bearing-taper roller-man trans mainshaft/primary shaft	1		
16	FTC3584	Ring-baulk	1		
		REVERSE GEAR			
12	FTC4242	Gear-reverse manual transmission	1	Note (1)	
12	FTC5070	Gear-reverse manual transmission	1	Note (2)	
13	FTC1311	Bearing-needle roller manual transmission	1		Suffix J
13	TUK100340	Bearing-needle roller manual transmission	2		Suffix K
14	FTC2005	Bush-1st gear selective-30.905 to 30.955mm	NLA		Use TUJ100050. Suffix J
14	TUJ100050	Bush-1st gear selective-30.905 to 30.955mm	1		Manual Suffix J
14	TUJ100050	Bush-1st gear selective-30.905 to 30.955mm	1		Suffix K

CHANGE POINTS:
(1) From (V) MA 081992 To (G) 0417152J
(2) From (G) 0417153J

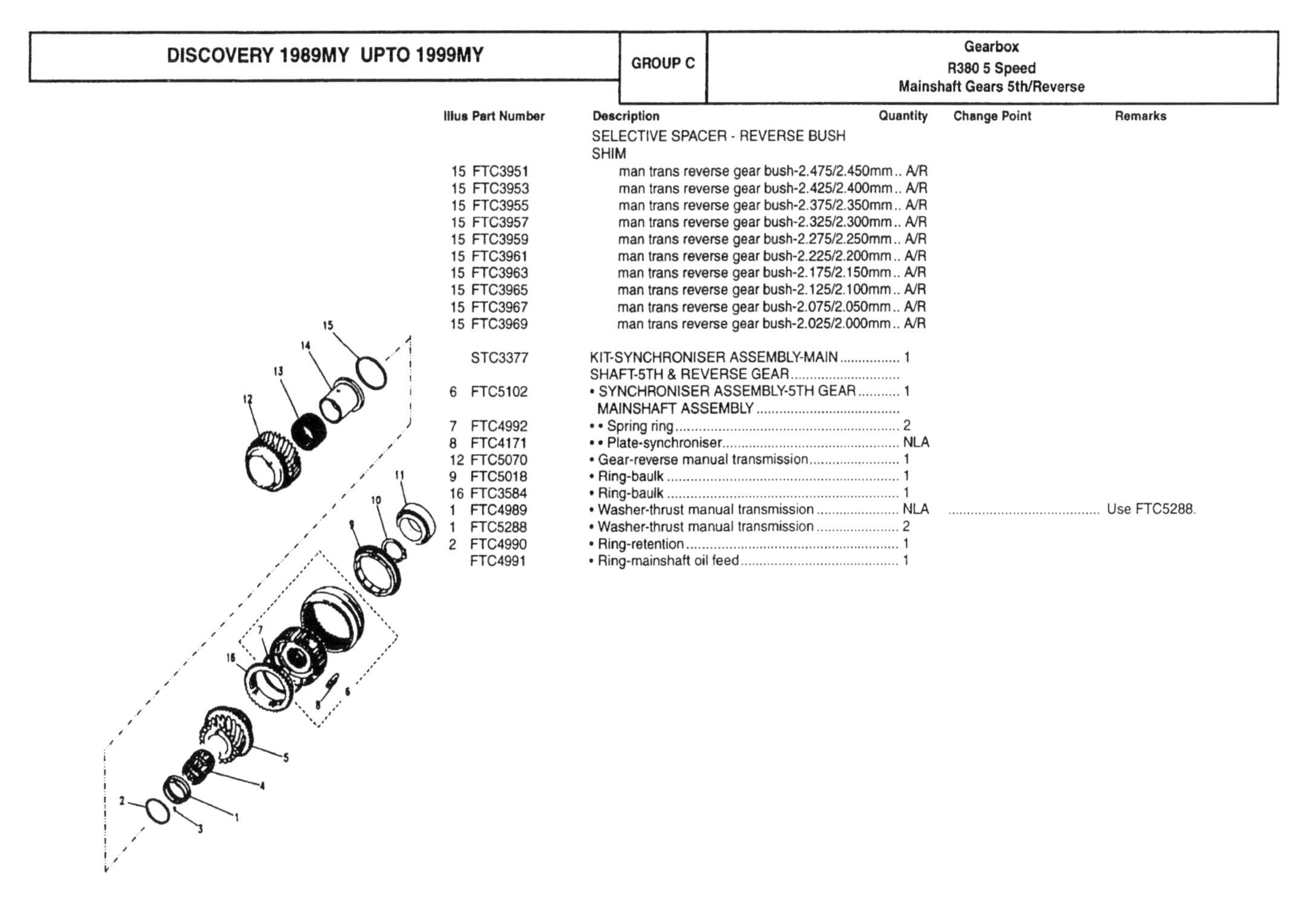

Illus	Part Number	Description	Quantity	Change Point	Remarks
		SELECTIVE SPACER - REVERSE BUSH SHIM			
15	FTC3951	man trans reverse gear bush-2.475/2.450mm	A/R		
15	FTC3953	man trans reverse gear bush-2.425/2.400mm	A/R		
15	FTC3955	man trans reverse gear bush-2.375/2.350mm	A/R		
15	FTC3957	man trans reverse gear bush-2.325/2.300mm	A/R		
15	FTC3959	man trans reverse gear bush-2.275/2.250mm	A/R		
15	FTC3961	man trans reverse gear bush-2.225/2.200mm	A/R		
15	FTC3963	man trans reverse gear bush-2.175/2.150mm	A/R		
15	FTC3965	man trans reverse gear bush-2.125/2.100mm	A/R		
15	FTC3967	man trans reverse gear bush-2.075/2.050mm	A/R		
15	FTC3969	man trans reverse gear bush-2.025/2.000mm	A/R		
	STC3377	KIT-SYNCHRONISER ASSEMBLY-MAIN SHAFT-5TH & REVERSE GEAR	1		
6	FTC5102	• SYNCHRONISER ASSEMBLY-5TH GEAR MAINSHAFT ASSEMBLY	1		
7	FTC4992	• • Spring ring	2		
8	FTC4171	• • Plate-synchroniser	NLA		
12	FTC5070	• Gear-reverse manual transmission	1		
9	FTC5018	• Ring-baulk	1		
16	FTC3584	• Ring-baulk	1		
1	FTC4989	• Washer-thrust manual transmission	NLA		Use FTC5288.
1	FTC5288	• Washer-thrust manual transmission	2		
2	FTC4990	• Ring-retention	1		
	FTC4991	• Ring-mainshaft oil feed	1		

Illus	Part Number	Description	Quantity	Change Point	Remarks
		FROM CHASSIS MA-MANUAL			
1	FTC2941	Shaft-lay	NLA	Note (1)	Fit layshaft FTC4982 and 5th gear kit STC3378..
1	FTC4982	Shaft-lay	1	Note (2)	Diesel Suffix J
1	FTC2723	Shaft-lay	1	Note (1)	
1	FTC4980	Shaft-lay	1	Note (2)	Petrol Suffix J
1	TUO100260	Shaft-lay	1		Petrol Suffix K
1	TUO100240	Shaft-lay	1		Diesel Suffix K
2	FTC317	Bearing-taper roller-man trans countershaft/centre plate	1		Diesel Suffix J
2	TZZ100200	Bearing-taper roller-man trans countershaft/centre plate	1		Suffix K
3	FTC248	Bearing-taper roller-man trans countershaft/main case	1		

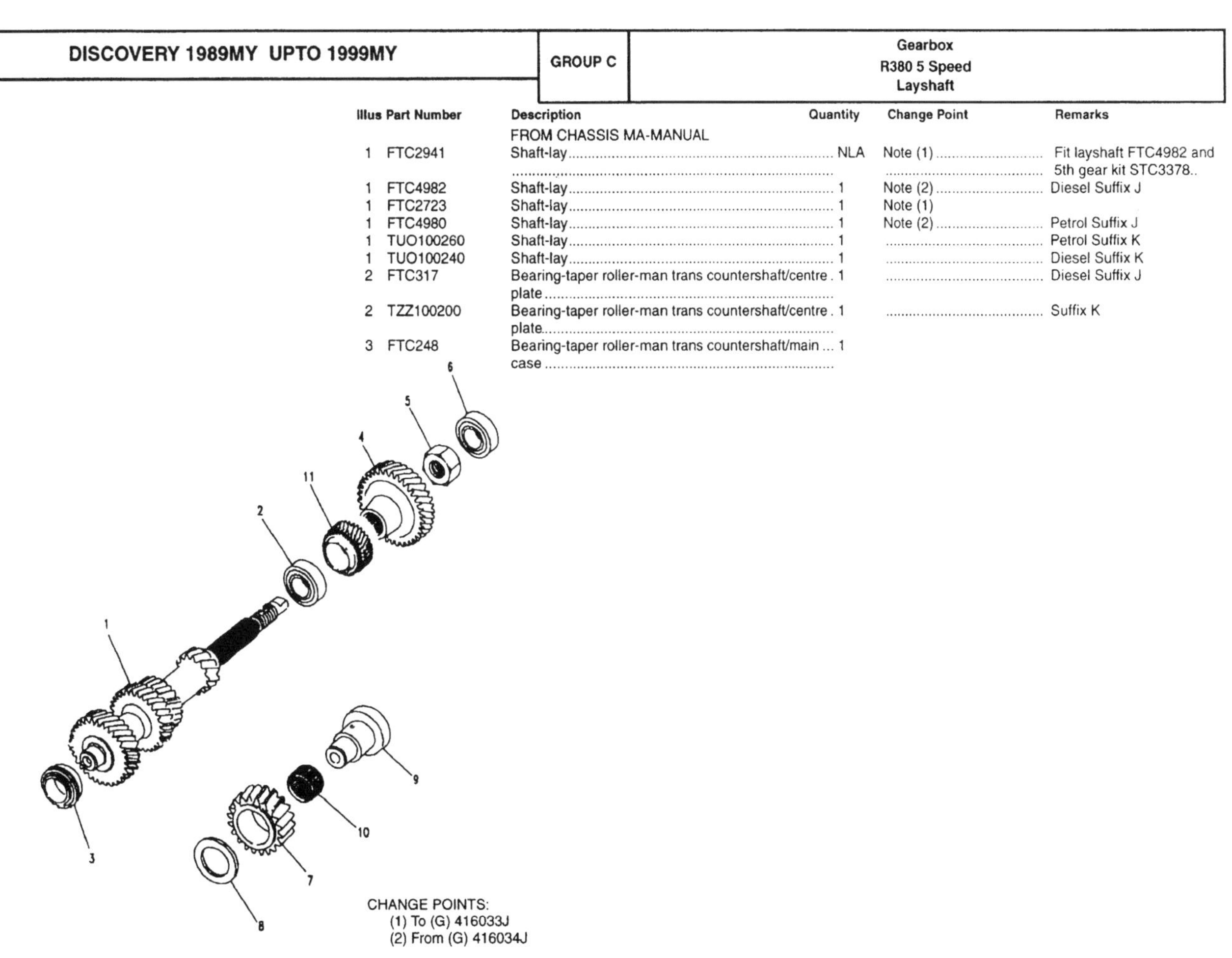

CHANGE POINTS:
(1) To (G) 416033J
(2) From (G) 416034J

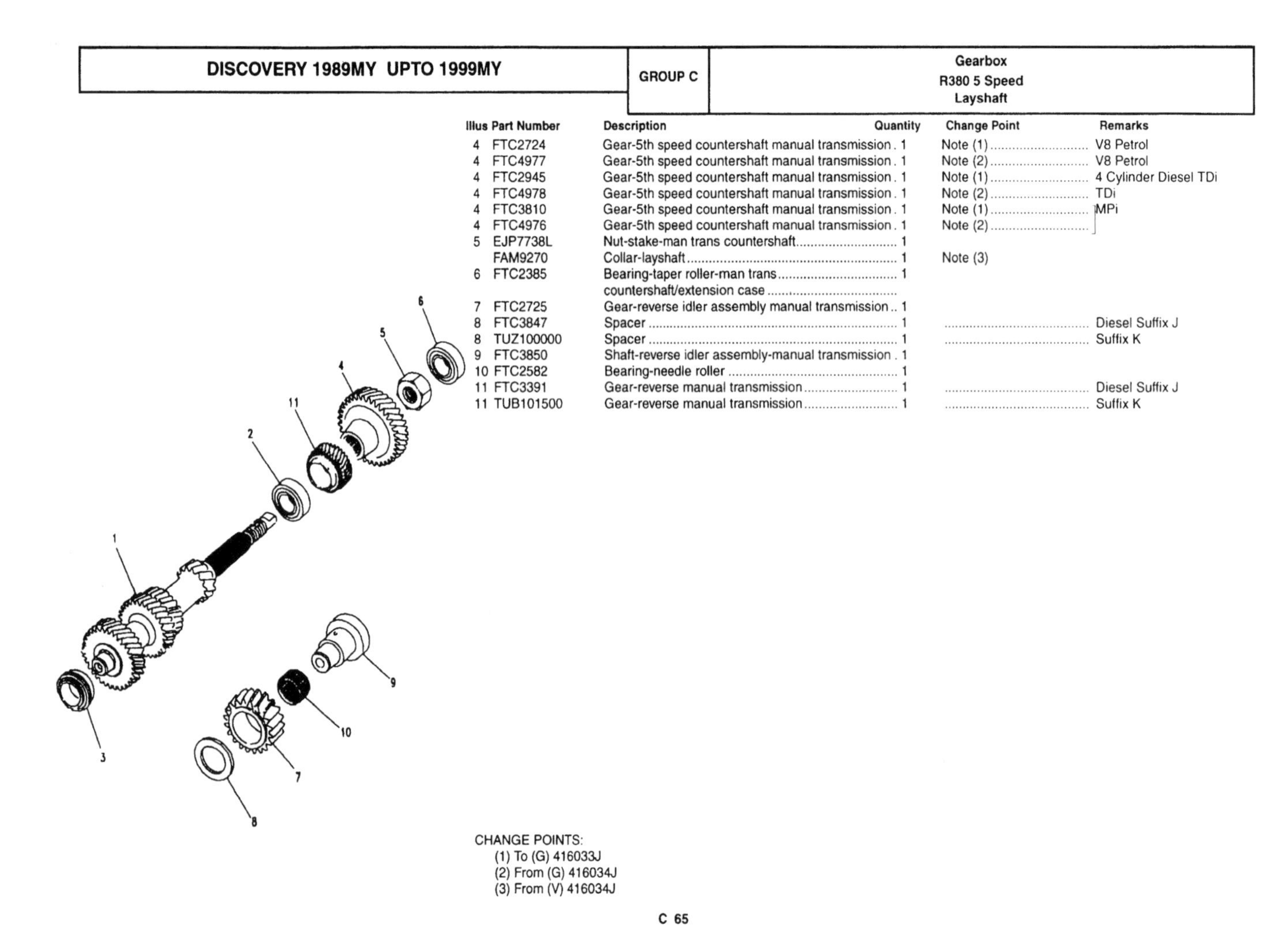

DISCOVERY 1989MY UPTO 1999MY

GROUP C

Gearbox
R380 5 Speed
Layshaft

Illus	Part Number	Description	Quantity	Change Point	Remarks
4	FTC2724	Gear-5th speed countershaft manual transmission	1	Note (1)	V8 Petrol
4	FTC4977	Gear-5th speed countershaft manual transmission	1	Note (2)	V8 Petrol
4	FTC2945	Gear-5th speed countershaft manual transmission	1	Note (1)	4 Cylinder Diesel TDi
4	FTC4978	Gear-5th speed countershaft manual transmission	1	Note (2)	TDi
4	FTC3810	Gear-5th speed countershaft manual transmission	1	Note (1)	MPi
4	FTC4976	Gear-5th speed countershaft manual transmission	1	Note (2)	
5	EJP7738L	Nut-stake-man trans countershaft	1		
	FAM9270	Collar-layshaft	1	Note (3)	
6	FTC2385	Bearing-taper roller-man trans countershaft/extension case	1		
7	FTC2725	Gear-reverse idler assembly manual transmission	1		
8	FTC3847	Spacer	1		Diesel Suffix J
8	TUZ100000	Spacer	1		Suffix K
9	FTC3850	Shaft-reverse idler assembly-manual transmission	1		
10	FTC2582	Bearing-needle roller	1		
11	FTC3391	Gear-reverse manual transmission	1		Diesel Suffix J
11	TUB101500	Gear-reverse manual transmission	1		Suffix K

CHANGE POINTS:
(1) To (G) 416033J
(2) From (G) 416034J
(3) From (V) 416034J

C 65

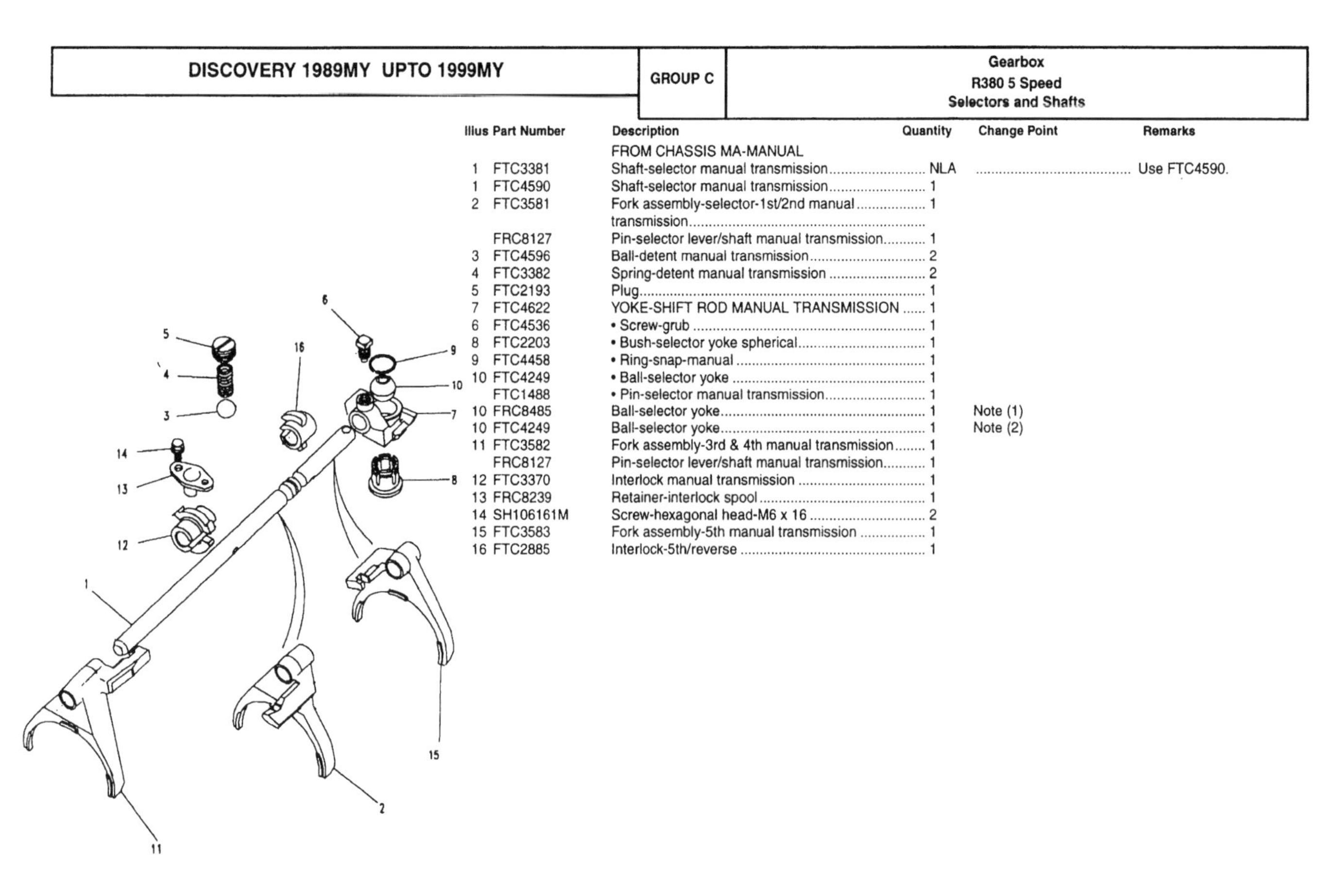

DISCOVERY 1989MY UPTO 1999MY

GROUP C

Gearbox
R380 5 Speed
Selectors and Shafts

Illus	Part Number	Description	Quantity	Change Point	Remarks
		FROM CHASSIS MA-MANUAL			
1	FTC3381	Shaft-selector manual transmission	NLA		Use FTC4590.
1	FTC4590	Shaft-selector manual transmission	1		
2	FTC3581	Fork assembly-selector-1st/2nd manual transmission	1		
	FRC8127	Pin-selector lever/shaft manual transmission	1		
3	FTC4596	Ball-detent manual transmission	2		
4	FTC3382	Spring-detent manual transmission	2		
5	FTC2193	Plug	1		
7	FTC4622	YOKE-SHIFT ROD MANUAL TRANSMISSION	1		
6	FTC4536	• Screw-grub	1		
8	FTC2203	• Bush-selector yoke spherical	1		
9	FTC4458	• Ring-snap-manual	1		
10	FTC4249	• Ball-selector yoke	1		
	FTC1488	• Pin-selector manual transmission	1		
10	FRC8485	Ball-selector yoke	1	Note (1)	
10	FTC4249	Ball-selector yoke	1	Note (2)	
11	FTC3582	Fork assembly-3rd & 4th manual transmission	1		
	FRC8127	Pin-selector lever/shaft manual transmission	1		
12	FTC3370	Interlock manual transmission	1		
13	FRC8239	Retainer-interlock spool	1		
14	SH106161M	Screw-hexagonal head-M6 x 16	2		
15	FTC3583	Fork assembly-5th manual transmission	1		
16	FTC2885	Interlock-5th/reverse	1		

CHANGE POINTS:
(1) To (V)53A 0269703J; To (V)55A 0269111J
(2) From (V)53A 0269704J; From (V)55A 0269112J

C 66

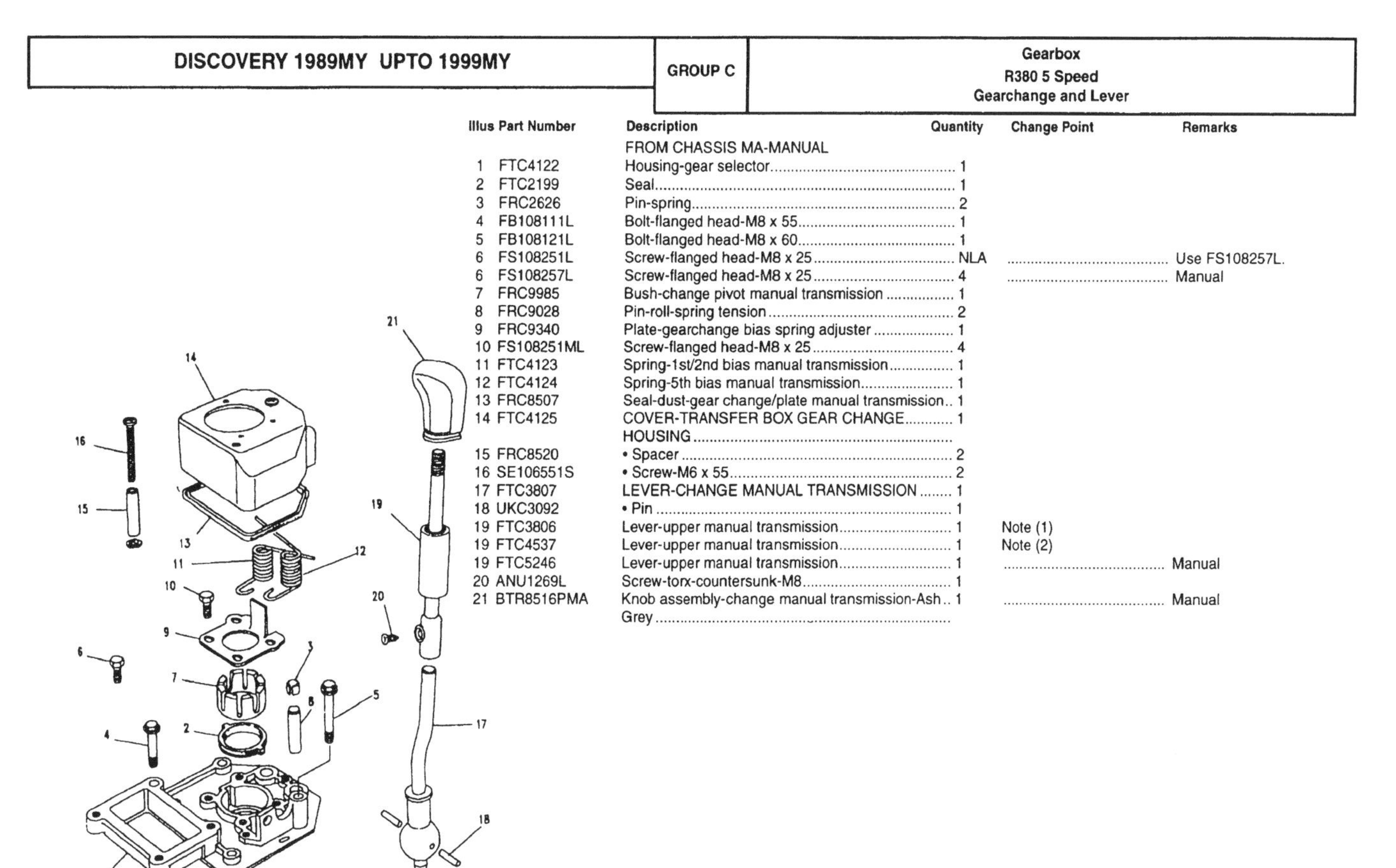

DISCOVERY 1989MY UPTO 1999MY

GROUP C

Gearbox
R380 5 Speed
Gearchange and Lever

Illus	Part Number	Description	Quantity	Change Point	Remarks
		FROM CHASSIS MA-MANUAL			
1	FTC4122	Housing-gear selector	1		
2	FTC2199	Seal	1		
3	FRC2626	Pin-spring	2		
4	FB108111L	Bolt-flanged head-M8 x 55	1		
5	FB108121L	Bolt-flanged head-M8 x 60	1		
6	FS108251L	Screw-flanged head-M8 x 25	NLA		Use FS108257L.
6	FS108257L	Screw-flanged head-M8 x 25	4		Manual
7	FRC9985	Bush-change pivot manual transmission	1		
8	FRC9028	Pin-roll-spring tension	2		
9	FRC9340	Plate-gearchange bias spring adjuster	1		
10	FS108251ML	Screw-flanged head-M8 x 25	4		
11	FTC4123	Spring-1st/2nd bias manual transmission	1		
12	FTC4124	Spring-5th bias manual transmission	1		
13	FRC8507	Seal-dust-gear change/plate manual transmission	1		
14	FTC4125	COVER-TRANSFER BOX GEAR CHANGE HOUSING	1		
15	FRC8520	• Spacer	2		
16	SE106551S	• Screw-M6 x 55	2		
17	FTC3807	LEVER-CHANGE MANUAL TRANSMISSION	1		
18	UKC3092	• Pin	1		
19	FTC3806	Lever-upper manual transmission	1	Note (1)	
19	FTC4537	Lever-upper manual transmission	1	Note (2)	
19	FTC5246	Lever-upper manual transmission	1		Manual
20	ANU1269L	Screw-torx-countersunk-M8	1		
21	BTR8516PMA	Knob assembly-change manual transmission-Ash Grey	1		Manual

CHANGE POINTS:
(1) To (V) MA 118209
(2) From (V) MA 118210

C 67

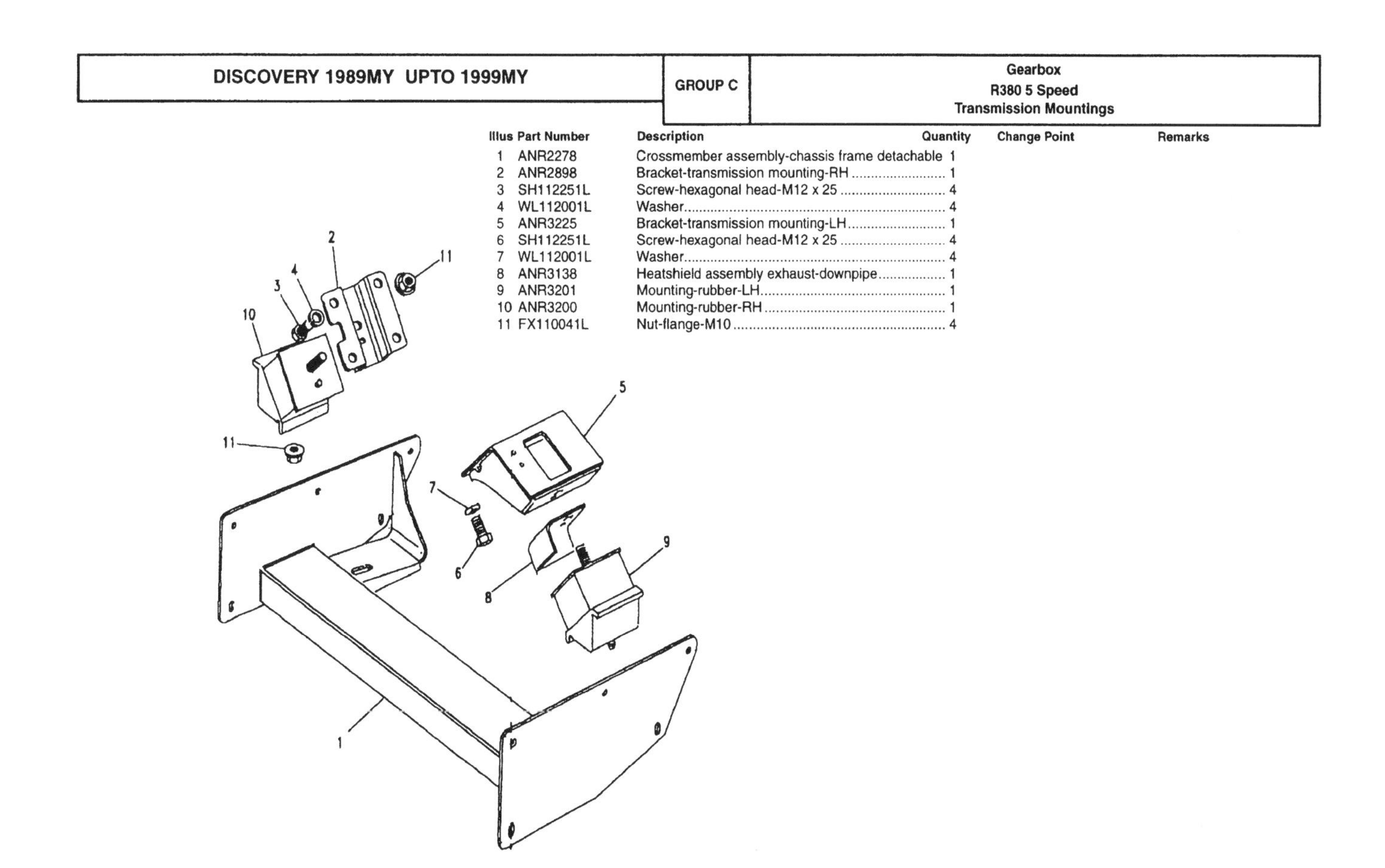

DISCOVERY 1989MY UPTO 1999MY

GROUP C

Gearbox
R380 5 Speed
Transmission Mountings

Illus	Part Number	Description	Quantity	Change Point	Remarks
1	ANR2278	Crossmember assembly-chassis frame detachable	1		
2	ANR2898	Bracket-transmission mounting-RH	1		
3	SH112251L	Screw-hexagonal head-M12 x 25	4		
4	WL112001L	Washer	4		
5	ANR3225	Bracket-transmission mounting-LH	1		
6	SH112251L	Screw-hexagonal head-M12 x 25	4		
7	WL112001L	Washer	4		
8	ANR3138	Heatshield assembly exhaust-downpipe	1		
9	ANR3201	Mounting-rubber-LH	1		
10	ANR3200	Mounting-rubber-RH	1		
11	FX110041L	Nut-flange-M10	4		

C 68

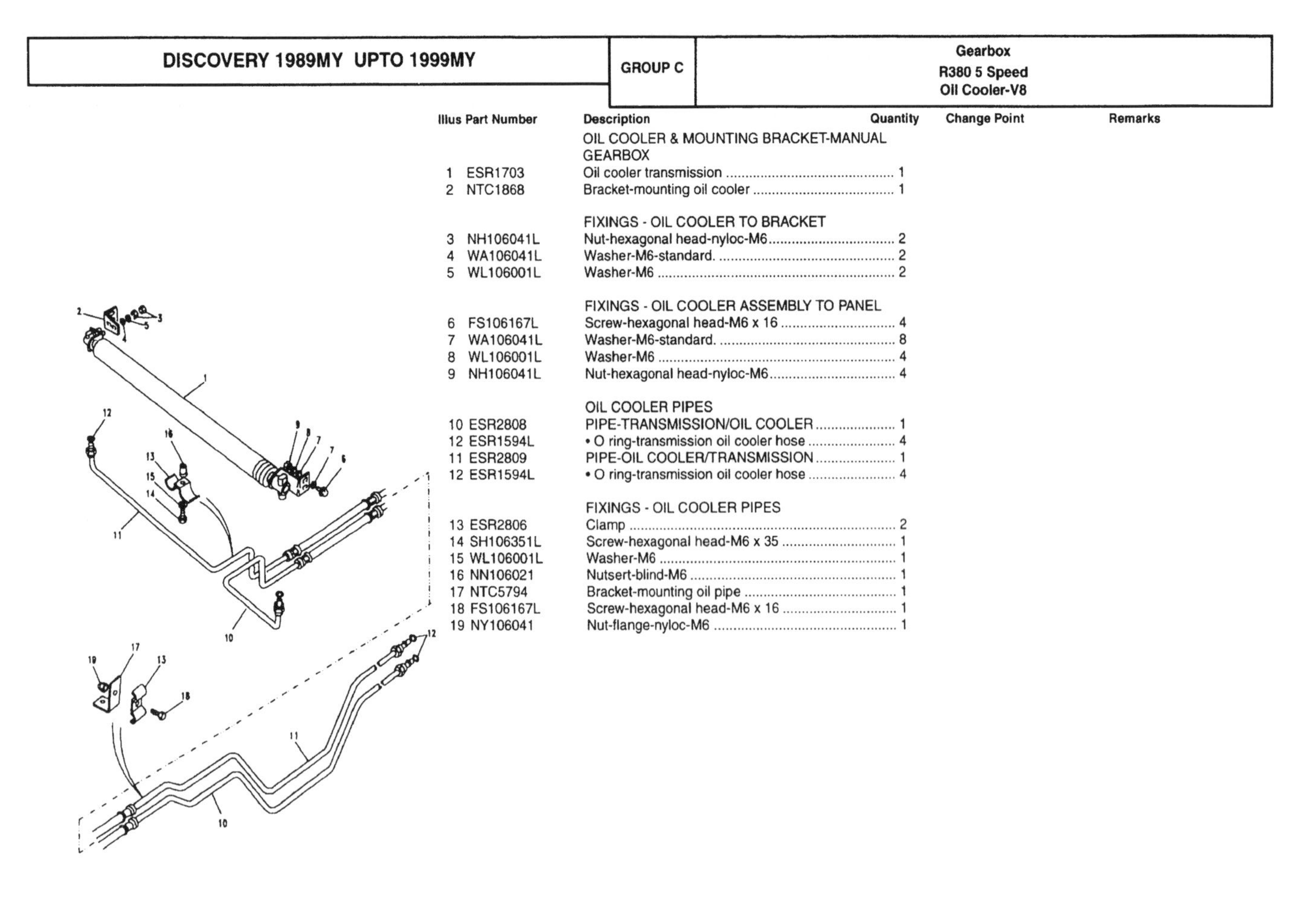

Illus	Part Number	Description	Quantity	Change Point	Remarks
		OIL COOLER & MOUNTING BRACKET-MANUAL GEARBOX			
1	ESR1703	Oil cooler transmission	1		
2	NTC1868	Bracket-mounting oil cooler	1		
		FIXINGS - OIL COOLER TO BRACKET			
3	NH106041L	Nut-hexagonal head-nyloc-M6	2		
4	WA106041L	Washer-M6-standard.	2		
5	WL106001L	Washer-M6	2		
		FIXINGS - OIL COOLER ASSEMBLY TO PANEL			
6	FS106167L	Screw-hexagonal head-M6 x 16	4		
7	WA106041L	Washer-M6-standard.	8		
8	WL106001L	Washer-M6	4		
9	NH106041L	Nut-hexagonal head-nyloc-M6	4		
		OIL COOLER PIPES			
10	ESR2808	PIPE-TRANSMISSION/OIL COOLER	1		
12	ESR1594L	• O ring-transmission oil cooler hose	4		
11	ESR2809	PIPE-OIL COOLER/TRANSMISSION	1		
12	ESR1594L	• O ring-transmission oil cooler hose	4		
		FIXINGS - OIL COOLER PIPES			
13	ESR2806	Clamp	2		
14	SH106351L	Screw-hexagonal head-M6 x 35	1		
15	WL106001L	Washer-M6	1		
16	NN106021	Nutsert-blind-M6	1		
17	NTC5794	Bracket-mounting oil pipe	1		
18	FS106167L	Screw-hexagonal head-M6 x 16	1		
19	NY106041	Nut-flange-nyloc-M6	1		

Illus	Part Number	Description	Quantity	Change Point	Remarks
		NEW			
1	FTC2258N	Transmission assembly automatic -new	1		3500 cc V8
1	FTC3099	Transmission assembly automatic -new	1	Note (1)	3900 cc V8
1	FTC2756	Transmission assembly automatic -new	1	Note (2)	TDi
1	FTC4969	Transmission assembly automatic -new	1	Note (3)	3900 cc V8
1	FTC4552	Transmission assembly automatic -new	1	Note (4)	TDi
		AUSTRALIA ONLY			
1	FTC2756	Transmission assembly automatic -new	1	Note (5)	TDi Australia
1	FTC2756E	Transmission assembly automatic -exchange	1	Note (5)	TDi Australia
		EXCHANGE			
1	FTC2258E	Transmission assembly automatic -exchange	1		3500 cc V8
1	FTC3099R	Transmission assembly automatic -reconditioned	NLA		Use FTC3099E. 3900 cc
1	FTC3099E	Transmission assembly automatic -exchange	1	Note (1)	3900 cc
1	FTC2756R	Transmission assembly automatic -reconditioned	NLA		Use FTC2756E. TDi
1	FTC2756E	Transmission assembly automatic -exchange	1		Automatic 4 Cylinder TDi except EDC
1	FTC4552E	Transmission assembly automatic -exchange	1	Note (4)	Automatic 4 Cylinder TDi EDC
1	FTC4969E	Transmission assembly automatic -exchange	1	Note (3)	3900 cc V8

CHANGE POINTS:
(1) To (E)36D 52139C
(2) To (E)19L 00409
(3) From (E)36D 52140C
(4) From (E)19L 00410A
(5) From (E)20L

Illus	Part Number	Description	Quantity	Change Point	Remarks
1	FRC5007	Adaptor-transmission mounting plate	1	Note (1)	V8
1	ERR4612	Adaptor-transmission mounting plate	1		TDi
2	FRC5054	Dowel	2		
3	FRC5024	Dowel	2		
4	BH506131L	Bolt-3/8UNC x 1 7/8	8		
5	BH608181L	Bolt	1		
6	594087	Seal-bottom	1		
7	FRC2859	Cover-bell housing-bottom	1		
8	FS106255L	Screw-cover bottom to plate-flanged head-M6 x 25	7		
9	WM106001L	Washer-spring-M6-square	7		
10	594134	Bolt-cover bottom to plate	2		
11	WM108001L	Washer-spring-M8	2		

CHANGE POINTS:
(1) To (V) KA 064754

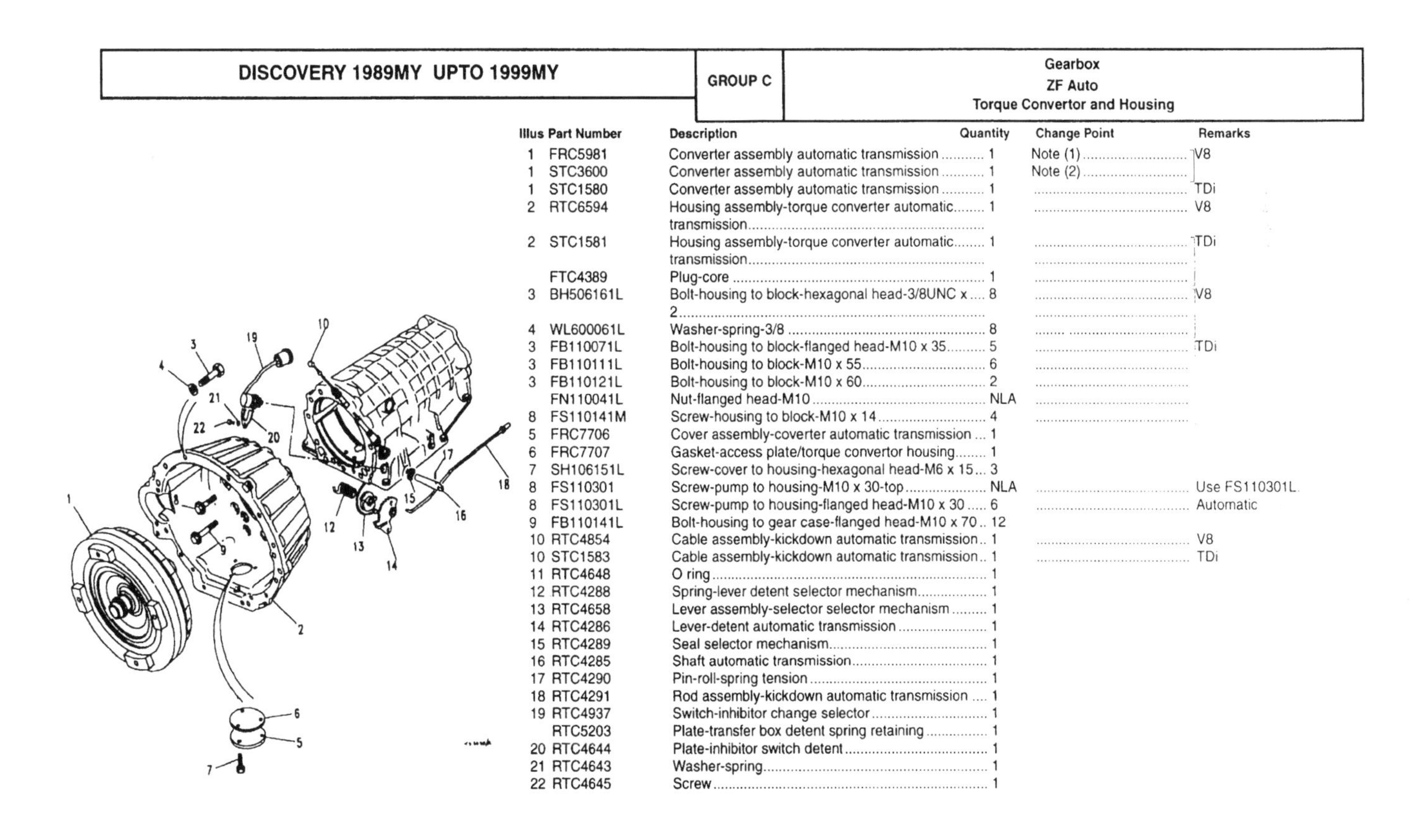

Illus	Part Number	Description	Quantity	Change Point	Remarks
1	FRC5981	Converter assembly automatic transmission	1	Note (1)	V8
1	STC3600	Converter assembly automatic transmission	1	Note (2)	
1	STC1580	Converter assembly automatic transmission	1		TDi
2	RTC6594	Housing assembly-torque converter automatic transmission	1		V8
2	STC1581	Housing assembly-torque converter automatic transmission	1		TDi
	FTC4389	Plug-core	1		
3	BH506161L	Bolt-housing to block-hexagonal head-3/8UNC x 2	8		V8
4	WL600061L	Washer-spring-3/8	8		
3	FB110071L	Bolt-housing to block-flanged head-M10 x 35	5		TDi
3	FB110111L	Bolt-housing to block-M10 x 55	6		
3	FB110121L	Bolt-housing to block-M10 x 60	2		
	FN110041L	Nut-flanged head-M10	NLA		
8	FS110141M	Screw-housing to block-M10 x 14	4		
5	FRC7706	Cover assembly-coverter automatic transmission	1		
6	FRC7707	Gasket-access plate/torque convertor housing	1		
7	SH106151L	Screw-cover to housing-hexagonal head-M6 x 15	3		
8	FS110301	Screw-pump to housing-M10 x 30-top	NLA		Use FS110301L
8	FS110301L	Screw-pump to housing-flanged head-M10 x 30	6		Automatic
9	FB110141L	Bolt-housing to gear case-flanged head-M10 x 70	12		
10	RTC4854	Cable assembly-kickdown automatic transmission	1		V8
10	STC1583	Cable assembly-kickdown automatic transmission	1		TDi
11	RTC4648	O ring	1		
12	RTC4288	Spring-lever detent selector mechanism	1		
13	RTC4658	Lever assembly-selector selector mechanism	1		
14	RTC4286	Lever-detent automatic transmission	1		
15	RTC4289	Seal selector mechanism	1		
16	RTC4285	Shaft automatic transmission	1		
17	RTC4290	Pin-roll-spring tension	1		
18	RTC4291	Rod assembly-kickdown automatic transmission	1		
19	RTC4937	Switch-inhibitor change selector	1		
	RTC5203	Plate-transfer box detent spring retaining	1		
20	RTC4644	Plate-inhibitor switch detent	1		
21	RTC4643	Washer-spring	1		
22	RTC4645	Screw	1		

CHANGE POINTS:
(1) To (E)36D 52139C
(2) From (E)36D 52140C

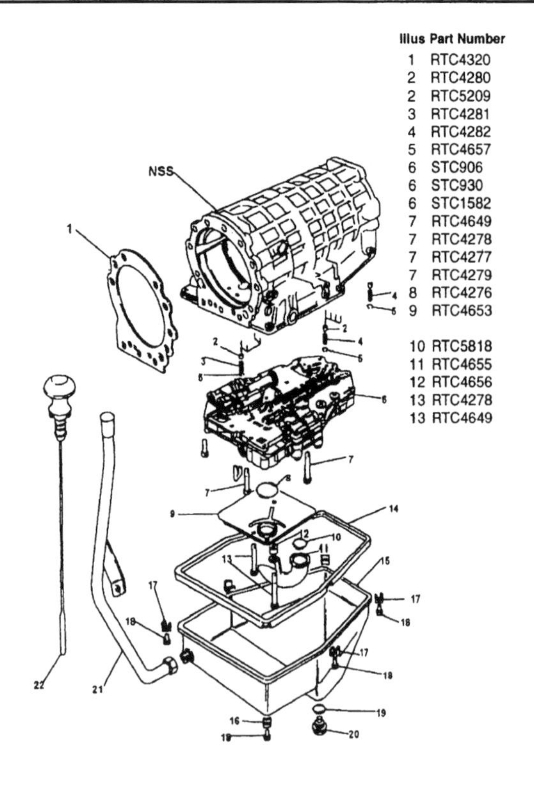

Illus	Part Number	Description	Quantity	Change Point	Remarks
1	RTC4320	Gasket-end cover automatic transmission	1		
2	RTC4280	Sleeve-short	8		
2	RTC5209	Sleeve-long	1		
3	RTC4281	Spring-short	4		
4	RTC4282	Spring-long	5		
5	RTC4657	Circlip	9		
6	STC906	Valve-block	1		3500 cc V8
6	STC930	Valve-block	1		3900 cc V8
6	STC1582	Valve-block	1		TDi
7	RTC4649	Screw-block to gearbox-83mm-long	1		
7	RTC4278	Screw-block to gearbox-65mm-long	4		
7	RTC4277	Screw-block to gearbox-60mm-long	3		
7	RTC4279	Screw-block to gearbox-30mm-long	8		
8	RTC4276	O ring-auto trans strainer to control unit	1		
9	RTC4653	Strainer-oil-automatic transmission fluid automatic transmission	1		
10	RTC5818	O ring-auto trans fluid suction pipe	1		
11	RTC4655	Pipe-suction automatic transmission	1		
12	RTC4656	Sleeve	1		
13	RTC4278	Screw-oil screen to block-65mm-long	2		
13	RTC4649	Screw-oil strainer to block-83mm-long	1		

DISCOVERY 1989MY UPTO 1999MY | GROUP C | Gearbox ZF Auto Casing and Sump

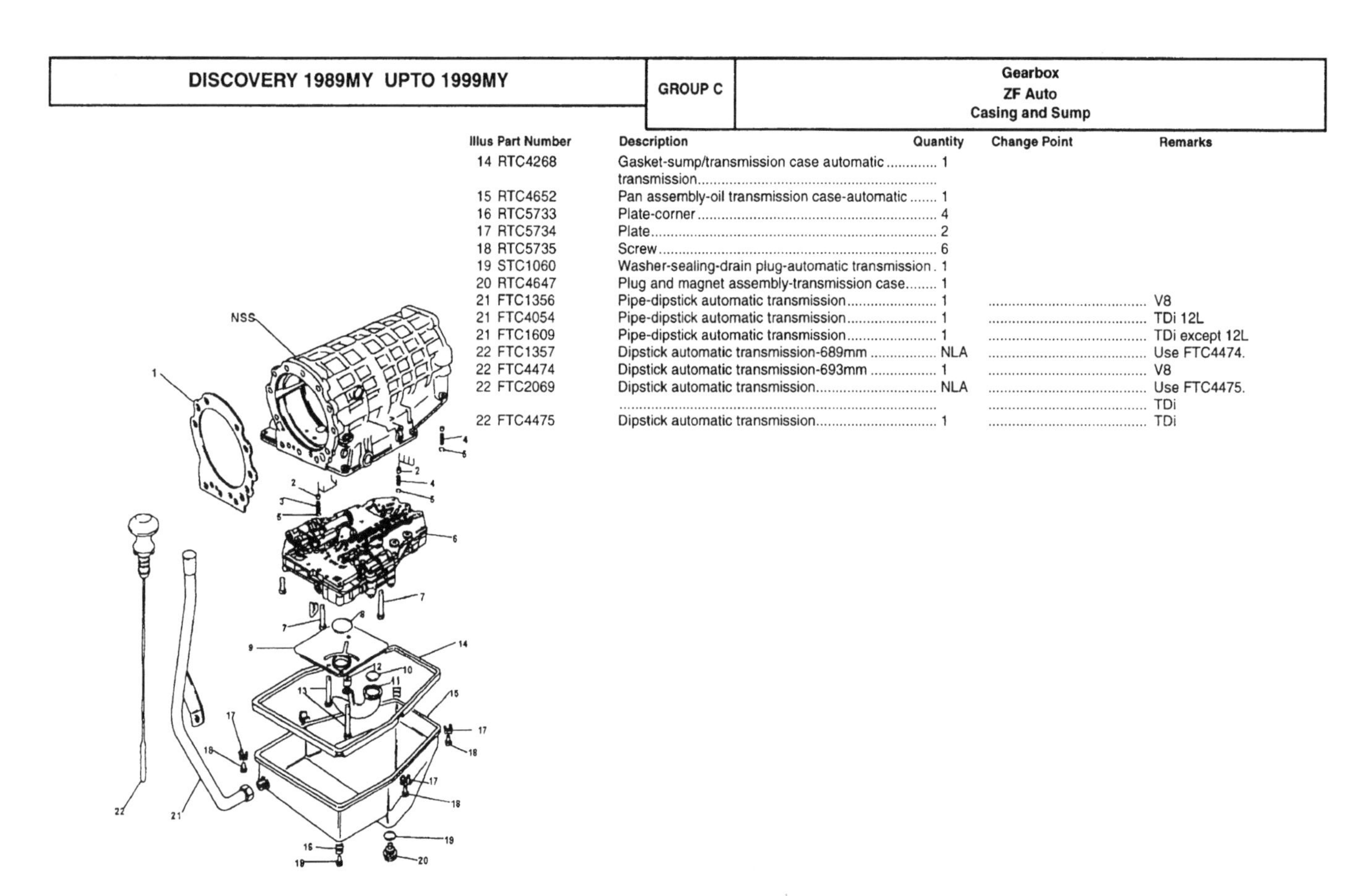

Illus	Part Number	Description	Quantity	Change Point	Remarks
14	RTC4268	Gasket-sump/transmission case automatic transmission	1		
15	RTC4652	Pan assembly-oil transmission case-automatic	1		
16	RTC5733	Plate-corner	4		
17	RTC5734	Plate	2		
18	RTC5735	Screw	6		
19	STC1060	Washer-sealing-drain plug-automatic transmission	1		
20	RTC4647	Plug and magnet assembly-transmission case	1		
21	FTC1356	Pipe-dipstick automatic transmission	1		V8
21	FTC4054	Pipe-dipstick automatic transmission	1		TDi 12L
21	FTC1609	Pipe-dipstick automatic transmission	1		TDi except 12L
22	FTC1357	Dipstick automatic transmission-689mm	NLA		Use FTC4474.
22	FTC4474	Dipstick automatic transmission-693mm	1		V8
22	FTC2069	Dipstick automatic transmission	NLA		Use FTC4475. TDi
22	FTC4475	Dipstick automatic transmission	1		TDi

Gearbox
ZF Auto
Oil Pump

Illus	Part Number	Description	Quantity	Change Point	Remarks
1	RTC5102	Seal-auto trans oil pump	1		
2	STC1836	Pump-transmission oil	1		
3	RTC5105	Screw-plate mounting to pump-M6 x 28	8		
4	RTC5106	Dowel-pin	1		
5	RTC5103	O ring	1		
6	RTC5107	Plate-intermediate	1		
7	RTC4339	Washer-sealing	1		
8	RTC4335	Plug-blanking	1		
9	RTC4338	Washer-sealing	1		
10	RTC4336	Plug-blanking	1		
		SHIM			
11	RTC5108	2.8mm	A/R		
11	RTC5109	2.6mm	A/R		
11	RTC5110	2.4mm	A/R		
11	RTC5111	2.20mm	A/R		
11	RTC5112	2.00mm	A/R		
11	RTC5113	1.80mm	A/R		
11	RTC5114	1.6mm	A/R		
11	RTC5115	1.40mm	A/R		
11	RTC5116	1.20mm	A/R		
11	RTC5117	1.0mm	A/R		

C 75

DISCOVERY 1989MY UPTO 1999MY | GROUP C

Gearbox
ZF Auto
Clutch 'A'

Illus	Part Number	Description	Quantity	Change Point	Remarks
1	RTC5143	Retainer-bearing automatic transmission	1		
2	RTC5141	Shaft-counter automatic transmission	1		
3	RTC5142	Kit-clutch-automatic	1		
4	RTC5144	Carrier-gear automatic transmission	1		

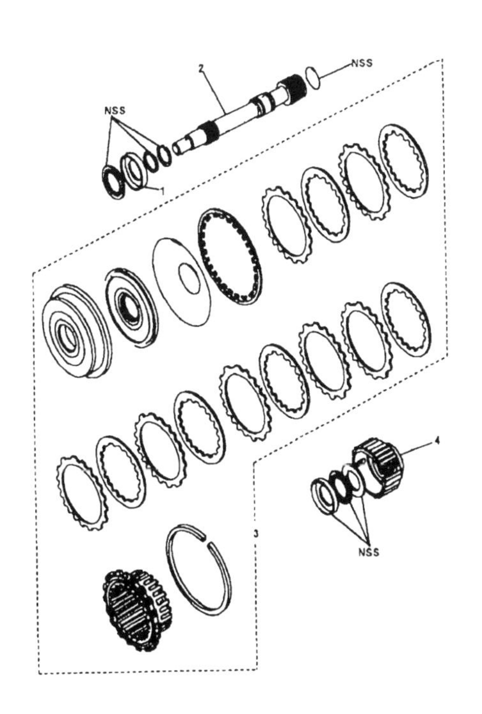

C 76

Illus	Part Number	Description	Quantity	Change Point	Remarks
1	RTC5124	Ring-snap automatic transmission	1		
2	RTC5123	Disc-clutch automatic transmission-end	1		
3	RTC5122	Disc-clutch automatic transmission-lined	4		
4	RTC5121	Disc-clutch automatic transmission-outer-steel	4		
5	RTC5148	Ring-snap automatic transmission	1		
6	RTC5147	Spring	1		
7	RTC5146	Piston-clutch-automatic	1		
8	RTC5149	Ring-support	1		
9	RTC5145	Guide automatic transmission	1		

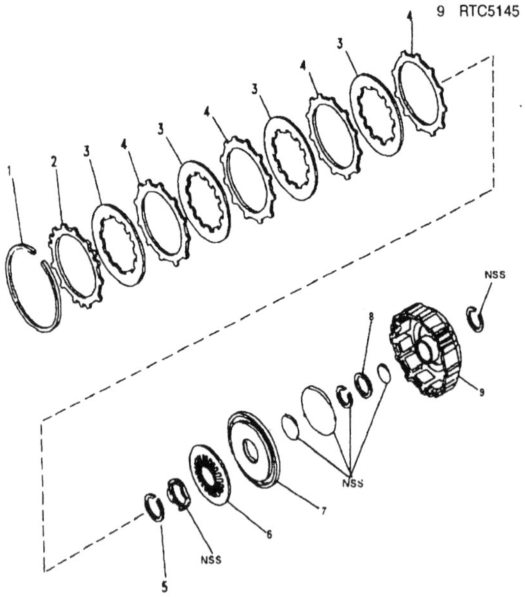

Illus	Part Number	Description	Quantity	Change Point	Remarks
1	RTC5137	Ring-snap automatic transmission	1		
2	RTC5139	Disc-clutch automatic transmission-end	1		
3	RTC5135	Disc-clutch automatic transmission-lined	4		
4	RTC5138	Disc-clutch automatic transmission-outer	4		
5	RTC5164	Spring	1		
6	RTC5163	Plate-forward clutch pressure	1		
7	RTC5162	Piston-clutch-automatic	1		
8	RTC5161	Cylinder	1		
9	RTC5171	Ring-snap automatic transmission-small	1		

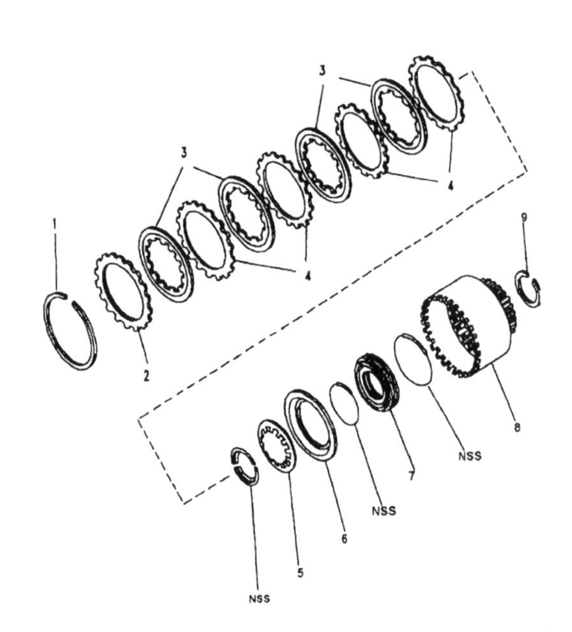

DISCOVERY 1989MY UPTO 1999MY

GROUP C

Gearbox
ZF Auto
Brakes 'F'

Illus	Part Number	Description	Quantity	Change Point	Remarks
1	RTC5137	Ring-snap automatic transmission	1		
2	RTC5136	Plate-automatic transmission clutch end	1		
3	RTC5135	Disc-clutch automatic transmission-lined	4		
4	RTC5134	Disc-clutch automatic transmission	4		
5	RTC5160	Ring-backing automatic transmission	1		
6	RTC5159	Disc-brake spring	1		
7	RTC5158	Piston-clutch-automatic	1		
8	RTC5157	Ring-piston-forward shaft automatic transmission	2		
9	RTC5155	Cylinder	1		
10	RTC5156	Screw	10		
11	RTC4660	Ring set assembly-piston	3		

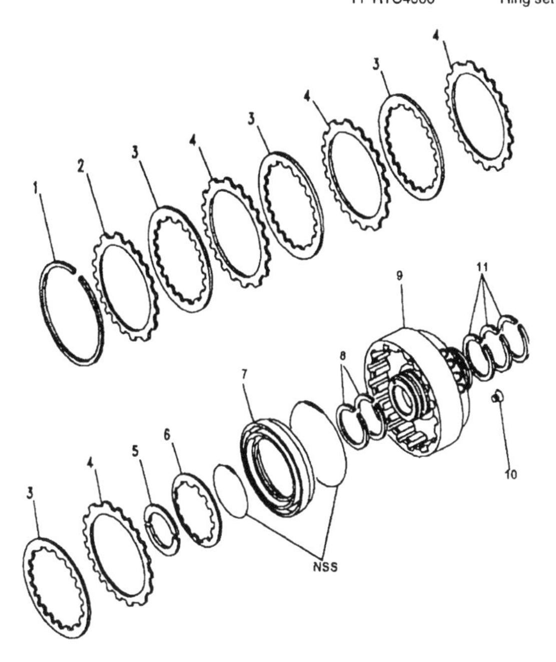

C 79

DISCOVERY 1989MY UPTO 1999MY

GROUP C

Gearbox
ZF Auto
Brake 'C-C'

Illus	Part Number	Description	Quantity	Change Point	Remarks
1	RTC5173	Ring-snap automatic transmission-large	1		
2	RTC5174	Plate-centre	1		
3	RTC5175	Piston-clutch-automatic	1		
4	RTC5176	Spring	1		
5	RTC5177	Ring-backing automatic transmission	1		
6	RTC5122	Disc-clutch automatic transmission-lined	4		
7	RTC5126	Disc-clutch automatic transmission-outer	4		
8	RTC5123	Disc-clutch automatic transmission-end	1		
9	RTC5178	Drive assembly-freewheel-automatic	1		
10	RTC5177	Ring-backing automatic transmission	1		
11	RTC5176	Spring	1		
12	RTC5179	Piston-clutch-automatic	1		

C 80

GROUP C	Gearbox ZF Auto Brake 'D'

Illus	Part Number	Description	Quantity	Change Point	Remarks
1	RTC5151	Cylinder	1		
2	RTC5152	Piston-clutch-automatic	1		
3	RTC5153	Spring	1		
4	RTC5154	Connector fit	1		
5	RTC5126	Disc-clutch automatic transmission-outer	4		
6	RTC5127	Disc-clutch automatic transmission-lined	4		
7	RTC5125	Disc-clutch automatic transmission-outer	1		
8	RTC5128	Ring-snap automatic transmission	1		

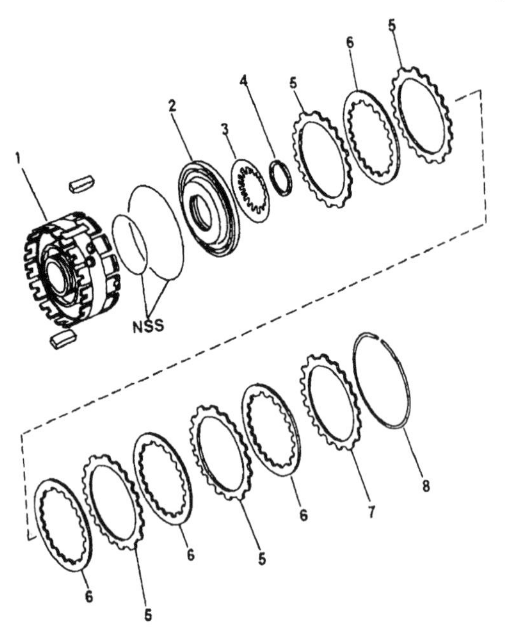

DISCOVERY 1989MY UPTO 1999MY

GROUP C	Gearbox ZF Auto Web Shaft and Fourth Gear

Illus	Part Number	Description	Quantity	Change Point	Remarks
1	RTC5183	Set-ring automatic transmission	2		
2	RTC5182	Shaft-sun gear	1		
	STC334	O ring	3		
3	RTC5185	Gear assembly-planetary automatic transmission-front	1		
4	RTC5181	Ring-snap automatic transmission	1		
5	RTC5180	Gear-ring automatic transmission-front	1		
6	RTC5188	Gear assembly-planetary automatic transmission-front	1		
7	RTC5190	Ring-snap automatic transmission-large-rear	1		
8	RTC5181	Ring-snap automatic transmission	1		
9	RTC5184	Gear-ring automatic transmission-centre	1		
10	RTC5186	Shaft-intermediate transmission	1		
11	RTC5187	Ring-snap automatic transmission-rear	1		
12	RTC5189	Shaft-spider	NLA		Use STC996.
12	STC996	Shaft-spider	1		
13	RTC5191	Gear-automatic transmission sun	1		
14	RTC5192	Gear assembly-planetary automatic transmission-rear	1		
15	RTC5193	Gear-ring automatic transmission-rear	1		

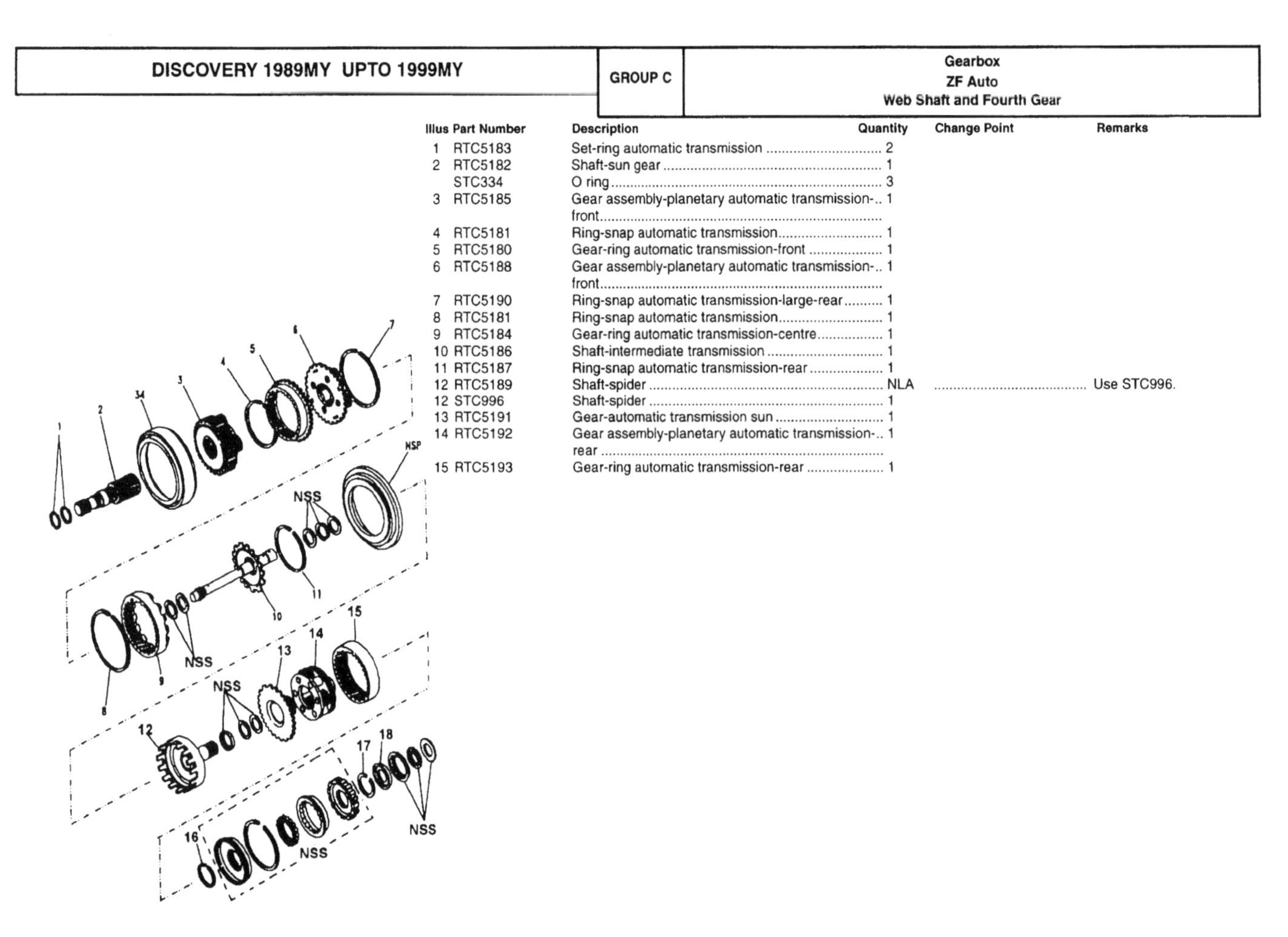

Illus	Part Number	Description	Quantity	Change Point	Remarks
16	RTC5171	Ring-snap automatic transmission-small	1		
17	RTC5172	Ring-snap automatic transmission-small	1		
18	RTC5129	Ring-backing automatic transmission	1		

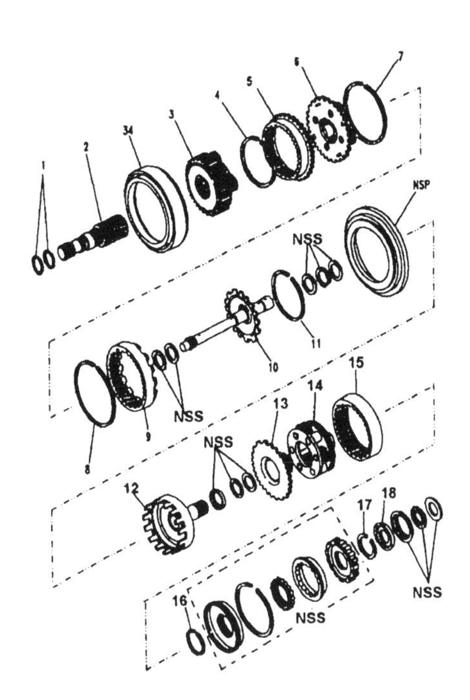

Illus	Part Number	Description	Quantity	Change Point	Remarks
1	RTC4314	Hub-governor	1		V8
1	STC1576	Hub-governor	1		TDi
	RTC5792	Weight-governor	1		V8
	STC1579	Weight-governor	1		TDi
2	RTC4313	Screw	2		
3	RTC4315	Clamp	1		
4	RTC4662	Gear-parking automatic transmission	1		
5	RTC4313	Screw-gear to hub	4		
6	RTC4661	Bolt-pawl to housing	1		
7	RTC4308	Spring-parking pawl automatic transmission	1		
8	RTC4307	Pawl-parklock automatic transmission	1		
9	RTC5212	Plate- guide-guide-pawl	1		
10	RTC4663	Plate-guide-pawl	1		
11	RTC5211	Screw-plate to housing	1		
12	RTC4295	Gasket-rear extension housing	1		
13	FTC3981	Shaft automatic transmission	NLA	Note (1)	Use FTC5090.
13	FTC5090	Shaft automatic transmission	1	Note (2)	
14	RTC4650	Seal-auto trans output shaft	1		
15	FRC5575	O ring	1		

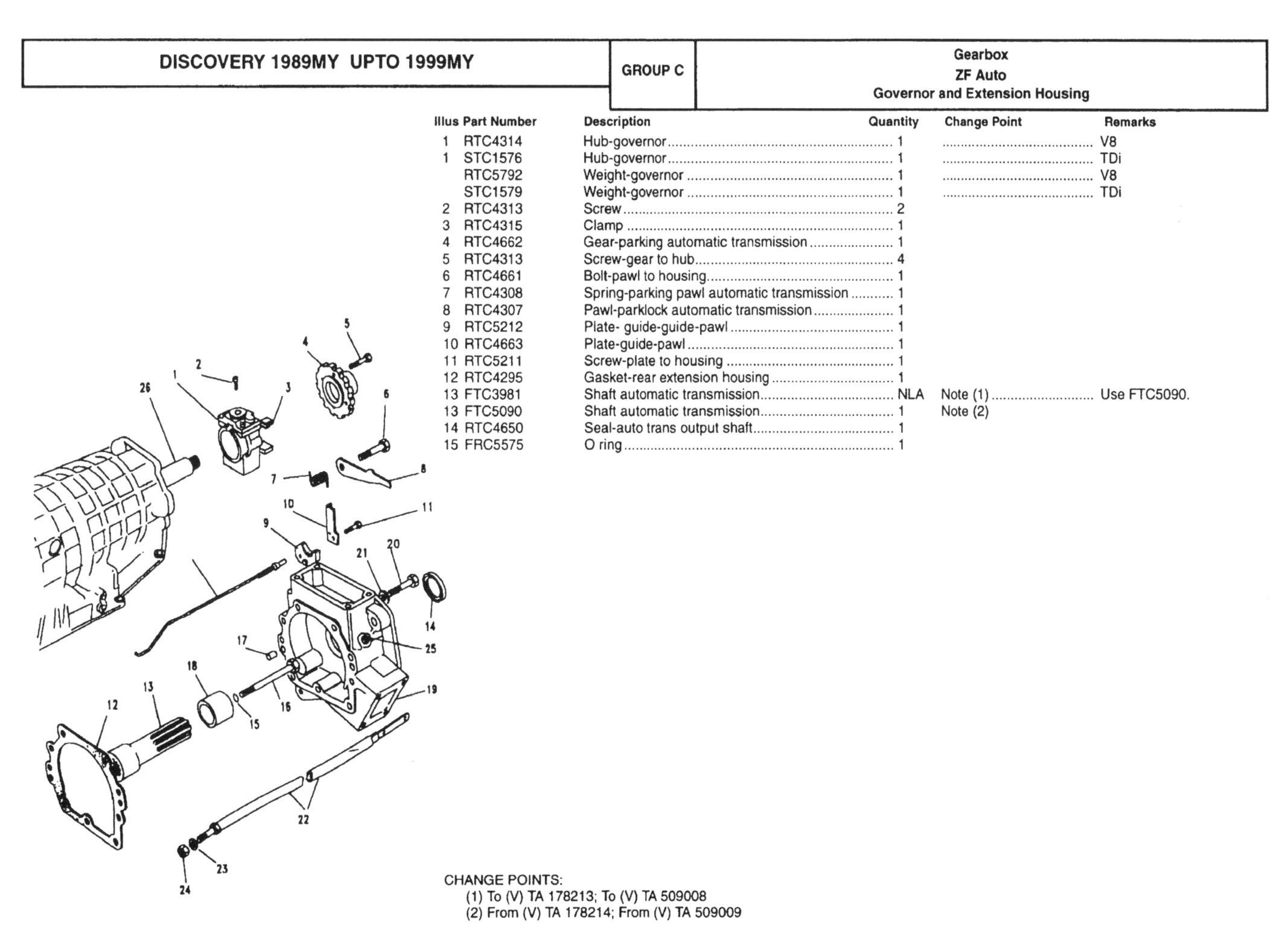

CHANGE POINTS:
(1) To (V) TA 178213; To (V) TA 509008
(2) From (V) TA 178214; From (V) TA 509009

Illus	Part Number	Description	Quantity	Change Point	Remarks
16	BH110261L	Bolt-output shaft to mainshaft-M10 x 130	1		
17	RTC4301	Dowel	2		
18	FTC3275	Spacer-output shaft	1		
19	RTC4659	Housing-governor extension	1		V8
19	STC1578	Housing-governor extension	1		4 Cylinder TDi
20	SH106121L	Screw-governer housing to gearbox-hexagonal head-M6 x 12	9		
21	WA106041L	Washer-M6-standard.	9		
22	FRC7500	Rod-transmission support	1		
23	WA110061L	Washer-plain-M10-standard	1		
24	NY110047L	Nut-hexagonal head-nyloc-M10	1		Automatic
25	FRC9427	Tube-breather transmission case	1		V8
25	FRC9430	Tube-breather transmission case	1		Automatic 4 Cylinder TDi
26	RTC5150	Shaft assembly-main automatic transmission	1		Automatic V8
26	STC1584	Shaft assembly-main automatic transmission	1		Automatic 4 Cylinder TDi
	FTC890	Stud-lifting bracket	1		

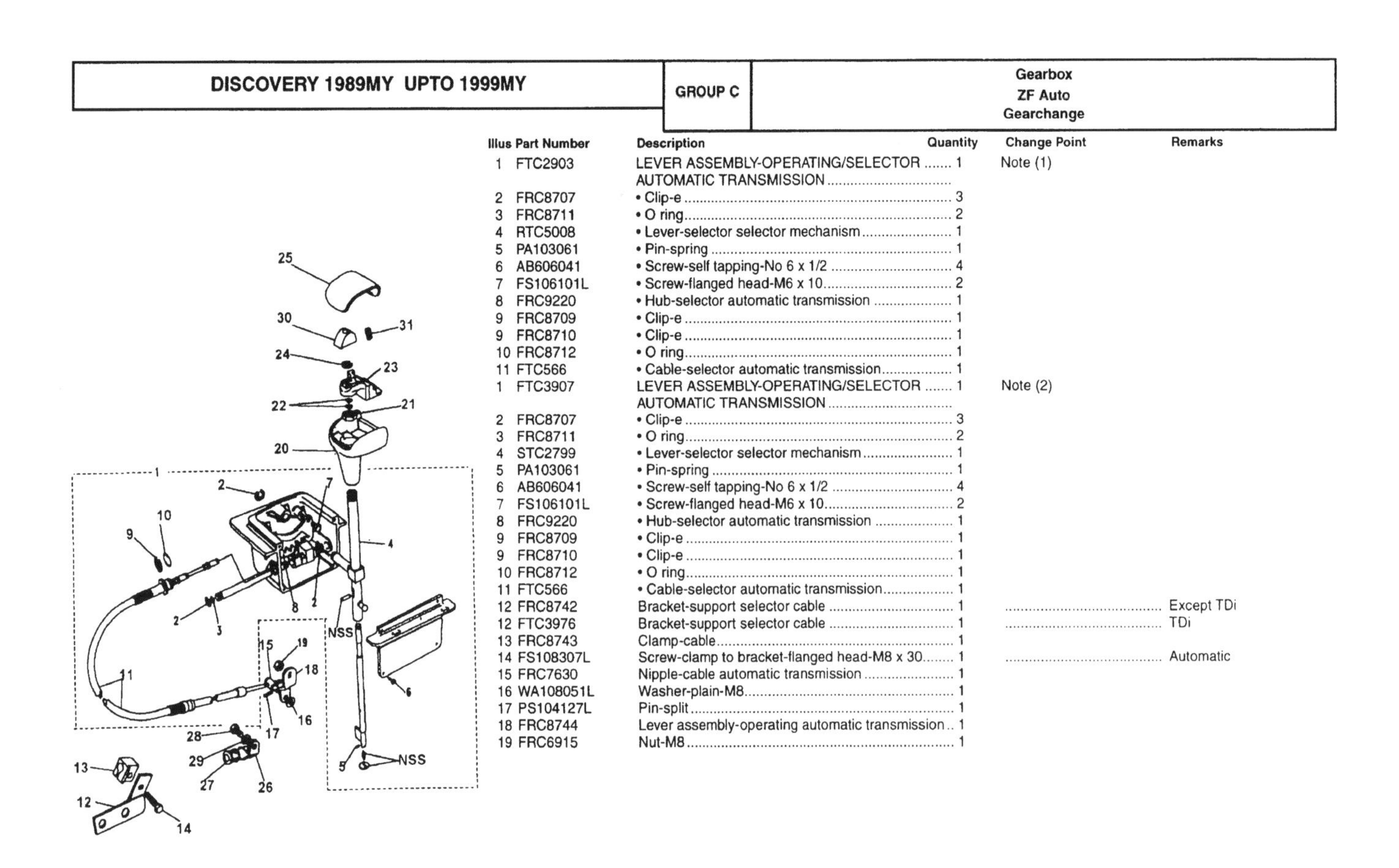

Illus	Part Number	Description	Quantity	Change Point	Remarks
1	FTC2903	LEVER ASSEMBLY-OPERATING/SELECTOR AUTOMATIC TRANSMISSION	1	Note (1)	
2	FRC8707	• Clip-e	3		
3	FRC8711	• O ring	2		
4	RTC5008	• Lever-selector selector mechanism	1		
5	PA103061	• Pin-spring	1		
6	AB606041	• Screw-self tapping-No 6 x 1/2	4		
7	FS106101L	• Screw-flanged head-M6 x 10	2		
8	FRC9220	• Hub-selector automatic transmission	1		
9	FRC8709	• Clip-e	1		
9	FRC8710	• Clip-e	1		
10	FRC8712	• O ring	1		
11	FTC566	• Cable-selector automatic transmission	1		
1	FTC3907	LEVER ASSEMBLY-OPERATING/SELECTOR AUTOMATIC TRANSMISSION	1	Note (2)	
2	FRC8707	• Clip-e	3		
3	FRC8711	• O ring	2		
4	STC2799	• Lever-selector selector mechanism	1		
5	PA103061	• Pin-spring	1		
6	AB606041	• Screw-self tapping-No 6 x 1/2	4		
7	FS106101L	• Screw-flanged head-M6 x 10	2		
8	FRC9220	• Hub-selector automatic transmission	1		
9	FRC8709	• Clip-e	1		
9	FRC8710	• Clip-e	1		
10	FRC8712	• O ring	1		
11	FTC566	• Cable-selector automatic transmission	1		
12	FRC8742	Bracket-support selector cable	1		Except TDi
12	FTC3976	Bracket-support selector cable	1		TDi
13	FRC8743	Clamp-cable	1		
14	FS108307L	Screw-clamp to bracket-flanged head-M8 x 30	1		Automatic
15	FRC7630	Nipple-cable automatic transmission	1		
16	WA108051L	Washer-plain-M8	1		
17	PS104127L	Pin-split	1		
18	FRC8744	Lever assembly-operating automatic transmission	1		
19	FRC6915	Nut-M8	1		

CHANGE POINTS:
(1) To (V) MA 081990
(2) From (V) MA 081991

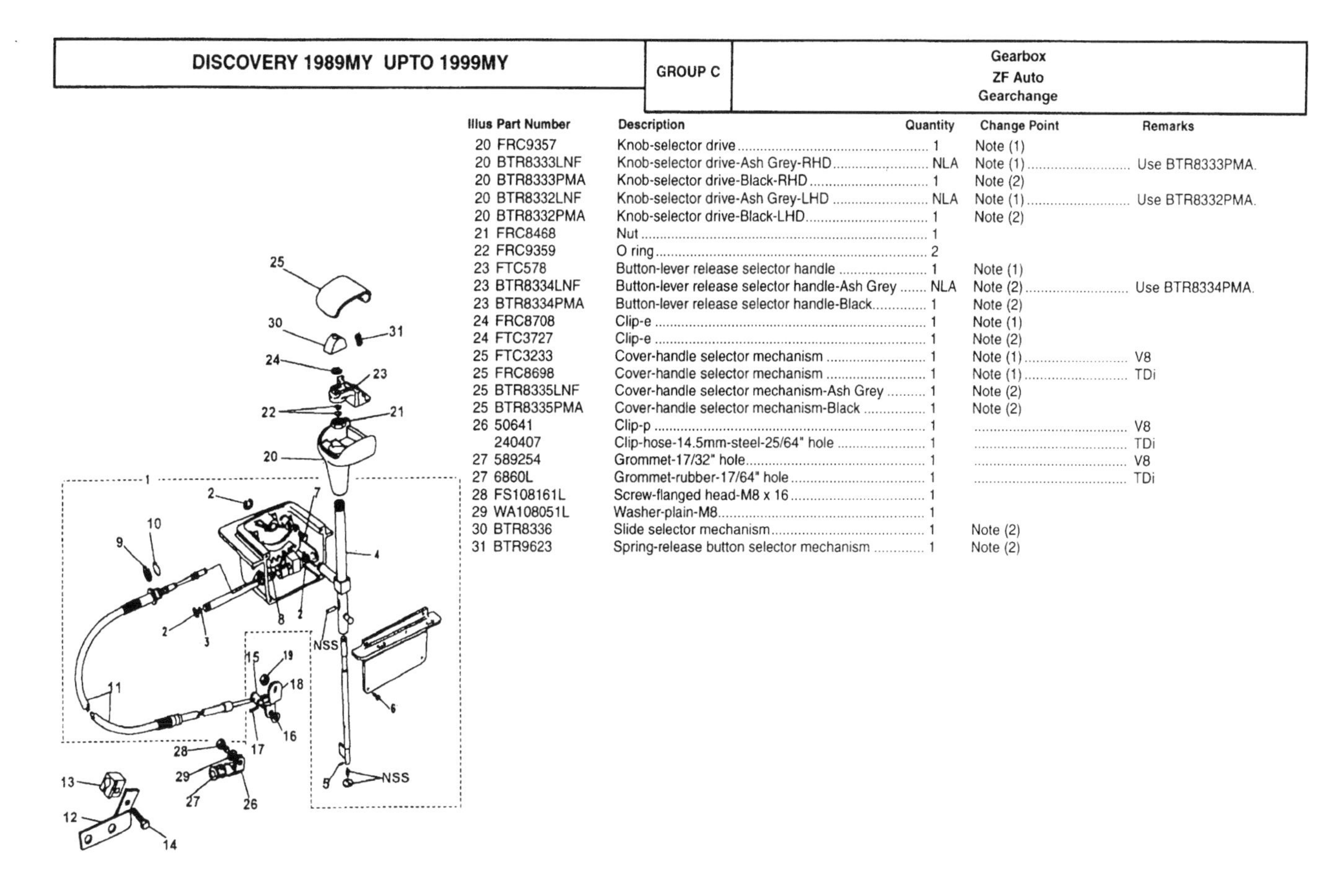

Illus	Part Number	Description	Quantity	Change Point	Remarks
20	FRC9357	Knob-selector drive	1	Note (1)	
20	BTR8333LNF	Knob-selector drive-Ash Grey-RHD	NLA	Note (1)	Use BTR8333PMA.
20	BTR8333PMA	Knob-selector drive-Black-RHD	1	Note (2)	
20	BTR8332LNF	Knob-selector drive-Ash Grey-LHD	NLA	Note (1)	Use BTR8332PMA.
20	BTR8332PMA	Knob-selector drive-Black-LHD	1	Note (2)	
21	FRC8468	Nut	1		
22	FRC9359	O ring	2		
23	FTC578	Button-lever release selector handle	1	Note (1)	
23	BTR8334LNF	Button-lever release selector handle-Ash Grey	NLA	Note (2)	Use BTR8334PMA.
23	BTR8334PMA	Button-lever release selector handle-Black	1	Note (2)	
24	FRC8708	Clip-e	1	Note (1)	
24	FTC3727	Clip-e	1	Note (2)	
25	FTC3233	Cover-handle selector mechanism	1	Note (1)	V8
25	FRC8698	Cover-handle selector mechanism	1	Note (1)	TDi
25	BTR8335LNF	Cover-handle selector mechanism-Ash Grey	1	Note (2)	
25	BTR8335PMA	Cover-handle selector mechanism-Black	1	Note (2)	
26	50641	Clip-p	1		V8
	240407	Clip-hose-14.5mm-steel-25/64" hole	1		TDi
27	589254	Grommet-17/32" hole	1		V8
27	6860L	Grommet-rubber-17/64" hole	1		TDi
28	FS108161L	Screw-flanged head-M8 x 16	1		
29	WA108051L	Washer-plain-M8	1		
30	BTR8336	Slide selector mechanism	1	Note (2)	
31	BTR9623	Spring-release button selector mechanism	1	Note (2)	

CHANGE POINTS:
(1) To (V) LA 081990
(2) From (V) MA 081991

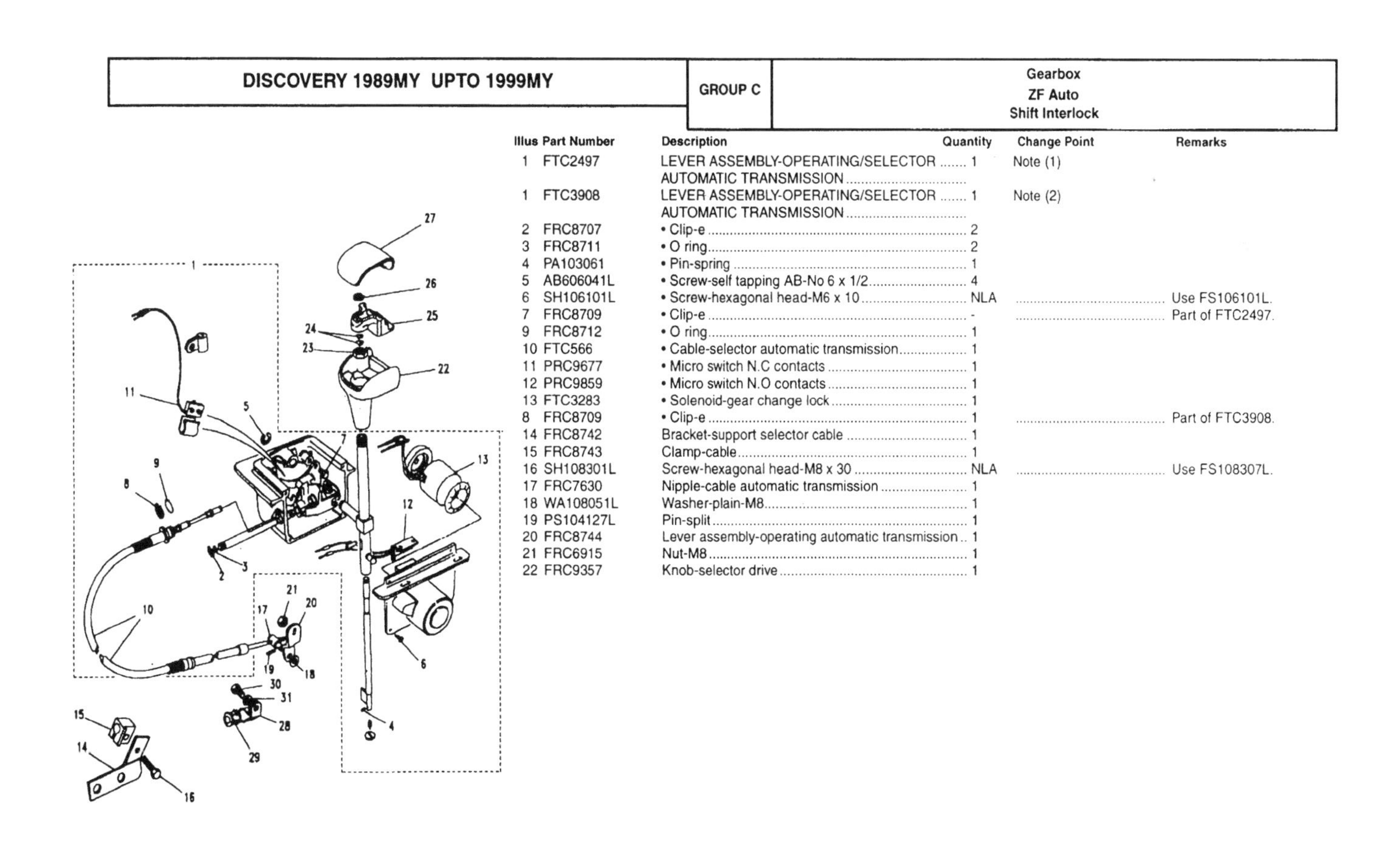

Illus	Part Number	Description	Quantity	Change Point	Remarks
1	FTC2497	LEVER ASSEMBLY-OPERATING/SELECTOR AUTOMATIC TRANSMISSION	1	Note (1)	
1	FTC3908	LEVER ASSEMBLY-OPERATING/SELECTOR AUTOMATIC TRANSMISSION	1	Note (2)	
2	FRC8707	• Clip-e	2		
3	FRC8711	• O ring	2		
4	PA103061	• Pin-spring	1		
5	AB606041L	• Screw-self tapping AB-No 6 x 1/2	4		
6	SH106101L	• Screw-hexagonal head-M6 x 10	NLA		Use FS106101L.
7	FRC8709	• Clip-e	-		Part of FTC2497.
9	FRC8712	• O ring	1		
10	FTC566	• Cable-selector automatic transmission	1		
11	PRC9677	• Micro switch N.C contacts	1		
12	PRC9859	• Micro switch N.O contacts	1		
13	FTC3283	• Solenoid-gear change lock	1		
8	FRC8709	• Clip-e	1		Part of FTC3908.
14	FRC8742	Bracket-support selector cable	1		
15	FRC8743	Clamp-cable	1		
16	SH108301L	Screw-hexagonal head-M8 x 30	NLA		Use FS108307L.
17	FRC7630	Nipple-cable automatic transmission	1		
18	WA108051L	Washer-plain-M8	1		
19	PS104127L	Pin-split	1		
20	FRC8744	Lever assembly-operating automatic transmission	1		
21	FRC6915	Nut-M8	1		
22	FRC9357	Knob-selector drive	1		

CHANGE POINTS:
(1) To (V) HA 012332
(2) From (V) JA 012333

Illus	Part Number	Description	Quantity	Change Point	Remarks
23	STC1128	Nut	1		
24	FRC9359	O ring	1		
25	FTC578	Button-lever release selector handle	1		
26	FRC8708	Clip-e	1		
27	FRC8698	Cover-handle selector mechanism	1		
28	50641	Clip-p	1		
29	589254	Grommet-17/32" hole	1		
30	SH108161L	Screw-hexagonal head-M8 x 16	NLA		Use FS108161L.
31	WA108051L	Washer-plain-M8	1		

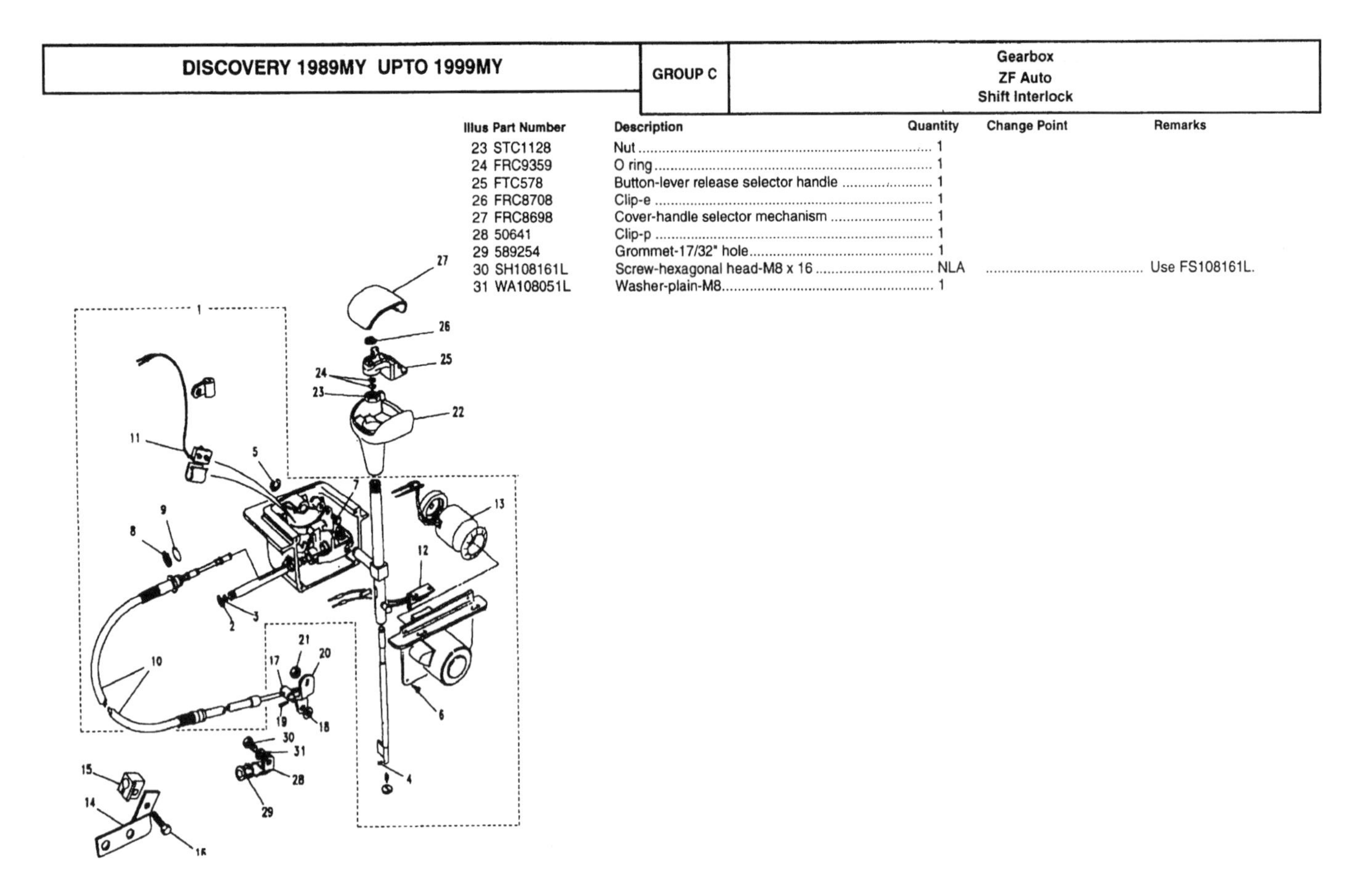

Illus	Part Number	Description	Quantity	Change Point	Remarks
		TRANSMISSION MOUNTING-4CYL			
1	ANR2278	Crossmember assembly-chassis frame detachable	1		
2	ANR2898	Bracket-transmission mounting-RH	1		
3	SH112251L	Screw-bracket to gearbox-hexagonal head-M12 x 25	4		
4	WL112001L	Washer	4		
5	ANR3225	Bracket-transmission mounting-LH	1		
6	SH112251L	Screw-bracket to gearbox-hexagonal head-M12 x 25	4		
7	WL112001L	Washer	4		
8	ANR3138	Heatshield assembly exhaust-downpipe	1		
9	ANR3201	Mounting-rubber-LH	1		
10	ANR3200	Mounting-rubber-RH	1		
11	FX110041L	Nut-flange-M10	4		

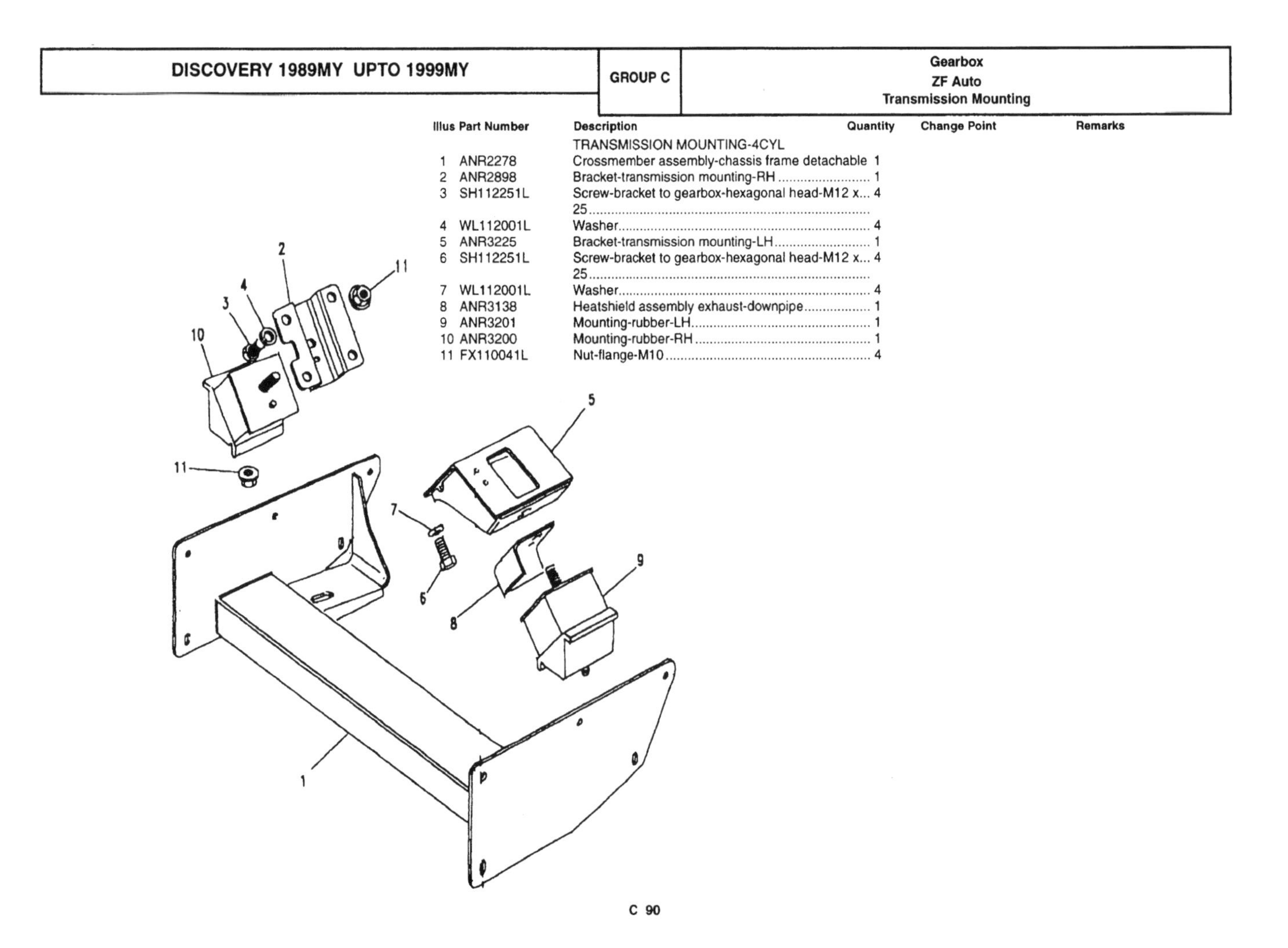

Illus	Part Number	Description	Quantity	Change Point	Remarks
1	RTC5100	Kit-gasket automatic transmission	1		
2	RTC5101	Kit-automatic transmission bearing overhaul	1		

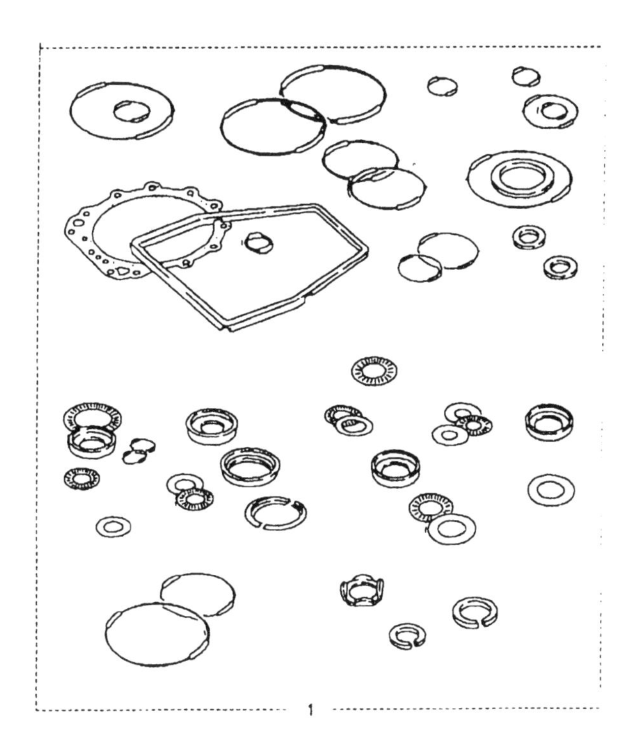

C 91

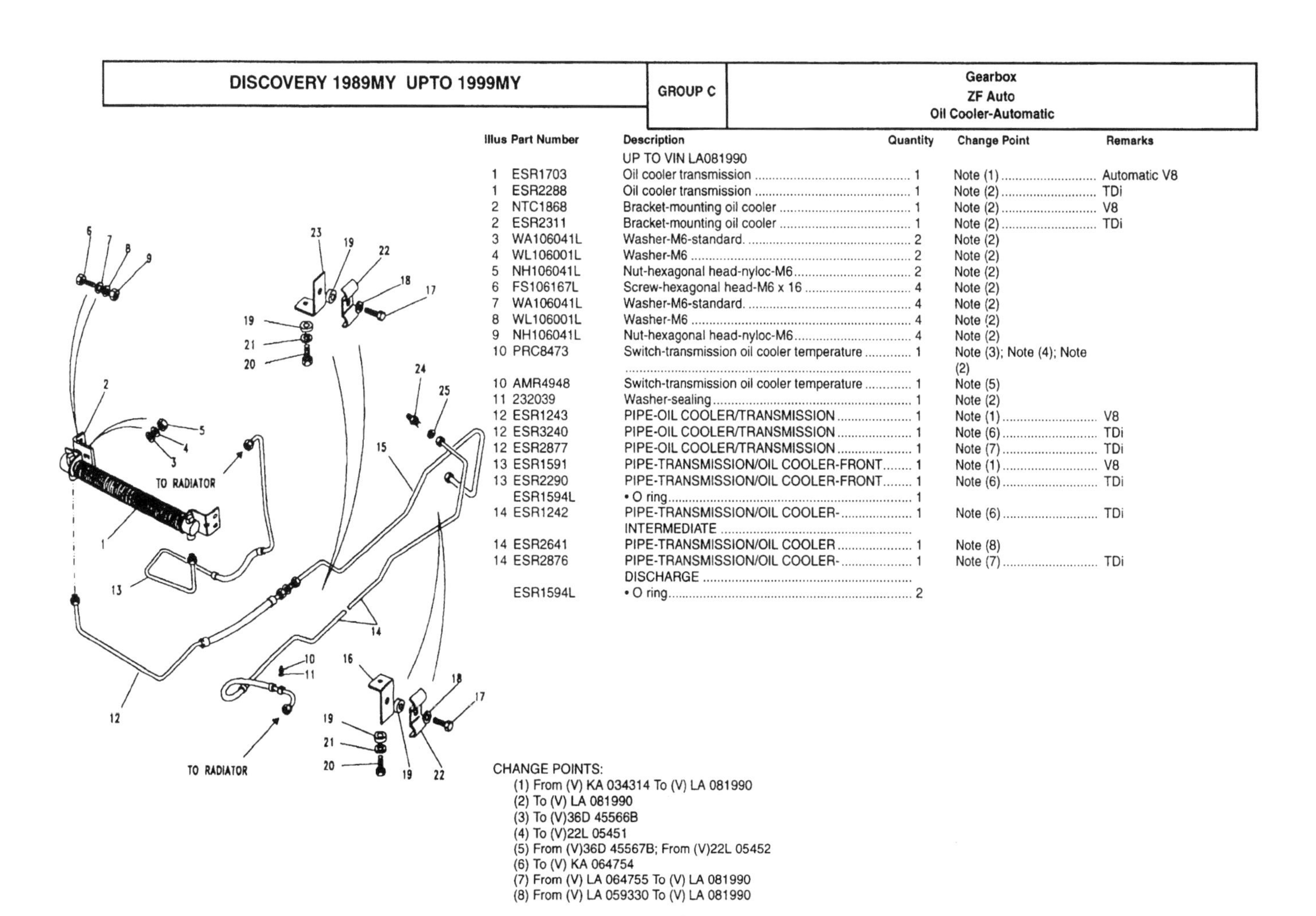

Illus	Part Number	Description	Quantity	Change Point	Remarks
		UP TO VIN LA081990			
1	ESR1703	Oil cooler transmission	1	Note (1)	Automatic V8
1	ESR2288	Oil cooler transmission	1	Note (2)	TDi
2	NTC1868	Bracket-mounting oil cooler	1	Note (2)	V8
2	ESR2311	Bracket-mounting oil cooler	1	Note (2)	TDi
3	WA106041L	Washer-M6-standard.	2	Note (2)	
4	WL106001L	Washer-M6	2	Note (2)	
5	NH106041L	Nut-hexagonal head-nyloc-M6	2	Note (2)	
6	FS106167L	Screw-hexagonal head-M6 x 16	4	Note (2)	
7	WA106041L	Washer-M6-standard.	4	Note (2)	
8	WL106001L	Washer-M6	4	Note (2)	
9	NH106041L	Nut-hexagonal head-nyloc-M6	4	Note (2)	
10	PRC8473	Switch-transmission oil cooler temperature	1	Note (3); Note (4); Note (2)	
10	AMR4948	Switch-transmission oil cooler temperature	1	Note (5)	
11	232039	Washer-sealing	1	Note (2)	
12	ESR1243	PIPE-OIL COOLER/TRANSMISSION	1	Note (1)	V8
12	ESR3240	PIPE-OIL COOLER/TRANSMISSION	1	Note (6)	TDi
12	ESR2877	PIPE-OIL COOLER/TRANSMISSION	1	Note (7)	TDi
13	ESR1591	PIPE-TRANSMISSION/OIL COOLER-FRONT	1	Note (1)	V8
13	ESR2290	PIPE-TRANSMISSION/OIL COOLER-FRONT	1	Note (6)	TDi
	ESR1594L	• O ring	1		
14	ESR1242	PIPE-TRANSMISSION/OIL COOLER-INTERMEDIATE	1	Note (6)	TDi
14	ESR2641	PIPE-TRANSMISSION/OIL COOLER	1	Note (8)	
14	ESR2876	PIPE-TRANSMISSION/OIL COOLER-DISCHARGE	1	Note (7)	TDi
	ESR1594L	• O ring	2		

CHANGE POINTS:
(1) From (V) KA 034314 To (V) LA 081990
(2) To (V) LA 081990
(3) To (V)36D 45566B
(4) To (V)22L 05451
(5) From (V)36D 45567B; From (V)22L 05452
(6) To (V) KA 064754
(7) From (V) LA 064755 To (V) LA 081990
(8) From (V) LA 059330 To (V) LA 081990

C 92

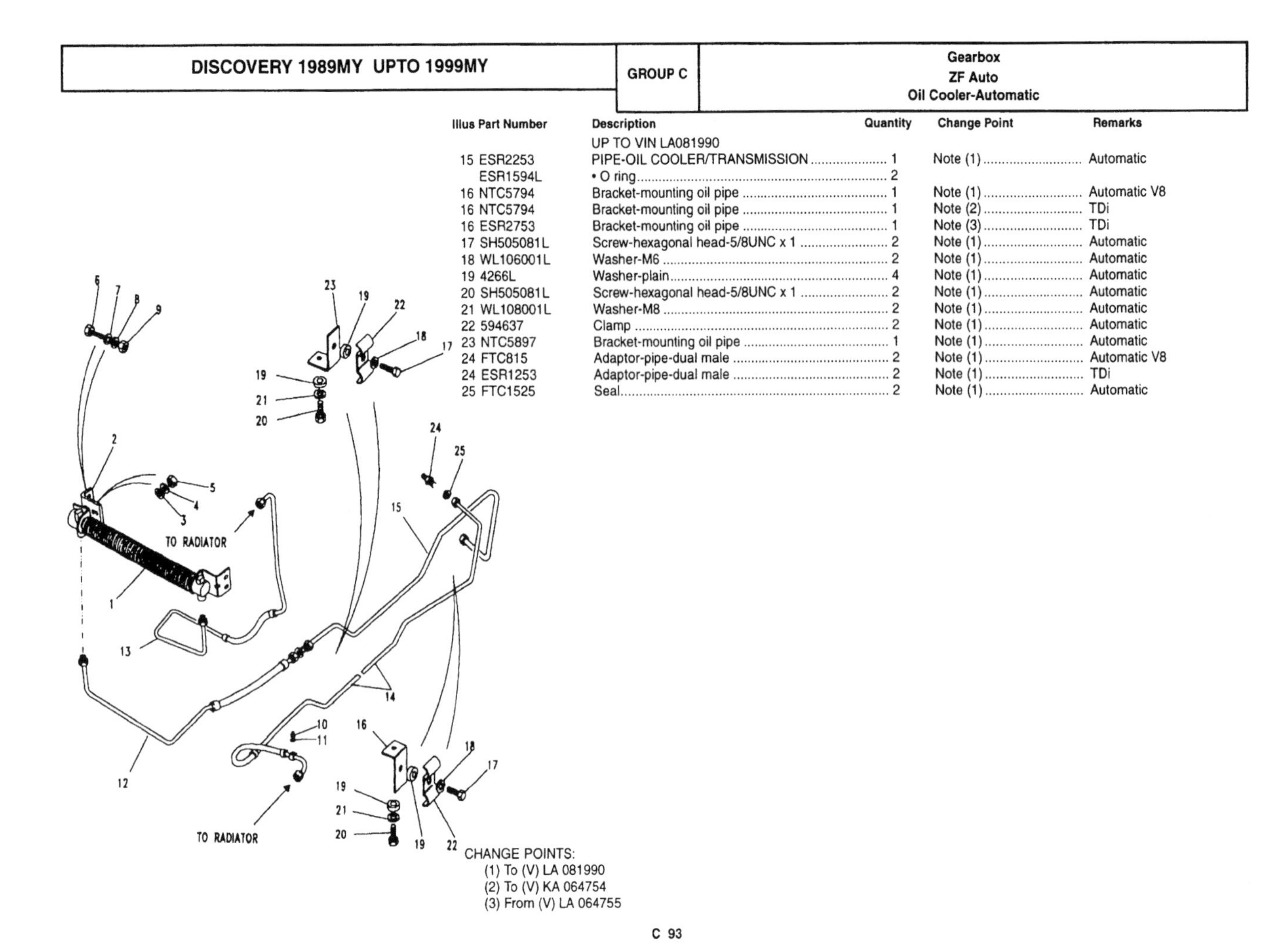

Illus	Part Number	Description	Quantity	Change Point	Remarks
		UP TO VIN LA081990			
15	ESR2253	PIPE-OIL COOLER/TRANSMISSION	1	Note (1)	Automatic
	ESR1594L	• O ring	2		
16	NTC5794	Bracket-mounting oil pipe	1	Note (1)	Automatic V8
16	NTC5794	Bracket-mounting oil pipe	1	Note (2)	TDi
16	ESR2753	Bracket-mounting oil pipe	1	Note (3)	TDi
17	SH505081L	Screw-hexagonal head-5/8UNC x 1	2	Note (1)	Automatic
18	WL106001L	Washer-M6	2	Note (1)	Automatic
19	4266L	Washer-plain	4	Note (1)	Automatic
20	SH505081L	Screw-hexagonal head-5/8UNC x 1	2	Note (1)	Automatic
21	WL108001L	Washer-M8	2	Note (1)	Automatic
22	594637	Clamp	2	Note (1)	Automatic
23	NTC5897	Bracket-mounting oil pipe	1	Note (1)	Automatic
24	FTC815	Adaptor-pipe-dual male	2	Note (1)	Automatic V8
24	ESR1253	Adaptor-pipe-dual male	2	Note (1)	TDi
25	FTC1525	Seal	2	Note (1)	Automatic

CHANGE POINTS:
(1) To (V) LA 081990
(2) To (V) KA 064754
(3) From (V) LA 064755

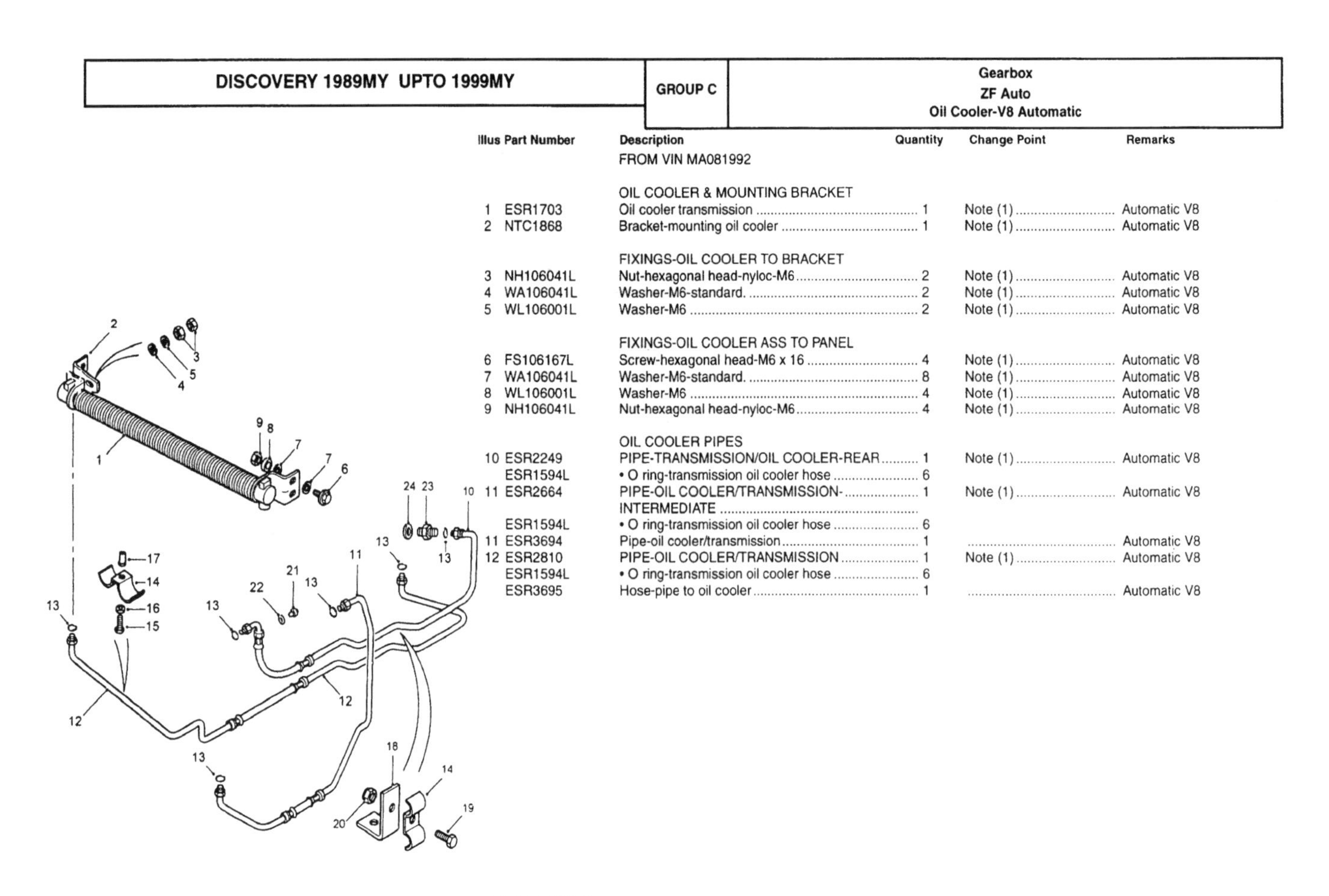

Illus	Part Number	Description	Quantity	Change Point	Remarks
		FROM VIN MA081992			
		OIL COOLER & MOUNTING BRACKET			
1	ESR1703	Oil cooler transmission	1	Note (1)	Automatic V8
2	NTC1868	Bracket-mounting oil cooler	1	Note (1)	Automatic V8
		FIXINGS-OIL COOLER TO BRACKET			
3	NH106041L	Nut-hexagonal head-nyloc-M6	2	Note (1)	Automatic V8
4	WA106041L	Washer-M6-standard.	2	Note (1)	Automatic V8
5	WL106001L	Washer-M6	2	Note (1)	Automatic V8
		FIXINGS-OIL COOLER ASS TO PANEL			
6	FS106167L	Screw-hexagonal head-M6 x 16	4	Note (1)	Automatic V8
7	WA106041L	Washer-M6-standard.	8	Note (1)	Automatic V8
8	WL106001L	Washer-M6	4	Note (1)	Automatic V8
9	NH106041L	Nut-hexagonal head-nyloc-M6	4	Note (1)	Automatic V8
		OIL COOLER PIPES			
10	ESR2249	PIPE-TRANSMISSION/OIL COOLER-REAR	1	Note (1)	Automatic V8
	ESR1594L	• O ring-transmission oil cooler hose	6		
11	ESR2664	PIPE-OIL COOLER/TRANSMISSION-INTERMEDIATE	1	Note (1)	Automatic V8
	ESR1594L	• O ring-transmission oil cooler hose	6		
11	ESR3694	Pipe-oil cooler/transmission	1		Automatic V8
12	ESR2810	PIPE-OIL COOLER/TRANSMISSION	1	Note (1)	Automatic V8
	ESR1594L	• O ring-transmission oil cooler hose	6		
	ESR3695	Hose-pipe to oil cooler	1		Automatic V8

CHANGE POINTS:
(1) From (V) MA 081992

Illus	Part Number	Description	Quantity	Change Point	Remarks
		FIXINGS-OIL COOLER PIPES			
14	ESR2806	Clamp	2	Note (1)	
15	SH106351L	Screw-hexagonal head-M6 x 35	1	Note (1)	
16	WL106001L	Washer-M6	1	Note (1)	
17	NN106021	Nutsert-blind-M6	1	Note (1)	
18	NTC5897	Bracket-mounting oil pipe	1	Note (1)	
19	FS106167L	Screw-hexagonal head-M6 x 16	1	Note (1)	
20	NY106041	Nut-flange-nyloc-M6	1	Note (1)	
		OIL TEMPERATURE SWITCH			
21	PRC8473	Switch-transmission oil cooler temperature	1	Note (2)	
21	AMR4948	Switch-transmission oil cooler temperature	1	Note (3)	
22	232039	Washer-sealing	1		
		PIPE ADAPTORS & SEALS			
23	ESR1253	Adaptor-pipe-dual male	2		
24	FTC1525	Seal	2		

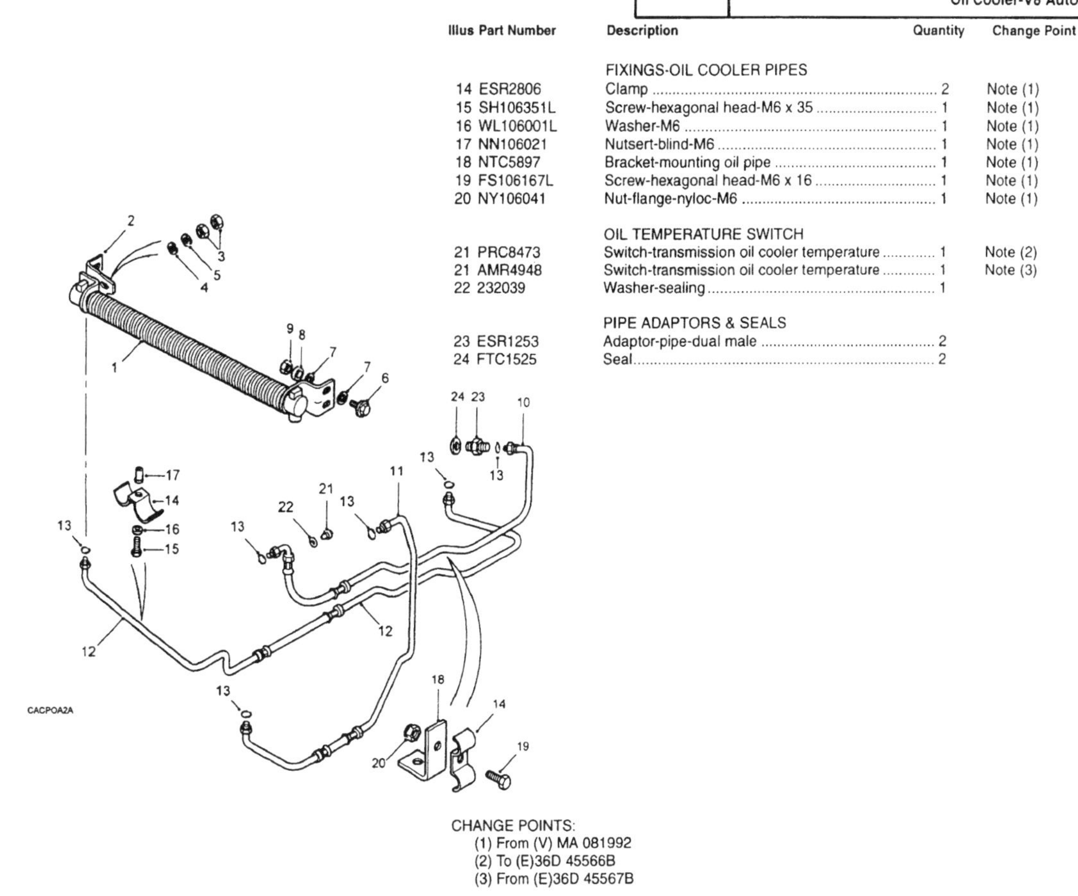

CHANGE POINTS:
(1) From (V) MA 081992
(2) To (E)36D 45566B
(3) From (E)36D 45567B

Illus	Part Number	Description	Quantity	Change Point	Remarks
		FROM VIN MA018992			
		OIL COOLER & MOUNTING BRACKET			
1	ESR3228	Oil cooler transmission	1	Note (1)	Automatic TDi
2	ESR2311	Bracket-mounting oil cooler	2	Note (1)	Automatic TDi
		FIXINGS-OIL COOLER TO BRACKETS			
3	FS106167L	Screw-hexagonal head-M6 x 16	2	Note (1)	Automatic TDi
4	WA106041L	Washer-M6-standard	4	Note (1)	Automatic TDi
5	WL106001L	Washer-M6	2	Note (1)	Automatic TDi
6	NH106041L	Nut-hexagonal head-nyloc-M6	2	Note (1)	Automatic TDi

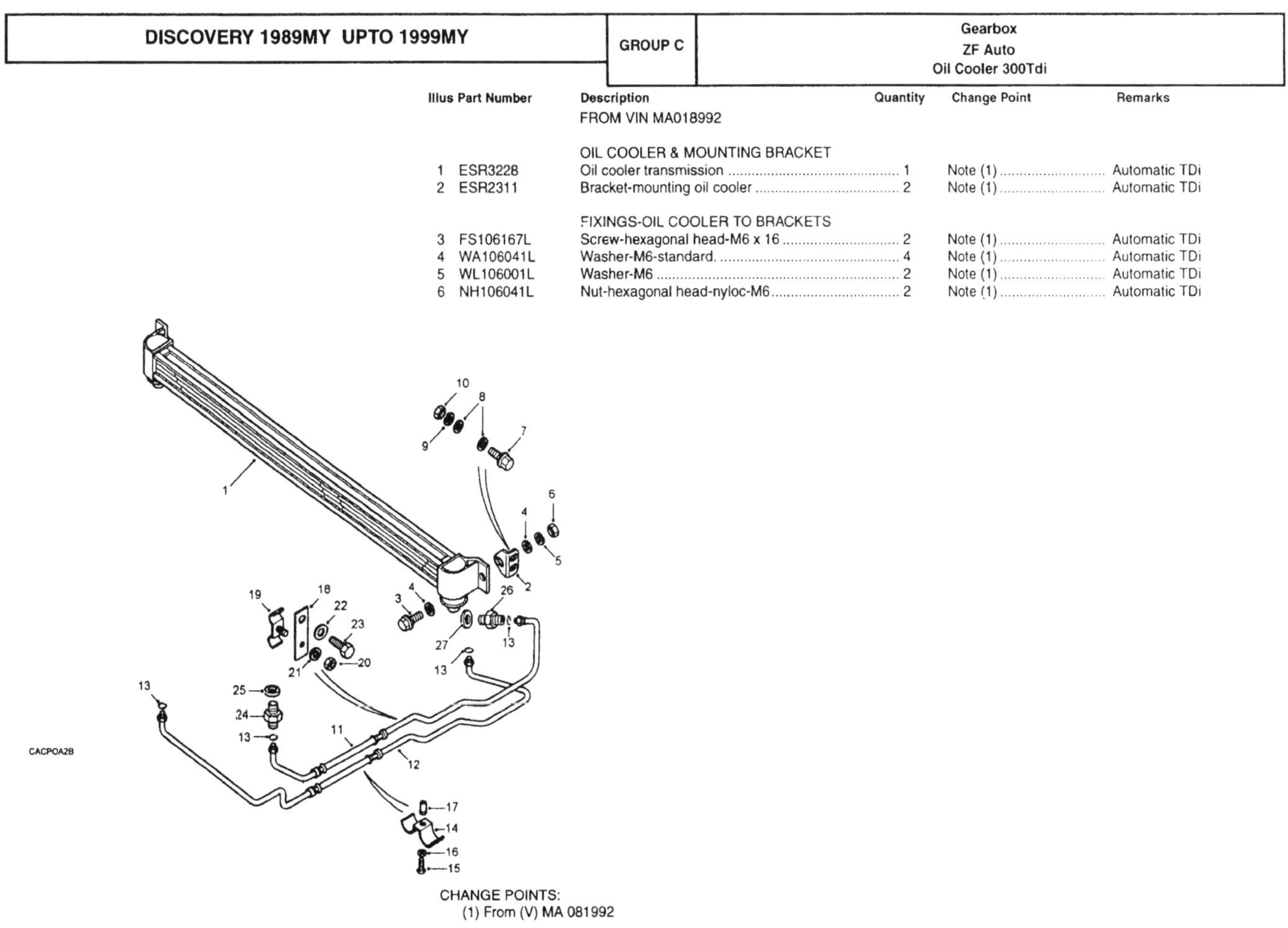

CHANGE POINTS:
(1) From (V) MA 081992

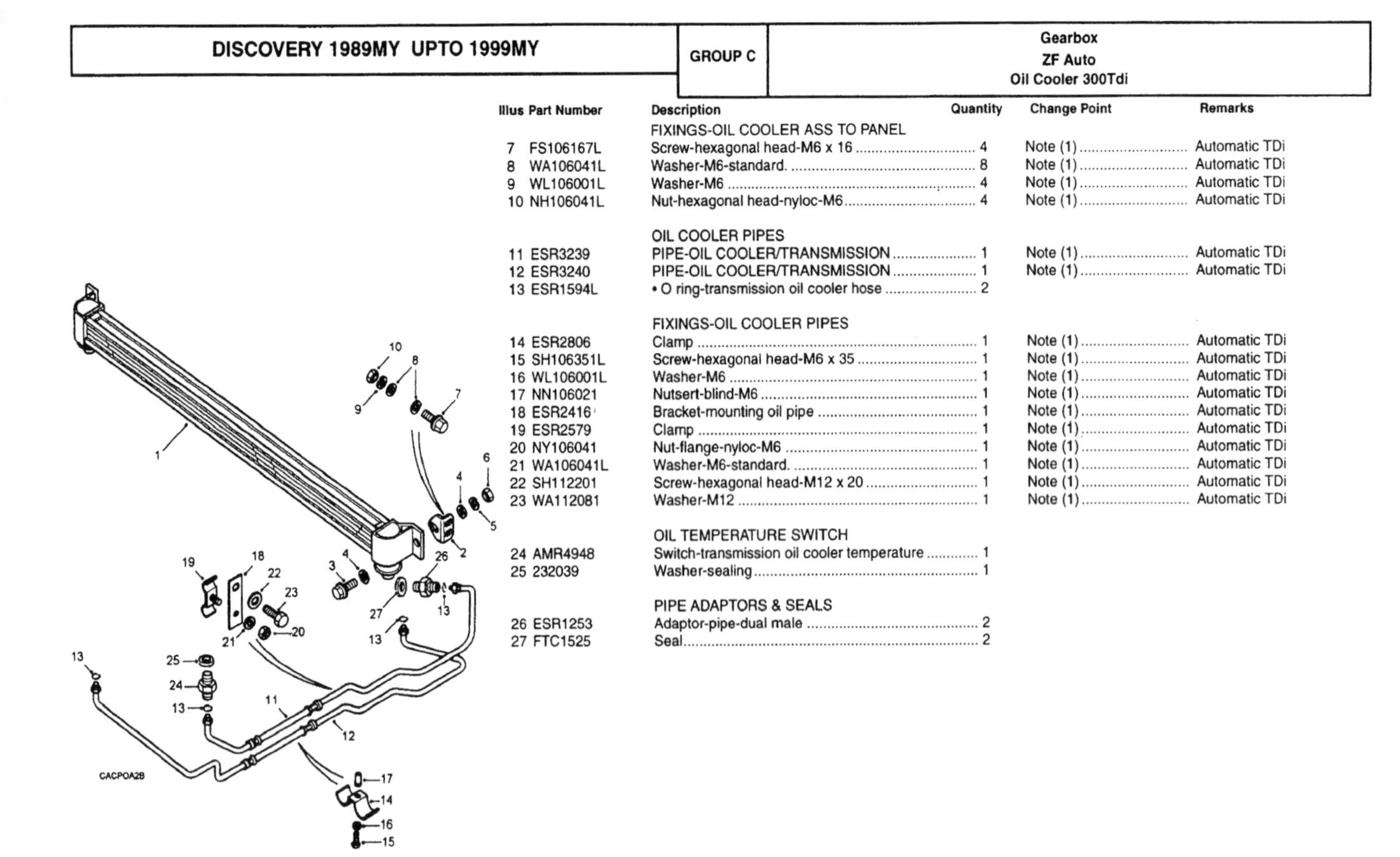

Illus	Part Number	Description	Quantity	Change Point	Remarks
		FIXINGS-OIL COOLER ASS TO PANEL			
7	FS106167L	Screw-hexagonal head-M6 x 16	4	Note (1)	Automatic TDi
8	WA106041L	Washer-M6-standard.	8	Note (1)	Automatic TDi
9	WL106001L	Washer-M6	4	Note (1)	Automatic TDi
10	NH106041L	Nut-hexagonal head-nyloc-M6	4	Note (1)	Automatic TDi
		OIL COOLER PIPES			
11	ESR3239	PIPE-OIL COOLER/TRANSMISSION	1	Note (1)	Automatic TDi
12	ESR3240	PIPE-OIL COOLER/TRANSMISSION	1	Note (1)	Automatic TDi
13	ESR1594L	• O ring-transmission oil cooler hose	2		
		FIXINGS-OIL COOLER PIPES			
14	ESR2806	Clamp	1	Note (1)	Automatic TDi
15	SH106351L	Screw-hexagonal head-M6 x 35	1	Note (1)	Automatic TDi
16	WL106001L	Washer-M6	1	Note (1)	Automatic TDi
17	NN106021	Nutsert-blind-M6	1	Note (1)	Automatic TDi
18	ESR2416	Bracket-mounting oil pipe	1	Note (1)	Automatic TDi
19	ESR2579	Clamp	1	Note (1)	Automatic TDi
20	NY106041	Nut-flange-nyloc-M6	1	Note (1)	Automatic TDi
21	WA106041L	Washer-M6-standard.	1	Note (1)	Automatic TDi
22	SH112201	Screw-hexagonal head-M12 x 20	1	Note (1)	Automatic TDi
23	WA112081	Washer-M12	1	Note (1)	Automatic TDi
		OIL TEMPERATURE SWITCH			
24	AMR4948	Switch-transmission oil cooler temperature	1		
25	232039	Washer-sealing	1		
		PIPE ADAPTORS & SEALS			
26	ESR1253	Adaptor-pipe-dual male	2		
27	FTC1525	Seal	2		

CHANGE POINTS:
(1) From (V) MA 081992

LT230 .... D 1
Differential .... D 7
Front Flange .... D 11
Front Housing .... D 3
Front Ouput Shaft .... D 13
Gasket Kit .... D 21
Gear Change-Automatic .... D 18
Gear Change-Manual .... D 16
High/Low Gear .... D 15
Housing Automatic .... D 4
Intermediate Shaft .... D 10
Mainshaft Gear .... D 8
Rear Flange .... D 12
Speedo Housing .... D 5
Transfer Assembly .... D 1
Transmission Brake-Direct Entry Cable .... D 23
Transmission Brake-Rod Operated .... D 22

Illus	Part Number	Description	Quantity	Change Point	Remarks
		SUFFIX C			
1	FRC9467N	Transfer gearbox assembly-1.2 : 1-new	1		TDi
1	FRC9467N	Transfer gearbox assembly-1.2 : 1-new	1		V8
		SUFFIX D			
		TRANSFER GEARBOX ASSEMBLY			
1	STC1010N	1.2 : 1-new	1		Twin Carburettor
1	STC1010N	1.2 : 1-new	1		EFi
1	STC1010N	1.2 : 1-new	1		TDi
		SUFFIX E			
		TRANSFER GEARBOX ASSEMBLY			
1	STC1475	1.2 : 1-new	1	To (T) 308626E	Automatic TDi
1	STC1596	1.2 : 1-new	1		Automatic EFi
1	STC1811	new	1		Manual MPi
1	STC1144	1.2 : 1-new	1	From (T) 309143E	Manual EFi
1	STC1143	1.2 : 1-new	1		Manual TDi
1	STC1108	1.2 : 1-new	1		Automatic EFi
1	STC1794	1.2 : 1-new	1		Automatic TDi
1	STC1072	1.2 : 1-new	1		Automatic EFi Japan
1	STC1830	1.2 : 1-new	1		Automatic TDi Japan
		SUFFIX F			
		TRANSFER GEARBOX ASSEMBLY			
1	STC1811	new	1		Manual MPi
1	STC1144	1.2 : 1-new	1		Manual Petrol EFi
1	STC1143	1.2 : 1-new	1		Manual TDi
1	STC1108	1.2 : 1-new	1		Automatic EFi
1	STC1794	1.2 : 1-new	1		Automatic TDi
1	STC1072	1.2 : 1-new	1		Automatic EFi Japan
1	STC1830	1.2 : 1-new	1		Automatic TDi Japan

1

D 1

Illus	Part Number	Description	Quantity	Change Point	Remarks
		SUFFIX G			
1	STC3608	Transfer gearbox assembly-new-1.2 : 1	1		Manual EFi
1	STC3608	Transfer gearbox assembly-new-1.2 : 1	1		Manual TDi
1	STC3608	Transfer gearbox assembly-new-1.2 : 1	1		Manual Twin Carburettor
1	STC3609	Transfer gearbox assembly-new-1.4 : 1	1		Manual MPi
1	STC3608	Transfer gearbox assembly-new-1.2 : 1	1		Automatic EFi
1	STC3608	Transfer gearbox assembly-new-1.2 : 1	1		Automatic TDi
1	STC3611	Transfer gearbox assembly-new-1.2 : 1	1		Automatic EFi Japan
1	STC3611	Transfer gearbox assembly-new-1.2 : 1	1		Automatic TDi Japan
		EXCHANGE BOXES-NOT AUTO JAPAN			
1	STC8682E	Transfer gearbox assembly-exchange-1.4 : 1	1	Note (1)	MPi Note (1)
1	STC8959E	Transfer gearbox assembly-exchange-1.2 : 1	1		Manual Note (1)
1	STC8960E	Transfer gearbox assembly-exchange-1.2 : 1	1		Automatic Note (1)

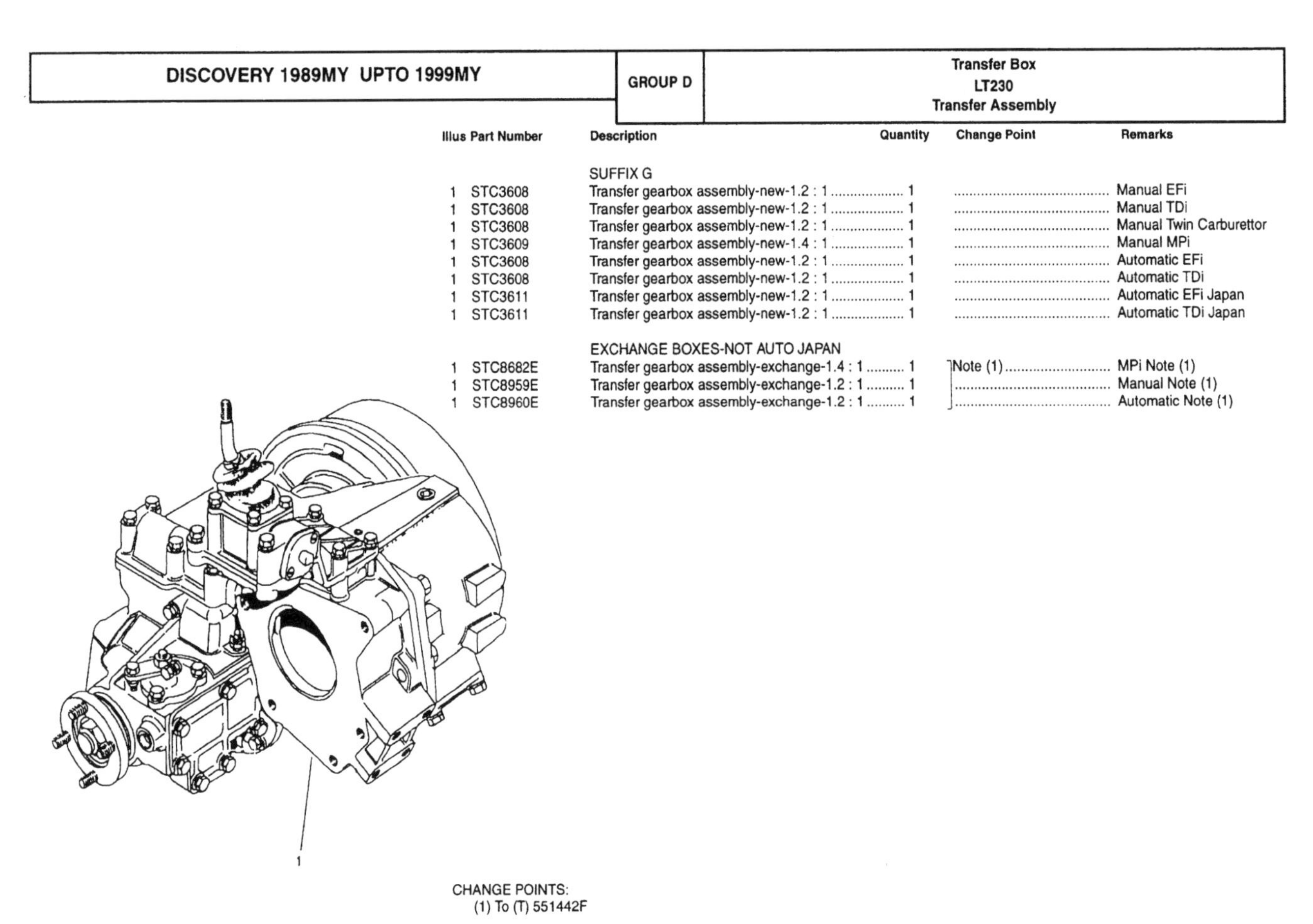

CHANGE POINTS:
(1) To (T) 551442F

Remarks:
(1) Remanufactured transfer boxes do not come with transmission brake assembly, any sensors, gear linkage or speedo pinion

D 2

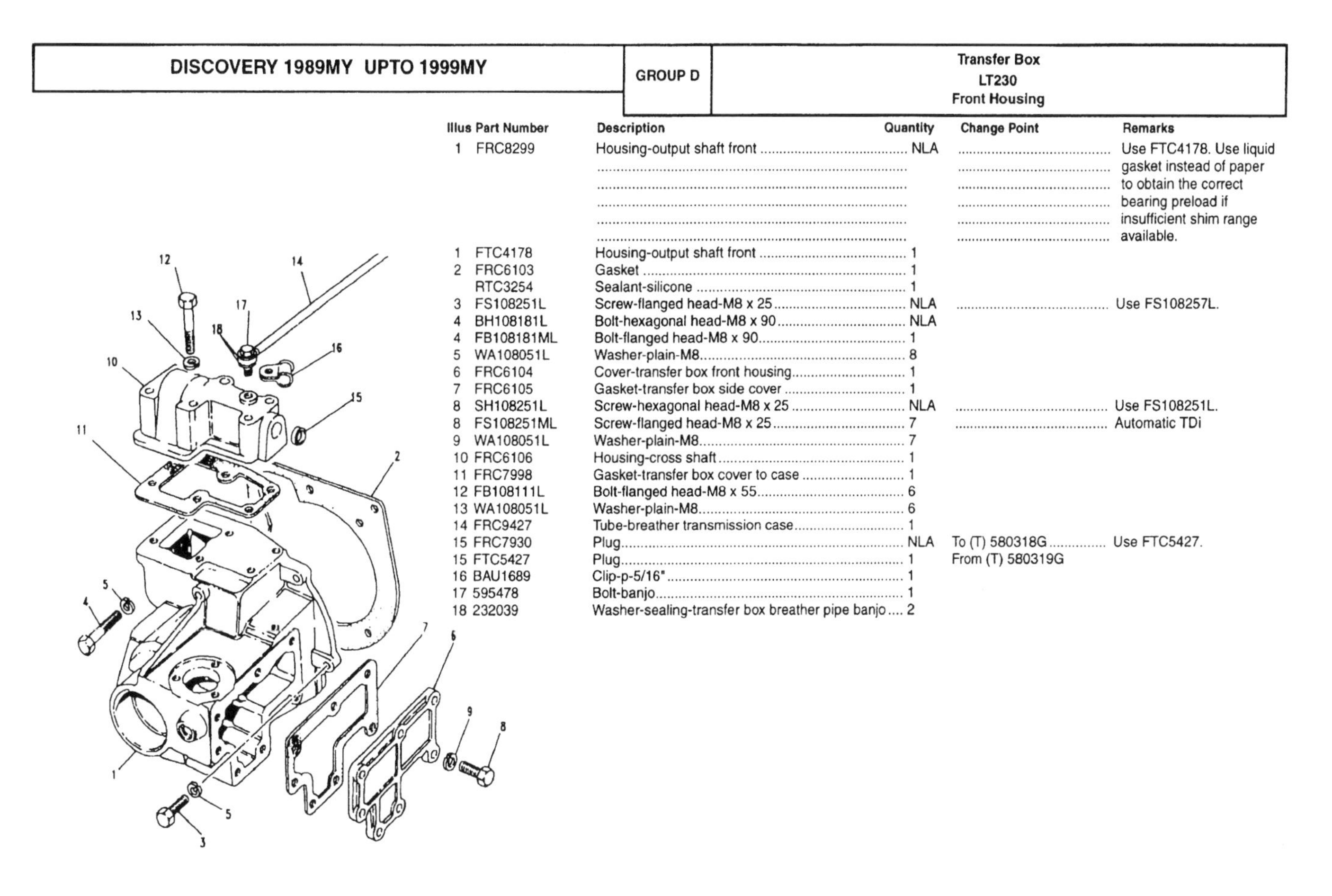

Illus	Part Number	Description	Quantity	Change Point	Remarks
1	FRC8299	Housing-output shaft front	NLA		Use FTC4178. Use liquid gasket instead of paper to obtain the correct bearing preload if insufficient shim range available.
1	FTC4178	Housing-output shaft front	1		
2	FRC6103	Gasket	1		
	RTC3254	Sealant-silicone	1		
3	FS108251L	Screw-flanged head-M8 x 25	NLA		Use FS108257L.
4	BH108181L	Bolt-hexagonal head-M8 x 90	NLA		
4	FB108181ML	Bolt-flanged head-M8 x 90	1		
5	WA108051L	Washer-plain-M8	8		
6	FRC6104	Cover-transfer box front housing	1		
7	FRC6105	Gasket-transfer box side cover	1		
8	SH108251L	Screw-hexagonal head-M8 x 25	NLA		Use FS108251L.
8	FS108251ML	Screw-flanged head-M8 x 25	7		Automatic TDi
9	WA108051L	Washer-plain-M8	7		
10	FRC6106	Housing-cross shaft	1		
11	FRC7998	Gasket-transfer box cover to case	1		
12	FB108111L	Bolt-flanged head-M8 x 55	6		
13	WA108051L	Washer-plain-M8	6		
14	FRC9427	Tube-breather transmission case	1		
15	FRC7930	Plug	NLA	To (T) 580318G	Use FTC5427.
15	FTC5427	Plug	1	From (T) 580319G	
16	BAU1689	Clip-p-5/16"	1		
17	595478	Bolt-banjo	1		
18	232039	Washer-sealing-transfer box breather pipe banjo	2		

D 3

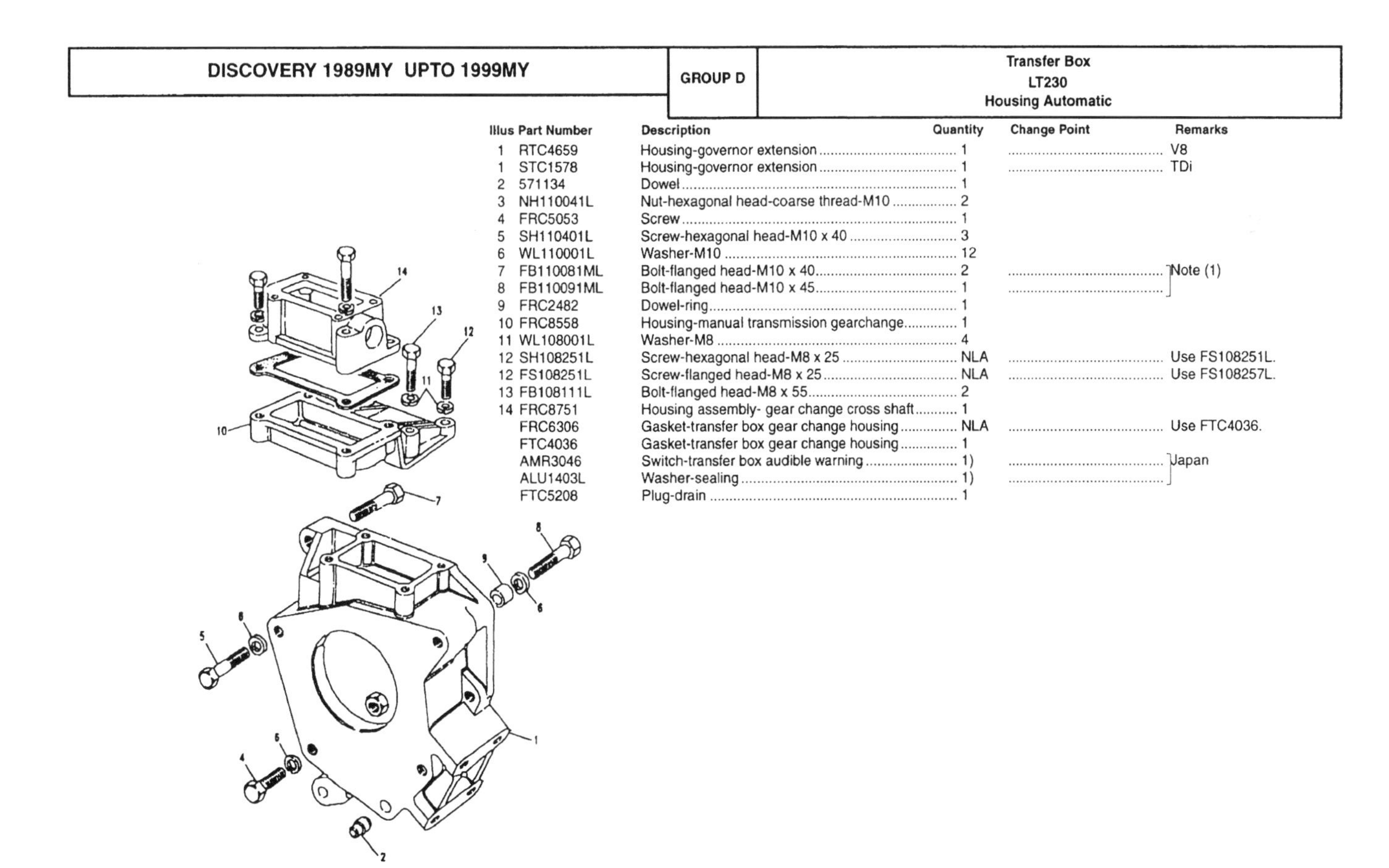

Illus	Part Number	Description	Quantity	Change Point	Remarks
1	RTC4659	Housing-governor extension	1		V8
1	STC1578	Housing-governor extension	1		TDi
2	571134	Dowel	1		
3	NH110041L	Nut-hexagonal head-coarse thread-M10	2		
4	FRC5053	Screw	1		
5	SH110401L	Screw-hexagonal head-M10 x 40	3		
6	WL110001L	Washer-M10	12		
7	FB110081ML	Bolt-flanged head-M10 x 40	2		Note (1)
8	FB110091ML	Bolt-flanged head-M10 x 45	1		Note (1)
9	FRC2482	Dowel-ring	1		
10	FRC8558	Housing-manual transmission gearchange	1		
11	WL108001L	Washer-M8	4		
12	SH108251L	Screw-hexagonal head-M8 x 25	NLA		Use FS108251L.
12	FS108251L	Screw-flanged head-M8 x 25	NLA		Use FS108257L.
13	FB108111L	Bolt-flanged head-M8 x 55	2		
14	FRC8751	Housing assembly- gear change cross shaft	1		
	FRC6306	Gasket-transfer box gear change housing	NLA		Use FTC4036.
	FTC4036	Gasket-transfer box gear change housing	1		
	AMR3046	Switch-transfer box audible warning	1)		Japan
	ALU1403L	Washer-sealing	1)		Japan
	FTC5208	Plug-drain	1		

Remarks:
(1) Pre-impregnated with adhesive

D 4

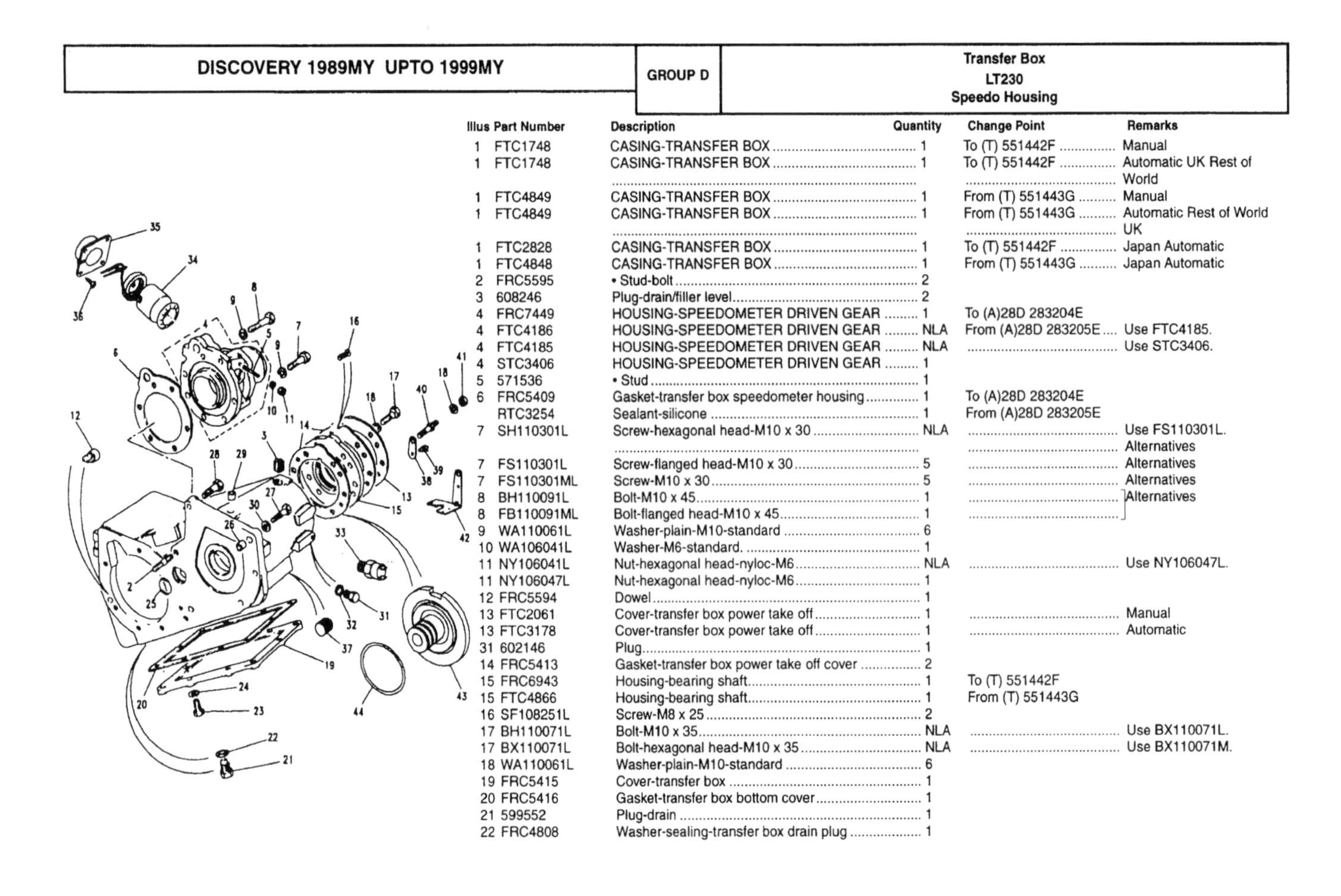

Illus	Part Number	Description	Quantity	Change Point	Remarks
1	FTC1748	CASING-TRANSFER BOX	1	To (T) 551442F	Manual
1	FTC1748	CASING-TRANSFER BOX	1	To (T) 551442F	Automatic UK Rest of World
1	FTC4849	CASING-TRANSFER BOX	1	From (T) 551443G	Manual
1	FTC4849	CASING-TRANSFER BOX	1	From (T) 551443G	Automatic Rest of World UK
1	FTC2828	CASING-TRANSFER BOX	1	To (T) 551442F	Japan Automatic
1	FTC4848	CASING-TRANSFER BOX	1	From (T) 551443G	Japan Automatic
2	FRC5595	• Stud-bolt	2		
3	608246	Plug-drain/filler level	2		
4	FRC7449	HOUSING-SPEEDOMETER DRIVEN GEAR	1	To (A)28D 283204E	
4	FTC4186	HOUSING-SPEEDOMETER DRIVEN GEAR	NLA	From (A)28D 283205E	Use FTC4185.
4	FTC4185	HOUSING-SPEEDOMETER DRIVEN GEAR	NLA		Use STC3406.
4	STC3406	HOUSING-SPEEDOMETER DRIVEN GEAR	1		
5	571536	• Stud	1		
6	FRC5409	Gasket-transfer box speedometer housing	1	To (A)28D 283204E	
	RTC3254	Sealant-silicone	1	From (A)28D 283205E	
7	SH110301L	Screw-hexagonal head-M10 x 30	NLA		Use FS110301L. Alternatives
7	FS110301L	Screw-flanged head-M10 x 30	5		Alternatives
7	FS110301ML	Screw-M10 x 30	5		Alternatives
8	BH110091L	Bolt-M10 x 45	1		Alternatives
8	FB110091ML	Bolt-flanged head-M10 x 45	1		
9	WA110061L	Washer-plain-M10-standard	6		
10	WA106041L	Washer-M6-standard.	1		
11	NY106041L	Nut-hexagonal head-nyloc-M6	NLA		Use NY106047L.
11	NY106047L	Nut-hexagonal head-nyloc-M6	1		
12	FRC5594	Dowel	1		
13	FTC2061	Cover-transfer box power take off	1		Manual
13	FTC3178	Cover-transfer box power take off	1		Automatic
31	602146	Plug	1		
14	FRC5413	Gasket-transfer box power take off cover	2		
15	FRC6943	Housing-bearing shaft	1	To (T) 551442F	
15	FTC4866	Housing-bearing shaft	1	From (T) 551443G	
16	SF108251L	Screw-M8 x 25	2		
17	BH110071L	Bolt-M10 x 35	NLA		Use BX110071L.
17	BX110071L	Bolt-hexagonal head-M10 x 35	NLA		Use BX110071M.
18	WA110061L	Washer-plain-M10-standard	6		
19	FRC5415	Cover-transfer box	1		
20	FRC5416	Gasket-transfer box bottom cover	1		
21	599552	Plug-drain	1		
22	FRC4808	Washer-sealing-transfer box drain plug	1		

DISCOVERY 1989MY UPTO 1999MY

GROUP D

Transfer Box
LT230
Speedo Housing

Illus	Part Number	Description	Quantity	Change Point	Remarks
23	BH108061L	Bolt-M8 x 30	NLA	To (V) LA 081991	Use FB108061.
23	FB108061	Bolt-flanged head-M8 x 30	NLA		Use FS108301L.
23	FS108301L	Screw-flanged head-M8 x 30	NLA	From (V) MA 081992	Use FS108307L.
23	FS108307L	Screw-flanged head-M8 x 30	10	Note (1)	
24	WA108051L	Washer-plain-M8	10		
25	FRC8292	O ring	1	To (V) LA 081991	
25	571665	O ring	1	From (V) MA 081992	
26	FRC2482	Dowel-ring	1		
27	BH110091L	Bolt-M10 x 45	1		
28	BH110081L	Bolt-hexagonal head-M10 x 40	3		Alternatives
28	BH110081L	Bolt-hexagonal head-M10 x 40	2		Alternatives
40	FTC890	Stud-lifting bracket	1		
41	NH110041L	Nut-hexagonal head-coarse thread-M10	1		
29	UKC24L	Plug-core-6mm	1		
30	WL110001L	Washer-M10	1		
31	SH214141	Plug	1	To (V) LA 081991	Alternatives
32	232042	Washer-sealing-transfer box drain plug	1		
33	PRC7984	Switch-transfer box temperature sensor	1	From (V) MA 081992	
34	FTC3283	Solenoid-gear change lock	1		Japan
35	FTC2829	Cover-interlock solenoid	1		
36	FS106201M	Screw-M6 x 20-flanged head-patch lock	4		
37	608246	Plug-drain/filler level	1		
38	ANR3625	Bracket-transmission retaining	NLA		Use ANR3227.
	ANR3227	Heatshield-bracket	1		
39	PRC3180	Clip-pipe-single	1		
40	FTC1919	Stud-heatshield attachment	1		
41	FN110041L	Nut-flanged head-M10	NLA		Use FN110047.
41	FN110047	Nut-flanged head-M10	1		
42	ANR3625	Bracket-transmission retaining	NLA		Use ANR3227.
	ANR3227	Heatshield-bracket	1		
43	FTC4993	Plate-oil guide transmission case	NLA	Note (2)	Use STC3615K.
	STC3615K	Kit-plate & seal-oil guide	1		
44	FTC4994	O ring	1	Note (2)	

CHANGE POINTS:
(1) From (V) MA 081992
(2) From (T) 445132E To (T) 461123E

DISCOVERY 1989MY UPTO 1999MY | GROUP D | Transfer Box LT230 Differential

Illus	Part Number	Description	Quantity	Change Point	Remarks
1	FRC7926	DIFFERENTIAL ASSEMBLY-TRANSFER BOX	1	To (T) 575041G	
1	FTC5207	DIFFERENTIAL ASSEMBLY-TRANSFER BOX	1	From (T) 575042G	
2	RTC3397	• Kit-shaft-cross	1		
3	RTC4490	• KIT-DIFFERENTIAL GEARS	1		Part of FRC7926.
4	FRC6968	• • Washer-thrust	4		
		• WASHER-THRUST			
5	FRC9845	• 1.05mm	A/R		
5	FRC9847	• 1.15mm	A/R		
5	FRC9849	• 1.25mm	A/R		
5	FRC9851	• 1.35mm	A/R		
5	FRC9853	• 1.45mm	A/R		
6	BH110121L	• Bolt-M10 x 60-long	8		
7	FRC7499	• Ring-transfer box differential retaining	1		
3	STC2940	• KIT-DIFFERENTIAL GEARS	1		Part of FTC5207.
4	FRC6968	• • Washer-thrust	4		

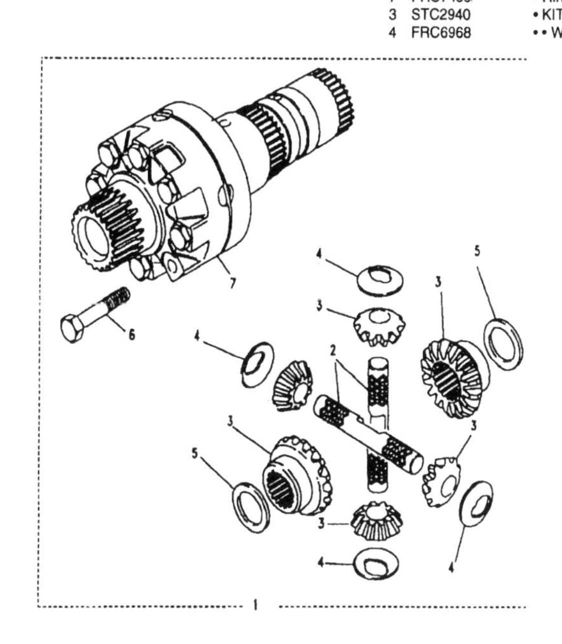

D 7

DISCOVERY 1989MY UPTO 1999MY | GROUP D | Transfer Box LT230 Mainshaft Gear

Illus	Part Number	Description	Quantity	Change Point	Remarks
		GEARS UP TO 551442F			
1	FRC8917	Gear-mainshaft-26 teeth	NLA	Note (1)	Use FTC5089. Except MPi
1	FTC5089	Gear-mainshaft-26 teeth	NLA		Except MPi
1	FTC5089	Gear-mainshaft-26 teeth	1	Note (2)	Except MPi
1	FTC4188	Gear-mainshaft-28 teeth	NLA	Note (3); Note (2)	Manual use FTC5087; auto use FTC5087 & FTC5090; later gear has longer internal spline, lubricating holes & dog teeth which old may not have.
1	FTC4818	Gear-mainshaft-28 teeth	NLA	Note (4); Note (5); Note (2)	FTC4818 is FTC4188 with dog teeth for power take off removed. FTC4818 is only used on Disco. FTC4188 is suitable for both Disco and Defender.
1	FTC5087	Gear-mainshaft-28 teeth	1	Note (6); Note (7)	
		GEARS FROM 551443G			
1	FTC4850	Gear-mainshaft-38 teeth-1.2 : 1	1	Note (8)	
1	FTC4962	Gear-mainshaft-26 teeth-1.4 : 1	1		MPi

CHANGE POINTS:
(1) To (V) LA 081991; To (T) 551442F
(2) To (T) 551442F
(3) From (V) MA 081992 To (V) MA 149447
(4) From (V) MA 149448 To (V) TA 178213
(5) From (V) TA 500000 To (V) TA 509009
(6) From (V) TA 178214
(7) From (V) TA 509010
(8) From (T) 551443G

D 8

Illus	Part Number	Description	Quantity	Change Point	Remarks
2	FRC2365	Seal-transfer box input	1		
3	FRC5564	Mainshaft bearing	2		
		SHIM			
4	FRC9926	3.15mm	A/R		
4	FRC9928	3.20mm	A/R		
4	FRC9930	3.25mm	A/R		
4	FRC9932	3.30mm	A/R		
4	FRC9934	3.35mm	A/R		
4	FRC9936	3.40mm	A/R		
4	FRC9938	3.45mm	A/R		
4	FRC9940	3.50mm	A/R		
4	FRC9942	3.55mm	A/R		
4	FRC9944	3.60mm	A/R		
4	FRC9946	3.65mm	A/R		
4	FRC9948	3.70mm	A/R		
4	FRC9950	3.75mm	A/R		
4	FRC9952	3.80mm	A/R		
4	FRC9954	3.85mm	A/R		
4	FRC9956	3.90mm	A/R		
4	FRC9958	3.95mm	A/R		
4	FRC9960	4.00mm	A/R		

D 9

Illus	Part Number	Description	Quantity	Change Point	Remarks
1	FRC8291	Shaft-transfer box-intermediate	1		
2	FRC8292	O ring	1		
		GEAR-INTERMEDIATE SHAFT			
3	FRC9552	1.2 : 1	1	Note (1)	
3	FTC4190	1.2 : 1	1	Note (2)	
3	FTC4846	1.2 : 1	1	Note (3)	
3	FRC9460	1.4 : 1	1		MPi
4	FRC7810	Bearing intermediate shaft assembly	NLA		Use STC3185.
4	STC3185	Bearing intermediate shaft assembly	2		
5	FRC7437	Spacer	1		
6	FRC7454	Circlip	2		
7	FRC7452	Plate-transfer box anti rotation	1		
8	FRC7453	Locknut	1		
9	FS108251ML	Screw-flanged head-M8 x 25	1		
10	WA108051L	Washer-plain-M8	1		
11	FRC7439	O ring	1		

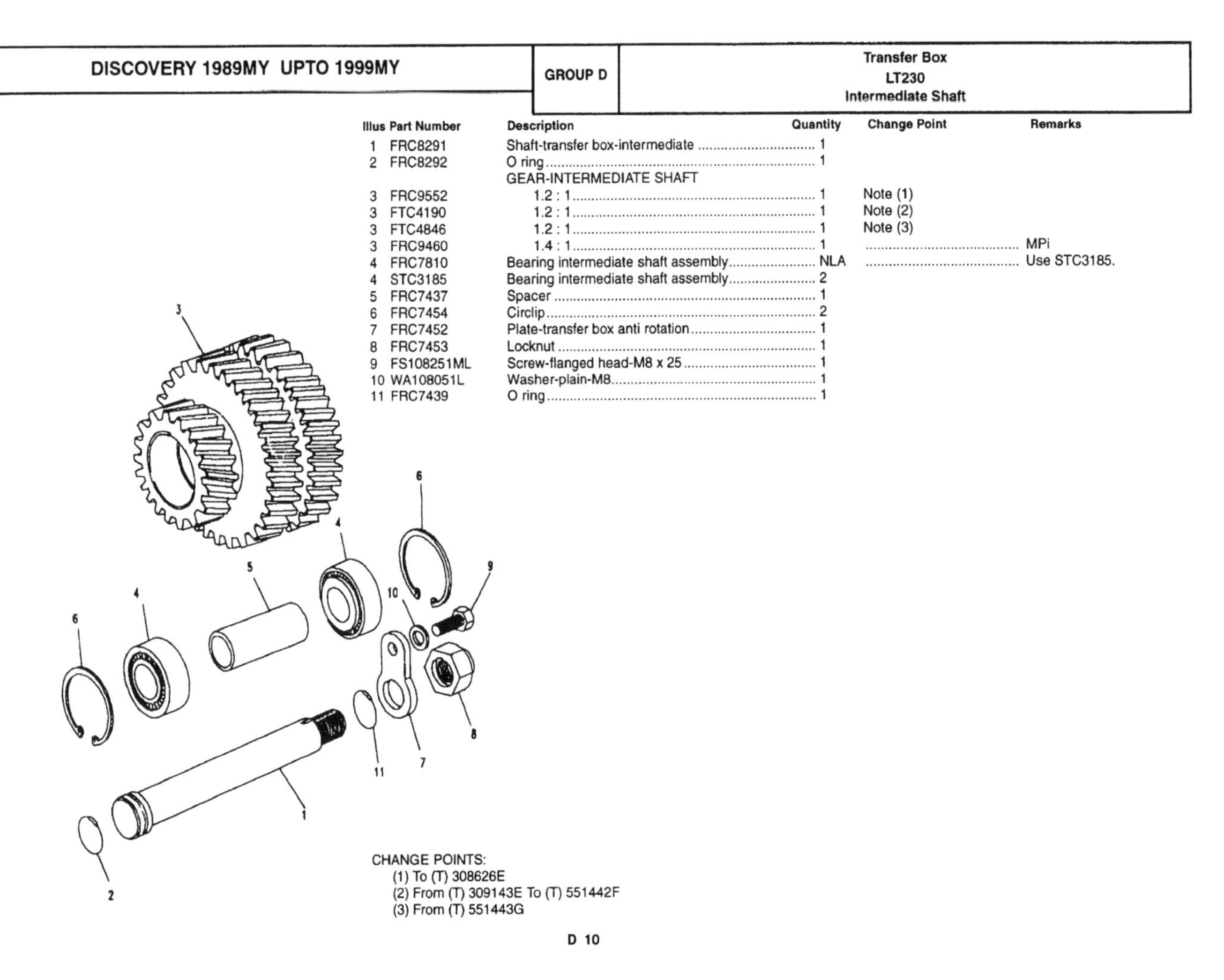

CHANGE POINTS:
(1) To (T) 308626E
(2) From (T) 309143E To (T) 551442F
(3) From (T) 551443G

D 10

DISCOVERY 1989MY UPTO 1999MY

GROUP D

Transfer Box LT230 Front Flange

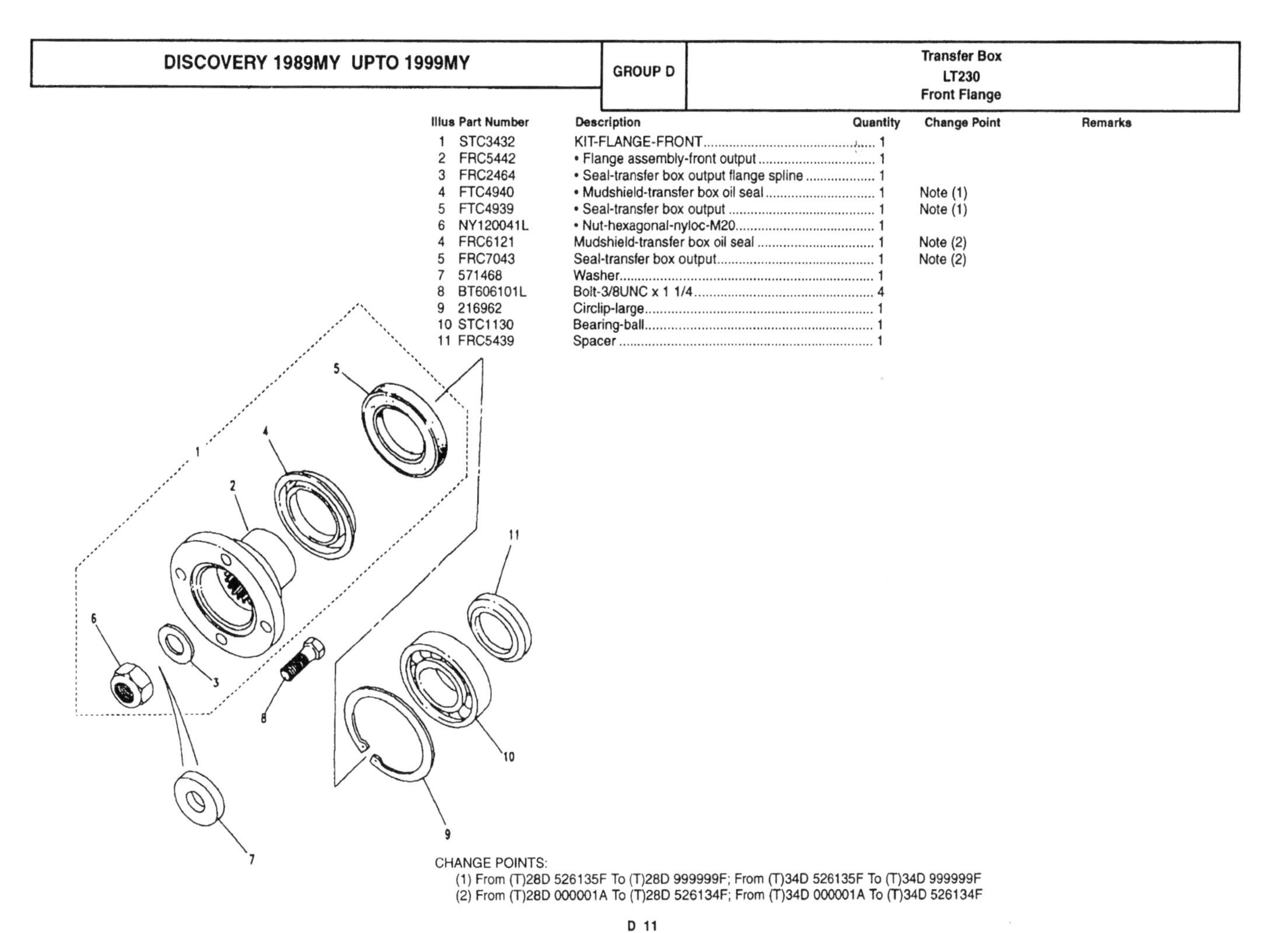

Illus	Part Number	Description	Quantity	Change Point	Remarks
1	STC3432	KIT-FLANGE-FRONT	1		
2	FRC5442	• Flange assembly-front output	1		
3	FRC2464	• Seal-transfer box output flange spline	1		
4	FTC4940	• Mudshield-transfer box oil seal	1	Note (1)	
5	FTC4939	• Seal-transfer box output	1	Note (1)	
6	NY120041L	• Nut-hexagonal-nyloc-M20	1		
4	FRC6121	Mudshield-transfer box oil seal	1	Note (2)	
5	FRC7043	Seal-transfer box output	1	Note (2)	
7	571468	Washer	1		
8	BT606101L	Bolt-3/8UNC x 1 1/4	4		
9	216962	Circlip-large	1		
10	STC1130	Bearing-ball	1		
11	FRC5439	Spacer	1		

CHANGE POINTS:
(1) From (T)28D 526135F To (T)28D 999999F; From (T)34D 526135F To (T)34D 999999F
(2) From (T)28D 000001A To (T)28D 526134F; From (T)34D 000001A To (T)34D 526134F

D 11

DISCOVERY 1989MY UPTO 1999MY

GROUP D

Transfer Box LT230 Rear Flange

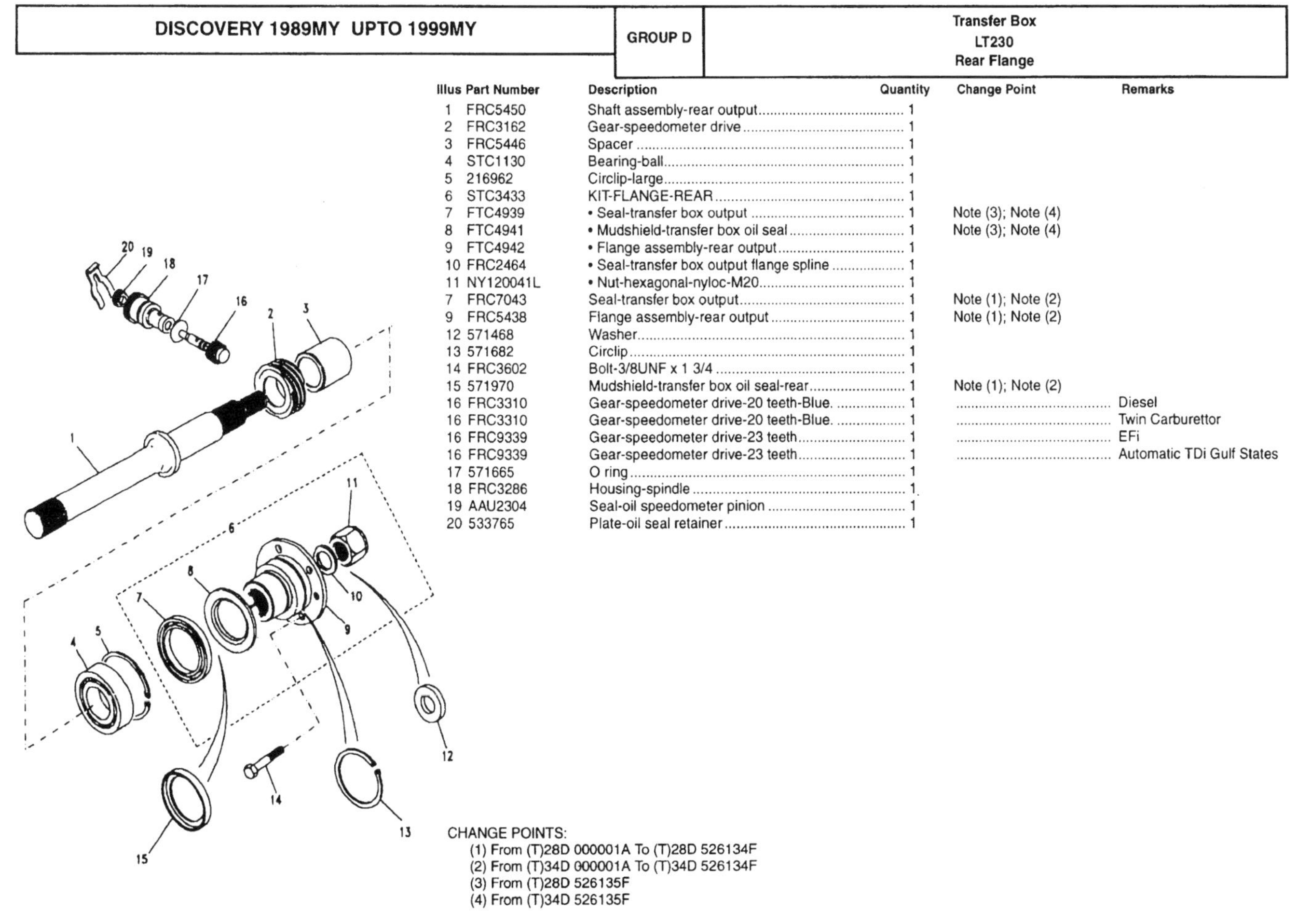

Illus	Part Number	Description	Quantity	Change Point	Remarks
1	FRC5450	Shaft assembly-rear output	1		
2	FRC3162	Gear-speedometer drive	1		
3	FRC5446	Spacer	1		
4	STC1130	Bearing-ball	1		
5	216962	Circlip-large	1		
6	STC3433	KIT-FLANGE-REAR	1		
7	FTC4939	• Seal-transfer box output	1	Note (3); Note (4)	
8	FTC4941	• Mudshield-transfer box oil seal	1	Note (3); Note (4)	
9	FTC4942	• Flange assembly-rear output	1		
10	FRC2464	• Seal-transfer box output flange spline	1		
11	NY120041L	• Nut-hexagonal-nyloc-M20	1		
7	FRC7043	Seal-transfer box output	1	Note (1); Note (2)	
9	FRC5438	Flange assembly-rear output	1	Note (1); Note (2)	
12	571468	Washer	1		
13	571682	Circlip	1		
14	FRC3602	Bolt-3/8UNF x 1 3/4	1		
15	571970	Mudshield-transfer box oil seal-rear	1	Note (1); Note (2)	
16	FRC3310	Gear-speedometer drive-20 teeth-Blue.	1		Diesel
16	FRC3310	Gear-speedometer drive-20 teeth-Blue.	1		Twin Carburettor
16	FRC9339	Gear-speedometer drive-23 teeth	1		EFi
16	FRC9339	Gear-speedometer drive-23 teeth	1		Automatic TDi Gulf States
17	571665	O ring	1		
18	FRC3286	Housing-spindle	1		
19	AAU2304	Seal-oil speedometer pinion	1		
20	533765	Plate-oil seal retainer	1		

CHANGE POINTS:
(1) From (T)28D 000001A To (T)28D 526134F
(2) From (T)34D 000001A To (T)34D 526134F
(3) From (T)28D 526135F
(4) From (T)34D 526135F

D 12

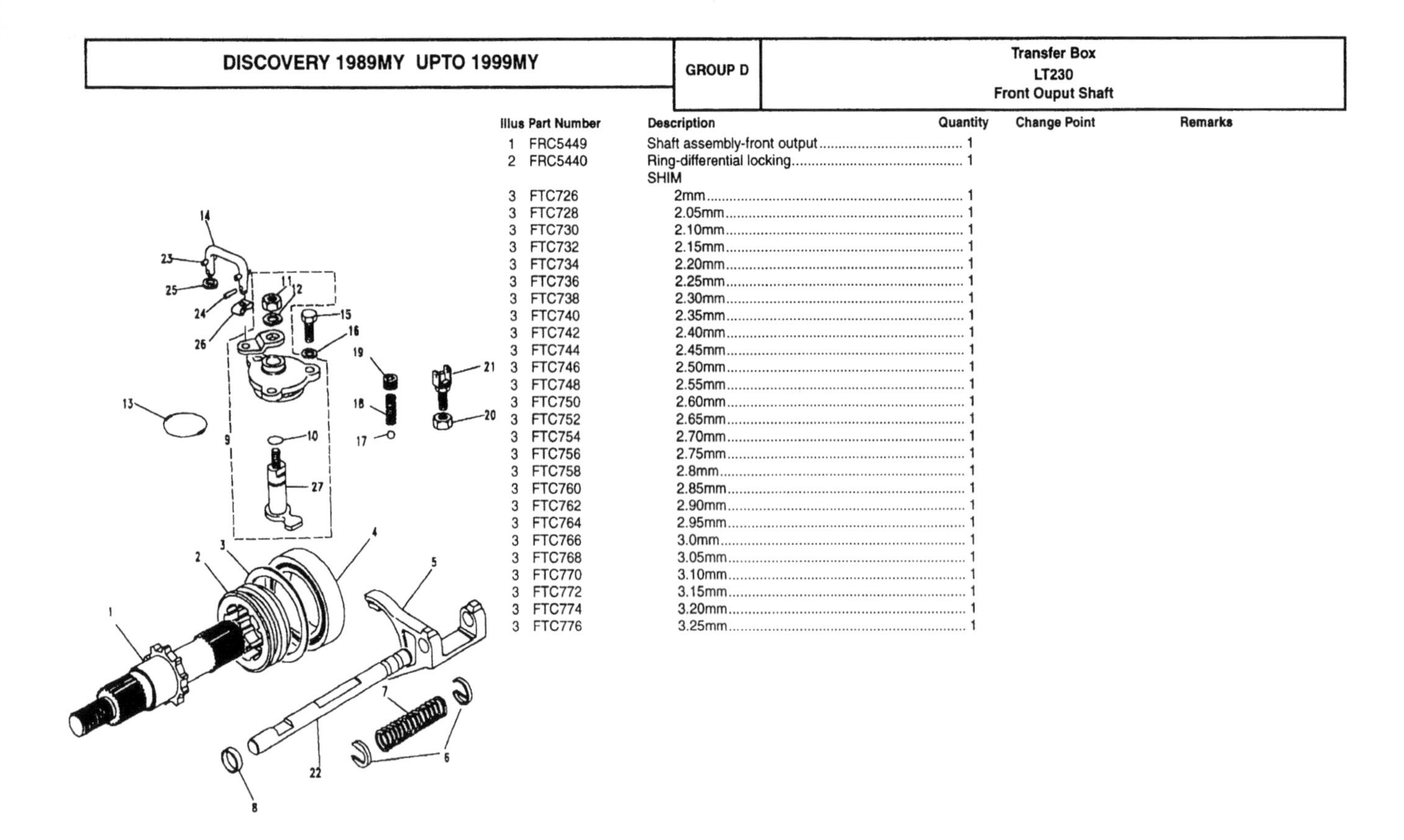

Illus	Part Number	Description	Quantity	Change Point	Remarks
1	FRC5449	Shaft assembly-front output	1		
2	FRC5440	Ring-differential locking	1		
		SHIM			
3	FTC726	2mm	1		
3	FTC728	2.05mm	1		
3	FTC730	2.10mm	1		
3	FTC732	2.15mm	1		
3	FTC734	2.20mm	1		
3	FTC736	2.25mm	1		
3	FTC738	2.30mm	1		
3	FTC740	2.35mm	1		
3	FTC742	2.40mm	1		
3	FTC744	2.45mm	1		
3	FTC746	2.50mm	1		
3	FTC748	2.55mm	1		
3	FTC750	2.60mm	1		
3	FTC752	2.65mm	1		
3	FTC754	2.70mm	1		
3	FTC756	2.75mm	1		
3	FTC758	2.8mm	1		
3	FTC760	2.85mm	1		
3	FTC762	2.90mm	1		
3	FTC764	2.95mm	1		
3	FTC766	3.0mm	1		
3	FTC768	3.05mm	1		
3	FTC770	3.10mm	1		
3	FTC772	3.15mm	1		
3	FTC774	3.20mm	1		
3	FTC776	3.25mm	1		

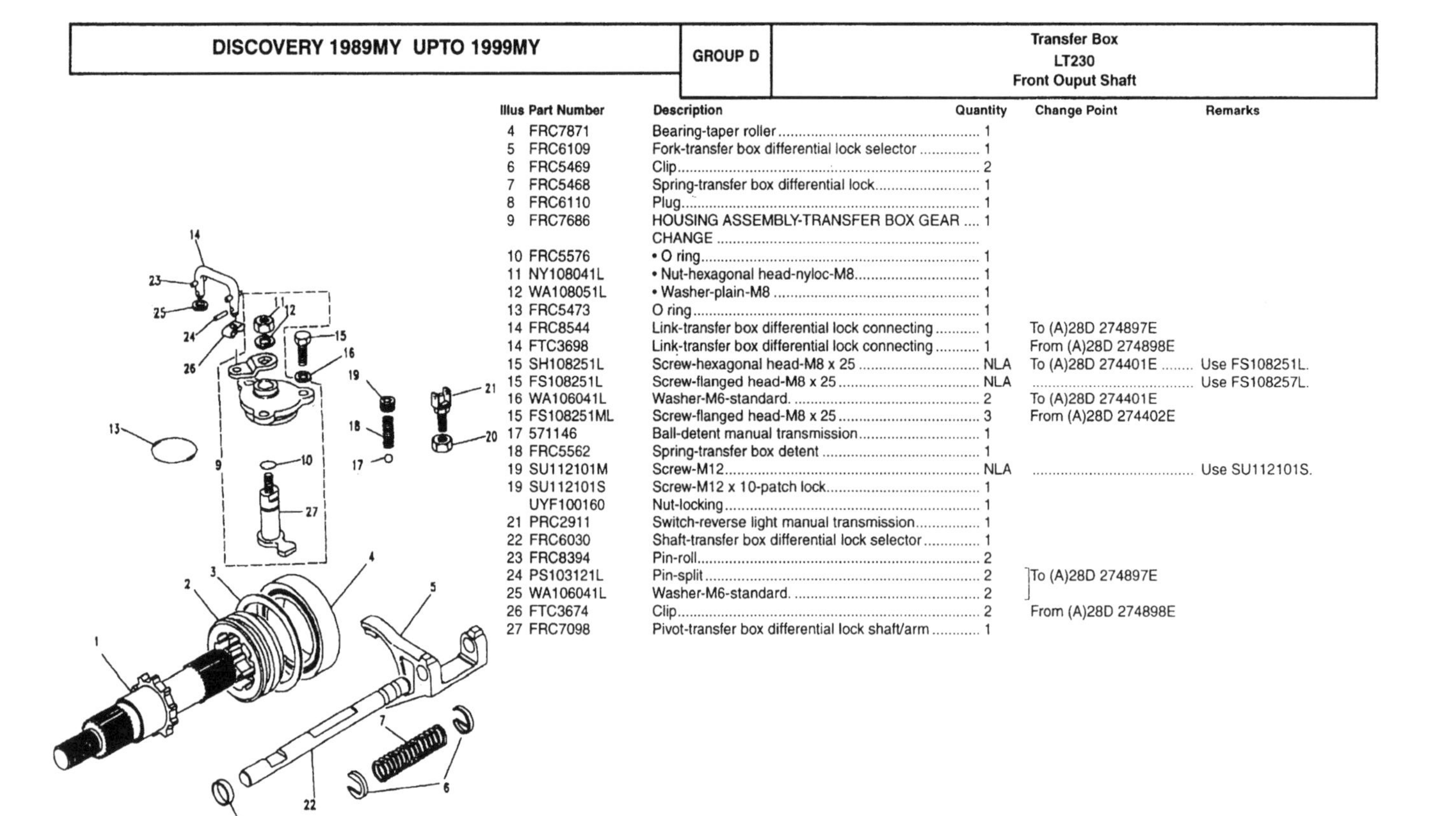

Illus	Part Number	Description	Quantity	Change Point	Remarks
4	FRC7871	Bearing-taper roller	1		
5	FRC6109	Fork-transfer box differential lock selector	1		
6	FRC5469	Clip	2		
7	FRC5468	Spring-transfer box differential lock	1		
8	FRC6110	Plug	1		
9	FRC7686	HOUSING ASSEMBLY-TRANSFER BOX GEAR CHANGE	1		
10	FRC5576	• O ring	1		
11	NY108041L	• Nut-hexagonal head-nyloc-M8	1		
12	WA108051L	• Washer-plain-M8	1		
13	FRC5473	O ring	1		
14	FRC8544	Link-transfer box differential lock connecting	1	To (A)28D 274897E	
14	FTC3698	Link-transfer box differential lock connecting	1	From (A)28D 274898E	
15	SH108251L	Screw-hexagonal head-M8 x 25	NLA	To (A)28D 274401E	Use FS108251L.
15	FS108251L	Screw-flanged head-M8 x 25	NLA		Use FS108257L.
16	WA106041L	Washer-M6-standard.	2	To (A)28D 274401E	
15	FS108251ML	Screw-flanged head-M8 x 25	3	From (A)28D 274402E	
17	571146	Ball-detent manual transmission	1		
18	FRC5562	Spring-transfer box detent	1		
19	SU112101M	Screw-M12	NLA		Use SU112101S.
19	SU112101S	Screw-M12 x 10-patch lock	1		
	UYF100160	Nut-locking	1		
21	PRC2911	Switch-reverse light manual transmission	1		
22	FRC6030	Shaft-transfer box differential lock selector	1		
23	FRC8394	Pin-roll	2		
24	PS103121L	Pin-split	2	To (A)28D 274897E	
25	WA106041L	Washer-M6-standard.	2		
26	FTC3674	Clip	2	From (A)28D 274898E	
27	FRC7098	Pivot-transfer box differential lock shaft/arm	1		

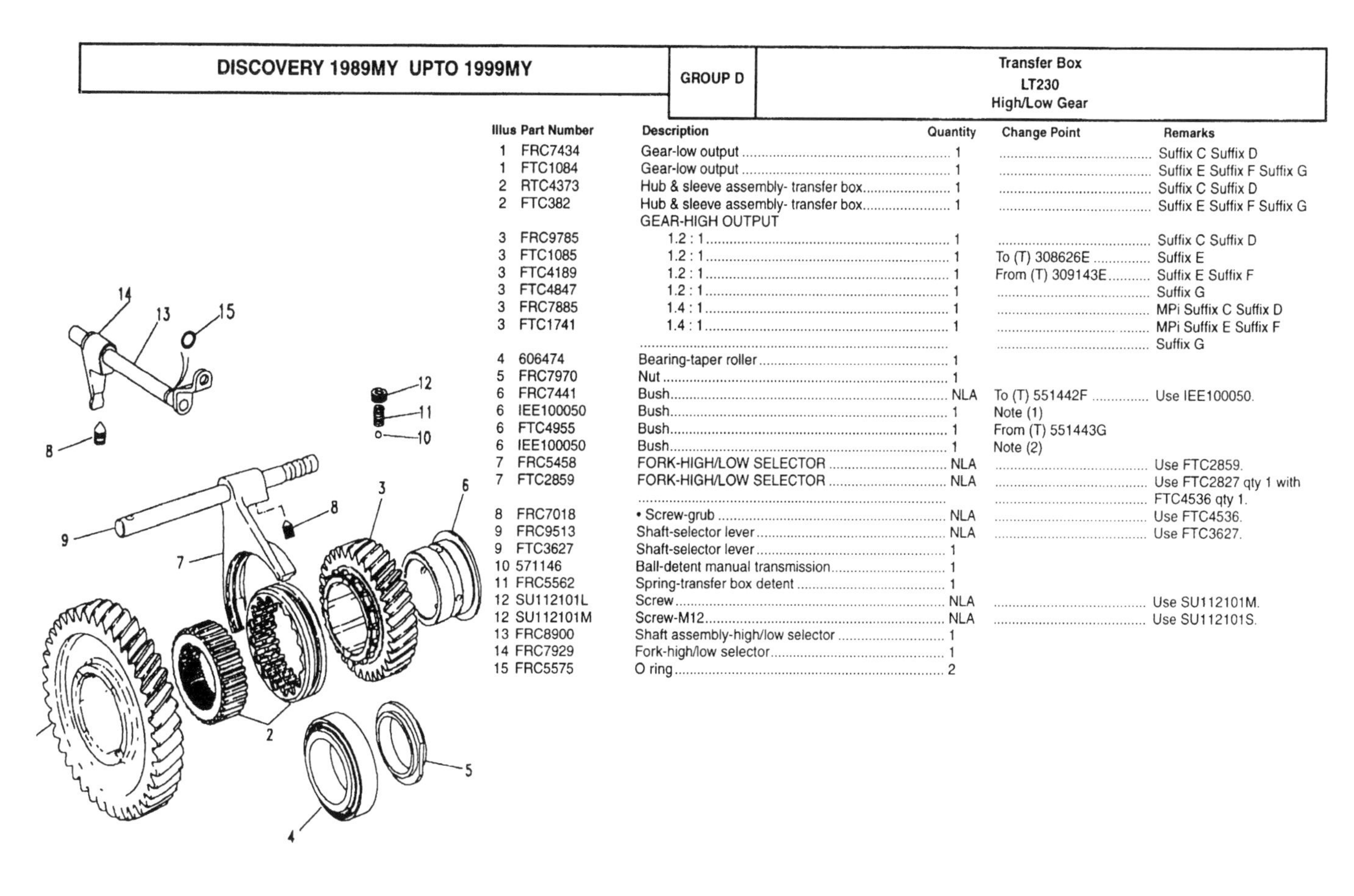

DISCOVERY 1989MY UPTO 1999MY

GROUP D

Transfer Box
LT230
High/Low Gear

Illus	Part Number	Description	Quantity	Change Point	Remarks
1	FRC7434	Gear-low output	1		Suffix C Suffix D
1	FTC1084	Gear-low output	1		Suffix E Suffix F Suffix G
2	RTC4373	Hub & sleeve assembly- transfer box	1		Suffix C Suffix D
2	FTC382	Hub & sleeve assembly- transfer box	1		Suffix E Suffix F Suffix G
		GEAR-HIGH OUTPUT			
3	FRC9785	1.2 : 1	1		Suffix C Suffix D
3	FTC1085	1.2 : 1	1	To (T) 308626E	Suffix E
3	FTC4189	1.2 : 1	1	From (T) 309143E	Suffix E Suffix F
3	FTC4847	1.2 : 1	1		Suffix G
3	FRC7885	1.4 : 1	1		MPi Suffix C Suffix D
3	FTC1741	1.4 : 1	1		MPi Suffix E Suffix F Suffix G
4	606474	Bearing-taper roller	1		
5	FRC7970	Nut	1		
6	FRC7441	Bush	NLA	To (T) 551442F	Use IEE100050.
6	IEE100050	Bush	1	Note (1)	
6	FTC4955	Bush	1	From (T) 551443G	
6	IEE100050	Bush	1	Note (2)	
7	FRC5458	FORK-HIGH/LOW SELECTOR	NLA		Use FTC2859.
7	FTC2859	FORK-HIGH/LOW SELECTOR	NLA		Use FTC2827 qty 1 with FTC4536 qty 1.
8	FRC7018	• Screw-grub	NLA		Use FTC4536.
9	FRC9513	Shaft-selector lever	NLA		Use FTC3627.
9	FTC3627	Shaft-selector lever	1		
10	571146	Ball-detent manual transmission	1		
11	FRC5562	Spring-transfer box detent	1		
12	SU112101L	Screw	NLA		Use SU112101M.
12	SU112101M	Screw-M12	NLA		Use SU112101S.
13	FRC8900	Shaft assembly-high/low selector	1		
14	FRC7929	Fork-high/low selector	1		
15	FRC5575	O ring	2		

CHANGE POINTS:
(1) To (T) 551442F
(2) From (T) 551443G

D 15

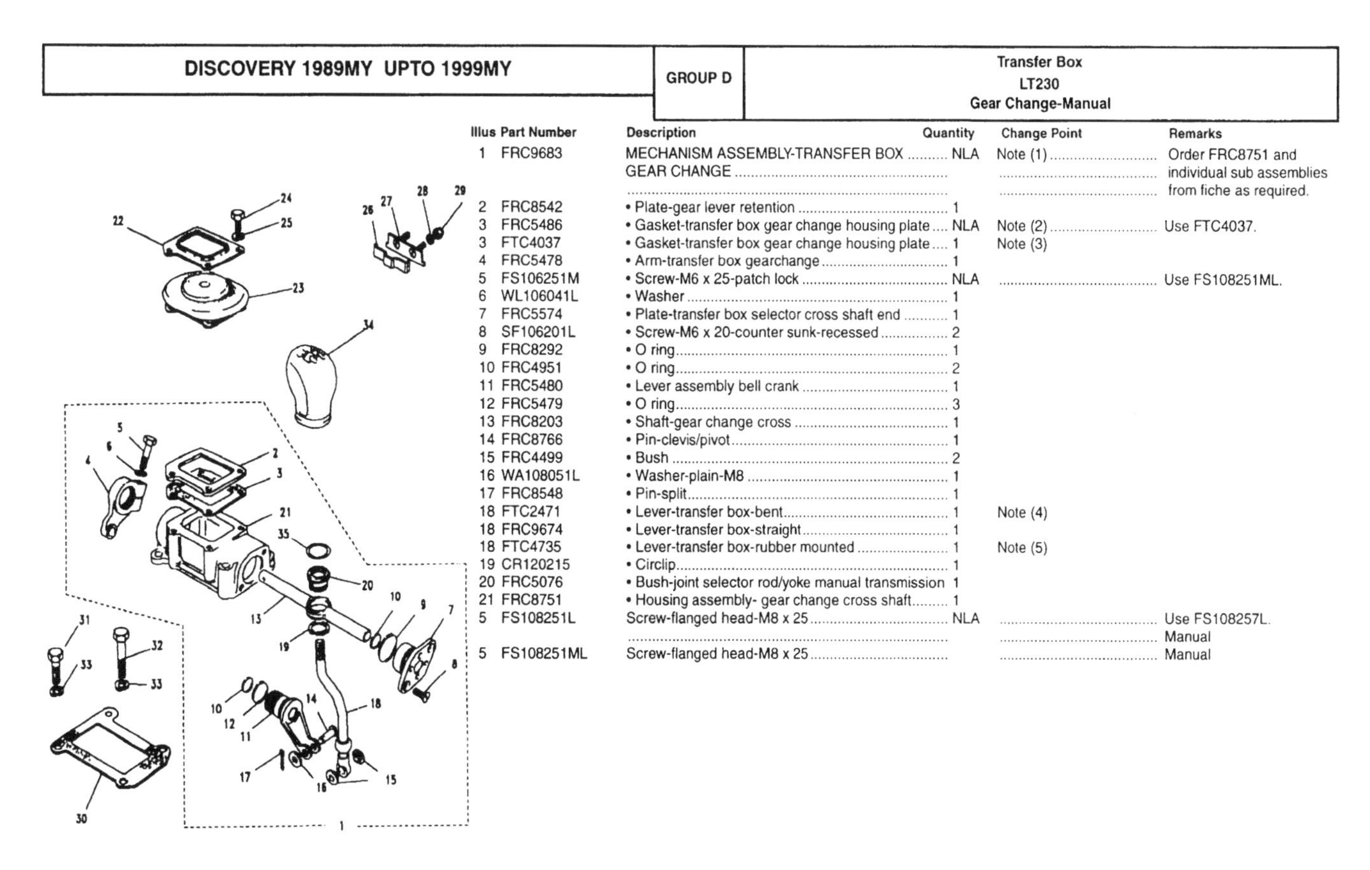

DISCOVERY 1989MY UPTO 1999MY

GROUP D

Transfer Box
LT230
Gear Change-Manual

Illus	Part Number	Description	Quantity	Change Point	Remarks
1	FRC9683	MECHANISM ASSEMBLY-TRANSFER BOX GEAR CHANGE	NLA	Note (1)	Order FRC8751 and individual sub assemblies from fiche as required.
2	FRC8542	• Plate-gear lever retention	1		
3	FRC5486	• Gasket-transfer box gear change housing plate	NLA	Note (2)	Use FTC4037.
3	FTC4037	• Gasket-transfer box gear change housing plate	1	Note (3)	
4	FRC5478	• Arm-transfer box gearchange	1		
5	FS106251M	• Screw-M6 x 25-patch lock	NLA		Use FS108251ML.
6	WL106041L	• Washer	1		
7	FRC5574	• Plate-transfer box selector cross shaft end	1		
8	SF106201L	• Screw-M6 x 20-counter sunk-recessed	2		
9	FRC8292	• O ring	1		
10	FRC4951	• O ring	2		
11	FRC5480	• Lever assembly bell crank	1		
12	FRC5479	• O ring	3		
13	FRC8203	• Shaft-gear change cross	1		
14	FRC8766	• Pin-clevis/pivot	1		
15	FRC4499	• Bush	2		
16	WA108051L	• Washer-plain-M8	1		
17	FRC8548	• Pin-split	1		
18	FTC2471	• Lever-transfer box-bent	1	Note (4)	
18	FRC9674	• Lever-transfer box-straight	1		
18	FTC4735	• Lever-transfer box-rubber mounted	1	Note (5)	
19	CR120215	• Circlip	1		
20	FRC5076	• Bush-joint selector rod/yoke manual transmission	1		
21	FRC8751	• Housing assembly- gear change cross shaft	1		
5	FS108251L	Screw-flanged head-M8 x 25	NLA		Use FS108257L. Manual
5	FS108251ML	Screw-flanged head-M8 x 25			Manual

CHANGE POINTS:
(1) From (V) JA 027886 To (V) MA 160601
(2) From (V) JA 027886 To (V) LA 081991
(3) From (V) MA 081992 To (V) MA 160601
(4) From (V) JA 027886 To (V) JA 027887
(5) From (V) MA 160602

D 16

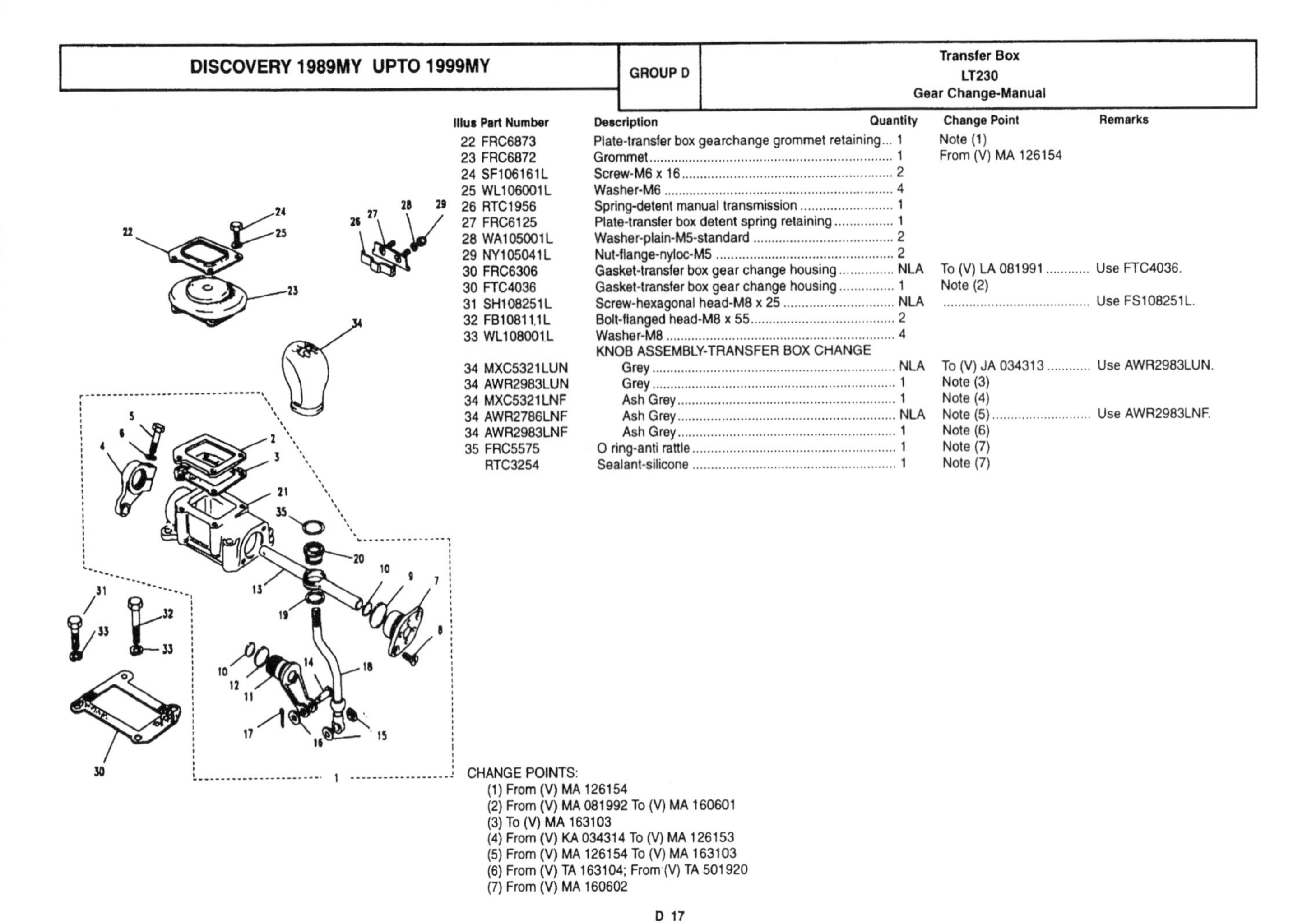

Illus	Part Number	Description	Quantity	Change Point	Remarks
22	FRC6873	Plate-transfer box gearchange grommet retaining	1	Note (1)	
23	FRC6872	Grommet	1	From (V) MA 126154	
24	SF106161L	Screw-M6 x 16	2		
25	WL106001L	Washer-M6	4		
26	RTC1956	Spring-detent manual transmission	1		
27	FRC6125	Plate-transfer box detent spring retaining	1		
28	WA105001L	Washer-plain-M5-standard	2		
29	NY105041L	Nut-flange-nyloc-M5	2		
30	FRC6306	Gasket-transfer box gear change housing	NLA	To (V) LA 081991	Use FTC4036.
30	FTC4036	Gasket-transfer box gear change housing	1	Note (2)	
31	SH108251L	Screw-hexagonal head-M8 x 25	NLA		Use FS108251L.
32	FB108111L	Bolt-flanged head-M8 x 55	2		
33	WL108001L	Washer-M8	4		
		KNOB ASSEMBLY-TRANSFER BOX CHANGE			
34	MXC5321LUN	Grey	NLA	To (V) JA 034313	Use AWR2983LUN.
34	AWR2983LUN	Grey	1	Note (3)	
34	MXC5321LNF	Ash Grey	1	Note (4)	
34	AWR2786LNF	Ash Grey	NLA	Note (5)	Use AWR2983LNF.
34	AWR2983LNF	Ash Grey	1	Note (6)	
35	FRC5575	O ring-anti rattle	1	Note (7)	
	RTC3254	Sealant-silicone	1	Note (7)	

CHANGE POINTS:

(1) From (V) MA 126154
(2) From (V) MA 081992 To (V) MA 160601
(3) To (V) MA 163103
(4) From (V) KA 034314 To (V) MA 126153
(5) From (V) MA 126154 To (V) MA 163103
(6) From (V) TA 163104; From (V) TA 501920
(7) From (V) MA 160602

D 17

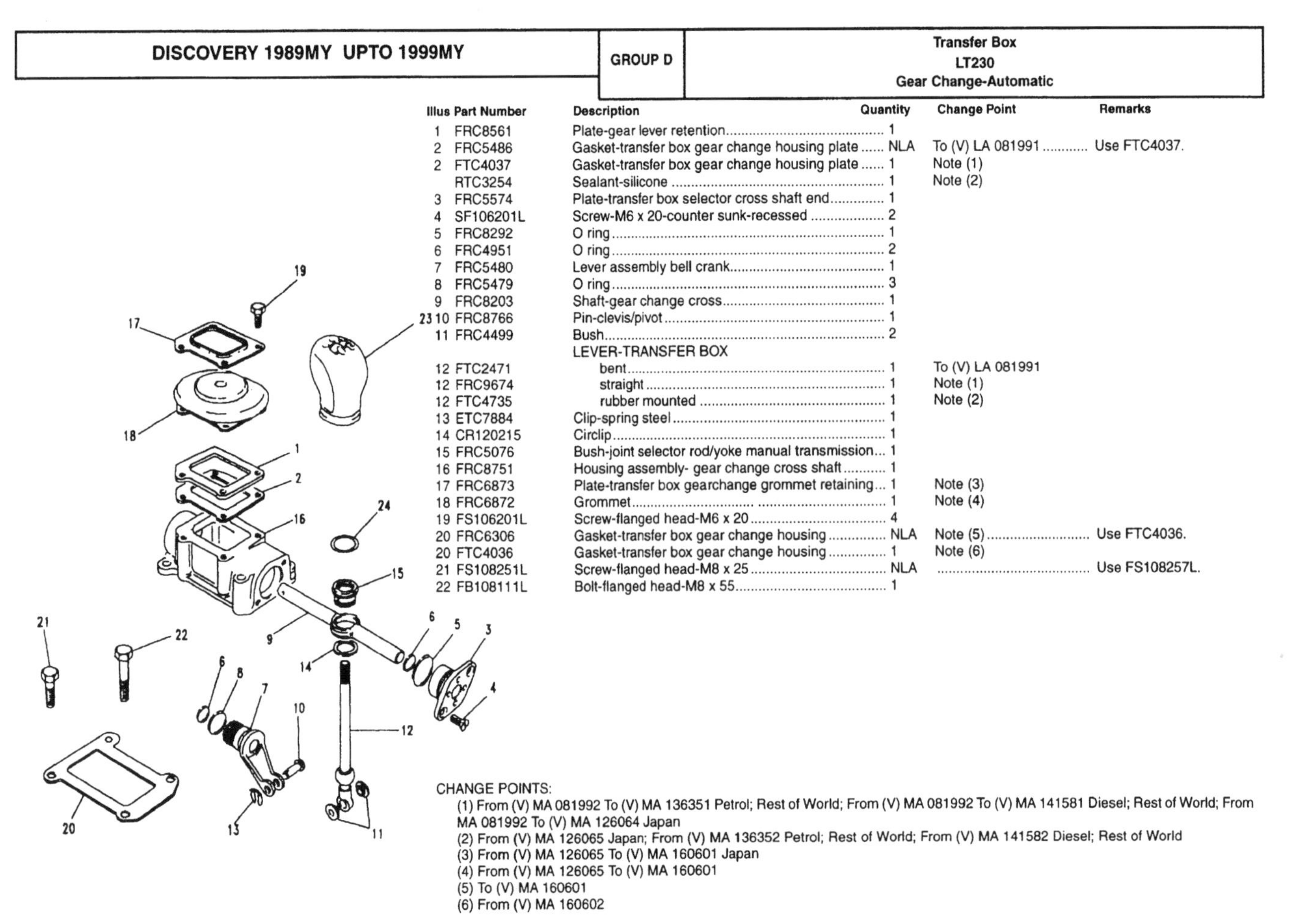

Illus	Part Number	Description	Quantity	Change Point	Remarks
1	FRC8561	Plate-gear lever retention	1		
2	FRC5486	Gasket-transfer box gear change housing plate	NLA	To (V) LA 081991	Use FTC4037.
2	FTC4037	Gasket-transfer box gear change housing plate	1	Note (1)	
	RTC3254	Sealant-silicone	1	Note (2)	
3	FRC5574	Plate-transfer box selector cross shaft end	1		
4	SF106201L	Screw-M6 x 20-counter sunk-recessed	2		
5	FRC8292	O ring	1		
6	FRC4951	O ring	2		
7	FRC5480	Lever assembly bell crank	1		
8	FRC5479	O ring	3		
9	FRC8203	Shaft-gear change cross	1		
10	FRC8766	Pin-clevis/pivot	1		
11	FRC4499	Bush	2		
		LEVER-TRANSFER BOX			
12	FTC2471	bent	1	To (V) LA 081991	
12	FRC9674	straight	1	Note (1)	
12	FTC4735	rubber mounted	1	Note (2)	
13	ETC7884	Clip-spring steel	1		
14	CR120215	Circlip	1		
15	FRC5076	Bush-joint selector rod/yoke manual transmission	1		
16	FRC8751	Housing assembly- gear change cross shaft	1		
17	FRC6873	Plate-transfer box gearchange grommet retaining	1	Note (3)	
18	FRC6872	Grommet	1	Note (4)	
19	FS106201L	Screw-flanged head-M6 x 20	4		
20	FRC6306	Gasket-transfer box gear change housing	NLA	Note (5)	Use FTC4036.
20	FTC4036	Gasket-transfer box gear change housing	1	Note (6)	
21	FS108251L	Screw-flanged head-M8 x 25	NLA		Use FS108257L.
22	FB108111L	Bolt-flanged head-M8 x 55	1		

CHANGE POINTS:

(1) From (V) MA 081992 To (V) MA 136351 Petrol; Rest of World; From (V) MA 081992 To (V) MA 141581 Diesel; Rest of World; From MA 081992 To (V) MA 126064 Japan
(2) From (V) MA 126065 Japan; From (V) MA 136352 Petrol; Rest of World; From (V) MA 141582 Diesel; Rest of World
(3) From (V) MA 126065 To (V) MA 160601 Japan
(4) From (V) MA 126065 To (V) MA 160601
(5) To (V) MA 160601
(6) From (V) MA 160602

D 18

Illus	Part Number	Description	Quantity	Change Point	Remarks
		KNOB ASSEMBLY-TRANSFER BOX CHANGE			
23	MXC5321LNF	Ash Grey	1	To (V) MA 126153	
23	AWR2786LNF	Ash Grey	NLA	Note (1)	Use AWR2983LNF.
23	AWR2983LNF	Ash Grey	1	Note (2)	Automatic
24	FRC5575	O ring	1	Note (3)	

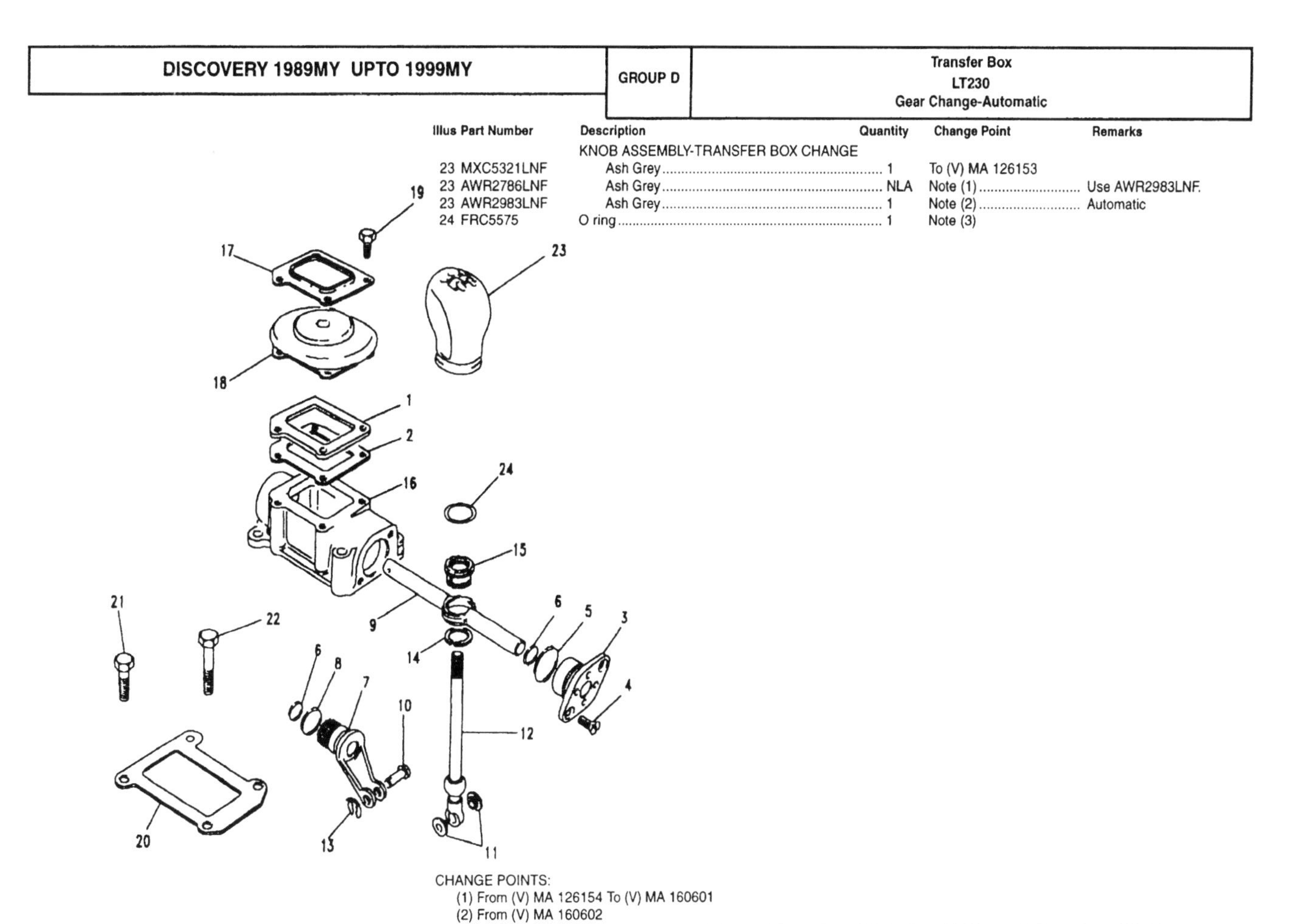

CHANGE POINTS:
(1) From (V) MA 126154 To (V) MA 160601
(2) From (V) MA 160602
(3) From (V) MA 126065 Japan; From (V) MA 136352 Petrol; Rest of World; From (V) MA 141582 Diesel; Rest of World

D 19

Illus	Part Number	Description	Quantity	Change Point	Remarks
1	FRC6000	Rod assembly-linkage connecting	1	Note (1)	
1	FTC3675	Rod assembly-linkage connecting	1	Note (2)	
2	FRC4499	Bush	4		
3	FRC8075	Arm-transfer box differential lock connecting rod adjusting	1		
4	FRC8767	Pin-clevis/pivot	2		
5	ANR1832	Clip-spring steel	2		
6	FRC8204	Arm-transfer box differential lock	1		Automatic
6	FRC8204	Arm-transfer box differential lock	1	To (V) LA 081991 Manual	Manual
7	FTC4095	Arm-transfer box differential lock	1	From (V) MA 081992 Manual	
8	FRC8202	Pin	1		
9	NH108041L	Nut-hexagonal head-M8	1		
10	ANR1832	Clip-spring steel	1		
11	FRC8547	Pin-clevis/pivot	1		
12	NY108041L	Nut-hexagonal head-nyloc-M8	1		
13	NT108041L	Nut-hexagonal head	1		
14	FRC8768	Pin-clevis/pivot	1		
15	FRC8769	Clip	1		
16	FB105051L	Bolt-M5	1	From (V) MA 081992	Manual
17	NY105041L	Nut-flange-nyloc-M5	1	Manual	
18	WA106041L	Washer-M6-standard.	1		

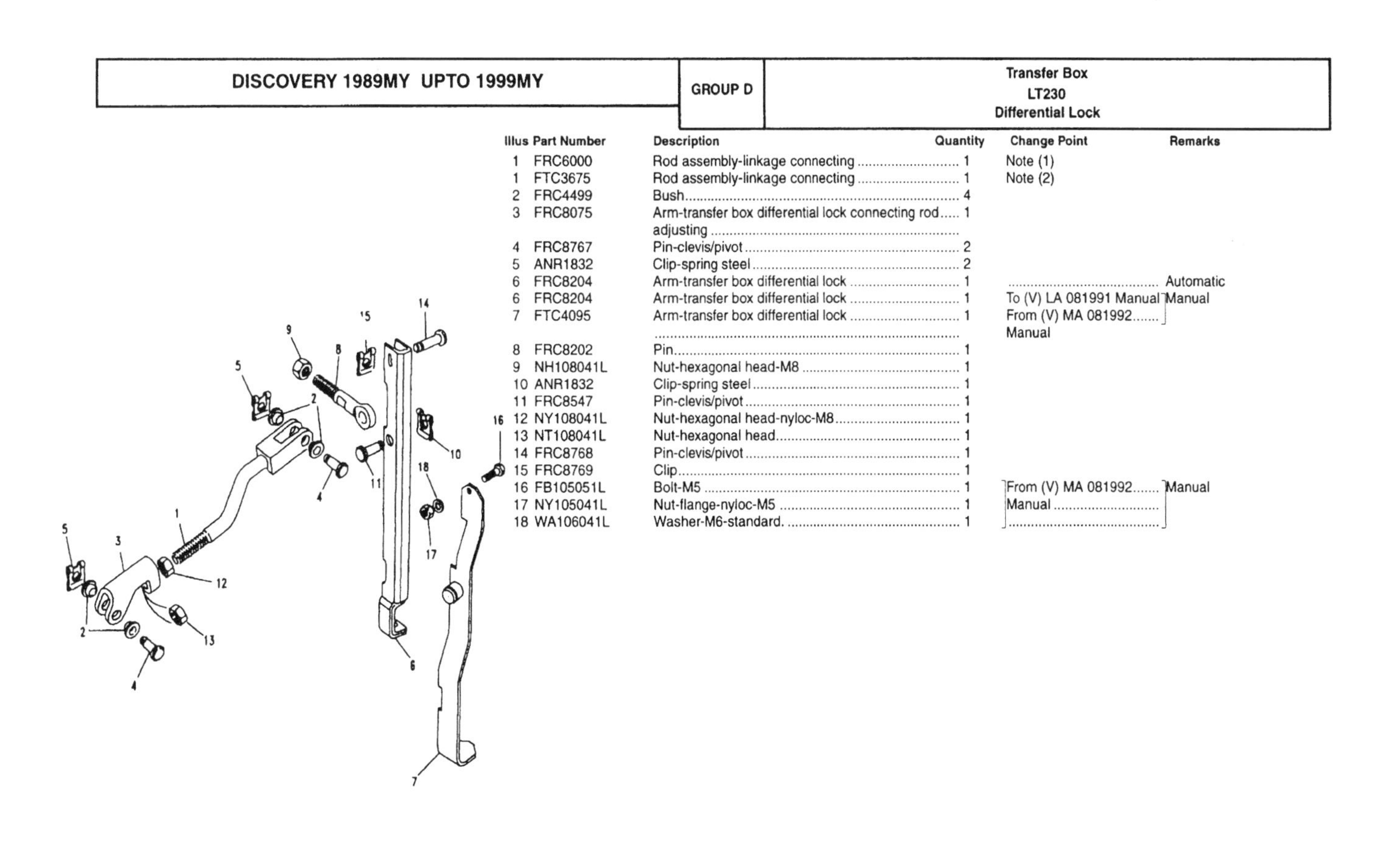

CHANGE POINTS:
(1) To (T)28D 274897E
(2) From (T)28D 274898E

D 20

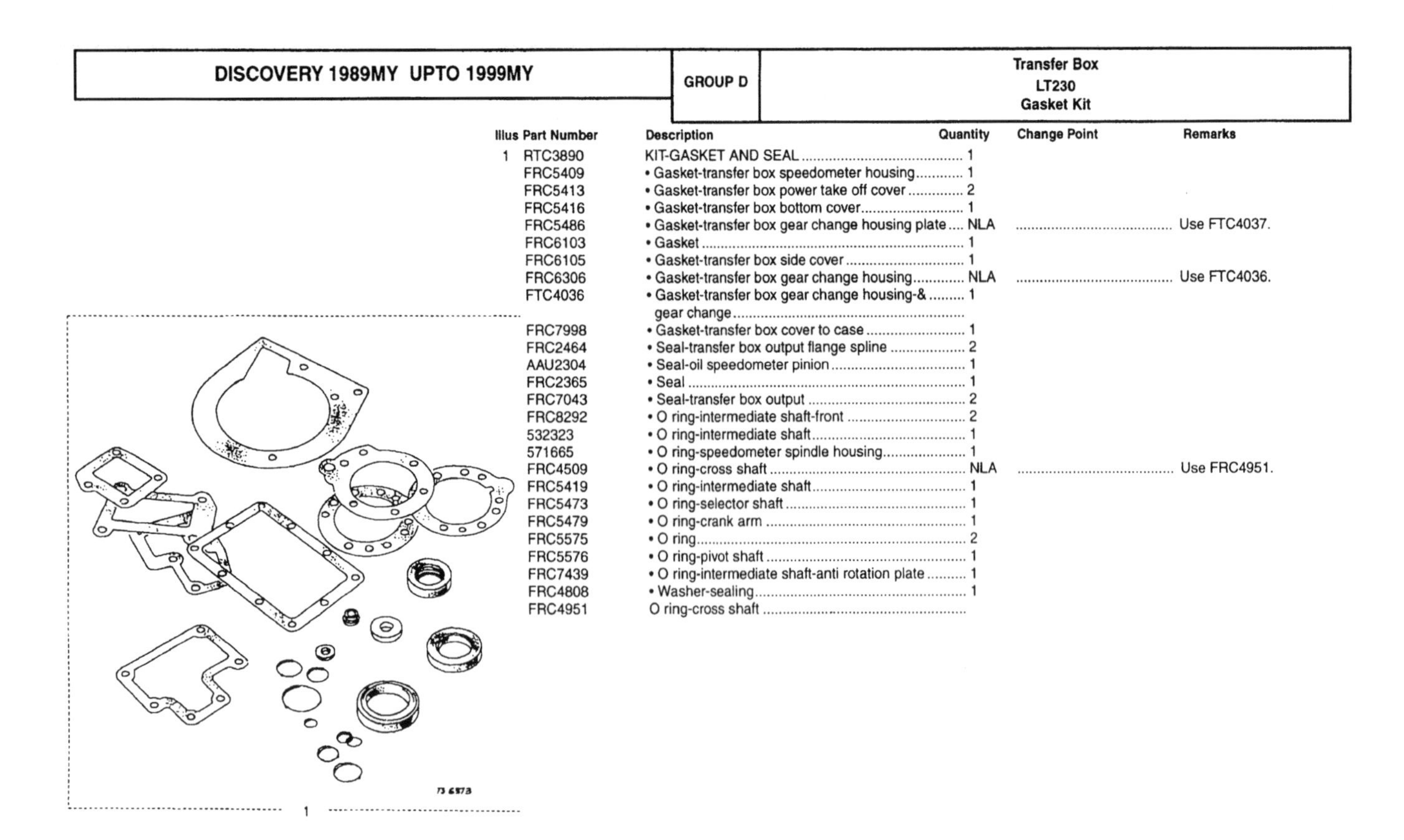

DISCOVERY 1989MY UPTO 1999MY

GROUP D — **Transfer Box LT230 Gasket Kit**

Illus	Part Number	Description	Quantity	Change Point	Remarks
1	RTC3890	KIT-GASKET AND SEAL	1		
	FRC5409	• Gasket-transfer box speedometer housing	1		
	FRC5413	• Gasket-transfer box power take off cover	2		
	FRC5416	• Gasket-transfer box bottom cover	1		
	FRC5486	• Gasket-transfer box gear change housing plate	NLA		Use FTC4037.
	FRC6103	• Gasket	1		
	FRC6105	• Gasket-transfer box side cover	1		
	FRC6306	• Gasket-transfer box gear change housing	NLA		Use FTC4036.
	FTC4036	• Gasket-transfer box gear change housing-& gear change	1		
	FRC7998	• Gasket-transfer box cover to case	1		
	FRC2464	• Seal-transfer box output flange spline	2		
	AAU2304	• Seal-oil speedometer pinion	1		
	FRC2365	• Seal	1		
	FRC7043	• Seal-transfer box output	2		
	FRC8292	• O ring-intermediate shaft-front	2		
	532323	• O ring-intermediate shaft	1		
	571665	• O ring-speedometer spindle housing	1		
	FRC4509	• O ring-cross shaft	NLA		Use FRC4951.
	FRC5419	• O ring-intermediate shaft	1		
	FRC5473	• O ring-selector shaft	1		
	FRC5479	• O ring-crank arm	1		
	FRC5575	• O ring	2		
	FRC5576	• O ring-pivot shaft	1		
	FRC7439	• O ring-intermediate shaft-anti rotation plate	1		
	FRC4808	• Washer-sealing	1		
	FRC4951	O ring-cross shaft			

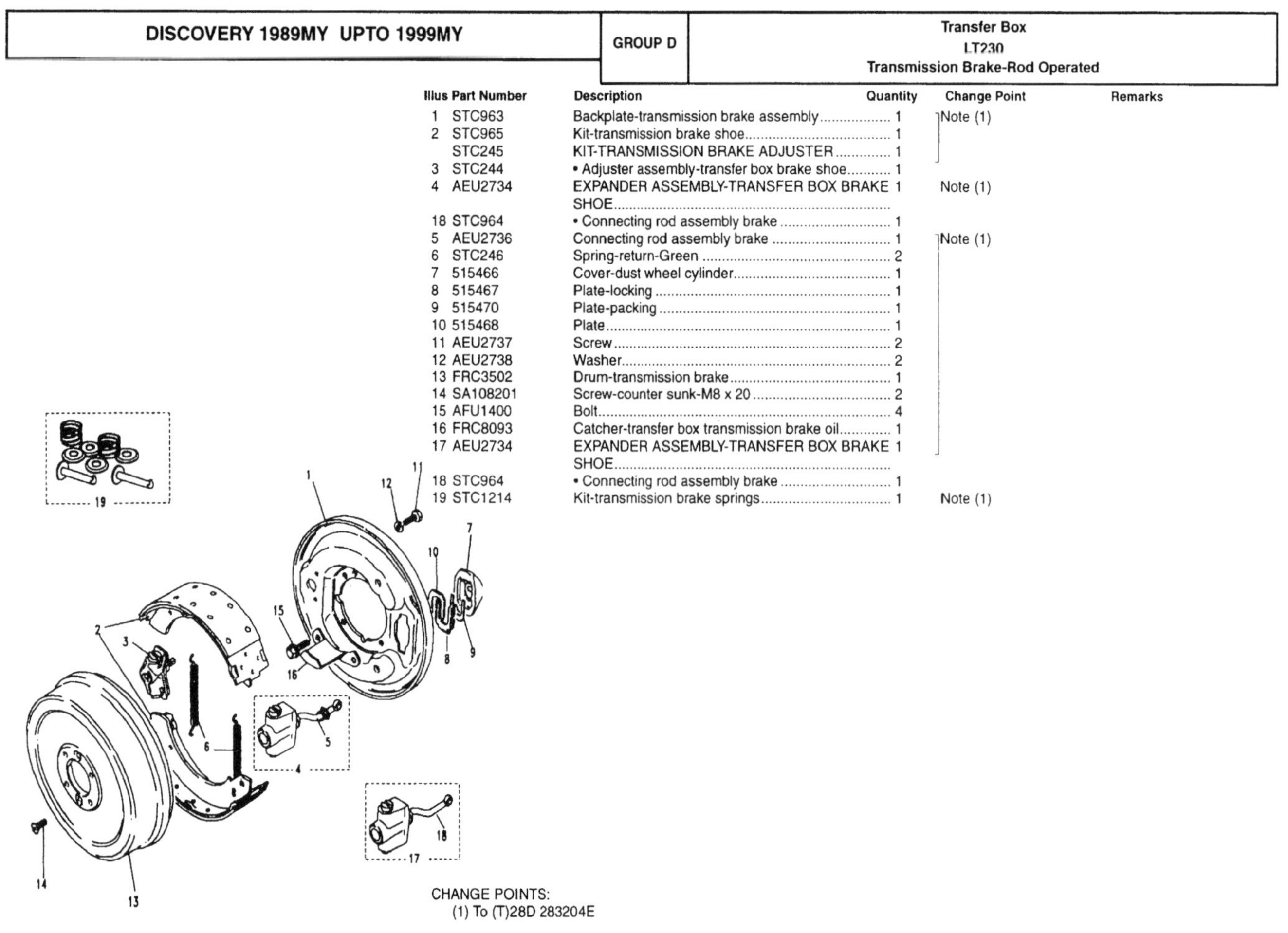

DISCOVERY 1989MY UPTO 1999MY

GROUP D — **Transfer Box LT230 Transmission Brake-Rod Operated**

Illus	Part Number	Description	Quantity	Change Point	Remarks
1	STC963	Backplate-transmission brake assembly	1	Note (1)	
2	STC965	Kit-transmission brake shoe	1		
	STC245	KIT-TRANSMISSION BRAKE ADJUSTER	1		
3	STC244	• Adjuster assembly-transfer box brake shoe	1		
4	AEU2734	EXPANDER ASSEMBLY-TRANSFER BOX BRAKE SHOE	1	Note (1)	
18	STC964	• Connecting rod assembly brake	1		
5	AEU2736	Connecting rod assembly brake	1	Note (1)	
6	STC246	Spring-return-Green	2		
7	515466	Cover-dust wheel cylinder	1		
8	515467	Plate-locking	1		
9	515470	Plate-packing	1		
10	515468	Plate	1		
11	AEU2737	Screw	2		
12	AEU2738	Washer	2		
13	FRC3502	Drum-transmission brake	1		
14	SA108201	Screw-counter sunk-M8 x 20	2		
15	AFU1400	Bolt	4		
16	FRC8093	Catcher-transfer box transmission brake oil	1		
17	AEU2734	EXPANDER ASSEMBLY-TRANSFER BOX BRAKE SHOE	1		
18	STC964	• Connecting rod assembly brake	1		
19	STC1214	Kit-transmission brake springs	1	Note (1)	

CHANGE POINTS:
(1) To (T)28D 283204E

Illus	Part Number	Description	Quantity	Change Point	Remarks
1	STC1533	Backplate-transmission brake assembly	1	Note (1)	
2	STC1526	Kit-transmission brake springs	1	Note (1)	
3	STC1527	Kit-transmission brake adjuster	1	Note (1)	
4	STC1532	KIT-TRANSMISSION BRAKE SHOE RETENTION	1	Note (1)	
	ICW100000	• Washer	1		
5	STC1525	Kit-transmission brake shoe	1	Note (1)	
6	STC1536	Shaft-cross-transmission brake	1	Note (1)	
7	STC1538	Lever-transmission brake	1	Note (1)	
8	AFU1400	Bolt	4	Note (1)	
9	FRC3502	Drum-transmission brake	1	Note (1)	
10	SA108201	Screw-counter sunk-M8 x 20	1	Note (1)	

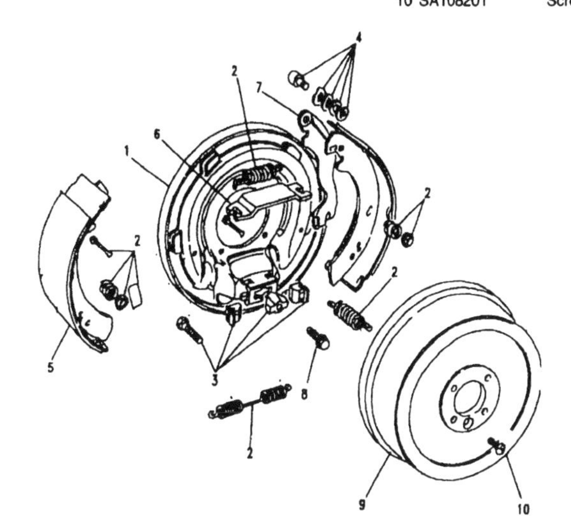

CHANGE POINTS:
(1) From (T)28D 283205E

D 23

Notes

Front Axle .... E 3
  Differential Front .... E 7
  Drive Shafts-ABS .... E 16
  Drive Shafts-Non ABS-From (V) JA032851 .... E 14
  Drive Shafts-Non ABS-To (V) JA032850 .... E 13
  Front Axle Assembly .... E 3
  Front Hubs-From (V) JA032851 .... E 24
  Front Hubs-To (V) JA032850 .... E 23
  Swivel Pin Housing-From (V) JA032850 .... E 20
  Swivel Pin Housing-To (V) JA032849 .... E 17
Front Suspension .... E 43
  Anti Roll Bar .... E 47
  Radius Arms & Links .... E 43
  Shock Absorbers & Springs .... E 45
Miscellaneous .... E 41
  Locking Wheel Nuts .... E 42
  Road Wheels .... E 41
Propshafts .... E 1
Rear Axle .... E 25
  Axle Assembly Rear-From (V) MA018992 .... E 26
  Axle Assembly Rear-To (V) LA081991 .... E 25
  Differential Rear .... E 28
  Hubs & Drive Shafts-From (V) JA032851 .... E 40
  Hubs & Drive Shafts-To (V) JA032850 .... E 39
Rear Suspension .... E 48
  Anti Roll Bar .... E 53
  Bottom Link .... E 48
  Shock Absorbers & Springs .... E 50
  Top Link, Fulcrum & Ball Joint .... E 49

Propshaft Front

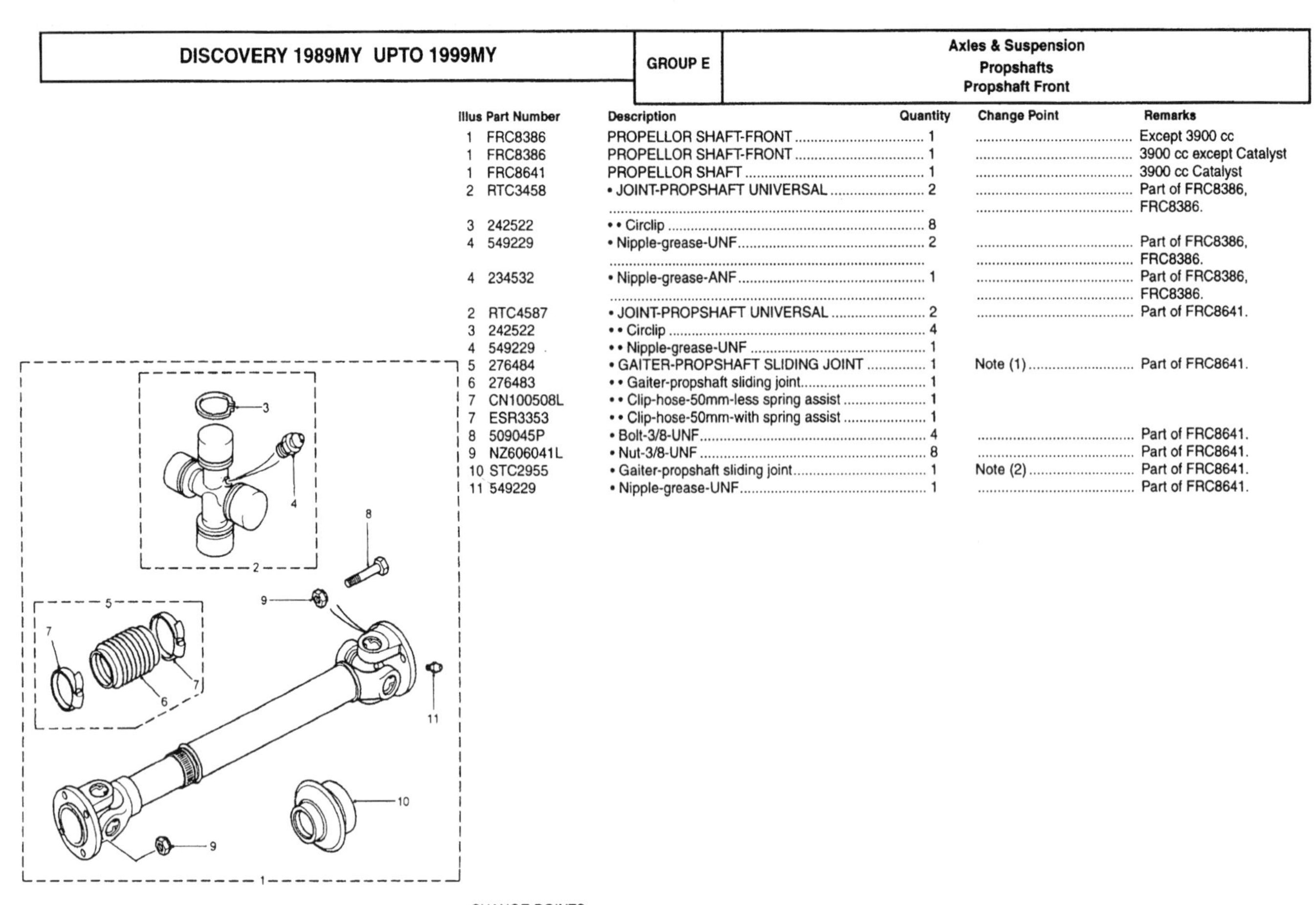

Illus	Part Number	Description	Quantity	Change Point	Remarks
1	FRC8386	PROPELLOR SHAFT-FRONT	1		Except 3900 cc
1	FRC8386	PROPELLOR SHAFT-FRONT	1		3900 cc except Catalyst
1	FRC8641	PROPELLOR SHAFT	1		3900 cc Catalyst
2	RTC3458	• JOINT-PROPSHAFT UNIVERSAL	2		Part of FRC8386, FRC8386.
3	242522	• • Circlip	8		
4	549229	• Nipple-grease-UNF	2		Part of FRC8386, FRC8386.
4	234532	• Nipple-grease-ANF	1		Part of FRC8386, FRC8386.
2	RTC4587	• JOINT-PROPSHAFT UNIVERSAL	2		Part of FRC8641.
3	242522	• • Circlip	4		
4	549229	• • Nipple-grease-UNF	1		
5	276484	• GAITER-PROPSHAFT SLIDING JOINT	1	Note (1)	Part of FRC8641.
6	276483	• • Gaiter-propshaft sliding joint	1		
7	CN100508L	• • Clip-hose-50mm-less spring assist	1		
7	ESR3353	• • Clip-hose-50mm-with spring assist	1		
8	509045P	• Bolt-3/8-UNF	4		Part of FRC8641.
9	NZ606041L	• Nut-3/8-UNF	8		Part of FRC8641.
10	STC2955	• Gaiter-propshaft sliding joint	1	Note (2)	Part of FRC8641.
11	549229	• Nipple-grease-UNF	1		Part of FRC8641.

CHANGE POINTS:
(1) To (V) October 1995
(2) From (V) October 1995

E 1

DISCOVERY 1989MY UPTO 1999MY | GROUP E | Axles & Suspension
Propshafts

Propshaft Rear

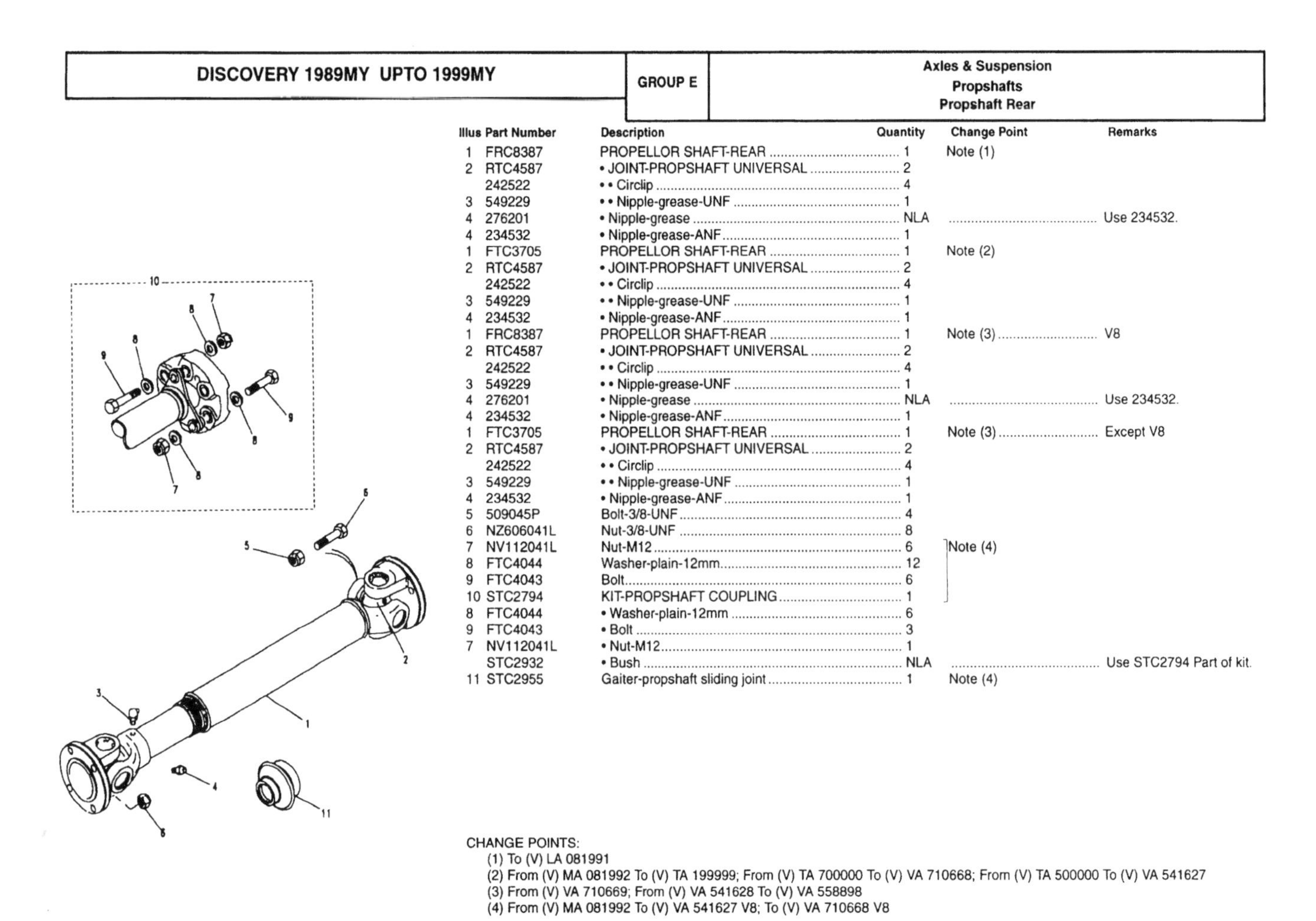

Illus	Part Number	Description	Quantity	Change Point	Remarks
1	FRC8387	PROPELLOR SHAFT-REAR	1	Note (1)	
2	RTC4587	• JOINT-PROPSHAFT UNIVERSAL	2		
	242522	• • Circlip	4		
3	549229	• • Nipple-grease-UNF	1		
4	276201	• Nipple-grease	NLA		Use 234532.
4	234532	• Nipple-grease-ANF	1		
1	FTC3705	PROPELLOR SHAFT-REAR	1	Note (2)	
2	RTC4587	• JOINT-PROPSHAFT UNIVERSAL	2		
	242522	• • Circlip	4		
3	549229	• • Nipple-grease-UNF	1		
4	234532	• Nipple-grease-ANF	1		
1	FRC8387	PROPELLOR SHAFT-REAR	1	Note (3)	V8
2	RTC4587	• JOINT-PROPSHAFT UNIVERSAL	2		
	242522	• • Circlip	4		
3	549229	• • Nipple-grease-UNF	1		
4	276201	• Nipple-grease	NLA		Use 234532.
4	234532	• Nipple-grease-ANF	1		
1	FTC3705	PROPELLOR SHAFT-REAR	1	Note (3)	Except V8
2	RTC4587	• JOINT-PROPSHAFT UNIVERSAL	2		
	242522	• • Circlip	4		
3	549229	• • Nipple-grease-UNF	1		
4	234532	• Nipple-grease-ANF	1		
5	509045P	Bolt-3/8-UNF	4		
6	NZ606041L	Nut-3/8-UNF	8		
7	NV112041L	Nut-M12	6	Note (4)	
8	FTC4044	Washer-plain-12mm	12		
9	FTC4043	Bolt	6		
10	STC2794	KIT-PROPSHAFT COUPLING	1		
8	FTC4044	• Washer-plain-12mm	6		
9	FTC4043	• Bolt	3		
7	NV112041L	• Nut-M12	1		
	STC2932	• Bush	NLA		Use STC2794 Part of kit.
11	STC2955	Gaiter-propshaft sliding joint	1	Note (4)	

CHANGE POINTS:
(1) To (V) LA 081991
(2) From (V) MA 081992 To (V) TA 199999; From (V) TA 700000 To (V) VA 710668; From (V) TA 500000 To (V) VA 541627
(3) From (V) VA 710669; From (V) VA 541628 To (V) VA 558898
(4) From (V) MA 081992 To (V) VA 541627 V8; To (V) VA 710668 V8

E 2

Illus	Part Number	Description	Quantity	Change Point	Remarks
		ASBESTOS-NON ABS-SOLID DISCS			
		V8			
		AXLE/DRIVE UNIT ASSEMBLY 4X4-FRONT			
1	FTC664	RHD-asbestos	NLA	Note (1)	Use axle FTC4410 and use old calipers with 4 x FTC3375 bolts. Securing holes in the calipers need to be enlarged to take the new 12mm bolts.
1	FTC3077	RHD-asbestos	1	Note (2)	
1	FTC665	LHD-asbestos	1	Note (3)	
1	FTC3076	LHD-asbestos	1	Note (4)	
		4 CYLINDER			
		AXLE/DRIVE UNIT ASSEMBLY 4X4-FRONT			
1	FTC790	RHD-asbestos	1	Note (5)	
1	FTC3078	RHD-asbestos	1	Note (6)	
1	FTC791	LHD-asbestos	1	Note (7)	
1	FTC3079	LHD-asbestos	NLA	Note (8)	Use FTC4403 and retain the calipers and pads from the replaced axle.
1	FTC4403	LHD-asbestos free	1	Note (9)	
		ASBESTOS FREE-NON ABS-VENTED DISCS			
		AXLE/DRIVE UNIT ASSEMBLY 4X4-FRONT			
1	FTC1464	RHD-asbestos free	1		V8
1	FTC1465	LHD-asbestos free	1		
1	FTC1466	RHD-asbestos free	1		4 Cylinder
1	FTC1467	LHD-asbestos free	1		

CHANGE POINTS:
(1) From (A)18L 00001C To (A)18L 12348C
(2) From (A)18L 12349C To (A)18L 99999C
(3) From (A)19L 00001C To (A)19L 13729C
(4) From (A)19L 13730C To (A)19L 99999C
(5) From (A)34L 00001C To (A)34L 23102C
(6) From (A)34L 23103C To (A)34L 99999C
(7) From (A)35L 00001C To (A)35L 22150C
(8) From (A)35L 22151C To (A)35L 99999C
(9) From (A)72L 04675A To (A)72L 99999A

Illus	Part Number	Description	Quantity	Change Point	Remarks
		ASBESTOS FREE-NON ABS-SOLID			
		V8			
		AXLE/DRIVE UNIT ASSEMBLY 4X4-FRONT			
1	FTC3237	RHD-asbestos free	1	Note (1)	
1	FTC3838	RHD-asbestos free	NLA	Note (2)	Use FTC4410.
1	FTC4410	RHD-asbestos free	1	Note (3)	
1	FTC3238	LHD-asbestos free	1	Note (4)	
1	FTC3839	LHD-asbestos free	1	Note (5)	
1	FTC4411	LHD-asbestos free	1	Note (6)	
		4 CYLINDER			
		AXLE/DRIVE UNIT ASSEMBLY 4X4-FRONT			
1	FTC3235	RHD-asbestos free	1	Note (7)	
1	FTC3840	RHD-asbestos free	NLA	Note (8)	Use FTC4402.
1	FTC4402	RHD-asbestos free	1	Note (9)	
1	FTC3236	LHD-asbestos free	1	Note (10)	
1	FTC3841	LHD-asbestos free	NLA	Note (11)	Use FTC4403.
1	FTC4403	LHD-asbestos free	1	Note (12)	

CHANGE POINTS:
(1) From (A)38L 02987C To (A)38L 99999C
(2) From (A)73L 00001A To (A)73L 01431A
(3) From (A)73L 01432A To (A)73L 99999A
(4) From (A)39L 05098C To (A)39L 99999C
(5) From (A)74L 00001A To (A)74L 00535A
(6) From (A)74L 00563A To (A)74L 99999A
(7) From (A)42L 00003A To (A)42L 99999A
(8) From (A)71L 00001A To (A)71L 06549A
(9) From (A)71L 06550A To (A)71L 99999A
(10) From (A)43L 16684A To (A)43L 99999A
(11) From (A)72L 00001A To (A)72L 04674A
(12) From (A)72L 04675A To (A)72L 99999A

DISCOVERY 1989MY UPTO 1999MY

GROUP E

Axles & Suspension
Front Axle
Front Axle Assembly

Illus	Part Number	Description	Quantity	Change Point	Remarks
		ASBESTOS FREE-ANTI LOCK BRAKES-SOLID			
		V8			
		AXLE/DRIVE UNIT ASSEMBLY 4X4-FRONT			
1	FTC3404	RHD-asbestos free	1	Note (1)	
1	FTC4408	RHD-asbestos free	1	Note (2)	
1	FTC3405	LHD-asbestos free	NLA	Note (3)	Use FTC4409.
1	FTC4409	LHD-asbestos free	1	Note (4)	
		4 CYLINDER			
		AXLE/DRIVE UNIT ASSEMBLY 4X4-FRONT			
1	FTC3406	RHD-asbestos free	1	Note (5)	
1	FTC4400	RHD-asbestos free	1	Note (6)	
1	FTC3407	LHD-asbestos free	1	Note (7)	
1	FTC4401	LHD-asbestos free	1	Note (8)	

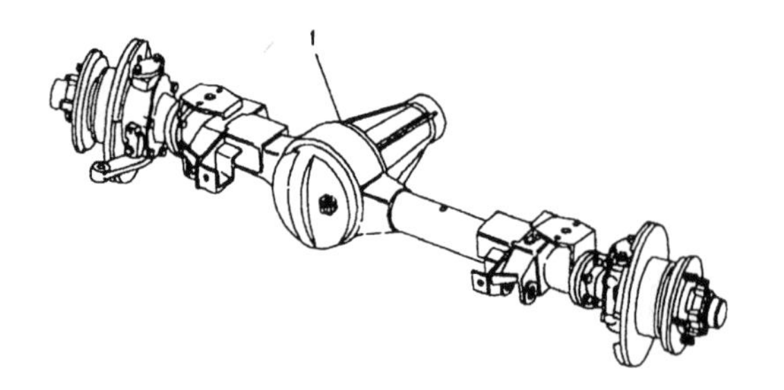

CHANGE POINTS:
(1) From (A)77L 00001A To (A)77L 00873A
(2) From (A)77L 00874A To (A)77L 99999A
(3) From (A)78L 00001A To (A)78L 03270A
(4) From (A)78L 03271A To (A)78L 99999A
(5) From (A)75L 00001A To (A)75L 01093A
(6) From (A)75L 01094A To (A)75L 99999A
(7) From (A)76L 00001A To (A)76L 00343A
(8) From (A)76L 00344A To (A)76L 99999A

E 5

DISCOVERY 1989MY UPTO 1999MY

GROUP E

Axles & Suspension
Front Axle
Case Assembly Front

Illus	Part Number	Description	Quantity	Change Point	Remarks
		EXCEPT ANTI-ROLL BAR			
1	FTC823	CASE-FRONT AXLE	NLA	To (V) August 1994	Use FTC4417.
				approx	V8
1	FTC825	CASE-FRONT AXLE	NLA		Use FTC2223.
					Diesel
1	FTC2223	CASE-FRONT AXLE	NLA		Use FTC4413.
					MPi
		WITH ANTI-ROLL BAR			
1	FTC2509	CASE-FRONT AXLE	NLA	To (V) August 1994	Use FTC4417.
				approx	V8
1	FTC2223	CASE-FRONT AXLE	NLA		Use FTC4413.
					Diesel
1	FTC2223	CASE-FRONT AXLE	NLA		Use FTC4413.
					MPi
		EXCEPT ANTI-ROLL BAR			
1	FTC4417	CASE-FRONT AXLE	1	From (V) August 1994	V8
1	FTC4413	CASE-FRONT AXLE	1	approx	Diesel
1	FTC4413	CASE-FRONT AXLE	1		MPi
		WITH ANTI-ROLL BAR			
1	FTC4417	CASE-FRONT AXLE	1	From (V) August 1994	V8
1	FTC4413	CASE-FRONT AXLE	1	approx	Diesel
1	FTC4413	CASE-FRONT AXLE	1		MPi
2	561196	• Stud-3/8UNF x 1 3/4	6		
2	561196	• Stud-3/8UNF x 1 3/4	4		
3	561195	• Stud-3/8UNF x 1 1/4	4		
4	608246	• Plug-drain/filler level	2		
5	FTC2175	PIPE-FRONT AXLE BREATHER	1		
6	595478	• Bolt-banjo	1		
7	232039	• Washer-sealing-axle drain plug	2		
8	NRC9246	Clip-pipe	3		
9	568680	Tie-cable-4.8 x 270mm-inside serated	3		
10	79123	Clip-pipe-single	1		

CFEXCC1A

E 6

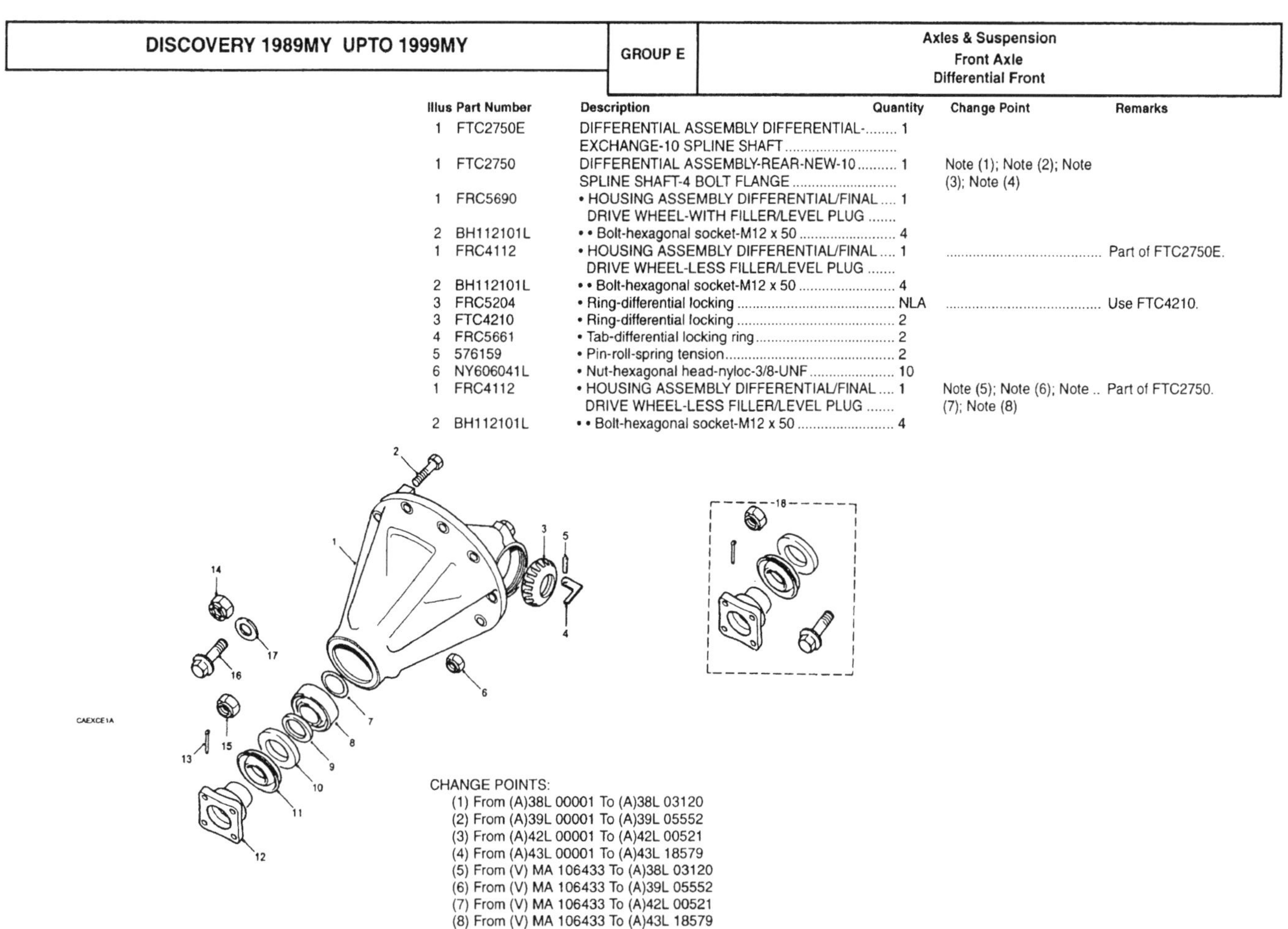

Illus	Part Number	Description	Quantity	Change Point	Remarks
1	FTC2750E	DIFFERENTIAL ASSEMBLY DIFFERENTIAL-EXCHANGE-10 SPLINE SHAFT	1		
1	FTC2750	DIFFERENTIAL ASSEMBLY-REAR-NEW-10 SPLINE SHAFT-4 BOLT FLANGE	1	Note (1); Note (2); Note (3); Note (4)	
1	FRC5690	• HOUSING ASSEMBLY DIFFERENTIAL/FINAL DRIVE WHEEL-WITH FILLER/LEVEL PLUG	1		
2	BH112101L	• • Bolt-hexagonal socket-M12 x 50	4		
1	FRC4112	• HOUSING ASSEMBLY DIFFERENTIAL/FINAL DRIVE WHEEL-LESS FILLER/LEVEL PLUG	1		Part of FTC2750E.
2	BH112101L	• • Bolt-hexagonal socket-M12 x 50	4		
3	FRC5204	• Ring-differential locking	NLA		Use FTC4210.
3	FTC4210	• Ring-differential locking	2		
4	FRC5661	• Tab-differential locking ring	2		
5	576159	• Pin-roll-spring tension	2		
6	NY606041L	• Nut-hexagonal head-nyloc-3/8-UNF	10		
1	FRC4112	• HOUSING ASSEMBLY DIFFERENTIAL/FINAL DRIVE WHEEL-LESS FILLER/LEVEL PLUG	1	Note (5); Note (6); Note (7); Note (8)	Part of FTC2750.
2	BH112101L	• • Bolt-hexagonal socket-M12 x 50	4		

CHANGE POINTS:
(1) From (A)38L 00001 To (A)38L 03120
(2) From (A)39L 00001 To (A)39L 05552
(3) From (A)42L 00001 To (A)42L 00521
(4) From (A)43L 00001 To (A)43L 18579
(5) From (V) MA 106433 To (A)38L 03120
(6) From (V) MA 106433 To (A)39L 05552
(7) From (V) MA 106433 To (A)42L 00521
(8) From (V) MA 106433 To (A)43L 18579

E 7

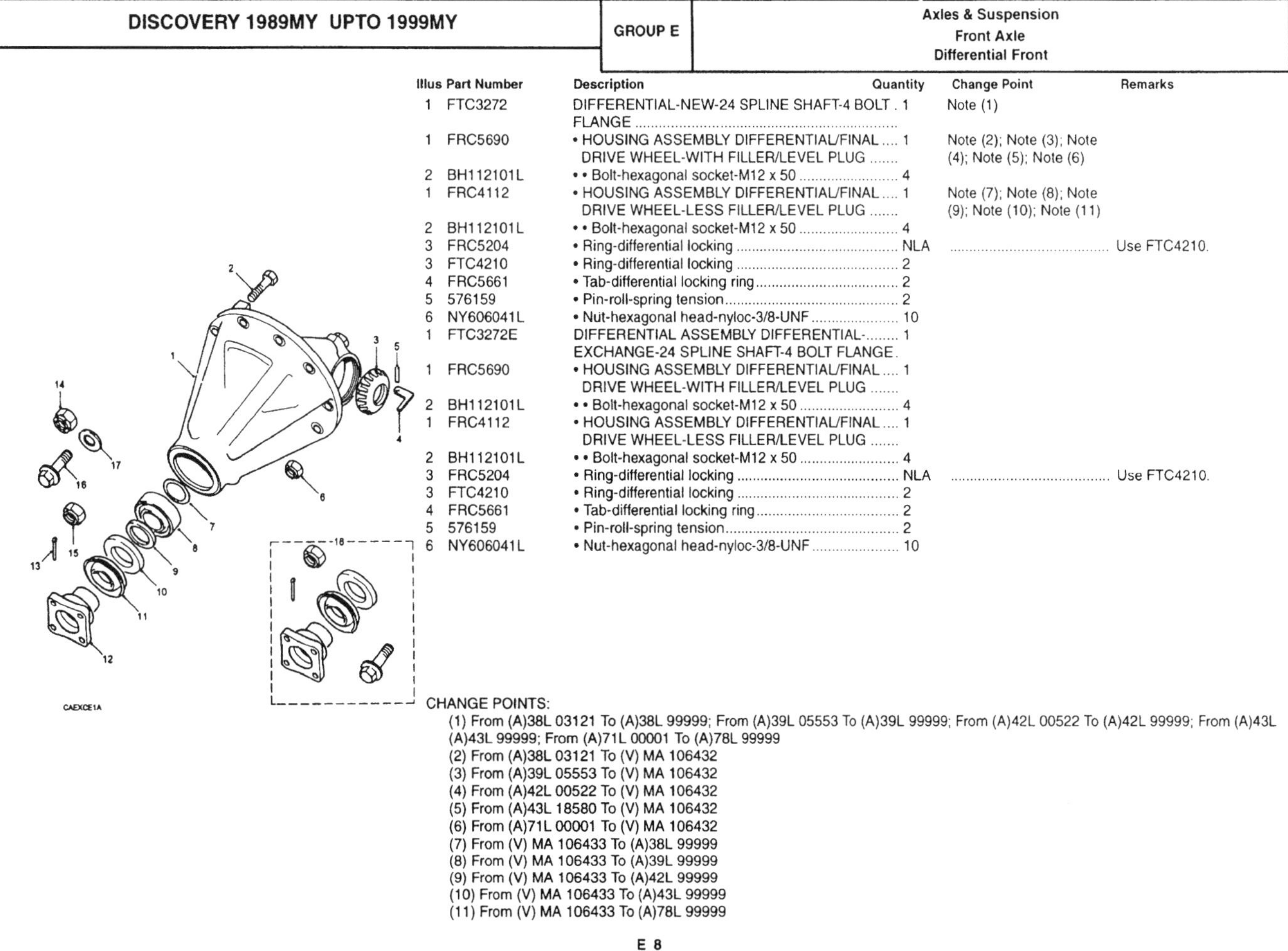

Illus	Part Number	Description	Quantity	Change Point	Remarks
1	FTC3272	DIFFERENTIAL-NEW-24 SPLINE SHAFT-4 BOLT FLANGE	1	Note (1)	
1	FRC5690	• HOUSING ASSEMBLY DIFFERENTIAL/FINAL DRIVE WHEEL-WITH FILLER/LEVEL PLUG	1	Note (2); Note (3); Note (4); Note (5); Note (6)	
2	BH112101L	• • Bolt-hexagonal socket-M12 x 50	4		
1	FRC4112	• HOUSING ASSEMBLY DIFFERENTIAL/FINAL DRIVE WHEEL-LESS FILLER/LEVEL PLUG	1	Note (7); Note (8); Note (9); Note (10); Note (11)	
2	BH112101L	• • Bolt-hexagonal socket-M12 x 50	4		
3	FRC5204	• Ring-differential locking	NLA		Use FTC4210.
3	FTC4210	• Ring-differential locking	2		
4	FRC5661	• Tab-differential locking ring	2		
5	576159	• Pin-roll-spring tension	2		
6	NY606041L	• Nut-hexagonal head-nyloc-3/8-UNF	10		
1	FTC3272E	DIFFERENTIAL ASSEMBLY DIFFERENTIAL-EXCHANGE-24 SPLINE SHAFT-4 BOLT FLANGE	1		
1	FRC5690	• HOUSING ASSEMBLY DIFFERENTIAL/FINAL DRIVE WHEEL-WITH FILLER/LEVEL PLUG	1		
2	BH112101L	• • Bolt-hexagonal socket-M12 x 50	4		
1	FRC4112	• HOUSING ASSEMBLY DIFFERENTIAL/FINAL DRIVE WHEEL-LESS FILLER/LEVEL PLUG	1		
2	BH112101L	• • Bolt-hexagonal socket-M12 x 50	4		
3	FRC5204	• Ring-differential locking	NLA		Use FTC4210.
3	FTC4210	• Ring-differential locking	2		
4	FRC5661	• Tab-differential locking ring	2		
5	576159	• Pin-roll-spring tension	2		
6	NY606041L	• Nut-hexagonal head-nyloc-3/8-UNF	10		

CHANGE POINTS:
(1) From (A)38L 03121 To (A)38L 99999; From (A)39L 05553 To (A)39L 99999; From (A)42L 00522 To (A)42L 99999; From (A)43L (A)43L 99999; From (A)71L 00001 To (A)78L 99999
(2) From (A)38L 03121 To (V) MA 106432
(3) From (A)39L 05553 To (V) MA 106432
(4) From (A)42L 00522 To (V) MA 106432
(5) From (A)43L 18580 To (V) MA 106432
(6) From (A)71L 00001 To (V) MA 106432
(7) From (V) MA 106433 To (A)38L 99999
(8) From (V) MA 106433 To (A)39L 99999
(9) From (V) MA 106433 To (A)42L 99999
(10) From (V) MA 106433 To (A)43L 99999
(11) From (V) MA 106433 To (A)78L 99999

E 8

Illus	Part Number	Description	Quantity	Change Point	Remarks
		SHIM			
7	FTC3869	2.155mm	A/R		
7	FTC3871	2.105mm	A/R		
7	FTC3873	2.055mm	A/R		
7	FTC3875	2.005mm	A/R		
7	FTC3877	1.955mm	A/R		
7	FTC3879	1.905mm	A/R		
7	FTC3881	1.855mm	A/R		
7	FTC3883	1.805mm	A/R		
7	FTC3885	1.755mm	A/R		
7	FTC3887	1.705mm	A/R		
7	FTC3889	1.655mm	A/R		
7	FTC3891	1.605mm	A/R		
7	FTC3893	1.555mm	A/R		
7	FTC3895	1.505mm	A/R		

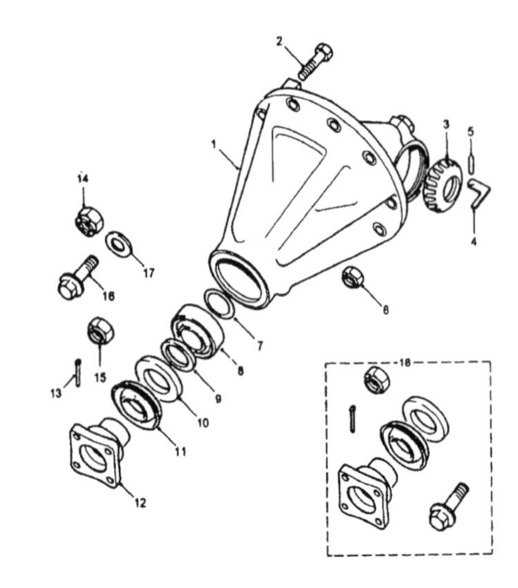

CAEXCE1A

DISCOVERY 1989MY UPTO 1999MY | GROUP E | Axles & Suspension
Front Axle
Differential Front

Illus	Part Number	Description	Quantity	Change Point	Remarks
8	539707	Bearing-taper roller	1		
9	539745	Spacer-pinion bearing	1		
10	FRC8220	Seal-differential final drive pinion	1	Note (1)	
10	FTC5258	Seal-differential final drive pinion	1	Note (2)	
11	FRC8154	Mudshield-front & rear differential	1	Note (1)	
11	FTC5317	Mudshield-front & rear differential	1	Note (2)	
12	FRC3002	Flange-driveshaft coupling differential-4 bolt flange	1	Note (1)	
13	PS608101L	Pin-split	1	Note (3)	Note (1)
14	3259	Nut-castle	1		Note (1)
15	NY116041L	Nut-hexagonal head-nyloc-M16	1	Note (4)	Note (2)
16	FS112301P	Screw-M12 x 30	1	Note (5)	Note (3)
17	90513454	Washer-plain	1	Note (4)	Nut securing flange
17	FTC5413	Washer-plain	1	Note (6)	Bolt securing flange
18	STC3722	KIT-FLANGE	1		
10	FTC5258	• Seal-stub axle	1		
12	FTC5322	• FLANGE-DRIVESHAFT COUPLING DIFFERENTIAL	1		
11	FTC5317	• • Mudshield-front & rear differential	1		
15	NY116041L	• Nut-hexagonal head-nyloc-M16	1		
13	PS608101L	• Pin-split	1		
16	FS112301P	• Screw-M12 x 30	1		

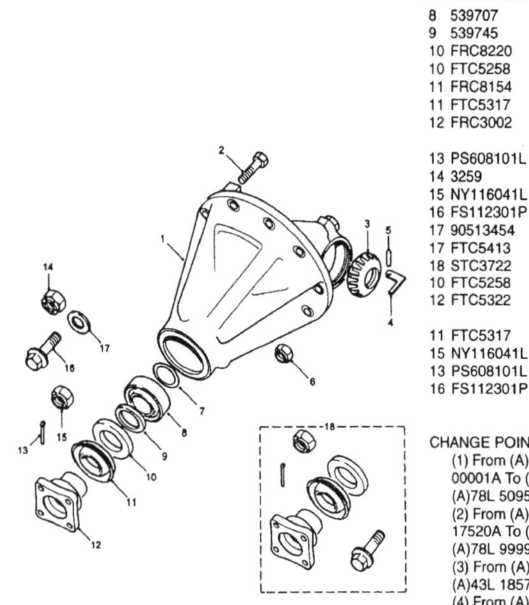

CAEXCE1A

CHANGE POINTS:

(1) From (A)71L 00001A To (A)71L 49101A; From (A)72L 00001A To (A)72L 34502A; From (A)74L 00001A To (A)74L 05336A; From 00001A To (A)75L 17519A; From (A)76L 00001A To (A)76L 05465A; From (A)77L 00001A To (A)77L 11316A; From (A)78L 00001A (A)78L 50959A

(2) From (A)71L 49102A To (A)71L 99999A; From (A)72L 34503A To (A)72L 99999A; From (A)74L 05337A To (A)74L 99999A; From 17520A To (A)75L 99999A; From (A)76L 05466A To (A)76L 99999A; From (A)77L 11317A To (A)77L 99999A; From (A)78L 50960A (A)78L 99999A

(3) From (A)38L 00001 To (A)38L 03120; From (A)39L 00001 To (A)39L 05552; From (A)42L 00001 To (A)42L 00521; From (A)43L (A)43L 18579

(4) From (A)71L 00001A To (A)71L 55169A; From (A)72L 00001A To (A)72L 40518A; From (A)73L 00001A To (A)73L 10454A; From 00001A To (A)74L 06441A; From (A)75L 00001A To (A)75L 20434A; From (A)76L 00001A To (A)76L 06647A; From (A)77L 00001A (A)77L 14993A; From (A)78L 00001A To (A)78L 61051A

(5) From (A)71L 55170A To (A)71L 99999A; From (A)72L 40519A To (A)72L 99999A; From (A)73L 10455A To (A)73L 99999A; From 06442A To (A)74L 99999A; From (A)75L 20435A To (A)75L 99999A; From (A)76L 06648A To (A)76L 99999A; From (A)77L 14994A (A)77L 99999A; From (A)78L 61052A To (A)78L 99999A

(6) From (A)72L 40519A To (A)72L 99999A; From (A)73L 10455A To (A)73L 99999A; From (A)74L 06442A To (A)74L 99999A; From 20435A To (A)75L 99999A; From (A)76L 06648A To (A)76L 99999A; From (A)77L 14994A To (A)77L 99999A; From (A)78L 61052A (A)78L 99999A; From (A)71L 55170A To (A)71L 99999A

Remarks:

(1) Use on differential assembly FTC2750 nut securing flange
(2) Use on differential assembly FTC3272 nut securing flange
(3) Use on differential assembly FTC3272 bolt securing flange

Illus	Part Number	Description	Quantity	Change Point	Remarks
1	FRC2933	Case differential	1	Note (1)	
1	FTC3269	Case differential	1	Note (2)	
2	STC851	KIT-DIFFERENTIAL GEARS-10 SPLINE SHAFT	NLA	Note (1)	Use STC1768.
3	599945	• Cross shaft differential	1		
4	CCN110L	• Circlip	2		
2	STC1768	KIT-DIFFERENTIAL GEARS-10 SPLINE SHAFT	1		
3	599945	• Cross shaft differential	1		
4	CCN110L	• Circlip	2		
2	FTC3267	Kit-differential gears	NLA	Note (5)	Use STC1846.
2	FTC3268	Kit-differential gears	NLA	Note (2)	
3	599945	Cross shaft differential	1		
4	CCN110L	Circlip	2		
5	FTC781	Crownwheel and pinion assembly	NLA	Note (1)	Use FTC3620 qty 1 with NY116041L qty 1.
5	FTC3620	Crownwheel and pinion assembly	1	Note (3)	Nut securing flange
5	TBH100040	Crownwheel and pinion assembly	1	Note (4)	Bolt securing flange
6	593692	Bolt	NLA	Note (1)	Use FTC3586.
6	FTC3586	Bolt-3/8UNF	10	Note (2)	
7	593693	Washer-plain	10		
8	RTC2726	Bearing-taper roller differential	1	Note (1)	
8	STC1602	Bearing-taper roller differential	NLA	Note (2)	Use RTC3095
8	RTC3095	Bearing-taper roller differential	1		
9	539706	Bearing-taper roller	1		

CHANGE POINTS:

(1) From (A)38L 00001 To (A)38L 03120; From (A)39L 00001 To (A)39L 05552; From (A)42L 00001 To (A)42L 00521; From (A)43L (A)43L 18579

(2) From (A)38L 03121 To (A)38L 99999; From (A)39L 05553 To (A)39L 99999; From (A)42L 00522 To (A)42L 99999; From (A)43L (A)43L 99999; From (A)71L 00001 To (A)78L 99999

(3) From (A)38L 03121 To (A)38L 99999; From (A)39L 05553 To (A)39L 99999; From (A)42L 00522 To (A)42L 99999; From (A)43L (A)43L 99999; From (A)71L 00001A To (A)71L 55169A; From (A)72L 00001A To (A)72L 40518A; From (A)73L 00001A To (A)73L 10454A; From (A)74L 00001A To (A)74L 06441A; From (A)75L 00001A To (A)75L 20434A; From (A)76L 00001A To (A)76L 06647A; (A)77L 00001A To (A)77L 14993A; From (A)78L 00001A To (A)78L 61051A

(4) From (A)71L 55170A To (A)71L 99999A; From (A)72L 40519A To (A)72L 99999A; From (A)73L 10455A To (A)73L 99999A; From 06442A To (A)74L 99999A; From (A)75L 20435A To (A)75L 99999A; From (A)76L 06648A To (A)76L 99999A; From (A)77L 14994A (A)77L 99999A; From (A)78L 61052A To (A)78L 99999A

(5) From (A)38L 03121 To (A)38L 99999; From (A)39L 05553 To (A)39L 99999; From (A)42L 00522 To (A)42L 99999; From (A)43L (A)43L 99999; From (A)71L 00001; From (A)72L 00001; From (A)73L 00001; From (A)74L 00001; From (A)75L 00001; From (A)76L 00001; From (A)77L 00001; From (A)78L 00001

Illus	Part Number	Description	Quantity	Change Point	Remarks
		SHIM			
10	FTC3853	1.548mm	A/R		
10	FTC3855	1.498mm	A/R		
10	FTC3857	1.448mm	A/R		
10	FTC3859	1.398mm	A/R		
10	FTC3861	1.348mm	A/R		
10	FTC3863	1.298mm	A/R		
10	FTC3865	1.248mm	A/R		
10	FTC3867	1.198mm	A/R		
11	7316	Gasket-differential housing to axle case	1		

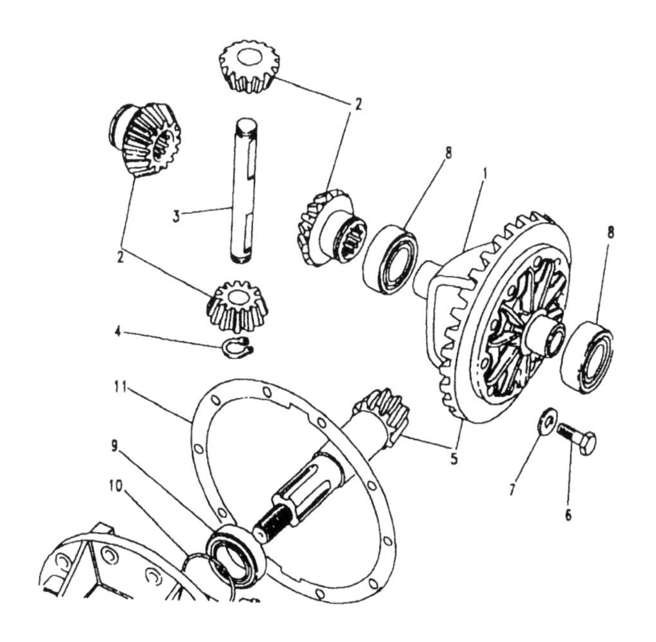

Drive Shafts-Non ABS-To (V) JA032850

Illus	Part Number	Description	Quantity	Change Point	Remarks
	FRC9518	SHAFT & JOINT ASSEMBLY- OUTER-RH	1		
	FRC9519	SHAFT & JOINT ASSEMBLY- OUTER-LH	1		
1	RTC5840	• Shaft-front axle half-RH	1		Part of FRC9518.
2	RTC5843	• Joint-constant velocity	2		
3	RTC5841	• Spacer	2		
	STC579	• Circlip-front axle shaft to CV joint	2		
1	RTC5842	• Shaft-front axle half-LH	1		Part of FRC9519.
4	571822	Sleeve	2		
5	FRC7065	Housing-swivel pin bearing	2		
6	FTC3456	Bolt-M10-hexagonal	12		
6	FTC3454	Bolt-dowel-hexagonal-M10	2		
7	FRC8530	Washer-plain	14		
8	571718	Seal-front driveshaft	2		
9	FTC3646	Gasket-swivel pin bearing housing to axle case	2		
10	571890	Seal-swivel pin housing-12.5mm	2		
10	FTC3401	Seal-swivel pin housing-9mm	2		
11	571755	Plate-oil seal retainer	2		
12	255205	Bolt-1/4UNF x 9/16	12		
13	WM600041L	Washer-spring-1/4 dia-square	12		
14	WA106041L	Washer-M6-standard.	12		

DISCOVERY 1989MY UPTO 1999MY | GROUP E | Axles & Suspension
Front Axle

Drive Shafts-Non ABS-From (V) JA032851

Illus	Part Number	Description	Quantity	Change Point	Remarks
1	FTC2880	DRIVESHAFT-FRONT-RH-10 SPLINE SHAFT	1	Note (1)	
1	FTC2881	DRIVESHAFT-FRONT-LH-10 SPLINE SHAFT	1	Note (1)	
1	FTC3148	DRIVESHAFT-FRONT-RH-24 SPLINE SHAFT	NLA	Note (2)	Use FTC3146.
1	FTC3146	DRIVESHAFT-FRONT-RH	1		Except Antilock Brakes
1	FTC3149	DRIVESHAFT-FRONT-LH-24 SPLINE SHAFT	NLA	Note (2)	Use FTC3147.
1	FTC3147	DRIVESHAFT-FRONT-LH	1		Except Antilock Brakes
2	RTC6754	• Shaft-front axle half-RH	1		Part of FTC2880.
3	STC3046	• Joint-constant velocity-10 spline shaft RH	2		Part of FTC2880, FTC2881.
4	RTC5841	• Spacer	2		Part of FTC2880, FTC2881.
5	STC579	• Circlip-front axle shaft to CV joint	1		
2	RTC6755	• Shaft-front axle half-LH	1		Part of FTC2881.
3	STC3051	• Joint-constant velocity-24 spline shaft RH	1		Part of FTC3148, FTC3146, FTC3149, FTC3147.
2	STC3049	• Shaft-front axle half-RH	1		Part of FTC3148, FTC3146.
4	RTC5841	• Spacer	1		Part of FTC3148, FTC3146, FTC3149, FTC3147.
2	STC3050	• Shaft-front axle half-LH	1		Part of FTC3149, FTC3147.
6	FRC7065	Housing-swivel pin bearing	2		
7	FTC3456	Bolt-M10-hexagonal	12	Note (3)	
7	TYG100590	Bolt-M10-double hex	12	Note (4)	
7	FTC3454	Bolt-dowel-hexagonal-M10	2	Note (3)	
7	TYG100580	Bolt-dowel-M10-double hex	2	Note (4)	
8	FRC8530	Washer-plain	14		
9	571718	Seal-front driveshaft	2	Note (1)	
9	FTC3276	Seal-front driveshaft	2	Note (2)	
10	FTC3646	Gasket-swivel pin bearing housing to axle case	2		
11	FTC3401	Seal-swivel pin housing-9mm	2		
12	549473	Circlip-axle shaft to hub driving member	2		
13	FTC943	Cap-hub front suspension-without logo	2	Note (5)	
13	FTC5414	Cap-hub front suspension-with Land Rover logo	2	Note (6)	

CHANGE POINTS:

(1) From (A)38L 00001B To (A)38L 03120B Petrol; From (A)39L 00001B To (A)39L 05552B Petrol; From (A)42L 00001B To (A)42L Diesel; From (A)43L 00001B To (A)43L 18579B Diesel

(2) From (A)38L 03121B To (A)38L 99999B Petrol; From (A)39L 05553B To (A)39L 99999B Petrol; From (A)43L 18580B To (A)43L Diesel; From (A)42L 00522B To (A)42L 99999B Diesel; From (A)71L 00001 To (A)78L 99999

(3) To (V) VA 729200

(4) From (V) VA 729201

(5) To (V) VA 720263

(6) From (V) VA 720264

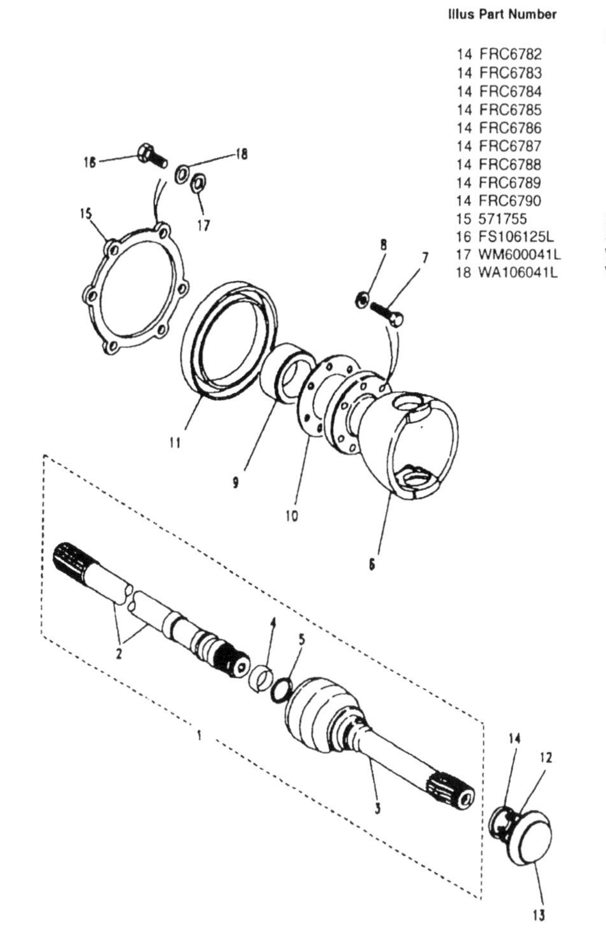

Illus	Part Number	Description	Quantity	Change Point	Remarks
		SHIM-FRONT AXLE HALF SHAFT			
14	FRC6782	0.45mm	A/R		
14	FRC6783	0.60mm	A/R		
14	FRC6784	0.75mm	A/R		
14	FRC6785	0.90mm	A/R		
14	FRC6786	1.05mm	A/R		
14	FRC6787	1.20mm	A/R		
14	FRC6788	1.35mm	A/R		
14	FRC6789	1.50mm	A/R		
14	FRC6790	1.65mm	A/R		
15	571755	Plate-oil seal retainer	2		
16	FS106125L	Screw-flanged head-M6 x 12	12		
17	WM600041L	Washer-spring-1/4 dia-square	12		
18	WA106041L	Washer-M6-standard.	12		

E 15

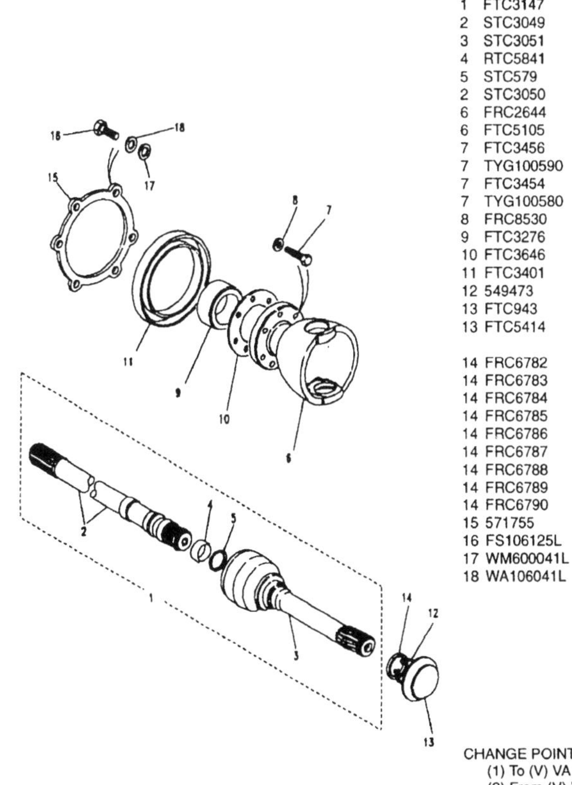

Illus	Part Number	Description	Quantity	Change Point	Remarks
1	FTC3146	DRIVESHAFT-FRONT-RH	1		
1	FTC3147	DRIVESHAFT-FRONT-LH	1		
2	STC3049	• Shaft-front axle half-RH	1		Part of FTC3146.
3	STC3051	• Joint-constant velocity	1		
4	RTC5841	• Spacer	1		
5	STC579	• Circlip-front axle shaft to CV joint	1		
2	STC3050	• Shaft-front axle half-LH	1		Part of FTC3147.
6	FRC2644	Housing-swivel pin bearing	NLA		Use FTC5105.
6	FTC5105	Housing-swivel pin bearing	2		
7	FTC3456	Bolt-M10-hexagonal	12	Note (1)	
7	TYG100590	Bolt-M10-double hex	12	Note (2)	
7	FTC3454	Bolt-dowel-hexagonal-M10	2	Note (1)	
7	TYG100580	Bolt-dowel-M10-double hex	2	Note (2)	
8	FRC8530	Washer-plain	14		
9	FTC3276	Seal-front driveshaft	2		
10	FTC3646	Gasket-swivel pin bearing housing to axle case	2		
11	FTC3401	Seal-swivel pin housing-9mm	2		
12	549473	Circlip-axle shaft to hub driving member	2		
13	FTC943	Cap-hub front suspension-without logo	2		
13	FTC5414	Cap-hub front suspension-with Land Rover logo	2	Note (3)	
		SHIM-FRONT AXLE HALF SHAFT			
14	FRC6782	0.45mm	A/R		
14	FRC6783	0.60mm	A/R		
14	FRC6784	0.75mm	A/R		
14	FRC6785	0.90mm	A/R		
14	FRC6786	1.05mm	A/R		
14	FRC6787	1.20mm	A/R		
14	FRC6788	1.35mm	A/R		
14	FRC6789	1.50mm	A/R		
14	FRC6790	1.65mm	A/R		
15	571755	Plate-oil seal retainer	2		
16	FS106125L	Screw-flanged head-M6 x 12	12		
17	WM600041L	Washer-spring-1/4 dia-square	12		
18	WA106041L	Washer-M6-standard.	12		

CHANGE POINTS:
(1) To (V) VA 729200
(2) From (V) VA 729201
(3) From (V) VA 720264

E 16

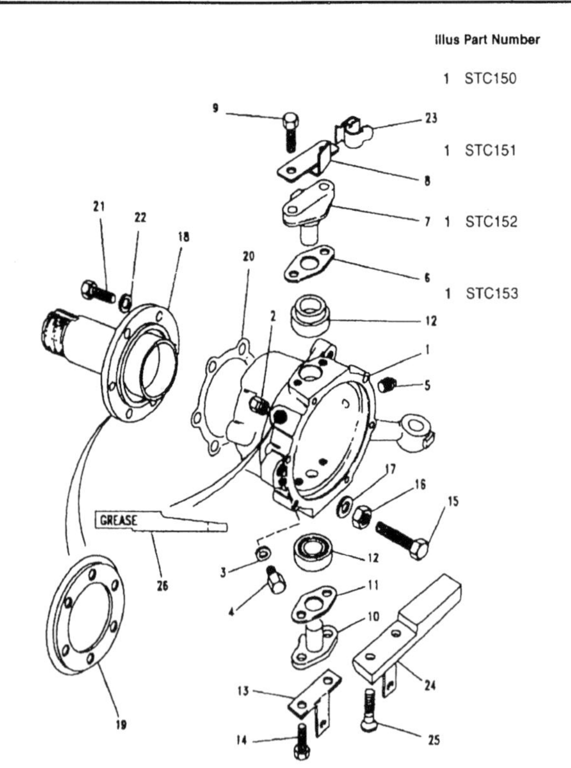

Illus	Part Number	Description	Quantity	Change Point	Remarks
		HOUSING-SWIVEL PIN			
1	STC150	RH-RHD	NLA	To (V) JA 018174	Use FTC2520 but cross rod needs to be adjusted to lengthen link by 5mm - see fitting instructions.
1	STC151	LH-RHD	NLA	To (V) JA 018174	Use FTC2521 but cross rod needs to be adjusted to lengthen link by 5mm - see fitting instructions.
1	STC152	RH-LHD	NLA	To (V) JA 032849	Use FTC2522 but cross rod needs to be adjusted to lengthen link by 5mm - see fitting instructions.
1	STC153	LH-LHD	NLA	To (V) JA 032849	Use FTC2523 but cross rod needs to be adjusted to lengthen link by 5mm - see fitting instructions.

DISCOVERY 1989MY UPTO 1999MY | GROUP E | Axles & Suspension
Front Axle
Swivel Pin Housing-To (V) JA032849

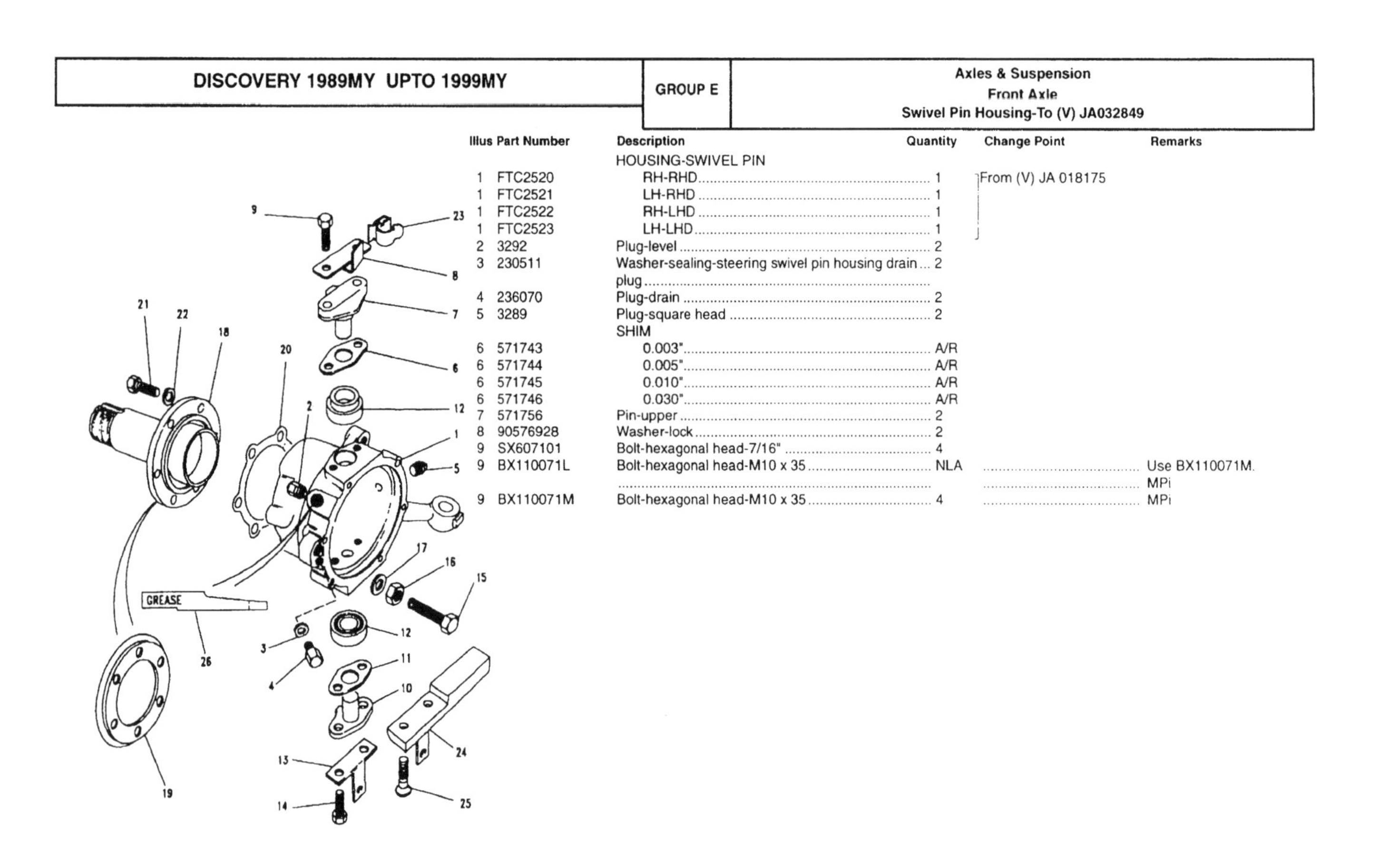

Illus	Part Number	Description	Quantity	Change Point	Remarks
		HOUSING-SWIVEL PIN			
1	FTC2520	RH-RHD	1	From (V) JA 018175	
1	FTC2521	LH-RHD	1		
1	FTC2522	RH-LHD	1		
1	FTC2523	LH-LHD	1		
2	3292	Plug-level	2		
3	230511	Washer-sealing-steering swivel pin housing drain plug	2		
4	236070	Plug-drain	2		
5	3289	Plug-square head	2		
		SHIM			
6	571743	0.003"	A/R		
6	571744	0.005"	A/R		
6	571745	0.010"	A/R		
6	571746	0.030"	A/R		
7	571756	Pin-upper	2		
8	90576928	Washer-lock	2		
9	SX607101	Bolt-hexagonal head-7/16"	4		
9	BX110071L	Bolt-hexagonal head-M10 x 35	NLA		Use BX110071M. MPi
9	BX110071M	Bolt-hexagonal head-M10 x 35	4		MPi

Swivel Pin Housing-To (V) JA032849

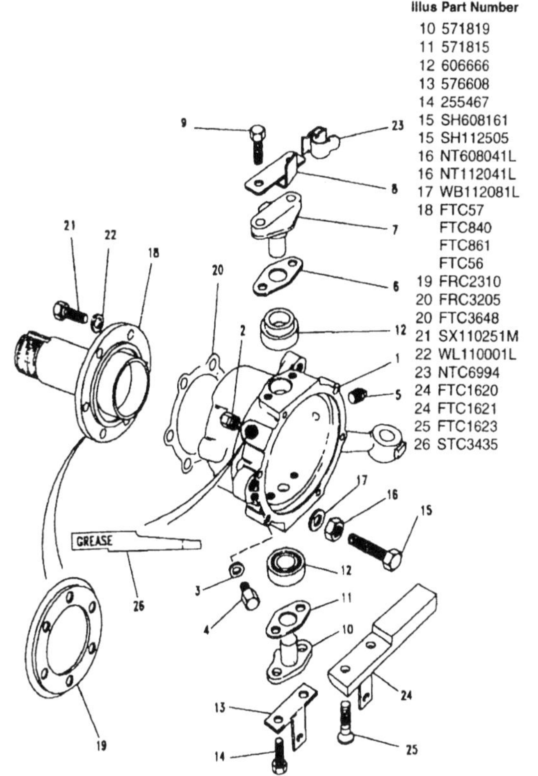

Illus	Part Number	Description	Quantity	Change Point	Remarks
10	571819	Pin	2		
11	571815	Gasket-lower swivel pin	2		
12	606666	Bearing-taper roller-lower	4		
13	576608	Washer-lock	2		
14	255467	Bolt-flanged head-7/16UNF x 1 1/8	4		
15	SH608161	Screw-hexagonal head-1/2UNF x 2	2		
15	SH112505	Bolt-shouldered-hexagonal head-M12 x 50	1		
16	NT608041L	Nut-hexagonal-thin-1/2"	2		
16	NT112041L	Nut-hexagonal-thin-M12	1		
17	WB112081L	Washer-plain-M12 x 22	4		
18	FTC57	AXLE-STUB-HUB FRONT SUSPENSION	2		
	FTC840	• Seal assembly-front/rear hub-outer	2		
	FTC861	• Bearing-front/rear axle hub	2		
	FTC56	• Washer-front hub thrust	2		
19	FRC2310	Plate-front & rear stub axle locking	2		
20	FRC3205	Gasket-stub axle	NLA		Use FTC3648.
20	FTC3648	Gasket-stub axle-asbestos free	2		
21	SX110251M	Screw-hexagon socket-M10 x 25	12		
22	WL110001L	Washer-M10	12		
23	NTC6994	Clip-cable	2		
24	FTC1620	Damper-swivel pin-RH	1		
24	FTC1621	Damper-swivel pin-LH	1		
25	FTC1623	Screw-torx-countersunk	1		
26	STC3435	Grease-375cc-one shot	1		Note (1)

Remarks:
(1) can be used instead of oil in swivel pin housing

Swivel Pin Housing-From (V) JA032850

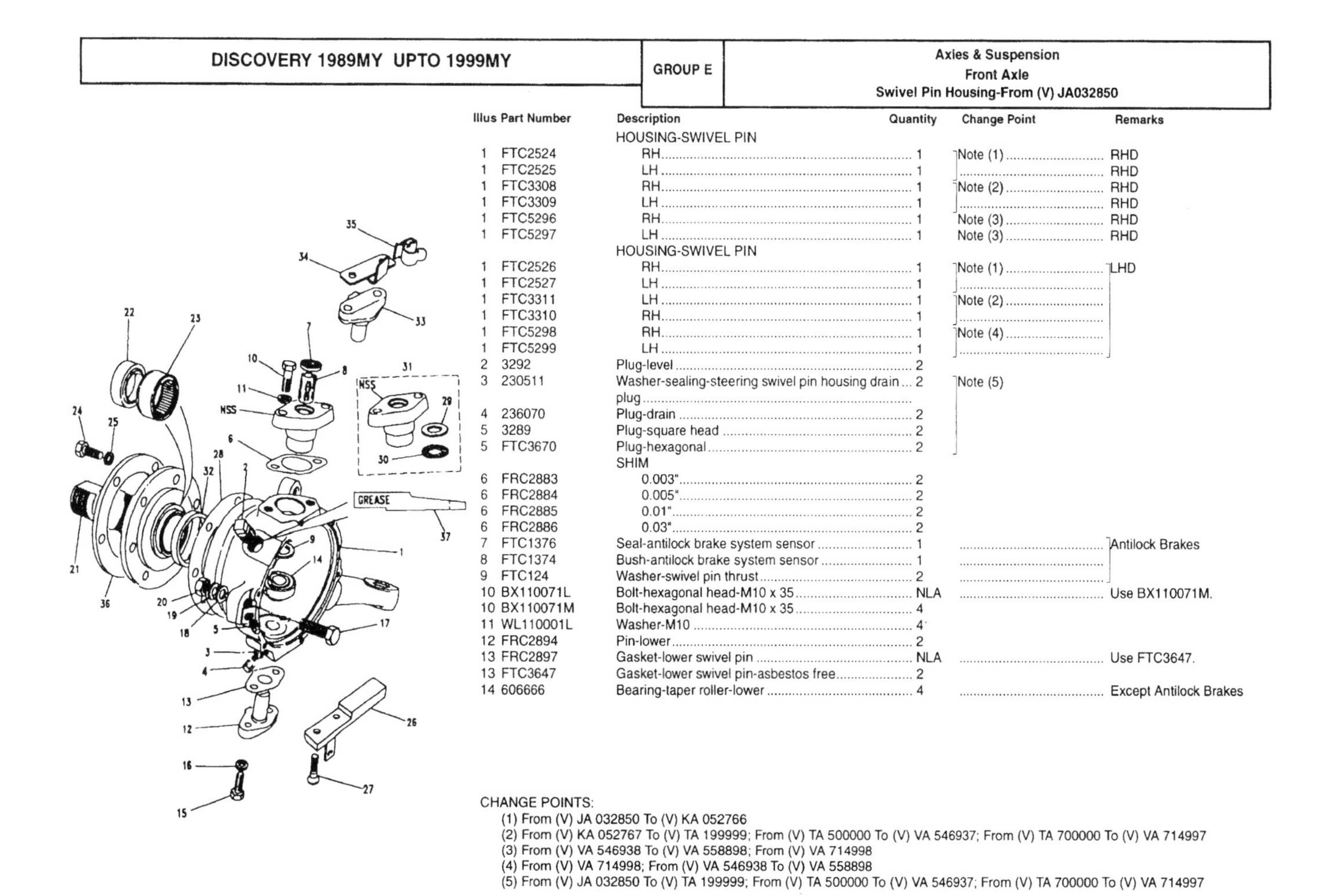

Illus	Part Number	Description	Quantity	Change Point	Remarks
		HOUSING-SWIVEL PIN			
1	FTC2524	RH	1	Note (1)	RHD
1	FTC2525	LH	1		RHD
1	FTC3308	RH	1	Note (2)	RHD
1	FTC3309	LH	1		RHD
1	FTC5296	RH	1	Note (3)	RHD
1	FTC5297	LH	1	Note (3)	RHD
		HOUSING-SWIVEL PIN			
1	FTC2526	RH	1	Note (1)	LHD
1	FTC2527	LH	1		
1	FTC3311	LH	1	Note (2)	
1	FTC3310	RH	1		
1	FTC5298	RH	1	Note (4)	
1	FTC5299	LH	1		
2	3292	Plug-level	2		
3	230511	Washer-sealing-steering swivel pin housing drain plug	2	Note (5)	
4	236070	Plug-drain	2		
5	3289	Plug-square head	2		
5	FTC3670	Plug-hexagonal	2		
		SHIM			
6	FRC2883	0.003"	2		
6	FRC2884	0.005"	2		
6	FRC2885	0.01"	2		
6	FRC2886	0.03"	2		
7	FTC1376	Seal-antilock brake system sensor	1		Antilock Brakes
8	FTC1374	Bush-antilock brake system sensor	1		
9	FTC124	Washer-swivel pin thrust	2		
10	BX110071L	Bolt-hexagonal head-M10 x 35	NLA		Use BX110071M.
10	BX110071M	Bolt-hexagonal head-M10 x 35	4		
11	WL110001L	Washer-M10	4		
12	FRC2894	Pin-lower	2		
13	FRC2897	Gasket-lower swivel pin	NLA		Use FTC3647.
13	FTC3647	Gasket-lower swivel pin-asbestos free	2		
14	606666	Bearing-taper roller-lower	4		Except Antilock Brakes

CHANGE POINTS:
(1) From (V) JA 032850 To (V) KA 052766
(2) From (V) KA 052767 To (V) TA 199999; From (V) TA 500000 To (V) VA 546937; From (V) TA 700000 To (V) VA 714997
(3) From (V) VA 546938 To (V) VA 558898; From (V) VA 714998
(4) From (V) VA 714998; From (V) VA 546938 To (V) VA 558898
(5) From (V) JA 032850 To (V) TA 199999; From (V) TA 500000 To (V) VA 546937; From (V) TA 700000 To (V) VA 714997

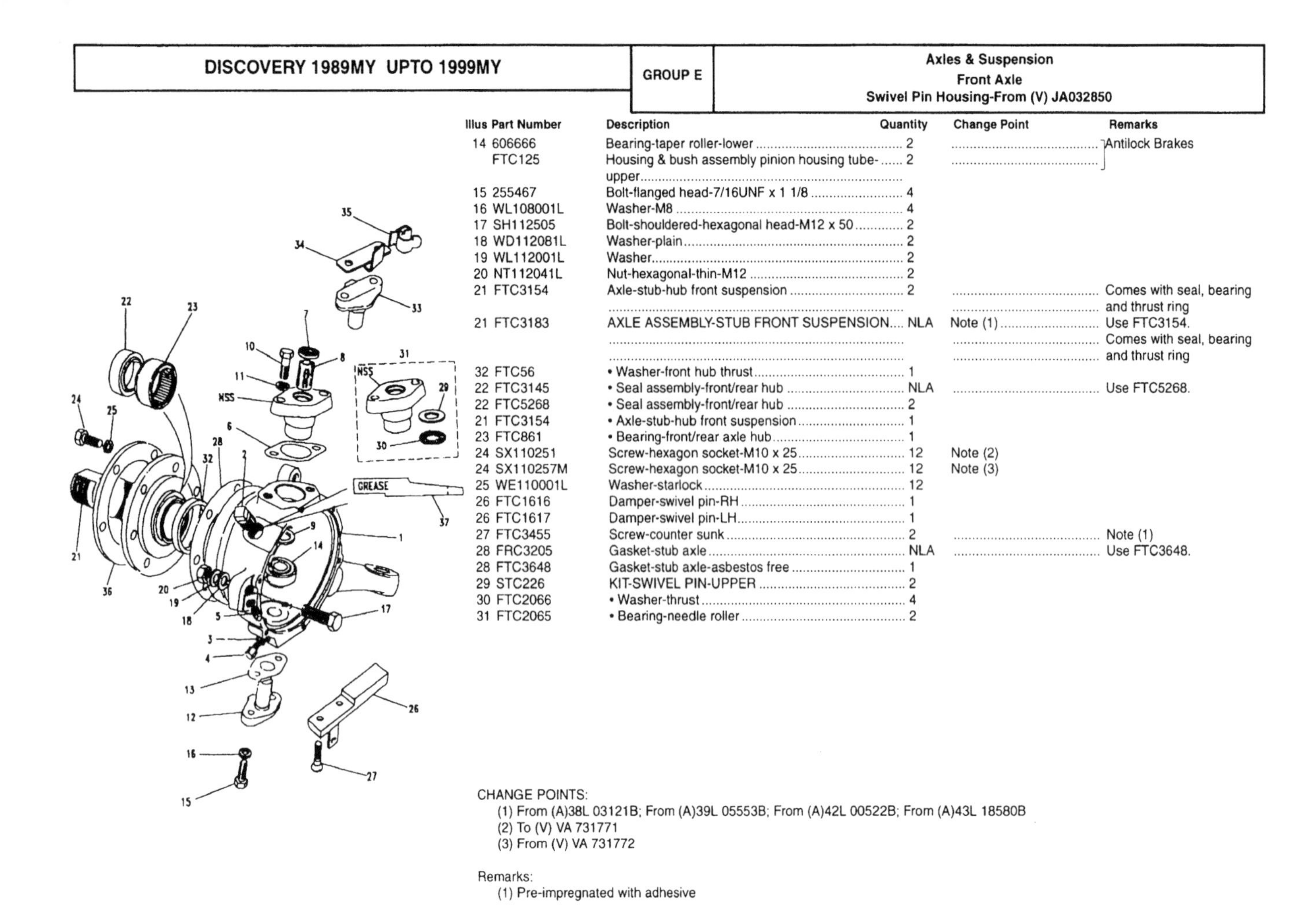

Illus	Part Number	Description	Quantity	Change Point	Remarks
14	606666	Bearing-taper roller-lower	2		Antilock Brakes
	FTC125	Housing & bush assembly pinion housing tube-upper	2		
15	255467	Bolt-flanged head-7/16UNF x 1 1/8	4		
16	WL108001L	Washer-M8	4		
17	SH112505	Bolt-shouldered-hexagonal head-M12 x 50	2		
18	WD112081L	Washer-plain	2		
19	WL112001L	Washer	2		
20	NT112041L	Nut-hexagonal-thin-M12	2		
21	FTC3154	Axle-stub-hub front suspension	2		Comes with seal, bearing and thrust ring
21	FTC3183	AXLE ASSEMBLY-STUB FRONT SUSPENSION	NLA	Note (1)	Use FTC3154. Comes with seal, bearing and thrust ring
32	FTC56	• Washer-front hub thrust	1		
22	FTC3145	• Seal assembly-front/rear hub	NLA		Use FTC5268.
22	FTC5268	• Seal assembly-front/rear hub	2		
21	FTC3154	• Axle-stub-hub front suspension	1		
23	FTC861	• Bearing-front/rear axle hub	1		
24	SX110251	Screw-hexagon socket-M10 x 25	12	Note (2)	
24	SX110257M	Screw-hexagon socket-M10 x 25	12	Note (3)	
25	WE110001L	Washer-starlock	12		
26	FTC1616	Damper-swivel pin-RH	1		
26	FTC1617	Damper-swivel pin-LH	1		
27	FTC3455	Screw-counter sunk	2		Note (1)
28	FRC3205	Gasket-stub axle	NLA		Use FTC3648.
28	FTC3648	Gasket-stub axle-asbestos free	1		
29	STC226	KIT-SWIVEL PIN-UPPER	2		
30	FTC2066	• Washer-thrust	4		
31	FTC2065	• Bearing-needle roller	2		

CHANGE POINTS:
(1) From (A)38L 03121B; From (A)39L 05553B; From (A)42L 00522B; From (A)43L 18580B
(2) To (V) VA 731771
(3) From (V) VA 731772

Remarks:
(1) Pre-impregnated with adhesive

DISCOVERY 1989MY UPTO 1999MY | GROUP E | Axles & Suspension
Front Axle
Swivel Pin Housing-From (V) JA032850

Illus	Part Number	Description	Quantity	Change Point	Remarks
33	FTC2882	Pin	1		
34	90576928	Washer-lock	2		
	FTC4045	Bracket-hose brake-RH-front	1	From (V) MA 081992	
	FTC4046	Bracket-hose brake-LH-front	1		
35	NTC6994	Clip-cable	2		
36	FRC2310	Plate-front & rear stub axle locking	2		
37	STC3435	Grease-375cc-one shot	1		Can be used instead of oil in swivel pin housing

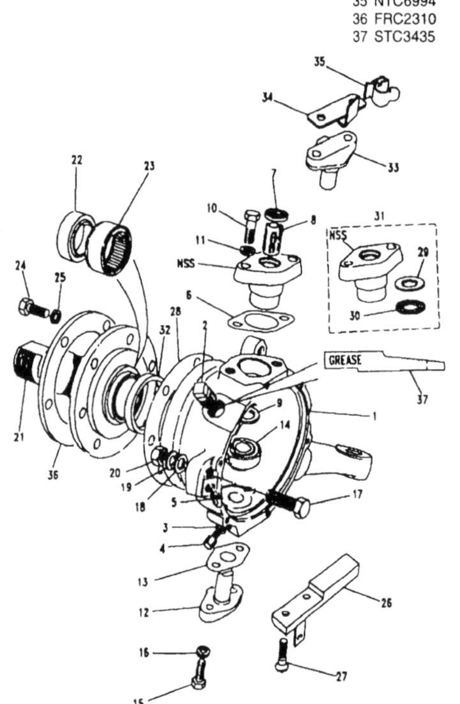

Illus	Part Number	Description	Quantity	Change Point	Remarks
1	571761	Driveshaft-front hub	2		
2	FRC8227	Spacer	2		
3	571752	Gasket-axle shaft drive member	2		
4	FRC8700	Locknut	4		
5	FRC8002	Washer-front & rear hub nut tab	2		
6	FTC2783	Seal assembly-front/rear hub-outer	NLA		Use FTC4785.
6	FTC4785	Seal assembly-front/rear hub-outer	2		
7	RTC3429	Bearing-front/rear axle hub	2		
8	FTC1457	HUB ASSEMBLY FRONT	2		
10	FRC8222	Seal assembly-front/rear hub-inner	2		
9	FRC5926	• Stud	10		
11	279146	Bolt	10		
12	WM600061L	Washer-3/8"-square	10		

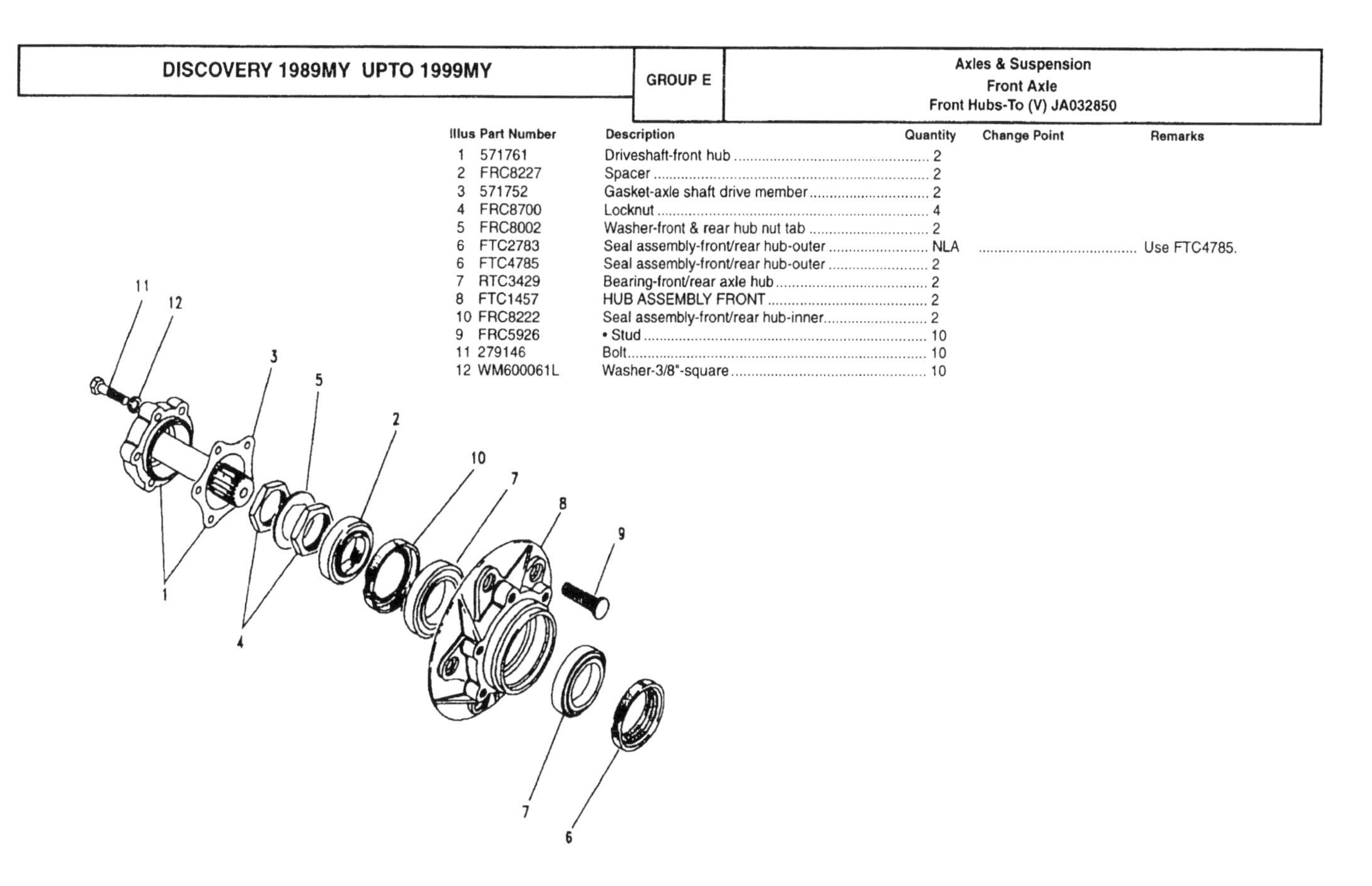

Illus	Part Number	Description	Quantity	Change Point	Remarks
1	FTC859	Hub assembly front	2		
2	571752	Gasket-axle shaft drive member	2		
3	FRC8700	Locknut	4		
4	FTC3179	Washer-lock-outer	2		
5	FTC3185	Washer-front & rear hub nut tab	NLA	Note (1)	Use FTC5241.
5	FTC5241	Washer-front & rear hub nut tab	2	Note (2); Note (3)	
6	FTC2783	Seal assembly-front/rear hub-outer	NLA		Use FTC4785.
6	FTC4785	Seal assembly-front/rear hub-outer	2		
7	RTC3429	Bearing-front/rear axle hub	4		
8	FTC860	Hub assembly-front & rear	NLA		Use FTC942.
8	FTC942	HUB ASSEMBLY-FRONT & REAR	2		
9	FRC5926	• Stud	10		
10	BX110095M	Bolt-M10 x 45	10		
11	WM600061L	Washer-3/8"-square	10		

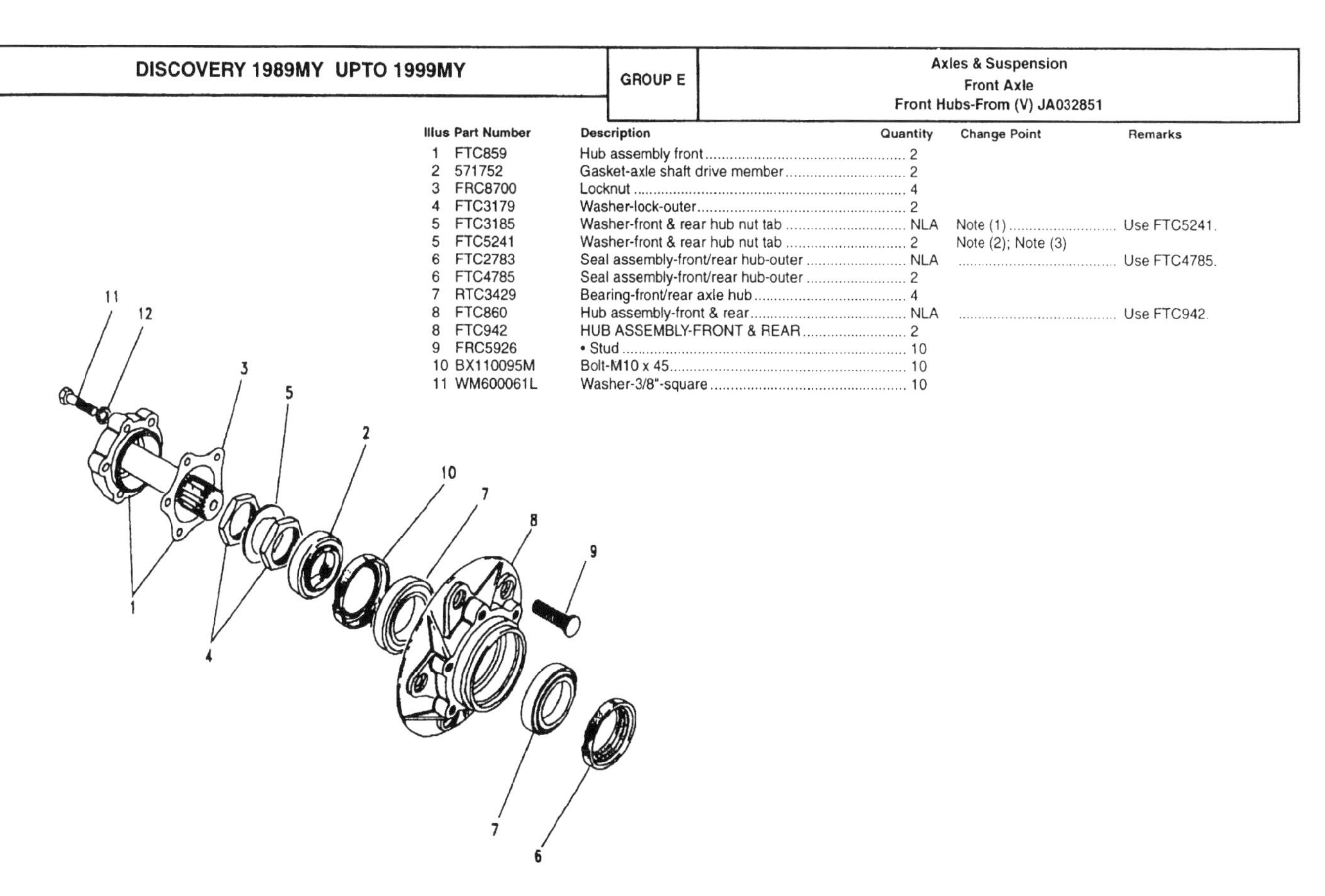

CHANGE POINTS:
(1) From (V) JA 032851 To (V) TA 199999; From (V) TA 700000 To (V) TA 703235; From (V) TA 500000 To (V) TA 534102
(2) From (V) VA 703236
(3) From (V) VA 534103 To (V) VA 558898

Illus	Part Number	Description	Quantity	Change Point	Remarks
		AXLE/DRIVE UNIT-4X4 REAR			
1	FTC490	asbestos-less splash shield	1	Note (1)	
1	FTC613	asbestos-splash shield	NLA		Use STC3405.
1	STC3405	asbestos-less splash shield	1		
1	FTC1468	asbestos free-less splash shield	1	Note (2)	
1	FTC1478	asbestos free-splash shield	1		
1	FTC3090	asbestos free	NLA	Note (3); Note (4)	Use STC3596.
1	STC3596	asbestos free	1		
1	FTC3438	asbestos free	NLA	Note (5); Note (6)	Use STC3269.
1	STC3269	asbestos free	1		

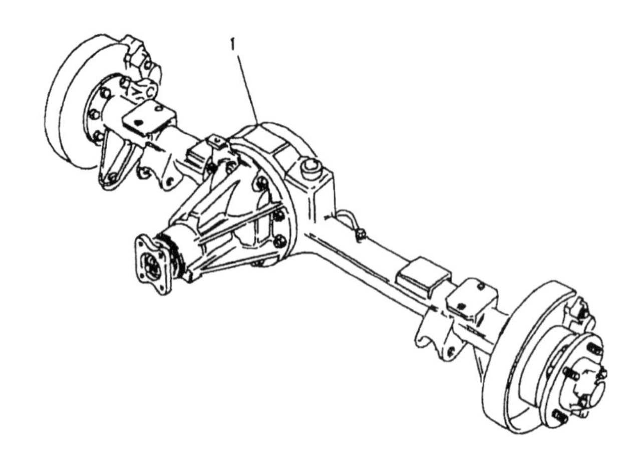

CHANGE POINTS:
(1) From (A)28S 00001C To (A)28S 28134C
(2) From (A)32S 00001A To (A)32S 25429A
(3) From (A)28S 28135C To (A)28S 54417C
(4) From (A)32S 25430A To (A)32S 29118B
(5) From (A)32S 29119B To (A)32S 99999B
(6) From (A)28S 54417D To (A)28S 99999D

E 25

Illus	Part Number	Description	Quantity	Change Point	Remarks
1	FTC3724	Axle/drive unit-4x4 rear-asbestos free	1	Note (1); Note (2); Note (3); Note (4)	3-bolt propshaft flange except Antilock Brakes
1	FTC5410	Axle/drive unit-4x4 rear-asbestos free	1	Note (5); Note (6)	V8 4-bolt propshaft flange except Antilock Brakes
1	FTC3725	Axle/drive unit-4x4 rear-asbestos free	1	Note (1); Note (2); Note (3); Note (4)	Antilock Brakes 3-bolt propshaft flange
1	FTC5411	Axle/drive unit-4x4 rear-asbestos free	1	Note (5); Note (6)	V8 Antilock Brakes 4-bolt propshaft flange

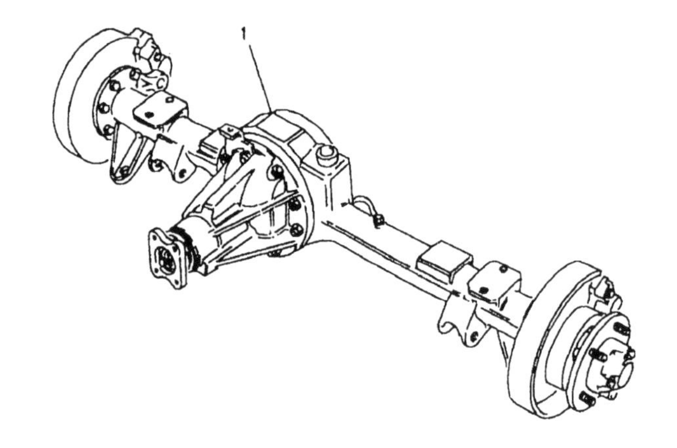

CHANGE POINTS:
(1) From (V) MA 081992 Except V8
(2) From (V) MA 081992 To (V) TA 199999 V8
(3) From (V) TA 500000 To (V) VA 541627 V8
(4) From (V) TA 700000 To (V) VA 710668 V8
(5) From (V) VA 710669
(6) From (V) VA 541628 To (V) VA 558898

E 26

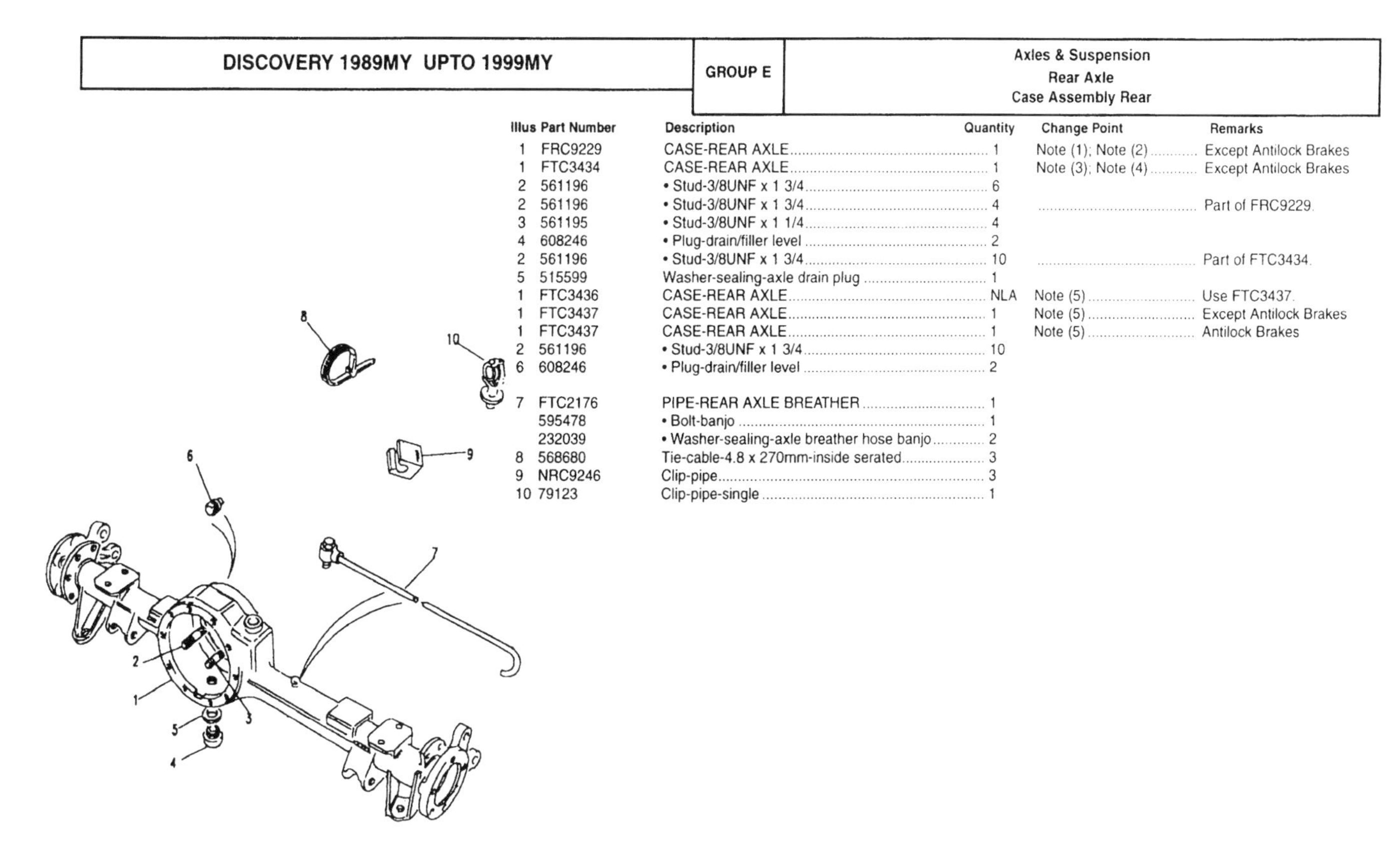

Illus	Part Number	Description	Quantity	Change Point	Remarks
1	FRC9229	CASE-REAR AXLE	1	Note (1); Note (2)	Except Antilock Brakes
1	FTC3434	CASE-REAR AXLE	1	Note (3); Note (4)	Except Antilock Brakes
2	561196	• Stud-3/8UNF x 1 3/4	6		
2	561196	• Stud-3/8UNF x 1 3/4	4		Part of FRC9229.
3	561195	• Stud-3/8UNF x 1 1/4	4		
4	608246	• Plug-drain/filler level	2		
2	561196	• Stud-3/8UNF x 1 3/4	10		Part of FTC3434.
5	515599	Washer-sealing-axle drain plug	1		
1	FTC3436	CASE-REAR AXLE	NLA	Note (5)	Use FTC3437.
1	FTC3437	CASE-REAR AXLE	1	Note (5)	Except Antilock Brakes
1	FTC3437	CASE-REAR AXLE	1	Note (5)	Antilock Brakes
2	561196	• Stud-3/8UNF x 1 3/4	10		
6	608246	• Plug-drain/filler level	2		
7	FTC2176	PIPE-REAR AXLE BREATHER	1		
	595478	• Bolt-banjo	1		
	232039	• Washer-sealing-axle breather hose banjo	2		
8	568680	Tie-cable-4.8 x 270mm-inside serated	3		
9	NRC9246	Clip-pipe	3		
10	79123	Clip-pipe-single	1		

CHANGE POINTS:
(1) From (A)32S 00001 To (A)32S 29121
(2) From (A)28S 00001 To (A)28S 44171
(3) From (A)32S 29122 To (A)32S 99999
(4) From (A)28S 44172 To (A)28S 99999
(5) From (A)46S 00001

Illus	Part Number	Description	Quantity	Change Point	Remarks
1	FTC2750	DIFFERENTIAL ASSEMBLY-REAR-NEW-10 SPLINE SHAFT-4 BOLT FLANGE	1	Note (1)	
	FRC5690	• HOUSING ASSEMBLY DIFFERENTIAL/FINAL DRIVE WHEEL-WITH FILLER/LEVEL PLUG	1		
	BH112101L	• • Bolt-hexagonal socket-M12 x 50	4		
	FRC4112	• HOUSING ASSEMBLY DIFFERENTIAL/FINAL DRIVE WHEEL-LESS FILLER/LEVEL PLUG	1	Note (2)	
	BH112101L	• • Bolt-hexagonal socket-M12 x 50	4		
3	FRC5204	• Ring-differential locking-new-10 spline shaft	NLA		Use FTC4210.
3	FTC4210	• Ring-differential locking-new-10 spline shaft	2		
4	FRC5661	• Tab-differential locking ring-new-10 spline shaft	2		
5	576159	• Pin-roll-spring tension-new-10 spline shaft	2		
6	NY606041L	• Nut-new-10 spline shaft-hexagonal head-nyloc-3/8-UNF	10		
1	FTC2750R	DIFFERENTIAL ASSEMBLY-REAR-RECONDITIONED-10 SPLINE SHAFT-4 BOLT FLANGE	NLA	From (A)32S 00001 To (A)32S 29121 From (A)28S 00001 To (A)28S 44171	Use FTC2750E.
	FRC5690	• HOUSING ASSEMBLY DIFFERENTIAL/FINAL DRIVE WHEEL-WITH FILLER/LEVEL PLUG	1		
	BH112101L	• • Bolt-hexagonal socket-M12 x 50	4		
	FRC4112	• HOUSING ASSEMBLY DIFFERENTIAL/FINAL DRIVE WHEEL-LESS FILLER/LEVEL PLUG	1	Note (2)	
	BH112101L	• • Bolt-hexagonal socket-M12 x 50	4		
3	FRC5204	• Ring-differential locking-reconditioned-10 spline shaft	NLA		Use FTC4210.
3	FTC4210	• Ring-differential locking-reconditioned-10 spline shaft	2		
4	FRC5661	• Tab-differential locking ring-reconditioned-10 spline shaft	2		
5	576159	• Pin-roll-spring tension-reconditioned-10 spline shaft	2		
6	NY606041L	• Nut-reconditioned-10 spline shaft-hexagonal head-nyloc-3/8-UNF	10		
1	FTC3272	Differential-new-24 spline shaft-4 bolt flange	1	Note (3)	

CAEXEE1A

CHANGE POINTS:
(1) From (A)32S 00001 To (A)32S 29121; From (A)28S 00001 To (A)28S 44171
(2) From (V) MA 118844
(3) From (V) VA 541628 To (V) VA 558898 V8; From (V) VA 710669 V8; From (A)32S 29122 To (A)32S 99999; From (A)28S 44172 To (A)28S 99999

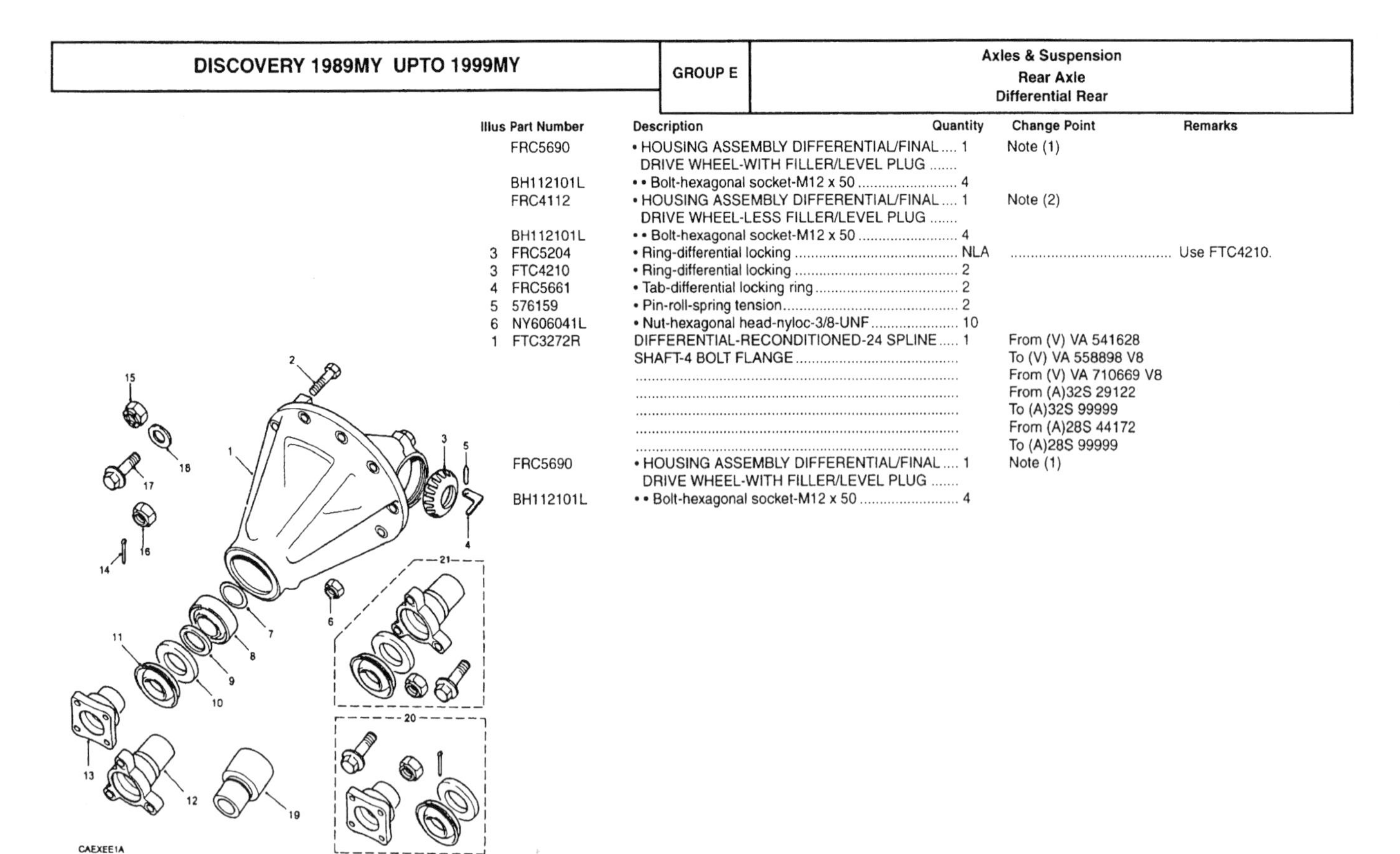

Illus	Part Number	Description	Quantity	Change Point	Remarks
	FRC5690	• HOUSING ASSEMBLY DIFFERENTIAL/FINAL DRIVE WHEEL-WITH FILLER/LEVEL PLUG	1	Note (1)	
	BH112101L	• • Bolt-hexagonal socket-M12 x 50	4		
	FRC4112	• HOUSING ASSEMBLY DIFFERENTIAL/FINAL DRIVE WHEEL-LESS FILLER/LEVEL PLUG	1	Note (2)	
	BH112101L	• • Bolt-hexagonal socket-M12 x 50	4		
3	FRC5204	• Ring-differential locking	NLA		Use FTC4210.
3	FTC4210	• Ring-differential locking	2		
4	FRC5661	• Tab-differential locking ring	2		
5	576159	• Pin-roll-spring tension	2		
6	NY606041L	• Nut-hexagonal head-nyloc-3/8-UNF	10		
1	FTC3272R	DIFFERENTIAL-RECONDITIONED-24 SPLINE SHAFT-4 BOLT FLANGE	1	From (V) VA 541628 To (V) VA 558898 V8 From (V) VA 710669 V8 From (A)32S 29122 To (A)32S 99999 From (A)28S 44172 To (A)28S 99999	
	FRC5690	• HOUSING ASSEMBLY DIFFERENTIAL/FINAL DRIVE WHEEL-WITH FILLER/LEVEL PLUG	1	Note (1)	
	BH112101L	• • Bolt-hexagonal socket-M12 x 50	4		

CHANGE POINTS:
(1) To (V) MA 118843
(2) From (V) MA 118844

E 29

DISCOVERY 1989MY UPTO 1999MY | GROUP E | Axles & Suspension
Rear Axle
Differential Rear

Illus	Part Number	Description	Quantity	Change Point	Remarks
	FRC4112	• HOUSING ASSEMBLY DIFFERENTIAL/FINAL DRIVE WHEEL-LESS FILLER/LEVEL PLUG	1	Note (1)	
	BH112101L	• • Bolt-hexagonal socket-M12 x 50	4		
3	FRC5204	• Ring-differential locking	NLA		Use FTC4210.
3	FTC4210	• Ring-differential locking	2		
4	FRC5661	• Tab-differential locking ring	2		
5	576159	• Pin-roll-spring tension	2		
6	NY606041L	• Nut-hexagonal head-nyloc-3/8-UNF	10		
1	FTC3723	Differential assembly-rear-new-24 spline shaft-3 bolt flange	1	From (V) MA 081992 To (V) TA 199999 From (V) TA 500000 To (V) VA 541627 From (V) TA 700000 To (V) VA 710668 From (V) VA 541628 To (V) VA 558898 Except V8, V8 From (V) VA 710669 Except V8	

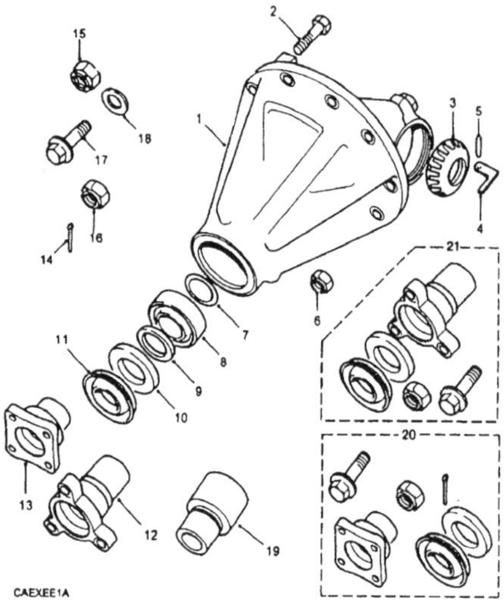

CAEXEE1A

CHANGE POINTS:
(1) From (V) MA 118844

E 30

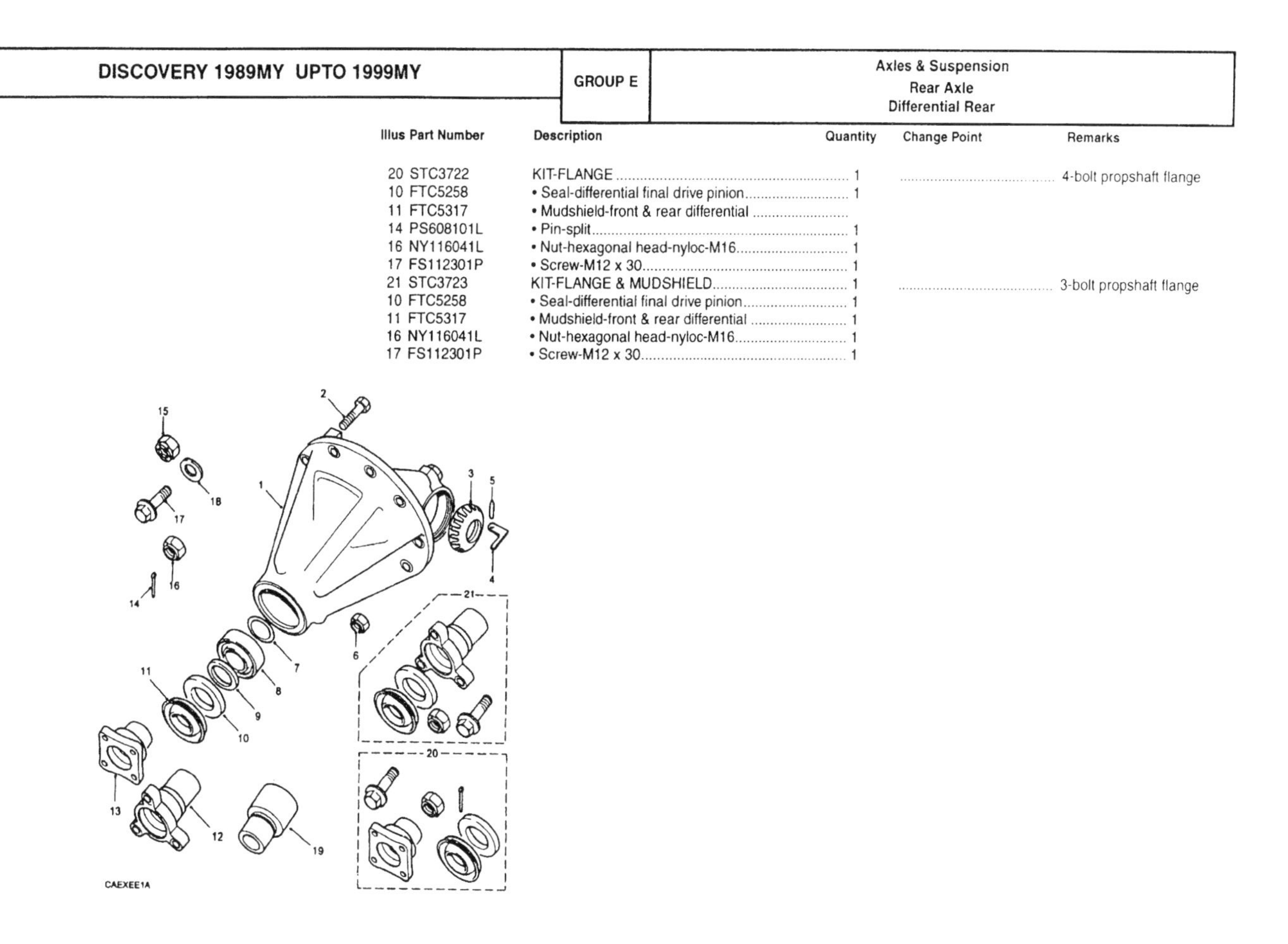

Illus	Part Number	Description	Quantity	Change Point	Remarks
20	STC3722	KIT-FLANGE	1		4-bolt propshaft flange
10	FTC5258	• Seal-differential final drive pinion	1		
11	FTC5317	• Mudshield-front & rear differential			
14	PS608101L	• Pin-split	1		
16	NY116041L	• Nut-hexagonal head-nyloc-M16	1		
17	FS112301P	• Screw-M12 x 30	1		
21	STC3723	KIT-FLANGE & MUDSHIELD	1		3-bolt propshaft flange
10	FTC5258	• Seal-differential final drive pinion	1		
11	FTC5317	• Mudshield-front & rear differential	1		
16	NY116041L	• Nut-hexagonal head-nyloc-M16	1		
17	FS112301P	• Screw-M12 x 30	1		

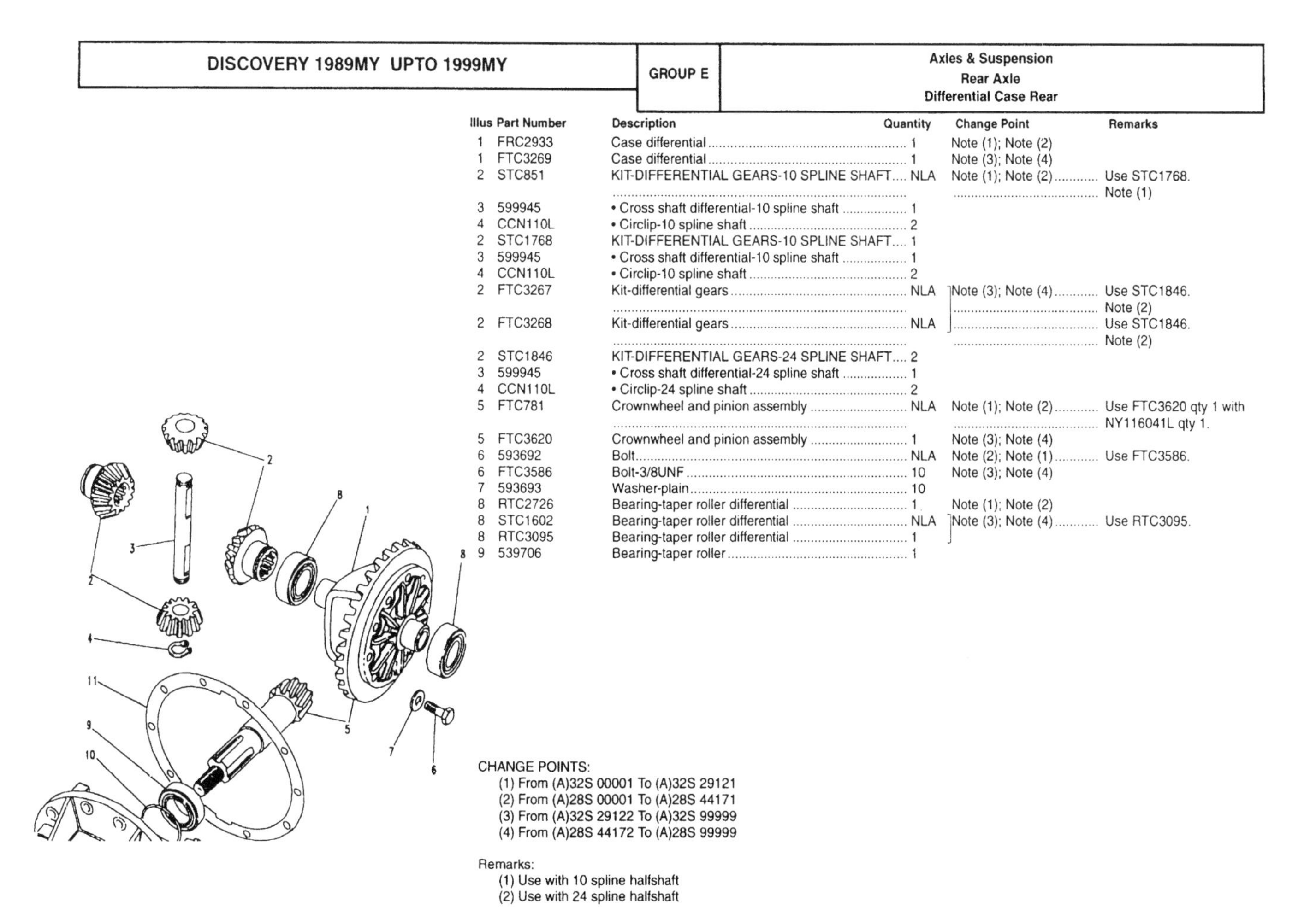

Illus	Part Number	Description	Quantity	Change Point	Remarks
1	FRC2933	Case differential	1	Note (1); Note (2)	
1	FTC3269	Case differential	1	Note (3); Note (4)	
2	STC851	KIT-DIFFERENTIAL GEARS-10 SPLINE SHAFT	NLA	Note (1); Note (2)	Use STC1768. Note (1)
3	599945	• Cross shaft differential-10 spline shaft	1		
4	CCN110L	• Circlip-10 spline shaft	2		
2	STC1768	KIT-DIFFERENTIAL GEARS-10 SPLINE SHAFT	1		
3	599945	• Cross shaft differential-10 spline shaft	1		
4	CCN110L	• Circlip-10 spline shaft	2		
2	FTC3267	Kit-differential gears	NLA	Note (3); Note (4)	Use STC1846. Note (2)
2	FTC3268	Kit-differential gears	NLA		Use STC1846. Note (2)
2	STC1846	KIT-DIFFERENTIAL GEARS-24 SPLINE SHAFT	2		
3	599945	• Cross shaft differential-24 spline shaft	1		
4	CCN110L	• Circlip-24 spline shaft	2		
5	FTC781	Crownwheel and pinion assembly	NLA	Note (1); Note (2)	Use FTC3620 qty 1 with NY116041L qty 1.
5	FTC3620	Crownwheel and pinion assembly	1	Note (3); Note (4)	
6	593692	Bolt	NLA	Note (2); Note (1)	Use FTC3586.
6	FTC3586	Bolt-3/8UNF	10	Note (3); Note (4)	
7	593693	Washer-plain	10		
8	RTC2726	Bearing-taper roller differential	1	Note (1); Note (2)	
8	STC1602	Bearing-taper roller differential	NLA	Note (3); Note (4)	Use RTC3095.
8	RTC3095	Bearing-taper roller differential	1		
9	539706	Bearing-taper roller	1		

CHANGE POINTS:
(1) From (A)32S 00001 To (A)32S 29121
(2) From (A)28S 00001 To (A)28S 44171
(3) From (A)32S 29122 To (A)32S 99999
(4) From (A)28S 44172 To (A)28S 99999

Remarks:
(1) Use with 10 spline halfshaft
(2) Use with 24 spline halfshaft

Illus	Part Number	Description	Quantity	Change Point	Remarks
		SHIM			
10	549230	0.038"	A/R	Note (1); Note (2)	
10	549232	0.04"	A/R		
10	549234	0.042"	A/R		
10	549236	0.044"	A/R		
10	549238	0.046"	A/R		
10	549240	0.048"	A/R		
10	549242	0.050"	A/R		
10	549244	0.052"	A/R		
10	549246	0.054"	A/R		
10	549248	0.056"	A/R		
10	549250	0.058"	A/R		
10	549252	0.060"	A/R		
10	576236	0.062"	A/R		
10	576237	0.063"	A/R		
10	576238	0.064"	A/R		
10	576239	0.065"	A/R		

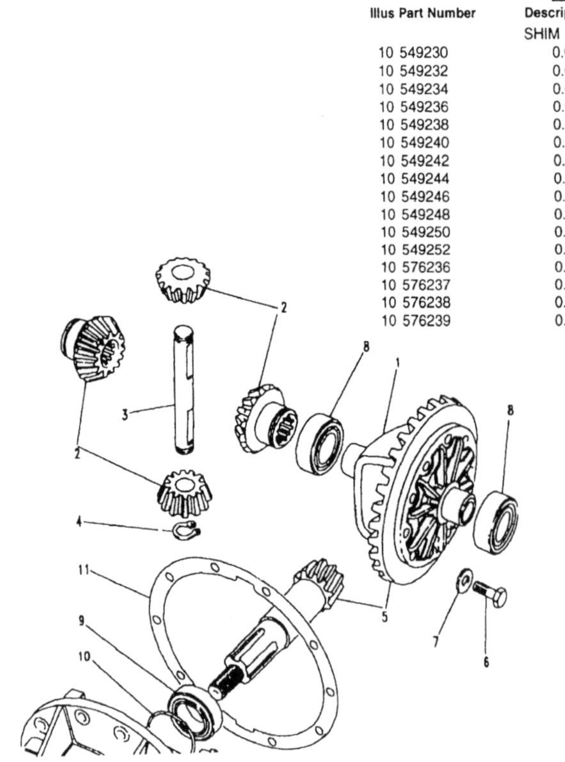

CHANGE POINTS:
(1) From (A)28S 00001 To (A)28S 44171
(2) From (A)32S 00001 To (A)32S 29118

Illus	Part Number	Description	Quantity	Change Point	Remarks
		SHIM			
10	FTC3853	1.548mm	A/R	Note (1); Note (2)	
10	FTC3855	1.498mm	A/R		
10	FTC3857	1.448mm	A/R		
10	FTC3859	1.398mm	A/R		
10	FTC3861	1.348mm	A/R		
10	FTC3863	1.298mm	A/R		
10	FTC3865	1.248mm	A/R		
10	FTC3867	1.198mm	A/R		
11	7316	Gasket-differential housing to axle case	1		

CHANGE POINTS:
(1) From (A)32S 29119 To (A)32S 99999
(2) From (A)28S 44172 To (A)28S 99999

Hubs & Drive Shafts-To (V) JA032850

Illus	Part Number	Description	Quantity	Change Point	Remarks
1	571882	Shaft-rear axle half-RH-10 spline shaft	1		
2	571883	Shaft-rear axle half-LH-10 spline shaft	1		
3	571752	Gasket-axle shaft drive member	2		
4	279146	Bolt	10		
5	WM600061L	Washer-3/8"-square	10		
6	FTC3648	Gasket-stub axle-asbestos free	2		
7	FRC8005	Shaft-stub rear	2		
8	SX110251M	Screw-hexagon socket-M10 x 25	12		
9	WL110001L	Washer-M10	12		
10	FRC2310	Plate-front & rear stub axle locking	2		
11	FTC2783	Seal assembly-front/rear hub-outer	NLA		Use FTC4785.
12	RTC3429	Bearing-front/rear axle hub	4		
11	FTC4785	Seal assembly-front/rear hub-outer	2		
13	FRC8532	HUB ASSEMBLY- REAR	2		
14	FRC5926	• Stud	10		
15	FRC8222	Seal assembly-front/rear hub-inner	2		
16	FRC8227	Spacer	2		
17	FRC8700	Locknut	4		
18	FRC8002	Washer-front & rear hub nut tab	2		

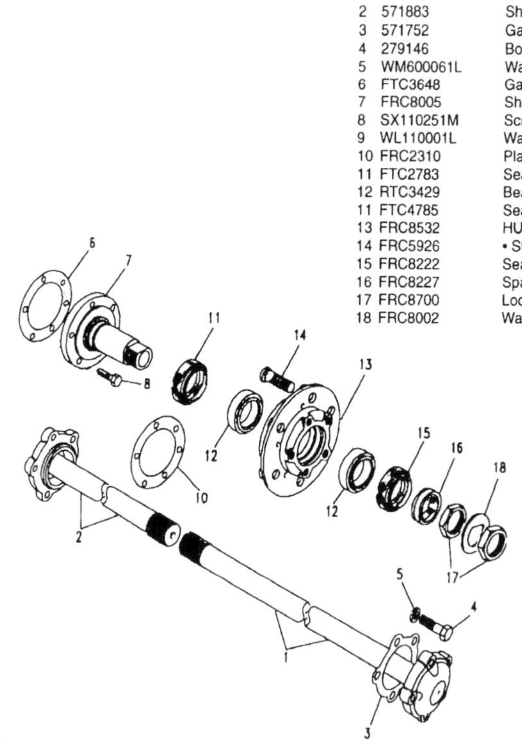

Hubs & Drive Shafts-From (V) JA032851

Illus	Part Number	Description	Quantity	Change Point	Remarks
		SHAFT-REAR AXLE HALF			
1	FTC3187	LH-10 spline shaft	1	Note (1); Note (2)	
1	FTC3186	RH-10 spline shaft	1		
1	FTC3271	LH-24 spline shaft	1	Note (3); Note (4)	
1	FTC3270	RH-24 spline shaft	1		
2	571752	Gasket-axle shaft drive member	2		
3	BX110095M	Bolt-M10 x 45	10		
4	WM600061L	Washer-3/8"-square	10		
5	FTC3650	Gasket-stub axle-asbestos free	2		Antilock Brakes
5	FTC3648	Gasket-stub axle-asbestos free	2		Except Antilock Brakes
6	FTC3188	Axle-rear suspension hub stub	2		
7	FTC3145	Seal assembly-front/rear hub	NLA		Use FTC5268.
7	FTC5268	Seal assembly-front/rear hub	2		
8	SX110251M	Screw-hexagon socket-M10 x 25	12	To (V) VA 731771	
8	SX110257M	Screw-hexagon socket-M10 x 25	12	From (V) VA 731772	
9	WL110001L	Washer-M10	12		
10	FTC1378	Plate-front & rear stub axle locking	2		Antilock Brakes
10	FRC2310	Plate-front & rear stub axle locking	2		Except Antilock Brakes
11	FTC2783	Seal assembly-front/rear hub-outer	NLA		Use FTC4785.
11	FTC4785	Seal assembly-front/rear hub-outer	2		
12	RTC3429	Bearing-front/rear axle hub	4		
13	FTC860	Hub assembly-front & rear	NLA		Use FTC942.
13	FTC942	HUB ASSEMBLY-FRONT & REAR	2		
15	FRC5926	• Stud			
14	FRC8700	Locknut	4		
15	FRC5926	Stud	10		
16	FTC3185	Washer-front & rear hub nut tab	NLA	Note (5)	Use FTC5241.
16	FTC5241	Washer-front & rear hub nut tab	2	Note (6)	
17	FTC3179	Washer-lock-outer	2		

CHANGE POINTS:

(1) From (A)28S 00001 To (A)28S 44171
(2) From (A)32S 00001 To (A)32S 29121
(3) From (A)28S 44172 To (A)28S 99999
(4) From (A)32S 29122 To (A)32S 99999
(5) From (V) JA 032851 To (V) TA 199999; From (V) TA 500000 To (V) TA 534102; From (V) TA 700000 To (V) TA 703235
(6) From (V) VA 534103 To (V) VA 558898; From (V) VA 703236

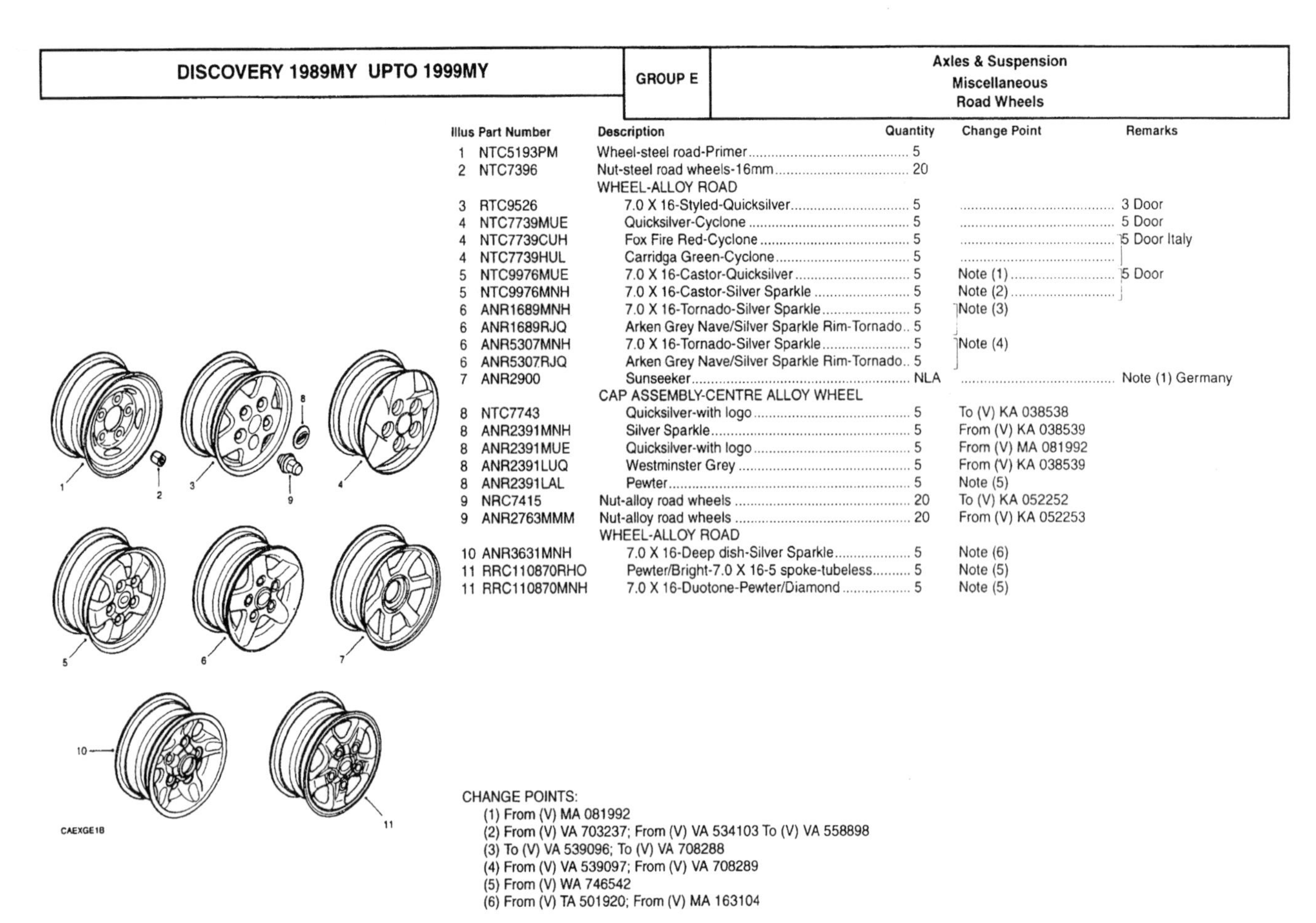

Illus	Part Number	Description	Quantity	Change Point	Remarks
1	NTC5193PM	Wheel-steel road-Primer	5		
2	NTC7396	Nut-steel road wheels-16mm	20		
		WHEEL-ALLOY ROAD			
3	RTC9526	7.0 X 16-Styled-Quicksilver	5		3 Door
4	NTC7739MUE	Quicksilver-Cyclone	5		5 Door
4	NTC7739CUH	Fox Fire Red-Cyclone	5		5 Door Italy
4	NTC7739HUL	Carridga Green-Cyclone	5		
5	NTC9976MUE	7.0 X 16-Castor-Quicksilver	5	Note (1)	5 Door
5	NTC9976MNH	7.0 X 16-Castor-Silver Sparkle	5	Note (2)	
6	ANR1689MNH	7.0 X 16-Tornado-Silver Sparkle	5	Note (3)	
6	ANR1689RJQ	Arken Grey Nave/Silver Sparkle Rim-Tornado	5		
6	ANR5307MNH	7.0 X 16-Tornado-Silver Sparkle	5	Note (4)	
6	ANR5307RJQ	Arken Grey Nave/Silver Sparkle Rim-Tornado	5		
7	ANR2900	Sunseeker	NLA		Note (1) Germany
		CAP ASSEMBLY-CENTRE ALLOY WHEEL			
8	NTC7743	Quicksilver-with logo	5	To (V) KA 038538	
8	ANR2391MNH	Silver Sparkle	5	From (V) KA 038539	
8	ANR2391MUE	Quicksilver-with logo	5	From (V) MA 081992	
8	ANR2391LUQ	Westminster Grey	5	From (V) KA 038539	
8	ANR2391LAL	Pewter	5	Note (5)	
9	NRC7415	Nut-alloy road wheels	20	To (V) KA 052252	
9	ANR2763MMM	Nut-alloy road wheels	20	From (V) KA 052253	
		WHEEL-ALLOY ROAD			
10	ANR3631MNH	7.0 X 16-Deep dish-Silver Sparkle	5	Note (6)	
11	RRC110870RHO	Pewter/Bright-7.0 X 16-5 spoke-tubeless	5	Note (5)	
11	RRC110870MNH	7.0 X 16-Duotone-Pewter/Diamond	5	Note (5)	

CHANGE POINTS:
(1) From (V) MA 081992
(2) From (V) VA 703237; From (V) VA 534103 To (V) VA 558898
(3) To (V) VA 539096; To (V) VA 708288
(4) From (V) VA 539097; From (V) VA 708289
(5) From (V) WA 746542
(6) From (V) TA 501920; From (V) MA 163104

Remarks:
(1) The centre cap for this wheel is fitted at port of entry and can only be bought from German dealers

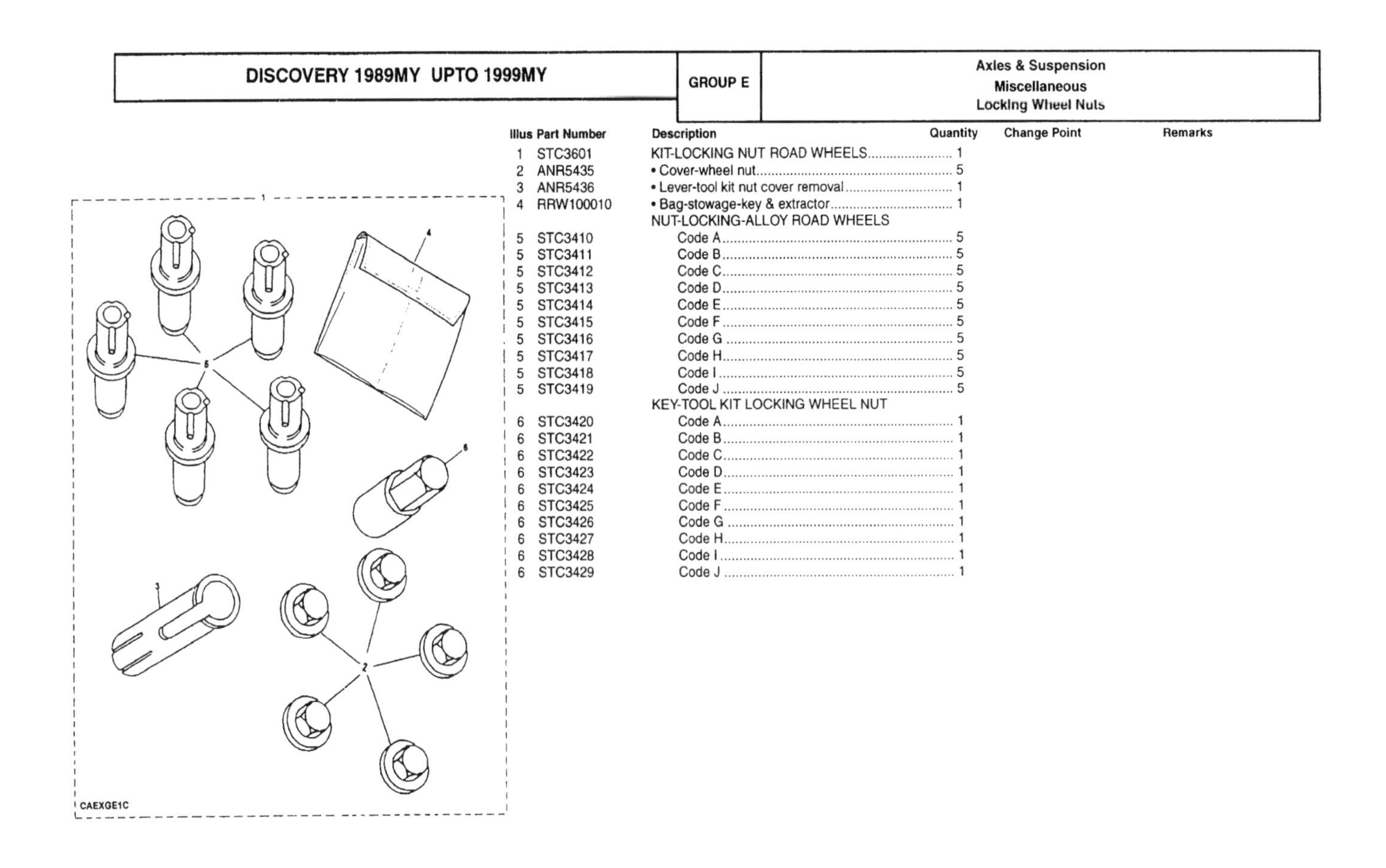

Illus	Part Number	Description	Quantity	Change Point	Remarks
1	STC3601	KIT-LOCKING NUT ROAD WHEELS	1		
2	ANR5435	• Cover-wheel nut	5		
3	ANR5436	• Lever-tool kit nut cover removal	1		
4	RRW100010	• Bag-stowage-key & extractor	1		
		NUT-LOCKING-ALLOY ROAD WHEELS			
5	STC3410	Code A	5		
5	STC3411	Code B	5		
5	STC3412	Code C	5		
5	STC3413	Code D	5		
5	STC3414	Code E	5		
5	STC3415	Code F	5		
5	STC3416	Code G	5		
5	STC3417	Code H	5		
5	STC3418	Code I	5		
5	STC3419	Code J	5		
		KEY-TOOL KIT LOCKING WHEEL NUT			
6	STC3420	Code A	1		
6	STC3421	Code B	1		
6	STC3422	Code C	1		
6	STC3423	Code D	1		
6	STC3424	Code E	1		
6	STC3425	Code F	1		
6	STC3426	Code G	1		
6	STC3427	Code H	1		
6	STC3428	Code I	1		
6	STC3429	Code J	1		

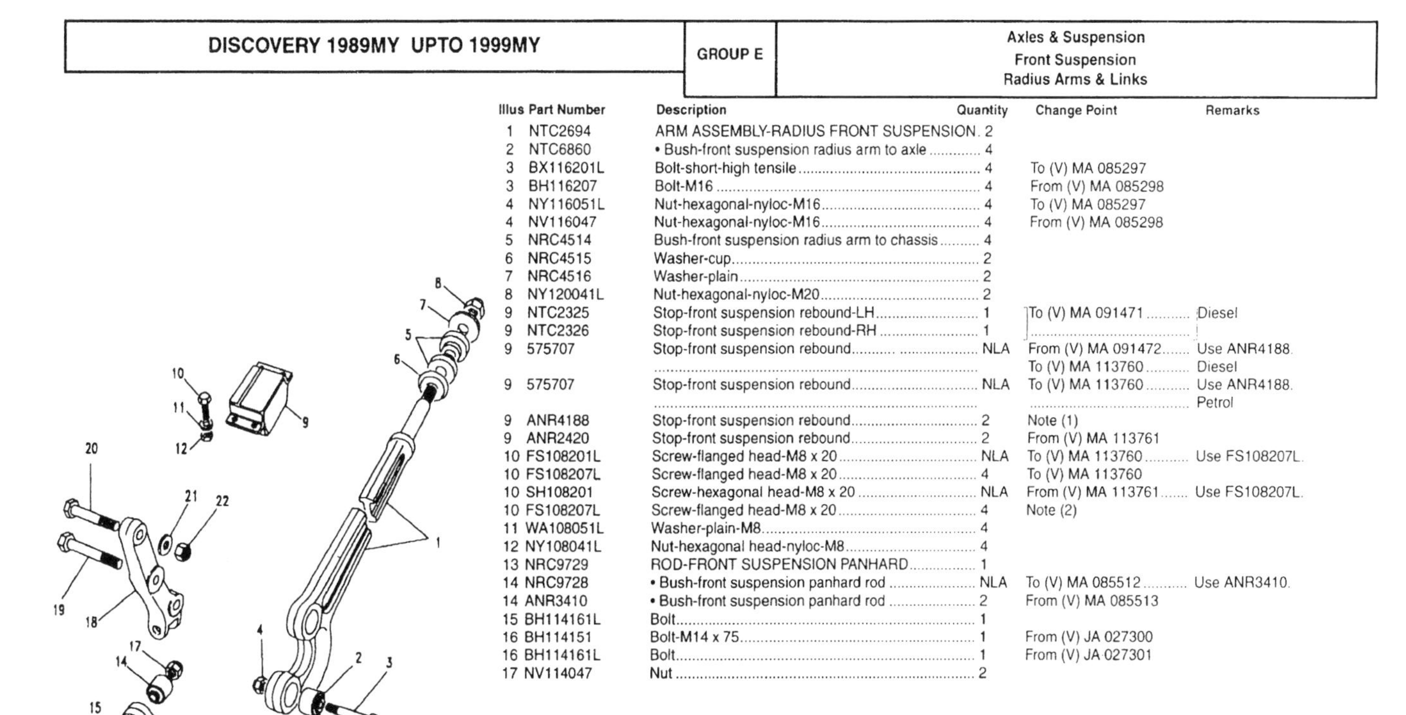

Illus	Part Number	Description	Quantity	Change Point	Remarks
1	NTC2694	ARM ASSEMBLY-RADIUS FRONT SUSPENSION	2		
2	NTC6860	• Bush-front suspension radius arm to axle	4		
3	BX116201L	Bolt-short-high tensile	4	To (V) MA 085297	
3	BH116207	Bolt-M16	4	From (V) MA 085298	
4	NY116051L	Nut-hexagonal-nyloc-M16	4	To (V) MA 085297	
4	NV116047	Nut-hexagonal-nyloc-M16	4	From (V) MA 085298	
5	NRC4514	Bush-front suspension radius arm to chassis	4		
6	NRC4515	Washer-cup	2		
7	NRC4516	Washer-plain	2		
8	NY120041L	Nut-hexagonal-nyloc-M20	2		
9	NTC2325	Stop-front suspension rebound-LH	1	To (V) MA 091471	Diesel
9	NTC2326	Stop-front suspension rebound-RH	1		
9	575707	Stop-front suspension rebound	NLA	From (V) MA 091472 To (V) MA 113760	Use ANR4188. Diesel
9	575707	Stop-front suspension rebound	NLA	To (V) MA 113760	Use ANR4188. Petrol
9	ANR4188	Stop-front suspension rebound	2	Note (1)	
9	ANR2420	Stop-front suspension rebound	2	From (V) MA 113761	
10	FS108201L	Screw-flanged head-M8 x 20	NLA	To (V) MA 113760	Use FS108207L.
10	FS108207L	Screw-flanged head-M8 x 20	4	To (V) MA 113760	
10	SH108201	Screw-hexagonal head-M8 x 20	NLA	From (V) MA 113761	Use FS108207L.
10	FS108207L	Screw-flanged head-M8 x 20	4	Note (2)	
11	WA108051L	Washer-plain-M8	4		
12	NY108041L	Nut-hexagonal head-nyloc-M8	4		
13	NRC9729	ROD-FRONT SUSPENSION PANHARD	1		
14	NRC9728	• Bush-front suspension panhard rod	NLA	To (V) MA 085512	Use ANR3410.
14	ANR3410	• Bush-front suspension panhard rod	2	From (V) MA 085513	
15	BH114161L	Bolt	1		
16	BH114151	Bolt-M14 x 75	1	From (V) JA 027300	
16	BH114161L	Bolt	1	From (V) JA 027301	
17	NV114047	Nut	2		

CHANGE POINTS:
(1) From (V) MA 091472 To (V) MA 113760
(2) From (V) MA 113761

DISCOVERY 1989MY UPTO 1999MY | GROUP E | Axles & Suspension
Front Suspension
Radius Arms & Links

Illus	Part Number	Description	Quantity	Change Point	Remarks
18	NTC9462	Arm-radius front suspension-RHD	1		
18	NTC9461	Arm-radius front suspension-LHD	1		
19	253952	Bolt	1		
20	BH608461	Bolt-hexagonal head	1		
21	WC112081L	Washer-plain-M12 x 28	2		
22	252164	Locknut	NLA		Use NY608041L.

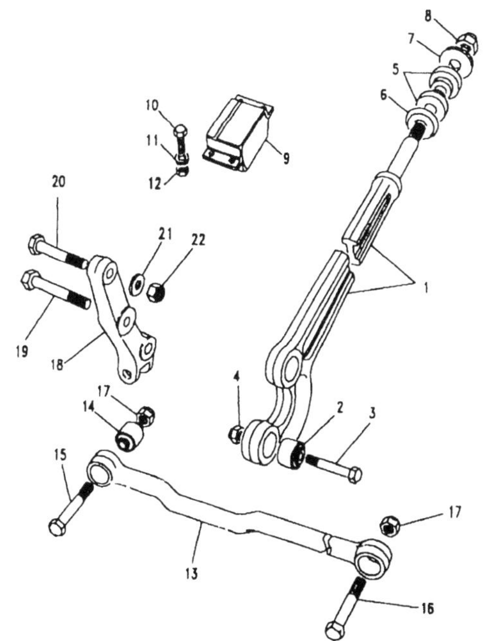

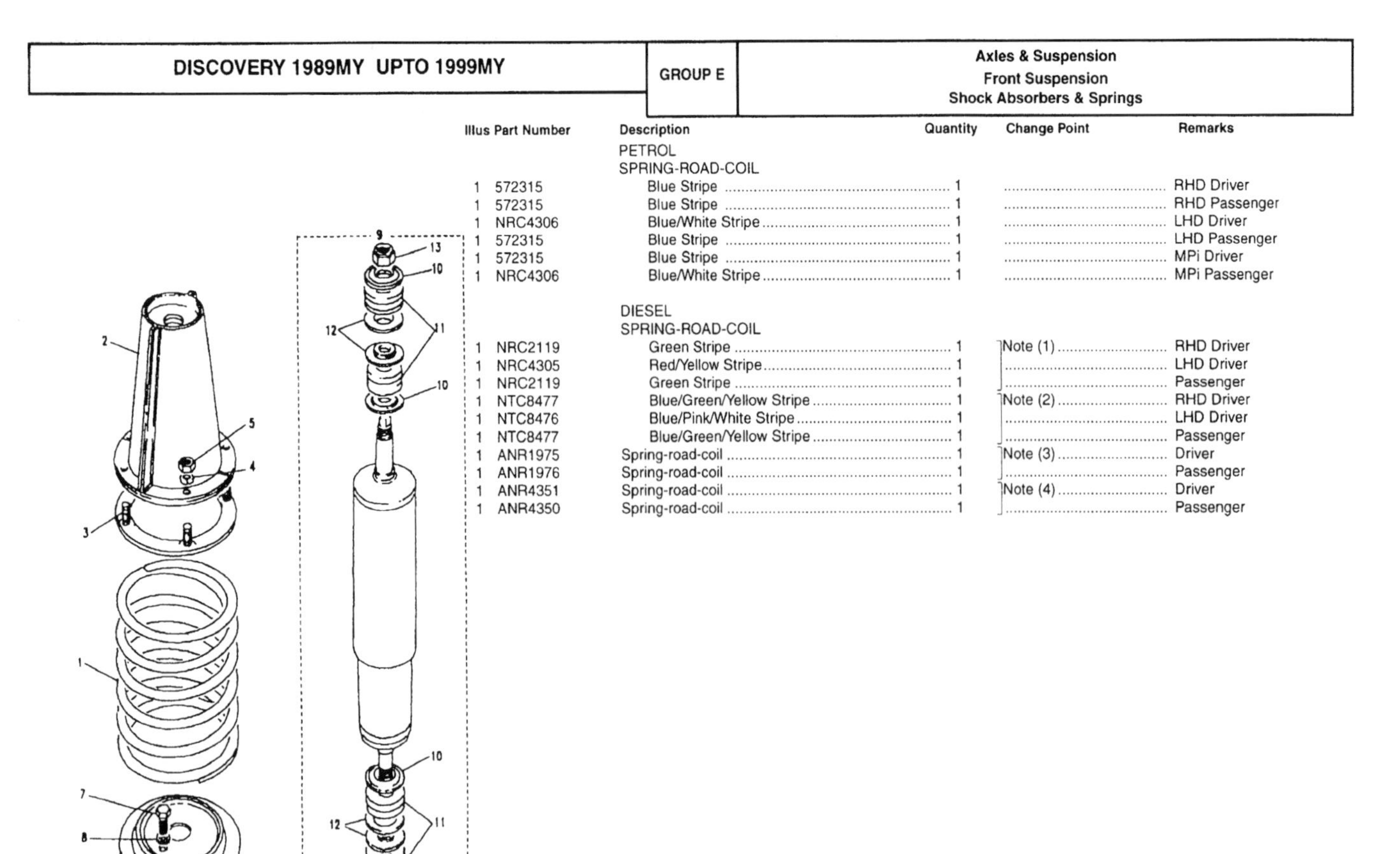

Illus	Part Number	Description	Quantity	Change Point	Remarks
		PETROL			
		SPRING-ROAD-COIL			
1	572315	Blue Stripe	1		RHD Driver
1	572315	Blue Stripe	1		RHD Passenger
1	NRC4306	Blue/White Stripe	1		LHD Driver
1	572315	Blue Stripe	1		LHD Passenger
1	572315	Blue Stripe	1		MPi Driver
1	NRC4306	Blue/White Stripe	1		MPi Passenger
		DIESEL			
		SPRING-ROAD-COIL			
1	NRC2119	Green Stripe	1	Note (1)	RHD Driver
1	NRC4305	Red/Yellow Stripe	1		LHD Driver
1	NRC2119	Green Stripe	1		Passenger
1	NTC8477	Blue/Green/Yellow Stripe	1	Note (2)	RHD Driver
1	NTC8476	Blue/Pink/White Stripe	1		LHD Driver
1	NTC8477	Blue/Green/Yellow Stripe	1		Passenger
1	ANR1975	Spring-road-coil	1	Note (3)	Driver
1	ANR1976	Spring-road-coil	1		Passenger
1	ANR4351	Spring-road-coil	1	Note (4)	Driver
1	ANR4350	Spring-road-coil	1		Passenger

CHANGE POINTS:
(1) From (V) GA 394352 To (V) GA 456980; From (V) HA 456981 To (V) HA 475972
(2) From (V) HA 475973 To (V) HA 486551; From (V) HA 000001 To (V) HA 012332; From (V) JA 012333 To (V) KA 038877
(3) From (V) KA 038878 To (V) MA 116460
(4) From (V) MA 116461

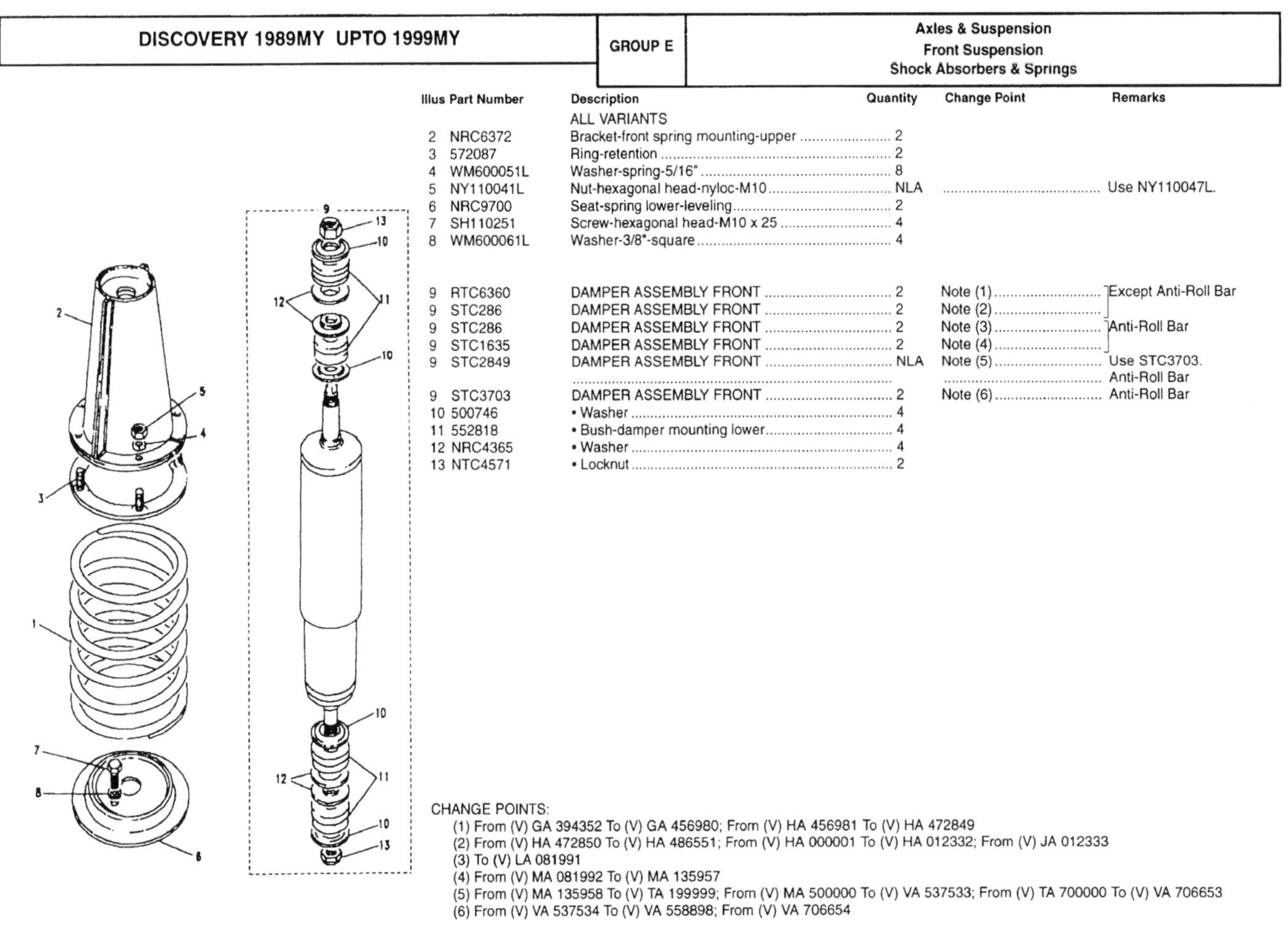

Illus	Part Number	Description	Quantity	Change Point	Remarks
		ALL VARIANTS			
2	NRC6372	Bracket-front spring mounting-upper	2		
3	572087	Ring-retention	2		
4	WM600051L	Washer-spring-5/16"	8		
5	NY110041L	Nut-hexagonal head-nyloc-M10	NLA		Use NY110047L.
6	NRC9700	Seat-spring lower-leveling	2		
7	SH110251	Screw-hexagonal head-M10 x 25	4		
8	WM600061L	Washer-3/8"-square	4		
9	RTC6360	DAMPER ASSEMBLY FRONT	2	Note (1)	Except Anti-Roll Bar
9	STC286	DAMPER ASSEMBLY FRONT	2	Note (2)	
9	STC286	DAMPER ASSEMBLY FRONT	2	Note (3)	Anti-Roll Bar
9	STC1635	DAMPER ASSEMBLY FRONT	2	Note (4)	
9	STC2849	DAMPER ASSEMBLY FRONT	NLA	Note (5)	Use STC3703. Anti-Roll Bar
9	STC3703	DAMPER ASSEMBLY FRONT	2	Note (6)	Anti-Roll Bar
10	500746	• Washer	4		
11	552818	• Bush-damper mounting lower	4		
12	NRC4365	• Washer	4		
13	NTC4571	• Locknut	2		

CHANGE POINTS:
(1) From (V) GA 394352 To (V) GA 456980; From (V) HA 456981 To (V) HA 472849
(2) From (V) HA 472850 To (V) HA 486551; From (V) HA 000001 To (V) HA 012332; From (V) JA 012333
(3) To (V) LA 081991
(4) From (V) MA 081992 To (V) MA 135957
(5) From (V) MA 135958 To (V) TA 199999; From (V) MA 500000 To (V) VA 537533; From (V) TA 700000 To (V) VA 706653
(6) From (V) VA 537534 To (V) VA 558898; From (V) VA 706654

Illus	Part Number	Description	Quantity	Change Point	Remarks
1	NTC6836	Bar-anti-roll front suspension	1	To (V) LA 081991	
1	NTC6837	Bar-anti-roll front suspension	1	From (V) MA 081992	
2	NTC6828	Bush-anti-roll bar-front	2		
3	NTC6776	Bracket-clamp roll bar mounting	2		
4	NY110041L	Nut-hexagonal head-nyloc-M10	NLA		Use NY110047L.
5	SH110351L	Screw-hexagonal head-M10 x 35	4		
6	WA110061L	Washer-plain-M10-standard	4		
7	NTC1888	Joint-ball-lower arm front/rear suspension	2		
8	NV116041L	Locknut-M16	2		
9	552819	Bush-lower arm front/rear suspension	4		
10	264024	Washer	2		
11	NTC8202	Pin	2		
12	WC112081L	Washer-plain-M12 x 28	2		
13	NC112041	Nut-castle-M12	2		
14	PS106321L	Pin-split	1		

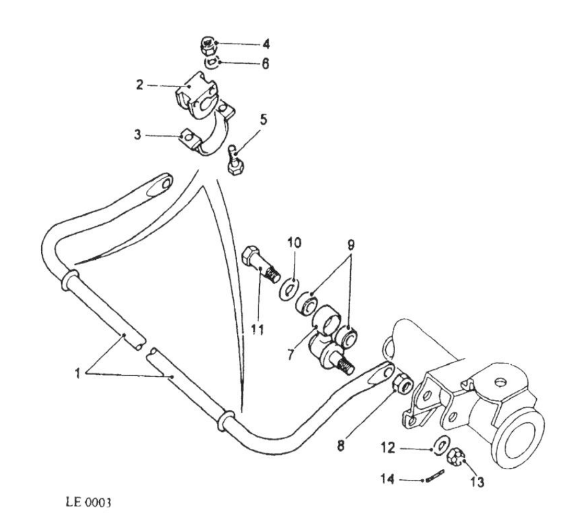

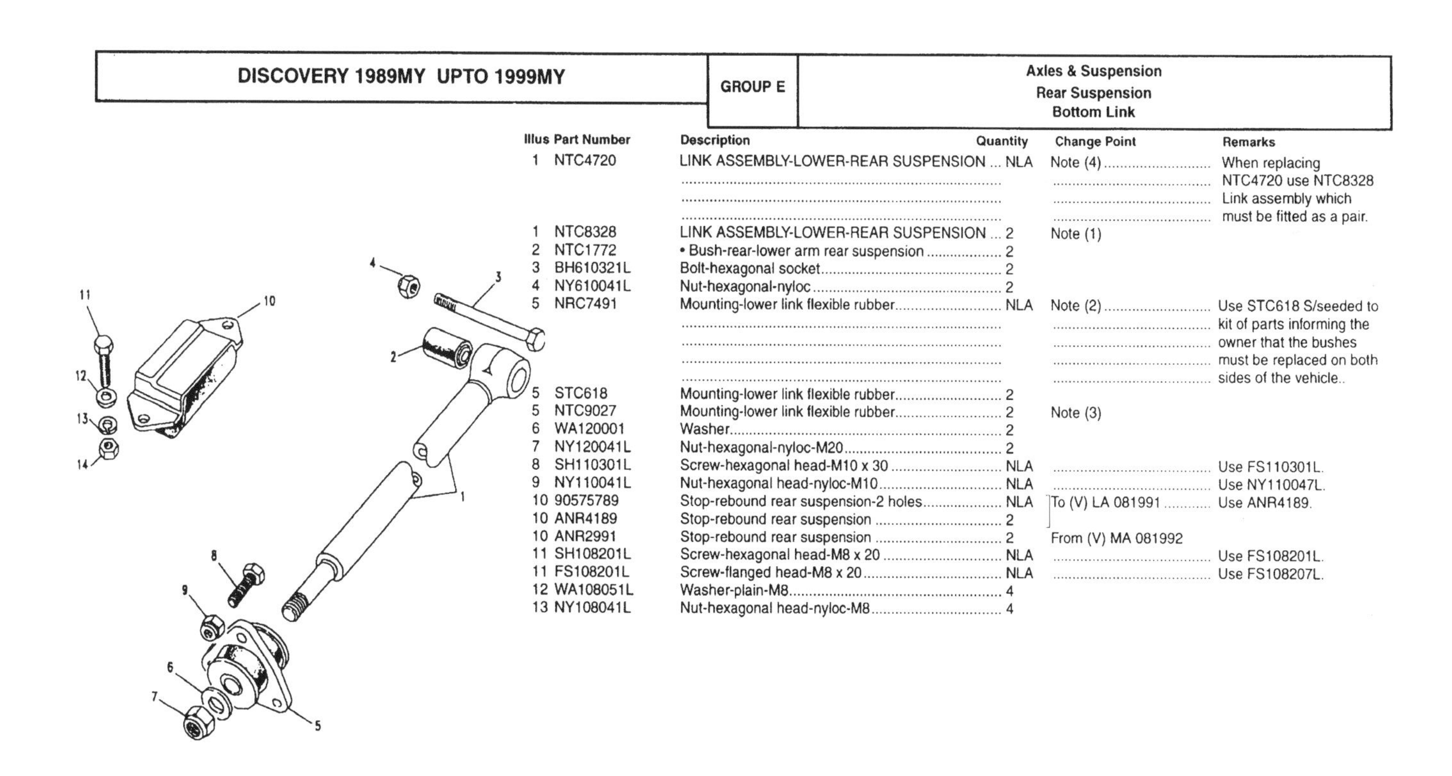

Illus	Part Number	Description	Quantity	Change Point	Remarks
1	NTC4720	LINK ASSEMBLY-LOWER-REAR SUSPENSION	NLA	Note (4)	When replacing NTC4720 use NTC8328 Link assembly which must be fitted as a pair.
1	NTC8328	LINK ASSEMBLY-LOWER-REAR SUSPENSION	2	Note (1)	
2	NTC1772	• Bush-rear-lower arm rear suspension	2		
3	BH610321L	Bolt-hexagonal socket	2		
4	NY610041L	Nut-hexagonal-nyloc	2		
5	NRC7491	Mounting-lower link flexible rubber	NLA	Note (2)	Use STC618 S/seeded to kit of parts informing the owner that the bushes must be replaced on both sides of the vehicle..
5	STC618	Mounting-lower link flexible rubber	2		
5	NTC9027	Mounting-lower link flexible rubber	2	Note (3)	
6	WA120001	Washer	2		
7	NY120041L	Nut-hexagonal-nyloc-M20	2		
8	SH110301L	Screw-hexagonal head-M10 x 30	NLA		Use FS110301L.
9	NY110041L	Nut-hexagonal head-nyloc-M10	NLA		Use NY110047L.
10	90575789	Stop-rebound rear suspension-2 holes	NLA	To (V) LA 081991	Use ANR4189.
10	ANR4189	Stop-rebound rear suspension	2		
10	ANR2991	Stop-rebound rear suspension	2	From (V) MA 081992	
11	SH108201L	Screw-hexagonal head-M8 x 20	NLA		Use FS108201L.
11	FS108201L	Screw-flanged head-M8 x 20	NLA		Use FS108207L.
12	WA108051L	Washer-plain-M8	4		
13	NY108041L	Nut-hexagonal head-nyloc-M8	4		

CHANGE POINTS:
(1) From (V) HA 477040 To (V) HA 486551; From (V) HA 000001 To (V) HA 012332; From (V) JA 012333
(2) To (V) JA 021208
(3) From (V) JA 021209
(4) From (V) GA 394352 To (V) GA 456980; From (V) HA 456981 To (V) HA 477039

DISCOVERY 1989MY UPTO 1999MY

GROUP E

Axles & Suspension
Rear Suspension
Top Link, Fulcrum & Ball Joint

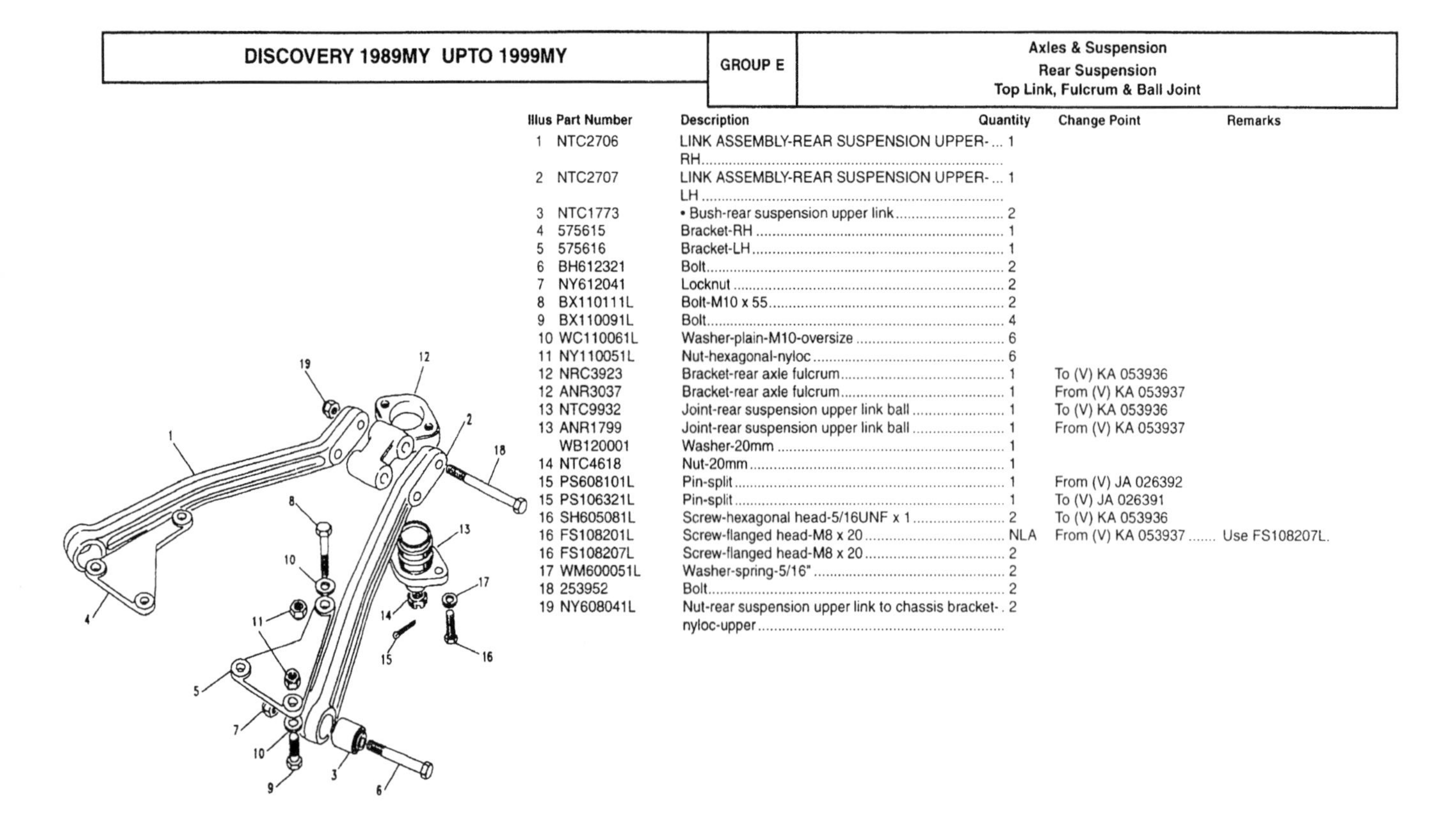

Illus	Part Number	Description	Quantity	Change Point	Remarks
1	NTC2706	LINK ASSEMBLY-REAR SUSPENSION UPPER-RH	1		
2	NTC2707	LINK ASSEMBLY-REAR SUSPENSION UPPER-LH	1		
3	NTC1773	• Bush-rear suspension upper link	2		
4	575615	Bracket-RH	1		
5	575616	Bracket-LH	1		
6	BH612321	Bolt	2		
7	NY612041	Locknut	2		
8	BX110111L	Bolt-M10 x 55	2		
9	BX110091L	Bolt	4		
10	WC110061L	Washer-plain-M10-oversize	6		
11	NY110051L	Nut-hexagonal-nyloc	6		
12	NRC3923	Bracket-rear axle fulcrum	1	To (V) KA 053936	
12	ANR3037	Bracket-rear axle fulcrum	1	From (V) KA 053937	
13	NTC9932	Joint-rear suspension upper link ball	1	To (V) KA 053936	
13	ANR1799	Joint-rear suspension upper link ball	1	From (V) KA 053937	
	WB120001	Washer-20mm	1		
14	NTC4618	Nut-20mm	1		
15	PS608101L	Pin-split	1	From (V) JA 026392	
15	PS106321L	Pin-split	1	To (V) JA 026391	
16	SH605081L	Screw-hexagonal head-5/16UNF x 1	2	To (V) KA 053936	
16	FS108201L	Screw-flanged head-M8 x 20	NLA	From (V) KA 053937	Use FS108207L.
16	FS108207L	Screw-flanged head-M8 x 20	2		
17	WM600051L	Washer-spring-5/16"	2		
18	253952	Bolt	2		
19	NY608041L	Nut-rear suspension upper link to chassis bracket-nyloc-upper	2		

E 49

DISCOVERY 1989MY UPTO 1999MY

GROUP E

Axles & Suspension
Rear Suspension
Shock Absorbers & Springs

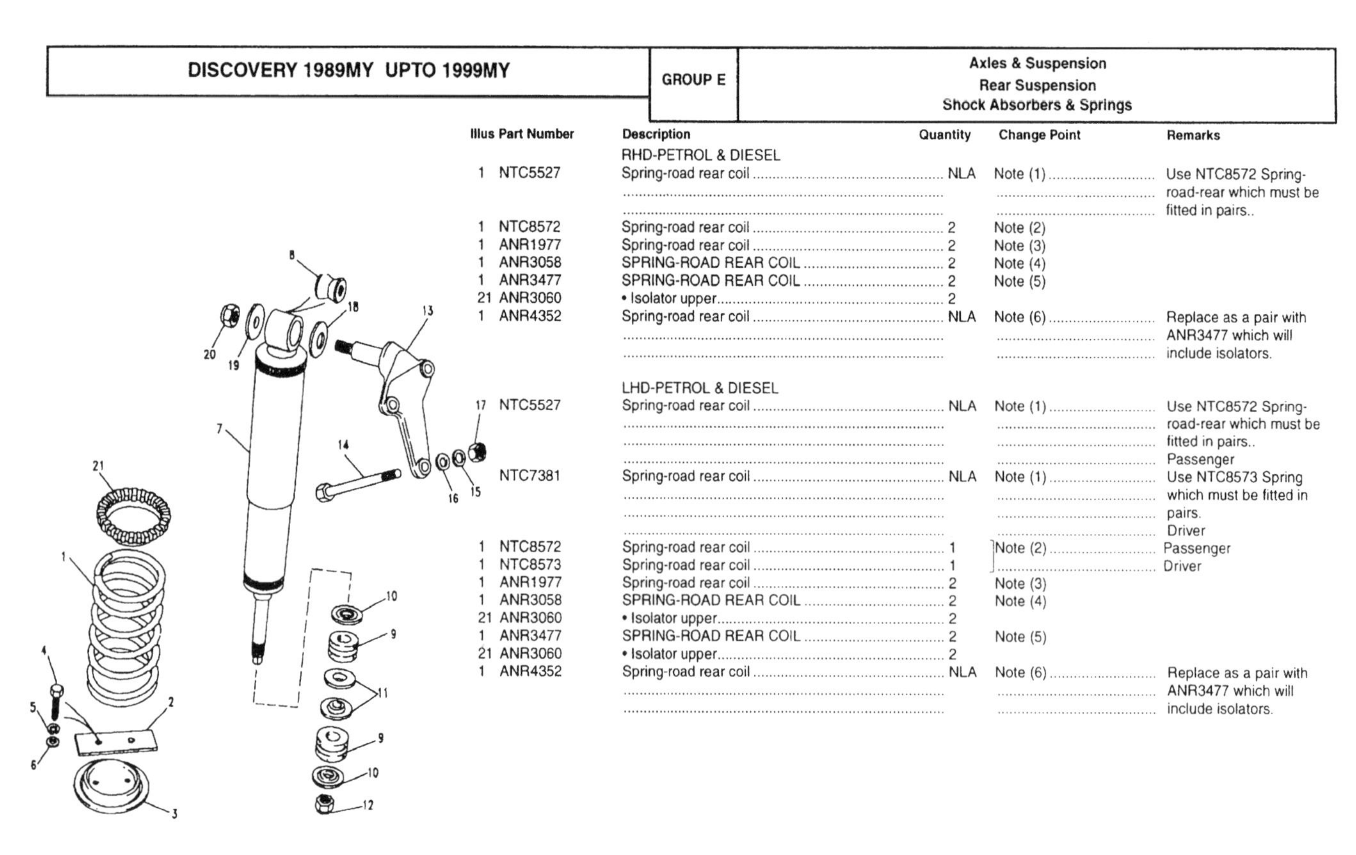

Illus	Part Number	Description	Quantity	Change Point	Remarks
		RHD-PETROL & DIESEL			
1	NTC5527	Spring-road rear coil	NLA	Note (1)	Use NTC8572 Spring-road-rear which must be fitted in pairs..
1	NTC8572	Spring-road rear coil	2	Note (2)	
1	ANR1977	Spring-road rear coil	2	Note (3)	
1	ANR3058	SPRING-ROAD REAR COIL	2	Note (4)	
1	ANR3477	SPRING-ROAD REAR COIL	2	Note (5)	
21	ANR3060	• Isolator upper	2		
1	ANR4352	Spring-road rear coil	NLA	Note (6)	Replace as a pair with ANR3477 which will include isolators.
		LHD-PETROL & DIESEL			
17	NTC5527	Spring-road rear coil	NLA	Note (1)	Use NTC8572 Spring-road-rear which must be fitted in pairs.. Passenger
	NTC7381	Spring-road rear coil	NLA	Note (1)	Use NTC8573 Spring which must be fitted in pairs. Driver
1	NTC8572	Spring-road rear coil	1	Note (2)	Passenger
1	NTC8573	Spring-road rear coil	1		Driver
1	ANR1977	Spring-road rear coil	2	Note (3)	
1	ANR3058	SPRING-ROAD REAR COIL	2	Note (4)	
21	ANR3060	• Isolator upper	2		
1	ANR3477	SPRING-ROAD REAR COIL	2	Note (5)	
21	ANR3060	• Isolator upper	2		
1	ANR4352	Spring-road rear coil	NLA	Note (6)	Replace as a pair with ANR3477 which will include isolators.

CHANGE POINTS:
(1) To (V) JA 015478
(2) From (V) JA 015479 To (V) KA 038877
(3) From (V) KA 038878 To (V) LA 066328
(4) From (V) LA 066329 To (V) LA 081991
(5) From (V) MA 081992 To (V) MA 119533
(6) From (V) MA 119534

E 50

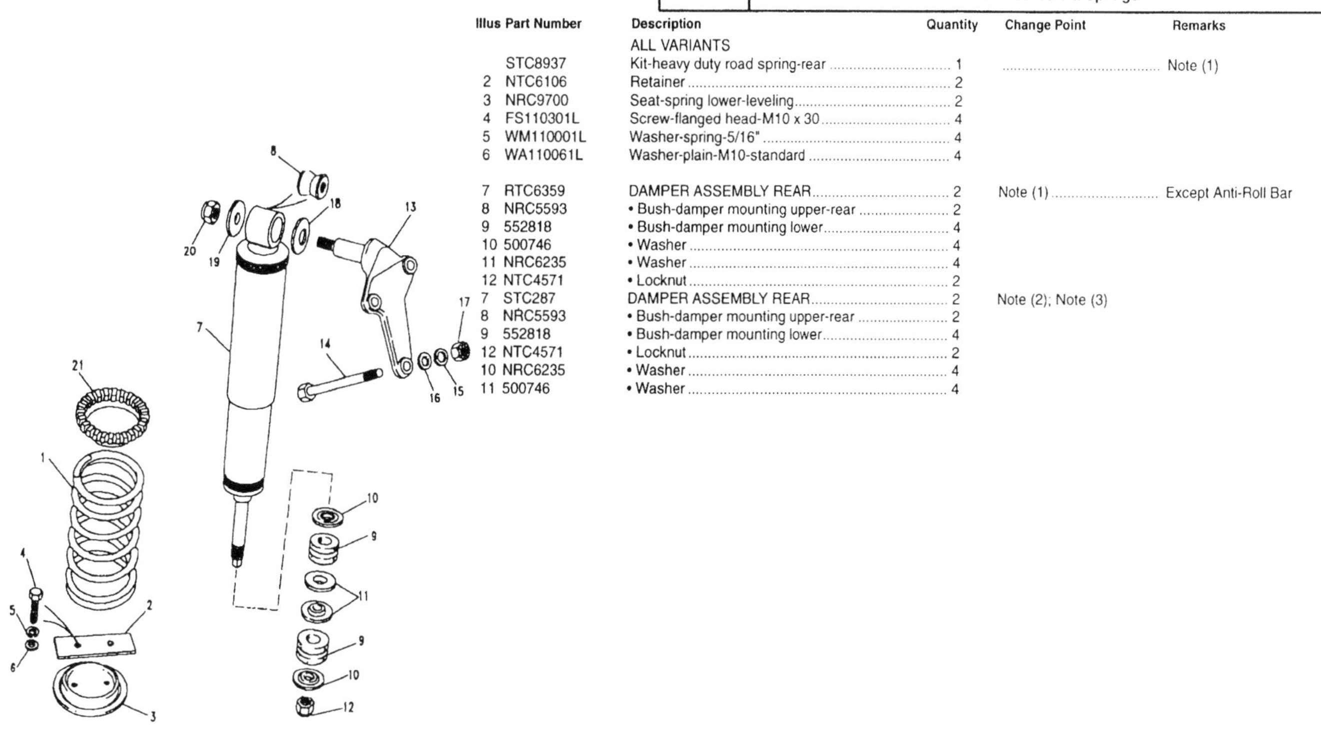

Illus	Part Number	Description	Quantity	Change Point	Remarks
		ALL VARIANTS			
	STC8937	Kit-heavy duty road spring-rear	1		Note (1)
2	NTC6106	Retainer	2		
3	NRC9700	Seat-spring lower-leveling	2		
4	FS110301L	Screw-flanged head-M10 x 30	4		
5	WM110001L	Washer-spring-5/16"	4		
6	WA110061L	Washer-plain-M10-standard	4		
7	RTC6359	DAMPER ASSEMBLY REAR	2	Note (1)	Except Anti-Roll Bar
8	NRC5593	• Bush-damper mounting upper-rear	2		
9	552818	• Bush-damper mounting lower	4		
10	500746	• Washer	4		
11	NRC6235	• Washer	4		
12	NTC4571	• Locknut	2		
7	STC287	DAMPER ASSEMBLY REAR	2	Note (2); Note (3)	
8	NRC5593	• Bush-damper mounting upper-rear	2		
9	552818	• Bush-damper mounting lower	4		
12	NTC4571	• Locknut	2		
10	NRC6235	• Washer	4		
11	500746	• Washer	4		

CHANGE POINTS:
(1) To (V) HA 472849
(2) From (V) HA 472850 To (V) LA 081991 Except Anti-Roll Bar, Anti-Roll Bar
(3) To (V) LA 081991 Anti-Roll Bar

Remarks:
(1) Vehicles fitted with heavy duty rear springs must have a front anti-roll bar fitted and have the rear anti-roll bar removed

DISCOVERY 1989MY UPTO 1999MY | GROUP E | Axles & Suspension
Rear Suspension
Shock Absorbers & Springs

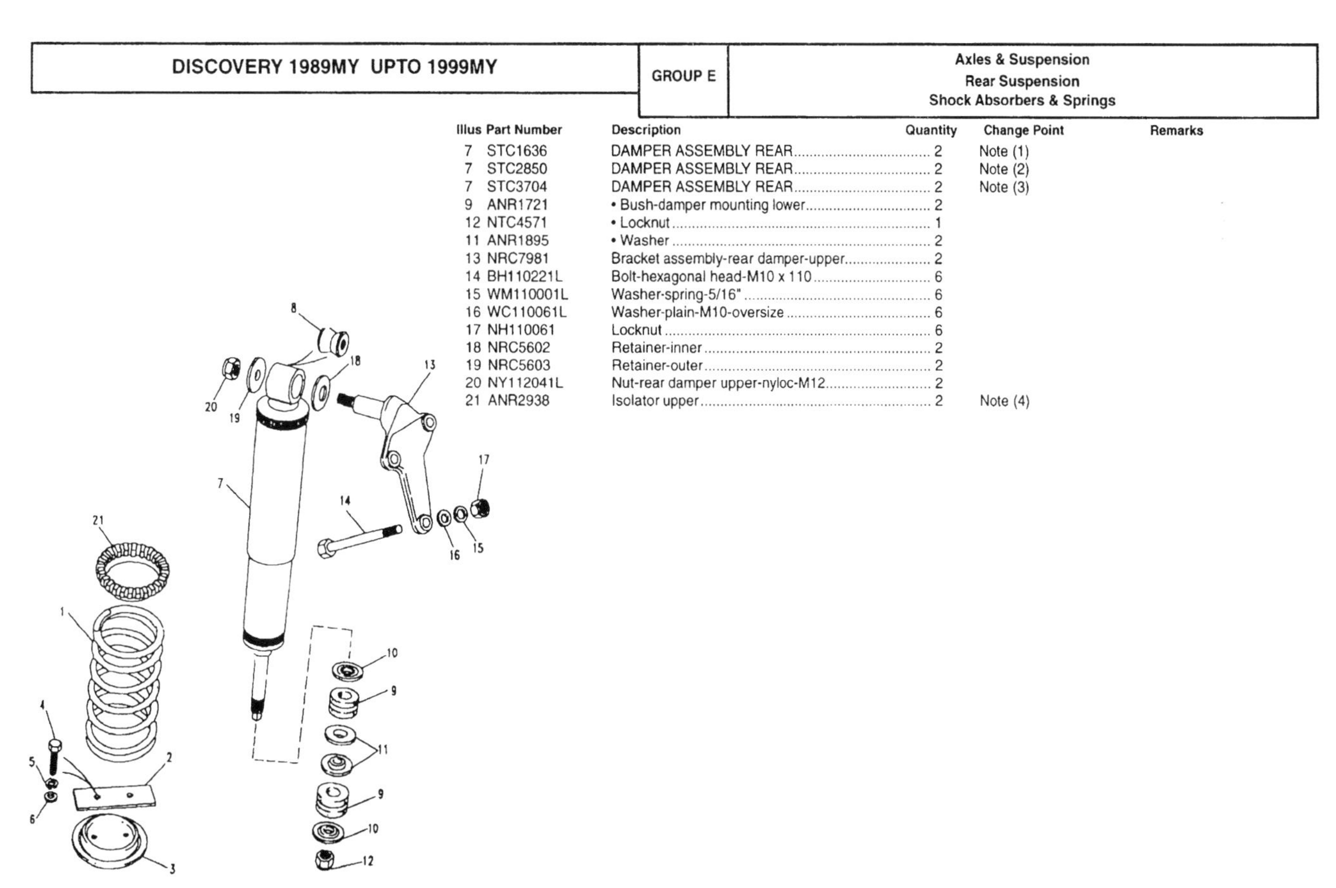

Illus	Part Number	Description	Quantity	Change Point	Remarks
7	STC1636	DAMPER ASSEMBLY REAR	2	Note (1)	
7	STC2850	DAMPER ASSEMBLY REAR	2	Note (2)	
7	STC3704	DAMPER ASSEMBLY REAR	2	Note (3)	
9	ANR1721	• Bush-damper mounting lower	2		
12	NTC4571	• Locknut	1		
11	ANR1895	• Washer	2		
13	NRC7981	Bracket assembly-rear damper-upper	2		
14	BH110221L	Bolt-hexagonal head-M10 x 110	6		
15	WM110001L	Washer-spring-5/16"	6		
16	WC110061L	Washer-plain-M10-oversize	6		
17	NH110061	Locknut	6		
18	NRC5602	Retainer-inner	2		
19	NRC5603	Retainer-outer	2		
20	NY112041L	Nut-rear damper upper-nyloc-M12	2		
21	ANR2938	Isolator upper	2	Note (4)	

CHANGE POINTS:
(1) From (V) MA 081992 To (V) MA 1359578
(2) From (V) MA 135958 To (V) TA 199999; From (V) TA 700000 To (V) VA 706653; From (V) TA 500000 To (V) VA 537533
(3) From (V) VA 537534 To (V) VA 558898; From (V) VA 706654
(4) From (V) MA 081992

Illus	Part Number	Description	Quantity	Change Point	Remarks
1	NRC6221	Bar assembly-anti-roll rear suspension	1	Note (1)	
1	ANR4344	Bar assembly-anti-roll rear suspension	1	Note (2)	
2	NTC7394	Bush-anti-roll bar-rear	2		
3	NTC6776	Bracket-clamp roll bar mounting	2		
4	SH110351L	Screw-hexagonal head-M10 x 35	4		
5	WA110061L	Washer-plain-M10-standard	4		
6	NY110041L	Nut-hexagonal head-nyloc-M10	NLA		Use NY110047L.
7	NTC1888	Joint-ball-lower arm front/rear suspension	2		
8	552819	Bush-lower arm front/rear suspension	2		
9	NTC8202	Pin	2		
10	264024	Washer	2		
11	WC116101	Washer	2		
12	NV116041L	Locknut-M16	2		
13	WC112081L	Washer-plain-M12 x 28	2		
14	NC112041	Nut-castle-M12	2		
15	PS106321L	Pin-split	1		

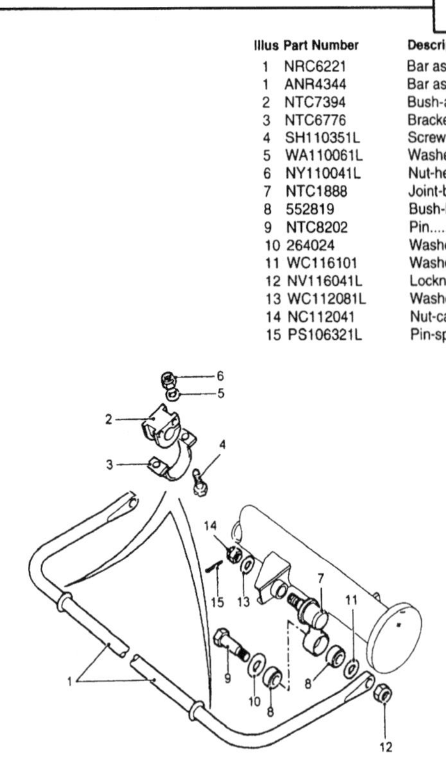

LE0004

CHANGE POINTS:
(1) To (V) MA 139686
(2) From (V) MA 139687

E 53

Steering Lower .... F 7
- Power Steering Hoses .... F 19
- Power Steering Reservoir-Diesel .... F 17
- Power Steering Reservoir-Petrol .... F 18
- Shaft-From (V) MA081992 .... F 9
- Shaft-From(V)HA002272to(V)LA081991 .... F 8
- Shaft-To (V) HA002271 .... F 7
- Steering Box-Power .... F 10
- Track Rods & Damper .... F 15

Steering Upper .... F 1
- Steering Column Lock .... F 6
- Steering Column Shroud .... F 4
- Steering Wheel & Column-From (V) .... F 2
- Steering Wheel & Column-To (V) LA081991 .... F 1

DISCOVERY 1989MY UPTO 1999MY

GROUP F

Steering
Steering Upper
Steering Wheel & Column-To (V) LA081991

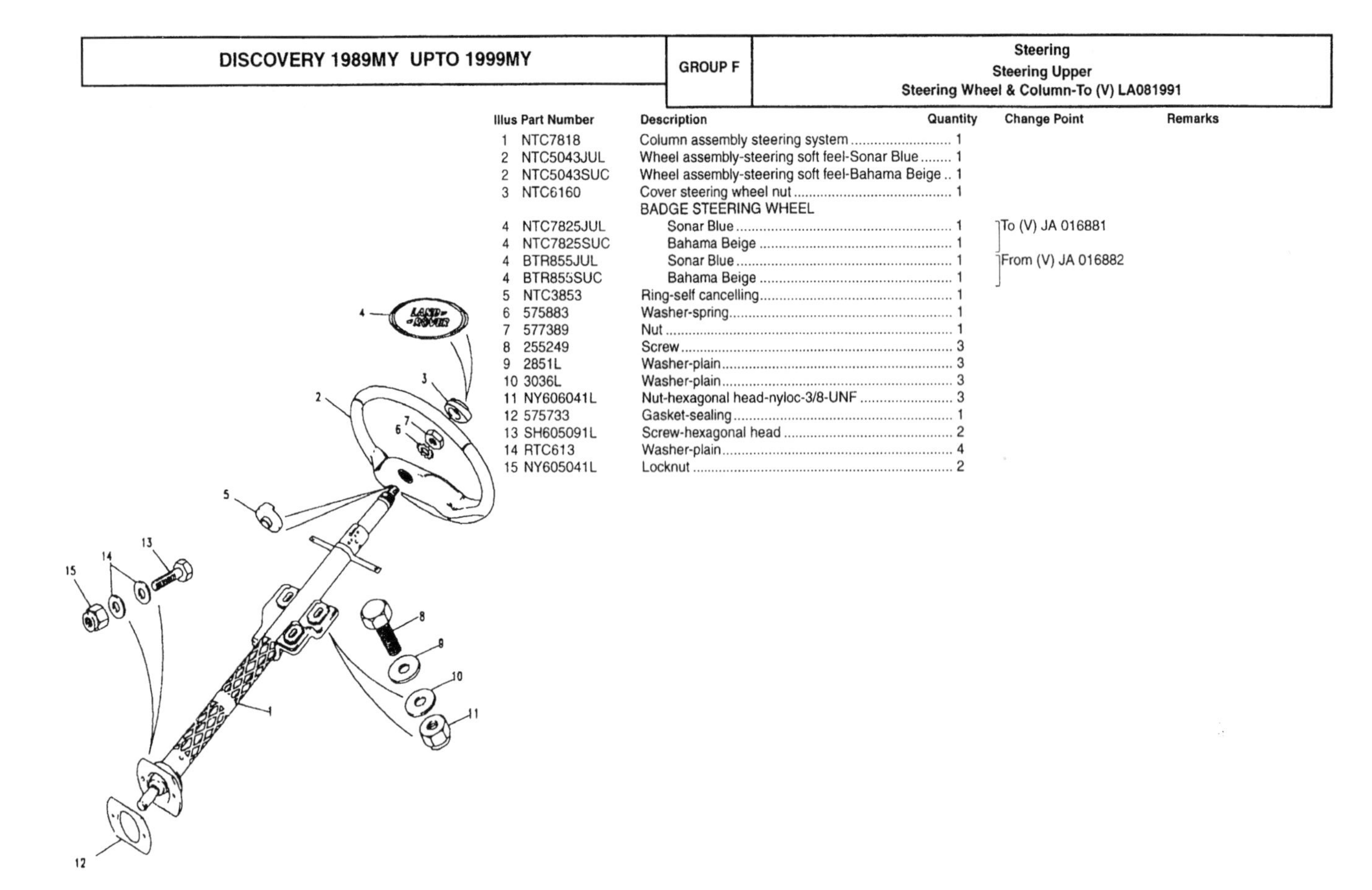

Illus	Part Number	Description	Quantity	Change Point	Remarks
1	NTC7818	Column assembly steering system	1		
2	NTC5043JUL	Wheel assembly-steering soft feel-Sonar Blue	1		
2	NTC5043SUC	Wheel assembly-steering soft feel-Bahama Beige	1		
3	NTC6160	Cover steering wheel nut	1		
		BADGE STEERING WHEEL			
4	NTC7825JUL	Sonar Blue	1	To (V) JA 016881	
4	NTC7825SUC	Bahama Beige	1		
4	BTR855JUL	Sonar Blue	1	From (V) JA 016882	
4	BTR855SUC	Bahama Beige	1		
5	NTC3853	Ring-self cancelling	1		
6	575883	Washer-spring	1		
7	577389	Nut	1		
8	255249	Screw	3		
9	2851L	Washer-plain	3		
10	3036L	Washer-plain	3		
11	NY606041L	Nut-hexagonal head-nyloc-3/8-UNF	3		
12	575733	Gasket-sealing	1		
13	SH605091L	Screw-hexagonal head	2		
14	RTC613	Washer-plain	4		
15	NY605041L	Locknut	2		

DISCOVERY 1989MY UPTO 1999MY

GROUP F

Steering
Steering Upper
Steering Wheel & Column-From (V) MA081992

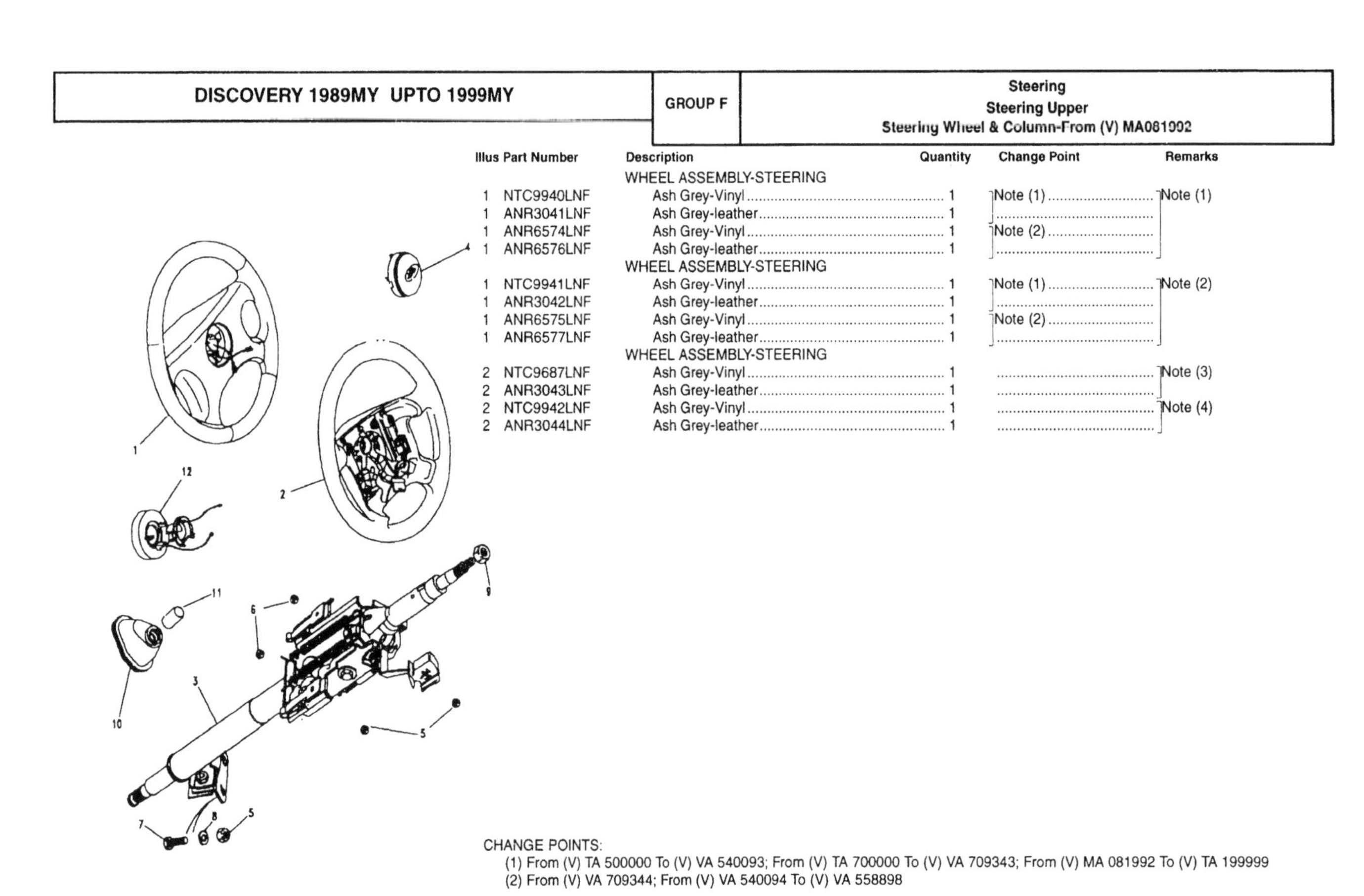

Illus	Part Number	Description	Quantity	Change Point	Remarks
		WHEEL ASSEMBLY-STEERING			
1	NTC9940LNF	Ash Grey-Vinyl	1	Note (1)	Note (1)
1	ANR3041LNF	Ash Grey-leather	1		
1	ANR6574LNF	Ash Grey-Vinyl	1	Note (2)	
1	ANR6576LNF	Ash Grey-leather	1		
		WHEEL ASSEMBLY-STEERING			
1	NTC9941LNF	Ash Grey-Vinyl	1	Note (1)	Note (2)
1	ANR3042LNF	Ash Grey-leather	1		
1	ANR6575LNF	Ash Grey-Vinyl	1	Note (2)	
1	ANR6577LNF	Ash Grey-leather	1		
		WHEEL ASSEMBLY-STEERING			
2	NTC9687LNF	Ash Grey-Vinyl	1		Note (3)
2	ANR3043LNF	Ash Grey-leather	1		
2	NTC9942LNF	Ash Grey-Vinyl	1		Note (4)
2	ANR3044LNF	Ash Grey-leather	1		

CHANGE POINTS:
(1) From (V) TA 500000 To (V) VA 540093; From (V) TA 700000 To (V) VA 709343; From (V) MA 081992 To (V) TA 199999
(2) From (V) VA 709344; From (V) VA 540094 To (V) VA 558898

Remarks:
(1) Except cruise control And drivers air bag
(2) cruise control Except drivers air bag
(3) drivers air bag Except cruise control
(4) cruise control drivers air bag

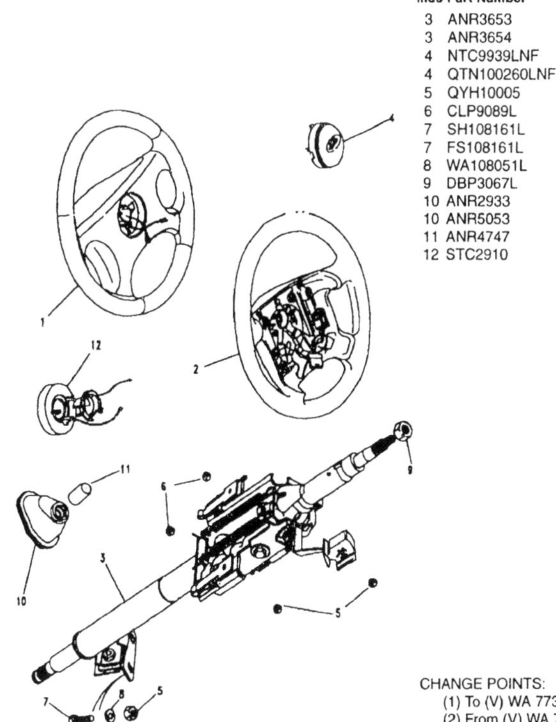

Illus	Part Number	Description	Quantity	Change Point	Remarks
3	ANR3653	Column assembly-tilt steering system	1		Note (1)
3	ANR3654	Column assembly-tilt steering system	1		Note (2)
4	NTC9939LNF	Horn push assembly steering wheel-Ash Grey	1	Note (1)	
4	QTN100260LNF	Horn push assembly steering wheel	1	Note (2)	
5	QYH10005	Nut	4		
6	CLP9089L	Washer-plain	2		
7	SH108161L	Screw-hexagonal head-M8 x 16	NLA		Use FS108161L.
7	FS108161L	Screw-flanged head-M8 x 16	2		
8	WA108051L	Washer-plain-M8	2		
9	DBP3067L	Nut-M12	1		
10	ANR2933	Grommet-sealing	1	Note (3)	
10	ANR5053	GROMMET-SEALING	1	Note (4)	
11	ANR4747	• Bush	1		
12	STC2910	Kit-indicator trip/slip ring	1		Note (3)

CHANGE POINTS:
(1) To (V) WA 773562
(2) From (V) WA 773563
(3) From (V) MA 081992 To (V) TA 165165; From (V) TA 500000 To (V) TA 502446
(4) From (V) TA 165166 To (V) TA 199999; From (V) TA 502447 To (V) VA 558898; From (V) TA 700000

Remarks:
(1) Except cruise control And passive restraint system
(2) cruise control passive restraint system
(3) For steering wheels only with horn push

F 3

DISCOVERY 1989MY UPTO 1999MY | GROUP F | Steering
Steering Upper
Steering Column Shroud

Illus	Part Number	Description	Quantity	Change Point	Remarks
1	MWC8326JUL	Cowl-upper half left hand indicator system-Sonar Blue	1	To (V) LA 081991	
1	MWC8326SUC	Cowl-upper half left hand indicator system-Bahama Beige	1		
		COWL-LOWER HALF LEFT HAND INDICATOR SYSTEM			
2	MXC2858JUL	Sonar Blue	1		
2	MXC2858SUC	Bahama Beige	1		
2	QRB10006LNF	Ash Grey	1	From (V) MA 081992	
3	DA606031	Screw	2		
4	DA606061L	Screw	2		
5	MWC1847	Spacer-lower	2		
6	SE104161L	Screw-M8 x 16	4		
6	QYP10007	Screw	3	From (V) MA 081992	

F 4

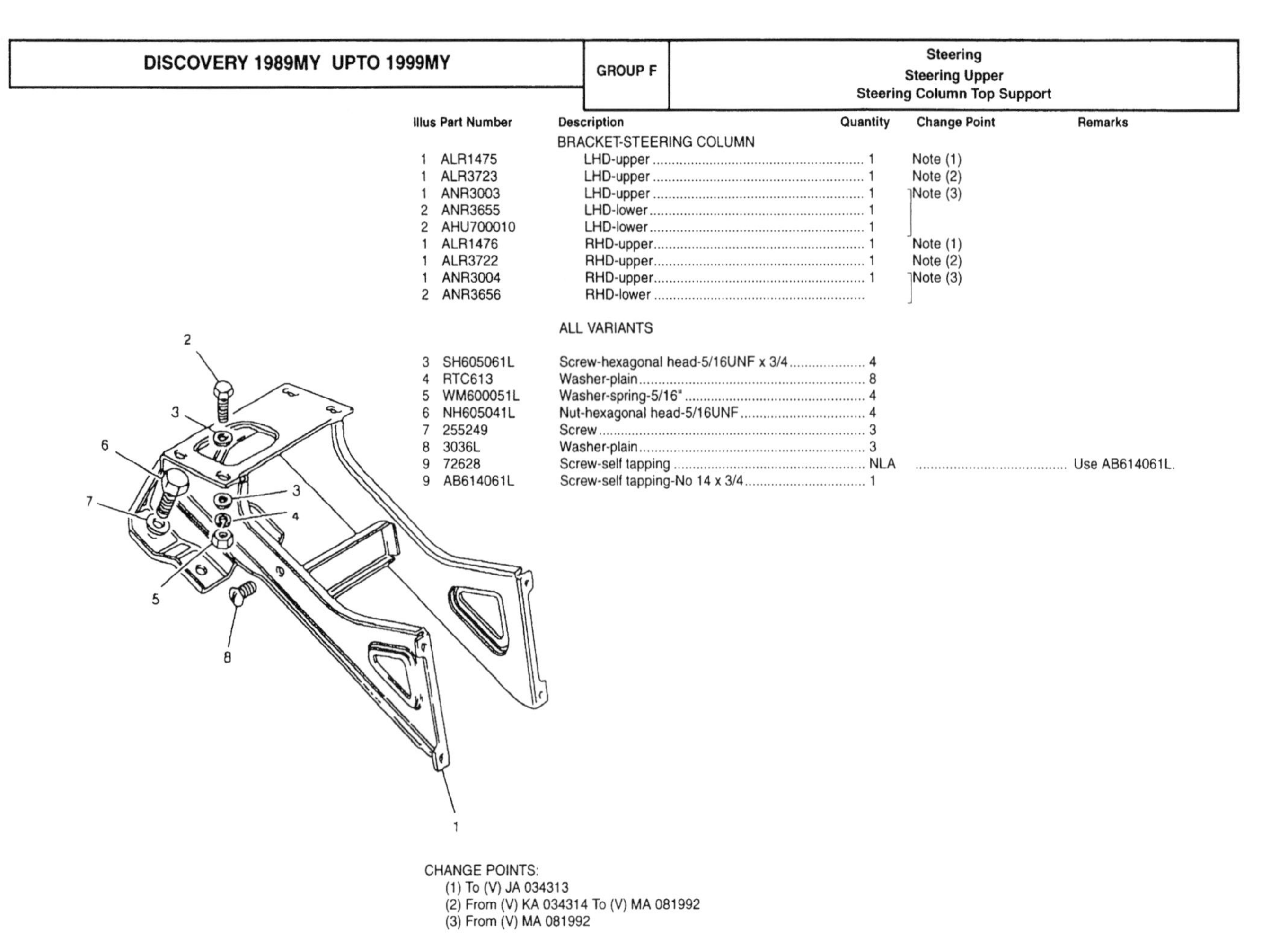

Illus	Part Number	Description	Quantity	Change Point	Remarks
		BRACKET-STEERING COLUMN			
1	ALR1475	LHD-upper	1	Note (1)	
1	ALR3723	LHD-upper	1	Note (2)	
1	ANR3003	LHD-upper	1	Note (3)	
2	ANR3655	LHD-lower	1		
2	AHU700010	LHD-lower	1		
1	ALR1476	RHD-upper	1	Note (1)	
1	ALR3722	RHD-upper	1	Note (2)	
1	ANR3004	RHD-upper	1	Note (3)	
2	ANR3656	RHD-lower			
		ALL VARIANTS			
3	SH605061L	Screw-hexagonal head-5/16UNF x 3/4	4		
4	RTC613	Washer-plain	8		
5	WM600051L	Washer-spring-5/16"	4		
6	NH605041L	Nut-hexagonal head-5/16UNF	4		
7	255249	Screw	3		
8	3036L	Washer-plain	3		
9	72628	Screw-self tapping	NLA		Use AB614061L.
9	AB614061L	Screw-self tapping-No 14 x 3/4	1		

CHANGE POINTS:
(1) To (V) JA 034313
(2) From (V) KA 034314 To (V) MA 081992
(3) From (V) MA 081992

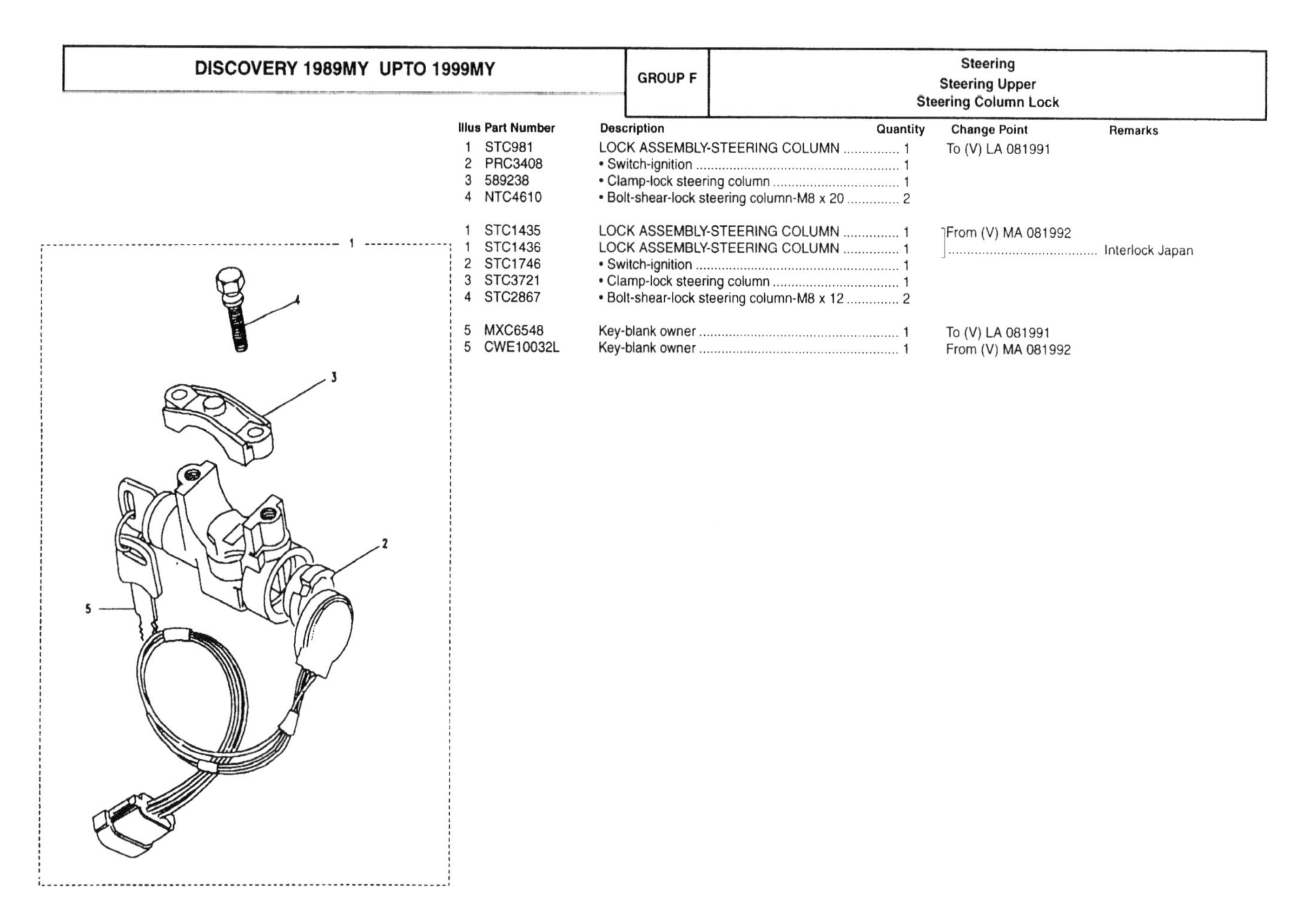

Illus	Part Number	Description	Quantity	Change Point	Remarks
1	STC981	LOCK ASSEMBLY-STEERING COLUMN	1	To (V) LA 081991	
2	PRC3408	• Switch-ignition	1		
3	589238	• Clamp-lock steering column	1		
4	NTC4610	• Bolt-shear-lock steering column-M8 x 20	2		
1	STC1435	LOCK ASSEMBLY-STEERING COLUMN	1	From (V) MA 081992	
1	STC1436	LOCK ASSEMBLY-STEERING COLUMN	1		Interlock Japan
2	STC1746	• Switch-ignition	1		
3	STC3721	• Clamp-lock steering column	1		
4	STC2867	• Bolt-shear-lock steering column-M8 x 12	2		
5	MXC6548	Key-blank owner	1	To (V) LA 081991	
5	CWE10032L	Key-blank owner	1	From (V) MA 081992	

Illus	Part Number	Description	Quantity	Change Point	Remarks
1	RTC5876	SHAFT-UNIVERSAL JOINT STEERING	1	Note (1)	Except LHD and TDi
2	NRC7863	• Joint-universal steering linkage	1		
1	NTC7370	Shaft-universal joint steering	NLA	Note (1)	Use RTC5876 qty 1 with NY108051L qty 2 with NT108041L qty 2 with NTC7369 qty 1 with FB108141 qty 2. LHD TDi
2	NRC7704	Joint-universal steering linkage-lower	1	Note (1)	
3	BX108081	Bolt-steering shaft-hexagonal head-M8	3		
4	NY108041L	Nut-hexagonal head-nyloc-M8	3		
	NTC7369	Heatshield assembly steering gear	1		
3	FB108141L	Bracket- compressor/tensioner assembly air conditioning-flanged head-M8 x 70	2		
4	NT108041L	Nut-hexagonal head	2		
	NY108051L	Locknut-M8	2		

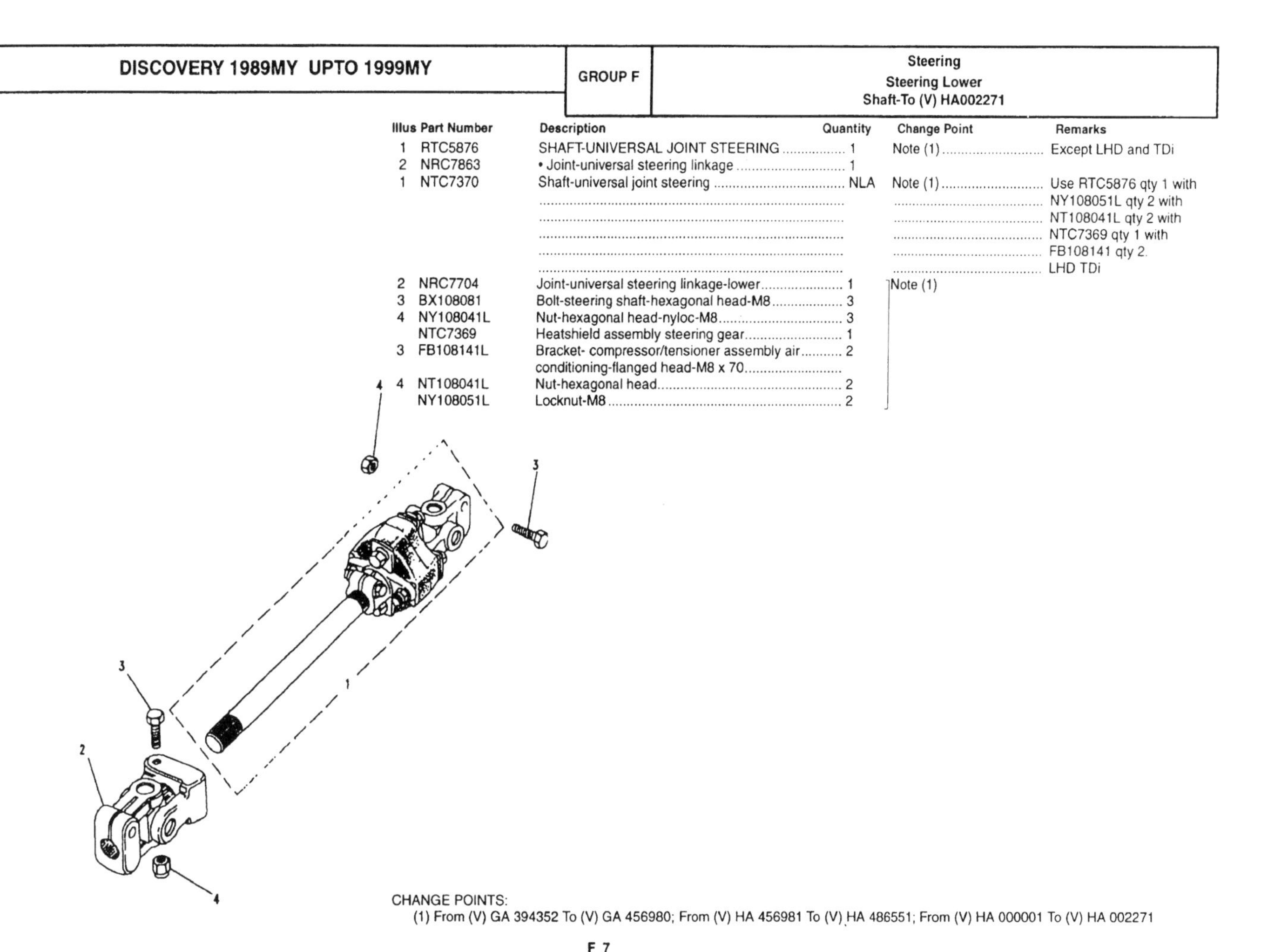

CHANGE POINTS:
(1) From (V) GA 394352 To (V) GA 456980; From (V) HA 456981 To (V) HA 486551; From (V) HA 000001 To (V) HA 002271

F 7

Illus	Part Number	Description	Quantity	Change Point	Remarks
1	NTC8478	SHAFT-UNIVERSAL JOINT STEERING	1		
2	NRC7387	• Joint-universal steering linkage-upper	1		
3	BX108081	Bolt-hexagonal head-M8	3		
4	NY108041L	Nut-hexagonal head-nyloc-M8	3		

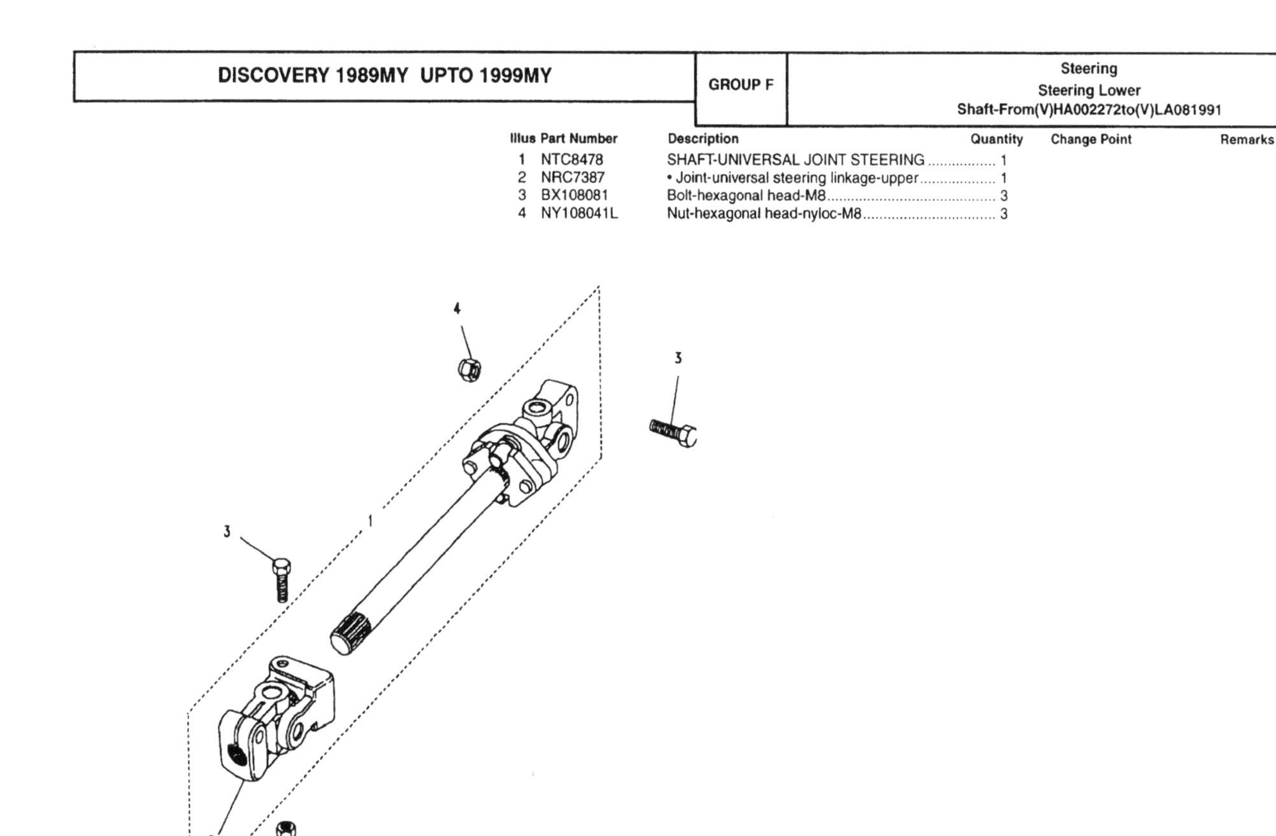

F 8

Illus	Part Number	Description	Quantity	Change Point	Remarks
1	ANR3171	SHAFT-UNIVERSAL JOINT STEERING	1		
2	STC2800	• Joint-universal steering linkage	1		
3	BH108077	• Bolt-M8 x 35	3		
4	ANR3518	Heatshield steering gear	1		

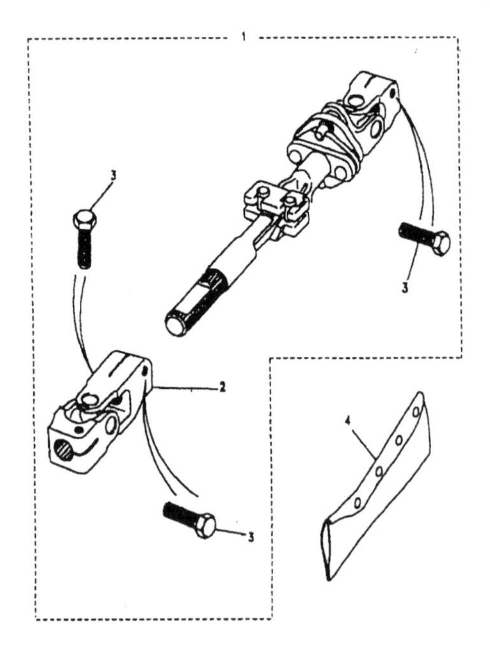

F 9

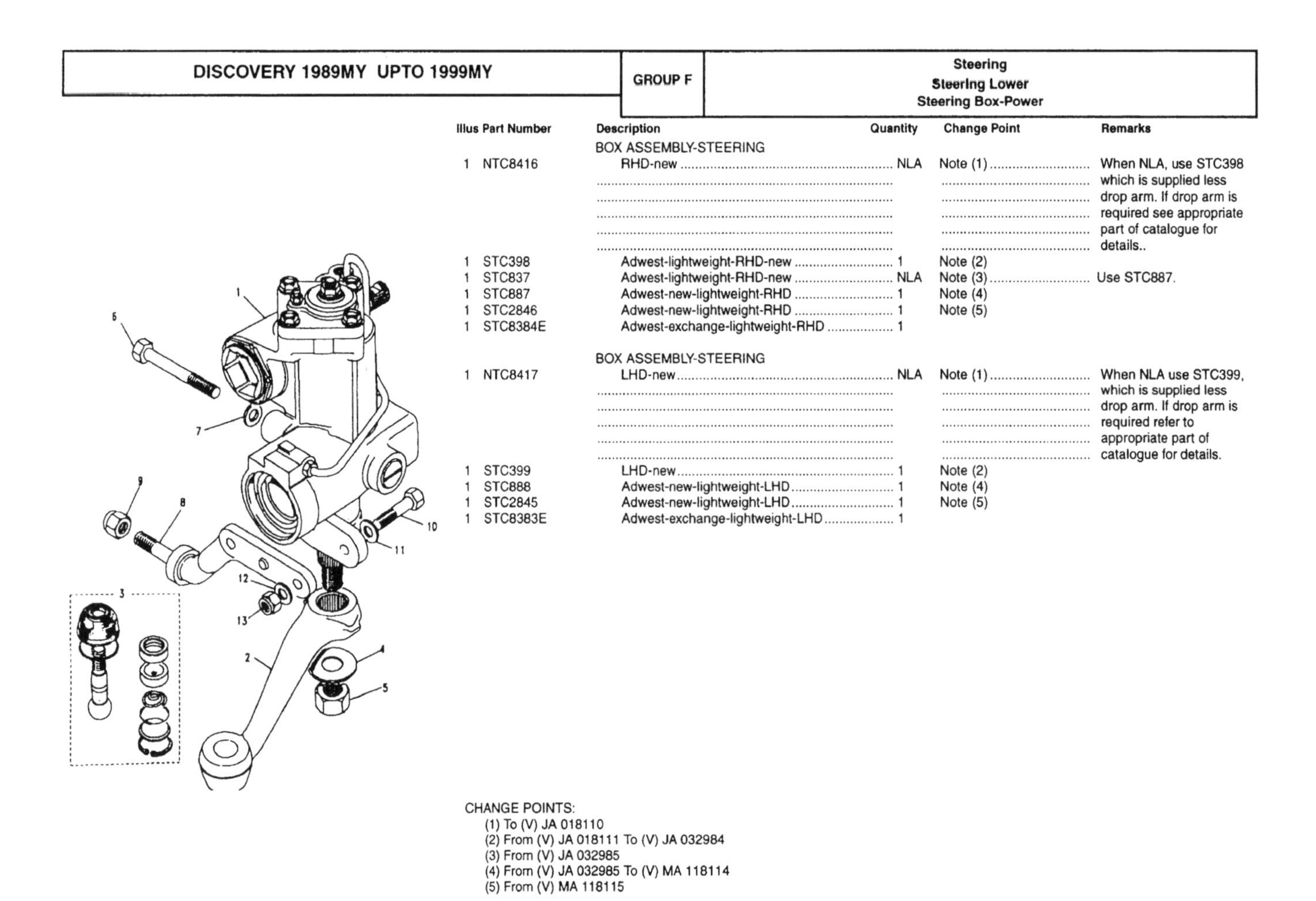

Illus	Part Number	Description	Quantity	Change Point	Remarks
		BOX ASSEMBLY-STEERING			
1	NTC8416	RHD-new	NLA	Note (1)	When NLA, use STC398 which is supplied less drop arm. If drop arm is required see appropriate part of catalogue for details..
1	STC398	Adwest-lightweight-RHD-new	1	Note (2)	
1	STC837	Adwest-lightweight-RHD-new	NLA	Note (3)	Use STC887.
1	STC887	Adwest-new-lightweight-RHD	1	Note (4)	
1	STC2846	Adwest-new-lightweight-RHD	1	Note (5)	
1	STC8384E	Adwest-exchange-lightweight-RHD	1		
		BOX ASSEMBLY-STEERING			
1	NTC8417	LHD-new	NLA	Note (1)	When NLA use STC399, which is supplied less drop arm. If drop arm is required refer to appropriate part of catalogue for details.
1	STC399	LHD-new	1	Note (2)	
1	STC888	Adwest-new-lightweight-LHD	1	Note (4)	
1	STC2845	Adwest-new-lightweight-LHD	1	Note (5)	
1	STC8383E	Adwest-exchange-lightweight-LHD	1		

CHANGE POINTS:
(1) To (V) JA 018110
(2) From (V) JA 018111 To (V) JA 032984
(3) From (V) JA 032985
(4) From (V) JA 032985 To (V) MA 118114
(5) From (V) MA 118115

F 10

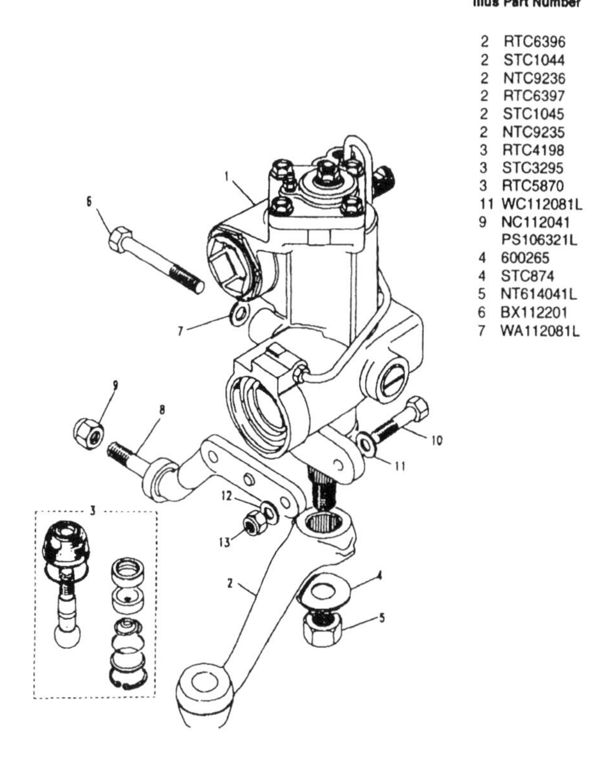

Illus	Part Number	Description	Quantity	Change Point	Remarks
		LEVER-DROP ARM STEERING			
2	RTC6396	RHD	NLA	Note (1)	Use STC1044.
2	STC1044	RHD	1		
2	NTC9236	RHD	1	Note (2)	
2	RTC6397	LHD	NLA	Note (1)	Use STC1045.
2	STC1045	LHD	1		
2	NTC9235	LHD	1	Note (2)	
3	RTC4198	Joint-ball-outer steering gear	NLA	Note (1)	Use STC3295.
3	STC3295	Joint-ball-outer steering gear	1		
3	RTC5870	KIT-BALL JOINT-LH-L/H THREAD	1	Note (2)	
11	WC112081L	• Washer-plain-M12 x 28	1		
9	NC112041	• Nut-castle-M12	1		
	PS106321L	• Pin-split	1		
4	600265	Washer-steering box drop arm nut tab-48mm	1	Note (1)	
4	STC874	Washer-steering box drop arm nut tab-60mm	1	Note (2)	
5	NT614041L	Nut	1		
6	BX112201	Bolt-hexagonal socket-M12 x 100	4		
7	WA112081L	Washer-plain-M12-standard	4		

CHANGE POINTS:
(1) To (V) JA 018109
(2) From (V) JA 018110

Illus	Part Number	Description	Quantity	Change Point	Remarks
		BAR-TIE-BENT FORGING			
8	594947	RHD	NLA	To (V) LA 081464	Use ANR2994 qty 1 with BH607201L qty 2.
8	594946	LHD	NLA	To (V) LA 081464	Use ANR2994 qty 1 with BH607201L qty 2.
8	ANR2994	Bar-tie-bent forging	1	From (V) LA 081465	
9	NY608041L	Nut-nyloc-upper	1		
	4905	Washer-1/2"	1		
10	BH607161L	Bolt-7/16UNF x 2	2	To (V) LA 081464	
11	3261L	Washer-plain-7/16"	2		
12	217245	Washer-plain-No 10	2		
13	NY607041L	Nut-nyloc-7/16UNF	2		
10	BH112121L	Bolt-M12 x 60	2	From (V) LA 081465	Use with ANR2994
11	WA112081L	Washer-plain-M12-standard	2		
12	WA112081L	Washer-plain-M12-standard	2		
13	NY112041L	Nut-nyloc-M12	1		
	STC875	Kit-steering box sector shaft seal	NLA	From (V) JA 027294	Use STC889.
	STC889	Kit-steering box sector shaft seal	1		
10	AFU2798L	Bolt-steering box centralise	1		

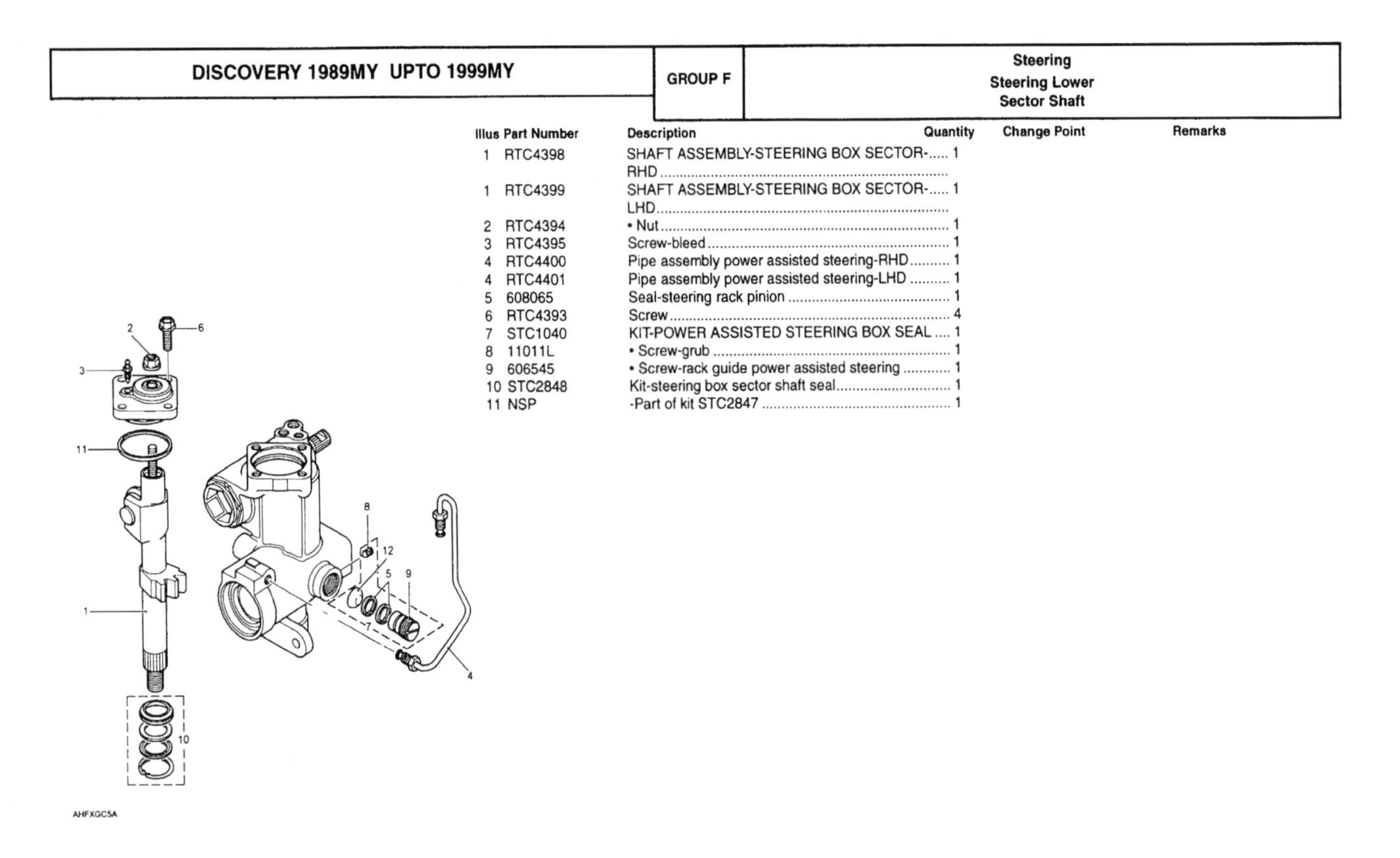

Illus	Part Number	Description	Quantity	Change Point	Remarks
1	RTC4398	SHAFT ASSEMBLY-STEERING BOX SECTOR-RHD	1		
1	RTC4399	SHAFT ASSEMBLY-STEERING BOX SECTOR-LHD	1		
2	RTC4394	• Nut	1		
3	RTC4395	Screw-bleed	1		
4	RTC4400	Pipe assembly power assisted steering-RHD	1		
4	RTC4401	Pipe assembly power assisted steering-LHD	1		
5	608065	Seal-steering rack pinion	1		
6	RTC4393	Screw	4		
7	STC1040	KIT-POWER ASSISTED STEERING BOX SEAL	1		
8	11011L	• Screw-grub	1		
9	606545	• Screw-rack guide power assisted steering	1		
10	STC2848	Kit-steering box sector shaft seal	1		
11	NSP	-Part of kit STC2847	1		

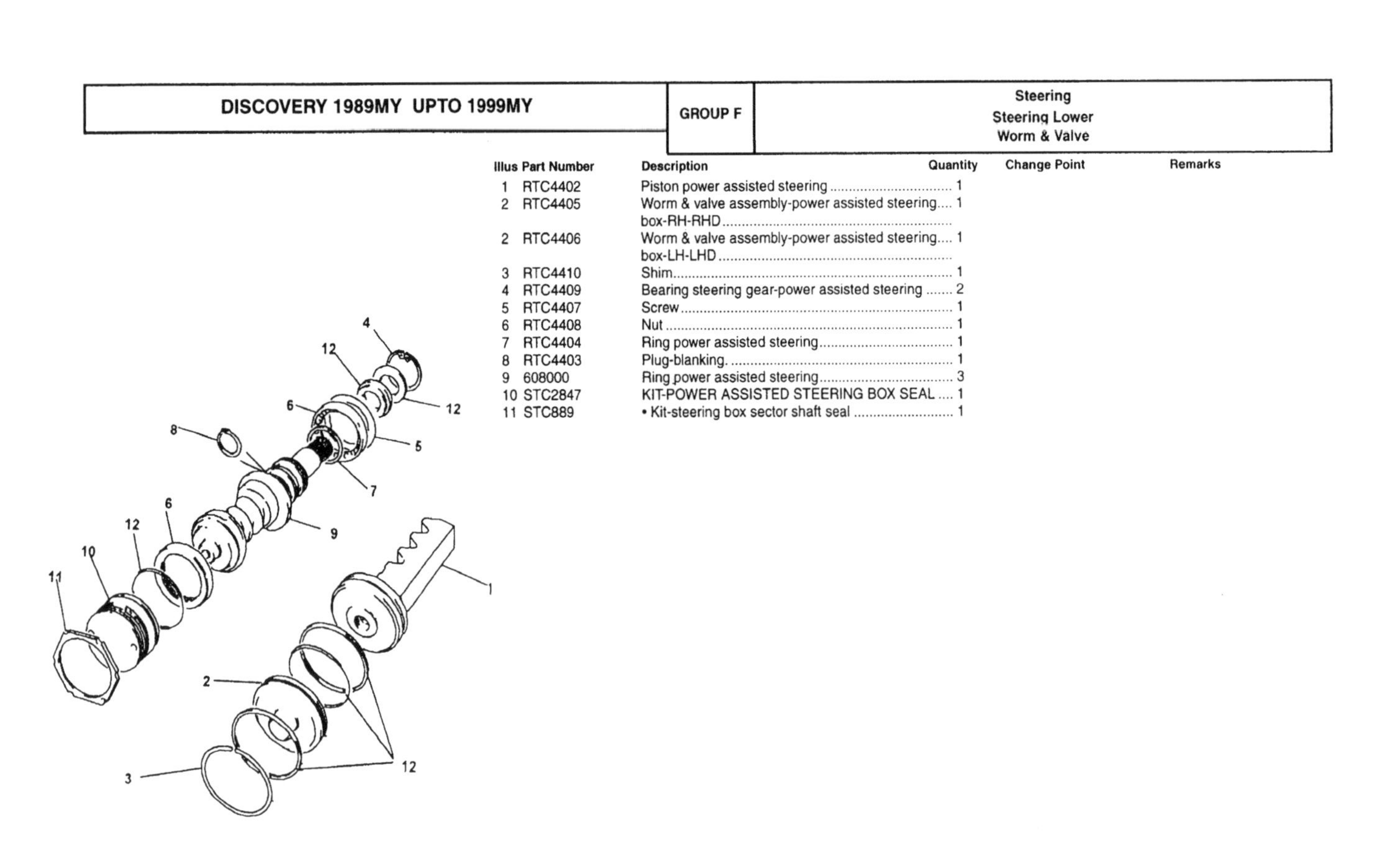

Illus	Part Number	Description	Quantity	Change Point	Remarks
1	RTC4402	Piston power assisted steering	1		
2	RTC4405	Worm & valve assembly-power assisted steering box-RH-RHD	1		
2	RTC4406	Worm & valve assembly-power assisted steering box-LH-LHD	1		
3	RTC4410	Shim	1		
4	RTC4409	Bearing steering gear-power assisted steering	2		
5	RTC4407	Screw	1		
6	RTC4408	Nut	1		
7	RTC4404	Ring power assisted steering	1		
8	RTC4403	Plug-blanking.	1		
9	608000	Ring power assisted steering	3		
10	STC2847	KIT-POWER ASSISTED STEERING BOX SEAL	1		
11	STC889	• Kit-steering box sector shaft seal	1		

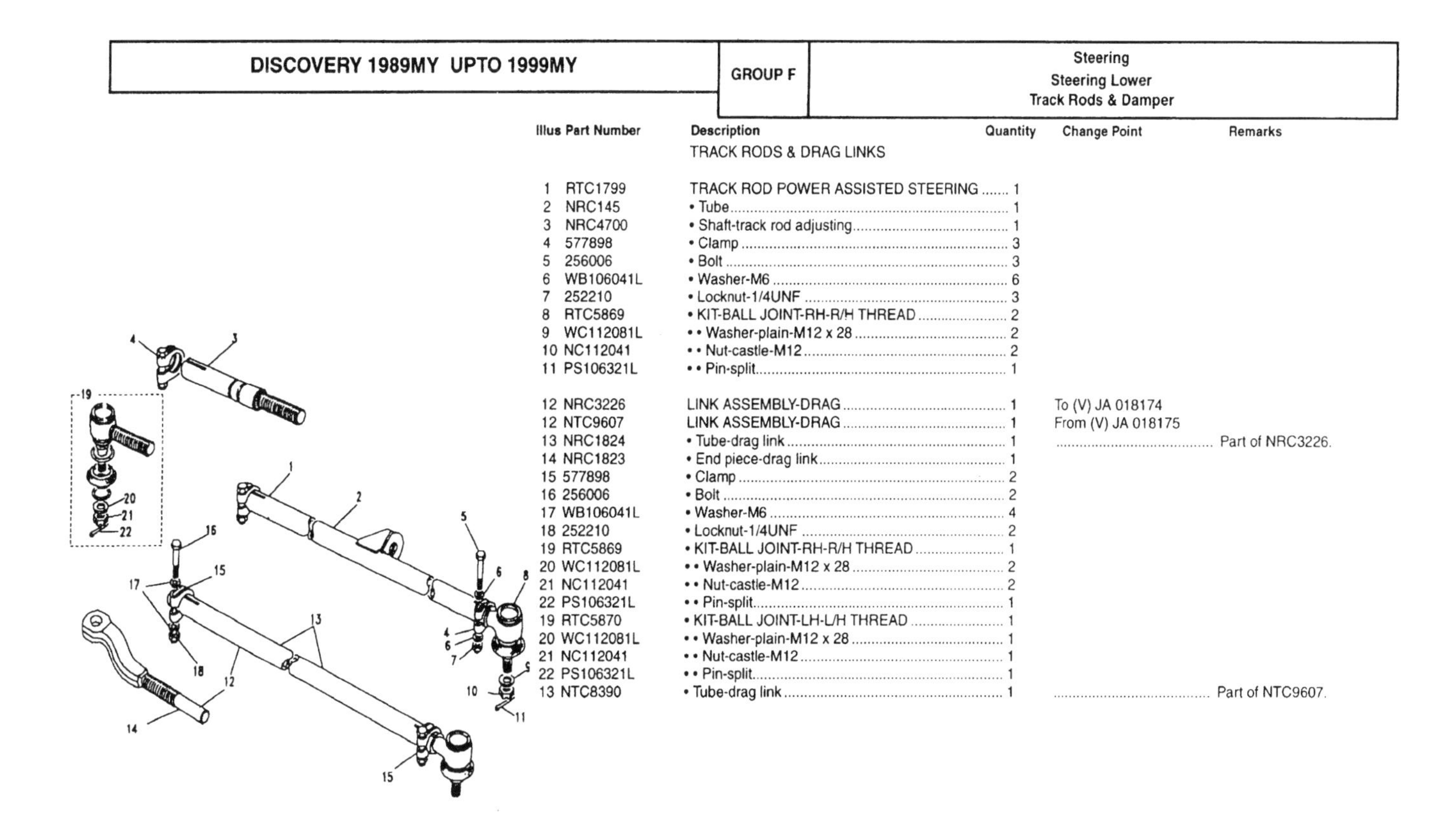

Illus	Part Number	Description	Quantity	Change Point	Remarks
		TRACK RODS & DRAG LINKS			
1	RTC1799	TRACK ROD POWER ASSISTED STEERING	1		
2	NRC145	• Tube	1		
3	NRC4700	• Shaft-track rod adjusting	1		
4	577898	• Clamp	3		
5	256006	• Bolt	3		
6	WB106041L	• Washer-M6	6		
7	252210	• Locknut-1/4UNF	3		
8	RTC5869	• KIT-BALL JOINT-RH-R/H THREAD	2		
9	WC112081L	• • Washer-plain-M12 x 28	2		
10	NC112041	• • Nut-castle-M12	2		
11	PS106321L	• • Pin-split	1		
12	NRC3226	LINK ASSEMBLY-DRAG	1	To (V) JA 018174	
12	NTC9607	LINK ASSEMBLY-DRAG	1	From (V) JA 018175	
13	NRC1824	• Tube-drag link	1		Part of NRC3226.
14	NRC1823	• End piece-drag link	1		
15	577898	• Clamp	2		
16	256006	• Bolt	2		
17	WB106041L	• Washer-M6	4		
18	252210	• Locknut-1/4UNF	2		
19	RTC5869	• KIT-BALL JOINT-RH-R/H THREAD	1		
20	WC112081L	• • Washer-plain-M12 x 28	2		
21	NC112041	• • Nut-castle-M12	2		
22	PS106321L	• • Pin-split	1		
19	RTC5870	• KIT-BALL JOINT-LH-L/H THREAD	1		
20	WC112081L	• • Washer-plain-M12 x 28	1		
21	NC112041	• • Nut-castle-M12	1		
22	PS106321L	• • Pin-split	1		
13	NTC8390	• Tube-drag link	1		Part of NTC9607.

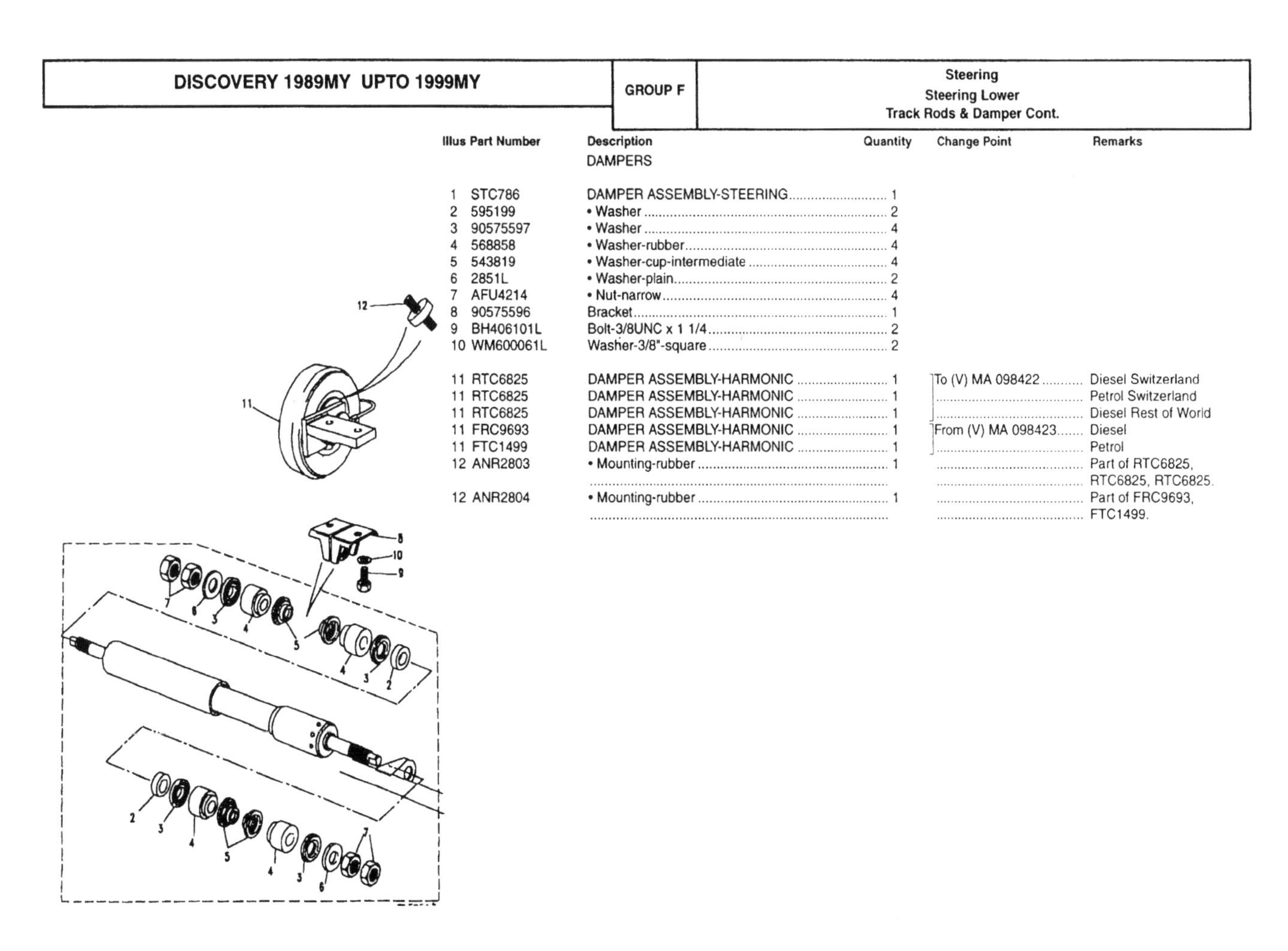

Illus	Part Number	Description	Quantity	Change Point	Remarks
		DAMPERS			
1	STC786	DAMPER ASSEMBLY-STEERING	1		
2	595199	• Washer	2		
3	90575597	• Washer	4		
4	568858	• Washer-rubber	4		
5	543819	• Washer-cup-intermediate	4		
6	2851L	• Washer-plain	2		
7	AFU4214	• Nut-narrow	4		
8	90575596	Bracket	1		
9	BH406101L	Bolt-3/8UNC x 1 1/4	2		
10	WM600061L	Washer-3/8"-square	2		
11	RTC6825	DAMPER ASSEMBLY-HARMONIC	1	To (V) MA 098422	Diesel Switzerland
11	RTC6825	DAMPER ASSEMBLY-HARMONIC	1		Petrol Switzerland
11	RTC6825	DAMPER ASSEMBLY-HARMONIC	1		Diesel Rest of World
11	FRC9693	DAMPER ASSEMBLY-HARMONIC	1	From (V) MA 098423	Diesel
11	FTC1499	DAMPER ASSEMBLY-HARMONIC	1		Petrol
12	ANR2803	• Mounting-rubber	1		Part of RTC6825, RTC6825, RTC6825.
12	ANR2804	• Mounting-rubber	1		Part of FRC9693, FTC1499.

Power Steering Reservoir-Diesel

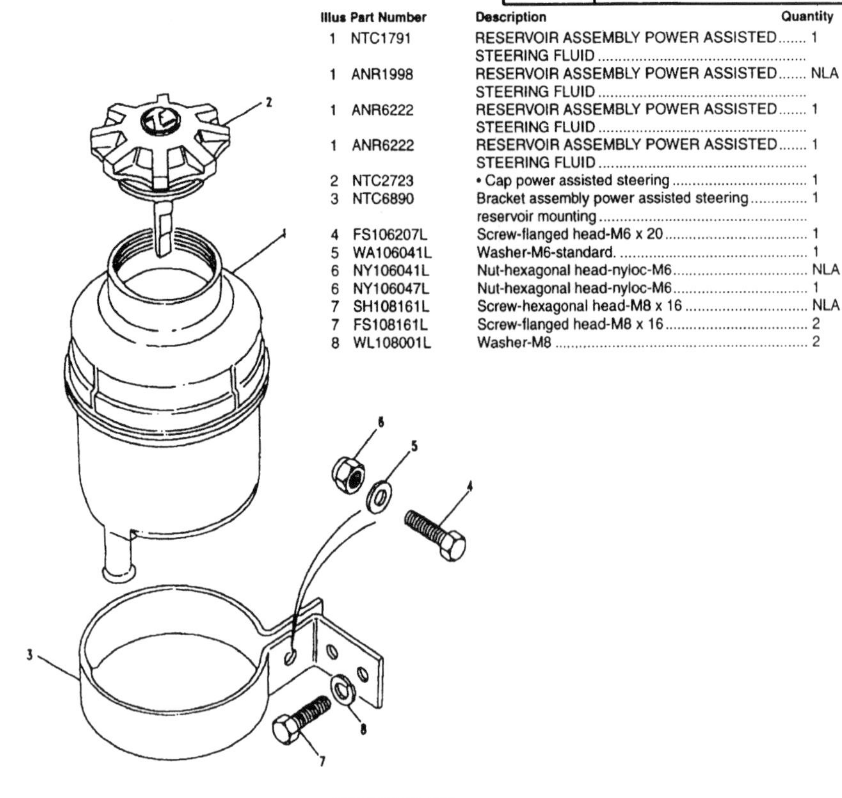

Illus	Part Number	Description	Quantity	Change Point	Remarks
1	NTC1791	RESERVOIR ASSEMBLY POWER ASSISTED STEERING FLUID	1	Note (1)	
1	ANR1998	RESERVOIR ASSEMBLY POWER ASSISTED STEERING FLUID	NLA	Note (2)	Use ANR6222.
1	ANR6222	RESERVOIR ASSEMBLY POWER ASSISTED STEERING FLUID	1		Diesel power steering
1	ANR6222	RESERVOIR ASSEMBLY POWER ASSISTED STEERING FLUID	1	Note (3)	
2	NTC2723	• Cap power assisted steering	1		
3	NTC6890	Bracket assembly power assisted steering reservoir mounting	1		
4	FS106207L	Screw-flanged head-M6 x 20	1		
5	WA106041L	Washer-M6-standard.	1		
6	NY106041L	Nut-hexagonal head-nyloc-M6	NLA		Use NY106047L.
6	NY106047L	Nut-hexagonal head-nyloc-M6	1		Diesel power steering
7	SH108161L	Screw-hexagonal head-M8 x 16	NLA		Use FS108161L.
7	FS108161L	Screw-flanged head-M8 x 16	2		Diesel power steering
8	WL108001L	Washer-M8	2		

CHANGE POINTS:
(1) To (V) KA 052813
(2) From (V) KA 052814 To (V) TA 199999; From (V) TA 700000 To (V) VA 711215; From (V) TA 500000 To (V) VA 541191
(3) From (V) VA 711216; From (V) VA 541192 To (V) VA 558898

F 17

Power Steering Reservoir-Petrol

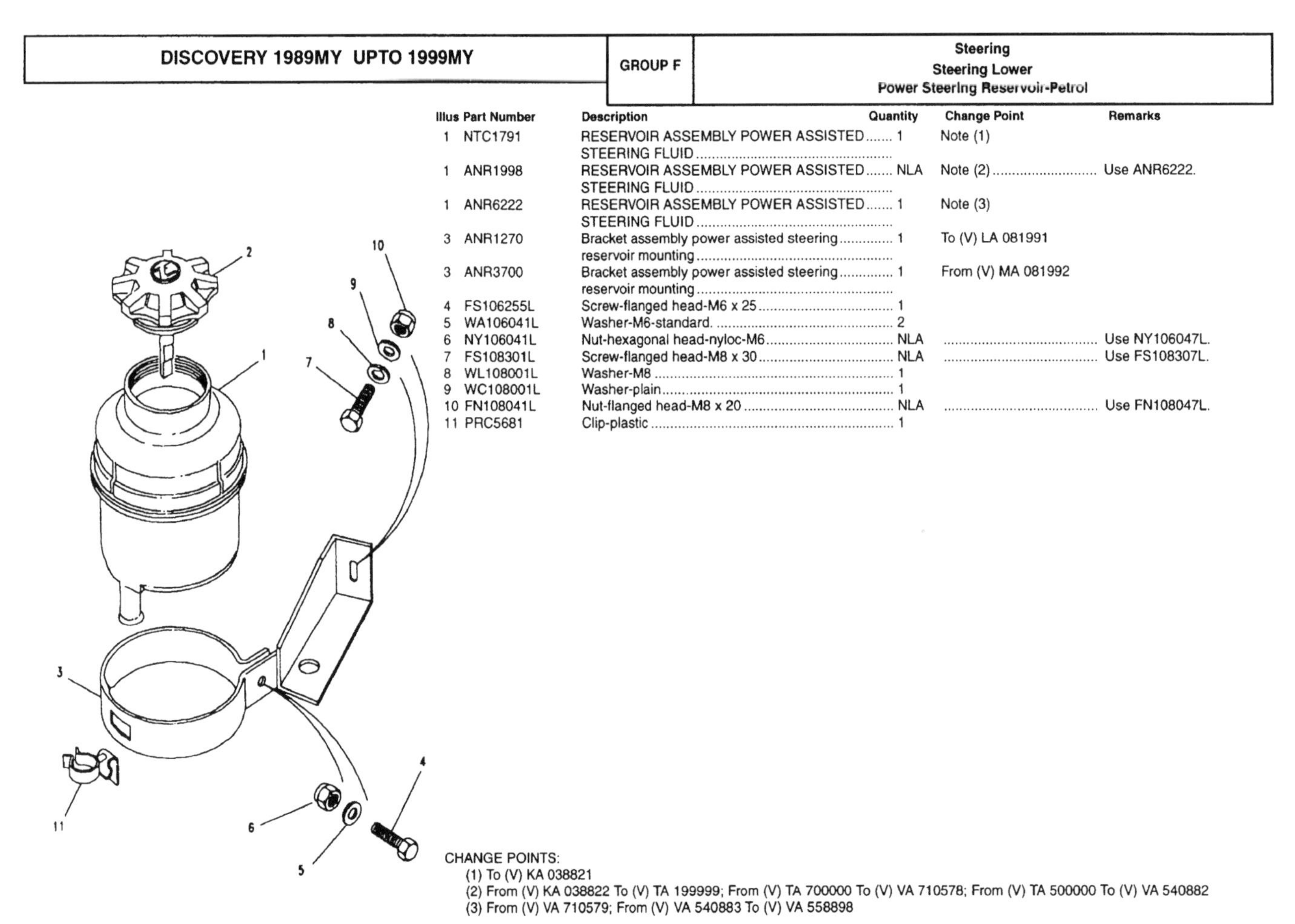

Illus	Part Number	Description	Quantity	Change Point	Remarks
1	NTC1791	RESERVOIR ASSEMBLY POWER ASSISTED STEERING FLUID	1	Note (1)	
1	ANR1998	RESERVOIR ASSEMBLY POWER ASSISTED STEERING FLUID	NLA	Note (2)	Use ANR6222.
1	ANR6222	RESERVOIR ASSEMBLY POWER ASSISTED STEERING FLUID	1	Note (3)	
3	ANR1270	Bracket assembly power assisted steering reservoir mounting	1	To (V) LA 081991	
3	ANR3700	Bracket assembly power assisted steering reservoir mounting	1	From (V) MA 081992	
4	FS106255L	Screw-flanged head-M6 x 25	1		
5	WA106041L	Washer-M6-standard.	2		
6	NY106041L	Nut-hexagonal head-nyloc-M6	NLA		Use NY106047L.
7	FS108301L	Screw-flanged head-M8 x 30	NLA		Use FS108307L.
8	WL108001L	Washer-M8	1		
9	WC108001L	Washer-plain	1		
10	FN108041L	Nut-flanged head-M8 x 20	NLA		Use FN108047L.
11	PRC5681	Clip-plastic	1		

CHANGE POINTS:
(1) To (V) KA 038821
(2) From (V) KA 038822 To (V) TA 199999; From (V) TA 700000 To (V) VA 710578; From (V) TA 500000 To (V) VA 540882
(3) From (V) VA 710579; From (V) VA 540883 To (V) VA 558898

F 18

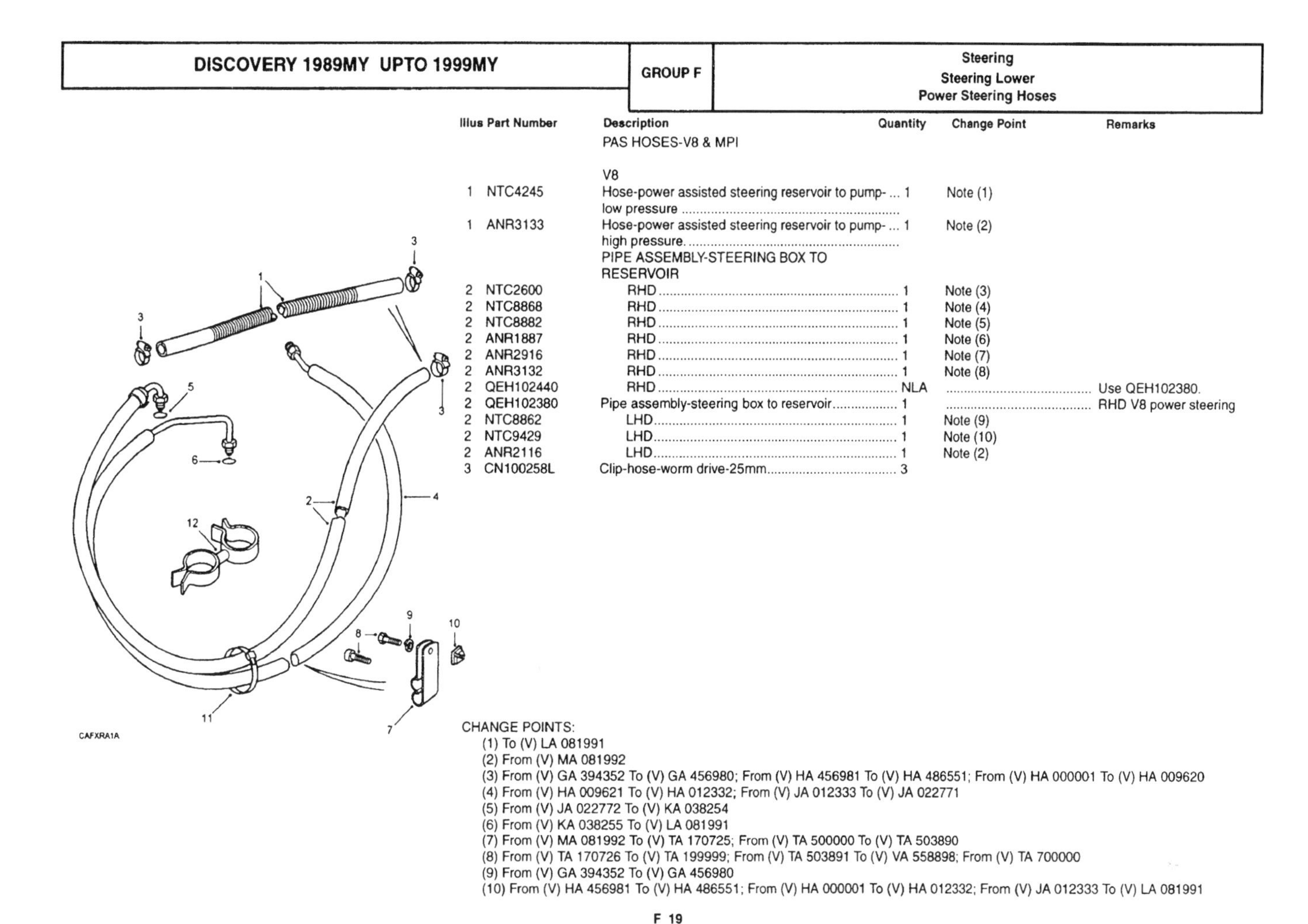

Illus	Part Number	Description	Quantity	Change Point	Remarks
		PAS HOSES-V8 & MPI			
		V8			
1	NTC4245	Hose-power assisted steering reservoir to pump-low pressure	1	Note (1)	
1	ANR3133	Hose-power assisted steering reservoir to pump-high pressure	1	Note (2)	
		PIPE ASSEMBLY-STEERING BOX TO RESERVOIR			
2	NTC2600	RHD	1	Note (3)	
2	NTC8868	RHD	1	Note (4)	
2	NTC8882	RHD	1	Note (5)	
2	ANR1887	RHD	1	Note (6)	
2	ANR2916	RHD	1	Note (7)	
2	ANR3132	RHD	1	Note (8)	
2	QEH102440	RHD	NLA		Use QEH102380.
2	QEH102380	Pipe assembly-steering box to reservoir	1		RHD V8 power steering
2	NTC8862	LHD	1	Note (9)	
2	NTC9429	LHD	1	Note (10)	
2	ANR2116	LHD	1	Note (2)	
3	CN100258L	Clip-hose-worm drive-25mm	3		

CHANGE POINTS:
(1) To (V) LA 081991
(2) From (V) MA 081992
(3) From (V) GA 394352 To (V) GA 456980; From (V) HA 456981 To (V) HA 486551; From (V) HA 000001 To (V) HA 009620
(4) From (V) HA 009621 To (V) HA 012332; From (V) JA 012333 To (V) JA 022771
(5) From (V) JA 022772 To (V) KA 038254
(6) From (V) KA 038255 To (V) LA 081991
(7) From (V) MA 081992 To (V) TA 170725; From (V) TA 500000 To (V) TA 503890
(8) From (V) TA 170726 To (V) TA 199999; From (V) TA 503891 To (V) VA 558898; From (V) TA 700000
(9) From (V) GA 394352 To (V) GA 456980
(10) From (V) HA 456981 To (V) HA 486551; From (V) HA 000001 To (V) HA 012332; From (V) JA 012333 To (V) LA 081991

F 19

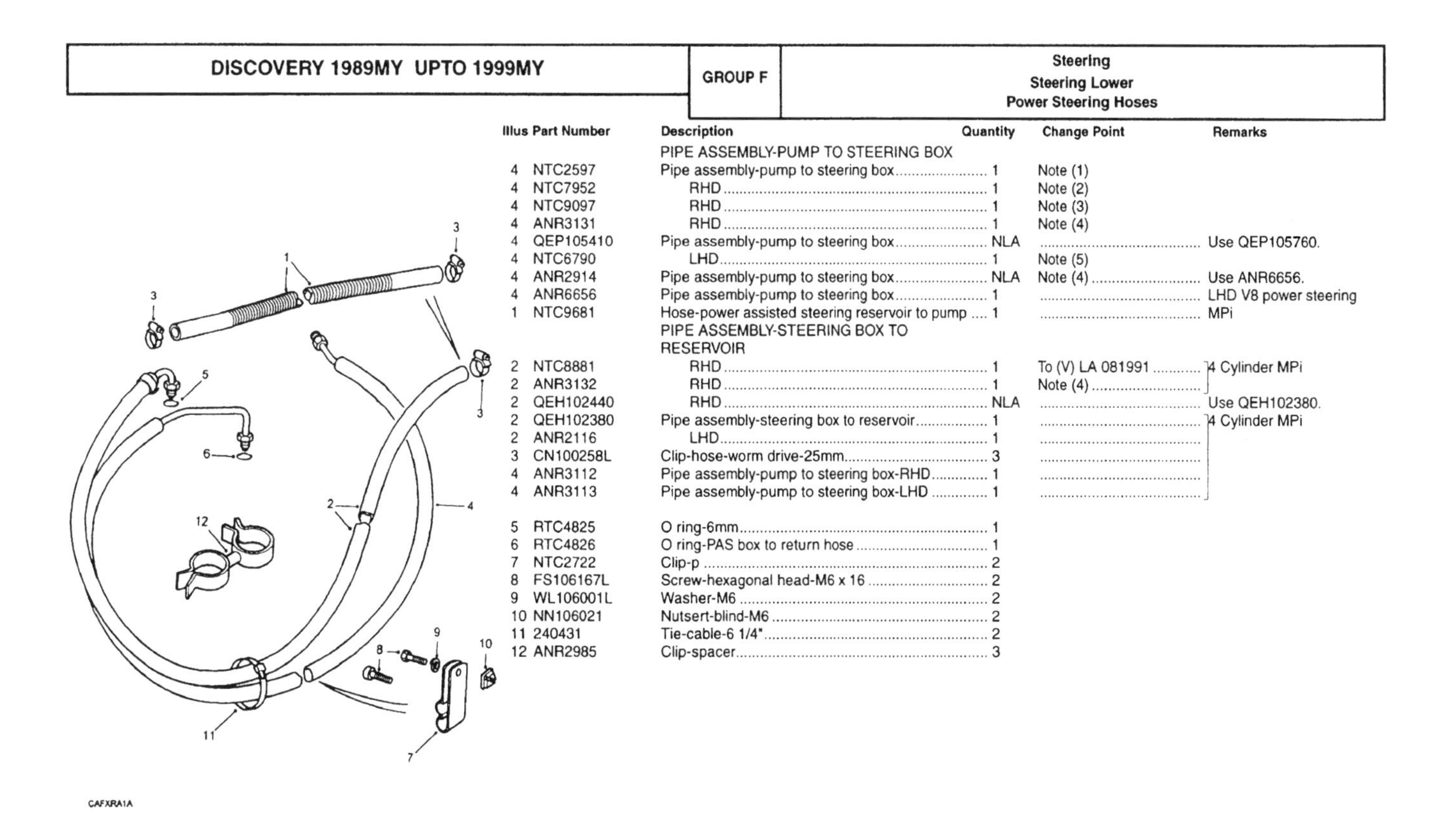

Illus	Part Number	Description	Quantity	Change Point	Remarks
		PIPE ASSEMBLY-PUMP TO STEERING BOX			
4	NTC2597	Pipe assembly-pump to steering box	1	Note (1)	
4	NTC7952	RHD	1	Note (2)	
4	NTC9097	RHD	1	Note (3)	
4	ANR3131	RHD	1	Note (4)	
4	QEP105410	Pipe assembly-pump to steering box	NLA		Use QEP105760.
4	NTC6790	LHD	1	Note (5)	
4	ANR2914	Pipe assembly-pump to steering box	NLA	Note (4)	Use ANR6656.
4	ANR6656	Pipe assembly-pump to steering box	1		LHD V8 power steering
1	NTC9681	Hose-power assisted steering reservoir to pump	1		MPi
		PIPE ASSEMBLY-STEERING BOX TO RESERVOIR			
2	NTC8881	RHD	1	To (V) LA 081991	4 Cylinder MPi
2	ANR3132	RHD	1	Note (4)	
2	QEH102440	RHD	NLA		Use QEH102380.
2	QEH102380	Pipe assembly-steering box to reservoir	1		4 Cylinder MPi
2	ANR2116	LHD	1		
3	CN100258L	Clip-hose-worm drive-25mm	3		
4	ANR3112	Pipe assembly-pump to steering box-RHD	1		
4	ANR3113	Pipe assembly-pump to steering box-LHD	1		
5	RTC4825	O ring-6mm	1		
6	RTC4826	O ring-PAS box to return hose	1		
7	NTC2722	Clip-p	2		
8	FS106167L	Screw-hexagonal head-M6 x 16	2		
9	WL106001L	Washer-M6	2		
10	NN106021	Nutsert-blind-M6	2		
11	240431	Tie-cable-6 1/4"	2		
12	ANR2985	Clip-spacer	3		

CHANGE POINTS:
(1) To (V) HA 546980
(2) From (V) HA 456981 To (V) KA 034313
(3) From (V) KA 034314 To (V) LA 081991
(4) From (V) MA 081992
(5) To (V) LA 081991

F 20

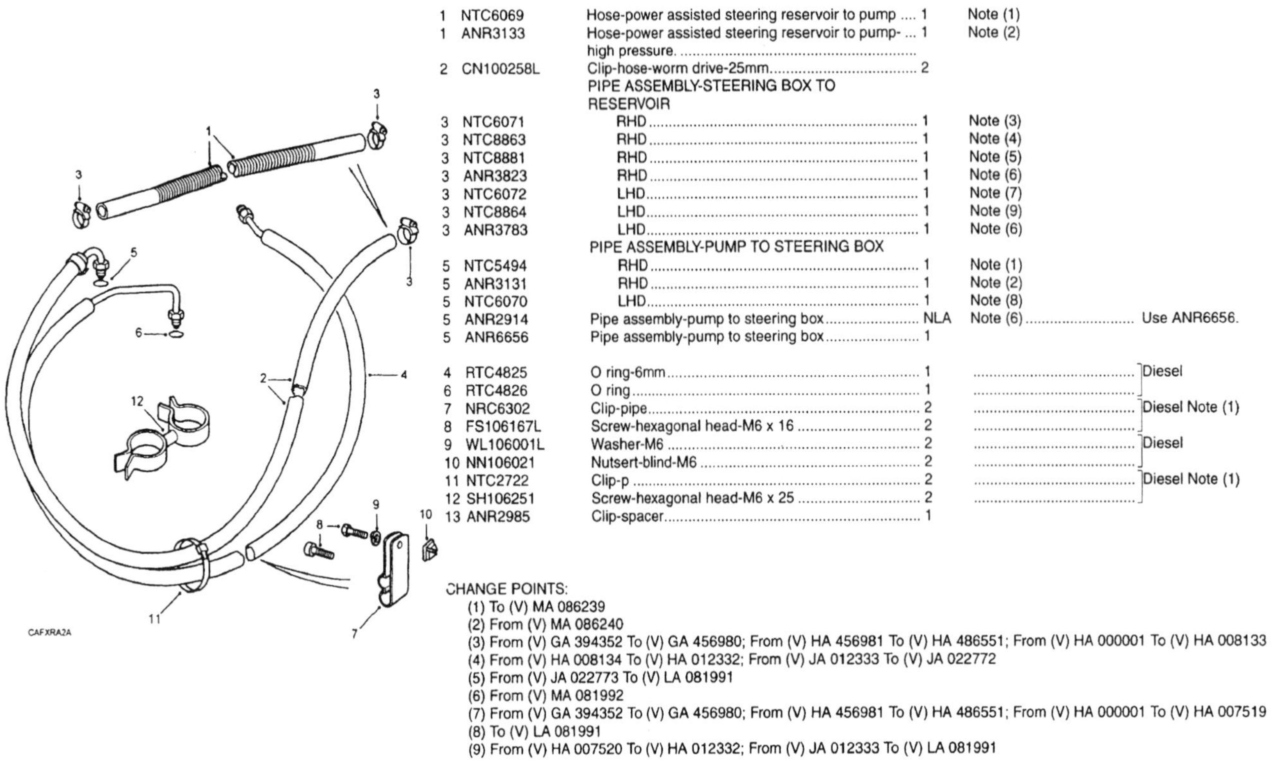

Illus	Part Number	Description	Quantity	Change Point	Remarks
		PAS HOSES-DIESEL			
1	NTC6069	Hose-power assisted steering reservoir to pump	1	Note (1)	
1	ANR3133	Hose-power assisted steering reservoir to pump-high pressure.	1	Note (2)	
2	CN100258L	Clip-hose-worm drive-25mm	2		
		PIPE ASSEMBLY-STEERING BOX TO RESERVOIR			
3	NTC6071	RHD	1	Note (3)	
3	NTC8863	RHD	1	Note (4)	
3	NTC8881	RHD	1	Note (5)	
3	ANR3823	RHD	1	Note (6)	
3	NTC6072	LHD	1	Note (7)	
3	NTC8864	LHD	1	Note (9)	
3	ANR3783	LHD	1	Note (6)	
		PIPE ASSEMBLY-PUMP TO STEERING BOX			
5	NTC5494	RHD	1	Note (1)	
5	ANR3131	RHD	1	Note (2)	
5	NTC6070	LHD	1	Note (8)	
5	ANR2914	Pipe assembly-pump to steering box	NLA	Note (6)	Use ANR6656.
5	ANR6656	Pipe assembly-pump to steering box	1		
4	RTC4825	O ring-6mm	1		Diesel
6	RTC4826	O ring	1		
7	NRC6302	Clip-pipe	2		Diesel Note (1)
8	FS106167L	Screw-hexagonal head-M6 x 16	2		
9	WL106001L	Washer-M6	2		Diesel
10	NN106021	Nutsert-blind-M6	2		
11	NTC2722	Clip-p	2		Diesel Note (1)
12	SH106251	Screw-hexagonal head-M6 x 25	2		
13	ANR2985	Clip-spacer	1		

CHANGE POINTS:
(1) To (V) MA 086239
(2) From (V) MA 086240
(3) From (V) GA 394352 To (V) GA 456980; From (V) HA 456981 To (V) HA 486551; From (V) HA 000001 To (V) HA 008133
(4) From (V) HA 008134 To (V) HA 012332; From (V) JA 012333 To (V) JA 022772
(5) From (V) JA 022773 To (V) LA 081991
(6) From (V) MA 081992
(7) From (V) GA 394352 To (V) GA 456980; From (V) HA 456981 To (V) HA 486551; From (V) HA 000001 To (V) HA 007519
(8) To (V) LA 081991
(9) From (V) HA 007520 To (V) HA 012332; From (V) JA 012333 To (V) LA 081991

Remarks:
(1) Both types used - check with vehicle

Engine Controls .... G 1
- Accelerator Pedal .... G 1
- Choke Cable .... G 14
- Clutch Master Cylinder & Piping Diesel .... G 12
- Clutch Master Cylinder & Piping Petrol .... G 11
- Pedalbox Assembly Automatic .... G 7
- Pedalbox Assembly Manual .... G 5

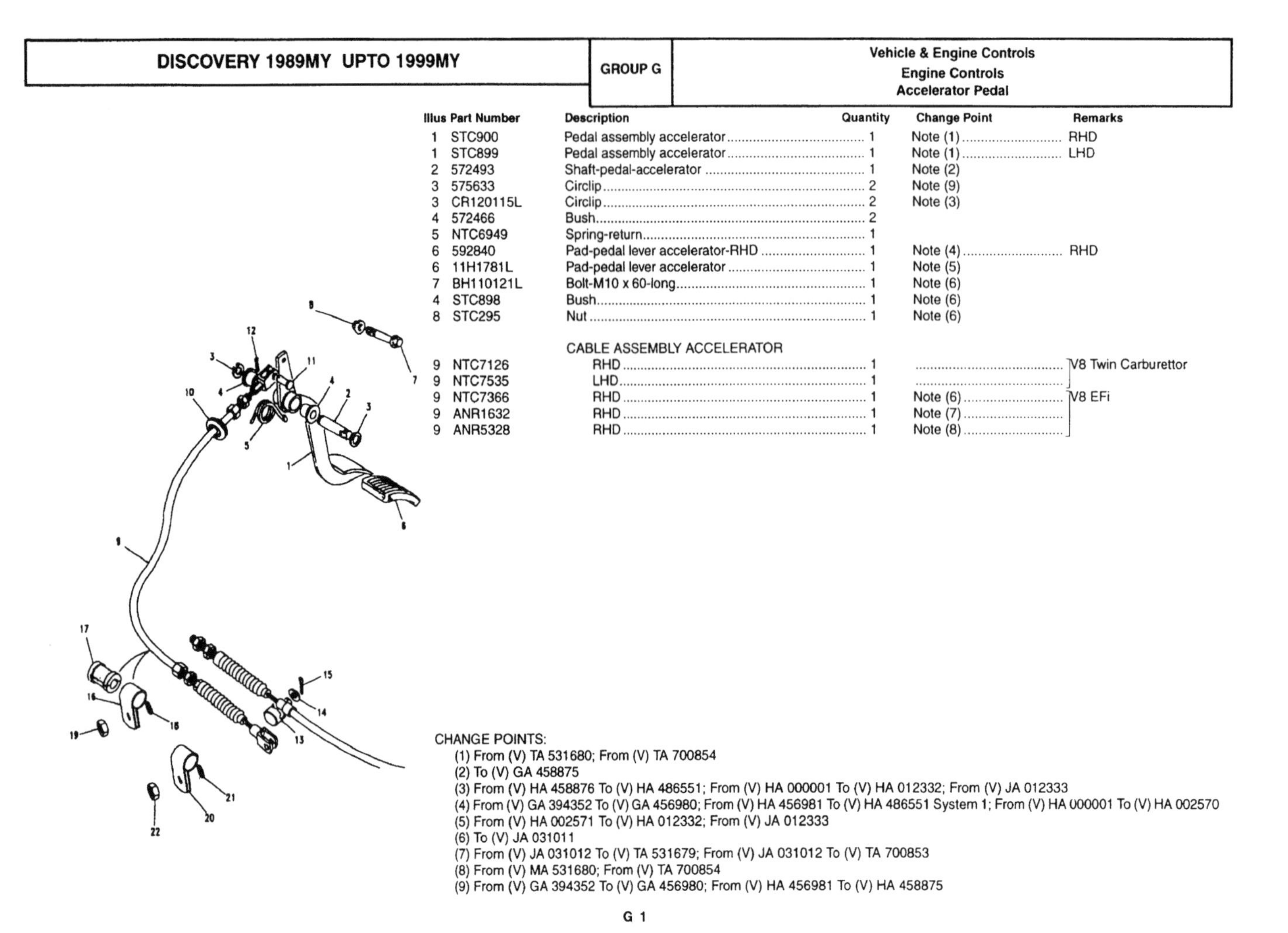

Illus	Part Number	Description	Quantity	Change Point	Remarks
1	STC900	Pedal assembly accelerator	1	Note (1)	RHD
1	STC899	Pedal assembly accelerator	1	Note (1)	LHD
2	572493	Shaft-pedal-accelerator	1	Note (2)	
3	575633	Circlip	2	Note (9)	
3	CR120115L	Circlip	2	Note (3)	
4	572466	Bush	2		
5	NTC6949	Spring-return	1		
6	592840	Pad-pedal lever accelerator-RHD	1	Note (4)	RHD
6	11H1781L	Pad-pedal lever accelerator	1	Note (5)	
7	BH110121L	Bolt-M10 x 60-long	1	Note (6)	
4	STC898	Bush	1	Note (6)	
8	STC295	Nut	1	Note (6)	
		CABLE ASSEMBLY ACCELERATOR			
9	NTC7126	RHD	1		V8 Twin Carburettor
9	NTC7535	LHD	1		V8 Twin Carburettor
9	NTC7366	RHD	1	Note (6)	V8 EFi
9	ANR1632	RHD	1	Note (7)	V8 EFi
9	ANR5328	RHD	1	Note (8)	V8 EFi

CHANGE POINTS:
(1) From (V) TA 531680; From (V) TA 700854
(2) To (V) GA 458875
(3) From (V) HA 458876 To (V) HA 486551; From (V) HA 000001 To (V) HA 012332; From (V) JA 012333
(4) From (V) GA 394352 To (V) GA 456980; From (V) HA 456981 To (V) HA 486551 System 1; From (V) HA 000001 To (V) HA 002570
(5) From (V) HA 002571 To (V) HA 012332; From (V) JA 012333
(6) To (V) JA 031011
(7) From (V) JA 031012 To (V) TA 531679; From (V) JA 031012 To (V) TA 700853
(8) From (V) MA 531680; From (V) TA 700854
(9) From (V) GA 394352 To (V) GA 456980; From (V) HA 456981 To (V) HA 458875

G 1

DISCOVERY 1989MY UPTO 1999MY | GROUP G | Vehicle & Engine Controls
Engine Controls
Accelerator Pedal

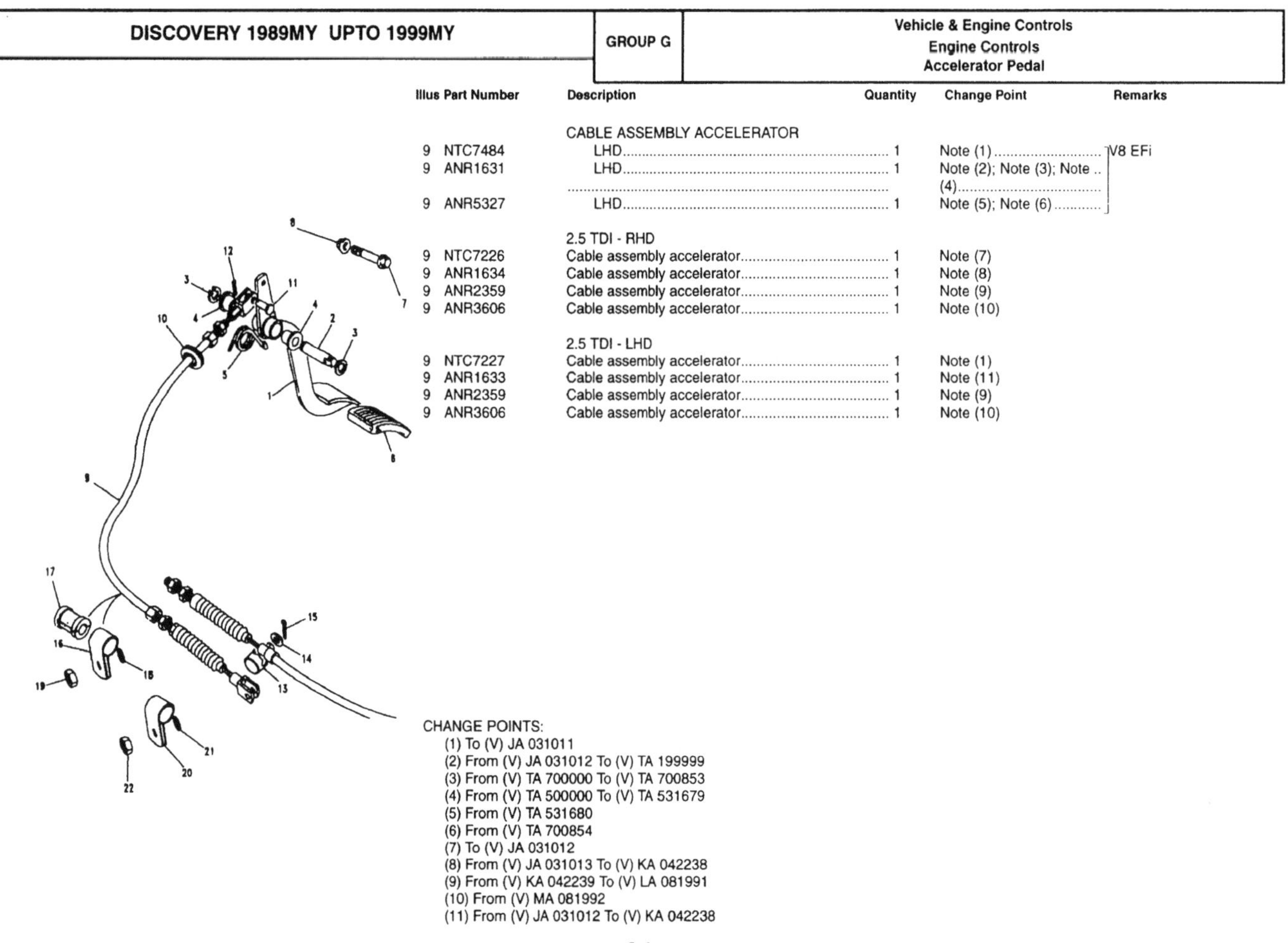

Illus	Part Number	Description	Quantity	Change Point	Remarks
		CABLE ASSEMBLY ACCELERATOR			
9	NTC7484	LHD	1	Note (1)	V8 EFi
9	ANR1631	LHD	1	Note (2); Note (3); Note (4)	V8 EFi
9	ANR5327	LHD	1	Note (5); Note (6)	V8 EFi
		2.5 TDI - RHD			
9	NTC7226	Cable assembly accelerator	1	Note (7)	
9	ANR1634	Cable assembly accelerator	1	Note (8)	
9	ANR2359	Cable assembly accelerator	1	Note (9)	
9	ANR3606	Cable assembly accelerator	1	Note (10)	
		2.5 TDI - LHD			
9	NTC7227	Cable assembly accelerator	1	Note (1)	
9	ANR1633	Cable assembly accelerator	1	Note (11)	
9	ANR2359	Cable assembly accelerator	1	Note (9)	
9	ANR3606	Cable assembly accelerator	1	Note (10)	

CHANGE POINTS:
(1) To (V) JA 031011
(2) From (V) JA 031012 To (V) TA 199999
(3) From (V) TA 700000 To (V) TA 700853
(4) From (V) TA 500000 To (V) TA 531679
(5) From (V) TA 531680
(6) From (V) TA 700854
(7) To (V) JA 031012
(8) From (V) JA 031013 To (V) KA 042238
(9) From (V) KA 042239 To (V) LA 081991
(10) From (V) MA 081992
(11) From (V) JA 031012 To (V) KA 042238

G 2

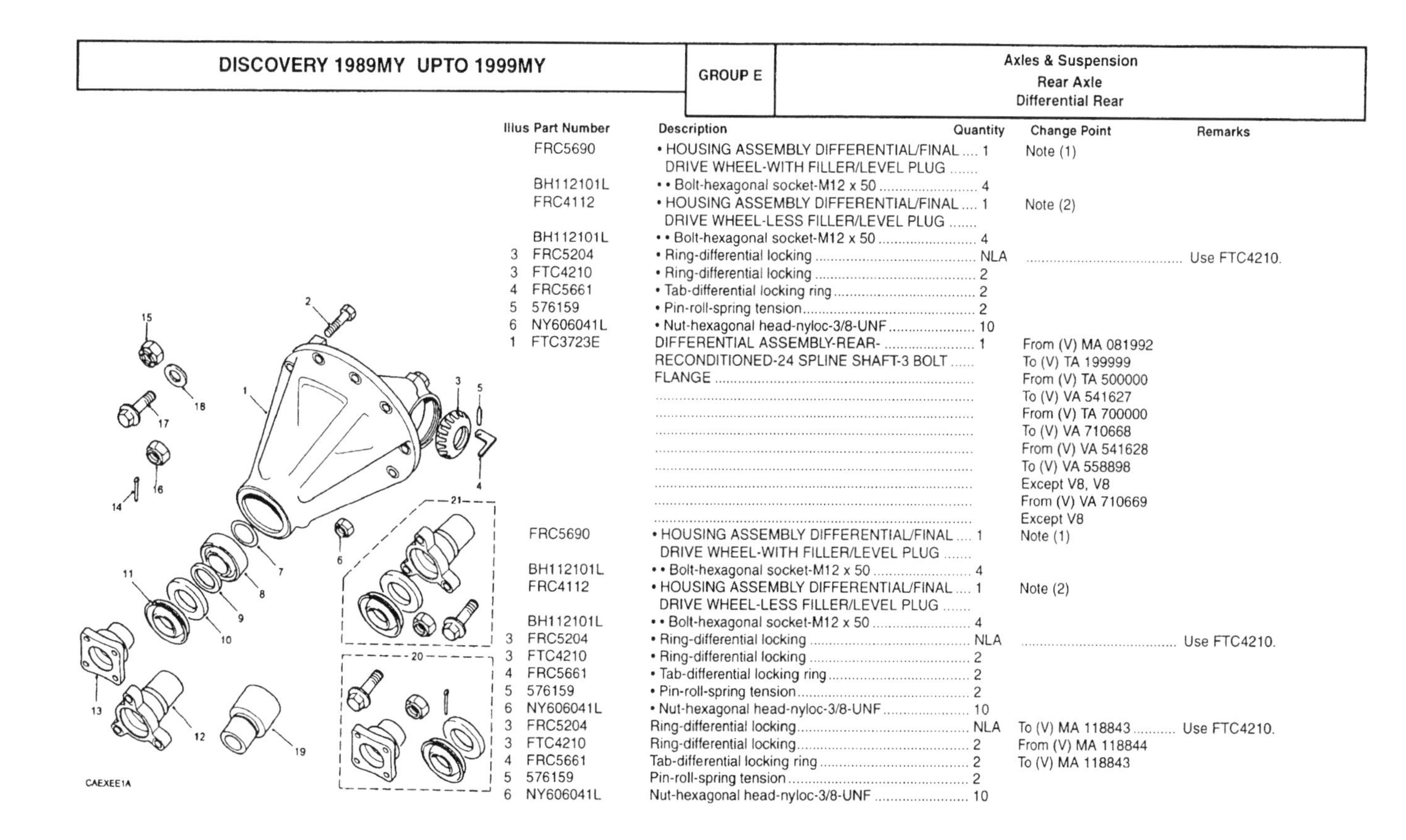

Illus	Part Number	Description	Quantity	Change Point	Remarks
	FRC5690	• HOUSING ASSEMBLY DIFFERENTIAL/FINAL DRIVE WHEEL-WITH FILLER/LEVEL PLUG	1	Note (1)	
	BH112101L	• • Bolt-hexagonal socket-M12 x 50	4		
	FRC4112	• HOUSING ASSEMBLY DIFFERENTIAL/FINAL DRIVE WHEEL-LESS FILLER/LEVEL PLUG	1	Note (2)	
	BH112101L	• • Bolt-hexagonal socket-M12 x 50	4		
3	FRC5204	• Ring-differential locking	NLA		Use FTC4210.
3	FTC4210	• Ring-differential locking	2		
4	FRC5661	• Tab-differential locking ring	2		
5	576159	• Pin-roll-spring tension	2		
6	NY606041L	• Nut-hexagonal head-nyloc-3/8-UNF	10		
1	FTC3723E	DIFFERENTIAL ASSEMBLY-REAR-RECONDITIONED-24 SPLINE SHAFT-3 BOLT FLANGE	1	From (V) MA 081992 To (V) TA 199999 From (V) TA 500000 To (V) VA 541627 From (V) TA 700000 To (V) VA 710668 From (V) VA 541628 To (V) VA 558898 Except V8, V8 From (V) VA 710669 Except V8	
	FRC5690	• HOUSING ASSEMBLY DIFFERENTIAL/FINAL DRIVE WHEEL-WITH FILLER/LEVEL PLUG	1	Note (1)	
	BH112101L	• • Bolt-hexagonal socket-M12 x 50	4		
	FRC4112	• HOUSING ASSEMBLY DIFFERENTIAL/FINAL DRIVE WHEEL-LESS FILLER/LEVEL PLUG	1	Note (2)	
	BH112101L	• • Bolt-hexagonal socket-M12 x 50	4		
3	FRC5204	• Ring-differential locking	NLA		Use FTC4210.
3	FTC4210	• Ring-differential locking	2		
4	FRC5661	• Tab-differential locking ring	2		
5	576159	• Pin-roll-spring tension	2		
6	NY606041L	• Nut-hexagonal head-nyloc-3/8-UNF	10		
3	FRC5204	Ring-differential locking	NLA	To (V) MA 118843	Use FTC4210.
3	FTC4210	Ring-differential locking	2	From (V) MA 118844	
4	FRC5661	Tab-differential locking ring	2	To (V) MA 118843	
5	576159	Pin-roll-spring tension	2		
6	NY606041L	Nut-hexagonal head-nyloc-3/8-UNF	10		

CHANGE POINTS:
(1) To (V) MA 118843
(2) From (V) MA 118844

CAEXEE1A

Illus	Part Number	Description	Quantity	Change Point	Remarks
		PINION BEARING (10 SPLINE)			
		SHIM			
7	FRC1193	0.060"	A/R	Note (1)	
7	FRC1195	0.062"	A/R		
7	FRC1197	0.064"	A/R		
7	FRC1199	0.066"	A/R		
7	FRC1201	0.068"	A/R		
7	FRC1203	0.070"	A/R		
7	539718	0.072"	A/R		
7	539720	0.074"	A/R		
7	539722	0.076"	A/R		
7	539724	0.080"	A/R		
		PINION BEARING (24 SPLINE)			
		SHIM-BEARING DIFFERENTIAL			
7	FTC3869	2.155mm	A/R	Note (2)	
7	FTC3871	2.105mm	A/R		
7	FTC3873	2.055mm	A/R		
7	FTC3875	2.005mm	A/R		
7	FTC3877	1.955mm	A/R		
7	FTC3879	1.905mm	A/R		
7	FTC3881	1.855mm	A/R		
7	FTC3883	1.805mm	A/R		
7	FTC3885	1.755mm	A/R		
7	FTC3887	1.705mm	A/R		
7	FTC3889	1.655mm	A/R		
7	FTC3891	1.605mm	A/R		
7	FTC3893	1.555mm	A/R		
7	FTC3895	1.505mm	A/R		
8	539707	Bearing-taper roller	1		
9	539745	Spacer-pinion bearing	1		

CHANGE POINTS:
(1) From (A)32S 00001 To (A)32S 29121; From (A)28S 00001 To (A)28S 44171
(2) From (A)32S 29122 To (A)32S 99999; From (A)28S 44172 To (A)28S 99999

Illus	Part Number	Description	Quantity	Change Point	Remarks
		SHIM-BEARING DIFFERENTIAL			
7	TYS101040	2.155mm	A/R		
7	TYS101060	2.105mm	A/R		
7	TYS101080	2.055mm	A/R		
7	TYS101100	2.005mm	A/R		
7	TYS101120	1.955mm	A/R		
7	TYS101140	1.905mm	A/R		
7	TYS101160	1.855mm	A/R		
7	TYS101180	1.805mm	A/R		
7	TYS101200	1.755mm	A/R		
7	TYS101220	1.705mm	A/R		
7	TYS101240	1.655mm	A/R		
7	TYS101260	1.605mm	A/R		
7	TYS101280	1.555mm	A/R		
7	TYS101300	1.505mm	A/R		

CAEXEE1A

DISCOVERY 1989MY UPTO 1999MY | GROUP E | Axles & Suspension
Rear Axle
Differential Rear

Illus	Part Number	Description	Quantity	Change Point	Remarks
10	FRC8220	Seal-differential final drive pinion	1	Note (1)	
10	FTC5258	Seal-differential final drive pinion	1	Note (2)	
11	FRC8154	Mudshield-front & rear differential	1	Note (1)	
11	FTC5317	Mudshield-front & rear differential	1	Note (2)	
		FLANGE-DRIVESHAFT COUPLING DIFFERENTIAL			
12	FTC3706	3 bolt flange	NLA	From (V) MA 081992	Use FTC3729.
12	FTC3729	3 bolt flange	NLA		Use STC3723.
13	FRC3002	4 bolt flange	1		
14	PS608101L	Pin-split	1	Note (3)	
15	3259	Nut-castle	1		
16	NY116041L	Nut-hexagonal head-nyloc-M16	1	Note (4)	
17	FS112301P	Screw-M12 x 30	1	Note (6)	
18	90513454	Washer-plain	1	Note (5)	Nut securing flange
18	FTC5413	Washer-plain	1	Note (6)	Bolt securing flange
19	FTC3707	Pin-rear axle drive flange centralising	1	Note (7)	

CAEXEE1A

CHANGE POINTS:

(1) To (V) VA 544244; From (V) TA 700000 To (V) VA 712973
(2) From (V) VA 544245 To (V) VA 558898; From (V) VA 712974
(3) To (A)32S 29121; To (A)28S 44171
(4) From (A)32S 29122 To (A)32S 99999; From (A)28S 44172 To (A)28S 99999; From (A)46S 00001B To (A)46S 10361B; From 00001A To (A)47S 90103A; From (A)47S 00001A To (A)56S 13055A; From (A)57S 00001A To (A)57S 03162A
(5) From (A)46S 00001B To (A)46S 10361B; From (A)47S 00001A To (A)47S 90103A; From (A)56S 00001A To (A)56S 13055A; From (A)57S 00001A To (A)57S 03162A
(6) From (A)46S 10362B To (A)46S 99999B; From (A)47S 90104A To (A)47S 99999A; From (A)56S 13056A To (A)56S 99999A; From (A)57S 03163A To (A)57S 99999A
(7) From (V) MA 081992 To (V) TA 199999; From (V) VA 541628 To (V) VA 558898; From (V) VA 710669

Illus	Part Number	Description	Quantity	Change Point	Remarks
		2.5 TDI - EDC DRIVE BY WIRE			
9	ANR5096	Cable assembly accelerator	1	Note (1)	RHD
9	ANR4617	Cable assembly accelerator	1		LHD
		2.5 TDI - CRUISE CONTROL			
9	ANR4513	Cable assembly accelerator	1	Note (1)	LHD
9	ANR4514	Cable assembly accelerator	1		RHD
		ALL VARIANTS			
10	562106	Grommet	1		
11	NTC6916	Pin-clevis/pivot	2	Note (2)	
11	ANR3660	Pin-clevis/pivot	2	Note (3)	
12	PS603041L	Pin-split	2		
	ANR3661	Clip-spring-push on	2	Note (3)	
		PETROL LHS			
13	90577449	Pin-clevis/pivot	1		
14	WA702101L	Washer	1		
15	PS603041	Pin-split	1		
		PETROL RHS			
16	240407	Clip-hose-14.5mm-steel-25/64" hole	1		
17	6860L	Grommet-rubber-17/64" hole	1		
18	WA108051L	Washer-plain-M8	1		
19	NH108041L	Nut-hexagonal head-M8	1		
		DIESEL			
20	CP108101L	Clip-p	1		
21	WA108051L	Washer-plain-M8	1		
22	NH108041L	Nut-hexagonal head-M8	1		
23	ANR1902	Link-lever accelerator	1	Note (4)	
23	ANR5093	Link-lever accelerator	1	Note (5)	

CHANGE POINTS:
(1) From (V) TA 163104; From (V) TA 501919
(2) To (V) MA 087363
(3) From (V) MA 087364
(4) To (E)19L 01106A
(5) From (E)19L 01107A

G 3

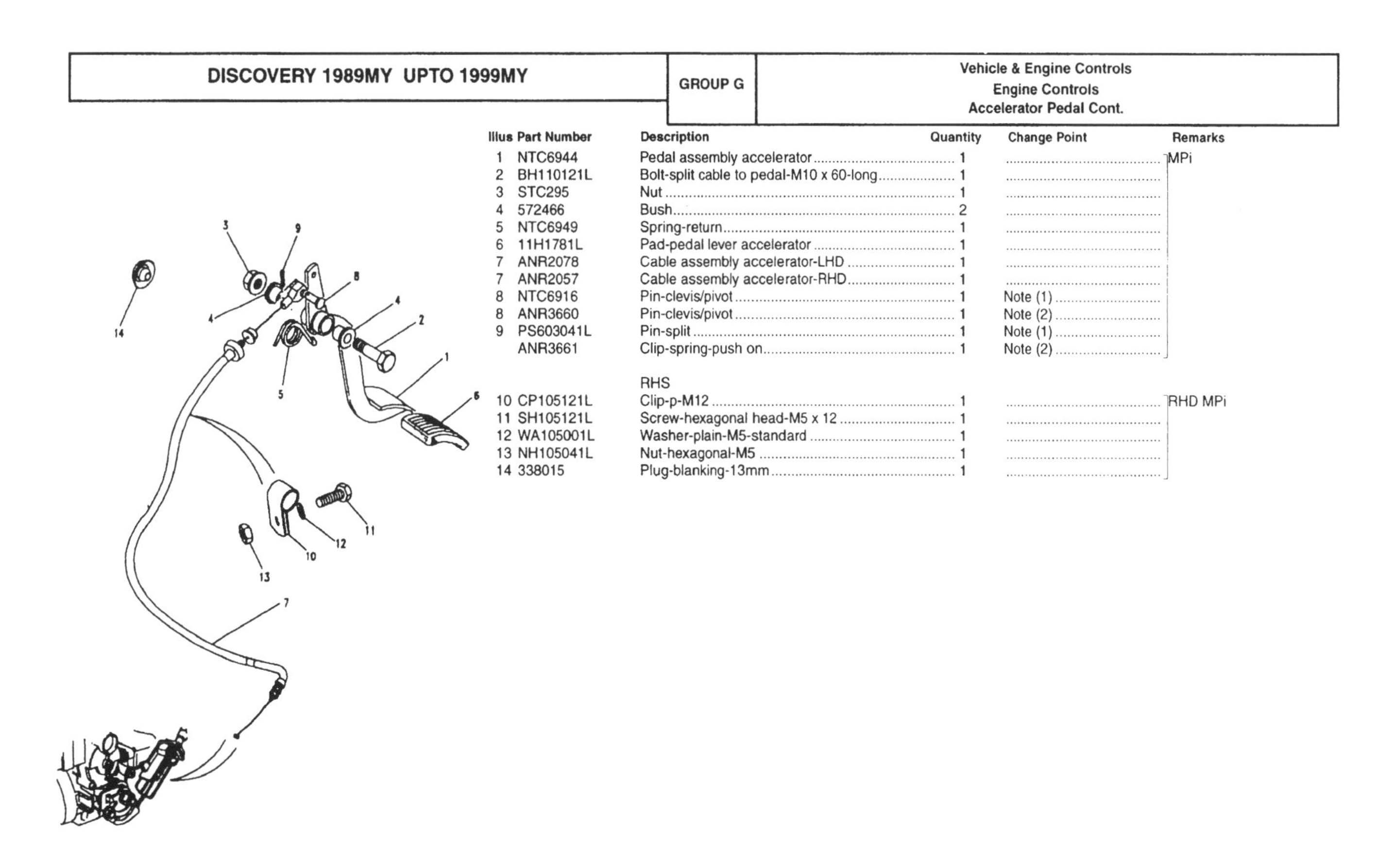

Illus	Part Number	Description	Quantity	Change Point	Remarks
1	NTC6944	Pedal assembly accelerator	1		MPi
2	BH110121L	Bolt-split cable to pedal-M10 x 60-long	1		
3	STC295	Nut	1		
4	572466	Bush	2		
5	NTC6949	Spring-return	1		
6	11H1781L	Pad-pedal lever accelerator	1		
7	ANR2078	Cable assembly accelerator-LHD	1		
7	ANR2057	Cable assembly accelerator-RHD	1		
8	NTC6916	Pin-clevis/pivot	1	Note (1)	
8	ANR3660	Pin-clevis/pivot	1	Note (2)	
9	PS603041L	Pin-split	1	Note (1)	
	ANR3661	Clip-spring-push on	1	Note (2)	
		RHS			
10	CP105121L	Clip-p-M12	1		RHD MPi
11	SH105121L	Screw-hexagonal head-M5 x 12	1		
12	WA105001L	Washer-plain-M5-standard	1		
13	NH105041L	Nut-hexagonal-M5	1		
14	338015	Plug-blanking-13mm	1		

CHANGE POINTS:
(1) To (V) LA 081991
(2) From (V) MA 081992

G 4

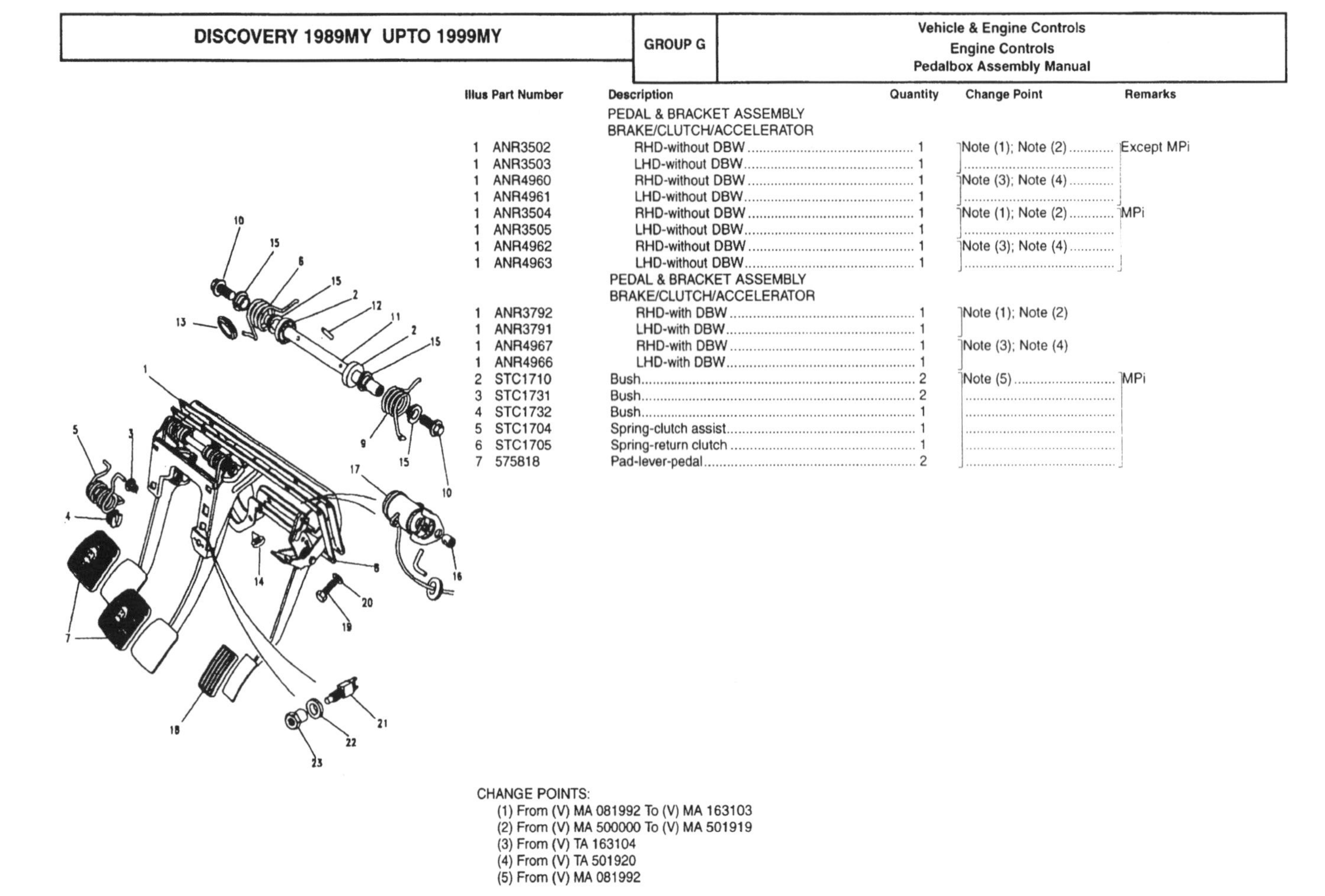

Illus	Part Number	Description	Quantity	Change Point	Remarks
		PEDAL & BRACKET ASSEMBLY BRAKE/CLUTCH/ACCELERATOR			
1	ANR3502	RHD-without DBW	1	Note (1); Note (2)	Except MPi
1	ANR3503	LHD-without DBW	1		
1	ANR4960	RHD-without DBW	1	Note (3); Note (4)	
1	ANR4961	LHD-without DBW	1		
1	ANR3504	RHD-without DBW	1	Note (1); Note (2)	MPi
1	ANR3505	LHD-without DBW	1		
1	ANR4962	RHD-without DBW	1	Note (3); Note (4)	
1	ANR4963	LHD-without DBW	1		
		PEDAL & BRACKET ASSEMBLY BRAKE/CLUTCH/ACCELERATOR			
1	ANR3792	RHD-with DBW	1	Note (1); Note (2)	
1	ANR3791	LHD-with DBW	1		
1	ANR4967	RHD-with DBW	1	Note (3); Note (4)	
1	ANR4966	LHD-with DBW	1		
2	STC1710	Bush	2	Note (5)	MPi
3	STC1731	Bush	2		
4	STC1732	Bush	1		
5	STC1704	Spring-clutch assist	1		
6	STC1705	Spring-return clutch	1		
7	575818	Pad-lever-pedal	2		

CHANGE POINTS:
(1) From (V) MA 081992 To (V) MA 163103
(2) From (V) MA 500000 To (V) MA 501919
(3) From (V) TA 163104
(4) From (V) TA 501920
(5) From (V) MA 081992

G 5

DISCOVERY 1989MY UPTO 1999MY | GROUP G | Vehicle & Engine Controls
Engine Controls
Pedalbox Assembly Manual

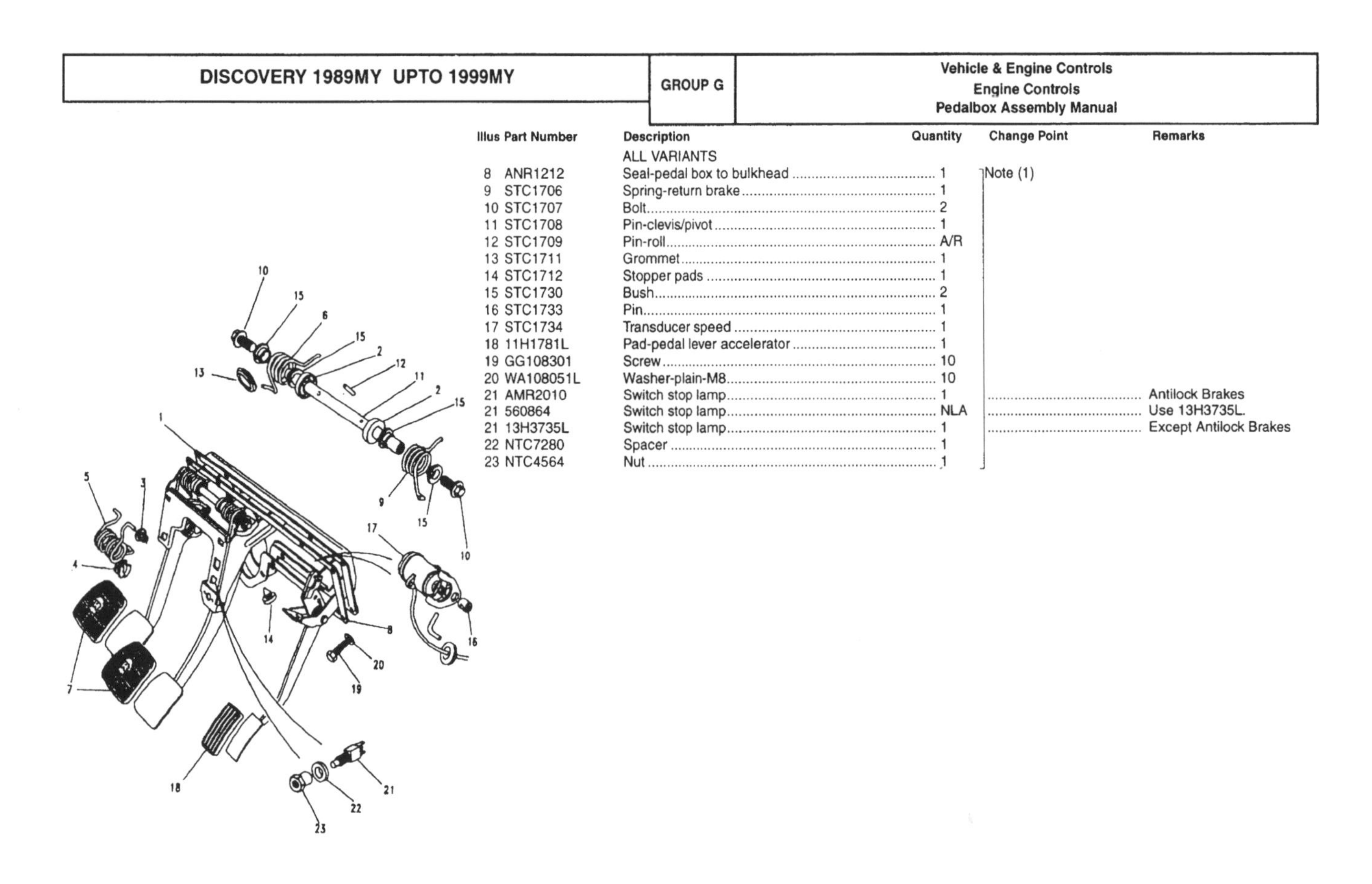

Illus	Part Number	Description	Quantity	Change Point	Remarks
		ALL VARIANTS			
8	ANR1212	Seal-pedal box to bulkhead	1	Note (1)	
9	STC1706	Spring-return brake	1		
10	STC1707	Bolt	2		
11	STC1708	Pin-clevis/pivot	1		
12	STC1709	Pin-roll	A/R		
13	STC1711	Grommet	1		
14	STC1712	Stopper pads	1		
15	STC1730	Bush	2		
16	STC1733	Pin	1		
17	STC1734	Transducer speed	1		
18	11H1781L	Pad-pedal lever accelerator	1		
19	GG108301	Screw	10		
20	WA108051L	Washer-plain-M8	10		
21	AMR2010	Switch stop lamp	1		Antilock Brakes
21	560864	Switch stop lamp	NLA		Use 13H3735L.
21	13H3735L	Switch stop lamp	1		Except Antilock Brakes
22	NTC7280	Spacer	1		
23	NTC4564	Nut	1		

CHANGE POINTS:
(1) From (V) MA 081992

G 6

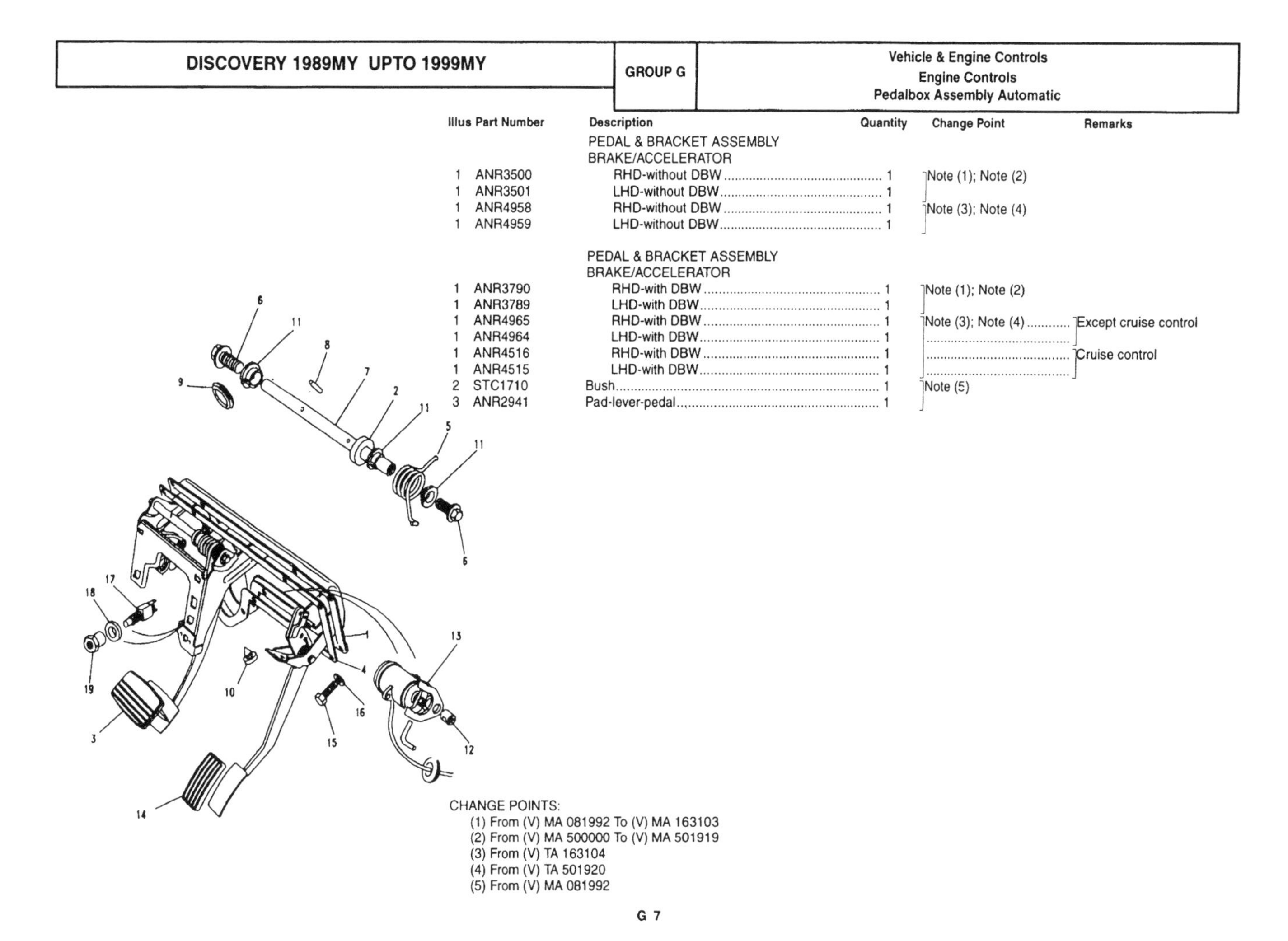

Illus	Part Number	Description	Quantity	Change Point	Remarks
		PEDAL & BRACKET ASSEMBLY BRAKE/ACCELERATOR			
1	ANR3500	RHD-without DBW	1	Note (1); Note (2)	
1	ANR3501	LHD-without DBW	1		
1	ANR4958	RHD-without DBW	1	Note (3); Note (4)	
1	ANR4959	LHD-without DBW	1		
		PEDAL & BRACKET ASSEMBLY BRAKE/ACCELERATOR			
1	ANR3790	RHD-with DBW	1	Note (1); Note (2)	
1	ANR3789	LHD-with DBW	1		
1	ANR4965	RHD-with DBW	1	Note (3); Note (4)	Except cruise control
1	ANR4964	LHD-with DBW	1		
1	ANR4516	RHD-with DBW	1		Cruise control
1	ANR4515	LHD-with DBW	1		
2	STC1710	Bush	1	Note (5)	
3	ANR2941	Pad-lever-pedal	1		

CHANGE POINTS:
(1) From (V) MA 081992 To (V) MA 163103
(2) From (V) MA 500000 To (V) MA 501919
(3) From (V) TA 163104
(4) From (V) TA 501920
(5) From (V) MA 081992

G 7

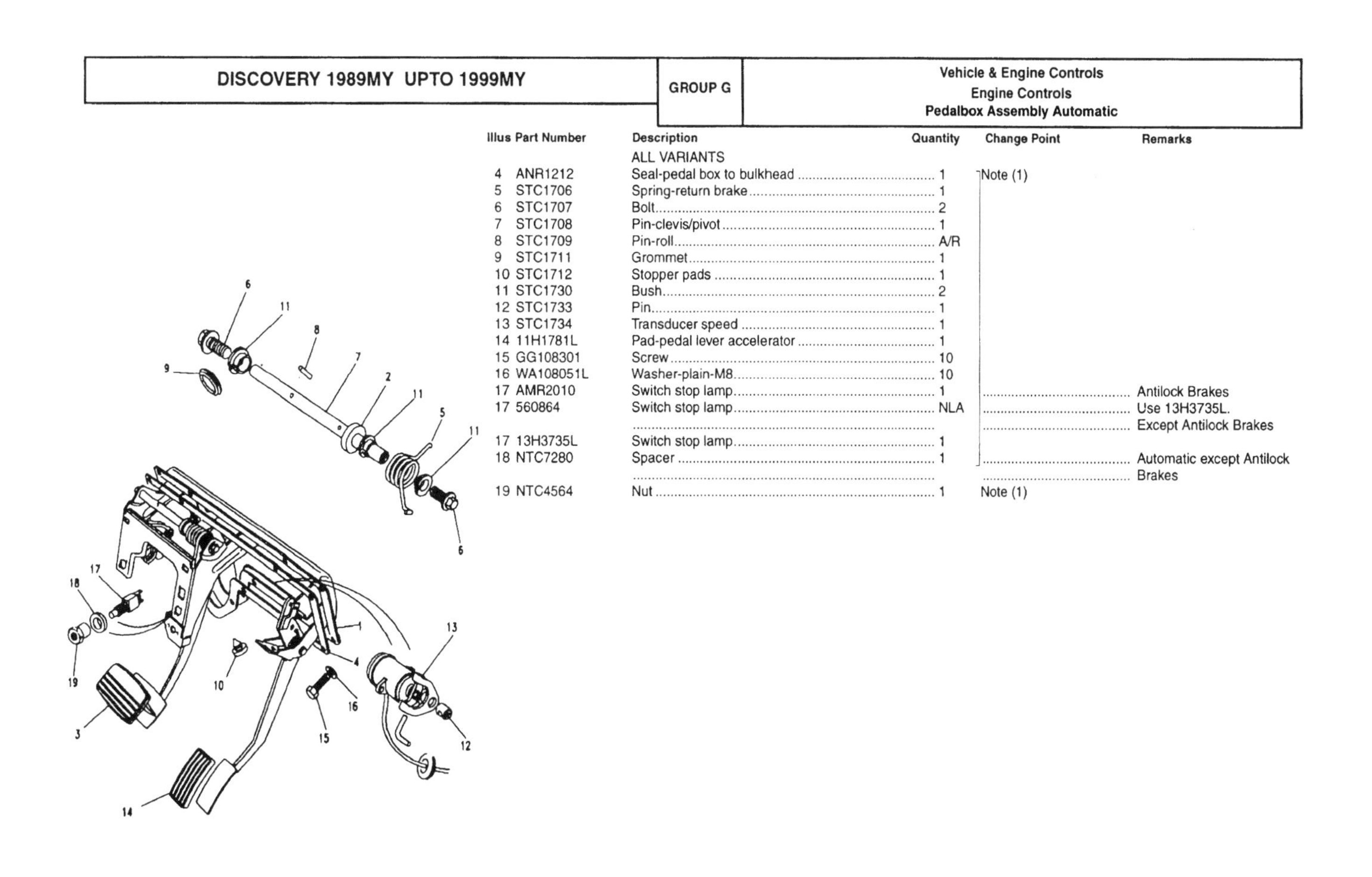

Illus	Part Number	Description	Quantity	Change Point	Remarks
		ALL VARIANTS			
4	ANR1212	Seal-pedal box to bulkhead	1	Note (1)	
5	STC1706	Spring-return brake	1		
6	STC1707	Bolt	2		
7	STC1708	Pin-clevis/pivot	1		
8	STC1709	Pin-roll	A/R		
9	STC1711	Grommet	1		
10	STC1712	Stopper pads	1		
11	STC1730	Bush	2		
12	STC1733	Pin	1		
13	STC1734	Transducer speed	1		
14	11H1781L	Pad-pedal lever accelerator	1		
15	GG108301	Screw	10		
16	WA108051L	Washer-plain-M8	10		
17	AMR2010	Switch stop lamp	1		Antilock Brakes
17	560864	Switch stop lamp	NLA		Use 13H3735L. Except Antilock Brakes
17	13H3735L	Switch stop lamp	1		Automatic except Antilock Brakes
18	NTC7280	Spacer	1		
19	NTC4564	Nut	1	Note (1)	

CHANGE POINTS:
(1) From (V) MA 081992

G 8

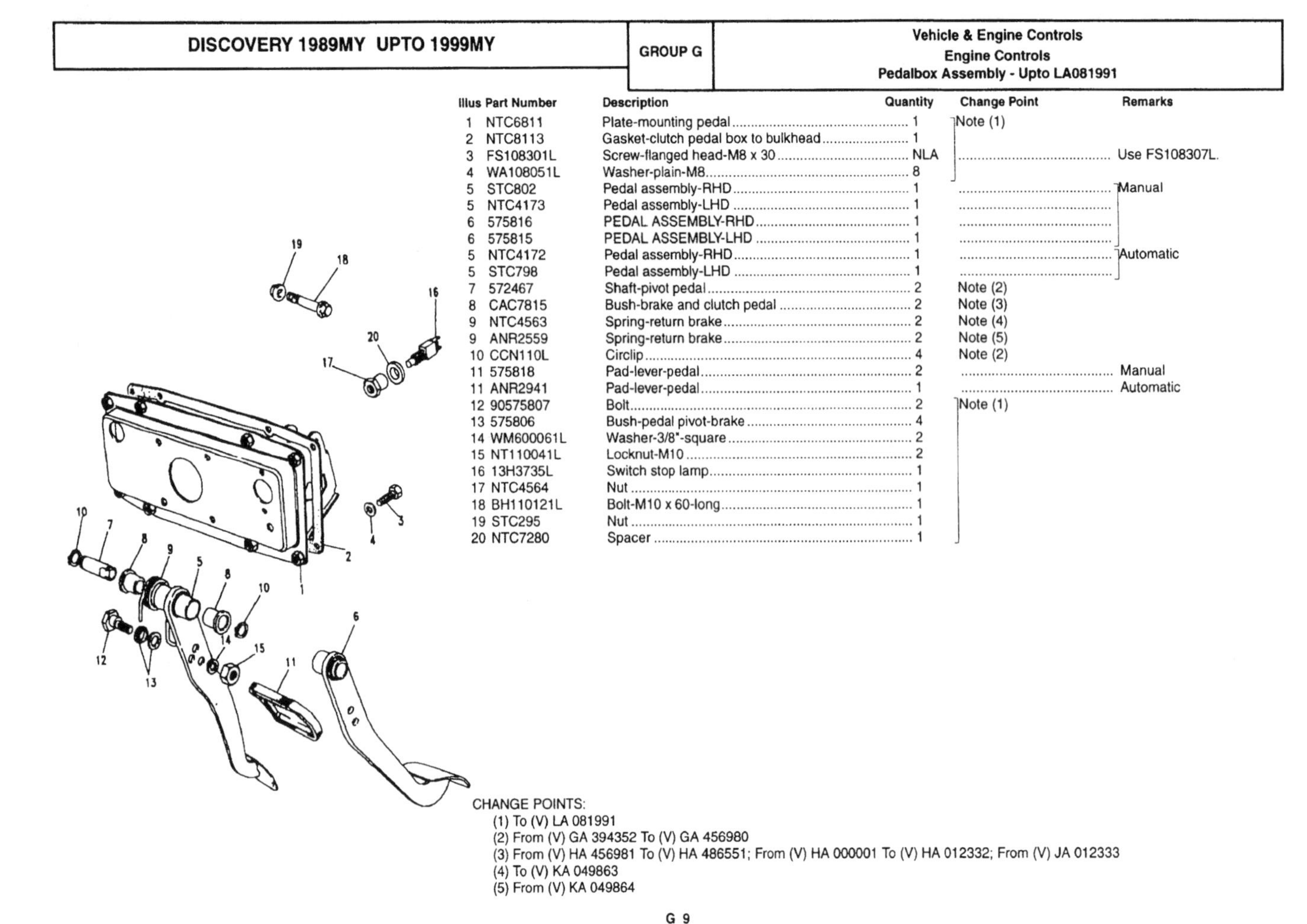

Illus	Part Number	Description	Quantity	Change Point	Remarks
1	NTC6811	Plate-mounting pedal	1	Note (1)	
2	NTC8113	Gasket-clutch pedal box to bulkhead	1		
3	FS108301L	Screw-flanged head-M8 x 30	NLA		Use FS108307L.
4	WA108051L	Washer-plain-M8	8		
5	STC802	Pedal assembly-RHD	1		Manual
5	NTC4173	Pedal assembly-LHD	1		
6	575816	PEDAL ASSEMBLY-RHD	1		
6	575815	PEDAL ASSEMBLY-LHD	1		
5	NTC4172	Pedal assembly-RHD	1		Automatic
5	STC798	Pedal assembly-LHD	1		
7	572467	Shaft-pivot pedal	2	Note (2)	
8	CAC7815	Bush-brake and clutch pedal	2	Note (3)	
9	NTC4563	Spring-return brake	2	Note (4)	
9	ANR2559	Spring-return brake	2	Note (5)	
10	CCN110L	Circlip	4	Note (2)	
11	575818	Pad-lever-pedal	2		Manual
11	ANR2941	Pad-lever-pedal	1		Automatic
12	90575807	Bolt	2	Note (1)	
13	575806	Bush-pedal pivot-brake	4		
14	WM600061L	Washer-3/8"-square	2		
15	NT110041L	Locknut-M10	2		
16	13H3735L	Switch stop lamp	1		
17	NTC4564	Nut	1		
18	BH110121L	Bolt-M10 x 60-long	1		
19	STC295	Nut	1		
20	NTC7280	Spacer	1		

CHANGE POINTS:
(1) To (V) LA 081991
(2) From (V) GA 394352 To (V) GA 456980
(3) From (V) HA 456981 To (V) HA 486551; From (V) HA 000001 To (V) HA 012332; From (V) JA 012333
(4) To (V) KA 049863
(5) From (V) KA 049864

Illus	Part Number	Description	Quantity	Change Point	Remarks
1	ERR2622	Switch-cruise control pedal vent	2	Note (1)	
2	PRC6260	Pump-cruise control vacuum	1		
3	AMR1114	T piece	1		
		HOSE-VACUUM CRUISE CONTROL			
4	AMR1886	pump to T-piece	1	Note (2); Note (3)	
5	NTC8232	T-piece to actuator	1		
6	AMR1888	T-piece to vent valve	1		
4	ANR4907	pump to T-piece	1	Note (4); Note (5)	
5	ANR4906	T-piece to actuator	1		
6	ANR4905	T-piece to vent valve	1		
7	CP105121L	Clip-p-M12	2	Note (2); Note (3)	
7	ANR4981	Clip-p	3	Note (4); Note (5)	
8	NTC4238	Grommet-Black-11 x 19mm	1	Note (1)	
	338017	Plug-blanking-14mm	1		

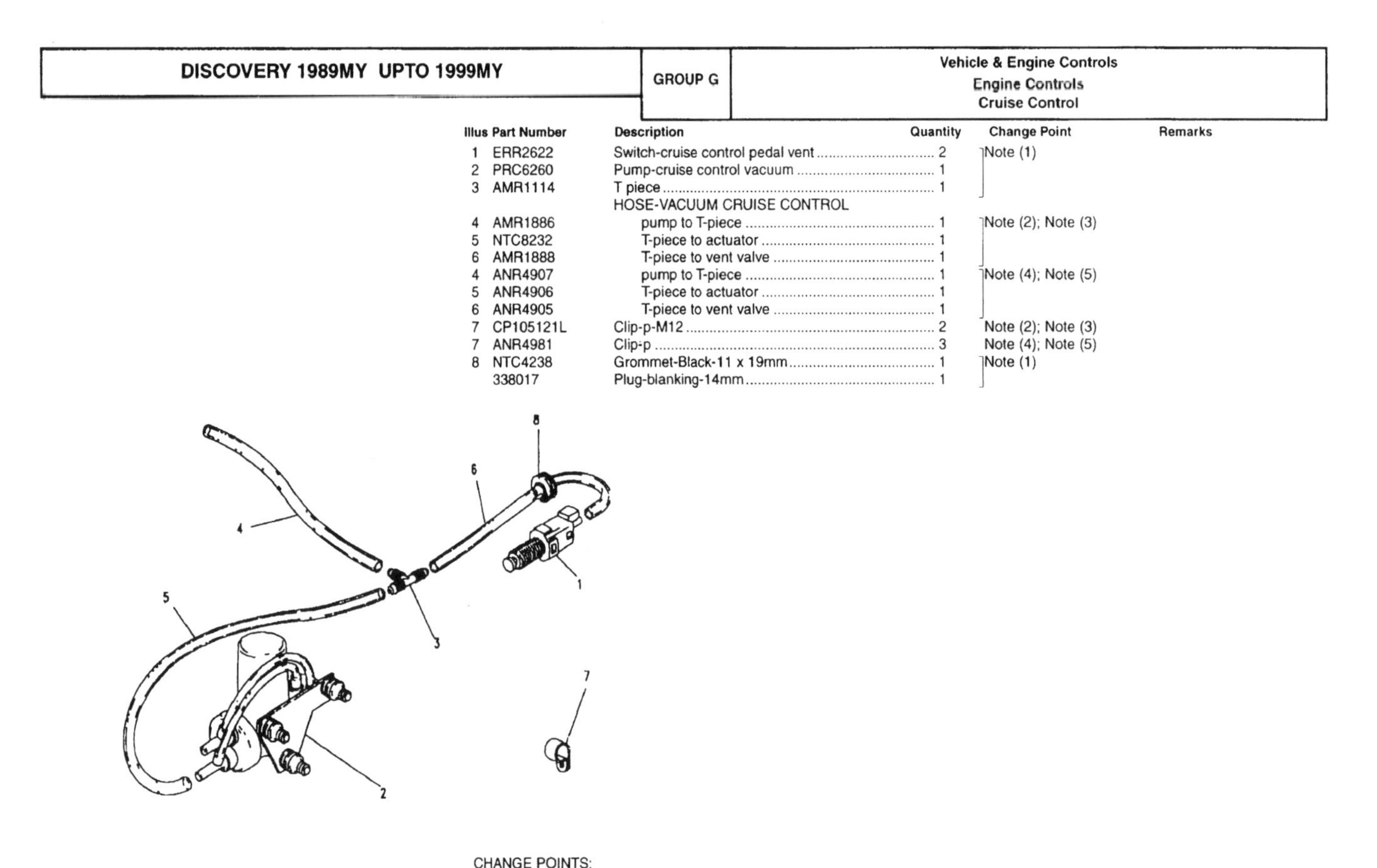

CHANGE POINTS:
(1) From (V) MA 081992
(2) From (V) MA 081992 To (V) MA 163103
(3) From (V) MA 500000 To (V) MA 501919
(4) From (V) TA 163104
(5) From (V) TA 501920

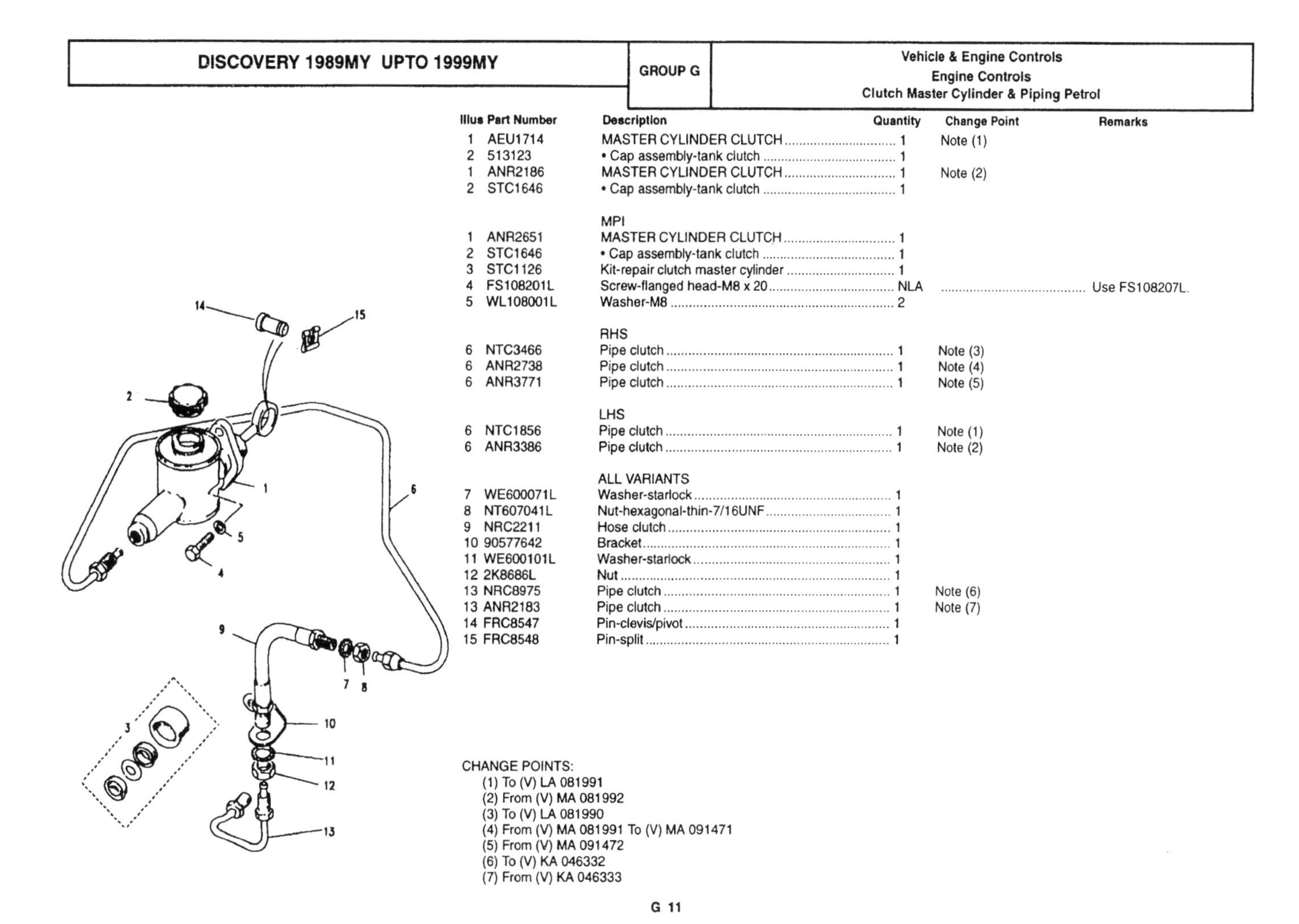

DISCOVERY 1989MY UPTO 1999MY

GROUP G — Vehicle & Engine Controls / Engine Controls / Clutch Master Cylinder & Piping Petrol

Illus	Part Number	Description	Quantity	Change Point	Remarks
1	AEU1714	MASTER CYLINDER CLUTCH	1	Note (1)	
2	513123	• Cap assembly-tank clutch	1		
1	ANR2186	MASTER CYLINDER CLUTCH	1	Note (2)	
2	STC1646	• Cap assembly-tank clutch	1		
		MPI			
1	ANR2651	MASTER CYLINDER CLUTCH	1		
2	STC1646	• Cap assembly-tank clutch	1		
3	STC1126	Kit-repair clutch master cylinder	1		
4	FS108201L	Screw-flanged head-M8 x 20	NLA		Use FS108207L.
5	WL108001L	Washer-M8	2		
		RHS			
6	NTC3466	Pipe clutch	1	Note (3)	
6	ANR2738	Pipe clutch	1	Note (4)	
6	ANR3771	Pipe clutch	1	Note (5)	
		LHS			
6	NTC1856	Pipe clutch	1	Note (1)	
6	ANR3386	Pipe clutch	1	Note (2)	
		ALL VARIANTS			
7	WE600071L	Washer-starlock	1		
8	NT607041L	Nut-hexagonal-thin-7/16UNF	1		
9	NRC2211	Hose clutch	1		
10	90577642	Bracket	1		
11	WE600101L	Washer-starlock	1		
12	2K8686L	Nut	1		
13	NRC8975	Pipe clutch	1	Note (6)	
13	ANR2183	Pipe clutch	1	Note (7)	
14	FRC8547	Pin-clevis/pivot	1		
15	FRC8548	Pin-split	1		

CHANGE POINTS:
(1) To (V) LA 081991
(2) From (V) MA 081992
(3) To (V) LA 081990
(4) From (V) MA 081991 To (V) MA 091471
(5) From (V) MA 091472
(6) To (V) KA 046332
(7) From (V) KA 046333

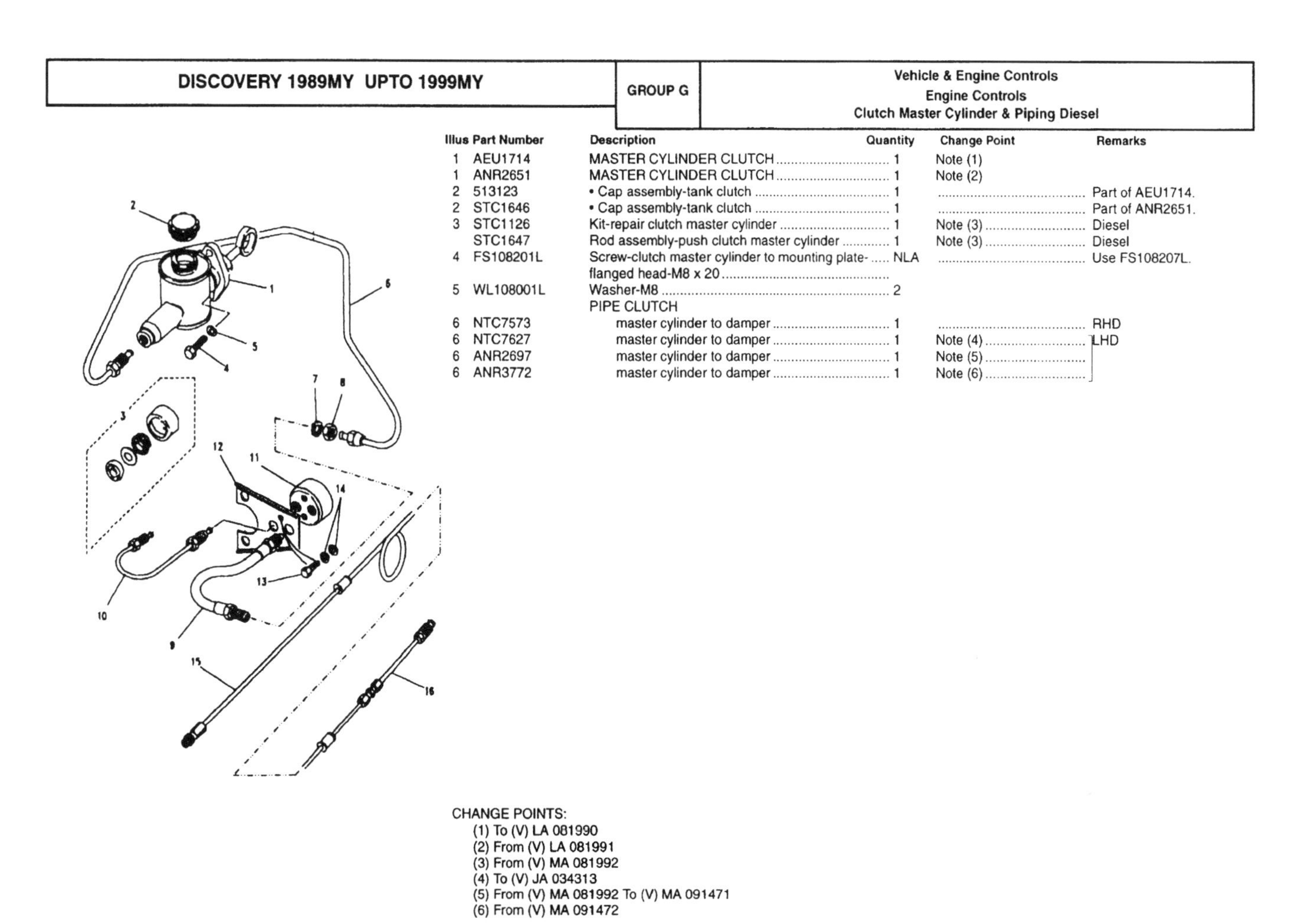

DISCOVERY 1989MY UPTO 1999MY

GROUP G — Vehicle & Engine Controls / Engine Controls / Clutch Master Cylinder & Piping Diesel

Illus	Part Number	Description	Quantity	Change Point	Remarks
1	AEU1714	MASTER CYLINDER CLUTCH	1	Note (1)	
1	ANR2651	MASTER CYLINDER CLUTCH	1	Note (2)	
2	513123	• Cap assembly-tank clutch	1		Part of AEU1714.
2	STC1646	• Cap assembly-tank clutch	1		Part of ANR2651.
3	STC1126	Kit-repair clutch master cylinder	1	Note (3)	Diesel
	STC1647	Rod assembly-push clutch master cylinder	1	Note (3)	Diesel
4	FS108201L	Screw-clutch master cylinder to mounting plate-flanged head-M8 x 20	NLA		Use FS108207L.
5	WL108001L	Washer-M8	2		
		PIPE CLUTCH			
6	NTC7573	master cylinder to damper	1		RHD
6	NTC7627	master cylinder to damper	1	Note (4)	LHD
6	ANR2697	master cylinder to damper	1	Note (5)	LHD
6	ANR3772	master cylinder to damper	1	Note (6)	LHD

CHANGE POINTS:
(1) To (V) LA 081990
(2) From (V) LA 081991
(3) From (V) MA 081992
(4) To (V) JA 034313
(5) From (V) MA 081992 To (V) MA 091471
(6) From (V) MA 091472

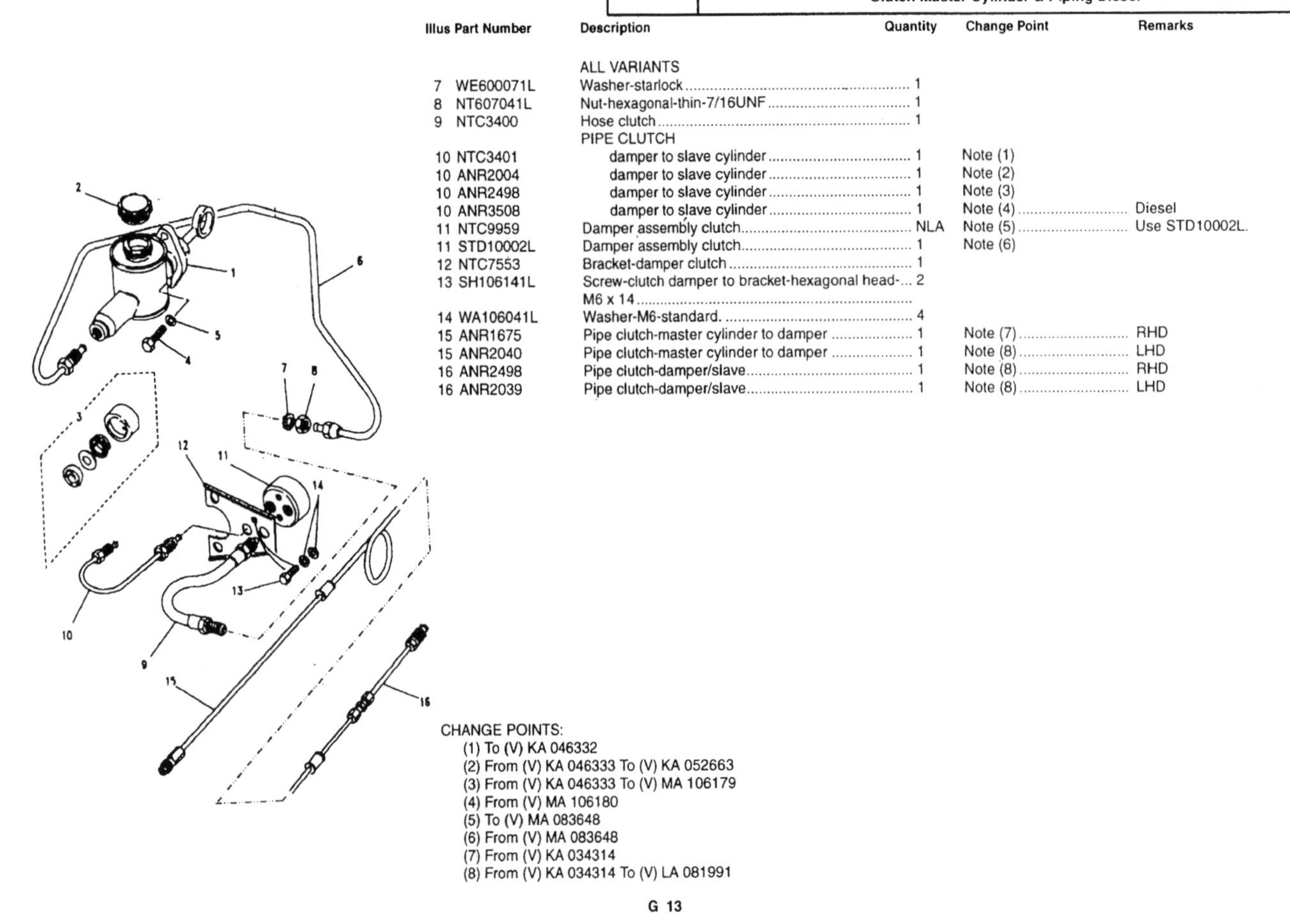

Illus	Part Number	Description	Quantity	Change Point	Remarks
		ALL VARIANTS			
7	WE600071L	Washer-starlock	1		
8	NT607041L	Nut-hexagonal-thin-7/16UNF	1		
9	NTC3400	Hose clutch	1		
		PIPE CLUTCH			
10	NTC3401	damper to slave cylinder	1	Note (1)	
10	ANR2004	damper to slave cylinder	1	Note (2)	
10	ANR2498	damper to slave cylinder	1	Note (3)	
10	ANR3508	damper to slave cylinder	1	Note (4)	Diesel
11	NTC9959	Damper assembly clutch	NLA	Note (5)	Use STD10002L.
11	STD10002L	Damper assembly clutch	1	Note (6)	
12	NTC7553	Bracket-damper clutch	1		
13	SH106141L	Screw-clutch damper to bracket-hexagonal head-M6 x 14	2		
14	WA106041L	Washer-M6-standard.	4		
15	ANR1675	Pipe clutch-master cylinder to damper	1	Note (7)	RHD
15	ANR2040	Pipe clutch-master cylinder to damper	1	Note (8)	LHD
16	ANR2498	Pipe clutch-damper/slave	1	Note (8)	RHD
16	ANR2039	Pipe clutch-damper/slave	1	Note (8)	LHD

CHANGE POINTS:
(1) To (V) KA 046332
(2) From (V) KA 046333 To (V) KA 052663
(3) From (V) KA 046333 To (V) MA 106179
(4) From (V) MA 106180
(5) To (V) MA 083648
(6) From (V) MA 083648
(7) From (V) KA 034314
(8) From (V) KA 034314 To (V) LA 081991

Illus	Part Number	Description	Quantity	Change Point	Remarks
1	NTC7723	Cable assembly choke	1		
2	ADU5065L	Clip-cable	1		
3	538890L	Clip	1		
4	MXC3176	Clip	1		
5	235113	Grommet	1		

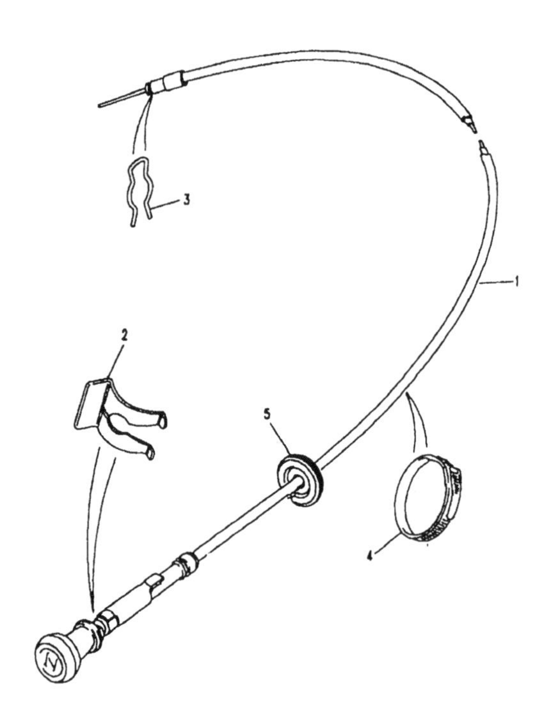

Brake Actuation .... H 1
  ABS Modulator .... H 3
  Master Cylinder & Servo .... H 1
Front & Rear .... H 14
  Front Disc Brakes .... H 14
  Rear Disc Brakes .... H 17
Handbrake .... H 20
  Handbrake Assembly .... H 20
Hoses & Pipes .... H 4
  Front-ABS-From (V) MA081992 .... H 9
  Front-LHD-To (V) LA081991 .... H 7
  Front-Non ABS-From (V) MA081992 .... H 10
  Front-RHD-To (V) LA081991 .... H 5
  Rear-ABS .... H 13
  Rear-Non ABS .... H 11
  Vacuum Hose .... H 4

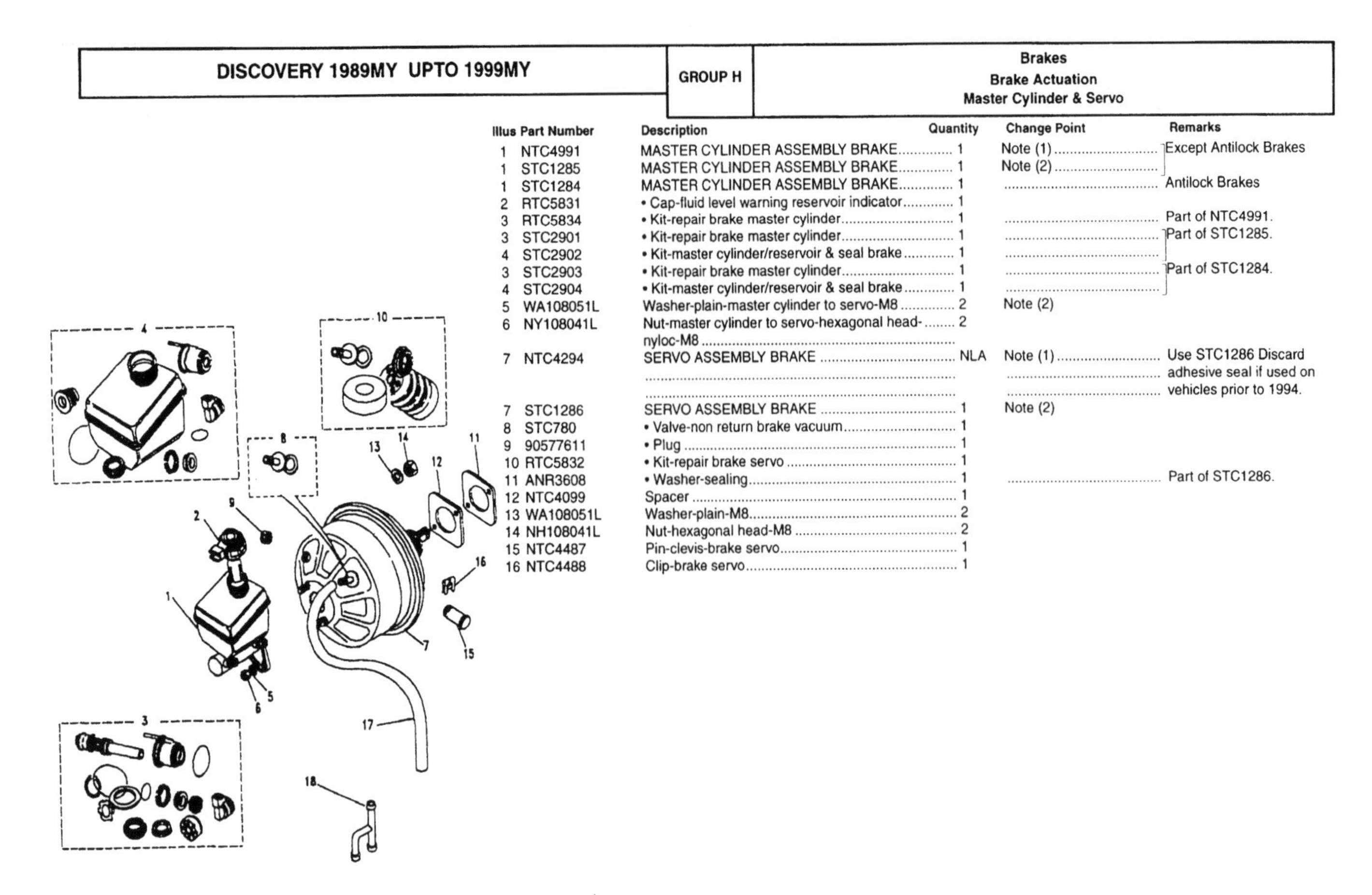

Illus	Part Number	Description	Quantity	Change Point	Remarks
1	NTC4991	MASTER CYLINDER ASSEMBLY BRAKE	1	Note (1)	Except Antilock Brakes
1	STC1285	MASTER CYLINDER ASSEMBLY BRAKE	1	Note (2)	
1	STC1284	MASTER CYLINDER ASSEMBLY BRAKE	1		Antilock Brakes
2	RTC5831	• Cap-fluid level warning reservoir indicator	1		
3	RTC5834	• Kit-repair brake master cylinder	1		Part of NTC4991.
3	STC2901	• Kit-repair brake master cylinder	1		Part of STC1285.
4	STC2902	• Kit-master cylinder/reservoir & seal brake	1		
3	STC2903	• Kit-repair brake master cylinder	1		Part of STC1284.
4	STC2904	• Kit-master cylinder/reservoir & seal brake	1		
5	WA108051L	Washer-plain-master cylinder to servo-M8	2	Note (2)	
6	NY108041L	Nut-master cylinder to servo-hexagonal head-nyloc-M8	2		
7	NTC4294	SERVO ASSEMBLY BRAKE	NLA	Note (1)	Use STC1286 Discard adhesive seal if used on vehicles prior to 1994.
7	STC1286	SERVO ASSEMBLY BRAKE	1	Note (2)	
8	STC780	• Valve-non return brake vacuum	1		
9	90577611	• Plug	1		
10	RTC5832	• Kit-repair brake servo	1		
11	ANR3608	• Washer-sealing	1		Part of STC1286.
12	NTC4099	Spacer	1		
13	WA108051L	Washer-plain-M8	2		
14	NH108041L	Nut-hexagonal head-M8	2		
15	NTC4487	Pin-clevis-brake servo	1		
16	NTC4488	Clip-brake servo	1		

CHANGE POINTS:
(1) To (V) LA 081991
(2) From (V) MA 081992

DISCOVERY 1989MY UPTO 1999MY | GROUP H | Brakes
Brake Actuation
Master Cylinder & Servo

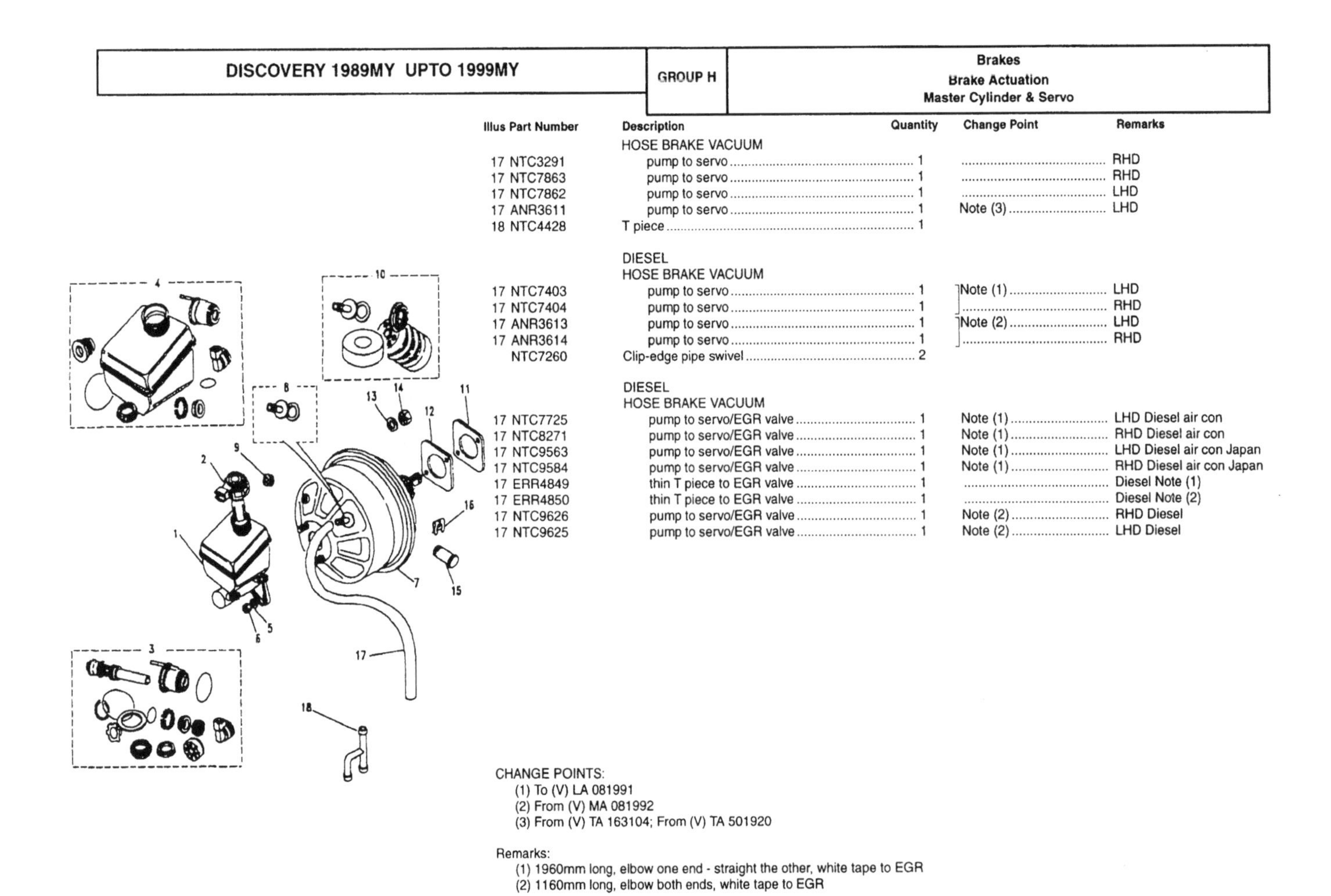

Illus	Part Number	Description	Quantity	Change Point	Remarks
		HOSE BRAKE VACUUM			
17	NTC3291	pump to servo	1		RHD
17	NTC7863	pump to servo	1		RHD
17	NTC7862	pump to servo	1		LHD
17	ANR3611	pump to servo	1	Note (3)	LHD
18	NTC4428	T piece	1		
		DIESEL			
		HOSE BRAKE VACUUM			
17	NTC7403	pump to servo	1	Note (1)	LHD
17	NTC7404	pump to servo	1		RHD
17	ANR3613	pump to servo	1	Note (2)	LHD
17	ANR3614	pump to servo	1		RHD
	NTC7260	Clip-edge pipe swivel	2		
		DIESEL			
		HOSE BRAKE VACUUM			
17	NTC7725	pump to servo/EGR valve	1	Note (1)	LHD Diesel air con
17	NTC8271	pump to servo/EGR valve	1	Note (1)	RHD Diesel air con
17	NTC9563	pump to servo/EGR valve	1	Note (1)	LHD Diesel air con Japan
17	NTC9584	pump to servo/EGR valve	1	Note (1)	RHD Diesel air con Japan
17	ERR4849	thin T piece to EGR valve	1		Diesel Note (1)
17	ERR4850	thin T piece to EGR valve	1		Diesel Note (2)
17	NTC9626	pump to servo/EGR valve	1	Note (2)	RHD Diesel
17	NTC9625	pump to servo/EGR valve	1	Note (2)	LHD Diesel

CHANGE POINTS:
(1) To (V) LA 081991
(2) From (V) MA 081992
(3) From (V) TA 163104; From (V) TA 501920

Remarks:
(1) 1960mm long, elbow one end - straight the other, white tape to EGR
(2) 1160mm long, elbow both ends, white tape to EGR

DISCOVERY 1989MY UPTO 1999MY — GROUP H

Brakes
Brake Actuation
ABS Modulator

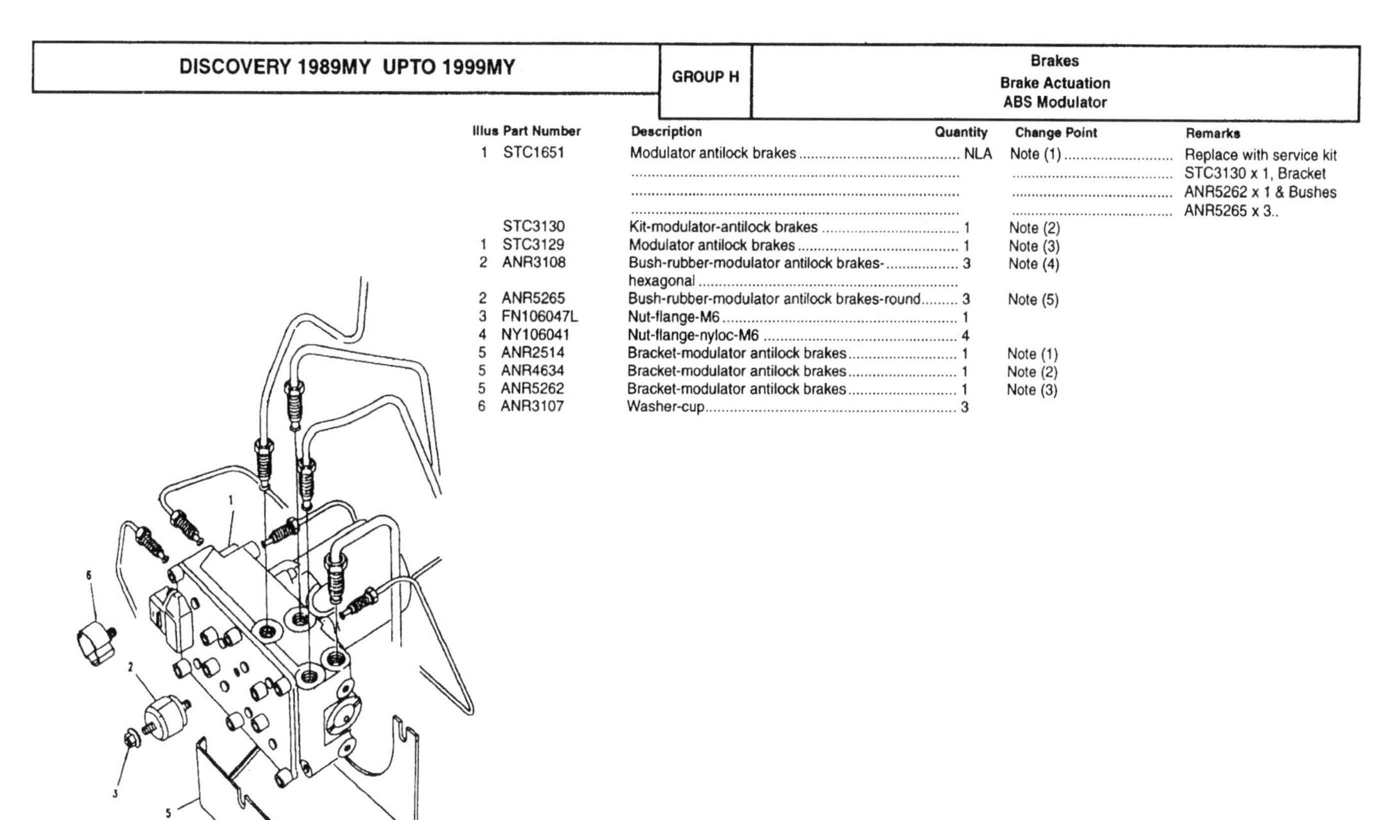

Illus	Part Number	Description	Quantity	Change Point	Remarks
1	STC1651	Modulator antilock brakes	NLA	Note (1)	Replace with service kit STC3130 x 1, Bracket ANR5262 x 1 & Bushes ANR5265 x 3..
	STC3130	Kit-modulator-antilock brakes	1	Note (2)	
1	STC3129	Modulator antilock brakes	1	Note (3)	
2	ANR3108	Bush-rubber-modulator antilock brakes-hexagonal	3	Note (4)	
2	ANR5265	Bush-rubber-modulator antilock brakes-round	3	Note (5)	
3	FN106047L	Nut-flange-M6	1		
4	NY106041	Nut-flange-nyloc-M6	4		
5	ANR2514	Bracket-modulator antilock brakes	1	Note (1)	
5	ANR4634	Bracket-modulator antilock brakes	1	Note (2)	
5	ANR5262	Bracket-modulator antilock brakes	1	Note (3)	
6	ANR3107	Washer-cup	3		

CHANGE POINTS:
(1) To (V) MA 163103
(2) From (V) TA 163104 To (V) TA 180925; From (V) TA 501920 To (V) TA 508779
(3) From (V) TA 180926 To (V) TA 199999; From (V) TA 508780 To (V) VA 558898; From (V) TA 700000
(4) To (V) TA 180925; To (V) TA 508779
(5) From (V) TA 180926; From (V) TA 508780

H 3

DISCOVERY 1989MY UPTO 1999MY — GROUP H

Brakes
Hoses & Pipes
Vacuum Hose

Illus	Part Number	Description	Quantity	Change Point	Remarks
1	NTC9677	Hose brake vacuum-to inlet manifold	1		
2	NTC7260	Clip-edge pipe swivel	2		
3	572548	Clip-hose	NLA		Use CN100168L.
3	CN100168L	Clip-hose	1		
4	RTC5907	Valve-inlet manifold non return	1		
5	232043	Washer-sealing	1		

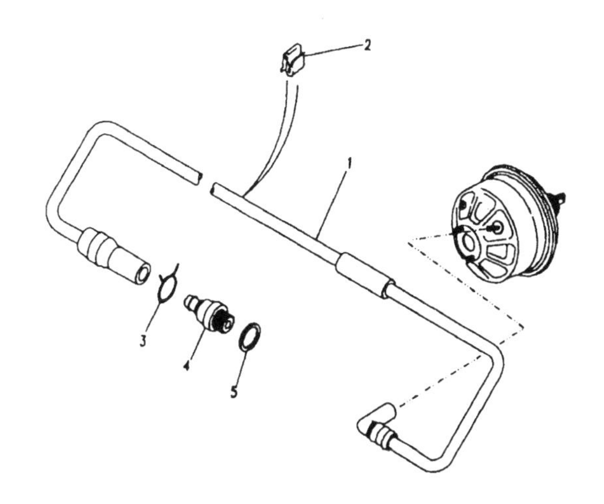

H 4

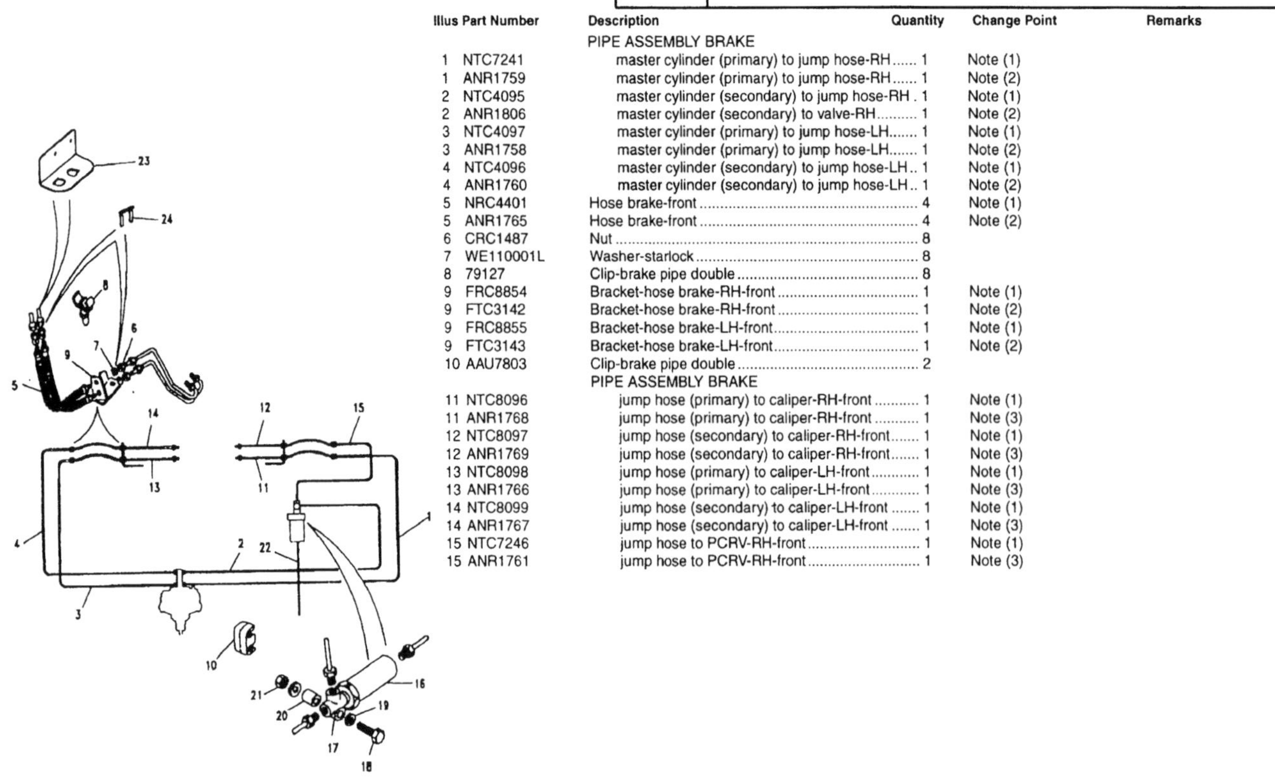

Illus	Part Number	Description	Quantity	Change Point	Remarks
		PIPE ASSEMBLY BRAKE			
1	NTC7241	master cylinder (primary) to jump hose-RH	1	Note (1)	
1	ANR1759	master cylinder (primary) to jump hose-RH	1	Note (2)	
2	NTC4095	master cylinder (secondary) to jump hose-RH	1	Note (1)	
2	ANR1806	master cylinder (secondary) to valve-RH	1	Note (2)	
3	NTC4097	master cylinder (primary) to jump hose-LH	1	Note (1)	
3	ANR1758	master cylinder (primary) to jump hose-LH	1	Note (2)	
4	NTC4096	master cylinder (secondary) to jump hose-LH	1	Note (1)	
4	ANR1760	master cylinder (secondary) to jump hose-LH	1	Note (2)	
5	NRC4401	Hose brake-front	4	Note (1)	
5	ANR1765	Hose brake-front	4	Note (2)	
6	CRC1487	Nut	8		
7	WE110001L	Washer-starlock	8		
8	79127	Clip-brake pipe double	8		
9	FRC8854	Bracket-hose brake-RH-front	1	Note (1)	
9	FTC3142	Bracket-hose brake-RH-front	1	Note (2)	
9	FRC8855	Bracket-hose brake-LH-front	1	Note (1)	
9	FTC3143	Bracket-hose brake-LH-front	1	Note (2)	
10	AAU7803	Clip-brake pipe double	2		
		PIPE ASSEMBLY BRAKE			
11	NTC8096	jump hose (primary) to caliper-RH-front	1	Note (1)	
11	ANR1768	jump hose (primary) to caliper-RH-front	1	Note (3)	
12	NTC8097	jump hose (secondary) to caliper-RH-front	1	Note (1)	
12	ANR1769	jump hose (secondary) to caliper-RH-front	1	Note (3)	
13	NTC8098	jump hose (primary) to caliper-LH-front	1	Note (1)	
13	ANR1766	jump hose (primary) to caliper-LH-front	1	Note (3)	
14	NTC8099	jump hose (secondary) to caliper-LH-front	1	Note (1)	
14	ANR1767	jump hose (secondary) to caliper-LH-front	1	Note (3)	
15	NTC7246	jump hose to PCRV-RH-front	1	Note (1)	
15	ANR1761	jump hose to PCRV-RH-front	1	Note (3)	

CHANGE POINTS:
(1) To (V) JA 034313
(2) From (V) KA 034314 To (V) LA 081991
(3) From (V) KA 034313 To (V) LA 081991

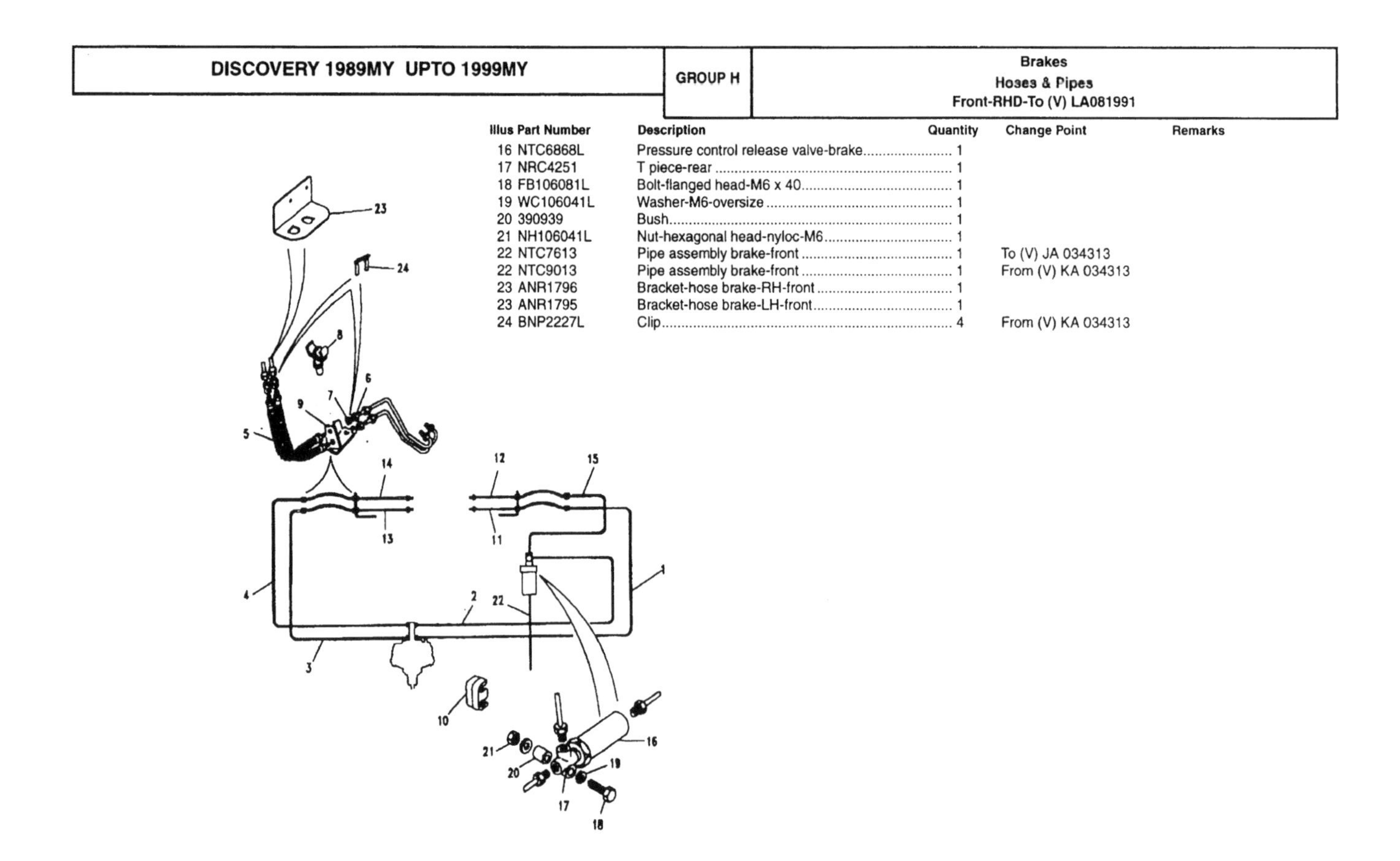

Illus	Part Number	Description	Quantity	Change Point	Remarks
16	NTC6868L	Pressure control release valve-brake	1		
17	NRC4251	T piece-rear	1		
18	FB106081L	Bolt-flanged head-M6 x 40	1		
19	WC106041L	Washer-M6-oversize	1		
20	390939	Bush	1		
21	NH106041L	Nut-hexagonal head-nyloc-M6	1		
22	NTC7613	Pipe assembly brake-front	1	To (V) JA 034313	
22	NTC9013	Pipe assembly brake-front	1	From (V) KA 034313	
23	ANR1796	Bracket-hose brake-RH-front	1		
23	ANR1795	Bracket-hose brake-LH-front	1		
24	BNP2227L	Clip	4	From (V) KA 034313	

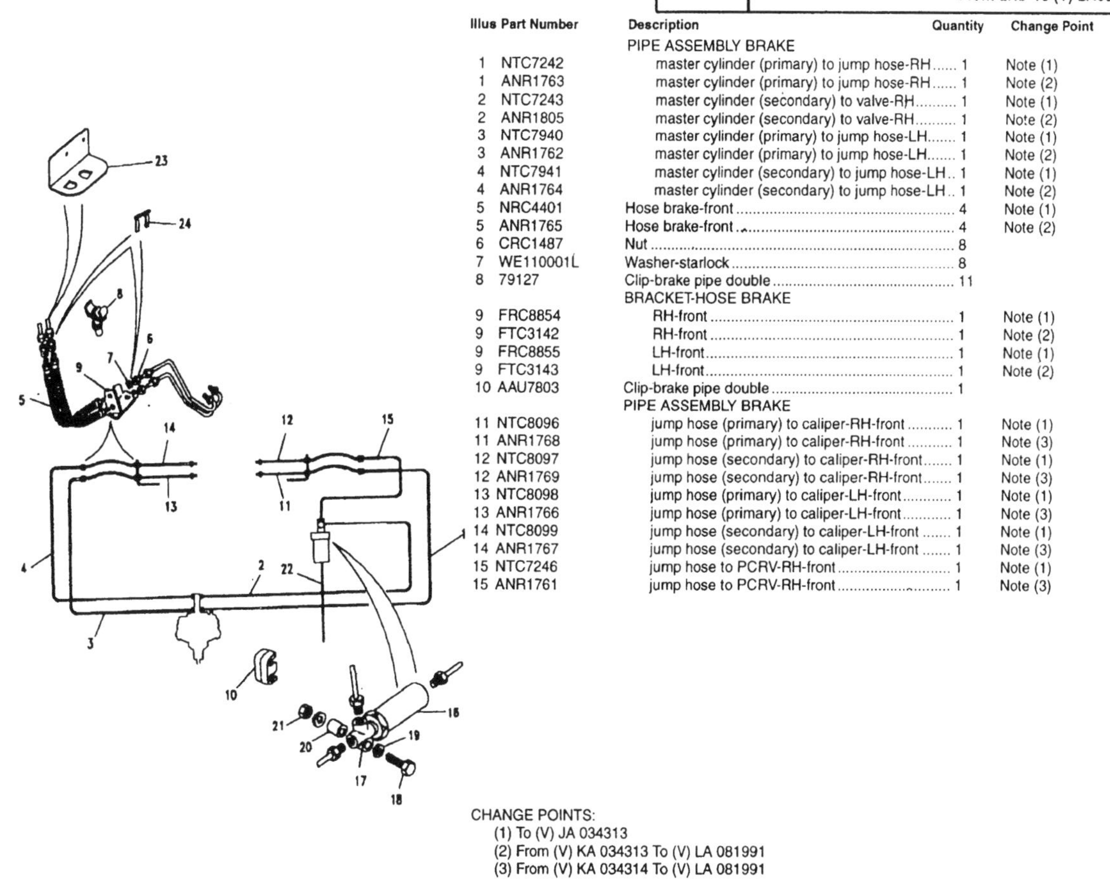

Illus	Part Number	Description	Quantity	Change Point	Remarks
		PIPE ASSEMBLY BRAKE			
1	NTC7242	master cylinder (primary) to jump hose-RH	1	Note (1)	
1	ANR1763	master cylinder (primary) to jump hose-RH	1	Note (2)	
2	NTC7243	master cylinder (secondary) to valve-RH	1	Note (1)	
2	ANR1805	master cylinder (secondary) to valve-RH	1	Note (2)	
3	NTC7940	master cylinder (primary) to jump hose-LH	1	Note (1)	
3	ANR1762	master cylinder (primary) to jump hose-LH	1	Note (2)	
4	NTC7941	master cylinder (secondary) to jump hose-LH	1	Note (1)	
4	ANR1764	master cylinder (secondary) to jump hose-LH	1	Note (2)	
5	NRC4401	Hose brake-front	4	Note (1)	
5	ANR1765	Hose brake-front	4	Note (2)	
6	CRC1487	Nut	8		
7	WE110001L	Washer-starlock	8		
8	79127	Clip-brake pipe double	11		
		BRACKET-HOSE BRAKE			
9	FRC8854	RH-front	1	Note (1)	
9	FTC3142	RH-front	1	Note (2)	
9	FRC8855	LH-front	1	Note (1)	
9	FTC3143	LH-front	1	Note (2)	
10	AAU7803	Clip-brake pipe double	1		
		PIPE ASSEMBLY BRAKE			
11	NTC8096	jump hose (primary) to caliper-RH-front	1	Note (1)	
11	ANR1768	jump hose (primary) to caliper-RH-front	1	Note (3)	
12	NTC8097	jump hose (secondary) to caliper-RH-front	1	Note (1)	
12	ANR1769	jump hose (secondary) to caliper-RH-front	1	Note (3)	
13	NTC8098	jump hose (primary) to caliper-LH-front	1	Note (1)	
13	ANR1766	jump hose (primary) to caliper-LH-front	1	Note (3)	
14	NTC8099	jump hose (secondary) to caliper-LH-front	1	Note (1)	
14	ANR1767	jump hose (secondary) to caliper-LH-front	1	Note (3)	
15	NTC7246	jump hose to PCRV-RH-front	1	Note (1)	
15	ANR1761	jump hose to PCRV-RH-front	1	Note (3)	

CHANGE POINTS:
(1) To (V) JA 034313
(2) From (V) KA 034313 To (V) LA 081991
(3) From (V) KA 034314 To (V) LA 081991

H 7

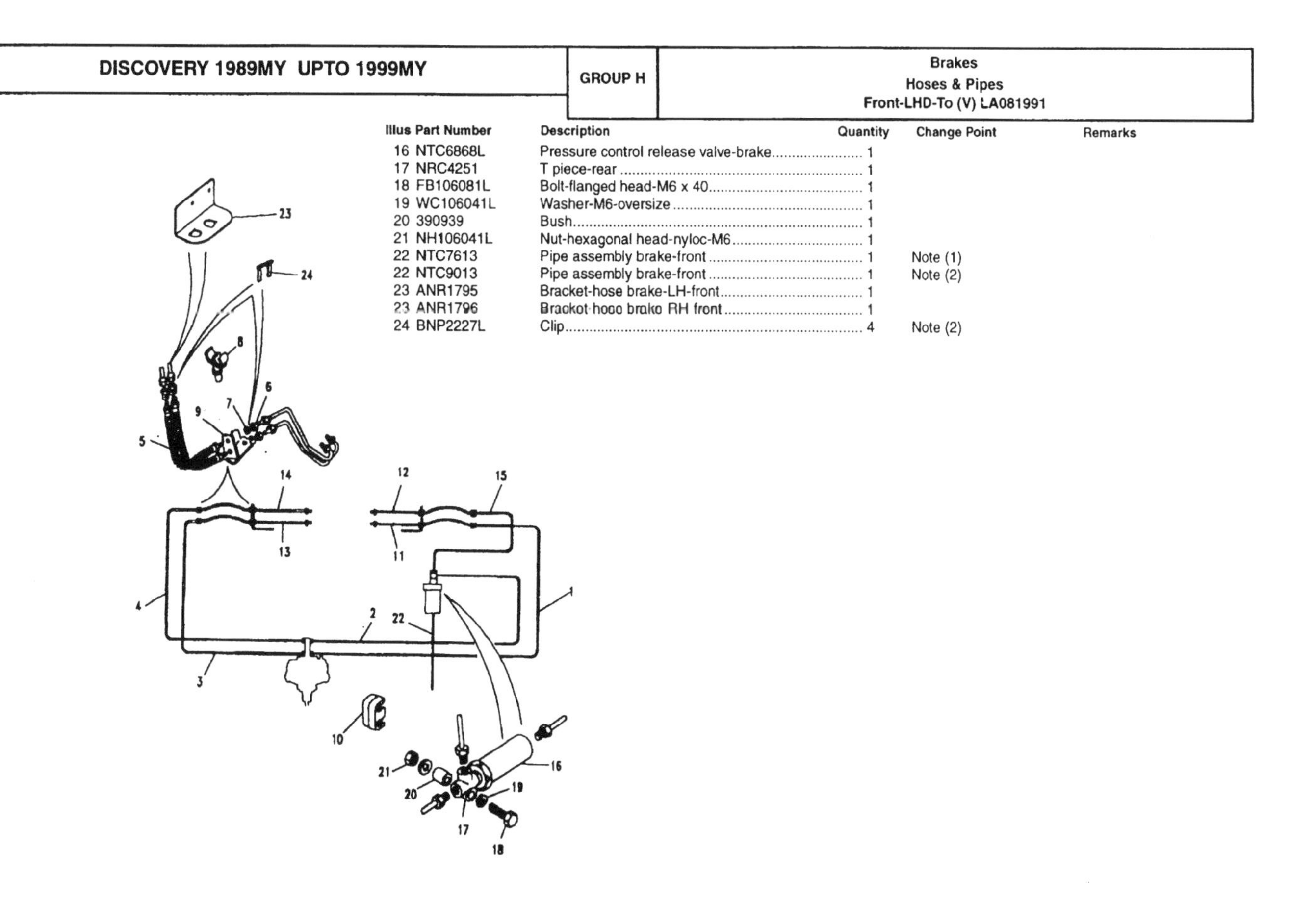

Illus	Part Number	Description	Quantity	Change Point	Remarks
16	NTC6868L	Pressure control release valve-brake	1		
17	NRC4251	T piece-rear	1		
18	FB106081L	Bolt-flanged head-M6 x 40	1		
19	WC106041L	Washer-M6-oversize	1		
20	390939	Bush	1		
21	NH106041L	Nut-hexagonal head-nyloc-M6	1		
22	NTC7613	Pipe assembly brake-front	1	Note (1)	
22	NTC9013	Pipe assembly brake-front	1	Note (2)	
23	ANR1795	Bracket-hose brake-LH-front	1		
23	ANR1796	Bracket-hose brake-RH front	1		
24	BNP2227L	Clip	4	Note (2)	

CHANGE POINTS:
(1) To (V) JA 034313
(2) From (V) KA 034314 To (V) LA 081991

H 8

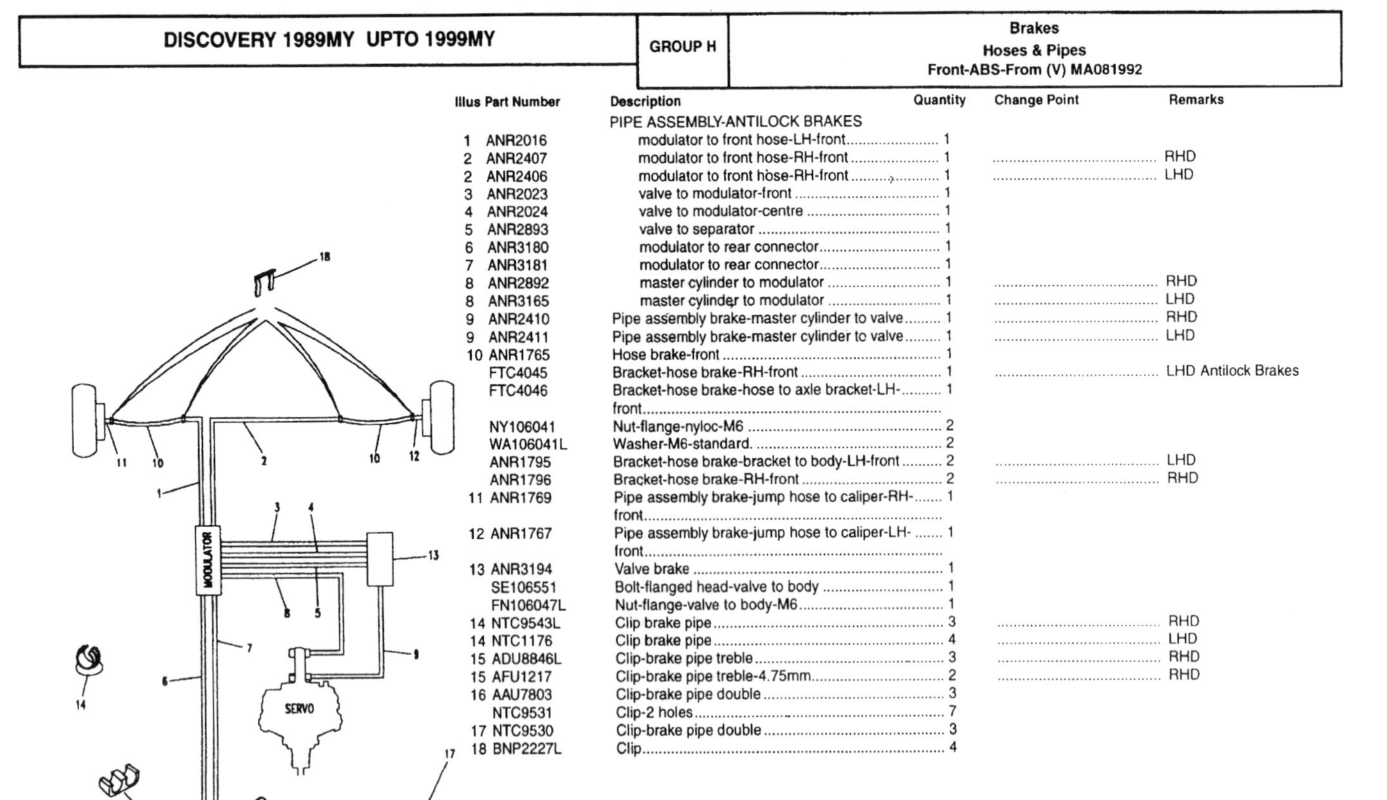

Illus	Part Number	Description	Quantity	Change Point	Remarks
		PIPE ASSEMBLY-ANTILOCK BRAKES			
1	ANR2016	modulator to front hose-LH-front	1		
2	ANR2407	modulator to front hose-RH-front	1		RHD
2	ANR2406	modulator to front hose-RH-front	1		LHD
3	ANR2023	valve to modulator-front	1		
4	ANR2024	valve to modulator-centre	1		
5	ANR2893	valve to separator	1		
6	ANR3180	modulator to rear connector	1		
7	ANR3181	modulator to rear connector	1		
8	ANR2892	master cylinder to modulator	1		RHD
8	ANR3165	master cylinder to modulator	1		LHD
9	ANR2410	Pipe assembly brake-master cylinder to valve	1		RHD
9	ANR2411	Pipe assembly brake-master cylinder to valve	1		LHD
10	ANR1765	Hose brake-front	1		
	FTC4045	Bracket-hose brake-RH-front	1		LHD Antilock Brakes
	FTC4046	Bracket-hose brake-hose to axle bracket-LH-front	1		
	NY106041	Nut-flange-nyloc-M6	2		
	WA106041L	Washer-M6-standard.	2		
	ANR1795	Bracket-hose brake-bracket to body-LH-front	2		LHD
	ANR1796	Bracket-hose brake-RH-front	2		RHD
11	ANR1769	Pipe assembly brake-jump hose to caliper-RH-front	1		
12	ANR1767	Pipe assembly brake-jump hose to caliper-LH-front	1		
13	ANR3194	Valve brake	1		
	SE106551	Bolt-flanged head-valve to body	1		
	FN106047L	Nut-flange-valve to body-M6	1		
14	NTC9543L	Clip brake pipe	3		RHD
14	NTC1176	Clip brake pipe	4		LHD
15	ADU8846L	Clip-brake pipe treble	3		RHD
15	AFU1217	Clip-brake pipe treble-4.75mm	2		RHD
16	AAU7803	Clip-brake pipe double	3		
	NTC9531	Clip-2 holes	7		
17	NTC9530	Clip-brake pipe double	3		
18	BNP2227L	Clip	4		

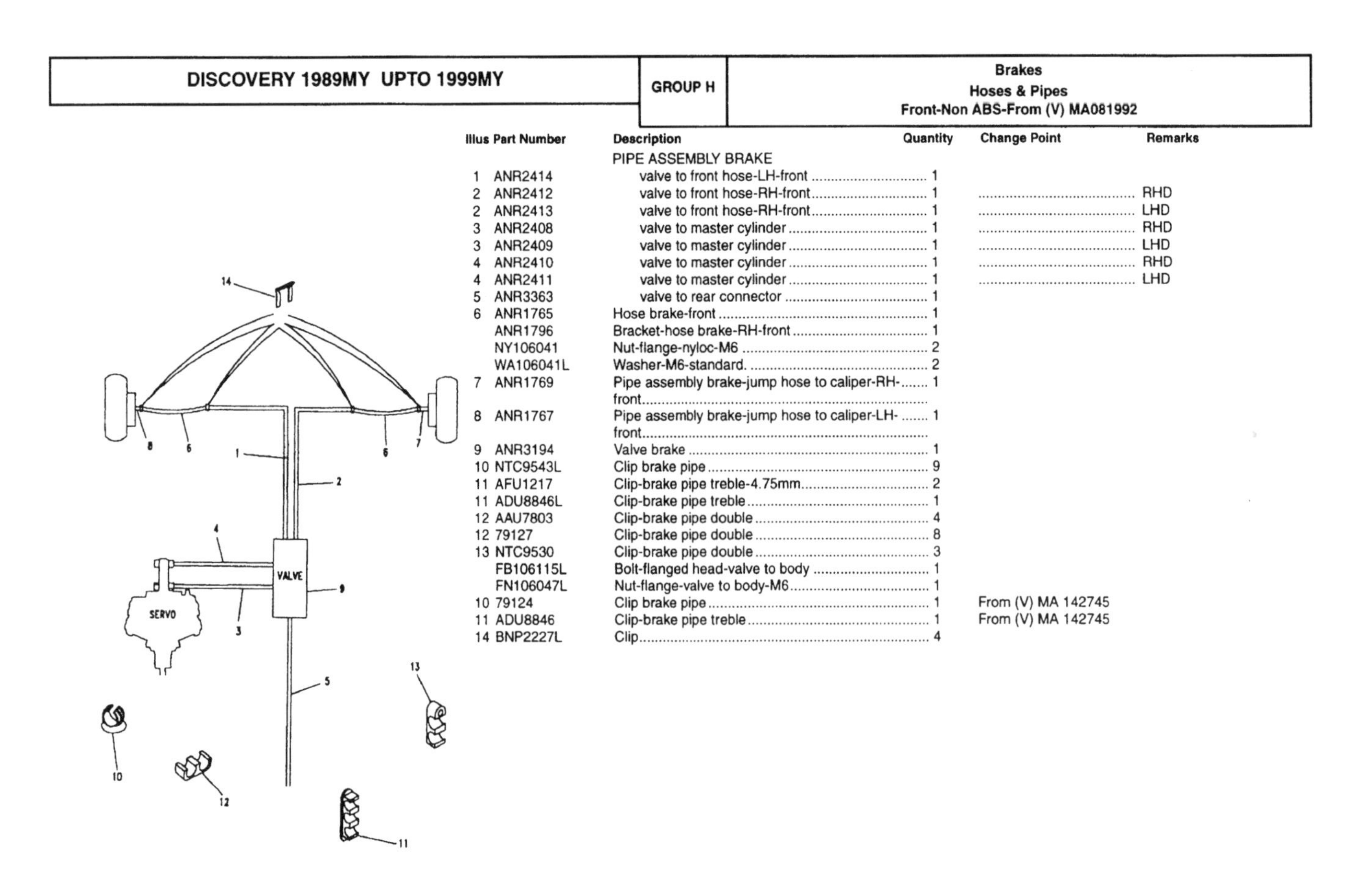

Illus	Part Number	Description	Quantity	Change Point	Remarks
		PIPE ASSEMBLY BRAKE			
1	ANR2414	valve to front hose-LH-front	1		
2	ANR2412	valve to front hose-RH-front	1		RHD
2	ANR2413	valve to front hose-RH-front	1		LHD
3	ANR2408	valve to master cylinder	1		RHD
3	ANR2409	valve to master cylinder	1		LHD
4	ANR2410	valve to master cylinder	1		RHD
4	ANR2411	valve to master cylinder	1		LHD
5	ANR3363	valve to rear connector	1		
6	ANR1765	Hose brake-front	1		
	ANR1796	Bracket-hose brake-RH-front	1		
	NY106041	Nut-flange-nyloc-M6	2		
	WA106041L	Washer-M6-standard.	2		
7	ANR1769	Pipe assembly brake-jump hose to caliper-RH-front	1		
8	ANR1767	Pipe assembly brake-jump hose to caliper-LH-front	1		
9	ANR3194	Valve brake	1		
10	NTC9543L	Clip brake pipe	9		
11	AFU1217	Clip-brake pipe treble-4.75mm	2		
11	ADU8846L	Clip-brake pipe treble	1		
12	AAU7803	Clip-brake pipe double	4		
12	79127	Clip-brake pipe double	8		
13	NTC9530	Clip-brake pipe double	3		
	FB106115L	Bolt-flanged head-valve to body	1		
	FN106047L	Nut-flange-valve to body-M6	1		
10	79124	Clip brake pipe	1	From (V) MA 142745	
11	ADU8846	Clip-brake pipe treble	1	From (V) MA 142745	
14	BNP2227L	Clip	4		

DISCOVERY 1989MY UPTO 1999MY

GROUP H | **Brakes** | **Hoses & Pipes** | **Rear-Non ABS**

Illus	Part Number	Description	Quantity	Change Point	Remarks
		PIPE ASSEMBLY BRAKE			
1	NTC2003	rear	1	Note (1)	
1	NTC8028	rear	NLA	Note (2)	Use NTC8028K qty 1 with STC589 qty 1.
1	NTC9273	rear	1	Note (3)	
1	NTC9293	rear	1	Note (4)	
1	NTC8874	rear-RH	1	Note (5)	
2	NRC4403	Connector brake-rear	1	Note (6)	
2	BMK2466	Connector brake	1	Note (7)	
3	WE110001L	Washer-starlock	2		
4	CRC1487	Nut	2		
5	79125	Clip-brake pipe double	4		
5	NTC9530	Clip-brake pipe double-stud fixing	3		MPi
5	NTC8242	Clip-brake pipe double	2		MPi
6	CRC1250L	Clip brake pipe	1		
6	NTC9543L	Clip brake pipe	9		MPi
7	NTC3458	Hose brake-rear	1		
8	NRC4251	T piece-rear	1		
9	BH106061L	Bolt-hexagonal head-M6 x 30	1		
10	WA106041L	Washer-M6-standard.	1		
11	NH106041L	Nut-hexagonal head-nyloc-M6	1		

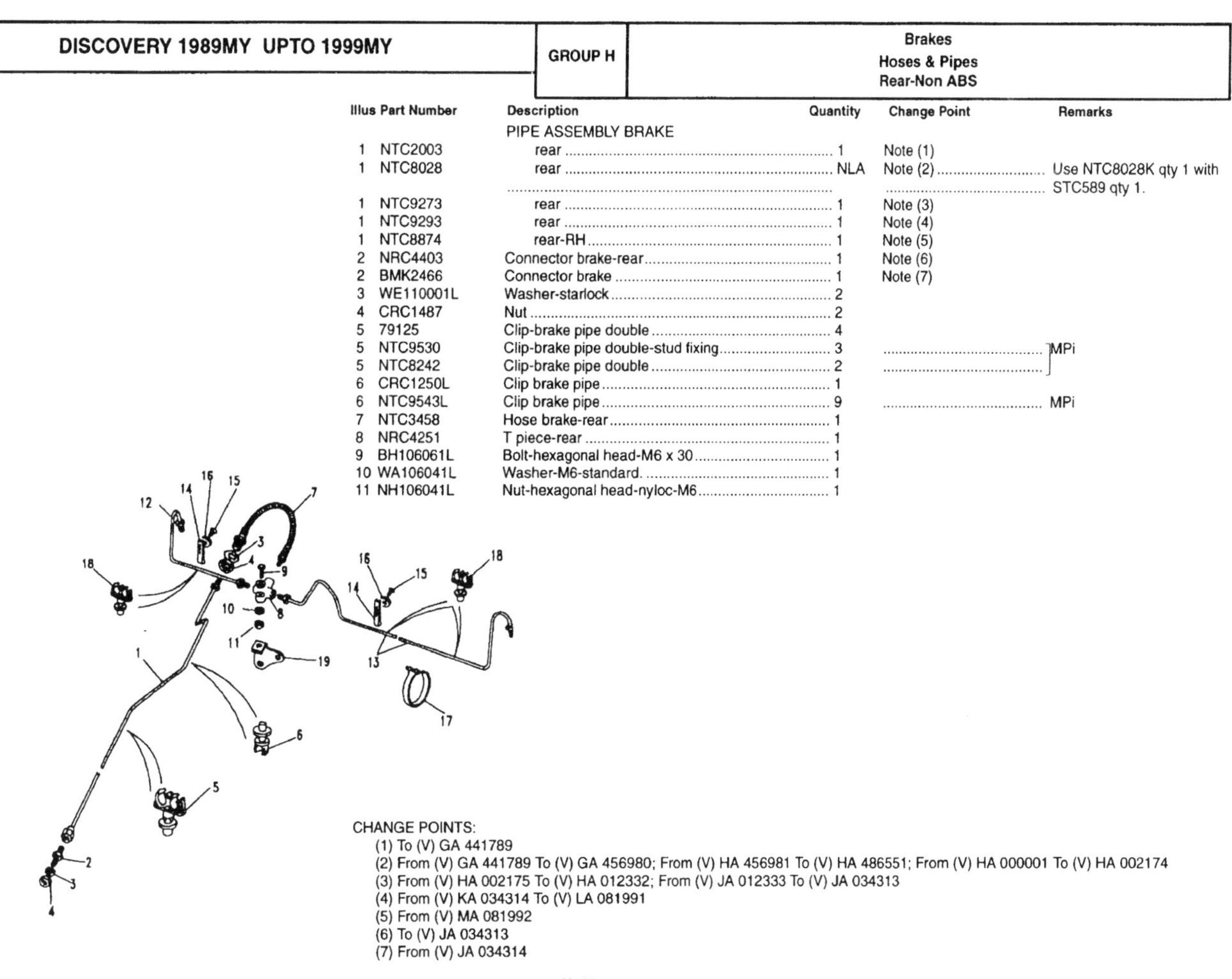

CHANGE POINTS:

(1) To (V) GA 441789
(2) From (V) GA 441789 To (V) GA 456980; From (V) HA 456981 To (V) HA 486551; From (V) HA 000001 To (V) HA 002174
(3) From (V) HA 002175 To (V) HA 012332; From (V) JA 012333 To (V) JA 034313
(4) From (V) KA 034314 To (V) LA 081991
(5) From (V) MA 081992
(6) To (V) JA 034313
(7) From (V) JA 034314

H 11

DISCOVERY 1989MY UPTO 1999MY

GROUP H | **Brakes** | **Hoses & Pipes** | **Rear-Non ABS**

Illus	Part Number	Description	Quantity	Change Point	Remarks
		PIPE ASSEMBLY BRAKE			
12	NRC4419	rear-RH	1	Note (1)	
12	NTC7676	rear-RH	1	Note (2)	
12	NTC7688	rear-RH	1	Note (3)	
13	NTC3071	rear-LH	1	Note (4)	
13	NTC7689	rear-LH	1	Note (5)	
14	577873	Clip-p	2		
15	SL510031	Screw	2		
16	WC702101L	Washer-plain	2		
17	11820L	Clip	1		
18	NTC8242	Clip-brake pipe double	3		
	NTC9603	Clip-brake pipe treble	1		
19	FRC8775	Bracket-brake pipe tee connector	NLA		Use FTC4889.
19	FTC4889	Bracket-brake pipe tee connector	1		

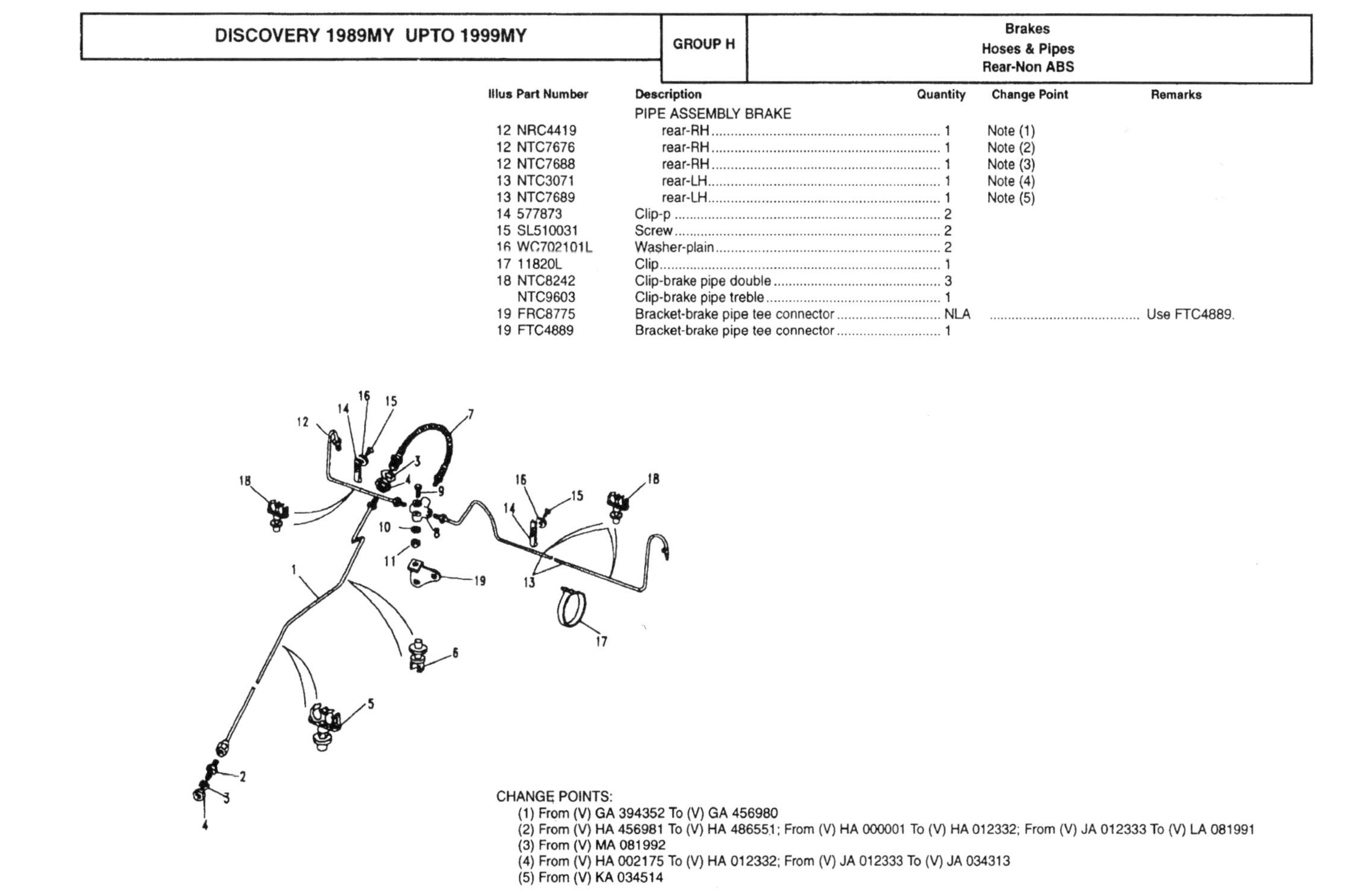

CHANGE POINTS:

(1) From (V) GA 394352 To (V) GA 456980
(2) From (V) HA 456981 To (V) HA 486551; From (V) HA 000001 To (V) HA 012332; From (V) JA 012333 To (V) LA 081991
(3) From (V) MA 081992
(4) From (V) HA 002175 To (V) HA 012332; From (V) JA 012333 To (V) JA 034313
(5) From (V) KA 034514

H 12

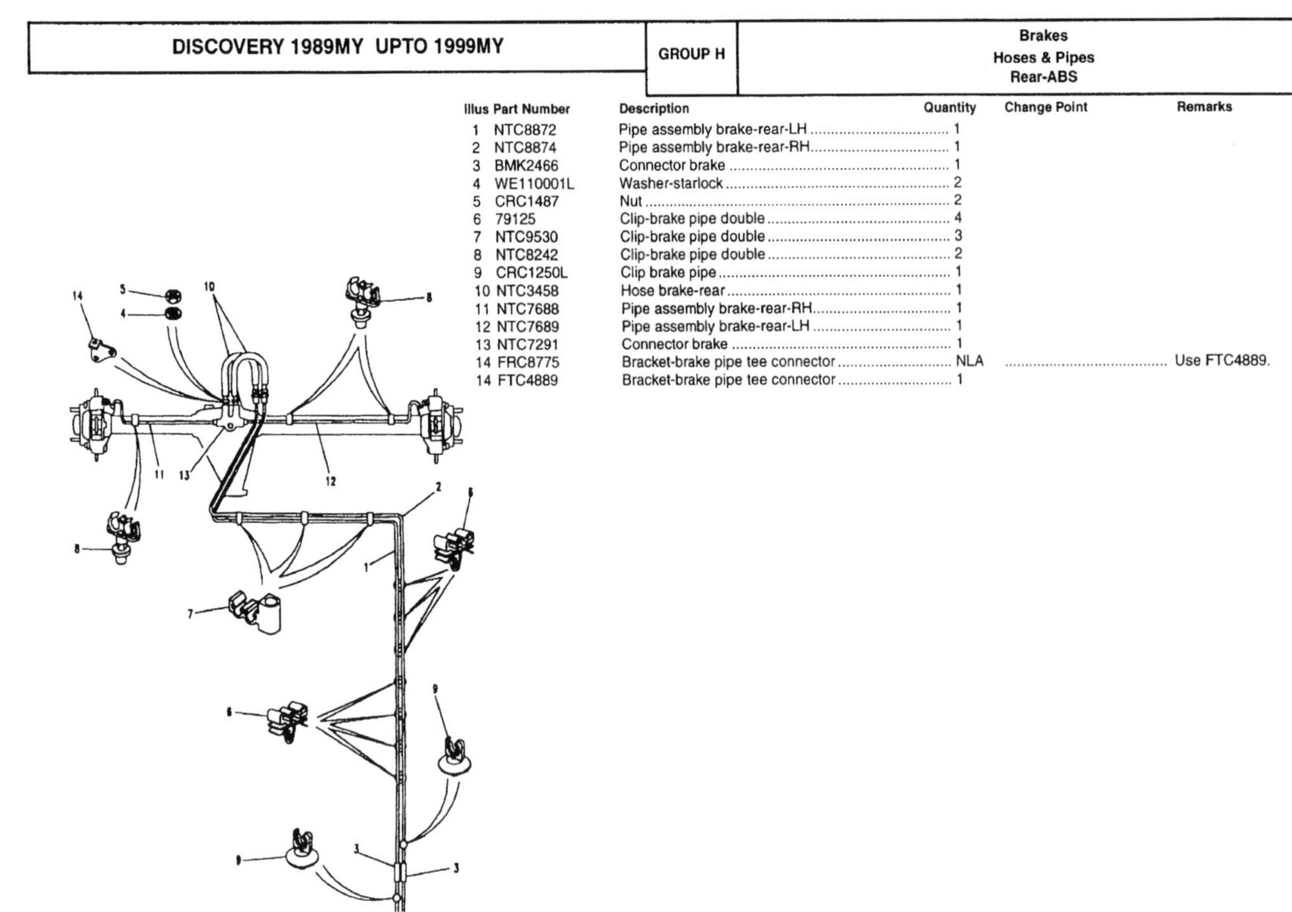

Illus	Part Number	Description	Quantity	Change Point	Remarks
1	NTC8872	Pipe assembly brake-rear-LH	1		
2	NTC8874	Pipe assembly brake-rear-RH	1		
3	BMK2466	Connector brake	1		
4	WE110001L	Washer-starlock	2		
5	CRC1487	Nut	2		
6	79125	Clip-brake pipe double	4		
7	NTC9530	Clip-brake pipe double	3		
8	NTC8242	Clip-brake pipe double	2		
9	CRC1250L	Clip brake pipe	1		
10	NTC3458	Hose brake-rear	1		
11	NTC7688	Pipe assembly brake-rear-RH	1		
12	NTC7689	Pipe assembly brake-rear-LH	1		
13	NTC7291	Connector brake	1		
14	FRC8775	Bracket-brake pipe tee connector	NLA		Use FTC4889.
14	FTC4889	Bracket-brake pipe tee connector	1		

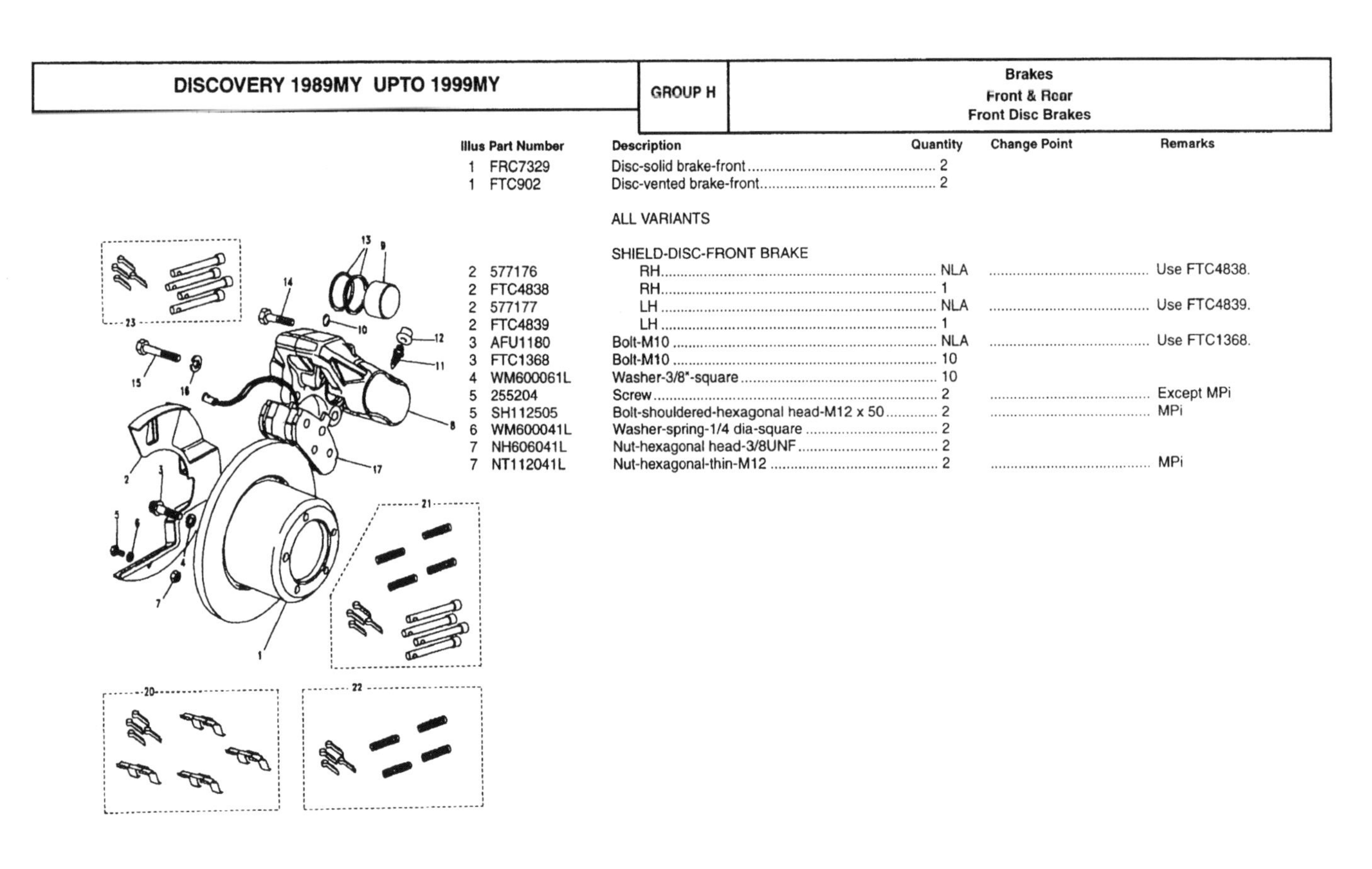

Illus	Part Number	Description	Quantity	Change Point	Remarks
1	FRC7329	Disc-solid brake-front	2		
1	FTC902	Disc-vented brake-front	2		
		ALL VARIANTS			
		SHIELD-DISC-FRONT BRAKE			
2	577176	RH	NLA		Use FTC4838.
2	FTC4838	RH	1		
2	577177	LH	NLA		Use FTC4839.
2	FTC4839	LH	1		
3	AFU1180	Bolt-M10	NLA		Use FTC1368.
3	FTC1368	Bolt-M10	10		
4	WM600061L	Washer-3/8"-square	10		
5	255204	Screw	2		Except MPi
5	SH112505	Bolt-shouldered-hexagonal head-M12 x 50	2		MPi
6	WM600041L	Washer-spring-1/4 dia-square	2		
7	NH606041L	Nut-hexagonal head-3/8UNF	2		
7	NT112041L	Nut-hexagonal-thin-M12	2		MPi

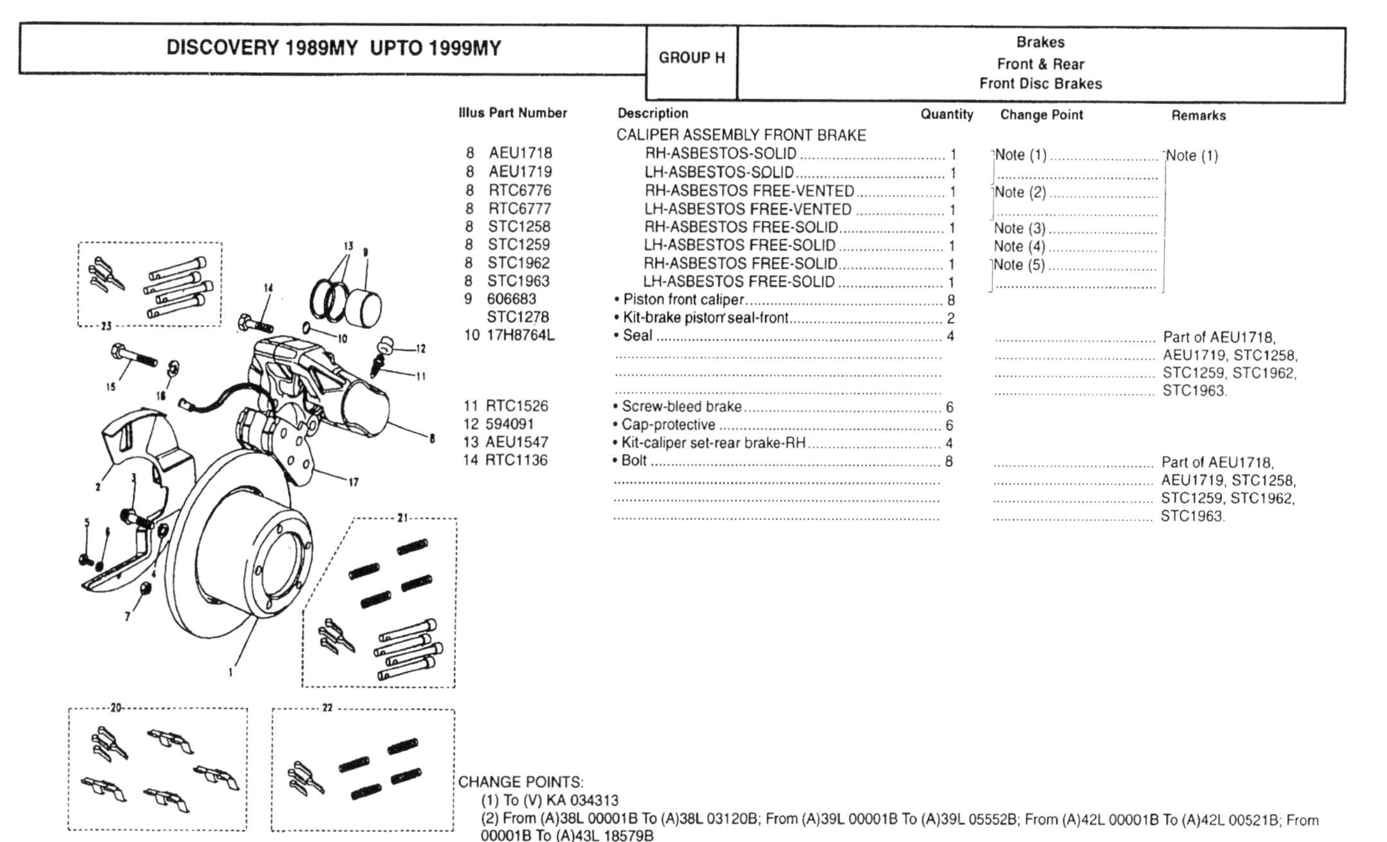

Illus	Part Number	Description	Quantity	Change Point	Remarks
		CALIPER ASSEMBLY FRONT BRAKE			
8	AEU1718	RH-ASBESTOS-SOLID	1	Note (1)	Note (1)
8	AEU1719	LH-ASBESTOS-SOLID	1		
8	RTC6776	RH-ASBESTOS FREE-VENTED	1	Note (2)	
8	RTC6777	LH-ASBESTOS FREE-VENTED	1		
8	STC1258	RH-ASBESTOS FREE-SOLID	1	Note (3)	
8	STC1259	LH-ASBESTOS FREE-SOLID	1	Note (4)	
8	STC1962	RH-ASBESTOS FREE-SOLID	1	Note (5)	
8	STC1963	LH-ASBESTOS FREE-SOLID	1		
9	606683	• Piston front caliper	8		
	STC1278	• Kit-brake piston seal-front	2		
10	17H8764L	• Seal	4		Part of AEU1718, AEU1719, STC1258, STC1259, STC1962, STC1963.
11	RTC1526	• Screw-bleed brake	6		
12	594091	• Cap-protective	6		
13	AEU1547	• Kit-caliper set-rear brake-RH	4		
14	RTC1136	• Bolt	8		Part of AEU1718, AEU1719, STC1258, STC1259, STC1962, STC1963.

CHANGE POINTS:
(1) To (V) KA 034313
(2) From (A)38L 00001B To (A)38L 03120B; From (A)39L 00001B To (A)39L 05552B; From (A)42L 00001B To (A)42L 00521B; From 00001B To (A)43L 18579B
(3) From (A)38L 03121B To (V) LA 081991; From (A)39L 05553B To (V) LA 081991; From (A)42L 00522B To (V) LA 081991; From 18580B To (V) LA 081991
(4) From (A)38L 03121B; From (A)39L 05553B; From (A)42L 00522B; From (A)43L 18580B To (V) LA 081991
(5) From (V) MA 081992

Remarks:
(1) Caliper seals are not a servicable item - replace caliper

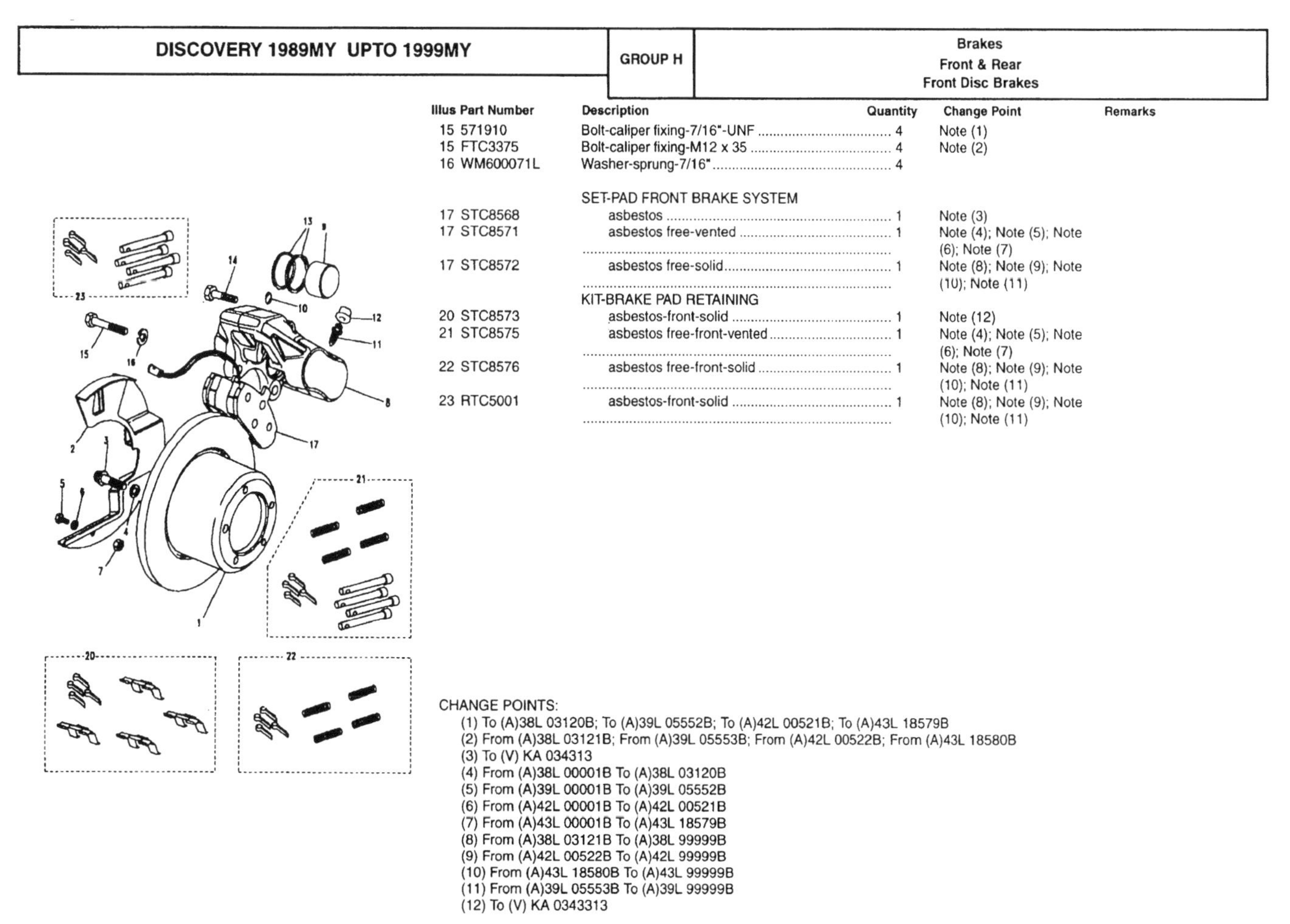

Illus	Part Number	Description	Quantity	Change Point	Remarks
15	571910	Bolt-caliper fixing-7/16"-UNF	4	Note (1)	
15	FTC3375	Bolt-caliper fixing-M12 x 35	4	Note (2)	
16	WM600071L	Washer-sprung-7/16"	4		
		SET-PAD FRONT BRAKE SYSTEM			
17	STC8568	asbestos	1	Note (3)	
17	STC8571	asbestos free-vented	1	Note (4); Note (5); Note (6); Note (7)	
17	STC8572	asbestos free-solid	1	Note (8); Note (9); Note (10); Note (11)	
		KIT-BRAKE PAD RETAINING			
20	STC8573	asbestos-front-solid	1	Note (12)	
21	STC8575	asbestos free-front-vented	1	Note (4); Note (5); Note (6); Note (7)	
22	STC8576	asbestos free-front-solid	1	Note (8); Note (9); Note (10); Note (11)	
23	RTC5001	asbestos-front-solid	1	Note (8); Note (9); Note (10); Note (11)	

CHANGE POINTS:
(1) To (A)38L 03120B; To (A)39L 05552B; To (A)42L 00521B; To (A)43L 18579B
(2) From (A)38L 03121B; From (A)39L 05553B; From (A)42L 00522B; From (A)43L 18580B
(3) To (V) KA 034313
(4) From (A)38L 00001B To (A)38L 03120B
(5) From (A)39L 00001B To (A)39L 05552B
(6) From (A)42L 00001B To (A)42L 00521B
(7) From (A)43L 00001B To (A)43L 18579B
(8) From (A)38L 03121B To (A)38L 99999B
(9) From (A)42L 00522B To (A)42L 99999B
(10) From (A)43L 18580B To (A)43L 99999B
(11) From (A)39L 05553B To (A)39L 99999B
(12) To (V) KA 0343313

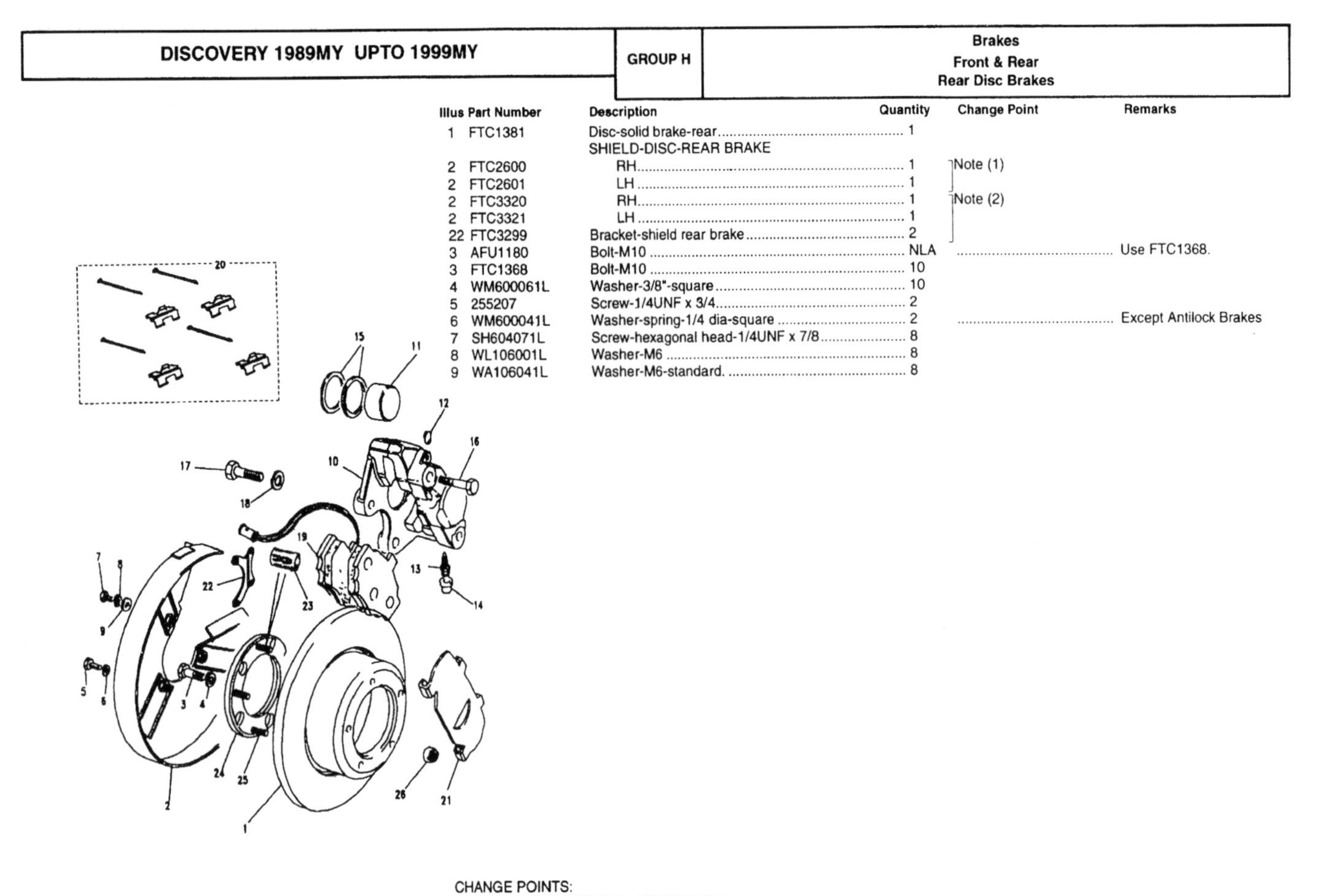

Illus	Part Number	Description	Quantity	Change Point	Remarks
1	FTC1381	Disc-solid brake-rear	1		
		SHIELD-DISC-REAR BRAKE			
2	FTC2600	RH	1	Note (1)	
2	FTC2601	LH	1		
2	FTC3320	RH	1	Note (2)	
2	FTC3321	LH	1		
22	FTC3299	Bracket-shield rear brake	2		
3	AFU1180	Bolt-M10	NLA		Use FTC1368.
3	FTC1368	Bolt-M10	10		
4	WM600061L	Washer-3/8"-square	10		
5	255207	Screw-1/4UNF x 3/4	2		
6	WM600041L	Washer-spring-1/4 dia-square	2		Except Antilock Brakes
7	SH604071L	Screw-hexagonal head-1/4UNF x 7/8	8		
8	WL106001L	Washer-M6	8		
9	WA106041L	Washer-M6-standard.	8		

CHANGE POINTS:
(1) To (A)28S 44171D; To (A)32S 2118B
(2) From (A)28S 44172D; From (A)32S 2119B

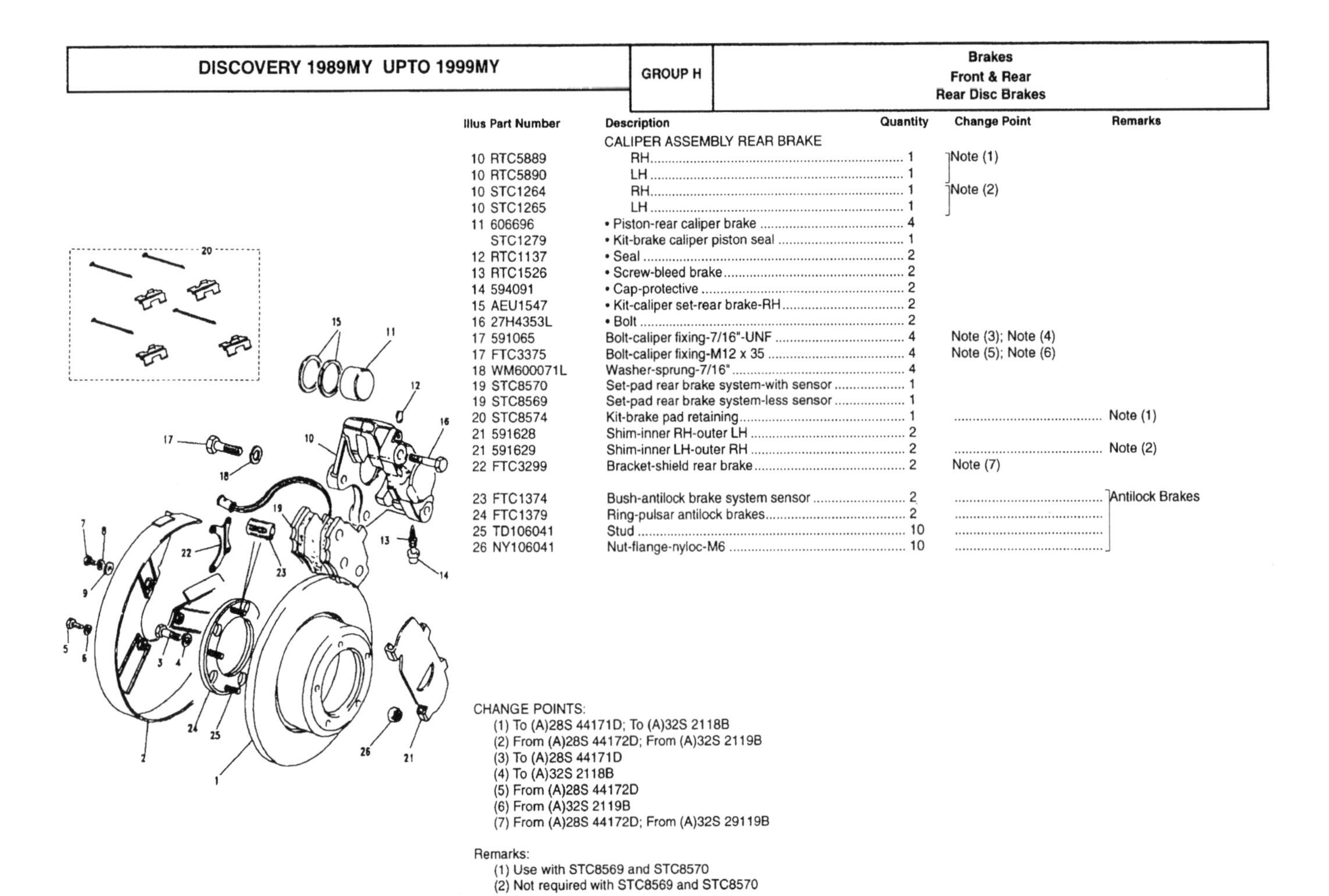

Illus	Part Number	Description	Quantity	Change Point	Remarks
		CALIPER ASSEMBLY REAR BRAKE			
10	RTC5889	RH	1	Note (1)	
10	RTC5890	LH	1		
10	STC1264	RH	1	Note (2)	
10	STC1265	LH	1		
11	606696	• Piston-rear caliper brake	4		
	STC1279	• Kit-brake caliper piston seal	1		
12	RTC1137	• Seal	2		
13	RTC1526	• Screw-bleed brake	2		
14	594091	• Cap-protective	2		
15	AEU1547	• Kit-caliper set-rear brake-RH	2		
16	27H4353L	• Bolt	2		
17	591065	Bolt-caliper fixing-7/16"-UNF	4	Note (3); Note (4)	
17	FTC3375	Bolt-caliper fixing-M12 x 35	4	Note (5); Note (6)	
18	WM600071L	Washer-sprung-7/16"	4		
19	STC8570	Set-pad rear brake system-with sensor	1		
19	STC8569	Set-pad rear brake system-less sensor	1		
20	STC8574	Kit-brake pad retaining	1		Note (1)
21	591628	Shim-inner RH-outer LH	2		
21	591629	Shim-inner LH-outer RH	2		Note (2)
22	FTC3299	Bracket-shield rear brake	2	Note (7)	
23	FTC1374	Bush-antilock brake system sensor	2		Antilock Brakes
24	FTC1379	Ring-pulsar antilock brakes	2		
25	TD106041	Stud	10		
26	NY106041	Nut-flange-nyloc-M6	10		

CHANGE POINTS:
(1) To (A)28S 44171D; To (A)32S 2118B
(2) From (A)28S 44172D; From (A)32S 2119B
(3) To (A)28S 44171D
(4) To (A)32S 2118B
(5) From (A)28S 44172D
(6) From (A)32S 2119B
(7) From (A)28S 44172D; From (A)32S 29119B

Remarks:
(1) Use with STC8569 and STC8570
(2) Not required with STC8569 and STC8570

DISCOVERY 1989MY UPTO 1999MY

GROUP H

Brakes
Front & Rear
Splash Shield

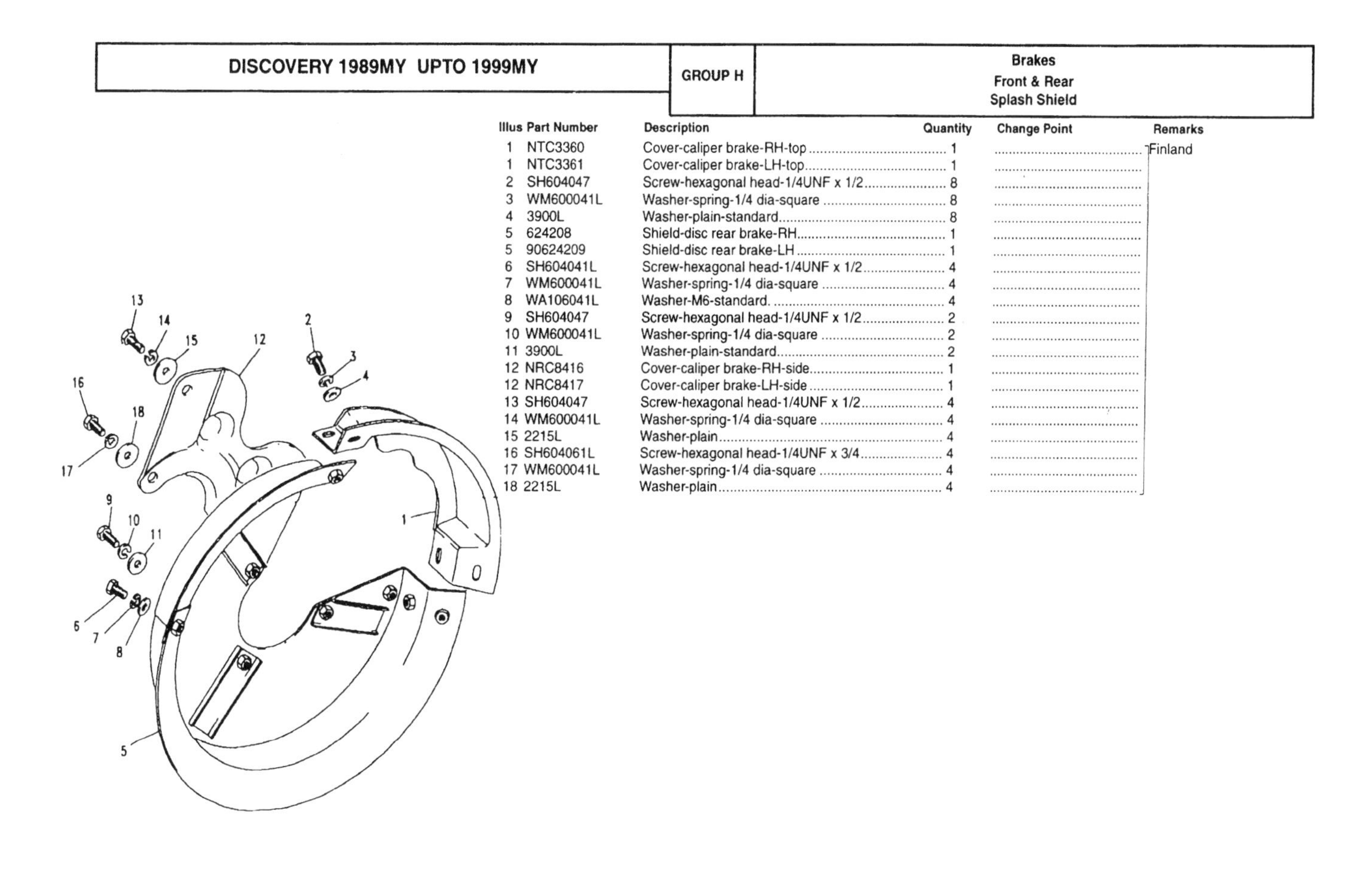

Illus	Part Number	Description	Quantity	Change Point	Remarks
1	NTC3360	Cover-caliper brake-RH-top	1		Finland
1	NTC3361	Cover-caliper brake-LH-top	1		
2	SH604047	Screw-hexagonal head-1/4UNF x 1/2	8		
3	WM600041L	Washer-spring-1/4 dia-square	8		
4	3900L	Washer-plain-standard	8		
5	624208	Shield-disc rear brake-RH	1		
5	90624209	Shield-disc rear brake-LH	1		
6	SH604041L	Screw-hexagonal head-1/4UNF x 1/2	4		
7	WM600041L	Washer-spring-1/4 dia-square	4		
8	WA106041L	Washer-M6-standard.	4		
9	SH604047	Screw-hexagonal head-1/4UNF x 1/2	2		
10	WM600041L	Washer-spring-1/4 dia-square	2		
11	3900L	Washer-plain-standard	2		
12	NRC8416	Cover-caliper brake-RH-side	1		
12	NRC8417	Cover-caliper brake-LH-side	1		
13	SH604047	Screw-hexagonal head-1/4UNF x 1/2	4		
14	WM600041L	Washer-spring-1/4 dia-square	4		
15	2215L	Washer-plain	4		
16	SH604061L	Screw-hexagonal head-1/4UNF x 3/4	4		
17	WM600041L	Washer-spring-1/4 dia-square	4		
18	2215L	Washer-plain	4		

H 19

DISCOVERY 1989MY UPTO 1999MY

GROUP H

Brakes
Handbrake
Handbrake Assembly

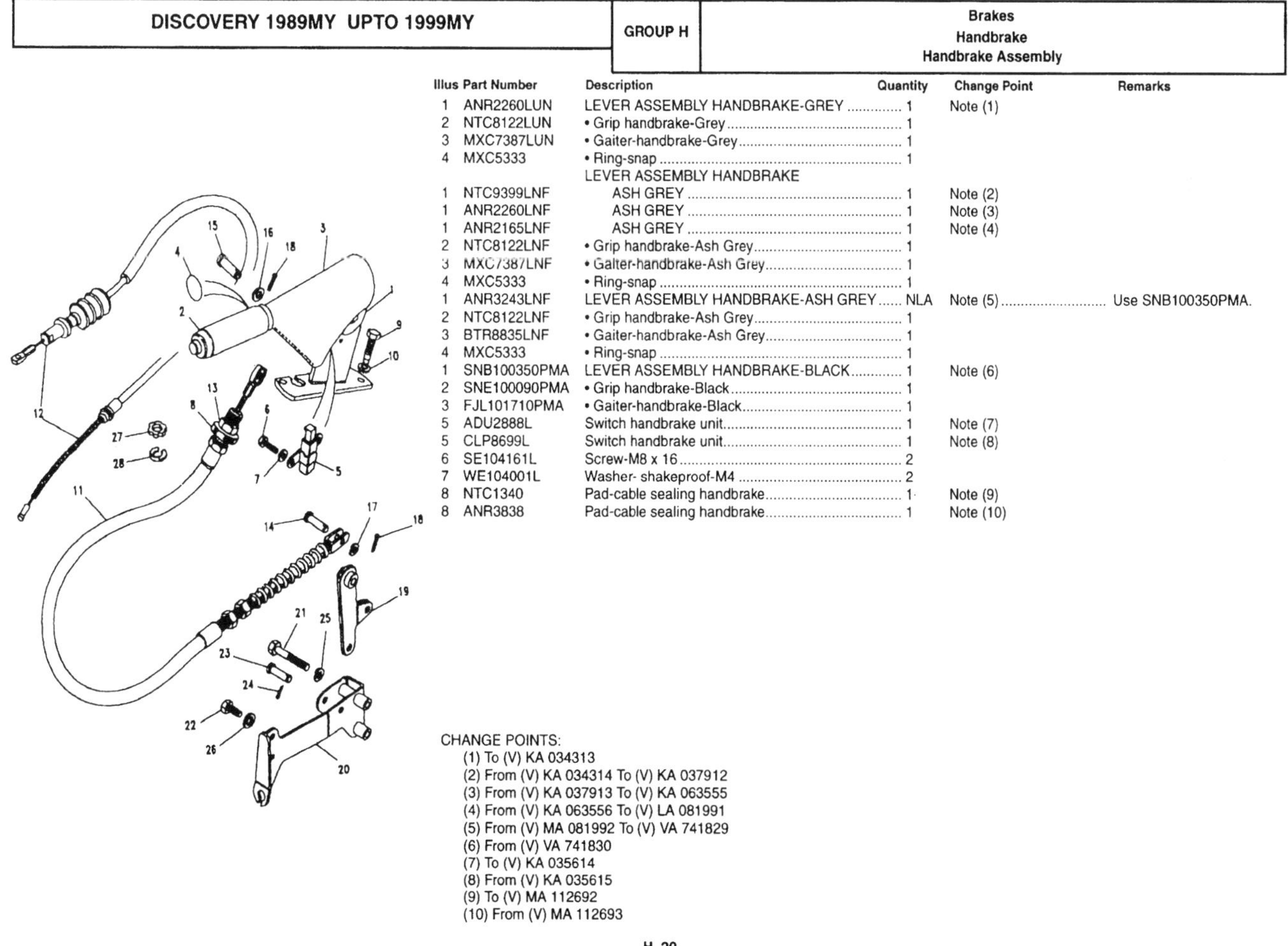

Illus	Part Number	Description	Quantity	Change Point	Remarks
1	ANR2260LUN	LEVER ASSEMBLY HANDBRAKE-GREY	1	Note (1)	
2	NTC8122LUN	• Grip handbrake-Grey	1		
3	MXC7387LUN	• Gaiter-handbrake-Grey	1		
4	MXC5333	• Ring-snap	1		
		LEVER ASSEMBLY HANDBRAKE			
1	NTC9399LNF	ASH GREY	1	Note (2)	
1	ANR2260LNF	ASH GREY	1	Note (3)	
1	ANR2165LNF	ASH GREY	1	Note (4)	
2	NTC8122LNF	• Grip handbrake-Ash Grey	1		
3	MXC7387LNF	• Gaiter-handbrake-Ash Grey	1		
4	MXC5333	• Ring-snap	1		
1	ANR3243LNF	LEVER ASSEMBLY HANDBRAKE-ASH GREY	NLA	Note (5)	Use SNB100350PMA.
2	NTC8122LNF	• Grip handbrake-Ash Grey	1		
3	BTR8835LNF	• Gaiter-handbrake-Ash Grey	1		
4	MXC5333	• Ring-snap	1		
1	SNB100350PMA	LEVER ASSEMBLY HANDBRAKE-BLACK	1	Note (6)	
2	SNE100090PMA	• Grip handbrake-Black	1		
3	FJL101710PMA	• Gaiter-handbrake-Black	1		
5	ADU2888L	Switch handbrake unit	1	Note (7)	
5	CLP8699L	Switch handbrake unit	1	Note (8)	
6	SE104161L	Screw-M8 x 16	2		
7	WE104001L	Washer- shakeproof-M4	2		
8	NTC1340	Pad-cable sealing handbrake	1	Note (9)	
8	ANR3838	Pad-cable sealing handbrake	1	Note (10)	

CHANGE POINTS:
(1) To (V) KA 034313
(2) From (V) KA 034314 To (V) KA 037912
(3) From (V) KA 037913 To (V) KA 063555
(4) From (V) KA 063556 To (V) LA 081991
(5) From (V) MA 081992 To (V) VA 741829
(6) From (V) VA 741830
(7) To (V) KA 035614
(8) From (V) KA 035615
(9) To (V) MA 112692
(10) From (V) MA 112693

H 20

Handbrake Assembly

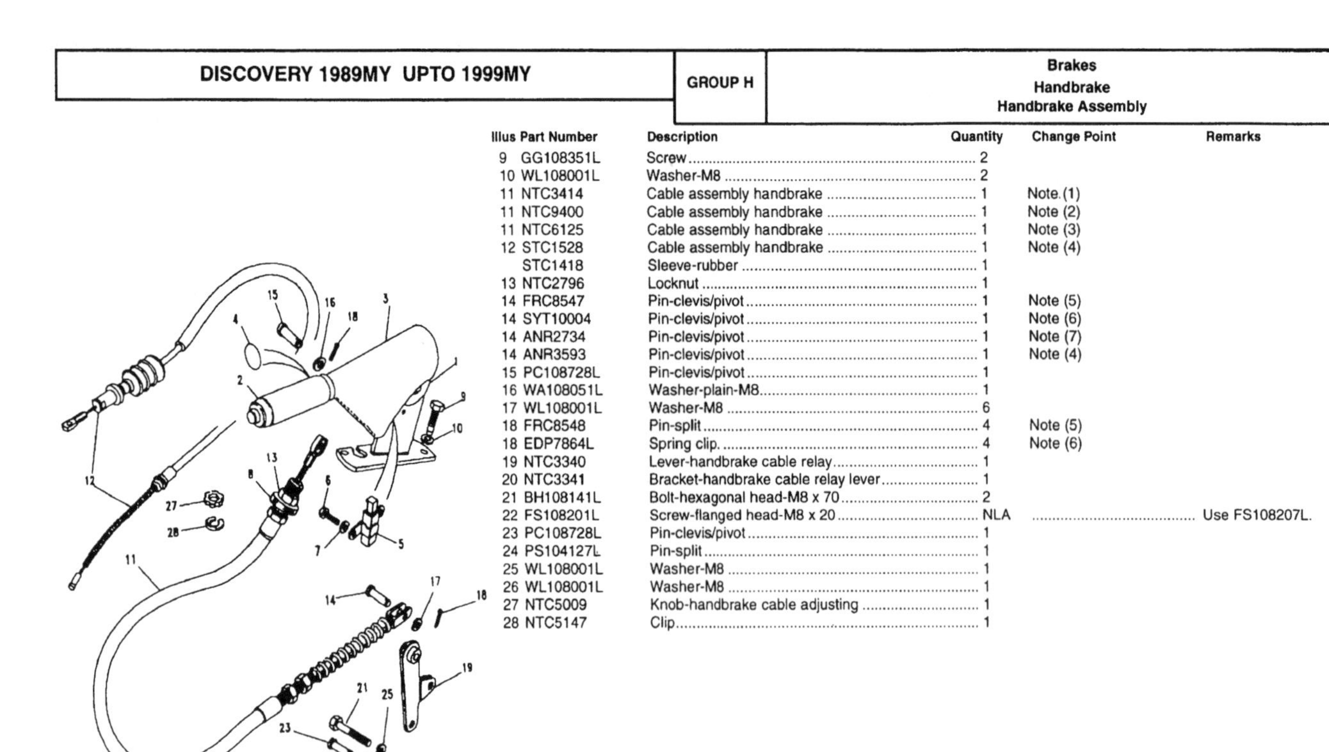

Illus	Part Number	Description	Quantity	Change Point	Remarks
9	GG108351L	Screw	2		
10	WL108001L	Washer-M8	2		
11	NTC3414	Cable assembly handbrake	1	Note.(1)	
11	NTC9400	Cable assembly handbrake	1	Note (2)	
11	NTC6125	Cable assembly handbrake	1	Note (3)	
12	STC1528	Cable assembly handbrake	1	Note (4)	
	STC1418	Sleeve-rubber	1		
13	NTC2796	Locknut	1		
14	FRC8547	Pin-clevis/pivot	1	Note (5)	
14	SYT10004	Pin-clevis/pivot	1	Note (6)	
14	ANR2734	Pin-clevis/pivot	1	Note (7)	
14	ANR3593	Pin-clevis/pivot	1	Note (4)	
15	PC108728L	Pin-clevis/pivot	1		
16	WA108051L	Washer-plain-M8	1		
17	WL108001L	Washer-M8	6		
18	FRC8548	Pin-split	4	Note (5)	
18	EDP7864L	Spring clip.	4	Note (6)	
19	NTC3340	Lever-handbrake cable relay	1		
20	NTC3341	Bracket-handbrake cable relay lever	1		
21	BH108141L	Bolt-hexagonal head-M8 x 70	2		
22	FS108201L	Screw-flanged head-M8 x 20	NLA		Use FS108207L.
23	PC108728L	Pin-clevis/pivot	1		
24	PS104127L	Pin-split	1		
25	WL108001L	Washer-M8	1		
26	WL108001L	Washer-M8	1		
27	NTC5009	Knob-handbrake cable adjusting	1		
28	NTC5147	Clip	1		

CHANGE POINTS:
(1) To (V) JA 020500
(2) From (V) JA 020501 To (V) KA 055568
(3) From (V) KA 055569 To (V) LA 081991
(4) From (V) MA 081992
(5) To (V) JA 017305
(6) From (V) JA 017306 To (V) KA 049862
(7) From (V) KA 048963 To (V) LA 081991

H 21

Handbrake Assembly-MPi-Up To KA038821

Illus	Part Number	Description	Quantity	Change Point	Remarks
1	ANR2260LNF	LEVER ASSEMBLY HANDBRAKE-ASH GREY	1		
2	NTC8122LNF	• Grip handbrake-Ash Grey	1		
3	MXC7387LNF	• Gaiter-handbrake-Ash Grey	1		
4	MXC5333	• Ring-snap	1		
5	NTC6125	Cable assembly handbrake	1		
6	STC1157	Bracket-handbrake cable relay lever	1		
7	ANR3054	Lever-handbrake cable relay	1		
8	PC108648	Pin assembly-clevis	1		

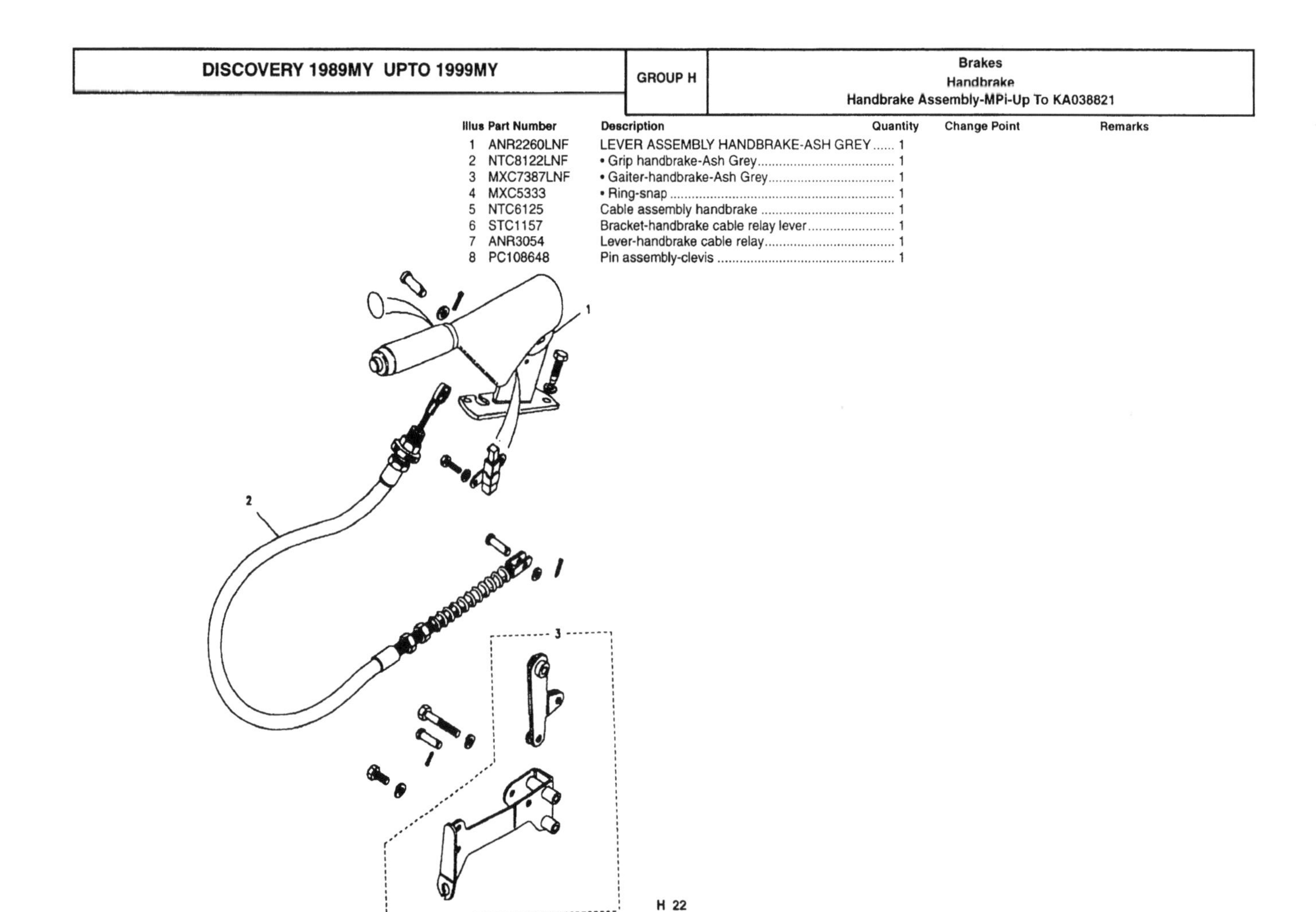

H 22

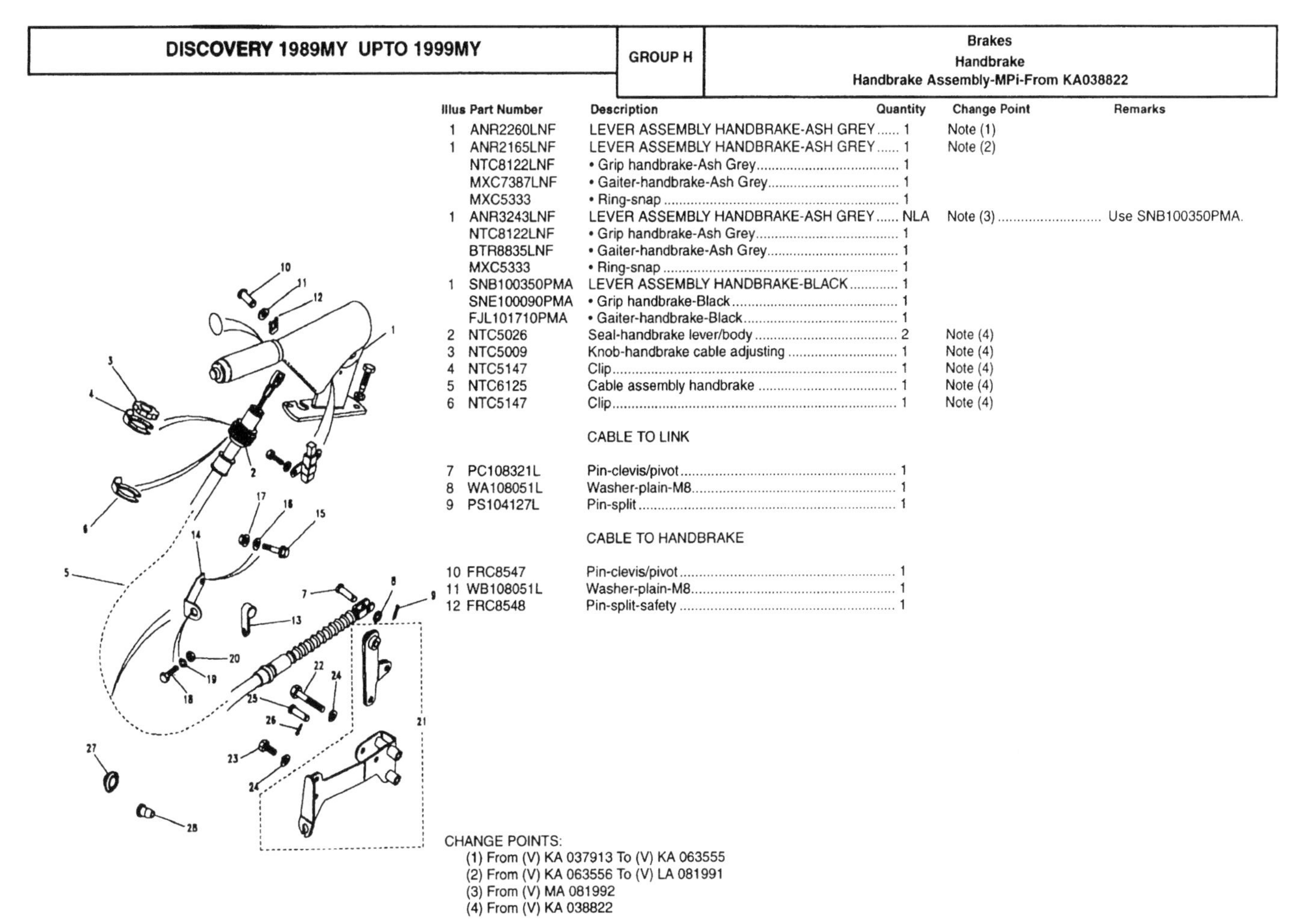

Illus	Part Number	Description	Quantity	Change Point	Remarks
1	ANR2260LNF	LEVER ASSEMBLY HANDBRAKE-ASH GREY	1	Note (1)	
1	ANR2165LNF	LEVER ASSEMBLY HANDBRAKE-ASH GREY	1	Note (2)	
	NTC8122LNF	• Grip handbrake-Ash Grey	1		
	MXC7387LNF	• Gaiter-handbrake-Ash Grey	1		
	MXC5333	• Ring-snap	1		
1	ANR3243LNF	LEVER ASSEMBLY HANDBRAKE-ASH GREY	NLA	Note (3)	Use SNB100350PMA.
	NTC8122LNF	• Grip handbrake-Ash Grey	1		
	BTR8835LNF	• Gaiter-handbrake-Ash Grey	1		
	MXC5333	• Ring-snap	1		
1	SNB100350PMA	LEVER ASSEMBLY HANDBRAKE-BLACK	1		
	SNE100090PMA	• Grip handbrake-Black	1		
	FJL101710PMA	• Gaiter-handbrake-Black	1		
2	NTC5026	Seal-handbrake lever/body	2	Note (4)	
3	NTC5009	Knob-handbrake cable adjusting	1	Note (4)	
4	NTC5147	Clip	1	Note (4)	
5	NTC6125	Cable assembly handbrake	1	Note (4)	
6	NTC5147	Clip	1	Note (4)	
		CABLE TO LINK			
7	PC108321L	Pin-clevis/pivot	1		
8	WA108051L	Washer-plain-M8	1		
9	PS104127L	Pin-split	1		
		CABLE TO HANDBRAKE			
10	FRC8547	Pin-clevis/pivot	1		
11	WB108051L	Washer-plain-M8	1		
12	FRC8548	Pin-split-safety	1		

CHANGE POINTS:
(1) From (V) KA 037913 To (V) KA 063555
(2) From (V) KA 063556 To (V) LA 081991
(3) From (V) MA 081992
(4) From (V) KA 038822

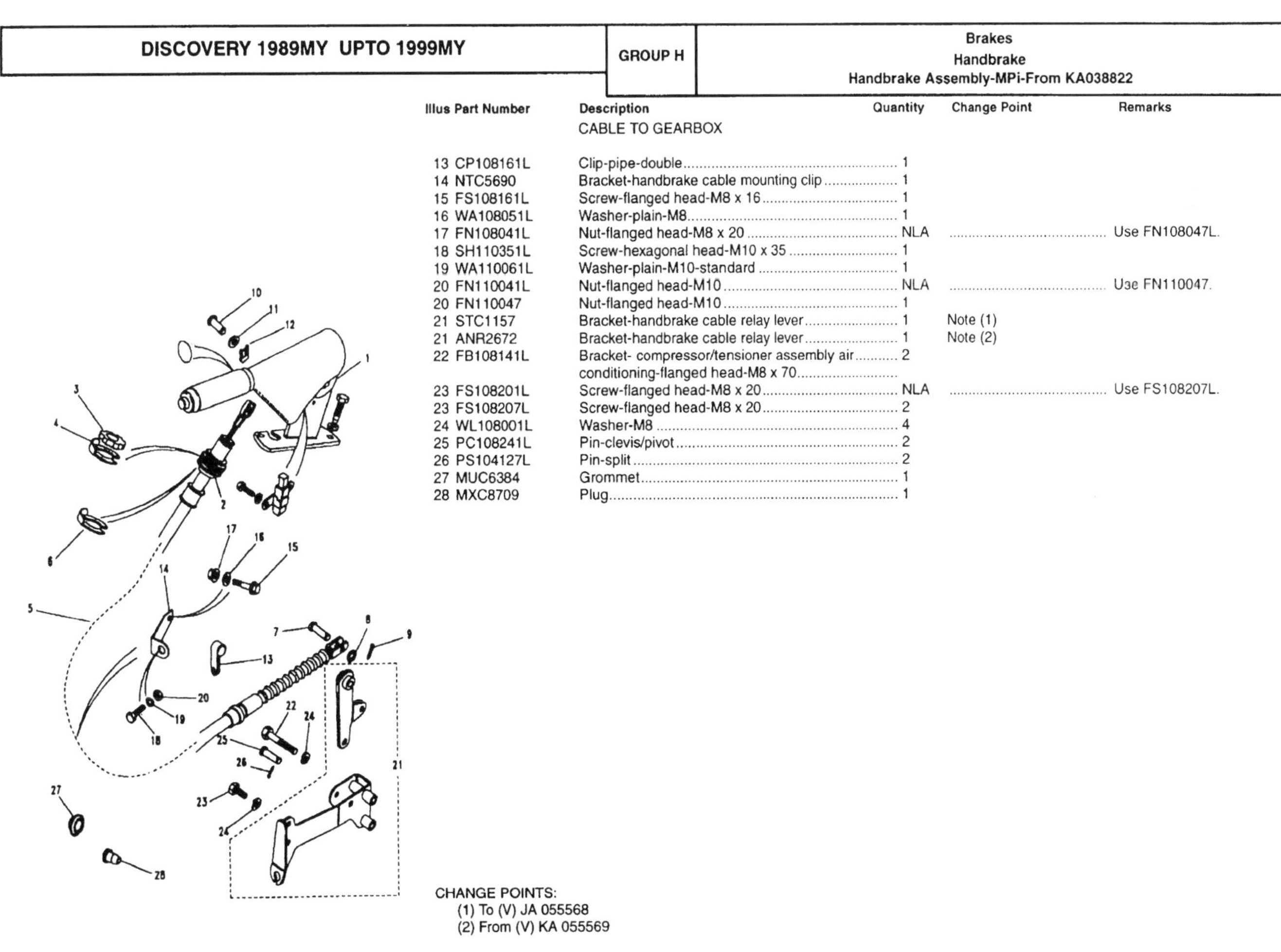

Illus	Part Number	Description	Quantity	Change Point	Remarks
		CABLE TO GEARBOX			
13	CP108161L	Clip-pipe-double	1		
14	NTC5690	Bracket-handbrake cable mounting clip	1		
15	FS108161L	Screw-flanged head-M8 x 16	1		
16	WA108051L	Washer-plain-M8	1		
17	FN108041L	Nut-flanged head-M8 x 20	NLA		Use FN108047L.
18	SH110351L	Screw-hexagonal head-M10 x 35	1		
19	WA110061L	Washer-plain-M10-standard	1		
20	FN110041L	Nut-flanged head-M10	NLA		Use FN110047.
20	FN110047	Nut-flanged head-M10	1		
21	STC1157	Bracket-handbrake cable relay lever	1	Note (1)	
21	ANR2672	Bracket-handbrake cable relay lever	1	Note (2)	
22	FB108141L	Bracket- compressor/tensioner assembly air conditioning-flanged head-M8 x 70	2		
23	FS108201L	Screw-flanged head-M8 x 20	NLA		Use FS108207L.
23	FS108207L	Screw-flanged head-M8 x 20	2		
24	WL108001L	Washer-M8	4		
25	PC108241L	Pin-clevis/pivot	2		
26	PS104127L	Pin-split	2		
27	MUC6384	Grommet	1		
28	MXC8709	Plug	1		

CHANGE POINTS:
(1) To (V) JA 055568
(2) From (V) KA 055569

Notes

Fuel-Air Cleaner-Evaporative Loss.......... J 1
Air Cleaner & Hoses Diesel.......... J 18
Air Cleaner & Hoses Petrol.......... J 14
Charcoal Canister.......... J 20
Evap Loss Control System.......... J 23
Filter & Sedimentor - Diesel.......... J 12
Fuel Filler.......... J 5
Fuel Pipes Diesel.......... J 10
Fuel Pipes Petrol.......... J 7
Fuel Tank Pump & Mountings Petrol.......... J 1
Fuel Tank Sender & Mountings Diesel.......... J 3

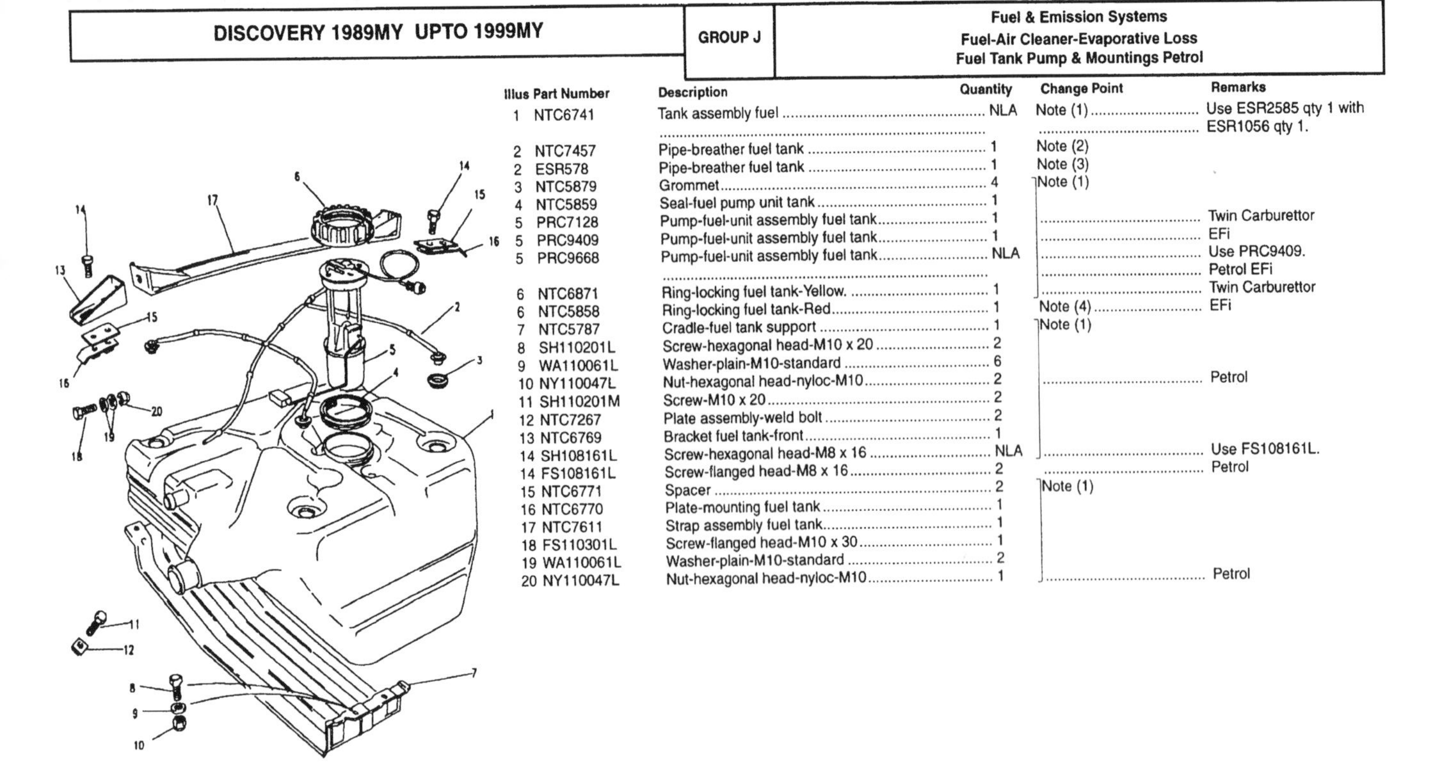

DISCOVERY 1989MY UPTO 1999MY

GROUP J

Fuel & Emission Systems
Fuel-Air Cleaner-Evaporative Loss
Fuel Tank Pump & Mountings Petrol

Illus	Part Number	Description	Quantity	Change Point	Remarks
1	NTC6741	Tank assembly fuel	NLA	Note (1)	Use ESR2585 qty 1 with ESR1056 qty 1.
2	NTC7457	Pipe-breather fuel tank	1	Note (2)	
2	ESR578	Pipe-breather fuel tank	1	Note (3)	
3	NTC5879	Grommet	4	Note (1)	
4	NTC5859	Seal-fuel pump unit tank	1		
5	PRC7128	Pump-fuel-unit assembly fuel tank	1		Twin Carburettor
5	PRC9409	Pump-fuel-unit assembly fuel tank	1		EFi
5	PRC9668	Pump-fuel-unit assembly fuel tank	NLA		Use PRC9409. Petrol EFi
6	NTC6871	Ring-locking fuel tank-Yellow.	1		Twin Carburettor
6	NTC5858	Ring-locking fuel tank-Red	1	Note (4)	EFi
7	NTC5787	Cradle-fuel tank support	1	Note (1)	
8	SH110201L	Screw-hexagonal head-M10 x 20	2		
9	WA110061L	Washer-plain-M10-standard	6		
10	NY110047L	Nut-hexagonal head-nyloc-M10	2		Petrol
11	SH110201M	Screw-M10 x 20	2		
12	NTC7267	Plate assembly-weld bolt	2		
13	NTC6769	Bracket fuel tank-front	1		
14	SH108161L	Screw-hexagonal head-M8 x 16	NLA		Use FS108161L.
14	FS108161L	Screw-flanged head-M8 x 16	2		Petrol
15	NTC6771	Spacer	2	Note (1)	
16	NTC6770	Plate-mounting fuel tank	1		
17	NTC7611	Strap assembly fuel tank	1		
18	FS110301L	Screw-flanged head-M10 x 30	1		
19	WA110061L	Washer-plain-M10-standard	2		
20	NY110047L	Nut-hexagonal head-nyloc-M10	1		Petrol

CHANGE POINTS:
(1) To (V) KA 053801
(2) From (V) GA 394352 To (V) GA 456980; From (V) HA 456981 To (V) HA 486551; From (V) HA 000001 To (V) HA 003613
(3) From (V) HA 003614 To (V) HA 012332; From (V) JA 012333 To (V) KA 053801
(4) From (V) HA 038200 To (V) KA 053801

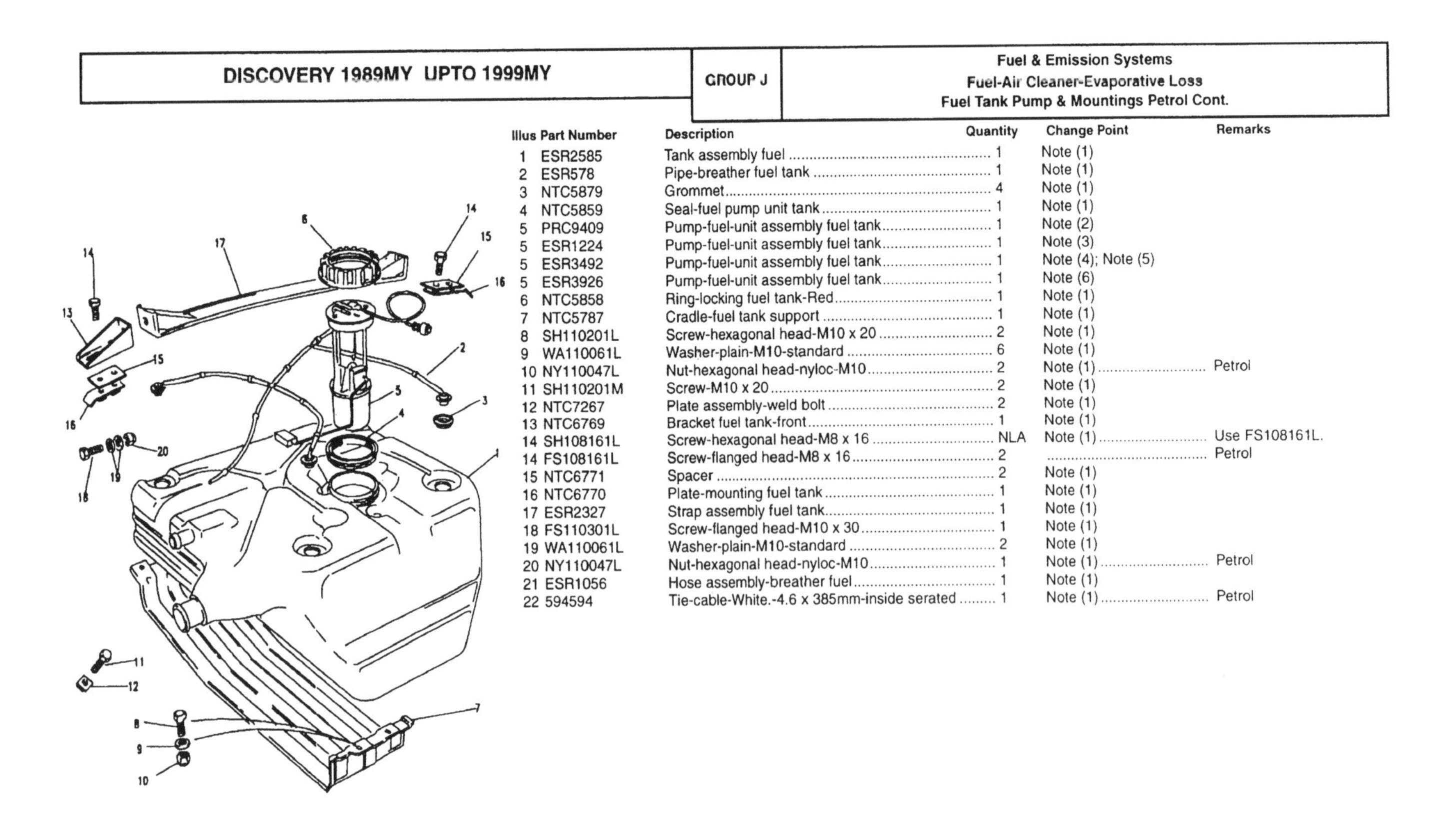

DISCOVERY 1989MY UPTO 1999MY

GROUP J

Fuel & Emission Systems
Fuel-Air Cleaner-Evaporative Loss
Fuel Tank Pump & Mountings Petrol Cont.

Illus	Part Number	Description	Quantity	Change Point	Remarks
1	ESR2585	Tank assembly fuel	1	Note (1)	
2	ESR578	Pipe-breather fuel tank	1	Note (1)	
3	NTC5879	Grommet	4	Note (1)	
4	NTC5859	Seal-fuel pump unit tank	1	Note (1)	
5	PRC9409	Pump-fuel-unit assembly fuel tank	1	Note (2)	
5	ESR1224	Pump-fuel-unit assembly fuel tank	1	Note (3)	
5	ESR3492	Pump-fuel-unit assembly fuel tank	1	Note (4); Note (5)	
5	ESR3926	Pump-fuel-unit assembly fuel tank	1	Note (6)	
6	NTC5858	Ring-locking fuel tank-Red	1	Note (1)	
7	NTC5787	Cradle-fuel tank support	1	Note (1)	
8	SH110201L	Screw-hexagonal head-M10 x 20	2	Note (1)	
9	WA110061L	Washer-plain-M10-standard	6	Note (1)	
10	NY110047L	Nut-hexagonal head-nyloc-M10	2	Note (1)	Petrol
11	SH110201M	Screw-M10 x 20	2	Note (1)	
12	NTC7267	Plate assembly-weld bolt	2	Note (1)	
13	NTC6769	Bracket fuel tank-front	1	Note (1)	
14	SH108161L	Screw-hexagonal head-M8 x 16	NLA	Note (1)	Use FS108161L.
14	FS108161L	Screw-flanged head-M8 x 16	2		Petrol
15	NTC6771	Spacer	2	Note (1)	
16	NTC6770	Plate-mounting fuel tank	1	Note (1)	
17	ESR2327	Strap assembly fuel tank	1	Note (1)	
18	FS110301L	Screw-flanged head-M10 x 30	1	Note (1)	
19	WA110061L	Washer-plain-M10-standard	2	Note (1)	
20	NY110047L	Nut-hexagonal head-nyloc-M10	1	Note (1)	Petrol
21	ESR1056	Hose assembly-breather fuel	1	Note (1)	
22	594594	Tie-cable-White.-4.6 x 385mm-inside serated	1	Note (1)	Petrol

CHANGE POINTS:
(1) From (V) KA 053802
(2) From (V) KA 053802 To (V) KA 065556
(3) From (V) KA 065557 To (V) MA 140197; From (V) KA 053802 To (V) MA 140197
(4) From (V) MA 140198 To (V) TA 172381
(5) From (V) KA 053802 To (V) TA 504321
(6) From (V) KA 053802; From (V) TA 172382; From (V) TA 504322

DISCOVERY 1989MY UPTO 1999MY

GROUP J

Fuel & Emission Systems
Fuel-Air Cleaner-Evaporative Loss
Fuel Tank Sender & Mountings Diesel

Illus	Part Number	Description	Quantity	Change Point	Remarks
1	NTC6741	Tank assembly fuel	NLA	Note (1)	Use ESR2585 qty 1 with ESR1056 qty 1.
1	ESR460	Tank assembly fuel	1	Note (2)	
2	NTC7487	Pipe-breather fuel tank	1	Note (3)	
2	ESR461	Pipe-breather fuel tank	1	Note (4)	
3	NTC5879	Grommet	4	Note (5)	
4	NTC5859	Seal-fuel pump unit tank	1		
5	PRC7129	Sensor unit fuel tank	1		
6	NTC6872	Ring-locking fuel tank-Black	1		
7	NTC5787	Cradle-fuel tank support	1		
8	SH110201L	Screw-hexagonal head-M10 x 20	2		
9	WA110061L	Washer-plain-M10-standard	6		
10	NY110047L	Nut-hexagonal head-nyloc-M10	2		
11	SH110201M	Screw-M10 x 20	2		
12	NTC7267	Plate assembly-weld bolt	2		
13	NTC6769	Bracket fuel tank-front	1		
14	SH108161L	Screw-hexagonal head-M8 x 16	NLA		Use FS108161L.
14	FS108161L	Screw-flanged head-M8 x 16	2		
15	NTC6771	Spacer	2	Note (5)	
16	NTC6770	Plate-mounting fuel tank	1		
17	NTC7611	Strap assembly fuel tank	1		
18	FS110301L	Screw-flanged head-M10 x 30	1		
19	WA110061L	Washer-plain-M10-standard	2		
20	NY110047L	Nut-hexagonal head-nyloc-M10	1		

CHANGE POINTS:
(1) To (V) KA 053845; To (V) JA 013791
(2) From (V) JA 013792 To (V) KA 053845
(3) From (V) GA 394352 To (V) GA 456980; From (V) HA 456981 To (V) HA 486551; From (V) HA 000001 To (V) HA 003613
(4) From (V) HA 003614 To (V) HA 012332; From (V) JA 012333 To (V) KA 053845
(5) To (V) KA 053845

J 3

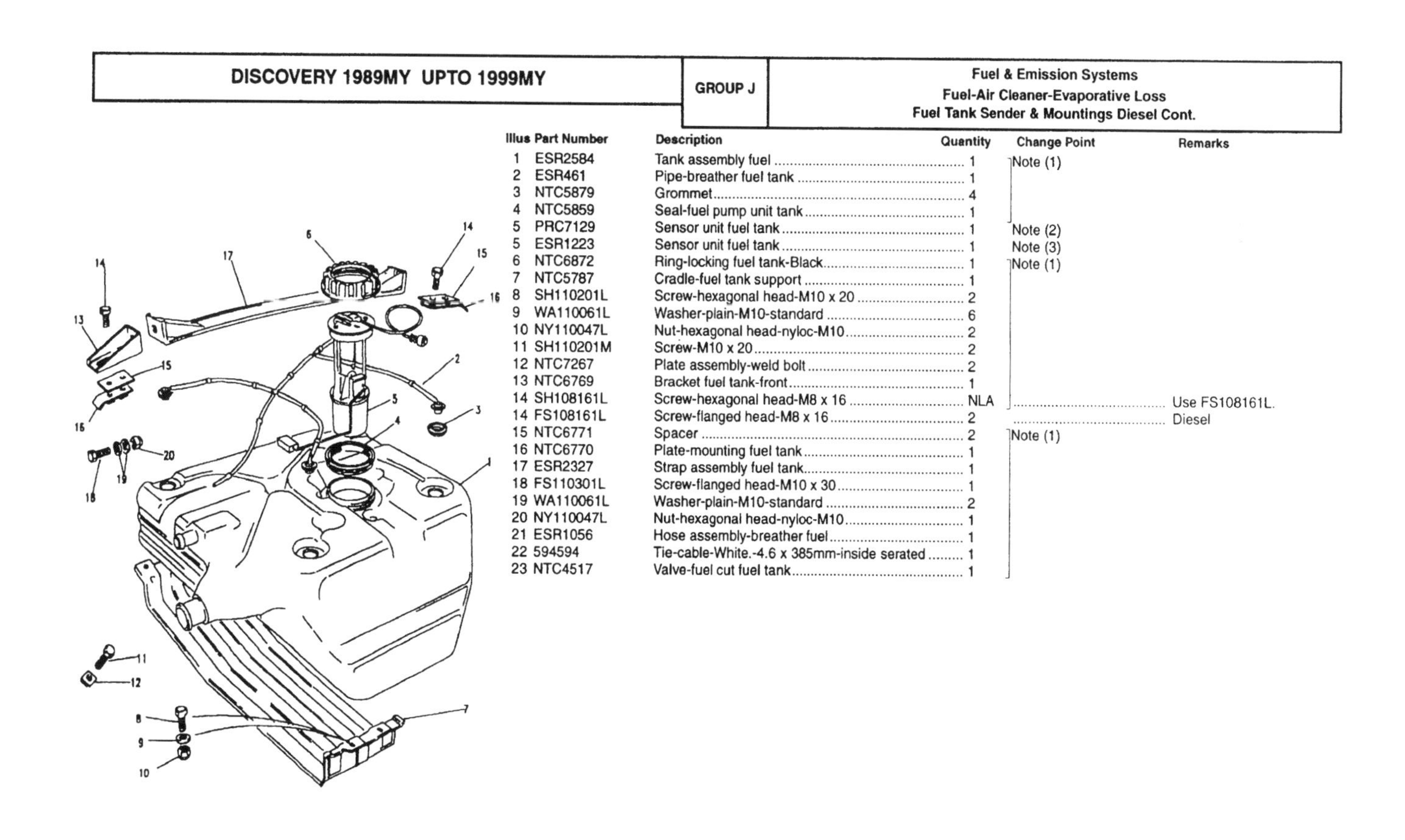

DISCOVERY 1989MY UPTO 1999MY

GROUP J

Fuel & Emission Systems
Fuel-Air Cleaner-Evaporative Loss
Fuel Tank Sender & Mountings Diesel Cont.

Illus	Part Number	Description	Quantity	Change Point	Remarks
1	ESR2584	Tank assembly fuel	1	Note (1)	
2	ESR461	Pipe-breather fuel tank	1		
3	NTC5879	Grommet	4		
4	NTC5859	Seal-fuel pump unit tank	1		
5	PRC7129	Sensor unit fuel tank	1	Note (2)	
5	ESR1223	Sensor unit fuel tank	1	Note (3)	
6	NTC6872	Ring-locking fuel tank-Black	1	Note (1)	
7	NTC5787	Cradle-fuel tank support	1		
8	SH110201L	Screw-hexagonal head-M10 x 20	2		
9	WA110061L	Washer-plain-M10-standard	6		
10	NY110047L	Nut-hexagonal head-nyloc-M10	2		
11	SH110201M	Screw-M10 x 20	2		
12	NTC7267	Plate assembly-weld bolt	2		
13	NTC6769	Bracket fuel tank-front	1		
14	SH108161L	Screw-hexagonal head-M8 x 16	NLA		Use FS108161L.
14	FS108161L	Screw-flanged head-M8 x 16	2		Diesel
15	NTC6771	Spacer	2	Note (1)	
16	NTC6770	Plate-mounting fuel tank	1		
17	ESR2327	Strap assembly fuel tank	1		
18	FS110301L	Screw-flanged head-M10 x 30	1		
19	WA110061L	Washer-plain-M10-standard	2		
20	NY110047L	Nut-hexagonal head-nyloc-M10	1		
21	ESR1056	Hose assembly-breather fuel	1		
22	594594	Tie-cable-White.-4.6 x 385mm-inside serated	1		
23	NTC4517	Valve-fuel cut fuel tank	1		

CHANGE POINTS:
(1) From (V) KA 053846
(2) From (V) KA 053846 To (V) KA 065556
(3) From (V) KA 053846; From (V) KA 065557

J 4

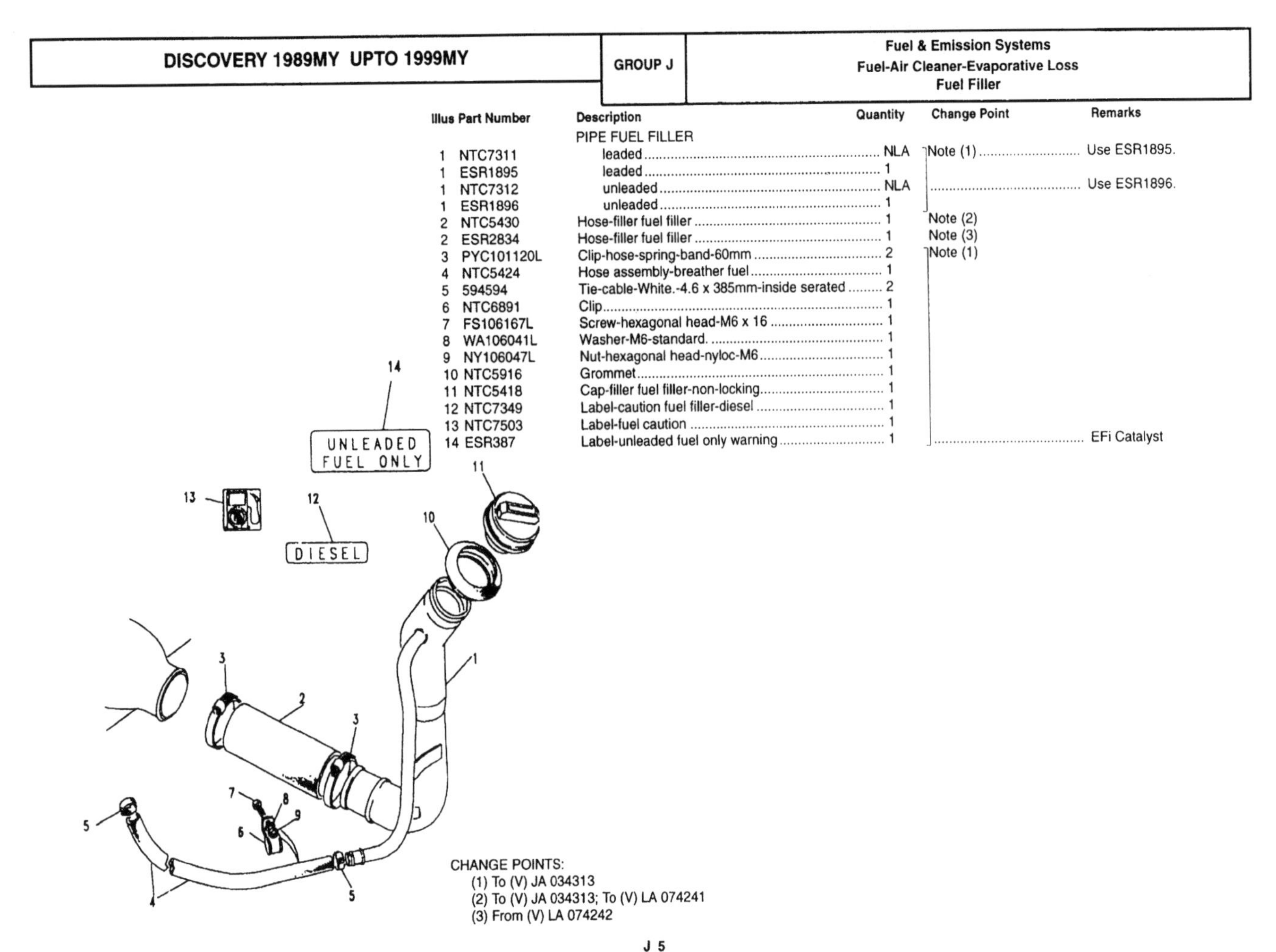

Illus	Part Number	Description	Quantity	Change Point	Remarks
		PIPE FUEL FILLER			
1	NTC7311	leaded	NLA	Note (1)	Use ESR1895.
1	ESR1895	leaded	1	Note (1)	
1	NTC7312	unleaded	NLA	Note (1)	Use ESR1896.
1	ESR1896	unleaded	1	Note (1)	
2	NTC5430	Hose-filler fuel filler	1	Note (2)	
2	ESR2834	Hose-filler fuel filler	1	Note (3)	
3	PYC101120L	Clip-hose-spring-band-60mm	2	Note (1)	
4	NTC5424	Hose assembly-breather fuel	1	Note (1)	
5	594594	Tie-cable-White.-4.6 x 385mm-inside serated	2	Note (1)	
6	NTC6891	Clip	1	Note (1)	
7	FS106167L	Screw-hexagonal head-M6 x 16	1	Note (1)	
8	WA106041L	Washer-M6-standard.	1	Note (1)	
9	NY106047L	Nut-hexagonal head-nyloc-M6	1	Note (1)	
10	NTC5916	Grommet	1	Note (1)	
11	NTC5418	Cap-filler fuel filler-non-locking	1	Note (1)	
12	NTC7349	Label-caution fuel filler-diesel	1	Note (1)	
13	NTC7503	Label-fuel caution	1	Note (1)	
14	ESR387	Label-unleaded fuel only warning	1	Note (1)	EFi Catalyst

CHANGE POINTS:
(1) To (V) JA 034313
(2) To (V) JA 034313; To (V) LA 074241
(3) From (V) LA 074242

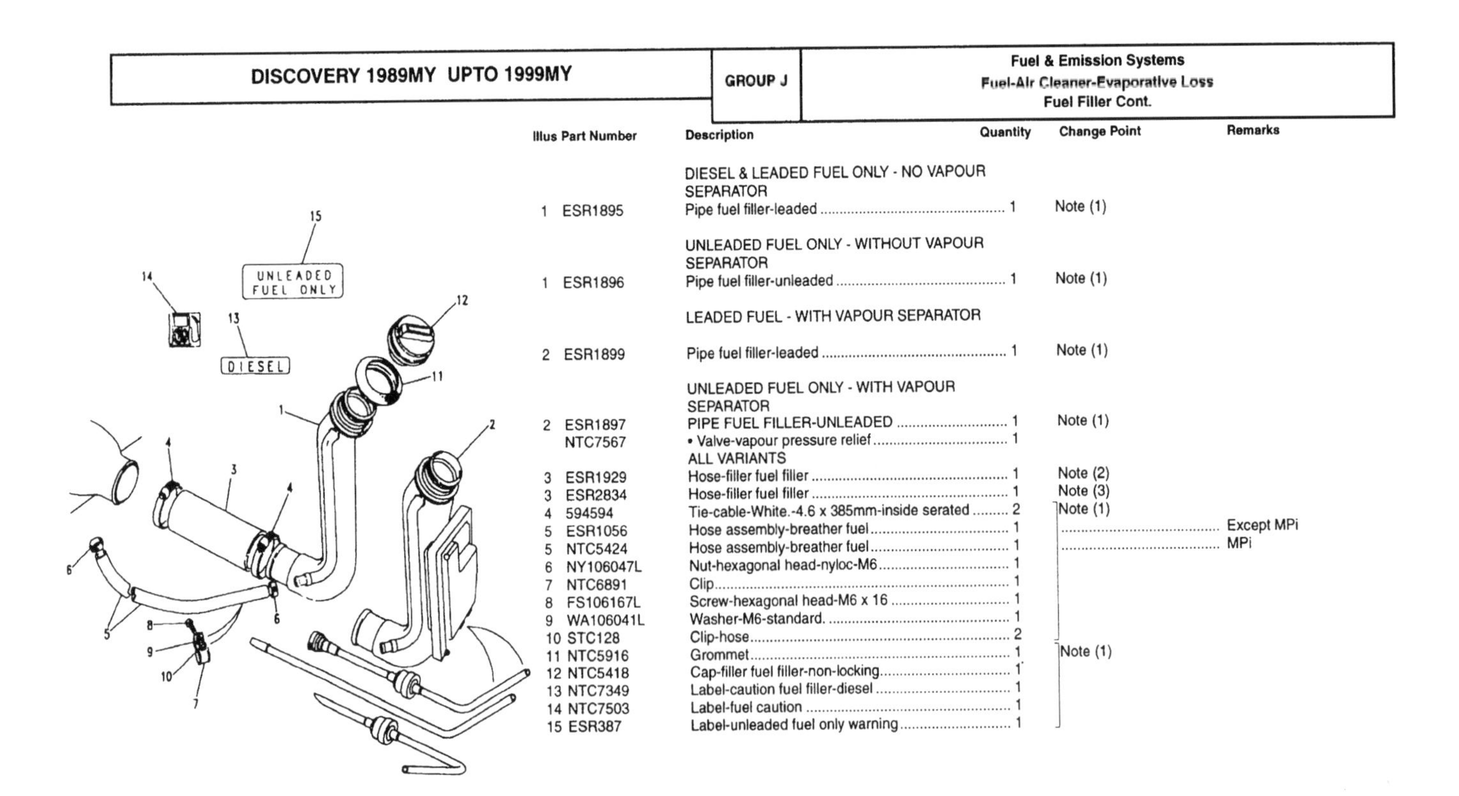

Illus	Part Number	Description	Quantity	Change Point	Remarks
		DIESEL & LEADED FUEL ONLY - NO VAPOUR SEPARATOR			
1	ESR1895	Pipe fuel filler-leaded	1	Note (1)	
		UNLEADED FUEL ONLY - WITHOUT VAPOUR SEPARATOR			
1	ESR1896	Pipe fuel filler-unleaded	1	Note (1)	
		LEADED FUEL - WITH VAPOUR SEPARATOR			
2	ESR1899	Pipe fuel filler-leaded	1	Note (1)	
		UNLEADED FUEL ONLY - WITH VAPOUR SEPARATOR			
2	ESR1897	PIPE FUEL FILLER-UNLEADED	1	Note (1)	
	NTC7567	• Valve-vapour pressure relief	1		
		ALL VARIANTS			
3	ESR1929	Hose-filler fuel filler	1	Note (2)	
3	ESR2834	Hose-filler fuel filler	1	Note (3)	
4	594594	Tie-cable-White.-4.6 x 385mm-inside serated	2	Note (1)	
5	ESR1056	Hose assembly-breather fuel	1	Note (1)	Except MPi
5	NTC5424	Hose assembly-breather fuel	1	Note (1)	MPi
6	NY106047L	Nut-hexagonal head-nyloc-M6	1	Note (1)	
7	NTC6891	Clip	1	Note (1)	
8	FS106167L	Screw-hexagonal head-M6 x 16	1	Note (1)	
9	WA106041L	Washer-M6-standard.	1	Note (1)	
10	STC128	Clip-hose	2	Note (1)	
11	NTC5916	Grommet	1	Note (1)	
12	NTC5418	Cap-filler fuel filler-non-locking	1	Note (1)	
13	NTC7349	Label-caution fuel filler-diesel	1	Note (1)	
14	NTC7503	Label-fuel caution	1	Note (1)	
15	ESR387	Label-unleaded fuel only warning	1	Note (1)	

CHANGE POINTS:
(1) From (V) KA 034314
(2) From (V) KA 034314 To (V) LA 074241
(3) From (V) LA 074242

Illus	Part Number	Description	Quantity	Change Point	Remarks
1	NTC7336	Pipe fuel-tank to filter	1		V8 Twin Carburettor
2	NTC7720	Pipe-filter to hose feed-tank to filter	1		
3	NTC7337	Pipe-return fuel lines-tank to filter	1		
4	CP108161L	Clip-pipe-tank to filter-double	2		
5	NRC9770	Nut-tube-female-tank to filter-M12	2		
6	NRC9771	Olive-tank to filter	2		
7	NTC2585	Clip-pipe-tank to filter-double	9		
8	79123	Clip-pipe-tank to filter-single	1		
9	NRC9786	Filter-in line fuel lines	1		V8 Petrol Twin Carburettor
	JS660L	Element-inline fuel filter	1		V8 Twin Carburettor

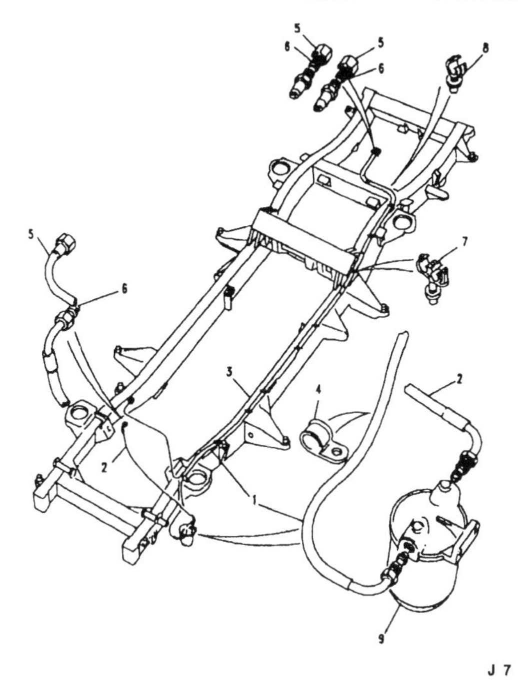

J 7

Illus	Part Number	Description	Quantity	Change Point	Remarks
		V8 & MPI			
1	ESR500	Pipe assembly-tank to filter	1	Note (1)	
1	ESR2364	Pipe assembly-tank to filter	1	Note (2)	
1	ESR2364	Pipe assembly-tank to filter	1	Note (2)	
		V8			
2	ESR501	Pipe assembly-filter to engine feed	1	Note (3)	
2	ESR3250	Pipe assembly-filter to engine feed	1	Note (4)	
		MPI			
2	ESR1673	Pipe assembly-filter to engine feed	1	Note (1)	
2	ESR2635	Pipe assembly-filter to engine feed	1	Note (5)	
		V8			
3	NTC6143	Pipe-return fuel lines	1	Note (6)	
3	ESR2188	Pipe-return fuel lines	1	Note (7)	
3	ESR2637	Pipe-return fuel lines	1	Note (8)	
3	ESR3267	Pipe-return fuel lines	1	Note (4)	
		MPI			
3	ESR1675	Pipe-return fuel lines	1	Note (1)	
3	ESR2636	Pipe-return fuel lines	1	Note (2)	

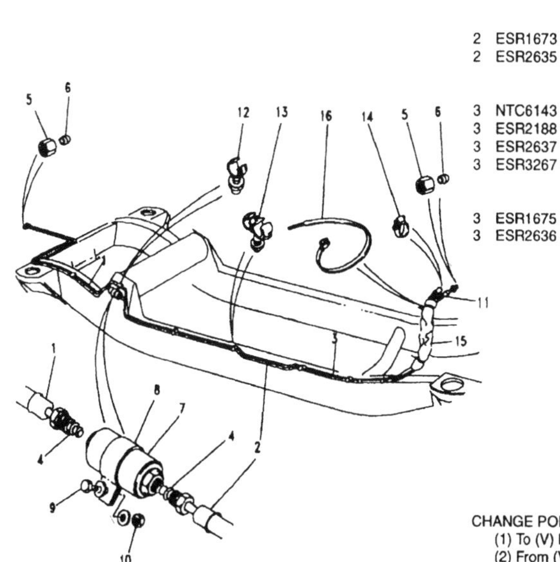

CHANGE POINTS:
(1) To (V) KA 081990
(2) From (V) LA 081991
(3) To (V) MA 091471
(4) From (V) MA 091472
(5) To (V) LA 081991
(6) From (V) GA 394352 To (V) GA 456980; From (V) HA 456981 To (V) HA 471372
(7) From (V) HA 471373 To (V) HA 486551; From (V) HA 000001 To (V) HA 012332; From (V) JA 012333 To (V) LA 081991
(8) From (V) LA 081991 To (V) MA 091471

J 8

Fuel Pipes Petrol Cont.

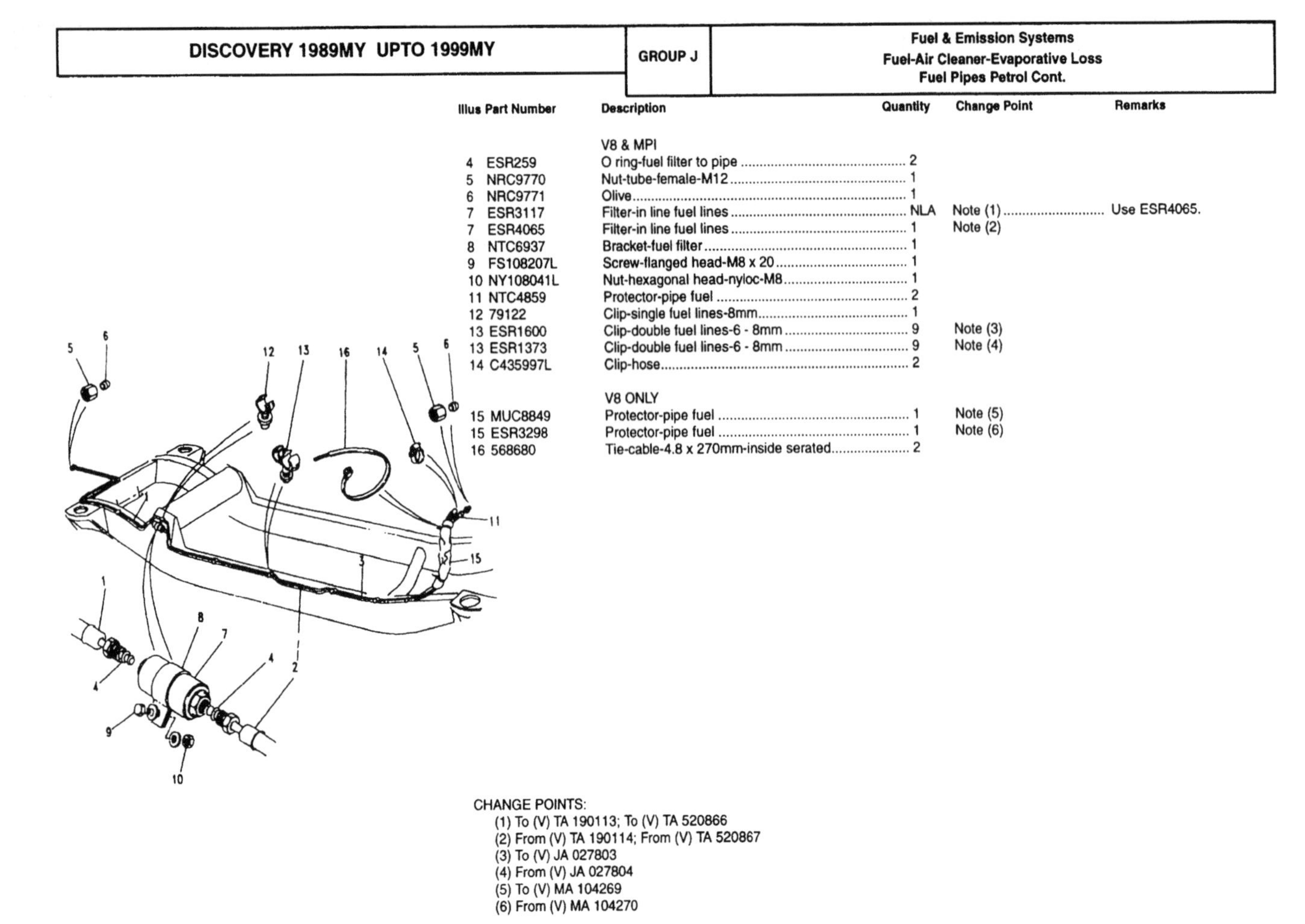

Illus	Part Number	Description	Quantity	Change Point	Remarks
		V8 & MPI			
4	ESR259	O ring-fuel filter to pipe	2		
5	NRC9770	Nut-tube-female-M12	1		
6	NRC9771	Olive	1		
7	ESR3117	Filter-in line fuel lines	NLA	Note (1)	Use ESR4065.
7	ESR4065	Filter-in line fuel lines	1	Note (2)	
8	NTC6937	Bracket-fuel filter	1		
9	FS108207L	Screw-flanged head-M8 x 20	1		
10	NY108041L	Nut-hexagonal head-nyloc-M8	1		
11	NTC4859	Protector-pipe fuel	2		
12	79122	Clip-single fuel lines-8mm	1		
13	ESR1600	Clip-double fuel lines-6 - 8mm	9	Note (3)	
13	ESR1373	Clip-double fuel lines-6 - 8mm	9	Note (4)	
14	C435997L	Clip-hose	2		
		V8 ONLY			
15	MUC8849	Protector-pipe fuel	1	Note (5)	
15	ESR3298	Protector-pipe fuel	1	Note (6)	
16	568680	Tie-cable-4.8 x 270mm-inside serated	2		

CHANGE POINTS:
(1) To (V) TA 190113; To (V) TA 520866
(2) From (V) TA 190114; From (V) TA 520867
(3) To (V) JA 027803
(4) From (V) JA 027804
(5) To (V) MA 104269
(6) From (V) MA 104270

Fuel Pipes Diesel

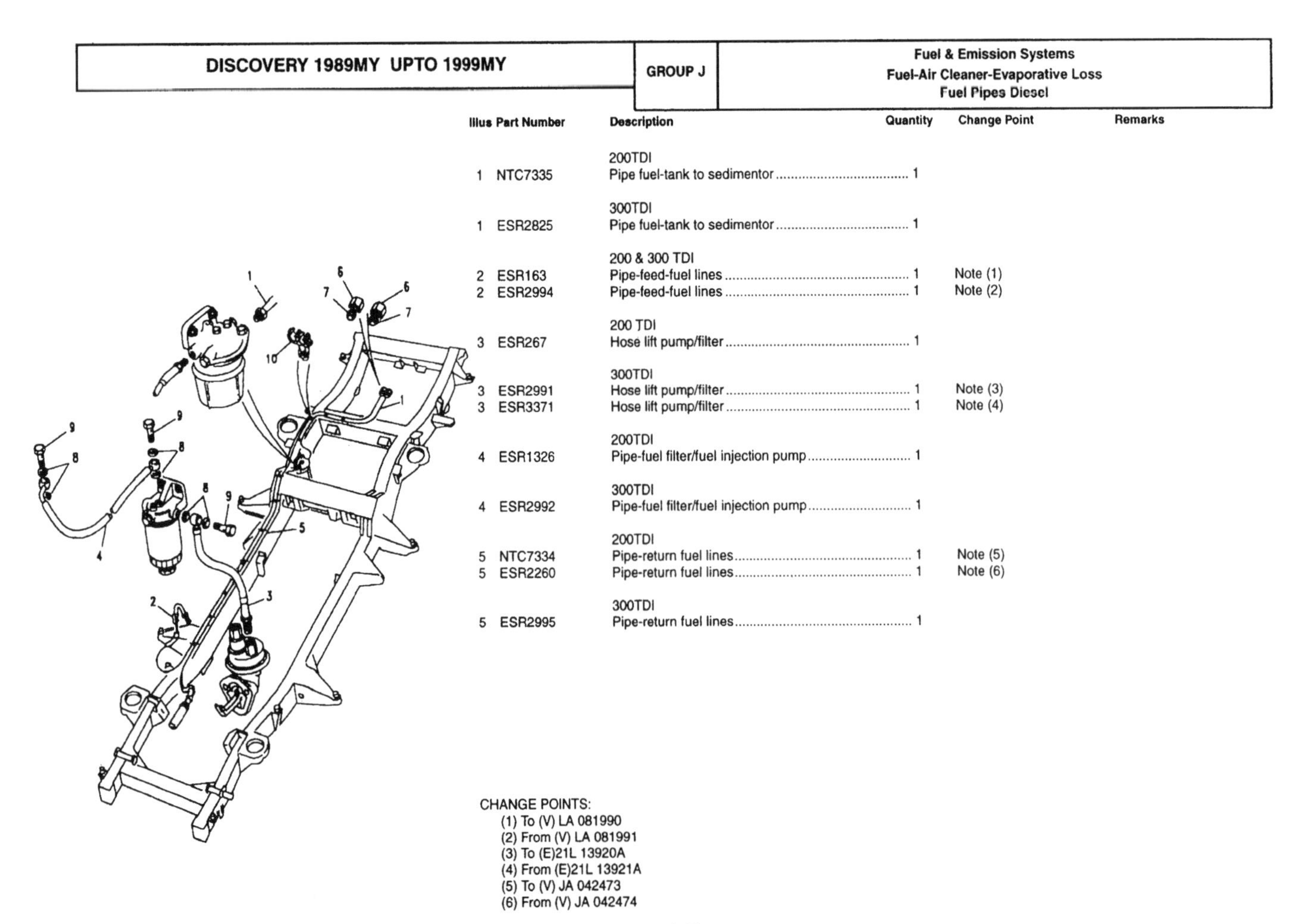

Illus	Part Number	Description	Quantity	Change Point	Remarks
		200TDI			
1	NTC7335	Pipe fuel-tank to sedimentor	1		
		300TDI			
1	ESR2825	Pipe fuel-tank to sedimentor	1		
		200 & 300 TDI			
2	ESR163	Pipe-feed-fuel lines	1	Note (1)	
2	ESR2994	Pipe-feed-fuel lines	1	Note (2)	
		200 TDI			
3	ESR267	Hose lift pump/filter	1		
		300TDI			
3	ESR2991	Hose lift pump/filter	1	Note (3)	
3	ESR3371	Hose lift pump/filter	1	Note (4)	
		200TDI			
4	ESR1326	Pipe-fuel filter/fuel injection pump	1		
		300TDI			
4	ESR2992	Pipe-fuel filter/fuel injection pump	1		
		200TDI			
5	NTC7334	Pipe-return fuel lines	1	Note (5)	
5	ESR2260	Pipe-return fuel lines	1	Note (6)	
		300TDI			
5	ESR2995	Pipe-return fuel lines	1		

CHANGE POINTS:
(1) To (V) LA 081990
(2) From (V) LA 081991
(3) To (E)21L 13920A
(4) From (E)21L 13921A
(5) To (V) JA 042473
(6) From (V) JA 042474

Illus	Part Number	Description	Quantity	Change Point	Remarks
		200 & 300 TDI FUEL PIPE FIXINGS			
6	NRC9770	Nut-tube-female-M12	3		
7	NRC9771	Olive	3		
8	ESR354	Washer-copper	4		
9	NTC3346	Bolt-banjo-pipe fuel to fuel filter	2		
10	ESR1601	Clip-double fuel lines	NLA	Note (1)	Use ESR1373.
10	ESR1373	Clip-double fuel lines-6 - 8mm	10	Note (2)	
	ERR943	Clip-swivel	1		

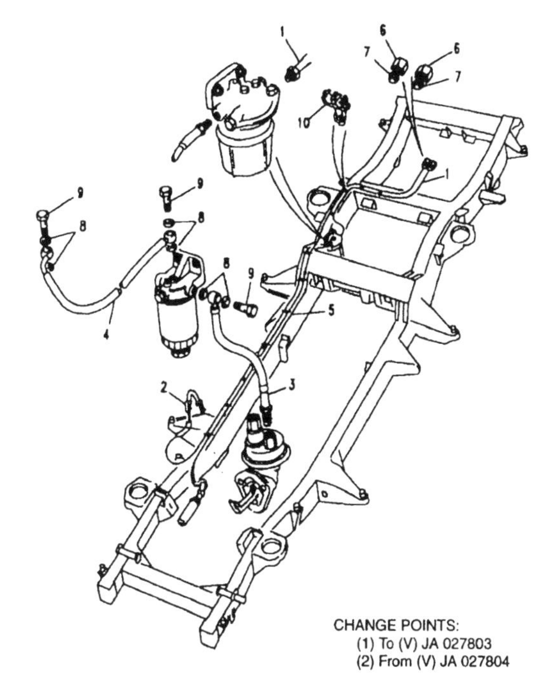

CHANGE POINTS:
(1) To (V) JA 027803
(2) From (V) JA 027804

J 11

Illus	Part Number	Description	Quantity	Change Point	Remarks
		HEATED			
1	RTC4939	FILTER-IN LINE FUEL LINES-HEATED	1		Cold Climate
2	RTC5938	• Cartridge-diesel fuel filter-heated	1		
3	STC612	• Switch-heater fuel filter-heated	1		
		NON HEATED			
4	NTC1518	FILTER ASSEMBLY-IN LINE FUEL LINES-UNHEATED	1		
5	AEU2147L	• Cartridge-diesel fuel filter-unheated	1		
6	AEU2148L	• Screw-bleed-unheated	1		
7	AEU2149L	• Washer-sealing-unheated	1		
		BRACKETRY			
8	NTC2873	Bracket-fuel filter	1		
9	SH110251L	Screw-hexagonal head-M10 x 25	2		
10	WA110061L	Washer-plain-M10-standard	2		
11	WL110001L	Washer-M10	2		
12	NH110041L	Nut-hexagonal head-coarse thread-M10	2		

J 12

Illus	Part Number	Description	Quantity	Change Point	Remarks
1	NRC9708	FILTER/SEDIMENTOR ASSEMBLY-DIESEL	1		
2	37H8119L	• Bolt	1		
3	522940	• Washer	1		
4	37H770	• O ring	1		
6	37H7920	• Plug-drain	1		
5	605011	• Seal-rubber	1		
7	517689	• Plug-blanking	2		
8	517706	• Washer-aluminium	2		
9	AAU9903	• Seal-diesel fuel sedimentor upper-top	1		
10	AAU9902	• Seal-diesel fuel sedimentor lower-bottom	1		
11	NTC7646	Bracket-fuel filter support	1		
12	SH110251L	Screw-sedimentor to bracket mounting-hexagonal head-M10 x 25	2		
13	WA110051	Washer-plain	2		
14	NH110041L	Nut-hexagonal head-coarse thread-M10	2		

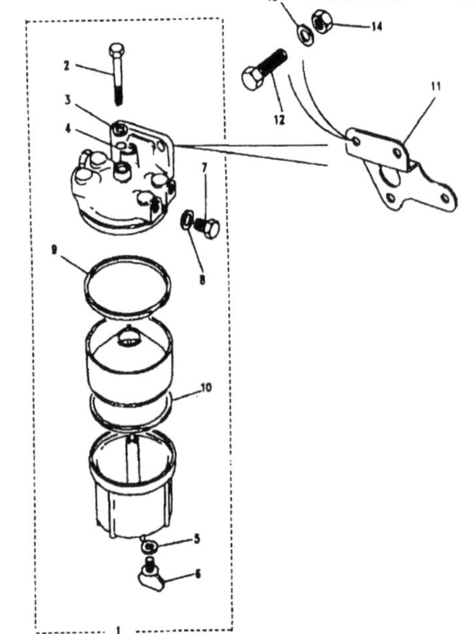

J 13

Illus	Part Number	Description	Quantity	Change Point	Remarks
1	ETC8830	AIR CLEANER ASSEMBLY	1		V8 Twin Carburettor
2	605191	• Element air cleaner	2		
3	RTC5888	• Seal	4		
4	RTC609	• Washer-plain-standard	2		
5	STC1120	• Valve assembly air intake	1		
6	606247	• Clip	1		
7	603670	Elbow-carburettor air inlet	2		V8 Twin Carburettor
8	603069	Seal induction system	2		
9	RTC3519	Clip-hose	2		

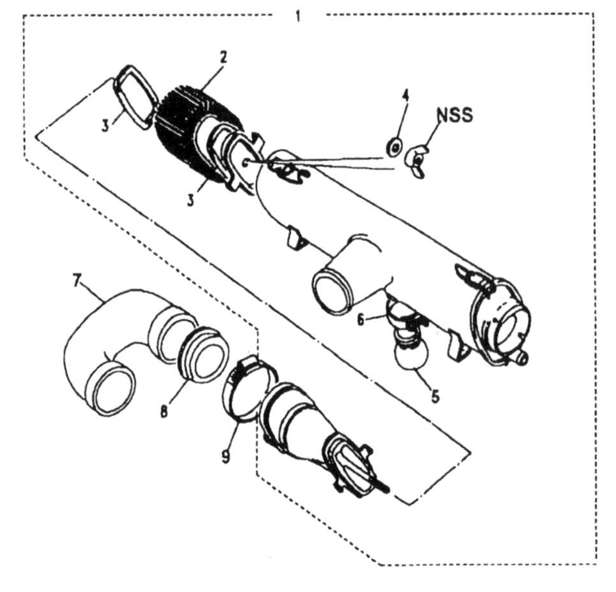

J 14

Illus	Part Number	Description	Quantity	Change Point	Remarks
1	NTC4408	AIR CLEANER ASSEMBLY	1	Note (1)	V8 Petrol EFi
2	RTC4683	• Element air cleaner	1		
3	NTC3751	• Valve assembly air intake	1		
4	RTC4819	• Grommet	2		
5	ESR274	Bracket-mounting air cleaner	1	Note (1)	V8 EFi
6	SH108161L	Screw-hexagonal head-M8 x 16	NLA		
6	FS108161L	Screw-flanged head-M8 x 16	4		
7	WL108001L	Washer-M8	6	Note (1)	
8	WC108051L	Washer-plain-M8	2		
9	FN108041L	Nut-flanged head-M8 x 20	NLA		
10	ESR1057L	Sensor-airflow-multi point inject	1		
11	NTC3354	O ring	1		
12	ERC5829	Bush	2		
13	ERC5830	Grommet	2		
14	ESR273	Bracket-mounting air cleaner-large	1		
15	ESR1611L	Hose-flexible	1		
16	CN100908L	Clip-hose	2		
	RTC5770	Plug-Red	1		

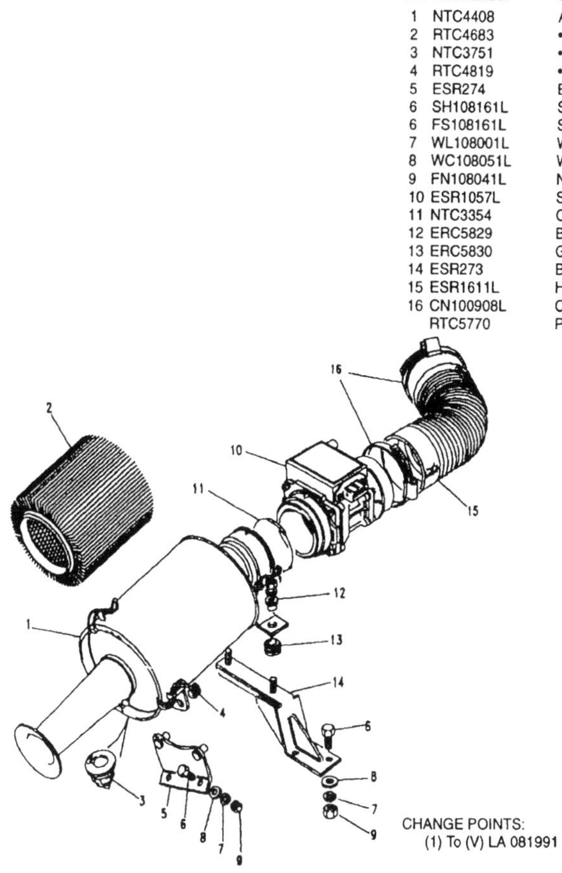

CHANGE POINTS:
(1) To (V) LA 081991

Illus	Part Number	Description	Quantity	Change Point	Remarks
1	ESR1804	AIR CLEANER ASSEMBLY	1	Note (1)	V8 Petrol EFi
1	ESR3011	AIR CLEANER ASSEMBLY	1	Note (2); Note (3)	V8 Petrol EFi
2	ESR1445	• Element air cleaner	1		
3	ESR3248	• Lid-air cleaner	1		Part of ESR1804.
4	ESR1803	• Grommet	2		
3	ESR3273	• Lid-air cleaner	1		Part of ESR3011.
5	ESR2067	Grommet-mounting air cleaner-front	2	Note (4)	V8 EFi
6	NTC8960	Grommet	2		
7	ESR2406	Grommet-blind	2		
8	ESR1057L	Sensor-airflow-multi point inject	1	Note (5)	
8	ERR5198	Sensor-airflow-multi point inject	1	Note (6)	
9	ESR1807	Hose-duct to throttle body induction system	1	Note (7)	
10	CN100908L	Clip-hose	2	Note (4)	
11	NTC3354	O ring	1		

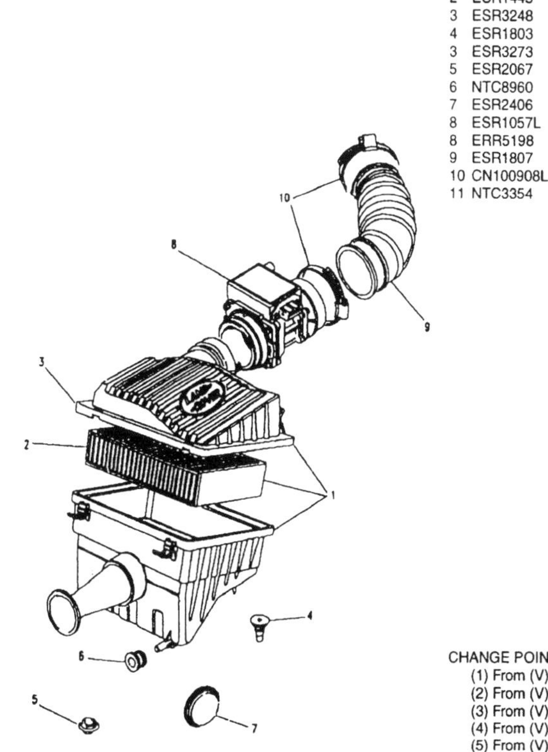

CHANGE POINTS:
(1) From (V) MA 081991 To (V) MA 163103
(2) From (V) TA 163104
(3) From (V) TA 501920
(4) From (V) MA 081991
(5) From (V) MA 081991 To (V) MA 112194
(6) From (V) MA 112195
(7) From (V) MA 081991; From (V) MA 112195

Illus	Part Number	Description	Quantity	Change Point	Remarks
1	ESR1630	AIR CLEANER ASSEMBLY	1		Petrol MPi
2	ESR1445	• Element air cleaner	1		
3	ESR2067	Grommet-mounting air cleaner-front	1	Note (1)	MPi
4	ESR1631	Hose-duct to throttle body induction system	1	Note (2)	MPi
4	ESR3056	Hose-duct to throttle body induction system	1	Note (3)	MPi
5	STC128	Clip-hose	1		MPi
6	CN100908L	Clip-hose	1		MPi
7	ESR2588	Bracket-mounting-duct induction system	2	Note (4)	MPi
	NTC8960	Grommet	2	Note (4)	MPi

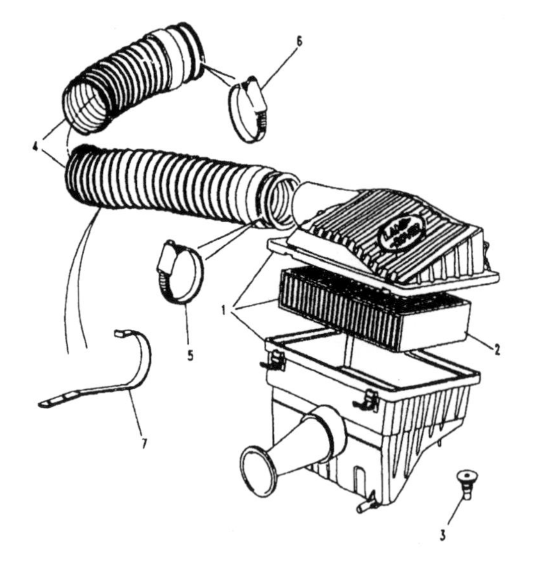

CHANGE POINTS:
(1) From (V) KA 037352
(2) To (V) MA 081990
(3) From (V) MA 081991
(4) From (V) KA 038822

J 17

Illus	Part Number	Description	Quantity	Change Point	Remarks
		AIR CLEANER - 200TDI			
1	ESR1050	AIR CLEANER ASSEMBLY	1		
2	NTC1435	• Element air cleaner	1	Note (1)	
2	ESR1049	• Element air cleaner	1	Note (2)	
18	NRC9239	• Nut-wing	1		
3	NRC8955	• Valve assembly air intake	1		
4	NTC5652	Clamp air intake/duct-rear	1		
5	MRC8388	Catch overcentre	1		
6	RA608176L	Rivet	2		
7	NH106041L	Nut-hexagonal head-nyloc-M6	2		
8	NRC9773	Mounting-rubber flexible support	2	Note (3)	
8	ESR1579	Mounting-rubber flexible support	2	Note (4)	
9	WA106041L	Washer-M6-standard.	2		
10	NTC5654	Clamp air intake/duct-front	1	Note (1)	
10	ESR1051	Clamp air intake/duct-front	1	Note (2)	
11	NTC5803	Elbow assembly-duct air cleaner/throttle body	1		
11	NTC7313	Elbow assembly-duct air cleaner/throttle body	1		Switzerland
12	CN100908L	Clip-hose	1		
13	NTC7083	Hose-air inlet turbocharger	1		
14	CN100608L	Clip-hose	NLA		Use PYC101120.
14	PYC101120	Clip-hose	NLA		Use PYC101120L.
15	NTC7086	Seal induction system	1		
16	ESR151	Duct-elbow induction system	1	Note (5)	
17	NTC7088	Resonator induction system	1	Note (5)	
18	NRC9239	Nut-wing	1		

CHANGE POINTS:
(1) To (V) JA 018272
(2) From (V) JA 018273
(3) To (V) JA 020994
(4) From (V) JA 020995
(5) To (V) JA 018278

J 18

Illus	Part Number	Description	Quantity	Change Point	Remarks
		AIR CLEANER 300TDI			
	STC7051	Kit-air intake-air cleaner	1		LHD
1	ESR1444	AIR CLEANER ASSEMBLY	1		
2	ESR1445	• Element air cleaner	1		
3	ESR3247	• Lid-air cleaner	1	Note (1)	
4	ESR2965	• Cap-air box blanking	1		
5	ESR1621	• Duct-cold air intake/air cleaner induction system	1		
6	ESR1446	• Grommet-mounting air cleaner	1		
7	ESR2392	• Grommet-blind	1		
8	ESR1803	• Grommet	2		
9	ESR2067	Grommet-mounting air cleaner-front	2		
10	NTC8960	Grommet	2		
		MECHANICAL EGR (WITHOUT AIR FLOW METER)			
11	ESR2150	Hose-air outlet turbocharger	1		
		EDC (WITH AIR FLOW METER)			
12	ESR2755	Duct-air cleaner/air flow meter induction system	1		
13	ESR1451	Hose-air flow meter to turbocharger induction system	1		
14	CN100808L	Clip-hose-worm drive	2		
15	CN100258L	Clip-hose-worm drive-25mm	2		

CHANGE POINTS:
(1) From (V) MA 086239

Illus	Part Number	Description	Quantity	Change Point	Remarks
1	NTC5179	Canister charcoal	1	Note (1)	V8
2	NTC5567	Bracket-charcoal canister	1		
3	NRC2268	Plate-charcoal canister bracket stiffening	1		
4	FS106167L	Screw-hexagonal head-M6 x 16	2		
5	WA106051L	Washer-M6	4		
6	NH106041L	Nut-hexagonal head-nyloc-M6	2		
7	NTC5415	Soleniod-charcoal canister	1		
8	NTC5454	Pipe fuel	1		
9	UKC3793L	Clip-hose	1		
10	UKC3795L	Clip-hose-12.7mm	1		
11	NTC5455	Pipe fuel	1		
12	NTC5562	Clip	1		

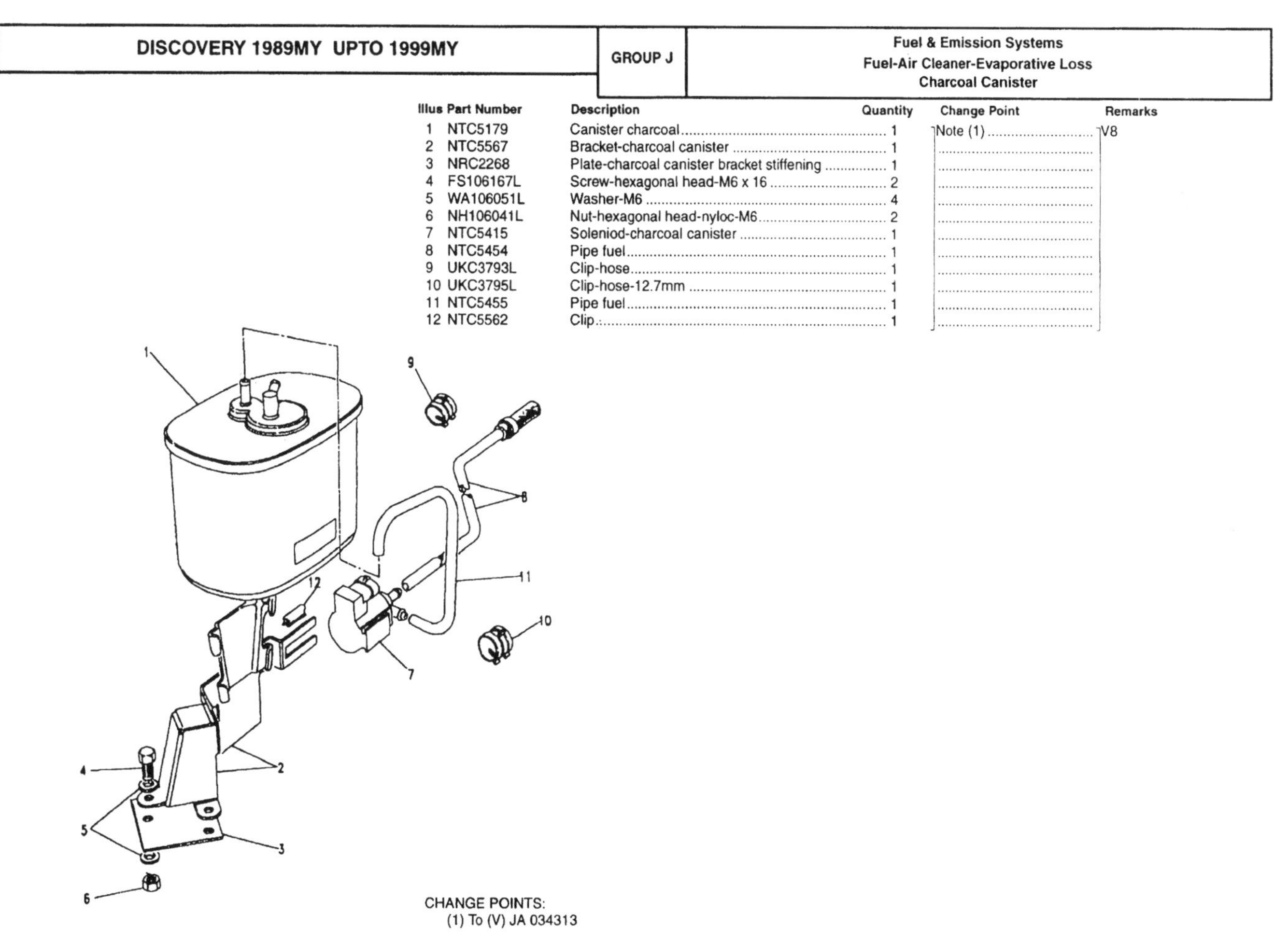

CHANGE POINTS:
(1) To (V) JA 034313

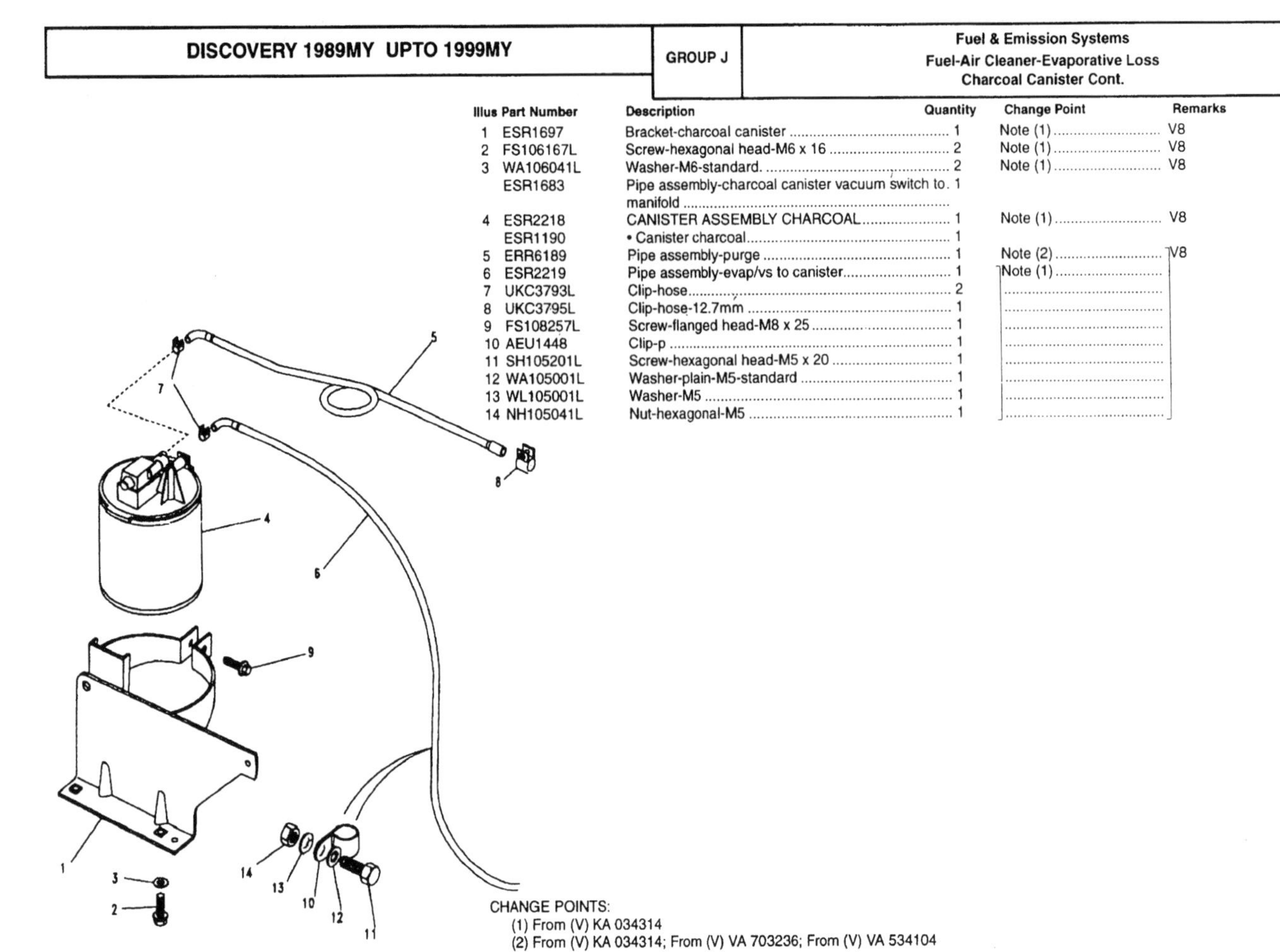

Illus	Part Number	Description	Quantity	Change Point	Remarks
1	ESR1697	Bracket-charcoal canister	1	Note (1)	V8
2	FS106167L	Screw-hexagonal head-M6 x 16	2	Note (1)	V8
3	WA106041L	Washer-M6-standard.	2	Note (1)	V8
	ESR1683	Pipe assembly-charcoal canister vacuum switch to. manifold	1		
4	ESR2218	CANISTER ASSEMBLY CHARCOAL	1	Note (1)	V8
	ESR1190	• Canister charcoal	1		
5	ERR6189	Pipe assembly-purge	1	Note (2)	V8
6	ESR2219	Pipe assembly-evap/vs to canister	1	Note (1)	
7	UKC3793L	Clip-hose	2		
8	UKC3795L	Clip-hose-12.7mm	1		
9	FS108257L	Screw-flanged head-M8 x 25	1		
10	AEU1448	Clip-p	1		
11	SH105201L	Screw-hexagonal head-M5 x 20	1		
12	WA105001L	Washer-plain-M5-standard	1		
13	WL105001L	Washer-M5	1		
14	NH105041L	Nut-hexagonal-M5	1		

CHANGE POINTS:
(1) From (V) KA 034314
(2) From (V) KA 034314; From (V) VA 703236; From (V) VA 534104

J 21

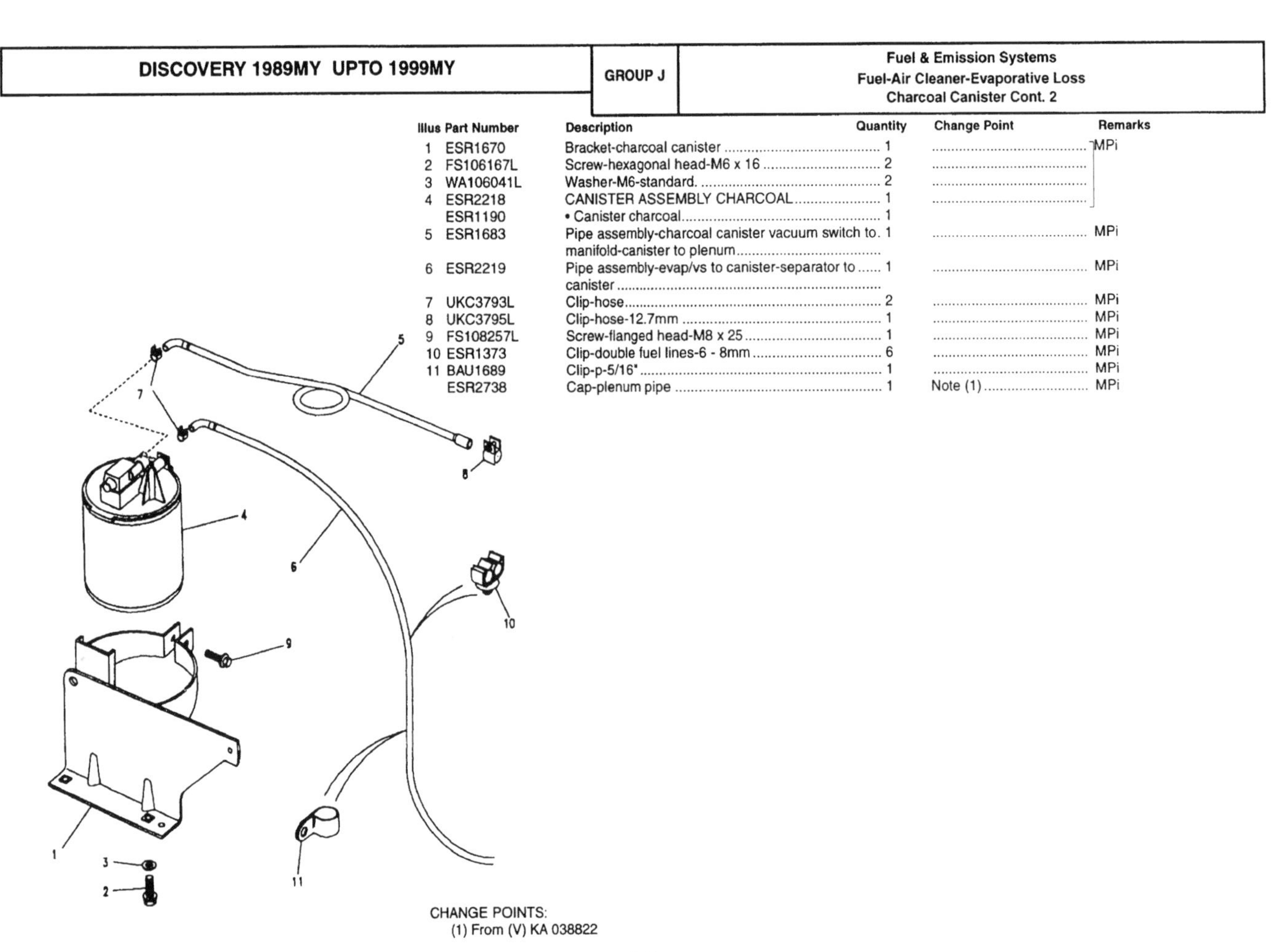

Illus	Part Number	Description	Quantity	Change Point	Remarks
1	ESR1670	Bracket-charcoal canister	1		MPi
2	FS106167L	Screw-hexagonal head-M6 x 16	2		
3	WA106041L	Washer-M6-standard.	2		
4	ESR2218	CANISTER ASSEMBLY CHARCOAL	1		
	ESR1190	• Canister charcoal	1		
5	ESR1683	Pipe assembly-charcoal canister vacuum switch to. manifold-canister to plenum	1		MPi
6	ESR2219	Pipe assembly-evap/vs to canister-separator to canister	1		MPi
7	UKC3793L	Clip-hose	2		MPi
8	UKC3795L	Clip-hose-12.7mm	1		MPi
9	FS108257L	Screw-flanged head-M8 x 25	1		MPi
10	ESR1373	Clip-double fuel lines-6 - 8mm	6		MPi
11	BAU1689	Clip-p-5/16"	1		MPi
	ESR2738	Cap-plenum pipe	1	Note (1)	MPi

CHANGE POINTS:
(1) From (V) KA 038822

J 22

DISCOVERY 1989MY UPTO 1999MY

GROUP J

Fuel & Emission Systems
Fuel-Air Cleaner-Evaporative Loss
Evap Loss Control System

Illus	Part Number	Description	Quantity	Change Point	Remarks
1	ESR167	Separator-vapour fuel filler	1	Note (1)	V8
2	ESR161	Seal	1		
3	SE105121L	Screw-pan head-M5 x 12	4		
4	WL105001L	Washer-M5	4		
5	NTC7491	Pipe assembly-canister to separator	1		
6	NTC7493	Pipe assembly-seperator to atmosphere	1		
7	NTC7492	Pipe assembly-tank to separator	1		
8	NTC4517	Valve-fuel cut fuel tank	1		
9	NTC4627	Grommet	1		
10	568680	Tie-cable-4.8 x 270mm-inside serated	2		
11	79122	Clip-single fuel lines-8mm	1		
12	78177	Screw	1		
13	WC702101L	Washer-plain	1		
14	AEU1448	Clip-p	4		
15	WM702001L	Washer-spring-3/16 dia-square	1		
16	RTC608	Nut	1		
17	NTC7497	Pipe fuel	1		
18	CN100128L	Clip-hose	1		
19	NTC7487	Pipe-breather fuel tank	1		

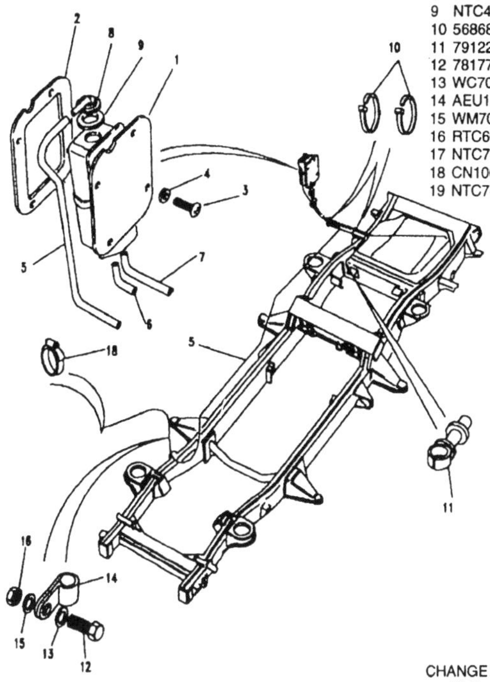

CHANGE POINTS:
(1) To (V) JA 034313

J 23

DISCOVERY 1989MY UPTO 1999MY

GROUP J

Fuel & Emission Systems
Fuel-Air Cleaner-Evaporative Loss
Evap Loss Control System Cont.

Illus	Part Number	Description	Quantity	Change Point	Remarks
1	ESR2219	Pipe assembly-evap/vs to canister-separator to canister	1	Note (1)	V8
2	ESR1897	PIPE FUEL FILLER-SEPARATOR TO CANISTER-. UNLEADED	1	Note (1)	V8
2	ESR1899	PIPE FUEL FILLER-SEPARATOR TO CANISTER-. LEADED	1	Note (1)	V8
3	NTC7567	• Valve-vapour pressure relief	1		Part of ESR1897.
5	568680	Tie-cable-separator to canister-4.8 x 270mm- inside serated	2		V8
6	78177	Screw	l		
7	WC702101L	Washer-plain	1		
8	AEU1448	Clip-p	1		
9	WM702001L	Washer-spring-3/16 dia-square	1		
10	RTC608	Nut	1		
11	79122	Clip-single fuel lines-separator to canister-8mm	1	Note (1)	V8

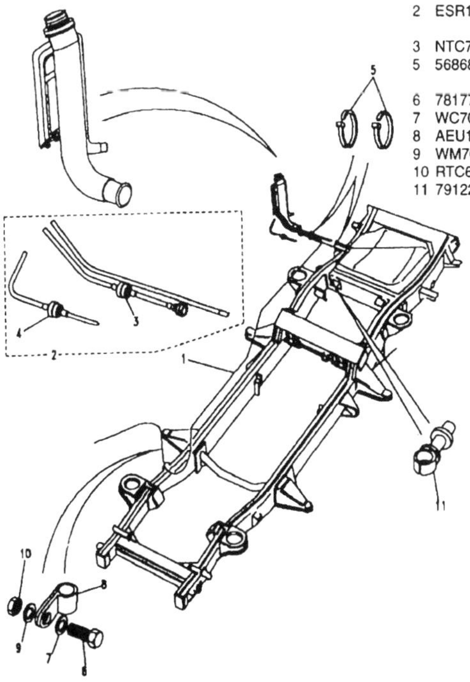

CHANGE POINTS:
(1) From (V) KA 034314

J 24

Notes

Exhaust Diesel .... K 13

Turbo Direct Injection .... K 13

Exhaust Petrol .... K 1

Carburetter Catalyst .... K 1

Downpipe Petrol .... K 10

Injection Catalyst MPi .... K 6

Injection Catalyst Petrol V8 .... K 3

Injection Non-Catalyst MPi To (V) KA038821 .... K 8

Injection Non-Catalyst Petrol V8 .... K 2

Intermediate Silencer Petrol .... K 11

Tailpipe Petrol .... K 12

V8 Service Condition .... K 5

DISCOVERY 1989MY UPTO 1999MY

GROUP K

Exhaust System
Exhaust Petrol
Carburetter Catalyst

Illus	Part Number	Description	Quantity	Change Point	Remarks
	ESR58	Catalytic converter assembly engine exhaust	1		V8 Twin Carburettor
	ESR59	Intermediate/rear assembly exhaust system	1		
	ESR67	Bracket-exhaust mounting/lashing hanger	1		
	NTC5615	Clamp	1		
	NTC7312	Pipe fuel filler-unleaded	NLA		Use ESR1896. V8 Twin Carburettor
	ESR1896	Pipe fuel filler-unleaded	1		V8 Twin Carburettor
	MWC2698	Label-catalyst-fuel filler neck	1		
	BAC10053	Label-catalyst-underbonnet	1		
	MWC2698	Label-catalyst	1		
	NTC1030	Olive-exhaust pipe joint flange	1		
	NTC5582	Mounting-rubber exhaust system	1		
	BH108101L	Bolt-M8 x 50	3		
	WL108001L	Washer-M8	3		
	NY108041L	Nut-hexagonal head-nyloc-M8	3		

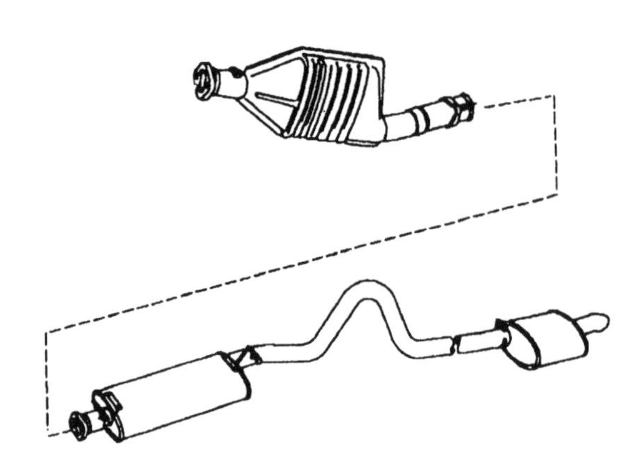

DISCOVERY 1989MY UPTO 1999MY

GROUP K

Exhaust System
Exhaust Petrol
Injection Non-Catalyst Petrol V8

Illus	Part Number	Description	Quantity	Change Point	Remarks
1	ESR225	Downpipe assembly exhaust system	1		V8 EFi
2	ETC4524	Gasket exhaust system-manifold to downpipe	2		
3	ESR2033	Nut-M8	6	Note (1)	
3	ESR2033	Nut-M8	6	Note (2)	
4	NTC4602	Heatshield assembly exhaust-downpipe	1		
5	FS106255L	Screw-flanged head-M6 x 25	2		
6	NTC4205	Heatshield-front engine mounting-LH	1		
7	SH505091L	Screw-hexagonal head-5/16UNC x 1 1/8	1		
8	4594L	Washer-plain	1		
9	ESR109	Heatshield-bracket	1	Note (3)	
9	ANR3227	Heatshield-bracket	1	Note (4)	
10	WA110061L	Washer-plain-M10-standard	1		
11	NH110041L	Nut-hexagonal head-coarse thread-M10	1		
12	ESR230	Intermediate/rear assembly exhaust system	1		3500 cc V8 EFi
13	ESR2917	Intermediate/rear assembly exhaust system	1		3900 cc V8 EFi
14	NV110041L	Nut-M10	2		V8 EFi
15	NRC7649	Heatshield-tunnel exhaust system	1		
16	FS106167L	Screw-hexagonal head-M6 x 16	4		
17	WA106041L	Washer-M6-standard.	4		
18	WL106001L	Washer-M6	4		
19	NH106041L	Nut-hexagonal head-nyloc-M6	4		
20	NTC5648	Bracket exhaust system-front	2		
21	NTC5634	Bracket exhaust system-centre	1		
22	SH108161L	Screw-hexagonal head-M8 x 16	NLA		
22	FS108161L	Screw-flanged head-M8 x 16	2		
23	SH108401L	Screw-hexagonal head-M8 x 40	1		
24	WP105L	Washer-plain	6		
25	NV108041L	Nut-M8	3		
26	NTC5582	Mounting-rubber exhaust system	3		

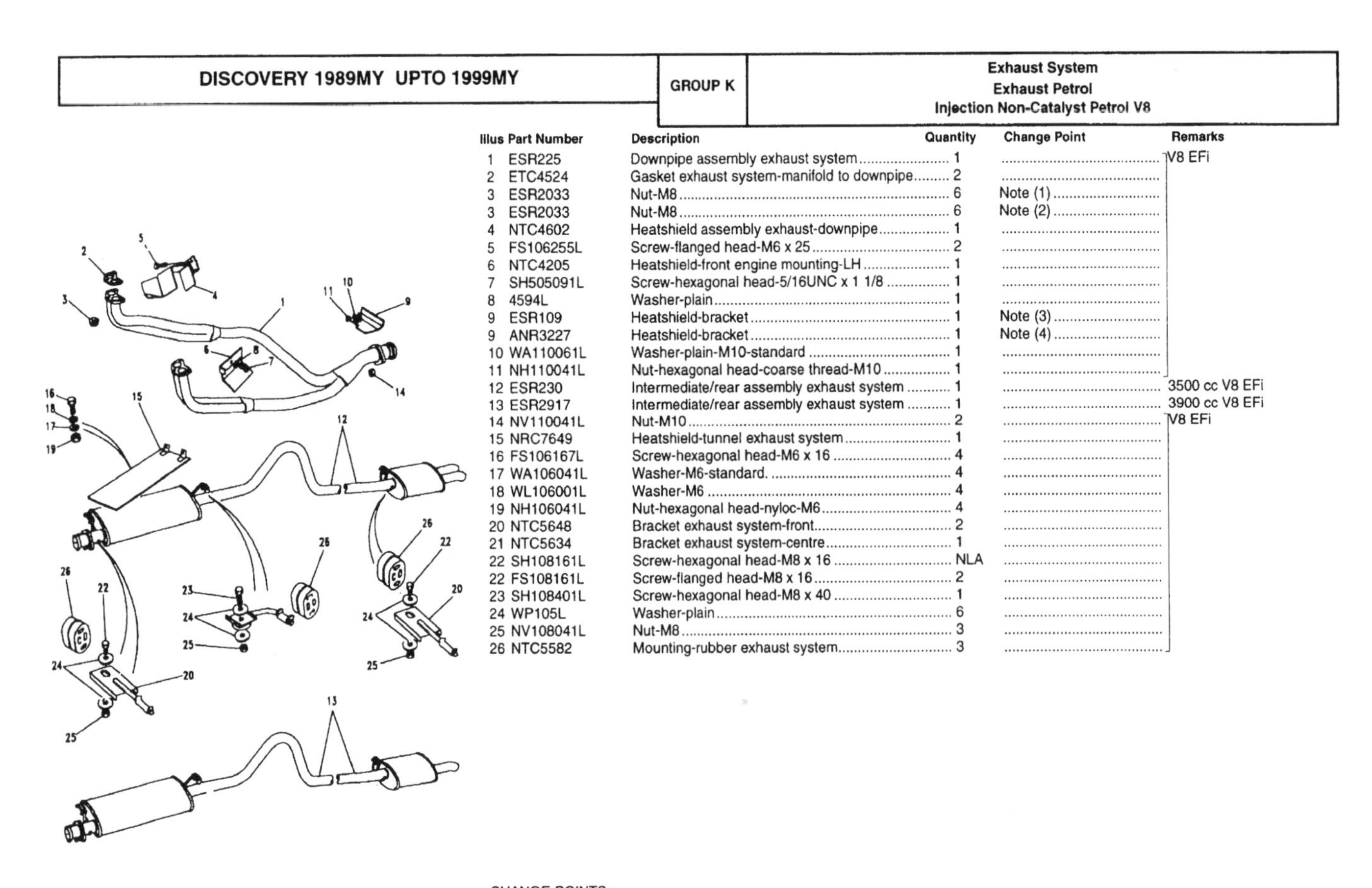

CHANGE POINTS:
(1) To (V) JA 030871
(2) From (V) JA 030872
(3) To (V) KA 064261
(4) From (V) KA 064262

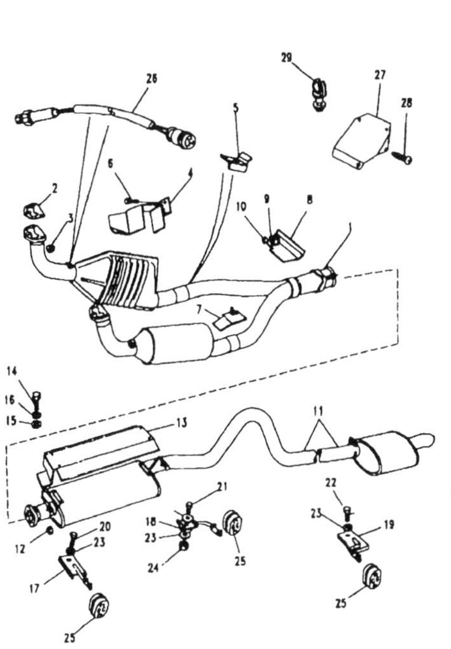

Illus	Part Number	Description	Quantity	Change Point	Remarks
1	NTC7319	Catalytic converter assembly engine exhaust	1	Note (1)	EFi
1	ESR3063	Catalytic converter assembly engine exhaust	1	Note (2)	
1	ESR3059	Catalytic converter assembly engine exhaust	1	Note (3)	
1	ESR3730	Downpipe assembly exhaust system	1	Note (4)	
1	ESR2254	Downpipe assembly exhaust system	1	Note (1)	EFi Japan
1	ESR3062	Downpipe assembly exhaust system	1	Note (5)	
2	ETC4524	Gasket exhaust system-manifold to downpipe	2		EFi
3	ESR2033	Nut-M8	6	Note (6)	
3	ESR2033	Nut-M8	6	Note (7)	
4	NTC4602	Heatshield assembly exhaust-downpipe	1		
5	ERR3601	Heatshield-catalyst bulkhead	1	Note (8)	
6	FS106255L	Screw-flanged head-M6 x 25	2		
7	NTC7345	Heatshield-clutch slave cylinder	1		
7	WEB101550	Heatshield-clutch slave cylinder	1	Note (9); Note (10)	
8	ESR109	Heatshield-bracket	1	Note (11)	
8	ANR3227	Heatshield-bracket	1	Note (12)	
9	WA110061L	Washer-plain-M10-standard	1		
10	FN110041L	Nut-flanged head-M10	NLA		Use FN110047.
10	FN110047	Nut-flanged head-M10	1		V8 EFi

CHANGE POINTS:
(1) To (V) LA 073567
(2) From (V) LA 073568 To (V) MA 135999
(3) From (V) MA 136000
(4) From (V) VA 703237; From (V) VA 534104
(5) From (V) LA 073568
(6) To (V) JA 030871
(7) From (V) JA 030872
(8) From (E)23D 06759D
(9) From (V) VA 703237
(10) From (V) VA 534104
(11) To (V) KA 064262
(12) From (V) KA 064262

K 3

DISCOVERY 1989MY UPTO 1999MY | GROUP K | Exhaust System
Exhaust Petrol
Injection Catalyst Petrol V8

Illus	Part Number	Description	Quantity	Change Point	Remarks
		PRE 97MY & 97MY AUTOMATIC			
11	NTC7426	Rear assembly exhaust system-rear	1	Note (1)	V8 EFi
11	ESR3670	Rear assembly exhaust system	1	Note (2); Note (3)	Automatic V8 EFi
11	ESR3969	Rear assembly exhaust system	1	Note (4)	EFi
		MANUAL 97MY ONLY			
	ESR3735	Silencer assembly exhaust system	1	Note (5)	EFi
	ESR3736	Tailpipe exhaust system	1	Note (5)	
2	ESR3737	Gasket exhaust system	1	Note (5)	
12	NV110041L	Nut-M10	2		
13	NTC7196	Heatshield assembly exhaust-intermediate	1	Note (1)	
13	ESR3671	Heatshield assembly exhaust-intermediate	1	Note (4)	
	ESR2928	Clip	1	Note (4)	
	ESR4117	Heatshield engine exhaust	1	Note (5)	
14	FS106161L	Screw-flanged head-M6 x 16	NLA	Note (4)	Use FS106167L.
14	FS106167L	Screw-hexagonal head-M6 x 16	1		V8 EFi
14	SH106167L	Screw-hexagonal head-M6 x 16	4		EFi
15	WA106041L	Washer-M6-standard.	4		EFi
16	WL106001L	Washer-M6	4		
17	ESR2421	Bracket-exhaust mounting	1		
18	NTC5634	Bracket exhaust system-centre	1		
19	ESR101	Bracket exhaust system-rear	1		
20	SH108161L	Screw-hexagonal head-M8 x 16	NLA		Use FS108161L.
20	FS108161L	Screw-flanged head-M8 x 16	1		V8 EFi
21	SH108401L	Screw-hexagonal head-M8 x 40	1		EFi
22	SH108251	Screw-hexagonal head-M8 x 25	NLA		Use FS108257.
22	FS108257	Screw-flanged head	1		V8 EFi
23	WP105L	Washer-plain	4		EFi
24	NV108041L	Nut-M8	1		
25	NTC5582	Mounting-rubber exhaust system	1		
26	PRC7062	Sensor-oxygen multi point injection	2		
27	ERR1608	Heatshield alternator-65 amp	1		
28	AB610031L	Screw-self tapping AB-10 x 3/8	1		
29	PRC3180	Clip-pipe-single	1		

CHANGE POINTS:
(1) To (V) MA 163104
(2) From (V) TA 501921 To (V) TA 502742
(3) From (V) TA 163105 To (V) TA 166329
(4) From (V) TA 502743; From (V) TA 166330
(5) From (V) VA 703237; From (V) VA 534104

K 4

Illus	Part Number	Description	Quantity	Change Point	Remarks
	STC3719	Adaptor-exhaust	1		
	STC3717	Tailpipe assembly-exhaust system	1		
	STC3716	Silencer assembly exhaust system	1		
	STC3619	Tailpipe exhaust system	1		
	NH110041L	Nut-hexagonal head-coarse thread-M10	3		L/R V8 Petrol
	ESR3737	Gasket exhaust system	1		

NOT ILLUSTRATED

ND 0001

Illus	Part Number	Description	Quantity	Change Point	Remarks
1	ESR1698	Downpipe assembly exhaust system	1	Note (1)	
2	WCM10027	Gasket exhaust system-manifold to downpipe	1		
3	ESR2034	Nut-flange-nyloc-M10	4		
4	ESR109	Heatshield-bracket	1		
5	WA110061L	Washer-plain-M10-standard	1		
6	FN110041L	Nut-flanged head-M10	NLA		Use FN110047.
6	FN110047	Nut-flanged head-M10	1		MPi
7	ESR1699	Intermediate/rear assembly exhaust system	1		
8	NV108041L	Nut-M8	2		
9	NRC7649	Heatshield-tunnel exhaust system	1		
10	FS106167L	Screw-hexagonal head-M6 x 16	2		
11	WA106041L	Washer-M6-standard.	4		
12	WL106001L	Washer-M6	4		
13	NH106041L	Nut-hexagonal head-nyloc-M6	4		
14	NTC5634	Bracket exhaust system-centre	1		
15	SH108401L	Screw-hexagonal head-M8 x 40	1		
16	WP105L	Washer-plain	2		
17	NV108041L	Nut-M8	1		
18	NTC5648	Bracket exhaust system-front	2		
19	GG108251L	Screw-hexagonal head-M8 x 125	2		
20	WP105L	Washer-plain	2		
21	NTC5582	Mounting-rubber exhaust system	3		
22	ESR2607	Hanger-exhaust front sub frame	1		
	ANR1892	Heatshield-front engine mounting	1	Note (2)	

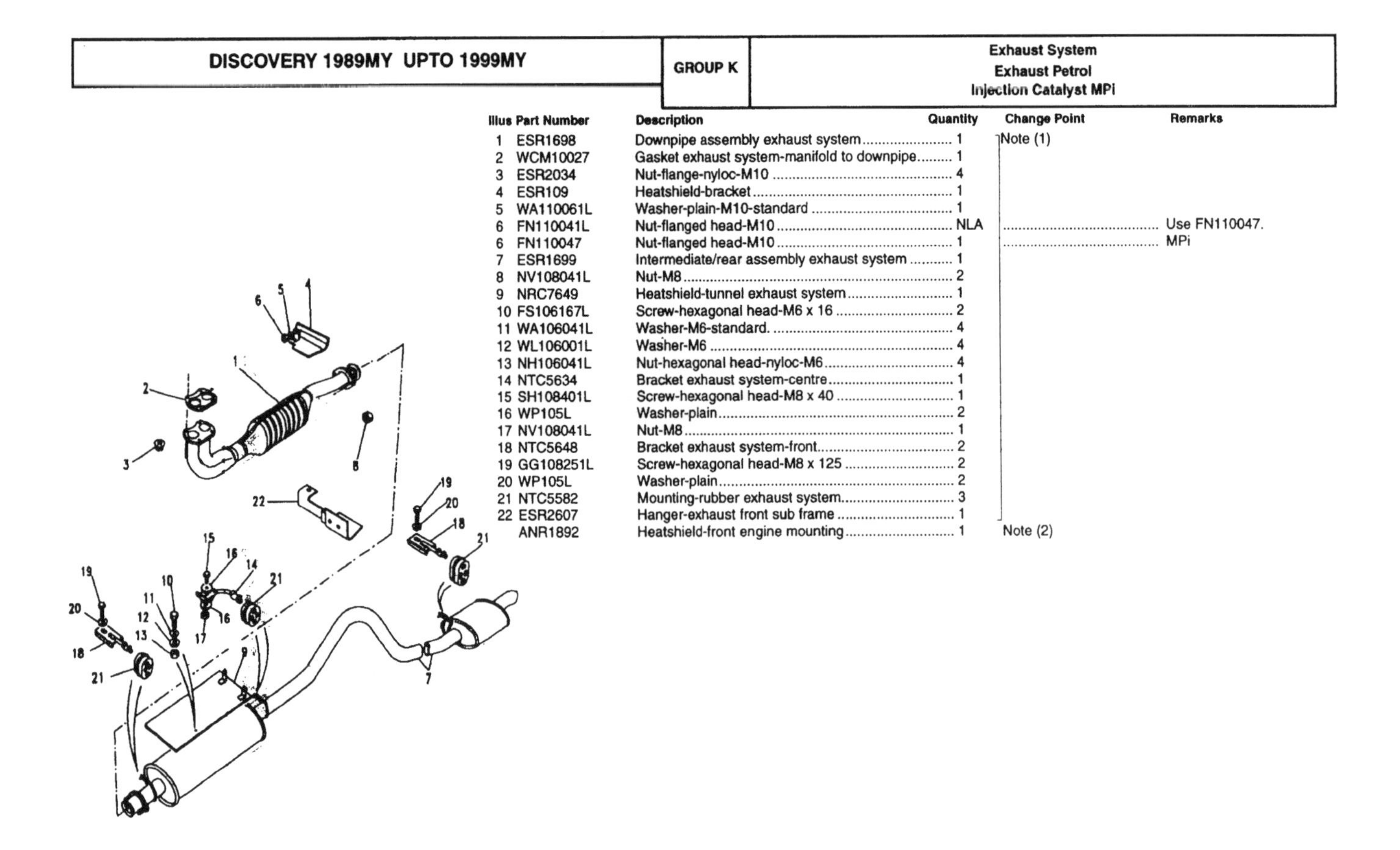

CHANGE POINTS:
(1) To (V) KA 038821
(2) From (V) KA 037352 To (V) KA 038821

Exhaust System
Exhaust Petrol
Injection Catalyst MPi

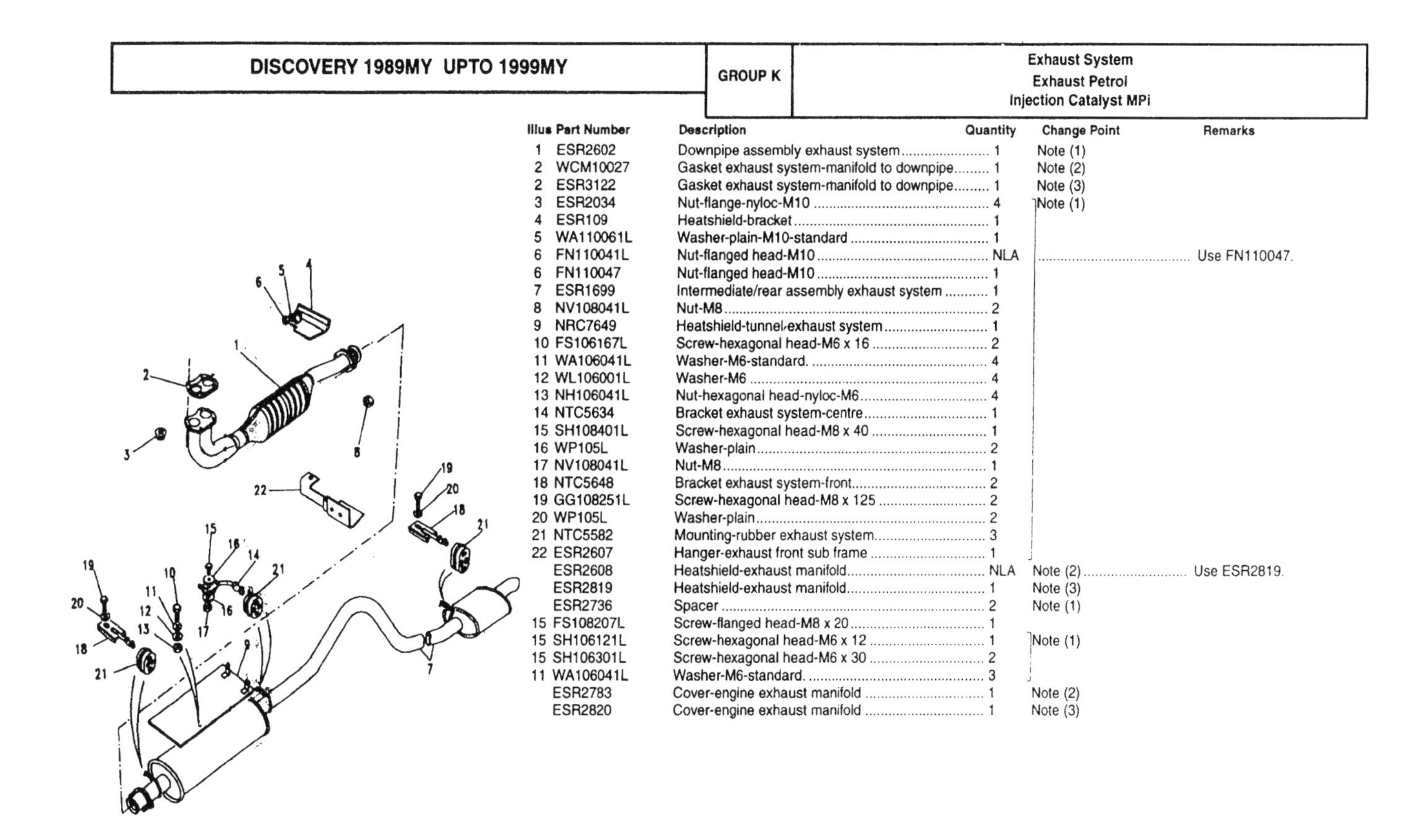

Illus	Part Number	Description	Quantity	Change Point	Remarks
1	ESR2602	Downpipe assembly exhaust system	1	Note (1)	
2	WCM10027	Gasket exhaust system-manifold to downpipe	1	Note (2)	
2	ESR3122	Gasket exhaust system-manifold to downpipe	1	Note (3)	
3	ESR2034	Nut-flange-nyloc-M10	4	Note (1)	
4	ESR109	Heatshield-bracket	1		
5	WA110061L	Washer-plain-M10-standard	1		
6	FN110041L	Nut-flanged head-M10	NLA		Use FN110047.
6	FN110047	Nut-flanged head-M10	1		
7	ESR1699	Intermediate/rear assembly exhaust system	1		
8	NV108041L	Nut-M8	2		
9	NRC7649	Heatshield-tunnel-exhaust system	1		
10	FS106167L	Screw-hexagonal head-M6 x 16	2		
11	WA106041L	Washer-M6-standard.	4		
12	WL106001L	Washer-M6	4		
13	NH106041L	Nut-hexagonal head-nyloc-M6	4		
14	NTC5634	Bracket exhaust system-centre	1		
15	SH108401L	Screw-hexagonal head-M8 x 40	1		
16	WP105L	Washer-plain	2		
17	NV108041L	Nut-M8	1		
18	NTC5648	Bracket exhaust system-front	2		
19	GG108251L	Screw-hexagonal head-M8 x 125	2		
20	WP105L	Washer-plain	2		
21	NTC5582	Mounting-rubber exhaust system	3		
22	ESR2607	Hanger-exhaust front sub frame	1		
	ESR2608	Heatshield-exhaust manifold	NLA	Note (2)	Use ESR2819.
	ESR2819	Heatshield-exhaust manifold	1	Note (3)	
	ESR2736	Spacer	2	Note (1)	
15	FS108207L	Screw-flanged head-M8 x 20	1		
15	SH106121L	Screw-hexagonal head-M6 x 12	1	Note (1)	
15	SH106301L	Screw-hexagonal head-M6 x 30	2		
11	WA106041L	Washer-M6-standard.	3		
	ESR2783	Cover-engine exhaust manifold	1	Note (2)	
	ESR2820	Cover-engine exhaust manifold	1	Note (3)	

CHANGE POINTS:
(1) From (V) KA 038822
(2) From (V) KA 038822 To (V) LA 081991
(3) From (V) KA 038822; From (V) MA 081992

DISCOVERY 1989MY UPTO 1999MY | GROUP K

Exhaust System
Exhaust Petrol
Injection Non-Catalyst MPi To (V) KA038821

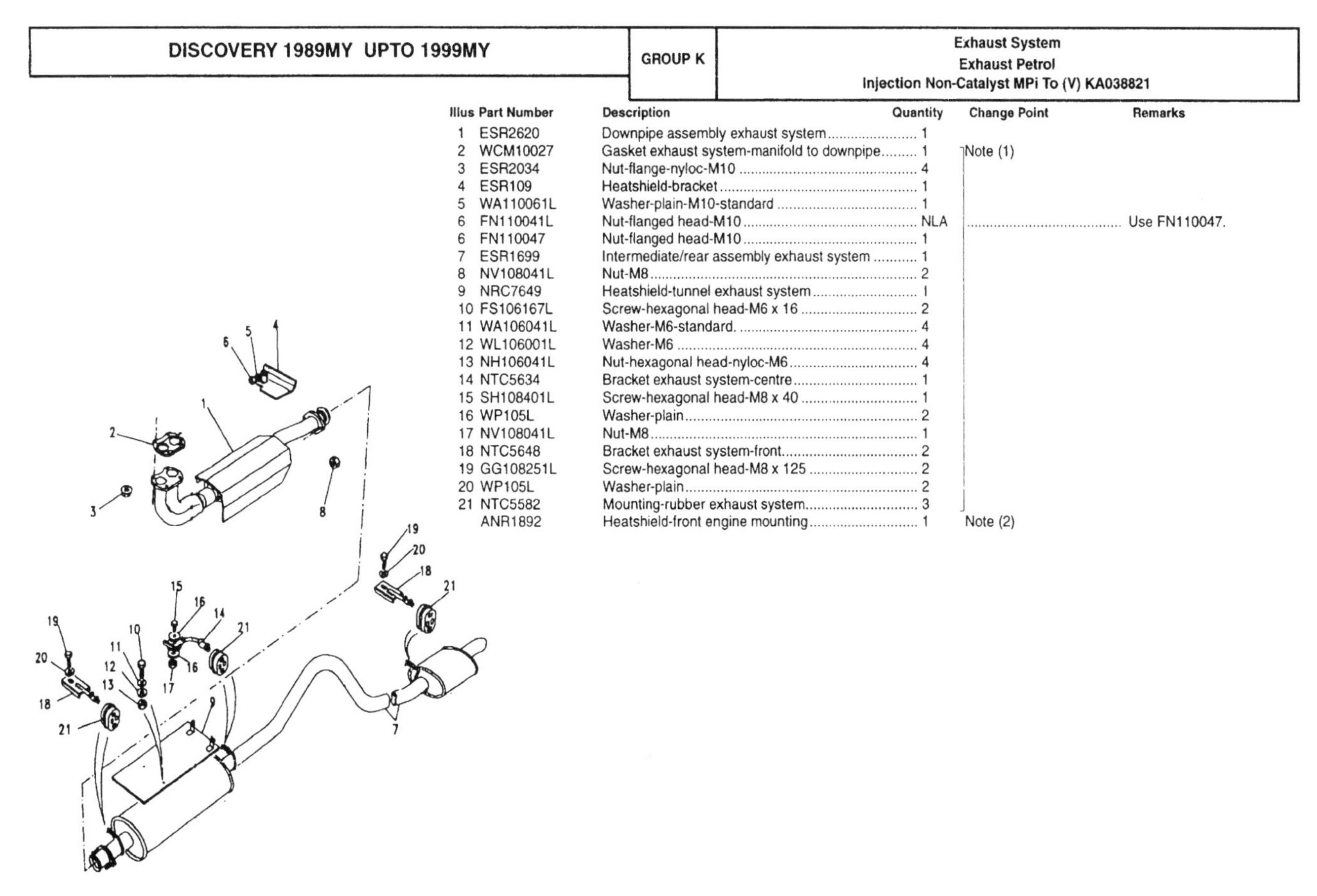

Illus	Part Number	Description	Quantity	Change Point	Remarks
1	ESR2620	Downpipe assembly exhaust system	1		
2	WCM10027	Gasket exhaust system-manifold to downpipe	1	Note (1)	
3	ESR2034	Nut-flange-nyloc-M10	4		
4	ESR109	Heatshield-bracket	1		
5	WA110061L	Washer-plain-M10-standard	1		
6	FN110041L	Nut-flanged head-M10	NLA		Use FN110047.
6	FN110047	Nut-flanged head-M10	1		
7	ESR1699	Intermediate/rear assembly exhaust system	1		
8	NV108041L	Nut-M8	2		
9	NRC7649	Heatshield-tunnel exhaust system	1		
10	FS106167L	Screw-hexagonal head-M6 x 16	2		
11	WA106041L	Washer-M6-standard.	4		
12	WL106001L	Washer-M6	4		
13	NH106041L	Nut-hexagonal head-nyloc-M6	4		
14	NTC5634	Bracket exhaust system-centre	1		
15	SH108401L	Screw-hexagonal head-M8 x 40	1		
16	WP105L	Washer-plain	2		
17	NV108041L	Nut-M8	1		
18	NTC5648	Bracket exhaust system-front	2		
19	GG108251L	Screw-hexagonal head-M8 x 125	2		
20	WP105L	Washer-plain	2		
21	NTC5582	Mounting-rubber exhaust system	3		
	ANR1892	Heatshield-front engine mounting	1	Note (2)	

CHANGE POINTS:
(1) To (V) KA 038821
(2) From (V) KA 037352 To (V) KA 038821

Exhaust System
Exhaust Petrol
Injection Non-Catalyst MPi From (V) KA038822

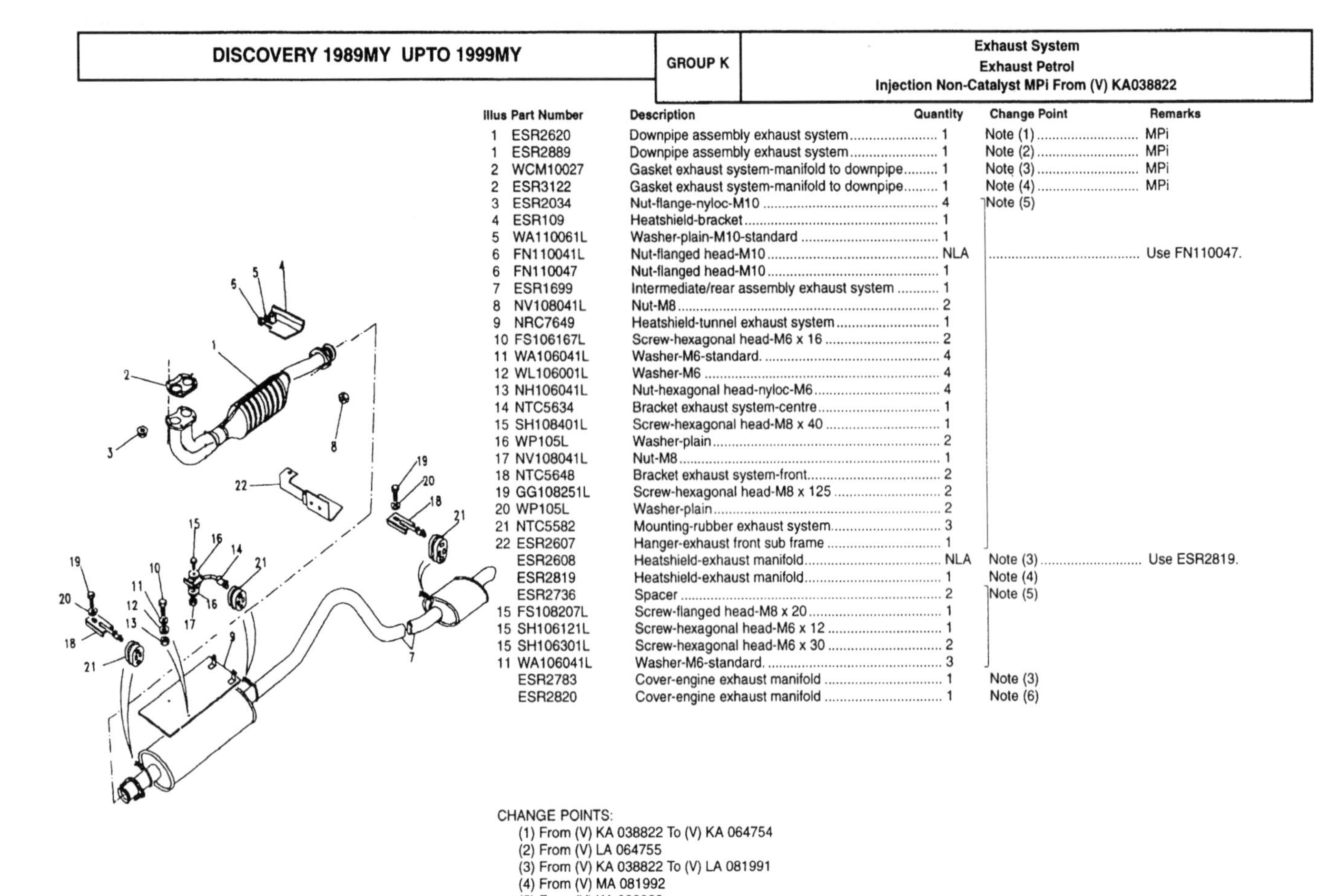

Illus	Part Number	Description	Quantity	Change Point	Remarks
1	ESR2620	Downpipe assembly exhaust system	1	Note (1)	MPi
1	ESR2889	Downpipe assembly exhaust system	1	Note (2)	MPi
2	WCM10027	Gasket exhaust system-manifold to downpipe	1	Note (3)	MPi
2	ESR3122	Gasket exhaust system-manifold to downpipe	1	Note (4)	MPi
3	ESR2034	Nut-flange-nyloc-M10	4	Note (5)	
4	ESR109	Heatshield-bracket	1		
5	WA110061L	Washer-plain-M10-standard	1		
6	FN110041L	Nut-flanged head-M10	NLA		Use FN110047.
6	FN110047	Nut-flanged head-M10	1		
7	ESR1699	Intermediate/rear assembly exhaust system	1		
8	NV108041L	Nut-M8	2		
9	NRC7649	Heatshield-tunnel exhaust system	1		
10	FS106167L	Screw-hexagonal head-M6 x 16	2		
11	WA106041L	Washer-M6-standard.	4		
12	WL106001L	Washer-M6	4		
13	NH106041L	Nut-hexagonal head-nyloc-M6	4		
14	NTC5634	Bracket exhaust system-centre	1		
15	SH108401L	Screw-hexagonal head-M8 x 40	1		
16	WP105L	Washer-plain	2		
17	NV108041L	Nut-M8	1		
18	NTC5648	Bracket exhaust system-front	2		
19	GG108251L	Screw-hexagonal head-M8 x 125	2		
20	WP105L	Washer-plain	2		
21	NTC5582	Mounting-rubber exhaust system	3		
22	ESR2607	Hanger-exhaust front sub frame	1		
	ESR2608	Heatshield-exhaust manifold	NLA	Note (3)	Use ESR2819.
	ESR2819	Heatshield-exhaust manifold	1	Note (4)	
	ESR2736	Spacer	2	Note (5)	
15	FS108207L	Screw-flanged head-M8 x 20	1		
15	SH106121L	Screw-hexagonal head-M6 x 12	1		
15	SH106301L	Screw-hexagonal head-M6 x 30	2		
11	WA106041L	Washer-M6-standard.	3		
	ESR2783	Cover-engine exhaust manifold	1	Note (3)	
	ESR2820	Cover-engine exhaust manifold	1	Note (6)	

CHANGE POINTS:
(1) From (V) KA 038822 To (V) KA 064754
(2) From (V) LA 064755
(3) From (V) KA 038822 To (V) LA 081991
(4) From (V) MA 081992
(5) From (V) KA 038822
(6) From (V) KA 038822; From (V) MA 081992

DISCOVERY 1989MY UPTO 1999MY | GROUP K

Exhaust System
Exhaust Petrol
Downpipe Petrol

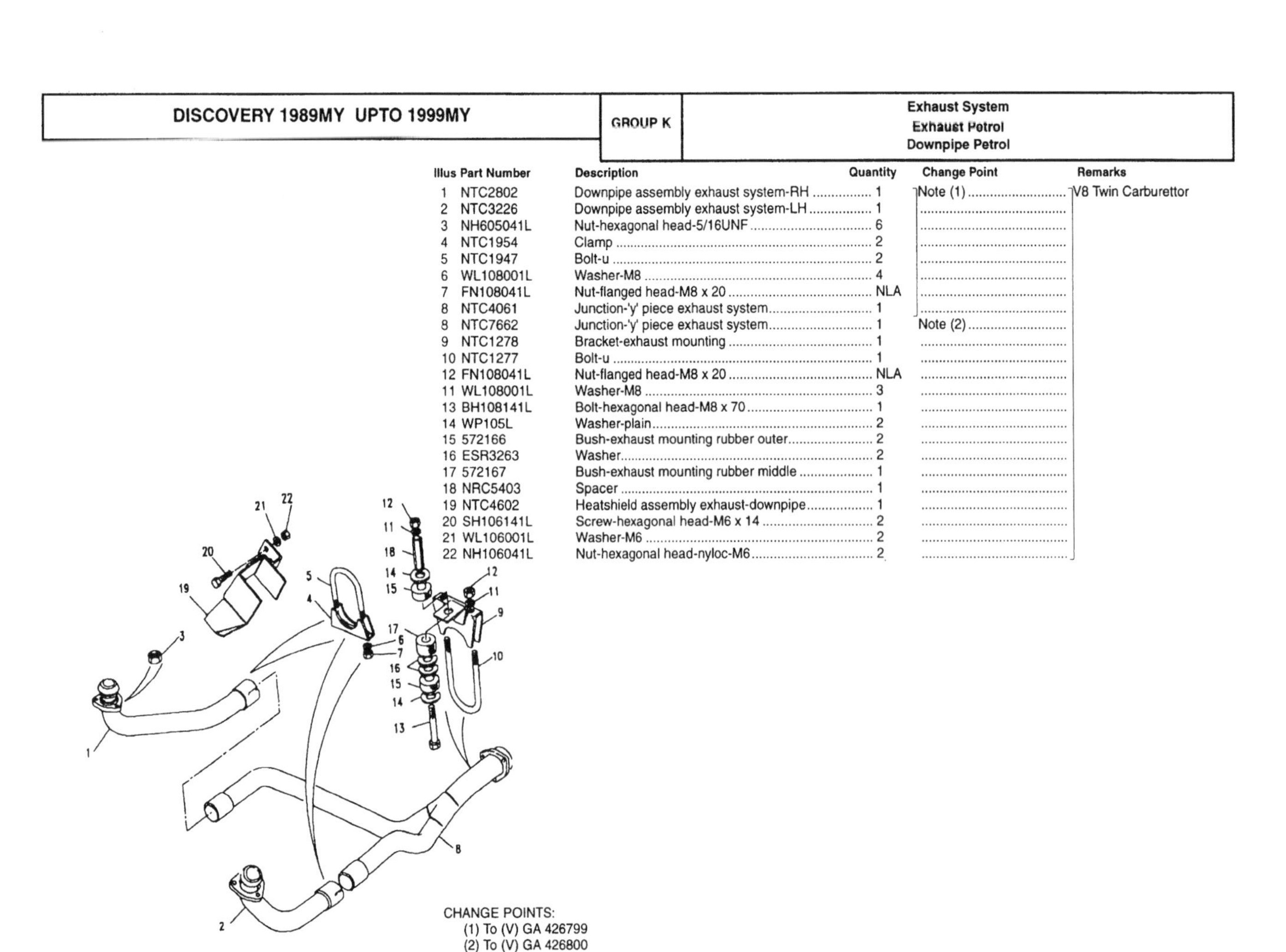

Illus	Part Number	Description	Quantity	Change Point	Remarks
1	NTC2802	Downpipe assembly exhaust system-RH	1	Note (1)	V8 Twin Carburettor
2	NTC3226	Downpipe assembly exhaust system-LH	1		
3	NH605041L	Nut-hexagonal head-5/16UNF	6		
4	NTC1954	Clamp	2		
5	NTC1947	Bolt-u	2		
6	WL108001L	Washer-M8	4		
7	FN108041L	Nut-flanged head-M8 x 20	NLA		
8	NTC4061	Junction-'y' piece exhaust system	1		
8	NTC7662	Junction-'y' piece exhaust system	1	Note (2)	
9	NTC1278	Bracket-exhaust mounting	1		
10	NTC1277	Bolt-u	1		
12	FN108041L	Nut-flanged head-M8 x 20	NLA		
11	WL108001L	Washer-M8	3		
13	BH108141L	Bolt-hexagonal head-M8 x 70	1		
14	WP105L	Washer-plain	2		
15	572166	Bush-exhaust mounting rubber outer	2		
16	ESR3263	Washer	2		
17	572167	Bush-exhaust mounting rubber middle	1		
18	NRC5403	Spacer	1		
19	NTC4602	Heatshield assembly exhaust-downpipe	1		
20	SH106141L	Screw-hexagonal head-M6 x 14	2		
21	WL106001L	Washer-M6	2		
22	NH106041L	Nut-hexagonal head-nyloc-M6	2		

CHANGE POINTS:
(1) To (V) GA 426799
(2) To (V) GA 426800

Exhaust Petrol
Intermediate Silencer Petrol

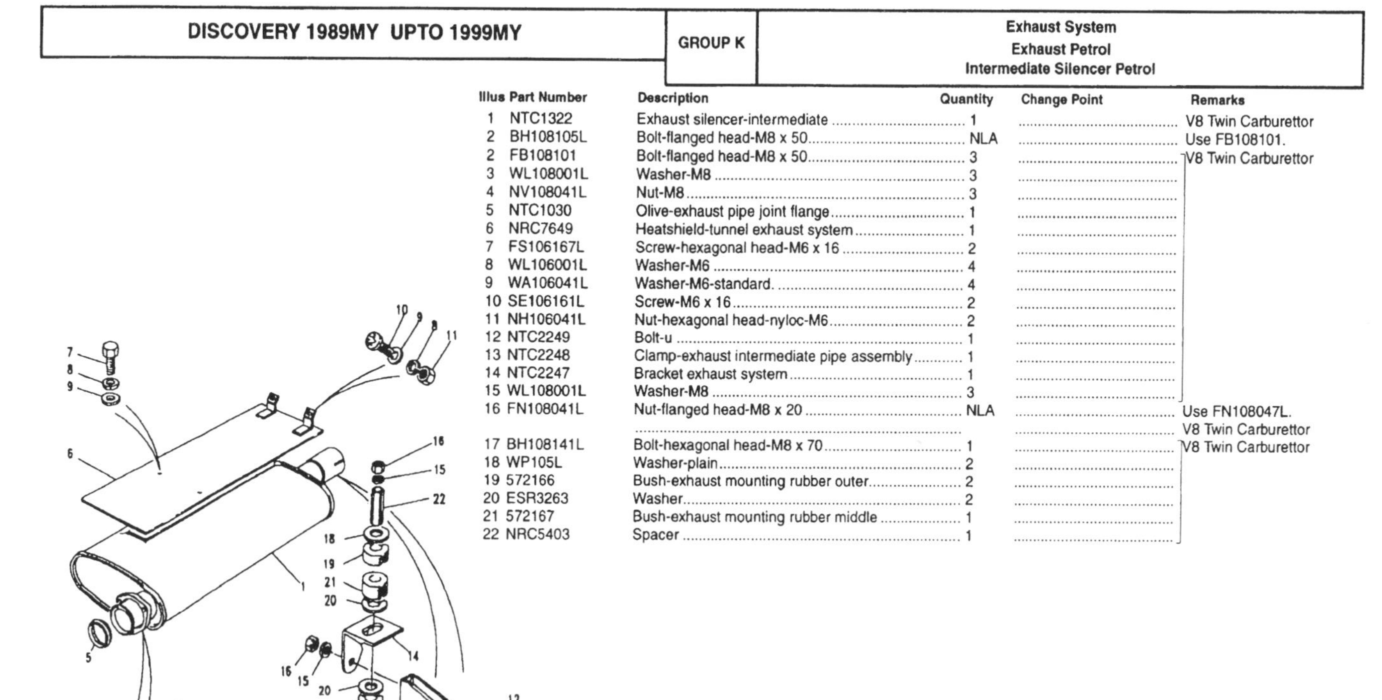

Illus	Part Number	Description	Quantity	Change Point	Remarks
1	NTC1322	Exhaust silencer-intermediate	1		V8 Twin Carburettor
2	BH108105L	Bolt-flanged head-M8 x 50	NLA		Use FB108101.
2	FB108101	Bolt-flanged head-M8 x 50	3		V8 Twin Carburettor
3	WL108001L	Washer-M8	3		
4	NV108041L	Nut-M8	3		
5	NTC1030	Olive-exhaust pipe joint flange	1		
6	NRC7649	Heatshield-tunnel exhaust system	1		
7	FS106167L	Screw-hexagonal head-M6 x 16	2		
8	WL106001L	Washer-M6	4		
9	WA106041L	Washer-M6-standard.	4		
10	SE106161L	Screw-M6 x 16	2		
11	NH106041L	Nut-hexagonal head-nyloc-M6	2		
12	NTC2249	Bolt-u	1		
13	NTC2248	Clamp-exhaust intermediate pipe assembly	1		
14	NTC2247	Bracket exhaust system	1		
15	WL108001L	Washer-M8	3		
16	FN108041L	Nut-flanged head-M8 x 20	NLA		Use FN108047L.
					V8 Twin Carburettor
17	BH108141L	Bolt-hexagonal head-M8 x 70	1		V8 Twin Carburettor
18	WP105L	Washer-plain	2		
19	572166	Bush-exhaust mounting rubber outer	2		
20	ESR3263	Washer	2		
21	572167	Bush-exhaust mounting rubber middle	1		
22	NRC5403	Spacer	1		

Exhaust Petrol
Tailpipe Petrol

Illus	Part Number	Description	Quantity	Change Point	Remarks
1	STC1428	Tailpipe assembly-exhaust system	1		V8 Twin Carburettor
2	NTC1637	Bracket exhaust system	1		
3	NTC1284	Bolt-u	1		
4	BH108141L	Bolt-hexagonal head-M8 x 70	1		
5	WP105L	Washer-plain	2		
6	572166	Bush-exhaust mounting rubber outer	2		
7	ESR3263	Washer	2		
8	NRC5403	Spacer	1		
9	572167	Bush-exhaust mounting rubber middle	1		
10	WL108001L	Washer-M8	3		
11	FN108041L	Nut-flanged head-M8 x 20	NLA		

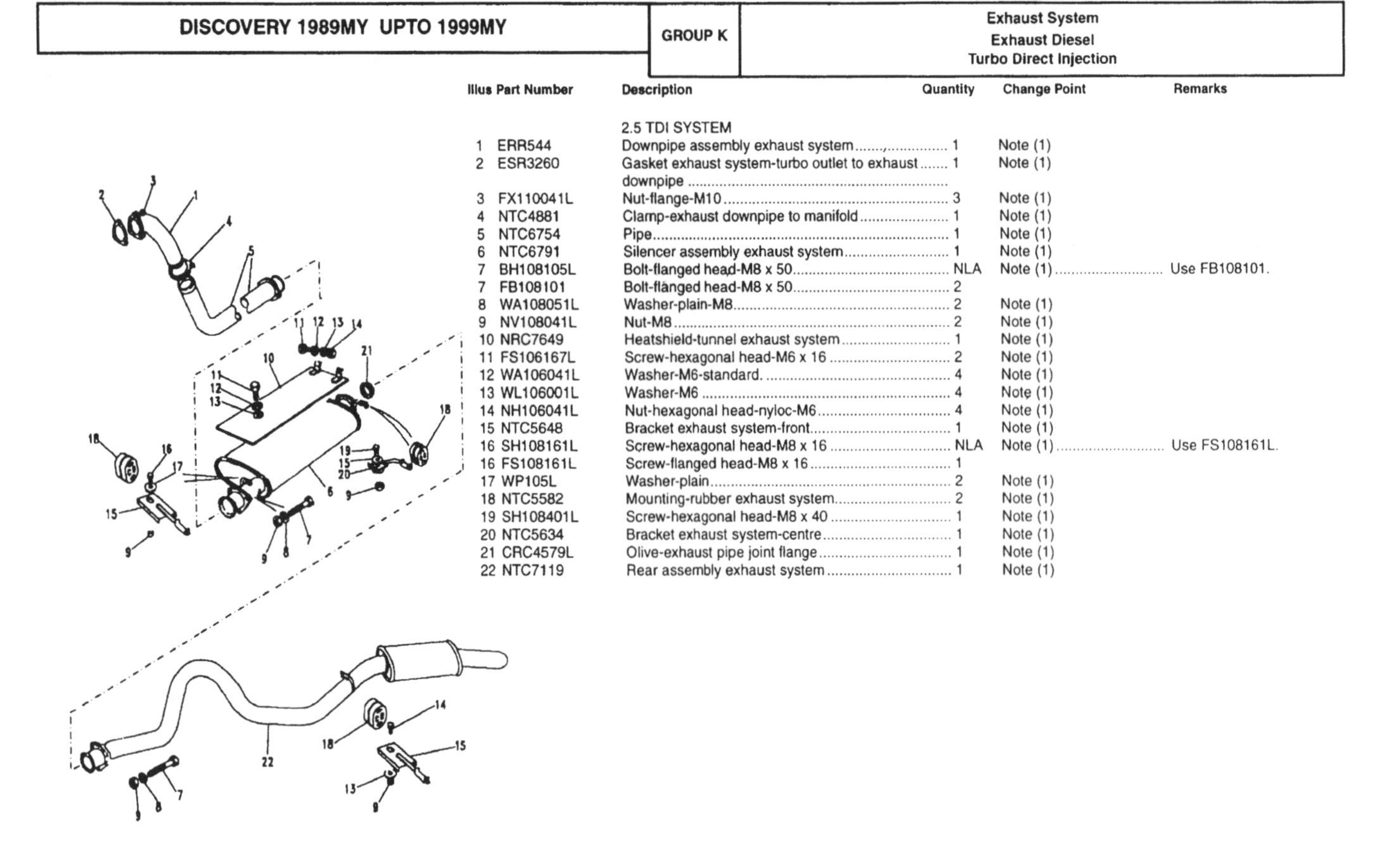

Illus	Part Number	Description	Quantity	Change Point	Remarks
		2.5 TDI SYSTEM			
1	ERR544	Downpipe assembly exhaust system	1	Note (1)	
2	ESR3260	Gasket exhaust system-turbo outlet to exhaust downpipe	1	Note (1)	
3	FX110041L	Nut-flange-M10	3	Note (1)	
4	NTC4881	Clamp-exhaust downpipe to manifold	1	Note (1)	
5	NTC6754	Pipe	1	Note (1)	
6	NTC6791	Silencer assembly exhaust system	1	Note (1)	
7	BH108105L	Bolt-flanged head-M8 x 50	NLA	Note (1)	Use FB108101.
7	FB108101	Bolt-flanged head-M8 x 50	2		
8	WA108051L	Washer-plain-M8	2	Note (1)	
9	NV108041L	Nut-M8	2	Note (1)	
10	NRC7649	Heatshield-tunnel exhaust system	1	Note (1)	
11	FS106167L	Screw-hexagonal head-M6 x 16	2	Note (1)	
12	WA106041L	Washer-M6-standard.	4	Note (1)	
13	WL106001L	Washer-M6	4	Note (1)	
14	NH106041L	Nut-hexagonal head-nyloc-M6	4	Note (1)	
15	NTC5648	Bracket exhaust system-front	1	Note (1)	
16	SH108161L	Screw-hexagonal head-M8 x 16	NLA	Note (1)	Use FS108161L.
16	FS108161L	Screw-flanged head-M8 x 16	1		
17	WP105L	Washer-plain	2	Note (1)	
18	NTC5582	Mounting-rubber exhaust system	2	Note (1)	
19	SH108401L	Screw-hexagonal head-M8 x 40	1	Note (1)	
20	NTC5634	Bracket exhaust system-centre	1	Note (1)	
21	CRC4579L	Olive-exhaust pipe joint flange	1	Note (1)	
22	NTC7119	Rear assembly exhaust system	1	Note (1)	

CHANGE POINTS:
(1) To (V) GA 460229

K 13

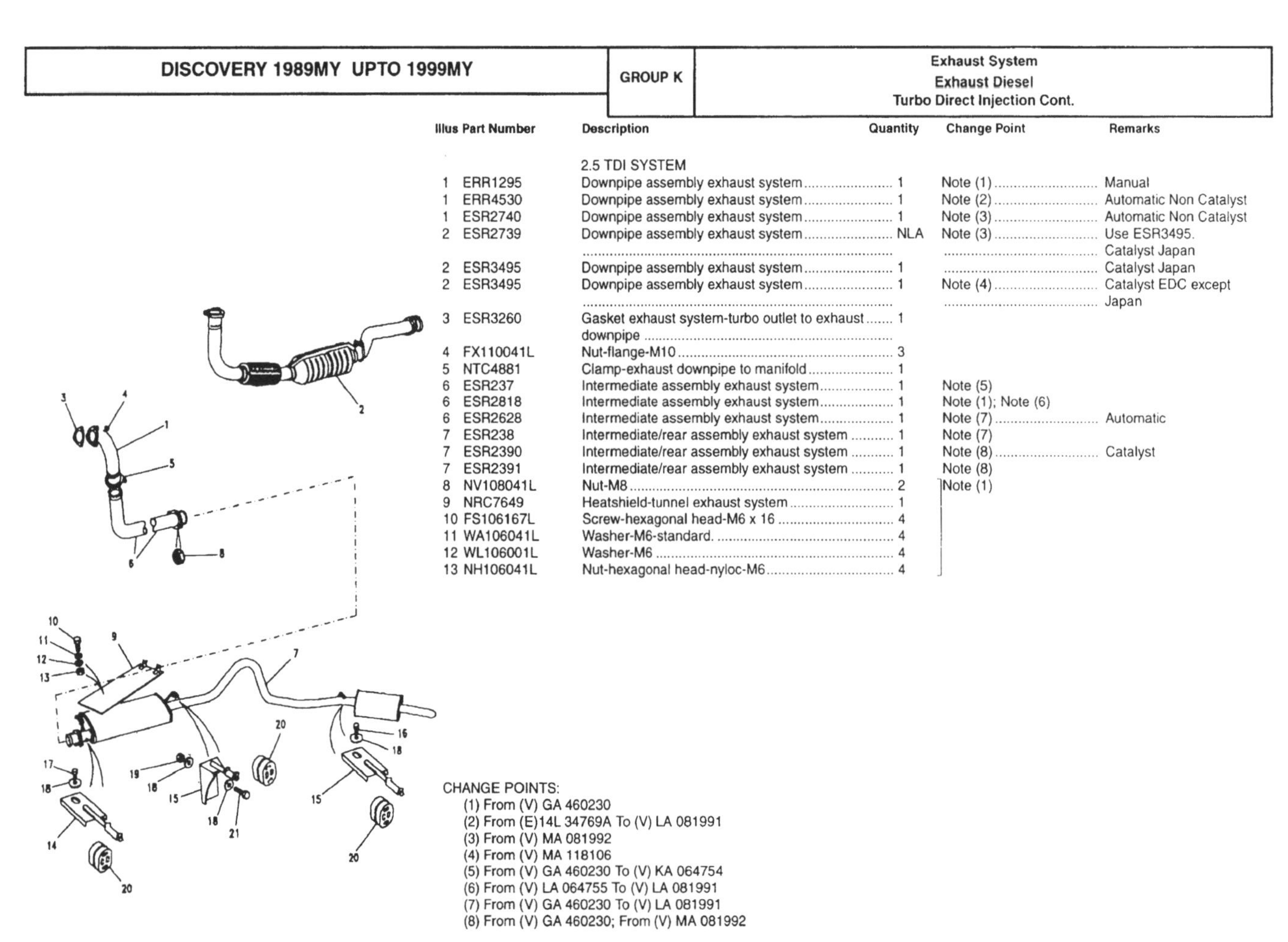

Illus	Part Number	Description	Quantity	Change Point	Remarks
		2.5 TDI SYSTEM			
1	ERR1295	Downpipe assembly exhaust system	1	Note (1)	Manual
1	ERR4530	Downpipe assembly exhaust system	1	Note (2)	Automatic Non Catalyst
1	ESR2740	Downpipe assembly exhaust system	1	Note (3)	Automatic Non Catalyst
2	ESR2739	Downpipe assembly exhaust system	NLA	Note (3)	Use ESR3495. Catalyst Japan
2	ESR3495	Downpipe assembly exhaust system	1		Catalyst Japan
2	ESR3495	Downpipe assembly exhaust system	1	Note (4)	Catalyst EDC except Japan
3	ESR3260	Gasket exhaust system-turbo outlet to exhaust downpipe	1		
4	FX110041L	Nut-flange-M10	3		
5	NTC4881	Clamp-exhaust downpipe to manifold	1		
6	ESR237	Intermediate assembly exhaust system	1	Note (5)	
6	ESR2818	Intermediate assembly exhaust system	1	Note (1); Note (6)	
6	ESR2628	Intermediate assembly exhaust system	1	Note (7)	Automatic
7	ESR238	Intermediate/rear assembly exhaust system	1	Note (7)	
7	ESR2390	Intermediate/rear assembly exhaust system	1	Note (8)	Catalyst
7	ESR2391	Intermediate/rear assembly exhaust system	1	Note (8)	
8	NV108041L	Nut-M8	2	Note (1)	
9	NRC7649	Heatshield-tunnel exhaust system	1		
10	FS106167L	Screw-hexagonal head-M6 x 16	4		
11	WA106041L	Washer-M6-standard.	4		
12	WL106001L	Washer-M6	4		
13	NH106041L	Nut-hexagonal head-nyloc-M6	4		

CHANGE POINTS:
(1) From (V) GA 460230
(2) From (E)14L 34769A To (V) LA 081991
(3) From (V) MA 081992
(4) From (V) MA 118106
(5) From (V) GA 460230 To (V) KA 064754
(6) From (V) LA 064755 To (V) LA 081991
(7) From (V) GA 460230 To (V) LA 081991
(8) From (V) GA 460230; From (V) MA 081992

K 14

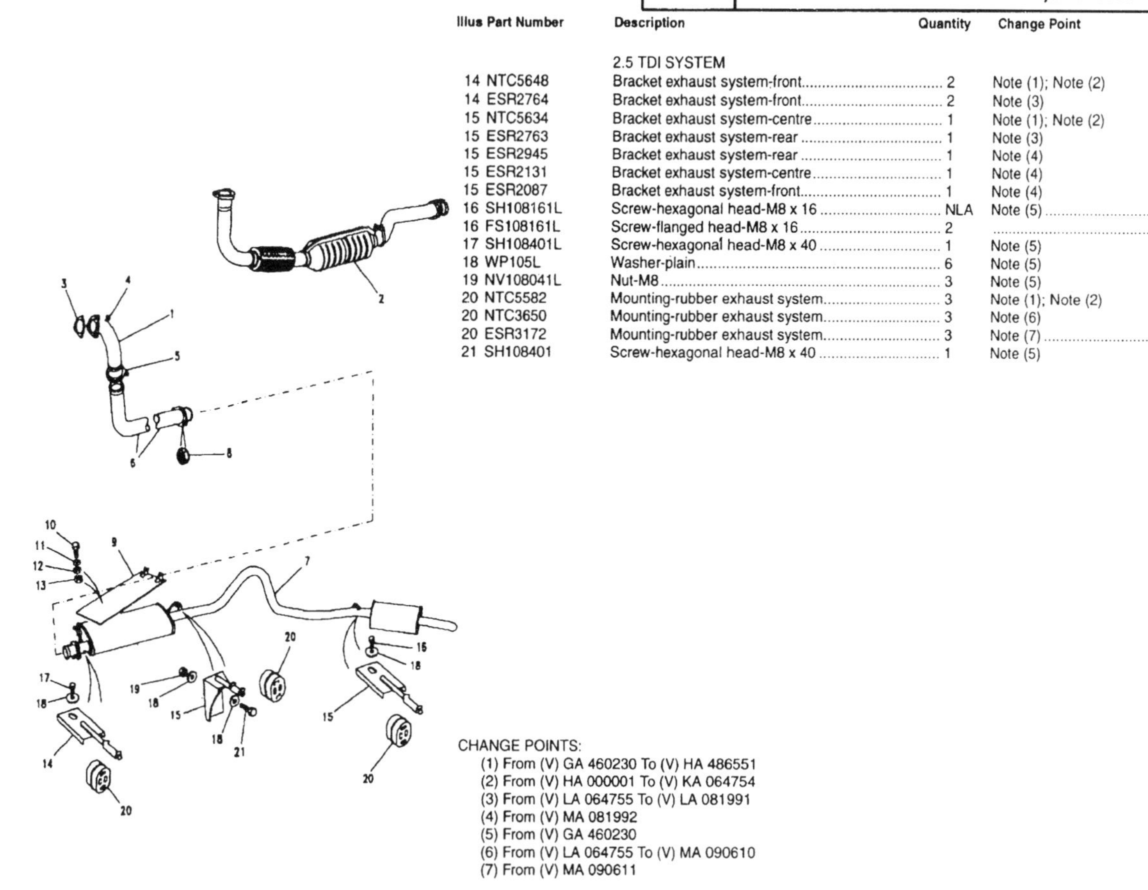

Illus	Part Number	Description	Quantity	Change Point	Remarks
		2.5 TDI SYSTEM			
14	NTC5648	Bracket exhaust system-front	2	Note (1); Note (2)	
14	ESR2764	Bracket exhaust system-front	2	Note (3)	
15	NTC5634	Bracket exhaust system-centre	1	Note (1); Note (2)	
15	ESR2763	Bracket exhaust system-rear	1	Note (3)	
15	ESR2945	Bracket exhaust system-rear	1	Note (4)	
15	ESR2131	Bracket exhaust system-centre	1	Note (4)	
15	ESR2087	Bracket exhaust system-front	1	Note (4)	
16	SH108161L	Screw-hexagonal head-M8 x 16	NLA	Note (5)	Use FS108161L.
16	FS108161L	Screw-flanged head-M8 x 16	2		2500 cc TDi
17	SH108401L	Screw-hexagonal head-M8 x 40	1	Note (5)	
18	WP105L	Washer-plain	6	Note (5)	
19	NV108041L	Nut-M8	3	Note (5)	
20	NTC5582	Mounting-rubber exhaust system	3	Note (1); Note (2)	
20	NTC3650	Mounting-rubber exhaust system	3	Note (6)	
20	ESR3172	Mounting-rubber exhaust system	3	Note (7)	2500 cc TDi
21	SH108401	Screw-hexagonal head-M8 x 40	1	Note (5)	

CHANGE POINTS:

(1) From (V) GA 460230 To (V) HA 486551
(2) From (V) HA 000001 To (V) KA 064754
(3) From (V) LA 064755 To (V) LA 081991
(4) From (V) MA 081992
(5) From (V) GA 460230
(6) From (V) LA 064755 To (V) MA 090610
(7) From (V) MA 090611

Notes

Air Conditioning .... L 38
  Ancillary Components Rear Air-Con (Cont) .... L 49
  Condenser Assembly .... L 47
  Control Assembly Rear .... L 50
  Evaporator and Blower Motor .... L 45
  Front Air Con-Assembly .... L 39
  Front Air Con-Blower Motor .... L 38
  Front Air Con-Control Assembly .... L 48
  Hose Layout-LHD .... L 55
  Hose Layout-RHD .... L 53
  Hoses Rear Under Bonnet .... L 51
  Hoses-Rear In Car .... L 52
  Rear Air Con-Ducts and Vents .... L 43
Cooling System .... L 1
  Cooling Unit & Fan Cowl - Diesel .... L 2
  Cooling Unit & Fan Cowl - Petrol .... L 1
  E.G.R. Hoses .... L 16
  Expansion Tank-Tdi & EFI .... L 12
  Oil Cooler Hoses-300TDi .... L 18
  Radiator Hoses-Petrol .... L 19
  Radiator Seals .... L 11
Heating System .... L 20
  Ancillary Components-Blower Assembly .... L 24

Ancillary Components-Front Air-Con .... L 25
Ancillary Components-Rear Air-Con .... L 26
Ancillary Components-Under Bonnet .... L 23
Blower Motor Assembly .... L 27
Control Assembly Front .... L 28
Ducting Front .... L 33
Ducting Rear .... L 34
Heater Assembly .... L 20
Hoses & Pipes .... L 29
Vacuum Tank and Hoses .... L 36

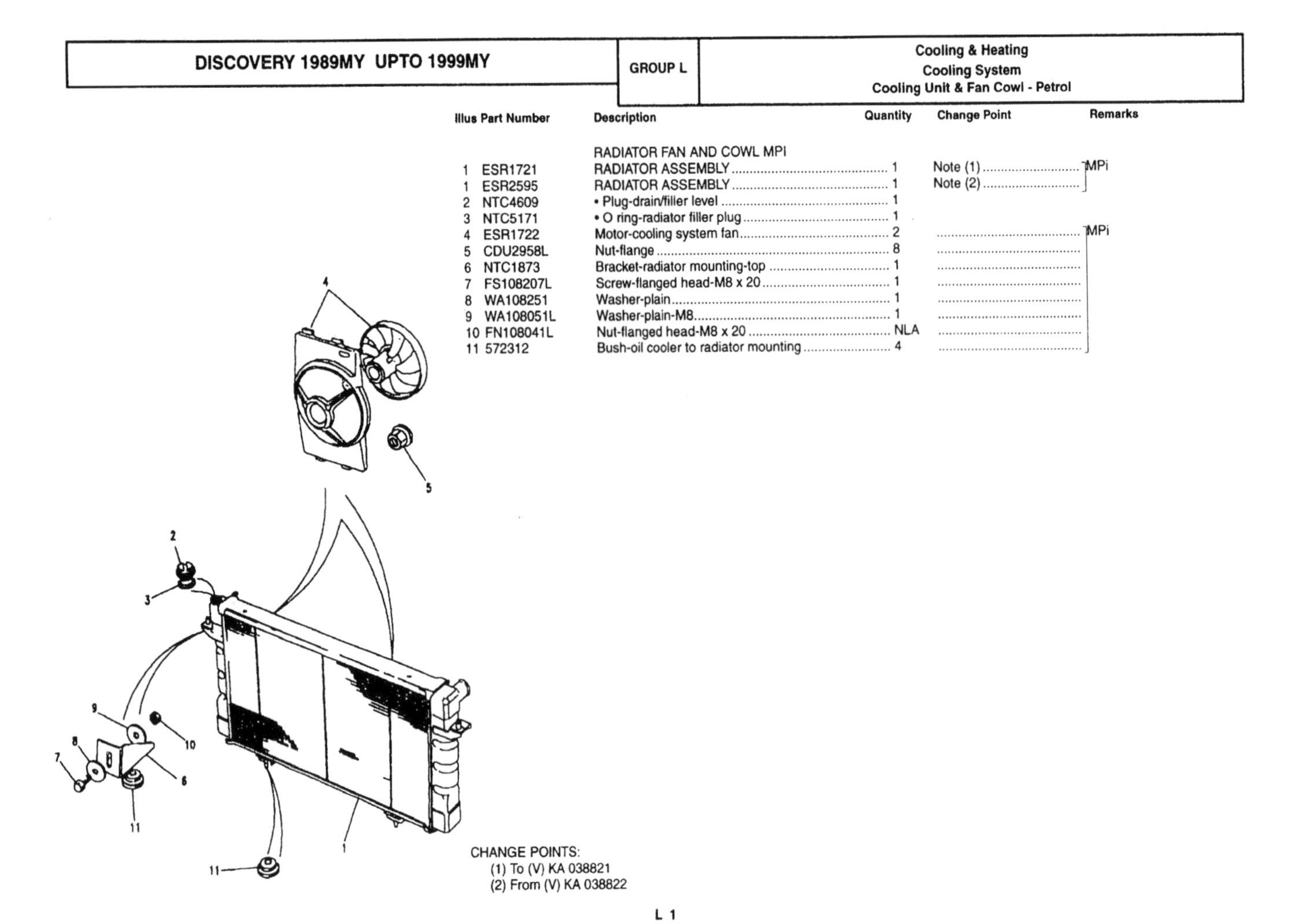

Illus	Part Number	Description	Quantity	Change Point	Remarks
		RADIATOR FAN AND COWL MPI			
1	ESR1721	RADIATOR ASSEMBLY	1	Note (1)	MPi
1	ESR2595	RADIATOR ASSEMBLY	1	Note (2)	
2	NTC4609	• Plug-drain/filler level	1		
3	NTC5171	• O ring-radiator filler plug	1		
4	ESR1722	Motor-cooling system fan	2		MPi
5	CDU2958L	Nut-flange	8		
6	NTC1873	Bracket-radiator mounting-top	1		
7	FS108207L	Screw-flanged head-M8 x 20	1		
8	WA108251	Washer-plain	1		
9	WA108051L	Washer-plain-M8	1		
10	FN108041L	Nut-flanged head-M8 x 20	NLA		
11	572312	Bush-oil cooler to radiator mounting	4		

CHANGE POINTS:
(1) To (V) KA 038821
(2) From (V) KA 038822

L 1

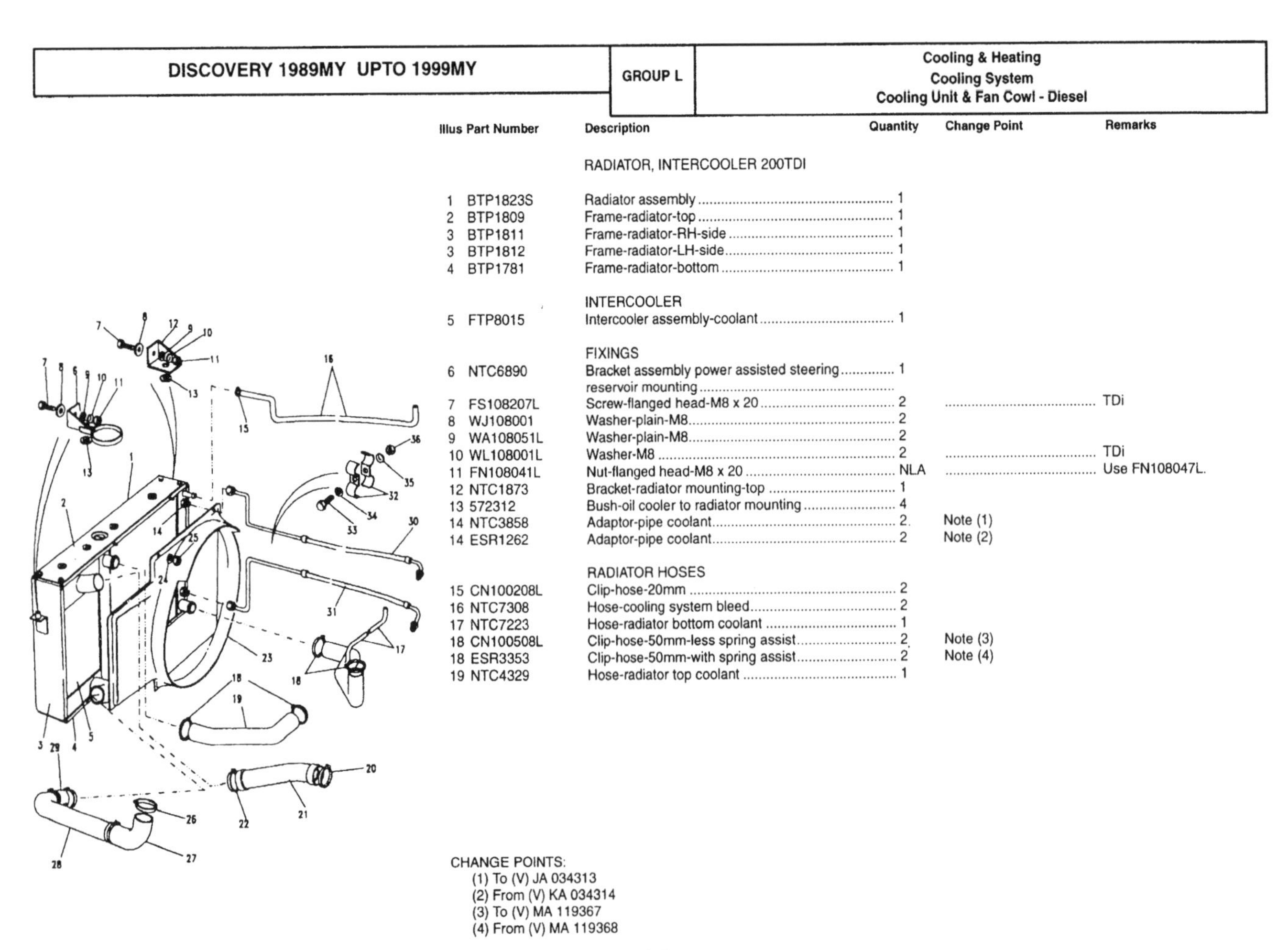

Illus	Part Number	Description	Quantity	Change Point	Remarks
		RADIATOR, INTERCOOLER 200TDI			
1	BTP1823S	Radiator assembly	1		
2	BTP1809	Frame-radiator-top	1		
3	BTP1811	Frame-radiator-RH-side	1		
3	BTP1812	Frame-radiator-LH-side	1		
4	BTP1781	Frame-radiator-bottom	1		
		INTERCOOLER			
5	FTP8015	Intercooler assembly-coolant	1		
		FIXINGS			
6	NTC6890	Bracket assembly power assisted steering reservoir mounting	1		
7	FS108207L	Screw-flanged head-M8 x 20	2		TDi
8	WJ108001	Washer-plain-M8	2		
9	WA108051L	Washer-plain-M8	2		
10	WL108001L	Washer-M8	2		TDi
11	FN108041L	Nut-flanged head-M8 x 20	NLA		Use FN108047L.
12	NTC1873	Bracket-radiator mounting-top	1		
13	572312	Bush-oil cooler to radiator mounting	4		
14	NTC3858	Adaptor-pipe coolant	2	Note (1)	
14	ESR1262	Adaptor-pipe coolant	2	Note (2)	
		RADIATOR HOSES			
15	CN100208L	Clip-hose-20mm	2		
16	NTC7308	Hose-cooling system bleed	2		
17	NTC7223	Hose-radiator bottom coolant	1		
18	CN100508L	Clip-hose-50mm-less spring assist	2	Note (3)	
18	ESR3353	Clip-hose-50mm-with spring assist	2	Note (4)	
19	NTC4329	Hose-radiator top coolant	1		

CHANGE POINTS:
(1) To (V) JA 034313
(2) From (V) KA 034314
(3) To (V) MA 119367
(4) From (V) MA 119368

L 2

DISCOVERY 1989MY UPTO 1999MY

GROUP L

Cooling & Heating
Cooling System
Cooling Unit & Fan Cowl - Diesel

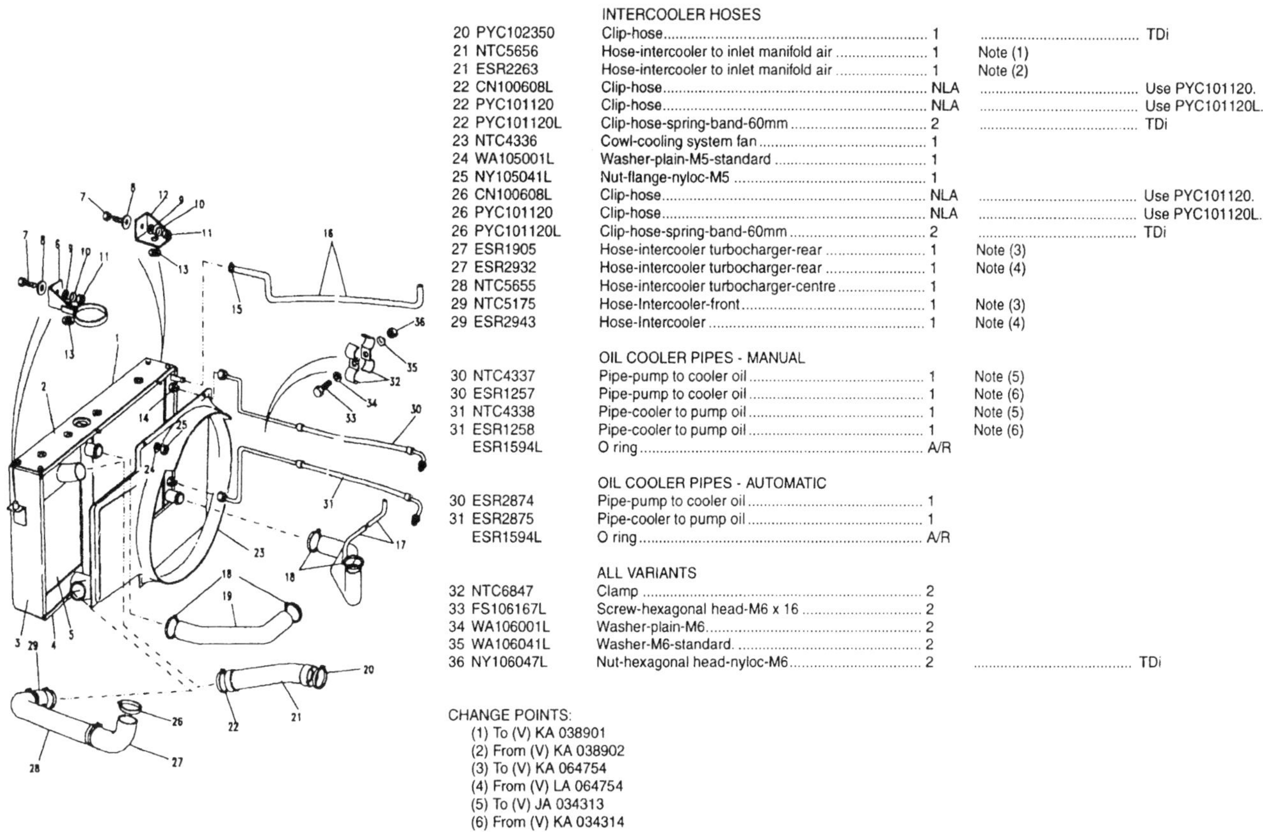

Illus	Part Number	Description	Quantity	Change Point	Remarks
		INTERCOOLER HOSES			
20	PYC102350	Clip-hose	1		TDi
21	NTC5656	Hose-intercooler to inlet manifold air	1	Note (1)	
21	ESR2263	Hose-intercooler to inlet manifold air	1	Note (2)	
22	CN100608L	Clip-hose	NLA		Use PYC101120.
22	PYC101120	Clip-hose	NLA		Use PYC101120L.
22	PYC101120L	Clip-hose-spring-band-60mm	2		TDi
23	NTC4336	Cowl-cooling system fan	1		
24	WA105001L	Washer-plain-M5-standard	1		
25	NY105041L	Nut-flange-nyloc-M5	1		
26	CN100608L	Clip-hose	NLA		Use PYC101120.
26	PYC101120	Clip-hose	NLA		Use PYC101120L.
26	PYC101120L	Clip-hose-spring-band-60mm	2		TDi
27	ESR1905	Hose-intercooler turbocharger-rear	1	Note (3)	
27	ESR2932	Hose-intercooler turbocharger-rear	1	Note (4)	
28	NTC5655	Hose-intercooler turbocharger-centre	1		
29	NTC5175	Hose-Intercooler-front	1	Note (3)	
29	ESR2943	Hose-Intercooler	1	Note (4)	
		OIL COOLER PIPES - MANUAL			
30	NTC4337	Pipe-pump to cooler oil	1	Note (5)	
30	ESR1257	Pipe-pump to cooler oil	1	Note (6)	
31	NTC4338	Pipe-cooler to pump oil	1	Note (5)	
31	ESR1258	Pipe-cooler to pump oil	1	Note (6)	
	ESR1594L	O ring	A/R		
		OIL COOLER PIPES - AUTOMATIC			
30	ESR2874	Pipe-pump to cooler oil	1		
31	ESR2875	Pipe-cooler to pump oil	1		
	ESR1594L	O ring	A/R		
		ALL VARIANTS			
32	NTC6847	Clamp	2		
33	FS106167L	Screw-hexagonal head-M6 x 16	2		
34	WA106001L	Washer-plain-M6	2		
35	WA106041L	Washer-M6-standard.	2		
36	NY106047L	Nut-hexagonal head-nyloc-M6	2		TDi

CHANGE POINTS:
(1) To (V) KA 038901
(2) From (V) KA 038902
(3) To (V) KA 064754
(4) From (V) LA 064754
(5) To (V) JA 034313
(6) From (V) KA 034314

L 3

DISCOVERY 1989MY UPTO 1999MY

GROUP L

Cooling & Heating
Cooling System
Cooling Unit, Fan Cowl & Hoses-V8-To (V) LA081991

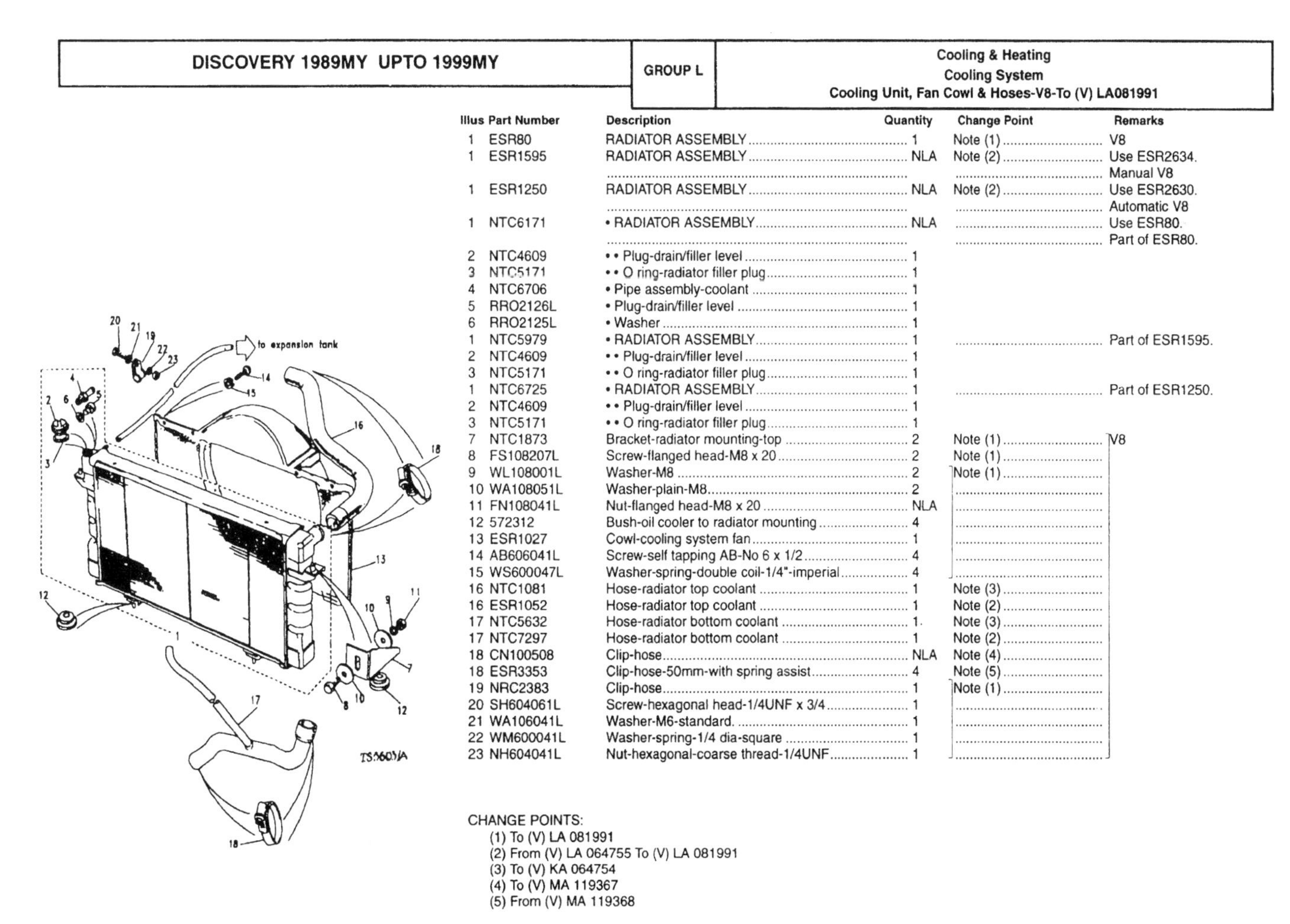

Illus	Part Number	Description	Quantity	Change Point	Remarks
1	ESR80	RADIATOR ASSEMBLY	1	Note (1)	V8
1	ESR1595	RADIATOR ASSEMBLY	NLA	Note (2)	Use ESR2634. Manual V8
1	ESR1250	RADIATOR ASSEMBLY	NLA	Note (2)	Use ESR2630. Automatic V8
1	NTC6171	• RADIATOR ASSEMBLY	NLA		Use ESR80. Part of ESR80.
2	NTC4609	• • Plug-drain/filler level	1		
3	NTC5171	• • O ring-radiator filler plug	1		
4	NTC6706	• Pipe assembly-coolant	1		
5	RRO2126L	• Plug-drain/filler level	1		
6	RRO2125L	• Washer	1		
1	NTC5979	• RADIATOR ASSEMBLY	1		Part of ESR1595.
2	NTC4609	• • Plug-drain/filler level	1		
3	NTC5171	• • O ring-radiator filler plug	1		
1	NTC6725	• RADIATOR ASSEMBLY	1		Part of ESR1250.
2	NTC4609	• • Plug-drain/filler level	1		
3	NTC5171	• • O ring-radiator filler plug	1		
7	NTC1873	Bracket-radiator mounting-top	2	Note (1)	V8
8	FS108207L	Screw-flanged head-M8 x 20	2	Note (1)	
9	WL108001L	Washer-M8	2	Note (1)	
10	WA108051L	Washer-plain-M8	2		
11	FN108041L	Nut-flanged head-M8 x 20	NLA		
12	572312	Bush-oil cooler to radiator mounting	4		
13	ESR1027	Cowl-cooling system fan	1		
14	AB606041L	Screw-self tapping AB-No 6 x 1/2	4		
15	WS600047L	Washer-spring-double coil-1/4"-imperial	4		
16	NTC1081	Hose-radiator top coolant	1	Note (3)	
16	ESR1052	Hose-radiator top coolant	1	Note (2)	
17	NTC5632	Hose-radiator bottom coolant	1	Note (3)	
17	NTC7297	Hose-radiator bottom coolant	1	Note (2)	
18	CN100508	Clip-hose	NLA	Note (4)	
18	ESR3353	Clip-hose-50mm-with spring assist	4	Note (5)	
19	NRC2383	Clip-hose	1	Note (1)	
20	SH604061L	Screw-hexagonal head-1/4UNF x 3/4	1		
21	WA106041L	Washer-M6-standard.	1		
22	WM600041L	Washer-spring-1/4 dia-square	1		
23	NH604041L	Nut-hexagonal-coarse thread-1/4UNF	1		

CHANGE POINTS:
(1) To (V) LA 081991
(2) From (V) LA 064755 To (V) LA 081991
(3) To (V) KA 064754
(4) To (V) MA 119367
(5) From (V) MA 119368

L 4

Cooling Unit, Fan Cowl & Hoses-V8-From (V) MA081992

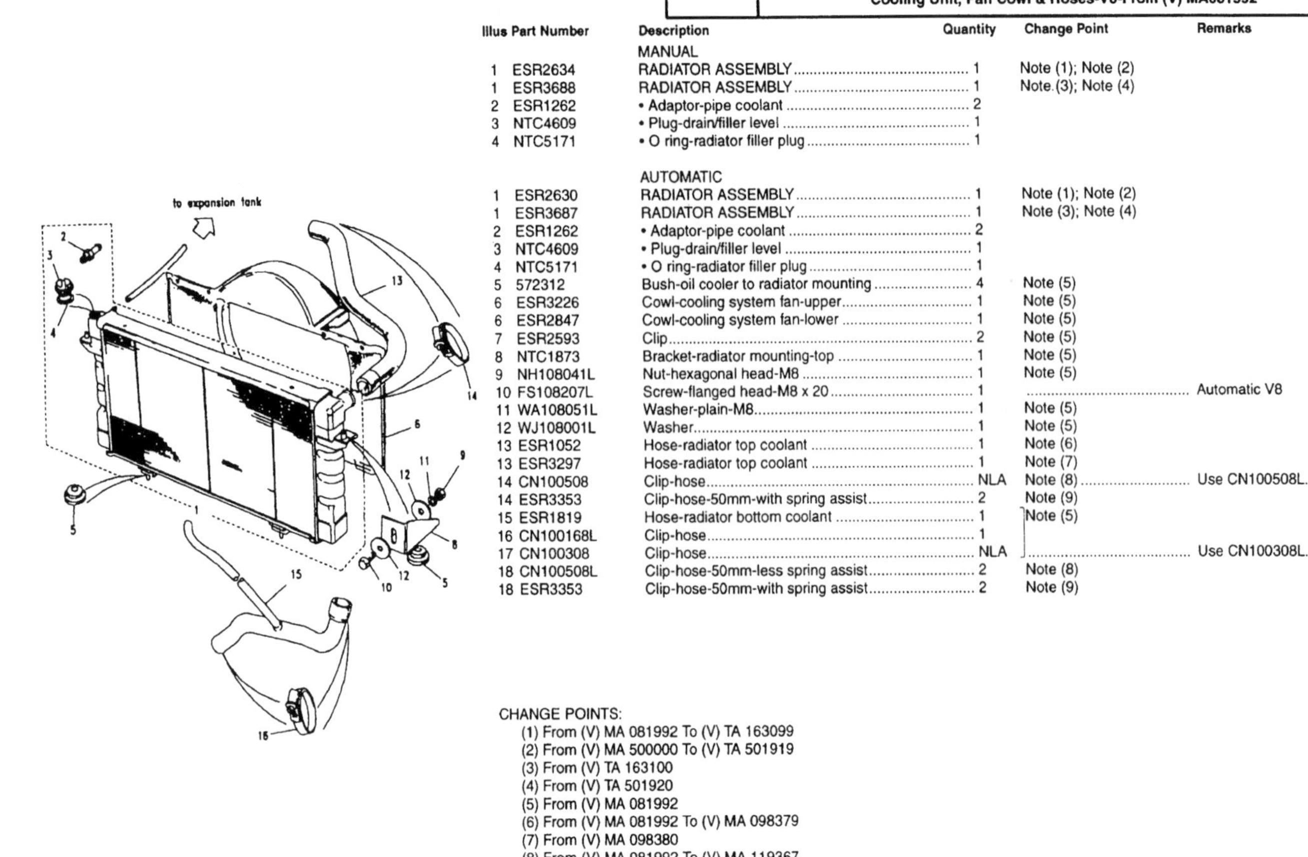

Illus	Part Number	Description	Quantity	Change Point	Remarks
		MANUAL			
1	ESR2634	RADIATOR ASSEMBLY	1	Note (1); Note (2)	
1	ESR3688	RADIATOR ASSEMBLY	1	Note (3); Note (4)	
2	ESR1262	• Adaptor-pipe coolant	2		
3	NTC4609	• Plug-drain/filler level	1		
4	NTC5171	• O ring-radiator filler plug	1		
		AUTOMATIC			
1	ESR2630	RADIATOR ASSEMBLY	1	Note (1); Note (2)	
1	ESR3687	RADIATOR ASSEMBLY	1	Note (3); Note (4)	
2	ESR1262	• Adaptor-pipe coolant	2		
3	NTC4609	• Plug-drain/filler level	1		
4	NTC5171	• O ring-radiator filler plug	1		
5	572312	Bush-oil cooler to radiator mounting	4	Note (5)	
6	ESR3226	Cowl-cooling system fan-upper	1	Note (5)	
6	ESR2847	Cowl-cooling system fan-lower	1	Note (5)	
7	ESR2593	Clip	2	Note (5)	
8	NTC1873	Bracket-radiator mounting-top	1	Note (5)	
9	NH108041L	Nut-hexagonal head-M8	1	Note (5)	
10	FS108207L	Screw-flanged head-M8 x 20	1		Automatic V8
11	WA108051L	Washer-plain-M8	1	Note (5)	
12	WJ108001L	Washer	1	Note (5)	
13	ESR1052	Hose-radiator top coolant	1	Note (6)	
13	ESR3297	Hose-radiator top coolant	1	Note (7)	
14	CN100508	Clip-hose	NLA	Note (8)	Use CN100508L.
14	ESR3353	Clip-hose-50mm-with spring assist	2	Note (9)	
15	ESR1819	Hose-radiator bottom coolant	1	Note (5)	
16	CN100168L	Clip-hose	1		
17	CN100308	Clip-hose	NLA		Use CN100308L.
18	CN100508L	Clip-hose-50mm-less spring assist	2	Note (8)	
18	ESR3353	Clip-hose-50mm-with spring assist	2	Note (9)	

CHANGE POINTS:
(1) From (V) MA 081992 To (V) TA 163099
(2) From (V) MA 500000 To (V) TA 501919
(3) From (V) TA 163100
(4) From (V) TA 501920
(5) From (V) MA 081992
(6) From (V) MA 081992 To (V) MA 098379
(7) From (V) MA 098380
(8) From (V) MA 081992 To (V) MA 119367
(9) From (V) MA 119368

Radiator Hoses - V8

Illus	Part Number	Description	Quantity	Change Point	Remarks
1	ESR2814	Pipe-engine to oil cooler	1	Note (1)	
1	ESR3693	Pipe-engine to oil cooler	1	Note (2); Note (3)	
2	ESR1594L	O ring-oil cooler hose	2	Note (2)	
3	ESR3098	O ring-small	1		
4	ESR2813	Pipe-oil cooler to engine	1	Note (1)	
4	ESR3692	Pipe-oil cooler to engine	NLA	Note (2); Note (3)	Use PBP101000.
4	PBP101000	Pipe-oil cooler to engine	1		
5	ESR1594L	O ring-engine oil cooler hose	2	Note (2)	
6	NTC6847	Clamp	2	Note (2)	
7	SH106161	Screw-hexagonal head-M6 x 16	NLA	Note (2)	Use FS106167L.
7	FS106167L	Screw-hexagonal head-M6 x 16	1		
8	WA106001	Washer-plain-M6	2	Note (2)	
9	NY106041	Nut-flange-nyloc-M6	1	Note (2)	

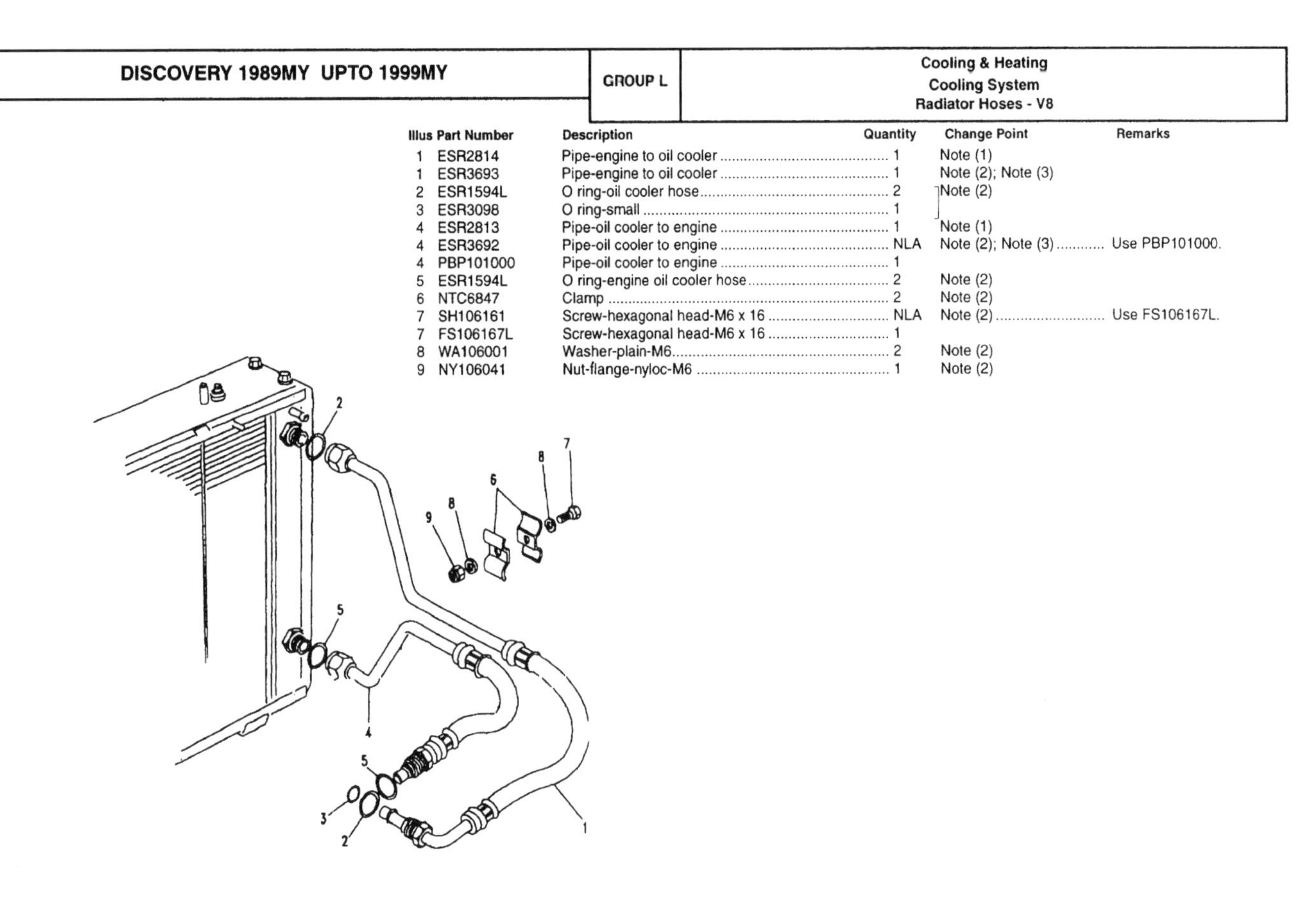

CHANGE POINTS:
(1) From (V) MA 081992 To (E)36D 52437C
(2) From (V) MA 081992
(3) From (E)36D 52438C

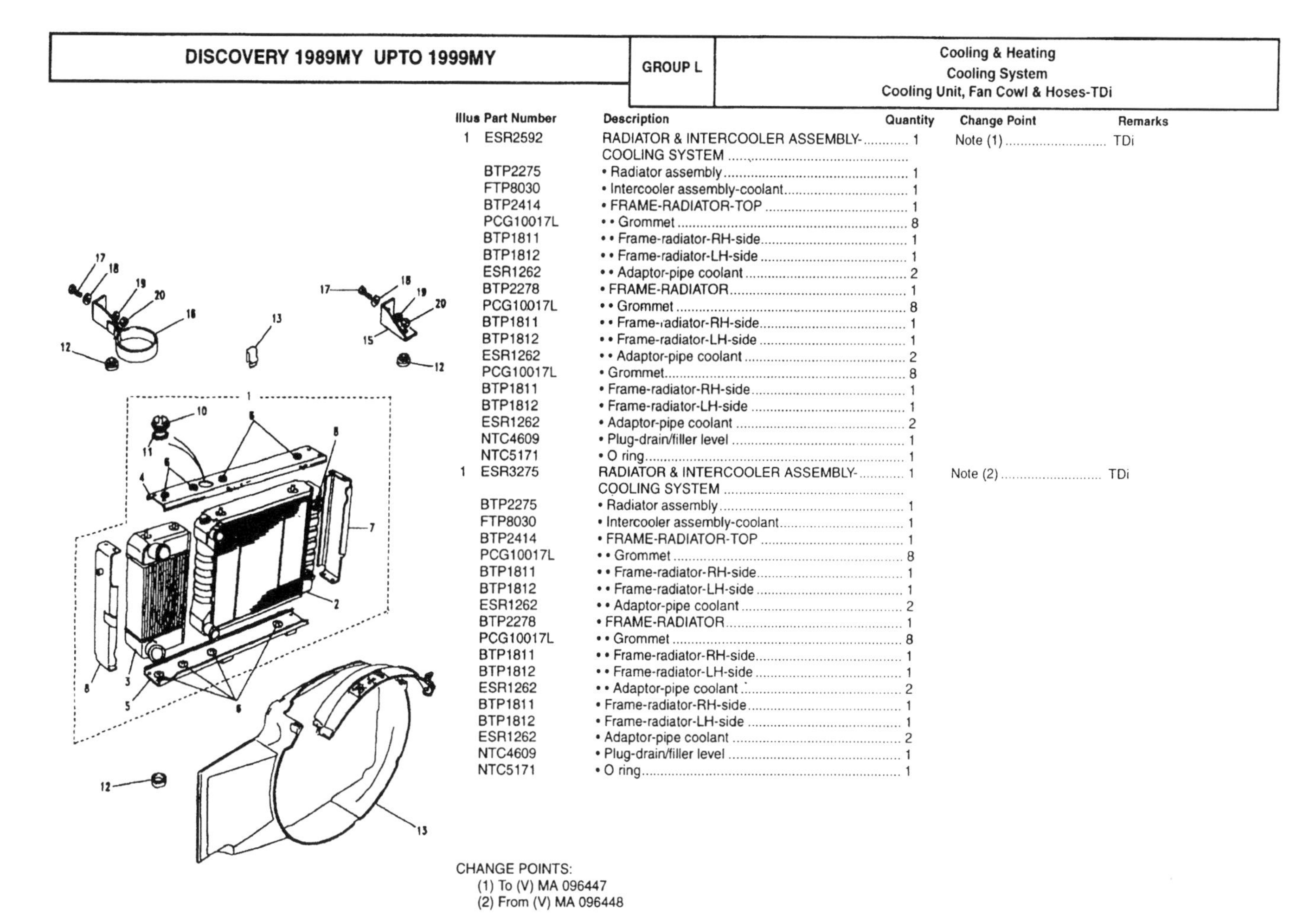

Illus	Part Number	Description	Quantity	Change Point	Remarks
1	ESR2592	RADIATOR & INTERCOOLER ASSEMBLY-COOLING SYSTEM	1	Note (1)	TDi
	BTP2275	• Radiator assembly	1		
	FTP8030	• Intercooler assembly-coolant	1		
	BTP2414	• FRAME-RADIATOR-TOP	1		
	PCG10017L	• • Grommet	8		
	BTP1811	• • Frame-radiator-RH-side	1		
	BTP1812	• • Frame-radiator-LH-side	1		
	ESR1262	• • Adaptor-pipe coolant	2		
	BTP2278	• FRAME-RADIATOR	1		
	PCG10017L	• • Grommet	8		
	BTP1811	• • Frame-radiator-RH-side	1		
	BTP1812	• • Frame-radiator-LH-side	1		
	ESR1262	• • Adaptor-pipe coolant	2		
	PCG10017L	• Grommet	8		
	BTP1811	• Frame-radiator-RH-side	1		
	BTP1812	• Frame-radiator-LH-side	1		
	ESR1262	• Adaptor-pipe coolant	2		
	NTC4609	• Plug-drain/filler level	1		
	NTC5171	• O ring	1		
1	ESR3275	RADIATOR & INTERCOOLER ASSEMBLY-COOLING SYSTEM	1	Note (2)	TDi
	BTP2275	• Radiator assembly	1		
	FTP8030	• Intercooler assembly-coolant	1		
	BTP2414	• FRAME-RADIATOR-TOP	1		
	PCG10017L	• • Grommet	8		
	BTP1811	• • Frame-radiator-RH-side	1		
	BTP1812	• • Frame-radiator-LH-side	1		
	ESR1262	• • Adaptor-pipe coolant	2		
	BTP2278	• FRAME-RADIATOR	1		
	PCG10017L	• • Grommet	8		
	BTP1811	• • Frame-radiator-RH-side	1		
	BTP1812	• • Frame-radiator-LH-side	1		
	ESR1262	• • Adaptor-pipe coolant	2		
	BTP1811	• Frame-radiator-RH-side	1		
	BTP1812	• Frame-radiator-LH-side	1		
	ESR1262	• Adaptor-pipe coolant	2		
	NTC4609	• Plug-drain/filler level	1		
	NTC5171	• O ring	1		

CHANGE POINTS:
(1) To (V) MA 096447
(2) From (V) MA 096448

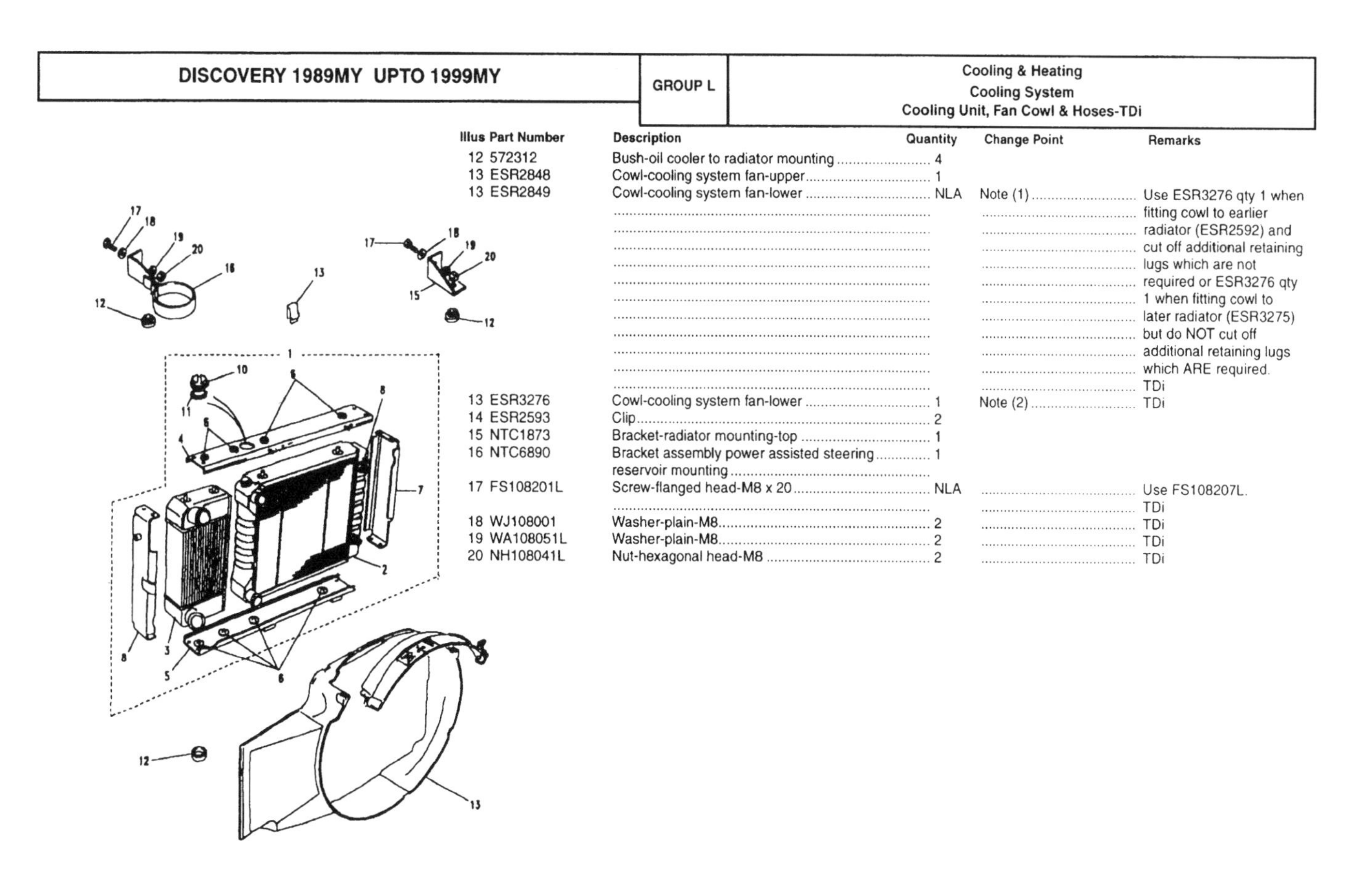

DISCOVERY 1989MY UPTO 1999MY | GROUP L | Cooling & Heating
Cooling System
Cooling Unit, Fan Cowl & Hoses-TDi

Illus	Part Number	Description	Quantity	Change Point	Remarks
12	572312	Bush-oil cooler to radiator mounting	4		
13	ESR2848	Cowl-cooling system fan-upper	1		
13	ESR2849	Cowl-cooling system fan-lower	NLA	Note (1)	Use ESR3276 qty 1 when fitting cowl to earlier radiator (ESR2592) and cut off additional retaining lugs which are not required or ESR3276 qty 1 when fitting cowl to later radiator (ESR3275) but do NOT cut off additional retaining lugs which ARE required. TDi
13	ESR3276	Cowl-cooling system fan-lower	1	Note (2)	TDi
14	ESR2593	Clip	2		
15	NTC1873	Bracket-radiator mounting-top	1		
16	NTC6890	Bracket assembly power assisted steering reservoir mounting	1		
17	FS108201L	Screw-flanged head-M8 x 20	NLA		Use FS108207L. TDi
18	WJ108001	Washer-plain-M8	2		TDi
19	WA108051L	Washer-plain-M8	2		TDi
20	NH108041L	Nut-hexagonal head-M8	2		TDi

CHANGE POINTS:
(1) To (V) MA 096447
(2) From (V) MA 096448

Illus	Part Number	Description	Quantity	Change Point	Remarks
		RADIATOR, INTERCOOLER 300TDI			
		RADIATOR HOSES			
1	ESR2491	Hose-radiator top coolant	1		
2	CN100508L	Clip-hose-50mm-less spring assist	1		
3	CN100608	Clip-hose	NLA		Use CN100608L.
3	CN100608L	Clip-hose	NLA		Use PYC101120.
3	PYC101120	Clip-hose	NLA		Use PYC101120L.
3	PYC101120L	Clip-hose-spring-band-60mm	1		
4	ESR3147	Hose-radiator bottom coolant	1	Note (1)	
4	ESR3296	Hose-radiator bottom coolant	1	Note (2)	
		CLIP-HOSE			
5	CN100308L	30mm	2		
5	ESR3607	26mm	2		
6	CN100508L	50mm-less spring assist	2	Note (3)	
6	ESR3353	50mm-with spring assist	2	Note (4)	
7	ESR2662	Clip-hose	1		
8	ESR1800	Clip-hose	2		
	594594	Tie-cable-White.-4.6 x 385mm-inside serated	2		

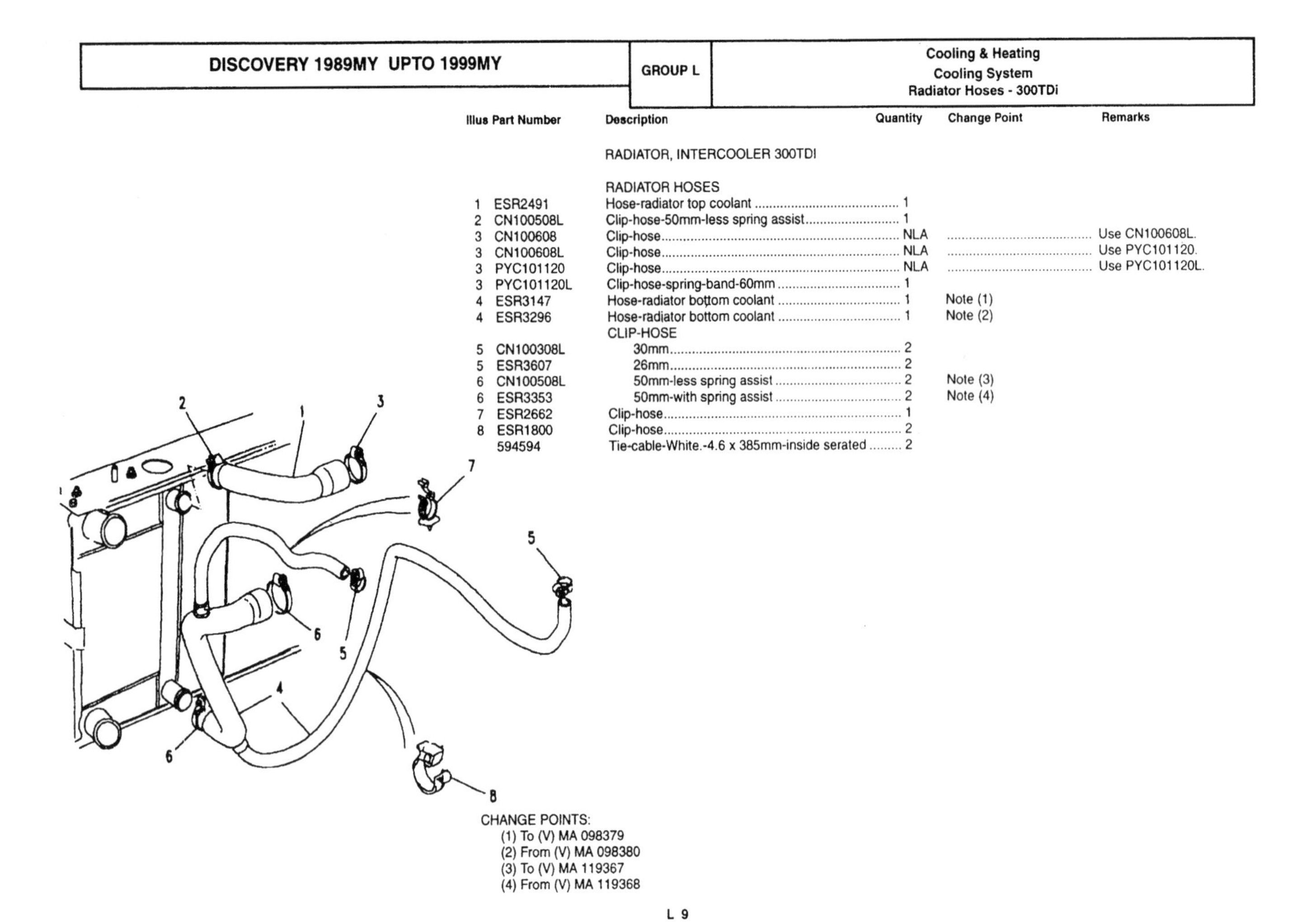

CHANGE POINTS:
(1) To (V) MA 098379
(2) From (V) MA 098380
(3) To (V) MA 119367
(4) From (V) MA 119368

Illus	Part Number	Description	Quantity	Change Point	Remarks
		RADIATOR, INTERCOOLER 300TDI			
		INTERCOOLER HOSES			
1	ESR3025	Hose-intercooler to inlet manifold air	1		
2	STC128	Clip-hose	2		TDi
3	ESR2910	HOSE ASSEMBLY-INTERCOOLER TURBOCHARGER	1		
4	ESR2911	• Pipe assembly-intercooler turbocharger	1		
5	ESR2912	• Hose-turbocharger to pipe	1		
6	ESR2913	• Hose-Intercooler	1		
7	CN100608	• Clip-hose	NLA		Use CN100608L.
7	CN100608L	Clip-hose	NLA		Use PYC101120.
7	PYC101120	Clip-hose	NLA		Use PYC101120L.
7	PYC101120L	Clip-hose-spring-band-60mm			TDi
8	CN100608	Clip-hose	NLA		Use CN100608L.
8	CN100608L	Clip-hose	NLA		Use PYC101120.
8	PYC101120	Clip-hose	NLA		Use PYC101120L.
8	PYC101120L	Clip-hose-spring-band-60mm	2		TDi

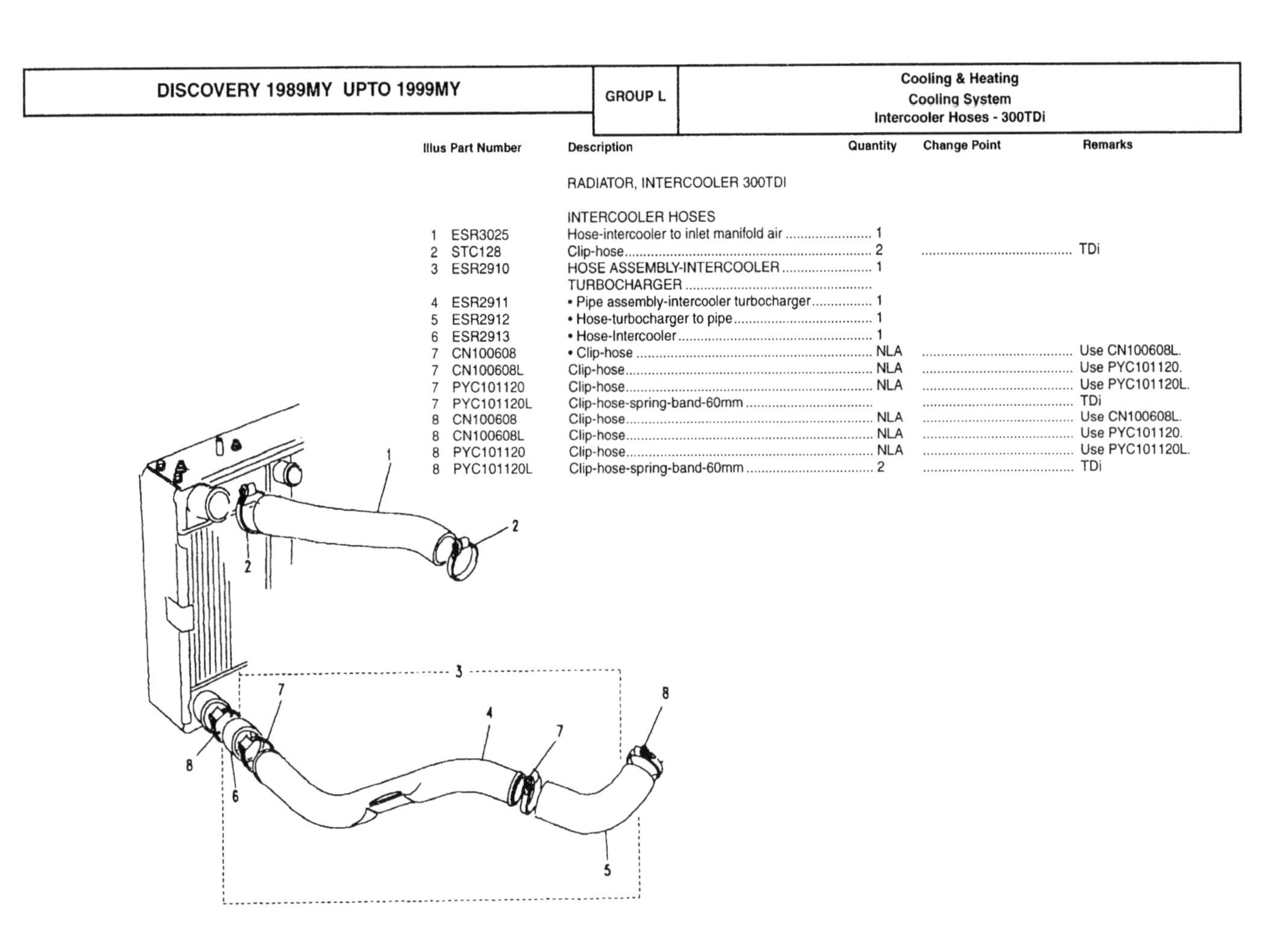

DISCOVERY 1989MY UPTO 1999MY

GROUP L

Cooling & Heating
Cooling System
Radiator Seals

Illus	Part Number	Description	Quantity	Change Point	Remarks
1	MUC6324	Seal-body to cooling pack air-top-bottom	2		
2	MRC218	Bracket-radiator upper mounting	1		
3	MRC206	Plate-radiator bottom	1		
4	MUC4177	Seal-body to cooling pack air-side	2		
		BRACKET-RADIATOR MOUNTING			
5	MUC6320	RH	1		Except air con
6	MUC6319	LH	1		Except air con
5	MUC6382	RH	1	Note (1)	Air con
5	BTR8266	RH	1	Note (2)	Air con
6	MUC6383	LH	1		Air con
7	RA608126L	Rivet-pop-steel	19		

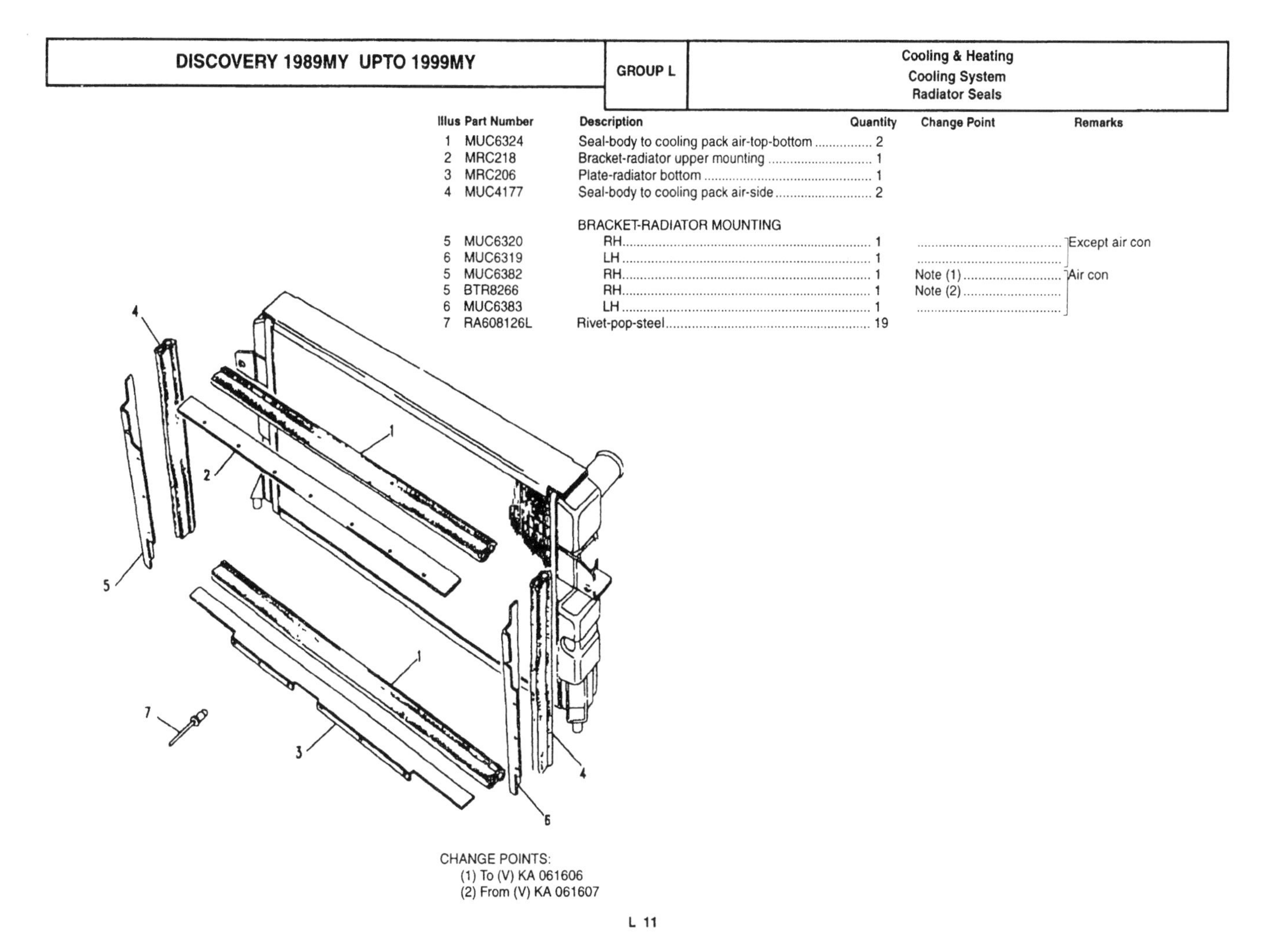

CHANGE POINTS:
(1) To (V) KA 061606
(2) From (V) KA 061607

L 11

DISCOVERY 1989MY UPTO 1999MY

GROUP L

Cooling & Heating
Cooling System
Expansion Tank-Tdi & EFI

Illus	Part Number	Description	Quantity	Change Point	Remarks
1	ESR63	Tank-radiator expansion	1		
2	NTC7161	Cap-expansion tank pressure-15psi	1		
3	NTC6874	Bracket-expansion tank	1		
		SCREW			
4	FS106161L	flanged head-M6 x 16	NLA		Use FS106167L.
4	FS106167L	hexagonal head-M6 x 16	4		V8
4	FS106167L	hexagonal head-M6 x 16	4		TDi
4	FS106161L	flanged head-M6 x 16	NLA		Use FS106167L.
5	WA106041L	Washer-M6-standard.	4		
6	NH106041L	Nut-hexagonal head-nyloc-M6	4		
7	NTC5844	Bracket-expansion tank	1		
8	NTC7308	Hose-cooling system bleed	1		TDi
9	CN100208L	Clip-hose-20mm	2		TDi
8	NTC7308	Hose-cooling system bleed	1	Note (1)	V8
9	CN100208L	Clip-hose-20mm	2	Note (1)	V8
8	ESR2344	Hose-cooling system bleed	1	From (V) MA 081992 V8	
9	CN100168L	Clip-hose	2	From (V) MA 081992 V8	
				Note (2)	

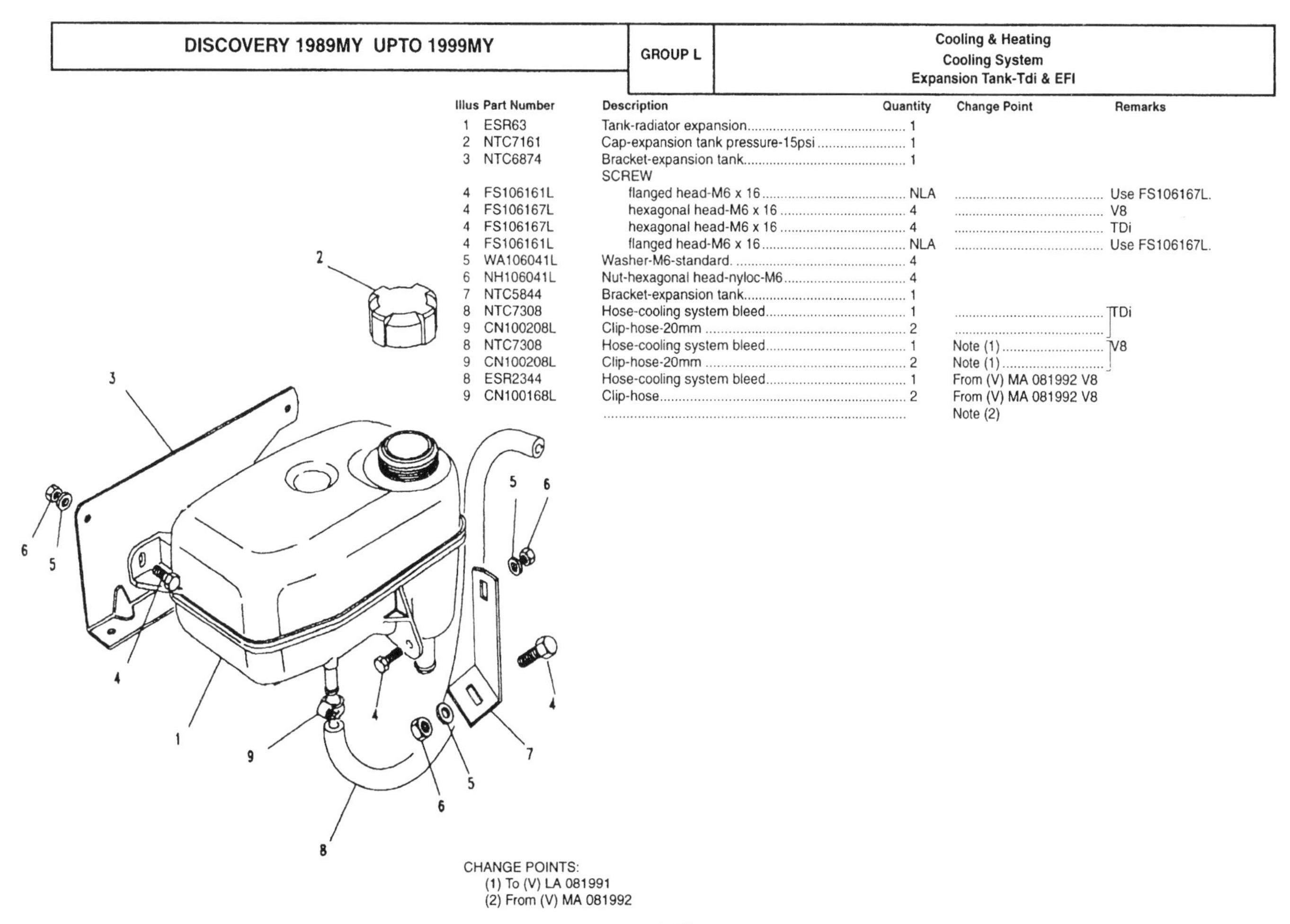

CHANGE POINTS:
(1) To (V) LA 081991
(2) From (V) MA 081992

L 12

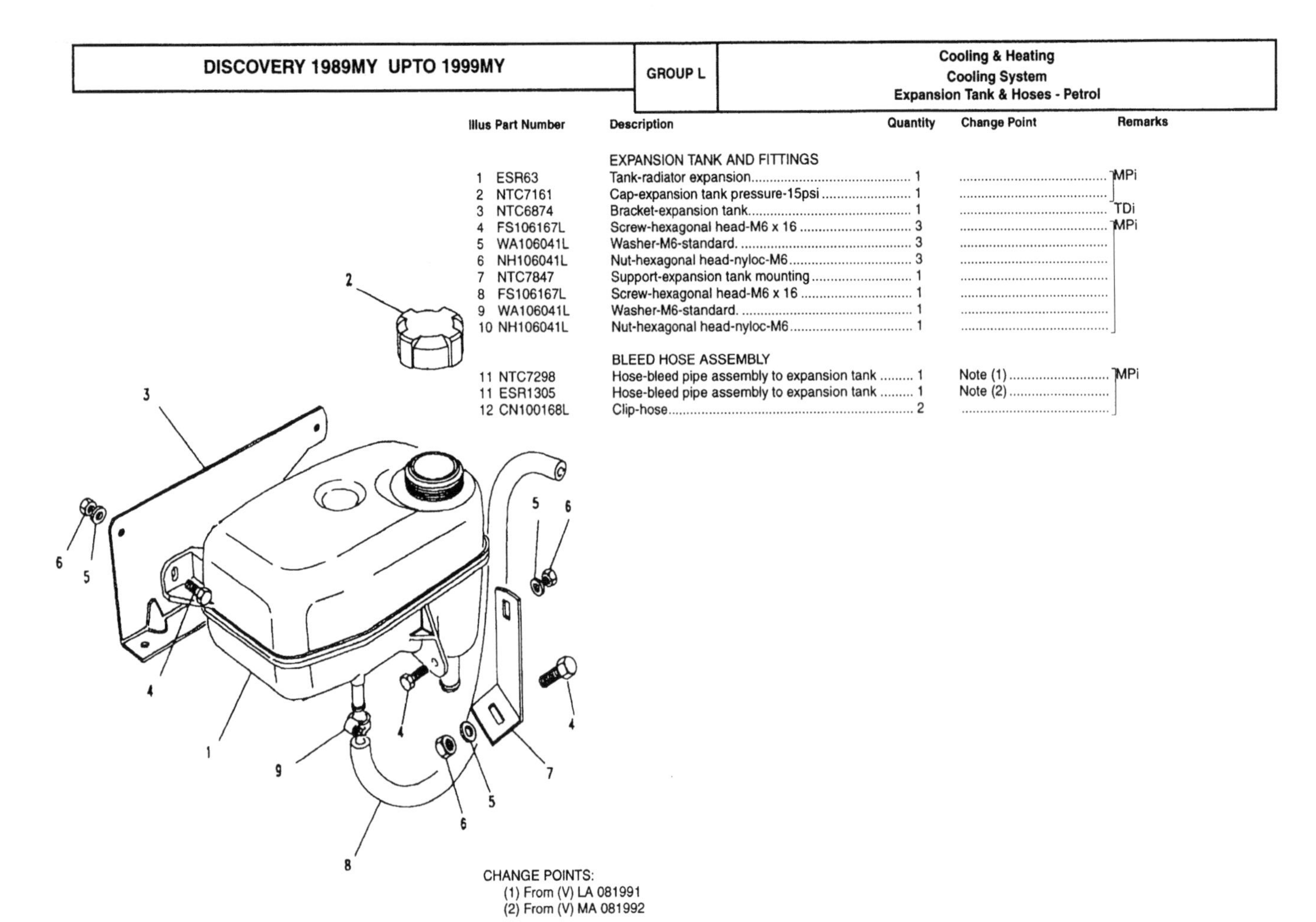

Illus	Part Number	Description	Quantity	Change Point	Remarks
		EXPANSION TANK AND FITTINGS			
1	ESR63	Tank-radiator expansion	1		MPi
2	NTC7161	Cap-expansion tank pressure-15psi	1		
3	NTC6874	Bracket-expansion tank	1		TDi
4	FS106167L	Screw-hexagonal head-M6 x 16	3		MPi
5	WA106041L	Washer-M6-standard.	3		
6	NH106041L	Nut-hexagonal head-nyloc-M6	3		
7	NTC7847	Support-expansion tank mounting	1		
8	FS106167L	Screw-hexagonal head-M6 x 16	1		
9	WA106041L	Washer-M6-standard.	1		
10	NH106041L	Nut-hexagonal head-nyloc-M6	1		
		BLEED HOSE ASSEMBLY			
11	NTC7298	Hose-bleed pipe assembly to expansion tank	1	Note (1)	MPi
11	ESR1305	Hose-bleed pipe assembly to expansion tank	1	Note (2)	
12	CN100168L	Clip-hose	2		

CHANGE POINTS:
(1) From (V) LA 081991
(2) From (V) MA 081992

Illus	Part Number	Description	Quantity	Change Point	Remarks
1	NTC7252	TANK-RADIATOR EXPANSION	1		V8 Twin Carburettor
2	565540	• Cap-expansion tank pressure	NLA		Use PCD100150.
2	PCD100150	Cap-expansion tank pressure			V8 Twin Carburettor
		BLEED HOSE ASSEMBLY			
3	NTC6858	Hose-radiator to expansion tank	1		V8 Twin Carburettor
4	CN100208L	Clip-hose-20mm	3		
		SUPPORT BRACKET & FITTINGS			
5	564724	Hose-expansion tank overflow	1		V8 Twin Carburettor
6	NTC7124	Bracket-expansion tank	1		
7	NRC5294	Tape-self adhesive-foam	2		
8	NRC5295	Tape-self adhesive-foam	2		
9	SH604071L	Screw-hexagonal head-1/4UNF x 7/8	1		
10	NY604041L	Nut-hexagonal-nyloc-1/4"	1		
11	SH108161L	Screw-hexagonal head-M8 x 16	NLA		Use FS108161L.
11	FS108161L	Screw-flanged head-M8 x 16	2		V8 Twin Carburettor
12	NV108041L	Nut-M8	2		
13	FS106167L	Screw-hexagonal head-M6 x 16	1		
14	NV106041L	Nut-M6	1		

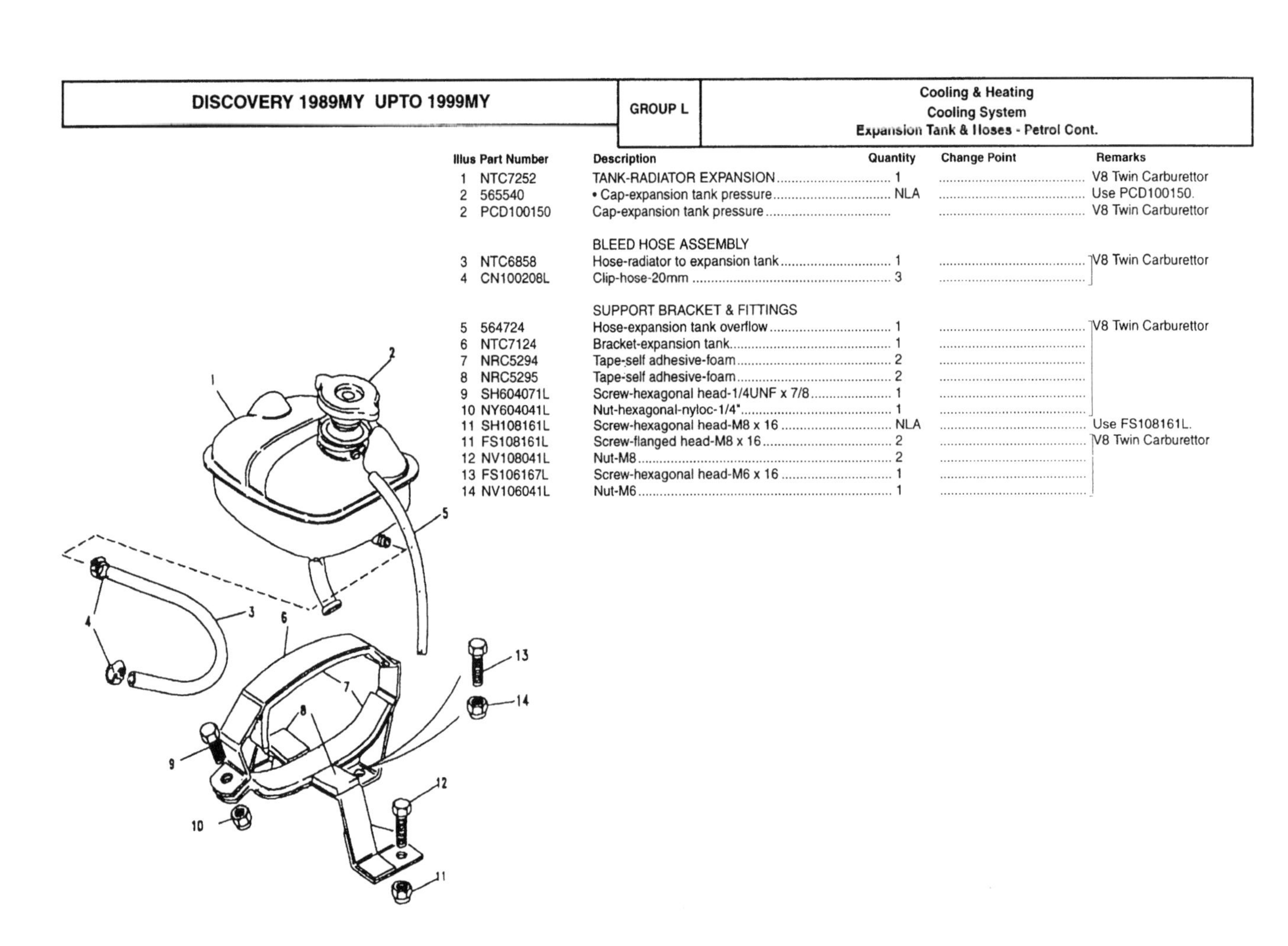

Illus	Part Number	Description	Quantity	Change Point	Remarks
		EXPANSION TANK - 300TDI			
		EXPANSION TANK & FITTINGS			
1	ESR63	Tank-radiator expansion	1		
2	NTC7161	Cap-expansion tank pressure-15psi	1		
3	NTC6874	Bracket-expansion tank	1		
4	FS106161L	Screw-flanged head-M6 x 16	NLA		Use FS106167L.
4	FS106167L	Screw-hexagonal head-M6 x 16	4		
5	WA106041L	Washer-M6-standard.	4		
6	NH106041L	Nut-hexagonal head-nyloc-M6	4		
7	NTC5844	Bracket-expansion tank	1		
8	ESR3436	HOSE-BLEED PIPE ASSEMBLY TO EXPANSION TANK	1		
9	ESR2348	• Pump-expansion tank ejector	1		
10	CN100168L	Clip-hose	3		

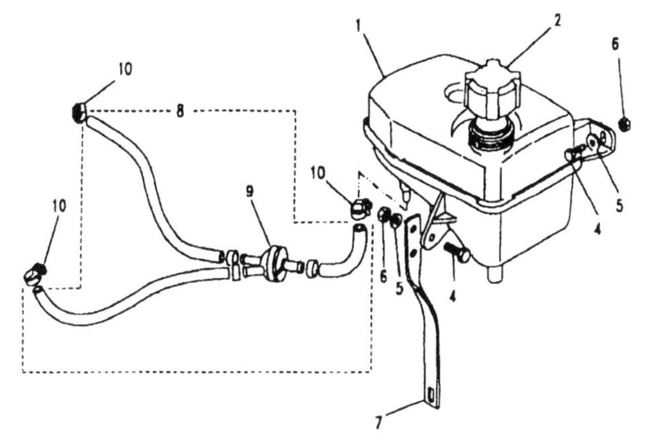

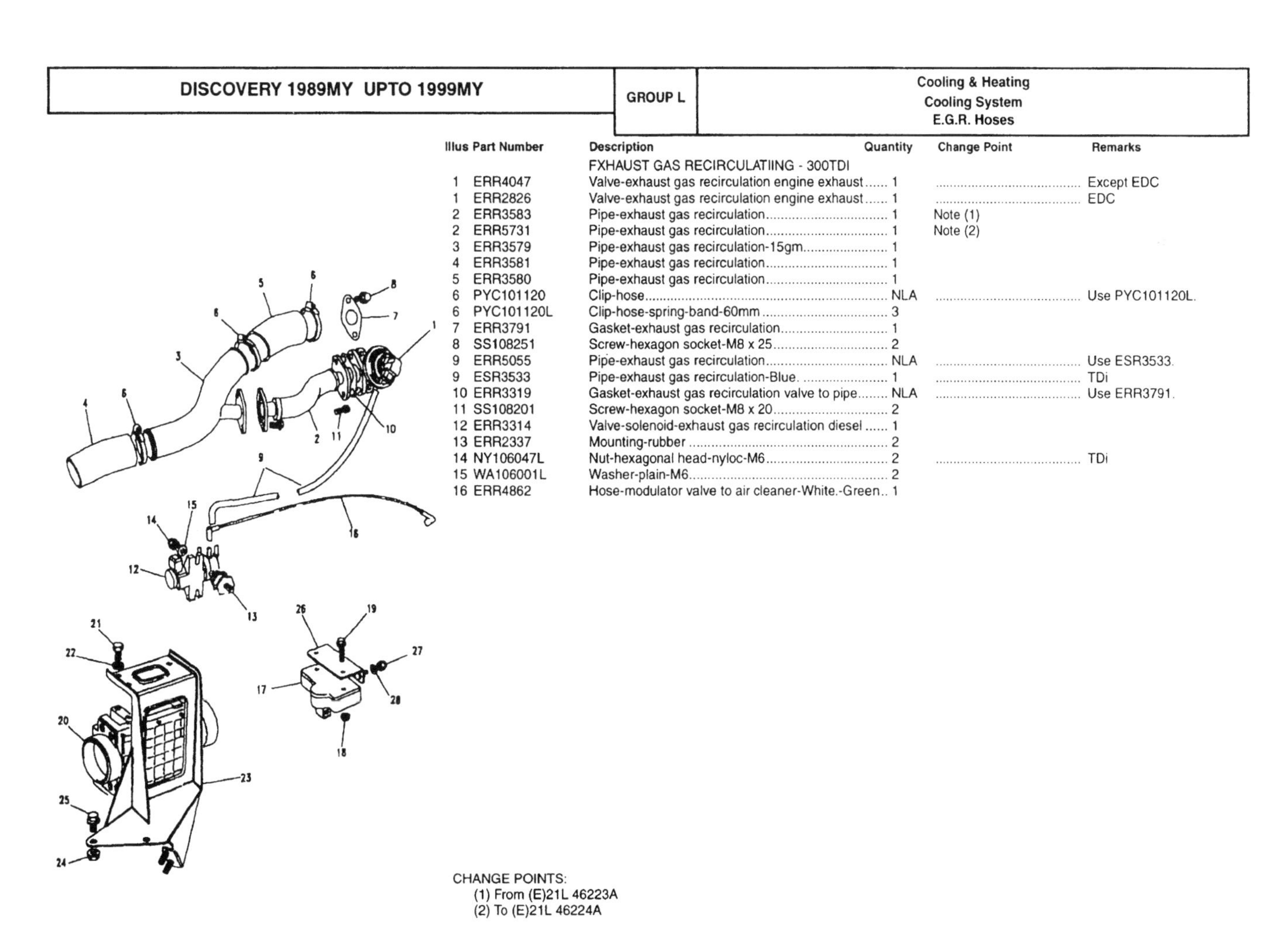

Illus	Part Number	Description	Quantity	Change Point	Remarks
		FXHAUST GAS RECIRCULATIING - 300TDI			
1	ERR4047	Valve-exhaust gas recirculation engine exhaust	1		Except EDC
1	ERR2826	Valve-exhaust gas recirculation engine exhaust	1		EDC
2	ERR3583	Pipe-exhaust gas recirculation	1	Note (1)	
2	ERR5731	Pipe-exhaust gas recirculation	1	Note (2)	
3	ERR3579	Pipe-exhaust gas recirculation-15gm	1		
4	ERR3581	Pipe-exhaust gas recirculation	1		
5	ERR3580	Pipe-exhaust gas recirculation	1		
6	PYC101120	Clip-hose	NLA		Use PYC101120L.
6	PYC101120L	Clip-hose-spring-band-60mm	3		
7	ERR3791	Gasket-exhaust gas recirculation	1		
8	SS108251	Screw-hexagon socket-M8 x 25	2		
9	ERR5055	Pipe-exhaust gas recirculation	NLA		Use ESR3533.
9	ESR3533	Pipe-exhaust gas recirculation-Blue.	1		TDi
10	ERR3319	Gasket-exhaust gas recirculation valve to pipe	NLA		Use ERR3791.
11	SS108201	Screw-hexagon socket-M8 x 20	2		
12	ERR3314	Valve-solenoid-exhaust gas recirculation diesel	1		
13	ERR2337	Mounting-rubber	2		
14	NY106047L	Nut-hexagonal head-nyloc-M6	2		TDi
15	WA106001L	Washer-plain-M6	2		
16	ERR4862	Hose-modulator valve to air cleaner-White.-Green.	1		

CHANGE POINTS:
(1) From (E)21L 46223A
(2) To (E)21L 46224A

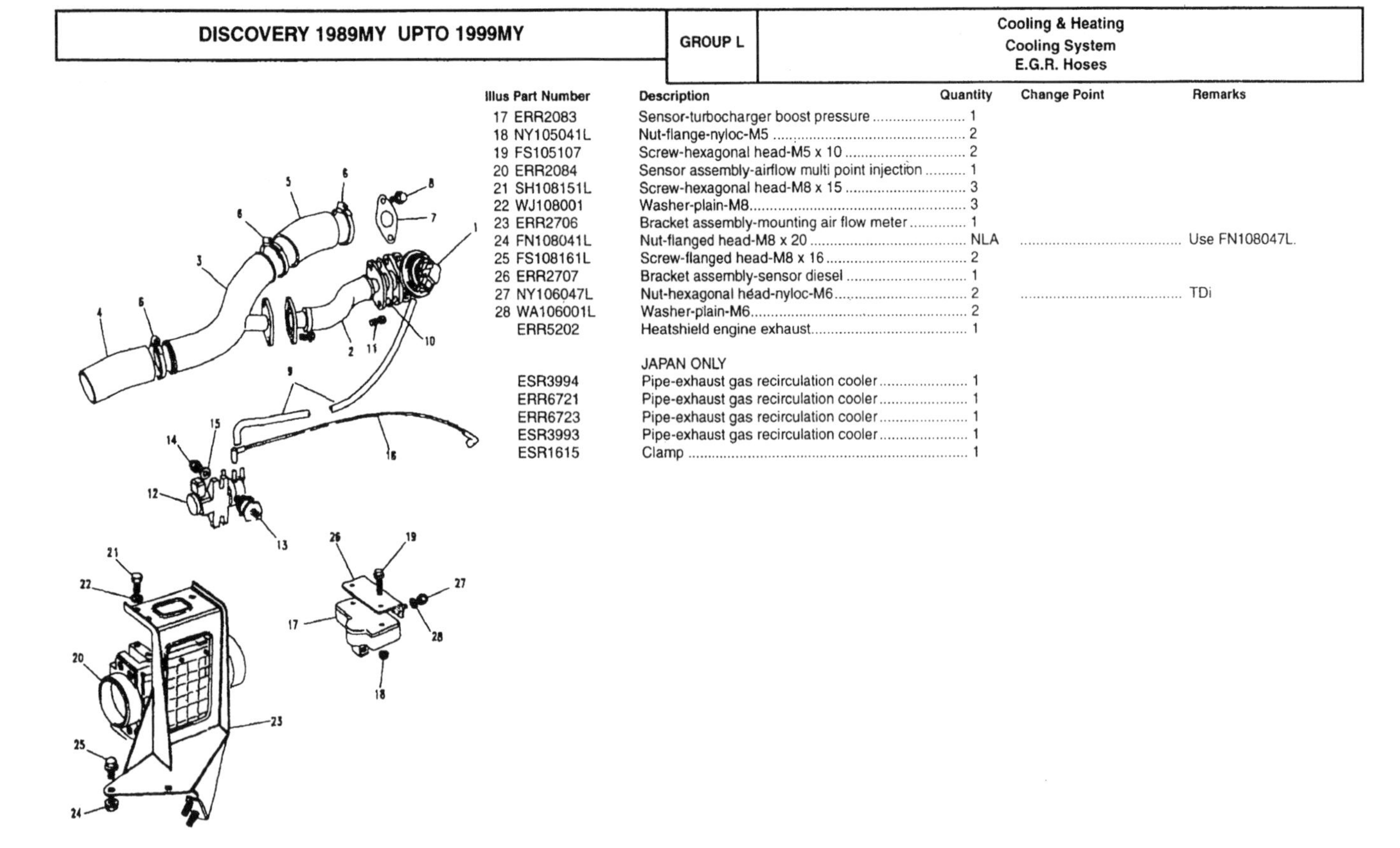

Illus	Part Number	Description	Quantity	Change Point	Remarks
17	ERR2083	Sensor-turbocharger boost pressure	1		
18	NY105041L	Nut-flange-nyloc-M5	2		
19	FS105107	Screw-hexagonal head-M5 x 10	2		
20	ERR2084	Sensor assembly-airflow multi point injection	1		
21	SH108151L	Screw-hexagonal head-M8 x 15	3		
22	WJ108001	Washer-plain-M8	3		
23	ERR2706	Bracket assembly-mounting air flow meter	1		
24	FN108041L	Nut-flanged head-M8 x 20	NLA		Use FN108047L.
25	FS108161L	Screw-flanged head-M8 x 16	2		
26	ERR2707	Bracket assembly-sensor diesel	1		
27	NY106047L	Nut-hexagonal head-nyloc-M6	2		TDi
28	WA106001L	Washer-plain-M6	2		
	ERR5202	Heatshield engine exhaust	1		
		JAPAN ONLY			
	ESR3994	Pipe-exhaust gas recirculation cooler	1		
	ERR6721	Pipe-exhaust gas recirculation cooler	1		
	ERR6723	Pipe-exhaust gas recirculation cooler	1		
	ESR3993	Pipe-exhaust gas recirculation cooler	1		
	ESR1615	Clamp	1		

Illus	Part Number	Description	Quantity	Change Point	Remarks
1	ESR1931	PIPE-OIL COOLER TO ENGINE	1		
2	ESR2844	PIPE-ENGINE TO OIL COOLER	1	Note (1); Note (2)	
2	ESR3691	PIPE-ENGINE TO OIL COOLER	1	Note (3); Note (4)	
	ESR1594L	• O ring	4		
3	NTC6847	Clamp	2		
4	SH106161	Screw-hexagonal head-M6 x 16	NLA		Use FS106167L.
4	FS106167L	Screw-hexagonal head-M6 x 16	1		
5	WA106041L	Washer-M6-standard.	2		
6	NY106041	Nut-flange-nyloc-M6	1		

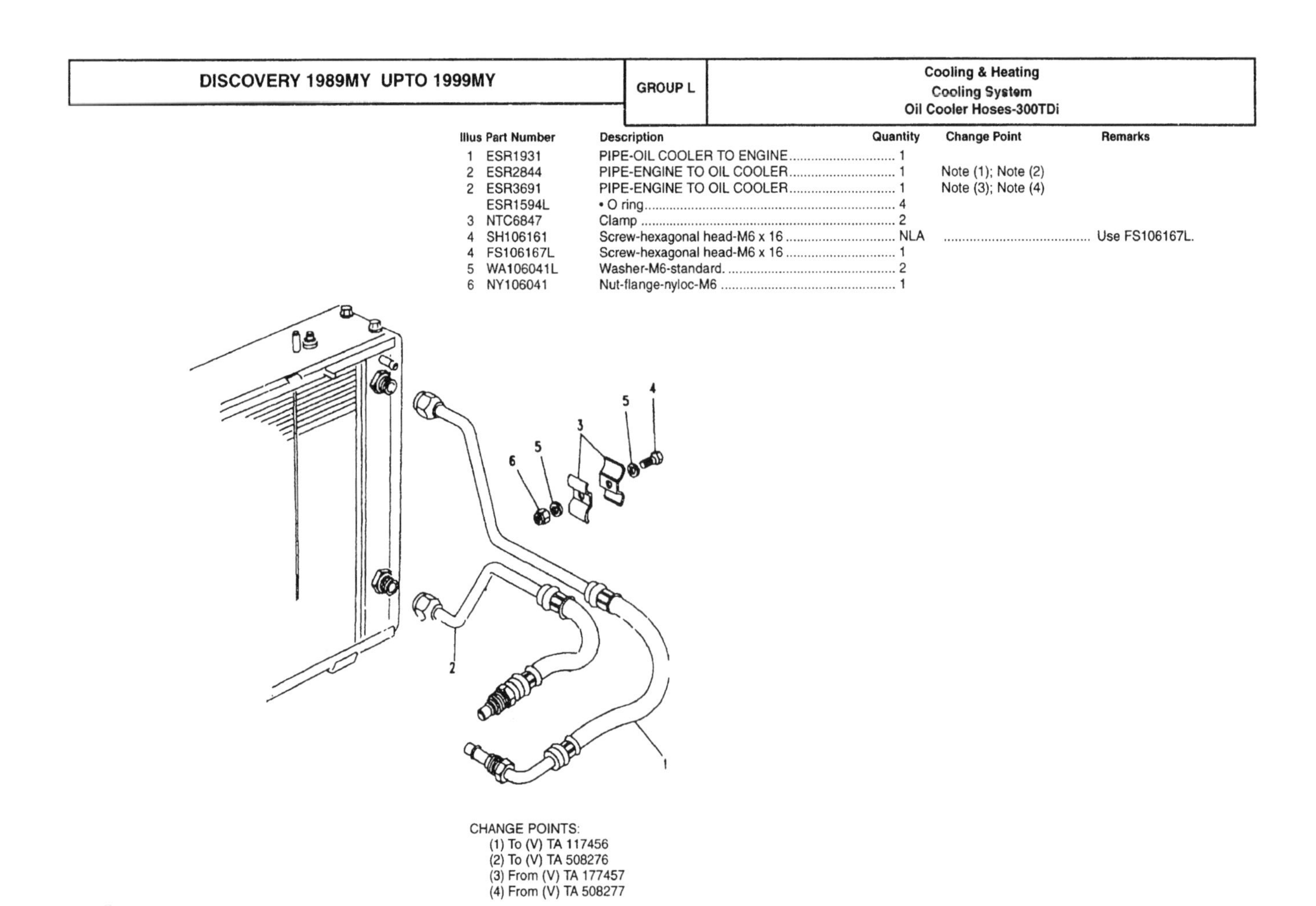

CHANGE POINTS:
(1) To (V) TA 117456
(2) To (V) TA 508276
(3) From (V) TA 177457
(4) From (V) TA 508277

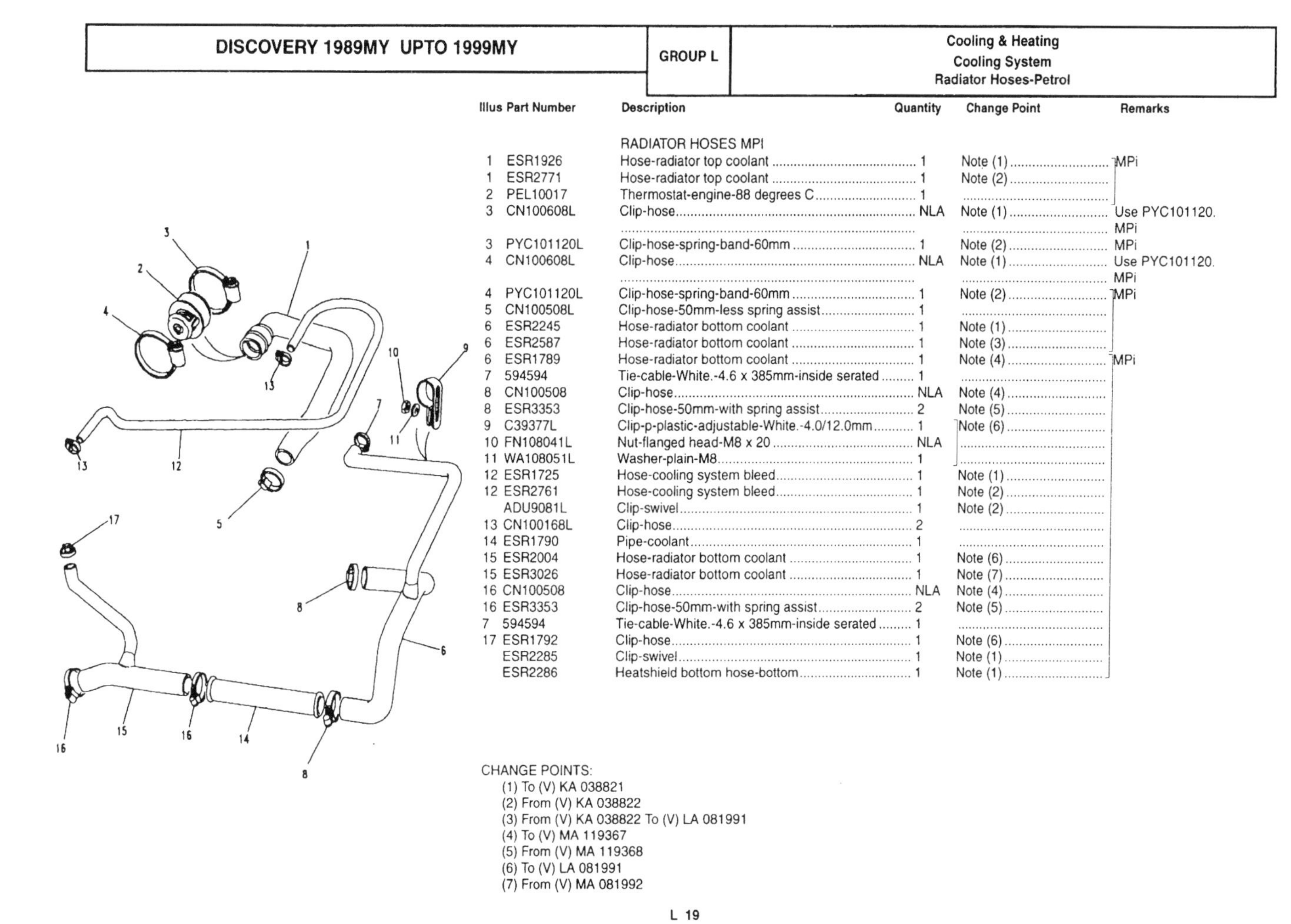

Illus	Part Number	Description	Quantity	Change Point	Remarks
		RADIATOR HOSES MPI			
1	ESR1926	Hose-radiator top coolant	1	Note (1)	MPi
1	ESR2771	Hose-radiator top coolant	1	Note (2)	
2	PEL10017	Thermostat-engine-88 degrees C	1		
3	CN100608L	Clip-hose	NLA	Note (1)	Use PYC101120. MPi
3	PYC101120L	Clip-hose-spring-band-60mm	1	Note (2)	MPi
4	CN100608L	Clip-hose	NLA	Note (1)	Use PYC101120. MPi
4	PYC101120L	Clip-hose-spring-band-60mm	1	Note (2)	MPi
5	CN100508L	Clip-hose-50mm-less spring assist	1		
6	ESR2245	Hose-radiator bottom coolant	1	Note (1)	
6	ESR2587	Hose-radiator bottom coolant	1	Note (3)	
6	ESR1789	Hose-radiator bottom coolant	1	Note (4)	MPi
7	594594	Tie-cable-White.-4.6 x 385mm-inside serated	1		
8	CN100508	Clip-hose	NLA	Note (4)	
8	ESR3353	Clip-hose-50mm-with spring assist	2	Note (5)	
9	C39377L	Clip-p-plastic-adjustable-White.-4.0/12.0mm	1	Note (6)	
10	FN108041L	Nut-flanged head-M8 x 20	NLA		
11	WA108051L	Washer-plain-M8	1		
12	ESR1725	Hose-cooling system bleed	1	Note (1)	
12	ESR2761	Hose-cooling system bleed	1	Note (2)	
	ADU9081L	Clip-swivel	1	Note (2)	
13	CN100168L	Clip-hose	2		
14	ESR1790	Pipe-coolant	1		
15	ESR2004	Hose-radiator bottom coolant	1	Note (6)	
15	ESR3026	Hose-radiator bottom coolant	1	Note (7)	
16	CN100508	Clip-hose	NLA	Note (4)	
16	ESR3353	Clip-hose-50mm-with spring assist	2	Note (5)	
7	594594	Tie-cable-White.-4.6 x 385mm-inside serated	1		
17	ESR1792	Clip-hose	1	Note (6)	
	ESR2285	Clip-swivel	1	Note (1)	
	ESR2286	Heatshield bottom hose-bottom	1	Note (1)	

CHANGE POINTS:
(1) To (V) KA 038821
(2) From (V) KA 038822
(3) From (V) KA 038822 To (V) LA 081991
(4) To (V) MA 119367
(5) From (V) MA 119368
(6) To (V) LA 081991
(7) From (V) MA 081992

L 19

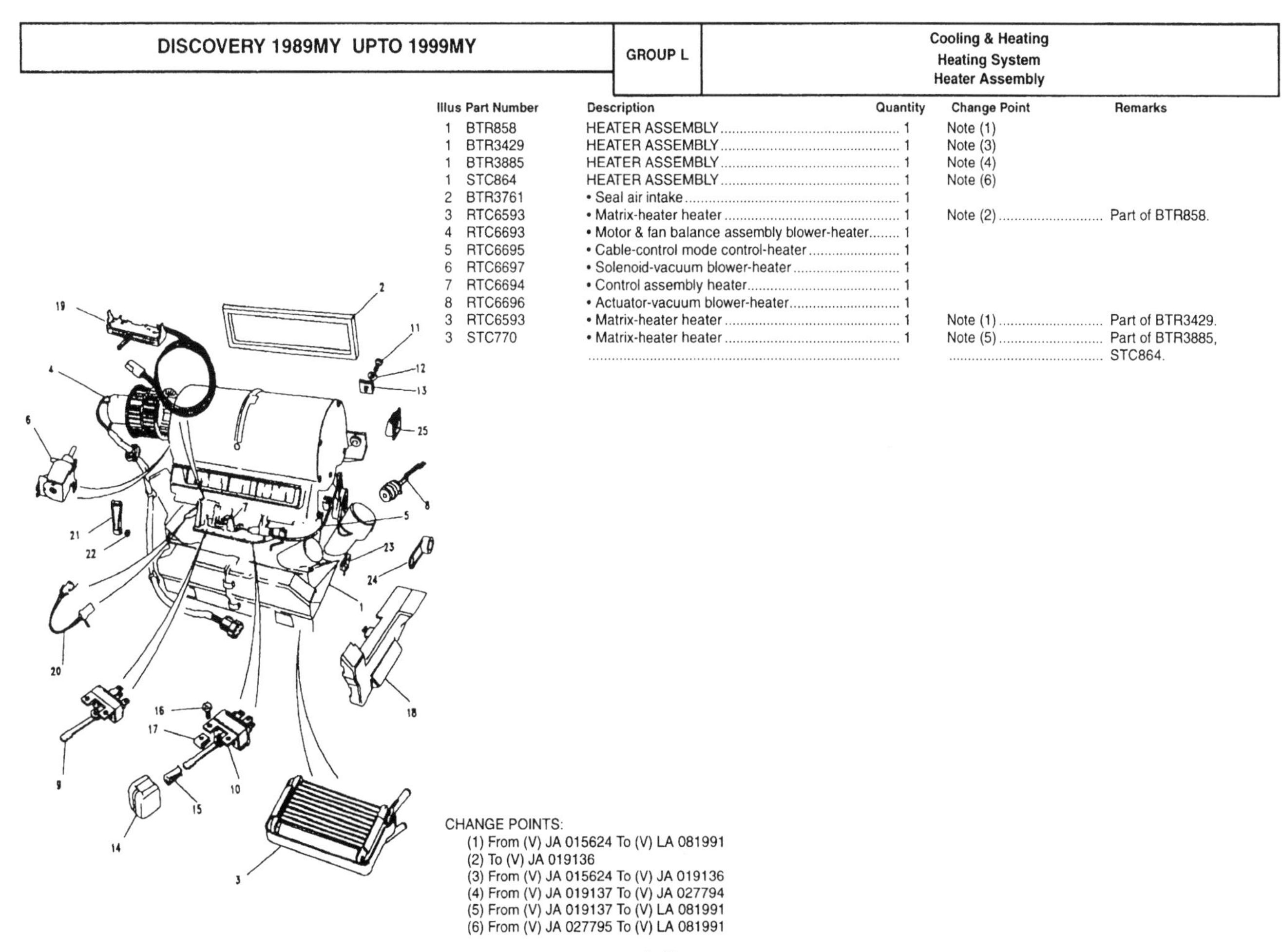

Illus	Part Number	Description	Quantity	Change Point	Remarks
1	BTR858	HEATER ASSEMBLY	1	Note (1)	
1	BTR3429	HEATER ASSEMBLY	1	Note (3)	
1	BTR3885	HEATER ASSEMBLY	1	Note (4)	
1	STC864	HEATER ASSEMBLY	1	Note (6)	
2	BTR3761	• Seal air intake	1		
3	RTC6593	• Matrix-heater heater	1	Note (2)	Part of BTR858.
4	RTC6693	• Motor & fan balance assembly blower-heater	1		
5	RTC6695	• Cable-control mode control-heater	1		
6	RTC6697	• Solenoid-vacuum blower-heater	1		
7	RTC6694	• Control assembly heater	1		
8	RTC6696	• Actuator-vacuum blower-heater	1		
3	RTC6593	• Matrix-heater heater	1	Note (1)	Part of BTR3429.
3	STC770	• Matrix-heater heater	1	Note (5)	Part of BTR3885, STC864.

CHANGE POINTS:
(1) From (V) JA 015624 To (V) LA 081991
(2) To (V) JA 019136
(3) From (V) JA 015624 To (V) JA 019136
(4) From (V) JA 019137 To (V) JA 027794
(5) From (V) JA 019137 To (V) LA 081991
(6) From (V) JA 027795 To (V) LA 081991

L 20

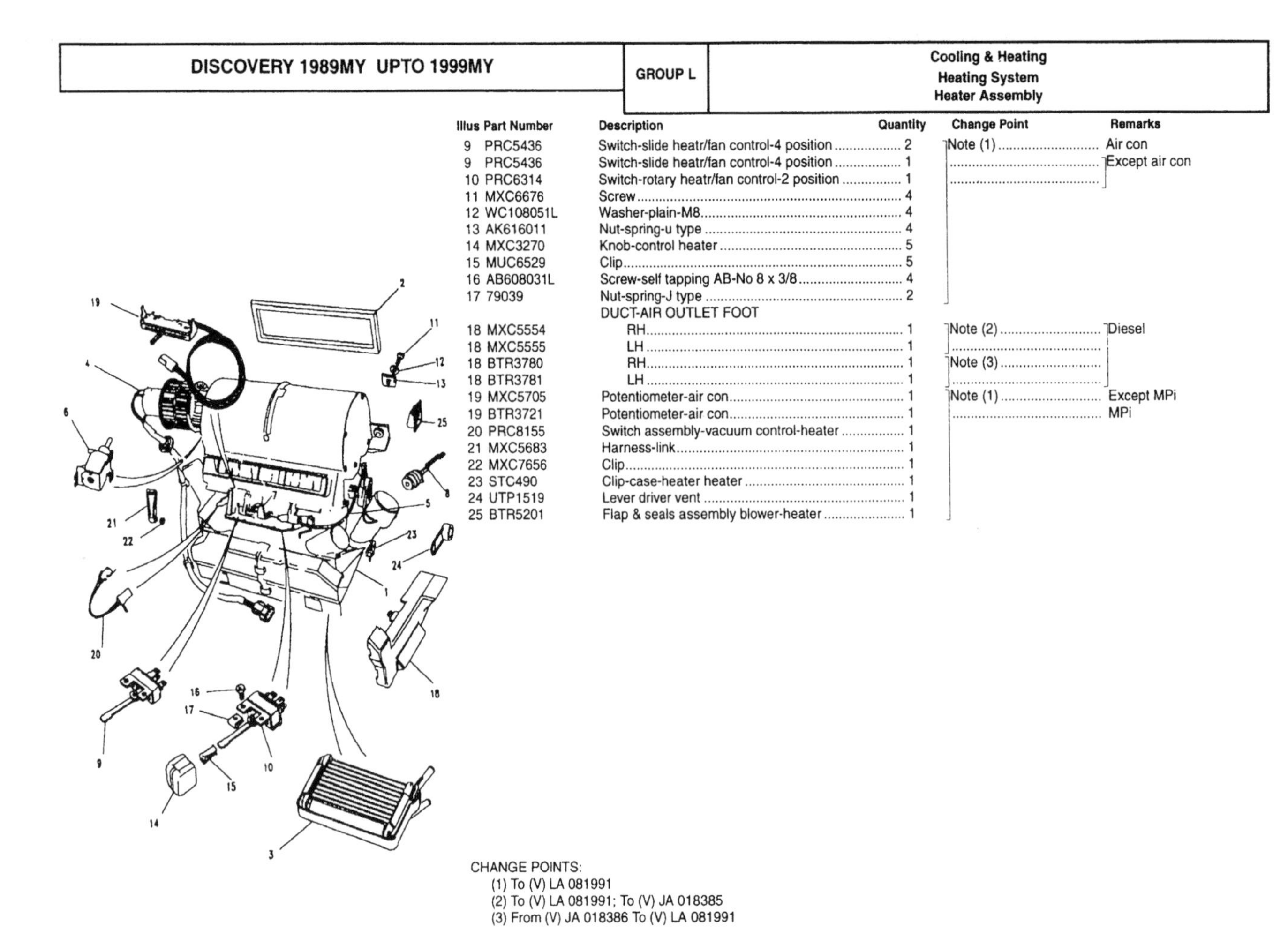

Illus	Part Number	Description	Quantity	Change Point	Remarks
9	PRC5436	Switch-slide heatr/fan control-4 position	2	Note (1)	Air con
9	PRC5436	Switch-slide heatr/fan control-4 position	1		Except air con
10	PRC6314	Switch-rotary heatr/fan control-2 position	1		
11	MXC6676	Screw	4		
12	WC108051L	Washer-plain-M8	4		
13	AK616011	Nut-spring-u type	4		
14	MXC3270	Knob-control heater	5		
15	MUC6529	Clip	5		
16	AB608031L	Screw-self tapping AB-No 8 x 3/8	4		
17	79039	Nut-spring-J type	2		
		DUCT-AIR OUTLET FOOT			
18	MXC5554	RH	1	Note (2)	Diesel
18	MXC5555	LH	1		
18	BTR3780	RH	1	Note (3)	
18	BTR3781	LH	1		
19	MXC5705	Potentiometer-air con	1	Note (1)	Except MPi
19	BTR3721	Potentiometer-air con	1		MPi
20	PRC8155	Switch assembly-vacuum control-heater	1		
21	MXC5683	Harness-link	1		
22	MXC7656	Clip	1		
23	STC490	Clip-case-heater heater	1		
24	UTP1519	Lever driver vent	1		
25	BTR5201	Flap & seals assembly blower-heater	1		

CHANGE POINTS:
(1) To (V) LA 081991
(2) To (V) LA 081991; To (V) JA 018385
(3) From (V) JA 018386 To (V) LA 081991

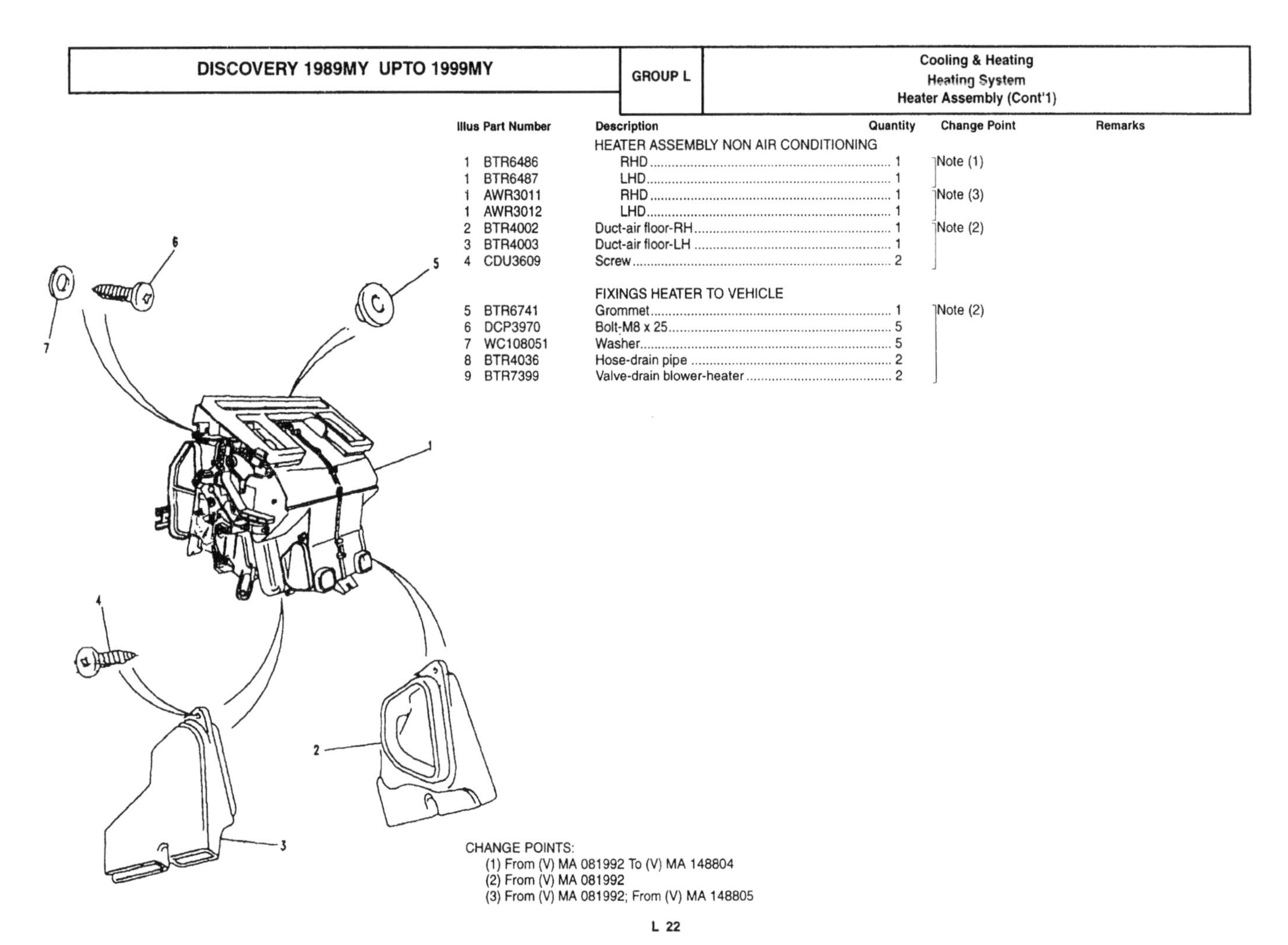

Illus	Part Number	Description	Quantity	Change Point	Remarks
		HEATER ASSEMBLY NON AIR CONDITIONING			
1	BTR6486	RHD	1	Note (1)	
1	BTR6487	LHD	1		
1	AWR3011	RHD	1	Note (3)	
1	AWR3012	LHD	1		
2	BTR4002	Duct-air floor-RH	1	Note (2)	
3	BTR4003	Duct-air floor-LH	1		
4	CDU3609	Screw	2		
		FIXINGS HEATER TO VEHICLE			
5	BTR6741	Grommet	1	Note (2)	
6	DCP3970	Bolt-M8 x 25	5		
7	WC108051	Washer	5		
8	BTR4036	Hose-drain pipe	2		
9	BTR7399	Valve-drain blower-heater	2		

CHANGE POINTS:
(1) From (V) MA 081992 To (V) MA 148804
(2) From (V) MA 081992
(3) From (V) MA 081992; From (V) MA 148805

Illus	Part Number	Description	Quantity	Change Point	Remarks
		UNDER BONNET - AIR CON			
1	STC3148	Receiver dryer assembly air conditioning	1	Note (1)	
2	STC3147	Fan-cooling/air conditioning	1		
3	STC3136	Switch-pressure air conditioning control	1		

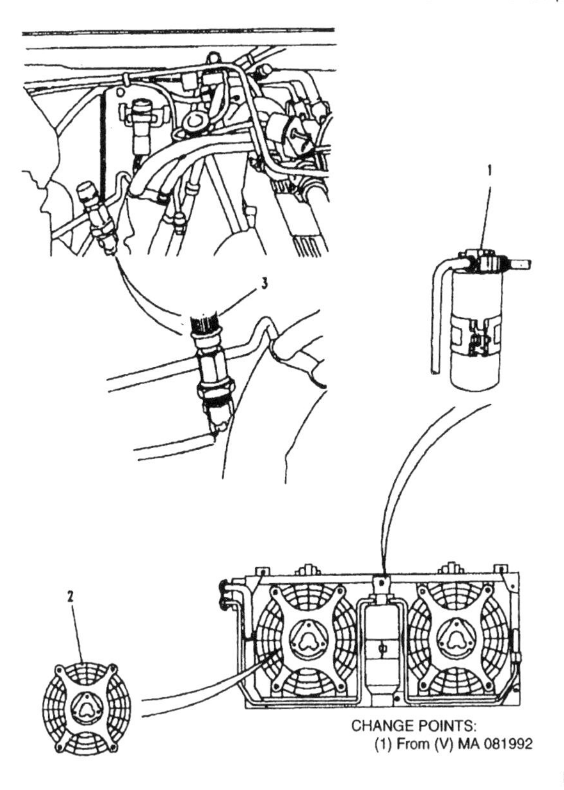

CHANGE POINTS:
(1) From (V) MA 081992

L 23

Illus	Part Number	Description	Quantity	Change Point	Remarks
		BLOWER UNIT - AIR CON			
1	STC3133	Resistor pack-speed control-blower-heater	1	Note (1)	
2	STC3141	Relay blower-heater	1		
3	STC3152	Motor assembly blower-air conditioning	1		RHD
3	STC3153	Motor assembly blower-air conditioning	1		LHD

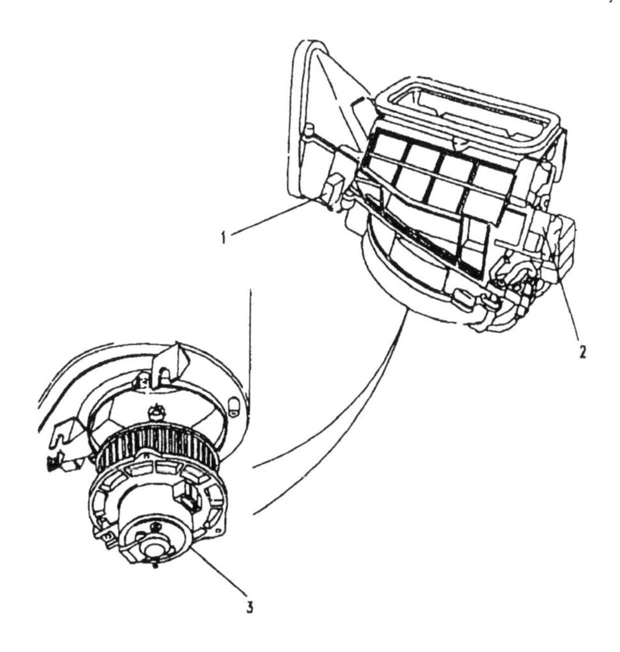

CHANGE POINTS:
(1) From (V) MA 081992

L 24

Illus	Part Number	Description	Quantity	Change Point	Remarks
		HEATER UNIT - AIR CON			
1	STC3139	Evaporator air conditioning	1	Note (1)	
2	STC3140	Valve-expansion evaporator-air conditioning	1		
3	STC3134	Thermostat evaporator-air conditioning	1		
4	STC3135	Matrix-heater heater	1		
5	STC3132	Pipe kit	1		
6	STC3166	O ring-17mm ID-18.2mm OD	2		
7	STC3154	Foam	1		
8	STC3917	Kit-pipes etc heater-heater inlet	1		
9	STC3917	Kit-pipes etc heater-heater outlet	1		
10	STC3917	Kit-pipes etc heater	1		
11	STC3917	Kit-pipes etc heater	1		

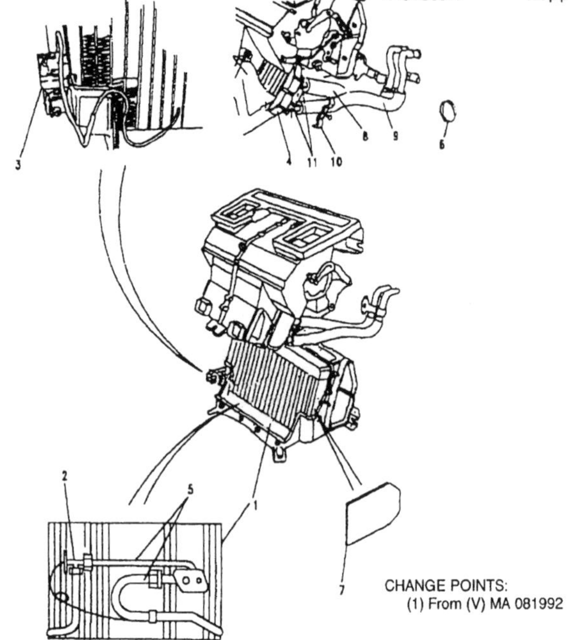

CHANGE POINTS:
(1) From (V) MA 081992

L 25

Illus	Part Number	Description	Quantity	Change Point	Remarks
		REAR AIR CONDITIONING UNIT			
1	STC3142	Evaporator assembly-air conditioning	1	Note (1)	
2	STC3143	Valve-expansion evaporator-air conditioning	1		
3	STC3144	Thermostat evaporator-air conditioning	1		
4	STC3145	Motor assembly blower-air conditioning	1		
5	STC3146	Fan-cooling/air conditioning	1		
6	STC3813	Relay-Black	1		
7	STC3815	Relay	1		
8	STC3814	Harness air conditioning-rear	1		
9	STC3816	Resistor-pack heater	1		
10	STC3817	Valve assembly-solenoid	1		
11	STC3818	Electric control unit-air conditioning	1		

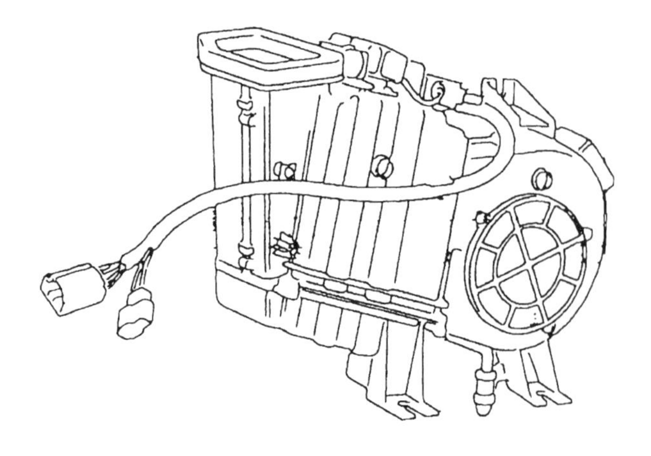

FULL ILLUSTRATION TO FOLLOW

CHANGE POINTS:
(1) From (V) MA 081992

L 26

Illus	Part Number	Description	Quantity	Change Point	Remarks
1	BTR6484	Blower assembly heater-RHD	1	Note (1)	
1	BTR6485	Blower assembly heater-LHD	1		
2	BTR4884	Bracket-heater blower mounting	1		
3	DCP3970	Bolt-M8 x 25	2		
4	WC108051	Washer	2		
5	DCP3970	Bolt-M8 x 25	1		
6	ADU7048	Nut-flange-M8	2		

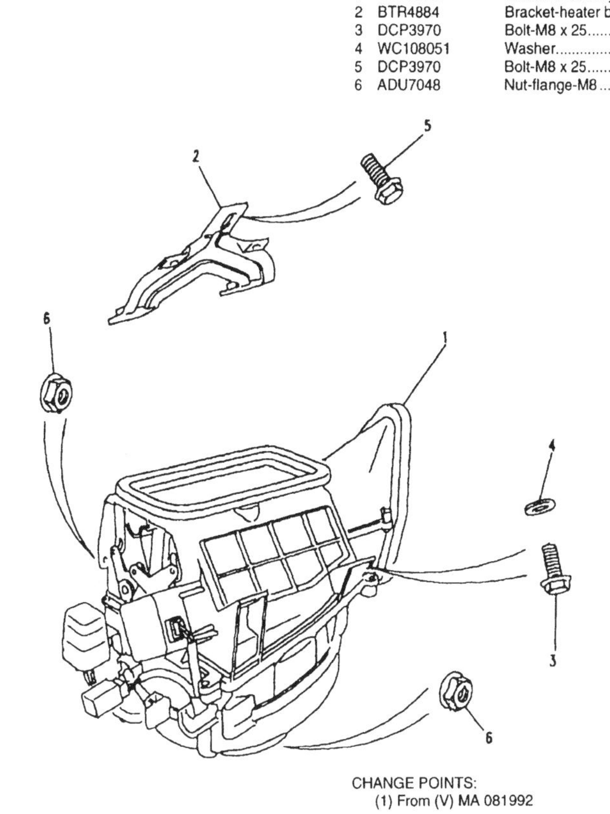

CHANGE POINTS:
(1) From (V) MA 081992

Illus	Part Number	Description	Quantity	Change Point	Remarks
		HEATER CONTROLS			
1	BTR6490	Control assembly heater-less cables	NLA	Note (1)	Use AWR1378.
1	AWR1378	Control assembly heater-with cables	NLA		Use AWR7155.
1	AWR7155	Control assembly heater	1		
2	AB608051	Screw-self tapping	4		
3	STC1863	SWITCH-PUSH PUSH AIR CONDITIONING RECIRCULATING	1		
	STC1877	• Bulb & holder assembly-Orange	A/R		
	STC1878	• Bulb & holder assembly-Green	A/R		
4	BTR6492	Plate-blanking air conditioning switch	2	Note (1)	
5	BTR8930	Knob-control heater	3		
6	BTR6494	Knob-slider control-heater	1		
7	BTR6495	Plate-graphics control-heater	1		
8	AB606031	Screw-self tapping-No 6 x 3/8	NLA	Note (1)	Use AB606031L.
8	AB608031	Screw-self tapping AB-No 8 x 3/8	2		
9	RTC3635	Bulb-286-w5/1.2-1.2 Watt-12V-1.2 Watt	1	Note (1)	

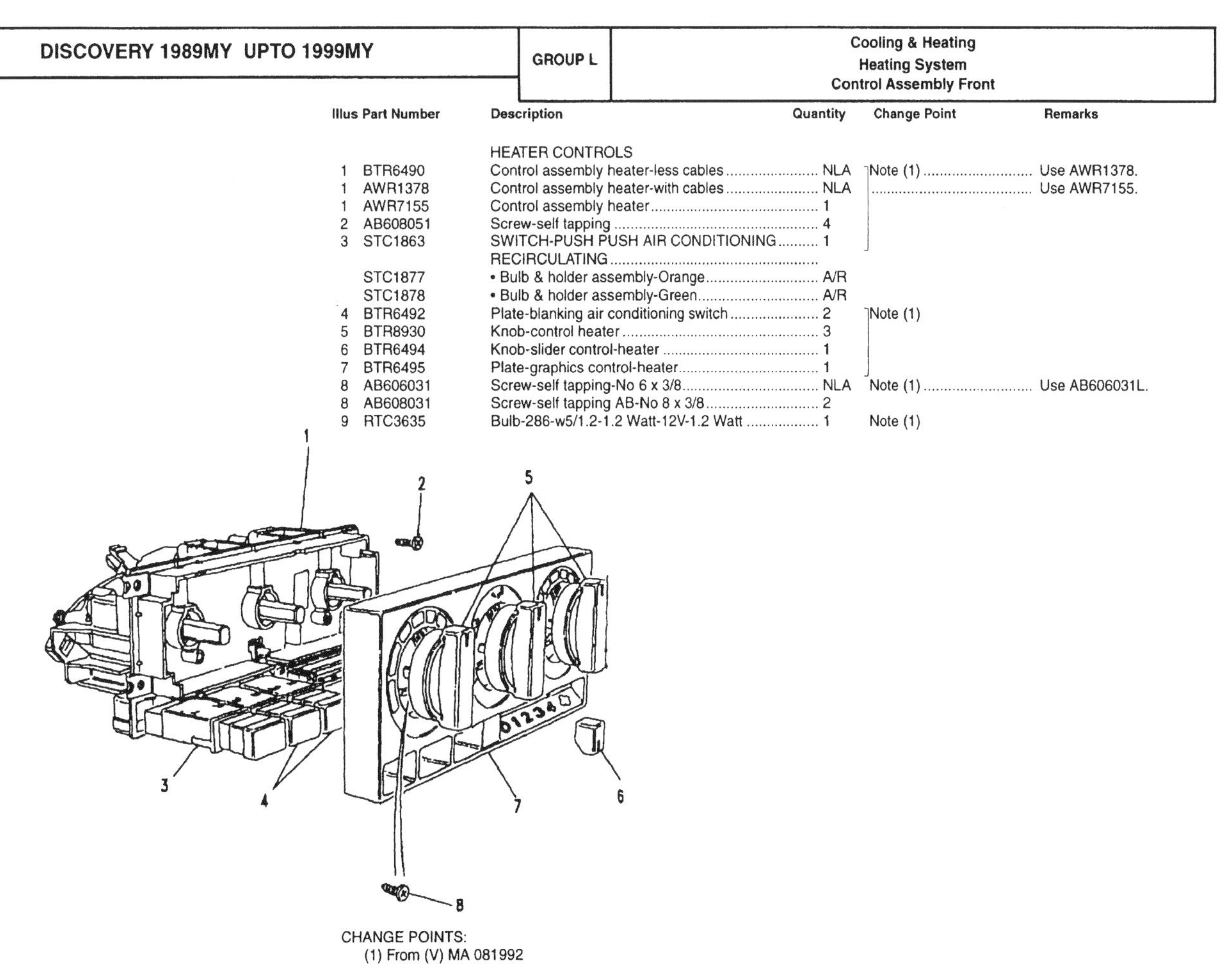

CHANGE POINTS:
(1) From (V) MA 081992

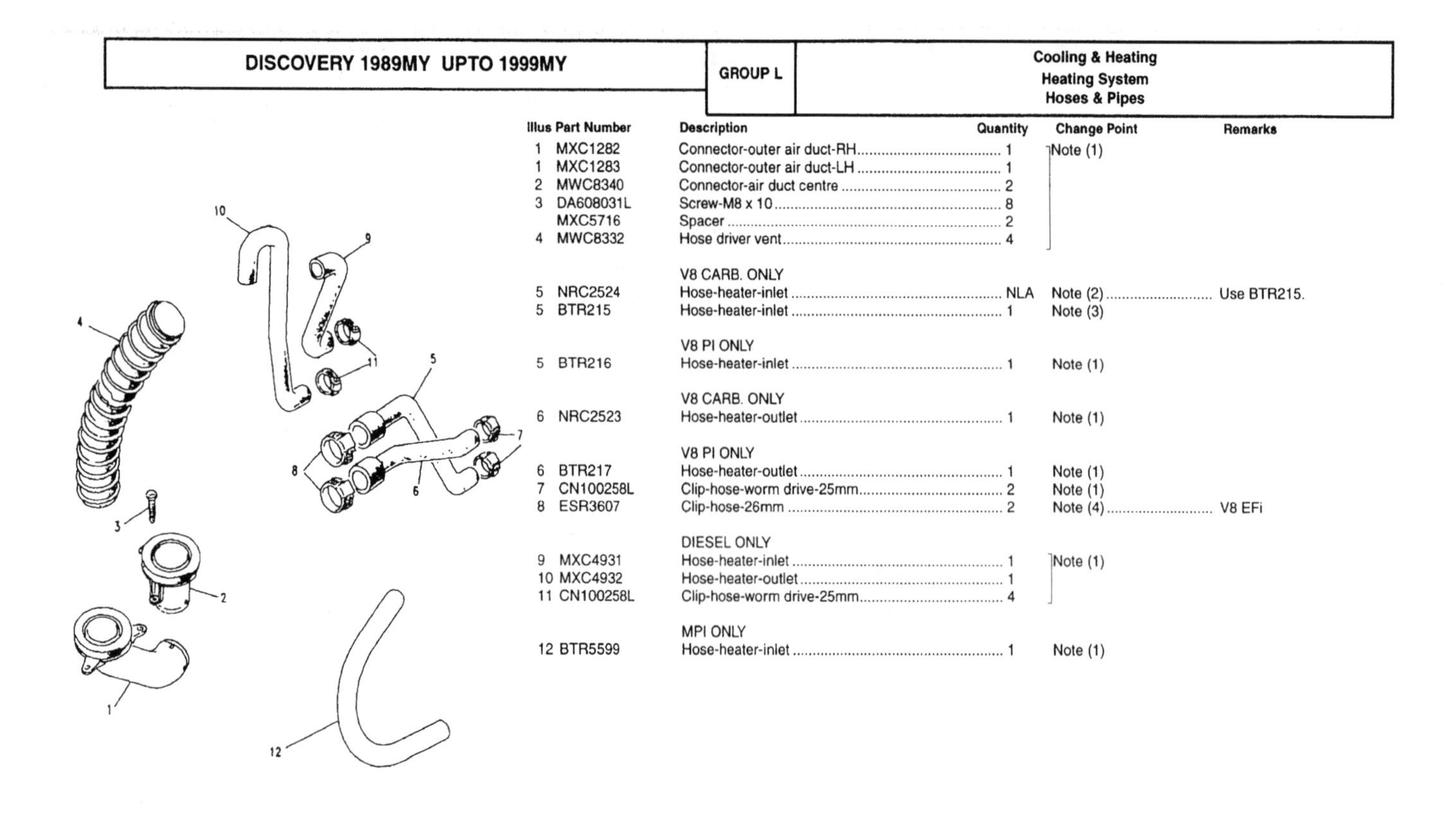

Illus	Part Number	Description	Quantity	Change Point	Remarks
1	MXC1282	Connector-outer air duct-RH	1	Note (1)	
1	MXC1283	Connector-outer air duct-LH	1	Note (1)	
2	MWC8340	Connector-air duct centre	2	Note (1)	
3	DA608031L	Screw-M8 x 10	8	Note (1)	
	MXC5716	Spacer	2	Note (1)	
4	MWC8332	Hose driver vent	4	Note (1)	
		V8 CARB. ONLY			
5	NRC2524	Hose-heater-inlet	NLA	Note (2)	Use BTR215.
5	BTR215	Hose-heater-inlet	1	Note (3)	
		V8 PI ONLY			
5	BTR216	Hose-heater-inlet	1	Note (1)	
		V8 CARB. ONLY			
6	NRC2523	Hose-heater-outlet	1	Note (1)	
		V8 PI ONLY			
6	BTR217	Hose-heater-outlet	1	Note (1)	
7	CN100258L	Clip-hose-worm drive-25mm	2	Note (1)	
8	ESR3607	Clip-hose-26mm	2	Note (4)	V8 EFi
		DIESEL ONLY			
9	MXC4931	Hose-heater-inlet	1	Note (1)	
10	MXC4932	Hose-heater-outlet	1	Note (1)	
11	CN100258L	Clip-hose-worm drive-25mm	4	Note (1)	
		MPI ONLY			
12	BTR5599	Hose-heater-inlet	1	Note (1)	

CHANGE POINTS:
(1) To (V) LA 081991
(2) To (V) GA 449987
(3) From (V) GA 449988 To (V) LA 081991
(4) To (V) MA 081991

L 29

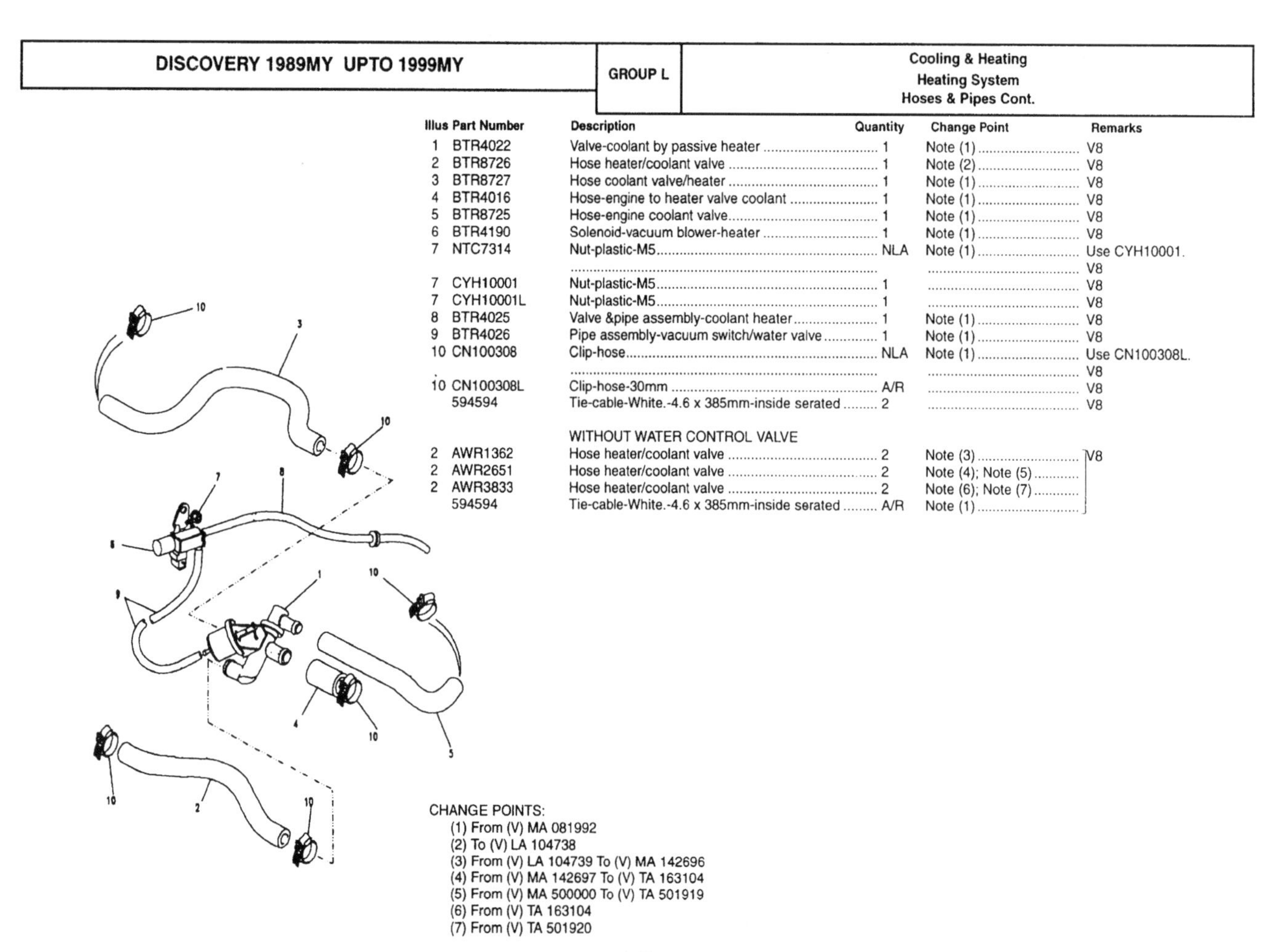

Illus	Part Number	Description	Quantity	Change Point	Remarks
1	BTR4022	Valve-coolant by passive heater	1	Note (1)	V8
2	BTR8726	Hose heater/coolant valve	1	Note (2)	V8
3	BTR8727	Hose coolant valve/heater	1	Note (1)	V8
4	BTR4016	Hose-engine to heater valve coolant	1	Note (1)	V8
5	BTR8725	Hose-engine coolant valve	1	Note (1)	V8
6	BTR4190	Solenoid-vacuum blower-heater	1	Note (1)	V8
7	NTC7314	Nut-plastic-M5	NLA	Note (1)	Use CYH10001. V8
7	CYH10001	Nut-plastic-M5	1		V8
7	CYH10001L	Nut-plastic-M5	1		V8
8	BTR4025	Valve &pipe assembly-coolant heater	1	Note (1)	V8
9	BTR4026	Pipe assembly-vacuum switch/water valve	1	Note (1)	V8
10	CN100308	Clip-hose	NLA	Note (1)	Use CN100308L. V8
10	CN100308L	Clip-hose-30mm	A/R		V8
	594594	Tie-cable-White.-4.6 x 385mm-inside serated	2		V8
		WITHOUT WATER CONTROL VALVE			
2	AWR1362	Hose heater/coolant valve	2	Note (3)	V8
2	AWR2651	Hose heater/coolant valve	2	Note (4); Note (5)	V8
2	AWR3833	Hose heater/coolant valve	2	Note (6); Note (7)	V8
	594594	Tie-cable-White.-4.6 x 385mm-inside serated	A/R	Note (1)	V8

CHANGE POINTS:
(1) From (V) MA 081992
(2) To (V) LA 104738
(3) From (V) LA 104739 To (V) MA 142696
(4) From (V) MA 142697 To (V) TA 163104
(5) From (V) MA 500000 To (V) TA 501919
(6) From (V) TA 163104
(7) From (V) TA 501920

L 30

Illus	Part Number	Description	Quantity	Change Point	Remarks
		MPI			
1	BTR4023	Valve-coolant by passive heater	1	Note (1)	
2	BTR4195	Hose heater/coolant valve	1		
3	BTR4196	Hose coolant valve/heater	1		
4	BTR4016	Hose-engine to heater valve coolant	1	Note (2)	
4	AWR1364	Hose-engine to heater valve coolant	1	Note (3)	
5	BTR4008	Hose assembly-engine coolant valve	1	Note (2)	
5	AWR1363	Hose assembly-engine coolant valve	1	Note (3)	
6	BTR4190	Solenoid-vacuum blower-heater	1	Note (1)	
7	NTC7314	Nut-plastic-M5	NLA		Use CYH10001.
8	BTR4027	Valve &pipe assembly-coolant heater	1	Note (1)	
9	BTR4026	Pipe assembly-vacuum switch/water valve	1		
10	ESR3607	Clip-hose-26mm	A/R		MPi
	594594	Tie-cable-White.-4.6 x 385mm-inside serated	A/R		MPi

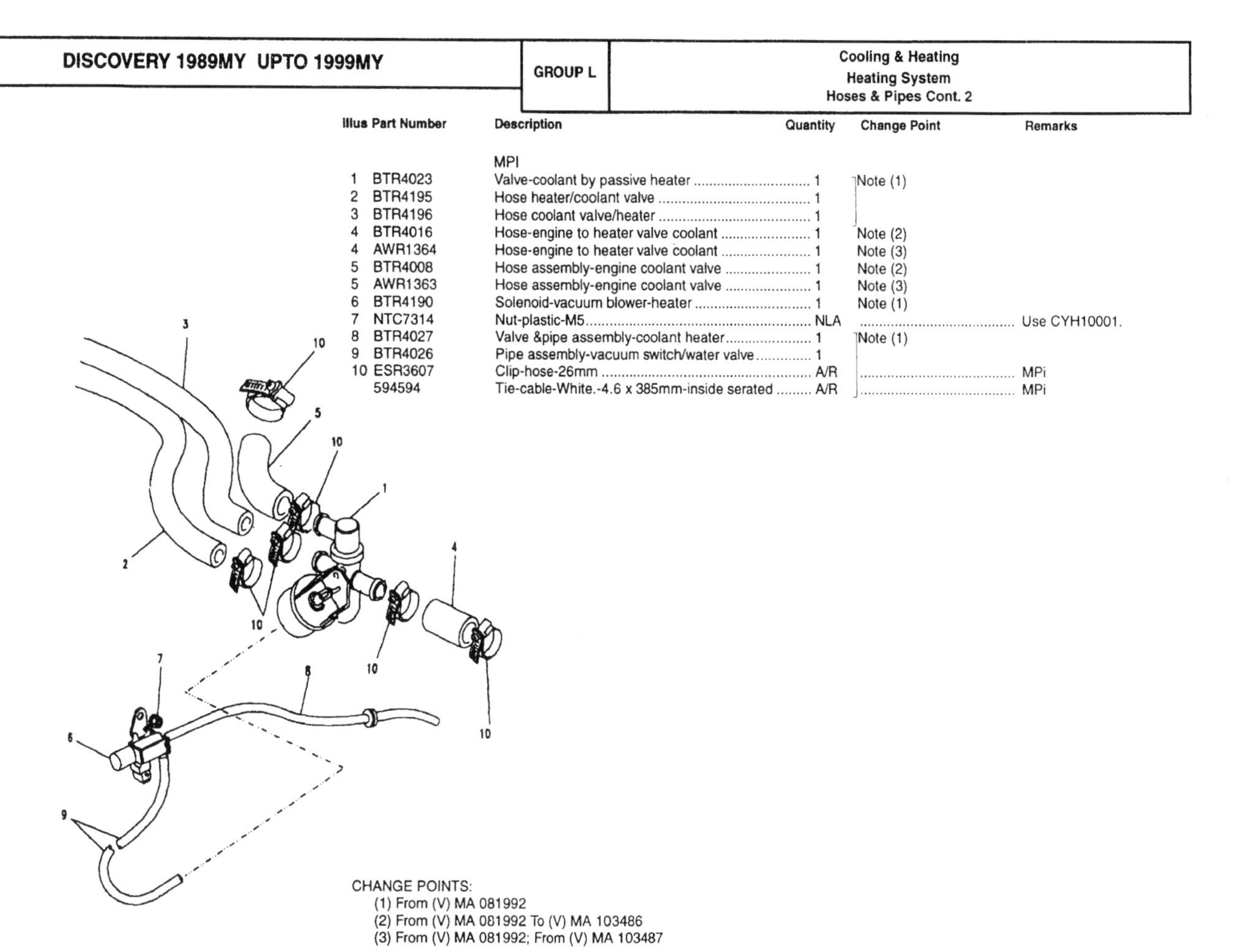

CHANGE POINTS:
(1) From (V) MA 081992
(2) From (V) MA 081992 To (V) MA 103486
(3) From (V) MA 081992; From (V) MA 103487

L 31

Illus	Part Number	Description	Quantity	Change Point	Remarks
		300TDI			
1	BTR9616	Hose-heater-inlet	1	Note (1)	
2	BTR9617	Hose-heater-outlet	1		
3	594594	Tie-cable-White.-4.6 x 385mm-inside serated	A/R		TDi
4	ERR4531	Pipe-heater to hose coolant	1		

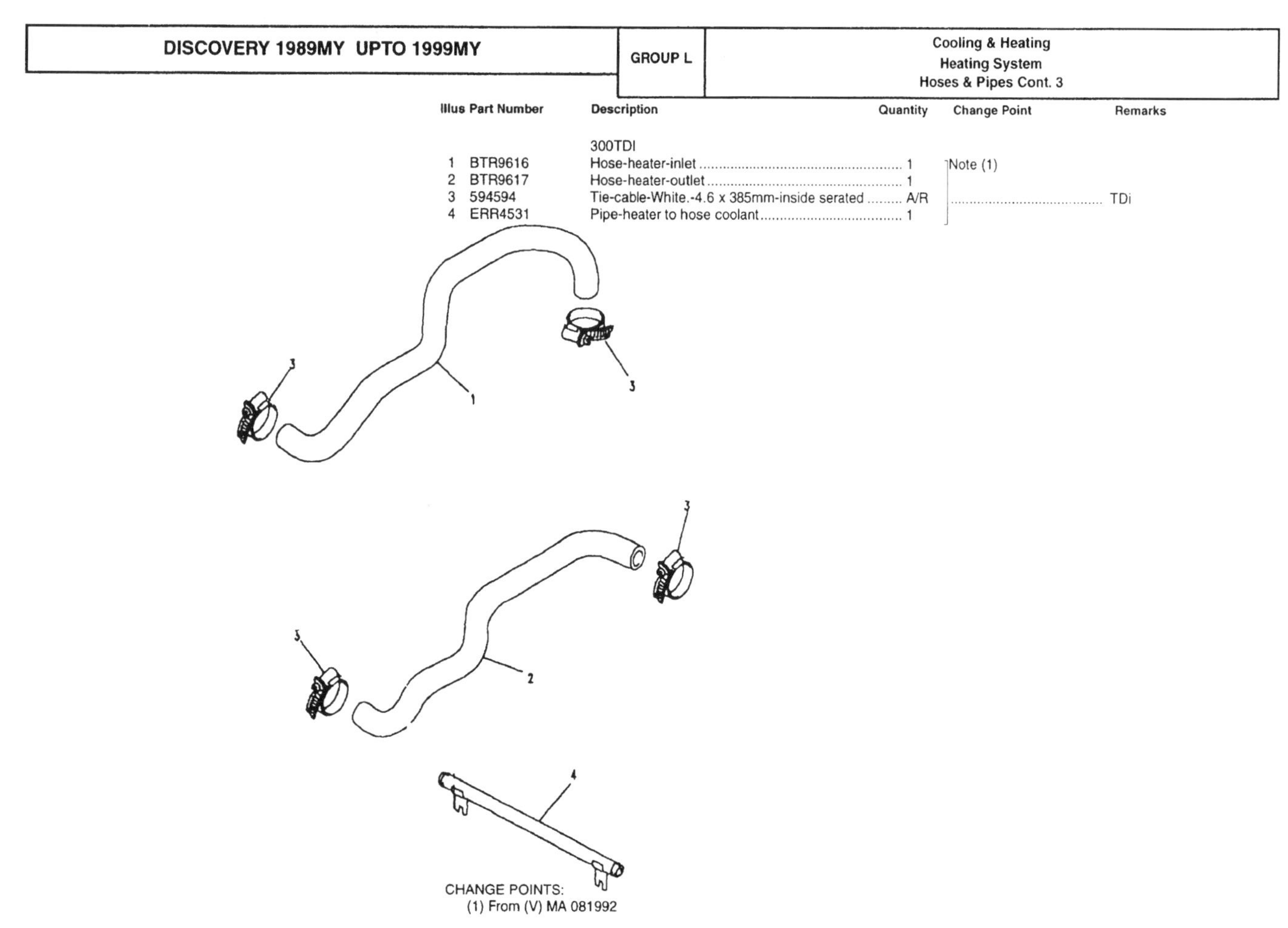

CHANGE POINTS:
(1) From (V) MA 081992

L 32

Illus	Part Number	Description	Quantity	Change Point	Remarks
		DUCT-AIR INLET HEATER			
1	BTR3998	RHD	1	Note (1)	
1	BTR3999	LHD	1		
1	AWR4125	RHD	1	Note (2)	
1	AWR4126	LHD	1		
2	GL106251	Screw-M6	5	Note (3)	
3	WJ106001L	Washer-plain-M6	5		

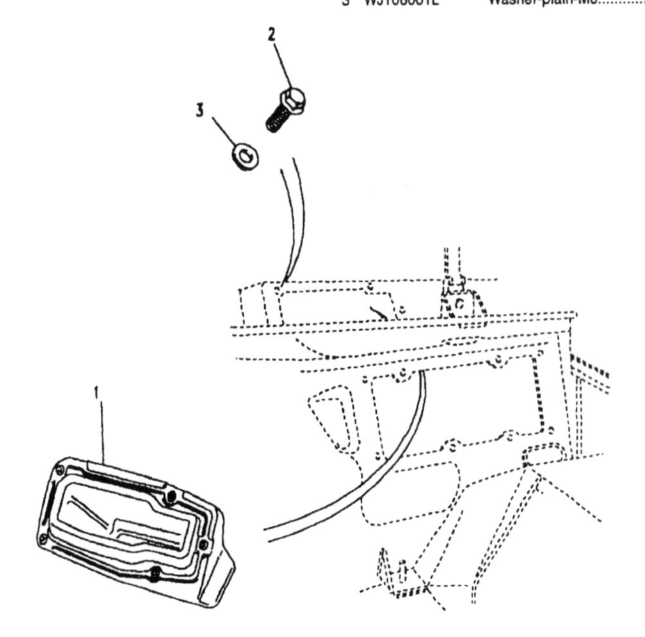

CHANGE POINTS:
(1) From (V) MA 081992 To (V) MA 163103
(2) From (V) MA 081992; From (V) TA 163104; From (V) TA 501920
(3) From (V) MA 081992

L 33

Illus	Part Number	Description	Quantity	Change Point	Remarks
1	MWC8366	Duct-air rear floor	2	Note (1)	
1	BTR5027	Duct-air rear floor	2	Note (2)	
2	BTR276	Joint-air duct floor	2	Note (3)	

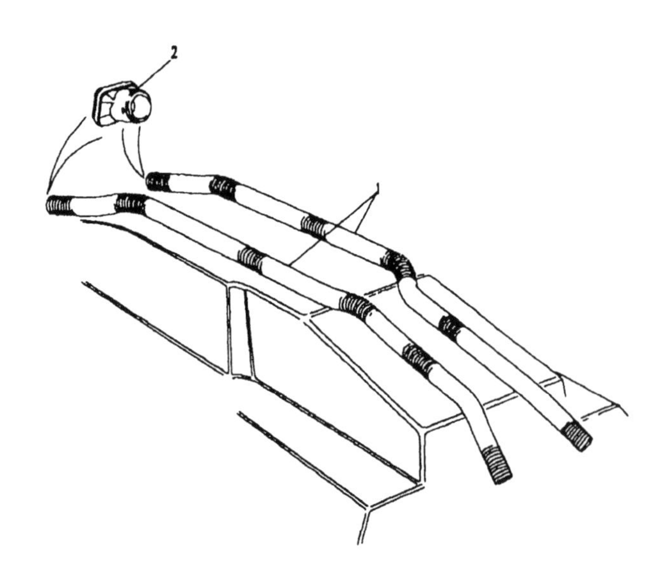

CHANGE POINTS:
(1) To (V) LA 081991; To (V) JA 031313
(2) From (V) KA 031314 To (V) LA 081991
(3) To (V) LA 081991

L 34

Illus	Part Number	Description	Quantity	Change Point	Remarks
1	BTR4430	Duct-air floor-RH	1	Note (1)	
2	BTR4431	Duct-air floor-LH	1		
3	BTR4494	Duct-air outlet foot-RH	1		
4	BTR4495	Duct-air outlet foot-LH	1		
5	DA610044	Screw-flanged head-No 10 x 1/2	4		
6	BTR4649	Bracket-tunnel console mounting	1	Note (2)	
6	FHC100410	Bracket-tunnel console mounting	1	Note (3)	
7	SE106101L	Screw-M6 x 10	NLA		Use FS106101L.
7	FS106101L	Screw-flanged head-M6 x 10	4		

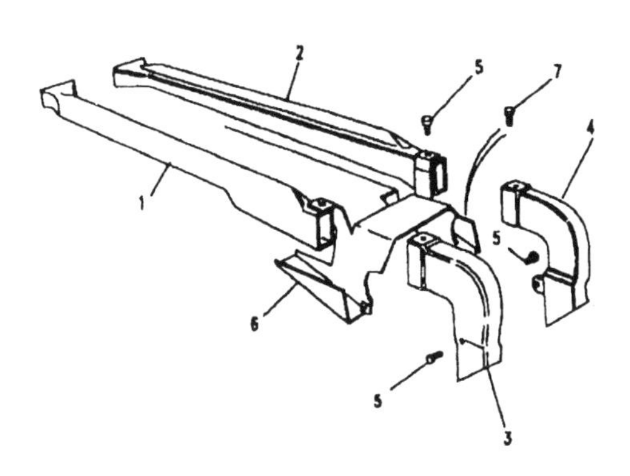

CHANGE POINTS:
(1) From (V) MA 647645
(2) To (V) VA 744994
(3) From (V) VA 744995

Illus	Part Number	Description	Quantity	Change Point	Remarks
1	MUC1679	Vacuum tank	1	Note (1)	
2	MWC3086	Pipe-vacuum heater	1		
3	MWC3088	Pipe-vacuum heater	1		
4	FS106167L	Screw-hexagonal head-M6 x 16	2		
5	WA106041L	Washer-M6-standard.	4		
6	WL106041	Washer-spring	2		
7	NH106041L	Nut-hexagonal head-nyloc-M6	2		
8	233243	Grommet-Black-rubber-3 x 25.5mm	1		
	MWC4340	Pipe-vacuum heater/air conditioning	1		

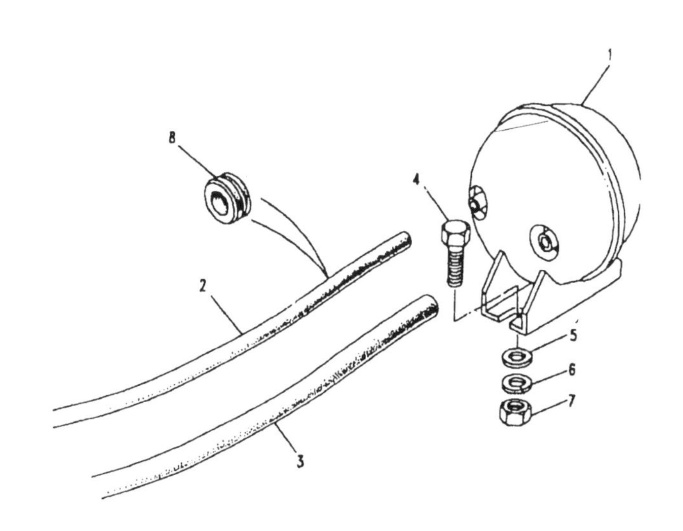

CHANGE POINTS:
(1) To (V) LA 067123

Illus	Part Number	Description	Quantity	Change Point	Remarks
1	MWC4340	Pipe-vacuum heater/air conditioning	1	Note (1)	
2	233243	Grommet-Black-rubber-3 x 25.5mm	1		
3	BTR5598	Valve-check-vacuum control-heater	1		

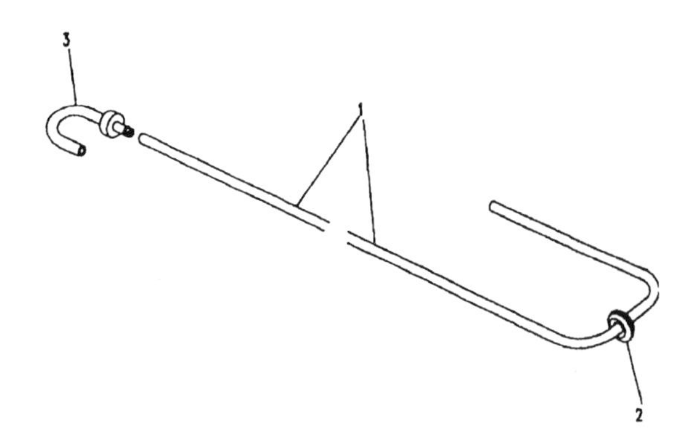

CHANGE POINTS:
(1) From (V) LA 067124 To (V) MA 081992

L 37

Illus	Part Number	Description	Quantity	Change Point	Remarks
		BLOWER UNIT			
1	BTR6484	Blower assembly heater-RHD	1	Note (1)	
1	BTR6485	Blower assembly heater-LHD	1		
		FIXINGS BLOWER TO VEHICLE			
2	BTR4884	Bracket-heater blower mounting	1	Note (1)	
3	DCP3970	Bolt-M8 x 25	2		
4	WC108051	Washer	2		
5	DCP3970	Bolt-M8 x 25	1		
6	ADU7048	Nut-flange-M8	2		

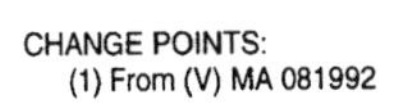

CHANGE POINTS:
(1) From (V) MA 081992

L 38

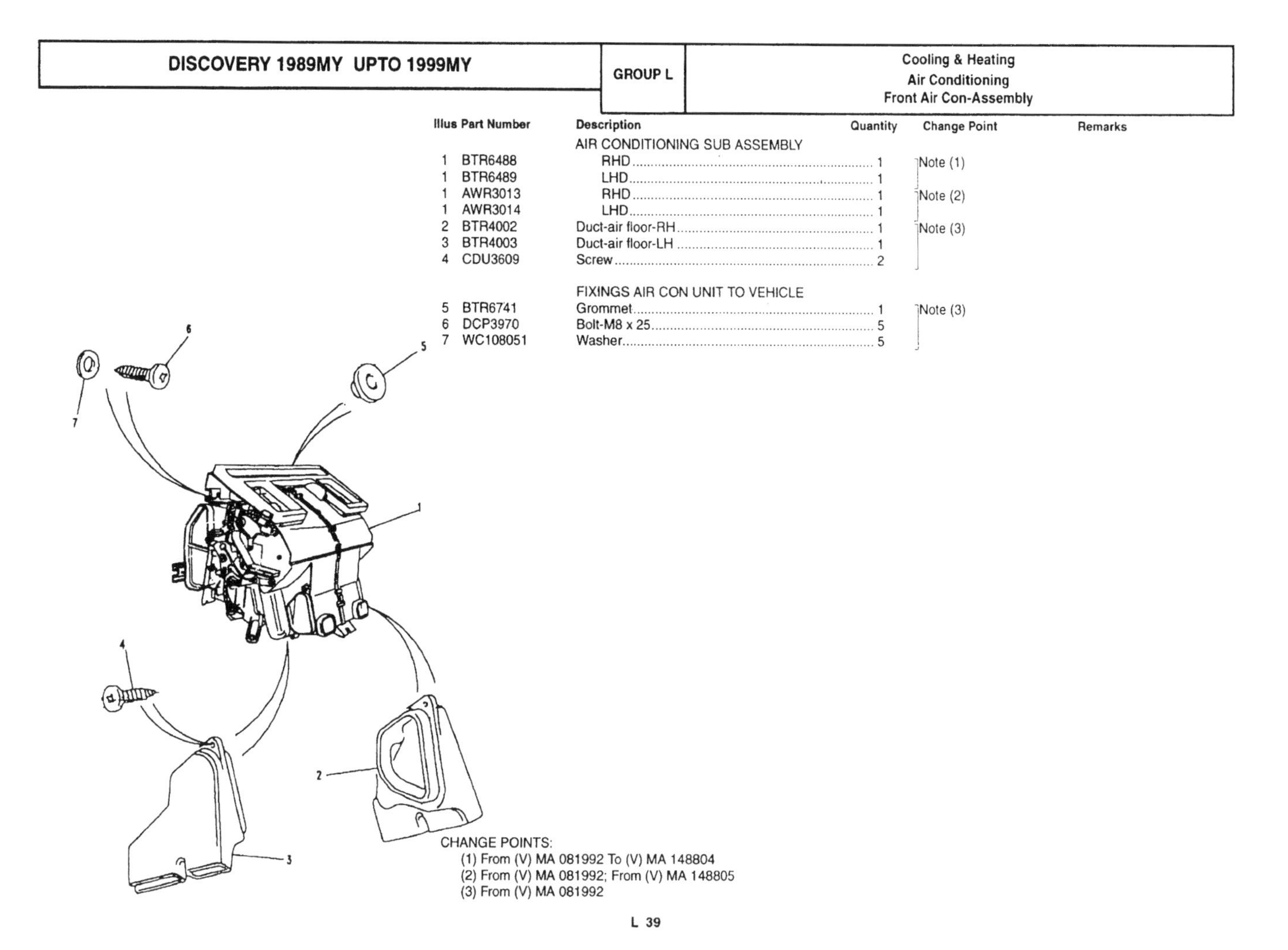

Illus	Part Number	Description	Quantity	Change Point	Remarks
		AIR CONDITIONING SUB ASSEMBLY			
1	BTR6488	RHD	1	Note (1)	
1	BTR6489	LHD	1		
1	AWR3013	RHD	1	Note (2)	
1	AWR3014	LHD	1		
2	BTR4002	Duct-air floor-RH	1	Note (3)	
3	BTR4003	Duct-air floor-LH	1		
4	CDU3609	Screw	2		
		FIXINGS AIR CON UNIT TO VEHICLE			
5	BTR6741	Grommet	1	Note (3)	
6	DCP3970	Bolt-M8 x 25	5		
7	WC108051	Washer	5		

CHANGE POINTS:
(1) From (V) MA 081992 To (V) MA 148804
(2) From (V) MA 081992; From (V) MA 148805
(3) From (V) MA 081992

L 39

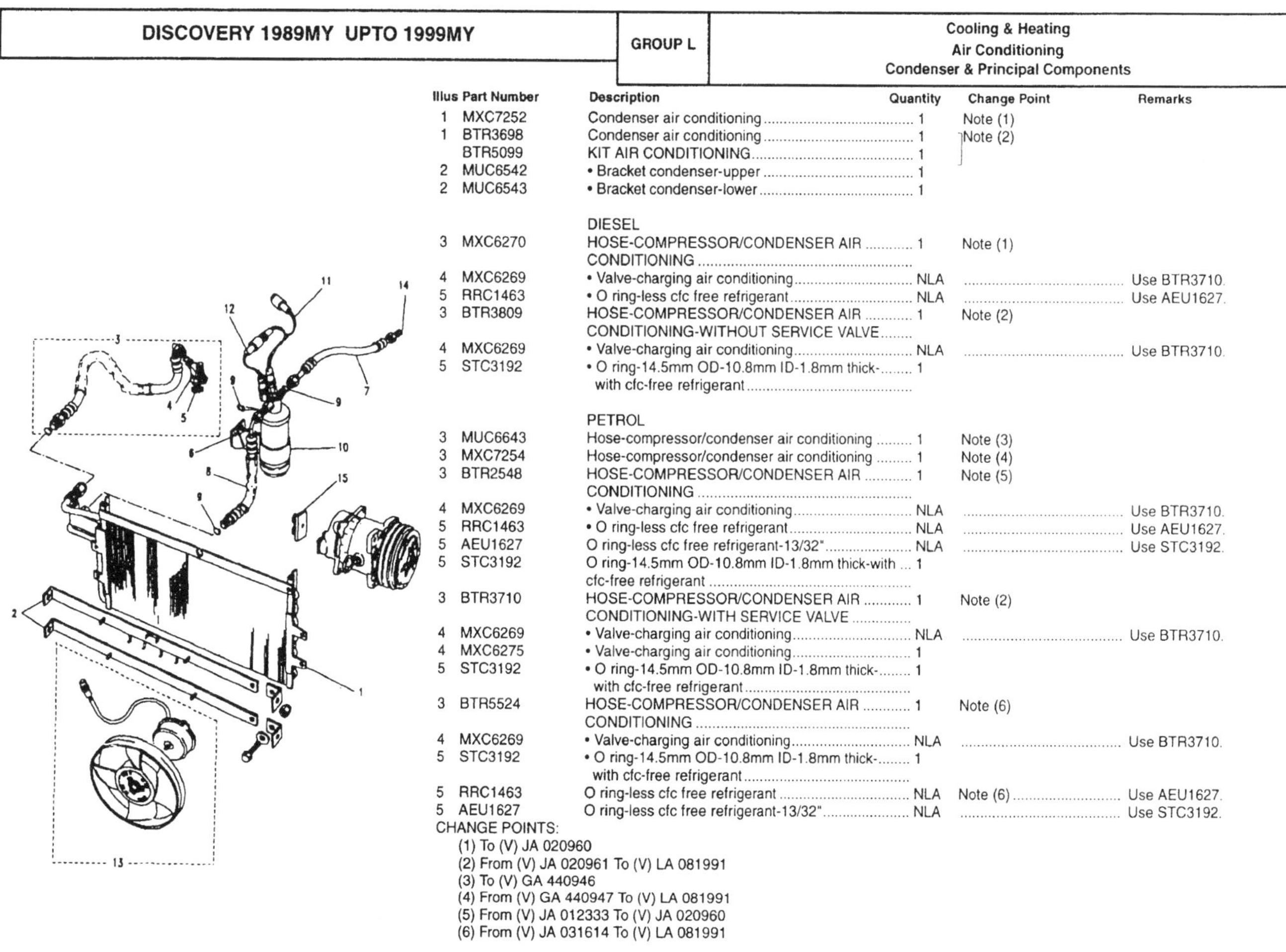

Illus	Part Number	Description	Quantity	Change Point	Remarks
1	MXC7252	Condenser air conditioning	1	Note (1)	
1	BTR3698	Condenser air conditioning	1	Note (2)	
	BTR5099	KIT AIR CONDITIONING	1		
2	MUC6542	• Bracket condenser-upper	1		
2	MUC6543	• Bracket condenser-lower	1		
		DIESEL			
3	MXC6270	HOSE-COMPRESSOR/CONDENSER AIR CONDITIONING	1	Note (1)	
4	MXC6269	• Valve-charging air conditioning	NLA		Use BTR3710.
5	RRC1463	• O ring-less cfc free refrigerant	NLA		Use AEU1627.
3	BTR3809	HOSE-COMPRESSOR/CONDENSER AIR CONDITIONING-WITHOUT SERVICE VALVE	1	Note (2)	
4	MXC6269	• Valve-charging air conditioning	NLA		Use BTR3710.
5	STC3192	• O ring-14.5mm OD-10.8mm ID-1.8mm thick-with cfc-free refrigerant	1		
		PETROL			
3	MUC6643	Hose-compressor/condenser air conditioning	1	Note (3)	
3	MXC7254	Hose-compressor/condenser air conditioning	1	Note (4)	
3	BTR2548	HOSE-COMPRESSOR/CONDENSER AIR CONDITIONING	1	Note (5)	
4	MXC6269	• Valve-charging air conditioning	NLA		Use BTR3710.
5	RRC1463	• O ring-less cfc free refrigerant	NLA		Use AEU1627.
5	AEU1627	O ring-less cfc free refrigerant-13/32"	NLA		Use STC3192.
5	STC3192	O ring-14.5mm OD-10.8mm ID-1.8mm thick-with cfc-free refrigerant	1		
3	BTR3710	HOSE-COMPRESSOR/CONDENSER AIR CONDITIONING-WITH SERVICE VALVE	1	Note (2)	
4	MXC6269	• Valve-charging air conditioning	NLA		Use BTR3710.
4	MXC6275	• Valve-charging air conditioning	1		
5	STC3192	• O ring-14.5mm OD-10.8mm ID-1.8mm thick-with cfc-free refrigerant	1		
3	BTR5524	HOSE-COMPRESSOR/CONDENSER AIR CONDITIONING	1	Note (6)	
4	MXC6269	• Valve-charging air conditioning	NLA		Use BTR3710.
5	STC3192	• O ring-14.5mm OD-10.8mm ID-1.8mm thick-with cfc-free refrigerant	1		
5	RRC1463	O ring-less cfc free refrigerant	NLA	Note (6)	Use AEU1627.
5	AEU1627	O ring-less cfc free refrigerant-13/32"	NLA		Use STC3192.

CHANGE POINTS:
(1) To (V) JA 020960
(2) From (V) JA 020961 To (V) LA 081991
(3) To (V) GA 440946
(4) From (V) GA 440947 To (V) LA 081991
(5) From (V) JA 012333 To (V) JA 020960
(6) From (V) JA 031614 To (V) LA 081991

L 40

Illus	Part Number	Description	Quantity	Change Point	Remarks
4	MXC6269	Valve-charging air conditioning	NLA	Note (1)	Use BTR3710.
3	BTR3710	HOSE-COMPRESSOR/CONDENSER AIR CONDITIONING-WITH SERVICE VALVE	1		
4	MXC6269	• Valve-charging air conditioning	NLA		Use BTR3710.
4	MXC6275	• Valve-charging air conditioning	1		
5	STC3192	• O ring-14.5mm OD-10.8mm ID-1.8mm thick-with cfc-free refrigerant	1		
5	RTC7452	O ring	1	Note (1)	
		PETROL - LHS			
6	MXC7257	Hose-receiver dryer/evaporator air conditioning	1	Note (2)	
6	BTR3630	Hose-receiver dryer/evaporator air conditioning	1	Note (3)	
		HOSE-RECEIVER DRYER/EVAPORATOR-AIR CONDITIONING			
7	BTR5527	LHD	1	Note (1)	
7	MXC7256	petrol-RHD	1		
7	BTR2564	diesel-LHD	1		
7	STC221	diesel-RHD	1		
8	AEU3064	Hose-condenser/receiver dryer air conditioning	NLA		Use BTR2564. LHD Diesel
8	AEU3064	Hose-condenser/receiver dryer air conditioning	NLA	Note (1)	Use BTR2564. Petrol
7	BTR2564	Hose-condenser/evaporator air conditioning	1	Note (1)	
9	STC3191	O ring-11.2mm OD-7.6mm ID-1.8mm thick-with cfc-free refrigerant	2	Note (1)	LHD Diesel
		ALL VARIANTS			
10	AEU1220	Receiver dryer assembly R134a-air conditioning	1	Note (4)	
10	MXC7258	Receiver dryer assembly R134a-air conditioning	NLA	Note (5)	Use BTR3717 qty 1, BTR3720 qty 1 & STC2992 qty 1. Discard old twin switch lead as new drier has a single connector.
10	BTR3717	Receiver dryer assembly R134a-air conditioning	1	Note (3)	

CHANGE POINTS:
(1) To (V) LA 081991
(2) To (V) JA 020961
(3) From (V) JA 020961 To (V) LA 081991
(4) To (V) HA 012332
(5) From (V) JA 012333 To (V) JA 020960

L 41

DISCOVERY 1989MY UPTO 1999MY | GROUP L | Cooling & Heating
Air Conditioning
Condenser & Principal Components

Illus	Part Number	Description	Quantity	Change Point	Remarks
		SWITCH-PRESSURE AIR CONDITIONING CONTROL			
11	MXC7261	low pressure	1	Note (1)	
12	MXC7260	high pressure	1		
12	BTR3720	dual pressure	1	Note (2)	
13	RTC6746	Motor assembly blower-air conditioning	2	Note (3)	
13	MXC7269	Motor assembly blower-air conditioning	2	Note (4)	
13	BTR3700	Motor assembly blower-air conditioning	2	Note (2)	
14	RTC5905	Adaptor-pipe-dual male	1	Note (5)	
15	MUC4358	Adaptor-air conditioning compressor hose	1		
	BTR3814	Resistor	1		

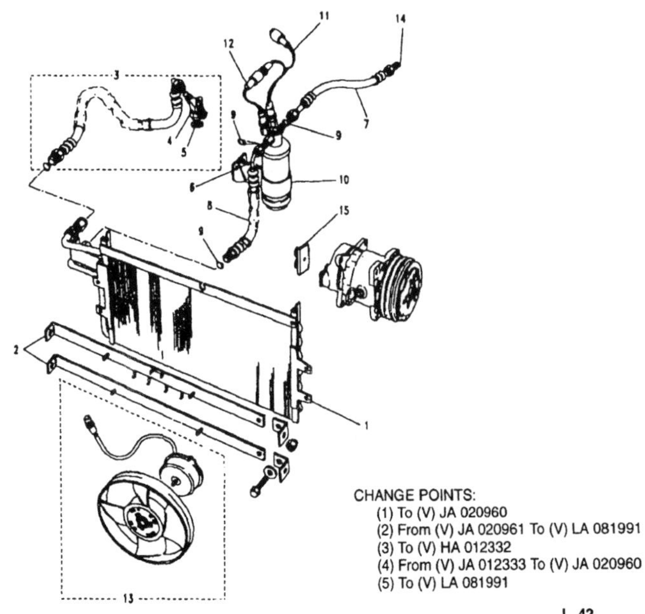

CHANGE POINTS:
(1) To (V) JA 020960
(2) From (V) JA 020961 To (V) LA 081991
(3) To (V) HA 012332
(4) From (V) JA 012333 To (V) JA 020960
(5) To (V) LA 081991

L 42

DISCOVERY 1989MY UPTO 1999MY	GROUP L	Cooling & Heating Air Conditioning Rear Air Con-Ducts and Vents

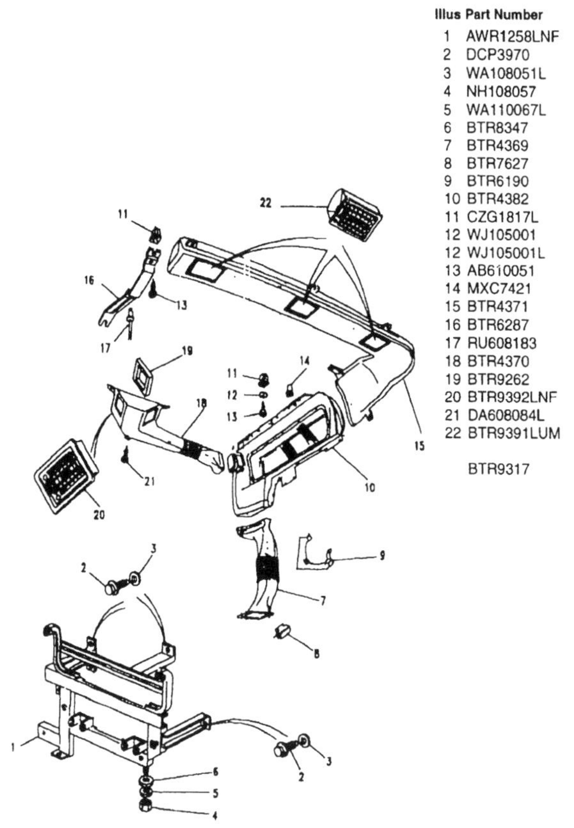

Illus	Part Number	Description	Quantity	Change Point	Remarks
1	AWR1258LNF	Frame-rear air conditioning unit	1	Note (1)	
2	DCP3970	Bolt-7 seats-M8 x 25	4	Note (1)	
3	WA108051L	Washer-plain-7 seats-M8	4	Note (1)	
4	NH108057	Locknut-7 seats	1	Note (1)	
5	WA110067L	Washer-7 seats-M10	1	Note (1)	
6	BTR8347	Washer-rubber-7 seats	1	Note (1)	
7	BTR4369	Duct-rear air conditioning-7 seats	1	Note (1)	
8	BTR7627	Clip-7 seats	3	Note (1)	
9	BTR6190	Bracket-speaker/E post-LH	1	Note (1)	
10	BTR4382	Panel-alpine light trim-7 seats	1	Note (1)	
11	CZG1817L	Nutsert-blind-7 seats	A/R	Note (1)	
12	WJ105001	Washer-plain-7 seats-M5-oversize-metric	NLA	Note (1)	Use WJ105001L.
12	WJ105001L	Washer-plain-7 seats-M5-oversize	A/R	Note (1)	
13	AB610051	Screw-7 seats	A/R	Note (1)	
14	MXC7421	Retainer-trim-7 seats	A/R	Note (1)	
15	BTR4371	Duct-rear air conditioning-7 seats	1	Note (1)	
16	BTR6287	Strap-air conditioning duct-7 seats	1	Note (1)	
17	RU608183	Rivet-7 seats	8	Note (1)	
18	BTR4370	Duct-rear air conditioning-7 seats	1	Note (1)	
19	BTR9262	Seal air conditioning-7 seats	2	Note (1)	
20	BTR9392LNF	Vent-rear air conditioning-7 seats-Ash Grey	2	Note (1)	
21	DA608084L	Screw-7 seats	1	Note (1)	
22	BTR9391LUM	Vent assembly-air conditioning duct-7 seats-Mist Grey	3	Note (1)	
	BTR9317	Bracket-air conditioning duct-7 seats	2	Note (1)	

CHANGE POINTS:
(1) From (V) MA 081992

L 43

DISCOVERY 1989MY UPTO 1999MY	GROUP L	Cooling & Heating Air Conditioning Hoses

Illus	Part Number	Description	Quantity	Change Point	Remarks
1	BTR6727	PIPE-EVAPORATOR/COMPRESSOR AIR CONDITIONING	1	Note (1)	
	STC3151	• O ring-5/8" diameter	2		
2	BTR8921	Clip-pipe	1	Note (1)	
3	BTR4041	HOSE-EVAPR/COMPRESSOR AIR CONDITIONING	1		V8 Petrol
3	BTR4042	HOSE-EVAPR/COMPRESSOR AIR CONDITIONING	1		4 Cylinder Petrol MPi
3	BTR4040	HOSE-EVAPR/COMPRESSOR AIR CONDITIONING	1		4 Cylinder Diesel TDi
	STC3151	• O ring-5/8" diameter	2		
4	BTR4045	HOSE-COMPRESSOR/CONDENSER AIR CONDITIONING	1	Note (1)	V8 Petrol
4	BTR4046	HOSE-COMPRESSOR/CONDENSER AIR CONDITIONING	1		4 Cylinder Petrol MPi
4	BTR4044	HOSE-COMPRESSOR/CONDENSER AIR CONDITIONING	1		4 Cylinder Diesel TDi
	STC3150	• O ring-1/2"diameter	2		
5	BTR7230	PIPE-FLUID AIR CONDITIONING	1	Note (1)	
	STC3149	• O ring-8mm diameter	1		
6	BTR4038	PIPE-FLUID AIR CONDITIONING	1	Note (1)	
	STC3149	• O ring-8mm diameter	1		
7	BTR9613	Bracket-air charge/body air cooler	1	Note (1)	
8	BTR7301	Clip-pipe	2		

CHANGE POINTS:
(1) From (V) MA 081992

L 44

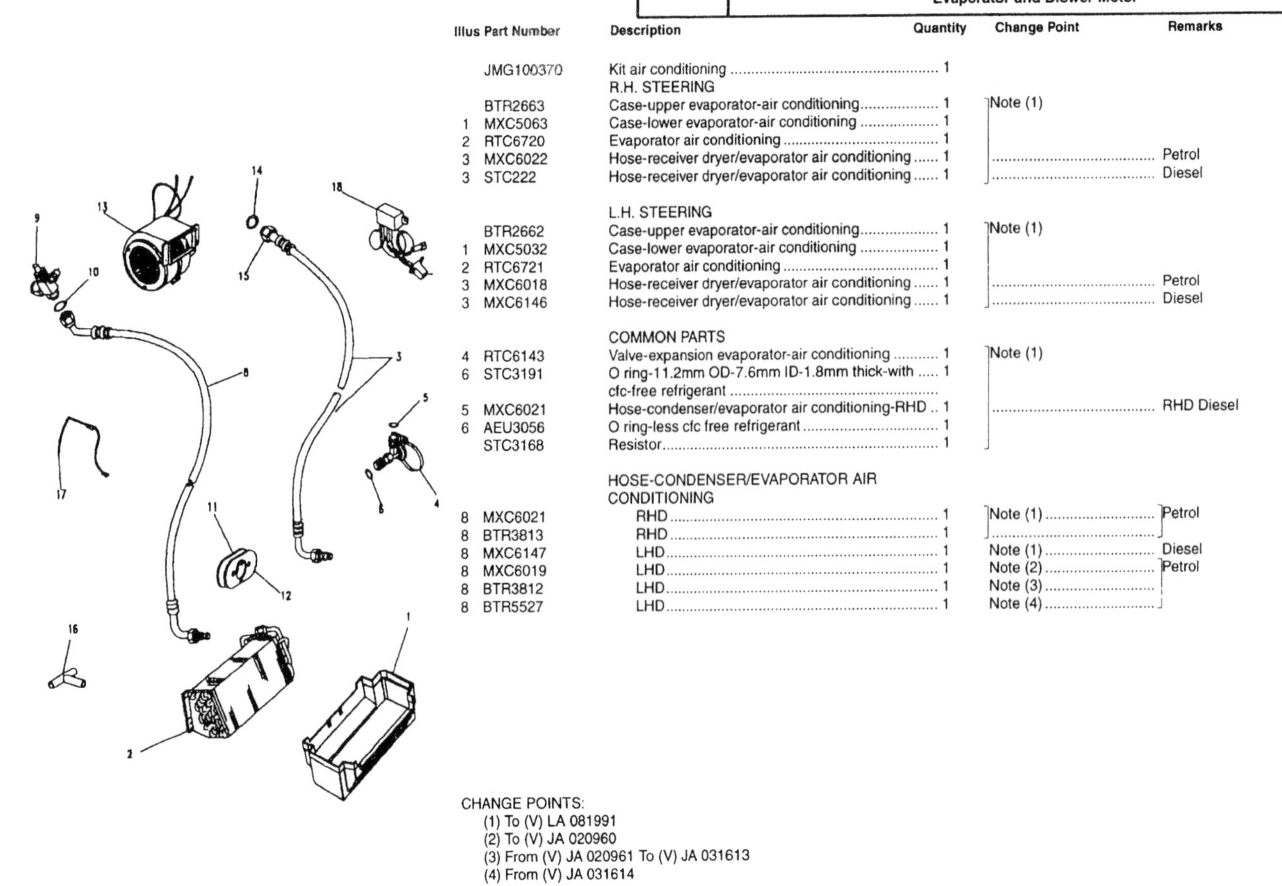

Illus	Part Number	Description	Quantity	Change Point	Remarks
	JMG100370	Kit air conditioning	1		
		R.H. STEERING			
	BTR2663	Case-upper evaporator-air conditioning	1	Note (1)	
1	MXC5063	Case-lower evaporator-air conditioning	1		
2	RTC6720	Evaporator air conditioning	1		
3	MXC6022	Hose-receiver dryer/evaporator air conditioning	1		Petrol
3	STC222	Hose-receiver dryer/evaporator air conditioning	1		Diesel
		L.H. STEERING			
	BTR2662	Case-upper evaporator-air conditioning	1	Note (1)	
1	MXC5032	Case-lower evaporator-air conditioning	1		
2	RTC6721	Evaporator air conditioning	1		
3	MXC6018	Hose-receiver dryer/evaporator air conditioning	1		Petrol
3	MXC6146	Hose-receiver dryer/evaporator air conditioning	1		Diesel
		COMMON PARTS			
4	RTC6143	Valve-expansion evaporator-air conditioning	1	Note (1)	
6	STC3191	O ring-11.2mm OD-7.6mm ID-1.8mm thick-with cfc-free refrigerant	1		
5	MXC6021	Hose-condenser/evaporator air conditioning-RHD	1		RHD Diesel
6	AEU3056	O ring-less cfc free refrigerant	1		
	STC3168	Resistor	1		
		HOSE-CONDENSER/EVAPORATOR AIR CONDITIONING			
8	MXC6021	RHD	1	Note (1)	Petrol
8	BTR3813	RHD	1		
8	MXC6147	LHD	1	Note (1)	Diesel
8	MXC6019	LHD	1	Note (2)	Petrol
8	BTR3812	LHD	1	Note (3)	
8	BTR5527	LHD	1	Note (4)	

CHANGE POINTS:
(1) To (V) LA 081991
(2) To (V) JA 020960
(3) From (V) JA 020961 To (V) JA 031613
(4) From (V) JA 031614

DISCOVERY 1989MY UPTO 1999MY | GROUP L | Cooling & Heating Air Conditioning Evaporator and Blower Motor

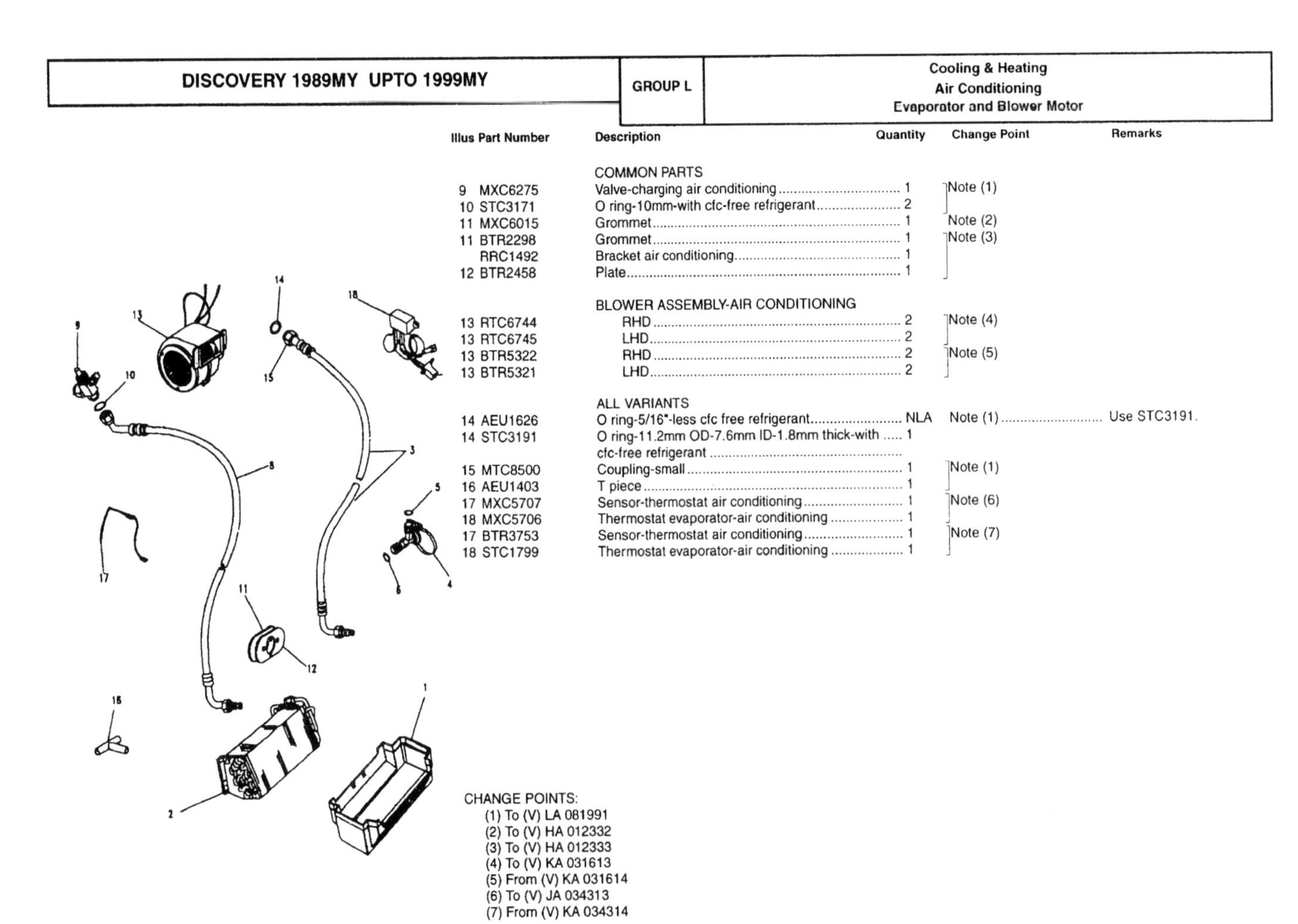

Illus	Part Number	Description	Quantity	Change Point	Remarks
		COMMON PARTS			
9	MXC6275	Valve-charging air conditioning	1	Note (1)	
10	STC3171	O ring-10mm-with cfc-free refrigerant	2		
11	MXC6015	Grommet	1	Note (2)	
11	BTR2298	Grommet	1	Note (3)	
	RRC1492	Bracket air conditioning	1		
12	BTR2458	Plate	1		
		BLOWER ASSEMBLY-AIR CONDITIONING			
13	RTC6744	RHD	2	Note (4)	
13	RTC6745	LHD	2		
13	BTR5322	RHD	2	Note (5)	
13	BTR5321	LHD	2		
		ALL VARIANTS			
14	AEU1626	O ring-5/16"-less cfc free refrigerant	NLA	Note (1)	Use STC3191.
14	STC3191	O ring-11.2mm OD-7.6mm ID-1.8mm thick-with cfc-free refrigerant	1		
15	MTC8500	Coupling-small	1	Note (1)	
16	AEU1403	T piece	1		
17	MXC5707	Sensor-thermostat air conditioning	1	Note (6)	
18	MXC5706	Thermostat evaporator-air conditioning	1		
17	BTR3753	Sensor-thermostat air conditioning	1	Note (7)	
18	STC1799	Thermostat evaporator-air conditioning	1		

CHANGE POINTS:
(1) To (V) LA 081991
(2) To (V) HA 012332
(3) To (V) HA 012333
(4) To (V) KA 031613
(5) From (V) KA 031614
(6) To (V) JA 034313
(7) From (V) KA 034314

Illus	Part Number	Description	Quantity	Change Point	Remarks
1	BTR6631	CONDENSER ASSEMBLY AIR CONDITIONING	1	Note (1)	4 Cylinder Petrol MPi
1	BTR6632	CONDENSER ASSEMBLY AIR CONDITIONING	1	Note (1)	4 Cylinder Diesel TDi
1	BTR6632	Condenser assembly air conditioning	1		V8 Petrol
2	HYG10014	Bolt-flanged head-M8	2	Note (2)	
3	CLP6235L	Grommet-pipe evaporator-air conditioning	1		
4	CLP6234L	Grommet	1		
5	AWR1322	Label-warning-CFC free	1	Note (2); Note (3)	

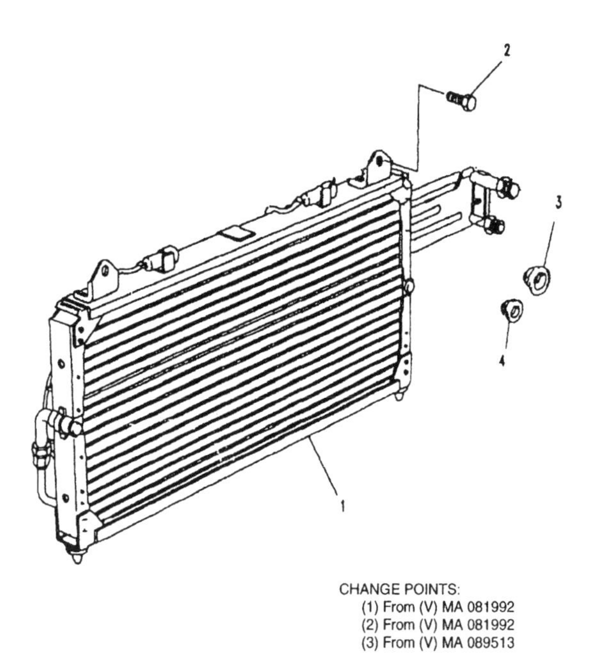

CHANGE POINTS:
(1) From (V) MA 081992
(2) From (V) MA 081992
(3) From (V) MA 089513

Illus	Part Number	Description	Quantity	Change Point	Remarks
		AIR-CONDITIONING CONTROLS			
1	AWR1379	CONTROL ASSEMBLY HEATER-AIR CONDITIONING-WITH CABLES	1	Note (1)	
1	BTR6490	• Control assembly heater-Air conditioning-less cables	NLA		Use AWR1378.
2	AB608051	• Screw-self tapping	4		
3	STC1863	• SWITCH-PUSH PUSH AIR CONDITIONING RECIRCULATING	1		
	STC1877	• • Bulb & holder assembly-Orange	A/R		
	STC1878	• • Bulb & holder assembly-Green	A/R		
4	BTR6491	• SWITCH-PUSH PUSH AIR CONDITIONING-FRONT	1		
	STC1877	• • Bulb & holder assembly-Orange	A/R		
	STC1878	• • Bulb & holder assembly-Green	A/R		
5	BTR4050	• SWITCH-PUSH PUSH AIR CONDITIONING-REAR	1		
	STC1877	• • Bulb & holder assembly-Orange	A/R		
	STC1878	• • Bulb & holder assembly-Green	A/R		
6	BTR6492	• Plate-blanking air conditioning switch	1		
7	BTR8930	• Knob-control heater	3		
8	BTR6494	• Knob-slider control-heater	1		
	JFE100640	• Plate-graphics control-heater	1		
10	AB606031	• Screw-self tapping-No 6 x 3/8	NLA		Use AB606031L.
11	RTC3635	• Bulb-286-w5/1.2-1.2 Watt-12V-1.2 Watt	1		
10	AB606031L	Screw-self tapping-No 6 x 3/8			

CHANGE POINTS:
(1) From (V) MA 081992

Illus	Part Number	Description	Quantity	Change Point	Remarks
1	BTR5258	Evaporator air conditioning	1	Note (1)	
2	BTR5263	Blower assembly-air conditioning	1		
3	BTR5250	Case-lower evaporator-air conditioning	1		
4	BTR5255	Case-upper evaporator-air conditioning-rear	1		
5	BTR5269	Valve-expansion evaporator-air conditioning-rear	1		
6	AEU1626	O ring-5/16"-less cfc free refrigerant	NLA	Note (1)	Use STC3191.
6	STC3191	O ring-11.2mm OD-7.6mm ID-1.8mm thick-with cfc-free refrigerant	1		
7	BTR6100	Case-heater with air conditioning	1	Note (1)	
		SUCTION PIPE - EVAPORATOR TO SUCTION HOSE			
8	BTR5272	Pipe-suction air conditioning-rear	1	Note (1)	
		LIQUID PIPE - LIQUID HOSE TO EXPANSION TANK			
9	BTR5274	Pipe-fluid air conditioning-rear	1	Note (1)	
		SUCTION HOSE - SUCTION PIPE TO COMPRESSOR			
10	BTR6032	Hose-suction	1	Note (1)	RHD Petrol
10	BTR6031	Hose-suction	1		LHD Petrol
10	BTR6092	Hose-suction	1		RHD Diesel
10	BTR6093	Hose-suction	1		LHD Diesel
		LIQUID HOSE - FROM CONDENSER TO LIQUID PIPE			
11	BTR6034	Hose-liquid	1	Note (1)	RHD
11	BTR6033	Hose-liquid	1		LHD
		ALL VARIANTS			
12	ANR1583	Nut-spring-u type	2	Note (1)	
13	BTR6087	Sleeve-air conditioning pipe protection-liquid	1		
13	BTR6088	Sleeve-air conditioning pipe protection-suction	1		
14	BTR6078	Coupling-female	1		
15	BTR6079	Coupling-large-male	1		
16	BTR6080	O ring-large	1		
17	BTR6081	Coupling-female	1		
18	BTR6082	Coupling-male-small	1		
19	BTR6095	O ring-large	1		

CHANGE POINTS:
(1) To (V) LA 081991

Illus	Part Number	Description	Quantity	Change Point	Remarks
		AIR CONDITIONING PRINCIPAL COMPONENTS			
1	BTR4048	Air conditioning sub assembly	1	Note (1)	
		CONTROL UNIT REAR AIR-CONDTIONING			
2	BTR4049	Switch pack-air conditioning	1	Note (1)	

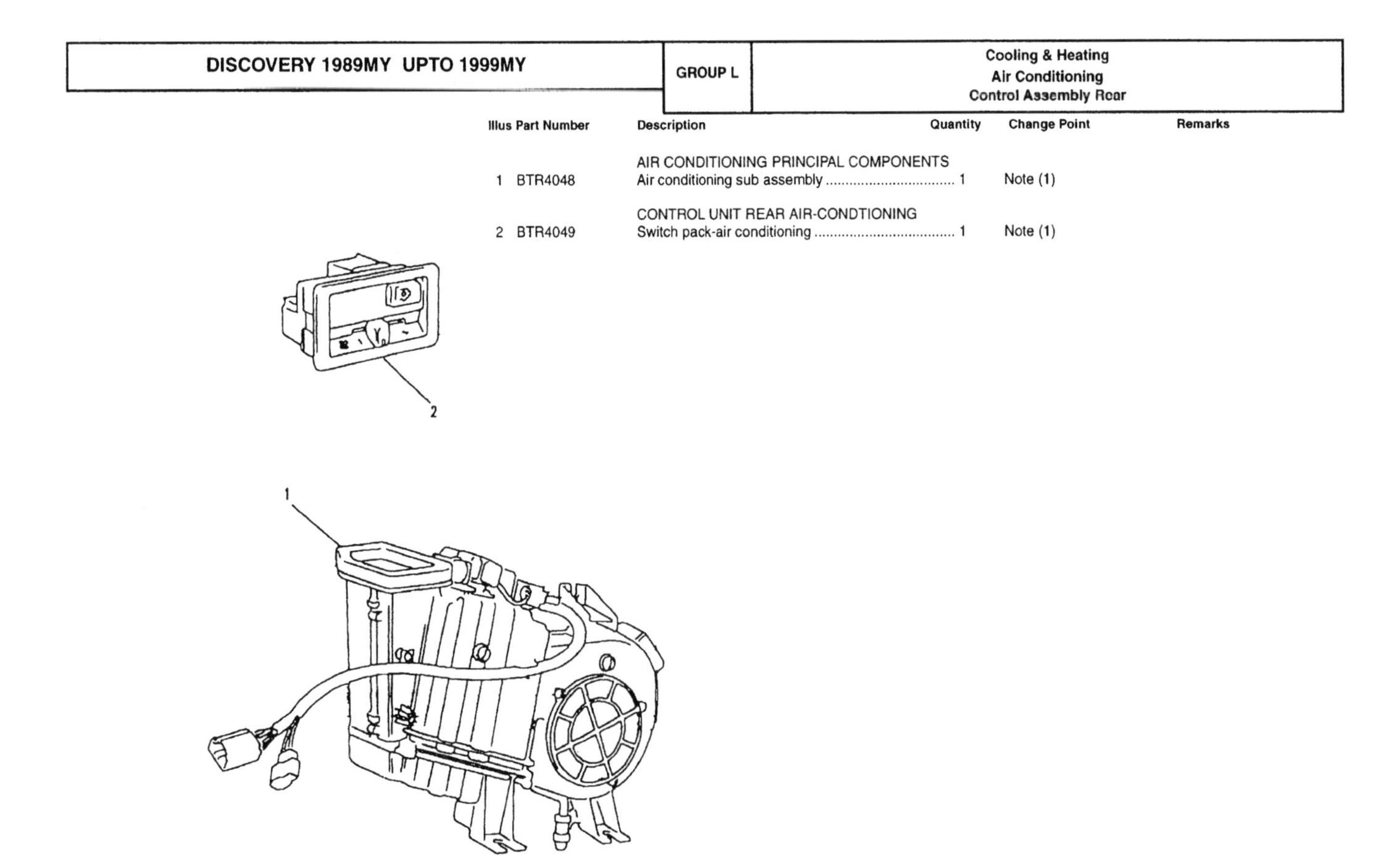

CHANGE POINTS:
(1) From (V) MA 081992

Hoses Rear Under Bonnet

Illus	Part Number	Description	Quantity	Change Point	Remarks
		EVAPORATOR TO COMPRESSOR			
1	BTR7260	Hose-evapr/compressor air conditioning-rear	1	Note (1)	
2	BTR8921	Clip-pipe	1		
3	BTR4041	HOSE-EVAPR/COMPRESSOR AIR CONDITIONING	1		Except TDi
3	BTR4040	HOSE-EVAPR/COMPRESSOR AIR CONDITIONING	1	Note (1)	TDi
	STC3151	• O ring-5/8" diameter	2		
		COMPRESSOR TO CONDENSER			
4	BTR4045	HOSE-COMPRESSOR/CONDENSER AIR CONDITIONING	1	Note (1)	V8 Petrol
4	BTR4044	HOSE-COMPRESSOR/CONDENSER AIR CONDITIONING	1		Diesel TDi
	STC3150	• O ring-1/2"diameter	2		
		CONDENSER TO EVAPORATOR			
5	BTR7230	PIPE-FLUID AIR CONDITIONING	1	Note (1)	
6	BTR7259	PIPE-FLUID AIR CONDITIONING-REAR	1		
	STC3149	• O ring-8mm diameter	1		
7	BTR9613	Bracket-air charge/body air cooler	1	Note (1)	
8	BTR7301	Clip-pipe	2		
		HOSE FIXINGS			
	FB106061M	Bolt-flanged head	2	Note (1)	

CHANGE POINTS:
(1) From (V) MA 081992

L 51

DISCOVERY 1989MY UPTO 1999MY | GROUP L | Cooling & Heating / Air Conditioning

Hoses-Rear In Car

Illus	Part Number	Description	Quantity	Change Point	Remarks
		LIQUID SUPPLY TO REAR EVAPORATOR			
1	BTR7359	Hose-receiver dryer/evaporator air conditioning	1	Note (1)	
2	BTR7360	Hose-receiver dryer/evaporator air conditioning-rear	1		
3	BTR7361	Hose-receiver dryer/evaporator air conditioning-rear	1		
4	BTR7362	Hose-receiver dryer/evaporator air conditioning-rear	1		
5	BTR8586	Hose-receiver dryer/evaporator air conditioning	1		
6	BTR7304	Clip-pipe	3		
7	BTR7302	Clip-pipe	4		
		VAPOUR RETURN FROM REAR EVAPORATOR			
8	BTR8587	Hose-evapr/compressor air conditioning	1	Note (1)	
9	BTR7366	Hose-evapr/compressor air conditioning-rear	1		
10	BTR7365	Hose-evapr/compressor air conditioning-rear	1		
11	BTR7364	Hose-evapr/compressor air conditioning-rear	1		
12	BTR7363	Hose-evapr/compressor air conditioning-rear	1		
13	BTR7303	Clip-pipe	3		

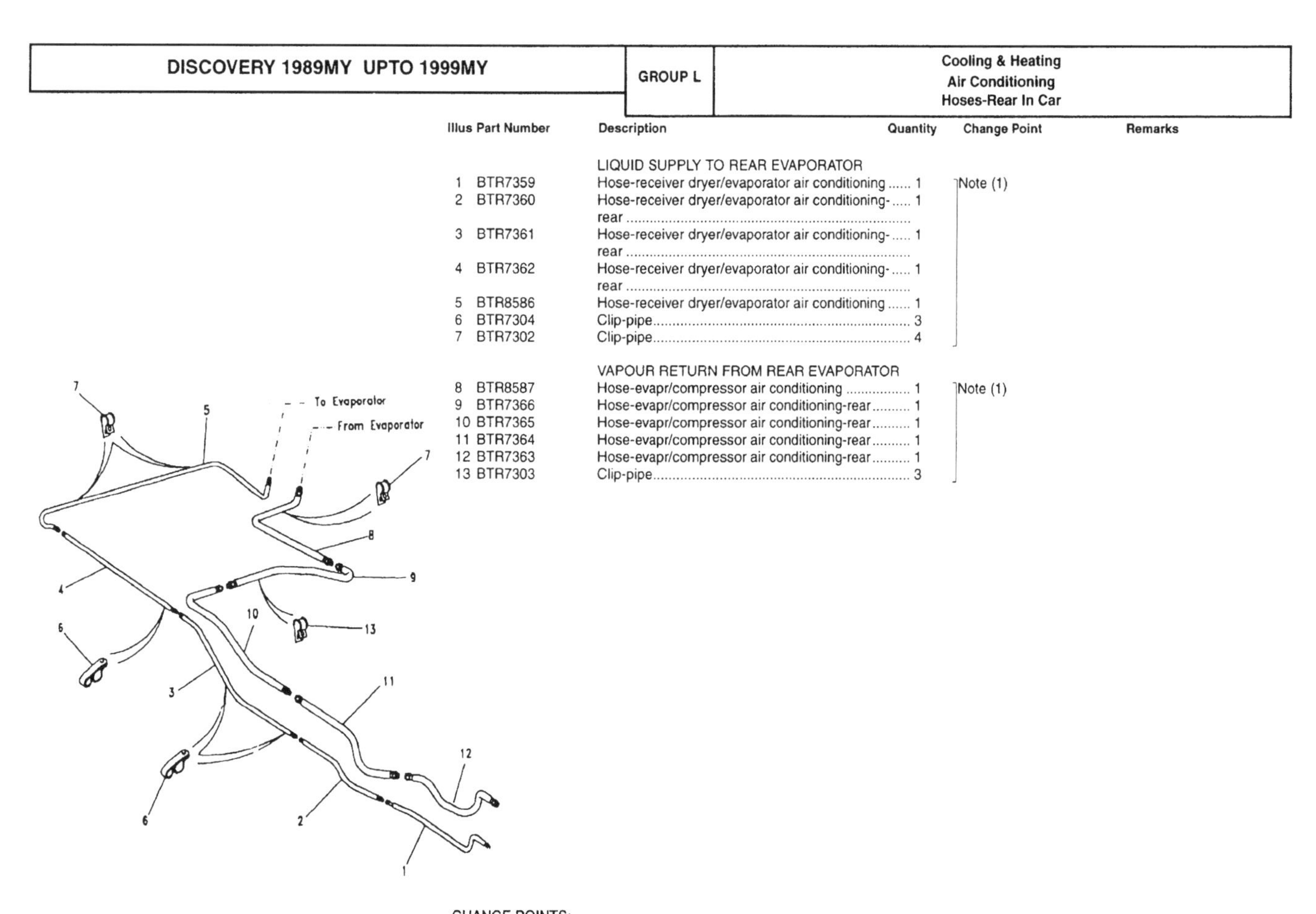

CHANGE POINTS:
(1) From (V) MA 081992

L 52

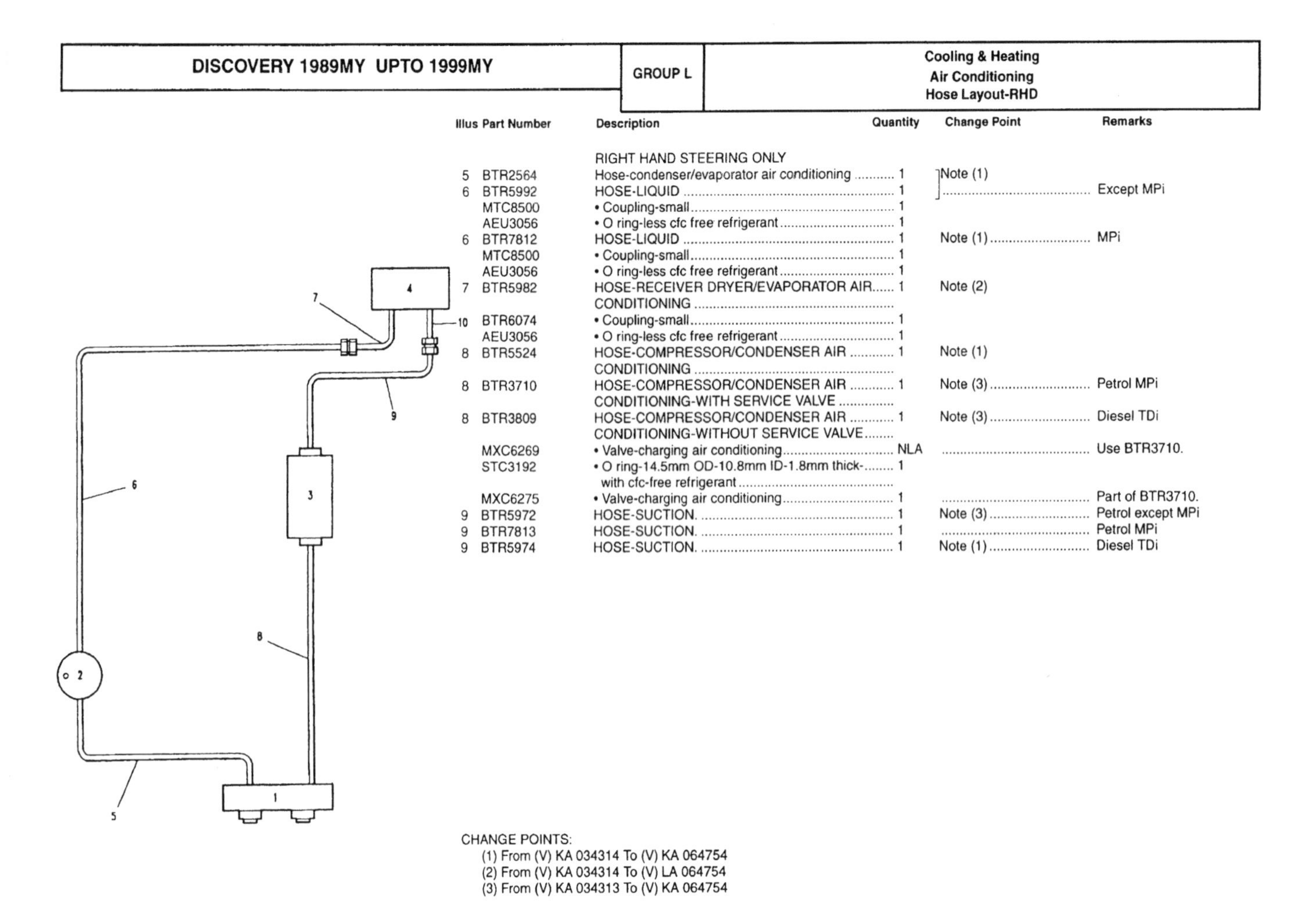

Illus	Part Number	Description	Quantity	Change Point	Remarks
		RIGHT HAND STEERING ONLY			
5	BTR2564	Hose-condenser/evaporator air conditioning	1	Note (1)	Except MPi
6	BTR5992	HOSE-LIQUID	1	Note (1)	Except MPi
	MTC8500	• Coupling-small	1		
	AEU3056	• O ring-less cfc free refrigerant	1		
6	BTR7812	HOSE-LIQUID	1	Note (1)	MPi
	MTC8500	• Coupling-small	1		
	AEU3056	• O ring-less cfc free refrigerant	1		
7	BTR5982	HOSE-RECEIVER DRYER/EVAPORATOR AIR CONDITIONING	1	Note (2)	
	BTR6074	• Coupling-small	1		
	AEU3056	• O ring-less cfc free refrigerant	1		
8	BTR5524	HOSE-COMPRESSOR/CONDENSER AIR CONDITIONING	1	Note (1)	
8	BTR3710	HOSE-COMPRESSOR/CONDENSER AIR CONDITIONING-WITH SERVICE VALVE	1	Note (3)	Petrol MPi
8	BTR3809	HOSE-COMPRESSOR/CONDENSER AIR CONDITIONING-WITHOUT SERVICE VALVE	1	Note (3)	Diesel TDi
	MXC6269	• Valve-charging air conditioning	NLA		Use BTR3710.
	STC3192	• O ring-14.5mm OD-10.8mm ID-1.8mm thick-with cfc-free refrigerant	1		
	MXC6275	• Valve-charging air conditioning	1		Part of BTR3710.
9	BTR5972	HOSE-SUCTION.	1	Note (3)	Petrol except MPi
9	BTR7813	HOSE-SUCTION.	1		Petrol MPi
9	BTR5974	HOSE-SUCTION.	1	Note (1)	Diesel TDi

CHANGE POINTS:
(1) From (V) KA 034314 To (V) KA 064754
(2) From (V) KA 034314 To (V) LA 064754
(3) From (V) KA 034313 To (V) KA 064754

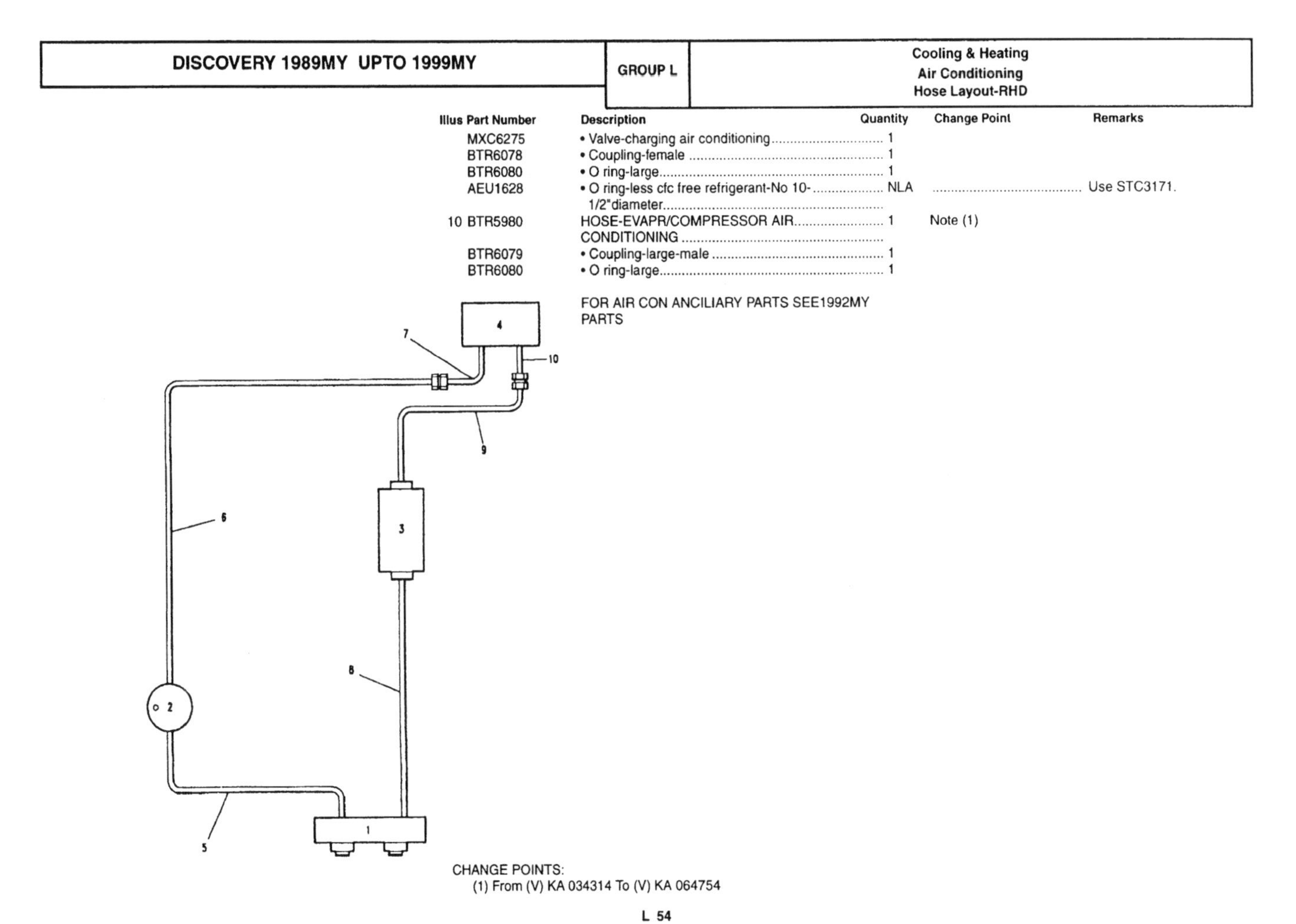

Illus	Part Number	Description	Quantity	Change Point	Remarks
	MXC6275	• Valve-charging air conditioning	1		
	BTR6078	• Coupling-female	1		
	BTR6080	• O ring-large	1		
	AEU1628	• O ring-less cfc free refrigerant-No 10-1/2"diameter	NLA		Use STC3171.
10	BTR5980	HOSE-EVAPR/COMPRESSOR AIR CONDITIONING	1	Note (1)	
	BTR6079	• Coupling-large-male	1		
	BTR6080	• O ring-large	1		

FOR AIR CON ANCILIARY PARTS SEE1992MY PARTS

CHANGE POINTS:
(1) From (V) KA 034314 To (V) KA 064754

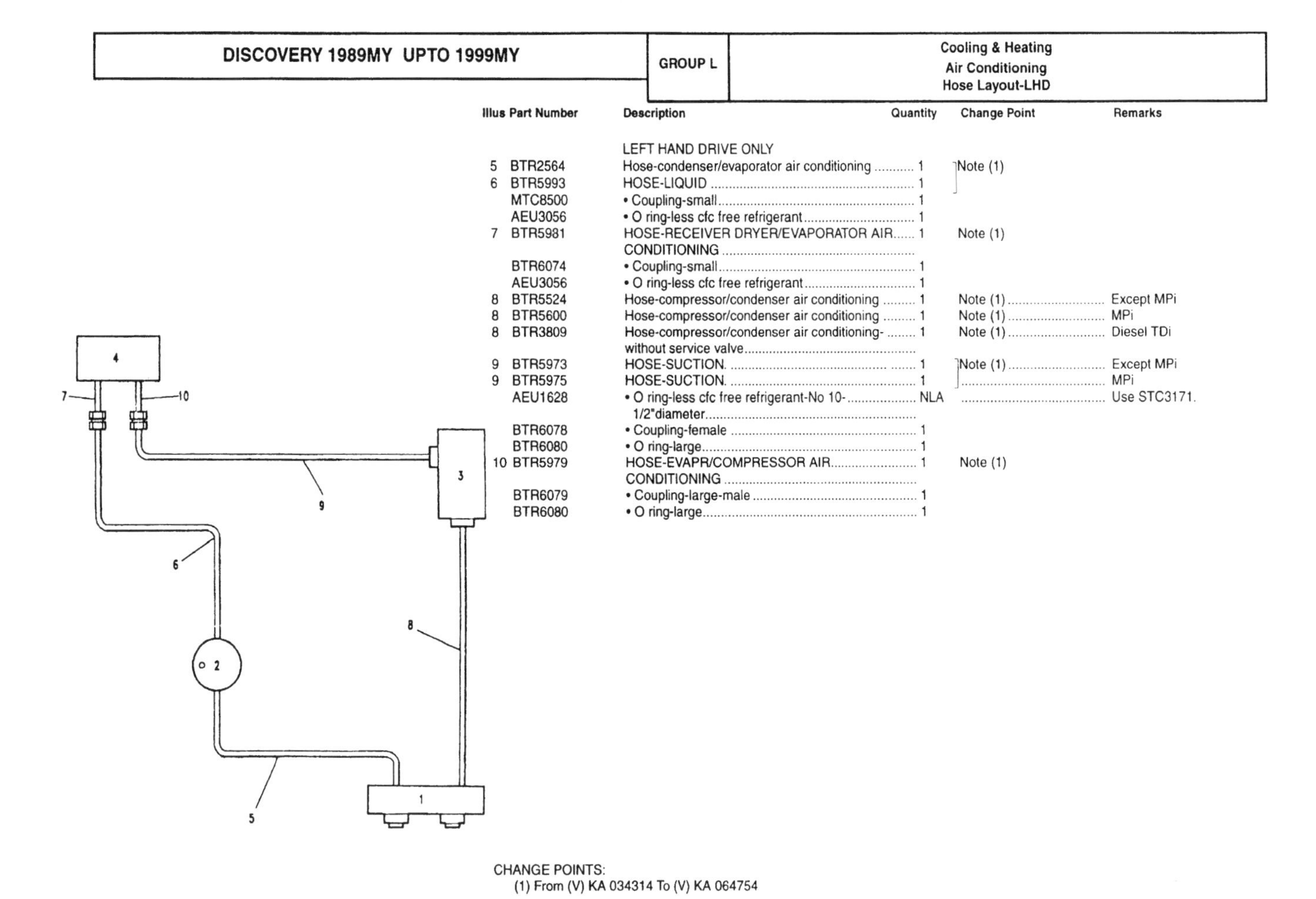

Illus	Part Number	Description	Quantity	Change Point	Remarks
		LEFT HAND DRIVE ONLY			
5	BTR2564	Hose-condenser/evaporator air conditioning	1	Note (1)	
6	BTR5993	HOSE-LIQUID	1		
	MTC8500	• Coupling-small	1		
	AEU3056	• O ring-less cfc free refrigerant	1		
7	BTR5981	HOSE-RECEIVER DRYER/EVAPORATOR AIR CONDITIONING	1	Note (1)	
	BTR6074	• Coupling-small	1		
	AEU3056	• O ring-less cfc free refrigerant	1		
8	BTR5524	Hose-compressor/condenser air conditioning	1	Note (1)	Except MPi
8	BTR5600	Hose-compressor/condenser air conditioning	1	Note (1)	MPi
8	BTR3809	Hose-compressor/condenser air conditioning-without service valve	1	Note (1)	Diesel TDi
9	BTR5973	HOSE-SUCTION.	1	Note (1)	Except MPi
9	BTR5975	HOSE-SUCTION.	1		MPi
	AEU1628	• O ring-less cfc free refrigerant-No 10-1/2"diameter	NLA		Use STC3171.
	BTR6078	• Coupling-female	1		
	BTR6080	• O ring-large	1		
10	BTR5979	HOSE-EVAPR/COMPRESSOR AIR CONDITIONING	1	Note (1)	
	BTR6079	• Coupling-large-male	1		
	BTR6080	• O ring-large	1		

CHANGE POINTS:
(1) From (V) KA 034314 To (V) KA 064754

L 55

Notes

Ancillaries ..... M 101
- ABS ..... M 111
- Alarm / Remote Locking System ..... M 102
- Cruise Control ..... M 112
- Heated Front Screen ..... M 116
- Horn ..... M 101
- Speedo Cable - V8 Twin Carb & 200TDi-To ..... M 113
- Speedo Transducer ..... M 115

Batteries & Harnesses ..... M 16
- Battery Diesel-To (V) LA081991 ..... M 20
- Battery Petrol - MPI From (V) MA081992 ..... M 19
- Battery Petrol - MPI To (V) LA081991 ..... M 18
- Battery Petrol - V8 Pi From (V) MA081992 ..... M 17
- Battery Petrol - V8 Pi To (V) LA081991 ..... M 16
- Harness - Air Conditioning ..... M 58
- Harness - Brake Pad Wear ..... M 59
- Harness - Fuel Pump-V8 EFI, 200 & 300 TDI ..... M 57
- Harness - Headlamp Levelling/Powerwash & ..... M 60
- Harness - S R S (Air Bag) ..... M 61
- Harness - Transmission-From (V) GA394352 ..... M 54
- Harness-Doors & 'A' Post-To (V) LA081991 ..... M 33
- Harness-Engine-Diesel Turbo-To (V) ..... M 50
- Harness-Engine-Petrol-To (V) LA081991 ..... M 44
- Harness-Front Wing-To (V) LA081991 ..... M 22
- Harness-Main-Without Wing Harness-To (V) ..... M 24

Cable Clips, Ties & Grommets ..... M 153
- Cable Clips ..... M 153
- Grommets ..... M 160

Electronic Control Units ..... M 91
- ECU - EGR-To (V) LA081991 ..... M 91
- ECU - Fuel V8-To (V) LA081991 ..... M 94
- ECU - MPi-To (V) LA081991 ..... M 98
- ECU - Sunroof Control Unit ..... M 97
- ECU - Window Lift ..... M 93
- Electronic Diesel Control ..... M 100

Fuse Box ..... M 81

Fuse Box & Relays ..... M 79

In-Car Entertainment ..... M 132
- Radio Systems-From (V) KA034314 To ..... M 133
- Radio Systems-From (V) MA081992 To ..... M 136
- Radio Systems-From (V) TA163104 & ..... M 138
- Radio Systems-From (V) VA703237 & ..... M 141
- Radio Systems-To (V) JA034313 ..... M 132

Instruments & Clock ..... M 117
- Clock & Cigar Lighter-To (V) LA081991 ..... M 130

Lighting ..... M 1
- Auxiliary ..... M 6
- Front Indicator-To (V) LA081991 ..... M 4
- Headlamps - Wipac To (V) LA081991 ..... M 1
- Interior Lamps-To (V) LA081991 ..... M 12
- Rear Lamps-To (V) LA081991 ..... M 7

Relays ..... M 83

Switches ..... M 63
- Binnacle - To (V) LA081991 ..... M 67
- Column - From (V) MA081992 ..... M 64
- Column - To (V) LA081991 ..... M 63
- Console Front ..... M 73
- Facia assembly ..... M 65
- Sunroof ..... M 76
- Switches Console & Rear Door ..... M 77

Wash Wipe Systems ..... M 144
- Headlamp Wash - To (V) LA081991 ..... M 150
- Rear Screen Wiper/Wash - To (V) LA081991 ..... M 151
- Windscreen & Headlamp Wash - To (V) ..... M 148
- Windscreen Wash - To (V) LA081991 ..... M 146
- Windscreen Wiper - To (V) LA081991 ..... M 144

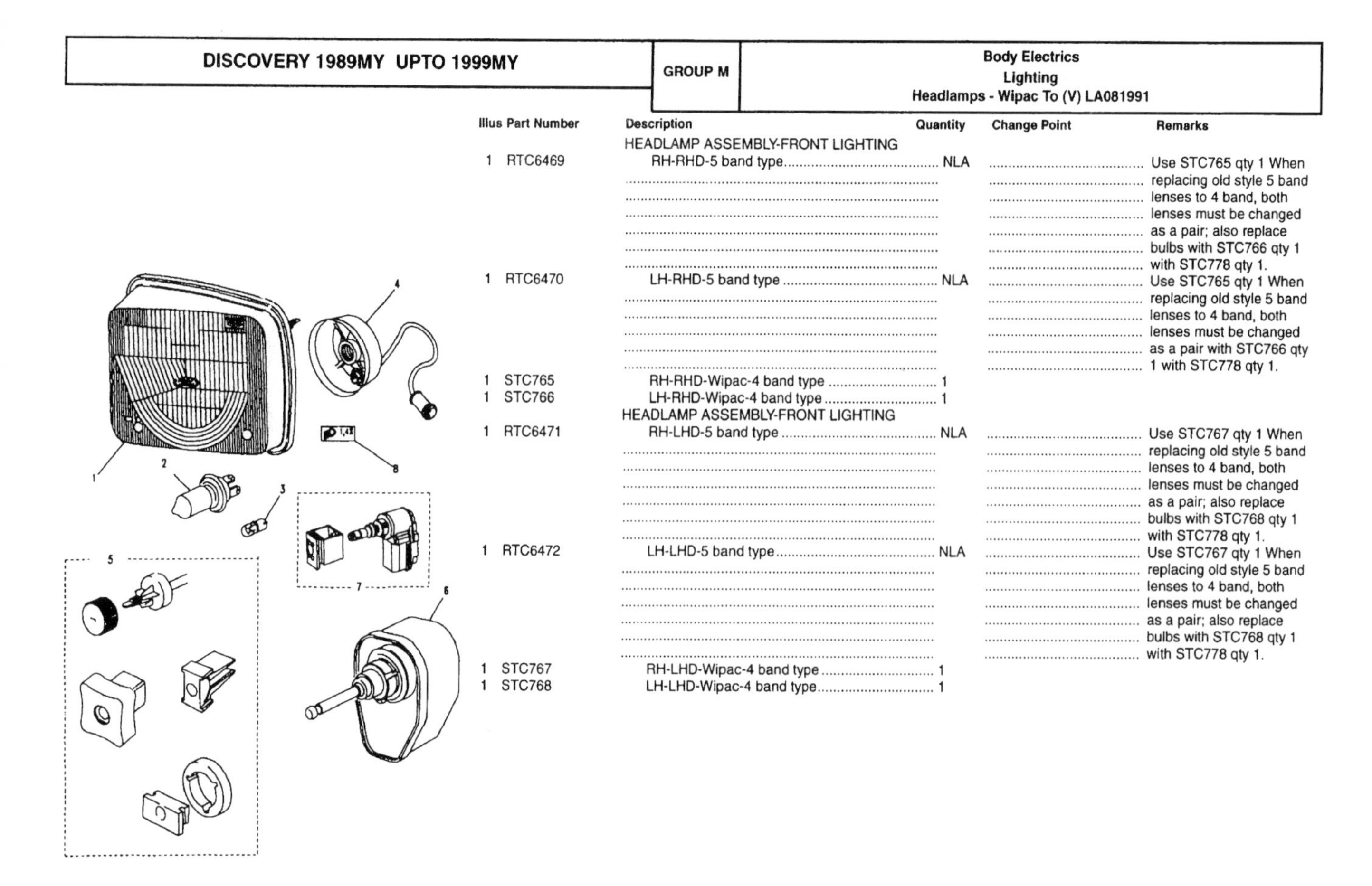

Illus	Part Number	Description	Quantity	Change Point	Remarks
		HEADLAMP ASSEMBLY-FRONT LIGHTING			
1	RTC6469	RH-RHD-5 band type	NLA		Use STC765 qty 1 When replacing old style 5 band lenses to 4 band, both lenses must be changed as a pair; also replace bulbs with STC766 qty 1 with STC778 qty 1.
1	RTC6470	LH-RHD-5 band type	NLA		Use STC765 qty 1 When replacing old style 5 band lenses to 4 band, both lenses must be changed as a pair with STC766 qty 1 with STC778 qty 1.
1	STC765	RH-RHD-Wipac-4 band type	1		
1	STC766	LH-RHD-Wipac-4 band type	1		
		HEADLAMP ASSEMBLY-FRONT LIGHTING			
1	RTC6471	RH-LHD-5 band type	NLA		Use STC767 qty 1 When replacing old style 5 band lenses to 4 band, both lenses must be changed as a pair; also replace bulbs with STC768 qty 1 with STC778 qty 1.
1	RTC6472	LH-LHD-5 band type	NLA		Use STC767 qty 1 When replacing old style 5 band lenses to 4 band, both lenses must be changed as a pair; also replace bulbs with STC768 qty 1 with STC778 qty 1.
1	STC767	RH-LHD-Wipac-4 band type	1		
1	STC768	LH-LHD-Wipac-4 band type	1		

DISCOVERY 1989MY UPTO 1999MY | GROUP M | Body Electrics Lighting Headlamps - Wipac To (V) LA081991

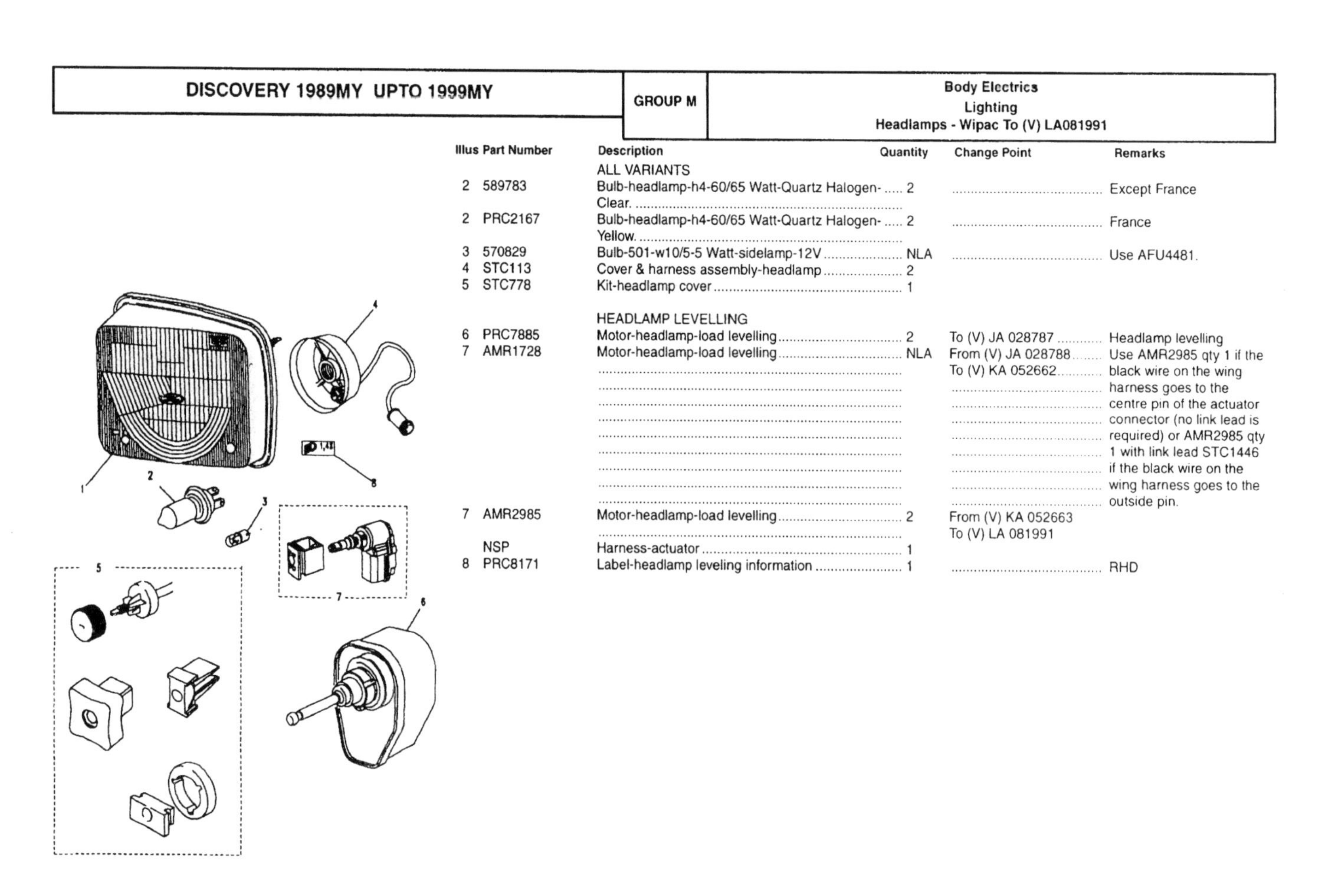

Illus	Part Number	Description	Quantity	Change Point	Remarks
		ALL VARIANTS			
2	589783	Bulb-headlamp-h4-60/65 Watt-Quartz Halogen-Clear.	2		Except France
2	PRC2167	Bulb-headlamp-h4-60/65 Watt-Quartz Halogen-Yellow.	2		France
3	570829	Bulb-501-w10/5-5 Watt-sidelamp-12V	NLA		Use AFU4481.
4	STC113	Cover & harness assembly-headlamp	2		
5	STC778	Kit-headlamp cover	1		
		HEADLAMP LEVELLING			
6	PRC7885	Motor-headlamp-load levelling	2	To (V) JA 028787	Headlamp levelling
7	AMR1728	Motor-headlamp-load levelling	NLA	From (V) JA 028788 To (V) KA 052662	Use AMR2985 qty 1 if the black wire on the wing harness goes to the centre pin of the actuator connector (no link lead is required) or AMR2985 qty 1 with link lead STC1446 if the black wire on the wing harness goes to the outside pin.
7	AMR2985	Motor-headlamp-load levelling	2	From (V) KA 052663 To (V) LA 081991	
	NSP	Harness-actuator	1		
8	PRC8171	Label-headlamp leveling information	1		RHD

Illus	Part Number	Description	Quantity	Change Point	Remarks
		CARELLO-FROM (V) MA081992			
		HEADLAMP ASSEMBLY-LOAD LEVELING			
1	STC1233	RH-RHD-Carello	1		
1	STC1234	LH-RHD-Carello	1		
1	STC1235	RH-LHD-Carello	1		
1	STC1236	LH-LHD-Carello	1		
		ALL VARIANTS			
2	589783	Bulb-headlamp-h4-60/65 Watt-main & dipped front light-Quartz Halogen-Clear	2		
3	570829	Bulb-501-w10/5-5 Watt-sidelamp-12V	NLA		Use AFU4481.
4	STC1240	Holder-front lighting bulb	2		
5	STC1239	Cover-headlamp bulb insulation	2		
6	STC1713	Gasket-headlamp lens-RH	1		
6	STC1714	Gasket-headlamp lens-LH	1		
7	STC1232	Kit-headlamp fittings	2		
11	STC3368	Adjuster unit-headlamp	4		
		HEADLAMP LEVELLING			
8	AMR2706	MOTOR-HEADLAMP-LOAD LEVELLING	2		
9	STC8646	• Grommet	2		
10	PRC8171	Label-headlamp leveling information	1		

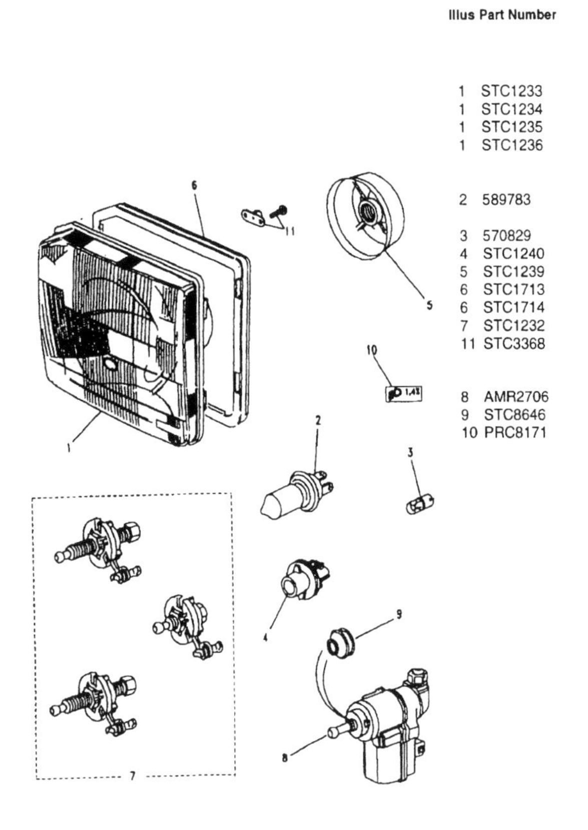

NOTE 1: THE ADJUSTER UNIT COMPLETE WITH FIXINGS IS FITTED AT THE REAR OF THE LIGHT UNIT AND CAN BE USED WITH ITEM 7

Illus	Part Number	Description	Quantity	Change Point	Remarks
		LAMP-FRONT DIRECTION INDICATOR			
1	PRC6471	RH	1	Note (1)	
1	PRC6472	LH	1		
1	PRC9306	RH	1	Note (2)	
1	PRC9307	LH	1		
2	264591	• Bulb-382-p25/1-21 Watt	2		
3	PRC7445	Screw	4		
4	WJ105001L	Washer-plain-M5-oversize	4		
		SIDE REPEATER LAMPS			
5	DCP8048	Lens-auxiliary lighting side repeater	2	To (V) JA 031114	
5	PRC9916	Lens-auxiliary lighting side repeater	2	From (V) JA 031115	
6	570829	Bulb-501-w10/5-5 Watt-side repeater-12V	NLA	From (V) KA 046106	Use AFU4481.
6	AFU4481	Bulb-501-w10/5-5 Watt-side repeater	2	From (V) KA 046107	
7	AFU3112L	O ring-side repeater	2		
8	STC966	Kit-side repeater harness	2		

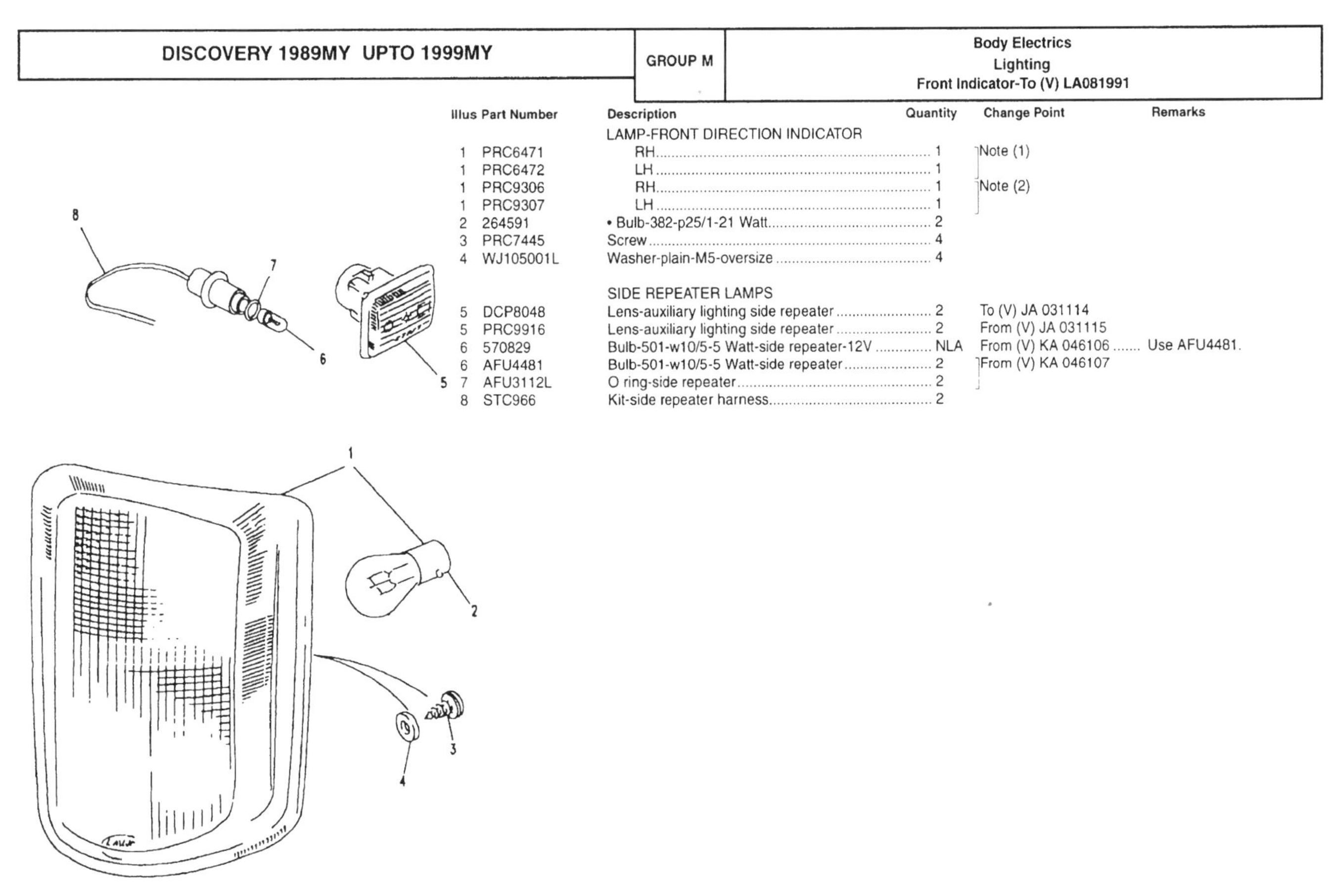

CHANGE POINTS:
(1) From (V) HA 000001 To (V) HA 009406; From (V) FA 393361 To (V) GA 456980; From (V) HA 456981 To (V) HA 486551
(2) From (V) HA 009407 To (V) HA 012332; From (V) JA 012333 To (V) LA 081991

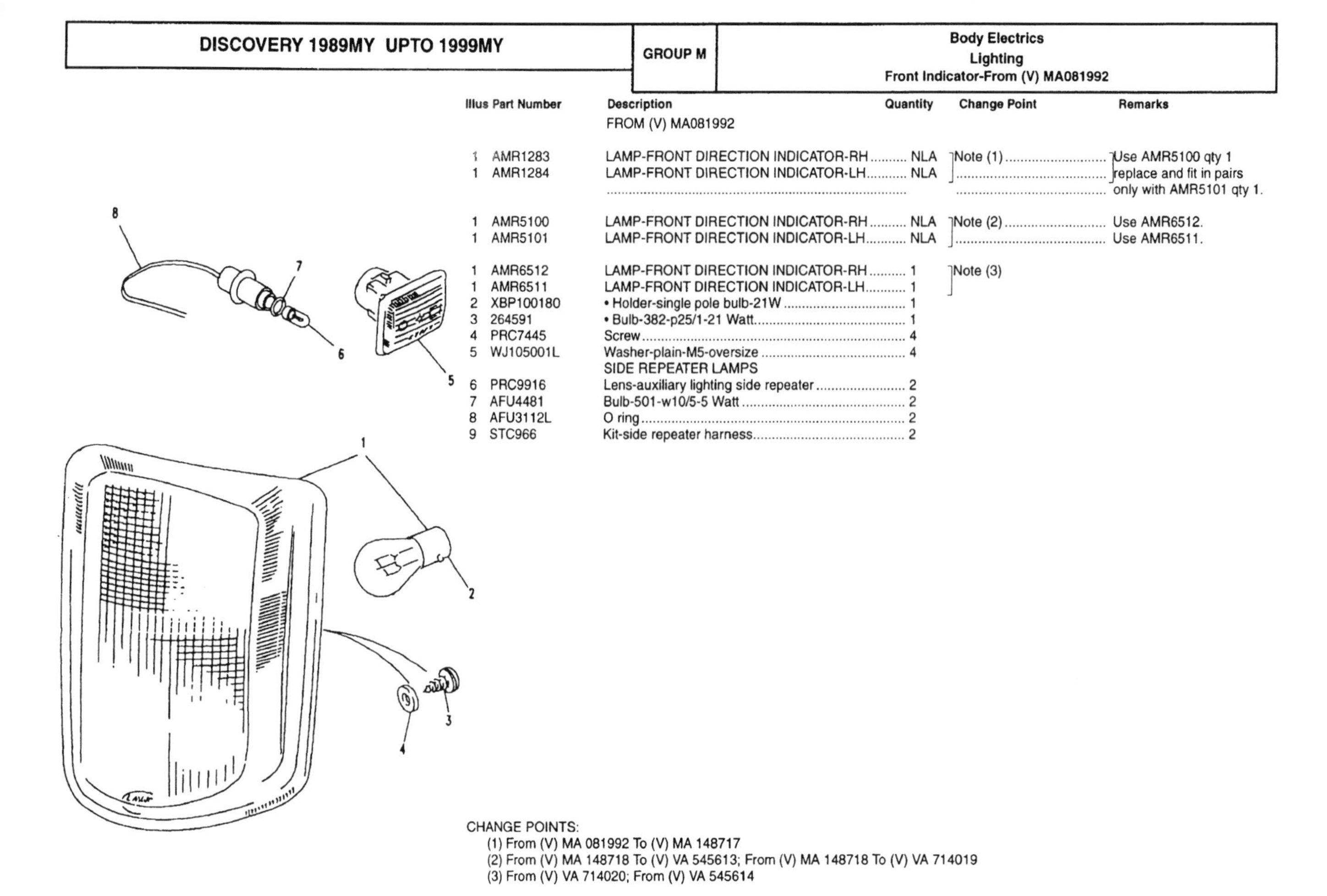

Illus	Part Number	Description	Quantity	Change Point	Remarks
		FROM (V) MA081992			
1	AMR1283	LAMP-FRONT DIRECTION INDICATOR-RH	NLA	Note (1)	Use AMR5100 qty 1
1	AMR1284	LAMP-FRONT DIRECTION INDICATOR-LH	NLA		replace and fit in pairs only with AMR5101 qty 1.
1	AMR5100	LAMP-FRONT DIRECTION INDICATOR-RH	NLA	Note (2)	Use AMR6512.
1	AMR5101	LAMP-FRONT DIRECTION INDICATOR-LH	NLA		Use AMR6511.
1	AMR6512	LAMP-FRONT DIRECTION INDICATOR-RH	1	Note (3)	
1	AMR6511	LAMP-FRONT DIRECTION INDICATOR-LH	1		
2	XBP100180	• Holder-single pole bulb-21W	1		
3	264591	• Bulb-382-p25/1-21 Watt	1		
4	PRC7445	Screw	4		
5	WJ105001L	Washer-plain-M5-oversize	4		
		SIDE REPEATER LAMPS			
6	PRC9916	Lens-auxiliary lighting side repeater	2		
7	AFU4481	Bulb-501-w10/5-5 Watt	2		
8	AFU3112L	O ring	2		
9	STC966	Kit-side repeater harness	2		

CHANGE POINTS:
(1) From (V) MA 081992 To (V) MA 148717
(2) From (V) MA 148718 To (V) VA 545613; From (V) MA 148718 To (V) VA 714019
(3) From (V) VA 714020; From (V) VA 545614

M 5

Illus	Part Number	Description	Quantity	Change Point	Remarks
		FROM (V) TA163104/TA501920			
		FRONT FOG LAMPS			
1	AMR4137	LAMP-FRONT LIGHTING FOG-RH	1		
1	AMR4136	LAMP-FRONT LIGHTING FOG-LH	1		
2	STC3085	• Bulb-headlamp-h3-55 Watt	2		
		FIXINGS			
3	SE105168L	Screw-M5	8		
4	WC105048L	Washer-M5	8		

NOTE: IMPORTANT-SPECIAL LOW UV EMISSION BULBS MUST BE USED FOR SERVICE
NOTE: THESE FOG LAMPS CANNOT BE FITTED TO VEHICLES PRIOR TO TA VIN PREFIX
NOTE: FOR VEHICLES TO TA VIN PREFIX, CONTACT LAND ROVER ACCESSORIES
NOTE: FOR VEHICLES FROM TA ONWARDS, LAMPS CAN BE ADDED WITH THE ADDITION OF A REPLACEMENT BELLY PANEL, RELAYS & SWITCH-SEE RELEVANT SECTIONS

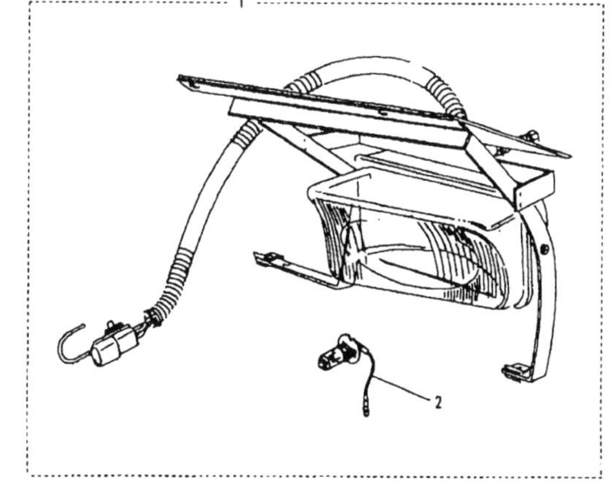

M 6

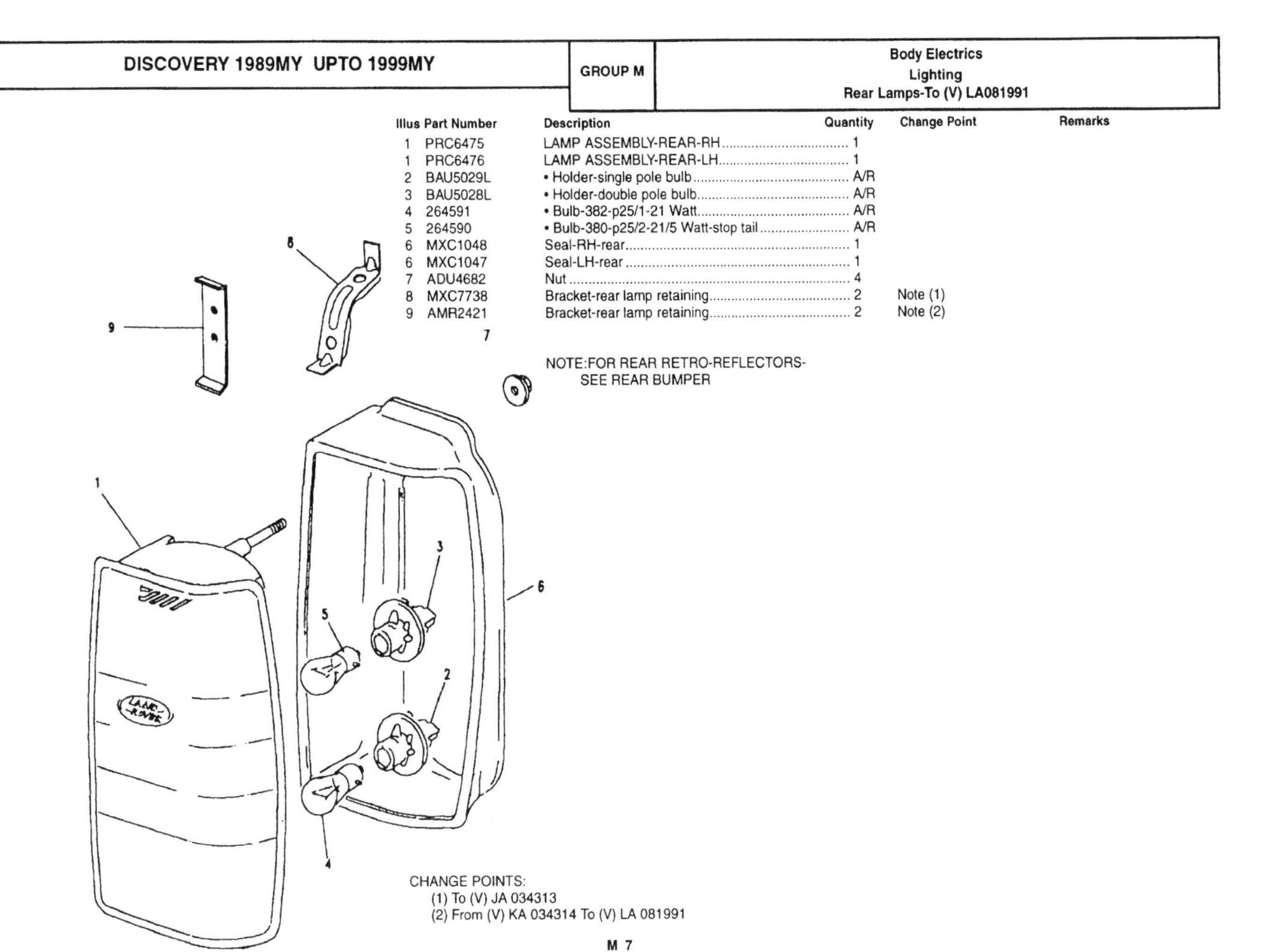

Illus	Part Number	Description	Quantity	Change Point	Remarks
1	PRC6475	LAMP ASSEMBLY-REAR-RH	1		
1	PRC6476	LAMP ASSEMBLY-REAR-LH	1		
2	BAU5029L	• Holder-single pole bulb	A/R		
3	BAU5028L	• Holder-double pole bulb	A/R		
4	264591	• Bulb-382-p25/1-21 Watt	A/R		
5	264590	• Bulb-380-p25/2-21/5 Watt-stop tail	A/R		
6	MXC1048	Seal-RH-rear	1		
6	MXC1047	Seal-LH-rear	1		
7	ADU4682	Nut	4		
8	MXC7738	Bracket-rear lamp retaining	2	Note (1)	
9	AMR2421	Bracket-rear lamp retaining	2	Note (2)	

NOTE:FOR REAR RETRO-REFLECTORS-
SEE REAR BUMPER

CHANGE POINTS:
(1) To (V) JA 034313
(2) From (V) KA 034314 To (V) LA 081991

M 7

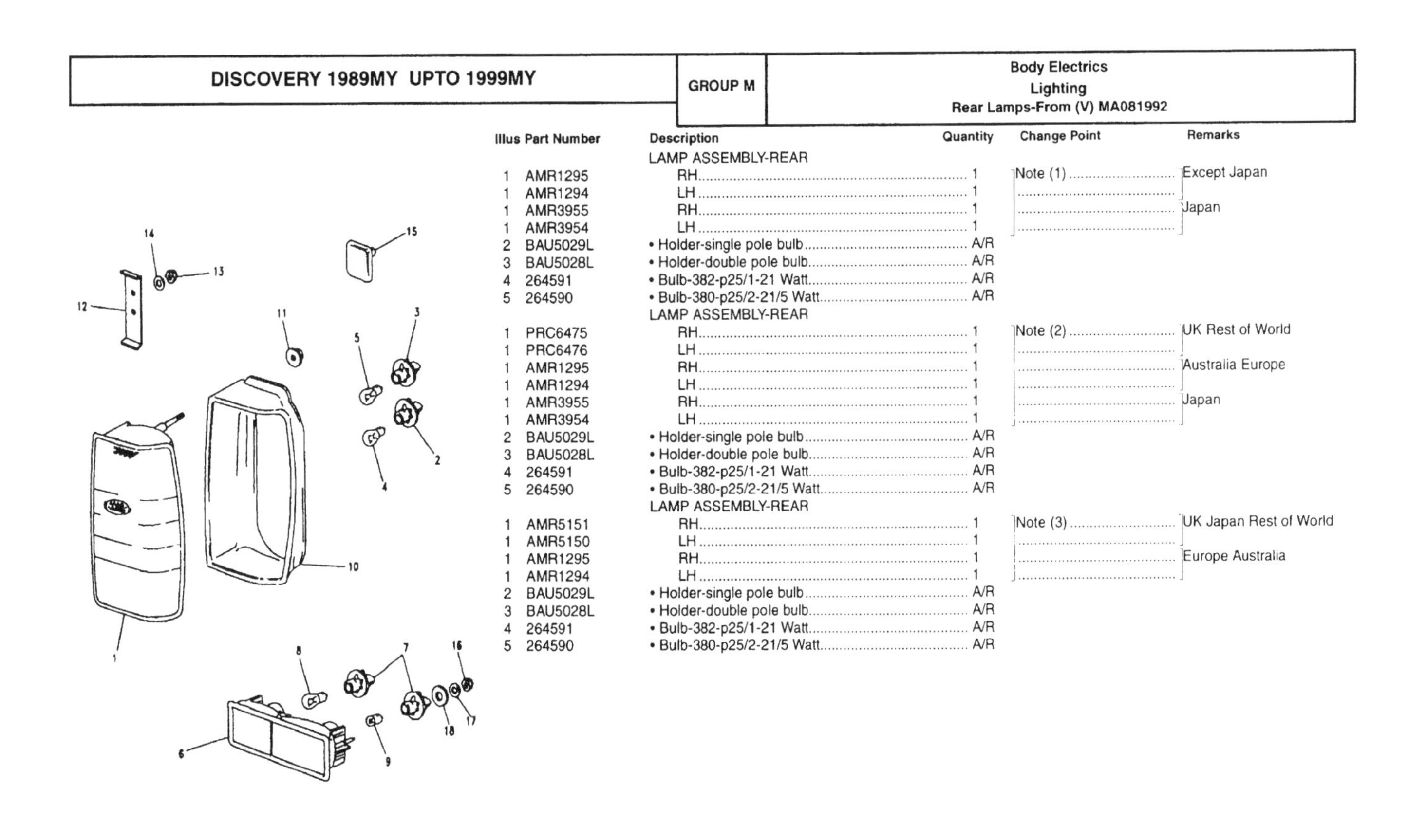

Illus	Part Number	Description	Quantity	Change Point	Remarks
		LAMP ASSEMBLY-REAR			
1	AMR1295	RH	1	Note (1)	Except Japan
1	AMR1294	LH	1		
1	AMR3955	RH	1		Japan
1	AMR3954	LH	1		
2	BAU5029L	• Holder-single pole bulb	A/R		
3	BAU5028L	• Holder-double pole bulb	A/R		
4	264591	• Bulb-382-p25/1-21 Watt	A/R		
5	264590	• Bulb-380-p25/2-21/5 Watt	A/R		
		LAMP ASSEMBLY-REAR			
1	PRC6475	RH	1	Note (2)	UK Rest of World
1	PRC6476	LH	1		
1	AMR1295	RH	1		Australia Europe
1	AMR1294	LH	1		
1	AMR3955	RH	1		Japan
1	AMR3954	LH	1		
2	BAU5029L	• Holder-single pole bulb	A/R		
3	BAU5028L	• Holder-double pole bulb	A/R		
4	264591	• Bulb-382-p25/1-21 Watt	A/R		
5	264590	• Bulb-380-p25/2-21/5 Watt	A/R		
		LAMP ASSEMBLY-REAR			
1	AMR5151	RH	1	Note (3)	UK Japan Rest of World
1	AMR5150	LH	1		
1	AMR1295	RH	1		Europe Australia
1	AMR1294	LH	1		
2	BAU5029L	• Holder-single pole bulb	A/R		
3	BAU5028L	• Holder-double pole bulb	A/R		
4	264591	• Bulb-382-p25/1-21 Watt	A/R		
5	264590	• Bulb-380-p25/2-21/5 Watt	A/R		

CHANGE POINTS:
(1) From (V) MA 081992 To (V) MA 135907
(2) From (V) MA 135908 To (V) MA 162389; From (V) MA 135908 To (V) MA 501704
(3) From (V) MA 162390 To (V) WA 746541; From (V) MA 501705 To (V) WA 746541

M 8

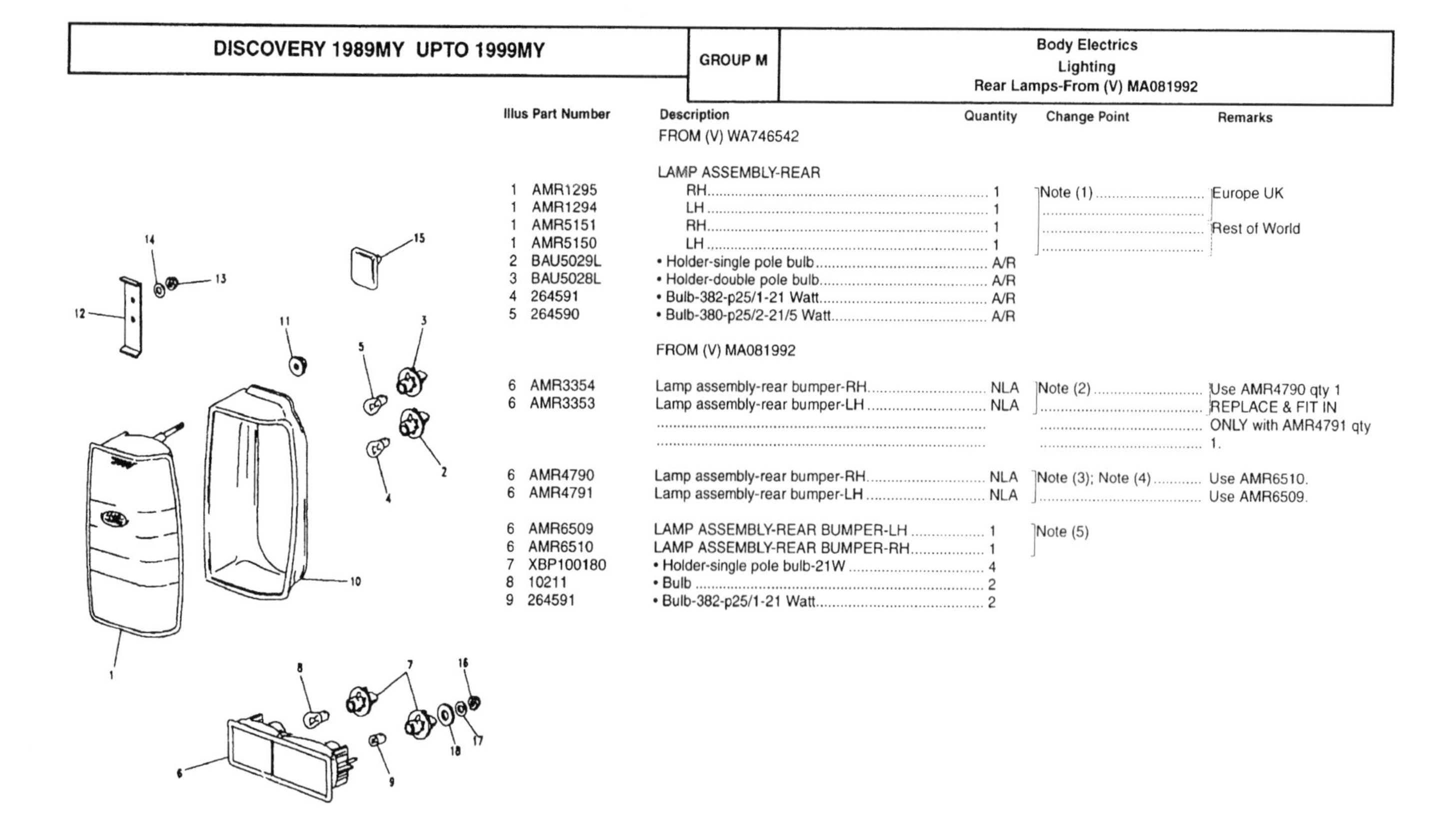

Illus	Part Number	Description	Quantity	Change Point	Remarks
		FROM (V) WA746542			
		LAMP ASSEMBLY-REAR			
1	AMR1295	RH	1	Note (1)	Europe UK
1	AMR1294	LH	1		
1	AMR5151	RH	1		Rest of World
1	AMR5150	LH	1		
2	BAU5029L	• Holder-single pole bulb	A/R		
3	BAU5028L	• Holder-double pole bulb	A/R		
4	264591	• Bulb-382-p25/1-21 Watt	A/R		
5	264590	• Bulb-380-p25/2-21/5 Watt	A/R		
		FROM (V) MA081992			
6	AMR3354	Lamp assembly-rear bumper-RH	NLA	Note (2)	Use AMR4790 qty 1 REPLACE & FIT IN ONLY with AMR4791 qty 1.
6	AMR3353	Lamp assembly-rear bumper-LH	NLA		
6	AMR4790	Lamp assembly-rear bumper-RH	NLA	Note (3); Note (4)	Use AMR6510.
6	AMR4791	Lamp assembly-rear bumper-LH	NLA		Use AMR6509.
6	AMR6509	LAMP ASSEMBLY-REAR BUMPER-LH	1	Note (5)	
6	AMR6510	LAMP ASSEMBLY-REAR BUMPER-RH	1		
7	XBP100180	• Holder-single pole bulb-21W	4		
8	10211	• Bulb	2		
9	264591	• Bulb-382-p25/1-21 Watt	2		

CHANGE POINTS:
(1) From (V) WA 746542
(2) From (V) MA 081992 To (V) MA 110580
(3) From (V) MA 110581 To (V) VA 714019
(4) From (V) MA 110581 To (V) VA 545613
(5) From (V) VA 545614; From (V) VA 714020

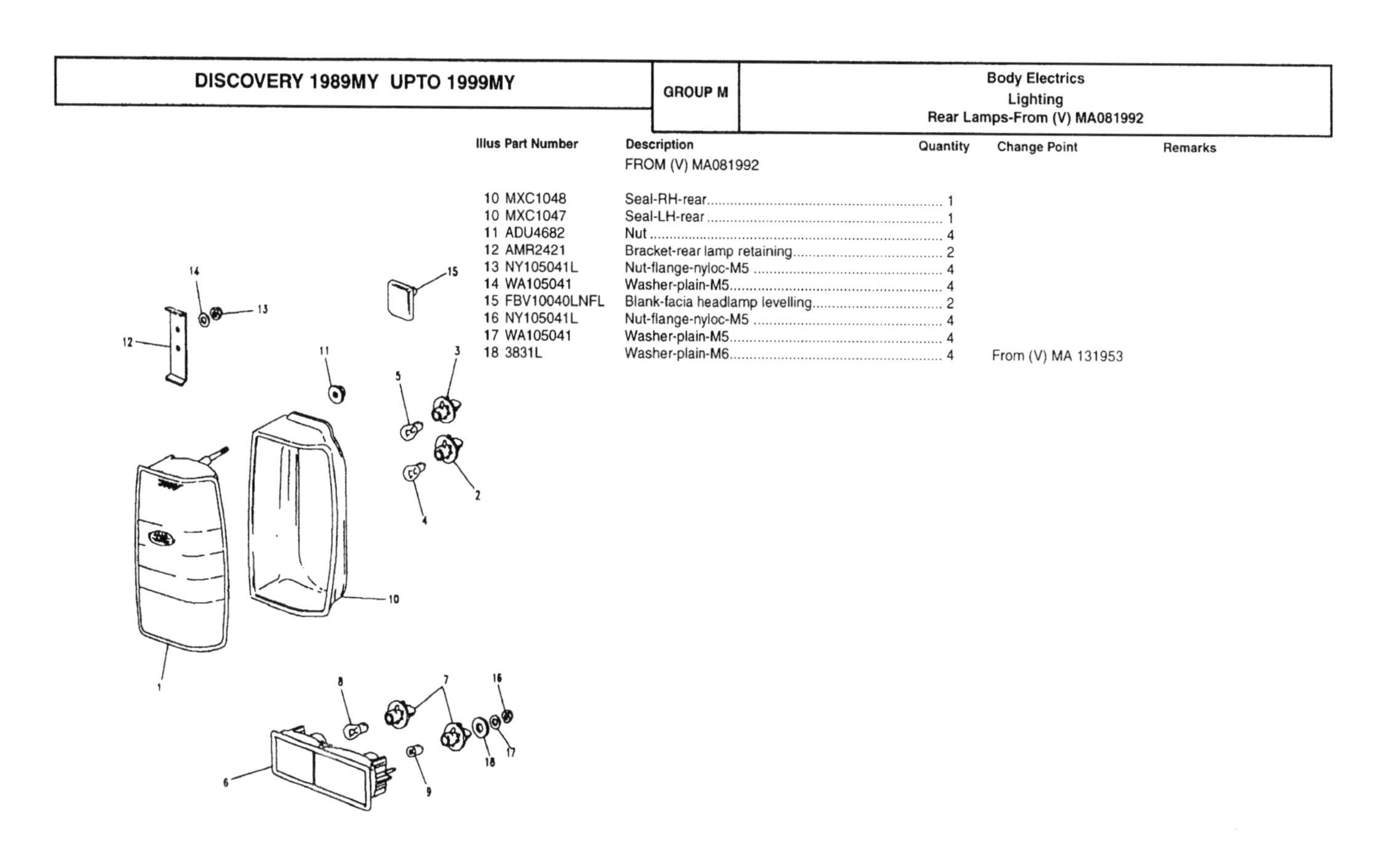

Illus	Part Number	Description	Quantity	Change Point	Remarks
		FROM (V) MA081992			
10	MXC1048	Seal-RH-rear	1		
10	MXC1047	Seal-LH-rear	1		
11	ADU4682	Nut	4		
12	AMR2421	Bracket-rear lamp retaining	2		
13	NY105041L	Nut-flange-nyloc-M5	4		
14	WA105041	Washer-plain-M5	4		
15	FBV10040LNFL	Blank-facia headlamp levelling	2		
16	NY105041L	Nut-flange-nyloc-M5	4		
17	WA105041	Washer-plain-M5	4		
18	3831L	Washer-plain-M6	4	From (V) MA 131953	

DISCOVERY 1989MY UPTO 1999MY

GROUP M — Body Electrics / Lighting / High Level Stop Lamp

Illus	Part Number	Description	Quantity	Change Point	Remarks
		FROM (V) MA081992			
1	AMR1285	LAMP-REAR HIGH MOUNTED STOP	1	Note (1)	
1	XFG100330	LAMP-REAR HIGH MOUNTED STOP	1	Note (2)	
2	264591	• Bulb-382-p25/1-21 Watt	1		
3	BTR2940	Screw-3.5 x 16	2		
4	BTR4414	Cover-high mounted stop lamp intergal	1		

NOTE: IF FITTING TO VEHICLE W/O HMSL A NEW HMSL REAR SCREEN IS REQUIRED

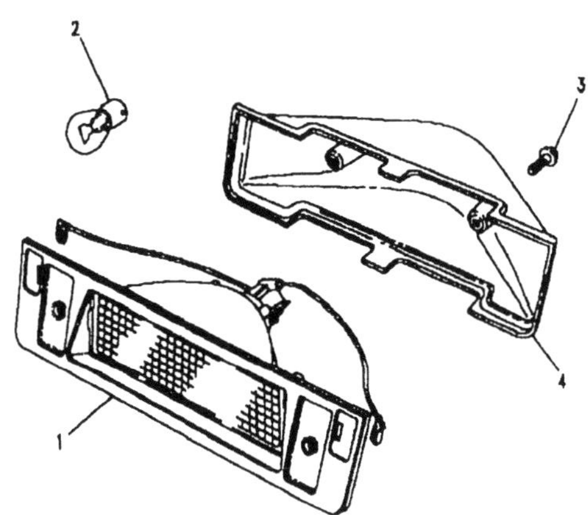

CHANGE POINTS:
(1) To (V) VA 724059; To (V) VA 588233
(2) From (V) VA 724060; From (V) VA 558234

M 11

DISCOVERY 1989MY UPTO 1999MY

GROUP M — Body Electrics / Lighting / Interior Lamps-To (V) LA081991

Illus	Part Number	Description	Quantity	Change Point	Remarks
1	AMR1183LUP	LAMP-CENTRE HEADLINING INTERIOR COURTESY-ASPEN GREY	2		
2	586438	• Bulb-272-10 Watt-festoon-36x10.5mm	2		
3	DA608061	Screw	8		
4	PRC6970	Switch-contact courtesy light	5	Note (1)	
4	PRC8548	Switch-contact courtesy light	5	Note (2)	
5	PRC6518L	Screw	3		
6	MXC2015	Bracket-courtesy lamp	1		
		LAMP REAR LOADSPACE			
7	AFU4092L	LAMP-INTERIOR COURTESY-WITH ON/OFF SWITCH	1	To (V) LA 064754	
	AFU4091	LAMP-TRUNK/LOADSPACE INTERIOR COURTESY-REAR-WITHOUT SWITCH	1	From (V) LA 064755	
8	STC1203	• Bulb-245-r19/10-10 Watt	1		

CHANGE POINTS:
(1) From (V) GA 394352 To (V) GA 456980; From (V) HA 456981 To (V) HA 486551; From (V) HA 000001 To (V) HA 008255
(2) From (V) HA 008256 To (V) HA 012332; From (V) JA 012333 To (V) LA 081991

M 12

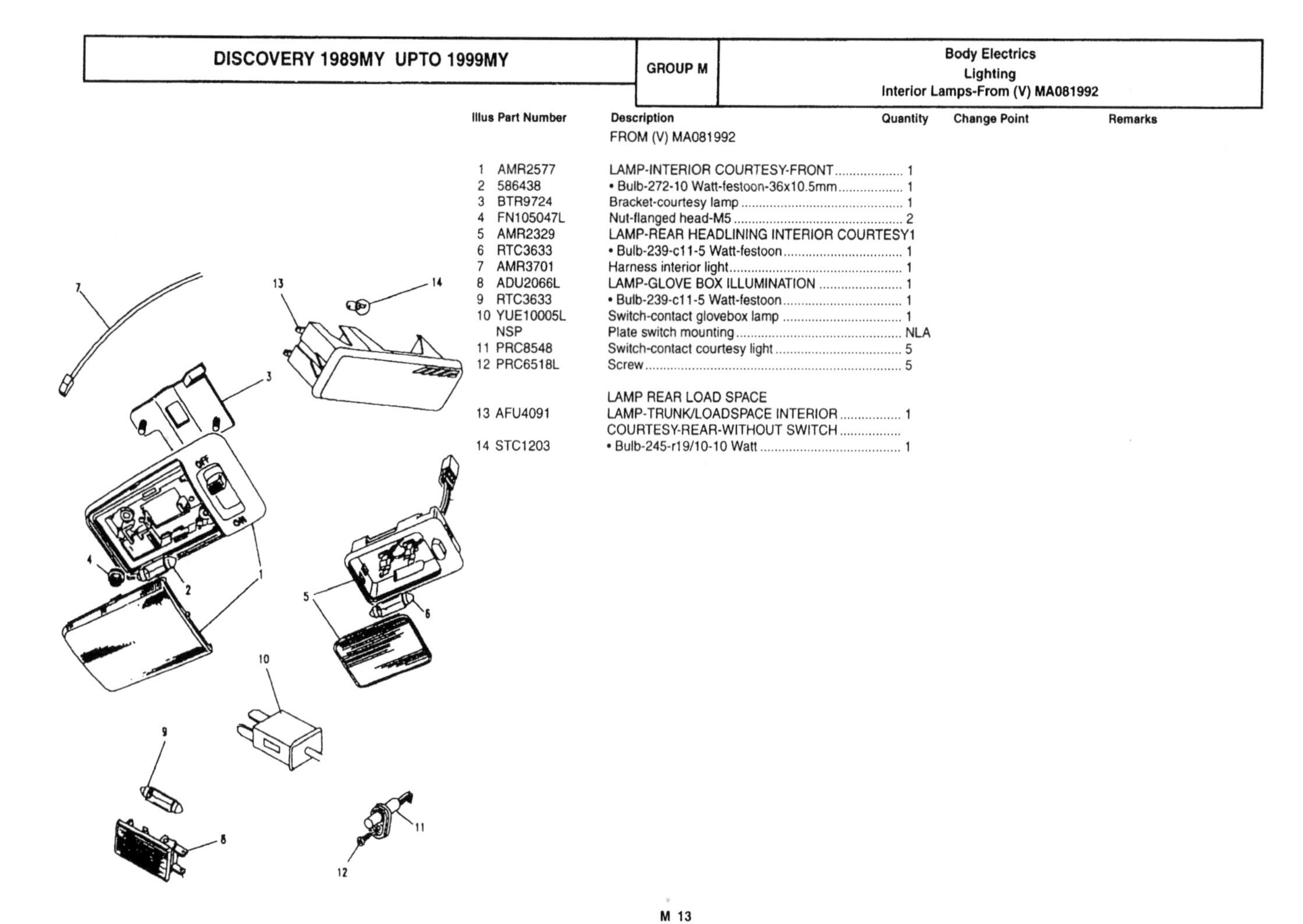

DISCOVERY 1989MY UPTO 1999MY

GROUP M

Body Electrics
Lighting
Interior Lamps-From (V) MA081992

Illus	Part Number	Description	Quantity	Change Point	Remarks
		FROM (V) MA081992			
1	AMR2577	LAMP-INTERIOR COURTESY-FRONT	1		
2	586438	• Bulb-272-10 Watt-festoon-36x10.5mm	1		
3	BTR9724	Bracket-courtesy lamp	1		
4	FN105047L	Nut-flanged head-M5	2		
5	AMR2329	LAMP-REAR HEADLINING INTERIOR COURTESY	1		
6	RTC3633	• Bulb-239-c11-5 Watt-festoon	1		
7	AMR3701	Harness interior light	1		
8	ADU2066L	LAMP-GLOVE BOX ILLUMINATION	1		
9	RTC3633	• Bulb-239-c11-5 Watt-festoon	1		
10	YUE10005L	Switch-contact glovebox lamp	1		
	NSP	Plate switch mounting	NLA		
11	PRC8548	Switch-contact courtesy light	5		
12	PRC6518L	Screw	5		
		LAMP REAR LOAD SPACE			
13	AFU4091	LAMP-TRUNK/LOADSPACE INTERIOR COURTESY-REAR-WITHOUT SWITCH	1		
14	STC1203	• Bulb-245-r19/10-10 Watt	1		

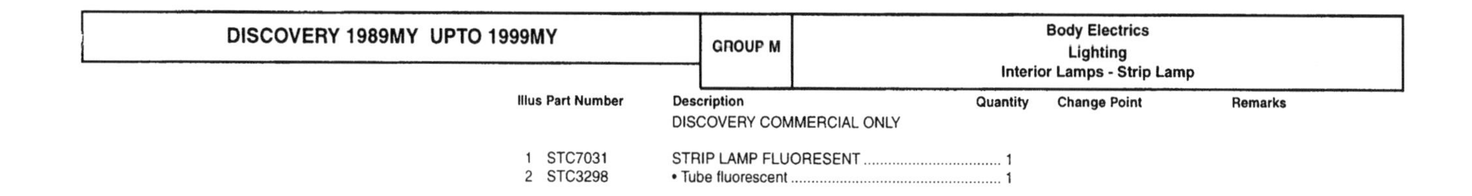

DISCOVERY 1989MY UPTO 1999MY

GROUP M

Body Electrics
Lighting
Interior Lamps - Strip Lamp

Illus	Part Number	Description	Quantity	Change Point	Remarks
		DISCOVERY COMMERCIAL ONLY			
1	STC7031	STRIP LAMP FLUORESENT	1		
2	STC3298	• Tube fluorescent	1		

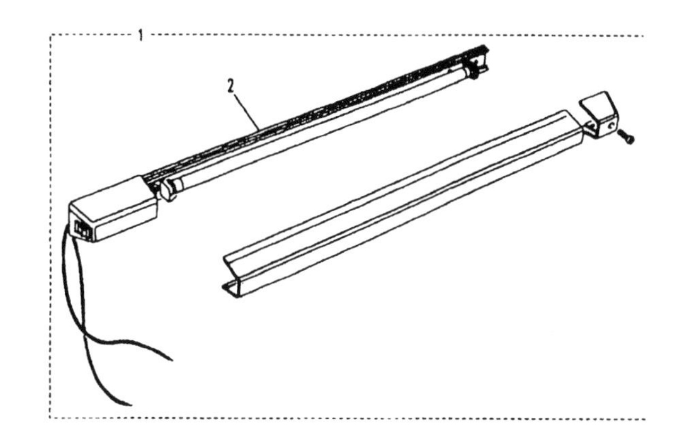

Body Electrics
Lighting
Lamp Number Plate

Illus	Part Number	Description	Quantity	Change Point	Remarks
1	PRC5838	LAMP-REAR LICENCE PLATE	2		
2	RTC3633	• Bulb-239-c11-5 Watt-festoon	2		
3	PRC7246	Gasket-rear lamp assembly	2		
4	DA606041	Screw-self tapping AB	4		
5	AK606021L	Nut-spring-u type	4		
6	PRC7266	Harness-licence plate lamp-rear	1		

NOTE: REAR DOOR MARKER LAMP SEE-
HARNESS BODY

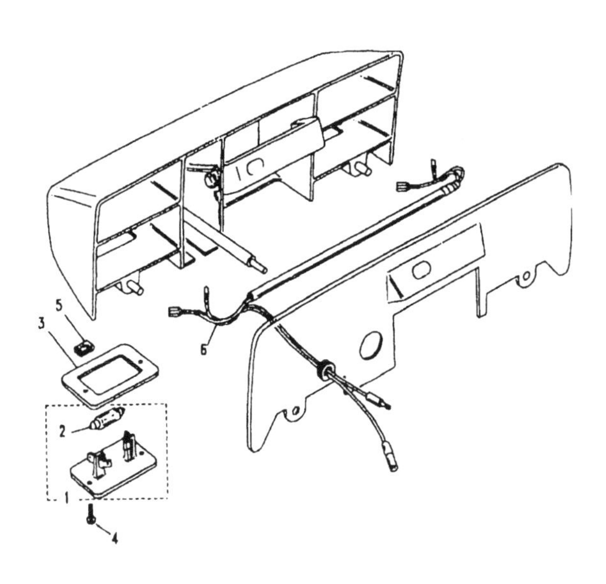

Body Electrics
Batteries & Harnesses
Battery Petrol - V8 Pi To (V) LA081991

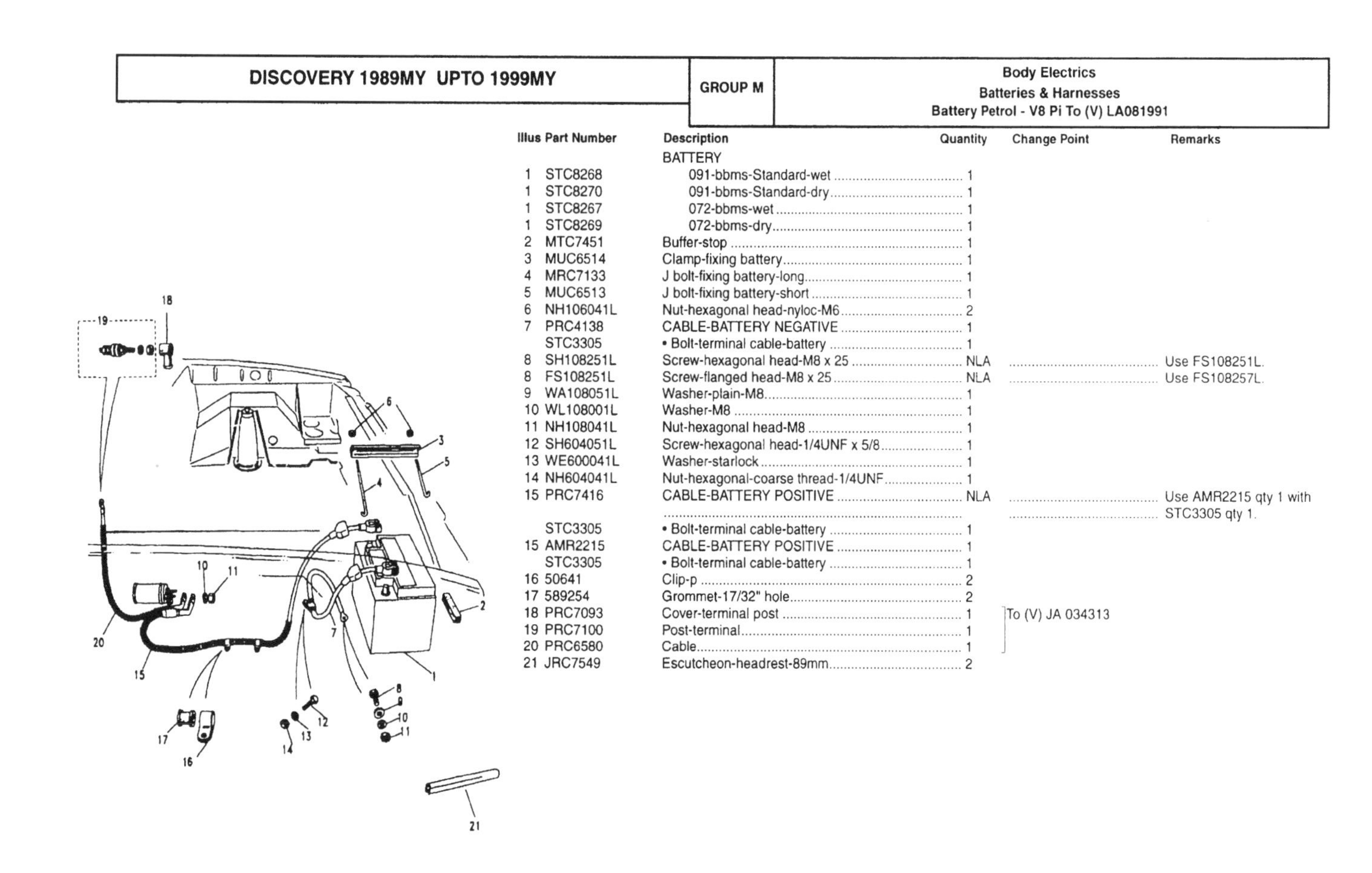

Illus	Part Number	Description	Quantity	Change Point	Remarks
		BATTERY			
1	STC8268	091-bbms-Standard-wet	1		
1	STC8270	091-bbms-Standard-dry	1		
1	STC8267	072-bbms-wet	1		
1	STC8269	072-bbms-dry	1		
2	MTC7451	Buffer-stop	1		
3	MUC6514	Clamp-fixing battery	1		
4	MRC7133	J bolt-fixing battery-long	1		
5	MUC6513	J bolt-fixing battery-short	1		
6	NH106041L	Nut-hexagonal head-nyloc-M6	2		
7	PRC4138	CABLE-BATTERY NEGATIVE	1		
	STC3305	• Bolt-terminal cable-battery	1		
8	SH108251L	Screw-hexagonal head-M8 x 25	NLA		Use FS108251L.
8	FS108251L	Screw-flanged head-M8 x 25	NLA		Use FS108257L.
9	WA108051L	Washer-plain-M8	1		
10	WL108001L	Washer-M8	1		
11	NH108041L	Nut-hexagonal head-M8	1		
12	SH604051L	Screw-hexagonal head-1/4UNF x 5/8	1		
13	WE600041L	Washer-starlock	1		
14	NH604041L	Nut-hexagonal-coarse thread-1/4UNF	1		
15	PRC7416	CABLE-BATTERY POSITIVE	NLA		Use AMR2215 qty 1 with STC3305 qty 1.
	STC3305	• Bolt-terminal cable-battery	1		
15	AMR2215	CABLE-BATTERY POSITIVE	1		
	STC3305	• Bolt-terminal cable-battery	1		
16	50641	Clip-p	2		
17	589254	Grommet-17/32" hole	2		
18	PRC7093	Cover-terminal post	1	To (V) JA 034313	
19	PRC7100	Post-terminal	1	To (V) JA 034313	
20	PRC6580	Cable	1	To (V) JA 034313	
21	JRC7549	Escutcheon-headrest-89mm	2		

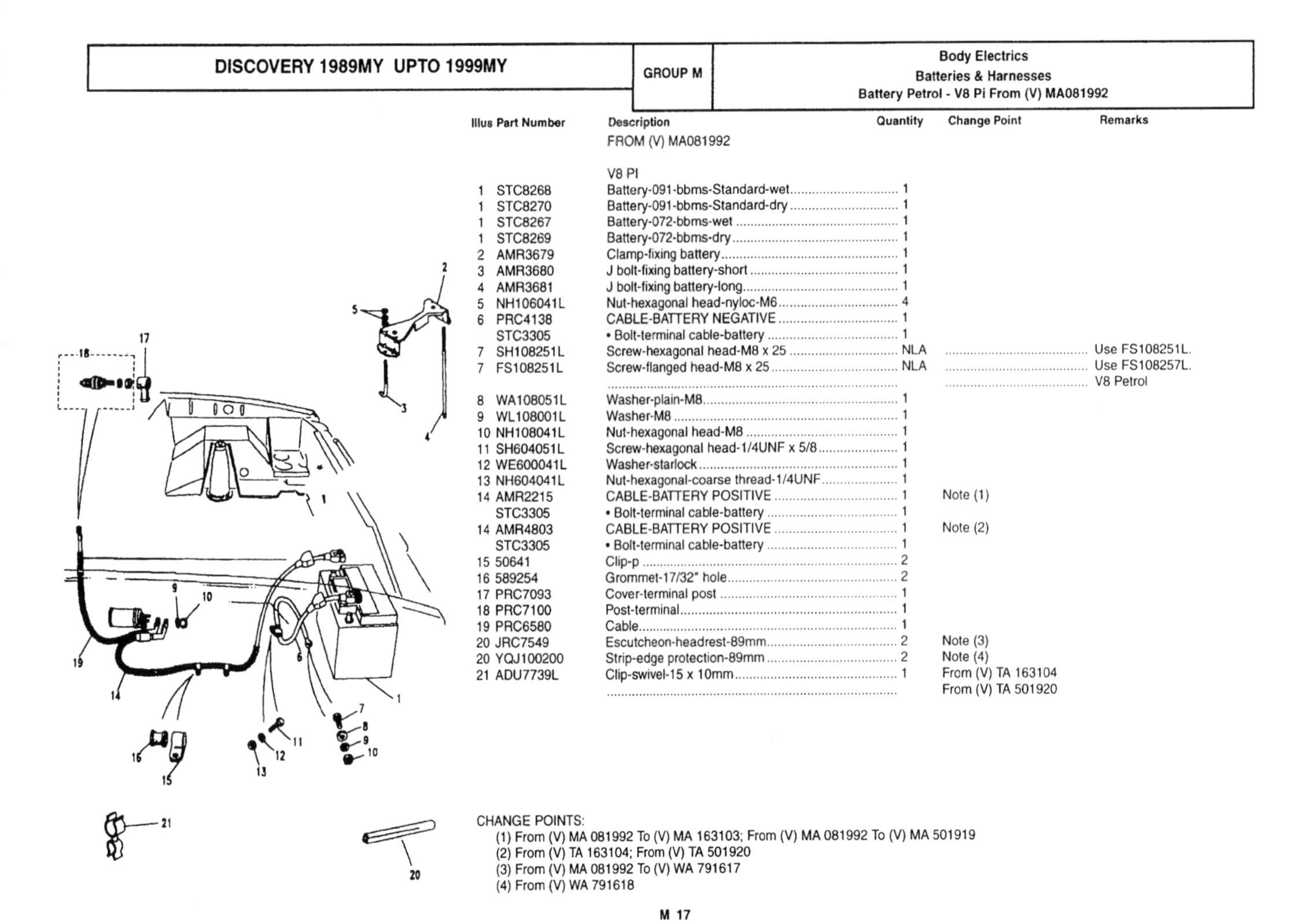

Illus	Part Number	Description	Quantity	Change Point	Remarks
		FROM (V) MA081992			
		V8 PI			
1	STC8268	Battery-091-bbms-Standard-wet	1		
1	STC8270	Battery-091-bbms-Standard-dry	1		
1	STC8267	Battery-072-bbms-wet	1		
1	STC8269	Battery-072-bbms-dry	1		
2	AMR3679	Clamp-fixing battery	1		
3	AMR3680	J bolt-fixing battery-short	1		
4	AMR3681	J bolt-fixing battery-long	1		
5	NH106041L	Nut-hexagonal head-nyloc-M6	4		
6	PRC4138	CABLE-BATTERY NEGATIVE	1		
	STC3305	• Bolt-terminal cable-battery	1		
7	SH108251L	Screw-hexagonal head-M8 x 25	NLA		Use FS108251L.
7	FS108251L	Screw-flanged head-M8 x 25	NLA		Use FS108257L.
					V8 Petrol
8	WA108051L	Washer-plain-M8	1		
9	WL108001L	Washer-M8	1		
10	NH108041L	Nut-hexagonal head-M8	1		
11	SH604051L	Screw-hexagonal head-1/4UNF x 5/8	1		
12	WE600041L	Washer-starlock	1		
13	NH604041L	Nut-hexagonal-coarse thread-1/4UNF	1		
14	AMR2215	CABLE-BATTERY POSITIVE	1	Note (1)	
	STC3305	• Bolt-terminal cable-battery	1		
14	AMR4803	CABLE-BATTERY POSITIVE	1	Note (2)	
	STC3305	• Bolt-terminal cable-battery	1		
15	50641	Clip-p	2		
16	589254	Grommet-17/32" hole	2		
17	PRC7093	Cover-terminal post	1		
18	PRC7100	Post-terminal	1		
19	PRC6580	Cable	1		
20	JRC7549	Escutcheon-headrest-89mm	2	Note (3)	
20	YQJ100200	Strip-edge protection-89mm	2	Note (4)	
21	ADU7739L	Clip-swivel-15 x 10mm	1	From (V) TA 163104	
				From (V) TA 501920	

CHANGE POINTS:
(1) From (V) MA 081992 To (V) MA 163103; From (V) MA 081992 To (V) MA 501919
(2) From (V) TA 163104; From (V) TA 501920
(3) From (V) MA 081992 To (V) WA 791617
(4) From (V) WA 791618

Illus	Part Number	Description	Quantity	Change Point	Remarks
		TO (V) LA081991			
		MPI			
		BATTERY			
1	STC8268	091-bbms-Standard-wet	1		
1	STC8270	091-bbms-Standard-dry	1		
1	STC8267	072-bbms-wet	1		
1	STC8269	072-bbms-dry	1		
2	MTC7451	Buffer-stop	1		
3	MUC6514	Clamp-fixing battery	1		
4	MRC7133	J bolt-fixing battery-long	1		
5	MUC6513	J bolt-fixing battery-short	1		
6	NH106041L	Nut-hexagonal head-nyloc-M6	2		
7	PRC4138	CABLE-BATTERY NEGATIVE	1		
	STC3305	• Bolt-terminal cable-battery	1		
8	SH108251L	Screw-hexagonal head-M8 x 25	NLA		Use FS108251L.
8	FS108251L	Screw-flanged head-M8 x 25	NLA		Use FS108257L.
9	WA108051L	Washer-plain-M8	1		
10	WL108001L	Washer-M8	1		
11	NH108041L	Nut-hexagonal head-M8	1		
12	SH604051L	Screw-hexagonal head-1/4UNF x 5/8	1		
13	WE600041L	Washer-starlock	1		
14	NH604041L	Nut-hexagonal-coarse thread-1/4UNF	1		
15	AMR2202	CABLE-BATTERY POSITIVE	1	To (V) KA 038821	
	STC3305	• Bolt-terminal cable-battery	1		
15	AMR2933	CABLE-BATTERY POSITIVE	1	From (V) KA 038822	
	STC3305	• Bolt-terminal cable-battery	1		
16	50641	Clip-p	2		
17	589254	Grommet-17/32" hole	2		
18	PRC7093	Cover-terminal post	1		
19	PRC7100	Post-terminal	1		
20	PRC6580	Cable.-T post to starter	1		
21	JRC7549	Escutcheon-headrest-89mm	2		
22	PRC3623	Cable assembly starter & earth	1		
23	SH106161	Screw-hexagonal head-M6 x 16	NLA		Use FS106167L.
23	FS106167L	Screw-hexagonal head-M6 x 16	1		
24	NH106041	Nut-hexagonal head-M6	NLA		Use NH106041L.
24	NH106041L	Nut-hexagonal head-nyloc-M6	1		
25	WL106041	Washer-spring	1		

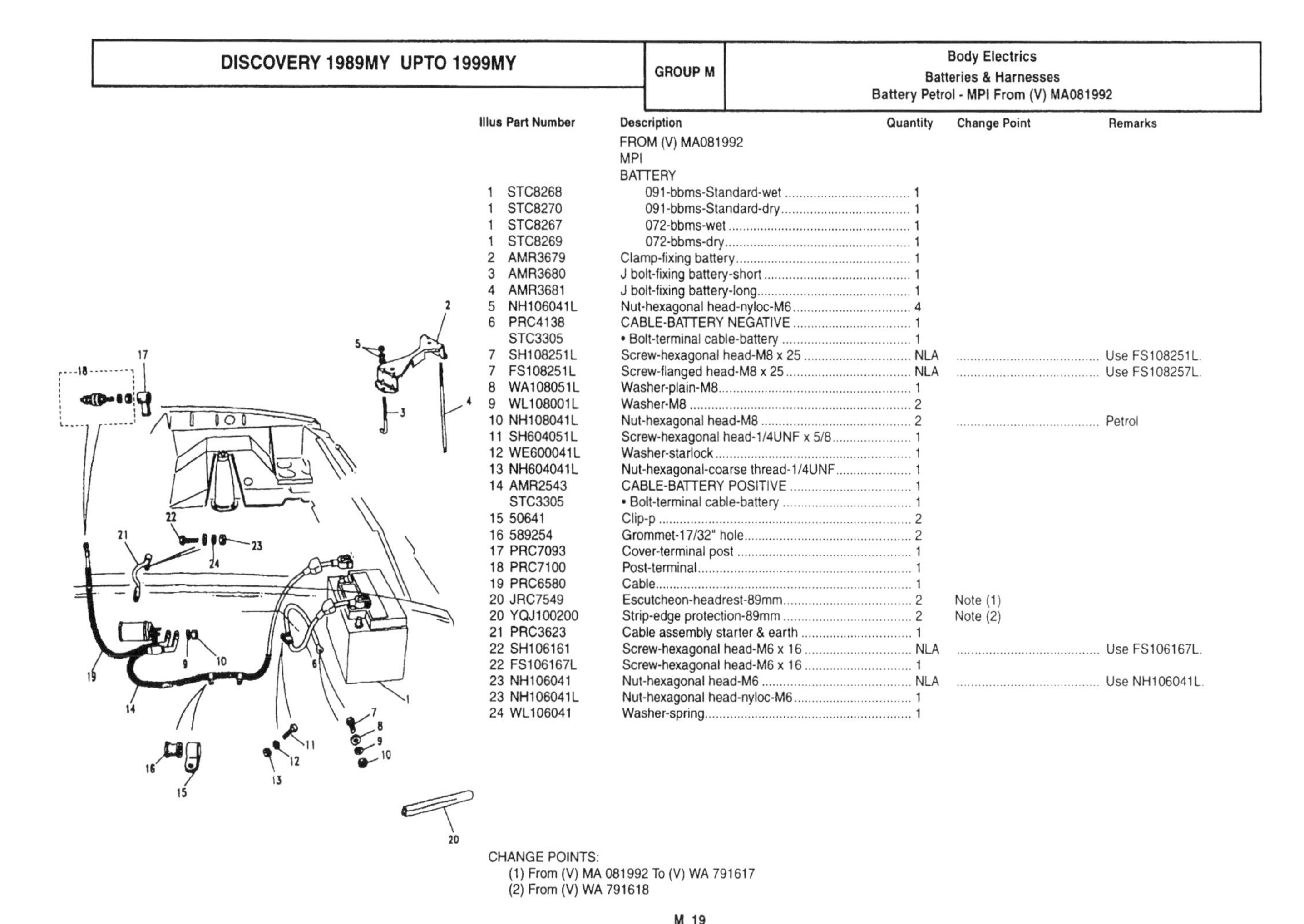

Illus	Part Number	Description	Quantity	Change Point	Remarks
		FROM (V) MA081992			
		MPI			
		BATTERY			
1	STC8268	091-bbms-Standard-wet	1		
1	STC8270	091-bbms-Standard-dry	1		
1	STC8267	072-bbms-wet	1		
1	STC8269	072-bbms-dry	1		
2	AMR3679	Clamp-fixing battery	1		
3	AMR3680	J bolt-fixing battery-short	1		
4	AMR3681	J bolt-fixing battery-long	1		
5	NH106041L	Nut-hexagonal head-nyloc-M6	4		
6	PRC4138	CABLE-BATTERY NEGATIVE	1		
	STC3305	• Bolt-terminal cable-battery	1		
7	SH108251L	Screw-hexagonal head-M8 x 25	NLA		Use FS108251L.
7	FS108251L	Screw-flanged head-M8 x 25	NLA		Use FS108257L.
8	WA108051L	Washer-plain-M8	1		
9	WL108001L	Washer-M8	2		
10	NH108041L	Nut-hexagonal head-M8	2		Petrol
11	SH604051L	Screw-hexagonal head-1/4UNF x 5/8	1		
12	WE600041L	Washer-starlock	1		
13	NH604041L	Nut-hexagonal-coarse thread-1/4UNF	1		
14	AMR2543	CABLE-BATTERY POSITIVE	1		
	STC3305	• Bolt-terminal cable-battery	1		
15	50641	Clip-p	2		
16	589254	Grommet-17/32" hole	2		
17	PRC7093	Cover-terminal post	1		
18	PRC7100	Post-terminal	1		
19	PRC6580	Cable	1		
20	JRC7549	Escutcheon-headrest-89mm	2	Note (1)	
20	YQJ100200	Strip-edge protection-89mm	2	Note (2)	
21	PRC3623	Cable assembly starter & earth	1		
22	SH106161	Screw-hexagonal head-M6 x 16	NLA		Use FS106167L.
22	FS106167L	Screw-hexagonal head-M6 x 16	1		
23	NH106041	Nut-hexagonal head-M6	NLA		Use NH106041L.
23	NH106041L	Nut-hexagonal head-nyloc-M6	1		
24	WL106041	Washer-spring	1		

CHANGE POINTS:
(1) From (V) MA 081992 To (V) WA 791617
(2) From (V) WA 791618

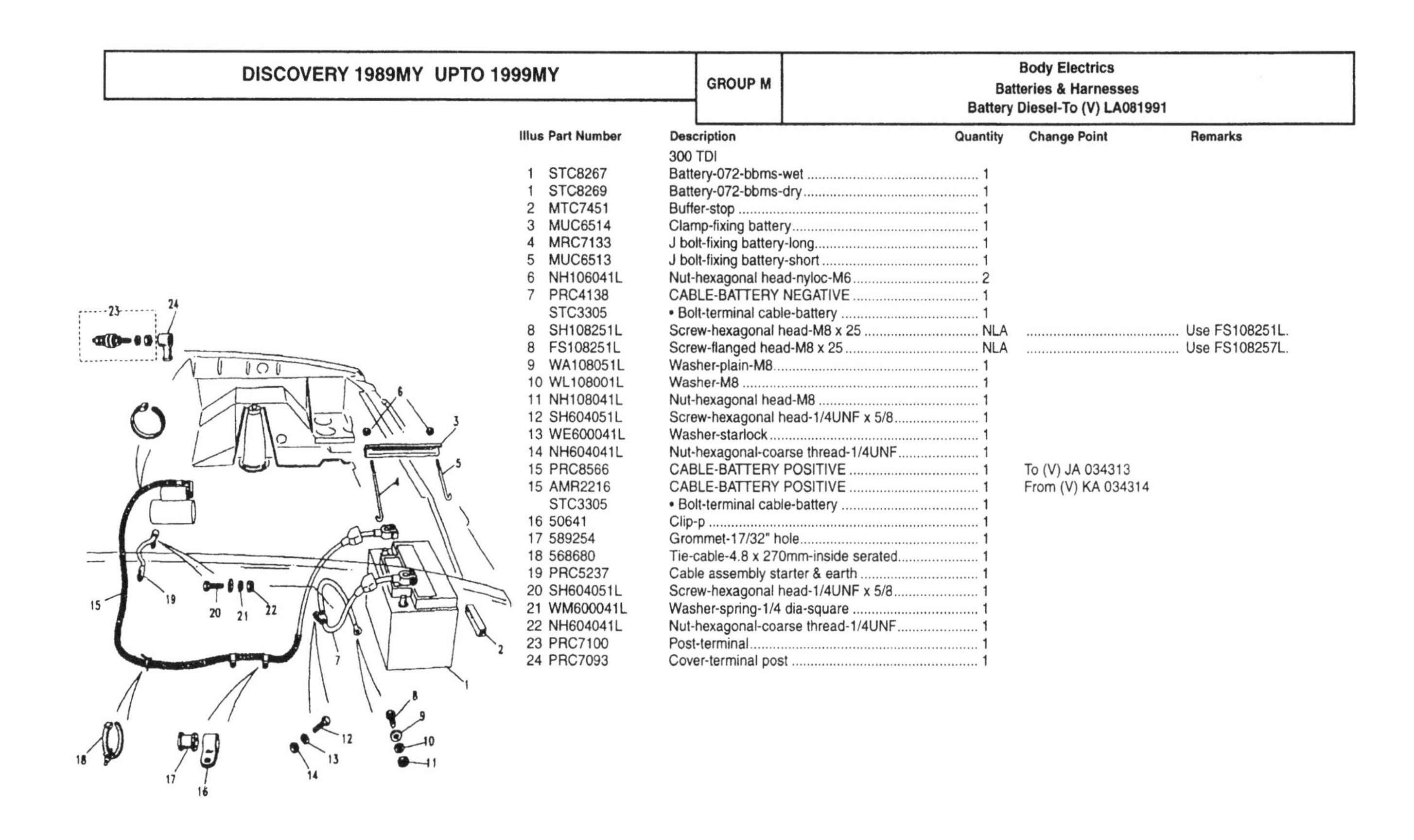

Illus	Part Number	Description	Quantity	Change Point	Remarks
		300 TDI			
1	STC8267	Battery-072-bbms-wet	1		
1	STC8269	Battery-072-bbms-dry	1		
2	MTC7451	Buffer-stop	1		
3	MUC6514	Clamp-fixing battery	1		
4	MRC7133	J bolt-fixing battery-long	1		
5	MUC6513	J bolt-fixing battery-short	1		
6	NH106041L	Nut-hexagonal head-nyloc-M6	2		
7	PRC4138	CABLE-BATTERY NEGATIVE	1		
	STC3305	• Bolt-terminal cable-battery	1		
8	SH108251L	Screw-hexagonal head-M8 x 25	NLA		Use FS108251L.
8	FS108251L	Screw-flanged head-M8 x 25	NLA		Use FS108257L.
9	WA108051L	Washer-plain-M8	1		
10	WL108001L	Washer-M8	1		
11	NH108041L	Nut-hexagonal head-M8	1		
12	SH604051L	Screw-hexagonal head-1/4UNF x 5/8	1		
13	WE600041L	Washer-starlock	1		
14	NH604041L	Nut-hexagonal-coarse thread-1/4UNF	1		
15	PRC8566	CABLE-BATTERY POSITIVE	1	To (V) JA 034313	
15	AMR2216	CABLE-BATTERY POSITIVE	1	From (V) KA 034314	
	STC3305	• Bolt-terminal cable-battery	1		
16	50641	Clip-p	1		
17	589254	Grommet-17/32" hole	1		
18	568680	Tie-cable-4.8 x 270mm-inside serated	1		
19	PRC5237	Cable assembly starter & earth	1		
20	SH604051L	Screw-hexagonal head-1/4UNF x 5/8	1		
21	WM600041L	Washer-spring-1/4 dia-square	1		
22	NH604041L	Nut-hexagonal-coarse thread-1/4UNF	1		
23	PRC7100	Post-terminal	1		
24	PRC7093	Cover-terminal post	1		

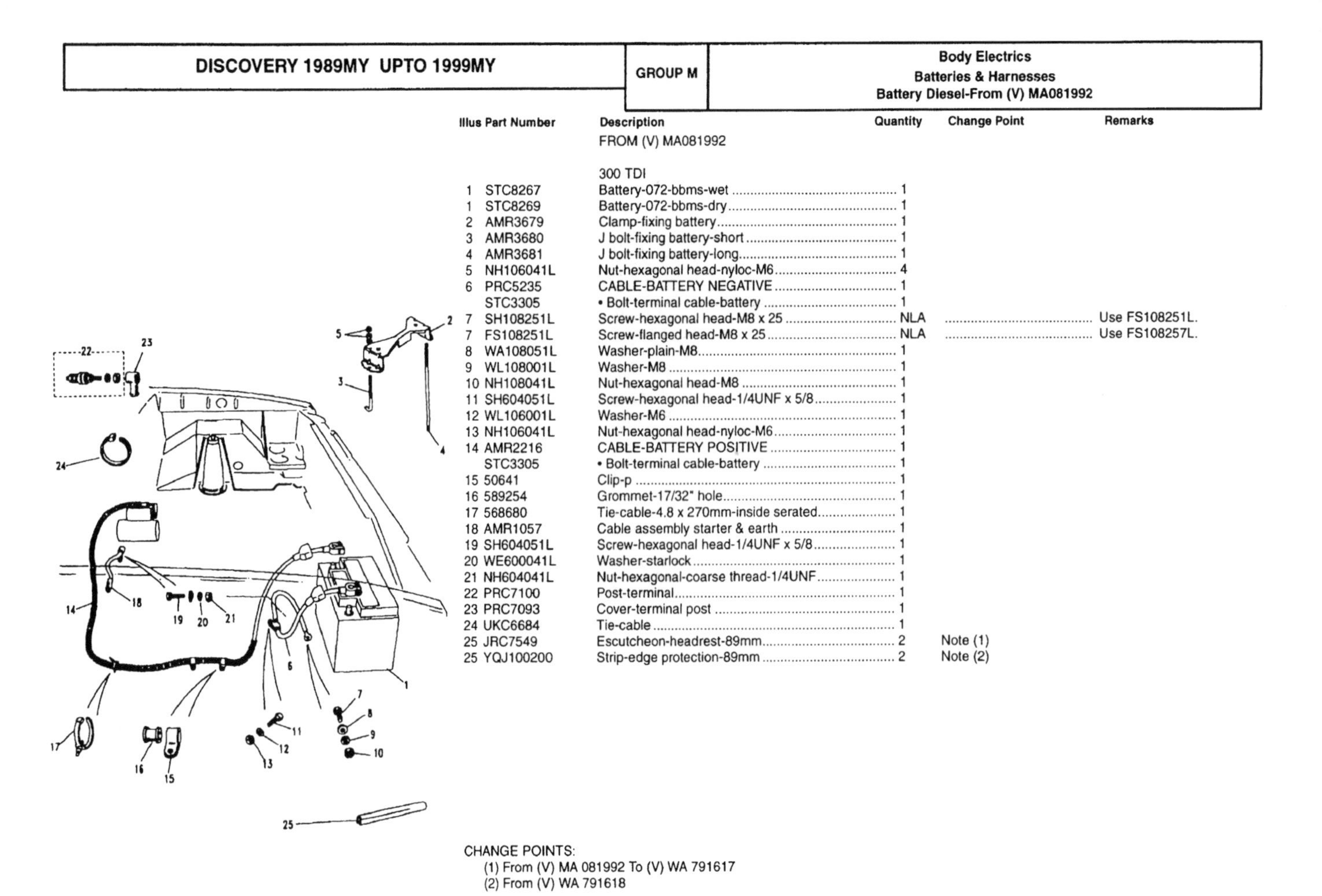

DISCOVERY 1989MY UPTO 1999MY	GROUP M	Body Electrics Batteries & Harnesses Battery Diesel-From (V) MA081992

Illus	Part Number	Description	Quantity	Change Point	Remarks
		FROM (V) MA081992			
		300 TDI			
1	STC8267	Battery-072-bbms-wet	1		
1	STC8269	Battery-072-bbms-dry	1		
2	AMR3679	Clamp-fixing battery	1		
3	AMR3680	J bolt-fixing battery-short	1		
4	AMR3681	J bolt-fixing battery-long	1		
5	NH106041L	Nut-hexagonal head-nyloc-M6	4		
6	PRC5235	CABLE-BATTERY NEGATIVE	1		
	STC3305	• Bolt-terminal cable-battery	1		
7	SH108251L	Screw-hexagonal head-M8 x 25	NLA		Use FS108251L.
7	FS108251L	Screw-flanged head-M8 x 25	NLA		Use FS108257L.
8	WA108051L	Washer-plain-M8	1		
9	WL108001L	Washer-M8	1		
10	NH108041L	Nut-hexagonal head-M8	1		
11	SH604051L	Screw-hexagonal head-1/4UNF x 5/8	1		
12	WL106001L	Washer-M6	1		
13	NH106041L	Nut-hexagonal head-nyloc-M6	1		
14	AMR2216	CABLE-BATTERY POSITIVE	1		
	STC3305	• Bolt-terminal cable-battery	1		
15	50641	Clip-p	1		
16	589254	Grommet-17/32" hole	1		
17	568680	Tie-cable-4.8 x 270mm-inside serated	1		
18	AMR1057	Cable assembly starter & earth	1		
19	SH604051L	Screw-hexagonal head-1/4UNF x 5/8	1		
20	WE600041L	Washer-starlock	1		
21	NH604041L	Nut-hexagonal-coarse thread-1/4UNF	1		
22	PRC7100	Post-terminal	1		
23	PRC7093	Cover-terminal post	1		
24	UKC6684	Tie-cable	1		
25	JRC7549	Escutcheon-headrest-89mm	2	Note (1)	
25	YQJ100200	Strip-edge protection-89mm	2	Note (2)	

CHANGE POINTS:
(1) From (V) MA 081992 To (V) WA 791617
(2) From (V) WA 791618

M 21

DISCOVERY 1989MY UPTO 1999MY	GROUP M	Body Electrics Batteries & Harnesses Harness-Front Wing-To (V) LA081991

Illus	Part Number	Description	Quantity	Change Point	Remarks
		UP TO (V) LA081991			
		HARNESS-FRONT WING			
1	PRC8080	RH-RHD	1	To (V) JA 034313	Except headlamp levelling
1	PRC9447	RH-RHD	1	From (V) KA 034314 To (V) KA 052662	
1	AMR2712	RH-RHD	1	From (V) KA 052663 To (V) LA 081991	
1	PRC8731	RH-RHD	1	To (V) JA 028787	Headlamp levelling
1	AMR1732	RH-RHD	1	From (V) JA 028788 To (V) JA 034313	
1	PRC9449	RH-RHD	1	From (V) KA 034314 To (V) KA 052662	
1	AMR2714	RH-RHD	1	From (V) KA 052663 To (V) LA 081991	
2	PRC8078	RH-LHD	1	To (V) JA 034313	Except headlamp levelling
2	PRC9448	RH-LHD	1	From (V) KA 034314 To (V) KA 052662	
2	AMR2713	RH-LHD	1	From (V) KA 052663 To (V) LA 081991	
2	PRC8089	RH-LHD	1	To (V) JA 028919	Headlamp levelling
2	AMR1729	RH-LHD	1	From (V) JA 028920 To (V) JA 034313	
2	PRC9450	RH-LHD	1	From (V) KA 034314 To (V) KA 052662	
2	AMR2715	RH-LHD	1	From (V) KA 052663 To (V) LA 081991	

NOTE: FROM (V) MA081992 FRONT WING HARNESS IS
INCORPORATED INTO THE MAIN HARNESS

M 22

Harness-Front Wing-To (V) LA081991

Illus	Part Number	Description	Quantity	Change Point	Remarks
		UP TO (V) LA081991			
		HARNESS-FRONT WING			
1	PRC8081	LH-RHD	1	To (V) JA 034313	Except headlamp levelling
1	PRC9045	LH-RHD	1	From (V) KA 034314 To (V) LA 081991	
1	PRC8692	LH-RHD	1	To (V) JA 028919	Headlamp levelling
1	AMR1730	LH-RHD	1	From (V) JA 028920 To (V) JA 034313	
1	PRC9047	LH-RHD	1	From (V) KA 034314 To (V) KA 052662	
1	AMR2986	LH-RHD	1	From (V) KA 052663 To (V) LA 081991	
2	PRC8079	LH-LHD	1	To (V) JA 034313	Except headlamp levelling
2	PRC9046	LH-LHD	1	From (V) JA 034314 To (V) LA 081991	
2	PRC8730	LH-LHD	1	To (V) JA 028787	Headlamp levelling
2	AMR1731	LH-LHD	1	From (V) JA 028788 To (V) KA 034313	
2	PRC9048	LH-LHD	1	From (V) KA 034314 To (V) KA 052662	
2	AMR2987	LH-LHD	1	From (V) KA 052663 To (V) LA 081991	
		NOTE: FROM MA081992 FRONT WING HARNESS IS INCORPORATED INTO THE MAIN HARNESS			
		ALL VARIANTS			
3	AFU1090L	Clip-cable-8mm hole	2		
4	AAU3686	Tie-cable-6.5mm hole-8.0 x 155mm	6		
5	DBP8169L	Clip-pivot-snap-13-17mm-7mm hole-Grey	5		
6	594594	Tie-cable-White.-4.6 x 385mm-inside serated	2	From (V) KA 034314	
7	79122	Clip-single fuel lines-8mm	5		
8	AFU1090L	Clip-cable-8mm hole	4		
9	UKC6684L	Tie-cable-Black-3.5 x 150mm-inside serated	4		

Harness-Main-Without Wing Harness-To (V) LA081991

Illus	Part Number	Description	Quantity	Change Point	Remarks
		V8 TWIN CARB & TDI			
1	PRC8181	Harness main	1	Note (1)	V8 TDi
		WITH ELECTRIC PACK &/OR AIR CON			
1	PRC8182	Harness main	1	Note (1)	V8 TDi
		ALL VARIANTS WITH HEADLAMP LEVELLING			
1	PRC8183	Harness main	1	Note (1)	
		V8 EFI			
1	PRC8694	Harness main	1	Note (1)	
1	PRC8695	Harness main	1	Note (2)	
		TDI-WITHOUT AIR CONDITIONING			
1	PRC8696	Harness main	1	Note (2)	
		TDI-WITH AIR CONDITIONING			
1	PRC8182	Harness main	1	Note (2)	

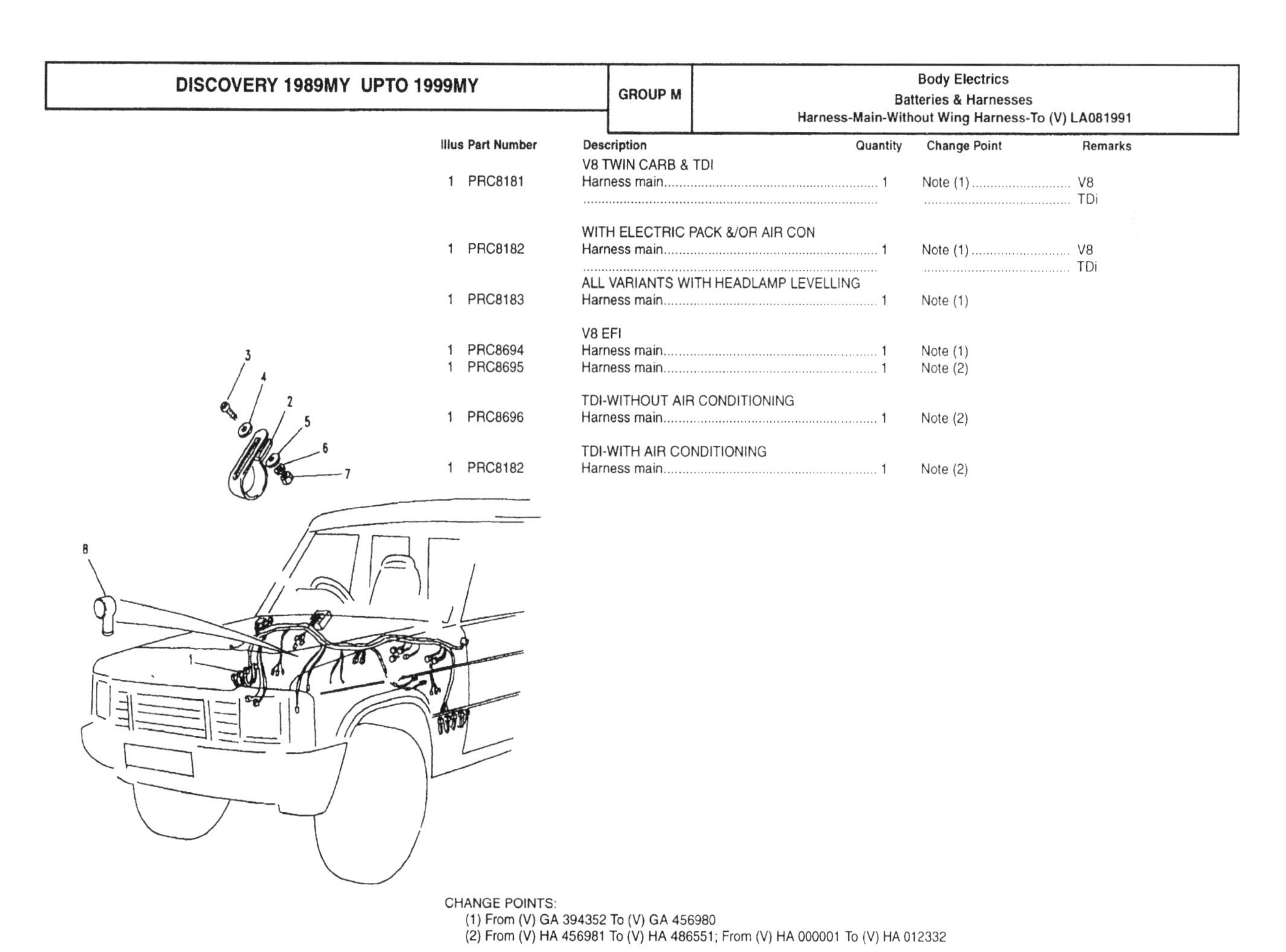

CHANGE POINTS:
(1) From (V) GA 394352 To (V) GA 456980
(2) From (V) HA 456981 To (V) HA 486551; From (V) HA 000001 To (V) HA 012332

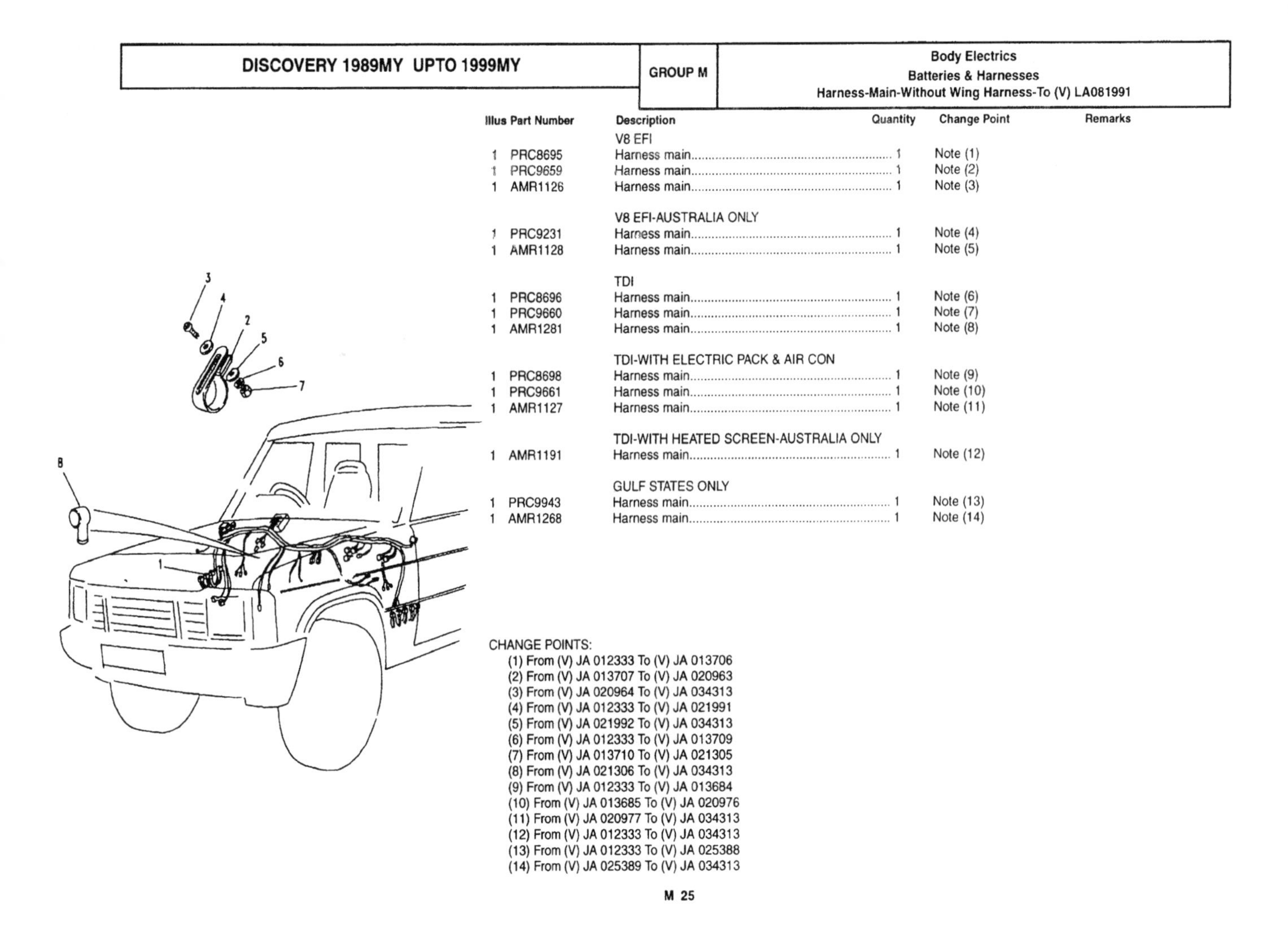

Illus	Part Number	Description	Quantity	Change Point	Remarks
		V8 EFI			
1	PRC8695	Harness main	1	Note (1)	
1	PRC9659	Harness main	1	Note (2)	
1	AMR1126	Harness main	1	Note (3)	
		V8 EFI-AUSTRALIA ONLY			
1	PRC9231	Harness main	1	Note (4)	
1	AMR1128	Harness main	1	Note (5)	
		TDI			
1	PRC8696	Harness main	1	Note (6)	
1	PRC9660	Harness main	1	Note (7)	
1	AMR1281	Harness main	1	Note (8)	
		TDI-WITH ELECTRIC PACK & AIR CON			
1	PRC8698	Harness main	1	Note (9)	
1	PRC9661	Harness main	1	Note (10)	
1	AMR1127	Harness main	1	Note (11)	
		TDI-WITH HEATED SCREEN-AUSTRALIA ONLY			
1	AMR1191	Harness main	1	Note (12)	
		GULF STATES ONLY			
1	PRC9943	Harness main	1	Note (13)	
1	AMR1268	Harness main	1	Note (14)	

CHANGE POINTS:
(1) From (V) JA 012333 To (V) JA 013706
(2) From (V) JA 013707 To (V) JA 020963
(3) From (V) JA 020964 To (V) JA 034313
(4) From (V) JA 012333 To (V) JA 021991
(5) From (V) JA 021992 To (V) JA 034313
(6) From (V) JA 012333 To (V) JA 013709
(7) From (V) JA 013710 To (V) JA 021305
(8) From (V) JA 021306 To (V) JA 034313
(9) From (V) JA 012333 To (V) JA 013684
(10) From (V) JA 013685 To (V) JA 020976
(11) From (V) JA 020977 To (V) JA 034313
(12) From (V) JA 012333 To (V) JA 034313
(13) From (V) JA 012333 To (V) JA 025388
(14) From (V) JA 025389 To (V) JA 034313

M 25

DISCOVERY 1989MY UPTO 1999MY | GROUP M | Body Electrics
Batteries & Harnesses
Harness-Main-Without Wing Harness-To (V) LA081991

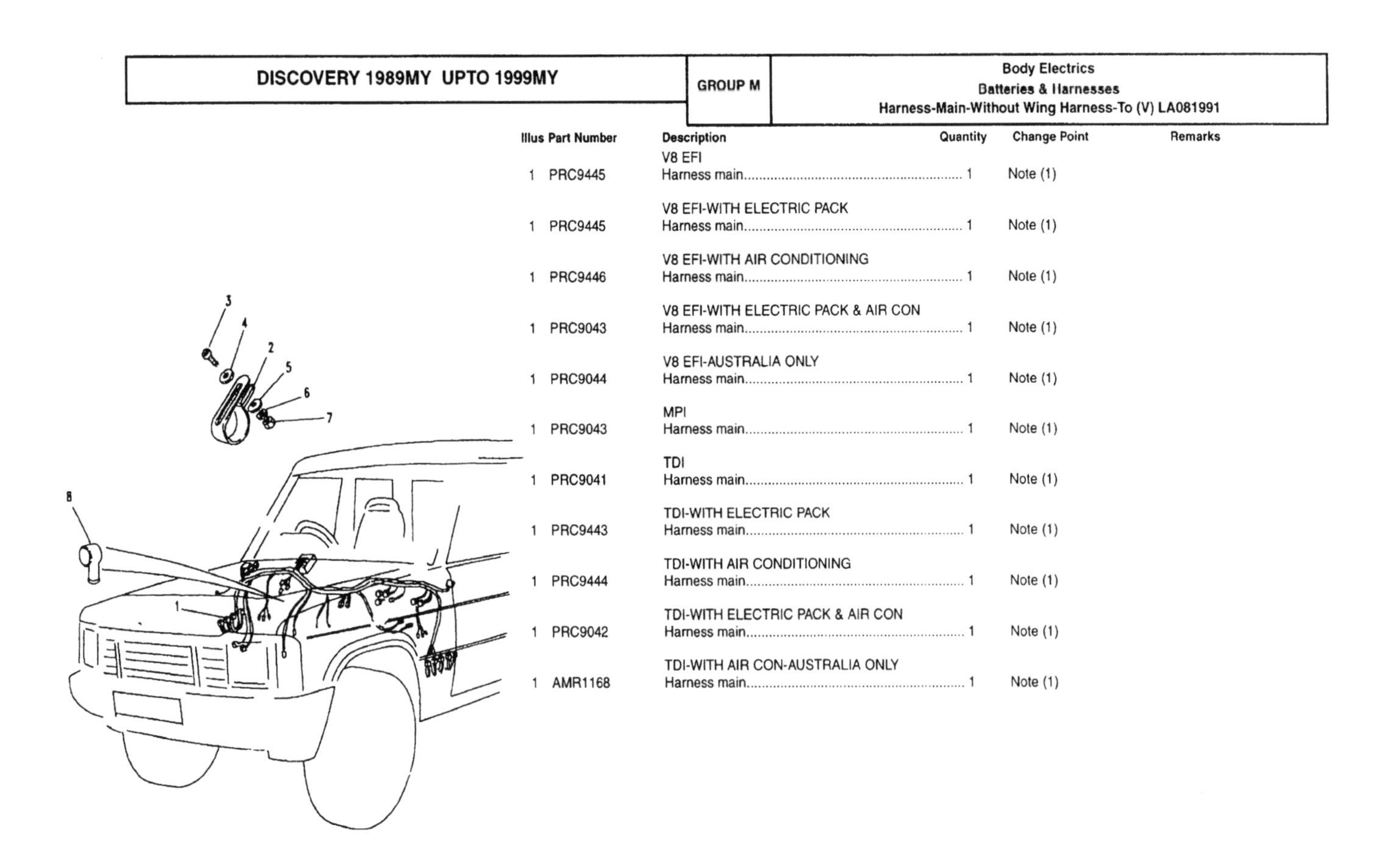

Illus	Part Number	Description	Quantity	Change Point	Remarks
		V8 EFI			
1	PRC9445	Harness main	1	Note (1)	
		V8 EFI-WITH ELECTRIC PACK			
1	PRC9445	Harness main	1	Note (1)	
		V8 EFI-WITH AIR CONDITIONING			
1	PRC9446	Harness main	1	Note (1)	
		V8 EFI-WITH ELECTRIC PACK & AIR CON			
1	PRC9043	Harness main	1	Note (1)	
		V8 EFI-AUSTRALIA ONLY			
1	PRC9044	Harness main	1	Note (1)	
		MPI			
1	PRC9043	Harness main	1	Note (1)	
		TDI			
1	PRC9041	Harness main	1	Note (1)	
		TDI-WITH ELECTRIC PACK			
1	PRC9443	Harness main	1	Note (1)	
		TDI-WITH AIR CONDITIONING			
1	PRC9444	Harness main	1	Note (1)	
		TDI-WITH ELECTRIC PACK & AIR CON			
1	PRC9042	Harness main	1	Note (1)	
		TDI-WITH AIR CON-AUSTRALIA ONLY			
1	AMR1168	Harness main	1	Note (1)	

CHANGE POINTS:
(1) From (V) KA 034314 To (V) KA 064754

M 26

DISCOVERY 1989MY UPTO 1999MY

GROUP M

Body Electrics
Batteries & Harnesses
Harness-Main-Without Wing Harness-To (V) LA081991

Illus	Part Number	Description	Quantity	Change Point	Remarks
		V8 EFI			
1	PRC9445	Harness main	1	Note (1)	
1	AMR3440	Harness main	1	Note (2)	
		V8 EFI-WITH ELECTRIC PACK			
1	PRC9445	Harness main	1	Note (1)	
1	AMR3440	Harness main	1	Note (2)	
		V8 EFI-WITH AIR CONDITIONING			
1	PRC9446	Harness main	1	Note (1)	
1	AMR3441	Harness main	1	Note (2)	
		V8 EFI-WITH ELECTRIC PACK & AIR CON			
1	PRC9043	Harness main	1	Note (1)	
1	AMR3438	Harness main	1	Note (2)	
		V8 EFI-AUSTRALIA ONLY			
1	PRC9044	Harness main	1	Note (1)	
1	AMR3439	Harness main	1	Note (2)	
		MPI			
	PRC9043	Harness main	1	Note (3)	
		TDI			
1	AMR3079	Harness main	1	Note (3)	
		TDI-WITH ELECTRIC PACK			
1	AMR3081	Harness main	1	Note (3)	
		TDI-WITH AIR CONDITIONING			
1	AMR3082	Harness main	1	Note (3)	
		TDI-WITH ELECTRIC PACK & AIR CON			
1	AMR3080	Harness main	1	Note (3)	
		TDI-WITH AIR CON-AUSTRALIA ONLY			
1	AMR3083	Harness main	1	Note (3)	

CHANGE POINTS:
(1) From (V) LA 064755 To (V) LA 065556
(2) From (V) LA 065557 To (V) LA 081991
(3) From (V) LA 064755 To (V) LA 081991

M 27

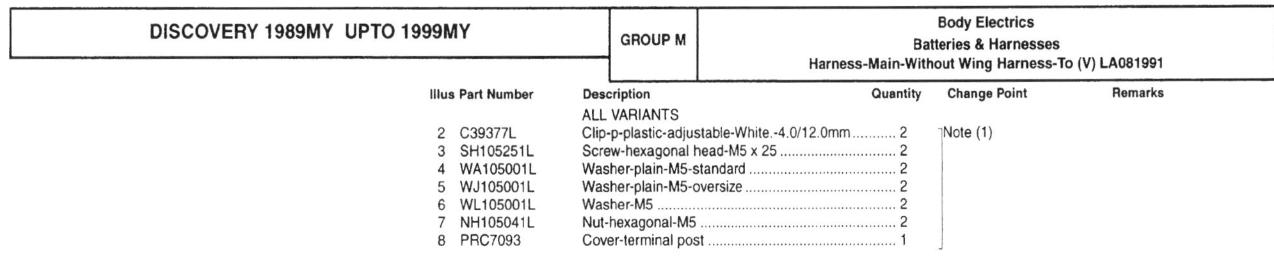

DISCOVERY 1989MY UPTO 1999MY

GROUP M

Body Electrics
Batteries & Harnesses
Harness-Main-Without Wing Harness-To (V) LA081991

Illus	Part Number	Description	Quantity	Change Point	Remarks
		ALL VARIANTS			
2	C39377L	Clip-p-plastic-adjustable-White.-4.0/12.0mm	2	Note (1)	
3	SH105251L	Screw-hexagonal head-M5 x 25	2		
4	WA105001L	Washer-plain-M5-standard	2		
5	WJ105001L	Washer-plain-M5-oversize	2		
6	WL105001L	Washer-M5	2		
7	NH105041L	Nut-hexagonal-M5	2		
8	PRC7093	Cover-terminal post	1		

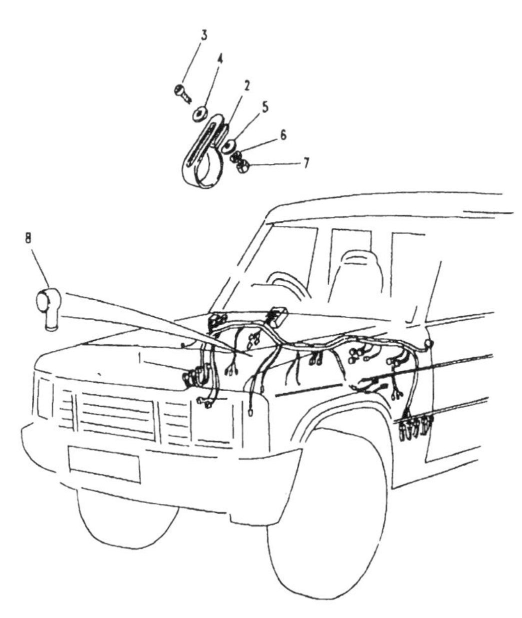

CHANGE POINTS:
(1) From (V) LA 064755 To (V) LA 081991

M 28

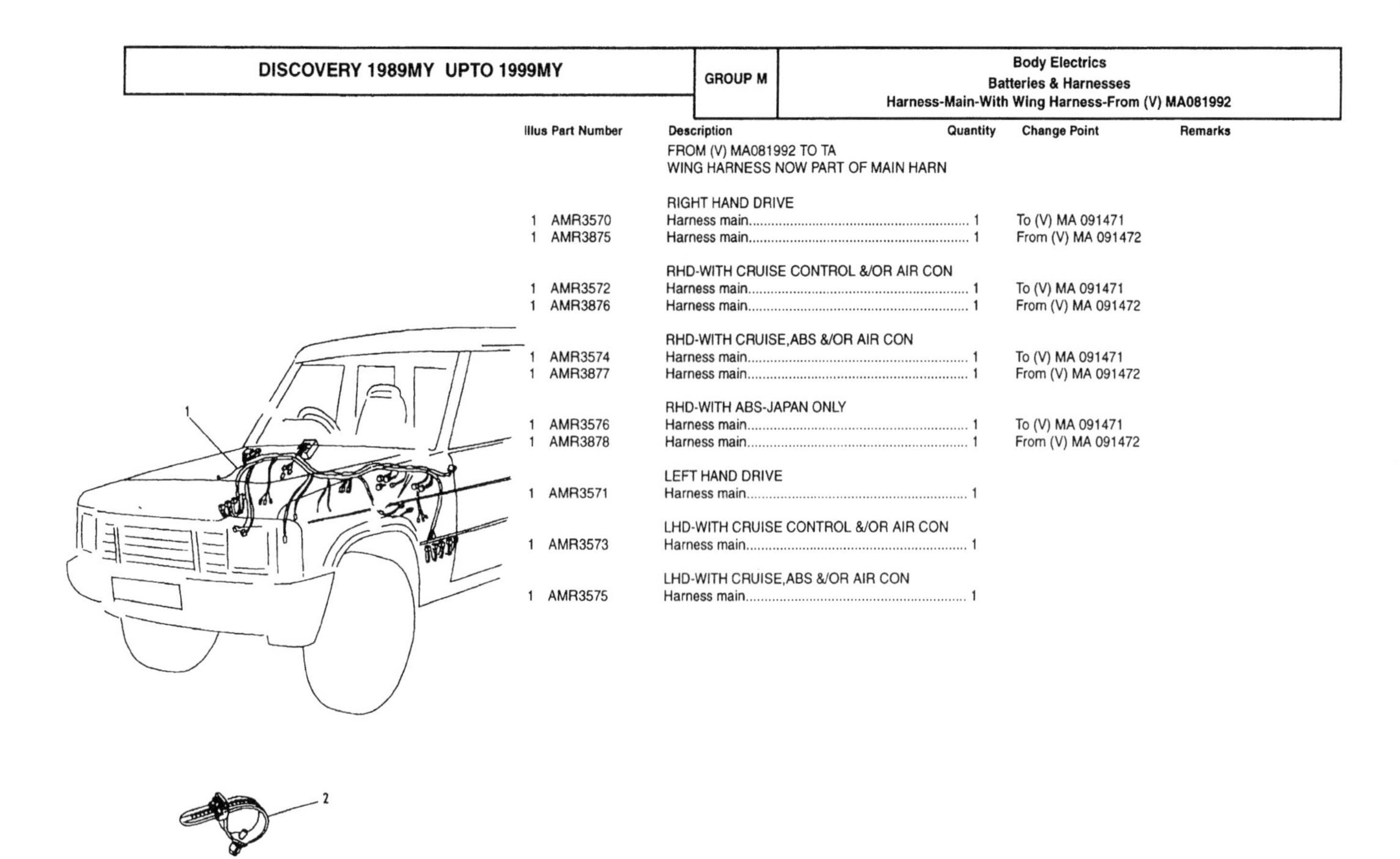

Illus	Part Number	Description	Quantity	Change Point	Remarks
		FROM (V) MA081992 TO TA WING HARNESS NOW PART OF MAIN HARN			
		RIGHT HAND DRIVE			
1	AMR3570	Harness main	1	To (V) MA 091471	
1	AMR3875	Harness main	1	From (V) MA 091472	
		RHD-WITH CRUISE CONTROL &/OR AIR CON			
1	AMR3572	Harness main	1	To (V) MA 091471	
1	AMR3876	Harness main	1	From (V) MA 091472	
		RHD-WITH CRUISE,ABS &/OR AIR CON			
1	AMR3574	Harness main	1	To (V) MA 091471	
1	AMR3877	Harness main	1	From (V) MA 091472	
		RHD-WITH ABS-JAPAN ONLY			
1	AMR3576	Harness main	1	To (V) MA 091471	
1	AMR3878	Harness main	1	From (V) MA 091472	
		LEFT HAND DRIVE			
1	AMR3571	Harness main	1		
		LHD-WITH CRUISE CONTROL &/OR AIR CON			
1	AMR3573	Harness main	1		
		LHD-WITH CRUISE,ABS &/OR AIR CON			
1	AMR3575	Harness main	1		

DISCOVERY 1989MY UPTO 1999MY | GROUP M | Body Electrics
Batteries & Harnesses
Harness-Main-With Wing Harness-From (V) MA081992

Illus	Part Number	Description	Quantity	Change Point	Remarks
		FROM (V) TA163104/TA501920 TO VA			
		RIGHT HAND DRIVE			
1	AMR4203	Harness main	1		
		RHD-WITH CRUISE &/OR AIR CON			
1	AMR4204	Harness main	1		
		RHD-WITH CRUISE,ABS &/OR AIR CON			
1	AMR4205	Harness main	1		
		RHD-WITH CRUISE,ABS &/OR AIR CON & ELECTRIC SEATS			
1	AMR4207	Harness main	1		
		RHD-WITH ABS-JAPAN ONLY			
1	AMR3682	Harness main	1		
		RHD-WITH ELECTRIC SEATS-JAPAN ONLY			
1	AMR4206	Harness main	1		
		LEFT HAND DRIVE			
1	AMR4208	Harness main	1		
		LHD-WITH CRUISE CONTROL &/OR AIR CON			
1	AMR4209	Harness main	1		
		LHD-WITH CRUISE &/OR AIR CON & ELECTRIC SEATS			
1	AMR4210	Harness main	1		
		LHD-WITH CRUISE,ABS &/OR AIR CON & ELECTRIC SEATS			
1	AMR4211	Harness main	1		

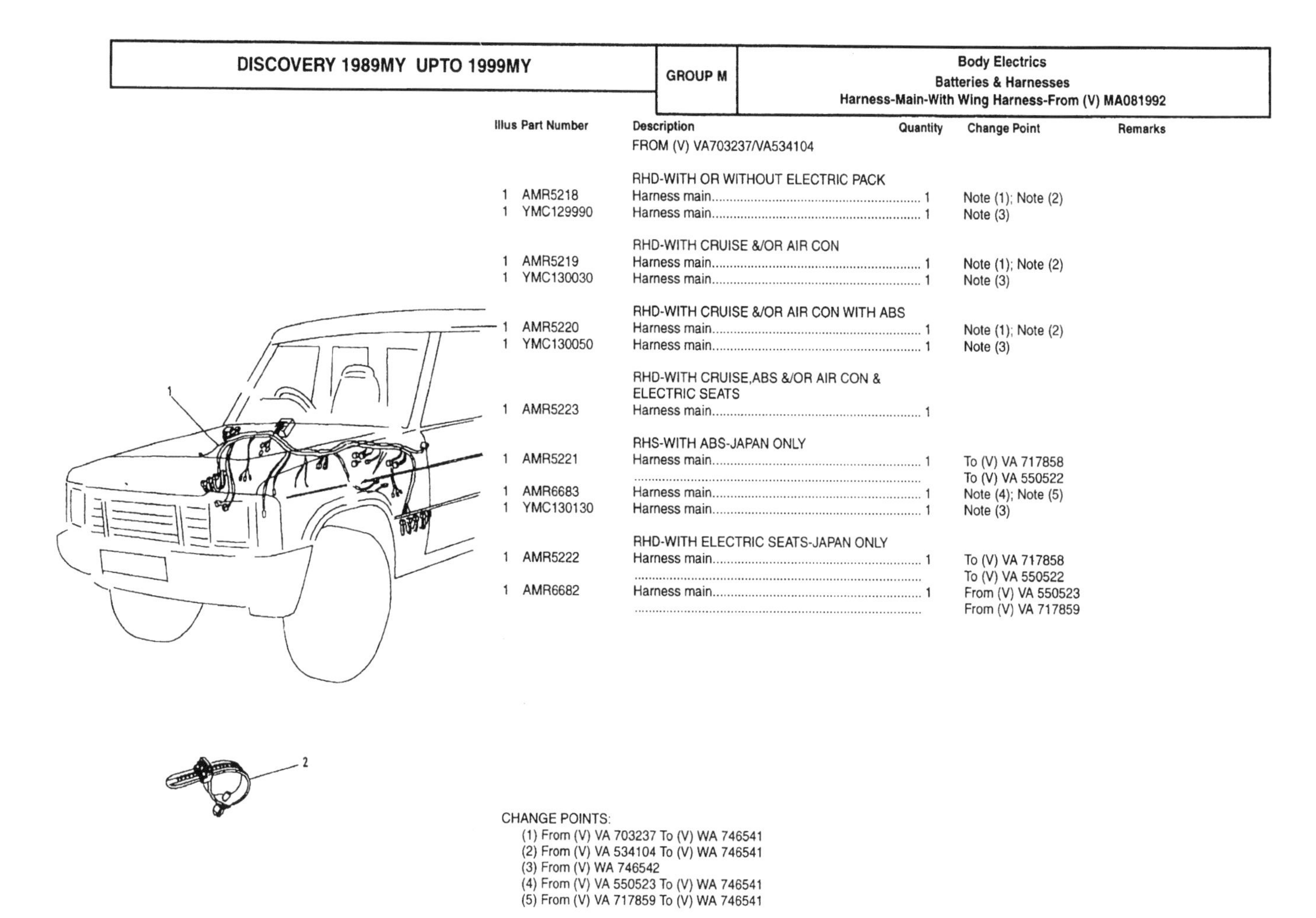

Illus	Part Number	Description	Quantity	Change Point	Remarks
		FROM (V) VA703237/VA534104			
		RHD-WITH OR WITHOUT ELECTRIC PACK			
1	AMR5218	Harness main	1	Note (1); Note (2)	
1	YMC129990	Harness main	1	Note (3)	
		RHD-WITH CRUISE &/OR AIR CON			
1	AMR5219	Harness main	1	Note (1); Note (2)	
1	YMC130030	Harness main	1	Note (3)	
		RHD-WITH CRUISE &/OR AIR CON WITH ABS			
1	AMR5220	Harness main	1	Note (1); Note (2)	
1	YMC130050	Harness main	1	Note (3)	
		RHD-WITH CRUISE,ABS &/OR AIR CON & ELECTRIC SEATS			
1	AMR5223	Harness main	1		
		RHS-WITH ABS-JAPAN ONLY			
1	AMR5221	Harness main	1	To (V) VA 717858	
				To (V) VA 550522	
1	AMR6683	Harness main	1	Note (4); Note (5)	
1	YMC130130	Harness main	1	Note (3)	
		RHD-WITH ELECTRIC SEATS-JAPAN ONLY			
1	AMR5222	Harness main	1	To (V) VA 717858	
				To (V) VA 550522	
1	AMR6682	Harness main	1	From (V) VA 550523	
				From (V) VA 717859	

CHANGE POINTS:
(1) From (V) VA 703237 To (V) WA 746541
(2) From (V) VA 534104 To (V) WA 746541
(3) From (V) WA 746542
(4) From (V) VA 550523 To (V) WA 746541
(5) From (V) VA 717859 To (V) WA 746541

M 31

DISCOVERY 1989MY UPTO 1999MY | GROUP M | Body Electrics
Batteries & Harnesses
Harness-Main-With Wing Harness-From (V) MA081992

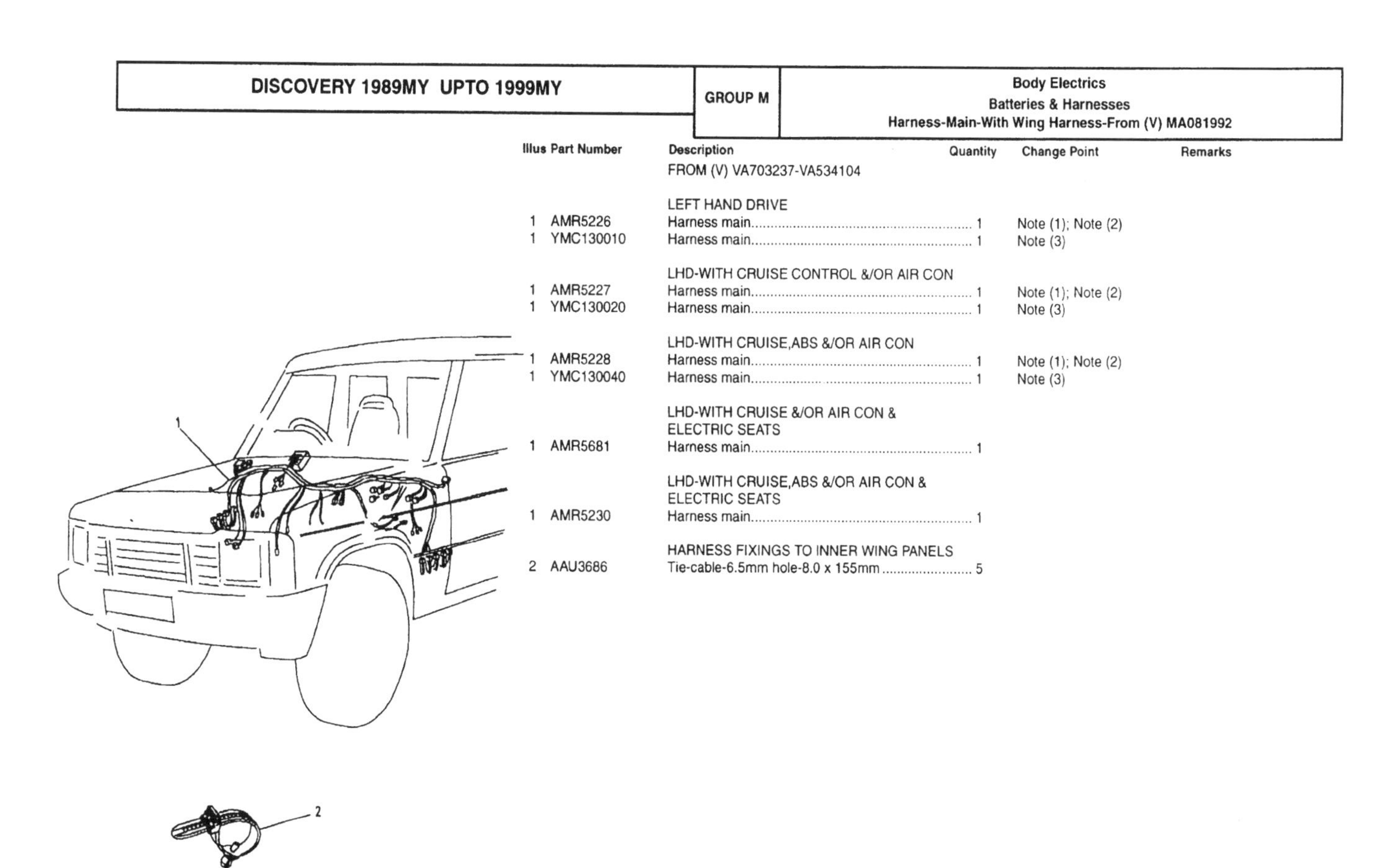

Illus	Part Number	Description	Quantity	Change Point	Remarks
		FROM (V) VA703237-VA534104			
		LEFT HAND DRIVE			
1	AMR5226	Harness main	1	Note (1); Note (2)	
1	YMC130010	Harness main	1	Note (3)	
		LHD-WITH CRUISE CONTROL &/OR AIR CON			
1	AMR5227	Harness main	1	Note (1); Note (2)	
1	YMC130020	Harness main	1	Note (3)	
		LHD-WITH CRUISE,ABS &/OR AIR CON			
1	AMR5228	Harness main	1	Note (1); Note (2)	
1	YMC130040	Harness main	1	Note (3)	
		LHD-WITH CRUISE &/OR AIR CON & ELECTRIC SEATS			
1	AMR5681	Harness main	1		
		LHD-WITH CRUISE,ABS &/OR AIR CON & ELECTRIC SEATS			
1	AMR5230	Harness main	1		
		HARNESS FIXINGS TO INNER WING PANELS			
2	AAU3686	Tie-cable-6.5mm hole-8.0 x 155mm	5		

CHANGE POINTS:
(1) From (V) VA 703237 To (V) WA 746541
(2) From (V) VA 534104 To (V) WA 746541
(3) From (V) WA 746542

M 32

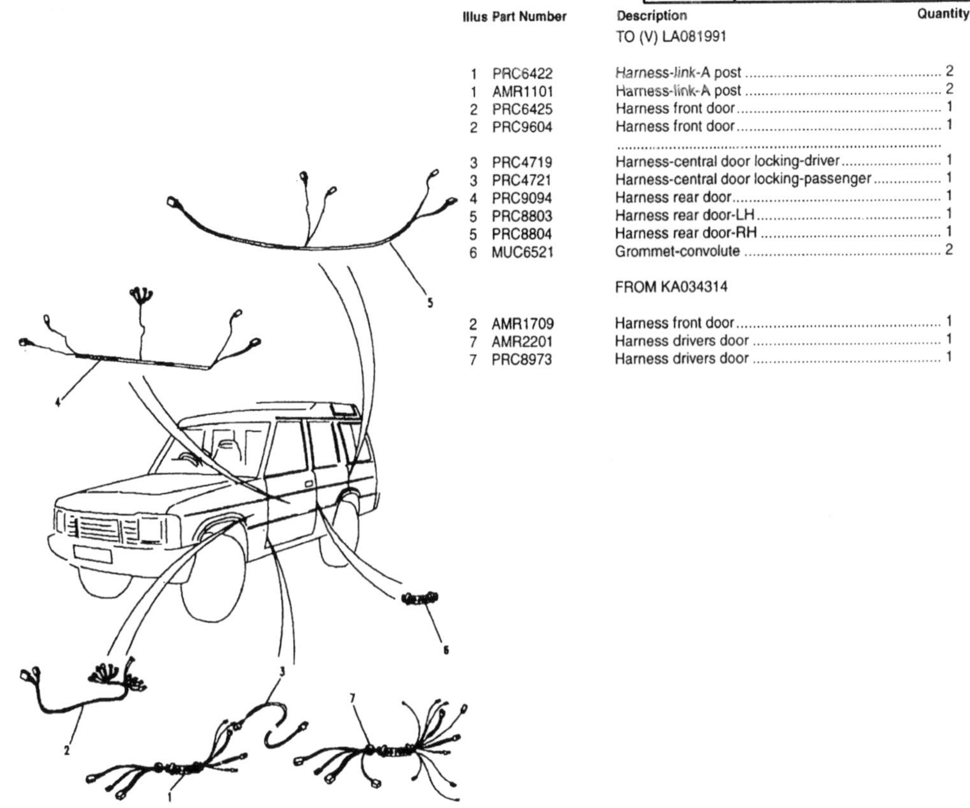

Illus	Part Number	Description	Quantity	Change Point	Remarks
		TO (V) LA081991			
1	PRC6422	Harness-link-A post	2	To (V) JA 034313	
1	AMR1101	Harness-link-A post	2	From (V) KA 034314	
2	PRC6425	Harness front door	1	From (V) JA 013672	Note (1)
2	PRC9604	Harness front door	1	From (V) JA 013673 To (V) JA 034313	Note (1)
3	PRC4719	Harness-central door locking-driver	1		
3	PRC4721	Harness-central door locking-passenger	1		
4	PRC9094	Harness rear door	1		
5	PRC8803	Harness rear door-LH	1		
5	PRC8804	Harness rear door-RH	1		
6	MUC6521	Grommet-convolute	2		
		FROM KA034314			
2	AMR1709	Harness front door	1		Electric windows
7	AMR2201	Harness drivers door	1		Note (2)
7	PRC8973	Harness drivers door	1		Except burglar alarm

Remarks:
(1) CDL electric windows electric door mirrors
(2) burglar alarm electric windows CDL electric door mirrors

M 33

DISCOVERY 1989MY UPTO 1999MY | **GROUP M** | **Body Electrics** **Batteries & Harnesses** **Harness-Body & Tail Door-To (V) LA081991**

Illus	Part Number	Description	Quantity	Change Point	Remarks
		V8 & TDI UP TO (V) LA081991			
1	PRC6413	Harness body-RH	1	To (V) JA 030856	High line
1	PRC8978	Harness body-RH	1	From (V) JA 030857	High line
2	PRC8185	Harness body-LH	1	To (V) JA 030856	High line
2	PRC8979	Harness body-LH	1	From (V) JA 030857 To (V) KA 064754	High line
2	AMR3437	Harness body-LH	1	From (V) LA 064755	High line
	AB608031L	Screw-self tapping AB-No 8 x 3/8	1		
3	AFU1090L	Clip-cable-8mm hole	8		
4	PRC6420	Harness- tail door	1	To (V) JA 030856	
4	PRC8977	Harness- tail door	1	From (V) JA 030857	
5	PRC7145	Grommet	1	To (V) JA 030856	
5	AMR1040	Grommet	1	From (V) JA 030857	
		LOWLINE			
1	PRC7423	Harness body-RH	1	To (V) JA 030856	
1	PRC8981	Harness body-RH	1	From (V) JA 030857	
2	PRC8184	Harness body-LH	1	To (V) JA 030856	
2	PRC8982	Harness body-LH	1	From (V) JA 030857 To (V) KA 064754	
2	AMR3436	Harness body-LH	1	From (V) KA 064755	
	AB608031L	Screw-self tapping AB-No 8 x 3/8	1		
3	AFU1090L	Clip-cable-8mm hole	8		
4	PRC7425	Harness- tail door	1		
5	PRC7145	Grommet	1	To (V) JA 030856	
5	AMR1040	Grommet	1	From (V) JA 030857	

M 34

Illus	Part Number	Description	Quantity	Change Point	Remarks
		MPI-TO (V) LA081991			
		TO (V) KA038821			
1	PRC8978	Harness body-RH	1		
2	PRC8979	Harness body-LH	1		
4	PRC8977	Harness- tail door	1		
		FROM (V) GA450690			
1	AMR1527	Harness body	1		Low line
1	AMR1528	Harness body	1		High line
4	AMR2304	Harness- tail door	1		Low line
4	AMR1540	Harness- tail door	1		High line
5	PRC7145	Grommet-for harness rear end door	1	To (V) JA 030856	
5	AMR1040	Grommet-for harness rear end door	1	From (V) JA 030857	
		REAR DOOR OPEN MARKER LAMP			
6	PRC8749	LAMP-STAND ALONE WARNING-MARKER REAR END DOOR	1		
7	AAU5034	• Diode-Pektron	1		
8	PRC8077	Lamp-stand alone warning-door open rear end door-Red	1		
9	MWC1722	Reflector-rear-rectangular	1		
10	AB608035	Screw-fixing reflector	2		

Illus	Part Number	Description	Quantity	Change Point	Remarks
		FROM (V) MA081992			
		LOWLINE-NOTE (1)			
1	AMR3556	Harness body	1	Note (1)	2 speaker system
					4 speaker system
		MIDLINE & HIGHLINE-NOTE (2&3)			
1	AMR3557	Harness body	1	Note (1)	Mid line 5 speaker system
					Mid line 7 speaker system
					CD Player high line
		ALL VARIANTS WITH REAR AIR CON.			
1	AMR3558	Harness body-with rear air-con	1	Note (1)	Rear air conditioning
		ALL VARIANTS WITHOUT REAR AIR CON.			
		FROM (V) MA135908			
		HARNESS BODY			
1	AMR4877	less rear air-con	1	Note (2)	
1	AMR4219	less rear air-con	1	Note (3)	
1	AMR5548	less rear air-con	1	Note (4)	
1	AMR5238	less rear air-con	1	Note (5)	

CHANGE POINTS:
(1) From (V) MA 081992 To (V) MA 135907
(2) From (V) MA 135908 To (V) MA 163103; From (V) MA 135908 To (V) TA 501919
(3) From (V) TA 163104 To (V) TA 172381; From (V) TA 501920 To (V) TA 504321
(4) From (V) TA 172382; From (V) TA 504322
(5) From (V) VA 703237; From (V) VA 534104

Illus	Part Number	Description	Quantity	Change Point	Remarks
		ALL VARIANTS WITH REAR AIR CON.			
		HARNESS BODY			
1	AMR4879	with rear air-con	1	Note (1)	
1	AMR3921	with rear air-con	1	Note (2)	
1	AMR5549	with rear air-con	1	Note (3)	
1	AMR5239	with rear air-con	1	Note (4)	

NOTE (1) LOWLINE-VEHICLES WITH BASE SPEC 2/4 SPEAKER RADIO SYSTEM
NOTE (2) MIDLINE-VEHICLES WITH 5/7 SPEAKER & REMOTE RADIO CONTROL
NOTE (3) HIGHLINE-VEHICLES WITH REMOTE CONTROL CD & RADIO SYSTEM

CHANGE POINTS:
(1) From (V) MA 135908 To (V) MA 163103; From (V) MA 135908 To (V) MA 501919
(2) From (V) TA 163104 To (V) TA 172381; From (V) TA 501920 To (V) TA 504321
(3) From (V) TA 172382; From (V) TA 504322
(4) From (V) VA 703237; From (V) VA 534104

DISCOVERY 1989MY UPTO 1999MY | GROUP M | Body Electrics
Batteries & Harnesses
Harness-Body,Facia & Strg Column-From (V) MA081992

Illus	Part Number	Description	Quantity	Change Point	Remarks
		FROM (V) MA081992			
		LOWLINE: 2/4 SPEAKER RADIO/CASSETTE			
2	AMR3608	Harness facia	1	Note (1)	Low line
2	AMR4214	Harness facia	1	Note (2)	
2	AMR5241	Harness facia	1	Note (3)	
		MIDLINE: 5/7 SPEAKER REMOTE R/CASSETTE			
2	AMR3609	Harness facia	1	Note (4)	Mid line
2	AMR4215	Harness facia	1	Note (2)	
2	AMR5242	Harness facia	1	Note (3)	

CHANGE POINTS:
(1) From (V) MA 081992 To (V) MA 163103; From (V) MA 081992 To (V) MA 501919
(2) From (V) TA 163104; From (V) TA 501920
(3) From (V) VA 703237; From (V) VA 534104
(4) From (V) MA 081992 To (V) MA 163104; From (V) MA 081992 To (V) MA 501919

Illus	Part Number	Description	Quantity	Change Point	Remarks
		HIGHLINE: REMOTE R/CASSETTE & CD			
2	AMR3610	Harness facia	4	Note (1)	High line
2	AMR4216	Harness facia	1	Note (2)	
2	AMR5243	Harness facia	1	Note (3)	
		JAPAN			
2	AMR3611	Harness facia	1	Note (1)	Japan
2	AMR3685	Harness facia	1	Note (2)	
2	AMR5245	Harness facia	1	Note (4)	
2	AMR6681	Harness facia	1	Note (5)	

NOTE: LOW/MID/HIGHLINE-SEE PREVIOUS PAGE

CHANGE POINTS:
(1) From (V) MA 081992 To (V) MA 163103; From (V) MA 081992 To (V) MA 501919
(2) From (V) TA 163104; From (V) TA 501920
(3) From (V) VA 703237; From (V) VA 534104
(4) From (V) VA 534104 To (V) VA 550522; From (V) VA 703237 To (V) VA 717858
(5) From (V) VA 717859; From (V) VA 550523

M 39

Illus	Part Number	Description	Quantity	Change Point	Remarks
		FROM (V) MA081992			
		STEERING COLUMN HARNESS (SEE NOTE)			
3	AMR3579	Harness steering column-with horn push only	1	Note (1)	
3	AMR3580	Harness steering column	1	Note (1)	Cruise control Rover airbag system
3	AMR3581	Harness steering column	1	Note (1)	Shift interlock
3	AMR3582	Harness steering column	1	Note (1)	
3	AMR3374	Harness steering column	1	Note (1)	Japan except Rover airbag system
		STEERING COLUMN BRACKET			
4	ANR3202	Bracket harness	1	Note (1)	
5	ANR5110	Bracket-steering column-without harness support	1	Note (2)	

NOTE: FROM (V) TA163104/TA501920-INCORPORATED INTO MAIN HARNESS

CHANGE POINTS:
(1) From (V) MA 081992 To (V) MA 163103; From (V) MA 081992 To (V) MA 501919
(2) From (V) TA 163104; From (V) TA 501920

M 40

Illus	Part Number	Description	Quantity	Change Point	Remarks
		FROM (V) MA081992			
		HARNESS FRONT DOOR			
1	AMR3617	driver	1	To (V) MA 137189	Except burglar alarm
1	AMR4239	driver	1	From (V) MA 137190	
1	AMR3618	driver	1	To (V) MA 137189	Burglar alarm
1	AMR4240	driver	1	From (V) MA 137190	
1	AMR3619	passenger	1	To (V) MA 137189	
1	AMR4241	passenger	1	From (V) MA 137190	
2	AMR3620	Harness-supplementary audio system	2		
		HARNESS REAR DOOR			
3	AMR3638	rear door service	1	Note (1)	
3	AMR5517	rear door service	1	Note (2)	
4	AMR3639	RH	1	To (V) MA 137189	
4	AMR4243	LH	1	From (V) MA 137190	
4	AMR3640	LH	1	To (V) MA 137189	
4	AMR4242	RH	1	From (V) MA 137190	
5	MUC6521	Grommet-convolute	2		

CHANGE POINTS:
(1) From (V) MA 081992 To (V) MA 163103; From (V) MA 081992 To (V) MA 501919
(2) From (V) TA 163104; From (V) TA 501920

M 41

DISCOVERY 1989MY UPTO 1999MY | GROUP M | Body Electrics
Batteries & Harnesses
Harness-Doors & Taildoor-From (V) MA081992

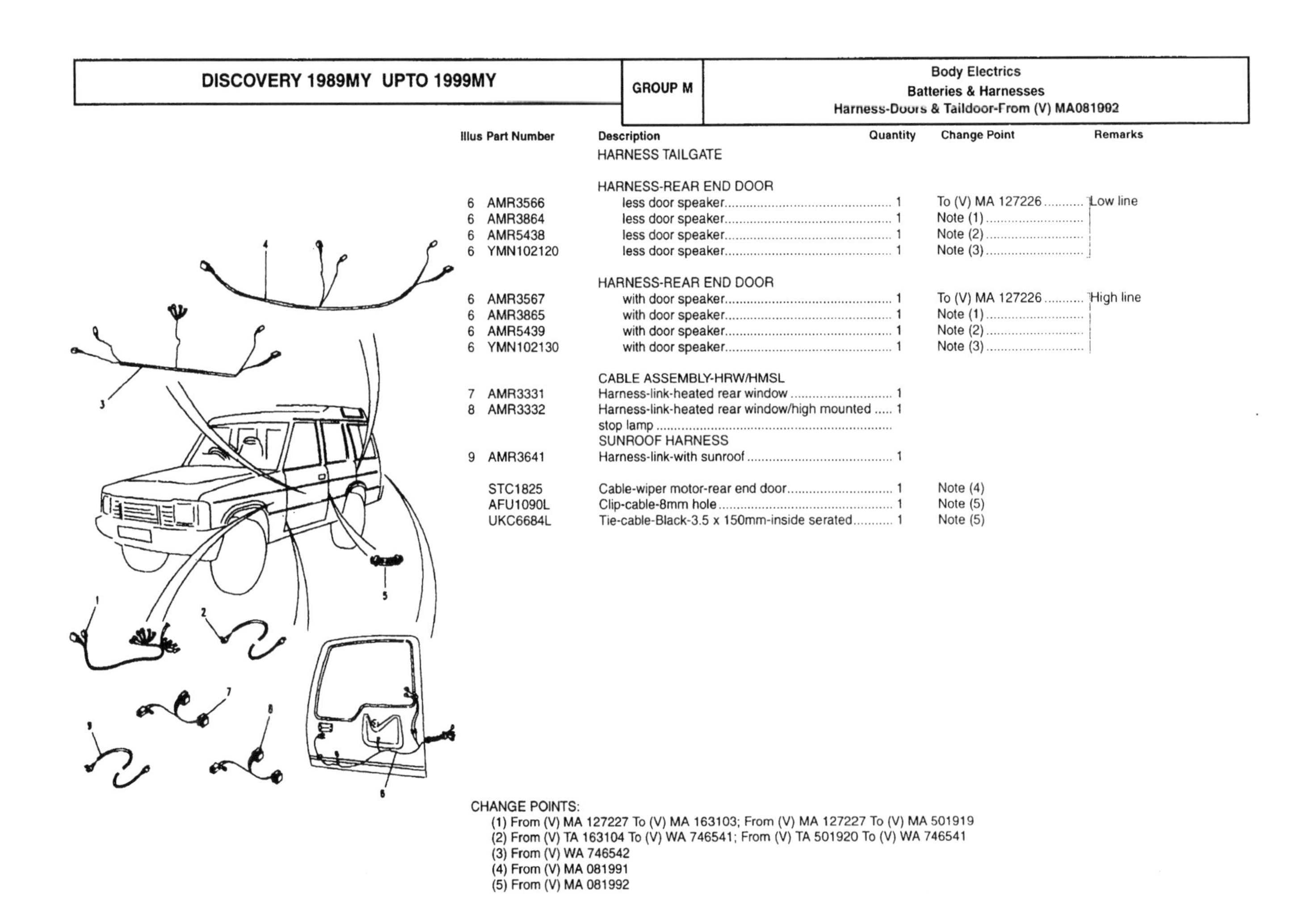

Illus	Part Number	Description	Quantity	Change Point	Remarks
		HARNESS TAILGATE			
		HARNESS-REAR END DOOR			
6	AMR3566	less door speaker	1	To (V) MA 127226	Low line
6	AMR3864	less door speaker	1	Note (1)	
6	AMR5438	less door speaker	1	Note (2)	
6	YMN102120	less door speaker	1	Note (3)	
		HARNESS-REAR END DOOR			
6	AMR3567	with door speaker	1	To (V) MA 127226	High line
6	AMR3865	with door speaker	1	Note (1)	
6	AMR5439	with door speaker	1	Note (2)	
6	YMN102130	with door speaker	1	Note (3)	
		CABLE ASSEMBLY-HRW/HMSL			
7	AMR3331	Harness-link-heated rear window	1		
8	AMR3332	Harness-link-heated rear window/high mounted stop lamp	1		
		SUNROOF HARNESS			
9	AMR3641	Harness-link-with sunroof	1		
	STC1825	Cable-wiper motor-rear end door	1	Note (4)	
	AFU1090L	Clip-cable-8mm hole	1	Note (5)	
	UKC6684L	Tie-cable-Black-3.5 x 150mm-inside serated	1	Note (5)	

CHANGE POINTS:
(1) From (V) MA 127227 To (V) MA 163103; From (V) MA 127227 To (V) MA 501919
(2) From (V) TA 163104 To (V) WA 746541; From (V) TA 501920 To (V) WA 746541
(3) From (V) WA 746542
(4) From (V) MA 081991
(5) From (V) MA 081992

M 42

Illus	Part Number	Description	Quantity	Change Point	Remarks
		FROM (V) MA081992			
		HEATER & AIR CON. SWITCH LINKS			
	AMR3613	Harness blower-heater	1		
1	AMR3614	Harness blower-air conditioning-front	1		Air con
1	AMR3615	Harness blower-air conditioning-front & rear	1		Rear air conditioning
		TRANS/BOX OIL TEMPERATURE SWITCH LINK LEAD AUTO ONLY			
2	PRC7292	Harness-link-oil-temp warning-automatic	1	To (V) MA 152906	
2	AMR4952	Harness-link-oil-temp warning	1	From (V) MA 152907	

M 43

Illus	Part Number	Description	Quantity	Change Point	Remarks
		ALL VARIANTS			
1	AMR2082	Harness engine	1	To (V) JA 055568	
		WITHOUT AIR CONDITIONING			
1	AMR2886	Harness engine	1	Note (1)	
		WITH AIR CONDITIONING			
1	AMR2082	Harness engine	1	Note (1)	
		HARNESS FIXINGS			
2	WL106001L	Washer-M6	1		
3	NH106041L	Nut-hexagonal head-nyloc-M6	1		
4	UKC6684L	Tie-cable-Black-3.5 x 150mm-inside serated	4		
5	AMR2294	Bracket-harness protector	1	Note (1)	
6	FS106101L	Screw-flanged head-M6 x 10	3	Note (1)	

CHANGE POINTS:
(1) From (V) KA 055569 To (V) LA 081991

M 44

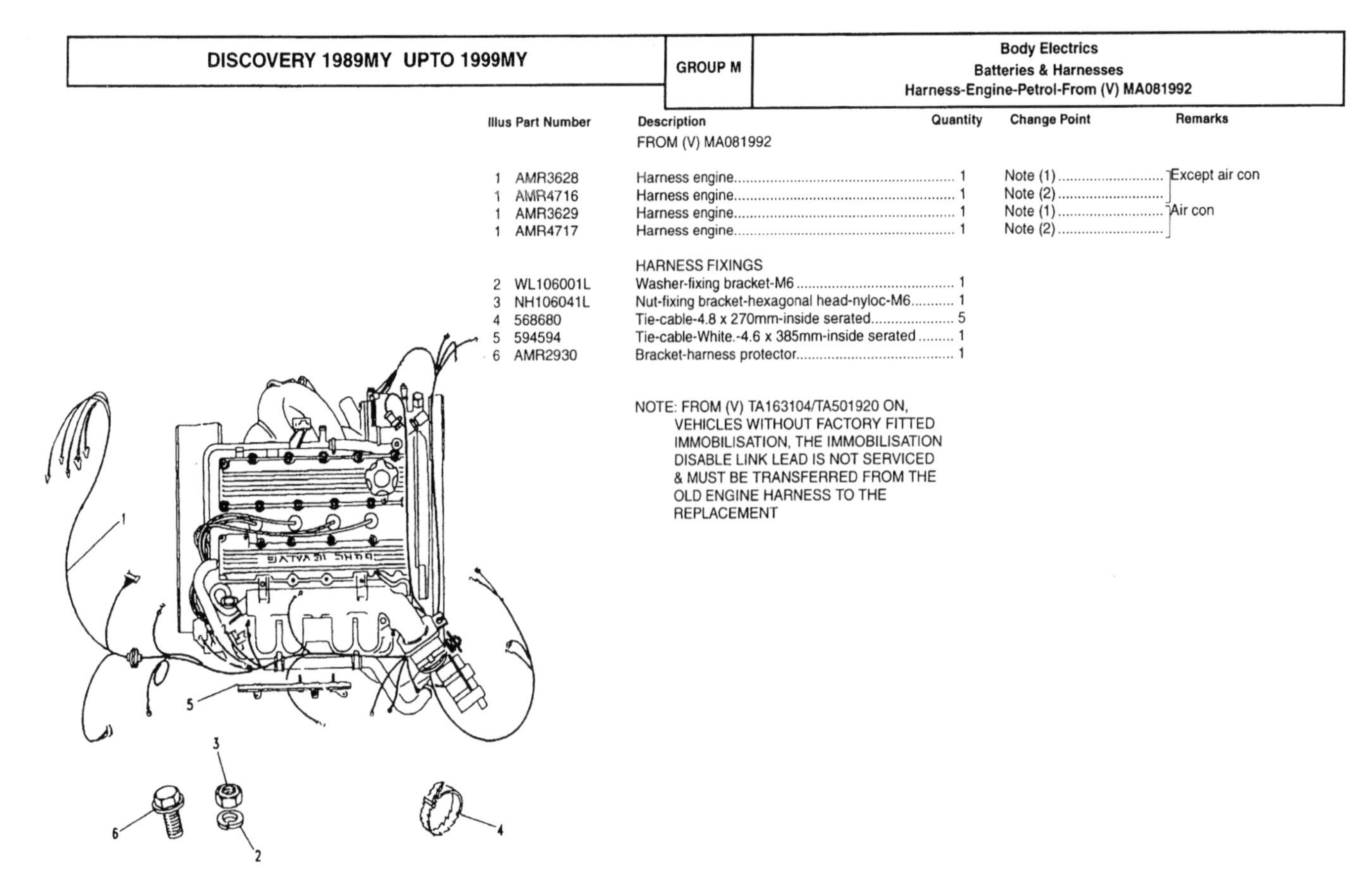

Illus	Part Number	Description	Quantity	Change Point	Remarks
		FROM (V) MA081992			
1	AMR3628	Harness engine	1	Note (1)	Except air con
1	AMR4716	Harness engine	1	Note (2)	
1	AMR3629	Harness engine	1	Note (1)	Air con
1	AMR4717	Harness engine	1	Note (2)	
		HARNESS FIXINGS			
2	WL106001L	Washer-fixing bracket-M6	1		
3	NH106041L	Nut-fixing bracket-hexagonal head-nyloc-M6	1		
4	568680	Tie-cable-4.8 x 270mm-inside serated	5		
5	594594	Tie-cable-White.-4.6 x 385mm-inside serated	1		
6	AMR2930	Bracket-harness protector	1		

NOTE: FROM (V) TA163104/TA501920 ON, VEHICLES WITHOUT FACTORY FITTED IMMOBILISATION, THE IMMOBILISATION DISABLE LINK LEAD IS NOT SERVICED & MUST BE TRANSFERRED FROM THE OLD ENGINE HARNESS TO THE REPLACEMENT

CHANGE POINTS:
(1) From (V) MA 081992 To (V) MA 163103; From (V) MA 081992 To (V) MA 501919
(2) From (V) TA 163104; From (V) TA 501920

M 45

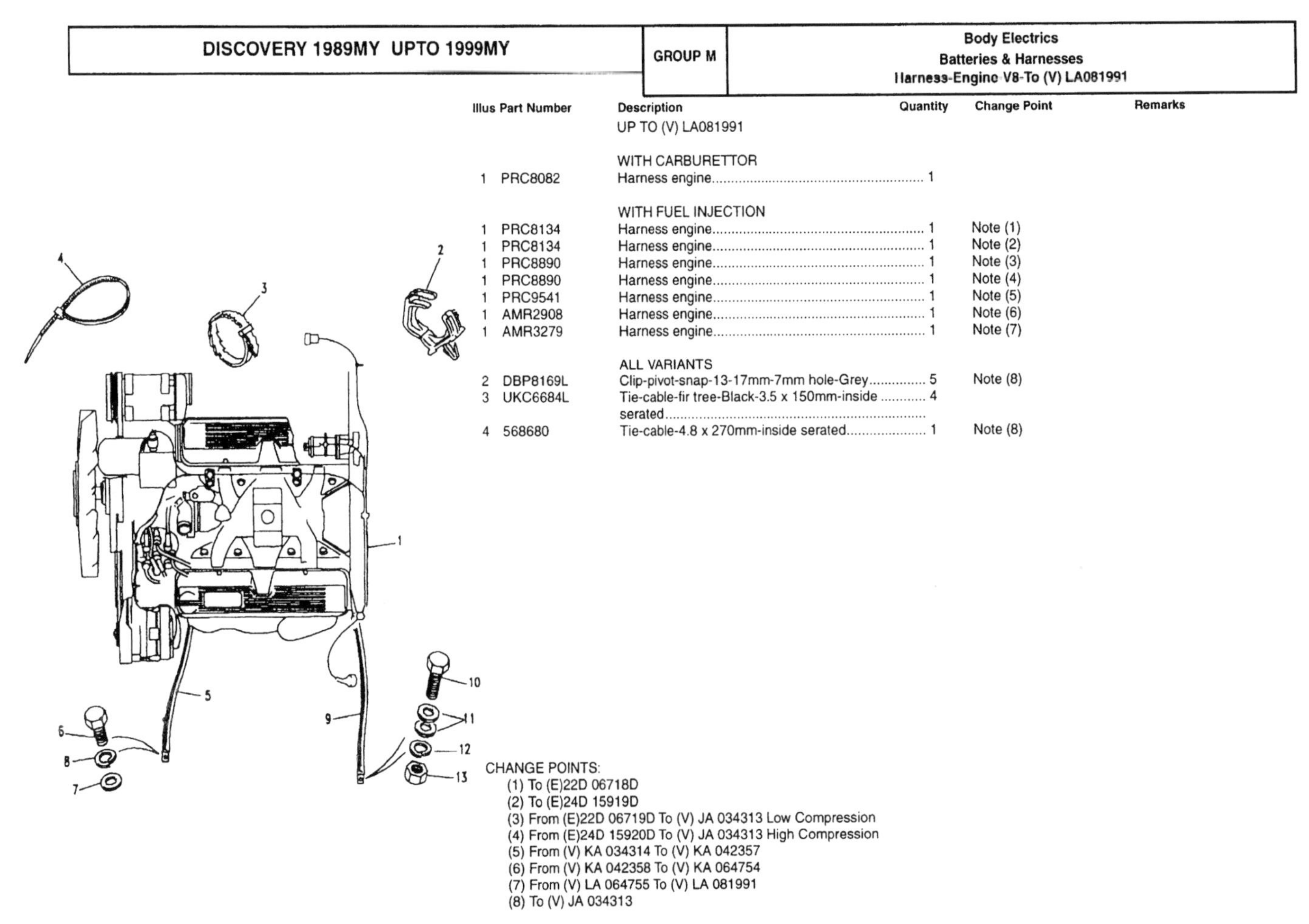

Illus	Part Number	Description	Quantity	Change Point	Remarks
		UP TO (V) LA081991			
		WITH CARBURETTOR			
1	PRC8082	Harness engine	1		
		WITH FUEL INJECTION			
1	PRC8134	Harness engine	1	Note (1)	
1	PRC8134	Harness engine	1	Note (2)	
1	PRC8890	Harness engine	1	Note (3)	
1	PRC8890	Harness engine	1	Note (4)	
1	PRC9541	Harness engine	1	Note (5)	
1	AMR2908	Harness engine	1	Note (6)	
1	AMR3279	Harness engine	1	Note (7)	
		ALL VARIANTS			
2	DBP8169L	Clip-pivot-snap-13-17mm-7mm hole-Grey	5	Note (8)	
3	UKC6684L	Tie-cable-fir tree-Black-3.5 x 150mm-inside serated	4		
4	568680	Tie-cable-4.8 x 270mm-inside serated	1	Note (8)	

CHANGE POINTS:
(1) To (E)22D 06718D
(2) To (E)24D 15919D
(3) From (E)22D 06719D To (V) JA 034313 Low Compression
(4) From (E)24D 15920D To (V) JA 034313 High Compression
(5) From (V) KA 034314 To (V) KA 042357
(6) From (V) KA 042358 To (V) KA 064754
(7) From (V) LA 064755 To (V) LA 081991
(8) To (V) JA 034313

M 46

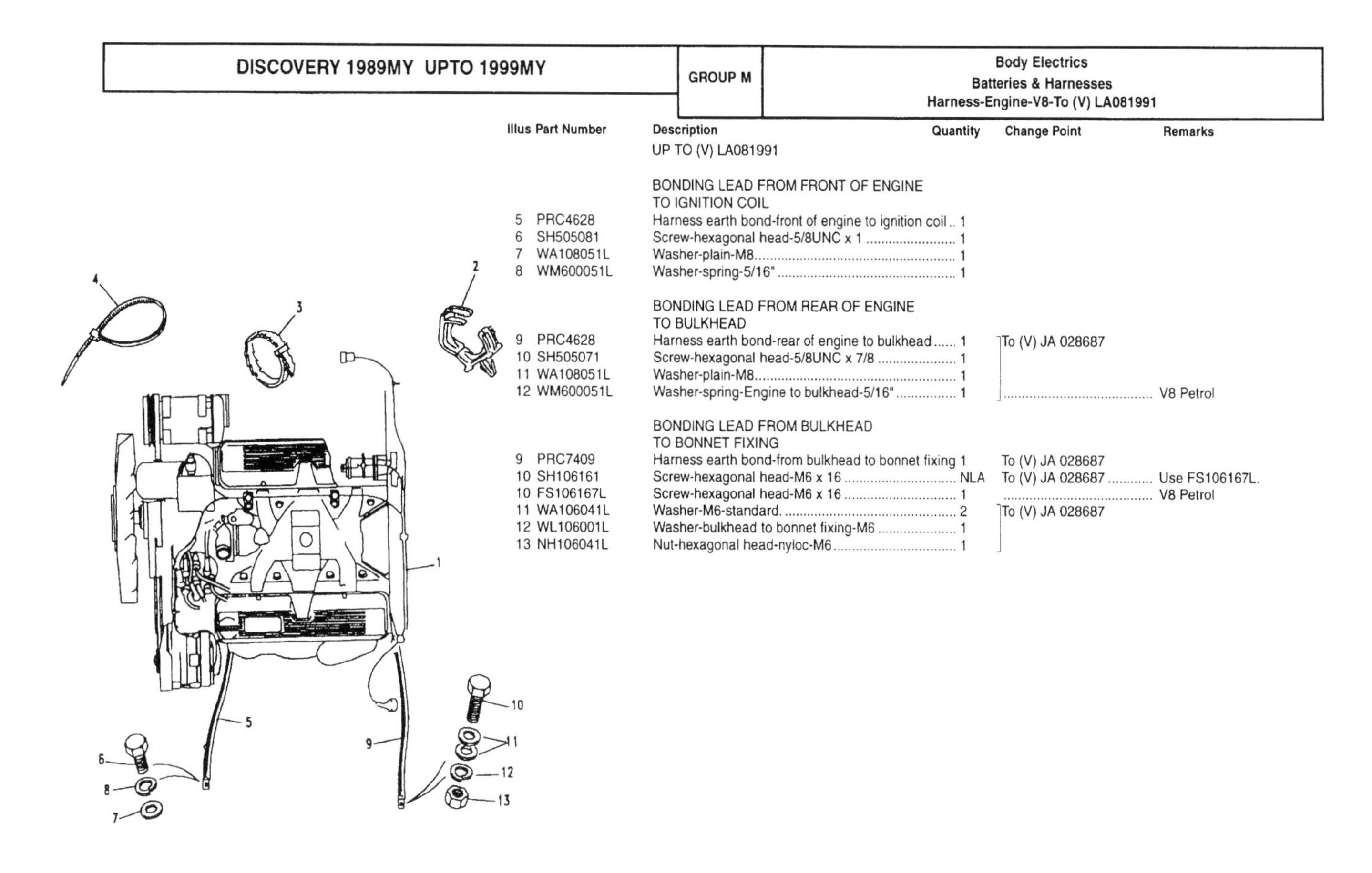

Illus	Part Number	Description	Quantity	Change Point	Remarks
		UP TO (V) LA081991			
		BONDING LEAD FROM FRONT OF ENGINE TO IGNITION COIL			
5	PRC4628	Harness earth bond-front of engine to ignition coil	1		
6	SH505081	Screw-hexagonal head-5/8UNC x 1	1		
7	WA108051L	Washer-plain-M8	1		
8	WM600051L	Washer-spring-5/16"	1		
		BONDING LEAD FROM REAR OF ENGINE TO BULKHEAD			
9	PRC4628	Harness earth bond-rear of engine to bulkhead	1	To (V) JA 028687	
10	SH505071	Screw-hexagonal head-5/8UNC x 7/8	1		
11	WA108051L	Washer-plain-M8	1		
12	WM600051L	Washer-spring-Engine to bulkhead-5/16"	1		V8 Petrol
		BONDING LEAD FROM BULKHEAD TO BONNET FIXING			
9	PRC7409	Harness earth bond-from bulkhead to bonnet fixing	1	To (V) JA 028687	
10	SH106161	Screw-hexagonal head-M6 x 16	NLA	To (V) JA 028687	Use FS106167L.
10	FS106167L	Screw-hexagonal head-M6 x 16	1		V8 Petrol
11	WA106041L	Washer-M6-standard.	2	To (V) JA 028687	
12	WL106001L	Washer-bulkhead to bonnet fixing-M6	1		
13	NH106041L	Nut-hexagonal head-nyloc-M6	1		

M 47

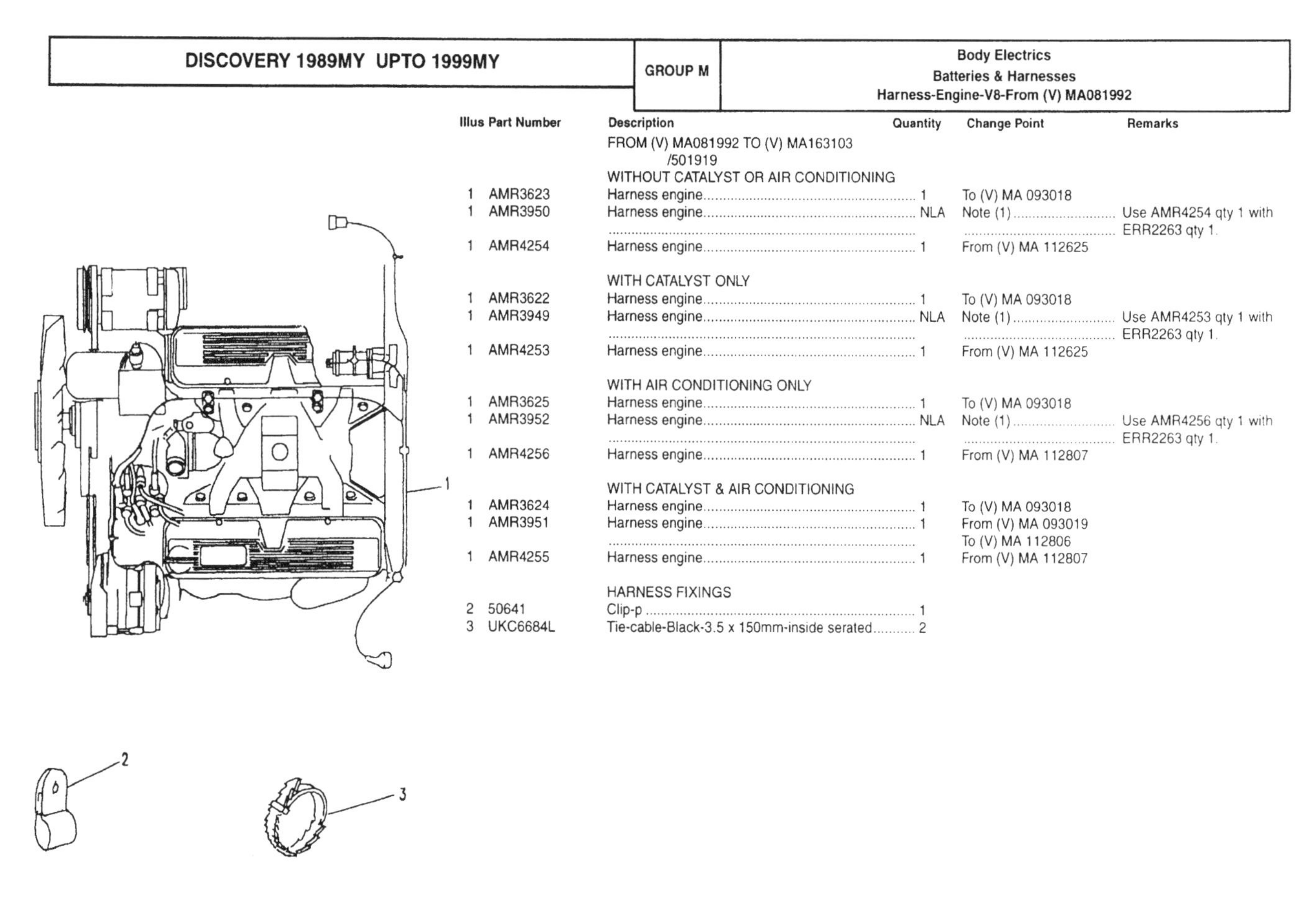

Illus	Part Number	Description	Quantity	Change Point	Remarks
		FROM (V) MA081992 TO (V) MA163103 /501919			
		WITHOUT CATALYST OR AIR CONDITIONING			
1	AMR3623	Harness engine	1	To (V) MA 093018	
1	AMR3950	Harness engine	NLA	Note (1)	Use AMR4254 qty 1 with ERR2263 qty 1.
1	AMR4254	Harness engine	1	From (V) MA 112625	
		WITH CATALYST ONLY			
1	AMR3622	Harness engine	1	To (V) MA 093018	
1	AMR3949	Harness engine	NLA	Note (1)	Use AMR4253 qty 1 with ERR2263 qty 1.
1	AMR4253	Harness engine	1	From (V) MA 112625	
		WITH AIR CONDITIONING ONLY			
1	AMR3625	Harness engine	1	To (V) MA 093018	
1	AMR3952	Harness engine	NLA	Note (1)	Use AMR4256 qty 1 with ERR2263 qty 1.
1	AMR4256	Harness engine	1	From (V) MA 112807	
		WITH CATALYST & AIR CONDITIONING			
1	AMR3624	Harness engine	1	To (V) MA 093018	
1	AMR3951	Harness engine	1	From (V) MA 093019 To (V) MA 112806	
1	AMR4255	Harness engine	1	From (V) MA 112807	
		HARNESS FIXINGS			
2	50641	Clip-p	1		
3	UKC6684L	Tie-cable-Black-3.5 x 150mm-inside serated	2		

CHANGE POINTS:
(1) From (V) MA 093019 To (V) MA 112624

M 48

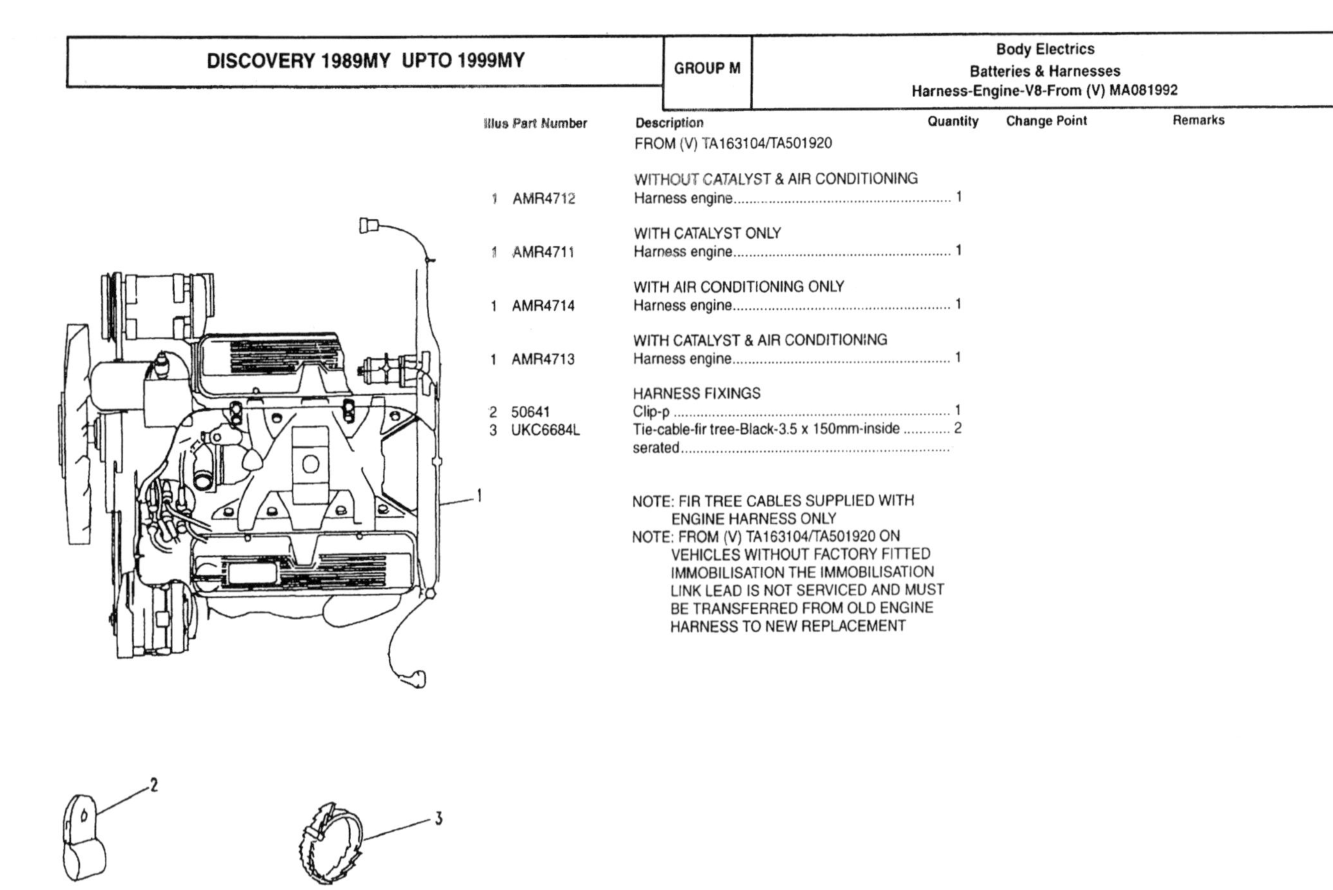

Illus	Part Number	Description	Quantity	Change Point	Remarks
		FROM (V) TA163104/TA501920			
		WITHOUT CATALYST & AIR CONDITIONING			
1	AMR4712	Harness engine	1		
		WITH CATALYST ONLY			
1	AMR4711	Harness engine	1		
		WITH AIR CONDITIONING ONLY			
1	AMR4714	Harness engine	1		
		WITH CATALYST & AIR CONDITIONING			
1	AMR4713	Harness engine	1		
		HARNESS FIXINGS			
2	50641	Clip-p	1		
3	UKC6684L	Tie-cable-fir tree-Black-3.5 x 150mm-inside serated	2		

NOTE: FIR TREE CABLES SUPPLIED WITH ENGINE HARNESS ONLY
NOTE: FROM (V) TA163104/TA501920 ON VEHICLES WITHOUT FACTORY FITTED IMMOBILISATION THE IMMOBILISATION LINK LEAD IS NOT SERVICED AND MUST BE TRANSFERRED FROM OLD ENGINE HARNESS TO NEW REPLACEMENT

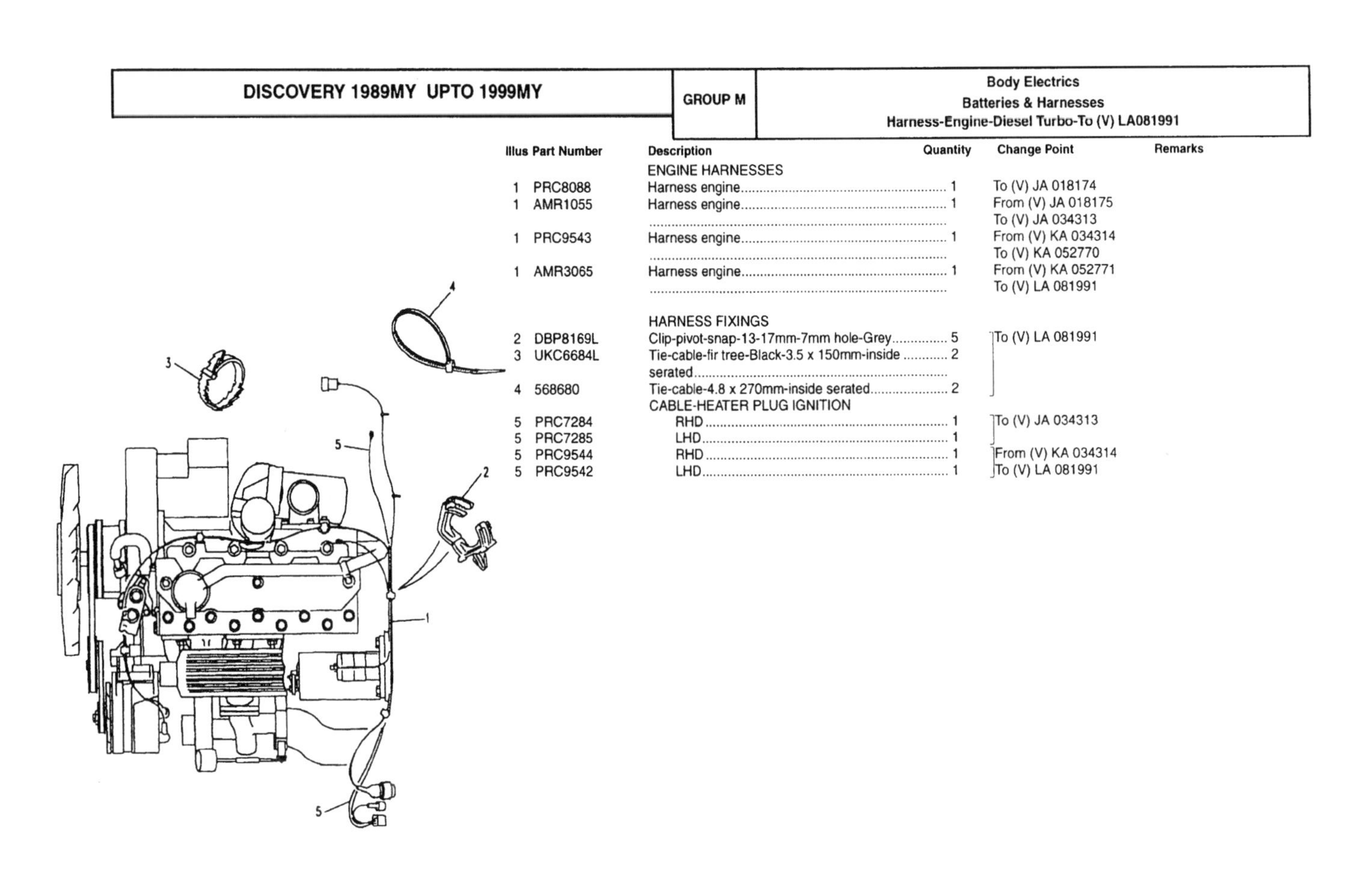

Illus	Part Number	Description	Quantity	Change Point	Remarks
		ENGINE HARNESSES			
1	PRC8088	Harness engine	1	To (V) JA 018174	
1	AMR1055	Harness engine	1	From (V) JA 018175 To (V) JA 034313	
1	PRC9543	Harness engine	1	From (V) KA 034314 To (V) KA 052770	
1	AMR3065	Harness engine	1	From (V) KA 052771 To (V) LA 081991	
		HARNESS FIXINGS			
2	DBP8169L	Clip-pivot-snap-13-17mm-7mm hole-Grey	5	To (V) LA 081991	
3	UKC6684L	Tie-cable-fir tree-Black-3.5 x 150mm-inside serated	2		
4	568680	Tie-cable-4.8 x 270mm-inside serated	2		
		CABLE-HEATER PLUG IGNITION			
5	PRC7284	RHD	1	To (V) JA 034313	
5	PRC7285	LHD	1		
5	PRC9544	RHD	1	From (V) KA 034314	
5	PRC9542	LHD	1	To (V) LA 081991	

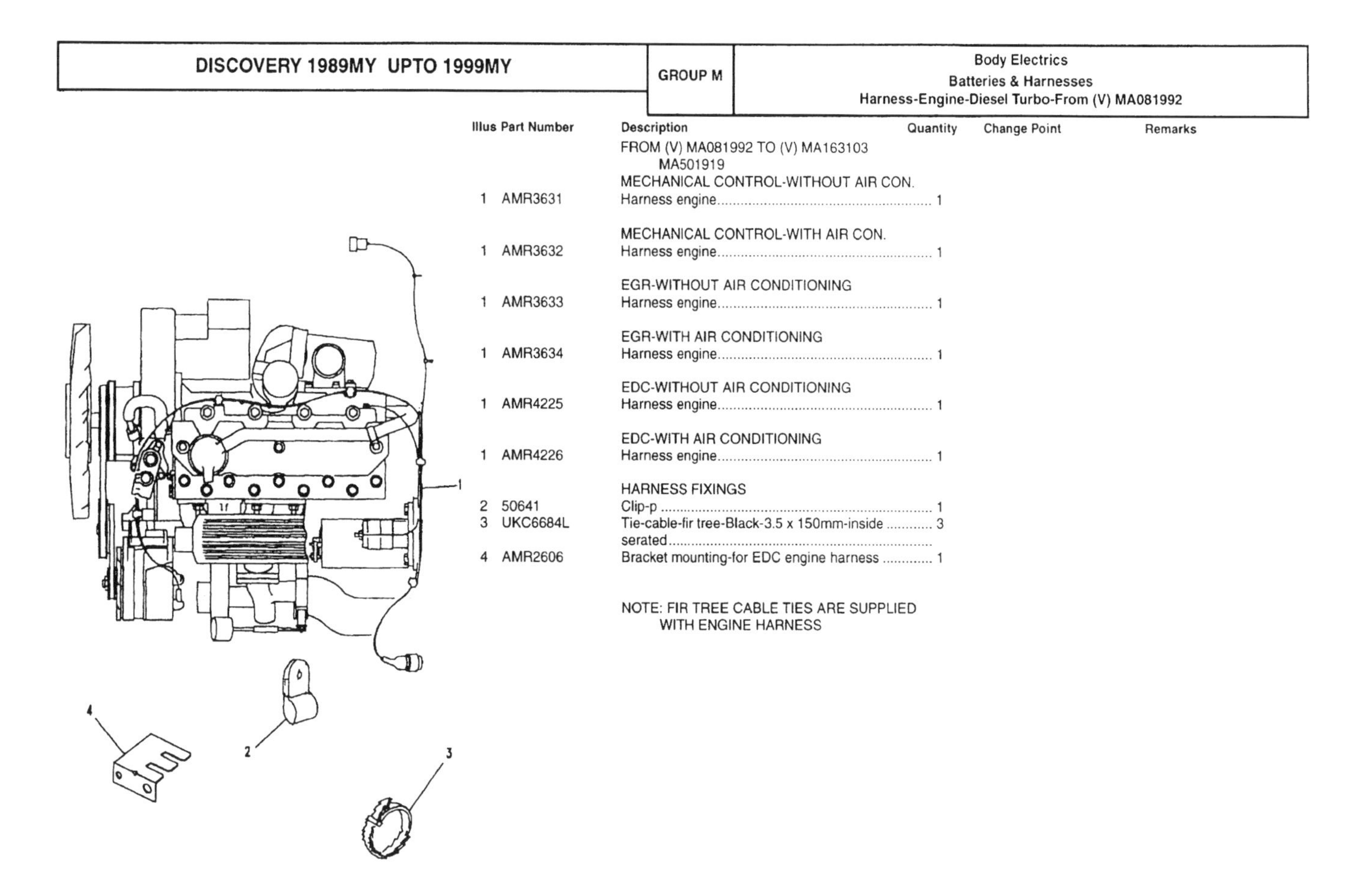

Illus	Part Number	Description	Quantity	Change Point	Remarks
		FROM (V) MA081992 TO (V) MA163103 MA501919			
		MECHANICAL CONTROL-WITHOUT AIR CON.			
1	AMR3631	Harness engine	1		
		MECHANICAL CONTROL-WITH AIR CON.			
1	AMR3632	Harness engine	1		
		EGR-WITHOUT AIR CONDITIONING			
1	AMR3633	Harness engine	1		
		EGR-WITH AIR CONDITIONING			
1	AMR3634	Harness engine	1		
		EDC-WITHOUT AIR CONDITIONING			
1	AMR4225	Harness engine	1		
		EDC-WITH AIR CONDITIONING			
1	AMR4226	Harness engine	1		
		HARNESS FIXINGS			
2	50641	Clip-p	1		
3	UKC6684L	Tie-cable-fir tree-Black-3.5 x 150mm-inside serated	3		
4	AMR2606	Bracket mounting-for EDC engine harness	1		

NOTE: FIR TREE CABLE TIES ARE SUPPLIED WITH ENGINE HARNESS

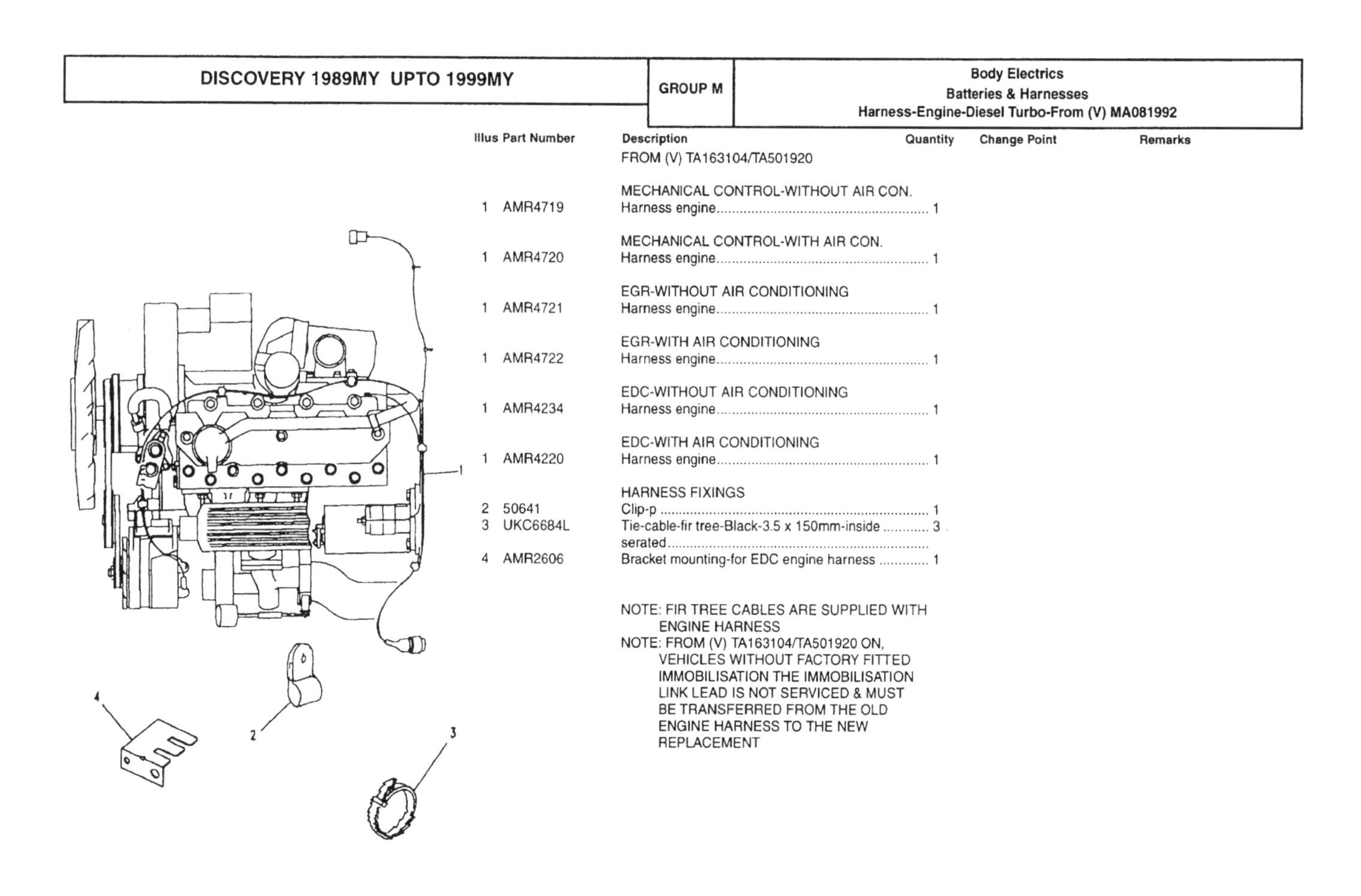

Illus	Part Number	Description	Quantity	Change Point	Remarks
		FROM (V) TA163104/TA501920			
		MECHANICAL CONTROL-WITHOUT AIR CON.			
1	AMR4719	Harness engine	1		
		MECHANICAL CONTROL-WITH AIR CON.			
1	AMR4720	Harness engine	1		
		EGR-WITHOUT AIR CONDITIONING			
1	AMR4721	Harness engine	1		
		EGR-WITH AIR CONDITIONING			
1	AMR4722	Harness engine	1		
		EDC-WITHOUT AIR CONDITIONING			
1	AMR4234	Harness engine	1		
		EDC-WITH AIR CONDITIONING			
1	AMR4220	Harness engine	1		
		HARNESS FIXINGS			
2	50641	Clip-p	1		
3	UKC6684L	Tie-cable-fir tree-Black-3.5 x 150mm-inside serated	3		
4	AMR2606	Bracket mounting-for EDC engine harness	1		

NOTE: FIR TREE CABLES ARE SUPPLIED WITH ENGINE HARNESS

NOTE: FROM (V) TA163104/TA501920 ON, VEHICLES WITHOUT FACTORY FITTED IMMOBILISATION THE IMMOBILISATION LINK LEAD IS NOT SERVICED & MUST BE TRANSFERRED FROM THE OLD ENGINE HARNESS TO THE NEW REPLACEMENT

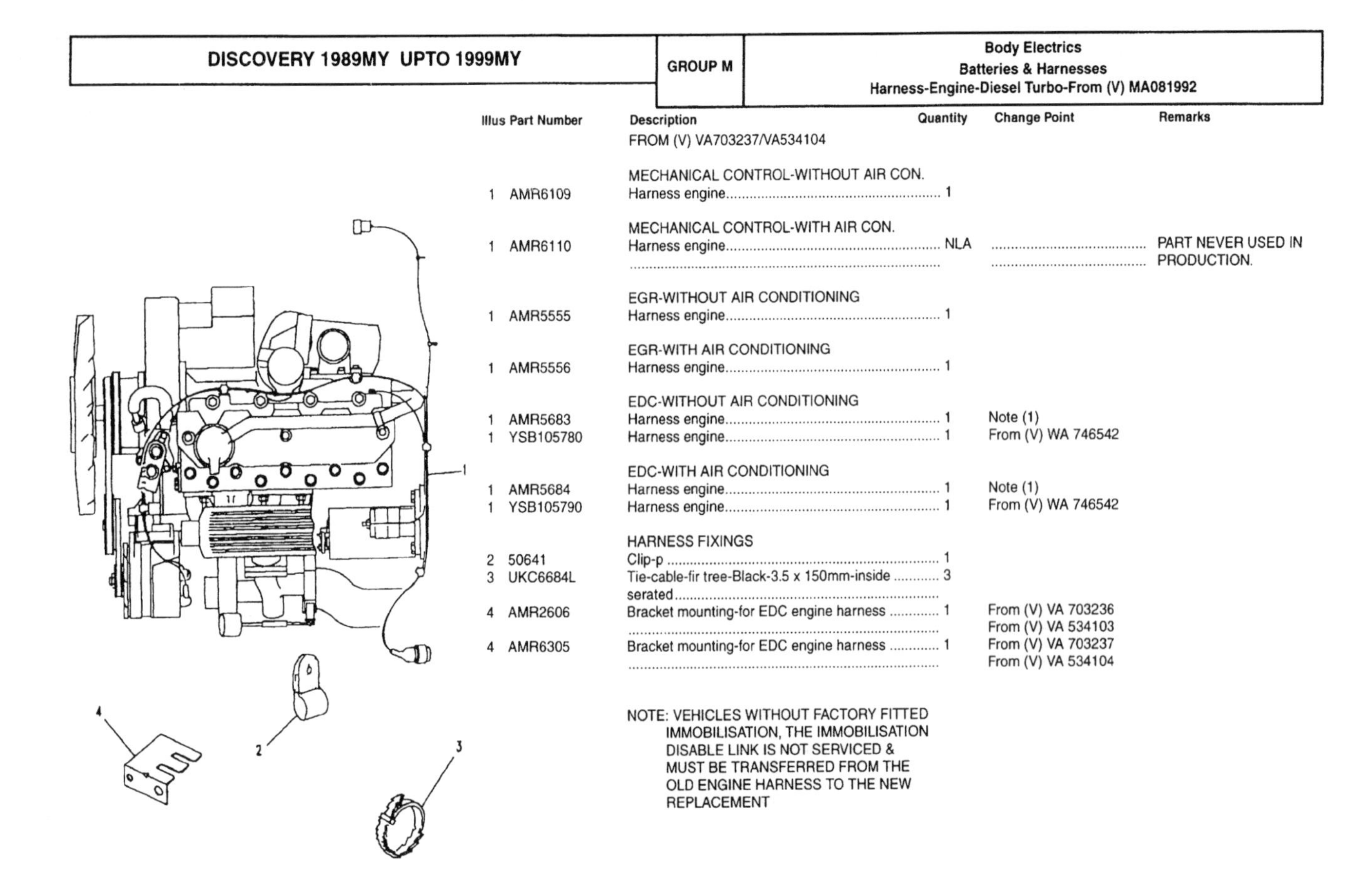

Illus	Part Number	Description	Quantity	Change Point	Remarks
		FROM (V) VA703237/VA534104			
		MECHANICAL CONTROL-WITHOUT AIR CON.			
1	AMR6109	Harness engine	1		
		MECHANICAL CONTROL-WITH AIR CON.			
1	AMR6110	Harness engine	NLA		PART NEVER USED IN PRODUCTION.
		EGR-WITHOUT AIR CONDITIONING			
1	AMR5555	Harness engine	1		
		EGR-WITH AIR CONDITIONING			
1	AMR5556	Harness engine	1		
		EDC-WITHOUT AIR CONDITIONING			
1	AMR5683	Harness engine	1	Note (1)	
1	YSB105780	Harness engine	1	From (V) WA 746542	
		EDC-WITH AIR CONDITIONING			
1	AMR5684	Harness engine	1	Note (1)	
1	YSB105790	Harness engine	1	From (V) WA 746542	
		HARNESS FIXINGS			
2	50641	Clip-p	1		
3	UKC6684L	Tie-cable-fir tree-Black-3.5 x 150mm-inside serated	3		
4	AMR2606	Bracket mounting-for EDC engine harness	1	From (V) VA 703236 From (V) VA 534103	
4	AMR6305	Bracket mounting-for EDC engine harness	1	From (V) VA 703237 From (V) VA 534104	

NOTE: VEHICLES WITHOUT FACTORY FITTED IMMOBILISATION, THE IMMOBILISATION DISABLE LINK IS NOT SERVICED & MUST BE TRANSFERRED FROM THE OLD ENGINE HARNESS TO THE NEW REPLACEMENT

CHANGE POINTS:
(1) From (V) VA 703237 To (V) WA 746541; From (V) VA 534104 To (V) WA 746541

Illus	Part Number	Description	Quantity	Change Point	Remarks
		FROM (V) GA394352 TO (V) JA034313			
		V8 CARB & TDI			
1	PRC6455	Harness-transmission	1		
		V8 PI			
1	PRC8147	Harness-transmission	1		
		GULF STATES TDI ONLY			
1	PRC9944	Harness-transmission	1		
		ALL VARIANTS			
2	568680	Tie-cable-4.8 x 270mm-inside serated	2		
		FROM (V) KA034314 TO (V) KA064754			
		V8 PI-MANUAL			
1	PRC9480	Harness-transmission	1		
		V8 PI-AUTOMATIC			
1	PRC8969	Harness-transmission	1		
		MPI			
1	PRC9480	Harness-transmission	1	To (V) KA 038821	
1	AMR1533	Harness-transmission	1	From (V) KA 038822	
	PRC6525	Clip-p	1		
		TDI-MANUAL			
1	PRC9481	Harness-transmission	1		
		GULF STATES TDI ONLY			
1	AMR2200	Harness-transmission	1		
		ALL VARIANTS			
2	UKC6684L	Tie-cable-fir tree-Black-3.5 x 150mm-inside serated	2		

1
2

DISCOVERY 1989MY UPTO 1999MY

GROUP M

Body Electrics
Batteries & Harnesses
Harness - Transmission-From (V) LA064755 To (V) LA081991

Illus	Part Number	Description	Quantity	Change Point	Remarks
		FROM (V) LA064755 TO LA081991			
		V8 PI-MANUAL			
1	PRC9480	Harness-transmission	1		
		V8 PI-AUTOMATIC			
1	PRC8969	Harness-transmission	1		
		MPI			
1	AMR1533	Harness-transmission	1		
	PRC6525	Clip-p	1		
		TDI-MANUAL			
1	AMR3077	Harness-transmission	1		
		TDI-AUTOMATIC			
1	AMR3078	Harness-transmission	1		
		TDI-GULF STATES ONLY			
1	AMR3084	Harness-transmission	1		Manual
1	AMR3235	Harness-transmission	1		Automatic
		ALL VARIANTS			
2	UKC6684L	Tie-cable-fir tree-Black-3.5 x 150mm-inside serated	2		
		TDI AUTOMATIC ONLY			
3	AMR3275	Harness-link-gearbox-temperature	1		

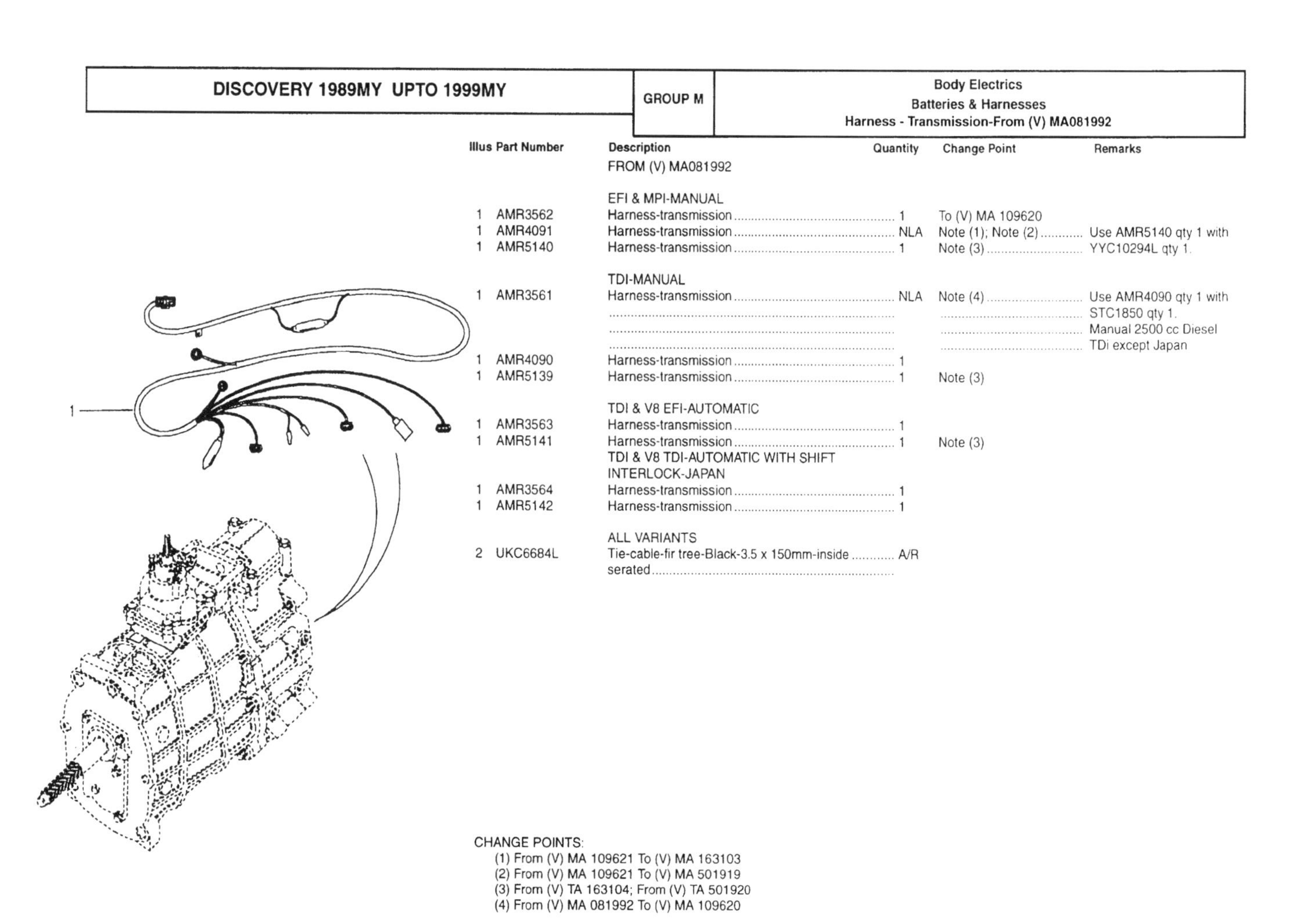

DISCOVERY 1989MY UPTO 1999MY

GROUP M

Body Electrics
Batteries & Harnesses
Harness - Transmission-From (V) MA081992

Illus	Part Number	Description	Quantity	Change Point	Remarks
		FROM (V) MA081992			
		EFI & MPI-MANUAL			
1	AMR3562	Harness-transmission	1	To (V) MA 109620	
1	AMR4091	Harness-transmission	NLA	Note (1); Note (2)	Use AMR5140 qty 1 with
1	AMR5140	Harness-transmission	1	Note (3)	YYC10294L qty 1.
		TDI-MANUAL			
1	AMR3561	Harness-transmission	NLA	Note (4)	Use AMR4090 qty 1 with STC1850 qty 1. Manual 2500 cc Diesel TDi except Japan
1	AMR4090	Harness-transmission	1		
1	AMR5139	Harness-transmission	1	Note (3)	
		TDI & V8 EFI-AUTOMATIC			
1	AMR3563	Harness-transmission	1		
1	AMR5141	Harness-transmission	1	Note (3)	
		TDI & V8 TDI-AUTOMATIC WITH SHIFT INTERLOCK-JAPAN			
1	AMR3564	Harness-transmission	1		
1	AMR5142	Harness-transmission	1		
		ALL VARIANTS			
2	UKC6684L	Tie-cable-fir tree-Black-3.5 x 150mm-inside serated	A/R		

CHANGE POINTS:
(1) From (V) MA 109621 To (V) MA 163103
(2) From (V) MA 109621 To (V) MA 501919
(3) From (V) TA 163104; From (V) TA 501920
(4) From (V) MA 081992 To (V) MA 109620

Harness - Fuel Pump-V8 EFI, 200 & 300 TDI

Illus	Part Number	Description	Quantity	Change Point	Remarks
		V8 EFI, 200 & 300 TDI			
1	PRC7943	Harness-link-fuel pump	1	Note (1)	
1	AMR2164	Harness-link-fuel pump	1	Note (2)	
1	AMR2587	Harness-link-fuel pump	1	Note (3)	
1	AMR4806	Harness-link-fuel pump	1	Note (4)	
1	AMR5670	Harness-link-fuel pump	1	Note (5)	
		MPI PETROL			
1	AMR2164	Harness-link-fuel pump	1	Note (6)	
1	AMR2587	Harness-link-fuel pump	1	Note (7)	
1	AMR4806	Harness-link-fuel pump	1	Note (4)	
1	AMR5670	Harness-link-fuel pump	1	Note (5)	
		ALL VARIANTS			
2	AFU1090L	Clip-cable-8mm hole	2		
3	SH105161L	Screw-hexagonal head-M5 x 16	1		
4	WL105001L	Washer-M5	1		
5	WA105001L	Washer-plain-M5-standard	1		

NOTE: SOME HARNESSES HAVE BRASS LUCAR BLADES INSTEAD OF TIN. SIKAFLEX MUST BE APPLIED TO BRASS LUCAR & ASTROLAN TO TIN

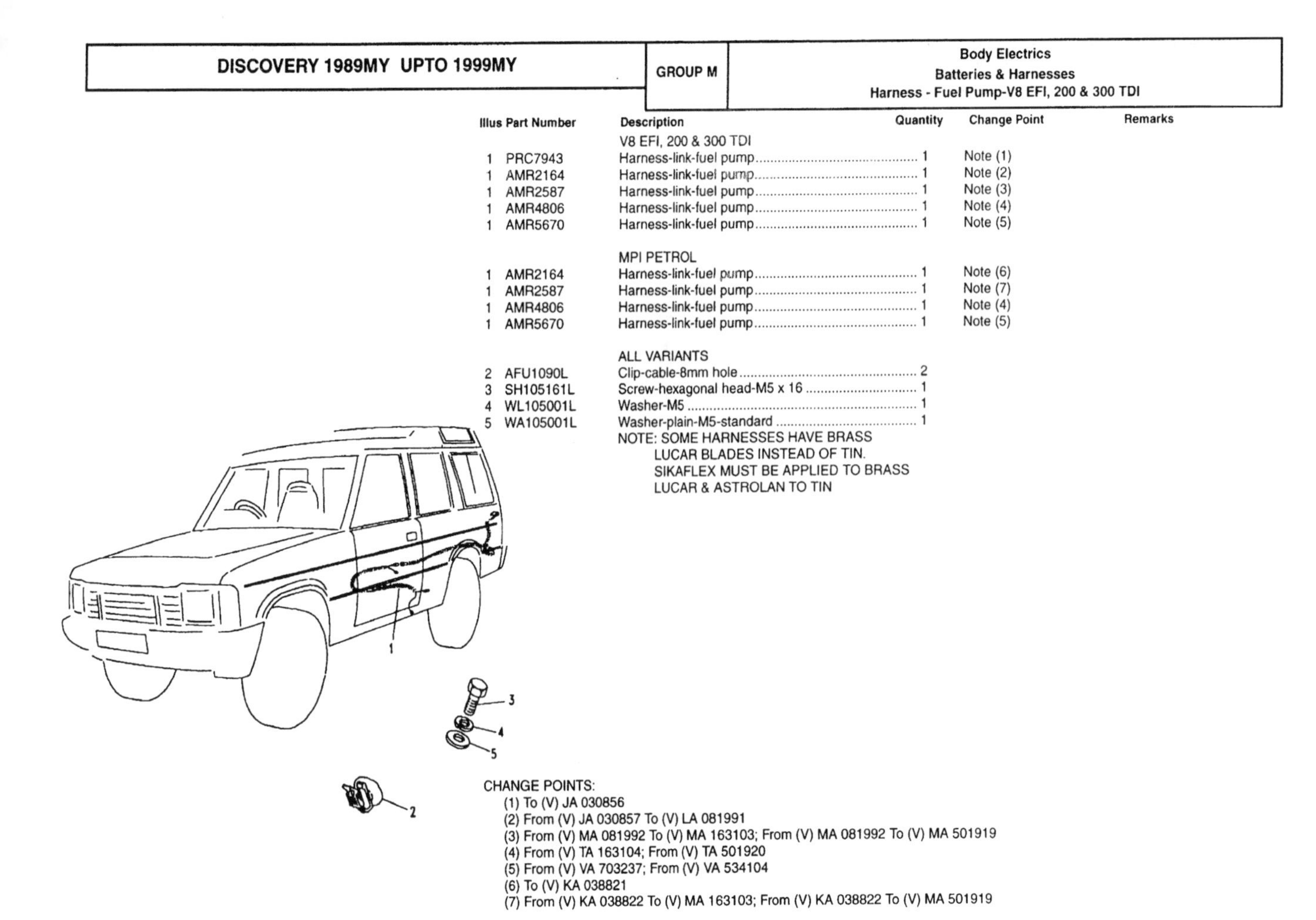

CHANGE POINTS:
(1) To (V) JA 030856
(2) From (V) JA 030857 To (V) LA 081991
(3) From (V) MA 081992 To (V) MA 163103; From (V) MA 081992 To (V) MA 501919
(4) From (V) TA 163104; From (V) TA 501920
(5) From (V) VA 703237; From (V) VA 534104
(6) To (V) KA 038821
(7) From (V) KA 038822 To (V) MA 163103; From (V) KA 038822 To (V) MA 501919

Harness - Air Conditioning

Illus	Part Number	Description	Quantity	Change Point	Remarks
1	PRC6453	Harness air conditioning	1		RHD V8 Twin Carburettor
1	PRC6454	Harness air conditioning	1		LHD V8 Twin Carburettor
1	PRC8534	Harness air conditioning	1	To (V) JA 034313	RHD V8 EFi
1	PRC9451	Harness air conditioning	1	From (V) KA 034314	
				To (V) KA 064754	
1	AMR3277	Harness air conditioning	1	From (V) LA 064755	
1	PRC8535	Harness air conditioning	1	To (V) JA 034313	LHD V8 EFi
1	PRC9452	Harness air conditioning	1	From (V) KA 034314	
				To (V) KA 064754	
1	AMR3277	Harness air conditioning	1	From (V) LA 064755	
1	PRC8572	Harness air conditioning	1	To (V) JA 034313	RHD TDi
1	PRC9451	Harness air conditioning	1	From (V) KA 034314	
				To (V) KA 064754	
1	AMR3277	Harness air conditioning	1	From (V) LA 064755	
1	PRC7282	Harness air conditioning	1	To (V) JA 034313	LHD TDi
1	PRC9452	Harness air conditioning	1	From (V) KA 034314	
				To (V) KA 064754	
1	AMR3278	Harness air conditioning	1	From (V) LA 064755	
1	AMR3273	Harness air conditioning	1		Automatic RHD TDi
1	AMR3274	Harness air conditioning	1		Automatic LHD TDi

NOTE: AIR CON. HARNESS INCORPORATED INTO MAIN HARNESS FROM (V) MA081992

1

Illus	Part Number	Description	Quantity	Change Point	Remarks
		HARNESS-PAD WEAR			
1	PRC8211	front	1	To (V) JA 034313	
1	AMR2299	front	1	From (V) KA 034314	
2	PRC8322	rear	1		
3	ADU8363L	Clip-harness-14.5mm-6.5mm hole	2		

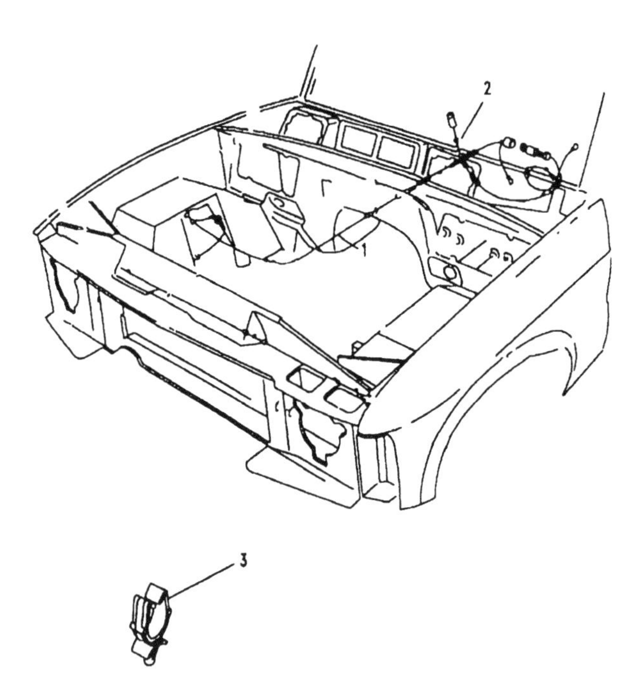

Illus	Part Number	Description	Quantity	Change Point	Remarks
1	PRC7941	Harness-link-headlamp levelling	1	To (V) LA 081991	
	PRC8414	Harness powerwash	1		EFi
		SEAT BELT WARNING CABLE-GULF STATES			
2	PRC8826	Harness-link-seatbelt warning	1		
3	AFU2881	Clip-harness-seat buckle lead	1		

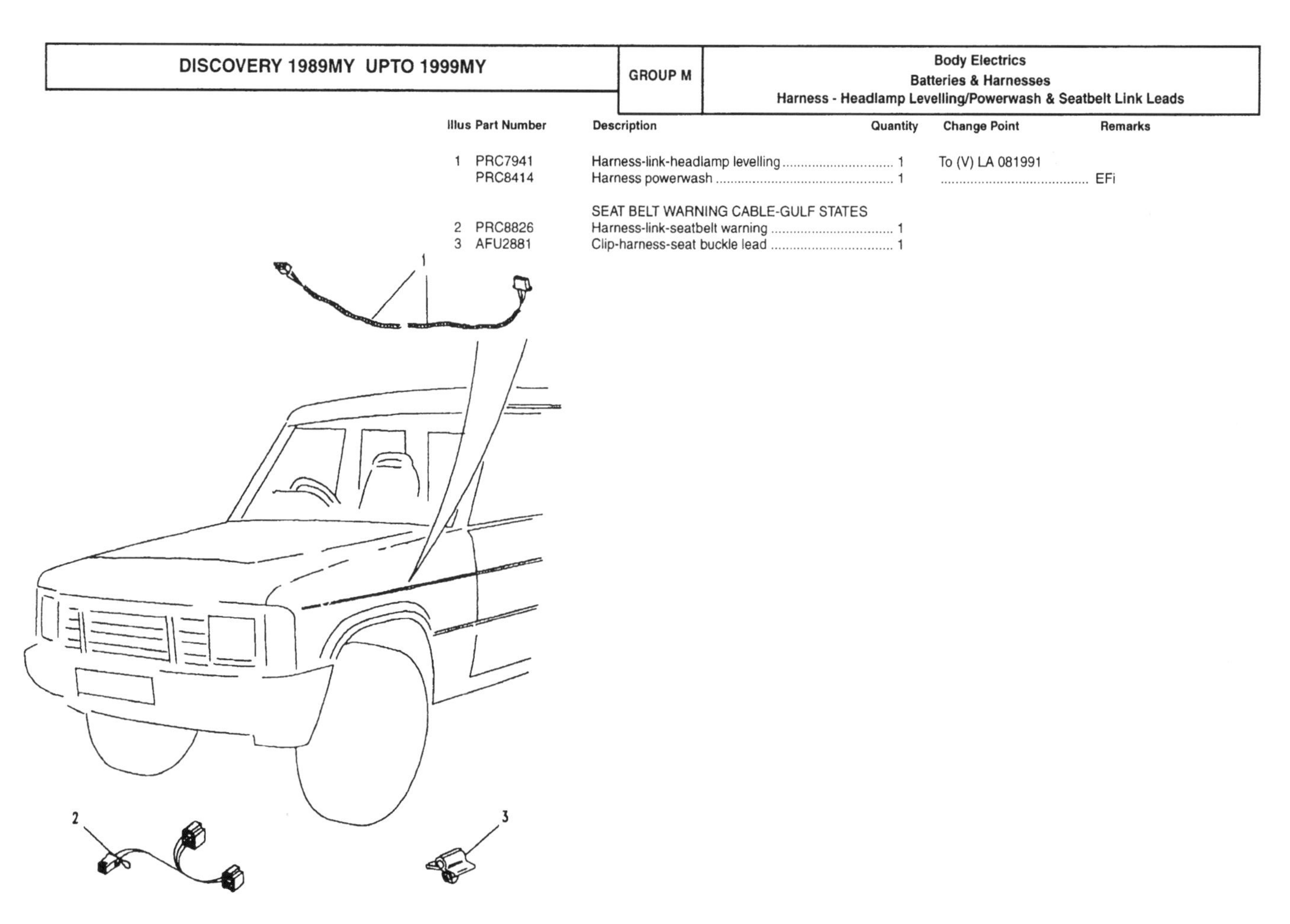

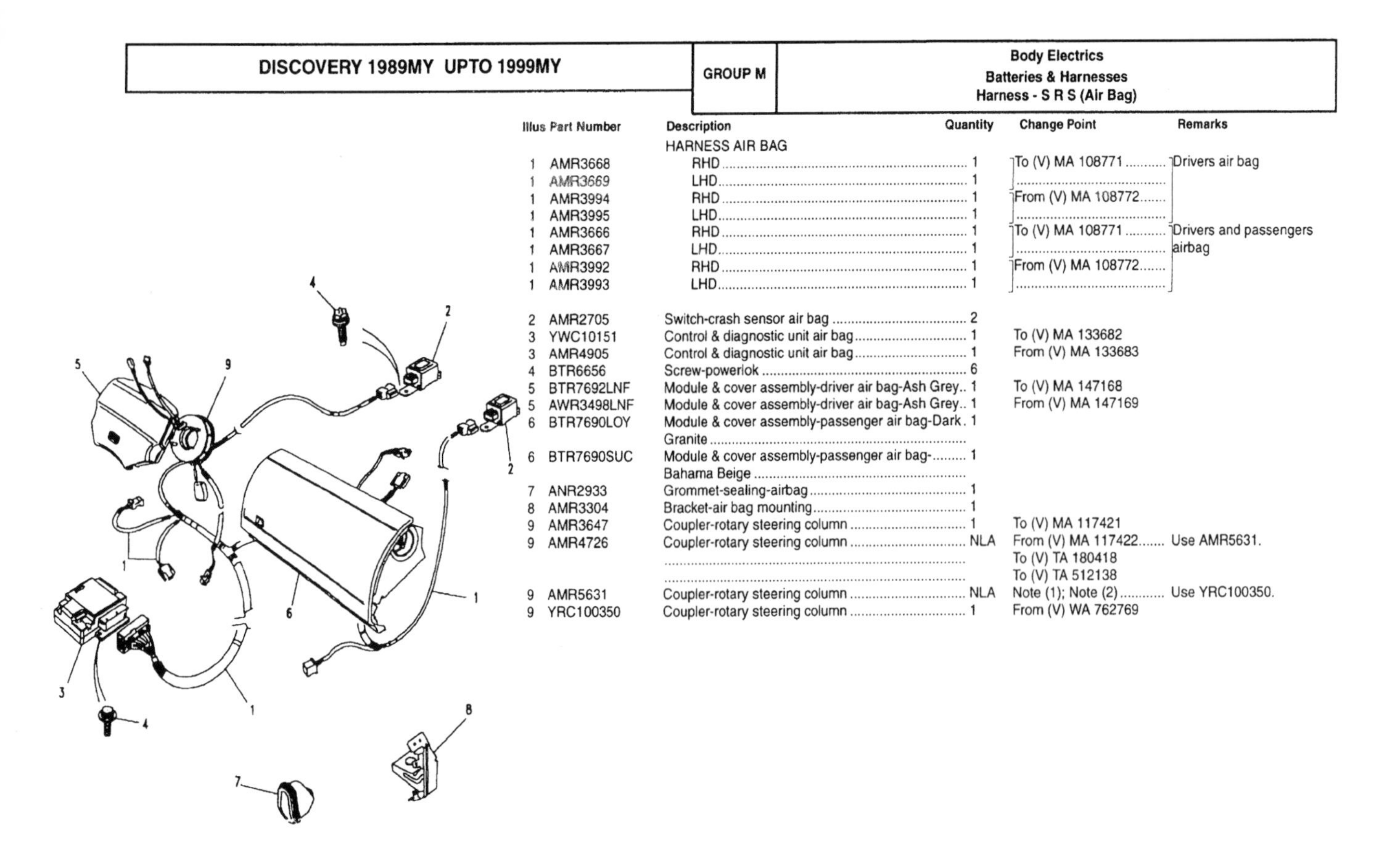

Illus	Part Number	Description	Quantity	Change Point	Remarks
		HARNESS AIR BAG			
1	AMR3668	RHD	1	To (V) MA 108771	Drivers air bag
1	AMR3669	LHD	1		
1	AMR3994	RHD	1	From (V) MA 108772	
1	AMR3995	LHD	1		
1	AMR3666	RHD	1	To (V) MA 108771	Drivers and passengers airbag
1	AMR3667	LHD	1		
1	AMR3992	RHD	1	From (V) MA 108772	
1	AMR3993	LHD	1		
2	AMR2705	Switch-crash sensor air bag	2		
3	YWC10151	Control & diagnostic unit air bag	1	To (V) MA 133682	
3	AMR4905	Control & diagnostic unit air bag	1	From (V) MA 133683	
4	BTR6656	Screw-powerlok	6		
5	BTR7692LNF	Module & cover assembly-driver air bag-Ash Grey	1	To (V) MA 147168	
5	AWR3498LNF	Module & cover assembly-driver air bag-Ash Grey	1	From (V) MA 147169	
6	BTR7690LOY	Module & cover assembly-passenger air bag-Dark Granite	1		
6	BTR7690SUC	Module & cover assembly-passenger air bag-Bahama Beige	1		
7	ANR2933	Grommet-sealing-airbag	1		
8	AMR3304	Bracket-air bag mounting	1		
9	AMR3647	Coupler-rotary steering column	1	To (V) MA 117421	
9	AMR4726	Coupler-rotary steering column	NLA	From (V) MA 117422 To (V) TA 180418 To (V) TA 512138	Use AMR5631.
9	AMR5631	Coupler-rotary steering column	NLA	Note (1); Note (2)	Use YRC100350.
9	YRC100350	Coupler-rotary steering column	1	From (V) WA 762769	

CHANGE POINTS:
(1) From (V) TA 512139 To (V) WA 762768
(2) From (V) TA 180419 To (V) WA 762768

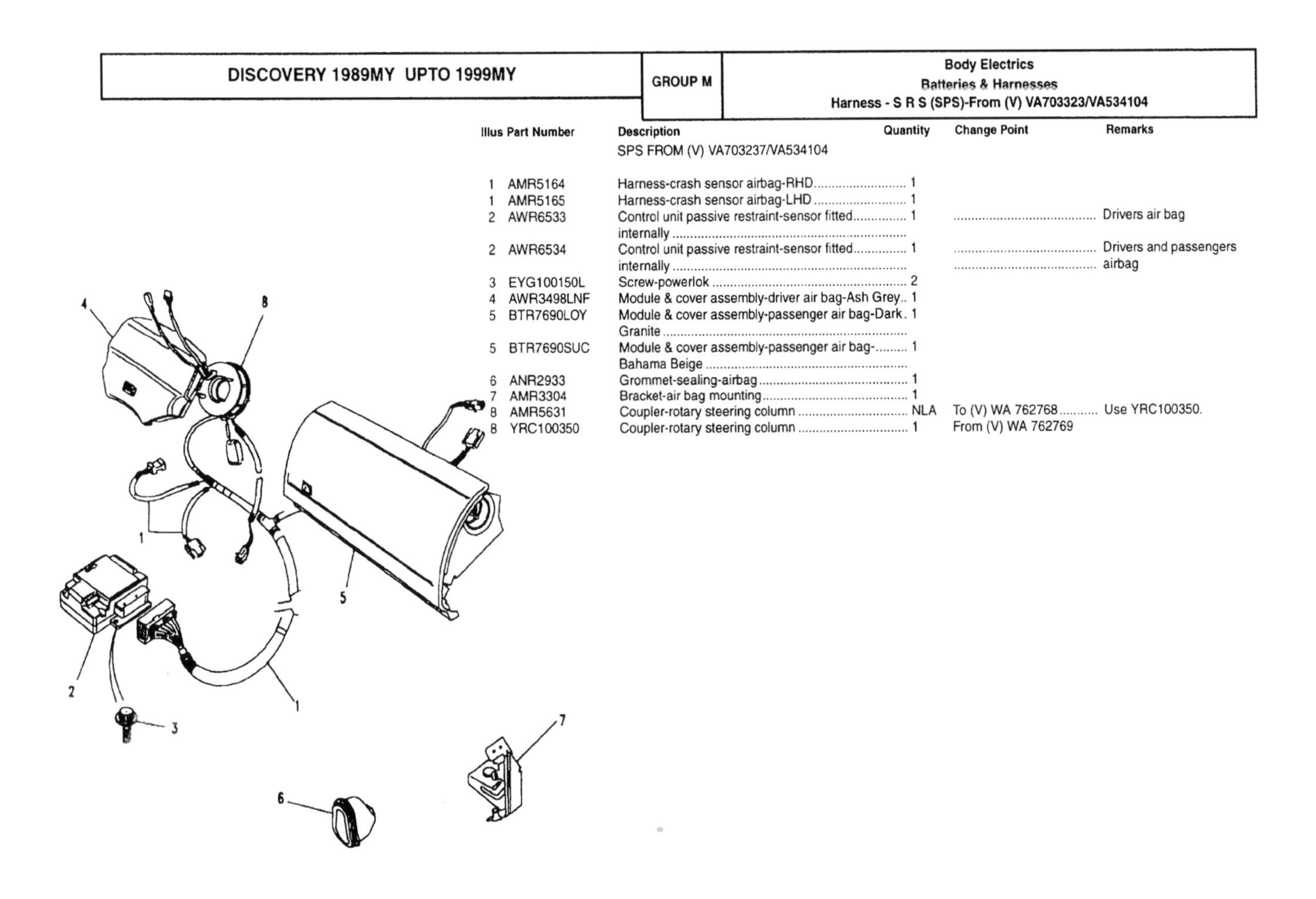

Illus	Part Number	Description	Quantity	Change Point	Remarks
		SPS FROM (V) VA703237/VA534104			
1	AMR5164	Harness-crash sensor airbag-RHD	1		
1	AMR5165	Harness-crash sensor airbag-LHD	1		
2	AWR6533	Control unit passive restraint-sensor fitted internally	1		Drivers air bag
2	AWR6534	Control unit passive restraint-sensor fitted internally	1		Drivers and passengers airbag
3	EYG100150L	Screw-powerlok	2		
4	AWR3498LNF	Module & cover assembly-driver air bag-Ash Grey	1		
5	BTR7690LOY	Module & cover assembly-passenger air bag-Dark Granite	1		
5	BTR7690SUC	Module & cover assembly-passenger air bag-Bahama Beige	1		
6	ANR2933	Grommet-sealing-airbag	1		
7	AMR3304	Bracket-air bag mounting	1		
8	AMR5631	Coupler-rotary steering column	NLA	To (V) WA 762768	Use YRC100350.
8	YRC100350	Coupler-rotary steering column	1	From (V) WA 762769	

Illus	Part Number	Description	Quantity	Change Point	Remarks
1	PRC8724	Switch-wash/wipe windscreen	1	Note (1)	
1	STC357	Switch-wash/wipe windscreen	1	Note (2)	
1	STC866	Switch-wash/wipe windscreen	1	Note (3)	
		HORN,LIGHTING,INDICATOR,HAZARD SWITCH			
2	RTC6618	Switch-master lighting /indicator/headlamp dip	1	Note (1)	
2	STC358	Switch-master lighting /indicator/headlamp dip	1	Note (2)	
2	STC865	Switch-master lighting /indicator/headlamp dip	1	Note (3)	
3	RTC3635	Bulb-286-w5/1.2-1.2 Watt-12V-1.2 Watt	1	To (V) LA 081991	
4	RTC5691	Knob-switch-hazard warning switch	1		
		SWITCH HOUSING			
5	PRC5438	Housing-centre column switches	1	To (V) JA 027299	
5	STC1597	Housing-centre column switches	1	Note (3)	

CHANGE POINTS:
(1) From (V) GA 394352 To (V) GA 456980
(2) From (V) HA 456981 To (V) HA 486551; From (V) HA 000001 To (V) HA 012332; From (V) JA 012333 To (V) JA 027299
(3) From (V) JA 027300 To (V) LA 081991

M 63

Illus	Part Number	Description	Quantity	Change Point	Remarks
		WITHOUT CRUISE CONTROL &/OR AIR BAG			
1	AMR3842	SWITCH ASSEMBLY STEERING COLUMN	1		
2	STC1790	• Switch-wash/wipe windscreen	1		
3	STC1791	• Switch-master lighting /indicator/headlamp dip	1		
4	STC2834	• Housing-centre column switches	1		
		WITH CRUISE CONTROL &/OR AIR BAG			
1	AMR3841	SWITCH ASSEMBLY STEERING COLUMN	1		
2	STC1790	• Switch-wash/wipe windscreen	1		
3	STC1791	• Switch-master lighting /indicator/headlamp dip	1		
4	STC2835	• Housing-centre column switches	1		
		WITHOUT SRS (AIR BAG)			
5	AMR3649	Coupler-rotary steering column	1	Note (1)	
5	AMR4727	Coupler-rotary steering column	1	Note (2)	
5	AMR5632	Coupler-rotary steering column	1	Note (3); Note (4)	
5	YRC100360	Coupler-rotary steering column	1	Note (5)	
6	PRC5797	Switch-cruise control set/resume	1		
7	PRC6590	Switch-cruise control-resume/cancel	1		
		WITH SRS (AIR BAG)			
5	AMR3647	Coupler-rotary steering column	1	Note (1)	Rover airbag system
5	AMR4726	Coupler-rotary steering column	NLA	Note (6)	Use AMR5631.
5	AMR5631	Coupler-rotary steering column	NLA	Note (7); Note (8)	Use YRC100350.
5	YRC100350	Coupler-rotary steering column	1	Note (5)	Rover airbag system
8	AMR3361	Switch-cruise control set/resume	1		Rover airbag system

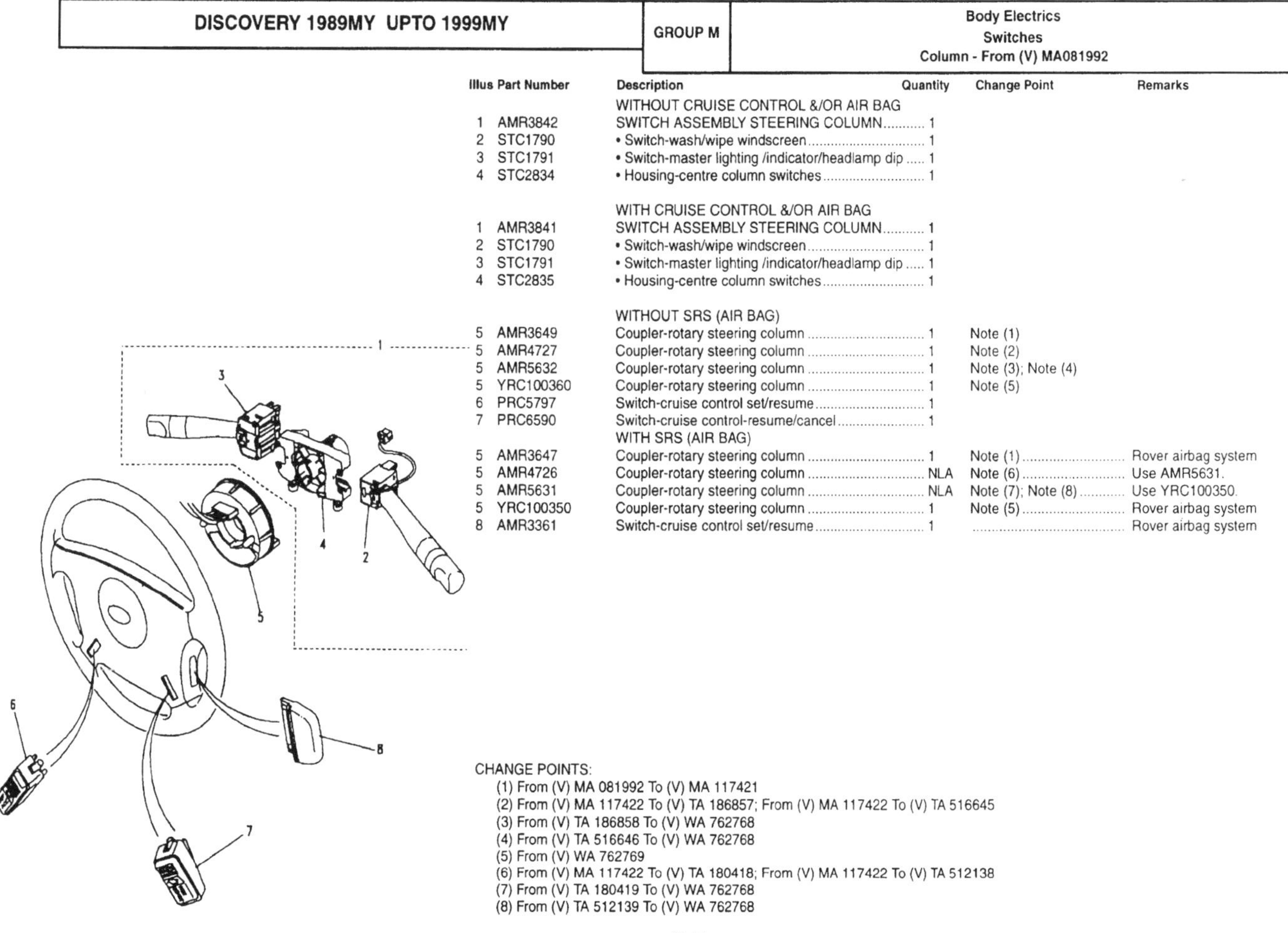

CHANGE POINTS:
(1) From (V) MA 081992 To (V) MA 117421
(2) From (V) MA 117422 To (V) TA 186857; From (V) MA 117422 To (V) TA 516645
(3) From (V) TA 186858 To (V) WA 762768
(4) From (V) TA 516646 To (V) WA 762768
(5) From (V) WA 762769
(6) From (V) MA 117422 To (V) TA 180418; From (V) MA 117422 To (V) TA 512138
(7) From (V) TA 180419 To (V) WA 762768
(8) From (V) TA 512139 To (V) WA 762768

M 64

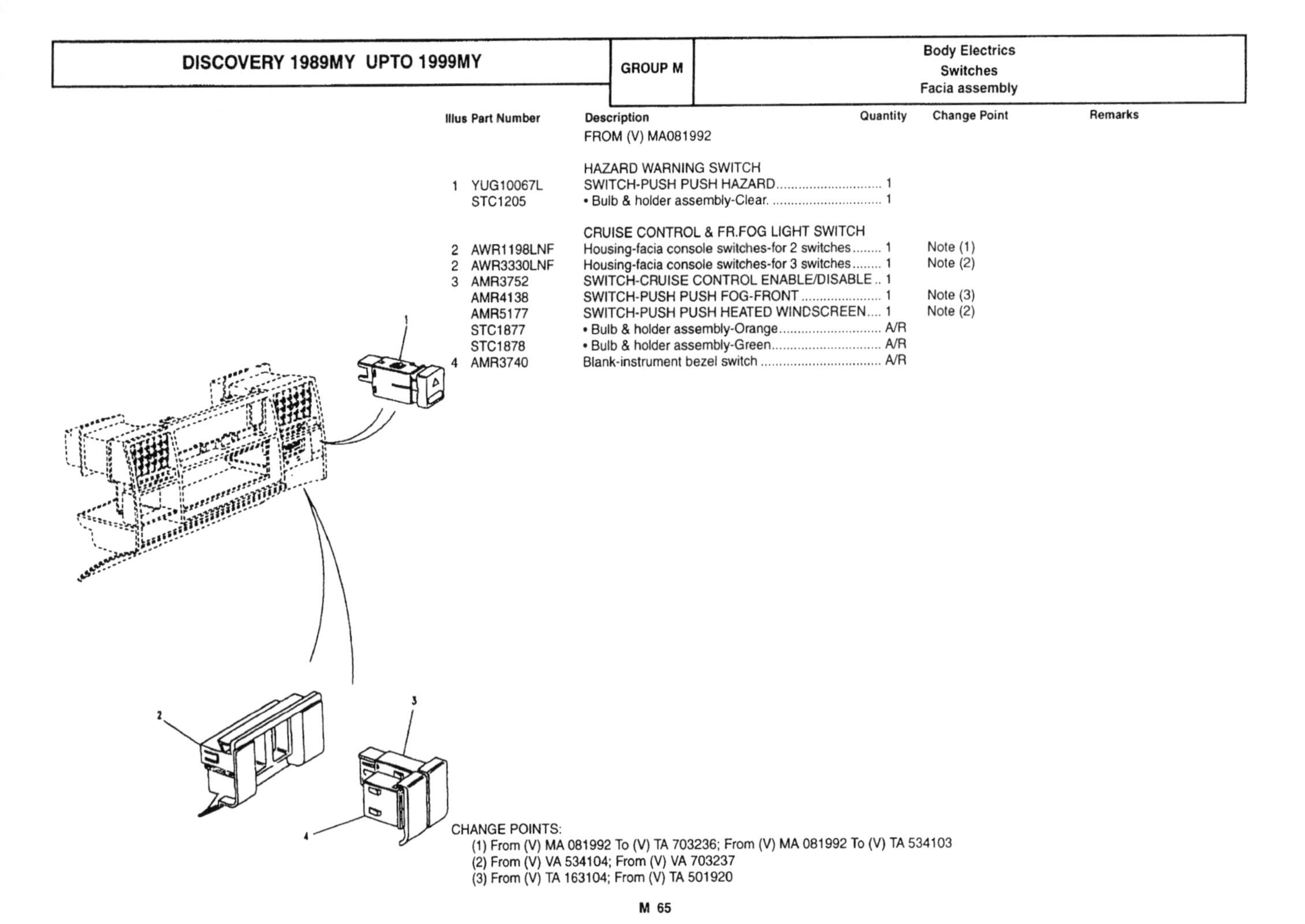

Illus	Part Number	Description	Quantity	Change Point	Remarks
		FROM (V) MA081992			
		HAZARD WARNING SWITCH			
1	YUG10067L	SWITCH-PUSH PUSH HAZARD	1		
	STC1205	• Bulb & holder assembly-Clear.	1		
		CRUISE CONTROL & FR.FOG LIGHT SWITCH			
2	AWR1198LNF	Housing-facia console switches-for 2 switches	1	Note (1)	
2	AWR3330LNF	Housing-facia console switches-for 3 switches	1	Note (2)	
3	AMR3752	SWITCH-CRUISE CONTROL ENABLE/DISABLE	1		
	AMR4138	SWITCH-PUSH PUSH FOG-FRONT	1	Note (3)	
	AMR5177	SWITCH-PUSH PUSH HEATED WINDSCREEN	1	Note (2)	
	STC1877	• Bulb & holder assembly-Orange	A/R		
	STC1878	• Bulb & holder assembly-Green	A/R		
4	AMR3740	Blank-instrument bezel switch	A/R		

CHANGE POINTS:
(1) From (V) MA 081992 To (V) TA 703236; From (V) MA 081992 To (V) TA 534103
(2) From (V) VA 534104; From (V) VA 703237
(3) From (V) TA 163104; From (V) TA 501920

M 65

Illus	Part Number	Description	Quantity	Change Point	Remarks
		FROM (V) MA081992			
		BEZEL-FACIA HEADLAMP LEVELLING SWITCH			
1	AMR2781LNF	RHD-Ash Grey-electric	1	Note (1)	Electric door mirrors
1	AMR2783LNF	LHD-Ash Grey-electric	1		
1	AMR2782LNF	RHD-Ash Grey-manual	1		Except electric door mirrors
1	AMR2784LNF	LHD-Ash Grey-manual	1		
1	FAV101160LNF	RHD-Ash Grey-electric	1	Note (2)	Electric door mirrors
1	FAV101180LNF	LHD-Ash Grey-electric	1		
1	FAV101170LNF	RHD-Ash Grey-manual	1		Except electric door mirrors
1	FAV101190LNF	LHD-Ash Grey-manual	1		
2	AMR2498	Switch-mirror	1		Except memory
3	AMR3935	Switch-rotary headlamp leveling	1	Note (1)	
3	AMR6424	Switch-rotary headlamp leveling	1	Note (2)	
4	FBV10040LNFL	Blank-facia headlamp levelling	1		

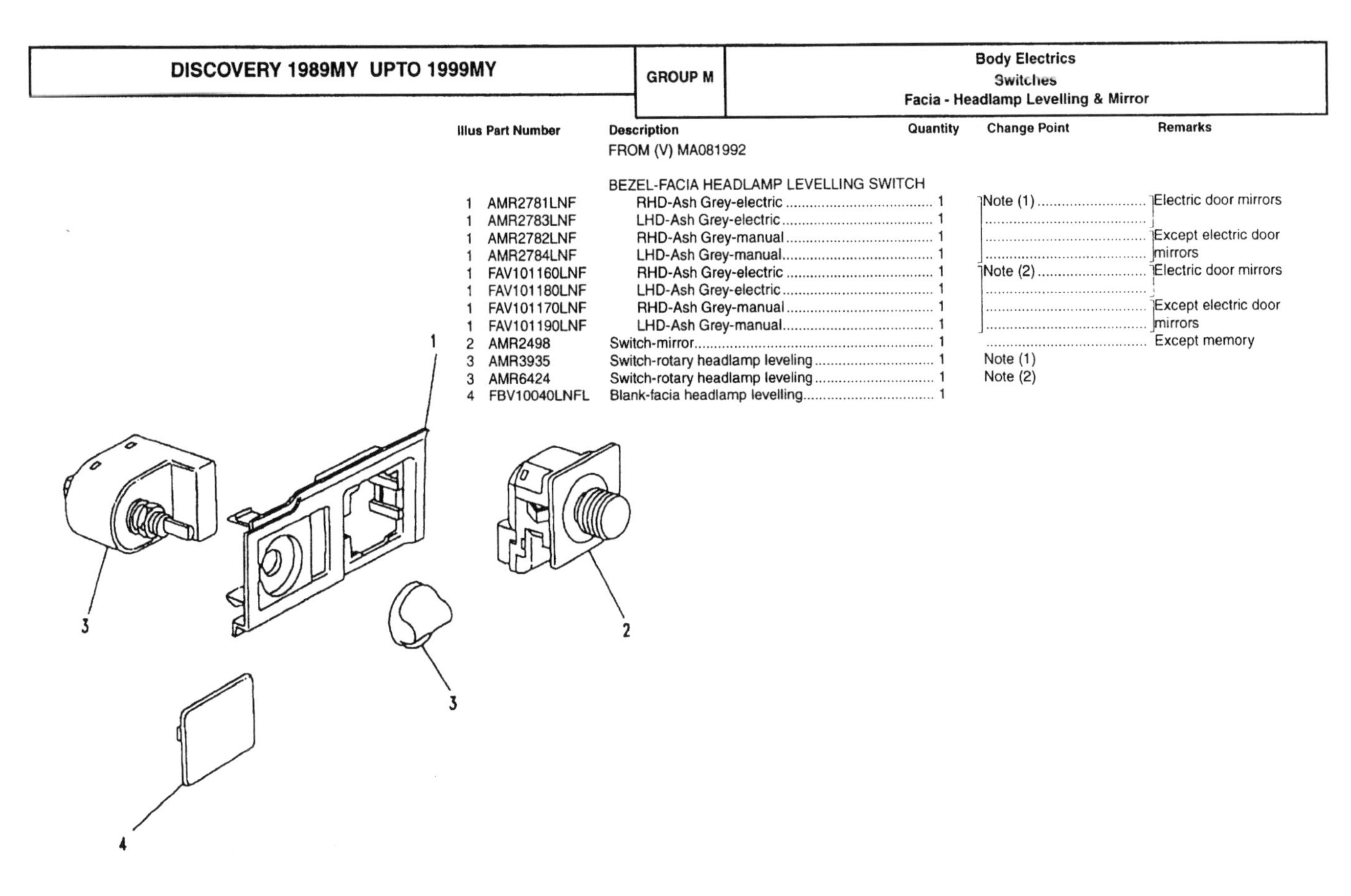

CHANGE POINTS:
(1) From (V) MA 081992 To (V) VA 745654
(2) From (V) VA 745655

M 66

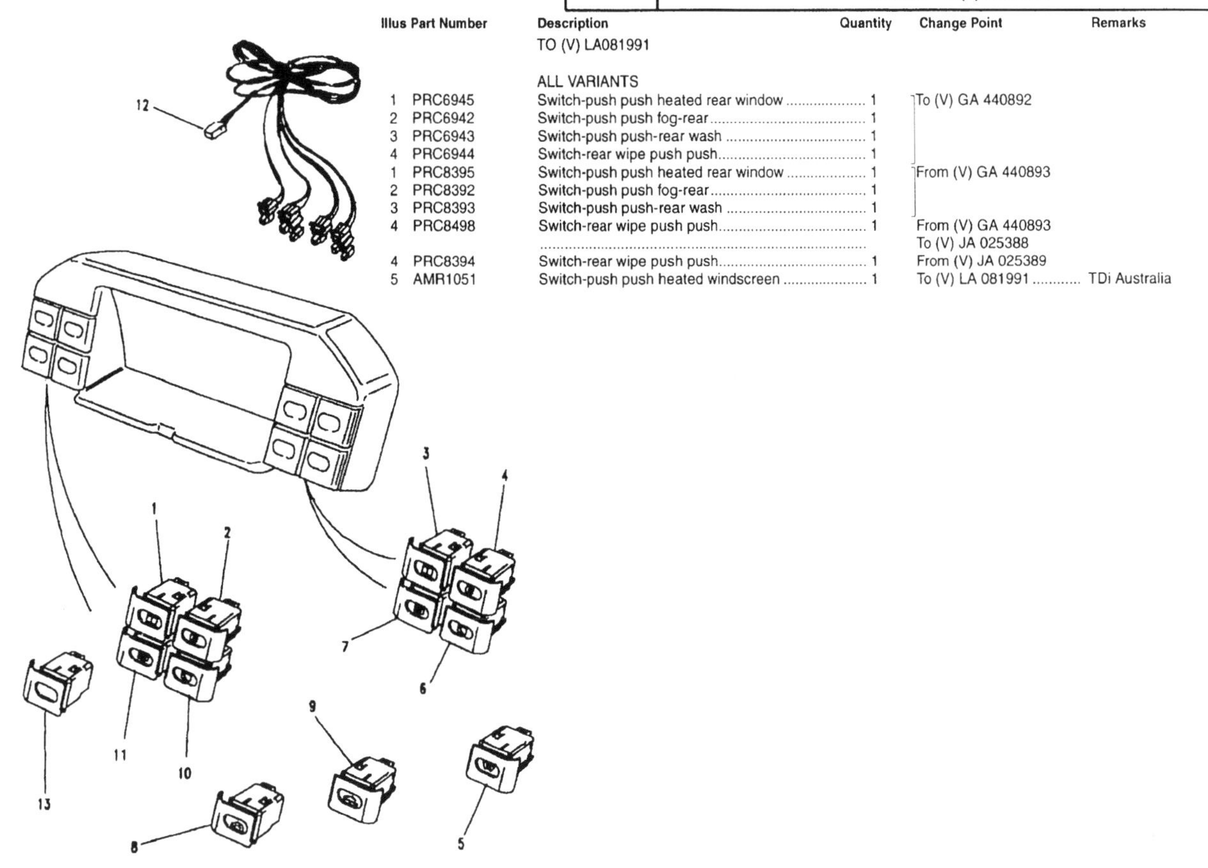

Illus	Part Number	Description	Quantity	Change Point	Remarks
		TO (V) LA081991			
		ALL VARIANTS			
1	PRC6945	Switch-push push heated rear window	1	To (V) GA 440892	
2	PRC6942	Switch-push push fog-rear	1		
3	PRC6943	Switch-push push-rear wash	1		
4	PRC6944	Switch-rear wipe push push	1		
1	PRC8395	Switch-push push heated rear window	1	From (V) GA 440893	
2	PRC8392	Switch-push push fog-rear	1		
3	PRC8393	Switch-push push-rear wash	1		
4	PRC8498	Switch-rear wipe push push	1	From (V) GA 440893 To (V) JA 025388	
4	PRC8394	Switch-rear wipe push push	1	From (V) JA 025389	
5	AMR1051	Switch-push push heated windscreen	1	To (V) LA 081991	TDi Australia

M 67

DISCOVERY 1989MY UPTO 1999MY | GROUP M | Body Electrics
Switches
Binnacle - To (V) LA081991

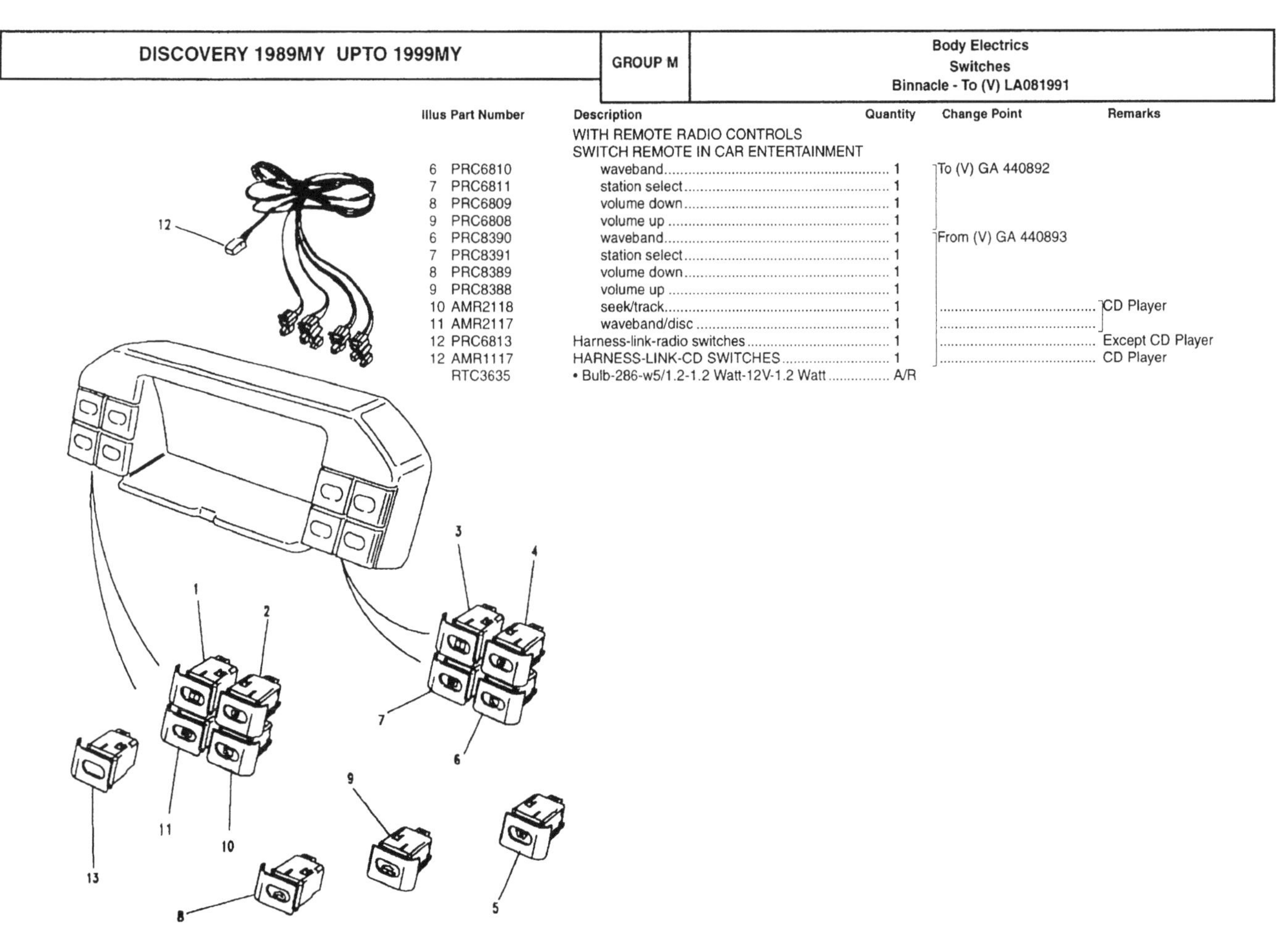

Illus	Part Number	Description	Quantity	Change Point	Remarks
		WITH REMOTE RADIO CONTROLS			
		SWITCH REMOTE IN CAR ENTERTAINMENT			
6	PRC6810	waveband	1	To (V) GA 440892	
7	PRC6811	station select	1		
8	PRC6809	volume down	1		
9	PRC6808	volume up	1		
6	PRC8390	waveband	1	From (V) GA 440893	
7	PRC8391	station select	1		
8	PRC8389	volume down	1		
9	PRC8388	volume up	1		
10	AMR2118	seek/track	1		CD Player
11	AMR2117	waveband/disc	1		
12	PRC6813	Harness-link-radio switches	1		Except CD Player
12	AMR1117	HARNESS-LINK-CD SWITCHES	1		CD Player
	RTC3635	• Bulb-286-w5/1.2-1.2 Watt-12V-1.2 Watt	A/R		

M 68

Illus	Part Number	Description	Quantity	Change Point	Remarks
		WITHOUT REMOTE RADIO CONTROLS			
13	PRC8396	Blank-facia switch	4		

NOTE: FOR BINNACLE HOUSING SEE-
TRIM VENT ASSEMBLY FACIA

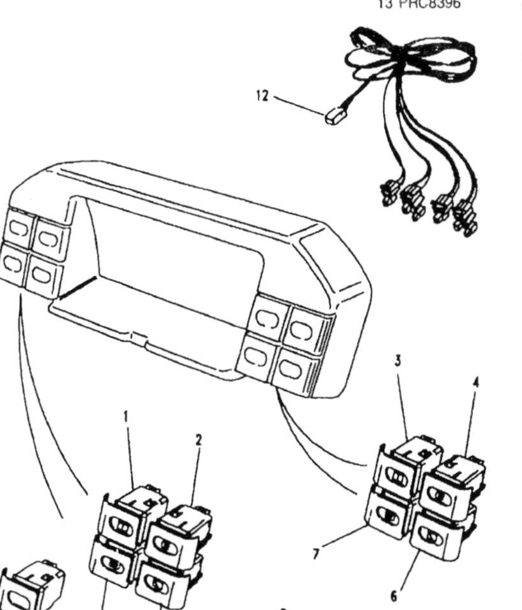

Illus	Part Number	Description	Quantity	Change Point	Remarks
		FROM (V) MA081992			
		ALL VARIANTS			
1	AMR3750	SWITCH-PUSH PUSH HEATED REAR WINDOW	1		
2	AMR3751	SWITCH-PUSH PUSH FOG-REAR	1		
2	AMR5178	SWITCH-PUSH PUSH FOG-REAR	1	From (V) VA 703237 From (V) VA 534104	
3	AMR3748	SWITCH-PUSH PUSH-REAR WASH	1		
4	AMR3749	SWITCH-REAR WIPE PUSH PUSH	1		
	STC1877	• Bulb & holder assembly-Orange	A/R		
	STC1878	• Bulb & holder assembly-Green	A/R		

Illus	Part Number	Description	Quantity	Change Point	Remarks
		WITH REMOTE RADIO CONTROLS-SEE NOTES			
		SWITCH REMOTE IN CAR ENTERTAINMENT			
5	AMR3741	VOLUME UP	1	Note (1)	
6	AMR3742	VOLUME DOWN	1		
7	AMR3743	NEXT STATION/TRACK	1		
8	AMR3744	BAND/CD SELECT	1		
	STC1877	• Bulb & holder assembly-Orange	A/R		
	STC1878	• Bulb & holder assembly-Green	A/R		
		SWITCH REMOTE IN CAR ENTERTAINMENT			
5	AMR6462	VOLUME UP	1	Note (2)	
6	AMR6463	VOLUME DOWN	1		
7	AMR6464	NEXT STATION/TRACK	1		
8	AMR6465	BAND/CD SELECT	1		
	STC1877	• Bulb & holder assembly-Orange	A/R		
	STC1878	• Bulb & holder assembly-Green	A/R		

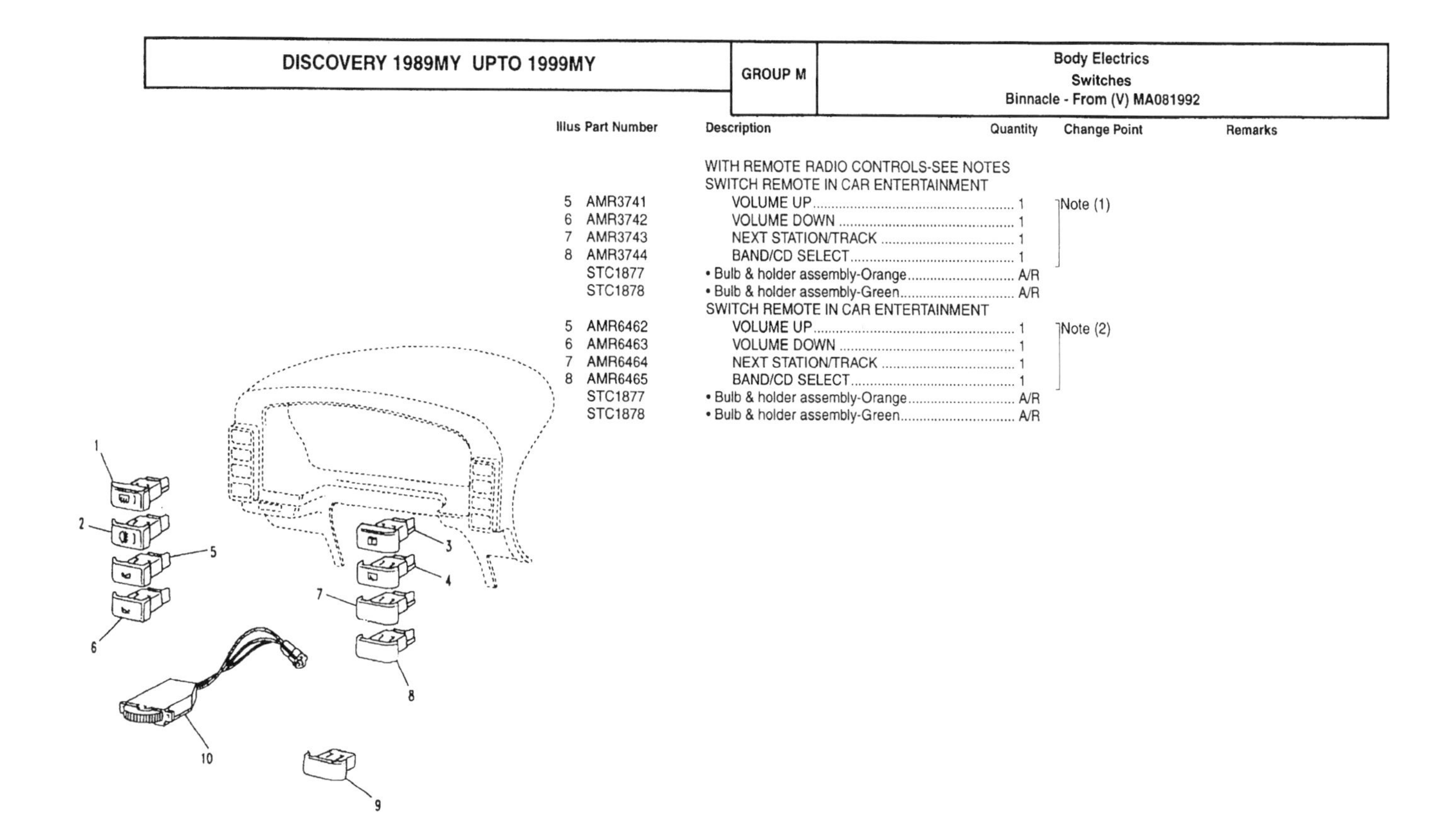

CHANGE POINTS:
(1) From (V) MA 081992 To (V) VA 543813; From (V) MA 081992 To (V) VA 712444
(2) From (V) VA 543814; From (V) VA 712445

Illus	Part Number	Description	Quantity	Change Point	Remarks
		WITHOUT REMOTE RADIO CONTROLS			
9	AMR3740	Blank-instrument bezel switch	4		
10	AMR2745	SWITCH-THUMBWHEEL ILLUMINATION CONTROL	1		
	STC1877	• Bulb & holder assembly-Orange	A/R		
	STC1878	• Bulb & holder assembly-Green	A/R		

NOTE: RADIO SWITCH HARNESS INCORPORATED INTO FACIA HARNESS
NOTE: FOR FACIA HARNESS SEE- HARNESS BODY

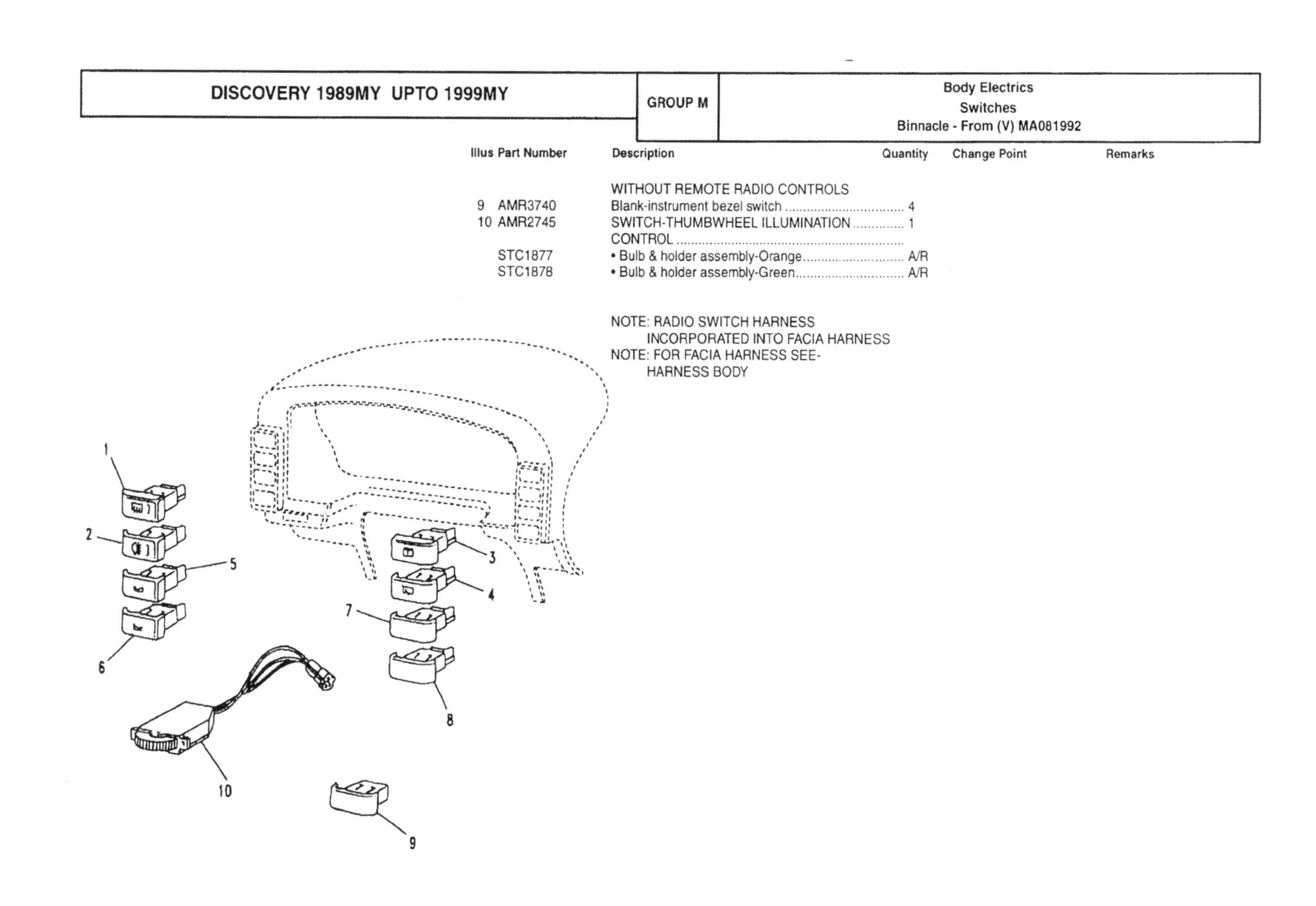

Illus	Part Number	Description	Quantity	Change Point	Remarks
		TO (V) LA081991			
		SWITCHES FOR 3 DOOR VEHICLES			
		HOUSING-TUNNEL CONSOLE SWITCH			
1	MWC9140LUN	RHD-Ash Grey-less headlamp levelling	1	To (V) JA 034313	
1	MWC9141LUN	LHD-Ash Grey-less headlamp levelling	1		
1	MWC9140LNF	RHD-Ash Grey-less headlamp levelling	1	From (V) KA 034314	
1	MWC9141LNF	LHD-Ash Grey-less headlamp levelling	1		
		HOUSING-TUNNEL CONSOLE SWITCH			
1	BTR1528LUN	RHD-Ash Grey-with headlamp levelling	1	To (V) JA 034313	Hand set door mirrors
1	BTR1529LUN	LHD-Ash Grey-with headlamp levelling	1		
1	BTR1528LNF	RHD-Ash Grey-with headlamp levelling	1	From (V) KA 034314	
1	BTR1529LNF	LHD-Ash Grey-with headlamp levelling	1		
		HOUSING-TUNNEL CONSOLE SWITCH			
1	BTR733LUN	RHD-Ash Grey-with headlamp levelling	1	To (V) JA 016651	Electric door mirrors
1	MXC7810LUN	LHD-Ash Grey-with headlamp levelling	1		
1	BTR1524LUN	RHD-Ash Grey-with headlamp levelling	1	From (V) JA 016652	
1	BTR1525LUN	LHD-Ash Grey-with headlamp levelling	1	To (V) JA 034313	
1	BTR1524LNF	RHD-Ash Grey-with headlamp levelling	1	From (V) KA 034314	
1	BTR1525LNF	LHD-Ash Grey-with headlamp levelling	1		

M 73

Illus	Part Number	Description	Quantity	Change Point	Remarks
		TO (V) LA081991			
		SWITCHES FOR 5 DOOR VEHICLES			
		HOUSING-TUNNEL CONSOLE SWITCH			
1	MWC9288LUN	RHD-Ash Grey-less headlamp levelling	1	To (V) JA 034313	
1	MWC9289LUN	LHD-Ash Grey-less headlamp levelling	1		
1	MWC9288LNF	RHD-Ash Grey-less headlamp levelling	1	From (V) KA 034314	
1	MWC9289LNF	LHD-Ash Grey-less headlamp levelling	1		
		HOUSING-TUNNEL CONSOLE SWITCH			
1	BTR1528LUN	RHD-Ash Grey-with headlamp levelling	1	To (V) JA 034313	Hand set door mirrors
1	BTR1529LUN	LHD-Ash Grey-with headlamp levelling	1		
1	BTR1528LNF	RHD-Ash Grey-with headlamp levelling	1	From (V) KA 034314	
1	BTR1529LNF	LHD-Ash Grey-with headlamp levelling	1		
		HOUSING-TUNNEL CONSOLE SWITCH			
1	BTR733LUN	RHD-Ash Grey-with headlamp levelling	1	To (V) JA 016651	Electric door mirrors
1	MXC7810LUN	LHD-Ash Grey-with headlamp levelling	1		
1	BTR1526LUN	RHD-Ash Grey-with headlamp levelling	1	From (V) JA 016652	
1	BTR1527LUN	LHD-Ash Grey-with headlamp levelling	1	To (V) JA 034313	
1	BTR1526LNF	RHD-Ash Grey-with headlamp levelling	1	From (V) KA 034314	
1	BTR1527LNF	LHD-Ash Grey-with headlamp levelling	1		

M 74

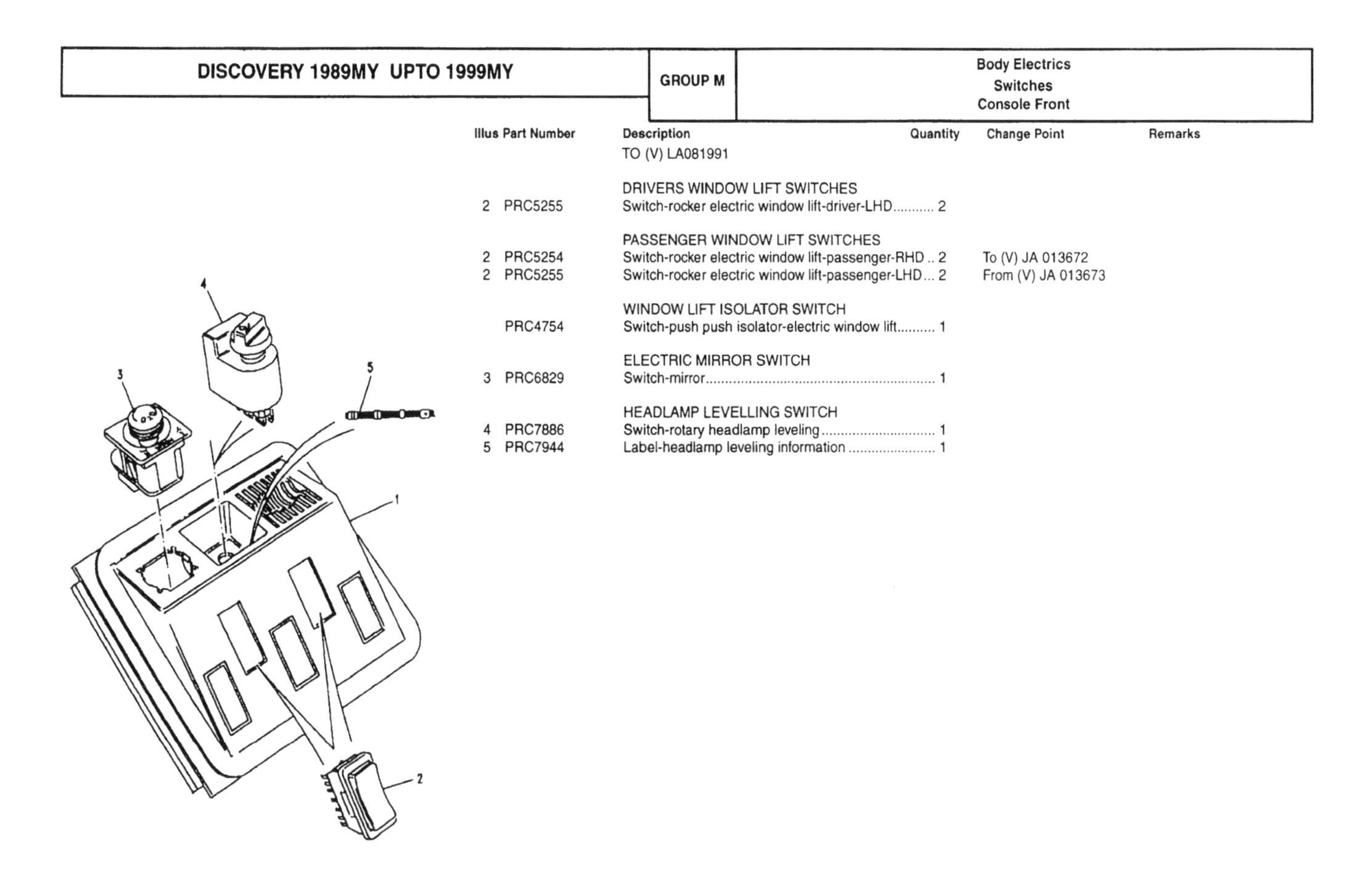

Illus	Part Number	Description	Quantity	Change Point	Remarks
		TO (V) LA081991			
		DRIVERS WINDOW LIFT SWITCHES			
2	PRC5255	Switch-rocker electric window lift-driver-LHD	2		
		PASSENGER WINDOW LIFT SWITCHES			
2	PRC5254	Switch-rocker electric window lift-passenger-RHD	2	To (V) JA 013672	
2	PRC5255	Switch-rocker electric window lift-passenger-LHD	2	From (V) JA 013673	
		WINDOW LIFT ISOLATOR SWITCH			
	PRC4754	Switch-push push isolator-electric window lift	1		
		ELECTRIC MIRROR SWITCH			
3	PRC6829	Switch-mirror	1		
		HEADLAMP LEVELLING SWITCH			
4	PRC7886	Switch-rotary headlamp leveling	1		
5	PRC7944	Label-headlamp leveling information	1		

Illus	Part Number	Description	Quantity	Change Point	Remarks
		FROM (V) MA081992			
		ELECTRIC SUNROOF CONTROL PANEL-FRONT			
1	AMR2471	Switch-rocker sunroof-front	2		
2	AMR3653	Switch-sunroof isolator push push-front	1		
3	BTR6984LUM	Bezel-sunroof switch-Mist Grey	1		
		ELECTRIC SUNROOF CONTROL PANEL-REAR			
4	AMR3652	Switch-rocker sunroof-rear	1		

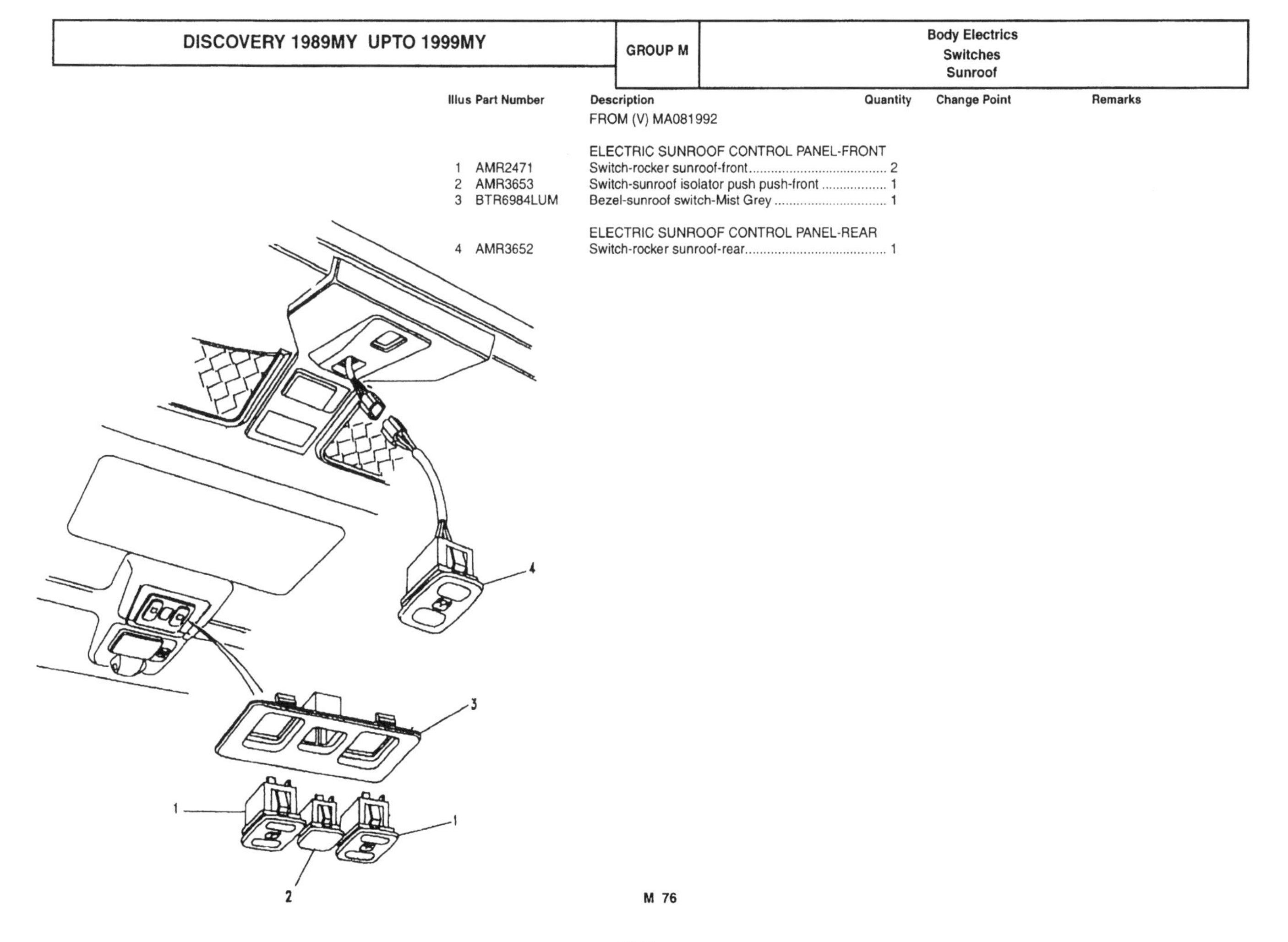

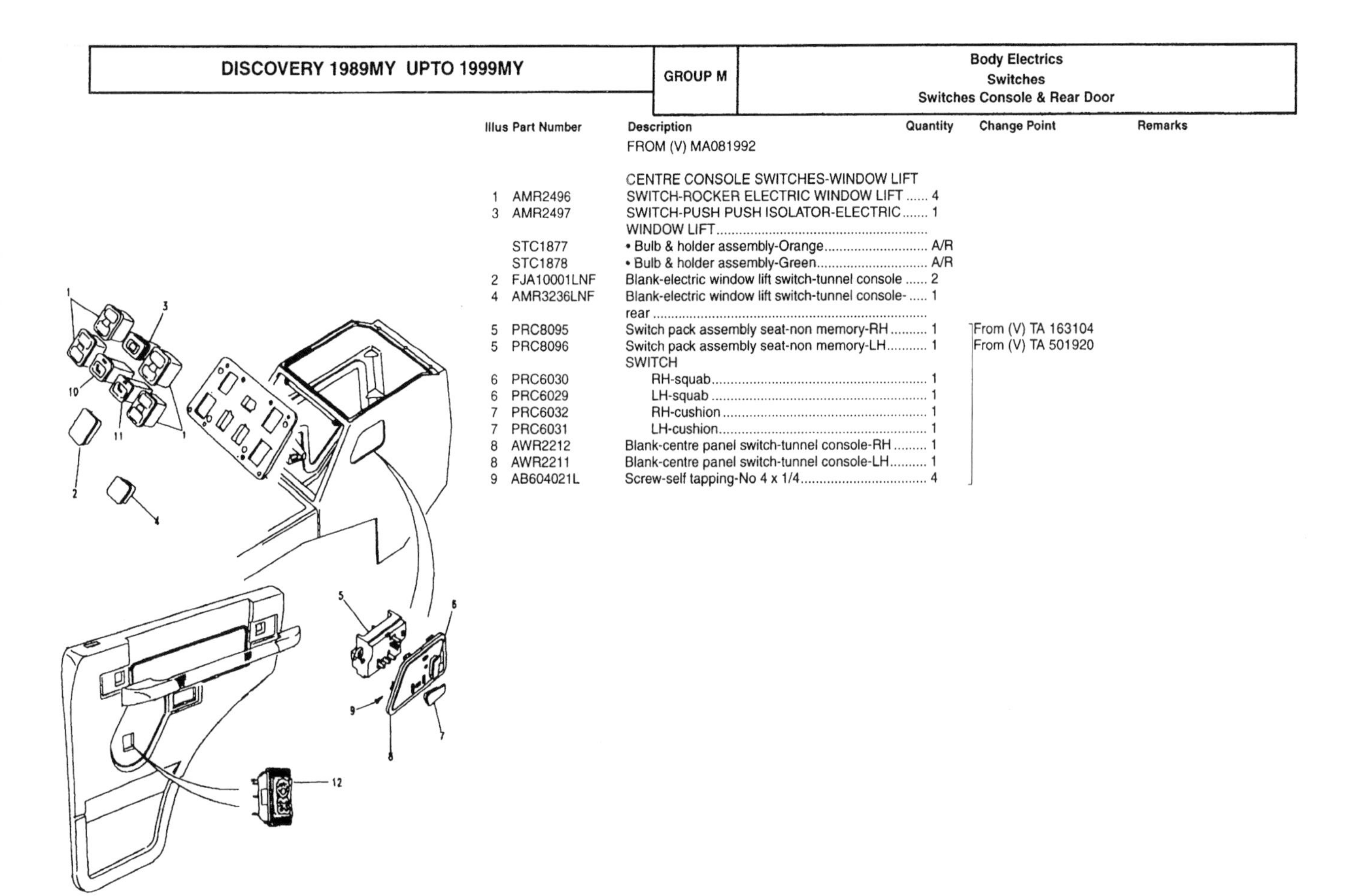

Illus	Part Number	Description	Quantity	Change Point	Remarks
		FROM (V) MA081992			
		CENTRE CONSOLE SWITCHES-WINDOW LIFT			
1	AMR2496	SWITCH-ROCKER ELECTRIC WINDOW LIFT	4		
3	AMR2497	SWITCH-PUSH PUSH ISOLATOR-ELECTRIC WINDOW LIFT	1		
	STC1877	• Bulb & holder assembly-Orange	A/R		
	STC1878	• Bulb & holder assembly-Green	A/R		
2	FJA10001LNF	Blank-electric window lift switch-tunnel console	2		
4	AMR3236LNF	Blank-electric window lift switch-tunnel console-rear	1		
5	PRC8095	Switch pack assembly seat-non memory-RH	1	From (V) TA 163104	
5	PRC8096	Switch pack assembly seat-non memory-LH	1	From (V) TA 501920	
		SWITCH			
6	PRC6030	RH-squab	1		
6	PRC6029	LH-squab	1		
7	PRC6032	RH-cushion	1		
7	PRC6031	LH-cushion	1		
8	AWR2212	Blank-centre panel switch-tunnel console-RH	1		
8	AWR2211	Blank-centre panel switch-tunnel console-LH	1		
9	AB604021L	Screw-self tapping-No 4 x 1/4	4		

M 77

DISCOVERY 1989MY UPTO 1999MY | GROUP M | Body Electrics
Switches
Switches Console & Rear Door

Illus	Part Number	Description	Quantity	Change Point	Remarks
10	AMR3602	SWITCH HEATED SEATS-RH	1	From (V) VA 703237 From (V) VA 534104	Heated seats
11	AMR3601	SWITCH HEATED SEATS-LH	1	From (V) VA 703237 From (V) VA 534104	Heated seats
	STC1877	• Bulb & holder assembly-Orange	A/R		
	STC1878	• Bulb & holder assembly-Green	A/R		
		REAR DOOR SWITCHES-WINDOW LIFT			
12	PRC5254	Switch-rocker electric window lift-rear door-RHD	2		
12	PRC5255	Switch-rocker electric window lift-rear door-LHD	2		

NOTE: FOR COVER PLATE & FIXINGS SEE-TRIM CUBBY BOX/CENTRE CONSOLE

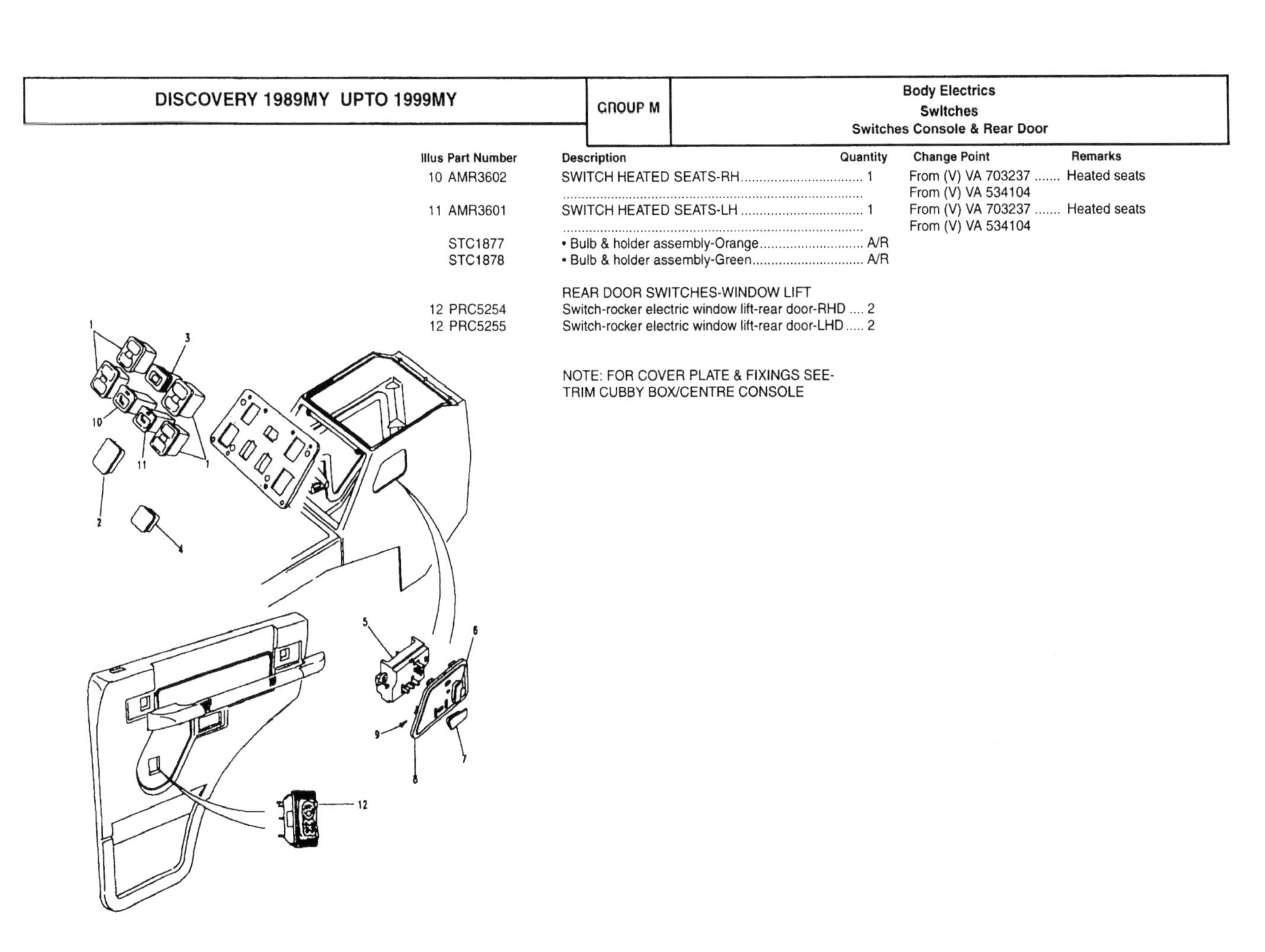

M 78

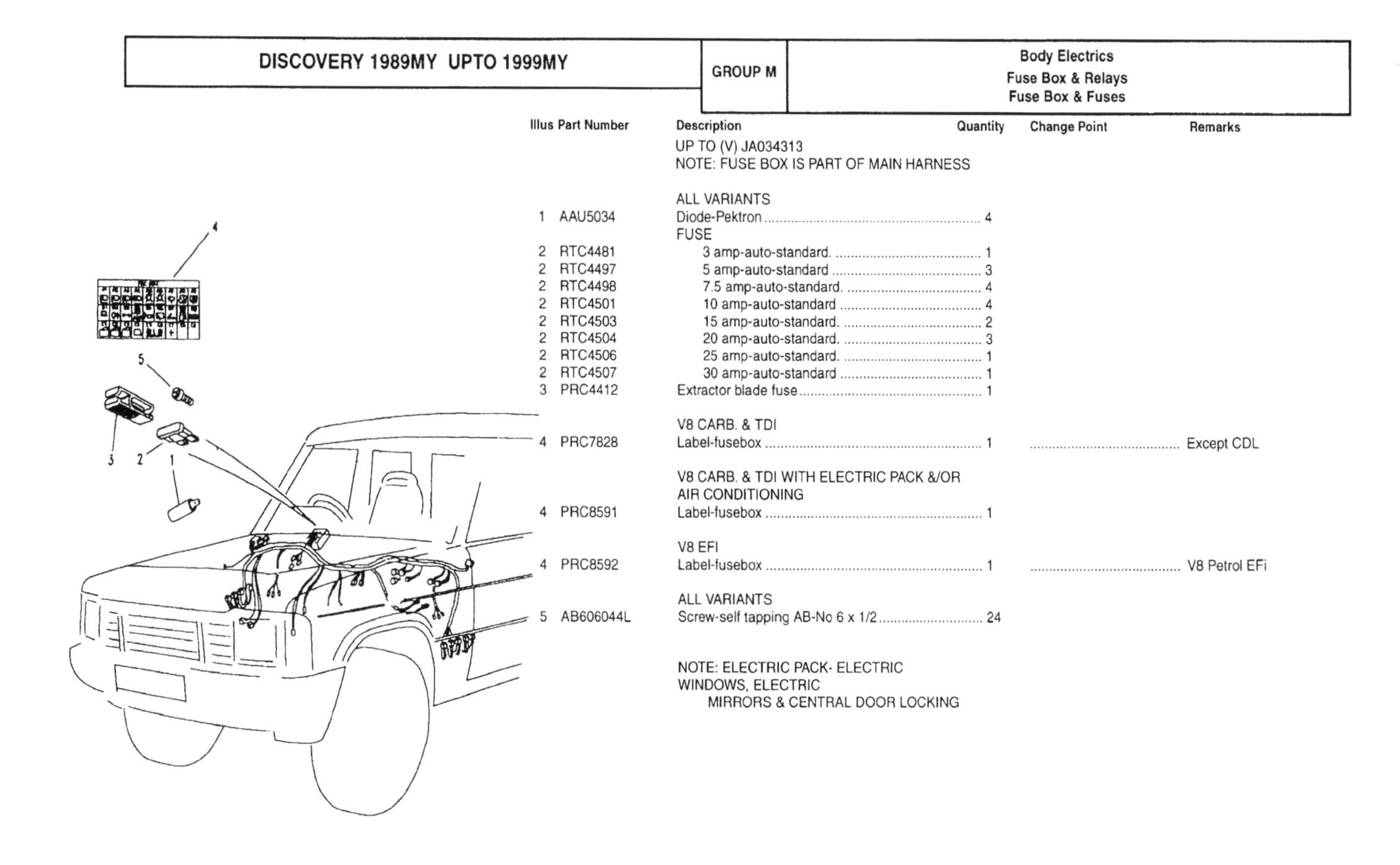

Illus	Part Number	Description	Quantity	Change Point	Remarks
		UP TO (V) JA034313			
		NOTE: FUSE BOX IS PART OF MAIN HARNESS			
		ALL VARIANTS			
1	AAU5034	Diode-Pektron	4		
		FUSE			
2	RTC4481	3 amp-auto-standard.	1		
2	RTC4497	5 amp-auto-standard	3		
2	RTC4498	7.5 amp-auto-standard.	4		
2	RTC4501	10 amp-auto-standard	4		
2	RTC4503	15 amp-auto-standard.	2		
2	RTC4504	20 amp-auto-standard.	3		
2	RTC4506	25 amp-auto-standard.	1		
2	RTC4507	30 amp-auto-standard	1		
3	PRC4412	Extractor blade fuse	1		
		V8 CARB. & TDI			
4	PRC7828	Label-fusebox	1		Except CDL
		V8 CARB. & TDI WITH ELECTRIC PACK &/OR AIR CONDITIONING			
4	PRC8591	Label-fusebox	1		
		V8 EFI			
4	PRC8592	Label-fusebox	1		V8 Petrol EFi
		ALL VARIANTS			
5	AB606044L	Screw-self tapping AB-No 6 x 1/2	24		

NOTE: ELECTRIC PACK- ELECTRIC WINDOWS, ELECTRIC MIRRORS & CENTRAL DOOR LOCKING

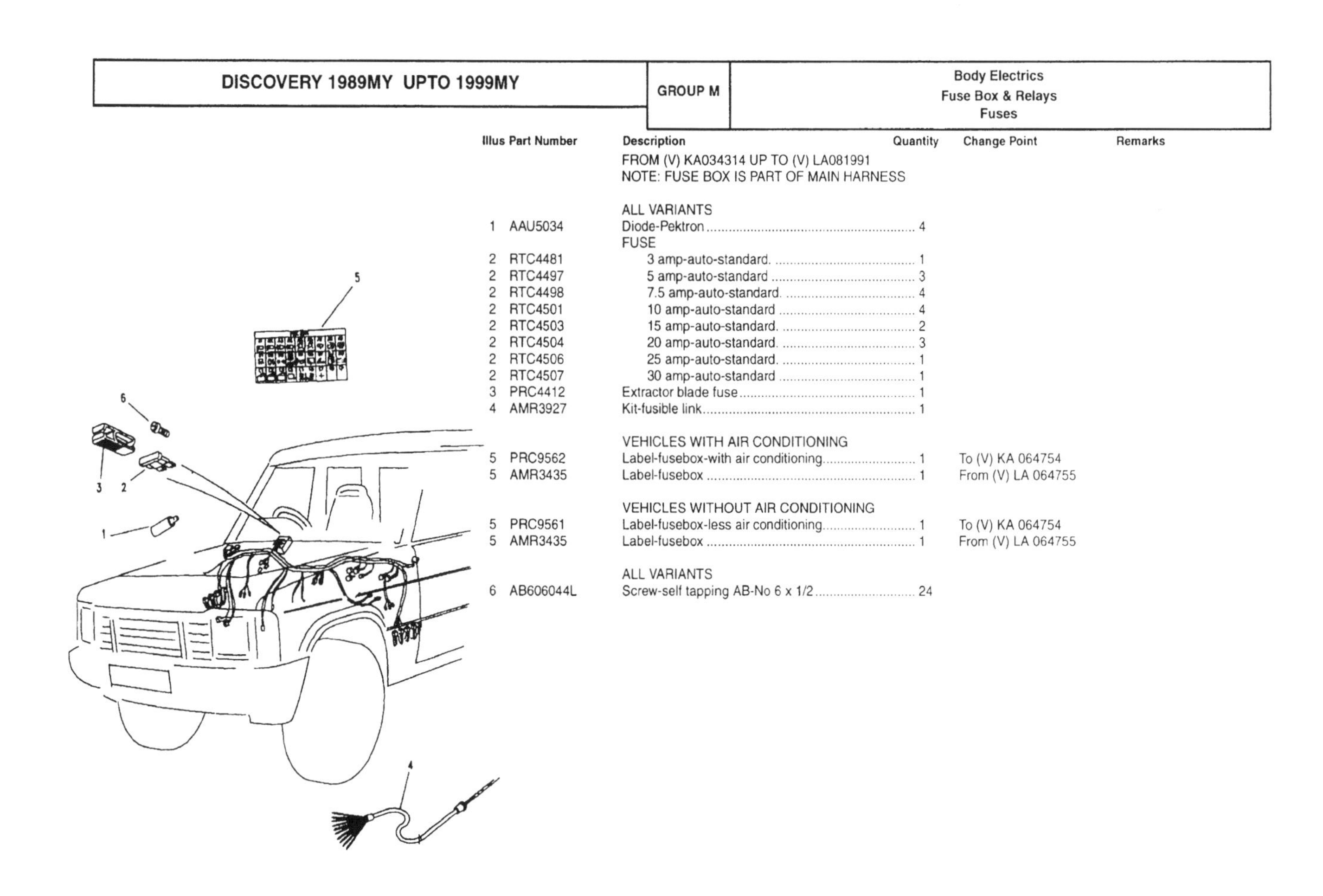

Illus	Part Number	Description	Quantity	Change Point	Remarks
		FROM (V) KA034314 UP TO (V) LA081991			
		NOTE: FUSE BOX IS PART OF MAIN HARNESS			
		ALL VARIANTS			
1	AAU5034	Diode-Pektron	4		
		FUSE			
2	RTC4481	3 amp-auto-standard.	1		
2	RTC4497	5 amp-auto-standard	3		
2	RTC4498	7.5 amp-auto-standard.	4		
2	RTC4501	10 amp-auto-standard	4		
2	RTC4503	15 amp-auto-standard.	2		
2	RTC4504	20 amp-auto-standard.	3		
2	RTC4506	25 amp-auto-standard.	1		
2	RTC4507	30 amp-auto-standard	1		
3	PRC4412	Extractor blade fuse	1		
4	AMR3927	Kit-fusible link	1		
		VEHICLES WITH AIR CONDITIONING			
5	PRC9562	Label-fusebox-with air conditioning	1	To (V) KA 064754	
5	AMR3435	Label-fusebox	1	From (V) LA 064755	
		VEHICLES WITHOUT AIR CONDITIONING			
5	PRC9561	Label-fusebox-less air conditioning	1	To (V) KA 064754	
5	AMR3435	Label-fusebox	1	From (V) LA 064755	
		ALL VARIANTS			
6	AB606044L	Screw-self tapping AB-No 6 x 1/2	24		

DISCOVERY 1989MY UPTO 1999MY	GROUP M	Body Electrics Fuse Box Interior Main

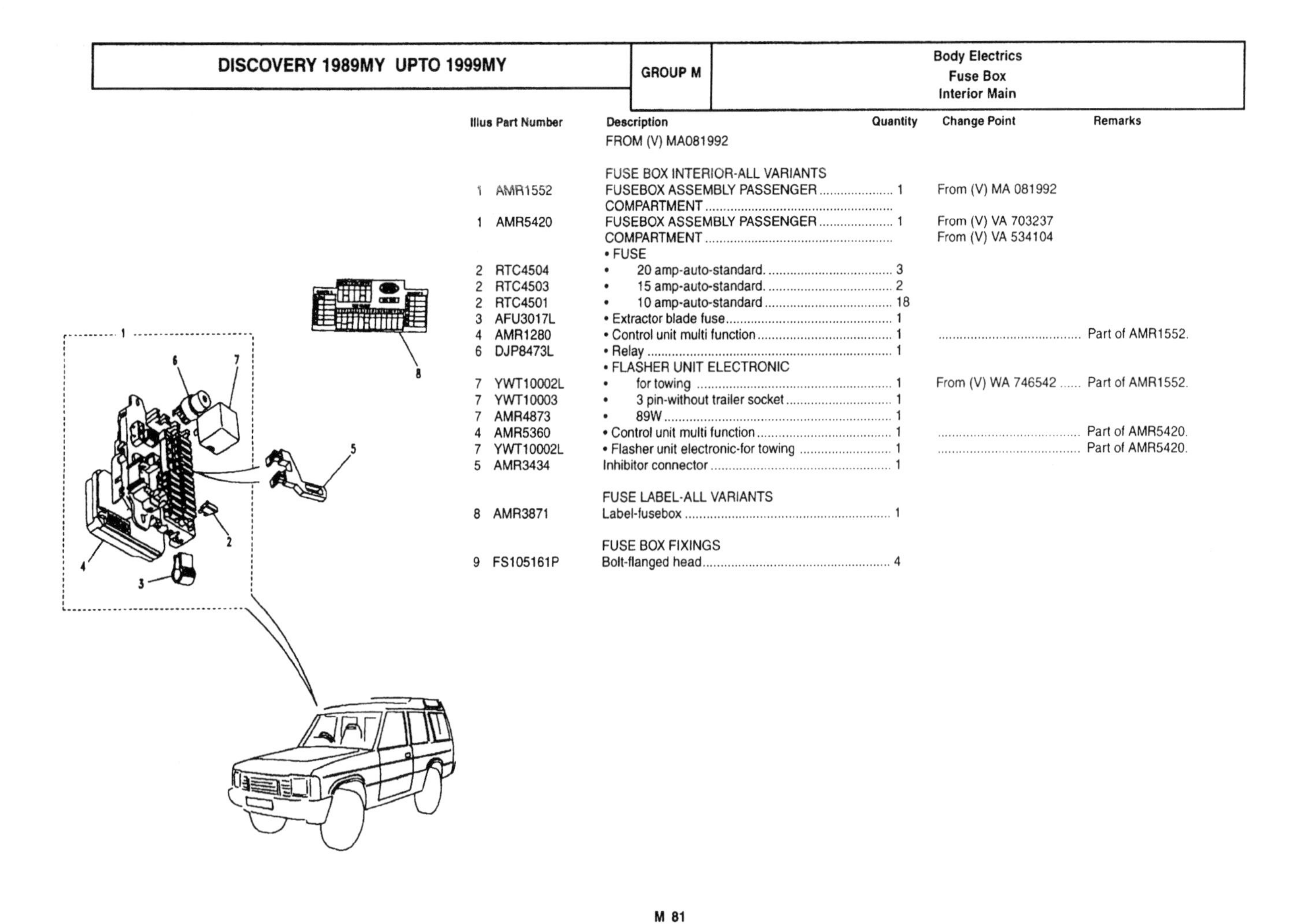

Illus	Part Number	Description	Quantity	Change Point	Remarks
		FROM (V) MA081992			
		FUSE BOX INTERIOR-ALL VARIANTS			
1	AMR1552	FUSEBOX ASSEMBLY PASSENGER COMPARTMENT	1	From (V) MA 081992	
1	AMR5420	FUSEBOX ASSEMBLY PASSENGER COMPARTMENT	1	From (V) VA 703237 From (V) VA 534104	
		• FUSE			
2	RTC4504	• 20 amp-auto-standard.	3		
2	RTC4503	• 15 amp-auto-standard.	2		
2	RTC4501	• 10 amp-auto-standard	18		
3	AFU3017L	• Extractor blade fuse	1		
4	AMR1280	• Control unit multi function	1		Part of AMR1552.
6	DJP8473L	• Relay	1		
		• FLASHER UNIT ELECTRONIC			
7	YWT10002L	• for towing	1	From (V) WA 746542	Part of AMR1552.
7	YWT10003	• 3 pin-without trailer socket	1		
7	AMR4873	• 89W	1		
4	AMR5360	• Control unit multi function	1		Part of AMR5420.
7	YWT10002L	• Flasher unit electronic-for towing	1		Part of AMR5420.
5	AMR3434	Inhibitor connector	1		
		FUSE LABEL-ALL VARIANTS			
8	AMR3871	Label-fusebox	1		
		FUSE BOX FIXINGS			
9	FS105161P	Bolt-flanged head	4		

M 81

DISCOVERY 1989MY UPTO 1999MY	GROUP M	Body Electrics Fuse Box Under Bonnet

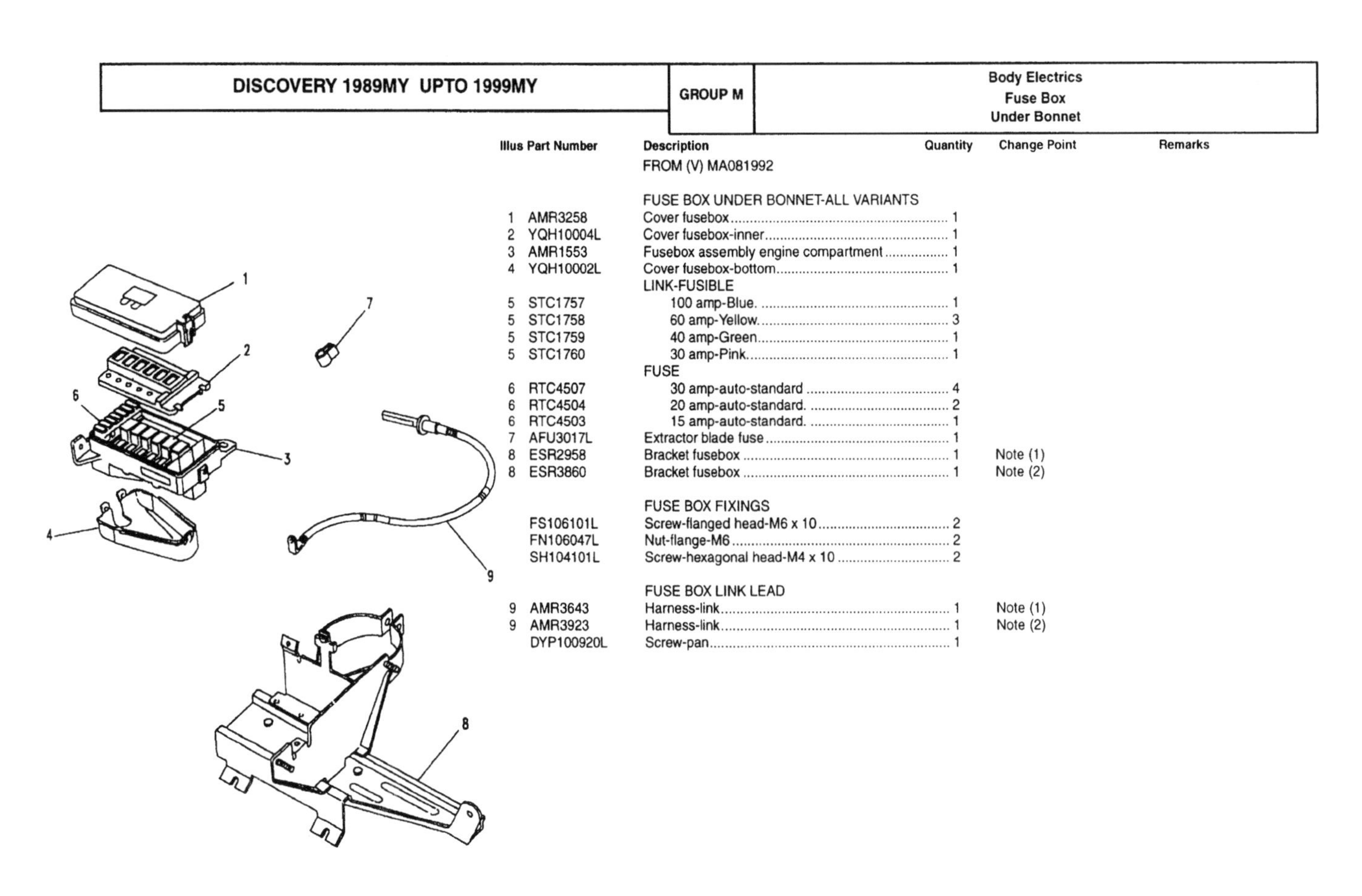

Illus	Part Number	Description	Quantity	Change Point	Remarks
		FROM (V) MA081992			
		FUSE BOX UNDER BONNET-ALL VARIANTS			
1	AMR3258	Cover fusebox	1		
2	YQH10004L	Cover fusebox-inner	1		
3	AMR1553	Fusebox assembly engine compartment	1		
4	YQH10002L	Cover fusebox-bottom	1		
		LINK-FUSIBLE			
5	STC1757	100 amp-Blue.	1		
5	STC1758	60 amp-Yellow.	3		
5	STC1759	40 amp-Green	1		
5	STC1760	30 amp-Pink	1		
		FUSE			
6	RTC4507	30 amp-auto-standard	4		
6	RTC4504	20 amp-auto-standard.	2		
6	RTC4503	15 amp-auto-standard.	1		
7	AFU3017L	Extractor blade fuse	1		
8	ESR2958	Bracket fusebox	1	Note (1)	
8	ESR3860	Bracket fusebox	1	Note (2)	
		FUSE BOX FIXINGS			
	FS106101L	Screw-flanged head-M6 x 10	2		
	FN106047L	Nut-flange-M6	2		
	SH104101L	Screw-hexagonal head-M4 x 10	2		
		FUSE BOX LINK LEAD			
9	AMR3643	Harness-link	1	Note (1)	
9	AMR3923	Harness-link	1	Note (2)	
	DYP100920L	Screw-pan	1		

CHANGE POINTS:
(1) From (V) MA 081992 To (V) MA 163103; From (V) MA 081992 To (V) MA 501919
(2) From (V) TA 163104; From (V) TA 501920

M 82

DISCOVERY 1989MY UPTO 1999MY

GROUP M — Body Electrics — Relays — Relays - To (V) LA081991

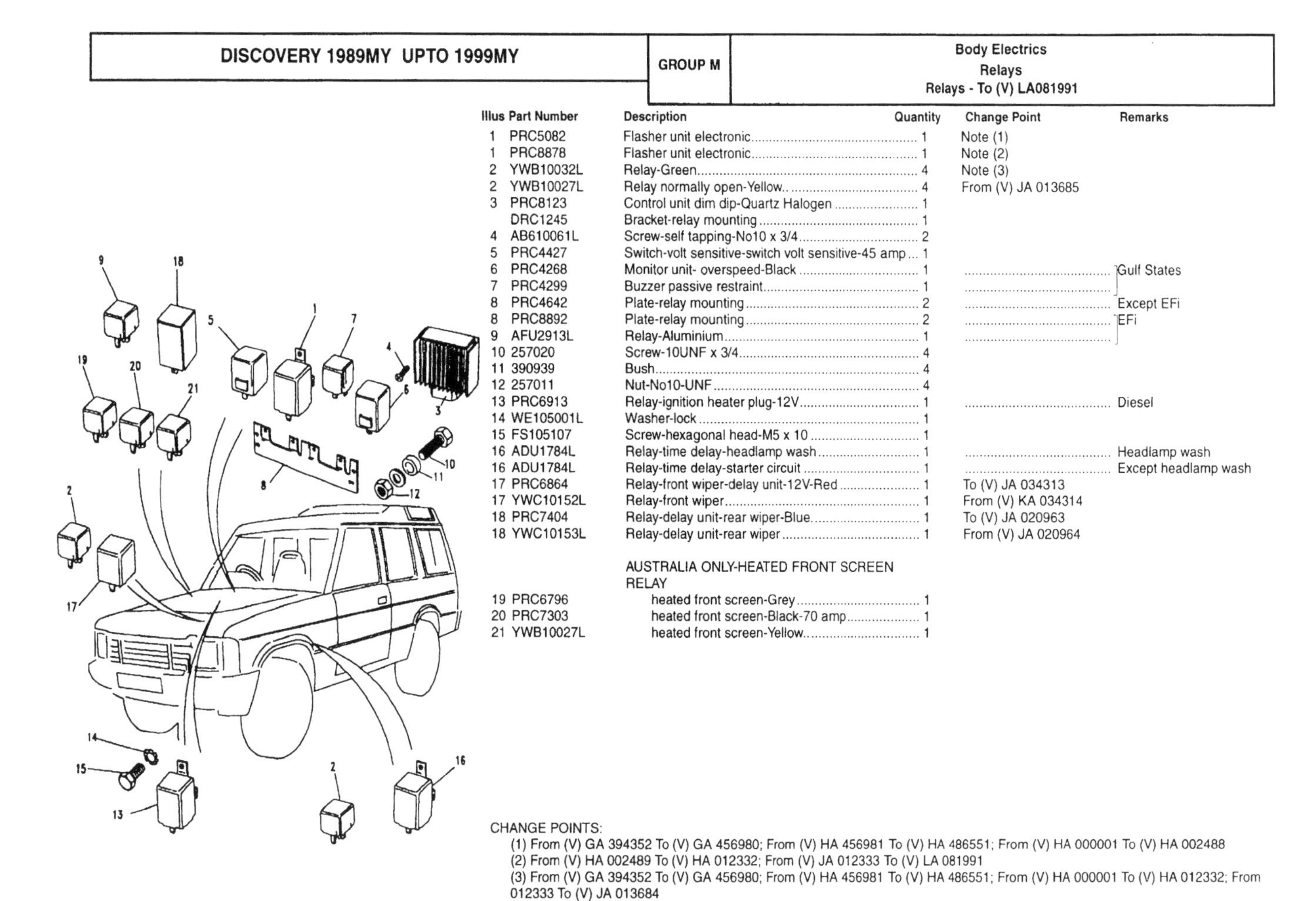

Illus	Part Number	Description	Quantity	Change Point	Remarks
1	PRC5082	Flasher unit electronic	1	Note (1)	
1	PRC8878	Flasher unit electronic	1	Note (2)	
2	YWB10032L	Relay-Green	4	Note (3)	
2	YWB10027L	Relay normally open-Yellow	4	From (V) JA 013685	
3	PRC8123	Control unit dim dip-Quartz Halogen	1		
	DRC1245	Bracket-relay mounting	1		
4	AB610061L	Screw-self tapping-No10 x 3/4	2		
5	PRC4427	Switch-volt sensitive-switch volt sensitive-45 amp	1		
6	PRC4268	Monitor unit- overspeed-Black	1		Gulf States
7	PRC4299	Buzzer passive restraint	1		Gulf States
8	PRC4642	Plate-relay mounting	2		Except EFi
8	PRC8892	Plate-relay mounting	2		EFi
9	AFU2913L	Relay-Aluminium	1		EFi
10	257020	Screw-10UNF x 3/4	4		
11	390939	Bush	4		
12	257011	Nut-No10-UNF	4		
13	PRC6913	Relay-ignition heater plug-12V	1		Diesel
14	WE105001L	Washer-lock	1		
15	FS105107	Screw-hexagonal head-M5 x 10	1		
16	ADU1784L	Relay-time delay-headlamp wash	1		Headlamp wash
16	ADU1784L	Relay-time delay-starter circuit	1		Except headlamp wash
17	PRC6864	Relay-front wiper-delay unit-12V-Red	1	To (V) JA 034313	
17	YWC10152L	Relay-front wiper	1	From (V) KA 034314	
18	PRC7404	Relay-delay unit-rear wiper-Blue	1	To (V) JA 020963	
18	YWC10153L	Relay-delay unit-rear wiper	1	From (V) JA 020964	
		AUSTRALIA ONLY-HEATED FRONT SCREEN RELAY			
19	PRC6796	heated front screen-Grey	1		
20	PRC7303	heated front screen-Black-70 amp	1		
21	YWB10027L	heated front screen-Yellow	1		

CHANGE POINTS:

(1) From (V) GA 394352 To (V) GA 456980; From (V) HA 456981 To (V) HA 486551; From (V) HA 000001 To (V) HA 002488
(2) From (V) HA 002489 To (V) HA 012332; From (V) JA 012333 To (V) LA 081991
(3) From (V) GA 394352 To (V) GA 456980; From (V) HA 456981 To (V) HA 486551; From (V) HA 000001 To (V) HA 012332; From 012333 To (V) JA 013684

M 83

DISCOVERY 1989MY UPTO 1999MY

GROUP M — Body Electrics — Relays — Relays - From (V) MA081992

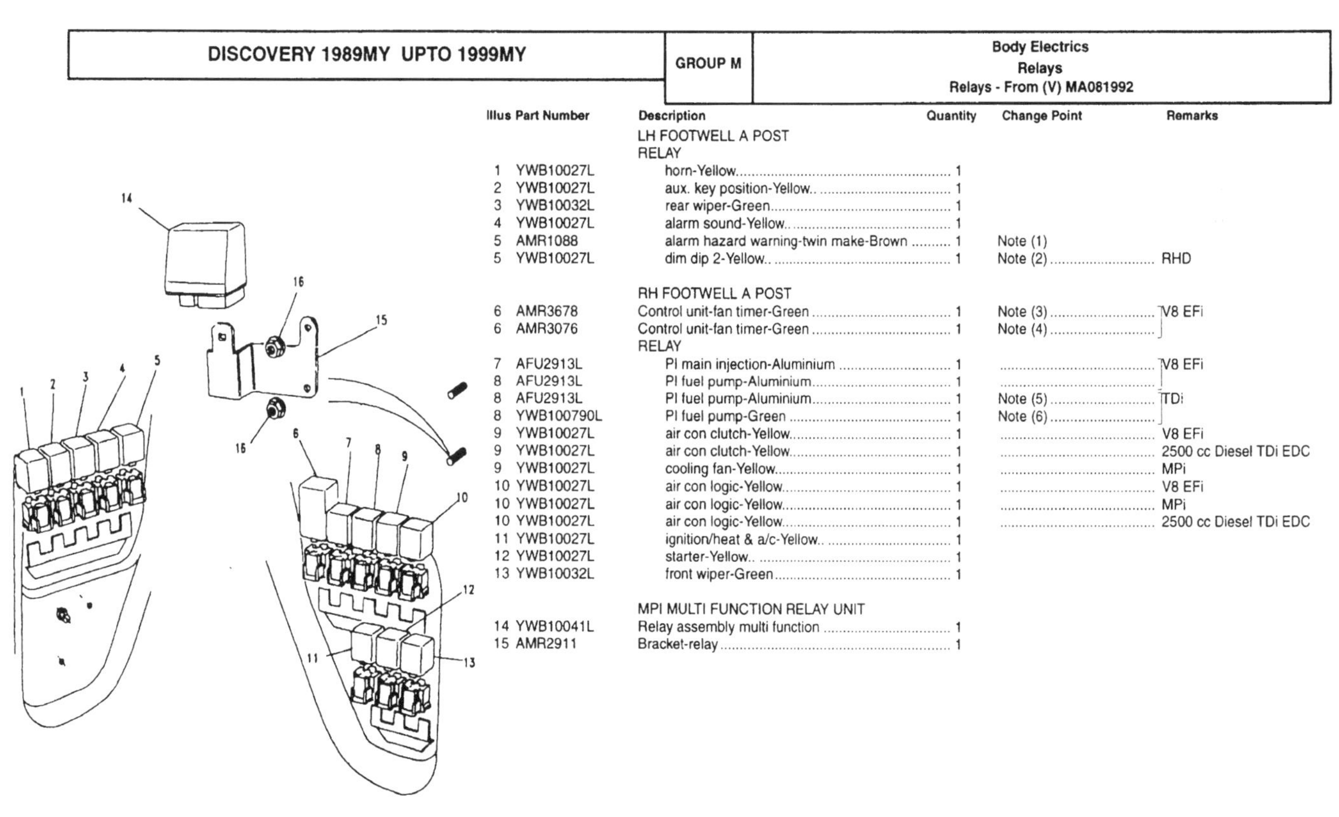

Illus	Part Number	Description	Quantity	Change Point	Remarks
		LH FOOTWELL A POST RELAY			
1	YWB10027L	horn-Yellow	1		
2	YWB10027L	aux. key position-Yellow	1		
3	YWB10032L	rear wiper-Green	1		
4	YWB10027L	alarm sound-Yellow	1		
5	AMR1088	alarm hazard warning-twin make-Brown	1	Note (1)	
5	YWB10027L	dim dip 2-Yellow	1	Note (2)	RHD
		RH FOOTWELL A POST			
6	AMR3678	Control unit-fan timer-Green	1	Note (3)	V8 EFi
6	AMR3076	Control unit-fan timer-Green	1	Note (4)	V8 EFi
		RELAY			
7	AFU2913L	PI main injection-Aluminium	1		V8 EFi
8	AFU2913L	PI fuel pump-Aluminium	1		V8 EFi
8	AFU2913L	PI fuel pump-Aluminium	1	Note (5)	TDi
8	YWB100790L	PI fuel pump-Green	1	Note (6)	TDi
9	YWB10027L	air con clutch-Yellow	1		V8 EFi
9	YWB10027L	air con clutch-Yellow	1		2500 cc Diesel TDi EDC
9	YWB10027L	cooling fan-Yellow	1		MPi
10	YWB10027L	air con logic-Yellow	1		V8 EFi
10	YWB10027L	air con logic-Yellow	1		MPi
10	YWB10027L	air con logic-Yellow	1		2500 cc Diesel TDi EDC
11	YWB10027L	ignition/heat & a/c-Yellow	1		
12	YWB10027L	starter-Yellow	1		
13	YWB10032L	front wiper-Green	1		
		MPI MULTI FUNCTION RELAY UNIT			
14	YWB10041L	Relay assembly multi function	1		
15	AMR2911	Bracket-relay	1		

CHANGE POINTS:

(1) From (V) MA 081992 To (V) MA 163104; From (V) MA 081992 To (V) MA 501920
(2) From (V) TA 163104; From (V) TA 501920
(3) From (V) MA 081992 To (V) TA 173886; From (V) MA 081992 To (V) TA 504867
(4) From (V) TA 173887; From (V) TA 504868
(5) From (V) MA 081992 To (V) WA 746541
(6) From (V) WA 746542

M 84

Illus	Part Number	Description	Quantity	Change Point	Remarks
		FROM (V) MA081992			
		PASSENGER FOOTWELL HINGED BRACKET -RIGHT HAND DRIVE VEHICLES			
1	AMR3066	Bracket-relay	1	To (V) MA 110669	
1	AMR4139	Bracket-relay	1	From (V) MA 110670	
		RELAY			
2	YWB10027L	dim dip-Yellow..	1	Note (1)	
2	YWB10027L	dim dip 1-Yellow..	1	Note (2)	
3	YWB10027L	power wash-Yellow	1		
4	YWB10032L	ABS warning-Green	1		
5	AMR3773	ABS pump-Black	1		
6	YWB10027L	a/c condenser fans-Yellow	1		
7	YWB10027L	ABS valve-Yellow..	1		
8	PRC9666	Switch-volt sensitive-Black	1		Manual
		RELAY			
9	YWB10032L	cruise control-Green	1		Automatic
9	YWB10027L	fog lamps front-Yellow..	1	Note (2)	
9	YWB10032L	fog lamps rear-Green	1		

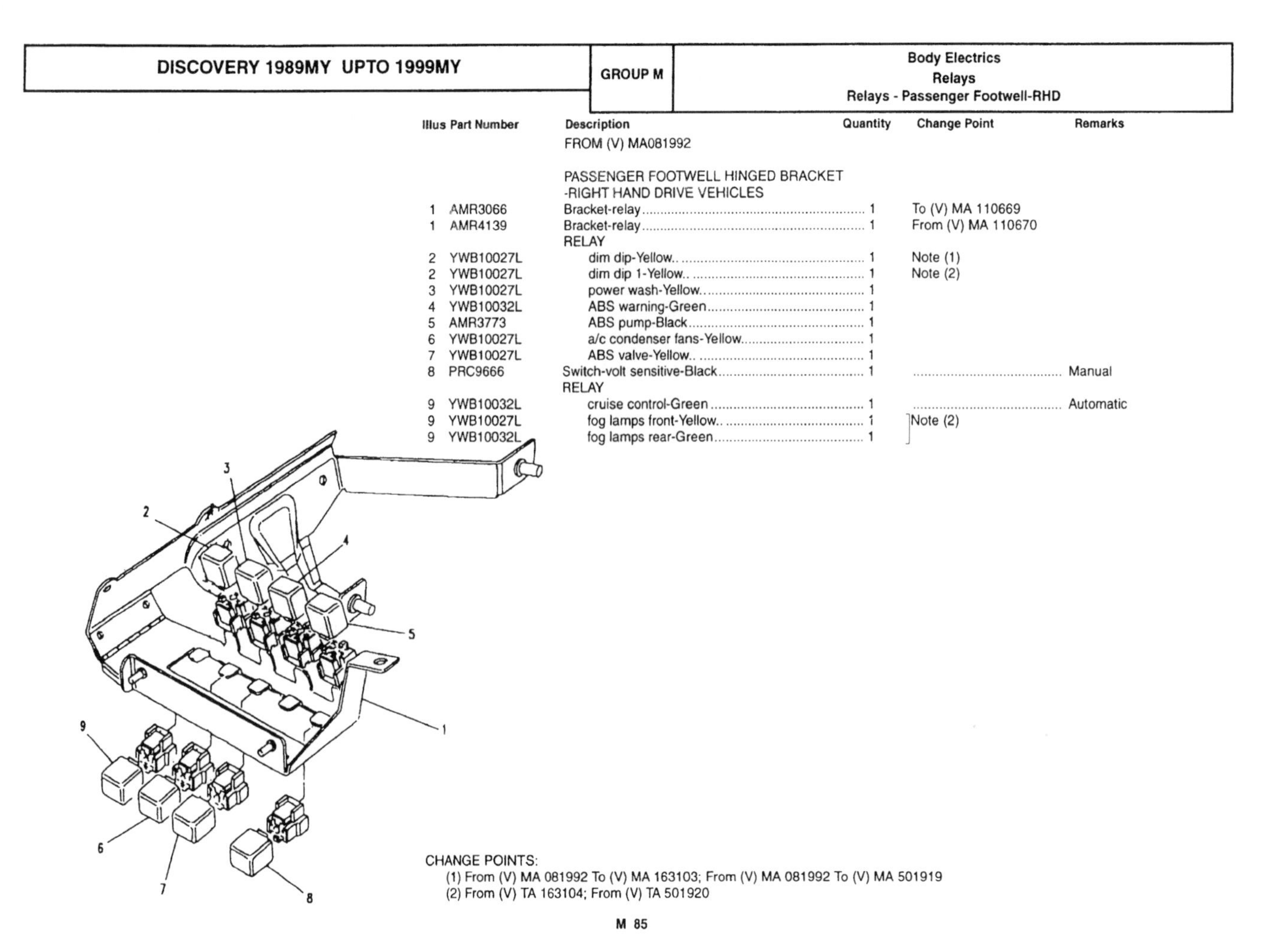

CHANGE POINTS:
(1) From (V) MA 081992 To (V) MA 163103; From (V) MA 081992 To (V) MA 501919
(2) From (V) TA 163104; From (V) TA 501920

Illus	Part Number	Description	Quantity	Change Point	Remarks
		FROM (V) MA081992			
		PASSENGER FOOTWELL HINGED BRACKET -LEFT HAND DRIVE VEHICLES			
1	AMR3067	Bracket-relay	1	To (V) MA 110669	
1	AMR4140	Bracket-relay	1	From (V) MA 110670	
2	YWB10027L	Relay normally open-power wash-Yellow..	1		
3	YWB10032L	Relay-ABS warning-Green	1		
4	AMR3773	Relay normally open-ABS pump-Black	1		
5	YWB10027L	Relay normally open-a/c condenser fans-Yellow	1		
7	YWB10027L	Relay normally open-ABS valve-Yellow..	1		
8	PRC9666	Switch-volt sensitive-cruise speed trip-Black	1		Manual
8	YWB10032L	Relay-cruise control-Green	1		Automatic
9	YWB10027L	Relay normally open-fog lamps front-Yellow..	1	Note (1)	
9	YWB10032L	Relay-fog lamps rear-Green	1		

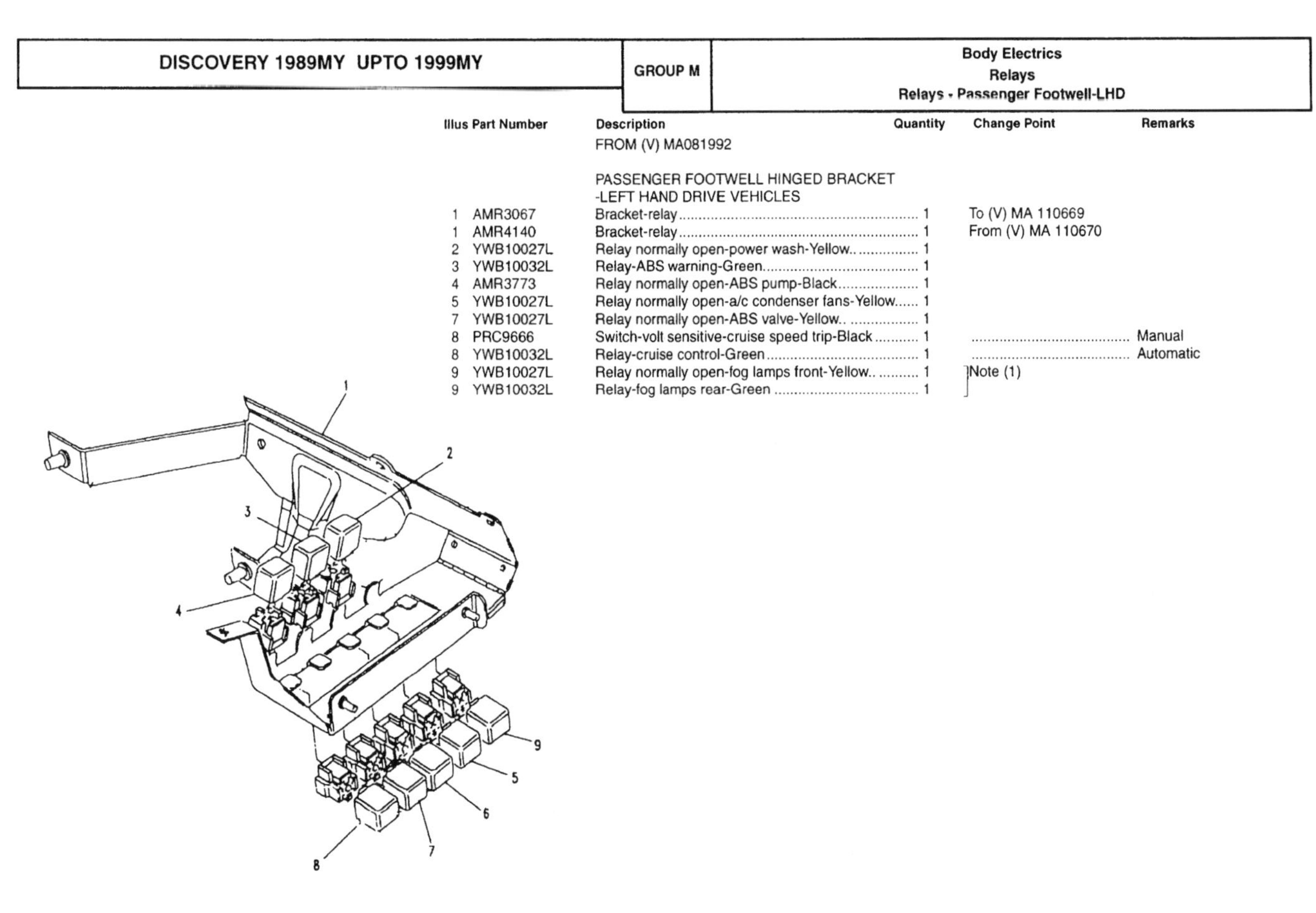

CHANGE POINTS:
(1) From (V) TA 163104; From (V) TA 501920

DISCOVERY 1989MY UPTO 1999MY | GROUP M | Body Electrics / Relays / Relays - Behind Glove Box

Illus	Part Number	Description	Quantity	Change Point	Remarks
		FROM (V) MA081992			
		BRACKET BEHIND GLOVE BOX			
1	AMR3134	Bracket-relay mounting	1		RHD Japan
		RELAY			
2	YWB10032L	shift interlock-Green	1		RHD Japan
3	YWB10027L	shift interlock-Yellow	1		RHD Japan
4	YWB10027L	catalyst warning-Yellow	1		RHD Japan
5	YWB10027L	power amplifier-Yellow	1		RHD Japan
		BRACKET BEHIND GLOVE BOX-WITH HEATED WINDSCREEN			
1	AMR3134	Bracket-relay mounting	1		Except Japan
6	AMR3773	Relay normally open-heated windscreen-Black	1	From (V) VA 534104	
				From (V) VA 703237	

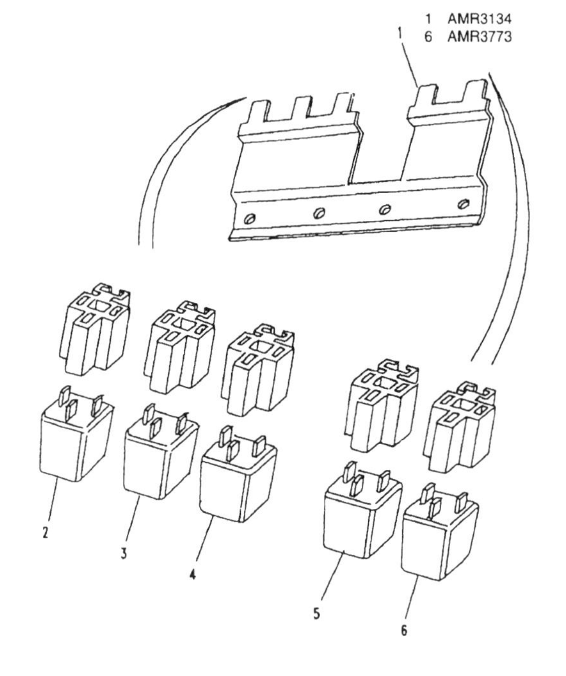

M 87

DISCOVERY 1989MY UPTO 1999MY | GROUP M | Body Electrics / Relays / Relays - Under LH Front Seat

Illus	Part Number	Description	Quantity	Change Point	Remarks
		FROM (V) MA081992			
		ELECTRIC SEAT RELAYS UNDER LH FRONT SEAT			
1	YWB10027L	Relay normally open-electric seat-Yellow	2	From (V) TA 163104	
2	RTC4507	Fuse-electric seat-30 amp-auto-standard	4	From (V) TA 501920	
3	AMR5485	Bracket-relay mounting-electric seat	1		
		HEATED SEAT RELAYS UNDER LH FRONT SEAT-FROM VA			
3	AMR5697	Bracket-relay mounting	1		
4	YWB10027L	Relay normally open-heated seat-Yellow	2		

M 88

Illus	Part Number	Description	Quantity	Change Point	Remarks
		MPI UP TO (V) LA081991			
1	YWB10041L	Relay assembly multi function	1		
	AMR2885	Bracket-relay	1	From (V) KA 038822	
2	AMR3678	Control unit-fan timer-Green	1		
3	DRC1245	Bracket-relay mounting	1		
4	AB608031L	Screw-self tapping AB-No 8 x 3/8	1		

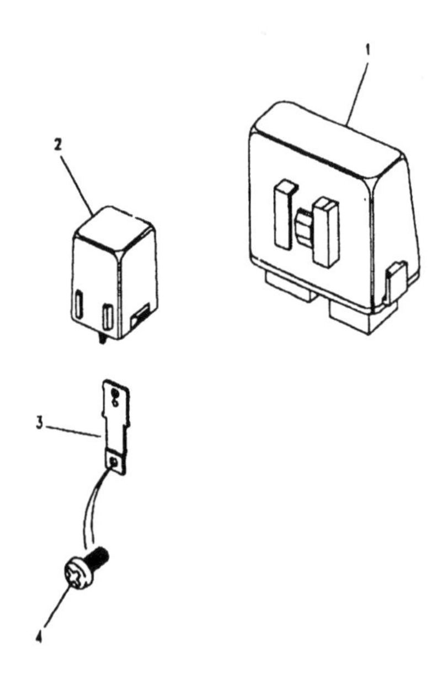

Illus	Part Number	Description	Quantity	Change Point	Remarks
		FROM (V) MA081992			
		DIM DIP SYSTEM			
	XBL10004L	Resistor-dim dip	1		
1	YWB10027L	Relay normally open-dim dip-Yellow	1		
		GLOW PLUG TIMER			
2	PRC6913	Relay-ignition heater plug-12V	1		2500 cc Diesel TDi
	SH910241L	Screw-fixing ignition heater plug-hexagonal head-10UNF x 3/4	1		2500 cc Diesel TDi
	WE105001L	Washer-lock-fixing ignition heater plug	1		2500 cc Diesel TDi
		CATALYST OVERHEAT WARNING SYSTEM			
3	AMR2608	Amplifier-sounder unit catalyst overheat	2		Japan
4	AMR2447	Harness-link-catalyst overheat	1		Japan
5	AMR2463	Thermocouple assembly-catalyst overheat	1		Japan

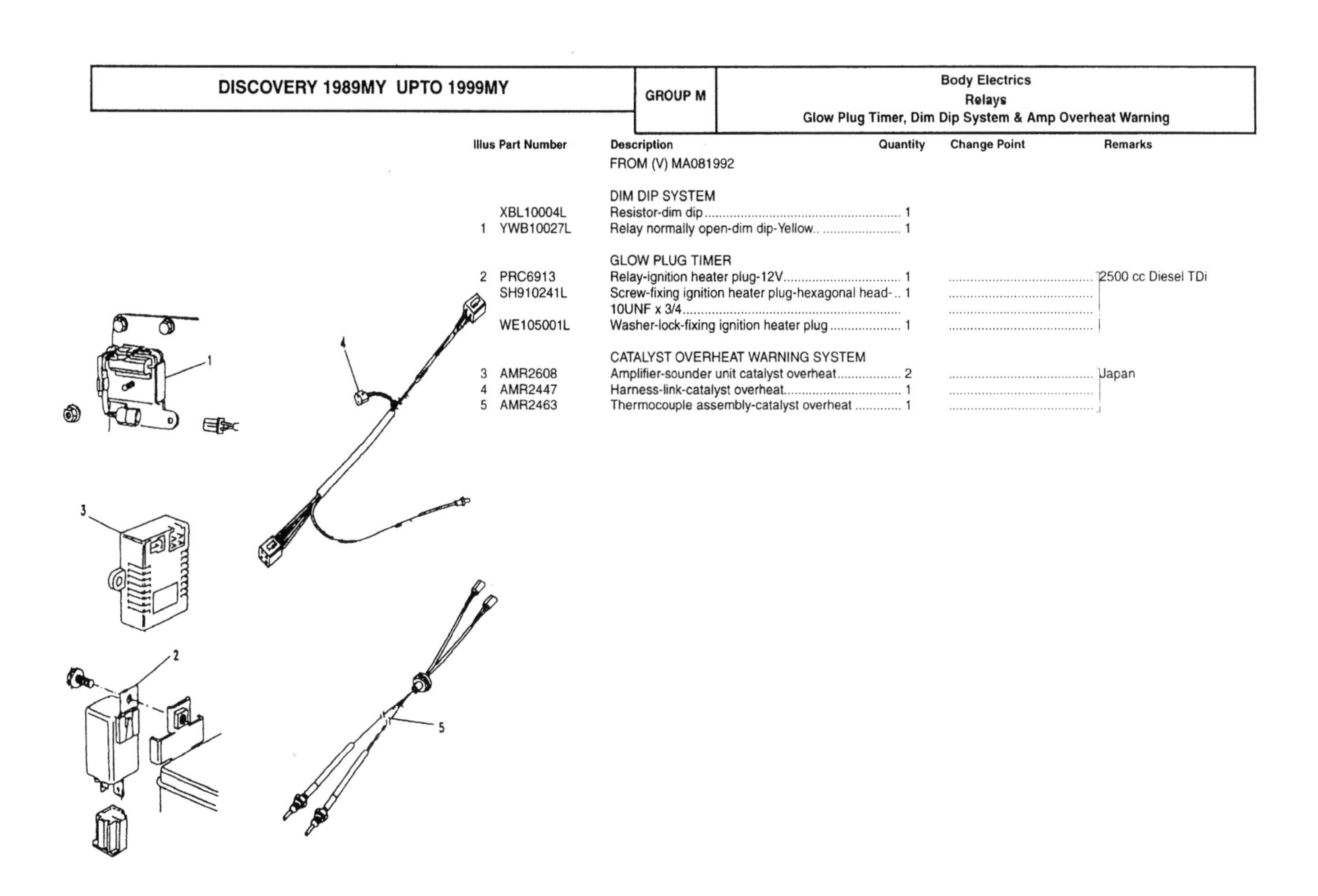

Illus	Part Number	Description	Quantity	Change Point	Remarks
1	PRC9147	Electronic control unit-exhaust gas recirculation diesel	1		
2	PRC9148	Bracket-electronic control unit mounting	1		
3	SH105101	Screw-fixing bracket ECU-hexagonal head	2		
4	WJ105001L	Washer-plain-fixing bracket ECU-M5-oversize	2		
5	WL105001L	Washer-fixing bracket ECU-M5	2		
6	PRC9579	Harness-link-exhaust gas recirculation	1		
7	PRC9146	Valve-solenoid-exhaust gas recirculation diesel	1		
8	AK608141L	Nut-spring-u type-fixing solenoid EGR	2		
9	DA608031L	Screw-fixing solenoid EGR-M8 x 10	2		
10	ERR2381	Bracket-mounting	1		
11	FN108041L	Nut-fixing bracket EGR-flanged head-M8 x 20	NLA		Use FN108047L.
12	SH108161L	Screw-fixing bracket EGR-hexagonal head-M8 x 16	NLA		Use FS108161L.
12	FS108161L	Screw-fixing bracket EGR-flanged head-M8 x 16	3		
13	WC108051L	Washer-plain-fixing bracket EGR-M8	3		

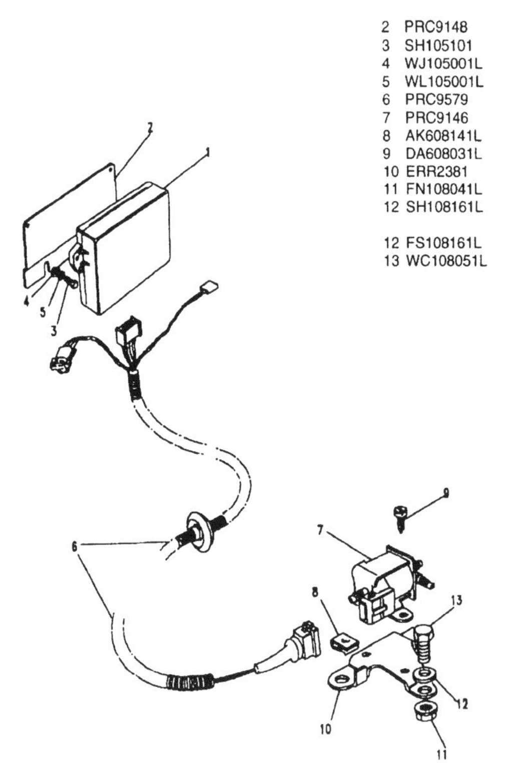

M 91

Illus	Part Number	Description	Quantity	Change Point	Remarks
		FROM (V) MA081992			
		EXHAUST GAS RECIRCULATION (EGR) LOCATED ON RH FOOTWELL A POST PANEL			
1	ERR4196	Electronic control unit-exhaust gas recirculation diesel	1	Note (1)	
1	ERR6234	Electronic control unit-exhaust gas recirculation diesel	1	Note (2)	
		CONTROL UNIT FIXINGS			
		NUT-PLASTIC			
2	NTC7314	push on-M5	NLA		Use CYH10001.
2	CYH10001	push on-M5	2		

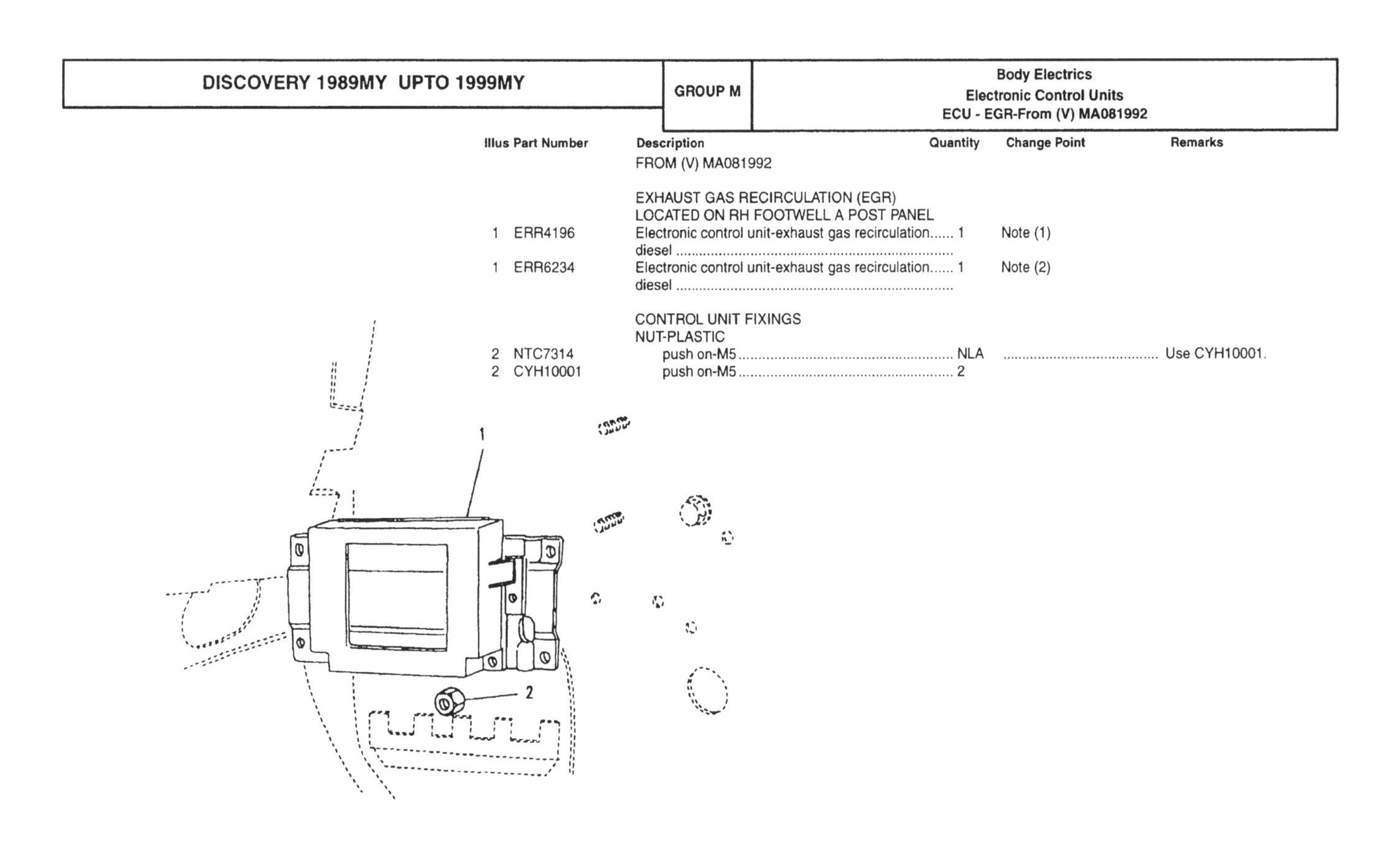

CHANGE POINTS:
(1) From (V) MA 081992 To (V) MA 163103; From (V) MA 081992 To (V) MA 501919
(2) From (V) TA 163104; From (V) TA 501920

M 92

Illus	Part Number	Description	Quantity	Change Point	Remarks
		FROM (V) MA081992			
		ELECTRONIC CONTROL UNIT			
1	AMR1282	ECU-electric windows	1		
2	NH104041	Nut-fixing ECU-M4	3		

NOTE: FOR MOUNTING BRACKET SEE
ABS SYSTEM

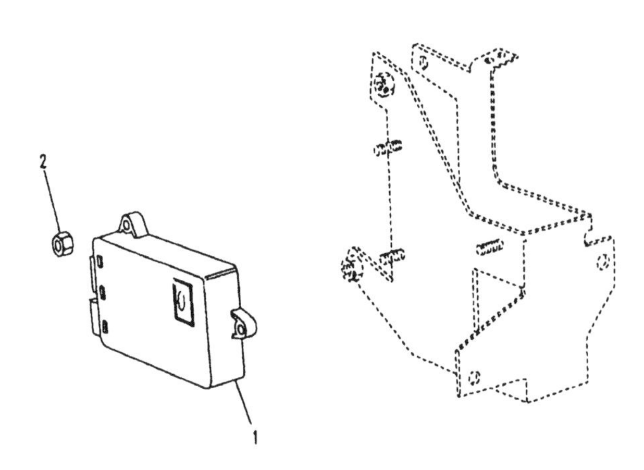

Illus	Part Number	Description	Quantity	Change Point	Remarks
		V8 TO (V) LA081991			
		ELECTRIC CONTROL UNIT-FUEL/IGNITION			
1	PRC9059	new	1	To (V) JA 018541	
1	PRC9612	new	1	From (V) JA 018542	
1	PRC9612E	exchange	1	To (V) JA 034313	
1	PRC9235	new	1	From (V) KA 034314	
1	PRC9235E	exchange	1	To (V) KA 064754	
1	AMR3276	new	1	From (V) LA 064755	
1	AMR3276E	exchange	1		
2	FS105107	Screw-fixing ECU-hexagonal head-M5 x 10	2		
3	WJ105001L	Washer-plain-fixing ECU-M5-oversize	2		
4	WL105001L	Washer-fixing ECU-M5	2		
		NON CATALYST-V8			
5	PRC8617	HARNESS-FUEL INJECTOR ENGINE	1	Note (1)	
5	PRC8889	HARNESS-FUEL INJECTOR ENGINE	1	Note (2)	
5	PRC9663	HARNESS-FUEL INJECTOR ENGINE	1	Note (3)	

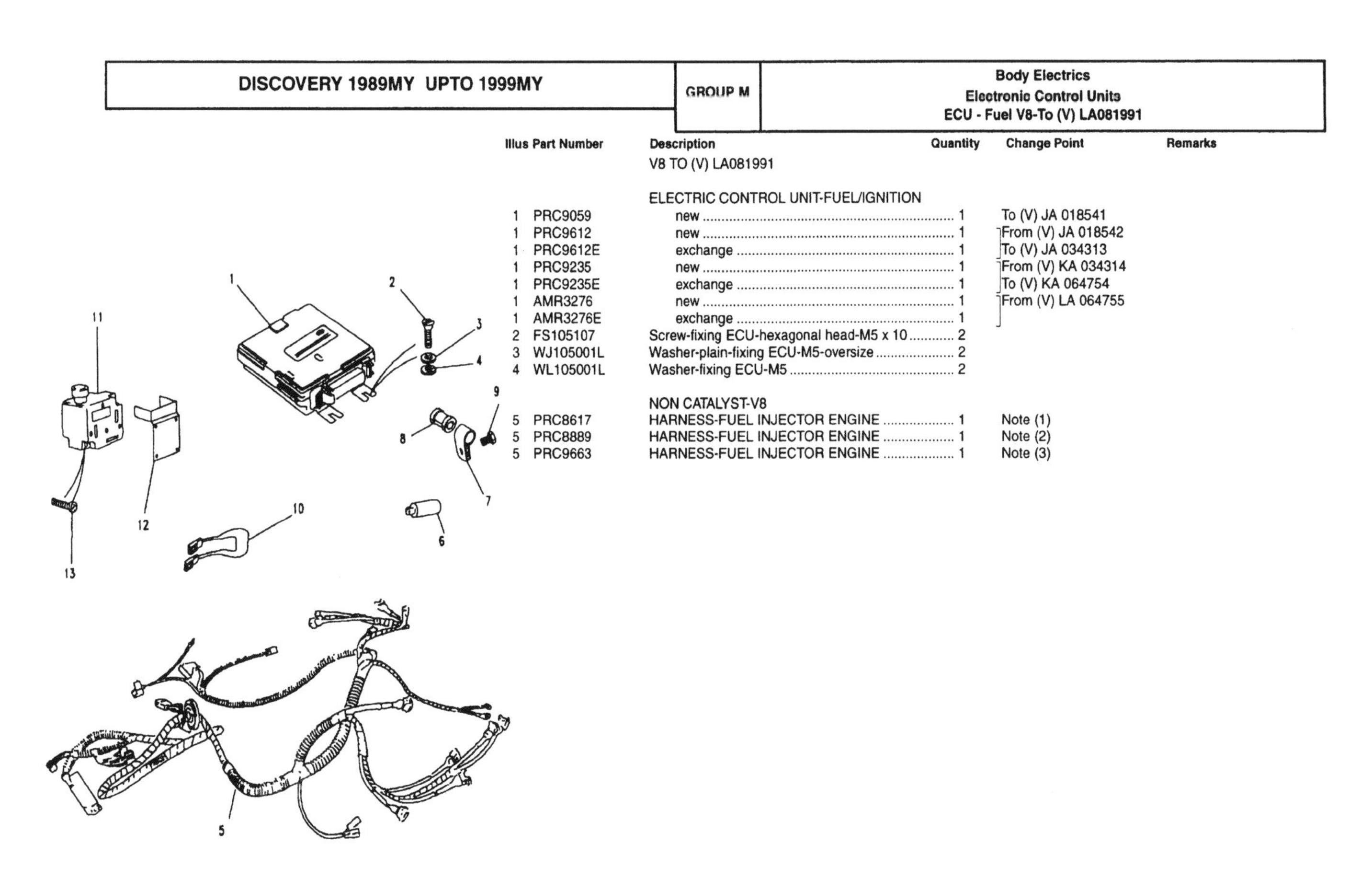

CHANGE POINTS:
(1) From (V) GA 394352 To (V) GA 456980; From (V) HA 456981 To (V) HA 486551; From (V) HA 000001 To (V) HA 008625
(2) From (V) HA 008626 To (V) HA 012332; From (V) JA 012333 To (V) JA 034313
(3) From (V) KA 034314 To (V) LA 081991

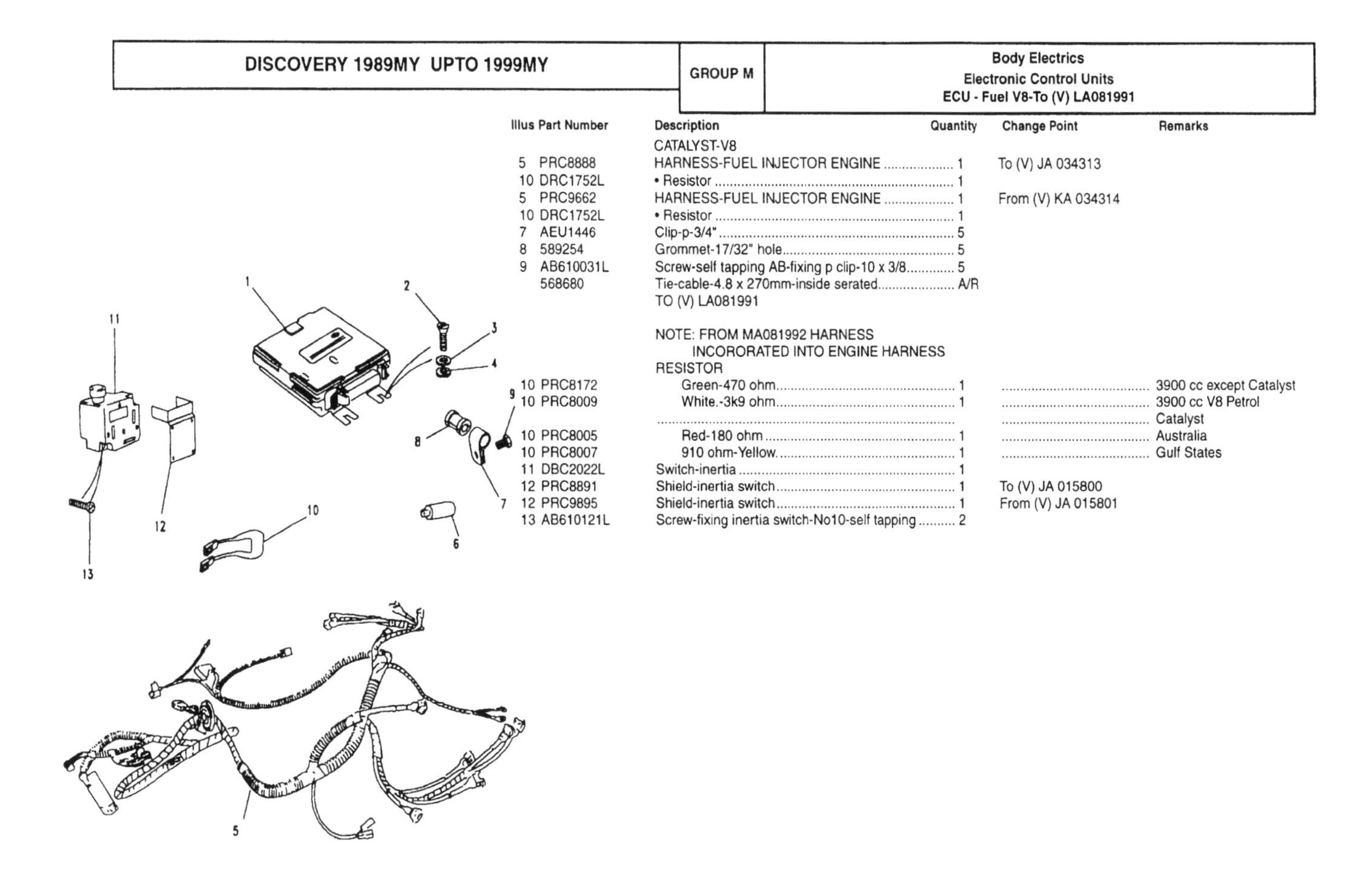

Illus	Part Number	Description	Quantity	Change Point	Remarks
		CATALYST-V8			
5	PRC8888	HARNESS-FUEL INJECTOR ENGINE	1	To (V) JA 034313	
10	DRC1752L	• Resistor	1		
5	PRC9662	HARNESS-FUEL INJECTOR ENGINE	1	From (V) KA 034314	
10	DRC1752L	• Resistor	1		
7	AEU1446	Clip-p-3/4"	5		
8	589254	Grommet-17/32" hole	5		
9	AB610031L	Screw-self tapping AB-fixing p clip-10 x 3/8	5		
	568680	Tie-cable-4.8 x 270mm-inside serated	A/R		
		TO (V) LA081991			
		NOTE: FROM MA081992 HARNESS INCORORATED INTO ENGINE HARNESS			
		RESISTOR			
10	PRC8172	Green-470 ohm	1		3900 cc except Catalyst
10	PRC8009	White.-3k9 ohm	1		3900 cc V8 Petrol Catalyst
10	PRC8005	Red-180 ohm	1		Australia
10	PRC8007	910 ohm-Yellow.	1		Gulf States
11	DBC2022L	Switch-inertia	1		
12	PRC8891	Shield-inertia switch	1	To (V) JA 015800	
12	PRC9895	Shield-inertia switch	1	From (V) JA 015801	
13	AB610121L	Screw-fixing inertia switch-No10-self tapping	2		

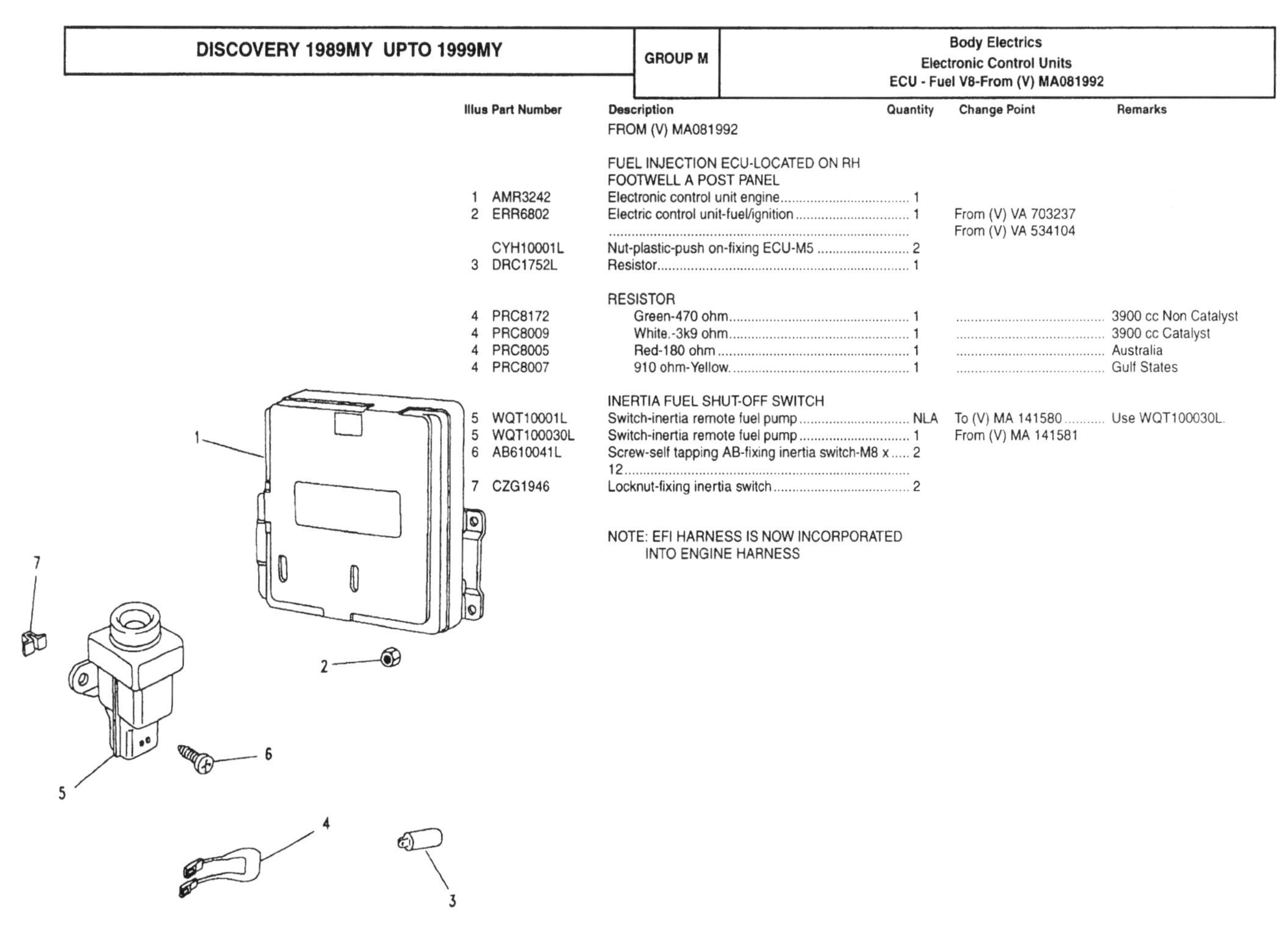

Illus	Part Number	Description	Quantity	Change Point	Remarks
		FROM (V) MA081992			
		FUEL INJECTION ECU-LOCATED ON RH FOOTWELL A POST PANEL			
1	AMR3242	Electronic control unit engine	1		
2	ERR6802	Electric control unit-fuel/ignition	1	From (V) VA 703237 From (V) VA 534104	
	CYH10001L	Nut-plastic-push on-fixing ECU-M5	2		
3	DRC1752L	Resistor	1		
		RESISTOR			
4	PRC8172	Green-470 ohm	1		3900 cc Non Catalyst
4	PRC8009	White.-3k9 ohm	1		3900 cc Catalyst
4	PRC8005	Red-180 ohm	1		Australia
4	PRC8007	910 ohm-Yellow.	1		Gulf States
		INERTIA FUEL SHUT-OFF SWITCH			
5	WQT10001L	Switch-inertia remote fuel pump	NLA	To (V) MA 141580	Use WQT100030L.
5	WQT100030L	Switch-inertia remote fuel pump	1	From (V) MA 141581	
6	AB610041L	Screw-self tapping AB-fixing inertia switch-M8 x 12	2		
7	CZG1946	Locknut-fixing inertia switch	2		

NOTE: EFI HARNESS IS NOW INCORPORATED INTO ENGINE HARNESS

Illus	Part Number	Description	Quantity	Change Point	Remarks
		ELECTRONIC CONTROL UNIT			
1	AMR2128	Control unit sunroof	1	Note (1)	
2	BTR9725	Bracket-electric control unit mounting	1		
3	AB606051	Screw-self tapping-fixing control unit-sunroof	2		
4	BTR475	Nut-spring-u type-fixing control unit-sunroof	2		

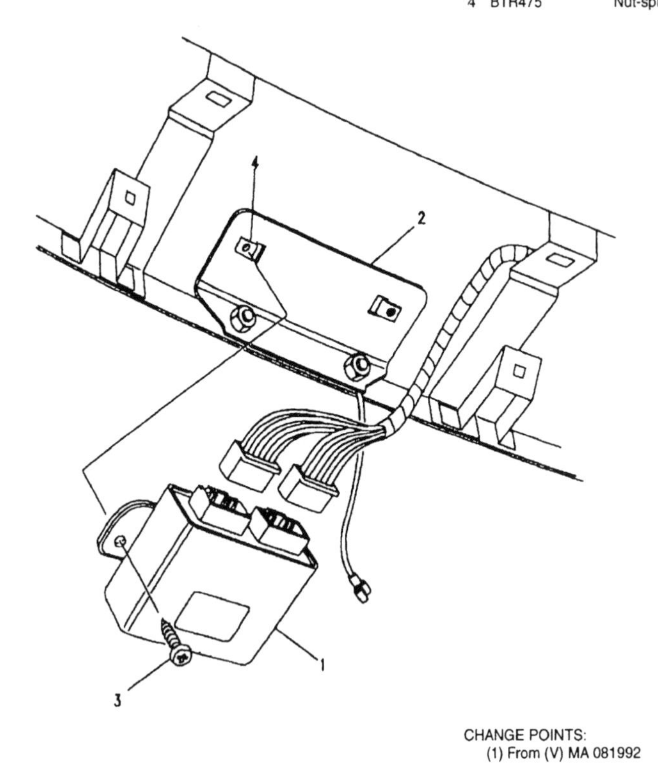

CHANGE POINTS:
(1) From (V) MA 081992

Illus	Part Number	Description	Quantity	Change Point	Remarks
		MPI TO (V) LA081991			
		NOTE: LOCATED ON RH INNER WING			
1	MKC10121	Electric control unit-fuel/ignition	1	To (V) KA 038821	
1	MKC101720	Electric control unit-fuel/ignition	1	From (V) KA 038822	
2	SH105201L	Screw-fixing ECU-hexagonal head-M5 x 20	3	To (V) KA 038821	
2	SH105201L	Screw-fixing ECU-hexagonal head-M5 x 20	3	From (V) KA 038822	LHD Catalyst
2	SH105201L	Screw-hexagonal head-M5 x 20	3	From (V) KA 038822	Non Catalyst
2	SH105121L	Screw-fixing ECU-hexagonal head-M5 x 12	3	From (V) KA 038822	RHD Catalyst
3	WA105001L	Washer-plain-fixing ECU-M5-standard	3		
4	WL105001L	Washer-fixing ECU-M5	3		
		BUSH			
5	390939	fixing ECU	3	To (V) KA 038821	
5	390939	fixing ECU	3	From (V) KA 038822	LHD Catalyst
5	390939	fixing ECU	3	From (V) KA 038822	Non Catalyst
6	MXC7420	Nutsert-blind-fixing ECU	3		
7	ERR2950	Pipe-inlet manifold vacuum	1	To (V) KA 038821	
7	ERR4641	Pipe-inlet manifold vacuum	NLA	From (V) KA 038822	Use AFU3048 qty 1 to vacuum pipe with ESR3428 qty 1.

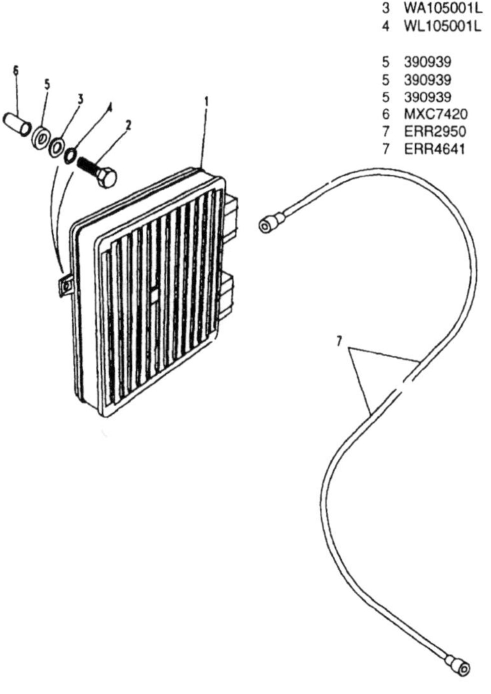

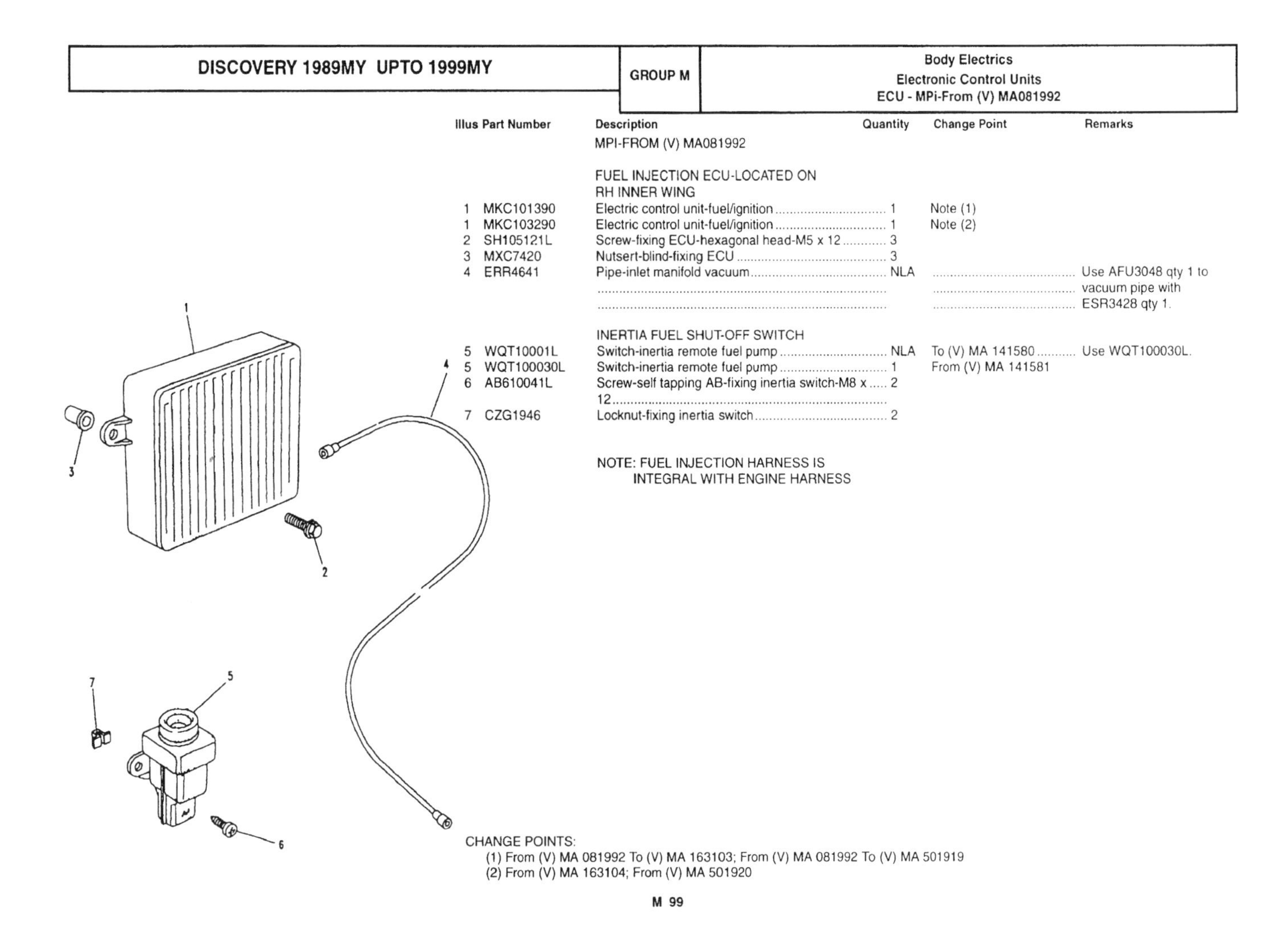

Illus	Part Number	Description	Quantity	Change Point	Remarks
		MPI-FROM (V) MA081992			
		FUEL INJECTION ECU-LOCATED ON RH INNER WING			
1	MKC101390	Electric control unit-fuel/ignition	1	Note (1)	
1	MKC103290	Electric control unit-fuel/ignition	1	Note (2)	
2	SH105121L	Screw-fixing ECU-hexagonal head-M5 x 12	3		
3	MXC7420	Nutsert-blind-fixing ECU	3		
4	ERR4641	Pipe-inlet manifold vacuum	NLA		Use AFU3048 qty 1 to vacuum pipe with ESR3428 qty 1.
		INERTIA FUEL SHUT-OFF SWITCH			
5	WQT10001L	Switch-inertia remote fuel pump	NLA	To (V) MA 141580	Use WQT100030L.
5	WQT100030L	Switch-inertia remote fuel pump	1	From (V) MA 141581	
6	AB610041L	Screw-self tapping AB-fixing inertia switch-M8 x 12	2		
7	CZG1946	Locknut-fixing inertia switch	2		

NOTE: FUEL INJECTION HARNESS IS INTEGRAL WITH ENGINE HARNESS

CHANGE POINTS:
(1) From (V) MA 081992 To (V) MA 163103; From (V) MA 081992 To (V) MA 501919
(2) From (V) MA 163104; From (V) MA 501920

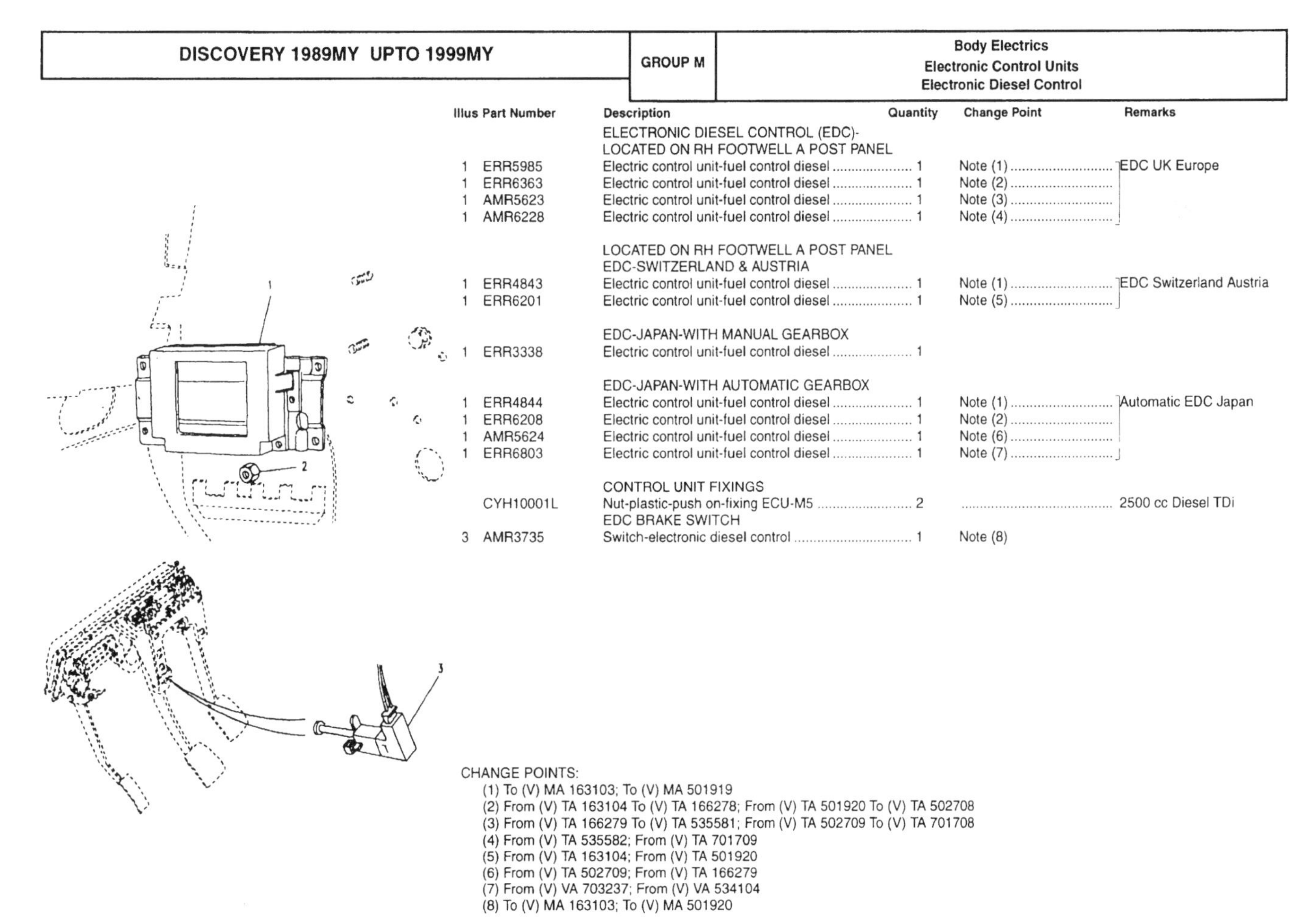

Illus	Part Number	Description	Quantity	Change Point	Remarks
		ELECTRONIC DIESEL CONTROL (EDC)-LOCATED ON RH FOOTWELL A POST PANEL			
1	ERR5985	Electric control unit-fuel control diesel	1	Note (1)	EDC UK Europe
1	ERR6363	Electric control unit-fuel control diesel	1	Note (2)	
1	AMR5623	Electric control unit-fuel control diesel	1	Note (3)	
1	AMR6228	Electric control unit-fuel control diesel	1	Note (4)	
		LOCATED ON RH FOOTWELL A POST PANEL EDC-SWITZERLAND & AUSTRIA			
1	ERR4843	Electric control unit-fuel control diesel	1	Note (1)	EDC Switzerland Austria
1	ERR6201	Electric control unit-fuel control diesel	1	Note (5)	
		EDC-JAPAN-WITH MANUAL GEARBOX			
1	ERR3338	Electric control unit-fuel control diesel	1		
		EDC-JAPAN-WITH AUTOMATIC GEARBOX			
1	ERR4844	Electric control unit-fuel control diesel	1	Note (1)	Automatic EDC Japan
1	ERR6208	Electric control unit-fuel control diesel	1	Note (2)	
1	AMR5624	Electric control unit-fuel control diesel	1	Note (6)	
1	ERR6803	Electric control unit-fuel control diesel	1	Note (7)	
		CONTROL UNIT FIXINGS			
	CYH10001L	Nut-plastic-push on-fixing ECU-M5	2		2500 cc Diesel TDi
		EDC BRAKE SWITCH			
3	AMR3735	Switch-electronic diesel control	1	Note (8)	

CHANGE POINTS:
(1) To (V) MA 163103; To (V) MA 501919
(2) From (V) TA 163104 To (V) TA 166278; From (V) TA 501920 To (V) TA 502708
(3) From (V) TA 166279 To (V) TA 535581; From (V) TA 502709 To (V) TA 701708
(4) From (V) TA 535582; From (V) TA 701709
(5) From (V) TA 163104; From (V) TA 501920
(6) From (V) TA 502709; From (V) TA 166279
(7) From (V) VA 703237; From (V) VA 534104
(8) To (V) MA 163103; To (V) MA 501920

Illus	Part Number	Description	Quantity	Change Point	Remarks
1	YEB10026	Horn assembly-high note	1	To (V) LA 081991	
	YEB10027	Horn low note	1		
1	AMR3466	Horn assembly-high note	1	From (V) MA 081992	
	AMR3467	Horn assembly-low note	1		

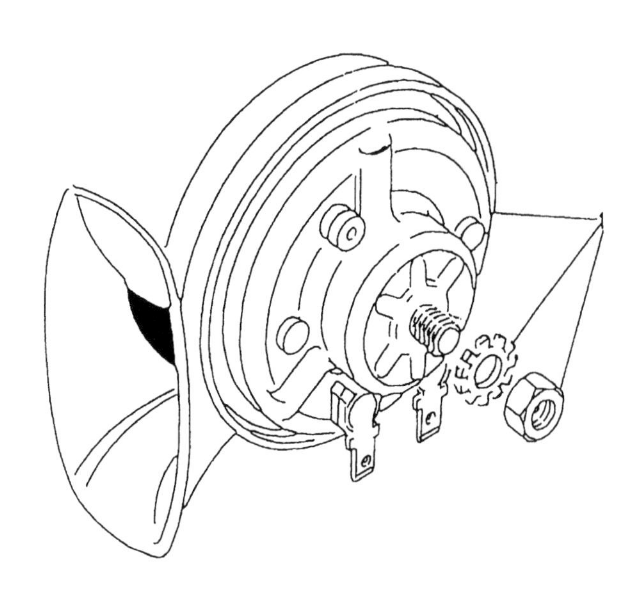

M 101

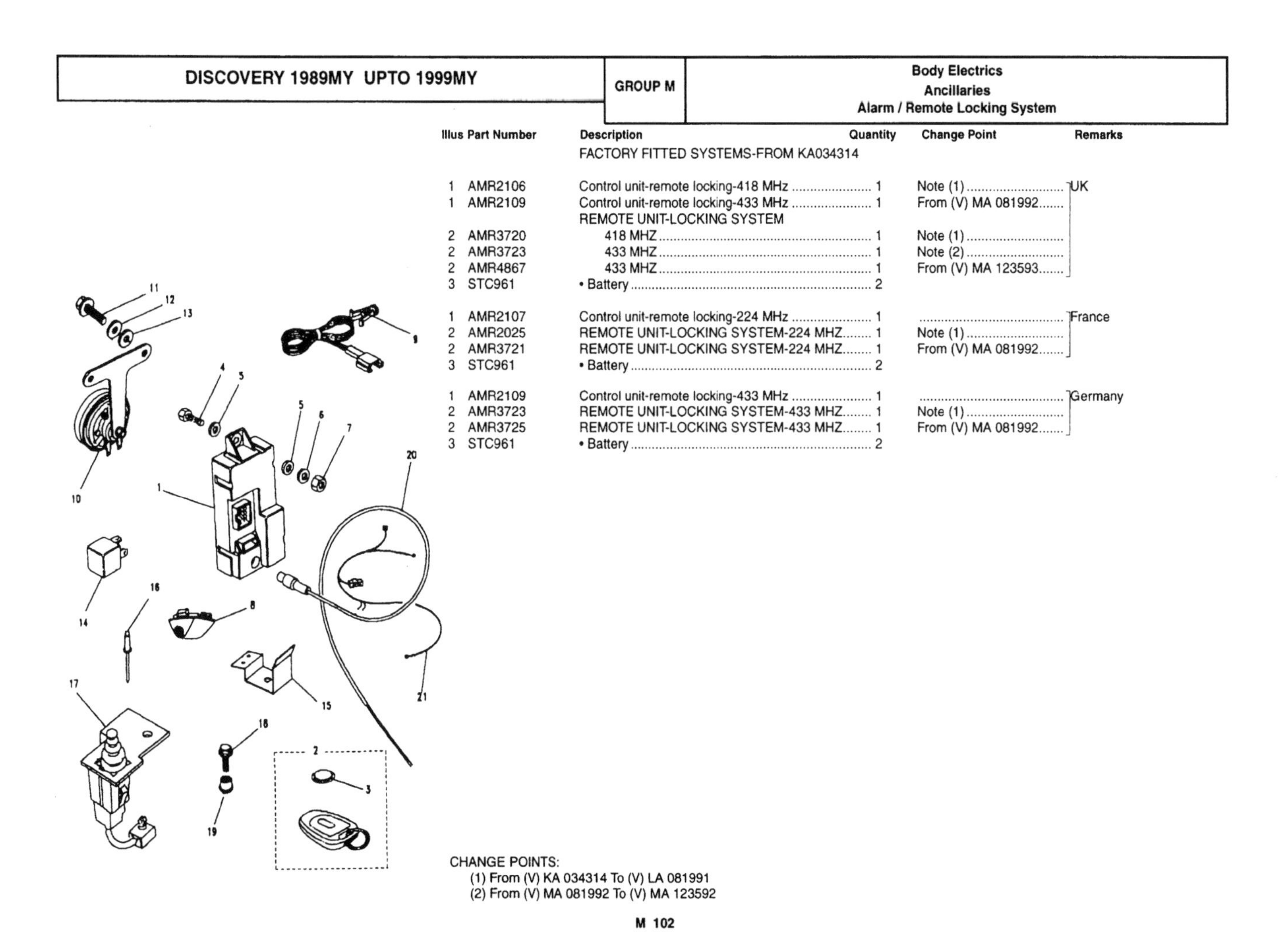

Illus	Part Number	Description	Quantity	Change Point	Remarks
		FACTORY FITTED SYSTEMS-FROM KA034314			
1	AMR2106	Control unit-remote locking-418 MHz	1	Note (1)	UK
1	AMR2109	Control unit-remote locking-433 MHz	1	From (V) MA 081992	
		REMOTE UNIT-LOCKING SYSTEM			
2	AMR3720	418 MHZ	1	Note (1)	
2	AMR3723	433 MHZ	1	Note (2)	
2	AMR4867	433 MHZ	1	From (V) MA 123593	
3	STC961	• Battery	2		
1	AMR2107	Control unit-remote locking-224 MHz	1		France
2	AMR2025	REMOTE UNIT-LOCKING SYSTEM-224 MHZ	1	Note (1)	
2	AMR3721	REMOTE UNIT-LOCKING SYSTEM-224 MHZ	1	From (V) MA 081992	
3	STC961	• Battery	2		
1	AMR2109	Control unit-remote locking-433 MHz	1		Germany
2	AMR3723	REMOTE UNIT-LOCKING SYSTEM-433 MHZ	1	Note (1)	
2	AMR3725	REMOTE UNIT-LOCKING SYSTEM-433 MHZ	1	From (V) MA 081992	
3	STC961	• Battery	2		

CHANGE POINTS:
(1) From (V) KA 034314 To (V) LA 081991
(2) From (V) MA 081992 To (V) MA 123592

M 102

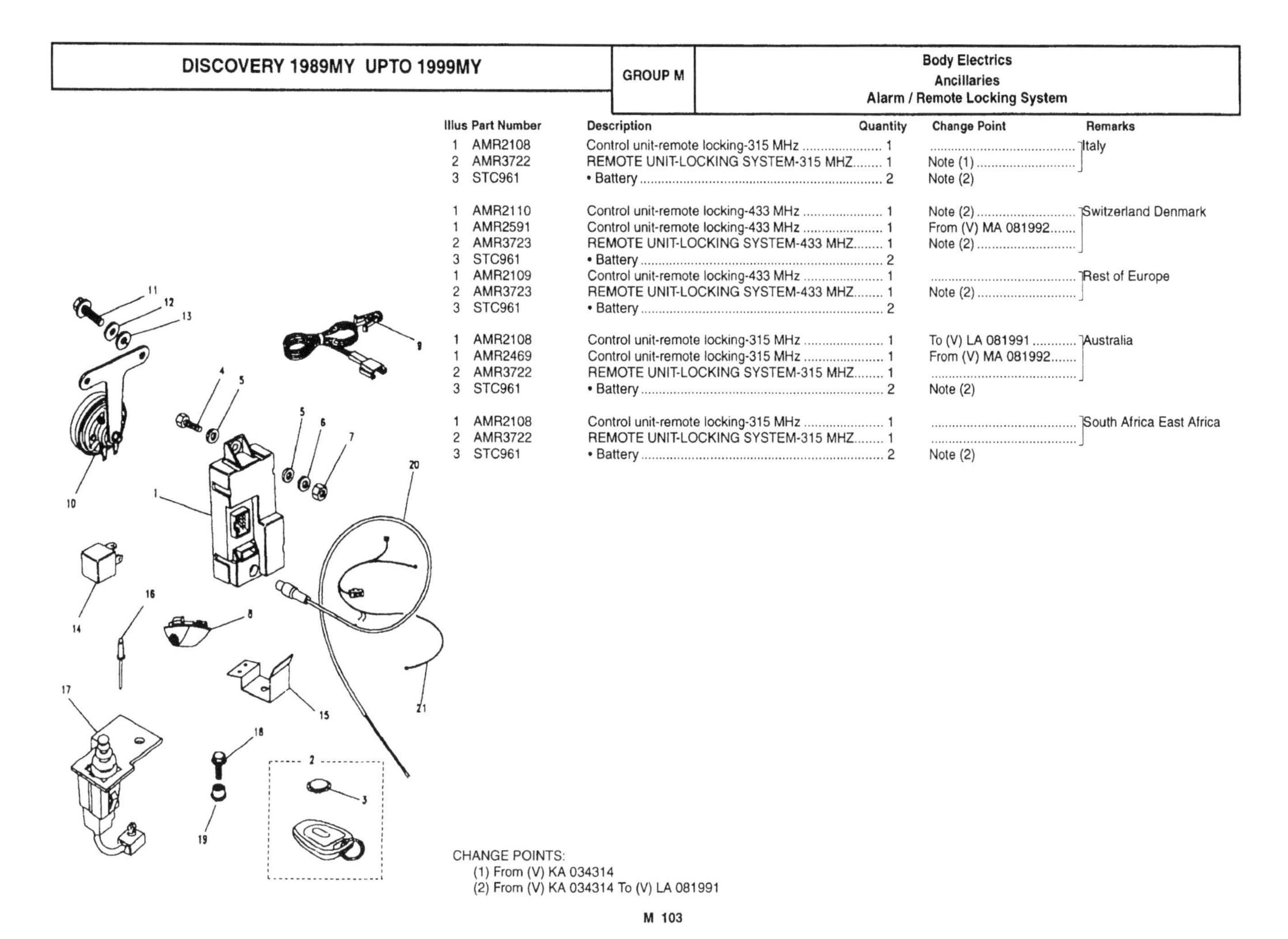

Illus	Part Number	Description	Quantity	Change Point	Remarks
1	AMR2108	Control unit-remote locking-315 MHz	1		Italy
2	AMR3722	REMOTE UNIT-LOCKING SYSTEM-315 MHZ	1	Note (1)	
3	STC961	• Battery	2	Note (2)	
1	AMR2110	Control unit-remote locking-433 MHz	1	Note (2)	Switzerland Denmark
1	AMR2591	Control unit-remote locking-433 MHz	1	From (V) MA 081992	
2	AMR3723	REMOTE UNIT-LOCKING SYSTEM-433 MHZ	1	Note (2)	
3	STC961	• Battery	2		
1	AMR2109	Control unit-remote locking-433 MHz	1		Rest of Europe
2	AMR3723	REMOTE UNIT-LOCKING SYSTEM-433 MHZ	1	Note (2)	
3	STC961	• Battery	2		
1	AMR2108	Control unit-remote locking-315 MHz	1	To (V) LA 081991	Australia
1	AMR2469	Control unit-remote locking-315 MHz	1	From (V) MA 081992	
2	AMR3722	REMOTE UNIT-LOCKING SYSTEM-315 MHZ	1		
3	STC961	• Battery	2	Note (2)	
1	AMR2108	Control unit-remote locking-315 MHz	1		South Africa East Africa
2	AMR3722	REMOTE UNIT-LOCKING SYSTEM-315 MHZ	1		
3	STC961	• Battery	2	Note (2)	

CHANGE POINTS:
(1) From (V) KA 034314
(2) From (V) KA 034314 To (V) LA 081991

M 103

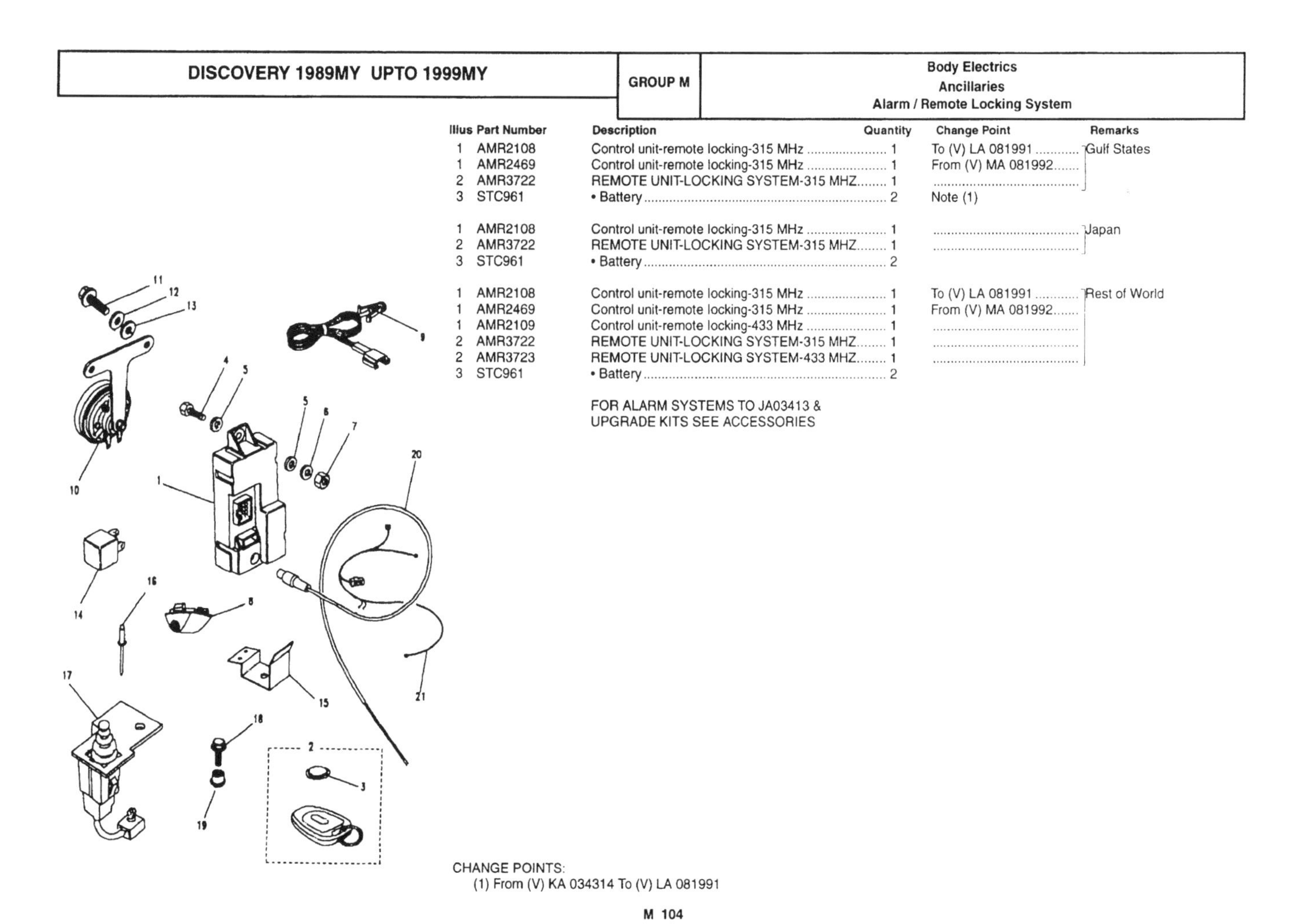

Illus	Part Number	Description	Quantity	Change Point	Remarks
1	AMR2108	Control unit-remote locking-315 MHz	1	To (V) LA 081991	Gulf States
1	AMR2469	Control unit-remote locking-315 MHz	1	From (V) MA 081992	
2	AMR3722	REMOTE UNIT-LOCKING SYSTEM-315 MHZ	1		
3	STC961	• Battery	2	Note (1)	
1	AMR2108	Control unit-remote locking-315 MHz	1		Japan
2	AMR3722	REMOTE UNIT-LOCKING SYSTEM-315 MHZ	1		
3	STC961	• Battery	2		
1	AMR2108	Control unit-remote locking-315 MHz	1	To (V) LA 081991	Rest of World
1	AMR2469	Control unit-remote locking-315 MHz	1	From (V) MA 081992	
1	AMR2109	Control unit-remote locking-433 MHz	1		
2	AMR3722	REMOTE UNIT-LOCKING SYSTEM-315 MHZ	1		
2	AMR3723	REMOTE UNIT-LOCKING SYSTEM-433 MHZ	1		
3	STC961	• Battery	2		

FOR ALARM SYSTEMS TO JA03413 &
UPGRADE KITS SEE ACCESSORIES

CHANGE POINTS:
(1) From (V) KA 034314 To (V) LA 081991

M 104

Illus	Part Number	Description	Quantity	Change Point	Remarks
		FACTORY FITTED SYSTEMS FROM KA034314			
		ALL VARIANTS			
4	SH105201L	Screw-fixing control unit-hexagonal head-M5 x 20	2	To (V) LA 081991	
5	WA105001L	Washer-plain-fixing control unit-M5-standard	2		
6	WL105001L	Washer-fixing control unit-M5	2		
7	NH105041L	Nut-hexagonal-fixing control unit-M5	2		

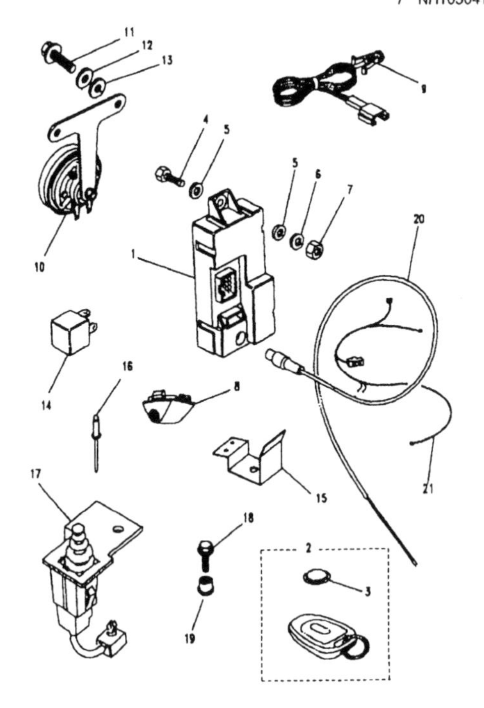

M 105

DISCOVERY 1989MY UPTO 1999MY | GROUP M | Body Electrics
Ancillaries
Alarm / Remote Locking System

Illus	Part Number	Description	Quantity	Change Point	Remarks
		ALARM EQUIPMENT			
		ALARM SYSTEM			
8	AMR2417LUP	Sensor ultrasonic-Aspen Grey	1	To (V) KA 051050	
8	YWC10227LUP	Sensor ultrasonic-Aspen Grey	1	Note (1)	
8	YWC102900LUP	Sensor ultrasonic-Aspen Grey	1	Note (2)	
8	YWC102890LUM	Sensor ultrasonic-Mink Grey	1	Note (3)	
8	YWC103640LUM	Sensor ultrasonic-Mink Grey	1	From (V) MA 097317	
9	AMR1891	Indicator-light emitting diode burglar alarm	1		
10	AMR1427	Horn assembly	1		
11	FS108201L	Screw-fixing horn-flanged head-M8 x 20	NLA	From (V) KA 034314	Use FS108207L.
12	WA108051L	Washer-plain-fixing horn-M8	4		
13	WL108001L	Washer-fixing horn-M8	2		
14	AMR1088	Relay-twin make-Brown	1		
14	YWB10027L	Relay normally open-Yellow.	1		
15	PRC9868	Bracket-alarm switch mounting	1	To (V) LA 081991	
15	AMR3490	Bracket-alarm switch mounting	1	From (V) MA 081992	
16	RA608123L	Rivet-fixing bracket	2		
17	YUE10015L	Switch-contact-bonnet burglar alarm	1		
17	AMR2022	Switch-contact-bonnet burglar alarm	1		
18	FS106167L	Screw-fixing switch bonnet-hexagonal head-M6 x 16	1		
18	FS106167L	Screw-fixing switch bonnet-hexagonal head-M6 x 16	2		
19	NN106021	Nutsert-blind-fixing switch bonnet-M6	1		
20	AMR2034	Aerial-receiver central door locking & alarm	1		
21	AMR2214	Harness-link-with remote locking	1	To (V) LA 081991	Japan Gulf States
21	PRC9527	Harness-link-burglar alarm-RHD	1		RHD
21	PRC9572	Harness-link-burglar alarm-LHD	1		LHD

FOR ALARM SYSTEMS TO JA034313 &
UPGRADE KITS SEE ACCESSORIES

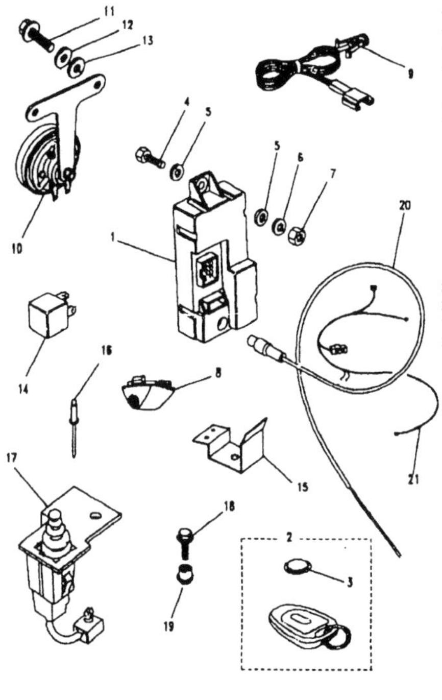

CHANGE POINTS:
(1) From (V) KA 051051 To (V) LA 072552
(2) From (V) LA 072553 To (V) LA 081991
(3) From (V) MA 081992 To (V) MA 097316

M 106

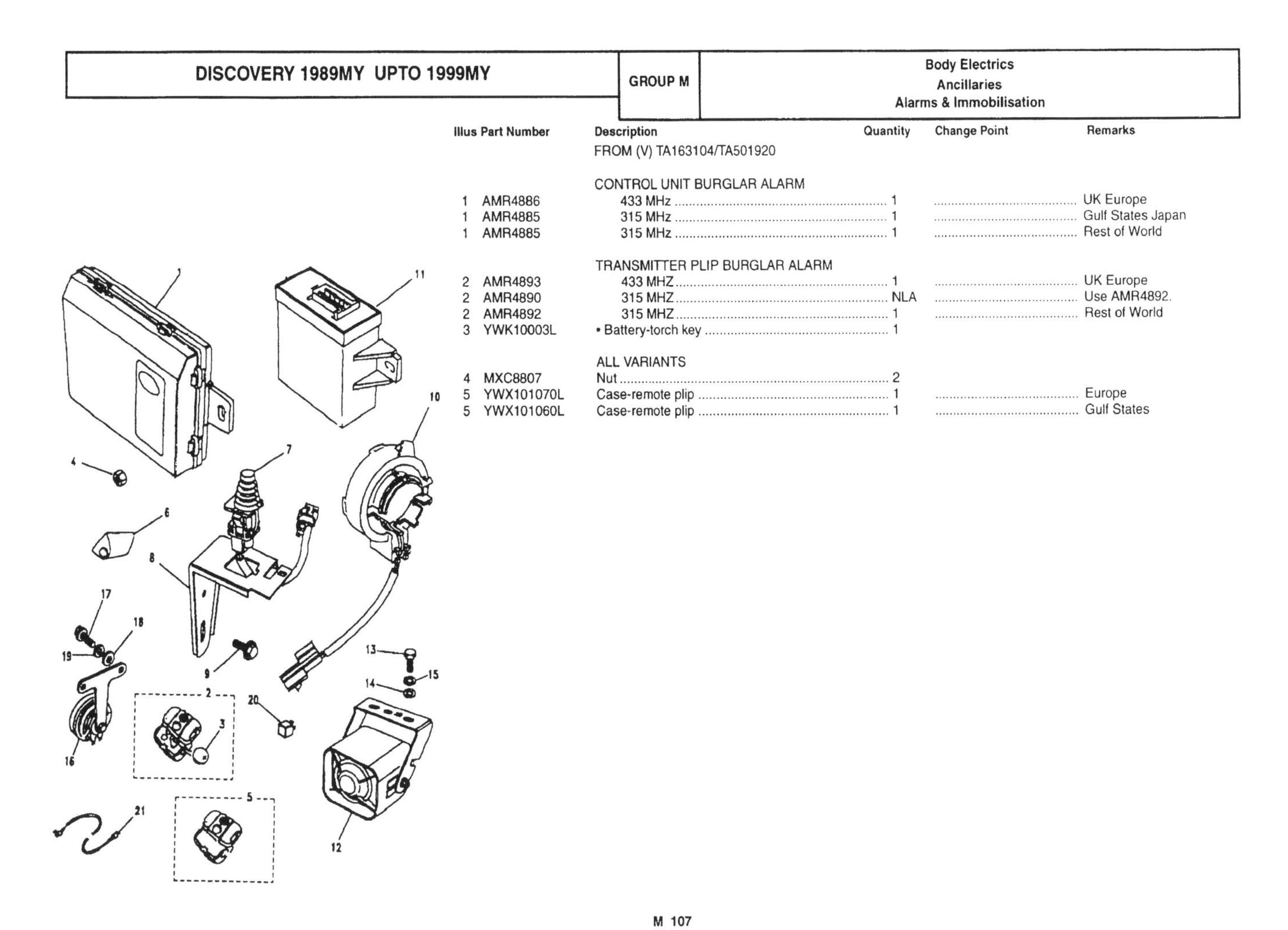

Illus	Part Number	Description	Quantity	Change Point	Remarks
		FROM (V) TA163104/TA501920			
		CONTROL UNIT BURGLAR ALARM			
1	AMR4886	433 MHz	1		UK Europe
1	AMR4885	315 MHz	1		Gulf States Japan
1	AMR4885	315 MHz	1		Rest of World
		TRANSMITTER PLIP BURGLAR ALARM			
2	AMR4893	433 MHZ	1		UK Europe
2	AMR4890	315 MHZ	NLA		Use AMR4892.
2	AMR4892	315 MHZ	1		Rest of World
3	YWK10003L	• Battery-torch key	1		
		ALL VARIANTS			
4	MXC8807	Nut	2		
5	YWX101070L	Case-remote plip	1		Europe
5	YWX101060L	Case-remote plip	1		Gulf States

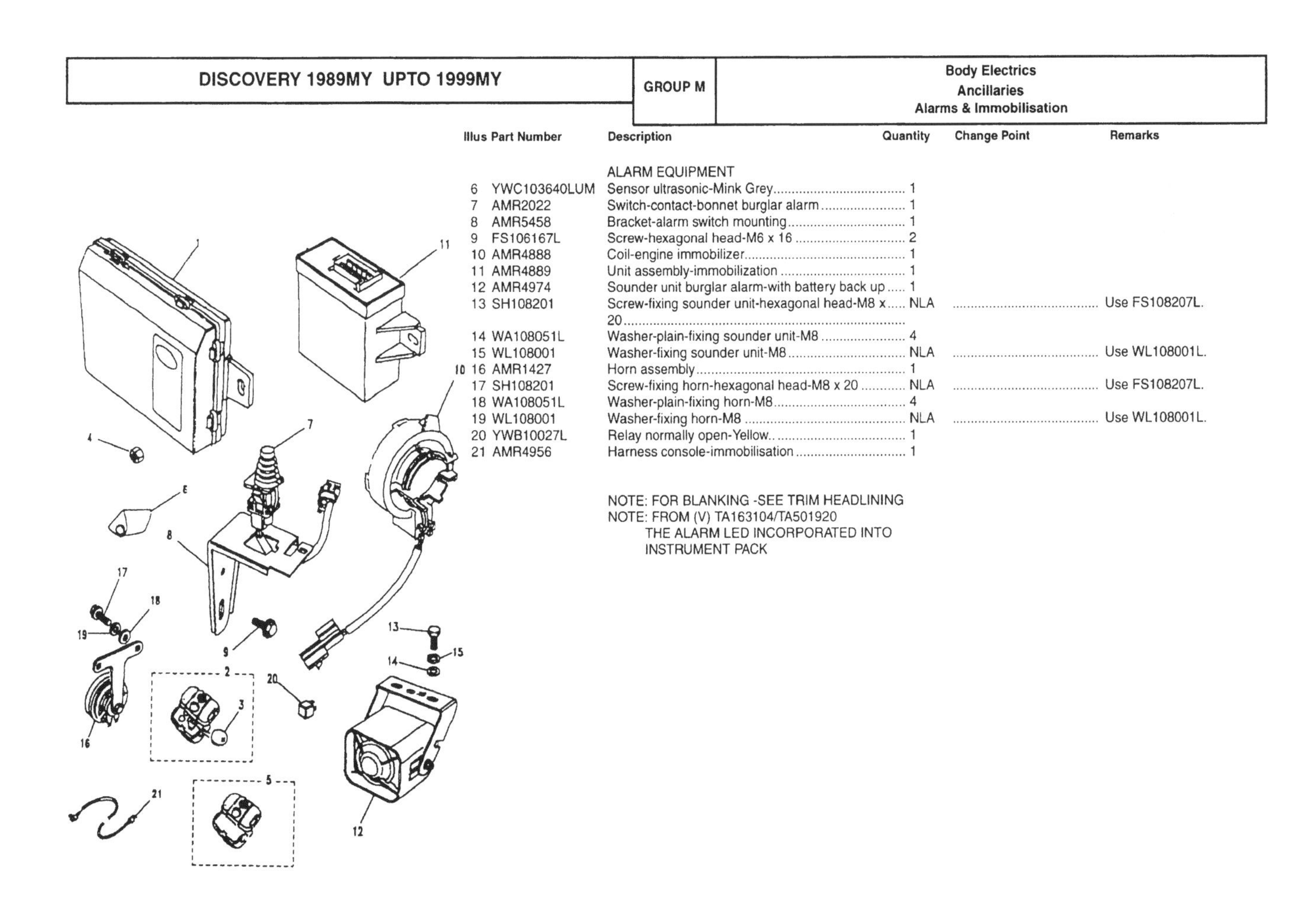

Illus	Part Number	Description	Quantity	Change Point	Remarks
		ALARM EQUIPMENT			
6	YWC103640LUM	Sensor ultrasonic-Mink Grey	1		
7	AMR2022	Switch-contact-bonnet burglar alarm	1		
8	AMR5458	Bracket-alarm switch mounting	1		
9	FS106167L	Screw-hexagonal head-M6 x 16	2		
10	AMR4888	Coil-engine immobilizer	1		
11	AMR4889	Unit assembly-immobilization	1		
12	AMR4974	Sounder unit burglar alarm-with battery back up	1		
13	SH108201	Screw-fixing sounder unit-hexagonal head-M8 x 20	NLA		Use FS108207L.
14	WA108051L	Washer-plain-fixing sounder unit-M8	4		
15	WL108001	Washer-fixing sounder unit-M8	NLA		Use WL108001L.
16	AMR1427	Horn assembly	1		
17	SH108201	Screw-fixing horn-hexagonal head-M8 x 20	NLA		Use FS108207L.
18	WA108051L	Washer-plain-fixing horn-M8	4		
19	WL108001	Washer-fixing horn-M8	NLA		Use WL108001L.
20	YWB10027L	Relay normally open-Yellow.	1		
21	AMR4956	Harness console-immobilisation	1		

NOTE: FOR BLANKING -SEE TRIM HEADLINING
NOTE: FROM (V) TA163104/TA501920
THE ALARM LED INCORPORATED INTO
INSTRUMENT PACK

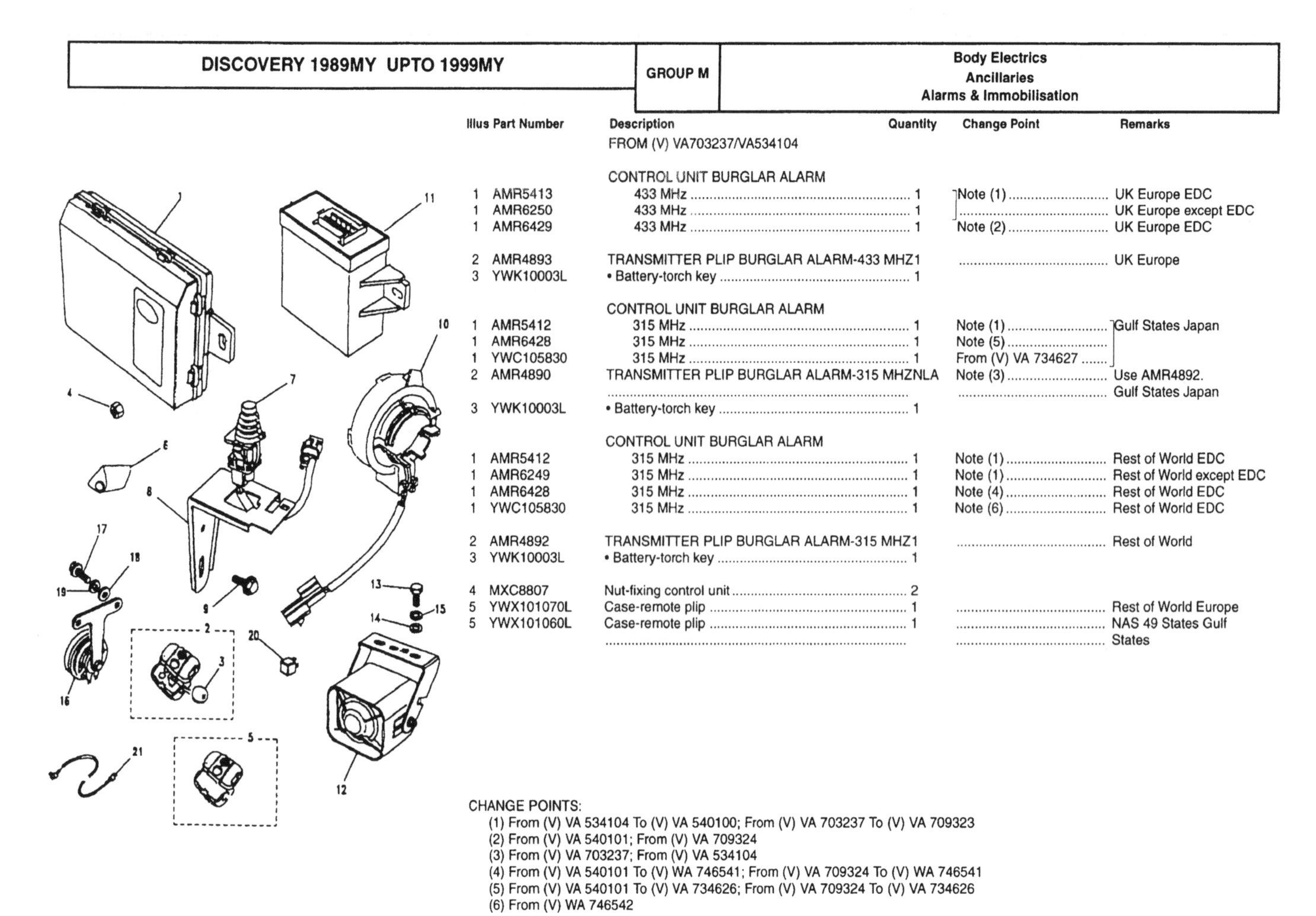

Illus	Part Number	Description	Quantity	Change Point	Remarks
		FROM (V) VA703237/VA534104			
		CONTROL UNIT BURGLAR ALARM			
1	AMR5413	433 MHz	1	Note (1)	UK Europe EDC
1	AMR6250	433 MHz	1		UK Europe except EDC
1	AMR6429	433 MHz	1	Note (2)	UK Europe EDC
2	AMR4893	TRANSMITTER PLIP BURGLAR ALARM-433 MHZ	1		UK Europe
3	YWK10003L	• Battery-torch key	1		
		CONTROL UNIT BURGLAR ALARM			
1	AMR5412	315 MHz	1	Note (1)	Gulf States Japan
1	AMR6428	315 MHz	1	Note (5)	
1	YWC105830	315 MHz	1	From (V) VA 734627	
2	AMR4890	TRANSMITTER PLIP BURGLAR ALARM-315 MHZ	NLA	Note (3)	Use AMR4892. Gulf States Japan
3	YWK10003L	• Battery-torch key	1		
		CONTROL UNIT BURGLAR ALARM			
1	AMR5412	315 MHz	1	Note (1)	Rest of World EDC
1	AMR6249	315 MHz	1	Note (1)	Rest of World except EDC
1	AMR6428	315 MHz	1	Note (4)	Rest of World EDC
1	YWC105830	315 MHz	1	Note (6)	Rest of World EDC
2	AMR4892	TRANSMITTER PLIP BURGLAR ALARM-315 MHZ	1		Rest of World
3	YWK10003L	• Battery-torch key	1		
4	MXC8807	Nut-fixing control unit	2		
5	YWX101070L	Case-remote plip	1		Rest of World Europe
5	YWX101060L	Case-remote plip	1		NAS 49 States Gulf States

CHANGE POINTS:
(1) From (V) VA 534104 To (V) VA 540100; From (V) VA 703237 To (V) VA 709323
(2) From (V) VA 540101; From (V) VA 709324
(3) From (V) VA 703237; From (V) VA 534104
(4) From (V) VA 540101 To (V) WA 746541; From (V) VA 709324 To (V) WA 746541
(5) From (V) VA 540101 To (V) VA 734626; From (V) VA 709324 To (V) VA 734626
(6) From (V) WA 746542

M 109

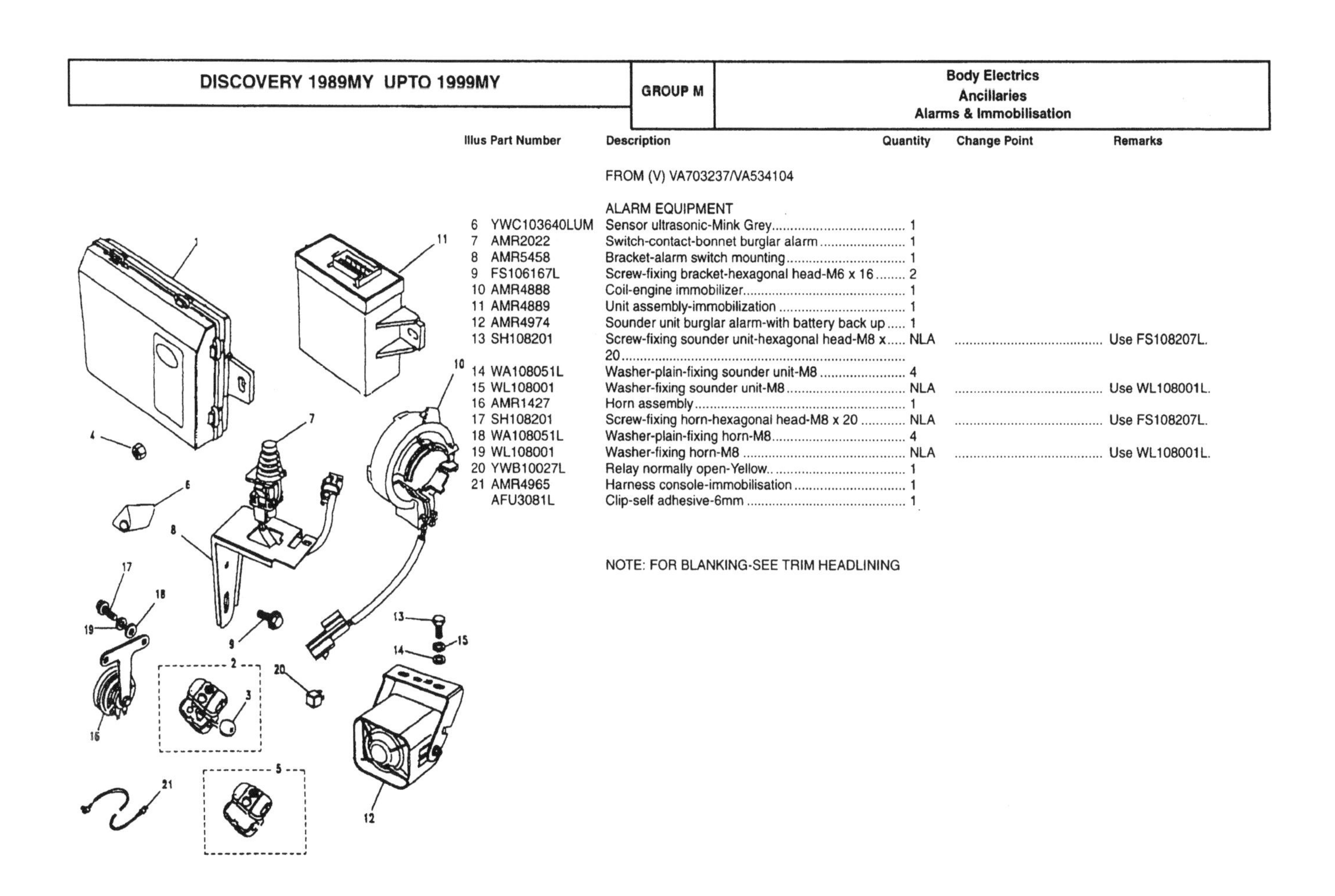

Illus	Part Number	Description	Quantity	Change Point	Remarks
		FROM (V) VA703237/VA534104			
		ALARM EQUIPMENT			
6	YWC103640LUM	Sensor ultrasonic-Mink Grey	1		
7	AMR2022	Switch-contact-bonnet burglar alarm	1		
8	AMR5458	Bracket-alarm switch mounting	1		
9	FS106167L	Screw-fixing bracket-hexagonal head-M6 x 16	2		
10	AMR4888	Coil-engine immobilizer	1		
11	AMR4889	Unit assembly-immobilization	1		
12	AMR4974	Sounder unit burglar alarm-with battery back up	1		
13	SH108201	Screw-fixing sounder unit-hexagonal head-M8 x 20	NLA		Use FS108207L.
14	WA108051L	Washer-plain-fixing sounder unit-M8	4		
15	WL108001	Washer-fixing sounder unit-M8	NLA		Use WL108001L.
16	AMR1427	Horn assembly	1		
17	SH108201	Screw-fixing horn-hexagonal head-M8 x 20	NLA		Use FS108207L.
18	WA108051L	Washer-plain-fixing horn-M8	4		
19	WL108001	Washer-fixing horn-M8	NLA		Use WL108001L.
20	YWB10027L	Relay normally open-Yellow	1		
21	AMR4965	Harness console-immobilisation	1		
	AFU3081L	Clip-self adhesive-6mm	1		

NOTE: FOR BLANKING-SEE TRIM HEADLINING

M 110

Illus	Part Number	Description	Quantity	Change Point	Remarks
1	AMR1097	Electric control unit antilock brakes	1	To (V) TA 180925 To (V) TA 508779	
1	AMR5557	Electric control unit antilock brakes	1	From (V) TA 180926 From (V) TA 508780	
2	BH106131	Bolt-fixing ECU-M6 x 65	2		
3	FN106047	Nut-flange-fixing ECU-M6	NLA		Use FN106047L.
3	FN106047L	Nut-flange-fixing ECU-M6	2		
4	AMR2741	Bracket assembly-electric control unit mounting-RHD	1		
4	AMR2742	Bracket assembly-electric control unit mounting-LHD	1		
5	FN108041	Nut-fixing bracket-flanged head-M8	NLA		Use FN108041L.
6	STC1749	Sensor assembly antilock brakes-front	2		
6	STC1750	Sensor assembly antilock brakes-rear	1		

NOTE: FOR ABS RELAYS SEE RELAY SECTION

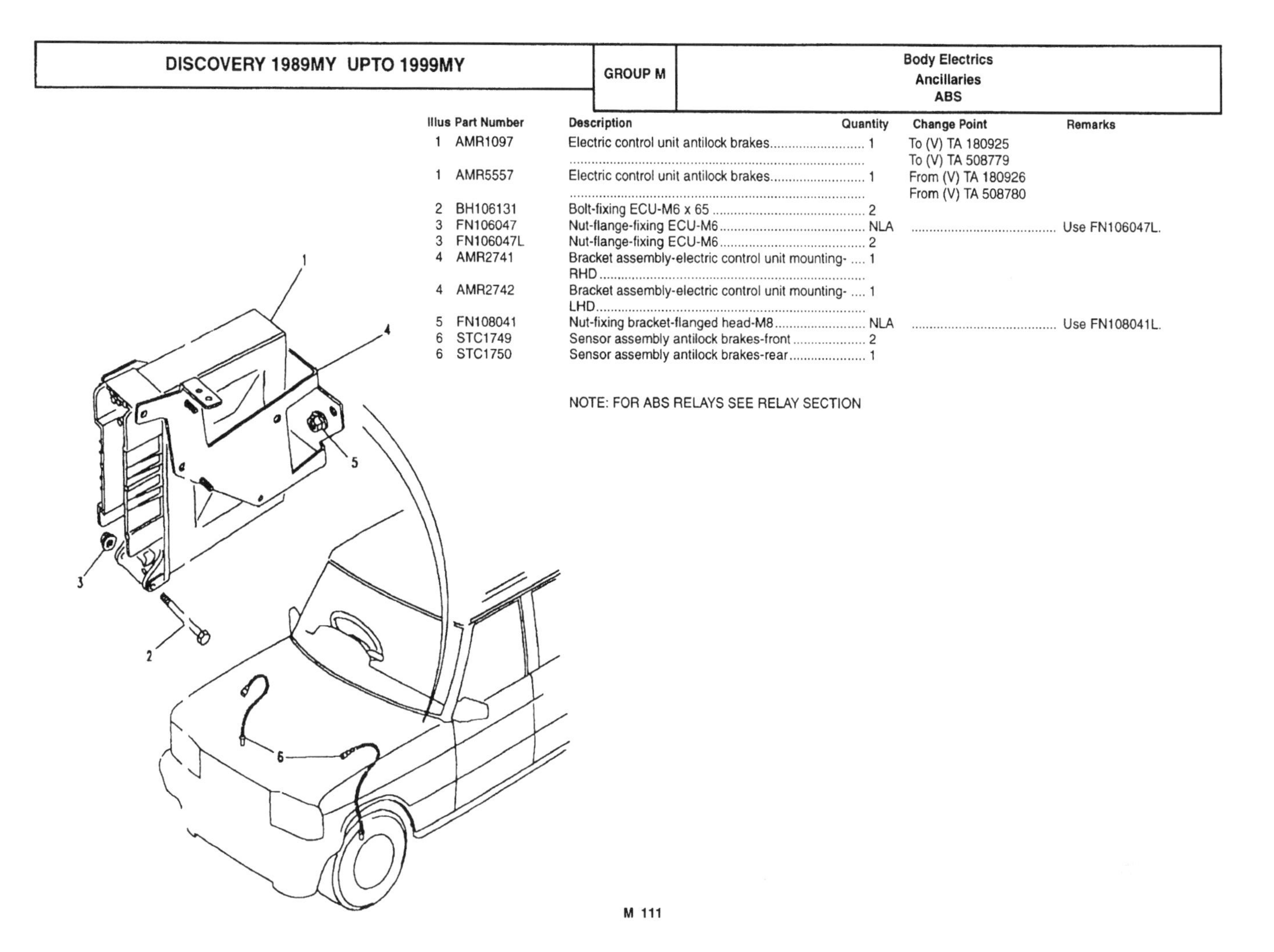

Illus	Part Number	Description	Quantity	Change Point	Remarks
		FROM MA081992			
		ALL VARIANTS			
4	AMR1173	Electric control unit cruise control	1		Petrol
4	AMR5441	Electric control unit cruise control	1	From (V) TA 163104 From (V) TA 501920	Diesel
5	NH104041	Nut-M4	2		
6	AMR1835	Cable assembly cruise control	1		Manual
6	PRC9991	Cable assembly cruise control	1		Automatic

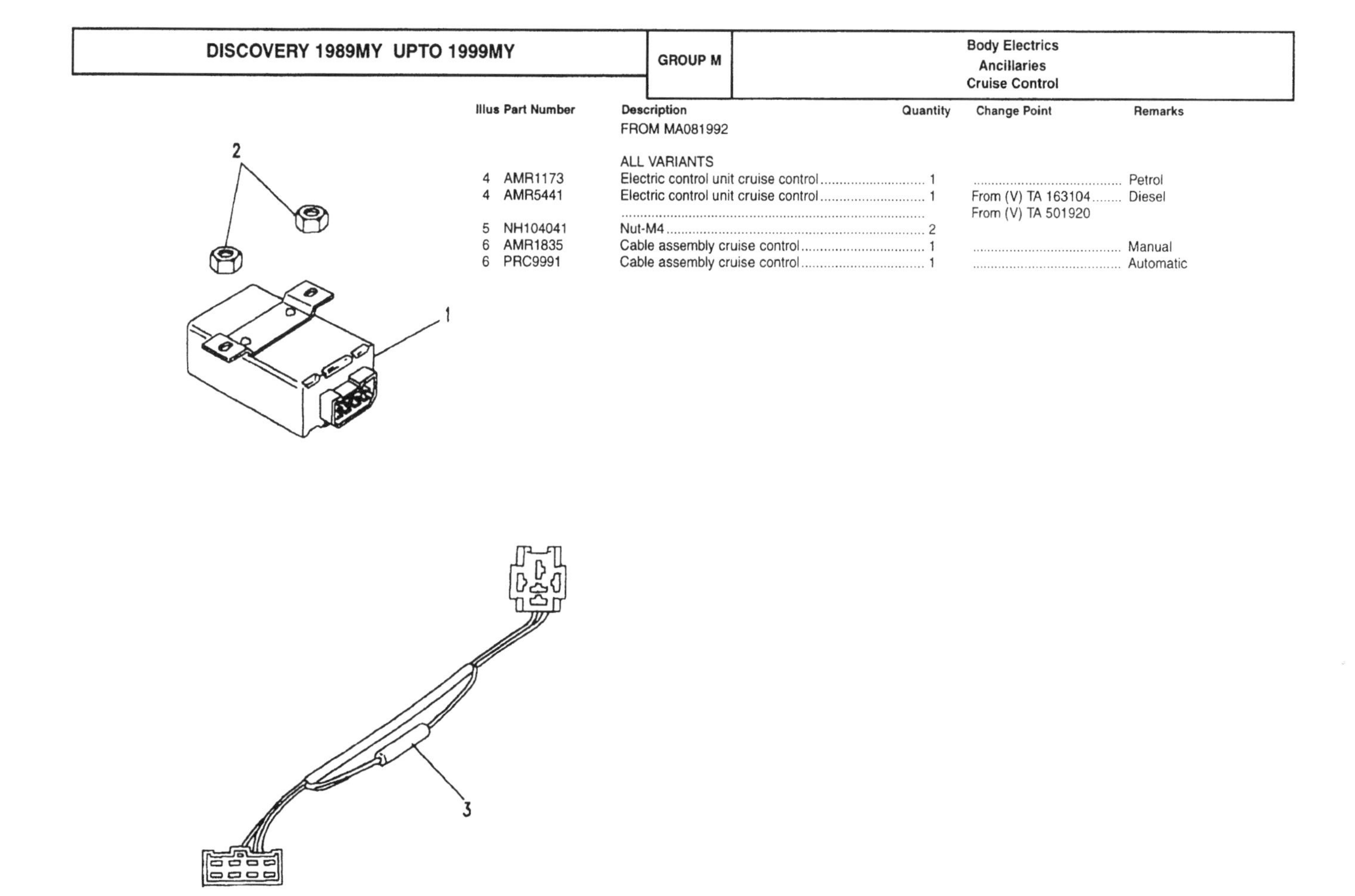

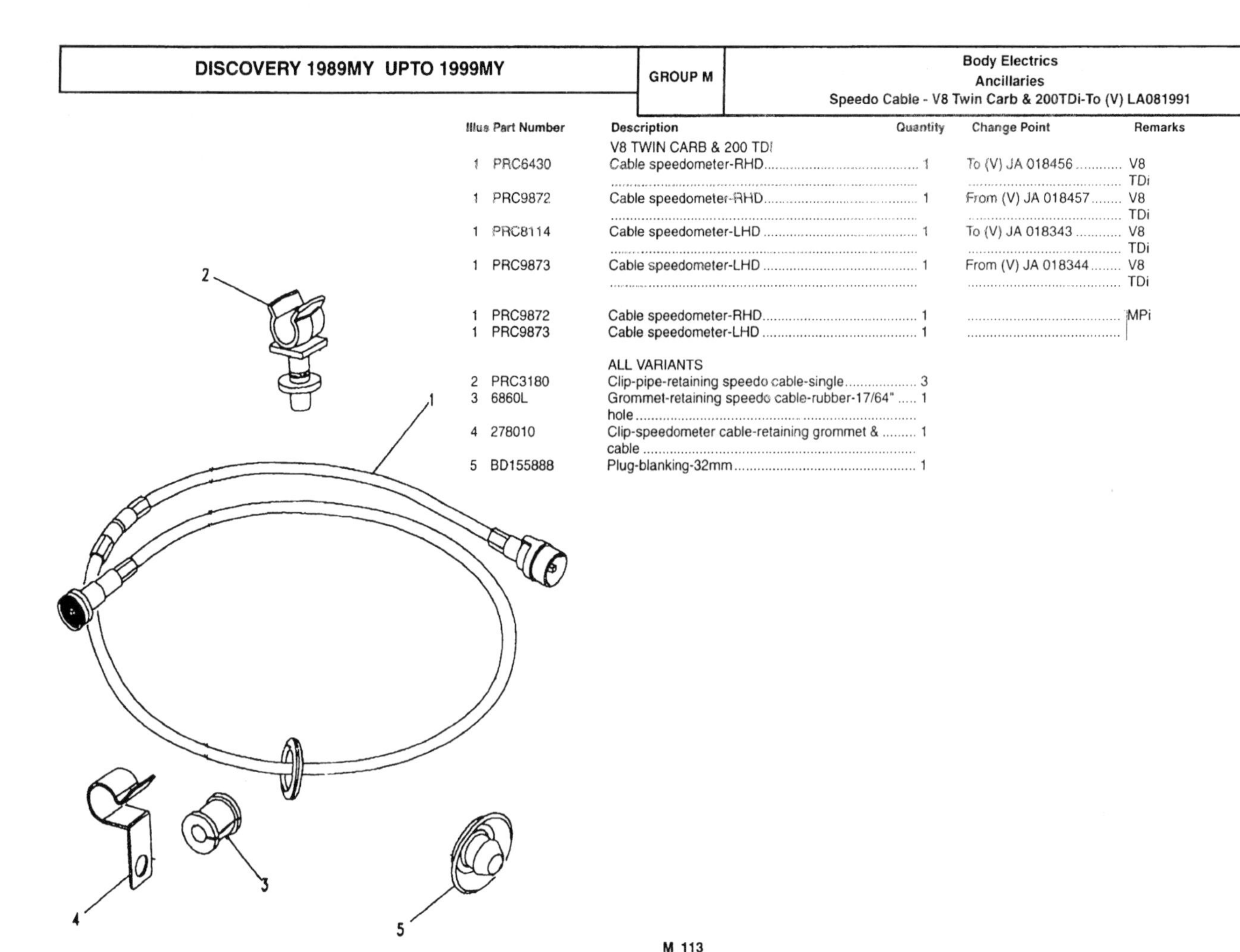

Illus	Part Number	Description	Quantity	Change Point	Remarks
		V8 TWIN CARB & 200 TDI			
1	PRC6430	Cable speedometer-RHD	1	To (V) JA 018456	V8 TDi
1	PRC9872	Cable speedometer-RHD	1	From (V) JA 018457	V8 TDi
1	PRC8114	Cable speedometer-LHD	1	To (V) JA 018343	V8 TDi
1	PRC9873	Cable speedometer-LHD	1	From (V) JA 018344	V8 TDi
1	PRC9872	Cable speedometer-RHD	1		MPi
1	PRC9873	Cable speedometer-LHD	1		
		ALL VARIANTS			
2	PRC3180	Clip-pipe-retaining speedo cable-single	3		
3	6860L	Grommet-retaining speedo cable-rubber-17/64" hole	1		
4	278010	Clip-speedometer cable-retaining grommet & cable	1		
5	BD155888	Plug-blanking-32mm	1		

M 113

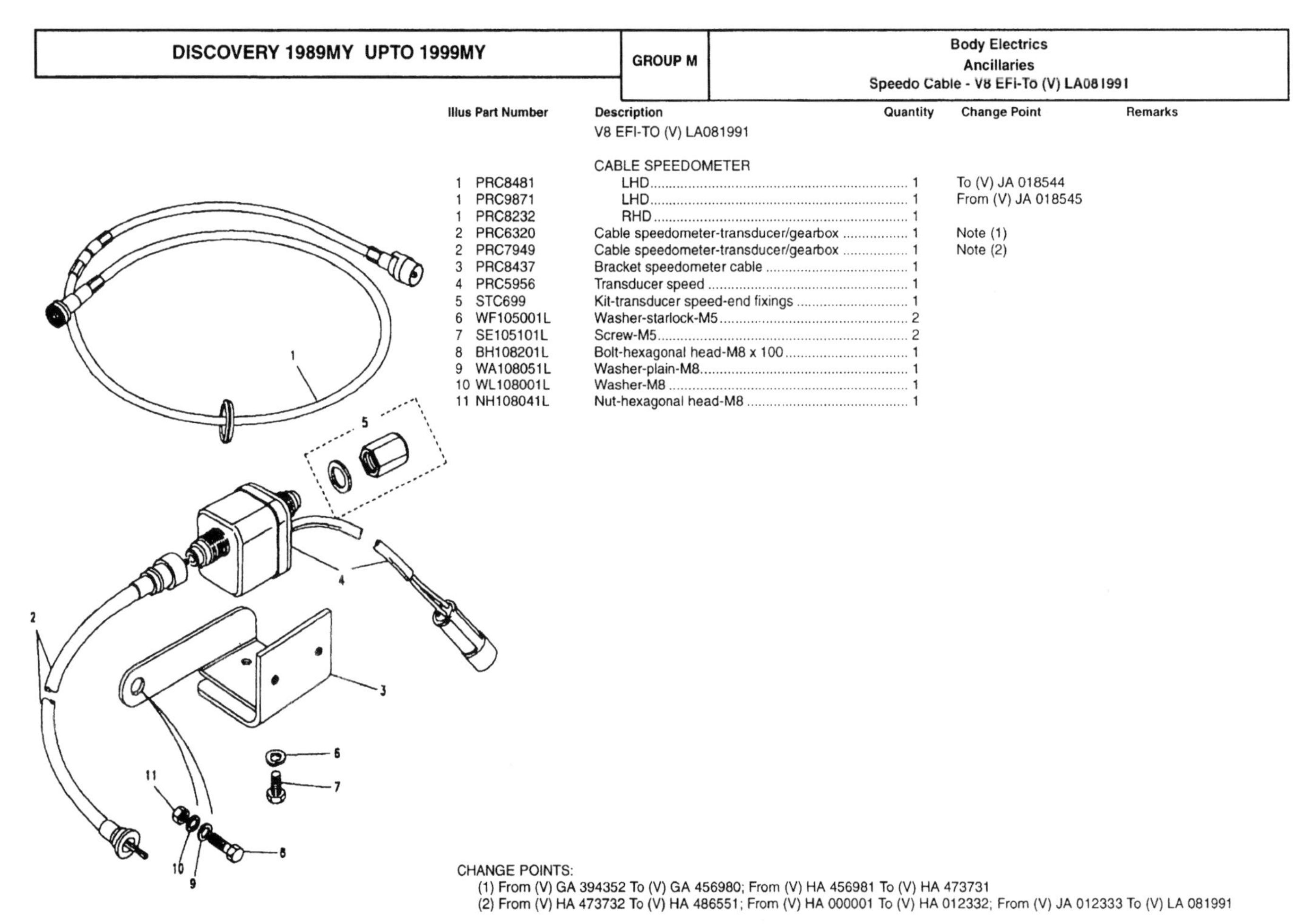

Illus	Part Number	Description	Quantity	Change Point	Remarks
		V8 EFI-TO (V) LA081991			
		CABLE SPEEDOMETER			
1	PRC8481	LHD	1	To (V) JA 018544	
1	PRC9871	LHD	1	From (V) JA 018545	
1	PRC8232	RHD	1		
2	PRC6320	Cable speedometer-transducer/gearbox	1	Note (1)	
2	PRC7949	Cable speedometer-transducer/gearbox	1	Note (2)	
3	PRC8437	Bracket speedometer cable	1		
4	PRC5956	Transducer speed	1		
5	STC699	Kit-transducer speed-end fixings	1		
6	WF105001L	Washer-starlock-M5	2		
7	SE105101L	Screw-M5	2		
8	BH108201L	Bolt-hexagonal head-M8 x 100	1		
9	WA108051L	Washer-plain-M8	1		
10	WL108001L	Washer-M8	1		
11	NH108041L	Nut-hexagonal head-M8	1		

CHANGE POINTS:
(1) From (V) GA 394352 To (V) GA 456980; From (V) HA 456981 To (V) HA 473731
(2) From (V) HA 473732 To (V) HA 486551; From (V) HA 000001 To (V) HA 012332; From (V) JA 012333 To (V) LA 081991

M 114

Illus	Part Number	Description	Quantity	Change Point	Remarks
		FROM (V) MA081992			
		SPEED SENSOR FITTED TO TRANSFER BOX			
1	AMR1253	Transducer speed	1		
2	SS106351	Screw-hexagon socket-fixing transducer	1		
3	WA106041L	Washer-fixing transducer-M6-standard	1		
4	571665	O ring-seal transducer	1		

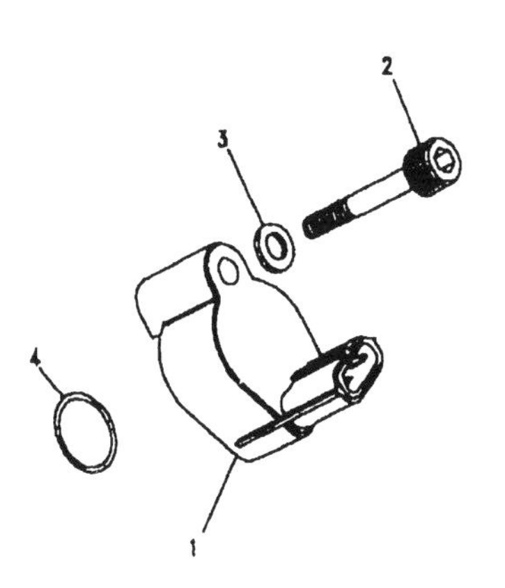

M 115

Illus	Part Number	Description	Quantity	Change Point	Remarks
		FROM (V) VA703237/VA534104			
		HEATED FRONT SCREEN OPTION			
1	AMR6248	Lead-heated screen fuse	1		
2	YQG10011L	Fuse-heated windscreen-60 amp-maxi-large-Blue	1		
3	AMR3464L	Cover-fuse	1		
5	RTC4507	Fuse-30 amp-auto-standard	2		
7	YQU10017L	Clip-harness	1		
8	AMR1935	Bracket-mounting-steel	1		
9	AMR5619	Grommet-cable blanking	2		

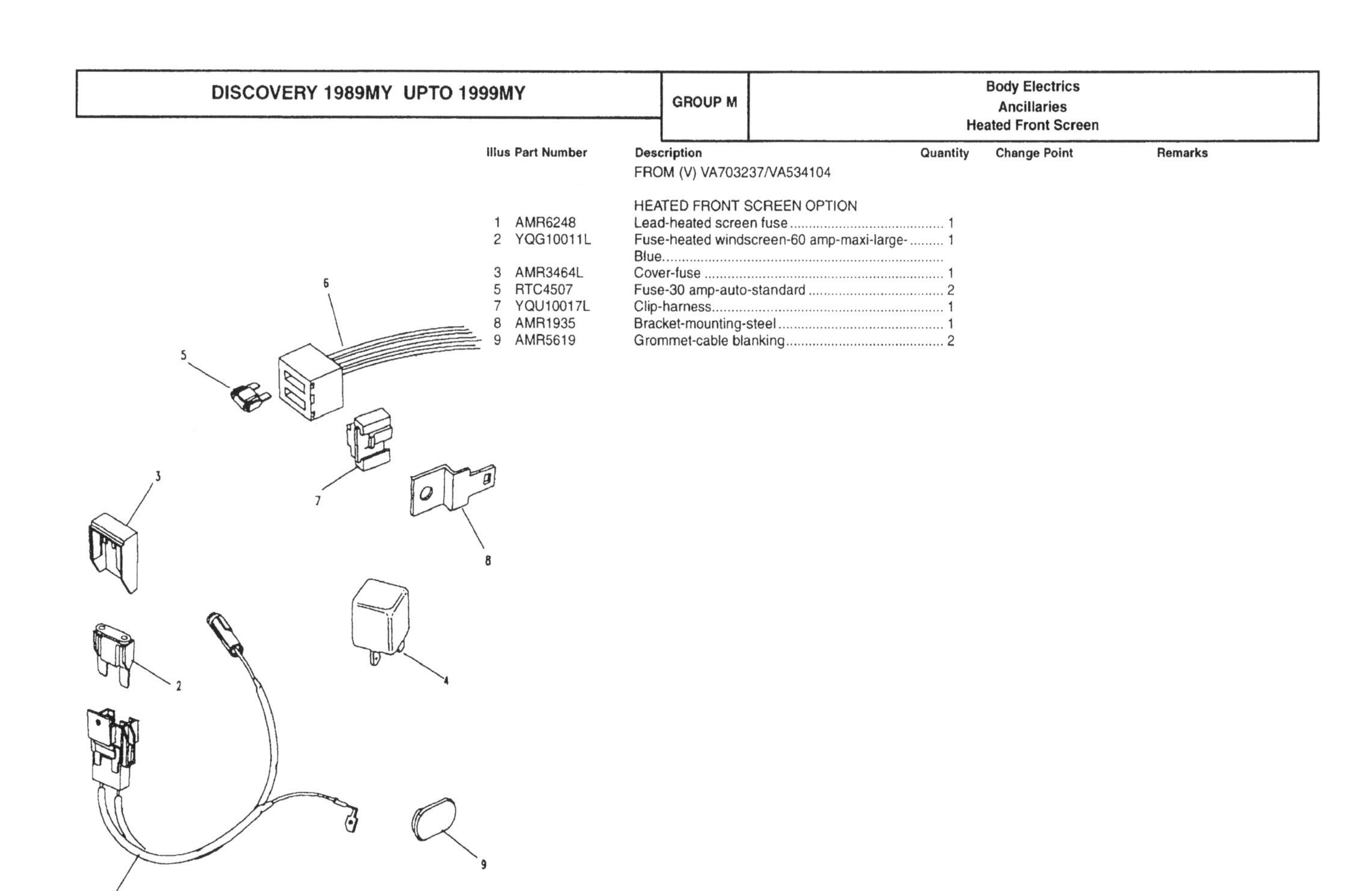

M 116

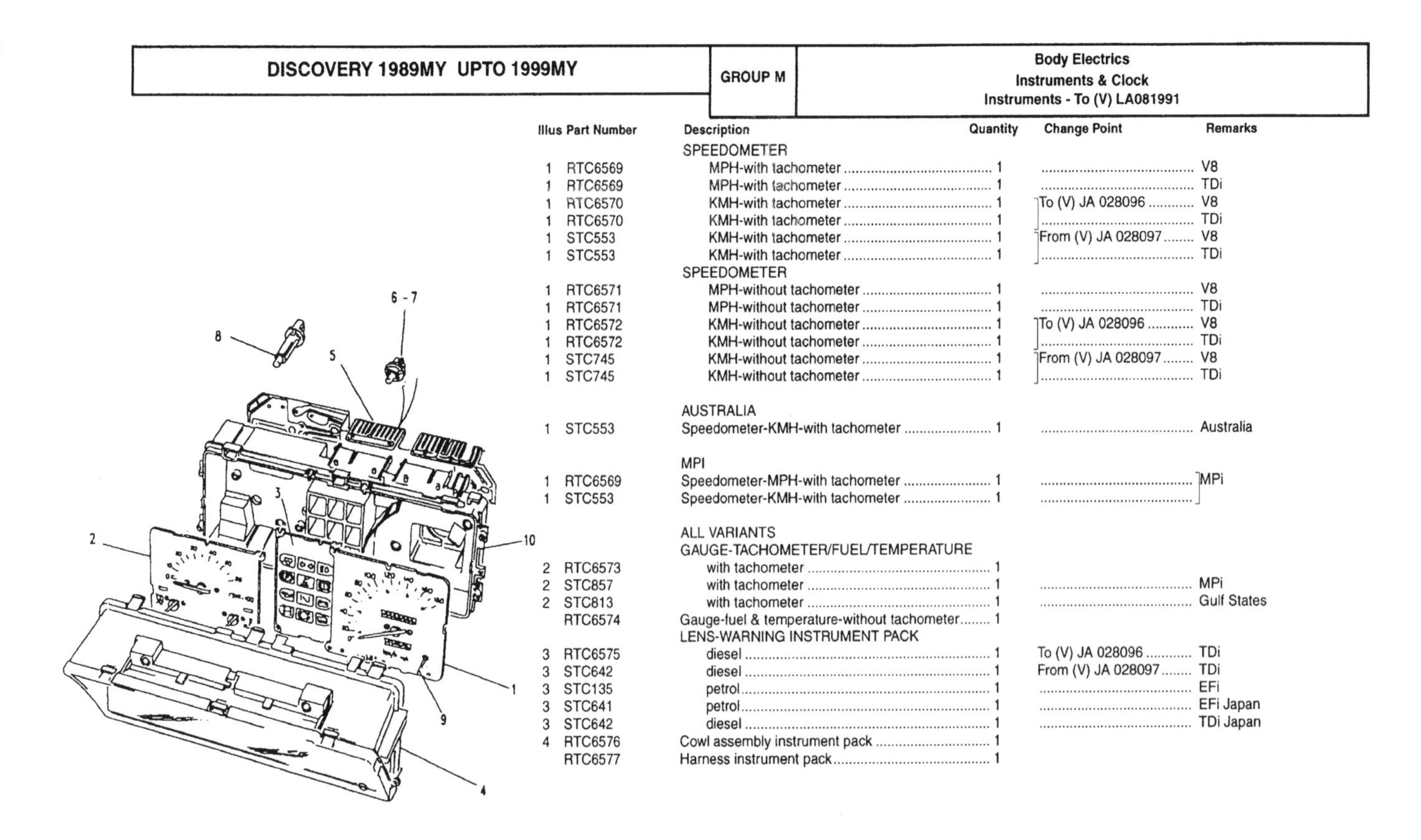

Illus	Part Number	Description	Quantity	Change Point	Remarks
		SPEEDOMETER			
1	RTC6569	MPH-with tachometer	1		V8
1	RTC6569	MPH-with tachometer	1		TDi
1	RTC6570	KMH-with tachometer	1	To (V) JA 028096	V8
1	RTC6570	KMH-with tachometer	1		TDi
1	STC553	KMH-with tachometer	1	From (V) JA 028097	V8
1	STC553	KMH-with tachometer	1		TDi
		SPEEDOMETER			
1	RTC6571	MPH-without tachometer	1		V8
1	RTC6571	MPH-without tachometer	1		TDi
1	RTC6572	KMH-without tachometer	1	To (V) JA 028096	V8
1	RTC6572	KMH-without tachometer	1		TDi
1	STC745	KMH-without tachometer	1	From (V) JA 028097	V8
1	STC745	KMH-without tachometer	1		TDi
		AUSTRALIA			
1	STC553	Speedometer-KMH-with tachometer	1		Australia
		MPI			
1	RTC6569	Speedometer-MPH-with tachometer	1		MPi
1	STC553	Speedometer-KMH-with tachometer	1		
		ALL VARIANTS			
		GAUGE-TACHOMETER/FUEL/TEMPERATURE			
2	RTC6573	with tachometer	1		
2	STC857	with tachometer	1		MPi
2	STC813	with tachometer	1		Gulf States
	RTC6574	Gauge-fuel & temperature-without tachometer	1		
		LENS-WARNING INSTRUMENT PACK			
3	RTC6575	diesel	1	To (V) JA 028096	TDi
3	STC642	diesel	1	From (V) JA 028097	TDi
3	STC135	petrol	1		EFi
3	STC641	petrol	1		EFi Japan
3	STC642	diesel	1		TDi Japan
4	RTC6576	Cowl assembly instrument pack	1		
	RTC6577	Harness instrument pack	1		

Illus	Part Number	Description	Quantity	Change Point	Remarks
5	RTC6578	Printed circuit board instrument pack	1	Note (1)	V8 Twin Carburettor
					TDi
5	STC136	Printed circuit board instrument pack	1	Note (2)	V8 EFi
					TDi
		ALL VARIANTS			
		BULB & HOLDER ASSEMBLY			
6	RTC6579	Grey	11		
7	RTC6580	Blue	1		
8	PRC7451	with tachometer-Brown-short	1		
8	RTC6582	with tachometer-Brown-long	3		
9	RTC6583	Knob trip reset	1		
10	STC983	Case assembly instrument pack	1		

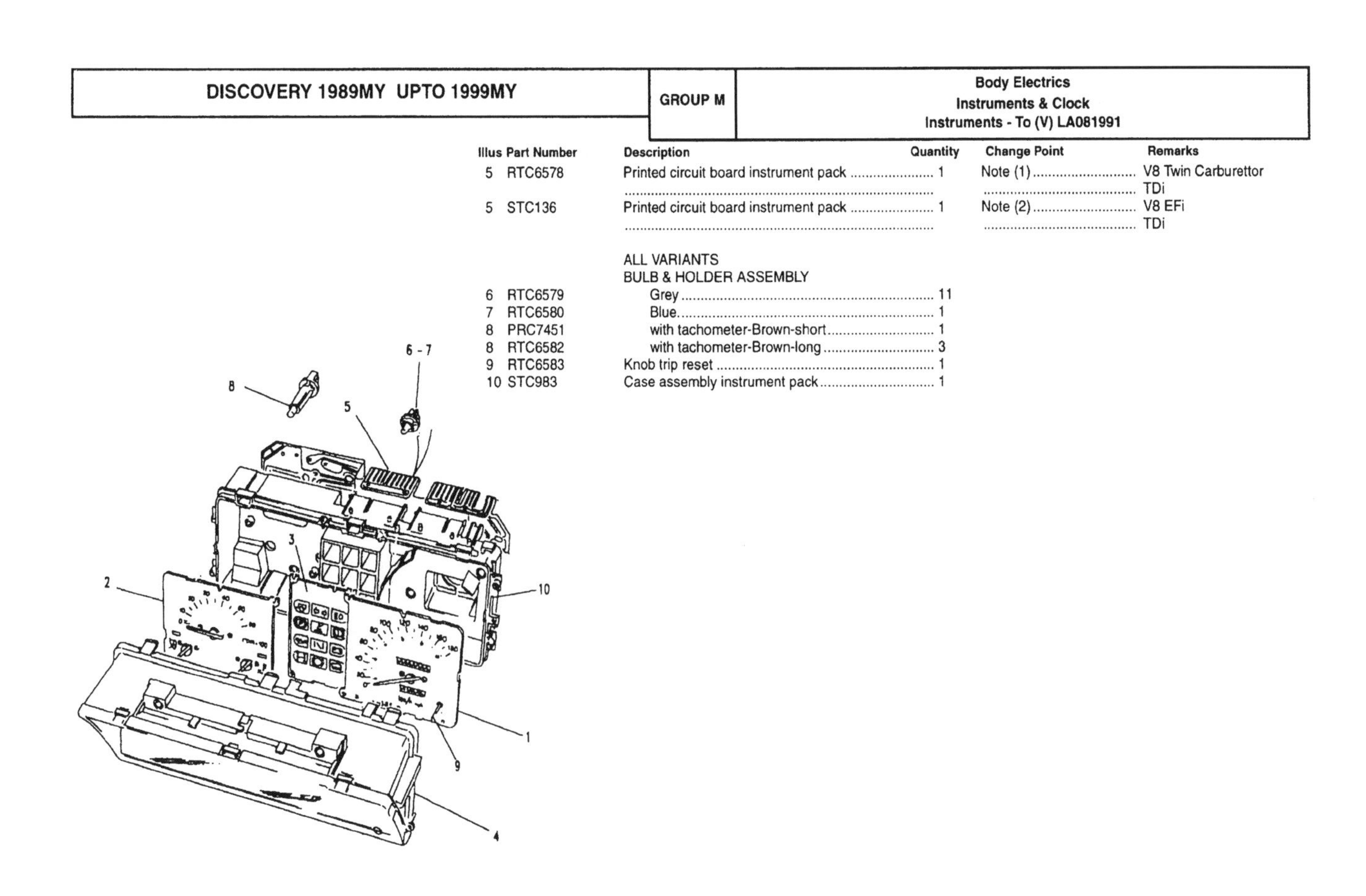

CHANGE POINTS:
(1) From (V) GA 394352 To (V) GA 440844
(2) From (V) GA 440845 To (V) GA 456980; From (V) HA 456981 To (V) HA 486551; From (V) HA 000001 To (V) HA 012332; From 012333 To (V) LA 081991

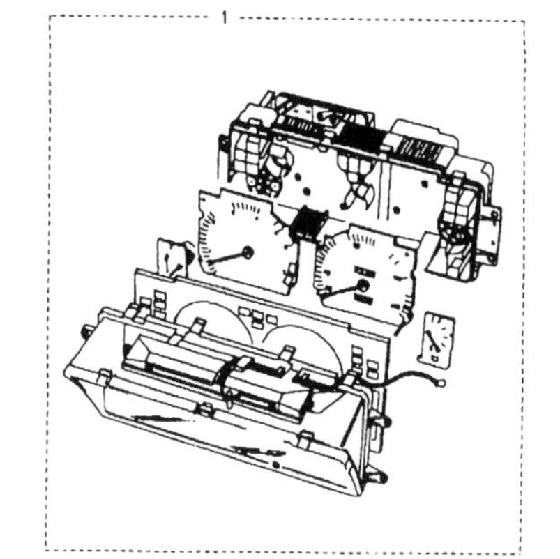

Illus	Part Number	Description	Quantity	Change Point	Remarks
		FROM MA081992 TO MA163103/MA501919			
		V8 PI & 300TDI- WITH SRS (AIR BAG) INSTRUMENT PACK			
1	AMR1339	with drivers air bag	1		UK
1	AMR1339	with drivers air bag	1		
1	AMR1340	with drivers air bag	1		Europe Gulf States Rest
1	AMR1340	with drivers air bag	1		of World
1	AMR1345	with drivers air bag	1		Australia
1	AMR1345	with drivers air bag	1		
1	AMR1349	with drivers air bag	1		Japan
1	AMR1349	with drivers air bag	1		
		MPI-WITH SRS (AIR BAG)			
1	AMR1343	Instrument pack-with tachometer-with drivers air bag	1		UK
1	AMR1344	Instrument pack-with tachometer-with drivers air bag	1		Europe Rest of World
		V8 EFI & 300TDI-WITHOUT SRS (AIR BAG) INSTRUMENT PACK			
1	AMR1233	less airbag	1		UK
1	AMR1233	less airbag	1		
1	AMR1234	less airbag	1		Europe Gulf States Rest
1	AMR1234	less airbag	1		of World
1	AMR1237	less airbag	1		Australia
1	AMR1237	less airbag	1		
1	AMR1262	less airbag	1		Japan
1	AMR1262	less airbag	1		
		MPI-WITHOUT SRS (AIR BAG)			
1	AMR1256	Instrument pack-with tachometer-less airbag	1		UK
1	AMR1257	Instrument pack-with tachometer-less airbag	1		Europe Rest of World

Illus	Part Number	Description	Quantity	Change Point	Remarks
		FROM MA081992 UP TO MA163103/MA501919			
		SPEEDOMETER			
1	STC1310	Speedometer-135mph	1		
1	STC1311	Speedometer-210km/h	1		
		TACHOMETER-V8 EFI & TDI TACHOMETER			
2	STC1314	less airbag	1		
2	STC1314	less airbag	1		
2	STC1316	with drivers air bag	1		
2	STC1316	with drivers air bag	1		
		TACHOMETER-MPI			
2	STC1317	Tachometer-with drivers air bag	1		
2	STC1315	Tachometer-less airbag	1		
		FUEL & WATER TEMPERATURE GAUGES-ALL VARIANTS			
3	STC1318	Gauge-fuel	1		
4	STC1319	Gauge-temperature	1		
5	STC1321	Mask assembly instrument pack	1		EFi UK
					TDi UK
5	STC1322	Mask assembly instrument pack	1		EFi Europe Gulf States
					Rest of World
					TDi
5	STC1323	Mask assembly instrument pack	1		EFi Australia
					Diesel Australia
5	STC1327	Mask assembly instrument pack	1		EFi Japan
					TDi Japan
		PRINTED MASK-MPI			
5	STC1324	Mask assembly instrument pack	1		UK
5	STC1325	Mask assembly instrument pack	1		Europe Rest of World

Illus	Part Number	Description	Quantity	Change Point	Remarks
		FROM MA081992 UP TO MA163103/MA501919			
		CASE ASSEMBLY-V8 EFI, TDI & MPI			
		CASE ASSEMBLY INSTRUMENT PACK			
6	STC1437	less airbag	1		EFi
6	STC1437	less airbag	1		TDi
6	STC1440	with drivers air bag	1		EFi
6	STC1440	with drivers air bag	1		TDi
6	STC1438	less airbag	1		MPi
6	STC1441	with drivers air bag	1		
7	STC1443	ILLUMINATION ASSEMBLY-INSTRUMENT PACK	1		
8	RTC6607	• Bulb & holder assembly-Clear.-black base	2		
9	STC1442	Bulb instrument pack	1		

DISCOVERY 1989MY UPTO 1999MY | GROUP M | Body Electrics
Instruments & Clock
Instrument Comp'ts-From (V) MA081992 To (V) MA163103/MA501919

Illus	Part Number	Description	Quantity	Change Point	Remarks
		PRINTED CIRCUIT BOARD INSTRUMENT PACK			
10	STC1332	less airbag	1		EFi UK Europe Rest of World
10	STC1332	less airbag	1		TDi
10	STC1333	less airbag	1		EFi Australia
10	STC1333	less airbag	1		TDi Australia
10	STC1337	less airbag	1		EFi Japan
10	STC1337	less airbag	1		
		PRINTED CIRCUIT BOARD INSTRUMENT PACK			
10	STC1335	with drivers air bag	1		EFi UK Europe Rest of World
10	STC1335	with drivers air bag	1		TDi
10	STC1336	with drivers air bag	1		EFi Australia
10	STC1336	with drivers air bag	1		TDi Australia
10	STC1338	with drivers air bag	1		EFi Japan
10	STC1338	with drivers air bag	1		TDi Japan
		PRINTED CIRCUIT BOARD INSTRUMENT PACK			
10	STC1332	less airbag	1		MPi UK
10	STC1337	less airbag	1		MPi Rest of World Europe
10	STC1335	with drivers air bag	1		MPi UK
10	STC1338	with drivers air bag	1		MPi Rest of World Europe
		PCB-ALL UK, V8 & TDI EUROPE & GULF			
10	STC1344	Printed circuit board instrument pack-without bulb check-Low fuel warning	1		UK
					EFi Europe Gulf States
					TDi Europe Gulf States
		PCB-ALL AUSTRALIA & JAPAN, & MPI EUROPE			
10	STC1345	Printed circuit board instrument pack-with bulb check-Low fuel warning	1		Australia Japan
					Europe

Illus	Part Number	Description	Quantity	Change Point	Remarks
		ALL VARIANTS			
11	STC1320	Cowl assembly instrument pack	1		
	STC2895	Kit-foam gasket-light seals	1		
12	YAL10001L	Knob trip reset	1		
13	YYP10011L	Screw-fixing instrument cowl	1		
		ALL VARIANTS			
		HOLDER ASSEMBLY AND BULBS-INSTRUMENT PACK			
14	RTC6607	black base-Clear.-black base	A/R		
14	STC1339	blue base-Clear.-blue base	1		
14	STC1341	black base-Green-black base	2		
15	STC1340	black base-Green-black base	1		
16	STC1342	grey base-Green-grey base	1		

Illus	Part Number	Description	Quantity	Change Point	Remarks
		FROM TA163104/TA501920			
		INSTRUMENT PACK WITH WARNING LIGHTS-INCLUDING TACHOMETER			
1	AMR4749	less airbag-RHD	1		EFi UK Rest of World
1	AMR4749	less airbag-RHD	1		TDi
1	AMR4747	less airbag-LHD	1		EFi Europe Rest of World
1	AMR4747	less airbag-LHD	1		TDi
1	AMR4845	less airbag	1		EFi Gulf States
1	AMR4845	less airbag	1		TDi
1	AMR4751	less airbag	1		EFi Australia
1	AMR4751	less airbag	1		TDi
1	AMR4753	less airbag	1		EFi Japan
1	AMR4753	less airbag	1		TDi
		MPI-WITHOUT SRS (AIR BAG)			
1	AMR4745	Instrument pack with warning lights-including tachometer-less airbag	1		UK
1	AMR4743	Instrument pack with warning lights-including tachometer-less airbag	1		Europe

Illus	Part Number	Description	Quantity	Change Point	Remarks
		FROM (V) TA163104/TA501920			
		INSTRUMENT PACK WITH WARNING LIGHTS-INCLUDING TACHOMETER			
1	AMR4750	with drivers air bag-RHD	1		EFi UK Rest of World
1	AMR4750	with drivers air bag-RHD	1		TDi
1	AMR4748	with drivers air bag-LHD	1		EFi Europe Rest of World
1	AMR4748	with drivers air bag-LHD	1		TDi
1	AMR4846	with drivers air bag	1		EFi Gulf States
1	AMR4846	with drivers air bag	1		TDi
1	AMR4752	with drivers air bag	1		EFi Australia
1	AMR4752	with drivers air bag	1		TDi
1	AMR4754	with drivers air bag	1		EFi Japan
1	AMR4754	with drivers air bag	1		
		MPI-WITH SRS (AIR BAG)			
1	AMR4746	Instrument pack with warning lights-including tachometer-with drivers air bag	1		UK
1	AMR4744	Instrument pack with warning lights-including tachometer-with drivers air bag	1		Europe

Illus	Part Number	Description	Quantity	Change Point	Remarks
		FROM TA163104/TA501920			
		SPEEDOMETER			
1	STC2965	Speedometer-135mph	1		
1	STC2966	Speedometer-210km/h	1		
		TACHOMETER			
2	STC2968	less airbag	1		EFi
2	STC2968	less airbag	1		TDi
2	STC2969	with drivers air bag	1		EFi
2	STC2969	with drivers air bag	1		TDi
2	STC1315	less airbag	1		MPi
2	STC1317	with drivers air bag	1		
		FUEL & WATER TEMPERATURE GAUGES-ALL VARIANTS			
3	STC1318	Gauge-fuel	1		
4	STC1319	Gauge-temperature	1		
		ITEM 5 PRINTED MASK IS A NSP			
		CASE ASSEMBLY INSTRUMENT PACK			
6	STC2970	less airbag	1		EFi
6	STC2970	less airbag	1		TDi
6	STC2971	with drivers air bag	1		EFi
6	STC2971	with drivers air bag	1		TDi
6	STC2972	less airbag	1		MPi
6	STC2973	with drivers air bag	1		

Illus	Part Number	Description	Quantity	Change Point	Remarks
		FROM (V) TA163104/TA501920			
		ILLUMINATION UNIT ASSY-ALL VARIANTS			
7	STC1443	ILLUMINATION ASSEMBLY-INSTRUMENT PACK	1		
8	RTC6607	• Bulb & holder assembly-Clear.-black base	2		
9	STC1442	Bulb instrument pack	1		
		FROM TA163104/TA501920			
		PRINTED CIRCUIT BOARD INSTRUMENT PACK			
10	STC2974	Printed circuit board instrument pack	1		EFi UK Europe Gulf States
10	STC2974	low fuel warning	1		TDi
10	STC2977	low fuel warning	1		EFi Australia
10	STC2977	low fuel warning	1		TDi
10	STC2975	low fuel warning	1		EFi Japan
10	STC2975	low fuel warning	1		TDi
10	STC2974	low fuel warning	1		MPi UK
10	STC2975	low fuel warning	1		MPi Europe
		PRINTED CIRCUIT BOARD INSTRUMENT PACK			
10	STC1344	without bulb check-Low fuel warning	1		EFi UK Europe Gulf States
10	STC1344	without bulb check-Low fuel warning	1		TDi
10	STC1344	without bulb check-Low fuel warning	1		Twin Carburettor
10	STC1345	with bulb check-Low fuel warning	1		Australia Japan
10	STC1345	with bulb check-Low fuel warning	1		MPi Europe

Illus	Part Number	Description	Quantity	Change Point	Remarks
		INSTRUMENT COWL-ALL VARIANTS			
11	STC1320	Cowl assembly instrument pack	1		
12	YAL10001L	Knob trip reset	1		
13	YYP10011L	Screw-cowl fixing	4		
		FROM (V) TA163104/TA501920			
		BULB & HOLDER ASSEMBLY			
14	RTC6607	black base-Clear.-black base	A/R		
14	STC1339	blue base-Clear.-blue base	1		
14	STC1341	black base-Green-black base	2		
15	STC1340	black base-Green-black base	1		
16	STC1342	grey base-Green-grey base	1		
17	YAJ100320L	Indicator-light emitting diode burglar alarm	1		

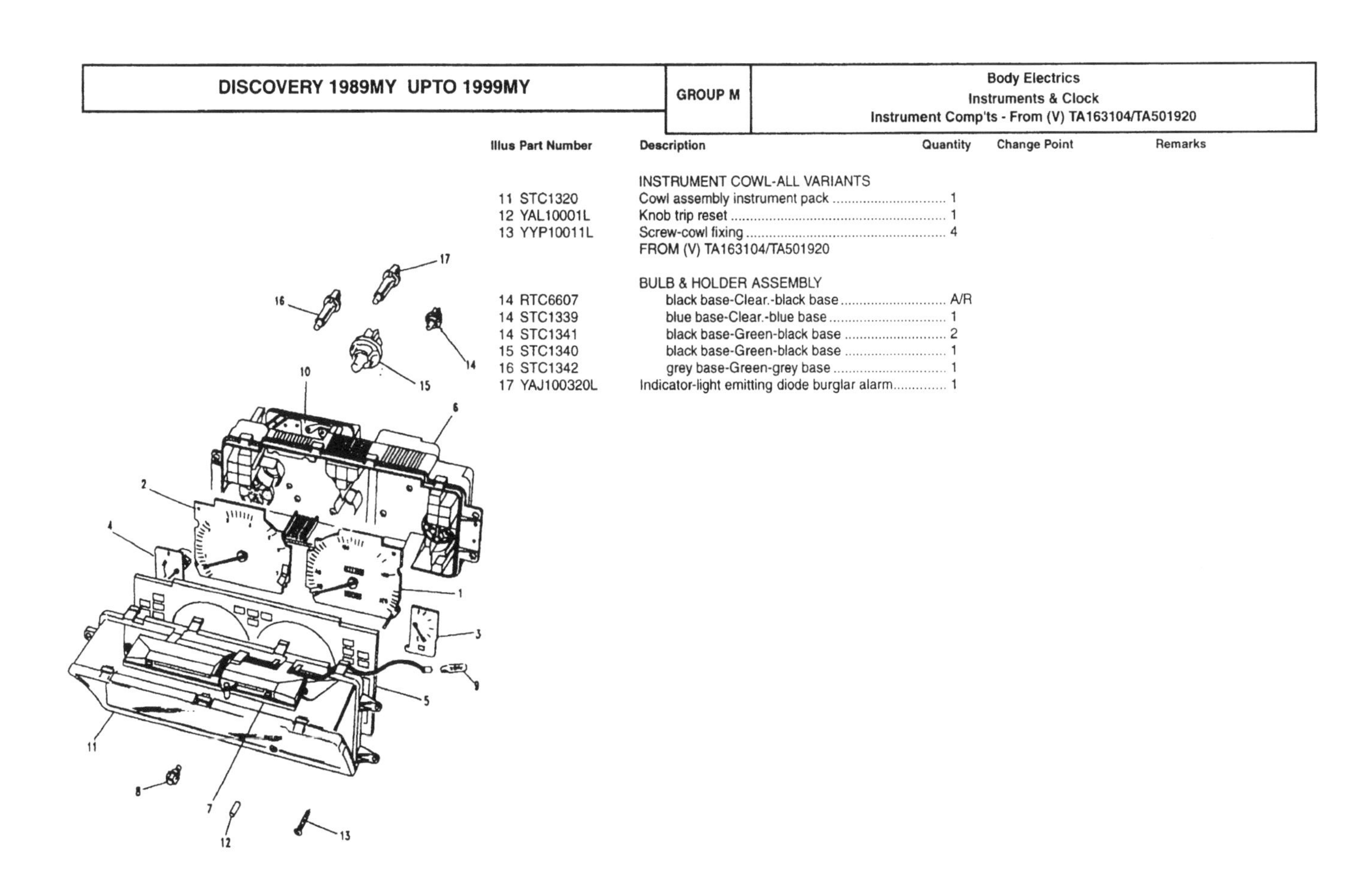

Illus	Part Number	Description	Quantity	Change Point	Remarks
		INSTRUMENT PACK			
1	AMR3929	UK police specification-less airbag	1	Note (1)	
1	AMR3930	UK police specification-with drivers air bag	1		
1	AMR4757	UK police specification-less airbag	1	From (V) TA 163104	
1	AMR4758	UK police specification-with drivers air bag	1		

NOTE:FOR SERVICE PARTS SEE PREVIOUS PAGES

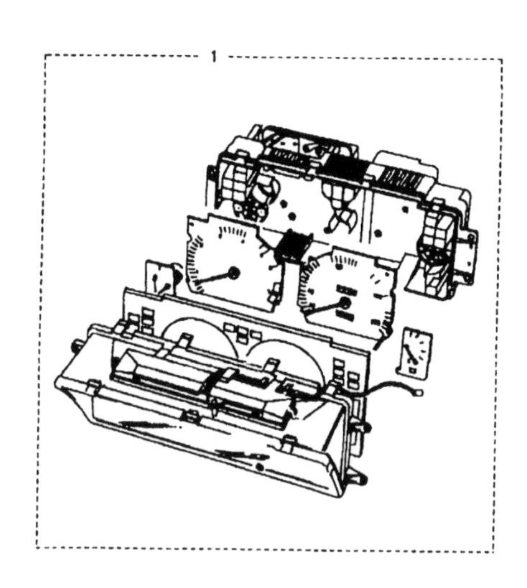

CHANGE POINTS:
(1) From (V) MA 081992 To (V) MA 163103; From (V) MA 081992 To (V) MA 501919

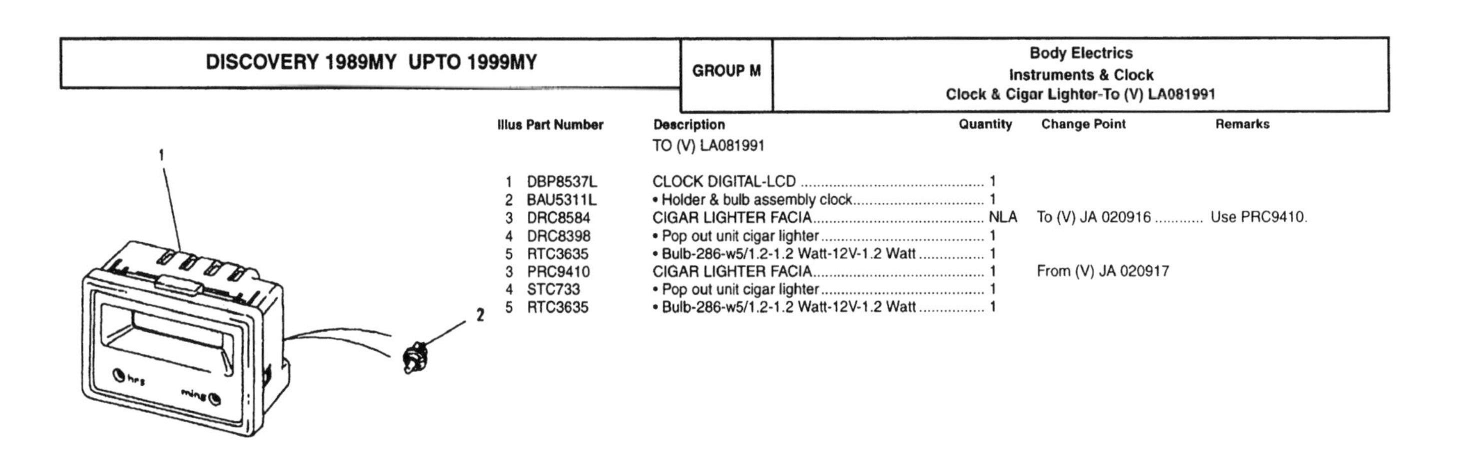

Illus	Part Number	Description	Quantity	Change Point	Remarks
		TO (V) LA081991			
1	DBP8537L	CLOCK DIGITAL-LCD	1		
2	BAU5311L	• Holder & bulb assembly clock	1		
3	DRC8584	CIGAR LIGHTER FACIA	NLA	To (V) JA 020916	Use PRC9410.
4	DRC8398	• Pop out unit cigar lighter	1		
5	RTC3635	• Bulb-286-w5/1.2-1.2 Watt-12V-1.2 Watt	1		
3	PRC9410	CIGAR LIGHTER FACIA	1	From (V) JA 020917	
4	STC733	• Pop out unit cigar lighter	1		
5	RTC3635	• Bulb-286-w5/1.2-1.2 Watt-12V-1.2 Watt	1		

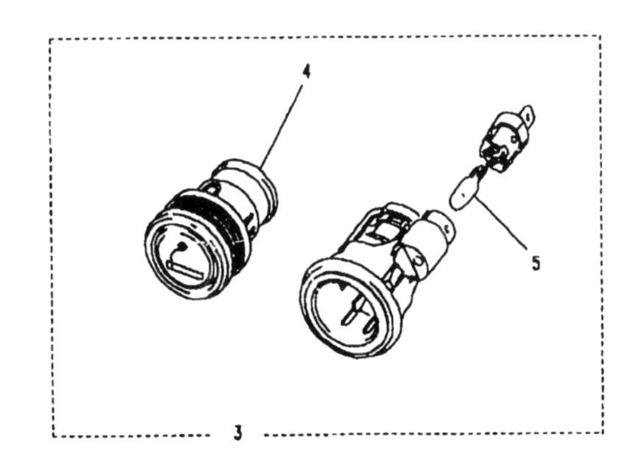

Illus	Part Number	Description	Quantity	Change Point	Remarks
		FROM (V) MA081992			
1	STC1800	CLOCK DIGITAL-LCD-WITHOUT HAZARD SWITCH - SILVER	1		
1	STC4113	CLOCK DIGITAL-LCD-WITHOUT HAZARD SWITCH - GOLD	1		
2	STC1983	• Holder & bulb assembly clock	1		
3	YUJ10037L	CIGAR LIGHTER FACIA	1		
4	RTC3635	• Bulb-286-w5/1.2-1.2 Watt-12V-1.2 Watt	1		

NOTE: HAZARD SWITCH-SEE AUXILIARY SWITCHES
NOTE: LAND-ROVER LOGO IS NOT SERVICED SEPARATELY
NOTE: POP-OUT UNIT IS NOT SERVICED SEPARATELY

CDMXIA3B

Illus	Part Number	Description	Quantity	Change Point	Remarks
1	PRC7410LUN	Blank-facia radio aperture-Ash	1		
		RADIO CASSETTE ELECTRONIC-AUTO/REVERSAL			
2	PRC6492LUN	Ash	NLA		Use PRC9466LUN qty 1 with STC910 qty 1.
2	PRC9466LUN	Ash	1		2 speaker system low line
2	PRC6499LUN	Ash	NLA		Use PRC9463LNF qty 1 with STC910 qty 1.
2	PRC9463LNF	Ash Grey	1		Remote control 6 speaker system high line UK
2	PRC7379LUN	Ash	1		Remote control 6 speaker system high line Germany
		LOWLINE-TWO SPEAKER SYSTEM			
3	PRC6848	Harness-link-front door speaker	2		2 speaker system low line
4	PRC9484LUN	Speaker-front single cone-door-Ash	2		
5	AB606074	Screw-front door speaker fixing-self tapping-M6	8		
		HIGHLINE 6 SPEAKER SYSTEM & GERMANY			
3	PRC6848	Harness-link-front door speaker	2		
4	PRC8301LUN	Speaker-front single cone-door-bass-Ash	2		
5	AB606124	Screw-fixing bass speaker-self tapping-No 6 x 40	8		
	PRC8496LUN	Tweeter audio system-front door-Ash	2		
5	AB606074	Screw-fixing tweeter-self tapping-M6	8		
6	PRC8564	Speaker assembly rear-E-post-RH	1		
6	PRC8565	Speaker assembly rear-E-post-LH	1		
7	RA608127	Rivet-pop-steel-fixing speaker rear E post	6		
		ALL VARIANTS			
8	PRC7471	Aerial front fender	1		
9	UKC6684L	Tie-cable-Black-3.5 x 150mm-inside serated	3		

Illus	Part Number	Description	Quantity	Change Point	Remarks
1	PRC7410LNF	Blank-facia radio aperture	1		
		LOWLINE - (NON RDS)			
2	AMR1029LNF	Radio cassette electronic-auto/reversal-Ash Grey	1		2 speaker system
		MIDLINE - TRAVELLER AUDIO CHOICE (RDS)			
2	PRC9463LNF	Radio cassette electronic-auto/reversal-Ash Grey	1		Remote control 4 speaker plus subwoofer system
		HIGHLINE - PREMIER AUDIO CHOICE (RDS)			
2	AMR1782LNF	Radio cassette electronic-auto/reversal-Ash Grey	1		CD Player remote control
		ALL VARIANTS			
3	AMR2014	Aerial front fender	NLA		Use AMR3461 qty 1 with AMR3462 qty 1.
4	UKC6684L	Tie-cable-Black-3.5 x 150mm-inside serated	3		
5	AMR1102	Harness-passenger door-door speakers	2		
6	AMR1030LNF	Speaker-front single cone-front door-Ash Grey	2		
7	AB606074	Screw-fixing speaker front door-self tapping-M6	8		
8	WA104004L	Washer-plain-fixing speaker front door-M4	8		

Illus	Part Number	Description	Quantity	Change Point	Remarks
		MIDLINE & HIGHLINE			
9	PRC8930LNF	Speaker assembly rear-E-post-Ash Grey	2		
7	AB608074	Screw-fixing speaker rear E post-self tapping-No8 x 7/8	8		
10	STC947	Speaker assembly rear-rear end door	1		
	BTR6580	Hexsert-fixing speaker to rear end door	8		
8	WJ105001L	Washer-plain-fixing speaker to rear end door-M5-oversize	8		
7	SE105301L	Screw-fixing speaker to rear end door-recessed-pan head-M5 x 30	8		
11	STC1824	Amplifier audio power-rear end door	1		
7	SP105121	Screw-amplifier to rear end door-pan head-M5 x 12	2		
12	BTR5498LNF	Spacer-radio-Ash Grey	1		

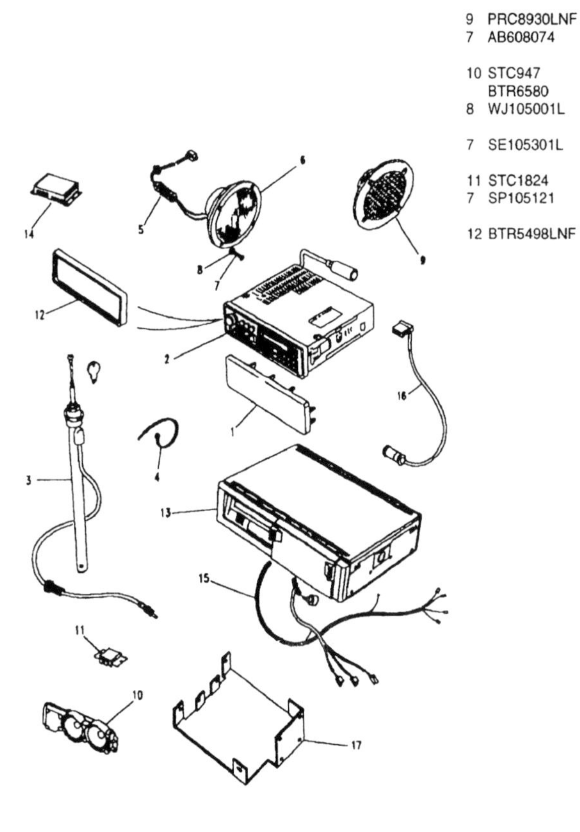

Radio Systems-From (V) KA034314 To LA081991

Illus	Part Number	Description	Quantity	Change Point	Remarks
13	AMR3154	CD AUTOMATIC CHANGER-PHILIPS	1		
	STC1624	• Magazine-automatic changer compact disc player	1		
7	SH104101L	Screw-fixing CD autochanger-hexagonal head-M4 x 10	4		
	WC104001	Washer-fixing CD autochanger-M4	4		
14	AMR3853	Amplifier assembly-audio power	1		
15	AMR1117	Harness-link-amplifier remote	1		
16	AMR3460	Harness-link-CD data lead	1		
17	AMR1785	Bracket-compact disc mounting	1		
7	SH105201	Screw-fixing CD autochanger bracket-hexagonal head-M5 x 20	4		
8	WJ105001	Washer-plain-fixing CD autochanger bracket-M5-oversize-metric	NLA		Use WJ105001L.
8	WJ105001L	Washer-plain-fixing CD autochanger bracket-M5-oversize	4		
	ADU3325	Nut-fixing CD autochanger bracket	4		

Radio Systems-From (V) MA081992 To MA163103 & MA501919

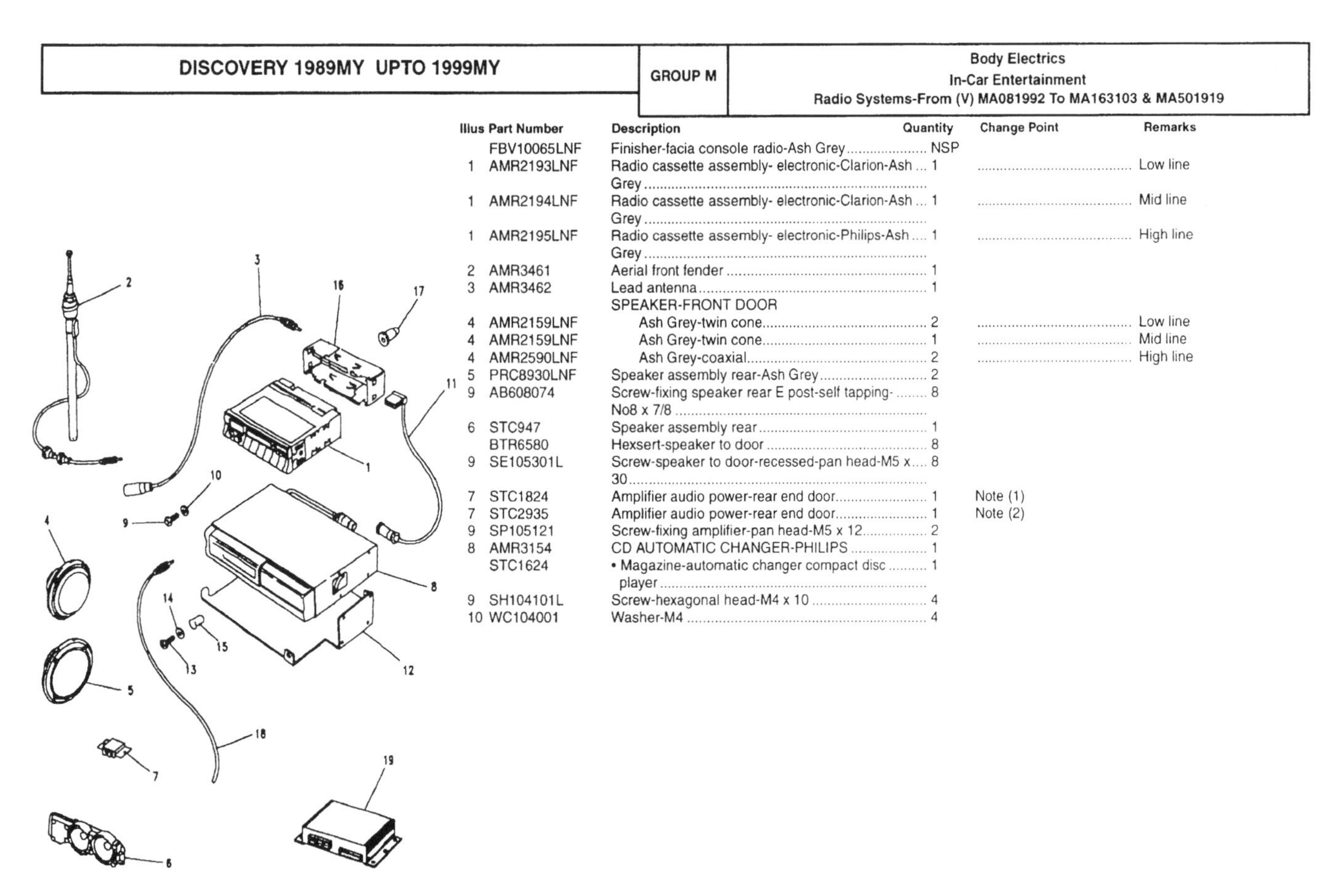

Illus	Part Number	Description	Quantity	Change Point	Remarks
	FBV10065LNF	Finisher-facia console radio-Ash Grey	NSP		
1	AMR2193LNF	Radio cassette assembly- electronic-Clarion-Ash Grey	1		Low line
1	AMR2194LNF	Radio cassette assembly- electronic-Clarion-Ash Grey	1		Mid line
1	AMR2195LNF	Radio cassette assembly- electronic-Philips-Ash Grey	1		High line
2	AMR3461	Aerial front fender	1		
3	AMR3462	Lead antenna	1		
		SPEAKER-FRONT DOOR			
4	AMR2159LNF	Ash Grey-twin cone	2		Low line
4	AMR2159LNF	Ash Grey-twin cone	1		Mid line
4	AMR2590LNF	Ash Grey-coaxial	2		High line
5	PRC8930LNF	Speaker assembly rear-Ash Grey	2		
9	AB608074	Screw-fixing speaker rear E post-self tapping-No8 x 7/8	8		
6	STC947	Speaker assembly rear	1		
	BTR6580	Hexsert-speaker to door	8		
9	SE105301L	Screw-speaker to door-recessed-pan head-M5 x 30	8		
7	STC1824	Amplifier audio power-rear end door	1	Note (1)	
7	STC2935	Amplifier audio power-rear end door	1	Note (2)	
9	SP105121	Screw-fixing amplifier-pan head-M5 x 12	2		
8	AMR3154	CD AUTOMATIC CHANGER-PHILIPS	1		
	STC1624	• Magazine-automatic changer compact disc player	1		
9	SH104101L	Screw-hexagonal head-M4 x 10	4		
10	WC104001	Washer-M4	4		

CHANGE POINTS:
(1) From (V) MA 081992 To (V) MA 156059; From (V) MA 500000 To (V) MA 503294
(2) From (V) MA 156060 To (V) MA 163103; From (V) MA 156060 To (V) MA 501919

DISCOVERY 1989MY UPTO 1999MY

GROUP M

Body Electrics
In-Car Entertainment
Radio Systems-From (V) MA081992 To MA163103 & MA501919

Illus	Part Number	Description	Quantity	Change Point	Remarks
11	AMR3460	Harness-link-CD data lead	1		Except Japan
12	AMR1785	Bracket-compact disc mounting	1		
13	SH105201	Screw-hexagonal head-M5 x 20	4		
14	WJ105001	Washer-plain-M5-oversize-metric	NLA		Use WJ105001L.
14	WJ105001L	Washer-plain-M5-oversize	4		
15	ADU3325	Nut	4		
16	AMR2125	Bracket-radio housing	1		
17	XQJ10002L	Bush-retention	1		
11	AMR2619	Harness-link-CD data lead	1		Japan
12	AMR3029	Bracket-compact disc mounting	1		
18	AMR2776	Lead-link radio	1		Local supply Japan
19	AMR3853	Amplifier assembly-audio power	1	Note (1)	Japan
19	AMR5117	Amplifier assembly-audio power	1	Note (2)	

CHANGE POINTS:
(1) From (V) MA 081992 To (V) MA 156059
(2) From (V) MA 156060 To (V) MA 163103; From (V) MA 156060 To (V) MA 501919

M 137

DISCOVERY 1989MY UPTO 1999MY

GROUP M

Body Electrics
In-Car Entertainment
Radio Systems-From (V) TA163104 & TA501920 To TA703236 & TA534103

Illus	Part Number	Description	Quantity	Change Point	Remarks
	FBV10065LNF	Finisher-facia console radio-Ash Grey	1		
1	AMR4794LNF	RADIO CASSETTE ASSEMBLY- ELECTRONIC-PHILIPS-ASH GREY	1	To (V) TA 532127 To (V) TA 701283	Low line
1	AMR5696LNF	RADIO CASSETTE ASSEMBLY- ELECTRONIC-PHILIPS-ASH GREY	1	From (V) TA 532128 From (V) TA 701284	
7	STC3078	• Pad-detachable-key radio	1		Part of AMR4794LNF.
7	XQI100040L	• Pad-detachable-key radio	1		Part of AMR5696LNF.
1	AMR2194LNF	Radio cassette assembly- electronic-Clarion-Ash Grey	1		Mid line
1	AMR4188LNF	Radio cassette assembly- electronic-Philips-Ash Grey	NLA	Note (1)	Use XQD101170LNF. High line
1	XQD101170LNF	Radio cassette assembly- electronic-Philips-Ash Grey	1		
2	AMR3461	Aerial front fender	1		
3	AMR3462	Lead antenna	1		

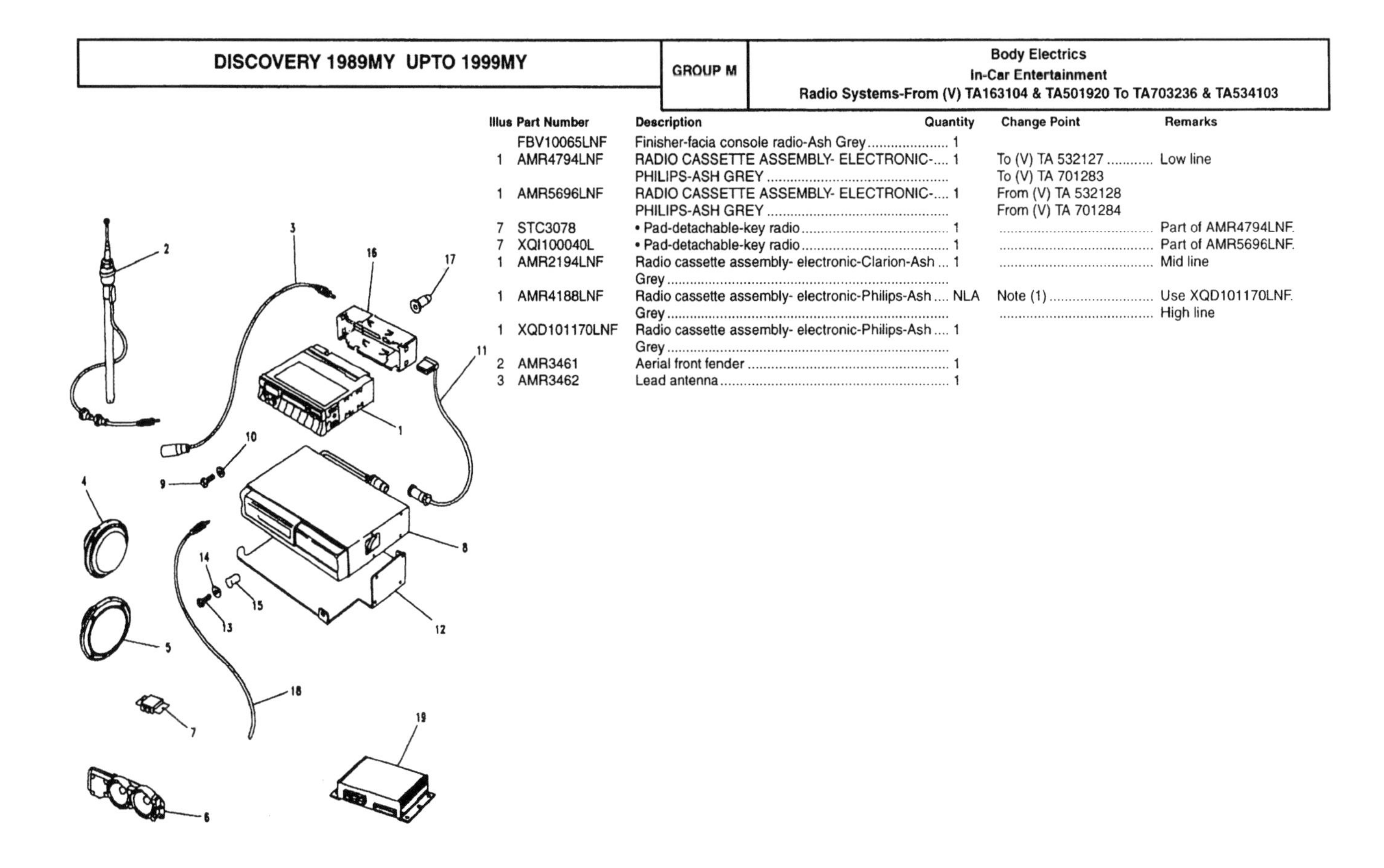

CHANGE POINTS:
(1) From (V) TA 163104; From (V) TA 501920

M 138

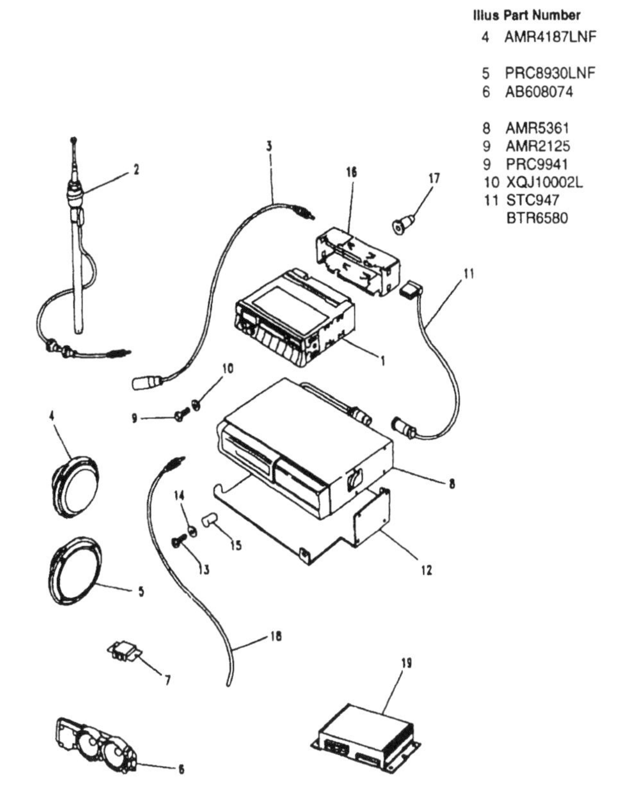

Illus	Part Number	Description	Quantity	Change Point	Remarks
4	AMR4187LNF	Speaker assembly-front single cone-door-Ash Grey	2		
5	PRC8930LNF	Speaker assembly rear-E-post-Ash Grey	2		
6	AB608074	Screw-fixing speaker rear E post-self tapping-No8 x 7/8	8		
8	AMR5361	Box-key pad radio	1		Low line
9	AMR2125	Bracket-radio housing	1	Note (1)	
9	PRC9941	Bracket-radio housing-Philips	1	Note (2)	
10	XQJ10002L	Bush-retention	1		
11	STC947	Speaker assembly rear-end door	1		
	BTR6580	Hexsert-fixing speaker to rear end door	8		

CHANGE POINTS:
(1) From (V) TA 163104 To (V) TA 191911; From (V) TA 501920 To (V) TA 522424
(2) From (V) TA 191912; From (V) TA 522424

Illus	Part Number	Description	Quantity	Change Point	Remarks
	WJ105001L	Washer-plain-fixing speaker to rear end door-M5-oversize	8		
6	SE105301L	Screw-fixing speaker to rear end door-recessed-pan head-M5 x 30	8		
12	STC2935	Amplifier audio power-rear end door	1		
6	SP105121	Screw-amplifier to rear end door-pan head-M5 x 12	2		
14	AMR3154	CD AUTOMATIC CHANGER-PHILIPS	1		
	STC1624	• Magazine-automatic changer compact disc player	1		
15	SH104101L	Screw-fixing CD autochanger-hexagonal head-M4 x 10	4		
16	WC104001	Washer-fixing CD autochanger-M4	4		
17	AMR4831L	Link-compact disc automatic changer-Philips	1		Except Japan
17	AMR4832	Link-compact disc automatic changer-Pioneer	1		Japan
18	AMR4184	Bracket-compact disc mounting-RH	1		
18	AMR4183	Bracket-compact disc mounting-LH	1		
19	FS105107	Screw-fixing CD bracket-hexagonal head-M5 x 10	4		
20	AMR5117	Amplifier assembly-audio power-footwell	1		Japan

Illus	Part Number	Description	Quantity	Change Point	Remarks
	FBV10065LNF	Finisher-facia console radio-Ash Grey	1		
1	AMR5696LNF	RADIO CASSETTE ASSEMBLY- ELECTRONIC-PHILIPS-ASH GREY	1		Low line
8	XQI100040L	• Pad-detachable-key radio	1		
9	AMR5361	Box-key pad radio	1		
1	AMR2194LNF	Radio cassette assembly- electronic-Clarion-Ash Grey	1		RDS EON mid line
1	AMR4188LNF	Radio cassette assembly- electronic-Philips-Ash Grey	NLA	Note (1); Note (2)	Use XQD101170LNF. CD Player high line
1	XQD101170LNF	Radio cassette assembly- electronic-Philips-Ash Grey	1	Note (3)	CD Player high line
2	AMR5270	Amplifier-audio antenna-AM/FM-Pioneer	1		
3	SN106127	Screw-thread forming-fixing amplifier-M6 x 12	NLA		Use SN106127L.
3	SN106127L	Screw-thread forming-fixing amplifier-M6 x 12	1		
4	AMR5362	Lead antenna-AM/FM	1		Except Japan
4	AMR5363	Lead antenna-AM/FM	1		Japan

CHANGE POINTS:
(1) From (V) VA 703237 To (V) WA 756115
(2) From (V) VA 534104 To (V) WA 756115
(3) From (V) WA 756116

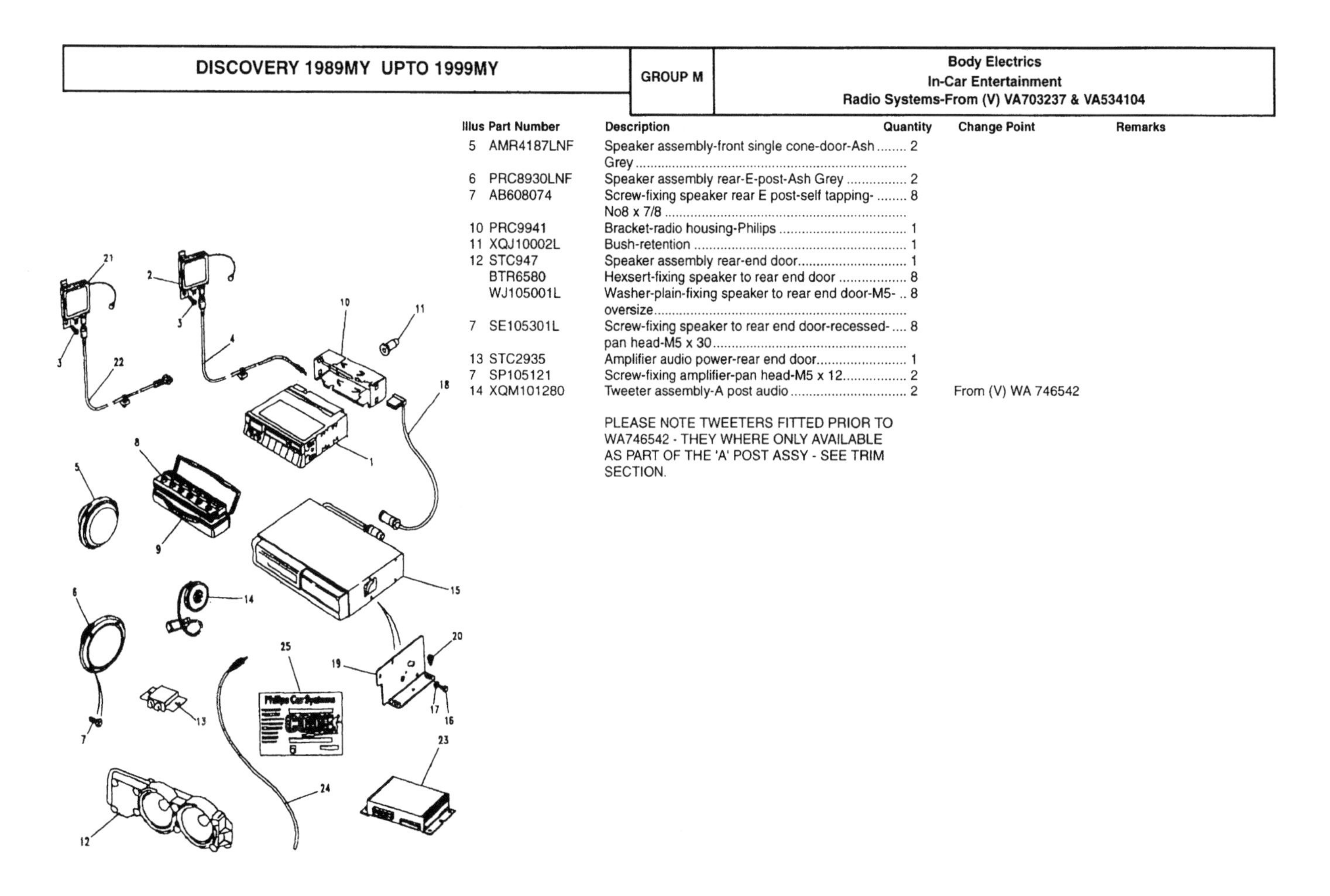

Illus	Part Number	Description	Quantity	Change Point	Remarks
5	AMR4187LNF	Speaker assembly-front single cone-door-Ash Grey	2		
6	PRC8930LNF	Speaker assembly rear-E-post-Ash Grey	2		
7	AB608074	Screw-fixing speaker rear E post-self tapping-No8 x 7/8	8		
10	PRC9941	Bracket-radio housing-Philips	1		
11	XQJ10002L	Bush-retention	1		
12	STC947	Speaker assembly rear-end door	1		
	BTR6580	Hexsert-fixing speaker to rear end door	8		
	WJ105001L	Washer-plain-fixing speaker to rear end door-M5-oversize	8		
7	SE105301L	Screw-fixing speaker to rear end door-recessed-pan head-M5 x 30	8		
13	STC2935	Amplifier audio power-rear end door	1		
7	SP105121	Screw-fixing amplifier-pan head-M5 x 12	2		
14	XQM101280	Tweeter assembly-A post audio	2	From (V) WA 746542	

PLEASE NOTE TWEETERS FITTED PRIOR TO WA746542 - THEY WHERE ONLY AVAILABLE AS PART OF THE 'A' POST ASSY - SEE TRIM SECTION.

Illus	Part Number	Description	Quantity	Change Point	Remarks
15	AMR3154	CD AUTOMATIC CHANGER-PHILIPS	1		
	STC1624	• Magazine-automatic changer compact disc player	1		
16	SH104101L	Screw-fixing CD autochanger-hexagonal head-M4 x 10	4		
17	WC104001	Washer-fixing CD autochanger-M4	4		
18	AMR4831L	Link-compact disc automatic changer-Philips	1		Except Japan
18	AMR4832	Link-compact disc automatic changer-Pioneer	1		Japan
19	AMR4184	Bracket-compact disc mounting-RH	1		
19	AMR4183	Bracket-compact disc mounting-LH	1		
20	FS105107	Screw-fixing CD bracket-hexagonal head-M5 x 10	4		
21	AMR5271	Amplifier-audio antenna-LH rear quarter window-FM	1		Japan
22	AMR5538	Lead antenna-FM	1		Japan
23	AMR5117	Amplifier assembly-audio power-footwell	1		Japan
25	STC3709	Label-warning	2		

Illus	Part Number	Description	Quantity	Change Point	Remarks
1	PRC7096	Motor & bracket assembly-windscreen wiper	1		
	STC1308	Seal-wiper linkage joint	1		
3	STC1306	Link assembly-windscreen wiper-RHD	NLA		Use AK612014 qty 6 with AMR3869 qty 1 with STC1883 qty 1 with WJ106001L qty 6 with 79131 qty 6 with PRC7781 qty 1.
3	STC1307	Link assembly-windscreen wiper-RHD	NLA		Use AK612014 qty 6 with AMR1515 qty 1 with AMR1882 qty 1 with STC1883 qty 1 with WJ106001L qty 6 with 79131 qty 6.
4	78759	Screw	1		
5	WP8019L	Washer	1		
6	PRC1352	Spacer	1		
7	79240	Nutsert-blind	1		
8	FS106167L	Screw-hexagonal head-M6 x 16	1		
9	WA106041L	Washer-M6-standard.	1		
10	WL106001L	Washer-M6	1		
11	PRC7781	Seal-windscreen wiper motor mounting bracket	1		
12	AB612081L	Screw-self tapping B-No 12 x 1"	4		
13	WJ106001L	Washer-plain-M6	4		
14	AK612011L	Nut	4		
15	PRC8235	Arm-windscreen wiper-RHD	2		
15	PRC8236	Arm-windscreen wiper-LHD	2		
16	NH108041L	Nut-hexagonal head-M8	2		
17	PRC8253	Cap-windscreen wiper spindle	2		
18	AMR1805	Blade-windscreen wiper	2		

Windscreen Wiper - From (V) MA081992

Illus	Part Number	Description	Quantity	Change Point	Remarks
		FROM (V) MA081992			
		MOTOR & BRACKET ASSEMBLY-WINDSCREEN WIPER			
1	AMR1514	RHD	1	To (V) MA 090726	
1	AMR3869	RHD	1	From (V) MA 090727	
1	AMR1515	LHD	1		
		WIPER ASSEMBLY FIXINGS			
2	AMR1882	Seal-wiper linkage joint	1		
3	79131	Screw	6		
4	NSP	Washer	NLA		
5	AK612011L	Nut	6		
6	SH106201	Screw-hexagon socket-M6 x 20-long-zinc	NLA		Use FS106201L.
6	FS106201L	Screw-flanged head-M6 x 20	1		
7	PRC1352	Spacer	1		
8	FM106041	Nut-flange-M6	1		
		ARM-WINDSCREEN WIPER			
9	PRC8235	RHD	2	Note (1)	
9	PRC8236	LHD	2		
9	DKB102720	RHD	2	Note (2)	
9	DKB102710	LHD	2		
10	NH108041L	Nut-hexagonal head-M8	2		
11	PRC8253	Cap-windscreen wiper spindle	2		
12	AMR1805	Blade-windscreen wiper	2	Note (1)	
12	DKC100920	Blade-windscreen wiper-passenger	2	Note (2)	
12	DKC100900	Blade-windscreen wiper-with spoiler-driver-RHD	1		
12	DKC100910	Blade-windscreen wiper-with spoiler-driver-LHD	1		

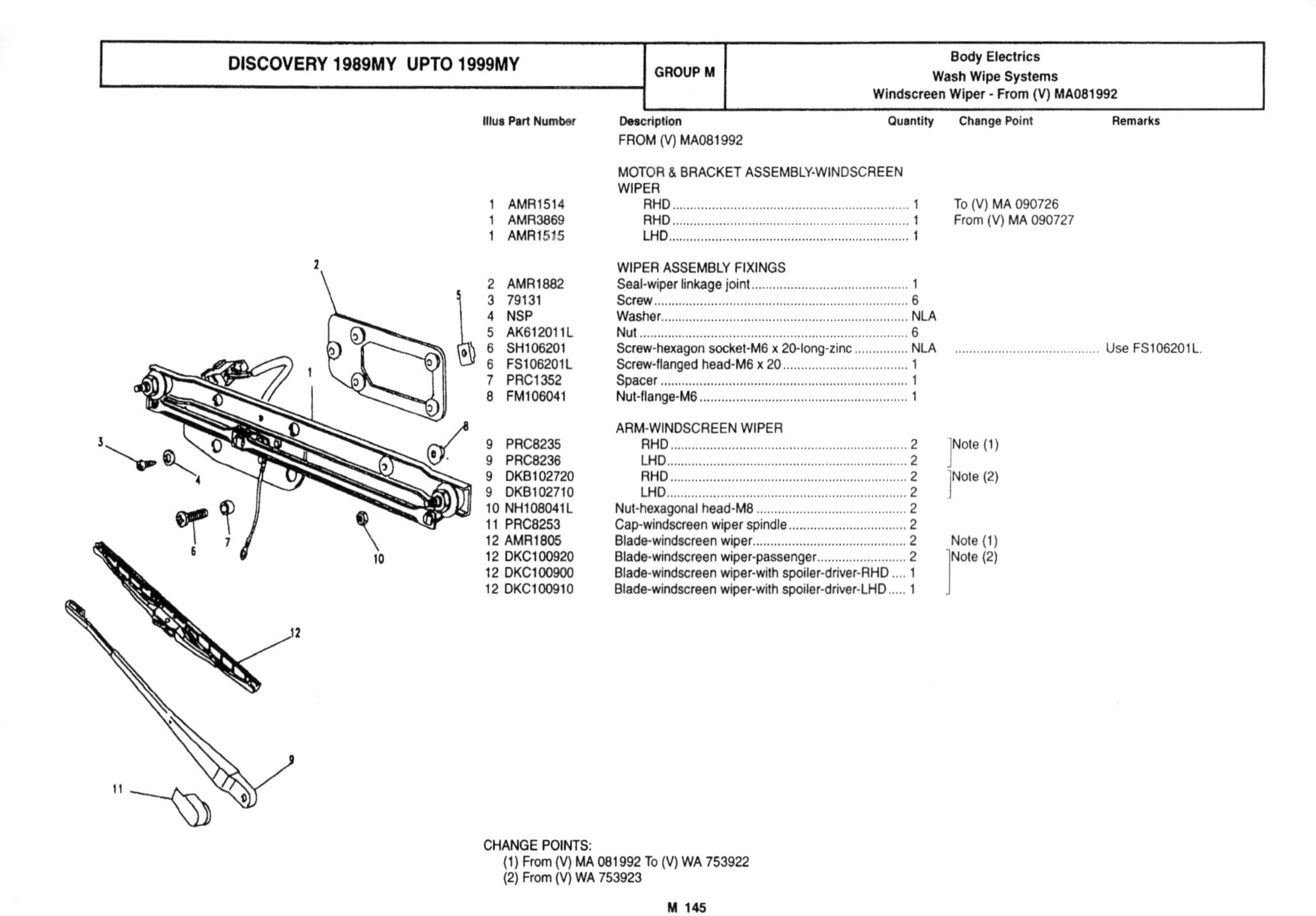

CHANGE POINTS:
(1) From (V) MA 081992 To (V) WA 753922
(2) From (V) WA 753923

Windscreen Wash - To (V) LA081991

Illus	Part Number	Description	Quantity	Change Point	Remarks
	PRC7307	CONTAINER ASSEMBLY-WINDSCREEN WASH	NLA		Use AMR3090 qty 1 with AMR3849 qty 1.
1	PRC7460	• Container-windscreen wash	1		
2	RTC3961	• Cap-washer system	1		
3	STC206	• Filter-container assembly filler neck	1		
4	PRC7045	• Pump assembly-windscreen wash-front	NLA		Use AMR3849 qty 1 with STC1454 qty 1.
5	ADU3905	• Pump assembly-backlight wash	NLA		Use AMR3849 qty 1 with STC1453 qty 1.
6	RTC3959	• Grommet-container assembly pump	2		
7	AMR3658	• Hose-wash system pump to non return valve-5mm	1		
8	AMR4060	• Clip-hose-5mm/ID	1		
9	AMR3659	• Hose-wash system pump to non return valve-4mm	1		
10	PRC6857	• Valve-wash system non return-5mm-front	1		
11	ACU5037	• Valve-wash system non return-4mm	1		
12	SH604061L	Screw-hexagonal head-1/4UNF x 3/4	3		
13	WF600041L	Washer- shakeproof	3		
14	PRC6852	Hose-wash system-720mm/long	A/R		
15	C15644L	T piece	1		
16	RTC3641	Hose-wash system jet to tee valve	A/R		
17	PRC5273	Clip-omega-Grey-10.5 x 8mm-6.5mm hole	1		
18	PRC7304	Jet assembly-nozzle-wash system	2		

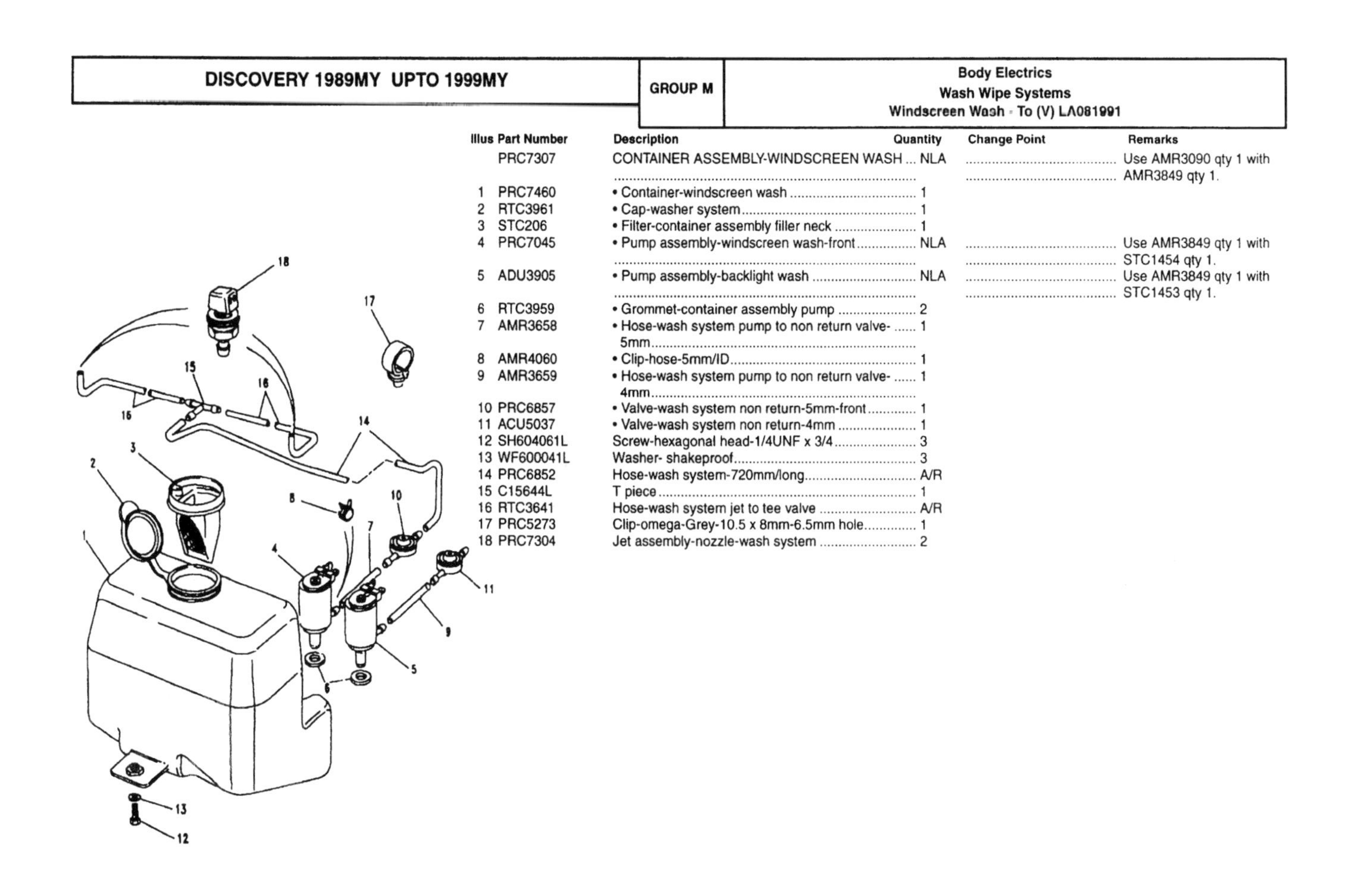

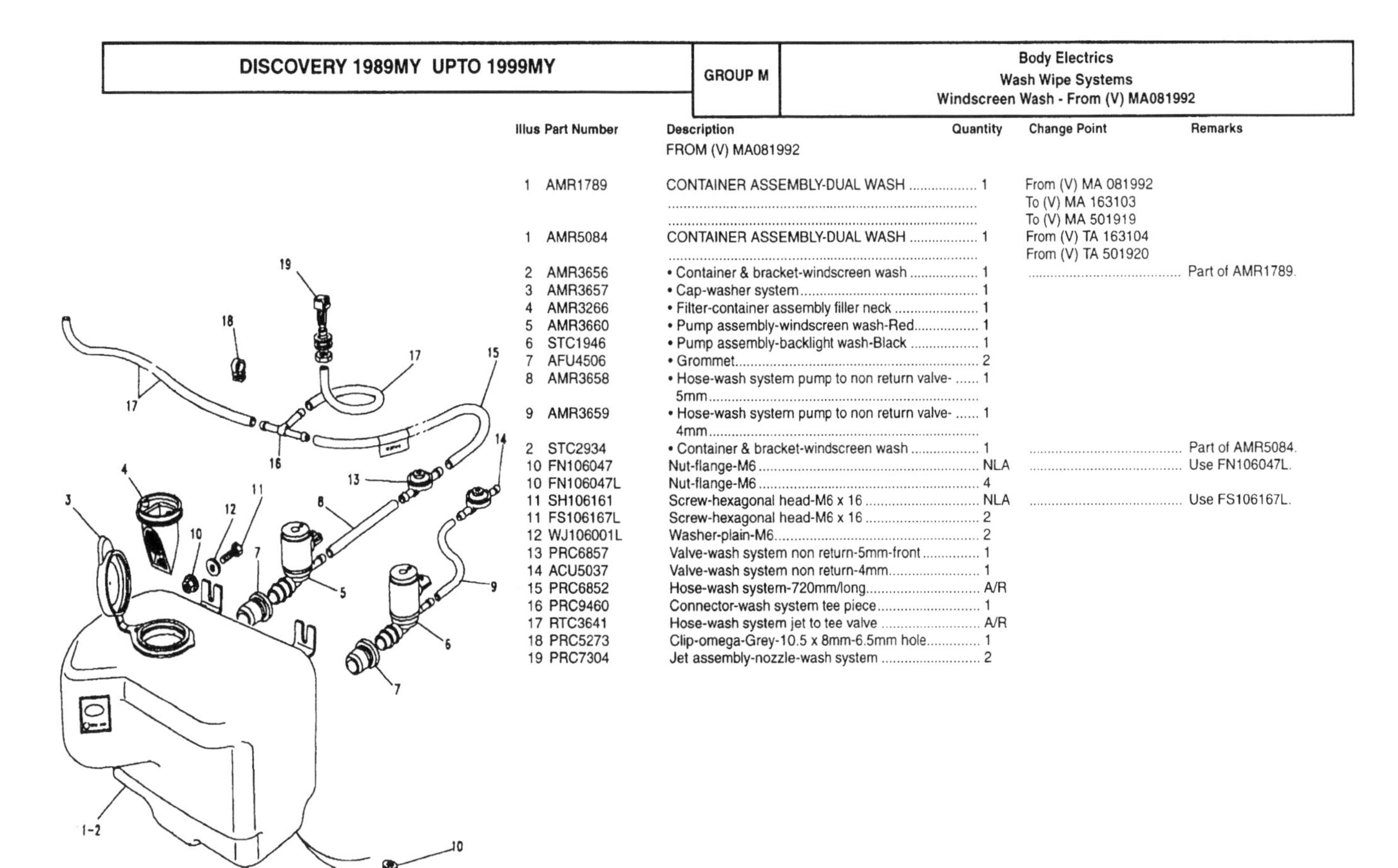

Illus	Part Number	Description	Quantity	Change Point	Remarks
		FROM (V) MA081992			
1	AMR1789	CONTAINER ASSEMBLY-DUAL WASH	1	From (V) MA 081992 To (V) MA 163103 To (V) MA 501919	
1	AMR5084	CONTAINER ASSEMBLY-DUAL WASH	1	From (V) TA 163104 From (V) TA 501920	
2	AMR3656	• Container & bracket-windscreen wash	1		Part of AMR1789.
3	AMR3657	• Cap-washer system	1		
4	AMR3266	• Filter-container assembly filler neck	1		
5	AMR3660	• Pump assembly-windscreen wash-Red	1		
6	STC1946	• Pump assembly-backlight wash-Black	1		
7	AFU4506	• Grommet	2		
8	AMR3658	• Hose-wash system pump to non return valve-5mm	1		
9	AMR3659	• Hose-wash system pump to non return valve-4mm	1		
2	STC2934	• Container & bracket-windscreen wash	1		Part of AMR5084.
10	FN106047	Nut-flange-M6	NLA		Use FN106047L.
10	FN106047L	Nut-flange-M6	4		
11	SH106161	Screw-hexagonal head-M6 x 16	NLA		Use FS106167L.
11	FS106167L	Screw-hexagonal head-M6 x 16	2		
12	WJ106001L	Washer-plain-M6	2		
13	PRC6857	Valve-wash system non return-5mm-front	1		
14	ACU5037	Valve-wash system non return-4mm	1		
15	PRC6852	Hose-wash system-720mm/long	A/R		
16	PRC9460	Connector-wash system tee piece	1		
17	RTC3641	Hose-wash system jet to tee valve	A/R		
18	PRC5273	Clip-omega-Grey-10.5 x 8mm-6.5mm hole	1		
19	PRC7304	Jet assembly-nozzle-wash system	2		

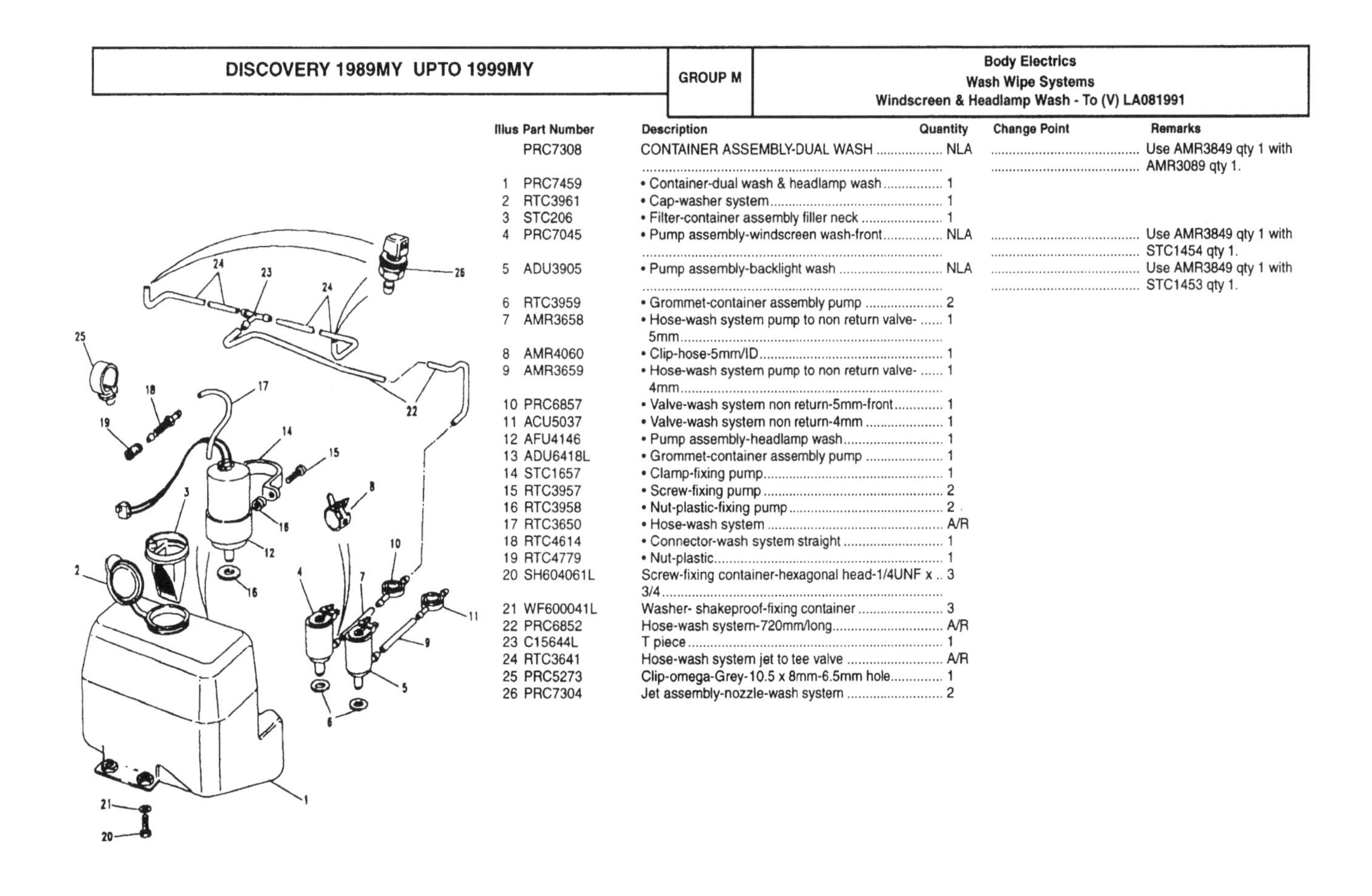

Illus	Part Number	Description	Quantity	Change Point	Remarks
	PRC7308	CONTAINER ASSEMBLY-DUAL WASH	NLA		Use AMR3849 qty 1 with AMR3089 qty 1.
1	PRC7459	• Container-dual wash & headlamp wash	1		
2	RTC3961	• Cap-washer system	1		
3	STC206	• Filter-container assembly filler neck	1		
4	PRC7045	• Pump assembly-windscreen wash-front	NLA		Use AMR3849 qty 1 with STC1454 qty 1.
5	ADU3905	• Pump assembly-backlight wash	NLA		Use AMR3849 qty 1 with STC1453 qty 1.
6	RTC3959	• Grommet-container assembly pump	2		
7	AMR3658	• Hose-wash system pump to non return valve-5mm	1		
8	AMR4060	• Clip-hose-5mm/ID	1		
9	AMR3659	• Hose-wash system pump to non return valve-4mm	1		
10	PRC6857	• Valve-wash system non return-5mm-front	1		
11	ACU5037	• Valve-wash system non return-4mm	1		
12	AFU4146	• Pump assembly-headlamp wash	1		
13	ADU6418L	• Grommet-container assembly pump	1		
14	STC1657	• Clamp-fixing pump	1		
15	RTC3957	• Screw-fixing pump	2		
16	RTC3958	• Nut-plastic-fixing pump	2		
17	RTC3650	• Hose-wash system	A/R		
18	RTC4614	• Connector-wash system straight	1		
19	RTC4779	• Nut-plastic	1		
20	SH604061L	Screw-fixing container-hexagonal head-1/4UNF x 3/4	3		
21	WF600041L	Washer- shakeproof-fixing container	3		
22	PRC6852	Hose-wash system-720mm/long	A/R		
23	C15644L	T piece	1		
24	RTC3641	Hose-wash system jet to tee valve	A/R		
25	PRC5273	Clip-omega-Grey-10.5 x 8mm-6.5mm hole	1		
26	PRC7304	Jet assembly-nozzle-wash system	2		

DISCOVERY 1989MY UPTO 1999MY

GROUP M

Body Electrics
Wash Wipe Systems
Windscreen & Headlamp Wash - From (V) MA081992

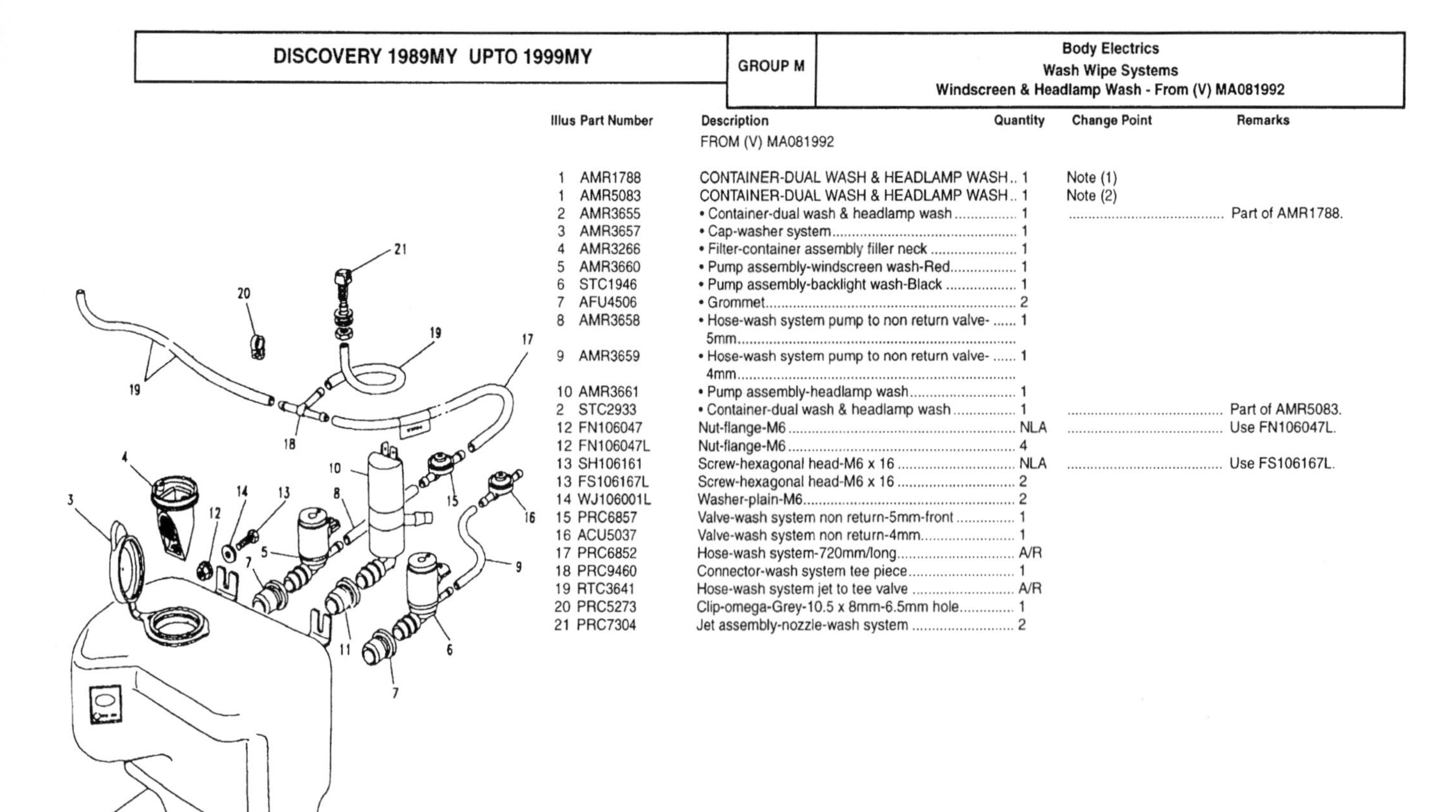

Illus	Part Number	Description	Quantity	Change Point	Remarks
		FROM (V) MA081992			
1	AMR1788	CONTAINER-DUAL WASH & HEADLAMP WASH	1	Note (1)	
1	AMR5083	CONTAINER-DUAL WASH & HEADLAMP WASH	1	Note (2)	
2	AMR3655	• Container-dual wash & headlamp wash	1		Part of AMR1788.
3	AMR3657	• Cap-washer system	1		
4	AMR3266	• Filter-container assembly filler neck	1		
5	AMR3660	• Pump assembly-windscreen wash-Red	1		
6	STC1946	• Pump assembly-backlight wash-Black	1		
7	AFU4506	• Grommet	2		
8	AMR3658	• Hose-wash system pump to non return valve-5mm	1		
9	AMR3659	• Hose-wash system pump to non return valve-4mm	1		
10	AMR3661	• Pump assembly-headlamp wash	1		
2	STC2933	• Container-dual wash & headlamp wash	1		Part of AMR5083.
12	FN106047	Nut-flange-M6	NLA		Use FN106047L.
12	FN106047L	Nut-flange-M6	4		
13	SH106161	Screw-hexagonal head-M6 x 16	NLA		Use FS106167L.
13	FS106167L	Screw-hexagonal head-M6 x 16	2		
14	WJ106001L	Washer-plain-M6	2		
15	PRC6857	Valve-wash system non return-5mm-front	1		
16	ACU5037	Valve-wash system non return-4mm	1		
17	PRC6852	Hose-wash system-720mm/long	A/R		
18	PRC9460	Connector-wash system tee piece	1		
19	RTC3641	Hose-wash system jet to tee valve	A/R		
20	PRC5273	Clip-omega-Grey-10.5 x 8mm-6.5mm hole	1		
21	PRC7304	Jet assembly-nozzle-wash system	2		

CHANGE POINTS:
(1) From (V) MA 081992 To (V) MA 163103; From (V) MA 081992 To (V) MA 501919
(2) From (V) TA 163104; From (V) TA 501920

M 149

DISCOVERY 1989MY UPTO 1999MY

GROUP M

Body Electrics
Wash Wipe Systems
Headlamp Wash - To (V) LA081991

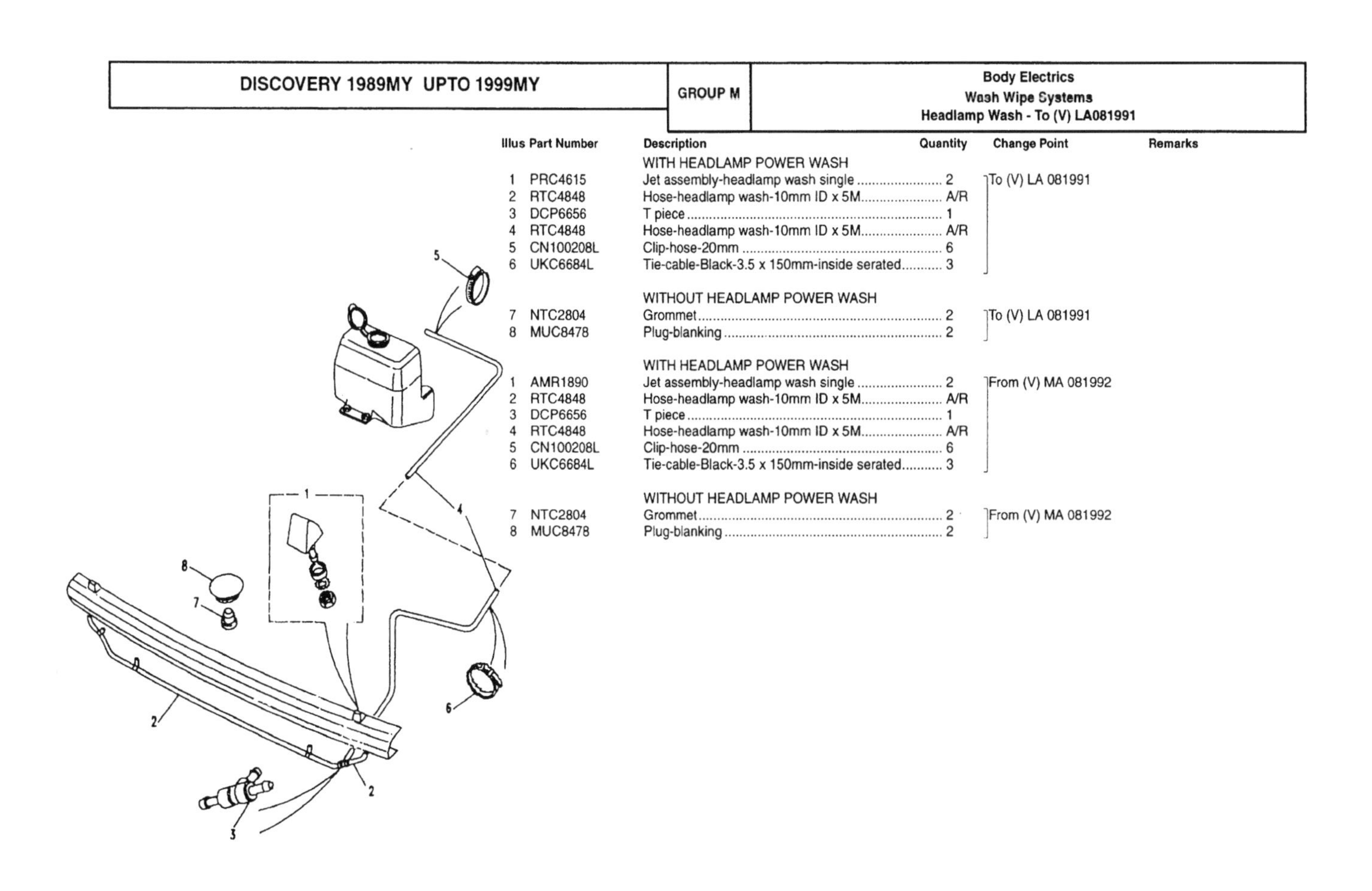

Illus	Part Number	Description	Quantity	Change Point	Remarks
		WITH HEADLAMP POWER WASH			
1	PRC4615	Jet assembly-headlamp wash single	2	To (V) LA 081991	
2	RTC4848	Hose-headlamp wash-10mm ID x 5M	A/R		
3	DCP6656	T piece	1		
4	RTC4848	Hose-headlamp wash-10mm ID x 5M	A/R		
5	CN100208L	Clip-hose-20mm	6		
6	UKC6684L	Tie-cable-Black-3.5 x 150mm-inside serated	3		
		WITHOUT HEADLAMP POWER WASH			
7	NTC2804	Grommet	2	To (V) LA 081991	
8	MUC8478	Plug-blanking	2		
		WITH HEADLAMP POWER WASH			
1	AMR1890	Jet assembly-headlamp wash single	2	From (V) MA 081992	
2	RTC4848	Hose-headlamp wash-10mm ID x 5M	A/R		
3	DCP6656	T piece	1		
4	RTC4848	Hose-headlamp wash-10mm ID x 5M	A/R		
5	CN100208L	Clip-hose-20mm	6		
6	UKC6684L	Tie-cable-Black-3.5 x 150mm-inside serated	3		
		WITHOUT HEADLAMP POWER WASH			
7	NTC2804	Grommet	2	From (V) MA 081992	
8	MUC8478	Plug-blanking	2		

M 150

Illus	Part Number	Description	Quantity	Change Point	Remarks
		TO (V) LA081991			
2	PRC6482	Motor-backlight wiper	NLA	To (V) JA 021175	Use STC1384.
1	STC1384	Motor-backlight wiper	NLA		Use STC3316 For vehicles prior to MA a later wiper arm will be needed - AMR3873.
1	STC3316	Motor-backlight wiper	1	From (V) MA 081992	
3	SH105201L	Screw-hexagonal head-M5 x 20	5		
4	WJ105001L	Washer-plain-M5-oversize	5		
5	WL105001L	Washer-M5	5		
6	NH105041L	Nut-hexagonal-M5	5	From (V) JA 021176	
7	PRC6484	Arm assembly-backlight wiper	1	To (V) JA 033862	
7	AMR2306	Arm assembly-backlight wiper	1	From (V) JA 033863	
8	PRC8160	Cap-backlight wiper spindle	1		
9	PRC6447	Blade-backlight wiper	1	To (V) JA 033862	
9	AMR1806	Blade-backlight wiper	1	From (V) JA 033863	
10	PRC6496	Jet assembly-backlight wash single nozzle	1	To (V) JA 034313	
10	PRC9008	Jet assembly-backlight wash single nozzle	1	From (V) KA 034314	
11	RTC3650	Hose-wash system	A/R		
12	233243	Grommet-Black-rubber-3 x 25.5mm	1		
13	PRC7417	Grommet-wash hose connector	1		
14	369092	Clip-pipe-rear	3		
15	AFU1090L	Clip-cable-8mm hole	2		

M 151

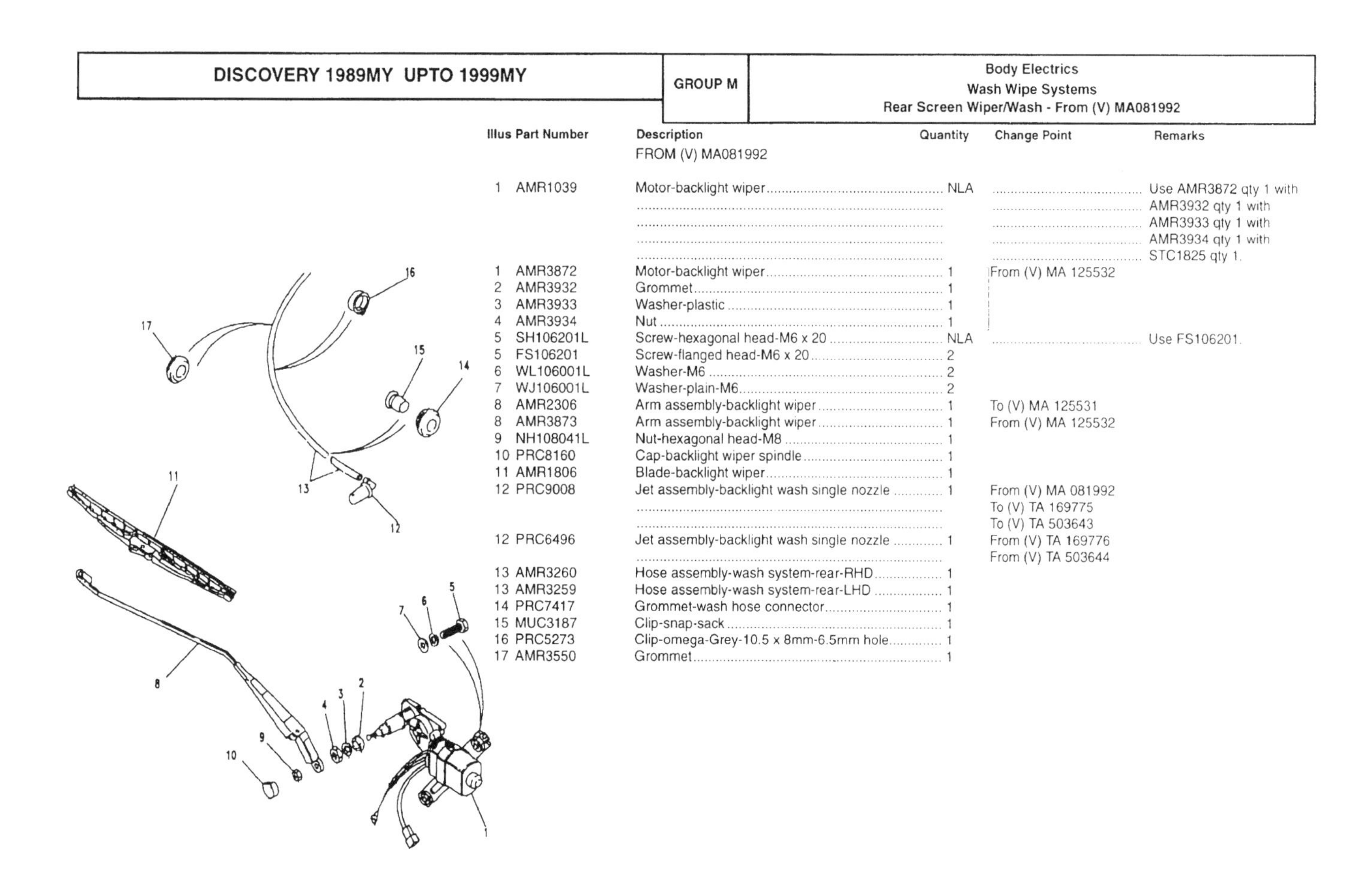

Illus	Part Number	Description	Quantity	Change Point	Remarks
		FROM (V) MA081992			
1	AMR1039	Motor-backlight wiper	NLA		Use AMR3872 qty 1 with AMR3932 qty 1 with AMR3933 qty 1 with AMR3934 qty 1 with STC1825 qty 1.
1	AMR3872	Motor-backlight wiper	1	From (V) MA 125532	
2	AMR3932	Grommet	1		
3	AMR3933	Washer-plastic	1		
4	AMR3934	Nut	1		
5	SH106201L	Screw-hexagonal head-M6 x 20	NLA		Use FS106201.
5	FS106201	Screw-flanged head-M6 x 20	2		
6	WL106001L	Washer-M6	2		
7	WJ106001L	Washer-plain-M6	2		
8	AMR2306	Arm assembly-backlight wiper	1	To (V) MA 125531	
8	AMR3873	Arm assembly-backlight wiper	1	From (V) MA 125532	
9	NH108041L	Nut-hexagonal head-M8	1		
10	PRC8160	Cap-backlight wiper spindle	1		
11	AMR1806	Blade-backlight wiper	1		
12	PRC9008	Jet assembly-backlight wash single nozzle	1	From (V) MA 081992 To (V) TA 169775 To (V) TA 503643	
12	PRC6496	Jet assembly-backlight wash single nozzle	1	From (V) TA 169776 From (V) TA 503644	
13	AMR3260	Hose assembly-wash system-rear-RHD	1		
13	AMR3259	Hose assembly-wash system-rear-LHD	1		
14	PRC7417	Grommet-wash hose connector	1		
15	MUC3187	Clip-snap-sack	1		
16	PRC5273	Clip-omega-Grey-10.5 x 8mm-6.5mm hole	1		
17	AMR3550	Grommet	1		

M 152

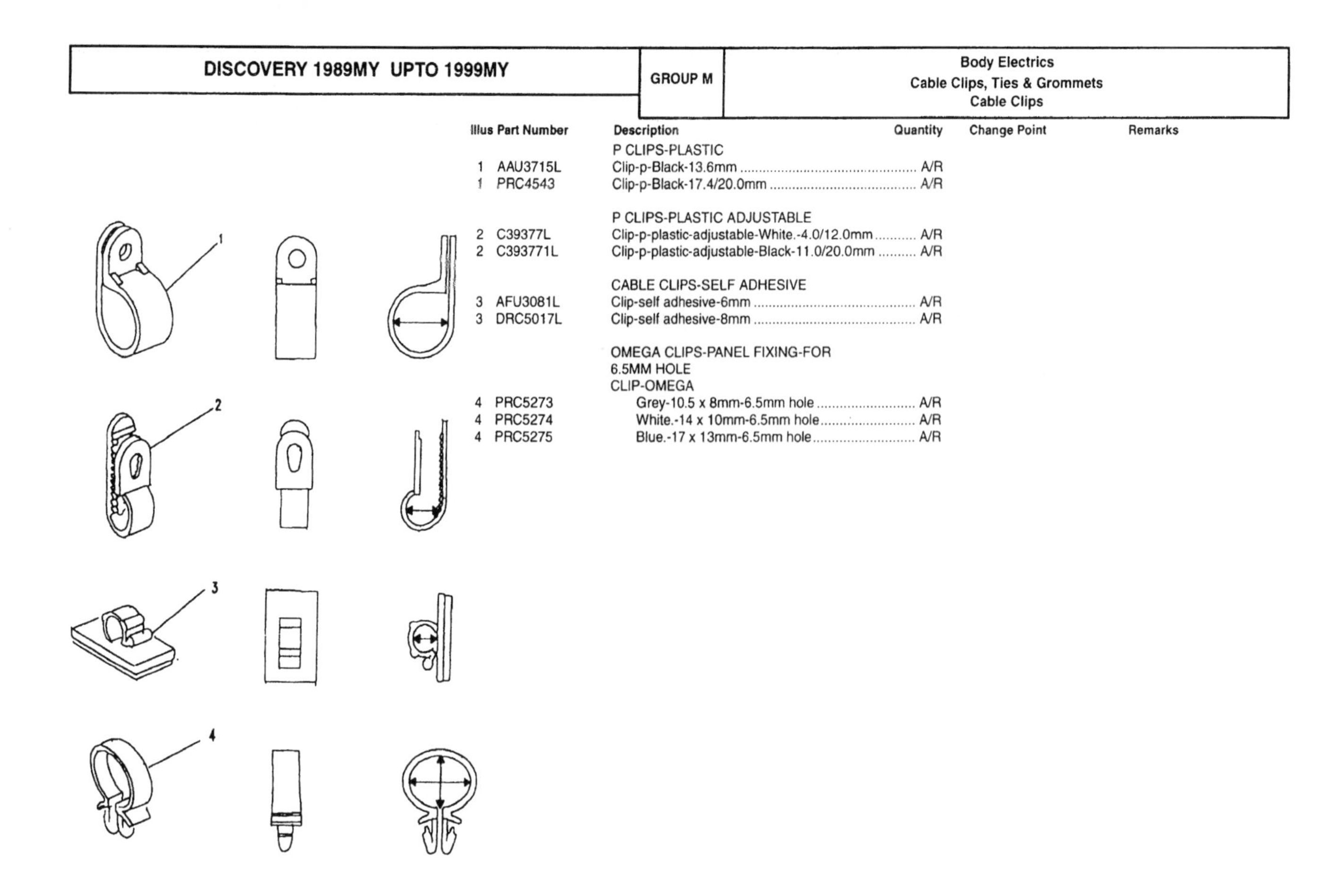

Illus	Part Number	Description	Quantity	Change Point	Remarks
		P CLIPS-PLASTIC			
1	AAU3715L	Clip-p-Black-13.6mm	A/R		
1	PRC4543	Clip-p-Black-17.4/20.0mm	A/R		
		P CLIPS-PLASTIC ADJUSTABLE			
2	C39377L	Clip-p-plastic-adjustable-White.-4.0/12.0mm	A/R		
2	C393771L	Clip-p-plastic-adjustable-Black-11.0/20.0mm	A/R		
		CABLE CLIPS-SELF ADHESIVE			
3	AFU3081L	Clip-self adhesive-6mm	A/R		
3	DRC5017L	Clip-self adhesive-8mm	A/R		
		OMEGA CLIPS-PANEL FIXING-FOR 6.5MM HOLE			
		CLIP-OMEGA			
4	PRC5273	Grey-10.5 x 8mm-6.5mm hole	A/R		
4	PRC5274	White.-14 x 10mm-6.5mm hole	A/R		
4	PRC5275	Blue.-17 x 13mm-6.5mm hole	A/R		

M 153

Illus	Part Number	Description	Quantity	Change Point	Remarks
		HARNESS SUPPORT CLIPS-STEEL			
1	240407	Clip-hose-14.5mm-steel-25/64" hole	A/R		
		HARNESS SUPPORT BRACKETS-STEEL			
2	PRC2979	Bracket-clipping-steel	A/R		
3	ERC9404	Bracket harness	A/R		
		CONNECTOR MOUNTING BRACKETS-STEEL			
4	AMR1935	Bracket-mounting-steel	A/R		

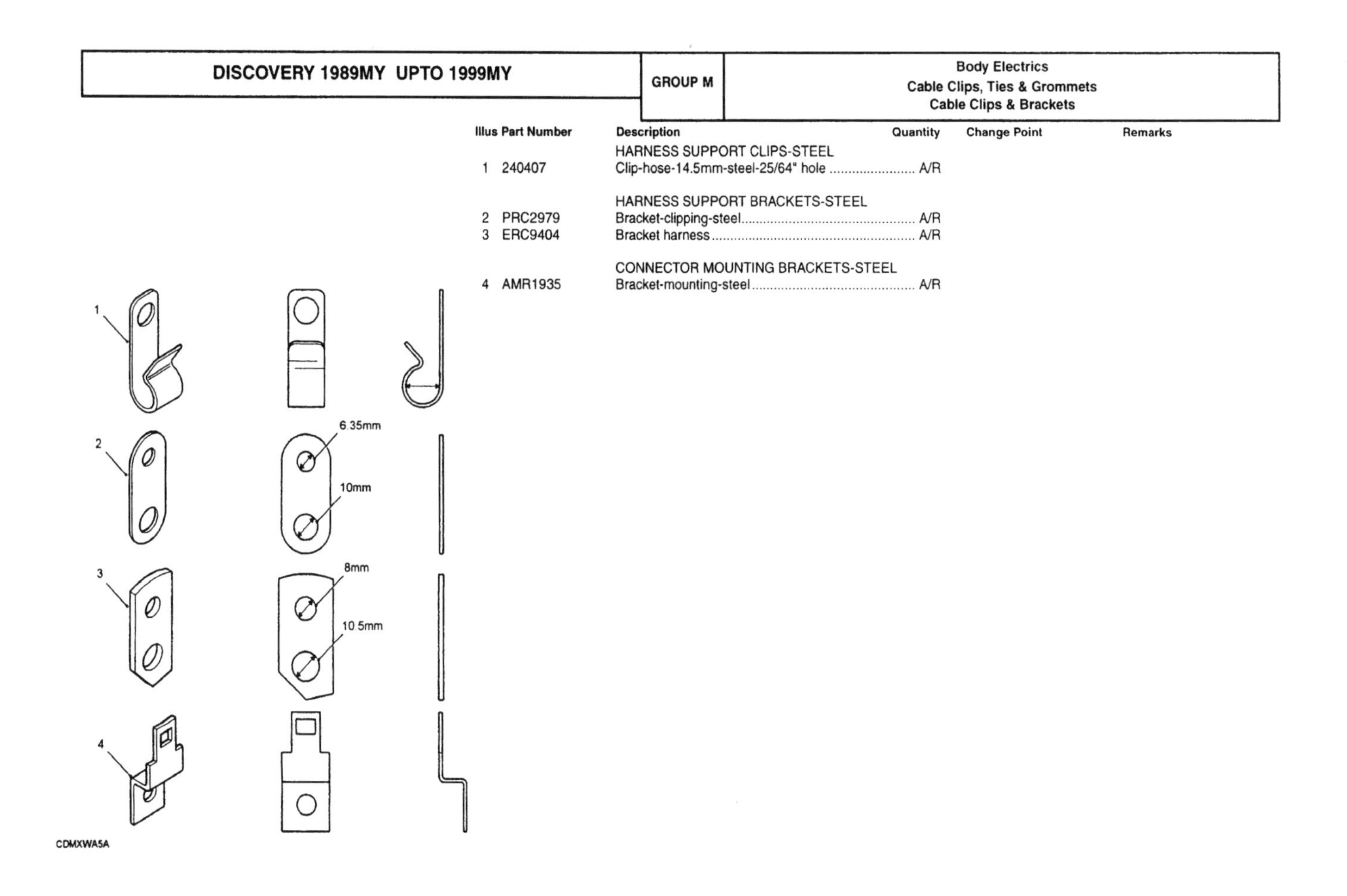

M 154

Illus	Part Number	Description	Quantity	Change Point	Remarks
		CABLE TIES-STANDARD-OUTSIDE SERRATED			
1	C43640	Tie-cable-Black-2.5 x 100mm-outside serated	A/R		
		CABLE TIES-STANDARD-INSIDE SERRATED			
		TIE-CABLE			
2	DRC1538	Black-2.5 x 100mm-inside serated	NLA		Use C43640.
2	UKC6684L	Black-3.5 x 150mm-inside serated	A/R		
2	NTC1154	3.5 x 300mm-inside serated	A/R		
2	594594	White.-4.6 x 385mm-inside serated	A/R		
2	573246	Black-4.8 x 115mm-inside serated	A/R		
2	RTC3772	Black-4.8 x 135mm-inside serated	A/R		
2	568680	4.8 x 270mm-inside serated	A/R		
2	ESR3301	Black-13.0 x 540mm-inside serated	A/R		
		CABLE TIES-RELEASABLE			
		TIE-CABLE			
3	13H9704L	Black-6.5 x 100mm-releasable	A/R		
3	13H9727L	Black-7.5 x 145mm-releasable	A/R		
3	AFU4173	Black-8.8 x 770mm-releasable	A/R		
		CABLE TIES-PANEL FIXING TYPE FOR PANELS UP TO 4MM THICK			
4	YQR10016L	Tie-cable-for panels up to 4mm thick-7mm hole-4.6 x 165mm-4.50mm	A/R		
		CABLE TIES-PANEL FIXING TYPE FOR PANELS UP TO 2.5MM THICK			
		TIE-CABLE			
4	YQR10075	for panels up to 2.5mm thick-Grey-4.6 x 165mm	A/R		
4	YYD10003	8mm hole-4.6 x 215mm	A/R		
5	AAU3686	6.5mm hole-8.0 x 155mm	A/R		

Illus	Part Number	Description	Quantity	Change Point	Remarks
		CABLE STRAPS-PLASTIC ADJUSTABLE-FOR 6.4MM HOLE			
1	ADU7981L	Tie-cable-plastic-adjustable-24mm diameter-97mm-6.4mm hole	A/R		
1	ADU8981	Tie-cable-plastic-adjustable-122mm-32mm diameter-6.4mm hole	A/R		
		CABLE TIES-STUD FIXING			
1	YYC10294L	Tie-cable-stud fixing-fixed contact-Clear.	A/R		

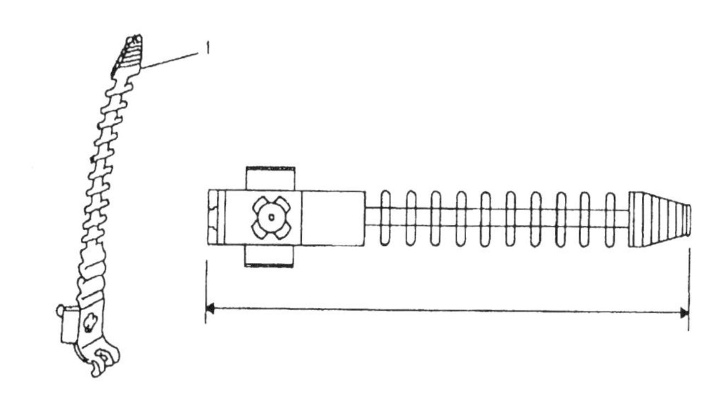

Illus	Part Number	Description	Quantity	Change Point	Remarks
		CRADLE CLIPS-PANEL FIXING			
1	CLP8934	Clip-cradle-13mm-8mm hole	A/R		
		CRADLE CLIPS-STUD FIXING			
2	AMR2323	Clip-cradle-13mm-5mm	A/R		
		SWIVEL CLIPS			
3	ESR301	Clip-swivel-10 x 10mm	A/R		
4	ADU7739L	Clip-swivel-15 x 10mm	A/R		

M 157

Illus	Part Number	Description	Quantity	Change Point	Remarks
		CABLE CLIPS-PANEL FIXING			
		CLIP-CABLE			
1	PRC3702	5mm-4.8mm hole	A/R		
2	AFU2879	5mm-6.5mm hole	A/R		
3	AFU1090L	8mm hole	A/R		
		CABLE CLIPS-EDGE FIXING			
4	AFU1296	Clip-edge harness-12mm-Clear	A/R		

M 158

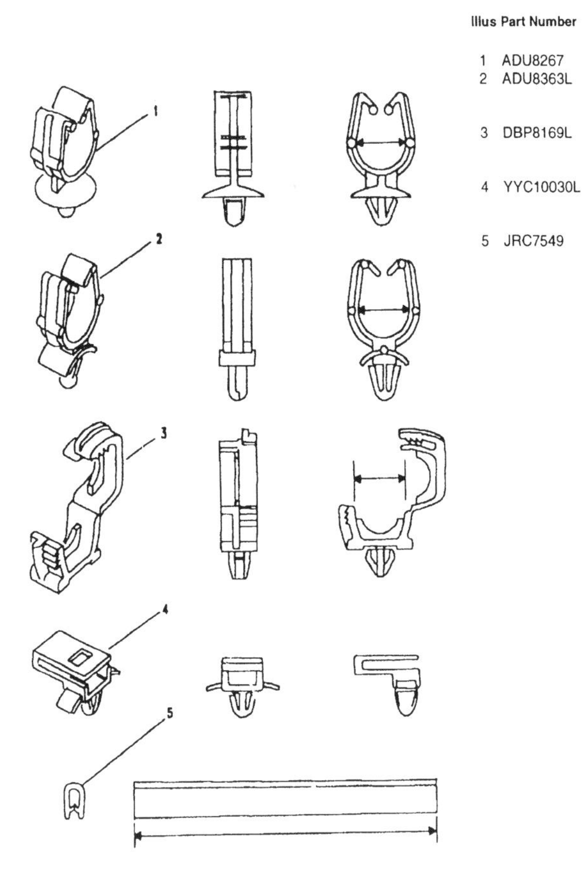

Illus	Part Number	Description	Quantity	Change Point	Remarks
		HARNESS CLIPS-PANEL FIXING			
1	ADU8267	Clip-harness-14mm-6mm hole	A/R		
2	ADU8363L	Clip-harness-14.5mm-6.5mm hole	A/R		
		SNAP CLIPS-PANEL FIXING-FOR 7MM HOLE			
3	DBP8169L	Clip-pivot-snap-13-17mm-7mm hole-Grey	A/R		
		HARNESS ADAPTOR CLIPS-PANEL FIXING			
4	YYC10030L	Clip-harness-adaptor-6.5mm hole	A/R		
		EDGE PROTECTION STRIP			
5	JRC7549	Escutcheon-headrest-89mm	A/R		

M 159

Illus	Part Number	Description	Quantity	Change Point	Remarks
		PLUG-BLANKING			
1	338029	8mm	A/R		
1	338013	9.5mm	A/R		
1	338014	11mm	A/R		
1	338015	13mm	A/R		
1	ANR1369	13.5mm	A/R		
1	338017	14mm	A/R		
1	338018	16mm	A/R		
1	338019	17.5mm	A/R		
1	338020	19mm	A/R		
1	338021	20.5mm	A/R		
1	338023	25.5mm	A/R		
1	BD155888	32mm	A/R		
1	338025	35mm	A/R		
1	338026	38mm	A/R		
1	AFU2627	41.5mm	A/R		
1	338009	47.5mm	A/R		
1	338028	50mm	A/R		
		GROMMET			
2	233244	cable-3 x 19mm-Black	A/R		
2	233243	Black-rubber-3 x 25.5mm	A/R		
2	269257	cable-6 x 33.5mm-Black	A/R		
2	555711	6 x 33.5mm-Red	A/R		
2	276054	cable-6 x 41.5mm-Black	A/R		
2	589452	cable-7 x 53mm-Black	A/R		
2	233566	cable-12.5 x 48mm-Red	A/R		

M 160

Notes

Body .... N 12
Body Colour Labels .... N 45
Body Tapes .... N 106
Body Tapes - 3 Door .... N 99
Body Tapes - FXi XS - Gulf States .... N 104
Body Tapes - FXi2 - Gulf States .... N 102
Body Tapes - FXi3 - Gulf States .... N 103
Body Tapes - FXi4 - Gulf States .... N 105
Body Tapes FXi 1996 .... N 107
Body Tapes XS 1996 .... N 108
Body Warning Labels .... N 44
Bodyshell - 3 Door .... N 16
Bodyshell - 5 Door .... N 17
Bodyside Mouldings - XS Only .... N 92
Bodyside Outer Panels .... N 29
Bonnet Assembly .... N 41
Bonnet Locking & Release .... N 69
Dog Guard - Commercial Only .... N 123
Drip Rail Mouldings .... N 90
Dust Proofing Seals .... N 26
Exterior Mirrors - Manual .... N 112
Exterior Mirrors-Electric-From (V) MA081991 .... N 117
Exterior Mirrors-Electric-To (V) LA081990 .... N 115
Front Bumper Assembly .... N 12
Front Door Assembly .... N 46
Front Wing Assembly .... N 27
Fuel Filler Flap .... N 33
Grille & Headlamp Surround From MA081991 .... N 43
Grille & Headlamp Surround Up To LA081990 .... N 42
Hard Top & Glass - Pick Up - Denmark .... N 120
Inner Bodyside Panels Rear End .... N 22
Locks and Handles - Front Door .... N 76
Locksets .... N 71
Mudflap Assembly .... N 110
Plenum Moulding .... N 89
Rear Bumper Assembly .... N 14
Rear Door Locks .... N 80
Rear End Door Assembly .... N 53
Rear End Door Lock .... N 83
Rear End Panels .... N 32
Rear Side Door Assembly .... N 50
Roof Assembly - Plain .... N 34
Roof Rack .... N 40
Rubbing Strips - Except XS .... N 94
Rubbing Strips - Italy SE Only .... N 93
Side Opening Windows .... N 56
Spare Wheel Mounting - Rear End Door - .... N 121
Styled Bodytapes .... N 87
Sunroof Assembly - Electric .... N 35
Sunroof Assembly Front-From (V) KA034314 .... N 38
Sunroof Assembly Front-To (V) JA034313 .... N 37
Sunroof Assembly Rear .... N 39
Wheelarch & Valance .... N 19
Wheelarch Extension-Germany"Sunseeker" .... N 109
Window Regulators - Front - Manual .... N 60
Windscreens - Alpine Lights & Interior Mirror .... N 66
Chassis .... N 1
Engine Mounting - 200Tdi .... N 4
Engine Mounting - 300Tdi .... N 5
Engine Mounting - MPi .... N 7
Engine Mountings - V8 .... N 6
Frame Assembly .... N 1
Replacement Outriggers .... N 3
Transmission Mounts .... N 8
Under Ride Protection Bar .... N 11

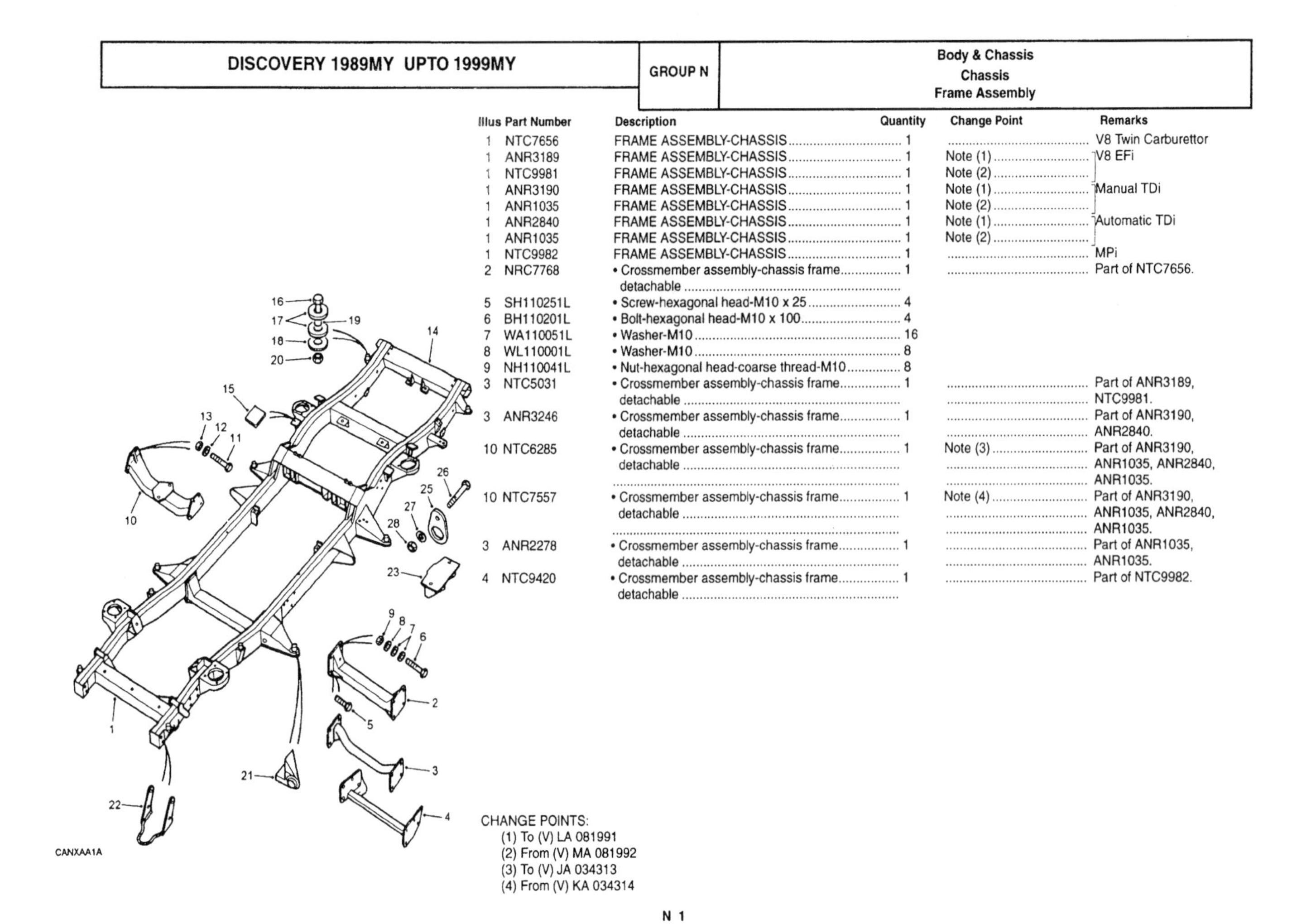

Illus	Part Number	Description	Quantity	Change Point	Remarks
1	NTC7656	FRAME ASSEMBLY-CHASSIS	1		V8 Twin Carburettor
1	ANR3189	FRAME ASSEMBLY-CHASSIS	1	Note (1)	V8 EFi
1	NTC9981	FRAME ASSEMBLY-CHASSIS	1	Note (2)	
1	ANR3190	FRAME ASSEMBLY-CHASSIS	1	Note (1)	Manual TDi
1	ANR1035	FRAME ASSEMBLY-CHASSIS	1	Note (2)	
1	ANR2840	FRAME ASSEMBLY-CHASSIS	1	Note (1)	Automatic TDi
1	ANR1035	FRAME ASSEMBLY-CHASSIS	1	Note (2)	
1	NTC9982	FRAME ASSEMBLY-CHASSIS	1		MPi
2	NRC7768	• Crossmember assembly-chassis frame detachable	1		Part of NTC7656.
5	SH110251L	• Screw-hexagonal head-M10 x 25	4		
6	BH110201L	• Bolt-hexagonal head-M10 x 100	4		
7	WA110051L	• Washer-M10	16		
8	WL110001L	• Washer-M10	8		
9	NH110041L	• Nut-hexagonal head-coarse thread-M10	8		
3	NTC5031	• Crossmember assembly-chassis frame detachable	1		Part of ANR3189, NTC9981.
3	ANR3246	• Crossmember assembly-chassis frame detachable	1		Part of ANR3190, ANR2840.
10	NTC6285	• Crossmember assembly-chassis frame detachable	1	Note (3)	Part of ANR3190, ANR1035, ANR2840, ANR1035.
10	NTC7557	• Crossmember assembly-chassis frame detachable	1	Note (4)	Part of ANR3190, ANR1035, ANR2840, ANR1035.
3	ANR2278	• Crossmember assembly-chassis frame detachable	1		Part of ANR1035, ANR1035.
4	NTC9420	• Crossmember assembly-chassis frame detachable	1		Part of NTC9982.

CHANGE POINTS:
(1) To (V) LA 081991
(2) From (V) MA 081992
(3) To (V) JA 034313
(4) From (V) KA 034314

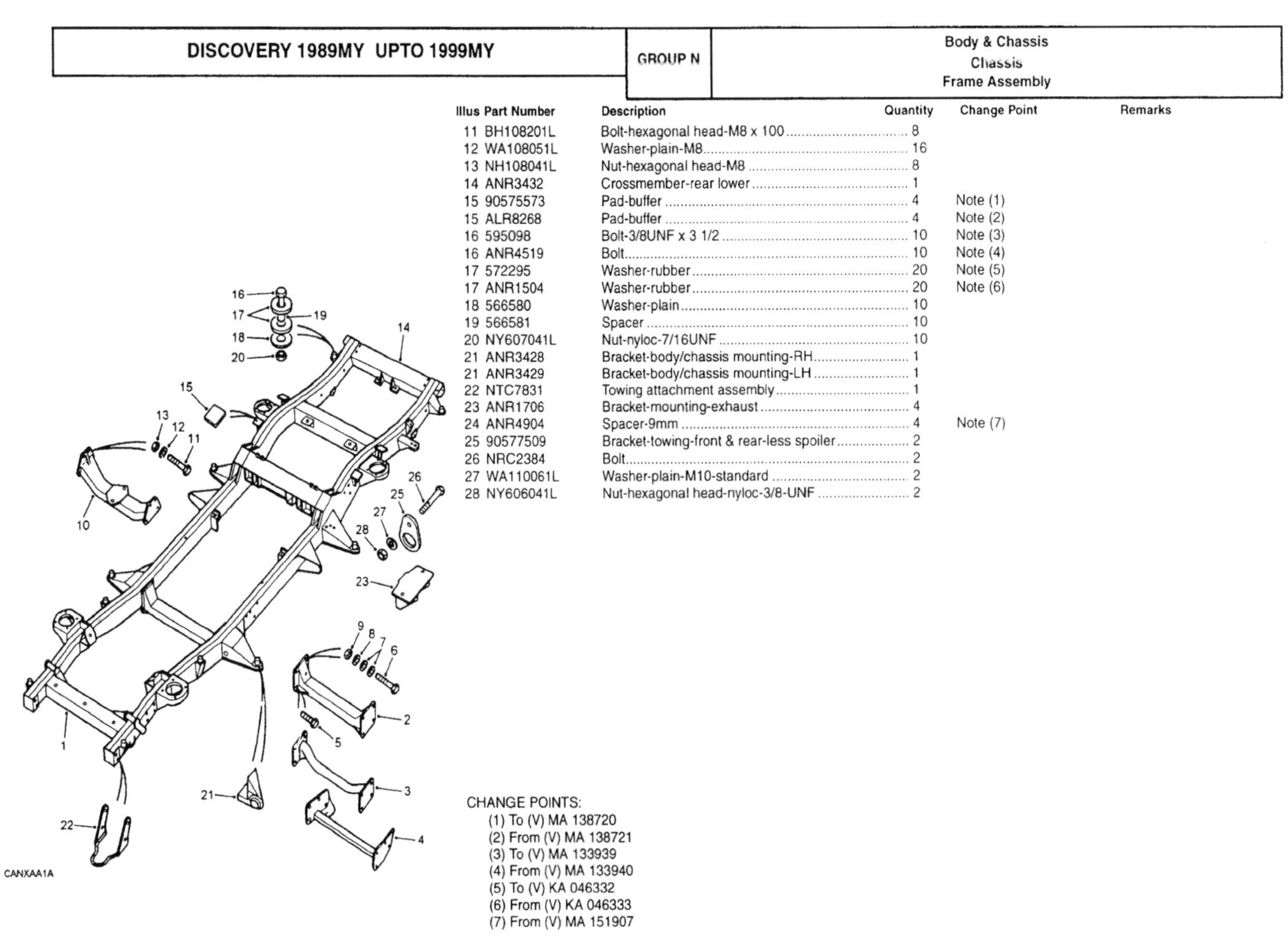

Illus	Part Number	Description	Quantity	Change Point	Remarks
11	BH108201L	Bolt-hexagonal head-M8 x 100	8		
12	WA108051L	Washer-plain-M8	16		
13	NH108041L	Nut-hexagonal head-M8	8		
14	ANR3432	Crossmember-rear lower	1		
15	90575573	Pad-buffer	4	Note (1)	
15	ALR8268	Pad-buffer	4	Note (2)	
16	595098	Bolt-3/8UNF x 3 1/2	10	Note (3)	
16	ANR4519	Bolt	10	Note (4)	
17	572295	Washer-rubber	20	Note (5)	
17	ANR1504	Washer-rubber	20	Note (6)	
18	566580	Washer-plain	10		
19	566581	Spacer	10		
20	NY607041L	Nut-nyloc-7/16UNF	10		
21	ANR3428	Bracket-body/chassis mounting-RH	1		
21	ANR3429	Bracket-body/chassis mounting-LH	1		
22	NTC7831	Towing attachment assembly	1		
23	ANR1706	Bracket-mounting-exhaust	4		
24	ANR4904	Spacer-9mm	4	Note (7)	
25	90577509	Bracket-towing-front & rear-less spoiler	2		
26	NRC2384	Bolt	2		
27	WA110061L	Washer-plain-M10-standard	2		
28	NY606041L	Nut-hexagonal head-nyloc-3/8-UNF	2		

CHANGE POINTS:
(1) To (V) MA 138720
(2) From (V) MA 138721
(3) To (V) MA 133939
(4) From (V) MA 133940
(5) To (V) KA 046332
(6) From (V) KA 046333
(7) From (V) MA 151907

DISCOVERY 1989MY UPTO 1999MY

GROUP N

Body & Chassis
Chassis
Replacement Outriggers

Illus	Part Number	Description	Quantity	Change Point	Remarks
1	STC9127	Chassis-fnt body mtg	1		
2	STC9125	Chassis-front end-LH	1		
2	STC9126	Chassis-front end-RH	1		
3	STC9128	Chassis-no2 body mtg	2		
4	STC9129	Chassis-no3 b/mtg-LH	1		
4	STC9130	Chassis-no3 b/mtg-RH	1		
5	STC9131	Chassis-no4 b/mtg-LH	1		
5	STC9132	Chassis-no4 b/mtg-RH	1		
6	STC9133	Chassis-no5 b/mtg-LH	1		
6	STC9134	Chassis-no5 b/mtg-RH	1		
7	STC9135	Crossmember assembly-chassis frame-front	1		
8	STC9137	Chassis-guss f/c/mem	1		
9	STC9136	Crossmember	1		
10	STC9138	Chassis-btm l/brk-LH	1		
10	STC9139	Chassis-btm l/brk-RH	1		
11	STC9140	Chassis-b/mtg-short	8	To (V) LA 081991	
11	STC9141	Chassis-b/mtg-long	8	From (V) MA 081992	

CANXAA1D

N 3

DISCOVERY 1989MY UPTO 1999MY

GROUP N

Body & Chassis
Chassis
Engine Mounting - 200Tdi

Illus	Part Number	Description	Quantity	Change Point	Remarks
		200TDI			
1	NTC6280	Bracket-engine mounting-LH	1		
2	NTC6281	Bracket-engine mounting-RH	1		
3	NTC6279	Mounting-engine rubber	2	Note (1)	
3	ANR2488	Mounting-engine rubber	2	Note (2)	
4	BH112261L	Bolt-bracket mounting to mounting engine rubber	2		
5	WA112081L	Washer-plain-bracket mounting to mounting engine rubber-M12-standard	4		
6	NY110041L	Nut-bracket mounting to mounting engine rubber-hexagonal head-nyloc-M10	NLA		Use NY110047L.
6	NY110047L	Nut-bracket mounting to mounting engine rubber-hexagonal head-nyloc-M10	2		
7	NY112041L	Nut-bracket mounting to mounting engine rubber-nyloc-M12	2		
8	WA110061L	Washer-plain-M10-standard	2		
9	SH112251L	Screw-hexagonal head-M12 x 25	4		
10	WA112081L	Washer-plain-bracket mounting to chassis-M12-standard	4		
11	SH110201L	Screw-bracket mounting to chassis-hexagonal head-M10 x 20	2		
12	WA110061L	Washer-plain-bracket mounting to chassis-M10-standard	2		

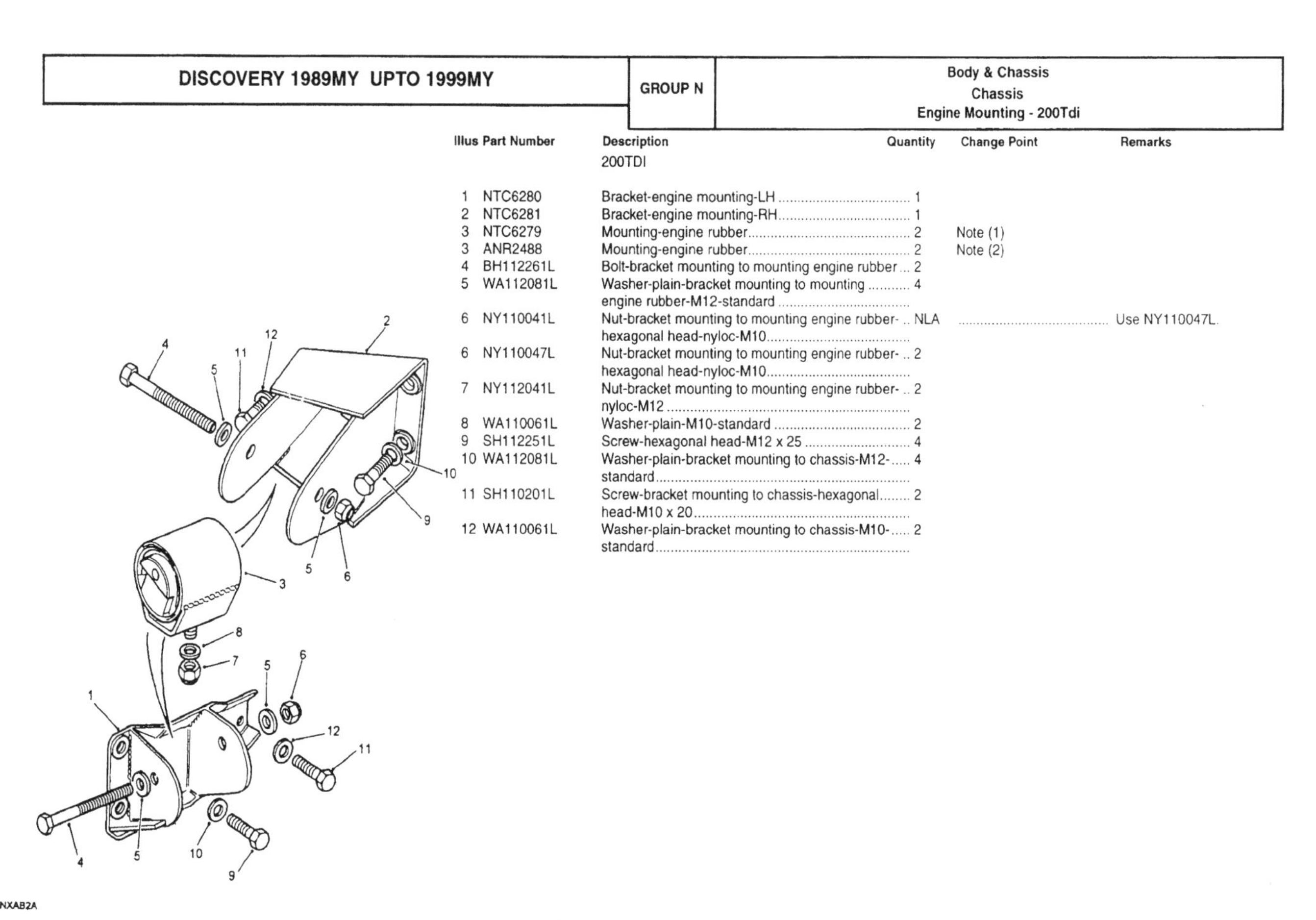

CANXAB2A

CHANGE POINTS:
(1) To (V) KA 064754
(2) From (V) LA 064755

N 4

Illus	Part Number	Description	Quantity	Change Point	Remarks
		300TDI			
1	ANR2868	Bracket-engine mounting-RH	1		
2	NTC9415	Bracket-engine mounting-LH	1		
3	SH112201	Screw-bracket mounting to chassis-hexagonal head-M12 x 20	8		
4	NTC9416	Bracket-ancillary mounting	2		
5	FX112041L	Nut-flange-bracket mounting to mounting rubber-M12	2		
6	FX110041L	Nut-flange-M10	2		
7	FX112041L	Nut-flange-bracket mounting to chassis-M12	2		

CANXAB3A

Illus	Part Number	Description	Quantity	Change Point	Remarks
1	NRC1302	Bracket-engine mounting-RH	NLA	Note (1)	Use ANR4697.
1	ANR2815	Bracket-engine mounting-RH	NLA	Note (2)	Use ANR4697.
1	ANR4697	Bracket-engine mounting-RH	1	Note (3)	
		BRACKET-ENGINE MOUNTING			
2	NRC3314	LH	1	Note (1)	
2	ANR2816	LH	1	Note (2)	
2	ANR4696	LH	1	Note (3)	
		ALL VARIANTS			
3	NTC4205	Heatshield-front engine mounting-LH	1		
4	STC434	MOUNTING-ENGINE RUBBER-BODY TO ENGINE BLOCK	2		
5	WC110061L	• Washer-plain-M10-oversize	2		
6	WL110001L	• Washer-M10	4		
7	NH110041L	• Nut-hexagonal head-coarse thread-M10	4		
8	SH505071L	Screw-body to engine block-hexagonal head-5/16UNC x 7/8	2		
9	WM600051L	Washer-spring-body to engine block-5/16"	2		
10	SH507091L	Screw-body to engine block-hexagonal head-7/16UNC x 1 1/8	4		
11	WM600071L	Washer-body to engine block-sprung-7/16"	4		
12	577846	Heatshield engine exhaust-RH	1		
13	SH604071L	Screw-bracket mounting to heatshield exhaust-hexagonal head-1/4UNF x 7/8	2		
14	WA106041L	Washer-bracket mounting to heatshield exhaust-M6-standard	2		
15	WM600041L	Washer-spring-bracket mounting to heatshield exhaust-1/4 dia-square	2		
16	NH604041L	Nut-hexagonal-coarse thread-1/4UNF	2		

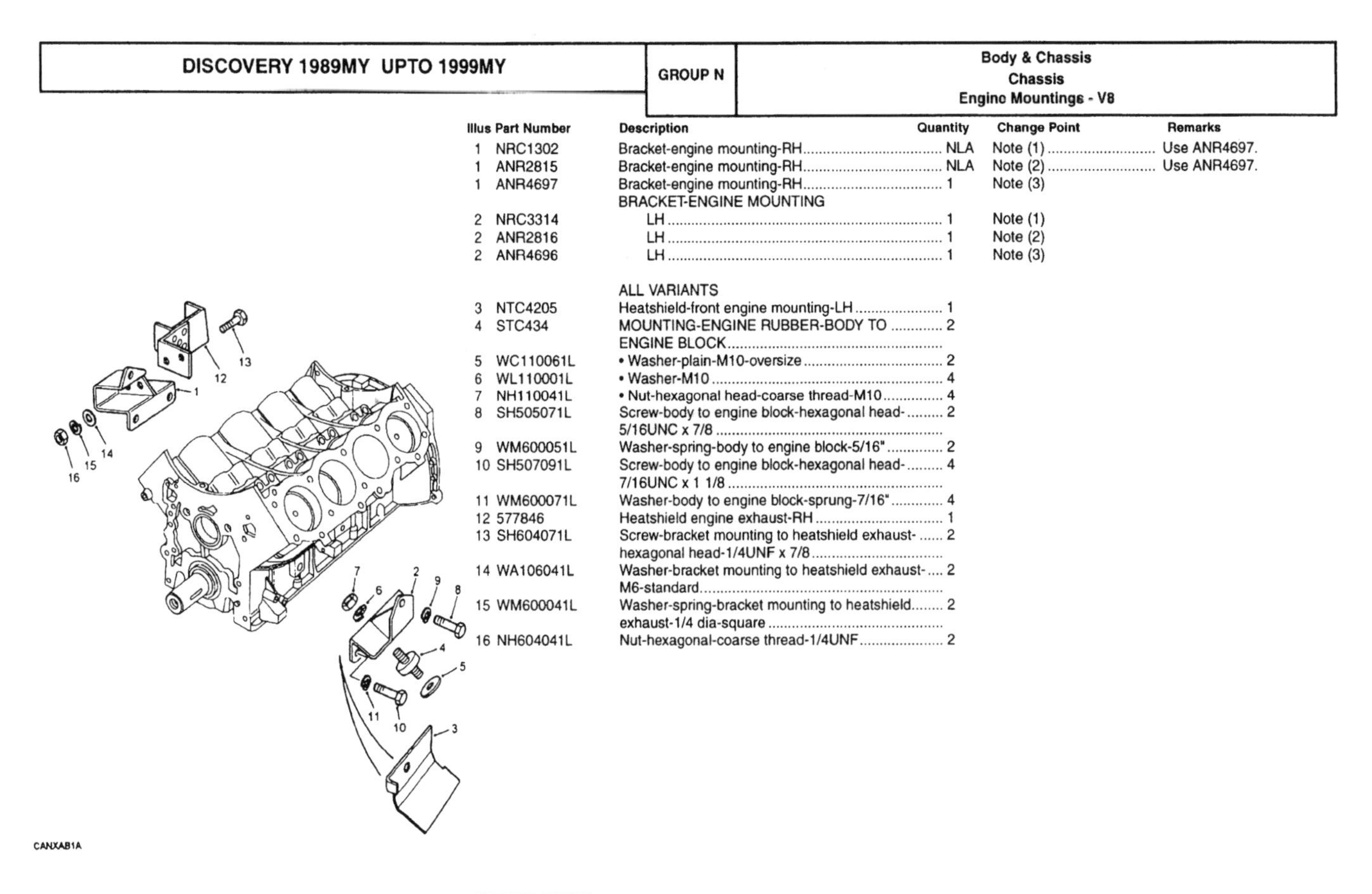

CANXAB1A

CHANGE POINTS:
(1) To (V) LA 081990
(2) From (V) MA 081991 To (V) MA 163103; From (V) MA 081991 To (V) MA 501919
(3) From (V) TA 163104; From (V) TA 501920

Engine Mounting - MPi

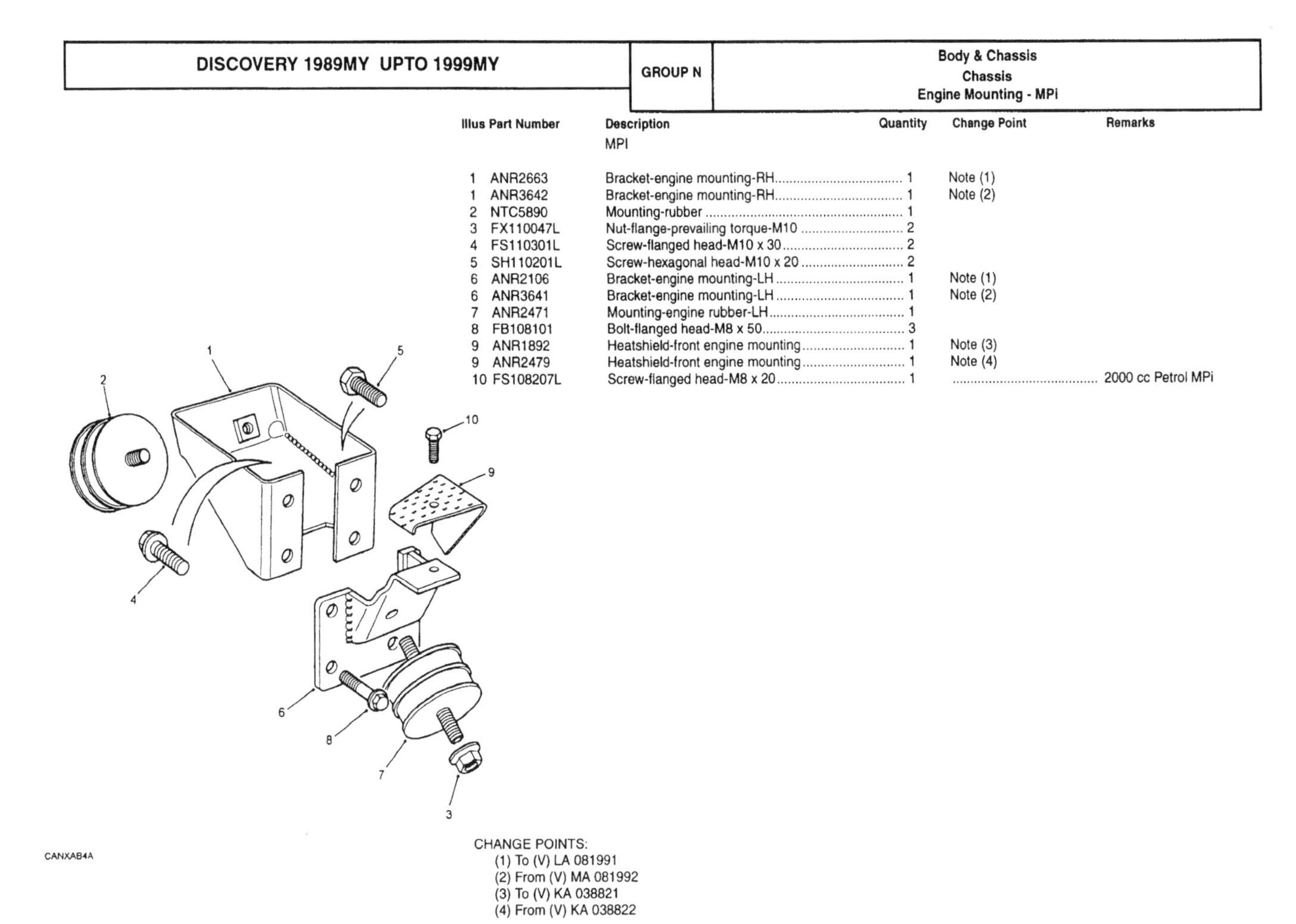

Illus	Part Number	Description	Quantity	Change Point	Remarks
		MPI			
1	ANR2663	Bracket-engine mounting-RH	1	Note (1)	
1	ANR3642	Bracket-engine mounting-RH	1	Note (2)	
2	NTC5890	Mounting-rubber	1		
3	FX110047L	Nut-flange-prevailing torque-M10	2		
4	FS110301L	Screw-flanged head-M10 x 30	2		
5	SH110201L	Screw-hexagonal head-M10 x 20	2		
6	ANR2106	Bracket-engine mounting-LH	1	Note (1)	
6	ANR3641	Bracket-engine mounting-LH	1	Note (2)	
7	ANR2471	Mounting-engine rubber-LH	1		
8	FB108101	Bolt-flanged head-M8 x 50	3		
9	ANR1892	Heatshield-front engine mounting	1	Note (3)	
9	ANR2479	Heatshield-front engine mounting	1	Note (4)	
10	FS108207L	Screw-flanged head-M8 x 20	1		2000 cc Petrol MPi

CHANGE POINTS:
(1) To (V) LA 081991
(2) From (V) MA 081992
(3) To (V) KA 038821
(4) From (V) KA 038822

Transmission Mounts

CANXAB5A

Illus	Part Number	Description	Quantity	Change Point	Remarks
		V8 ONLY			
1	NRC8204	Plate mounting-RH	NLA		Use ANR2819 qty 1 with WC110061L qty 1.
1	ANR2819	Plate mounting-RH	1		
		TDI / MPI			
2	ANR2898	Bracket-transmission mounting-RH	1		
2	ANR2898	Bracket-transmission mounting-RH	1		
		V8 ONLY			
2	NRC6528	Bracket-transmission mounting-LH-rear	1	Note (1)	
2	ANR2817	Bracket-transmission mounting-LH-rear	1	Note (2)	
2	ANR4615	Bracket-transmission mounting-LH	1	Note (3)	
		TDI / MPI			
2	ANR2897	Bracket-transmission mounting-LH	1	Note (4)	
2	ANR2897	Bracket-transmission mounting-LH	1	Note (4)	
2	ANR3225	Bracket-transmission mounting-LH	1	Note (5)	
2	ANR3225	Bracket-transmission mounting-LH	1	Note (5)	
		V8 ONLY			
3	NTC5890	Mounting-rubber-body to chassis frame	2	Note (1)	
3	ANR2805	Mounting-rubber-body to chassis frame	2	Note (6)	
		TDI / MPI			
3	ANR2128	Mounting-rubber-body to chassis frame-RH	1	Note (7)	
3	ANR2127	Mounting-rubber-body to chassis frame-LH	1	Note (7)	
3	ANR3200	Mounting-rubber-body to chassis frame-RH	1	Note (8)	
3	ANR3201	Mounting-rubber-body to chassis frame-LH	1	Note (8)	

CHANGE POINTS:
(1) To (V) LA 081990
(2) From (V) MA 081991 To (V) MA 163103; From (V) MA 081991 To (V) MA 501919
(3) From (V) MA 163104; From (V) MA 501920
(4) To (V) LA 075278
(5) From (V) LA 075279
(6) From (V) MA 081991
(7) To (V) MA 086333
(8) From (V) MA 086334

Illus	Part Number	Description	Quantity	Change Point	Remarks
		V8 ONLY			
4	ANR2818	Bracket-transmission mounting-RH-rear	1		
5	90575585	Bracket-transmission mounting-rear-LH	NLA		Use ANR2820 qty 1 with WC110061L qty 1.
5	ANR2820	Bracket-transmission mounting-rear-LH	1		
		ALL VARIANTS			
6	WA110061L	Washer-plain-M10-standard	2		
7	WM600061L	Washer-3/8"-square	4		
8	NH606041L	Nut-hexagonal head-3/8UNF	6		
9	SH112251L	Screw-hexagonal head-M12 x 25	6		
10	WL112001L	Washer	6		
11	BH108201L	Bolt-hexagonal head-M8 x 100	6		
12	WA108051L	Washer-plain-M8	12		
13	WL108001L	Washer-M8	6		
14	NH108041L	Nut-hexagonal head-M8	6		

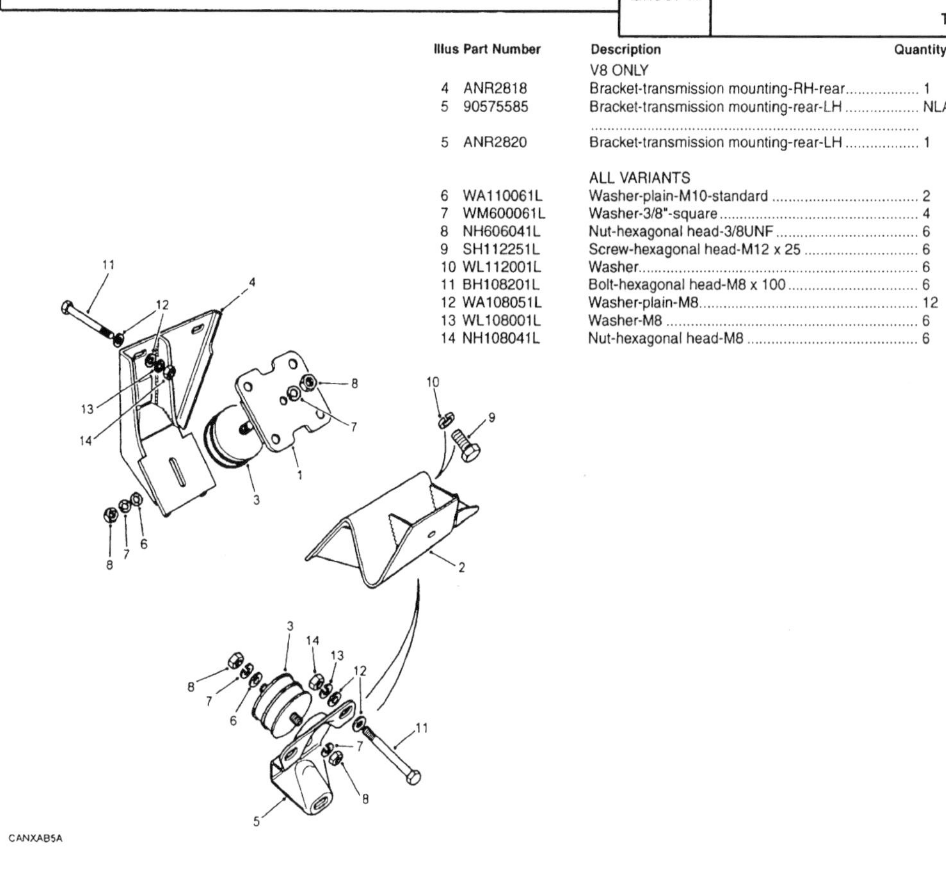

Transmission Mounts - MPi Greece Only

Illus	Part Number	Description	Quantity	Change Point	Remarks
		ONLY UP TO KA038821			
1	NTC9424	Bracket-transmission mounting-RH	1	Note (1)	
2	ANR2128	Mounting-rubber-RH	1	Note (2)	
2	ANR3200	Mounting-rubber-RH	1	Note (3)	
3	SH112251L	Screw-hexagonal head-M12 x 25	8	Note (1)	
4	WL112001L	Washer	8	Note (1)	
5	NTC9425	Bracket-transmission mounting-LH	1	Note (1)	
6	ANR2127	Mounting-rubber-LH	1	Note (2)	
6	ANR3200	Mounting-rubber-RH	1	Note (3)	
7	FX110041L	Nut-flange-bracket mounting to mounting rubber-M10	2	Note (1)	
8	ANR3138	Heatshield assembly exhaust-downpipe	1	Note (3)	

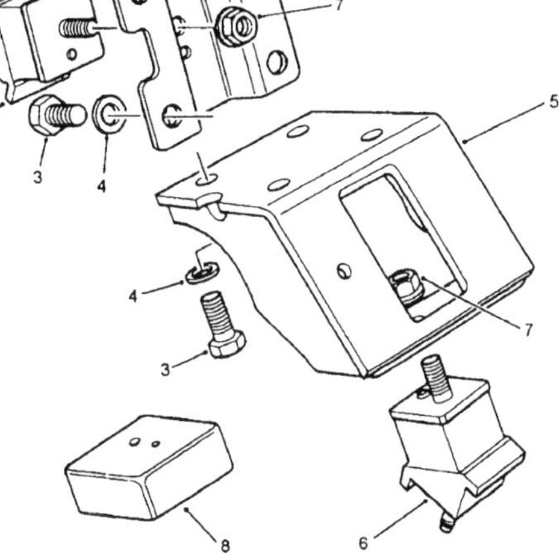

CHANGE POINTS:
(1) To (V) KA 038821
(2) To (V) KA 064754
(3) From (V) LA 064755

Under Ride Protection Bar

Illus	Part Number	Description	Quantity	Change Point	Remarks
1	NTC8239	Bar-underride protection	1		
2	NRC5886	Plug-blanking	2		
3	253948	Bolt-1/2UNF x 6 1/4	1		
4	WC112081L	Washer-plain-M12 x 28	3		
5	NY608041L	Nut-nyloc-upper	1		
6	BH606401	Bolt-3/8UNF x 5	1		
7	BH606361	Bolt	1		
8	3036L	Washer-plain	8		
9	NH604041L	Nut-hexagonal-coarse thread-1/4UNF	4		

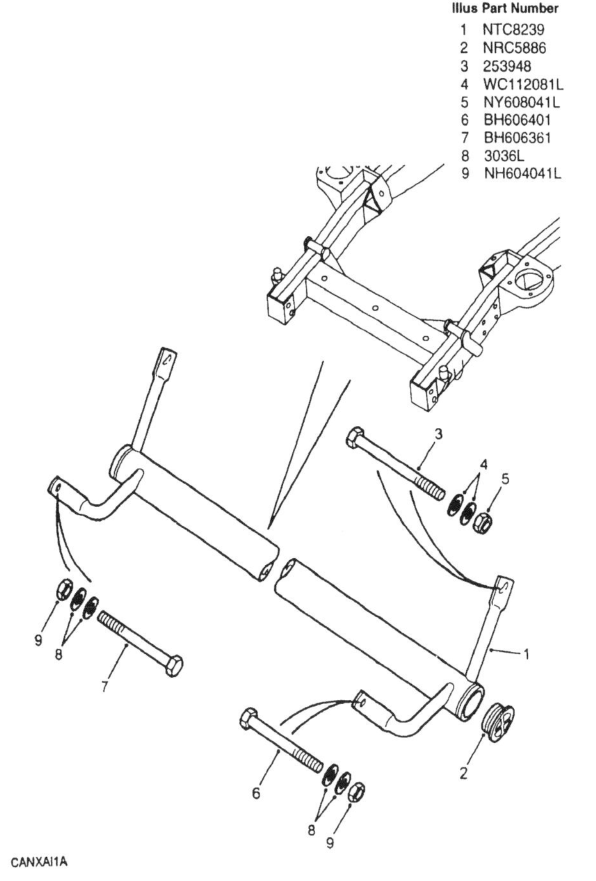

CANXAI1A

Front Bumper Assembly

Illus	Part Number	Description	Quantity	Change Point	Remarks
1	NTC5077	Bumper assembly-front-painted	1	Note (1)	
1	ANR2029	Bumper assembly-front-painted	1	Note (2)	
2	ANR3255	Hook-front sub frame towing	1		
3	BH606361L	Bolt-3/8UNF	2		
4	NH606041L	Nut-tow loop to bumper-hexagonal head-3/8UNF	4		
5	NRC2384	Bolt	2	Note (3)	
5	BH606361L	Bolt-3/8UNF	2	Note (4)	
6	3036L	Washer-plain	8		
		BELLY PANEL - NO DRIVING LAMPS			
7	MWC8004PUB	Valance front-lower	1	Note (5)	
7	AWR1479PMD	Valance front-lower	1	Note (6)	
		BELLY PANEL - SUITABLE FOR DRIVING LAMPS			
7	AWR2438PMD	Valance front-lower	1	Note (7)	
	DYP10035L	Screw-washer-M5 x 20	4		
8	WC108051L	Washer-plain-M8	6		
9	FN108047L	Nut-flanged head-M8	6		
		END CAP-FRONT BUMPER			
10	NTC5082PUB	RH-Black	1	Note (5)	
10	NTC5083PUB	LH-Black	1		
10	AWR2988PMD	RH-Black	1	Note (6)	
10	AWR2987PMD	LH-Black	1		

CANXAC1A

CHANGE POINTS:

(1) To (V) LA 081990
(2) From (V) MA 081991
(3) To (V) KA 040246
(4) From (V) KA 040247
(5) To (V) LA 081991
(6) From (V) MA 081992
(7) From (V) TA 163104; From (V) TA 501920

Illus	Part Number	Description	Quantity	Change Point	Remarks
		ALL VARIANTS			
11	FS106201L	Screw-flanged head-M6 x 20	2		
11	FS106207L	Screw-flanged head-M6 x 20	2		
12	WA106041L	Washer-M6-standard.	4		
13	AL610011L	Nut-spring-special-No10	8		
14	DA610051L	Screw-flanged head-No10	8		
		PLINTH-LICENCE PLATE			
15	STC8836AA	Black-with bull bar	1	Note (1)	
15	AWR6883PMD	Black-without bull bar	1		
15	NTC5486PUB	Black-without bull bar	1	Note (2)	
		FIXINGS FOR AWR6883			
16	BTR9688	Screw-poly top	2		
17	BTR9634	Washer-retaining	2		
18	WJ105001L	Washer-plain-M5-oversize	2		
19	WS108001L	Washer-spring	2		
		ALL VARIANTS			
18	WA106041L	Washer-M6-standard.	2		
19	NY106041L	Nut-hexagonal head-nyloc-M6	NLA		Use NY106047L.
19	NY106047L	Nut-hexagonal head-nyloc-M6	2		
20	NTC7482	Bracket-front bumper support	2	Note (2)	
21	ANR2741	Bracket-front bumper support	2	Note (1)	
22	WA106041L	Washer-M6-standard.	2		
23	NH106041L	Nut-bracket support to front bumper-hexagonal head-nyloc-M6	2		

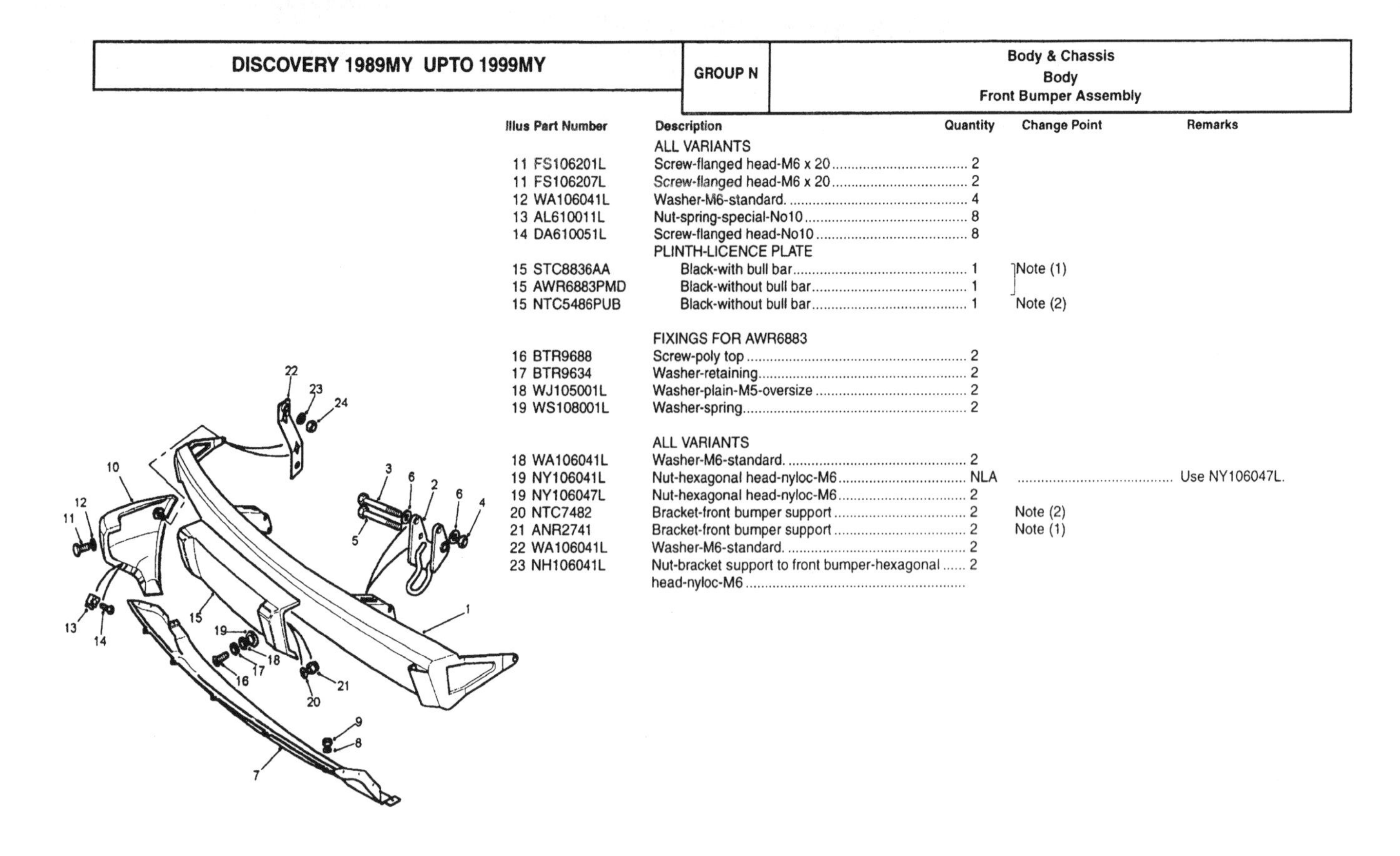

CHANGE POINTS:
(1) From (V) MA 081992
(2) To (V) LA 081991

N 13

Illus	Part Number	Description	Quantity	Change Point	Remarks
1	NTC5097	Bumper assembly-painted rear	NLA		Use BTR3542PUB qty 1 with BTR3543PUB qty 6 with NTC9665PUB qty 1.
1	NTC9665PUB	Bumper assembly-painted rear-Black	1	Note (1)	
1	ANR2743	Bumper assembly-painted rear	1	Note (2)	
2	NTC5106	Bracket-rear bumper mounting	2		
3	ANR3026	Bracket-rear bumper mounting-LH	1	Note (3)	
3	ANR3025	Bracket-rear bumper mounting-RH	1		
4	FS108257L	Screw-flanged head-M8 x 25	2		
5	WA108001L	Washer-8mm	4	Note (3)	
6	SH110251L	Screw-hexagonal head-M10 x 25	4		
7	WL110001L	Washer-M10	4		
8	SH110251L	Screw-hexagonal head-M10 x 25	2		
9	WC110061L	Washer-plain-M10-oversize	2		
10	WL110001L	Washer-M10	2		
	NTC8830	Shim	2		
		END CAP ASSEMBLY-REAR BUMPER			
11	NTC5098PUB	RH-Black	1	Note (1)	
11	NTC5099PUB	LH-Black	1		
11	AWR1002PMD	RH-Black	1	Note (4); Note (5)	
11	AWR1001PMD	LH-Black	1		
11	AWR2984PMD	RH-Black	1	Note (6)	
11	AWR2985PMD	LH-Black	1		
12	BTR9689	Screw-poly top-M5 x 10	2	Note (2)	
12	SH106121L	Screw-hexagonal head-M6 x 12	2		
13	WA106041L	Washer-M6-standard.	4		

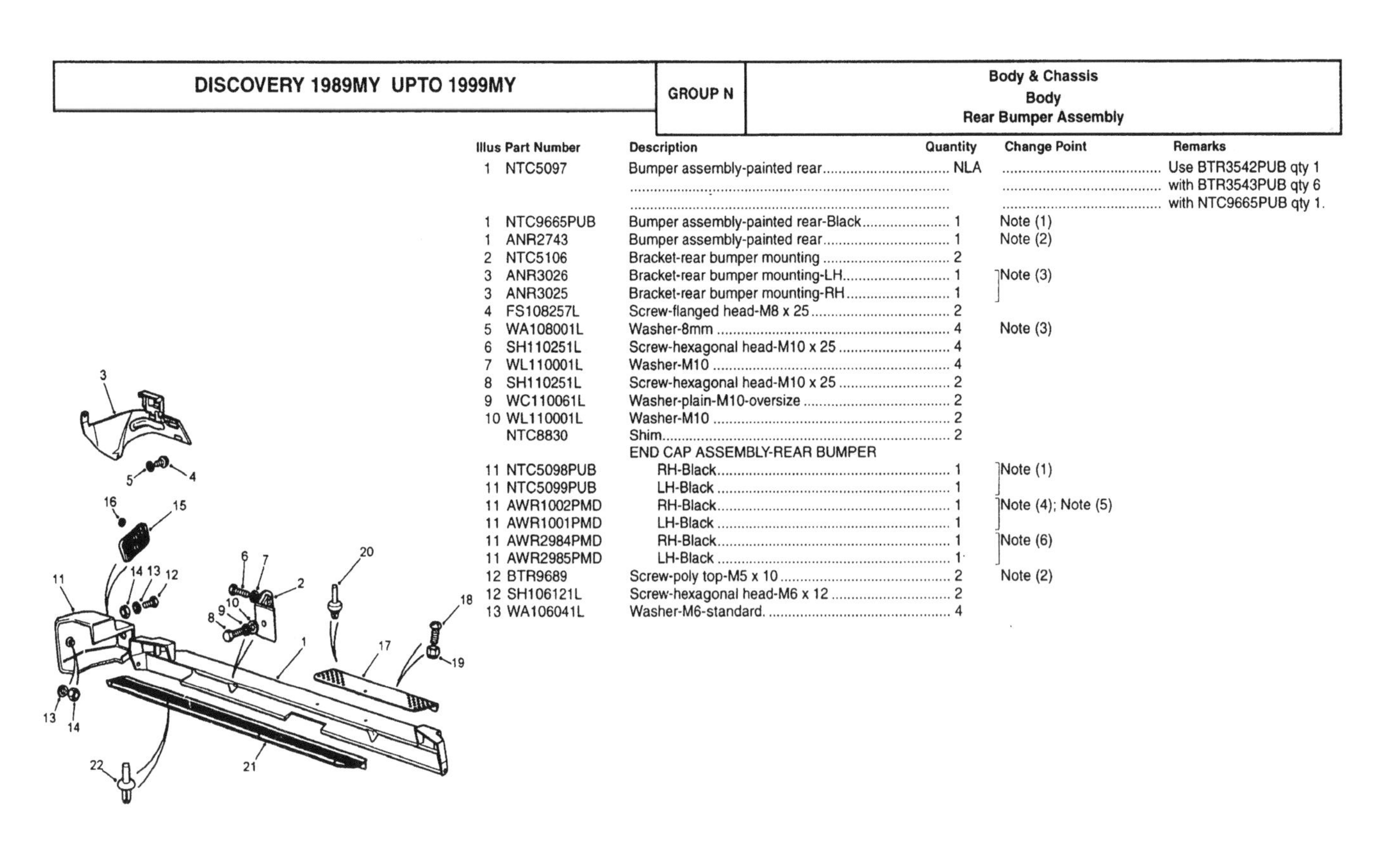

CHANGE POINTS:
(1) To (V) LA 081991
(2) From (V) MA 081992
(3) From (V) MA 081991
(4) From (V) MA 081992 To (V) TA 164023
(5) From (V) MA 500000 To (V) TA 502127
(6) From (V) TA 164024; From (V) TA 502128

N 14

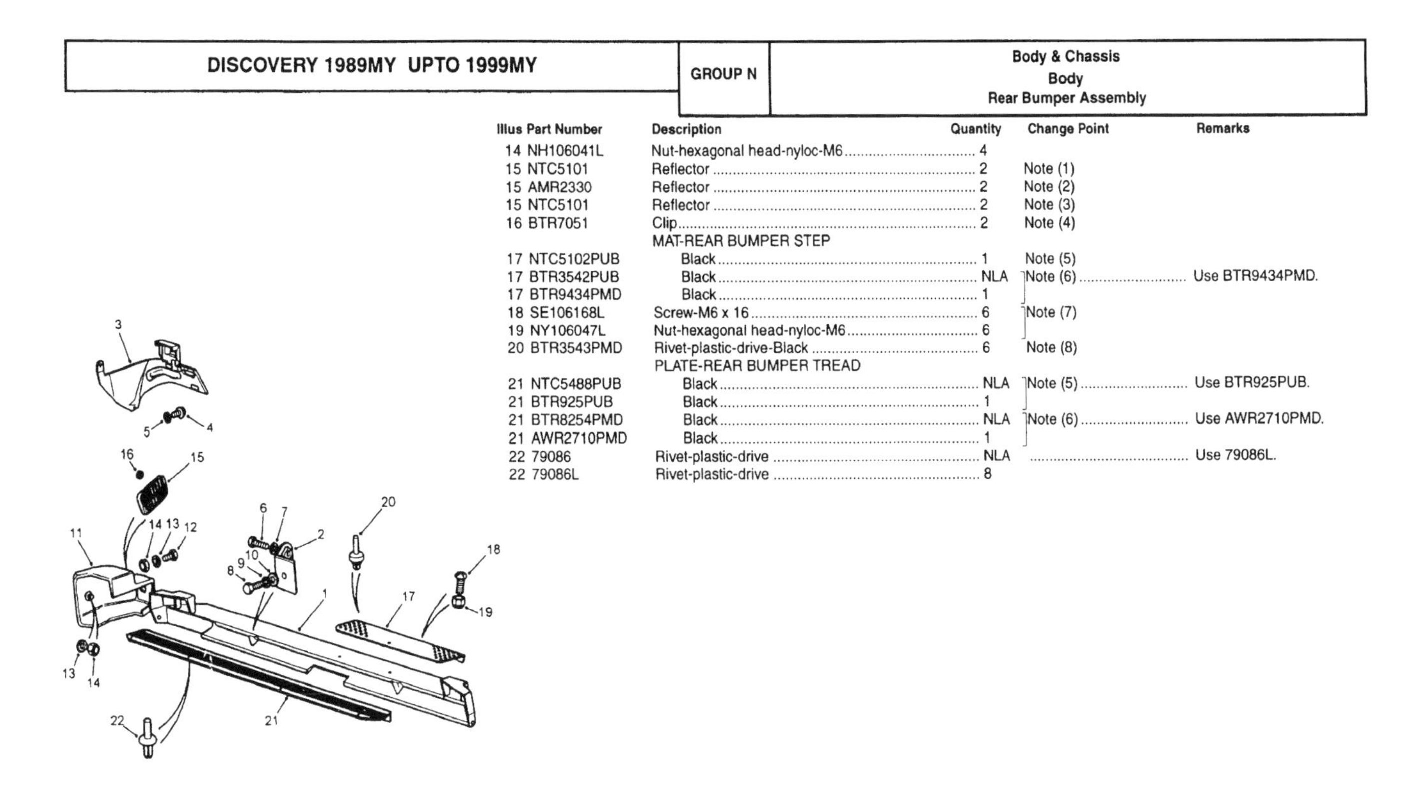

Illus	Part Number	Description	Quantity	Change Point	Remarks
14	NH106041L	Nut-hexagonal head-nyloc-M6	4		
15	NTC5101	Reflector	2	Note (1)	
15	AMR2330	Reflector	2	Note (2)	
15	NTC5101	Reflector	2	Note (3)	
16	BTR7051	Clip	2	Note (4)	
		MAT-REAR BUMPER STEP			
17	NTC5102PUB	Black	1	Note (5)	
17	BTR3542PUB	Black	NLA	Note (6)	Use BTR9434PMD.
17	BTR9434PMD	Black	1		
18	SE106168L	Screw-M6 x 16	6	Note (7)	
19	NY106047L	Nut-hexagonal head-nyloc-M6	6		
20	BTR3543PMD	Rivet-plastic-drive-Black	6	Note (8)	
		PLATE-REAR BUMPER TREAD			
21	NTC5488PUB	Black	NLA	Note (5)	Use BTR925PUB.
21	BTR925PUB	Black	1		
21	BTR8254PMD	Black	NLA	Note (6)	Use AWR2710PMD.
21	AWR2710PMD	Black	1		
22	79086	Rivet-plastic-drive	NLA		Use 79086L.
22	79086L	Rivet-plastic-drive	8		

CANXAE1B

CHANGE POINTS:
(1) To (V) KA 055950
(2) From (V) KA 055951 To (V) LA 067187
(3) From (V) LA 067188
(4) From (V) KA 055950
(5) To (V) LA 081991
(6) From (V) MA 081992
(7) To (V) JA 034313
(8) From (V) KA 034314; From (V) MA 081992

Illus	Part Number	Description	Quantity	Change Point	Remarks
		3 DOOR			
1	STC1817	Body shell	1	Note (1)	
2	STC1822	Template-sunroof	1		
		3 DOOR - NON SUNROOF OR ROOF RACK			
1	ALR7696	Body shell	1	Note (2)	
1	ALR9856	Body shell	1	Note (3)	
		3 DOOR - WITH TWIN SUNROOF - NO ROOF RACK			
1	ALR7697	Body shell-with sunroof	1	Note (2)	
1	ALR9857	Body shell-with sunroof	1	Note (3)	
1	ASR2337	Body shell	1	Note (3)	
		3 DOOR - NON SUNROOF WITH ROOFRACK			
1	ALR7694	Body shell	1	Note (2)	
1	ALR9854	Body shell	1	Note (3)	
1	ASR2335	Body shell	1	Note (3)	
		3 DOOR - WITH TWIN SUNROOF & ROOF RACK			
1	ALR7695	Body shell-with sunroof	1	Note (2)	
1	ALR9855	Body shell-with sunroof	1	Note (3)	

CANXBA1A

CHANGE POINTS:
(1) To (V) LA 081990
(2) From (V) MA 081991 To (V) TA 703236; From (V) MA 081991 To (V) TA 534103
(3) From (V) VA 703237; From (V) VA 534104

Illus	Part Number	Description	Quantity	Change Point	Remarks
		NON SUNROOF			
1	STC1833	Body shell	1	To (V) LA 081990	
2	STC1822	Template-sunroof	1		
		NON SUNROOF OR ROOF RACK			
		BODY SHELL			
1	ALR3752	less sunroof	1	Note (1)	
1	ALR9852	less sunroof	1	Note (2); Note (3)	
1	ASR2332	less sunroof	1	Note (4); Note (5)	
1	AAB700030	less sunroof	1	Note (6)	
		TWIN SUNROOF - NO ROOF RACK			
1	ALR3753	Body shell	1	Note (7); Note (8)	
1	ALR7693	Body shell	1	Note (9); Note (10)	
1	ALR9853	Body shell	1	Note (11); Note (12)	
1	ASR2333	Body shell	1	Note (2); Note (3); Note (13)	
1	AAB700040	Body shell	1	Note (6)	

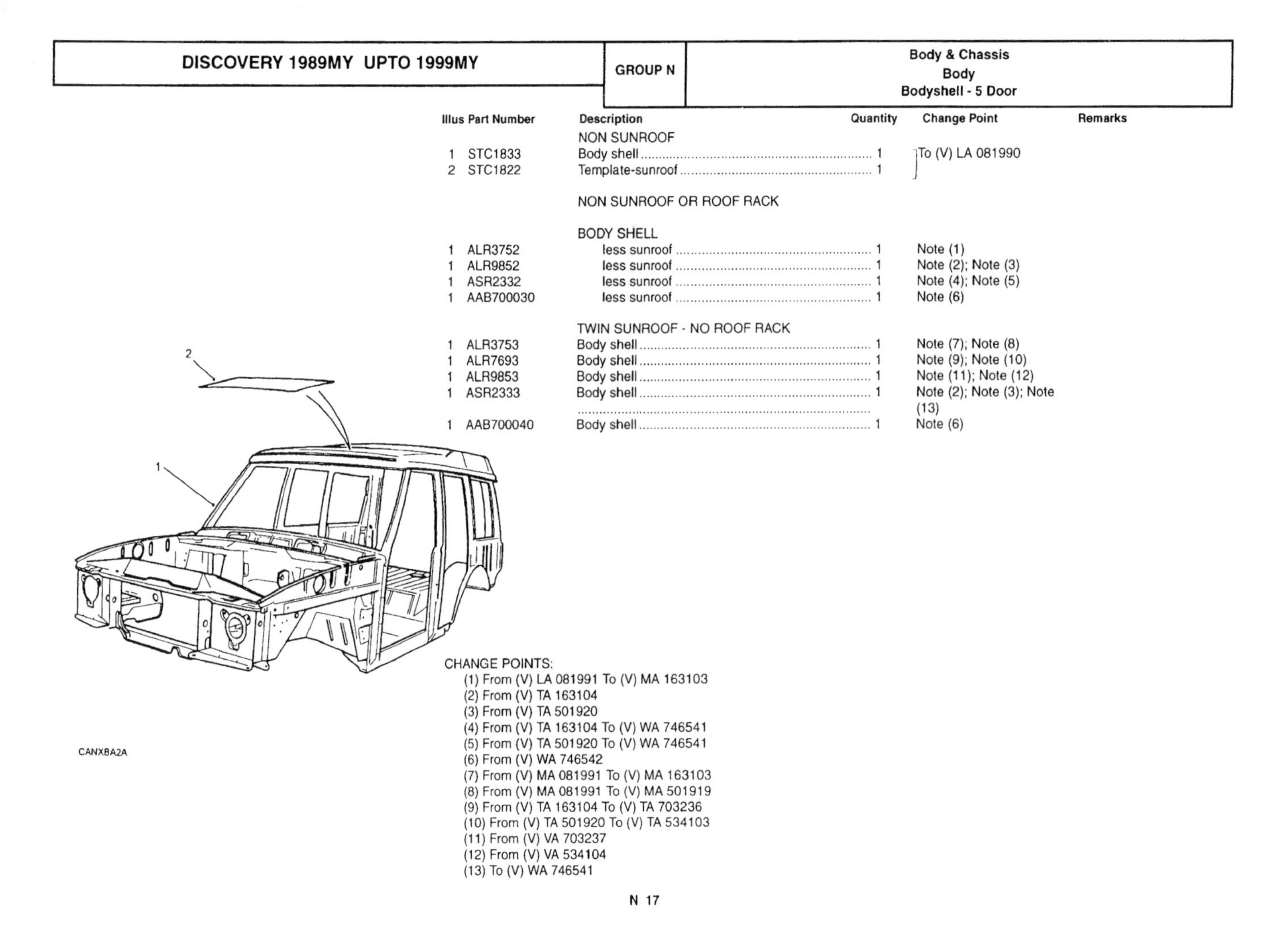

CHANGE POINTS:
(1) From (V) LA 081991 To (V) MA 163103
(2) From (V) TA 163104
(3) From (V) TA 501920
(4) From (V) TA 163104 To (V) WA 746541
(5) From (V) TA 501920 To (V) WA 746541
(6) From (V) WA 746542
(7) From (V) MA 081991 To (V) MA 163103
(8) From (V) MA 081991 To (V) MA 501919
(9) From (V) TA 163104 To (V) TA 703236
(10) From (V) TA 501920 To (V) TA 534103
(11) From (V) VA 703237
(12) From (V) VA 534104
(13) To (V) WA 746541

Illus	Part Number	Description	Quantity	Change Point	Remarks
		NON SUNROOF WITH ROOF RACK			
1	ALR3740	Body shell	1	Note (1)	
1	ALR9851	Body shell	NLA	Note (2); Note (3)	Use ASR2331.
1	ASR2331	Body shell	1	Note (2); Note (3); Note (4)	
1	AAB700020	Body shell	1	Note (5)	
		TWIN SUNROOF & ROOF RACK			
1	ALR3751	Body shell	1	Note (1)	
1	ALR9850	Body shell	NLA	Note (2); Note (3)	Use ASR2330.
1	ASR2330	Body shell	1	Note (2); Note (3); Note (4)	
1	AAB700010	Body shell	1	Note (5)	

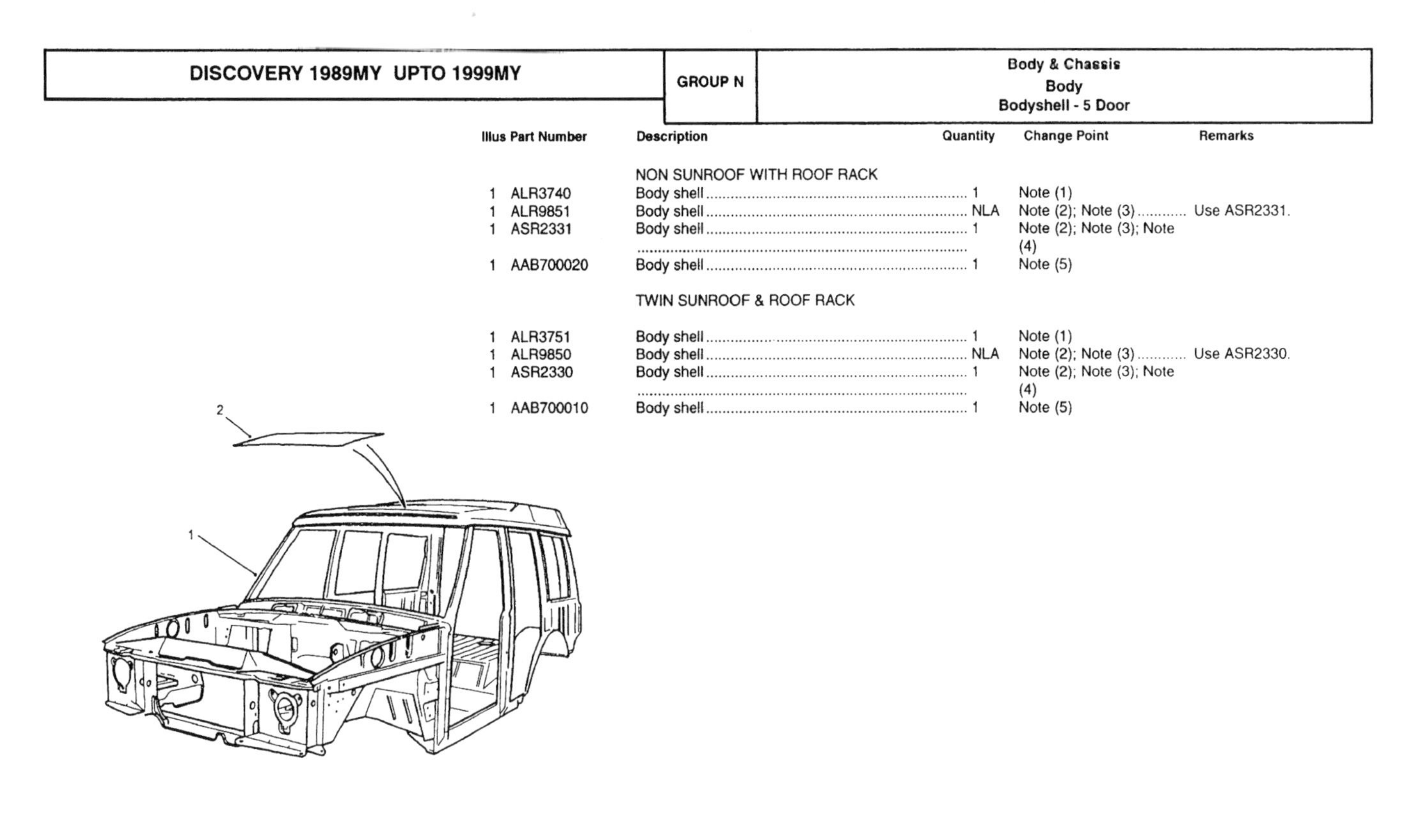

CHANGE POINTS:
(1) From (V) MA 081991 To (V) MA 163103; From (V) MA 081991 To (V) MA 501919
(2) From (V) TA 163104
(3) From (V) TA 501920
(4) To (V) WA 746541
(5) From (V) WA 746542

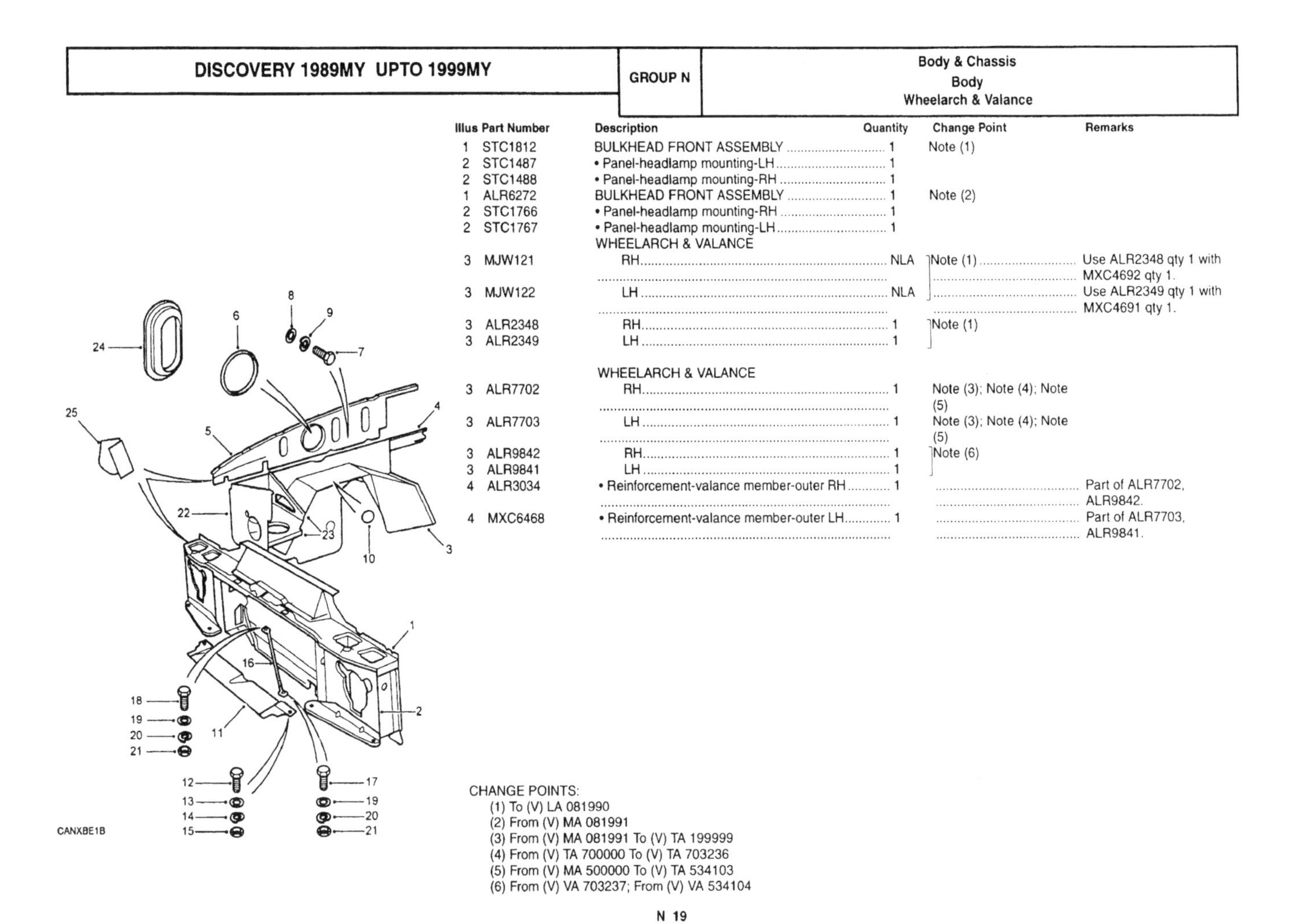

Illus	Part Number	Description	Quantity	Change Point	Remarks
1	STC1812	BULKHEAD FRONT ASSEMBLY	1	Note (1)	
2	STC1487	• Panel-headlamp mounting-LH	1		
2	STC1488	• Panel-headlamp mounting-RH	1		
1	ALR6272	BULKHEAD FRONT ASSEMBLY	1	Note (2)	
2	STC1766	• Panel-headlamp mounting-RH	1		
2	STC1767	• Panel-headlamp mounting-LH	1		
		WHEELARCH & VALANCE			
3	MJW121	RH	NLA	Note (1)	Use ALR2348 qty 1 with MXC4692 qty 1.
3	MJW122	LH	NLA		Use ALR2349 qty 1 with MXC4691 qty 1.
3	ALR2348	RH	1	Note (1)	
3	ALR2349	LH	1		
		WHEELARCH & VALANCE			
3	ALR7702	RH	1	Note (3); Note (4); Note (5)	
3	ALR7703	LH	1	Note (3); Note (4); Note (5)	
3	ALR9842	RH	1	Note (6)	
3	ALR9841	LH	1		
4	ALR3034	• Reinforcement-valance member-outer RH	1		Part of ALR7702, ALR9842.
4	MXC6468	• Reinforcement-valance member-outer LH	1		Part of ALR7703, ALR9841.

CHANGE POINTS:
(1) To (V) LA 081990
(2) From (V) MA 081991
(3) From (V) MA 081991 To (V) TA 199999
(4) From (V) TA 700000 To (V) TA 703236
(5) From (V) MA 500000 To (V) TA 534103
(6) From (V) VA 703237; From (V) VA 534104

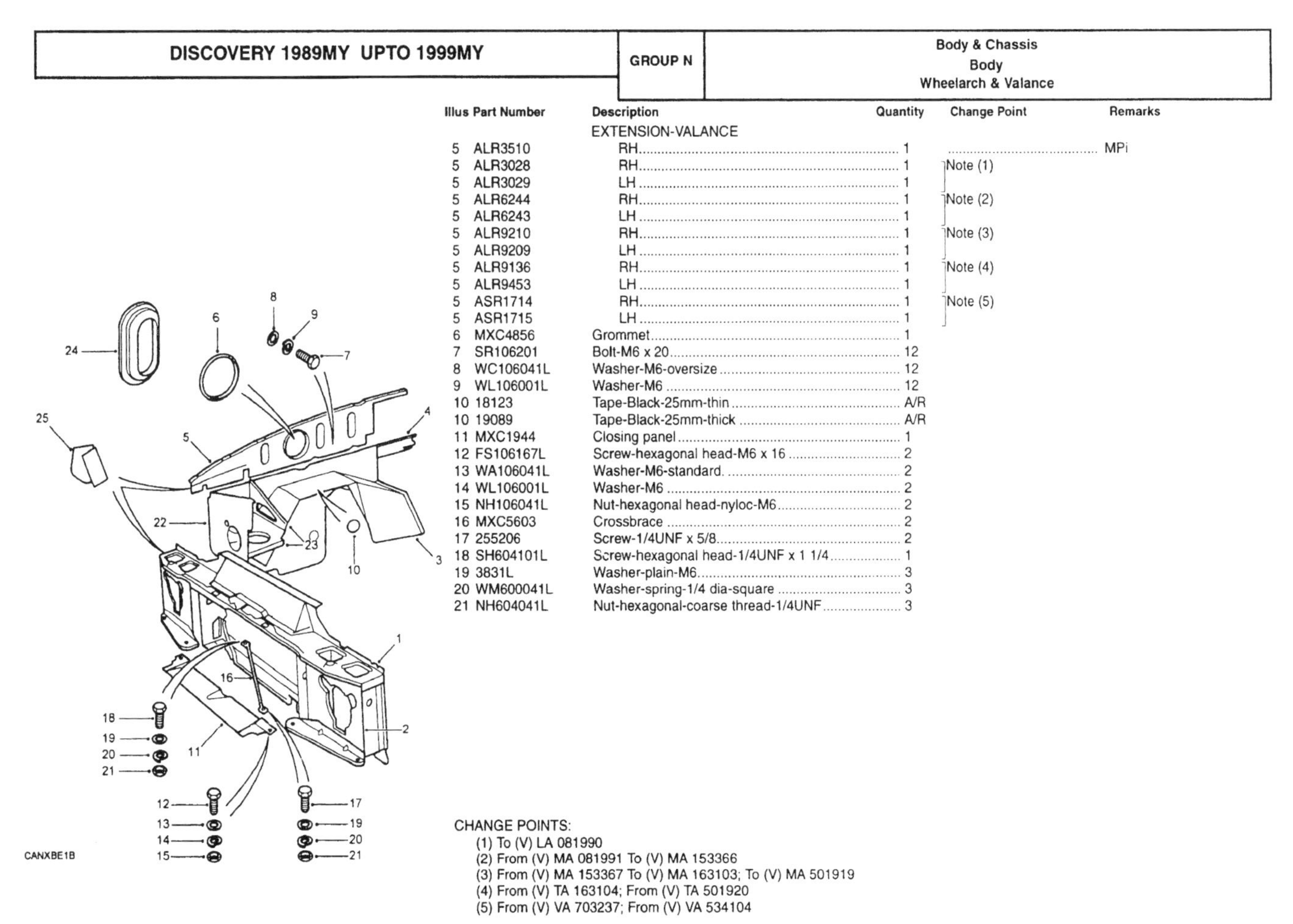

Illus	Part Number	Description	Quantity	Change Point	Remarks
		EXTENSION-VALANCE			
5	ALR3510	RH	1		MPi
5	ALR3028	RH	1	Note (1)	
5	ALR3029	LH	1		
5	ALR6244	RH	1	Note (2)	
5	ALR6243	LH	1		
5	ALR9210	RH	1	Note (3)	
5	ALR9209	LH	1		
5	ALR9136	RH	1	Note (4)	
5	ALR9453	LH	1		
5	ASR1714	RH	1	Note (5)	
5	ASR1715	LH	1		
6	MXC4856	Grommet	1		
7	SR106201	Bolt-M6 x 20	12		
8	WC106041L	Washer-M6-oversize	12		
9	WL106001L	Washer-M6	12		
10	18123	Tape-Black-25mm-thin	A/R		
10	19089	Tape-Black-25mm-thick	A/R		
11	MXC1944	Closing panel	1		
12	FS106167L	Screw-hexagonal head-M6 x 16	2		
13	WA106041L	Washer-M6-standard.	2		
14	WL106001L	Washer-M6	2		
15	NH106041L	Nut-hexagonal head-nyloc-M6	2		
16	MXC5603	Crossbrace	2		
17	255206	Screw-1/4UNF x 5/8	2		
18	SH604101L	Screw-hexagonal head-1/4UNF x 1 1/4	1		
19	3831L	Washer-plain-M6	3		
20	WM600041L	Washer-spring-1/4 dia-square	3		
21	NH604041L	Nut-hexagonal-coarse thread-1/4UNF	3		

CHANGE POINTS:
(1) To (V) LA 081990
(2) From (V) MA 081991 To (V) MA 153366
(3) From (V) MA 153367 To (V) MA 163103; To (V) MA 501919
(4) From (V) TA 163104; From (V) TA 501920
(5) From (V) VA 703237; From (V) VA 534104

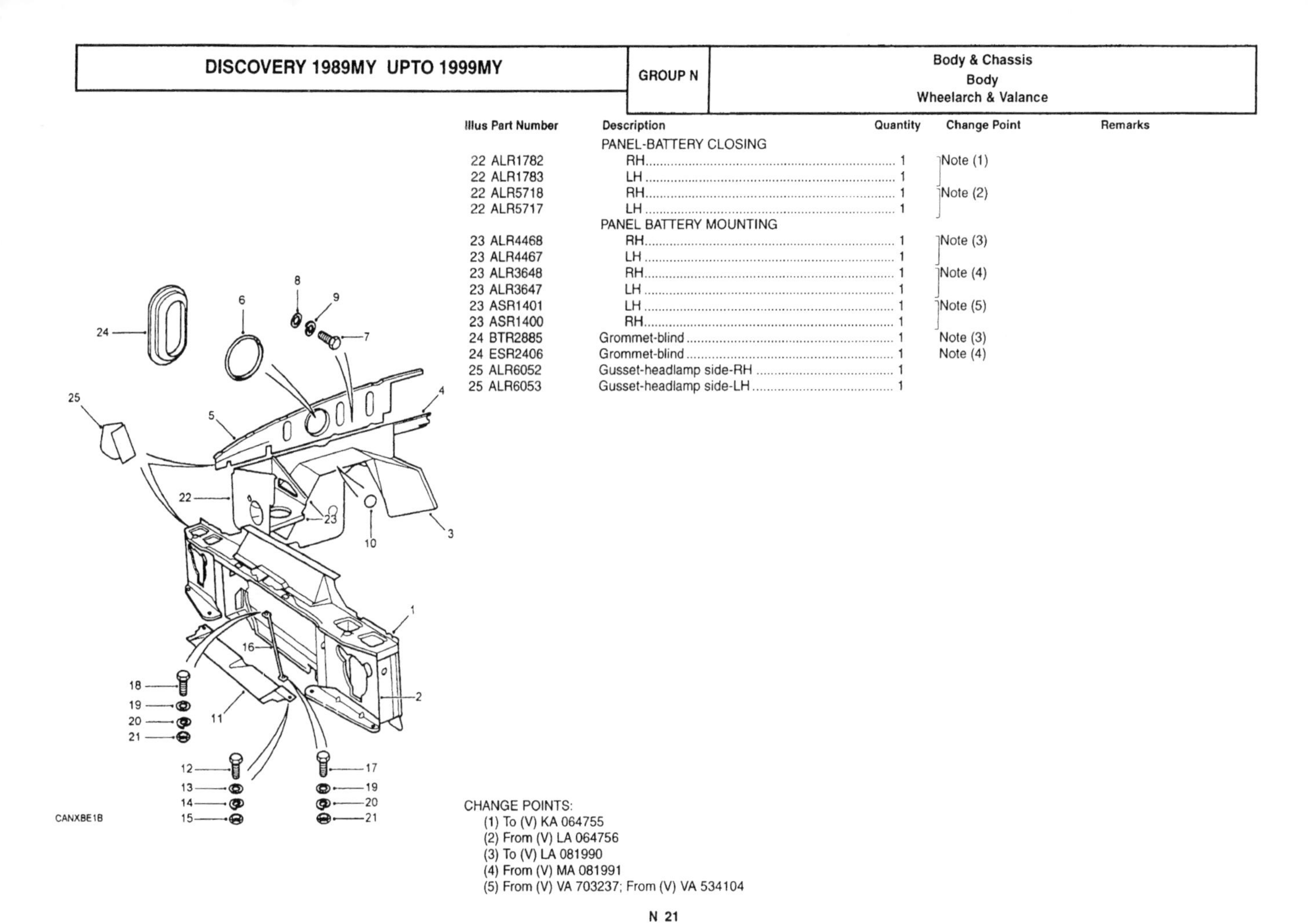

Illus	Part Number	Description	Quantity	Change Point	Remarks
		PANEL-BATTERY CLOSING			
22	ALR1782	RH	1	Note (1)	
22	ALR1783	LH	1		
22	ALR5718	RH	1	Note (2)	
22	ALR5717	LH	1		
		PANEL BATTERY MOUNTING			
23	ALR4468	RH	1	Note (3)	
23	ALR4467	LH	1		
23	ALR3648	RH	1	Note (4)	
23	ALR3647	LH	1		
23	ASR1401	LH	1	Note (5)	
23	ASR1400	RH	1		
24	BTR2885	Grommet-blind	1	Note (3)	
24	ESR2406	Grommet-blind	1	Note (4)	
25	ALR6052	Gusset-headlamp side-RH	1		
25	ALR6053	Gusset-headlamp side-LH	1		

CHANGE POINTS:
(1) To (V) KA 064755
(2) From (V) LA 064756
(3) To (V) LA 081990
(4) From (V) MA 081991
(5) From (V) VA 703237; From (V) VA 534104

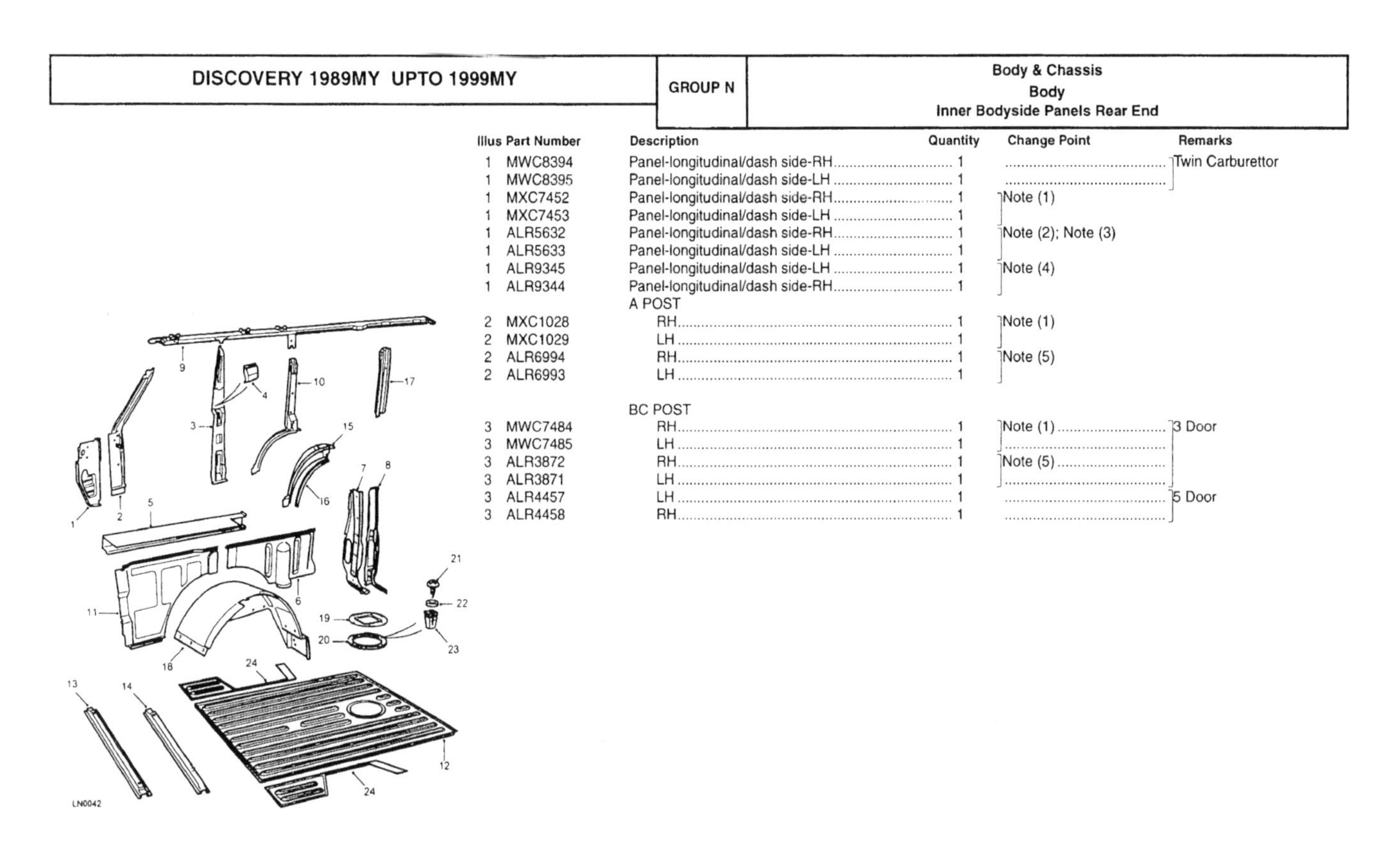

Illus	Part Number	Description	Quantity	Change Point	Remarks
1	MWC8394	Panel-longitudinal/dash side-RH	1		Twin Carburettor
1	MWC8395	Panel-longitudinal/dash side-LH	1		
1	MXC7452	Panel-longitudinal/dash side-RH	1	Note (1)	
1	MXC7453	Panel-longitudinal/dash side-LH	1		
1	ALR5632	Panel-longitudinal/dash side-RH	1	Note (2); Note (3)	
1	ALR5633	Panel-longitudinal/dash side-LH	1		
1	ALR9345	Panel-longitudinal/dash side-LH	1	Note (4)	
1	ALR9344	Panel-longitudinal/dash side-RH	1		
		A POST			
2	MXC1028	RH	1	Note (1)	
2	MXC1029	LH	1		
2	ALR6994	RH	1	Note (5)	
2	ALR6993	LH	1		
		BC POST			
3	MWC7484	RH	1	Note (1)	3 Door
3	MWC7485	LH	1		
3	ALR3872	RH	1	Note (5)	
3	ALR3871	LH	1		
3	ALR4457	LH	1		5 Door
3	ALR4458	RH	1		

CHANGE POINTS:
(1) To (V) LA 081990
(2) From (V) MA 081991 To (V) MA 163103
(3) From (V) MA 500000 To (V) TA 501919
(4) From (V) TA 163104; From (V) TA 501920
(5) From (V) MA 081991

Illus	Part Number	Description	Quantity	Change Point	Remarks
		ALL VARIANTS			
4	BTR1684	Escutcheon- BC post lower	1		
		3 DOOR			
		SILL			
5	MWC7502	outer-RH	1	Note (1)	
5	MWC7503	outer-LH	1		
5	ALR3784	outer-RH	NLA	Note (2)	Use STC2814 qty 1 with ALR2098 qty 1.
5	ALR3783	outer-LH	NLA		Use STC2813 qty 1 with ALR2098 qty 1.
		5 DOOR			
5	ALR3282	Sill-outer-RH	NLA		Use STC2816 qty 1 with ALR2098 qty 1.
5	ALR3283	Sill-outer-LH	NLA		Use STC2815 qty 1 with ALR2098 qty 1.

CHANGE POINTS:
(1) To (V) JA 034313
(2) From (V) KA 034314

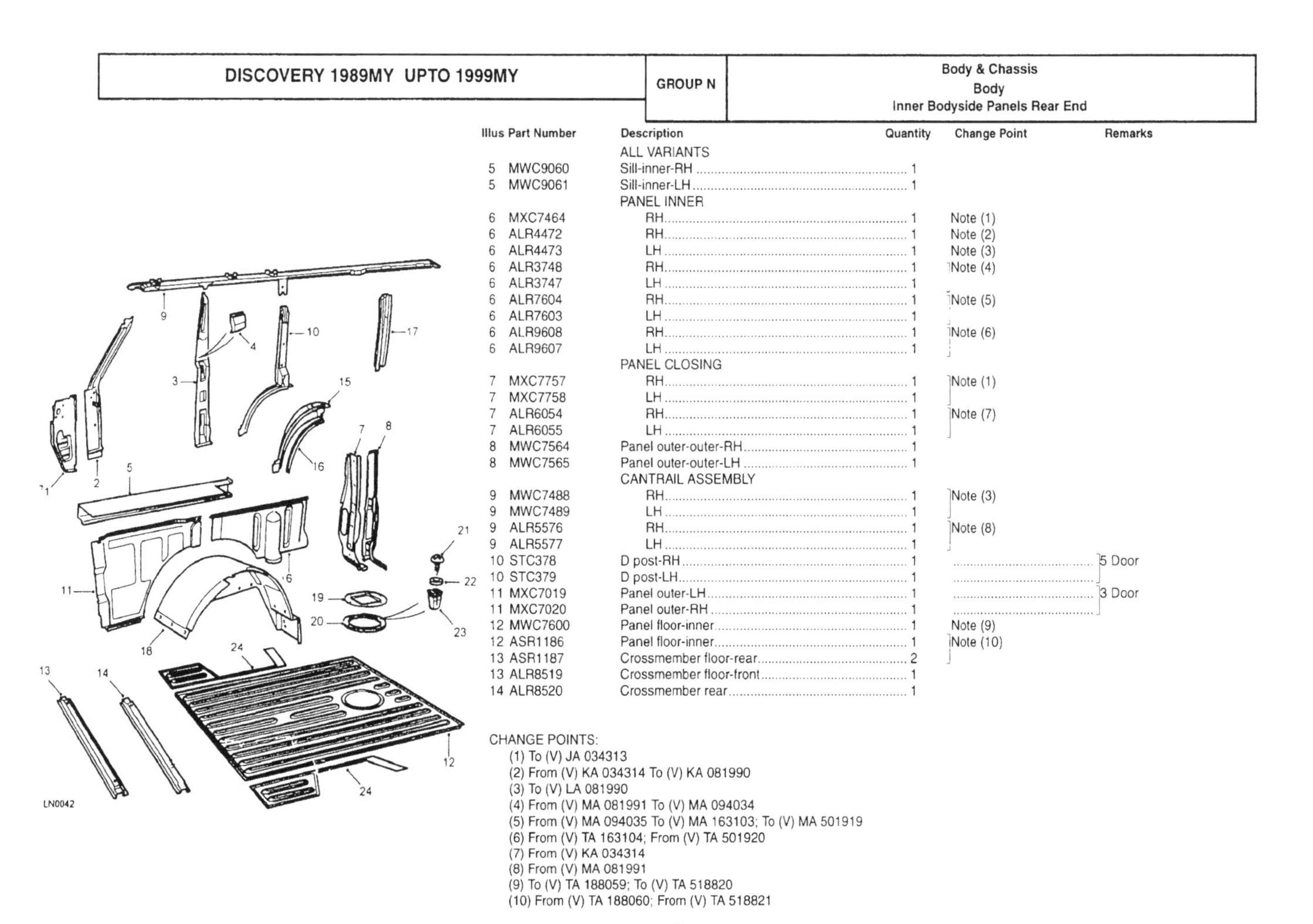

Illus	Part Number	Description	Quantity	Change Point	Remarks
		ALL VARIANTS			
5	MWC9060	Sill-inner-RH	1		
5	MWC9061	Sill-inner-LH	1		
		PANEL INNER			
6	MXC7464	RH	1	Note (1)	
6	ALR4472	RH	1	Note (2)	
6	ALR4473	LH	1	Note (3)	
6	ALR3748	RH	1	Note (4)	
6	ALR3747	LH	1		
6	ALR7604	RH	1	Note (5)	
6	ALR7603	LH	1		
6	ALR9608	RH	1	Note (6)	
6	ALR9607	LH	1		
		PANEL CLOSING			
7	MXC7757	RH	1	Note (1)	
7	MXC7758	LH	1		
7	ALR6054	RH	1	Note (7)	
7	ALR6055	LH	1		
8	MWC7564	Panel outer-outer-RH	1		
8	MWC7565	Panel outer-outer-LH	1		
		CANTRAIL ASSEMBLY			
9	MWC7488	RH	1	Note (3)	
9	MWC7489	LH	1		
9	ALR5576	RH	1	Note (8)	
9	ALR5577	LH	1		
10	STC378	D post-RH	1		5 Door
10	STC379	D post-LH	1		
11	MXC7019	Panel outer-LH	1		3 Door
11	MXC7020	Panel outer-RH	1		
12	MWC7600	Panel floor-inner	1	Note (9)	
12	ASR1186	Panel floor-inner	1	Note (10)	
13	ASR1187	Crossmember floor-rear	2		
13	ALR8519	Crossmember floor-front	1		
14	ALR8520	Crossmember rear	1		

CHANGE POINTS:
(1) To (V) JA 034313
(2) From (V) KA 034314 To (V) KA 081990
(3) To (V) LA 081990
(4) From (V) MA 081991 To (V) MA 094034
(5) From (V) MA 094035 To (V) MA 163103; To (V) MA 501919
(6) From (V) TA 163104; From (V) TA 501920
(7) From (V) KA 034314
(8) From (V) MA 081991
(9) To (V) TA 188059; To (V) TA 518820
(10) From (V) TA 188060; From (V) TA 518821

Inner Bodyside Panels Rear End

Illus	Part Number	Description	Quantity	Change Point	Remarks
15	MWC4841	Panel wheelarch-LH	1		5 Door
15	MWC4840	Panel wheelarch-RH	1		
16	MXC9553	Panel make up piece-LH	1		
16	MXC9552	Panel make up piece-RH	1		
17	MXC8662	Panel inner D post-RH	1		3 Door
17	MXC8663	Panel inner D post-LH	1		
18	ALR3789	Wheelarch assembly-rear-LH	1		
18	ALR3790	Wheelarch assembly-rear-RH	1		
19	MXC1420	PANEL-ACCESS-RR FLOOR	1		
20	MXC1421	• Seal	1		
21	AB610041L	Screw-self tapping AB-M8 x 12	6		
22	WC702101L	Washer-plain	6		
23	RTC3743	Nut-lokut	6		
24	ASR1594	Panel floor-side-RH	1		
24	ASR1595	Panel floor-side-LH	1		
		DISCOVERY COMMERCIAL ONLY			
	STC7019	Panel-floor extension	1		

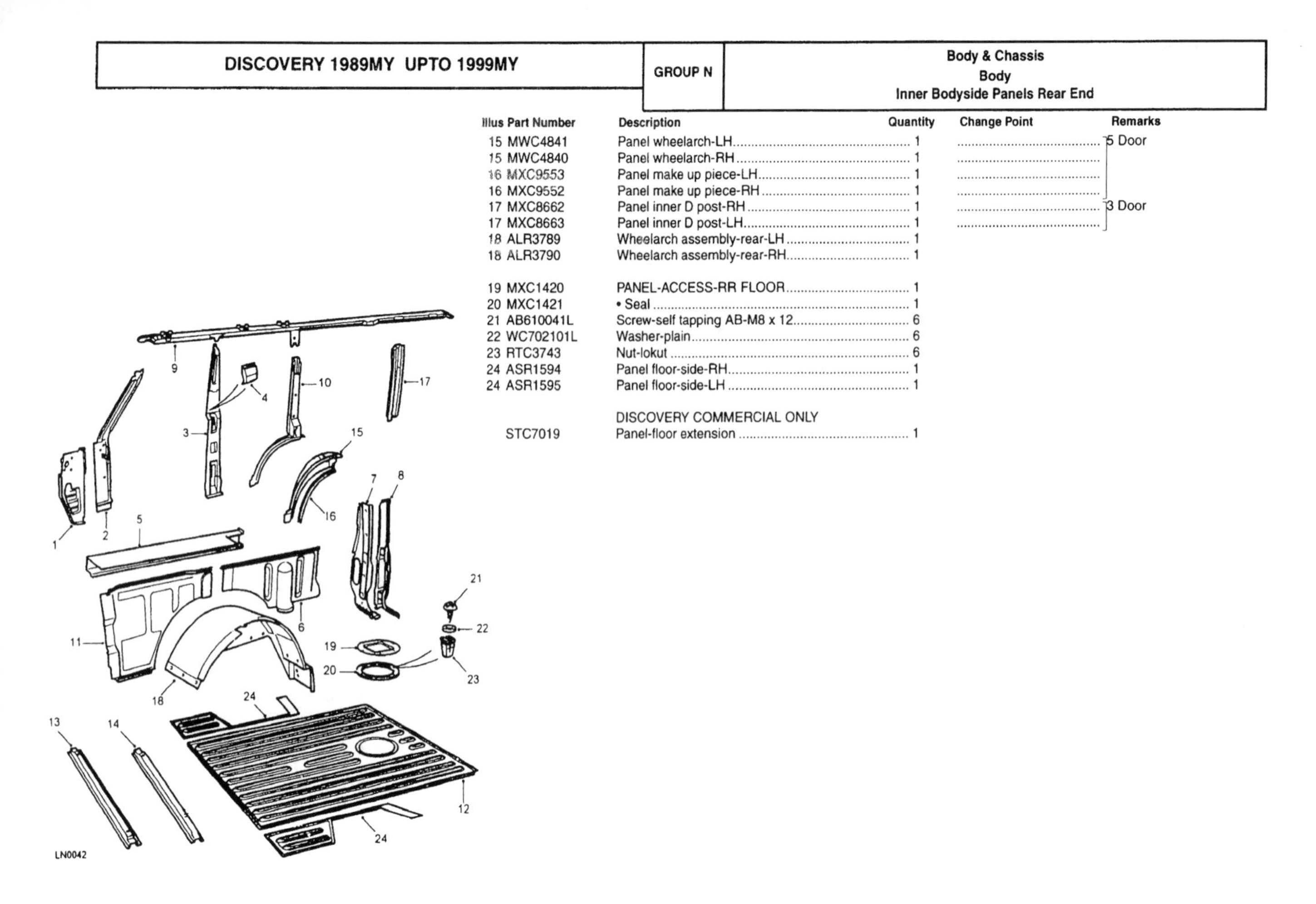

Dust Proofing Seals

Illus	Part Number	Description	Quantity	Change Point	Remarks
1	MUC1818	Seal-front	4		
2	MUC1819	Seal-rear	2		
3	MUC1822	Seal	4		
4	MUC1820	Seal	2		

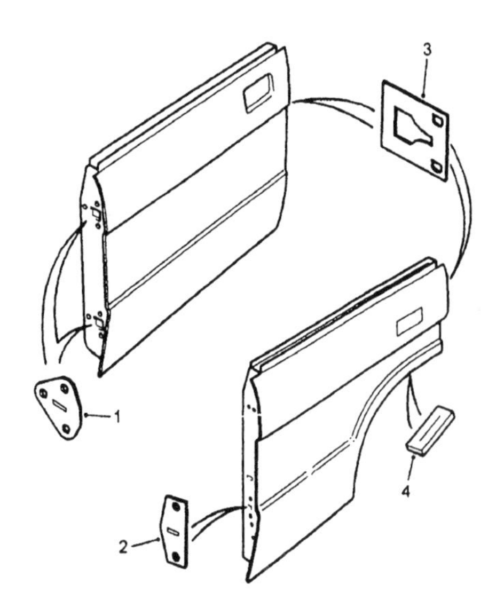

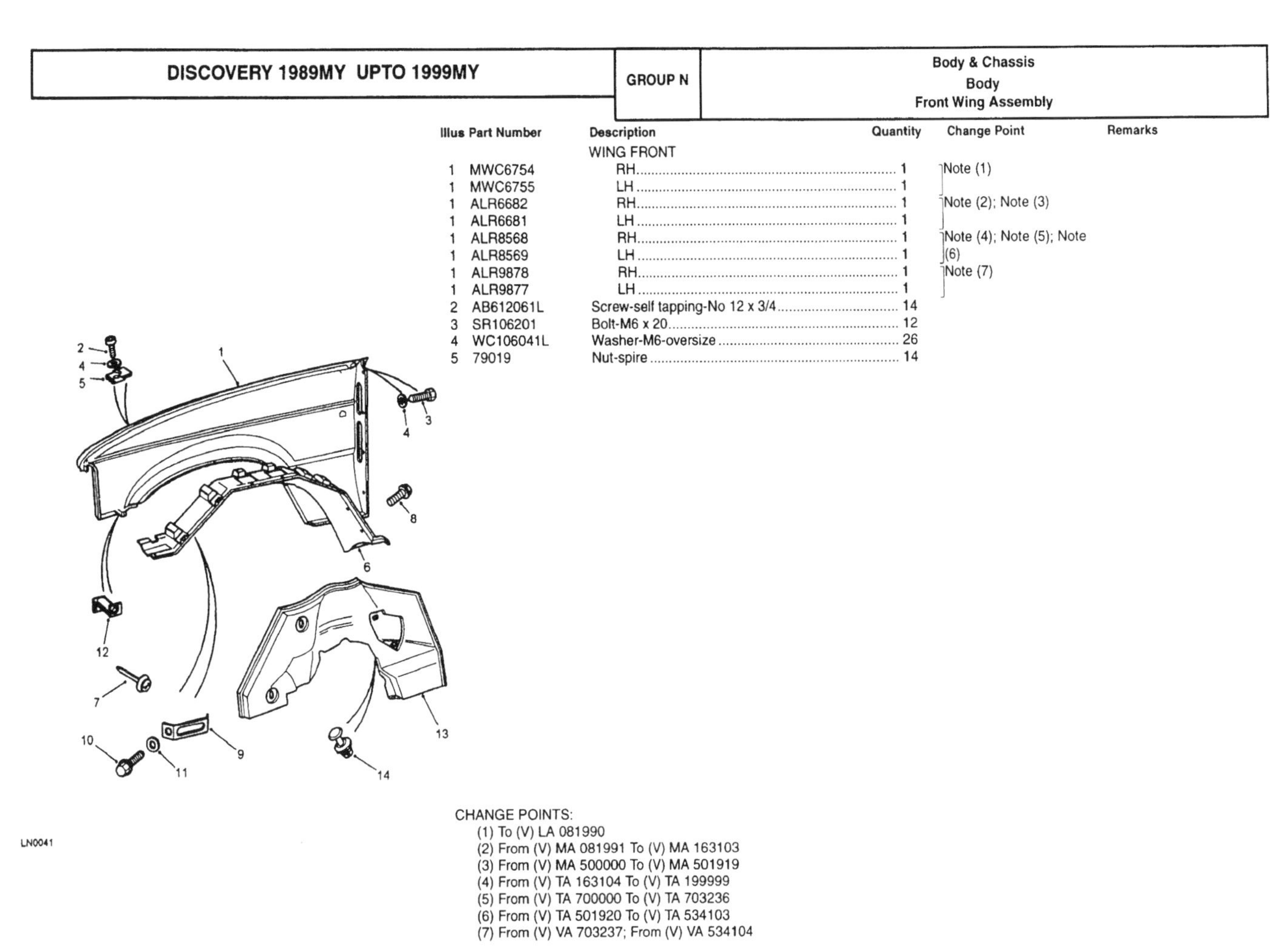

Illus	Part Number	Description	Quantity	Change Point	Remarks
		WING FRONT			
1	MWC6754	RH	1	Note (1)	
1	MWC6755	LH	1		
1	ALR6682	RH	1	Note (2); Note (3)	
1	ALR6681	LH	1		
1	ALR8568	RH	1	Note (4); Note (5); Note (6)	
1	ALR8569	LH	1		
1	ALR9878	RH	1	Note (7)	
1	ALR9877	LH	1		
2	AB612061L	Screw-self tapping-No 12 x 3/4	14		
3	SR106201	Bolt-M6 x 20	12		
4	WC106041L	Washer-M6-oversize	26		
5	79019	Nut-spire	14		

CHANGE POINTS:
(1) To (V) LA 081990
(2) From (V) MA 081991 To (V) MA 163103
(3) From (V) MA 500000 To (V) MA 501919
(4) From (V) TA 163104 To (V) TA 199999
(5) From (V) TA 700000 To (V) TA 703236
(6) From (V) TA 501920 To (V) TA 534103
(7) From (V) VA 703237; From (V) VA 534104

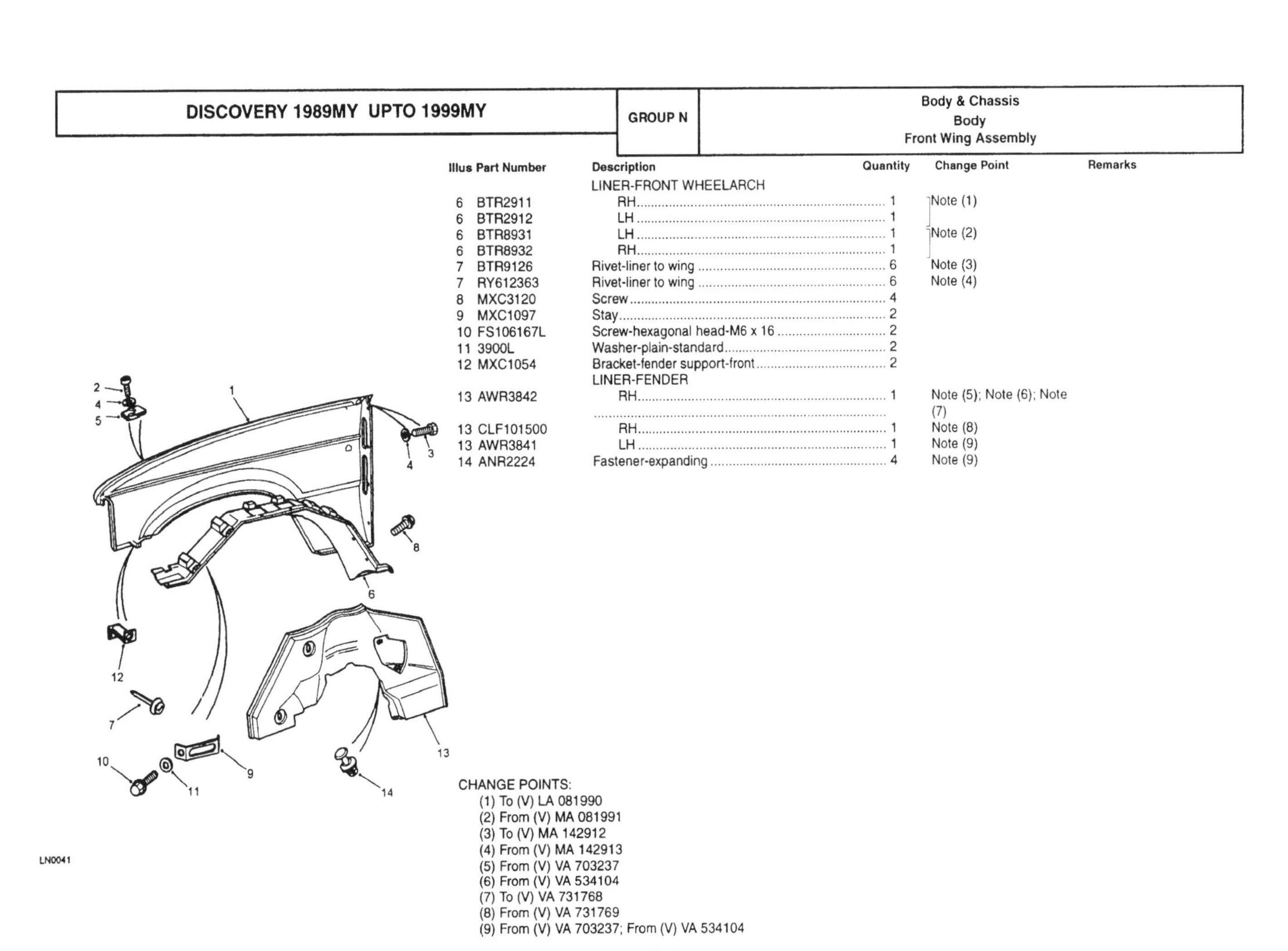

Illus	Part Number	Description	Quantity	Change Point	Remarks
		LINER-FRONT WHEELARCH			
6	BTR2911	RH	1	Note (1)	
6	BTR2912	LH	1		
6	BTR8931	LH	1	Note (2)	
6	BTR8932	RH	1		
7	BTR9126	Rivet-liner to wing	6	Note (3)	
7	RY612363	Rivet-liner to wing	6	Note (4)	
8	MXC3120	Screw	4		
9	MXC1097	Stay	2		
10	FS106167L	Screw-hexagonal head-M6 x 16	2		
11	3900L	Washer-plain-standard	2		
12	MXC1054	Bracket-fender support-front	2		
		LINER-FENDER			
13	AWR3842	RH	1	Note (5); Note (6); Note (7)	
13	CLF101500	RH	1	Note (8)	
13	AWR3841	LH	1	Note (9)	
14	ANR2224	Fastener-expanding	4	Note (9)	

CHANGE POINTS:
(1) To (V) LA 081990
(2) From (V) MA 081991
(3) To (V) MA 142912
(4) From (V) MA 142913
(5) From (V) VA 703237
(6) From (V) VA 534104
(7) To (V) VA 731768
(8) From (V) VA 731769
(9) From (V) VA 703237; From (V) VA 534104

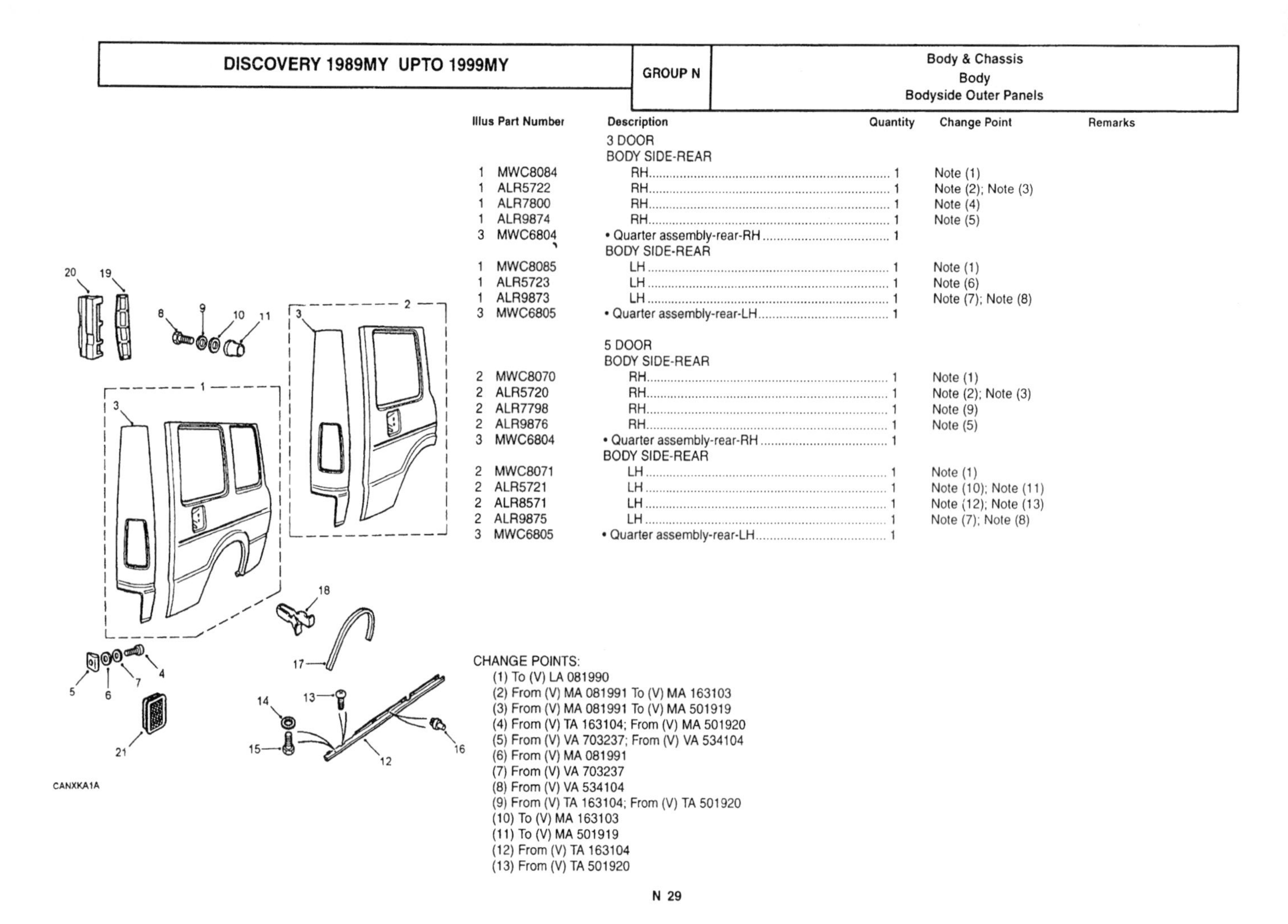

Illus	Part Number	Description	Quantity	Change Point	Remarks
		3 DOOR			
		BODY SIDE-REAR			
1	MWC8084	RH	1	Note (1)	
1	ALR5722	RH	1	Note (2); Note (3)	
1	ALR7800	RH	1	Note (4)	
1	ALR9874	RH	1	Note (5)	
3	MWC6804	• Quarter assembly-rear-RH	1		
		BODY SIDE-REAR			
1	MWC8085	LH	1	Note (1)	
1	ALR5723	LH	1	Note (6)	
1	ALR9873	LH	1	Note (7); Note (8)	
3	MWC6805	• Quarter assembly-rear-LH	1		
		5 DOOR			
		BODY SIDE-REAR			
2	MWC8070	RH	1	Note (1)	
2	ALR5720	RH	1	Note (2); Note (3)	
2	ALR7798	RH	1	Note (9)	
2	ALR9876	RH	1	Note (5)	
3	MWC6804	• Quarter assembly-rear-RH	1		
		BODY SIDE-REAR			
2	MWC8071	LH	1	Note (1)	
2	ALR5721	LH	1	Note (10); Note (11)	
2	ALR8571	LH	1	Note (12); Note (13)	
2	ALR9875	LH	1	Note (7); Note (8)	
3	MWC6805	• Quarter assembly-rear-LH	1		

CHANGE POINTS:
(1) To (V) LA 081990
(2) From (V) MA 081991 To (V) MA 163103
(3) From (V) MA 081991 To (V) MA 501919
(4) From (V) TA 163104; From (V) MA 501920
(5) From (V) VA 703237; From (V) VA 534104
(6) From (V) MA 081991
(7) From (V) VA 703237
(8) From (V) VA 534104
(9) From (V) TA 163104; From (V) TA 501920
(10) To (V) MA 163103
(11) To (V) MA 501919
(12) From (V) TA 163104
(13) From (V) TA 501920

N 29

CANXKA1A

Illus	Part Number	Description	Quantity	Change Point	Remarks
		ALL VARIANTS			
4	78322	Screw-self tapping-No 8 x 1/2	NLA		Use AB608047L.
4	AB608047L	Screw-self tapping-No 8 x 1/2	14		
5	AK608045	Nut-spring-self thread	14		
6	WA105001L	Washer-plain-M5-standard	14		
7	WL105001L	Washer-M5	14		
8	SR106201	Bolt-M6 x 20	52		
9	WL106001L	Washer-M6	52		
10	WJ106001L	Washer-plain-M6	38		
11	MXC1089	Hexsert-Black	34		
		FINISHER-BODY SIDE SILL			
12	MXC2676PUB	RH-Black	1	Note (1)	3 Door
12	MXC2677PUB	LH-Black	1		
12	BTR5706PUB	RH-Black	1	Note (2)	
12	BTR5705PUB	LH-Black	1		
12	BTR5704PUB	RH-Black	1		5 Door
12	BTR5703PUB	LH-Black	1		
		ALL VARIANTS			
13	AB612061L	Screw-self tapping-No 12 x 3/4	2		
14	RTC613	Washer-plain	2		
15	SH604081L	Screw-hexagonal head-M5-6mm	2		
16	79086	Rivet-plastic-drive	NLA		Use 79086L.
16	79086L	Rivet-plastic-drive	40		

CANXKA1A

CHANGE POINTS:
(1) To (V) JA 031297
(2) From (V) JA 031298

N 30

Illus	Part Number	Description	Quantity	Change Point	Remarks
17	MXC5752	Seal-wheelarch flare	2		
18	MXC3177L	Clip	10		
19	MXC7580	Moulding-E post upper air vent	1	Note (1)	
19	BTR1691	Moulding-E post upper air vent	1	Note (2)	
20	BTR8110	Baffle-air vent-rear	1	Note (3)	
21	BTR3996	Vent assembly-air extractor	2	Note (4)	
		AUSTRALIA & GULF STATES ONLY			
	BTR4383	Seal air intake	1		

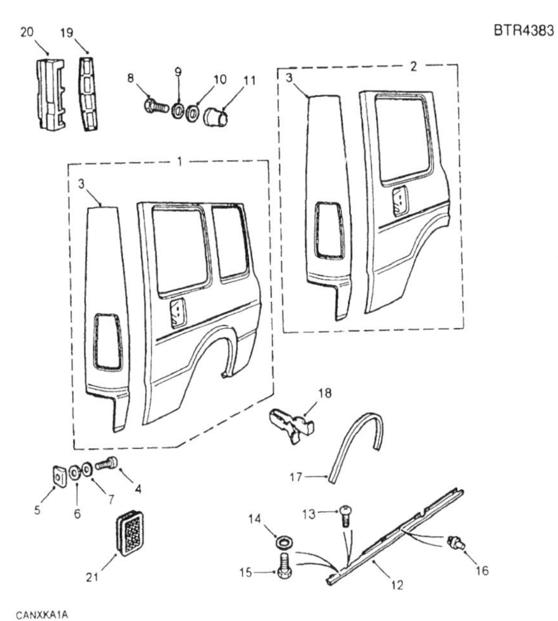

CHANGE POINTS:
(1) To (V) JA 019457
(2) From (V) JA 019458
(3) From (V) KA 038822
(4) From (V) MA 081991

Illus	Part Number	Description	Quantity	Change Point	Remarks
		PANEL HEADER ROOF			
1	ALR6238	outer	1		
2	MXC8657	rear	1	To (V) VA 737406	
2	ASR1862	rear	1	From (V) VA 737407	
3	MXC1280	Finisher-headlining-rear	1		
4	ALR6176	Panel closing	1	To (V) VA 737406	
4	AQH700010	Panel closing	1	From (V) VA 737407	
		AUSTRALIA ONLY			
4	MXC9314	Panel closing	1	To (V) LA 081990	
4	ALR6749	Panel closing	1	From (V) MA 081991	
		ALL VARIANTS			
5	STC1066	CROSSMEMBER ASSEMBLY-REAR FLOOR	1		
6	MWC7584	• Panel-rear end-lower	1		
7	MWC7562	• Closing-rear crossmember	1		
8	ALR1773	Panel-floor to rear end	1		
9	MXC5278	Support bracket	A/R		

CDNXBR1A

Illus	Part Number	Description	Quantity	Change Point	Remarks
1	ALR5404	FUEL FILLER DOOR ASSEMBLY	1		
	CCP2208	• Spring-fuel fill door	1		
2	MXC6366	Shim	A/R		
3	SE104161L	Screw-M8 x 16	2		
4	WC104001L	Washer-plain	2		
5	RTC6687	LOCK-FUEL FILLER FLAP	1		
6	CDP6702L	• Clip	1		
7	CDP1568L	Washer-sealing	1		
8	CDP1567L	Striker-fuel filler door	1		
9	MWC9919	Buffer filler flap	2		

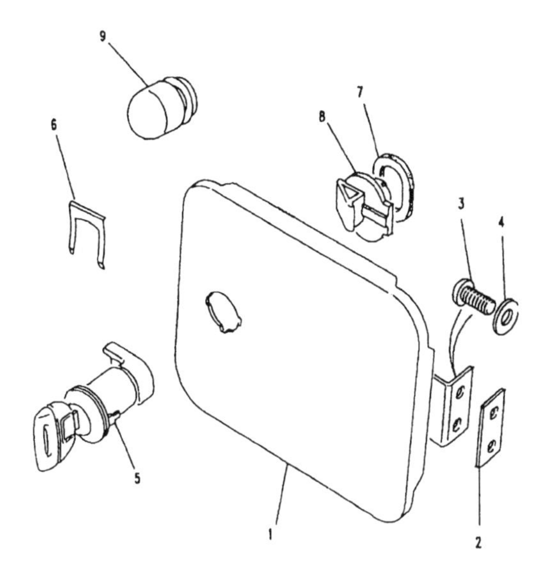

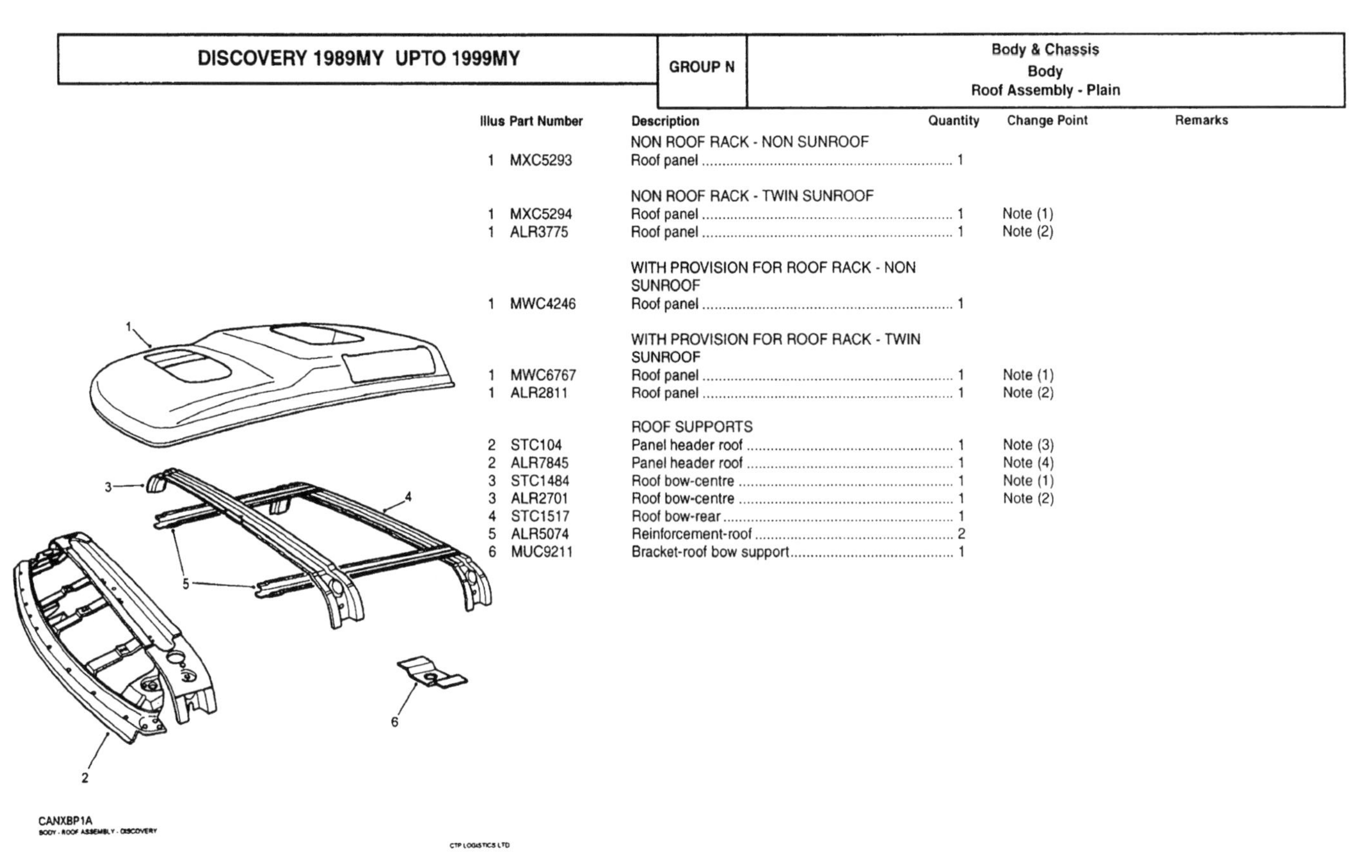

Illus	Part Number	Description	Quantity	Change Point	Remarks
		NON ROOF RACK - NON SUNROOF			
1	MXC5293	Roof panel	1		
		NON ROOF RACK - TWIN SUNROOF			
1	MXC5294	Roof panel	1	Note (1)	
1	ALR3775	Roof panel	1	Note (2)	
		WITH PROVISION FOR ROOF RACK - NON SUNROOF			
1	MWC4246	Roof panel	1		
		WITH PROVISION FOR ROOF RACK - TWIN SUNROOF			
1	MWC6767	Roof panel	1	Note (1)	
1	ALR2811	Roof panel	1	Note (2)	
		ROOF SUPPORTS			
2	STC104	Panel header roof	1	Note (3)	
2	ALR7845	Panel header roof	1	Note (4)	
3	STC1484	Roof bow-centre	1	Note (1)	
3	ALR2701	Roof bow-centre	1	Note (2)	
4	STC1517	Roof bow-rear	1		
5	ALR5074	Reinforcement-roof	2		
6	MUC9211	Bracket-roof bow support	1		

CANXBP1A

CHANGE POINTS:
(1) To (V) LA 081990
(2) From (V) MA 081991
(3) To (V) LA 081991
(4) From (V) MA 081992

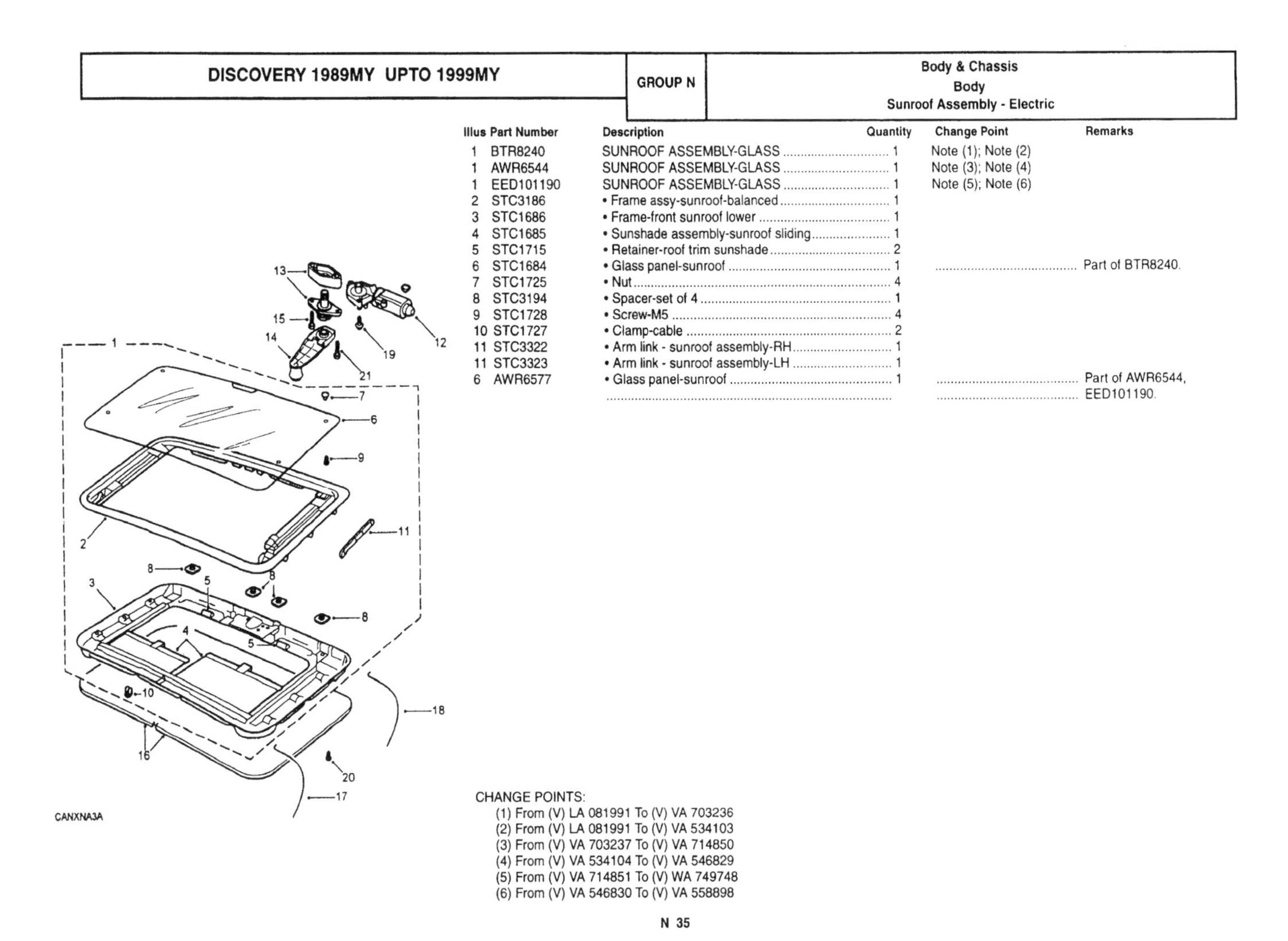

Illus	Part Number	Description	Quantity	Change Point	Remarks
1	BTR8240	SUNROOF ASSEMBLY-GLASS	1	Note (1); Note (2)	
1	AWR6544	SUNROOF ASSEMBLY-GLASS	1	Note (3); Note (4)	
1	EED101190	SUNROOF ASSEMBLY-GLASS	1	Note (5); Note (6)	
2	STC3186	• Frame assy-sunroof-balanced	1		
3	STC1686	• Frame-front sunroof lower	1		
4	STC1685	• Sunshade assembly-sunroof sliding	1		
5	STC1715	• Retainer-roof trim sunshade	2		
6	STC1684	• Glass panel-sunroof	1		Part of BTR8240.
7	STC1725	• Nut	4		
8	STC3194	• Spacer-set of 4	1		
9	STC1728	• Screw-M5	4		
10	STC1727	• Clamp-cable	2		
11	STC3322	• Arm link - sunroof assembly-RH	1		
11	STC3323	• Arm link - sunroof assembly-LH	1		
6	AWR6577	• Glass panel-sunroof	1		Part of AWR6544, EED101190.

CHANGE POINTS:
(1) From (V) LA 081991 To (V) VA 703236
(2) From (V) LA 081991 To (V) VA 534103
(3) From (V) VA 703237 To (V) VA 714850
(4) From (V) VA 534104 To (V) VA 546829
(5) From (V) VA 714851 To (V) WA 749748
(6) From (V) VA 546830 To (V) VA 558898

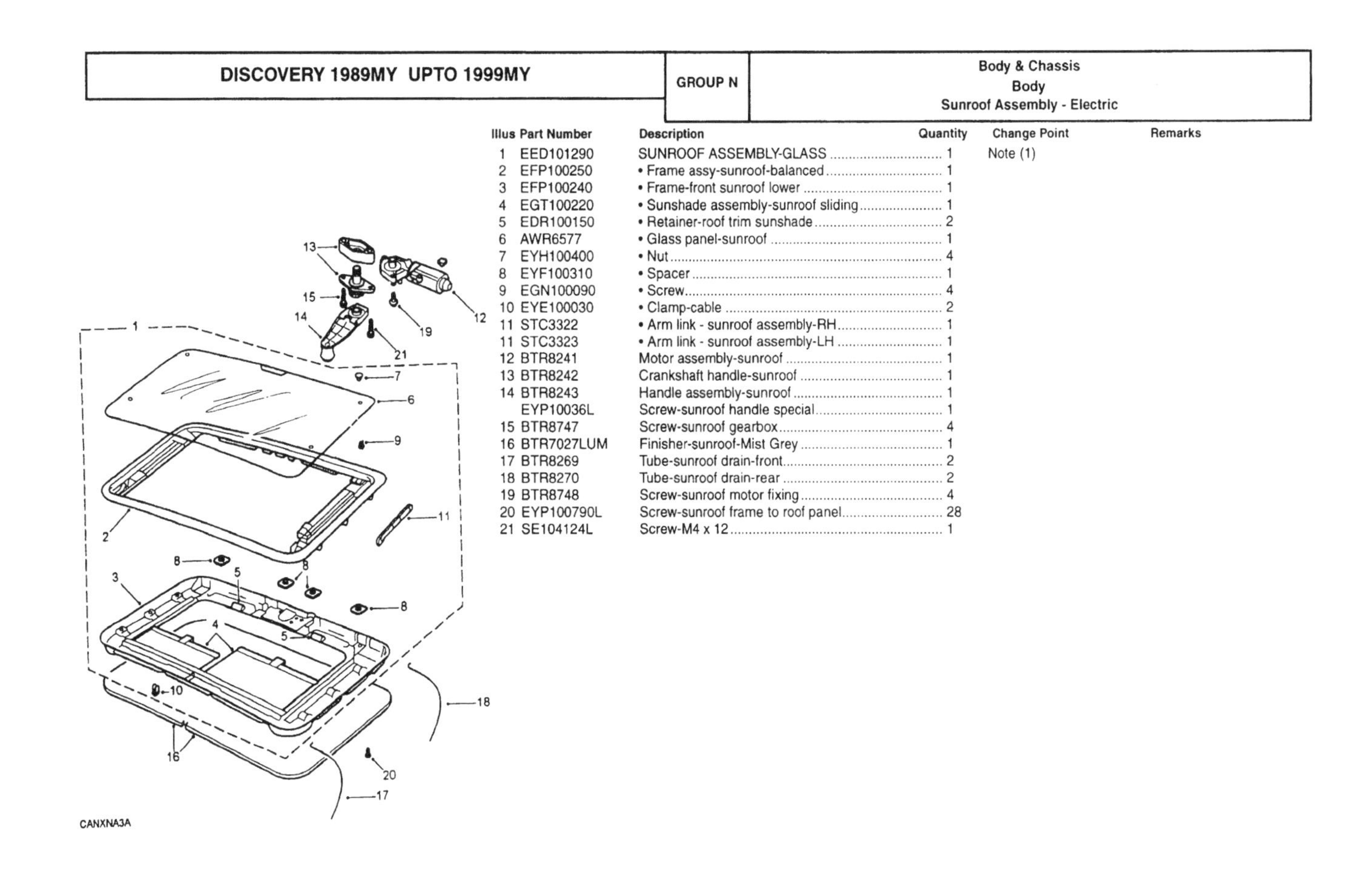

Illus	Part Number	Description	Quantity	Change Point	Remarks
1	EED101290	SUNROOF ASSEMBLY-GLASS	1	Note (1)	
2	EFP100250	• Frame assy-sunroof-balanced	1		
3	EFP100240	• Frame-front sunroof lower	1		
4	EGT100220	• Sunshade assembly-sunroof sliding	1		
5	EDR100150	• Retainer-roof trim sunshade	2		
6	AWR6577	• Glass panel-sunroof	1		
7	EYH100400	• Nut	4		
8	EYF100310	• Spacer	1		
9	EGN100090	• Screw	4		
10	EYE100030	• Clamp-cable	2		
11	STC3322	• Arm link - sunroof assembly-RH	1		
11	STC3323	• Arm link - sunroof assembly-LH	1		
12	BTR8241	Motor assembly-sunroof	1		
13	BTR8242	Crankshaft handle-sunroof	1		
14	BTR8243	Handle assembly-sunroof	1		
	EYP10036L	Screw-sunroof handle special	1		
15	BTR8747	Screw-sunroof gearbox	4		
16	BTR7027LUM	Finisher-sunroof-Mist Grey	1		
17	BTR8269	Tube-sunroof drain-front	2		
18	BTR8270	Tube-sunroof drain-rear	2		
19	BTR8748	Screw-sunroof motor fixing	4		
20	EYP100790L	Screw-sunroof frame to roof panel	28		
21	SE104124L	Screw-M4 x 12	1		

CHANGE POINTS:
(1) From (V) WA 749749

DISCOVERY 1989MY UPTO 1999MY	GROUP N	Body & Chassis Body Sunroof Assembly Front-To (V) JA034313

Illus	Part Number	Description	Quantity	Change Point	Remarks
1	MXC3414	Frame-front sunroof upper	1		
2	RTC6481	Seal-rubber	1		
3	RTC6482	Seal-sunroof frame	1		
4	MXC3413	Sunroof assembly-glass	1		
5	MXC3415	Frame-sunroof-lower	1		
6	AR606041L	Screw	18		
7	MXC7881LUP	Finisher-sunroof-Grey	1		
8	STC982	Handle assembly-sunroof	1		

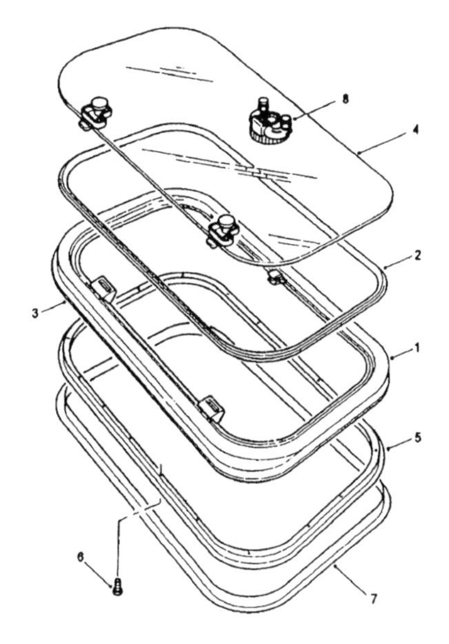

CANXNA1B

N 37

DISCOVERY 1989MY UPTO 1999MY	GROUP N	Body & Chassis Body Sunroof Assembly Front-From (V) KA034314

Illus	Part Number	Description	Quantity	Change Point	Remarks
1	BTR3072	SUNROOF ASSEMBLY-GLASS	1		
2	STC773	• Seal-rubber	1		
3	RTC6482	• Seal-sunroof frame	1		
4	MXC3415	Frame-sunroof-lower	1		
5	AR606041L	Screw	18		
6	MXC7882LUP	Finisher-sunroof-Grey	1		
7	STC982	Handle assembly-sunroof	1		
8	STC3682	Hinge-tilt & remove sunroof	2		

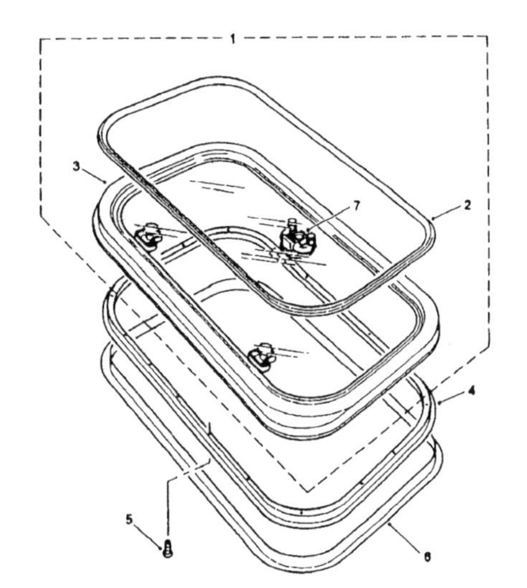

CANXNA2A

N 38

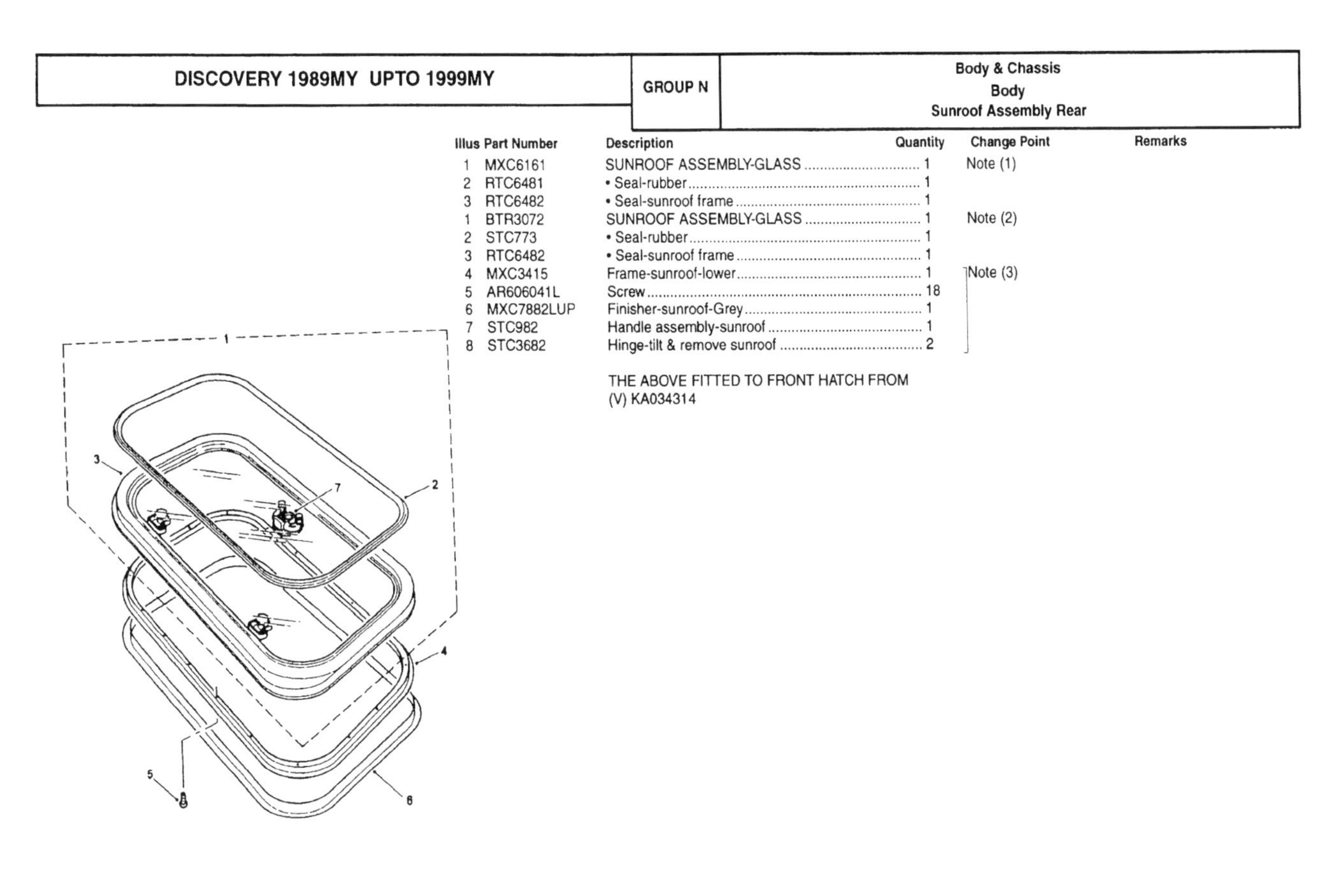

Illus	Part Number	Description	Quantity	Change Point	Remarks
1	MXC6161	SUNROOF ASSEMBLY-GLASS	1	Note (1)	
2	RTC6481	• Seal-rubber	1		
3	RTC6482	• Seal-sunroof frame	1		
1	BTR3072	SUNROOF ASSEMBLY-GLASS	1	Note (2)	
2	STC773	• Seal-rubber	1		
3	RTC6482	• Seal-sunroof frame	1		
4	MXC3415	Frame-sunroof-lower	1	Note (3)	
5	AR606041L	Screw	18		
6	MXC7882LUP	Finisher-sunroof-Grey	1		
7	STC982	Handle assembly-sunroof	1		
8	STC3682	Hinge-tilt & remove sunroof	2		

THE ABOVE FITTED TO FRONT HATCH FROM (V) KA034314

CHANGE POINTS:
(1) To (V) JA 019144
(2) From (V) JA 019145
(3) From (V) KA 034314

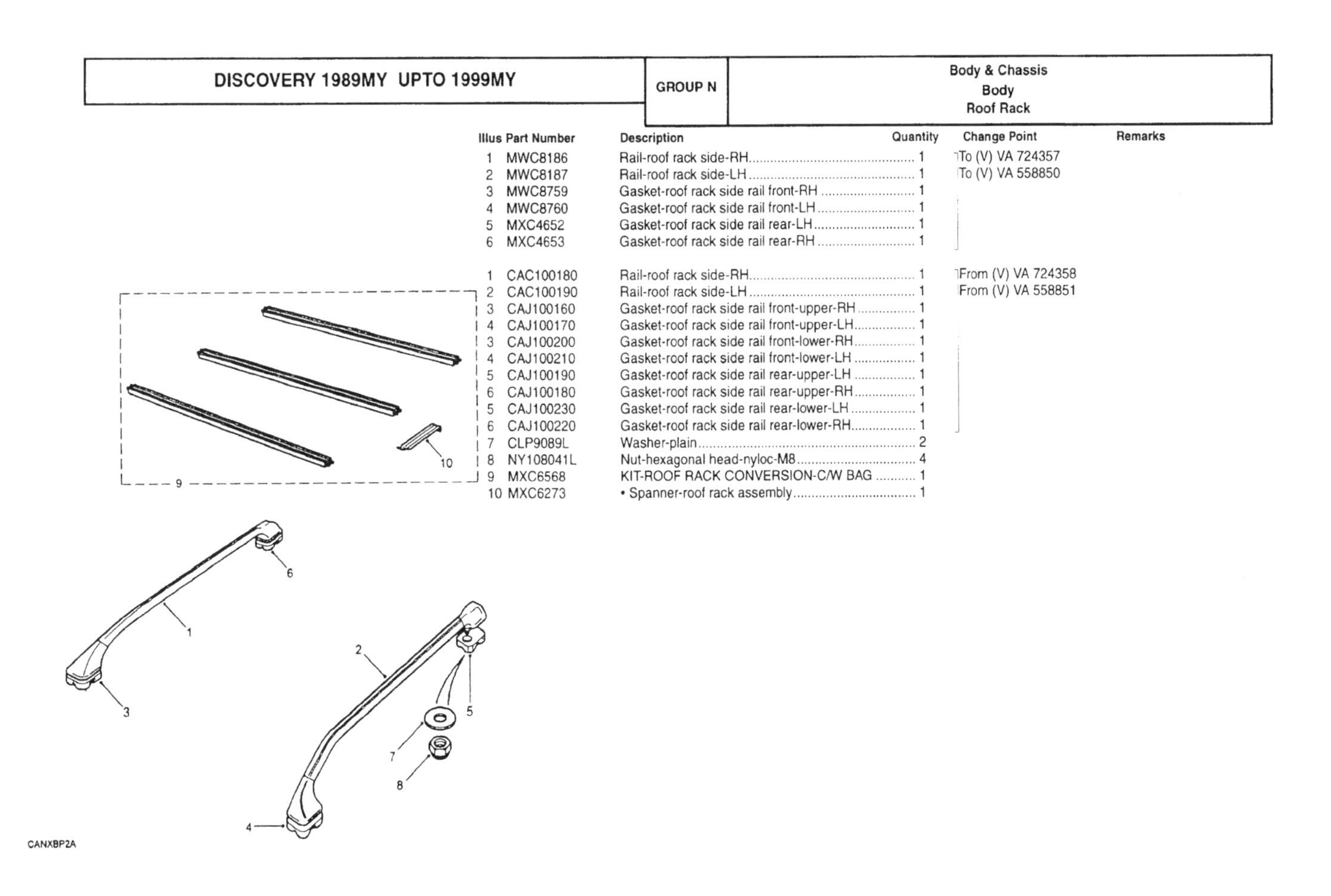

Illus	Part Number	Description	Quantity	Change Point	Remarks
1	MWC8186	Rail-roof rack side-RH	1	To (V) VA 724357	
2	MWC8187	Rail-roof rack side-LH	1	To (V) VA 558850	
3	MWC8759	Gasket-roof rack side rail front-RH	1		
4	MWC8760	Gasket-roof rack side rail front-LH	1		
5	MXC4652	Gasket-roof rack side rail rear-LH	1		
6	MXC4653	Gasket-roof rack side rail rear-RH	1		
1	CAC100180	Rail-roof rack side-RH	1	From (V) VA 724358	
2	CAC100190	Rail-roof rack side-LH	1	From (V) VA 558851	
3	CAJ100160	Gasket-roof rack side rail front-upper-RH	1		
4	CAJ100170	Gasket-roof rack side rail front-upper-LH	1		
3	CAJ100200	Gasket-roof rack side rail front-lower-RH	1		
4	CAJ100210	Gasket-roof rack side rail front-lower-LH	1		
5	CAJ100190	Gasket-roof rack side rail rear-upper-LH	1		
6	CAJ100180	Gasket-roof rack side rail rear-upper-RH	1		
5	CAJ100230	Gasket-roof rack side rail rear-lower-LH	1		
6	CAJ100220	Gasket-roof rack side rail rear-lower-RH	1		
7	CLP9089L	Washer-plain	2		
8	NY108041L	Nut-hexagonal head-nyloc-M8	4		
9	MXC6568	KIT-ROOF RACK CONVERSION-C/W BAG	1		
10	MXC6273	• Spanner-roof rack assembly	1		

DISCOVERY 1989MY UPTO 1999MY	GROUP N	Body & Chassis Body Bonnet Assembly

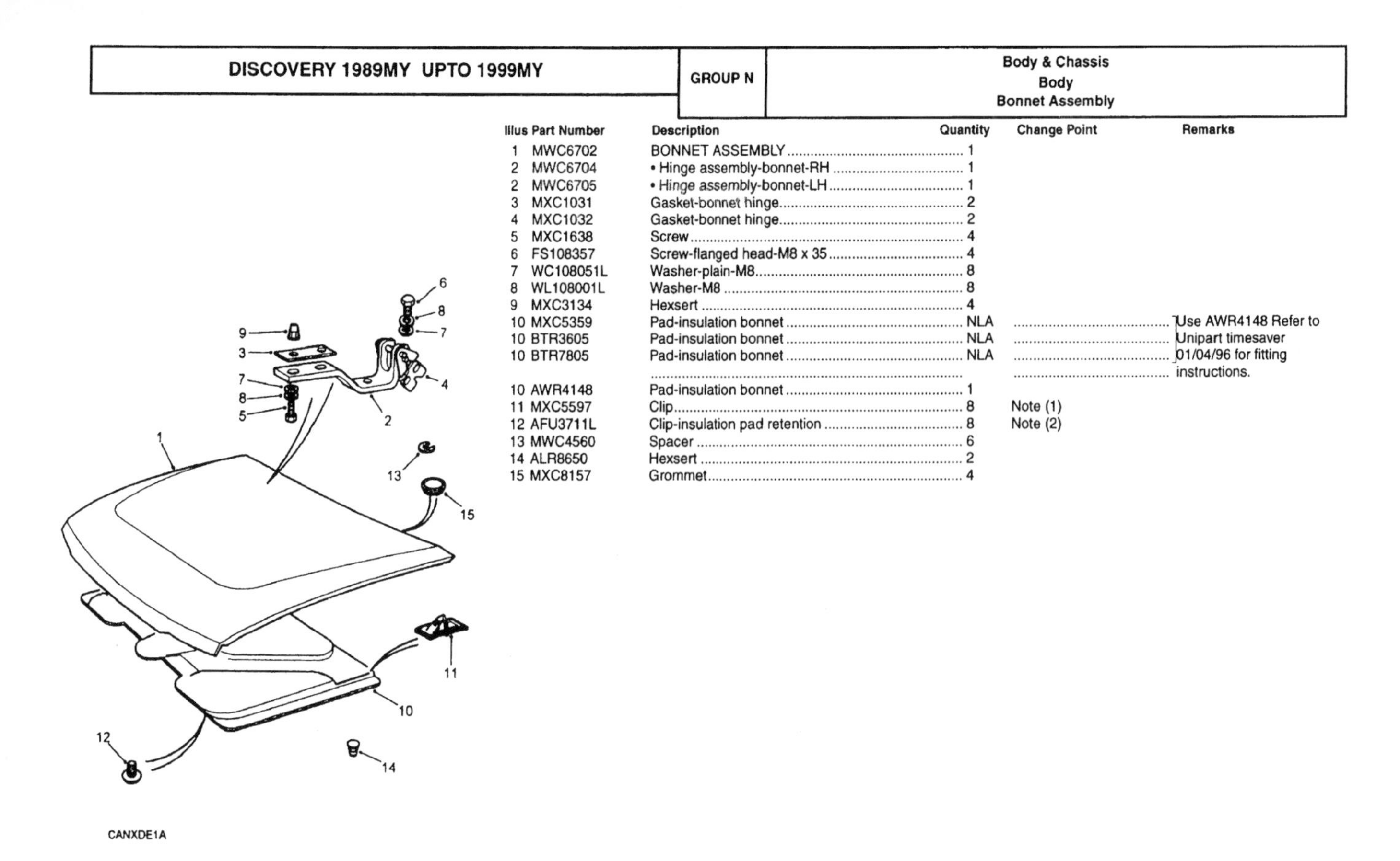

CANXDE1A

Illus	Part Number	Description	Quantity	Change Point	Remarks
1	MWC6702	BONNET ASSEMBLY	1		
2	MWC6704	• Hinge assembly-bonnet-RH	1		
2	MWC6705	• Hinge assembly-bonnet-LH	1		
3	MXC1031	Gasket-bonnet hinge	2		
4	MXC1032	Gasket-bonnet hinge	2		
5	MXC1638	Screw	4		
6	FS108357	Screw-flanged head-M8 x 35	4		
7	WC108051L	Washer-plain-M8	8		
8	WL108001L	Washer-M8	8		
9	MXC3134	Hexsert	4		
10	MXC5359	Pad-insulation bonnet	NLA		Use AWR4148 Refer to
10	BTR3605	Pad-insulation bonnet	NLA		Unipart timesaver
10	BTR7805	Pad-insulation bonnet	NLA		01/04/96 for fitting
					instructions.
10	AWR4148	Pad-insulation bonnet	1		
11	MXC5597	Clip	8	Note (1)	
12	AFU3711L	Clip-insulation pad retention	8	Note (2)	
13	MWC4560	Spacer	6		
14	ALR8650	Hexsert	2		
15	MXC8157	Grommet	4		

CHANGE POINTS:
(1) To (V) KA 054484
(2) From (V) KA 054485

N 41

DISCOVERY 1989MY UPTO 1999MY	GROUP N	Body & Chassis Body Grille & Headlamp Surround Up To LA081990

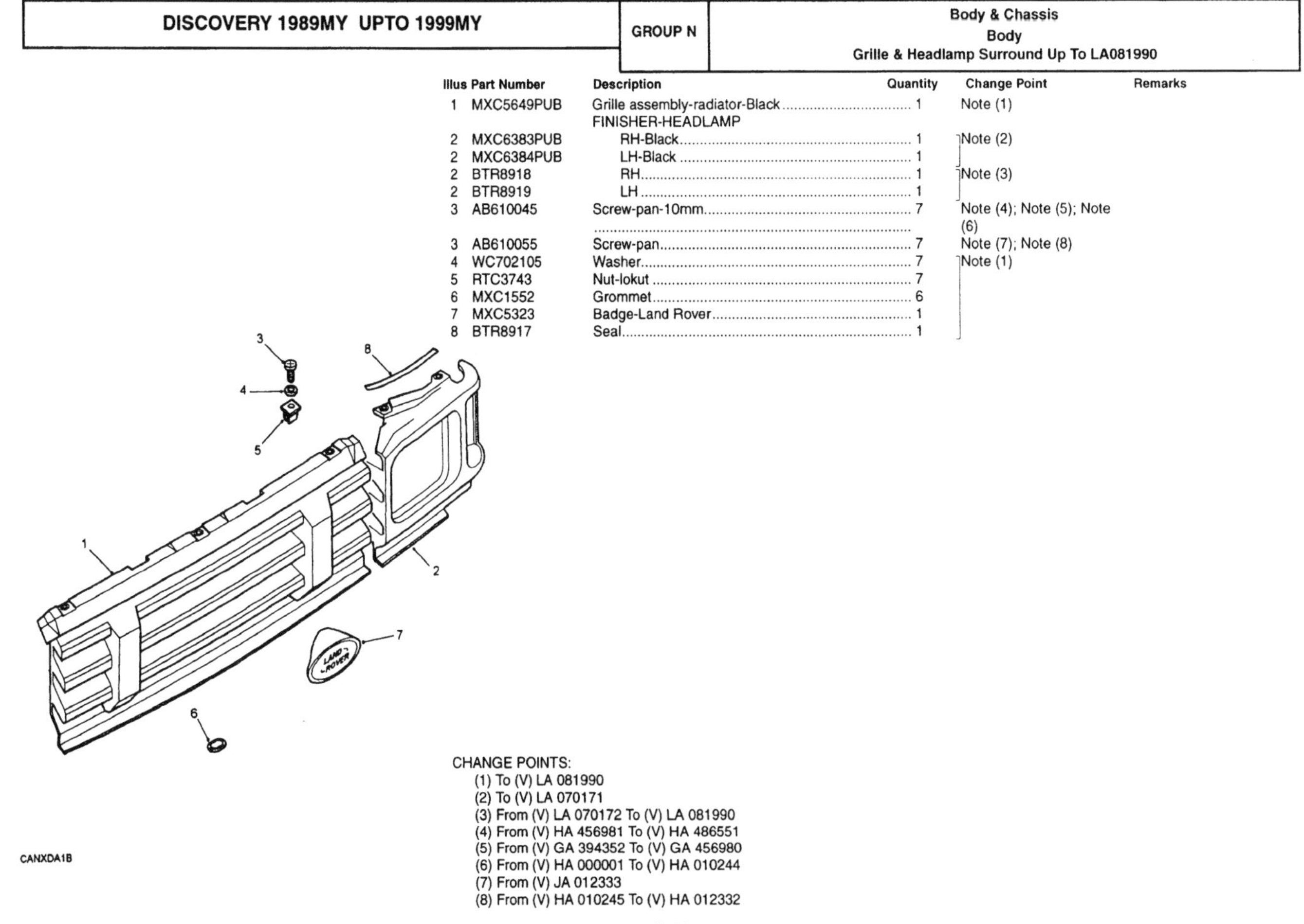

CANXDA1B

Illus	Part Number	Description	Quantity	Change Point	Remarks
1	MXC5649PUB	Grille assembly-radiator-Black	1	Note (1)	
		FINISHER-HEADLAMP			
2	MXC6383PUB	RH-Black	1	Note (2)	
2	MXC6384PUB	LH-Black	1		
2	BTR8918	RH	1	Note (3)	
2	BTR8919	LH	1		
3	AB610045	Screw-pan-10mm	7	Note (4); Note (5); Note (6)	
3	AB610055	Screw-pan	7	Note (7); Note (8)	
4	WC702105	Washer	7	Note (1)	
5	RTC3743	Nut-lokut	7		
6	MXC1552	Grommet	6		
7	MXC5323	Badge-Land Rover	1		
8	BTR8917	Seal	1		

CHANGE POINTS:
(1) To (V) LA 081990
(2) To (V) LA 070171
(3) From (V) LA 070172 To (V) LA 081990
(4) From (V) HA 456981 To (V) HA 486551
(5) From (V) GA 394352 To (V) GA 456980
(6) From (V) HA 000001 To (V) HA 010244
(7) From (V) JA 012333
(8) From (V) HA 010245 To (V) HA 012332

N 42

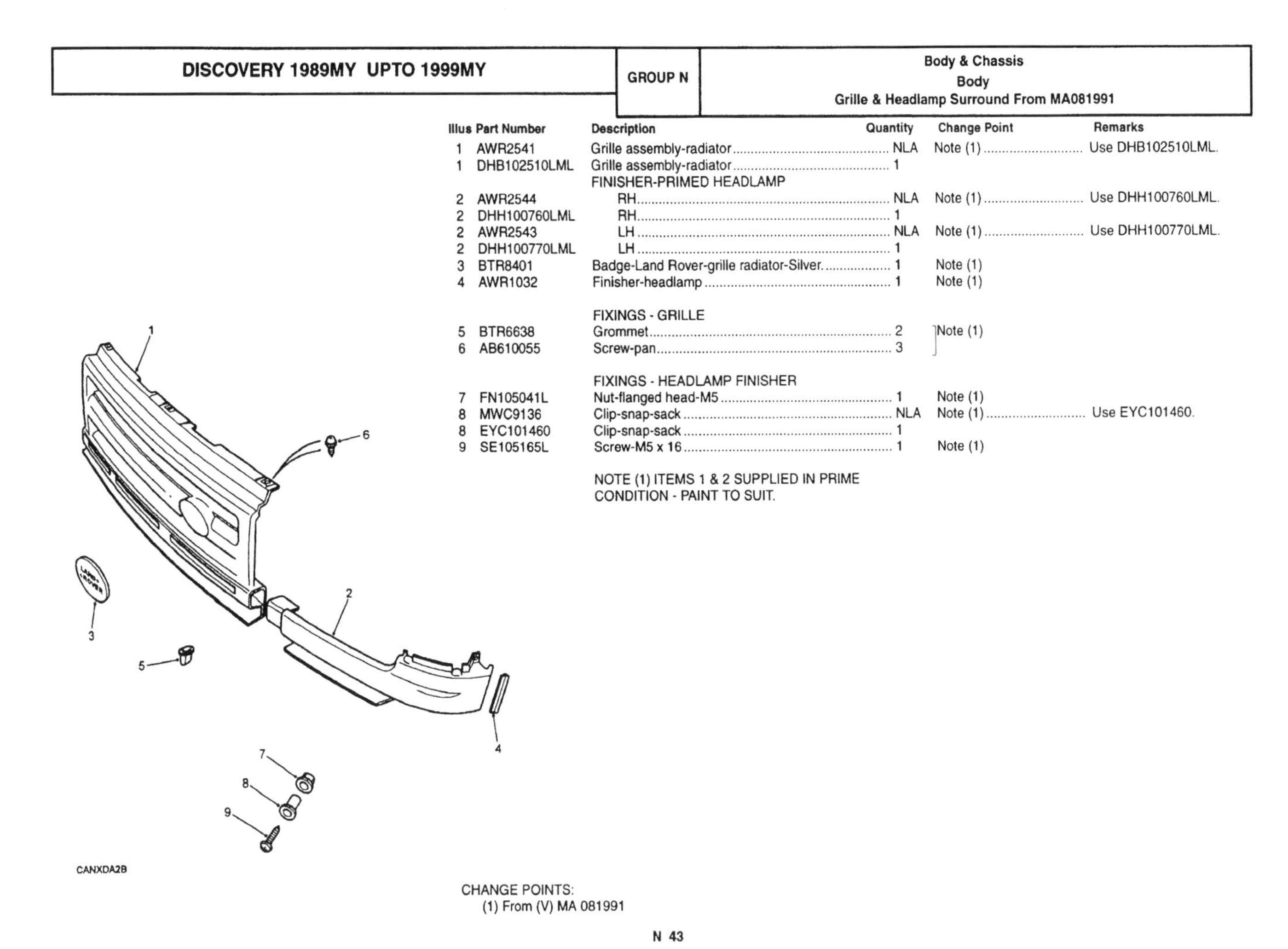

Illus	Part Number	Description	Quantity	Change Point	Remarks
1	AWR2541	Grille assembly-radiator	NLA	Note (1)	Use DHB102510LML.
1	DHB102510LML	Grille assembly-radiator	1		
		FINISHER-PRIMED HEADLAMP			
2	AWR2544	RH	NLA	Note (1)	Use DHH100760LML.
2	DHH100760LML	RH	1		
2	AWR2543	LH	NLA	Note (1)	Use DHH100770LML.
2	DHH100770LML	LH	1		
3	BTR8401	Badge-Land Rover-grille radiator-Silver.	1	Note (1)	
4	AWR1032	Finisher-headlamp	1	Note (1)	
		FIXINGS - GRILLE			
5	BTR6638	Grommet	2	Note (1)	
6	AB610055	Screw-pan	3		
		FIXINGS - HEADLAMP FINISHER			
7	FN105041L	Nut-flanged head-M5	1	Note (1)	
8	MWC9136	Clip-snap-sack	NLA	Note (1)	Use EYC101460.
8	EYC101460	Clip-snap-sack	1		
9	SE105165L	Screw-M5 x 16	1	Note (1)	

NOTE (1) ITEMS 1 & 2 SUPPLIED IN PRIME CONDITION - PAINT TO SUIT.

CHANGE POINTS:
(1) From (V) MA 081991

N 43

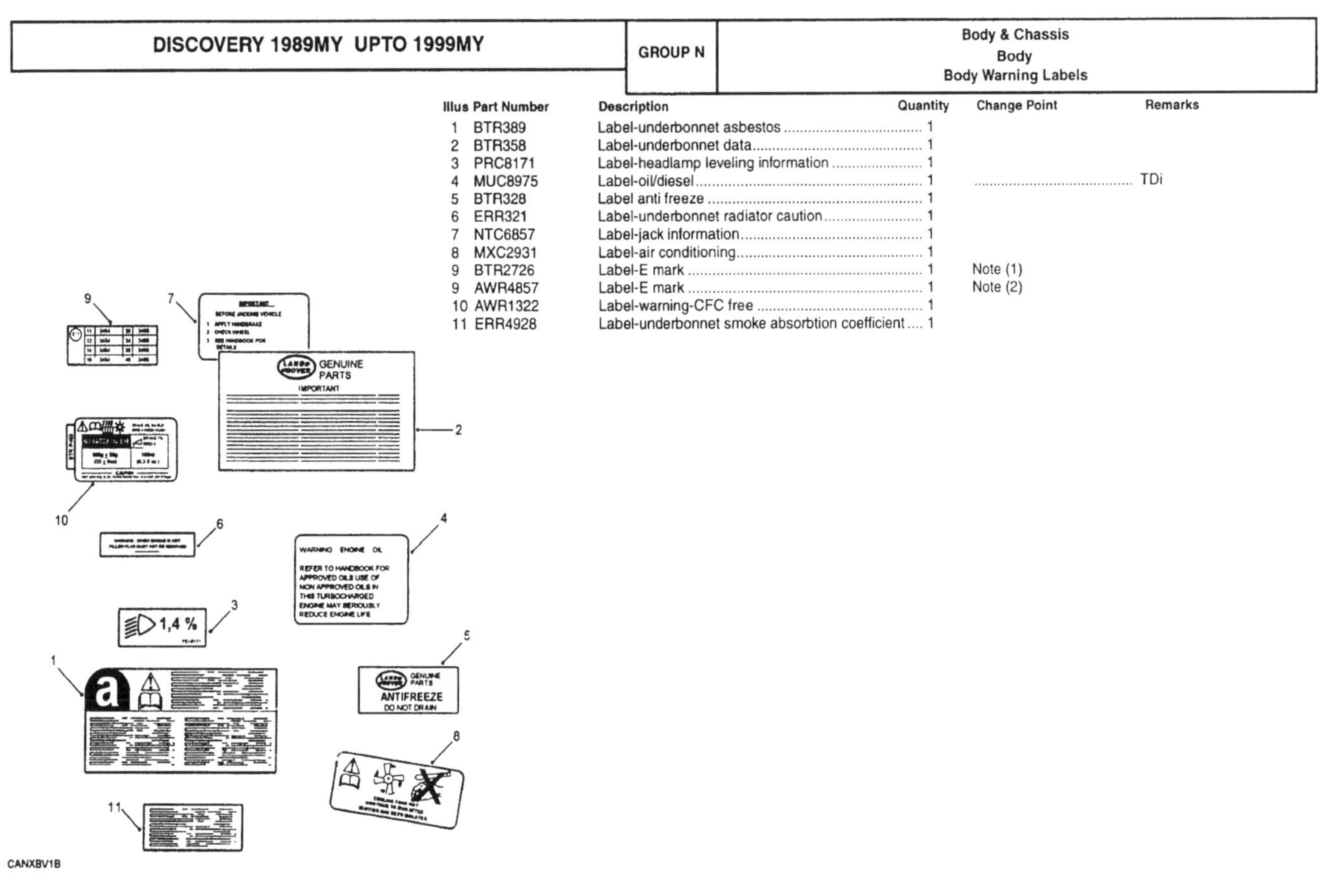

Illus	Part Number	Description	Quantity	Change Point	Remarks
1	BTR389	Label-underbonnet asbestos	1		
2	BTR358	Label-underbonnet data	1		
3	PRC8171	Label-headlamp leveling information	1		
4	MUC8975	Label-oil/diesel	1		TDi
5	BTR328	Label anti freeze	1		
6	ERR321	Label-underbonnet radiator caution	1		
7	NTC6857	Label-jack information	1		
8	MXC2931	Label-air conditioning	1		
9	BTR2726	Label-E mark	1	Note (1)	
9	AWR4857	Label-E mark	1	Note (2)	
10	AWR1322	Label-warning-CFC free	1		
11	ERR4928	Label-underbonnet smoke absorbtion coefficient	1		

CHANGE POINTS:
(1) To (V) MA 163103; To (V) MA 501919
(2) From (V) TA 163104; From (V) TA 501920

N 44

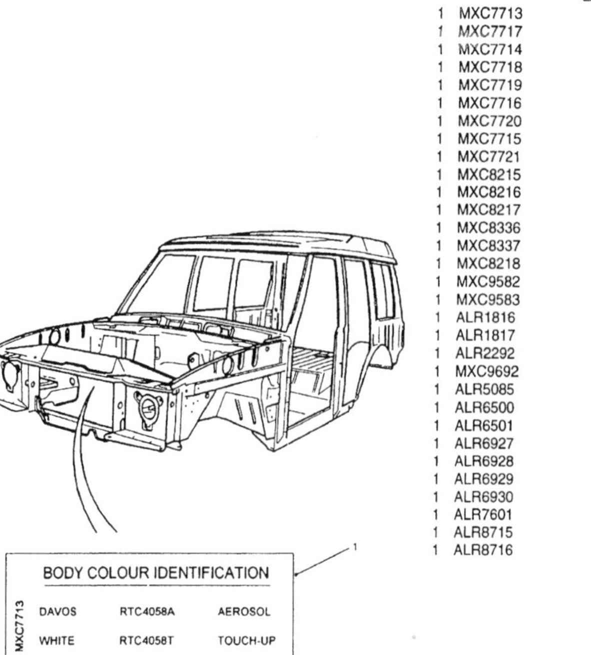

Illus	Part Number	Description	Quantity	Change Point	Remarks
		LABEL-BODY COLOUR ADHESIVE			
1	MXC7713	Davas White	1		
1	MXC7717	Zanzibar Silver	1		
1	MXC7714	Corallin Red	1		
1	MXC7718	Mistrale Light Blue	1		
1	MXC7719	Foxfire Red	1		
1	MXC7716	Caracal Black	1		
1	MXC7720	Marseilles Blue	1		
1	MXC7715	Windjammer Blue	1		
1	MXC7721	Arken Grey	1		
1	MXC8215	Eskdale Green	1		
1	MXC8216	Peddlestone Beige	1		
1	MXC8217	Carridga Green	1		
1	MXC8336	Savarin White	1		
1	MXC8337	Pennine Grey	1		
1	MXC8218	Armada Gold	1		
1	MXC9582	Tintern	1		
1	MXC9583	Ionian Blue	1		
1	ALR1816	Cairngorm Brown	1		
1	ALR1817	Aegean Blue/Bright	1		
1	ALR2292	Aspen Silver	1		
1	MXC9692	Sonoran Brown	1		
1	ALR5085	Sandglow	1		
1	ALR6500	Montpellier Red	1		
1	ALR6501	Coniston Green	1		
1	ALR6927	Plymouth Blue	1		
1	ALR6928	Niagara Grey	1		
1	ALR6929	Avalon Blue	1		
1	ALR6930	Slate Grey	1		
1	ALR7601	Caprice Turquoise	1		
1	ALR8715	Biaritz Blue	1		
1	ALR8716	Epsom Green	1		

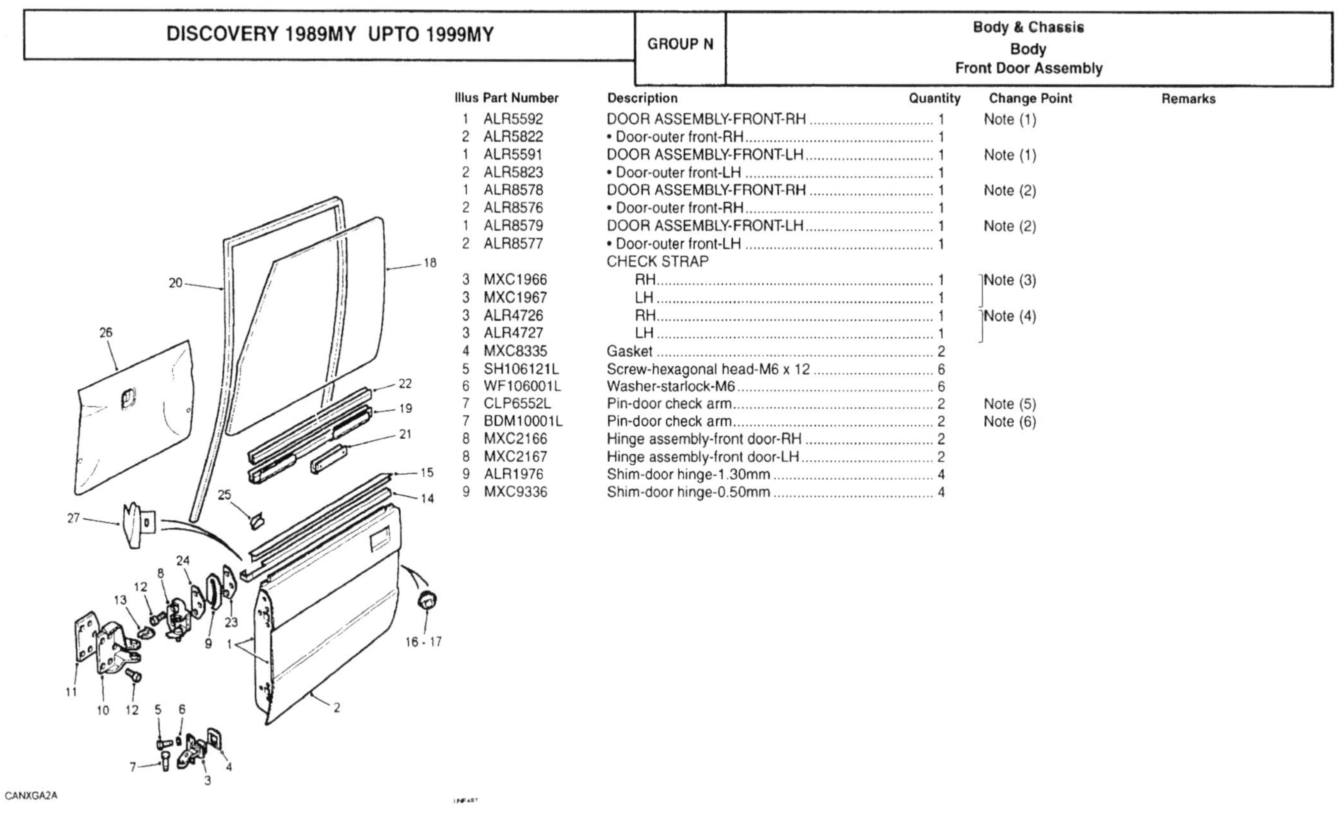

Illus	Part Number	Description	Quantity	Change Point	Remarks
1	ALR5592	DOOR ASSEMBLY-FRONT-RH	1	Note (1)	
2	ALR5822	• Door-outer front-RH	1		
1	ALR5591	DOOR ASSEMBLY-FRONT-LH	1	Note (1)	
2	ALR5823	• Door-outer front-LH	1		
1	ALR8578	DOOR ASSEMBLY-FRONT-RH	1	Note (2)	
2	ALR8576	• Door-outer front-RH	1		
1	ALR8579	DOOR ASSEMBLY-FRONT-LH	1	Note (2)	
2	ALR8577	• Door-outer front-LH	1		
		CHECK STRAP			
3	MXC1966	RH	1	Note (3)	
3	MXC1967	LH	1		
3	ALR4726	RH	1	Note (4)	
3	ALR4727	LH	1		
4	MXC8335	Gasket	2		
5	SH106121L	Screw-hexagonal head-M6 x 12	6		
6	WF106001L	Washer-starlock-M6	6		
7	CLP6552L	Pin-door check arm	2	Note (5)	
7	BDM10001L	Pin-door check arm	2	Note (6)	
8	MXC2166	Hinge assembly-front door-RH	2		
8	MXC2167	Hinge assembly-front door-LH	2		
9	ALR1976	Shim-door hinge-1.30mm	4		
9	MXC9336	Shim-door hinge-0.50mm	4		

CHANGE POINTS:
(1) To (V) MA 163103; To (V) MA 501919
(2) From (V) TA 163104; From (V) TA 501920
(3) To (V) KA 048963
(4) From (V) KA 048964
(5) To (V) MA 129951
(6) From (V) MA 129952

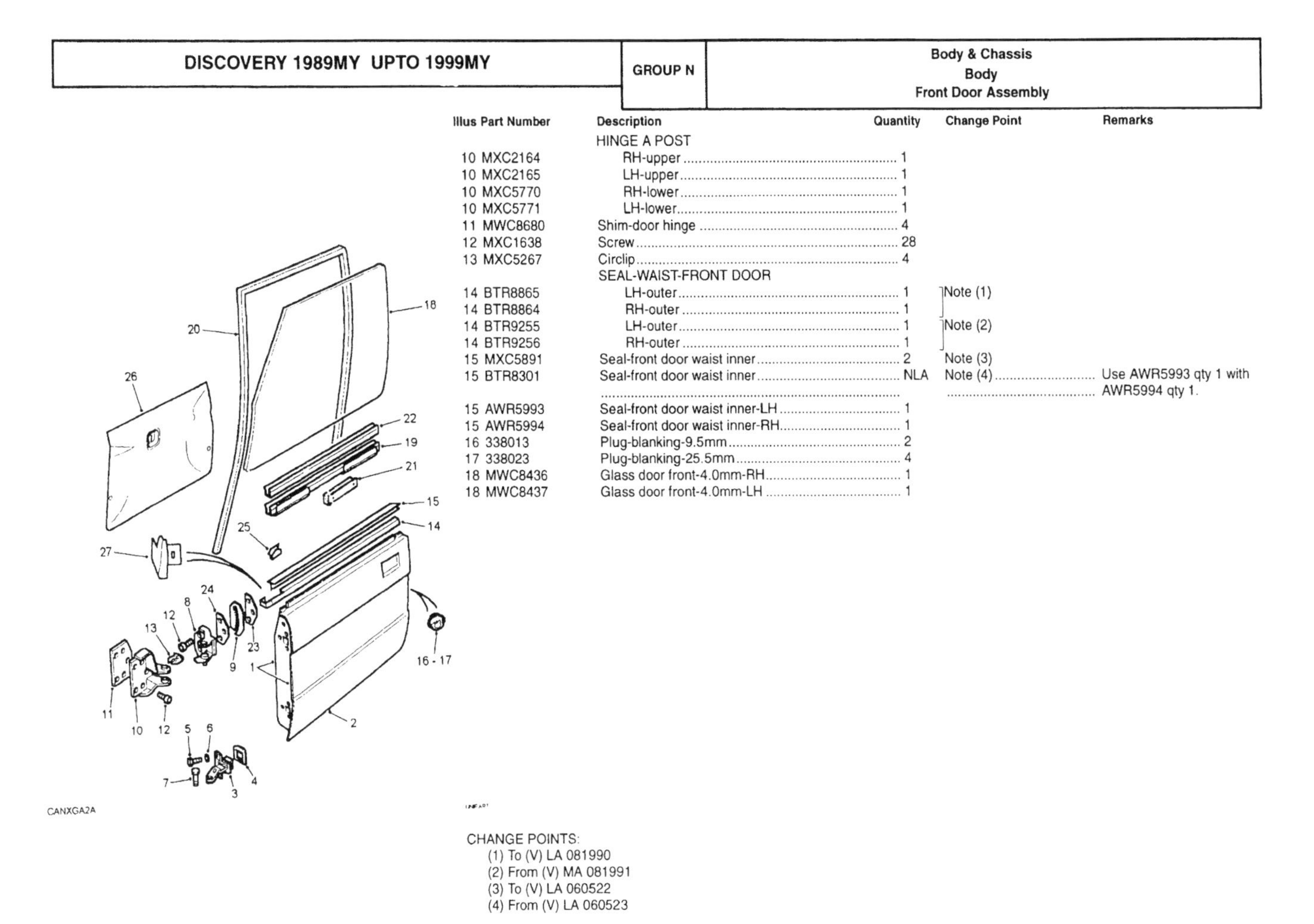

Illus	Part Number	Description	Quantity	Change Point	Remarks
		HINGE A POST			
10	MXC2164	RH-upper	1		
10	MXC2165	LH-upper	1		
10	MXC5770	RH-lower	1		
10	MXC5771	LH-lower	1		
11	MWC8680	Shim-door hinge	4		
12	MXC1638	Screw	28		
13	MXC5267	Circlip	4		
		SEAL-WAIST-FRONT DOOR			
14	BTR8865	LH-outer	1	Note (1)	
14	BTR8864	RH-outer	1		
14	BTR9255	LH-outer	1	Note (2)	
14	BTR9256	RH-outer	1		
15	MXC5891	Seal-front door waist inner	2	Note (3)	
15	BTR8301	Seal-front door waist inner	NLA	Note (4)	Use AWR5993 qty 1 with AWR5994 qty 1.
15	AWR5993	Seal-front door waist inner-LH	1		
15	AWR5994	Seal-front door waist inner-RH	1		
16	338013	Plug-blanking-9.5mm	2		
17	338023	Plug-blanking-25.5mm	4		
18	MWC8436	Glass door front-4.0mm-RH	1		
18	MWC8437	Glass door front-4.0mm-LH	1		

CHANGE POINTS:
(1) To (V) LA 081990
(2) From (V) MA 081991
(3) To (V) LA 060522
(4) From (V) LA 060523

N 47

DISCOVERY 1989MY UPTO 1999MY | GROUP N | Body & Chassis Body Front Door Assembly

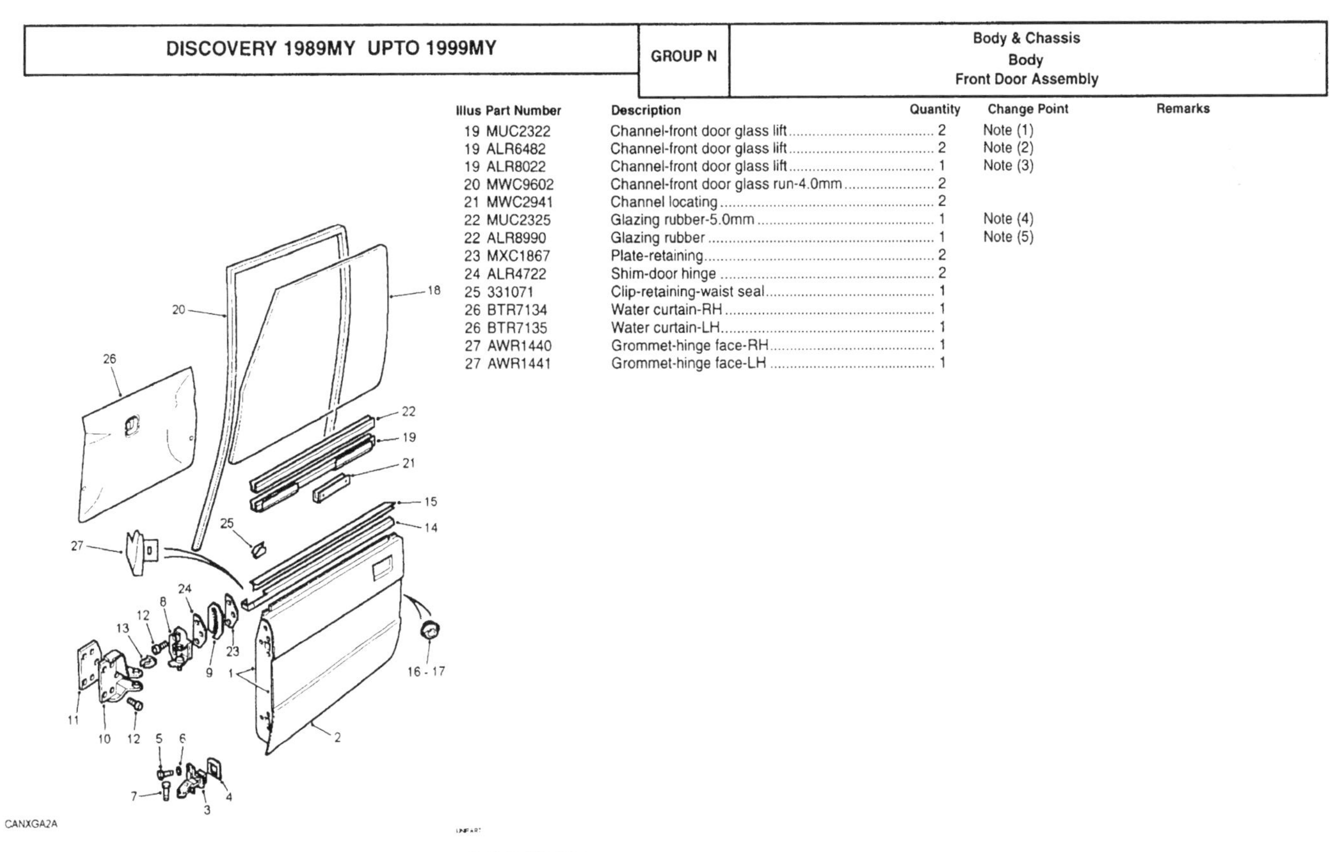

Illus	Part Number	Description	Quantity	Change Point	Remarks
19	MUC2322	Channel-front door glass lift	2	Note (1)	
19	ALR6482	Channel-front door glass lift	2	Note (2)	
19	ALR8022	Channel-front door glass lift	1	Note (3)	
20	MWC9602	Channel-front door glass run-4.0mm	2		
21	MWC2941	Channel locating	2		
22	MUC2325	Glazing rubber-5.0mm	1	Note (4)	
22	ALR8990	Glazing rubber	1	Note (5)	
23	MXC1867	Plate-retaining	2		
24	ALR4722	Shim-door hinge	2		
25	331071	Clip-retaining-waist seal	1		
26	BTR7134	Water curtain-RH	1		
26	BTR7135	Water curtain-LH	1		
27	AWR1440	Grommet-hinge face-RH	1		
27	AWR1441	Grommet-hinge face-LH	1		

CHANGE POINTS:
(1) To (V) LA 081990
(2) From (V) MA 081991 To (V) MA 137189
(3) From (V) MA 137190
(4) To (V) MA 139664
(5) From (V) MA 139665

N 48

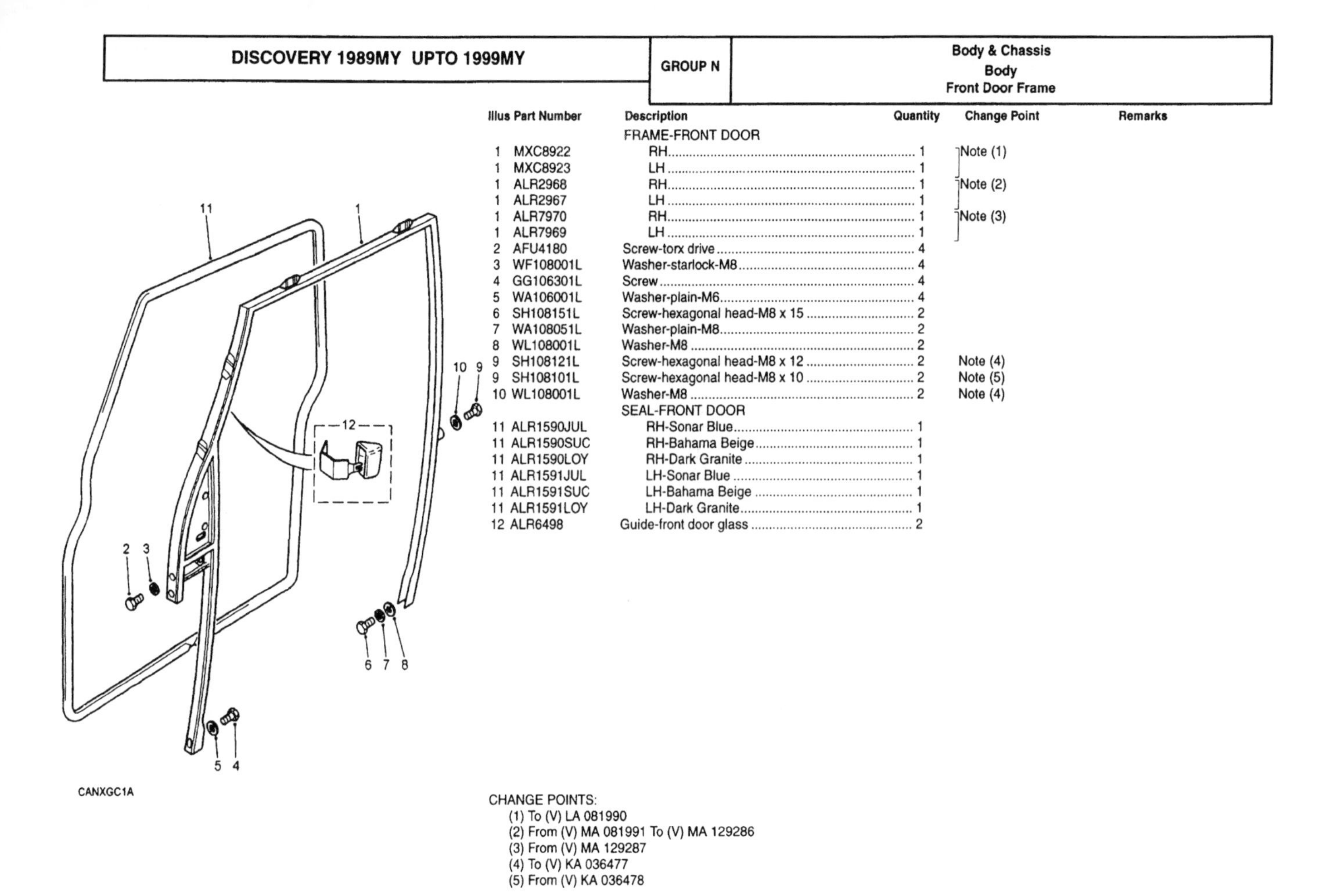

Illus	Part Number	Description	Quantity	Change Point	Remarks
		FRAME-FRONT DOOR			
1	MXC8922	RH	1	Note (1)	
1	MXC8923	LH	1		
1	ALR2968	RH	1	Note (2)	
1	ALR2967	LH	1		
1	ALR7970	RH	1	Note (3)	
1	ALR7969	LH	1		
2	AFU4180	Screw-torx drive	4		
3	WF108001L	Washer-starlock-M8	4		
4	GG106301L	Screw	4		
5	WA106001L	Washer-plain-M6	4		
6	SH108151L	Screw-hexagonal head-M8 x 15	2		
7	WA108051L	Washer-plain-M8	2		
8	WL108001L	Washer-M8	2		
9	SH108121L	Screw-hexagonal head-M8 x 12	2	Note (4)	
9	SH108101L	Screw-hexagonal head-M8 x 10	2	Note (5)	
10	WL108001L	Washer-M8	2	Note (4)	
		SEAL-FRONT DOOR			
11	ALR1590JUL	RH-Sonar Blue	1		
11	ALR1590SUC	RH-Bahama Beige	1		
11	ALR1590LOY	RH-Dark Granite	1		
11	ALR1591JUL	LH-Sonar Blue	1		
11	ALR1591SUC	LH-Bahama Beige	1		
11	ALR1591LOY	LH-Dark Granite	1		
12	ALR6498	Guide-front door glass	2		

CHANGE POINTS:
(1) To (V) LA 081990
(2) From (V) MA 081991 To (V) MA 129286
(3) From (V) MA 129287
(4) To (V) KA 036477
(5) From (V) KA 036478

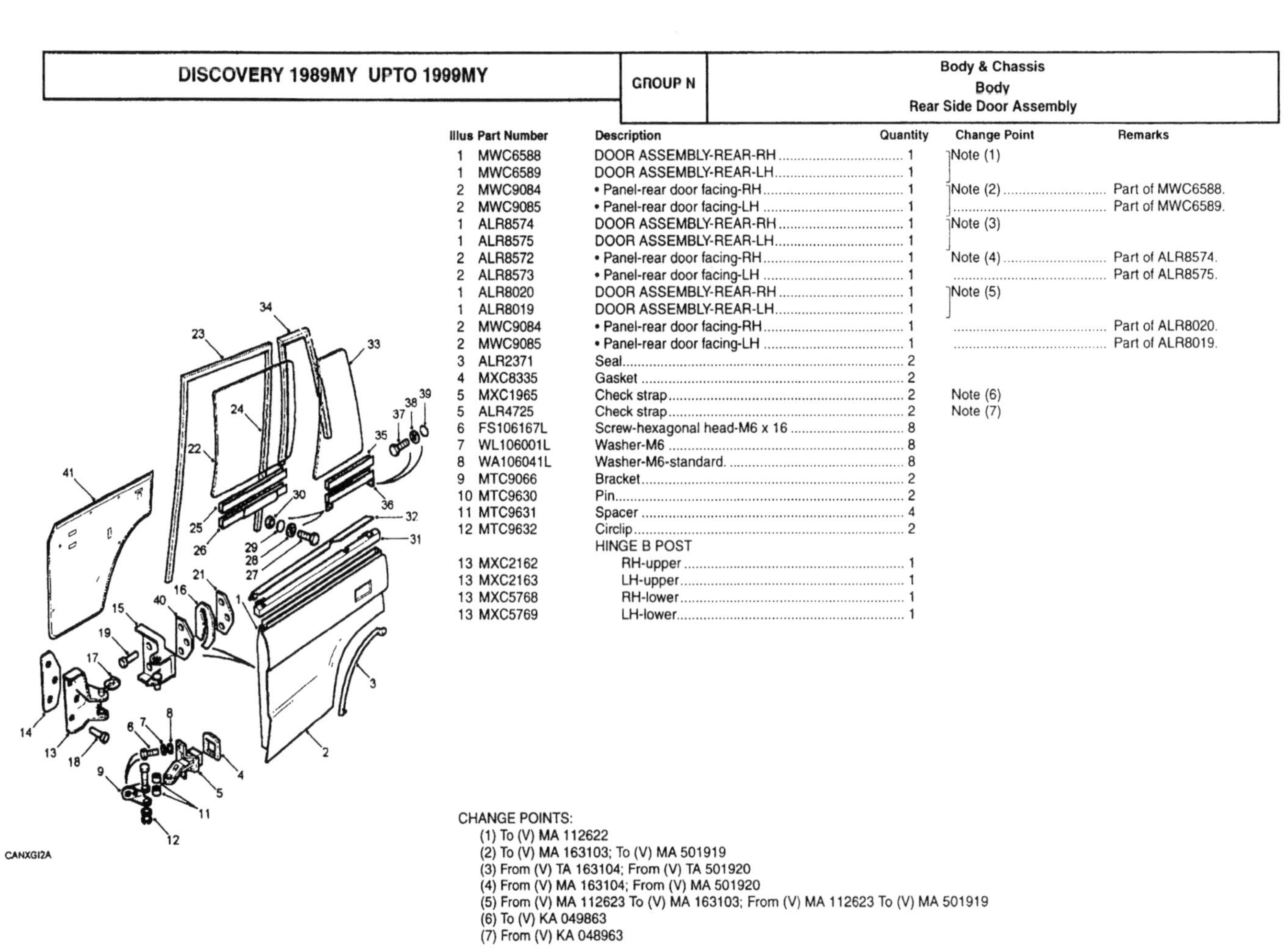

Illus	Part Number	Description	Quantity	Change Point	Remarks
1	MWC6588	DOOR ASSEMBLY-REAR-RH	1	Note (1)	
1	MWC6589	DOOR ASSEMBLY-REAR-LH	1		
2	MWC9084	• Panel-rear door facing-RH	1	Note (2)	Part of MWC6588.
2	MWC9085	• Panel-rear door facing-LH	1		Part of MWC6589.
1	ALR8574	DOOR ASSEMBLY-REAR-RH	1	Note (3)	
1	ALR8575	DOOR ASSEMBLY-REAR-LH	1		
2	ALR8572	• Panel-rear door facing-RH	1	Note (4)	Part of ALR8574.
2	ALR8573	• Panel-rear door facing-LH	1		Part of ALR8575.
1	ALR8020	DOOR ASSEMBLY-REAR-RH	1	Note (5)	
1	ALR8019	DOOR ASSEMBLY-REAR-LH	1		
2	MWC9084	• Panel-rear door facing-RH	1		Part of ALR8020.
2	MWC9085	• Panel-rear door facing-LH	1		Part of ALR8019.
3	ALR2371	Seal	2		
4	MXC8335	Gasket	2		
5	MXC1965	Check strap	2	Note (6)	
5	ALR4725	Check strap	2	Note (7)	
6	FS106167L	Screw-hexagonal head-M6 x 16	8		
7	WL106001L	Washer-M6	8		
8	WA106041L	Washer-M6-standard.	8		
9	MTC9066	Bracket	2		
10	MTC9630	Pin	2		
11	MTC9631	Spacer	4		
12	MTC9632	Circlip	2		
		HINGE B POST			
13	MXC2162	RH-upper	1		
13	MXC2163	LH-upper	1		
13	MXC5768	RH-lower	1		
13	MXC5769	LH-lower	1		

CHANGE POINTS:
(1) To (V) MA 112622
(2) To (V) MA 163103; To (V) MA 501919
(3) From (V) TA 163104; From (V) TA 501920
(4) From (V) MA 163104; From (V) MA 501920
(5) From (V) MA 112623 To (V) MA 163103; From (V) MA 112623 To (V) MA 501919
(6) To (V) KA 049863
(7) From (V) KA 048963

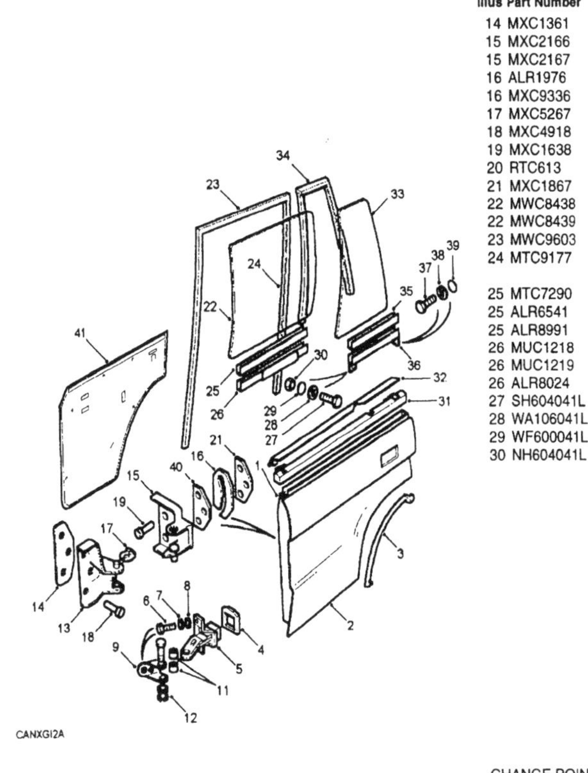

Illus	Part Number	Description	Quantity	Change Point	Remarks
14	MXC1361	Shim-door hinge-lower	4		
15	MXC2166	Hinge assembly-front door-RH	2		
15	MXC2167	Hinge assembly-front door-LH	2		
16	ALR1976	Shim-door hinge-1.30mm	2		
16	MXC9336	Shim-door hinge-0.50mm	2		
17	MXC5267	Circlip	4		
18	MXC4918	Screw-pan	12		
19	MXC1638	Screw	12		
20	RTC613	Washer-plain	12		
21	MXC1867	Plate-retaining	4		
22	MWC8438	Glass-rear-door-4.0mm-RH	1		
22	MWC8439	Glass-rear-door-4.0mm-LH	1		
23	MWC9603	Channel-rear door glass run-front-top-4.0mm	2		
24	MTC9177	Channel-rear door glass run	2		
		CHANNEL-REAR DOOR GLASS RUN			
25	MTC7290	rubber	2	Note (1)	
25	ALR6541	rubber	1	Note (2)	
25	ALR8991	rubber	1	Note (3)	
26	MUC1218	Channel-rear door glass lift-LH	1	Note (4)	
26	MUC1219	Channel-rear door glass lift-RH	1		
26	ALR8024	Channel-rear door glass lift	1	Note (3)	
27	SH604041L	Screw-hexagonal head-1/4UNF x 1/2	2		
28	WA106041L	Washer-M6-standard.	2		
29	WF600041L	Washer- shakeproof	2		
30	NH604041L	Nut-hexagonal-coarse thread-1/4UNF	2		

CHANGE POINTS:
(1) To (V) LA 081990
(2) From (V) MA 081991 To (V) MA 137189
(3) From (V) MA 137190
(4) From (V) LA 081991 To (V) MA 137189

N 51

DISCOVERY 1989MY UPTO 1999MY | GROUP N | Body & Chassis
Body
Rear Side Door Assembly

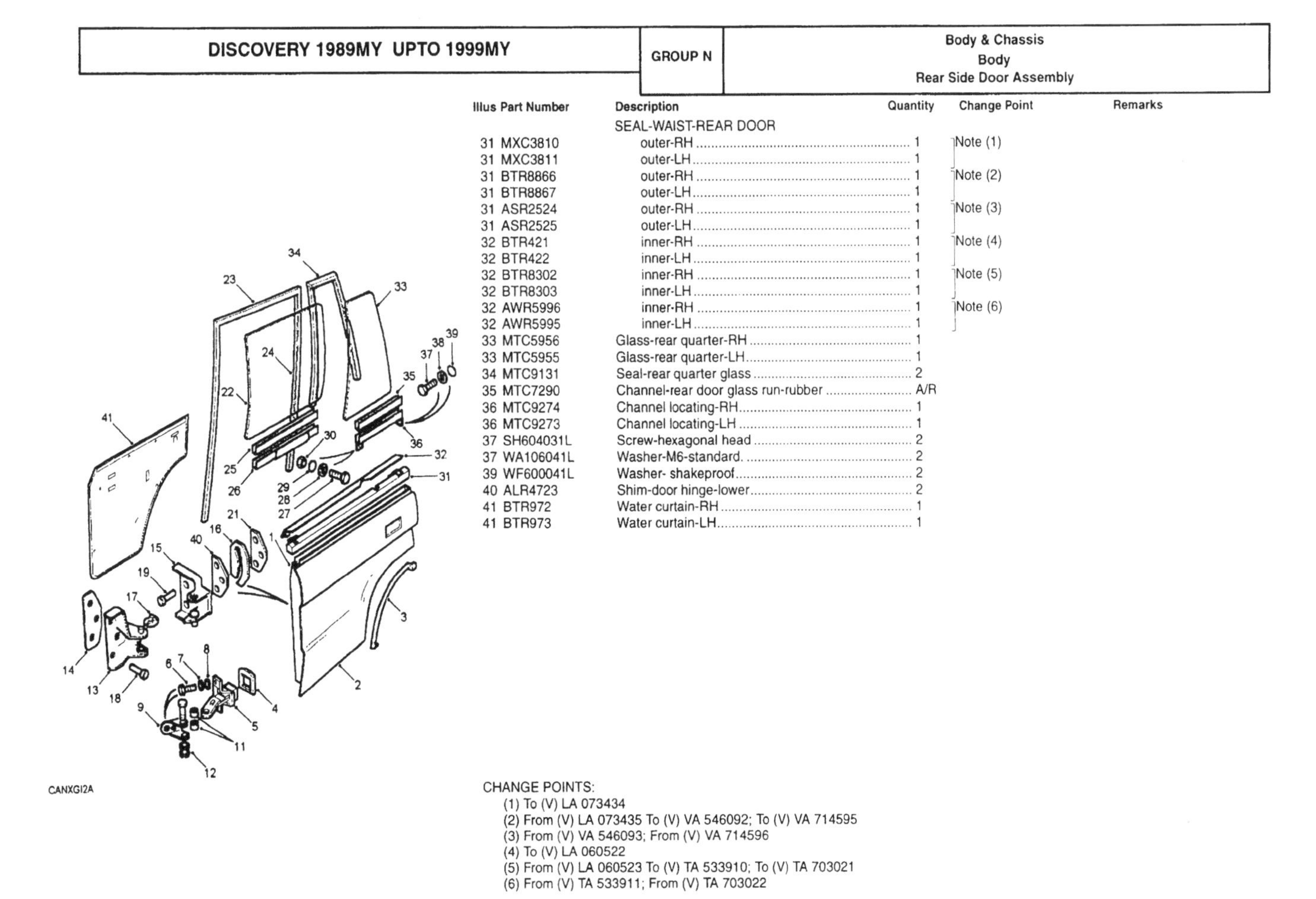

Illus	Part Number	Description	Quantity	Change Point	Remarks
		SEAL-WAIST-REAR DOOR			
31	MXC3810	outer-RH	1	Note (1)	
31	MXC3811	outer-LH	1		
31	BTR8866	outer-RH	1	Note (2)	
31	BTR8867	outer-LH	1		
31	ASR2524	outer-RH	1	Note (3)	
31	ASR2525	outer-LH	1		
32	BTR421	inner-RH	1	Note (4)	
32	BTR422	inner-LH	1		
32	BTR8302	inner-RH	1	Note (5)	
32	BTR8303	inner-LH	1		
32	AWR5996	inner-RH	1	Note (6)	
32	AWR5995	inner-LH	1		
33	MTC5956	Glass-rear quarter-RH	1		
33	MTC5955	Glass-rear quarter-LH	1		
34	MTC9131	Seal-rear quarter glass	2		
35	MTC7290	Channel-rear door glass run-rubber	A/R		
36	MTC9274	Channel locating-RH	1		
36	MTC9273	Channel locating-LH	1		
37	SH604031L	Screw-hexagonal head	2		
37	WA106041L	Washer-M6-standard.	2		
39	WF600041L	Washer- shakeproof	2		
40	ALR4723	Shim-door hinge-lower	2		
41	BTR972	Water curtain-RH	1		
41	BTR973	Water curtain-LH	1		

CHANGE POINTS:
(1) To (V) LA 073434
(2) From (V) LA 073435 To (V) VA 546092; To (V) VA 714595
(3) From (V) VA 546093; From (V) VA 714596
(4) To (V) LA 060522
(5) From (V) LA 060523 To (V) TA 533910; To (V) TA 703021
(6) From (V) TA 533911; From (V) TA 703022

N 52

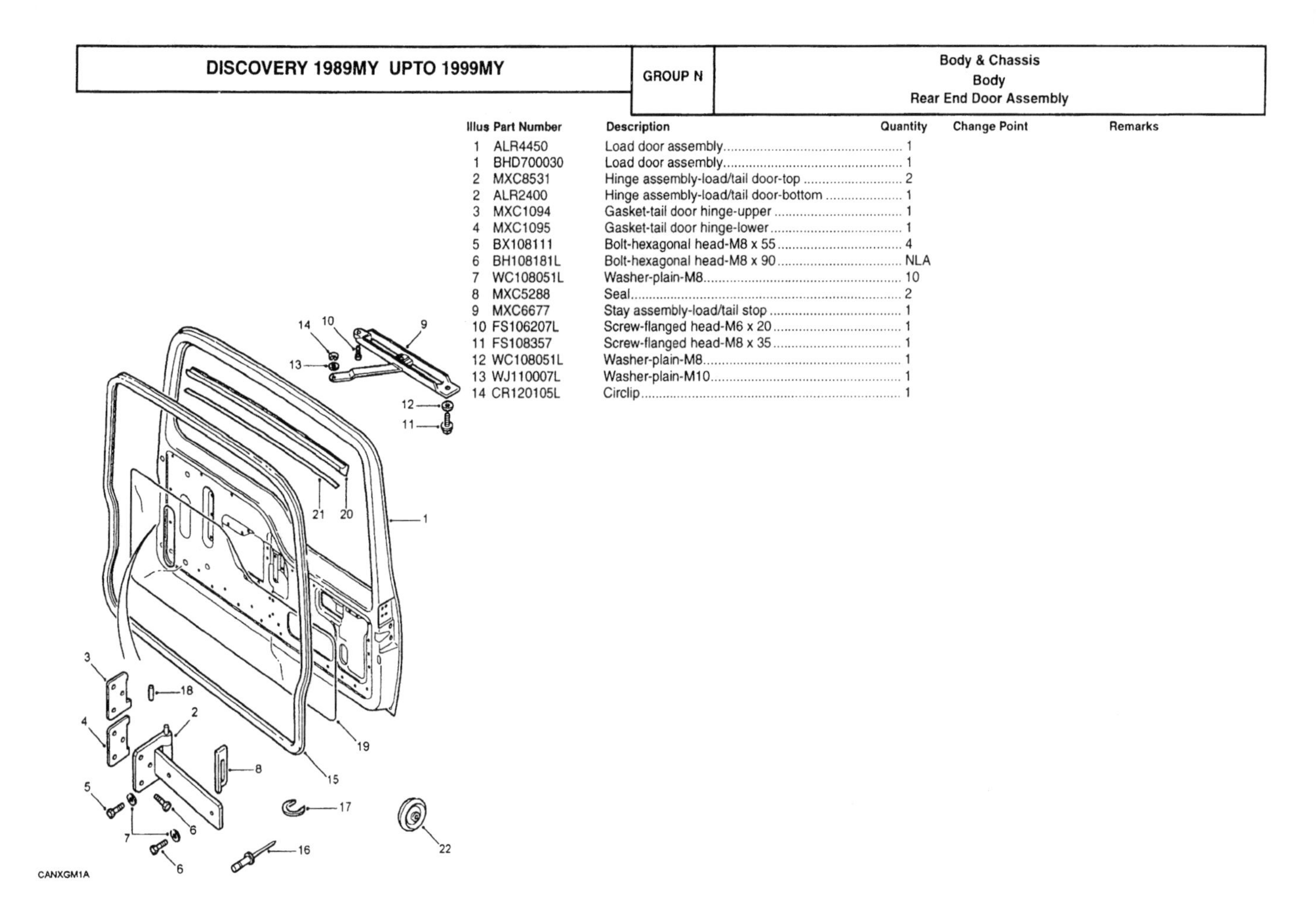

Illus	Part Number	Description	Quantity	Change Point	Remarks
1	ALR4450	Load door assembly	1		
1	BHD700030	Load door assembly	1		
2	MXC8531	Hinge assembly-load/tail door-top	2		
2	ALR2400	Hinge assembly-load/tail door-bottom	1		
3	MXC1094	Gasket-tail door hinge-upper	1		
4	MXC1095	Gasket-tail door hinge-lower	1		
5	BX108111	Bolt-hexagonal head-M8 x 55	4		
6	BH108181L	Bolt-hexagonal head-M8 x 90	NLA		
7	WC108051L	Washer-plain-M8	10		
8	MXC5288	Seal	2		
9	MXC6677	Stay assembly-load/tail stop	1		
10	FS106207L	Screw-flanged head-M6 x 20	1		
11	FS108357	Screw-flanged head-M8 x 35	1		
12	WC108051L	Washer-plain-M8	1		
13	WJ110007L	Washer-plain-M10	1		
14	CR120105L	Circlip	1		

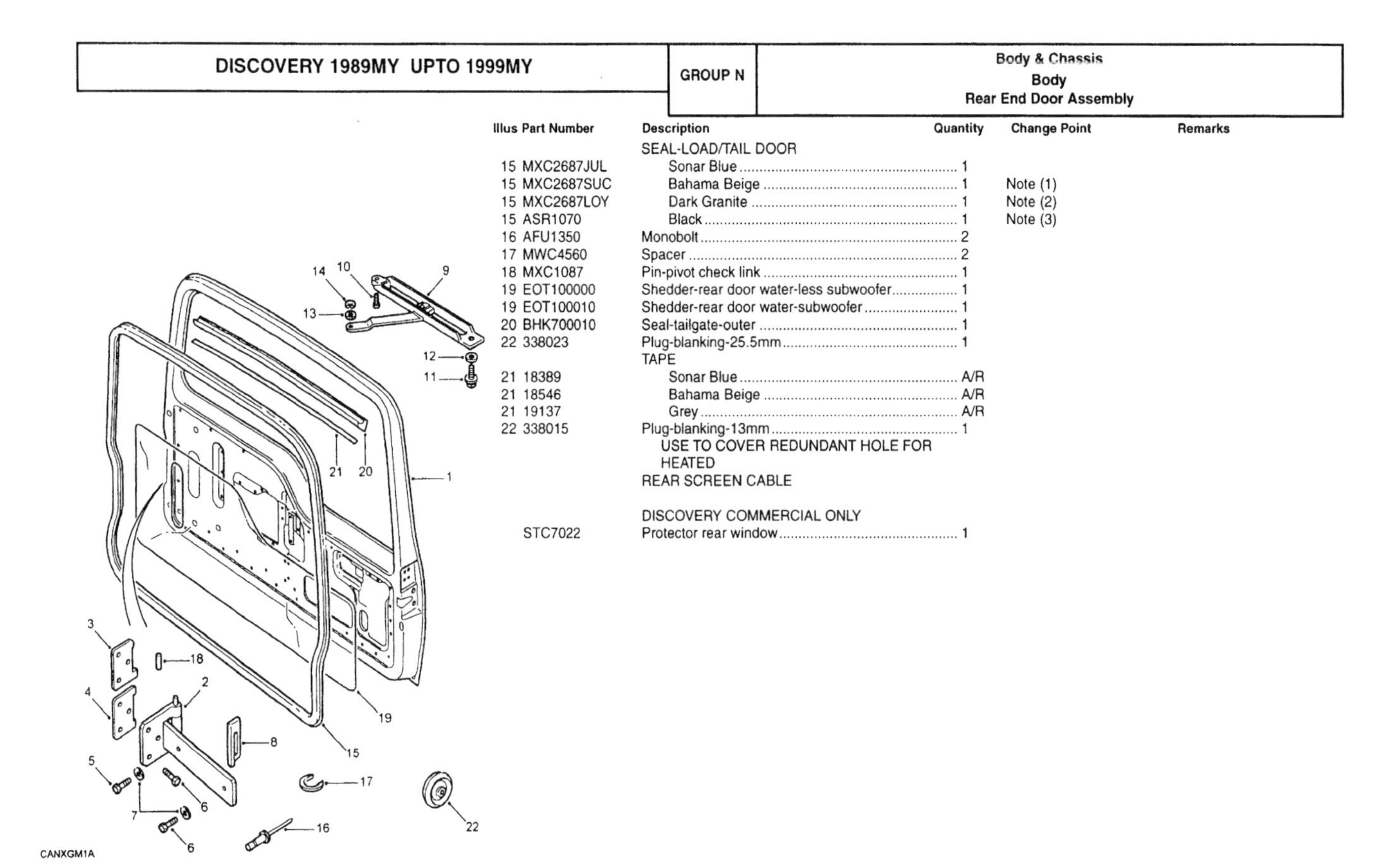

Illus	Part Number	Description	Quantity	Change Point	Remarks
		SEAL-LOAD/TAIL DOOR			
15	MXC2687JUL	Sonar Blue	1		
15	MXC2687SUC	Bahama Beige	1	Note (1)	
15	MXC2687LOY	Dark Granite	1	Note (2)	
15	ASR1070	Black	1	Note (3)	
16	AFU1350	Monobolt	2		
17	MWC4560	Spacer	2		
18	MXC1087	Pin-pivot check link	1		
19	EOT100000	Shedder-rear door water-less subwoofer	1		
19	EOT100010	Shedder-rear door water-subwoofer	1		
20	BHK700010	Seal-tailgate-outer	1		
22	338023	Plug-blanking-25.5mm	1		
		TAPE			
21	18389	Sonar Blue	A/R		
21	18546	Bahama Beige	A/R		
21	19137	Grey	A/R		
22	338015	Plug-blanking-13mm	1		
		USE TO COVER REDUNDANT HOLE FOR HEATED REAR SCREEN CABLE			
		DISCOVERY COMMERCIAL ONLY			
	STC7022	Protector rear window	1		

CHANGE POINTS:
(1) To (V) VA 533121; To (V) VA 702245
(2) To (V) VA 532188; To (V) VA 702298
(3) From (V) VA 533122 Beige; From (V) VA 702246 Beige; From (V) VA 532189 Grey; From (V) VA 702299 Grey

DISCOVERY 1989MY UPTO 1999MY

GROUP N — Body & Chassis / Body / Rear Door Frame

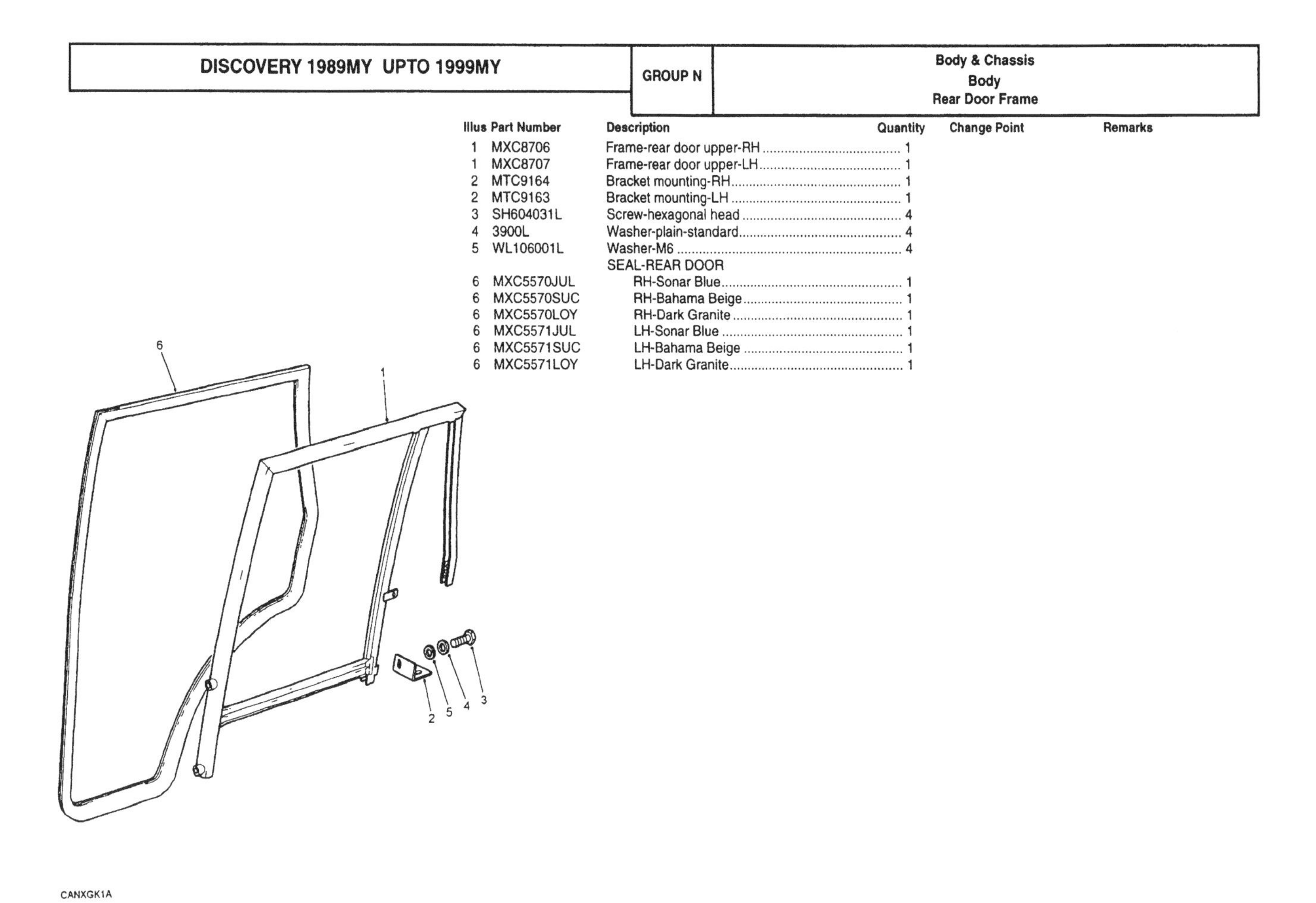

Illus	Part Number	Description	Quantity	Change Point	Remarks
1	MXC8706	Frame-rear door upper-RH	1		
1	MXC8707	Frame-rear door upper-LH	1		
2	MTC9164	Bracket mounting-RH	1		
2	MTC9163	Bracket mounting-LH	1		
3	SH604031L	Screw-hexagonal head	4		
4	3900L	Washer-plain-standard	4		
5	WL106001L	Washer-M6	4		
		SEAL-REAR DOOR			
6	MXC5570JUL	RH-Sonar Blue	1		
6	MXC5570SUC	RH-Bahama Beige	1		
6	MXC5570LOY	RH-Dark Granite	1		
6	MXC5571JUL	LH-Sonar Blue	1		
6	MXC5571SUC	LH-Bahama Beige	1		
6	MXC5571LOY	LH-Dark Granite	1		

N 55

DISCOVERY 1989MY UPTO 1999MY

GROUP N — Body & Chassis / Body / Side Opening Windows

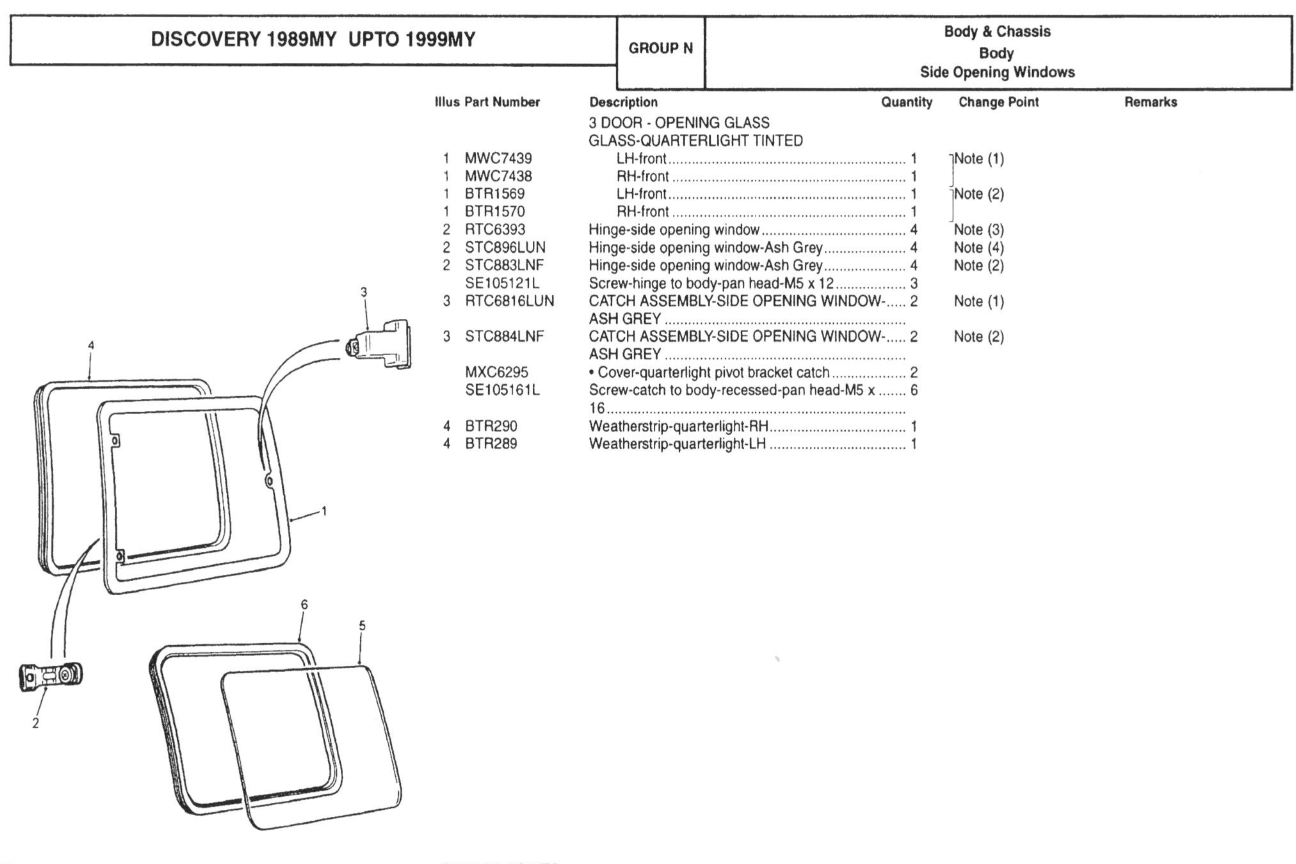

Illus	Part Number	Description	Quantity	Change Point	Remarks
		3 DOOR - OPENING GLASS			
		GLASS-QUARTERLIGHT TINTED			
1	MWC7439	LH-front	1	Note (1)	
1	MWC7438	RH-front	1		
1	BTR1569	LH-front	1	Note (2)	
1	BTR1570	RH-front	1		
2	RTC6393	Hinge-side opening window	4	Note (3)	
2	STC896LUN	Hinge-side opening window-Ash Grey	4	Note (4)	
2	STC883LNF	Hinge-side opening window-Ash Grey	4	Note (2)	
	SE105121L	Screw-hinge to body-pan head-M5 x 12	3		
3	RTC6816LUN	CATCH ASSEMBLY-SIDE OPENING WINDOW-ASH GREY	2	Note (1)	
3	STC884LNF	CATCH ASSEMBLY-SIDE OPENING WINDOW-ASH GREY	2	Note (2)	
	MXC6295	• Cover-quarterlight pivot bracket catch	2		
	SE105161L	Screw-catch to body-recessed-pan head-M5 x 16	6		
4	BTR290	Weatherstrip-quarterlight-RH	1		
4	BTR289	Weatherstrip-quarterlight-LH	1		

CHANGE POINTS:
(1) To (V) JA 034313
(2) From (V) KA 034314
(3) To (V) JA 032325
(4) From (V) JA 032326 To (V) JA 034313

N 56

Illus	Part Number	Description	Quantity	Change Point	Remarks
		3 DOOR			
		GLASS-REAR QUARTER			
5	MXC6813	LH-no aerial	1	Note (1)	Except rear air conditioning
5	AWR3290	RH-with aerial	1	Note (2)	
5	MXC6813	LH-no aerial	1		Except rear air conditioning
5	MXC6812	RH-no aerial	1	Note (1)	
5	BTR9643	LH-no aerial	1		Rear air conditioning
5	BTR9643	LH-no aerial	1	Note (2)	

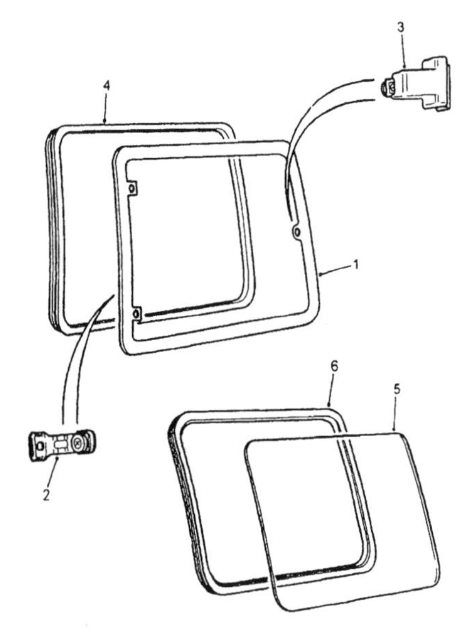

CANXLA1A

CHANGE POINTS:
(1) To (V) TA 534103; To (V) TA 703236
(2) From (V) VA 534104; From (V) VA 703237

N 57

Illus	Part Number	Description	Quantity	Change Point	Remarks
		5 DOOR			
		GLASS-REAR QUARTER			
5	MXC1210	RH-no aerial	1	Note (1)	
5	AWR3292	RH-with aerial	1	Note (2)	
5	AWR3292	RH-with aerial	1		
5	MXC1211	LH-no aerial	1		Except rear air conditioning
5	AWR3291	LH-with aerial	1	Note (3)	
5	BTR9644	LH-no aerial	1		Rear air conditioning
5	AWR3294	LH-with aerial	1	Note (2)	

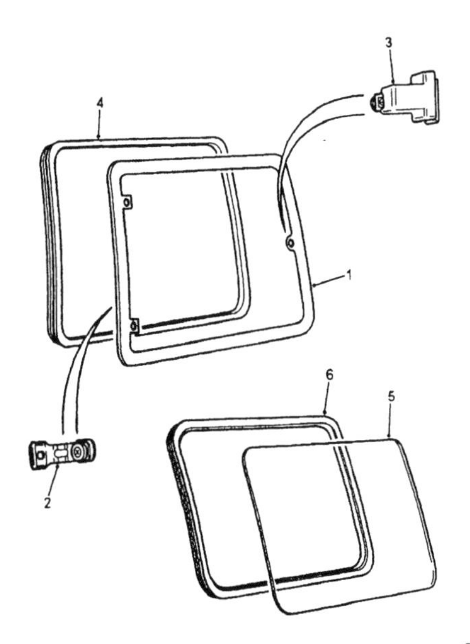

CANXLA1A

CHANGE POINTS:
(1) To (V) TA 534103; To (V) TA 703236
(2) From (V) VA 534104; From (V) VA 703237
(3) From (V) VA 534104; From (V) TA 703237

N 58

Illus	Part Number	Description	Quantity	Change Point	Remarks
		WEATHERSTRIP-REAR QUARTER			
6	MXC7877	RH	1	Note (1)	3 Door
6	MXC7878	LH	1		
6	AWR5386	RH	1	Note (2)	
6	AWR5387	LH	1		
		WEATHERSTRIP-REAR QUARTER			
6	MXC7272	RH	1	Note (1)	5 Door
6	MXC7273	LH	1		
6	AWR5388	RH	1	Note (2)	
6	AWR5389	LH	1		

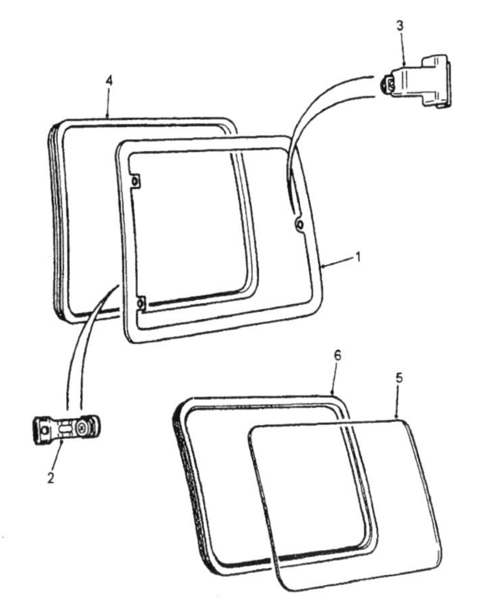

CANXLA1A

CHANGE POINTS:
(1) To (V) TA 534103; To (V) TA 703236
(2) From (V) VA 534104; From (V) VA 703237

N 59

Illus	Part Number	Description	Quantity	Change Point	Remarks
		REGULATOR-FRONT DOOR GLASS			
1	MUC2038	manual-RH	1	Note (1)	
1	MUC2039	manual-LH	1		
1	ALR6474	manual-RH	1	Note (2)	
1	ALR6473	manual-LH	1		
2	SE106201L	Screw-pan head-M6 x 20	8		
3	WA106041L	Washer-M6-standard.	8		
4	WE106001L	Washer-lock	8		
		HANDLE-FRONT/REAR DOOR WINDOW REGULATOR			
5	MWC6512LUN	Ash Grey	2	Note (4)	
5	MWC6512LNF	Ash Grey	2	Note (6)	
5	BTR9275LNF	Ash Grey	2	Note (3)	
6	MXC2049LUN	Escutcheon-window regulator-Ash Grey	2	Note (4)	
6	MXC2049LNF	Escutcheon-window regulator-Ash Grey	2	Note (5)	
7	SE105161L	Screw-recessed-pan head-M5 x 16	2		
8	WE105001L	Washer-lock	2		
9	MWC8021LUN	Cover plate-Ash Grey	2	Note (4)	
9	MWC8021LNF	Cover plate-Ash Grey	2	Note (5)	

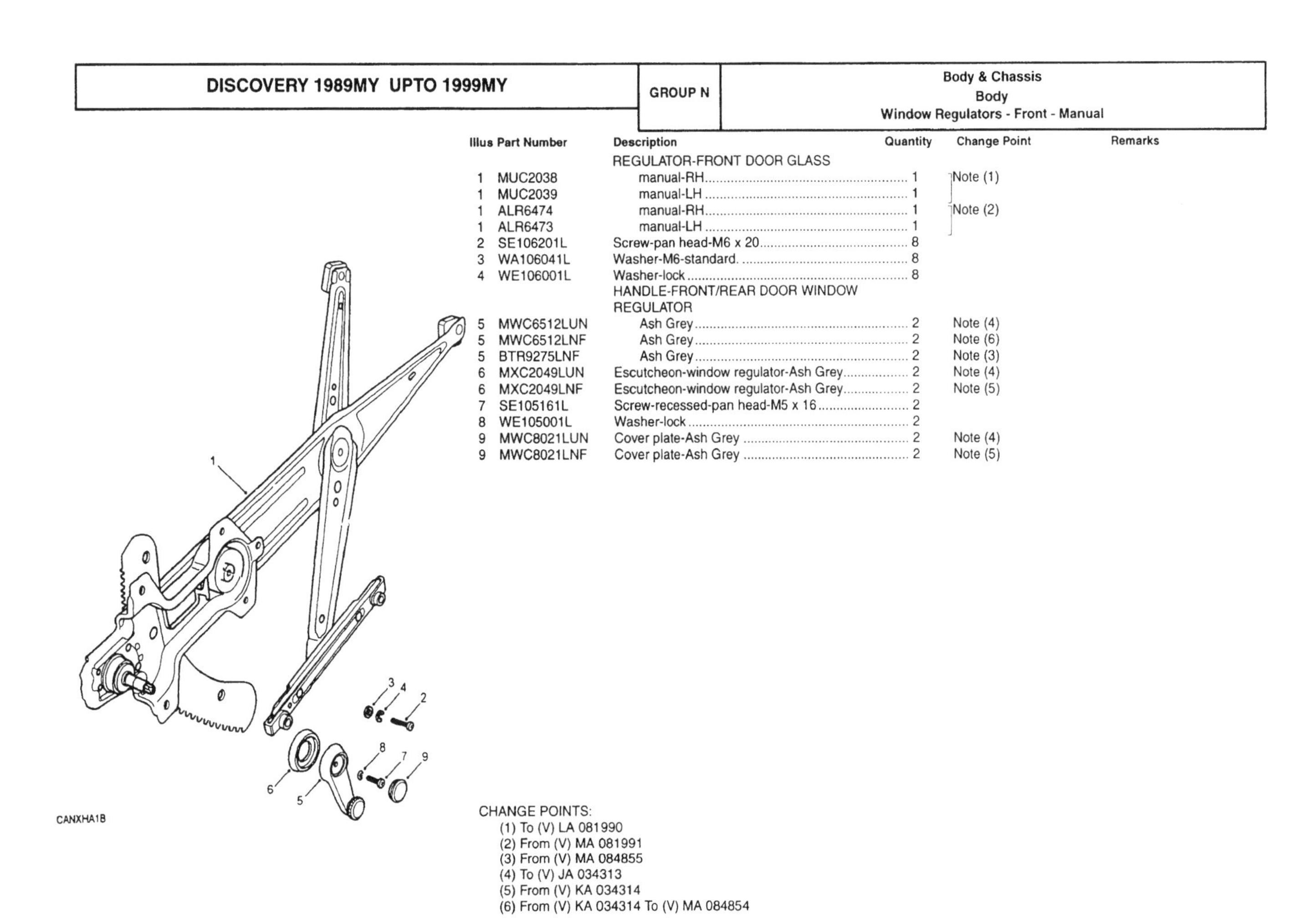

CANXHA1B

CHANGE POINTS:
(1) To (V) LA 081990
(2) From (V) MA 081991
(3) From (V) MA 084855
(4) To (V) JA 034313
(5) From (V) KA 034314
(6) From (V) KA 034314 To (V) MA 084854

N 60

DISCOVERY 1989MY UPTO 1999MY

GROUP N

Body & Chassis
Body
Window Regulator-Front-Electric-Up To MA137189

Illus	Part Number	Description	Quantity	Change Point	Remarks
1	RTC3814	Regulator-front door glass-electric-RH	1	Note (1)	
1	RTC3815	Regulator-front door glass-electric-LH	1		
		MOTOR-FRONT DOOR WINDOW REGULATOR			
2	RTC3820	RH	1	Note (2); Note (3); Note (4)	
2	RTC3821	LH	1	Note (2); Note (3); Note (4)	
2	RTC6641	RH	1	Note (5); Note (6)	
2	RTC6640	LH	1	Note (5); Note (6)	
3	SE106201L	Screw-pan head-M6 x 20	8		
4	WA106041L	Washer-M6-standard.	8		
5	WE106001L	Washer-lock	8		

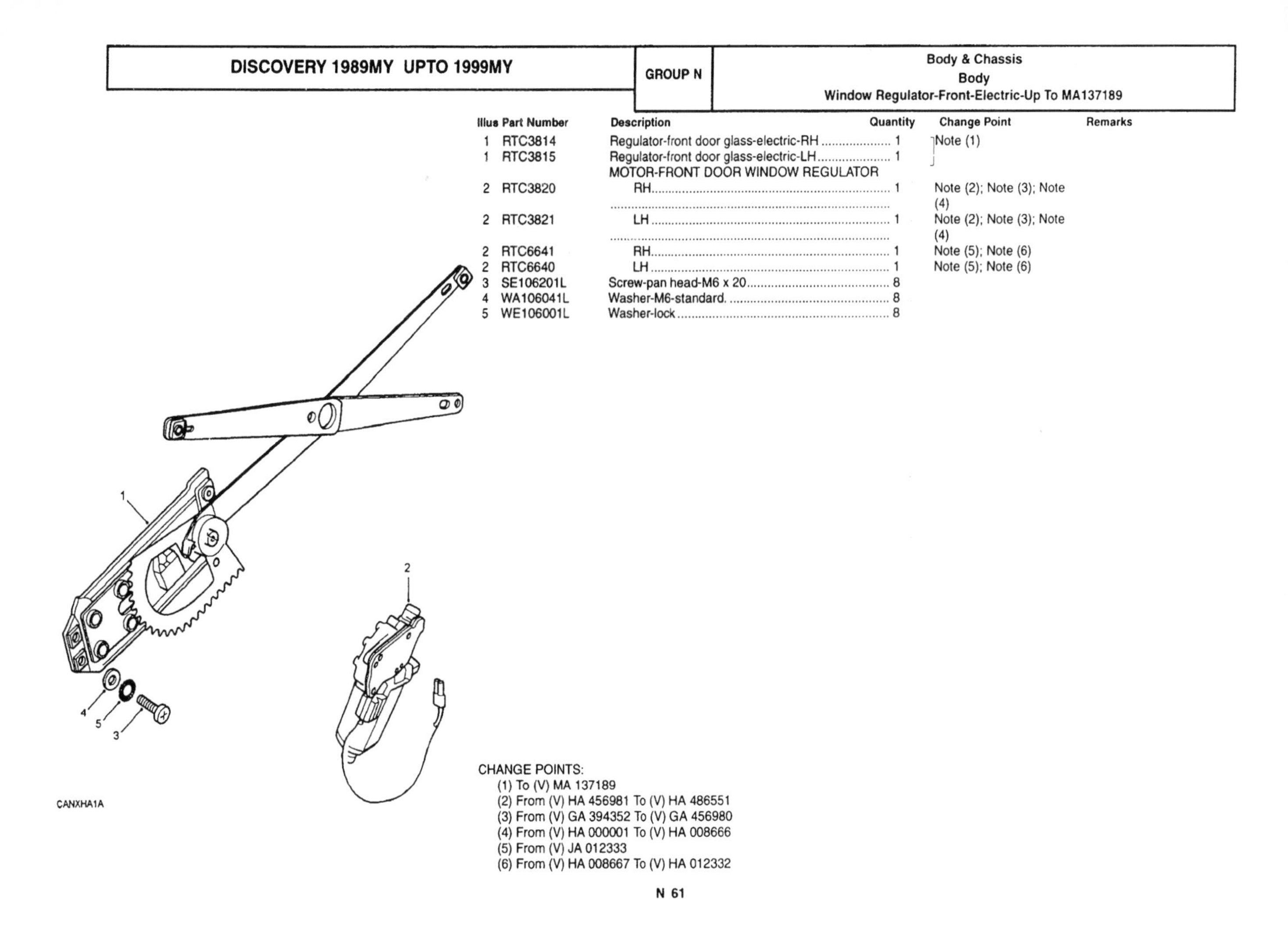

CHANGE POINTS:
(1) To (V) MA 137189
(2) From (V) HA 456981 To (V) HA 486551
(3) From (V) GA 394352 To (V) GA 456980
(4) From (V) HA 000001 To (V) HA 008666
(5) From (V) JA 012333
(6) From (V) HA 008667 To (V) HA 012332

N 61

DISCOVERY 1989MY UPTO 1999MY

GROUP N

Body & Chassis
Body
Window Regulator-Front-Electric-From MA137190

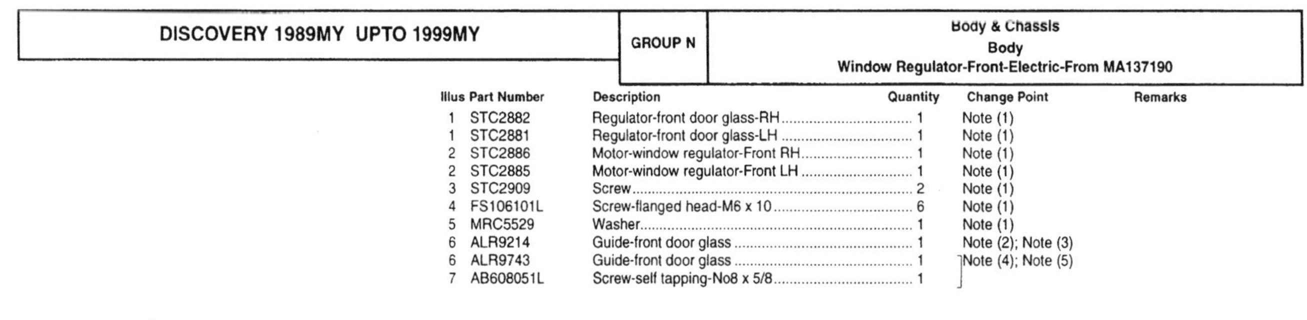

Illus	Part Number	Description	Quantity	Change Point	Remarks
1	STC2882	Regulator-front door glass-RH	1	Note (1)	
1	STC2881	Regulator-front door glass-LH	1	Note (1)	
2	STC2886	Motor-window regulator-Front RH	1	Note (1)	
2	STC2885	Motor-window regulator-Front LH	1	Note (1)	
3	STC2909	Screw	2	Note (1)	
4	FS106101L	Screw-flanged head-M6 x 10	6	Note (1)	
5	MRC5529	Washer	1	Note (1)	
6	ALR9214	Guide-front door glass	1	Note (2); Note (3)	
6	ALR9743	Guide-front door glass	1	Note (4); Note (5)	
7	AB608051L	Screw-self tapping-No8 x 5/8	1		

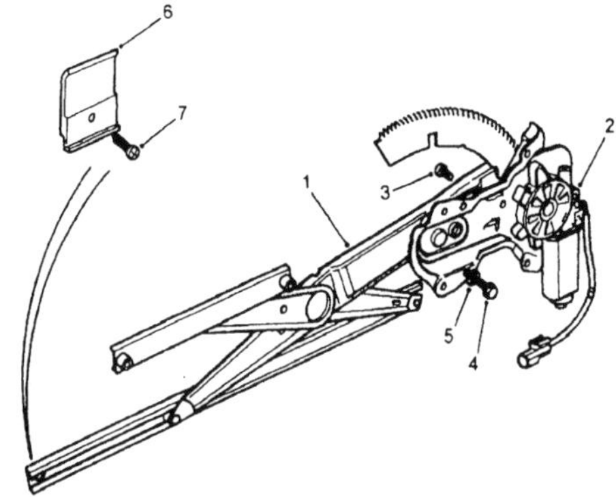

CANXHA2A

CHANGE POINTS:
(1) From (V) MA 137190
(2) From (V) MA 137190 To (V) TA 172958
(3) From (V) MA 137190 To (V) TA 504505
(4) From (V) TA 172959
(5) From (V) TA 504506

N 62

Illus	Part Number	Description	Quantity	Change Point	Remarks
1	MTC8906	Regulator-rear door glass-manual-RH	1		
1	MTC8907	Regulator-rear door glass-manual-LH	1		
2	SE106121L	Screw-pan head-M6 x 12	4		
3	WE106001L	Washer-lock	4		
4	WA106041L	Washer-M6-standard.	4		
5	MWC6512LUN	Handle-front/rear door window regulator-Ash Grey	2	Note (1)	
6	MXC2049LUN	Escutcheon-window regulator-Ash Grey	2		
7	MWC8021LUN	Cover plate-Ash Grey	2		
5	MWC6512LNF	Handle-front/rear door window regulator-Ash Grey	2	Note (2)	
6	MXC2049LNF	Escutcheon-window regulator-Ash Grey	2		
7	MWC8021LNF	Cover plate-Ash Grey	2		
8	SE105161L	Screw-recessed-pan head-M5 x 16	2		
9	WE105001L	Washer-lock	2		

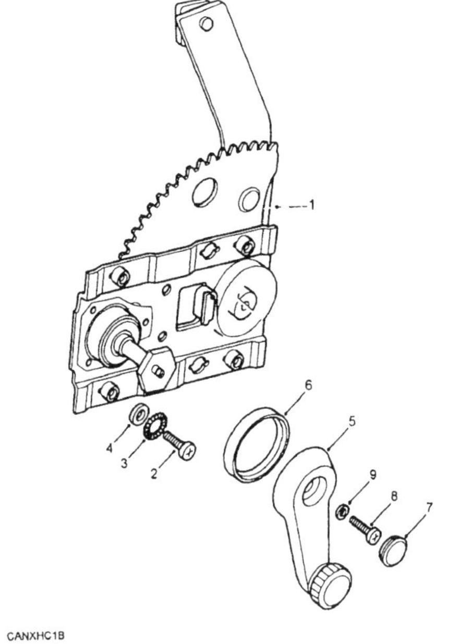

CHANGE POINTS:
(1) To (V) JA 034313
(2) From (V) KA 034314

Illus	Part Number	Description	Quantity	Change Point	Remarks
1	RTC3816	Regulator assembly-rear door electric glass-RH	1		
1	RTC3817	Regulator assembly-rear door electric glass-LH	1		
2	RTC6643	Motor-rear door window regulator-RH	1		
2	RTC6642	Motor-rear door window regulator-LH	1		
3	SE106121L	Screw-pan head-M6 x 12	4		
4	WE106001L	Washer-lock	4		
5	WA106041L	Washer-M6-standard.	4		

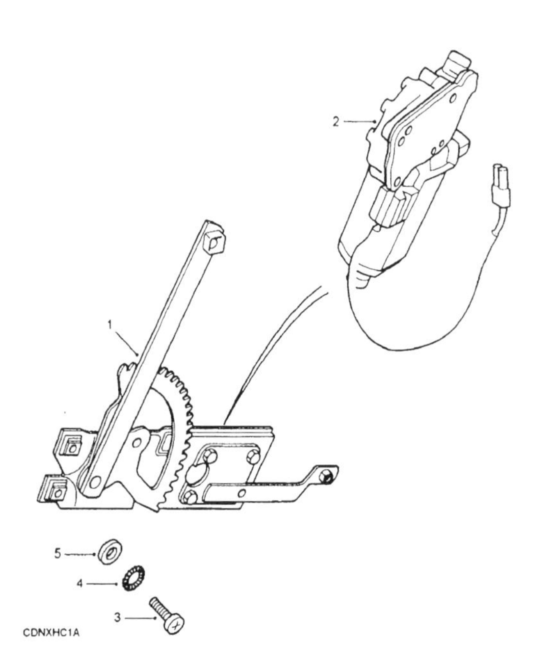

Illus	Part Number	Description	Quantity	Change Point	Remarks
1	STC2884	Regulator-rear door glass-RH	1		
1	STC2883	Regulator-rear door glass-LH	1		
2	STC2885	Motor-window regulator-Rear RH	1		
2	STC2886	Motor-window regulator-Rear LH	1		
3	STC2909	Screw	2		
4	FS106161L	Screw-flanged head-M6 x 16	NLA		Use FS106167L.
4	FS106167L	Screw-hexagonal head-M6 x 16	4		

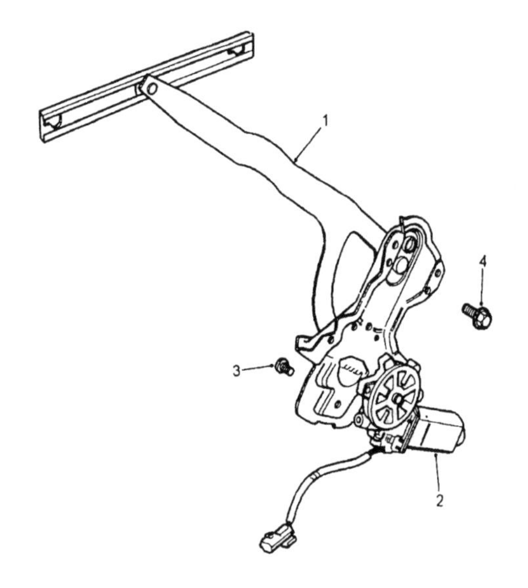

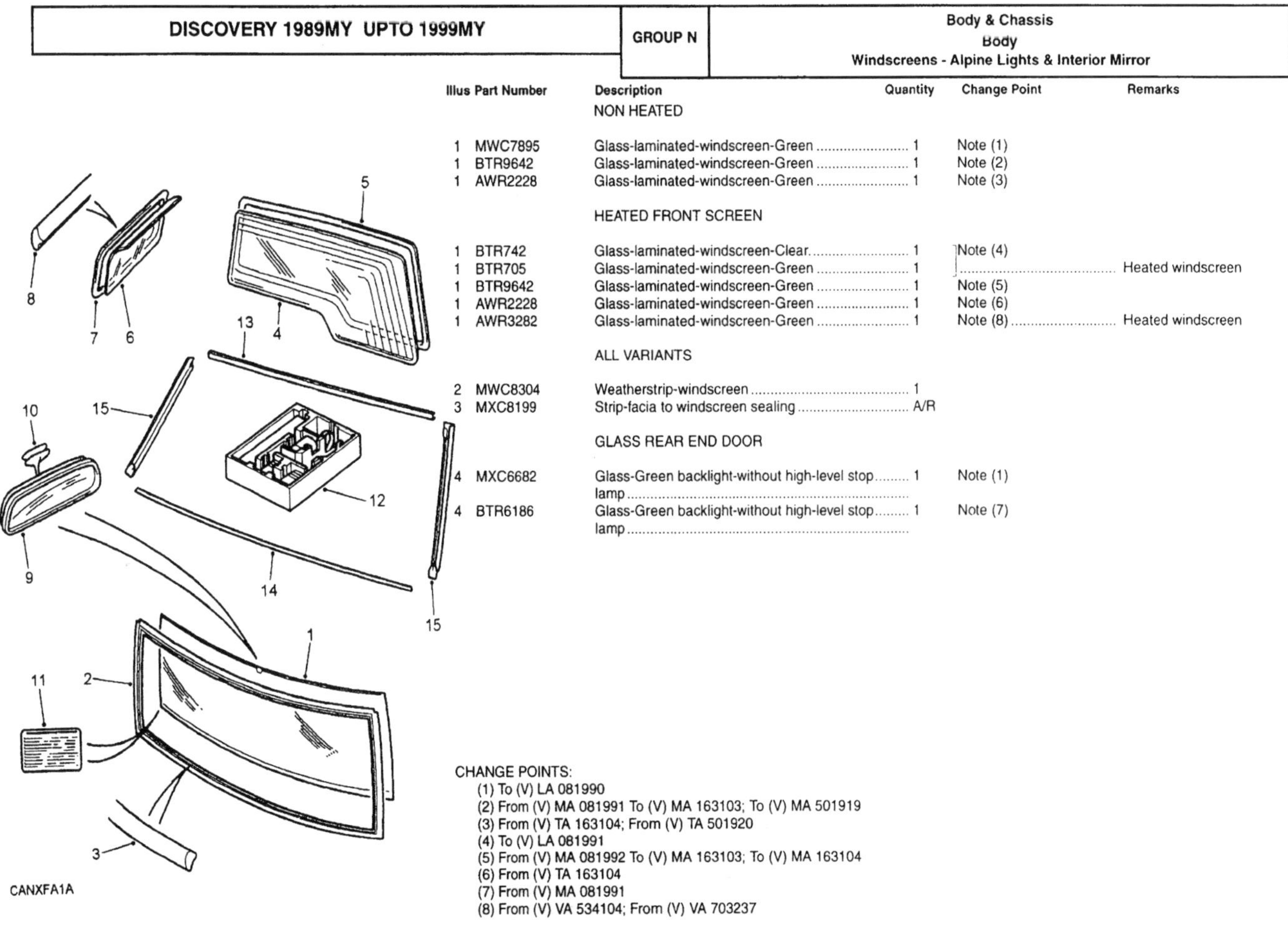

Illus	Part Number	Description	Quantity	Change Point	Remarks
		NON HEATED			
1	MWC7895	Glass-laminated-windscreen-Green	1	Note (1)	
1	BTR9642	Glass-laminated-windscreen-Green	1	Note (2)	
1	AWR2228	Glass-laminated-windscreen-Green	1	Note (3)	
		HEATED FRONT SCREEN			
1	BTR742	Glass-laminated-windscreen-Clear	1	Note (4)	
1	BTR705	Glass-laminated-windscreen-Green	1		Heated windscreen
1	BTR9642	Glass-laminated-windscreen-Green	1	Note (5)	
1	AWR2228	Glass-laminated-windscreen-Green	1	Note (6)	
1	AWR3282	Glass-laminated-windscreen-Green	1	Note (8)	Heated windscreen
		ALL VARIANTS			
2	MWC8304	Weatherstrip-windscreen	1		
3	MXC8199	Strip-facia to windscreen sealing	A/R		
		GLASS REAR END DOOR			
4	MXC6682	Glass-Green backlight-without high-level stop lamp	1	Note (1)	
4	BTR6186	Glass-Green backlight-without high-level stop lamp	1	Note (7)	

CHANGE POINTS:
(1) To (V) LA 081990
(2) From (V) MA 081991 To (V) MA 163103; To (V) MA 501919
(3) From (V) TA 163104; From (V) TA 501920
(4) To (V) LA 081991
(5) From (V) MA 081992 To (V) MA 163103; To (V) MA 163104
(6) From (V) TA 163104
(7) From (V) MA 081991
(8) From (V) VA 534104; From (V) VA 703237

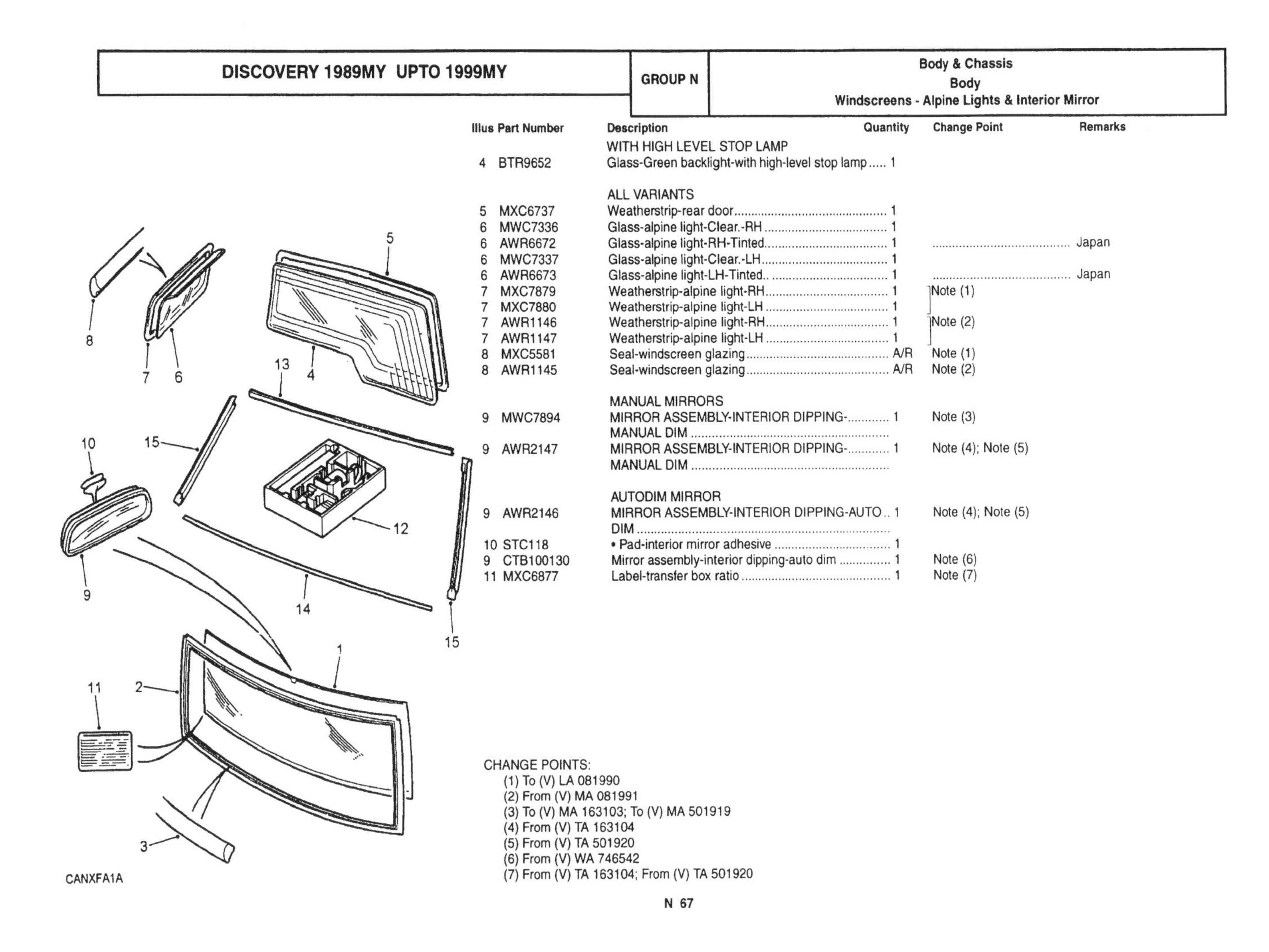

Illus	Part Number	Description	Quantity	Change Point	Remarks
		WITH HIGH LEVEL STOP LAMP			
4	BTR9652	Glass-Green backlight-with high-level stop lamp	1		
		ALL VARIANTS			
5	MXC6737	Weatherstrip-rear door	1		
6	MWC7336	Glass-alpine light-Clear.-RH	1		
6	AWR6672	Glass-alpine light-RH-Tinted	1		Japan
6	MWC7337	Glass-alpine light-Clear.-LH	1		
6	AWR6673	Glass-alpine light-LH-Tinted.	1		Japan
7	MXC7879	Weatherstrip-alpine light-RH	1	Note (1)	
7	MXC7880	Weatherstrip-alpine light-LH	1		
7	AWR1146	Weatherstrip-alpine light-RH	1	Note (2)	
7	AWR1147	Weatherstrip-alpine light-LH	1		
8	MXC5581	Seal-windscreen glazing	A/R	Note (1)	
8	AWR1145	Seal-windscreen glazing	A/R	Note (2)	
		MANUAL MIRRORS			
9	MWC7894	MIRROR ASSEMBLY-INTERIOR DIPPING-MANUAL DIM	1	Note (3)	
9	AWR2147	MIRROR ASSEMBLY-INTERIOR DIPPING-MANUAL DIM	1	Note (4); Note (5)	
		AUTODIM MIRROR			
9	AWR2146	MIRROR ASSEMBLY-INTERIOR DIPPING-AUTO DIM	1	Note (4); Note (5)	
10	STC118	• Pad-interior mirror adhesive	1		
9	CTB100130	Mirror assembly-interior dipping-auto dim	1	Note (6)	
11	MXC6877	Label-transfer box ratio	1	Note (7)	

CHANGE POINTS:
(1) To (V) LA 081990
(2) From (V) MA 081991
(3) To (V) MA 163103; To (V) MA 501919
(4) From (V) TA 163104
(5) From (V) TA 501920
(6) From (V) WA 746542
(7) From (V) TA 163104; From (V) TA 501920

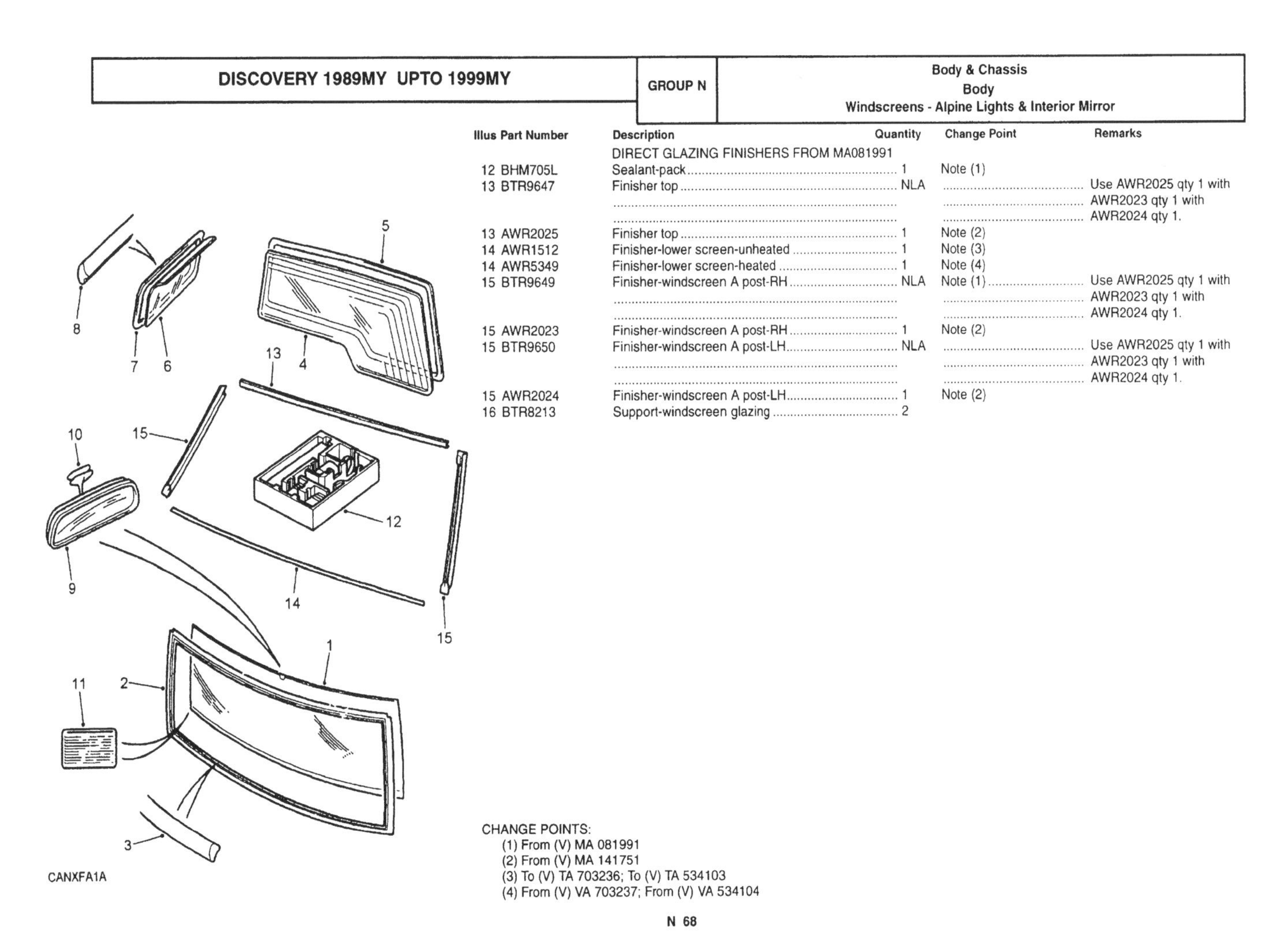

Illus	Part Number	Description	Quantity	Change Point	Remarks
		DIRECT GLAZING FINISHERS FROM MA081991			
12	BHM705L	Sealant-pack	1	Note (1)	
13	BTR9647	Finisher top	NLA		Use AWR2025 qty 1 with AWR2023 qty 1 with AWR2024 qty 1.
13	AWR2025	Finisher top	1	Note (2)	
14	AWR1512	Finisher-lower screen-unheated	1	Note (3)	
14	AWR5349	Finisher-lower screen-heated	1	Note (4)	
15	BTR9649	Finisher-windscreen A post-RH	NLA	Note (1)	Use AWR2025 qty 1 with AWR2023 qty 1 with AWR2024 qty 1.
15	AWR2023	Finisher-windscreen A post-RH	1	Note (2)	
15	BTR9650	Finisher-windscreen A post-LH	NLA		Use AWR2025 qty 1 with AWR2023 qty 1 with AWR2024 qty 1.
15	AWR2024	Finisher-windscreen A post-LH	1	Note (2)	
16	BTR8213	Support-windscreen glazing	2		

CHANGE POINTS:
(1) From (V) MA 081991
(2) From (V) MA 141751
(3) To (V) TA 703236; To (V) TA 534103
(4) From (V) VA 703237; From (V) VA 534104

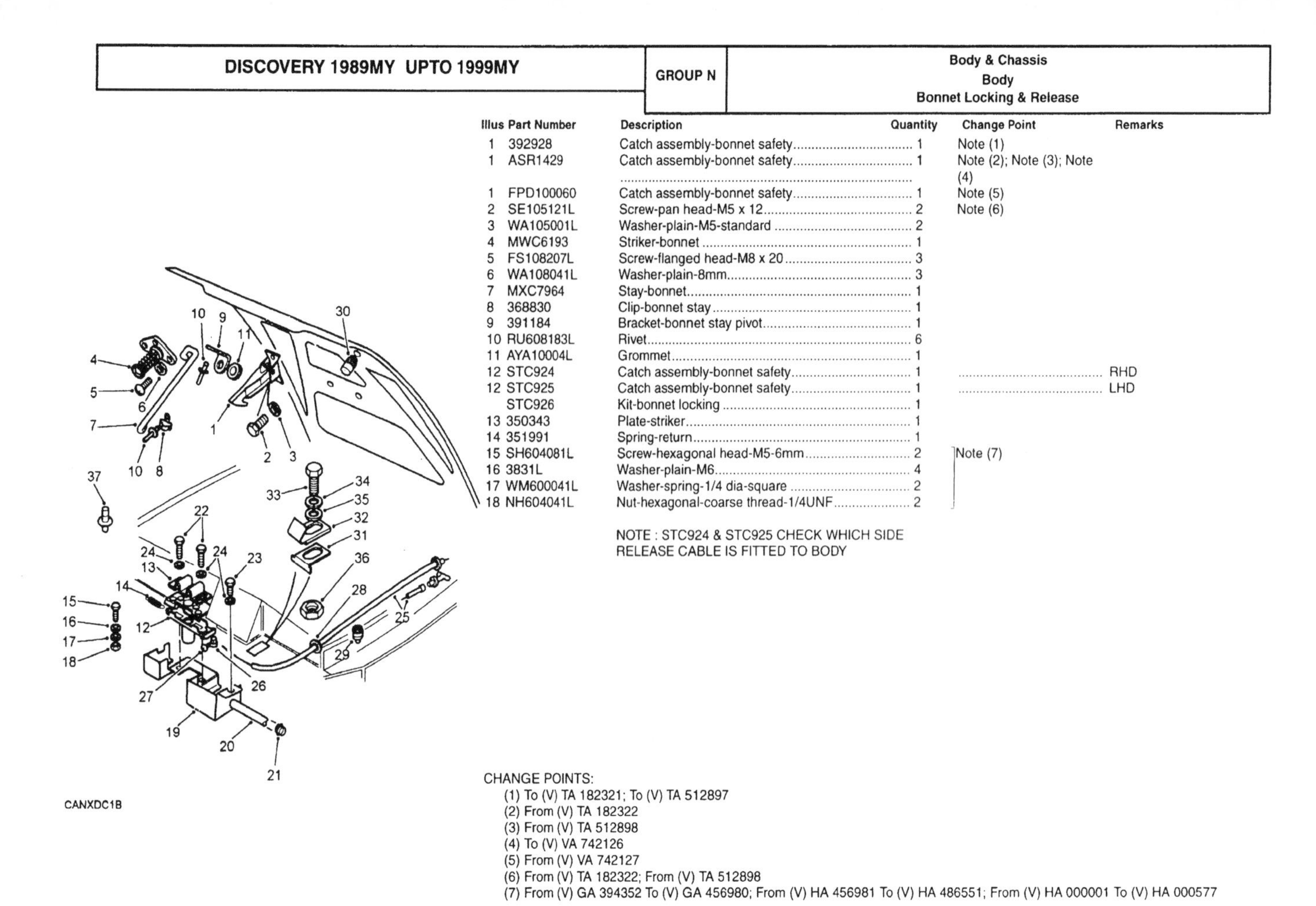

Illus	Part Number	Description	Quantity	Change Point	Remarks
1	392928	Catch assembly-bonnet safety	1	Note (1)	
1	ASR1429	Catch assembly-bonnet safety	1	Note (2); Note (3); Note (4)	
1	FPD100060	Catch assembly-bonnet safety	1	Note (5)	
2	SE105121L	Screw-pan head-M5 x 12	2	Note (6)	
3	WA105001L	Washer-plain-M5-standard	2		
4	MWC6193	Striker-bonnet	1		
5	FS108207L	Screw-flanged head-M8 x 20	3		
6	WA108041L	Washer-plain-8mm	3		
7	MXC7964	Stay-bonnet	1		
8	368830	Clip-bonnet stay	1		
9	391184	Bracket-bonnet stay pivot	1		
10	RU608183L	Rivet	6		
11	AYA10004L	Grommet	1		
12	STC924	Catch assembly-bonnet safety	1		RHD
12	STC925	Catch assembly-bonnet safety	1		LHD
	STC926	Kit-bonnet locking	1		
13	350343	Plate-striker	1		
14	351991	Spring-return	1		
15	SH604081L	Screw-hexagonal head-M5-6mm	2	Note (7)	
16	3831L	Washer-plain-M6	4	Note (7)	
17	WM600041L	Washer-spring-1/4 dia-square	2	Note (7)	
18	NH604041L	Nut-hexagonal-coarse thread-1/4UNF	2	Note (7)	

NOTE : STC924 & STC925 CHECK WHICH SIDE RELEASE CABLE IS FITTED TO BODY

CHANGE POINTS:

(1) To (V) TA 182321; To (V) TA 512897
(2) From (V) TA 182322
(3) From (V) TA 512898
(4) To (V) VA 742126
(5) From (V) VA 742127
(6) From (V) TA 182322; From (V) TA 512898
(7) From (V) GA 394352 To (V) GA 456980; From (V) HA 456981 To (V) HA 486551; From (V) HA 000001 To (V) HA 000577

N 69

DISCOVERY 1989MY UPTO 1999MY | GROUP N | Body & Chassis
Body
Bonnet Locking & Release

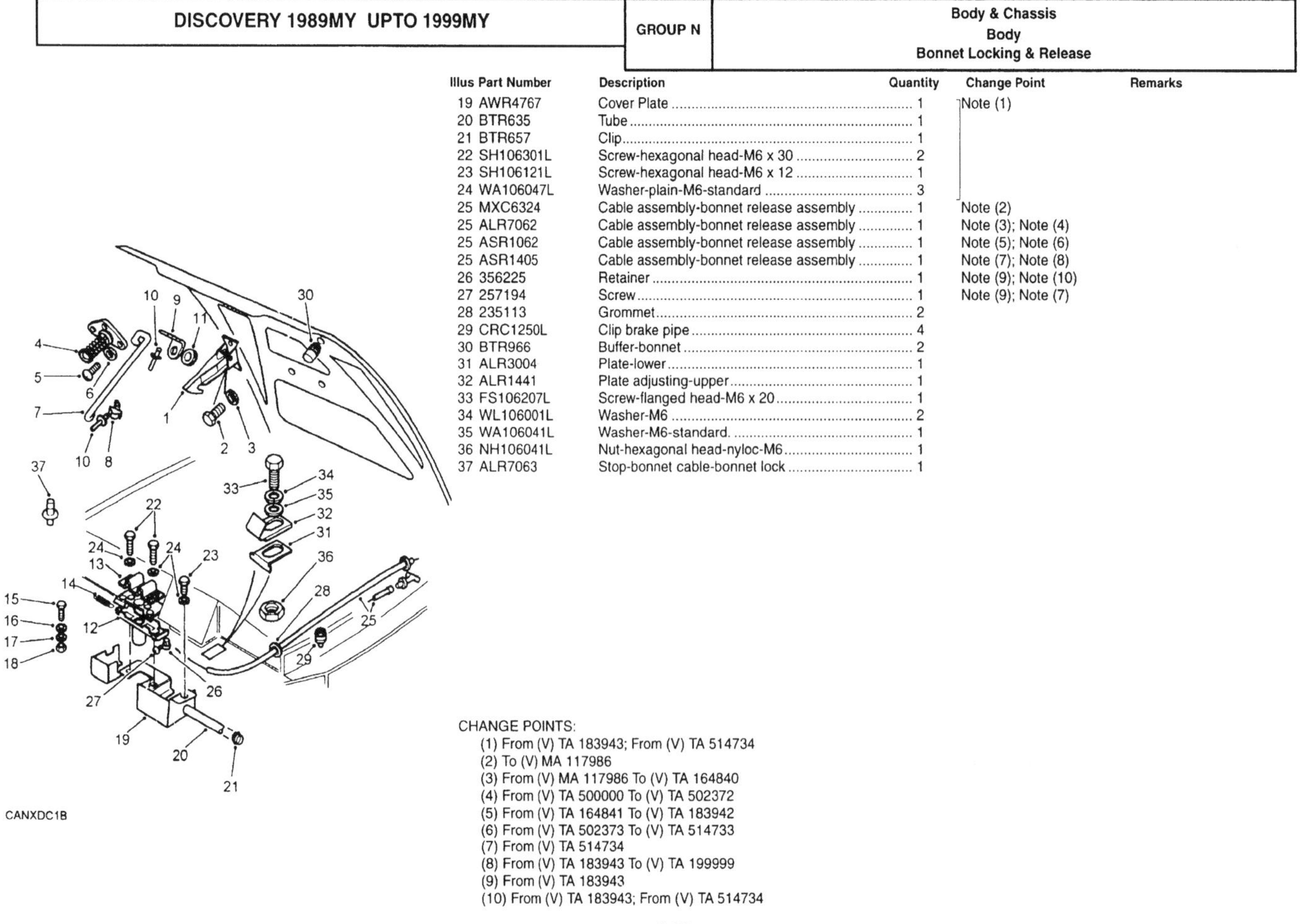

Illus	Part Number	Description	Quantity	Change Point	Remarks
19	AWR4767	Cover Plate	1	Note (1)	
20	BTR635	Tube	1	Note (1)	
21	BTR657	Clip	1	Note (1)	
22	SH106301L	Screw-hexagonal head-M6 x 30	2	Note (1)	
23	SH106121L	Screw-hexagonal head-M6 x 12	1	Note (1)	
24	WA106047L	Washer-plain-M6-standard	3	Note (1)	
25	MXC6324	Cable assembly-bonnet release assembly	1	Note (2)	
25	ALR7062	Cable assembly-bonnet release assembly	1	Note (3); Note (4)	
25	ASR1062	Cable assembly-bonnet release assembly	1	Note (5); Note (6)	
25	ASR1405	Cable assembly-bonnet release assembly	1	Note (7); Note (8)	
26	356225	Retainer	1	Note (9); Note (10)	
27	257194	Screw	1	Note (9); Note (7)	
28	235113	Grommet	2		
29	CRC1250L	Clip brake pipe	4		
30	BTR966	Buffer-bonnet	2		
31	ALR3004	Plate-lower	1		
32	ALR1441	Plate adjusting-upper	1		
33	FS106207L	Screw-flanged head-M6 x 20	1		
34	WL106001L	Washer-M6	2		
35	WA106041L	Washer-M6-standard.	1		
36	NH106041L	Nut-hexagonal head-nyloc-M6	1		
37	ALR7063	Stop-bonnet cable-bonnet lock	1		

CHANGE POINTS:

(1) From (V) TA 183943; From (V) TA 514734
(2) To (V) MA 117986
(3) From (V) MA 117986 To (V) TA 164840
(4) From (V) TA 500000 To (V) TA 502372
(5) From (V) TA 164841 To (V) TA 183942
(6) From (V) TA 502373 To (V) TA 514733
(7) From (V) TA 514734
(8) From (V) TA 183943 To (V) TA 199999
(9) From (V) TA 183943
(10) From (V) TA 183943; From (V) TA 514734

N 70

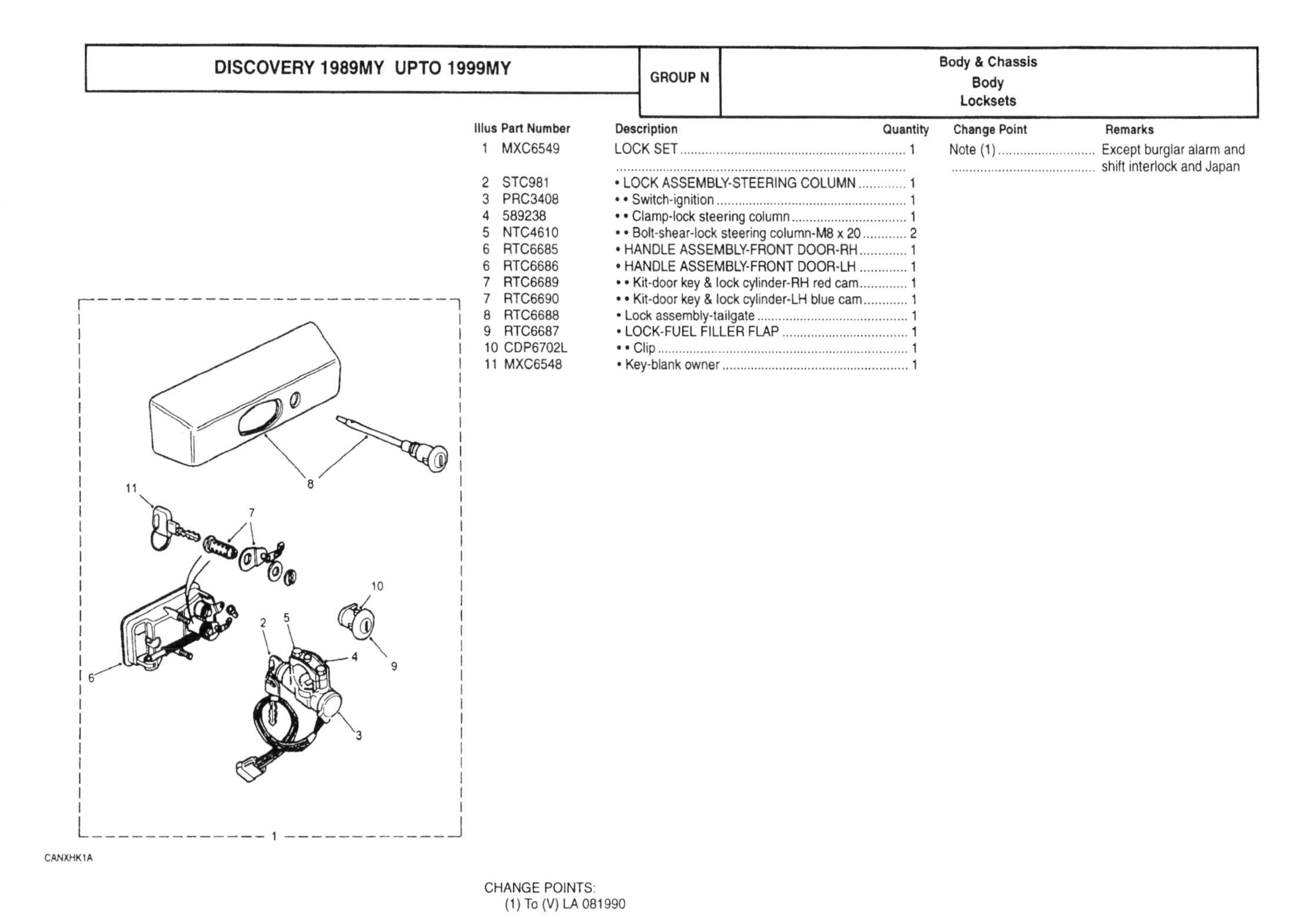

Illus	Part Number	Description	Quantity	Change Point	Remarks
1	MXC6549	LOCK SET	1	Note (1)	Except burglar alarm and shift interlock and Japan
2	STC981	• LOCK ASSEMBLY-STEERING COLUMN	1		
3	PRC3408	• • Switch-ignition	1		
4	589238	• • Clamp-lock steering column	1		
5	NTC4610	• • Bolt-shear-lock steering column-M8 x 20	2		
6	RTC6685	• HANDLE ASSEMBLY-FRONT DOOR-RH	1		
6	RTC6686	• HANDLE ASSEMBLY-FRONT DOOR-LH	1		
7	RTC6689	• • Kit-door key & lock cylinder-RH red cam	1		
7	RTC6690	• • Kit-door key & lock cylinder-LH blue cam	1		
8	RTC6688	• Lock assembly-tailgate	1		
9	RTC6687	• LOCK-FUEL FILLER FLAP	1		
10	CDP6702L	• • Clip	1		
11	MXC6548	• Key-blank owner	1		

CHANGE POINTS:
(1) To (V) LA 081990

N 71

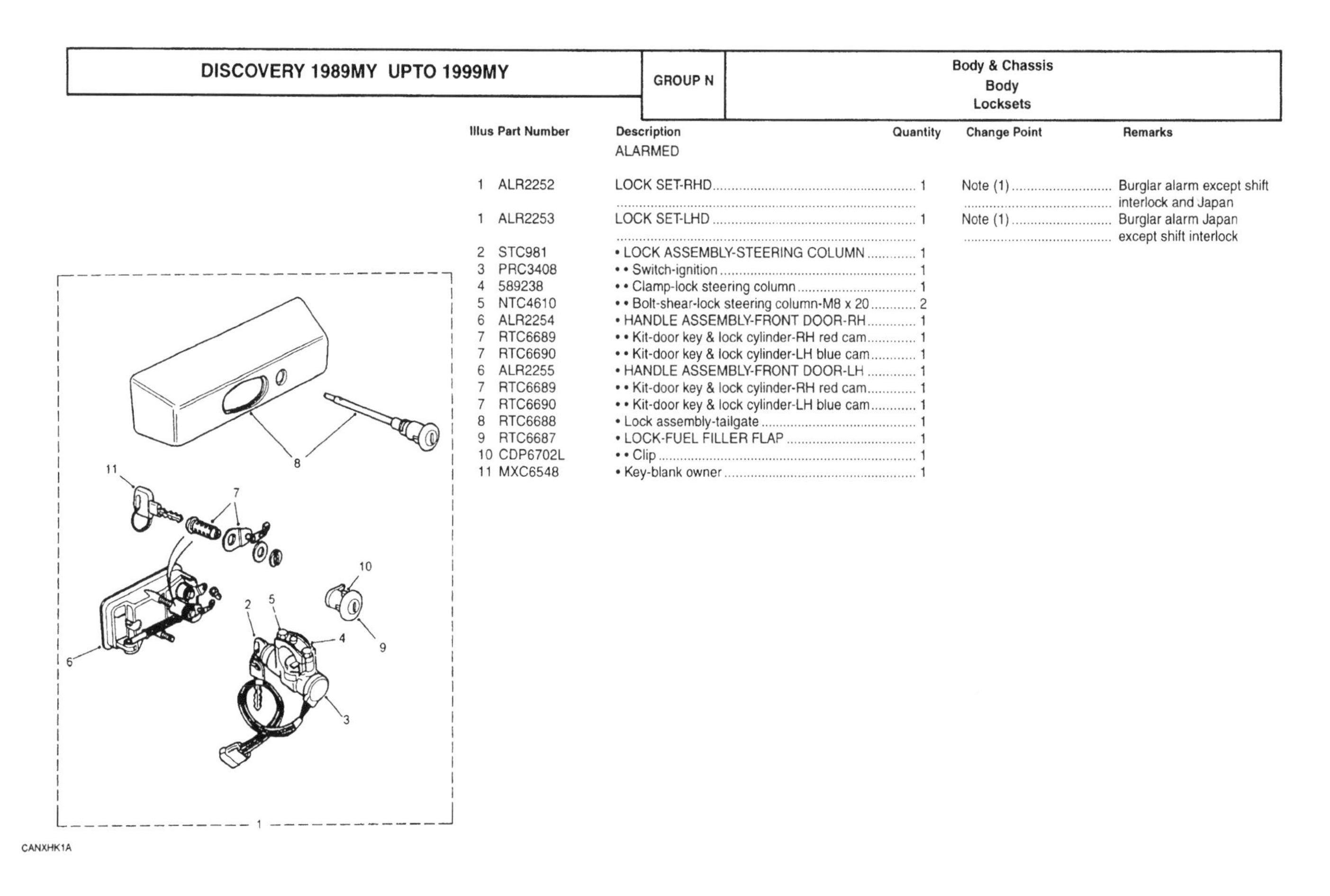

Illus	Part Number	Description	Quantity	Change Point	Remarks
		ALARMED			
1	ALR2252	LOCK SET-RHD	1	Note (1)	Burglar alarm except shift interlock and Japan
1	ALR2253	LOCK SET-LHD	1	Note (1)	Burglar alarm Japan except shift interlock
2	STC981	• LOCK ASSEMBLY-STEERING COLUMN	1		
3	PRC3408	• • Switch-ignition	1		
4	589238	• • Clamp-lock steering column	1		
5	NTC4610	• • Bolt-shear-lock steering column-M8 x 20	2		
6	ALR2254	• HANDLE ASSEMBLY-FRONT DOOR-RH	1		
7	RTC6689	• • Kit-door key & lock cylinder-RH red cam	1		
7	RTC6690	• • Kit-door key & lock cylinder-LH blue cam	1		
6	ALR2255	• HANDLE ASSEMBLY-FRONT DOOR-LH	1		
7	RTC6689	• • Kit-door key & lock cylinder-RH red cam	1		
7	RTC6690	• • Kit-door key & lock cylinder-LH blue cam	1		
8	RTC6688	• Lock assembly-tailgate	1		
9	RTC6687	• LOCK-FUEL FILLER FLAP	1		
10	CDP6702L	• • Clip	1		
11	MXC6548	• Key-blank owner	1		

CHANGE POINTS:
(1) To (V) LA 081990

N 72

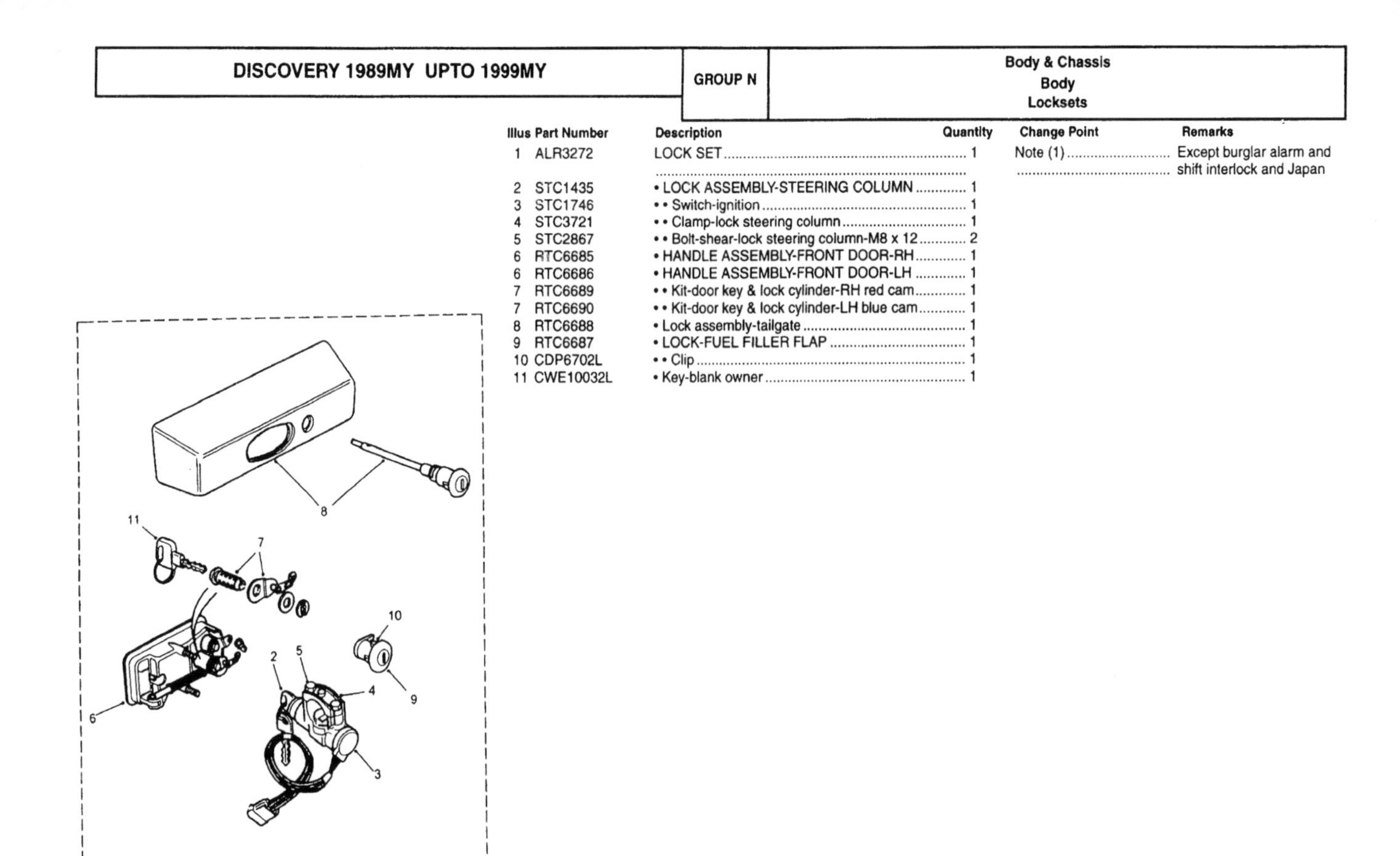

Illus	Part Number	Description	Quantity	Change Point	Remarks
1	ALR3272	LOCK SET	1	Note (1)	Except burglar alarm and shift interlock and Japan
2	STC1435	• LOCK ASSEMBLY-STEERING COLUMN	1		
3	STC1746	• • Switch-ignition	1		
4	STC3721	• • Clamp-lock steering column	1		
5	STC2867	• • Bolt-shear-lock steering column-M8 x 12	2		
6	RTC6685	• HANDLE ASSEMBLY-FRONT DOOR-RH	1		
6	RTC6686	• HANDLE ASSEMBLY-FRONT DOOR-LH	1		
7	RTC6689	• • Kit-door key & lock cylinder-RH red cam	1		
7	RTC6690	• • Kit-door key & lock cylinder-LH blue cam	1		
8	RTC6688	• Lock assembly-tailgate	1		
9	RTC6687	• LOCK-FUEL FILLER FLAP	1		
10	CDP6702L	• • Clip	1		
11	CWE10032L	• Key-blank owner	1		

CHANGE POINTS:
(1) From (V) MA 081991

N 73

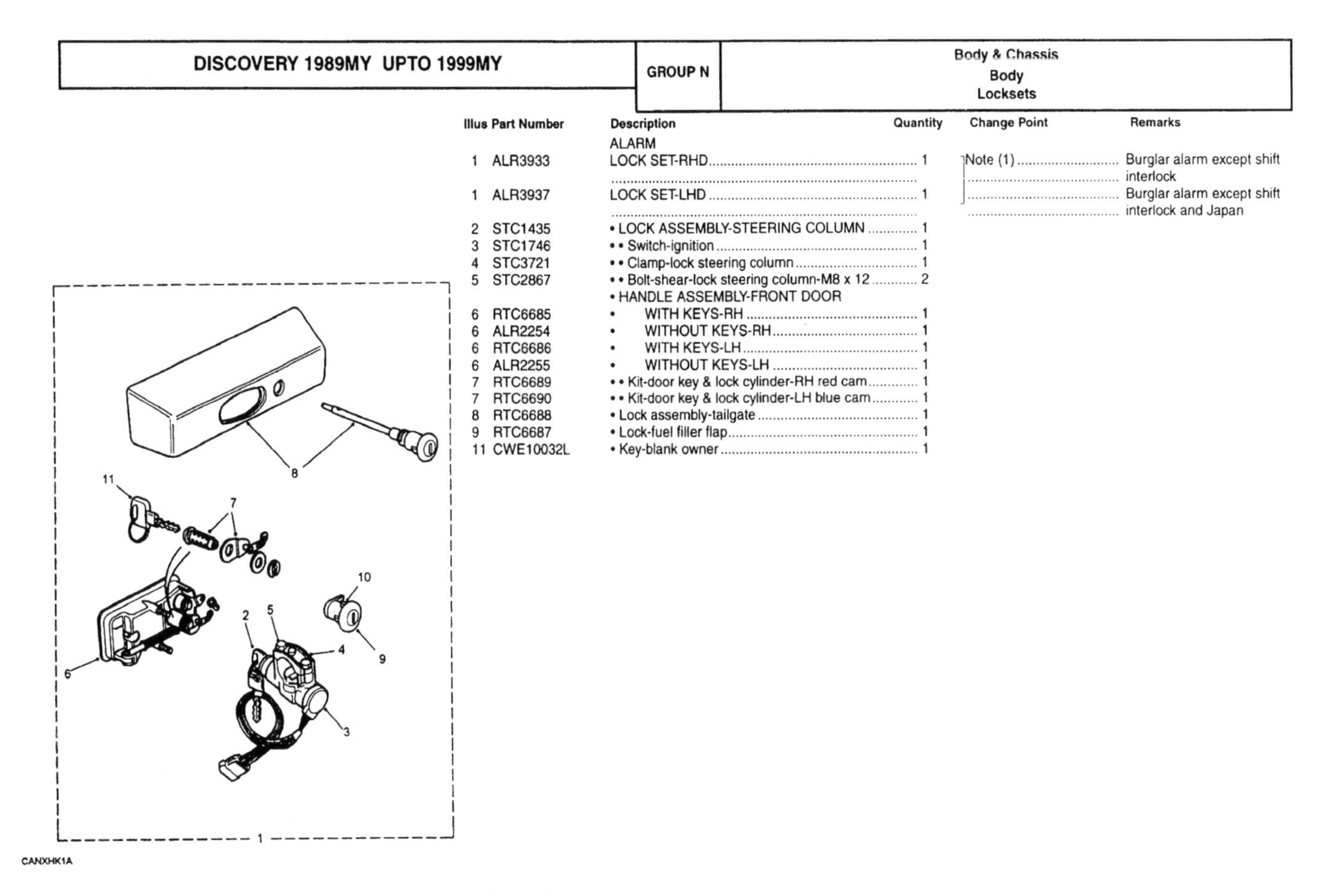

Illus	Part Number	Description	Quantity	Change Point	Remarks
		ALARM			
1	ALR3933	LOCK SET-RHD	1	Note (1)	Burglar alarm except shift interlock
1	ALR3937	LOCK SET-LHD	1		Burglar alarm except shift interlock and Japan
2	STC1435	• LOCK ASSEMBLY-STEERING COLUMN	1		
3	STC1746	• • Switch-ignition	1		
4	STC3721	• • Clamp-lock steering column	1		
5	STC2867	• • Bolt-shear-lock steering column-M8 x 12	2		
		• HANDLE ASSEMBLY-FRONT DOOR			
6	RTC6685	• WITH KEYS-RH	1		
6	ALR2254	• WITHOUT KEYS-RH	1		
6	RTC6686	• WITH KEYS-LH	1		
6	ALR2255	• WITHOUT KEYS-LH	1		
7	RTC6689	• • Kit-door key & lock cylinder-RH red cam	1		
7	RTC6690	• • Kit-door key & lock cylinder-LH blue cam	1		
8	RTC6688	• Lock assembly-tailgate	1		
9	RTC6687	• Lock-fuel filler flap	1		
11	CWE10032L	• Key-blank owner	1		

CHANGE POINTS:
(1) From (V) MA 081991

N 74

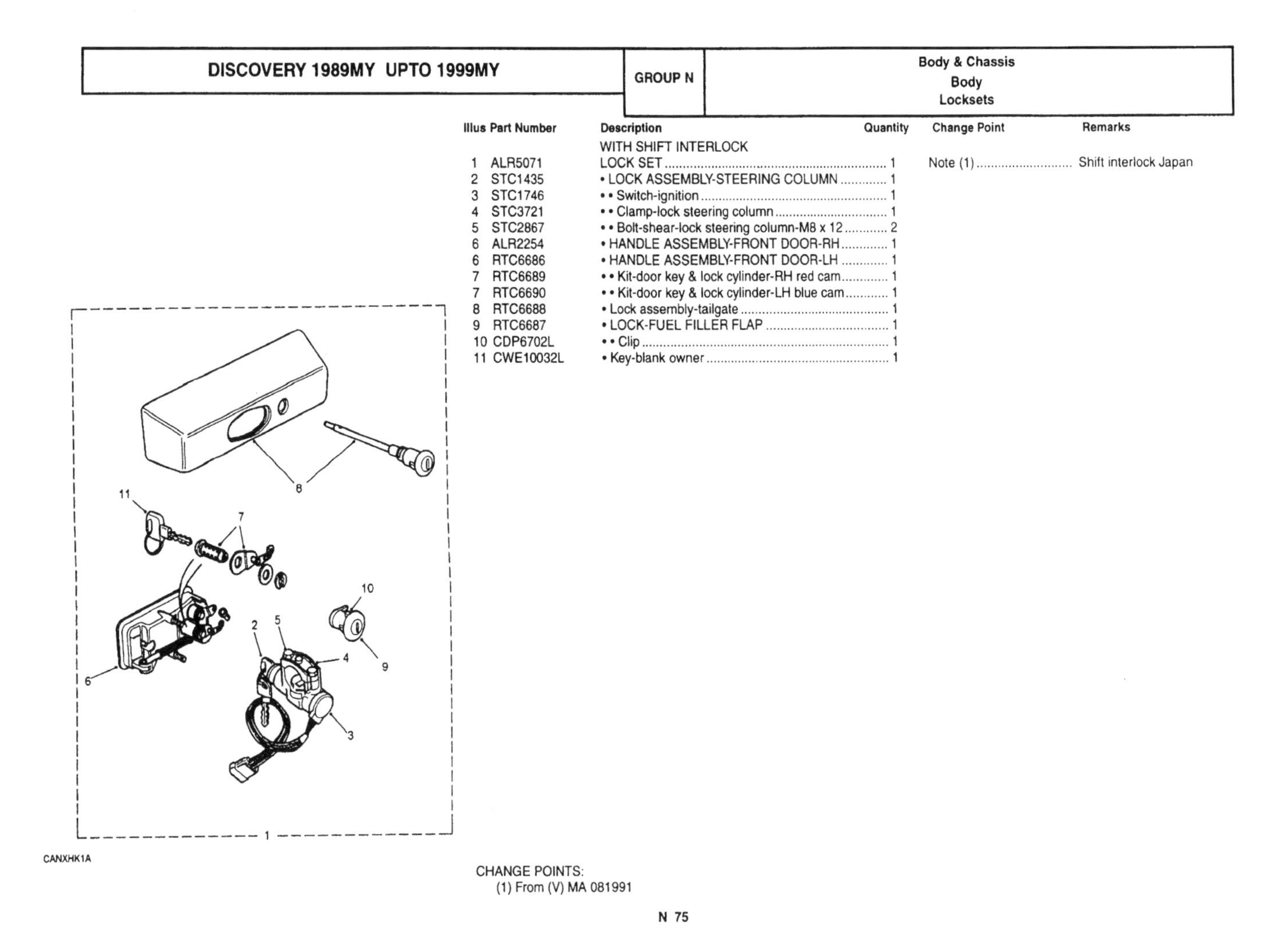

Illus	Part Number	Description	Quantity	Change Point	Remarks
		WITH SHIFT INTERLOCK			
1	ALR5071	LOCK SET	1	Note (1)	Shift interlock Japan
2	STC1435	• LOCK ASSEMBLY-STEERING COLUMN	1		
3	STC1746	• • Switch-ignition	1		
4	STC3721	• • Clamp-lock steering column	1		
5	STC2867	• • Bolt-shear-lock steering column-M8 x 12	2		
6	ALR2254	• HANDLE ASSEMBLY-FRONT DOOR-RH	1		
6	RTC6686	• HANDLE ASSEMBLY-FRONT DOOR-LH	1		
7	RTC6689	• • Kit-door key & lock cylinder-RH red cam	1		
7	RTC6690	• • Kit-door key & lock cylinder-LH blue cam	1		
8	RTC6688	• Lock assembly-tailgate	1		
9	RTC6687	• LOCK-FUEL FILLER FLAP	1		
10	CDP6702L	• • Clip	1		
11	CWE10032L	• Key-blank owner	1		

CHANGE POINTS:
(1) From (V) MA 081991

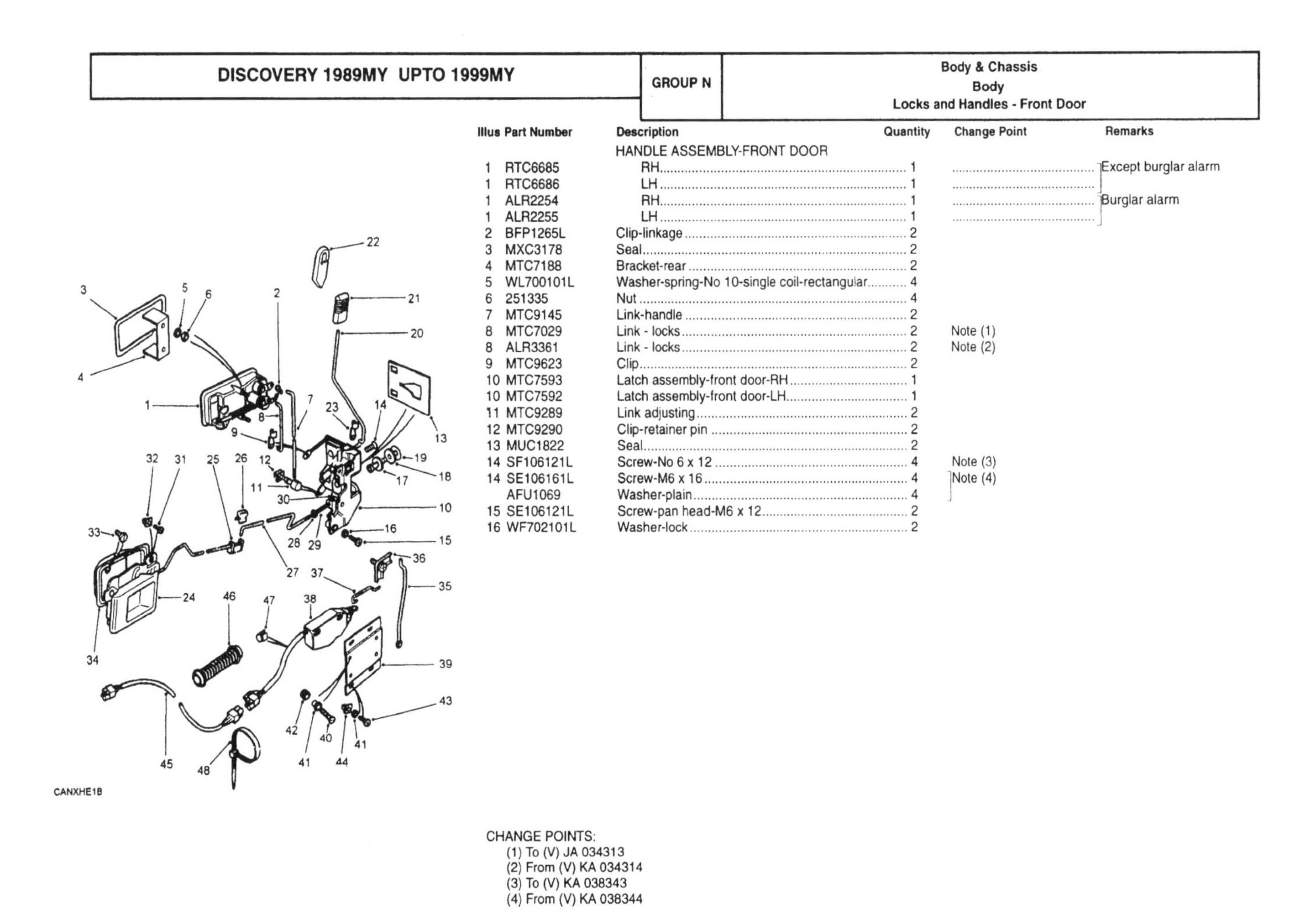

Illus	Part Number	Description	Quantity	Change Point	Remarks
		HANDLE ASSEMBLY-FRONT DOOR			
1	RTC6685	RH	1		Except burglar alarm
1	RTC6686	LH	1		
1	ALR2254	RH	1		Burglar alarm
1	ALR2255	LH	1		
2	BFP1265L	Clip-linkage	2		
3	MXC3178	Seal	2		
4	MTC7188	Bracket-rear	2		
5	WL700101L	Washer-spring-No 10-single coil-rectangular	4		
6	251335	Nut	4		
7	MTC9145	Link-handle	2		
8	MTC7029	Link - locks	2	Note (1)	
8	ALR3361	Link - locks	2	Note (2)	
9	MTC9623	Clip	2		
10	MTC7593	Latch assembly-front door-RH	1		
10	MTC7592	Latch assembly-front door-LH	1		
11	MTC9289	Link adjusting	2		
12	MTC9290	Clip-retainer pin	2		
13	MUC1822	Seal	2		
14	SF106121L	Screw-No 6 x 12	4	Note (3)	
14	SE106161L	Screw-M6 x 16	4	Note (4)	
	AFU1069	Washer-plain	4		
15	SE106121L	Screw-pan head-M6 x 12	2		
16	WF702101L	Washer-lock	2		

CHANGE POINTS:
(1) To (V) JA 034313
(2) From (V) KA 034314
(3) To (V) KA 038343
(4) From (V) KA 038344

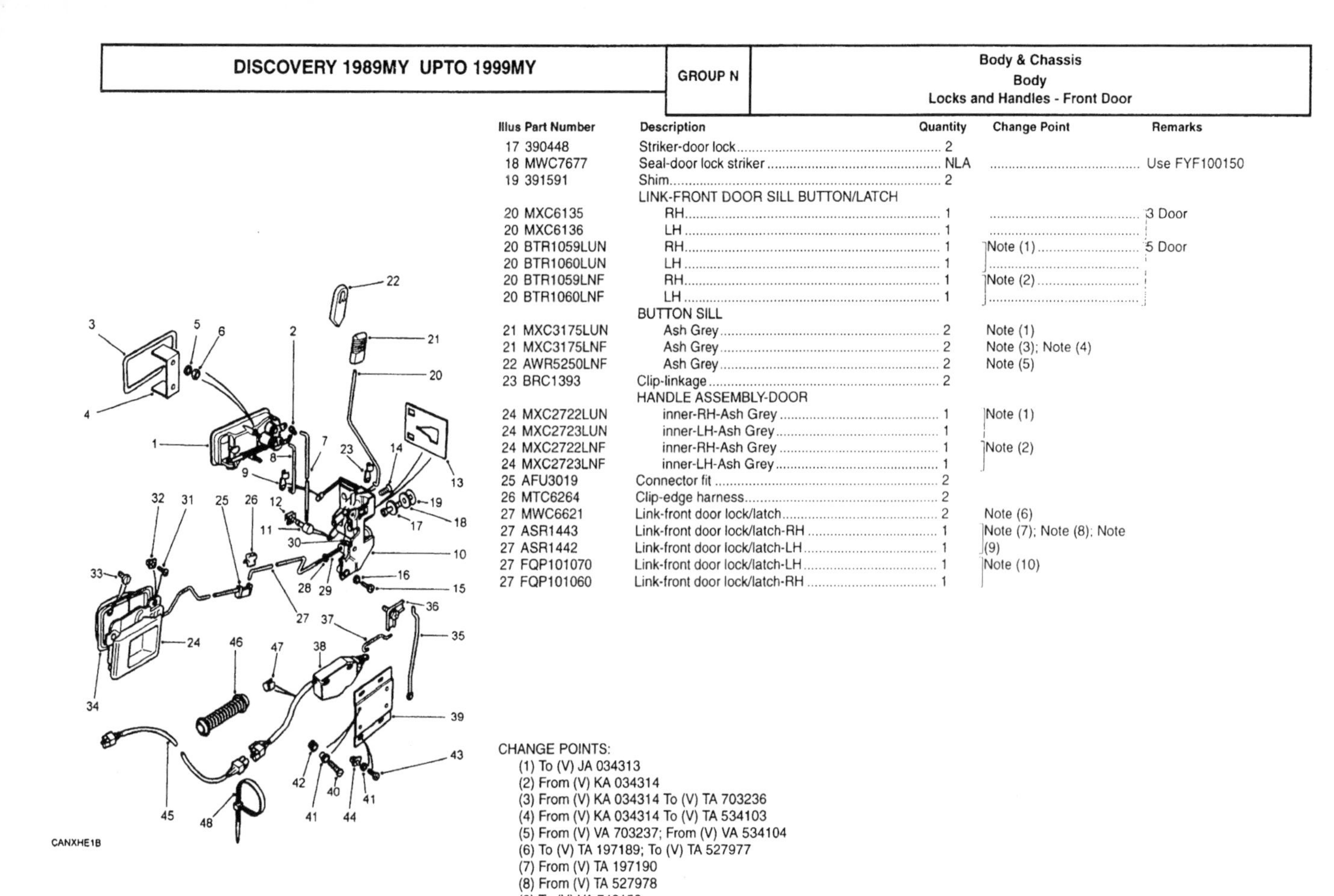

Illus	Part Number	Description	Quantity	Change Point	Remarks
17	390448	Striker-door lock	2		
18	MWC7677	Seal-door lock striker	NLA		Use FYF100150
19	391591	Shim	2		
		LINK-FRONT DOOR SILL BUTTON/LATCH			
20	MXC6135	RH	1		3 Door
20	MXC6136	LH	1		
20	BTR1059LUN	RH	1	Note (1)	5 Door
20	BTR1060LUN	LH	1		
20	BTR1059LNF	RH	1	Note (2)	
20	BTR1060LNF	LH	1		
		BUTTON SILL			
21	MXC3175LUN	Ash Grey	2	Note (1)	
21	MXC3175LNF	Ash Grey	2	Note (3); Note (4)	
22	AWR5250LNF	Ash Grey	2	Note (5)	
23	BRC1393	Clip-linkage	2		
		HANDLE ASSEMBLY-DOOR			
24	MXC2722LUN	inner-RH-Ash Grey	1	Note (1)	
24	MXC2723LUN	inner-LH-Ash Grey	1		
24	MXC2722LNF	inner-RH-Ash Grey	1	Note (2)	
24	MXC2723LNF	inner-LH-Ash Grey	1		
25	AFU3019	Connector fit	2		
26	MTC6264	Clip-edge harness	2		
27	MWC6621	Link-front door lock/latch	2	Note (6)	
27	ASR1443	Link-front door lock/latch-RH	1	Note (7); Note (8); Note (9)	
27	ASR1442	Link-front door lock/latch-LH	1		
27	FQP101070	Link-front door lock/latch-LH	1	Note (10)	
27	FQP101060	Link-front door lock/latch-RH	1		

CHANGE POINTS:
(1) To (V) JA 034313
(2) From (V) KA 034314
(3) From (V) KA 034314 To (V) TA 703236
(4) From (V) KA 034314 To (V) TA 534103
(5) From (V) VA 703237; From (V) VA 534104
(6) To (V) TA 197189; To (V) TA 527977
(7) From (V) TA 197190
(8) From (V) TA 527978
(9) To (V) VA 743153
(10) From (V) VA 743154

DISCOVERY 1989MY UPTO 1999MY | GROUP N | Body & Chassis
Body
Locks and Handles - Front Door

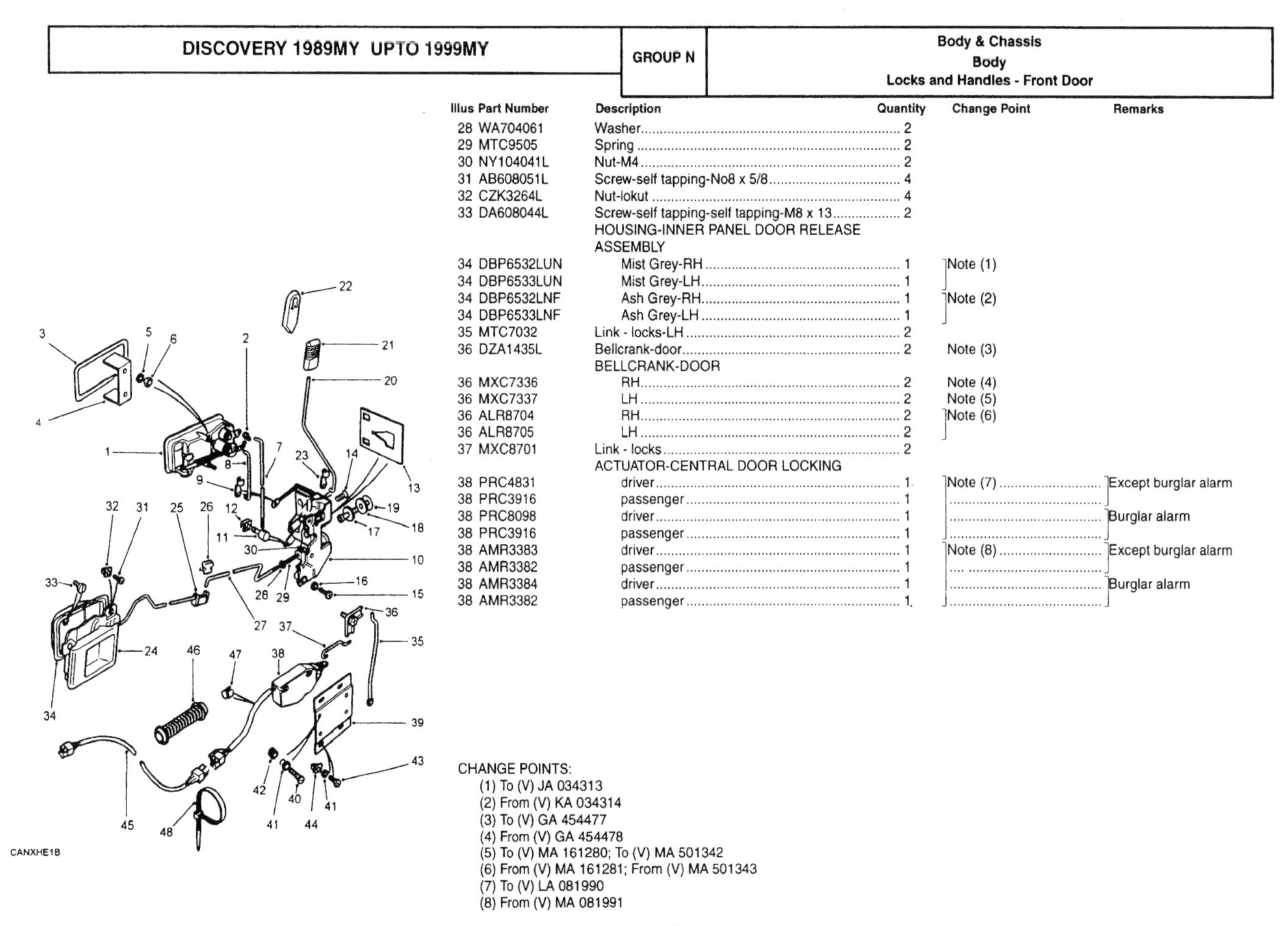

Illus	Part Number	Description	Quantity	Change Point	Remarks
28	WA704061	Washer	2		
29	MTC9505	Spring	2		
30	NY104041L	Nut-M4	2		
31	AB608051L	Screw-self tapping-No8 x 5/8	4		
32	CZK3264L	Nut-lokut	4		
33	DA608044L	Screw-self tapping-self tapping-M8 x 13	2		
		HOUSING-INNER PANEL DOOR RELEASE ASSEMBLY			
34	DBP6532LUN	Mist Grey-RH	1	Note (1)	
34	DBP6533LUN	Mist Grey-LH	1		
34	DBP6532LNF	Ash Grey-RH	1	Note (2)	
34	DBP6533LNF	Ash Grey-LH	1		
35	MTC7032	Link - locks-LH	2		
36	DZA1435L	Bellcrank-door	2	Note (3)	
		BELLCRANK-DOOR			
36	MXC7336	RH	2	Note (4)	
36	MXC7337	LH	2	Note (5)	
36	ALR8704	RH	2	Note (6)	
36	ALR8705	LH	2		
37	MXC8701	Link - locks	2		
		ACTUATOR-CENTRAL DOOR LOCKING			
38	PRC4831	driver	1	Note (7)	Except burglar alarm
38	PRC3916	passenger	1		
38	PRC8098	driver	1		Burglar alarm
38	PRC3916	passenger	1		
38	AMR3383	driver	1	Note (8)	Except burglar alarm
38	AMR3382	passenger	1		
38	AMR3384	driver	1		Burglar alarm
38	AMR3382	passenger	1		

CHANGE POINTS:
(1) To (V) JA 034313
(2) From (V) KA 034314
(3) To (V) GA 454477
(4) From (V) GA 454478
(5) To (V) MA 161280; To (V) MA 501342
(6) From (V) MA 161281; From (V) MA 501343
(7) To (V) LA 081990
(8) From (V) MA 081991

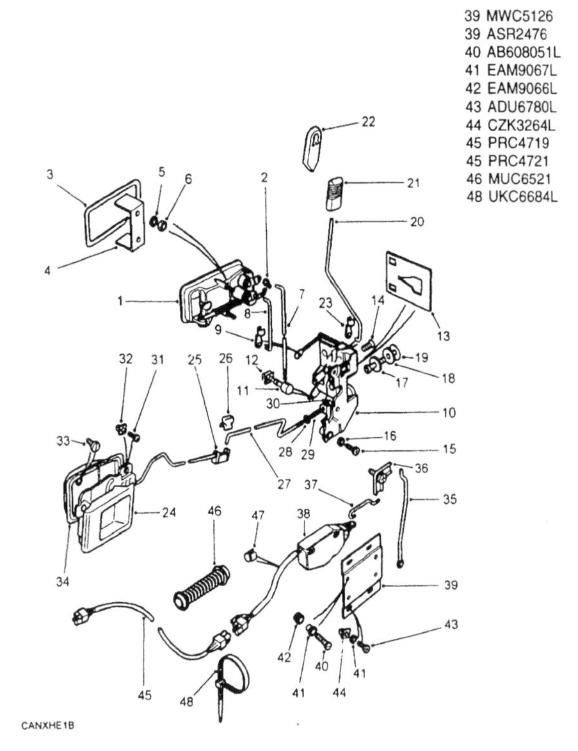

Illus	Part Number	Description	Quantity	Change Point	Remarks
		ALL VARIANTS			
39	MWC5126	Plate-door actuator mounting	2	Note (1)	
39	ASR2476	Plate-door actuator mounting	2	Note (2)	
40	AB608051L	Screw-self tapping-No8 x 5/8	4		
41	EAM9067L	Bush	8		
42	EAM9066L	Grommet	4		
43	ADU6780L	Screw	8		
44	CZK3264L	Nut-lokut	8		
45	PRC4719	Harness-central door locking-driver	1		
45	PRC4721	Harness-central door locking-passenger	1		
46	MUC6521	Grommet-convolute	2		
48	UKC6684L	Tie-cable-Black-3.5 x 150mm-inside serated	2	Note (3)	

CHANGE POINTS:
(1) To (V) VA 542991; To (V) VA 711791
(2) From (V) VA 542992; From (V) VA 711792
(3) From (V) GA 022853

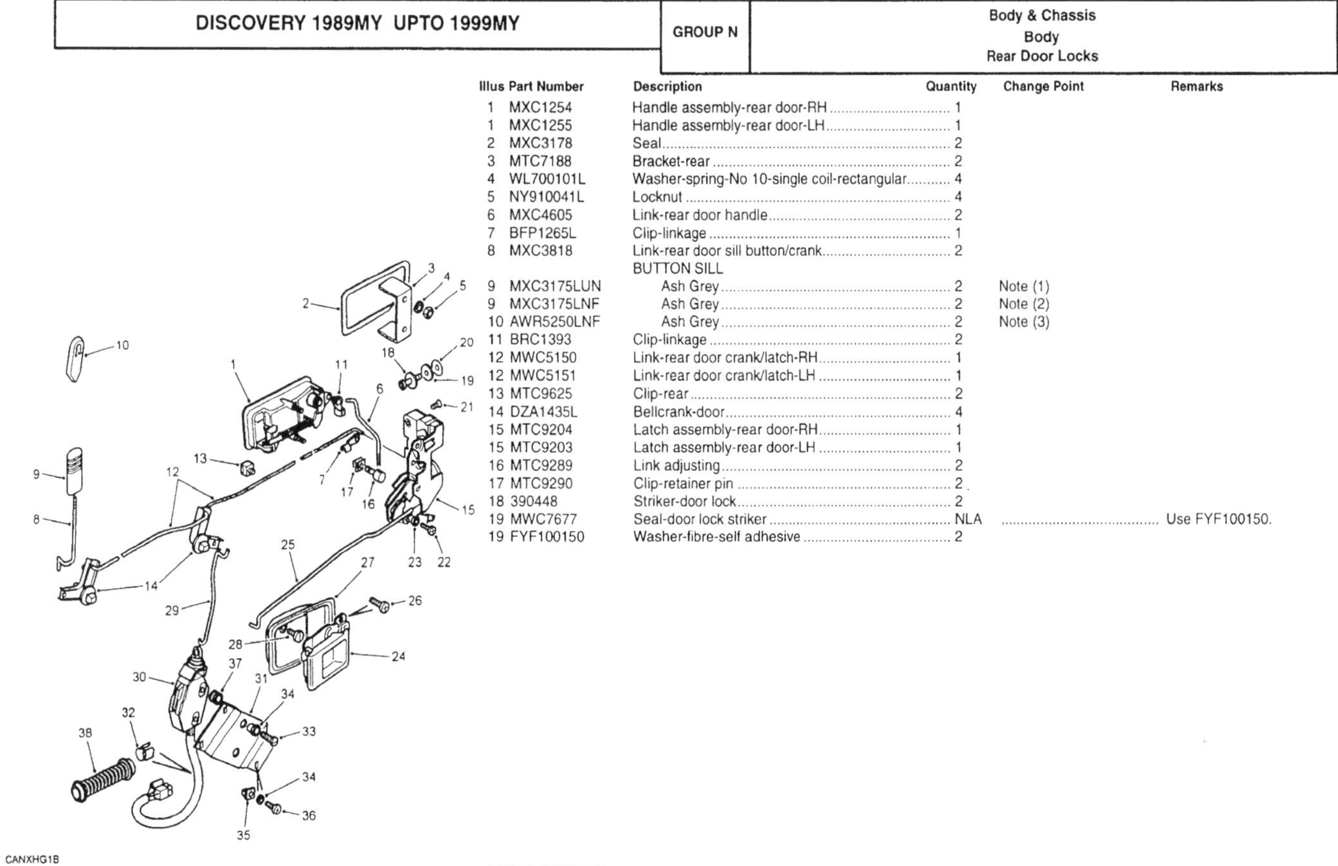

Illus	Part Number	Description	Quantity	Change Point	Remarks
1	MXC1254	Handle assembly-rear door-RH	1		
1	MXC1255	Handle assembly-rear door-LH	1		
2	MXC3178	Seal	2		
3	MTC7188	Bracket-rear	2		
4	WL700101L	Washer-spring-No 10-single coil-rectangular	4		
5	NY910041L	Locknut	4		
6	MXC4605	Link-rear door handle	2		
7	BFP1265L	Clip-linkage	1		
8	MXC3818	Link-rear door sill button/crank	2		
		BUTTON SILL			
9	MXC3175LUN	Ash Grey	2	Note (1)	
9	MXC3175LNF	Ash Grey	2	Note (2)	
10	AWR5250LNF	Ash Grey	2	Note (3)	
11	BRC1393	Clip-linkage	2		
12	MWC5150	Link-rear door crank/latch-RH	1		
12	MWC5151	Link-rear door crank/latch-LH	1		
13	MTC9625	Clip-rear	2		
14	DZA1435L	Bellcrank-door	4		
15	MTC9204	Latch assembly-rear door-RH	1		
15	MTC9203	Latch assembly-rear door-LH	1		
16	MTC9289	Link adjusting	2		
17	MTC9290	Clip-retainer pin	2		
18	390448	Striker-door lock	2		
19	MWC7677	Seal-door lock striker	NLA		Use FYF100150.
19	FYF100150	Washer-fibre-self adhesive	2		

CHANGE POINTS:
(1) To (V) JA 034313
(2) From (V) KA 034314
(3) From (V) VA 703237; From (V) VA 534104

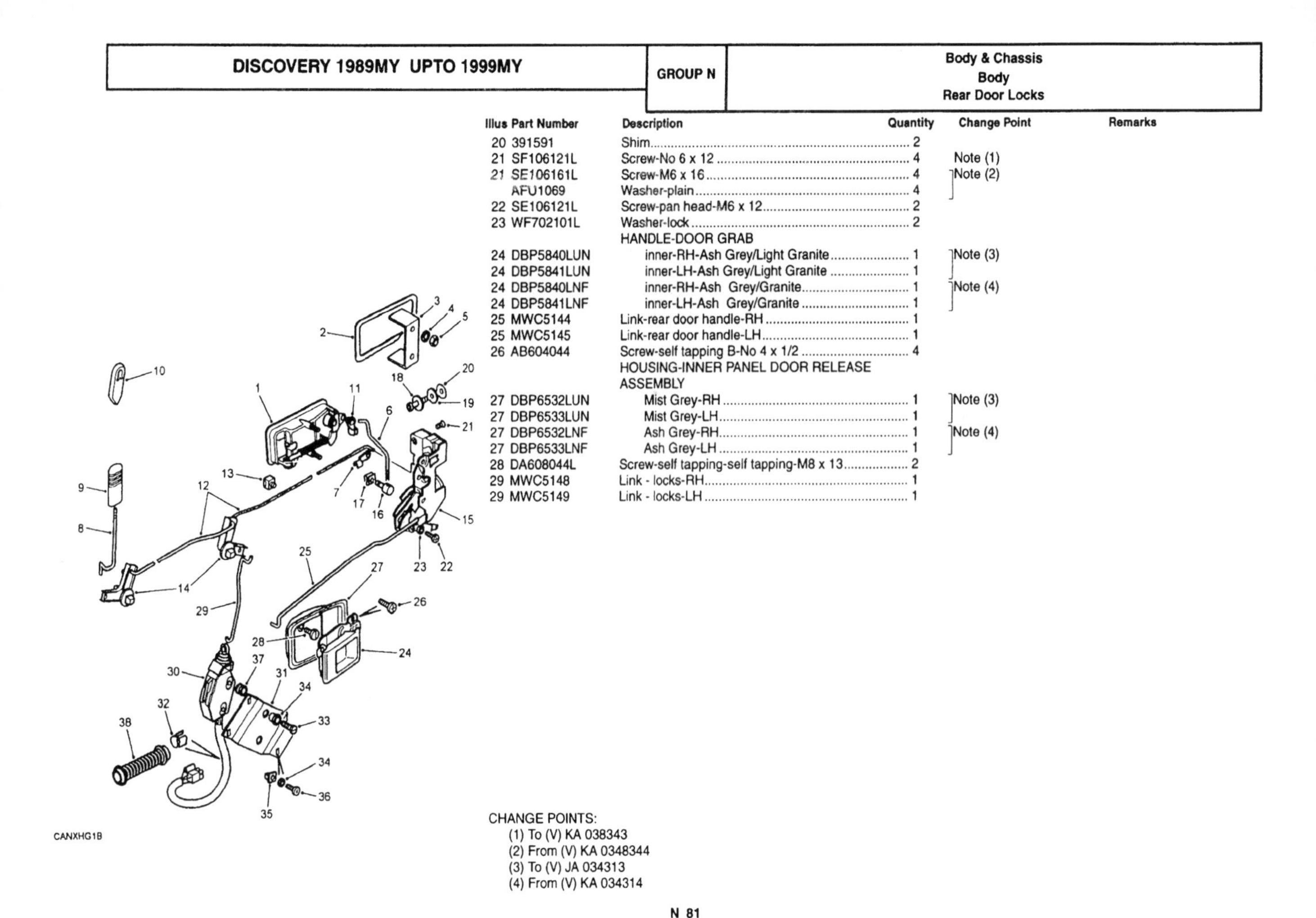

Illus	Part Number	Description	Quantity	Change Point	Remarks
20	391591	Shim	2		
21	SF106121L	Screw-No 6 x 12	4	Note (1)	
21	SE106161L	Screw-M6 x 16	4	Note (2)	
	AFU1069	Washer-plain	4		
22	SE106121L	Screw-pan head-M6 x 12	2		
23	WF702101L	Washer-lock	2		
		HANDLE-DOOR GRAB			
24	DBP5840LUN	inner-RH-Ash Grey/Light Granite	1	Note (3)	
24	DBP5841LUN	inner-LH-Ash Grey/Light Granite	1		
24	DBP5840LNF	inner-RH-Ash Grey/Granite	1	Note (4)	
24	DBP5841LNF	inner-LH-Ash Grey/Granite	1		
25	MWC5144	Link-rear door handle-RH	1		
25	MWC5145	Link-rear door handle-LH	1		
26	AB604044	Screw-self tapping B-No 4 x 1/2	4		
		HOUSING-INNER PANEL DOOR RELEASE ASSEMBLY			
27	DBP6532LUN	Mist Grey-RH	1	Note (3)	
27	DBP6533LUN	Mist Grey-LH	1		
27	DBP6532LNF	Ash Grey-RH	1	Note (4)	
27	DBP6533LNF	Ash Grey-LH	1		
28	DA608044L	Screw-self tapping-self tapping-M8 x 13	2		
29	MWC5148	Link - locks-RH	1		
29	MWC5149	Link - locks-LH	1		

CHANGE POINTS:
(1) To (V) KA 038343
(2) From (V) KA 0348344
(3) To (V) JA 034313
(4) From (V) KA 034314

Illus	Part Number	Description	Quantity	Change Point	Remarks
30	PRC3916	Actuator-central door locking-passenger	2	Note (1)	
30	AMR3382	Actuator-central door locking-passenger	2	Note (2)	
31	MWC6432	Plate-rear door actuator-RH	1		
31	MWC6433	Plate-rear door actuator-LH	1		
32	345662	Clip-cable-rear	A/R		
33	AB608051L	Screw-self tapping-No8 x 5/8	8		
34	EAM9067L	Bush	8		
35	CZK3264	Nut-lokut	NLA		Use CZK3264L.
35	CZK3264L	Nut-lokut	8		
36	ADU6780	Screw-plastite	8		
37	EAM9066L	Grommet	4		
38	MUC6521	Grommet-convolute	2		

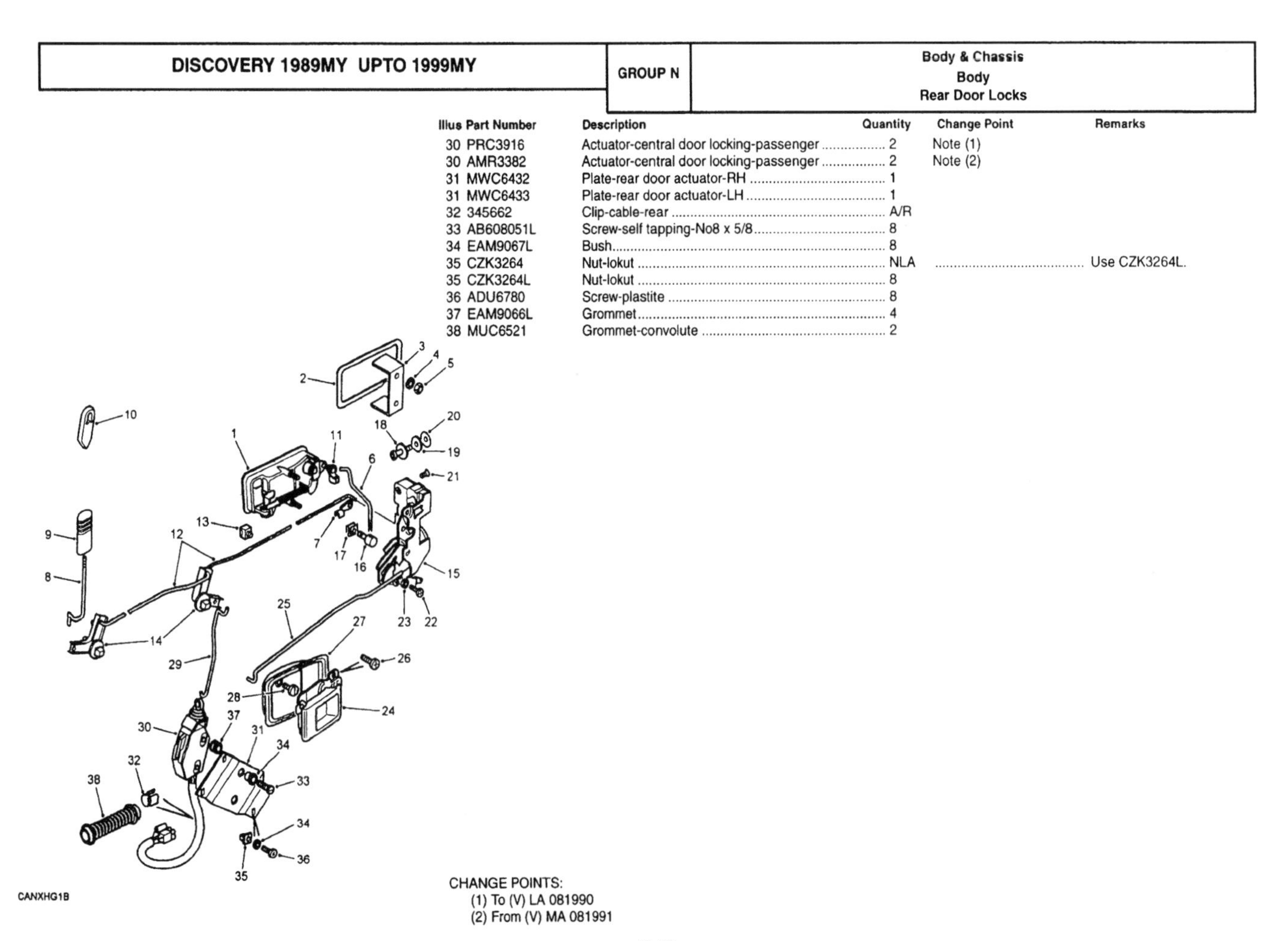

CHANGE POINTS:
(1) To (V) LA 081990
(2) From (V) MA 081991

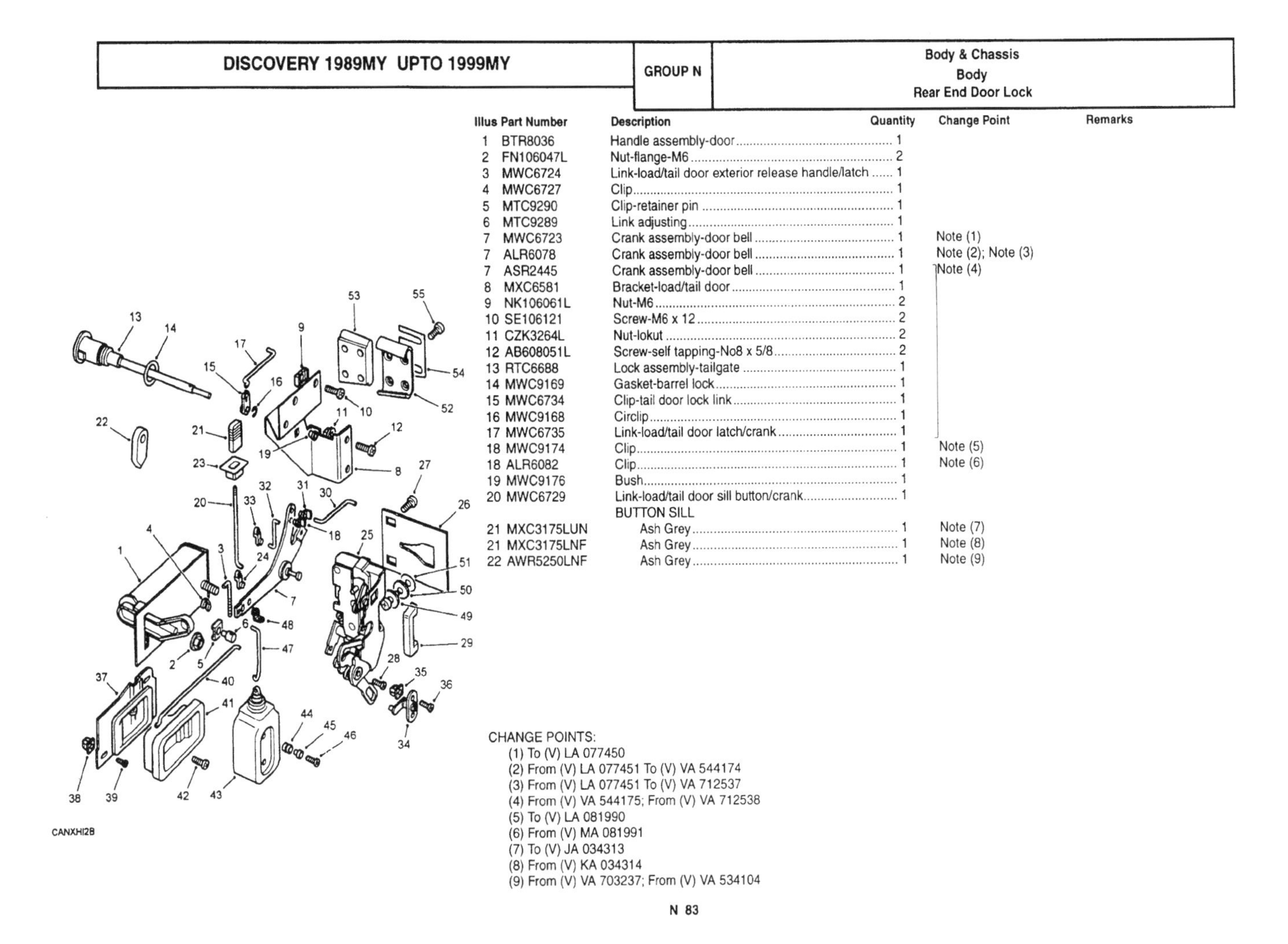

Illus	Part Number	Description	Quantity	Change Point	Remarks
1	BTR8036	Handle assembly-door	1		
2	FN106047L	Nut-flange-M6	2		
3	MWC6724	Link-load/tail door exterior release handle/latch	1		
4	MWC6727	Clip	1		
5	MTC9290	Clip-retainer pin	1		
6	MTC9289	Link adjusting	1		
7	MWC6723	Crank assembly-door bell	1	Note (1)	
7	ALR6078	Crank assembly-door bell	1	Note (2); Note (3)	
7	ASR2445	Crank assembly-door bell	1	Note (4)	
8	MXC6581	Bracket-load/tail door	1		
9	NK106061L	Nut-M6	2		
10	SE106121	Screw-M6 x 12	2		
11	CZK3264L	Nut-lokut	2		
12	AB608051L	Screw-self tapping-No8 x 5/8	2		
13	RTC6688	Lock assembly-tailgate	1		
14	MWC9169	Gasket-barrel lock	1		
15	MWC6734	Clip-tail door lock link	1		
16	MWC9168	Circlip	1		
17	MWC6735	Link-load/tail door latch/crank	1		
18	MWC9174	Clip	1	Note (5)	
18	ALR6082	Clip	1	Note (6)	
19	MWC9176	Bush	1		
20	MWC6729	Link-load/tail door sill button/crank	1		
		BUTTON SILL			
21	MXC3175LUN	Ash Grey	1	Note (7)	
21	MXC3175LNF	Ash Grey	1	Note (8)	
22	AWR5250LNF	Ash Grey	1	Note (9)	

CANXHI2B

CHANGE POINTS:
(1) To (V) LA 077450
(2) From (V) LA 077451 To (V) VA 544174
(3) From (V) LA 077451 To (V) VA 712537
(4) From (V) VA 544175; From (V) VA 712538
(5) To (V) LA 081990
(6) From (V) MA 081991
(7) To (V) JA 034313
(8) From (V) KA 034314
(9) From (V) VA 703237; From (V) VA 534104

N 83

DISCOVERY 1989MY UPTO 1999MY | GROUP N | Body & Chassis / Body / Rear End Door Lock

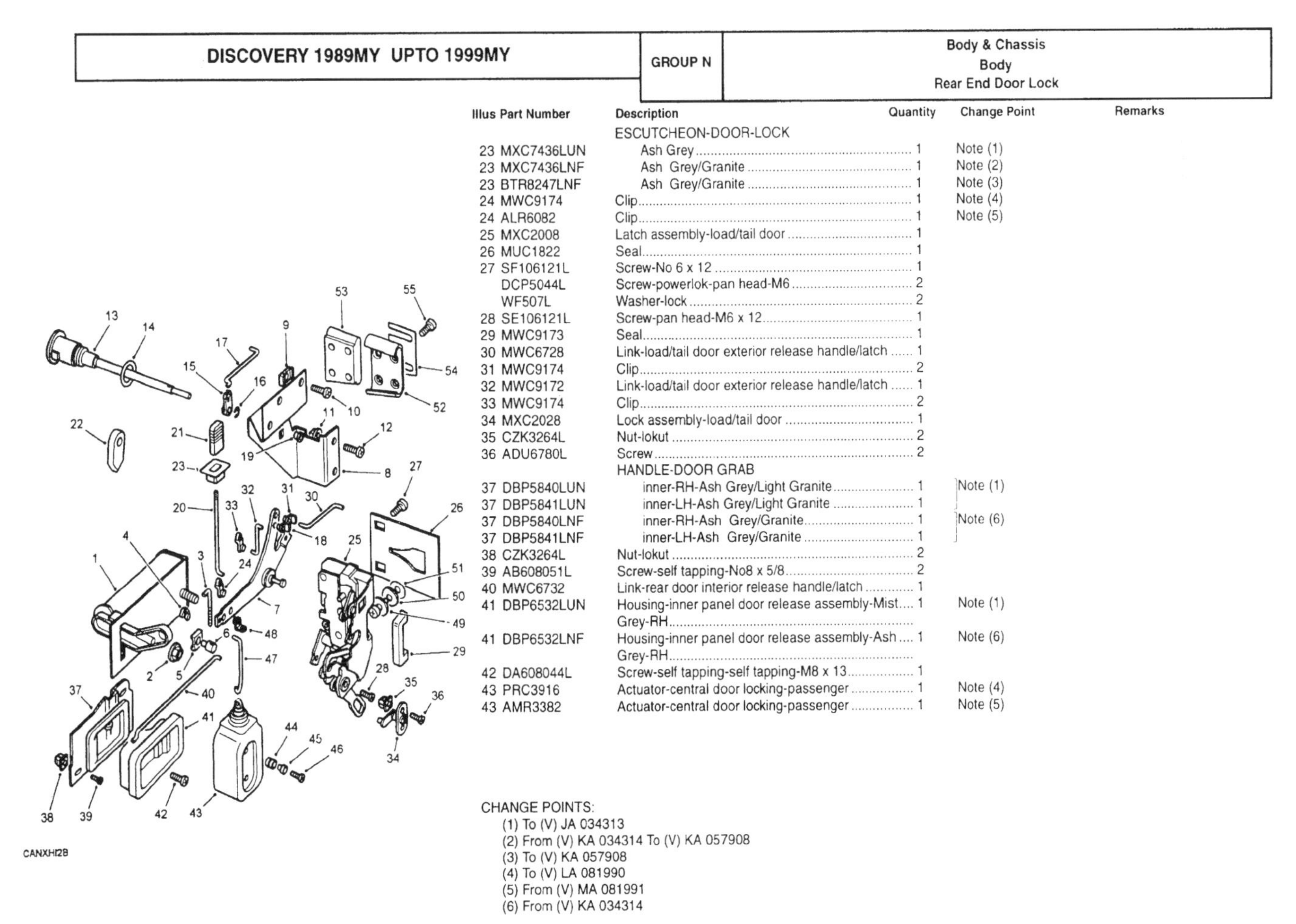

Illus	Part Number	Description	Quantity	Change Point	Remarks
		ESCUTCHEON-DOOR-LOCK			
23	MXC7436LUN	Ash Grey	1	Note (1)	
23	MXC7436LNF	Ash Grey/Granite	1	Note (2)	
23	BTR8247LNF	Ash Grey/Granite	1	Note (3)	
24	MWC9174	Clip	1	Note (4)	
24	ALR6082	Clip	1	Note (5)	
25	MXC2008	Latch assembly-load/tail door	1		
26	MUC1822	Seal	1		
27	SF106121L	Screw-No 6 x 12	1		
	DCP5044L	Screw-powerlok-pan head-M6	2		
	WF507L	Washer-lock	2		
28	SE106121L	Screw-pan head-M6 x 12	1		
29	MWC9173	Seal	1		
30	MWC6728	Link-load/tail door exterior release handle/latch	1		
31	MWC9174	Clip	2		
32	MWC9172	Link-load/tail door exterior release handle/latch	1		
33	MWC9174	Clip	2		
34	MXC2028	Lock assembly-load/tail door	1		
35	CZK3264L	Nut-lokut	2		
36	ADU6780L	Screw	2		
		HANDLE-DOOR GRAB			
37	DBP5840LUN	inner-RH-Ash Grey/Light Granite	1	Note (1)	
37	DBP5841LUN	inner-LH-Ash Grey/Light Granite	1		
37	DBP5840LNF	inner-RH-Ash Grey/Granite	1	Note (6)	
37	DBP5841LNF	inner-LH-Ash Grey/Granite	1		
38	CZK3264L	Nut-lokut	2		
39	AB608051L	Screw-self tapping-No8 x 5/8	2		
40	MWC6732	Link-rear door interior release handle/latch	1		
41	DBP6532LUN	Housing-inner panel door release assembly-Mist Grey-RH	1	Note (1)	
41	DBP6532LNF	Housing-inner panel door release assembly-Ash Grey-RH	1	Note (6)	
42	DA608044L	Screw-self tapping-self tapping-M8 x 13	1		
43	PRC3916	Actuator-central door locking-passenger	1	Note (4)	
43	AMR3382	Actuator-central door locking-passenger	1	Note (5)	

CANXHI2B

CHANGE POINTS:
(1) To (V) JA 034313
(2) From (V) KA 034314 To (V) KA 057908
(3) To (V) KA 057908
(4) To (V) LA 081990
(5) From (V) MA 081991
(6) From (V) KA 034314

N 84

DISCOVERY 1989MY UPTO 1999MY	GROUP N	Body & Chassis Body Rear End Door Lock

Illus	Part Number	Description	Quantity	Change Point	Remarks
44	EAM9066L	Grommet	2		
45	EAM9067L	Bush	2		
46	AB608051L	Screw-self tapping-No8 x 5/8	2		
47	BTR157	Link-central door locking motor to latch	1		
48	MWC9174	Clip	1		
49	390448	Striker-door lock	1		
50	391591	Shim	1		
51	MWC7677	Seal-door lock striker	NLA		Use FYF100150.
51	FYF100150	Washer-fibre-self adhesive	2		
52	MXC2630	Dovetail female-rear door	1	Note (1)	
53	MXC2631	Dovetail male-rear door	1		
52	ALR5691	Dovetail female-rear door	1	Note (2)	
53	ALR6306	Dovetail male-rear door	1		
54	MXC3132	Shim	A/R		
55	SF106201L	Screw-M6 x 20-counter sunk-recessed	8		

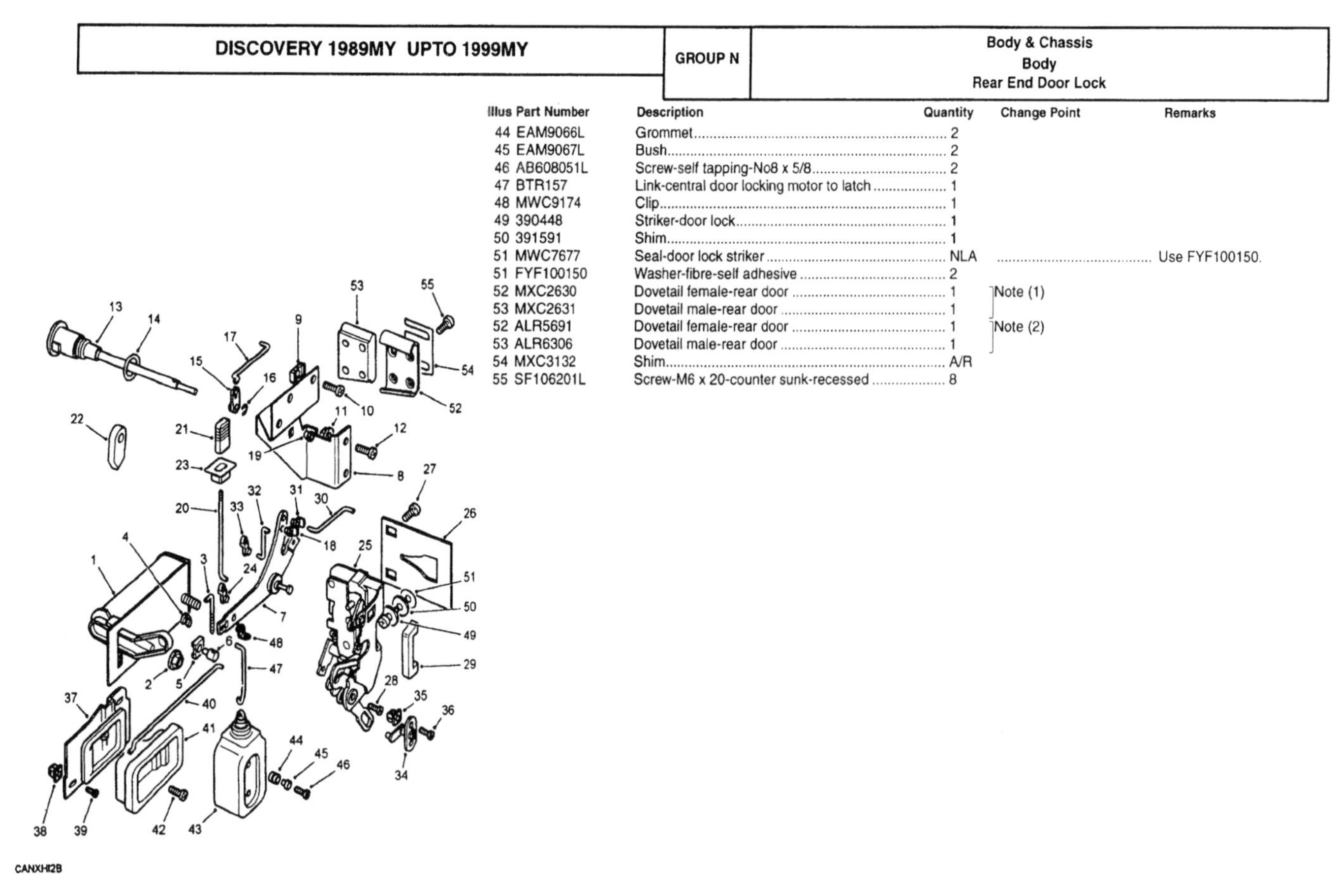

CHANGE POINTS:
(1) To (V) MA 099941
(2) From (V) MA 099942

N 85

DISCOVERY 1989MY UPTO 1999MY	GROUP N	Body & Chassis Body Rear Door Lock Cover

Illus	Part Number	Description	Quantity	Change Point	Remarks
1	MWC6733	Cover-tail door handle-exterior	1		
2	MWC9171	Gasket-door handle	1		
3	FN106047L	Nut-flange-M6	2		
4	SH106121L	Screw-hexagonal head-M6 x 12	3		
5	WA106041L	Washer-M6-standard.	3		
6	MXC1245	Bracket-load/tail door-centre	1		
7	MXC1256	Bracket-load/tail door	1		
8	MXC2519	Badge-Land Rover	1		

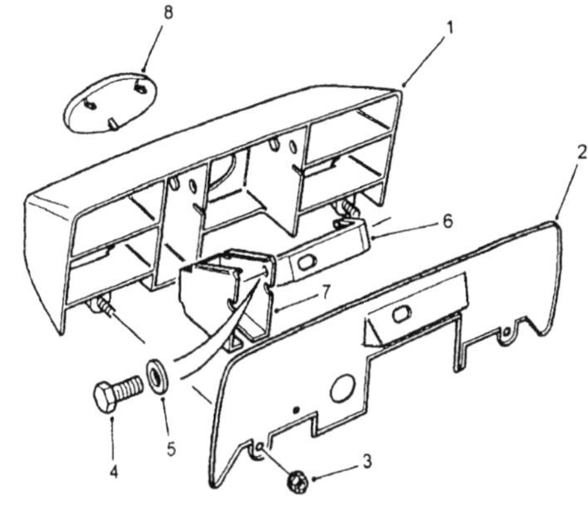

CANXHI1A

N 86

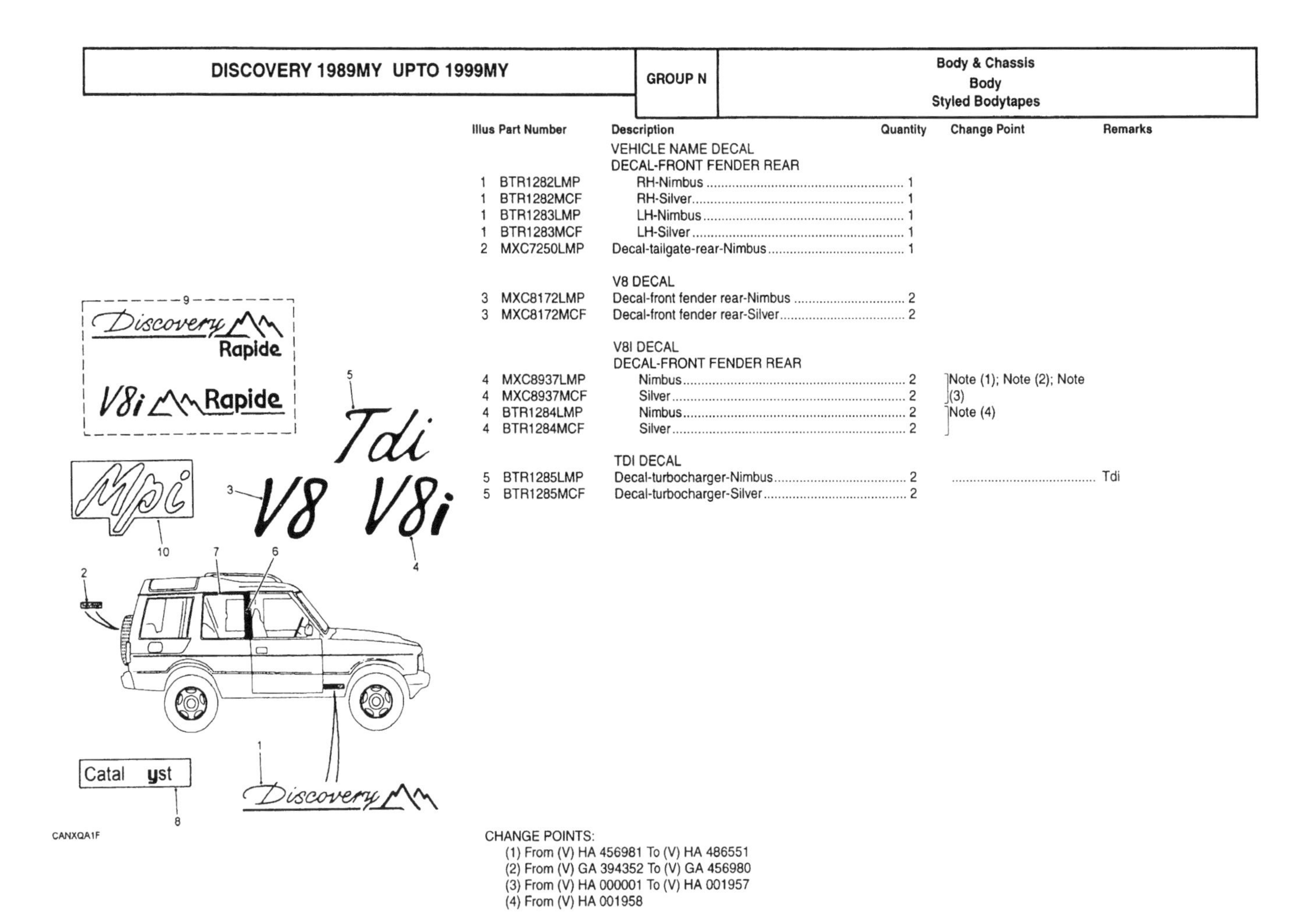

Illus	Part Number	Description	Quantity	Change Point	Remarks
		VEHICLE NAME DECAL			
		DECAL-FRONT FENDER REAR			
1	BTR1282LMP	RH-Nimbus	1		
1	BTR1282MCF	RH-Silver	1		
1	BTR1283LMP	LH-Nimbus	1		
1	BTR1283MCF	LH-Silver	1		
2	MXC7250LMP	Decal-tailgate-rear-Nimbus	1		
		V8 DECAL			
3	MXC8172LMP	Decal-front fender rear-Nimbus	2		
3	MXC8172MCF	Decal-front fender rear-Silver	2		
		V8I DECAL			
		DECAL-FRONT FENDER REAR			
4	MXC8937LMP	Nimbus	2	Note (1); Note (2); Note (3)	
4	MXC8937MCF	Silver	2		
4	BTR1284LMP	Nimbus	2	Note (4)	
4	BTR1284MCF	Silver	2		
		TDI DECAL			
5	BTR1285LMP	Decal-turbocharger-Nimbus	2		Tdi
5	BTR1285MCF	Decal-turbocharger-Silver	2		

CHANGE POINTS:
(1) From (V) HA 456981 To (V) HA 486551
(2) From (V) GA 394352 To (V) GA 456980
(3) From (V) HA 000001 To (V) HA 001957
(4) From (V) HA 001958

N 87

Illus	Part Number	Description	Quantity	Change Point	Remarks
6	MXC8366	Decal-B Post blackout-RH	1		
6	MXC8367	Decal-B Post blackout-LH	1		
7	MXC8368	Decal-door blackout-rear-upper-RH	1		
7	MXC8369	Decal-door blackout-rear-upper-LH	1		
		CATALYST DECAL			
8	MXC8938LMP	Decal catalyst-Nimbus	2		
8	MXC8938MCF	Decal catalyst-Silver	2		
		ITALY ONLY			
9	STC622	Kit decal	1		
		2.0L MPI ONLY			
10	BTR7815MCF	Decal-front fender rear-Silver	1		
10	BTR7815LMP	Decal-front fender rear-Grey	1		

NOTE: STC622 ONLY AVAILABLE THROUGH ROVER ITALY

Discovery
Rapide
V8i Rapide
Tdi
Mpi
V8 V8i
Catalyst
CANXQA1F

N 88

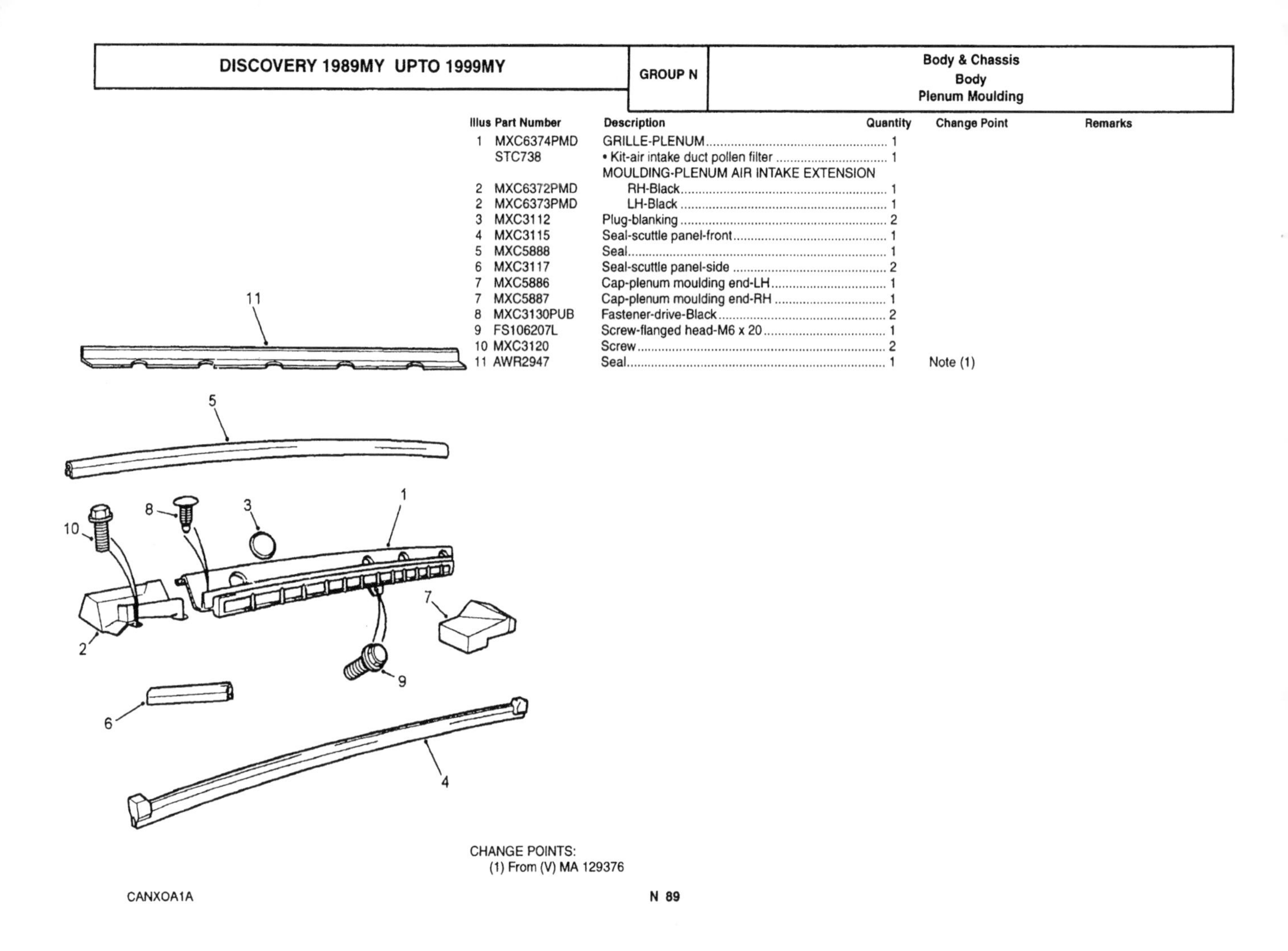

Illus	Part Number	Description	Quantity	Change Point	Remarks
1	MXC6374PMD	GRILLE-PLENUM	1		
	STC738	• Kit-air intake duct pollen filter	1		
		MOULDING-PLENUM AIR INTAKE EXTENSION			
2	MXC6372PMD	RH-Black	1		
2	MXC6373PMD	LH-Black	1		
3	MXC3112	Plug-blanking	2		
4	MXC3115	Seal-scuttle panel-front	1		
5	MXC5888	Seal	1		
6	MXC3117	Seal-scuttle panel-side	2		
7	MXC5886	Cap-plenum moulding end-LH	1		
7	MXC5887	Cap-plenum moulding end-RH	1		
8	MXC3130PUB	Fastener-drive-Black	2		
9	FS106207L	Screw-flanged head-M6 x 20	1		
10	MXC3120	Screw	2		
11	AWR2947	Seal	1	Note (1)	

CHANGE POINTS:
(1) From (V) MA 129376

CANXOA1A

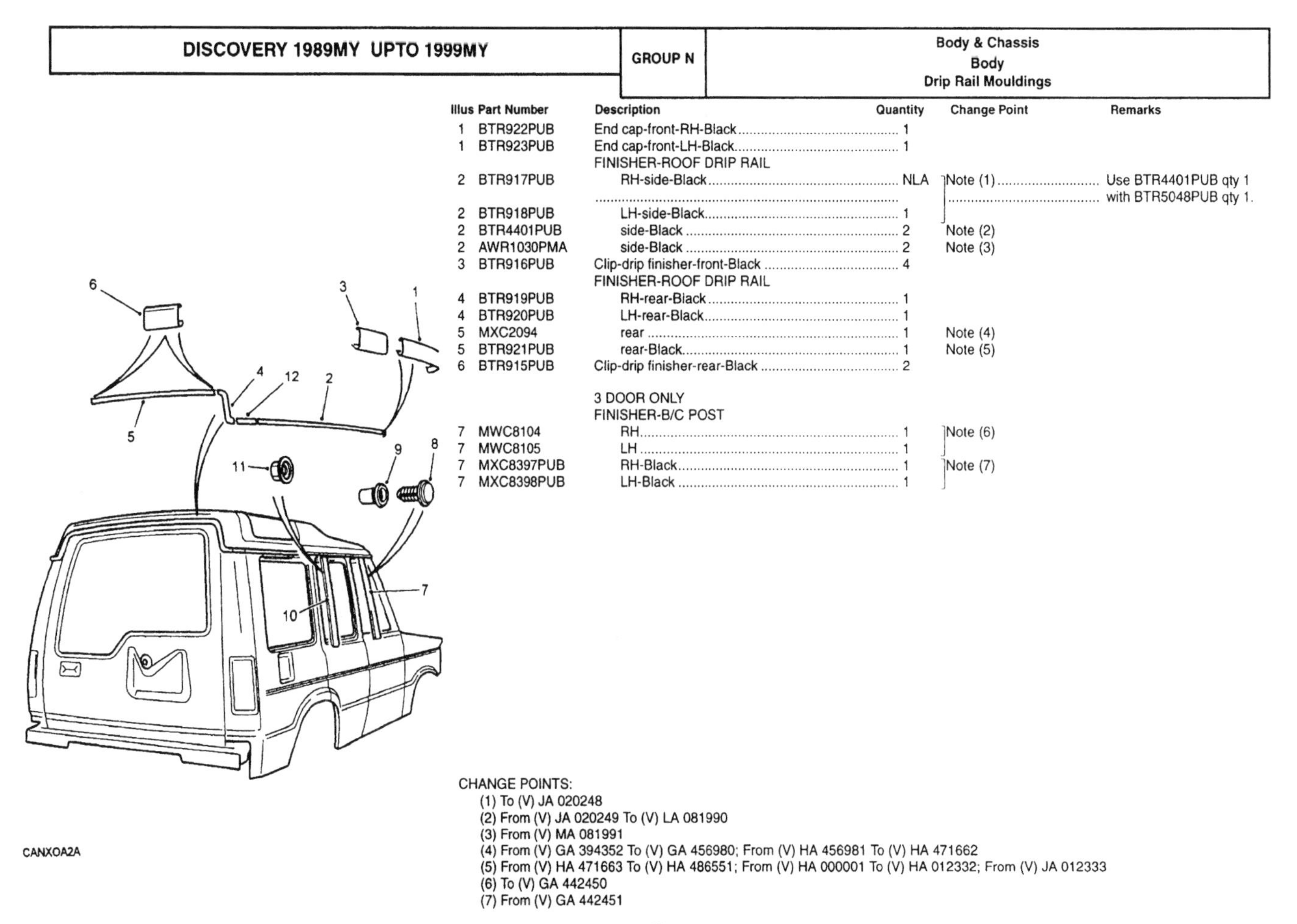

Illus	Part Number	Description	Quantity	Change Point	Remarks
1	BTR922PUB	End cap-front-RH-Black	1		
1	BTR923PUB	End cap-front-LH-Black	1		
		FINISHER-ROOF DRIP RAIL			
2	BTR917PUB	RH-side-Black	NLA	Note (1)	Use BTR4401PUB qty 1 with BTR5048PUB qty 1.
2	BTR918PUB	LH-side-Black	1		
2	BTR4401PUB	side-Black	2	Note (2)	
2	AWR1030PMA	side-Black	2	Note (3)	
3	BTR916PUB	Clip-drip finisher-front-Black	4		
		FINISHER-ROOF DRIP RAIL			
4	BTR919PUB	RH-rear-Black	1		
4	BTR920PUB	LH-rear-Black	1		
5	MXC2094	rear	1	Note (4)	
5	BTR921PUB	rear-Black	1	Note (5)	
6	BTR915PUB	Clip-drip finisher-rear-Black	2		
		3 DOOR ONLY			
		FINISHER-B/C POST			
7	MWC8104	RH	1	Note (6)	
7	MWC8105	LH	1		
7	MXC8397PUB	RH-Black	1	Note (7)	
7	MXC8398PUB	LH-Black	1		

CHANGE POINTS:
(1) To (V) JA 020248
(2) From (V) JA 020249 To (V) LA 081990
(3) From (V) MA 081991
(4) From (V) GA 394352 To (V) GA 456980; From (V) HA 456981 To (V) HA 471662
(5) From (V) HA 471663 To (V) HA 486551; From (V) HA 000001 To (V) HA 012332; From (V) JA 012333
(6) To (V) GA 442450
(7) From (V) GA 442451

CANXOA2A

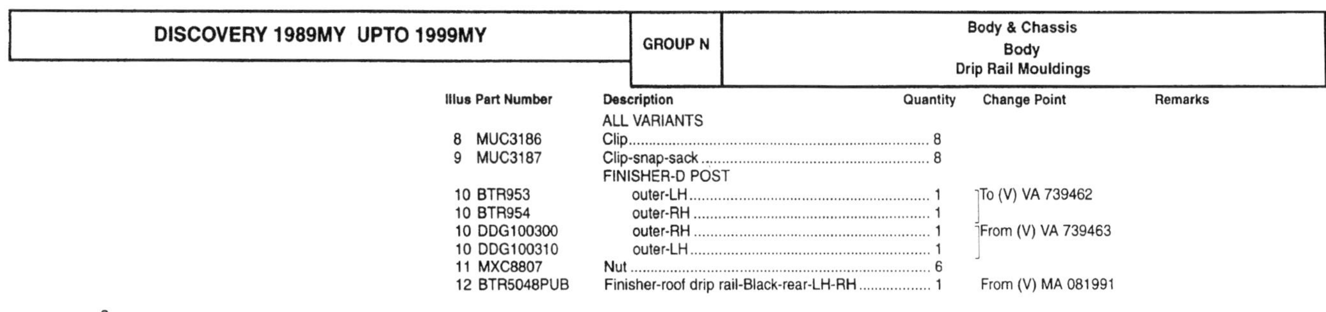

Illus	Part Number	Description	Quantity	Change Point	Remarks
		ALL VARIANTS			
8	MUC3186	Clip	8		
9	MUC3187	Clip-snap-sack	8		
		FINISHER-D POST			
10	BTR953	outer-LH	1	To (V) VA 739462	
10	BTR954	outer-RH	1		
10	DDG100300	outer-RH	1	From (V) VA 739463	
10	DDG100310	outer-LH	1		
11	MXC8807	Nut	6		
12	BTR5048PUB	Finisher-roof drip rail-Black-rear-LH-RH	1	From (V) MA 081991	

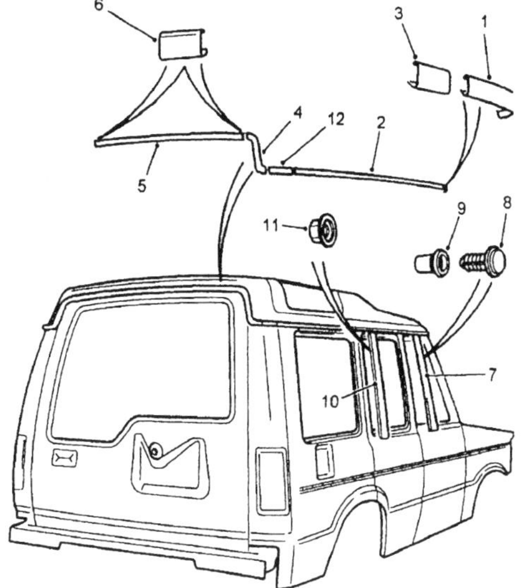

CANXOA2A

Illus	Part Number	Description	Quantity	Change Point	Remarks
1	AWR2098PMD	Rubbing strip assembly-front fender-RH-Black	1	Note (1)	
1	AWR2099PMD	Rubbing strip assembly-front fender-LH-Black	1	Note (1)	
2	AWR2100PMD	Rubbing strip assembly-front door-RH-Black	1	Note (1)	
2	AWR2101PMD	Rubbing strip assembly-front door-LH-Black	1	Note (1)	
3	AWR2102PMD	Rubbing strip-rear door-RH-Black	1	Note (1)	
3	AWR2103PMD	Rubbing strip-rear door-LH-Black	1	Note (1)	
4	AWR2104PMD	Rubbing strip-rear fender-RH-Black	1	Note (1)	
4	AWR2105PMD	Rubbing strip-rear fender-LH-Black	1	Note (1)	

FOR 3 DOOR XS USE ACCESSORIE KIT

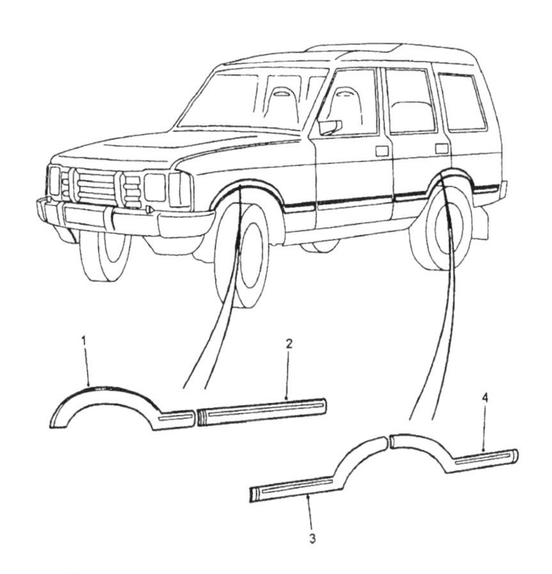

CANXQA1L

CHANGE POINTS:
(1) From (V) TA 163104; From (V) TA 501920

Illus	Part Number	Description	Quantity	Change Point	Remarks
1	BTR503PUB	Rubbing strip-front fender-RH-Black	1		
1	BTR504PUB	Rubbing strip-front fender-LH-Black	1		
2	BTR505PUB	Rubbing strip-front door-RH-Black	1		
2	BTR506PUB	Rubbing strip-front door-LH-Black	1		
3	BTR507PUB	Rubbing strip-rear door-RH-Black	1		
3	BTR508PUB	Rubbing strip-rear door-LH-Black	1		
4	BTR509PUB	Rubbing strip-rear fender-RH-Black	1		
4	BTR510PUB	Rubbing strip-rear fender-LH-Black	1		

NOTE (1) THE ABOVE PARTS ARE ONLY AVAILABLE IN
THE ITALIAN MARKET THEY CANNOT BE OBTAINED IN ANY
COUNTRY OTHER THAN ITALY

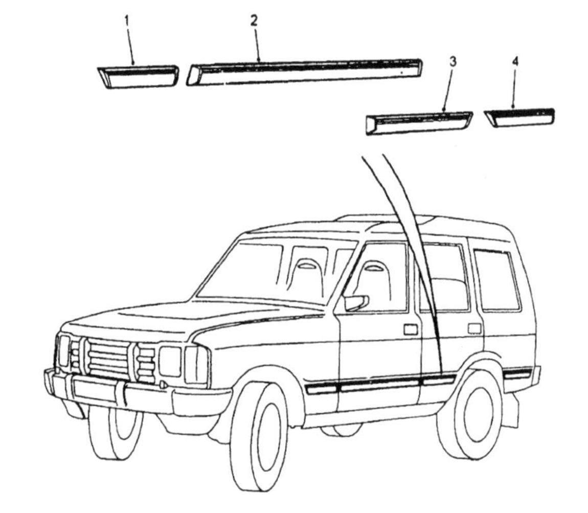

CANXOA3A

N 93

Illus	Part Number	Description	Quantity	Change Point	Remarks
1	BTR503LUZ	Rubbing strip-front fender-RH-with chrome inserts	1	Note (1)	
1	BTR504LUZ	Rubbing strip-front fender-LH-with chrome inserts	1		
2	BTR505LUZ	Rubbing strip-front door-RH-with chrome inserts	1		
2	BTR506LUZ	Rubbing strip-front door-LH-with chrome inserts	1		
3	BTR507LUZ	Rubbing strip-rear door-RH-with chrome inserts	1		
3	BTR508LUZ	Rubbing strip-rear door-LH-with chrome inserts	1		

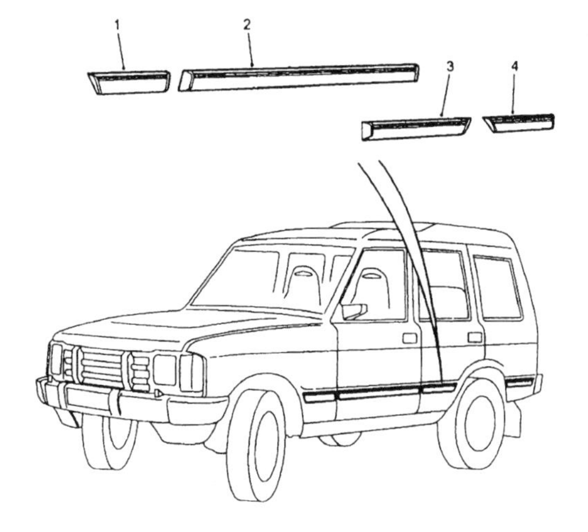

CANXOA4A

CHANGE POINTS:
(1) To (V) MA 163103; To (V) MA 501919

N 94

Illus	Part Number	Description	Quantity	Change Point	Remarks
		RUBBING STRIP-REAR QUARTER			
4	BTR509LUZ	RH-with chrome inserts	1	Note (1)	
4	BTR510LUZ	LH-with chrome inserts	1		
4	BTR8238LUZ	RH-with chrome inserts	1	Note (2); Note (3)	
4	BTR8237LUZ	LH-with chrome inserts	1		

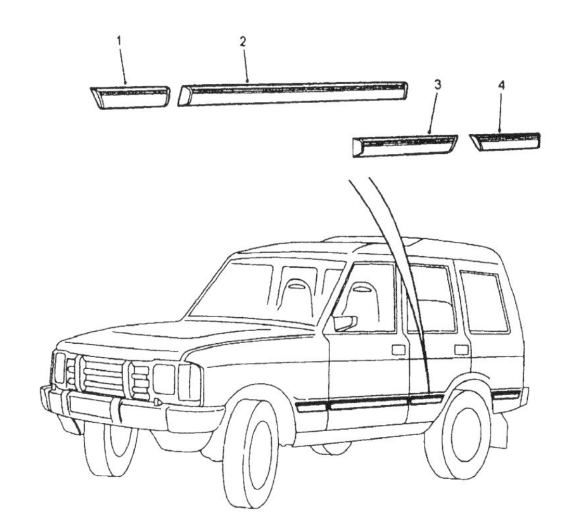

CANXOA4A

CHANGE POINTS:
(1) To (V) LA 081990
(2) From (V) MA 081991 To (V) MA 163103
(3) From (V) MA 081991 To (V) MA 501919

N 95

Illus	Part Number	Description	Quantity	Change Point	Remarks
		5 DOOR - UP TO MA163103 & MA501919			
1	BTR8680PMD	Rubbing strip-front fender-Black insert-RH	1	Note (1)	
1	BTR8681PMD	Rubbing strip-front fender-Black insert-LH	1		
2	BTR8682PMD	Rubbing strip-front door-Black insert-RH	1		
2	BTR8683PMD	Rubbing strip-front door-Black insert-LH	1		
3	BTR8684PMD	Rubbing strip-rear door-Black insert-RH	1		
3	BTR8685PMD	Rubbing strip-rear door-Black insert-LH	1		
		RUBBING STRIP-REAR QUARTER			
4	BTR8686PMD	Black insert-RH	1	Note (2)	
4	BTR8687PMD	Black insert-LH	1		
4	BTR8700PMD	Black insert-RH	1	Note (3)	
4	BTR8701PMD	Black insert-LH	1		

CANXOA4A

CHANGE POINTS:
(1) To (V) MA 163103; To (V) MA 501919
(2) To (V) LA 081990
(3) From (V) MA 081991 To (V) MA 163103; To (V) MA 501919

N 96

Illus	Part Number	Description	Quantity	Change Point	Remarks
1	AWR4334LUZ	Rubbing strip-front fender-RH-with chrome inserts	1	Note (1)	
1	AWR4335LUZ	Rubbing strip-front fender-LH-with chrome inserts	1		
2	AWR4336LUZ	Rubbing strip-front door-RH-with chrome inserts	1		
2	AWR4337LUZ	Rubbing strip-front door-LH-with chrome inserts	1		
3	AWR4338LUZ	Rubbing strip-rear door-RH-with chrome inserts	1		
3	AWR4339LUZ	Rubbing strip-rear door-LH-with chrome inserts	1		
4	AWR4340LUZ	Rubbing strip-rear quarter-RH-with chrome inserts	1		
4	AWR4341LUZ	Rubbing strip-rear quarter-LH-with chrome inserts	1		

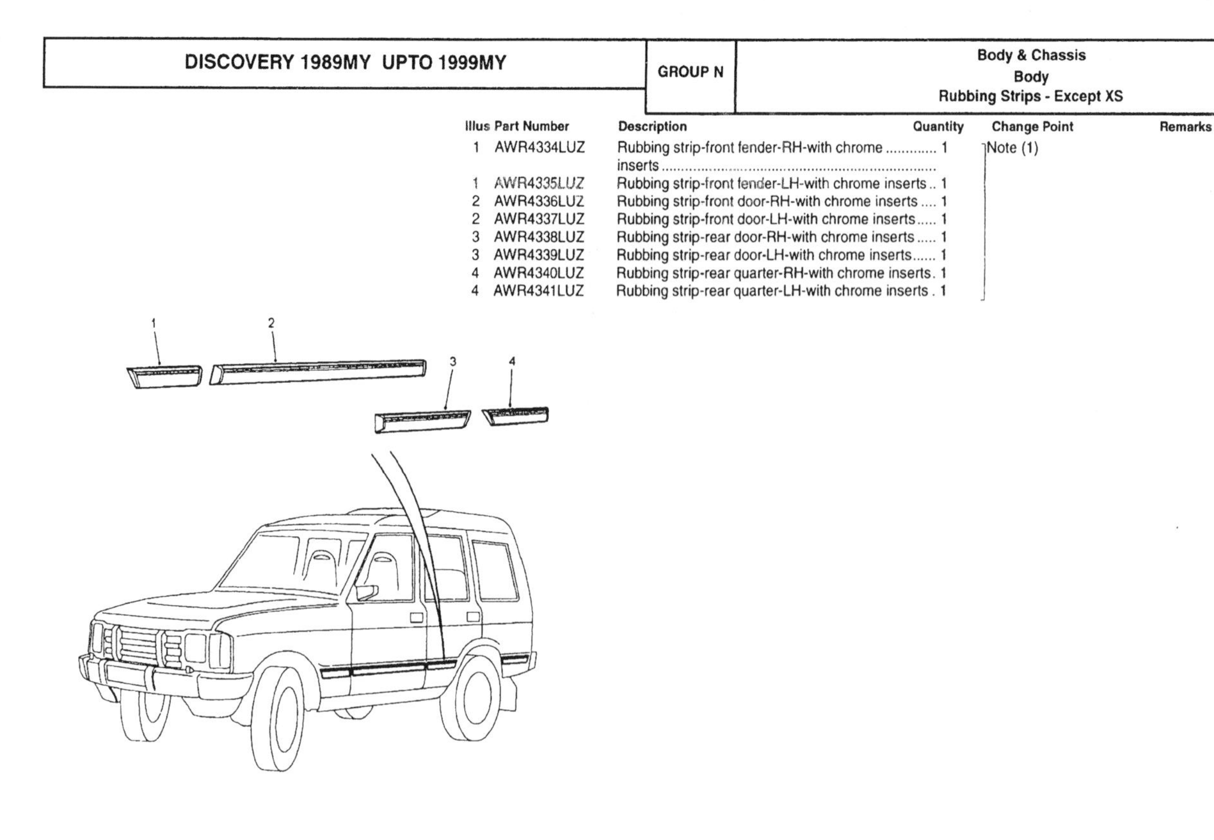

CHANGE POINTS:
(1) From (V) TA 163104; From (V) TA 501920

Illus	Part Number	Description	Quantity	Change Point	Remarks
		RUBBING STRIP-FRONT FENDER			
1	AWR4334PMD	Black insert-RH	1	Note (1)	
1	AWR4334PMD	Black insert-RH	1		
1	AWR4335PMD	Black insert-LH	1		
1	AWR4335PMD	Black insert-LH	1		
		RUBBING STRIP-FRONT DOOR			
2	AWR4336PMD	Black insert-RH	1		
2	AWR4336PMD	Black insert-RH	1		
2	AWR4337PMD	Black insert-LH	1		
2	AWR4337PMD	Black insert-LH	1		
		RUBBING STRIP-REAR DOOR			
3	AWR4338PMD	Black insert-RH	1		
3	AWR4338PMD	Black insert-RH	1		
3	AWR4339PMD	Black insert-LH	1		
3	AWR4339PMD	Black insert-LH	1		
		RUBBING STRIP-REAR QUARTER			
4	AWR4340PMD	Black insert-RH	1		
4	AWR4340PMD	Black insert-RH	1		
4	AWR4341PMD	Black insert-LH	1		
4	AWR4341PMD	Black insert-LH	1		

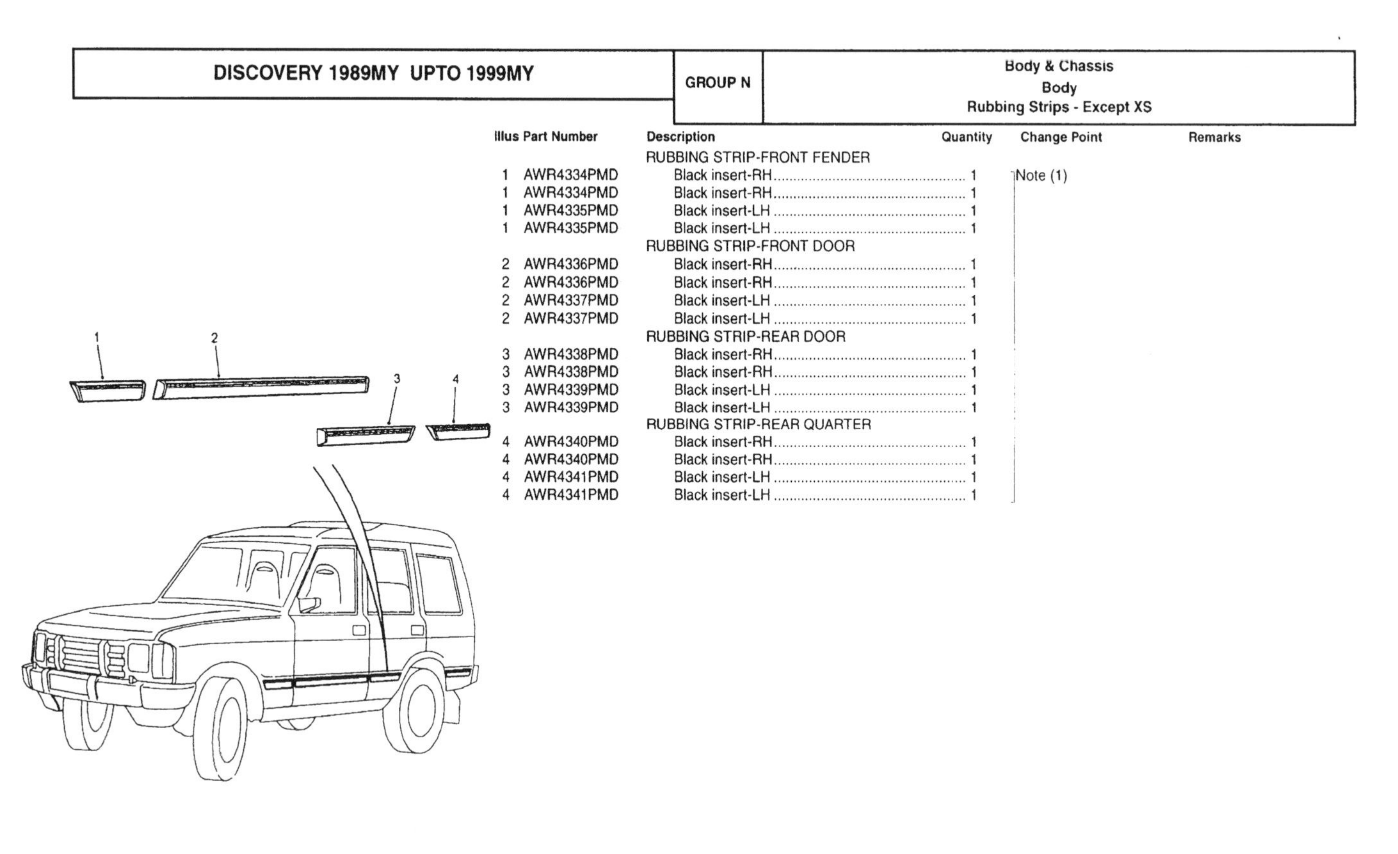

CHANGE POINTS:
(1) From (V) TA 163104; From (V) TA 501920

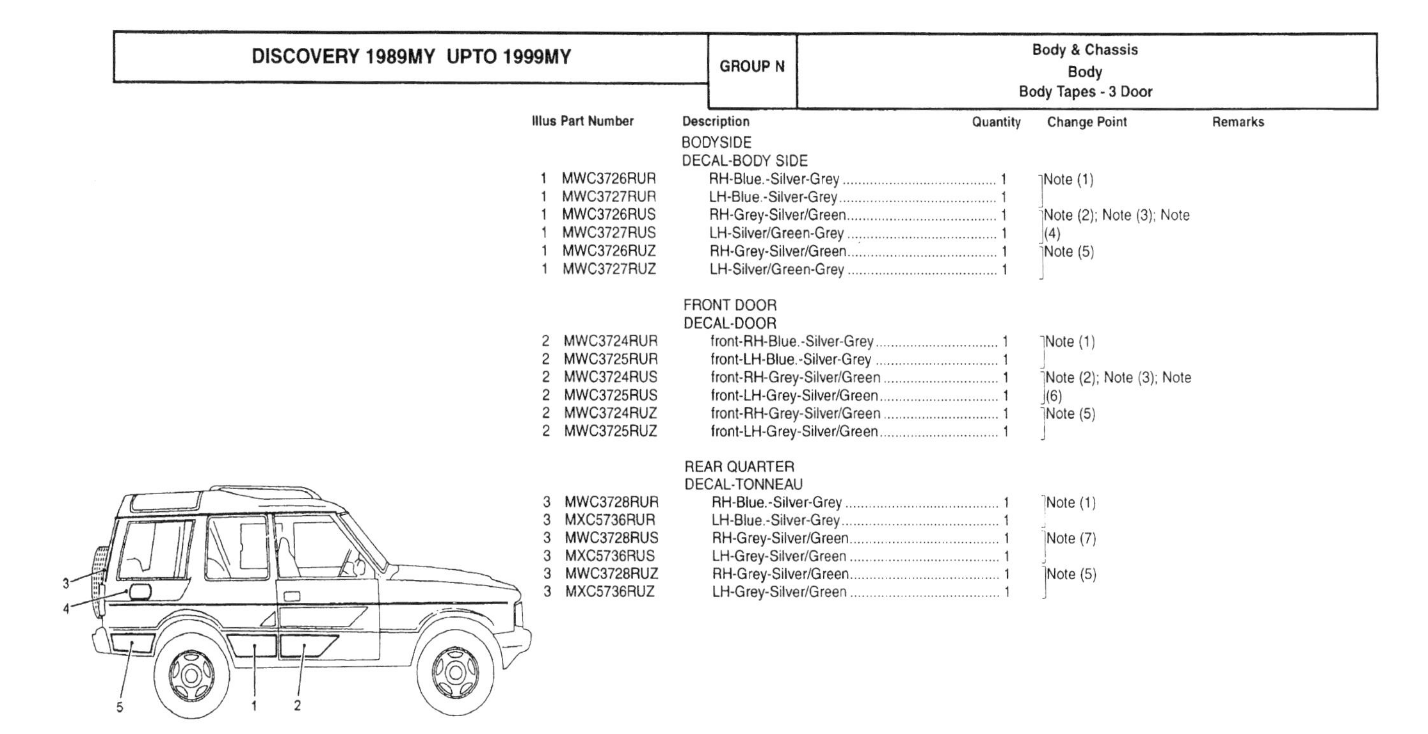

Illus	Part Number	Description	Quantity	Change Point	Remarks
		BODYSIDE			
		DECAL-BODY SIDE			
1	MWC3726RUR	RH-Blue.-Silver-Grey	1	Note (1)	
1	MWC3727RUR	LH-Blue.-Silver-Grey	1		
1	MWC3726RUS	RH-Grey-Silver/Green	1	Note (2); Note (3); Note (4)	
1	MWC3727RUS	LH-Silver/Green-Grey	1		
1	MWC3726RUZ	RH-Grey-Silver/Green	1	Note (5)	
1	MWC3727RUZ	LH-Silver/Green-Grey	1		
		FRONT DOOR			
		DECAL-DOOR			
2	MWC3724RUR	front-RH-Blue.-Silver-Grey	1	Note (1)	
2	MWC3725RUR	front-LH-Blue.-Silver-Grey	1		
2	MWC3724RUS	front-RH-Grey-Silver/Green	1	Note (2); Note (3); Note (6)	
2	MWC3725RUS	front-LH-Grey-Silver/Green	1		
2	MWC3724RUZ	front-RH-Grey-Silver/Green	1	Note (5)	
2	MWC3725RUZ	front-LH-Grey-Silver/Green	1		
		REAR QUARTER			
		DECAL-TONNEAU			
3	MWC3728RUR	RH-Blue.-Silver-Grey	1	Note (1)	
3	MXC5736RUR	LH-Blue.-Silver-Grey	1		
3	MWC3728RUS	RH-Grey-Silver/Green	1	Note (7)	
3	MXC5736RUS	LH-Grey-Silver/Green	1		
3	MWC3728RUZ	RH-Grey-Silver/Green	1	Note (5)	
3	MXC5736RUZ	LH-Grey-Silver/Green	1		

CANXQA1D

CHANGE POINTS:
(1) To (V) HA 012333
(2) From (V) HA 456981 To (V) HA 486551
(3) From (V) GA 394352 To (V) GA 456980
(4) From (V) HA 000001 To (V) HA 002104
(5) From (V) HA 002105
(6) From (V) HA 000001 To (V) HA 001957
(7) To (V) HA 002104

N 99

DISCOVERY 1989MY UPTO 1999MY | GROUP N | Body & Chassis
Body
Body Tapes - 3 Door

Illus	Part Number	Description	Quantity	Change Point	Remarks
		DECAL-FUEL FILLER DOOR			
4	MXC6748RUR	Blue.-Grey-Silver	1	Note (1)	
4	MXC6748RUS	Grey-Silver/Green	1	Note (2)	
4	MXC6748RUZ	Grey-Silver/Green	1	Note (3)	
		DECAL-BODY SIDE			
5	MXC5734RUR	RH-Blue.-Grey-Silver	1	Note (1)	
5	MXC5735RUR	LH-Blue.-Grey-Silver	1		
5	MXC5734RUS	RH-Grey-Silver/Green	1	Note (2)	
5	MXC5735RUS	LH-Grey-Silver/Green	1		
5	MXC5734RUZ	RH-Grey-Silver/Green	1	Note (3)	
5	MXC5735RUZ	LH-Grey-Silver/Green	1		

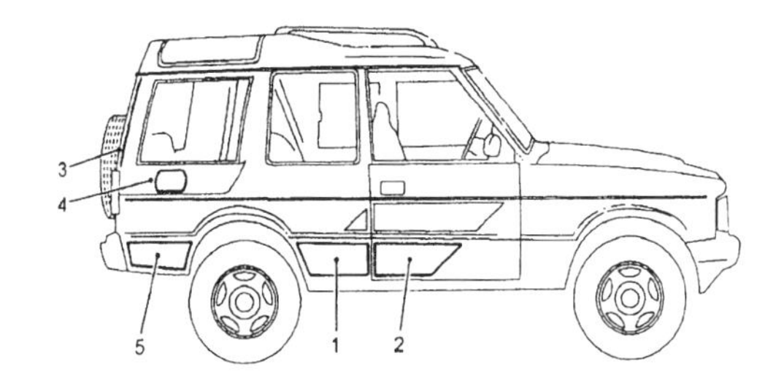

CANXQA1D

CHANGE POINTS:
(1) To (V) HA 012333
(2) To (V) HA 002104
(3) From (V) HA 002105

N 100

Illus	Part Number	Description	Quantity	Change Point	Remarks
		BODYSIDE			
1	BTR2078RWE	Decal-body side-RH-Grey-Silver/Green	1	Note (1)	
1	BTR2079RWE	Decal-body side-LH-Grey-Silver/Green	1		
		FRONT DOOR			
2	BTR2074RWE	Decal-door-RH-Grey-Silver/Green	1	Note (1)	
2	BTR2075RWE	Decal-door-LH-Grey-Silver/Green	1		
		BODYSIDE			
3	BTR2082RWE	Decal-body side-RH-Grey-Silver/Green	1	Note (1)	
3	BTR2083RWE	Decal-body side-LH-Grey-Silver/Green	1		

NOTE: UK VEHICLES MAY BE FITTED WITH PRE JA BODY TAPES

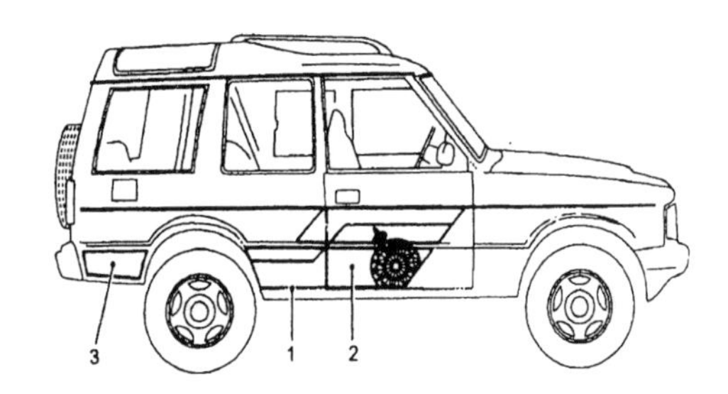

CANXQA1E

CHANGE POINTS:
(1) From (V) JA 012334

Illus	Part Number	Description	Quantity	Change Point	Remarks
1	STC4225RCH	DECAL-FRONT FENDER REAR-LH-BLUE	1		
2	BTR8998RCH	• Decal-front fender rear-LH-Blue.	1		
3	BTR9008RCJ	• Decal-front fender rear-LH-Silver	1		
1	STC4226RCH	DECAL-FRONT FENDER REAR-RH-BLUE	1		
2	BTR8999RCH	• Decal-front fender rear-RH-Blue	1		
3	BTR9009RCJ	• Decal-front fender rear-RH-Silver	1		
1	STC4225RCJ	DECAL-FRONT FENDER REAR-LH-SILVER	1		
2	BTR8998RCJ	• Decal-front fender rear-LH-Silver	1		
3	BTR9008RCH	• Decal-front fender rear-LH-Blue.	1		
1	STC4226RCJ	DECAL-FRONT FENDER REAR-RH-SILVER	1		
2	BTR8999RCJ	• Decal-front fender rear-RH-Silver	1		
3	BTR9009RCH	• Decal-front fender rear-RH-Blue	1		
		DECAL-DOOR			
4	BTR9002RCH	LH-rear-Blue.	1		
4	BTR9003RCH	RH-rear-Blue.	1		
4	BTR9002RCJ	LH-rear-Silver	1		
4	BTR9003RCJ	RH-rear-Silver	1		
		TAPE REAR QUARTER			
5	BTR9004RCH	LH-Blue.	1		
5	BTR9005RCH	RH-Blue.	1		
5	BTR9004RCJ	LH-Silver	1		
5	BTR9005RCJ	RH-Silver	1		
		DECAL-DOOR			
6	BTR9000RCH	LH-front-Blue.	1		
6	BTR9001RCH	RH-front-Blue.	1		
6	BTR9000RCJ	LH-front-Silver	1		
6	BTR9001RCJ	RH-front-Silver	1		
		DECAL-DOOR			
7	BTR9006RCH	LH-rear-Blue.	1		
7	BTR9007RCH	RH-rear-Blue.	1		
7	BTR9006RCJ	LH-rear-Silver	1		
7	BTR9007RCJ	RH-rear-Silver	1		

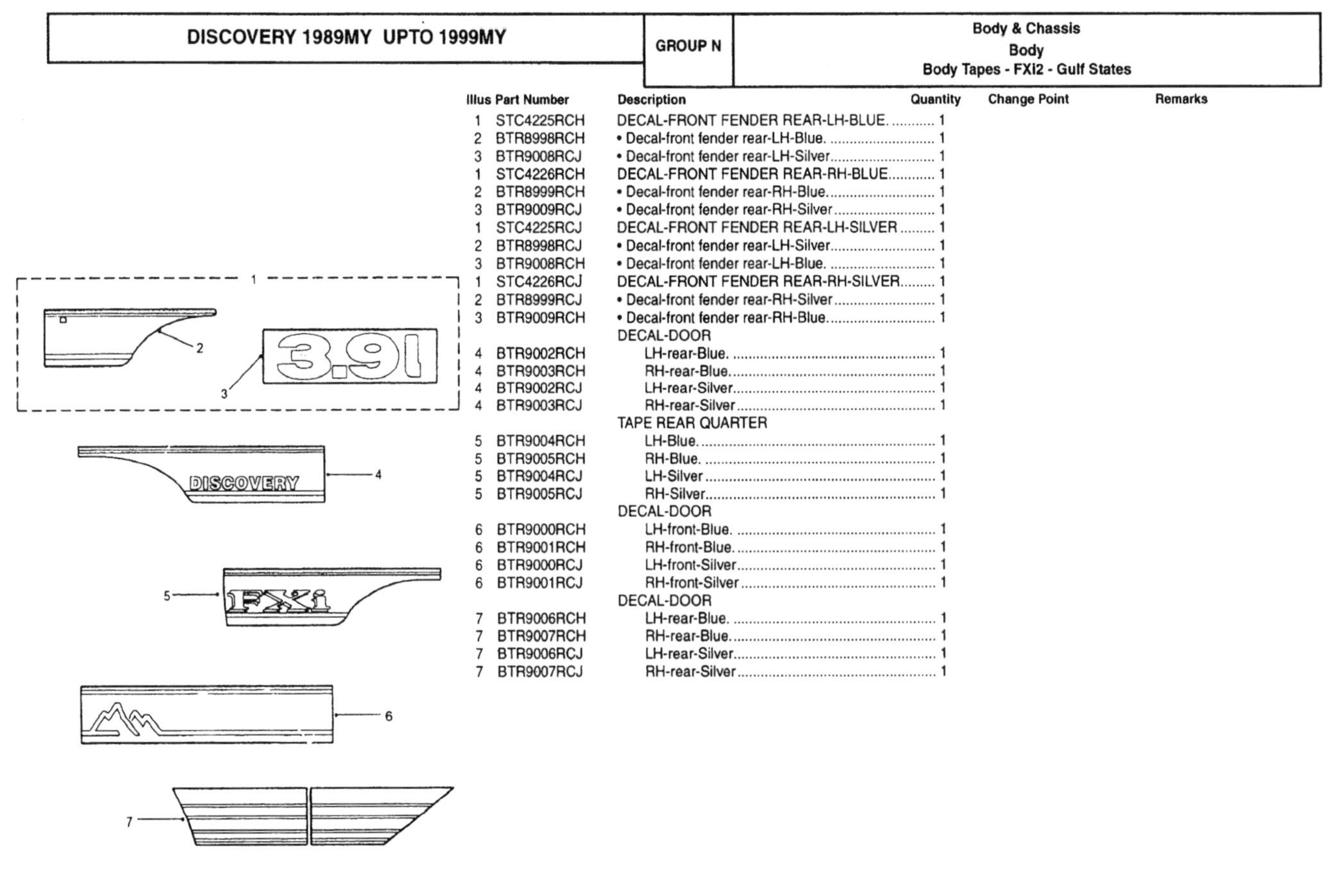

LN0013

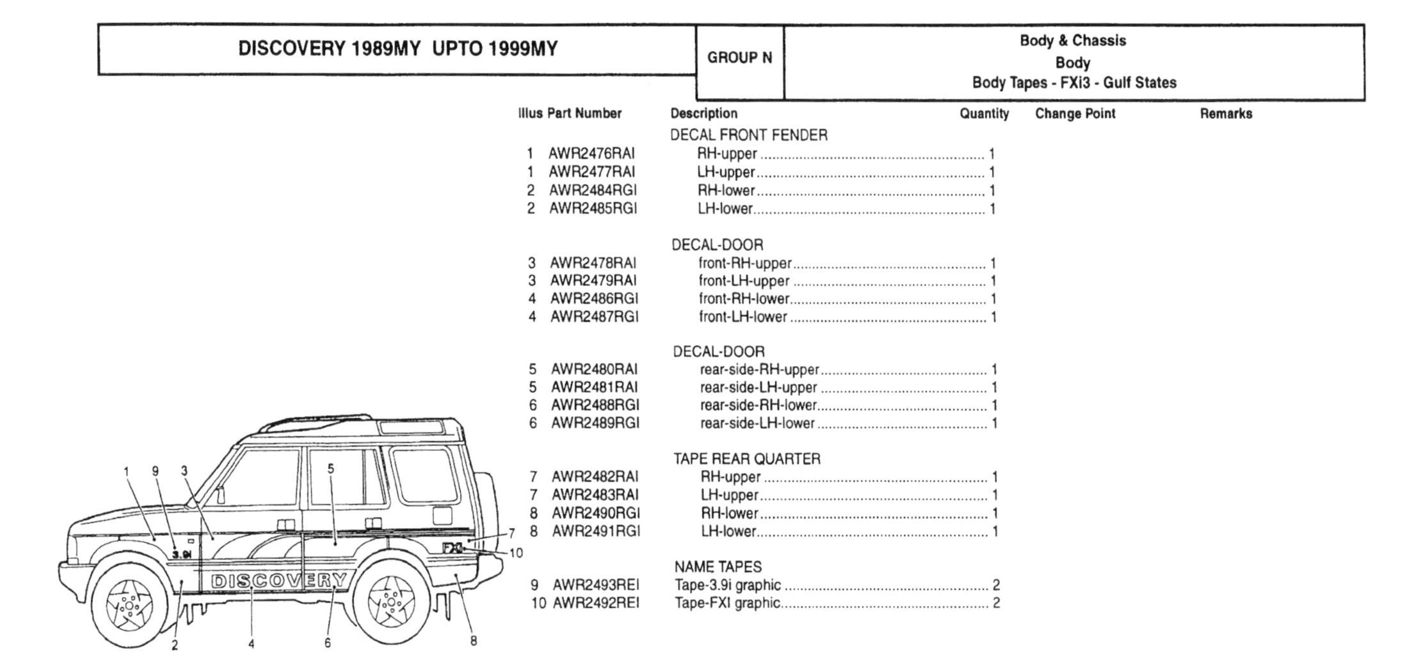

Illus	Part Number	Description	Quantity	Change Point	Remarks
		DECAL FRONT FENDER			
1	AWR2476RAI	RH-upper	1		
1	AWR2477RAI	LH-upper	1		
2	AWR2484RGI	RH-lower	1		
2	AWR2485RGI	LH-lower	1		
		DECAL-DOOR			
3	AWR2478RAI	front-RH-upper	1		
3	AWR2479RAI	front-LH-upper	1		
4	AWR2486RGI	front-RH-lower	1		
4	AWR2487RGI	front-LH-lower	1		
		DECAL-DOOR			
5	AWR2480RAI	rear-side-RH-upper	1		
5	AWR2481RAI	rear-side-LH-upper	1		
6	AWR2488RGI	rear-side-RH-lower	1		
6	AWR2489RGI	rear-side-LH-lower	1		
		TAPE REAR QUARTER			
7	AWR2482RAI	RH-upper	1		
7	AWR2483RAI	LH-upper	1		
8	AWR2490RGI	RH-lower	1		
8	AWR2491RGI	LH-lower	1		
		NAME TAPES			
9	AWR2493REI	Tape-3.9i graphic	2		
10	AWR2492REI	Tape-FXI graphic	2		

CANXQA1J

Illus	Part Number	Description	Quantity	Change Point	Remarks
		FRONT WING			
1	AWR2484RGI	Decal front fender-RH-lower	1		
1	AWR2485RGI	Decal front fender-LH-lower	1		
		FRONT DOOR			
2	AWR2486RGI	Decal-door-front-RH-lower	1		
2	AWR2487RGI	Decal-door-front-LH-lower	1		
		REAR DOOR			
3	AWR2488RGI	Decal-door-rear-side-RH-lower	1		
3	AWR2489RGI	Decal-door-rear-side-LH-lower	1		
		REAR QUARTER			
4	AWR2490RGI	Tape rear quarter-RH-lower	1		
4	AWR2491RGI	Tape rear quarter-LH-lower	1		
		NAME TAPES			
5	AWR2494REI	Tape-XS graphic	2		
6	AWR2493REI	Tape-3.9i graphic	2		
7	AWR2492REI	Tape-FXI graphic	2		

CANXQA1K

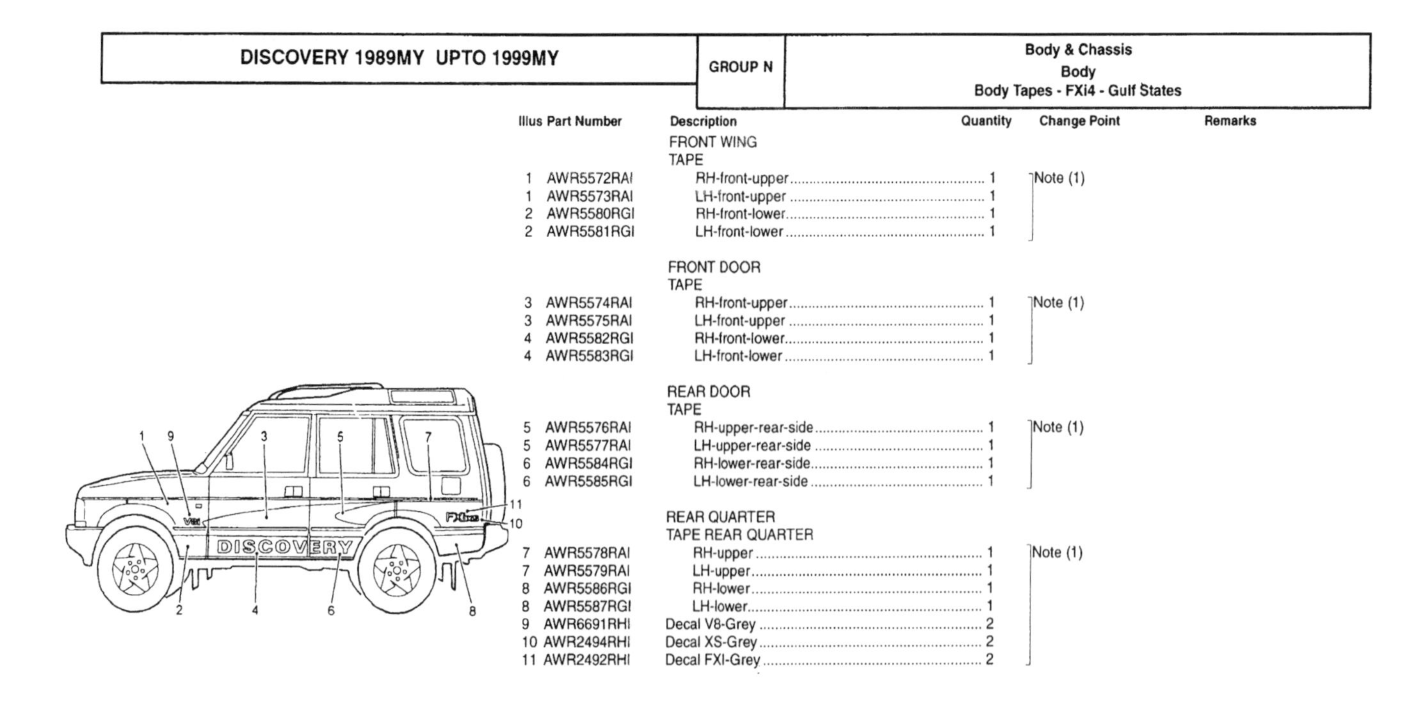

Illus	Part Number	Description	Quantity	Change Point	Remarks
		FRONT WING			
		TAPE			
1	AWR5572RAI	RH-front-upper	1	Note (1)	
1	AWR5573RAI	LH-front-upper	1		
2	AWR5580RGI	RH-front-lower	1		
2	AWR5581RGI	LH-front-lower	1		
		FRONT DOOR			
		TAPE			
3	AWR5574RAI	RH-front-upper	1	Note (1)	
3	AWR5575RAI	LH-front-upper	1		
4	AWR5582RGI	RH-front-lower	1		
4	AWR5583RGI	LH-front-lower	1		
		REAR DOOR			
		TAPE			
5	AWR5576RAI	RH-upper-rear-side	1	Note (1)	
5	AWR5577RAI	LH-upper-rear-side	1		
6	AWR5584RGI	RH-lower-rear-side	1		
6	AWR5585RGI	LH-lower-rear-side	1		
		REAR QUARTER			
		TAPE REAR QUARTER			
7	AWR5578RAI	RH-upper	1	Note (1)	
7	AWR5579RAI	LH-upper	1		
8	AWR5586RGI	RH-lower	1		
8	AWR5587RGI	LH-lower	1		
9	AWR6691RHI	Decal V8-Grey	2		
10	AWR2494RHI	Decal XS-Grey	2		
11	AWR2492RHI	Decal FXI-Grey	2		

CANXQA10

CHANGE POINTS:
(1) From (V) VA 703237; From (V) VA 534104

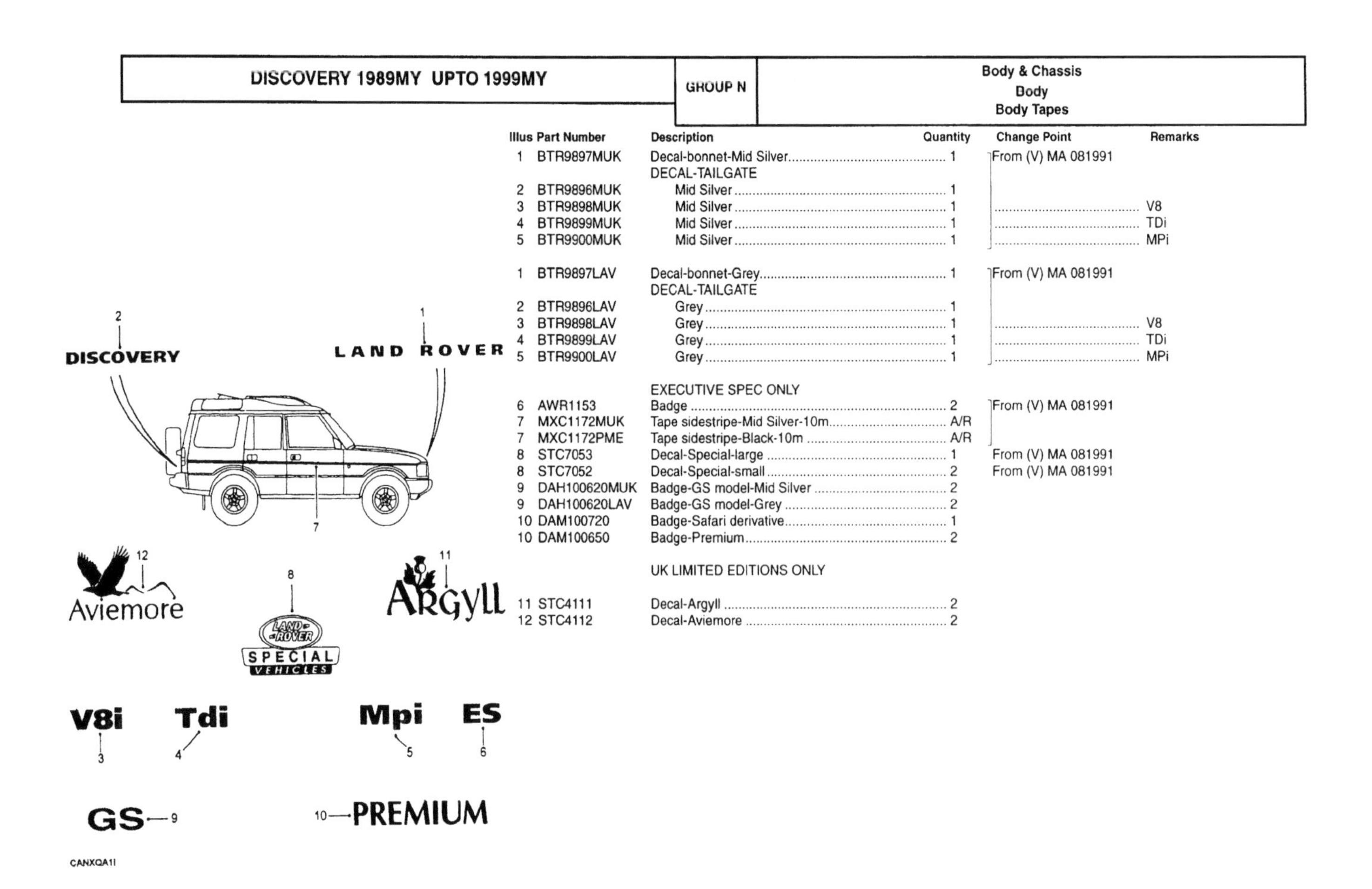

Illus	Part Number	Description	Quantity	Change Point	Remarks
1	BTR9897MUK	Decal-bonnet-Mid Silver	1	From (V) MA 081991	
		DECAL-TAILGATE			
2	BTR9896MUK	Mid Silver	1		
3	BTR9898MUK	Mid Silver	1		V8
4	BTR9899MUK	Mid Silver	1		TDi
5	BTR9900MUK	Mid Silver	1		MPi
1	BTR9897LAV	Decal-bonnet-Grey	1	From (V) MA 081991	
		DECAL-TAILGATE			
2	BTR9896LAV	Grey	1		
3	BTR9898LAV	Grey	1		V8
4	BTR9899LAV	Grey	1		TDi
5	BTR9900LAV	Grey	1		MPi
		EXECUTIVE SPEC ONLY			
6	AWR1153	Badge	2	From (V) MA 081991	
7	MXC1172MUK	Tape sidestripe-Mid Silver-10m	A/R		
7	MXC1172PME	Tape sidestripe-Black-10m	A/R		
8	STC7053	Decal-Special-large	1	From (V) MA 081991	
8	STC7052	Decal-Special-small	2	From (V) MA 081991	
9	DAH100620MUK	Badge-GS model-Mid Silver	2		
9	DAH100620LAV	Badge-GS model-Grey	2		
10	DAM100720	Badge-Safari derivative	1		
10	DAM100650	Badge-Premium	2		
		UK LIMITED EDITIONS ONLY			
11	STC4111	Decal-Argyll	2		
12	STC4112	Decal-Aviemore	2		

CANXQA11

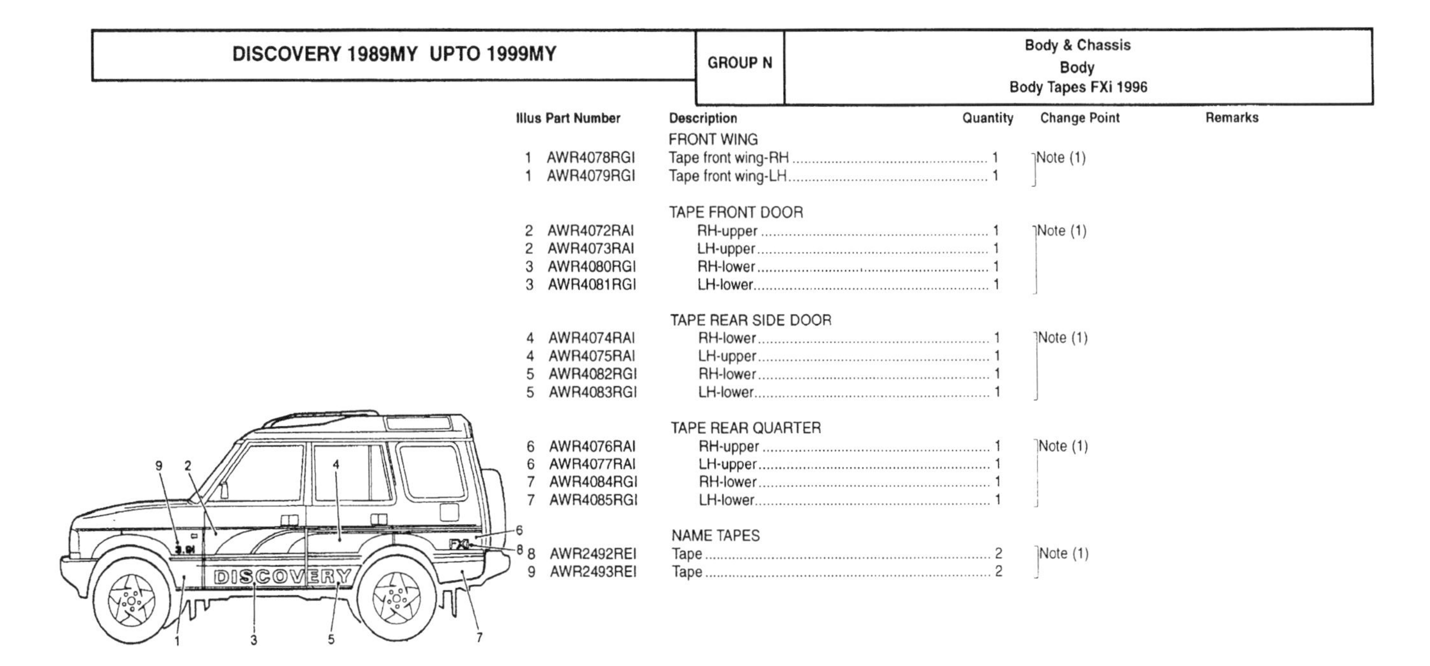

Illus	Part Number	Description	Quantity	Change Point	Remarks
		FRONT WING			
1	AWR4078RGI	Tape front wing-RH	1	Note (1)	
1	AWR4079RGI	Tape front wing-LH	1		
		TAPE FRONT DOOR			
2	AWR4072RAI	RH-upper	1	Note (1)	
2	AWR4073RAI	LH-upper	1		
3	AWR4080RGI	RH-lower	1		
3	AWR4081RGI	LH-lower	1		
		TAPE REAR SIDE DOOR			
4	AWR4074RAI	RH-lower	1	Note (1)	
4	AWR4075RAI	LH-upper	1		
5	AWR4082RGI	RH-lower	1		
5	AWR4083RGI	LH-lower	1		
		TAPE REAR QUARTER			
6	AWR4076RAI	RH-upper	1	Note (1)	
6	AWR4077RAI	LH-upper	1		
7	AWR4084RGI	RH-lower	1		
7	AWR4085RGI	LH-lower	1		
		NAME TAPES			
8	AWR2492REI	Tape	2	Note (1)	
9	AWR2493REI	Tape	2		

CANXQA1N

CHANGE POINTS:
(1) From (V) TA 163104; From (V) TA 501920

Illus	Part Number	Description	Quantity	Change Point	Remarks
		FRONT WING			
1	AWR2800RCI	Tape front wing-RH-Dark Grey	1	Note (1)	
1	AWR2801RCI	Tape front wing-LH-Dark Grey	1		
		FRONT DOOR			
2	AWR2802RCI	Tape front door-RH-Dark Grey	1	Note (1)	
2	AWR2803RCI	Tape front door-LH-Dark Grey	1		
		REAR DOOR			
3	AWR2804RCI	Tape rear side door-RH-Dark Grey	1	Note (1)	
3	AWR2805RCI	Tape rear side door-LH-Dark Grey	1		
		REAR DOOR			
4	AWR2806RCI	Tape rear quarter-RH-Dark Grey	1	Note (1)	
4	AWR2807RCI	Tape rear quarter-LH-Dark Grey	1		

CANXQA1M

CHANGE POINTS:
(1) From (V) TA 163104; From (V) TA 501920

Illus	Part Number	Description	Quantity	Change Point	Remarks
		3 DOOR ONLY			
1	STC8140	Kit-wheelarch flare	1		
		5 DOOR ONLY			
1	STC8139	Kit-wheelarch flare	1		

CANXQA1G

Illus	Part Number	Description	Quantity	Change Point	Remarks
1	RTC6820	Mudflaps-kit front	1		
2	392678	Bracket-front-RH	1		
2	392679	Bracket-front-LH	1		
3	MXC6094	Bracket	NSS		
4	SH106121L	Screw-hexagonal head-M6 x 12	3		
5	WA106041L	Washer-M6-standard.	3		
6	WL106001L	Washer-M6	3		
7	NH106041L	Nut-hexagonal head-nyloc-M6	3		
8	SE106161L	Screw-M6 x 16	6		
9	WA106041L	Washer-M6-standard.	6		
10	WL106001L	Washer-M6	6		
11	NH106041L	Nut-hexagonal head-nyloc-M6	6		

NOTE: FOR PART MXC6094 USE MUDFLAP ASSEMBLY
IF BRACKET REQUIRED

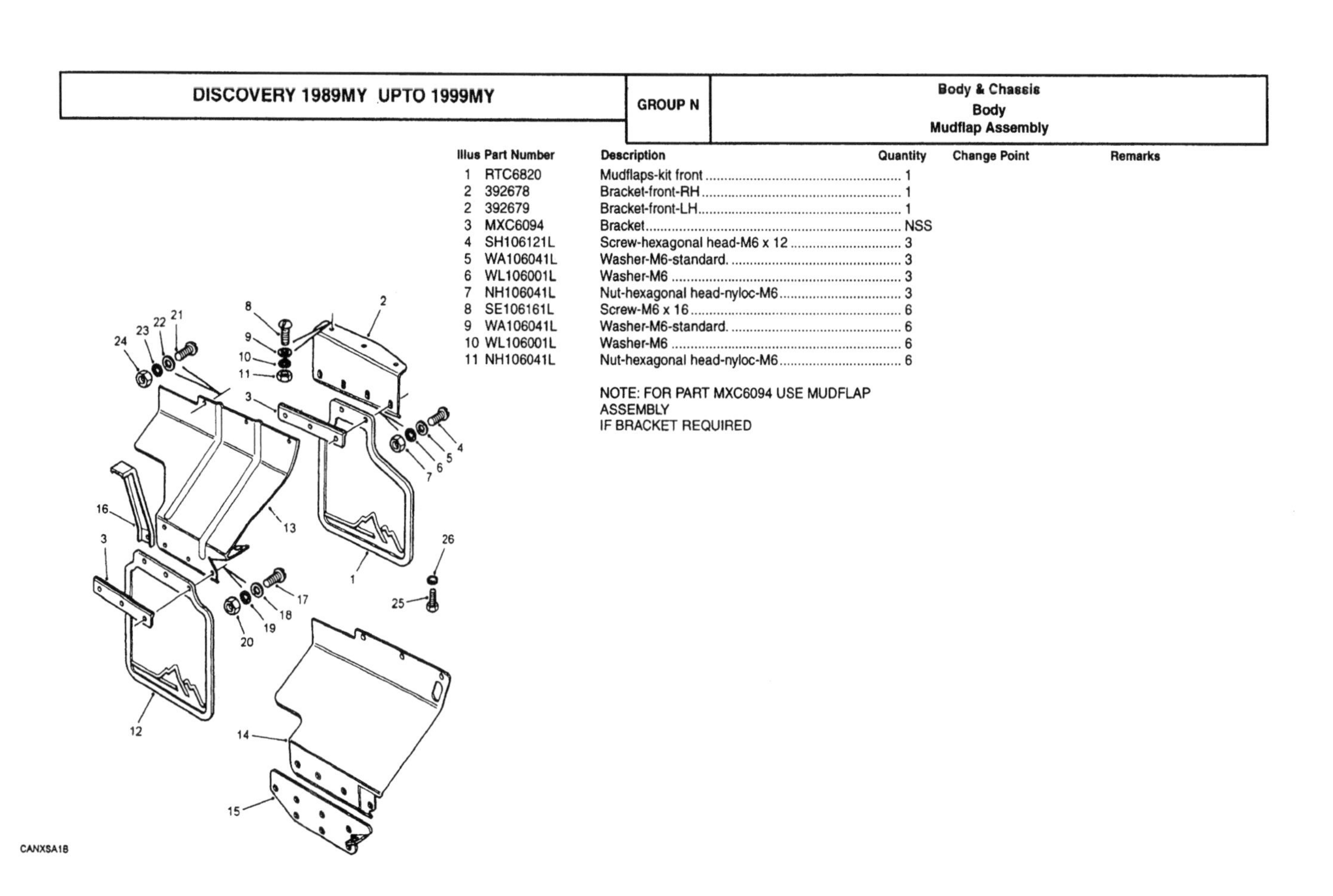

CANXSA1B

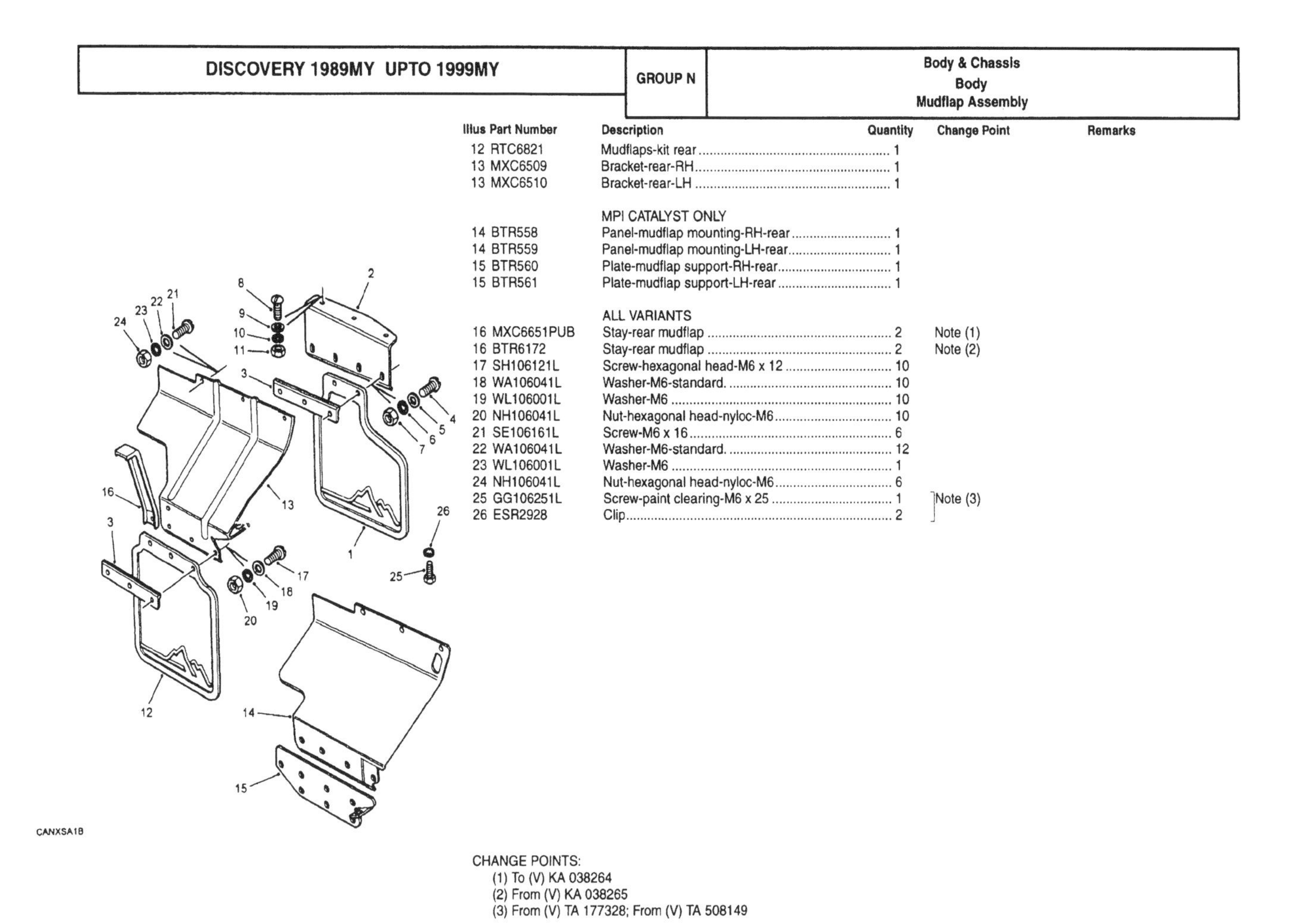

Illus	Part Number	Description	Quantity	Change Point	Remarks
12	RTC6821	Mudflaps-kit rear	1		
13	MXC6509	Bracket-rear-RH	1		
13	MXC6510	Bracket-rear-LH	1		
		MPI CATALYST ONLY			
14	BTR558	Panel-mudflap mounting-RH-rear	1		
14	BTR559	Panel-mudflap mounting-LH-rear	1		
15	BTR560	Plate-mudflap support-RH-rear	1		
15	BTR561	Plate-mudflap support-LH-rear	1		
		ALL VARIANTS			
16	MXC6651PUB	Stay-rear mudflap	2	Note (1)	
16	BTR6172	Stay-rear mudflap	2	Note (2)	
17	SH106121L	Screw-hexagonal head-M6 x 12	10		
18	WA106041L	Washer-M6-standard.	10		
19	WL106001L	Washer-M6	10		
20	NH106041L	Nut-hexagonal head-nyloc-M6	10		
21	SE106161L	Screw-M6 x 16	6		
22	WA106041L	Washer-M6-standard.	12		
23	WL106001L	Washer-M6	1		
24	NH106041L	Nut-hexagonal head-nyloc-M6	6		
25	GG106251L	Screw-paint clearing-M6 x 25	1	Note (3)	
26	ESR2928	Clip	2		

CANXSA1B

CHANGE POINTS:
(1) To (V) KA 038264
(2) From (V) KA 038265
(3) From (V) TA 177328; From (V) TA 508149

N 111

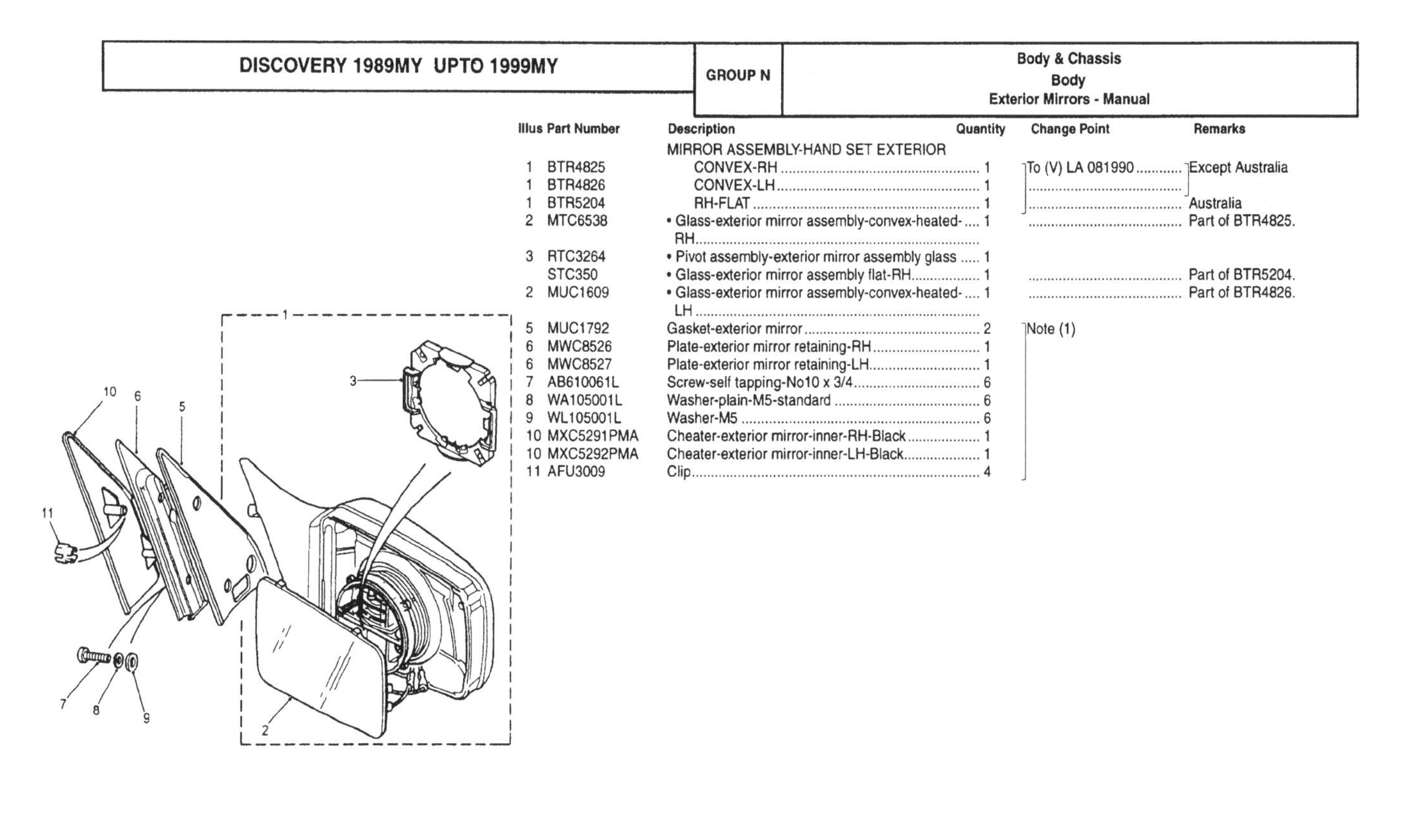

Illus	Part Number	Description	Quantity	Change Point	Remarks
		MIRROR ASSEMBLY-HAND SET EXTERIOR			
1	BTR4825	CONVEX-RH	1	To (V) LA 081990	Except Australia
1	BTR4826	CONVEX-LH	1		
1	BTR5204	RH-FLAT	1		Australia
2	MTC6538	• Glass-exterior mirror assembly-convex-heated-RH	1		Part of BTR4825.
3	RTC3264	• Pivot assembly-exterior mirror assembly glass	1		
	STC350	• Glass-exterior mirror assembly flat-RH	1		Part of BTR5204.
2	MUC1609	• Glass-exterior mirror assembly-convex-heated-LH	1		Part of BTR4826.
5	MUC1792	Gasket-exterior mirror	2	Note (1)	
6	MWC8526	Plate-exterior mirror retaining-RH	1		
6	MWC8527	Plate-exterior mirror retaining-LH	1		
7	AB610061L	Screw-self tapping-No10 x 3/4	6		
8	WA105001L	Washer-plain-M5-standard	6		
9	WL105001L	Washer-M5	6		
10	MXC5291PMA	Cheater-exterior mirror-inner-RH-Black	1		
10	MXC5292PMA	Cheater-exterior mirror-inner-LH-Black	1		
11	AFU3009	Clip	4		

CANXRA1A

CHANGE POINTS:
(1) To (V) LA 081990

N 112

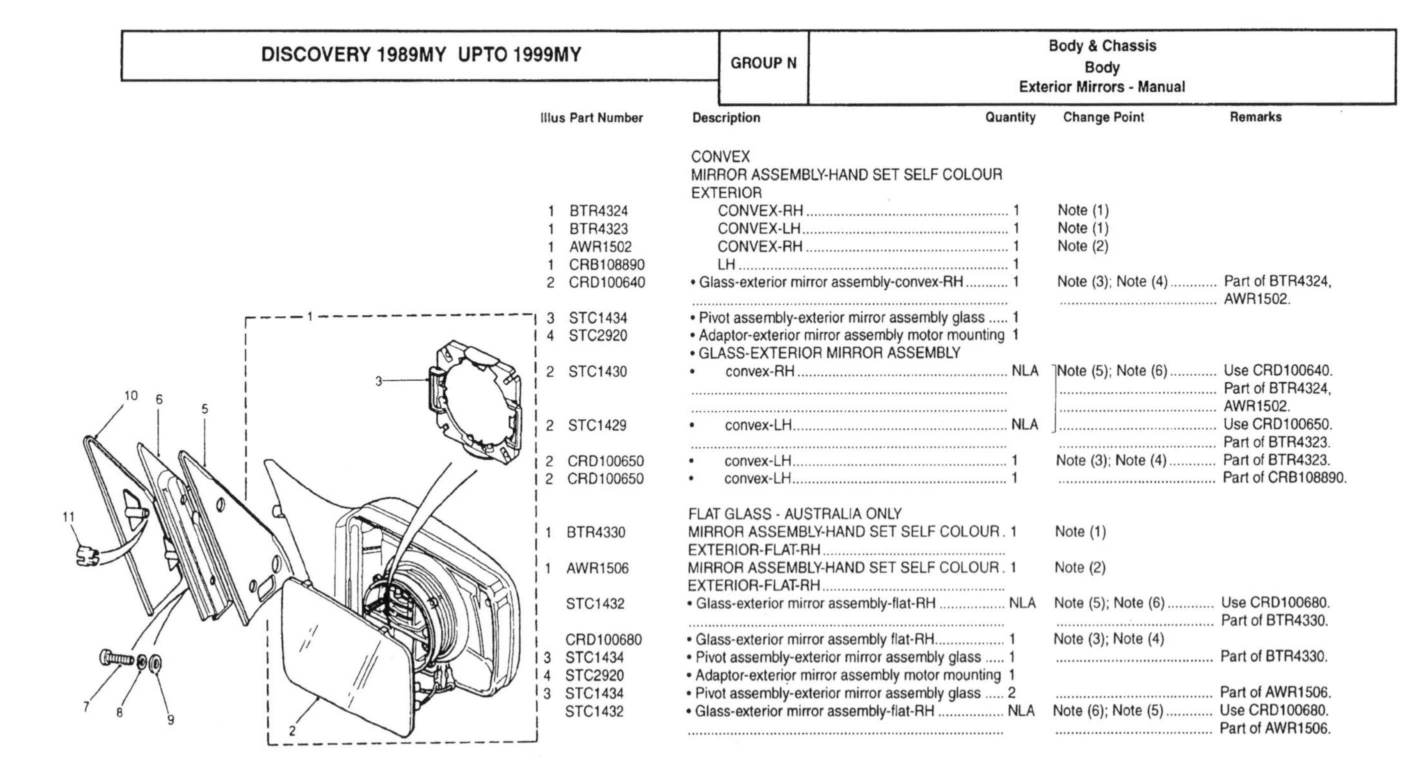

Illus	Part Number	Description	Quantity	Change Point	Remarks
		CONVEX			
		MIRROR ASSEMBLY-HAND SET SELF COLOUR EXTERIOR			
1	BTR4324	CONVEX-RH	1	Note (1)	
1	BTR4323	CONVEX-LH	1	Note (1)	
1	AWR1502	CONVEX-RH	1	Note (2)	
1	CRB108890	LH	1		
2	CRD100640	• Glass-exterior mirror assembly-convex-RH	1	Note (3); Note (4)	Part of BTR4324, AWR1502.
3	STC1434	• Pivot assembly-exterior mirror assembly glass	1		
4	STC2920	• Adaptor-exterior mirror assembly motor mounting	1		
		• GLASS-EXTERIOR MIRROR ASSEMBLY			
2	STC1430	• convex-RH	NLA	Note (5); Note (6)	Use CRD100640. Part of BTR4324, AWR1502.
2	STC1429	• convex-LH	NLA		Use CRD100650. Part of BTR4323.
2	CRD100650	• convex-LH	1	Note (3); Note (4)	Part of BTR4323.
2	CRD100650	• convex-LH	1		Part of CRB108890.
		FLAT GLASS - AUSTRALIA ONLY			
1	BTR4330	MIRROR ASSEMBLY-HAND SET SELF COLOUR EXTERIOR-FLAT-RH	1	Note (1)	
1	AWR1506	MIRROR ASSEMBLY-HAND SET SELF COLOUR EXTERIOR-FLAT-RH	1	Note (2)	
	STC1432	• Glass-exterior mirror assembly-flat-RH	NLA	Note (5); Note (6)	Use CRD100680. Part of BTR4330.
	CRD100680	• Glass-exterior mirror assembly flat-RH	1	Note (3); Note (4)	
3	STC1434	• Pivot assembly-exterior mirror assembly glass	1		Part of BTR4330.
4	STC2920	• Adaptor-exterior mirror assembly motor mounting	1		
3	STC1434	• Pivot assembly-exterior mirror assembly glass	2		Part of AWR1506.
	STC1432	• Glass-exterior mirror assembly-flat-RH	NLA	Note (6); Note (5)	Use CRD100680. Part of AWR1506.

CANXRA1A

CHANGE POINTS:
(1) From (V) MA 081991 To (V) MA 129286
(2) From (V) MA 129287
(3) From (V) VA 723455
(4) From (V) VA 557506
(5) To (V) VA 723454
(6) To (V) VA 557505

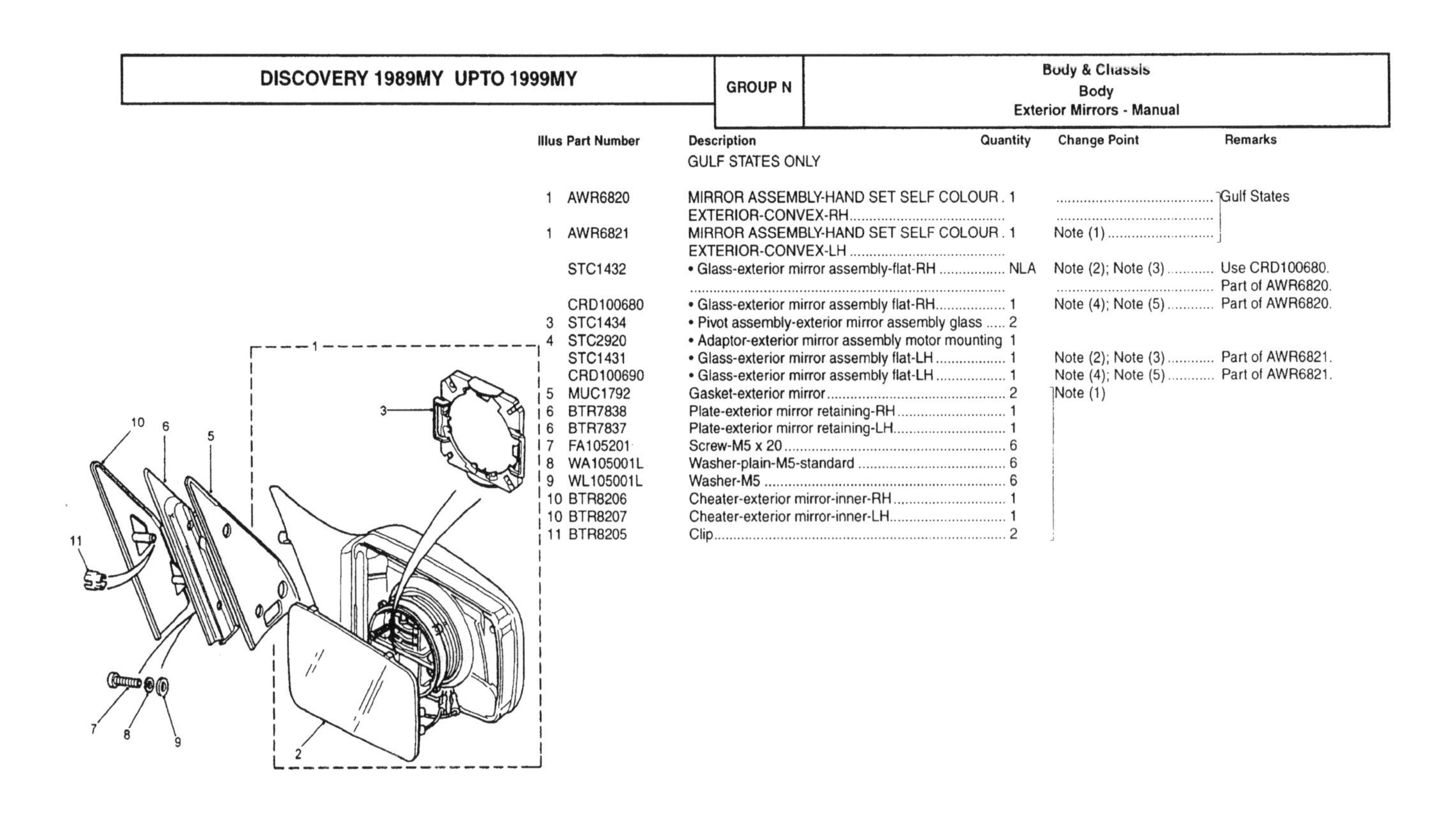

Illus	Part Number	Description	Quantity	Change Point	Remarks
		GULF STATES ONLY			
1	AWR6820	MIRROR ASSEMBLY-HAND SET SELF COLOUR EXTERIOR-CONVEX-RH	1		Gulf States
1	AWR6821	MIRROR ASSEMBLY-HAND SET SELF COLOUR EXTERIOR-CONVEX-LH	1	Note (1)	
	STC1432	• Glass-exterior mirror assembly-flat-RH	NLA	Note (2); Note (3)	Use CRD100680. Part of AWR6820.
	CRD100680	• Glass-exterior mirror assembly flat-RH	1	Note (4); Note (5)	Part of AWR6820.
3	STC1434	• Pivot assembly-exterior mirror assembly glass	2		
4	STC2920	• Adaptor-exterior mirror assembly motor mounting	1		
	STC1431	• Glass-exterior mirror assembly flat-LH	1	Note (2); Note (3)	Part of AWR6821.
	CRD100690	• Glass-exterior mirror assembly flat-LH	1	Note (4); Note (5)	Part of AWR6821.
5	MUC1792	Gasket-exterior mirror	2	Note (1)	
6	BTR7838	Plate-exterior mirror retaining-RH	1		
6	BTR7837	Plate-exterior mirror retaining-LH	1		
7	FA105201	Screw-M5 x 20	6		
8	WA105001L	Washer-plain-M5-standard	6		
9	WL105001L	Washer-M5	6		
10	BTR8206	Cheater-exterior mirror-inner-RH	1		
10	BTR8207	Cheater-exterior mirror-inner-LH	1		
11	BTR8205	Clip	2		

CANXRA1A

CHANGE POINTS:
(1) From (V) MA 081991
(2) To (V) VA 723454
(3) To (V) VA 557505
(4) From (V) VA 723455
(5) From (V) VA 557506

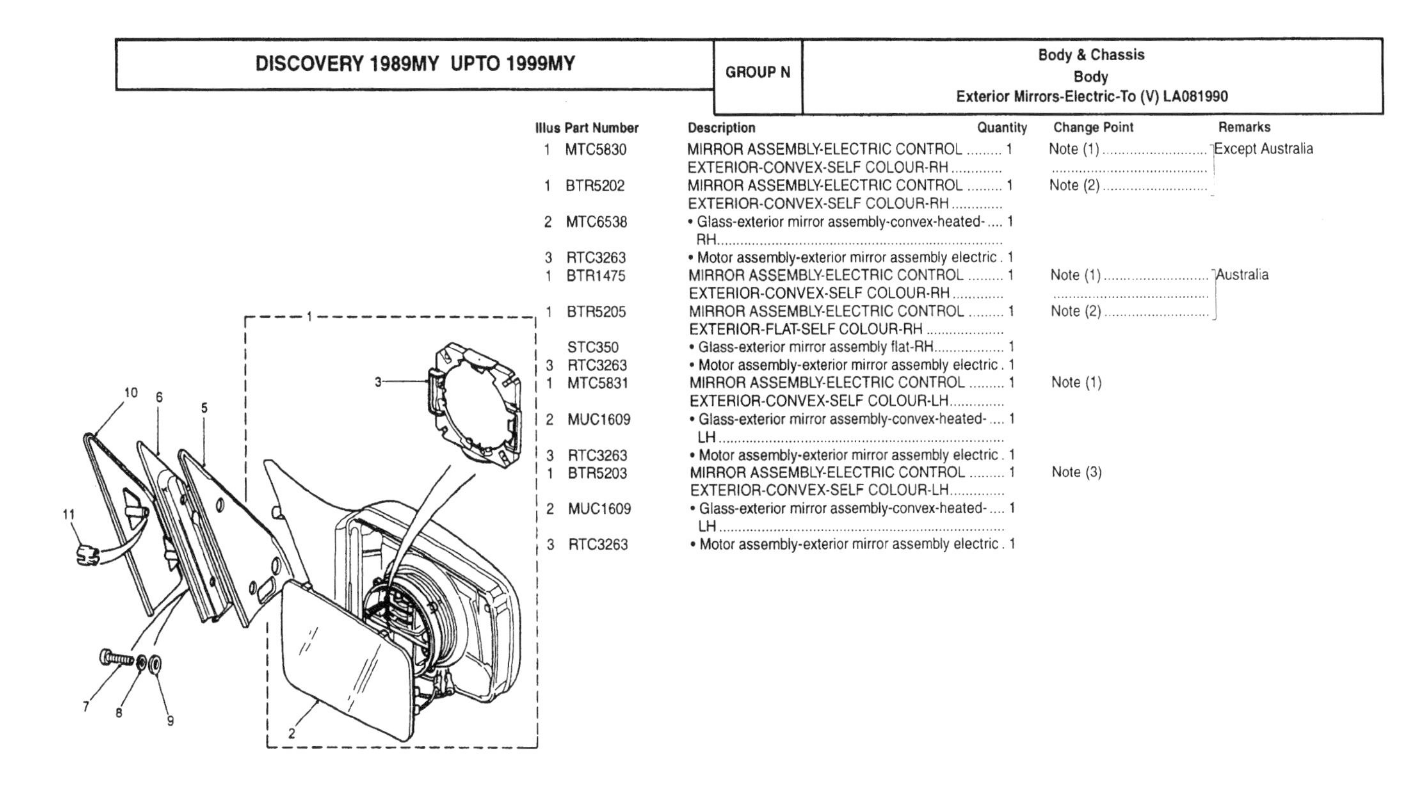

Illus	Part Number	Description	Quantity	Change Point	Remarks
1	MTC5830	MIRROR ASSEMBLY-ELECTRIC CONTROL EXTERIOR-CONVEX-SELF COLOUR-RH	1	Note (1)	Except Australia
1	BTR5202	MIRROR ASSEMBLY-ELECTRIC CONTROL EXTERIOR-CONVEX-SELF COLOUR-RH	1	Note (2)	
2	MTC6538	• Glass-exterior mirror assembly-convex-heated-RH	1		
3	RTC3263	• Motor assembly-exterior mirror assembly electric	1		
1	BTR1475	MIRROR ASSEMBLY-ELECTRIC CONTROL EXTERIOR-CONVEX-SELF COLOUR-RH	1	Note (1)	Australia
1	BTR5205	MIRROR ASSEMBLY-ELECTRIC CONTROL EXTERIOR-FLAT-SELF COLOUR-RH	1	Note (2)	
	STC350	• Glass-exterior mirror assembly flat-RH	1		
3	RTC3263	• Motor assembly-exterior mirror assembly electric	1		
1	MTC5831	MIRROR ASSEMBLY-ELECTRIC CONTROL EXTERIOR-CONVEX-SELF COLOUR-LH	1	Note (1)	
2	MUC1609	• Glass-exterior mirror assembly-convex-heated-LH	1		
3	RTC3263	• Motor assembly-exterior mirror assembly electric	1		
1	BTR5203	MIRROR ASSEMBLY-ELECTRIC CONTROL EXTERIOR-CONVEX-SELF COLOUR-LH	1	Note (3)	
2	MUC1609	• Glass-exterior mirror assembly-convex-heated-LH	1		
3	RTC3263	• Motor assembly-exterior mirror assembly electric	1		

CANXRA1A

CHANGE POINTS:
(1) To (V) JA 033904
(2) From (V) JA 033905 To (V) LA 081990
(3) From (V) JA 033905 To (V) LA 081880

N 115

DISCOVERY 1989MY UPTO 1999MY | GROUP N | Body & Chassis
Body
Exterior Mirrors-Electric-To (V) LA081990

Illus	Part Number	Description	Quantity	Change Point	Remarks
6	MUC1792	Gasket-exterior mirror	2	Note (1)	
5	MWC8526	Plate-exterior mirror retaining-RH	1		
5	MWC8527	Plate-exterior mirror retaining-LH	1		
	AB610061L	Screw-self tapping-No10 x 3/4	6		
7	WA105001L	Washer-plain-M5-standard	6		
8	WL105001L	Washer-M5	6		
9	MXC5291PMA	Cheater-exterior mirror-inner-RH-Black	1		
9	MXC5292PMA	Cheater-exterior mirror-inner-LH-Black	1		
10	AFU3009	Clip	4		

CANXRA1A

CHANGE POINTS:
(1) To (V) LA 081990
(2) To (V) LA 081990

N 116

Illus	Part Number	Description	Quantity	Change Point	Remarks
		MIRROR ASSEMBLY-ELECTRIC CONTROL EXTERIOR			
1	BTR4325	SELF COLOUR-LH	1	Note (1)	UK
1	BTR4326	SELF COLOUR-RH	1		
1	AWR1505	SELF COLOUR-LH	1	Note (2)	
1	AWR1504	SELF COLOUR-RH	1		
2	STC1429	• Glass-exterior mirror assembly-convex-LH	NLA		Use CRD100650. Part of BTR4325, AWR1505.
3	STC2920	• Adaptor-exterior mirror assembly motor mounting	1		
2	STC1430	• Glass-exterior mirror assembly-convex-RH	NLA		Use CRD100640. Part of BTR4326, AWR1504.
2	CRD100650	Glass-exterior mirror assembly-convex-LH			UK
2	CRD100640	Glass-exterior mirror assembly-convex-RH			
		MIRROR ASSEMBLY-ELECTRIC CONTROL EXTERIOR			
1	BTR4328	SELF COLOUR-RH	1	Note (1)	Australia
1	BTR4325	SELF COLOUR-LH	1		
1	AWR1507	SELF COLOUR-RH	1	Note (2)	
1	AWR1505	SELF COLOUR-LH	1		
2	STC1432	• Glass-exterior mirror assembly-flat-RH	NLA		Use CRD100680. Part of BTR4328, AWR1507.
3	STC2920	• Adaptor-exterior mirror assembly motor mounting	1		
2	STC1429	• Glass-exterior mirror assembly-convex-LH	NLA		Use CRD100650. Part of BTR4325, AWR1505.
2	CRD100680	Glass-exterior mirror assembly flat-RH			Australia

CANXRA1A

CHANGE POINTS:
(1) To (V) MA 129286
(2) From (V) MA 129287

Illus	Part Number	Description	Quantity	Change Point	Remarks
1	AWR6338	MIRROR ASSEMBLY-ELECTRIC CONTROL EXTERIOR-SELF COLOUR-RH	1	Note (1); Note (2)	Gulf States
2	STC1430	• Glass-exterior mirror assembly-convex-RH	NLA		Use CRD100640.
3	STC2920	• Adaptor-exterior mirror assembly motor mounting	1		
1	CRB108860	MIRROR ASSEMBLY-ELECTRIC CONTROL EXTERIOR-SELF COLOUR-RH	1	Note (3); Note (4)	Gulf States
2	AWR6861	• Glass-exterior mirror assembly-convex-RH	1		
	CTT100030	• Foam-pad-self adhesive	1		
1	AWR6339	MIRROR ASSEMBLY-ELECTRIC CONTROL EXTERIOR-SELF COLOUR-LH	1	Note (1); Note (2)	Gulf States
2	STC1429	• Glass-exterior mirror assembly-convex-LH	NLA		Use CRD100650.
3	STC2920	• Adaptor-exterior mirror assembly motor mounting	1		
1	CRB108870	MIRROR ASSEMBLY-ELECTRIC CONTROL EXTERIOR-SELF COLOUR-LH	1	Note (3); Note (4)	Gulf States
2	AWR6862	• Glass-exterior mirror assembly-convex-LH	1		
	CTT100030	• Foam-pad-self adhesive	1		

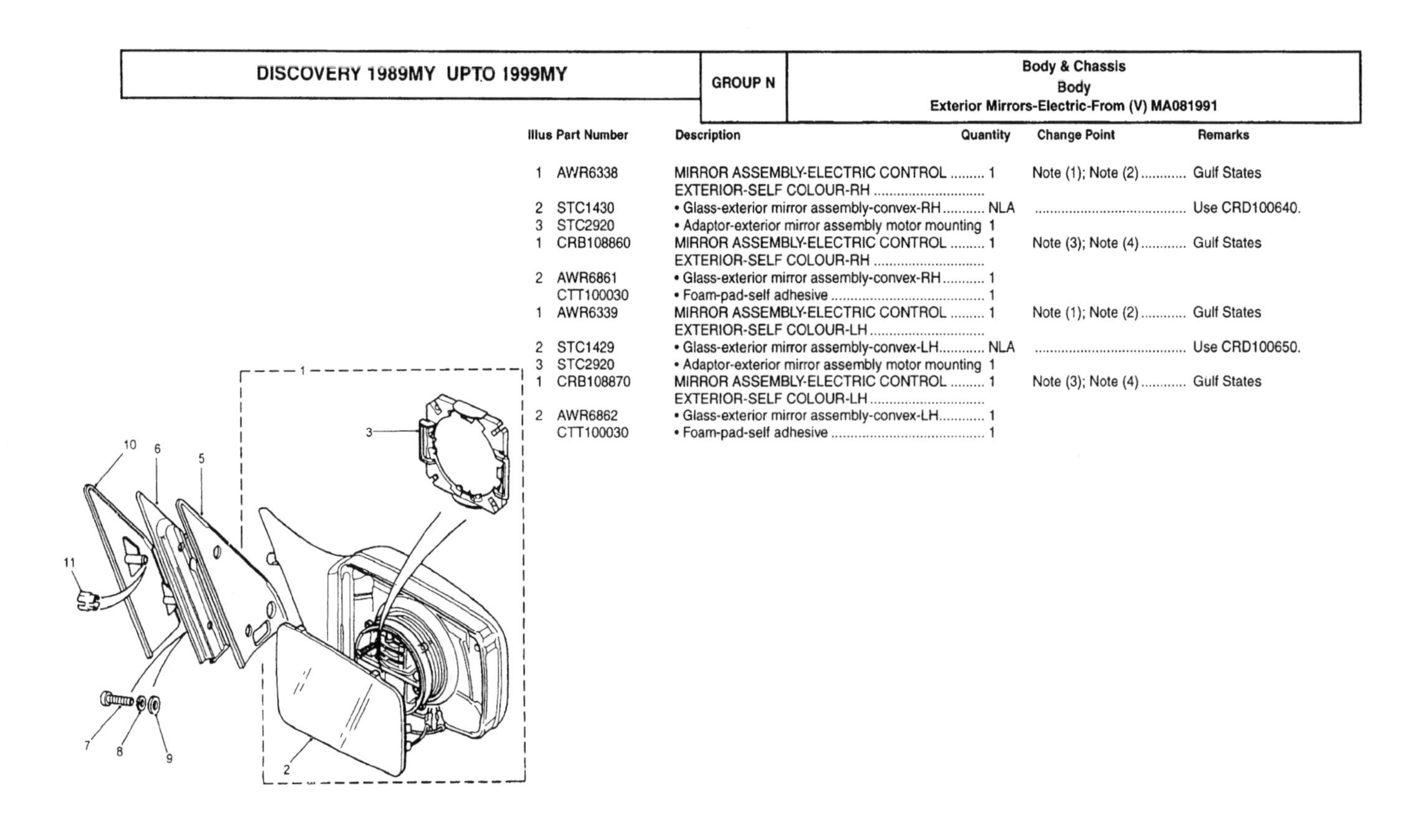

CANXRA1A

CHANGE POINTS:
(1) From (V) VA 703237 To (V) VA 723454
(2) From (V) VA 534104 To (V) VA 557505
(3) From (V) VA 723455
(4) From (V) VA 557506

Illus	Part Number	Description	Quantity	Change Point	Remarks
		NOTE: NOT POSSIBLE TO CHANGE MOTOR ASSY - USE COMPLETE MIRROR			
5	MUC1792	Gasket-exterior mirror	2	Note (1)	
6	BTR7838	Plate-exterior mirror retaining-RH	1		
6	BTR7837	Plate-exterior mirror retaining-LH	1		
7	FA105201	Screw-M5 x 20	6		
8	WA105001L	Washer-plain-M5-standard	6		
9	WL105001L	Washer-M5	6		
10	BTR8206	Cheater-exterior mirror-inner-RH	1		
10	BTR8207	Cheater-exterior mirror-inner-LH	1		
11	BTR8205	Clip	2		

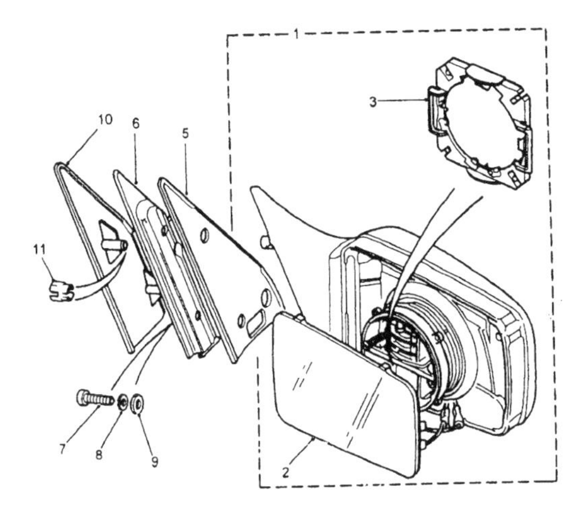

CANXRA1A

CHANGE POINTS:
(1) From (V) MA 081991
(2) From (V) MA 081991

Illus	Part Number	Description	Quantity	Change Point	Remarks
1	STC7156	HARDTOP ASSEMBLY-C/W TAILGATE	1		
2	STC60092	• Tailgate assembly-upper-glass reinforced plastic	1		
3	STC7158	• Hinge assembly-tailgate-upper	2		
4	STC7155	Glass-rear door clear	1		
5	STC7157	Strut-tailgate gas	2		
		GLASS-REAR QUARTER			
6	STC7154	RH	1		
6	STC7153	LH	1		
7	STC7152	RH	1		
7	STC7151	LH	1		

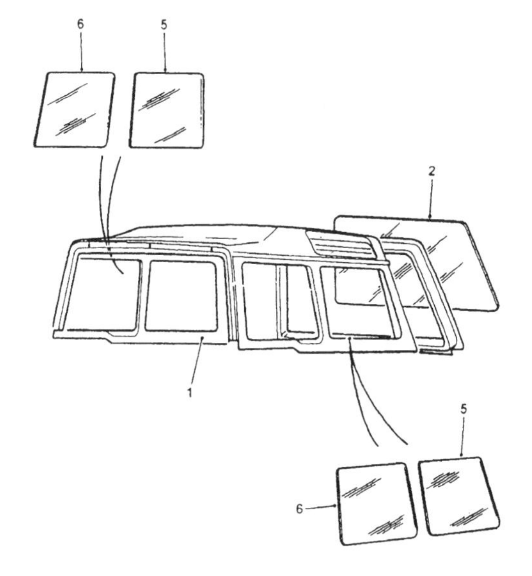

CANXVC1A

Illus	Part Number	Description	Quantity	Change Point	Remarks
		205 TYRES ONLY			
1	ALR5693	Carrier-load/tail door spare wheel mounting	1		
		235 GOODYEAR TYRES ONLY			
1	ANR4743	Carrier-load/tail door spare wheel mounting	NLA	Note (1)	Use ANR5040 qty 2 with FB108161L qty 5 with FB108171L qty 1 with FN108041L qty 6 with ALR5693 qty 01.
2	ANR5040	Spacer	2		
		ALL 205 TYRE VARIANTS			
3	BH108141L	Bolt-hexagonal head-M8 x 70	5		
4	BH108161L	Bolt-hexagonal head-M8 x 80	1		
5	WC108051L	Washer-plain-M8	6		
6	NV108041L	Nut-M8	6		
		ALL 235 TYRE VARIANTS			
3	FB108161L	Bolt-flanged head-M8 x 80	5		
4	FB108171L	Bolt-flanged head-M8 x 85-long	1		
6	FN108041L	Nut-flanged head-M8 x 20	NLA		Use FN108047L.
		COVER ASSEMBLY-SPARE WHEEL			
7	BTR924MUE	Quicksilver	1	Note (2)	
7	MWC9071NCM	Davas White	1	Note (3)	
7	BTR924NUC	Savarin White	1	Note (4)	
7	BTR9438MUE	Quicksilver	1	Note (5)	
7	BTR9438NUC	Savarin White	1	Note (6)	
7	BTR9438MNH	Silver Sparkle	1	Note (6)	
		235 GOODYEAR TYRES ONLY			
7	ANR4751	Cover assembly-spare wheel-with L/R logo	1		

CANXTA2A

CHANGE POINTS:
(1) From (V) MA 141111
(2) To (V) LA 081991
(3) From (V) GA 394352 To (V) GA 415597
(4) From (V) GA 415598 To (V) GA 456980; From (V) HA 456981 To (V) HA 486551; From (V) HA 000001 To (V) HA 012332; From 012333
(5) From (V) MA 081992
(6) From (V) MA 081991

N 121

Illus	Part Number	Description	Quantity	Change Point	Remarks
		SPARE WHEEL MOUNTING			
8	MXC7335	Washer-plastic	3		
9	MXC7887	Nut-M16	3		
		ALLOY WHEEL MOUNTING			
8	MXC7335	Washer-plastic	3		
10	NTC8240	Nut	2	Note (1)	
10	ANR4914	Nut	2	Note (2)	
		STEEL WHEEL MOUNTING			
10	ANR3116	Nut	2		
		ALLOY WHEEL MOUNTING			
11	NTC8278	NUT-LOCKING	1		
12	STC463	• Key-blank owner	1		
13	BTR9901LAV	Badge-Land Rover-Grey	1	Note (3)	

NOTE : NTC8240 AND NTC8278 ARE TO BE USED ON
SPARE WHEEL MOUNTING ONLY. THEY MUST NOT
BE USED FOR RETAINING ROAD WHEELS ONTO AXLES.

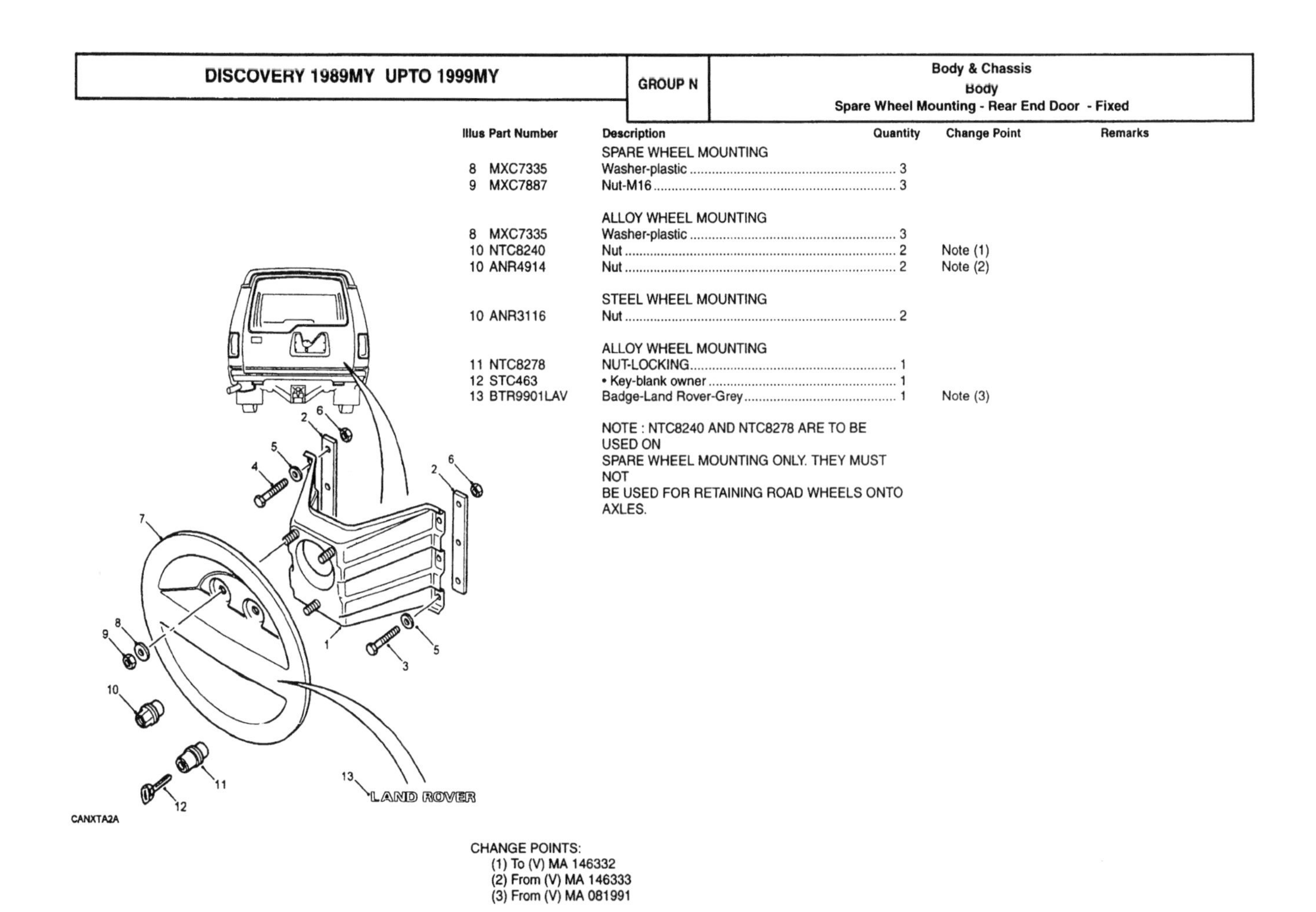

CHANGE POINTS:
(1) To (V) MA 146332
(2) From (V) MA 146333
(3) From (V) MA 081991

N 122

Illus	Part Number	Description	Quantity	Change Point	Remarks
1	STC7017	Guard-dog-loadspace	1		
2	STC7018	Bracket	2		

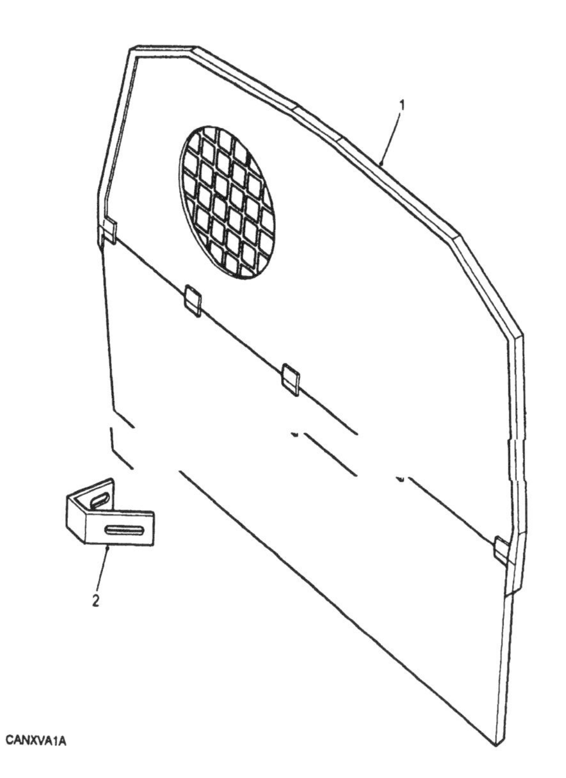

N 123

Notes

Trim .... P 1
Auto Select Mounting .... P 17
Auto and Manual Select Mounting Assembly .... P 16
Body Trim Rear (Cont) .... P 63
Bodyside Front & Rear .... P 57
Bodyside Trim Front - 3 Door .... P 43
Bodyside Trim Front - 5 Door .... P 51
Carpets and Finishers .... P 76
Centre Console Bag .... P 18
Centre Console & Cubby Box-From (V) .... P 19
Centre Console & Cubby Box-From (V) .... P 20
Facia Assembly .... P 1
Facia Centre Console .... P 12
Footrest and Toebox Trim .... P 64
Front Door Casing .... P 24
Grab Handle Facia .... P 6
Headlining .... P 66
Instrument Binnacle Mounting .... P 5
Insulation - Front .... P 74
Loadspace Cover .... P 22
Lower Facia & Centre Console-To (V) .... P 13
Rear Air Conditioning .... P 15
Rear End Door Casing .... P 39
Rear Side Door Casing .... P 35
Sunroof Roller Blind .... P 73

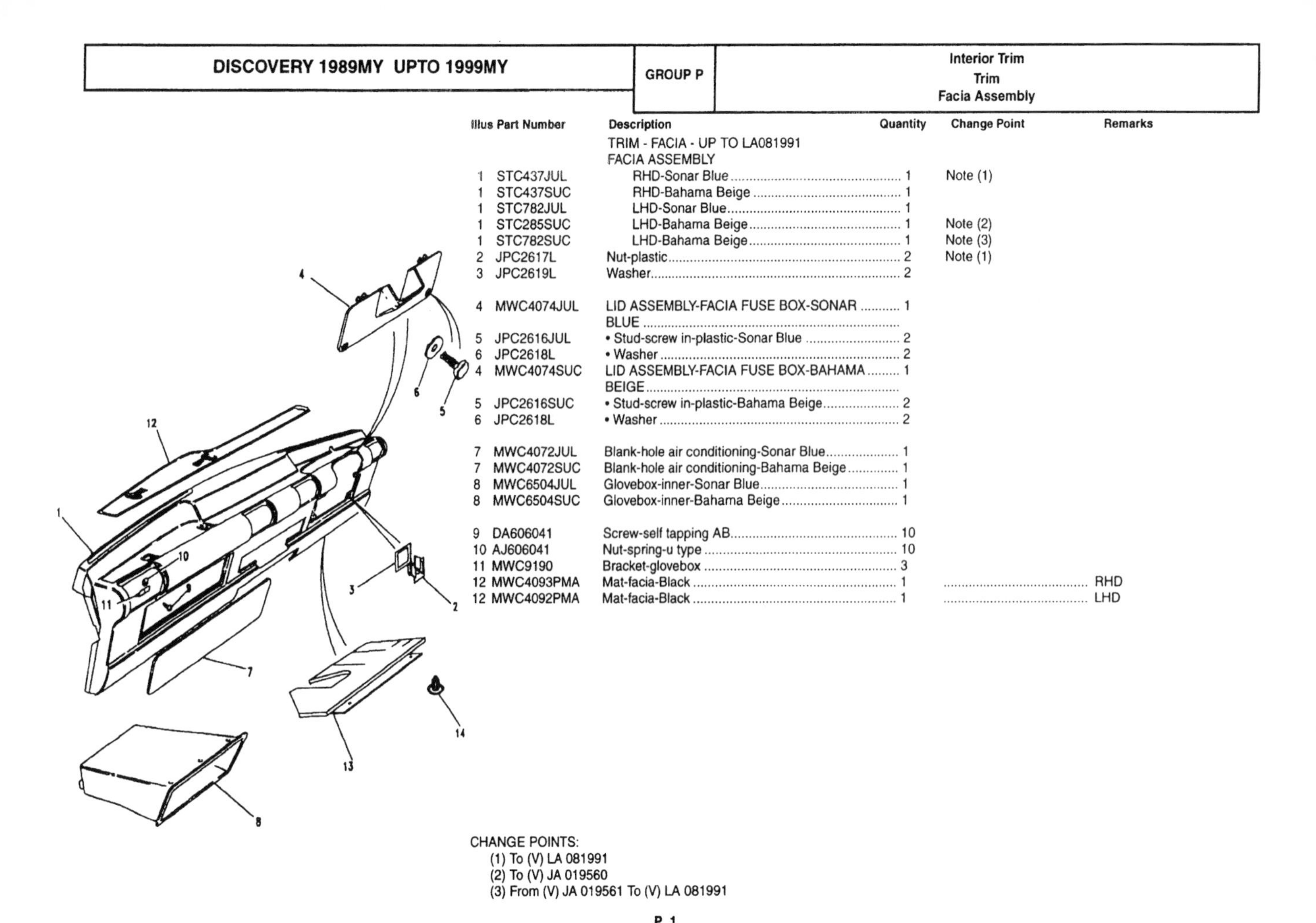

Illus	Part Number	Description	Quantity	Change Point	Remarks
		TRIM - FACIA - UP TO LA081991			
		FACIA ASSEMBLY			
1	STC437JUL	RHD-Sonar Blue	1	Note (1)	
1	STC437SUC	RHD-Bahama Beige	1		
1	STC782JUL	LHD-Sonar Blue	1		
1	STC285SUC	LHD-Bahama Beige	1	Note (2)	
1	STC782SUC	LHD-Bahama Beige	1	Note (3)	
2	JPC2617L	Nut-plastic	2	Note (1)	
3	JPC2619L	Washer	2		
4	MWC4074JUL	LID ASSEMBLY-FACIA FUSE BOX-SONAR BLUE	1		
5	JPC2616JUL	• Stud-screw in-plastic-Sonar Blue	2		
6	JPC2618L	• Washer	2		
4	MWC4074SUC	LID ASSEMBLY-FACIA FUSE BOX-BAHAMA BEIGE	1		
5	JPC2616SUC	• Stud-screw in-plastic-Bahama Beige	2		
6	JPC2618L	• Washer	2		
7	MWC4072JUL	Blank-hole air conditioning-Sonar Blue	1		
7	MWC4072SUC	Blank-hole air conditioning-Bahama Beige	1		
8	MWC6504JUL	Glovebox-inner-Sonar Blue	1		
8	MWC6504SUC	Glovebox-inner-Bahama Beige	1		
9	DA606041	Screw-self tapping AB	10		
10	AJ606041	Nut-spring-u type	10		
11	MWC9190	Bracket-glovebox	3		
12	MWC4093PMA	Mat-facia-Black	1		RHD
12	MWC4092PMA	Mat-facia-Black	1		LHD

CHANGE POINTS:
(1) To (V) LA 081991
(2) To (V) JA 019560
(3) From (V) JA 019561 To (V) LA 081991

P 1

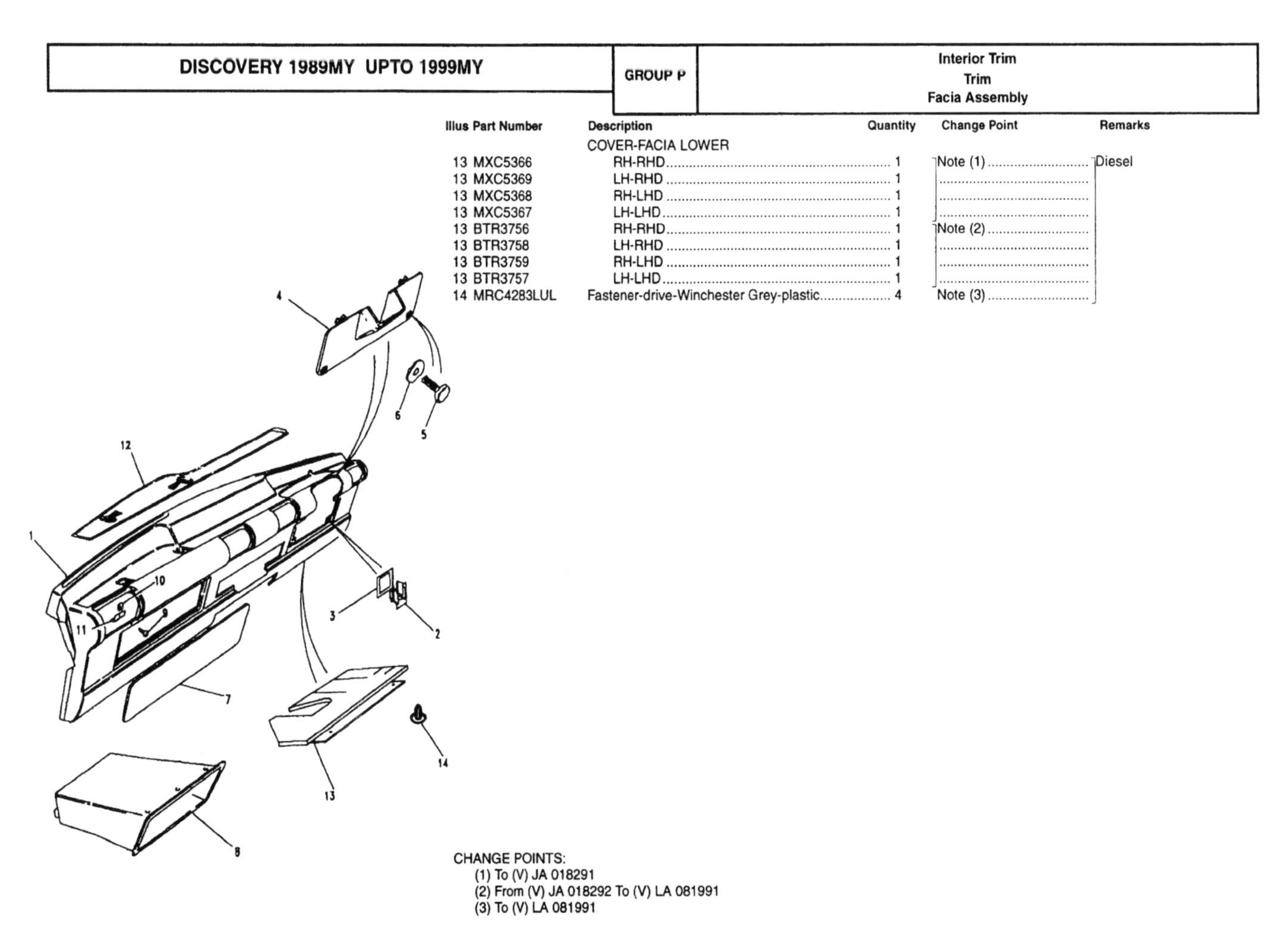

Illus	Part Number	Description	Quantity	Change Point	Remarks
		COVER-FACIA LOWER			
13	MXC5366	RH-RHD	1	Note (1)	Diesel
13	MXC5369	LH-RHD	1		
13	MXC5368	RH-LHD	1		
13	MXC5367	LH-LHD	1		
13	BTR3756	RH-RHD	1	Note (2)	
13	BTR3758	LH-RHD	1		
13	BTR3759	RH-LHD	1		
13	BTR3757	LH-LHD	1		
14	MRC4283LUL	Fastener-drive-Winchester Grey-plastic	4	Note (3)	

CHANGE POINTS:
(1) To (V) JA 018291
(2) From (V) JA 018292 To (V) LA 081991
(3) To (V) LA 081991

P 2

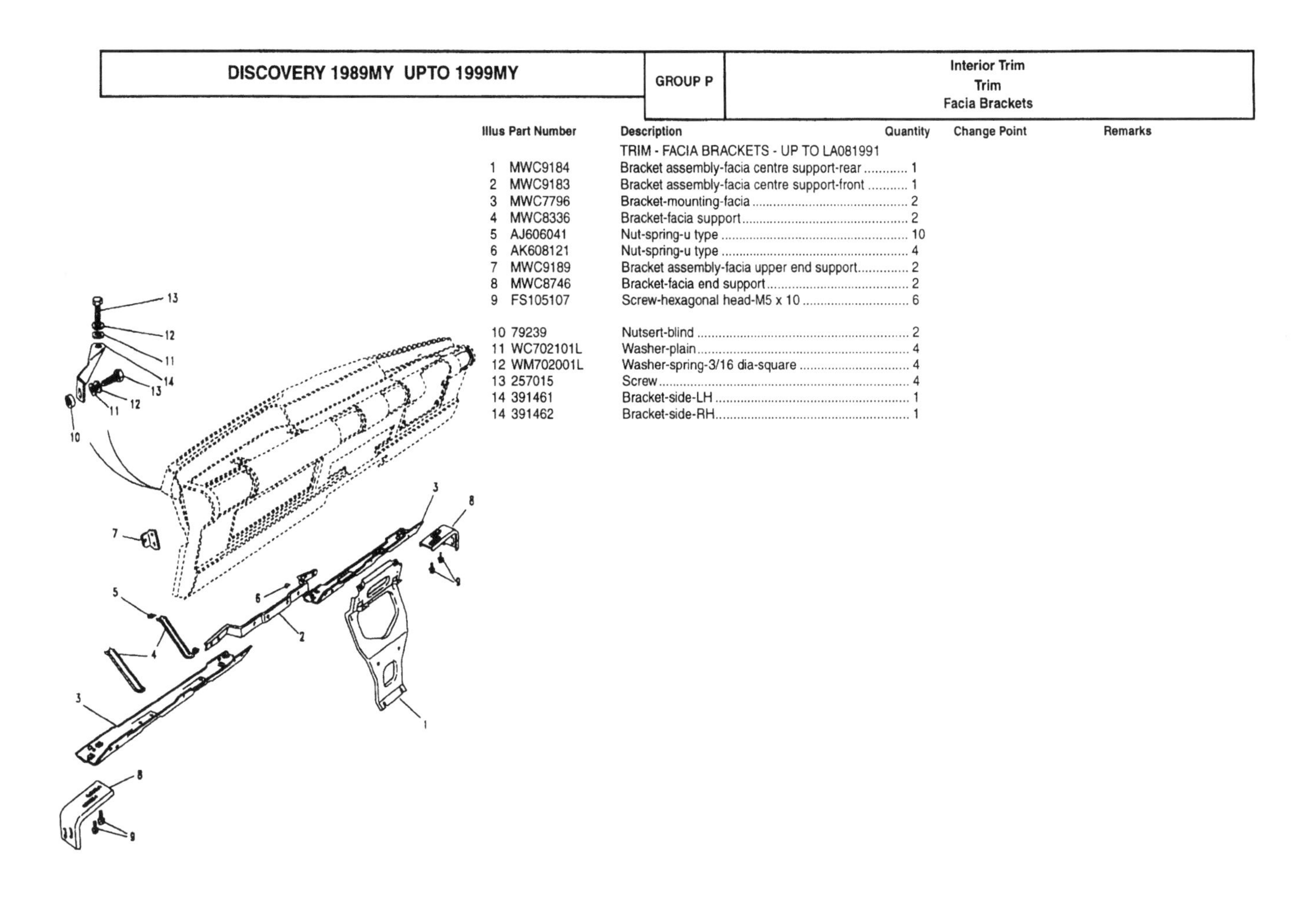

Illus	Part Number	Description	Quantity	Change Point	Remarks
		TRIM - FACIA BRACKETS - UP TO LA081991			
1	MWC9184	Bracket assembly-facia centre support-rear	1		
2	MWC9183	Bracket assembly-facia centre support-front	1		
3	MWC7796	Bracket-mounting-facia	2		
4	MWC8336	Bracket-facia support	2		
5	AJ606041	Nut-spring-u type	10		
6	AK608121	Nut-spring-u type	4		
7	MWC9189	Bracket assembly-facia upper end support	2		
8	MWC8746	Bracket-facia end support	2		
9	FS105107	Screw-hexagonal head-M5 x 10	6		
10	79239	Nutsert-blind	2		
11	WC702101L	Washer-plain	4		
12	WM702001L	Washer-spring-3/16 dia-square	4		
13	257015	Screw	4		
14	391461	Bracket-side-LH	1		
14	391462	Bracket-side-RH	1		

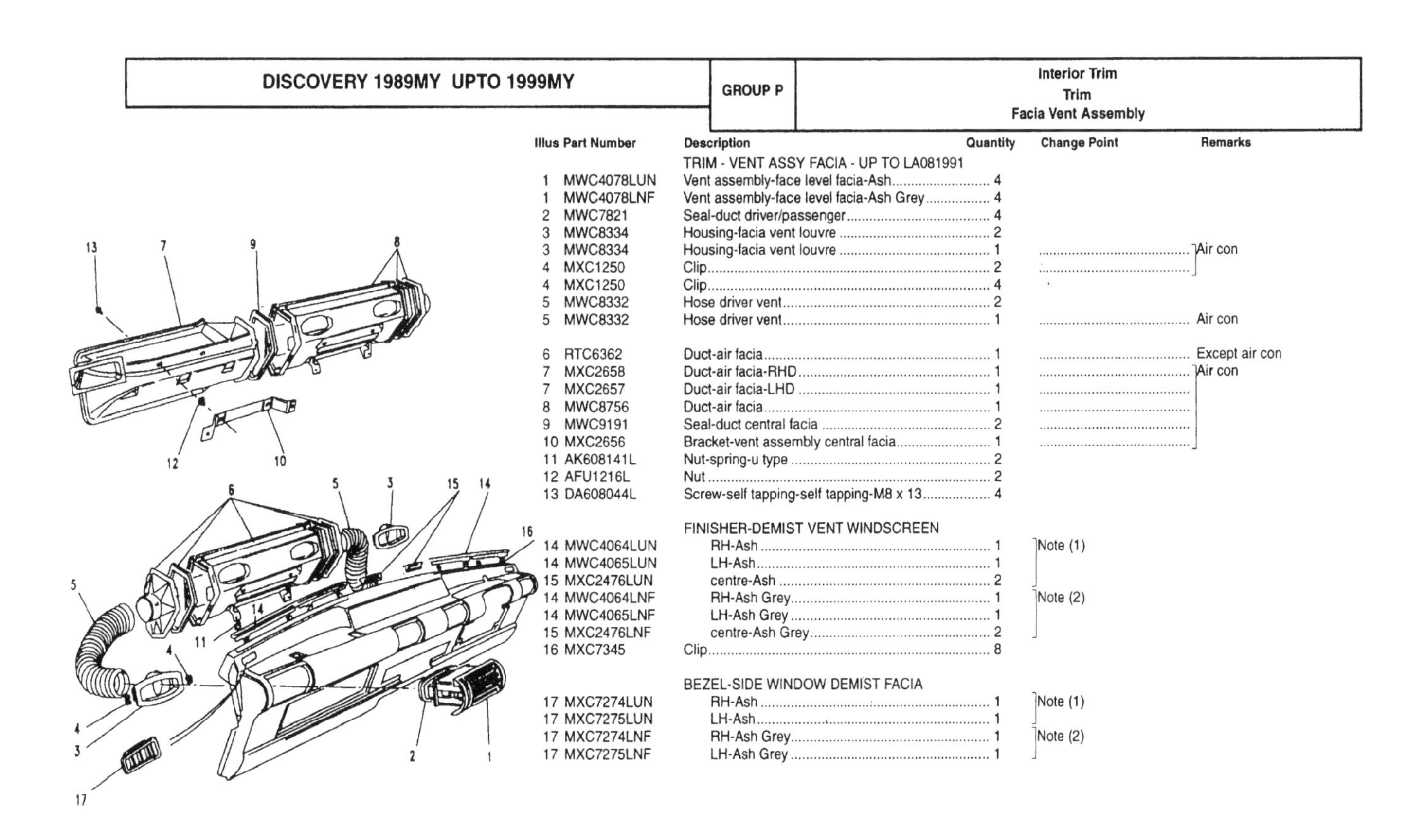

Illus	Part Number	Description	Quantity	Change Point	Remarks
		TRIM - VENT ASSY FACIA - UP TO LA081991			
1	MWC4078LUN	Vent assembly-face level facia-Ash	4		
1	MWC4078LNF	Vent assembly-face level facia-Ash Grey	4		
2	MWC7821	Seal-duct driver/passenger	4		
3	MWC8334	Housing-facia vent louvre	2		
3	MWC8334	Housing-facia vent louvre	1		Air con
4	MXC1250	Clip	2		
4	MXC1250	Clip	4		
5	MWC8332	Hose driver vent	2		
5	MWC8332	Hose driver vent	1		Air con
6	RTC6362	Duct-air facia	1		Except air con
7	MXC2658	Duct-air facia-RHD	1		Air con
7	MXC2657	Duct-air facia-LHD	1		
8	MWC8756	Duct-air facia	1		
9	MWC9191	Seal-duct central facia	2		
10	MXC2656	Bracket-vent assembly central facia	1		
11	AK608141L	Nut-spring-u type	2		
12	AFU1216L	Nut	2		
13	DA608044L	Screw-self tapping-self tapping-M8 x 13	4		
		FINISHER-DEMIST VENT WINDSCREEN			
14	MWC4064LUN	RH-Ash	1	Note (1)	
14	MWC4065LUN	LH-Ash	1		
15	MXC2476LUN	centre-Ash	2		
14	MWC4064LNF	RH-Ash Grey	1	Note (2)	
14	MWC4065LNF	LH-Ash Grey	1		
15	MXC2476LNF	centre-Ash Grey	2		
16	MXC7345	Clip	8		
		BEZEL-SIDE WINDOW DEMIST FACIA			
17	MXC7274LUN	RH-Ash	1	Note (1)	
17	MXC7275LUN	LH-Ash	1		
17	MXC7274LNF	RH-Ash Grey	1	Note (2)	
17	MXC7275LNF	LH-Ash Grey	1		

CHANGE POINTS:
(1) To (V) JA 034313
(2) From (V) KA 034314 To (V) LA 081991

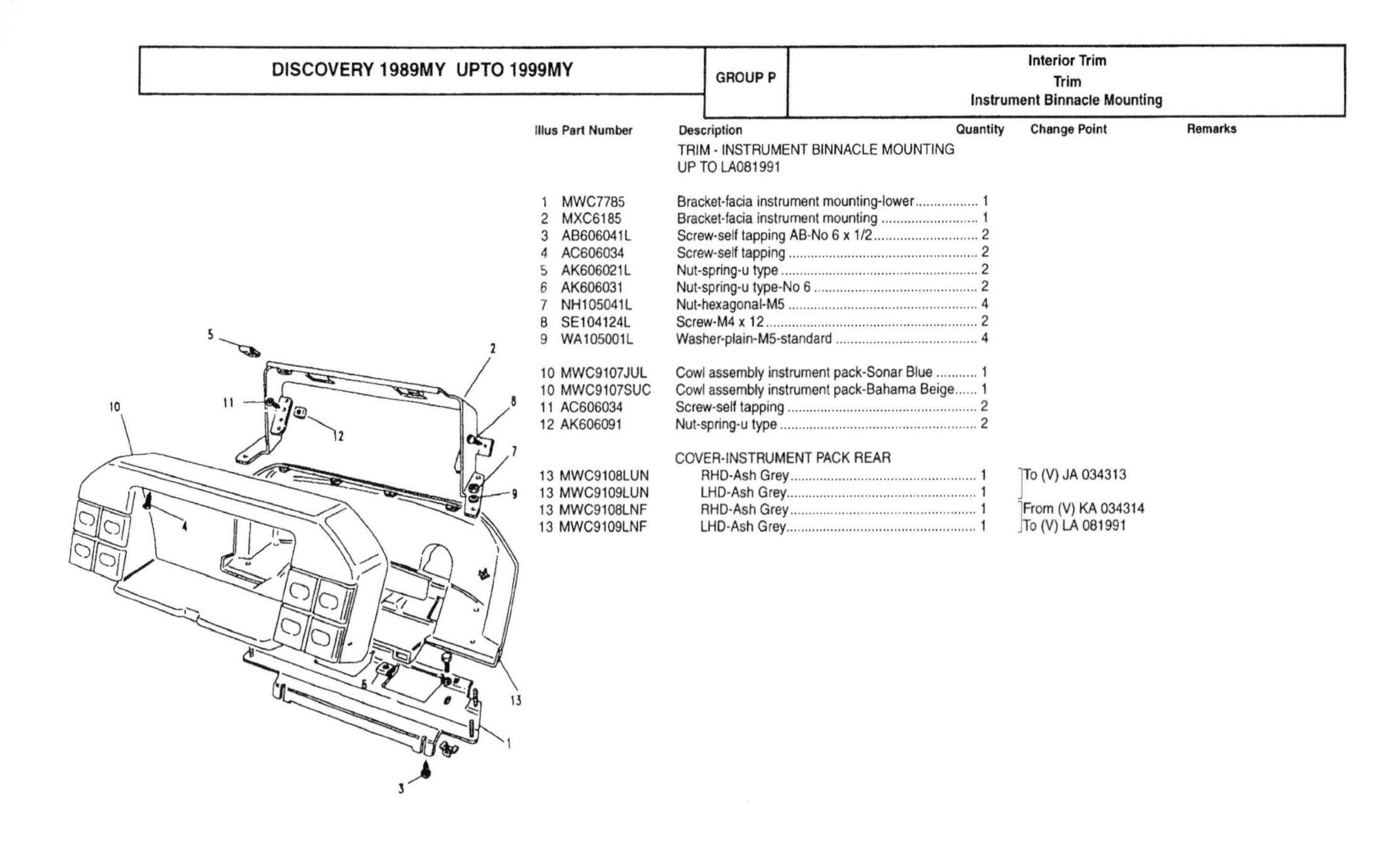

Illus	Part Number	Description	Quantity	Change Point	Remarks
		TRIM - INSTRUMENT BINNACLE MOUNTING UP TO LA081991			
1	MWC7785	Bracket-facia instrument mounting-lower	1		
2	MXC6185	Bracket-facia instrument mounting	1		
3	AB606041L	Screw-self tapping AB-No 6 x 1/2	2		
4	AC606034	Screw-self tapping	2		
5	AK606021L	Nut-spring-u type	2		
6	AK606031	Nut-spring-u type-No 6	2		
7	NH105041L	Nut-hexagonal-M5	4		
8	SE104124L	Screw-M4 x 12	2		
9	WA105001L	Washer-plain-M5-standard	4		
10	MWC9107JUL	Cowl assembly instrument pack-Sonar Blue	1		
10	MWC9107SUC	Cowl assembly instrument pack-Bahama Beige	1		
11	AC606034	Screw-self tapping	2		
12	AK606091	Nut-spring-u type	2		
		COVER-INSTRUMENT PACK REAR			
13	MWC9108LUN	RHD-Ash Grey	1	To (V) JA 034313	
13	MWC9109LUN	LHD-Ash Grey	1		
13	MWC9108LNF	RHD-Ash Grey	1	From (V) KA 034314	
13	MWC9109LNF	LHD-Ash Grey	1	To (V) LA 081991	

Illus	Part Number	Description	Quantity	Change Point	Remarks
		TRIM - GRAB HANDLE FACIA - UP TO LA081991			
		HANDLE ASSEMBLY-GRAB			
1	MWC4098JUL	Sonar Blue	1	Note (1)	
1	MWC4098SUC	Bahama Beige	1		
1	BTR1180JUL	Sonar Blue	1	Note (2)	
1	BTR1180SUC	Bahama Beige	1		
2	SF106251L	Screw	8		
3	MWC9149	Bracket-facia grab handle reinforcement-RHD	1		
3	MWC9148	Bracket-facia grab handle reinforcement-LHD	1		
4	NK106061L	Nut-M6	2		
5	NK106081	Nut-lokut	2		
6	WA105001L	Washer-plain-M5-standard	4		
7	NH105041L	Nut-hexagonal-M5	4		
8	MWC8744	Bracket-facia grab handle reinforcement-RHD	1		
8	MWC8745	Bracket-facia grab handle reinforcement-LHD	1		
9	WA106041L	Washer-M6-standard.	2		
10	NH106041L	Nut-hexagonal head-nyloc-M6	2		
11	MWC9202	Bracket-facia grab handle reinforcement-RHD	1		
11	MWC9203	Bracket-facia grab handle reinforcement-LHD	1		

CHANGE POINTS:
(1) To (V) HA 482391
(2) From (V) HA 482392 To (V) LA 081991

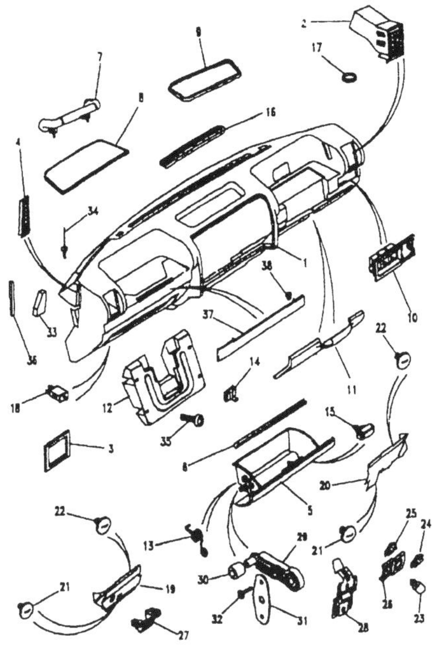

Illus	Part Number	Description	Quantity	Change Point	Remarks
		TRIM- FACIA ASSEMBLY - MA0819912 ON			
		FACIA ASSEMBLY			
1	STC1287SUC	RHD-Bahama Beige	1		Except passenger air bag
1	STC1287LOY	RHD-Dark Granite	1		
1	STC1289SUC	LHD-Bahama Beige	1		
1	STC1288LOY	LHD-Dark Granite	1		
1	STC1290SUC	RHD-Bahama Beige	1		Passenger air bag
1	STC1290LOY	RHD-Dark Granite	1		
1	STC1291SUC	LHD-Bahama Beige	1		
1	STC1291LOY	LHD-Dark Granite	1		
2	AWR1194LNF	Vent assembly-face level facia-Ash Grey-outer	2		
3	BTR7174	Seal-duct driver/passenger-outer	2		
4	BTR3313LNF	Bezel-side window demist facia-Ash Grey-LH	1		
4	BTR3312LNF	Bezel-side window demist facia-Ash Grey-RH	1		
5	BTR3351LOY	Glovebox assembly-facia-Dark Granite	1		
5	FFB101460SUC	Glovebox assembly-facia-Bahama Beige	1		
6	BTR3730LOY	Finisher-glovebox-Dark Granite	1		
6	BTR3730SUC	Finisher-glovebox-Bahama Beige	1		
7	BTR3896LOY	Handle assembly-grab-Dark Granite	1		
7	BTR3896SUC	Handle assembly-grab-Bahama Beige	1		
8	BTR4307PMA	Mat-facia-Black	1		
9	BTR9334PMA	Mat-facia-middle-Black	1		

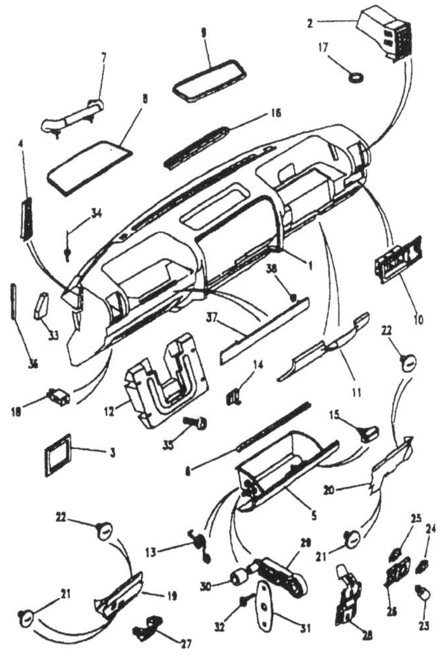

Illus	Part Number	Description	Quantity	Change Point	Remarks
		FOR ITEM 10 SEE ELECTRICAL SECTION			
11	BTR3723SUC	LID-FACIA FUSE BOX-BAHAMA BEIGE-RH	1		
11	BTR3722SUC	LID-FACIA FUSE BOX-BAHAMA BEIGE-LH	1		
	AWR7009	• Washer-rubber	1		
	AWR7010	• Washer-foam	2		
	FBO100260SUC	• Cover fusebox	1		
	BTR4830	• Hinge assembly-fusebox lid	2		
24	MUC6523	• Washer	2		
23	FBR100060SUC	• Turnbuckle-fuse box lid-Bahama Beige	1		
11	BTR3723LOY	Lid-facia fuse box-LH-Dark Granite	1		
11	BTR3722LOY	Lid-facia fuse box-RH-Dark Granite	1		
12	BTR7281	Foam-facia finisher assembly	2		
13	BTR9908	Spring-glovebox return-RH	1	Note (1)	
13	BTR9909	Spring-glovebox return-LH	1		
14	BTR9515	Clip	8		
15	BTR3367	Latch-glovebox	1	Note (2); Note (3)	
15	FNC100080L	Latch-glovebox	1	Note (4); Note (5)	
16	BTR9333LNF	Bezel assembly windscreen demist-Ash Grey	1		
17	BTR7175	Plug-facia blanking	2		

CHANGE POINTS:
(1) From (V) MA 081992
(2) From (V) MA 081992 To (V) VA 539099
(3) From (V) MA 081992 To (V) VA 708283
(4) From (V) VA 708284
(5) From (V) VA 539100

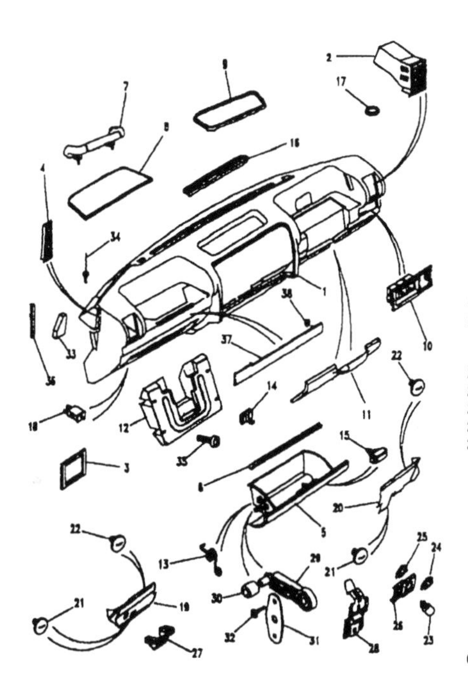

Illus	Part Number	Description	Quantity	Change Point	Remarks
		TRIM - FACIA ASSEMBLY - MA081992 ON			
19	BTR9888	Panel-passenger closure-RHD	1		
		PANEL-DRIVER CLOSING			
20	AWR3331	RH-RHD	1	Note (2)	
20	BTR9889	LH-RHD	1	Note (1)	
20	AWR6200	RH-RHD	1	Note (3)	
20	FBS100760	LH-RHD	1	Note (4)	
21	BTR5200	Clip-trim	1	Note (5)	
20	FBS100700	Panel-driver closing-RH	1	Note (4)	
22	BTR9437	Fastener-fir tree	3		
	MUC1820	Seal	A/R		RHD
19	BTR9890	Panel-passenger closure-LHD	1		
		PANEL-DRIVER CLOSING			
20	BTR9891	LHD	1	Note (1)	
20	AWR3332	LH-LHD	1	Note (2)	
20	AWR6199	LH-LHD	1	Note (3)	
20	FBS100770	LH-LHD	1	Note (4)	
21	BTR5200	Clip-trim	1	Note (5)	
22	BTR9437	Fastener-fir tree	3		
	STC2911	Panel-closing-facia-RHD	1		
	STC2912	Panel-closing-facia-LHD	1		
21	BTR5200	Clip-trim	1	Note (5)	
22	BTR9437	Fastener-fir tree	3		
		TURNBUCKLE-FUSE BOX LID			
23	FBR10001LOY	Dark Granite	2		
23	FBR10001SUC	Bahama Beige	2		
24	MUC6523	Washer	2		
25	AFU4131	Fastener-quarter turn	2		
26	BTR4264	Receiver-turnbuckle	2		

CHANGE POINTS:
(1) From (V) MA 081992 To (V) MA 162178
(2) From (V) MA 162179 To (V) VA 711412; From (V) MA 162179 To (V) VA 542549
(3) From (V) VA 542005; From (V) VA 711413; To (V) WA 763561
(4) From (V) WA 763562
(5) From (V) MA 081992

Illus	Part Number	Description	Quantity	Change Point	Remarks
27	BTR3731	Striker-facia glovebox	1		
28	BTR6836	Clip	2	Note (1)	
28	AWR4781	Clip	2	Note (2)	
29	FFV10002	Arm-glovebox stay	2		
30	FXB10002	Roller	2		
31	FFU10004	Plate-glovebox stay retaining	2		
32	FYP10003	Screw	4		
33	MUC6426	Foam	2		
34	BTR9284	Rivet	2		
35	BTR6656	Screw-powerlok	4		
36	STC3096	Strip-anti rattle	A/R		
	BTR2913	Seal-E post	1		
	MUC1820	Seal-child lock	A/R	Note (4)	RHD
		NEW WOOD CAPPING ABOVE GLOVEBOX FROM 98MY			
37	FAE101350	Capping-wood-facia-RHD	1	Note (3)	
37	FAE101340	Capping-wood-facia-LHD	1		
38	FYH100280	Nut	2		
	VUR100260	Template-drilling	1		

CHANGE POINTS:
(1) From (V) MA 081992 To (V) TA 195772; From (V) MA 081992 To (V) TA 526563
(2) From (V) TA 195773; From (V) TA 526564
(3) From (V) WA 746542
(4) From (V) MA 081992

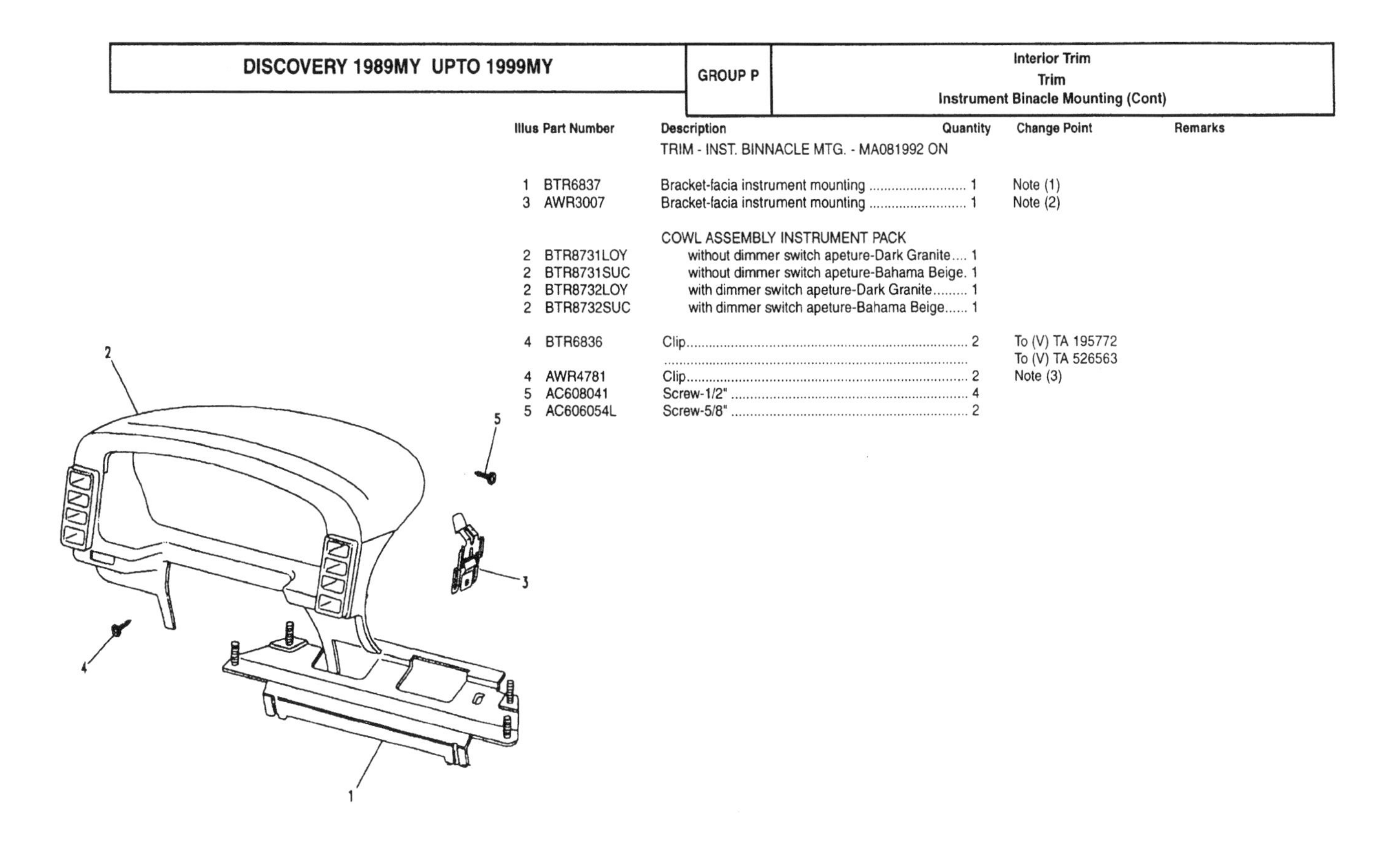

Illus	Part Number	Description	Quantity	Change Point	Remarks
		TRIM - INST. BINNACLE MTG. - MA081992 ON			
1	BTR6837	Bracket-facia instrument mounting	1	Note (1)	
3	AWR3007	Bracket-facia instrument mounting	1	Note (2)	
		COWL ASSEMBLY INSTRUMENT PACK			
2	BTR8731LOY	without dimmer switch apeture-Dark Granite	1		
2	BTR8731SUC	without dimmer switch apeture-Bahama Beige.	1		
2	BTR8732LOY	with dimmer switch apeture-Dark Granite	1		
2	BTR8732SUC	with dimmer switch apeture-Bahama Beige	1		
4	BTR6836	Clip	2	To (V) TA 195772	
				To (V) TA 526563	
4	AWR4781	Clip	2	Note (3)	
5	AC608041	Screw-1/2"	4		
5	AC606054L	Screw-5/8"	2		

CHANGE POINTS:
(1) From (V) MA 081992 To (V) MA 146407
(2) From (V) MA 146408
(3) From (V) TA 195773; From (V) TA 526564

P 11

DISCOVERY 1989MY UPTO 1999MY | GROUP P | Interior Trim
Trim
Facia Centre Console

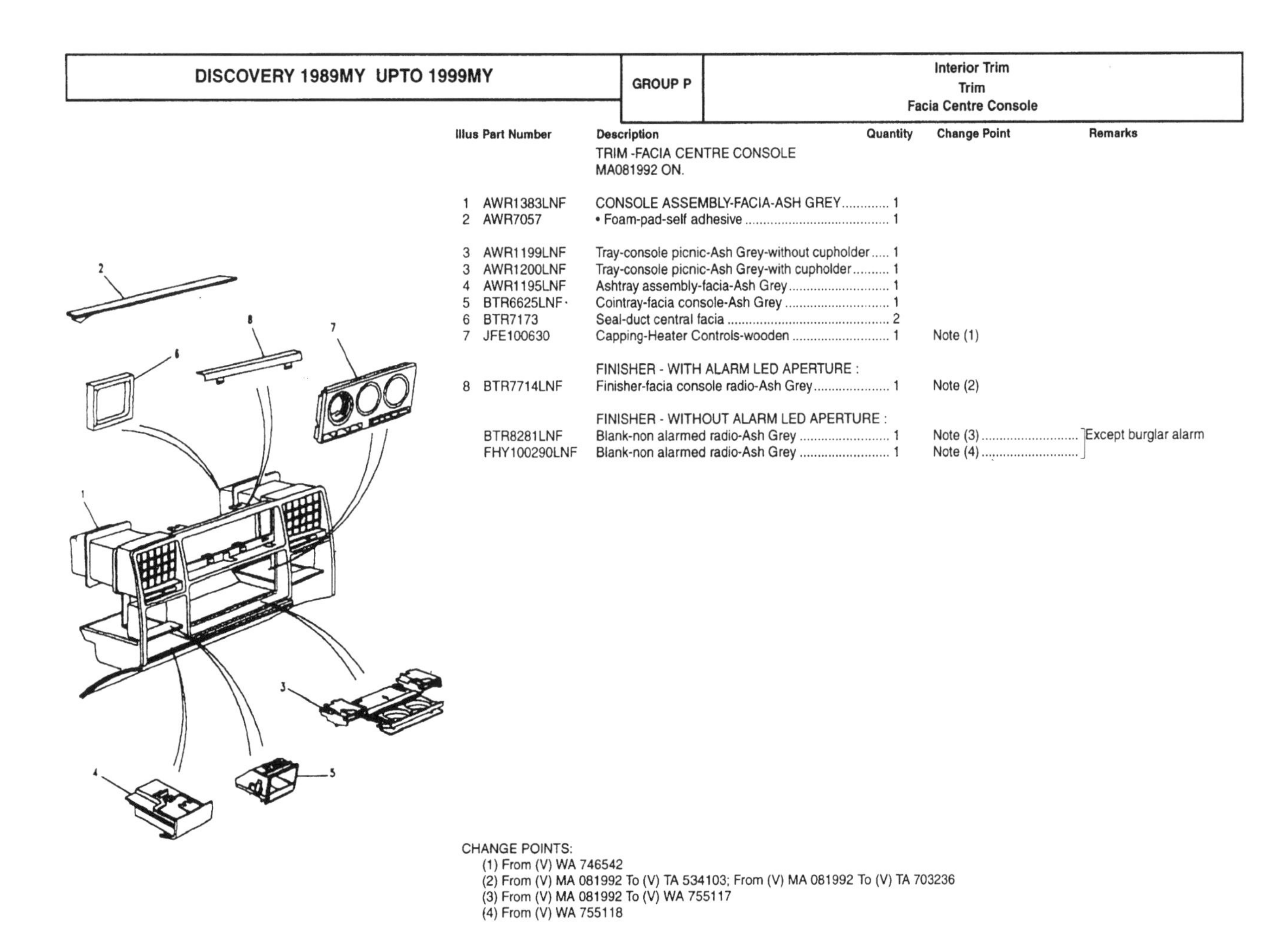

Illus	Part Number	Description	Quantity	Change Point	Remarks
		TRIM -FACIA CENTRE CONSOLE MA081992 ON.			
1	AWR1383LNF	CONSOLE ASSEMBLY-FACIA-ASH GREY	1		
2	AWR7057	• Foam-pad-self adhesive	1		
3	AWR1199LNF	Tray-console picnic-Ash Grey-without cupholder	1		
3	AWR1200LNF	Tray-console picnic-Ash Grey-with cupholder	1		
4	AWR1195LNF	Ashtray assembly-facia-Ash Grey	1		
5	BTR6625LNF	Cointray-facia console-Ash Grey	1		
6	BTR7173	Seal-duct central facia	2		
7	JFE100630	Capping-Heater Controls-wooden	1	Note (1)	
		FINISHER - WITH ALARM LED APERTURE :			
8	BTR7714LNF	Finisher-facia console radio-Ash Grey	1	Note (2)	
		FINISHER - WITHOUT ALARM LED APERTURE :			
	BTR8281LNF	Blank-non alarmed radio-Ash Grey	1	Note (3)	Except burglar alarm
	FHY100290LNF	Blank-non alarmed radio-Ash Grey	1	Note (4)	

CHANGE POINTS:
(1) From (V) WA 746542
(2) From (V) MA 081992 To (V) TA 534103; From (V) MA 081992 To (V) TA 703236
(3) From (V) MA 081992 To (V) WA 755117
(4) From (V) WA 755118

P 12

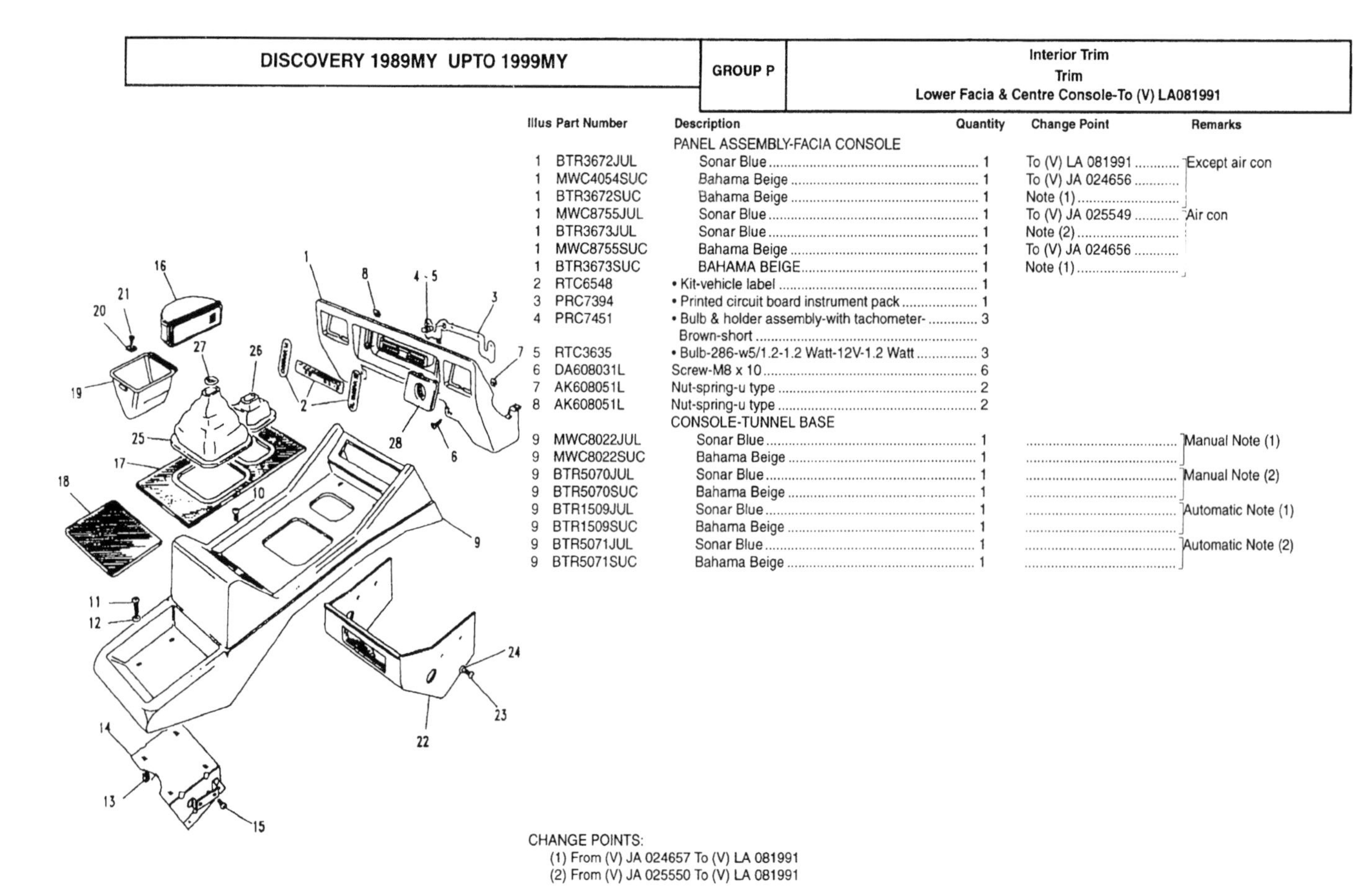

Illus	Part Number	Description	Quantity	Change Point	Remarks
		PANEL ASSEMBLY-FACIA CONSOLE			
1	BTR3672JUL	Sonar Blue	1	To (V) LA 081991	Except air con
1	MWC4054SUC	Bahama Beige	1	To (V) JA 024656	
1	BTR3672SUC	Bahama Beige	1	Note (1)	
1	MWC8755JUL	Sonar Blue	1	To (V) JA 025549	Air con
1	BTR3673JUL	Sonar Blue	1	Note (2)	
1	MWC8755SUC	Bahama Beige	1	To (V) JA 024656	
1	BTR3673SUC	BAHAMA BEIGE	1	Note (1)	
2	RTC6548	• Kit-vehicle label	1		
3	PRC7394	• Printed circuit board instrument pack	1		
4	PRC7451	• Bulb & holder assembly-with tachometer-Brown-short	3		
5	RTC3635	• Bulb-286-w5/1.2-1.2 Watt-12V-1.2 Watt	3		
6	DA608031L	Screw-M8 x 10	6		
7	AK608051L	Nut-spring-u type	2		
8	AK608051L	Nut-spring-u type	2		
		CONSOLE-TUNNEL BASE			
9	MWC8022JUL	Sonar Blue	1		Manual Note (1)
9	MWC8022SUC	Bahama Beige	1		
9	BTR5070JUL	Sonar Blue	1		Manual Note (2)
9	BTR5070SUC	Bahama Beige	1		
9	BTR1509JUL	Sonar Blue	1		Automatic Note (1)
9	BTR1509SUC	Bahama Beige	1		
9	BTR5071JUL	Sonar Blue	1		Automatic Note (2)
9	BTR5071SUC	Bahama Beige	1		

CHANGE POINTS:
(1) From (V) JA 024657 To (V) LA 081991
(2) From (V) JA 025550 To (V) LA 081991

Remarks:
(1) Except rear air conditioning
(2) rear air conditioning

P 13

DISCOVERY 1989MY UPTO 1999MY | GROUP P | Interior Trim
Trim
Lower Facia & Centre Console-To (V) LA081991

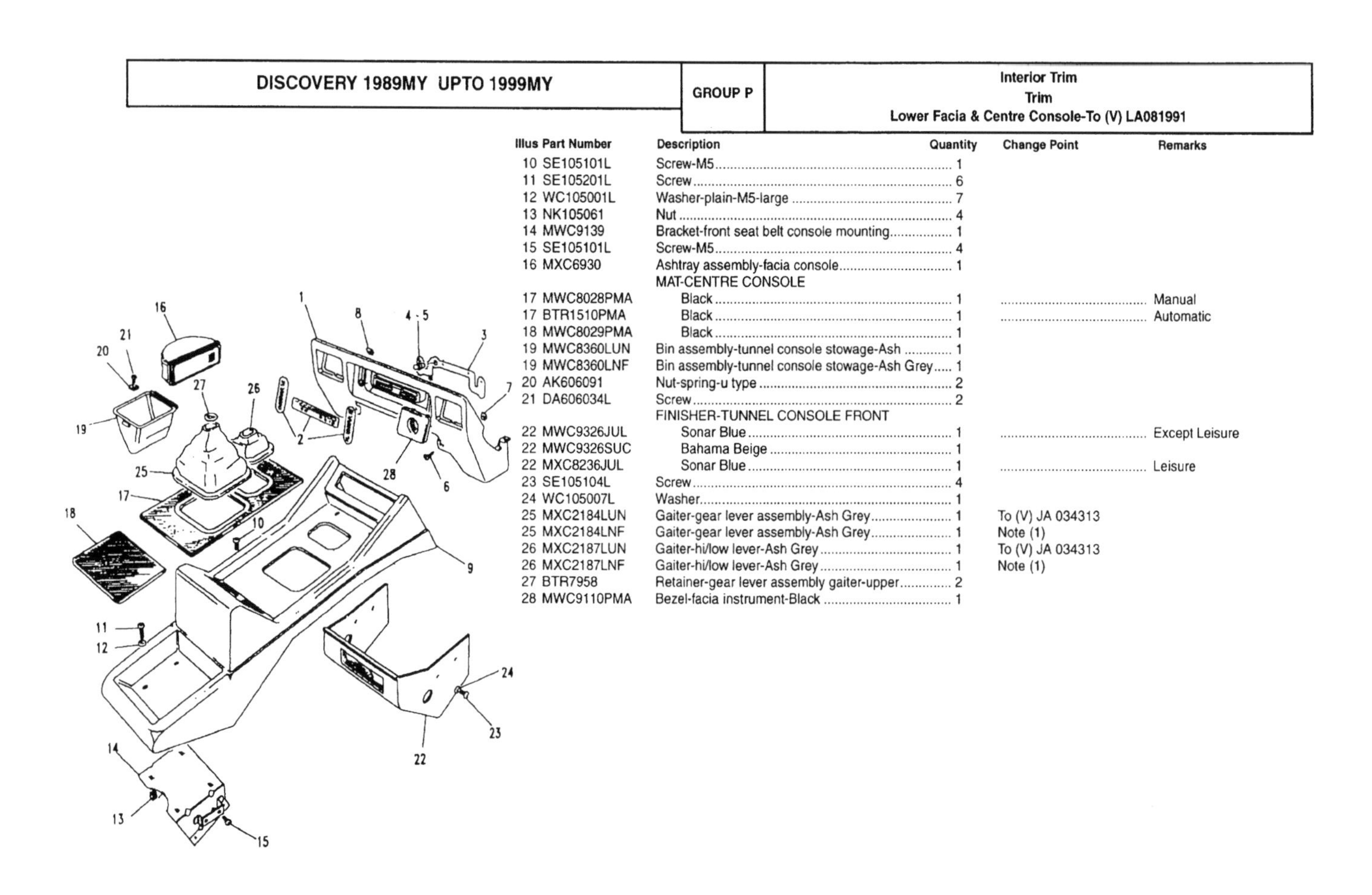

Illus	Part Number	Description	Quantity	Change Point	Remarks
10	SE105101L	Screw-M5	1		
11	SE105201L	Screw	6		
12	WC105001L	Washer-plain-M5-large	7		
13	NK105061	Nut	4		
14	MWC9139	Bracket-front seat belt console mounting	1		
15	SE105101L	Screw-M5	4		
16	MXC6930	Ashtray assembly-facia console	1		
		MAT-CENTRE CONSOLE			
17	MWC8028PMA	Black	1		Manual
17	BTR1510PMA	Black	1		Automatic
18	MWC8029PMA	Black	1		
19	MWC8360LUN	Bin assembly-tunnel console stowage-Ash	1		
19	MWC8360LNF	Bin assembly-tunnel console stowage-Ash Grey	1		
20	AK606091	Nut-spring-u type	2		
21	DA606034L	Screw	2		
		FINISHER-TUNNEL CONSOLE FRONT			
22	MWC9326JUL	Sonar Blue	1		Except Leisure
22	MWC9326SUC	Bahama Beige	1		
22	MXC8236JUL	Sonar Blue	1		Leisure
23	SE105104L	Screw	4		
24	WC105007L	Washer	1		
25	MXC2184LUN	Gaiter-gear lever assembly-Ash Grey	1	To (V) JA 034313	
25	MXC2184LNF	Gaiter-gear lever assembly-Ash Grey	1	Note (1)	
26	MXC2187LUN	Gaiter-hi/low lever-Ash Grey	1	To (V) JA 034313	
26	MXC2187LNF	Gaiter-hi/low lever-Ash Grey	1	Note (1)	
27	BTR7958	Retainer-gear lever assembly gaiter-upper	2		
28	MWC9110PMA	Bezel-facia instrument-Black	1		

CHANGE POINTS:
(1) From (V) KA 034314 To (V) LA 081991

P 14

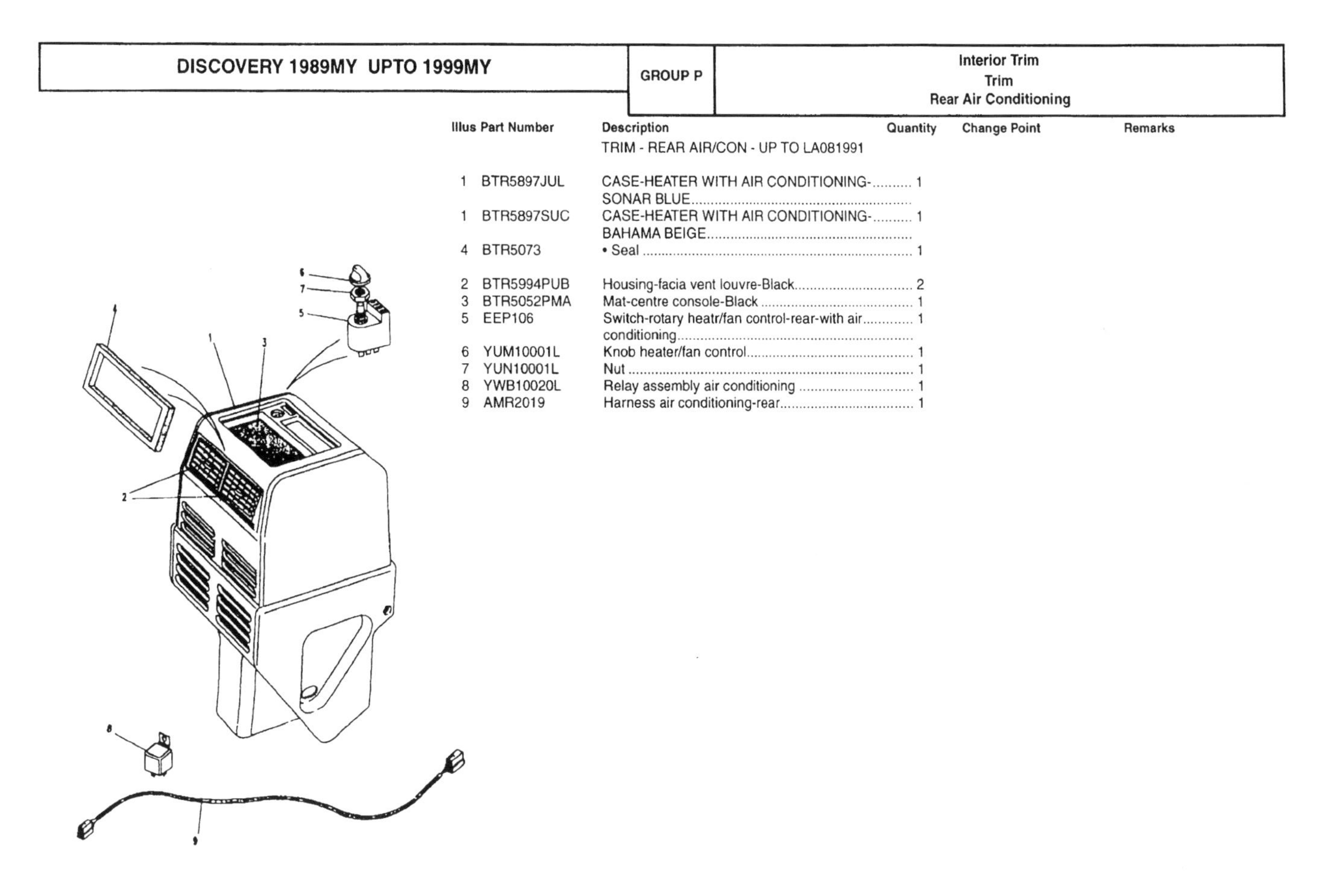

Illus	Part Number	Description	Quantity	Change Point	Remarks
		TRIM - REAR AIR/CON - UP TO LA081991			
1	BTR5897JUL	CASE-HEATER WITH AIR CONDITIONING-SONAR BLUE	1		
1	BTR5897SUC	CASE-HEATER WITH AIR CONDITIONING-BAHAMA BEIGE	1		
4	BTR5073	• Seal	1		
2	BTR5994PUB	Housing-facia vent louvre-Black	2		
3	BTR5052PMA	Mat-centre console-Black	1		
5	EEP106	Switch-rotary heatr/fan control-rear-with air conditioning	1		
6	YUM10001L	Knob heater/fan control	1		
7	YUN10001L	Nut	1		
8	YWB10020L	Relay assembly air conditioning	1		
9	AMR2019	Harness air conditioning-rear	1		

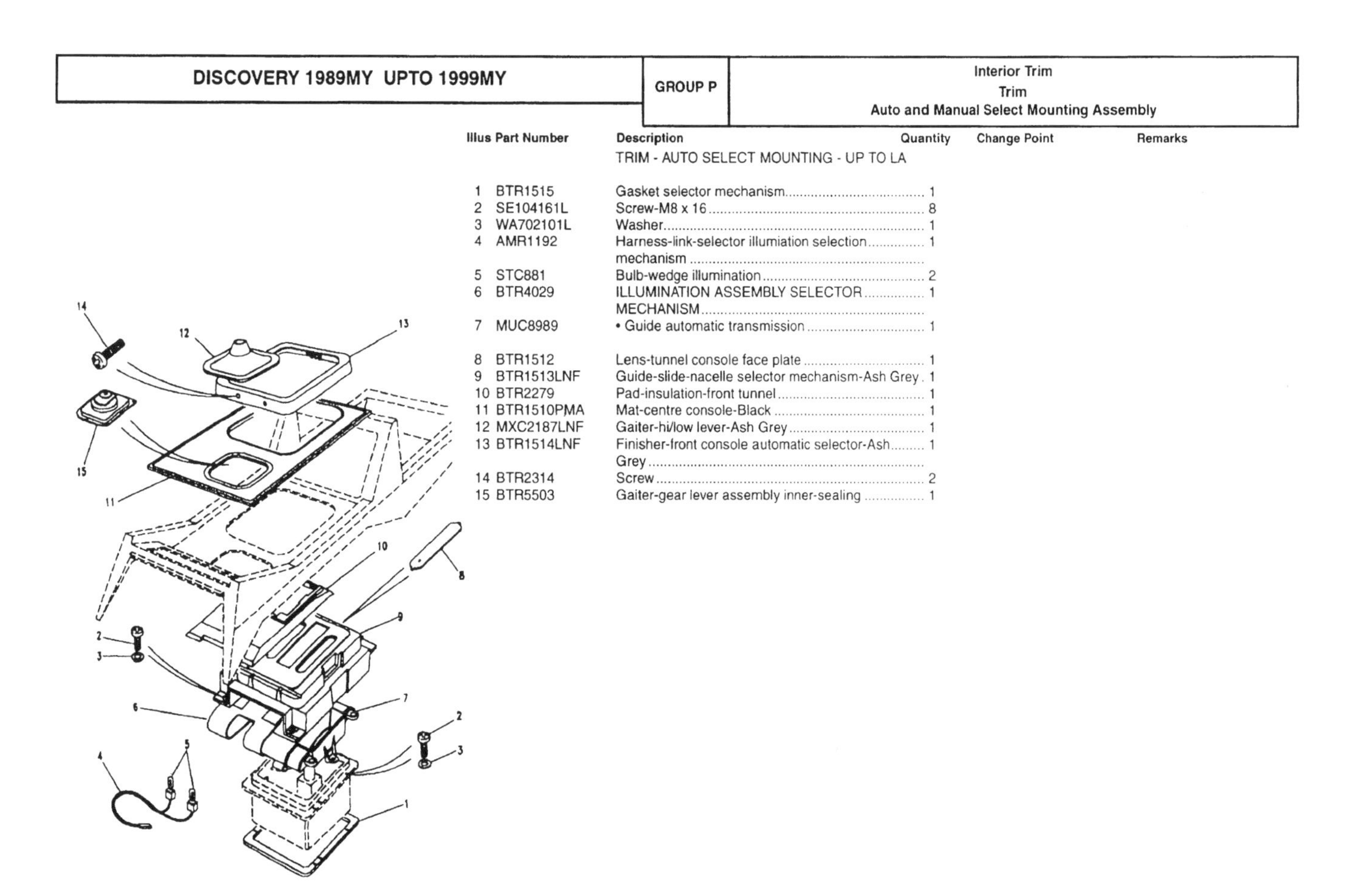

Illus	Part Number	Description	Quantity	Change Point	Remarks
		TRIM - AUTO SELECT MOUNTING - UP TO LA			
1	BTR1515	Gasket selector mechanism	1		
2	SE104161L	Screw-M8 x 16	8		
3	WA702101L	Washer	1		
4	AMR1192	Harness-link-selector illumiation selection mechanism	1		
5	STC881	Bulb-wedge illumination	2		
6	BTR4029	ILLUMINATION ASSEMBLY SELECTOR MECHANISM	1		
7	MUC8989	• Guide automatic transmission	1		
8	BTR1512	Lens-tunnel console face plate	1		
9	BTR1513LNF	Guide-slide-nacelle selector mechanism-Ash Grey	1		
10	BTR2279	Pad-insulation-front tunnel	1		
11	BTR1510PMA	Mat-centre console-Black	1		
12	MXC2187LNF	Gaiter-hi/low lever-Ash Grey	1		
13	BTR1514LNF	Finisher-front console automatic selector-Ash Grey	1		
14	BTR2314	Screw	2		
15	BTR5503	Gaiter-gear lever assembly inner-sealing	1		

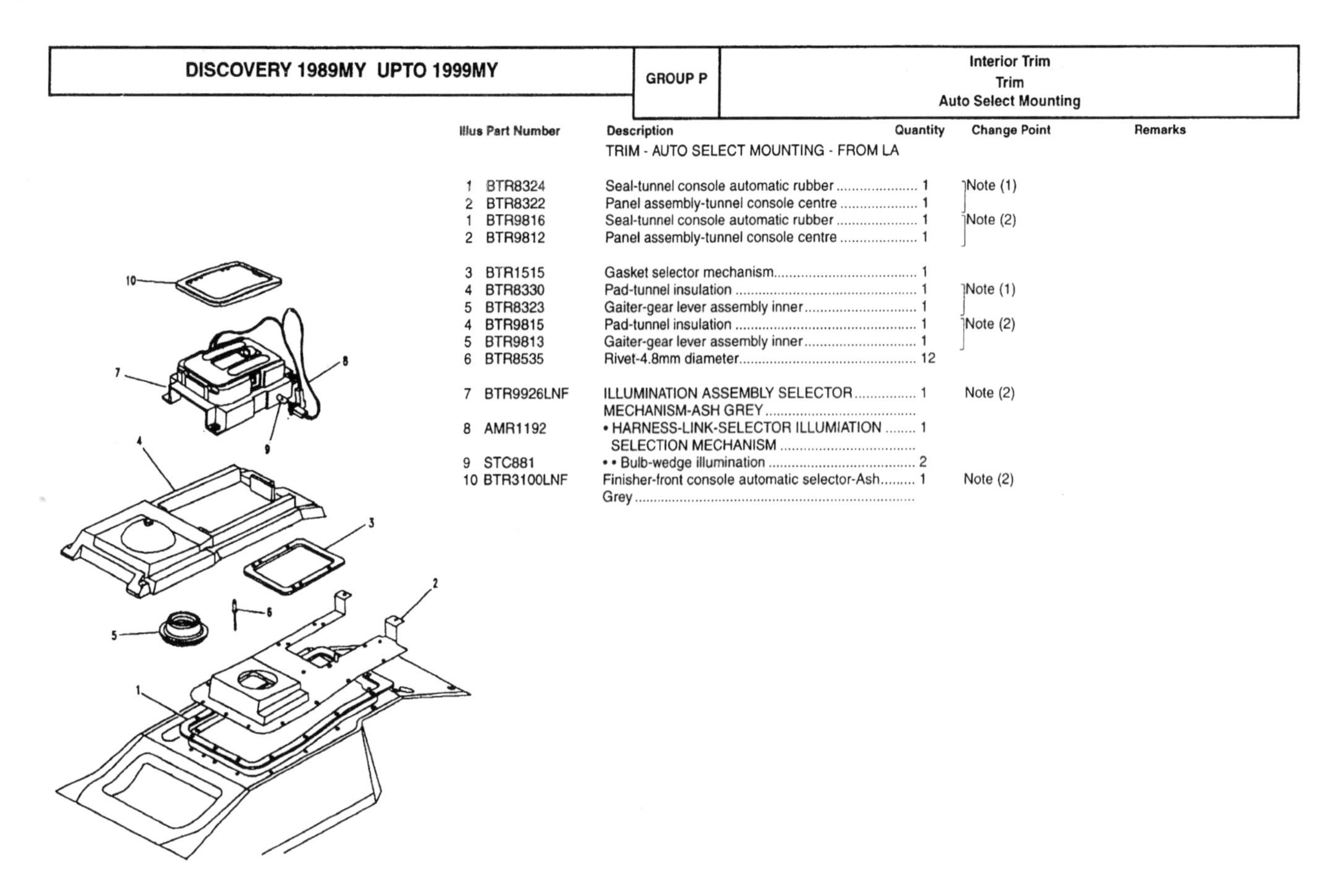

TRIM - AUTO SELECT MOUNTING - FROM LA

Illus	Part Number	Description	Quantity	Change Point	Remarks
1	BTR8324	Seal-tunnel console automatic rubber	1	Note (1)	
2	BTR8322	Panel assembly-tunnel console centre	1		
1	BTR9816	Seal-tunnel console automatic rubber	1	Note (2)	
2	BTR9812	Panel assembly-tunnel console centre	1		
3	BTR1515	Gasket selector mechanism	1		
4	BTR8330	Pad-tunnel insulation	1	Note (1)	
5	BTR8323	Gaiter-gear lever assembly inner	1		
4	BTR9815	Pad-tunnel insulation	1	Note (2)	
5	BTR9813	Gaiter-gear lever assembly inner	1		
6	BTR8535	Rivet-4.8mm diameter	12		
7	BTR9926LNF	ILLUMINATION ASSEMBLY SELECTOR MECHANISM-ASH GREY	1	Note (2)	
8	AMR1192	• HARNESS-LINK-SELECTOR ILLUMIATION SELECTION MECHANISM	1		
9	STC881	• • Bulb-wedge illumination	2		
10	BTR3100LNF	Finisher-front console automatic selector-Ash Grey	1	Note (2)	

CHANGE POINTS:
(1) From (V) LA 064755 To (V) LA 081991
(2) From (V) MA 081992

Illus	Part Number	Description	Quantity	Change Point	Remarks
1	STC1445	KIT-CONSOLE BAG-BAHAMA BEIGE	1	Note (1)	
2	BTR3063SUC	• Bag-console-Bahama Beige	1		
3	MXC6292LUN	• Fastener-quarter turn-Grey	2		
4	MXC6569LUN	• Rivet-Grey	4		
1	STC1444	KIT-CONSOLE BAG-SONAR BLUE	NLA	Note (1)	Use STC1574.
2	BTR3063JUL	• Bag-console-Sonar Blue	1		
3	MXC6292LNF	• Fastener-quarter turn-Ash Grey	2		
4	MXC6569LUN	• Rivet-Grey	4		
1	STC1575	KIT-CONSOLE BAG-BAHAMA BEIGE	1	Note (2)	
2	NSP	• Bag-console-Bahama Beige	NLA		Use STC1575 Bag no longer serviced separate- - use kit.
3	MXC6292LNF	• Fastener-quarter turn-Ash Grey	2		
4	MXC6569LUN	• Rivet-Grey	4		
1	STC1574	KIT-CONSOLE BAG-SONAR BLUE	1	Note (2)	
2	NSP	• Bag-console-Sonar Blue	NLA		Use STC1574 Bag no longer serviced separate- - use kit.
3	MXC6292LNF	• Fastener-quarter turn-Ash Grey	2		
4	MXC6569LNF	• Rivet-Ash Grey	4		

CHANGE POINTS:
(1) To (V) KA 064754
(2) From (V) LA 064755

Centre Console & Cubby Box-From (V) MA081992 To MA163103

Illus	Part Number	Description	Quantity	Change Point	Remarks
		CONSOLE ASSEMBLY-TUNNEL			
1	BTR4285LOY	Dark Granite	1		Manual
1	BTR4285SUC	Bahama Beige	1		
1	BTR4286SUC	Bahama Beige	1		Automatic
1	BTR4286LOY	Dark Granite	1		
2	STC1806LOY	Lid assembly-tunnel console-Dark Granite	1		
2	STC1807SUC	Lid assembly-tunnel console-Bahama Beige	1		
3	BTR9910LNF	Hinge assembly-tunnel console lid-Ash Grey	1		
4	BTR3089LNF	Finisher-tunnel console centre-Ash Grey	1		
6	AWR1439	Housing-tunnel console switch	NLA		Use FHR100500PMA.
7	AWR1399	Ashtray assembly-tunnel console	1		
8	AWR1183	Label-warning-manual	1		Manual
8	BTR9547	Label-warning-automatic	1		Automatic
9	BTR8901LNF	Latch assembly-tunnel console lid-Ash Grey	1		
10	BTR8902LNF	Striker-centre console stowage box lid-Ash Grey	1		
11	BTR3644LNF	Housing-tunnel console switch-Ash Grey	1		
12	BTR3087PMA	Mat-centre console-Black-manual	1		Manual
12	BTR3088PMA	Mat-centre console-automatic-Black	1		Automatic
13	BTR9217LNF	Gaiter-hi/low lever-Ash Grey	1	Note (1)	
14	BTR9840LNF	Gaiter-gear lever assembly-Ash Grey	1		
13	AWR3247LNF	Gaiter-hi/low lever-Ash Grey	1	Note (2)	
14	AWR3246LNF	Gaiter-gear lever assembly-Ash Grey	1		
15	BTR7958	Retainer-gear lever assembly gaiter-upper	1		
16	BTR8842	Fastener-expanding	4		
17	NK105061	Nut	2		
18	SE105164L	Screw-M5 x 16	4		
19	ADU1670L	Clip-spring steel	4		
20	AC606044	Screw-self tapping	4		
21	MXC5713	Pad-anti rattle	A/R		

FOR CIGAR LIGHTER (5) SEE ELECTRICS

CAPXCC2A

CHANGE POINTS:
(1) From (V) MA 081992 To (V) MA 152977
(2) From (V) MA 152978 To (V) MA 163103

Centre Console & Cubby Box-From (V) TA163104 To TA501920

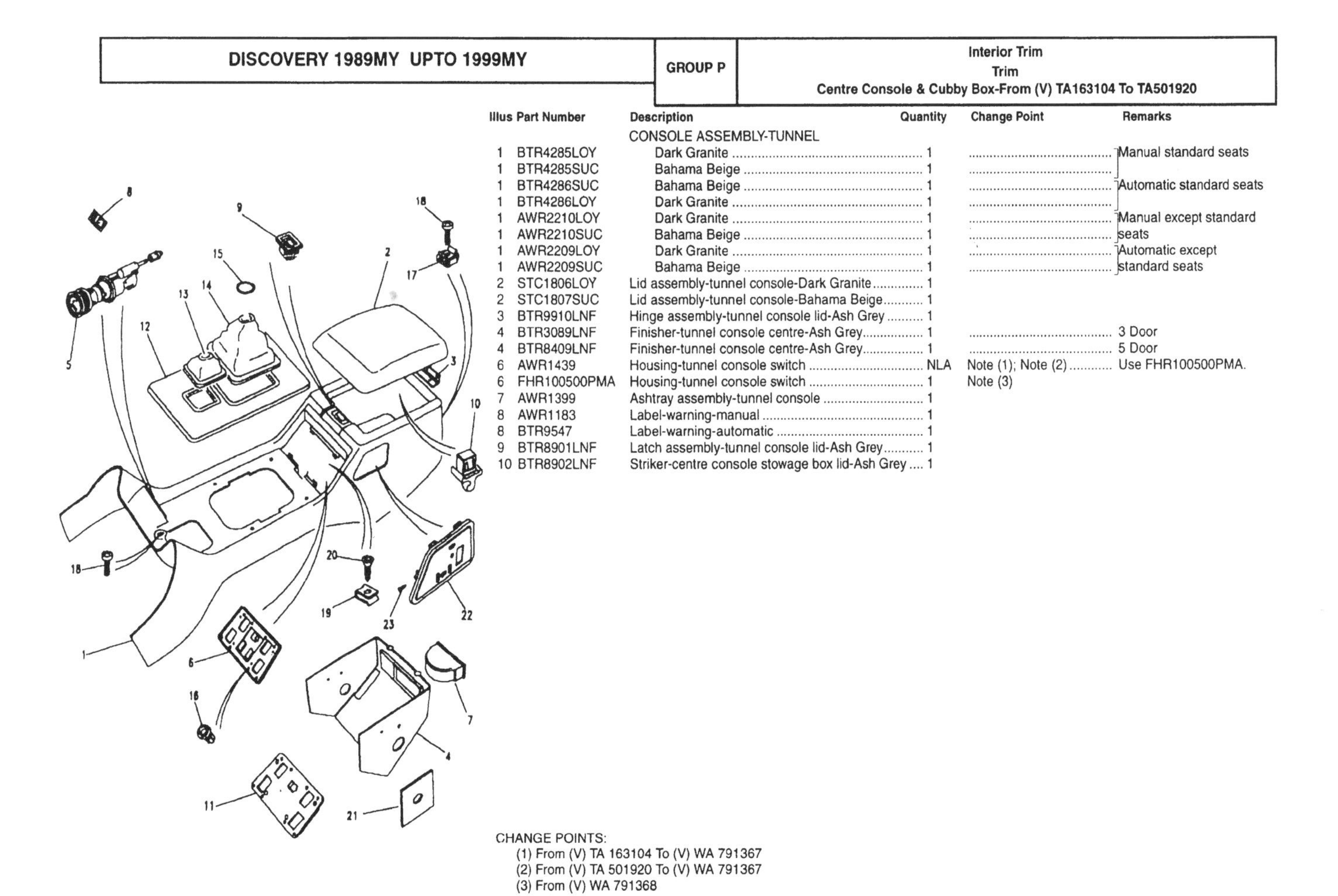

Illus	Part Number	Description	Quantity	Change Point	Remarks
		CONSOLE ASSEMBLY-TUNNEL			
1	BTR4285LOY	Dark Granite	1		Manual standard seats
1	BTR4285SUC	Bahama Beige	1		
1	BTR4286SUC	Bahama Beige	1		Automatic standard seats
1	BTR4286LOY	Dark Granite	1		
1	AWR2210LOY	Dark Granite	1		Manual except standard seats
1	AWR2210SUC	Bahama Beige	1		
1	AWR2209LOY	Dark Granite	1		Automatic except standard seats
1	AWR2209SUC	Bahama Beige	1		
2	STC1806LOY	Lid assembly-tunnel console-Dark Granite	1		
2	STC1807SUC	Lid assembly-tunnel console-Bahama Beige	1		
3	BTR9910LNF	Hinge assembly-tunnel console lid-Ash Grey	1		
4	BTR3089LNF	Finisher-tunnel console centre-Ash Grey	1		3 Door
4	BTR8409LNF	Finisher-tunnel console centre-Ash Grey	1		5 Door
6	AWR1439	Housing-tunnel console switch	NLA	Note (1); Note (2)	Use FHR100500PMA.
6	FHR100500PMA	Housing-tunnel console switch	1	Note (3)	
7	AWR1399	Ashtray assembly-tunnel console	1		
8	AWR1183	Label-warning-manual	1		
8	BTR9547	Label-warning-automatic	1		
9	BTR8901LNF	Latch assembly-tunnel console lid-Ash Grey	1		
10	BTR8902LNF	Striker-centre console stowage box lid-Ash Grey	1		

CHANGE POINTS:
(1) From (V) TA 163104 To (V) WA 791367
(2) From (V) TA 501920 To (V) WA 791367
(3) From (V) WA 791368

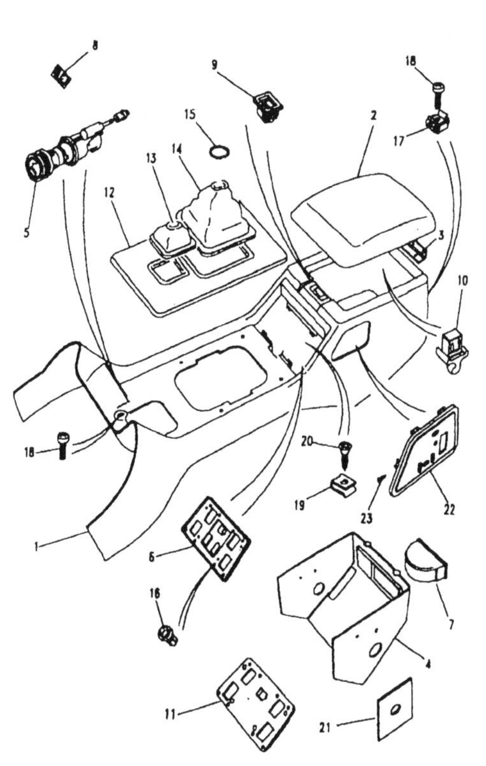

Illus	Part Number	Description	Quantity	Change Point	Remarks
11	BTR3644LNF	Housing-tunnel console switch-Ash Grey	1		Except heated seats
11	BTR3655LNF	Housing-tunnel console switch-Ash Grey	1	Note (1); Note (2)	Heated seats
12	BTR3088PMA	Mat-centre console-automatic-Black	1		Automatic heated seats
12	BTR3087PMA	Mat-centre console-Black-manual	1		Manual heated seats
13	AWR3247LNF	Gaiter-hi/low lever-Ash Grey	1	From (V) TA 163104	Heated seats
14	AWR3246LNF	Gaiter-gear lever assembly-Ash Grey	1	To (V) VA 741829	
				From (V) TA 501920	
13	FJL101530PMA	Gaiter-hi/low lever-Ash Grey	1	From (V) VA 741830	
14	FJL101680PMA	Gaiter-gear lever assembly-Ash Grey	1		
15	BTR7958	Retainer-gear lever assembly gaiter-upper	1		
16	BTR8842	Fastener-expanding	4		
17	NK105061	Nut-console to floor	2		
18	SE105164L	Screw-console to floor-M5 x 16	4		
19	ADU1670L	Clip-spring steel-switch panel to console	4		
20	AC606044	Screw-self tapping-switch panel to console	4		
21	MXC5713	Pad-anti rattle	A/R		
22	AWR2212	Blank-centre panel switch-tunnel console-RH	1		
22	AWR2211	Blank-centre panel switch-tunnel console-LH	1		
23	AB604021L	Screw-self tapping-No 4 x 1/4	4		
	STC3095	Strip-anti rattle	A/R		

CHANGE POINTS:
(1) From (V) VA 703237
(2) From (V) VA 534104

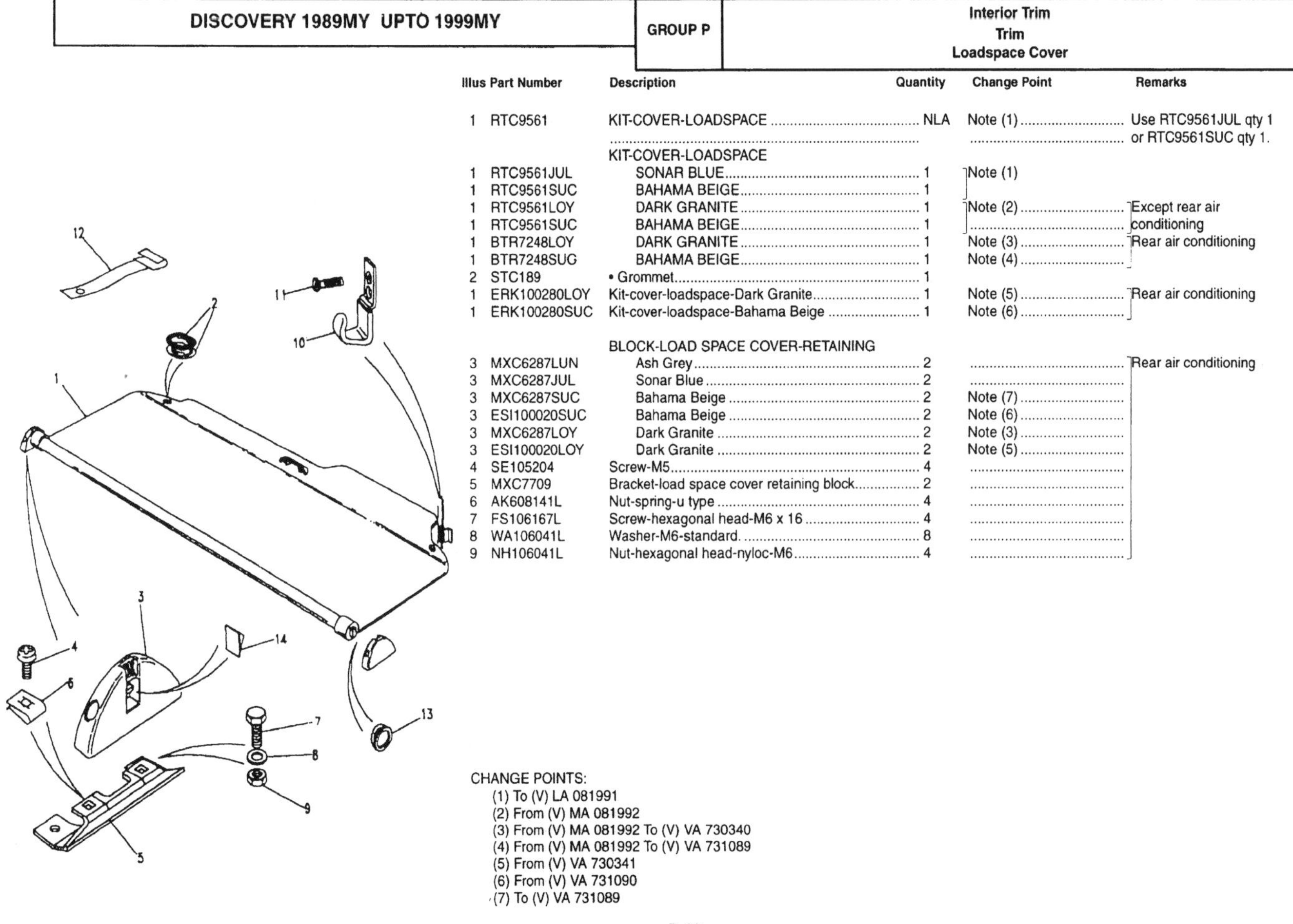

Illus	Part Number	Description	Quantity	Change Point	Remarks
1	RTC9561	KIT-COVER-LOADSPACE	NLA	Note (1)	Use RTC9561JUL qty 1 or RTC9561SUC qty 1.
		KIT-COVER-LOADSPACE			
1	RTC9561JUL	SONAR BLUE	1	Note (1)	
1	RTC9561SUC	BAHAMA BEIGE	1		
1	RTC9561LOY	DARK GRANITE	1	Note (2)	Except rear air conditioning
1	RTC9561SUC	BAHAMA BEIGE	1		
1	BTR7248LOY	DARK GRANITE	1	Note (3)	Rear air conditioning
1	BTR7248SUG	BAHAMA BEIGE	1	Note (4)	
2	STC189	• Grommet	1		
1	ERK100280LOY	Kit-cover-loadspace-Dark Granite	1	Note (5)	Rear air conditioning
1	ERK100280SUC	Kit-cover-loadspace-Bahama Beige	1	Note (6)	
		BLOCK-LOAD SPACE COVER-RETAINING			
3	MXC6287LUN	Ash Grey	2		Rear air conditioning
3	MXC6287JUL	Sonar Blue	2		
3	MXC6287SUC	Bahama Beige	2	Note (7)	
3	ESI100020SUC	Bahama Beige	2	Note (6)	
3	MXC6287LOY	Dark Granite	2	Note (3)	
3	ESI100020LOY	Dark Granite	2	Note (5)	
4	SE105204	Screw-M5	4		
5	MXC7709	Bracket-load space cover retaining block	2		
6	AK608141L	Nut-spring-u type	4		
7	FS106167L	Screw-hexagonal head-M6 x 16	4		
8	WA106041L	Washer-M6-standard.	8		
9	NH106041L	Nut-hexagonal head-nyloc-M6	4		

CHANGE POINTS:
(1) To (V) LA 081991
(2) From (V) MA 081992
(3) From (V) MA 081992 To (V) VA 730340
(4) From (V) MA 081992 To (V) VA 731089
(5) From (V) VA 730341
(6) From (V) VA 731090
(7) To (V) VA 731089

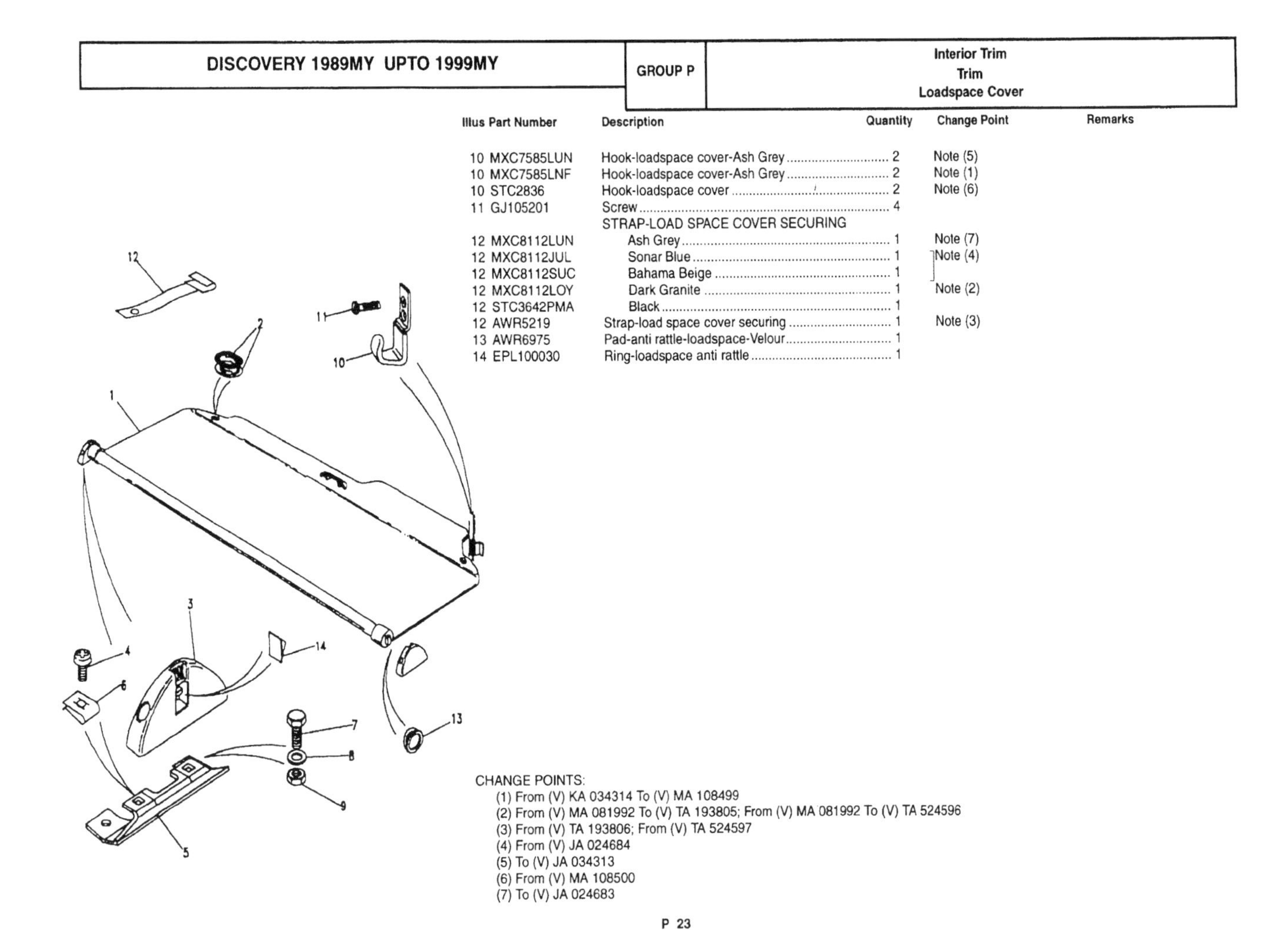

Illus	Part Number	Description	Quantity	Change Point	Remarks
10	MXC7585LUN	Hook-loadspace cover-Ash Grey	2	Note (5)	
10	MXC7585LNF	Hook-loadspace cover-Ash Grey	2	Note (1)	
10	STC2836	Hook-loadspace cover	2	Note (6)	
11	GJ105201	Screw	4		
		STRAP-LOAD SPACE COVER SECURING			
12	MXC8112LUN	Ash Grey	1	Note (7)	
12	MXC8112JUL	Sonar Blue	1	Note (4)	
12	MXC8112SUC	Bahama Beige	1		
12	MXC8112LOY	Dark Granite	1	Note (2)	
12	STC3642PMA	Black	1		
12	AWR5219	Strap-load space cover securing	1	Note (3)	
13	AWR6975	Pad-anti rattle-loadspace-Velour	1		
14	EPL100030	Ring-loadspace anti rattle	1		

CHANGE POINTS:
(1) From (V) KA 034314 To (V) MA 108499
(2) From (V) MA 081992 To (V) TA 193805; From (V) MA 081992 To (V) TA 524596
(3) From (V) TA 193806; From (V) TA 524597
(4) From (V) JA 024684
(5) To (V) JA 034313
(6) From (V) MA 108500
(7) To (V) JA 024683

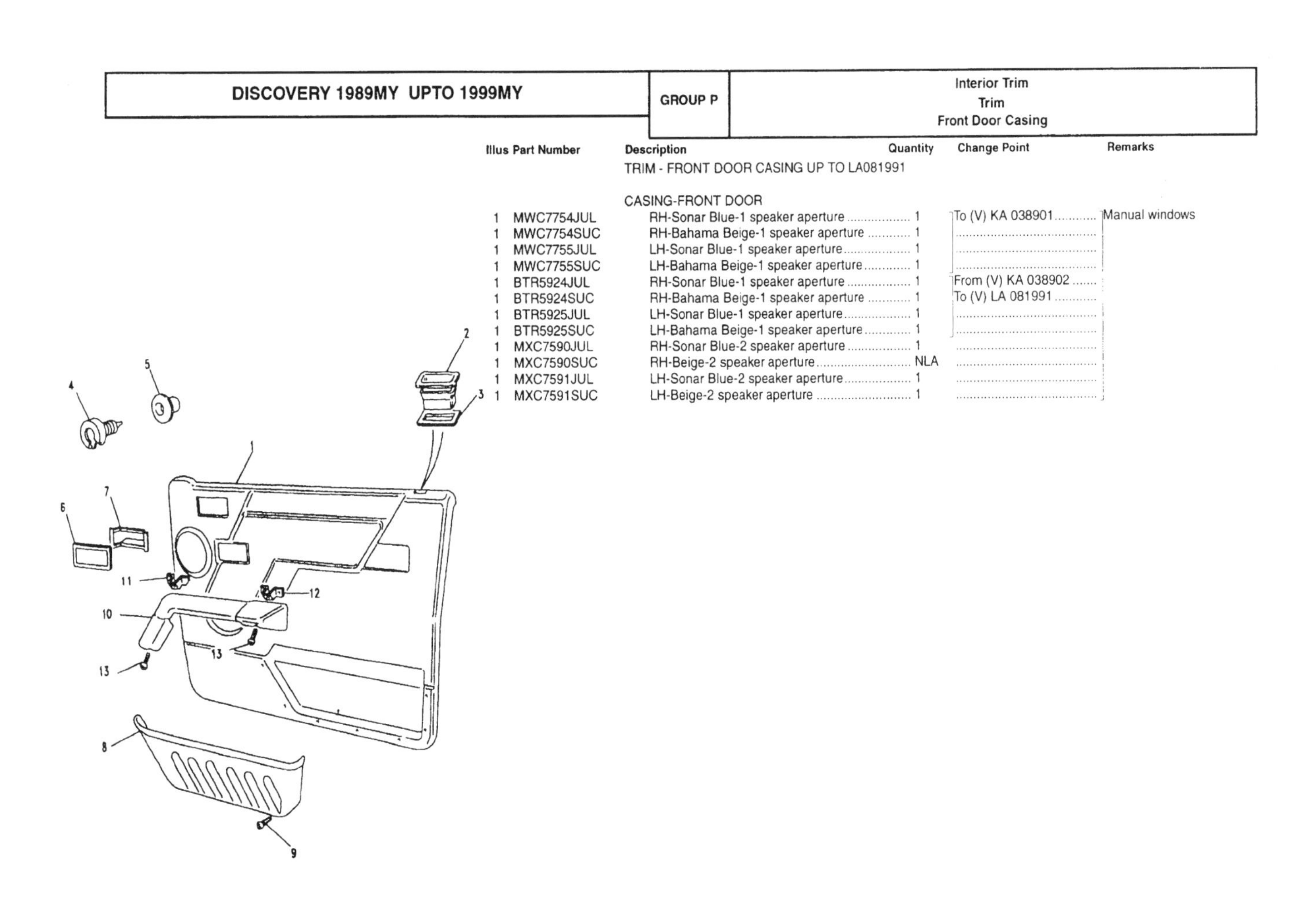

TRIM - FRONT DOOR CASING UP TO LA081991

Illus	Part Number	Description	Quantity	Change Point	Remarks
		CASING-FRONT DOOR			
1	MWC7754JUL	RH-Sonar Blue-1 speaker aperture	1	To (V) KA 038901	Manual windows
1	MWC7754SUC	RH-Bahama Beige-1 speaker aperture	1		
1	MWC7755JUL	LH-Sonar Blue-1 speaker aperture	1		
1	MWC7755SUC	LH-Bahama Beige-1 speaker aperture	1		
1	BTR5924JUL	RH-Sonar Blue-1 speaker aperture	1	From (V) KA 038902	
1	BTR5924SUC	RH-Bahama Beige-1 speaker aperture	1	To (V) LA 081991	
1	BTR5925JUL	LH-Sonar Blue-1 speaker aperture	1		
1	BTR5925SUC	LH-Bahama Beige-1 speaker aperture	1		
1	MXC7590JUL	RH-Sonar Blue-2 speaker aperture	1		
1	MXC7590SUC	RH-Beige-2 speaker aperture	NLA		
1	MXC7591JUL	LH-Sonar Blue-2 speaker aperture	1		
1	MXC7591SUC	LH-Beige-2 speaker aperture	1		

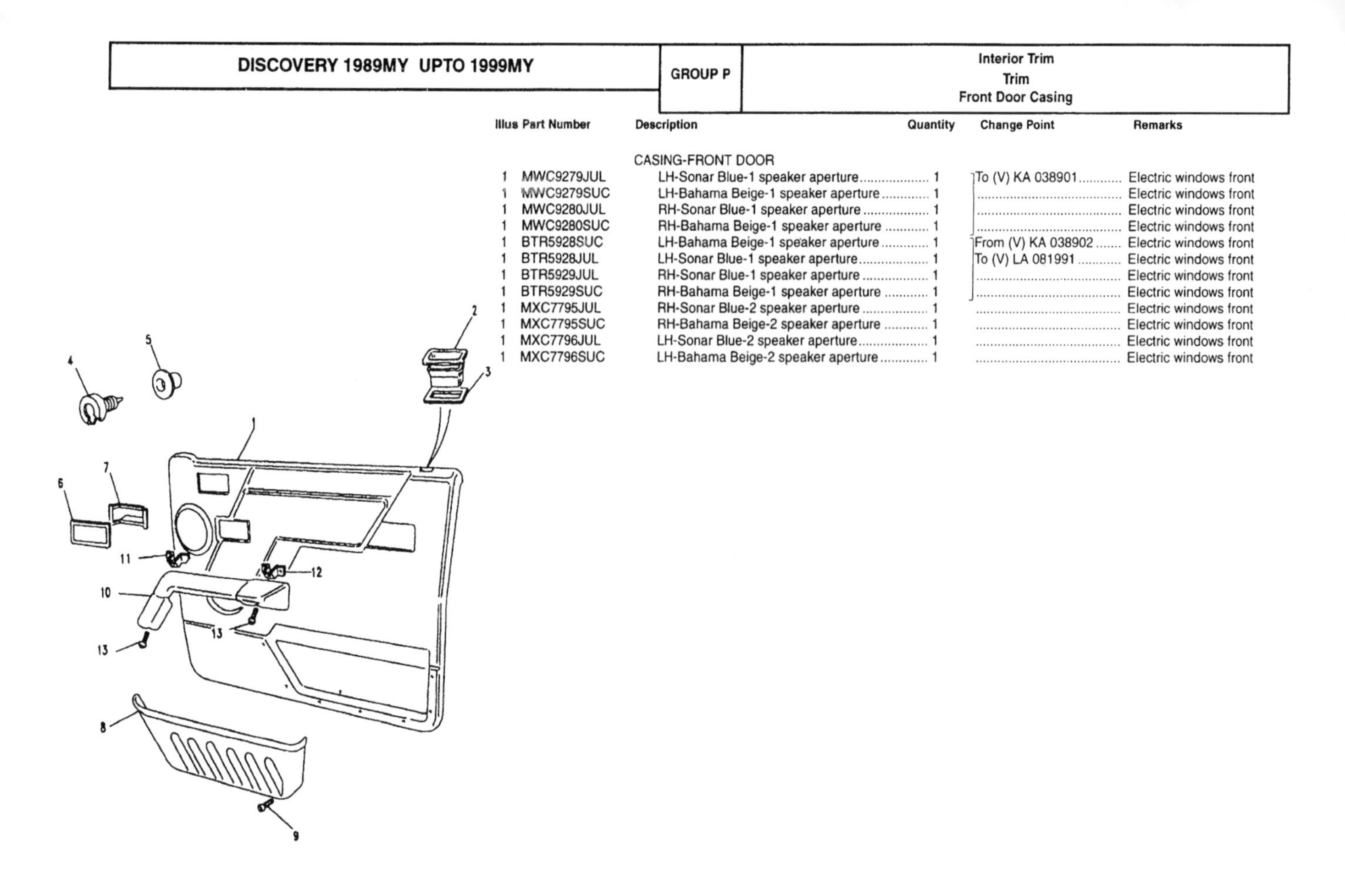

Illus	Part Number	Description	Quantity	Change Point	Remarks
		CASING-FRONT DOOR			
1	MWC9279JUL	LH-Sonar Blue-1 speaker aperture	1	To (V) KA 038901	Electric windows front
1	MWC9279SUC	LH-Bahama Beige-1 speaker aperture	1		Electric windows front
1	MWC9280JUL	RH-Sonar Blue-1 speaker aperture	1		Electric windows front
1	MWC9280SUC	RH-Bahama Beige-1 speaker aperture	1		Electric windows front
1	BTR5928SUC	LH-Bahama Beige-1 speaker aperture	1	From (V) KA 038902	Electric windows front
1	BTR5928JUL	LH-Sonar Blue-1 speaker aperture	1	To (V) LA 081991	Electric windows front
1	BTR5929JUL	RH-Sonar Blue-1 speaker aperture	1		Electric windows front
1	BTR5929SUC	RH-Bahama Beige-1 speaker aperture	1		Electric windows front
1	MXC7795JUL	RH-Sonar Blue-2 speaker aperture	1		Electric windows front
1	MXC7795SUC	RH-Bahama Beige-2 speaker aperture	1		Electric windows front
1	MXC7796JUL	LH-Sonar Blue-2 speaker aperture	1		Electric windows front
1	MXC7796SUC	LH-Bahama Beige-2 speaker aperture	1		Electric windows front

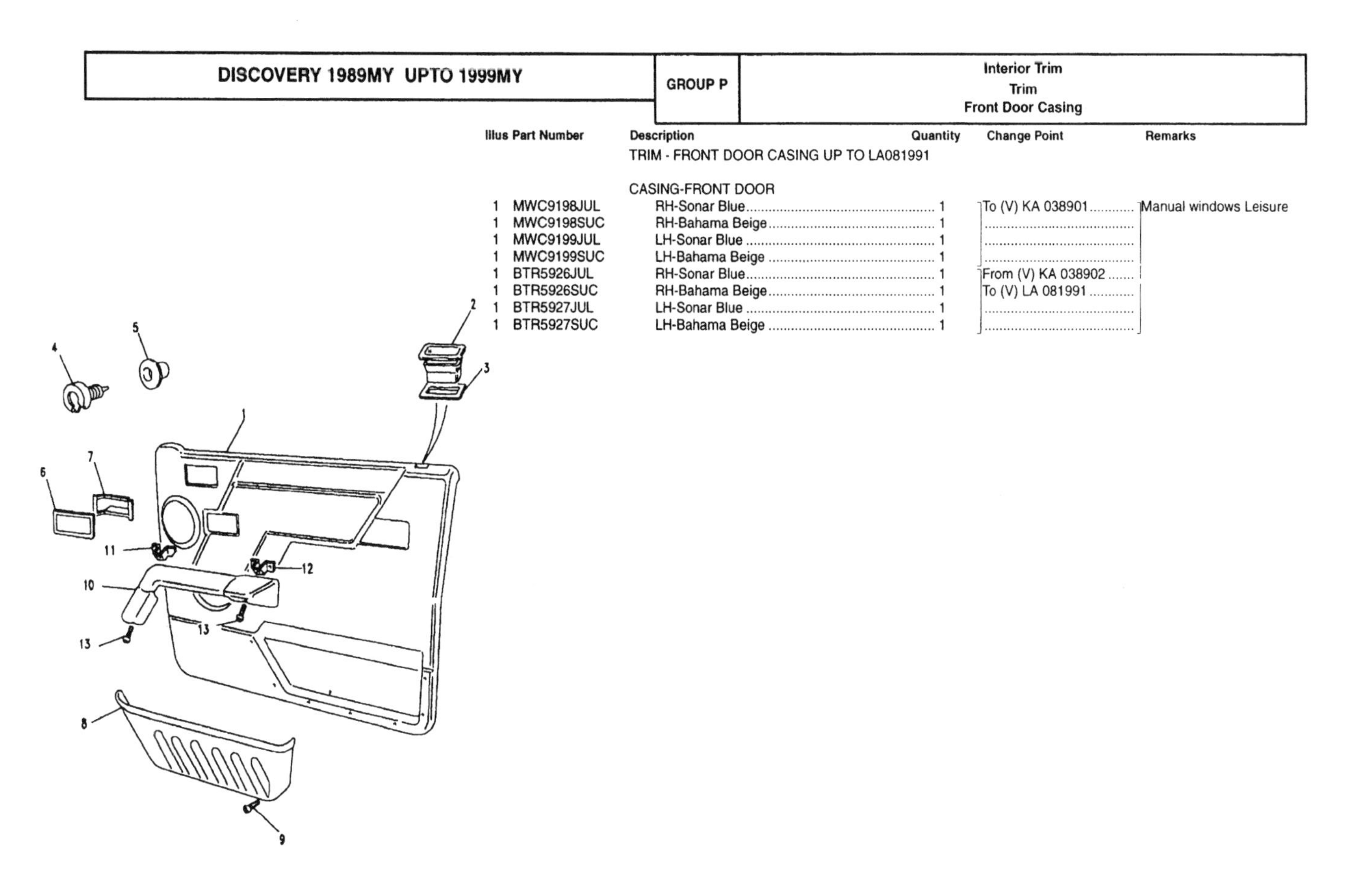

TRIM - FRONT DOOR CASING UP TO LA081991

Illus	Part Number	Description	Quantity	Change Point	Remarks
		CASING-FRONT DOOR			
1	MWC9198JUL	RH-Sonar Blue	1	To (V) KA 038901	Manual windows Leisure
1	MWC9198SUC	RH-Bahama Beige	1		
1	MWC9199JUL	LH-Sonar Blue	1		
1	MWC9199SUC	LH-Bahama Beige	1		
1	BTR5926JUL	RH-Sonar Blue	1	From (V) KA 038902	
1	BTR5926SUC	RH-Bahama Beige	1	To (V) LA 081991	
1	BTR5927JUL	LH-Sonar Blue	1		
1	BTR5927SUC	LH-Bahama Beige	1		

Illus	Part Number	Description	Quantity	Change Point	Remarks
		CASING-FRONT DOOR			
1	BTR703JUL	LH-Sonar Blue	1	To (V) KA 038901	Electric windows Leisure
1	BTR703SUC	LH-Bahama Beige	1		
1	BTR704JUL	RH-Sonar Blue	1		
1	BTR704SUC	RH-Bahama Beige	1		
1	BTR5935JUL	LH-Sonar Blue	1	From (V) KA 038902	
1	BTR5935SUC	LH-Bahama Beige	1	To (V) LA 081991	
1	BTR5934JUL	RH-Sonar Blue	1		
1	BTR5934SUC	RH-Bahama Beige	1		

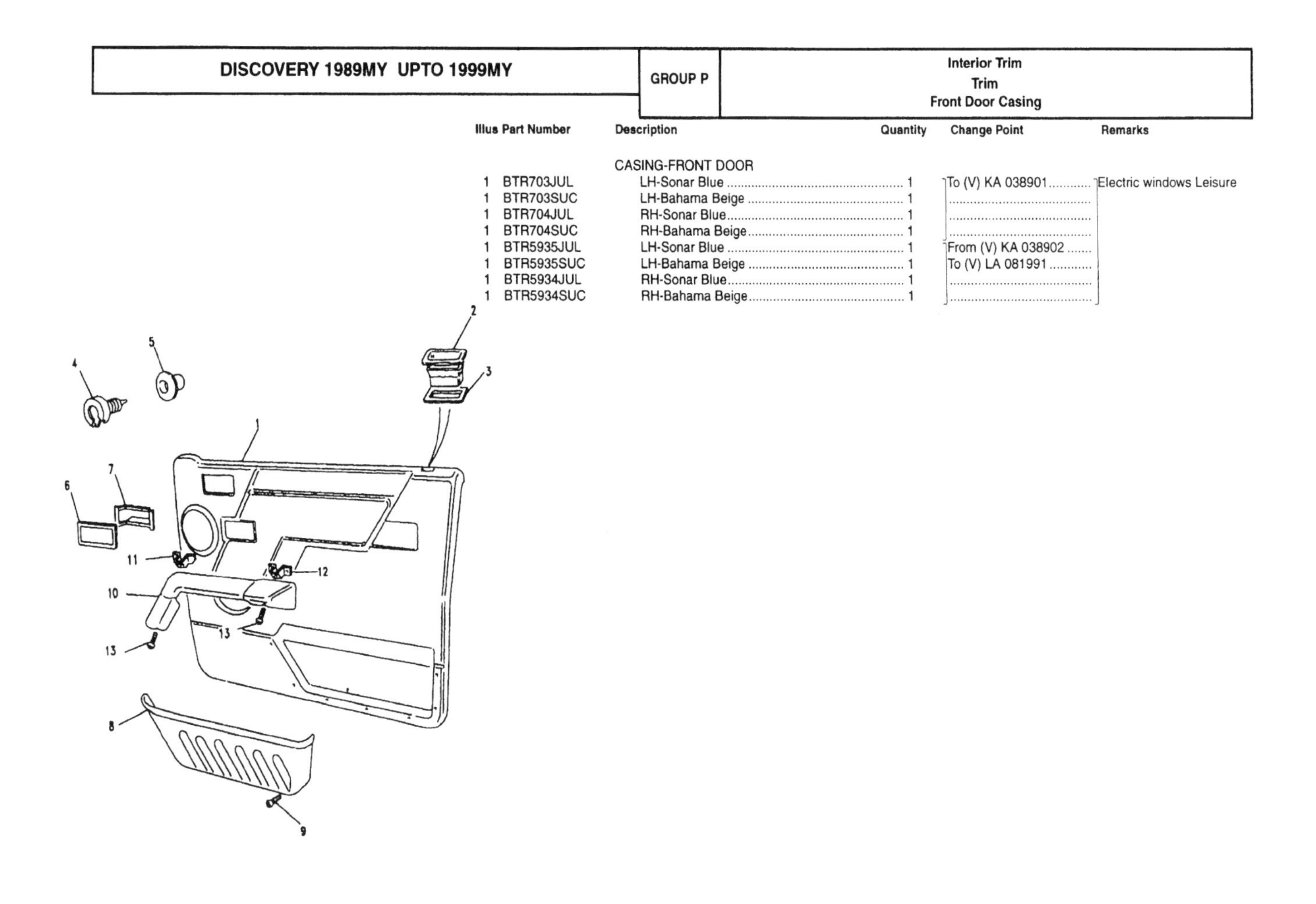

Illus	Part Number	Description	Quantity	Change Point	Remarks
		TRIM - FRONT DOOR CASING UP TO LA081991			
2	CDP1582LUN	Grommet-Grey	2	Note (1)	
2	CDP1582LNF	Grommet-Ash Grey	2	Note (2)	
3	BEP6073	Clip	2		
4	MWC9134	Stud-plastic	14	Note (3)	
5	MWC9136	Clip-snap-sack	NLA		Use EYC101460.
5	EYC101460	Clip-snap-sack	14		
6	YOO4340LUN	Ashtray-door-Ash Grey	2	Note (1)	
6	YOO4340LNF	Ashtray-door-Ash Grey	2	Note (2)	
7	CZA4736L	Clip	2		
		BIN ASSEMBLY-FRONT DOOR			
8	MWC7756JUL	RH-Sonar Blue	1		
8	MWC7756SUC	RH-Bahama Beige	1		
8	MWC7757JUL	LH-Sonar Blue	1		
8	MWC7757SUC	LH-Bahama Beige	1		
9	AB604051L	Screw	14		

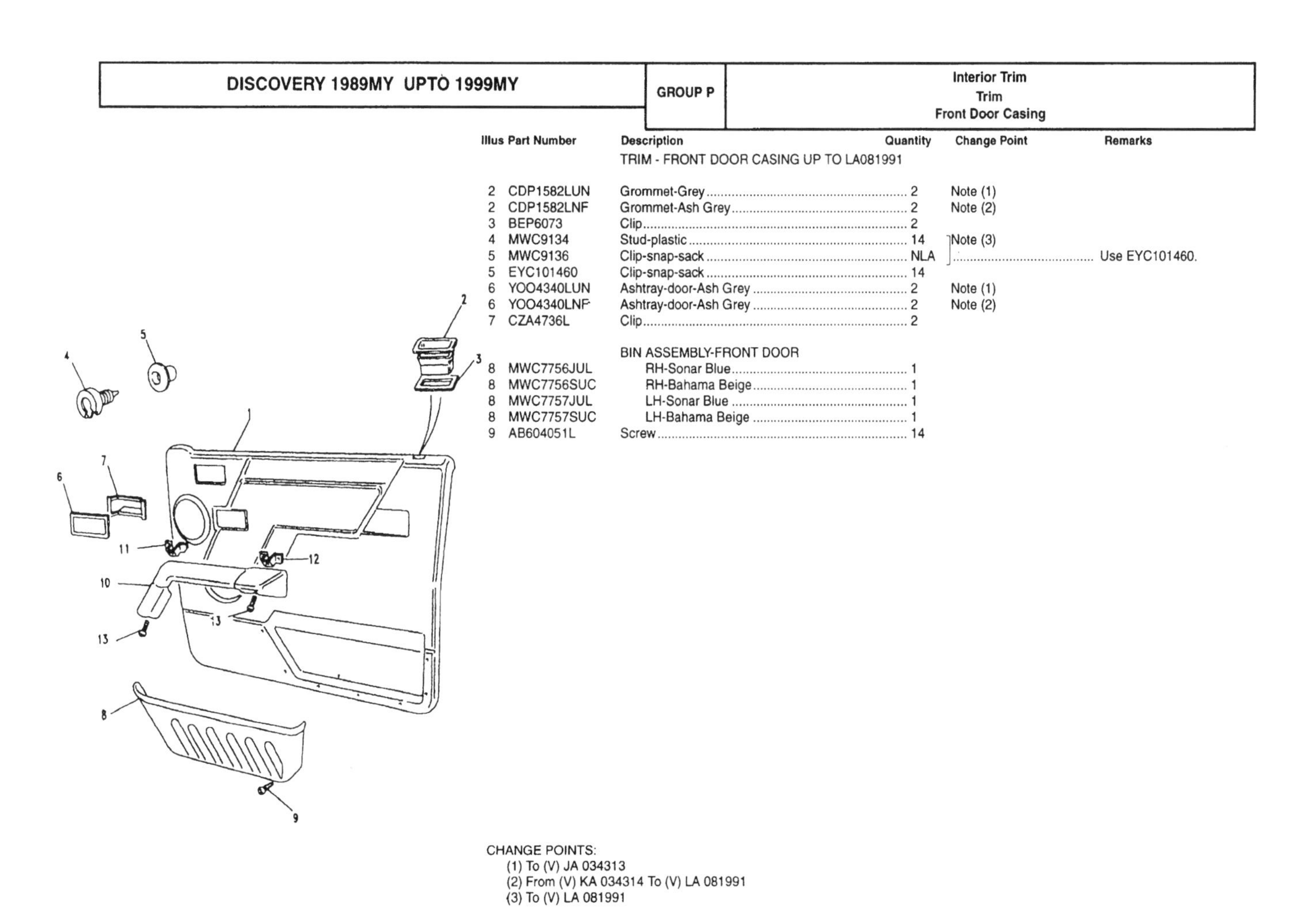

CHANGE POINTS:
(1) To (V) JA 034313
(2) From (V) KA 034314 To (V) LA 081991
(3) To (V) LA 081991

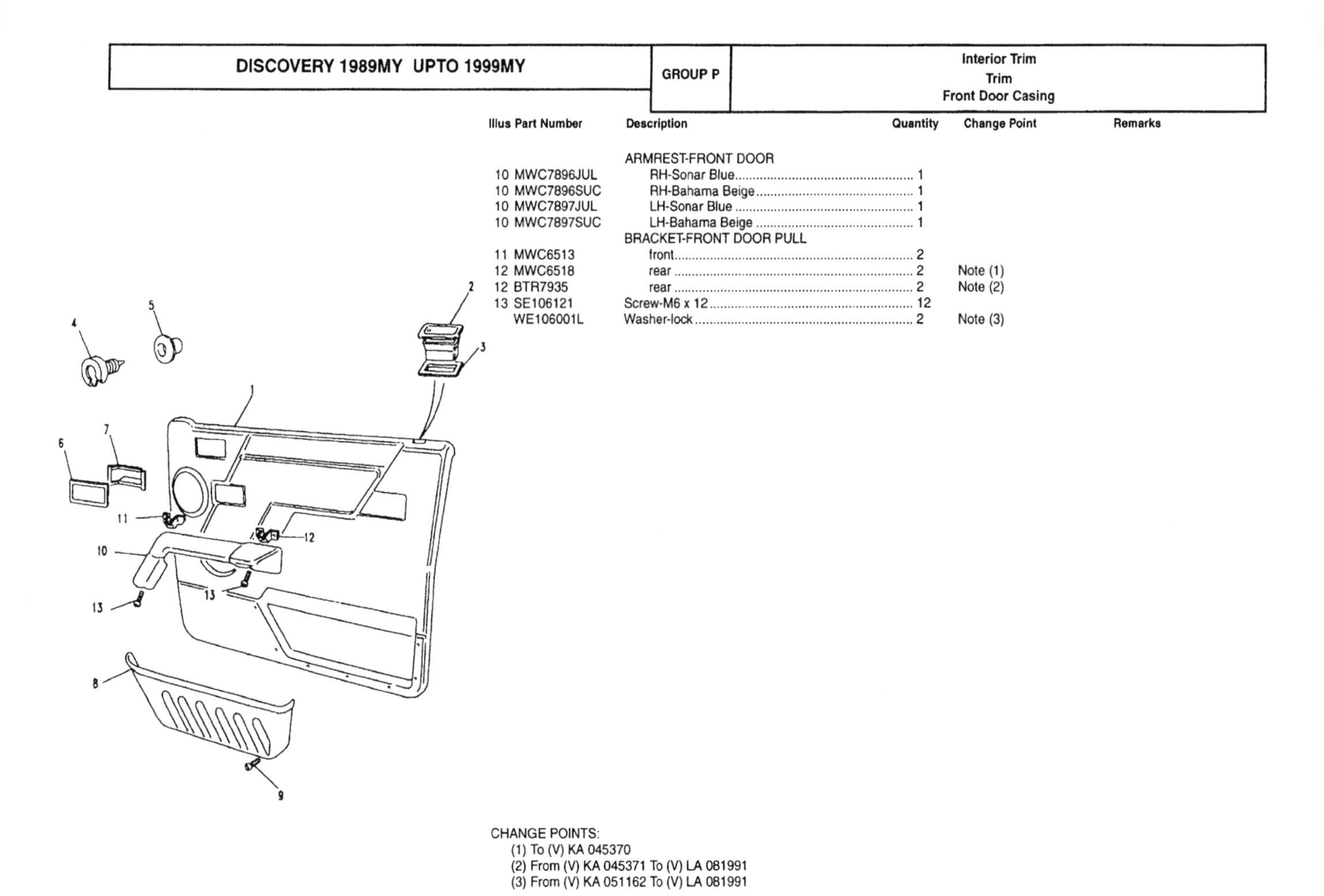

Illus	Part Number	Description	Quantity	Change Point	Remarks
		ARMREST-FRONT DOOR			
10	MWC7896JUL	RH-Sonar Blue	1		
10	MWC7896SUC	RH-Bahama Beige	1		
10	MWC7897JUL	LH-Sonar Blue	1		
10	MWC7897SUC	LH-Bahama Beige	1		
		BRACKET-FRONT DOOR PULL			
11	MWC6513	front	2		
12	MWC6518	rear	2	Note (1)	
12	BTR7935	rear	2	Note (2)	
13	SE106121	Screw-M6 x 12	12		
	WE106001L	Washer-lock	2	Note (3)	

CHANGE POINTS:
(1) To (V) KA 045370
(2) From (V) KA 045371 To (V) LA 081991
(3) From (V) KA 051162 To (V) LA 081991

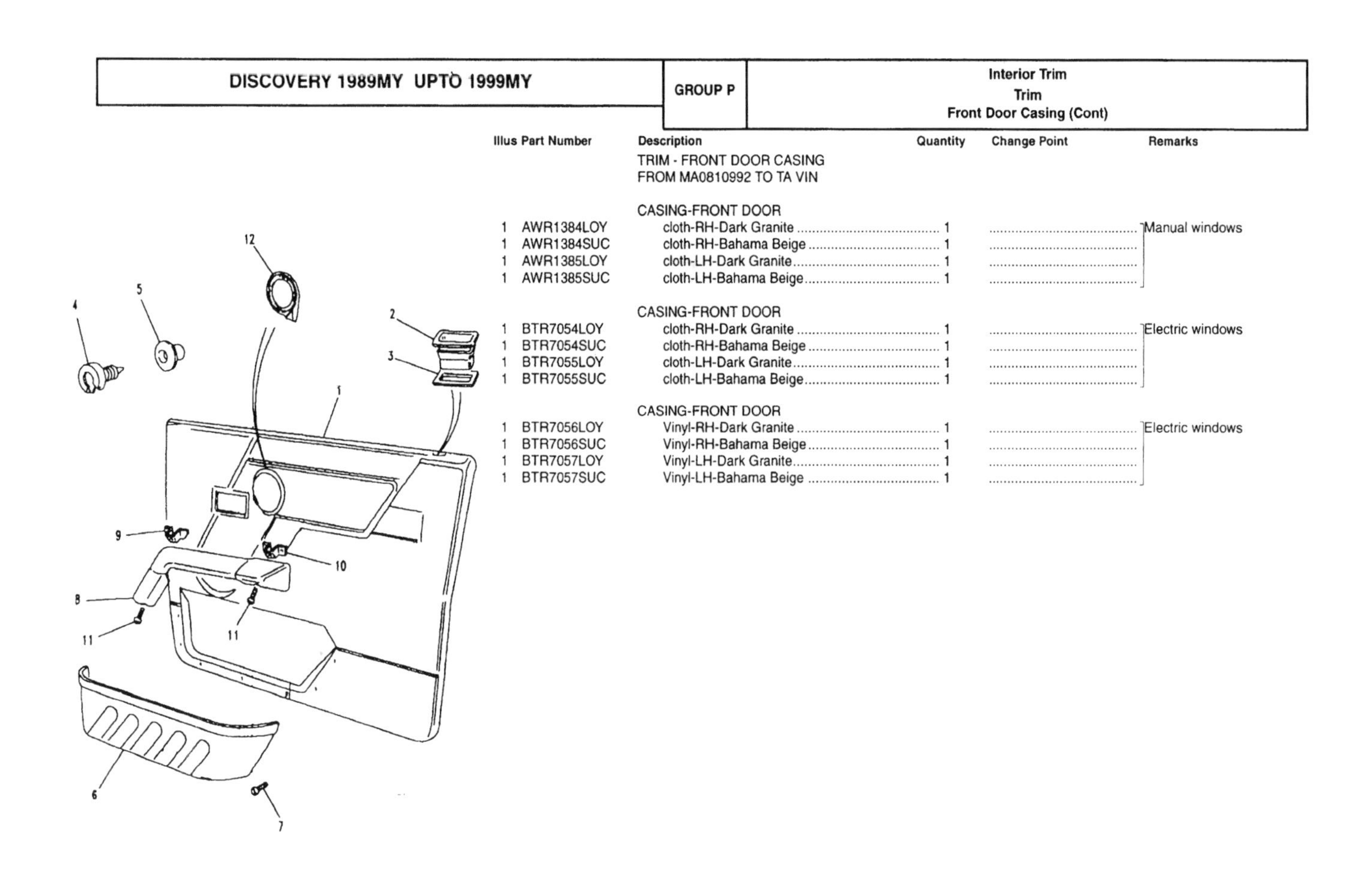

Illus	Part Number	Description	Quantity	Change Point	Remarks
		TRIM - FRONT DOOR CASING FROM MA0810992 TO TA VIN			
		CASING-FRONT DOOR			
1	AWR1384LOY	cloth-RH-Dark Granite	1		Manual windows
1	AWR1384SUC	cloth-RH-Bahama Beige	1		
1	AWR1385LOY	cloth-LH-Dark Granite	1		
1	AWR1385SUC	cloth-LH-Bahama Beige	1		
		CASING-FRONT DOOR			
1	BTR7054LOY	cloth-RH-Dark Granite	1		Electric windows
1	BTR7054SUC	cloth-RH-Bahama Beige	1		
1	BTR7055LOY	cloth-LH-Dark Granite	1		
1	BTR7055SUC	cloth-LH-Bahama Beige	1		
		CASING-FRONT DOOR			
1	BTR7056LOY	Vinyl-RH-Dark Granite	1		Electric windows
1	BTR7056SUC	Vinyl-RH-Bahama Beige	1		
1	BTR7057LOY	Vinyl-LH-Dark Granite	1		
1	BTR7057SUC	Vinyl-LH-Bahama Beige	1		

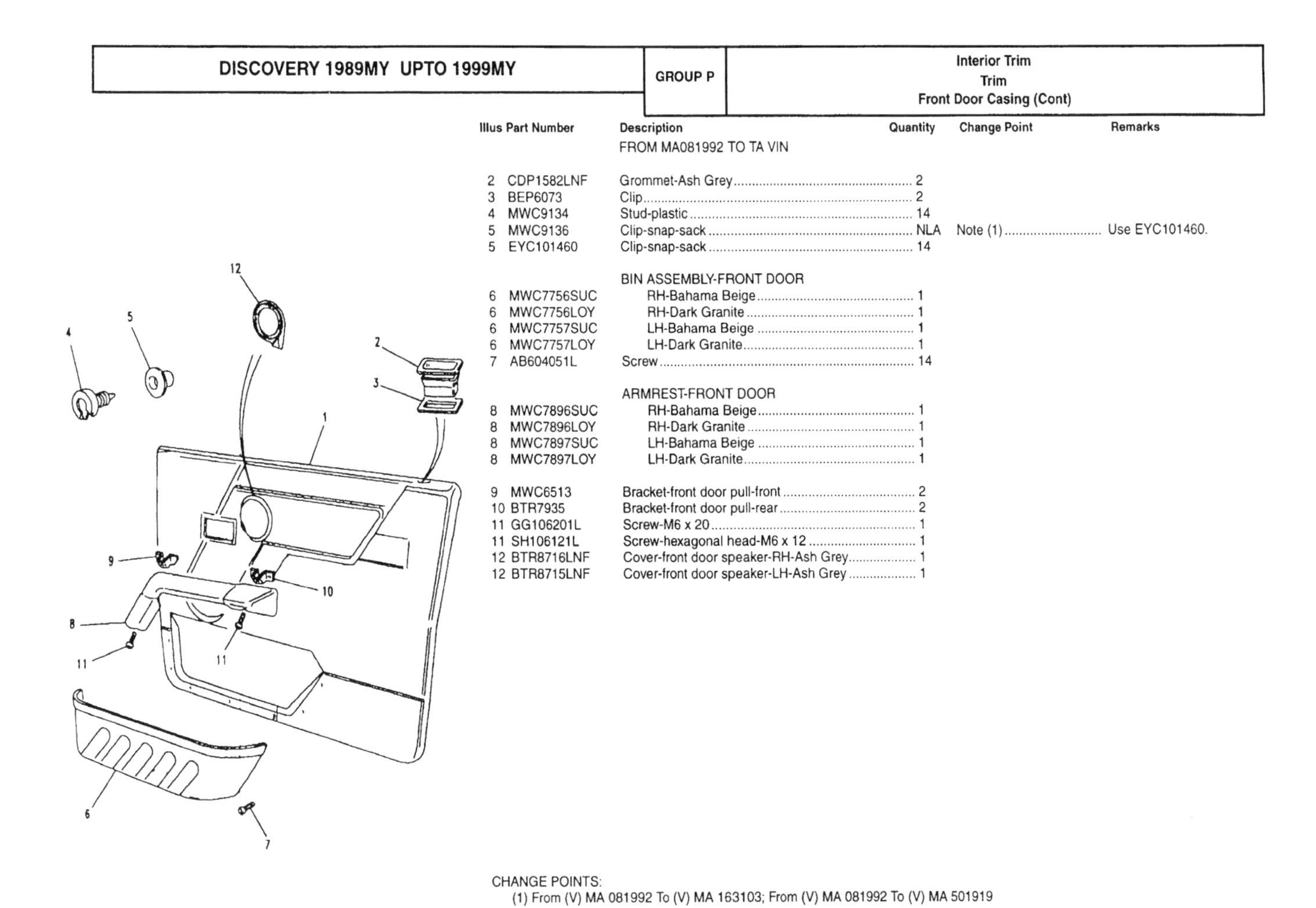

Illus	Part Number	Description	Quantity	Change Point	Remarks
		FROM MA081992 TO TA VIN			
2	CDP1582LNF	Grommet-Ash Grey	2		
3	BEP6073	Clip	2		
4	MWC9134	Stud-plastic	14		
5	MWC9136	Clip-snap-sack	NLA	Note (1)	Use EYC101460.
5	EYC101460	Clip-snap-sack	14		
		BIN ASSEMBLY-FRONT DOOR			
6	MWC7756SUC	RH-Bahama Beige	1		
6	MWC7756LOY	RH-Dark Granite	1		
6	MWC7757SUC	LH-Bahama Beige	1		
6	MWC7757LOY	LH-Dark Granite	1		
7	AB604051L	Screw	14		
		ARMREST-FRONT DOOR			
8	MWC7896SUC	RH-Bahama Beige	1		
8	MWC7896LOY	RH-Dark Granite	1		
8	MWC7897SUC	LH-Bahama Beige	1		
8	MWC7897LOY	LH-Dark Granite	1		
9	MWC6513	Bracket-front door pull-front	2		
10	BTR7935	Bracket-front door pull-rear	2		
11	GG106201L	Screw-M6 x 20	1		
11	SH106121L	Screw-hexagonal head-M6 x 12	1		
12	BTR8716LNF	Cover-front door speaker-RH-Ash Grey	1		
12	BTR8715LNF	Cover-front door speaker-LH-Ash Grey	1		

CHANGE POINTS:
(1) From (V) MA 081992 To (V) MA 163103; From (V) MA 081992 To (V) MA 501919

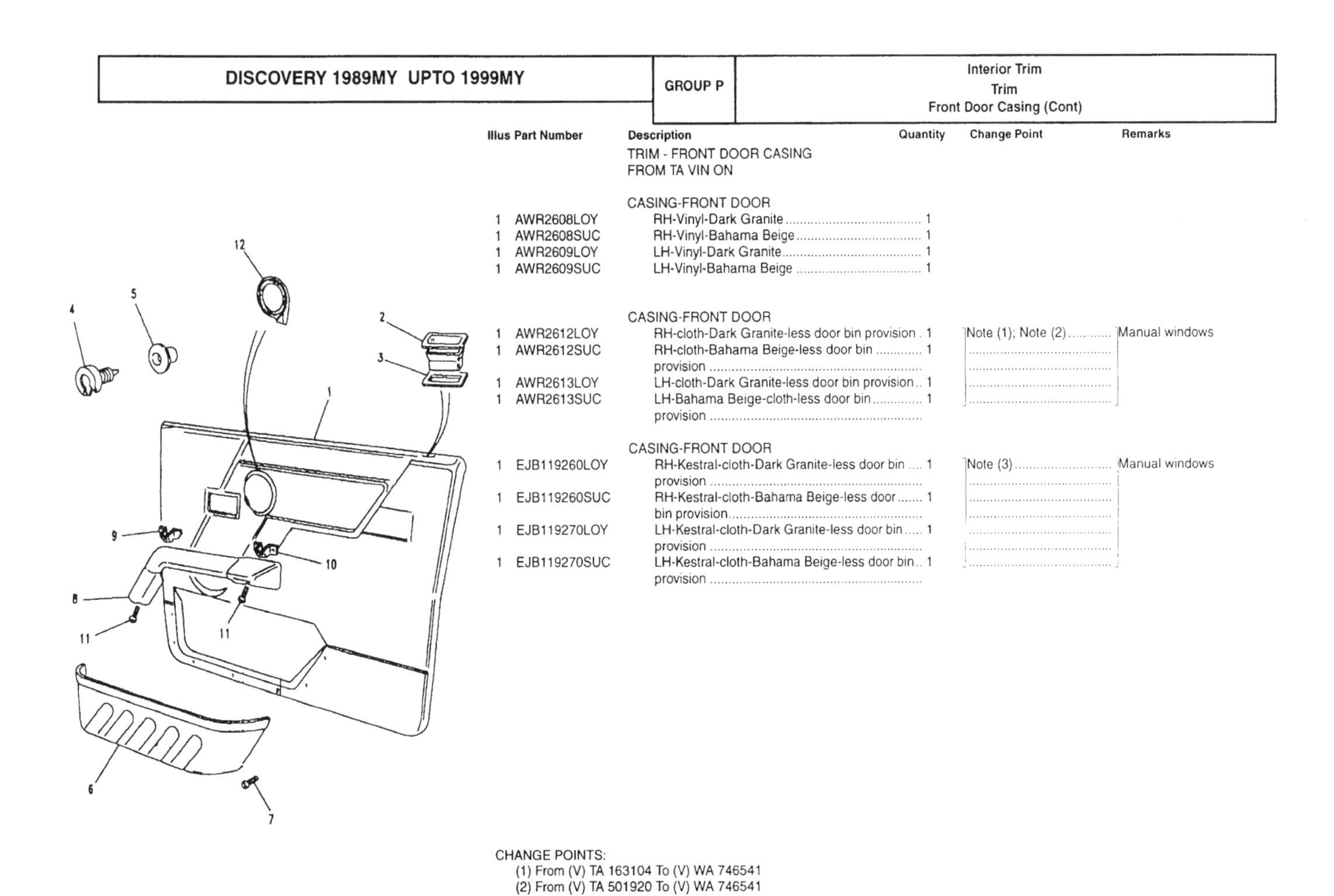

Illus	Part Number	Description	Quantity	Change Point	Remarks
		TRIM - FRONT DOOR CASING FROM TA VIN ON			
		CASING-FRONT DOOR			
1	AWR2608LOY	RH-Vinyl-Dark Granite	1		
1	AWR2608SUC	RH-Vinyl-Bahama Beige	1		
1	AWR2609LOY	LH-Vinyl-Dark Granite	1		
1	AWR2609SUC	LH-Vinyl-Bahama Beige	1		
		CASING-FRONT DOOR			
1	AWR2612LOY	RH-cloth-Dark Granite-less door bin provision	1	Note (1); Note (2)	Manual windows
1	AWR2612SUC	RH-cloth-Bahama Beige-less door bin provision	1		
1	AWR2613LOY	LH-cloth-Dark Granite-less door bin provision	1		
1	AWR2613SUC	LH-Bahama Beige-cloth-less door bin provision	1		
		CASING-FRONT DOOR			
1	EJB119260LOY	RH-Kestral-cloth-Dark Granite-less door bin provision	1	Note (3)	Manual windows
1	EJB119260SUC	RH-Kestral-cloth-Bahama Beige-less door bin provision	1		
1	EJB119270LOY	LH-Kestral-cloth-Dark Granite-less door bin provision	1		
1	EJB119270SUC	LH-Kestral-cloth-Bahama Beige-less door bin provision	1		

CHANGE POINTS:
(1) From (V) TA 163104 To (V) WA 746541
(2) From (V) TA 501920 To (V) WA 746541
(3) From (V) WA 746542

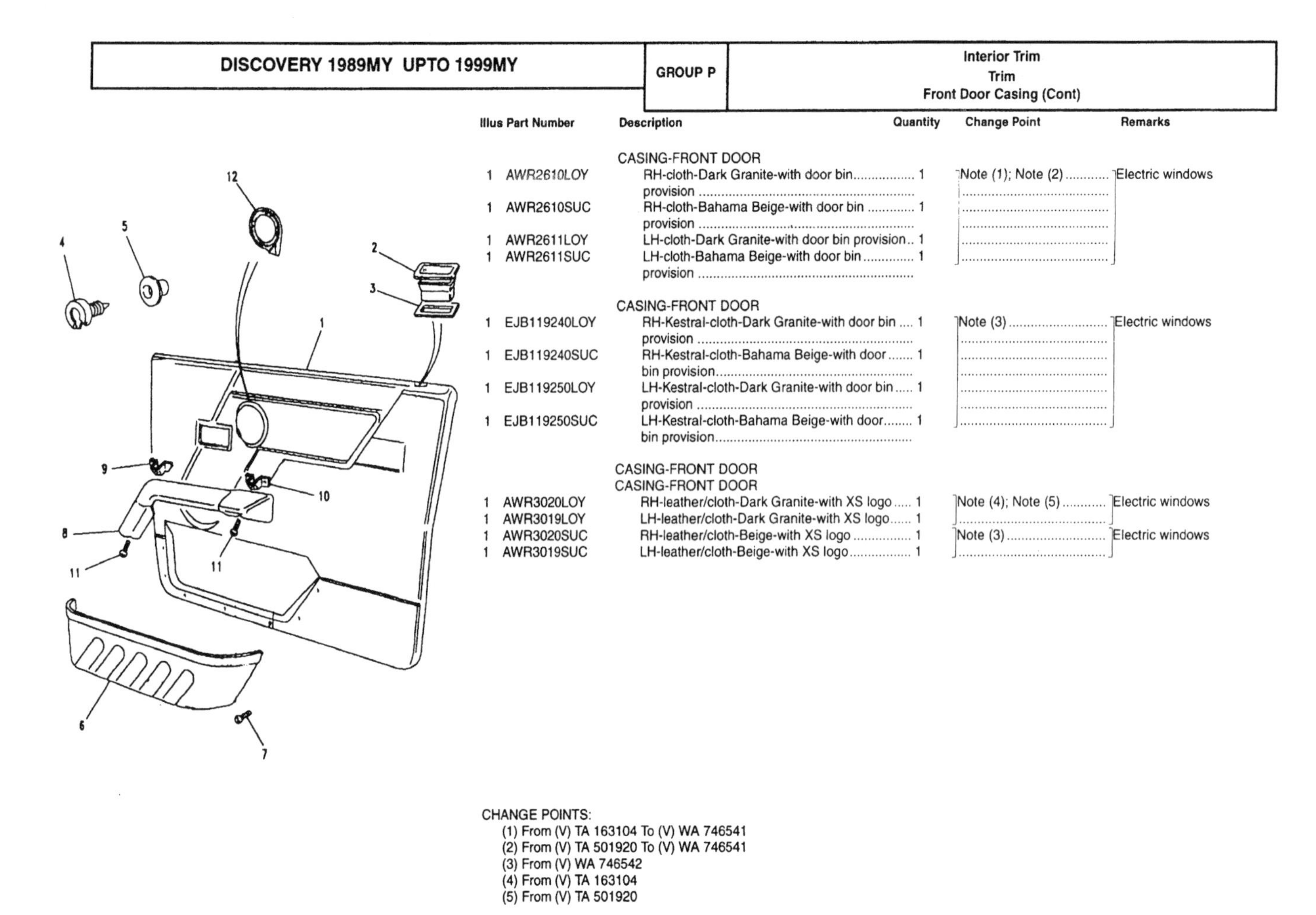

Illus	Part Number	Description	Quantity	Change Point	Remarks
		CASING-FRONT DOOR			
1	AWR2610LOY	RH-cloth-Dark Granite-with door bin provision	1	Note (1); Note (2)	Electric windows
1	AWR2610SUC	RH-cloth-Bahama Beige-with door bin provision	1		
1	AWR2611LOY	LH-cloth-Dark Granite-with door bin provision	1		
1	AWR2611SUC	LH-cloth-Bahama Beige-with door bin provision	1		
		CASING-FRONT DOOR			
1	EJB119240LOY	RH-Kestral-cloth-Dark Granite-with door bin provision	1	Note (3)	Electric windows
1	EJB119240SUC	RH-Kestral-cloth-Bahama Beige-with door bin provision	1		
1	EJB119250LOY	LH-Kestral-cloth-Dark Granite-with door bin provision	1		
1	EJB119250SUC	LH-Kestral-cloth-Bahama Beige-with door bin provision	1		
		CASING-FRONT DOOR			
		CASING-FRONT DOOR			
1	AWR3020LOY	RH-leather/cloth-Dark Granite-with XS logo	1	Note (4); Note (5)	Electric windows
1	AWR3019LOY	LH-leather/cloth-Dark Granite-with XS logo	1		
1	AWR3020SUC	RH-leather/cloth-Beige-with XS logo	1	Note (3)	Electric windows
1	AWR3019SUC	LH-leather/cloth-Beige-with XS logo	1		

CHANGE POINTS:
(1) From (V) TA 163104 To (V) WA 746541
(2) From (V) TA 501920 To (V) WA 746541
(3) From (V) WA 746542
(4) From (V) TA 163104
(5) From (V) TA 501920

P 33

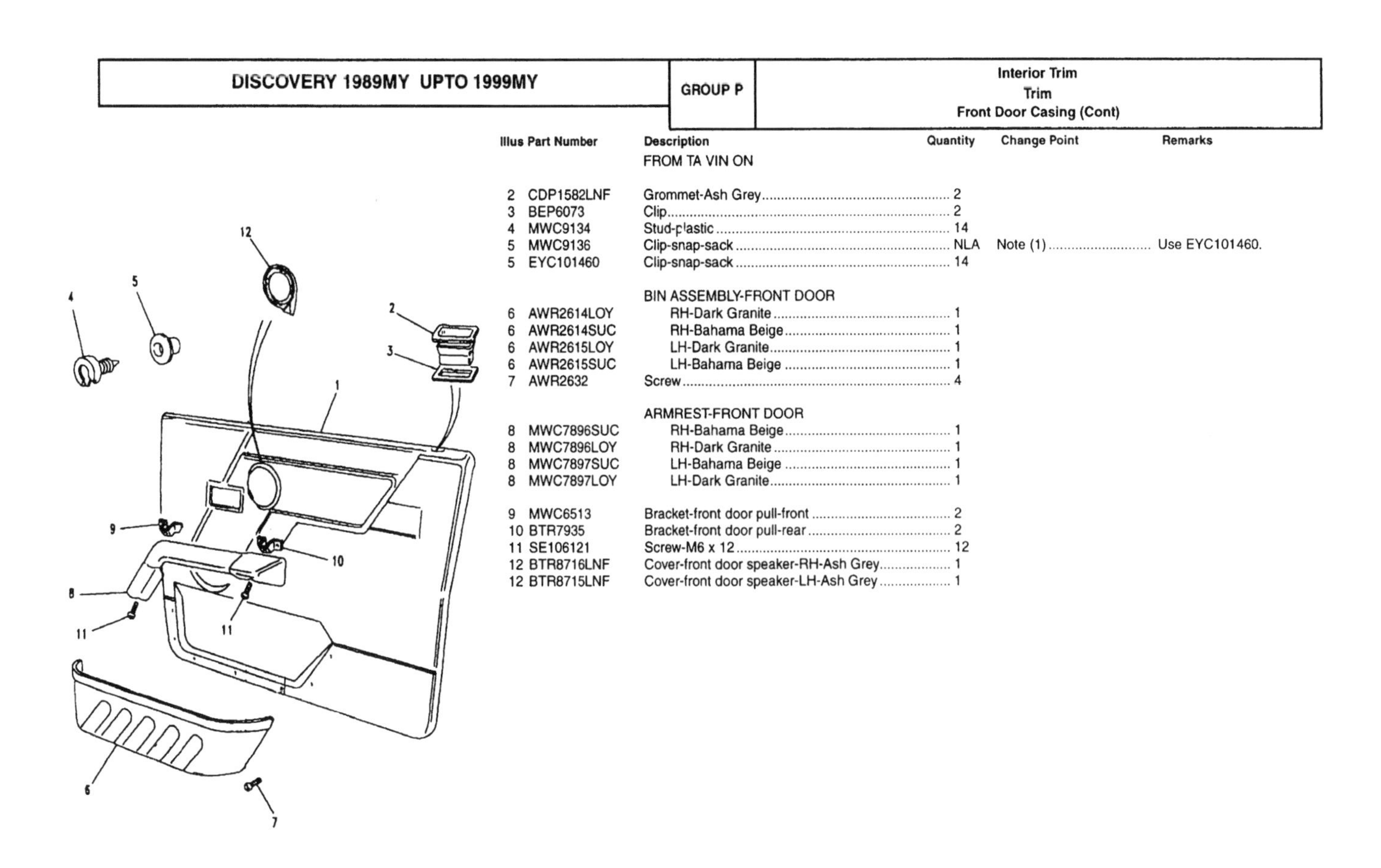

Illus	Part Number	Description	Quantity	Change Point	Remarks
		FROM TA VIN ON			
2	CDP1582LNF	Grommet-Ash Grey	2		
3	BEP6073	Clip	2		
4	MWC9134	Stud-plastic	14		
5	MWC9136	Clip-snap-sack	NLA	Note (1)	Use EYC101460.
5	EYC101460	Clip-snap-sack	14		
		BIN ASSEMBLY-FRONT DOOR			
6	AWR2614LOY	RH-Dark Granite	1		
6	AWR2614SUC	RH-Bahama Beige	1		
6	AWR2615LOY	LH-Dark Granite	1		
6	AWR2615SUC	LH-Bahama Beige	1		
7	AWR2632	Screw	4		
		ARMREST-FRONT DOOR			
8	MWC7896SUC	RH-Bahama Beige	1		
8	MWC7896LOY	RH-Dark Granite	1		
8	MWC7897SUC	LH-Bahama Beige	1		
8	MWC7897LOY	LH-Dark Granite	1		
9	MWC6513	Bracket-front door pull-front	2		
10	BTR7935	Bracket-front door pull-rear	2		
11	SE106121	Screw-M6 x 12	12		
12	BTR8716LNF	Cover-front door speaker-RH-Ash Grey	1		
12	BTR8715LNF	Cover-front door speaker-LH-Ash Grey	1		

CHANGE POINTS:
(1) From (V) TA 163104; From (V) TA 501920

P 34

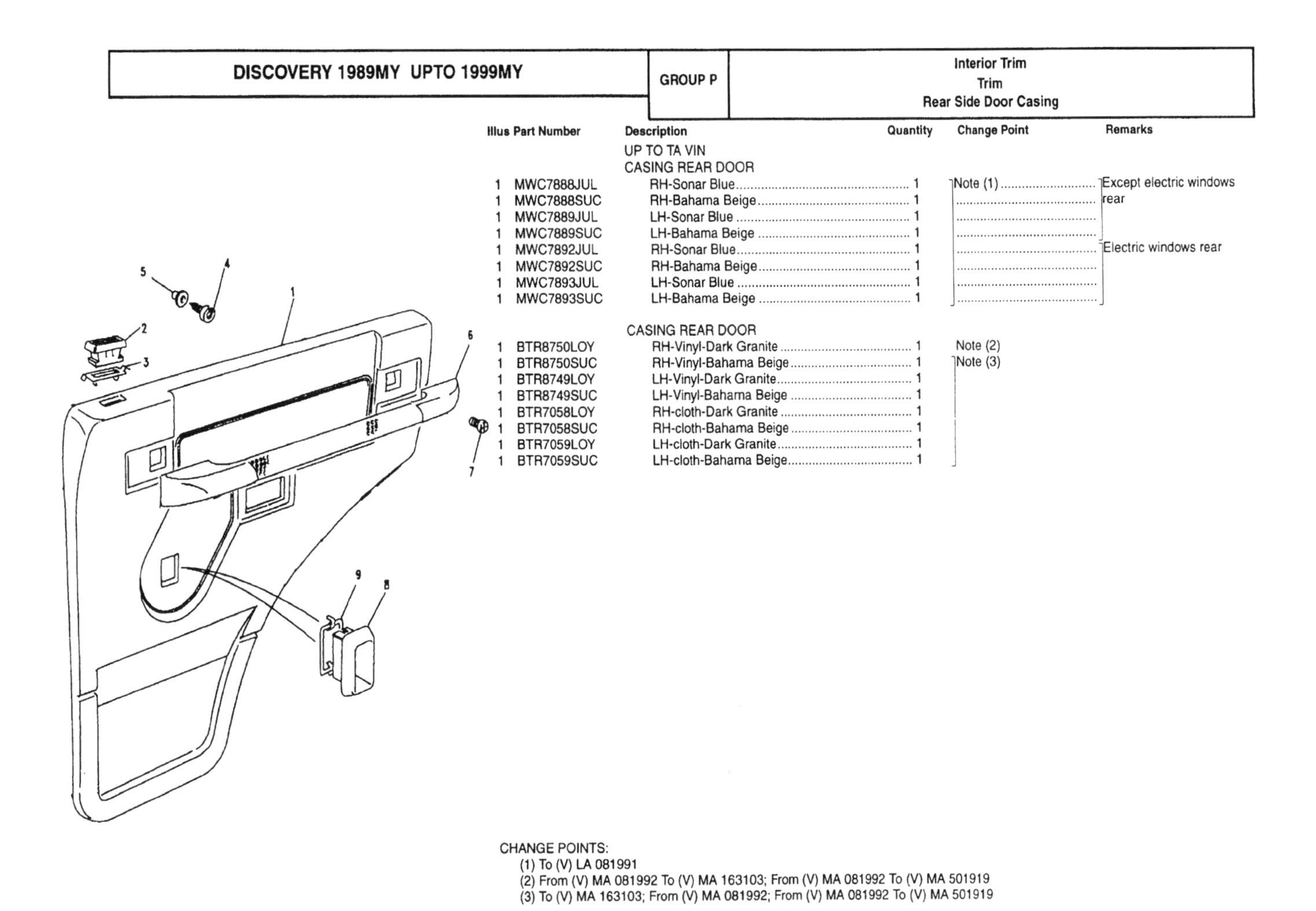

Illus	Part Number	Description	Quantity	Change Point	Remarks
		UP TO TA VIN			
		CASING REAR DOOR			
1	MWC7888JUL	RH-Sonar Blue	1	Note (1)	Except electric windows rear
1	MWC7888SUC	RH-Bahama Beige	1		
1	MWC7889JUL	LH-Sonar Blue	1		
1	MWC7889SUC	LH-Bahama Beige	1		
1	MWC7892JUL	RH-Sonar Blue	1		Electric windows rear
1	MWC7892SUC	RH-Bahama Beige	1		
1	MWC7893JUL	LH-Sonar Blue	1		
1	MWC7893SUC	LH-Bahama Beige	1		
		CASING REAR DOOR			
1	BTR8750LOY	RH-Vinyl-Dark Granite	1	Note (2)	
1	BTR8750SUC	RH-Vinyl-Bahama Beige	1	Note (3)	
1	BTR8749LOY	LH-Vinyl-Dark Granite	1		
1	BTR8749SUC	LH-Vinyl-Bahama Beige	1		
1	BTR7058LOY	RH-cloth-Dark Granite	1		
1	BTR7058SUC	RH-cloth-Bahama Beige	1		
1	BTR7059LOY	LH-cloth-Dark Granite	1		
1	BTR7059SUC	LH-cloth-Bahama Beige	1		

CHANGE POINTS:
(1) To (V) LA 081991
(2) From (V) MA 081992 To (V) MA 163103; From (V) MA 081992 To (V) MA 501919
(3) To (V) MA 163103; From (V) MA 081992; From (V) MA 081992 To (V) MA 501919

P 35

Illus	Part Number	Description	Quantity	Change Point	Remarks
		UP TO TA VIN			
2	CDP1582LUN	Grommet-Grey	2	Note (1)	
2	CDP1582LNF	Grommet-Ash Grey	2	Note (2)	
3	BEP6073	Clip	2	Note (3)	
4	MWC9134	Stud-plastic	14		
5	MWC9136	Clip-snap-sack	NLA		Use EYC101460.
5	EYC101460	Clip-snap-sack	14		
		ARMREST ASSEMBLY-REAR DOOR			
6	MWC7898JUL	Sonar Blue	2	Note (3)	
6	MWC7898SUC	Bahama Beige-7 seats	2		
6	MWC7898LOY	Dark Granite	2	Note (4)	
	STC8495JUL	Kit-armrest-Sonar Blue	2	Note (5); Note (6)	
	STC8495SUC	Kit-armrest-Bahama Beige	2		
7	SE106121L	Screw-pan head-M6 x 12	4	Note (3)	
7	FS106125L	Screw-flanged head-M6 x 12	4	Note (4)	
8	MXC7628LUN	Escutcheon-rear door window regulator-Ash Grey	2	Note (1)	
8	MXC7628LNF	Escutcheon-rear door window regulator-Ash Grey	2	Note (2)	
9	BTR404	Clip	2	Note (3)	

CHANGE POINTS:
(1) To (V) JA 034313
(2) From (V) KA 034314 To (V) MA 163103; From (V) KA 034314 To (V) MA 501919
(3) To (V) MA 163103; To (V) MA 501919
(4) From (V) MA 081992 To (V) MA 163103; From (V) MA 081992 To (V) MA 501919
(5) To (V) MA 163103
(6) To (V) MA 501919

P 36

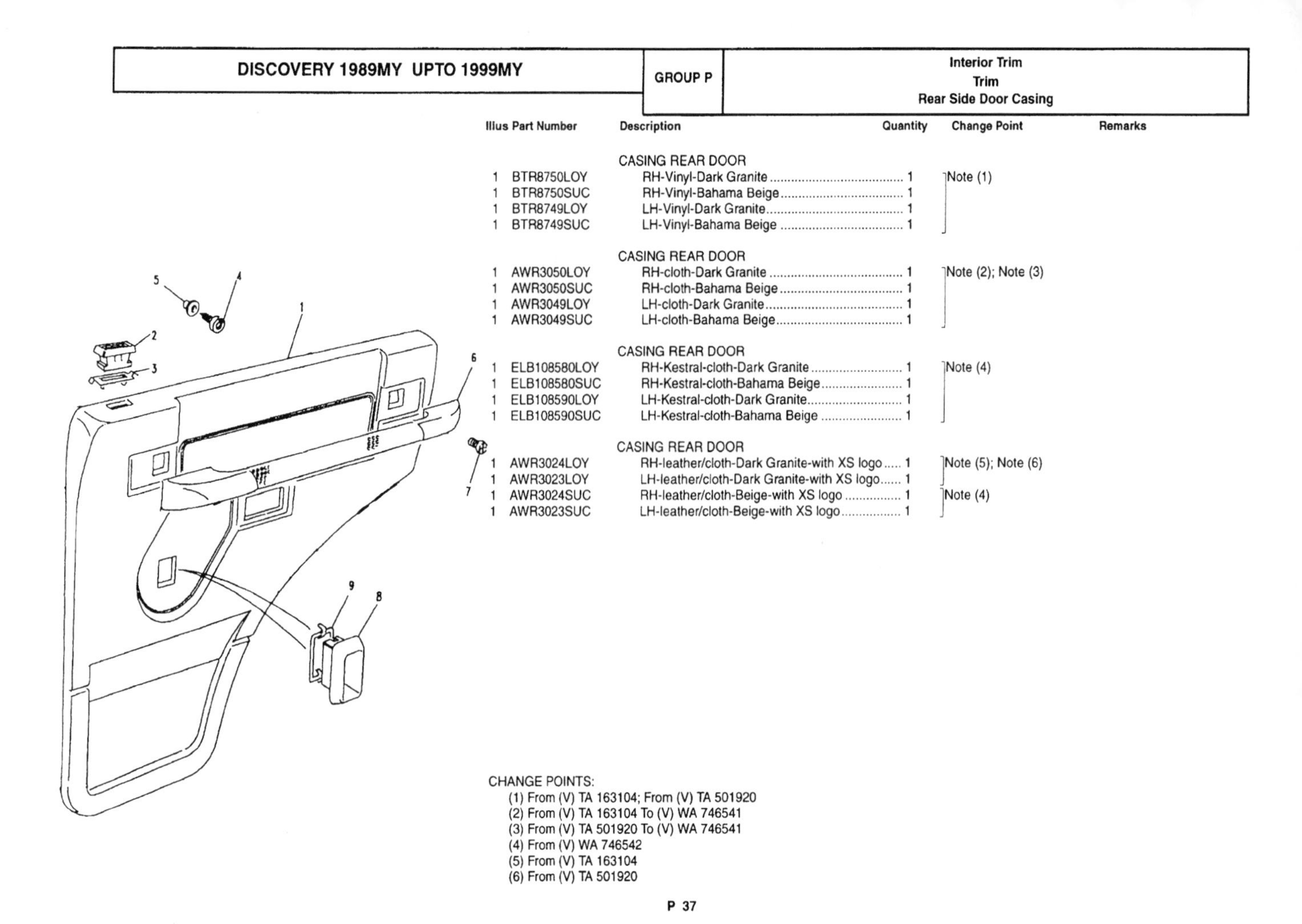

Illus	Part Number	Description	Quantity	Change Point	Remarks
		CASING REAR DOOR			
1	BTR8750LOY	RH-Vinyl-Dark Granite	1	Note (1)	
1	BTR8750SUC	RH-Vinyl-Bahama Beige	1		
1	BTR8749LOY	LH-Vinyl-Dark Granite	1		
1	BTR8749SUC	LH-Vinyl-Bahama Beige	1		
		CASING REAR DOOR			
1	AWR3050LOY	RH-cloth-Dark Granite	1	Note (2); Note (3)	
1	AWR3050SUC	RH-cloth-Bahama Beige	1		
1	AWR3049LOY	LH-cloth-Dark Granite	1		
1	AWR3049SUC	LH-cloth-Bahama Beige	1		
		CASING REAR DOOR			
1	ELB108580LOY	RH-Kestral-cloth-Dark Granite	1	Note (4)	
1	ELB108580SUC	RH-Kestral-cloth-Bahama Beige	1		
1	ELB108590LOY	LH-Kestral-cloth-Dark Granite	1		
1	ELB108590SUC	LH-Kestral-cloth-Bahama Beige	1		
		CASING REAR DOOR			
1	AWR3024LOY	RH-leather/cloth-Dark Granite-with XS logo	1	Note (5); Note (6)	
1	AWR3023LOY	LH-leather/cloth-Dark Granite-with XS logo	1		
1	AWR3024SUC	RH-leather/cloth-Beige-with XS logo	1	Note (4)	
1	AWR3023SUC	LH-leather/cloth-Beige-with XS logo	1		

CHANGE POINTS:
(1) From (V) TA 163104; From (V) TA 501920
(2) From (V) TA 163104 To (V) WA 746541
(3) From (V) TA 501920 To (V) WA 746541
(4) From (V) WA 746542
(5) From (V) TA 163104
(6) From (V) TA 501920

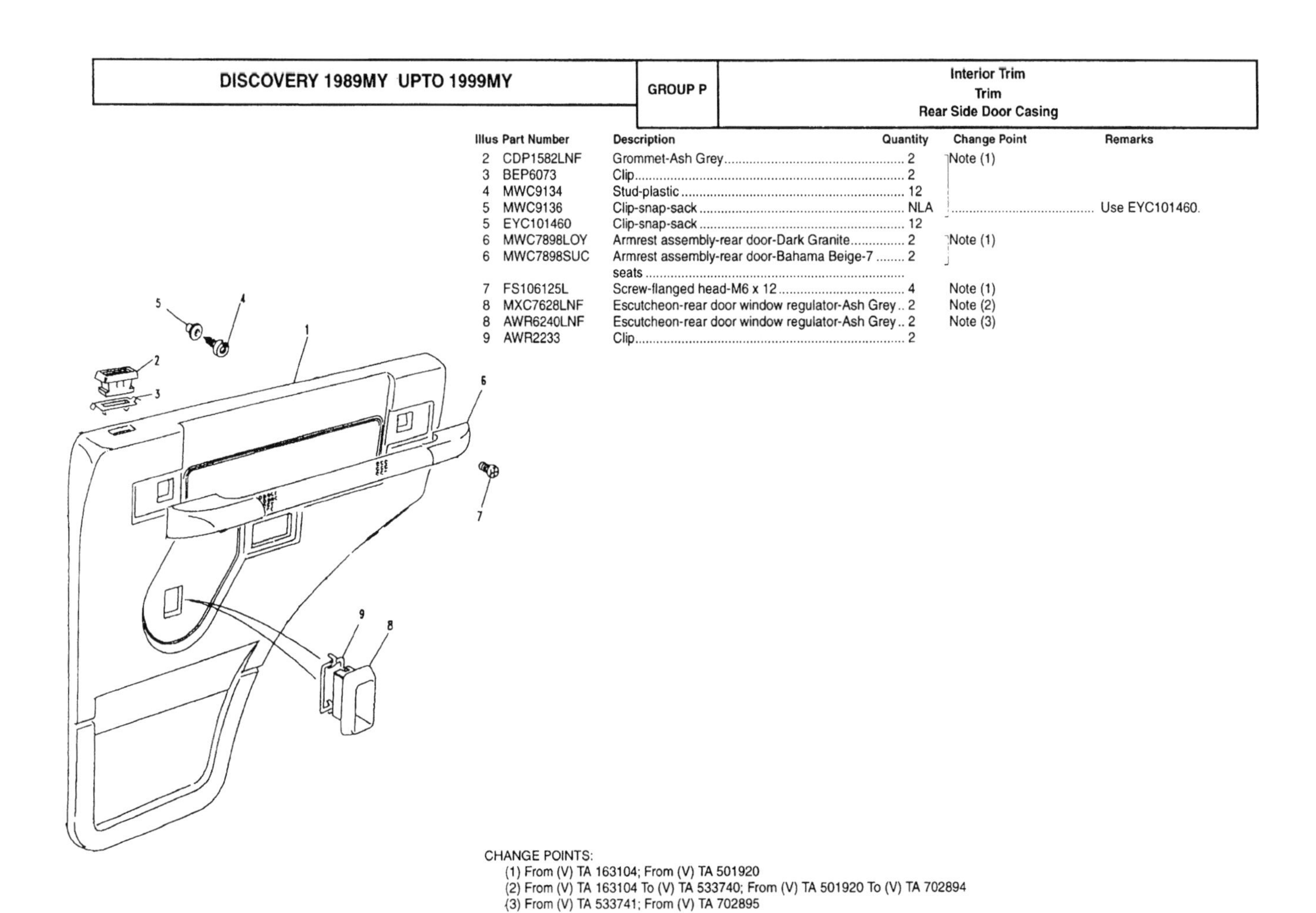

Illus	Part Number	Description	Quantity	Change Point	Remarks
2	CDP1582LNF	Grommet-Ash Grey	2	Note (1)	
3	BEP6073	Clip	2		
4	MWC9134	Stud-plastic	12		
5	MWC9136	Clip-snap-sack	NLA		Use EYC101460.
5	EYC101460	Clip-snap-sack	12		
6	MWC7898LOY	Armrest assembly-rear door-Dark Granite	2	Note (1)	
6	MWC7898SUC	Armrest assembly-rear door-Bahama Beige-7 seats	2		
7	FS106125L	Screw-flanged head-M6 x 12	4	Note (1)	
8	MXC7628LNF	Escutcheon-rear door window regulator-Ash Grey	2	Note (2)	
8	AWR6240LNF	Escutcheon-rear door window regulator-Ash Grey	2	Note (3)	
9	AWR2233	Clip	2		

CHANGE POINTS:
(1) From (V) TA 163104; From (V) TA 501920
(2) From (V) TA 163104 To (V) TA 533740; From (V) TA 501920 To (V) TA 702894
(3) From (V) TA 533741; From (V) TA 702895

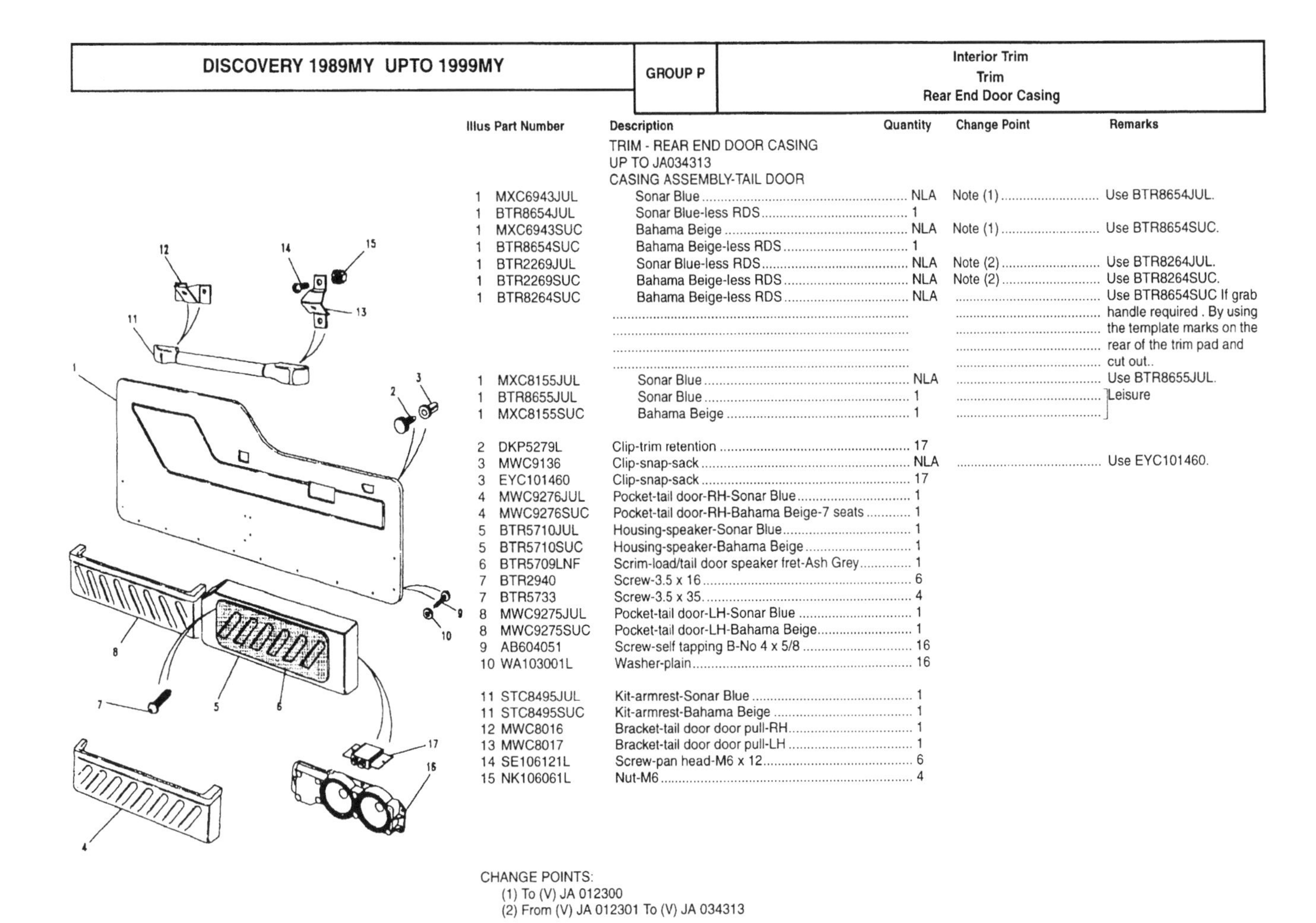

Illus	Part Number	Description	Quantity	Change Point	Remarks
		TRIM - REAR END DOOR CASING UP TO JA034313			
		CASING ASSEMBLY-TAIL DOOR			
1	MXC6943JUL	Sonar Blue	NLA	Note (1)	Use BTR8654JUL.
1	BTR8654JUL	Sonar Blue-less RDS	1		
1	MXC6943SUC	Bahama Beige	NLA	Note (1)	Use BTR8654SUC.
1	BTR8654SUC	Bahama Beige-less RDS	1		
1	BTR2269JUL	Sonar Blue-less RDS	NLA	Note (2)	Use BTR8264JUL.
1	BTR2269SUC	Bahama Beige-less RDS	NLA	Note (2)	Use BTR8264SUC.
1	BTR8264SUC	Bahama Beige-less RDS	NLA		Use BTR8654SUC If grab handle required . By using the template marks on the rear of the trim pad and cut out..
1	MXC8155JUL	Sonar Blue	NLA		Use BTR8655JUL.
1	BTR8655JUL	Sonar Blue	1		Leisure
1	MXC8155SUC	Bahama Beige	1		
2	DKP5279L	Clip-trim retention	17		
3	MWC9136	Clip-snap-sack	NLA		Use EYC101460.
3	EYC101460	Clip-snap-sack	17		
4	MWC9276JUL	Pocket-tail door-RH-Sonar Blue	1		
4	MWC9276SUC	Pocket-tail door-RH-Bahama Beige-7 seats	1		
5	BTR5710JUL	Housing-speaker-Sonar Blue	1		
5	BTR5710SUC	Housing-speaker-Bahama Beige	1		
6	BTR5709LNF	Scrim-load/tail door speaker fret-Ash Grey	1		
7	BTR2940	Screw-3.5 x 16	6		
7	BTR5733	Screw-3.5 x 35	4		
8	MWC9275JUL	Pocket-tail door-LH-Sonar Blue	1		
8	MWC9275SUC	Pocket-tail door-LH-Bahama Beige	1		
9	AB604051	Screw-self tapping B-No 4 x 5/8	16		
10	WA103001L	Washer-plain	16		
11	STC8495JUL	Kit-armrest-Sonar Blue	1		
11	STC8495SUC	Kit-armrest-Bahama Beige	1		
12	MWC8016	Bracket-tail door door pull-RH	1		
13	MWC8017	Bracket-tail door door pull-LH	1		
14	SE106121L	Screw-pan head-M6 x 12	6		
15	NK106061L	Nut-M6	4		

CHANGE POINTS:
(1) To (V) JA 012300
(2) From (V) JA 012301 To (V) JA 034313

P 39

Illus	Part Number	Description	Quantity	Change Point	Remarks
		TRIM - REAR END DOOR CASING FROM KA034314			
		CASING ASSEMBLY-TAIL DOOR			
1	BTR2269JUL	Sonar Blue-less RDS	NLA	Note (1)	Use BTR8264JUL.
1	BTR2269SUC	Bahama Beige-less RDS	NLA		Use BTR8264SUC.
1	BTR8264SUC	Bahama Beige-less RDS	NLA	Note (2)	Use BTR8654SUC If grab handle required . By using the template marks on the rear of the trim pad and cut out..
1	BTR8264JUL	Sonar Blue-less RDS	1	Note (2)	
1	BTR8654JUL	Sonar Blue-less RDS	1	Note (3)	
1	BTR8654SUC	Bahama Beige-less RDS	1		Except Leisure
1	BTR8654LOY	Dark Granite-less RDS	1	Note (4)	
		CASING ASSEMBLY-TAIL DOOR			
1	BTR2615JUL	Sonar Blue-with RDS	1	Note (1)	
1	BTR2615SUC	Bahama Beige-with RDS	1		
1	BTR8263SUC	Bahama Beige-with RDS	1	Note (2)	
1	BTR8263JUL	Sonar Blue-with RDS	1		
1	BTR8653SUC	Bahama Beige-with RDS	1	Note (3)	
1	BTR8653JUL	Sonar Blue-with RDS	1		
1	BTR8653LOY	Dark Granite-with RDS	1	Note (4)	
		CASING ASSEMBLY-TAIL DOOR			
1	MXC8155JUL	Sonar Blue	NLA	Note (5)	Use BTR8655JUL.
1	MXC8155SUC	Bahama Beige	1		Leisure
1	BTR8655JUL	Sonar Blue	1	Note (3)	
1	BTR8655SUC	Bahama Beige	1		
1	BTR8655LOY	Dark Granite	1	Note (4)	

CAPXEG2B

CHANGE POINTS:
(1) From (V) KA 034314 To (V) KA 063712
(2) From (V) KA 063713 To (V) LA 072327
(3) From (V) LA 072328
(4) From (V) MA 081992
(5) From (V) KA 034314 To (V) LA 072327

P 40

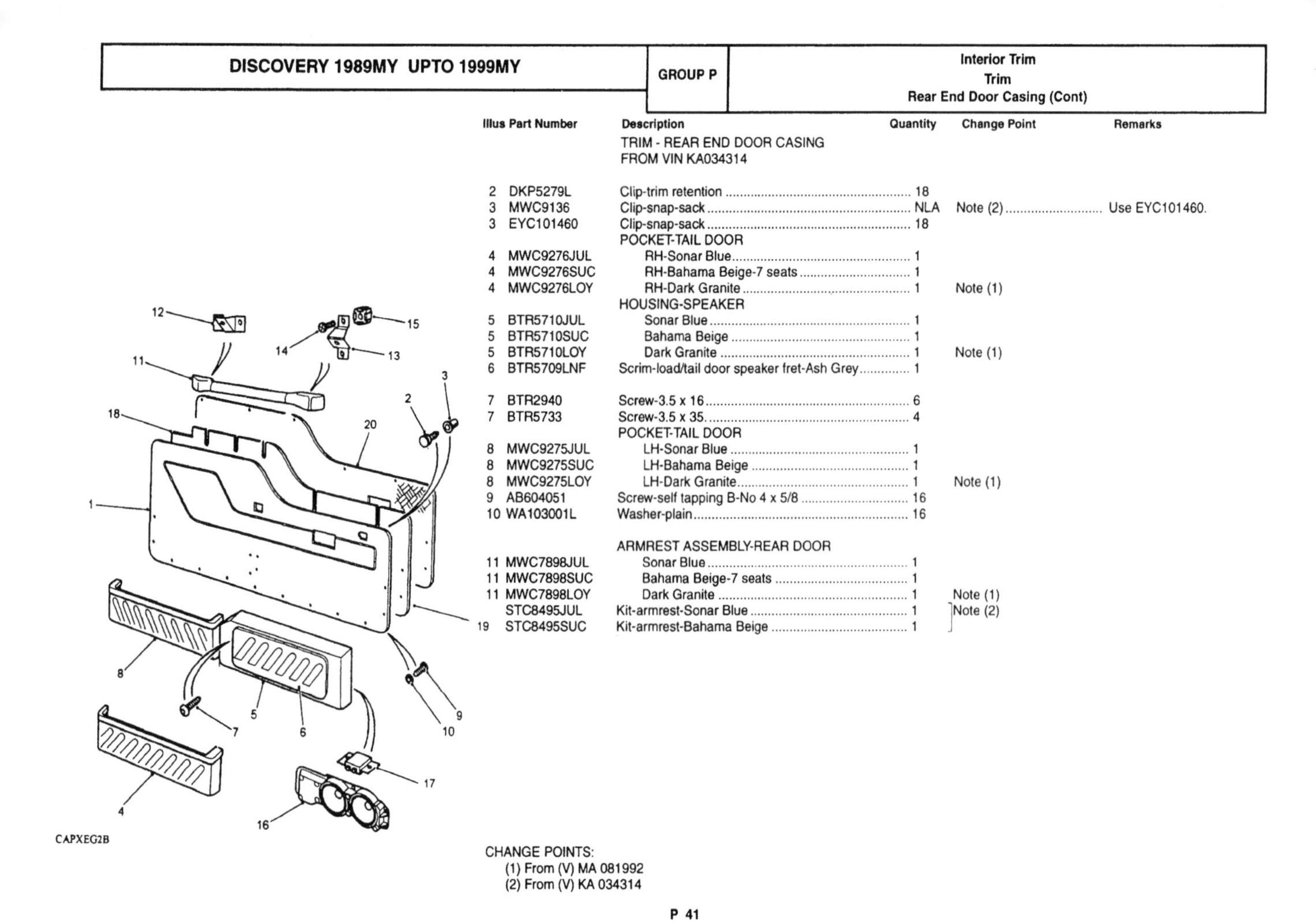

Illus	Part Number	Description	Quantity	Change Point	Remarks
		TRIM - REAR END DOOR CASING FROM VIN KA034314			
2	DKP5279L	Clip-trim retention	18		
3	MWC9136	Clip-snap-sack	NLA	Note (2)	Use EYC101460.
3	EYC101460	Clip-snap-sack	18		
		POCKET-TAIL DOOR			
4	MWC9276JUL	RH-Sonar Blue	1		
4	MWC9276SUC	RH-Bahama Beige-7 seats	1		
4	MWC9276LOY	RH-Dark Granite	1	Note (1)	
		HOUSING-SPEAKER			
5	BTR5710JUL	Sonar Blue	1		
5	BTR5710SUC	Bahama Beige	1		
5	BTR5710LOY	Dark Granite	1	Note (1)	
6	BTR5709LNF	Scrim-load/tail door speaker fret-Ash Grey	1		
7	BTR2940	Screw-3.5 x 16	6		
7	BTR5733	Screw-3.5 x 35	4		
		POCKET-TAIL DOOR			
8	MWC9275JUL	LH-Sonar Blue	1		
8	MWC9275SUC	LH-Bahama Beige	1		
8	MWC9275LOY	LH-Dark Granite	1	Note (1)	
9	AB604051	Screw-self tapping B-No 4 x 5/8	16		
10	WA103001L	Washer-plain	16		
		ARMREST ASSEMBLY-REAR DOOR			
11	MWC7898JUL	Sonar Blue	1		
11	MWC7898SUC	Bahama Beige-7 seats	1		
11	MWC7898LOY	Dark Granite	1	Note (1)	
	STC8495JUL	Kit-armrest-Sonar Blue	1	Note (2)	
	STC8495SUC	Kit-armrest-Bahama Beige	1	Note (2)	

CHANGE POINTS:
(1) From (V) MA 081992
(2) From (V) KA 034314

Illus	Part Number	Description	Quantity	Change Point	Remarks
12	MWC8016	Bracket-tail door door pull-RH	1		
13	MWC8017	Bracket-tail door door pull-LH	1		
14	SE106121L	Screw-pan head-M6 x 12	6		
15	NK106061L	Nut-M6	4		
		ITEMES 16 & 17 SEE ELECTRICAL			
18	ESQ100090	Pad-anti rattle-small-LH	1		
19	ESQ100080	Pad-anti rattle-large-RH	1		
20	STC7030	Protector-tail door-aluminium	1		

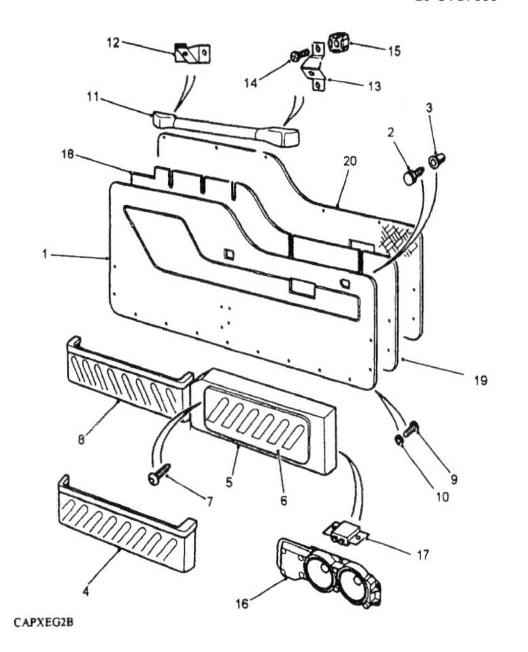

Illus	Part Number	Description	Quantity	Change Point	Remarks
		FINISHER-A POST UPPER			
1	MWC8920JUL	RH-Sonar Blue	1	Note (1)	
1	MWC8920SUC	RH-Bahama Beige	1		
1	MWC8921JUL	LH-Sonar Blue	1		
1	MWC8921SUC	LH-Bahama Beige	1		
1	AWR1442LUM	RH-Mist Grey-without speaker	1	Note (2)	
1	AWR1443LUM	LH-Mist Grey-without speaker	1		
	AWR1450LUM	FINISHER ASSEMBLY-A POST UPPER-RH-MIST GREY-WITH SPEAKER	1	Note (3)	
1	AWR1444LUM	• FINISHER-A POST UPPER-RH-WITH TWEETERS	1		
	EYC10008L	• • Clip-trim retention	3		
	AWR1449LUM	FINISHER ASSEMBLY-A POST UPPER-LH-MIST GREY-WITH SPEAKER	1	Note (3)	
1	AWR1445LUM	• FINISHER-A POST UPPER-LH-WITH TWEETERS	1		
	EYC10008L	• • Clip-trim retention	3		

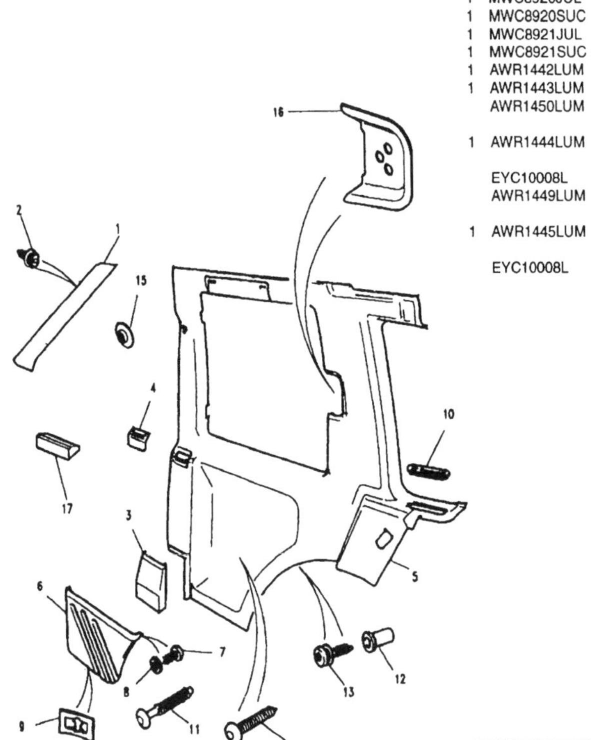

CHANGE POINTS:
(1) To (V) LA 081991
(2) From (V) MA 081992
(3) From (V) TA 163104; From (V) TA 501920

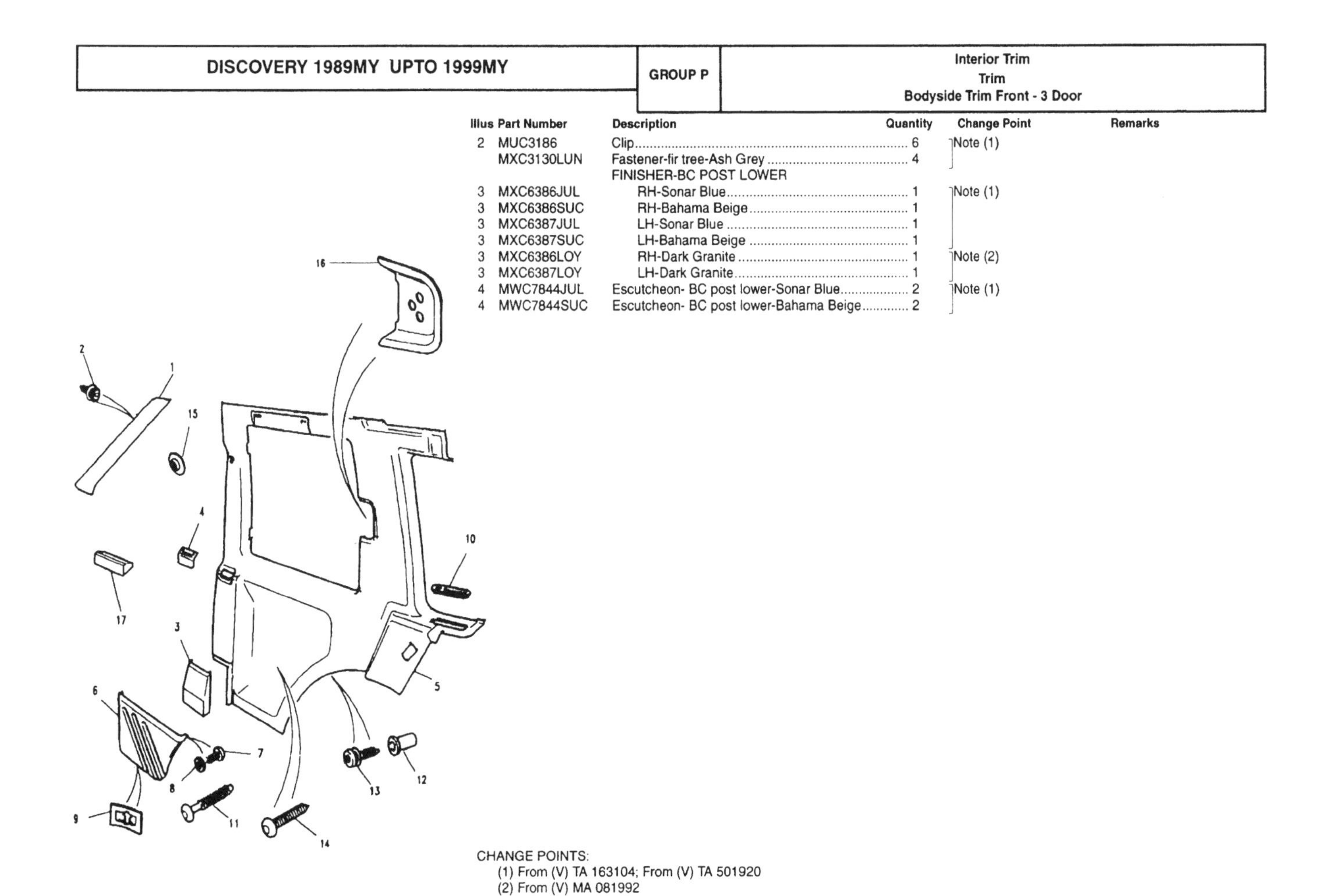

Illus	Part Number	Description	Quantity	Change Point	Remarks
2	MUC3186	Clip	6	Note (1)	
	MXC3130LUN	Fastener-fir tree-Ash Grey	4		
		FINISHER-BC POST LOWER			
3	MXC6386JUL	RH-Sonar Blue	1	Note (1)	
3	MXC6386SUC	RH-Bahama Beige	1		
3	MXC6387JUL	LH-Sonar Blue	1		
3	MXC6387SUC	LH-Bahama Beige	1		
3	MXC6386LOY	RH-Dark Granite	1	Note (2)	
3	MXC6387LOY	LH-Dark Granite	1		
4	MWC7844JUL	Escutcheon- BC post lower-Sonar Blue	2	Note (1)	
4	MWC7844SUC	Escutcheon- BC post lower-Bahama Beige	2		

CHANGE POINTS:
(1) From (V) TA 163104; From (V) TA 501920
(2) From (V) MA 081992

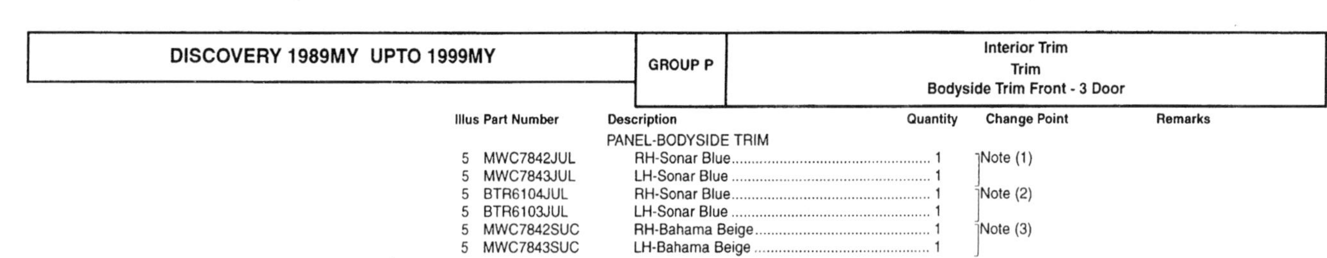

Illus	Part Number	Description	Quantity	Change Point	Remarks
		PANEL-BODYSIDE TRIM			
5	MWC7842JUL	RH-Sonar Blue	1	Note (1)	
5	MWC7843JUL	LH-Sonar Blue	1		
5	BTR6104JUL	RH-Sonar Blue	1	Note (2)	
5	BTR6103JUL	LH-Sonar Blue	1		
5	MWC7842SUC	RH-Bahama Beige	1	Note (3)	
5	MWC7843SUC	LH-Bahama Beige	1		

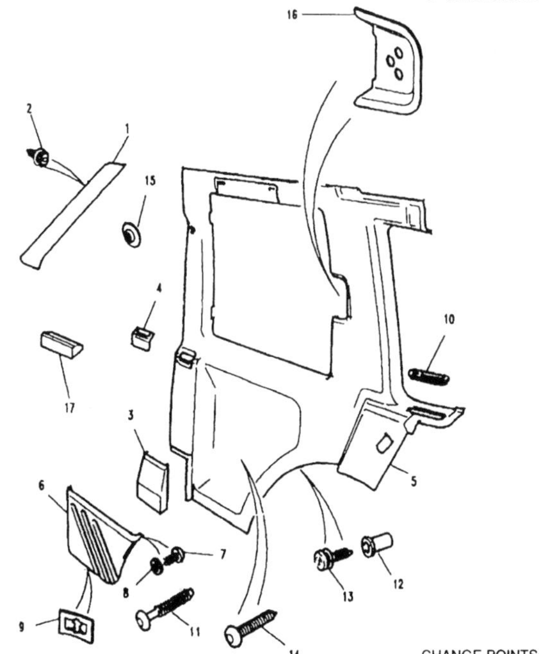

CHANGE POINTS:
(1) To (V) KA 044131
(2) From (V) KA 044132 To (V) LA 081991
(3) To (V) KA 043967

P 45

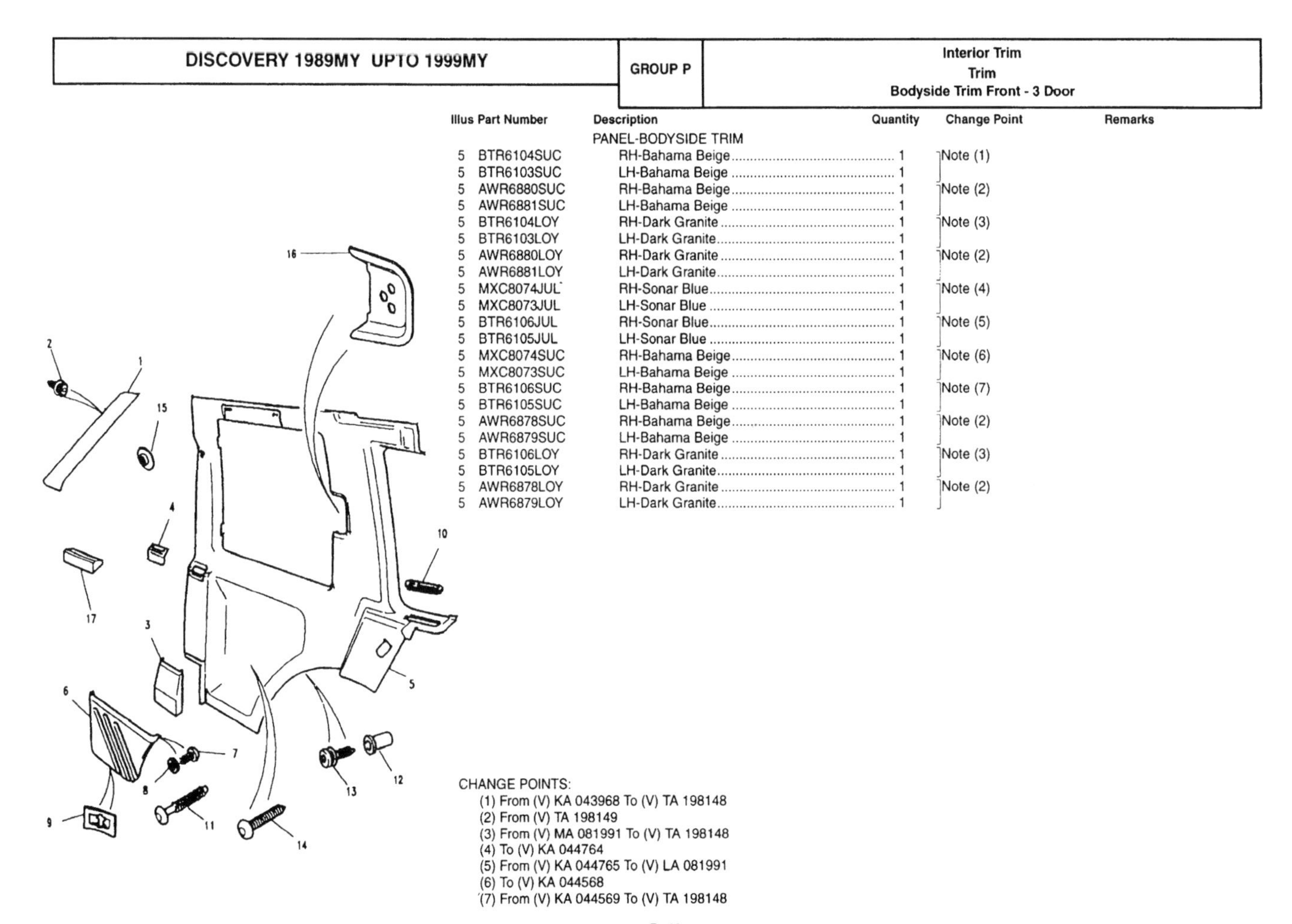

DISCOVERY 1989MY UPTO 1999MY

GROUP P

Interior Trim
Trim
Bodyside Trim Front - 3 Door

Illus	Part Number	Description	Quantity	Change Point	Remarks
		PANEL-BODYSIDE TRIM			
5	BTR6104SUC	RH-Bahama Beige	1	Note (1)	
5	BTR6103SUC	LH-Bahama Beige	1		
5	AWR6880SUC	RH-Bahama Beige	1	Note (2)	
5	AWR6881SUC	LH-Bahama Beige	1		
5	BTR6104LOY	RH-Dark Granite	1	Note (3)	
5	BTR6103LOY	LH-Dark Granite	1		
5	AWR6880LOY	RH-Dark Granite	1	Note (2)	
5	AWR6881LOY	LH-Dark Granite	1		
5	MXC8074JUL	RH-Sonar Blue	1	Note (4)	
5	MXC8073JUL	LH-Sonar Blue	1		
5	BTR6106JUL	RH-Sonar Blue	1	Note (5)	
5	BTR6105JUL	LH-Sonar Blue	1		
5	MXC8074SUC	RH-Bahama Beige	1	Note (6)	
5	MXC8073SUC	LH-Bahama Beige	1		
5	BTR6106SUC	RH-Bahama Beige	1	Note (7)	
5	BTR6105SUC	LH-Bahama Beige	1		
5	AWR6878SUC	RH-Bahama Beige	1	Note (2)	
5	AWR6879SUC	LH-Bahama Beige	1		
5	BTR6106LOY	RH-Dark Granite	1	Note (3)	
5	BTR6105LOY	LH-Dark Granite	1		
5	AWR6878LOY	RH-Dark Granite	1	Note (2)	
5	AWR6879LOY	LH-Dark Granite	1		

CHANGE POINTS:
(1) From (V) KA 043968 To (V) TA 198148
(2) From (V) TA 198149
(3) From (V) MA 081991 To (V) TA 198148
(4) To (V) KA 044764
(5) From (V) KA 044765 To (V) LA 081991
(6) To (V) KA 044568
(7) From (V) KA 044569 To (V) TA 198148

P 46

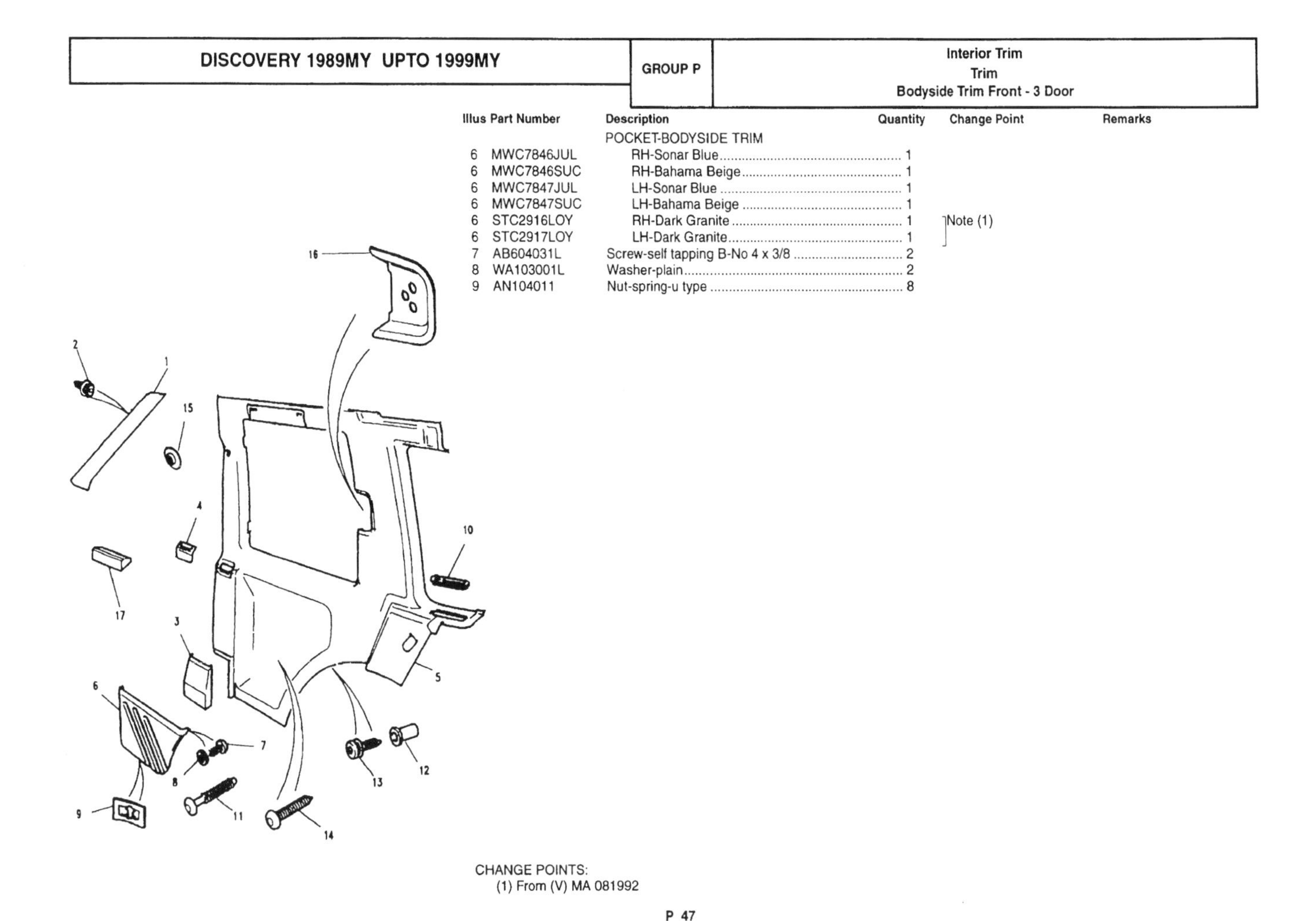

Illus	Part Number	Description	Quantity	Change Point	Remarks
		POCKET-BODYSIDE TRIM			
6	MWC7846JUL	RH-Sonar Blue	1		
6	MWC7846SUC	RH-Bahama Beige	1		
6	MWC7847JUL	LH-Sonar Blue	1		
6	MWC7847SUC	LH-Bahama Beige	1		
6	STC2916LOY	RH-Dark Granite	1	Note (1)	
6	STC2917LOY	LH-Dark Granite	1		
7	AB604031L	Screw-self tapping B-No 4 x 3/8	2		
8	WA103001L	Washer-plain	2		
9	AN104011	Nut-spring-u type	8		

CHANGE POINTS:
(1) From (V) MA 081992

DISCOVERY 1989MY UPTO 1999MY | GROUP P | Interior Trim
Trim
Bodyside Trim Front - 3 Door

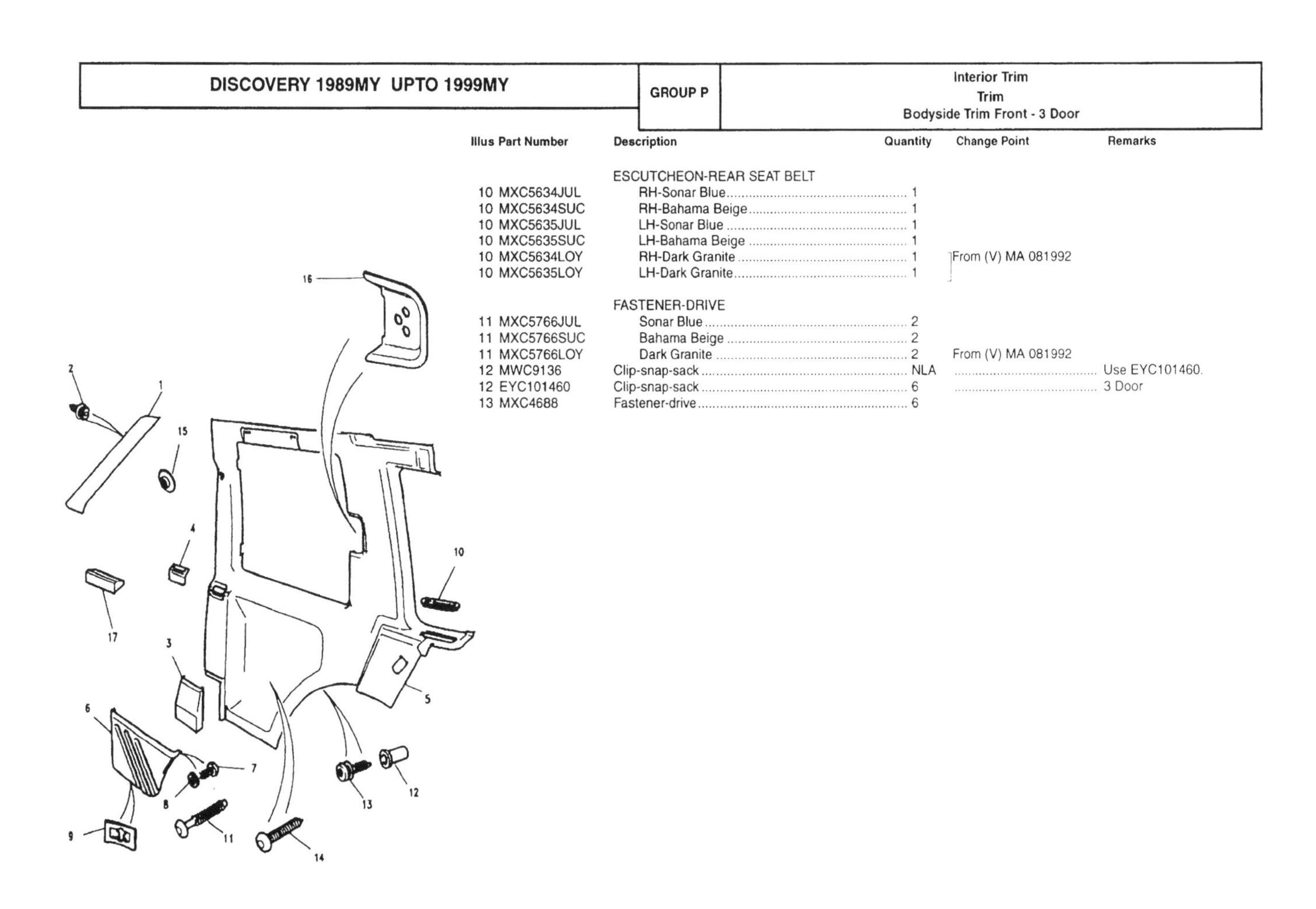

Illus	Part Number	Description	Quantity	Change Point	Remarks
		ESCUTCHEON-REAR SEAT BELT			
10	MXC5634JUL	RH-Sonar Blue	1		
10	MXC5634SUC	RH-Bahama Beige	1		
10	MXC5635JUL	LH-Sonar Blue	1		
10	MXC5635SUC	LH-Bahama Beige	1		
10	MXC5634LOY	RH-Dark Granite	1	From (V) MA 081992	
10	MXC5635LOY	LH-Dark Granite	1		
		FASTENER-DRIVE			
11	MXC5766JUL	Sonar Blue	2		
11	MXC5766SUC	Bahama Beige	2		
11	MXC5766LOY	Dark Granite	2	From (V) MA 081992	
12	MWC9136	Clip-snap-sack	NLA		Use EYC101460.
12	EYC101460	Clip-snap-sack	6		3 Door
13	MXC4688	Fastener-drive	6		

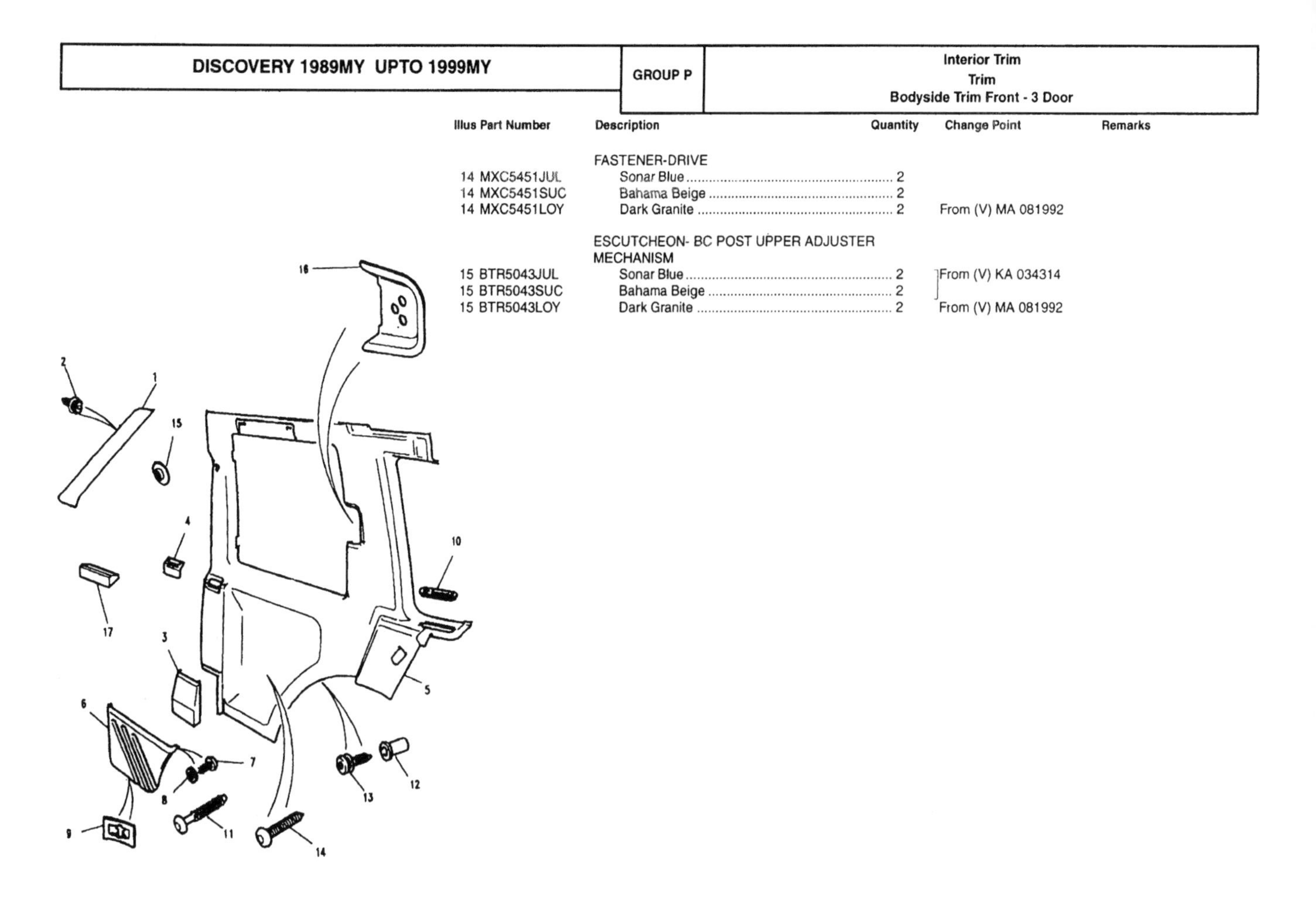

Illus	Part Number	Description	Quantity	Change Point	Remarks
		FASTENER-DRIVE			
14	MXC5451JUL	Sonar Blue	2		
14	MXC5451SUC	Bahama Beige	2		
14	MXC5451LOY	Dark Granite	2	From (V) MA 081992	
		ESCUTCHEON- BC POST UPPER ADJUSTER MECHANISM			
15	BTR5043JUL	Sonar Blue	2	From (V) KA 034314	
15	BTR5043SUC	Bahama Beige	2		
15	BTR5043LOY	Dark Granite	2	From (V) MA 081992	

P 49

DISCOVERY 1989MY UPTO 1999MY | GROUP P | Interior Trim
Trim
Bodyside Trim Front - 3 Door

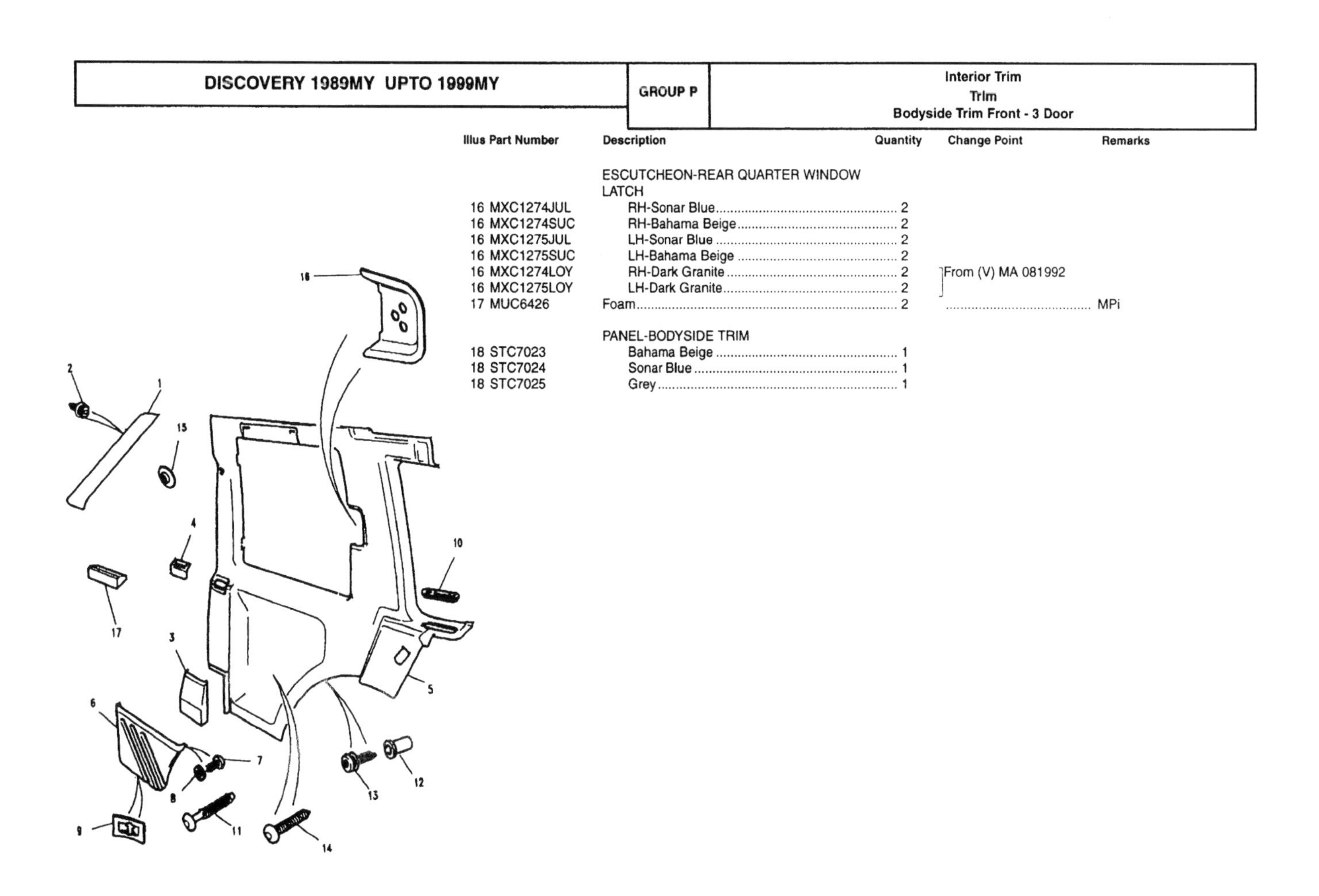

Illus	Part Number	Description	Quantity	Change Point	Remarks
		ESCUTCHEON-REAR QUARTER WINDOW LATCH			
16	MXC1274JUL	RH-Sonar Blue	2		
16	MXC1274SUC	RH-Bahama Beige	2		
16	MXC1275JUL	LH-Sonar Blue	2		
16	MXC1275SUC	LH-Bahama Beige	2		
16	MXC1274LOY	RH-Dark Granite	2	From (V) MA 081992	
16	MXC1275LOY	LH-Dark Granite	2		
17	MUC6426	Foam	2		MPi
		PANEL-BODYSIDE TRIM			
18	STC7023	Bahama Beige	1		
18	STC7024	Sonar Blue	1		
18	STC7025	Grey	1		

P 50

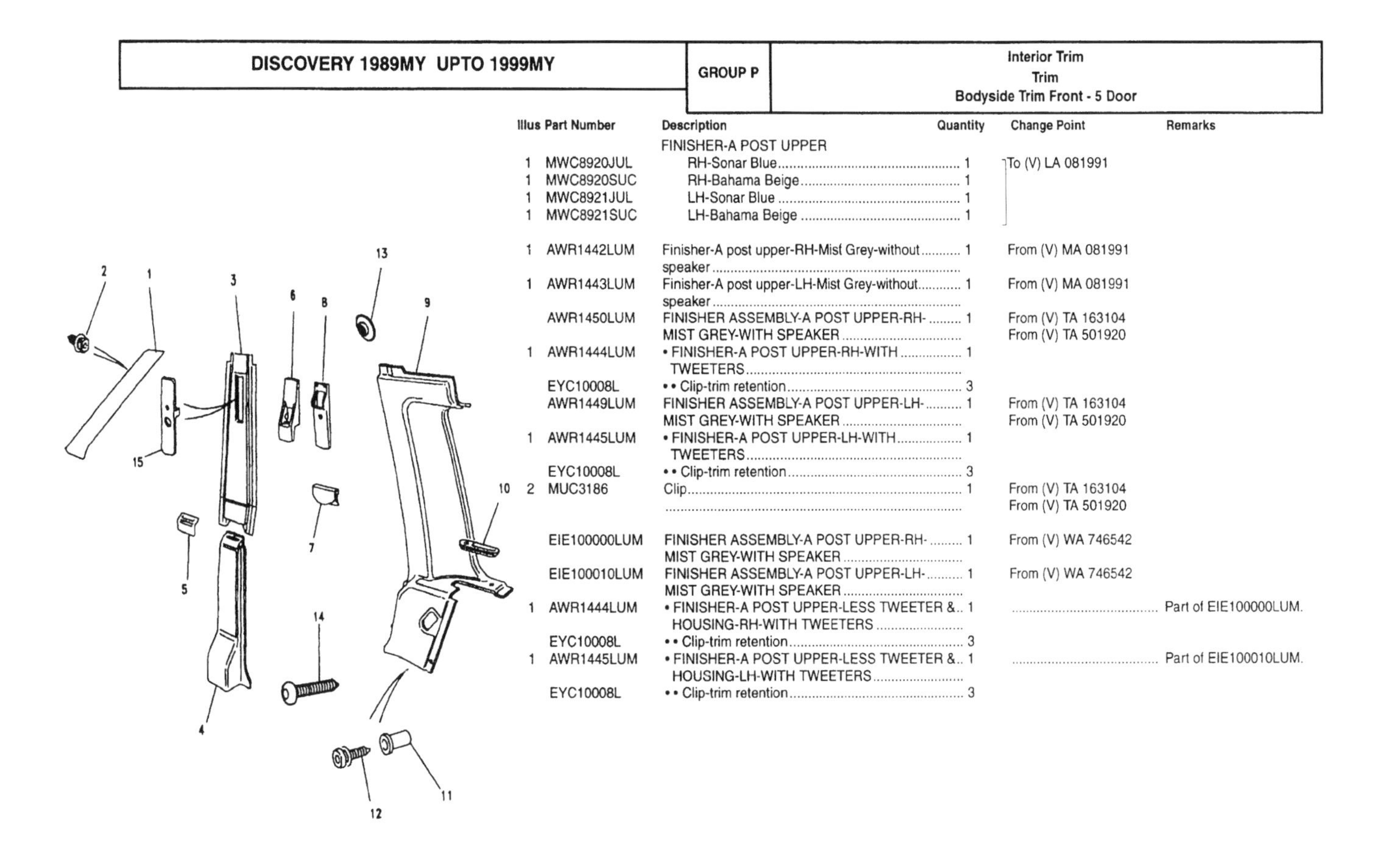

Illus	Part Number	Description	Quantity	Change Point	Remarks
		FINISHER-A POST UPPER			
1	MWC8920JUL	RH-Sonar Blue	1	To (V) LA 081991	
1	MWC8920SUC	RH-Bahama Beige	1		
1	MWC8921JUL	LH-Sonar Blue	1		
1	MWC8921SUC	LH-Bahama Beige	1		
1	AWR1442LUM	Finisher-A post upper-RH-Mist Grey-without speaker	1	From (V) MA 081991	
1	AWR1443LUM	Finisher-A post upper-LH-Mist Grey-without speaker	1	From (V) MA 081991	
	AWR1450LUM	FINISHER ASSEMBLY-A POST UPPER-RH-MIST GREY-WITH SPEAKER	1	From (V) TA 163104 From (V) TA 501920	
1	AWR1444LUM	• FINISHER-A POST UPPER-RH-WITH TWEETERS	1		
	EYC10008L	• • Clip-trim retention	3		
	AWR1449LUM	FINISHER ASSEMBLY-A POST UPPER-LH-MIST GREY-WITH SPEAKER	1	From (V) TA 163104 From (V) TA 501920	
1	AWR1445LUM	• FINISHER-A POST UPPER-LH-WITH TWEETERS	1		
	EYC10008L	• • Clip-trim retention	3		
2	MUC3186	Clip	1	From (V) TA 163104 From (V) TA 501920	
	EIE100000LUM	FINISHER ASSEMBLY-A POST UPPER-RH-MIST GREY-WITH SPEAKER	1	From (V) WA 746542	
	EIE100010LUM	FINISHER ASSEMBLY-A POST UPPER-LH-MIST GREY-WITH SPEAKER	1	From (V) WA 746542	
1	AWR1444LUM	• FINISHER-A POST UPPER-LESS TWEETER & HOUSING-RH-WITH TWEETERS	1		Part of EIE100000LUM.
	EYC10008L	• • Clip-trim retention	3		
1	AWR1445LUM	• FINISHER-A POST UPPER-LESS TWEETER & HOUSING-LH-WITH TWEETERS	1		Part of EIE100010LUM.
	EYC10008L	• • Clip-trim retention	3		

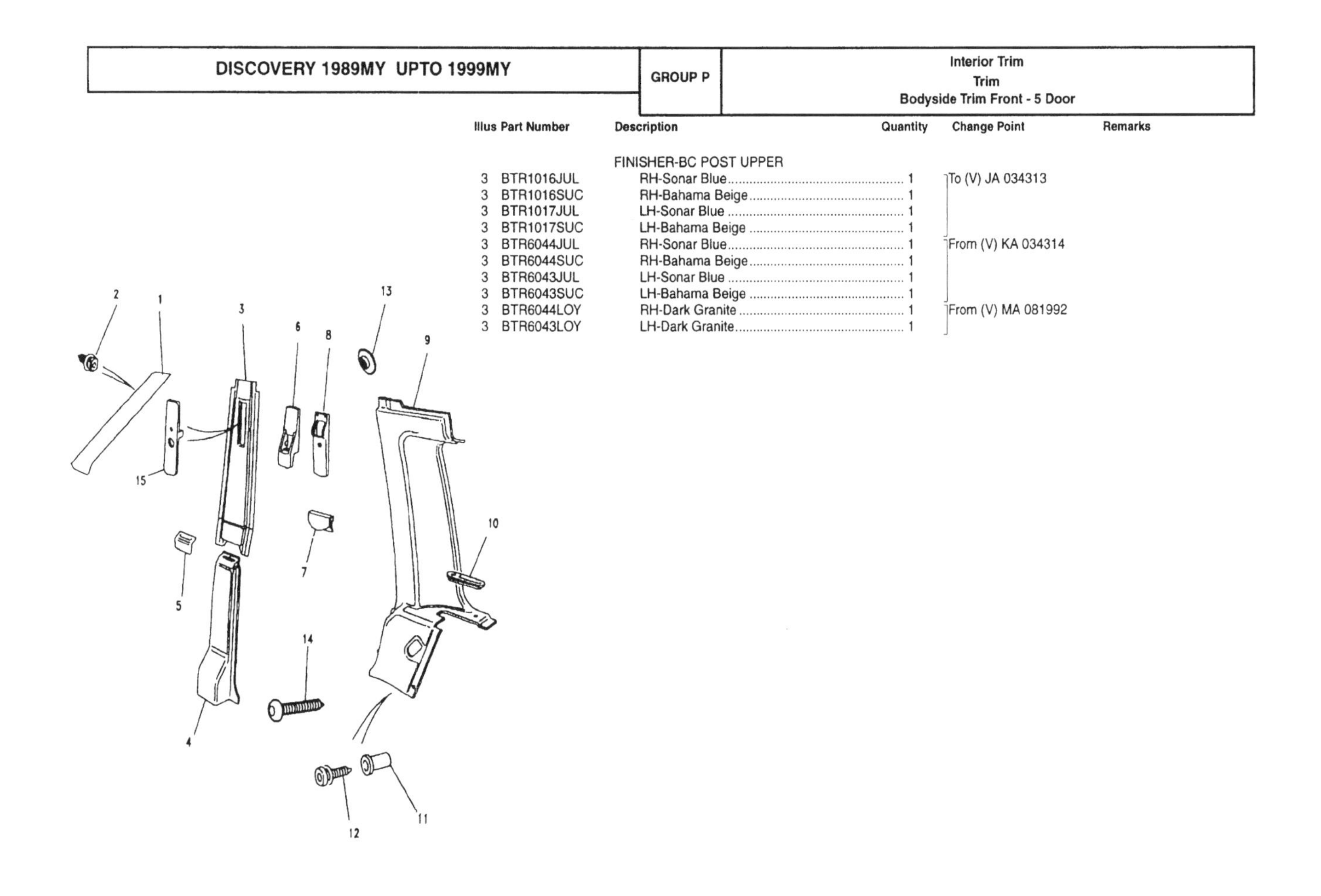

Illus	Part Number	Description	Quantity	Change Point	Remarks
		FINISHER-BC POST UPPER			
3	BTR1016JUL	RH-Sonar Blue	1	To (V) JA 034313	
3	BTR1016SUC	RH-Bahama Beige	1		
3	BTR1017JUL	LH-Sonar Blue	1		
3	BTR1017SUC	LH-Bahama Beige	1		
3	BTR6044JUL	RH-Sonar Blue	1	From (V) KA 034314	
3	BTR6044SUC	RH-Bahama Beige	1		
3	BTR6043JUL	LH-Sonar Blue	1		
3	BTR6043SUC	LH-Bahama Beige	1		
3	BTR6044LOY	RH-Dark Granite	1	From (V) MA 081992	
3	BTR6043LOY	LH-Dark Granite	1		

DISCOVERY 1989MY UPTO 1999MY

GROUP P — **Interior Trim / Trim / Bodyside Trim Front - 5 Door**

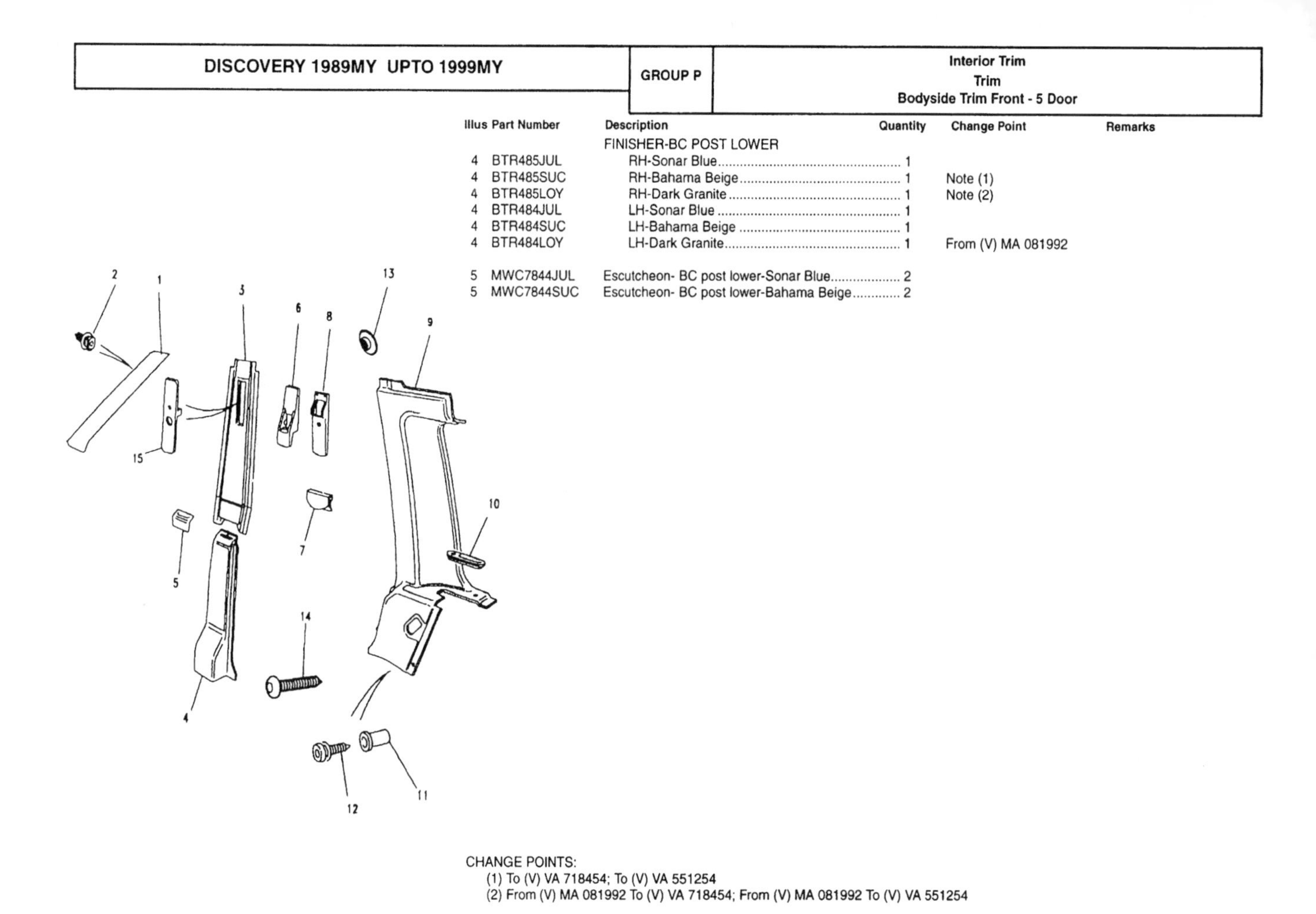

Illus	Part Number	Description	Quantity	Change Point	Remarks
		FINISHER-BC POST LOWER			
4	BTR485JUL	RH-Sonar Blue	1		
4	BTR485SUC	RH-Bahama Beige	1	Note (1)	
4	BTR485LOY	RH-Dark Granite	1	Note (2)	
4	BTR484JUL	LH-Sonar Blue	1		
4	BTR484SUC	LH-Bahama Beige	1		
4	BTR484LOY	LH-Dark Granite	1	From (V) MA 081992	
5	MWC7844JUL	Escutcheon- BC post lower-Sonar Blue	2		
5	MWC7844SUC	Escutcheon- BC post lower-Bahama Beige	2		

CHANGE POINTS:
(1) To (V) VA 718454; To (V) VA 551254
(2) From (V) MA 081992 To (V) VA 718454; From (V) MA 081992 To (V) VA 551254

P 53

DISCOVERY 1989MY UPTO 1999MY

GROUP P — **Interior Trim / Trim / Bodyside Trim Front - 5 Door**

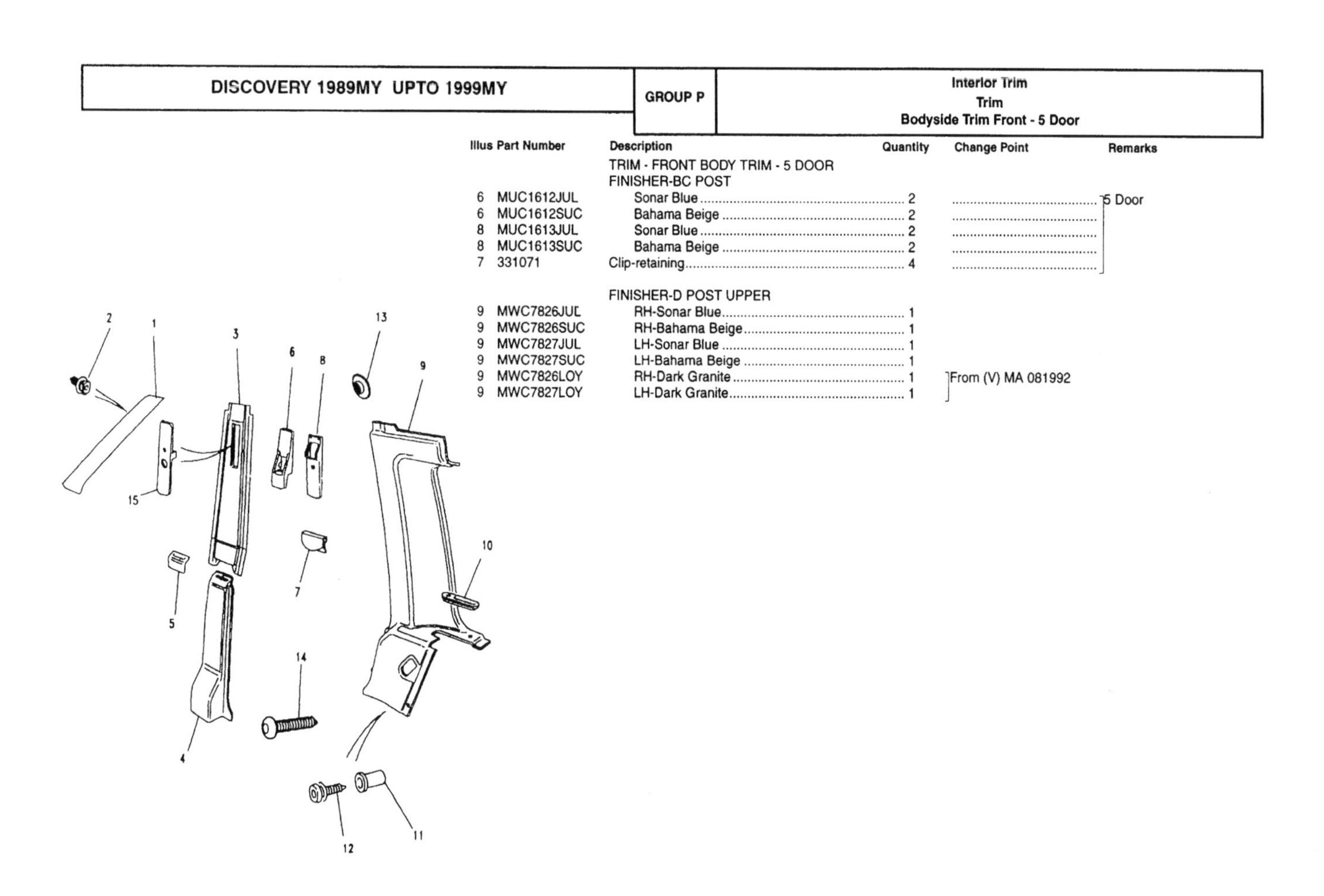

Illus	Part Number	Description	Quantity	Change Point	Remarks
		TRIM - FRONT BODY TRIM - 5 DOOR			
		FINISHER-BC POST			
6	MUC1612JUL	Sonar Blue	2		5 Door
6	MUC1612SUC	Bahama Beige	2		
8	MUC1613JUL	Sonar Blue	2		
8	MUC1613SUC	Bahama Beige	2		
7	331071	Clip-retaining	4		
		FINISHER-D POST UPPER			
9	MWC7826JUL	RH-Sonar Blue	1		
9	MWC7826SUC	RH-Bahama Beige	1		
9	MWC7827JUL	LH-Sonar Blue	1		
9	MWC7827SUC	LH-Bahama Beige	1		
9	MWC7826LOY	RH-Dark Granite	1	From (V) MA 081992	
9	MWC7827LOY	LH-Dark Granite	1		

P 54

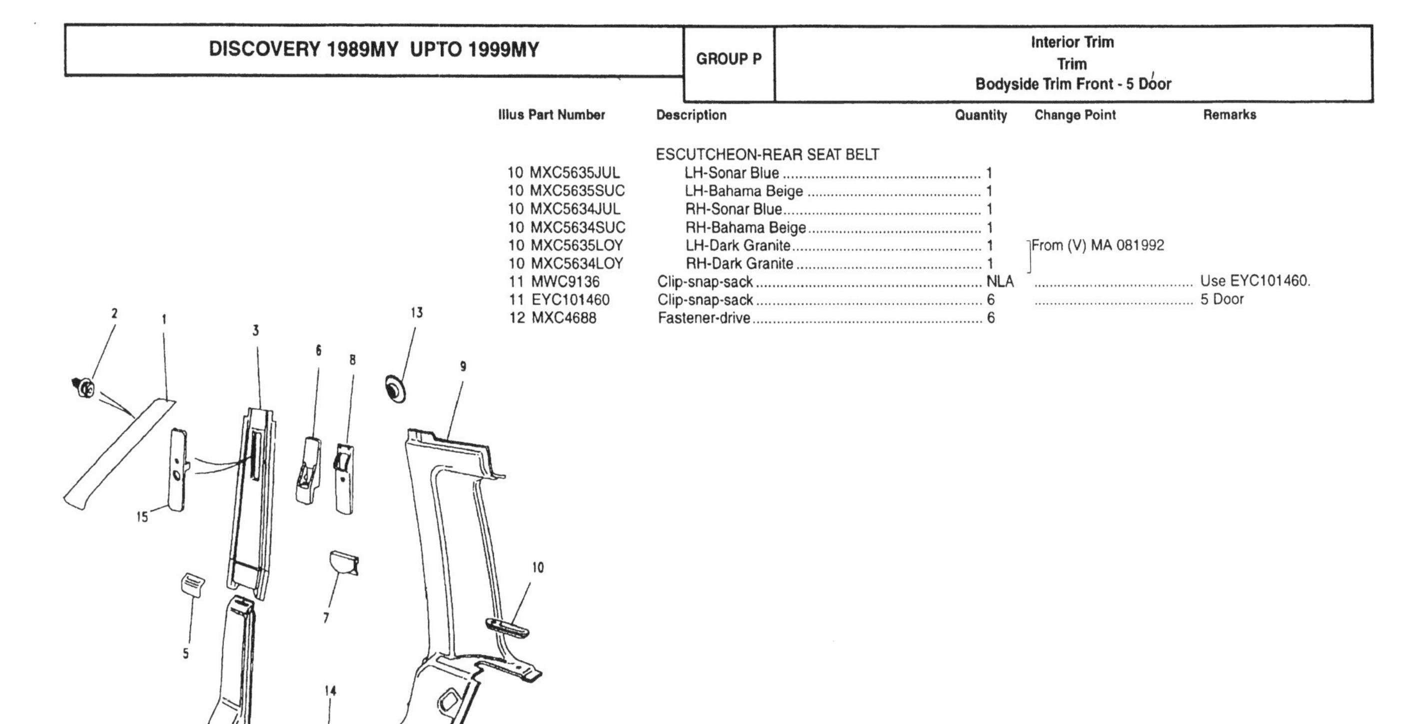

Illus	Part Number	Description	Quantity	Change Point	Remarks
		ESCUTCHEON-REAR SEAT BELT			
10	MXC5635JUL	LH-Sonar Blue	1		
10	MXC5635SUC	LH-Bahama Beige	1		
10	MXC5634JUL	RH-Sonar Blue	1		
10	MXC5634SUC	RH-Bahama Beige	1		
10	MXC5635LOY	LH-Dark Granite	1	From (V) MA 081992	
10	MXC5634LOY	RH-Dark Granite	1		
11	MWC9136	Clip-snap-sack	NLA		Use EYC101460.
11	EYC101460	Clip-snap-sack	6		5 Door
12	MXC4688	Fastener-drive	6		

P 55

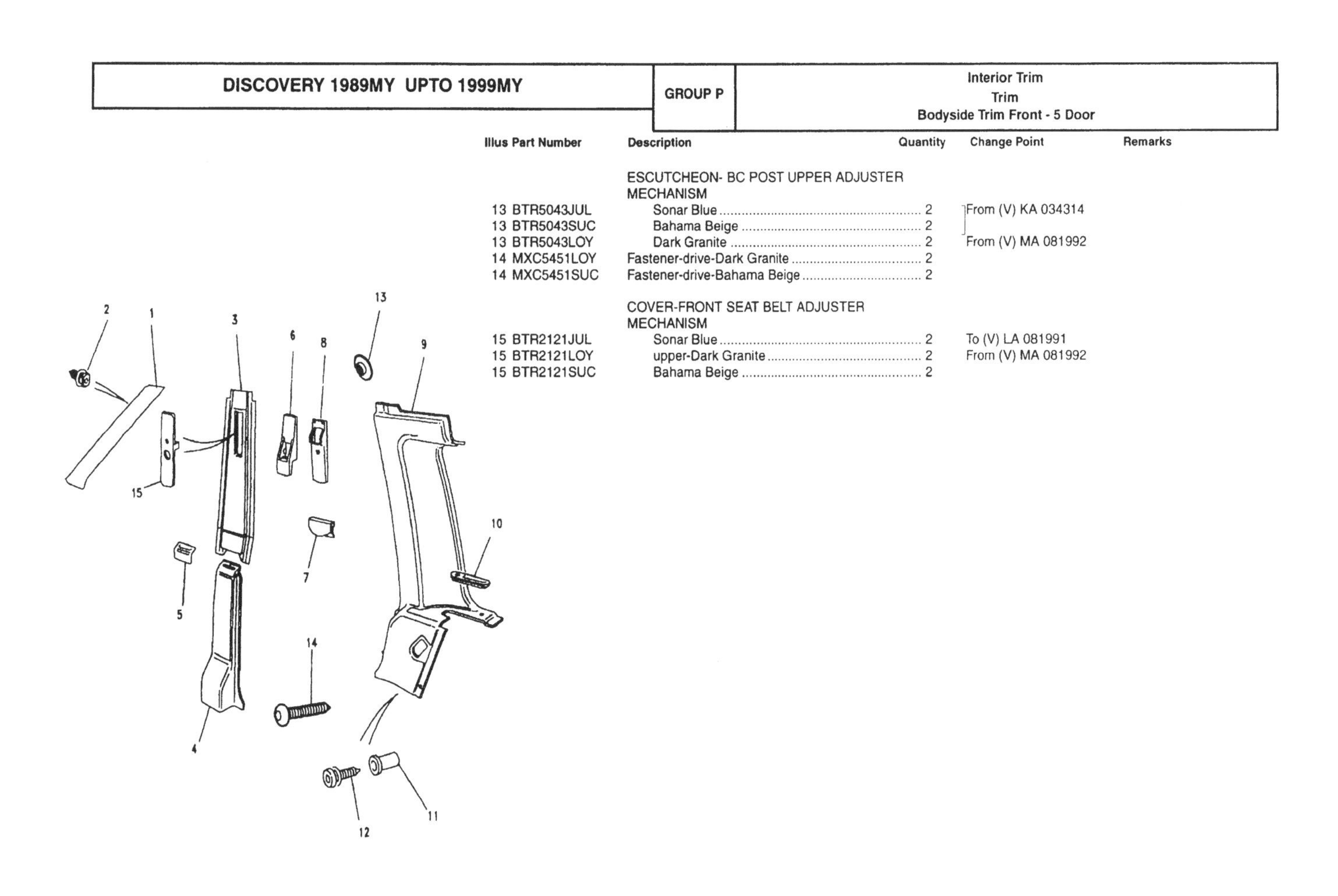

Illus	Part Number	Description	Quantity	Change Point	Remarks
		ESCUTCHEON- BC POST UPPER ADJUSTER MECHANISM			
13	BTR5043JUL	Sonar Blue	2	From (V) KA 034314	
13	BTR5043SUC	Bahama Beige	2		
13	BTR5043LOY	Dark Granite	2	From (V) MA 081992	
14	MXC5451LOY	Fastener-drive-Dark Granite	2		
14	MXC5451SUC	Fastener-drive-Bahama Beige	2		
		COVER-FRONT SEAT BELT ADJUSTER MECHANISM			
15	BTR2121JUL	Sonar Blue	2	To (V) LA 081991	
15	BTR2121LOY	upper-Dark Granite	2	From (V) MA 081992	
15	BTR2121SUC	Bahama Beige	2		

P 56

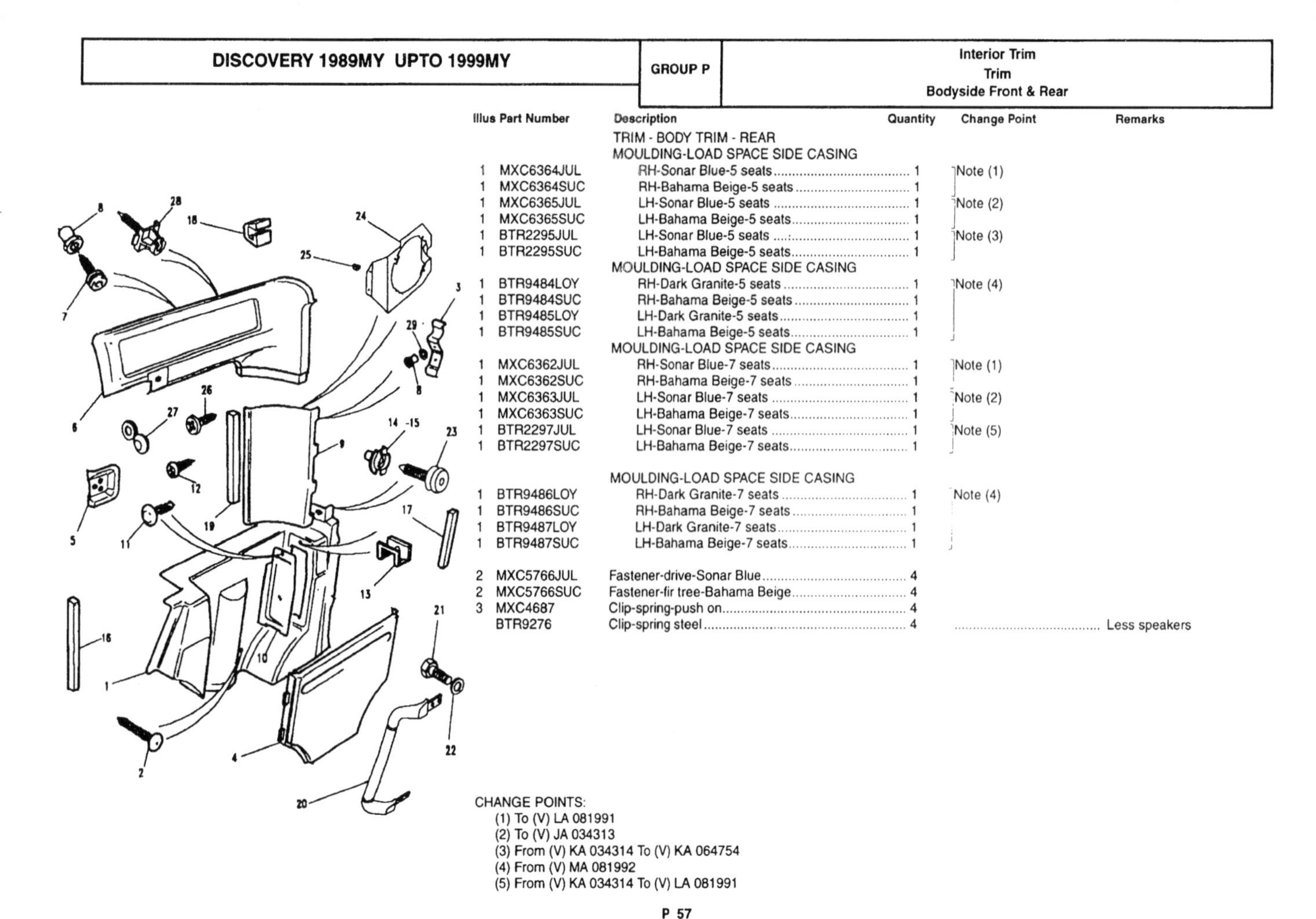

Illus	Part Number	Description	Quantity	Change Point	Remarks
		TRIM - BODY TRIM - REAR			
		MOULDING-LOAD SPACE SIDE CASING			
1	MXC6364JUL	RH-Sonar Blue-5 seats	1	Note (1)	
1	MXC6364SUC	RH-Bahama Beige-5 seats	1		
1	MXC6365JUL	LH-Sonar Blue-5 seats	1	Note (2)	
1	MXC6365SUC	LH-Bahama Beige-5 seats	1		
1	BTR2295JUL	LH-Sonar Blue-5 seats	1	Note (3)	
1	BTR2295SUC	LH-Bahama Beige-5 seats	1		
		MOULDING-LOAD SPACE SIDE CASING			
1	BTR9484LOY	RH-Dark Granite-5 seats	1	Note (4)	
1	BTR9484SUC	RH-Bahama Beige-5 seats	1		
1	BTR9485LOY	LH-Dark Granite-5 seats	1		
1	BTR9485SUC	LH-Bahama Beige-5 seats	1		
		MOULDING-LOAD SPACE SIDE CASING			
1	MXC6362JUL	RH-Sonar Blue-7 seats	1	Note (1)	
1	MXC6362SUC	RH-Bahama Beige-7 seats	1		
1	MXC6363JUL	LH-Sonar Blue-7 seats	1	Note (2)	
1	MXC6363SUC	LH-Bahama Beige-7 seats	1		
1	BTR2297JUL	LH-Sonar Blue-7 seats	1	Note (5)	
1	BTR2297SUC	LH-Bahama Beige-7 seats	1		
		MOULDING-LOAD SPACE SIDE CASING			
1	BTR9486LOY	RH-Dark Granite-7 seats	1	Note (4)	
1	BTR9486SUC	RH-Bahama Beige-7 seats	1		
1	BTR9487LOY	LH-Dark Granite-7 seats	1		
1	BTR9487SUC	LH-Bahama Beige-7 seats	1		
2	MXC5766JUL	Fastener-drive-Sonar Blue	4		
2	MXC5766SUC	Fastener-fir tree-Bahama Beige	4		
3	MXC4687	Clip-spring-push on	4		
	BTR9276	Clip-spring steel	4		Less speakers

CHANGE POINTS:
(1) To (V) LA 081991
(2) To (V) JA 034313
(3) From (V) KA 034314 To (V) KA 064754
(4) From (V) MA 081992
(5) From (V) KA 034314 To (V) LA 081991

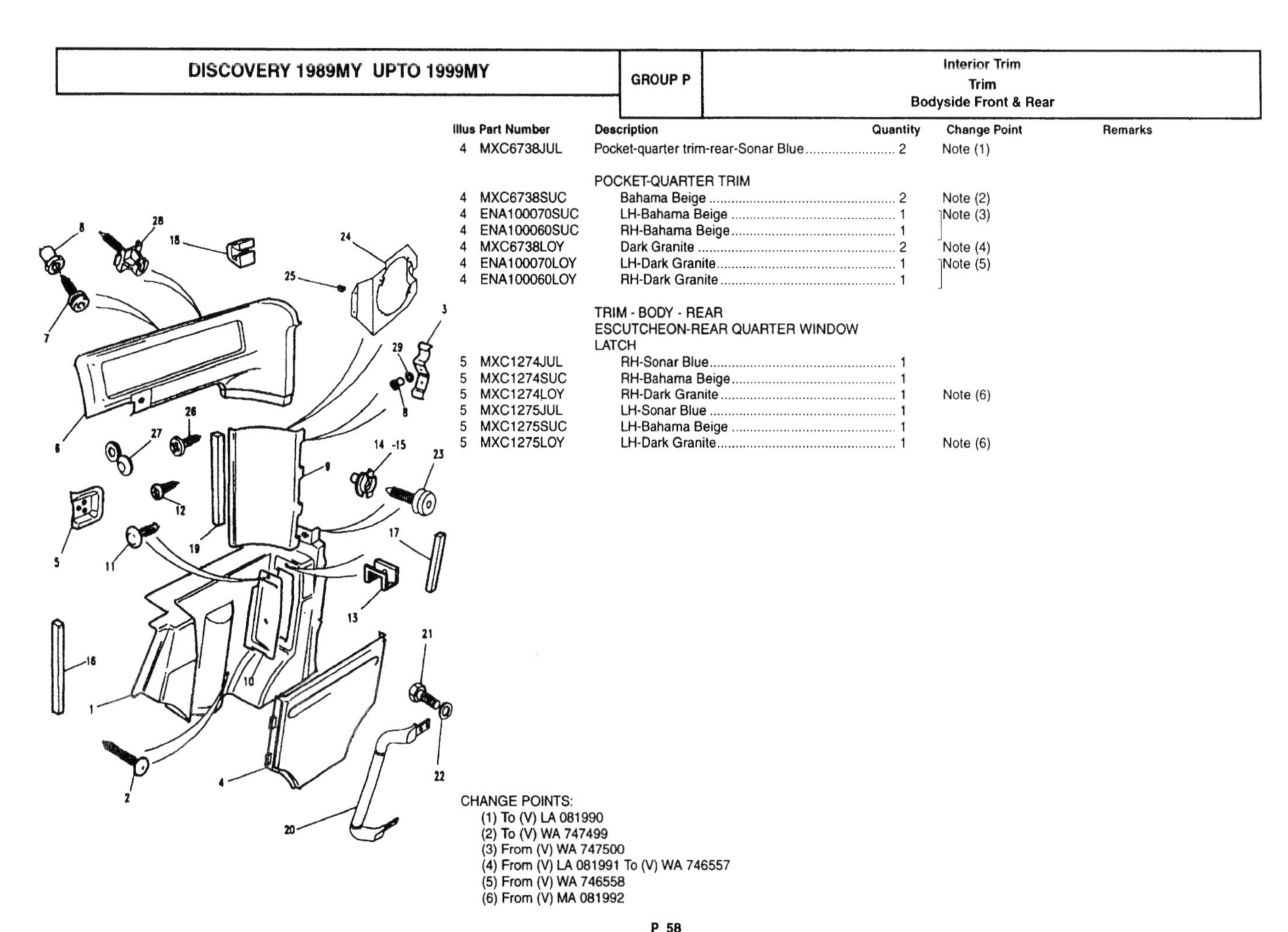

Illus	Part Number	Description	Quantity	Change Point	Remarks
4	MXC6738JUL	Pocket-quarter trim-rear-Sonar Blue	2	Note (1)	
		POCKET-QUARTER TRIM			
4	MXC6738SUC	Bahama Beige	2	Note (2)	
4	ENA100070SUC	LH-Bahama Beige	1	Note (3)	
4	ENA100060SUC	RH-Bahama Beige	1		
4	MXC6738LOY	Dark Granite	2	Note (4)	
4	ENA100070LOY	LH-Dark Granite	1	Note (5)	
4	ENA100060LOY	RH-Dark Granite	1		
		TRIM - BODY - REAR			
		ESCUTCHEON-REAR QUARTER WINDOW LATCH			
5	MXC1274JUL	RH-Sonar Blue	1		
5	MXC1274SUC	RH-Bahama Beige	1		
5	MXC1274LOY	RH-Dark Granite	1	Note (6)	
5	MXC1275JUL	LH-Sonar Blue	1		
5	MXC1275SUC	LH-Bahama Beige	1		
5	MXC1275LOY	LH-Dark Granite	1	Note (6)	

CHANGE POINTS:
(1) To (V) LA 081990
(2) To (V) WA 747499
(3) From (V) WA 747500
(4) From (V) LA 081991 To (V) WA 746557
(5) From (V) WA 746558
(6) From (V) MA 081992

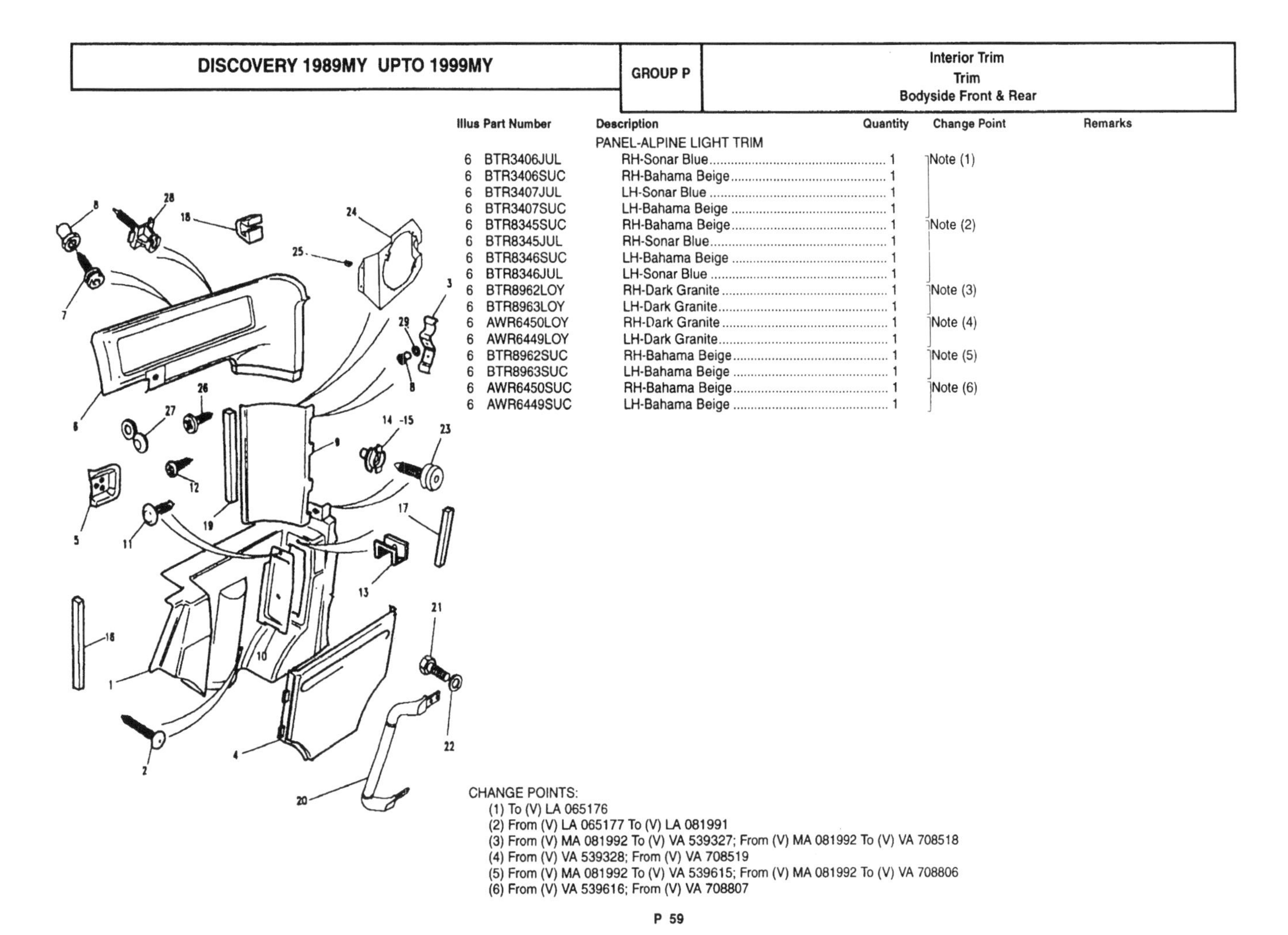

Illus	Part Number	Description	Quantity	Change Point	Remarks
		PANEL-ALPINE LIGHT TRIM			
6	BTR3406JUL	RH-Sonar Blue	1	Note (1)	
6	BTR3406SUC	RH-Bahama Beige	1		
6	BTR3407JUL	LH-Sonar Blue	1		
6	BTR3407SUC	LH-Bahama Beige	1		
6	BTR8345SUC	RH-Bahama Beige	1	Note (2)	
6	BTR8345JUL	RH-Sonar Blue	1		
6	BTR8346SUC	LH-Bahama Beige	1		
6	BTR8346JUL	LH-Sonar Blue	1		
6	BTR8962LOY	RH-Dark Granite	1	Note (3)	
6	BTR8963LOY	LH-Dark Granite	1		
6	AWR6450LOY	RH-Dark Granite	1	Note (4)	
6	AWR6449LOY	LH-Dark Granite	1		
6	BTR8962SUC	RH-Bahama Beige	1	Note (5)	
6	BTR8963SUC	LH-Bahama Beige	1		
6	AWR6450SUC	RH-Bahama Beige	1	Note (6)	
6	AWR6449SUC	LH-Bahama Beige	1		

CHANGE POINTS:
(1) To (V) LA 065176
(2) From (V) LA 065177 To (V) LA 081991
(3) From (V) MA 081992 To (V) VA 539327; From (V) MA 081992 To (V) VA 708518
(4) From (V) VA 539328; From (V) VA 708519
(5) From (V) MA 081992 To (V) VA 539615; From (V) MA 081992 To (V) VA 708806
(6) From (V) VA 539616; From (V) VA 708807

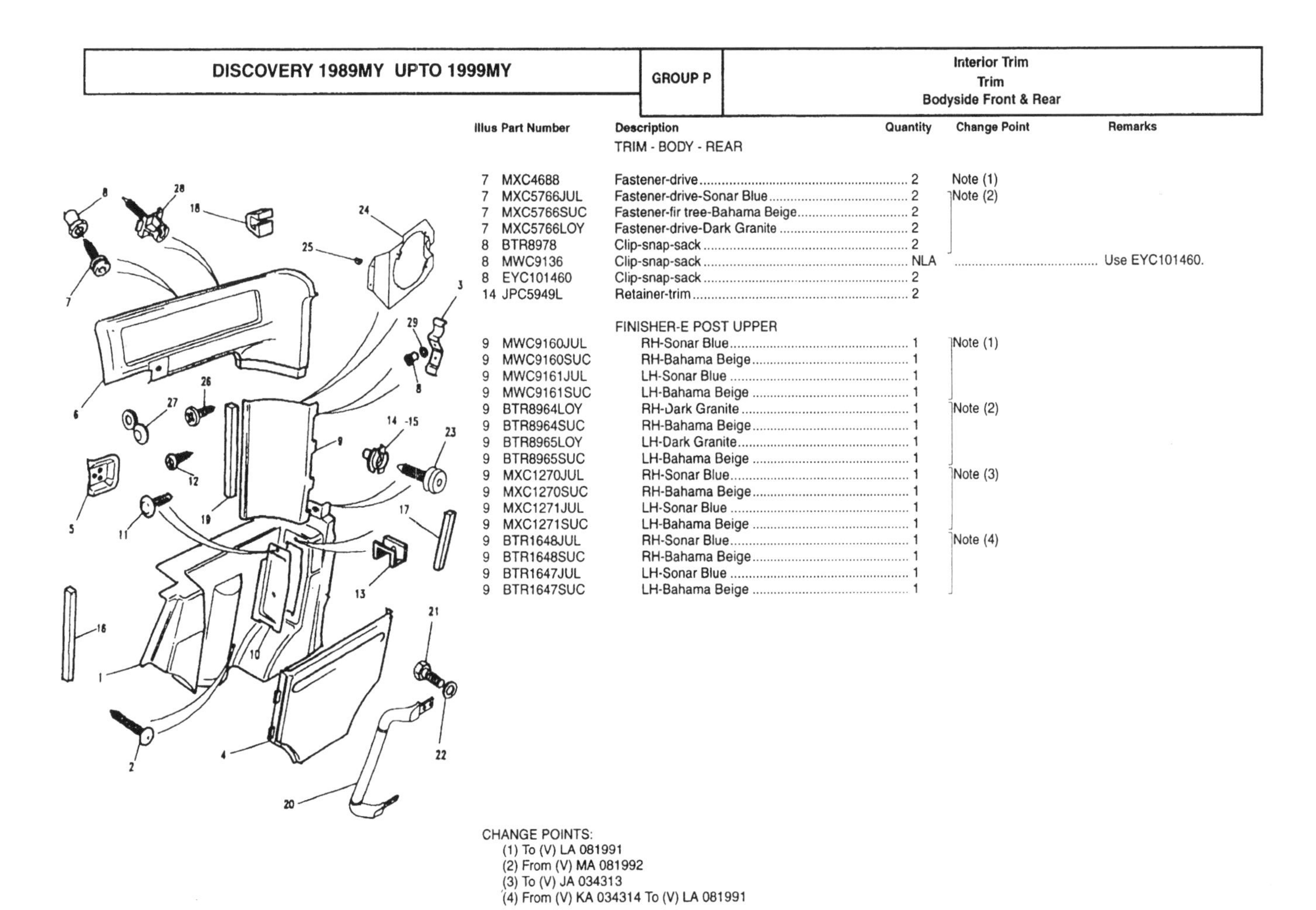

Illus	Part Number	Description	Quantity	Change Point	Remarks
		TRIM - BODY - REAR			
7	MXC4688	Fastener-drive	2	Note (1)	
7	MXC5766JUL	Fastener-drive-Sonar Blue	2	Note (2)	
7	MXC5766SUC	Fastener-fir tree-Bahama Beige	2		
7	MXC5766LOY	Fastener-drive-Dark Granite	2		
8	BTR8978	Clip-snap-sack	2		
8	MWC9136	Clip-snap-sack	NLA		Use EYC101460.
8	EYC101460	Clip-snap-sack	2		
14	JPC5949L	Retainer-trim	2		
		FINISHER-E POST UPPER			
9	MWC9160JUL	RH-Sonar Blue	1	Note (1)	
9	MWC9160SUC	RH-Bahama Beige	1		
9	MWC9161JUL	LH-Sonar Blue	1		
9	MWC9161SUC	LH-Bahama Beige	1		
9	BTR8964LOY	RH-Dark Granite	1	Note (2)	
9	BTR8964SUC	RH-Bahama Beige	1		
9	BTR8965LOY	LH-Dark Granite	1		
9	BTR8965SUC	LH-Bahama Beige	1		
9	MXC1270JUL	RH-Sonar Blue	1	Note (3)	
9	MXC1270SUC	RH-Bahama Beige	1		
9	MXC1271JUL	LH-Sonar Blue	1		
9	MXC1271SUC	LH-Bahama Beige	1		
9	BTR1648JUL	RH-Sonar Blue	1	Note (4)	
9	BTR1648SUC	RH-Bahama Beige	1		
9	BTR1647JUL	LH-Sonar Blue	1		
9	BTR1647SUC	LH-Bahama Beige	1		

CHANGE POINTS:
(1) To (V) LA 081991
(2) From (V) MA 081992
(3) To (V) JA 034313
(4) From (V) KA 034314 To (V) LA 081991

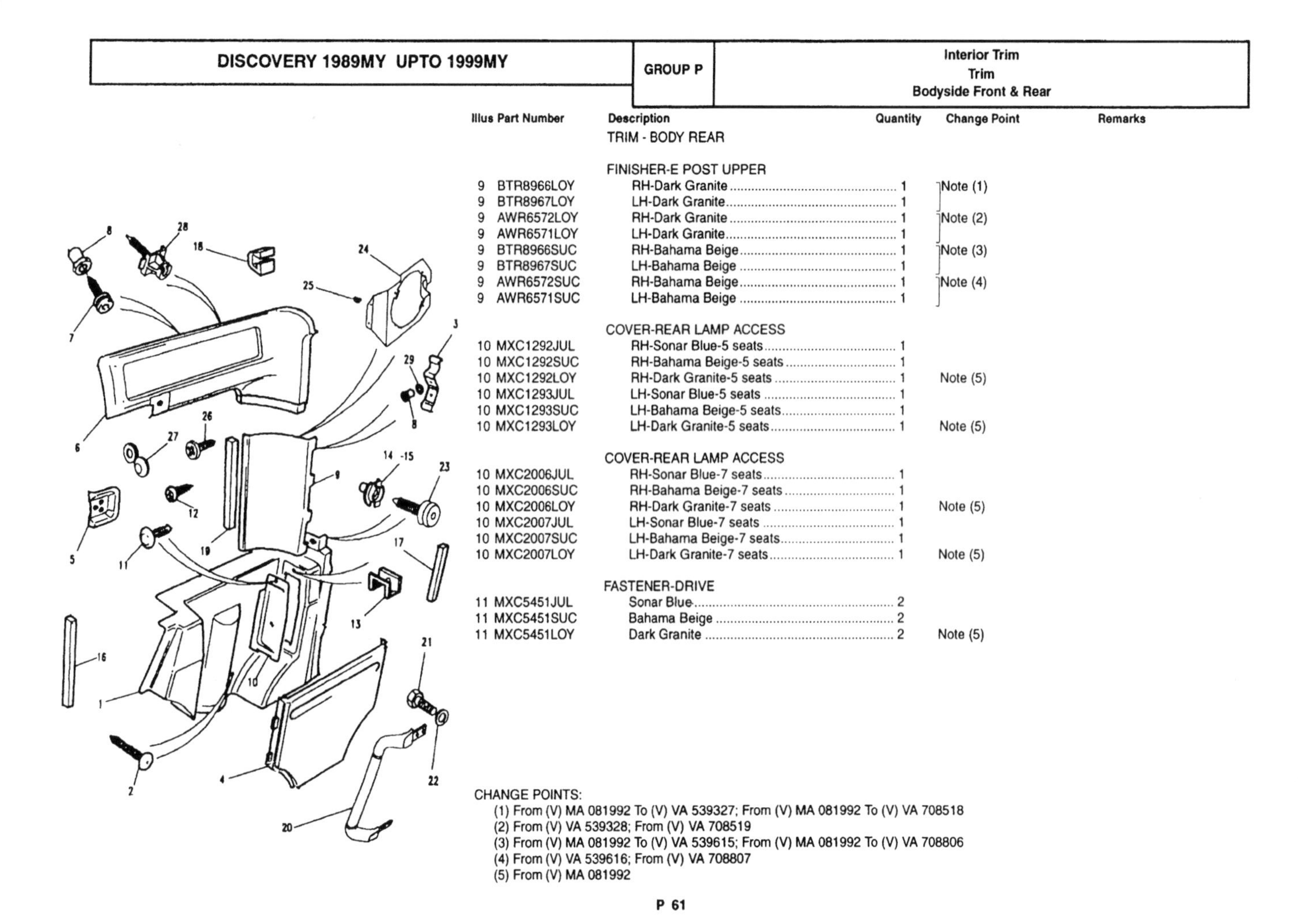

Illus	Part Number	Description	Quantity	Change Point	Remarks
		TRIM - BODY REAR			
		FINISHER-E POST UPPER			
9	BTR8966LOY	RH-Dark Granite	1	Note (1)	
9	BTR8967LOY	LH-Dark Granite	1		
9	AWR6572LOY	RH-Dark Granite	1	Note (2)	
9	AWR6571LOY	LH-Dark Granite	1		
9	BTR8966SUC	RH-Bahama Beige	1	Note (3)	
9	BTR8967SUC	LH-Bahama Beige	1		
9	AWR6572SUC	RH-Bahama Beige	1	Note (4)	
9	AWR6571SUC	LH-Bahama Beige	1		
		COVER-REAR LAMP ACCESS			
10	MXC1292JUL	RH-Sonar Blue-5 seats	1		
10	MXC1292SUC	RH-Bahama Beige-5 seats	1		
10	MXC1292LOY	RH-Dark Granite-5 seats	1	Note (5)	
10	MXC1293JUL	LH-Sonar Blue-5 seats	1		
10	MXC1293SUC	LH-Bahama Beige-5 seats	1		
10	MXC1293LOY	LH-Dark Granite-5 seats	1	Note (5)	
		COVER-REAR LAMP ACCESS			
10	MXC2006JUL	RH-Sonar Blue-7 seats	1		
10	MXC2006SUC	RH-Bahama Beige-7 seats	1		
10	MXC2006LOY	RH-Dark Granite-7 seats	1	Note (5)	
10	MXC2007JUL	LH-Sonar Blue-7 seats	1		
10	MXC2007SUC	LH-Bahama Beige-7 seats	1		
10	MXC2007LOY	LH-Dark Granite-7 seats	1	Note (5)	
		FASTENER-DRIVE			
11	MXC5451JUL	Sonar Blue	2		
11	MXC5451SUC	Bahama Beige	2		
11	MXC5451LOY	Dark Granite	2	Note (5)	

CHANGE POINTS:
(1) From (V) MA 081992 To (V) VA 539327; From (V) MA 081992 To (V) VA 708518
(2) From (V) VA 539328; From (V) VA 708519
(3) From (V) MA 081992 To (V) VA 539615; From (V) MA 081992 To (V) VA 708806
(4) From (V) VA 539616; From (V) VA 708807
(5) From (V) MA 081992

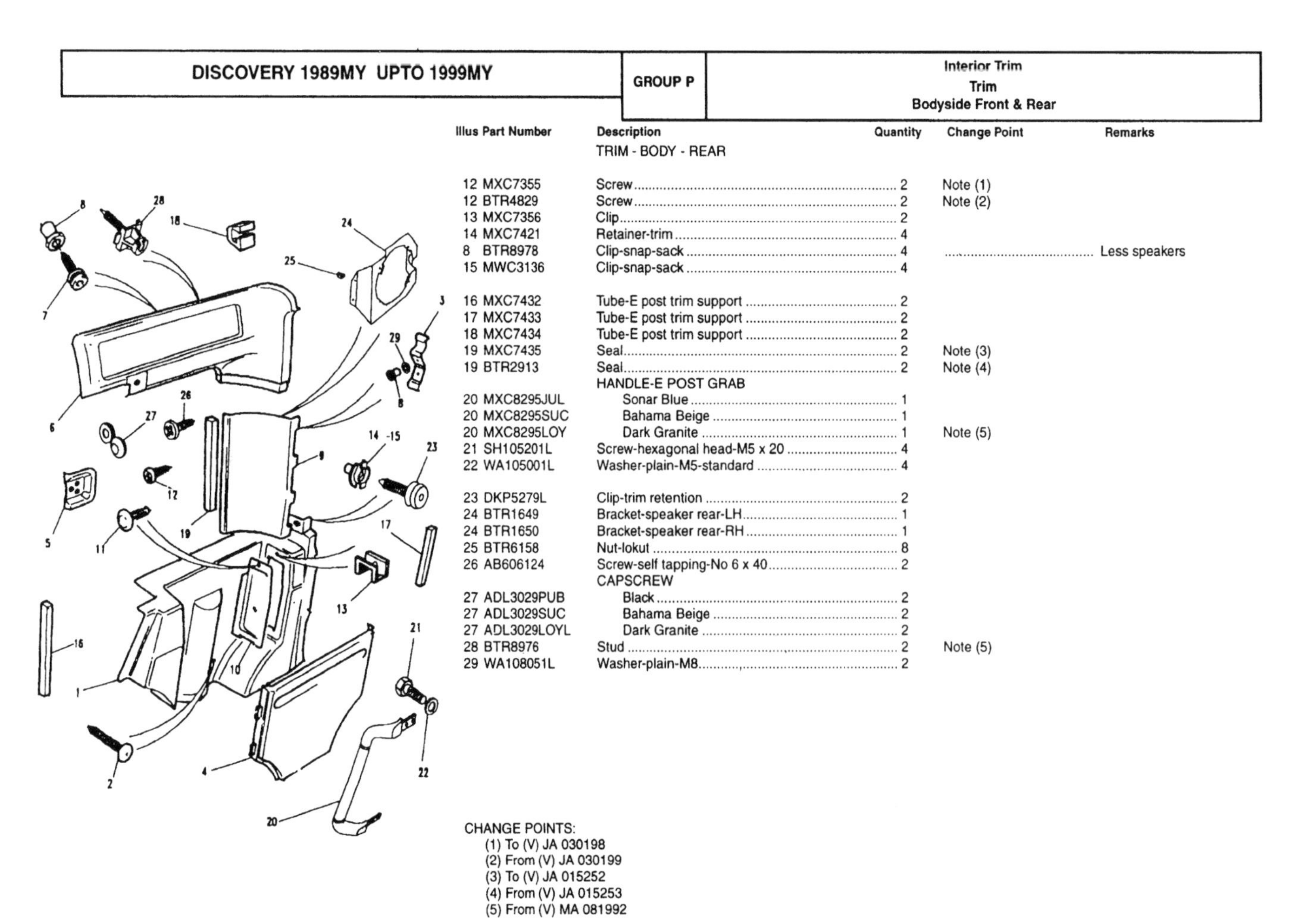

Illus	Part Number	Description	Quantity	Change Point	Remarks
		TRIM - BODY - REAR			
12	MXC7355	Screw	2	Note (1)	
12	BTR4829	Screw	2	Note (2)	
13	MXC7356	Clip	2		
14	MXC7421	Retainer-trim	4		
8	BTR8978	Clip-snap-sack	4		Less speakers
15	MWC3136	Clip-snap-sack	4		
16	MXC7432	Tube-E post trim support	2		
17	MXC7433	Tube-E post trim support	2		
18	MXC7434	Tube-E post trim support	2		
19	MXC7435	Seal	2	Note (3)	
19	BTR2913	Seal	2	Note (4)	
		HANDLE-E POST GRAB			
20	MXC8295JUL	Sonar Blue	1		
20	MXC8295SUC	Bahama Beige	1		
20	MXC8295LOY	Dark Granite	1	Note (5)	
21	SH105201L	Screw-hexagonal head-M5 x 20	4		
22	WA105001L	Washer-plain-M5-standard	4		
23	DKP5279L	Clip-trim retention	2		
24	BTR1649	Bracket-speaker rear-LH	1		
24	BTR1650	Bracket-speaker rear-RH	1		
25	BTR6158	Nut-lokut	8		
26	AB606124	Screw-self tapping-No 6 x 40	2		
		CAPSCREW			
27	ADL3029PUB	Black	2		
27	ADL3029SUC	Bahama Beige	2		
27	ADL3029LOYL	Dark Granite	2		
28	BTR8976	Stud	2	Note (5)	
29	WA108051L	Washer-plain-M8	2		

CHANGE POINTS:
(1) To (V) JA 030198
(2) From (V) JA 030199
(3) To (V) JA 015252
(4) From (V) JA 015253
(5) From (V) MA 081992

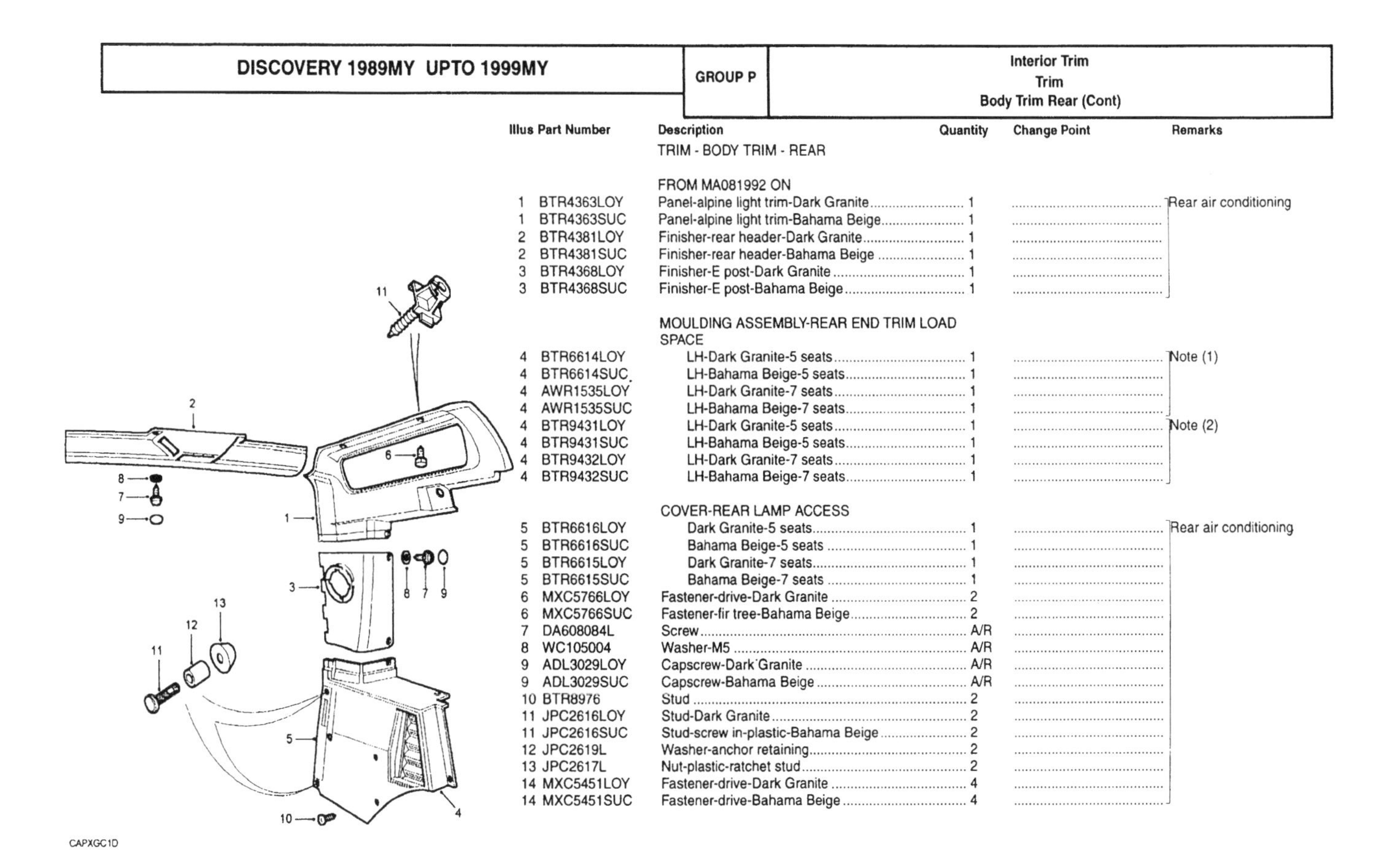

Illus	Part Number	Description	Quantity	Change Point	Remarks
		TRIM - BODY TRIM - REAR			
		FROM MA081992 ON			
1	BTR4363LOY	Panel-alpine light trim-Dark Granite	1		Rear air conditioning
1	BTR4363SUC	Panel-alpine light trim-Bahama Beige	1		
2	BTR4381LOY	Finisher-rear header-Dark Granite	1		
2	BTR4381SUC	Finisher-rear header-Bahama Beige	1		
3	BTR4368LOY	Finisher-E post-Dark Granite	1		
3	BTR4368SUC	Finisher-E post-Bahama Beige	1		
		MOULDING ASSEMBLY-REAR END TRIM LOAD SPACE			
4	BTR6614LOY	LH-Dark Granite-5 seats	1		Note (1)
4	BTR6614SUC	LH-Bahama Beige-5 seats	1		
4	AWR1535LOY	LH-Dark Granite-7 seats	1		
4	AWR1535SUC	LH-Bahama Beige-7 seats	1		
4	BTR9431LOY	LH-Dark Granite-5 seats	1		Note (2)
4	BTR9431SUC	LH-Bahama Beige-5 seats	1		
4	BTR9432LOY	LH-Dark Granite-7 seats	1		
4	BTR9432SUC	LH-Bahama Beige-7 seats	1		
		COVER-REAR LAMP ACCESS			
5	BTR6616LOY	Dark Granite-5 seats	1		Rear air conditioning
5	BTR6616SUC	Bahama Beige-5 seats	1		
5	BTR6615LOY	Dark Granite-7 seats	1		
5	BTR6615SUC	Bahama Beige-7 seats	1		
6	MXC5766LOY	Fastener-drive-Dark Granite	2		
6	MXC5766SUC	Fastener-fir tree-Bahama Beige	2		
7	DA608084L	Screw	A/R		
8	WC105004	Washer-M5	A/R		
9	ADL3029LOY	Capscrew-Dark Granite	A/R		
9	ADL3029SUC	Capscrew-Bahama Beige	A/R		
10	BTR8976	Stud	2		
11	JPC2616LOY	Stud-Dark Granite	2		
11	JPC2616SUC	Stud-screw in-plastic-Bahama Beige	2		
12	JPC2619L	Washer-anchor retaining	2		
13	JPC2617L	Nut-plastic-ratchet stud	2		
14	MXC5451LOY	Fastener-drive-Dark Granite	4		
14	MXC5451SUC	Fastener-drive-Bahama Beige	4		

Remarks:
(1) rear air conditioning Except Australia And New Zealand
(2) rear air conditioning Australia New Zealand

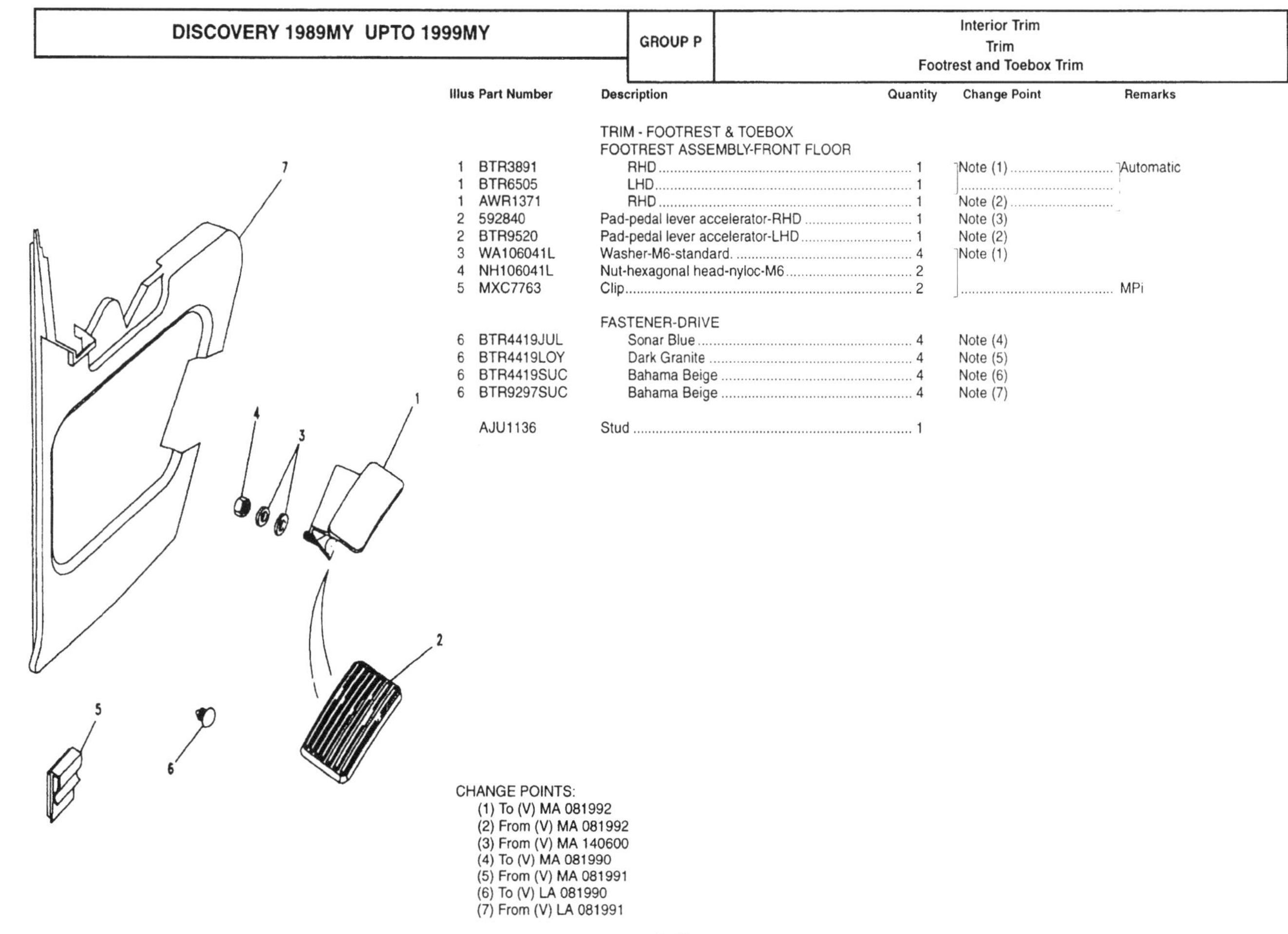

Illus	Part Number	Description	Quantity	Change Point	Remarks
		TRIM - FOOTREST & TOEBOX			
		FOOTREST ASSEMBLY-FRONT FLOOR			
1	BTR3891	RHD	1	Note (1)	Automatic
1	BTR6505	LHD	1		
1	AWR1371	RHD	1	Note (2)	
2	592840	Pad-pedal lever accelerator-RHD	1	Note (3)	
2	BTR9520	Pad-pedal lever accelerator-LHD	1	Note (2)	
3	WA106041L	Washer-M6-standard.	4	Note (1)	
4	NH106041L	Nut-hexagonal head-nyloc-M6	2		
5	MXC7763	Clip	2		MPi
		FASTENER-DRIVE			
6	BTR4419JUL	Sonar Blue	4	Note (4)	
6	BTR4419LOY	Dark Granite	4	Note (5)	
6	BTR4419SUC	Bahama Beige	4	Note (6)	
6	BTR9297SUC	Bahama Beige	4	Note (7)	
	AJU1136	Stud	1		

CHANGE POINTS:
(1) To (V) MA 081992
(2) From (V) MA 081992
(3) From (V) MA 140600
(4) To (V) MA 081990
(5) From (V) MA 081991
(6) To (V) LA 081990
(7) From (V) LA 081991

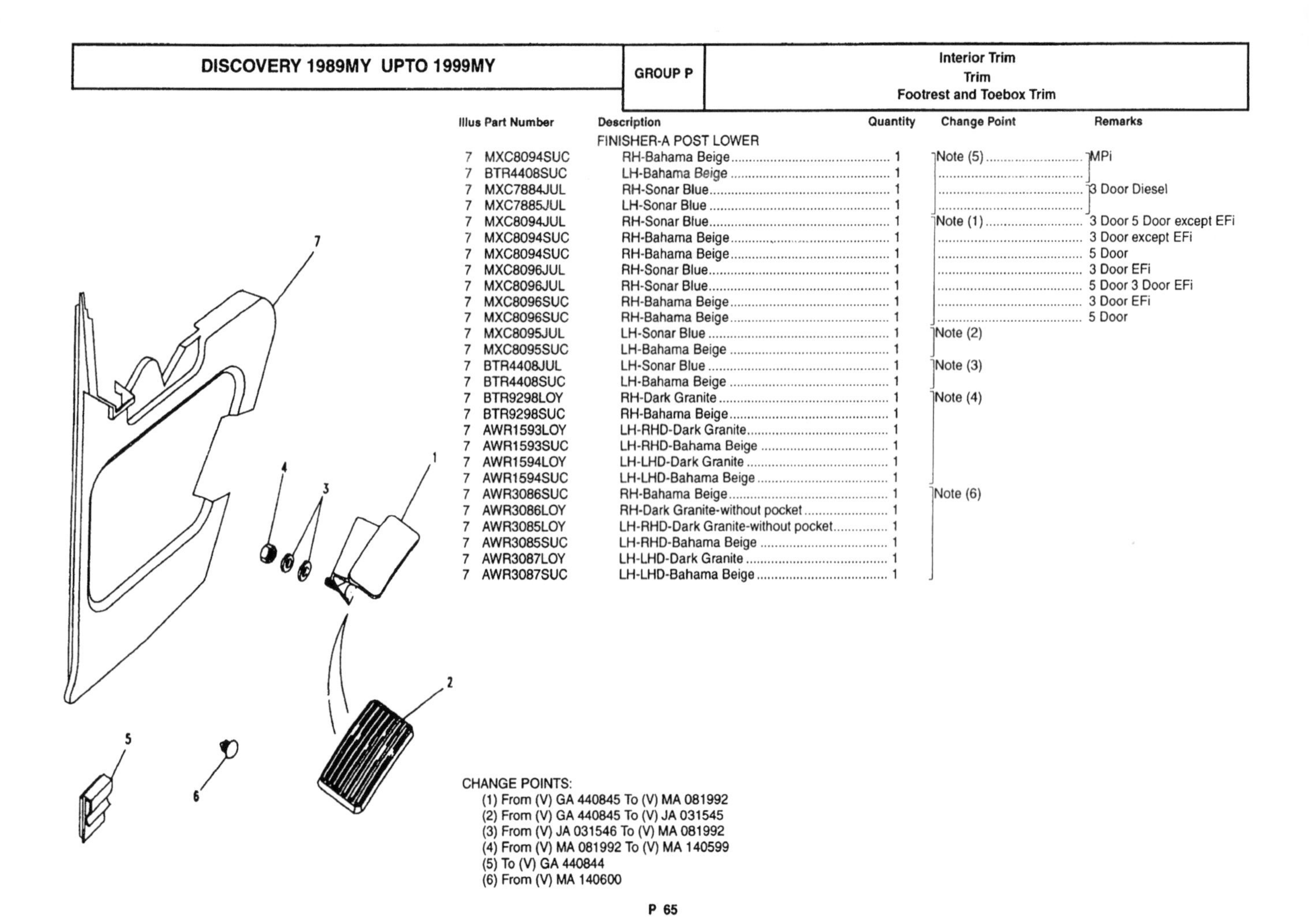

Illus	Part Number	Description	Quantity	Change Point	Remarks
		FINISHER-A POST LOWER			
7	MXC8094SUC	RH-Bahama Beige	1	Note (5)	MPi
7	BTR4408SUC	LH-Bahama Beige	1		
7	MXC7884JUL	RH-Sonar Blue	1		3 Door Diesel
7	MXC7885JUL	LH-Sonar Blue	1		
7	MXC8094JUL	RH-Sonar Blue	1	Note (1)	3 Door 5 Door except EFi
7	MXC8094SUC	RH-Bahama Beige	1		3 Door except EFi
7	MXC8094SUC	RH-Bahama Beige	1		5 Door
7	MXC8096JUL	RH-Sonar Blue	1		3 Door EFi
7	MXC8096JUL	RH-Sonar Blue	1		5 Door 3 Door EFi
7	MXC8096SUC	RH-Bahama Beige	1		3 Door EFi
7	MXC8096SUC	RH-Bahama Beige	1		5 Door
7	MXC8095JUL	LH-Sonar Blue	1	Note (2)	
7	MXC8095SUC	LH-Bahama Beige	1		
7	BTR4408JUL	LH-Sonar Blue	1	Note (3)	
7	BTR4408SUC	LH-Bahama Beige	1		
7	BTR9298LOY	RH-Dark Granite	1	Note (4)	
7	BTR9298SUC	RH-Bahama Beige	1		
7	AWR1593LOY	LH-RHD-Dark Granite	1		
7	AWR1593SUC	LH-RHD-Bahama Beige	1		
7	AWR1594LOY	LH-LHD-Dark Granite	1		
7	AWR1594SUC	LH-LHD-Bahama Beige	1		
7	AWR3086SUC	RH-Bahama Beige	1	Note (6)	
7	AWR3086LOY	RH-Dark Granite-without pocket	1		
7	AWR3085LOY	LH-RHD-Dark Granite-without pocket	1		
7	AWR3085SUC	LH-RHD-Bahama Beige	1		
7	AWR3087LOY	LH-LHD-Dark Granite	1		
7	AWR3087SUC	LH-LHD-Bahama Beige	1		

CHANGE POINTS:
(1) From (V) GA 440845 To (V) MA 081992
(2) From (V) GA 440845 To (V) JA 031545
(3) From (V) JA 031546 To (V) MA 081992
(4) From (V) MA 081992 To (V) MA 140599
(5) To (V) GA 440844
(6) From (V) MA 140600

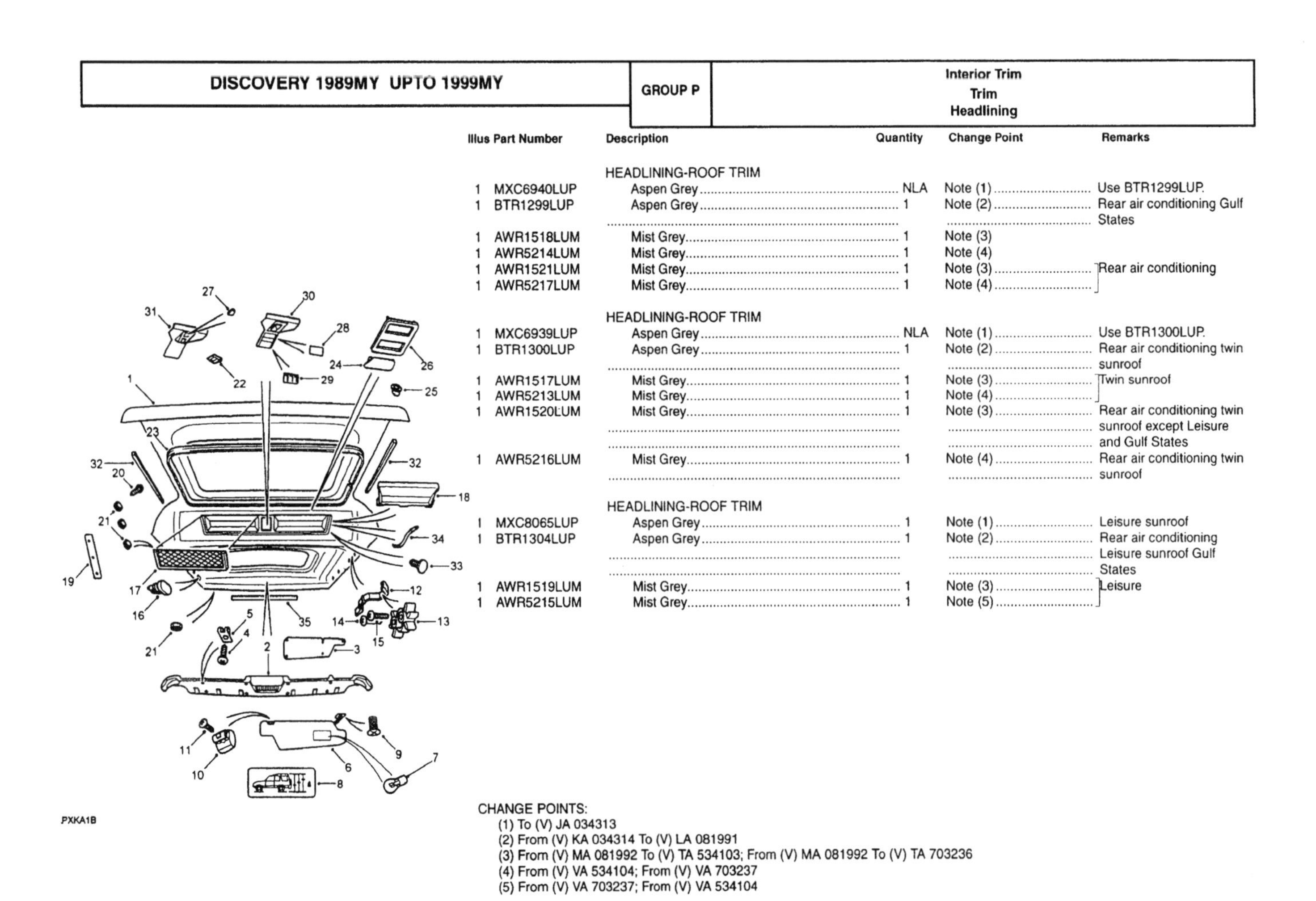

Illus	Part Number	Description	Quantity	Change Point	Remarks
		HEADLINING-ROOF TRIM			
1	MXC6940LUP	Aspen Grey	NLA	Note (1)	Use BTR1299LUP.
1	BTR1299LUP	Aspen Grey	1	Note (2)	Rear air conditioning Gulf States
1	AWR1518LUM	Mist Grey	1	Note (3)	
1	AWR5214LUM	Mist Grey	1	Note (4)	
1	AWR1521LUM	Mist Grey	1	Note (3)	Rear air conditioning
1	AWR5217LUM	Mist Grey	1	Note (4)	
		HEADLINING-ROOF TRIM			
1	MXC6939LUP	Aspen Grey	NLA	Note (1)	Use BTR1300LUP.
1	BTR1300LUP	Aspen Grey	1	Note (2)	Rear air conditioning twin sunroof
1	AWR1517LUM	Mist Grey	1	Note (3)	Twin sunroof
1	AWR5213LUM	Mist Grey	1	Note (4)	
1	AWR1520LUM	Mist Grey	1	Note (3)	Rear air conditioning twin sunroof except Leisure and Gulf States
1	AWR5216LUM	Mist Grey	1	Note (4)	Rear air conditioning twin sunroof
		HEADLINING-ROOF TRIM			
1	MXC8065LUP	Aspen Grey	1	Note (1)	Leisure sunroof
1	BTR1304LUP	Aspen Grey	1	Note (2)	Rear air conditioning Leisure sunroof Gulf States
1	AWR1519LUM	Mist Grey	1	Note (3)	Leisure
1	AWR5215LUM	Mist Grey	1	Note (5)	

CHANGE POINTS:
(1) To (V) JA 034313
(2) From (V) KA 034314 To (V) LA 081991
(3) From (V) MA 081992 To (V) TA 534103; From (V) MA 081992 To (V) TA 703236
(4) From (V) VA 534104; From (V) VA 703237
(5) From (V) VA 703237; From (V) VA 534104

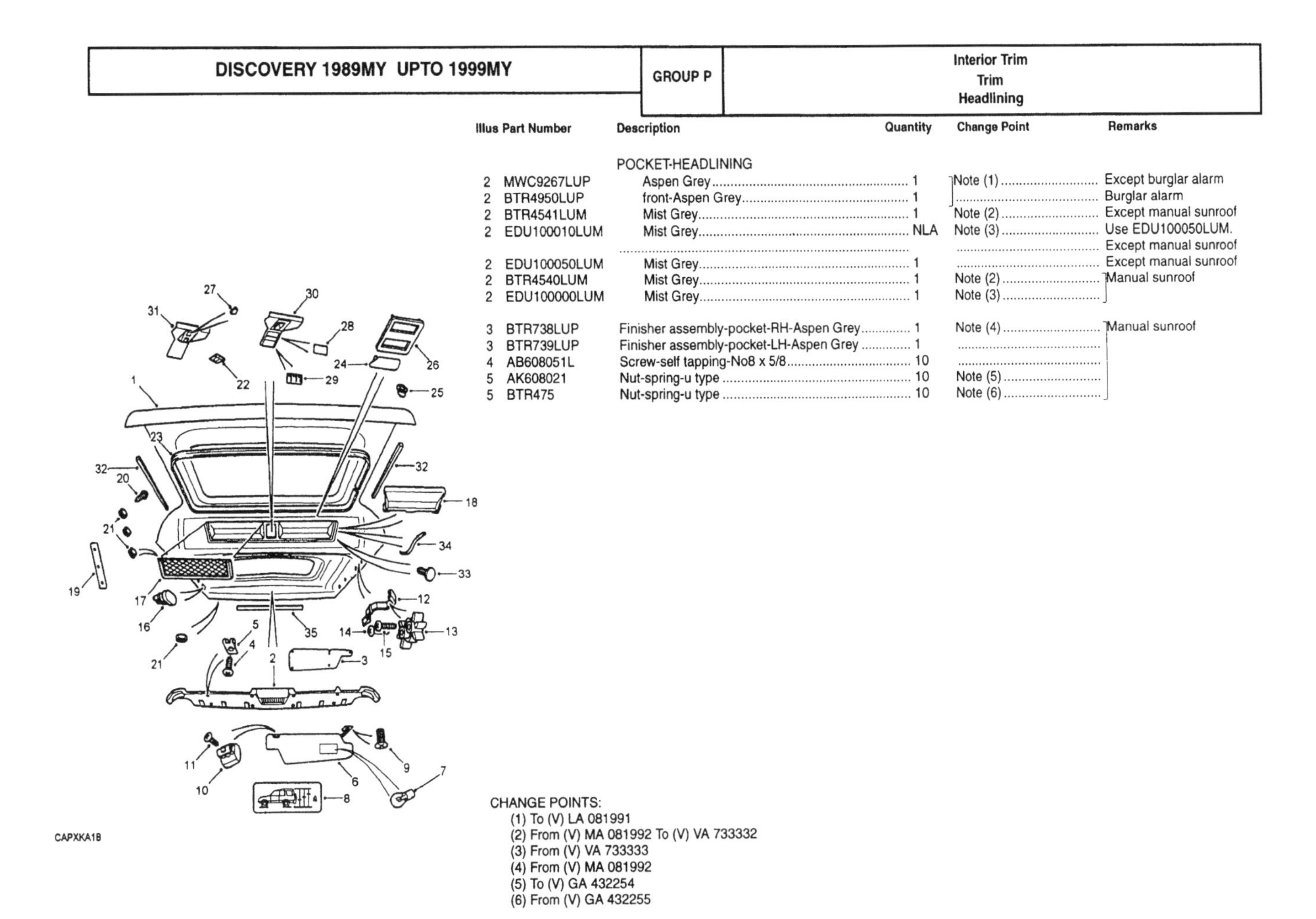

Illus	Part Number	Description	Quantity	Change Point	Remarks
		POCKET-HEADLINING			
2	MWC9267LUP	Aspen Grey	1	Note (1)	Except burglar alarm
2	BTR4950LUP	front-Aspen Grey	1		Burglar alarm
2	BTR4541LUM	Mist Grey	1	Note (2)	Except manual sunroof
2	EDU100010LUM	Mist Grey	NLA	Note (3)	Use EDU100050LUM. Except manual sunroof
2	EDU100050LUM	Mist Grey	1		Except manual sunroof
2	BTR4540LUM	Mist Grey	1	Note (2)	Manual sunroof
2	EDU100000LUM	Mist Grey	1	Note (3)	
3	BTR738LUP	Finisher assembly-pocket-RH-Aspen Grey	1	Note (4)	Manual sunroof
3	BTR739LUP	Finisher assembly-pocket-LH-Aspen Grey	1		
4	AB608051L	Screw-self tapping-No8 x 5/8	10		
5	AK608021	Nut-spring-u type	10	Note (5)	
5	BTR475	Nut-spring-u type	10	Note (6)	

CHANGE POINTS:
(1) To (V) LA 081991
(2) From (V) MA 081992 To (V) VA 733332
(3) From (V) VA 733333
(4) From (V) MA 081992
(5) To (V) GA 432254
(6) From (V) GA 432255

Illus	Part Number	Description	Quantity	Change Point	Remarks
		SUNVISORS UP TO LA081991 WITH MIRRORS			
6	MXC6097LUP	Sunvisor assembly-front header-with mirror-less height label-with airbag warning label-Aspen Grey-RH	1	Note (1)	Manual RHD except Australia
6	MXC6098LUP	Sunvisor assembly-front header-with mirror-less air bag warning label-less height label-Aspen Grey-LH	1	Note (1)	Manual RHD except Australia
		NO MIRRORS			
6	BTR806LUP	Sunvisor assembly-front header-less mirror-with height label-less air bag warning label-Aspen Grey-RH	1	Note (1)	Manual RHD except Australia
6	BTR807LUP	Sunvisor assembly-front header-less mirror-Aspen Grey-LH	1	Note (1)	Manual RHD except Australia
		AUSTRALIA			
6	BTR1013LUP	Sunvisor assembly-front header-less mirror-less air bag warning label-less height label-Aspen Grey-LH-RHD	1	Note (1)	Manual RHD Australia
		AUTOMATIC ONLY			
6	BTR2560LUP	Sunvisor assembly-front header-less mirror-with height label-less air bag warning label-Aspen Grey-RH	1	Note (1)	Automatic RHD except Australia
6	BTR2561LUP	Sunvisor assembly-front header-Aspen Grey-less mirror-LH	1	Note (1)	Automatic RHD except Australia

CAPXKA1B

CHANGE POINTS:
(1) To (V) LA 081991

Illus	Part Number	Description	Quantity	Change Point	Remarks
		SUNVISORS MA081992 ON RHS			
		WITH MIROR			
6	EDQ104820LUM	Sunvisor assembly-front header-with mirror-less air bag warning label-less height label-Mist Grey-LH	1	Note (1)	Manual RHD except Australia
		WITH HIGHT LABEL			
6	EDQ104810LUM	Sunvisor assembly-front header-less mirror-less air bag warning label-with height label-Mist Grey-RH	1	Note (1)	Manual RHD except Australia
6	EDQ104800LUM	Sunvisor assembly-front header-with height label-less air bag warning label-Mist Grey-LH	1	Note (1)	Manual RHD except Australia
		WITH SRS/HEIGHT LABEL			
6	EDQ104870LUM	Sunvisor assembly-front header-with airbag warning label-with height label-Mist Grey-RH	1	Note (1)	Manual RHD except Australia
		WITH SRS/HEIGHT LABEL & MIRROR			
6	EDQ104880LUM	Sunvisor assembly-front header-with mirror-with airbag warning label-with height label-Mist Grey-LH-RHD	1	Note (2)	Manual RHD except Australia
		ILLUMINATED			
6	EDQ104960LUM	Sunvisor assembly-front header-with illuminated mirror-with airbag warning label-with height label-Mist Grey-RH	1	Note (3)	Manual RHD except Australia
6	EDQ104970LUM	Sunvisor assembly-front header-with illuminated mirror-with airbag warning label-less height label-Mist Grey-LH	1	Note (3)	Manual RHD except Australia

CAPXKA1B

CHANGE POINTS:
(1) From (V) MA 081992
(2) From (V) MA 081992 To (V) 3509SV
(3) From (V) VA 742062

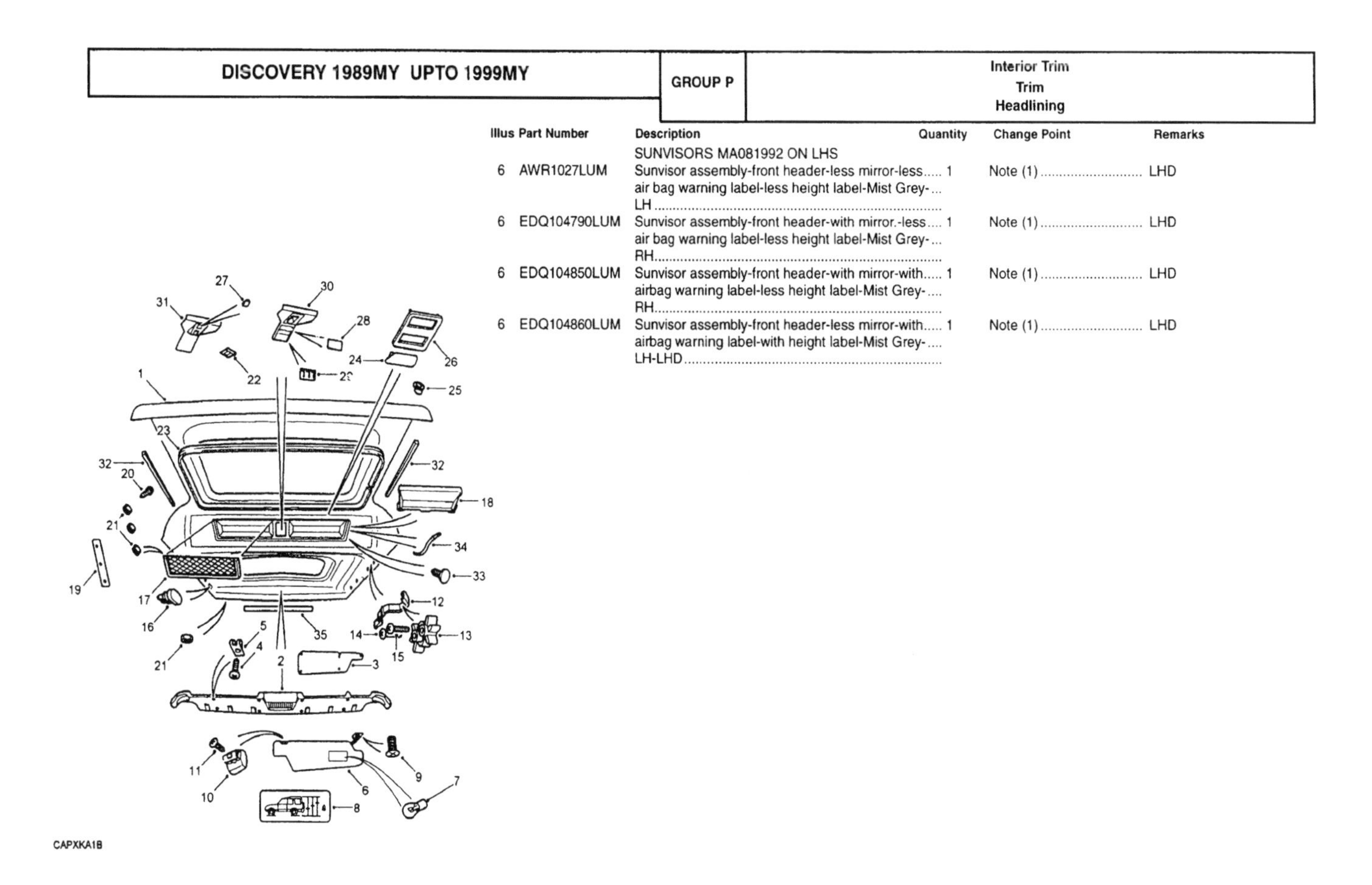

Illus	Part Number	Description	Quantity	Change Point	Remarks
		SUNVISORS MA081992 ON LHS			
6	AWR1027LUM	Sunvisor assembly-front header-less mirror-less air bag warning label-less height label-Mist Grey-LH	1	Note (1)	LHD
6	EDQ104790LUM	Sunvisor assembly-front header-with mirror.-less air bag warning label-less height label-Mist Grey-RH	1	Note (1)	LHD
6	EDQ104850LUM	Sunvisor assembly-front header-with mirror-with airbag warning label-less height label-Mist Grey-RH	1	Note (1)	LHD
6	EDQ104860LUM	Sunvisor assembly-front header-less mirror-with airbag warning label-with height label-Mist Grey-LH-LHD	1	Note (1)	LHD

CAPXKA1B

CHANGE POINTS:
(1) From (V) MA 081992

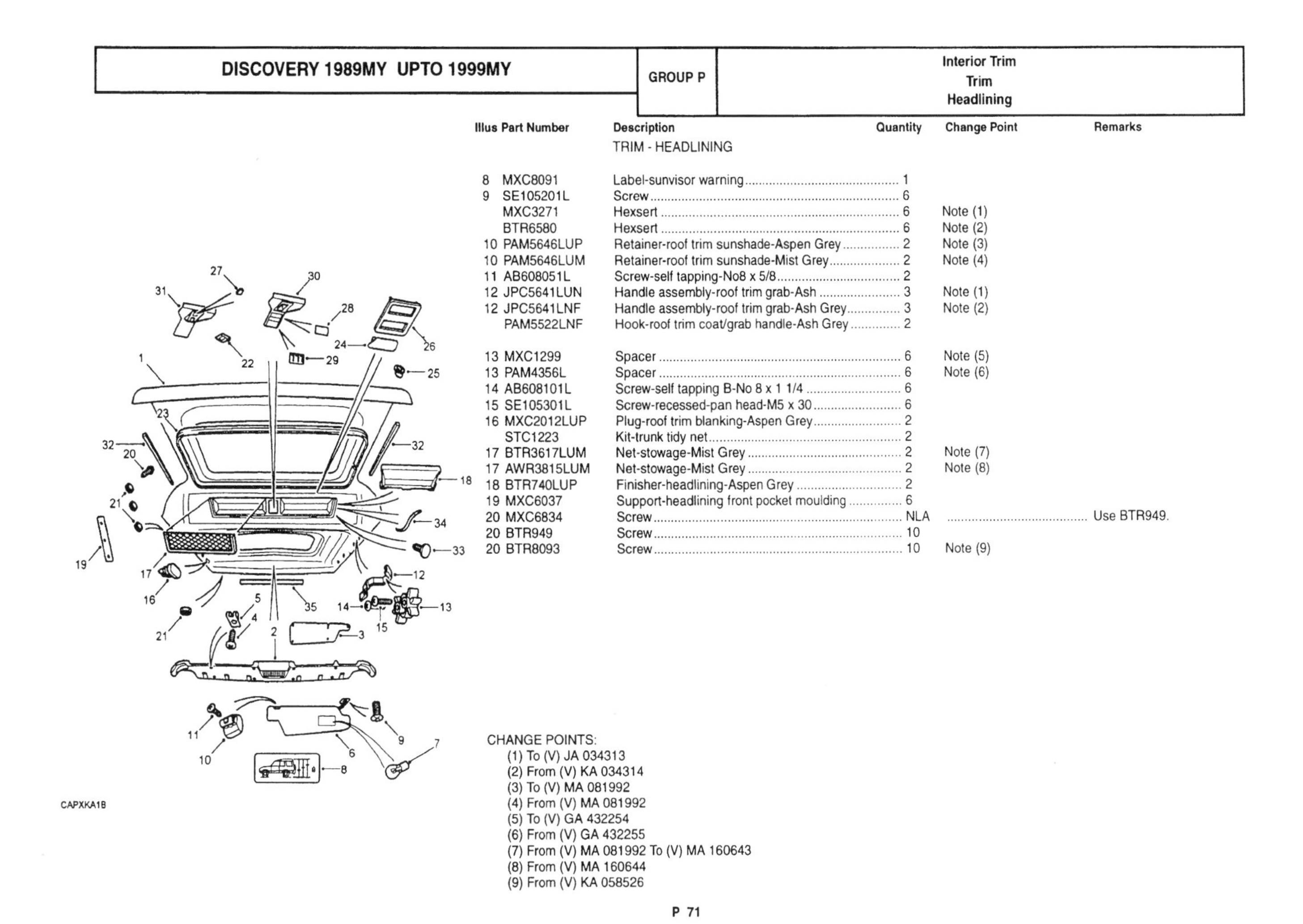

Illus	Part Number	Description	Quantity	Change Point	Remarks
		TRIM - HEADLINING			
8	MXC8091	Label-sunvisor warning	1		
9	SE105201L	Screw	6		
	MXC3271	Hexsert	6	Note (1)	
	BTR6580	Hexsert	6	Note (2)	
10	PAM5646LUP	Retainer-roof trim sunshade-Aspen Grey	2	Note (3)	
10	PAM5646LUM	Retainer-roof trim sunshade-Mist Grey	2	Note (4)	
11	AB608051L	Screw-self tapping-No8 x 5/8	2		
12	JPC5641LUN	Handle assembly-roof trim grab-Ash	3	Note (1)	
12	JPC5641LNF	Handle assembly-roof trim grab-Ash Grey	3	Note (2)	
	PAM5522LNF	Hook-roof trim coat/grab handle-Ash Grey	2		
13	MXC1299	Spacer	6	Note (5)	
13	PAM4356L	Spacer	6	Note (6)	
14	AB608101L	Screw-self tapping B-No 8 x 1 1/4	6		
15	SE105301L	Screw-recessed-pan head-M5 x 30	6		
16	MXC2012LUP	Plug-roof trim blanking-Aspen Grey	2		
	STC1223	Kit-trunk tidy net	2		
17	BTR3617LUM	Net-stowage-Mist Grey	2	Note (7)	
17	AWR3815LUM	Net-stowage-Mist Grey	2	Note (8)	
18	BTR740LUP	Finisher-headlining-Aspen Grey	2		
19	MXC6037	Support-headlining front pocket moulding	6		
20	MXC6834	Screw	NLA		Use BTR949.
20	BTR949	Screw	10		
20	BTR8093	Screw	10	Note (9)	

CHANGE POINTS:
(1) To (V) JA 034313
(2) From (V) KA 034314
(3) To (V) MA 081992
(4) From (V) MA 081992
(5) To (V) GA 432254
(6) From (V) GA 432255
(7) From (V) MA 081992 To (V) MA 160643
(8) From (V) MA 160644
(9) From (V) KA 058526

Illus	Part Number	Description	Quantity	Change Point	Remarks
		TRIM - HEADLINING - FROM MA081992			
21	BTR9905	Nut	A/R		
22	AWR1184	Bracket-sunroof mounting	1		
23	BTR7027LUM	Finisher-sunroof-Mist Grey	1		
24	BTR7006LUM	Plate-Mist Grey	1		
25	DRC5582	Nut-lokut	4		
26	BTR5351LUM	Moulding-centre roof lamp-less sunroof	2		Except sunroof
27	BTR6983LUM	Plug-roof trim blanking-Mist Grey	1		
28	BTR6985LUM	Plate-switch pack blanking-Mist Grey	2		Except electric sunroof
29	BTR6984LUM	Bezel-sunroof switch-Mist Grey	2		
30	BTR4542LUM	Bezel-manual sunroof-Mist Grey	1		
31	BTR4543LUM	Bezel-sunroof switch-Mist Grey	1		
32	AWR1400	Foam-pad-self adhesive	4		
33	JPC4651LUM	Fastener-headlining fir tree-Mist Grey	2		Except sunroof
34	BTR9611	Bracket-headlining support	2		
35	BTR2913	Seal	1		
	AWR4646LUM	Velcro-hook-Mist Grey	2		

CAPXKA1B

Illus	Part Number	Description	Quantity	Change Point	Remarks
		TRIM - SUNROOF ROLLER BLIND			
1	STC8158	Kit-sunroof roller blind	UP	Note (1)	
2	BTR4518	Spacer	6		
3	AB606071L	Screw	6		

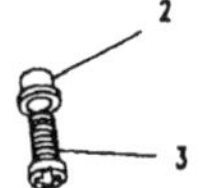

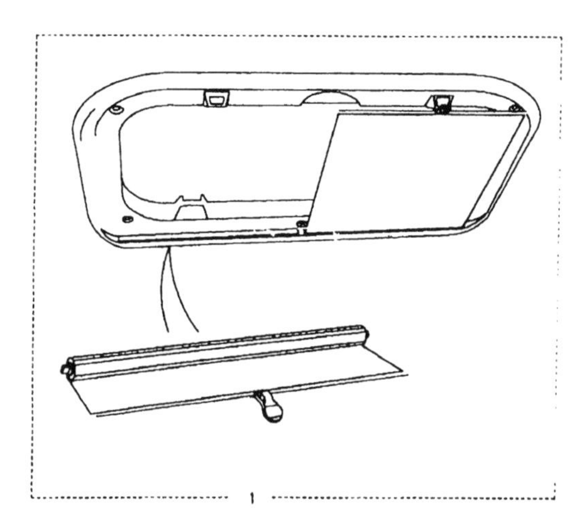

CHANGE POINTS:
(1) To (V) LA 081991

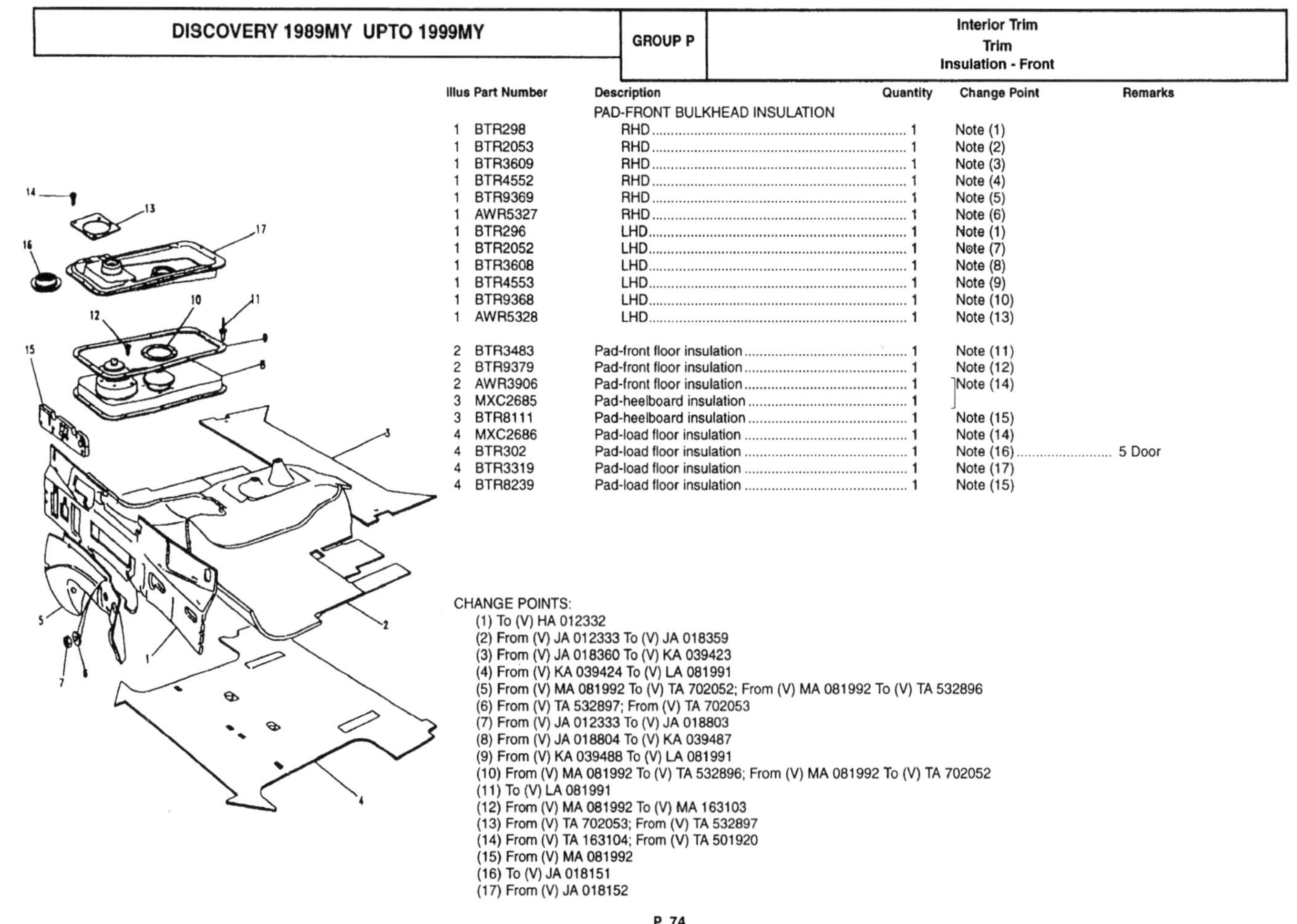

Illus	Part Number	Description	Quantity	Change Point	Remarks
		PAD-FRONT BULKHEAD INSULATION			
1	BTR298	RHD	1	Note (1)	
1	BTR2053	RHD	1	Note (2)	
1	BTR3609	RHD	1	Note (3)	
1	BTR4552	RHD	1	Note (4)	
1	BTR9369	RHD	1	Note (5)	
1	AWR5327	RHD	1	Note (6)	
1	BTR296	LHD	1	Note (1)	
1	BTR2052	LHD	1	Note (7)	
1	BTR3608	LHD	1	Note (8)	
1	BTR4553	LHD	1	Note (9)	
1	BTR9368	LHD	1	Note (10)	
1	AWR5328	LHD	1	Note (13)	
2	BTR3483	Pad-front floor insulation	1	Note (11)	
2	BTR9379	Pad-front floor insulation	1	Note (12)	
2	AWR3906	Pad-front floor insulation	1	Note (14)	
3	MXC2685	Pad-heelboard insulation	1		
3	BTR8111	Pad-heelboard insulation	1	Note (15)	
4	MXC2686	Pad-load floor insulation	1	Note (14)	
4	BTR302	Pad-load floor insulation	1	Note (16)	5 Door
4	BTR3319	Pad-load floor insulation	1	Note (17)	
4	BTR8239	Pad-load floor insulation	1	Note (15)	

CHANGE POINTS:
(1) To (V) HA 012332
(2) From (V) JA 012333 To (V) JA 018359
(3) From (V) JA 018360 To (V) KA 039423
(4) From (V) KA 039424 To (V) LA 081991
(5) From (V) MA 081992 To (V) TA 702052; From (V) MA 081992 To (V) TA 532896
(6) From (V) TA 532897; From (V) TA 702053
(7) From (V) JA 012333 To (V) JA 018803
(8) From (V) JA 018804 To (V) KA 039487
(9) From (V) KA 039488 To (V) LA 081991
(10) From (V) MA 081992 To (V) TA 532896; From (V) MA 081992 To (V) TA 702052
(11) To (V) LA 081991
(12) From (V) MA 081992 To (V) MA 163103
(13) From (V) TA 702053; From (V) TA 532897
(14) From (V) TA 163104; From (V) TA 501920
(15) From (V) MA 081992
(16) To (V) JA 018151
(17) From (V) JA 018152

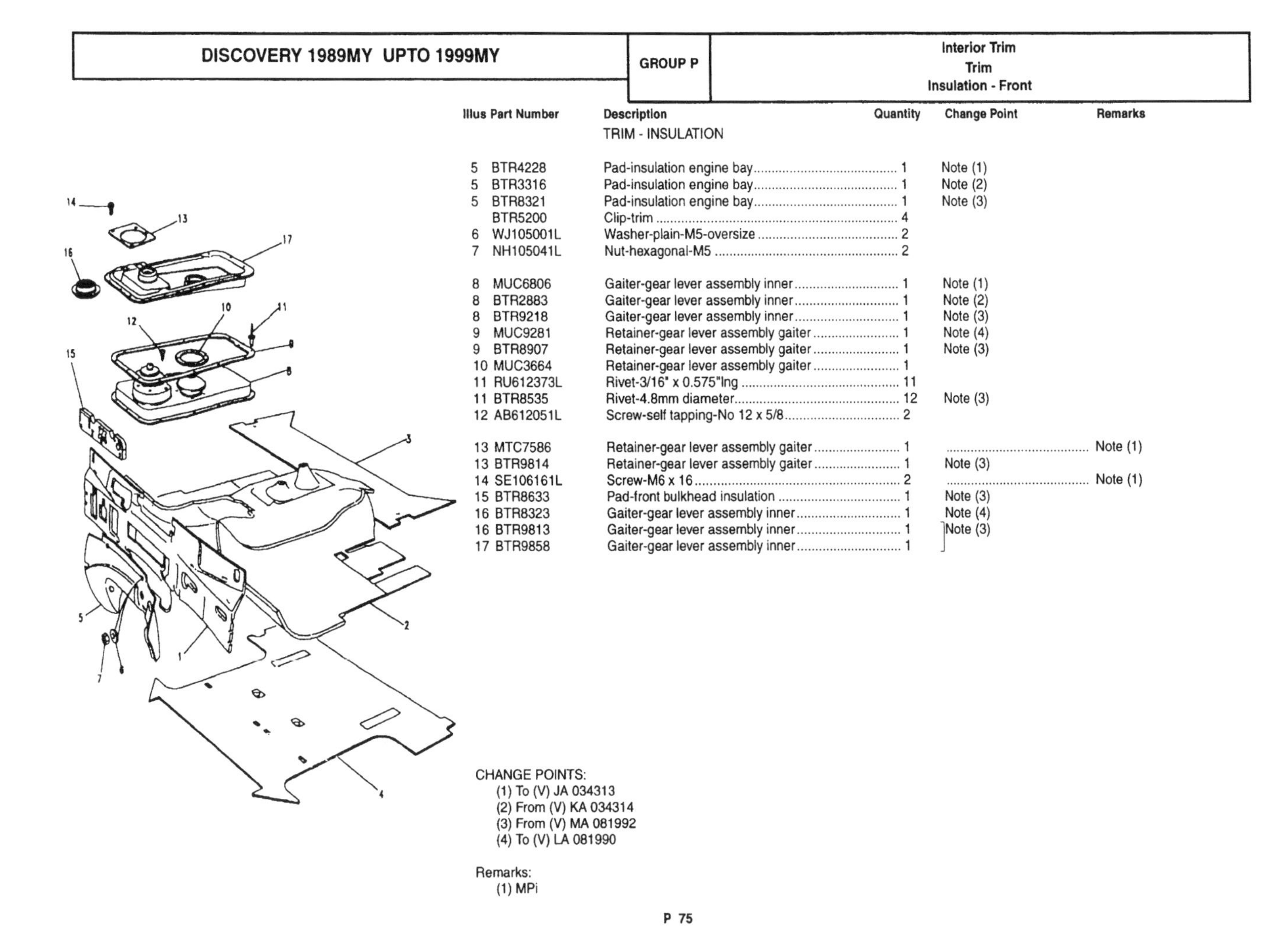

DISCOVERY 1989MY UPTO 1999MY	GROUP P	Interior Trim Trim Insulation - Front

Illus	Part Number	Description	Quantity	Change Point	Remarks
		TRIM - INSULATION			
5	BTR4228	Pad-insulation engine bay	1	Note (1)	
5	BTR3316	Pad-insulation engine bay	1	Note (2)	
5	BTR8321	Pad-insulation engine bay	1	Note (3)	
	BTR5200	Clip-trim	4		
6	WJ105001L	Washer-plain-M5-oversize	2		
7	NH105041L	Nut-hexagonal-M5	2		
8	MUC6806	Gaiter-gear lever assembly inner	1	Note (1)	
8	BTR2883	Gaiter-gear lever assembly inner	1	Note (2)	
8	BTR9218	Gaiter-gear lever assembly inner	1	Note (3)	
9	MUC9281	Retainer-gear lever assembly gaiter	1	Note (4)	
9	BTR8907	Retainer-gear lever assembly gaiter	1	Note (3)	
10	MUC3664	Retainer-gear lever assembly gaiter	1		
11	RU612373L	Rivet-3/16" x 0.575"lng	11		
11	BTR8535	Rivet-4.8mm diameter	12	Note (3)	
12	AB612051L	Screw-self tapping-No 12 x 5/8	2		
13	MTC7586	Retainer-gear lever assembly gaiter	1		Note (1)
13	BTR9814	Retainer-gear lever assembly gaiter	1	Note (3)	
14	SE106161L	Screw-M6 x 16	2		Note (1)
15	BTR8633	Pad-front bulkhead insulation	1	Note (3)	
16	BTR8323	Gaiter-gear lever assembly inner	1	Note (4)	
16	BTR9813	Gaiter-gear lever assembly inner	1	Note (3)	
17	BTR9858	Gaiter-gear lever assembly inner	1		

CHANGE POINTS:
(1) To (V) JA 034313
(2) From (V) KA 034314
(3) From (V) MA 081992
(4) To (V) LA 081990

Remarks:
(1) MPi

P 75

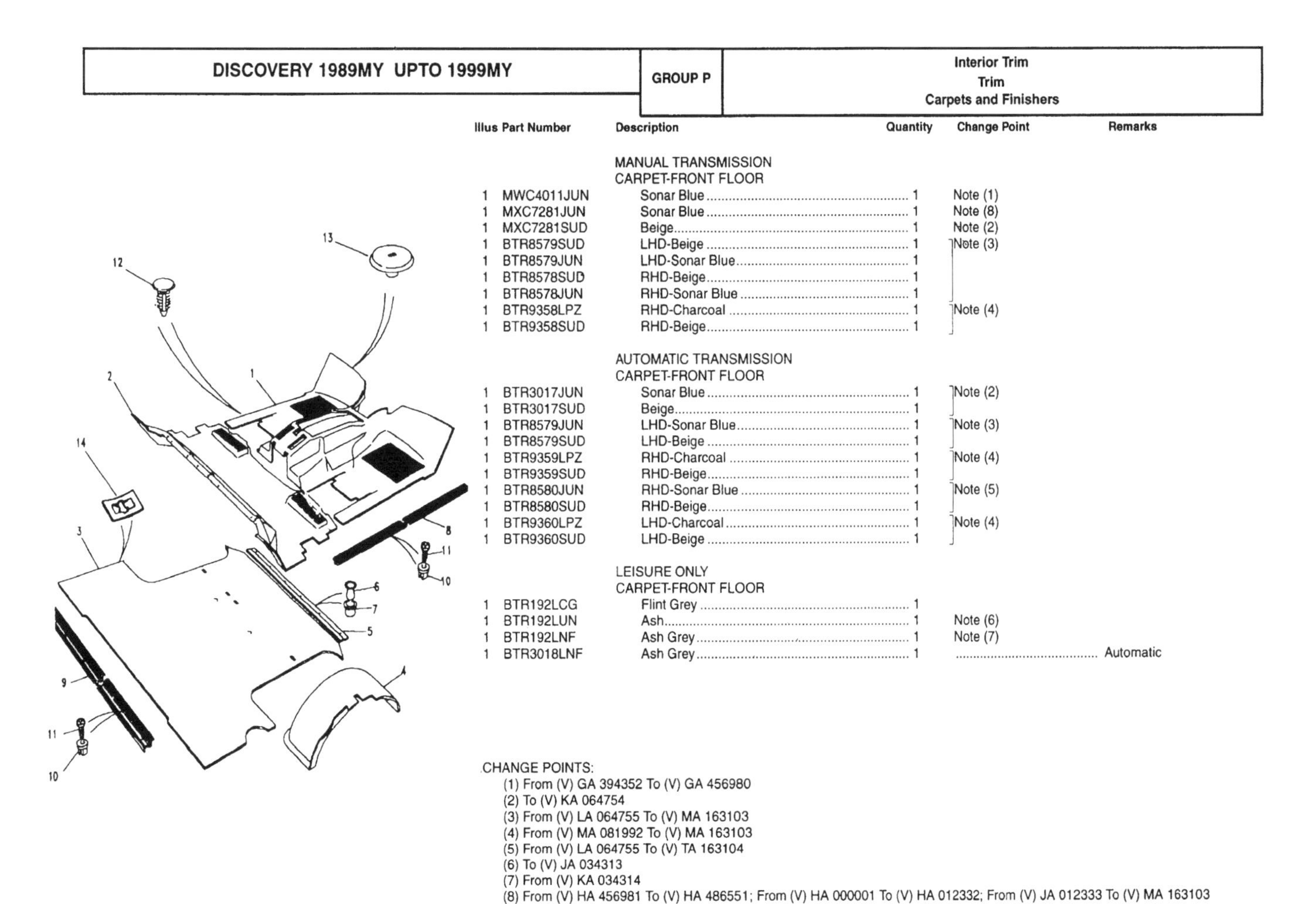

DISCOVERY 1989MY UPTO 1999MY	GROUP P	Interior Trim Trim Carpets and Finishers

Illus	Part Number	Description	Quantity	Change Point	Remarks
		MANUAL TRANSMISSION			
		CARPET-FRONT FLOOR			
1	MWC4011JUN	Sonar Blue	1	Note (1)	
1	MXC7281JUN	Sonar Blue	1	Note (8)	
1	MXC7281SUD	Beige	1	Note (2)	
1	BTR8579SUD	LHD-Beige	1	Note (3)	
1	BTR8579JUN	LHD-Sonar Blue	1		
1	BTR8578SUD	RHD-Beige	1		
1	BTR8578JUN	RHD-Sonar Blue	1		
1	BTR9358LPZ	RHD-Charcoal	1	Note (4)	
1	BTR9358SUD	RHD-Beige	1		
		AUTOMATIC TRANSMISSION			
		CARPET-FRONT FLOOR			
1	BTR3017JUN	Sonar Blue	1	Note (2)	
1	BTR3017SUD	Beige	1		
1	BTR8579JUN	LHD-Sonar Blue	1	Note (3)	
1	BTR8579SUD	LHD-Beige	1		
1	BTR9359LPZ	RHD-Charcoal	1	Note (4)	
1	BTR9359SUD	RHD-Beige	1		
1	BTR8580JUN	RHD-Sonar Blue	1	Note (5)	
1	BTR8580SUD	RHD-Beige	1		
1	BTR9360LPZ	LHD-Charcoal	1	Note (4)	
1	BTR9360SUD	LHD-Beige	1		
		LEISURE ONLY			
		CARPET-FRONT FLOOR			
1	BTR192LCG	Flint Grey	1		
1	BTR192LUN	Ash	1	Note (6)	
1	BTR192LNF	Ash Grey	1	Note (7)	
1	BTR3018LNF	Ash Grey	1		Automatic

CHANGE POINTS:
(1) From (V) GA 394352 To (V) GA 456980
(2) To (V) KA 064754
(3) From (V) LA 064755 To (V) MA 163103
(4) From (V) MA 081992 To (V) MA 163103
(5) From (V) LA 064755 To (V) TA 163104
(6) To (V) JA 034313
(7) From (V) KA 034314
(8) From (V) HA 456981 To (V) HA 486551; From (V) HA 000001 To (V) HA 012332; From (V) JA 012333 To (V) MA 163103

P 76

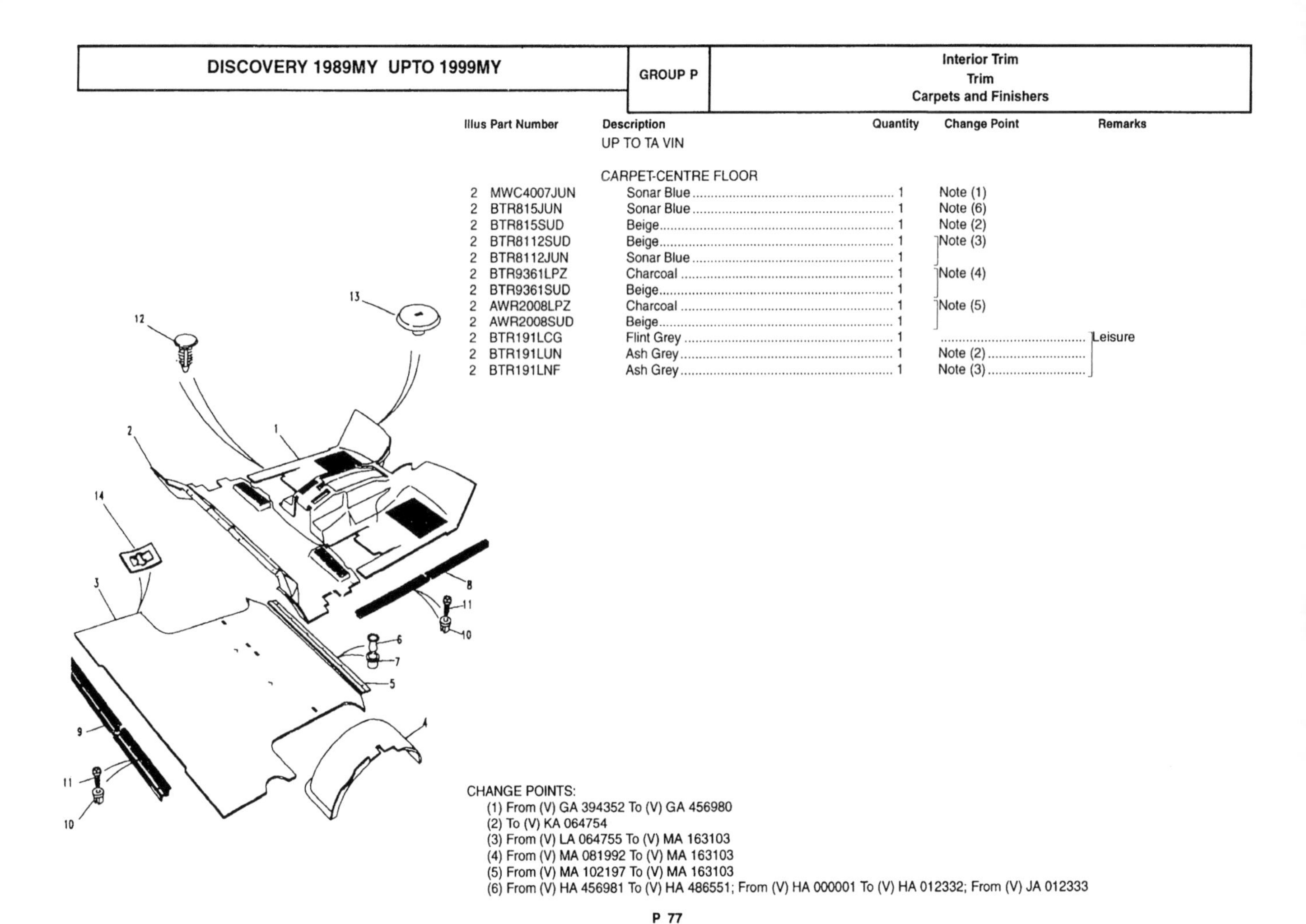

UP TO TA VIN

CARPET-CENTRE FLOOR

Illus	Part Number	Description	Quantity	Change Point	Remarks
2	MWC4007JUN	Sonar Blue	1	Note (1)	
2	BTR815JUN	Sonar Blue	1	Note (6)	
2	BTR815SUD	Beige	1	Note (2)	
2	BTR8112SUD	Beige	1	Note (3)	
2	BTR8112JUN	Sonar Blue	1		
2	BTR9361LPZ	Charcoal	1	Note (4)	
2	BTR9361SUD	Beige	1		
2	AWR2008LPZ	Charcoal	1	Note (5)	
2	AWR2008SUD	Beige	1		
2	BTR191LCG	Flint Grey	1		Leisure
2	BTR191LUN	Ash Grey	1	Note (2)	
2	BTR191LNF	Ash Grey	1	Note (3)	

CHANGE POINTS:
(1) From (V) GA 394352 To (V) GA 456980
(2) To (V) KA 064754
(3) From (V) LA 064755 To (V) MA 163103
(4) From (V) MA 081992 To (V) MA 163103
(5) From (V) MA 102197 To (V) MA 163103
(6) From (V) HA 456981 To (V) HA 486551; From (V) HA 000001 To (V) HA 012332; From (V) JA 012333

P 77

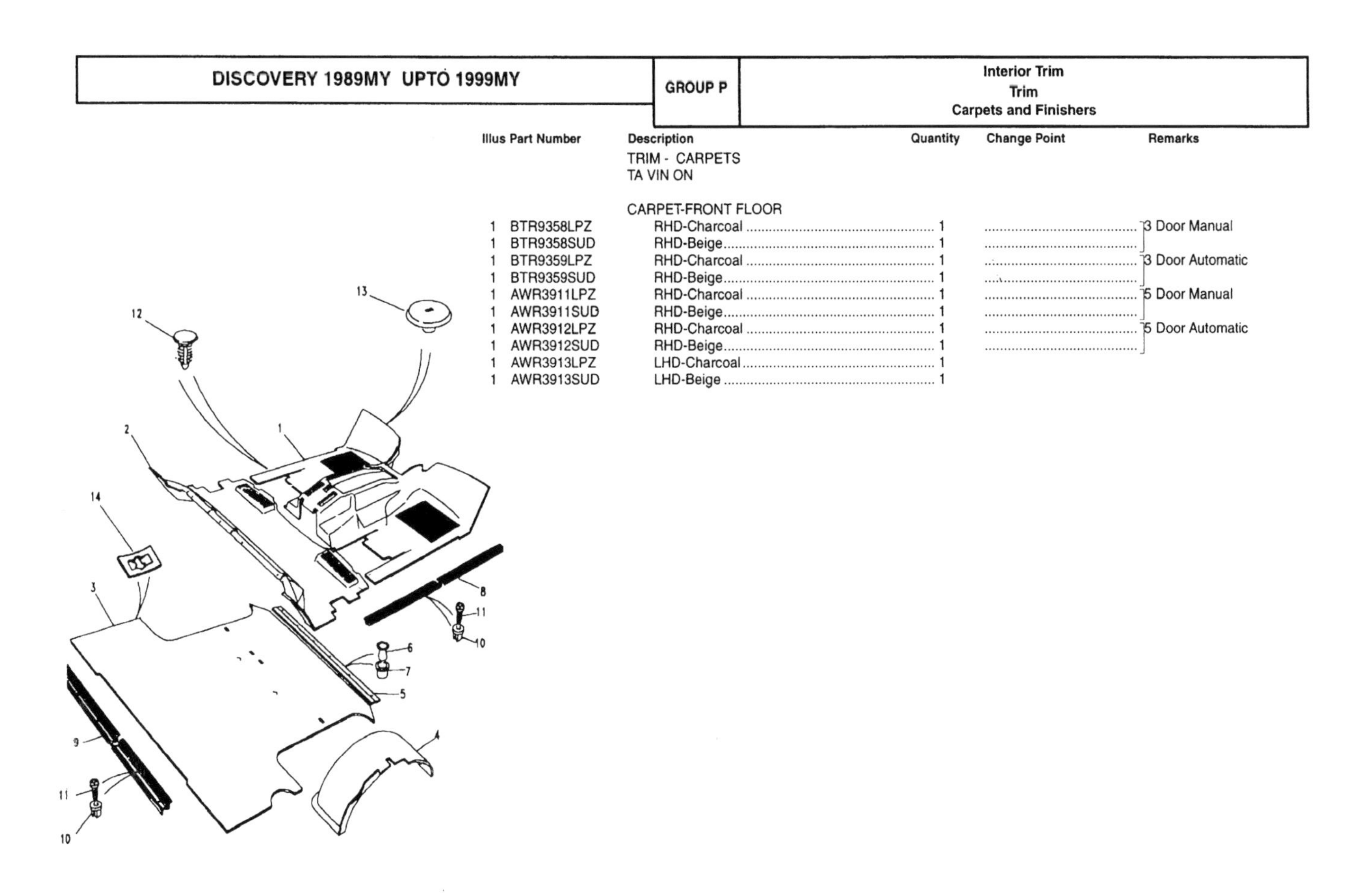

TRIM - CARPETS
TA VIN ON

CARPET-FRONT FLOOR

Illus	Part Number	Description	Quantity	Change Point	Remarks
1	BTR9358LPZ	RHD-Charcoal	1		3 Door Manual
1	BTR9358SUD	RHD-Beige	1		
1	BTR9359LPZ	RHD-Charcoal	1		3 Door Automatic
1	BTR9359SUD	RHD-Beige	1		
1	AWR3911LPZ	RHD-Charcoal	1		5 Door Manual
1	AWR3911SUD	RHD-Beige	1		
1	AWR3912LPZ	RHD-Charcoal	1		5 Door Automatic
1	AWR3912SUD	RHD-Beige	1		
1	AWR3913LPZ	LHD-Charcoal	1		
1	AWR3913SUD	LHD-Beige	1		

P 78

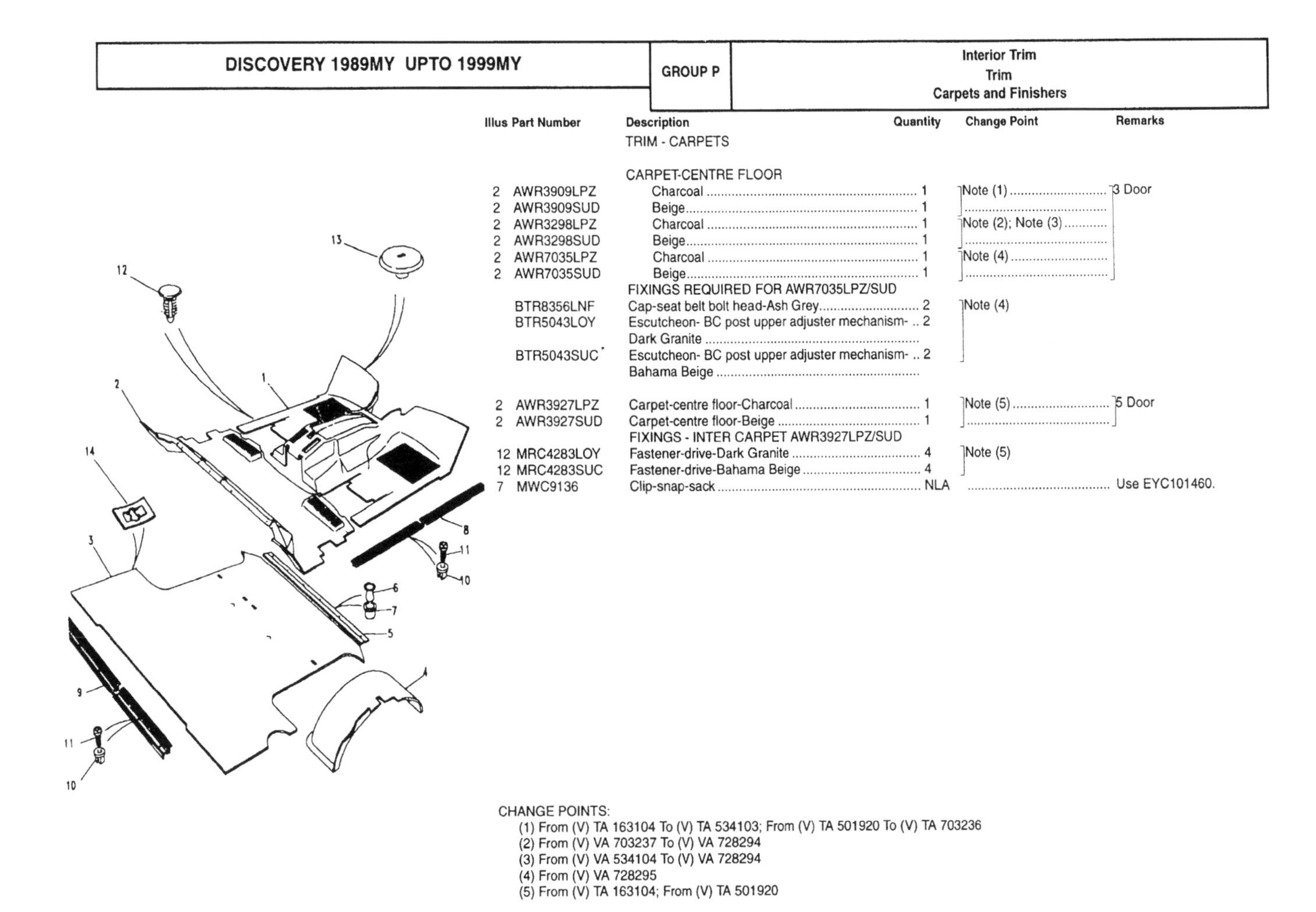

Illus	Part Number	Description	Quantity	Change Point	Remarks
		TRIM - CARPETS			
		CARPET-CENTRE FLOOR			
2	AWR3909LPZ	Charcoal	1	Note (1)	3 Door
2	AWR3909SUD	Beige	1		
2	AWR3298LPZ	Charcoal	1	Note (2); Note (3)	
2	AWR3298SUD	Beige	1		
2	AWR7035LPZ	Charcoal	1	Note (4)	
2	AWR7035SUD	Beige	1		
		FIXINGS REQUIRED FOR AWR7035LPZ/SUD			
	BTR8356LNF	Cap-seat belt bolt head-Ash Grey	2	Note (4)	
	BTR5043LOY	Escutcheon- BC post upper adjuster mechanism- Dark Granite	2		
	BTR5043SUC*	Escutcheon- BC post upper adjuster mechanism- Bahama Beige	2		
2	AWR3927LPZ	Carpet-centre floor-Charcoal	1	Note (5)	5 Door
2	AWR3927SUD	Carpet-centre floor-Beige	1		
		FIXINGS - INTER CARPET AWR3927LPZ/SUD			
12	MRC4283LOY	Fastener-drive-Dark Granite	4	Note (5)	
12	MRC4283SUC	Fastener-drive-Bahama Beige	4		
7	MWC9136	Clip-snap-sack	NLA		Use EYC101460.

CHANGE POINTS:
(1) From (V) TA 163104 To (V) TA 534103; From (V) TA 501920 To (V) TA 703236
(2) From (V) VA 703237 To (V) VA 728294
(3) From (V) VA 534104 To (V) VA 728294
(4) From (V) VA 728295
(5) From (V) TA 163104; From (V) TA 501920

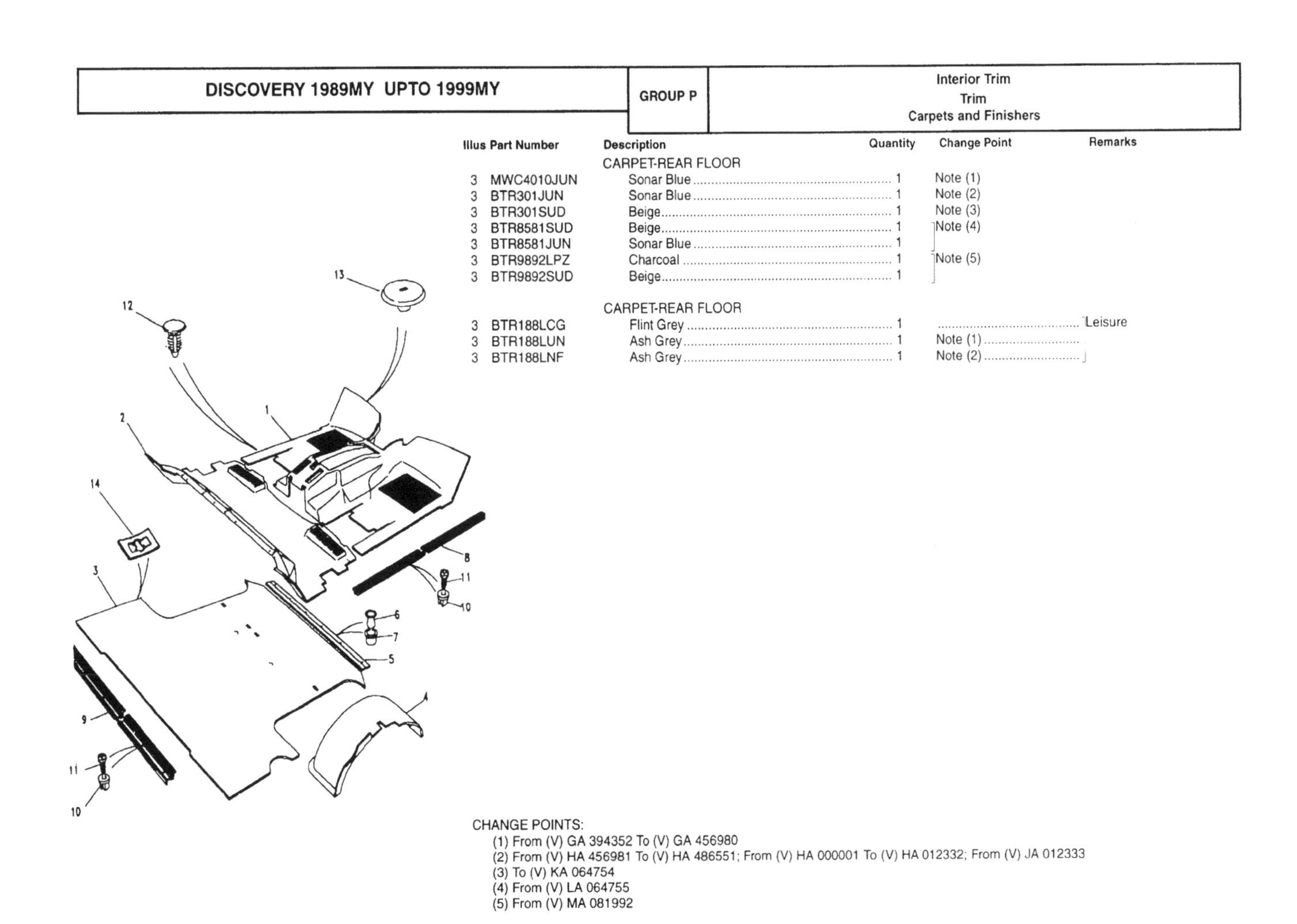

Illus	Part Number	Description	Quantity	Change Point	Remarks
		CARPET-REAR FLOOR			
3	MWC4010JUN	Sonar Blue	1	Note (1)	
3	BTR301JUN	Sonar Blue	1	Note (2)	
3	BTR301SUD	Beige	1	Note (3)	
3	BTR8581SUD	Beige	1	Note (4)	
3	BTR8581JUN	Sonar Blue	1		
3	BTR9892LPZ	Charcoal	1	Note (5)	
3	BTR9892SUD	Beige	1		
		CARPET-REAR FLOOR			
3	BTR188LCG	Flint Grey	1		Leisure
3	BTR188LUN	Ash Grey	1	Note (1)	
3	BTR188LNF	Ash Grey	1	Note (2)	

CHANGE POINTS:
(1) From (V) GA 394352 To (V) GA 456980
(2) From (V) HA 456981 To (V) HA 486551; From (V) HA 000001 To (V) HA 012332; From (V) JA 012333
(3) To (V) KA 064754
(4) From (V) LA 064755
(5) From (V) MA 081992

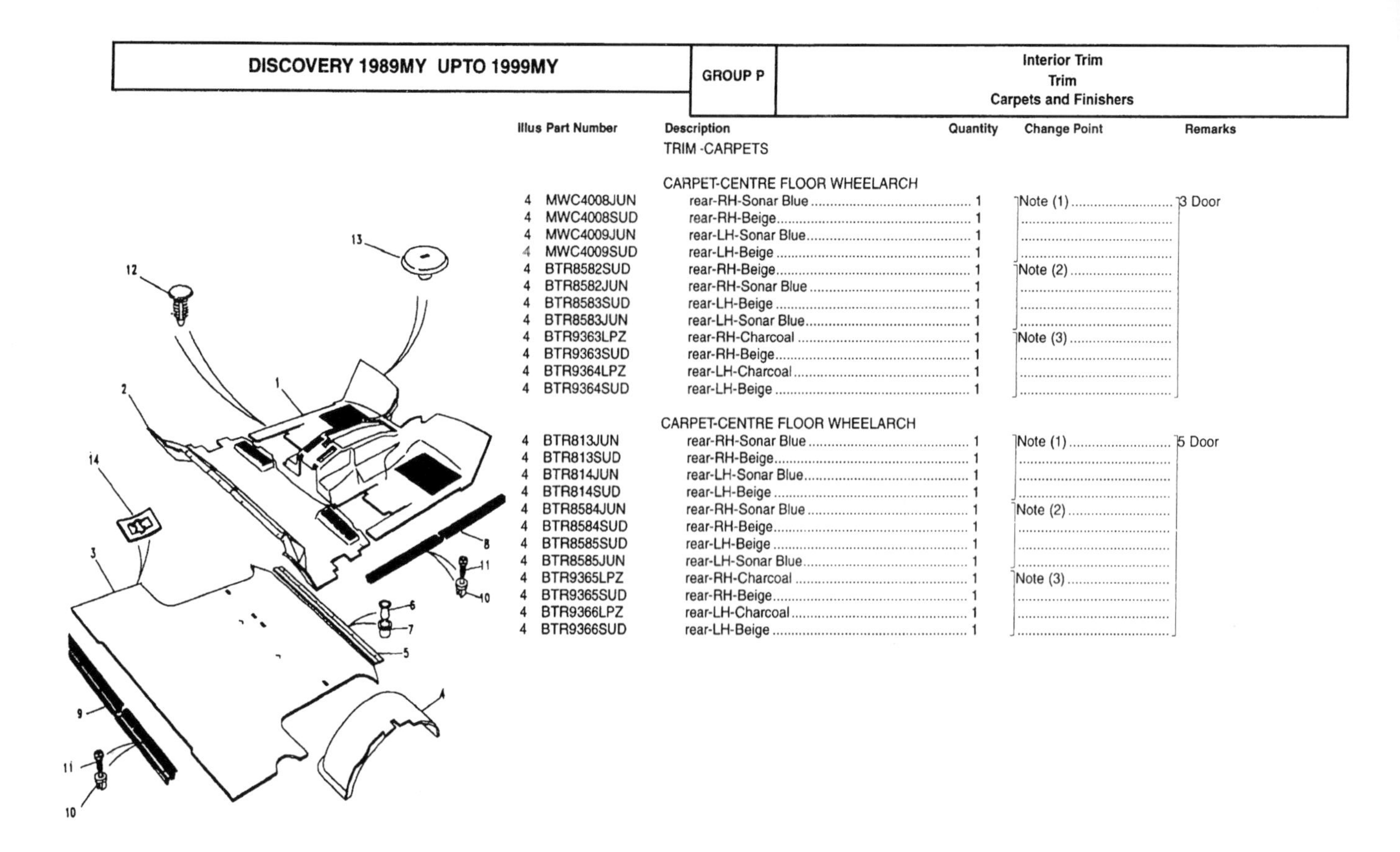

TRIM -CARPETS

Illus	Part Number	Description	Quantity	Change Point	Remarks
		CARPET-CENTRE FLOOR WHEELARCH			
4	MWC4008JUN	rear-RH-Sonar Blue	1	Note (1)	3 Door
4	MWC4008SUD	rear-RH-Beige	1		
4	MWC4009JUN	rear-LH-Sonar Blue	1		
4	MWC4009SUD	rear-LH-Beige	1		
4	BTR8582SUD	rear-RH-Beige	1	Note (2)	
4	BTR8582JUN	rear-RH-Sonar Blue	1		
4	BTR8583SUD	rear-LH-Beige	1		
4	BTR8583JUN	rear-LH-Sonar Blue	1		
4	BTR9363LPZ	rear-RH-Charcoal	1	Note (3)	
4	BTR9363SUD	rear-RH-Beige	1		
4	BTR9364LPZ	rear-LH-Charcoal	1		
4	BTR9364SUD	rear-LH-Beige	1		
		CARPET-CENTRE FLOOR WHEELARCH			
4	BTR813JUN	rear-RH-Sonar Blue	1	Note (1)	5 Door
4	BTR813SUD	rear-RH-Beige	1		
4	BTR814JUN	rear-LH-Sonar Blue	1		
4	BTR814SUD	rear-LH-Beige	1		
4	BTR8584JUN	rear-RH-Sonar Blue	1	Note (2)	
4	BTR8584SUD	rear-RH-Beige	1		
4	BTR8585SUD	rear-LH-Beige	1		
4	BTR8585JUN	rear-LH-Sonar Blue	1		
4	BTR9365LPZ	rear-RH-Charcoal	1	Note (3)	
4	BTR9365SUD	rear-RH-Beige	1		
4	BTR9366LPZ	rear-LH-Charcoal	1		
4	BTR9366SUD	rear-LH-Beige	1		

CHANGE POINTS:
(1) To (V) KA 064754
(2) From (V) LA 064755
(3) From (V) MA 081992

P 81

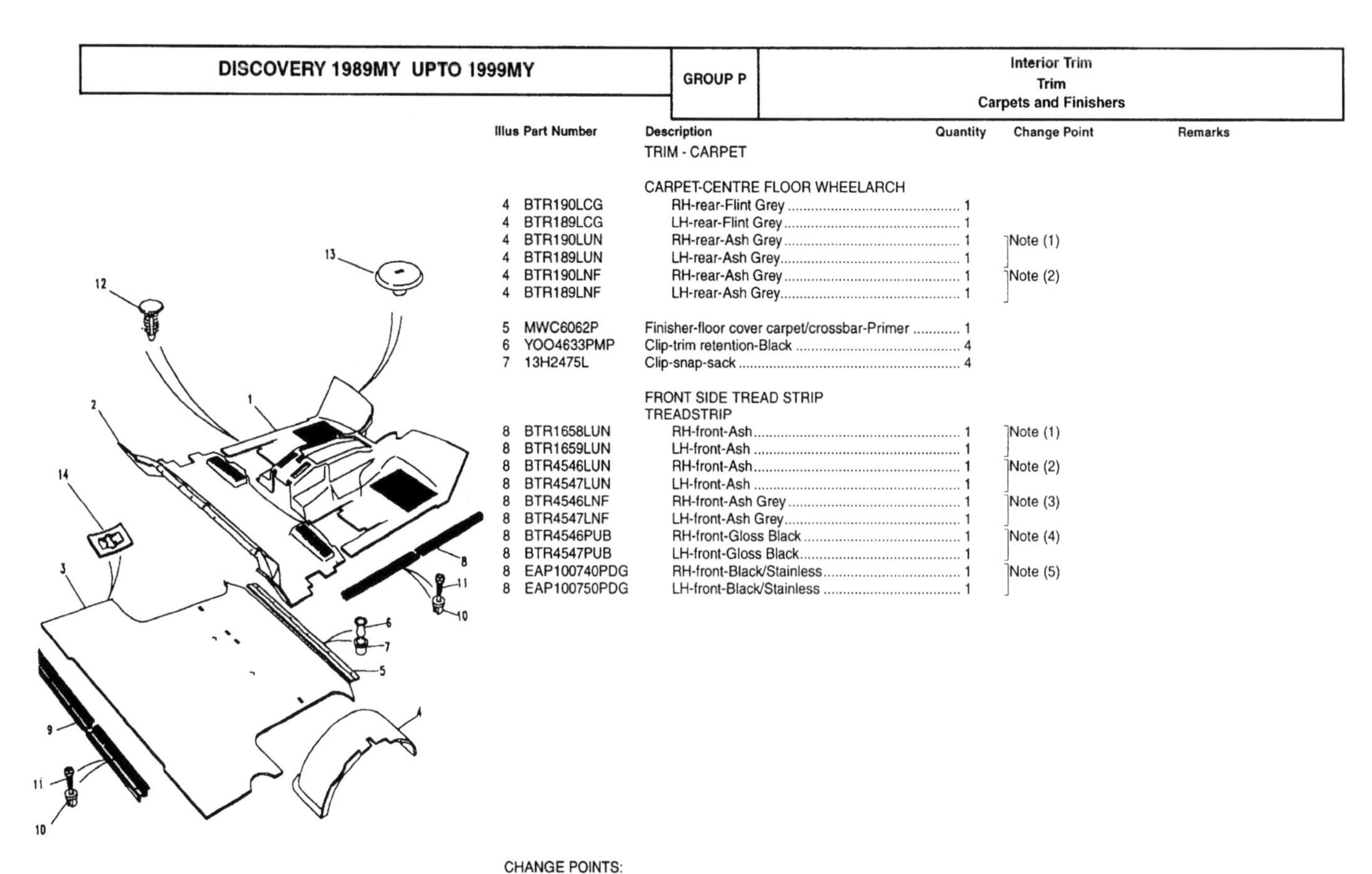

TRIM - CARPET

Illus	Part Number	Description	Quantity	Change Point	Remarks
		CARPET-CENTRE FLOOR WHEELARCH			
4	BTR190LCG	RH-rear-Flint Grey	1		
4	BTR189LCG	LH-rear-Flint Grey	1		
4	BTR190LUN	RH-rear-Ash Grey	1	Note (1)	
4	BTR189LUN	LH-rear-Ash Grey	1		
4	BTR190LNF	RH-rear-Ash Grey	1	Note (2)	
4	BTR189LNF	LH-rear-Ash Grey	1		
5	MWC6062P	Finisher-floor cover carpet/crossbar-Primer	1		
6	YOO4633PMP	Clip-trim retention-Black	4		
7	13H2475L	Clip-snap-sack	4		
		FRONT SIDE TREAD STRIP			
		TREADSTRIP			
8	BTR1658LUN	RH-front-Ash	1	Note (1)	
8	BTR1659LUN	LH-front-Ash	1		
8	BTR4546LUN	RH-front-Ash	1	Note (2)	
8	BTR4547LUN	LH-front-Ash	1		
8	BTR4546LNF	RH-front-Ash Grey	1	Note (3)	
8	BTR4547LNF	LH-front-Ash Grey	1		
8	BTR4546PUB	RH-front-Gloss Black	1	Note (4)	
8	BTR4547PUB	LH-front-Gloss Black	1		
8	EAP100740PDG	RH-front-Black/Stainless	1	Note (5)	
8	EAP100750PDG	LH-front-Black/Stainless	1		

CHANGE POINTS:
(1) To (V) JA 025549
(2) From (V) JA 025550 To (V) JA 034313
(3) From (V) KA 034314 To (V) LA 081991
(4) From (V) MA 081991 To (V) WA 746541
(5) From (V) WA 746542

P 82

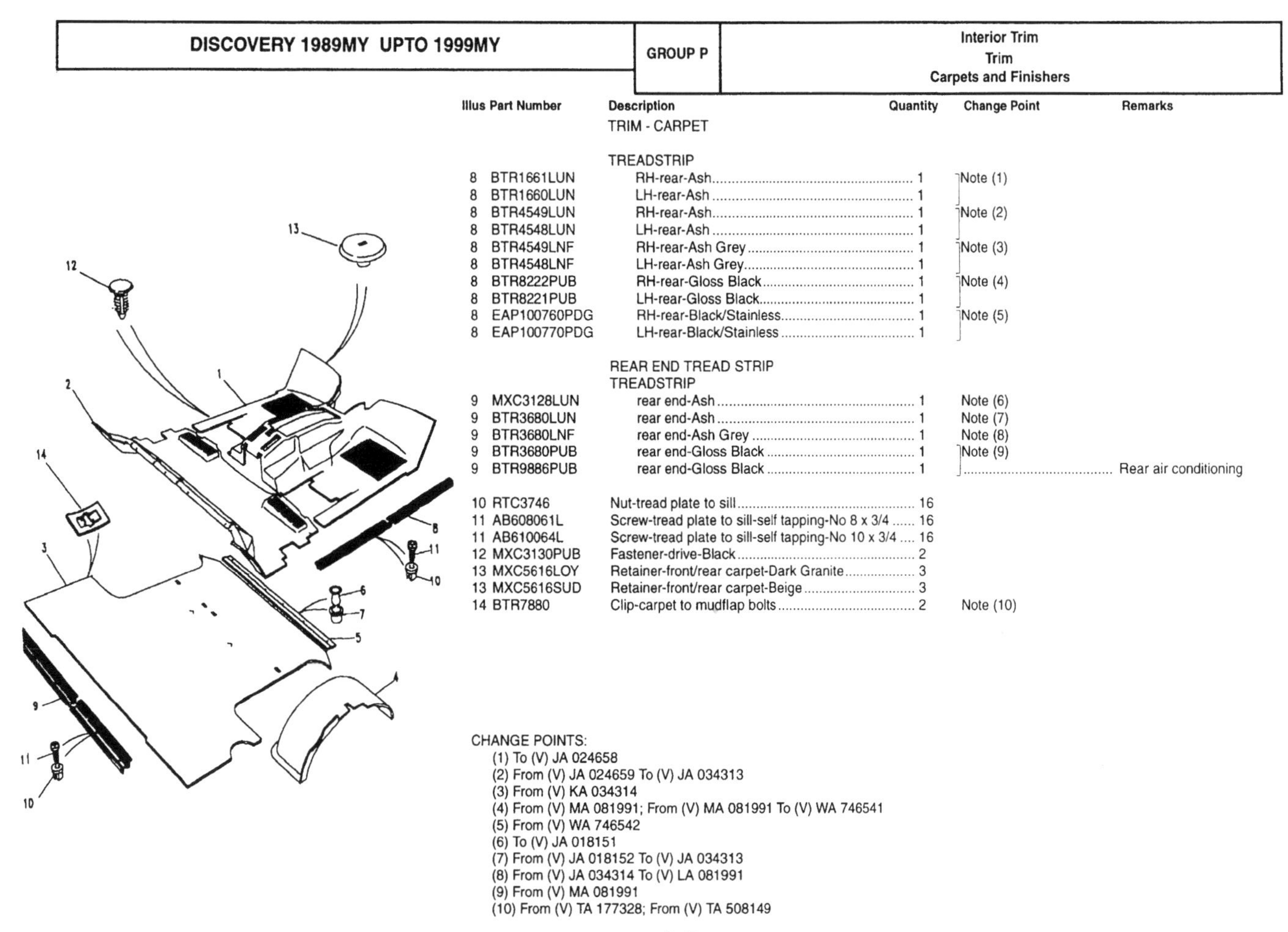

Illus	Part Number	Description	Quantity	Change Point	Remarks
		TRIM - CARPET			
		TREADSTRIP			
8	BTR1661LUN	RH-rear-Ash	1	Note (1)	
8	BTR1660LUN	LH-rear-Ash	1		
8	BTR4549LUN	RH-rear-Ash	1	Note (2)	
8	BTR4548LUN	LH-rear-Ash	1		
8	BTR4549LNF	RH-rear-Ash Grey	1	Note (3)	
8	BTR4548LNF	LH-rear-Ash Grey	1		
8	BTR8222PUB	RH-rear-Gloss Black	1	Note (4)	
8	BTR8221PUB	LH-rear-Gloss Black	1		
8	EAP100760PDG	RH-rear-Black/Stainless	1	Note (5)	
8	EAP100770PDG	LH-rear-Black/Stainless	1		
		REAR END TREAD STRIP			
		TREADSTRIP			
9	MXC3128LUN	rear end-Ash	1	Note (6)	
9	BTR3680LUN	rear end-Ash	1	Note (7)	
9	BTR3680LNF	rear end-Ash Grey	1	Note (8)	
9	BTR3680PUB	rear end-Gloss Black	1	Note (9)	
9	BTR9886PUB	rear end-Gloss Black	1		Rear air conditioning
10	RTC3746	Nut-tread plate to sill	16		
11	AB608061L	Screw-tread plate to sill-self tapping-No 8 x 3/4	16		
11	AB610064L	Screw-tread plate to sill-self tapping-No 10 x 3/4	16		
12	MXC3130PUB	Fastener-drive-Black	2		
13	MXC5616LOY	Retainer-front/rear carpet-Dark Granite	3		
13	MXC5616SUD	Retainer-front/rear carpet-Beige	3		
14	BTR7880	Clip-carpet to mudflap bolts	2	Note (10)	

CHANGE POINTS:
- (1) To (V) JA 024658
- (2) From (V) JA 024659 To (V) JA 034313
- (3) From (V) KA 034314
- (4) From (V) MA 081991; From (V) MA 081991 To (V) WA 746541
- (5) From (V) WA 746542
- (6) To (V) JA 018151
- (7) From (V) JA 018152 To (V) JA 034313
- (8) From (V) JA 034314 To (V) LA 081991
- (9) From (V) MA 081991
- (10) From (V) TA 177328; From (V) TA 508149

Notes

Seat Belts .... R 53
- Front Seat Belt Assembly .... R 53
- Rear Seat Belt .... R 56
- Rear Seat Belt - Inward Facing .... R 58
- Safety Harness - Rear Anchorage .... R 59

Seats .... R 1
- Front Seats Electrically Adjusted .... R 22
- Front Seats Manual .... R 2
- Inward Facing Seats .... R 47
- Plinths and Finishers .... R 1
- Rear Seat with Armrest .... R 41
- Rear Seat with Armrest and Logo .... R 44
- Rear Seats .... R 27
- Seat Covers-50LE .... R 52

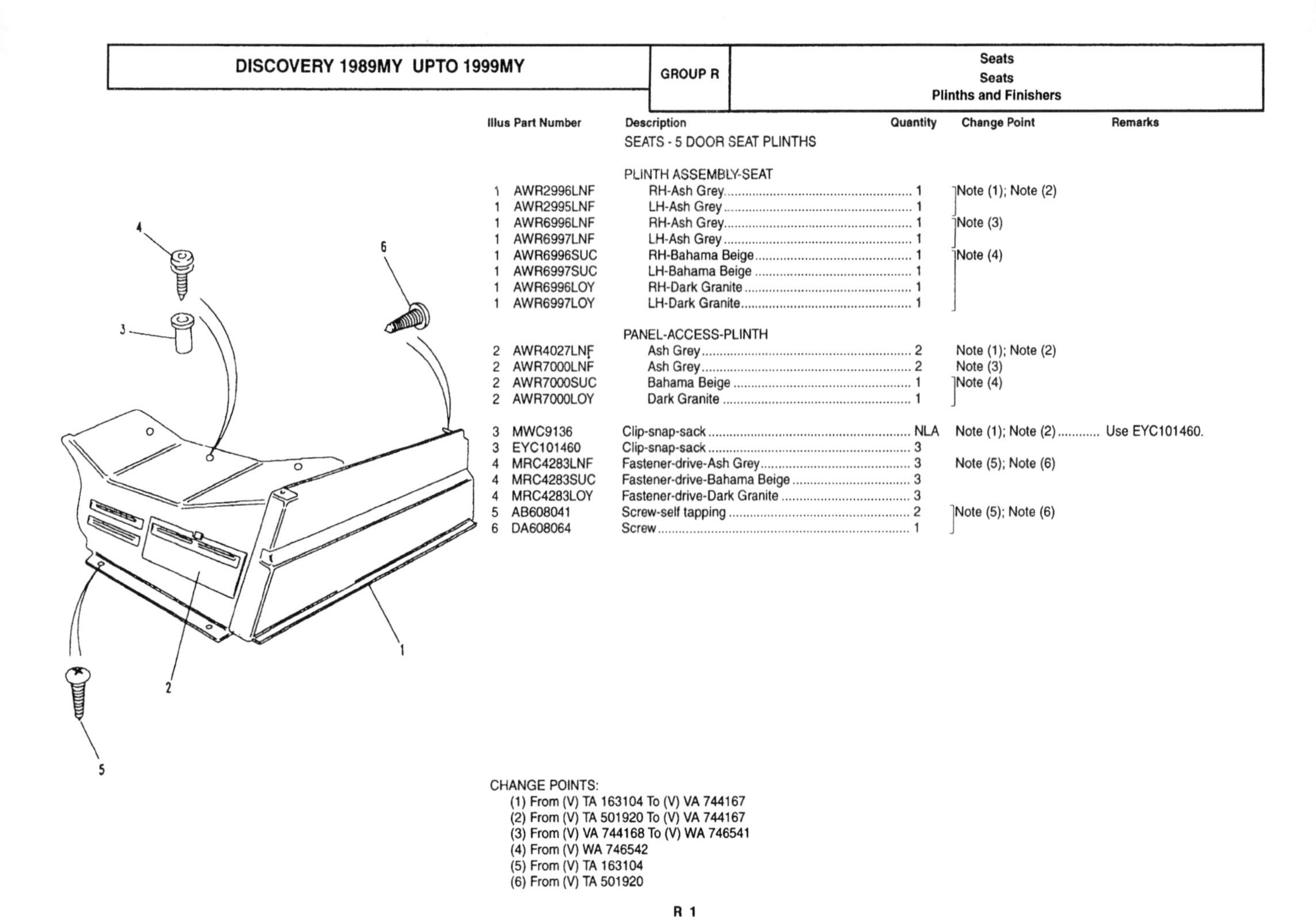

Illus	Part Number	Description	Quantity	Change Point	Remarks
		SEATS - 5 DOOR SEAT PLINTHS			
		PLINTH ASSEMBLY-SEAT			
1	AWR2996LNF	RH-Ash Grey	1	Note (1); Note (2)	
1	AWR2995LNF	LH-Ash Grey	1		
1	AWR6996LNF	RH-Ash Grey	1	Note (3)	
1	AWR6997LNF	LH-Ash Grey	1		
1	AWR6996SUC	RH-Bahama Beige	1	Note (4)	
1	AWR6997SUC	LH-Bahama Beige	1		
1	AWR6996LOY	RH-Dark Granite	1		
1	AWR6997LOY	LH-Dark Granite	1		
		PANEL-ACCESS-PLINTH			
2	AWR4027LNF	Ash Grey	2	Note (1); Note (2)	
2	AWR7000LNF	Ash Grey	2	Note (3)	
2	AWR7000SUC	Bahama Beige	1	Note (4)	
2	AWR7000LOY	Dark Granite	1		
3	MWC9136	Clip-snap-sack	NLA	Note (1); Note (2)	Use EYC101460.
3	EYC101460	Clip-snap-sack	3		
4	MRC4283LNF	Fastener-drive-Ash Grey	3	Note (5); Note (6)	
4	MRC4283SUC	Fastener-drive-Bahama Beige	3		
4	MRC4283LOY	Fastener-drive-Dark Granite	3		
5	AB608041	Screw-self tapping	2	Note (5); Note (6)	
6	DA608064	Screw	1		

CHANGE POINTS:
(1) From (V) TA 163104 To (V) VA 744167
(2) From (V) TA 501920 To (V) VA 744167
(3) From (V) VA 744168 To (V) WA 746541
(4) From (V) WA 746542
(5) From (V) TA 163104
(6) From (V) TA 501920

R 1

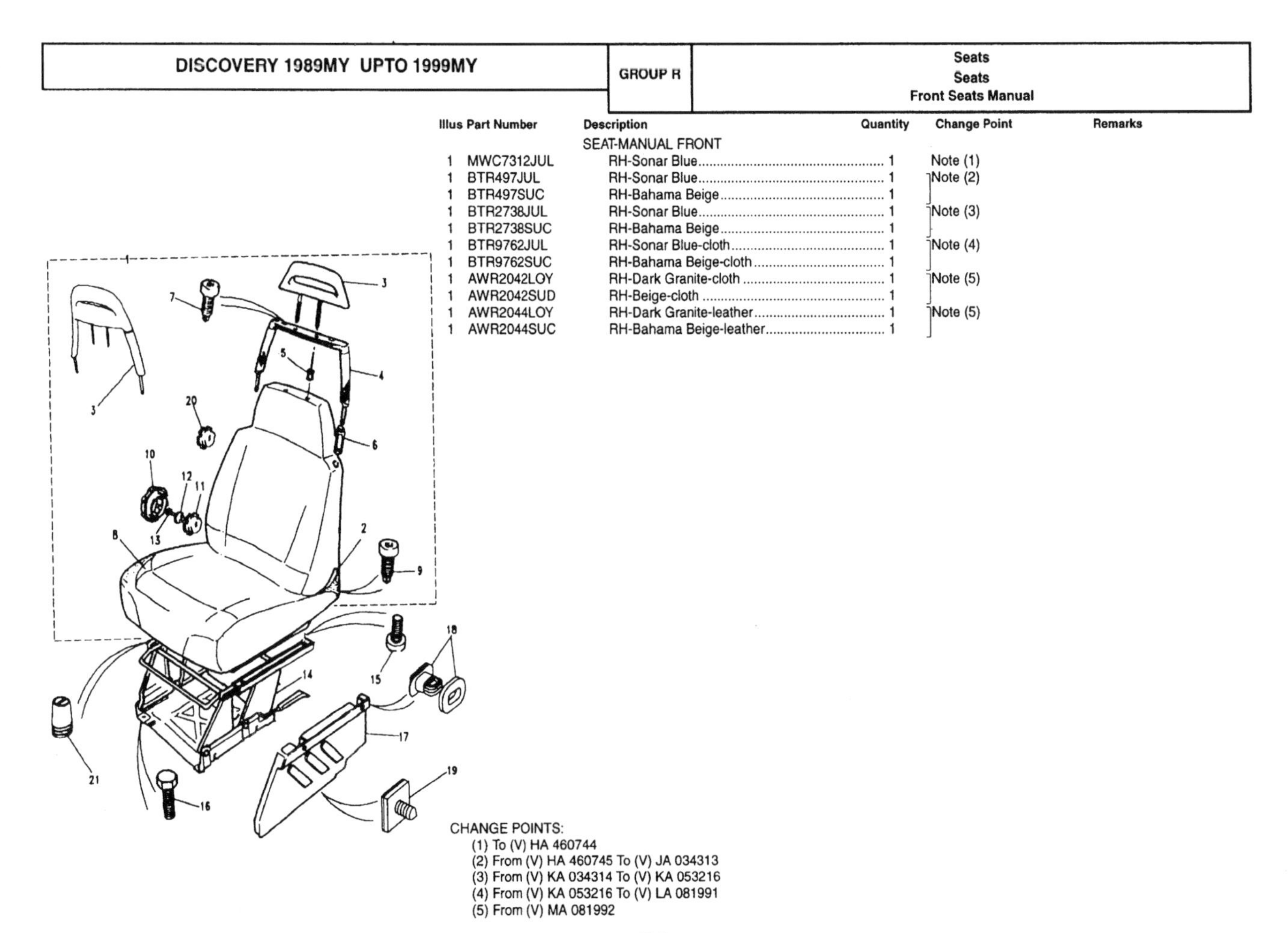

Illus	Part Number	Description	Quantity	Change Point	Remarks
		SEAT-MANUAL FRONT			
1	MWC7312JUL	RH-Sonar Blue	1	Note (1)	
1	BTR497JUL	RH-Sonar Blue	1	Note (2)	
1	BTR497SUC	RH-Bahama Beige	1		
1	BTR2738JUL	RH-Sonar Blue	1	Note (3)	
1	BTR2738SUC	RH-Bahama Beige	1		
1	BTR9762JUL	RH-Sonar Blue-cloth	1	Note (4)	
1	BTR9762SUC	RH-Bahama Beige-cloth	1		
1	AWR2042LOY	RH-Dark Granite-cloth	1	Note (5)	
1	AWR2042SUD	RH-Beige-cloth	1		
1	AWR2044LOY	RH-Dark Granite-leather	1	Note (5)	
1	AWR2044SUC	RH-Bahama Beige-leather	1		

CHANGE POINTS:
(1) To (V) HA 460744
(2) From (V) HA 460745 To (V) JA 034313
(3) From (V) KA 034314 To (V) KA 053216
(4) From (V) KA 053216 To (V) LA 081991
(5) From (V) MA 081992

R 2

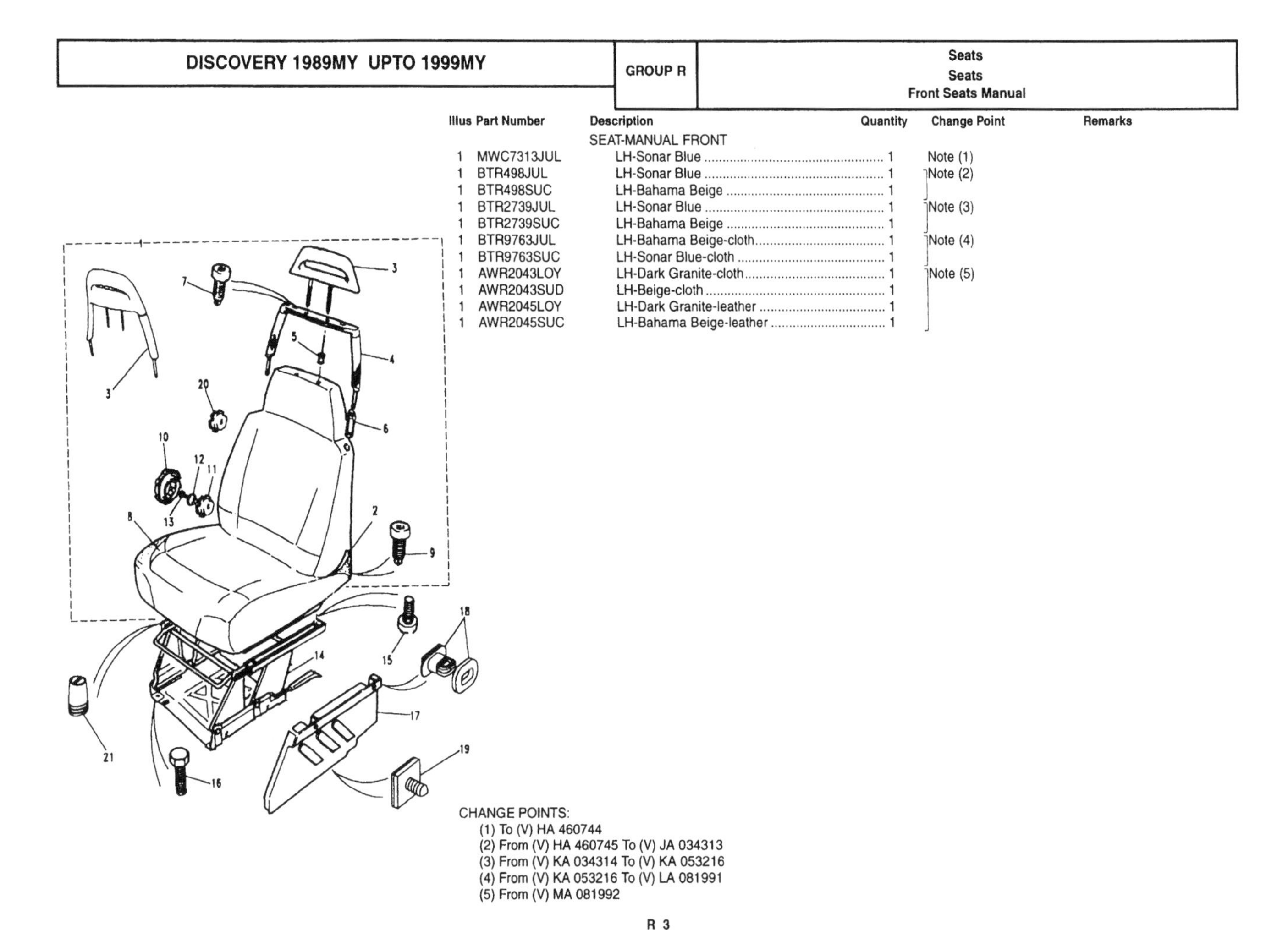

Illus	Part Number	Description	Quantity	Change Point	Remarks
		SEAT-MANUAL FRONT			
1	MWC7313JUL	LH-Sonar Blue	1	Note (1)	
1	BTR498JUL	LH-Sonar Blue	1	Note (2)	
1	BTR498SUC	LH-Bahama Beige	1		
1	BTR2739JUL	LH-Sonar Blue	1	Note (3)	
1	BTR2739SUC	LH-Bahama Beige	1		
1	BTR9763JUL	LH-Bahama Beige-cloth	1	Note (4)	
1	BTR9763SUC	LH-Sonar Blue-cloth	1		
1	AWR2043LOY	LH-Dark Granite-cloth	1	Note (5)	
1	AWR2043SUD	LH-Beige-cloth	1		
1	AWR2045LOY	LH-Dark Granite-leather	1		
1	AWR2045SUC	LH-Bahama Beige-leather	1		

CHANGE POINTS:
(1) To (V) HA 460744
(2) From (V) HA 460745 To (V) JA 034313
(3) From (V) KA 034314 To (V) KA 053216
(4) From (V) KA 053216 To (V) LA 081991
(5) From (V) MA 081992

R 3

DISCOVERY 1989MY UPTO 1999MY | GROUP R | Seats
Seats
Front Seats Manual

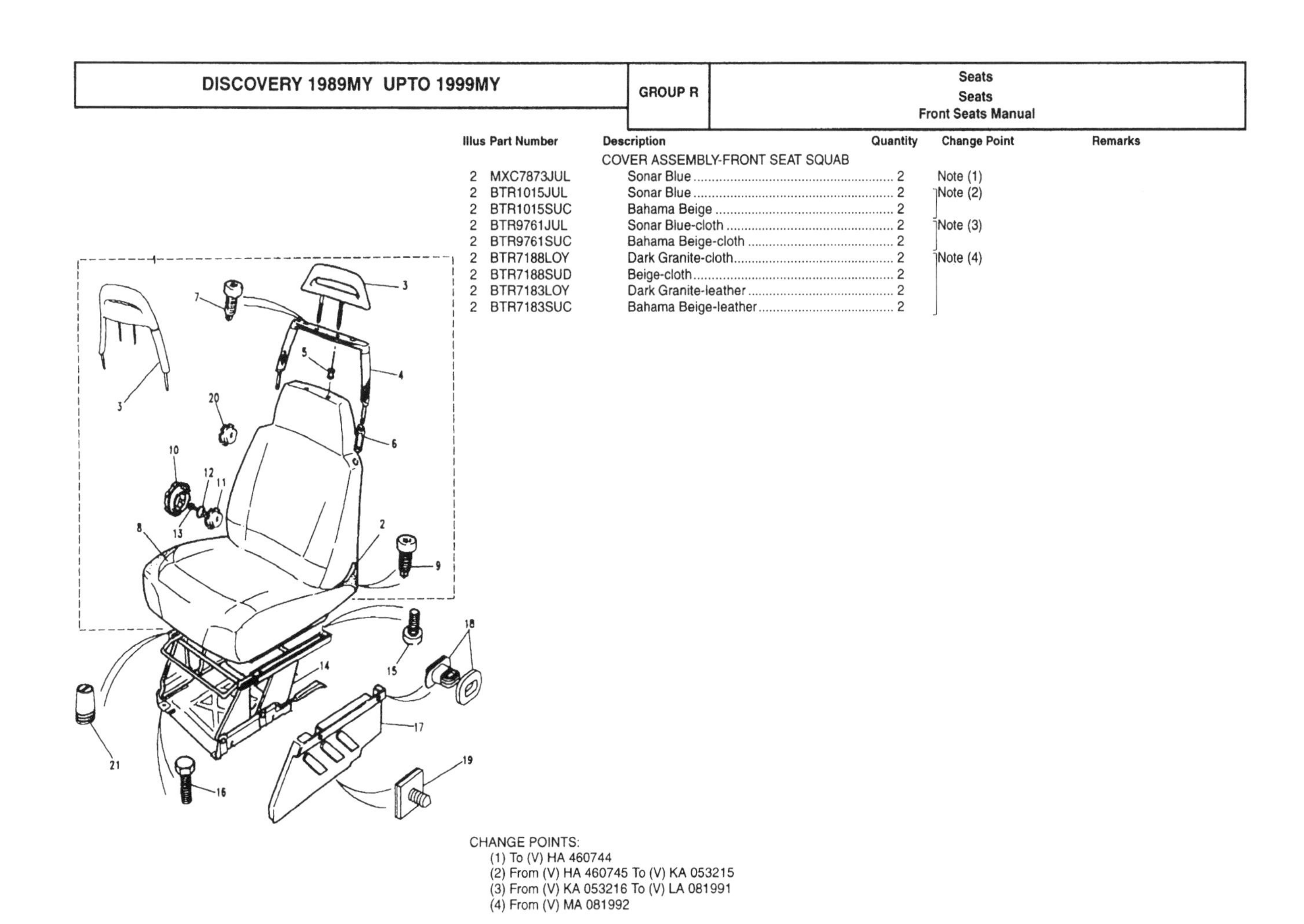

Illus	Part Number	Description	Quantity	Change Point	Remarks
		COVER ASSEMBLY-FRONT SEAT SQUAB			
2	MXC7873JUL	Sonar Blue	2	Note (1)	
2	BTR1015JUL	Sonar Blue	2	Note (2)	
2	BTR1015SUC	Bahama Beige	2		
2	BTR9761JUL	Sonar Blue-cloth	2	Note (3)	
2	BTR9761SUC	Bahama Beige-cloth	2		
2	BTR7188LOY	Dark Granite-cloth	2	Note (4)	
2	BTR7188SUD	Beige-cloth	2		
2	BTR7183LOY	Dark Granite-leather	2		
2	BTR7183SUC	Bahama Beige-leather	2		

CHANGE POINTS:
(1) To (V) HA 460744
(2) From (V) HA 460745 To (V) KA 053215
(3) From (V) KA 053216 To (V) LA 081991
(4) From (V) MA 081992

R 4

Illus	Part Number	Description	Quantity	Change Point	Remarks
		HEADRESTRAINT ASSEMBLY			
3	MWC8831JUN	Sonar Blue	2	Note (1)	
3	BTR437JUN	Sonar Blue	2	Note (2)	
3	BTR437SUC	Bahama Beige	2		
3	BTR437LOY	Dark Granite	2	Note (3)	
3	BTR437SUD	Beige	2		
3	BTR437LOY	Dark Granite	2		
3	BTR437SUC	Bahama Beige	2		

CHANGE POINTS:
(1) To (V) HA 460744
(2) From (V) HA 460745 To (V) LA 081991
(3) From (V) MA 081992

DISCOVERY 1989MY UPTO 1999MY | GROUP R | Seats
Seats
Front Seats Manual

Illus	Part Number	Description	Quantity	Change Point	Remarks
4	MWC9704JUN	Handle-front headrestraint grab-Sonar Blue	2	Note (1)	
		ESCUTCHEON-HEADREST			
5	MUC2338PMA	Black	4	Note (2)	
5	BTR3636LUL	Winchester Grey	4	Note (3)	
5	BTR3636SUA	Sorrel Brown	4	Note (3)	
5	BTR3636LOY	Dark Granite	4	Note (4)	
		INSERT-FRONT HEADRESTRAINT GRAB HANDLE			
6	MWC9719JUL	Sonar Blue	4		
6	MWC9719SUC	Bahama Beige	4		
6	MWC9719LOY	Dark Granite	4	Note (4)	
7	MWC9705	Screw-grab handle to squab-M8	4		
		FROM HA460745 GRAB HANDLE IS PART OF H/RESTRAINT			

CHANGE POINTS:
(1) To (V) HA 460744
(2) To (V) JA 031559
(3) From (V) JA 031560
(4) From (V) MA 081992

Illus	Part Number	Description	Quantity	Change Point	Remarks
		COVER ASSEMBLY-FRONT SEAT CUSHION			
8	MXC7872JUL	Sonar Blue	2	Note (1)	
8	MXC7872SUC	Bahama Beige	2		
8	BTR9753JUL	Sonar Blue	2	Note (2)	
8	BTR9753SUC	Bahama Beige	2		
8	BTR7189LOY	cloth-Dark Granite	2	Note (3)	
8	BTR7189SUD	cloth-Beige	2		
8	BTR7182LOY	leather-Dark Granite	2		
8	BTR7182SUC	leather-Bahama Beige	2		
	BTR2069	Clip-cover to cushion	6		

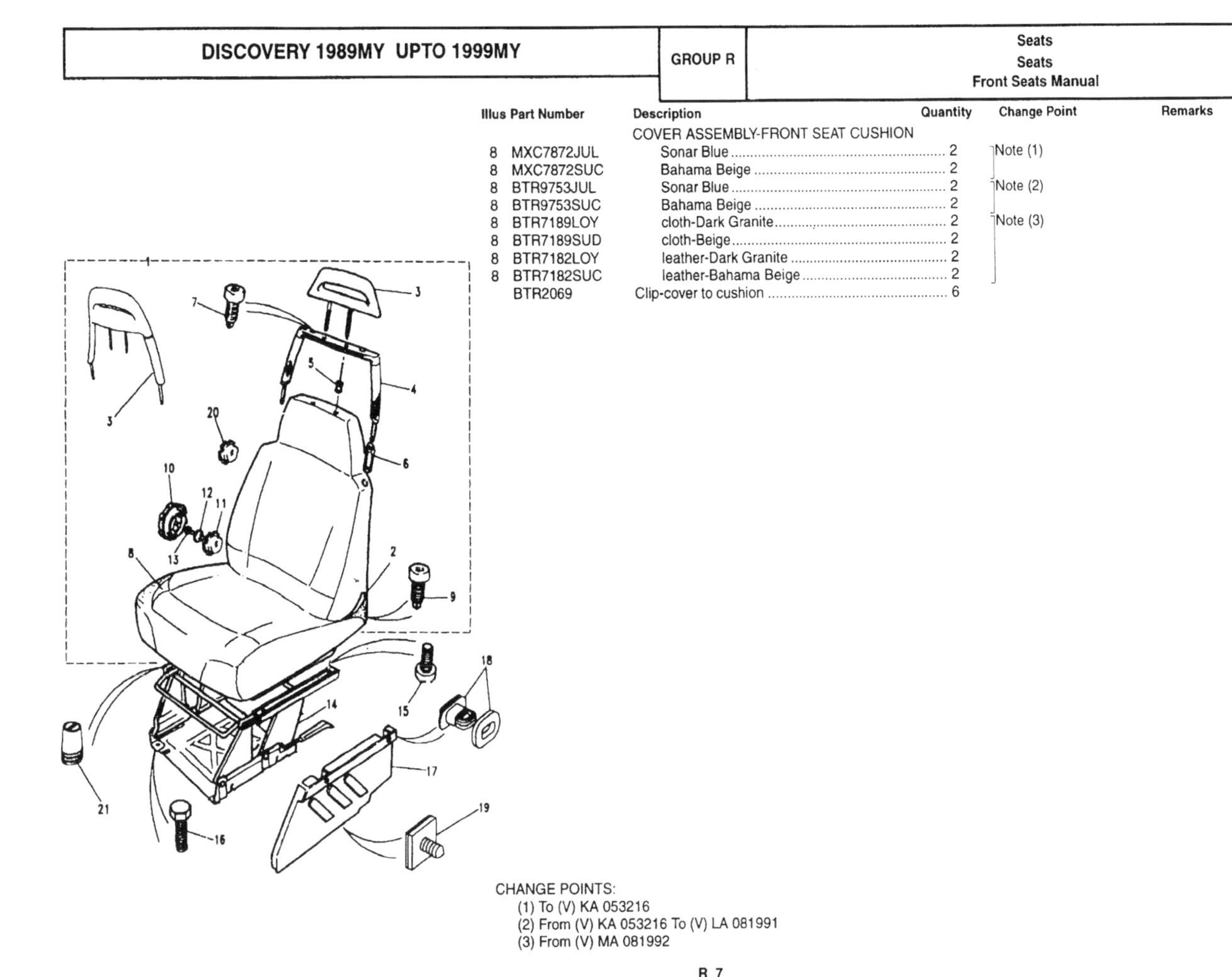

CHANGE POINTS:
(1) To (V) KA 053216
(2) From (V) KA 053216 To (V) LA 081991
(3) From (V) MA 081992

Illus	Part Number	Description	Quantity	Change Point	Remarks
9	MWC9705	Screw-squab to frame-M8	12	Note (1)	
9	HYG10014	Bolt-flanged head-squab to frame-M8	12	Note (2)	
		HANDWHEEL-FRONT SEAT SQUAB MANUAL RECLINE MECHANISM			
10	MWC9725LUN	Ash Grey	2	Note (3)	
10	MWC9725LNF	Ash Grey	2	Note (4)	
10	BTR6115LNF	Ash Grey	2	Note (5)	
	BTR6171LNF	Cap-front seat squab manual recline mechanism hand wheel-Ash Grey	2		
11	MWC9707	Disc support-recline handle-seat	2	Note (6)	
11	BTR6513	Disc support-recline handle-seat	2	Note (5)	
12	MWC9701	Washer-squab to frame	2	Note (6)	
13	MWC8894	Screw-seat recline mechanism	2		
13	BTR6512	Screw-seat recline mechanism	1	Note (5)	

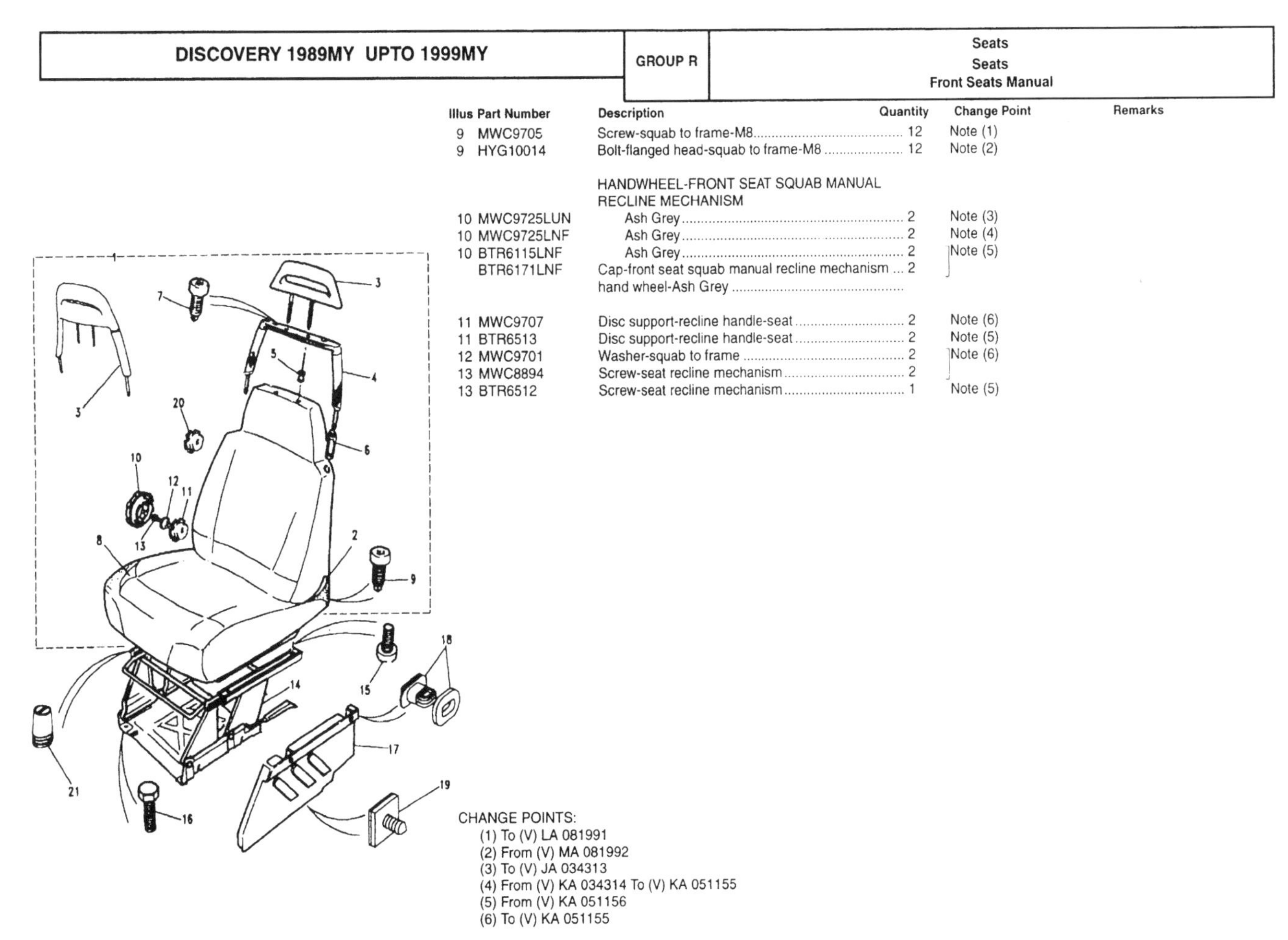

CHANGE POINTS:
(1) To (V) LA 081991
(2) From (V) MA 081992
(3) To (V) JA 034313
(4) From (V) KA 034314 To (V) KA 051155
(5) From (V) KA 051156
(6) To (V) KA 051155

Illus	Part Number	Description	Quantity	Change Point	Remarks
		NON-ELECTRIC SEAT ADJUSTMENT			
		3 DOOR ONLY TILT & SLIDE MECHANISM FRAME ASSEMBLY-MANUAL FRONT SEAT			
14	MXC6310LUN	RH-Ash Grey	1	Note (1)	
14	MXC6311LUN	LH-Ash Grey	1		
14	MXC6310LNF	RH-Ash Grey	1	Note (2)	
14	MXC6311LNF	LH-Ash Grey	1		
14	BTR9848PMA	RH-Black	1	Note (3)	
14	BTR9847PMA	LH-Black	1		
		5 DOOR & LEISURE ONLY PLINTH & SLIDES FRAME ASSEMBLY-MANUAL FRONT SEAT			
14	MXC7270LUN	RH-Ash Grey	1	Note (1)	
14	MXC7271LUN	LH-Ash Grey	1		
14	MXC7270LNF	RH-Ash Grey	1	Note (2)	
14	MXC7271LNF	LH-Ash Grey	1		
14	BTR9850PMA	RH-Black	1	Note (3)	
14	BTR9849PMA	LH-Black	1		

THE SEAT SLIDES ARE PART OF THE MECHANISM OR PLINTH

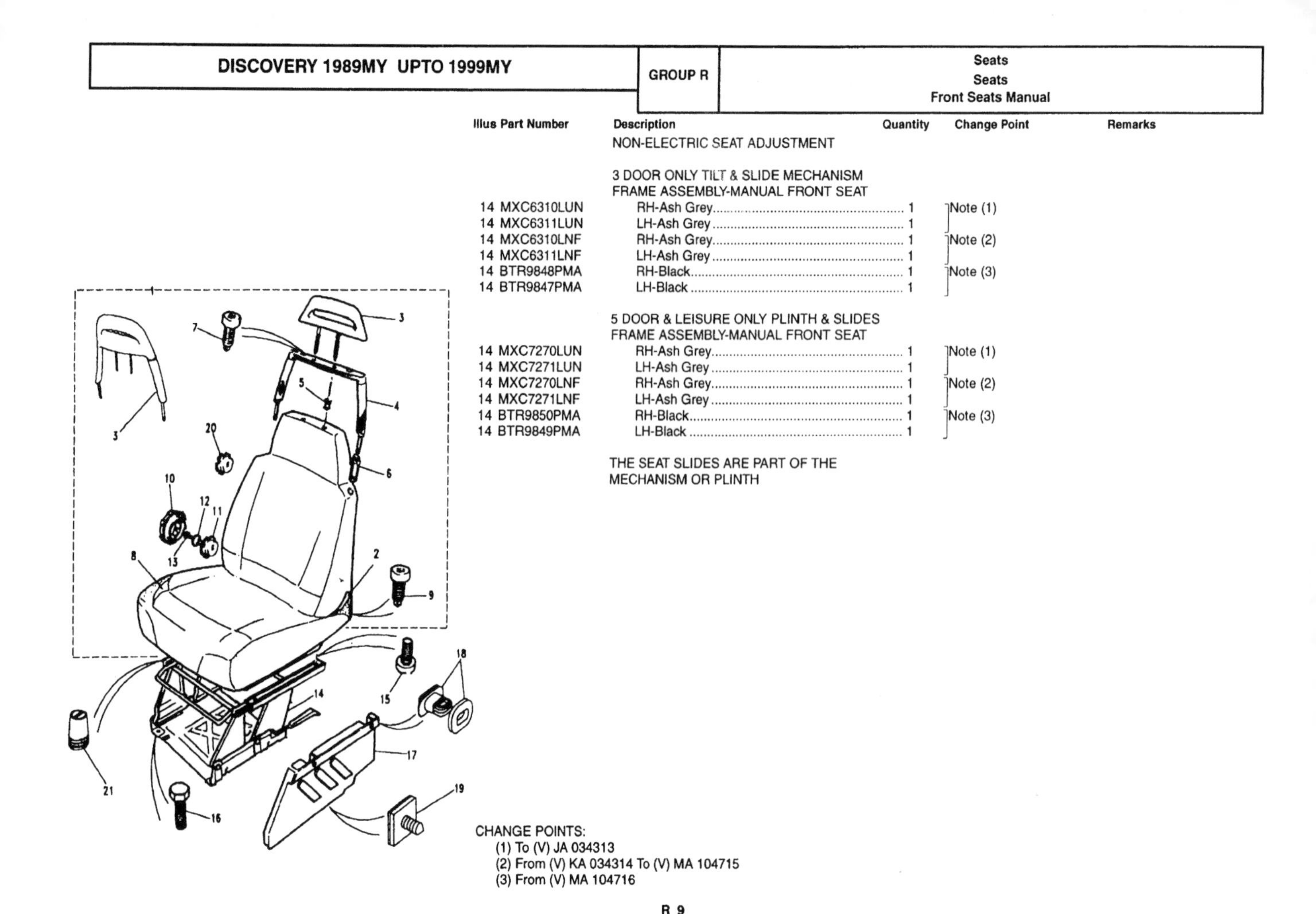

CHANGE POINTS:
(1) To (V) JA 034313
(2) From (V) KA 034314 To (V) MA 104715
(3) From (V) MA 104716

Illus	Part Number	Description	Quantity	Change Point	Remarks
15	MWC9706	Screw-seat base to plinth	8	Note (1)	
15	BTR5487	Screw-seat base to plinth-M8	8	Note (2)	
16	MWC9705	Screw-plinth to floor-M8	8	Note (3)	
16	DAM7844L	Screw-plinth to floor	8	Note (4)	
16	MWC9705	Screw-plinth to floor-M8	8	Note (5)	
16	BTR5487	Screw-plinth to floor-M8	8	Note (2)	

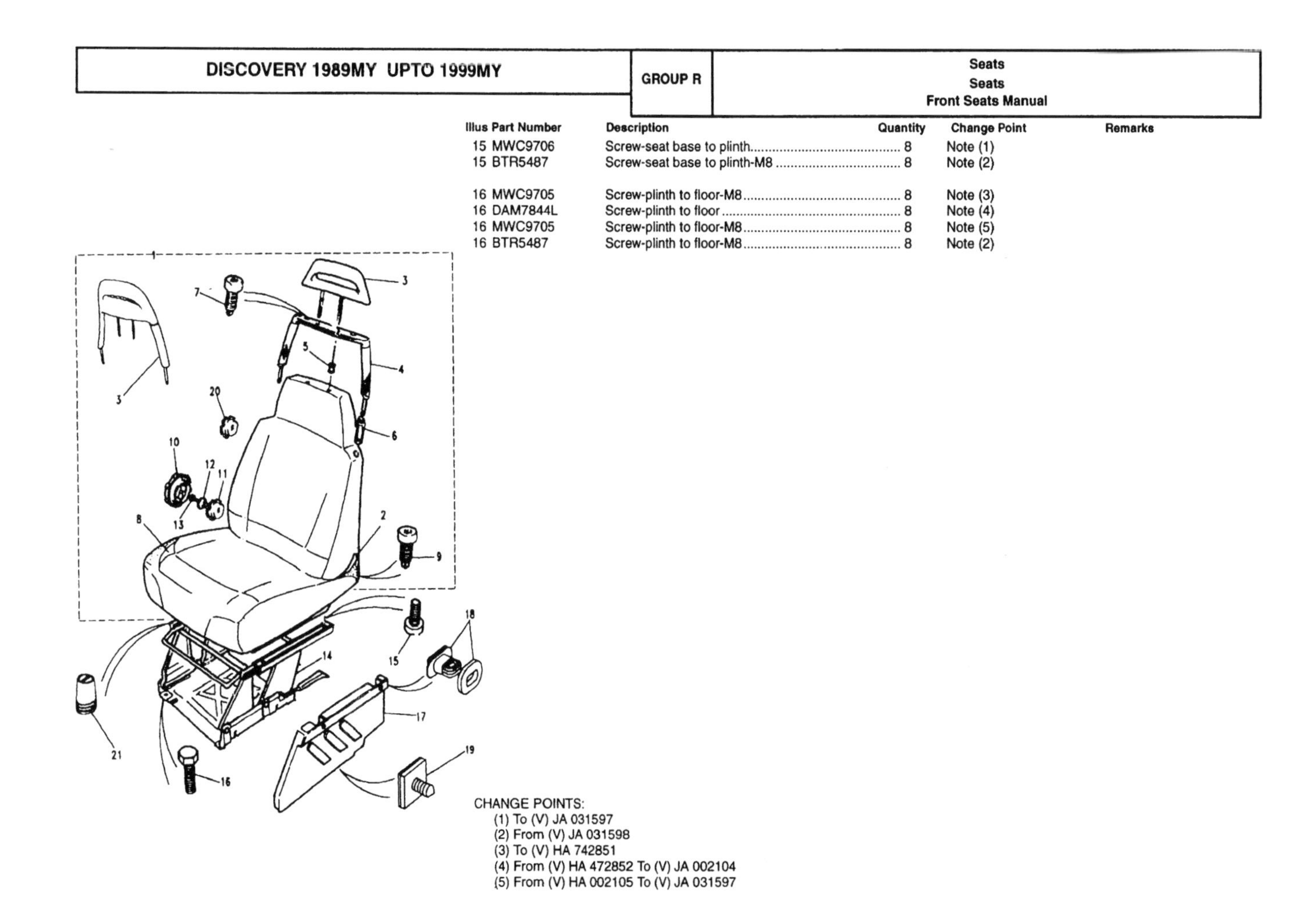

CHANGE POINTS:
(1) To (V) JA 031597
(2) From (V) JA 031598
(3) To (V) HA 742851
(4) From (V) HA 472852 To (V) JA 002104
(5) From (V) HA 002105 To (V) JA 031597

Illus	Part Number	Description	Quantity	Change Point	Remarks
		VALANCE-FRONT SEAT OUTER			
17	MXC8394LUN	RH-Ash Grey	1	Note (1)	
17	MXC8394LNF	RH-Ash Grey	1	Note (2)	
17	MXC8395LUN	LH-Ash Grey	1	Note (1)	
17	MXC8395LNF	LH-Ash Grey	1	Note (2)	
17	BTR5212LNF	LH-Ash Grey	1		Note (1)
	MXC3130LUN	Fastener-fir tree-Ash Grey	4	Note (1)	
	MXC3130LNF	Fastener-fir tree-Ash Grey	4	Note (2)	
18	BTR3264	Catch	2		Note (1)
19	BTR3269	Stud-plastic	2		Note (1)
20	AWR2157LNF	Handwheel-front seat squab manual recline mechanism-Ash Grey	2		
21	STC3184	Buffer rubber-seat front	4		

CHANGE POINTS:
(1) To (V) JA 034313
(2) From (V) KA 034314 To (V) MA 163103; From (V) KA 034314 To (V) MA 501919

Remarks:
(1) CD Player

Illus	Part Number	Description	Quantity	Change Point	Remarks
		3 AND 5 DOOR			
1	AWR2640PMA	Frame assembly-manual lumbar support front seat squab-RH-Black	1	Note (1)	
1	AWR2639PMA	Frame assembly-manual lumbar support front seat squab-LH-Black	1	Note (1)	
		CUSHION - RH & LH			
2	AWR2641PMA	Frame assembly-manual front seat cushion-Black	2	Note (1)	3 Door
2	AWR2642PMA	Frame assembly-manual front seat cushion-Black	2	Note (1)	5 Door
		3 AND 5 DOOR RH & LH			
3	AWR2814	Foam-squab	2	Note (1)	
4	MXC7870	Foam cushion-front	2	Note (1)	
5	AWR4320	Kit-lumber mechanism-RH	1	Note (1)	
5	AWR4321	Kit-lumber mechanism-LH	1	Note (1)	

CARXEA2B

CHANGE POINTS:
(1) From (V) TA 163104; From (V) TA 501920

Illus	Part Number	Description	Quantity	Change Point	Remarks
		SEATS - FRAMES & OVERLAYS - UP TO TA VIN			
1	AWR2028PMA	Frame sub assembly-front seat squab-front-RH-Black	1		Note (1)
1	AWR2029PMA	Frame sub assembly-front seat squab-front-LH-Black	1		Note (1)
2	BTR8762PMA	Frame assembly-manual front seat cushion-Black	2		
3	AWR2158	Foam-squab-front	2		Note (1)
4	MXC7870	Foam cushion-front	2		
5	AWR4320	Kit-lumber mechanism-RH	1		
5	AWR4321	Kit-lumber mechanism-LH	1		

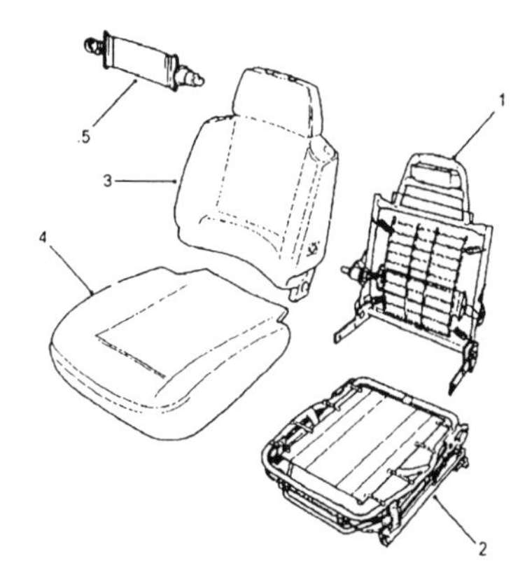

CARXEA2B

Remarks:
(1) lumbar adjustment

Illus	Part Number	Description	Quantity	Change Point	Remarks
		SEAT-MANUAL FRONT			
1	AWR3526LOY	RH-cloth-Dark Granite	1	Note (1)	3 Door except electric seats
1	AWR3526SUC	RH-cloth-Bahama Beige	1		
1	AWR3525LOY	LH-cloth-Dark Granite	1		
1	AWR3525SUC	LH-cloth-Bahama Beige	1		
		SEAT-MANUAL FRONT			
1	AWR3522LOY	RH-cloth-Dark Granite	1	Note (1)	5 Door except electric seats
1	AWR3522SUC	RH-cloth-Bahama Beige	1		
1	AWR3521LOY	LH-cloth-Dark Granite	1		
1	AWR3521SUC	LH-cloth-Bahama Beige	1		
		SEAT-MANUAL FRONT			
1	AWR2646LOY	RH-leather-Dark Granite	1	Note (2)	5 Door except electric seats
1	AWR6096LOY	RH-leather-Dark Granite	1	Note (3)	
1	AWR2646SUC	RH-leather-Bahama Beige	1	Note (4)	
1	AWR6096SUC	RH-leather-Bahama Beige	1	Note (5)	
1	AWR2647LOY	LH-leather-Dark Granite	1	Note (2)	
1	AWR6095LOY	LH-leather-Dark Granite	1	Note (6)	
1	AWR2647SUC	LH-leather-Bahama Beige	1	Note (4)	
1	AWR6095SUC	LH-leather-Bahama Beige	1	Note (5)	

CHANGE POINTS:
(1) From (V) TA 163104 To (V) WA 746541; From (V) TA 501920 To (V) WA 746541
(2) To (V) TA 530158; To (V) TA 700125
(3) From (V) TA 530159; From (V) TA 700126
(4) To (V) TA 703236; To (V) TA 534104
(5) From (V) VA 703237 To (V) WA 746541; From (V) VA 534104 To (V) WA 746541
(6) From (V) TA 530159 To (V) WA 746541; From (V) TA 700126 To (V) WA 746541

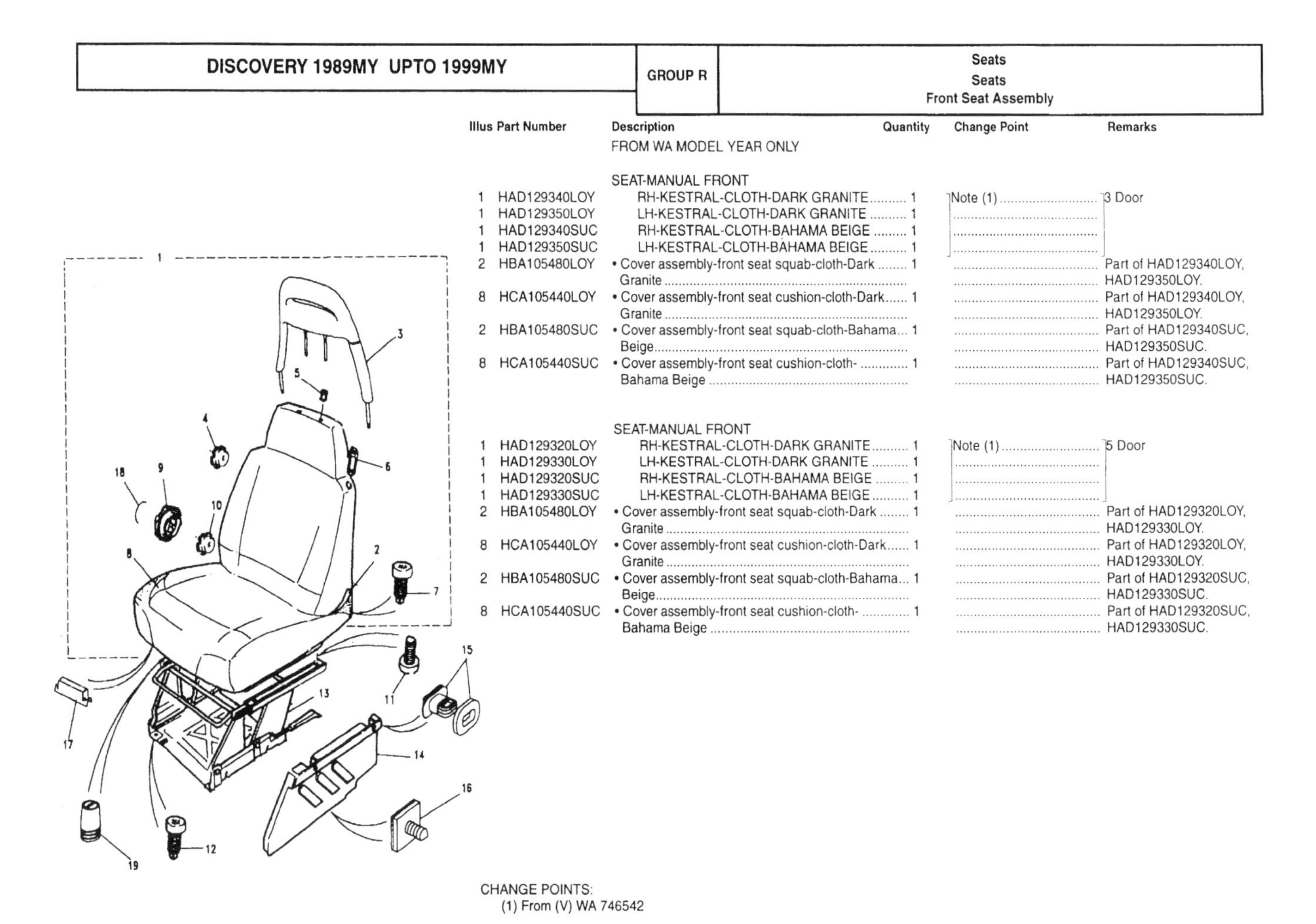

Illus	Part Number	Description	Quantity	Change Point	Remarks
		FROM WA MODEL YEAR ONLY			
		SEAT-MANUAL FRONT			
1	HAD129340LOY	RH-KESTRAL-CLOTH-DARK GRANITE	1	Note (1)	3 Door
1	HAD129350LOY	LH-KESTRAL-CLOTH-DARK GRANITE	1		
1	HAD129340SUC	RH-KESTRAL-CLOTH-BAHAMA BEIGE	1		
1	HAD129350SUC	LH-KESTRAL-CLOTH-BAHAMA BEIGE	1		
2	HBA105480LOY	• Cover assembly-front seat squab-cloth-Dark Granite	1		Part of HAD129340LOY, HAD129350LOY.
8	HCA105440LOY	• Cover assembly-front seat cushion-cloth-Dark Granite	1		Part of HAD129340LOY, HAD129350LOY.
2	HBA105480SUC	• Cover assembly-front seat squab-cloth-Bahama Beige	1		Part of HAD129340SUC, HAD129350SUC.
8	HCA105440SUC	• Cover assembly-front seat cushion-cloth-Bahama Beige	1		Part of HAD129340SUC, HAD129350SUC.
		SEAT-MANUAL FRONT			
1	HAD129320LOY	RH-KESTRAL-CLOTH-DARK GRANITE	1	Note (1)	5 Door
1	HAD129330LOY	LH-KESTRAL-CLOTH-DARK GRANITE	1		
1	HAD129320SUC	RH-KESTRAL-CLOTH-BAHAMA BEIGE	1		
1	HAD129330SUC	LH-KESTRAL-CLOTH-BAHAMA BEIGE	1		
2	HBA105480LOY	• Cover assembly-front seat squab-cloth-Dark Granite	1		Part of HAD129320LOY, HAD129330LOY.
8	HCA105440LOY	• Cover assembly-front seat cushion-cloth-Dark Granite	1		Part of HAD129320LOY, HAD129330LOY.
2	HBA105480SUC	• Cover assembly-front seat squab-cloth-Bahama Beige	1		Part of HAD129320SUC, HAD129330SUC.
8	HCA105440SUC	• Cover assembly-front seat cushion-cloth-Bahama Beige	1		Part of HAD129320SUC, HAD129330SUC.

CHANGE POINTS:
(1) From (V) WA 746542

R 15

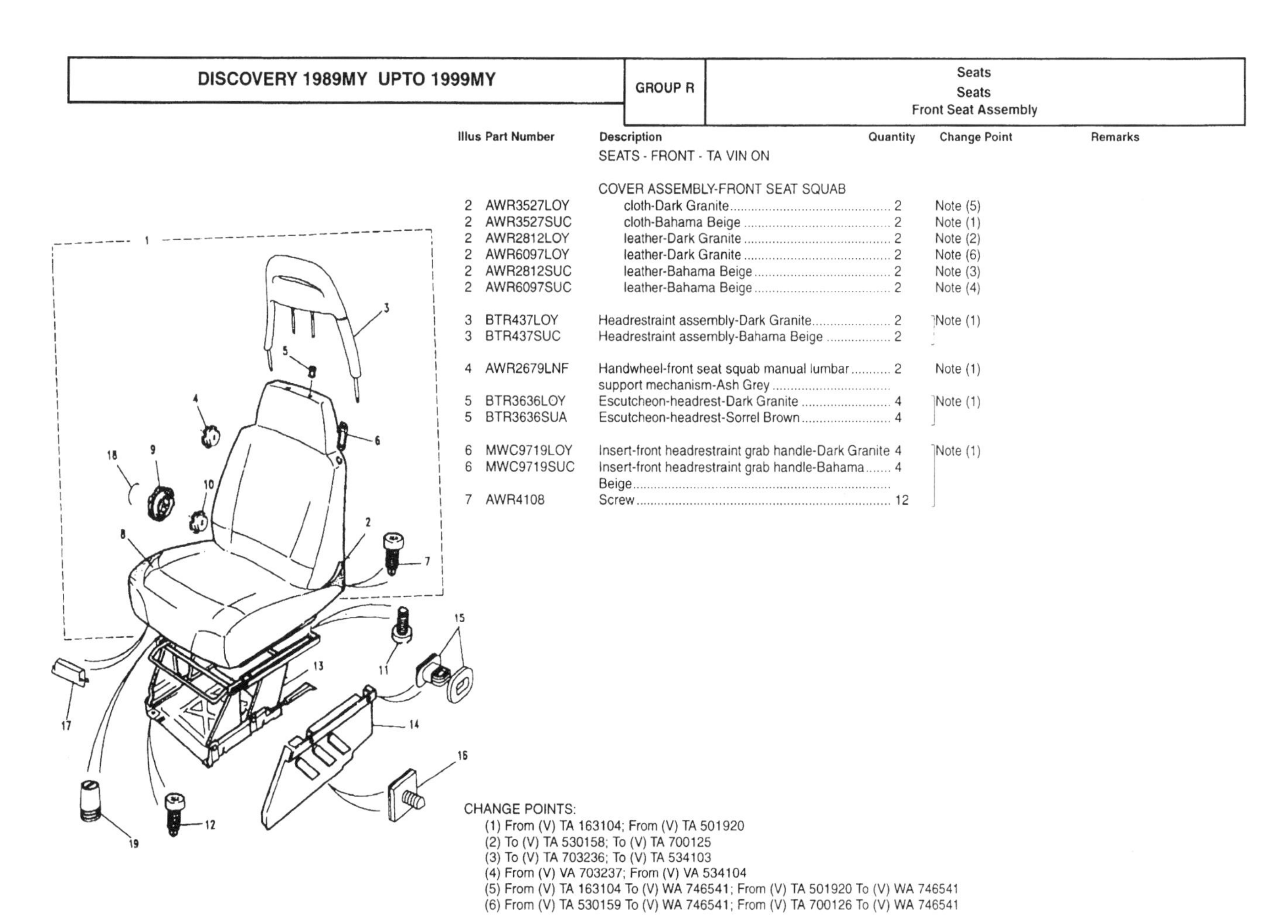

Illus	Part Number	Description	Quantity	Change Point	Remarks
		SEATS - FRONT - TA VIN ON			
		COVER ASSEMBLY-FRONT SEAT SQUAB			
2	AWR3527LOY	cloth-Dark Granite	2	Note (5)	
2	AWR3527SUC	cloth-Bahama Beige	2	Note (1)	
2	AWR2812LOY	leather-Dark Granite	2	Note (2)	
2	AWR6097LOY	leather-Dark Granite	2	Note (6)	
2	AWR2812SUC	leather-Bahama Beige	2	Note (3)	
2	AWR6097SUC	leather-Bahama Beige	2	Note (4)	
3	BTR437LOY	Headrestraint assembly-Dark Granite	2	Note (1)	
3	BTR437SUC	Headrestraint assembly-Bahama Beige	2		
4	AWR2679LNF	Handwheel-front seat squab manual lumbar support mechanism-Ash Grey	2	Note (1)	
5	BTR3636LOY	Escutcheon-headrest-Dark Granite	4	Note (1)	
5	BTR3636SUA	Escutcheon-headrest-Sorrel Brown	4		
6	MWC9719LOY	Insert-front headrestraint grab handle-Dark Granite	4	Note (1)	
6	MWC9719SUC	Insert-front headrestraint grab handle-Bahama Beige	4		
7	AWR4108	Screw	12		

CHANGE POINTS:
(1) From (V) TA 163104; From (V) TA 501920
(2) To (V) TA 530158; To (V) TA 700125
(3) To (V) TA 703236; To (V) TA 534103
(4) From (V) VA 703237; From (V) VA 534104
(5) From (V) TA 163104 To (V) WA 746541; From (V) TA 501920 To (V) WA 746541
(6) From (V) TA 530159 To (V) WA 746541; From (V) TA 700126 To (V) WA 746541

R 16

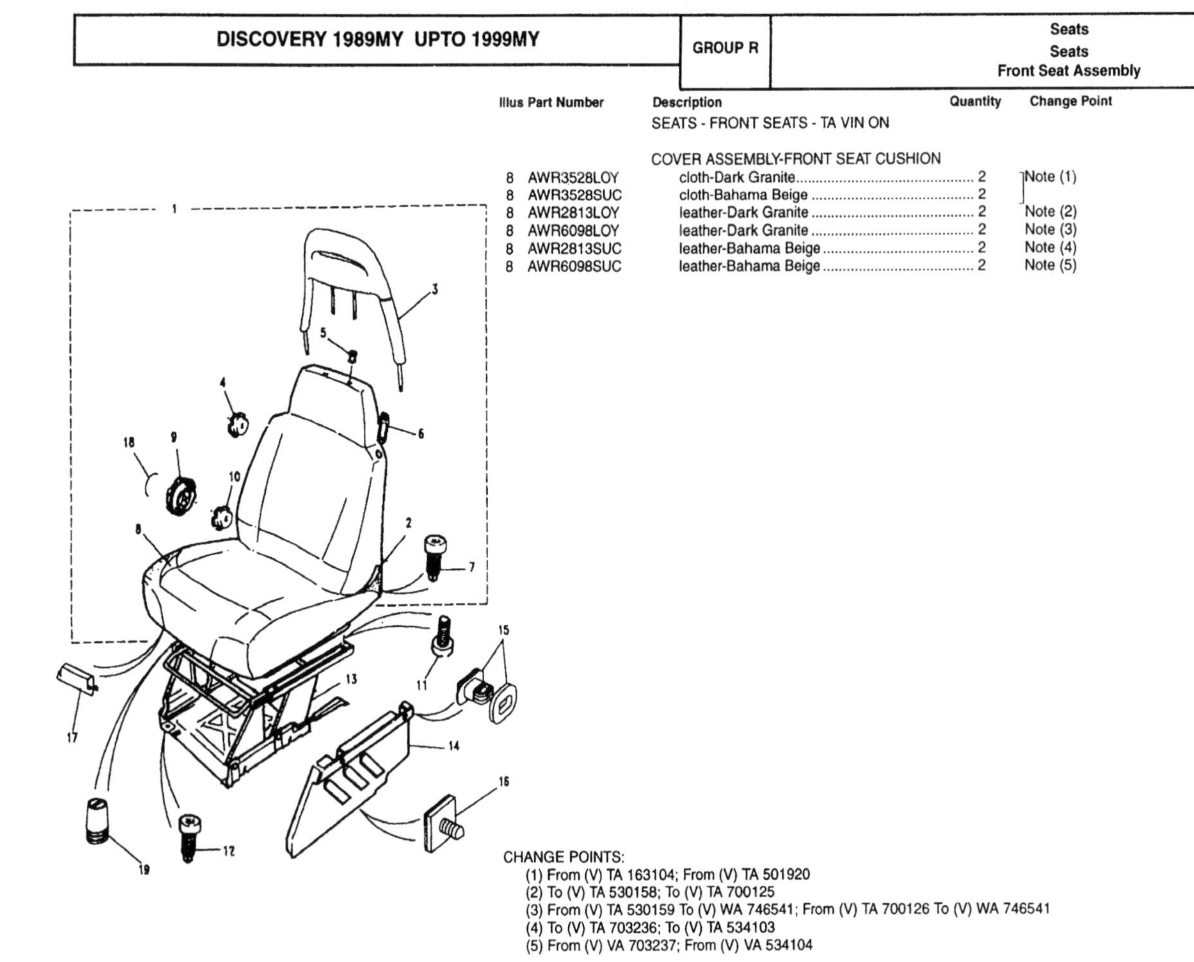

Illus	Part Number	Description	Quantity	Change Point	Remarks
		SEATS - FRONT SEATS - TA VIN ON			
		COVER ASSEMBLY-FRONT SEAT CUSHION			
8	AWR3528LOY	cloth-Dark Granite	2	Note (1)	
8	AWR3528SUC	cloth-Bahama Beige	2		
8	AWR2813LOY	leather-Dark Granite	2	Note (2)	
8	AWR6098LOY	leather-Dark Granite	2	Note (3)	
8	AWR2813SUC	leather-Bahama Beige	2	Note (4)	
8	AWR6098SUC	leather-Bahama Beige	2	Note (5)	

CHANGE POINTS:
(1) From (V) TA 163104; From (V) TA 501920
(2) To (V) TA 530158; To (V) TA 700125
(3) From (V) TA 530159 To (V) WA 746541; From (V) TA 700126 To (V) WA 746541
(4) To (V) TA 703236; To (V) TA 534103
(5) From (V) VA 703237; From (V) VA 534104

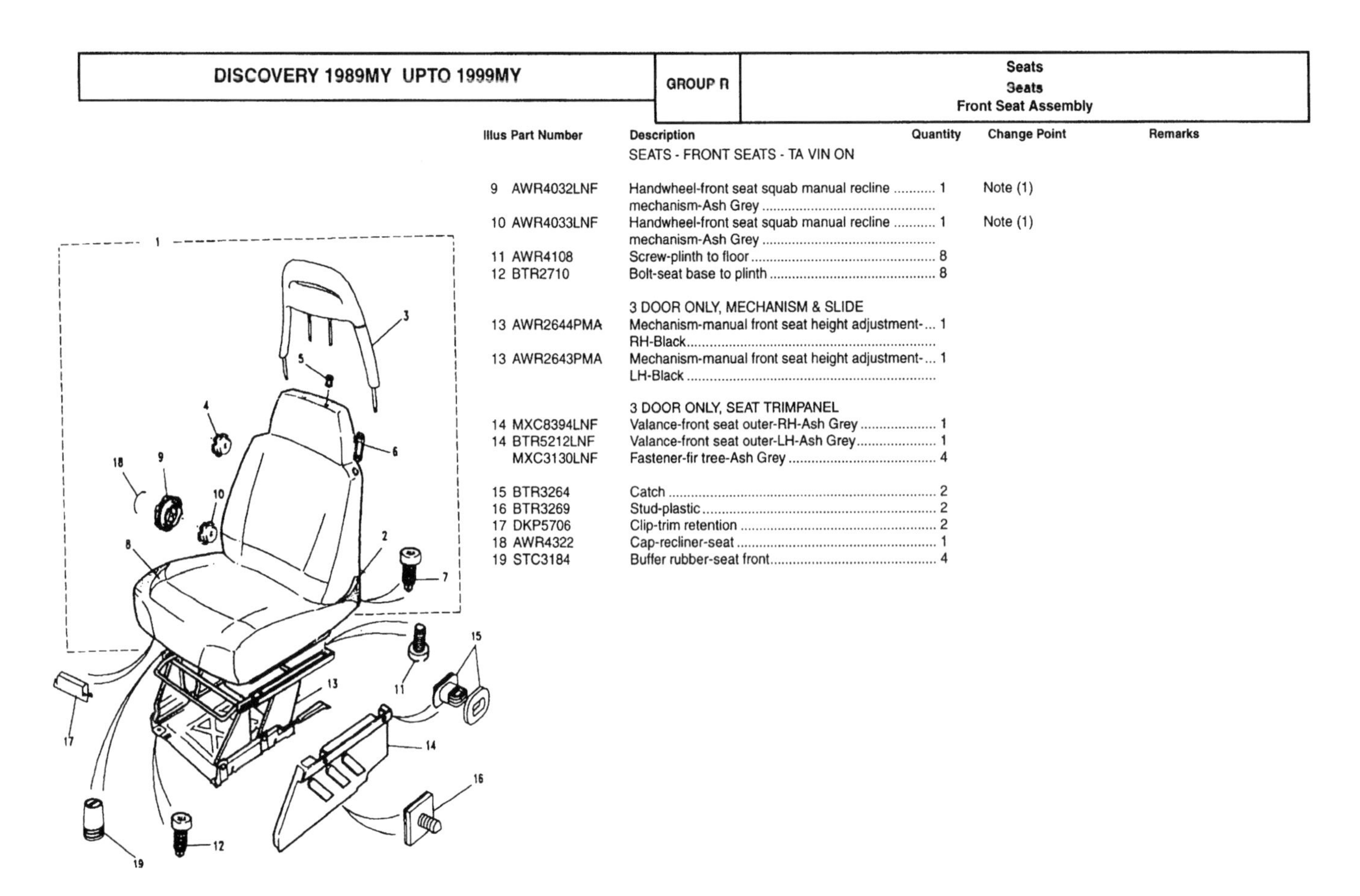

Illus	Part Number	Description	Quantity	Change Point	Remarks
		SEATS - FRONT SEATS - TA VIN ON			
9	AWR4032LNF	Handwheel-front seat squab manual recline mechanism-Ash Grey	1	Note (1)	
10	AWR4033LNF	Handwheel-front seat squab manual recline mechanism-Ash Grey	1	Note (1)	
11	AWR4108	Screw-plinth to floor	8		
12	BTR2710	Bolt-seat base to plinth	8		
		3 DOOR ONLY, MECHANISM & SLIDE			
13	AWR2644PMA	Mechanism-manual front seat height adjustment-RH-Black	1		
13	AWR2643PMA	Mechanism-manual front seat height adjustment-LH-Black	1		
		3 DOOR ONLY, SEAT TRIMPANEL			
14	MXC8394LNF	Valance-front seat outer-RH-Ash Grey	1		
14	BTR5212LNF	Valance-front seat outer-LH-Ash Grey	1		
	MXC3130LNF	Fastener-fir tree-Ash Grey	4		
15	BTR3264	Catch	2		
16	BTR3269	Stud-plastic	2		
17	DKP5706	Clip-trim retention	2		
18	AWR4322	Cap-recliner-seat	1		
19	STC3184	Buffer rubber-seat front	4		

CHANGE POINTS:
(1) From (V) TA 163104; From (V) TA 501920

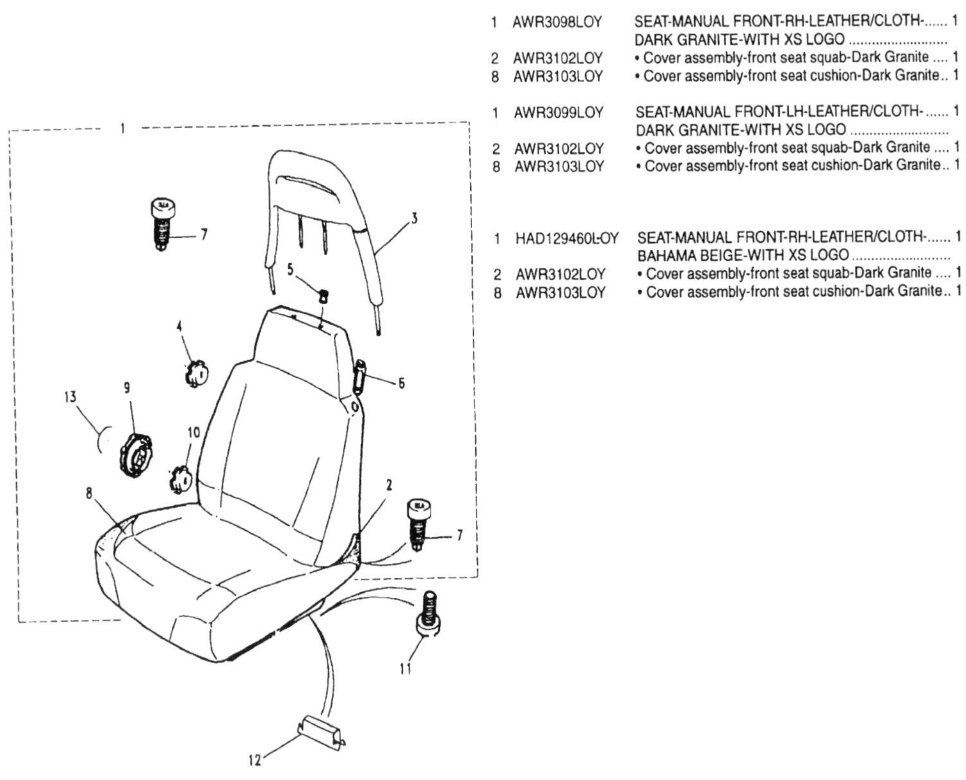

Illus	Part Number	Description	Quantity	Change Point	Remarks
1	AWR3098LOY	SEAT-MANUAL FRONT-RH-LEATHER/CLOTH-DARK GRANITE-WITH XS LOGO	1	Note (1)	5 Door
2	AWR3102LOY	• Cover assembly-front seat squab-Dark Granite	1		
8	AWR3103LOY	• Cover assembly-front seat cushion-Dark Granite	1		
1	AWR3099LOY	SEAT-MANUAL FRONT-LH-LEATHER/CLOTH-DARK GRANITE-WITH XS LOGO	1	Note (1)	5 Door
2	AWR3102LOY	• Cover assembly-front seat squab-Dark Granite	1		
8	AWR3103LOY	• Cover assembly-front seat cushion-Dark Granite	1		
1	HAD129460LOY	SEAT-MANUAL FRONT-RH-LEATHER/CLOTH-BAHAMA BEIGE-WITH XS LOGO	1	Note (2)	3 Door
2	AWR3102LOY	• Cover assembly-front seat squab-Dark Granite	1		
8	AWR3103LOY	• Cover assembly-front seat cushion-Dark Granite	1		

CHANGE POINTS:
(1) From (V) TA 163104; From (V) TA 501920
(2) From (V) WA 746542

Illus	Part Number	Description	Quantity	Change Point	Remarks
1	HAD129470LOY	SEAT-MANUAL FRONT-LH-LEATHER/CLOTH-BAHAMA BEIGE-WITH XS LOGO	1	Note (1)	3 Door
2	AWR3102LOY	• Cover assembly-front seat squab-Dark Granite	1		
8	AWR3103LOY	• Cover assembly-front seat cushion-Dark Granite	1		
3	BTR437LOY	Headrestraint assembly-Dark Granite	1		
4	AWR2679LNF	Handwheel-front seat squab manual lumbar support mechanism-Ash Grey	1		
5	BTR3636LOY	Escutcheon-headrest-Dark Granite	2		
5	BTR3636SUC	Escutcheon-headrest-Bahama Beige	1	Note (1)	
6	MWC9719LOY	Insert-front headrestraint grab handle-Dark Granite	2		
7	AWR4108	Screw	12		
9	AWR4032LNF	Handwheel-front seat squab manual recline mechanism-Ash Grey	1		
10	AWR4033LNF	Handwheel-front seat squab manual recline mechanism-Ash Grey	1		
11	BTR2710	Bolt-seat base to plinth	8		
12	DKP5706	Clip-trim retention	2		
13	AWR4322	Cap-recliner-seat	1		

CHANGE POINTS:
(1) From (V) WA 746542

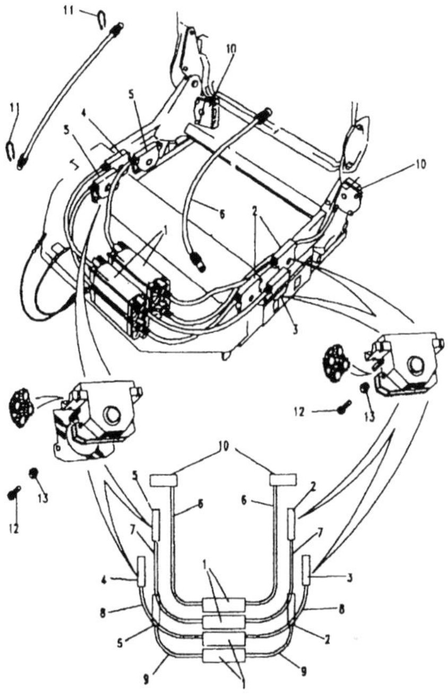

Illus	Part Number	Description	Quantity	Change Point	Remarks
		FROM (V) TA163104/TA501920			
1	STC2936	Motor-front seat	8		
		GEARBOX ASSEMBLY-FRONT SEAT FORE & AFT MECHANISM			
2	RTC5623	inner LH	4		
3	RTC6412	outer LH	2		
4	RTC6411	outer RH	2		
5	RTC5786	inner RH	4		
		CABLE-SLIDE ASSEMBLY FLEXIBLE DRIVE			
6	STC4230	524mm	4		
7	STC4227	281mm	NLA		
8	STC4228	230mm	4		
9	STC4229	255.5mm	4		
		GEARBOX RECLINE			
10	RTC5789	Gearbox assembly-recline-RH	1		
10	RTC5790	Gearbox assembly-recline-LH	1		
11	STC1888	Clip-spring steel	16		
12	RTC5797	Screw-40mm long	8		
12	RTC5798	Screw-30mm long	16		
13	NH104041L	Nut-hexagonal-M4	24		

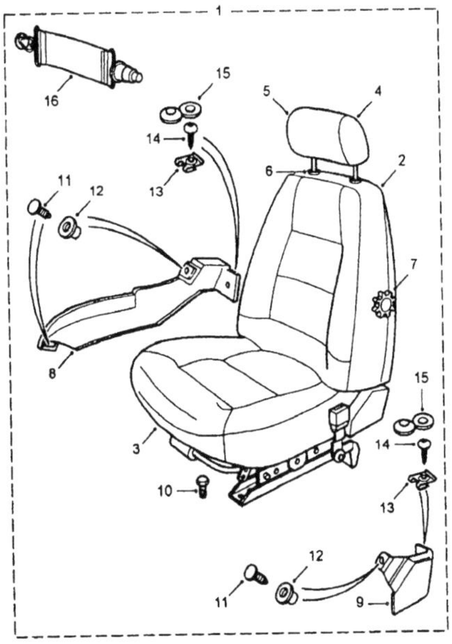

Illus	Part Number	Description	Quantity	Change Point	Remarks
		SEAT-ELECTRIC FRONT			
1	AWR2140LOY	RH-leather-Dark Granite	1	Note (1)	
1	AWR2141LOY	LH-leather-Dark Granite	1		
1	AWR6080LOY	RH-leather-Dark Granite	1	Note (2)	
1	AWR6079LOY	LH-leather-Dark Granite	1		
1	HAD128480LOY	RH-leather-Dark Granite	1	Note (3)	
1	HAD128490LOY	LH-leather-Dark Granite	1		
1	AWR2140SUC	RH-leather-Bahama Beige	1	Note (4)	
1	AWR2141SUC	LH-leather-Bahama Beige	1		
1	AWR6080SUC	RH-leather-Bahama Beige	1	Note (5)	
1	AWR6079SUC	LH-leather-Bahama Beige	1		
1	HAD128480SUC	RH-leather-Bahama Beige	1	Note (3)	
1	HAD128490SUC	LH-leather-Bahama Beige	1		
		LEATHER HEATED SEATS ELECTRIC SEAT-ELECTRIC FRONT			
1	AWR3884LOY	RH-leather-Dark Granite	1	Note (5)	
1	AWR3883LOY	LH-leather-Dark Granite	1		
1	HAD128520LOY	RH-leather-Dark Granite	1	Note (6); Note (7)	
1	HAD128530LOY	LH-leather-Dark Granite	1		
1	AWR3884SUC	RH-leather-Bahama Beige	1	Note (5)	
1	AWR3883SUC	LH-leather-Bahama Beige	1		
1	HAD128520SUC	RH-leather-Bahama Beige	1	Note (6); Note (7)	
1	HAD128530SUC	LH-leather-Bahama Beige	1		

CARXEE1A

CHANGE POINTS:
(1) To (V) TA 530158; To (V) TA 700125
(2) From (V) TA 530159; From (V) TA 700126
(3) From (V) VA 715218; From (V) VA 547236
(4) To (V) TA 703236; To (V) TA 534103
(5) From (V) VA 703237 To (V) VA 715217; From (V) VA 534104 To (V) VA 547235
(6) From (V) VA 715218
(7) From (V) VA 547236

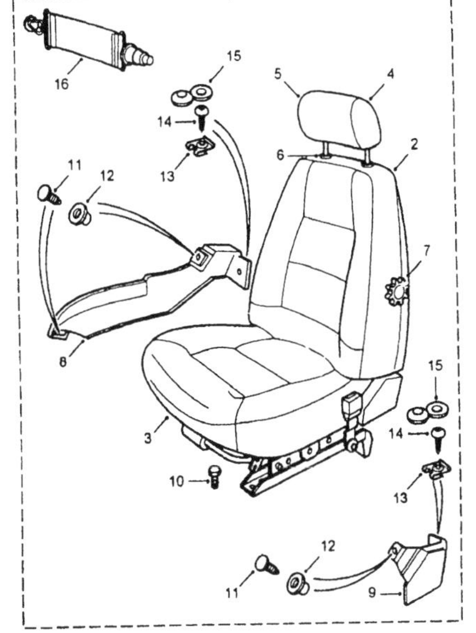

CARXEE1A

Illus	Part Number	Description	Quantity	Change Point	Remarks
		FROM 98MY ONLY			
1	HAD129480SMK	Seat-electric front-RH-leather-Light Stone Beige	1	Note (1)	Heated seats
1	HAD129490SMK	Seat-electric front-LH-leather-Light Stone Beige	1	Note (1)	
		COVER ASSEMBLY-FRONT SEAT SQUAB			
2	AWR2143LOY	leather-Dark Granite	2	Note (2)	
2	AWR6082LOY	leather-Dark Granite	2	Note (3)	
2	HBA105370LOY	leather-Dark Granite	2	Note (4)	
2	AWR2143SUC	leather-Bahama Beige	2	Note (5)	
2	AWR6082SUC	leather-Bahama Beige	2	Note (3)	
2	HBA105370SUC	leather-Bahama Beige	2	Note (4)	
		COVER ASSEMBLY-FRONT SEAT CUSHION			
3	AWR2142LOY	leather-Dark Granite	2	Note (2)	
3	AWR6081LOY	leather-Dark Granite	2	Note (6)	
3	HCA105330LOY	leather-Dark Granite	2	Note (4)	
3	AWR2142SUC	leather-Bahama Beige	2	Note (5)	
3	AWR6081SUC	leather-Bahama Beige	2	Note (3)	
3	HCA105330SUC	leather-Bahama Beige	2	Note (4)	
		HEADRESTRAINT ASSEMBLY			
4	AWR2144LOY	leather-Dark Granite	2	Note (2)	
4	AWR6083LOY	leather-Dark Granite	2	Note (6)	
4	HAH103190LOY	leather-Dark Granite	2	Note (4)	
4	AWR2144SUC	leather-Bahama Beige	2	Note (5)	
4	AWR6083SUC	leather-Bahama Beige	2	Note (3)	
4	HAH103190SUC	leather-Bahama Beige	2	Note (4)	

CHANGE POINTS:
(1) From (V) WA 746542
(2) To (V) TA 530158; To (V) TA 700125
(3) From (V) VA 703237 To (V) VA 715217; From (V) VA 534104 To (V) VA 547235
(4) From (V) VA 715218; From (V) VA 547236
(5) To (V) TA 703236; To (V) TA 534103
(6) From (V) TA 530159; From (V) TA 700126

R 23

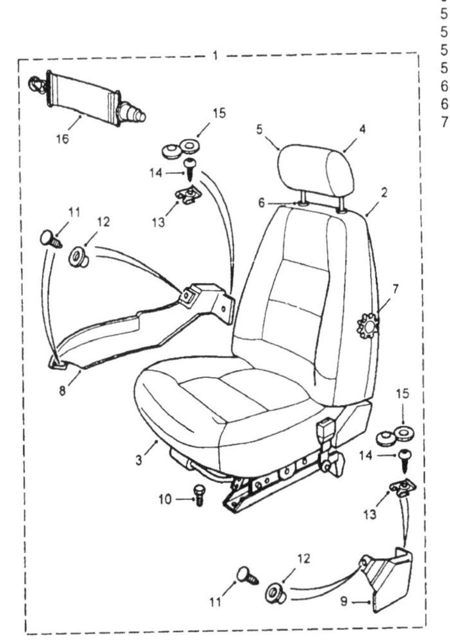

CARXEE1A

Illus	Part Number	Description	Quantity	Change Point	Remarks
		COVER ASSEMBLY-SEAT-FRONT-HEADRESTRAINT			
5	AWR2145LOY	leather-Dark Granite	2	Note (1)	
5	AWR6084LOY	leather-Dark Granite	2	Note (2)	
5	HDA102360LOY	leather-Dark Granite	2	Note (3)	
5	AWR2145SUC	leather-Bahama Beige	2	Note (4)	
5	AWR6084SUC	leather-Bahama Beige	2	Note (5)	
5	HDA102360SUC	leather-Bahama Beige	2	Note (3)	
6	BTR3636LOY	Escutcheon-headrest-Dark Granite	4		
6	BTR3636SUA	Escutcheon-headrest-Sorrel Brown	4		
7	AWR2679LNF	Handwheel-front seat squab manual lumbar support mechanism-Ash Grey	2		

CHANGE POINTS:
(1) To (V) TA 530158; To (V) TA 700125
(2) From (V) TA 530159; From (V) TA 700126
(3) From (V) VA 715218; From (V) VA 547236
(4) To (V) TA 703236; To (V) TA 534103
(5) From (V) VA 703237 To (V) VA 715217; From (V) VA 534104 To (V) VA 547235

R 24

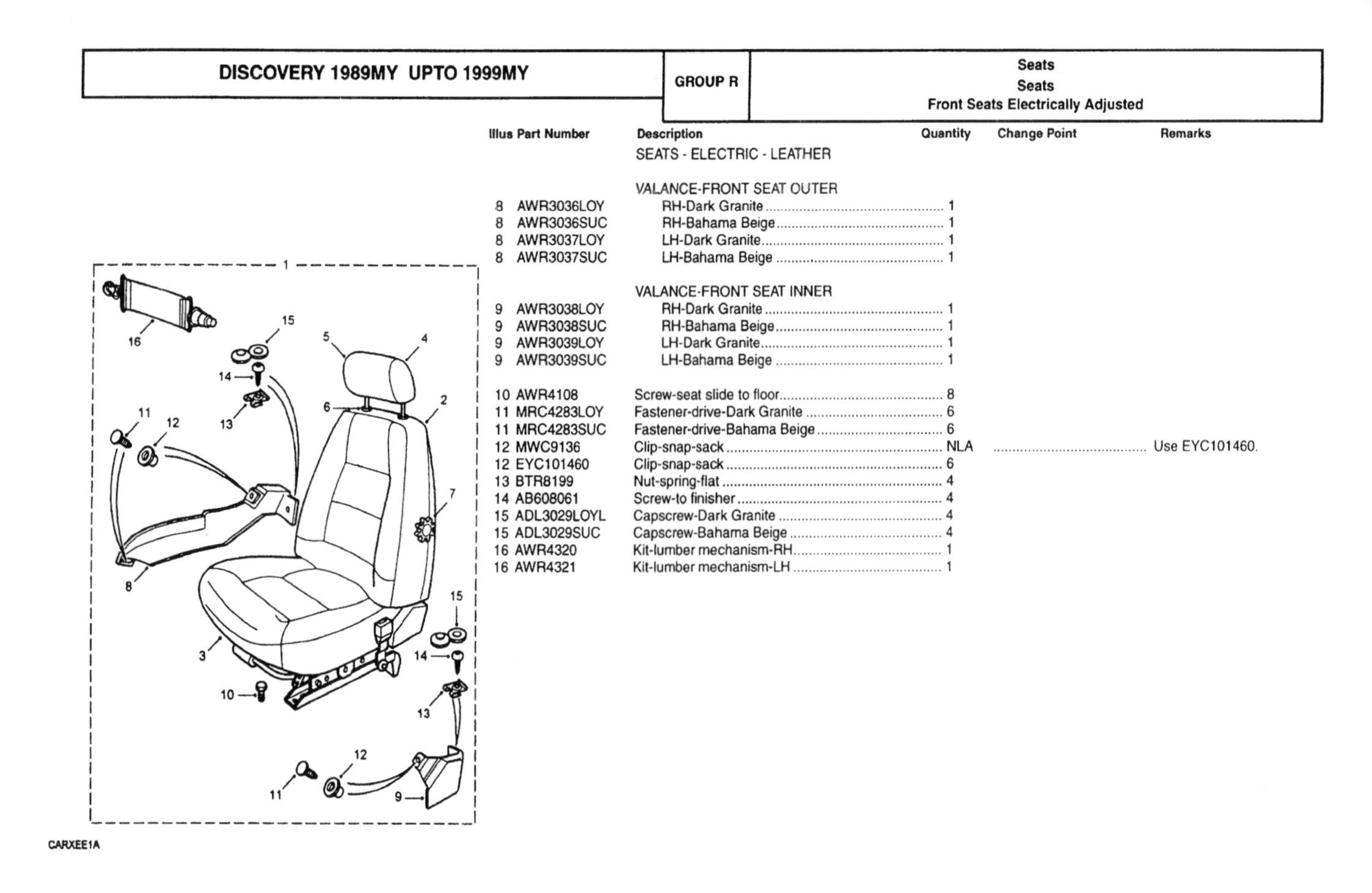

CARXEE1A

Illus	Part Number	Description	Quantity	Change Point	Remarks
		SEATS - ELECTRIC - LEATHER			
		VALANCE-FRONT SEAT OUTER			
8	AWR3036LOY	RH-Dark Granite	1		
8	AWR3036SUC	RH-Bahama Beige	1		
8	AWR3037LOY	LH-Dark Granite	1		
8	AWR3037SUC	LH-Bahama Beige	1		
		VALANCE-FRONT SEAT INNER			
9	AWR3038LOY	RH-Dark Granite	1		
9	AWR3038SUC	RH-Bahama Beige	1		
9	AWR3039LOY	LH-Dark Granite	1		
9	AWR3039SUC	LH-Bahama Beige	1		
10	AWR4108	Screw-seat slide to floor	8		
11	MRC4283LOY	Fastener-drive-Dark Granite	6		
11	MRC4283SUC	Fastener-drive-Bahama Beige	6		
12	MWC9136	Clip-snap-sack	NLA		Use EYC101460.
12	EYC101460	Clip-snap-sack	6		
13	BTR8199	Nut-spring-flat	4		
14	AB608061	Screw-to finisher	4		
15	ADL3029LOYL	Capscrew-Dark Granite	4		
15	ADL3029SUC	Capscrew-Bahama Beige	4		
16	AWR4320	Kit-lumber mechanism-RH	1		
16	AWR4321	Kit-lumber mechanism-LH	1		

R 25

DISCOVERY 1989MY UPTO 1999MY | GROUP R | Seats
Seats
Front Seats Electrically Adjusted - Frames

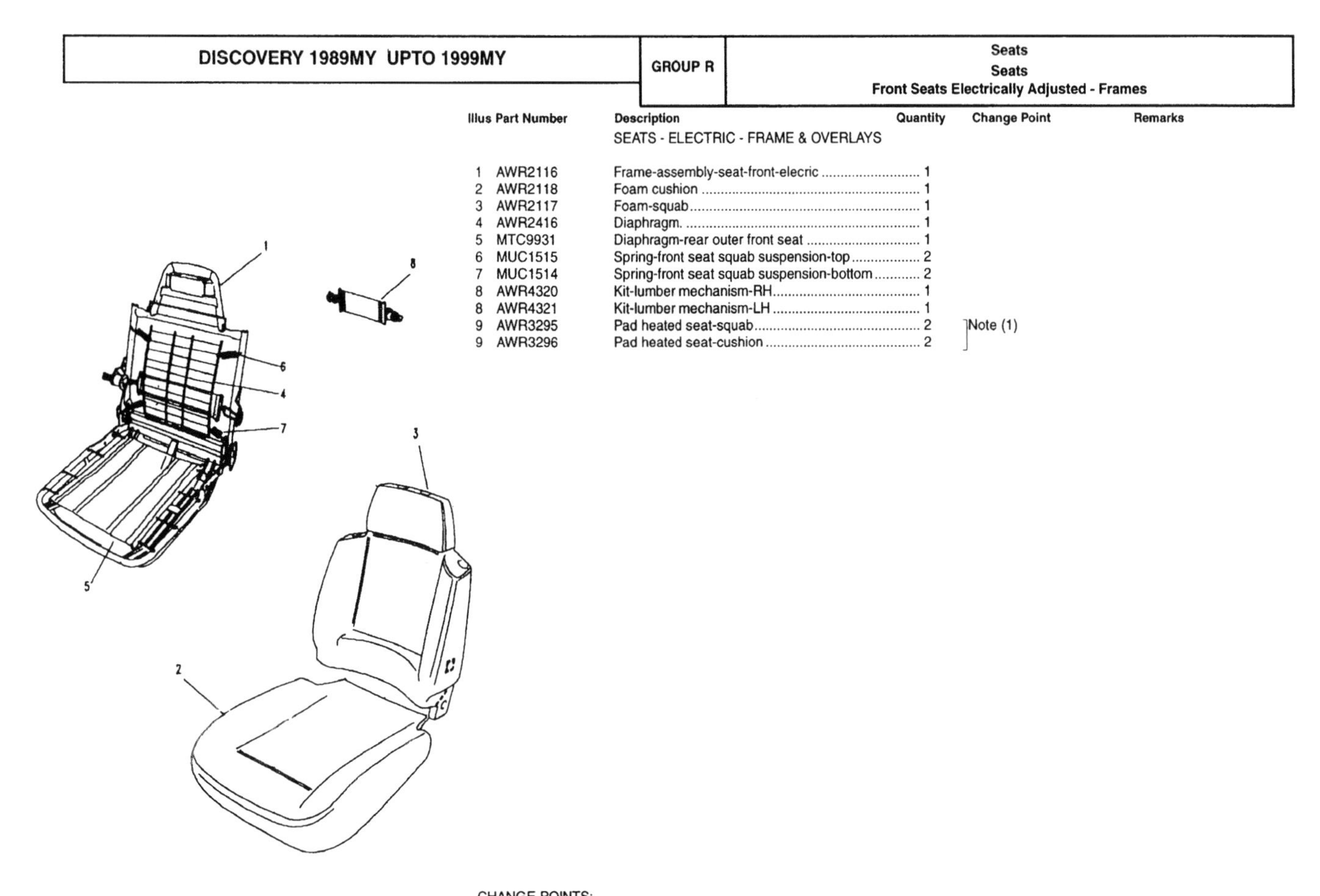

Illus	Part Number	Description	Quantity	Change Point	Remarks
		SEATS - ELECTRIC - FRAME & OVERLAYS			
1	AWR2116	Frame-assembly-seat-front-elecric	1		
2	AWR2118	Foam cushion	1		
3	AWR2117	Foam-squab	1		
4	AWR2416	Diaphragm.	1		
5	MTC9931	Diaphragm-rear outer front seat	1		
6	MUC1515	Spring-front seat squab suspension-top	2		
7	MUC1514	Spring-front seat squab suspension-bottom	2		
8	AWR4320	Kit-lumber mechanism-RH	1		
8	AWR4321	Kit-lumber mechanism-LH	1		
9	AWR3295	Pad heated seat-squab	2	Note (1)	
9	AWR3296	Pad heated seat-cushion	2	Note (1)	

CHANGE POINTS:
(1) From (V) VA 703237; From (V) VA 534104

R 26

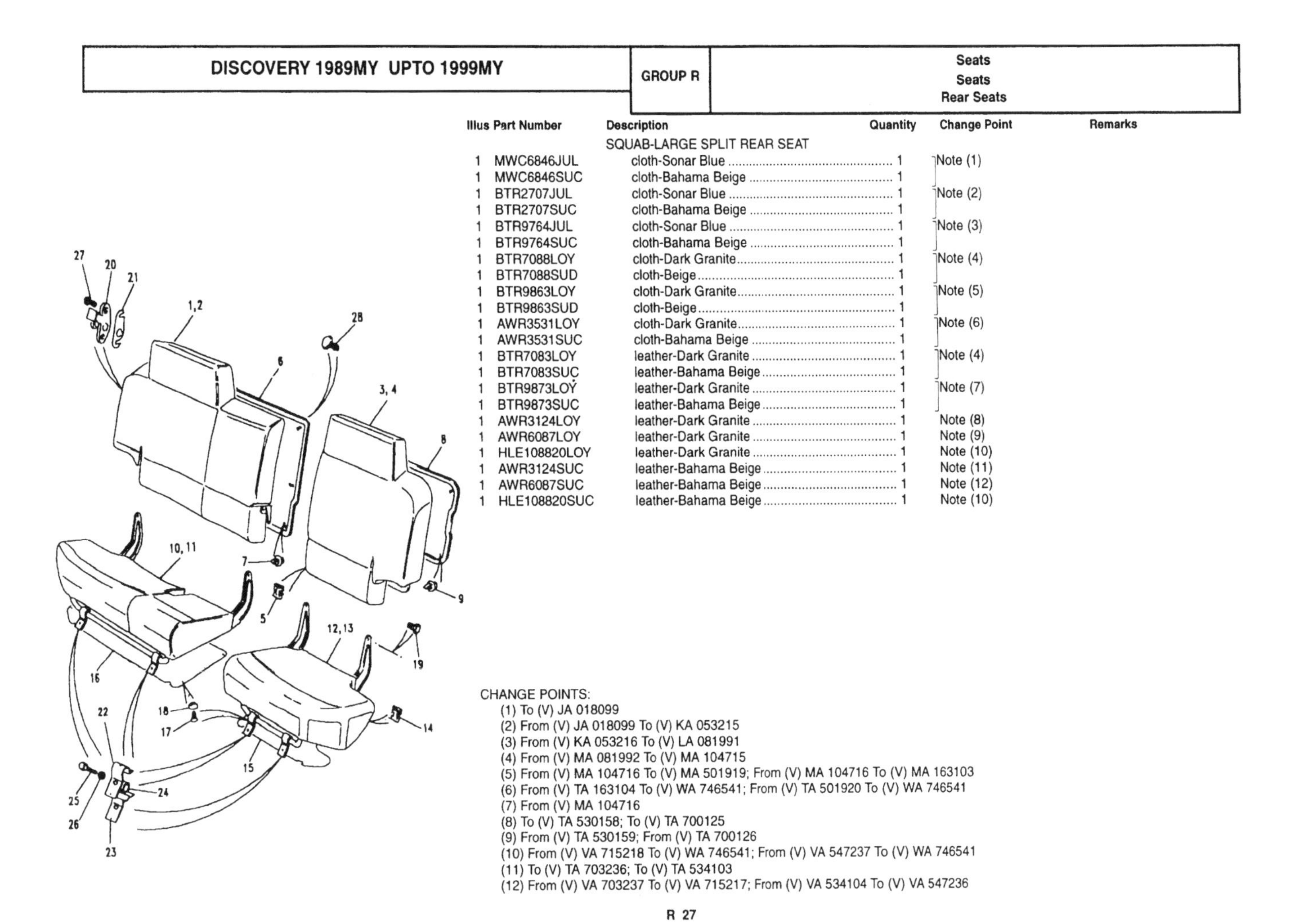

Illus	Part Number	Description	Quantity	Change Point	Remarks
		SQUAB-LARGE SPLIT REAR SEAT			
1	MWC6846JUL	cloth-Sonar Blue	1	Note (1)	
1	MWC6846SUC	cloth-Bahama Beige	1		
1	BTR2707JUL	cloth-Sonar Blue	1	Note (2)	
1	BTR2707SUC	cloth-Bahama Beige	1		
1	BTR9764JUL	cloth-Sonar Blue	1	Note (3)	
1	BTR9764SUC	cloth-Bahama Beige	1		
1	BTR7088LOY	cloth-Dark Granite	1	Note (4)	
1	BTR7088SUD	cloth-Beige	1		
1	BTR9863LOY	cloth-Dark Granite	1	Note (5)	
1	BTR9863SUD	cloth-Beige	1		
1	AWR3531LOY	cloth-Dark Granite	1	Note (6)	
1	AWR3531SUC	cloth-Bahama Beige	1		
1	BTR7083LOY	leather-Dark Granite	1	Note (4)	
1	BTR7083SUC	leather-Bahama Beige	1		
1	BTR9873LOY	leather-Dark Granite	1	Note (7)	
1	BTR9873SUC	leather-Bahama Beige	1		
1	AWR3124LOY	leather-Dark Granite	1	Note (8)	
1	AWR6087LOY	leather-Dark Granite	1	Note (9)	
1	HLE108820LOY	leather-Dark Granite	1	Note (10)	
1	AWR3124SUC	leather-Bahama Beige	1	Note (11)	
1	AWR6087SUC	leather-Bahama Beige	1	Note (12)	
1	HLE108820SUC	leather-Bahama Beige	1	Note (10)	

CHANGE POINTS:
(1) To (V) JA 018099
(2) From (V) JA 018099 To (V) KA 053215
(3) From (V) KA 053216 To (V) LA 081991
(4) From (V) MA 081992 To (V) MA 104715
(5) From (V) MA 104716 To (V) MA 501919; From (V) MA 104716 To (V) MA 163103
(6) From (V) TA 163104 To (V) WA 746541; From (V) TA 501920 To (V) WA 746541
(7) From (V) MA 104716
(8) To (V) TA 530158; To (V) TA 700125
(9) From (V) TA 530159; From (V) TA 700126
(10) From (V) VA 715218 To (V) WA 746541; From (V) VA 547237 To (V) WA 746541
(11) To (V) TA 703236; To (V) TA 534103
(12) From (V) VA 703237 To (V) VA 715217; From (V) VA 534104 To (V) VA 547236

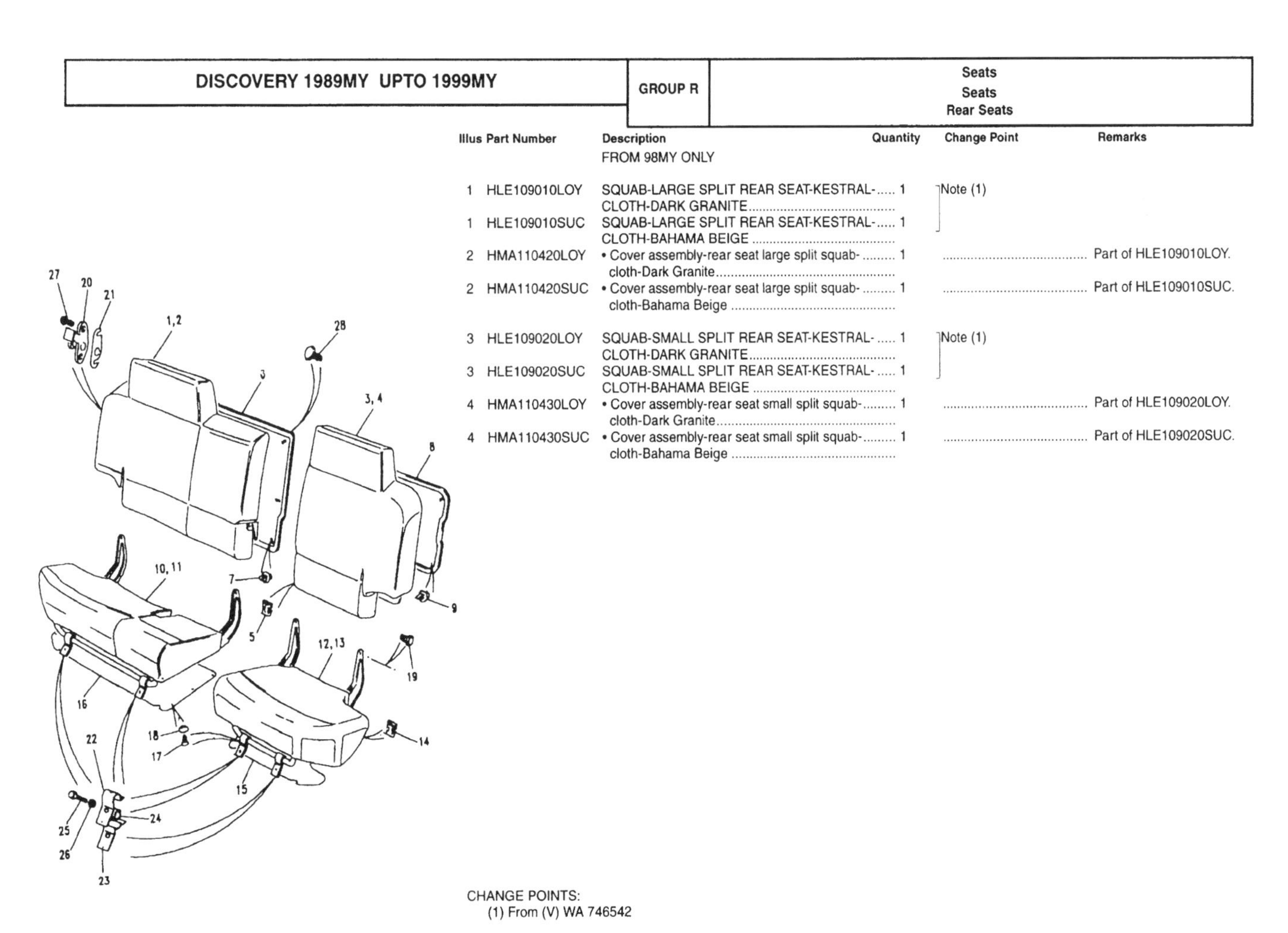

Illus	Part Number	Description	Quantity	Change Point	Remarks
		FROM 98MY ONLY			
1	HLE109010LOY	SQUAB-LARGE SPLIT REAR SEAT-KESTRAL-CLOTH-DARK GRANITE	1	Note (1)	
1	HLE109010SUC	SQUAB-LARGE SPLIT REAR SEAT-KESTRAL-CLOTH-BAHAMA BEIGE	1		
2	HMA110420LOY	• Cover assembly-rear seat large split squab-cloth-Dark Granite	1		Part of HLE109010LOY.
2	HMA110420SUC	• Cover assembly-rear seat large split squab-cloth-Bahama Beige	1		Part of HLE109010SUC.
3	HLE109020LOY	SQUAB-SMALL SPLIT REAR SEAT-KESTRAL-CLOTH-DARK GRANITE	1	Note (1)	
3	HLE109020SUC	SQUAB-SMALL SPLIT REAR SEAT-KESTRAL-CLOTH-BAHAMA BEIGE	1		
4	HMA110430LOY	• Cover assembly-rear seat small split squab-cloth-Dark Granite	1		Part of HLE109020LOY.
4	HMA110430SUC	• Cover assembly-rear seat small split squab-cloth-Bahama Beige	1		Part of HLE109020SUC.

CHANGE POINTS:
(1) From (V) WA 746542

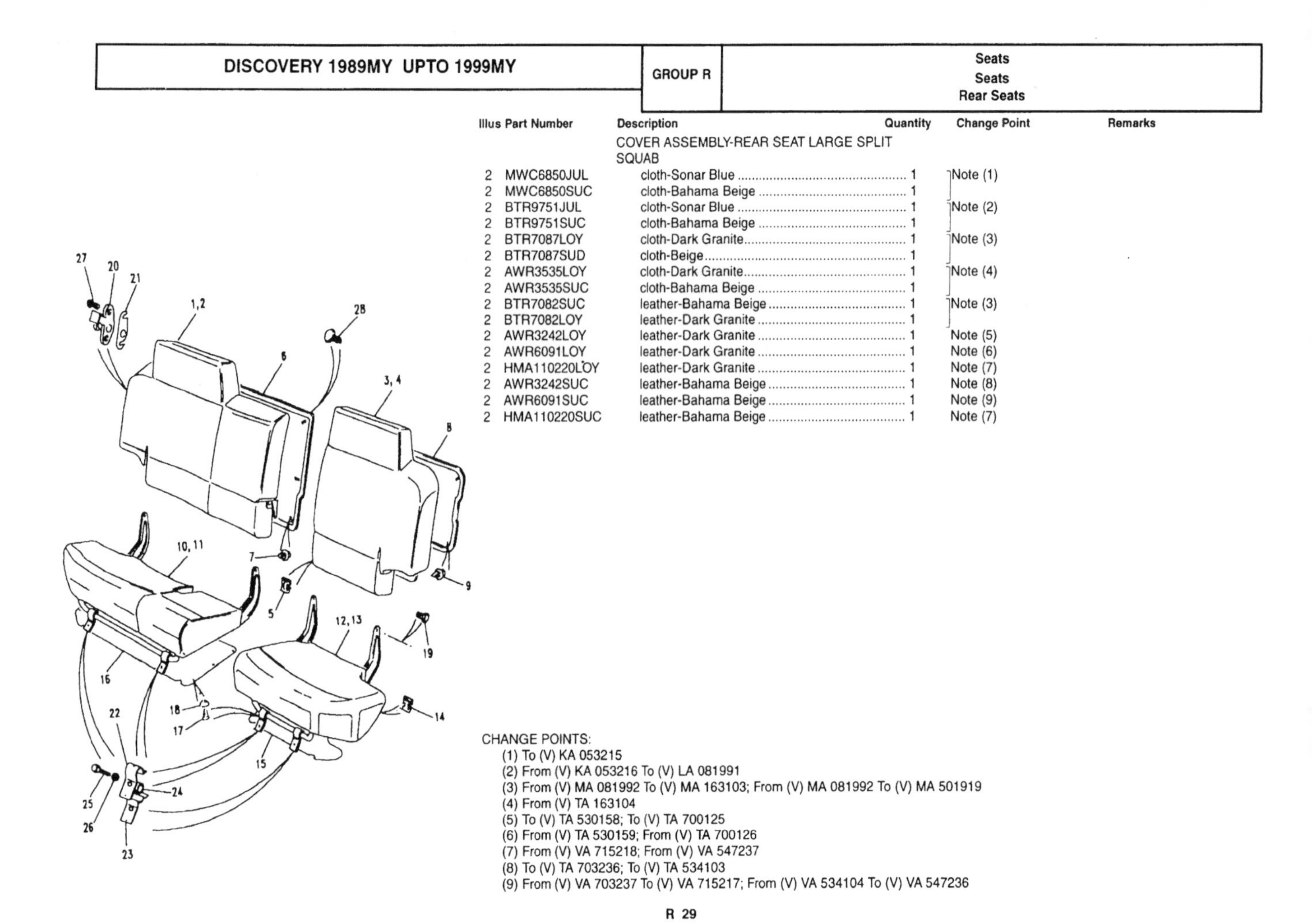

Illus	Part Number	Description	Quantity	Change Point	Remarks
		COVER ASSEMBLY-REAR SEAT LARGE SPLIT SQUAB			
2	MWC6850JUL	cloth-Sonar Blue	1	Note (1)	
2	MWC6850SUC	cloth-Bahama Beige	1		
2	BTR9751JUL	cloth-Sonar Blue	1	Note (2)	
2	BTR9751SUC	cloth-Bahama Beige	1		
2	BTR7087LOY	cloth-Dark Granite	1	Note (3)	
2	BTR7087SUD	cloth-Beige	1		
2	AWR3535LOY	cloth-Dark Granite	1	Note (4)	
2	AWR3535SUC	cloth-Bahama Beige	1		
2	BTR7082SUC	leather-Bahama Beige	1	Note (3)	
2	BTR7082LOY	leather-Dark Granite	1		
2	AWR3242LOY	leather-Dark Granite	1	Note (5)	
2	AWR6091LOY	leather-Dark Granite	1	Note (6)	
2	HMA110220LOY	leather-Dark Granite	1	Note (7)	
2	AWR3242SUC	leather-Bahama Beige	1	Note (8)	
2	AWR6091SUC	leather-Bahama Beige	1	Note (9)	
2	HMA110220SUC	leather-Bahama Beige	1	Note (7)	

CHANGE POINTS:
(1) To (V) KA 053215
(2) From (V) KA 053216 To (V) LA 081991
(3) From (V) MA 081992 To (V) MA 163103; From (V) MA 081992 To (V) MA 501919
(4) From (V) TA 163104
(5) To (V) TA 530158; To (V) TA 700125
(6) From (V) TA 530159; From (V) TA 700126
(7) From (V) VA 715218; From (V) VA 547237
(8) To (V) TA 703236; To (V) TA 534103
(9) From (V) VA 703237 To (V) VA 715217; From (V) VA 534104 To (V) VA 547236

R 29

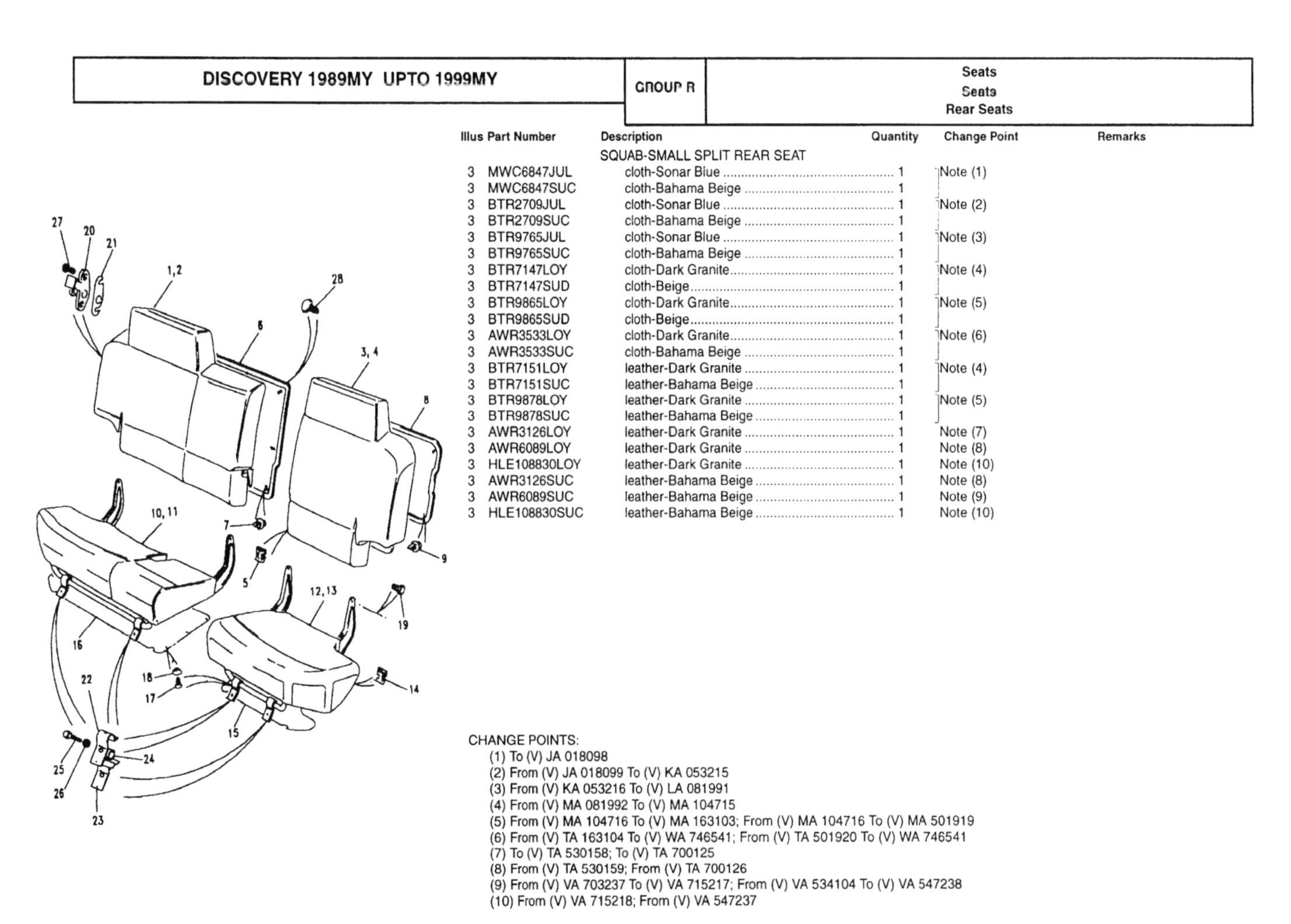

Illus	Part Number	Description	Quantity	Change Point	Remarks
		SQUAB-SMALL SPLIT REAR SEAT			
3	MWC6847JUL	cloth-Sonar Blue	1	Note (1)	
3	MWC6847SUC	cloth-Bahama Beige	1		
3	BTR2709JUL	cloth-Sonar Blue	1	Note (2)	
3	BTR2709SUC	cloth-Bahama Beige	1		
3	BTR9765JUL	cloth-Sonar Blue	1	Note (3)	
3	BTR9765SUC	cloth-Bahama Beige	1		
3	BTR7147LOY	cloth-Dark Granite	1	Note (4)	
3	BTR7147SUD	cloth-Beige	1		
3	BTR9865LOY	cloth-Dark Granite	1	Note (5)	
3	BTR9865SUD	cloth-Beige	1		
3	AWR3533LOY	cloth-Dark Granite	1	Note (6)	
3	AWR3533SUC	cloth-Bahama Beige	1		
3	BTR7151LOY	leather-Dark Granite	1	Note (4)	
3	BTR7151SUC	leather-Bahama Beige	1		
3	BTR9878LOY	leather-Dark Granite	1	Note (5)	
3	BTR9878SUC	leather-Bahama Beige	1		
3	AWR3126LOY	leather-Dark Granite	1	Note (7)	
3	AWR6089LOY	leather-Dark Granite	1	Note (8)	
3	HLE108830LOY	leather-Dark Granite	1	Note (10)	
3	AWR3126SUC	leather-Bahama Beige	1	Note (8)	
3	AWR6089SUC	leather-Bahama Beige	1	Note (9)	
3	HLE108830SUC	leather-Bahama Beige	1	Note (10)	

CHANGE POINTS:
(1) To (V) JA 018098
(2) From (V) JA 018099 To (V) KA 053215
(3) From (V) KA 053216 To (V) LA 081991
(4) From (V) MA 081992 To (V) MA 104715
(5) From (V) MA 104716 To (V) MA 163103; From (V) MA 104716 To (V) MA 501919
(6) From (V) TA 163104 To (V) WA 746541; From (V) TA 501920 To (V) WA 746541
(7) To (V) TA 530158; To (V) TA 700125
(8) From (V) TA 530159; From (V) TA 700126
(9) From (V) VA 703237 To (V) VA 715217; From (V) VA 534104 To (V) VA 547238
(10) From (V) VA 715218; From (V) VA 547237

R 30

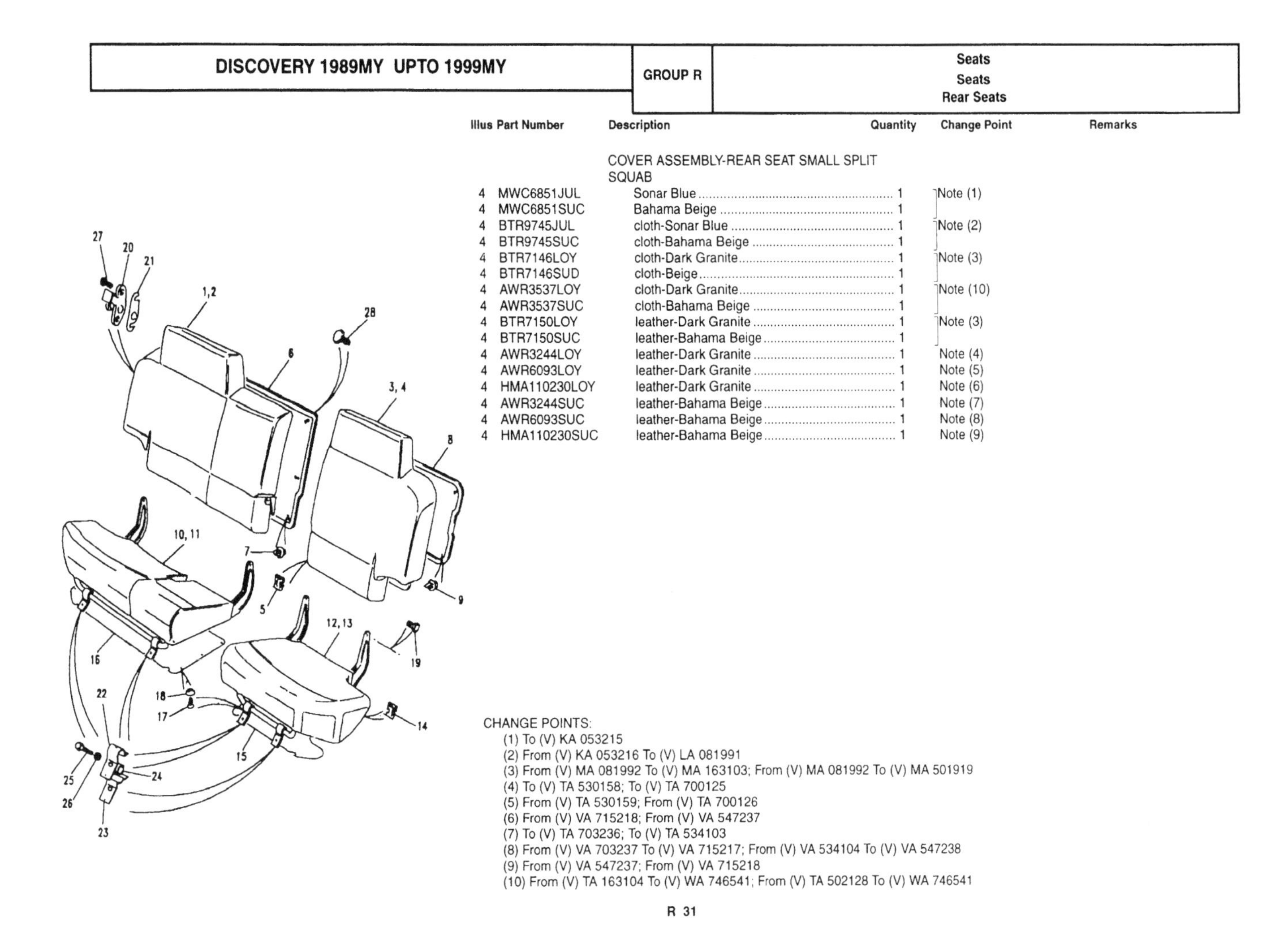

Illus	Part Number	Description	Quantity	Change Point	Remarks
		COVER ASSEMBLY-REAR SEAT SMALL SPLIT SQUAB			
4	MWC6851JUL	Sonar Blue	1	Note (1)	
4	MWC6851SUC	Bahama Beige	1		
4	BTR9745JUL	cloth-Sonar Blue	1	Note (2)	
4	BTR9745SUC	cloth-Bahama Beige	1		
4	BTR7146LOY	cloth-Dark Granite	1	Note (3)	
4	BTR7146SUD	cloth-Beige	1		
4	AWR3537LOY	cloth-Dark Granite	1	Note (10)	
4	AWR3537SUC	cloth-Bahama Beige	1		
4	BTR7150LOY	leather-Dark Granite	1	Note (3)	
4	BTR7150SUC	leather-Bahama Beige	1		
4	AWR3244LOY	leather-Dark Granite	1	Note (4)	
4	AWR6093LOY	leather-Dark Granite	1	Note (5)	
4	HMA110230LOY	leather-Dark Granite	1	Note (6)	
4	AWR3244SUC	leather-Bahama Beige	1	Note (7)	
4	AWR6093SUC	leather-Bahama Beige	1	Note (8)	
4	HMA110230SUC	leather-Bahama Beige	1	Note (9)	

CHANGE POINTS:
(1) To (V) KA 053215
(2) From (V) KA 053216 To (V) LA 081991
(3) From (V) MA 081992 To (V) MA 163103; From (V) MA 081992 To (V) MA 501919
(4) To (V) TA 530158; To (V) TA 700125
(5) From (V) TA 530159; From (V) TA 700126
(6) From (V) VA 715218; From (V) VA 547237
(7) To (V) TA 703236; To (V) TA 534103
(8) From (V) VA 703237 To (V) VA 715217; From (V) VA 534104 To (V) VA 547238
(9) From (V) VA 547237; From (V) VA 715218
(10) From (V) TA 163104 To (V) WA 746541; From (V) TA 502128 To (V) WA 746541

R 31

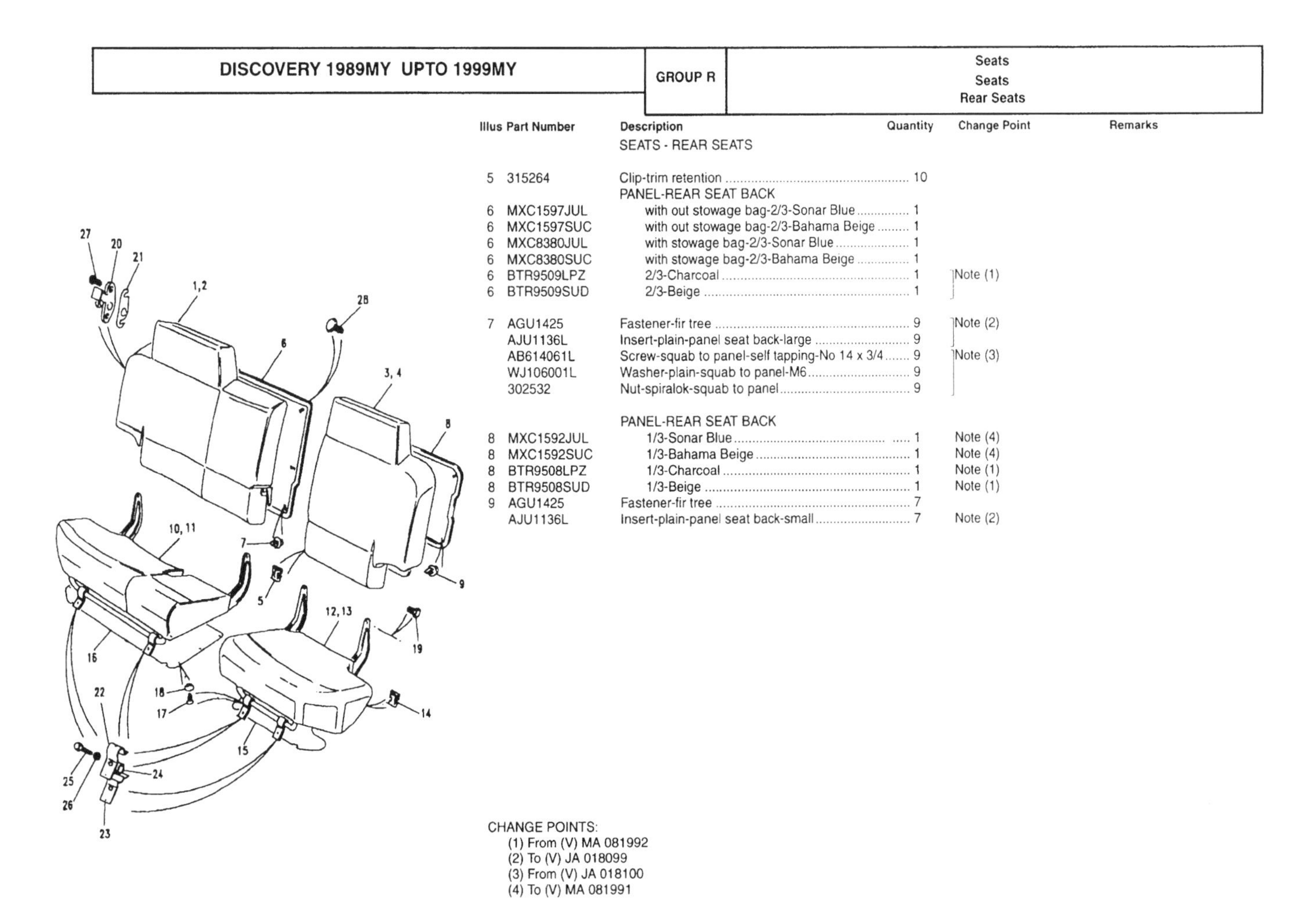

Illus	Part Number	Description	Quantity	Change Point	Remarks
		SEATS - REAR SEATS			
5	315264	Clip-trim retention	10		
		PANEL-REAR SEAT BACK			
6	MXC1597JUL	with out stowage bag-2/3-Sonar Blue	1		
6	MXC1597SUC	with out stowage bag-2/3-Bahama Beige	1		
6	MXC8380JUL	with stowage bag-2/3-Sonar Blue	1		
6	MXC8380SUC	with stowage bag-2/3-Bahama Beige	1		
6	BTR9509LPZ	2/3-Charcoal	1	Note (1)	
6	BTR9509SUD	2/3-Beige	1		
7	AGU1425	Fastener-fir tree	9	Note (2)	
	AJU1136L	Insert-plain-panel seat back-large	9		
	AB614061L	Screw-squab to panel-self tapping-No 14 x 3/4	9	Note (3)	
	WJ106001L	Washer-plain-squab to panel-M6	9		
	302532	Nut-spiralok-squab to panel	9		
		PANEL-REAR SEAT BACK			
8	MXC1592JUL	1/3-Sonar Blue	1	Note (4)	
8	MXC1592SUC	1/3-Bahama Beige	1	Note (4)	
8	BTR9508LPZ	1/3-Charcoal	1	Note (1)	
8	BTR9508SUD	1/3-Beige	1	Note (1)	
9	AGU1425	Fastener-fir tree	7		
	AJU1136L	Insert-plain-panel seat back-small	7	Note (2)	

CHANGE POINTS:
(1) From (V) MA 081992
(2) To (V) JA 018099
(3) From (V) JA 018100
(4) To (V) MA 081991

R 32

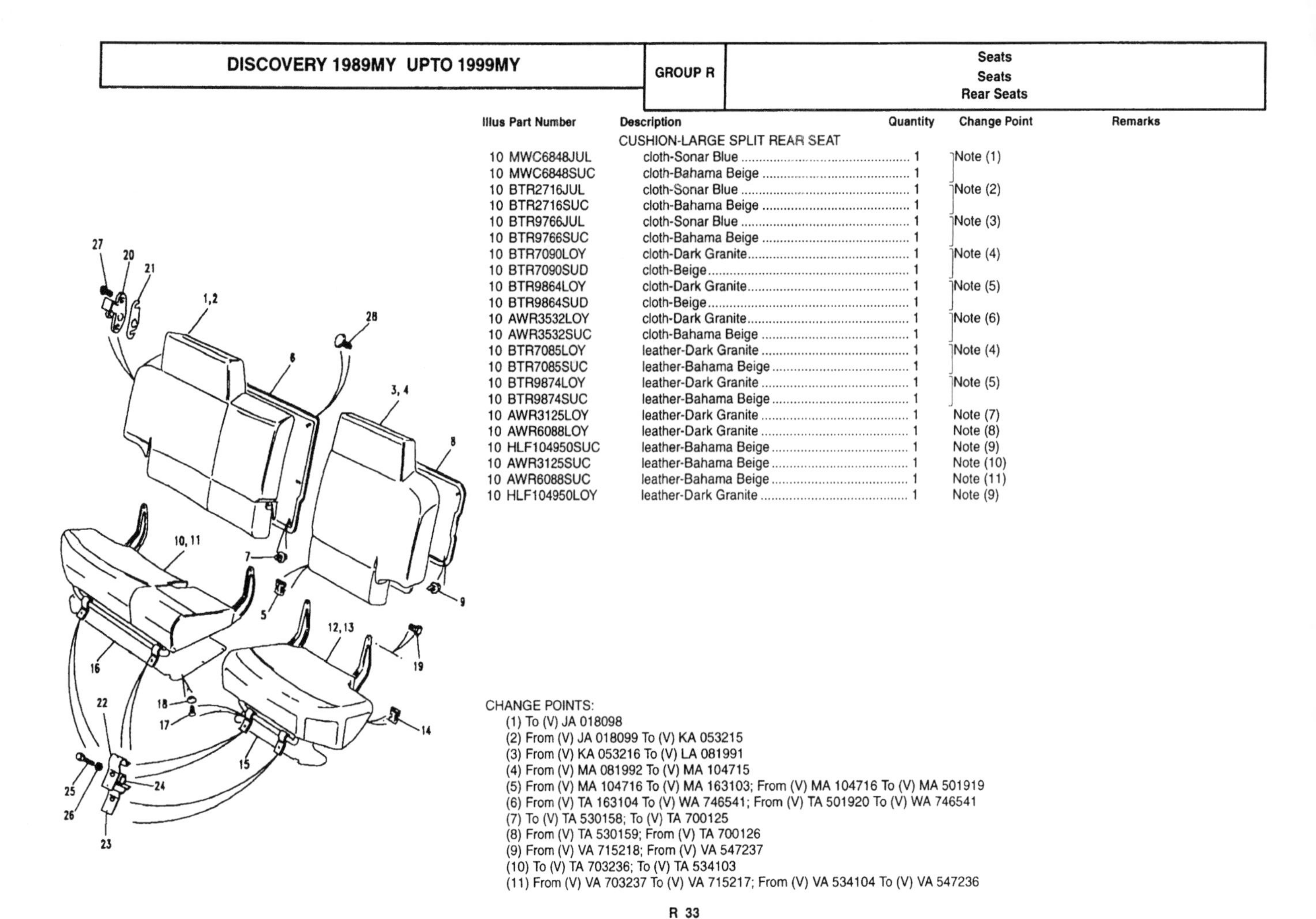

Illus	Part Number	Description	Quantity	Change Point	Remarks
		CUSHION-LARGE SPLIT REAR SEAT			
10	MWC6848JUL	cloth-Sonar Blue	1	Note (1)	
10	MWC6848SUC	cloth-Bahama Beige	1		
10	BTR2716JUL	cloth-Sonar Blue	1	Note (2)	
10	BTR2716SUC	cloth-Bahama Beige	1		
10	BTR9766JUL	cloth-Sonar Blue	1	Note (3)	
10	BTR9766SUC	cloth-Bahama Beige	1		
10	BTR7090LOY	cloth-Dark Granite	1	Note (4)	
10	BTR7090SUD	cloth-Beige	1		
10	BTR9864LOY	cloth-Dark Granite	1	Note (5)	
10	BTR9864SUD	cloth-Beige	1		
10	AWR3532LOY	cloth-Dark Granite	1	Note (6)	
10	AWR3532SUC	cloth-Bahama Beige	1		
10	BTR7085LOY	leather-Dark Granite	1	Note (4)	
10	BTR7085SUC	leather-Bahama Beige	1		
10	BTR9874LOY	leather-Dark Granite	1	Note (5)	
10	BTR9874SUC	leather-Bahama Beige	1		
10	AWR3125LOY	leather-Dark Granite	1	Note (7)	
10	AWR6088LOY	leather-Dark Granite	1	Note (8)	
10	HLF104950SUC	leather-Bahama Beige	1	Note (9)	
10	AWR3125SUC	leather-Bahama Beige	1	Note (10)	
10	AWR6088SUC	leather-Bahama Beige	1	Note (11)	
10	HLF104950LOY	leather-Dark Granite	1	Note (9)	

CHANGE POINTS:
(1) To (V) JA 018098
(2) From (V) JA 018099 To (V) KA 053215
(3) From (V) KA 053216 To (V) LA 081991
(4) From (V) MA 081992 To (V) MA 104715
(5) From (V) MA 104716 To (V) MA 163103; From (V) MA 104716 To (V) MA 501919
(6) From (V) TA 163104 To (V) WA 746541; From (V) TA 501920 To (V) WA 746541
(7) To (V) TA 530158; To (V) TA 700125
(8) From (V) TA 530159; From (V) TA 700126
(9) From (V) VA 715218; From (V) VA 547237
(10) To (V) TA 703236; To (V) TA 534103
(11) From (V) VA 703237 To (V) VA 715217; From (V) VA 534104 To (V) VA 547236

R 33

Illus	Part Number	Description	Quantity	Change Point	Remarks
10	HLF104980LOY	CUSHION-LARGE SPLIT REAR SEAT-KESTRAL-CLOTH-DARK GRANITE	1	Note (1)	
10	HLF104980SUC	CUSHION-LARGE SPLIT REAR SEAT-KESTRAL-CLOTH-BAHAMA BEIGE	1		
11	HPA106170LOY	• Cover assembly-rear seat large split cushion-cloth-Dark Granite	1		Part of HLF104980LOY.
11	HPA106170SUC	• Cover assembly-rear seat large split cushion-cloth-Bahama Beige	1		Part of HLF104980SUC.
12	HLF104990LOY	CUSHION-SMALL SPLIT REAR SEAT-KESTRAL-CLOTH-DARK GRANITE	1	Note (1)	
12	HLF104990SUC	CUSHION-SMALL SPLIT REAR SEAT-KESTRAL-CLOTH-BAHAMA BEIGE	1		
13	HPA106180LOY	• Cover assembly-rear seat small split cushion-cloth-Dark Granite	1		Part of HLF104990LOY.
13	HPA106180SUC	• Cover assembly-rear seat small split cushion-cloth-Bahama Beige	1		Part of HLF104990SUC.

CHANGE POINTS:
(1) From (V) WA 746542

R 34

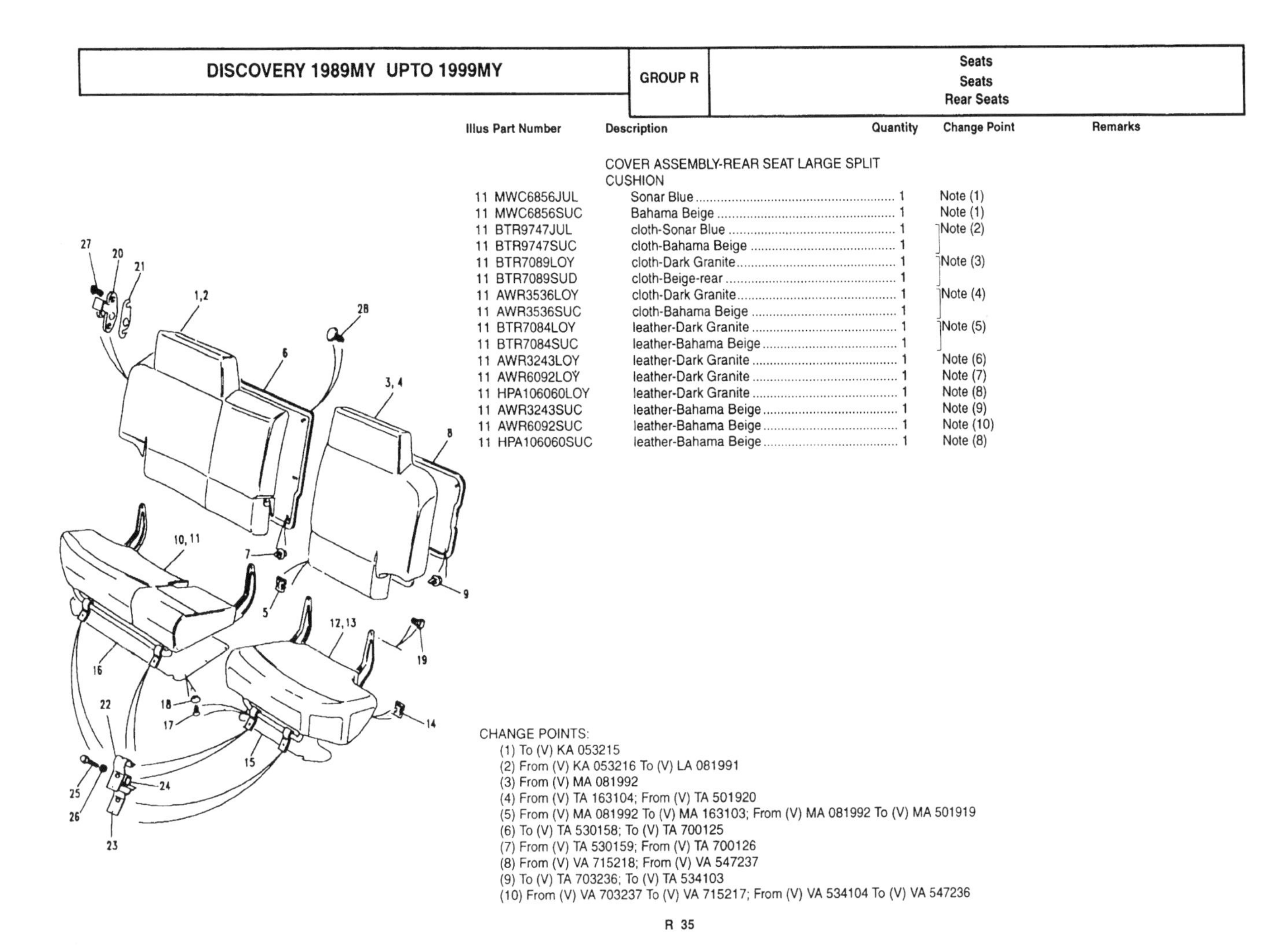

Illus	Part Number	Description	Quantity	Change Point	Remarks
		COVER ASSEMBLY-REAR SEAT LARGE SPLIT CUSHION			
11	MWC6856JUL	Sonar Blue	1	Note (1)	
11	MWC6856SUC	Bahama Beige	1	Note (1)	
11	BTR9747JUL	cloth-Sonar Blue	1	Note (2)	
11	BTR9747SUC	cloth-Bahama Beige	1		
11	BTR7089LOY	cloth-Dark Granite	1	Note (3)	
11	BTR7089SUD	cloth-Beige-rear	1		
11	AWR3536LOY	cloth-Dark Granite	1	Note (4)	
11	AWR3536SUC	cloth-Bahama Beige	1		
11	BTR7084LOY	leather-Dark Granite	1	Note (5)	
11	BTR7084SUC	leather-Bahama Beige	1		
11	AWR3243LOY	leather-Dark Granite	1	Note (6)	
11	AWR6092LOY	leather-Dark Granite	1	Note (7)	
11	HPA106060LOY	leather-Dark Granite	1	Note (8)	
11	AWR3243SUC	leather-Bahama Beige	1	Note (9)	
11	AWR6092SUC	leather-Bahama Beige	1	Note (10)	
11	HPA106060SUC	leather-Bahama Beige	1	Note (8)	

CHANGE POINTS:
(1) To (V) KA 053215
(2) From (V) KA 053216 To (V) LA 081991
(3) From (V) MA 081992
(4) From (V) TA 163104; From (V) TA 501920
(5) From (V) MA 081992 To (V) MA 163103; From (V) MA 081992 To (V) MA 501919
(6) To (V) TA 530158; To (V) TA 700125
(7) From (V) TA 530159; From (V) TA 700126
(8) From (V) VA 715218; From (V) VA 547237
(9) To (V) TA 703236; To (V) TA 534103
(10) From (V) VA 703237 To (V) VA 715217; From (V) VA 534104 To (V) VA 547236

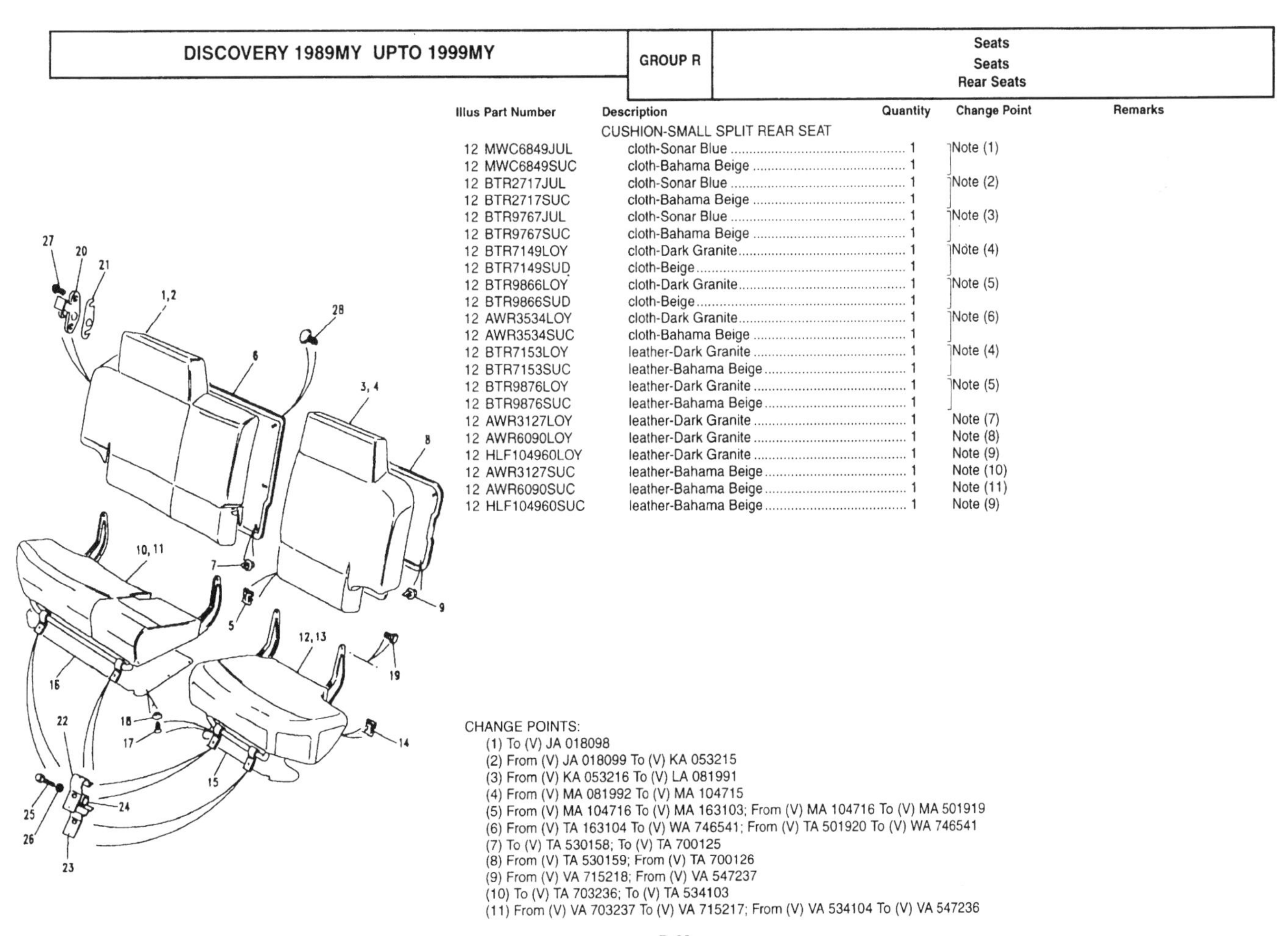

Illus	Part Number	Description	Quantity	Change Point	Remarks
		CUSHION-SMALL SPLIT REAR SEAT			
12	MWC6849JUL	cloth-Sonar Blue	1	Note (1)	
12	MWC6849SUC	cloth-Bahama Beige	1		
12	BTR2717JUL	cloth-Sonar Blue	1	Note (2)	
12	BTR2717SUC	cloth-Bahama Beige	1		
12	BTR9767JUL	cloth-Sonar Blue	1	Note (3)	
12	BTR9767SUC	cloth-Bahama Beige	1		
12	BTR7149LOY	cloth-Dark Granite	1	Note (4)	
12	BTR7149SUD	cloth-Beige	1		
12	BTR9866LOY	cloth-Dark Granite	1	Note (5)	
12	BTR9866SUD	cloth-Beige	1		
12	AWR3534LOY	cloth-Dark Granite	1	Note (6)	
12	AWR3534SUC	cloth-Bahama Beige	1		
12	BTR7153LOY	leather-Dark Granite	1	Note (4)	
12	BTR7153SUC	leather-Bahama Beige	1		
12	BTR9876LOY	leather-Dark Granite	1	Note (5)	
12	BTR9876SUC	leather-Bahama Beige	1		
12	AWR3127LOY	leather-Dark Granite	1	Note (7)	
12	AWR6090LOY	leather-Dark Granite	1	Note (8)	
12	HLF104960LOY	leather-Dark Granite	1	Note (9)	
12	AWR3127SUC	leather-Bahama Beige	1	Note (10)	
12	AWR6090SUC	leather-Bahama Beige	1	Note (11)	
12	HLF104960SUC	leather-Bahama Beige	1	Note (9)	

CHANGE POINTS:
(1) To (V) JA 018098
(2) From (V) JA 018099 To (V) KA 053215
(3) From (V) KA 053216 To (V) LA 081991
(4) From (V) MA 081992 To (V) MA 104715
(5) From (V) MA 104716 To (V) MA 163103; From (V) MA 104716 To (V) MA 501919
(6) From (V) TA 163104 To (V) WA 746541; From (V) TA 501920 To (V) WA 746541
(7) To (V) TA 530158; To (V) TA 700125
(8) From (V) TA 530159; From (V) TA 700126
(9) From (V) VA 715218; From (V) VA 547237
(10) To (V) TA 703236; To (V) TA 534103
(11) From (V) VA 703237 To (V) VA 715217; From (V) VA 534104 To (V) VA 547236

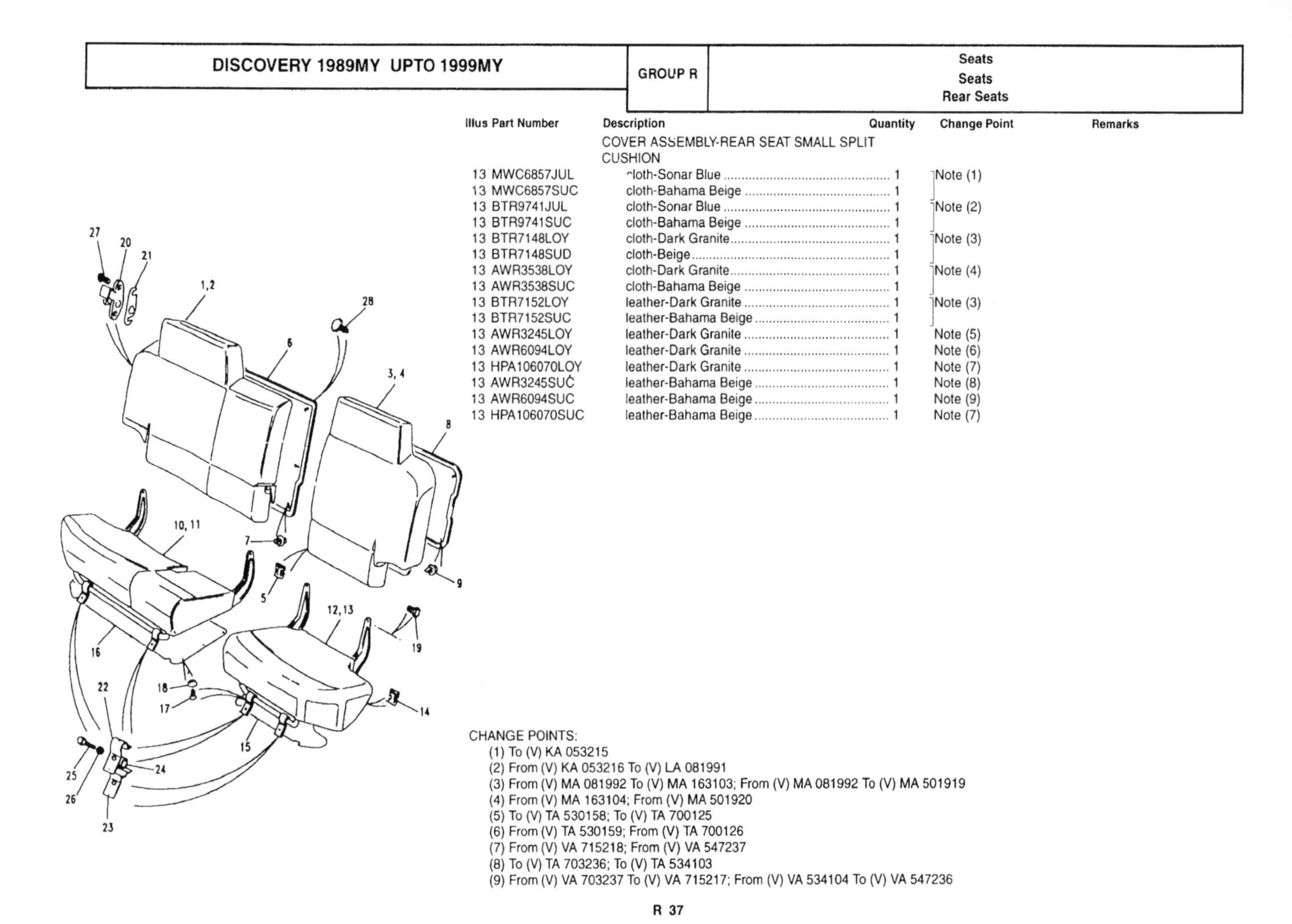

DISCOVERY 1989MY UPTO 1999MY

GROUP R

Seats
Seats
Rear Seats

Illus	Part Number	Description	Quantity	Change Point	Remarks
		COVER ASSEMBLY-REAR SEAT SMALL SPLIT CUSHION			
13	MWC6857JUL	cloth-Sonar Blue	1	Note (1)	
13	MWC6857SUC	cloth-Bahama Beige	1		
13	BTR9741JUL	cloth-Sonar Blue	1	Note (2)	
13	BTR9741SUC	cloth-Bahama Beige	1		
13	BTR7148LOY	cloth-Dark Granite	1	Note (3)	
13	BTR7148SUD	cloth-Beige	1		
13	AWR3538LOY	cloth-Dark Granite	1	Note (4)	
13	AWR3538SUC	cloth-Bahama Beige	1		
13	BTR7152LOY	leather-Dark Granite	1	Note (3)	
13	BTR7152SUC	leather-Bahama Beige	1		
13	AWR3245LOY	leather-Dark Granite	1	Note (5)	
13	AWR6094LOY	leather-Dark Granite	1	Note (6)	
13	HPA106070LOY	leather-Dark Granite	1	Note (7)	
13	AWR3245SUC	leather-Bahama Beige	1	Note (8)	
13	AWR6094SUC	leather-Bahama Beige	1	Note (9)	
13	HPA106070SUC	leather-Bahama Beige	1	Note (7)	

CHANGE POINTS:
(1) To (V) KA 053215
(2) From (V) KA 053216 To (V) LA 081991
(3) From (V) MA 081992 To (V) MA 163103; From (V) MA 081992 To (V) MA 501919
(4) From (V) MA 163104; From (V) MA 501920
(5) To (V) TA 530158; To (V) TA 700125
(6) From (V) TA 530159; From (V) TA 700126
(7) From (V) VA 715218; From (V) VA 547237
(8) To (V) TA 703236; To (V) TA 534103
(9) From (V) VA 703237 To (V) VA 715217; From (V) VA 534104 To (V) VA 547236

R 37

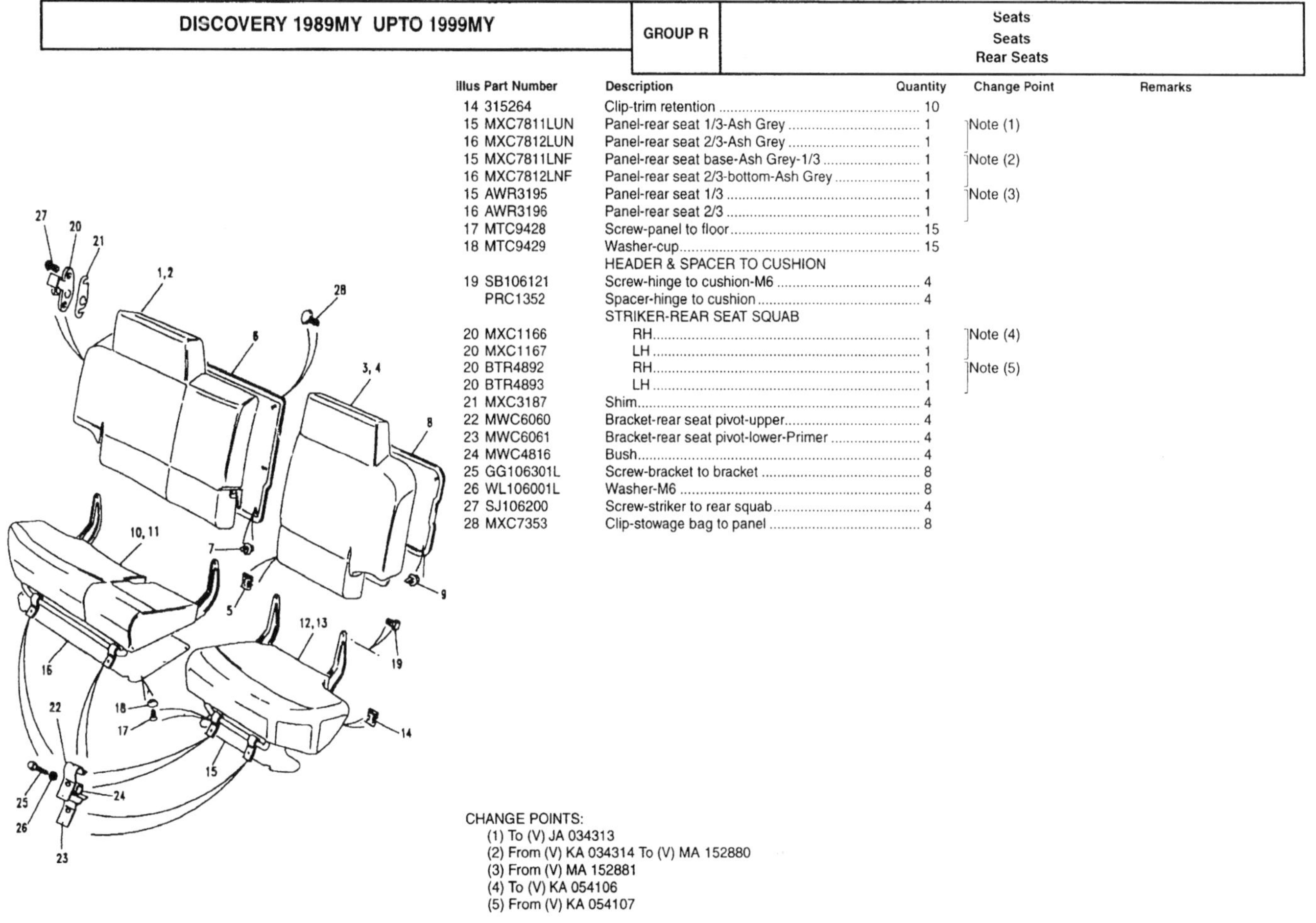

DISCOVERY 1989MY UPTO 1999MY

GROUP R

Seats
Seats
Rear Seats

Illus	Part Number	Description	Quantity	Change Point	Remarks
14	315264	Clip-trim retention	10		
15	MXC7811LUN	Panel-rear seat 1/3-Ash Grey	1	Note (1)	
16	MXC7812LUN	Panel-rear seat 2/3-Ash Grey	1		
15	MXC7811LNF	Panel-rear seat base-Ash Grey-1/3	1	Note (2)	
16	MXC7812LNF	Panel-rear seat 2/3-bottom-Ash Grey	1		
15	AWR3195	Panel-rear seat 1/3	1	Note (3)	
16	AWR3196	Panel-rear seat 2/3	1		
17	MTC9428	Screw-panel to floor	15		
18	MTC9429	Washer-cup	15		
		HEADER & SPACER TO CUSHION			
19	SB106121	Screw-hinge to cushion-M6	4		
	PRC1352	Spacer-hinge to cushion	4		
		STRIKER-REAR SEAT SQUAB			
20	MXC1166	RH	1	Note (4)	
20	MXC1167	LH	1		
20	BTR4892	RH	1	Note (5)	
20	BTR4893	LH	1		
21	MXC3187	Shim	4		
22	MWC6060	Bracket-rear seat pivot-upper	4		
23	MWC6061	Bracket-rear seat pivot-lower-Primer	4		
24	MWC4816	Bush	4		
25	GG106301L	Screw-bracket to bracket	8		
26	WL106001L	Washer-M6	8		
27	SJ106200	Screw-striker to rear squab	4		
28	MXC7353	Clip-stowage bag to panel	8		

CHANGE POINTS:
(1) To (V) JA 034313
(2) From (V) KA 034314 To (V) MA 152880
(3) From (V) MA 152881
(4) To (V) KA 054106
(5) From (V) KA 054107

R 38

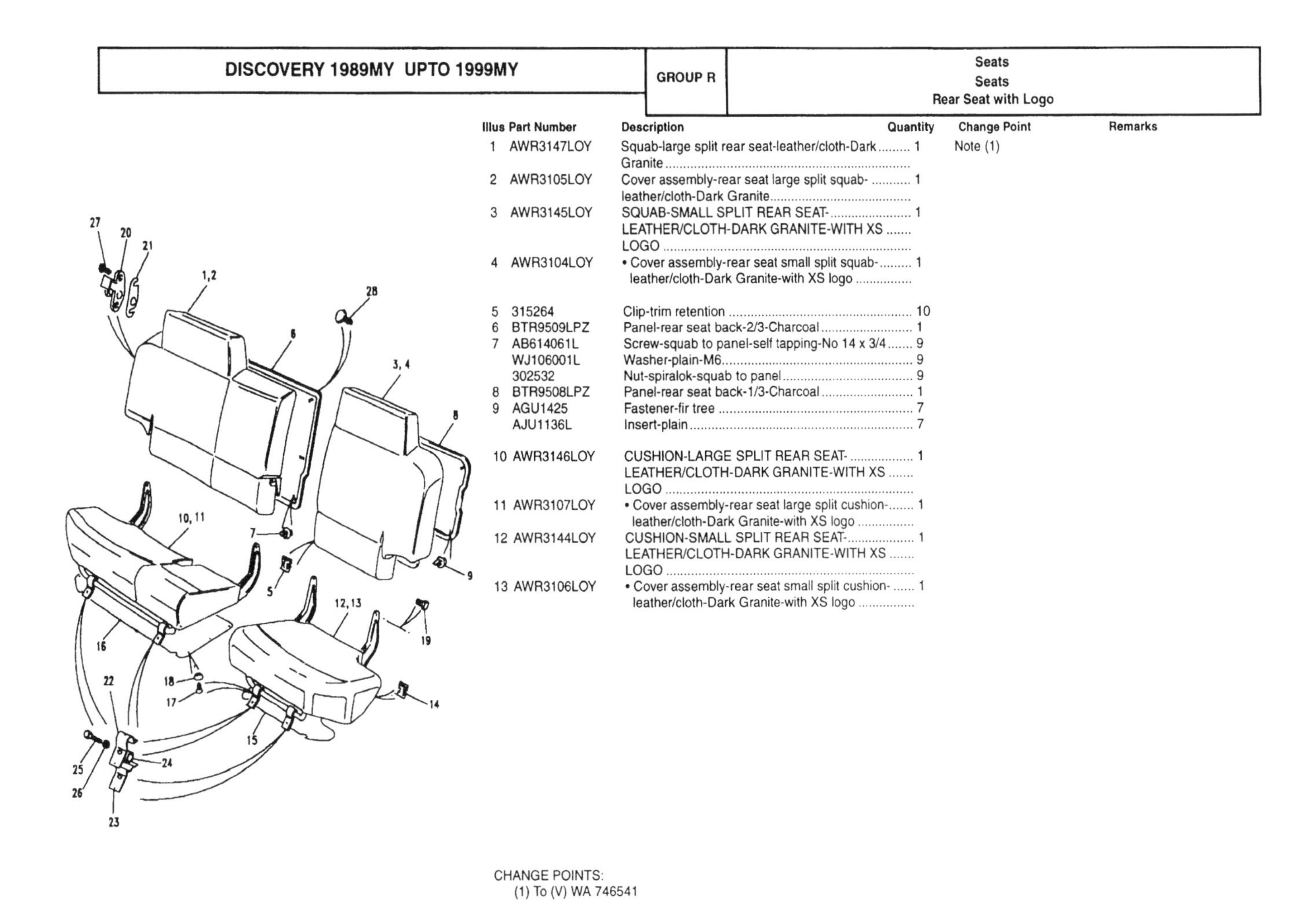

Illus	Part Number	Description	Quantity	Change Point	Remarks
1	AWR3147LOY	Squab-large split rear seat-leather/cloth-Dark Granite	1	Note (1)	
2	AWR3105LOY	Cover assembly-rear seat large split squab-leather/cloth-Dark Granite	1		
3	AWR3145LOY	SQUAB-SMALL SPLIT REAR SEAT-LEATHER/CLOTH-DARK GRANITE-WITH XS LOGO	1		
4	AWR3104LOY	• Cover assembly-rear seat small split squab-leather/cloth-Dark Granite-with XS logo	1		
5	315264	Clip-trim retention	10		
6	BTR9509LPZ	Panel-rear seat back-2/3-Charcoal	1		
7	AB614061L	Screw-squab to panel-self tapping-No 14 x 3/4	9		
	WJ106001L	Washer-plain-M6	9		
	302532	Nut-spiralok-squab to panel	9		
8	BTR9508LPZ	Panel-rear seat back-1/3-Charcoal	1		
9	AGU1425	Fastener-fir tree	7		
	AJU1136L	Insert-plain	7		
10	AWR3146LOY	CUSHION-LARGE SPLIT REAR SEAT-LEATHER/CLOTH-DARK GRANITE-WITH XS LOGO	1		
11	AWR3107LOY	• Cover assembly-rear seat large split cushion-leather/cloth-Dark Granite-with XS logo	1		
12	AWR3144LOY	CUSHION-SMALL SPLIT REAR SEAT-LEATHER/CLOTH-DARK GRANITE-WITH XS LOGO	1		
13	AWR3106LOY	• Cover assembly-rear seat small split cushion-leather/cloth-Dark Granite-with XS logo	1		

CHANGE POINTS:
(1) To (V) WA 746541

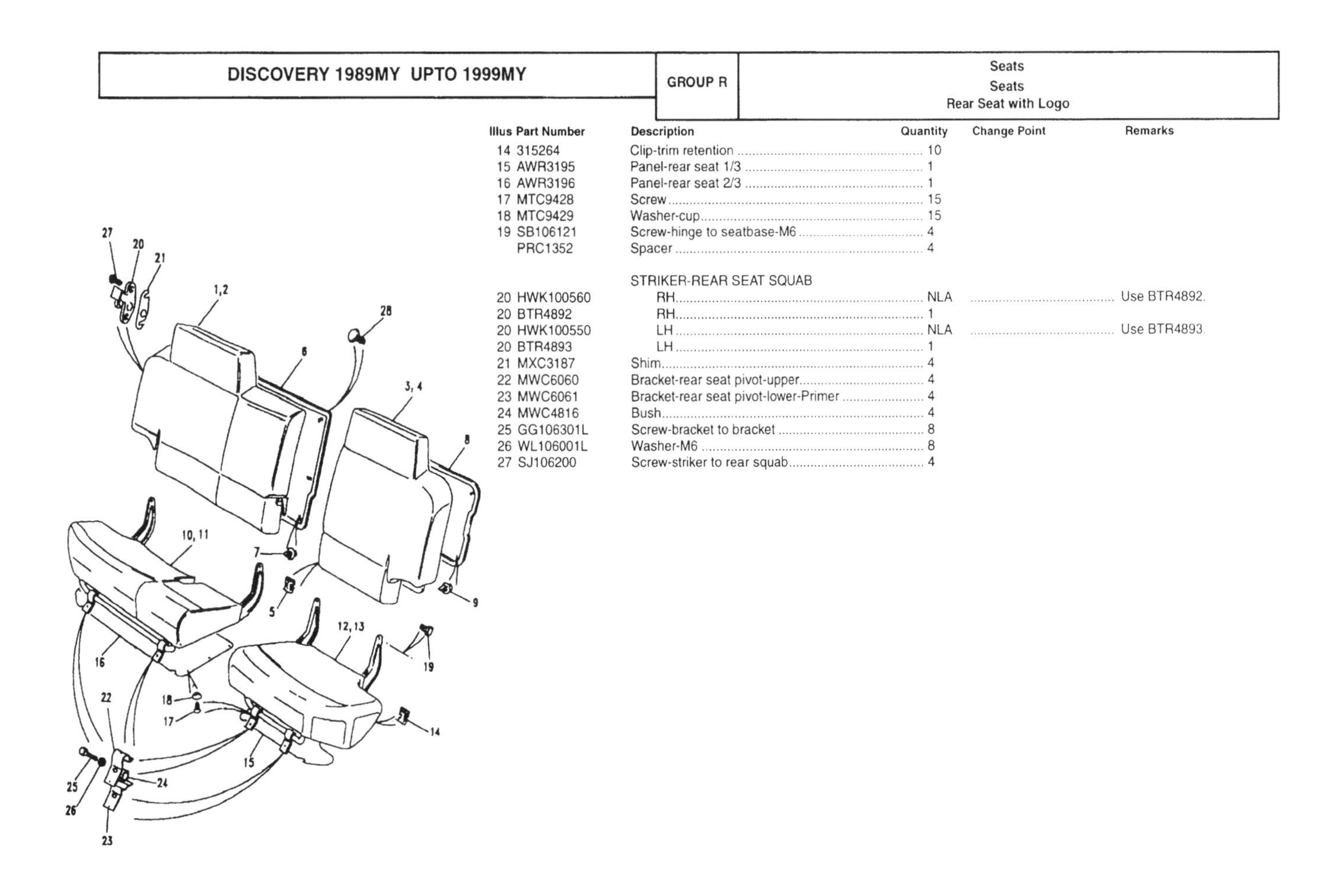

Illus	Part Number	Description	Quantity	Change Point	Remarks
14	315264	Clip-trim retention	10		
15	AWR3195	Panel-rear seat 1/3	1		
16	AWR3196	Panel-rear seat 2/3	1		
17	MTC9428	Screw	15		
18	MTC9429	Washer-cup	15		
19	SB106121	Screw-hinge to seatbase-M6	4		
	PRC1352	Spacer	4		
		STRIKER-REAR SEAT SQUAB			
20	HWK100560	RH	NLA		Use BTR4892.
20	BTR4892	RH	1		
20	HWK100550	LH	NLA		Use BTR4893.
20	BTR4893	LH	1		
21	MXC3187	Shim	4		
22	MWC6060	Bracket-rear seat pivot-upper	4		
23	MWC6061	Bracket-rear seat pivot-lower-Primer	4		
24	MWC4816	Bush	4		
25	GG106301L	Screw-bracket to bracket	8		
26	WL106001L	Washer-M6	8		
27	SJ106200	Screw-striker to rear squab	4		

Illus	Part Number	Description	Quantity	Change Point	Remarks
		SQUAB-LARGE SPLIT REAR SEAT			
1	HLE108990LOY	LEATHER-DARK GRANITE-WITH ARMREST	1	Note (1)	
1	HLE108990SUC	LEATHER-BAHAMA BEIGE-WITH ARMREST	1		
1	HLE109040SMK	LEATHER-LIGHT STONE BEIGE-WITH ARMREST	1		
		• COVER ASSEMBLY-REAR SEAT LARGE SPLIT SQUAB			
2	HMA110560LOY	• leather-Dark Granite	1		Part of HLE108990LOY.
2	HMA110560SUC	• leather-Bahama Beige	1		Part of HLE108990SUC.
2	HMA110630SMK	• leather-Light Stone Beige	1		Part of HLE109040SMK.
		SQUAB-SMALL SPLIT REAR SEAT			
3	HLE108830LOY	LEATHER-DARK GRANITE	1	Note (2); Note (3)	
3	HLE108830SUC	LEATHER-BAHAMA BEIGE	1		
3	HLE109000SMK	LEATHER-LIGHT STONE BEIGE	1	Note (1)	
		• COVER ASSEMBLY-REAR SEAT SMALL SPLIT SQUAB			
4	HMA110230LOY	• leather-Dark Granite	1		Part of HLE108830LOY.
4	HMA110230SUC	• leather-Bahama Beige	1		Part of HLE108830SUC.
4	HMA110570SMK	• leather-Light Stone Beige	1		Part of HLE109000SMK.

CHANGE POINTS:
(1) From (V) WA 746542
(2) From (V) VA 715218
(3) From (V) VA 547237

DISCOVERY 1989MY UPTO 1999MY | GROUP R | Seats
Seats
Rear Seat with Armrest

Illus	Part Number	Description	Quantity	Change Point	Remarks
		CUSHION-LARGE SPLIT REAR SEAT			
5	HLF104950LOY	LEATHER-DARK GRANITE	1	Note (1); Note (2)	
5	HLF104950SUC	LEATHER-BAHAMA BEIGE	1	Note (2); Note (1)	
5	HLF105070SMK	LEATHER-LIGHT STONE BEIGE	1	Note (5)	
		• COVER ASSEMBLY-REAR SEAT LARGE SPLIT CUSHION			
6	HPA106060LOY	• leather-Dark Granite	1		Part of HLF104950LOY.
6	HPA106060SUC	• leather-Bahama Beige	1		Part of HLF104950SUC.
6	HPA106290SMK	• leather-Light Stone Beige	1		Part of HLF105070SMK.
		CUSHION-SMALL SPLIT REAR SEAT			
7	HLF104960LOY	LEATHER-DARK GRANITE	1	Note (3); Note (4)	
7	HLF104960SUC	LEATHER-BAHAMA BEIGE	1	Note (2); Note (1)	
7	HLF105080SMK	LEATHER-LIGHT STONE BEIGE	1	Note (5)	
		• COVER ASSEMBLY-REAR SEAT SMALL SPLIT CUSHION			
8	HPA106070LOY	• leather-Dark Granite	1		Part of HLF104960LOY.
8	HPA106070SUC	• leather-Bahama Beige	1		Part of HLF104960SUC.
8	HPA106280SMK	• leather-Light Stone Beige	1		Part of HLF105080SMK.

CHANGE POINTS:
(1) From (V) VA 547237
(2) From (V) VA 715218
(3) From (V) VA 547237
(4) From (V) VA 715218
(5) From (V) WA 746542

Illus	Part Number	Description	Quantity	Change Point	Remarks
		ARMREST-SPLIT REAR SEAT			
9	HLJ101170LOY	LEATHER-DARK GRANITE	1	Note (1)	
9	HLJ101170SUC	LEATHER-BAHAMA BEIGE	1		
9	HLJ101170SMK	LEATHER-LIGHT STONE BEIGE	1		
		• COVER-ARMREST.			
10	HRA102750LOY	• leather-Dark Granite	1		Part of HLJ101170LOY.
10	HRA102750SUC	• leather-Bahama Beige	1		Part of HLJ101170SUC.
10	HRA102780SMK	• leather-Light Stone Beige	1		Part of HLJ101170SMK.
		BOARD-ARMREST.			
11	HXE100640LOY	leather-Dark Granite	1	Note (1)	
11	HXE100640SUC	leather-Bahama Beige	1		
11	HXE100640SMK	leather-Light Stone Beige	1		
12	HTL100840	Hinge-rear seat armrest-RH	1	Note (1)	
13	HTL100850	Hinge-rear seat armrest-LH	1		

CHANGE POINTS:
(1) From (V) WA 746542

Illus	Part Number	Description	Quantity	Change Point	Remarks
1	HLE109580LOY	Squab-large split rear seat-leather/cloth-Dark Granite-with XS logo	NLA		Use HLE108970LOY.
1	HLE108970LOY	SQUAB-LARGE SPLIT REAR SEAT-LEATHER/CLOTH-DARK GRANITE-WITH XS LOGO	1	Note (1)	
2	HMA110540LOY	• Cover assembly-rear seat large split squab-leather/cloth-Dark Granite-with XS logo	1		
1	HLE109580SUC	Squab-large split rear seat-leather/cloth-Bahama Beige-with XS logo	NLA		Use HLE108970SUC.
1	HLE108970SUC	SQUAB-LARGE SPLIT REAR SEAT-LEATHER/CLOTH-BAHAMA BEIGE-WITH XS LOGO	1		
2	HMA110540SUC	• Cover assembly-rear seat large split squab-leather/cloth-Bahama Beige-with XS logo	1		
3	HLE109530LOY	Squab-small split rear seat-leather/cloth-Dark Granite-with XS logo	NLA		Use AWR3145LOY.
3	AWR3145LOY	SQUAB-SMALL SPLIT REAR SEAT-LEATHER/CLOTH-DARK GRANITE-WITH XS LOGO	1		
4	AWR3104LOY	• Cover assembly-rear seat small split squab-leather/cloth-Dark Granite-with XS logo	1		
3	HLE109530SUC	Squab-small split rear seat-leather/cloth-Bahama Beige-with XS logo	NLA		Use AWR3145SUC.
3	AWR3145SUC	SQUAB-SMALL SPLIT REAR SEAT-LEATHER/CLOTH-BAHAMA BEIGE-WITH XS LOGO	1		
4	AWR3104SUC	• Cover assembly-rear seat small split squab-leather/cloth-Bahama Beige-with XS logo	1		
5	AWR3146LOY	CUSHION-LARGE SPLIT REAR SEAT-LEATHER/CLOTH-DARK GRANITE-WITH XS LOGO	1		
6	AWR3107LOY	• Cover assembly-rear seat large split cushion-leather/cloth-Dark Granite-with XS logo	1		
5	AWR3146SUC	CUSHION-LARGE SPLIT REAR SEAT-LEATHER/CLOTH-BAHAMA BEIGE-WITH XS LOGO	1	Note (1)	
6	AWR3107SUC	• Cover assembly-rear seat large split cushion-leather/cloth-Bahama Beige-with XS logo	1		

CHANGE POINTS:
(1) From (V) WA 746542

Illus	Part Number	Description	Quantity	Change Point	Remarks
7	AWR3144LOY	CUSHION-SMALL SPLIT REAR SEAT-LEATHER/CLOTH-DARK GRANITE-WITH XS LOGO	1		
8	AWR3106LOY	• Cover assembly-rear seat small split cushion-leather/cloth-Dark Granite-with XS logo	1		
9	HLJ101150LOY	ARMREST-SPLIT REAR SEAT-LEATHER/CLOTH-DARK GRANITE-WITH XS LOGO	1	Note (1)	
10	HRA102720LOY	• Cover-armrest.-leather/cloth-Dark Granite-with XS logo	1		
11	HXE100650LOY	Board-armrest.-leather-Dark Granite	1	Note (1)	
12	HTL100840	Hinge-rear seat armrest-RH	1	Note (1)	
13	HTL100850	Hinge-rear seat armrest-LH	1	Note (1)	

CHANGE POINTS:
(1) From (V) WA 746542

Illus	Part Number	Description	Quantity	Change Point	Remarks
		SEATS - SIDE SUPPORTS - REAR SEATS			
1	MXC6025	Bracket-rear seat release catch-RH	1		
1	MXC6026	Bracket-rear seat release catch-LH	1		
2	FS106167L	Screw-hexagonal head-M6 x 16	8		
3	WA106041L	Washer-M6-standard.	8		
4	MWC8299	Latch assembly-rear seat squab-LH	1		
4	MWC8300	Latch assembly-rear seat squab-RH	1		
5	SF106201L	Screw-support bracket-M6 x 20-counter sunk-recessed	6		
6	MXC6441	Rod-rear seat squab latch link-RH	1		
6	MXC6442	Rod-rear seat squab latch link-LH	1		
7	MXC1198	Clip	2		
8	MXC3073LUL	Button-rear seat squab release-Winchester Grey	2	Note (1)	
8	MXC3073LNF	Button-rear seat squab release-Ash Grey	2	Note (2)	

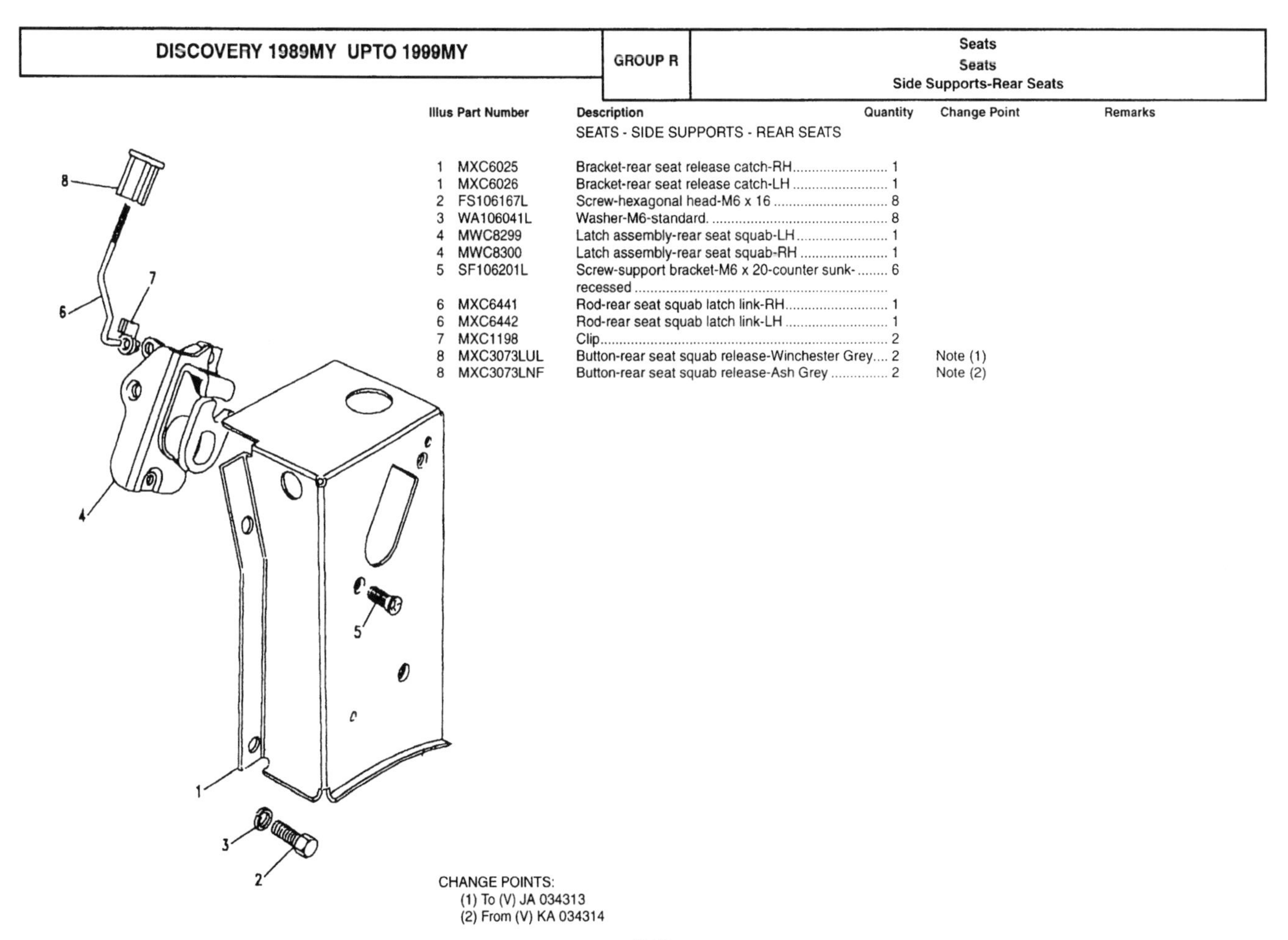

CHANGE POINTS:
(1) To (V) JA 034313
(2) From (V) KA 034314

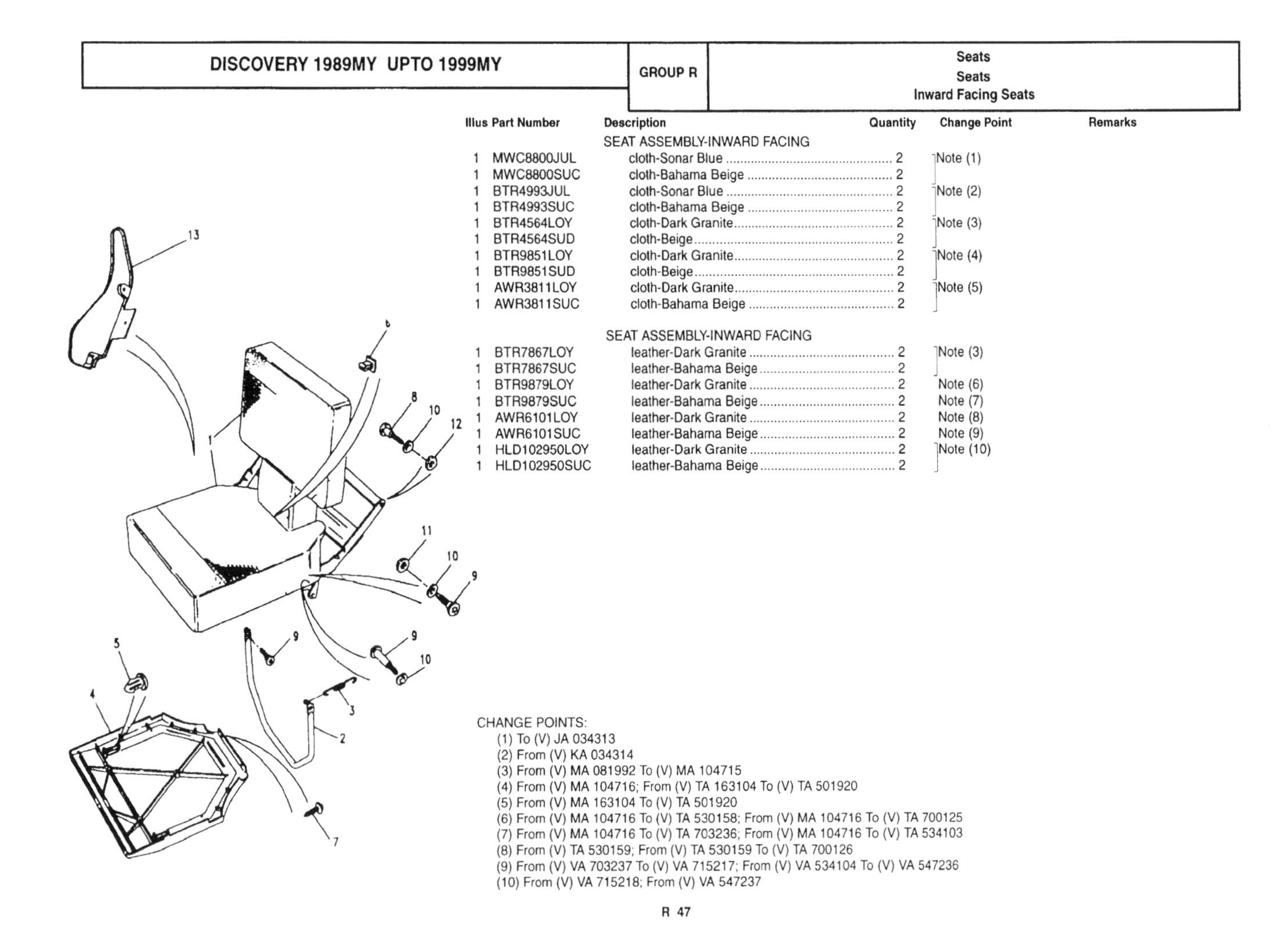

Illus	Part Number	Description	Quantity	Change Point	Remarks
		SEAT ASSEMBLY-INWARD FACING			
1	MWC8800JUL	cloth-Sonar Blue	2	Note (1)	
1	MWC8800SUC	cloth-Bahama Beige	2		
1	BTR4993JUL	cloth-Sonar Blue	2	Note (2)	
1	BTR4993SUC	cloth-Bahama Beige	2		
1	BTR4564LOY	cloth-Dark Granite	2	Note (3)	
1	BTR4564SUD	cloth-Beige	2		
1	BTR9851LOY	cloth-Dark Granite	2	Note (4)	
1	BTR9851SUD	cloth-Beige	2		
1	AWR3811LOY	cloth-Dark Granite	2	Note (5)	
1	AWR3811SUC	cloth-Bahama Beige	2		
		SEAT ASSEMBLY-INWARD FACING			
1	BTR7867LOY	leather-Dark Granite	2	Note (3)	
1	BTR7867SUC	leather-Bahama Beige	2		
1	BTR9879LOY	leather-Dark Granite	2	Note (6)	
1	BTR9879SUC	leather-Bahama Beige	2	Note (7)	
1	AWR6101LOY	leather-Dark Granite	2	Note (8)	
1	AWR6101SUC	leather-Bahama Beige	2	Note (9)	
1	HLD102950LOY	leather-Dark Granite	2	Note (10)	
1	HLD102950SUC	leather-Bahama Beige	2		

CHANGE POINTS:
(1) To (V) JA 034313
(2) From (V) KA 034314
(3) From (V) MA 081992 To (V) MA 104715
(4) From (V) MA 104716; From (V) TA 163104 To (V) TA 501920
(5) From (V) MA 163104 To (V) TA 501920
(6) From (V) MA 104716 To (V) TA 530158; From (V) MA 104716 To (V) TA 700125
(7) From (V) MA 104716 To (V) TA 703236; From (V) MA 104716 To (V) TA 534103
(8) From (V) TA 530159; From (V) TA 530159 To (V) TA 700126
(9) From (V) VA 703237 To (V) VA 715217; From (V) VA 534104 To (V) VA 547236
(10) From (V) VA 715218; From (V) VA 547237

R 47

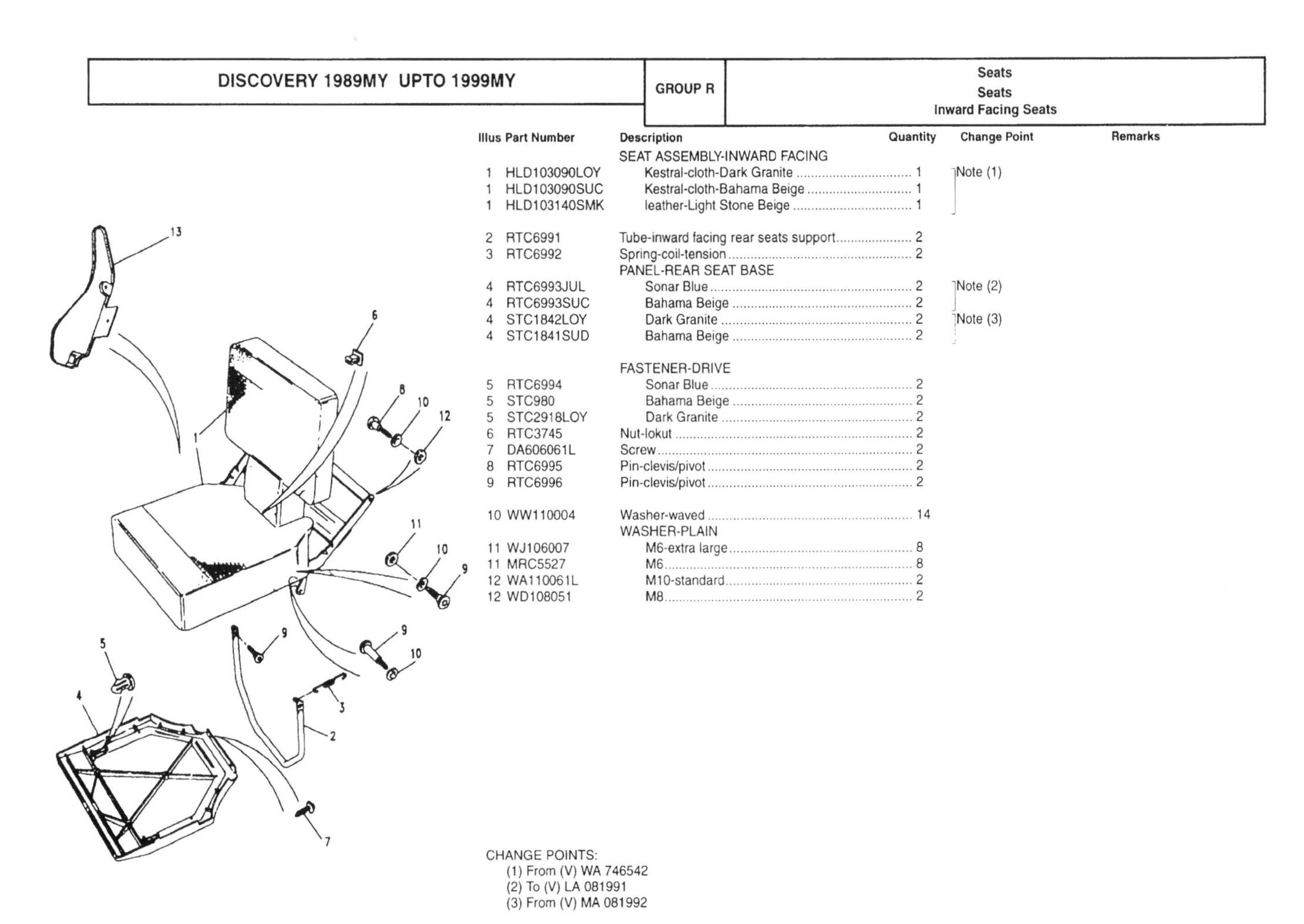

Illus	Part Number	Description	Quantity	Change Point	Remarks
		SEAT ASSEMBLY-INWARD FACING			
1	HLD103090LOY	Kestral-cloth-Dark Granite	1	Note (1)	
1	HLD103090SUC	Kestral-cloth-Bahama Beige	1		
1	HLD103140SMK	leather-Light Stone Beige	1		
2	RTC6991	Tube-inward facing rear seats support	2		
3	RTC6992	Spring-coil-tension	2		
		PANEL-REAR SEAT BASE			
4	RTC6993JUL	Sonar Blue	2	Note (2)	
4	RTC6993SUC	Bahama Beige	2		
4	STC1842LOY	Dark Granite	2	Note (3)	
4	STC1841SUD	Bahama Beige	2		
		FASTENER-DRIVE			
5	RTC6994	Sonar Blue	2		
5	STC980	Bahama Beige	2		
5	STC2918LOY	Dark Granite	2		
6	RTC3745	Nut-lokut	2		
7	DA606061L	Screw	2		
8	RTC6995	Pin-clevis/pivot	2		
9	RTC6996	Pin-clevis/pivot	2		
10	WW110004	Washer-waved	14		
		WASHER-PLAIN			
11	WJ106007	M6-extra large	8		
11	MRC5527	M6	8		
12	WA110061L	M10-standard	2		
12	WD108051	M8	2		

CHANGE POINTS:
(1) From (V) WA 746542
(2) To (V) LA 081991
(3) From (V) MA 081992

R 48

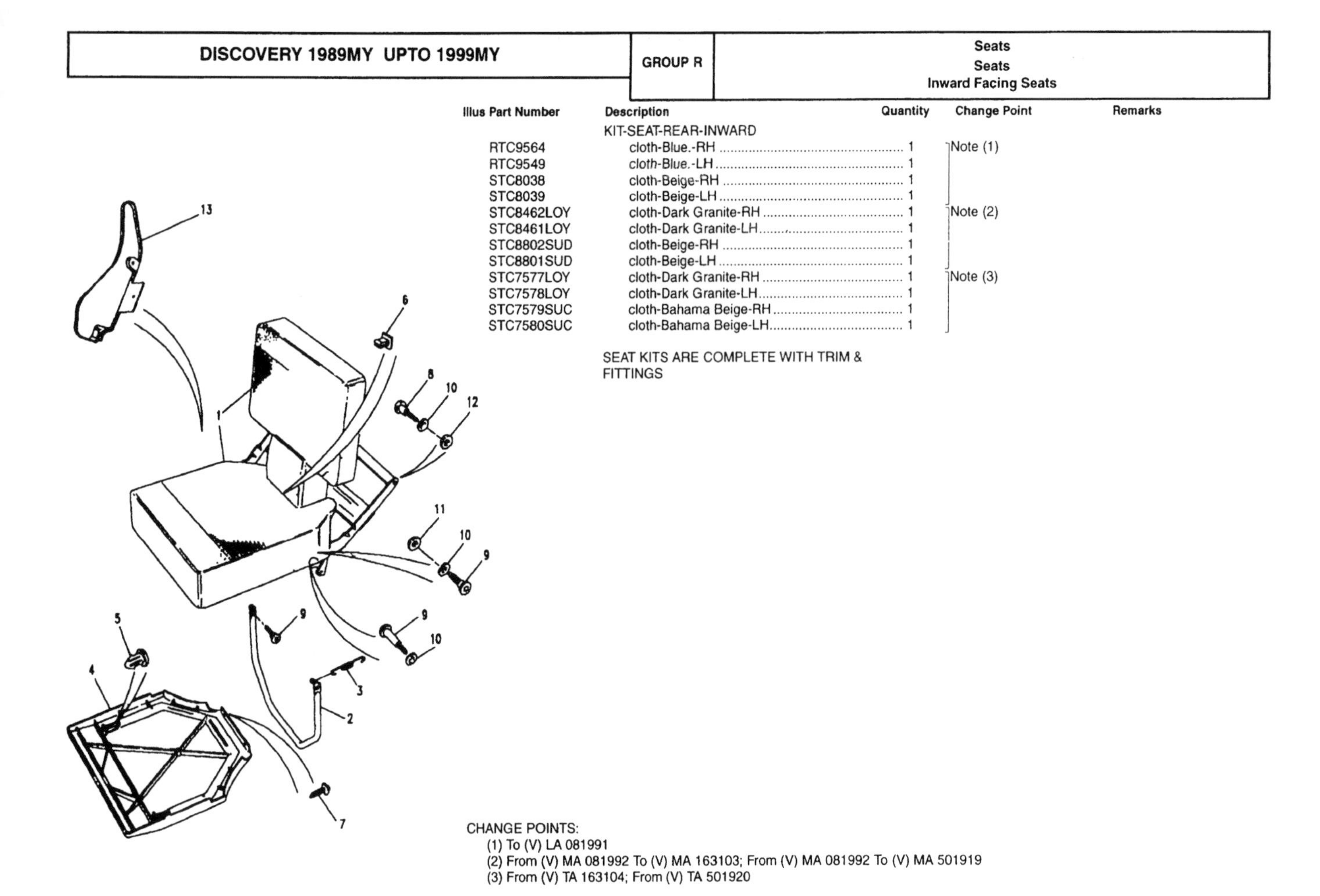

Illus	Part Number	Description	Quantity	Change Point	Remarks
		KIT-SEAT-REAR-INWARD			
	RTC9564	cloth-Blue.-RH	1	Note (1)	
	RTC9549	cloth-Blue.-LH	1		
	STC8038	cloth-Beige-RH	1		
	STC8039	cloth-Beige-LH	1		
	STC8462LOY	cloth-Dark Granite-RH	1	Note (2)	
	STC8461LOY	cloth-Dark Granite-LH	1		
	STC8802SUD	cloth-Beige-RH	1		
	STC8801SUD	cloth-Beige-LH	1		
	STC7577LOY	cloth-Dark Granite-RH	1	Note (3)	
	STC7578LOY	cloth-Dark Granite-LH	1		
	STC7579SUC	cloth-Bahama Beige-RH	1		
	STC7580SUC	cloth-Bahama Beige-LH	1		

SEAT KITS ARE COMPLETE WITH TRIM & FITTINGS

CHANGE POINTS:
(1) To (V) LA 081991
(2) From (V) MA 081992 To (V) MA 163103; From (V) MA 081992 To (V) MA 501919
(3) From (V) TA 163104; From (V) TA 501920

R 49

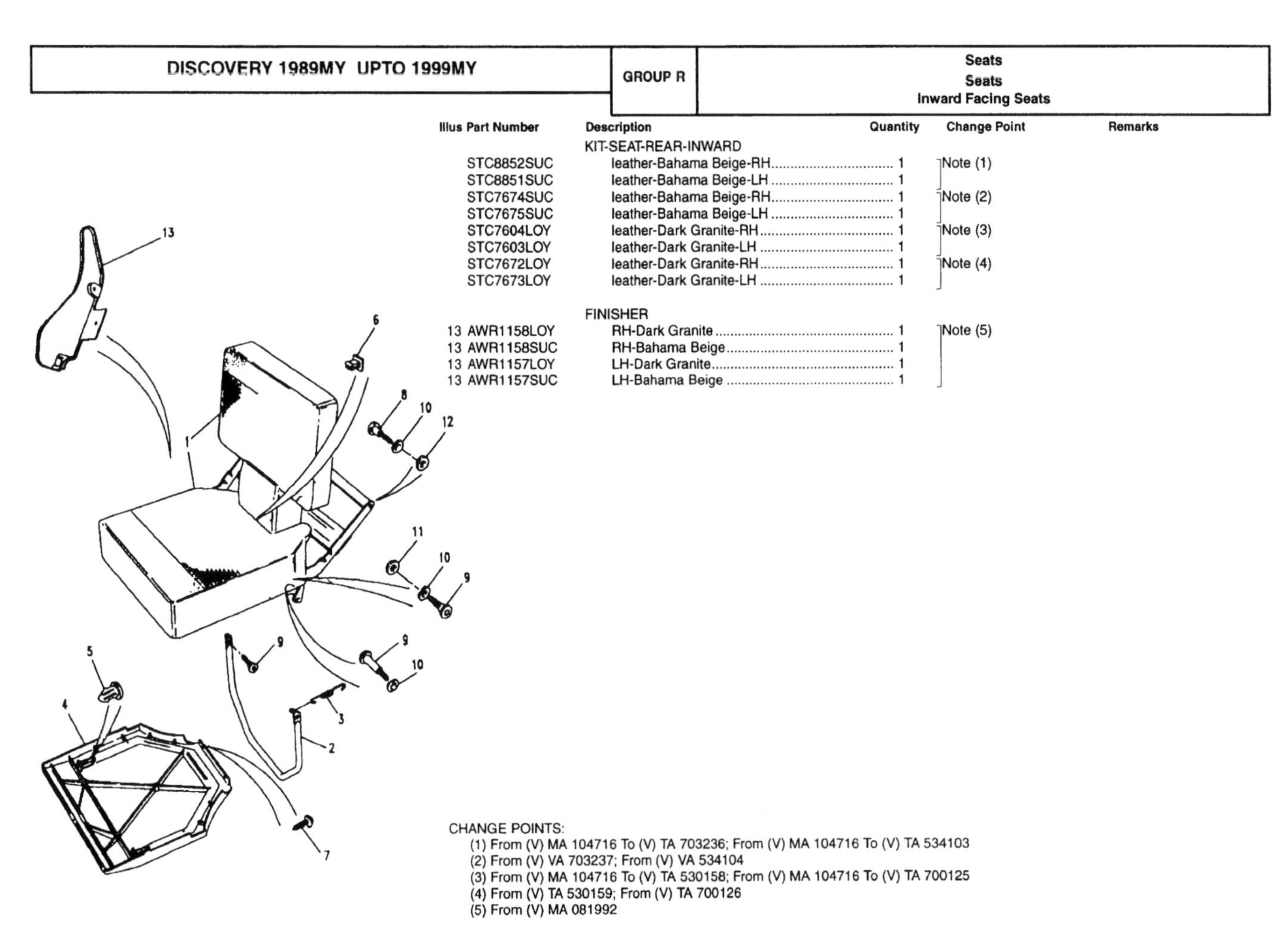

Illus	Part Number	Description	Quantity	Change Point	Remarks
		KIT-SEAT-REAR-INWARD			
	STC8852SUC	leather-Bahama Beige-RH	1	Note (1)	
	STC8851SUC	leather-Bahama Beige-LH	1		
	STC7674SUC	leather-Bahama Beige-RH	1	Note (2)	
	STC7675SUC	leather-Bahama Beige-LH	1		
	STC7604LOY	leather-Dark Granite-RH	1	Note (3)	
	STC7603LOY	leather-Dark Granite-LH	1		
	STC7672LOY	leather-Dark Granite-RH	1	Note (4)	
	STC7673LOY	leather-Dark Granite-LH	1		
		FINISHER			
13	AWR1158LOY	RH-Dark Granite	1	Note (5)	
13	AWR1158SUC	RH-Bahama Beige	1		
13	AWR1157LOY	LH-Dark Granite	1		
13	AWR1157SUC	LH-Bahama Beige	1		

CHANGE POINTS:
(1) From (V) MA 104716 To (V) TA 703236; From (V) MA 104716 To (V) TA 534103
(2) From (V) VA 703237; From (V) VA 534104
(3) From (V) MA 104716 To (V) TA 530158; From (V) MA 104716 To (V) TA 700125
(4) From (V) TA 530159; From (V) TA 700126
(5) From (V) MA 081992

R 50

Seats
Rear Inward Facing Seats - Fixings

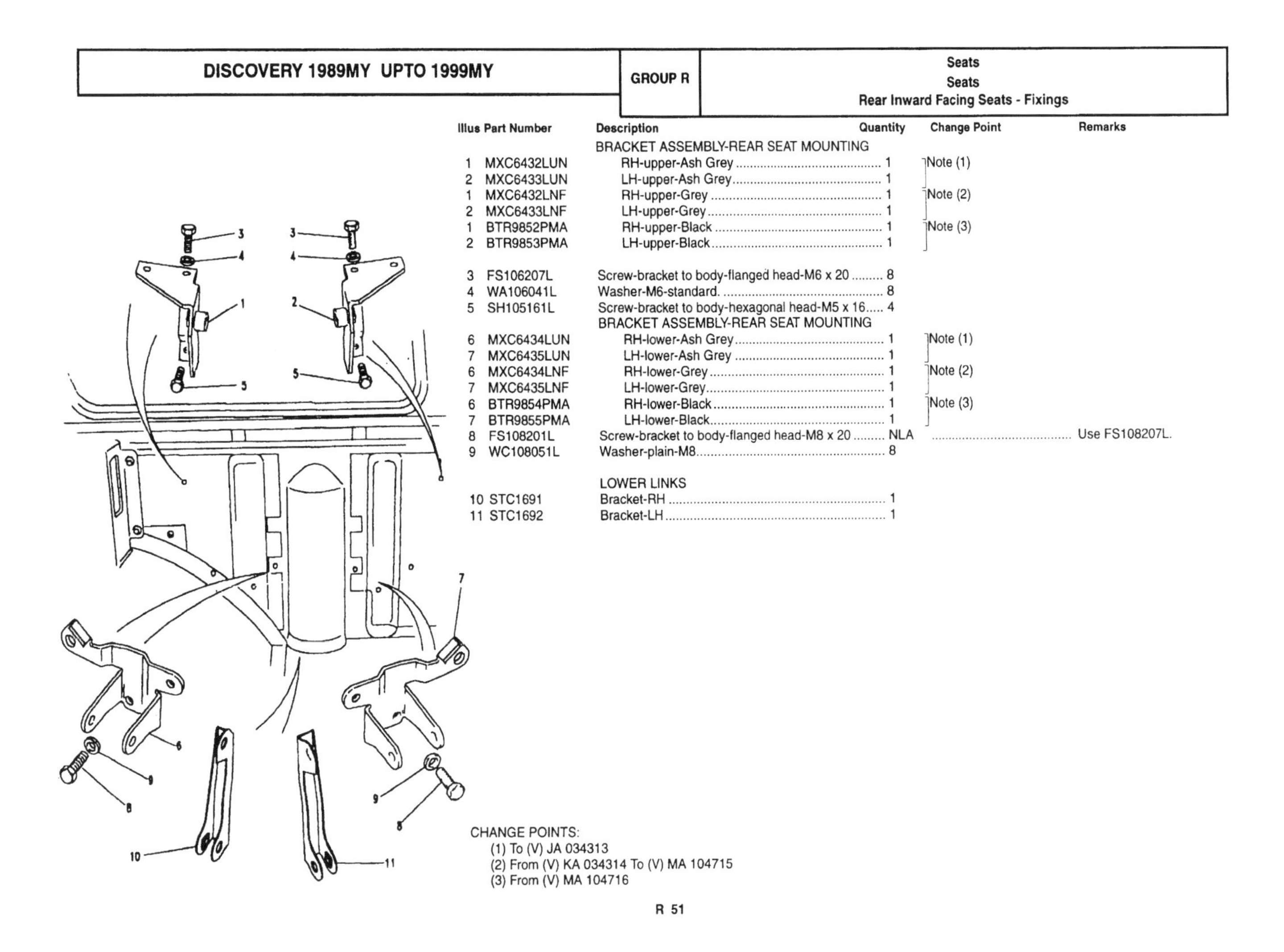

Illus	Part Number	Description	Quantity	Change Point	Remarks
		BRACKET ASSEMBLY-REAR SEAT MOUNTING			
1	MXC6432LUN	RH-upper-Ash Grey	1	Note (1)	
2	MXC6433LUN	LH-upper-Ash Grey	1		
1	MXC6432LNF	RH-upper-Grey	1	Note (2)	
2	MXC6433LNF	LH-upper-Grey	1		
1	BTR9852PMA	RH-upper-Black	1	Note (3)	
2	BTR9853PMA	LH-upper-Black	1		
3	FS106207L	Screw-bracket to body-flanged head-M6 x 20	8		
4	WA106041L	Washer-M6-standard.	8		
5	SH105161L	Screw-bracket to body-hexagonal head-M5 x 16	4		
		BRACKET ASSEMBLY-REAR SEAT MOUNTING			
6	MXC6434LUN	RH-lower-Ash Grey	1	Note (1)	
7	MXC6435LUN	LH-lower-Ash Grey	1		
6	MXC6434LNF	RH-lower-Grey	1	Note (2)	
7	MXC6435LNF	LH-lower-Grey	1		
6	BTR9854PMA	RH-lower-Black	1	Note (3)	
7	BTR9855PMA	LH-lower-Black	1		
8	FS108201L	Screw-bracket to body-flanged head-M8 x 20	NLA		Use FS108207L.
9	WC108051L	Washer-plain-M8	8		
		LOWER LINKS			
10	STC1691	Bracket-RH	1		
11	STC1692	Bracket-LH	1		

CHANGE POINTS:
(1) To (V) JA 034313
(2) From (V) KA 034314 To (V) MA 104715
(3) From (V) MA 104716

Seats
Seat Covers-50LE

Illus	Part Number	Description	Quantity	Change Point	Remarks
1	HBA105380SMK	Cover assembly-front seat squab-Light Smokestone	1		
2	HCA105340SMK	Cover assembly-front seat cushion-Light Smokestone	1		
3	HMA110560SMK	Cover assembly-squab-large-Light Smokestone	1		
4	HMA110230SMK	Cover assembly-squab-small-Light Smokestone	1		
5	HPA106060SMK	Cover assembly-cushion-large-Light Smokestone	1		
6	HPA106070SMK	Cover assembly-cushion-small-Light Smokestone	1		

ILLUSTRATION TO FOLLOW

TF 0001

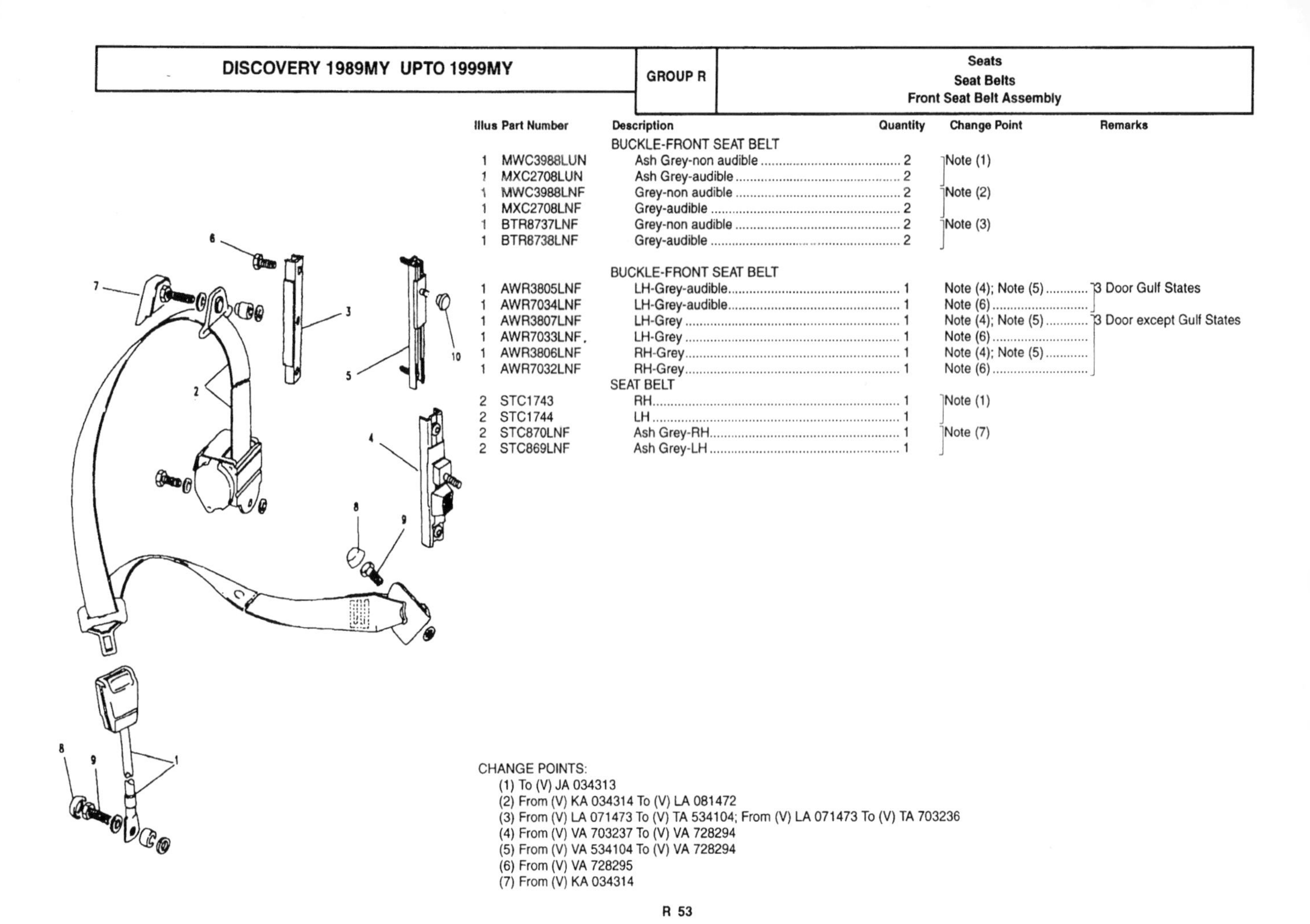

Illus	Part Number	Description	Quantity	Change Point	Remarks
		BUCKLE-FRONT SEAT BELT			
1	MWC3988LUN	Ash Grey-non audible	2	Note (1)	
1	MXC2708LUN	Ash Grey-audible	2		
1	MWC3988LNF	Grey-non audible	2	Note (2)	
1	MXC2708LNF	Grey-audible	2		
1	BTR8737LNF	Grey-non audible	2	Note (3)	
1	BTR8738LNF	Grey-audible	2		
		BUCKLE-FRONT SEAT BELT			
1	AWR3805LNF	LH-Grey-audible	1	Note (4); Note (5)	3 Door Gulf States
1	AWR7034LNF	LH-Grey-audible	1	Note (6)	
1	AWR3807LNF	LH-Grey	1	Note (4); Note (5)	3 Door except Gulf States
1	AWR7033LNF	LH-Grey	1	Note (6)	
1	AWR3806LNF	RH-Grey	1	Note (4); Note (5)	
1	AWR7032LNF	RH-Grey	1	Note (6)	
		SEAT BELT			
2	STC1743	RH	1	Note (1)	
2	STC1744	LH	1		
2	STC870LNF	Ash Grey-RH	1	Note (7)	
2	STC869LNF	Ash Grey-LH	1		

CHANGE POINTS:
(1) To (V) JA 034313
(2) From (V) KA 034314 To (V) LA 081472
(3) From (V) LA 071473 To (V) TA 534104; From (V) LA 071473 To (V) TA 703236
(4) From (V) VA 703237 To (V) VA 728294
(5) From (V) VA 534104 To (V) VA 728294
(6) From (V) VA 728295
(7) From (V) KA 034314

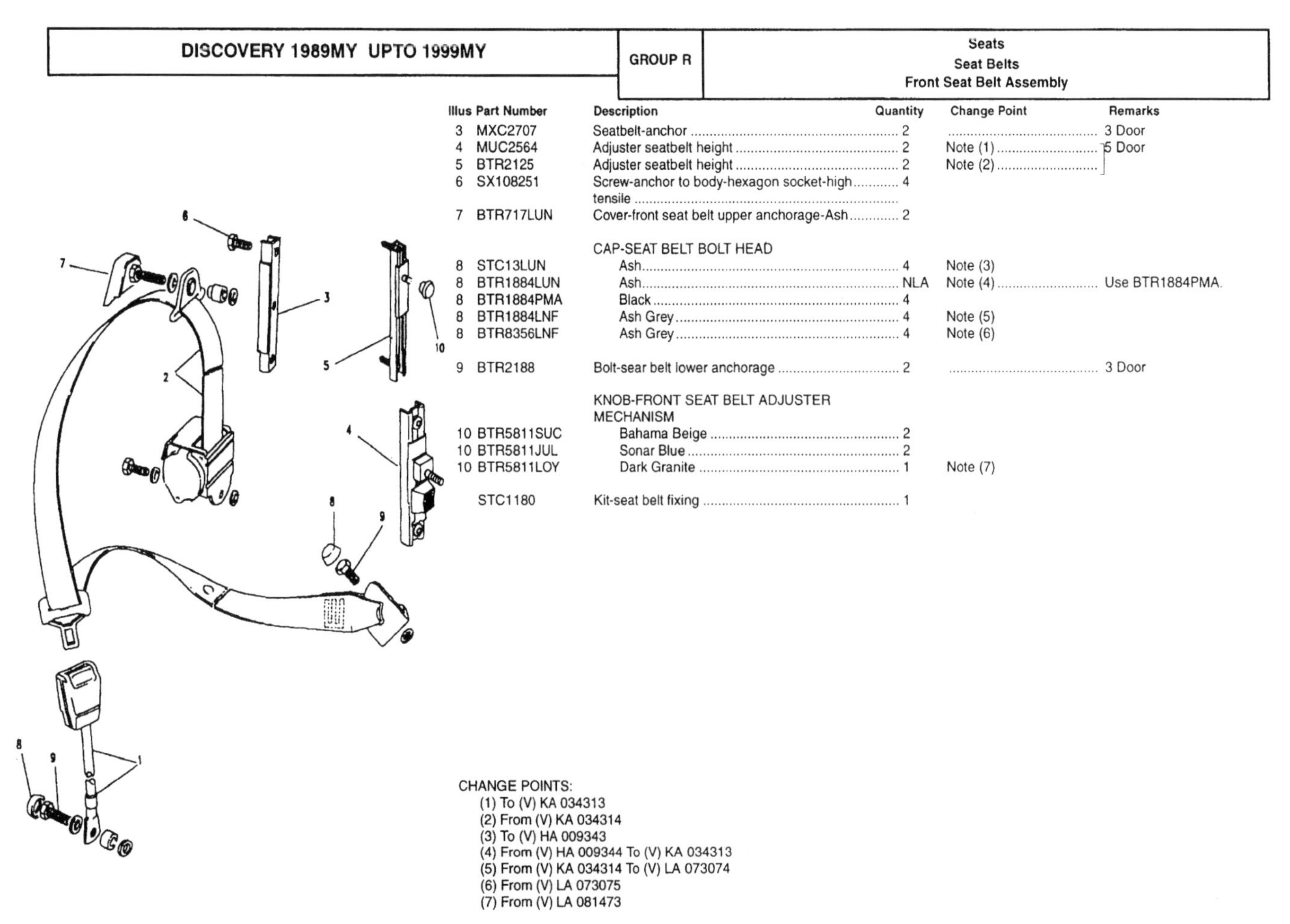

Illus	Part Number	Description	Quantity	Change Point	Remarks
3	MXC2707	Seatbelt-anchor	2		3 Door
4	MUC2564	Adjuster seatbelt height	2	Note (1)	5 Door
5	BTR2125	Adjuster seatbelt height	2	Note (2)	
6	SX108251	Screw-anchor to body-hexagon socket-high tensile	4		
7	BTR717LUN	Cover-front seat belt upper anchorage-Ash	2		
		CAP-SEAT BELT BOLT HEAD			
8	STC13LUN	Ash	4	Note (3)	
8	BTR1884LUN	Ash	NLA	Note (4)	Use BTR1884PMA.
8	BTR1884PMA	Black	4		
8	BTR1884LNF	Ash Grey	4	Note (5)	
8	BTR8356LNF	Ash Grey	4	Note (6)	
9	BTR2188	Bolt-sear belt lower anchorage	2		3 Door
		KNOB-FRONT SEAT BELT ADJUSTER MECHANISM			
10	BTR5811SUC	Bahama Beige	2		
10	BTR5811JUL	Sonar Blue	2		
10	BTR5811LOY	Dark Granite	1	Note (7)	
	STC1180	Kit-seat belt fixing	1		

CHANGE POINTS:
(1) To (V) KA 034313
(2) From (V) KA 034314
(3) To (V) HA 009343
(4) From (V) HA 009344 To (V) KA 034313
(5) From (V) KA 034314 To (V) LA 073074
(6) From (V) LA 073075
(7) From (V) LA 081473

Seats
Seat Belts
Front Seat Belt Assembly 5 Door

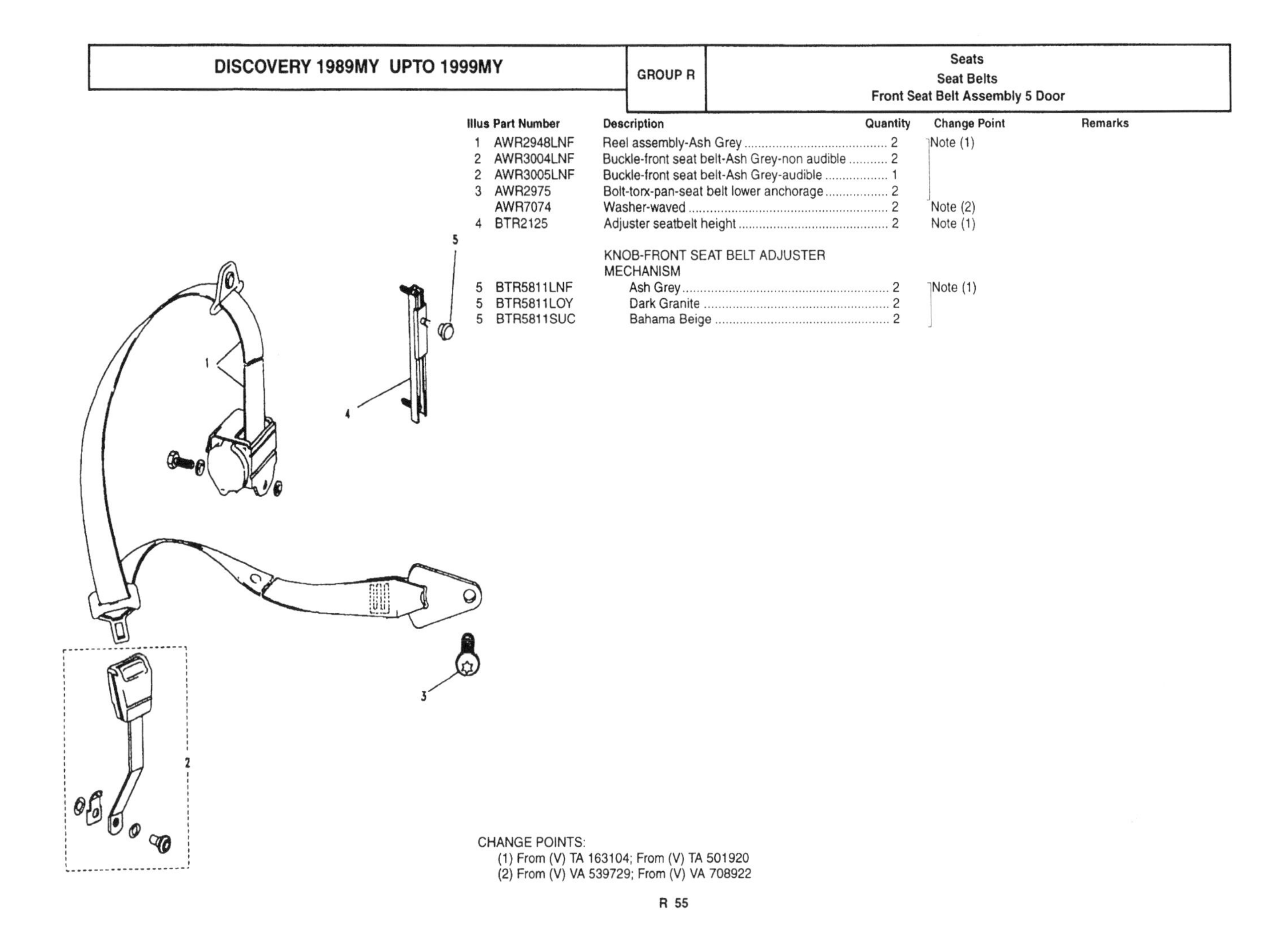

Illus	Part Number	Description	Quantity	Change Point	Remarks
1	AWR2948LNF	Reel assembly-Ash Grey	2	Note (1)	
2	AWR3004LNF	Buckle-front seat belt-Ash Grey-non audible	2		
2	AWR3005LNF	Buckle-front seat belt-Ash Grey-audible	1		
3	AWR2975	Bolt-torx-pan-seat belt lower anchorage	2		
	AWR7074	Washer-waved	2	Note (2)	
4	BTR2125	Adjuster seatbelt height	2	Note (1)	
		KNOB-FRONT SEAT BELT ADJUSTER MECHANISM			
5	BTR5811LNF	Ash Grey	2	Note (1)	
5	BTR5811LOY	Dark Granite	2		
5	BTR5811SUC	Bahama Beige	2		

CHANGE POINTS:
(1) From (V) TA 163104; From (V) TA 501920
(2) From (V) VA 539729; From (V) VA 708922

R 55

DISCOVERY 1989MY UPTO 1999MY | GROUP R

Seats
Seat Belts
Rear Seat Belt

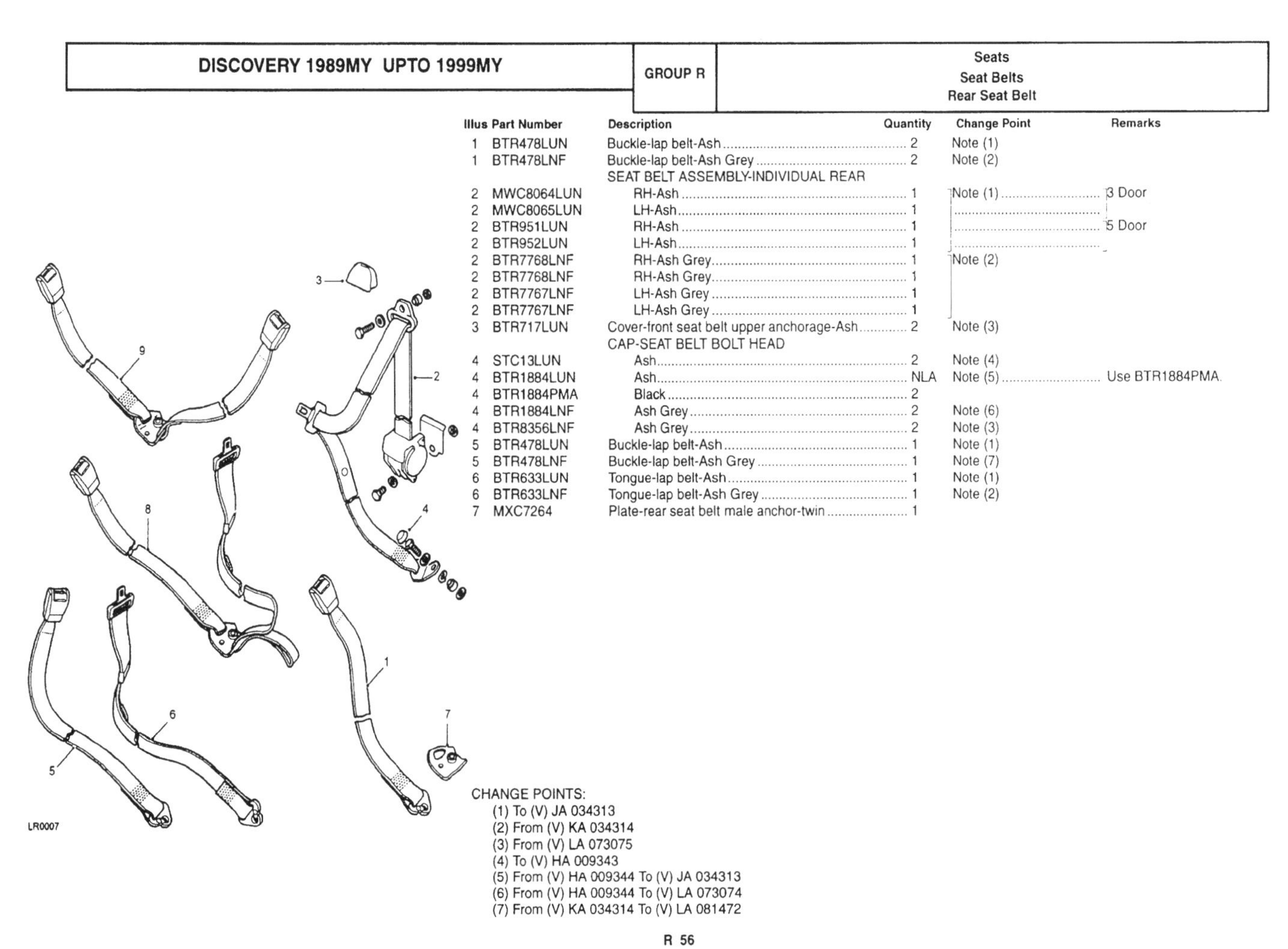

Illus	Part Number	Description	Quantity	Change Point	Remarks
1	BTR478LUN	Buckle-lap belt-Ash	2	Note (1)	
1	BTR478LNF	Buckle-lap belt-Ash Grey	2	Note (2)	
		SEAT BELT ASSEMBLY-INDIVIDUAL REAR			
2	MWC8064LUN	RH-Ash	1	Note (1)	3 Door
2	MWC8065LUN	LH-Ash	1		
2	BTR951LUN	RH-Ash	1		5 Door
2	BTR952LUN	LH-Ash	1		
2	BTR7768LNF	RH-Ash Grey	1	Note (2)	
2	BTR7768LNF	RH-Ash Grey	1		
2	BTR7767LNF	LH-Ash Grey	1		
2	BTR7767LNF	LH-Ash Grey	1		
3	BTR717LUN	Cover-front seat belt upper anchorage-Ash	2	Note (3)	
		CAP-SEAT BELT BOLT HEAD			
4	STC13LUN	Ash	2	Note (4)	
4	BTR1884LUN	Ash	NLA	Note (5)	Use BTR1884PMA.
4	BTR1884PMA	Black	2		
4	BTR1884LNF	Ash Grey	2	Note (6)	
4	BTR8356LNF	Ash Grey	2	Note (3)	
5	BTR478LUN	Buckle-lap belt-Ash	1	Note (1)	
5	BTR478LNF	Buckle-lap belt-Ash Grey	1	Note (7)	
6	BTR633LUN	Tongue-lap belt-Ash	1	Note (1)	
6	BTR633LNF	Tongue-lap belt-Ash Grey	1	Note (2)	
7	MXC7264	Plate-rear seat belt male anchor-twin	1		

CHANGE POINTS:
(1) To (V) JA 034313
(2) From (V) KA 034314
(3) From (V) LA 073075
(4) To (V) HA 009343
(5) From (V) HA 009344 To (V) JA 034313
(6) From (V) HA 009344 To (V) LA 073074
(7) From (V) KA 034314 To (V) LA 081472

R 56

DISCOVERY 1989MY UPTO 1999MY	GROUP R	Seats Seat Belts Rear Seat Belt

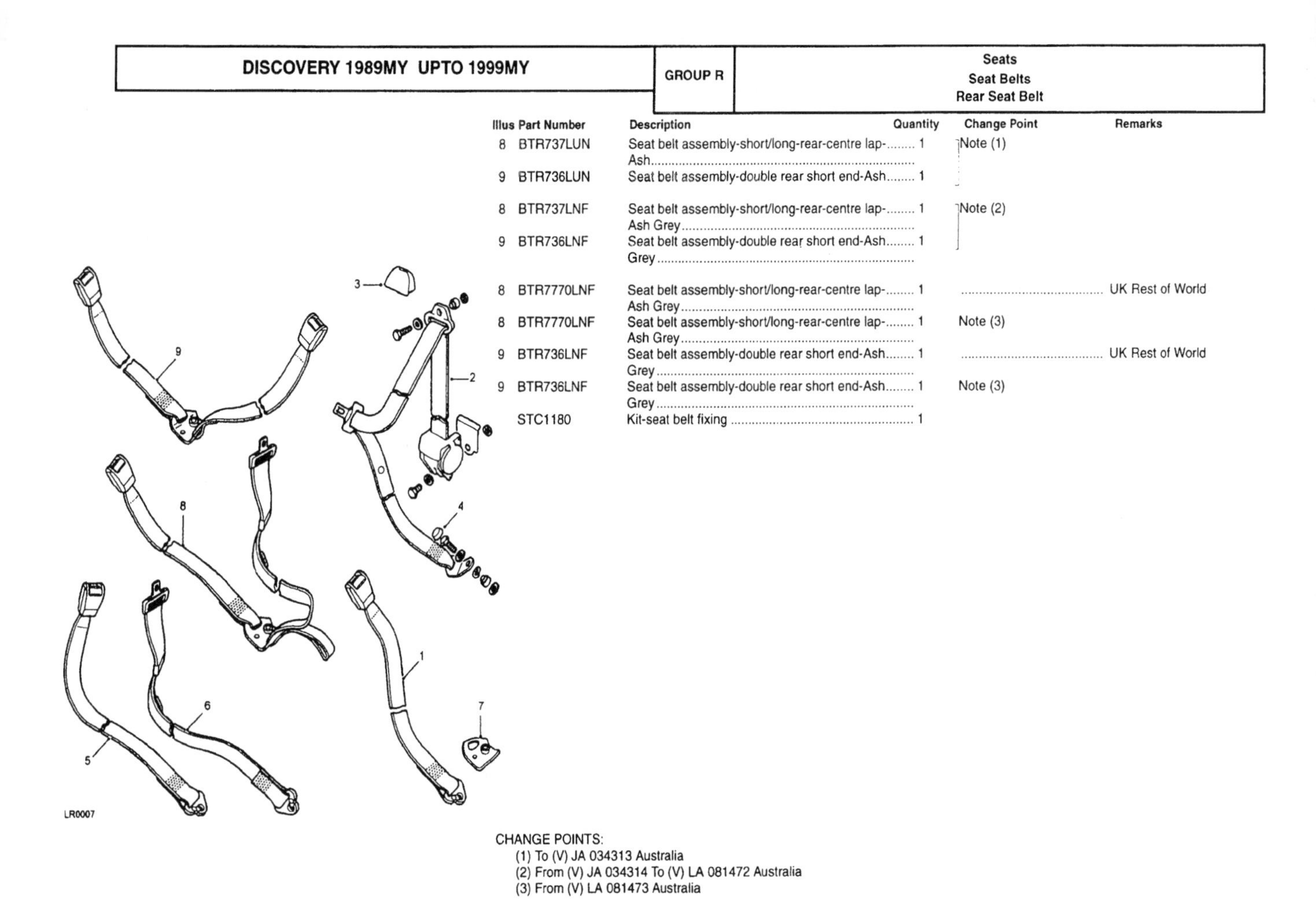

Illus	Part Number	Description	Quantity	Change Point	Remarks
8	BTR737LUN	Seat belt assembly-short/long-rear-centre lap-Ash	1	Note (1)	
9	BTR736LUN	Seat belt assembly-double rear short end-Ash	1		
8	BTR737LNF	Seat belt assembly-short/long-rear-centre lap-Ash Grey	1	Note (2)	
9	BTR736LNF	Seat belt assembly-double rear short end-Ash Grey	1		
8	BTR7770LNF	Seat belt assembly-short/long-rear-centre lap-Ash Grey	1		UK Rest of World
8	BTR7770LNF	Seat belt assembly-short/long-rear-centre lap-Ash Grey	1	Note (3)	
9	BTR736LNF	Seat belt assembly-double rear short end-Ash Grey	1		UK Rest of World
9	BTR736LNF	Seat belt assembly-double rear short end-Ash Grey	1	Note (3)	
	STC1180	Kit-seat belt fixing	1		

CHANGE POINTS:
(1) To (V) JA 034313 Australia
(2) From (V) JA 034314 To (V) LA 081472 Australia
(3) From (V) LA 081473 Australia

R 57

DISCOVERY 1989MY UPTO 1999MY	GROUP R	Seats Seat Belts Rear Seat Belt - Inward Facing

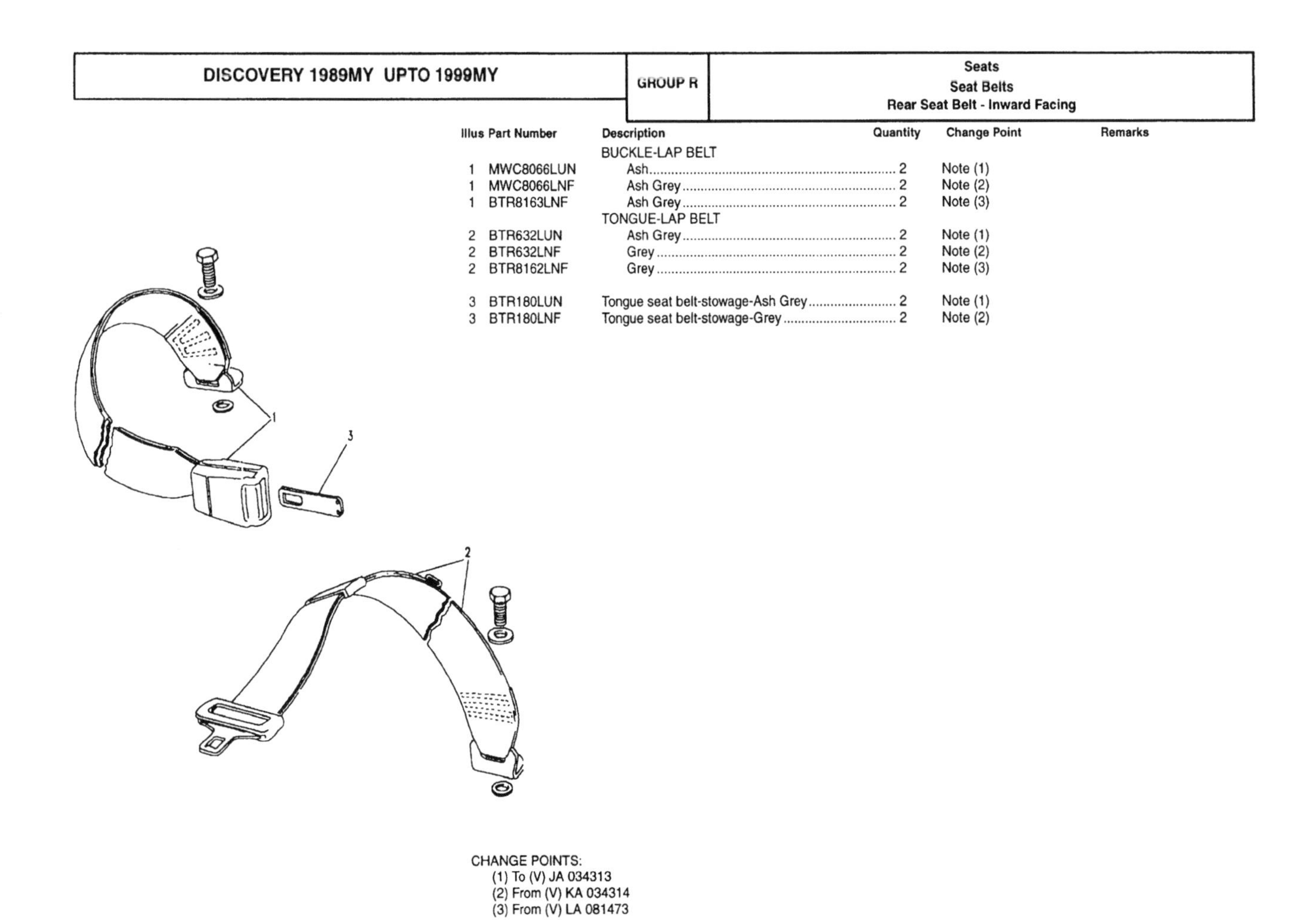

Illus	Part Number	Description	Quantity	Change Point	Remarks
		BUCKLE-LAP BELT			
1	MWC8066LUN	Ash	2	Note (1)	
1	MWC8066LNF	Ash Grey	2	Note (2)	
1	BTR8163LNF	Ash Grey	2	Note (3)	
		TONGUE-LAP BELT			
2	BTR632LUN	Ash Grey	2	Note (1)	
2	BTR632LNF	Grey	2	Note (2)	
2	BTR8162LNF	Grey	2	Note (3)	
3	BTR180LUN	Tongue seat belt-stowage-Ash Grey	2	Note (1)	
3	BTR180LNF	Tongue seat belt-stowage-Grey	2	Note (2)	

CHANGE POINTS:
(1) To (V) JA 034313
(2) From (V) KA 034314
(3) From (V) LA 081473

R 58

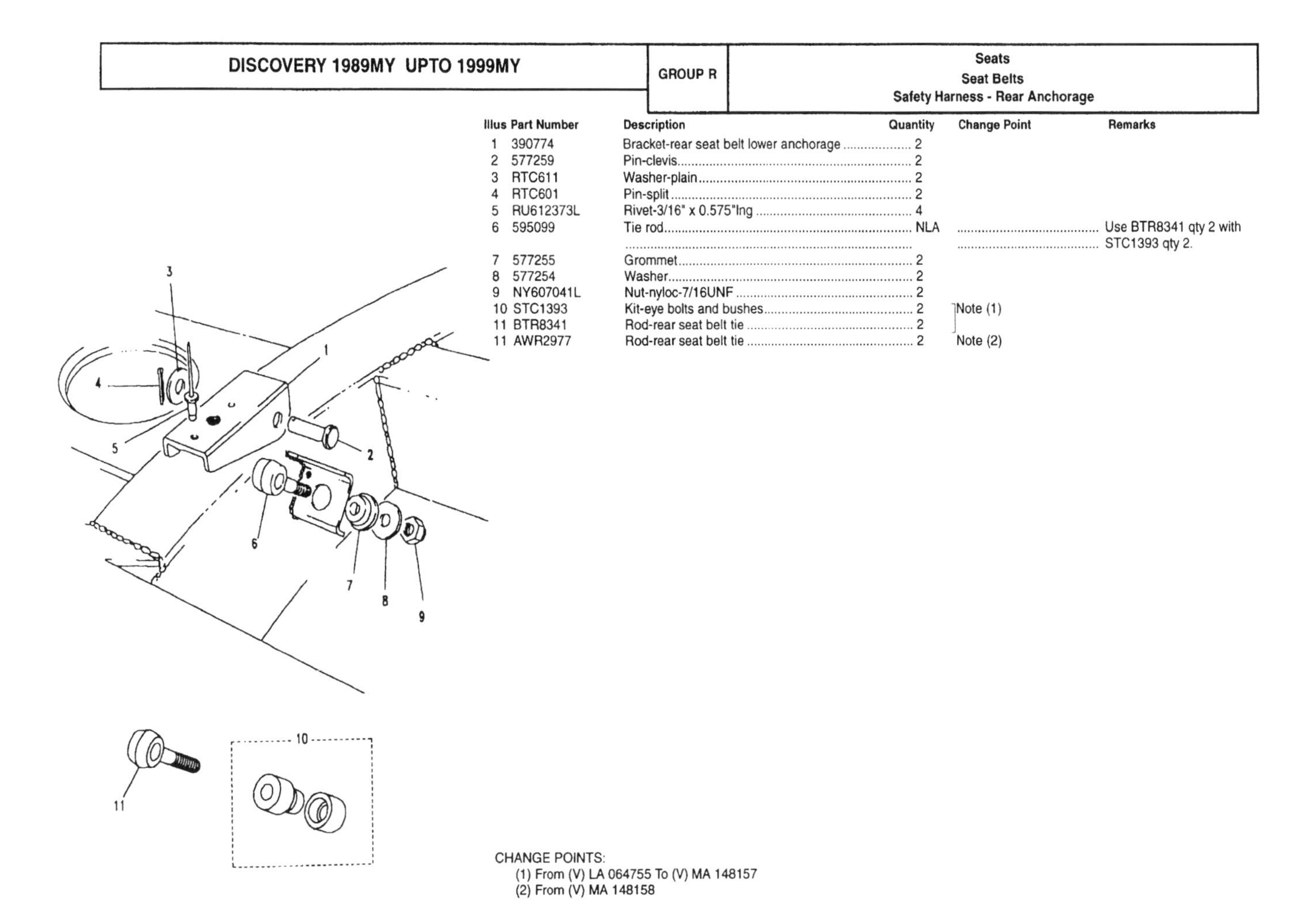

Illus	Part Number	Description	Quantity	Change Point	Remarks
1	390774	Bracket-rear seat belt lower anchorage	2		
2	577259	Pin-clevis	2		
3	RTC611	Washer-plain	2		
4	RTC601	Pin-split	2		
5	RU612373L	Rivet-3/16" x 0.575"lng	4		
6	595099	Tie rod	NLA		Use BTR8341 qty 2 with STC1393 qty 2.
7	577255	Grommet	2		
8	577254	Washer	2		
9	NY607041L	Nut-nyloc-7/16UNF	2		
10	STC1393	Kit-eye bolts and bushes	2	Note (1)	
11	BTR8341	Rod-rear seat belt tie	2	Note (1)	
11	AWR2977	Rod-rear seat belt tie	2	Note (2)	

CHANGE POINTS:
(1) From (V) LA 064755 To (V) MA 148157
(2) From (V) MA 148158

Notes

Stowage .... S 1

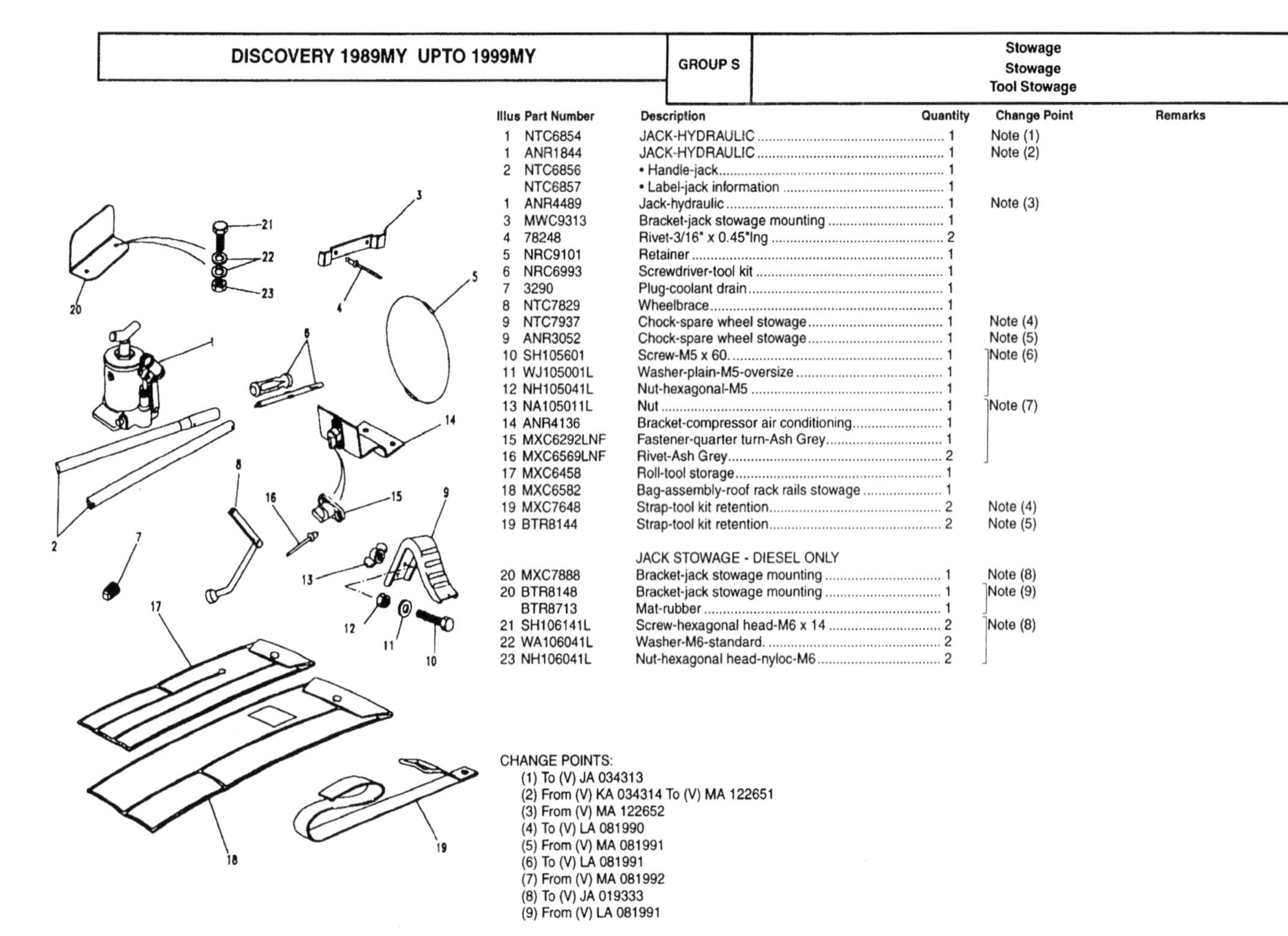

Illus	Part Number	Description	Quantity	Change Point	Remarks
1	NTC6854	JACK-HYDRAULIC	1	Note (1)	
1	ANR1844	JACK-HYDRAULIC	1	Note (2)	
2	NTC6856	• Handle-jack	1		
	NTC6857	• Label-jack information	1		
1	ANR4489	Jack-hydraulic	1	Note (3)	
3	MWC9313	Bracket-jack stowage mounting	1		
4	78248	Rivet-3/16" x 0.45"lng	2		
5	NRC9101	Retainer	1		
6	NRC6993	Screwdriver-tool kit	1		
7	3290	Plug-coolant drain	1		
8	NTC7829	Wheelbrace	1		
9	NTC7937	Chock-spare wheel stowage	1	Note (4)	
9	ANR3052	Chock-spare wheel stowage	1	Note (5)	
10	SH105601	Screw-M5 x 60	1	Note (6)	
11	WJ105001L	Washer-plain-M5-oversize	1		
12	NH105041L	Nut-hexagonal-M5	1		
13	NA105011L	Nut	1	Note (7)	
14	ANR4136	Bracket-compressor air conditioning	1		
15	MXC6292LNF	Fastener-quarter turn-Ash Grey	1		
16	MXC6569LNF	Rivet-Ash Grey	2		
17	MXC6458	Roll-tool storage	1		
18	MXC6582	Bag-assembly-roof rack rails stowage	1		
19	MXC7648	Strap-tool kit retention	2	Note (4)	
19	BTR8144	Strap-tool kit retention	2	Note (5)	
		JACK STOWAGE - DIESEL ONLY			
20	MXC7888	Bracket-jack stowage mounting	1	Note (8)	
20	BTR8148	Bracket-jack stowage mounting	1	Note (9)	
	BTR8713	Mat-rubber	1		
21	SH106141L	Screw-hexagonal head-M6 x 14	2	Note (8)	
22	WA106041L	Washer-M6-standard	2		
23	NH106041L	Nut-hexagonal head-nyloc-M6	2		

CHANGE POINTS:
(1) To (V) JA 034313
(2) From (V) KA 034314 To (V) MA 122651
(3) From (V) MA 122652
(4) To (V) LA 081990
(5) From (V) MA 081991
(6) To (V) LA 081991
(7) From (V) MA 081992
(8) To (V) JA 019333
(9) From (V) LA 081991

S 1

Accessories .... T 5
Alarm - Non Central Locking .... T 34
Alarm Systems .... T 33
Alarm-Thatcham Cat One Approved .... T 35
Body Styling Kit .... T 7
Body Tape Kit .... T 5
Box-Luggage/Ski .... T 41
Carpets & Mats .... T 32
Dog Guard .... T 13
Fog & Driving Lamps .... T 23
Headlamp Leveling Kit-95MY On .... T 25
Heavy Duty Rear Suspension .... T 38
Interior Protection - Seat Covers .... T 11
Interior Wood Trim .... T 15
Inward Facing Seat Kits .... T 29
Loadspace Protection-Rigid .... T 30
Loadspace Protection-Stowable .... T 31
Lubricants & Car Care .... T 28
Mudflaps .... T 45
Nudge Bar-Air Bag Vehicles Only .... T 9
Nudge Bar-Non Air Bag Vehicles Only .... T 8
Optional Extras .... T 46
Rear Bike/Ski Carrier .... T 42
Rear Step-Retractable .... T 43
Refrigerator .... T 14
Roof Racks .... T 39
Rubbing Strips .... T 6
Safety & Security - Child Seat .... T 16
Safety & Security - Fire Extinguisher .... T 17
Safety & Security - Security Box .... T 18
Side Step .... T 44
Ski Carrier .... T 40
Soft Protection Bar- Air Bag Vehicles Only .... T 10
Spare Wheel Covers .... T 37
Storage - Cubby Box .... T 19
Sunroof Glass & Sunblinds .... T 47
Towing .... T 20
Wheel-Accessories .... T 36
Winch-Husky .... T 26
Winch-Xd9000i .... T 27
Miscellaneous .... T 1
Paint .... T 1

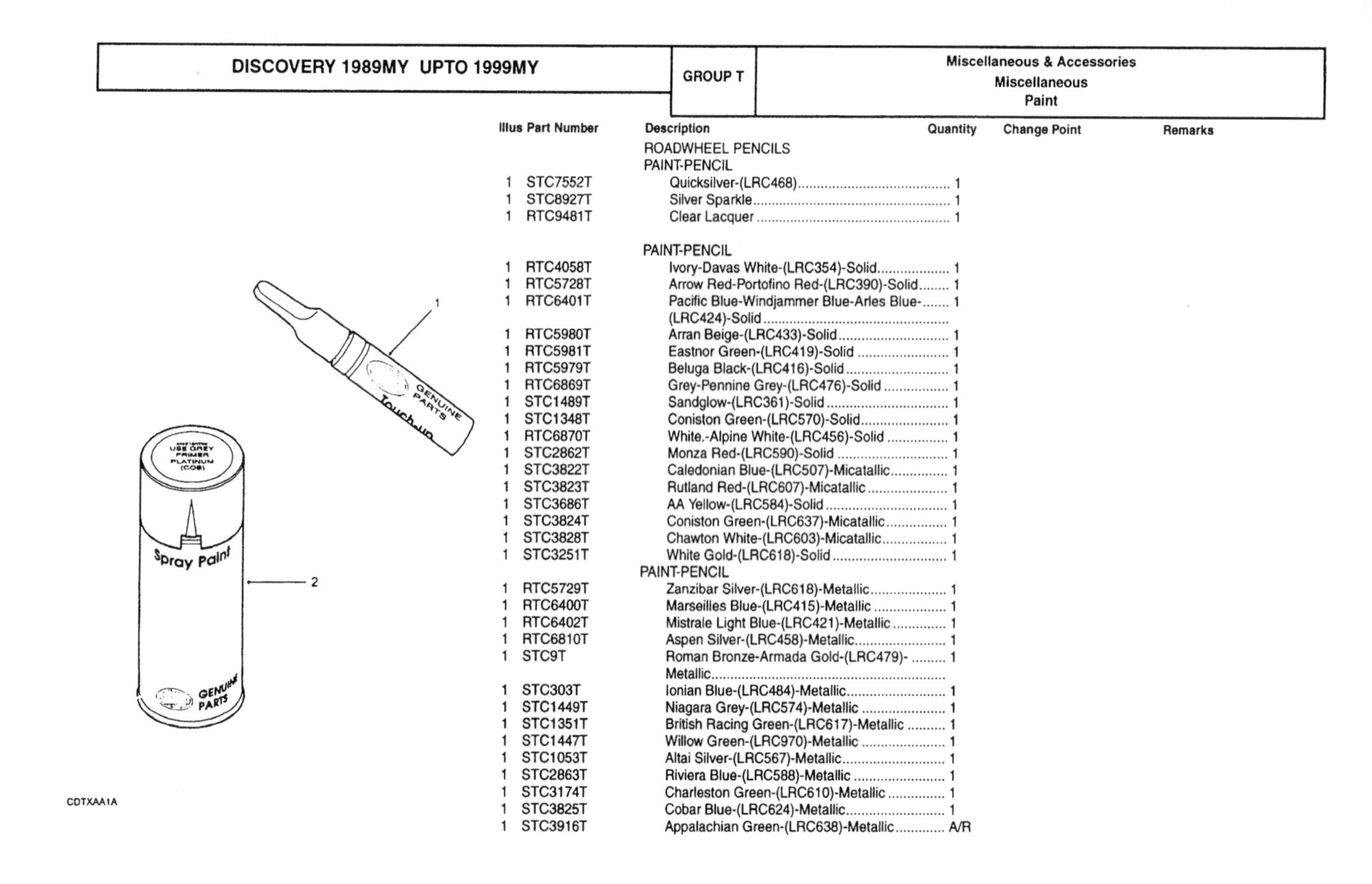

Illus	Part Number	Description	Quantity	Change Point	Remarks
		ROADWHEEL PENCILS			
		PAINT-PENCIL			
1	STC7552T	Quicksilver-(LRC468)	1		
1	STC8927T	Silver Sparkle	1		
1	RTC9481T	Clear Lacquer	1		
		PAINT-PENCIL			
1	RTC4058T	Ivory-Davas White-(LRC354)-Solid	1		
1	RTC5728T	Arrow Red-Portofino Red-(LRC390)-Solid	1		
1	RTC6401T	Pacific Blue-Windjammer Blue-Arles Blue-(LRC424)-Solid	1		
1	RTC5980T	Arran Beige-(LRC433)-Solid	1		
1	RTC5981T	Eastnor Green-(LRC419)-Solid	1		
1	RTC5979T	Beluga Black-(LRC416)-Solid	1		
1	RTC6869T	Grey-Pennine Grey-(LRC476)-Solid	1		
1	STC1489T	Sandglow-(LRC361)-Solid	1		
1	STC1348T	Coniston Green-(LRC570)-Solid	1		
1	RTC6870T	White.-Alpine White-(LRC456)-Solid	1		
1	STC2862T	Monza Red-(LRC590)-Solid	1		
1	STC3822T	Caledonian Blue-(LRC507)-Micatallic	1		
1	STC3823T	Rutland Red-(LRC607)-Micatallic	1		
1	STC3686T	AA Yellow-(LRC584)-Solid	1		
1	STC3824T	Coniston Green-(LRC637)-Micatallic	1		
1	STC3828T	Chawton White-(LRC603)-Micatallic	1		
1	STC3251T	White Gold-(LRC618)-Solid	1		
		PAINT-PENCIL			
1	RTC5729T	Zanzibar Silver-(LRC618)-Metallic	1		
1	RTC6400T	Marseilles Blue-(LRC415)-Metallic	1		
1	RTC6402T	Mistrale Light Blue-(LRC421)-Metallic	1		
1	RTC6810T	Aspen Silver-(LRC458)-Metallic	1		
1	STC9T	Roman Bronze-Armada Gold-(LRC479)-Metallic	1		
1	STC303T	Ionian Blue-(LRC484)-Metallic	1		
1	STC1449T	Niagara Grey-(LRC574)-Metallic	1		
1	STC1351T	British Racing Green-(LRC617)-Metallic	1		
1	STC1447T	Willow Green-(LRC970)-Metallic	1		
1	STC1053T	Altai Silver-(LRC567)-Metallic	1		
1	STC2863T	Riviera Blue-(LRC588)-Metallic	1		
1	STC3174T	Charleston Green-(LRC610)-Metallic	1		
1	STC3825T	Cobar Blue-(LRC624)-Metallic	1		
1	STC3916T	Appalachian Green-(LRC638)-Metallic	A/R		

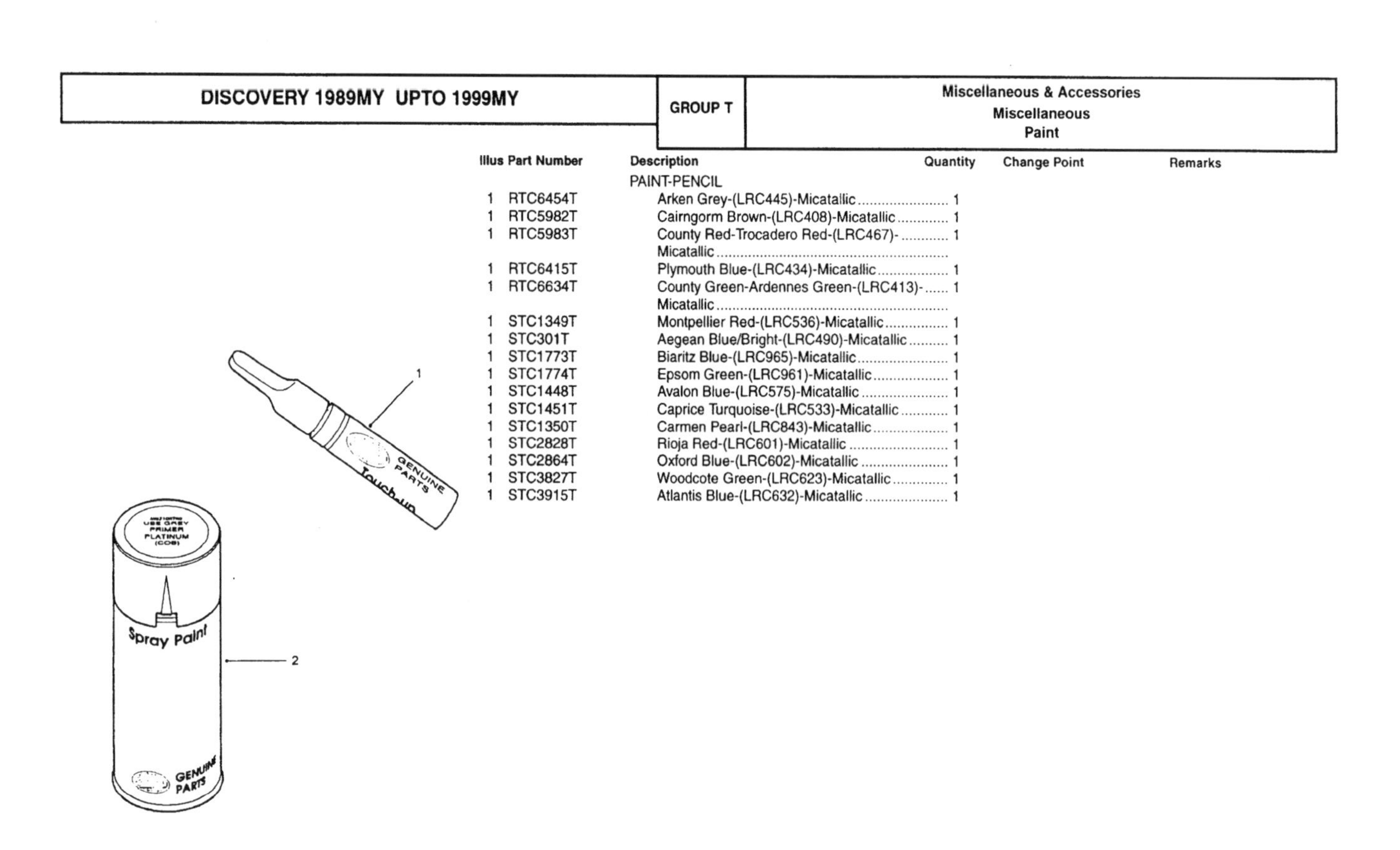

Illus	Part Number	Description	Quantity	Change Point	Remarks
		PAINT-PENCIL			
1	RTC6454T	Arken Grey-(LRC445)-Micatallic	1		
1	RTC5982T	Cairngorm Brown-(LRC408)-Micatallic	1		
1	RTC5983T	County Red-Trocadero Red-(LRC467)-Micatallic	1		
1	RTC6415T	Plymouth Blue-(LRC434)-Micatallic	1		
1	RTC6634T	County Green-Ardennes Green-(LRC413)-Micatallic	1		
1	STC1349T	Montpellier Red-(LRC536)-Micatallic	1		
1	STC301T	Aegean Blue/Bright-(LRC490)-Micatallic	1		
1	STC1773T	Biaritz Blue-(LRC965)-Micatallic	1		
1	STC1774T	Epsom Green-(LRC961)-Micatallic	1		
1	STC1448T	Avalon Blue-(LRC575)-Micatallic	1		
1	STC1451T	Caprice Turquoise-(LRC533)-Micatallic	1		
1	STC1350T	Carmen Pearl-(LRC843)-Micatallic	1		
1	STC2828T	Rioja Red-(LRC601)-Micatallic	1		
1	STC2864T	Oxford Blue-(LRC602)-Micatallic	1		
1	STC3827T	Woodcote Green-(LRC623)-Micatallic	1		
1	STC3915T	Atlantis Blue-(LRC632)-Micatallic	1		

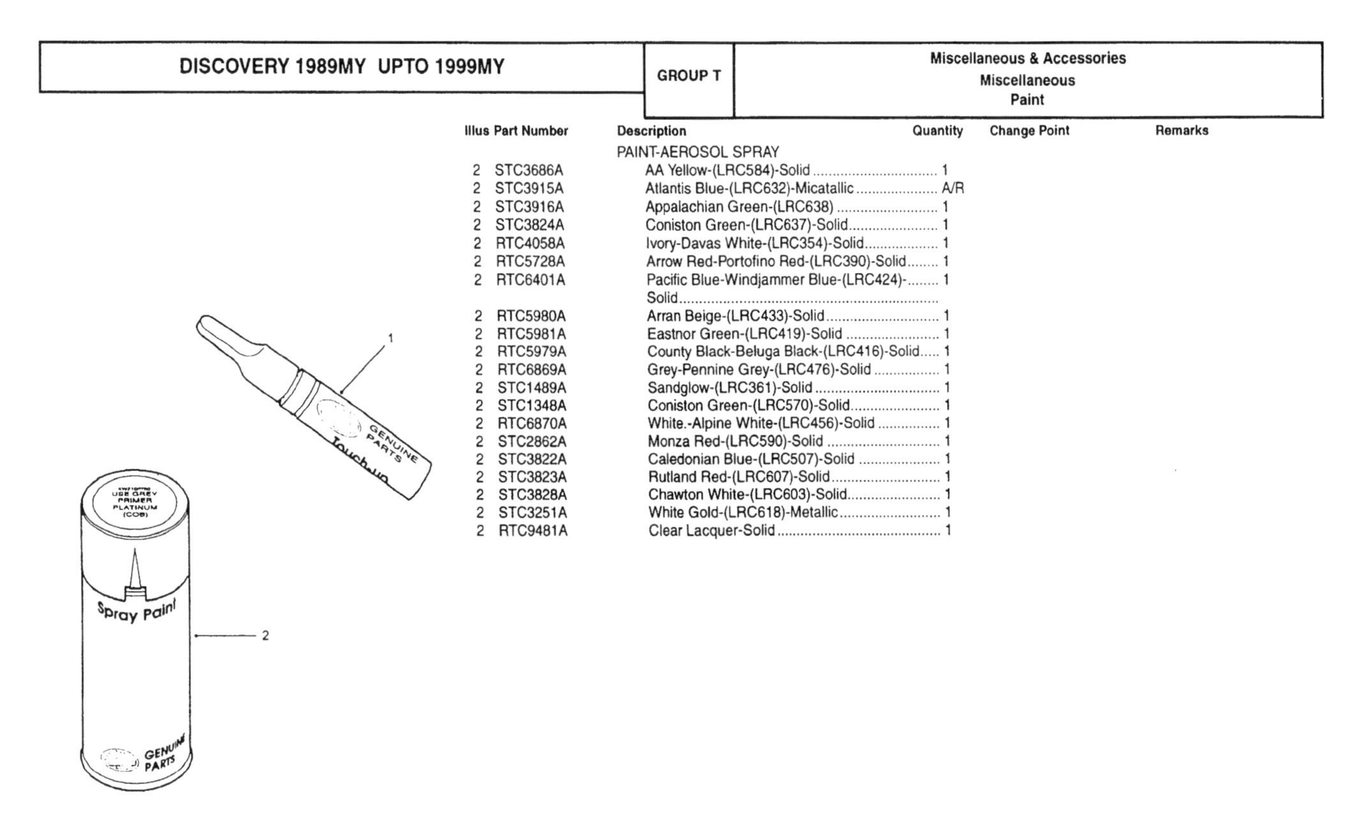

Illus	Part Number	Description	Quantity	Change Point	Remarks
		PAINT-AEROSOL SPRAY			
2	STC3686A	AA Yellow-(LRC584)-Solid	1		
2	STC3915A	Atlantis Blue-(LRC632)-Micatallic	A/R		
2	STC3916A	Appalachian Green-(LRC638)	1		
2	STC3824A	Coniston Green-(LRC637)-Solid	1		
2	RTC4058A	Ivory-Davas White-(LRC354)-Solid	1		
2	RTC5728A	Arrow Red-Portofino Red-(LRC390)-Solid	1		
2	RTC6401A	Pacific Blue-Windjammer Blue-(LRC424)-Solid	1		
2	RTC5980A	Arran Beige-(LRC433)-Solid	1		
2	RTC5981A	Eastnor Green-(LRC419)-Solid	1		
2	RTC5979A	County Black-Beluga Black-(LRC416)-Solid	1		
2	RTC6869A	Grey-Pennine Grey-(LRC476)-Solid	1		
2	STC1489A	Sandglow-(LRC361)-Solid	1		
2	STC1348A	Coniston Green-(LRC570)-Solid	1		
2	RTC6870A	White.-Alpine White-(LRC456)-Solid	1		
2	STC2862A	Monza Red-(LRC590)-Solid	1		
2	STC3822A	Caledonian Blue-(LRC507)-Solid	1		
2	STC3823A	Rutland Red-(LRC607)-Solid	1		
2	STC3828A	Chawton White-(LRC603)-Solid	1		
2	STC3251A	White Gold-(LRC618)-Metallic	1		
2	RTC9481A	Clear Lacquer-Solid	1		

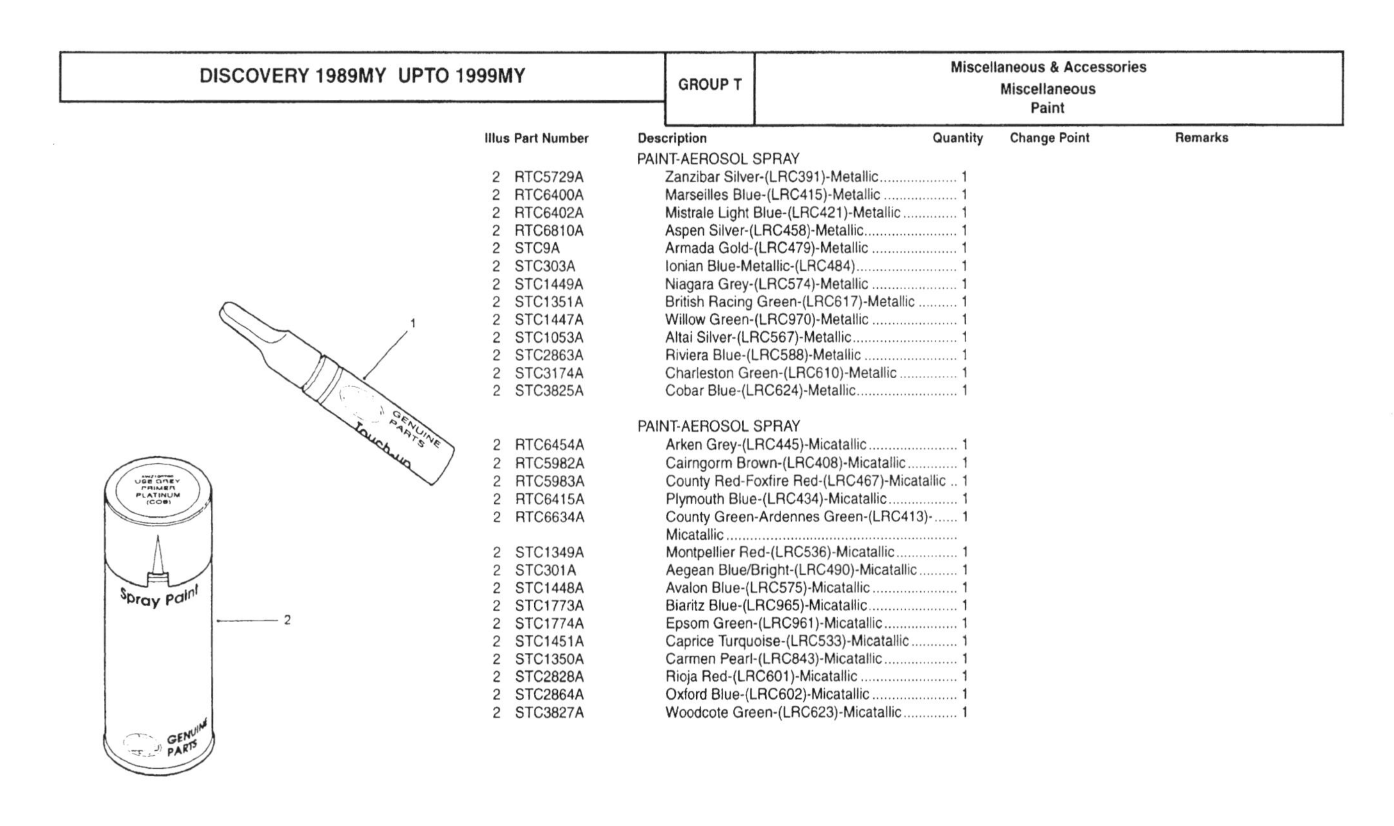

Illus	Part Number	Description	Quantity	Change Point	Remarks
		PAINT-AEROSOL SPRAY			
2	RTC5729A	Zanzibar Silver-(LRC391)-Metallic	1		
2	RTC6400A	Marseilles Blue-(LRC415)-Metallic	1		
2	RTC6402A	Mistrale Light Blue-(LRC421)-Metallic	1		
2	RTC6810A	Aspen Silver-(LRC458)-Metallic	1		
2	STC9A	Armada Gold-(LRC479)-Metallic	1		
2	STC303A	Ionian Blue-Metallic-(LRC484)	1		
2	STC1449A	Niagara Grey-(LRC574)-Metallic	1		
2	STC1351A	British Racing Green-(LRC617)-Metallic	1		
2	STC1447A	Willow Green-(LRC970)-Metallic	1		
2	STC1053A	Altai Silver-(LRC567)-Metallic	1		
2	STC2863A	Riviera Blue-(LRC588)-Metallic	1		
2	STC3174A	Charleston Green-(LRC610)-Metallic	1		
2	STC3825A	Cobar Blue-(LRC624)-Metallic	1		
		PAINT-AEROSOL SPRAY			
2	RTC6454A	Arken Grey-(LRC445)-Micatallic	1		
2	RTC5982A	Cairngorm Brown-(LRC408)-Micatallic	1		
2	RTC5983A	County Red-Foxfire Red-(LRC467)-Micatallic	1		
2	RTC6415A	Plymouth Blue-(LRC434)-Micatallic	1		
2	RTC6634A	County Green-Ardennes Green-(LRC413)-Micatallic	1		
2	STC1349A	Montpellier Red-(LRC536)-Micatallic	1		
2	STC301A	Aegean Blue/Bright-(LRC490)-Micatallic	1		
2	STC1448A	Avalon Blue-(LRC575)-Micatallic	1		
2	STC1773A	Biaritz Blue-(LRC965)-Micatallic	1		
2	STC1774A	Epsom Green-(LRC961)-Micatallic	1		
2	STC1451A	Caprice Turquoise-(LRC533)-Micatallic	1		
2	STC1350A	Carmen Pearl-(LRC843)-Micatallic	1		
2	STC2828A	Rioja Red-(LRC601)-Micatallic	1		
2	STC2864A	Oxford Blue-(LRC602)-Micatallic	1		
2	STC3827A	Woodcote Green-(LRC623)-Micatallic	1		

Illus	Part Number	Description	Quantity	Change Point	Remarks
		5 DOOR			
		KIT-BODY DECALS/BADGES			
1	STC8013RWB	Turquoise/Light Silver	1		
1	STC8014RWA	Orange/Dark Silver	1		
1	STC8085	Light Silver/Dark Silver	1		
1	STC8483	Comet	1		
1	STC8482	Grey	1		

Illus	Part Number	Description	Quantity	Change Point	Remarks
		3 DOOR ONLY - UP TO LA81990-NOTE(1)			
1	STC8037	Kit-body rubbing strip	1	Note (1)	3 Door
2	STC8140	Kit-wheelarch flare	1		
		3 DOOR ONLY - FROM MA081991-NOTE(2)			
1	STC8478	Kit-body rubbing strip	1	Note (2)	3 Door
2	STC8497	Kit-wheelarch flare	1		
		5 DOOR ONLY - UP TO LA081990-NOTE(1&3)			
1	STC8037	Kit-body rubbing strip	1	Note (3)	5 Door
2	STC8139	Kit-wheelarch flare	1		
		5 DOOR ONLY - FROM MA081991-NOTE(2)			
1	STC8463	Kit-body rubbing strip	1	Note (2)	5 Door
2	STC8498	Kit-wheelarch flare	1		

NOTE:1:KIT INCLUDES WHEELARCH COVERS AND SIDE
MOULDINGS.ONLY SUITABLE FOR VEHICLES WITHOUT BODY STRIPES.

NOTE(2) SUITABLE FOR NARROW RUBBING STRIPS.

NOTE(3)ONLY FOR FITMENT WITH BODY SIDE MOULDINGS.

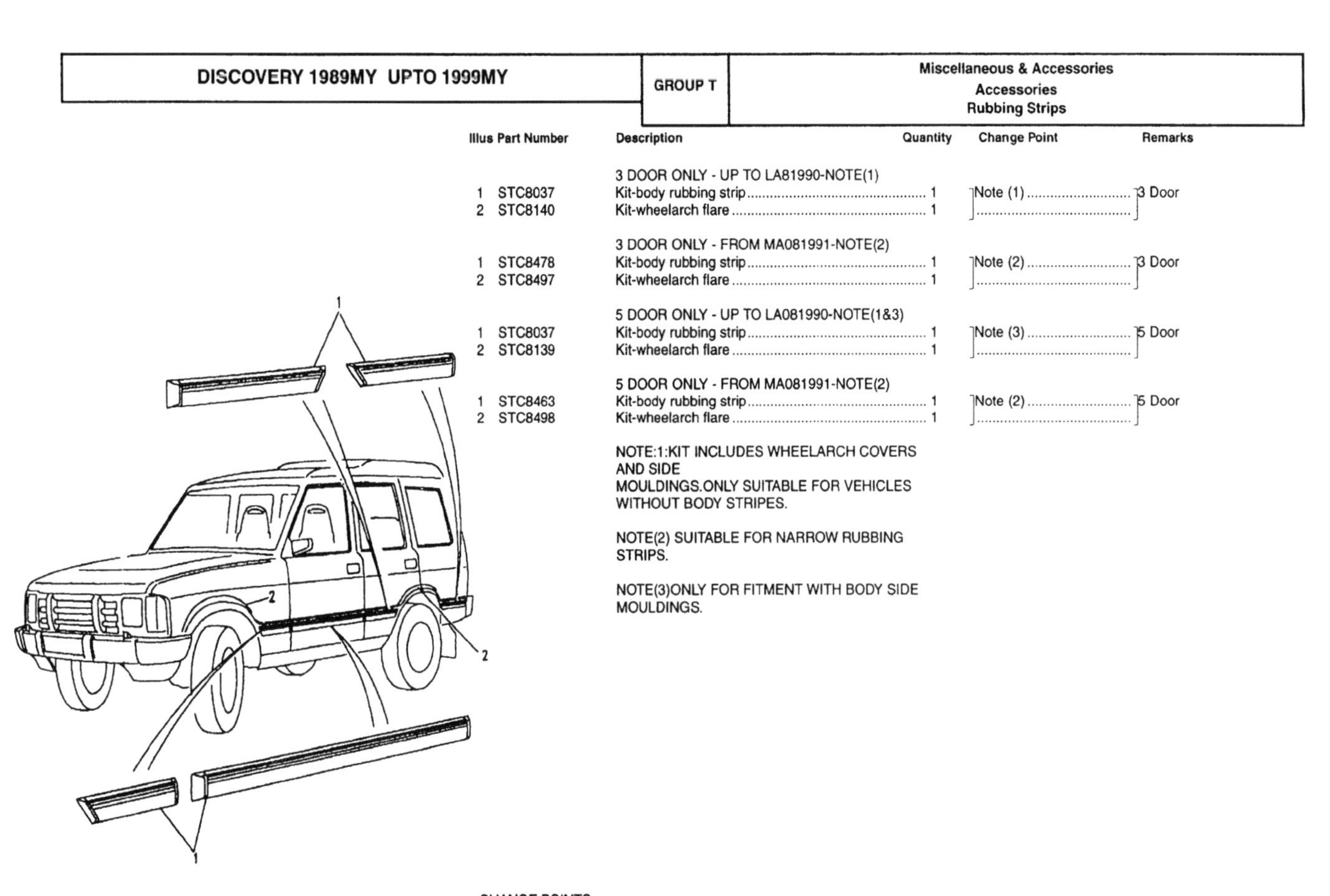

CHANGE POINTS:
(1) To (V) MA 081990
(2) From (V) MA 081991
(3) To (V) LA 081990

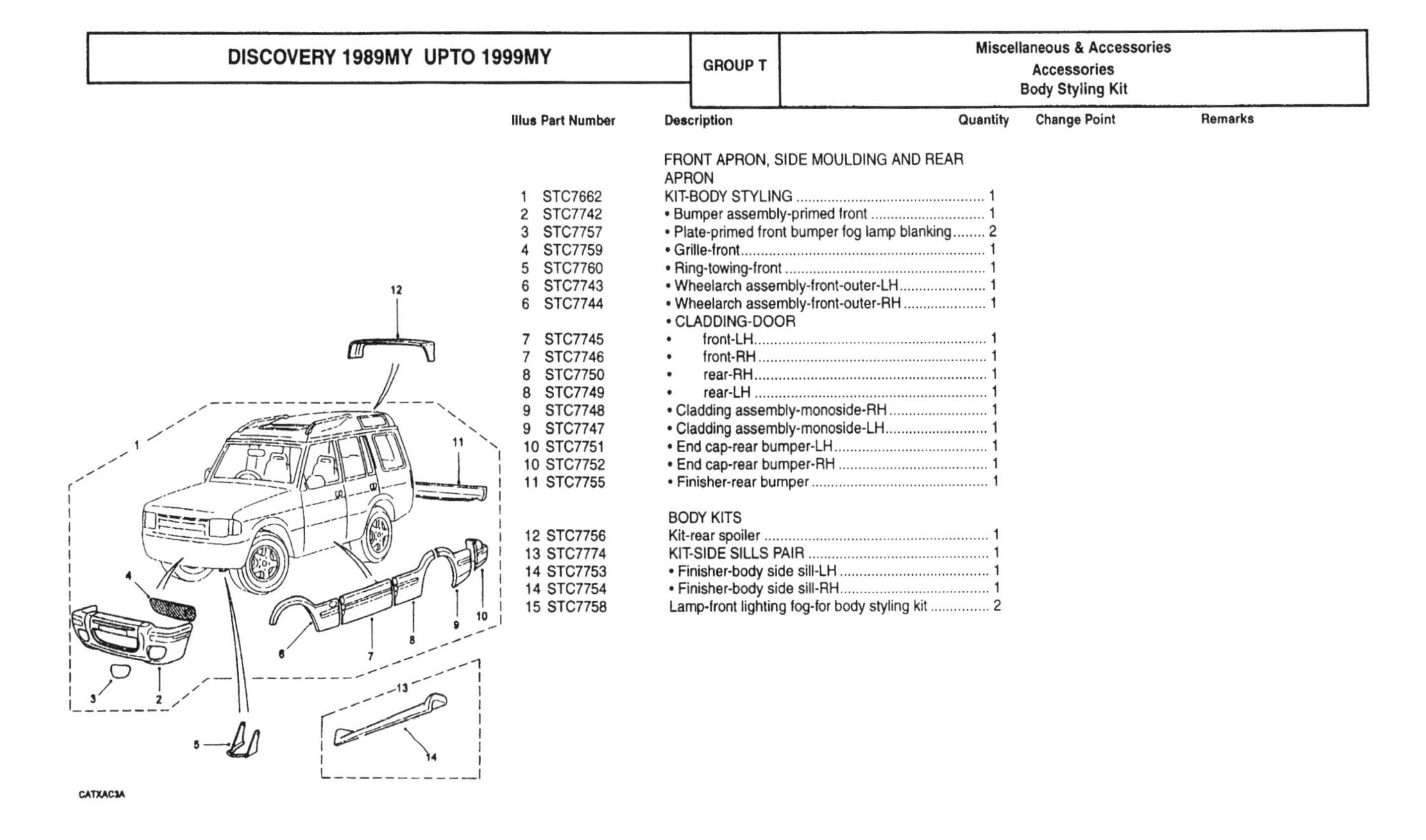

Illus	Part Number	Description	Quantity	Change Point	Remarks
		FRONT APRON, SIDE MOULDING AND REAR APRON			
1	STC7662	KIT-BODY STYLING	1		
2	STC7742	• Bumper assembly-primed front	1		
3	STC7757	• Plate-primed front bumper fog lamp blanking	2		
4	STC7759	• Grille-front	1		
5	STC7760	• Ring-towing-front	1		
6	STC7743	• Wheelarch assembly-front-outer-LH	1		
6	STC7744	• Wheelarch assembly-front-outer-RH	1		
		• CLADDING-DOOR			
7	STC7745	• front-LH	1		
7	STC7746	• front-RH	1		
8	STC7750	• rear-RH	1		
8	STC7749	• rear-LH	1		
9	STC7748	• Cladding assembly-monoside-RH	1		
9	STC7747	• Cladding assembly-monoside-LH	1		
10	STC7751	• End cap-rear bumper-LH	1		
10	STC7752	• End cap-rear bumper-RH	1		
11	STC7755	• Finisher-rear bumper	1		
		BODY KITS			
12	STC7756	Kit-rear spoiler	1		
13	STC7774	KIT-SIDE SILLS PAIR	1		
14	STC7753	• Finisher-body side sill-LH	1		
14	STC7754	• Finisher-body side sill-RH	1		
15	STC7758	Lamp-front lighting fog-for body styling kit	2		

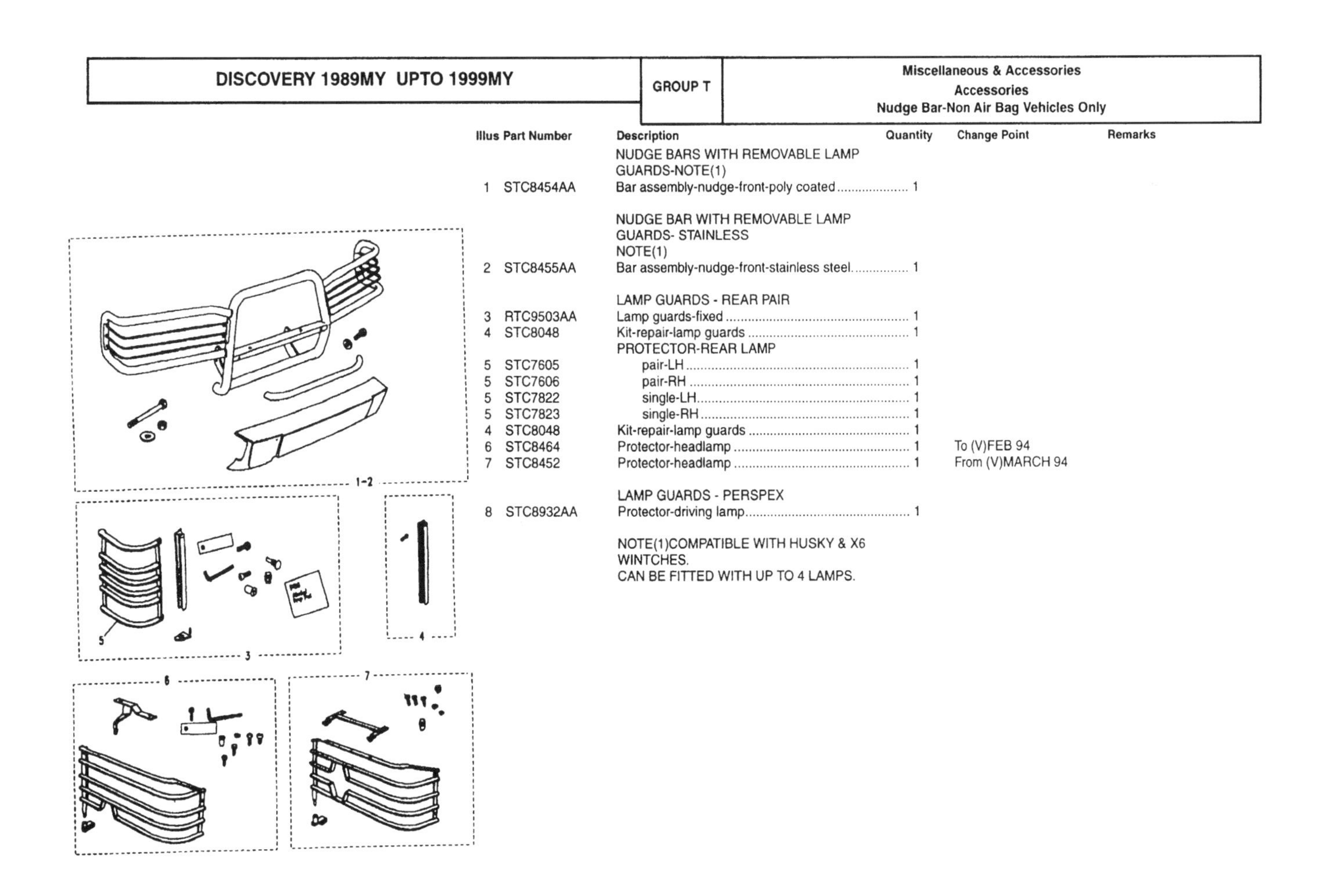

Illus	Part Number	Description	Quantity	Change Point	Remarks
		NUDGE BARS WITH REMOVABLE LAMP GUARDS-NOTE(1)			
1	STC8454AA	Bar assembly-nudge-front-poly coated	1		
		NUDGE BAR WITH REMOVABLE LAMP GUARDS- STAINLESS NOTE(1)			
2	STC8455AA	Bar assembly-nudge-front-stainless steel	1		
		LAMP GUARDS - REAR PAIR			
3	RTC9503AA	Lamp guards-fixed	1		
4	STC8048	Kit-repair-lamp guards	1		
		PROTECTOR-REAR LAMP			
5	STC7605	pair-LH	1		
5	STC7606	pair-RH	1		
5	STC7822	single-LH	1		
5	STC7823	single-RH	1		
4	STC8048	Kit-repair-lamp guards	1		
6	STC8464	Protector-headlamp	1	To (V)FEB 94	
7	STC8452	Protector-headlamp	1	From (V)MARCH 94	
		LAMP GUARDS - PERSPEX			
8	STC8932AA	Protector-driving lamp	1		

NOTE(1)COMPATIBLE WITH HUSKY & X6 WINTCHES.
CAN BE FITTED WITH UP TO 4 LAMPS.

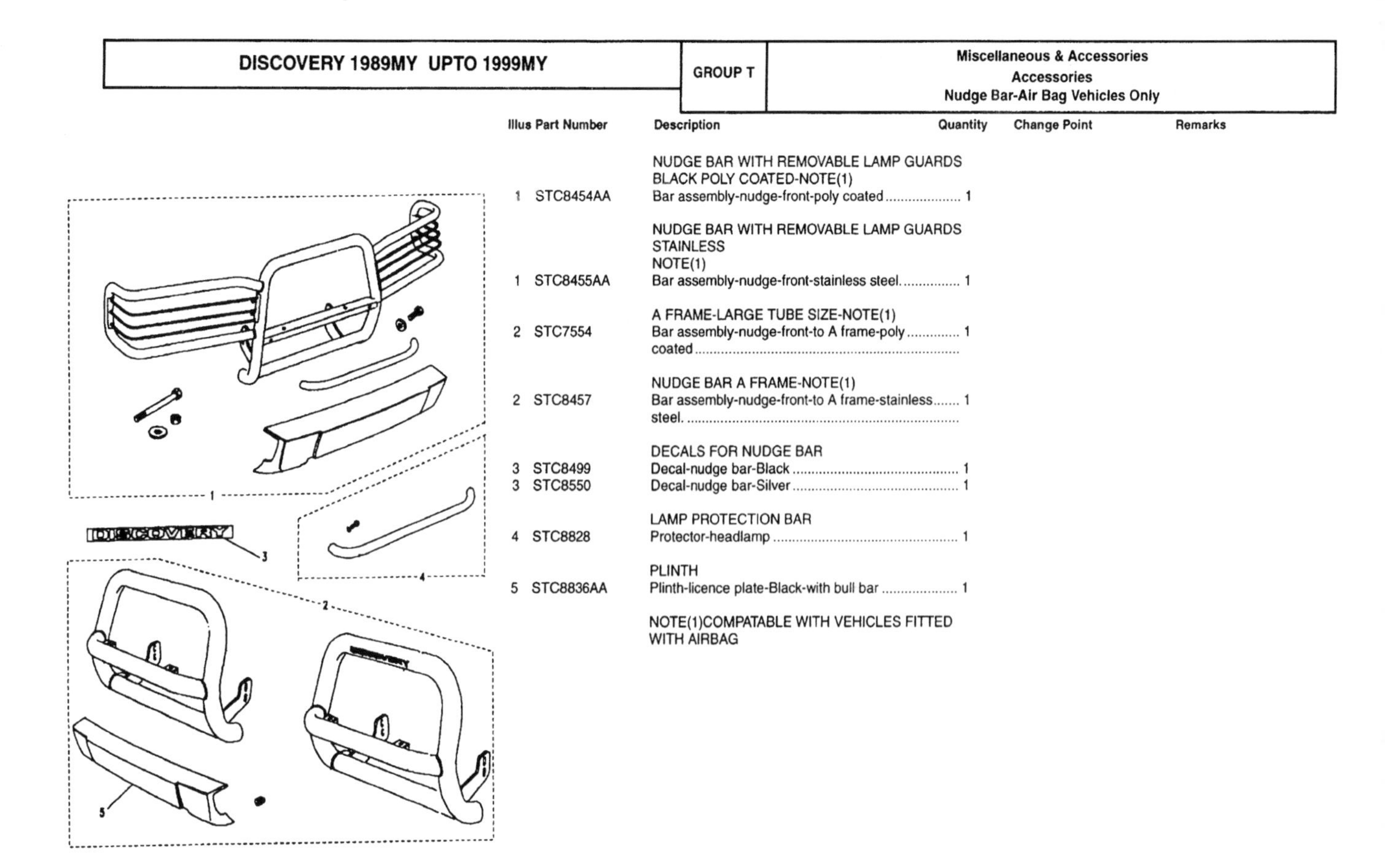

Illus	Part Number	Description	Quantity	Change Point	Remarks
		NUDGE BAR WITH REMOVABLE LAMP GUARDS BLACK POLY COATED-NOTE(1)			
1	STC8454AA	Bar assembly-nudge-front-poly coated	1		
		NUDGE BAR WITH REMOVABLE LAMP GUARDS STAINLESS NOTE(1)			
1	STC8455AA	Bar assembly-nudge-front-stainless steel.	1		
		A FRAME-LARGE TUBE SIZE-NOTE(1)			
2	STC7554	Bar assembly-nudge-front-to A frame-poly coated	1		
		NUDGE BAR A FRAME-NOTE(1)			
2	STC8457	Bar assembly-nudge-front-to A frame-stainless steel.	1		
		DECALS FOR NUDGE BAR			
3	STC8499	Decal-nudge bar-Black	1		
3	STC8550	Decal-nudge bar-Silver	1		
		LAMP PROTECTION BAR			
4	STC8828	Protector-headlamp	1		
		PLINTH			
5	STC8836AA	Plinth-licence plate-Black-with bull bar	1		

NOTE(1)COMPATABLE WITH VEHICLES FITTED WITH AIRBAG

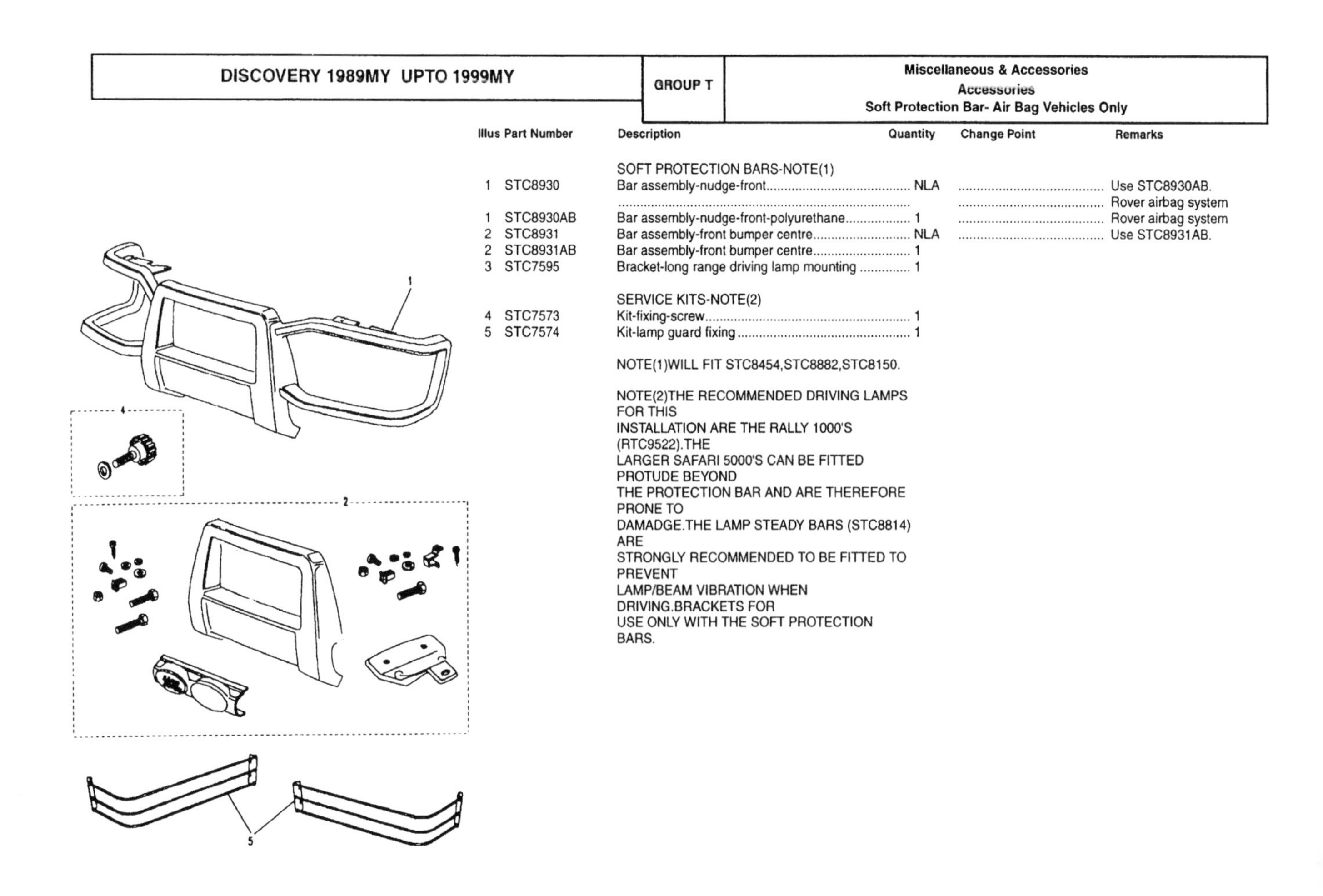

Illus	Part Number	Description	Quantity	Change Point	Remarks
		SOFT PROTECTION BARS-NOTE(1)			
1	STC8930	Bar assembly-nudge-front	NLA		Use STC8930AB.
					Rover airbag system
1	STC8930AB	Bar assembly-nudge-front-polyurethane	1		Rover airbag system
2	STC8931	Bar assembly-front bumper centre	NLA		Use STC8931AB.
2	STC8931AB	Bar assembly-front bumper centre	1		
3	STC7595	Bracket-long range driving lamp mounting	1		
		SERVICE KITS-NOTE(2)			
4	STC7573	Kit-fixing-screw	1		
5	STC7574	Kit-lamp guard fixing	1		

NOTE(1)WILL FIT STC8454,STC8882,STC8150.

NOTE(2)THE RECOMMENDED DRIVING LAMPS FOR THIS INSTALLATION ARE THE RALLY 1000'S (RTC9522).THE LARGER SAFARI 5000'S CAN BE FITTED PROTUDE BEYOND THE PROTECTION BAR AND ARE THEREFORE PRONE TO DAMADGE.THE LAMP STEADY BARS (STC8814) ARE STRONGLY RECOMMENDED TO BE FITTED TO PREVENT LAMP/BEAM VIBRATION WHEN DRIVING.BRACKETS FOR USE ONLY WITH THE SOFT PROTECTION BARS.

Illus	Part Number	Description	Quantity	Change Point	Remarks
	STC8170AC	CAR SET-SEAT COVER-GREY-WATERPROOF	1		Note (1)
		• COVER-SEAT-WATERPROOF			
	STC8171AC	• front-Grey	1		
	STC8173AA	• rear inward facing seat-Grey	1		
	STC8172AA	• rear forward facing seat-Grey	1		
	STC8174AC	CAR SET-SEAT COVER-BEIGE-WATERPROOF	1		Note (1)
	STC8175AC	• Cover-seat-waterproof-front seat-Beige	1		
	STC8177AA	• Cover-seat-waterproof-rear inward facing seat-Beige	1		
	STC8176AA	• Cover-seat-waterproof-rear forward facing seat-Beige	1		
	STC8944	CAR SET-SEAT COVER-GREY-WATERPROOF	1		Electric seats
	STC8945	• Cover-seat-front-waterproof-Grey	NLA		Use STC8945AA.
	STC8173	• Cover-waterproof-seat-rear inward facing seat-Grey	1		
	STC8172	• Cover-waterproof-seat-rear forward facing seat-Grey	1		
	STC8945AA	Cover-seat-front-waterproof-Grey			Electric seats

Remarks:
(1) Except electric seats

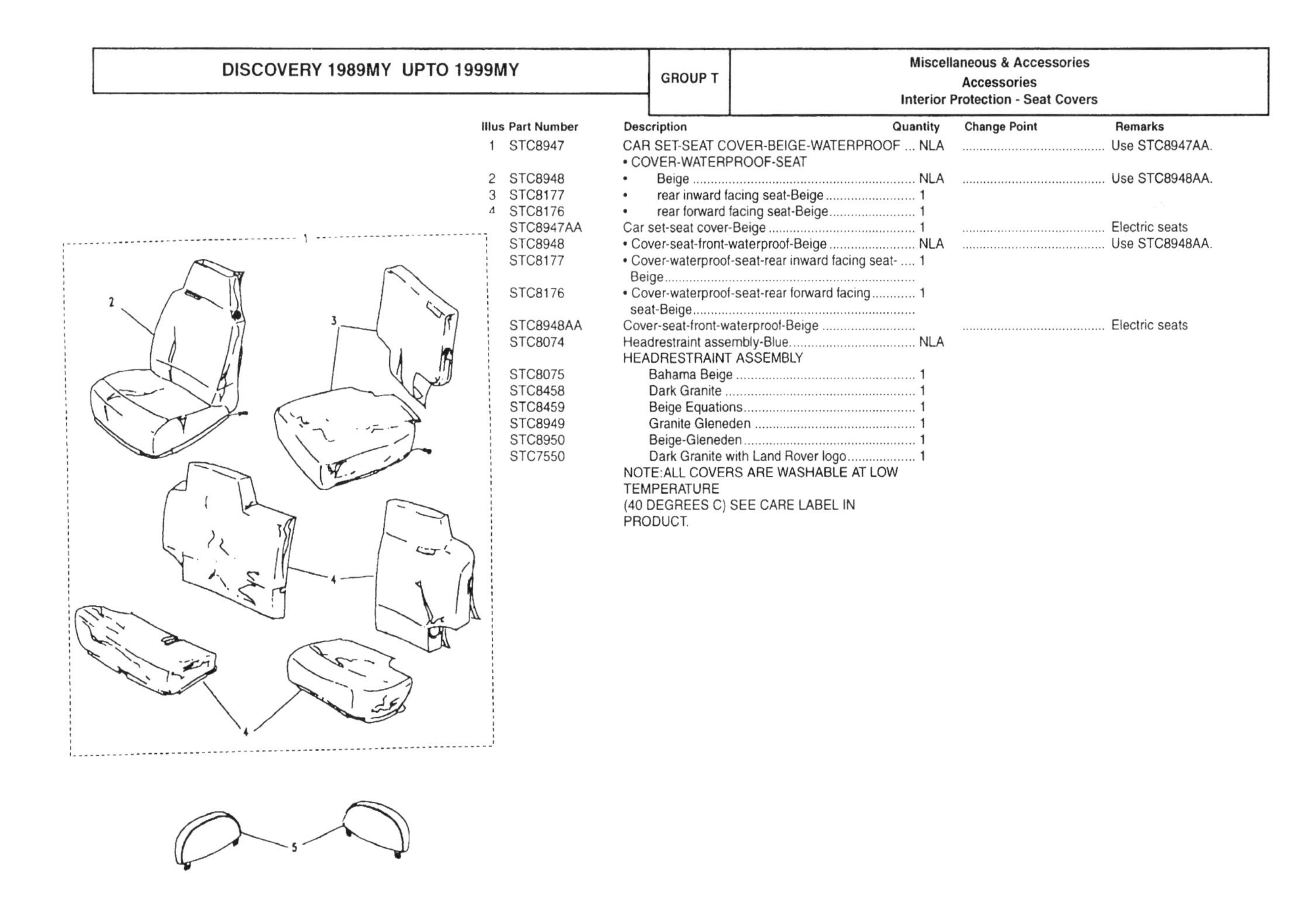

Illus	Part Number	Description	Quantity	Change Point	Remarks
1	STC8947	CAR SET-SEAT COVER-BEIGE-WATERPROOF	NLA		Use STC8947AA.
		• COVER-WATERPROOF-SEAT			
2	STC8948	• Beige	NLA		Use STC8948AA.
3	STC8177	• rear inward facing seat-Beige	1		
4	STC8176	• rear forward facing seat-Beige	1		
	STC8947AA	Car set-seat cover-Beige	1		Electric seats
	STC8948	• Cover-seat-front-waterproof-Beige	NLA		Use STC8948AA.
	STC8177	• Cover-waterproof-seat-rear inward facing seat-Beige	1		
	STC8176	• Cover-waterproof-seat-rear forward facing seat-Beige	1		
	STC8948AA	Cover-seat-front-waterproof-Beige			Electric seats
	STC8074	Headrestraint assembly-Blue	NLA		
		HEADRESTRAINT ASSEMBLY			
	STC8075	Bahama Beige	1		
	STC8458	Dark Granite	1		
	STC8459	Beige Equations	1		
	STC8949	Granite Gleneden	1		
	STC8950	Beige-Gleneden	1		
	STC7550	Dark Granite with Land Rover logo	1		
		NOTE:ALL COVERS ARE WASHABLE AT LOW TEMPERATURE (40 DEGREES C) SEE CARE LABEL IN PRODUCT.			

Illus	Part Number	Description	Quantity	Change Point	Remarks
		GUARD-DOG-LOADSPACE			
1	STC8414	MESH TYPE	1		Except rear air conditioning
2	STC8413	BAR TYPE	1		
2	STC7570	BAR TYPE	1		Rear air conditioning
4	STC8104	• Kit-dog guard fitting	1		
5	STC8017	Rack-riding tack	1		

NOTE: COMPATIBLE WITH ROLLER LOADSPACE COVER

Illus	Part Number	Description	Quantity	Change Point	Remarks
		REFRIGERATOR - CFC FREE			
1	STC8519	Refridgerator	NLA		Use STC8519AA fridge & STC7896 power lead for Freelander or STC8519AA fridge & STC7897 power lead for Discovery,Classic & New Range Rover. or STC8519AA.
1	STC8519AA	Refridgerator	1		
2	STC7897	Socket-accessory power-fridge	1		
	STC7814	Transformer-mains-240v to 12v	1		
	STC7809	Coolbox-20 litre	1		
	STC7809AA	Coolbox-20 litre	1		
2	STC7891	Socket-accessory power-coolbox	1		
2	STC8427AA	Socket-accessory power-twin top coolbox	1		

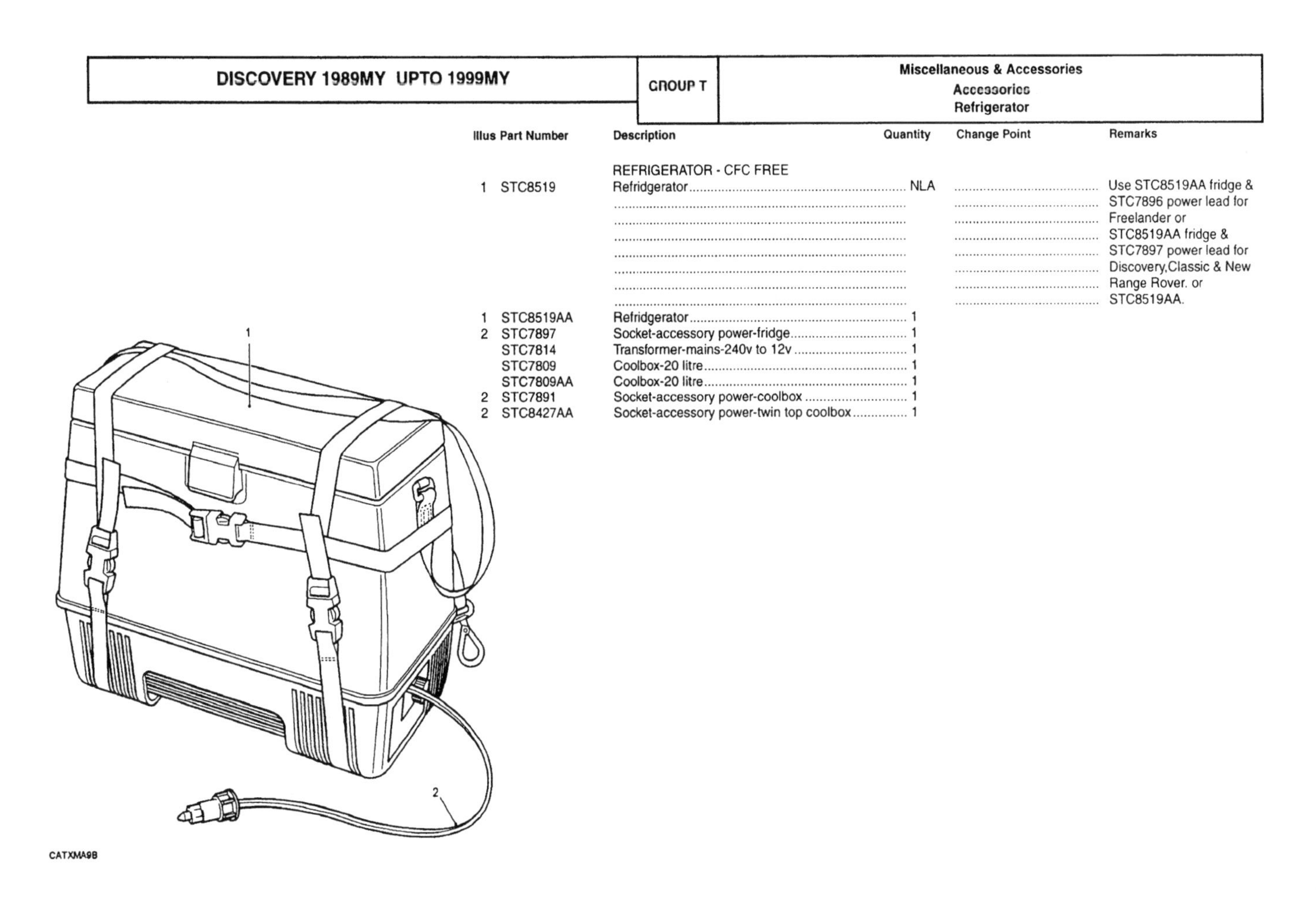

Illus	Part Number	Description	Quantity	Change Point	Remarks
		WITH AIR BAG			
1	STC7859	Kit interior wood trim-LHD			Drivers and passengers airbag

ILLUSTRATION TO FOLLOW
L'ILLUSTRATION A SUIVRE
DAS BILD IST ZU FOLGEN
IL DISEGNO SEGUIRA
EL DISENO SIGUE
A ILLUSTRACAO SEGUIRA MAIS TARDE

TF 0007

Illus	Part Number	Description	Quantity	Change Point	Remarks
1	STC50013	Seat-child restraint	1		
2	STC8154	Cushion-child restraint booster	1		

NOTE: NOT TO BE FITTED ON INWARD FACING REAR SEATS AND REARWARD FACING ON THE FRONT
PASSENGER SEAT IN VEHICLES WITH AIR BAG(SRS) FITTED.
NOTE: QUERIES RELATING TO FITTING, ALTERNATIVE
APPLICATIONS AND SERVICE LEVELS SHOULD BE
DIRECTED TO ACCESSORIES DEPARTMENT AT LAND ROVER.

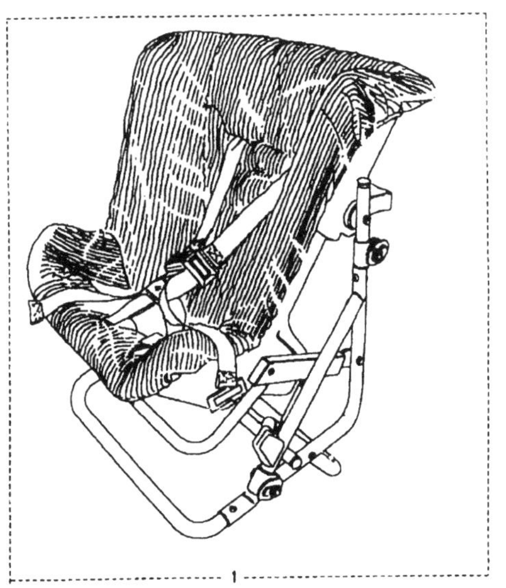

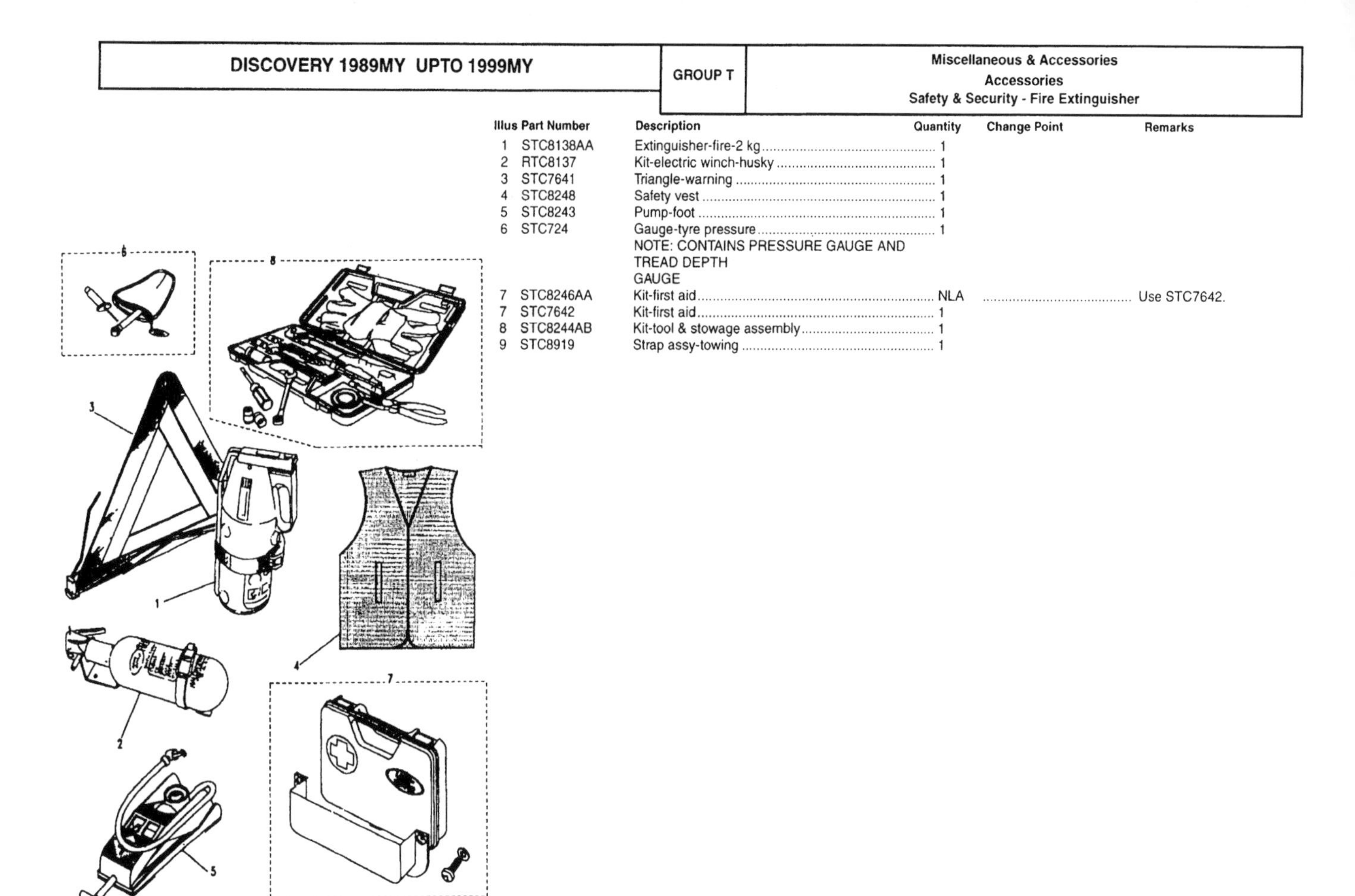

DISCOVERY 1989MY UPTO 1999MY — GROUP T

Miscellaneous & Accessories
Accessories
Safety & Security - Fire Extinguisher

Illus	Part Number	Description	Quantity	Change Point	Remarks
1	STC8138AA	Extinguisher-fire-2 kg	1		
2	RTC8137	Kit-electric winch-husky	1		
3	STC7641	Triangle-warning	1		
4	STC8248	Safety vest	1		
5	STC8243	Pump-foot	1		
6	STC724	Gauge-tyre pressure	1		
		NOTE: CONTAINS PRESSURE GAUGE AND TREAD DEPTH GAUGE			
7	STC8246AA	Kit-first aid	NLA		Use STC7642.
7	STC7642	Kit-first aid	1		
8	STC8244AB	Kit-tool & stowage assembly	1		
9	STC8919	Strap assy-towing	1		

T 17

DISCOVERY 1989MY UPTO 1999MY — GROUP T

Miscellaneous & Accessories
Accessories
Safety & Security - Security Box

Illus	Part Number	Description	Quantity	Change Point	Remarks
1	STC8018L	Box-gun	1		
2	STC8124	Kit-gun box fixing	1		

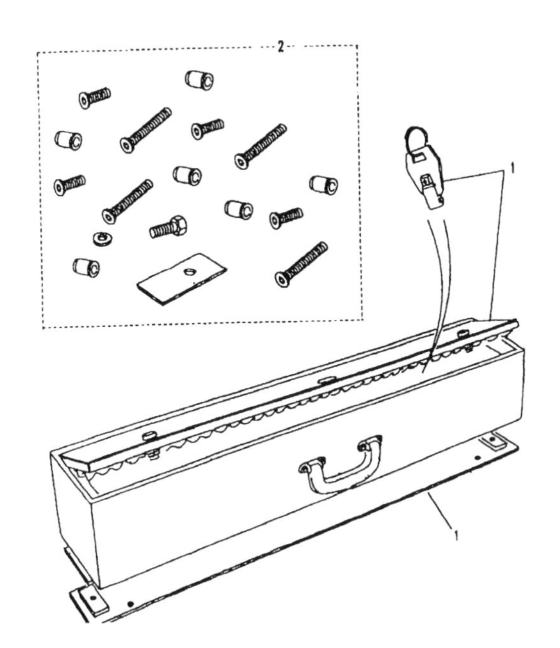

T 18

Illus	Part Number	Description	Quantity	Change Point	Remarks
		CUBBY BOX - UP TO FEB 94			
1	STC8089	BIN ASSEMBLY-TUNNEL CONSOLE STOWAGE-BEIGE	1		
1	STC8088	BIN ASSEMBLY-TUNNEL CONSOLE STOWAGE-BLUE.	1		
	STC7618SUC	• Hinges-plastic-Beige	1		Part of STC8089.
	STC7611	• Cubby box-stay	1		
	STC7612	• Lock & keys	1		
	STC7617JUL	• Hinges-plastic-Blue	2		Part of STC8088.
		LID CUBBY BOX - FROM MARCH 94			
2	STC8934LOY	Lid assembly-tunnel console-Granite	1		
2	STC8934SUC	Cubby box lid-Bahama Beige	1		

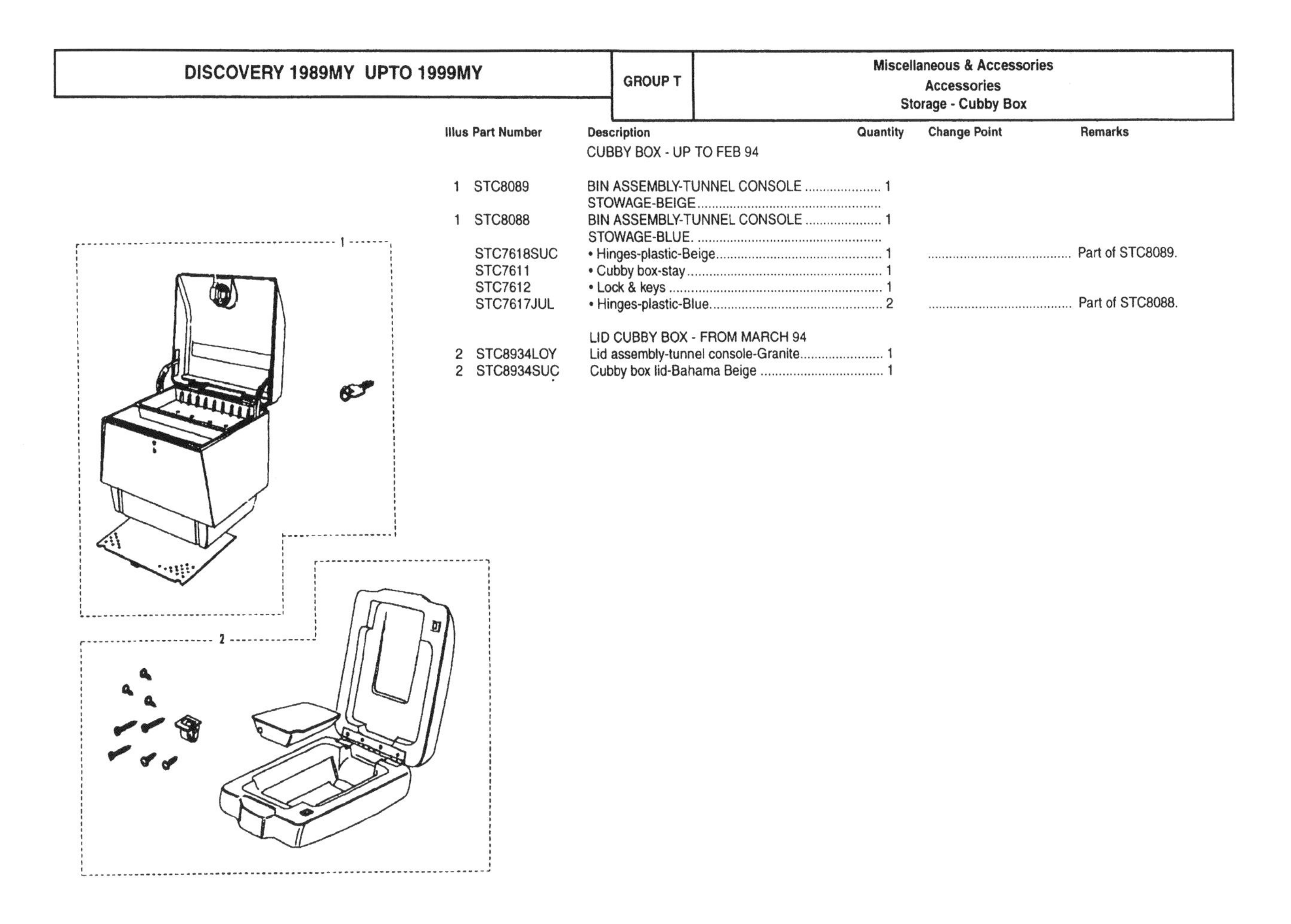

Illus	Part Number	Description	Quantity	Change Point	Remarks
	RTC9499	KIT-TOWING	1	Note (1)	
	STC4286D	• Socket-electrical towing-7 pin-12N	1		
1	NTC7080PUB	Plate towing	1	Note (1)	
2	NTC7634	Bar-vehicle recovery rear-RH	1		
2	NTC7635	Bar-vehicle recovery rear-LH	1		
3	BH608321L	Bolt-plate to body	2		
4	NTC7099PUB	Spacer	2		
5	WL600081L	Washer	2		
6	NH608041L	Nut	2		
7	SH606091L	Screw-towing plate to tie bar-hexagonal head-long	2		
8	NY606041L	Nut-hexagonal head-nyloc-3/8-UNF	2		
9	90577509	Bracket-towing-front & rear-less spoiler	2		
10	BH606361L	Bolt-3/8UNF	2		
11	WA110061L	Washer-plain-M10-standard	2		
12	NY606041L	Nut-hexagonal head-nyloc-3/8-UNF	2		
13	RTC8891	Ball-towing attachment-50mm	1		
	RTC8159	Jaw assembly-towing	1		
	STC8114	Cover-towing attachment ball	1		
14	BH610181L	Bolt-tow ball to plate	2	Note (1)	
15	WA116101L	Washer-plain	2		
16	NH610041L	Nut	4		
	STC8060AA	Tow bar-adjustable	1		
	STC8060AB	Tow bar-adjustable	1		
	RTC9581	Pin-towing pintle	1	Note (1)	
	RTC8831	Plate-tow bar bracket slider	1		
	STC8816	Tow bar kit	1		
	STC8875	Socket-accessory power-12 S	1		
		NOTE: INCLUDING WIRING AND SPLIT CHARGE.			
		NOTE: WITH STC8060AA USE RTC9581			

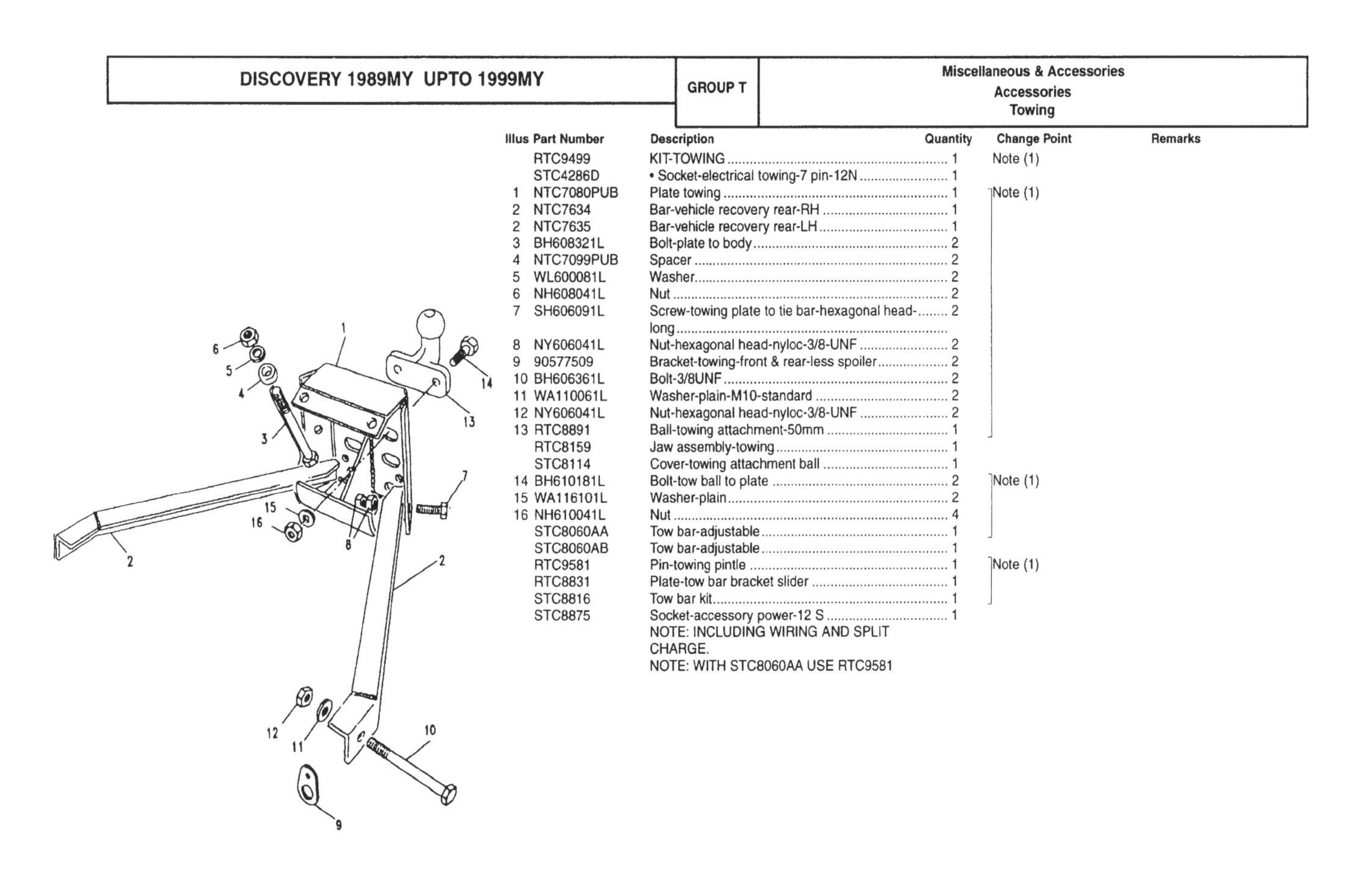

CHANGE POINTS:
(1) To (V) LA 081880

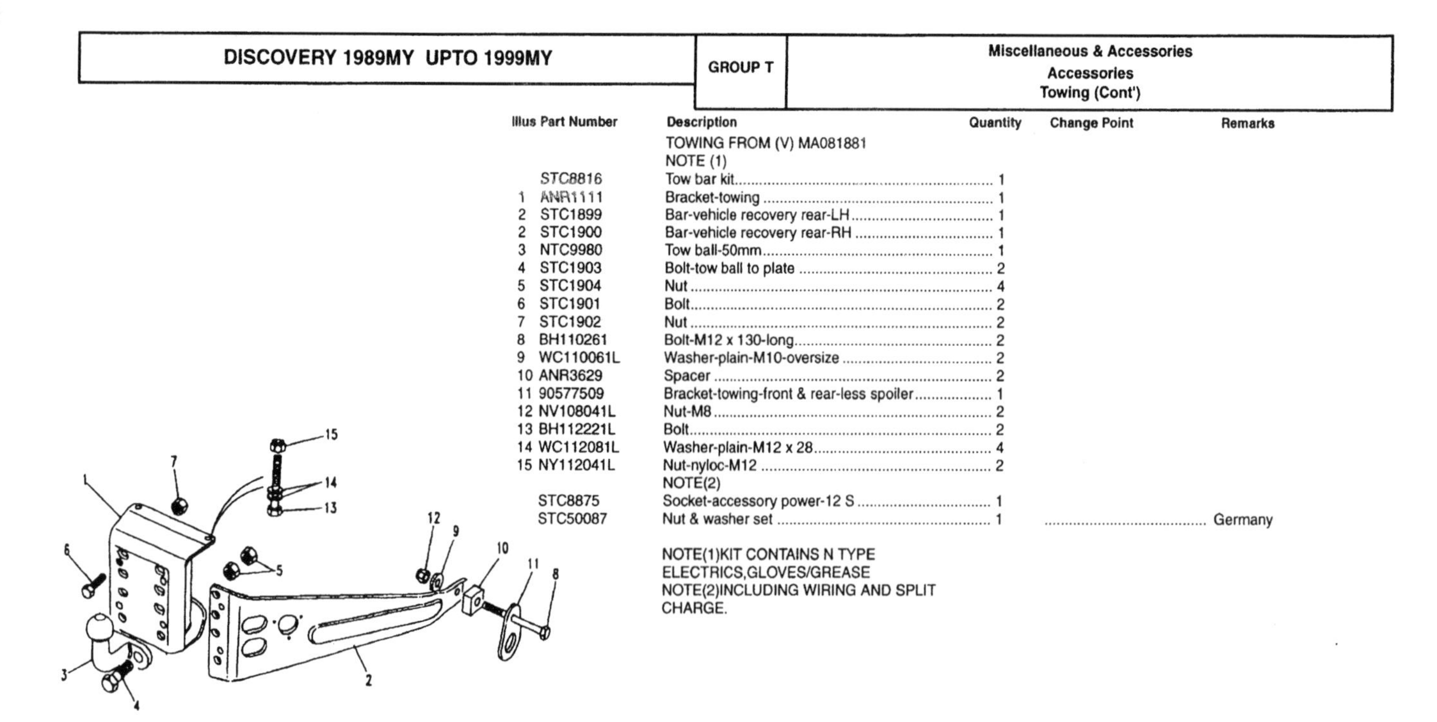

Illus	Part Number	Description	Quantity	Change Point	Remarks
		TOWING FROM (V) MA081881			
		NOTE (1)			
	STC8816	Tow bar kit	1		
1	ANR1111	Bracket-towing	1		
2	STC1899	Bar-vehicle recovery rear-LH	1		
2	STC1900	Bar-vehicle recovery rear-RH	1		
3	NTC9980	Tow ball-50mm	1		
4	STC1903	Bolt-tow ball to plate	2		
5	STC1904	Nut	4		
6	STC1901	Bolt	2		
7	STC1902	Nut	2		
8	BH110261	Bolt-M12 x 130-long	2		
9	WC110061L	Washer-plain-M10-oversize	2		
10	ANR3629	Spacer	2		
11	90577509	Bracket-towing-front & rear-less spoiler	1		
12	NV108041L	Nut-M8	2		
13	BH112221L	Bolt	2		
14	WC112081L	Washer-plain-M12 x 28	4		
15	NY112041L	Nut-nyloc-M12	2		
		NOTE(2)			
	STC8875	Socket-accessory power-12 S	1		
	STC50087	Nut & washer set	1		Germany

NOTE(1)KIT CONTAINS N TYPE ELECTRICS,GLOVES/GREASE
NOTE(2)INCLUDING WIRING AND SPLIT CHARGE.

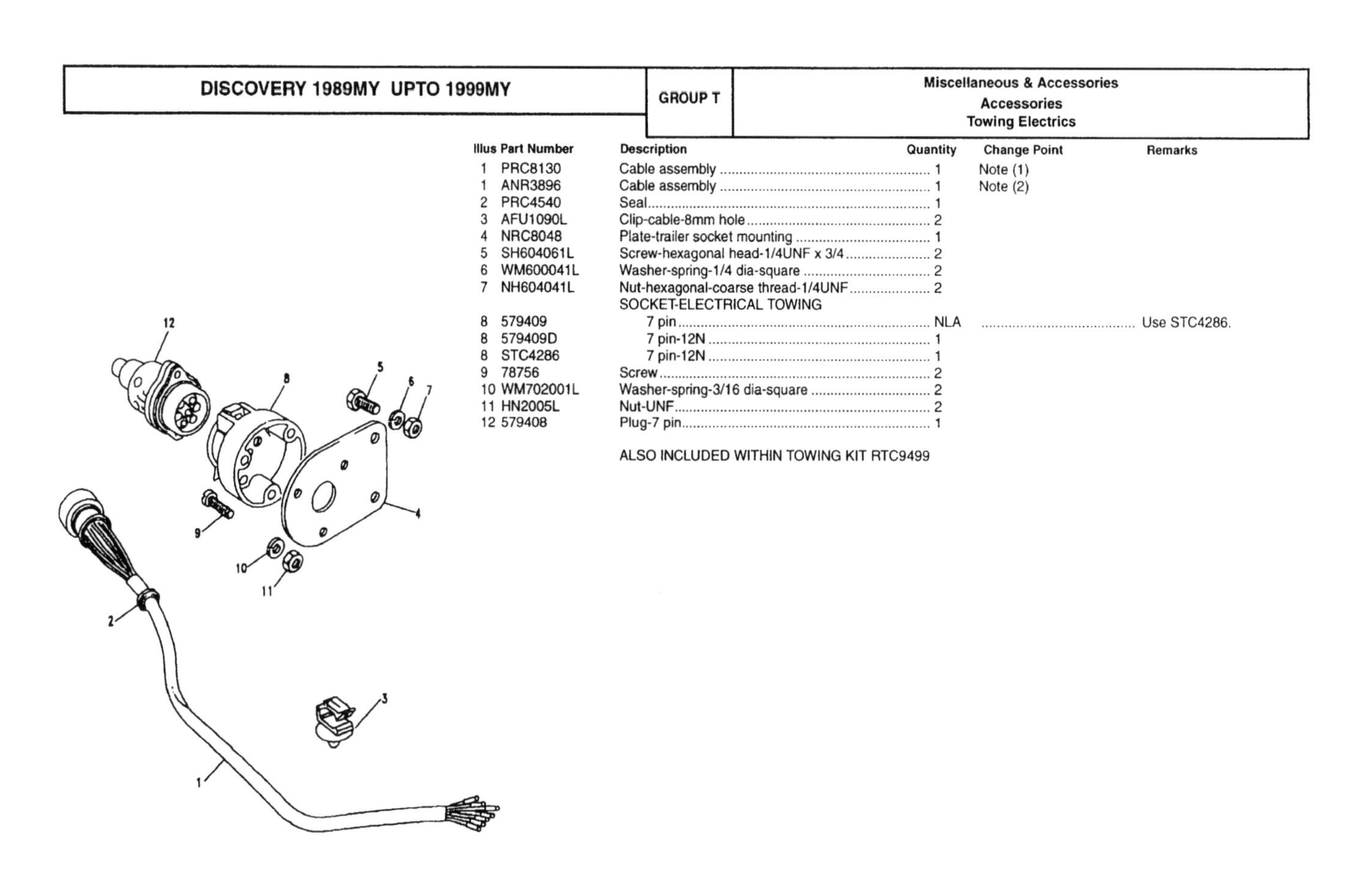

Illus	Part Number	Description	Quantity	Change Point	Remarks
1	PRC8130	Cable assembly	1	Note (1)	
1	ANR3896	Cable assembly	1	Note (2)	
2	PRC4540	Seal	1		
3	AFU1090L	Clip-cable-8mm hole	2		
4	NRC8048	Plate-trailer socket mounting	1		
5	SH604061L	Screw-hexagonal head-1/4UNF x 3/4	2		
6	WM600041L	Washer-spring-1/4 dia-square	2		
7	NH604041L	Nut-hexagonal-coarse thread-1/4UNF	2		
		SOCKET-ELECTRICAL TOWING			
8	579409	7 pin	NLA		Use STC4286.
8	579409D	7 pin-12N	1		
8	STC4286	7 pin-12N	1		
9	78756	Screw	2		
10	WM702001L	Washer-spring-3/16 dia-square	2		
11	HN2005L	Nut-UNF	2		
12	579408	Plug-7 pin	1		

ALSO INCLUDED WITHIN TOWING KIT RTC9499

CHANGE POINTS:
(1) To (V) MA 118309
(2) From (V) MA 118310

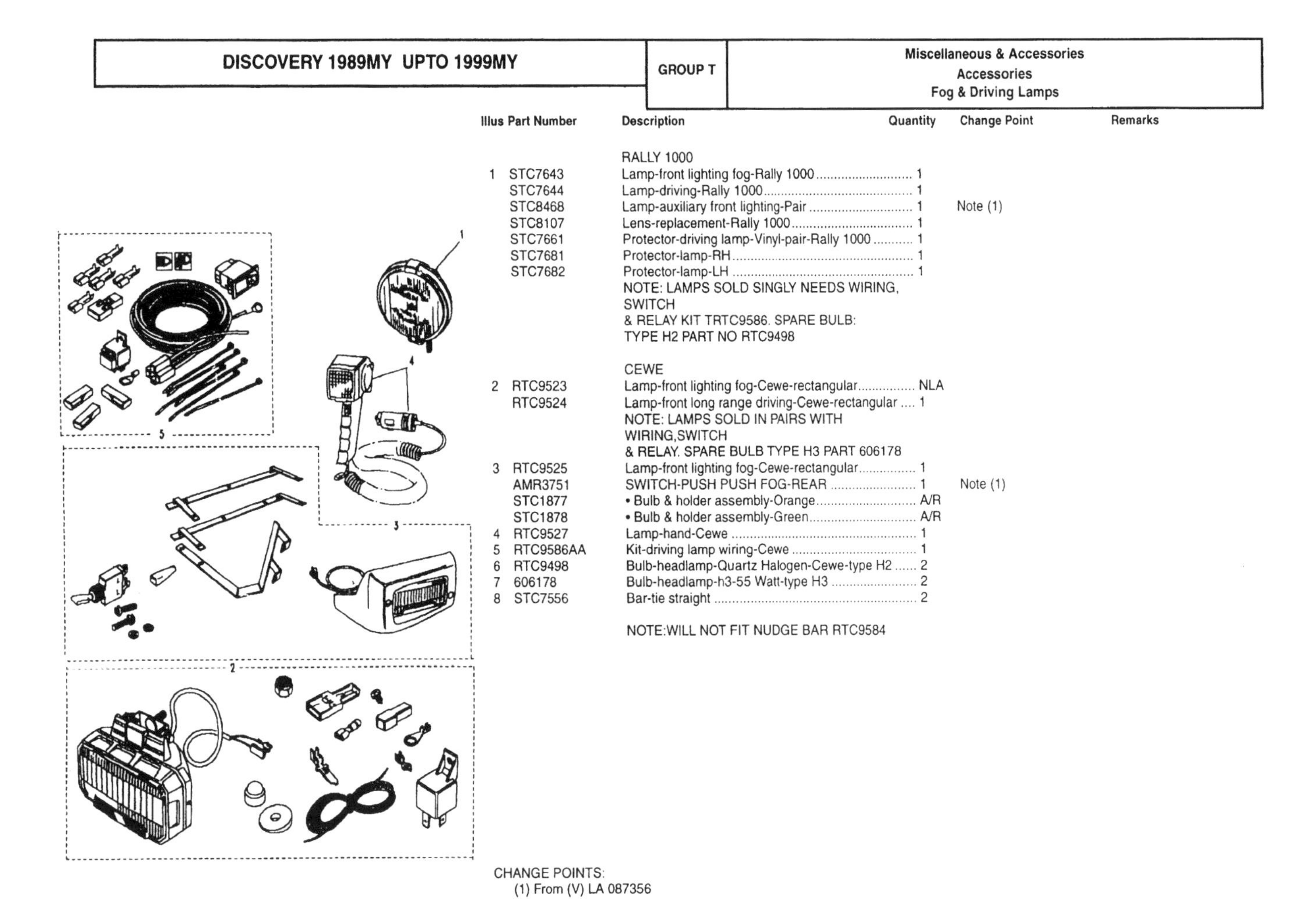

DISCOVERY 1989MY UPTO 1999MY	GROUP T	Miscellaneous & Accessories Accessories Fog & Driving Lamps

Illus	Part Number	Description	Quantity	Change Point	Remarks
		RALLY 1000			
1	STC7643	Lamp-front lighting fog-Rally 1000	1		
	STC7644	Lamp-driving-Rally 1000	1		
	STC8468	Lamp-auxiliary front lighting-Pair	1	Note (1)	
	STC8107	Lens-replacement-Rally 1000	1		
	STC7661	Protector-driving lamp-Vinyl-pair-Rally 1000	1		
	STC7681	Protector-lamp-RH	1		
	STC7682	Protector-lamp-LH	1		
		NOTE: LAMPS SOLD SINGLY NEEDS WIRING, SWITCH & RELAY KIT TRTC9586. SPARE BULB: TYPE H2 PART NO RTC9498			
		CEWE			
2	RTC9523	Lamp-front lighting fog-Cewe-rectangular	NLA		
	RTC9524	Lamp-front long range driving-Cewe-rectangular	1		
		NOTE: LAMPS SOLD IN PAIRS WITH WIRING,SWITCH & RELAY. SPARE BULB TYPE H3 PART 606178			
3	RTC9525	Lamp-front lighting fog-Cewe-rectangular	1		
	AMR3751	SWITCH-PUSH PUSH FOG-REAR	1	Note (1)	
	STC1877	• Bulb & holder assembly-Orange	A/R		
	STC1878	• Bulb & holder assembly-Green	A/R		
4	RTC9527	Lamp-hand-Cewe	1		
5	RTC9586AA	Kit-driving lamp wiring-Cewe	1		
6	RTC9498	Bulb-headlamp-Quartz Halogen-Cewe-type H2	2		
7	606178	Bulb-headlamp-h3-55 Watt-type H3	2		
8	STC7556	Bar-tie straight	2		

NOTE:WILL NOT FIT NUDGE BAR RTC9584

CHANGE POINTS:
(1) From (V) LA 087356

T 23

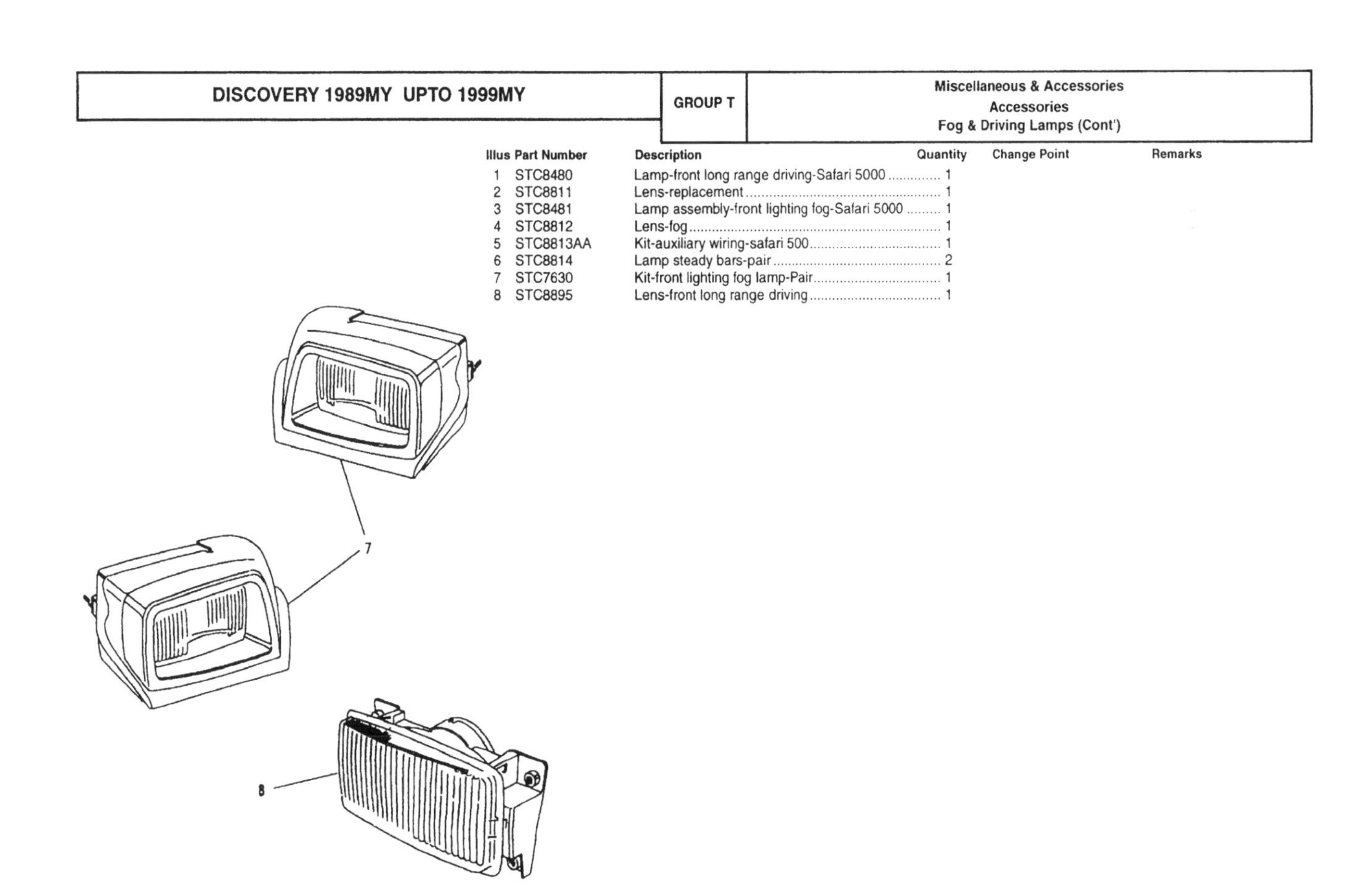

DISCOVERY 1989MY UPTO 1999MY	GROUP T	Miscellaneous & Accessories Accessories Fog & Driving Lamps (Cont')

Illus	Part Number	Description	Quantity	Change Point	Remarks
1	STC8480	Lamp-front long range driving-Safari 5000	1		
2	STC8811	Lens-replacement	1		
3	STC8481	Lamp assembly-front lighting fog-Safari 5000	1		
4	STC8812	Lens-fog	1		
5	STC8813AA	Kit-auxiliary wiring-safari 500	1		
6	STC8814	Lamp steady bars-pair	2		
7	STC7630	Kit-front lighting fog lamp-Pair	1		
8	STC8895	Lens-front long range driving	1		

T 24

Headlamp Leveling Kit-95MY On

Illus	Part Number	Description	Quantity	Change Point	Remarks
1	STC8938	Headlamp assembly-load levelling	1	Note (1)	
1	STC8938AA	Headlamp assembly-load levelling	1	Note (2)	

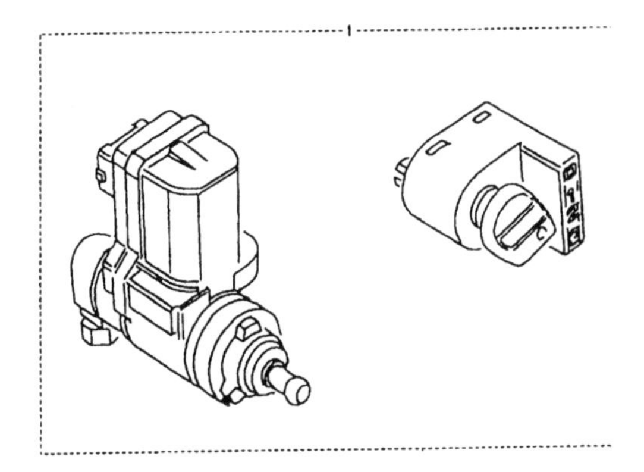

CHANGE POINTS:
(1) From (V) MA 081992 To (V) VA 745654
(2) From (V) VA 745655

Winch-Husky

Illus	Part Number	Description	Quantity	Change Point	Remarks
		X6 WINCH-NOTE(2&3)			
1	RTC9587AB	Winch-electric-husky	1		V8 except drivers and passengers airbag
		ALL VARIANTS			
2	RTC9520	Kit-winch accessories	1		
3	STC8123	Roller fairlead-winch	1		
4	STC8110	Control-winch remote	1		

NOTE(1)COMPATIBLE WITH NUDGE BARS.USE RTC9587 FOR V8 MODELS.

NOTE(2)NOT COMPATIBLE TO VEHICLES FITTED WITH AIR BAGS.

NOTE(3)NEW LEISURE WINCH.COMPATABLE WITH NUDGE BARS

Illus	Part Number	Description	Quantity	Change Point	Remarks
		XD9000I-NOTE(1)			
1	STC8896	Winch-electric-XD9000i	NLA		Use winch STC8896AA which is supplied without the A frame protection bar. If the A frame bar is also required, order STC7692. or STC8896AA.
1	STC8896AA	Winch-electric-XD9000i	1		
2	STC8893	Cover-winch	1		
		ALL VARIANTS			
3	RTC9520	Kit-winch accessories	1		

NOTE(1)COMPATIBLE TO VEHICLES WITH AIR BAGS.

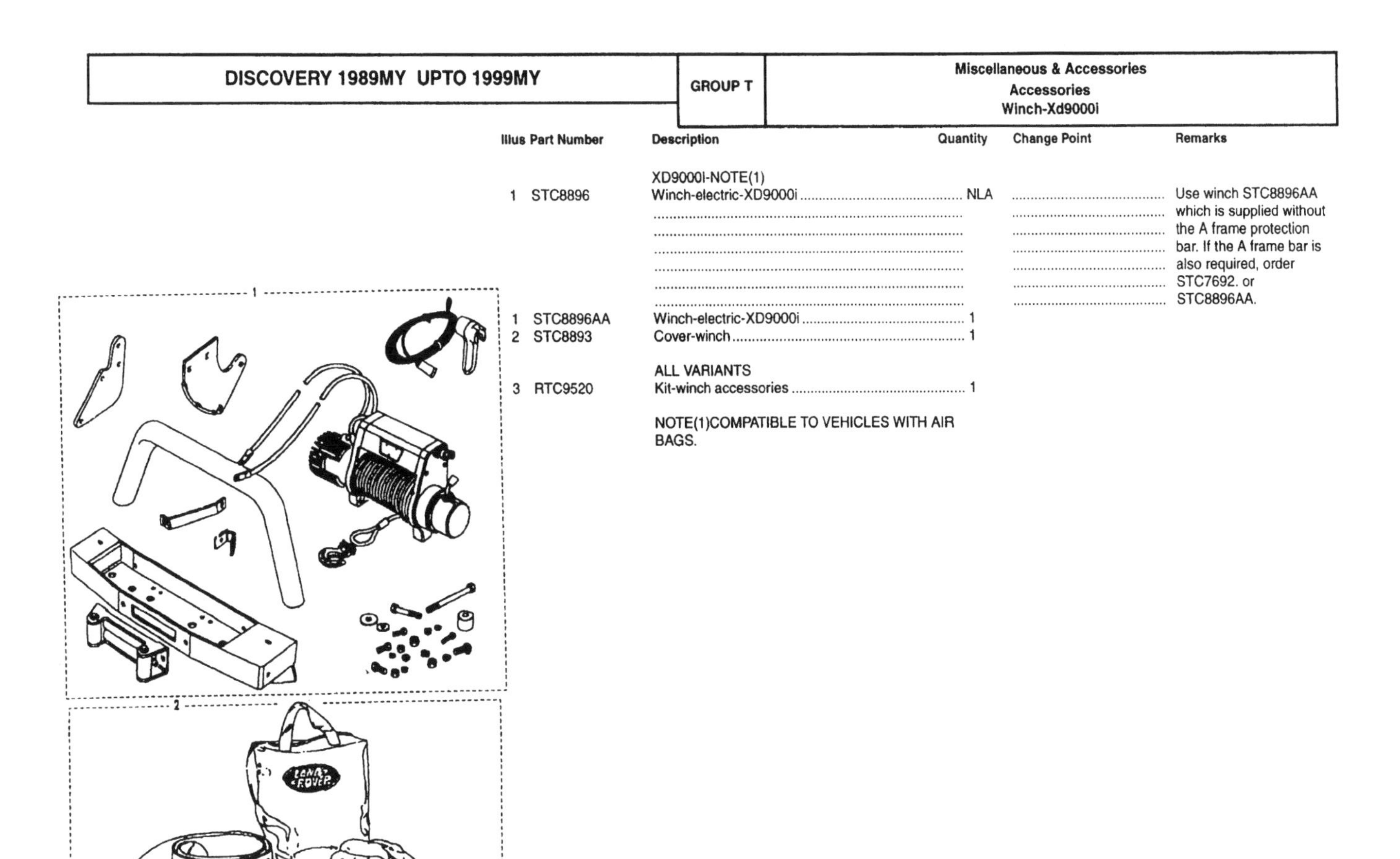

Illus	Part Number	Description	Quantity	Change Point	Remarks
		LUBRICANTS - UK ONLY			
	STC8955	Oil-1ltr-15W-40	A/R		
	STC9181	Oil-Engine-1ltr	A/R		UK
	STC8259	Gear oil-EP90	A/R		
	STC8261	Fluid-automatic transmission-DEXTRON 2	A/R		
	STC8263	Fluid-automatic transmission-Type G	A/R		
		BRAKE FLUID - UK ONLY			
	STC8292	Brake fluid	A/R		
		ANTI-FREEZE			
	RTC5782A	Antifreeze-205ltr	A/R		
	RTC5781A	Antifreeze-5ltr	A/R		
	RTC5779A	Antifreeze-1ltr	A/R		
		NOTE:NOT AVAILABLE IN ALL TERRITORIES.			
		EUROPE ONLY			
	STC8249	Cleaner-glazing-1ltr	A/R 1		
	STC717	De-icer-aerosol-500ml	A/R		
	STC718	Cleaner-alloy wheel-500ml	A/R		
	STC719	Cleaner-glazing-500ml	A/R		
	STC721	Bumper cleaner-black-500ml	A/R		
	STC722	Shampoo-500ml	A/R		
	STC723	Polish wax-500ml	A/R		

NOTE : NOT AVAILABLE IN ALL TERRITORIES

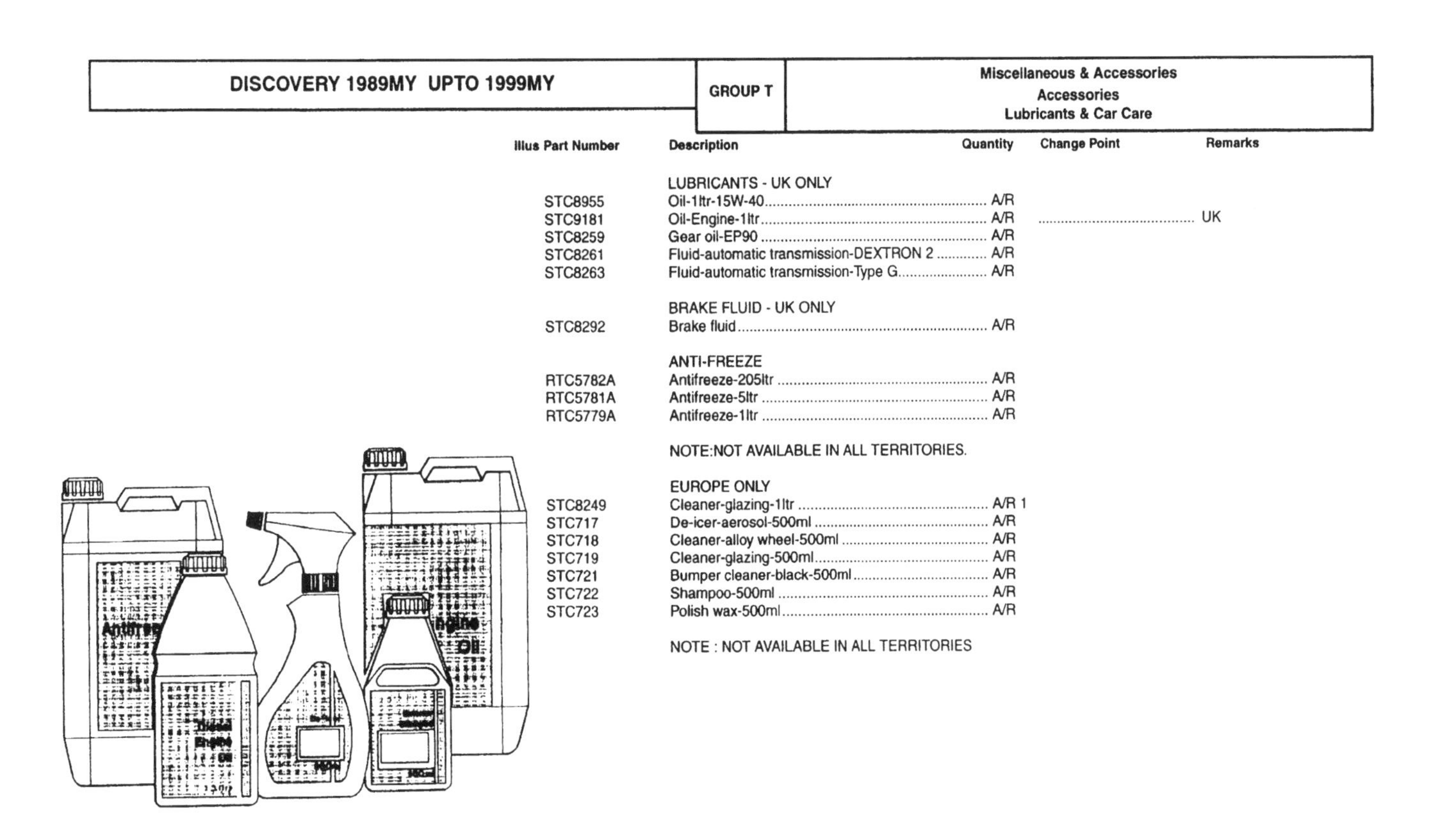

DISCOVERY 1989MY UPTO 1999MY

GROUP T

Miscellaneous & Accessories
Accessories
Inward Facing Seat Kits

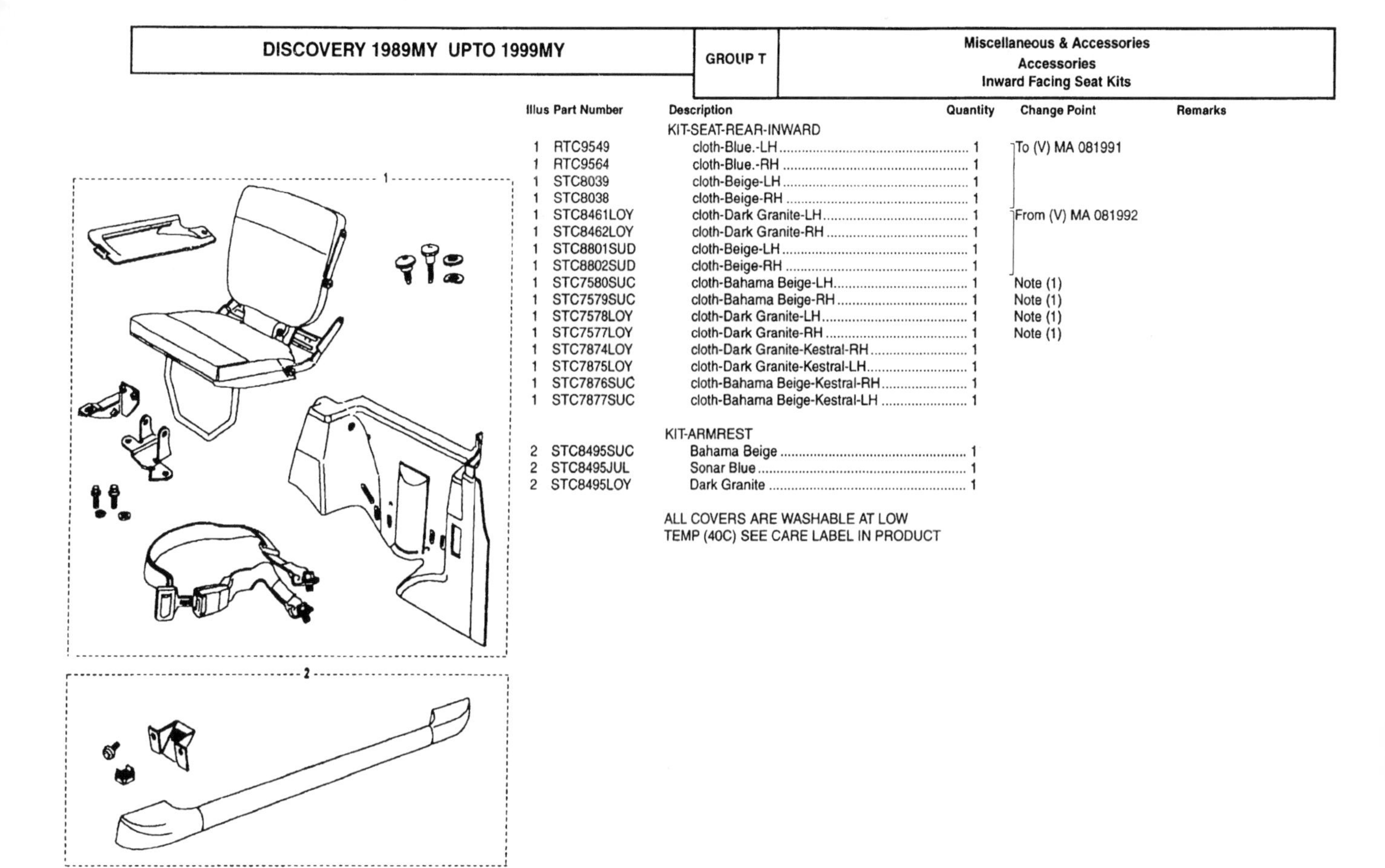

Illus	Part Number	Description	Quantity	Change Point	Remarks
		KIT-SEAT-REAR-INWARD			
1	RTC9549	cloth-Blue.-LH	1	To (V) MA 081991	
1	RTC9564	cloth-Blue.-RH	1		
1	STC8039	cloth-Beige-LH	1		
1	STC8038	cloth-Beige-RH	1		
1	STC8461LOY	cloth-Dark Granite-LH	1	From (V) MA 081992	
1	STC8462LOY	cloth-Dark Granite-RH	1		
1	STC8801SUD	cloth-Beige-LH	1		
1	STC8802SUD	cloth-Beige-RH	1		
1	STC7580SUC	cloth-Bahama Beige-LH	1	Note (1)	
1	STC7579SUC	cloth-Bahama Beige-RH	1	Note (1)	
1	STC7578LOY	cloth-Dark Granite-LH	1	Note (1)	
1	STC7577LOY	cloth-Dark Granite-RH	1	Note (1)	
1	STC7874LOY	cloth-Dark Granite-Kestral-RH	1		
1	STC7875LOY	cloth-Dark Granite-Kestral-LH	1		
1	STC7876SUC	cloth-Bahama Beige-Kestral-RH	1		
1	STC7877SUC	cloth-Bahama Beige-Kestral-LH	1		
		KIT-ARMREST			
2	STC8495SUC	Bahama Beige	1		
2	STC8495JUL	Sonar Blue	1		
2	STC8495LOY	Dark Granite	1		

ALL COVERS ARE WASHABLE AT LOW TEMP (40C) SEE CARE LABEL IN PRODUCT

CHANGE POINTS:
(1) From (V) TA 163104; From (V) TA 501920

DISCOVERY 1989MY UPTO 1999MY

GROUP T

Miscellaneous & Accessories
Accessories
Loadspace Protection-Rigid

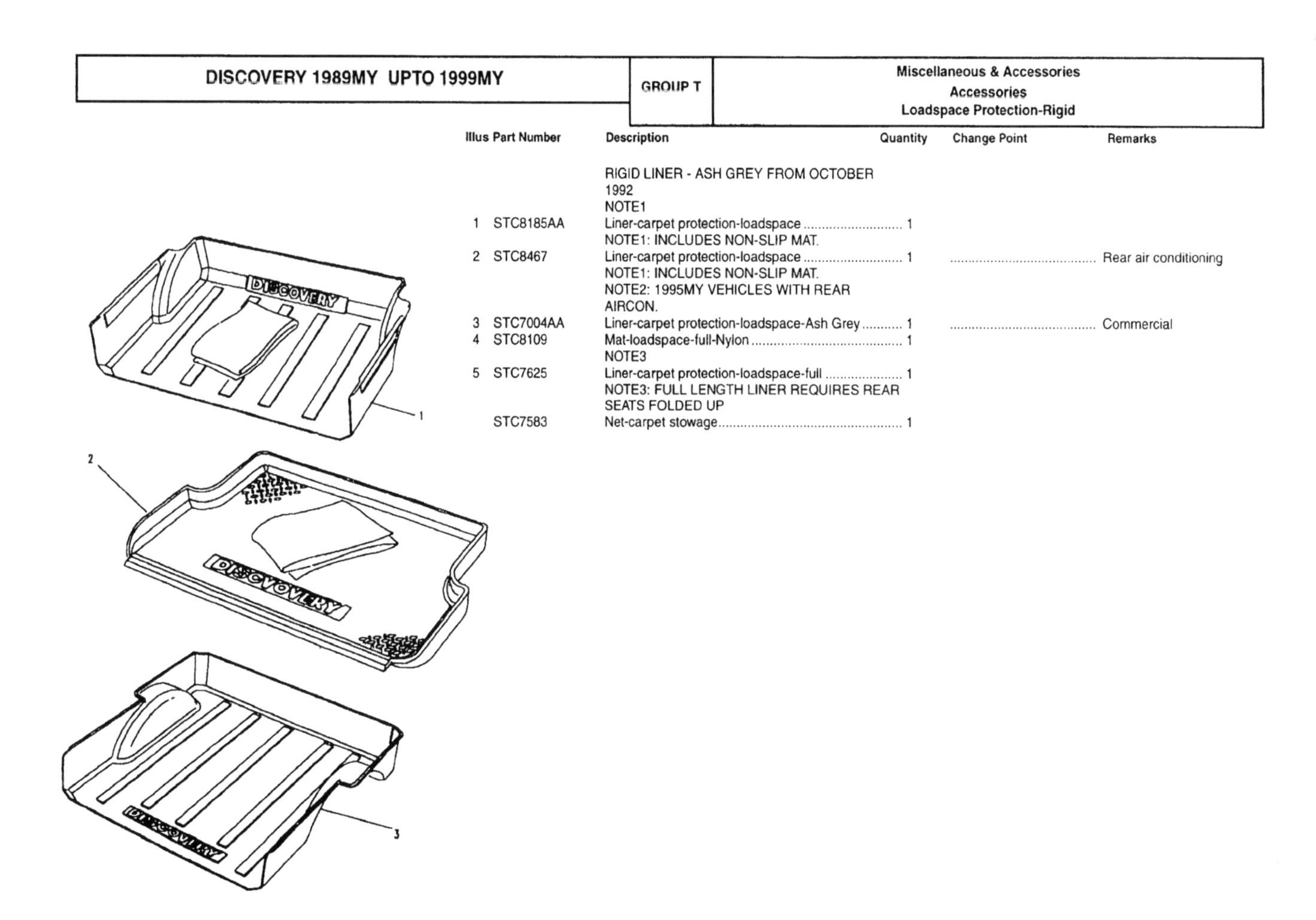

Illus	Part Number	Description	Quantity	Change Point	Remarks
		RIGID LINER - ASH GREY FROM OCTOBER 1992 NOTE1			
1	STC8185AA	Liner-carpet protection-loadspace NOTE1: INCLUDES NON-SLIP MAT.	1		
2	STC8467	Liner-carpet protection-loadspace NOTE1: INCLUDES NON-SLIP MAT. NOTE2: 1995MY VEHICLES WITH REAR AIRCON.	1		Rear air conditioning
3	STC7004AA	Liner-carpet protection-loadspace-Ash Grey	1		Commercial
4	STC8109	Mat-loadspace-full-Nylon NOTE3	1		
5	STC7625	Liner-carpet protection-loadspace-full NOTE3: FULL LENGTH LINER REQUIRES REAR SEATS FOLDED UP	1		
	STC7583	Net-carpet stowage	1		

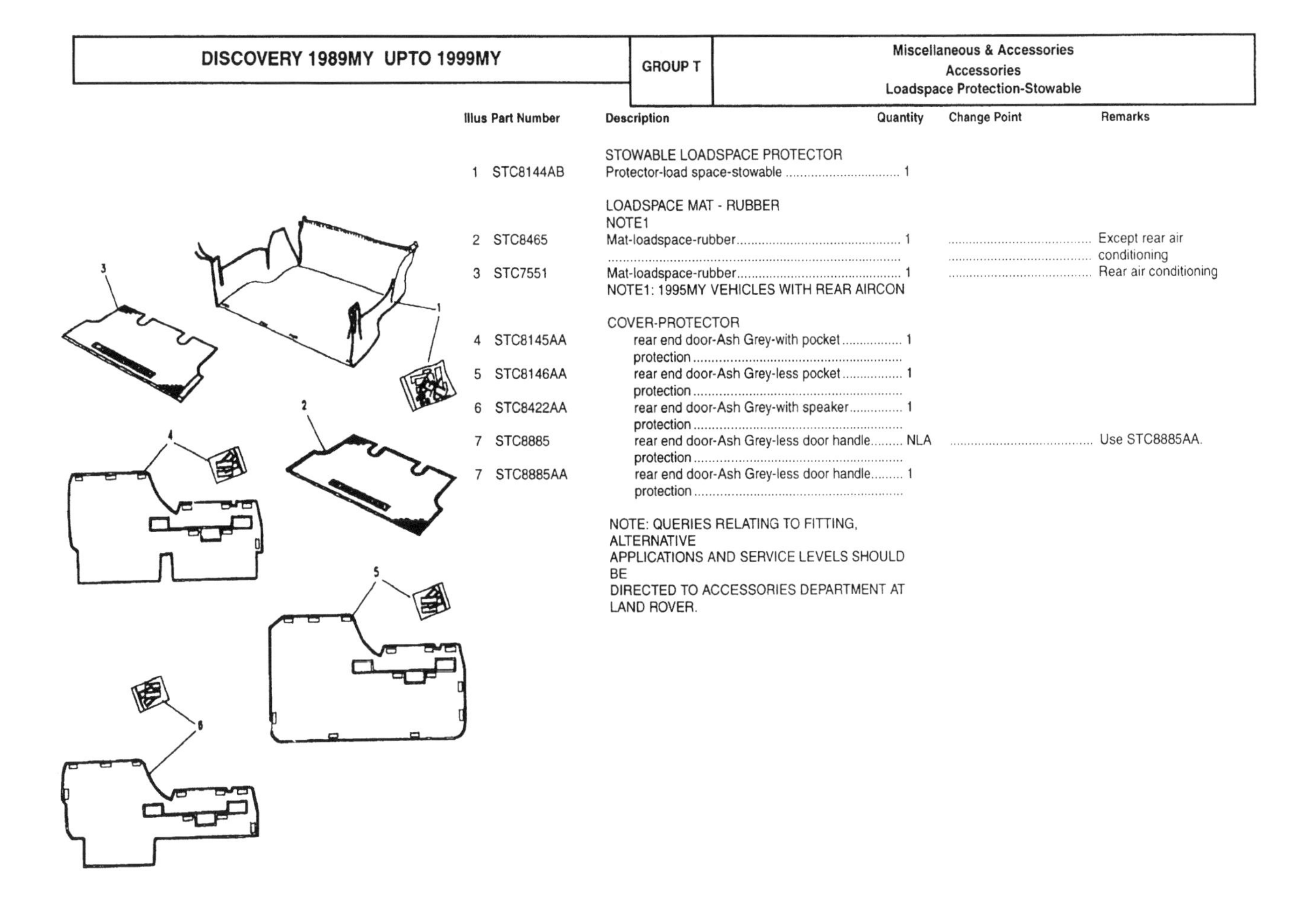

Illus	Part Number	Description	Quantity	Change Point	Remarks
		STOWABLE LOADSPACE PROTECTOR			
1	STC8144AB	Protector-load space-stowable	1		
		LOADSPACE MAT - RUBBER			
		NOTE1			
2	STC8465	Mat-loadspace-rubber	1		Except rear air conditioning
3	STC7551	Mat-loadspace-rubber	1		Rear air conditioning
		NOTE1: 1995MY VEHICLES WITH REAR AIRCON			
		COVER-PROTECTOR			
4	STC8145AA	rear end door-Ash Grey-with pocket protection	1		
5	STC8146AA	rear end door-Ash Grey-less pocket protection	1		
6	STC8422AA	rear end door-Ash Grey-with speaker protection	1		
7	STC8885	rear end door-Ash Grey-less door handle protection	NLA		Use STC8885AA.
7	STC8885AA	rear end door-Ash Grey-less door handle protection	1		

NOTE: QUERIES RELATING TO FITTING, ALTERNATIVE APPLICATIONS AND SERVICE LEVELS SHOULD BE DIRECTED TO ACCESSORIES DEPARTMENT AT LAND ROVER.

Illus	Part Number	Description	Quantity	Change Point	Remarks
		3 DOOR ONLY - NON AUTOMATIC VEHICLES			
1	RTC9512	Car set-drop in mat-Blue	NLA		Use STC8443. 3 Door Manual LHD
		CAR SET-DROP IN MAT			
1	STC8010AA	Ash Grey	NLA		Use STC8443.
1	STC8443	Sonar Blue	1		3 Door Manual LHD
1	STC8443	Sonar Blue	1		5 Door RHD
		3 & 5 DOOR			
		NOTE1			
2	STC8188AB	Car set-drop in mat-front-pair-Black	1		
2	STC8847	Car set-drop in mat-front-pair-Ash Grey	1		
		NOTE1: FRONT PAIR ONLY			
		UNIVERSAL CARPET MAT SETS			
		NOTE2			
		CAR SET-DROP IN MAT			
1	STC8445AB	Dark Granite	NLA		Use STC8445AC.
1	STC8445AC	Granite Grey	1	Note (1)	
1	STC8444AC	Beige	1		
1	STC7589	Beige	NLA		Use STC50088.
1	STC7590	Granite	1		5 Door
		UNIVERSAL CARPET MATS SETS - WITH REAR AIR CON PROVISION			
1	STC8420	Car set-drop in mat-Blue	1		Rear air conditioning
1	STC8421	Car set-drop in mat-Beige	1		Rear air conditioning
		CARPET-LOADSPACE			
	RTC9513	Blue	1		
	STC8007	Beige	1		
	STC8935	Granite/Ash Grey	1		
		CONCEPT 2000 MATS			
1	STC50088	Car set-drop in mat-Concept 2000-Bahama Beige	1		
1	STC50089	Car set-drop in mat-Concept 2000-Granite	1		

CHANGE POINTS:
(1) From (V) LA 087356

Illus	Part Number	Description	Quantity	Change Point	Remarks
1	STC7601	Kit burglar alarm	1	Note (1); Note (2)	
2	STC7560	Actuator-central door locking	1	Note (4)	CDL
3	STC7602	Kit burglar alarm	1	Note (3)	Ultrasonic sensor

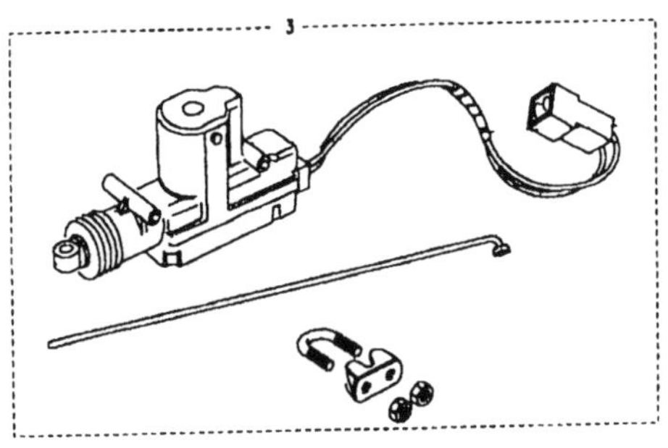

CHANGE POINTS:
(1) From (V) KA 034314
(2) To (V) LA 087355
(3) To (V) LA 072533
(4) To (V) LA 081991

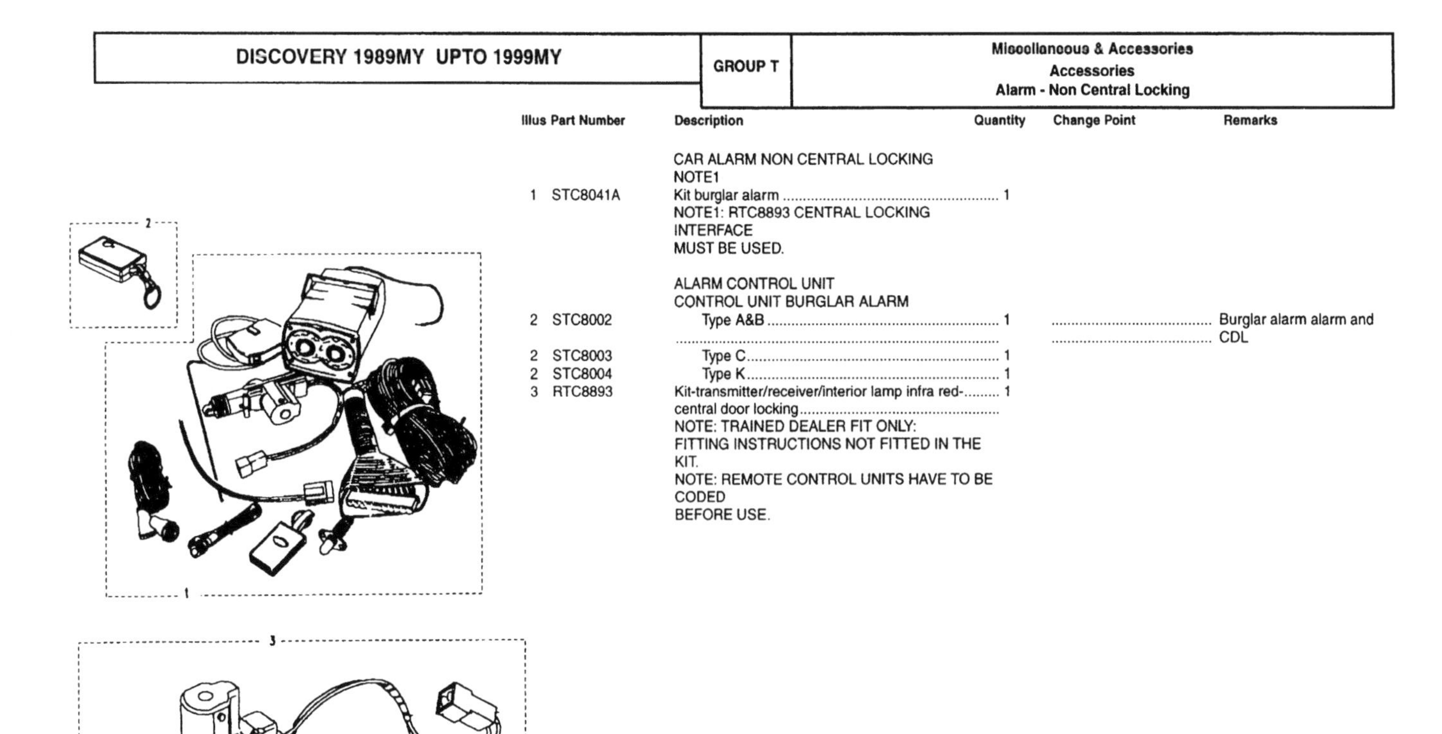

Illus	Part Number	Description	Quantity	Change Point	Remarks
		CAR ALARM NON CENTRAL LOCKING NOTE1			
1	STC8041A	Kit burglar alarm	1		
		NOTE1: RTC8893 CENTRAL LOCKING INTERFACE MUST BE USED.			
		ALARM CONTROL UNIT			
		CONTROL UNIT BURGLAR ALARM			
2	STC8002	Type A&B	1		Burglar alarm alarm and CDL
2	STC8003	Type C	1		
2	STC8004	Type K	1		
3	RTC8893	Kit-transmitter/receiver/interior lamp infra red-central door locking	1		
		NOTE: TRAINED DEALER FIT ONLY: FITTING INSTRUCTIONS NOT FITTED IN THE KIT.			
		NOTE: REMOTE CONTROL UNITS HAVE TO BE CODED BEFORE USE.			

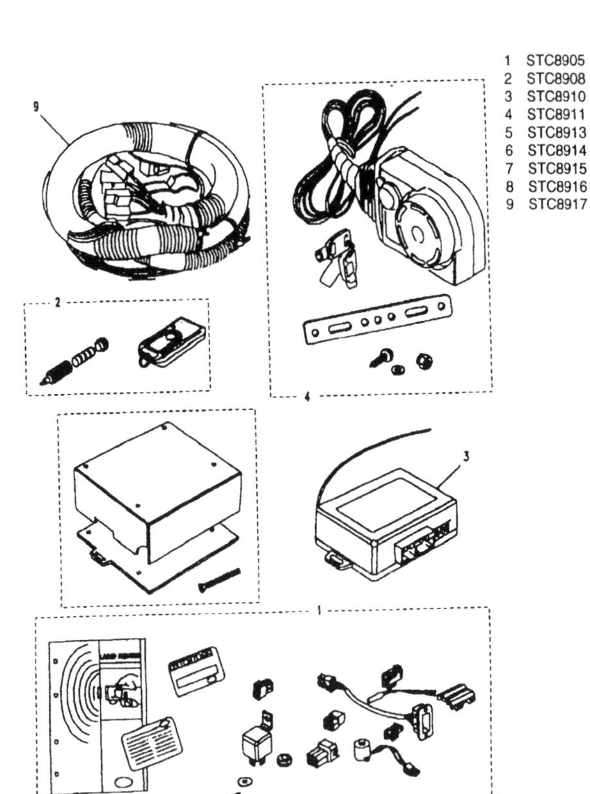

Illus	Part Number	Description	Quantity	Change Point	Remarks
		THATCHAM CATEGORY ONE APPROVED NOTE:1			
1	STC8905	Kit burglar alarm	1		
2	STC8908	Transmitter plip burglar alarm	1		
3	STC8910	Control unit burglar alarm	1		
4	STC8911	Sounder unit burglar alarm	1		
5	STC8913	Relay changeover	1		
6	STC8914	Switch-contact-bonnet burglar alarm	1		
7	STC8915	Cover-control unit burglar alarm	1		
8	STC8916	Battery-transmitter burglar alarm	1		
9	STC8917	Harness-link-with burglar alarm	1		

NOTE:1 TRAINED DEALER FIT ONLY:
FITTING INSTRUCTIONS NOT INCLUDED IN
THE KIT

Illus	Part Number	Description	Quantity	Change Point	Remarks
1	STC8843AA	Kit-locking nut road wheels-set of 5-alloy-except Deep dish wheel	1		
2	RTC9535	Kit-locking nut road wheels-set of 5-steel	1		
	STC7575	Key-tool kit locking wheel nut-pin type	1		When ordering a replacement key, the key number must be quoted in the customer field data
	STC7576	Key-tool kit locking wheel nut-fluted	1		When ordering a replacement key, the key number must be quoted in the customer field data
3	RTC9526	Wheel-alloy road-7.0 X 16-Styled-Quicksilver NOTE: REQUIRES WHEEL NUT KIT RTC9563AA KNAVE CENTRE SUPPLIED WITH WHEEL.	5		
3	RTC9526POL	Wheel-alloy road-7.0 X 16-Styled-polished	5		
4	RTC9589	Set-snow chains-205 x 16-600 x 16	1		
4	STC8518	Set-snow chains-235 x 16-rear	1		
		CAP-LOCKING-NUT ROAD WHEELS			
5	STC8112	plastic-Silver	A/R		
5	STC8113	plastic-Black	A/R		
5	STC8844	stainless steel.	A/R		

NOTE: APPLICABLE TO STEEL & ALLOY WHEELS.

NOTE: IT IS MOST IMPORTANT TO QUOTE THE IN THE KEY NUMBER IN THE CUSTOMER DATA FIELD WHEN ORDERING REPLACEMENTS.

NOTE: QUERIES RELATING TO FITTING, ALTERNATIVE
APPLICATIONS AND SERVICE LEVELS SHOULD BE
DIRECTED TO ACCESSORIES DEPARTMENT AT LAND ROVER

Spare Wheel Covers

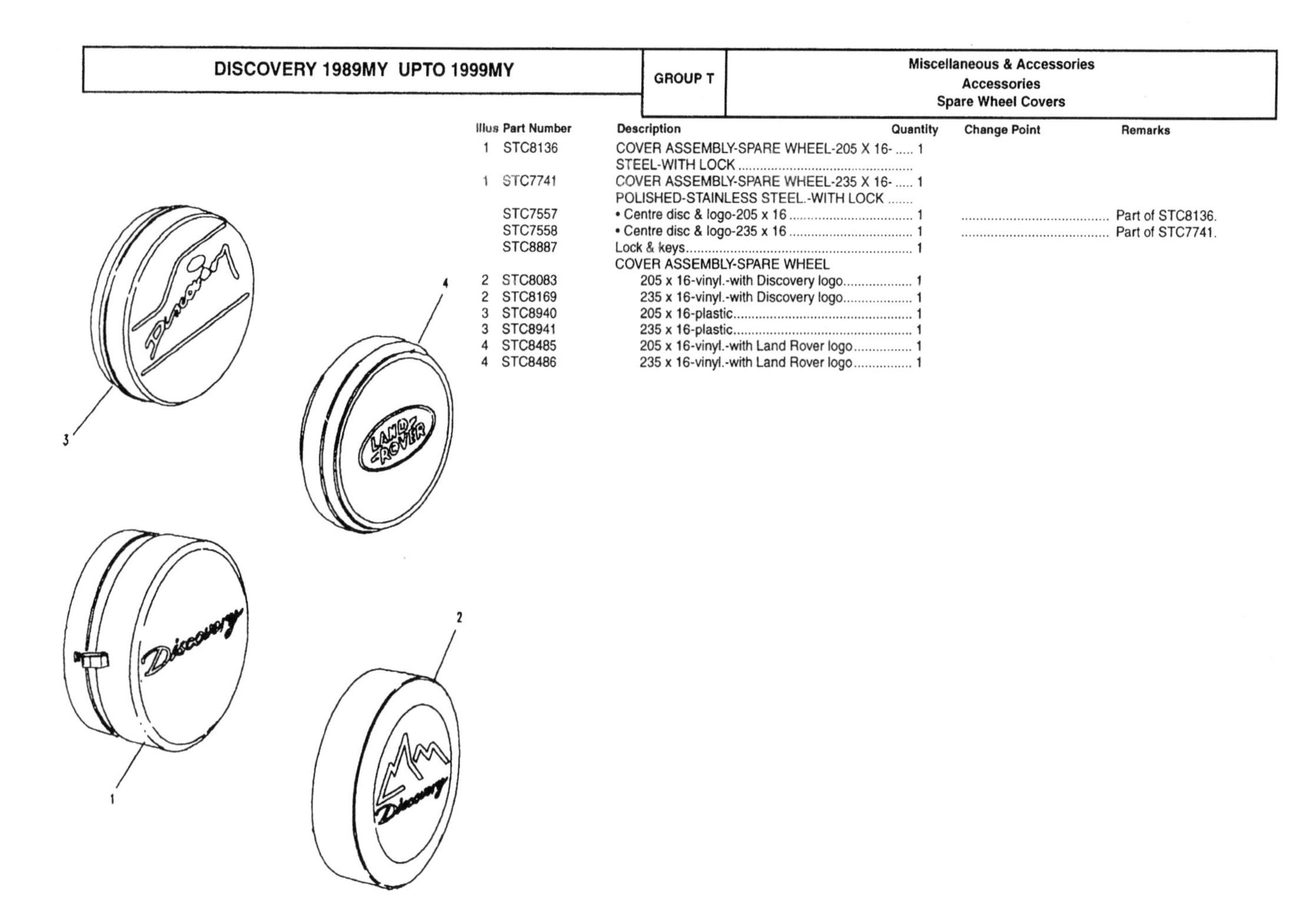

Illus	Part Number	Description	Quantity	Change Point	Remarks
1	STC8136	COVER ASSEMBLY-SPARE WHEEL-205 X 16-STEEL-WITH LOCK	1		
1	STC7741	COVER ASSEMBLY-SPARE WHEEL-235 X 16-POLISHED-STAINLESS STEEL.-WITH LOCK	1		
	STC7557	• Centre disc & logo-205 x 16	1		Part of STC8136.
	STC7558	• Centre disc & logo-235 x 16	1		Part of STC7741.
	STC8887	Lock & keys	1		
		COVER ASSEMBLY-SPARE WHEEL			
2	STC8083	205 x 16-vinyl.-with Discovery logo	1		
2	STC8169	235 x 16-vinyl.-with Discovery logo	1		
3	STC8940	205 x 16-plastic	1		
3	STC8941	235 x 16-plastic	1		
4	STC8485	205 x 16-vinyl.-with Land Rover logo	1		
4	STC8486	235 x 16-vinyl.-with Land Rover logo	1		

T 37

Heavy Duty Rear Suspension

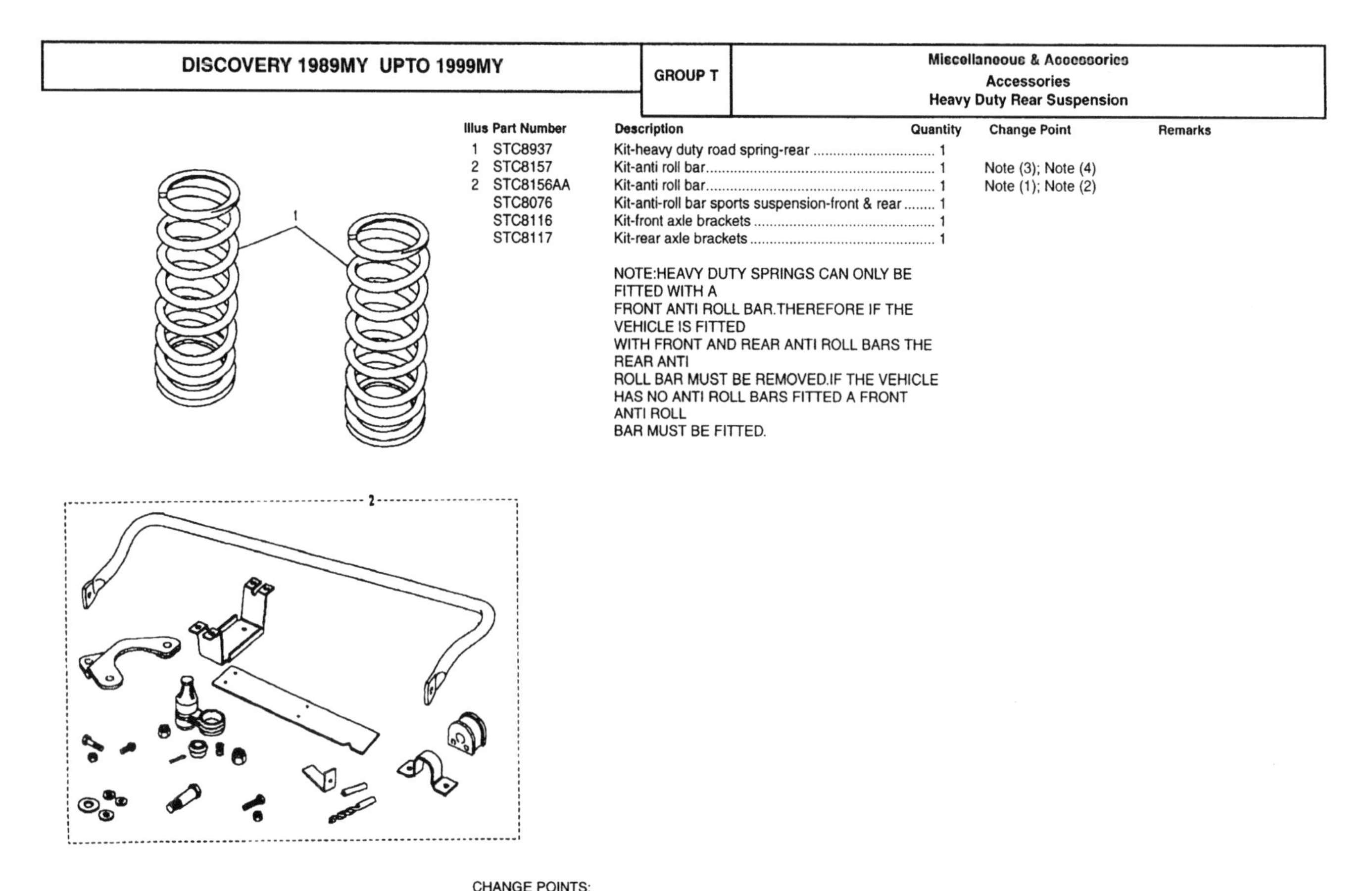

Illus	Part Number	Description	Quantity	Change Point	Remarks
1	STC8937	Kit-heavy duty road spring-rear	1		
2	STC8157	Kit-anti roll bar	1	Note (3); Note (4)	
2	STC8156AA	Kit-anti roll bar	1	Note (1); Note (2)	
	STC8076	Kit-anti-roll bar sports suspension-front & rear	1		
	STC8116	Kit-front axle brackets	1		
	STC8117	Kit-rear axle brackets	1		

NOTE:HEAVY DUTY SPRINGS CAN ONLY BE FITTED WITH A FRONT ANTI ROLL BAR.THEREFORE IF THE VEHICLE IS FITTED WITH FRONT AND REAR ANTI ROLL BARS THE REAR ANTI ROLL BAR MUST BE REMOVED.IF THE VEHICLE HAS NO ANTI ROLL BARS FITTED A FRONT ANTI ROLL BAR MUST BE FITTED.

CHANGE POINTS:
(1) From (V) KA 049523
(2) To (V) LA 083522
(3) From (V) GA 410786
(4) To (V) KA 047135

T 38

Roof Racks

Illus	Part Number	Description	Quantity	Change Point	Remarks
		NOTE:1&2			
1	RTC9539AB	Rack assembly-roof-standard	1		
1	STC8830	Rack assembly-roof-Full Length	1		
2	STC8125AA	Ladder-roof rack access	1		
3	STC8057AB	Bar-roof sports	1		

NOTE:1 MAXIMUM ROOF LOAD 50KG (114IB) CAN BE FITTED WITH OE ROOF RACK. USE FIXING KIT STC8878
NOTE:2 USE FIXINGS STC8878

NOTE:QUERIES RELATING TO FITTING,ALTERNATIVE APPLICATIONS AND SERVICE LEVELS SHOULD BE DIRECTED TO ACCESSORIES DEPARTMENT AT LAND ROVER

DISCOVERY 1989MY UPTO 1999MY | GROUP T | Miscellaneous & Accessories
Accessories

Ski Carrier

Illus	Part Number	Description	Quantity	Change Point	Remarks
		NOT SUITABLE FOR FITMENT TO O.E. ROOF RACKS			
1	STC7568	Carrier-ski roof rack-use with sports bars	1		
		NOTE: TAKES UP TO 4 PAIRS OF SKIS & WATERSKIS AND SNOWBOARDS.			
2	STC8435	Carrier-sailboard-use with sports bars	1		
3	RTC9477	Load stop adjustable-set of 4	1		
4	STC7566	Strap-roof rack lashing-pair-2 metres	1		
		NOTE: USEFUL LOAD RETENTION DEVICE			

NOTE: QUERIES RELATING TO FITTING, ALTERNATIVE APPLICATIONS AND SERVICE LEVELS SHOULD BE DIRECTED TO ACCESSORIES DEPARTMENT AT LAND ROVER

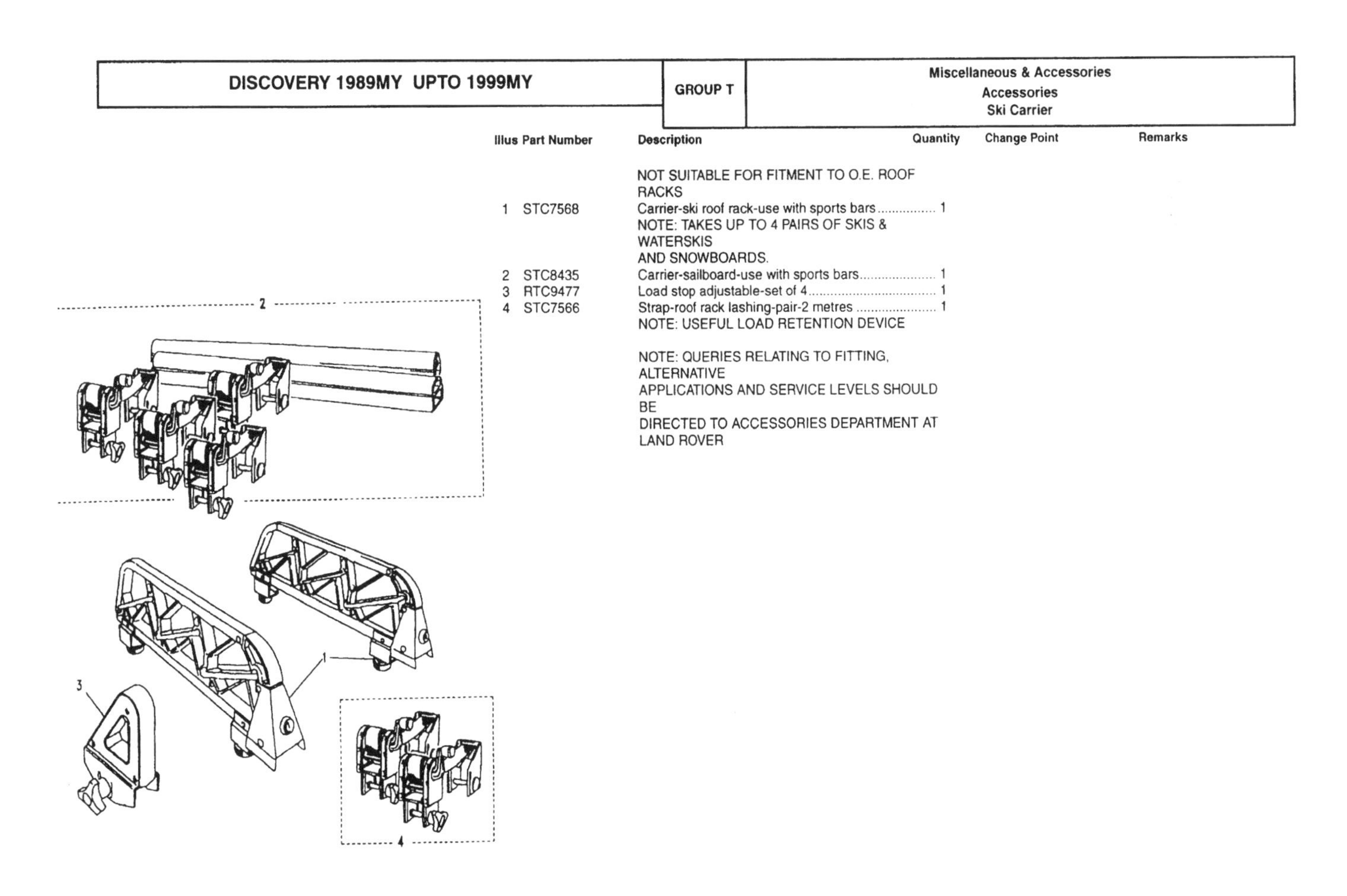

Illus	Part Number	Description	Quantity	Change Point	Remarks
1	STC7627	Box-luggage	1		
2	STC8511	Sportbox-roof rack assembly	1		

NOTE: LUGGAGE TOP BOX & SKI BOX REQUIRE SPORTS BARS (STC8057) FOR FITMENT TO VEHICLE ROOF.
NOT SUITABLE FOR FITMENT TO O.E ROOF BARS.

NOTE: QUERIES RELATING TO FITTING, ALTERNATIVE APPLICATIONS AND SERVICE LEVELS SHOULD BE DIRECTED TO ACCESSORIES DEPARTMENT AT LAND ROVER.

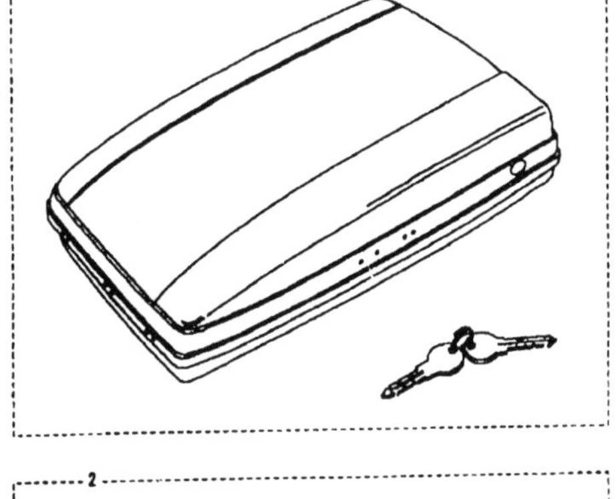

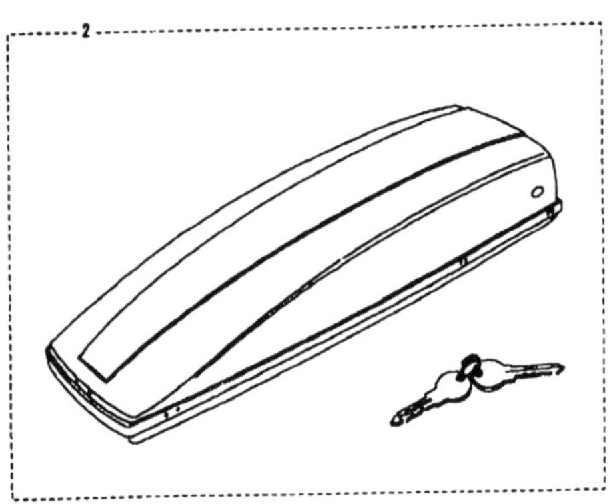

Illus	Part Number	Description	Quantity	Change Point	Remarks
		NOTE1			
1	STC8025AA	Carrier-trunk rack ski	1	Note (1)	
		NOTE2			
1	STC8817	Carrier-trunk rack ski	1	Note (2)	

NOTE1: VEHICLE CANNOT TOW CRARVAN OR TRAILER WITH THIS ITEM FITTED.

NOTE2: 1995MY VEHICLES ONWARDS- REVISED BUMPER END CAPS.

NOTE: QUERIES RELATING TO FITTING, ALTERNATIVE APPLICATIONS AND SERVICE LEVELS SHOULD BE DIRECTED TO ACCESSORIES DEPARTMENT AT LAND ROVER.

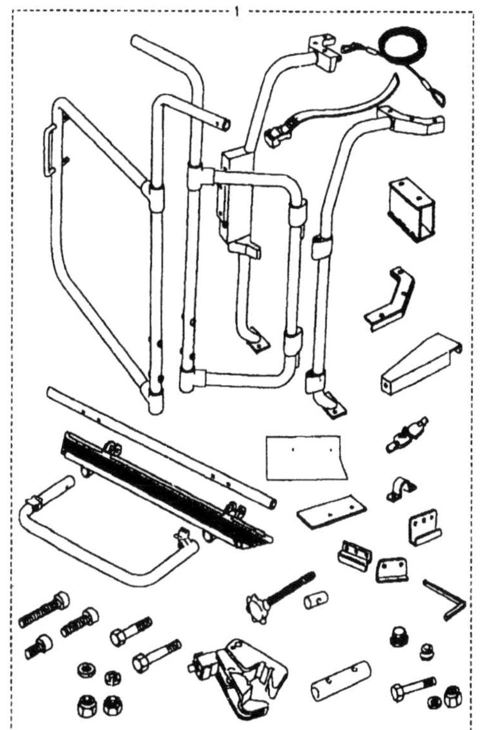

CHANGE POINTS:
(1) To (V) LA 081990
(2) From (V) MA 081991

Rear Step-Retractable

Illus	Part Number	Description	Quantity	Change Point	Remarks
1	RTC9505AC	STEP REAR RETRACTABLE	1		
2	RTC9593AA	• Kit-rear step fixing	1		

NOTE: WHEN FITTING REAR STEP INCONJUNCTION WITH TOW KIT STC8816 TOW PLATE STC8879 IS REQUIRED FOR MOUNTING S TYPE SOCKET.

NOTE: QUERIES RELATING TO FITTING, ALTERNATIVE APPLICATIONS AND SERVICE LEVELS SHOULD BE DIRECTED TO ACCESSORIES DEPARTMENT AT LAND ROVER

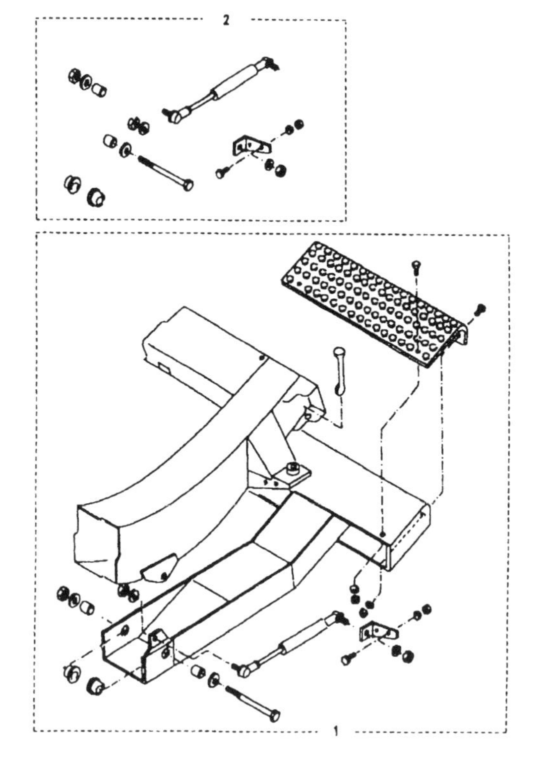

Side Step

Illus	Part Number	Description	Quantity	Change Point	Remarks
		NOTE1			
1	RTC9507AB	Kit-side step	1		
2	STC8087AA	Kit-side step	1		
3	STC8049	Kit service-side step	1		
4	STC8130AA	KIT-SIDE STEP	1		
5	STC8942	• Step-side runner-RH	1		
5	STC8943	• Step-side runner-LH	1		
	STC8807	Extrusion-side step	1		
	STC7808	Mudflaps	1		

NOTE1:USE BRACKET STC8819 FOR FITTING FRONT MUDFLAPS.

NOTE: QUERIES RELATING TO FITTING, ALTERNATIVE APPLICATIONS AND SERVICE LEVELS SHOULD BE DIRECTED TO ACCESSORIES DEPARTMENT AT LAND ROVER.

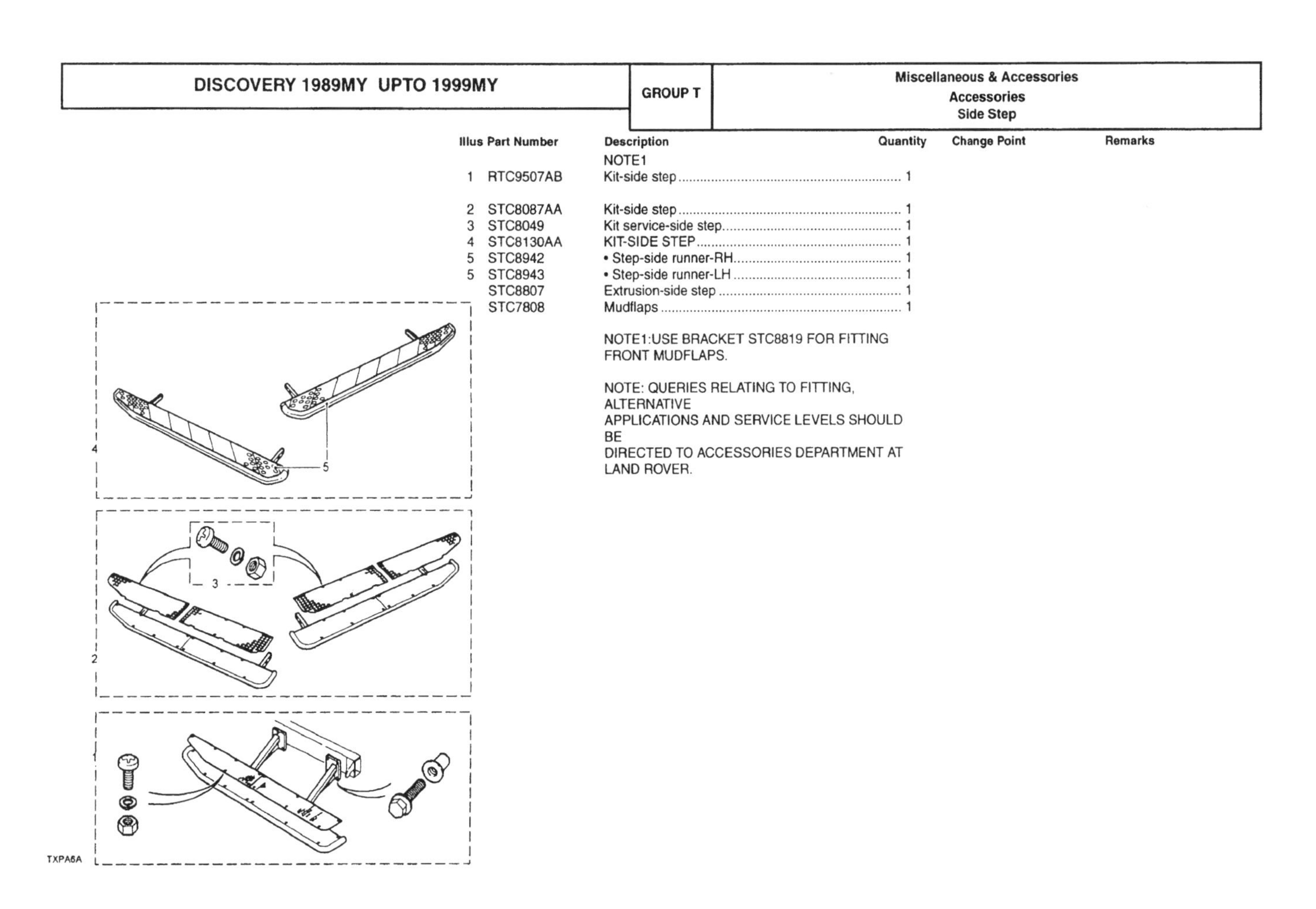

Illus	Part Number	Description	Quantity	Change Point	Remarks
	RTC9554	Mudflaps-kit front	1		
2	STC7857	Mudflaps-kit rear	1		

ILLUSTRATION TO FOLLOW
L'ILLUSTRATION A SUIVRE
DAS BILD IST ZU FOLGEN
IL DISEGNO SEGUIRA
EL DISENO SIGUE
A ILLUSTRACAO SEGUIRA MAIS TARDE

TF 0007

NOT ILLUSTRATED

Illus	Part Number	Description	Quantity	Change Point	Remarks
	STC8784	Road atlas	1		
	STC8247AA	Kit-bulb	1		
	STC8040	Aerial.-electric aerial	1		
		KIT-AIR CONDITIONING			
	STC8406	RHD	NLA	Note (1)	Use STC8833. TDi
	STC8833	RHD	1	Note (2)	TDi
	STC8407	LHD	NLA	Note (1)	Use STC8832. TDi
	STC8832	LHD	1	Note (2)	TDi
	STC8408	RHD	NLA	Note (1)	Use STC8831. V8
	STC8831	RHD	1	Note (2)	V8
	STC8409	LHD	NLA	Note (1)	Use STC8829. V8
	STC8829	LHD	1	Note (2)	V8
	STC8880	Harness air conditioning-RHD	1	Note (2)	Except Antilock Brakes
	STC8881	Harness air conditioning-LHD	1	Note (2)	and cruise control
	STC8401	Harness air bag-RHD	1		

ND 0001

CHANGE POINTS:
(1) To (V) LA 087355
(2) From (V) LA 087356

Illus	Part Number	Description	Quantity	Change Point	Remarks
	STC8834	Sunroof assembly-electric operating glass-pair	1	Note (1)	Electric sunroof
	STC8933	Sunroof complete assy-rear-front	1		Manual sunroof
	STC8096	KIT-GLASS SUNROOF ASSEMBLY-FRONT-REAR	1		
	BTR3074	• Frame-front sunroof upper	1		
	STC773	• Seal-rubber	1		
	RTC6482	• Seal-sunroof frame	1		
	BTR3075	• Sunroof assembly-glass	1		
	MXC3415	• Frame-sunroof-lower	1		
	AR606041L	• Screw	18		
	STC982	• Handle assembly-sunroof	1		
	STC8030	Kit-sunblind	1		5 Door except rear air conditioning
	STC7596	Kit-sunblind	1		5 Door rear air conditioning high mounted stop light
NOT ILLUSTRATED	STC7597	Kit-sunblind	1	Note (1)	5 Door rear air conditioning

ND 0001

CHANGE POINTS:
(1) From (V) LA 087356

T 47

Notes

SVO Options.. V 1
Commercial Bodyside Panels....................................... V 2
Police Spec Mudflaps .... V 1
Roof Rack - Camel LE Only .... V 4

Illus	Part Number	Description	Quantity	Change Point	Remarks
1	STC7334	Kit-mudflaps	4		Police

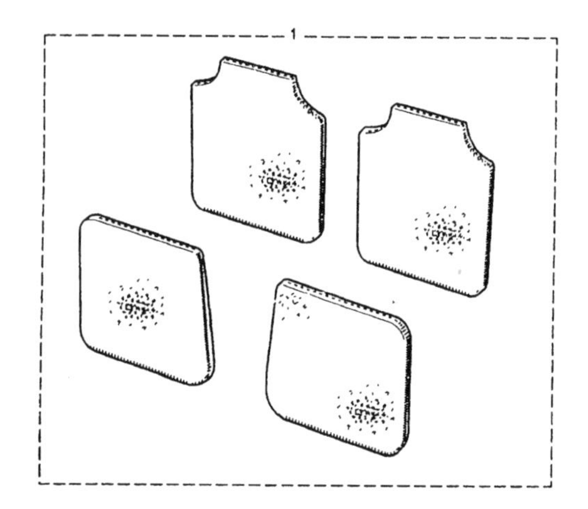

AJVXSA1A

Illus	Part Number	Description	Quantity	Change Point	Remarks
		COMPLETE WITH WINDOW BLANKING PANELS			
1	STC7020	Body side-rear-LH	1		Commercial
1	STC7021	Body side-rear-RH	1		Commercial

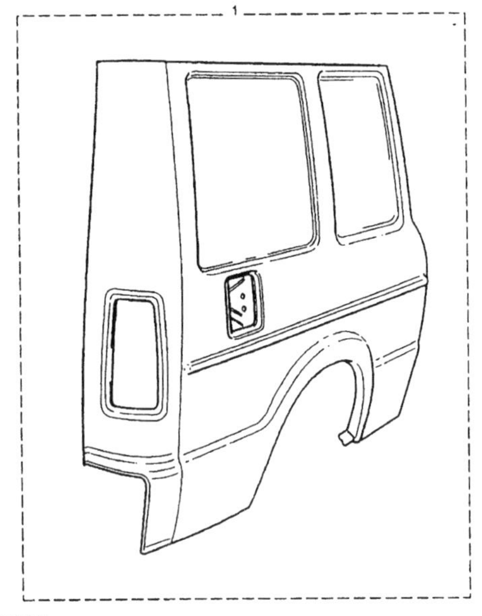

CAVXAA1A

Illus	Part Number	Description	Quantity	Change Point	Remarks
1	STC7337	Capping-wood-facia	1		
2	STC7338	Capping-wood-facia-driver-side-LH	1		
3	STC7339	Capping-wood-facia-driver-side-RH	1		

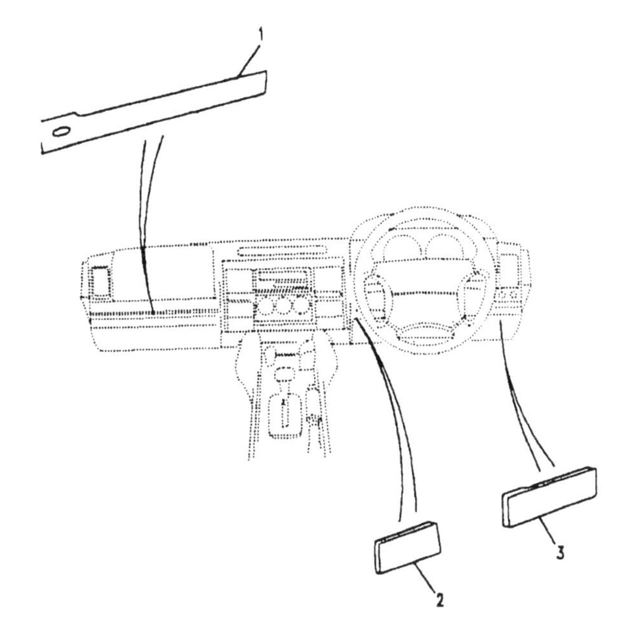

Illus	Part Number	Description	Quantity	Change Point	Remarks
1	STC7525	Rack assembly-roof-Camel L.E. only	1		

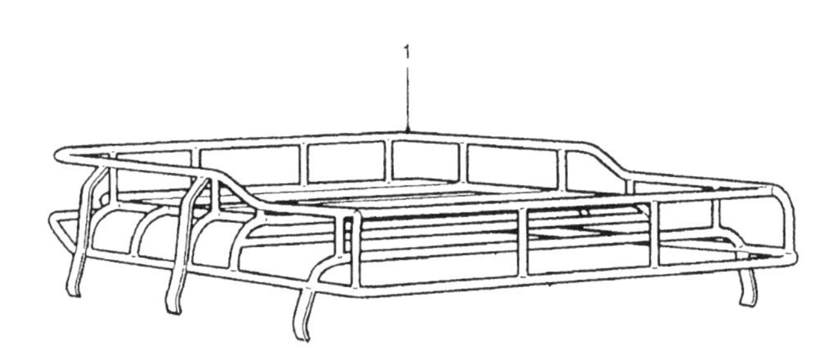

CAVXZA1A

Illus	Part Number	Description	Quantity	Change Point	Remarks
		BADGE-CAMEL TROPHY			
1	STC60123	large	1		
2	STC60124	small	1		
3	STC60125	front door	2		

LV0001

V 5

10211	M 09	232043	B138	256006	F 15	3289	E 20	390939	M 83	515599	E 27	549236	E 37	570829	M 02
11011L	F 13	232043	H 04	257002	B 98	3290	B 79	391184	N 69	517689	J 13	549238	E 37	570829	M 04
11820L	H 12	232044	B 14	257002	B139	3290	B111	391461	P 03	517706	J 13	549240	E 37	571134	C 52
11H1781L	G 06	233243	M160	257004	B139	3290	C 52	391462	P 03	521453	B139	549242	E 37	571134	C 04
11H1781L	G 08	233243	L 36	257011	B139	3290	C 26	391591	N 77	522932	B 94	549244	E 37	571134	D 04
11H1781L	G 01	233243	L 37	257011	M 83	3290	S 01	391591	N 81	522940	J 13	549246	E 37	571146	D 14
11H1781L	G 04	233243	M151	257015	P 03	3292	C 03	391591	N 85	524765	B 16	549248	E 37	571146	D 15
12H4547L	B164	233244	M160	257020	M 83	3292	C 25	392678	N110	524765	B 24	549250	E 37	571146	C 20
12H4636L	B164	233566	M160	257194	N 70	3292	E 18	392679	N110	524765	B 54	549252	E 37	571146	C 40
13H2023L	C 17	234532	E 01	264024	E 47	3292	E 20	392928	N 69	524765	B 62	549473	E 14	571160	C 24
13H2023L	C 37	234532	E 02	264024	E 53	331071	P 54	4075L	B 84	525497	B 16	549473	E 16	571161	C 46
13H2475L	P 82	235113	N 70	264590	M 08	331071	N 48	4075L	B118	525497	B 54	549911	B 80	571161	C 24
13H3735L	G 06	235113	G 14	264590	M 09	338009	M160	4266L	C 93	529364	B 40	549911	B112	571163	C 46
13H3735L	G 08	235770	B 07	264590	M 07	338013	N 47	4266L	C 42	529364	B 76	552818	E 46	571163	C 24
13H3735L	G 09	236070	E 18	264591	M 11	338013	M160	4478	B 99	529364	B161	552818	E 51	571468	D 11
13H9704L	M155	236070	E 20	264591	M 05	338014	M160	4589L	B 13	532323	D 21	552819	E 47	571468	D 12
13H9727L	M155	236257	B 14	264591	M 08	338015	N 54	4594L	B106	532943	C 19	552819	E 53	571536	D 05
154545	B 79	236408	B 09	264591	M 09	338015	M160	4594L	B156	532943	C 39	555711	M160	571665	D 12
154545	B 84	240407	C 87	264591	M 04	338015	G 04	4594L	K 02	533765	D 12	560864	G 06	571665	D 21
154545	B 89	240407	M154	264591	M 07	338017	M160	4905	F 12	534790	B 97	560864	G 08	571665	M115
154545	B111	240407	G 03	266945	B 08	338017	G 10	4910L	B 86	534897	B 94	561195	E 06	571665	D 06
154545	B117	240431	F 20	266945	B 46	338018	M160	500746	E 46	535781	B160	561195	E 27	571682	D 12
154545	B126	242522	E 01	267604	B 94	338019	M160	500746	E 51	535781	B161	561196	E 06	571718	E 13
17H8764L	H 15	242522	E 02	269257	M160	338020	M160	501593	B 04	535782	B160	561196	E 27	571718	E 14
18123	N 20	243959	B 17	273069	B 28	338021	M160	501593	B 43	535782	B161	562106	G 03	571743	E 18
18389	N 54	243967	B130	273069	B 62	338023	N 54	502116	B 39	538039	B 10	562481	B140	571744	E 18
18546	N 54	243968	B 87	273166	B 87	338023	N 47	502116	B 75	538131	B 07	564258	B 90	571745	E 18
19089	N 20	243968	B122	273166	B122	338023	M160	502473	B 19	538131	B 45	564258	B127	571746	E 18
19137	N 54	244487	B 14	276054	M160	338025	M160	502473	B 57	538132	B 07	564455	B 14	571752	E 23
1G2984	B189	244488	B 14	276201	E 02	338026	M160	50641	B157	538132	B 45	564456	B 14	571752	E 24
1K51L	B164	247127	B 04	276483	E 01	338028	M160	50641	C 87	538134	B 07	564724	L 14	571752	E 39
213961	B 09	247127	B 43	276484	E 01	338029	M160	50641	M 48	538890L	G 14	565540	L 14	571752	E 40
213961	B 82	247665	B 14	278010	M113	345662	N 82	50641	M 49	539706	E 11	566222	B 83	571755	E 13
216962	D 11	247766	B117	279146	E 23	350343	N 69	50641	M 51	539706	E 36	566580	N 02	571755	E 15
216962	D 12	251335	N 76	279146	E 39	351991	N 69	50641	M 52	539707	E 10	566581	N 02	571755	E 16
217245	F 12	252164	E 44	27H4353L	H 18	356225	N 70	50641	M 53	539707	E 32	566737	B 98	571756	E 18
2204L	B130	252210	F 15	2851L	F 16	368830	N 69	50641	M 17	539718	E 32	568431	B 39	571761	E 23
2204L	B 95	252513	B 79	2851L	F 01	369092	M151	50641	M 18	539720	E 32	568431	B 75	571815	E 19
2215L	H 19	252513	B111	2K8686L	G 11	37D2260L	B164	50641	M 19	539722	E 32	568680	B 26	571819	E 19
2266L	B142	252513	B118	302532	R 32	37H770	J 13	50641	M 21	539724	E 32	568680	B148	571822	E 13
22G2076L	B165	252514	B 94	302532	R 39	37H7920	J 13	50641	C 89	539745	E 10	568680	E 06	571882	E 39
230313	B 22	252623	B 96	3036L	B 96	37H8119L	J 13	50641	M 16	539745	E 32	568680	E 27	571883	E 39
230511	E 18	253023	C 05	3036L	B131	3831L	N 69	50641	M 20	541010	B 21	568680	J 23	571890	E 13
230511	E 20	253023	C 27	3036L	N 12	3831L	N 20	50642	B101	543819	F 16	568680	M155	571910	H 16
232038	C 05	253948	N 11	3036L	N 11	3831L	M 10	50642	B147	546194	B160	568680	M 46	571970	D 12
232038	C 27	253952	E 44	3036L	F 05	3900L	N 28	509045P	E 01	546194	B161	568680	J 09	572087	E 46
232039	E 06	253952	E 49	3036L	F 01	3900L	H 19	509045P	E 02	546198	B160	568680	J 24	572166	K 12
232039	E 27	255204	H 14	315264	R 32	3900L	N 55	513123	G 11	546198	B161	568680	M 54	572166	K 10
232039	D 03	255205	E 13	315264	R 38	390448	N 77	513123	G 12	546799	B 19	568680	M 95	572166	K 11
232039	C 92	255206	N 20	315264	R 39	390448	N 80	514244	C 02	546799	B 57	568680	M 21	572167	K 12
232039	C 42	255207	H 17	315264	R 40	390448	N 85	514244	C 24	549229	E 01	568680	M 45	572167	K 10
232039	C 95	255249	F 05	3259	E 10	390774	R 59	515466	D 22	549229	E 02	568680	M 20	572167	K 11
232039	C 97	255249	F 01	3259	E 34	390939	M 98	515467	D 22	549230	E 37	568680	M 50	572295	N 02
232042	D 06	255467	E 19	3261L	F 12	390939	H 06	515468	D 22	549232	E 37	568858	F 16	572312	L 01
232043	B 94	255467	E 21	3289	E 18	390939	H 08	515470	D 22	549234	E 37	570829	M 03	572312	L 02

Part No.	Ref	Part No.	Ref	Part No.	Ref	Part No.	Ref	Part No.	Ref	Part No.	Ref	Part No.	Ref	Part No.	Ref
572312	L 04	586438	M 13	594594	J 06	602146	B 79	602634	B 98	606545	F 13	613718L	B 91	79122	M 23
572312	L 05	586438	M 12	594594	L 09	602146	B111	602910	B 87	606666	E 19	613805	B 91	79122	J 23
572312	L 08	586440	B 90	594594	L 30	602146	D 05	602910	B122	606666	E 20	613857	B101	79122	J 09
572315	E 45	586440	C 52	594594	L 31	602147	B 79	602912	B 87	606666	E 21	613857	B147	79122	J 24
572466	G 01	587477L	B151	594594	L 32	602147	B111	602912	B122	606683	H 15	614037	B 87	79123	E 06
572466	G 04	589238	N 71	594594	M 45	602148	B 89	602913	B 87	606696	H 18	614037	B122	79123	E 27
572467	G 09	589238	N 72	594594	B 95	602148	B126	602913	B122	606733	C 44	614154	B119	79123	J 07
572493	G 01	589238	F 06	594637	C 93	602152	B 79	602915	B111	606733	C 02	614154	B144	79124	H 10
572548	B 94	589254	C 87	594637	C 42	602152	B 94	602953	B101	606733	C 45	614188	B 92	79125	H 11
572548	H 04	589254	M 95	594946	F 12	602152	B111	602953	B147	606733	C 46	614188	B128	79125	H 13
573246	B 26	589254	M 17	594947	F 12	602153	B 89	603069	J 14	606733	C 24	614202	B 87	79127	H 10
573246	M155	589254	M 18	595098	N 02	602153	B126	603127	B 90	608000	F 14	614202	B122	79127	H 05
575615	E 49	589254	M 19	595099	R 59	602154	B 89	603127	B127	608065	F 13	614443	B 96	79127	H 07
575616	E 49	589254	M 21	595199	F 16	602154	B126	603143	B 79	608246	E 06	614443	B131	79131	M145
575633	G 01	589254	C 89	595478	E 06	602172	B 89	603224	B 94	608246	E 27	614538L	B 98	79239	P 03
575707	E 43	589254	M 16	595478	E 27	602186	B 89	603224	B138	608246	D 05	614718	B 35	79240	M144
575733	F 01	589254	M 20	595478	D 03	602186	B126	603277	B 94	608246	D 06	614718	B 85	8566L	B 07
575806	G 09	589452	M160	597586	B 04	602191	B 88	603301	B 80	610025	B 91	614718	B100	8566L	B 45
575815	G 09	589783	M 03	599552	D 05	602191	B123	603301	B112	610178	B113	614718	B119	90513454	E 10
575816	G 09	589783	M 02	599945	E 11	602191	B125	603330	B 91	610289	B 92	622324	C 52	90513454	E 34
575818	G 05	591065	H 18	599945	E 36	602192	B 88	603340	B161	610289	B128	622324	C 04	90519055	B 04
575818	G 09	591628	H 18	600265	F 11	602192	B123	603376	B 91	610402	B 90	624091	B 17	90519055	B 43
575883	F 01	591629	H 18	602040	B 88	602192	B125	603378	B 89	610489	B116	624091	B 55	90568054	B130
576137	C 46	591988	C 02	602040	B123	602193	B 88	603378	B126	610735	B161	624208	H 19	90568054	B 95
576137	C 24	591988	C 24	602040	B125	602193	B124	603428	B129	610736	B161	625038	B 90	90571104	C 06
576159	E 07	592840	G 01	602061	B 81	602193	B125	603440	B 94	610833	B 98	625038	B127	90571104	C 07
576159	E 08	592840	P 64	602061	B114	602199	B 82	603446	B101	610846L	B140	6395L	B 10	90571104	C 28
576159	E 28	593692	E 11	602067	B 87	602199	B115	603446	B147	611092	B 91	6395L	B 48	90575573	N 02
576159	E 29	593692	E 36	602067	B122	602200	B 88	603535	B 80	611097	B 91	6395L	B108	90575585	N 09
576159	E 30	593693	E 11	602070	B 82	602200	B124	603535	B112	611109	B 91	6395L	B160	90575596	F 16
576159	E 31	593693	E 36	602070	B115	602200	B125	603535	B113	611110	B 91	6860L	B151	90575597	F 16
576203	C 46	594087	C 71	602071	B 87	602201	B 84	603554	B 88	611323	B108	6860L	C 87	90575789	E 48
576203	C 24	594087	C 05	602071	B122	602201	B117	603554	B123	611323	B160	6860L	G 03	90575807	G 09
576236	E 37	594087	C 27	602076	B130	602201	B129	603561	B 94	611324	B108	6860L	M113	90576928	E 18
576237	E 37	594087	C 53	602076	B 95	602212	B 79	603622	B 98	611351	B 91	72628	F 05	90576928	E 22
576238	E 37	594091	C 44	602082	B 81	602212	B111	603659	B 09	611612	B 86	7316	E 12	90577449	G 03
576239	E 37	594091	C 02	602082	B114	602227	B 92	603659	B 82	611612	B100	7316	E 38	90577509	N 02
576476	B160	594091	H 15	602087	B 82	602227	B128	603659	B115	611612	B121	78177	J 23	90577509	T 20
576557	B189	594091	H 18	602087	B115	602236	B130	603670	J 14	611612	B144	78177	J 24	90577509	T 21
576608	E 19	594091	C 45	602097	B 89	602236	B 95	603672	B 90	611659L	B 89	78248	S 01	90577611	H 01
576723	C 46	594091	C 46	602097	B126	602289	B 88	603672	B103	611659L	B126	78322	N 30	90577642	G 11
576723	C 24	594091	C 24	602098	B 88	602289	B123	603672	B151	611660	B 89	78756	T 22	90602025	B 80
577176	H 14	594134	C 71	602098	B124	602289	B137	603673	B151	611660	B126	78759	M144	90602025	B112
577177	H 14	594134	C 05	602098	B125	602289	B125	603734	B 89	611786	B 93	78782	B117	90602064	B 87
577254	R 59	594134	C 27	602099	B130	602388	B 84	603734	B126	611795	B135	78862	B 90	90602064	B122
577255	R 59	594134	C 53	602099	B 95	602388	B118	603796	B 88	611950	B135	79019	N 27	90602202	B 84
577259	R 59	594594	B 91	602123	B 88	602411	B113	603796	B124	612064	B 97	79026	B 21	90602202	B117
577389	F 01	594594	B130	602123	B123	602512	B 90	603920	B 94	612435	B 98	79027	B 59	90602372	B 80
577846	B 83	594594	B176	602123	B125	602512	B127	605011	J 13	612898	B 79	79027	B 90	90608178	B107
577846	N 06	594594	M 23	602130	B 79	602545	B 79	605191	J 14	612898	B111	79027	B127	90611504	B 95
577873	H 12	594594	J 05	602130	B111	602545	B116	605410	B153	612989	B 80	79039	L 21	90624209	H 19
577898	F 15	594594	L 19	602141	B 79	602587	B 80	606168	B 91	612989	B112	79086	N 15	AAB700010	N 18
579408	T 22	594594	M155	602141	B111	602587	B112	606178	T 23	613402	B 91	79086	N 30	AAB700020	N 18
579409	T 22	594594	J 02	602142	B 89	602609	B 81	606247	J 14	613671	B 80	79086L	N 15	AAB700030	N 17
579409D	T 22	594594	J 04	602142	B126	602609	B114	606474	D 15	613671	B113	79086L	N 30	AAB700040	N 17

AAU2249L .... B 36
AAU2249L B105
AAU2304 .... D 12
AAU2304 .... D 21
AAU3686 .... M 23
AAU3686 .... M155
AAU3686 .... M 32
AAU3715L M153
AAU5034 .... M 79
AAU5034 .... M 80
AAU5034 .... M 35
AAU7803 .... H 09
AAU7803 .... H 10
AAU7803 .... H 05
AAU7803 .... H 07
AAU9902 .... J 13
AAU9903 .... J 13
AB604021L .... M 77
AB604021L .... P 21
AB604031L .... P 47
AB604044 .... N 81
AB604051 .... P 39
AB604055 .... P 41
AB604051L .... P 28
AB604051L .... P 31
AB606031 .... L 28
AB606031 .... L 48
AB606031L .... L 48
AB606041 .... C 86
AB606041L .... L 04
AB606041L .... P 05
AB606041L .... C 88
AB606044L .... M 79
AB606044L .... M 80
AB606051 .... M 97
AB606071L .... P 73
AB606074 .... M132
AB606074 .... M133
AB606124 .... P 62
AB606124 .... M132
AB608031 .... L 28
AB608031L .... L 21
AB608031L .... M 34
AB608031L .... M 89
AB608035 .... M 35
AB608041 .... R 01
AB608047L .... N 30
AB608051 .... L 28
AB608051 .... L 48
AB608051L .... N 78
AB608051L .... N 79
AB608051L .... N 82
AB608051L .... N 83
AB608051L .... N 84
AB608051L .... N 85
AB608051L .... P 67
AB608051L .... P 71
AB608051L .... N 62
AB608061 .... R 25
AB608061L .... P 83
AB608074 .... M134
AB608074 .... M136
AB608074 .... M139
AB608074 .... M142
AB608101L .... P 71
AB610031L .... B 79
AB610031L .... B 91
AB610031L .... K 04
AB610031L .... M 95
AB610041L B103
AB610041L B116
AB610041L B151
AB610041L .... N 25
AB610041L .... M 96
AB610041L .... M 99
AB610045 .... N 42
AB610051 .... L 43
AB610055 .... N 42
AB610055 .... N 43
AB610061L .... B 91
AB610061L N112
AB610061L N116
AB610061L .... M 83
AB610064L .... P 83
AB610121L .... M 95
AB612051L .... P 75
AB612061L .... N 27
AB612061L .... N 30
AB612081L M144
AB614061L .... F 05
AB614061L .... R 32
AB614061L .... R 39
AC606034 .... P 05
AC606044 .... P 19
AC606044 .... P 21
AC606054L .... P 11
AC608041 .... P 11
ACU5037 .... M147
ACU5037 .... M149
ACU5037 .... M146
ACU5037 .... M148
ADL3029LOY .... P 63
ADL3029LOYL .... P 62
ADL3029LOYL .... R 25
ADL3029PUB .... P 62
ADL3029SUC .... P 62
ADL3029SUC .... R 25
ADL3029SUC .... P 63
ADU1670L .... P 19
ADU1670L .... P 21
ADU1784L .... M 83
ADU2066L .... M 13
ADU2888L .... H 20
ADU3325 .... M135
ADU3325 .... M137
ADU3905 .... M146
ADU3905 .... M148
ADU4682 .... M 10
ADU4682 .... M 07
ADU5065L .... G 14
ADU6418L M148
ADU6780 .... N 82
ADU6780L .... N 79
ADU6780L .... N 84
ADU7048 .... L 27
ADU7048 .... L 38
ADU7161 .... B176
ADU7242L B104
ADU7242L B148
ADU7340L B185
ADU7739L M157
ADU7739L .... M 17
ADU7981L M156
ADU8229L B164
ADU8267 .... M159
ADU8363L M159
ADU8363L .... M 59
ADU8846 .... H 10
ADU8846L .... H 09
ADU8846L .... H 10
ADU8891 .... B171
ADU8981 .... M156
ADU9081L .... L 19
AEA635L B172
AEU1220 .... L 41
AEU1403 .... L 46
AEU1446 .... B151
AEU1446 .... M 95
AEU1448 .... B 79
AEU1448 .... J 23
AEU1448 .... J 21
AEU1448 .... J 24
AEU1547 .... H 15
AEU1547 .... H 18
AEU1616 .... B153
AEU1616 .... B154
AEU1626 .... L 46
AEU1626 .... L 49
AEU1627 .... L 40
AEU1628 .... L 54
AEU1628 .... L 55
AEU1714 .... G 11
AEU1714 .... G 12
AEU1718 .... H 15
AEU1719 .... H 15
AEU2129L .... B 29
AEU2147L .... J 12
AEU2148L .... J 12
AEU2149L .... J 12
AEU2718 .... B 33
AEU2719 .... B 33
AEU2720 .... B 33
AEU2734 .... D 22
AEU2736 .... D 22
AEU2737 .... D 22
AEU2738 .... D 22
AEU3056 .... L 45
AEU3056 .... L 53
AEU3056 .... L 55
AEU3064 .... L 41
AFU1069 .... N 76
AFU1069 .... N 81
AFU1090L .... M 23
AFU1090L M158
AFU1090L .... T 22
AFU1090L .... M 34
AFU1090L .... M 42
AFU1090L .... M 57
AFU1090L M151
AFU1158 .... B 26
AFU1180 .... H 14
AFU1180 .... H 17
AFU1216L .... P 04
AFU1217 .... H 09
AFU1217 .... H 10
AFU1296 .... M158
AFU1350 .... N 54
AFU1400 .... D 23
AFU1400 .... D 22
AFU1879L .... B 05
AFU1882L .... B 05
AFU1882L .... B 44
AFU1887L .... B 05
AFU1887L .... B 53
AFU1890L .... B 23
AFU2627 .... M160
AFU2798L .... F 12
AFU2879 .... M158
AFU2881 .... M 60
AFU2913L .... M 84
AFU2913L .... M 83
AFU3009 .... N112
AFU3009 .... N116
AFU3017L .... M 81
AFU3017L .... M 82
AFU3019 .... N 77
AFU3081L M110
AFU3081L M153
AFU3112L .... M 05
AFU3112L .... M 04
AFU3711L .... N 41
AFU4091 .... M 13
AFU4091 .... M 12
AFU4092L .... M 12
AFU4131 .... P 09
AFU4146 .... M148
AFU4173 .... M155
AFU4180 .... N 49
AFU4214 .... F 16
AFU4481 .... M 05
AFU4481 .... M 04
AFU4506 .... M147
AFU4506 .... M149
AGU1425 .... R 32
AGU1425 .... R 39
AHU1026L B165
AHU700010 .... F 05
AJ606041 .... P 01
AJ606041 .... P 03
AJU1136 .... P 64
AJU1136L .... R 32
AJU1136L .... R 39
AK606021L .... M 15
AK606021L .... P 05
AK606031 .... P 05
AK606091 .... P 05
AK606091 .... P 14
AK608021 .... P 67
AK608045 .... N 30
AK608051L .... P 13
AK608121 .... P 03
AK608141L .... P 04
AK608141L .... P 22
AK608141L .... M 91
AK612011L M145
AK612011L M144
AK616011 .... L 21
AL610011L .... N 13
ALR1441 .... N 70
ALR1475 .... F 05
ALR1476 .... F 05
ALR1590JUL .... N 49
ALR1590LOY .... N 49
ALR1590SUC .... N 49
ALR1591JUL .... N 49
ALR1591LOY .... N 49
ALR1591SUC .... N 49
ALR1773 .... N 32
ALR1782 .... N 21
ALR1783 .... N 21
ALR1816 .... N 45
ALR1817 .... N 45
ALR1976 .... N 51
ALR1976 .... N 46
ALR2252 .... N 72
ALR2253 .... N 72
ALR2254 .... N 72
ALR2254 .... N 74
ALR2254 .... N 75
ALR2254 .... N 76
ALR2255 .... N 72
ALR2255 .... N 74
ALR2255 .... N 76
ALR2292 .... N 45
ALR2348 .... N 19
ALR2349 .... N 19
ALR2371 .... N 50
ALR2400 .... N 53
ALR2701 .... N 34
ALR2811 .... N 34
ALR2967 .... N 49
ALR2968 .... N 49
ALR3004 .... N 70
ALR3028 .... N 20
ALR3029 .... N 20
ALR3034 .... N 19
ALR3272 .... N 73
ALR3282 .... N 23
ALR3283 .... N 23
ALR3361 .... N 76
ALR3510 .... N 20
ALR3647 .... N 21
ALR3648 .... N 21
ALR3722 .... F 05
ALR3723 .... F 05
ALR3740 .... N 18
ALR3747 .... N 24
ALR3748 .... N 24
ALR3751 .... N 18
ALR3752 .... N 17
ALR3753 .... N 17
ALR3775 .... N 34
ALR3783 .... N 23
ALR3784 .... N 23
ALR3789 .... N 25
ALR3790 .... N 25
ALR3871 .... N 22
ALR3872 .... N 22
ALR3933 .... N 74
ALR3937 .... N 74
ALR4450 .... N 53
ALR4457 .... N 22
ALR4458 .... N 22
ALR4467 .... N 21
ALR4468 .... N 21
ALR4472 .... N 24
ALR4473 .... N 24
ALR4722 .... N 48
ALR4723 .... N 52
ALR4725 .... N 50
ALR4726 .... N 46
ALR4727 .... N 46
ALR5071 .... N 75
ALR5074 .... N 34
ALR5085 .... N 45
ALR5404 .... N 33
ALR5576 .... N 24
ALR5577 .... N 24
ALR5591 .... N 46
ALR5592 .... N 46
ALR5632 .... N 22
ALR5633 .... N 22
ALR5691 .... N 85
ALR5693 .... N121
ALR5717 .... N 21
ALR5718 .... N 21
ALR5720 .... N 29
ALR5721 .... N 29
ALR5722 .... N 29
ALR5723 .... N 29
ALR5822 .... N 46
ALR5823 .... N 46
ALR6052 .... N 21
ALR6053 .... N 21
ALR6054 .... N 24
ALR6055 .... N 24
ALR6078 .... N 83
ALR6082 .... N 83
ALR6082 .... N 84
ALR6176 .... N 32
ALR6238 .... N 32
ALR6243 .... N 20
ALR6244 .... N 20
ALR6272 .... N 19
ALR6306 .... N 85
ALR6473 .... N 60
ALR6474 .... N 60
ALR6482 .... N 48
ALR6498 .... N 49
ALR6500 .... N 45
ALR6501 .... N 45
ALR6541 .... N 51
ALR6681 .... N 27
ALR6682 .... N 27
ALR6749 .... N 32
ALR6927 .... N 45
ALR6928 .... N 45
ALR6929 .... N 45
ALR6930 .... N 45
ALR6993 .... N 22
ALR6994 .... N 22
ALR7062 .... N 70
ALR7063 .... N 70
ALR7601 .... N 45
ALR7603 .... N 24
ALR7604 .... N 24
ALR7693 .... N 17
ALR7694 .... N 16
ALR7695 .... N 16
ALR7696 .... N 16
ALR7697 .... N 16
ALR7702 .... N 19
ALR7703 .... N 19
ALR7798 .... N 29
ALR7800 .... N 29
ALR7845 .... N 34
ALR7969 .... N 49
ALR7970 .... N 49
ALR8019 .... N 50
ALR8020 .... N 50
ALR8022 .... N 48
ALR8024 .... N 51
ALR8268 .... N 02
ALR8519 .... N 24
ALR8520 .... N 24
ALR8568 .... N 27
ALR8569 .... N 27
ALR8571 .... N 29
ALR8572 .... N 50
ALR8573 .... N 50
ALR8574 .... N 50
ALR8575 .... N 50
ALR8576 .... N 46
ALR8577 .... N 46
ALR8578 .... N 46
ALR8579 .... N 46
ALR8650 .... N 41
ALR8704 .... N 78
ALR8705 .... N 78
ALR8715 .... N 45
ALR8716 .... N 45
ALR8990 .... N 48
ALR8991 .... N 51
ALR9136 .... N 20
ALR9209 .... N 20
ALR9210 .... N 20
ALR9214 .... N 62
ALR9344 .... N 22
ALR9345 .... N 22
ALR9453 .... N 20
ALR9607 .... N 24
ALR9608 .... N 24
ALR9743 .... N 62
ALR9841 .... N 19
ALR9842 .... N 19
ALR9850 .... N 18
ALR9851 .... N 18
ALR9852 .... N 17
ALR9853 .... N 17
ALR9854 .... N 16

Part	Ref
ALR9855	N 16
ALR9856	N 16
ALR9857	N 16
ALR9873	N 29
ALR9874	N 29
ALR9875	N 29
ALR9876	N 29
ALR9877	N 27
ALR9878	N 27
ALU1403	C 54
ALU1403L	B 47
ALU1403L	C 54
ALU1403L	D 04
AMR1029LNF	M133
AMR1030LNF	M133
AMR1039	M152
AMR1040	M 34
AMR1040	M 35
AMR1051	M 67
AMR1053	B 38
AMR1055	M 50
AMR1057	M 21
AMR1088	M106
AMR1088	M 84
AMR1097	M111
AMR1101	M 33
AMR1102	M133
AMR1114	G 10
AMR1117	M 68
AMR1117	M135
AMR1126	M 25
AMR1127	M 25
AMR1128	M 25
AMR1168	M 26
AMR1173	M112
AMR1183LUP	M 12
AMR1191	M 25
AMR1192	P 16
AMR1192	P 17
AMR1233	M119
AMR1234	M119
AMR1237	M119
AMR1253	M115
AMR1256	M119
AMR1257	M119
AMR1262	M119
AMR1268	M 25
AMR1280	M 81
AMR1281	M 25
AMR1282	M 93
AMR1283	M 05
AMR1284	M 05
AMR1285	M 11
AMR1294	M 08
AMR1294	M 09

Part	Ref
AMR1295	M 08
AMR1295	M 09
AMR1339	M119
AMR1340	M119
AMR1343	M119
AMR1344	M119
AMR1345	M119
AMR1349	M119
AMR1425	B 61
AMR1425	B130
AMR1425	B176
AMR1427	M108
AMR1427	M110
AMR1427	M106
AMR1514	M145
AMR1515	M145
AMR1527	M 35
AMR1528	M 35
AMR1533	M 54
AMR1533	M 55
AMR1540	M 35
AMR1552	M 81
AMR1553	M 82
AMR1709	M 33
AMR1728	M 02
AMR1729	M 22
AMR1730	M 23
AMR1731	M 23
AMR1732	M 22
AMR1782LNF	M133
AMR1785	M135
AMR1785	M137
AMR1788	M149
AMR1789	M147
AMR1805	M145
AMR1805	M144
AMR1806	M152
AMR1806	M151
AMR1835	M112
AMR1882	M145
AMR1886	G 10
AMR1888	G 10
AMR1890	M150
AMR1891	M106
AMR1935	M154
AMR1935	M116
AMR2010	G 06
AMR2010	G 08
AMR2014	M133
AMR2019	P 15
AMR2022	M108
AMR2022	M110
AMR2022	M106
AMR2025	M102
AMR2034	M106

Part	Ref
AMR2082	M 44
AMR2092	B122
AMR2106	M102
AMR2107	M102
AMR2108	M103
AMR2108	M104
AMR2109	M102
AMR2109	M103
AMR2109	M104
AMR2110	M103
AMR2117	M 68
AMR2118	M 68
AMR2125	M137
AMR2125	M139
AMR2128	M 97
AMR2159LNF	M136
AMR2164	M 57
AMR2193LNF	M136
AMR2194LNF	M136
AMR2194LNF	M138
AMR2194LNF	M141
AMR2195LNF	M136
AMR2200	M 54
AMR2201	M 33
AMR2202	M 18
AMR2214	M106
AMR2215	M 17
AMR2215	M 16
AMR2216	M 21
AMR2216	M 20
AMR2217	B157
AMR2218	B 73
AMR2218	B157
AMR2218	B155
AMR2263	B 73
AMR2263	B157
AMR2263	B155
AMR2294	M 44
AMR2299	M 59
AMR2304	M 35
AMR2306	M152
AMR2306	M151
AMR2323	M157
AMR2329	M 13
AMR2330	N 15
AMR2417LUP	M106
AMR2421	M 10
AMR2421	M 07
AMR2425	B 16
AMR2425	B 54
AMR2447	M 90
AMR2463	M 90
AMR2469	M103
AMR2469	M104
AMR2471	M 76

Part	Ref
AMR2496	M 77
AMR2497	M 77
AMR2498	M 66
AMR2543	M 19
AMR2577	M 13
AMR2587	M 57
AMR2590LNF	M136
AMR2591	M103
AMR2606	M 51
AMR2606	M 52
AMR2606	M 53
AMR2608	M 90
AMR2619	M137
AMR2705	M 61
AMR2706	M 03
AMR2712	M 22
AMR2713	M 22
AMR2714	M 22
AMR2715	M 22
AMR2741	M111
AMR2742	M111
AMR2745	M 72
AMR2776	M137
AMR2781LNF	M 66
AMR2782LNF	M 66
AMR2783LNF	M 66
AMR2784LNF	M 66
AMR2885	M 89
AMR2886	M 44
AMR2908	M 46
AMR2911	M 84
AMR2930	M 45
AMR2933	M 18
AMR2985	M 02
AMR2986	M 23
AMR2987	M 23
AMR3029	M137
AMR3046	D 04
AMR3065	M 50
AMR3066	M 85
AMR3067	M 86
AMR3076	M 84
AMR3077	M 55
AMR3078	M 55
AMR3079	M 27
AMR3080	M 27
AMR3081	M 27
AMR3082	M 27
AMR3083	M 27
AMR3084	M 55
AMR3107	B155
AMR3107E	B155
AMR3134	M 87
AMR3154	M135
AMR3154	M136

Part	Ref
AMR3154	M140
AMR3154	M143
AMR3235	M 55
AMR3236LNF	M 77
AMR3242	M 96
AMR3258	M 82
AMR3259	M152
AMR3260	M152
AMR3266	M147
AMR3266	M149
AMR3273	M 58
AMR3274	M 58
AMR3275	M 55
AMR3276	M 94
AMR3276E	M 94
AMR3277	M 58
AMR3278	M 58
AMR3279	M 46
AMR3304	M 61
AMR3304	M 62
AMR3331	M 42
AMR3332	M 42
AMR3353	M 09
AMR3354	M 09
AMR3361	M 64
AMR3374	M 40
AMR3382	N 78
AMR3382	N 82
AMR3382	N 84
AMR3383	N 78
AMR3384	N 78
AMR3434	M 81
AMR3435	M 80
AMR3436	M 34
AMR3437	M 34
AMR3438	M 27
AMR3439	M 27
AMR3440	M 27
AMR3441	M 27
AMR3460	M135
AMR3460	M137
AMR3461	M136
AMR3461	M138
AMR3462	M136
AMR3462	M138
AMR3464L	M116
AMR3466	M101
AMR3467	M101
AMR3490	M106
AMR3550	M152
AMR3556	M 36
AMR3557	M 36
AMR3558	M 36
AMR3561	M 56
AMR3562	M 56

Part	Ref
AMR3563	M 56
AMR3564	M 56
AMR3566	M 42
AMR3567	M 42
AMR3570	M 29
AMR3571	M 29
AMR3572	M 29
AMR3573	M 29
AMR3574	M 29
AMR3575	M 29
AMR3576	M 29
AMR3579	M 40
AMR3580	M 40
AMR3581	M 40
AMR3582	M 40
AMR3601	M 78
AMR3602	M 78
AMR3608	M 38
AMR3609	M 38
AMR3610	M 39
AMR3611	M 39
AMR3613	M 43
AMR3614	M 43
AMR3615	M 43
AMR3617	M 41
AMR3618	M 41
AMR3619	M 41
AMR3620	M 41
AMR3622	M 48
AMR3623	M 48
AMR3624	M 48
AMR3625	M 48
AMR3628	M 45
AMR3629	M 45
AMR3631	M 51
AMR3632	M 51
AMR3633	M 51
AMR3634	M 51
AMR3638	M 41
AMR3639	M 41
AMR3640	M 41
AMR3641	M 42
AMR3643	M 82
AMR3647	M 61
AMR3647	M 64
AMR3649	M 64
AMR3652	M 76
AMR3653	M 76
AMR3655	M149
AMR3656	M147
AMR3657	M147
AMR3657	M149
AMR3658	M147
AMR3658	M149
AMR3658	M146

Part	Ref
AMR3658	M148
AMR3659	M147
AMR3659	M149
AMR3659	M146
AMR3659	M148
AMR3660	M147
AMR3660	M149
AMR3661	M149
AMR3666	M 61
AMR3667	M 61
AMR3668	M 61
AMR3669	M 61
AMR3678	M 84
AMR3678	M 89
AMR3679	M 17
AMR3679	M 19
AMR3679	M 21
AMR3680	M 17
AMR3680	M 19
AMR3680	M 21
AMR3681	M 17
AMR3681	M 19
AMR3681	M 21
AMR3682	M 30
AMR3685	M 39
AMR3701	M 13
AMR3720	M102
AMR3721	M102
AMR3722	M103
AMR3722	M104
AMR3723	M102
AMR3723	M103
AMR3723	M104
AMR3725	M102
AMR3735	M100
AMR3740	M 65
AMR3740	M 72
AMR3741	M 71
AMR3742	M 71
AMR3743	M 71
AMR3744	M 71
AMR3748	M 70
AMR3749	M 70
AMR3750	M 70
AMR3751	T 23
AMR3751	M 70
AMR3752	M 65
AMR3754	B 76
AMR3755	B 75
AMR3773	M 85
AMR3773	M 86
AMR3773	M 87
AMR3841	M 64
AMR3842	M 64
AMR3853	M135

Part	Ref
AMR3853	M137
AMR3859	C 54
AMR3864	M 42
AMR3865	M 42
AMR3869	M145
AMR3871	M 81
AMR3872	M152
AMR3873	M152
AMR3875	M 29
AMR3876	M 29
AMR3877	M 29
AMR3878	M 29
AMR3921	M 37
AMR3923	M 82
AMR3927	M 80
AMR3929	M129
AMR3930	M129
AMR3932	M152
AMR3933	M152
AMR3934	M152
AMR3935	M 66
AMR3949	M 48
AMR3950	M 48
AMR3951	M 48
AMR3952	M 48
AMR3954	M 08
AMR3955	M 08
AMR3992	M 61
AMR3993	M 61
AMR3994	M 61
AMR3995	M 61
AMR4060	M146
AMR4060	M148
AMR4090	M 56
AMR4091	M 56
AMR4136	M 06
AMR4137	M 06
AMR4138	M 65
AMR4139	M 85
AMR4140	M 86
AMR4183	M140
AMR4183	M143
AMR4184	M140
AMR4184	M143
AMR4187LNF	M139
AMR4187LNF	M142
AMR4188LNF	M138
AMR4188LNF	M141
AMR4203	M 30
AMR4204	M 30
AMR4205	M 30
AMR4206	M 30
AMR4207	M 30
AMR4208	M 30
AMR4209	M 30

DISCOVERY 1989MY UPTO 1999MY | GROUP X | Numeric Index "AMR4210 " TO "ANR3838 "

AMR4210	M 30	AMR4831L	M140	AMR5221	M 31	AMR6228	M100	ANR1767	H 07	ANR2260LUN	H 20	ANR2914	F 20	ANR3202	M 40
AMR4211	M 30	AMR4831L	M143	AMR5222	M 31	AMR6248	M116	ANR1768	H 05	ANR2278	N 01	ANR2914	F 21	ANR3225	C 68
AMR4214	M 38	AMR4832	M140	AMR5223	M 31	AMR6249	M109	ANR1768	H 07	ANR2278	C 68	ANR2916	F 19	ANR3225	C 90
AMR4215	M 38	AMR4832	M143	AMR5226	M 32	AMR6250	M109	ANR1769	H 09	ANR2278	C 90	ANR2933	M 61	ANR3225	N 08
AMR4216	M 39	AMR4845	M124	AMR5227	M 32	AMR6305	M 53	ANR1769	H 10	ANR2359	G 02	ANR2933	M 62	ANR3227	K 02
AMR4219	M 36	AMR4846	M125	AMR5228	M 32	AMR6424	M 66	ANR1769	H 05	ANR2391LAL	E 41	ANR2933	F 03	ANR3227	K 03
AMR4220	M 52	AMR4867	M102	AMR5230	M 32	AMR6428	M109	ANR1769	H 07	ANR2391LUQ	E 41	ANR2938	E 52	ANR3227	D 06
AMR4225	M 51	AMR4873	M 81	AMR5238	M 36	AMR6429	M109	ANR1795	H 09	ANR2391MNH	E 41	ANR2941	G 07	ANR3243LNF	H 20
AMR4226	M 51	AMR4877	M 36	AMR5239	M 37	AMR6462	M 71	ANR1795	H 06	ANR2391MUE	E 41	ANR2941	G 09	ANR3243LNF	H 23
AMR4234	M 52	AMR4879	M 37	AMR5241	M 38	AMR6463	M 71	ANR1795	H 08	ANR2406	H 09	ANR2985	F 20	ANR3246	N 01
AMR4239	M 41	AMR4885	M107	AMR5242	M 38	AMR6464	M 71	ANR1796	H 09	ANR2407	H 09	ANR2985	F 21	ANR3255	N 12
AMR4240	M 41	AMR4886	M107	AMR5243	M 39	AMR6465	M 71	ANR1796	H 10	ANR2408	H 10	ANR2991	E 48	ANR3363	H 10
AMR4241	M 41	AMR4888	M108	AMR5245	M 39	AMR6509	M 09	ANR1796	H 06	ANR2409	H 10	ANR2994	F 12	ANR3386	G 11
AMR4242	M 41	AMR4888	M110	AMR5270	M141	AMR6510	M 09	ANR1796	H 08	ANR2410	H 09	ANR3003	F 05	ANR3410	E 43
AMR4243	M 41	AMR4889	M108	AMR5271	M143	AMR6511	M 05	ANR1799	E 49	ANR2410	H 10	ANR3004	F 05	ANR3428	N 02
AMR4247	B155	AMR4889	M110	AMR5360	M 81	AMR6512	M 05	ANR1805	H 07	ANR2411	H 09	ANR3025	N 14	ANR3429	N 02
AMR4247E	B155	AMR4890	M107	AMR5361	M139	AMR6681	M 39	ANR1806	H 05	ANR2411	H 10	ANR3026	N 14	ANR3432	N 02
AMR4248	B 73	AMR4890	M109	AMR5361	M141	AMR6682	M 31	ANR1832	D 20	ANR2412	H 10	ANR3037	E 49	ANR3477	E 50
AMR4248E	B 73	AMR4892	M107	AMR5362	M141	AMR6683	M 31	ANR1844	S 01	ANR2413	H 10	ANR3041LNF	F 02	ANR3500	G 07
AMR4253	M 48	AMR4892	M109	AMR5363	M141	AN104011	P 47	ANR1887	F 19	ANR2414	H 10	ANR3042LNF	F 02	ANR3501	G 07
AMR4254	M 48	AMR4893	M107	AMR5412	M109	ANR1035	N 01	ANR1892	K 06	ANR2420	E 43	ANR3043LNF	F 02	ANR3502	G 05
AMR4255	M 48	AMR4893	M109	AMR5413	M109	ANR1111	T 21	ANR1892	K 08	ANR2471	N 07	ANR3044LNF	F 02	ANR3503	G 05
AMR4256	M 48	AMR4905	M 61	AMR5420	M 81	ANR1212	G 06	ANR1892	N 07	ANR2479	N 07	ANR3052	S 01	ANR3504	G 05
AMR4711	M 49	AMR4939	B 73	AMR5425	B 73	ANR1212	G 08	ANR1895	E 52	ANR2488	N 04	ANR3054	H 22	ANR3505	G 05
AMR4712	M 49	AMR4939	B155	AMR5438	M 42	ANR1270	F 18	ANR1902	G 03	ANR2498	G 13	ANR3058	E 50	ANR3508	G 13
AMR4713	M 49	AMR4948	C 92	AMR5439	M 42	ANR1369	M160	ANR1975	E 45	ANR2514	H 03	ANR3060	E 50	ANR3518	F 09
AMR4714	M 49	AMR4948	C 95	AMR5441	M112	ANR1467	B182	ANR1976	E 45	ANR2559	G 09	ANR3107	H 03	ANR3593	H 21
AMR4716	M 45	AMR4948	C 97	AMR5458	M108	ANR1504	N 02	ANR1977	E 50	ANR2651	G 11	ANR3108	H 03	ANR3606	G 02
AMR4717	M 45	AMR4952	M 43	AMR5458	M110	ANR1583	L 49	ANR1998	F 18	ANR2651	G 12	ANR3112	F 20	ANR3608	H 01
AMR4719	M 52	AMR4956	M108	AMR5485	M 88	ANR1631	G 02	ANR1998	F 17	ANR2663	N 07	ANR3113	F 20	ANR3611	H 02
AMR4720	M 52	AMR4965	M110	AMR5517	M 41	ANR1632	G 01	ANR2004	G 13	ANR2672	H 24	ANR3116	N122	ANR3613	H 02
AMR4721	M 52	AMR4974	M108	AMR5538	M143	ANR1633	G 02	ANR2016	H 09	ANR2697	G 12	ANR3131	F 20	ANR3614	H 02
AMR4722	M 52	AMR4974	M110	AMR5548	M 36	ANR1634	G 02	ANR2023	H 09	ANR2734	H 21	ANR3131	F 21	ANR3625	D 06
AMR4726	M 61	AMR5083	M149	AMR5549	M 37	ANR1675	G 13	ANR2024	H 09	ANR2738	G 11	ANR3132	F 19	ANR3629	T 21
AMR4726	M 64	AMR5084	M147	AMR5555	M 53	ANR1689MNH	E 41	ANR2029	N 12	ANR2741	N 13	ANR3132	F 20	ANR3631MNH	E 41
AMR4727	M 64	AMR5100	M 05	AMR5556	M 53	ANR1689RJQ	E 41	ANR2039	G 13	ANR2743	N 14	ANR3133	F 19	ANR3641	N 07
AMR4743	M124	AMR5101	M 05	AMR5557	M111	ANR1706	N 02	ANR2040	G 13	ANR2763MMM	E 41	ANR3133	F 21	ANR3642	N 07
AMR4744	M125	AMR5117	M137	AMR5619	M116	ANR1721	E 52	ANR2057	G 04	ANR2803	F 16	ANR3138	C 68	ANR3653	F 03
AMR4745	M124	AMR5117	M140	AMR5623	M100	ANR1758	H 05	ANR2078	G 04	ANR2804	F 16	ANR3138	C 90	ANR3654	F 03
AMR4746	M125	AMR5117	M143	AMR5624	M100	ANR1759	H 05	ANR2106	N 07	ANR2805	N 08	ANR3138	N 10	ANR3655	F 05
AMR4747	M124	AMR5139	M 56	AMR5631	M 61	ANR1760	H 05	ANR2116	F 19	ANR2815	N 06	ANR3165	H 09	ANR3656	F 05
AMR4748	M125	AMR5140	M 56	AMR5631	M 64	ANR1761	H 05	ANR2116	F 20	ANR2816	N 06	ANR3171	F 09	ANR3660	G 03
AMR4749	M124	AMR5141	M 56	AMR5631	M 62	ANR1761	H 07	ANR2127	N 08	ANR2817	N 08	ANR3180	H 09	ANR3660	G 04
AMR4750	M125	AMR5142	M 56	AMR5632	M 64	ANR1762	H 07	ANR2127	N 10	ANR2818	N 09	ANR3181	H 09	ANR3661	G 03
AMR4751	M124	AMR5150	M 08	AMR5670	M 57	ANR1763	H 07	ANR2128	N 08	ANR2819	N 08	ANR3189	N 01	ANR3661	G 04
AMR4752	M125	AMR5150	M 09	AMR5681	M 32	ANR1764	H 07	ANR2128	N 10	ANR2820	N 09	ANR3190	N 01	ANR3700	F 18
AMR4753	M124	AMR5151	M 08	AMR5683	M 53	ANR1765	H 09	ANR2157	B 70	ANR2840	N 01	ANR3194	H 09	ANR3771	G 11
AMR4754	M125	AMR5151	M 09	AMR5684	M 53	ANR1765	H 10	ANR2165LNF	H 20	ANR2868	N 05	ANR3194	H 10	ANR3772	G 12
AMR4757	M129	AMR5164	M 62	AMR5696LNF	M138	ANR1765	H 05	ANR2165LNF	H 23	ANR2892	H 09	ANR3200	C 68	ANR3783	F 21
AMR4758	M129	AMR5165	M 62	AMR5696LNF	M141	ANR1765	H 07	ANR2183	G 11	ANR2893	H 09	ANR3200	C 90	ANR3789	G 07
AMR4790	M 09	AMR5177	M 65	AMR5697	M 88	ANR1766	H 05	ANR2186	G 11	ANR2897	N 08	ANR3200	N 08	ANR3790	G 07
AMR4791	M 09	AMR5178	M 70	AMR6109	M 53	ANR1766	H 07	ANR2224	N 28	ANR2898	C 68	ANR3200	N 10	ANR3791	G 05
AMR4794LNF	M138	AMR5218	M 31	AMR6110	M 53	ANR1767	H 09	ANR2260LNF	H 20	ANR2898	C 90	ANR3201	C 68	ANR3792	G 05
AMR4803	M 17	AMR5219	M 31	AMR6111	B 73	ANR1767	H 10	ANR2260LNF	H 22	ANR2898	N 08	ANR3201	C 90	ANR3823	F 21
AMR4806	M 57	AMR5220	M 31	AMR6111	B155	ANR1767	H 05	ANR2260LNF	H 23	ANR2900	E 41	ANR3201	N 08	ANR3838	H 20

ANR3896 T 22	ANR6222 F 18	AWR1158LOY R 50	AWR1535LOY P 63	AWR2212 P 21	AWR2615LOY P 34	AWR3020LOY P 33	AWR3147LOY R 39
ANR4136 S 01	ANR6222 F 17	AWR1158SUC R 50	AWR1535SUC P 63	AWR2228 N 66	AWR2615SUC P 34	AWR3020SUC P 33	AWR3195 R 38
ANR4188 E 43	ANR6574LNF F 02	AWR1183 P 19	AWR1593LOY P 65	AWR2233 P 38	AWR2632 P 34	AWR3023LOY P 37	AWR3195 R 40
ANR4189 E 48	ANR6575LNF F 02	AWR1183 P 20	AWR1593SUC P 65	AWR2416 R 26	AWR2639PMA R 12	AWR3023SUC P 37	AWR3196 R 38
ANR4344 E 53	ANR6576LNF F 02	AWR1184 P 72	AWR1594LOY P 65	AWR2438PMD N 12	AWR2640PMA R 12	AWR3024LOY P 37	AWR3196 R 40
ANR4350 E 45	ANR6577LNF F 02	AWR1194LNF P 07	AWR1594SUC P 65	AWR2476RAI N103	AWR2641PMA R 12	AWR3024SUC P 37	AWR3242LOY R 29
ANR4351 E 45	ANR6656 F 20	AWR1195LNF P 12	AWR2008LPZ P 77	AWR2477RAI N103	AWR2642PMA R 12	AWR3036LOY R 25	AWR3242SUC R 29
ANR4352 E 50	ANR6656 F 21	AWR1198LNF M 65	AWR2008SUD P 77	AWR2478RAI N103	AWR2643PMA R 18	AWR3036SUC R 25	AWR3243LOY R 35
ANR4489 S 01	ANU1269L C 67	AWR1199LNF P 12	AWR2023 N 68	AWR2479RAI N103	AWR2644PMA R 18	AWR3037LOY R 25	AWR3243SUC R 35
ANR4513 G 03	AQH700010 N 32	AWR1200LNF P 12	AWR2024 N 68	AWR2480RAI N103	AWR2646LOY R 14	AWR3037SUC R 25	AWR3244LOY R 31
ANR4514 G 03	AR606041L N 39	AWR1258LNF L 43	AWR2025 N 68	AWR2481RAI N103	AWR2646SUC R 14	AWR3038LOY R 25	AWR3244SUC R 31
ANR4515 G 07	AR606041L T 47	AWR1322 N 44	AWR2028PMA R 13	AWR2482RAI N103	AWR2647LOY R 14	AWR3038SUC R 25	AWR3245LOY R 37
ANR4516 G 07	AR606041L N 37	AWR1322 L 47	AWR2029PMA R 13	AWR2483RAI N103	AWR2647SUC R 14	AWR3039LOY R 25	AWR3245SUC R 37
ANR4517 B 65	AR606041L N 38	AWR1362 L 30	AWR2042LOY R 02	AWR2484RGI N103	AWR2651 L 30	AWR3039SUC R 25	AWR3246LNF P 19
ANR4518 B 65	ASR1062 N 70	AWR1363 L 31	AWR2042SUD R 02	AWR2484RGI N104	AWR2679LNF R 24	AWR3049LOY P 37	AWR3246LNF P 21
ANR4519 N 02	ASR1070 N 54	AWR1364 L 31	AWR2043LOY R 03	AWR2485RGI N103	AWR2679LNF R 16	AWR3049SUC P 37	AWR3247LNF P 19
ANR4615 N 08	ASR1186 N 24	AWR1371 P 64	AWR2043SUD R 03	AWR2485RGI N104	AWR2679LNF R 20	AWR3050LOY P 37	AWR3247LNF P 21
ANR4617 G 03	ASR1187 N 24	AWR1378 L 28	AWR2044LOY R 02	AWR2486RGI N103	AWR2710PMD N 15	AWR3050SUC P 37	AWR3282 N 66
ANR4634 H 03	ASR1400 N 21	AWR1379 L 48	AWR2044SUC R 02	AWR2486RGI N104	AWR2786LNF D 17	AWR3085LOY P 65	AWR3290 N 57
ANR4647 B182	ASR1401 N 21	AWR1383LNF P 12	AWR2045LOY R 03	AWR2487RGI N103	AWR2786LNF D 19	AWR3085SUC P 65	AWR3291 N 58
ANR4696 N 06	ASR1405 N 70	AWR1384LOY P 30	AWR2045SUC R 03	AWR2487RGI N104	AWR2800RCI N108	AWR3086LOY P 65	AWR3292 N 58
ANR4697 B 83	ASR1429 N 69	AWR1384SUC P 30	AWR2098PMD N 92	AWR2488RGI N103	AWR2801RCI N108	AWR3086SUC P 65	AWR3294 N 58
ANR4697 N 06	ASR1442 N 77	AWR1385LOY P 30	AWR2099PMD N 92	AWR2488RGI N104	AWR2802RCI N108	AWR3087LOY P 65	AWR3295 R 26
ANR4743 N121	ASR1443 N 77	AWR1385SUC P 30	AWR2100PMD N 92	AWR2489RGI N103	AWR2803RCI N108	AWR3087SUC P 65	AWR3296 R 26
ANR4747 F 03	ASR1594 N 25	AWR1399 P 19	AWR2101PMD N 92	AWR2489RGI N104	AWR2804RCI N108	AWR3098LOY R 19	AWR3298LPZ P 79
ANR4751 N121	ASR1595 N 25	AWR1399 P 20	AWR2102PMD N 92	AWR2490RGI N103	AWR2805RCI N108	AWR3099LOY R 19	AWR3298SUD P 79
ANR4904 N 02	ASR1714 N 20	AWR1400 P 72	AWR2103PMD N 92	AWR2490RGI N104	AWR2806RCI N108	AWR3102LOY R 19	AWR3330LNF M 65
ANR4905 G 10	ASR1715 N 20	AWR1439 P 19	AWR2104PMD N 92	AWR2491RGI N103	AWR2807RCI N108	AWR3102LOY R 20	AWR3331 P 09
ANR4906 G 10	ASR1862 N 32	AWR1439 P 20	AWR2105PMD N 92	AWR2491RGI N104	AWR2812LOY R 16	AWR3103LOY R 19	AWR3332 P 09
ANR4907 G 10	ASR2330 N 18	AWR1440 N 48	AWR2116 R 26	AWR2492REI N103	AWR2812SUC R 16	AWR3103LOY R 20	AWR3498LNF M 61
ANR4914 N122	ASR2331 N 18	AWR1441 N 48	AWR2117 R 26	AWR2492REI N104	AWR2813LOY R 17	AWR3104LOY R 39	AWR3498LNF M 62
ANR4958 G 07	ASR2332 N 17	AWR1442LUM P 43	AWR2118 R 26	AWR2492REI N107	AWR2813SUC R 17	AWR3104LOY R 44	AWR3521LOY R 14
ANR4959 G 07	ASR2333 N 17	AWR1442LUM P 51	AWR2140LOY R 22	AWR2492RHI N105	AWR2814 R 12	AWR3104SUC R 44	AWR3521SUC R 14
ANR4960 G 05	ASR2335 N 16	AWR1443LUM P 43	AWR2140SUC R 22	AWR2493REI N103	AWR2947 N 89	AWR3105LOY R 39	AWR3522LOY R 14
ANR4961 G 05	ASR2337 N 16	AWR1443LUM P 51	AWR2141LOY R 22	AWR2493REI N104	AWR2948LNF R 55	AWR3106LOY R 39	AWR3522SUC R 14
ANR4962 G 05	ASR2445 N 83	AWR1444LUM P 43	AWR2141SUC R 22	AWR2493REI N107	AWR2975 R 55	AWR3106LOY R 45	AWR3525LOY R 14
ANR4963 G 05	ASR2476 N 79	AWR1444LUM P 51	AWR2142LOY R 23	AWR2494REI N104	AWR2977 R 59	AWR3107LOY R 39	AWR3525SUC R 14
ANR4964 G 07	ASR2524 N 52	AWR1445LUM P 43	AWR2142SUC R 23	AWR2494RHI N105	AWR2983LNF D 17	AWR3107LOY R 44	AWR3526LOY R 14
ANR4965 G 07	ASR2525 N 52	AWR1445LUM P 51	AWR2143LOY R 23	AWR2541 N 43	AWR2983LNF D 19	AWR3107SUC R 44	AWR3526SUC R 14
ANR4966 G 05	AUD2437L B 98	AWR1449LUM P 43	AWR2143SUC R 23	AWR2543 N 43	AWR2983LUN D 17	AWR3124LOY R 27	AWR3527LOY R 16
ANR4967 G 05	AUD3306L B 97	AWR1449LUM P 51	AWR2144LOY R 23	AWR2544 N 43	AWR2984PMD N 14	AWR3124SUC R 27	AWR3527SUC R 16
ANR4981 G 10	AUD3577L B135	AWR1450LUM P 43	AWR2144SUC R 23	AWR2608LOY P 32	AWR2985PMD N 14	AWR3125LOY R 33	AWR3528LOY R 17
ANR5040 N121	AUD4398L B 97	AWR1450LUM P 51	AWR2145LOY R 24	AWR2608SUC P 32	AWR2987PMD N 12	AWR3125SUC R 33	AWR3528SUC R 17
ANR5053 F 03	AUD4771L B 97	AWR1479PMD N 12	AWR2145SUC R 24	AWR2609LOY P 32	AWR2988PMD N 12	AWR3126LOY R 30	AWR3531LOY R 27
ANR5093 G 03	AWR1001PMD N 14	AWR1502 N113	AWR2146 N 67	AWR2609SUC P 32	AWR2995LNF R 01	AWR3126SUC R 30	AWR3531SUC R 27
ANR5096 G 03	AWR1002PMD N 14	AWR1504 N117	AWR2147 N 67	AWR2610LOY P 33	AWR2996LNF R 01	AWR3127LOY R 36	AWR3532LOY R 33
ANR5110 M 40	AWR1027LUM P 70	AWR1505 N117	AWR2157LNF R 11	AWR2610SUC P 33	AWR3004LNF R 55	AWR3127SUC R 36	AWR3532SUC R 33
ANR5262 H 03	AWR1030PMA N 90	AWR1506 N113	AWR2158 R 13	AWR2611LOY P 33	AWR3005LNF R 55	AWR3144LOY R 39	AWR3533LOY R 30
ANR5265 H 03	AWR1032 N 43	AWR1507 N117	AWR2209LOY P 20	AWR2611SUC P 33	AWR3007 P 11	AWR3144LOY R 45	AWR3533SUC R 30
ANR5307MNH E 41	AWR1145 N 67	AWR1512 N 68	AWR2209SUC P 20	AWR2612LOY P 32	AWR3011 L 22	AWR3145LOY R 39	AWR3534LOY R 36
ANR5307RJQ E 41	AWR1146 N 67	AWR1517LUM P 66	AWR2210LOY P 20	AWR2612SUC P 32	AWR3012 L 22	AWR3145LOY R 44	AWR3534SUC R 36
ANR5327 G 02	AWR1147 N 67	AWR1518LUM P 66	AWR2210SUC P 20	AWR2613LOY P 32	AWR3013 L 39	AWR3145SUC R 44	AWR3535LOY R 29
ANR5328 G 01	AWR1153 N106	AWR1519LUM P 66	AWR2211 M 77	AWR2613SUC P 32	AWR3014 L 39	AWR3146LOY R 39	AWR3535SUC R 29
ANR5435 E 42	AWR1157LOY R 50	AWR1520LUM P 66	AWR2211 P 21	AWR2614LOY P 34	AWR3019LOY P 33	AWR3146LOY R 44	AWR3536LOY R 35
ANR5436 E 42	AWR1157SUC R 50	AWR1521LUM P 66	AWR2212 M 77	AWR2614SUC P 34	AWR3019SUC P 33	AWR3146SUC R 44	AWR3536SUC R 35

AWR3537LOY .... R 31
AWR3537SUC .... R 31
AWR3538LOY .... R 37
AWR3538SUC .... R 37
AWR3805LNF .... R 53
AWR3806LNF .... R 53
AWR3807LNF .... R 53
AWR3811LOY .... R 47
AWR3811SUC .... R 47
AWR3815LUM .... P 71
AWR3833 .... L 30
AWR3841 .... N 28
AWR3842 .... N 28
AWR3883LOY .... R 22
AWR3883SUC .... R 22
AWR3884LOY .... R 22
AWR3884SUC .... R 22
AWR3906 .... P 74
AWR3909LPZ .... P 79
AWR3909SUD .... P 79
AWR3911LPZ .... P 78
AWR3911SUD .... P 78
AWR3912LPZ .... P 78
AWR3912SUD .... P 78
AWR3913LPZ .... P 78
AWR3913SUD .... P 78
AWR3927LPZ .... P 79
AWR3927SUD .... P 79
AWR4027LNF .... R 01
AWR4032LNF .... R 18
AWR4032LNF .... R 20
AWR4033LNF .... R 18
AWR4033LNF .... R 20
AWR4072RAI N107
AWR4073RAI N107
AWR4074RAI N107
AWR4075RAI N107
AWR4076RAI N107
AWR4077RAI N107
AWR4078RGI N107
AWR4079RGI N107
AWR4080RGI N107
AWR4081RGI N107
AWR4082RGI N107
AWR4083RGI N107
AWR4084RGI N107
AWR4085RGI N107
AWR4108 .... R 25
AWR4108 .... R 16
AWR4108 .... R 18
AWR4108 .... R 20
AWR4125 .... L 33
AWR4126 .... L 33
AWR4148 .... N 41
AWR4320 .... R 13

AWR4320 .... R 25
AWR4320 .... R 26
AWR4320 .... R 12
AWR4321 .... R 13
AWR4321 .... R 25
AWR4321 .... R 26
AWR4321 .... R 12
AWR4322 .... R 18
AWR4322 .... R 20
AWR4334LUZ .... N 97
AWR4334PMD .... N 98
AWR4335LUZ .... N 97
AWR4335PMD .... N 98
AWR4336LUZ .... N 97
AWR4336PMD .... N 98
AWR4337LUZ .... N 97
AWR4337PMD .... N 98
AWR4338LUZ .... N 97
AWR4338PMD .... N 98
AWR4339LUZ .... N 97
AWR4339PMD .... N 98
AWR4340LUZ .... N 97
AWR4340PMD .... N 98
AWR4341LUZ .... N 97
AWR4341PMD .... N 98
AWR4646LUM .... P 72
AWR4767 .... N 70
AWR4781 .... P 10
AWR4781 .... P 11
AWR4857 .... N 44
AWR5213LUM .... P 66
AWR5214LUM .... P 66
AWR5215LUM .... P 66
AWR5216LUM .... P 66
AWR5217LUM .... P 66
AWR5219 .... P 23
AWR5250LNF .... N 77
AWR5250LNF .... N 80
AWR5250LNF .... N 83
AWR5327 .... P 74
AWR5328 .... P 74
AWR5349 .... N 68
AWR5386 .... N 59
AWR5387 .... N 59
AWR5388 .... N 59
AWR5389 .... N 59
AWR5572RAI N105
AWR5573RAI N105
AWR5574RAI N105
AWR5575RAI N105
AWR5576RAI N105
AWR5577RAI N105
AWR5578RAI N105
AWR5579RAI N105
AWR5580RGI N105

AWR5581RGI N105
AWR5582RGI N105
AWR5583RGI N105
AWR5584RGI N105
AWR5585RGI N105
AWR5586RGI N105
AWR5587RGI N105
AWR5993 .... N 47
AWR5994 .... N 47
AWR5995 .... N 52
AWR5996 .... N 52
AWR6079LOY .... R 22
AWR6079SUC .... R 22
AWR6080LOY .... R 22
AWR6080SUC .... R 22
AWR6081LOY .... R 23
AWR6081SUC .... R 23
AWR6082LOY .... R 23
AWR6082SUC .... R 23
AWR6083LOY .... R 23
AWR6083SUC .... R 23
AWR6084LOY .... R 24
AWR6084SUC .... R 24
AWR6087LOY .... R 27
AWR6087SUC .... R 27
AWR6088LOY .... R 33
AWR6088SUC .... R 33
AWR6089LOY .... R 30
AWR6089SUC .... R 30
AWR6090LOY .... R 36
AWR6090SUC .... R 36
AWR6091LOY .... R 29
AWR6091SUC .... R 29
AWR6092LOY .... R 35
AWR6092SUC .... R 35
AWR6093LOY .... R 31
AWR6093SUC .... R 31
AWR6094LOY .... R 37
AWR6094SUC .... R 37
AWR6095LOY .... R 14
AWR6095SUC .... R 14
AWR6096LOY .... R 14
AWR6096SUC .... R 14
AWR6097LOY .... R 16
AWR6097SUC .... R 16
AWR6098LOY .... R 17
AWR6098SUC .... R 17
AWR6101LOY .... R 47
AWR6101SUC .... R 47
AWR6199 .... P 09
AWR6200 .... P 09
AWR6240LNF .... P 38
AWR6338 .... N118
AWR6339 .... N118
AWR6449LOY .... P 59

AWR6449SUC .... P 59
AWR6450LOY .... P 59
AWR6450SUC .... P 59
AWR6533 .... M 62
AWR6534 .... M 62
AWR6544 .... N 35
AWR6571LOY .... P 61
AWR6571SUC .... P 61
AWR6572LOY .... P 61
AWR6572SUC .... P 61
AWR6577 .... N 35
AWR6577 .... N 36
AWR6672 .... N 67
AWR6673 .... N 67
AWR6691RHI N105
AWR6820 .... N114
AWR6821 .... N114
AWR6861 .... N118
AWR6862 .... N118
AWR6878LOY .... P 46
AWR6878SUC .... P 46
AWR6879LOY .... P 46
AWR6879SUC .... P 46
AWR6880LOY .... P 46
AWR6880SUC .... P 46
AWR6881LOY .... P 46
AWR6881SUC .... P 46
AWR6883PMD .... N 13
AWR6975 .... P 23
AWR6996LNF .... R 01
AWR6996LOY .... R 01
AWR6996SUC .... R 01
AWR6997LNF .... R 01
AWR6997LOY .... R 01
AWR6997SUC .... R 01
AWR7000LNF .... R 01
AWR7000LOY .... R 01
AWR7000SUC .... R 01
AWR7009 .... P 08
AWR7010 .... P 08
AWR7032LNF .... R 53
AWR7033LNF .... R 53
AWR7034LNF .... R 53
AWR7035LPZ .... P 79
AWR7035SUD .... P 79
AWR7057 .... P 12
AWR7074 .... R 55
AWR7155 .... L 28
AYA10004L .... N 69
BAC10053 .... K 01
BAU1689 .... D 03
BAU1689 .... C 03
BAU1689 .... C 25
BAU1689 .... J 22
BAU5028L .... M 08

BAU5028L .... M 09
BAU5028L .... M 07
BAU5029L .... M 08
BAU5029L .... M 09
BAU5029L .... M 07
BAU5311L M130
BAU5325L B132
BAU5325L B179
BAU5325L B133
BAU5325L B134
BD155888 .... M160
BD155888 .... M113
BDM10001L .... N 46
BDU1055L B165
BDU1431L B171
BDU1496L .... B 60
BDU1649L B164
BEP6073 .... P 36
BEP6073 .... P 38
BEP6073 .... P 28
BEP6073 .... P 31
BEP6073 .... P 34
BFP1265L .... N 76
BFP1265L .... N 80
BH106061L .... C 06
BH106061L .... H 11
BH106131 .... M111
BH108061L .... D 06
BH108077 .... F 09
BH108101L .... K 01
BH108105L .... K 11
BH108105L .... K 13
BH108141L N121
BH108141L .... K 12
BH108141L .... K 10
BH108141L .... K 11
BH108141L .... H 21
BH108161L N121
BH108181L .... D 03
BH108181L .... N 53
BH108201L .... N 02
BH108201L .... N 09
BH108201L M114
BH108281 .... B157
BH110071L .... D 05
BH110081L .... D 06
BH110091L .... B 37
BH110091L .... B 39
BH110091L .... D 05
BH110091L .... D 06
BH110111L B161
BH110121L .... G 01
BH110121L .... G 04
BH110121L .... G 09
BH110121L .... D 07

BH110201L .... N 01
BH110221L .... E 52
BH110221L .... C 04
BH110261 .... T 21
BH110261L .... C 85
BH112091L .... C 04
BH112091L .... C 26
BH112101L .... E 07
BH112101L .... E 08
BH112101L .... E 28
BH112101L .... E 29
BH112101L .... E 30
BH112101L .... E 31
BH112121L .... F 12
BH112221L .... T 21
BH112261L .... N 04
BH114151 .... E 43
BH114161L .... E 43
BH116207 .... E 43
BH406101L .... F 16
BH504101L .... B 85
BH504101L B120
BH504111L .... B 85
BH504111L B120
BH504121L .... B 85
BH504121L B120
BH504121L B142
BH504151L .... B 85
BH504151L B119
BH504161L .... B 85
BH504161L B119
BH504161L B120
BH504181L B119
BH505121L .... B 85
BH505121L B119
BH505181L B138
BH505241L .... B 84
BH505241L B117
BH505341L B100
BH505381L .... B 85
BH505381L B120
BH505381L B142
BH505401L .... B 85
BH505401L B120
BH505401L B142
BH505441L .... B 85
BH505441L B119
BH505485L .... B 85
BH505485L B119
BH506121L B130
BH506121L .... B 95
BH506131L B100
BH506131L B144
BH506131L .... C 71
BH506161L B130

BH506161L .... C 52
BH506161L .... C 72
BH506161L .... C 26
BH506161L .... B 95
BH506345 .... B144
BH604101L .... B 86
BH605101L .... B 35
BH605101L B100
BH605101L B119
BH605101L B144
BH605131L .... B 80
BH605131L B113
BH606361 .... N 11
BH606361L .... N 12
BH606361L .... T 20
BH606401 .... N 11
BH606401L .... B 35
BH606401L B100
BH606401L B144
BH607161L .... F 12
BH608181L .... C 71
BH608321L .... T 20
BH608461 .... E 44
BH610181L .... T 20
BH610321L .... E 48
BH612321 .... E 49
BHD700030 .... N 53
BHK700010 .... N 54
BHM1153L B165
BHM1505L B165
BHM705L .... N 68
BLS112L .... C 47
BLS112L .... C 18
BLS112L .... C 38
BMK2466 .... H 11
BMK2466 .... H 13
BNP2227L .... H 09
BNP2227L .... H 10
BNP2227L .... H 06
BNP2227L .... H 08
BRC1393 .... N 77
BRC1393 .... N 80
BT606101L .... D 11
BTP1781 .... L 02
BTP1809 .... L 02
BTP1811 .... L 02
BTP1811 .... L 07
BTP1812 .... L 02
BTP1812 .... L 07
BTP1823S .... L 02
BTP2275 .... L 07
BTP2278 .... L 07
BTP2414 .... L 07
BTR1013LUP .... P 68
BTR1015JUL .... R 04

BTR1015SUC .... R 04
BTR1016JUL .... P 52
BTR1016SUC .... P 52
BTR1017JUL .... P 52
BTR1017SUC .... P 52
BTR1059LNF .... N 77
BTR1059LUN .... N 77
BTR1060LNF .... N 77
BTR1060LUN .... N 77
BTR1180JUL .... P 06
BTR1180SUC .... P 06
BTR1282LMP .... N 87
BTR1282MCF .... N 87
BTR1283LMP .... N 87
BTR1283MCF .... N 87
BTR1284LMP .... N 87
BTR1284MCF .... N 87
BTR1285LMP .... N 87
BTR1285MCF .... N 87
BTR1299LUP .... P 66
BTR1300LUP .... P 66
BTR1304LUP .... P 66
BTR1475 .... N115
BTR1509JUL .... P 13
BTR1509SUC .... P 13
BTR1510PMA .... P 16
BTR1510PMA .... P 14
BTR1512 .... P 16
BTR1513LNF .... P 16
BTR1514LNF .... P 16
BTR1515 .... P 16
BTR1515 .... P 17
BTR1524LNF .... M 73
BTR1524LUN .... M 73
BTR1525LNF .... M 73
BTR1525LUN .... M 73
BTR1526LNF .... M 74
BTR1526LUN .... M 74
BTR1527LNF .... M 74
BTR1527LUN .... M 74
BTR1528LNF .... M 73
BTR1528LNF .... M 74
BTR1528LUN .... M 73
BTR1528LUN .... M 74
BTR1529LNF .... M 73
BTR1529LNF .... M 74
BTR1529LUN .... M 73
BTR1529LUN .... M 74
BTR1569 .... N 56
BTR157 .... N 85
BTR1570 .... N 56
BTR1647JUL .... P 60
BTR1647SUC .... P 60
BTR1648JUL .... P 60
BTR1648SUC .... P 60

BTR1649 P 62
BTR1650 P 62
BTR1658LUN P 82
BTR1659LUN P 82
BTR1660LUN P 83

BTR1661LUN P 83
BTR1684 N 23
BTR1691 N 31
BTR180LNF R 58
BTR180LUN R 58

BTR1884LNF R 54
BTR1884LNF R 56
BTR1884LUN R 54
BTR1884LUN R 56
BTR1884PMA R 54

BTR1884PMA R 56
BTR188LCG P 80
BTR188LNF P 80
BTR188LUN P 80
BTR189LCG P 82

BTR189LNF P 82
BTR189LUN P 82
BTR190LCG P 82
BTR190LNF P 82
BTR190LUN P 82

BTR191LCG P 77
BTR191LNF P 77
BTR191LUN P 77
BTR192LCG P 76
BTR192LNF P 76

BTR192LUN P 76
BTR2052 P 74
BTR2053 P 74
BTR2069 R 07
BTR2074RWE N101

BTR2075RWE N101
BTR2078RWE N101
BTR2079RWE N101
BTR2082RWE N101
BTR2083RWE N101

BTR2121JUL P 56
BTR2121LOY P 56
BTR2121SUC P 56
BTR2125 R 54
BTR2125 R 55

BTR215 L 29
BTR216 L 29
BTR217 L 29
BTR2188 R 54
BTR2269JUL P 39

BTR2269JUL P 40
BTR2269SUC P 39
BTR2269SUC P 40
BTR2279 P 16
BTR2295JUL P 57

BTR2295SUC P 57
BTR2297JUL P 57
BTR2297SUC P 57
BTR2298 L 46
BTR2314 P 16

BTR2458 L 46
BTR2548 L 40
BTR2560LUP P 68
BTR2561LUP P 68
BTR2564 L 41

BTR2564 L 53
BTR2564 L 55
BTR2615JUL P 40
BTR2615SUC P 40
BTR2662 L 45

BTR2663 L 45
BTR2707JUL R 27
BTR2707SUC R 27
BTR2709JUL R 30
BTR2709SUC R 30

BTR2710 R 18
BTR2710 R 20
BTR2716JUL R 33
BTR2716SUC R 33
BTR2717JUL R 36

BTR2717SUC R 36
BTR2726 N 44
BTR2738JUL R 02
BTR2738SUC R 02
BTR2739JUL R 03

BTR2739SUC R 03
BTR276 L 34
BTR2883 P 75
BTR2885 N 21
BTR289 N 56

BTR290 N 56
BTR2911 N 28
BTR2912 N 28
BTR2913 P 62
BTR2913 P 10

BTR2913 P 72
BTR2940 P 39
BTR2940 M 11
BTR2940 P 41
BTR296 P 74

BTR298 P 74
BTR3017JUN P 76
BTR3017SUD P 76
BTR3018LNF P 76
BTR301JUN P 80

BTR301SUD P 80
BTR302 P 74
BTR3063JUL P 18
BTR3063SUC P 18
BTR3072 N 39

BTR3072 N 38
BTR3074 T 47
BTR3075 T 47
BTR3087PMA P 19
BTR3087PMA P 21

BTR3088PMA P 19
BTR3088PMA P 21
BTR3089LNF P 19
BTR3089LNF P 20
BTR3100LNF P 17

BTR3264 R 11
BTR3264 R 18
BTR3269 R 11
BTR3269 R 18
BTR328 N 44

BTR3312LNF P 07
BTR3313LNF P 07
BTR3316 P 75
BTR3319 P 74
BTR3351LOY P 07

BTR3367 P 08
BTR3406JUL P 59
BTR3406SUC P 59
BTR3407JUL P 59
BTR3407SUC P 59

BTR3429 L 20
BTR3483 P 74
BTR3542PUB N 15
BTR3543PMD N 15
BTR358 N 44

BTR3605 N 41
BTR3608 P 74
BTR3609 P 74
BTR3617LUM P 71
BTR3630 L 41

BTR3636LOY R 24
BTR3636LOY R 06
BTR3636LOY R 16
BTR3636LOY R 20
BTR3636LUL R 06

BTR3636SUA R 24
BTR3636SUA R 06
BTR3636SUA R 16
BTR3636SUC R 20
BTR3644LNF P 19

BTR3644LNF P 21
BTR3655LNF P 21
BTR3672JUL P 13
BTR3672SUC P 13
BTR3673JUL P 13

BTR3673SUC P 13
BTR3680LNF P 83
BTR3680LUN P 83
BTR3680PUB P 83
BTR3698 L 40

BTR3700 L 42
BTR3710 L 40
BTR3710 L 41
BTR3710 L 53
BTR3717 L 41

BTR3720 L 42
BTR3721 L 21
BTR3722LOY P 08
BTR3722SUC P 08
BTR3723LOY P 08

BTR3723SUC P 08
BTR3730LOY P 07
BTR3730SUC P 07
BTR3731 P 10
BTR3753 L 46

BTR3756 P 02
BTR3757 P 02
BTR3758 P 02
BTR3759 P 02
BTR3761 L 20

BTR3780 L 21
BTR3781 L 21
BTR3809 L 40
BTR3809 L 53
BTR3809 L 55

BTR3812 L 45
BTR3813 L 45
BTR3814 L 42
BTR3885 L 20
BTR389 N 44

BTR3891 P 64
BTR3896LOY P 07
BTR3896SUC P 07
BTR3996 N 31
BTR3998 L 33

BTR3999 L 33
BTR4002 L 22
BTR4002 L 39
BTR4003 L 22
BTR4003 L 39

BTR4008 L 31
BTR4016 L 30
BTR4016 L 31
BTR4022 L 30
BTR4023 L 31

BTR4025 L 30
BTR4026 L 30
BTR4026 L 31
BTR4027 L 31
BTR4029 P 16

BTR4036 L 22
BTR4038 L 44
BTR404 P 36
BTR4040 L 44
BTR4040 L 51

BTR4041 L 44
BTR4041 L 51
BTR4042 L 44
BTR4044 L 44
BTR4044 L 51

BTR4045 L 44
BTR4045 L 51
BTR4046 L 44
BTR4048 L 50
BTR4049 L 50

BTR4050 L 48
BTR4187 B183
BTR4190 L 30
BTR4190 L 31
BTR4195 L 31

BTR4196 L 31
BTR421 N 52
BTR422 N 52
BTR4228 P 75
BTR4264 P 09

BTR4285LOY P 19
BTR4285LOY P 20
BTR4285SUC P 19
BTR4285SUC P 20
BTR4286LOY P 19

BTR4286LOY P 20
BTR4286SUC P 19
BTR4286SUC P 20
BTR4307PMA P 07
BTR4323 N113

BTR4324 N113
BTR4325 N117
BTR4326 N117
BTR4328 N117
BTR4330 N113

BTR4363LOY P 63
BTR4363SUC P 63
BTR4368LOY P 63
BTR4368SUC P 63
BTR4369 L 43

BTR4370 L 43
BTR4371 L 43
BTR437JUN R 05
BTR437LOY R 05
BTR437LOY R 16

BTR437LOY R 20
BTR437SUC R 05
BTR437SUC R 16
BTR437SUD R 05
BTR4381LOY P 63

BTR4381SUC P 63
BTR4382 L 43
BTR4383 N 31
BTR4401PUB N 90
BTR4408JUL P 65

BTR4408SUC P 65
BTR4414 M 11
BTR4419JUL P 64
BTR4419LOY P 64
BTR4419SUC P 64

BTR4430 L 35
BTR4431 L 35
BTR4494 L 35
BTR4495 L 35
BTR4518 P 73

BTR4540LUM P 67
BTR4541LUM P 67
BTR4542LUM P 72
BTR4543LUM P 72
BTR4546LNF P 82

BTR4546LUN P 82
BTR4546PUB P 82
BTR4547LNF P 82
BTR4547LUN P 82
BTR4547PUB P 82

BTR4548LNF P 83
BTR4548LUN P 83
BTR4549LNF P 83
BTR4549LUN P 83
BTR4552 P 74

BTR4553 P 74
BTR4564LOY R 47
BTR4564SUD R 47
BTR4649 L 35
BTR4717 B 71

BTR4719 B183
BTR475 P 67
BTR475 M 97
BTR478LNF R 56
BTR478LUN R 56

BTR4825 N112
BTR4826 N112
BTR4829 P 62
BTR4830 P 08
BTR484JUL P 53

BTR484LOY P 53
BTR484SUC P 53
BTR485JUL P 53
BTR485LOY P 53
BTR485SUC P 53

BTR4884 L 27
BTR4884 L 38
BTR4892 R 38
BTR4892 R 40
BTR4893 R 38

BTR4893 R 40
BTR4950LUP P 67
BTR497JUL R 02
BTR497SUC R 02
BTR498JUL R 03

BTR498SUC R 03
BTR4993JUL R 47
BTR4993SUC R 47
BTR5027 L 34
BTR503LUZ N 94

BTR503PUB N 93
BTR5043JUL P 49
BTR5043JUL P 56
BTR5043LOY P 49
BTR5043LOY P 56

BTR5043LOY P 79
BTR5043SUC P 49
BTR5043SUC P 56
BTR5043SUC P 79
BTR5048PUB N 91

BTR504LUZ N 94
BTR504PUB N 93
BTR5052PMA P 15
BTR505LUZ N 94
BTR505PUB N 93

BTR506LUZ N 94
BTR506PUB N 93
BTR5070JUL P 13
BTR5070SUC P 13
BTR5071JUL P 13

BTR5071SUC P 13
BTR5073 P 15
BTR507LUZ N 94
BTR507PUB N 93
BTR508LUZ N 94

BTR508PUB N 93
BTR5099 L 40
BTR509LUZ N 95
BTR509PUB N 93
BTR510LUZ N 95

BTR510PUB N 93
BTR5200 P 75
BTR5200 P 09
BTR5201 L 21
BTR5202 N115

BTR5203 N115
BTR5204 N112
BTR5205 N115
BTR5212LNF R 11
BTR5212LNF R 18

BTR5250 L 49
BTR5255 L 49
BTR5258 L 49
BTR5263 L 49
BTR5269 L 49

BTR5272 L 49
BTR5274 L 49
BTR5321 L 46
BTR5322 L 46
BTR5351LUM P 72

BTR5487 R 10
BTR5498LNF M134
BTR5503 P 16
BTR5524 L 40
BTR5524 L 53

BTR5524 L 55
BTR5527 L 41
BTR5527 L 45
BTR558 N111
BTR559 N111

BTR5598 L 37
BTR5599 L 29
BTR560 N111
BTR5600 L 55
BTR561 N111

BTR5703PUB N 30
BTR5704PUB N 30
BTR5705PUB N 30
BTR5706PUB N 30
BTR5709LNF P 39

BTR5709LNF P 41
BTR5710JUL P 39
BTR5710JUL P 41
BTR5710LOY P 41
BTR5710SUC P 39

BTR5710SUC P 41
BTR5733 P 39
BTR5733 P 41
BTR5750 B145
BTR5811JUL R 54

BTR5811LNF R 55
BTR5811LOY R 54
BTR5811LOY R 55
BTR5811SUC R 54
BTR5811SUC R 55

BTR5897JUL P 15
BTR5897SUC P 15
BTR5924JUL P 24
BTR5924SUC P 24
BTR5925JUL P 24

BTR5925SUC P 24
BTR5926JUL P 26
BTR5926SUC P 26
BTR5927JUL P 26
BTR5927SUC P 26

BTR5928JUL P 25
BTR5928SUC P 25
BTR5929JUL P 25
BTR5929SUC P 25
BTR5934JUL P 27

BTR5934SUC P 27
BTR5935JUL P 27
BTR5935SUC P 27
BTR5972 L 53
BTR5973 L 55

BTR5974 L 53
BTR5975 L 55
BTR5979 L 55
BTR5980 L 54
BTR5981 L 55
BTR5982 L 53
BTR5992 L 53
BTR5993 L 55
BTR5994PUB P 15
BTR6031 L 49
BTR6032 L 49
BTR6033 L 49
BTR6034 L 49
BTR6043JUL P 52
BTR6043LOY P 52
BTR6043SUC P 52
BTR6044JUL P 52
BTR6044LOY P 52
BTR6044SUC P 52
BTR6074 L 53
BTR6074 L 55
BTR6078 L 49
BTR6078 L 54
BTR6078 L 55
BTR6079 L 49
BTR6079 L 54
BTR6079 L 55
BTR6080 L 49
BTR6080 L 54
BTR6080 L 55
BTR6081 L 49
BTR6082 L 49
BTR6087 L 49
BTR6088 L 49
BTR6092 L 49
BTR6093 L 49
BTR6095 L 49
BTR6100 L 49
BTR6103JUL P 45
BTR6103LOY P 46
BTR6103SUC P 46
BTR6104JUL P 45
BTR6104LOY P 46
BTR6104SUC P 46
BTR6105JUL P 46
BTR6105LOY P 46
BTR6105SUC P 46
BTR6106JUL P 46
BTR6106LOY P 46
BTR6106SUC P 46
BTR6115LNF R 08
BTR6158 P 62
BTR6171LNF R 08
BTR6172 N111
BTR6186 N 66
BTR6190 L 43
BTR6287 L 43
BTR632LNF R 58
BTR632LUN R 58
BTR633LNF R 56
BTR633LUN R 56
BTR635 N 70
BTR6484 L 27
BTR6484 L 38
BTR6485 L 27
BTR6485 L 38
BTR6486 L 22
BTR6487 L 22
BTR6488 L 39
BTR6489 L 39
BTR6490 L 28
BTR6490 L 48
BTR6491 L 48
BTR6492 L 28
BTR6492 L 48
BTR6494 L 28
BTR6494 L 48
BTR6495 L 28
BTR6505 P 64
BTR6512 R 08
BTR6513 R 08
BTR657 N 70
BTR6580 P 71
BTR6580 M134
BTR6580 M136
BTR6580 M139
BTR6580 M142
BTR6614LOY P 63
BTR6614SUC P 63
BTR6615LOY P 63
BTR6615SUC P 63
BTR6616LOY P 63
BTR6616SUC P 63
BTR6625LNF P 12
BTR6631 L 47
BTR6632 L 47
BTR6638 N 43
BTR6656 M 61
BTR6656 P 10
BTR6727 L 44
BTR6741 L 22
BTR6741 L 39
BTR6836 P 10
BTR6836 P 11
BTR6837 P 11
BTR6983LUM P 72
BTR6984LUM M 76
BTR6984LUM P 72
BTR6985LUM P 72
BTR7006LUM P 72
BTR7027LUM N 36
BTR7027LUM P 72
BTR703JUL P 27
BTR703SUC P 27
BTR704JUL P 27
BTR704SUC P 27
BTR705 N 66
BTR7051 N 15
BTR7054LOY P 30
BTR7054SUC P 30
BTR7055LOY P 30
BTR7055SUC P 30
BTR7056LOY P 30
BTR7056SUC P 30
BTR7057LOY P 30
BTR7057SUC P 30
BTR7058LOY P 35
BTR7058SUC P 35
BTR7059LOY P 35
BTR7059SUC P 35
BTR7082LOY R 29
BTR7082SUC R 29
BTR7083LOY R 27
BTR7083SUC R 27
BTR7084LOY R 35
BTR7084SUC R 35
BTR7085LOY R 33
BTR7085SUC R 33
BTR7087LOY R 29
BTR7087SUD R 29
BTR7088LOY R 27
BTR7088SUD R 27
BTR7089LOY R 35
BTR7089SUD R 35
BTR7090LOY R 33
BTR7090SUD R 33
BTR7134 N 48
BTR7135 N 48
BTR7146LOY R 31
BTR7146SUD R 31
BTR7147LOY R 30
BTR7147SUD R 30
BTR7148LOY R 37
BTR7148SUD R 37
BTR7149LOY R 36
BTR7149SUD R 36
BTR7150LOY R 31
BTR7150SUC R 31
BTR7151LOY R 30
BTR7151SUC R 30
BTR7152LOY R 37
BTR7152SUC R 37
BTR7153LOY R 36
BTR7153SUC R 36
BTR7173 P 12
BTR7174 P 07
BTR7175 P 08
BTR717LUN R 54
BTR717LUN R 56
BTR7182LOY R 07
BTR7182SUC R 07
BTR7183LOY R 04
BTR7183SUC R 04
BTR7188LOY R 04
BTR7188SUD R 04
BTR7189LOY R 07
BTR7189SUD R 07
BTR7230 L 44
BTR7230 L 51
BTR7248LOY P 22
BTR7248SUC P 22
BTR7259 L 51
BTR7260 L 51
BTR7281 P 08
BTR7301 L 44
BTR7301 L 51
BTR7302 L 52
BTR7303 L 52
BTR7304 L 52
BTR733LUN M 73
BTR733LUN M 74
BTR7359 L 52
BTR7360 L 52
BTR7361 L 52
BTR7362 L 52
BTR7363 L 52
BTR7364 L 52
BTR7365 L 52
BTR7366 L 52
BTR736LNF R 57
BTR736LUN R 57
BTR737LNF R 57
BTR737LUN R 57
BTR738LUP P 67
BTR7399 L 22
BTR739LUP P 67
BTR740LUP P 71
BTR742 N 66
BTR7627 L 43
BTR7690LOY M 61
BTR7690LOY M 62
BTR7690SUC M 61
BTR7690SUC M 62
BTR7692LNF M 61
BTR7714LNF P 12
BTR7767LNF R 56
BTR7768LNF R 56
BTR7770LNF R 57
BTR7805 N 41
BTR7812 L 53
BTR7813 L 53
BTR7815LMP N 88
BTR7815MCF N 88
BTR7837 N114
BTR7837 N119
BTR7838 N114
BTR7838 N119
BTR7867LOY R 47
BTR7867SUC R 47
BTR7880 P 83
BTR7935 P 29
BTR7935 P 31
BTR7935 P 34
BTR7958 P 14
BTR7958 P 19
BTR7958 P 21
BTR8036 N 83
BTR806LUP P 68
BTR807LUP P 68
BTR8093 P 71
BTR8110 N 31
BTR8111 P 74
BTR8112JUN P 77
BTR8112SUD P 77
BTR813JUN P 81
BTR813SUD P 81
BTR8144 S 01
BTR8148 S 01
BTR814JUN P 81
BTR814SUD P 81
BTR815JUN P 77
BTR815SUD P 77
BTR8162LNF R 58
BTR8163LNF R 58
BTR8199 R 25
BTR8205 N114
BTR8205 N119
BTR8206 N114
BTR8206 N119
BTR8207 N114
BTR8207 N119
BTR8213 N 68
BTR8221PUB P 83
BTR8222PUB P 83
BTR8237LUZ N 95
BTR8238LUZ N 95
BTR8239 P 74
BTR8240 N 35
BTR8241 N 36
BTR8242 N 36
BTR8243 N 36
BTR8247LNF N 84
BTR8254PMD N 15
BTR8263JUL P 40
BTR8263SUC P 40
BTR8264JUL P 40
BTR8264SUC P 39
BTR8264SUC P 40
BTR8266 L 11
BTR8269 N 36
BTR8270 N 36
BTR8281LNF P 12
BTR8301 N 47
BTR8302 N 52
BTR8303 N 52
BTR8321 P 75
BTR8322 P 17
BTR8323 P 75
BTR8323 P 17
BTR8324 P 17
BTR8330 P 17
BTR8332LNF C 87
BTR8332PMA C 87
BTR8333LNF C 87
BTR8333PMA C 87
BTR8334LNF C 87
BTR8334PMA C 87
BTR8335LNF C 87
BTR8335PMA C 87
BTR8336 C 87
BTR8341 R 59
BTR8345JUL P 59
BTR8345SUC P 59
BTR8346JUL P 59
BTR8346SUC P 59
BTR8347 L 43
BTR8356LNF R 54
BTR8356LNF P 79
BTR8356LNF R 56
BTR8401 N 43
BTR8409LNF P 20
BTR8516LNF C 41
BTR8516PMA C 67
BTR8516PMA C 41
BTR8535 P 75
BTR8535 P 17
BTR855JUL F 01
BTR855SUC F 01
BTR8578JUN P 76
BTR8578SUD P 76
BTR8579JUN P 76
BTR8579SUD P 76
BTR858 L 20
BTR8580JUN P 76
BTR8580SUD P 76
BTR8581JUN P 80
BTR8581SUD P 80
BTR8582JUN P 81
BTR8582SUD P 81
BTR8583JUN P 81
BTR8583SUD P 81
BTR8584JUN P 81
BTR8584SUD P 81
BTR8585JUN P 81
BTR8585SUD P 81
BTR8586 L 52
BTR8587 L 52
BTR8633 P 75
BTR8653JUL P 40
BTR8653LOY P 40
BTR8653SUC P 40
BTR8654JUL P 39
BTR8654JUL P 40
BTR8654LOY P 40
BTR8654SUC P 39
BTR8654SUC P 40
BTR8655JUL P 39
BTR8655JUL P 40
BTR8655LOY P 40
BTR8655SUC P 40
BTR8680PMD N 96
BTR8681PMD N 96
BTR8682PMD N 96
BTR8683PMD N 96
BTR8684PMD N 96
BTR8685PMD N 96
BTR8686PMD N 96
BTR8687PMD N 96
BTR8700PMD N 96
BTR8701PMD N 96
BTR8713 S 01
BTR8715LNF P 31
BTR8715LNF P 34
BTR8716LNF P 31
BTR8716LNF P 34
BTR8725 L 30
BTR8726 L 30
BTR8727 L 30
BTR8731LOY P 11
BTR8731SUC P 11
BTR8732LOY P 11
BTR8732SUC P 11
BTR8737LNF R 53
BTR8738LNF R 53
BTR8747 N 36
BTR8748 N 36
BTR8749LOY P 35
BTR8749LOY P 37
BTR8749SUC P 35
BTR8749SUC P 37
BTR8750LOY P 35
BTR8750LOY P 37
BTR8750SUC P 35
BTR8750SUC P 37
BTR8762PMA R 13
BTR8835LNF H 20
BTR8835LNF H 23
BTR8842 P 19
BTR8842 P 21
BTR8864 N 47
BTR8865 N 47
BTR8866 N 52
BTR8867 N 52
BTR8901LNF P 19
BTR8901LNF P 20
BTR8902LNF P 19
BTR8902LNF P 20
BTR8907 P 75
BTR8917 N 42
BTR8918 N 42
BTR8919 N 42
BTR8921 L 44
BTR8921 L 51
BTR8930 L 28
BTR8930 L 48
BTR8931 N 28
BTR8932 N 28
BTR8962LOY P 59
BTR8962SUC P 59
BTR8963LOY P 59
BTR8963SUC P 59
BTR8964LOY P 60
BTR8964SUC P 60
BTR8965LOY P 60
BTR8965SUC P 60
BTR8966LOY P 61
BTR8966SUC P 61
BTR8967LOY P 61
BTR8967SUC P 61
BTR8976 P 62
BTR8976 P 63
BTR8978 P 60
BTR8978 P 62
BTR8998RCH N102
BTR8998RCJ N102
BTR8999RCH N102
BTR8999RCJ N102
BTR9000RCH N102
BTR9000RCJ N102
BTR9001RCH N102
BTR9001RCJ N102
BTR9002RCH N102
BTR9002RCJ N102
BTR9003RCH N102
BTR9003RCJ N102
BTR9004RCH N102
BTR9004RCJ N102
BTR9005RCH N102
BTR9005RCJ N102
BTR9006RCH N102

BTR9006RCJ N102
BTR9007RCH............N102
BTR9007RCJ N102
BTR9008RCH............N102
BTR9008RCJ N102

BTR9009RCH............N102
BTR9009RCJ N102
BTR9126 ....N 28
BTR915PUB ....N 90
BTR916PUB ....N 90

BTR917PUB ....N 90
BTR918PUB ....N 90
BTR919PUB ....N 90
BTR920PUB ....N 90
BTR9217LNF ....P 19

BTR9218 ....P 75
BTR921PUB ....N 90
BTR922PUB ....N 90
BTR923PUB ....N 90
BTR924MUE N121

BTR924NUC N121
BTR9255 ....N 47
BTR9256 ....N 47
BTR925PUB ....N 15
BTR9262 ....L 43

BTR9275LNF ....N 60
BTR9276 ....P 57
BTR9284 ....P 10
BTR9297SUC............P 64
BTR9298LOY ....P 65

BTR9298SUC............P 65
BTR9317 ....L 43
BTR9333LNF ....P 08
BTR9334PMA............P 07
BTR9358LPZ ....P 76

BTR9358LPZ ....P 78
BTR9358SUD............P 76
BTR9358SUD............P 78
BTR9359LPZ ....P 76
BTR9359LPZ ....P 78

BTR9359SUD............P 76
BTR9359SUD............P 78
BTR9360LPZ ....P 76
BTR9360SUD............P 76
BTR9361LPZ ....P 77

BTR9361SUD............P 77
BTR9363LPZ ....P 81
BTR9363SUD............P 81
BTR9364LPZ ....P 81
BTR9364SUD............P 81

BTR9365LPZ ....P 81
BTR9365SUD............P 81
BTR9366LPZ ....P 81
BTR9366SUD............P 81
BTR9368 ....P 74

BTR9369 ....P 74
BTR9379 ....P 74
BTR9391LUM............L 43
BTR9392LNF ....L 43
BTR9431LOY ....P 63

BTR9431SUC............P 63
BTR9432LOY ....P 63
BTR9432SUC............P 63
BTR9434PMD ....N 15
BTR9437 ....P 09

BTR9438MNH N121
BTR9438MUE N121
BTR9438NUC............N121
BTR9484LOY ....P 57
BTR9484SUC............P 57

BTR9485LOY ....P 57
BTR9485SUC............P 57
BTR9486LOY ....P 57
BTR9486SUC............P 57
BTR9487LOY ....P 57

BTR9487SUC............P 57
BTR949 ....P 71
BTR9508LPZ ....R 32
BTR9508LPZ ....R 39
BTR9508SUD............R 32

BTR9509LPZ ....R 32
BTR9509LPZ ....R 39
BTR9509SUD............R 32
BTR9515 ....P 08
BTR951LUN ....R 56

BTR9520 ....P 64
BTR952LUN ....R 56
BTR953 ....N 91
BTR954 ....N 91
BTR9547 ....P 19

BTR9547 ....P 20
BTR9611 ....P 72
BTR9613 ....L 44
BTR9613 ....L 51
BTR9616 ....L 32

BTR9617 ....L 32
BTR9623 ....C 87
BTR9634 ....N 13
BTR9642 ....N 66
BTR9643 ....N 57

BTR9644 ....N 58
BTR9647 ....N 68
BTR9649 ....N 68
BTR9650 ....N 68
BTR9652 ....N 67

BTR966 ....N 70
BTR9688 ....N 13
BTR9689 ....N 14
BTR972 ....N 52
BTR9724 ....M 13

BTR9725 ....M 97
BTR973 ....N 52
BTR9741JUL ....R 37
BTR9741SUC ....R 37
BTR9745JUL ....R 31

BTR9745SUC ....R 31
BTR9747JUL ....R 35
BTR9747SUC ....R 35
BTR9751JUL ....R 29
BTR9751SUC ....R 29

BTR9753JUL ....R 07
BTR9753SUC ....R 07
BTR9761JUL ....R 04
BTR9761SUC ....R 04
BTR9762JUL ....R 02

BTR9762SUC ....R 02
BTR9763JUL ....R 03
BTR9763SUC ....R 03
BTR9764JUL ....R 27
BTR9764SUC ....R 27

BTR9765JUL ....R 30
BTR9765SUC ....R 30
BTR9766JUL ....R 33
BTR9766SUC ....R 33
BTR9767JUL ....R 36

BTR9767SUC ....R 36
BTR9812 ....P 17
BTR9813 ....P 75
BTR9813 ....P 17
BTR9814 ....P 75

BTR9815 ....P 17
BTR9816 ....P 17
BTR9840LNF ....P 19
BTR9847PMA ....R 09
BTR9848PMA ....R 09

BTR9849PMA ....R 09
BTR9850PMA ....R 09
BTR9851LOY ....R 47
BTR9851SUD ....R 47
BTR9852PMA ....R 51

BTR9853PMA ....R 51
BTR9854PMA ....R 51
BTR9855PMA ....R 51
BTR9858 ....P 75
BTR9863LOY ....R 27

BTR9863SUD ....R 27
BTR9864LOY ....R 33
BTR9864SUD ....R 33
BTR9865LOY ....R 30
BTR9865SUD ....R 30

BTR9866LOY ....R 36
BTR9866SUD ....R 36
BTR9873LOY ....R 27
BTR9873SUC ....R 27
BTR9874LOY ....R 33

BTR9874SUC............R 33
BTR9876LOY ....R 36
BTR9876SUC............R 36
BTR9878LOY ....R 30
BTR9878SUC............R 30

BTR9879LOY ....R 47
BTR9879SUC............R 47
BTR9886PUB ....P 83
BTR9888 ....P 09
BTR9889 ....P 09

BTR9890 ....P 09
BTR9891 ....P 09
BTR9892LPZ ....P 80
BTR9892SUD............P 80
BTR9896LAV N106

BTR9896MUK N106
BTR9897LAV N106
BTR9897MUK N106
BTR9898LAV N106
BTR9898MUK N106

BTR9899LAV N106
BTR9899MUK N106
BTR9900LAV N106
BTR9900MUK N106
BTR9901LAV N122

BTR9905 ....P 72
BTR9908 ....P 08
BTR9909 ....P 08
BTR9910LNF ....P 19
BTR9910LNF ....P 20

BTR9926LNF ....P 17
BX108081 ....F 08
BX108081 ....F 07
BX108111 ....N 53
BX110071L ....D 05

BX110071L ....E 18
BX110071L ....E 20
BX110071M ....B 22
BX110071M ....E 18
BX110071M ....E 20

BX110091L ....E 49
BX110095M ....E 24
BX110095M ....E 40
BX110111L ....E 49
BX112091 ....B 60

BX112201 ....F 11
BX116201L ....E 43
BZV1051 ....C 54
C15644L M146
C15644L M148

C393771L M153
C39377L ....L 19
C39377L M153
C39377L ....M 28
C43231L B140

C435997L ....J 09
C43640 ....M155
C457593 ....B 93
C457593 ....B130
CAC100180 ....N 40

CAC100190 ....N 40
CAC7815 ....G 09
CAJ100160 ....N 40
CAJ100170 ....N 40
CAJ100180 ....N 40

CAJ100190 ....N 40
CAJ100200 ....N 40
CAJ100210 ....N 40
CAJ100220 ....N 40
CAJ100230 ....N 40

CAM1210L B172
CAM1539L B164
CAM4133L B172
CAM6263L B166
CCN110L ....G 09

CCN110L ....E 11
CCN110L ....E 36
CCP2208 ....N 33
CDP1567L ....N 33
CDP1568L ....N 33

CDP1582LNF ....P 36
CDP1582LNF ....P 38
CDP1582LNF ....P 28
CDP1582LNF ....P 31
CDP1582LNF ....P 34

CDP1582LUN ....P 36
CDP1582LUN ....P 28
CDP6702L ....N 71
CDP6702L ....N 72
CDP6702L ....N 73

CDP6702L ....N 75
CDP6702L ....N 33
CDU1344L B167
CDU1345L B171
CDU1346L B189

CDU2958L ....L 01
CDU3609 ....L 22
CDU3609 ....L 39
CDU4083 ....B167
CDU51 ....C 54

CJ600304L ....B 23
CLF101500 ....N 28
CLP6234L ....L 47
CLP6235L ....L 47
CLP6552L ....N 46

CLP8699L ....H 20
CLP8934 ....M157
CLP9089L ....N 40
CLP9089L ....F 03
CN100128L ....J 23

CN100168L ....B 94
CN100168L B180
CN100168L ....L 19
CN100168L ....L 13
CN100168L ....H 04

CN100168L ....L 05
CN100168L ....L 15
CN100168L ....L 12
CN100208L ....B 21
CN100208L ....B 59

CN100208L ....L 02
CN100208L ....L 14
CN100208L ....L 12
CN100208L M150
CN100258L ....B 21

CN100258L ....B 24
CN100258L ....B 59
CN100258L B176
CN100258L ....L 29
CN100258L ....J 19

CN100258L ....F 19
CN100258L ....F 20
CN100258L ....F 21
CN100308 ....L 05
CN100308 ....L 30

CN100308L B129
CN100308L B130
CN100308L B176
CN100308L ....L 09
CN100308L ....L 30

CN100308L ....B 95
CN100408L ....B 12
CN100508 ....L 19
CN100508 ....L 04
CN100508 ....L 05

CN100508L ....L 19
CN100508L ....E 01
CN100508L ....L 02
CN100508L ....L 05
CN100508L ....L 09

CN100608 ....L 09
CN100608 ....L 10
CN100608L ....J 18
CN100608L ....L 19
CN100608L ....L 03

CN100608L ....L 09
CN100608L ....L 10
CN100808L ....J 19
CN100908L ....J 18
CN100908L ....J 15

CN100908L ....J 16
CN100908L ....J 17
CP105121L ....G 04
CP105121L ....G 10
CP106221 ....B174

CP108101L ....G 03
CP108161L ....J 07
CP108161L ....H 24
CP108251L ....B 55
CP108251L ....B 59

CR120105L ....N 53
CR120115L ....G 01
CR120215 ....D 16
CR120215 ....D 18
CR120305 ....B 14

CRB108860 ....N118
CRB108870 ....N118
CRB108890 ....N113
CRC1250L ....N 70
CRC1250L ....H 11

CRC1250L ....H 13
CRC1487 ....H 11
CRC1487 ....H 13
CRC1487 ....H 05
CRC1487 ....H 07

CRC4579L ....K 13
CRD100640 ....N113
CRD100640 ....N117
CRD100650 ....N113
CRD100650 ....N117

CRD100680 ....N113
CRD100680 ....N114
CRD100680 ....N117
CRD100690 ....N114
CTB100130 ....N 67

CTT100030 ....N118
CUD2399L ....B 97
CUD2785L ....B 97
CUD2788L ....B 97
CWE10032L ....N 73

CWE10032L ....N 74
CWE10032L ....N 75
CWE10032L ....F 06
CYH10001 ....L 30
CYH10001 ....M 92

CYH10001L ....L 30
CYH10001L ....M 96
CYH10001L M100
CZA4736L ....P 28
CZG1817L ....L 43

CZG1946 ....M 96
CZG1946 ....M 99
CZK3264 ....N 82
CZK3264L ....N 78
CZK3264L ....N 79

CZK3264L ....N 82
CZK3264L ....N 83
CZK3264L ....N 84
DA606031 ....F 04
DA606034L ....P 14

DA606041 ....P 01
DA606041 ....M 15
DA606061L ....F 04
DA606061L ....R 48
DA608031L ....L 29

DA608031L ....M 91
DA608031L ....P 13
DA608044L ....P 04
DA608044L ....N 78
DA608044L ....N 81

DA608044L ....N 84
DA608061 ....M 12
DA608064 ....R 01
DA608084L ....L 43
DA608084L ....P 63

DA610044 ....L 35
DA610051L ....N 13
DAH100620LAV N106
DAH100620MUK............N106
DAM100650 ....N106

DAM100720 ....N106
DAM3653L B165
DAM7638 ....B189
DAM7844L ....R 10
DBC2022L ....M 95

DBP3067L ....F 03
DBP5840LNF............N 81
DBP5840LNF............N 84
DBP5840LUN ....N 81
DBP5840LUN ....N 84

DBP5841LNF............N 81
DBP5841LNF............N 84
DBP5841LUN ....N 81
DBP5841LUN ....N 84
DBP6532LNF............N 78

DBP6532LNF............N 81
DBP6532LNF............N 84
DBP6532LUN ....N 78
DBP6532LUN ....N 81
DBP6532LUN ....N 84

DBP6533LNF............N 78
DBP6533LNF............N 81
DBP6533LUN ....N 78
DBP6533LUN ....N 81
DBP8169L B120

DBP8169L ....M 23
DBP8169L M159
DBP8169L ....M 46
DBP8169L ....M 50
DBP8537L M130

DCP3970 ....L 27
DCP3970 ....L 22
DCP3970 ....L 39
DCP3970 ....L 38
DCP3970 ....L 43

DCP5044L ... N 84
DCP6656 ... M150
DCP7384L ... B 66
DCP8048 ... M 04
DDG100300 ... N 91

DDG100310 ... N 91
DHB102510LML ... N 43
DHH100760LML ... N 43
DHH100770LML ... N 43
DJP8473L ... M 81

DKB102710 ... M145
DKB102720 ... M145
DKC100900 ... M145
DKC100910 ... M145
DKC100920 ... M145

DKP5279L ... P 39
DKP5279L ... P 62
DKP5279L ... P 41
DKP5706 ... R 18
DKP5706 ... R 20

DRC1245 ... M 89
DRC1245 ... M 83
DRC1538 ... M155
DRC1752L ... B104
DRC1752L ... B148

DRC1752L ... M 95
DRC1752L ... M 96
DRC5017L ... M153
DRC5582 ... P 72
DRC8398 ... M130

DRC8584 ... M130
DYP10035L ... N 12
DYP100920L ... M 82
DZA1435L ... N 78
DZA1435L ... N 80

EAC22723 ... B136
EAC22723 ... B137
EAC2414L ... B132
EAC2414L ... B133
EAC32151 ... B136

EAM9066L ... N 79
EAM9066L ... N 82
EAM9066L ... N 85
EAM9067L ... N 79
EAM9067L ... N 82

EAM9067L ... N 85
EAP100740PDG ... P 82
EAP100750PDG ... P 82
EAP100760PDG ... P 83
EAP100770PDG ... P 83

EDP7864L ... H 21
EDQ104790LUM ... P 70
EDQ104800LUM ... P 69
EDQ104810LUM ... P 69
EDQ104820LUM ... P 69

EDQ104850LUM ... P 70
EDQ104860LUM ... P 70
EDQ104870LUM ... P 69
EDQ104880LUM ... P 69
EDQ104960LUM ... P 69

EDQ104970LUM ... P 69
EDR100150 ... N 36
EDU100000LUM ... P 67
EDU100010LUM ... P 67
EDU100050LUM ... P 67

EED101190 ... N 35
EED101290 ... N 36
EEP106 ... P 15
EFP100240 ... N 36
EFP100250 ... N 36

EGN100090 ... N 36
EGT100220 ... N 36
EIE100000LUM ... P 51
EIE100010LUM ... P 51
EJB119240LOY ... P 33

EJB119240SUC ... P 33
EJB119250LOY ... P 33
EJB119250SUC ... P 33
EJB119260LOY ... P 32
EJB119260SUC ... P 32

EJB119270LOY ... P 32
EJB119270SUC ... P 32
EJP7738L ... C 65
ELB108580LOY ... P 37
ELB108580SUC ... P 37

ELB108590LOY ... P 37
ELB108590SUC ... P 37
ENA100060LOY ... P 58
ENA100060SUC ... P 58
ENA100070LOY ... P 58

ENA100070SUC ... P 58
EOT100000 ... N 54
EOT100010 ... N 54
EPL100030 ... P 23
ERC1161L ... B 98

ERC1167 ... B 85
ERC1351 ... B 87
ERC1351 ... B122
ERC1353 ... B101
ERC1353 ... B147

ERC1561 ... B 22
ERC1637 ... B 88
ERC1637 ... B123
ERC1637 ... B125
ERC210 ... B 88

ERC211 ... B 88
ERC2143 ... B 95
ERC2154 ... B 98
ERC2226L ... B122
ERC224 ... B 88

ERC224 ... B123
ERC224 ... B125
ERC225 ... B 88
ERC2297 ... B 94
ERC2298 ... B 94

ERC2319 ... B 95
ERC2320 ... B 95
ERC248 ... B127
ERC255 ... B135
ERC256 ... B 94

ERC2838 ... B 92
ERC2838 ... B128
ERC2839 ... B 92
ERC2839 ... B128
ERC2989 ... B 90

ERC2989 ... B127
ERC3102 ... B 96
ERC3208 ... B127
ERC3209 ... B127
ERC3346L ... B 88

ERC3579L ... B140
ERC3606 ... B 96
ERC3606 ... B131
ERC3690 ... B 96
ERC3690 ... B131

ERC377L ... B138
ERC3792L ... B139
ERC3792L ... B140
ERC3853L ... B137
ERC3933 ... B127

ERC3990 ... B 95
ERC416 ... B 80
ERC416 ... B113
ERC417 ... B 80
ERC417 ... B113

ERC4211 ... B139
ERC4215 ... B139
ERC4626 ... B138
ERC4644 ... B 04
ERC4650 ... B 07

ERC4658 ... B 39
ERC4877 ... B 80
ERC4877 ... B113
ERC4878 ... B 80
ERC4878 ... B113

ERC4879 ... B 80
ERC4879 ... B113
ERC4880 ... B 80
ERC4880 ... B113
ERC4949 ... B 89

ERC4949 ... B126
ERC4996 ... B 04
ERC4996 ... B 43
ERC5068 ... B 96
ERC5708 ... B 86

ERC5709 ... B 13
ERC5709 ... B 86
ERC573 ... B 88
ERC573 ... B125
ERC5749 ... B128

ERC5829 ... J 15
ERC5830 ... B106
ERC5830 ... B156
ERC5830 ... J 15
ERC5875 ... B 96

ERC5875 ... B131
ERC5913 ... B 15
ERC5913 ... B 53
ERC5923 ... B 15
ERC5923 ... B 53

ERC614 ... B 90
ERC614 ... B124
ERC614 ... B125
ERC6551 ... B 39
ERC6552 ... B 92

ERC6552 ... B128
ERC675 ... B 99
ERC675 ... B142
ERC6878 ... B 91
ERC6934 ... B 79

ERC6997 ... B101
ERC7295 ... B 10
ERC7295 ... B 39
ERC7295 ... B 40
ERC7295 ... B 75

ERC7295 ... B 76
ERC7321 ... B 96
ERC7321 ... B131
ERC7508 ... B137
ERC7536 ... B137

ERC7538 ... B137
ERC7865 ... B 88
ERC7865 ... B123
ERC7929 ... B 92
ERC7929 ... B128

ERC8501 ... B122
ERC8598 ... B136
ERC8626 ... B 98
ERC8847 ... B 22
ERC8849 ... B 22

ERC8864 ... B 05
ERC9088 ... B 88
ERC9088 ... B123
ERC9110 ... B138
ERC9114L ... B137

ERC9115 ... B137
ERC9116 ... B137
ERC9117 ... B137
ERC9118 ... B137
ERC9187 ... B137

ERC9240 ... B 39
ERC9404 ... B 39
ERC9404 ... B 55
ERC9404 ... B 65
ERC9404 ... B 69

ERC9404 ... B120
ERC9404 ... B160
ERC9404 ... M154
ERC9410 ... B 05
ERC9410 ... B 44

ERC9480 ... B 06
ERC9480 ... B 55
ERC9706 ... B 14
ERK100280LOY ... P 22
ERK100280SUC ... P 22

ERR1004 ... B 24
ERR1019 ... B 17
ERR1019 ... B 55
ERR1040 ... B 48
ERR1041 ... B 48

ERR1063 ... B 14
ERR1084 ... B 48
ERR1085 ... B 48
ERR1086 ... B 48
ERR1088 ... B 14

ERR1092 ... B 50
ERR1094 ... B 44
ERR1103 ... B 09
ERR1107 ... B 18
ERR1108 ... B 10

ERR1111 ... B 10
ERR1115 ... B 34
ERR1119M ... B 12
ERR1120M ... B 12
ERR1121M ... B 10

ERR1121M ... B 12
ERR1122M ... B 10
ERR1125 ... B 28
ERR1125 ... B 31
ERR1125 ... B 62

ERR1125 ... B 65
ERR1125 ... B 67
ERR1132 ... B 26
ERR1133 ... B 26
ERR1134 ... B 26

ERR1148 ... B 26
ERR1149 ... B 25
ERR1156 ... B 19
ERR1156 ... B 57
ERR1157 ... B 19

ERR117 ... B 96
ERR117 ... B131
ERR1173 ... B 34
ERR1174 ... B 34
ERR1178 ... B 14

ERR1181 ... B 07
ERR1195 ... B 10
ERR1200 ... B 29
ERR1201 ... B 18
ERR1202 ... B 18

ERR1203 ... B 18
ERR1203 ... B 56
ERR1208 ... B 24
ERR1209 ... B 18
ERR1209 ... B 56

ERR1210 ... B 18
ERR1213 ... B 09
ERR1248 ... B 09
ERR1261 ... B136
ERR1262 ... B136

ERR1290 ... B138
ERR1291 ... B 09
ERR1293 ... B 31
ERR1295 ... B 31
ERR1295 ... K 14

ERR1299 ... B 15
ERR1304 ... B 29
ERR1304 ... B 66
ERR1351 ... B 59
ERR1361 ... B 12

ERR1390 ... B 08
ERR1391 ... B 08
ERR14 ... B 10
ERR1440 ... B 39
ERR1471 ... B 21

ERR1471 ... B 59
ERR1475 ... B 06
ERR1504 ... B 25
ERR1509 ... B 29
ERR1521 ... B 14

ERR1530 ... B 20
ERR1536 ... B 17
ERR1548 ... B 26
ERR1549 ... B 25
ERR1550 ... B 25

ERR1560 ... B 49
ERR1561 ... B 49
ERR1564 ... B 45
ERR1574 ... B156
ERR1589 ... B156

ERR1604 ... B 28
ERR1604 ... B 65
ERR1605 ... B 21
ERR1607 ... B 10
ERR1608 ... B153

ERR1608 ... K 04
ERR1616 ... B 76
ERR1617 ... B 76
ERR1618 ... B 76
ERR1619 ... B 76

ERR1630 ... B 07
ERR1630 ... B 45
ERR1632 ... B 10
ERR1632 ... B 84
ERR1632 ... B117

ERR1642 ... B 07
ERR1653 ... B 44
ERR1653 ... B 47
ERR1661 ... B 28
ERR1730 ... B 21

ERR1756 ... B135
ERR1780 ... B 88
ERR1780 ... B123
ERR1780 ... B125
ERR1781 ... B125

ERR1782 ... B 88
ERR1782 ... B123
ERR1782 ... B125
ERR1790 ... B 09
ERR1808 ... B146

ERR1809 ... B146
ERR1810 ... B146
ERR1885 ... B106
ERR1885 ... B156
ERR1919 ... B 43

ERR1921 ... B123
ERR1922L ... B 79
ERR1922L ... B116
ERR1939 ... B 54
ERR1972 ... B 50

ERR1973 ... B 50
ERR1985 ... B 28
ERR1986 ... B 26
ERR1990 ... B 87
ERR1990 ... B122

ERR2023 ... B 64
ERR2026 ... B 06
ERR2026 ... B 59
ERR2027 ... B 69
ERR2028 ... B 27

ERR2028 ... B 63
ERR2073 ... B105
ERR2073 ... B153
ERR2073 ... B154
ERR2079 ... B 76

ERR2081 ... B 55
ERR2082 ... B 62
ERR2083 ... L 17
ERR2084 ... L 17
ERR2100 ... B 45

ERR2109 ... B 67
ERR2112 ... B 45
ERR2215 ... B 71
ERR2216 ... B 60
ERR2216 ... B 64

ERR2220 ... B 45
ERR2228 ... B 70
ERR2241 ... B 15
ERR2241 ... B 53
ERR2263 ... B135

ERR2264 ... B160
ERR2266 ... B 52
ERR2284 ... B 62
ERR2317 ... B 53
ERR2337 ... L 16

ERR2344 ... B 49
ERR2351 ... B 07
ERR2352 ... B 07
ERR2354 ... B 26
ERR2378 ... B 51

ERR2381 ... B 26
ERR2381 ... M 91
ERR2386 ... B 25
ERR2393 ... B 20
ERR2396 ... B 54

ERR2405 ... B 56
ERR2409 ... B 58
ERR2410 ... B 46
ERR2418 ... B 46
ERR2419 ... B 46

ERR2428 ... B 85
ERR2428 ... B119
ERR2429 ... B130
ERR2443 ... B 26
ERR2490 ... B122

ERR25 ... B 10
ERR250 ... B 24
ERR2529 ... B 20
ERR2530 ... B 11
ERR2532 ... B 39

ERR2532 ... B 40
ERR2533 ... B 96
ERR2535 ... B 96
ERR2551 ... B 88
ERR2622 ... G 10

ERR2623 ... B 15
ERR2640 ... B 79
ERR2640 ... B111
ERR2673 ... B120
ERR2676 ... B156

ERR2677 ... B157
ERR2678 ... B153
ERR2692 ... B114
ERR2693 ... B114
ERR2706 ... L 17

ERR2707 ... L 17
ERR2711 ... B 15
ERR2732 ... B 56
ERR2767 ... B 43
ERR2789 ... B 52

ERR2798	B 72	ERR3368	B 20	ERR3735	B 52	ERR4247	B136	ERR4641	M 98	ERR4870	B127	ERR5261	B 54	ERR631	B 28
ERR2799	B112	ERR338	B 23	ERR3736	B 51	ERR4252	B146	ERR4641	M 99	ERR4883	B 56	ERR5262	B 16	ERR635	B 10
ERR2803	B 23	ERR3380	B 13	ERR3737	B 61	ERR4254	B146	ERR4645	B 75	ERR489	B 97	ERR5262	B 54	ERR6363	M100
ERR2826	L 16	ERR3417	B 47	ERR3738	B 61	ERR4256	B146	ERR4664	B 48	ERR4892	B129	ERR5263	B 16	ERR6382	B131
ERR2846	B 80	ERR3419	B 43	ERR3753	B 54	ERR4309	B129	ERR4672	B127	ERR4894	B 67	ERR5263	B 54	ERR6490	B 10
ERR2846	B112	ERR3424	B 58	ERR3754	B 60	ERR4323	B123	ERR4676	B147	ERR4895	B 67	ERR528	B 14	ERR6490	B 84
ERR2926	B136	ERR3439	B121	ERR3756	B 60	ERR4381	B 98	ERR4679	B143	ERR4897	B 28	ERR530	B 14	ERR6490	B117
ERR2943	B123	ERR3440	B145	ERR3764	B153	ERR4383	B 98	ERR4682	B 18	ERR4897	B 65	ERR531	B 14	ERR657	B 25
ERR2943	B125	ERR3442	B112	ERR3767	B119	ERR4388	B137	ERR4685	B 61	ERR490	B 97	ERR532	B 14	ERR663	B 20
ERR2944	B124	ERR3443	B121	ERR3777	B 57	ERR4419	B 64	ERR4686	B 61	ERR4912	B 25	ERR535	B 33	ERR666	B 22
ERR2944	B125	ERR3451	B 38	ERR3779	B 56	ERR4419E	B 64	ERR4687	B 56	ERR492	B 23	ERR537	B 24	ERR667	B 28
ERR2950	M 98	ERR3452	B 38	ERR3780	B 66	ERR4421	B117	ERR4689	B 59	ERR4923	B130	ERR541	B 14	ERR6700	B 64
ERR2958	B 80	ERR3453	B 38	ERR3785	B 62	ERR4427	B143	ERR4691	B 58	ERR4928	N 44	ERR542	B 25	ERR6704	B139
ERR2958	B112	ERR3454	B131	ERR3788	B 82	ERR4461	B121	ERR4696	B 47	ERR4935	B 94	ERR544	B 31	ERR6721	L 17
ERR296	B 25	ERR3455	B131	ERR3788	B115	ERR4513	B156	ERR4697	B 47	ERR4936	B 84	ERR544	K 13	ERR6723	L 17
ERR2961	B 26	ERR3457	B 56	ERR3791	L 16	ERR4513	B145	ERR4698	B 67	ERR4936	B117	ERR545	B 24	ERR6727	B 64
ERR2962	B 26	ERR3468	B 66	ERR3799	B103	ERR4521	B100	ERR4699	B 62	ERR4937	B133	ERR5473	B129	ERR6733	B131
ERR2966	B133	ERR3479	B 61	ERR3799	B151	ERR4521	B144	ERR4700	B 62	ERR4944	B135	ERR5474	B129	ERR675	B 39
ERR2966	B134	ERR3481	B 62	ERR3807	B 72	ERR4530	B 31	ERR4706	B 59	ERR4975	B 71	ERR551L	B 24	ERR6802	M 96
ERR297	B 25	ERR3490	B 61	ERR3809	B 47	ERR4530	K 14	ERR4707	B 45	ERR500	B 14	ERR551L	B 31	ERR6803	M100
ERR3	B 04	ERR3492	B 31	ERR388	B 12	ERR4531	L 32	ERR4708	B 51	ERR5008	B131	ERR551L	B131	ERR6811	B 43
ERR3062	B143	ERR3493	B 31	ERR39	B 12	ERR4541	B133	ERR4709	B 60	ERR5009	B 74	ERR5572	B117	ERR6835	B 64
ERR3062	B145	ERR3539	B 69	ERR390	B 10	ERR4543	B 72	ERR4710	B 45	ERR5009	B 38	ERR5575	B108	ERR690	B 25
ERR3063	B143	ERR3545	B 60	ERR3922	B 39	ERR455	B 20	ERR4722	B 75	ERR5009E	B 74	ERR5575	B160	ERR696	B 23
ERR309	B 14	ERR3545	B 64	ERR394	B 10	ERR4551	B 20	ERR4723	B 75	ERR5009E	B 38	ERR5579	B121	ERR7002	B 47
ERR319	B129	ERR3547	B 60	ERR395	B151	ERR4556	B 79	ERR4738	B146	ERR5010	B131	ERR5579	B145	ERR7012	B135
ERR321	B130	ERR355	B 25	ERR3987	B 45	ERR4556	B116	ERR4739	B146	ERR5023	B131	ERR559	B 18	ERR7013	B135
ERR321	N 44	ERR3562	B119	ERR4029	B 26	ERR4566	B138	ERR4740	B146	ERR5024	B131	ERR560	B 18	ERR7027	B125
ERR3283	B 53	ERR3579	L 16	ERR4029	B 30	ERR4574	B 39	ERR4753	B146	ERR5027	B 54	ERR560	B 56	ERR703	B 08
ERR3284	B 51	ERR3580	L 16	ERR4046	B 64	ERR4574	B 40	ERR4754	B146	ERR5034	B 04	ERR561	B 19	ERR703	B 46
ERR3286	B 49	ERR3581	L 16	ERR4046E	B 64	ERR4574	B 75	ERR4755	B146	ERR5034	B 43	ERR561	B 57	ERR705	B 07
ERR3287	B 52	ERR3583	L 16	ERR4047	L 16	ERR4574	B 76	ERR4762	B136	ERR5041	B 20	ERR5622	B164	ERR7154	B 54
ERR3291	B 61	ERR359	B 28	ERR4052	B145	ERR4575	B 48	ERR4763	B136	ERR5041	B 58	ERR5691	B 51	ERR7173	B 67
ERR330	B136	ERR3601	K 03	ERR4053	B143	ERR4576	B 49	ERR4764	B136	ERR5055	L 16	ERR5731	L 16	ERR719	B 39
ERR3306	B 66	ERR3605	B 21	ERR4054	B143	ERR4578	B 49	ERR4764	B137	ERR5057	B 63	ERR5795	B131	ERR719	B 75
ERR3314	L 16	ERR3606	B 14	ERR4060	B 80	ERR459	B 28	ERR4765	B136	ERR506	B 06	ERR5911	B 52	ERR7218	B124
ERR3319	L 16	ERR3607	B 15	ERR4060	B112	ERR459E	B 28	ERR4802	B 67	ERR506	B 21	ERR5924	B128	ERR722	B132
ERR3336	B 64	ERR3622	B 61	ERR4066	B143	ERR4602	B125	ERR4802E	B 67	ERR5087	B 45	ERR5926	B128	ERR722	B133
ERR3336E	B 64	ERR3633	B 47	ERR4077	B119	ERR4612	B 40	ERR4824	B 49	ERR509	B 21	ERR5985	M100	ERR722	B134
ERR3337	B 66	ERR3648	B 88	ERR4120	B 49	ERR4612	C 71	ERR4825	B 51	ERR5098	B 61	ERR605	B 07	ERR7283	B130
ERR3337E	B 66	ERR3648	B123	ERR4121	B150	ERR4621	B 29	ERR4827	B 40	ERR5099	B 51	ERR607	B 19	ERR7283	B 95
ERR3338	M100	ERR3648	B125	ERR4122	B150	ERR4621	B 66	ERR4829	B 76	ERR514	B110	ERR607	B 57	ERR7287	B 79
ERR3339	B 66	ERR3649	B123	ERR4123	B150	ERR4623	B121	ERR4834	B 58	ERR5189	B 73	ERR6072	B138	ERR7287	B111
ERR3339E	B 66	ERR3650	B 88	ERR4124	B150	ERR4623	B145	ERR4843	M100	ERR5198	J 16	ERR6087	B159	ERR7293	B 49
ERR3340	B 15	ERR3651	B 29	ERR4125	B150	ERR4627	B125	ERR4844	M100	ERR5199	B136	ERR6087E	B159	ERR7295	B 72
ERR3340	B 53	ERR3652	B 29	ERR4126	B150	ERR4628	B 88	ERR4848	B 56	ERR5202	L 17	ERR6184	B134	ERR7306	B130
ERR3340	B 87	ERR3671	B145	ERR4127	B151	ERR4628	B123	ERR4849	H 02	ERR5207	B146	ERR6185	B134	ERR7338	B123
ERR3340	B122	ERR3677	B 82	ERR4128	B151	ERR4628	B125	ERR4850	H 02	ERR5208	B146	ERR6186	B134	ERR7338	B125
ERR3342	B 56	ERR3677	B115	ERR4129	B151	ERR4632	B 58	ERR4859	B 73	ERR5209	B146	ERR6187	B134	ERR7340	B 59
ERR3343	B 56	ERR3682	B 23	ERR4165	B110	ERR4633	B 82	ERR4860	B 48	ERR5225	B111	ERR6189	J 21	ERR751	B 07
ERR3348	B 66	ERR3683	B 47	ERR4170	B110	ERR4633	B115	ERR4862	L 16	ERR5227	B111	ERR6191	B121	ERR765	B 17
ERR3348E	B 66	ERR371	B 24	ERR4179	B 43	ERR4635	B129	ERR4866	B112	ERR5238	B129	ERR6201	M100	ERR767	B154
ERR335	B 31	ERR372	B 24	ERR4194	B121	ERR4639	B 71	ERR4867	B119	ERR5258	B117	ERR6208	M100	ERR788	B 24
ERR335	B 67	ERR3733	B 70	ERR4196	M 92	ERR4640	B 19	ERR4867	B121	ERR5259	B 51	ERR6234	M 92	ERR789	B 24
ERR3356	B 48	ERR3734	B 51	ERR4245	B 31	ERR4640	B 57	ERR4868	B143	ERR5261	B 16	ERR630	B 23	ERR798	B 99

Part	Ref	Part	Ref	Part	Ref	Part	Ref	Part	Ref	Part	Ref	Part	Ref	Part	Ref
ERR798	B142	ESR1057L	J 16	ESR163	J 10	ESR2087	K 15	ESR2608	K 07	ESR2889	K 09	ESR3353	E 01	ETC4068	B 19
ERR807	B 37	ESR109	K 02	ESR1630	J 17	ESR2093	C 42	ESR2608	K 09	ESR2910	L 10	ESR3353	L 02	ETC4068	B 57
ERR808	B 37	ESR109	K 03	ESR1631	J 17	ESR2094	C 42	ESR2620	K 08	ESR2911	L 10	ESR3353	L 04	ETC4069	B 19
ERR809	B 23	ESR109	K 06	ESR167	J 23	ESR2131	K 15	ESR2620	K 09	ESR2912	L 10	ESR3353	L 05	ETC4069	B 57
ERR810	B 13	ESR109	K 07	ESR1670	J 22	ESR2150	J 19	ESR2628	K 14	ESR2913	L 10	ESR3353	L 09	ETC4076	B 22
ERR810	B 34	ESR109	K 08	ESR1673	J 08	ESR2188	J 08	ESR2630	L 05	ESR2917	K 02	ESR3371	J 10	ETC4154	B 10
ERR816	B 31	ESR109	K 09	ESR1675	J 08	ESR2218	J 21	ESR2634	L 05	ESR2928	N111	ESR3436	L 15	ETC4246	B 19
ERR817	B 98	ESR1190	J 21	ESR1683	J 21	ESR2218	J 22	ESR2635	J 08	ESR2928	K 04	ESR3492	J 02	ETC4246	B 57
ERR845	B 35	ESR1190	J 22	ESR1683	J 22	ESR2219	J 21	ESR2636	J 08	ESR2932	L 03	ESR3495	K 14	ETC4276	B 87
ERR845	B100	ESR1223	J 04	ESR1697	J 21	ESR2219	J 22	ESR2637	J 08	ESR2943	L 03	ESR3533	L 16	ETC4276	B122
ERR845	B144	ESR1224	J 02	ESR1698	K 06	ESR2219	J 24	ESR2641	C 92	ESR2945	K 15	ESR354	J 11	ETC4330	B 80
ERR848	B 14	ESR1242	C 92	ESR1699	K 06	ESR2245	L 19	ESR2662	L 09	ESR295	C 42	ESR3607	L 29	ETC4330	B112
ERR850	B 14	ESR1243	C 92	ESR1699	K 07	ESR2249	C 94	ESR2664	C 94	ESR2958	M 82	ESR3607	L 09	ETC4390	B 07
ERR852	B122	ESR1243	C 42	ESR1699	K 08	ESR225	K 02	ESR267	J 10	ESR2965	J 19	ESR3607	L 31	ETC4523	B140
ERR874	B 21	ESR1247	C 42	ESR1699	K 09	ESR2253	C 93	ESR269	C 42	ESR2991	J 10	ESR3670	K 04	ETC4524	K 02
ERR874	B 59	ESR1250	L 04	ESR1703	C 92	ESR2254	K 03	ESR270	C 42	ESR2992	J 10	ESR3671	K 04	ETC4524	K 03
ERR875	B 21	ESR1253	C 93	ESR1703	C 69	ESR2260	J 10	ESR273	J 15	ESR2994	J 10	ESR3687	L 05	ETC4529	B 04
ERR877	B 21	ESR1253	C 95	ESR1703	C 94	ESR2263	L 03	ESR2736	K 07	ESR2995	J 10	ESR3688	L 05	ETC4596	B 93
ERR878	B 21	ESR1253	C 97	ESR1721	L 01	ESR2285	L 19	ESR2736	K 09	ESR301	M157	ESR3691	L 18	ETC4596	B130
ERR879	B 21	ESR1256	C 42	ESR1722	L 01	ESR2286	L 19	ESR2738	J 22	ESR3011	J 16	ESR3692	L 06	ETC4616	B 33
ERR882	B 21	ESR1257	L 03	ESR1725	L 19	ESR2288	C 92	ESR2739	K 14	ESR3025	L 10	ESR3693	L 06	ETC4670	B 22
ERR886	B 28	ESR1258	L 03	ESR1789	L 19	ESR2290	C 92	ESR274	J 15	ESR3026	L 19	ESR3694	C 94	ETC4728L	B 98
ERR886	B 62	ESR1262	L 02	ESR1790	L 19	ESR230	K 02	ESR2740	K 14	ESR3056	J 17	ESR3695	C 94	ETC4765	B 93
ERR886	B 65	ESR1262	L 05	ESR1792	L 19	ESR2311	C 92	ESR2753	C 93	ESR3059	K 03	ESR3730	K 03	ETC4765	B129
ERR894	B 55	ESR1262	L 07	ESR1800	L 09	ESR2311	C 96	ESR2755	J 19	ESR3062	K 03	ESR3735	K 04	ETC4880	B 14
ERR894	B 62	ESR1305	L 13	ESR1803	J 16	ESR2327	J 02	ESR2761	L 19	ESR3063	K 03	ESR3736	K 04	ETC4922	B 04
ERR894	B 65	ESR1326	J 10	ESR1803	J 19	ESR2327	J 04	ESR2763	K 15	ESR3098	L 06	ESR3737	K 04	ETC4922	B 43
ERR894	B 75	ESR1373	J 11	ESR1804	J 16	ESR2344	L 12	ESR2764	K 15	ESR3117	J 09	ESR3737	K 05	ETC5064	B 10
ERR894	B 76	ESR1373	J 09	ESR1807	J 16	ESR2348	L 15	ESR2771	L 19	ESR3122	K 07	ESR3860	M 82	ETC5065	B 10
ERR896	B 31	ESR1373	J 22	ESR1819	L 05	ESR2364	J 08	ESR2783	K 07	ESR3122	K 09	ESR387	J 05	ETC5065	B 84
ERR900	B 11	ESR1444	J 19	ESR1895	J 05	ESR237	K 14	ESR2783	K 09	ESR3147	L 09	ESR387	J 06	ETC5065	B117
ERR900	B 50	ESR1445	J 16	ESR1895	J 06	ESR238	K 14	ESR2806	C 69	ESR3172	K 15	ESR3926	J 02	ETC5422	B 86
ERR904	B 98	ESR1445	J 17	ESR1896	J 05	ESR2390	K 14	ESR2806	C 95	ESR3226	L 05	ESR3969	K 04	ETC5422	B121
ERR913	B 04	ESR1445	J 19	ESR1896	K 01	ESR2391	K 14	ESR2806	C 97	ESR3228	C 96	ESR3993	L 17	ETC5423	B 86
ERR913	B 43	ESR1446	J 19	ESR1896	J 06	ESR2392	J 19	ESR2808	C 69	ESR3239	C 97	ESR3994	L 17	ETC5423	B121
ERR914	B154	ESR1451	J 19	ESR1897	J 06	ESR2406	J 16	ESR2809	C 69	ESR3240	C 92	ESR4065	J 09	ETC5499	B 13
ERR928	B 14	ESR151	J 18	ESR1897	J 24	ESR2406	N 21	ESR2810	C 94	ESR3240	C 97	ESR4117	K 04	ETC5499	B 86
ERR941	B135	ESR1579	J 18	ESR1899	J 06	ESR2416	C 97	ESR2813	L 06	ESR3247	J 19	ESR460	J 03	ETC5499	B121
ERR943	J 11	ESR1591	C 92	ESR1899	J 24	ESR2421	K 04	ESR2814	L 06	ESR3248	J 16	ESR461	J 03	ETC5524	B140
ERR978	B 17	ESR1594L	C 92	ESR1905	L 03	ESR2422	B 62	ESR2818	K 14	ESR3250	J 08	ESR461	J 04	ETC5576	B106
ESI100020LOY	P 22	ESR1594L	C 93	ESR1926	L 19	ESR2491	L 09	ESR2819	K 07	ESR3260	B 31	ESR500	J 08	ETC5576	B156
ESI100020SUC	P 22	ESR1594L	C 42	ESR1929	J 06	ESR2579	C 97	ESR2819	K 09	ESR3260	K 13	ESR501	J 08	ETC5577	B 24
ESQ100080	P 42	ESR1594L	L 03	ESR1931	L 18	ESR2584	J 04	ESR2820	K 07	ESR3260	K 14	ESR578	J 01	ETC5577	B 55
ESQ100090	P 42	ESR1594L	L 06	ESR1972	B179	ESR2585	J 02	ESR2820	K 09	ESR3263	K 12	ESR578	J 02	ETC5577	B 62
ESR101	K 04	ESR1594L	C 69	ESR2004	L 19	ESR2587	L 19	ESR2825	J 10	ESR3263	K 10	ESR58	K 01	ETC5592	B 05
ESR1027	L 04	ESR1594L	C 94	ESR2033	K 02	ESR2588	J 17	ESR2834	J 05	ESR3263	K 11	ESR59	K 01	ETC5592	B 44
ESR1049	J 18	ESR1594L	C 97	ESR2033	K 03	ESR259	J 09	ESR2834	J 06	ESR3267	J 08	ESR63	L 13	ETC5679	B 80
ESR1050	J 18	ESR1594L	L 18	ESR2034	B178	ESR2592	L 07	ESR2844	L 18	ESR3273	J 16	ESR63	L 15	ETC5679	B112
ESR1051	J 18	ESR1595	L 04	ESR2034	K 06	ESR2593	L 05	ESR2847	L 05	ESR3275	L 07	ESR63	L 12	ETC5797	B 98
ESR1052	L 04	ESR1600	J 09	ESR2034	K 07	ESR2593	L 08	ESR2848	L 08	ESR3276	L 08	ESR67	K 01	ETC5964	B124
ESR1052	L 05	ESR1601	J 11	ESR2034	K 08	ESR2595	L 01	ESR2849	L 08	ESR3296	L 09	ESR80	L 04	ETC5967	B 23
ESR1056	J 02	ESR161	J 23	ESR2034	K 09	ESR2602	K 07	ESR2874	L 03	ESR3297	L 05	ETC1260	B121	ETC6093	B103
ESR1056	J 04	ESR1611L	J 15	ESR2067	J 16	ESR2607	K 06	ESR2875	L 03	ESR3298	J 09	ETC1275	B121	ETC6093	B151
ESR1056	J 06	ESR1615	L 17	ESR2067	J 17	ESR2607	K 07	ESR2876	C 92	ESR3301	M155	ETC4021	B 15	ETC6099	B128
ESR1057L	J 15	ESR1621	J 19	ESR2067	J 19	ESR2607	K 09	ESR2877	C 92	ESR3353	L 19	ETC4022	B 15	ETC6104	B 85

ETC6104	B 86
ETC6104	B120
ETC6135	B 93
ETC6135	B130
ETC6148	B 85
ETC6148	B119
ETC6159	B144
ETC6214	B136
ETC6222	B135
ETC6223	B135
ETC6237	B103
ETC6291	B106
ETC6291	B156
ETC6375	B132
ETC6375	B133
ETC6375	B134
ETC6408	B 99
ETC6408	B142
ETC6444	B129
ETC6496	B 99
ETC6503	B129
ETC6505	B138
ETC6531	B 05
ETC6531	B 44
ETC6532	B 05
ETC6532	B 44
ETC6547	B129
ETC6607	B 99
ETC6607	B142
ETC6657	B139
ETC6660	B136
ETC6661	B132
ETC6661	B133
ETC6661	B134
ETC6850L	B 92
ETC6874	B135
ETC6874	B136
ETC6889	B129
ETC6890	B129
ETC6913	B132
ETC6976	B101
ETC7068	B 98
ETC7126	B 98
ETC7127	B 97
ETC7128	B 22
ETC7149	B140
ETC7150	B140
ETC7184	B 31
ETC7188	B 91
ETC7189	B 91
ETC7201	B 91
ETC7208	B 85
ETC7208	B106
ETC7208	B156
ETC7238	B 13
ETC7244	B151
ETC7286	B 08
ETC7313	B 25
ETC7313	B132
ETC7313	B133
ETC7313	B134
ETC7322	B147
ETC7329	B 25
ETC7329	B 26
ETC7340	B 80
ETC7340	B112
ETC7345	B 84
ETC7345	B118
ETC7353	B136
ETC7357	B 08
ETC7357	B 46
ETC7385	B 84
ETC7385	B117
ETC7398	B 09
ETC7398	B 15
ETC7398	B115
ETC7419	B113
ETC7427	B101
ETC7427	B147
ETC7442	B137
ETC7461	B 30
ETC7461E	B 30
ETC7463	B 31
ETC7468	B 34
ETC7469	B 36
ETC7513	B 31
ETC7514	B 31
ETC7519	B 36
ETC7528	B 34
ETC7530	B 18
ETC7532	B 34
ETC7553L	B 86
ETC7668ML	B140
ETC7713	B110
ETC7714	B 78
ETC7714	B110
ETC7769	B140
ETC7771	B135
ETC7773	B139
ETC7819	B125
ETC7869	B 27
ETC7870	B139
ETC7874	B136
ETC7875	B140
ETC7884	B139
ETC7884	D 18
ETC7886	B139
ETC7887	B139
ETC7899	B153
ETC7899	B154
ETC7915	B 94
ETC7929	B 06
ETC7971	B103
ETC7971	B151
ETC7994	B100
ETC7994	B144
ETC7996	B112
ETC8001	B 16
ETC8002	B 16
ETC8003	B 16
ETC8003	B 54
ETC8007	B 23
ETC8031	B 17
ETC8031	B 55
ETC8036	B 16
ETC8074	B 04
ETC8074	B 43
ETC8086	B 08
ETC8095	B 18
ETC8103	B 18
ETC8191	B 08
ETC8191	B 46
ETC8193	B 19
ETC8193	B 57
ETC8194	B 16
ETC8197	B 29
ETC8352	B 04
ETC8352	B 43
ETC8412	B 29
ETC8440	B156
ETC8441	B106
ETC8441	B156
ETC8442	B 04
ETC8442	B 43
ETC8470	B 29
ETC8494	B132
ETC8494	B133
ETC8495	B135
ETC8496	B130
ETC8550	B 11
ETC8559	B106
ETC8559	B156
ETC8560	B 11
ETC8560	B 50
ETC8579	B106
ETC8579	B156
ETC8591	B 91
ETC8596	B123
ETC8596	B125
ETC8610	B102
ETC8610	B149
ETC8611	B102
ETC8611	B149
ETC8612	B102
ETC8612	B149
ETC8613	B102
ETC8613	B149
ETC8614	B102
ETC8614	B149
ETC8615	B102
ETC8615	B149
ETC8616	B102
ETC8616	B149
ETC8617	B103
ETC8617	B149
ETC8618	B103
ETC8618	B149
ETC8622	B123
ETC8663	B 19
ETC8663	B 57
ETC8679	B127
ETC8680	B127
ETC8681	B 90
ETC8682	B 90
ETC8684	B123
ETC8686	B128
ETC8720	B103
ETC8765	B106
ETC8765	B156
ETC8808	B 16
ETC8808	B 54
ETC8809	B 16
ETC8810	B 16
ETC8810	B 54
ETC8820	B 31
ETC8820	B 67
ETC8830	J 14
ETC8833	B 87
ETC8833	B122
ETC8847	B 16
ETC8847	B 54
ETC8852	B 35
ETC8853	B 10
ETC9009	B 35
ETC9083	B105
ETC9083	B153
EYC10008L	P 43
EYC10008L	P 51
EYC101460	P 36
EYC101460	P 38
EYC101460	P 39
EYC101460	P 48
EYC101460	P 55
EYC101460	R 01
EYC101460	P 28
EYC101460	P 60
EYC101460	R 25
EYC101460	N 43
EYC101460	P 31
EYC101460	P 34
EYC101460	P 41
EYE100030	N 36
EYF100310	N 36
EYG100150L	M 62
EYH100400	N 36
EYP10036L	N 36
EYP100790L	N 36
FA105201	N114
FA105201	N119
FAE101340	P 10
FAE101350	P 10
FAM9270	C 65
FAV101160LNF	M 66
FAV101170LNF	M 66
FAV101180LNF	M 66
FAV101190LNF	M 66
FB105051L	D 20
FB106061M	L 51
FB106065L	B168
FB106071	B179
FB106081L	B 23
FB106081L	H 06
FB106081L	H 08
FB106105L	B165
FB106105L	B174
FB106111L	B 59
FB106115L	H 10
FB106121L	B 21
FB106121L	B 28
FB106255	B174
FB108051	B 65
FB108061	D 06
FB108061L	B148
FB108071L	B 10
FB108071L	B 48
FB108071L	B 49
FB108081L	B 21
FB108081L	B 49
FB108081L	C 44
FB108081L	C 55
FB108081L	C 45
FB108081L	C 46
FB108101	B 18
FB108101	B 49
FB108101	B 56
FB108101	B 64
FB108101	B172
FB108101	B176
FB108101	B187
FB108101	C 02
FB108101	K 11
FB108101	K 13
FB108101	N 07
FB108111L	D 03
FB108111L	C 55
FB108111L	C 67
FB108111L	D 04
FB108111L	B170
FB108111L	D 17
FB108111L	D 18
FB108121L	B 09
FB108121L	C 67
FB108131L	B 51
FB108141L	B 10
FB108141L	H 24
FB108141L	F 07
FB108141S	C 55
FB108161L	B 10
FB108161L	B 48
FB108161L	B182
FB108161L	N121
FB108171L	B 62
FB108171L	N121
FB108171L	C 55
FB108181	B 24
FB108181ML	B 10
FB108181ML	D 03
FB108197	B145
FB108201L	B 49
FB108221	3 49
FB108227	B155
FB108241L	B 51
FB108251	B 09
FB108251	B 34
FB108251	B 73
FB108251	B170
FB108261	B 51
FB110071L	B 75
FB110071L	C 52
FB110071L	C 72
FB110081L	B170
FB110081ML	D 04
FB110091L	B171
FB110091L	B184
FB110091L	B187
FB110091ML	D 04
FB110091ML	D 05
FB110101L	B170
FB110111L	B 40
FB110111L	C 72
FB110121L	B184
FB110121L	B187
FB110121L	C 72
FB110141L	B 11
FB110141L	B 50
FB110141L	B 75
FB110141L	C 72
FB110151L	B145
FB110151L	B170
FB110201L	B183
FB504117S	B136
FB505181L	B138
FB505251S	B117
FB505311S	B117
FB505381S	B117
FB506267	B143
FB506267	B145
FBO100260SUC	P 08
FBR10001LOY	P 09
FBR10001SUC	P 09
FBR100060SUC	P 08
FBS100700	P 09
FBS100760	P 09
FBS100770	P 09
FBV10040LNFL	M 66
FBV10040LNFL	M 10
FBV10065LNF	M136
FBV10065LNF	M138
FBV10065LNF	M141
FFB101460SUC	P 07
FFU10004	P 10
FFV10002	P 10
FHC100410	L 35
FHR100500PMA	P 20
FHY100290LNF	P 12
FJA10001LNF	M 77
FJL101530PMA	P 21
FJL101680PMA	P 21
FJL101710PMA	H 20
FJL101710PMA	H 23
FM106041	M145
FN105041L	N 43
FN105047L	B 73
FN105047L	B186
FN105047L	M 13
FN106047	M111
FN106047	M147
FN106047	M149
FN106047L	B180
FN106047L	B133
FN106047L	B134
FN106047L	M 82
FN106047L	C 53
FN106047L	M111
FN106047L	M147
FN106047L	H 03
FN106047L	H 09
FN106047L	H 10
FN106047L	N 86
FN106047L	N 83
FN106047L	M149
FN108041	M111
FN108041L	B 24
FN108041L	B 28
FN108041L	B 34
FN108041L	B 62
FN108041L	B 64
FN108041L	B 66
FN108041L	B 71
FN108041L	B 74
FN108041L	B 99
FN108041L	B104
FN108041L	B148
FN108041L	B157
FN108041L	B177
FN108041L	B155
FN108041L	N121
FN108041L	K 12
FN108041L	L 01
FN108041L	L 19
FN108041L	L 17
FN108041L	L 02
FN108041L	J 15
FN108041L	K 10
FN108041L	K 11
FN108041L	L 04
FN108041L	F 18
FN108041L	H 24
FN108041L	M 91
FN108045L	B 25
FN108047L	B 12
FN108047L	B 29
FN108047L	B 56
FN108047L	B104
FN108047L	B140
FN108047L	N 12
FN110041L	B 11
FN110041L	B 24
FN110041L	B 37
FN110041L	B 50
FN110041L	B 51
FN110041L	B 74
FN110041L	K 03
FN110041L	K 06
FN110041L	K 07
FN110041L	C 52
FN110041L	C 72
FN110041L	C 03
FN110041L	C 04
FN110041L	K 08
FN110041L	K 09
FN110041L	D 06
FN110041L	H 24
FN110047	B 11
FN110047	B 24
FN110047	B 50
FN110047	B 51
FN110047	B 74
FN110047	K 03
FN110047	K 06

FN110047 K 07	FRC2883 E 20	FRC4343 C 29	FRC4951 D 18	FRC5305 C 30	FRC5679 C 32	FRC6786 E 15	FRC7810 D 10
FN110047 C 52	FRC2884 E 20	FRC4345 C 09	FRC5007 C 71	FRC5409 D 21	FRC5690 E 07	FRC6786 E 16	FRC7851 B161
FN110047 C 03	FRC2885 E 20	FRC4345 C 29	FRC5024 C 71	FRC5409 D 05	FRC5690 E 08	FRC6787 E 15	FRC7855 C 54
FN110047 C 04	FRC2886 E 20	FRC4347 C 09	FRC5053 D 04	FRC5413 D 21	FRC5690 E 28	FRC6787 E 16	FRC7871 D 14
FN110047 K 08	FRC2894 E 20	FRC4347 C 29	FRC5054 C 71	FRC5413 D 05	FRC5690 E 29	FRC6788 E 15	FRC7885 D 15
FN110047 K 09	FRC2897 E 20	FRC4349 C 10	FRC5076 D 16	FRC5415 D 05	FRC5690 E 31	FRC6788 E 16	FRC7926 D 07
FN110047 D 06	FRC2933 E 11	FRC4349 C 29	FRC5076 D 18	FRC5416 D 21	FRC5926 E 23	FRC6789 E 15	FRC7929 D 15
FN110047 H 24	FRC2933 E 36	FRC4351 C 10	FRC5095 C 17	FRC5416 D 05	FRC5926 E 24	FRC6789 E 16	FRC7930 D 03
FNC100080L P 08	FRC2975 C 46	FRC4351 C 29	FRC5095 C 37	FRC5419 D 21	FRC5926 E 39	FRC6790 E 15	FRC7970 D 15
FPD100060 N 69	FRC2975 C 24	FRC4353 C 10	FRC5186 C 17	FRC5438 D 12	FRC5926 E 40	FRC6790 E 16	FRC7998 D 03
FQP101060 N 77	FRC3002 E 10	FRC4353 C 29	FRC5186 C 37	FRC5439 D 11	FRC5981 C 72	FRC6872 D 17	FRC7998 D 21
FQP101070 N 77	FRC3002 E 34	FRC4355 C 10	FRC5204 E 07	FRC5440 D 13	FRC6000 D 20	FRC6872 D 18	FRC8002 E 23
FRC1193 E 32	FRC3162 D 12	FRC4355 C 29	FRC5204 E 08	FRC5442 D 11	FRC6030 D 14	FRC6873 D 17	FRC8002 E 39
FRC1195 E 32	FRC3205 E 19	FRC4357 C 10	FRC5204 E 28	FRC5446 D 12	FRC6103 D 03	FRC6873 D 18	FRC8005 E 39
FRC1197 E 32	FRC3205 E 21	FRC4357 C 29	FRC5204 E 29	FRC5449 D 13	FRC6103 D 21	FRC6915 C 86	FRC8075 D 20
FRC1199 E 32	FRC3286 D 12	FRC4359 C 10	FRC5204 E 30	FRC5450 D 12	FRC6104 D 03	FRC6915 C 88	FRC8093 D 22
FRC1201 E 32	FRC3310 D 12	FRC4359 C 29	FRC5204 E 31	FRC5458 D 15	FRC6105 D 03	FRC6943 D 05	FRC8104 C 06
FRC1203 E 32	FRC3326 C 02	FRC4361 C 10	FRC5235 C 14	FRC5468 D 14	FRC6105 D 21	FRC6950 C 03	FRC8104 C 07
FRC2310 E 19	FRC3327 C 44	FRC4361 C 29	FRC5235 C 34	FRC5469 D 14	FRC6106 D 03	FRC6950 C 25	FRC8104 C 28
FRC2310 E 22	FRC3327 C 02	FRC4363 C 10	FRC5243 C 13	FRC5473 D 14	FRC6109 D 14	FRC6968 D 07	FRC8127 C 66
FRC2310 E 39	FRC3327 C 45	FRC4363 C 29	FRC5243 C 33	FRC5473 D 21	FRC6110 D 14	FRC7018 D 15	FRC8127 C 18
FRC2310 E 40	FRC3416 C 44	FRC4365 C 10	FRC5244 C 13	FRC5478 D 16	FRC6121 D 11	FRC7043 D 11	FRC8127 C 38
FRC2365 D 09	FRC3416 C 02	FRC4365 C 29	FRC5244 C 33	FRC5479 D 21	FRC6125 D 17	FRC7043 D 12	FRC8154 E 10
FRC2365 D 21	FRC3416 C 45	FRC4367 C 10	FRC5245 C 13	FRC5479 D 16	FRC6145 C 47	FRC7043 D 21	FRC8154 E 34
FRC2365 C 07	FRC3417 C 02	FRC4367 C 29	FRC5245 C 33	FRC5479 D 18	FRC6145 C 25	FRC7065 E 13	FRC8170 C 19
FRC2402 C 44	FRC3502 D 23	FRC4369 C 10	FRC5246 C 13	FRC5480 D 16	FRC6154 C 26	FRC7065 E 14	FRC8170 C 39
FRC2402 C 02	FRC3502 D 22	FRC4369 C 29	FRC5246 C 33	FRC5480 D 18	FRC6244 C 06	FRC7075 B 40	FRC8202 D 20
FRC2402 C 45	FRC3602 D 12	FRC4449 C 06	FRC5247 C 13	FRC5486 D 21	FRC6244 C 07	FRC7075 B161	FRC8203 D 16
FRC2402 C 46	FRC4078 B 39	FRC4449 C 28	FRC5247 C 33	FRC5486 D 16	FRC6244 C 28	FRC7080 B161	FRC8203 D 18
FRC2402 C 24	FRC4078 C 02	FRC4493 C 07	FRC5255 C 44	FRC5486 D 18	FRC6291 B 40	FRC7081 B 40	FRC8204 D 20
FRC2464 D 11	FRC4112 E 07	FRC4493 C 28	FRC5255 C 02	FRC5562 D 14	FRC6291 B 76	FRC7081 B161	FRC8220 E 10
FRC2464 D 12	FRC4112 E 08	FRC4494 C 07	FRC5255 C 45	FRC5562 D 15	FRC6293 B 40	FRC7098 D 14	FRC8220 E 34
FRC2464 D 21	FRC4112 E 28	FRC4494 C 28	FRC5280 C 61	FRC5564 D 09	FRC6293 B 76	FRC7192 C 19	FRC8222 E 23
FRC2481 C 44	FRC4112 E 29	FRC4499 D 20	FRC5280 C 14	FRC5574 D 16	FRC6295 B 40	FRC7192 C 39	FRC8222 E 39
FRC2481 C 02	FRC4112 E 30	FRC4499 D 16	FRC5280 C 34	FRC5574 D 18	FRC6295 B 76	FRC7195 C 18	FRC8227 E 23
FRC2481 C 19	FRC4112 E 31	FRC4499 D 18	FRC5284 C 14	FRC5575 D 15	FRC6297 B 40	FRC7195 C 38	FRC8227 E 39
FRC2481 C 39	FRC4282 C 07	FRC4501 C 07	FRC5284 C 35	FRC5575 D 21	FRC6297 B 76	FRC7214 C 17	FRC8232 C 12
FRC2481 C 45	FRC4282 C 28	FRC4501 C 28	FRC5286 C 14	FRC5575 C 84	FRC6299 B 40	FRC7214 C 37	FRC8232 C 31
FRC2481 C 46	FRC4327 C 09	FRC4509 D 21	FRC5286 C 35	FRC5575 D 17	FRC6299 B 76	FRC7329 H 14	FRC8232 C 13
FRC2481 C 24	FRC4327 C 29	FRC4803 C 46	FRC5288 C 14	FRC5575 D 19	FRC6301 B 40	FRC7434 D 15	FRC8232 C 14
FRC2482 B 40	FRC4329 C 09	FRC4803 C 24	FRC5288 C 35	FRC5576 D 14	FRC6301 B 76	FRC7437 D 10	FRC8232 C 32
FRC2482 B 43	FRC4329 C 29	FRC4808 B 31	FRC5290 C 14	FRC5576 D 21	FRC6306 D 21	FRC7439 D 10	FRC8232 C 34
FRC2482 D 04	FRC4331 C 09	FRC4808 B 55	FRC5290 C 35	FRC5594 D 05	FRC6306 D 04	FRC7439 D 21	FRC8239 C 66
FRC2482 D 06	FRC4331 C 29	FRC4808 B 67	FRC5292 C 14	FRC5595 D 05	FRC6306 D 17	FRC7441 D 15	FRC8239 C 39
FRC2528 C 46	FRC4333 C 09	FRC4808 D 21	FRC5292 C 35	FRC5661 E 07	FRC6306 D 18	FRC7449 D 05	FRC8246 C 17
FRC2528 C 24	FRC4333 C 29	FRC4808 D 05	FRC5294 C 14	FRC5661 E 08	FRC6318 C 20	FRC7452 D 10	FRC8246 C 37
FRC2626 C 54	FRC4335 C 09	FRC4810 C 06	FRC5294 C 35	FRC5661 E 28	FRC6318 C 40	FRC7453 D 10	FRC8285 C 17
FRC2626 C 67	FRC4335 C 29	FRC4810 C 07	FRC5296 C 14	FRC5661 E 29	FRC6782 E 15	FRC7454 D 10	FRC8285 C 37
FRC2626 C 21	FRC4337 C 09	FRC4810 C 28	FRC5296 C 35	FRC5661 E 30	FRC6782 E 16	FRC7499 D 07	FRC8291 D 10
FRC2626 C 41	FRC4337 C 29	FRC4905 C 19	FRC5298 C 14	FRC5661 E 31	FRC6783 E 15	FRC7500 C 85	FRC8292 D 10
FRC2644 E 16	FRC4339 C 09	FRC4905 C 39	FRC5298 C 35	FRC5678 C 12	FRC6783 E 16	FRC7630 C 86	FRC8292 D 21
FRC2859 C 71	FRC4339 C 29	FRC4947 C 17	FRC5300 C 14	FRC5678 C 31	FRC6784 E 15	FRC7630 C 88	FRC8292 D 06
FRC2859 C 05	FRC4341 C 09	FRC4947 C 37	FRC5300 C 35	FRC5678 C 13	FRC6784 E 16	FRC7686 D 14	FRC8292 D 16
FRC2859 C 27	FRC4341 C 29	FRC4951 D 21	FRC5302 C 14	FRC5678 C 32	FRC6785 E 15	FRC7706 C 72	FRC8292 D 18
FRC2859 C 53	FRC4343 C 09	FRC4951 D 16	FRC5302 C 35	FRC5679 C 13	FRC6785 E 16	FRC7707 C 72	FRC8299 D 03

FRC8382	C 17	FRC8709	C 86	FRC9359	C 87	FRC9946	D 09	FS106167L	B 21	FS106167L	N 65	FS108161L	C 87	FS108207L	B143
FRC8382	C 37	FRC8709	C 88	FRC9359	C 89	FRC9948	D 09	FS106167L	B 22	FS106167L	L 18	FS108161L	J 02	FS108207L	E 43
FRC8386	E 01	FRC8710	C 86	FRC9388	C 14	FRC9950	D 09	FS106167L	B 34	FS106181L	B167	FS108161L	J 04	FS108207L	E 49
FRC8387	E 02	FRC8711	C 86	FRC9388	C 34	FRC9952	D 09	FS106167L	B 53	FS106201	M152	FS108161L	J 15	FS108207L	N 69
FRC8394	D 14	FRC8711	C 88	FRC9390	C 40	FRC9954	D 09	FS106167L	B 60	FS106201L	B 28	FS108161L	K 13	FS108207L	K 07
FRC8468	C 87	FRC8712	C 86	FRC9427	D 03	FRC9956	D 09	FS106167L	B 65	FS106201L	B 65	FS108161L	K 15	FS108207L	L 01
FRC8481	C 19	FRC8712	C 88	FRC9427	C 85	FRC9958	D 09	FS106167L	B 99	FS106201L	N 13	FS108161L	L 14	FS108207L	L 02
FRC8481	C 39	FRC8742	C 86	FRC9427	C 03	FRC9960	D 09	FS106167L	B164	FS106201L	C 07	FS108161L	F 17	FS108207L	J 09
FRC8482	C 19	FRC8742	C 88	FRC9427	C 25	FRC9985	C 67	FS106167L	B177	FS106201L	M145	FS108161L	H 24	FS108207L	K 09
FRC8482	C 39	FRC8743	C 86	FRC9430	C 55	FRC9985	C 20	FS106167L	N 28	FS106201L	D 18	FS108161L	F 03	FS108207L	L 04
FRC8485	C 66	FRC8743	C 88	FRC9430	C 85	FRC9985	C 40	FS106167L	N 50	FS106201M	D 06	FS108161L	M 91	FS108207L	L 05
FRC8485	C 19	FRC8744	C 86	FRC9430	C 03	FRC9986	C 18	FS106167L	P 22	FS106207L	B 65	FS108161ML	B 40	FS108207L	N 07
FRC8485	C 38	FRC8744	C 88	FRC9430	C 25	FRC9986	C 38	FS106167L	J 05	FS106207L	B129	FS108161ML	B 76	FS108207L	H 24
FRC8490	C 20	FRC8751	D 04	FRC9460	D 10	FRC9987	C 20	FS106167L	K 02	FS106207L	B168	FS108201L	B 14	FS108251L	B 14
FRC8490	C 40	FRC8751	D 16	FRC9467N	D 01	FRC9987	C 40	FS106167L	K 04	FS106207L	N 13	FS108201L	B 24	FS108251L	B 49
FRC8496	C 21	FRC8751	D 18	FRC9513	D 15	FS105107	B 09	FS106167L	K 06	FS106207L	N 70	FS108201L	B 64	FS108251L	B 61
FRC8497	C 21	FRC8766	D 16	FRC9518	E 13	FS105107	B 37	FS106207L	K 07	FS106207L	N 53	FS108201L	B 69	FS108251L	B 72
FRC8497	C 41	FRC8766	D 18	FRC9519	E 13	FS105107	B 62	FS106167L	L 13	FS106207L	C 28	FS108201L	B 70	FS108251L	B 99
FRC8505	C 21	FRC8767	D 20	FRC9526	C 14	FS105107	B 38	FS106167L	C 92	FS106207L	R 51	FS108201L	B157	FS108251L	B142
FRC8505	C 41	FRC8768	D 20	FRC9526	C 35	FS105107	L 17	FS106167L	C 19	FS106207L	N 89	FS108201L	B169	FS108251L	B157
FRC8507	C 67	FRC8769	D 20	FRC9552	D 10	FS105107	M 94	FS106167L	C 39	FS106207L	F 17	FS108201L	B176	FS108251L	D 03
FRC8507	C 20	FRC8775	H 12	FRC9568	B 39	FS105107	P 03	FS106167L	C 42	FS106251M	D 16	FS108201L	B182	FS108251L	D 14
FRC8507	C 40	FRC8775	H 13	FRC9568	B108	FS105107	M140	FS106255L	M108	FS106255L	B 20	FS108201L	B187	FS108251L	C 67
FRC8517	C 21	FRC8776	C 20	FRC9568	B160	FS105107	M143	FS106167L	M110	FS106255L	B167	FS108201L	E 43	FS108251L	C 02
FRC8517	C 41	FRC8776	C 40	FRC9568	C 44	FS105107	M 83	FS106167L	L 03	FS106255L	B174	FS108201L	E 48	FS108251L	C 08
FRC8520	C 67	FRC8793	C 20	FRC9568	C 02	FS105161P	M 81	FS106167L	M106	FS106255L	B177	FS108201L	E 49	FS108251L	C 21
FRC8520	C 20	FRC8793	C 40	FRC9568	C 46	FS105201L	B185	FS106167L	M 47	FS106255L	K 02	FS108201L	G 11	FS108251L	C 29
FRC8520	C 40	FRC8827	C 20	FRC9568	C 24	FS106101L	B 10	FS106167L	M147	FS106255L	K 03	FS108201L	G 12	FS108251L	C 41
FRC8530	E 13	FRC8827	C 40	FRC9669	C 21	FS106101L	B179	FS106167L	J 06	FS106255L	C 71	FS108201L	M106	FS108251L	C 22
FRC8530	E 14	FRC8854	H 05	FRC9669	C 41	FS106101L	M 82	FS106167L	J 20	FS106255L	C 05	FS108201L	L 08	FS108251L	D 04
FRC8530	E 16	FRC8854	H 07	FRC9674	D 16	FS106101L	C 86	FS106167L	J 21	FS106255L	C 27	FS108201L	H 21	FS108251L	C 24
FRC8532	E 39	FRC8855	H 05	FRC9674	D 18	FS106101L	L 35	FS106167L	J 22	FS106255L	C 53	FS108201L	R 51	FS108251L	M 17
FRC8542	D 16	FRC8855	H 07	FRC9683	D 16	FS106101L	C 46	FS106167L	K 11	FS106255L	B170	FS108201L	H 24	FS108251L	M 18
FRC8544	D 14	FRC8882	C 18	FRC9693	F 16	FS106101L	C 24	FS106167L	K 08	FS106255L	F 18	FS108205L	B 25	FS108251L	M 19
FRC8547	D 20	FRC8882	C 38	FRC9706	C 36	FS106101L	M 44	FS106167L	K 09	FS108127	B 49	FS108207L	B 07	FS108251L	M 21
FRC8547	G 11	FRC8900	D 15	FRC9785	D 15	FS106101L	N 62	FS106167L	K 13	FS108127	B 51	FS108207L	B 24	FS108251L	D 16
FRC8547	H 21	FRC8915	C 20	FRC9792L	C 12	FS106121	B 28	FS106167L	K 14	FS108127	B 52	FS108207L	B 28	FS108251L	D 18
FRC8547	H 23	FRC8915	C 40	FRC9792L	C 31	FS106121	B 65	FS106167L	L 06	FS108127	B 70	FS108207L	B 34	FS108251L	M 16
FRC8548	G 11	FRC8917	D 08	FRC9812	C 14	FS106125L	B167	FS106167L	L 14	FS108127	B121	FS108207L	B 44	FS108251L	M 20
FRC8548	H 21	FRC9028	C 67	FRC9812	C 35	FS106125L	B173	FS106167L	L 15	FS108127	B185	FS108207L	B 47	FS108251ML	D 03
FRC8548	H 23	FRC9028	C 20	FRC9845	D 07	FS106125L	B177	FS106167L	F 20	FS108127	B143	FS108207L	B 55	FS108251ML	D 10
FRC8548	D 16	FRC9028	C 40	FRC9847	D 07	FS106125L	B180	FS106167L	F 21	FS108127L	B164	FS108207L	B 59	FS108251ML	D 14
FRC8558	D 04	FRC9203	B161	FRC9849	D 07	FS106125L	B182	FS106167L	C 69	FS108161L	B 04	FS108207L	B 64	FS108251ML	C 44
FRC8561	D 18	FRC9205	B161	FRC9851	D 07	FS106125L	P 36	FS106167L	L 36	FS108161L	B 24	FS108207L	B 67	FS108251ML	C 56
FRC8641	E 01	FRC9207	B161	FRC9853	D 07	FS106125L	P 38	FS106167L	N 20	FS108161L	B 47	FS108207L	B 69	FS108251ML	C 67
FRC8698	C 87	FRC9209	B161	FRC9926	D 09	FS106125L	E 15	FS106167L	C 94	FS108161L	B 52	FS108207L	B 70	FS108251ML	C 45
FRC8698	C 89	FRC9211	B161	FRC9928	D 09	FS106125L	E 16	FS106167L	C 95	FS108161L	B 55	FS108207L	B 76	FS108251ML	C 46
FRC8700	E 23	FRC9220	C 86	FRC9930	D 09	FS106127L	B133	FS106167L	C 96	FS108161L	B 60	FS108207L	B 99	FS108251ML	D 16
FRC8700	E 24	FRC9229	E 27	FRC9932	D 09	FS106127L	B134	FS106167L	C 97	FS108161L	B 65	FS108207L	B142	FS108257	K 04
FRC8700	E 39	FRC9339	D 12	FRC9934	D 09	FS106161L	B 65	FS106167L	R 46	FS108161L	B182	FS108207L	B156	FS108257L	B 06
FRC8700	E 40	FRC9340	C 67	FRC9936	D 09	FS106161L	K 04	FS106167L	M 18	FS108161L	J 01	FS108207L	B169	FS108257L	B 09
FRC8707	C 86	FRC9340	C 20	FRC9938	D 09	FS106161L	L 15	FS106167L	M 19	FS108161L	J 03	FS108207L	B176	FS108257L	B 10
FRC8707	C 88	FRC9340	C 40	FRC9940	D 09	FS106161L	L 12	FS106167L	M149	FS108161L	K 02	FS108207L	B182	FS108257L	B 16
FRC8708	C 87	FRC9357	C 87	FRC9942	D 09	FS106161L	N 65	FS106167L	L 12	FS108161L	K 04	FS108207L	B187	FS108257L	B 17
FRC8708	C 89	FRC9357	C 88	FRC9944	D 09	FS106161ML	C 47	FS106167L	M144	FS108161L	L 17	FS108207L	B189	FS108257L	B 21

FS108257L	B 23	FS110301L	J 03	FTC1356	C 74	FTC1616	E 21	FTC2203	C 66	FTC265	C 16	FTC287	C 36	FTC3185	E 40
FS108257L	B 27	FS110301L	C 72	FTC1357	C 74	FTC1617	E 21	FTC2203	C 18	FTC265	C 36	FTC2880	E 14	FTC3186	E 40
FS108257L	B 43	FS110301L	J 02	FTC1359	C 37	FTC1620	E 19	FTC2203	C 38	FTC267	C 16	FTC2881	E 14	FTC3187	E 40
FS108257L	B 48	FS110301L	J 04	FTC1360	C 34	FTC1621	E 19	FTC2210	C 06	FTC267	C 36	FTC2882	E 22	FTC3188	E 40
FS108257L	B 59	FS110301L	D 05	FTC1368	H 14	FTC1623	E 19	FTC2223	E 06	FTC2687	C 55	FTC2885	C 66	FTC3197	B160
FS108257L	B 61	FS110301L	N 07	FTC1368	H 17	FTC1741	D 15	FTC2258E	C 70	FTC2687	C 28	FTC2887	C 60	FTC3233	C 87
FS108257L	B 72	FS110301ML	D 05	FTC1374	H 18	FTC1748	D 05	FTC2258N	C 70	FTC269	C 16	FTC289	C 16	FTC3235	E 04
FS108257L	B142	FS112301P	E 10	FTC1374	E 20	FTC1752	C 14	FTC2282	C 13	FTC269	C 36	FTC289	C 36	FTC3236	E 04
FS108257L	B157	FS112301P	E 34	FTC1376	E 20	FTC1752	C 34	FTC2282	C 32	FTC271	C 16	FTC2903	C 86	FTC3237	E 04
FS108257L	N 14	FS112301P	E 35	FTC1378	E 40	FTC1791	C 14	FTC2283	C 14	FTC271	C 36	FTC291	C 16	FTC3238	E 04
FS108257L	C 54	FS112401	B 60	FTC1379	H 18	FTC1791	C 34	FTC2283	C 34	FTC2712	C 60	FTC291	C 36	FTC3267	E 11
FS108257L	C 67	FS112401L	B184	FTC1381	H 17	FTC1908	C 40	FTC2284	C 14	FTC2714	C 58	FTC293	C 16	FTC3267	E 36
FS108257L	C 02	FS112401L	B187	FTC1403	C 06	FTC1919	D 06	FTC2284	C 34	FTC2716	C 59	FTC293	C 36	FTC3268	E 11
FS108257L	C 08	FS112401L	C 52	FTC1403	C 28	FTC1982	C 13	FTC2285	C 13	FTC2723	C 64	FTC2941	C 64	FTC3268	E 36
FS108257L	J 21	FS504047	B 84	FTC1404	C 28	FTC1982	C 32	FTC2285	C 32	FTC2724	C 65	FTC2945	C 65	FTC3269	E 11
FS108257L	J 22	FS504047	B 98	FTC1408	C 32	FTC1988	C 20	FTC2383	C 54	FTC2725	C 65	FTC2948	C 59	FTC3269	E 36
FS108301L	C 55	FS504047	B132	FTC1426	C 11	FTC1988	C 40	FTC2385	C 65	FTC2727	C 62	FTC295	C 16	FTC3270	E 40
FS108301L	G 09	FS504047	B139	FTC1435	C 17	FTC1988	C 41	FTC2392	C 54	FTC273	C 16	FTC295	C 36	FTC3271	E 40
FS108301L	D 06	FS504047	B140	FTC1435	C 37	FTC1989	C 13	FTC2396	C 58	FTC273	C 36	FTC2957	C 44	FTC3272	E 08
FS108301L	F 18	FS504047	B 95	FTC1441	C 55	FTC2001	B160	FTC2396	C 59	FTC2731	C 60	FTC2957	C 02	FTC3272	E 28
FS108301S	C 56	FS504055	B115	FTC1441	C 28	FTC2002	B160	FTC2396	C 60	FTC2731	C 62	FTC2957	C 45	FTC3272E	E 08
FS108307L	B 09	FS504087S	B136	FTC1446	C 11	FTC2005	C 59	FTC2397	C 58	FTC2733	C 58	FTC2957	C 24	FTC3272R	E 29
FS108307L	B 34	FS505077	B130	FTC1446	C 30	FTC2005	C 62	FTC2397	C 60	FTC2737	C 57	FTC3076	E 03	FTC3275	C 85
FS108307L	B 51	FTC1073	C 15	FTC1455	C 34	FTC2005	C 13	FTC2448	C 20	FTC275	C 16	FTC3077	E 03	FTC3276	E 14
FS108307L	B 63	FTC1073	C 36	FTC1457	E 23	FTC2005	C 33	FTC2448	C 40	FTC275	C 36	FTC3078	E 03	FTC3276	E 16
FS108307L	B 73	FTC1074	C 15	FTC1464	E 03	FTC2006	C 13	FTC2450	C 18	FTC2750	E 07	FTC3079	E 03	FTC3283	D 06
FS108307L	B104	FTC1084	D 15	FTC1465	E 03	FTC2006	C 33	FTC2450	C 38	FTC2750	E 28	FTC3090	E 25	FTC3283	C 88
FS108307L	B177	FTC1085	D 15	FTC1466	E 03	FTC2007	C 13	FTC2461	C 13	FTC2750E	E 07	FTC3099	C 70	FTC3299	H 17
FS108307L	B187	FTC124	E 20	FTC1467	E 03	FTC2007	C 33	FTC2461	C 32	FTC2750R	E 28	FTC3099E	C 70	FTC3299	H 18
FS108307L	C 55	FTC125	E 21	FTC1468	E 25	FTC2008	C 13	FTC2462	C 13	FTC2756	C 70	FTC3099R	C 70	FTC3308	E 20
FS108307L	C 86	FTC1270	C 04	FTC1478	E 25	FTC2008	C 33	FTC2471	D 16	FTC2756E	C 70	FTC311	C 08	FTC3309	E 20
FS108307L	C 28	FTC1271	C 44	FTC1482	C 28	FTC2009	C 13	FTC2471	D 18	FTC2756R	C 70	FTC311	C 29	FTC3310	E 20
FS108307L	D 06	FTC1271	C 45	FTC1486	C 18	FTC2009	C 33	FTC248	C 64	FTC277	C 16	FTC3142	H 05	FTC3311	E 20
FS108357	N 41	FTC1282	C 11	FTC1486	C 38	FTC2061	D 05	FTC248	C 15	FTC277	C 36	FTC3142	H 07	FTC3317	B189
FS108357	N 53	FTC1282	C 30	FTC1488	C 66	FTC2065	E 21	FTC248	C 36	FTC2770	C 11	FTC3143	H 05	FTC3320	H 17
FS108357	C 39	FTC1293	C 30	FTC1488	C 18	FTC2066	E 21	FTC2497	C 88	FTC2772	C 02	FTC3143	H 07	FTC3321	H 17
FS110141M	B 76	FTC1301	C 14	FTC1488	C 38	FTC2069	C 74	FTC2498	C 02	FTC2772	C 45	FTC3145	E 21	FTC3370	C 66
FS110141M	C 72	FTC1301	C 34	FTC1489	C 19	FTC2081	B 40	FTC2498	C 24	FTC2783	E 23	FTC3145	E 40	FTC3371	C 62
FS110251L	B187	FTC1310	C 12	FTC1489	C 38	FTC2084	C 13	FTC2504	C 12	FTC2783	E 24	FTC3146	E 14	FTC3375	H 16
FS110301	C 72	FTC1310	C 31	FTC1490	C 18	FTC2084	C 32	FTC2504	C 31	FTC2783	E 39	FTC3146	E 16	FTC3375	H 18
FS110301L	B 15	FTC1311	C 58	FTC1490	C 38	FTC2089	C 32	FTC2505	C 12	FTC2783	E 40	FTC3147	E 14	FTC3381	C 66
FS110301L	B 35	FTC1311	C 60	FTC1491	C 18	FTC2090	C 13	FTC2509	E 06	FTC279	C 16	FTC3147	E 16	FTC3382	C 47
FS110301L	B 37	FTC1311	C 62	FTC1491	C 38	FTC2102	C 06	FTC2520	E 18	FTC279	C 36	FTC3148	E 14	FTC3382	C 66
FS110301L	B 39	FTC1311	C 13	FTC1494	C 06	FTC2149	B 39	FTC2521	E 18	FTC281	C 16	FTC3149	E 14	FTC3382	C 38
FS110301L	B 53	FTC1311	C 32	FTC1499	F 16	FTC2175	E 06	FTC2522	E 18	FTC281	C 36	FTC3154	E 21	FTC3387	C 54
FS110301L	B 71	FTC1312	C 13	FTC1512	C 19	FTC2176	E 27	FTC2523	E 18	FTC2822	C 08	FTC316	C 08	FTC3391	C 65
FS110301L	B 72	FTC1312	C 32	FTC1512	C 39	FTC2192	C 03	FTC2524	E 20	FTC2822	C 29	FTC316	C 29	FTC3401	E 13
FS110301L	B 74	FTC1313	C 12	FTC1525	C 93	FTC2192	C 25	FTC2525	E 20	FTC2828	D 05	FTC317	C 64	FTC3401	E 14
FS110301L	B 75	FTC1313	C 31	FTC1525	C 42	FTC2193	C 66	FTC2526	E 20	FTC2829	D 06	FTC317	C 15	FTC3401	E 16
FS110301L	B 76	FTC1316	C 18	FTC1525	C 95	FTC2193	C 18	FTC2527	E 20	FTC283	C 16	FTC317	C 36	FTC3404	E 05
FS110301L	B184	FTC1316	C 38	FTC1525	C 97	FTC2193	C 38	FTC2582	C 65	FTC283	C 36	FTC3178	D 05	FTC3405	E 05
FS110301L	B187	FTC1327	C 13	FTC1533	C 06	FTC2195	C 18	FTC2600	H 17	FTC285	C 16	FTC3179	E 24	FTC3406	E 05
FS110301L	B188	FTC1327	C 32	FTC1562	C 17	FTC2199	C 67	FTC2601	H 17	FTC285	C 36	FTC3179	E 40	FTC3407	E 05
FS110301L	E 51	FTC1329	C 14	FTC160	B160	FTC2199	C 20	FTC263	C 16	FTC2859	D 15	FTC3183	E 21	FTC3434	E 27
FS110301L	J 01	FTC1329	C 34	FTC1609	C 74	FTC2199	C 40	FTC263	C 36	FTC287	C 16	FTC3185	E 24	FTC3436	E 27

FTC3437 ....E 27
FTC3438 ....E 25
FTC3454 ....E 13
FTC3454 ....E 14
FTC3454 ....E 16

FTC3455 ....E 21
FTC3456 ....E 13
FTC3456 ....E 14
FTC3456 ....E 16
FTC346 ....C 37

FTC347 ....C 30
FTC357 ....C 13
FTC357 ....C 32
FTC358 ....C 12
FTC358 ....C 31

FTC3581 ....C 66
FTC3582 ....C 66
FTC3583 ....C 66
FTC3584 ....C 58
FTC3584 ....C 59

FTC3584 ....C 60
FTC3584 ....C 62
FTC3584 ....C 63
FTC3586 ....E 11
FTC3586 ....E 36

FTC3620 ....E 11
FTC3620 ....E 36
FTC3623 ....C 06
FTC3623 ....C 28
FTC3627 ....D 15

FTC3646 ....E 13
FTC3646 ....E 14
FTC3646 ....E 16
FTC3647 ....E 20
FTC3648 ....E 19

FTC3648 ....E 21
FTC3648 ....E 39
FTC3648 ....E 40
FTC3650 ....E 40
FTC3664 ....C 11

FTC3670 ....E 20
FTC3674 ....D 14
FTC3675 ....D 20
FTC3696 ....C 56
FTC3697 ....C 62

FTC3698 ....D 14
FTC3699 ....C 61
FTC3701 ....C 54
FTC3703 ....C 57
FTC3705 ....E 02

FTC3706 ....E 34
FTC3707 ....E 34
FTC3711 ....C 55
FTC3713 ....C 55
FTC3720 ....C 03

FTC3720 ....C 25
FTC3723 ....E 30
FTC3723E ....E 31
FTC3724 ....E 26
FTC3725 ....E 26

FTC3727 ....C 87
FTC3729 ....E 34
FTC3730 ....C 54
FTC3739 ....C 48
FTC3741 ....C 48

FTC3743 ....C 48
FTC3745 ....C 48
FTC3747 ....C 48
FTC3749 ....C 48
FTC3751 ....C 48

FTC3753 ....C 48
FTC3755 ....C 48
FTC3757 ....C 48
FTC3759 ....C 48
FTC3761 ....C 48

FTC3763 ....C 49
FTC3765 ....C 49
FTC3767 ....C 49
FTC3769 ....C 49
FTC3771 ....C 49

FTC3773 ....C 49
FTC3775 ....C 49
FTC3777 ....C 49
FTC3779 ....C 49
FTC3781 ....C 49

FTC3783 ....C 49
FTC3785 ....C 49
FTC3787 ....C 49
FTC3806 ....C 67
FTC3807 ....C 67

FTC3810 ....C 65
FTC382 ....D 15
FTC3838 ....E 04
FTC3839 ....E 04
FTC3840 ....E 04

FTC3841 ....E 04
FTC3847 ....C 65
FTC3848 ....C 47
FTC3850 ....C 65
FTC3853 ....E 12

FTC3853 ....E 38
FTC3855 ....E 12
FTC3855 ....E 38
FTC3857 ....E 12
FTC3857 ....E 38

FTC3859 ....E 12
FTC3859 ....E 38
FTC3861 ....E 12
FTC3861 ....E 38
FTC3863 ....E 12

FTC3863 ....E 38
FTC3865 ....E 12
FTC3865 ....E 38
FTC3867 ....E 12
FTC3867 ....E 38

FTC3869 ....E 09
FTC3869 ....E 32
FTC3871 ....E 09
FTC3871 ....E 32
FTC3873 ....E 09

FTC3873 ....E 32
FTC3875 ....E 09
FTC3875 ....E 32
FTC3877 ....E 09
FTC3877 ....E 32

FTC3879 ....E 09
FTC3879 ....E 32
FTC3881 ....E 09
FTC3881 ....E 32
FTC3883 ....E 09

FTC3883 ....E 32
FTC3885 ....E 09
FTC3885 ....E 32
FTC3887 ....E 09
FTC3887 ....E 32

FTC3889 ....E 09
FTC3889 ....E 32
FTC3891 ....E 09
FTC3891 ....E 32
FTC3893 ....E 09

FTC3893 ....E 32
FTC3895 ....E 09
FTC3895 ....E 32
FTC3907 ....C 86
FTC3908 ....C 88

FTC3911 ....C 44
FTC3911 ....C 45
FTC3911 ....C 46
FTC3912 ....C 44
FTC3912 ....C 45

FTC3913 ....C 46
FTC3921 ....C 52
FTC3922 ....C 52
FTC3951 ....C 63
FTC3953 ....C 63

FTC3955 ....C 63
FTC3957 ....C 63
FTC3959 ....C 63
FTC3961 ....C 63
FTC3963 ....C 63

FTC3965 ....C 63
FTC3967 ....C 63
FTC3969 ....C 63
FTC3976 ....C 86
FTC3981 ....C 84

FTC4007 ....C 13
FTC4007 ....C 32
FTC4008 ....C 14
FTC4008 ....C 34
FTC4009 ....C 13

FTC4009 ....C 32
FTC4010 ....C 14
FTC4010 ....C 34
FTC4011 ....C 13
FTC4011 ....C 32

FTC4021 ....C 54
FTC4026 ....C 21
FTC4026 ....C 41
FTC4034 ....B189
FTC4036 ....D 21

FTC4036 ....D 04
FTC4036 ....D 17
FTC4036 ....D 18
FTC4037 ....D 16
FTC4037 ....D 18

FTC4043 ....E 02
FTC4044 ....E 02
FTC4045 ....H 09
FTC4045 ....E 22
FTC4046 ....H 09

FTC4046 ....E 22
FTC4049 ....B 40
FTC4050 ....B 76
FTC4052 ....B 76
FTC4053 ....C 55

FTC4054 ....C 74
FTC4056 ....C 47
FTC4082 ....C 20
FTC4082 ....C 40
FTC4087 ....C 32

FTC4095 ....D 20
FTC4108 ....C 47
FTC4112 ....C 47
FTC4112 ....C 03
FTC4112 ....C 25

FTC4122 ....C 67
FTC4123 ....C 67
FTC4124 ....C 67
FTC4125 ....C 67
FTC4129 ....C 21

FTC4129 ....C 41
FTC4131 ....C 47
FTC4171 ....C 58
FTC4171 ....C 59
FTC4171 ....C 60

FTC4171 ....C 62
FTC4171 ....C 63
FTC4172 ....C 58
FTC4172 ....C 60
FTC4178 ....D 03

FTC4185 ....D 05
FTC4186 ....D 05
FTC4188 ....D 08
FTC4189 ....D 15
FTC419 ....C 17

FTC419 ....C 37
FTC4190 ....D 10
FTC4204 ....B 39
FTC4204 ....B 75
FTC4206 ....C 54

FTC4210 ....E 07
FTC4210 ....E 08
FTC4210 ....E 28
FTC4210 ....E 29
FTC4210 ....E 30

FTC4210 ....E 31
FTC4213 ....B 40
FTC4213 ....B 76
FTC4214 ....B 40
FTC4214 ....B 76

FTC4241 ....C 54
FTC4242 ....C 62
FTC4245 ....C 62
FTC4249 ....C 66
FTC4296 ....C 50

FTC4298 ....C 50
FTC4300 ....C 50
FTC4302 ....C 50
FTC4304 ....C 50
FTC4306 ....C 50

FTC4308 ....C 50
FTC4310 ....C 50
FTC4312 ....C 50
FTC4314 ....C 50
FTC4316 ....C 50

FTC4318 ....C 50
FTC4320 ....C 50
FTC4322 ....C 50
FTC4324 ....C 50
FTC4326 ....C 50

FTC4388 ....B161
FTC4389 ....C 72
FTC4400 ....E 05
FTC4401 ....E 05
FTC4402 ....E 04

FTC4403 ....E 03
FTC4403 ....E 04
FTC4408 ....E 05
FTC4409 ....E 05
FTC4410 ....E 04

FTC4411 ....E 04
FTC4413 ....E 06
FTC4417 ....E 06
FTC4449 ....C 54
FTC4458 ....C 66

FTC4474 ....C 74
FTC4475 ....C 74
FTC4483 ....C 55
FTC4489 ....C 23
FTC4494 ....C 23

FTC4495 ....C 01
FTC4497 ....C 01
FTC4522 ....C 54
FTC4524 ....C 61
FTC4536 ....C 66

FTC4536 ....C 18
FTC4536 ....C 38
FTC4537 ....C 67
FTC4538 ....C 61
FTC4539 ....C 61

FTC4552 ....C 70
FTC4552E ....C 70
FTC4557 ....B 40
FTC4590 ....C 66
FTC4596 ....C 66

FTC4622 ....C 66
FTC4662 ....B160
FTC4735 ....D 16
FTC4735 ....D 18
FTC4785 ....E 23

FTC4785 ....E 24
FTC4785 ....E 39
FTC4785 ....E 40
FTC4818 ....D 08
FTC4828 ....B160

FTC4829 ....C 52
FTC4830 ....C 52
FTC4838 ....H 14
FTC4839 ....H 14
FTC4846 ....D 10

FTC4847 ....D 15
FTC4848 ....D 05
FTC4849 ....D 05
FTC4850 ....D 08
FTC4866 ....D 05

FTC4889 ....H 12
FTC4889 ....H 13
FTC490 ....E 25
FTC4900 ....C 52
FTC4939 ....D 11

FTC4939 ....D 12
FTC4940 ....D 11
FTC4941 ....D 12
FTC4942 ....D 12
FTC4955 ....D 15

FTC4962 ....D 08
FTC4969 ....C 70
FTC4969E ....C 70
FTC4976 ....C 65
FTC4977 ....C 65

FTC4978 ....C 65
FTC4980 ....C 64
FTC4982 ....C 64
FTC4989 ....C 61
FTC4989 ....C 63

FTC4990 ....C 61
FTC4990 ....C 63
FTC4991 ....C 54
FTC4991 ....C 61
FTC4991 ....C 63

FTC4992 ....C 58
FTC4992 ....C 59
FTC4992 ....C 60
FTC4992 ....C 62
FTC4992 ....C 63

FTC4993 ....D 06
FTC4994 ....D 06
FTC5018 ....C 60
FTC5018 ....C 62
FTC5018 ....C 63

FTC5019 ....C 58
FTC5019 ....C 59
FTC5019 ....C 60
FTC5036 ....C 59
FTC5037 ....C 59

FTC5038 ....C 58
FTC5041 ....C 60
FTC5042 ....C 61
FTC5043 ....C 61
FTC5044 ....C 56

FTC5046 ....C 56
FTC5048 ....C 56
FTC5070 ....C 62
FTC5070 ....C 63
FTC5072 ....C 44

FTC5072 ....C 45
FTC5072 ....C 46
FTC5087 ....D 08
FTC5089 ....D 08
FTC5090 ....C 84

FTC5100 ....C 58
FTC5100 ....C 59
FTC5101 ....C 60
FTC5102 ....C 62
FTC5102 ....C 63

FTC5105 ....E 16
FTC5119 ....C 03
FTC5119 ....C 25
FTC5207 ....D 07
FTC5208 ....D 04

FTC5218 ....C 44
FTC5218 ....C 02
FTC5218 ....C 45
FTC5241 ....E 24
FTC5241 ....E 40

FTC5246 ....C 67
FTC5258 ....E 10
FTC5258 ....E 34
FTC5258 ....E 35
FTC5268 ....E 21

FTC5268 ....E 40
FTC5287 ....C 57
FTC5288 ....C 61
FTC5288 ....C 63
FTC5296 ....E 20

FTC5297 ....E 20
FTC5298 ....E 20
FTC5299 ....E 20
FTC5303 ....C 56
FTC5303 ....C 08

FTC5303 ....C 29
FTC5317 ....E 10
FTC5317 ....E 34
FTC5317 ....E 35
FTC5322 ....E 10

FTC5410 ....E 26
FTC5411 ....E 26
FTC5413 ....E 10
FTC5413 ....E 34
FTC5414 ....E 14

FTC5414 ....E 16
FTC5427 ....D 03
FTC56 ....E 19
FTC56 ....E 21
FTC566 ....C 86

FTC566 ....C 88
FTC57 ....E 19
FTC575 ....B 39
FTC578 ....C 87
FTC578 ....C 89

FTC613 ....E 25
FTC651 ....B161
FTC664 ....E 03
FTC665 ....E 03
FTC726 ....D 13

FTC728 ....D 13
FTC730 ....D 13
FTC732 ....D 13
FTC734 ....D 13
FTC736 ....D 13

FTC738 ....D 13
FTC740 ....D 13
FTC742 ....D 13
FTC744 ....D 13
FTC746 ....D 13

FTC748 ....D 13
FTC750 ....D 13
FTC752 ....D 13
FTC754 ....D 13
FTC756 ....D 13

FTC758	D 13	FX110047L	N 07	HLD102950LOY	R 47	HMA110540SUC	R 44	JPC4651LUM	P 72	LGH101060L	B173	LSF100040L	B 47	MKC10121	M 98
FTC760	D 13	FX112041L	N 05	HLD102950SUC	R 47	HMA110560LOY	R 41	JPC5641LNF	P 71	LGH101070L	B173	LSO100000	B 82	MKC101390	M 99
FTC762	D 13	FXB10002	P 10	HLD103090LOY	R 48	HMA110560SMK	R 52	JPC5641LUN	P 71	LGJ100880	B 16	LSO100000	B115	MKC101720	M 98
FTC764	D 13	FYF100150	N 80	HLD103090SUC	R 48	HMA110560SUC	R 41	JPC5949L	P 60	LGJ100880	B 54	LSP10037	B167	MKC103290	M 99
FTC766	D 13	FYF100150	N 85	HLD103140SMK	R 48	HMA110570SMK	R 41	JPD10016	B184	LGJ100900	B 88	LUB10009	B173	MKG10004L	B180
FTC768	D 13	FYH100280	P 10	HLE108820LOY	R 27	HMA110630SMK	R 41	JPD10026	B184	LGJ100900	B123	LUC100120	B173	MKW10011L	B179
FTC770	D 13	FYP10003	P 10	HLE108820SUC	R 27	HN2005L	T 22	JRC7549	M159	LGJ100900	B125	LUC100120	B175	MRC206	L 11
FTC772	D 13	GG106201L	P 31	HLE108830LOY	R 30	HPA106060LOY	R 35	JRC7549	M 17	LGK10005L	B172	LUD100040L	B174	MRC218	L 11
FTC774	D 13	GG106251L	N111	HLE108830LOY	R 41	HPA106060LOY	R 42	JRC7549	M 18	LGL10015	B173	LUF100430	B 43	MRC4283LNF	R 01
FTC776	D 13	GG106301L	N 49	HLE108830SUC	R 30	HPA106060SMK	R 52	JRC7549	M 19	LGL100390L	B173	LUL10007L	B164	MRC4283LOY	R 01
FTC781	E 11	GG106301L	R 38	HLE108830SUC	R 41	HPA106060SUC	R 35	JRC7549	M 21	LGR10002L	B173	LVB10045	B173	MRC4283LOY	P 79
FTC781	E 36	GG106301L	R 40	HLE108970LOY	R 44	HPA106060SUC	R 42	JRC7549	M 16	LGR100100L	B173	LVB101000	B173	MRC4283LOY	R 25
FTC790	E 03	GG108251L	K 06	HLE108970SUC	R 44	HPA106070LOY	R 37	JS660L	J 07	LHB10028L	B175	LVF10005	B167	MRC4283LUL	P 02
FTC791	E 03	GG108251L	K 07	HLE108990LOY	R 41	HPA106070LOY	R 42	JZV1085L	B180	LHG10036	B165	LVG10005	B171	MRC4283SUC	R 01
FTC813	B108	GG108251L	K 08	HLE108990SUC	R 41	HPA106070SMK	R 52	JZX1039L	B 97	LHH100660	B 45	LVL10001	B164	MRC4283SUC	P 79
FTC814	B108	GG108251L	K 09	HLE109000SMK	R 41	HPA106070SUC	R 37	JZX1181L	B 97	LHN10016	B169	LVP10005	B174	MRC4283SUC	R 25
FTC815	C 93	GG108301	G 06	HLE109010LOY	R 28	HPA106070SUC	R 42	JZX1303L	B 97	LHP10011L	B169	LVP10006	B174	MRC5527	R 48
FTC815	C 42	GG108301	G 08	HLE109010SUC	R 28	HPA106170LOY	R 34	JZX1394L	B 97	LHP10016L	B169	LVP100170	B174	MRC5529	N 62
FTC823	E 06	GG108351L	H 21	HLE109020LOY	R 28	HPA106170SUC	R 34	KSG10001	B189	LHP100860	B 50	LVP100320L	B174	MRC7133	M 18
FTC825	E 06	GJ105201	P 23	HLE109020SUC	R 28	HPA106180LOY	R 34	KSG10002	B189	LHR10025L	B169	LVQ10040	B181	MRC7133	M 16
FTC840	E 19	GL106251	L 33	HLE109040SMK	R 41	HPA106180SUC	R 34	KSP10093	B189	LJC10008L	B173	LYC10004L	B174	MRC7133	M 20
FTC859	E 24	HAD128480LOY	R 22	HLE109530LOY	R 44	HPA106280SMK	R 42	KSP10094	B189	LJC100270	B 19	LYC10005L	B174	MRC8388	J 18
FTC860	E 24	HAD128480SUC	R 22	HLE109530SUC	R 44	HPA106290SMK	R 42	KSP101360	B189	LJC100270	B 57	LYG10012L	B189	MTC5830	N115
FTC860	E 40	HAD128490LOY	R 22	HLE109580LOY	R 44	HRA102720LOY	R 45	KYG10050	B189	LJQ100560	B173	LYG10035L	B171	MTC5831	N115
FTC861	E 19	HAD128490SUC	R 22	HLE109580SUC	R 44	HRA102750LOY	R 43	LBB10468NL	B163	LJR10159L	B168	LYG10040L	B169	MTC5955	N 52
FTC861	E 21	HAD128520LOY	R 22	HLF104950LOY	R 33	HRA102750SUC	R 43	LCG10003	B169	LJR10161L	B168	LYG10046	B165	MTC5956	N 52
FTC890	C 85	HAD128520SUC	R 22	HLF104950LOY	R 42	HRA102780SMK	R 43	LCK10001L	B164	LJR10163L	B168	LYP10028L	B178	MTC6264	N 77
FTC890	C 06	HAD128530LOY	R 22	HLF104950SUC	R 33	HTL100840	R 43	LDF10331L	B172	LJR10165L	B168	LYQ100050	C 54	MTC6538	N112
FTC890	C 07	HAD128530SUC	R 22	HLF104950SUC	R 42	HTL100840	R 45	LDF105080L	B172	LJR10169L	B168	LYQ100090	C 47	MTC6538	N115
FTC890	C 28	HAD129320LOY	R 15	HLF104960LOY	R 36	HTL100850	R 43	LDF108090	B125	LJR102640	B168	LYQ100090	C 03	MTC7029	N 76
FTC890	D 06	HAD129320SUC	R 15	HLF104960LOY	R 42	HTL100850	R 45	LDM10003	B174	LJR102660	B168	LYQ100090	C 25	MTC7032	N 78
FTC902	H 14	HAD129330LOY	R 15	HLF104960SUC	R 36	HWK100550	R 40	LDN10028L	B173	LJR102670	B168	LYR100500L	B172	MTC7188	N 76
FTC916	C 14	HAD129330SUC	R 15	HLF104960SUC	R 42	HWK100560	R 40	LDR10135L	B174	LKB10473L	B177	LYU10011L	B164	MTC7188	N 80
FTC916	C 34	HAD129340LOY	R 15	HLF104980LOY	R 34	HXE100640LOY	R 43	LDR10140	B174	LKB105090	B177	LYU100300L	B164	MTC7290	N 51
FTC917	C 08	HAD129340SUC	R 15	HLF104980SUC	R 34	HXE100640SMK	R 43	LDR10144	B174	LKB105620	B177	LZB100270L	B173	MTC7290	N 52
FTC917	C 29	HAD129350LOY	R 15	HLF104990LOY	R 34	HXE100640SUC	R 43	LDR10152	B175	LKB107080	B177	LZN10001L	B177	MTC7451	M 18
FTC918	C 03	HAD129350SUC	R 15	HLF104990SUC	R 34	HXE100650LOY	R 45	LDR10159L	B174	LKC10060L	B178	LZX1505	B 97	MTC7451	M 16
FTC918	C 25	HAD129460LOY	R 19	HLF105070SMK	R 42	HYG10014	R 08	LDR102720L	B174	LKC10066	B178	LZX1600L	B 97	MTC7451	M 20
FTC942	E 24	HAD129470LOY	R 20	HLF105080SMK	R 42	HYG10014	L 47	LDR102840	B174	LKG10002	B178	LZX1988L	B 97	MTC7586	P 75
FTC942	E 40	HAD129480SMK	R 23	HLJ101150LOY	R 45	ICV100000	C 28	LEF10078	B165	LKJ10027	B177	LZX1989L	B 97	MTC7592	N 76
FTC943	E 14	HAD129490SMK	R 23	HLJ101170LOY	R 43	ICW100000	D 23	LEG10003	B165	LKJ10030	B177	LZX2107L	B 97	MTC7593	N 76
FTC943	E 16	HAH103190LOY	R 23	HLJ101170SMK	R 43	IEE100050	D 15	LEG10004	B165	LKU10034L	B177	MEK10002L	B179	MTC8500	L 46
FTP8015	L 02	HAH103190SUC	R 23	HLJ101170SUC	R 43	IGM100010	C 54	LEJ10002L	B165	LKX100440	B 94	MHB10085	B180	MTC8500	L 53
FTP8030	L 07	HBA105370LOY	R 23	HMA110220LOY	R 29	JFE100630	P 12	LFH10005L	B166	LLH10137	B174	MHB101140	B180	MTC8500	L 55
FX108041L	B 51	HBA105370SUC	R 23	HMA110220SUC	R 29	JFE100640	L 48	LFL10238	B166	LLH10138	B174	MHB101620	B180	MTC8906	N 63
FX108047	B 73	HBA105380SMK	R 52	HMA110230LOY	R 31	JMG100370	L 45	LFL10239	B166	LLO100000	B 59	MHH10020L	B179	MTC8907	N 63
FX110041L	B 24	HBA105480LOY	R 15	HMA110230LOY	R 41	JPC2616JUL	P 01	LFL102860L	B166	LPF10055	B171	MHK10004L	B178	MTC9066	N 50
FX110041L	B 31	HBA105480SUC	R 15	HMA110230SMK	R 52	JPC2616LOY	P 63	LFL102870L	B166	LPM10003	B171	MHN10004L	B179	MTC9131	N 52
FX110041L	B 62	HCA105330LOY	R 23	HMA110230SUC	R 31	JPC2616SUC	P 01	LFP10076L	B166	LPS10001L	B171	MHU10068	B179	MTC9145	N 76
FX110041L	C 68	HCA105330SUC	R 23	HMA110230SUC	R 41	JPC2616SUC	P 63	LFT10013L	B165	LQC10010	B174	MHU10069	B179	MTC9163	N 55
FX110041L	C 90	HCA105340SMK	R 52	HMA110420LOY	R 28	JPC2617L	P 01	LFT10014L	B165	LQC100190L	B174	MHX10002L	B180	MTC9164	N 55
FX110041L	K 13	HCA105440LOY	R 15	HMA110420SUC	R 28	JPC2617L	P 63	LGC10219L	B175	LQM10026	B167	MJN10026L	B179	MTC9177	N 51
FX110041L	K 14	HCA105440SUC	R 15	HMA110430LOY	R 28	JPC2618L	P 01	LGH10020L	B173	LQN10033L	B167	MJW121	N 19	MTC9203	N 80
FX110041L	N 05	HDA102360LOY	R 24	HMA110430SUC	R 28	JPC2619L	P 01	LGH10049L	B173	LSB10076	B167	MJW122	N 19	MTC9204	N 80
FX110041L	N 10	HDA102360SUC	R 24	HMA110540LOY	R 44	JPC2619L	P 63	LGH101000L	B173	LSB102610	B 47	MJY10029L	B179	MTC9273	N 52

MTC9274 N 52
MTC9289 N 76
MTC9289 N 80
MTC9289 N 83
MTC9290 N 76
MTC9290 N 80
MTC9290 N 83
MTC9428 R 38
MTC9428 R 40
MTC9429 R 38
MTC9429 R 40
MTC9505 N 78
MTC9623 N 76
MTC9625 N 80
MTC9630 N 50
MTC9631 N 50
MTC9632 N 50
MTC9931 R 26
MUC1218 N 51
MUC1219 N 51
MUC1514 R 26
MUC1515 R 26
MUC1609 N112
MUC1609 N115
MUC1612JUL P 54
MUC1612SUC P 54
MUC1613JUL P 54
MUC1613SUC P 54
MUC1679 L 36
MUC1792 N112
MUC1792 N114
MUC1792 N116
MUC1792 N119
MUC1818 N 26
MUC1819 N 26
MUC1820 N 26
MUC1820 P 09
MUC1820 P 10
MUC1822 N 26
MUC1822 N 76
MUC1822 N 84
MUC2038 N 60
MUC2039 N 60
MUC2322 N 48
MUC2325 N 48
MUC2338PMA R 06
MUC2564 R 54
MUC3186 P 44
MUC3186 P 51
MUC3186 N 91
MUC3187 N 91
MUC3187 M152
MUC3664 P 75
MUC4177 L 11
MUC4358 L 42

MUC6319 L 11
MUC6320 L 11
MUC6324 L 11
MUC6382 L 11
MUC6383 L 11
MUC6384 H 24
MUC6426 P 50
MUC6426 P 10
MUC6513 M 18
MUC6513 M 16
MUC6513 M 20
MUC6514 M 18
MUC6514 M 16
MUC6514 M 20
MUC6521 M 33
MUC6521 N 79
MUC6521 N 82
MUC6521 M 41
MUC6523 P 08
MUC6523 P 09
MUC6529 L 21
MUC6542 L 40
MUC6543 L 40
MUC6643 L 40
MUC6806 P 75
MUC8478 M150
MUC8849 J 09
MUC8975 N 44
MUC8989 P 16
MUC9211 N 34
MUC9281 P 75
MWC1722 M 35
MWC1847 F 04
MWC2698 K 01
MWC2941 N 48
MWC3086 L 36
MWC3088 L 36
MWC3136 P 62
MWC3724RUR N 99
MWC3724RUS N 99
MWC3724RUZ N 99
MWC3725RUR N 99
MWC3725RUS N 99
MWC3725RUZ N 99
MWC3726RUR N 99
MWC3726RUS N 99
MWC3726RUZ N 99
MWC3727RUR N 99
MWC3727RUS N 99
MWC3727RUZ N 99
MWC3728RUR N 99
MWC3728RUS N 99
MWC3728RUZ N 99
MWC3988LNF R 53
MWC3988LUN R 53

MWC4007JUN P 77
MWC4008JUN P 81
MWC4008SUD P 81
MWC4009JUN P 81
MWC4009SUD P 81
MWC4010JUN P 80
MWC4011JUN P 76
MWC4054SUC P 13
MWC4064LNF P 04
MWC4064LUN P 04
MWC4065LNF P 04
MWC4065LUN P 04
MWC4072JUL P 01
MWC4072SUC P 01
MWC4074JUL P 01
MWC4074SUC P 01
MWC4078LNF P 04
MWC4078LUN P 04
MWC4092PMA P 01
MWC4093PMA P 01
MWC4098JUL P 06
MWC4098SUC P 06
MWC4246 N 34
MWC4340 L 36
MWC4340 L 37
MWC4560 N 41
MWC4560 N 54
MWC4816 R 38
MWC4816 R 40
MWC4840 N 25
MWC4841 N 25
MWC5126 N 79
MWC5144 N 81
MWC5145 N 81
MWC5148 N 81
MWC5149 N 81
MWC5150 N 80
MWC5151 N 80
MWC6060 R 38
MWC6060 R 40
MWC6061 R 38
MWC6061 R 40
MWC6062P P 82
MWC6193 N 69
MWC6432 N 82
MWC6433 N 82
MWC6504JUL P 01
MWC6504SUC P 01
MWC6512LNF N 60
MWC6512LNF N 63
MWC6512LUN N 60
MWC6512LUN N 63
MWC6513 P 29
MWC6513 P 31
MWC6513 P 34

MWC6518 P 29
MWC6588 N 50
MWC6589 N 50
MWC6621 N 77
MWC6702 N 41
MWC6704 N 41
MWC6705 N 41
MWC6723 N 83
MWC6724 N 83
MWC6727 N 83
MWC6728 N 84
MWC6729 N 83
MWC6732 N 84
MWC6733 N 86
MWC6734 N 83
MWC6735 N 83
MWC6754 N 27
MWC6755 N 27
MWC6767 N 34
MWC6804 N 29
MWC6805 N 29
MWC6846JUL R 27
MWC6846SUC R 27
MWC6847JUL R 30
MWC6847SUC R 30
MWC6848JUL R 33
MWC6848SUC R 33
MWC6849JUL R 36
MWC6849SUC R 36
MWC6850JUL R 29
MWC6850SUC R 29
MWC6851JUL R 31
MWC6851SUC R 31
MWC6856JUL R 35
MWC6856SUC R 35
MWC6857JUL R 37
MWC6857SUC R 37
MWC7312JUL R 02
MWC7313JUL R 03
MWC7336 N 67
MWC7337 N 67
MWC7438 N 56
MWC7439 N 56
MWC7484 N 22
MWC7485 N 22
MWC7488 N 24
MWC7489 N 24
MWC7502 N 23
MWC7503 N 23
MWC7562 N 32
MWC7564 N 24
MWC7565 N 24
MWC7584 N 32
MWC7600 N 24
MWC7677 N 77

MWC7677 N 80
MWC7677 N 85
MWC7754JUL P 24
MWC7754SUC P 24
MWC7755JUL P 24
MWC7755SUC P 24
MWC7756JUL P 28
MWC7756LOY P 31
MWC7756SUC P 28
MWC7756SUC P 31
MWC7757JUL P 28
MWC7757LOY P 31
MWC7757SUC P 28
MWC7757SUC P 31
MWC7785 P 05
MWC7796 P 03
MWC7821 P 04
MWC7826JUL P 54
MWC7826LOY P 54
MWC7826SUC P 54
MWC7827JUL P 54
MWC7827LOY P 54
MWC7827SUC P 54
MWC7842JUL P 45
MWC7842SUC P 45
MWC7843JUL P 45
MWC7843SUC P 45
MWC7844JUL P 44
MWC7844JUL P 53
MWC7844SUC P 44
MWC7844SUC P 53
MWC7846JUL P 47
MWC7846SUC P 47
MWC7847JUL P 47
MWC7847SUC P 47
MWC7888JUL P 35
MWC7888SUC P 35
MWC7889JUL P 35
MWC7889SUC P 35
MWC7892JUL P 35
MWC7892SUC P 35
MWC7893JUL P 35
MWC7893SUC P 35
MWC7894 N 67
MWC7895 N 66
MWC7896JUL P 29
MWC7896LOY P 31
MWC7896LOY P 34
MWC7896SUC P 29
MWC7896SUC P 31
MWC7896SUC P 34
MWC7897JUL P 29
MWC7897LOY P 31
MWC7897LOY P 34
MWC7897SUC P 29

MWC7897SUC P 31
MWC7897SUC P 34
MWC7898JUL P 36
MWC7898JUL P 41
MWC7898LOY P 36
MWC7898LOY P 38
MWC7898LOY P 41
MWC7898SUC P 36
MWC7898SUC P 38
MWC7898SUC P 41
MWC8004PUB N 12
MWC8016 P 39
MWC8016 P 42
MWC8017 P 39
MWC8017 P 42
MWC8021LNF N 60
MWC8021LNF N 63
MWC8021LUN N 60
MWC8021LUN N 63
MWC8022JUL P 13
MWC8022SUC P 13
MWC8028PMA P 14
MWC8029PMA P 14
MWC8064LUN R 56
MWC8065LUN R 56
MWC8066LNF R 58
MWC8066LUN R 58
MWC8070 N 29
MWC8071 N 29
MWC8084 N 29
MWC8085 N 29
MWC8104 N 90
MWC8105 N 90
MWC8186 N 40
MWC8187 N 40
MWC8299 R 46
MWC8300 R 46
MWC8304 N 66
MWC8326JUL F 04
MWC8326SUC F 04
MWC8332 P 04
MWC8332 L 29
MWC8334 P 04
MWC8336 P 03
MWC8340 L 29
MWC8360LNF P 14
MWC8360LUN P 14
MWC8366 L 34
MWC8394 N 22
MWC8395 N 22
MWC8436 N 47
MWC8437 N 47
MWC8438 N 51
MWC8439 N 51
MWC8526 N112

MWC8526 N116
MWC8527 N112
MWC8527 N116
MWC8680 N 47
MWC8744 P 06
MWC8745 P 06
MWC8746 P 03
MWC8755JUL P 13
MWC8755SUC P 13
MWC8756 P 04
MWC8759 N 40
MWC8760 N 40
MWC8800JUL R 47
MWC8800SUC R 47
MWC8831JUN R 05
MWC8894 R 08
MWC8920JUL P 43
MWC8920JUL P 51
MWC8920SUC P 43
MWC8920SUC P 51
MWC8921JUL P 43
MWC8921JUL P 51
MWC8921SUC P 43
MWC8921SUC P 51
MWC9060 N 24
MWC9061 N 24
MWC9071NCM N121
MWC9084 N 50
MWC9085 N 50
MWC9107JUL P 05
MWC9107SUC P 05
MWC9108LNF P 05
MWC9108LUN P 05
MWC9109LNF P 05
MWC9109LUN P 05
MWC9110PMA P 14
MWC9134 P 36
MWC9134 P 38
MWC9134 P 28
MWC9134 P 31
MWC9134 P 34
MWC9136 P 36
MWC9136 P 38
MWC9136 P 39
MWC9136 P 48
MWC9136 P 55
MWC9136 R 01
MWC9136 P 28
MWC9136 P 60
MWC9136 P 79
MWC9136 R 25
MWC9136 N 43
MWC9136 P 31
MWC9136 P 34
MWC9136 P 41

MWC9139 P 14
MWC9140LNF M 73
MWC9140LUN M 73
MWC9141LNF M 73
MWC9141LUN M 73
MWC9148 P 06
MWC9149 P 06
MWC9160JUL P 60
MWC9160SUC P 60
MWC9161JUL P 60
MWC9161SUC P 60
MWC9168 N 83
MWC9169 N 83
MWC9171 N 86
MWC9172 N 84
MWC9173 N 84
MWC9174 N 83
MWC9174 N 84
MWC9174 N 85
MWC9176 N 83
MWC9183 P 03
MWC9184 P 03
MWC9189 P 03
MWC9190 P 01
MWC9191 P 04
MWC9198JUL P 26
MWC9198SUC P 26
MWC9199JUL P 26
MWC9199SUC P 26
MWC9202 P 06
MWC9203 P 06
MWC9267LUP P 67
MWC9275JUL P 39
MWC9275JUL P 41
MWC9275LOY P 41
MWC9275SUC P 39
MWC9275SUC P 41
MWC9276JUL P 39
MWC9276JUL P 41
MWC9276LOY P 41
MWC9276SUC P 39
MWC9276SUC P 41
MWC9279JUL P 25
MWC9279SUC P 25
MWC9280JUL P 25
MWC9280SUC P 25
MWC9288LNF M 74
MWC9288LUN M 74
MWC9289LNF M 74
MWC9289LUN M 74
MWC9313 S 01
MWC9326JUL P 14
MWC9326SUC P 14
MWC9602 N 48
MWC9603 N 51

MWC9701 R 08
MWC9704JUN R 06
MWC9705 R 06
MWC9705 R 08
MWC9705 R 10
MWC9706 R 10
MWC9707 R 08
MWC9719JUL R 06
MWC9719LOY R 06
MWC9719LOY R 16
MWC9719LOY R 20
MWC9719SUC R 06
MWC9719SUC R 16
MWC9725LNF R 08
MWC9725LUN R 08
MWC9919 N 33
MXC1028 N 22
MXC1029 N 22
MXC1031 N 41
MXC1032 N 41
MXC1047 M 10
MXC1047 M 07
MXC1048 M 10
MXC1048 M 07
MXC1054 N 28
MXC1087 N 54
MXC1089 N 30
MXC1094 N 53
MXC1095 N 53
MXC1097 N 28
MXC1166 R 38
MXC1167 R 38
MXC1172MUK N106
MXC1172PME N106
MXC1198 R 46
MXC1210 N 58
MXC1211 N 58
MXC1245 N 86
MXC1250 P 04
MXC1254 N 80
MXC1255 N 80
MXC1256 N 86
MXC1270JUL P 60
MXC1270SUC P 60
MXC1271JUL P 60
MXC1271SUC P 60
MXC1274JUL P 50
MXC1274JUL P 58
MXC1274LOY P 50
MXC1274LOY P 58
MXC1274SUC P 50
MXC1274SUC P 58
MXC1275JUL P 50
MXC1275JUL P 58
MXC1275LOY P 50
MXC1275LOY P 58
MXC1275SUC P 50
MXC1275SUC P 58
MXC1280 N 32
MXC1282 L 29
MXC1283 L 29
MXC1292JUL P 61
MXC1292LOY P 61
MXC1292SUC P 61
MXC1293JUL P 61
MXC1293LOY P 61
MXC1293SUC P 61
MXC1299 P 71
MXC1361 N 51
MXC1420 N 25
MXC1421 N 25
MXC1552 N 42
MXC1592JUL R 32
MXC1592SUC R 32
MXC1597JUL R 32
MXC1597SUC R 32
MXC1638 N 41
MXC1638 N 51
MXC1638 N 47
MXC1848 B133
MXC1848 B134
MXC1867 N 51
MXC1867 N 48
MXC1944 N 20
MXC1965 N 50
MXC1966 N 46
MXC1967 N 46
MXC2006JUL P 61
MXC2006LOY P 61
MXC2006SUC P 61
MXC2007JUL P 61
MXC2007LOY P 61
MXC2007SUC P 61
MXC2008 N 84
MXC2012LUP P 71
MXC2015 M 12
MXC2028 N 84
MXC2049LNF N 60
MXC2049LNF N 63
MXC2049LUN N 60
MXC2049LUN N 63
MXC2094 N 90
MXC2162 N 50
MXC2163 N 50
MXC2164 N 47
MXC2165 N 47
MXC2166 N 51
MXC2166 N 46
MXC2167 N 51
MXC2167 N 46
MXC2184LNF P 14
MXC2184LUN P 14
MXC2187LNF P 16
MXC2187LNF P 14
MXC2187LUN P 14
MXC2476LNF P 04
MXC2476LUN P 04
MXC2519 N 86
MXC2630 N 85
MXC2631 N 85
MXC2656 P 04
MXC2657 P 04
MXC2658 P 04
MXC2676PUB N 30
MXC2677PUB N 30
MXC2685 P 74
MXC2686 P 74
MXC2687JUL N 54
MXC2687LOY N 54
MXC2687SUC N 54
MXC2707 R 54
MXC2708LNF R 53
MXC2708LUN R 53
MXC2722LNF N 77
MXC2722LUN N 77
MXC2723LNF N 77
MXC2723LUN N 77
MXC2858JUL F 04
MXC2858SUC F 04
MXC2931 N 44
MXC3073LNF R 46
MXC3073LUL R 46
MXC3112 N 89
MXC3115 N 89
MXC3117 N 89
MXC3120 N 28
MXC3120 N 89
MXC3128LUN P 83
MXC3130LNF R 11
MXC3130LNF R 18
MXC3130LUN P 44
MXC3130LUN R 11
MXC3130PUB P 83
MXC3130PUB N 89
MXC3132 N 85
MXC3134 N 41
MXC3175LNF N 77
MXC3175LNF N 80
MXC3175LNF N 83
MXC3175LUN N 77
MXC3175LUN N 80
MXC3175LUN N 83
MXC3176 G 14
MXC3177L N 31
MXC3178 N 76
MXC3178 N 80
MXC3187 R 38
MXC3187 R 40
MXC3270 L 21
MXC3271 P 71
MXC3413 N 37
MXC3414 N 37
MXC3415 N 39
MXC3415 T 47
MXC3415 N 37
MXC3415 N 38
MXC3810 N 52
MXC3811 N 52
MXC3818 N 80
MXC4605 N 80
MXC4652 N 40
MXC4653 N 40
MXC4687 P 57
MXC4688 P 48
MXC4688 P 55
MXC4688 P 60
MXC4856 N 20
MXC4918 N 51
MXC4931 L 29
MXC4932 L 29
MXC5032 L 45
MXC5063 L 45
MXC5267 N 51
MXC5267 N 47
MXC5278 N 32
MXC5288 N 53
MXC5291PMA N112
MXC5291PMA N116
MXC5292PMA N112
MXC5292PMA N116
MXC5293 N 34
MXC5294 N 34
MXC5321LNF D 17
MXC5321LNF D 19
MXC5321LUN D 17
MXC5322LNF C 21
MXC5322LUN C 21
MXC5322LUN C 41
MXC5323 N 42
MXC5333 H 20
MXC5333 H 22
MXC5333 H 23
MXC5359 N 41
MXC5366 P 02
MXC5367 P 02
MXC5368 P 02
MXC5369 P 02
MXC5451JUL P 49
MXC5451JUL P 61
MXC5451LOY P 49
MXC5451LOY P 56
MXC5451LOY P 61
MXC5451LOY P 63
MXC5451SUC P 49
MXC5451SUC P 56
MXC5451SUC P 61
MXC5451SUC P 63
MXC5554 L 21
MXC5555 L 21
MXC5570JUL N 55
MXC5570LOY N 55
MXC5570SUC N 55
MXC5571JUL N 55
MXC5571LOY N 55
MXC5571SUC N 55
MXC5581 N 67
MXC5597 N 41
MXC5603 N 20
MXC5616LOY P 83
MXC5616SUD P 83
MXC5634JUL P 48
MXC5634JUL P 55
MXC5634LOY P 48
MXC5634LOY P 55
MXC5634SUC P 48
MXC5634SUC P 55
MXC5635JUL P 48
MXC5635JUL P 55
MXC5635LOY P 48
MXC5635LOY P 55
MXC5635SUC P 48
MXC5635SUC P 55
MXC5649PUB N 42
MXC5683 L 21
MXC5705 L 21
MXC5706 L 46
MXC5707 L 46
MXC5713 P 19
MXC5713 P 21
MXC5716 L 29
MXC5734RUR N100
MXC5734RUS N100
MXC5734RUZ N100
MXC5735RUR N100
MXC5735RUS N100
MXC5735RUZ N100
MXC5736RUR N 99
MXC5736RUS N 99
MXC5736RUZ N 99
MXC5752 N 31
MXC5766JUL P 48
MXC5766JUL P 57
MXC5766JUL P 60
MXC5766LOY P 48
MXC5766LOY P 60
MXC5766LOY P 63
MXC5766SUC P 48
MXC5766SUC P 57
MXC5766SUC P 60
MXC5766SUC P 63
MXC5768 N 50
MXC5769 N 50
MXC5770 N 47
MXC5771 N 47
MXC5886 N 89
MXC5887 N 89
MXC5888 N 89
MXC5891 N 47
MXC6015 L 46
MXC6018 L 45
MXC6019 L 45
MXC6021 L 45
MXC6022 L 45
MXC6025 R 46
MXC6026 R 46
MXC6037 P 71
MXC6094 N110
MXC6097LUP P 68
MXC6098LUP P 68
MXC6135 N 77
MXC6136 N 77
MXC6146 L 45
MXC6147 L 45
MXC6161 N 39
MXC6185 P 05
MXC6269 L 40
MXC6269 L 41
MXC6269 L 53
MXC6270 L 40
MXC6273 N 40
MXC6275 L 40
MXC6275 L 41
MXC6275 L 46
MXC6275 L 53
MXC6275 L 54
MXC6287JUL P 22
MXC6287LOY P 22
MXC6287LUN P 22
MXC6287SUC P 22
MXC6292LNF P 18
MXC6292LNF S 01
MXC6292LUN P 18
MXC6295 N 56
MXC6310LNF R 09
MXC6310LUN R 09
MXC6311LNF R 09
MXC6311LUN R 09
MXC6324 N 70
MXC6362JUL P 57
MXC6362SUC P 57
MXC6363JUL P 57
MXC6363SUC P 57
MXC6364JUL P 57
MXC6364SUC P 57
MXC6365JUL P 57
MXC6365SUC P 57
MXC6366 N 33
MXC6372PMD N 89
MXC6373PMD N 89
MXC6374PMD N 89
MXC6383PUB N 42
MXC6384PUB N 42
MXC6386JUL P 44
MXC6386LOY P 44
MXC6386SUC P 44
MXC6387JUL P 44
MXC6387LOY P 44
MXC6387SUC P 44
MXC6432LNF R 51
MXC6432LUN R 51
MXC6433LNF R 51
MXC6433LUN R 51
MXC6434LNF R 51
MXC6434LUN R 51
MXC6435LNF R 51
MXC6435LUN R 51
MXC6441 R 46
MXC6442 R 46
MXC6458 S 01
MXC6468 N 19
MXC6509 N111
MXC6510 N111
MXC6548 N 71
MXC6548 N 72
MXC6548 F 06
MXC6549 N 71
MXC6568 N 40
MXC6569LNF P 18
MXC6569LNF S 01
MXC6569LUN P 18
MXC6581 N 83
MXC6582 S 01
MXC6651PUB N111
MXC6676 L 21
MXC6677 N 53
MXC6682 N 66
MXC6737 N 67
MXC6738JUL P 58
MXC6738LOY P 58
MXC6738SUC P 58
MXC6748RUR N100
MXC6748RUS N100
MXC6748RUZ N100
MXC6812 N 57
MXC6813 N 57
MXC6834 P 71
MXC6877 N 67
MXC6930 P 14
MXC6939LUP P 66
MXC6940LUP P 66
MXC6943JUL P 39
MXC6943SUC P 39
MXC7019 N 24
MXC7020 N 24
MXC7250LMP N 87
MXC7252 L 40
MXC7254 L 40
MXC7256 L 41
MXC7257 L 41
MXC7258 L 41
MXC7260 L 42
MXC7261 L 42
MXC7264 R 56
MXC7269 L 42
MXC7270LNF R 09
MXC7270LUN R 09
MXC7271LNF R 09
MXC7271LUN R 09
MXC7272 N 59
MXC7273 N 59
MXC7274LNF P 04
MXC7274LUN P 04
MXC7275LNF P 04
MXC7275LUN P 04
MXC7281JUN P 76
MXC7281SUD P 76
MXC7335 N122
MXC7336 N 78
MXC7337 N 78
MXC7345 P 04
MXC7353 R 38
MXC7355 P 62
MXC7356 P 62
MXC7387LNF H 20
MXC7387LNF H 22
MXC7387LNF H 23
MXC7387LUN H 20
MXC7420 M 98
MXC7420 M 99
MXC7421 P 62
MXC7421 L 43
MXC7432 P 62
MXC7433 P 62
MXC7434 P 62
MXC7435 P 62
MXC7436LNF N 84
MXC7436LUN N 84
MXC7452 N 22
MXC7453 N 22
MXC7464 N 24

MXC7580 ....N 31
MXC7585LNF ....P 23
MXC7585LUN ....P 23
MXC7590JUL ....P 24
MXC7590SUC ....P 24
MXC7591JUL ....P 24
MXC7591SUC ....P 24
MXC7628LNF ....P 36
MXC7628LNF ....P 38
MXC7628LUN ....P 36
MXC7648 ....S 01
MXC7656 ....L 21
MXC7709 ....P 22
MXC7713 ....N 45
MXC7714 ....N 45
MXC7715 ....N 45
MXC7716 ....N 45
MXC7717 ....N 45
MXC7718 ....N 45
MXC7719 ....N 45
MXC7720 ....N 45
MXC7721 ....N 45
MXC7738 ....M 07
MXC7757 ....N 24
MXC7758 ....N 24
MXC7763 ....P 64
MXC7795JUL ....P 25
MXC7795SUC ....P 25
MXC7796JUL ....P 25
MXC7796SUC ....P 25
MXC7810LUN ....M 73
MXC7810LUN ....M 74
MXC7811LNF ....R 38
MXC7811LUN ....R 38
MXC7812LNF ....R 38
MXC7812LUN ....R 38
MXC7870 ....R 13
MXC7870 ....R 12
MXC7872JUL ....R 07
MXC7872SUC ....R 07
MXC7873JUL ....R 04
MXC7877 ....N 59
MXC7878 ....N 59
MXC7879 ....N 67
MXC7880 ....N 67
MXC7881LUP ....N 37
MXC7882LUP ....N 39
MXC7882LUP ....N 38
MXC7884JUL ....P 65
MXC7885JUL ....P 65
MXC7887 ....N122
MXC7888 ....S 01
MXC7964 ....N 69
MXC8065LUP ....P 66
MXC8073JUL ....P 46
MXC8073SUC ....P 46
MXC8074JUL ....P 46
MXC8074SUC ....P 46
MXC8091 ....P 71
MXC8094JUL ....P 65
MXC8094SUC ....P 65
MXC8095JUL ....P 65
MXC8095SUC ....P 65
MXC8096JUL ....P 65
MXC8096SUC ....P 65
MXC8112JUL ....P 23
MXC8112LOY ....P 23
MXC8112LUN ....P 23
MXC8112SUC ....P 23
MXC8155JUL ....P 39
MXC8155JUL ....P 40
MXC8155SUC ....P 39
MXC8155SUC ....P 40
MXC8157 ....N 41
MXC8172LMP ....N 87
MXC8172MCF ....N 87
MXC8199 ....N 66
MXC8215 ....N 45
MXC8216 ....N 45
MXC8217 ....N 45
MXC8218 ....N 45
MXC8236JUL ....P 14
MXC8295JUL ....P 62
MXC8295LOY ....P 62
MXC8295SUC ....P 62
MXC8335 ....N 50
MXC8335 ....N 46
MXC8336 ....N 45
MXC8337 ....N 45
MXC8366 ....N 88
MXC8367 ....N 88
MXC8368 ....N 88
MXC8369 ....N 88
MXC8380JUL ....R 32
MXC8380SUC ....R 32
MXC8394LNF ....R 11
MXC8394LNF ....R 18
MXC8394LUN ....R 11
MXC8395LNF ....R 11
MXC8395LUN ....R 11
MXC8397PUB ....N 90
MXC8398PUB ....N 90
MXC8531 ....N 53
MXC8657 ....N 32
MXC8662 ....N 25
MXC8663 ....N 25
MXC8701 ....N 78
MXC8706 ....N 55
MXC8707 ....N 55
MXC8709 ....H 24
MXC8807 ....M107
MXC8807 ....M109
MXC8807 ....N 91
MXC8922 ....N 49
MXC8923 ....N 49
MXC8937LMP ....N 87
MXC8937MCF ....N 87
MXC8938LMP ....N 88
MXC8938MCF ....N 88
MXC9314 ....N 32
MXC9336 ....N 51
MXC9336 ....N 46
MXC9552 ....N 25
MXC9553 ....N 25
MXC9582 ....N 45
MXC9583 ....N 45
MXC9692 ....N 45
MYB100040L B178
MYF10018L B179
MYF100730L B178
MYP10013L B179
NA105011L ....S 01
NAD10038E B159
NAD10039 ....B 37
NAD10048 ....B188
NAD101190 ....B159
NC112041 ....E 47
NC112041 ....E 53
NC112041 ....F 11
NC112041 ....F 15
NEC10047 ....B185
NEC10049 ....B185
NGC10223 ....B185
NGC10224 ....B185
NGC10225 ....B185
NGC10226 ....B185
NGD10016L B185
NH100041L ....C 25
NH104041 ....M112
NH104041 ....M 93
NH104041L ....B 54
NH104041L ....R 21
NH105041L ....B 36
NH105041L ....B 90
NH105041L ....C 52
NH105041L ....P 75
NH105041L ....S 01
NH105041L M105
NH105041L ....G 04
NH105041L ....J 21
NH105041L ....P 05
NH105041L ....P 06
NH105041L ....M 28
NH105041L M151
NH106041 ....M 18
NH106041 ....M 19
NH106041L ....B 38
NH106041L ....N 13
NH106041L ....N 15
NH106041L ....N 70
NH106041L N110
NH106041L N111
NH106041L ....P 22
NH106041L ....J 18
NH106041L ....K 02
NH106041L ....K 06
NH106041L ....K 07
NH106041L ....L 13
NH106041L ....C 92
NH106041L ....C 05
NH106041L ....C 27
NH106041L ....C 42
NH106041L ....S 01
NH106041L ....M 47
NH106041L ....J 20
NH106041L ....K 10
NH106041L ....K 11
NH106041L ....K 08
NH106041L ....K 09
NH106041L ....K 13
NH106041L ....K 14
NH106041L ....L 15
NH106041L ....C 69
NH106041L ....L 36
NH106041L ....N 20
NH106041L ....C 94
NH106041L ....C 96
NH106041L ....C 97
NH106041L ....P 06
NH106041L ....P 64
NH106041L ....M 17
NH106041L ....M 18
NH106041L ....M 19
NH106041L ....M 21
NH106041L ....M 45
NH106041L ....L 12
NH106041L ....H 11
NH106041L ....H 06
NH106041L ....H 08
NH106041L ....M 16
NH106041L ....M 20
NH106041L ....M 44
NH108041L ....B 65
NH108041L B106
NH108041L B140
NH108041L B142
NH108041L B156
NH108041L ....D 20
NH108041L ....N 02
NH108041L ....H 01
NH108041L ....G 03
NH108041L M145
NH108041L ....L 05
NH108041L ....L 08
NH108041L ....N 09
NH108041L ....M 17
NH108041L ....M 18
NH108041L ....M 19
NH108041L ....M 21
NH108041L M152
NH108041L M114
NH108041L ....M 16
NH108041L ....M 20
NH108041L M144
NH108057 ....L 43
NH110041L ....B 83
NH110041L ....N 01
NH110041L ....N 06
NH110041L ....J 12
NH110041L ....K 02
NH110041L ....J 13
NH110041L ....D 04
NH110041L ....D 06
NH110041L ....K 05
NH110061 ....E 52
NH504041L ....B 85
NH504041L B151
NH604041L ....B 83
NH604041L B132
NH604041L ....N 06
NH604041L ....N 11
NH604041L ....N 69
NH604041L ....N 51
NH604041L ....T 22
NH604041L ....L 04
NH604041L ....N 20
NH604041L ....M 17
NH604041L ....M 18
NH604041L ....M 19
NH604041L ....M 21
NH604041L ....M 16
NH604041L ....M 20
NH605041L ....B 80
NH605041L ....B 84
NH605041L ....B 98
NH605041L B113
NH605041L B118
NH605041L B137
NH605041L ....F 05
NH605041L ....K 10
NH606041L ....B 35
NH606041L B100
NH606041L B142
NH606041L B144
NH606041L ....N 12
NH606041L ....H 14
NH606041L ....N 09
NH608041L ....T 20
NH610041L ....T 20
NH910011L ....B 98
NH910011L B103
NH910011L B151
NK105061 ....P 14
NK105061 ....P 19
NK105061 ....P 21
NK106061L ....P 39
NK106061L ....N 83
NK106061L ....P 06
NK106061L ....P 42
NK106081 ....P 06
NLP10003 ....B185
NN106021 ....M106
NN106021 ....F 20
NN106021 ....F 21
NN106021 ....C 69
NN106021 ....C 95
NN106021 ....C 97
NNK10001L B177
NR604090 ....B 96
NR604090 ....B131
NRC1302 ....N 06
NRC145 ....F 15
NRC1823 ....F 15
NRC1824 ....F 15
NRC2119 ....E 45
NRC2211 ....G 11
NRC2268 ....J 20
NRC2383 ....L 04
NRC2384 ....N 02
NRC2384 ....N 12
NRC2523 ....L 29
NRC2524 ....L 29
NRC3226 ....F 15
NRC3314 ....B 83
NRC3314 ....N 06
NRC3923 ....E 49
NRC4251 ....H 11
NRC4251 ....H 06
NRC4251 ....H 08
NRC4305 ....E 45
NRC4306 ....E 45
NRC4365 ....E 46
NRC4401 ....H 05
NRC4401 ....H 07
NRC4403 ....H 11
NRC4419 ....H 12
NRC4514 ....E 43
NRC4515 ....E 43
NRC4516 ....E 43
NRC4700 ....F 15
NRC5294 ....L 14
NRC5295 ....L 14
NRC5403 ....K 12
NRC5403 ....K 10
NRC5403 ....K 11
NRC5593 ....E 51
NRC5602 ....E 52
NRC5603 ....E 52
NRC5886 ....N 11
NRC6221 ....E 53
NRC6235 ....E 51
NRC6302 ....F 21
NRC6372 ....E 46
NRC6528 ....N 08
NRC6993 ....S 01
NRC7387 ....F 08
NRC7415 ....E 41
NRC7491 ....E 48
NRC7649 ....K 02
NRC7649 ....K 06
NRC7649 ....K 07
NRC7649 ....K 11
NRC7649 ....K 08
NRC7649 ....K 09
NRC7649 ....K 13
NRC7649 ....K 14
NRC7704 ....F 07
NRC7768 ....N 01
NRC7863 ....F 07
NRC7981 ....E 52
NRC8048 ....T 22
NRC8204 ....N 08
NRC8416 ....H 19
NRC8417 ....H 19
NRC8618 ....B 15
NRC8955 ....J 18
NRC8975 ....G 11
NRC9101 ....S 01
NRC9239 ....J 18
NRC9246 ....E 06
NRC9246 ....E 27
NRC9700 ....E 46
NRC9700 ....E 51
NRC9708 ....J 13
NRC9728 ....E 43
NRC9729 ....E 43
NRC9770 ....B 29
NRC9770 ....B 63
NRC9770 ....J 07
NRC9770 ....J 11
NRC9770 ....J 09
NRC9771 ....B 29
NRC9771 ....B 63
NRC9771 ....J 07
NRC9771 ....J 11
NRC9771 ....J 09
NRC9773 ....J 18
NRC9786 ....J 07
NSC100390L B185
NT108041L ....D 20
NT108041L ....F 07
NT110041L ....C 20
NT110041L ....C 40
NT110041L ....G 09
NT112041L ....H 14
NT112041L ....E 19
NT112041L ....E 21
NT607041L ....G 11
NT607041L ....G 13
NT608041L ....E 19
NT614041L ....F 11
NTC1030 ....K 01
NTC1030 ....K 11
NTC1081 ....L 04
NTC1154 ....M155
NTC1176 ....H 09
NTC1277 ....K 10
NTC1278 ....K 10
NTC1284 ....K 12
NTC1322 ....K 11
NTC1340 ....H 20
NTC1435 ....J 18
NTC1518 ....J 12
NTC1637 ....K 12
NTC1772 ....E 48
NTC1773 ....E 49
NTC1791 ....F 18
NTC1791 ....F 17
NTC1856 ....G 11
NTC1868 ....C 92
NTC1868 ....C 42
NTC1868 ....C 69
NTC1868 ....C 94
NTC1873 ....L 01
NTC1873 ....L 02
NTC1873 ....L 04
NTC1873 ....L 05
NTC1873 ....L 08
NTC1888 ....E 47
NTC1888 ....E 53
NTC1947 ....K 10
NTC1954 ....K 10
NTC2003 ....H 11
NTC2247 ....K 11
NTC2248 ....K 11
NTC2249 ....K 11
NTC2325 ....E 43
NTC2326 ....E 43
NTC2585 ....J 07
NTC2597 ....F 20

Part No.	Ref.	Part No.	Ref.	Part No.	Ref.	Part No.	Ref.	Part No.	Ref.	Part No.	Ref.	Part No.	Ref.	Part No.	Ref.
NTC2600	F 19	NTC4564	G 06	NTC5454	J 20	NTC5859	J 04	NTC6837	E 47	NTC7267	J 03	NTC7611	J 03	NTC8242	H 13
NTC2694	E 43	NTC4564	G 08	NTC5455	J 20	NTC5879	J 01	NTC6847	L 03	NTC7267	J 02	NTC7613	H 06	NTC8271	H 02
NTC2706	E 49	NTC4564	G 09	NTC5486PUB	N 13	NTC5879	J 03	NTC6847	L 06	NTC7267	J 04	NTC7613	H 08	NTC8278	N122
NTC2707	E 49	NTC4571	E 46	NTC5488PUB	N 15	NTC5879	J 02	NTC6847	L 18	NTC7275	C 42	NTC7627	G 12	NTC8286	B 99
NTC2722	F 20	NTC4571	E 51	NTC5494	F 21	NTC5879	J 04	NTC6854	S 01	NTC7280	G 06	NTC7634	T 20	NTC8286	B142
NTC2722	F 21	NTC4571	E 52	NTC5527	E 50	NTC5890	N 07	NTC6856	S 01	NTC7280	G 08	NTC7635	T 20	NTC8288	B 34
NTC2723	F 17	NTC4602	K 02	NTC5562	J 20	NTC5890	N 08	NTC6857	N 44	NTC7280	G 09	NTC7646	J 13	NTC8328	E 48
NTC2796	H 21	NTC4602	K 03	NTC5567	J 20	NTC5897	C 93	NTC6857	S 01	NTC7291	H 13	NTC7656	N 01	NTC8390	F 15
NTC2802	K 10	NTC4602	K 10	NTC5582	K 02	NTC5897	C 42	NTC6858	L 14	NTC7297	L 04	NTC7662	K 10	NTC8416	F 10
NTC2804	M150	NTC4609	B129	NTC5582	K 04	NTC5897	C 95	NTC6860	E 43	NTC7298	L 13	NTC7676	H 12	NTC8417	F 10
NTC2873	J 12	NTC4609	L 01	NTC5582	K 01	NTC5900	B 83	NTC6868L	H 06	NTC7308	L 02	NTC7688	H 12	NTC8476	E 45
NTC3071	H 12	NTC4609	L 04	NTC5582	K 06	NTC5916	J 05	NTC6868L	H 08	NTC7308	L 12	NTC7688	H 13	NTC8477	E 45
NTC3226	K 10	NTC4609	L 05	NTC5582	K 07	NTC5916	J 06	NTC6871	J 01	NTC7311	J 05	NTC7689	H 12	NTC8478	F 08
NTC3291	H 02	NTC4609	L 07	NTC5582	K 08	NTC5979	L 04	NTC6872	J 03	NTC7312	J 05	NTC7689	H 13	NTC8572	E 50
NTC3340	H 21	NTC4610	N 71	NTC5582	K 09	NTC6069	F 21	NTC6872	J 04	NTC7312	K 01	NTC7720	J 07	NTC8573	E 50
NTC3341	H 21	NTC4610	N 72	NTC5582	K 13	NTC6070	F 21	NTC6874	L 13	NTC7313	J 18	NTC7723	G 14	NTC8830	N 14
NTC3346	J 11	NTC4610	F 06	NTC5582	K 15	NTC6071	F 21	NTC6874	L 15	NTC7314	L 30	NTC7725	H 02	NTC8862	F 19
NTC3354	J 15	NTC4618	E 49	NTC5615	K 01	NTC6072	F 21	NTC6874	L 12	NTC7314	L 31	NTC7739CUH	E 41	NTC8863	F 21
NTC3354	J 16	NTC4627	J 23	NTC5632	L 04	NTC6106	E 51	NTC6890	L 02	NTC7314	M 92	NTC7739HUL	E 41	NTC8864	F 21
NTC3360	H 19	NTC4720	E 48	NTC5634	K 02	NTC6125	H 21	NTC6890	L 08	NTC7319	K 03	NTC7739MUE	E 41	NTC8868	F 19
NTC3361	H 19	NTC4859	J 09	NTC5634	K 04	NTC6125	H 22	NTC6890	F 17	NTC7334	J 10	NTC7743	E 41	NTC8872	H 13
NTC3400	G 13	NTC4881	K 13	NTC5634	K 06	NTC6125	H 23	NTC6891	J 05	NTC7335	J 10	NTC7818	F 01	NTC8874	H 11
NTC3401	G 13	NTC4881	K 14	NTC5634	K 07	NTC6143	J 08	NTC6891	J 06	NTC7336	J 07	NTC7825JUL	F 01	NTC8874	H 13
NTC3414	H 21	NTC4991	H 01	NTC5634	K 08	NTC6160	F 01	NTC6916	G 03	NTC7337	J 07	NTC7825SUC	F 01	NTC8881	F 20
NTC3458	H 11	NTC5009	H 21	NTC5634	K 09	NTC6166	C 42	NTC6916	G 04	NTC7345	K 03	NTC7829	S 01	NTC8881	F 21
NTC3458	H 13	NTC5009	H 23	NTC5634	K 13	NTC6171	L 04	NTC6937	J 09	NTC7349	J 05	NTC7831	N 02	NTC8882	F 19
NTC3466	G 11	NTC5026	H 23	NTC5634	K 15	NTC6279	N 04	NTC6944	G 04	NTC7349	J 06	NTC7847	L 13	NTC8960	J 16
NTC3650	K 15	NTC5031	N 01	NTC5648	K 02	NTC6280	N 04	NTC6949	G 01	NTC7366	G 01	NTC7853	C 22	NTC8960	J 17
NTC3751	J 15	NTC5043JUL	F 01	NTC5648	K 06	NTC6281	N 04	NTC6949	G 04	NTC7369	F 07	NTC7862	H 02	NTC8960	J 19
NTC3853	F 01	NTC5043SUC	F 01	NTC5648	K 07	NTC6282	C 22	NTC6994	E 19	NTC7370	F 07	NTC7863	H 02	NTC9013	H 06
NTC3858	L 02	NTC5077	N 12	NTC5648	K 08	NTC6285	N 01	NTC6994	E 22	NTC7381	E 50	NTC7937	S 01	NTC9013	H 08
NTC4061	K 10	NTC5082PUB	N 12	NTC5648	K 09	NTC6706	L 04	NTC7080PUB	T 20	NTC7394	E 53	NTC7940	H 07	NTC9027	E 48
NTC4095	H 05	NTC5083PUB	N 12	NTC5648	K 13	NTC6725	L 04	NTC7083	J 18	NTC7396	E 41	NTC7941	H 07	NTC9097	F 20
NTC4096	H 05	NTC5097	N 14	NTC5648	K 15	NTC6741	J 01	NTC7086	J 18	NTC7403	H 02	NTC7952	F 20	NTC9198	B142
NTC4097	H 05	NTC5098PUB	N 14	NTC5652	J 18	NTC6741	J 03	NTC7088	J 18	NTC7404	H 02	NTC8028	H 11	NTC9235	F 11
NTC4099	H 01	NTC5099PUB	N 14	NTC5654	J 18	NTC6754	K 13	NTC7099PUB	T 20	NTC7426	K 04	NTC8096	H 05	NTC9236	F 11
NTC4172	G 09	NTC5101	N 15	NTC5655	L 03	NTC6769	J 01	NTC7119	K 13	NTC7457	J 01	NTC8096	H 07	NTC9273	H 11
NTC4173	G 09	NTC5102PUB	N 15	NTC5656	L 03	NTC6769	J 03	NTC7124	L 14	NTC7482	N 13	NTC8097	H 05	NTC9293	H 11
NTC4205	B 83	NTC5106	N 14	NTC5690	H 24	NTC6769	J 02	NTC7126	G 01	NTC7484	G 02	NTC8097	H 07	NTC9399LNF	H 20
NTC4205	N 06	NTC5147	H 21	NTC5787	J 01	NTC6769	J 04	NTC7161	L 13	NTC7487	J 03	NTC8098	H 05	NTC9400	H 21
NTC4205	K 02	NTC5147	H 23	NTC5787	J 03	NTC6770	J 01	NTC7161	L 15	NTC7487	J 23	NTC8098	H 07	NTC9415	N 05
NTC4238	G 10	NTC5171	B129	NTC5787	J 02	NTC6770	J 03	NTC7161	L 12	NTC7491	J 23	NTC8099	H 05	NTC9416	N 05
NTC4245	F 19	NTC5171	L 01	NTC5787	J 04	NTC6770	J 02	NTC7196	K 04	NTC7492	J 23	NTC8099	H 07	NTC9420	N 01
NTC4294	H 01	NTC5171	L 04	NTC5794	C 93	NTC6770	J 04	NTC7223	L 02	NTC7493	J 23	NTC8113	G 09	NTC9424	N 10
NTC4329	L 02	NTC5171	L 05	NTC5794	C 42	NTC6771	J 01	NTC7226	G 02	NTC7497	J 23	NTC8122LNF	H 20	NTC9425	N 10
NTC4336	L 03	NTC5171	L 07	NTC5794	C 69	NTC6771	J 03	NTC7227	G 02	NTC7503	J 05	NTC8122LNF	H 22	NTC9429	F 19
NTC4337	L 03	NTC5175	L 03	NTC5799	C 42	NTC6771	J 02	NTC7241	H 05	NTC7503	J 06	NTC8122LNF	H 23	NTC9461	E 44
NTC4338	L 03	NTC5179	J 20	NTC5803	J 18	NTC6771	J 04	NTC7242	H 07	NTC7535	G 01	NTC8122LUN	H 20	NTC9462	E 44
NTC4408	J 15	NTC5193PM	E 41	NTC5844	L 15	NTC6776	E 47	NTC7243	H 07	NTC7553	G 13	NTC8202	E 47	NTC9530	H 09
NTC4428	H 02	NTC5415	J 20	NTC5844	L 12	NTC6776	E 53	NTC7246	H 05	NTC7557	N 01	NTC8202	E 53	NTC9530	H 10
NTC4487	H 01	NTC5418	J 05	NTC5858	J 01	NTC6790	F 20	NTC7246	H 07	NTC7558	C 22	NTC8232	G 10	NTC9530	H 11
NTC4488	H 01	NTC5418	J 06	NTC5858	J 02	NTC6791	K 13	NTC7252	L 14	NTC7567	J 06	NTC8239	N 11	NTC9530	H 13
NTC4517	J 23	NTC5424	J 05	NTC5859	J 01	NTC6811	G 09	NTC7260	H 02	NTC7567	J 24	NTC8240	N122	NTC9531	H 09
NTC4517	J 04	NTC5424	J 06	NTC5859	J 03	NTC6828	E 47	NTC7260	H 04	NTC7573	G 12	NTC8242	H 11	NTC9543L	H 09
NTC4563	G 09	NTC5430	J 05	NTC5859	J 02	NTC6836	E 47	NTC7267	J 01	NTC7611	J 01	NTC8242	H 12	NTC9543L	H 10

NTC9543L ....H 11
NTC9563 ....H 02
NTC9584 ....H 02
NTC9603 ....H 12
NTC9607 ....F 15

NTC9625 ....H 02
NTC9626 ....H 02
NTC9665PUB............N 14
NTC9677 ....H 04
NTC9681 ....F 20

NTC9687LNF ....F 02
NTC9932 ....E 49
NTC9939LNF ....F 03
NTC9940LNF ....F 02
NTC9941LNF ....F 02

NTC9942LNF ....F 02
NTC9959 ....G 13
NTC9976MNH............E 41
NTC9976MUE............E 41
NTC9980 ....T 21

NTC9981 ....N 01
NTC9982 ....N 01
NUC10003 ....B171
NV106041L ....L 14
NV108041L N121

NV108041L ....K 02
NV108041L ....K 04
NV108041L ....K 06
NV108041L ....K 07
NV108041L ....K 11

NV108041L ....K 08
NV108041L ....K 09
NV108041L ....K 13
NV108041L ....K 14
NV108041L ....K 15

NV108041L ....L 14
NV108041L ....T 21
NV110041L ....K 02
NV110041L ....K 04
NV112041L ....E 02

NV114047 ....E 43
NV116041L ....E 47
NV116041L ....E 53
NV116047 ....E 43
NY104041L ....N 78

NY105041L ....D 20
NY105041L ....L 17
NY105041L ....L 03
NY105041L ....M 10
NY105041L ....D 17

NY106041 ....H 18
NY106041 ....L 06
NY106041 ....H 03
NY106041 ....H 09
NY106041 ....H 10

NY106041 ....C 69
NY106041 ....C 95
NY106041 ....C 97
NY106041 ....L 18
NY106041L ....N 13

NY106041L ....D 05
NY106041L ....F 18
NY106041L ....F 17
NY106047L ....N 13
NY106047L ....N 15

NY106047L ....J 05
NY106047L ....L 16
NY106047L ....L 17
NY106047L ....L 03
NY106047L ....J 06

NY106047L ....D 05
NY106047L ....F 17
NY108041L ....E 43
NY108041L ....E 48
NY108041L ....D 20

NY108041L ....D 14
NY108041L ....K 01
NY108041L ....H 01
NY108041L ....J 09
NY108041L ....N 40

NY108041L ....F 08
NY108041L ....F 07
NY108051L ....B 34
NY108051L ....F 07
NY110041L ....E 46

NY110041L ....E 47
NY110041L ....E 48
NY110041L ....E 53
NY110041L ....N 04
NY110047L ....J 01

NY110047L ....J 03
NY110047L ....C 85
NY110047L ....C 22
NY110047L ....J 02
NY110047L ....J 04

NY110047L ....N 04
NY110051L ....E 49
NY112041L ....E 52
NY112041L ....F 12
NY112041L ....N 04

NY112041L ....T 21
NY116041L ....E 10
NY116041L ....E 34
NY116041L ....E 35
NY116051L ....E 43

NY120041L ....E 43
NY120041L ....E 48
NY120041L ....D 11
NY120041L ....D 12
NY604041 ....B 86

NY604041L ....L 14
NY605041L ....F 01
NY606041L ....E 07
NY606041L ....E 08
NY606041L ....E 28

NY606041L ....E 29
NY606041L ....E 30
NY606041L ....E 31
NY606041L ....N 02
NY606041L ....T 20

NY606041L ....F 01
NY607041L ....N 02
NY607041L ....F 12
NY607041L ....R 59
NY608041L ....E 49

NY608041L ....N 11
NY608041L ....F 12
NY610041L ....E 48
NY612041 ....E 49
NY910041L ....N 80

NZ606041L ....E 01
NZ606041L ....E 02
PA103061 ....C 86
PA103061 ....C 88
PA108101L ....C 61

PAM4356L ....P 71
PAM5522LNF............P 71
PAM5646LUM............P 71
PAM5646LUP............P 71
PBP101000 ....L 06

PC108241L ....H 24
PC108321L ....H 23
PC108648 ....H 22
PC108728L ....H 21
PCD100150 ....L 14

PCG10017L ....L 07
PEB10095 ....B170
PEH10033 ....B176
PEH100570 ....B180
PEH101170 ....B129

PEL10017 ....B176
PEL10017 ....L 19
PEN10029L B170
PEP10124 ....B176
PEP10148 ....B176

PEP10150 ....B176
PEP101730 ....B176
PEP101950 ....B176
PEP102840 ....B 51
PEP102840 ....B 61

PEQ10036L B176
PEQ100580 ....B176
PEQ100600L B176
PET10014 ....B176
PEU101620 ....B170

PFQ10001L B170
PKX100030 ....B 98
PQG10008 ....B187
PQG10008L B184
PQR10041L B182

PQR10050 ....B187
PQR10051 ....B184
PQR10051 ....B187
PQR10052 ....B187
PQS10061 ....B186

PQS100900 ....B186
PQU10007 ....B187
PRC10001 ....B187
PRC10002 ....B187
PRC1352 ....R 38

PRC1352 ....M145
PRC1352 ....R 40
PRC1352 ....M144
PRC1360 ....B148
PRC2167 ....M 02

PRC2911 ....D 14
PRC2911 ....C 20
PRC2911 ....C 40
PRC2979 ....M154
PRC3139 ....C 42

PRC3180 ....K 04
PRC3180 ....D 06
PRC3180 ....M113
PRC3359 ....B 23
PRC3359 ....B 61

PRC3408 ....N 71
PRC3408 ....N 72
PRC3408 ....F 06
PRC3505 ....B 61
PRC3505 ....B 93

PRC3505 ....B130
PRC3623 ....B107
PRC3623 ....B158
PRC3623 ....B164
PRC3623 ....M 18

PRC3623 ....M 19
PRC3702 ....M158
PRC3916 ....N 78
PRC3916 ....N 82
PRC3916 ....N 84

PRC4138 ....M 17
PRC4138 ....M 18
PRC4138 ....M 19
PRC4138 ....M 16
PRC4138 ....M 20

PRC4268 ....M 83
PRC4299 ....M 83
PRC4412 ....M 79
PRC4412 ....M 80
PRC4427 ....M 83

PRC4540 ....T 22
PRC4543 ....M153
PRC4615 ....M150
PRC4628 ....B104
PRC4628 ....B148

PRC4628 ....M 47
PRC4642 ....M 83
PRC4719 ....M 33
PRC4719 ....N 79
PRC4721 ....M 33

PRC4721 ....N 79
PRC4754 ....M 75
PRC4831 ....N 78
PRC5082 ....M 83
PRC5086 ....B101

PRC5086 ....B147
PRC5109E ....B 37
PRC5109N ....B 37
PRC5208 ....B107
PRC5208 ....B158

PRC5235 ....M 21
PRC5237 ....M 20
PRC5254 ....M 75
PRC5254 ....M 78
PRC5255 ....M 75

PRC5255 ....M 78
PRC5273 ....B153
PRC5273 ....M153
PRC5273 ....M147
PRC5273 ....M152

PRC5273 ....M149
PRC5273 ....M146
PRC5273 ....M148
PRC5274 ....M153
PRC5275 ....M153

PRC5436 ....L 21
PRC5438 ....M 63
PRC5538 ....B 37
PRC5681 ....F 18
PRC5797 ....M 64

PRC5838 ....M 15
PRC5956 ....M114
PRC6029 ....M 77
PRC6030 ....M 77
PRC6031 ....M 77

PRC6032 ....M 77
PRC6141 ....B104
PRC6144 ....B104
PRC6144 ....B148
PRC6260 ....G 10

PRC6314 ....L 21
PRC6320 ....M114
PRC6387 ....B 15
PRC6387 ....B 53
PRC6413 ....M 34

PRC6420 ....M 34
PRC6422 ....M 33
PRC6425 ....M 33
PRC6430 ....M113
PRC6447 ....M151

PRC6453 ....M 58
PRC6454 ....M 58
PRC6455 ....M 54
PRC6471 ....M 04
PRC6472 ....M 04

PRC6475 ....M 08
PRC6475 ....M 07
PRC6476 ....M 08
PRC6476 ....M 07
PRC6482 ....M151

PRC6484 ....M151
PRC6492LUN............M132
PRC6496 ....M152
PRC6496 ....M151
PRC6499LUN............M132

PRC6518L ....M 13
PRC6518L ....M 12
PRC6525 ....M 54
PRC6525 ....M 55
PRC6574 ....B148

PRC6580 ....M 17
PRC6580 ....M 18
PRC6580 ....M 19
PRC6580 ....M 16
PRC6590 ....M 64

PRC6796 ....M 83
PRC6808 ....M 68
PRC6809 ....M 68
PRC6810 ....M 68
PRC6811 ....M 68

PRC6813 ....M 68
PRC6829 ....M 75
PRC6848 ....M132
PRC6852 ....M147
PRC6852 ....M149

PRC6852 ....M146
PRC6852 ....M148
PRC6857 ....M147
PRC6857 ....M149
PRC6857 ....M146

PRC6857 ....M148
PRC6864 ....M 83
PRC6913 ....M 90
PRC6913 ....M 83
PRC6942 ....M 67

PRC6943 ....M 67
PRC6944 ....M 67
PRC6945 ....M 67
PRC6947 ....B 16
PRC6970 ....M 12

PRC7045 ....M146
PRC7045 ....M148
PRC7062 ....K 04
PRC7093 ....M 17
PRC7093 ....M 18

PRC7093 ....M 19
PRC7093 ....M 21
PRC7093 ....M 16
PRC7093 ....M 20
PRC7093 ....M 28

PRC7096 ....M144
PRC7100 ....M 17
PRC7100 ....M 18
PRC7100 ....M 19
PRC7100 ....M 21

PRC7100 ....M 16
PRC7100 ....M 20
PRC7128 ....J 01
PRC7129 ....J 03
PRC7129 ....J 04

PRC7145 ....M 34
PRC7145 ....M 35
PRC7204 ....B122
PRC7246 ....M 15
PRC7266 ....M 15

PRC7282 ....M 58
PRC7284 ....M 50
PRC7285 ....M 50
PRC7292 ....M 43
PRC7303 ....M 83

PRC7304 ....M147
PRC7304 ....M149
PRC7304 ....M146
PRC7304 ....M148
PRC7307 ....M146

PRC7308 ....M148
PRC7379LUN............M132
PRC7394 ....P 13
PRC7404 ....M 83
PRC7409 ....M 47

PRC7410LNF............M133
PRC7410LUN............M132
PRC7416 ....M 16
PRC7417 ....M152
PRC7417 ....M151

PRC7423 ....M 34
PRC7425 ....M 34
PRC7445 ....M 05
PRC7445 ....M 04
PRC7451 ....M118

PRC7451 ....P 13
PRC7459 ....M148
PRC7460 ....M146
PRC7471 ....M132
PRC7781 ....M144

PRC7828 ....M 79
PRC7885 ....M 02
PRC7886 ....M 75
PRC7922 ....B158
PRC7941 ....M 60

PRC7943 ....M 57
PRC7944 ....M 75
PRC7949 ....M114
PRC7984 ....D 06
PRC8001 ....B 23

PRC8003 ....B130
PRC8003 ....B 95
PRC8005 ....M 95
PRC8005 ....M 96
PRC8007 ....M 95

PRC8007 ....M 96
PRC8009 ....M 95
PRC8009 ....M 96
PRC8077 ....M 35
PRC8078 ....M 22

PRC8079 ....M 23
PRC8080 ....M 22
PRC8081 ....M 23
PRC8082 ....M 46
PRC8088 ....M 50

PRC8089 ....M 22
PRC8095 ....M 77
PRC8096 ....M 77
PRC8098 ....N 78
PRC8114 ....M113

PRC8123 ....M 83
PRC8130 ....T 22
PRC8134 ....M 46
PRC8147 ....M 54
PRC8155 ....L 21

PRC8160 ....M152
PRC8160 ....M151
PRC8171 ....N 44
PRC8171 ....M 03
PRC8171 ....M 02

PRC8172 ....M 95
PRC8172 ....M 96
PRC8181 ....M 24
PRC8182 ....M 24
PRC8183 ....M 24

PRC8184 ....M 34
PRC8185 ....M 34
PRC8211 ....M 59
PRC8232 ....M114
PRC8235 ....M145

PRC8235 ....M144
PRC8236 ....M145
PRC8236 ....M144
PRC8253 ....M145
PRC8253 ....M144

PRC8301LUN M132
PRC8322 .... M 59
PRC8388 .... M 68
PRC8389 .... M 68
PRC8390 .... M 68

PRC8391 .... M 68
PRC8392 .... M 67
PRC8393 .... M 67
PRC8394 .... M 67
PRC8395 .... M 67

PRC8396 .... M 69
PRC8414 .... M 60
PRC8437 .... M114
PRC8473 .... C 92
PRC8473 .... C 42

PRC8473 .... C 95
PRC8481 .... M114
PRC8496LUN M132
PRC8498 .... M 67
PRC8534 .... M 58

PRC8535 .... M 58
PRC8548 .... M 13
PRC8548 .... M 12
PRC8564 .... M132
PRC8565 .... M132

PRC8566 .... M 20
PRC8572 .... M 58
PRC8591 .... M 79
PRC8592 .... M 79
PRC8617 .... M 94

PRC8692 .... M 23
PRC8694 .... M 24
PRC8695 .... M 24
PRC8695 .... M 25
PRC8696 .... M 24

PRC8696 .... M 25
PRC8698 .... M 25
PRC8724 .... M 63
PRC8730 .... M 23
PRC8731 .... M 22

PRC8749 .... M 35
PRC8783 .... B130
PRC8803 .... M 33
PRC8804 .... M 33
PRC8826 .... M 60

PRC8878 .... M 83
PRC8888 .... M 95
PRC8889 .... M 94
PRC8890 .... M 46
PRC8891 .... M 95

PRC8892 .... M 83
PRC8930LNF M134
PRC8930LNF M136
PRC8930LNF M139
PRC8930LNF M142

PRC8969 .... M 54
PRC8969 .... M 55
PRC8973 .... M 33
PRC8977 .... M 34
PRC8977 .... M 35

PRC8978 .... M 34
PRC8978 .... M 35
PRC8979 .... M 34
PRC8979 .... M 35
PRC8981 .... M 34

PRC8982 .... M 34
PRC9008 .... M152
PRC9008 .... M151
PRC9041 .... M 26
PRC9042 .... M 26

PRC9043 .... M 26
PRC9043 .... M 27
PRC9044 .... M 26
PRC9044 .... M 27
PRC9045 .... M 23

PRC9046 .... M 23
PRC9047 .... M 23
PRC9048 .... M 23
PRC9059 .... M 94
PRC9094 .... M 33

PRC9146 .... B 26
PRC9146 .... M 91
PRC9147 .... M 91
PRC9148 .... M 91
PRC9231 .... M 25

PRC9235 .... M 94
PRC9235E .... M 94
PRC9306 .... M 04
PRC9307 .... M 04
PRC9409 .... J 01

PRC9409 .... J 02
PRC9410 .... M130
PRC9443 .... M 26
PRC9444 .... M 26
PRC9445 .... M 26

PRC9445 .... M 27
PRC9446 .... M 26
PRC9446 .... M 27
PRC9447 .... M 22
PRC9448 .... M 22

PRC9449 .... M 22
PRC9450 .... M 22
PRC9451 .... M 58
PRC9452 .... M 58
PRC9460 .... M147

PRC9460 .... M149
PRC9463LNF M132
PRC9463LNF M133
PRC9466LUN M132
PRC9480 .... M 54

PRC9480 .... M 55
PRC9481 .... M 54
PRC9484LUN M132
PRC9527 .... M106
PRC9541 .... M 46

PRC9542 .... M 50
PRC9543 .... M 50
PRC9544 .... M 50
PRC9561 .... M 80
PRC9562 .... M 80

PRC9572 .... M106
PRC9579 .... M 91
PRC9604 .... M 33
PRC9612 .... M 94
PRC9612E .... M 94

PRC9659 .... M 25
PRC9660 .... M 25
PRC9661 .... M 25
PRC9662 .... M 95
PRC9663 .... M 94

PRC9666 .... M 85
PRC9666 .... M 86
PRC9668 .... J 01
PRC9677 .... C 88
PRC9859 .... C 88

PRC9868 .... M106
PRC9871 .... M114
PRC9872 .... M113
PRC9873 .... M113
PRC9895 .... M 95

PRC9916 .... M 05
PRC9916 .... M 04
PRC9917 .... B 23
PRC9941 .... M139
PRC9941 .... M142

PRC9942 .... B 23
PRC9943 .... M 25
PRC9944 .... M 54
PRC9991 .... M112
PS103121L .... D 14

PS104127L .... C 86
PS104127L .... H 21
PS104127L .... C 88
PS104127L .... H 23
PS104127L .... H 24

PS106321L .... E 47
PS106321L .... E 49
PS106321L .... E 53
PS106321L .... F 11
PS106321L .... F 15

PS603041 .... G 03
PS603041L B140
PS603041L .... G 03
PS603041L .... G 04
PS606101L .... B 89

PS606101L B126
PS608101L .... E 10
PS608101L .... E 34
PS608101L .... E 35
PS608101L .... E 49

PSD10102 .... B189
PSD102420 .... B189
PYC101120 .... J 18
PYC101120 .... L 16
PYC101120 .... L 03

PYC101120 .... L 09
PYC101120 .... L 10
PYC101120L .... J 05
PYC101120L .... L 19
PYC101120L .... L 16

PYC101120L .... L 03
PYC101120L .... L 09
PYC101120L .... L 10
PYC102350 .... L 03
PYG10006L B184

PYG10006L B187
QEH102380 .... F 19
QEH102380 .... F 20
QEH102440 .... F 19
QEH102440 .... F 20

QEP105410 .... F 20
QRB10006LNF .... F 04
QTN100260LNF .... F 03
QVU10037 .... B182
QYH10005 .... F 03

QYP10007 .... F 04
RA608123L M106
RA608126L .... L 11
RA608127 .... M132
RA608176L .... J 18

RPS2016L B175
RRC110870MNH .... E 41
RRC110870RHO .... E 41
RRC1463 .... L 40
RRC1492 .... L 46

RRO2125L .... L 04
RRO2126L .... L 04
RRW100010 .... E 42
RTC1136 .... H 15
RTC1137 .... H 18

RTC1526 .... H 15
RTC1526 .... H 18
RTC1718 .... B 80
RTC1718 .... B112
RTC171810 .... B 80

RTC171810 .... B112
RTC171820 .... B 80
RTC171820 .... B112
RTC1799 .... F 15
RTC1956 .... D 17

RTC2117 .... B 81
RTC2117 .... B114
RTC211710 .... B 81
RTC211710 .... B114
RTC211720 .... B 81

RTC211720 .... B114
RTC218620 .... B 81
RTC218620 .... B114
RTC2186S .... B 81
RTC2186S B114

RTC229520 .... B114
RTC2295S B114
RTC2408 .... B 81
RTC2408 .... B114
RTC240820 .... B 81

RTC240820 .... B114
RTC2726 .... E 11
RTC2726 .... E 36
RTC2825 .... B 07
RTC2825 .... B 45

RTC2914 .... C 14
RTC2914 .... C 35
RTC2993 .... B 08
RTC2993 .... B 46
RTC299310 .... B 08

RTC299310 .... B 46
RTC3095 .... E 11
RTC3095 .... E 36
RTC3197 .... B101
RTC3197 .... B147

RTC3201 .... B147
RTC3254 .... D 03
RTC3254 .... D 05
RTC3254 .... D 17
RTC3254 .... D 18

RTC3263 .... N115
RTC3264 .... N112
RTC3306 .... B141
RTC3397 .... D 07
RTC3429 .... E 23

RTC3429 .... E 24
RTC3429 .... E 39
RTC3429 .... E 40
RTC3458 .... E 01
RTC3519 .... J 14

RTC3566 .... B 97
RTC3633 .... M 15
RTC3633 .... M 13
RTC3635 .... M 68
RTC3635 .... L 28

RTC3635 .... M130
RTC3635 .... M131
RTC3635 .... L 48
RTC3635 .... M 63
RTC3635 .... P 13

RTC3641 .... M147
RTC3641 .... M149
RTC3641 .... M146
RTC3641 .... M148
RTC3650 .... M148

RTC3650 .... M151
RTC3725 .... B 91
RTC3743 .... N 25
RTC3743 .... N 42
RTC3745 .... R 48

RTC3746 .... P 83
RTC3772 .... M155
RTC3812 .... B151
RTC3814 .... N 61
RTC3815 .... N 61

RTC3816 .... N 64
RTC3817 .... N 64
RTC3820 .... N 61
RTC3821 .... N 61
RTC3890 .... D 21

RTC3942 .... B101
RTC3957 .... M148
RTC3958 .... M148
RTC3959 .... M146
RTC3959 .... M148

RTC3961 .... M146
RTC3961 .... M148
RTC4058A .... T 03
RTC4058T .... T 01
RTC4198 .... F 11

RTC4268 .... C 74
RTC4276 .... C 73
RTC4277 .... C 73
RTC4278 .... C 73
RTC4279 .... C 73

RTC4280 .... C 73
RTC4281 .... C 73
RTC4282 .... C 73
RTC4285 .... C 72
RTC4286 .... C 72

RTC4288 .... C 72
RTC4289 .... C 72
RTC4290 .... C 72
RTC4291 .... C 72
RTC4295 .... C 84

RTC4301 .... C 85
RTC4307 .... C 84
RTC4308 .... C 84
RTC4313 .... C 84
RTC4314 .... C 84

RTC4315 .... C 84
RTC4320 .... C 73
RTC4335 .... C 75
RTC4336 .... C 75
RTC4338 .... C 75

RTC4339 .... C 75
RTC4373 .... D 15
RTC4393 .... F 13
RTC4394 .... F 13
RTC4395 .... F 13

RTC4398 .... F 13
RTC4399 .... F 13
RTC4400 .... F 13
RTC4401 .... F 13
RTC4402 .... F 14

RTC4403 .... F 14
RTC4404 .... F 14
RTC4405 .... F 14
RTC4406 .... F 14
RTC4407 .... F 14

RTC4408 .... F 14
RTC4409 .... F 14
RTC4410 .... F 14
RTC4477 .... B 87
RTC4481 .... M 79

RTC4481 .... M 80
RTC4490 .... D 07
RTC4497 .... M 79
RTC4497 .... M 80
RTC4498 .... M 79

RTC4498 .... M 80
RTC4501 .... M 81
RTC4501 .... M 79
RTC4501 .... M 80
RTC4503 .... M 81

RTC4503 .... M 82
RTC4503 .... M 79
RTC4503 .... M 80
RTC4504 .... M 81
RTC4504 .... M 82

RTC4504 .... M 79
RTC4504 .... M 80
RTC4506 .... M 79
RTC4506 .... M 80
RTC4507 .... M 82

RTC4507 .... M 79
RTC4507 .... M 80
RTC4507 .... M 88
RTC4507 .... M116
RTC4587 .... E 01

RTC4587 .... E 02
RTC4614 .... M148
RTC4643 .... C 72
RTC4644 .... C 72
RTC4645 .... C 72

RTC4647 .... C 74
RTC4648 .... C 72
RTC4649 .... C 73
RTC4650 .... C 84
RTC4652 .... C 74

RTC4653 .... C 73
RTC4655 .... C 73
RTC4656 .... C 73
RTC4657 .... C 73
RTC4658 .... C 72

RTC4659 .... C 85
RTC4659 .... D 04
RTC4660 .... C 79
RTC4661 .... C 84
RTC4662 .... C 84

RTC4663 .... C 84
RTC4683 .... J 15
RTC4779 .... M148
RTC4783 .... B 07
RTC4783 .... B 45

RTC478310 .... B 07
RTC478310 .... B 45
RTC4819 .... J 15
RTC4825 .... F 20
RTC4825 .... F 21

RTC4826 .... F 20
RTC4826 .... F 21
RTC4848 .... M150
RTC4854 .... C 72
RTC4937 .... C 72

RTC4939 .... J 12
RTC4978 .... B 37
RTC4979 .... B 37
RTC4980 .... B 37
RTC4981 .... B 37

RTC4982 .... B 37
RTC4983 .... B 37
RTC5001 .... H 16
RTC5008 .... C 86
RTC5046 .... B105

RTC5046 .... B152
RTC5046 .... B153
RTC5047 .... B152
RTC5047 .... B153
RTC5048 .... B107

RTC5049 .... B107
RTC5050 .... B107
RTC5053 .... B154
RTC5053E B154
RTC5054 .... B154

RTC5089 .... B101
RTC5089 .... B147
RTC5090 .... B101
RTC5090 .... B147
RTC5091 .... B101

RTC5091 .... B147
RTC5092 .... B101
RTC5100 .... C 91
RTC5101 .... C 91
RTC5102 .... C 75

Part	Ref	Part	Ref	Part	Ref	Part	Ref	Part	Ref	Part	Ref	Part	Ref	Part	Ref
RTC5103	C 75	RTC5162	C 78	RTC5687	B105	RTC5979A	T 03	RTC6436	B 97	RTC6641	N 61	RTC6751	C 11	RTC9561	P 22
RTC5105	C 75	RTC5163	C 78	RTC5687	B152	RTC5979T	T 01	RTC6454A	T 04	RTC6642	N 64	RTC6751	C 30	RTC9561JUL	P 22
RTC5106	C 75	RTC5164	C 78	RTC5687	B153	RTC5980A	T 03	RTC6454T	T 02	RTC6643	N 64	RTC6754	E 14	RTC9561LOY	P 22
RTC5107	C 75	RTC5171	C 83	RTC5688	B 36	RTC5980T	T 01	RTC6457	B 08	RTC6682N	B 77	RTC6755	E 14	RTC9561SUC	P 22
RTC5108	C 75	RTC5171	C 78	RTC5689	B 36	RTC5981A	T 03	RTC645720	B 08	RTC6682R	B 77	RTC6760	B142	RTC9564	R 49
RTC5109	C 75	RTC5172	C 83	RTC5691	M 63	RTC5981T	T 01	RTC645740	B 08	RTC6685	N 71	RTC6776	H 15	RTC9564	T 29
RTC5110	C 75	RTC5173	C 80	RTC5728A	T 03	RTC5982A	T 04	RTC6469	M 01	RTC6685	N 73	RTC6777	H 15	RTC9581	T 20
RTC5111	C 75	RTC5174	C 80	RTC5728T	T 01	RTC5982T	T 02	RTC6470	M 01	RTC6685	N 74	RTC6810A	T 04	RTC9586AA	T 23
RTC5112	C 75	RTC5175	C 80	RTC5729A	T 04	RTC5983A	T 04	RTC6471	M 01	RTC6685	N 76	RTC6810T	T 01	RTC9587AB	T 26
RTC5113	C 75	RTC5176	C 80	RTC5729T	T 01	RTC5983T	T 02	RTC6472	M 01	RTC6686	N 71	RTC6816LUN	N 56	RTC9589	T 36
RTC5114	C 75	RTC5177	C 80	RTC5733	C 74	RTC601	R 59	RTC6481	N 39	RTC6686	N 73	RTC6820	N110	RTC9593AA	T 43
RTC5115	C 75	RTC5178	C 80	RTC5734	C 74	RTC6061N	B107	RTC6481	N 37	RTC6686	N 74	RTC6821	N111	RU608183	L 43
RTC5116	C 75	RTC5179	C 80	RTC5735	C 74	RTC6061N	B158	RTC6482	N 39	RTC6686	N 75	RTC6825	F 16	RU608183L	N 69
RTC5117	C 75	RTC5180	C 82	RTC5770	J 15	RTC6066S	B114	RTC6482	T 47	RTC6686	N 76	RTC6869A	T 03	RU612373L	P 75
RTC5121	C 77	RTC5181	C 82	RTC5779A	T 28	RTC6072	B 97	RTC6482	N 37	RTC6687	N 71	RTC6869T	T 01	RU612373L	R 59
RTC5122	C 77	RTC5182	C 82	RTC5781A	T 28	RTC6074	B 99	RTC6482	N 38	RTC6687	N 72	RTC6870A	T 03	RY612363	N 28
RTC5122	C 80	RTC5183	C 82	RTC5782A	T 28	RTC6074	B142	RTC6535	B 30	RTC6687	N 73	RTC6870T	T 01	SA108201	D 23
RTC5123	C 77	RTC5184	C 82	RTC5786	R 21	RTC6075	B 99	RTC6536	B 30	RTC6687	N 74	RTC6896	B 16	SA108201	D 22
RTC5123	C 80	RTC5185	C 82	RTC5789	R 21	RTC6075	B142	RTC6537	B 30	RTC6687	N 75	RTC6991	R 48	SB106121	R 38
RTC5124	C 77	RTC5186	C 82	RTC5790	R 21	RTC608	J 23	RTC6538	B 30	RTC6687	N 33	RTC6992	R 48	SB106121	R 40
RTC5125	C 81	RTC5187	C 82	RTC5792	C 84	RTC608	J 24	RTC6539	B 30	RTC6688	N 71	RTC6993JUL	R 48	SC104141L	B135
RTC5126	C 80	RTC5188	C 82	RTC5797	R 21	RTC609	B 84	RTC6548	P 13	RTC6688	N 72	RTC6993SUC	R 48	SE104124L	N 36
RTC5126	C 81	RTC5189	C 82	RTC5798	R 21	RTC609	B118	RTC6551	B102	RTC6688	N 73	RTC6994	R 48	SE104124L	P 05
RTC5127	C 81	RTC5190	C 82	RTC5818	C 73	RTC609	J 14	RTC6551	B149	RTC6688	N 74	RTC6995	R 48	SE104161L	N 33
RTC5128	C 81	RTC5191	C 82	RTC5831	H 01	RTC611	R 59	RTC6563R	C 01	RTC6688	N 75	RTC6996	R 48	SE104161L	P 16
RTC5129	C 83	RTC5192	C 82	RTC5832	H 01	RTC6114	B 36	RTC6564	B 19	RTC6688	N 83	RTC7452	L 41	SE104161L	F 04
RTC5134	C 79	RTC5193	C 82	RTC5834	H 01	RTC6114	B105	RTC6564	B 57	RTC6689	N 71	RTC8137	T 17	SE104161L	H 20
RTC5135	C 79	RTC5203	C 72	RTC5840	E 13	RTC6114	B152	RTC6569	M117	RTC6689	N 72	RTC8159	T 20	SE105101L	M114
RTC5135	C 78	RTC5209	C 73	RTC5841	E 13	RTC6114	B153	RTC6570	M117	RTC6689	N 73	RTC8831	T 20	SE105101L	P 14
RTC5136	C 79	RTC5211	C 84	RTC5841	E 14	RTC6115	B 83	RTC6571	M117	RTC6689	N 74	RTC8891	T 20	SE105104L	P 14
RTC5137	C 79	RTC5212	C 84	RTC5841	E 16	RTC613	B 96	RTC6572	M117	RTC6689	N 75	RTC8893	T 34	SE105105L	B174
RTC5137	C 78	RTC5228E	B158	RTC5842	E 13	RTC613	B131	RTC6573	M117	RTC6690	N 71	RTC9477	T 40	SE105121L	N 69
RTC5138	C 78	RTC5254	B141	RTC5843	E 13	RTC613	N 51	RTC6574	M117	RTC6690	N 72	RTC9481A	T 03	SE105121L	J 23
RTC5139	C 78	RTC5609	B154	RTC5847	B 36	RTC613	F 05	RTC6575	M117	RTC6690	N 73	RTC9481T	T 01	SE105121L	N 56
RTC5141	C 76	RTC5610	B154	RTC5869	F 15	RTC613	N 30	RTC6576	M117	RTC6690	N 74	RTC9498	T 23	SE105161L	N 56
RTC5142	C 76	RTC5611	B154	RTC5870	F 11	RTC613	F 01	RTC6577	M117	RTC6690	N 75	RTC9499	T 20	SE105161L	N 60
RTC5143	C 76	RTC5623	R 21	RTC5870	F 15	RTC6143	L 45	RTC6578	M118	RTC6693	L 20	RTC9503AA	T 08	SE105161L	N 63
RTC5144	C 76	RTC5628	B104	RTC5876	F 07	RTC6166	B154	RTC6579	M118	RTC6694	L 20	RTC9505AC	T 43	SE105164L	P 19
RTC5145	C 77	RTC5628	B148	RTC5888	J 14	RTC6359	E 51	RTC6580	M118	RTC6695	L 20	RTC9507AB	T 44	SE105164L	P 21
RTC5146	C 77	RTC5670	B 36	RTC5889	H 18	RTC6360	E 46	RTC6582	M118	RTC6696	L 20	RTC9512	T 32	SE105165L	N 43
RTC5147	C 77	RTC5670	B105	RTC5890	H 18	RTC6362	P 04	RTC6583	M118	RTC6697	L 20	RTC9513	T 32	SE105168L	M 06
RTC5148	C 77	RTC5670	B152	RTC5905	L 42	RTC6393	N 56	RTC6593	L 20	RTC6702	B 28	RTC9520	T 26	SE105201L	P 71
RTC5149	C 77	RTC5671	B 36	RTC5907	B 94	RTC6395	B 12	RTC6594	C 72	RTC6702	B 64	RTC9520	T 27	SE105201L	P 14
RTC5150	C 85	RTC5671	B105	RTC5907	B138	RTC6396	F 11	RTC6607	M121	RTC6720	L 45	RTC9523	T 23	SE105204	P 22
RTC5151	C 81	RTC5671	B152	RTC5907	H 04	RTC6397	F 11	RTC6607	M123	RTC6721	L 45	RTC9524	T 23	SE105301L	P 71
RTC5152	C 81	RTC5679	B132	RTC5918	B 79	RTC6400A	T 04	RTC6607	M127	RTC6722N	C 23	RTC9525	T 23	SE105301L	M134
RTC5153	C 81	RTC5679	B179	RTC5918	B111	RTC6400T	T 01	RTC6607	M128	RTC6722R	C 23	RTC9526	E 41	SE105301L	M136
RTC5154	C 81	RTC5679	B133	RTC5925	B 36	RTC6401A	T 03	RTC6618	M 63	RTC6732	B 29	RTC9526	T 36	SE105301L	M140
RTC5155	C 79	RTC5679	B134	RTC5926	B 36	RTC6401T	T 01	RTC6634A	T 04	RTC6733	B 29	RTC9526POL	T 36	SE105301L	M142
RTC5156	C 79	RTC5680E	B105	RTC5926	B105	RTC6402A	T 04	RTC6634T	T 02	RTC6734	B 29	RTC9527	T 23	SE106101L	B 10
RTC5157	C 79	RTC5680E	B152	RTC5926	B152	RTC6402T	T 01	RTC6635	B 03	RTC6735	B 29	RTC9535	T 36	SE106101L	L 35
RTC5158	C 79	RTC5680N	B105	RTC5928	B154	RTC6411	R 21	RTC6636R	B 02	RTC6744	L 46	RTC9539AB	T 39	SE106121	P 29
RTC5159	C 79	RTC5680N	B152	RTC5935	B 99	RTC6412	R 21	RTC6637	B 01	RTC6745	L 46	RTC9549	R 49	SE106121	N 83
RTC5160	C 79	RTC5681E	B 36	RTC5938	J 12	RTC6415A	T 04	RTC6638	B 01	RTC6746	L 42	RTC9549	T 29	SE106121	P 34
RTC5161	C 78	RTC5681N	B 36	RTC5965	B135	RTC6415T	T 02	RTC6640	N 61	RTC6751	C 56	RTC9554	T 45	SE106121L	P 36

Part	Ref	Part	Ref	Part	Ref	Part	Ref	Part	Ref	Part	Ref	Part	Ref	Part	Ref
SE106121L	P 39	SH105251L	M 28	SH108161L	K 15	SH110201L	N 04	SH506071L	B124	SJ106200	R 40	STC1102N	B109	STC1251	B 74
SE106121L	N 76	SH105601	S 01	SH108161L	L 14	SH110201L	N 07	SH506071L	B156	SL510031	H 12	STC1108	D 01	STC1251	B 38
SE106121L	N 81	SH106101L	C 24	SH108161L	C 89	SH110201M	J 01	SH506071L	B125	SN104101L	B185	STC1120	J 14	STC1252	B159
SE106121L	N 84	SH106101L	C 88	SH108161L	F 17	SH110201M	J 03	SH506095L	B 96	SN106127	M141	STC1122	B 88	STC1253	B159
SE106121L	P 42	SH106121L	B137	SH108161L	F 03	SH110201M	J 02	SH506095L	B131	SN106127L	M141	STC1126	G 11	STC1254	B 37
SE106121L	N 63	SH106121L	N 14	SH108161L	M 91	SH110201M	J 04	SH506095L	B156	SNB100350PMA	H 20	STC1126	G 12	STC1254	B 74
SE106121L	N 64	SH106121L	N 70	SH108161L	C 22	SH110251	E 46	SH506101L	B106	SNB100350PMA	H 23	STC1128	C 89	STC1254	B 38
SE106161L	N110	SH106121L	N110	SH108161M	B 76	SH110251L	N 01	SH506101L	B156	SNE100090PMA	H 20	STC113	M 02	STC1255	B 37
SE106161L	N111	SH106121L	N111	SH108201	E 43	SH110251L	N 14	SH506125	B101	SNE100090PMA	H 23	STC1130	D 11	STC1255	B 74
SE106161L	P 75	SH106121L	K 07	SH108201	M108	SH110251L	J 12	SH506125	B147	SP105121	M134	STC1130	D 12	STC1255	B 38
SE106161L	K 11	SH106121L	C 85	SH108201	M110	SH110251L	C 22	SH507091L	B 83	SP105121	M136	STC1143	D 01	STC1258	H 15
SE106161L	N 76	SH106121L	N 46	SH108201L	E 48	SH110251L	J 13	SH507091L	N 06	SP105121	M140	STC1144	D 01	STC1259	H 15
SE106161L	N 81	SH106121L	K 09	SH108251	K 04	SH110301L	E 48	SH604031L	N 52	SP105121	M142	STC1157	H 22	STC1264	H 18
SE106168L	N 15	SH106121L	N 86	SH108251	C 54	SH110301L	D 05	SH604031L	N 55	SR106201	N 27	STC1157	H 24	STC1265	H 18
SE106201L	N 60	SH106121L	P 31	SH108251L	B 99	SH110351L	E 47	SH604041L	B132	SR106201	N 20	STC1172	B 32	STC1278	H 15
SE106201L	N 61	SH106141L	B 15	SH108251L	D 03	SH110351L	E 53	SH604041L	B140	SR106201	N 30	STC118	N 67	STC1279	H 18
SE106251L	B131	SH106141L	G 13	SH108251L	D 14	SH110351L	H 24	SH604041L	N 51	SS106351	M115	STC1180	R 54	STC128	J 06
SE106551	H 09	SH106141L	S 01	SH108251L	C 21	SH110401L	D 04	SH604041L	H 19	SS108201	B 25	STC1180	R 57	STC128	J 17
SE106551S	C 67	SH106141L	K 10	SH108251L	C 29	SH112201	N 05	SH604047	H 19	SS108201	B 67	STC1184	B101	STC128	L 10
SE106601	C 20	SH106151L	C 72	SH108251L	C 41	SH112201	C 97	SH604051L	B 13	SS108201	L 16	STC1184	B147	STC1284	H 01
SE106601	C 40	SH106161	M 47	SH108251L	C 22	SH112251L	C 68	SH604051L	B 86	SS108251	L 16	STC1186	B147	STC1285	H 01
SE910161L	B103	SH106161	M147	SH108251L	D 04	SH112251L	C 90	SH604051L	B121	SS108251L	B 33	STC1190	B 27	STC1286	H 01
SE910161L	B151	SH106161	L 06	SH108251L	C 24	SH112251L	N 04	SH604051L	M 17	SS108251L	B 67	STC1203	M 13	STC1287LOY	P 07
SF106121L	N 76	SH106161	M 18	SH108251L	M 17	SH112251L	N 09	SH604051L	M 18	SS108301L	B 27	STC1203	M 12	STC1287SUC	P 07
SF106121L	N 81	SH106161	M 19	SH108251L	M 18	SH112251L	N 10	SH604051L	M 19	SS108801L	B135	STC1205	M 65	STC1288LOY	P 07
SF106121L	N 84	SH106161	M149	SH108251L	M 19	SH112301L	C 04	SH604051L	M 21	SS110801	B 39	STC1212	B101	STC1289SUC	P 07
SF106161L	D 17	SH106161	L 18	SH108251L	M 21	SH112301L	C 26	SH604051L	M 16	SS505061L	B129	STC1214	D 22	STC1290LOY	P 07
SF106201L	N 85	SH106161M	C 66	SH108251L	D 17	SH112505	H 14	SH604051L	M 20	SS505061L	B136	STC1223	P 71	STC1290SUC	P 07
SF106201L	R 46	SH106167L	K 04	SH108251L	M 16	SH112505	E 19	SH604061L	T 22	SS506060	B 99	STC1232	M 03	STC1291LOY	P 07
SF106201L	D 16	SH106201	M145	SH108251L	M 20	SH112505	E 21	SH604061L	L 04	SS506121L	B107	STC1233	M 03	STC1291SUC	P 07
SF106201L	D 18	SH106201L	M152	SH108301L	C 28	SH214141	D 06	SH604061L	H 19	SS506121L	B158	STC1234	M 03	STC1306	M144
SF106251L	P 06	SH106251	F 21	SH108301L	C 88	SH504051L	B115	SH604061L	M146	SS607061	B161	STC1235	M 03	STC1307	M144
SF108251L	D 05	SH106301L	N 70	SH108351L	C 39	SH504051L	B118	SH604061L	M148	ST606080L	B135	STC1236	M 03	STC1308	M144
SH104101L	M 82	SH106301L	K 07	SH108401	K 15	SH504061L	B 82	SH604071L	B 83	STC1000R	C 23	STC1239	M 03	STC1310	M120
SH104101L	M135	SH106301L	K 09	SH108401L	B106	SH504061L	B 94	SH604071L	N 06	STC1003R	C 23	STC1240	M 03	STC1311	M120
SH104101L	M136	SH106351L	C 69	SH108401L	B156	SH504061L	B115	SH604071L	H 17	STC1010N	D 01	STC1241	B159	STC1314	M120
SH104101L	M140	SH106351L	C 95	SH108401L	K 02	SH504061L	B129	SH604071L	L 14	STC1036	C 25	STC1242	B159	STC1315	M120
SH104101L	M143	SH106351L	C 97	SH108401L	K 04	SH504161L	B 85	SH604081L	N 69	STC104	N 34	STC1243	B159	STC1315	M126
SH105101	M 91	SH106451	C 05	SH108401L	K 06	SH504161L	B119	SH604081L	N 30	STC1040	F 13	STC1244	B 37	STC1316	M120
SH105121L	B140	SH106451L	C 27	SH108401L	K 07	SH505051L	B121	SH604101L	N 20	STC1044	F 11	STC1244	B 74	STC1317	M120
SH105121L	G 04	SH108101L	N 49	SH108401L	C 28	SH505061L	B128	SH605051L	B 96	STC1045	F 11	STC1244	B 38	STC1317	M126
SH105121L	M 98	SH108121L	N 49	SH108401L	K 08	SH505071	M 47	SH605051L	B131	STC1052020	B 08	STC1245	B 37	STC1318	M120
SH105121L	M 99	SH108141L	C 52	SH108401L	K 09	SH505071L	B 83	SH605061L	F 05	STC1052040	B 08	STC1245	B 74	STC1318	M126
SH105161L	B 26	SH108141L	C 04	SH108401L	K 13	SH505071L	N 06	SH605081L	E 49	STC1053A	T 04	STC1245	B 38	STC1319	M120
SH105161L	C 52	SH108151L	N 49	SH108401L	K 15	SH505081	M 47	SH605091L	B119	STC1053T	T 01	STC1246	B 37	STC1319	M126
SH105161L	R 51	SH108151L	L 17	SH108401L	C 24	SH505081L	B 93	SH605091L	F 01	STC1060	C 74	STC1246	B 74	STC1320	M123
SH105161L	M 57	SH108161L	B 52	SH108451L	B106	SH505081L	C 93	SH606061L	B108	STC1063	C 18	STC1246	B 38	STC1320	M128
SH105201	M135	SH108161L	J 01	SH108451L	B156	SH505081L	C 42	SH606061L	B160	STC1063	C 38	STC1247	B 37	STC1321	M120
SH105201	M137	SH108161L	J 03	SH110201L	B 21	SH505091L	B 84	SH606091L	T 20	STC1065N	B 02	STC1247	B 74	STC1322	M120
SH105201L	P 62	SH108161L	K 02	SH110201L	B161	SH505091L	B118	SH607081L	B108	STC1066	N 32	STC1247	B 38	STC1323	M120
SH105201L	M105	SH108161L	K 04	SH110201L	B184	SH505091L	B124	SH607081L	B160	STC1072	D 01	STC1250	B 37	STC1324	M120
SH105201L	J 21	SH108161L	J 02	SH110201L	J 01	SH505091L	B131	SH608161	E 19	STC1086	B 51	STC1250	B 74	STC1325	M120
SH105201L	M 98	SH108161L	J 04	SH110201L	J 03	SH505091L	K 02	SH910241L	B 98	STC1093R	C 01	STC1250	B159	STC1327	M120
SH105201L	M151	SH108161L	J 15	SH110201L	J 02	SH506071L	B 90	SH910241L	M 90	STC1101N	B109	STC1250	B 38	STC1332	M122
SH105251L	B 90	SH108161L	K 13	SH110201L	J 04	SH506071L	B106	SJ106200	R 38	STC1102E	B109	STC1251	B 37	STC1333	M122

STC1333 .... M122
STC1335 .... M122
STC1336 .... M122
STC1337 .... M122
STC1338 .... M122
STC1339 .... M123
STC1339 .... M128
STC1340 .... M123
STC1340 .... M128
STC1341 .... M123
STC1341 .... M128
STC1342 .... M123
STC1342 .... M128
STC1344 .... M122
STC1344 .... M127
STC1345 .... M122
STC1345 .... M127
STC1348A .... T 03
STC1348T .... T 01
STC1349A .... T 04
STC1349T .... T 02
STC135 .... M117
STC1350A .... T 04
STC1350T .... T 02
STC1351A .... T 04
STC1351T .... T 01
STC136 .... M118
STC1384 .... M151
STC1393 .... R 59
STC13LUN .... R 54
STC13LUN .... R 56
STC1418 .... H 21
STC1428 .... K 12
STC1429 .... N113
STC1429 .... N117
STC1429 .... N118
STC1430 .... N113
STC1430 .... N117
STC1430 .... N118
STC1431 .... N114
STC1432 .... N113
STC1432 .... N114
STC1432 .... N117
STC1434 .... N113
STC1434 .... N114
STC1435 .... N 73
STC1435 .... N 74
STC1435 .... N 75
STC1435 .... F 06
STC1436 .... F 06
STC1437 .... M121
STC1438 .... M121
STC1440 .... M121
STC1441 .... M121
STC1442 .... M121
STC1442 .... M127
STC1443 .... M121
STC1443 .... M127
STC1444 .... P 18
STC1445 .... P 18
STC1447A .... T 04
STC1447T .... T 01
STC1448A .... T 04
STC1448T .... T 02
STC1449A .... T 04
STC1449T .... T 01
STC1451A .... T 04
STC1451T .... T 02
STC1475 .... D 01
STC1484 .... N 34
STC1487 .... N 19
STC1488 .... N 19
STC1489A .... T 03
STC1489T .... T 01
STC150 .... E 17
STC151 .... E 17
STC1514R .... C 01
STC1517 .... N 34
STC152 .... E 17
STC1525 .... D 23
STC1526 .... D 23
STC1527 .... D 23
STC1528 .... H 21
STC153 .... E 17
STC1532 .... D 23
STC1533 .... D 23
STC1536 .... D 23
STC1538 .... D 23
STC1541E .... C 43
STC1543E .... C 43
STC1544E .... C 43
STC1557 .... B 32
STC155N B109
STC155R B109
STC156N B109
STC156R B109
STC1574 .... P 18
STC1575 .... P 18
STC1576 .... C 84
STC1578 .... C 85
STC1578 .... D 04
STC1579 .... C 84
STC157R .... C 23
STC1580 .... C 72
STC1581 .... C 72
STC1582 .... C 73
STC1583 .... C 72
STC1584 .... C 85
STC1596 .... D 01
STC1597 .... M 63
STC1600N B109
STC1602 .... E 11
STC1602 .... E 36
STC1604 .... B152
STC1605 .... B152
STC1624 .... M135
STC1624 .... M136
STC1624 .... M140
STC1624 .... M143
STC1628 .... C 47
STC1633 .... B 99
STC1633 .... B142
STC1635 .... E 46
STC1636 .... E 52
STC1642 .... B141
STC1646 .... G 11
STC1646 .... G 12
STC1647 .... G 12
STC1651 .... H 03
STC1657 .... M148
STC1675 .... B 42
STC1684 .... N 35
STC1685 .... N 35
STC1686 .... N 35
STC1691 .... R 51
STC1692 .... R 51
STC1693 .... B119
STC1694 .... B 66
STC1695 .... B 66
STC1696 .... B 66
STC1697 .... B 66
STC1698 .... B 66
STC1699 .... B 66
STC1700 .... B 66
STC1701 .... B 66
STC1704 .... G 05
STC1705 .... G 05
STC1706 .... G 06
STC1706 .... G 08
STC1707 .... G 06
STC1707 .... G 08
STC1708 .... G 06
STC1708 .... G 08
STC1709 .... G 06
STC1709 .... G 08
STC1710 .... G 05
STC1710 .... G 07
STC1711 .... G 06
STC1711 .... G 08
STC1712 .... G 06
STC1712 .... G 08
STC1713 .... M 03
STC1714 .... M 03
STC1715 .... N 35
STC1725 .... N 35
STC1727 .... N 35
STC1728 .... N 35
STC1730 .... G 06
STC1730 .... G 08
STC1731 .... G 05
STC1732 .... G 05
STC1733 .... G 06
STC1733 .... G 08
STC1734 .... G 06
STC1734 .... G 08
STC1736E .... B 41
STC1736N .... B 41
STC1738 .... B 66
STC1739 .... B 66
STC1743 .... R 53
STC1744 .... R 53
STC1746 .... N 73
STC1746 .... N 74
STC1746 .... N 75
STC1746 .... F 06
STC1749 .... M111
STC1750 .... M111
STC1753 .... B154
STC1757 .... M 82
STC1758 .... M 82
STC1759 .... M 82
STC1760 .... M 82
STC1766 .... N 19
STC1767 .... N 19
STC1768 .... E 11
STC1768 .... E 36
STC1773A .... T 04
STC1773T .... T 02
STC1774A .... T 04
STC1774T .... T 02
STC1790 .... M 64
STC1791 .... M 64
STC1794 .... D 01
STC1796 .... B 73
STC1799 .... L 46
STC1800 .... M131
STC1806LOY .... P 19
STC1806LOY .... P 20
STC1807SUC .... P 19
STC1807SUC .... P 20
STC1811 .... D 01
STC1812 .... N 19
STC1817 .... N 16
STC1822 .... N 17
STC1822 .... N 16
STC1824 .... M134
STC1824 .... M136
STC1825 .... M 42
STC1830 .... D 01
STC1833 .... N 17
STC1836 .... C 75
STC1841SUD .... R 48
STC1842LOY .... R 48
STC1846 .... E 36
STC1856 .... B147
STC1857 .... B101
STC1857 .... B147
STC1861E B109
STC1861N B109
STC1862E B109
STC1862N B109
STC1863 .... L 28
STC1863 .... L 48
STC1877 .... M 65
STC1877 .... L 28
STC1877 .... L 48
STC1877 .... T 23
STC1877 .... M 70
STC1877 .... M 71
STC1877 .... M 72
STC1877 .... M 77
STC1877 .... M 78
STC1878 .... M 65
STC1878 .... L 28
STC1878 .... L 48
STC1878 .... T 23
STC1878 .... M 70
STC1878 .... M 71
STC1878 .... M 72
STC1878 .... M 77
STC1878 .... M 78
STC1888 .... R 21
STC1889 .... C 11
STC1889 .... C 30
STC189 .... P 22
STC1897 .... B 30
STC1899 .... T 21
STC1900 .... T 21
STC1901 .... T 21
STC1902 .... T 21
STC1903 .... T 21
STC1904 .... T 21
STC1946 .... M147
STC1946 .... M149
STC1962 .... H 15
STC1963 .... H 15
STC1983 .... M131
STC205 .... B 97
STC206 .... M146
STC206 .... M148
STC221 .... L 41
STC222 .... L 45
STC226 .... E 21
STC234 .... B105
STC234 .... B152
STC244 .... D 22
STC245 .... D 22
STC246 .... D 22
STC262 .... B 28
STC2772 .... C 01
STC2772 .... C 23
STC2773 .... C 01
STC2773 .... C 23
STC2794 .... E 02
STC2799 .... C 86
STC2800 .... F 09
STC2801 .... B 68
STC2802 .... B 68
STC2818 .... C 44
STC2818 .... C 45
STC2818 .... C 46
STC2822 .... B141
STC2823 .... B141
STC2828A .... T 04
STC2828T .... T 02
STC2834 .... M 64
STC2835 .... M 64
STC2836 .... P 23
STC2845 .... F 10
STC2846 .... F 10
STC2847 .... F 14
STC2848 .... F 13
STC2849 .... E 46
STC2850 .... E 52
STC285SUC .... P 01
STC286 .... E 46
STC2862A .... T 03
STC2862T .... T 01
STC2863A .... T 04
STC2863T .... T 01
STC2864A .... T 04
STC2864T .... T 02
STC2867 .... N 73
STC2867 .... N 74
STC2867 .... N 75
STC2867 .... F 06
STC287 .... E 51
STC2881 .... N 62
STC2882 .... N 62
STC2883 .... N 65
STC2884 .... N 65
STC2885 .... N 62
STC2885 .... N 65
STC2886 .... N 62
STC2886 .... N 65
STC2895 .... M123
STC2898 .... B147
STC2901 .... H 01
STC2902 .... H 01
STC2903 .... H 01
STC2904 .... H 01
STC2909 .... N 62
STC2909 .... N 65
STC2910 .... F 03
STC2911 .... P 09
STC2912 .... P 09
STC2916LOY .... P 47
STC2917LOY .... P 47
STC2918LOY .... R 48
STC2920 .... N113
STC2920 .... N114
STC2920 .... N117
STC2920 .... N118
STC2932 .... E 02
STC2933 .... M149
STC2934 .... M147
STC2935 .... M136
STC2935 .... M140
STC2935 .... M142
STC2936 .... R 21
STC2940 .... D 07
STC295 .... G 01
STC295 .... G 04
STC295 .... G 09
STC2955 .... E 01
STC2955 .... E 02
STC2965 .... M126
STC2966 .... M126
STC2968 .... M126
STC2969 .... M126
STC2970 .... M126
STC2971 .... M126
STC2972 .... M126
STC2973 .... M126
STC2974 .... M127
STC2975 .... M127
STC2977 .... M127
STC298210 .... B 46
STC298220 .... B 46
STC301A .... T 04
STC301T .... T 02
STC303A .... T 04
STC303T .... T 01
STC3046 .... E 14
STC3049 .... E 14
STC3049 .... E 16
STC3050 .... E 14
STC3050 .... E 16
STC3051 .... E 14
STC3051 .... E 16
STC3078 .... M138
STC3084 .... B 67
STC3085 .... M 06
STC3095 .... P 21
STC3096 .... P 10
STC3129 .... H 03
STC313 .... B 31
STC3130 .... H 03
STC3131 .... B100
STC3131 .... B144
STC3132 .... L 25
STC3133 .... L 24
STC3134 .... L 25
STC3135 .... L 25
STC3136 .... L 23
STC3139 .... L 25
STC3140 .... L 25
STC3141 .... L 24
STC3142 .... L 26
STC3143 .... L 26
STC3144 .... L 26
STC3145 .... L 26
STC3146 .... L 26
STC3147 .... L 23
STC3148 .... L 23
STC3149 .... L 44
STC3149 .... L 51
STC3150 .... L 44
STC3150 .... L 51
STC3151 .... L 44
STC3151 .... L 51
STC3152 .... L 24
STC3153 .... L 24
STC3154 .... L 25
STC3159 .... B 67
STC3160 .... B181
STC3163 .... B 73
STC3166 .... L 25
STC3168 .... L 45
STC3171 .... L 46
STC3174A .... T 04
STC3174T .... T 01
STC3184 .... R 11
STC3184 .... R 18
STC3185 .... D 10
STC3186 .... N 35
STC3191 .... L 41
STC3191 .... L 45
STC3191 .... L 46
STC3191 .... L 49
STC3192 .... L 40
STC3192 .... L 41
STC3192 .... L 53
STC3194 .... N 35
STC3251A .... T 03
STC3251T .... T 01
STC3252 .... B 64
STC3253 .... B 64
STC3254 .... B 64
STC3269 .... E 25

STC3272	B160	STC3592	B 67	STC3915A	T 03	STC7020	V 02	STC7580SUC	R 49	STC7755	T 07	STC8083	T 37	STC8248	T 17
STC3272	B161	STC3596	E 25	STC3915T	T 02	STC7021	V 02	STC7580SUC	T 29	STC7756	T 07	STC8085	T 05	STC8249	T 28
STC3295	F 11	STC3597	B 64	STC3916A	T 03	STC7022	N 54	STC7583	T 30	STC7757	T 07	STC8087AA	T 44	STC8259	T 28
STC3297	B 29	STC3600	C 72	STC3916T	T 01	STC7023	P 50	STC7589	T 32	STC7758	T 07	STC8088	T 19	STC8261	T 28
STC3297	B 66	STC3601	E 42	STC3917	L 25	STC7024	P 50	STC7590	T 32	STC7759	T 07	STC8089	T 19	STC8263	T 28
STC3298	M 14	STC3608	D 02	STC398	F 10	STC7025	P 50	STC7595	T 10	STC7760	T 07	STC8096	T 47	STC8267	M 17
STC3305	M 17	STC3609	D 02	STC399	F 10	STC7030	P 42	STC7596	T 47	STC7774	T 07	STC8104	T 13	STC8267	M 18
STC3305	M 18	STC3611	D 02	STC4104	B 87	STC7031	M 14	STC7597	T 47	STC778	M 02	STC8107	T 23	STC8267	M 19
STC3305	M 19	STC3615K	D 06	STC4111	N106	STC7051	J 19	STC7601	T 33	STC779	C 28	STC8109	T 30	STC8267	M 21
STC3305	M 21	STC3619	K 05	STC4112	N106	STC7052	N106	STC7602	T 33	STC780	H 01	STC8110	T 26	STC8267	M 16
STC3305	M 16	STC3624	B 64	STC4113	M131	STC7053	N106	STC7603LOY	R 50	STC7808	T 44	STC8112	T 36	STC8267	M 20
STC3305	M 20	STC363	B 32	STC4225RCH	N102	STC7151	N120	STC7604LOY	R 50	STC7809	T 14	STC8113	T 36	STC8268	M 17
STC3316	M151	STC3642PMA	P 23	STC4225RCJ	N102	STC7152	N120	STC7605	T 08	STC7809AA	T 14	STC8114	T 20	STC8268	M 18
STC3322	N 35	STC3682	N 39	STC4226RCH	N102	STC7153	N120	STC7606	T 08	STC7814	T 14	STC8116	T 38	STC8268	M 19
STC3322	N 36	STC3682	N 38	STC4226RCJ	N102	STC7154	N120	STC7611	T 19	STC7822	T 08	STC8117	T 38	STC8268	M 16
STC3323	N 35	STC3686A	T 03	STC4227	R 21	STC7155	N120	STC7612	T 19	STC7823	T 08	STC8123	T 26	STC8269	M 17
STC3323	N 36	STC3686T	T 01	STC4228	R 21	STC7156	N120	STC7617JUL	T 19	STC782JUL	P 01	STC8124	T 18	STC8269	M 18
STC334	C 82	STC3691	B 73	STC4229	R 21	STC7157	N120	STC7618SUC	T 19	STC782SUC	P 01	STC8125AA	T 39	STC8269	M 19
STC3368	M 03	STC3703	E 46	STC4230	R 21	STC7158	N120	STC7625	T 30	STC7857	T 45	STC813	M117	STC8269	M 21
STC3375	C 59	STC3704	E 52	STC4286	T 22	STC717	T 28	STC7627	T 41	STC7859	T 15	STC8130AA	T 44	STC8269	M 16
STC3376	C 60	STC3709	M143	STC4286D	T 20	STC718	T 28	STC7630	T 24	STC786	F 16	STC8136	T 37	STC8269	M 20
STC3377	C 63	STC3715	B 37	STC434	B 83	STC719	T 28	STC7641	T 17	STC7874LOY	T 29	STC8138AA	T 17	STC8270	M 17
STC3395	B 07	STC3715	B 74	STC434	N 06	STC721	T 28	STC7642	T 17	STC7875LOY	T 29	STC8139	T 06	STC8270	M 18
STC3395	B 45	STC3715	B 38	STC437JUL	P 01	STC722	T 28	STC7643	T 23	STC7876SUC	T 29	STC8139	N109	STC8270	M 19
STC3405	E 25	STC3716	K 05	STC437SUC	P 01	STC723	T 28	STC7644	T 23	STC7877SUC	T 29	STC8140	T 06	STC8270	M 16
STC3406	D 05	STC3717	K 05	STC463	N122	STC724	T 17	STC765	M 01	STC7891	T 14	STC8140	N109	STC8292	T 28
STC3407	B 48	STC3719	K 05	STC483	B 85	STC733	M130	STC766	M 01	STC7897	T 14	STC8144AB	T 31	STC8358	B 39
STC3410	E 42	STC3721	N 73	STC483	B119	STC7334	V 01	STC7661	T 23	STC798	G 09	STC8145AA	T 31	STC8360	B108
STC3411	E 42	STC3721	N 74	STC490	L 21	STC7337	V 03	STC7662	T 07	STC8002	T 34	STC8146AA	T 31	STC8361	B160
STC3412	E 42	STC3721	N 75	STC50013	T 16	STC7338	V 03	STC767	M 01	STC8003	T 34	STC8154	T 16	STC837	F 10
STC3413	E 42	STC3721	F 06	STC50087	T 21	STC7339	V 03	STC7672LOY	R 50	STC8004	T 34	STC8156AA	T 38	STC8383E	F 10
STC3414	E 42	STC3722	E 10	STC50088	T 32	STC738	N 89	STC7673LOY	R 50	STC8007	T 32	STC8157	T 38	STC8384E	F 10
STC3415	E 42	STC3722	E 35	STC50089	T 32	STC745	M117	STC7674SUC	R 50	STC8010AA	T 32	STC8158	P 73	STC8401	T 46
STC3416	E 42	STC3723	E 35	STC509	B 97	STC7525	V 04	STC7675SUC	R 50	STC8013RWB	T 05	STC8169	T 37	STC8406	T 46
STC3417	E 42	STC3724	B112	STC510	B 97	STC7550	T 12	STC768	M 01	STC8014RWA	T 05	STC8170AC	T 11	STC8407	T 46
STC3418	E 42	STC378	N 24	STC553	M117	STC7551	T 31	STC7681	T 23	STC8017	T 13	STC8171AC	T 11	STC8408	T 46
STC3419	E 42	STC379	N 24	STC579	E 13	STC7552T	T 01	STC7682	T 23	STC8018L	T 18	STC8172	T 11	STC8409	T 46
STC3420	E 42	STC3813	L 26	STC579	E 14	STC7554	T 09	STC770	L 20	STC802	G 09	STC8172AA	T 11	STC8413	T 13
STC3421	E 42	STC3814	L 26	STC579	E 16	STC7556	T 23	STC773	N 39	STC8025AA	T 42	STC8173	T 11	STC8414	T 13
STC3422	E 42	STC3815	L 26	STC60092	N120	STC7557	T 37	STC773	T 47	STC8030	T 47	STC8173AA	T 11	STC8420	T 32
STC3423	E 42	STC3816	L 26	STC60123	V 05	STC7558	T 37	STC773	N 38	STC8037	T 06	STC8174AC	T 11	STC8421	T 32
STC3424	E 42	STC3817	L 26	STC60124	V 05	STC7560	T 33	STC7741	T 37	STC8038	R 49	STC8175AC	T 11	STC8422AA	T 31
STC3425	E 42	STC3818	L 26	STC60125	V 05	STC7566	T 40	STC7742	T 07	STC8038	T 29	STC8176	T 12	STC8427AA	T 14
STC3426	E 42	STC3822A	T 03	STC611	B 09	STC7568	T 40	STC7743	T 07	STC8039	R 49	STC8176AA	T 11	STC8435	T 40
STC3427	E 42	STC3822T	T 01	STC611	B 47	STC7570	T 13	STC7744	T 07	STC8039	T 29	STC8177	T 12	STC8443	T 32
STC3428	E 42	STC3823A	T 03	STC612	J 12	STC7573	T 10	STC7745	T 07	STC8040	T 46	STC8177AA	T 11	STC8444AC	T 32
STC3429	E 42	STC3823T	T 01	STC618	E 48	STC7574	T 10	STC7746	T 07	STC8041A	T 34	STC8185AA	T 30	STC8445AB	T 32
STC3432	D 11	STC3824A	T 03	STC622	N 88	STC7575	T 36	STC7747	T 07	STC8048	T 08	STC8188AB	T 32	STC8445AC	T 32
STC3433	D 12	STC3824T	T 01	STC641	M117	STC7576	T 36	STC7748	T 07	STC8049	T 44	STC822	C 06	STC8452	T 08
STC3435	E 19	STC3825A	T 04	STC642	M117	STC7577LOY	R 49	STC7749	T 07	STC8057AB	T 39	STC822	C 07	STC8454AA	T 08
STC3435	E 22	STC3825T	T 01	STC699	M114	STC7577LOY	T 29	STC7750	T 07	STC8060AA	T 20	STC822	C 28	STC8454AA	T 09
STC350	N112	STC3827A	T 04	STC7004AA	T 30	STC7578LOY	R 49	STC7751	T 07	STC8060AB	T 20	STC8243	T 17	STC8455AA	T 08
STC350	N115	STC3827T	T 02	STC7017	N123	STC7578LOY	T 29	STC7752	T 07	STC8074	T 12	STC8244AB	T 17	STC8455AA	T 09
STC357	M 63	STC3828A	T 03	STC7018	N123	STC7579SUC	R 49	STC7753	T 07	STC8075	T 12	STC8246AA	T 17	STC8457	T 09
STC358	M 63	STC3828T	T 01	STC7019	N 25	STC7579SUC	T 29	STC7754	T 07	STC8076	T 38	STC8247AA	T 46	STC8458	T 12

Code	Ref	Code	Ref	Code	Ref	Code	Ref	Code	Ref	Code	Ref	Code	Ref	Code	Ref
STC8459	T 12	STC8646	M 03	STC8905	T 35	STC9134	N 03	SX110351L	B175	TRC103040	C 43	TZZ100200	C 64	UKC3795L	B 91
STC8461LOY	R 49	STC865	M 63	STC8908	T 35	STC9135	N 03	SX607101	E 18	TRE100460	C 47	UAM1637L	B164	UKC3795L	J 20
STC8461LOY	T 29	STC8656	B151	STC8910	T 35	STC9136	N 03	SY504072L	B 90	TUB101500	C 65	UAM1737L	B172	UKC3795L	J 21
STC8462LOY	R 49	STC866	M 63	STC8911	T 35	STC9137	N 03	SY504072L	B127	TUB101800	C 59	UAM2663L	B171	UKC3795L	J 22
STC8462LOY	T 29	STC8682E	D 02	STC8913	T 35	STC9138	N 03	SYT10004	H 21	TUD101720	C 57	UAM2857L	B 47	UKC3799L	B 91
STC8463	T 06	STC869LNF	R 53	STC8914	T 35	STC9139	N 03	TBH100040	E 11	TUD101970	C 56	UAM2857L	B167	UKC37L	C 11
STC8464	T 08	STC870LNF	R 53	STC8915	T 35	STC9140	N 03	TD106041	H 18	TUD102340	C 56	UAM2957L	B 47	UKC37L	C 30
STC8465	T 31	STC874	F 11	STC8916	T 35	STC9141	N 03	TD108201	B 10	TUD102370	C 56	UAM2957L	B167	UKC37L	C 13
STC8467	T 30	STC875	F 12	STC8917	T 35	STC9181	T 28	TD108201	B 12	TUJ100050	C 59	UAM5029L	B174	UKC37L	C 32
STC8468	T 23	STC8784	T 46	STC8919	T 17	STC924	N 69	TD110041L	B 11	TUJ100050	C 62	UAM5038L	B164	UKC3803L	B 30
STC8478	T 06	STC8801SUD	R 49	STC8927T	T 01	STC925	N 69	TE108041L	B 06	TUJ100050	C 13	UAM5191L	B173	UKC3803L	B 62
STC848	B181	STC8801SUD	T 29	STC8930	T 10	STC926	N 69	TE108041L	B 24	TUJ100050	C 33	UAM5327L	B171	UKC6684	M 21
STC8480	T 24	STC8802SUD	R 49	STC8930AB	T 10	STC930	C 73	TE108041L	B 34	TUK10011L	C 13	UAM7218L	B172	UKC6684L	M 23
STC8481	T 24	STC8802SUD	T 29	STC8931	T 10	STC947	M134	TE108051L	B 16	TUK10011L	C 32	UAM7498L	B 36	UKC6684L	M155
STC8482	T 05	STC8807	T 44	STC8931AB	T 10	STC947	M136	TE108051L	B 24	TUK100340	C 58	UAM7987L	B165	UKC6684L	M 46
STC8483	T 05	STC881	P 16	STC8932AA	T 08	STC947	M139	TE108051L	B 54	TUK100340	C 60	UKC1060L	C 56	UKC6684L	M 48
STC8485	T 37	STC881	P 17	STC8933	T 47	STC947	M142	TE108051L	B 62	TUK100340	C 62	UKC1060L	C 08	UKC6684L	M 49
STC8486	T 37	STC8811	T 24	STC8934LOY	T 19	STC958	B 46	TE108051L	B 96	TUO100240	C 64	UKC1060L	C 29	UKC6684L	M 51
STC8495JUL	P 36	STC8812	T 24	STC8934SUC	T 19	STC961	M102	TE108051L	B172	TUO100260	C 64	UKC170L	C 47	UKC6684L	M 52
STC8495JUL	P 39	STC8813AA	T 24	STC8935	T 32	STC961	M103	TE108061L	B 10	TUZ100000	C 65	UKC170L	C 03	UKC6684L	M 53
STC8495JUL	T 29	STC8814	T 24	STC8937	E 51	STC961	M104	TE108061L	B 49	TUZ100020	C 51	UKC170L	C 25	UKC6684L	M 54
STC8495JUL	P 41	STC8816	T 20	STC8937	T 38	STC963	D 22	TE108181	B 71	TUZ100040	C 51	UKC18L	C 17	UKC6684L	M 55
STC8495LOY	T 29	STC8816	T 21	STC8938	T 25	STC964	D 22	TE110051L	B 16	TUZ100060	C 51	UKC18L	C 37	UKC6684L	N 79
STC8495SUC	P 36	STC8817	T 42	STC8938AA	T 25	STC965	D 22	TE110051L	B 24	TUZ100080	C 51	UKC2089L	C 19	UKC6684L	M 42
STC8495SUC	P 39	STC8828	T 09	STC8940	T 37	STC966	M 05	TE110051L	B 49	TUZ100100	C 51	UKC2089L	C 39	UKC6684L	M 56
STC8495SUC	T 29	STC8829	T 46	STC8941	T 37	STC966	M 04	TE110051L	B 75	TUZ100120	C 51	UKC2105L	C 19	UKC6684L	M132
STC8495SUC	P 41	STC8830	T 39	STC8942	T 44	STC974	B171	TE110051L	B178	TUZ100140	C 51	UKC2105L	C 39	UKC6684L	M133
STC8497	T 06	STC8831	T 46	STC8943	T 44	STC980	R 48	TE110061L	B 39	TUZ100160	C 51	UKC24L	C 03	UKC6684L	M 44
STC8498	T 06	STC8832	T 46	STC8944	T 11	STC981	N 71	TE110061L	B 75	TUZ100200	C 51	UKC24L	C 25	UKC6684L	M 50
STC8499	T 09	STC8833	T 46	STC8945	T 11	STC981	N 72	TE110061L	B189	TUZ100220	C 51	UKC24L	D 06	UKC6684L	M150
STC851	E 11	STC8834	T 47	STC8945AA	T 11	STC981	F 06	TE110061L	C 52	TUZ100240	C 51	UKC25L	C 52	UKC75	C 18
STC851	E 36	STC8836AA	N 13	STC8947	T 12	STC982	N 39	TE110061L	C 04	TUZ100270	C 51	UKC25L	C 04	UKC75	C 38
STC8511	T 41	STC8836AA	T 09	STC8947AA	T 12	STC982	T 47	TE110071L	B 16	TUZ100290	C 51	UKC25L	C 26	UKC8137L	B 88
STC8518	T 36	STC883LNF	N 56	STC8948	T 12	STC982	N 37	TE110071L	B 24	TUZ100310	C 51	UKC2738L	C 06	UKC8L	C 56
STC8519	T 14	STC8843AA	T 36	STC8948AA	T 12	STC982	N 38	TE110071L	B 40	TUZ100330	C 51	UKC3092	C 67	UKC8L	C 11
STC8519AA	T 14	STC8844	T 36	STC8949	T 12	STC983	M118	TE110071L	B 54	TVQ100130	C 56	UKC3092	C 21	UKC8L	C 30
STC852	B188	STC8847	T 32	STC8950	T 12	STC985	B153	TE110071L	B 62	TYG100580	E 14	UKC3092	C 41	URB100760	B 39
STC853	B188	STC884LNF	N 56	STC8955	T 28	STC986	B153	TE110071L	B 76	TYG100580	E 16	UKC30L	C 47	URB100760	B 75
STC854	B188	STC8851SUC	R 50	STC8959E	D 02	STC996	C 82	TE110071L	B172	TYG100590	E 14	UKC31L	C 12	UTP1519	L 21
STC8550	T 09	STC8852SUC	R 50	STC8960E	D 02	STC9A	T 04	TE110151	B 51	TYG100590	E 16	UKC31L	C 31	UYF100160	D 14
STC8568	H 16	STC887	F 10	STC896LUN	N 56	STC9T	T 01	TE505105L	B 98	TYS101040	E 33	UKC31L	C 13	VUR100260	P 10
STC8569	H 18	STC8875	T 20	STC898	G 01	STD10002L	G 13	TE506101L	B 99	TYS101060	E 33	UKC31L	C 14	WA103001L	P 39
STC857	M117	STC8875	T 21	STC899	G 01	SU112101L	D 15	TE506101L	B142	TYS101080	E 33	UKC31L	C 32	WA103001L	P 47
STC8570	H 18	STC888	F 10	STC900	G 01	SU112101M	D 14	TKC1229L	C 03	TYS101100	E 33	UKC31L	C 34	WA103001L	P 41
STC8571	H 16	STC8880	T 46	STC906	C 73	SU112101M	D 15	TKC1229L	C 25	TYS101120	E 33	UKC3530L	C 12	WA104001L	B 54
STC8572	H 16	STC8881	T 46	STC9124	B151	SU112101S	D 14	TKC1235L	C 03	TYS101140	E 33	UKC3530L	C 31	WA104001L	B135
STC8573	H 16	STC8885	T 31	STC9125	N 03	SX108201L	B 39	TKC1235L	C 25	TYS101160	E 33	UKC3530L	C 14	WA104004L	M133
STC8574	H 18	STC8885AA	T 31	STC9126	N 03	SX108201L	B 75	TKC2786L	C 02	TYS101180	E 33	UKC3530L	C 34	WA105001L	B 26
STC8575	H 16	STC8887	T 37	STC9127	N 03	SX108251	R 54	TKC2786L	C 24	TYS101200	E 33	UKC3531L	C 13	WA105001L	N 69
STC8576	H 16	STC889	F 12	STC9128	N 03	SX110251	E 21	TKC290L	C 19	TYS101220	E 33	UKC3531L	C 32	WA105001L	C 52
STC8577	C 13	STC889	F 14	STC9129	N 03	SX110251M	E 19	TKC290L	C 39	TYS101240	E 33	UKC3660L	C 19	WA105001L	P 62
STC8577	C 32	STC8893	T 27	STC9130	N 03	SX110251M	E 39	TKC5779L	C 03	TYS101260	E 33	UKC3660L	C 39	WA105001L	L 03
STC8617R	B162	STC8895	T 24	STC9131	N 03	SX110251M	E 40	TKC5779L	C 25	TYS101280	E 33	UKC3793L	J 20	WA105001L	M105
STC862	B153	STC8896	T 27	STC9132	N 03	SX110257M	E 21	TRC102940	C 43	TYS101300	E 33	UKC3793L	J 21	WA105001L	G 04
STC864	L 20	STC8896AA	T 27	STC9133	N 03	SX110257M	E 40	TRC102960	C 43	TZZ100190	C 56	UKC3793L	J 22	WA105001L	J 21

WA105001L	P 05	WA106041L	J 06	WA108051L	B106	WA108051L	C 88	WA112081L	N 04	WC112081L	F 15	WJ105001L	P 75	WL106001L	N111
WA105001L	P 06	WA106041L	J 21	WA108051L	B118	WA108051L	C 89	WA116101L	T 20	WC112081L	T 21	WJ105001L	S 01	WL106001L	K 02
WA105001L	N 30	WA106041L	J 22	WA108051L	B119	WA108051L	H 23	WA120001	E 48	WC116101	E 53	WJ105001L	M 94	WL106001L	K 04
WA105001L	N112	WA106041L	K 11	WA108051L	B131	WA108051L	H 24	WA702101L	B 98	WC702101L	N 25	WJ105001L	L 43	WL106001L	K 06
WA105001L	N114	WA106041L	K 08	WA108051L	B135	WA108051L	D 16	WA702101L	B139	WC702101L	J 23	WJ105001L	M 05	WL106001L	K 07
WA105001L	M 57	WA106041L	K 09	WA108051L	B138	WA108051L	F 03	WA702101L	P 16	WC702101L	J 24	WJ105001L	M134	WL106001L	C 92
WA105001L	M 98	WA106041L	K 13	WA108051L	B139	WA108051L	M 16	WA702101L	G 03	WC702101L	P 03	WJ105001L	M135	WL106001L	C 93
WA105001L	D 17	WA106041L	K 14	WA108051L	B142	WA108051L	M 20	WA704061	N 78	WC702101L	H 12	WJ105001L	M137	WL106001L	C 05
WA105001L	N116	WA106041L	L 04	WA108051L	B144	WA108251	L 01	WAM2578L	B173	WC702105	N 42	WJ105001L	M140	WL106001L	C 07
WA105001L	N119	WA106041L	L 15	WA108051L	B148	WA110051	J 13	WAM3885L	B169	WCM10027	K 06	WJ105001L	M142	WL106001L	C 19
WA105001L	M 28	WA106041L	D 05	WA108051L	B156	WA110051L	N 01	WB106041L	B 90	WCM10027	K 07	WJ105001L	M 04	WL106001L	C 27
WA105041	M 10	WA106041L	H 09	WA108051L	B157	WA110061L	B 24	WB106041L	B127	WCM10027	K 08	WJ105001L	M 28	WL106001L	C 28
WA106001	L 06	WA106041L	H 10	WA108051L	E 43	WA110061L	B 35	WB106041L	F 15	WCM10027	K 09	WJ105001L	M 91	WL106001L	C 39
WA106001L	N 49	WA106041L	H 19	WA108051L	E 48	WA110061L	B 39	WB108051L	B139	WD108051	R 48	WJ105001L	M151	WL106001L	C 42
WA106001L	L 16	WA106041L	C 69	WA108051L	D 03	WA110061L	B 75	WB108051L	H 23	WD112081L	E 21	WJ106001L	L 33	WL106001L	H 17
WA106001L	L 17	WA106041L	L 36	WA108051L	D 10	WA110061L	B100	WB112081L	E 19	WE104001L	B 54	WJ106001L	R 32	WL106001L	R 38
WA106001L	C 06	WA106041L	N 20	WA108051L	D 14	WA110061L	B106	WB120001	E 49	WE104001L	H 20	WJ106001L	M147	WL106001L	M 47
WA106001L	L 03	WA106041L	N 86	WA108051L	N 02	WA110061L	B107	WC104001	M135	WE105001L	M 90	WJ106001L	N 30	WL106001L	K 10
WA106041L	B 13	WA106041L	C 94	WA108051L	N 49	WA110061L	B144	WC104001	M136	WE105001L	M 83	WJ106001L	M152	WL106001L	K 11
WA106041L	B 15	WA106041L	C 96	WA108051L	G 06	WA110061L	B156	WC104001	M140	WE105001L	N 60	WJ106001L	M149	WL106001L	K 08
WA106041L	B 83	WA106041L	C 97	WA108051L	G 08	WA110061L	B158	WC104001	M143	WE105001L	N 63	WJ106001L	R 39	WL106001L	K 09
WA106041L	B 85	WA106041L	R 46	WA108051L	L 01	WA110061L	B161	WC104001L	N 33	WE106001L	P 29	WJ106001L	M144	WL106001L	K 13
WA106041L	B 86	WA106041L	R 51	WA108051L	L 19	WA110061L	E 47	WC105001L	P 14	WE106001L	N 60	WJ106007	R 48	WL106001L	K 14
WA106041L	B 98	WA106041L	P 06	WA108051L	C 86	WA110061L	E 51	WC105004	P 63	WE106001L	N 61	WJ108001	L 17	WL106001L	F 20
WA106041L	B119	WA106041L	P 64	WA108051L	C 87	WA110061L	E 53	WC105007L	P 14	WE106001L	N 63	WJ108001	L 02	WL106001L	F 21
WA106041L	B121	WA106041L	F 18	WA108051L	C 02	WA110061L	N 02	WC105048L	M 06	WE106001L	N 64	WJ108001	L 08	WL106001L	C 69
WA106041L	B132	WA106041L	F 17	WA108051L	C 08	WA110061L	J 01	WC106041L	B119	WE110001L	H 11	WJ108001L	L 05	WL106001L	N 20
WA106041L	D 20	WA106041L	L 12	WA108051L	C 21	WA110061L	J 03	WC106041L	N 27	WE110001L	H 13	WJ110007L	B175	WL106001L	N 55
WA106041L	D 14	WA106041L	H 11	WA108051L	C 28	WA110061L	J 12	WC106041L	N 20	WE110001L	E 21	WJ110007L	B187	WL106001L	C 94
WA106041L	N 06	WA106041L	E 13	WA108051L	C 29	WA110061L	K 02	WC106041L	H 06	WE110001L	H 05	WJ110007L	N 53	WL106001L	C 95
WA106041L	N 13	WA106041L	E 15	WA108051L	C 41	WA110061L	K 03	WC106041L	H 08	WE110001L	H 07	WJM10006L	B133	WL106001L	C 96
WA106041L	N 14	WA106041L	E 16	WA108051L	C 22	WA110061L	K 06	WC108001L	F 18	WE600041L	M 17	WJM10006L	B134	WL106001L	C 97
WA106041L	N 70	WA106041L	M144	WA108051L	H 01	WA110061L	K 07	WC108051	L 27	WE600041L	M 18	WL100001L	C 25	WL106001L	C 24
WA106041L	N 50	WA106041L	N 60	WA108051L	P 62	WA110061L	C 52	WC108051	L 22	WE600041L	M 19	WL105001L	J 23	WL106001L	N 30
WA106041L	N 51	WA106041L	N 61	WA108051L	M108	WA110061L	C 85	WC108051	L 39	WE600041L	M 21	WL105001L	C 52	WL106001L	M 21
WA106041L	N 52	WA106041L	N 63	WA108051L	M110	WA110061L	C 03	WC108051	L 38	WE600041L	M 16	WL105001L	M105	WL106001L	M 45
WA106041L	N110	WA106041L	N 64	WA108051L	L 02	WA110061L	C 04	WC108051L	N 12	WE600041L	M 20	WL105001L	J 21	WL106001L	M152
WA106041L	N111	WA106041L	L 18	WA108051L	M106	WA110061L	C 25	WC108051L	N 41	WE600071L	G 11	WL105001L	M 94	WL106001L	D 17
WA106041L	P 22	WA106047L	N 70	WA108051L	M 47	WA110061L	C 22	WC108051L	N 53	WE600071L	G 13	WL105001L	N 30	WL106001L	R 40
WA106041L	G 13	WA106051L	J 20	WA108051L	G 03	WA110061L	J 02	WC108051L	N121	WE600101L	G 11	WL105001L	N112	WL106001L	M 44
WA106041L	J 05	WA108001L	N 14	WA108051L	G 09	WA110061L	J 04	WC108051L	L 21	WEB101550	K 03	WL105001L	N114	WL106001L	M144
WA106041L	J 18	WA108041L	N 69	WA108051L	K 13	WA110061L	K 08	WC108051L	J 15	WF105001L	M114	WL105001L	M 57	WL106041	L 36
WA106041L	K 02	WA108051L	B 06	WA108051L	L 04	WA110061L	K 09	WC108051L	R 51	WF106001L	N 46	WL105001L	M 98	WL106041	M 18
WA106041L	K 04	WA108051L	B 14	WA108051L	L 05	WA110061L	D 05	WC108051L	M 91	WF108001L	N 49	WL105001L	N116	WL106041	M 19
WA106041L	K 06	WA108051L	B 24	WA108051L	L 08	WA110061L	N 04	WC110061L	B 83	WF507L	N 84	WL105001L	N119	WL106041L	D 16
WA106041L	K 07	WA108051L	B 25	WA108051L	D 06	WA110061L	N 09	WC110061L	E 49	WF600041L	N 51	WL105001L	M 28	WL108001	M108
WA106041L	L 13	WA108051L	B 27	WA108051L	H 21	WA110061L	T 20	WC110061L	E 52	WF600041L	N 52	WL105001L	M 91	WL108001	M110
WA106041L	C 85	WA108051L	B 28	WA108051L	N 09	WA110061L	R 48	WC110061L	N 06	WF600041L	M146	WL105001L	M151	WL108001L	B 06
WA106041L	C 92	WA108051L	B 31	WA108051L	B170	WA110061L	H 24	WC110061L	N 14	WF600041L	M148	WL106001L	B 22	WL108001L	B 14
WA106041L	C 42	WA108051L	B 34	WA108051L	L 43	WA110067L	L 43	WC110061L	T 21	WF702101L	N 76	WL106001L	B 33	WL108001L	B 24
WA106041L	H 17	WA108051L	B 84	WA108051L	M 17	WA112081	C 97	WC112081L	E 44	WF702101L	N 81	WL106001L	B137	WL108001L	B 39
WA106041L	S 01	WA108051L	B 85	WA108051L	M 18	WA112081L	C 04	WC112081L	E 47	WJ105001	L 43	WL106001L	B 38	WL108001L	B 75
WA106041L	M115	WA108051L	B 99	WA108051L	M 19	WA112081L	C 26	WC112081L	E 53	WJ105001	M135	WL106001L	N 70	WL108001L	B104
WA106041L	L 03	WA108051L	B100	WA108051L	M 21	WA112081L	F 11	WC112081L	N 11	WJ105001	M137	WL106001L	N 50	WL108001L	B148
WA106041L	M 47	WA108051L	B104	WA108051L	M114	WA112081L	F 12	WC112081L	F 11	WJ105001L	N 13	WL106001L	N110	WL108001L	B189

WL108001L ....N 41
WL108001L ....N 49
WL108001L ....G 11
WL108001L ....G 12
WL108001L ....K 12

WL108001L ....K 01
WL108001L ....C 93
WL108001L ....C 02
WL108001L ....C 05
WL108001L ....C 07

WL108001L ....C 19
WL108001L ....C 27
WL108001L ....C 28
WL108001L ....C 39
WL108001L ....C 42

WL108001L ....L 02
WL108001L M106
WL108001L ....J 15
WL108001L ....K 10
WL108001L ....K 11

WL108001L ....L 04
WL108001L ....D 04
WL108001L ....H 21
WL108001L ....N 09
WL108001L ....C 24

WL108001L ....M 17
WL108001L ....M 18
WL108001L ....M 19
WL108001L ....M 21
WL108001L M114

WL108001L ....F 18
WL108001L ....F 17
WL108001L ....H 24
WL108001L ....D 17
WL108001L ....E 21

WL108001L ....M 16
WL108001L ....M 20
WL110001L ....B 15
WL110001L ....B 37
WL110001L ....B 83

WL110001L B161
WL110001L ....N 01
WL110001L ....N 06
WL110001L ....N 14
WL110001L ....J 12

WL110001L ....C 03
WL110001L ....D 04
WL110001L ....D 06
WL110001L ....E 19
WL110001L ....E 20

WL110001L ....E 39
WL110001L ....E 40
WL112001L ....C 68
WL112001L ....C 90
WL112001L ....C 04

WL112001L ....C 26
WL112001L ....N 09
WL112001L ....N 10
WL112001L ....E 21
WL600041L B140

WL600061L ....C 72
WL600081L ....T 20
WL700101L B103
WL700101L B140
WL700101L B151

WL700101L ....N 76
WL700101L ....N 80
WM106001L ....C 71
WM106001L ....C 19
WM106001L ....C 39

WM108001L ....C 71
WM110001L ....E 51
WM110001L ....E 52
WM600041L ....B 82
WM600041L ....B 83

WM600041L ....B 84
WM600041L ....B 85
WM600041L ....B 90
WM600041L ....B 94
WM600041L B115

WM600041L B118
WM600041L B127
WM600041L B129
WM600041L B132
WM600041L B151

WM600041L ....N 06
WM600041L ....N 69
WM600041L ....H 14
WM600041L ....H 17
WM600041L ....T 22

WM600041L ....L 04
WM600041L ....H 19
WM600041L ....N 20
WM600041L ....E 13
WM600041L ....E 15

WM600041L ....E 16
WM600041L ....B 95
WM600041L ....M 20
WM600051L ....B 33
WM600051L ....B 35

WM600051L ....B 83
WM600051L ....B 93
WM600051L ....B 98
WM600051L ....B 99
WM600051L B119

WM600051L B124
WM600051L B130
WM600051L B136
WM600051L B137
WM600051L B142

WM600051L B144
WM600051L ....E 46
WM600051L ....E 49
WM600051L ....N 06
WM600051L ....F 05

WM600051L ....M 47
WM600061L ....B 35
WM600061L ....B 90
WM600061L B100
WM600061L B106

WM600061L B107
WM600061L B108
WM600061L B124
WM600061L B156
WM600061L B158

WM600061L B160
WM600061L B125
WM600061L ....E 46
WM600061L ....C 26
WM600061L ....H 14

WM600061L ....H 17
WM600061L ....G 09
WM600061L ....F 16
WM600061L ....N 09
WM600061L ....E 23

WM600061L ....E 24
WM600061L ....E 39
WM600061L ....E 40
WM600071L ....B 83
WM600071L ....N 06

WM600071L ....H 16
WM600071L ....H 18
WM702001L B151
WM702001L ....J 23
WM702001L ....J 24

WM702001L ....T 22
WM702001L ....P 03
WP105L ....K 02
WP105L ....K 04
WP105L ....K 12

WP105L ....K 06
WP105L ....K 07
WP105L ....K 10
WP105L ....K 11
WP105L ....K 08

WP105L ....K 09
WP105L ....K 13
WP105L ....K 15
WP8019L M144
WQT10001L ....M 96

WQT10001L ....M 99
WQT100030L............M 96
WQT100030L............M 99
WS108001L ....N 13
WS600047L ....L 04

WW110004 ....R 48
WZX1505 ....B 97
XBL10004L ....M 90
XBP100180 ....M 05
XBP100180 ....M 09

XFG100330 ....M 11
XQD101170LNF........M138
XQD101170LNF........M141
XQI100040L M138
XQI100040L M141

XQJ10002L M137
XQJ10002L M139
XQJ10002L M142
XQM101280 ....M142
YAJ100320L M128

YAL10001L M123
YAL10001L M128
YCB10019 ....B176
YEB10026 ....M101
YEB10027 ....M101

YFM4509 ....B162
YFM4510 ....B162
YFM4511 ....B162
YFM4571 ....B162
YFM4572 ....B162

YFM4590 ....B162
YFM4591 ....B162
YLE10088 ....B152
YLE10088E B152
YLE10099 ....B153

YLE10099E B153
YLE10113 ....B 73
YLE10113E ....B 73
YLE10124 ....B186
YLE10124E B186

YLE101880 ....B186
YLF10039 ....B187
YLP10033L B186
YLU10097 ....B187
YLU10127 ....B187

YLU101600 ....B187
YMC129990 ....M 31
YMC130010 ....M 32
YMC130020 ....M 32
YMC130030 ....M 31

YMC130040 ....M 32
YMC130050 ....M 31
YMC130130 ....M 31
YMN102120 ....M 42
YMN102130 ....M 42

YOO4340LNF............P 28
YOO4340LUN............P 28
YOO4633PMP............P 82
YQG10011L M116
YQH10002L ....M 82

YQH10004L ....M 82
YQJ100200 ....M 17
YQJ100200 ....M 19
YQJ100200 ....M 21
YQR10016L M155

YQR10075 ....M155
YQU10017L M116
YRC100350 ....M 61
YRC100350 ....M 64
YRC100350 ....M 62

YRC100360 ....M 64
YSB105780 ....M 53
YSB105790 ....M 53
YUE10005L ....M 13
YUE10015L M106

YUG10067L ....M 65
YUJ10037L M131
YUM10001L ....P 15
YUN10001L ....P 15
YWB10020L ....P 15

YWB10027L M108
YWB10027L M110
YWB10027L M106
YWB10027L ....M 84
YWB10027L ....M 85

YWB10027L ....M 86
YWB10027L ....M 87
YWB10027L ....M 88
YWB10027L ....M 90
YWB10027L ....M 83

YWB10032L ....M 84
YWB10032L ....M 85
YWB10032L ....M 86
YWB10032L ....M 87
YWB10032L ....M 83

YWB10041L ....M 84
YWB10041L ....M 89
YWB100790L............M 84
YWC10151 ....M 61
YWC10152L ....M 83

YWC10153L ....M 83
YWC10227LUP M106
YWC102890LUM.......M106
YWC102900LUP.......M106
YWC103640LUM.......M108

YWC103640LUM.......M110
YWC103640LUM.......M106
YWC105830 ....M109
YWK10003L M107
YWK10003L M109

YWT10002L ....M 81
YWT10003 ....M 81
YWX101060L............M107
YWX101060L............M109
YWX101070L............M107

YWX101070L............M109
YYC10030L M159
YYC10294L M156
YYD10003 ....B157
YYD10003 ....M155

YYP10011L M123
YYP10011L M128

Brooklands Land Rover & Range Rover Titles

Available from Amazon or, in case of difficulty from Land Rover distributors:

Brooklands Books Ltd., P.O. Box 146, Cobham, Surrey, KT11 1LG, England, UK
Phone: +44 (0) 1932 865051 info@brooklands-books.com www.brooklandsbooks.com

www.brooklandsbooks.com

LAND ROVER OFFICIAL FACTORY PUBLICATIONS

Title	Part No.
Land Rover Series 1 Workshop Manual	4291
Land Rover Series 1 1948-53 Parts Catalogue	4051
Land Rover Series 1 1954-58 Parts Catalogue	4107
Land Rover Series 1 Instruction Manual	4277
Land Rover Series 1 and II Diesel Instruction Manual	4343
Land Rover Series II and IIA Workshop Manual	AKM8159
Land Rover Series II and Early IIA Bonneted Control Parts Catalogue	605957
Land Rover Series IIA Bonneted Control Parts Catalogue	RTC9840CC
Land Rover Series IIA, III and 109 V8 Optional Equipment Parts Catalogue	RTC9842CE
Land Rover Series IIA/IIB Instruction Manual	LSM64IM
Land Rover Series 2A and 3 88 Parts Catalogue Supplement (USA Spec)	606494
Land Rover Series III Workshop Manual	AKM3648
Land Rover Series III Workshop Manual V8 Supplement (edn. 2)	AKM8022
Land Rover Series III 88, 109 and 109 V8 Parts Catalogue	RTC9841CE
Land Rover Series III Owners Manual 1971-1978	607324B
Land Rover Series III Owners Manual 1979-1985	AKM8155
Military Land Rover (Lightweight) Series III Parts Catalogue	61278
Military Land Rover Series III (L.W.B.) User Handbook	608179
Military Land Rover (Lightweight) Series III User Manual	608180
Land Rover 90/110 and Defender Workshop Manual 1983-1992	SLR621ENWM
Land Rover Defender Workshop Manual 1993-1995	LDAWMEN93
Land Rover Defender 300 Tdi and Supplements Workshop Manual 1996-1998	LRL0097ENGBB
Land Rover Defender Td5 Workshop Manual and Supplements 1999-2006	LRL0410BB
Land Rover Defender Electrical Manual Td5 1999-06 and 300Tdi 2002-2006	LRD5EHBB
Land Rover 110 Parts Catalogue 1983-1986	RTC9863CE
Land Rover Defender Parts Catalogue 1987-2006	STC9021CC
Land Rover 90 • 110 Handbook 1983-1990 MY	LSM0054
Land Rover Defender 90 • 110 • 130 Handbook 1991 MY - Feb. 1994	LHAHBEN93
Land Rover Defender 90 • 110 • 130 Handbook Mar. 1994 - 1998 MY	LRL0087ENG/2
Military Land Rover 90/110 All Variants (Excluding APV and SAS) User Manual	2320-D-122-201
Military Land Rover 90 and 110 2.5 Diesel Engine Versions User Handbook	SLR989WDHB
Military Land Rover Defender XD - Wolf Workshop Manual - 2320D128 -	302 522 523 524
Military Land Rover Defender XD - Wolf Parts Catalogue	2320D128711
Discovery Workshop Manual 1990-1994 (petrol 3.5, 3.9, Mpi and diesel 200 Tdi)	SJR900ENWM
Discovery Workshop Manual 1995-1998 (petrol 2.0 Mpi, 3.9, 4.0 V8 and diesel 300 Tdi)	LRL0079BB
Discovery Series II Workshop Manual 1999-2003 (petrol 4.0 V8 and diesel Td5 2.5)	VDR100090/6
Discovery Parts Catalogue 1989-1998 (2.0 Mpi, 3.5, 3.9 V8 and 200 Tdi and 300 Tdi)	RTC9947CF
Discovery Parts Catalogue 1999-2003 (petrol 4.0 V8 and diesel Td5 2.5)	STC9049CA
Discovery Owners Handbook 1990-1991 (petrol 3.5 V8 and diesel 200 Tdi)	SJR820ENHB90
Discovery Series II Handbook 1999-2004 MY (petrol 4.0 V8 and Td5 diesel)	LRL0459BB
Freelander Workshop Manual 1998-2000 (petrol 1.8 and diesel 2.0)	LRL0144
Freelander Workshop Manual 2001-2003 ON (petrol 1.8L, 2.5L and diesel Td4 2.0)	LRL0350ENG/4
Land Rover 101 1 Tonne Forward Control Workshop Manual	RTC9120
Land Rover 101 1 Tonne Forward Control Parts Catalogue	608294B
Land Rover 101 1 Tonne Forward Control User Manual	608239
Range Rover Workshop Manual 1970-1985 (petrol 3.5)	AKM3630
Range Rover Workshop Manual 1986-1989 (petrol 3.5 and diesel 2.4 Turbo VM)	SRR660ENWM & LSM180WS4/2
Range Rover Workshop Manual 1990-1994 (petrol 3.9, 4.2 and diesel 2.5 Turbo VM, 200 Tdi)	LHAWMENA02
Range Rover Workshop Manual 1995-2001 (petrol 4.0, 4.6 and BMW 2.5 diesel)	LRL0326ENGBB
Range Rover Workshop Manual 2002-2005 (BMW petrol 4.4 and BMW 3.0 diesel)	LRL0477
Range Rover Electrical Manual 2002-2005 UK version (petrol 4.4 and 3.0 diesel)	RR02KEMBB
Range Rover Electrical Manual 2002-2005 USA version (BMW petrol 4.4)	RR02AEMBB
Range Rover Parts Catalogue 1970-1985 (petrol 3.5)	RTC9846CH
Range Rover Parts Catalogue 1986-1991 (petrol 3.5, 3.9 and diesel 2.4 and 2.5 Turbo VM)	RTC9908CB
Range Rover Parts Catalogue 1992-1994 MY and 95 MY Classic (petrol 3.9, 4.2 and diesel 2.5 Turbo VM, 200 Tdi and 300 Tdi)	RTC9961CB
Range Rover Parts Catalogue 1995-2001 MY (petrol 4.0, 4.6 and BMW 2.5 diesel)	RTC9970CE
Range Rover Owners Handbook 1970-1980 (petrol 3.5)	606917
Range Rover Owners Handbook 1981-1982 (petrol 3.5)	AKM8139
Range Rover Owners Handbook 1983-1985 (petrol 3.5)	LSM0001HB
Range Rover Owners Handbook 1986-1987 (petrol 3.5 and diesel 2.4 Turbo VM)	LSM129HB

Engine Overhaul Manuals for Land Rover and Range Rover

Title	Part No.
300 Tdi Engine, R380 Manual Gearbox and LT230T Transfer Gearbox Overhaul Manuals	LRL003, 070 & 081
Petrol Engine V8 3.5, 3.9, 4.0, 4.2 and 4.6 Overhaul Manuals	LRL004 & 164
Land Rover/Range Rover Driving Techniques	LR369
Working in the Wild - Manual for Africa	SMR684MI
Winching in Safety - Complete guide to winching Land Rovers and Range Rovers	SMR699MI

Workshop Manual Owners Edition

Land Rover 2 / 2A / 3 Owners Workshop Manual 1959-1983
Land Rover 90, 110 and Defender Workshop Manual Owners Edition 1983-1995
Land Rover Discovery Workshop Manual Owners Edition 1990-1998

All titles available from Amazon or Land Rover specialists

Brooklands Books Ltd., P.O. Box 146, Cobham, Surrey, KT11 1LG, England, UK
Phone: +44 (0) 1932 865051 info@brooklands-books.com www.brooklands-books.com

www.brooklandsbooks.com

Printed in Great Britain
by Amazon